UNDERSTANDING NUTRITION

TENTH EDITION

Ellie Whitney

Sharon Rady Rolfes

THOMSON

WADSWORTH

Australia • Canada • Mexico • Singapore • Spain
United Kingdom • United States

THOMSON

WADSWORTH

Publisher: Peter Marshall
Development Editor: Elizabeth Howe
Assistant Editor: Elesha Feldman
Editorial Assistant: Lisa Michel
Technology Project Manager: Travis Metz
Marketing Manager: Jennifer Somerville
Marketing Assistant: Melanie Banfield
Advertising Project Manager: Shemika Britt
Project Manager, Editorial Production: Sandra Craig
Creative Director: Rob Hugel
Print/Media Buyer: Rebecca Cross

Permissions Editor: Kiely Sexton
Production: The Book Company
Text and Cover Designer: John Walker
Art Editor: Carolyn Deacy
Photo Researcher: Myrna Engler
Copy Editor: Patricia Lewis
Illustrations: Imagineering
Cover Image: ©FoodPix
Cover Printer: Phoenix Color Corp
Compositor: Parkwood Composition Service
Printer: R. R. Donnelley/Willard

Printed in the United States of America
3 4 5 6 7 08 07 06 05

For more information about our products, contact us at:
Thomson Learning Academic Resource Center
1-800-423-0563

For permission to use material from this text or product,
submit a request online at:
http://www.thomsonrights.com

Any additional questions about permissions can be
submitted by email to thomsonrights@thomson.com.

Library of Congress Control Number: 2003111919

Student Edition: ISBN 0-534-62226-7
Instructor's Edition: ISBN 0-534-62239-9
International Student Edition: ISBN 0-534-62242-9 (Not for sale in the United States)

Thomson Wadsworth
10 Davis Drive
Belmont, CA 94002–3098
USA

Asia
Thomson Learning
5 Shenton Way #01-01
UIC Building
Singapore 068808

Australia/New Zealand
Thomson Learning
102 Dodds Street
Southbank, Victoria 3006
Australia

Canada
Nelson
1120 Birchmount Road
Toronto, Ontario M1K 5G4
Canada

Europe/Middle East/Africa
Thomson Learning
High Holborn House
50/51 Bedford Row
London WC1R 4LR
United Kingdom

Latin America
Thomson Learning
Seneca, 53
Colonia Polanco
11560 Mexico D.F.
Mexico

Spain/Portugal
Paraninfo
Calle Magallanes, 25
28015 Madrid, Spain

To
*the memory of Gary
Woodruff, the editor who
first encouraged me to write.*

Ellie

To

*my daughter Marni Jay,
whose sparkle delights
my heart. The early years
of your life have been
wonderful, and I
look forward to watching
you pursue your dreams
as you venture into your
college years.*

**Mom
(Sharon)**

About the Authors

Ellie Whitney grew up in New York City and received her B.A. and Ph.D. degrees in English and Biology at Radcliffe/Harvard University and Washington University, respectively. She has lived in Tallahassee since 1970, has taught at both the Florida State University and Florida A&M University, has written newspaper columns on environmental matters for the *Tallahassee Democrat,* and has authored almost a dozen college textbooks on nutrition, health, and related topics, some of which are in their seventh (or later) editions. In addition to teaching and writing, she has spent the past three-plus decades exploring outdoor Florida and studying its ecology. Her latest book is *Priceless Florida: The Natural Ecosystems* (Pineapple Press, 2004).

Sharon Rady Rolfes received her M.S. in nutrition and food science from the Florida State University. She is a founding member of Nutrition and Health Associates, an information resource center that maintains a research database on over 1000 nutrition-related topics. Her other publications include the college textbooks *Understanding Normal and Clinical Nutrition* and *Nutrition for Health and Health Care* and a multimedia CD-ROM called *Nutrition Interactive.* In addition to writing, she occasionally lectures at universities and at professional conferences and serves as a consultant for various educational projects. Her volunteer activities include coordinating meals for the hungry and homeless and serving as a partner in Stepping Toward Health, a community initiative that encourages individuals "to be more active and eat more nutritionally." She maintains her registration as a dietitian and membership in the American Dietetic Association.

Brief Contents

Contents

Preface

In some ways, things haven't changed much since the first edition of this introductory nutrition text was written. This tenth edition maintains the same goals: to reveal the fascination of the science of nutrition and share the fun and excitement of nutrition with the reader. We have learned from the hundreds of professors and over a million students who have used this book over the years that readers want an *understanding* of the science of nutrition so that they can make healthy choices in their daily lives.

Yet much has changed in the world of nutrition and in our daily lives over the past three decades. The number of food options has increased dramatically—even as we spend less time in the kitchen preparing meals. The connections between diet and disease have become more apparent, stimulating interest in dietary choices that can help people enjoy longer and healthier lives. The science of nutrition has grown rapidly, with new "facts" emerging daily. Current hot topics—such as nutritional genomics, ghrelin, and functional foods—had yet to be explored and, consequently, weren't even mentioned in the first edition. This edition discusses each of these topics, and more. As with every previous edition, every chapter has been substantially revised to reflect the many changes that have occurred in the field of nutrition and in our daily lives over the years. We hope that this book serves you well.

The Chapters *Understanding Nutrition* presents the core information of an introductory nutrition course. Chapter 1 wastes no time in exploring why we eat the foods we do and continues with a brief overview of the nutrients, the science of nutrition, recommended nutrient intakes, assessment, and important relationships between diet and health. Chapter 2 describes the diet-planning principles and food guides used to create diets that support good health and includes instructions on how to read a food label. In Chapter 3, readers follow the journey of digestion and absorption as the body transforms foods into nutrients. Chapters 4 through 6 describe carbohydrates, fats, and proteins—their chemistry, roles in the body, and places in the diet. Then Chapter 7 shows how the body derives energy from these three nutrients. Chapters 8 and 9 continue the story with a look at energy balance, the factors associated with overweight and underweight, and the benefits and dangers of weight loss and weight gain. Chapters 10 through 13 complete the introductory lessons by describing the vitamins, the minerals, and water—their roles in the body, deficiency and toxicity symptoms, and sources.

The next seven chapters weave that basic information into practical applications, showing how nutrition influences people's lives. Chapter 14 describes how physical activity and nutrition work together to support fitness. Chapters 15, 16, and 17 present the special nutrient needs of people through the life cycle—pregnancy and lactation; infancy, childhood, and adolescence; and adulthood and the later years. Chapter 18 focuses on the dietary risk factors and recommendations associated with chronic diseases, and Chapter 19 addresses consumer concerns about the safety of the food and water supply. Chapter 20 closes the book by examining hunger and the global environment and exploring possible solutions for establishing sustainable foodways.

The Highlights Every chapter is followed by a highlight that provides readers with an in-depth look at a current, and often controversial, topic that relates to its companion chapter. New highlights in this edition feature a comparison of dietary guidelines from around the world (Highlight 2), the benefits of (some) high-fat foods (Highlight 5), and the health connections of functional foods and the phytochemicals they contain (Highlight 13).

Special Features The art and layout in this edition have been redesigned to add visual appeal and enhance learning. In addition, special features help readers identify key concepts and apply nutrition knowledge. For example, a **definition** is provided whenever a new term is introduced. These definitions often include pronunciations and derivations to facilitate understanding. A glossary at the end of the text includes all defined terms.

> **definition** (DEF-eh-NISH-en): the meaning of a word.
> • **de** = from
> • **finis** = boundary

Nutrition in Your Life

New to this edition are Nutrition in Your Life sections at the beginning and end of each chapter. The opening section introduces the essence of the chapter in a friendly and familiar scenario. Then the closing section revisits that message and prompts readers to consider whether their personal choices are meeting the dietary goals introduced in the chapter.

IN SUMMARY Each major section within a chapter concludes with a summary paragraph that reviews the key concepts. Similarly, summary tables cue readers to important reviews.

Also featured in this edition are the Healthy People 2010 nutrition-related priorities, which are presented whenever their subjects are discussed. Healthy People 2010 is a report developed by the U.S. Department of Health and Human Services that establishes national objectives in health promotion and disease prevention for the year 2010.

HEALTHY PEOPLE 2010

These nutrition-related priorities are presented throughout the text whenever their subjects are discussed.

HOW TO

Many chapters include "How to" sections that guide readers through problem-solving tasks. For example, the "How to" in Chapter 1 shows readers how to calculate energy intake from the grams of carbohydrate, fat, and protein in a food; another "How to" in Chapter 20 describes how to plan healthy meals on a tight budget.

NUTRITION CALCULATION

Several chapters close with a "Nutrition Calculation" section. These sections often reinforce the "How to" lessons and provide practice in doing nutrition-related calculations. The problems enable readers to apply their skills to hypothetical situations and then check their answers (found at the end of the chapter). Readers who successfully master these exercises will be well prepared for "real-life" nutrition-related problems.

NUTRITION ON THE NET

Each chapter and many highlights also conclude with Nutrition on the Net—a list of websites for further study of topics covered in the accompanying text. These listings do not imply an endorsement of the organizations or their programs. We have tried to provide reputable sources, but cannot be responsible for the content of these sites. (Read Highlight 1 to learn how to find reliable information on the Internet.)

STUDY QUESTIONS

Each chapter ends with study questions in essay and multiple-choice format. Study questions offer readers the opportunity to review the major concepts presented in the chapters in preparation for exams. The page numbers after each essay question refer readers to discussions that answer the question; multiple-choice answers appear at the end of the chapter.

The Appendixes The appendixes are valuable references for a number of purposes. Appendix A summarizes background information on the hormonal and nervous systems, complementing Appendixes B and C on basic chemistry, the chemical structures of nutrients, and major metabolic pathways. Appendix D describes measures of protein quality. Appendix E provides detailed coverage on nutrition assessment, and Appendix F presents the estimated energy requirements for men and women at various levels of physical activity. Appendix G presents the 2003 U.S. Exchange System. Appendix H is a 2000-item food composition table compiled from the latest nutrient database assembled by ESHA Research, Inc., of Salem, Oregon. Appendix I presents recommendations from the World Health Organization (WHO) and information for Canadians—the Choice System and guidelines to healthy eating and physical activities.

The Inside Covers The inside covers put commonly used information at your fingertips. The front covers (pp. A, B, and C) present the current nutrient recommendations; the inside back cover (p. Y on the left) features the Daily Values used on food labels and a glossary of nutrient measures; and the inside back cover (p. Z on the right) shows the suggested weight ranges for various heights. The pages just prior to the back cover (pp. W–X) assist readers with calculations and conversions.

Closing Comments We have tried to keep the number of references manageable. Many statements that have appeared in previous editions with references now appear without them, but every statement is backed by research, and the authors will supply references upon request. We have not provided a separate list of suggested readings, but have tried to include references that will provide readers with additional details or a good overview of the subject. Nutrition is a fascinating subject, and we hope our enthusiasm for it comes through on every page.

Acknowledgments

To produce a book requires the coordinated effort of a team of people—and, no doubt, each team member has another team of support people as well. We salute, with a big round of applause, everyone who has worked so diligently to ensure the quality of this book.

We thank our partners and friends, Linda DeBruyne and Fran Webb, for their valuable consultations and contributions; working together over the past 20 years has been a most wonderful experience. We especially appreciate Linda's research assistance on several chapters and Margaret Hedley's attention to the Canadian information throughout the text and in Appendix I. A million thank-yous to Lynn Earnest for her careful attention to manuscript preparation and a multitude of other daily tasks and to Marni Jay Rolfes for her assistance in proofreading and copying numerous pages. To Kiely Sexton, a special thanks for her assistance in obtaining permissions. We also thank the many people who have prepared the ancillaries that accompany this text: Harry Sitren for writing and enhancing the Test Bank; Lori Turner, Mary Rhiner, and Margaret Hedley for preparing the Instructor's Manual; Thomas Castonguay, Steven Nizielski, and Richard Morel for developing the Nutrition Explorer CD; Eugene Fenster for gathering and creating slides for the Multimedia Manager and creating material for the WebTutor; Charlene Hamilton for selecting the transparencies; and Lori Turner for organizing the Student Study Guide. A big thank-you to the folks at ESHA for creating the food composition appendix, verifying the data in figures and tables, and developing the computerized diet analysis program that accompanies this book. Our special thanks to Peter Marshall for his continued support and insightful ideas; to Beth Howe for her brilliant suggestions for improvements and efficient coordination of reviews; to Sandra Craig for her guidance of this revision from conception to conclusion; to Dusty Friedman for her diligent attention to the innumerable details involved in production; to Jennifer Somerville, Melanie Banfield, Joy Westberg, Brian Chaffee, and Shemika Britt for their enthusiastic efforts in marketing; to Travis Metz for his talented techno tango on our website and CD products; to Elesha Feldman for her competent management of ancillary development; and to Lisa Michel for her willingness to fill in the gaps whenever the need arose. We also thank John Walker for creatively designing these pages; Carolyn Deacy for styling the figures to add visual appeal and enhance learning; Lisa Sovran and the team of artists at Imagineering for creating accurate and attractive artwork to complement our writing; Myrna Engler for selecting photographs and coordinating photography sessions to deliver nutrition messages beautifully; Pat Lewis for sharing her grammar knowledge and copyediting over 2000 manuscript pages; Martha Ghent for proofreading close to 1000 final text pages; and Micki Taylor for composing a thorough and useful index. To the hundreds of others involved in production and sales, we tip our hats in appreciation.

We are especially grateful to our associates, friends, and families for their continued encouragement and support. We also thank our many reviewers for their comments and contributions.

Ellie Whitney
Sharon Rady Rolfes
March 2004

Reviewers of *Understanding Nutrition*

Fernando Agudelo-Silva
Laney College

Nancy Amy
*University of California
at Berkeley*

James Baily
*University of Tennessee
at Knoxville*

Kathleen D. Bauer
Montclair State University

Eugenia Bearden
*Clayton College and State
University*

Nancy Becker
Portland State University

Patricia Benarducci
Miami-Dade Community College

Margaret Ann Berry
University of Central Oklahoma

Sharleen J. Birkimer
University of Louisville

Debra Boardley
University of Toledo

Jeanne S. Boone
Palm Beach Community College

Ellen Brennan
San Antonio College

Dorothy A. Byrne
*University of Texas at
San Antonio*

Nancy Canolty
University of Georgia

Leah Carter
Bakersfield College

Mary Ann Cessna
*Indiana University
of Pennsylvania*

Jo Carol Chezum
Ball State University

Michele Ciccazzo
Florida International University

Donald D. Clarke
Fordham College of Fordham U.

Ava Craig
Sacramento City College

Tina Crook
Univ. of Central Arkansas

Wendy Cunningham
*California State University
Sacramento*

Jim Daugherty
Glendale Community College

Beth Ellen DiLuglio
Palm Beach Community College

Robert DiSilvestro
Ohio State University

Eugene J. Fenster
Longview Community College

Pam Fletcher
*Albuquerque Technical
Vocational Institute*

Betty Forbes
West Virginia University

Eileen Ford
University of Pennsylvania

William Forsythe
University of Southern Mississippi

Coni Francis
*University of Colorado Health
Sciences Center*

Jean Fremont
Simon Fraser University

Julie Rae Friedman
*State University of New York at
Farmingdale*

Trish Froehlich
Palm Beach Community College

Patricia Garrett
*University of Tennessee,
Chattanooga*

Francine Genta
Cabrillo College

Leonard E. Gerber
University of Rhode Island

Victoria Getty
Indiana University

Jill Golden
Orange Coast College

Sandra M. Gross
West Chester University

Deborah Gustafson
Utah State University

Leon Hageman
Burlington County College

Charlene Hamilton
University of Delaware

Shelley Hancock
The University of Alabama

Margaret Hedley
University of Guelph

Carol A. Heinz-Bennett
Mesa Community College

Nancy Hillquist
Elgin Community College

Carolyn Hoffman
Central Michigan University

Kim M. Hohol
Mesa Community College

Tracy Horton
*University of Colorado Health
Sciences Center*

Andie Hsueh
Texas Woman's University

Eleanor B. Huang
Orange Coast College

Donna-Jean Hunt
Stephen F. Austin University

Bernadette Janas
Rutgers University

Michael Jenkins
Kent State University

Carol Johnston
Arizona State University

Connie Jones
*Northwestern State University
of Louisiana*

Jayanthi Kandiah
Ball State University

Younghee Kim
Bowling Green State University

Beth Kitchin
*University of Alabama
at Birmingham*

Kim Kline
University of Texas, Austin

Vicki Kloosterhouse
Oakland Community College

Susan M. Krueger
*University of Wisconsin–
Eau Claire*

Joanne Kuchta
Texas A&M University

Michael LaFontaine
*Central Connecticut
State University*

Betty Larson
Concordia College

Chunhye Kim Lee
Northern Arizona University

Robert D. Lee
Central Michigan University

Anne Leftwich
University of Central Arkansas

Joseph Leichter
University of British Columbia

Alan Levine
Marywood University

Janet Levins
Pensacola Junior College

Samantha Logan
*University of Massachusetts,
Amherst*

Jack Logomarsino
Central Michigan University

Elaine M. Long
Boise State University

Swarna Mandali
Central Missouri State University

Laura McArthur
East Carolina University

Harriet McCoy
*University of Arkansas,
Fayetteville*

Bruce McDonald
University of Manitoba

Lisa McKee
New Mexico State University

Mary Mead
University of California Berkeley

Rhonda L. Meyers
Lower Columbia College

Lynn Monahan-Couch
West Chester University

Cynthia K. Moore
University of Montevallo

William Moore
*Wytheville Community
College*

Edith Moran
Chicago State University

Mithia Mukutmoni
Sierra College

Yasmin Neggers
*University of Alabama,
Tuscaloosa*

Paula Netherton
Tulsa Junior College

Steven Nizielski
Grand Valley State University

Amy Olson
*College of St. Benedict at
St. John's University*

Marvin Parent
Oakland Community College

Linda Peck
University of Findlay, Ohio

Susan S. Percival
University of Florida

Erwina Peterson
*Yakima Valley Community
College*

Roseanne L. Poole
Tallahassee Community College

Julie Priday
Centralia College

Stephanie Raach
Rock Valley College

Ann Raymon
Chemeketa Community College

Nuha F. Rice
*Portland Community College &
Clackamas Community College*

Ramona G. Rice
Georgia Military College,
Milledgeville, Georgia

Robin R. Roach
The University of Memphis

Christian K. Roberts
University of California,
Los Angeles

Janet Sass
Northern Virginia Community
College

Tammy Sakanashi
Santa Rosa Junior College

Padmini Shankar
Georgia Southern University

Nancy Shearer
Cape Cod Community College

Linda Shelton
California State University,
Fresno

Melissa Shock
University of Central Arkansas

Brenda J. Smith
Oklahoma State University

Diana-Marie Spillman
Miami University

Karen Stammen
Chapman University

Wendy Stuhldreher
Slippery Rock University of
Pennsylvania

Carla Taylor
University of Manitoba

Janet Thompson
University of Waterloo

Michele Trankina
St. Mary's University

Josephine Umoren
Northern Illinois University

Anne VanBeber
Texas Christian University

Michelle L. Vineyard
University of Tennessee,
Chattanooga

Ava Craig-Waite
Sacramento City College

Dana Wassmer
California State University,
Sacramento

Suzy Weems
Stephen F. Austin University

D. Katie Wiedman
University of Saint Francis

Garrison Wilkes
University of Massachusetts/
Boston

Richard A. Willis
University of Texas at Austin

Shahla M. Wunderlich
Montclair State University

Lisa Young
New York University

UNDERSTANDING NUTRITION

An Overview of Nutrition

© Lew Robertson/FoodPix/Getty Images

Nutrition in Your Life

Believe it or not, you have probably eaten at least 20,000 meals in your life. Without any conscious effort on your part, your body uses the nutrients from those foods to make all its components, fuel all its activities, and defend itself against diseases. How successfully your body handles these tasks depends, in part, on your food choices. Nutritious food choices support healthy bodies.

Welcome to the world of **nutrition.** Nutrition has played a significant role in your life, even from before your birth, although you may not always have been aware of it. And it will continue to affect you in major ways, depending on the **foods** you select.

Every day, several times a day, you make food choices that influence your body's health for better or worse. Each day's choices may benefit or harm your health only a little, but when these choices are repeated over years and decades, the rewards or consequences become major. That being the case, close attention to good eating habits now can bring health benefits later. Conversely, carelessness about food choices from youth on can contribute to many chronic diseases■ prevalent in later life, including heart disease and cancer. Of course, some people will become ill or die young no matter what choices they make, and others will live long lives despite making poor choices. For the large majority, however, the food choices they make each and every day will benefit or impair their health in proportion to the wisdom of those choices.

Although most people realize that their food habits affect their health, they often choose foods for other reasons. After all, foods bring to the table a variety of pleasures, traditions, and associations as well as nourishment. The challenge, then, is to combine favorite foods and fun times with a nutritionally balanced **diet.**

■ In general, a **chronic** disease progresses slowly or with little change and lasts a long time. By comparison, an **acute** disease develops quickly, produces sharp symptoms, and runs a short course.
- **chronos** = time
- **acute** = sharp

nutrition: the science of foods and the nutrients and other substances they contain, and of their actions within the body (including ingestion, digestion, absorption, transport, metabolism, and excretion). A broader definition includes the social, economic, cultural, and psychological implications of food and eating.

foods: products derived from plants or animals that can be taken into the body to yield energy and nutrients for the maintenance of life and the growth and repair of tissues.

diet: the foods and beverages a person eats and drinks.

3

Food Choices

People decide what to eat, when to eat, and even whether to eat in highly personal ways, often based on behavioral or social motives rather than on awareness of nutrition's importance to health. Fortunately, many different food choices can be healthy ones, but nutrition awareness helps to make them so.

Personal Preference As you might expect, the number one reason people choose foods is taste—they like certain flavors. Two widely shared preferences are for the sweetness of sugar and the savoriness of salt. Liking high-fat foods appears to be another universally common preference. Other preferences might be for the hot peppers common in Mexican cooking or the curry spices of Indian cuisine. Some research suggests that genetics may influence people's food preferences.[1]

An enjoyable way to learn about other cultures is to taste their ethnic foods.

© Michael Newman/PhotoEdit

Habit People sometimes select foods out of habit. They eat cereal every morning, for example, simply because they have always eaten cereal for breakfast. Eating a familiar food and not having to make any decisions can be comforting.

Ethnic Heritage or Tradition Among the strongest influences on food choices are ethnic heritage and tradition. People eat the foods they grew up eating. Every country, and in fact every region of a country, has its own typical foods and ways of combining them into meals. The "American diet" includes many ethnic foods from various countries, all adding variety to the diet. This is most evident when eating out: 60 percent of U.S. restaurants (excluding fast-food places) have an ethnic emphasis, most commonly Chinese, Italian, or Mexican.

Social Interactions Most people enjoy companionship while eating. It's fun to join friends when they are ordering pizza or going out for ice cream. Meals are social events, and the sharing of food is part of hospitality. Social customs almost compel people to accept food or drink offered by a host or shared by a group.

Availability, Convenience, and Economy People eat foods that are accessible, quick and easy to prepare, and within their financial means. Consumers today value convenience highly and are willing to spend over half of their food budget on meals that require little, if any, further preparation.[2] They frequently eat out, bring home ready-to-eat meals, or have food delivered. Even when they venture into the kitchen, they want to prepare a meal in 15 to 20 minutes, using less than a half dozen ingredients—and those "ingredients" are often semiprepared foods, such as canned soups. Such emphasis on convenience limits food choices to the selections offered on menus and products designed for quick preparation. Whether decisions based on convenience meet a person's nutrition needs depends on the choices made. Eating a banana or a candy bar may be equally convenient, but the fruit offers more vitamins and minerals and less sugar and fat.

Positive and Negative Associations People tend to like foods with happy associations—such as hot dogs at ball games or cake and ice cream at birthday parties. By the same token, people can attach intense and unalterable dislikes to foods that they ate when they felt sick or that were forced on them when they weren't hungry. Parents may teach their children to like and dislike certain foods by using those foods as rewards or punishments.

Emotional Comfort Some people cannot eat when they are emotionally upset. Others may eat in response to a variety of emotional stimuli—for example, to relieve boredom or depression or to calm anxiety.[3] A depressed person may choose to eat chocolates rather than to call a friend. A person who has returned home from an exciting evening out may unwind with a late-night sandwich. These people may find emotional comfort, in part, because foods can influence the brain's chemistry and the mind's response. Carbohydrates and alcohol, for example, tend to calm, whereas proteins and caffeine are more likely to activate.[4] Eating in response to emotions can easily lead to overeating and obesity, but may be appropriate at times. For example, sharing food at times of bereavement serves both the giver's need to provide comfort and the receiver's need to be cared for and to interact with others, as well as to take nourishment.

Values Food choices may reflect people's religious beliefs, political views, or environmental concerns. For example, many Christians forgo meat during Lent, the period prior to Easter, and Jewish law includes an extensive set of dietary rules that govern the use of foods derived from animals. Muslims fast between sunrise and sunset during Ramadan, the ninth month of the Islamic calendar. A concerned consumer may boycott fruit picked by migrant workers who have been exploited. People may buy vegetables from local farmers to save the fuel and environmental costs of foods shipped in from far away. They may also select foods packaged in containers that can be reused or recycled. Some consumers accept or reject foods that have been irradiated or genetically modified, depending on their approval of these processes (see Chapter and Highlight 19 for a complete discussion).

Body Weight and Image Sometimes people select certain foods and supplements that they believe will improve their physical appearance and avoid those they believe might be detrimental. Such decisions can be beneficial when based on sound nutrition and fitness knowledge, but undermine good health when based on faddism or carried to extremes, as pointed out in later discussions of eating disorders (Highlight 9) and supplements athletes commonly use (Highlight 14).

Nutrition and Health Benefits Finally, of course, many consumers make food choices that will benefit health. Food manufacturers and restaurant chefs have responded to scientific findings linking health with nutrition by offering an abundant selection of health-promoting foods and beverages. Foods that provide health benefits beyond their nutrient contributions are called **functional foods.** In some cases, functional foods are as natural and familiar as oatmeal or tomatoes. In other cases, the foods have been modified in a way that provides health benefits, perhaps by lowering the fat contents. In still other cases, manufacturers have fortified foods by adding nutrients or phytochemicals that provide health benefits (see Highlight 13). Examples of these functional foods include orange juice fortified with calcium to help build strong bones and margarine made with a plant sterol that lowers blood cholesterol.

Consumers welcome these new foods into their diets, provided that the foods are reasonably priced, clearly labeled, easy to find in the grocery store, and convenient to prepare. These foods must also taste good—as good as the traditional choices. Of course, a person need not eat any of these "special" foods to enjoy a healthy diet; many "regular" foods provide numerous health benefits as well. In fact, "regular" foods such as whole grains; vegetables and legumes; fruits; meats, fish, and poultry; and milk products are among the healthiest choices a person can make.

IN SUMMARY A person selects foods for a variety of reasons. Whatever those reasons may be, food choices influence health. Individual food selections neither make nor break a diet's healthfulness, but the balance of foods selected over time can make an important difference to health.[5] For this reason, people are wise to think "nutrition" when making their food choices.

© Ariel Skelley/CORBIS

To enhance your health, keep nutrition in mind when selecting foods.

functional foods: foods that provide health benefits beyond their nutrient contributions. Functional foods may include whole foods, fortified foods, and modified foods.

Foods bring pleasure—and nutrients.

The Nutrients

Biologically speaking, people eat to receive nourishment. Do you ever think of yourself as a biological being made of carefully arranged atoms, molecules, cells, tissues, and organs? Are you aware of the activity going on within your body even as you sit still? The atoms, molecules, and cells of your body continually move and change, even though the structures of your tissues and organs and your external appearance remain relatively constant. Your skin, which has covered you since your birth, is replaced entirely by new cells every seven years. The fat beneath your skin is not the same fat that was there a year ago. Your oldest red blood cell is only 120 days old, and the entire lining of your digestive tract is renewed every 3 days. To maintain your "self," you must continually replenish, from foods, the **energy** and the **nutrients** you deplete in maintaining your body.

Nutrients in Foods and in the Body

Amazingly, the body can derive all the energy, structural materials, and regulating agents that it needs from the foods we eat. This section introduces the nutrients that foods deliver and shows how they participate in the dynamic processes that keep people alive and well.

Composition of Foods Chemical analysis of a food such as a tomato shows that it is composed primarily of water (95 percent). Most of the solid materials are carbohydrates, lipids,■ and proteins. If you could remove these materials, you would find a tiny residue of vitamins, minerals, and other compounds. Water, carbohydrates, lipids, proteins, vitamins, and some of the minerals found in foods are nutrients—substances the body uses for the growth, maintenance, and repair of its tissues.

This book focuses mostly on the nutrients, but foods contain other compounds as well—fibers, **phytochemicals**, pigments, additives, alcohols, and others. Some are beneficial, some are neutral, and a few are harmful. Later sections of the book touch on these **nonnutrients** and their significance.

Composition of the Body A complete chemical analysis of your body would show that it is made of materials similar to those found in foods (see Figure 1-1). A healthy 150-pound body contains about 90 pounds of water and about 20 to 45 pounds of fat. The remaining pounds are mostly protein, carbohydrate, and the major minerals of the bones. Vitamins, other minerals, and incidental extras constitute a fraction of a pound.

Chemical Composition of Nutrients The simplest of the nutrients are the minerals. Each mineral is a chemical element; its atoms are all alike. As a result, its identity never changes. Iron may change its form, for example, but it remains iron when a food is cooked, when a person eats the food, when iron becomes part of a red blood cell, when the cell is broken down, and when the iron is lost from the body by excretion. The next simplest nutrient is water, a compound made of two elements—hydrogen and oxygen. Minerals and water are **inorganic** nutrients—they contain no carbon.

The other four classes of nutrients (carbohydrates, lipids, proteins, and vitamins) are more complex. In addition to hydrogen and oxygen, they all contain carbon, an element found in all living things. They are therefore called **organic** compounds (meaning, literally, "alive"). Protein and some vitamins also contain nitrogen and may contain other elements as well (see Table 1-1).

■ As Chapter 5 explains, most lipids are fats.

---

energy: the capacity to do work. The energy in food is chemical energy. The body can convert this chemical energy to mechanical, electrical, or heat energy.

nutrients: chemical substances obtained from food and used in the body to provide energy, structural materials, and regulating agents to support growth, maintenance, and repair of the body's tissues. Nutrients may also reduce the risks of some diseases.

phytochemicals (FIE-toe-KEM-ih-cals): nonnutrient compounds found in plant-derived foods that have biological activity in the body.
  • phyto = plant

nonnutrients: compounds in foods that do not fit within the six classes of nutrients.

inorganic: not containing carbon or pertaining to living things.
  • in = not

organic: in chemistry, a substance or molecule containing carbon-carbon bonds or carbon-hydrogen bonds.* In agriculture, organic means growing crops and raising livestock according to U.S. Department of Agriculture (USDA) standards (see Chapter 19).

* This definition excludes coal, diamonds, and a few carbon-containing compounds that contain only a single carbon and no hydrogen, such as carbon dioxide (CO_2), calcium carbonate ($CaCO_3$), magnesium carbonate ($MgCO_3$), and sodium cyanide ($NaCN$).

FIGURE 1-1 Body Composition of Healthy-Weight Men and Women

The human body is made of compounds similar to those found in foods—mostly water (60 percent) and some fat (13 to 21 percent for young men, 23 to 31 percent for young women), with carbohydrate, protein, vitamins, minerals, and other minor constituents making up the remainder. (Chapter 8 describes the health hazards of too little or too much body fat.)

Essential Nutrients The body can make some nutrients, but it cannot make all of them, and it makes some in insufficient quantities to meet its needs. It must obtain these nutrients from foods. The nutrients that foods must supply are **essential nutrients.** When used to refer to nutrients, the word *essential* means more than just "necessary"; it means "needed from outside the body"—normally, from foods.

The Energy-Yielding Nutrients

In the body, three of the organic nutrients can be used to provide energy: carbohydrate, fat, and protein.■ In contrast to these **energy-yielding nutrients,** vitamins, minerals, and water do not yield energy in the human body.

■ Carbohydrate, fat, and protein are sometimes called **macronutrients** because they are required by the body in relatively large amounts (many grams daily). In contrast, vitamins and minerals are **micronutrients,** required in small amounts (milligrams or micrograms daily).

TABLE 1-1 Elements in the Six Classes of Nutrients

Notice that organic nutrients contain carbon.

	Carbon	Hydrogen	Oxygen	Nitrogen	Minerals
Inorganic nutrients					
Minerals					✓
Water		✓	✓		
Organic nutrients					
Carbohydrates	✓	✓	✓		
Lipids (fats)	✓	✓	✓		
Proteins[a]	✓	✓	✓	✓	
Vitamins[b]	✓	✓	✓		

[a] Some proteins also contain the mineral sulfur.
[b] Some vitamins contain nitrogen; some contain minerals.

essential nutrients: nutrients a person must obtain from food because the body cannot make them for itself in sufficient quantity to meet physiological needs; also called **indispensable nutrients.** About 40 nutrients are currently known to be essential for human beings.

energy-yielding nutrients: the nutrients that break down to yield energy the body can use:
• Carbohydrate.
• Fat.
• Protein.

HOW TO | Think Metric

Like other scientists, nutrition scientists use metric units of measure. They measure food energy in kilocalories, people's height in centimeters, people's weight in kilograms, and the weights of foods and nutrients in grams, milligrams, or micrograms. For ease in using these measures, it helps to remember that the prefixes on the grams imply 1000. For example, a *kilo*gram is 1000 grams, a *milli*gram is 1/1000 of a gram, and a *micro*gram is 1/1000 of a milligram.

Most food labels and many recipe books provide "dual measures," listing both household measures, such as cups, quarts, and teaspoons, and metric measures, such as milliliters, liters, and grams. This practice gives people an opportunity to gradually learn to "think metric."

A person might begin to "think metric" by simply observing the measure—by noticing the amount of soda in a 2-liter bottle, for example. Through such experiences, a person can become familiar with a measure without having to do any conversions.

To facilitate communication, many members of the international scientific community have adopted a common system of measurement—the International System of Units (SI). In addition to using metric measures, the SI establishes common units of measurement. For example, the SI unit for measuring food energy is the joule (not the kcalorie). A joule is the amount of energy expended when 1 kilogram is moved 1 meter by a force of 1 newton. The joule is thus a measure of *work* energy, whereas the kcalorie is a measure of *heat* energy. While many scientists and journals report their findings in kilojoules (kJ), many others, particularly those in the United States, use kcalories. To convert energy measures from kcalories to kilojoules, multiply by 4.2. For example, a 50-kcalorie cookie provides 210 kilojoules:

$$50 \text{ kcal} \times 4.2 = 210 \text{ kJ}.$$

Exact conversion factors for these and other units of measure are in the Aids to Calculation section on the last two pages of the book.

Volume: Liters (L)

1L = 1000 milliliters (mL).
0.95 L = 1 quart.
1 mL = 0.03 fluid ounces.
240 mL = 1 cup.

A liter of liquid is approximately one U.S. quart. (Four liters are only about 5 percent more than a gallon.)

One cup is about 240 milliliters; a half-cup of liquid is about 120 milliliters.

Weight: Grams (g)

1 g = 1000 milligrams (mg).
1 g = 0.04 ounce (oz).
1 oz = 28.35 g or ≈ 30 g.
100 g ≈ 3½ oz.
1 kilogram (kg) = 1000 g.
1 kg = 2.2 pounds (lb).
454 g = 1 lb.

A kilogram is slightly more than 2 lb; conversely, a pound is about ½ kg.

A half-cup of vegetables weighs about 100 grams; one pea weighs about ½ gram.

A 5-pound bag of potatoes weighs about 2 kilograms, and a 176-pound person weighs 80 kilograms.

Energy Measured in kCalories The energy released from carbohydrates, fats, and proteins can be measured in **calories**—tiny units of energy so small that a single apple provides tens of thousands of them. To ease calculations, energy is expressed in 1000-calorie metric units known as kilocalories (shortened to kcalories, but commonly called "calories"). When you read in popular books or magazines that an apple provides "100 calories," understand that it means 100 kcalories. This book uses the term *kcalorie* and its abbreviation *kcal* throughout, as do other scientific books and journals. The accompanying "How to" provides a few tips on how to "think metric."

calories: units by which energy is measured. Food energy is measured in **kilocalories** (1000 calories equal 1 kilocalorie), abbreviated **kcalories** or **kcal.** One kcalorie is the amount of heat necessary to raise the temperature of 1 kilogram (kg) of water 1°C. The scientific use of the term *kcalorie* is the same as the popular use of the term *calorie*.

Energy from Foods The amount of energy a food provides depends on how much carbohydrate, fat, and protein it contains. When completely broken down

HOW TO Calculate the Energy Available from Foods

To calculate the energy available from a food, multiply the number of grams of carbohydrate, protein, and fat by 4, 4, and 9, respectively. Then add the results together. For example, 1 slice of bread with 1 tablespoon of peanut butter on it contains 16 grams carbohydrate, 7 grams protein, and 9 grams fat:

$$16 \text{ g carbohydrate} \times 4 \text{ kcal/g} = 64 \text{ kcal.}$$
$$7 \text{ g protein} \times 4 \text{ kcal/g} = 28 \text{ kcal.}$$
$$9 \text{ g fat} \times 9 \text{ kcal/g} = 81 \text{ kcal.}$$
$$\text{Total} = 173 \text{ kcal.}$$

From this information, you can calculate the percentage of kcalories each of the energy nutrients contributes to the total. To determine the percentage of kcalories from fat, for example, divide the 81 fat kcalories by the total 173 kcalories:

$$81 \text{ fat kcal} \div 173 \text{ total kcal} = 0.468$$
(rounded to 0.47).

Then multiply by 100 to get the percentage:

$$0.47 \times 100 = 47\%.$$

Dietary recommendations that urge people to limit fat intake to 20 to 35 percent of kcalories refer to the day's total energy intake, not to individual foods. Still, if the proportion of fat in each food choice throughout a day exceeds 35 percent of kcalories, then the day's total surely will, too. Knowing that this snack provides 47 percent of its kcalories from fat alerts a person to the need to make lower-fat selections at other times that day. ■

in the body, a gram of carbohydrate yields about 4 kcalories of energy; a gram of protein also yields 4 kcalories; and a gram of fat yields 9 kcalories (see Table 1-2). Fat, therefore, has a greater **energy density** than either carbohydrate or protein. Figure 1-2 (on p. 10) compares the energy density of two breakfast options, and later chapters describe how considering a food's energy density can help with weight management.■ The accompanying "How to" explains how to calculate the energy available from foods.

One other substance contributes energy: alcohol. Alcohol is not considered a nutrient because it interferes with the growth, maintenance, and repair of the body, but it does yield energy (7 kcalories per gram) when metabolized in the body. (Highlight 7 and Chapter 18 present the potential harms and possible benefits of alcohol consumption.)

Most foods contain all three energy-yielding nutrients, as well as water, vitamins, minerals, and other substances. For example, meat contains water, fat, vitamins, and minerals as well as protein. Bread contains water, a trace of fat, a little protein, and some vitamins and minerals in addition to its carbohydrate. Only a few foods are exceptions to this rule, the common ones being sugar (pure carbohydrate) and oil (essentially pure fat).

Energy in the Body The body uses the energy-yielding nutrients to fuel all its activities. When the body uses carbohydrate, fat, or protein for energy, the bonds between the nutrient's atoms break. As the bonds break, they release energy.■ Some of this energy is released as heat, but some is used to send electrical impulses through the brain and nerves, to synthesize body compounds, and to move muscles. Thus the energy from food supports every activity from quiet thought to vigorous sports.

If the body does not use these nutrients to fuel its current activities, it rearranges them into storage compounds (such as body fat), to be used between meals and overnight when fresh energy supplies run low. If more energy is consumed than expended, the result is an increase in energy stores and weight gain. Similarly, if less energy is consumed than expended, the result is a decrease in energy stores and weight loss.

When consumed in excess of energy need, alcohol, too, can be converted to body fat and stored. When alcohol contributes a substantial portion of the energy in a person's diet, the harm it does far exceeds the problems of excess body fat. (Highlight 7 describes the effects of alcohol on health and nutrition.)

■ Foods with a high energy density help with weight gain, whereas those with a low energy density help with weight loss.

■ The processes by which nutrients are broken down to yield energy or used to make body structures are known as **metabolism** (defined and described further in Chapter 7).

TABLE 1-2 kCalorie Values of Energy Nutrients

Energy Nutrients	kCalories[a] (per gram)
Carbohydrate	4 kcal/g
Fat	9 kcal/g
Protein	4 kcal/g

NOTE: Alcohol contributes 7 kcalories per gram that can be used for energy, but it is not considered a nutrient because it interferes with the body's growth, maintenance, and repair.
[a] For those using kilojoules: 1 g carbohydrate = 17 kJ; 1 g protein = 17 kJ; 1 g fat = 37 kJ; and 1 g alcohol = 29 kJ.

energy density: a measure of the energy a food provides relative to the amount of food (kcalories per gram).

FIGURE 1-2 | Energy Density of Two Breakfast Options Compared

Gram for gram, ounce for ounce, and bite for bite, foods with a high energy density deliver more kcalories than foods with a low energy density. Both of these breakfast options provide 500 kcalories, but the cereal with milk, fruit salad, scrambled egg, turkey sausage, and toast with jam offers three times as much food as the doughnuts (based on weight); it has a lower energy density than the doughnuts. Selecting a variety of foods also helps to ensure nutrient adequacy.

© Matthew Farruggio (both)

LOWER ENERGY DENSITY

This 450-gram breakfast delivers 500 kcalories, for an energy density of 1.1 (500 kcal ÷ 450 g = 1.1 kcal/g).

HIGHER ENERGY DENSITY

This 144-gram breakfast also delivers 500 kcalories, for an energy density of 3.5 (500 kcal ÷ 144 g = 3.5 kcal/g).

Other Roles of Energy-Yielding Nutrients In addition to providing energy, carbohydrates, fats, and proteins provide the raw materials for building the body's tissues and regulating its many activities. In fact, protein's role as a fuel source is relatively minor compared with both the other two nutrients and its other roles. Proteins are found in structures such as the muscles and skin and help to regulate activities such as digestion and energy metabolism.

The Vitamins

The **vitamins** are also organic, but they do not provide energy. Instead, they facilitate the release of energy from carbohydrate, fat, and protein and participate in numerous other activities throughout the body.

There are 13 different vitamins, each with its own special roles to play.* One vitamin enables the eyes to see in dim light, another helps protect the lungs from air pollution, and still another helps make the sex hormones—among other things. When you cut yourself, one vitamin helps stop the bleeding and another helps repair the skin. Vitamins busily help replace old red blood cells and the lining of the digestive tract. Almost every action in the body requires the assistance of vitamins.

Vitamins can function only if they are intact, but because they are complex organic molecules, they are vulnerable to destruction by heat, light, and chemical agents. This is why the body handles them carefully, and why nutrition-wise cooks do, too. The strategies of cooking vegetables at moderate temperatures, using small amounts of water, and for short times all help to preserve the vitamins.

vitamins: organic, essential nutrients required in small amounts by the body for health.

* The water-soluble vitamins are vitamin C and the eight B vitamins: thiamin, riboflavin, niacin, vitamins B_6 and B_{12}, folate, biotin, and pantothenic acid. The fat-soluble vitamins are vitamins A, D, E, and K. The water-soluble vitamins are the subject of Chapter 10 and the fat-soluble vitamins, of Chapter 11.

The Minerals

In the body, some **minerals** are put together in orderly arrays in such structures as bones and teeth. Minerals are also found in the fluids of the body and influence their properties. Whatever their roles, minerals do not yield energy.

Some 16 minerals are known to be essential in human nutrition.* Others are still being studied to determine whether they play significant roles in the human body. Still other minerals are *not* essential nutrients, but are important nevertheless because they are environmental contaminants that displace the nutrient minerals from their workplaces in the body, disrupting body functions. The problems caused by contaminant minerals are described in Chapter 13.

Because minerals are inorganic, they are indestructible and need not be handled with the special care that vitamins require. Minerals can, however, be bound by substances that interfere with the body's ability to absorb them. They can also be lost during food-refining processes or during cooking when they leach into water that is discarded.

© Ron Chapple/Taxi/Getty Images

Water itself is an essential nutrient and naturally carries many minerals.

Water

Water, indispensable and abundant, provides the environment in which nearly all the body's activities are conducted. It participates in many metabolic reactions and supplies the medium for transporting vital materials to cells and waste products away from them. Water is discussed fully in Chapter 12, but it is mentioned in every chapter. If you watch for it, you cannot help but be impressed by water's participation in all life processes.

IN SUMMARY Foods provide nutrients—substances that support the growth, maintenance, and repair of the body's tissues. The six classes of nutrients include:

- Carbohydrates.
- Lipids (fats).
- Proteins.
- Vitamins.
- Minerals.
- Water.

Foods rich in the energy-yielding nutrients (carbohydrates, fats, and proteins) provide the major materials for building the body's tissues and yield energy for the body's use or storage. Energy is measured in kcalories. Vitamins, minerals, and water facilitate a variety of activities in the body. Without exaggeration, nutrients provide the physical and metabolic basis for nearly all that we are and all that we do.

The Science of Nutrition

The science of nutrition is the study of the nutrients and other substances in foods and the body's handling of them. Its foundation depends on several other sciences, including biology, biochemistry, and physiology. As sciences go, nutrition is a young one, but as you can see from the size of this book, much has happened in nutrition's short life. And it is currently entering a tremendous growth spurt as

minerals: inorganic elements. Some minerals are essential nutrients required in small amounts by the body for health.

---

* The major minerals are calcium, phosphorus, potassium, sodium, chloride, magnesium, and sulfur. The trace minerals are iron, iodine, zinc, chromium, selenium, fluoride, molybdenum, copper, and manganese. Chapters 12 and 13 are devoted to the major and trace minerals, respectively.

FIGURE 1-3   The Scientific Method

In conducting research, scientists follow the scientific method. Most often, research generates additional problems and questions. Thus the sequence begins anew, and research continues in a never-ending, somewhat cyclical way.

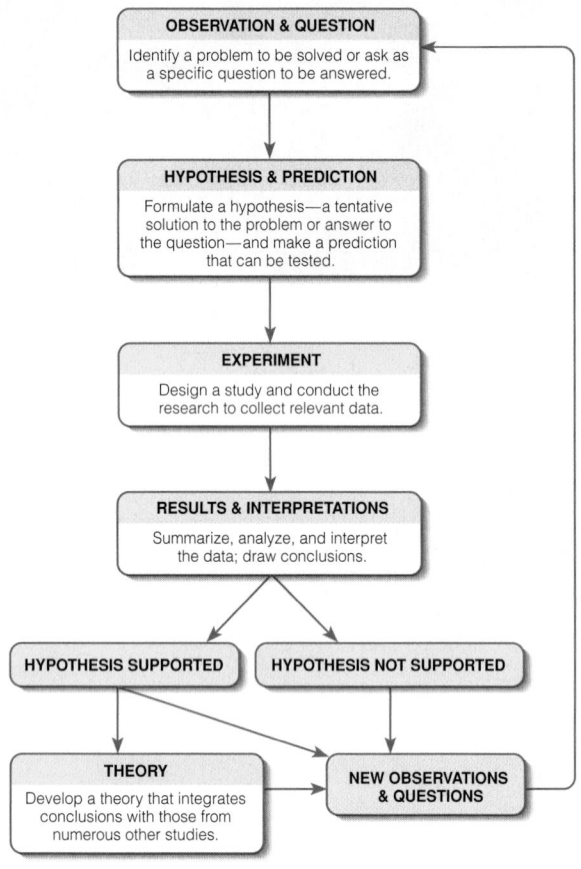

scientists apply knowledge gained from sequencing the human **genome**. The integration of nutrition, genomics, and molecular biology has opened up a whole new world of study called **nutritional genomics**—the science of how nutrients affect the activities of genes and how genes affect the activities of nutrients.[6] Look for examples of these interactions and of how nutritional genomics is shaping the science of nutrition in later sections of the book. This section introduces the research methods scientists have used in uncovering the wonders of nutrition.

Nutrition Research

Researchers use the scientific method to guide their work (see Figure 1-3). As the figure shows, research always begins with a problem or a question. For example, "What foods or nutrients might protect against the common cold?" In search of an answer, scientists make an educated guess (hypothesis), such as "foods rich in vitamin C reduce the number of common colds." Then they systematically conduct research studies to collect data that will test the **hypothesis** (see the glossary on p. 14 for definitions of research terms). Some examples of various types of research designs are presented in Figure 1-4. Each type of study has strengths and weaknesses (see Table 1-3 on p. 14). Consequently, some provide stronger evidence than others. Findings must be analyzed and interpreted with an awareness of each study's limitations. Importantly, scientists must be cautious about drawing any conclusions until they have accumulated a body of evidence from multiple studies that have used various types of research designs. As evidence accumulates, scientists begin to develop a **theory** that integrates the various findings and explains the complex relationships. (See Highlight 1 for a discussion of how to evaluate research findings.)

In attempting to discover whether a nutrient relieves symptoms or cures a disease, researchers deliberately manipulate one variable (for example, the amount of vitamin C in the diet) and measure any observed changes (perhaps the number of colds). As much as possible, all other conditions are held constant. The following paragraphs illustrate how this is accomplished using research on vitamin C and the common cold as an example.

Controls In studies examining the effectiveness of vitamin C, researchers typically divide the **subjects** into two groups. One group (the **experimental group**) receives a vitamin C supplement, and the other (the **control group**) does not. Researchers observe both groups to determine whether the vitamin C group has fewer or shorter colds than the control group. A number of pitfalls are inherent in an experiment of this kind and must be avoided.

First, each person must have an equal chance of being assigned to either the experimental group or the control group. This is accomplished by **randomization**; that is, the members are chosen from the same population by flipping a coin or some other method involving chance.

Importantly, the two groups of people must be similar and must have the same track record with respect to colds to rule out the possibility that observed differences in the rate, severity, or duration of colds might have occurred anyway. If, for example, the control group would normally catch twice as many colds as the experimental group, then the findings prove nothing.

In experiments involving a nutrient, the diets of both groups must also be similar, especially with respect to the nutrient being studied. If those in the experimental group were receiving less vitamin C from their diet, then the effects of the supplement may not be apparent.

genome (GEE-nome): the full complement of genetic material (DNA) in the chromosomes of a cell. In human beings, the genome consists of 23 pairs of chromosomes. The study of genomes is **genomics.**

nutritional genomics: the science of how nutrients affect the activities of genes and how genes affect the activities of nutrients.

FIGURE 1-4 Examples of Research Designs

EPIDEMIOLOGICAL STUDIES

CROSS-SECTIONAL

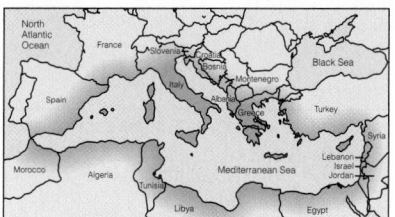

Researchers observe how much and what kinds of foods a group of people eat and how healthy those people are. Their findings identify factors that might influence the incidence of a disease in various populations.

Example. The people of the Mediterranean region drink lots of wine, eat plenty of fat from olive oil, and have a lower incidence of heart disease than northern Europeans and North Americans.

CASE-CONTROL

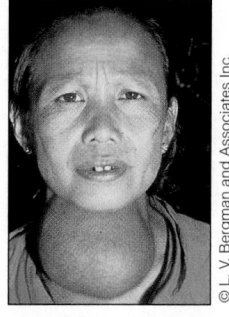

© L. V. Bergman and Associates Inc.

Researchers compare people who do and do not have a given condition such as a disease, closely matching them in age, gender, and other key variables so that differences in other factors will stand out. These differences may account for the condition in the group that has it.

Example. People with goiter lack iodine in their diets.

COHORT

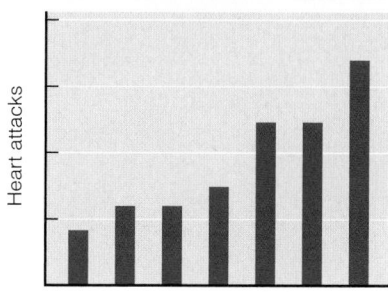

Researchers analyze data collected from a selected group of people (a cohort) at intervals over a certain period of time.

Example. Data collected periodically over the past several decades from over 5000 people randomly selected from the town of Framingham, Massachusetts, in 1948 have revealed that the risk of heart attack increases as blood cholesterol increases.

EXPERIMENTAL STUDIES

LABORATORY-BASED ANIMAL STUDIES

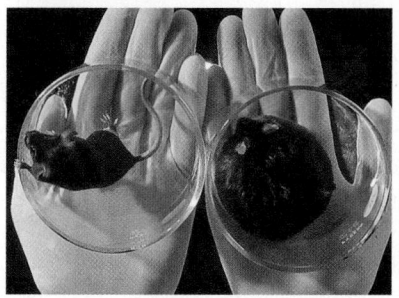

© R. Benali/Gamma

Researchers feed animals special diets that provide or omit specific nutrients and then observe any changes in health. Such studies test possible disease causes and treatments in a laboratory where all conditions can be controlled.

Example. Mice fed a high-fat diet eat less food than mice given a lower-fat diet, so they receive the same number of kcalories—but the mice eating the fat-rich diet become severely obese.

LABORATORY-BASED IN VITRO STUDIES

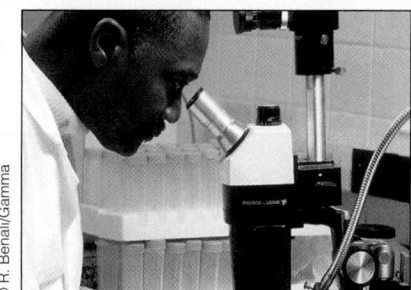

Researchers examine the effects of a specific variable on a tissue, cell, or molecule isolated from a living organism.

Example. Laboratory studies find that fish oils inhibit the growth and activity of the bacteria implicated in ulcer formation.

HUMAN INTERVENTION (OR CLINICAL) TRIALS

USDA Agricultural Research Service

© 2001 Photo Disc Inc.

Researchers ask people to adopt a new behavior (for example, eat a citrus fruit, take a vitamin C supplement, or exercise daily). These trials help determine the effectiveness of such interventions on the development or prevention of disease.

Example. Heart disease risk factors improve when men receive fresh-squeezed orange juice daily for two months compared with those on a diet low in vitamin C—even when both groups follow a diet high in saturated fat.

Sample Size To ensure that chance variation between the two groups does not influence the results, the groups must be large. If one member of a group of five people catches a bad cold by chance, he will pull the whole group's average toward bad colds; but if one member of a group of 500 catches a bad cold, she will not unduly affect the group average. Statistical methods are used to determine whether differences between groups of various sizes support a hypothesis.

TABLE 1-3   Strengths and Weaknesses of Research Designs

Type of Research	Strengths	Weaknesses
Epidemiological studies determine the incidence and distribution of diseases in a population. Epidemiological studies include cross-sectional, case-control, and cohort (see Figure 1-4).	• Can narrow down the list of possible causes • Can raise questions to pursue through other types of studies	• Cannot control variables that may influence the development or the prevention of a disease • Cannot prove cause and effect
Laboratory-based studies explore the effects of a specific variable on a tissue, cell, or molecule. Laboratory-based studies are often conducted in test tubes (in vitro) or on animals.	• Can control conditions • Can determine effects of a variable	• Cannot apply results from test tubes or animals to human beings
Human intervention or **clinical trials** involve human beings who follow a specified regimen.	• Can control conditions (for the most part) • Can apply findings to some groups of human beings	• Cannot generalize findings to all human beings • Cannot use certain treatments for clinical or ethical reasons

Placebos If people take vitamin C for colds and *believe* it will cure them, their chances of recovery may improve. Taking anything believed to be beneficial may hasten recovery. This phenomenon, the result of expectations, is known as the **placebo effect**. In experiments designed to determine vitamin C's effect on colds, this mind-body effect must be rigorously controlled. Severity of symptoms is often a subjective measure, and people who believe they are receiving treatment may report less severe symptoms.

GLOSSARY OF RESEARCH TERMS

blind experiment: an experiment in which the subjects do not know whether they are members of the experimental group or the control group.

control group: a group of individuals similar in all possible respects to the experimental group except for the treatment. Ideally, the control group receives a placebo while the experimental group receives a real treatment.

correlation (CORE-ee-LAY-shun): the simultaneous increase, decrease, or change in two variables. If A increases as B increases, or if A decreases as B decreases, the correlation is **positive**. (This does not mean that A causes B or vice versa.) If A increases as B decreases, or if A decreases as B increases, the correlation is **negative**. (This does not mean that A prevents B or vice versa.) Some third factor may account for both A and B.

double-blind experiment: an experiment in which neither the subjects nor the researchers know which subjects are members of the experimental group and which are serving as control subjects, until after the experiment is over.

experimental group: a group of individuals similar in all possible respects to the control group except for the treatment. The experimental group receives the real treatment.

hypothesis (hi-POTH-eh-sis): an unproven statement that tentatively explains the relationships between two or more variables.

peer review: a process in which a panel of scientists rigorously evaluates a research study to assure that the scientific method was followed.

placebo (pla-SEE-bo): an inert, harmless medication given to provide comfort and hope; a sham treatment used in controlled research studies.

placebo effect: the result of expectations in the effectiveness of a medicine, even medicine without pharmaceutical effects.

randomization (RAN-dom-ih-ZAY-shun): a process of choosing the members of the experimental and control groups without bias.

replication (REP-lee-KAY-shun): repeating an experiment and getting the same results. The skeptical scientist, on hearing of a new, exciting finding, will ask, "Has it been replicated yet?" If it hasn't, the scientist will withhold judgment regarding the finding's validity.

subjects: the people or animals participating in a research project.

theory: a tentative explanation that integrates many and diverse findings to further the understanding of a defined topic.

validity (va-LID-ih-tee): having the quality of being founded on fact or evidence.

variables: factors that change. A variable may depend on another variable (for example, a child's height depends on his age), or it may be independent (for example, a child's height does not depend on the color of her eyes). Sometimes both variables correlate with a third variable (a child's height and eye color both depend on genetics).

One way experimenters control for the placebo effect is to give pills to all participants; those in the experimental group receive pills containing vitamin C, and those in the control group receive a **placebo**, pills of similar appearance and taste containing an inactive ingredient. This way, the expectations of both groups will be equivalent. It is not necessary to convince all subjects that they are receiving vitamin C, but the extent of belief or unbelief must be the same in both groups. A study conducted under these conditions is called a **blind experiment**—that is, the subjects do not know (are blind to) whether they are members of the experimental group (receiving treatment) or the control group (receiving the placebo).

Double Blind When both the subjects and the researchers do not know which subjects are in which group, the study is called a **double-blind experiment.** Being fallible human beings and having an emotional and sometimes financial investment in a successful outcome, researchers might record and interpret results with a bias in the expected direction. To prevent such bias, the pills are coded by a third party, who does not reveal to the experimenters which subjects were in which group until all results have been recorded.

Correlations and Causes Researchers often examine the relationships between two or more **variables**—for example, daily vitamin C intake and the number of colds or the duration and severity of cold symptoms. Importantly, researchers must be able to observe, measure, or verify the variables selected. Findings sometimes suggest no **correlation** between the two variables (regardless of the amount of vitamin C consumed, the number of colds remains the same). Other times, studies find either a **positive correlation** (the more vitamin C, the more colds) or a **negative correlation** (the more vitamin C, the fewer colds). Correlational evidence proves only that two variables are associated, not that one is the cause of the other. People often jump to conclusions when they learn of correlations, but the conclusions are often wrong. To prove that A causes B, scientists have to find evidence of the *mechanism*—that is, to catch A in the act of causing B, so to speak. Furthermore, other scientists must confirm or disprove the findings through **replication** before the results are accepted into the body of nutrition knowledge. Before the findings are published, they are subjected to **peer review**—a process whereby a panel of scientists critically evaluates the study to confirm that it followed standard scientific methods.

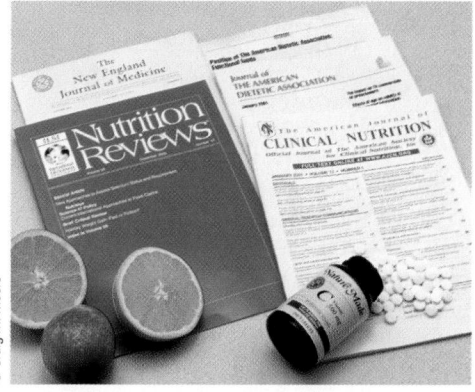

Knowledge about the nutrients and their effects on health comes from scientific study.

Research versus Rumors

In discussing these subtleties of experimental design, our intent is to show you what a far cry scientific **validity** is from the experience of your neighbor Sarah (sample size, one; no control group), who says she takes vitamin C when she feels a cold coming on and "it works every time."■ She knows what she is taking, she believes in its effectiveness, and she tends not to notice when it doesn't work. Before concluding that an experiment has shown that a nutrient cures a disease or alleviates a symptom, ask these questions:

- Who participated in the study, and how were participants selected?
- Were the control group and the experimental group similar?
- Was the sample size large enough to rule out chance variation?
- Was a placebo effectively administered (blind)?
- Was the experiment double blind?
- Were the variables selected appropriately and measured accurately?
- Do the data support the conclusions?

These characteristics of well-designed research have enabled scientists to study the actions of nutrients in the body. Such research has laid the foundation for quantifying how much of each nutrient the body needs.

■ A personal account of an experience or event is an **anecdote** and is not accepted as reliable scientific information.
- **anekdotos** = unpublished

IN SUMMARY Scientists learn about nutrition by conducting experiments that follow the protocol of scientific research. Researchers take care to establish similar control and experimental groups, large sample sizes, placebos, and blind treatments. Their findings must be reviewed and replicated by other scientists before being accepted as valid.

Dietary Reference Intakes

Nutrition experts have produced a set of standards that define the amounts of energy, nutrients, other dietary components, and physical activity that best support health. These recommendations are called **Dietary Reference Intakes (DRI)** and reflect the collaborative efforts of researchers in both the United States and Canada.[*7] The inside front covers provide a handy reference for DRI values.

Establishing Nutrient Recommendations

The DRI Committee consists of highly qualified scientists who base their estimates of nutrient needs on careful examination and interpretation of scientific evidence. These recommendations apply to healthy people and may not be appropriate for people with diseases that increase or decrease nutrient needs. The next several paragraphs discuss specific aspects of how the committee goes about establishing the values that make up the DRI:

- Estimated Average Requirements (EAR).
- Recommended Dietary Allowances (RDA).
- Adequate Intakes (AI).
- Tolerable Upper Intake Levels (UL).

Estimated Average Requirements (EAR) The committee reviews hundreds of research studies to determine the **requirement** for a nutrient—how much is needed in the diet. The committee selects a different criterion for each nutrient based on its roles both in performing activities in the body and in reducing disease risks.■ From this information, the committee determines an **Estimated Average Requirement (EAR)** for the nutrient—the average amount that appears sufficient to maintain a specific body function in half of the population.

An examination of all the available data reveals that each person's body is unique and has its own set of requirements. Men differ from women, and needs change as a person grows from infancy through old age. For this reason, the committee clusters its recommendations for people into groups by age and gender. Even so, the exact requirements of people the same age and gender are likely to be different. For example, person A might need 40 units of the nutrient each day; person B might need 35; person C, 57. A look at enough individuals might reveal that their requirements fall into a symmetrical distribution, with most near the midpoint (shown in Figure 1-5 as 45 units) and only a few at the extremes.

Recommended Dietary Allowances (RDA) Then the committee must decide what intake to recommend for everybody—the **Recommended Dietary Allowance (RDA)**. Assuming the distribution shown in Figure 1-5, the Estimated Average Requirement (shown in the figure as 45 units) for each nutrient is probably closest to everyone's need. But if people consumed exactly the average requirement of a given nutrient each day, half of the population would develop deficiencies of that nutrient; in Figure 1-5, person C would be among them. Recommendations should be set high enough above the Estimated Average Requirement to meet the needs of most healthy people.

Don't let the DRI's "alphabet soup" of nutrient intake standards confuse you. Their names make sense when you learn their purposes.

■ Research in nutritional genomics is expected to identify specific nutrient-gene interactions that will help estimate nutrient requirements more precisely.

Dietary Reference Intakes (DRI): a set of nutrient intake values for healthy people in the United States and Canada. These values are used for planning and assessing diets and include:
- Estimated Average Requirements (EAR).
- Recommended Dietary Allowances (RDA).
- Adequate Intakes (AI).
- Tolerable Upper Intake Levels (UL).

requirement: the lowest continuing intake of a nutrient that will maintain a specified criterion of adequacy.

Estimated Average Requirement (EAR): the average daily amount of a nutrient that will maintain a specific biochemical or physiological function in half the healthy people of a given age and gender group.

Recommended Dietary Allowance (RDA): the average daily amount of a nutrient considered adequate to meet the known nutrient needs of practically all healthy people; a goal for dietary intake by individuals.

* The DRI reports are produced by the Food and Nutrition Board, Institute of Medicine of the National Academies, with active involvement of scientists from Canada.

FIGURE 1-5 Estimated Average Requirements and Recommended Dietary Allowances Compared

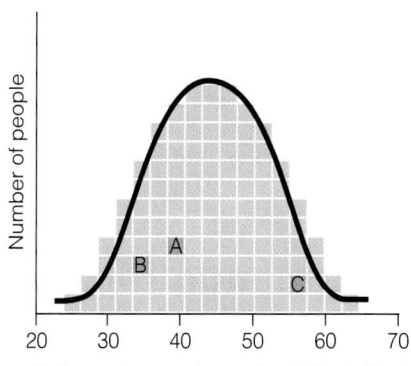

Each square represents a person. Some people require only a small amount of the nutrient, and some require a lot, but most fall somewhere near the middle. The text discusses three of these people: A, B, and C.

The RDA for a nutrient is set well above the Estimated Average Requirement. It covers about 98% of the population.

Small amounts above the daily requirement do no harm, whereas amounts below the requirement lead to health problems. When people's nutrient intakes are consistently **deficient** (less than the requirement), their nutrient stores decline, and over time this decline leads to poor health and deficiency symptoms. Therefore, to ensure that the nutrient RDA meet the needs of as many people as possible, the RDA are set near the top end of the range of the population's estimated requirements.

In this example, a reasonable RDA might be 63 units a day (see Figure 1-5). Such a point can be calculated mathematically so that it covers about 98 percent of a population. Almost everybody—including person C whose needs were higher than the average—would be covered if they met this dietary goal. Relatively few people's requirements would exceed this recommendation, and even then, they wouldn't exceed by much.

Adequate Intakes (AI) For some nutrients, there is insufficient scientific evidence to determine an Estimated Average Requirement (which is needed to set an RDA). In these cases, the committee establishes an **Adequate Intake (AI)** instead of an RDA. An AI reflects the average amount of a nutrient that a group of healthy people consumes. Like the RDA, the AI may be used as nutrient goals for individuals.

Although both the RDA and the AI serve as nutrient intake goals for individuals, their differences are noteworthy. An RDA for a given nutrient is based on enough scientific evidence to expect that the needs of almost all healthy people will be met. An AI, on the other hand, must rely more heavily on scientific judgments because sufficient evidence is lacking. The percentage of people covered by an AI is unknown; an AI is expected to exceed average requirements, but it may cover more or fewer people than an RDA would (if an RDA could be determined). For these reasons, AI values are more tentative than RDA. The table on the inside front cover identifies which nutrients have an RDA and which have an AI. Later chapters present the RDA and AI values for the vitamins and minerals.

Tolerable Upper Intake Levels (UL) As mentioned earlier, the recommended intakes for nutrients are generous, and although they do not necessarily cover every individual for every nutrient, they probably should not be exceeded by much. People's tolerances for high doses of nutrients vary, and somewhere above

deficient: the amount of a nutrient below which almost all healthy people can be expected, over time, to experience deficiency symptoms.

Adequate Intake (AI): the average daily amount of a nutrient that appears sufficient to maintain a specified criterion; a value used as a guide for nutrient intake when an RDA cannot be determined.

FIGURE 1-6 — Naive versus Accurate View of Nutrient Intakes

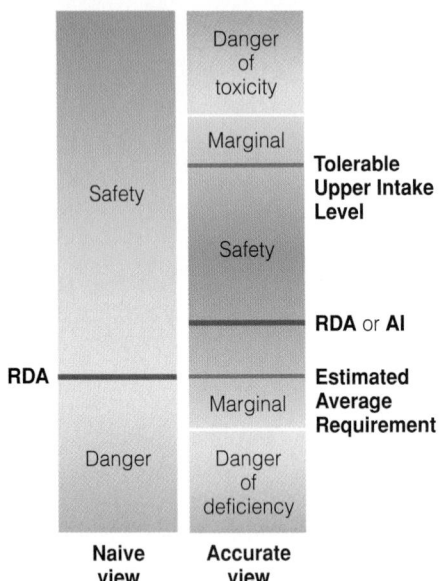

The RDA or AI for a given nutrient represents a point that lies within a range of appropriate and reasonable intakes between toxicity and deficiency. Both of these recommendations are high enough to provide reserves in times of short-term dietary inadequacies, but not so high as to approach toxicity. Nutrient intakes above or below this range may be equally harmful.

■ Reference adults:
  • Men: 19–30 yr, 5 ft 10 in, 154 lb.
  • Women: 19–30 yr, 5 ft 4 in, 126 lb.

Tolerable Upper Intake Level (UL): the maximum daily amount of a nutrient that appears safe for most healthy people and beyond which there is an increased risk of adverse health effects.

Estimated Energy Requirement (EER): the average dietary energy intake that maintains energy balance and good health in a person of a given age, gender, height, and level of physical activity.

Acceptable Macronutrient Distribution Ranges (AMDR): ranges of intakes for the energy nutrients that provide adequate energy and nutrients and reduce the risk of chronic diseases.

the recommended intake is a **Tolerable Upper Intake Level (UL)** beyond which a nutrient is likely to become toxic. It is naive—and inaccurate—to think of recommendations as minimum amounts. A more accurate view is to see a person's nutrient needs as falling within a range, with marginal and danger zones both below and above it (see Figure 1-6).

Upper levels are particularly useful in guarding against the overconsumption of nutrients, which may occur when people use large-dose supplements and fortified foods regularly. Later chapters discuss the dangers associated with excessively high intakes of vitamins and minerals, and the inside front cover presents a table that includes the upper-level values for selected nutrients.

Establishing Energy Recommendations

In contrast to the RDA and AI values for nutrients, the recommendation for energy is not generous. Energy's recommendation—called the **Estimated Energy Requirement (EER)**—is similar to the Estimated Average Requirement in that it is set at the *average* of the population's estimated requirements (see Figure 1-7).

Estimated Energy Requirement (EER) The Estimated Energy Requirement represents the average dietary energy intake (kcalories per day) that will maintain energy balance in a healthy person of a given age, gender, weight, height, and physical activity level.■ Balance is key to the energy recommendation. Enough energy is needed to sustain a healthy and active life, but too much energy can lead to weight gain and obesity. Because any amount in excess of needs results in weight gain, there is no upper level for energy.

Acceptable Macronutrient Distribution Ranges (AMDR) People don't eat energy directly; they derive energy from foods containing carbohydrate, fat, and protein. Each of these three energy-yielding nutrients contributes to the total energy intake, and those contributions vary in relation to each other. The DRI committee has determined that the composition of a diet that provides adequate energy and nutrients and reduces the risk of chronic diseases is:

• 45 to 65 percent from carbohydrate.

• 20 to 35 percent from fat.

• 10 to 35 percent from protein.

These values are known as **Acceptable Macronutrient Distribution Ranges (AMDR).**

FIGURE 1-7 — Recommended Intakes of Nutrients and Energy Compared

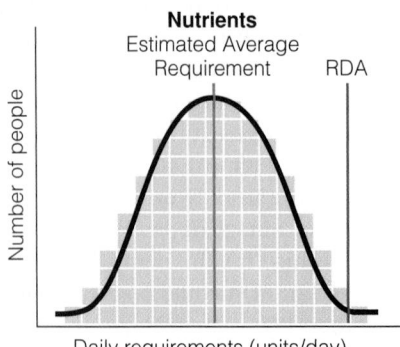

The nutrient intake recommendations are set high enough to cover nearly everyone's requirements (the boxes represent people).

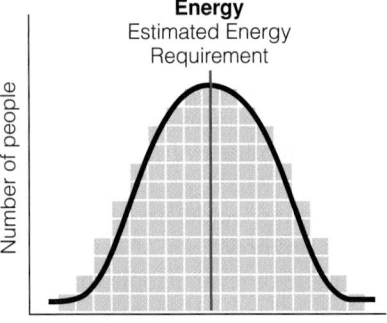

The recommended intake for energy is set at the average that will maintain energy balance in a healthy person of desirable body weight.

Using Nutrient Recommendations

Although the intent of nutrient recommendations may seem simple enough, they are the subject of much misunderstanding and controversy. Perhaps the following facts will help put them in perspective. First, estimates of adequate energy and nutrient intakes apply to *healthy* people. They need to be adjusted for malnourished people or those with medical problems who may require supplemented or restricted intakes.

Second, these *recommendations* are not minimum requirements, nor are they necessarily optimal intakes for all individuals. Recommendations can only target "most" of the people and cannot account for individual variations in nutrient needs—yet. Given the recent explosion of knowledge about genetics, the day may be fast approaching when nutrition scientists will be able to determine an individual's optimal nutrient needs.[8] Until then, registered dietitians■ and other qualified health professionals can help determine whether recommendations should be adjusted to meet individual needs.

Third, most nutrient goals are intended to be met through diets composed of a variety of *foods* whenever possible. Because foods contain mixtures of nutrients and nonnutrients, they deliver more than just those nutrients covered by the recommendations. Excess intakes of vitamins and minerals are unlikely when their sources are foods rather than supplements.

Fourth, recommendations apply to *average* daily intakes. Meeting recommendations for every nutrient every day is difficult and unnecessary. The length of time over which a person's intake can deviate from the average without risk of deficiency or overdose varies for each nutrient, depending on the body's use and storage of the nutrient. For most nutrients (such as thiamin and vitamin C), deprivation would lead to rapid development of deficiency symptoms (within days or weeks); for others (such as vitamin A and vitamin B_{12}), deficiencies would develop more slowly (over months or years).

Fifth, each of the DRI categories serves a unique purpose. For example, the Estimated Average Requirements are most appropriately used to develop and evaluate nutrition programs for *groups* such as schoolchildren or military personnel. The RDA (or AI if an RDA is not available) can be used to set goals for *individuals*. Tolerable Upper Intake Levels help to keep nutrient intakes below the amounts that increase the risk of toxicity. With these understandings, professionals can use the DRI for a variety of purposes.

Comparing Nutrient Recommendations

At least 40 different nations and international organizations have published nutrient standards similar to those used in the United States and Canada. Slight differences may be apparent, reflecting differences both in the interpretation of the data from which the standards were derived and in the food habits and physical activities of the populations they serve.

Many countries use the recommendations developed by two international groups: FAO (Food and Agriculture Organization) and WHO (World Health Organization).■ The FAO/WHO recommendations are considered sufficient to maintain health in nearly all healthy people worldwide.

■ A **registered dietitian** is a college-educated food and nutrition specialist who is qualified to evaluate people's nutritional health and needs. See Highlight 1 for more on what constitutes a nutrition expert.

■ Nutrient recommendations from FAO/WHO are provided in Appendix I.

IN SUMMARY The Dietary Reference Intakes (DRI) are a set of nutrient intake values that can be used to plan and evaluate diets for healthy people. The Estimated Average Requirement defines the amount of a nutrient that supports a specific function in the body for half of the population. The Recommended Dietary Allowance (RDA) is based on the Estimated Average Requirement and establishes a goal for dietary intake that will meet the needs of almost all healthy people. An Adequate Intake (AI) serves a similar purpose when an RDA cannot be determined. The Estimated Energy Requirement defines the average amount of energy intake needed to maintain energy balance, and the Acceptable Macronutrient Distribution Ranges define the proportions contributed by carbohydrate, fat, and protein to a healthy diet. The Tolerable Upper Intake Level establishes the highest amount that appears safe for regular consumption.

Nutrition Assessment

A peek inside the mouth provides clues to a person's nutrition status—an inflamed tongue indicating a B vitamin deficiency or mottled teeth revealing fluoride toxicity, for example.

What happens when a person doesn't get enough or gets too much of a nutrient or energy? If the deficiency or excess is significant over time, the person exhibits signs of **malnutrition.** With a deficiency of energy, the person may display the symptoms of **undernutrition** by becoming extremely thin, losing muscle tissue, and becoming prone to infection and disease. With a deficiency of a nutrient, the person may experience skin rashes, depression, hair loss, bleeding gums, muscle spasms, night blindness, or other symptoms. With an excess of energy, the person may become obese and vulnerable to diseases associated with **overnutrition** such as heart disease and diabetes. With a sudden nutrient overdose, the person may experience hot flashes, yellowing skin, a rapid heart rate, low blood pressure, or other symptoms. Similarly, regular intakes in excess of needs may also have adverse effects.

Malnutrition symptoms are easy to miss. They resemble the symptoms of other diseases: diarrhea, skin rashes, pain, and the like. But a person who has learned how to use assessment techniques to detect malnutrition can tell when these conditions are caused by poor nutrition and can take steps to correct it. This discussion presents the basics of nutrition assessment; many more details are offered in later chapters and in Appendix E.

Nutrition Assessment of Individuals

To prepare a **nutrition assessment,** a registered dietitian or other trained health care professional uses:

- Historical information.
- Anthropometric data.
- Physical examinations.
- Laboratory tests.

Each of these methods involves collecting data in various ways and interpreting each finding in relation to the others to create a total picture.

Historical Information One step in evaluating nutrition status is to obtain information about a person's history with respect to health status, socioeconomic status, drug use, and diet. The health history reflects a person's medical record and may reveal a disease that interferes with the person's ability to eat or the body's use of nutrients. The person's family history of major diseases is also noteworthy, especially for conditions such as heart disease that have a genetic tendency to run in families. Economic circumstances may show a financial inability to buy foods or inadequate kitchen facilities in which to prepare them. Social factors such as marital status, ethnic background, and educational level also influence food choices and nutrition status. A drug history may highlight possible diet-medication interactions that lead to nutrient deficiencies (as described in Highlight 17). A diet history can indicate whether the diet may be under- or oversupplying nutrients or energy.

malnutrition: any condition caused by excess or deficient food energy or nutrient intake or by an imbalance of nutrients.

- **mal** = bad

undernutrition: deficient energy or nutrients.

overnutrition: excess energy or nutrients.

nutrition assessment: a comprehensive analysis of a person's nutrition status that uses health, socioeconomic, drug, and diet histories; anthropometric measurements; physical examinations; and laboratory tests.

To take a diet history, the assessor collects data about the foods a person eats. The data may be collected by recording the foods the person has eaten over a period of 24 hours, three days, or a week or more or by asking what foods the person typically eats and how much of each. The days in the record have to be fairly typical of the person's diet, and portion sizes must be recorded accurately. To determine the amounts of nutrients consumed, the assessor usually enters the foods and their portion sizes into a computer using a diet analysis program. This step can also be done manually by looking up each food in a table of food composition such as Appendix H in this book. Then the assessor compares the calculated nutrient intakes with the DRI to determine the probability of adequacy (see Figure 1-8).[9] Alternatively, the diet history might be compared against standards such as the Food Guide Pyramid or *Dietary Guidelines* (described in Chapter 2).

An estimate of energy and nutrient intakes from a diet history, combined with other sources of information, can help confirm or rule out the *possibility* of suspected nutrition problems. A sufficient intake of a nutrient does not guarantee adequacy, and an insufficient intake does not always indicate a deficiency, but such findings warn of possible problems.

Anthropometric Data A second technique that may help to reveal nutrition problems is the taking of **anthropometric** measures such as height and weight. The assessor compares measurements taken on an individual with standards specific for gender and age or with previous measures on the same individual. (Chapter 8 presents information on body weight and its standards.)

Measurements taken periodically and compared with previous measurements reveal patterns and indicate trends in a person's overall nutrition status, but they provide little information about specific nutrients. Instead, measurements out of line with expectations may reveal such problems as growth failure in children, wasting or swelling of body tissues in adults, and obesity—conditions that may reflect energy or nutrient deficiencies or excesses.

Physical Examinations A third nutrition assessment technique is a physical examination that looks for clues to poor nutrition status. Every part of the body that can be inspected can offer such clues: the hair, eyes, skin, posture, tongue, fingernails, and others. The examination requires skill, for many physical signs can reflect more than one nutrient deficiency or toxicity or even nonnutrition conditions. Like the other assessment techniques, a physical examination does not by itself point to firm conclusions. Instead, it reveals possible nutrient imbalances for other assessment techniques to confirm, or it confirms data collected from other assessment measures.

Laboratory Tests A fourth way to detect a developing deficiency, imbalance, or toxicity is to take samples of blood or urine, analyze them in the laboratory, and compare the results with normal values for a similar population.■ A goal of nutrition assessment is to uncover early signs of malnutrition before symptoms appear, and laboratory tests are most useful for this purpose. In addition, they can confirm suspicions raised by other assessment methods.

Iron, for Example The mineral iron can be used to illustrate the stages in the development of a nutrient deficiency and the assessment techniques useful in detecting them. The **overt**, or outward, signs of an iron deficiency appear at the end of a long sequence of events. Figure 1-9 describes what happens in the body as a nutrient deficiency progresses and shows which assessment methods can reveal those changes.

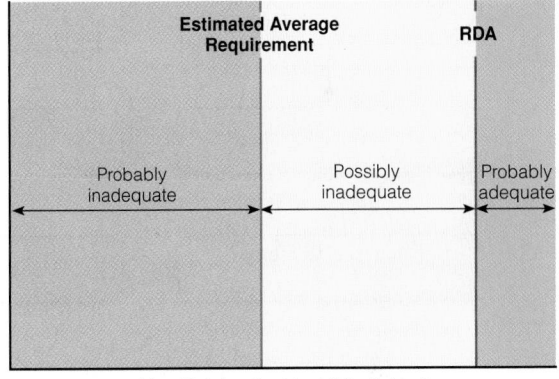

FIGURE 1-8 Using the DRI to Assess the Dietary Intake of a Healthy Individual

If a person's usual intake falls above the RDA, the intake is probably adequate because the RDA covers the needs of almost all people. A usual intake that falls between the RDA and the Estimated Average Requirement is more difficult to access; the intake may be adequate, but the chances are greater or equal that it is inadequate. If the usual intake falls below the Estimated Average Requirement, it is probably inadequate.

Usual intake of nutrient X (units/day)

■ Assessment may one day depend on measures of how a nutrient influences genetic activity within the cells, instead of quantities in the blood or other tissues.

anthropometric (AN-throw-poe-MET-rick): relating to measurement of the physical characteristics of the body, such as height and weight.
- **anthropos** = human
- **metric** = measuring

overt (oh-VERT): out in the open and easy to observe.
- **ouvrir** = to open

FIGURE 1-9 Stages in the Development of a Nutrient Deficiency

Internal changes precede outward signs of deficiencies. As a corollary, signs of sickness need not appear before a person takes corrective measures. Tests can either reveal the presence of problems in the early stages or confirm that nutrient stores are adequate.

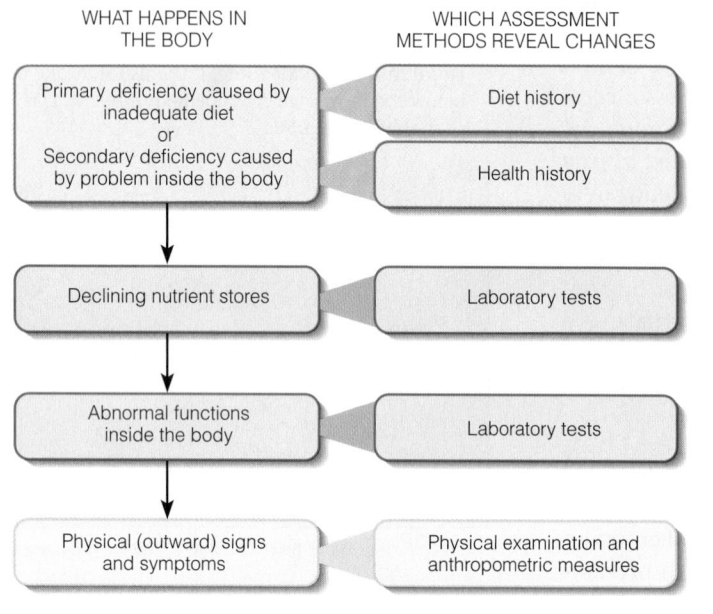

First, the body has too little iron—either because iron is lacking in the person's diet (a **primary deficiency**) or because the person's body doesn't absorb enough, excretes too much, or uses iron inefficiently (a **secondary deficiency**). A diet history provides clues to primary deficiencies; a health history provides clues to secondary deficiencies.

Then the body begins to use up its stores of iron. At this stage, the deficiency might be described as **subclinical**. It exists as a **covert** condition and might be detected by laboratory tests, but outward signs have not yet appeared.

Finally, iron stores are exhausted. Now, the body cannot make enough iron-containing red blood cells to replace those that are aging and dying. The iron in red blood cells normally carries oxygen to all the body's tissues. When iron is lacking, fewer red blood cells are made, the new ones are pale and small, and every part of the body feels the effects of an oxygen shortage. Now the overt symptoms of deficiency appear—weakness, fatigue, pallor, and headaches, reflecting the iron-deficient state of the blood. Physical examination would reveal these symptoms.

HEALTHY PEOPLE 2010

Increase the proportion of primary health care providers who provide nutrition assessment when appropriate and formulate a diet plan for those who need intervention.

■ Healthy People 2010 is described on p. 23.

primary deficiency: a nutrient deficiency caused by inadequate dietary intake of a nutrient.

secondary deficiency: a nutrient deficiency caused by something other than an inadequate intake such as a disease condition or drug interaction that reduces absorption, accelerates use, hastens excretion, or destroys the nutrient.

subclinical deficiency: a deficiency in the early stages, before the outward signs have appeared.

covert (KOH-vert): hidden, as if under covers.
• couvrir = to cover

Nutrition Assessment of Populations

To assess a population's nutrition status, researchers conduct surveys using techniques similar to those used on individuals. The data collected are then used by various agencies for numerous purposes, including the development of national health goals.

National Nutrition Surveys The National Nutrition Monitoring program coordinates the many nutrition-related activities of various federal agencies. One of its most recent projects is the integration of two major national surveys to provide comprehensive data efficiently.[10] One portion of the survey collects data on the kinds and amounts of foods people eat.* Then researchers calculate the energy and nutrients in the foods and compare the amounts consumed with a standard. The other portion examines the people themselves, using anthropometric measurements, physical examinations, and laboratory tests.†[11] The data provide valuable information on several nutrition-related conditions, such as growth retardation, heart disease, and nutrient deficiencies. National nutrition surveys often oversample high-risk groups (low-income families, pregnant women, adolescents, the elderly, African Americans, and Mexican Americans) in order to glean an accurate estimate of their health and nutrition status.

The resulting wealth of information from the national nutrition surveys is used for a variety of purposes. For example, Congress uses this information to establish

---

* This portion of the survey was formerly called the Continuing Survey of Food Intakes by Individuals (CSFII), popularly known as *What We Eat in America.*
† This portion of the survey is known as the National Health and Nutrition Examination Survey (NHANES).

public policy on nutrition education, food assistance programs, and the regulation of the food supply. Scientists use the information to establish research priorities. The food industry uses these data to guide decisions in public relations and product development.[12] The Dietary Reference Intakes and other major reports that examine the relationships between diet and health depend on information collected from these nutrition surveys. These data also provide the basis for developing and monitoring national health goals.

National Health Goals **Healthy People,** a program that identifies the nation's health priorities and guides policies that promote health and prevent disease, was initiated over 20 years ago. At the start of each decade, the program sets goals for improving the nation's health during the following ten years. The goals of Healthy People 2010 focus on "improving the quality of life and eliminating disparity in health among racial and ethnic groups."[13] Nutrition is one of many focus areas, each with numerous objectives. Table 1-4 lists the nutrition objectives for 2010. These and other nutrition-related objectives appear throughout the text where their subjects are discussed.

At the close of the twentieth century, the nation's progress toward meeting its Healthy People 2000 goals was mixed.[14] For almost 60 percent of the objectives, the population either met the target or was moving in the right direction. Successes included reductions in the incidence of food- and water-borne infections, oral and breast cancer, and infant mortality, for example. On the downside, the population was moving in the opposite direction of several key objectives, most notably for reducing overweight and increasing physical activity.

Surveys provide valuable information about the kinds of foods people are eating.

TABLE 1-4	Healthy People 2010 Nutrition and Overweight Objectives	HEALTHY PEOPLE 2010

- Increase the proportion of adults who are at a *healthy weight.*
- Reduce the proportion of adults who are *obese.*
- Reduce the proportion of children and adolescents who are *overweight* or *obese.*
- Reduce *growth retardation* among low-income children under age 5 years.
- Increase the proportion of persons aged 2 years and older who consume at least two daily servings of *fruit.*
- Increase the proportion of persons aged 2 years and older who consume at least three daily servings of *vegetables,* with at least one-third being dark green or orange vegetables.
- Increase the proportion of persons aged 2 years and older who consume at least six daily servings of *grain products,* with at least three being whole grains.
- Increase the proportion of persons aged 2 years and older who consume less than 10 percent of kcalories from *saturated fat.*
- Increase the proportion of persons aged 2 years and older who consume no more than 30 percent of kcalories from *total fat.*
- Increase the proportion of persons aged 2 years and older who consume 2400 mg or less of *sodium.*

- Increase the proportion of persons aged 2 years and older who meet dietary recommendations for *calcium.*
- Reduce *iron deficiency* among young children, females of childbearing age, and pregnant females.
- Reduce *anemia* among low-income pregnant females in their third trimester.
- Increase the proportion of children and adolescents aged 6 to 19 years whose intake of *meals and snacks at school* contributes to good overall dietary quality.
- Increase the proportion of schools that teach all essential *nutrition education* topics in one course.
- Increase the proportion of worksites that offer *nutrition or weight management classes or counseling.*
- Increase the proportion of primary care providers who provide nutrition *assessment* when appropriate and who formulate a diet plan for those who need *intervention.*
- Increase the proportion of physician office visits made by patients with a diagnosis of cardiovascular disease, diabetes, or hyperlipidemia that include *counseling or education related to diet and nutrition.*
- Increase *food security* among U.S. households and in so doing reduce hunger.

NOTE: "Nutrition and Overweight" is one of 28 focus areas, each with numerous objectives. Several of the other focus areas have nutrition-related objectives, and these are presented in later chapters.

SOURCE: Healthy People 2010, www.healthypeople.gov

Healthy People: a national public health initiative under the jurisdiction of the U.S. Department of Health and Human Services (DHHS) that identifies the most significant preventable threats to health and focuses efforts toward eliminating them.

IN SUMMARY People become malnourished when they get too little or too much energy or nutrients. Deficiencies, excesses, and imbalances of nutrients lead to malnutrition diseases. To detect malnutrition in individuals, health care professionals use four nutrition assessment methods. Reviewing dietary data and health information may suggest a nutrition problem in its earliest stages. Laboratory tests may detect it before it becomes overt, whereas anthropometrics and physical examinations pick up on the problem only after it is causing symptoms. Similar assessment methods are used in national surveys to measure people's food consumption and to evaluate the nutrition status of populations.

Diet and Health

Diet has always played a vital role in supporting health.■ Early nutrition research focused on identifying the nutrients in foods that would prevent such common diseases as rickets and scurvy, the vitamin D– and vitamin C–deficiency diseases. With this knowledge, developed countries have been successful in protecting against nutrient deficiency diseases. World hunger and nutrient deficiency diseases still pose a major health threat in developing countries, but not because of a lack of nutrition knowledge, as Chapter 20 explains. More recently, nutrition research has focused on **chronic diseases** associated with energy and nutrient excesses. Once thought to be "rich countries' problems," chronic diseases have become epidemic in developing countries as well—contributing to three out of five deaths worldwide.[15]

Chronic Diseases

Table 1-5 lists the ten leading causes of death in the United States. These "causes" are stated as if a single condition such as heart disease caused death, but most chronic diseases arise from multiple factors over many years. A person who died of heart disease may have been overweight, had high blood pressure, been a cigarette smoker, and spent years eating a diet high in saturated fat and getting too little exercise.

Of course, not all people who die of heart disease fit this description, nor do all people with these characteristics die of heart disease. People who are overweight might die from the complications of diabetes instead, or those who smoke might die of cancer. They might even die from something totally unrelated to any of these factors, such as an automobile accident. Still, statistical studies have shown that certain conditions and behaviors are linked to certain diseases.

Risk Factors for Chronic Diseases

Factors that increase or reduce the *risk* of developing chronic diseases are identified by analyzing statistical data. A strong association between a **risk factor** and a disease means that when the factor is present, the *likelihood* of developing the disease increases. It does not mean that all people with the risk factor will develop the disease. Similarly, a lack of risk factors does not guarantee freedom from a given disease. On the average, though, the more risk factors in a person's life, the greater that person's chances of developing the disease. Conversely, the fewer risk factors in a person's life, the better the chances for good health.

Risk Factors Persist Risk factors tend to persist over time. Without intervention, a young adult with high blood pressure will most likely continue to have

■ Nutritional genomics will provide a better understanding of the relationships among genes, foods, and health.

TABLE 1-5 Leading Causes of Death in the United States

	Percentage of Total Deaths
1. Heart disease	28.9
2. Cancers	22.9
3. Strokes	6.8
4. Chronic lung diseases	5.1
5. Accidents	4.0
6. Diabetes mellitus	2.9
7. Pneumonia and influenza	2.6
8. Alzheimer's disease	2.2
9. Kidney diseases	1.6
10. Blood infections	1.3

NOTE: The diseases highlighted in green have relationships with diet; yellow indicates a relationship with alcohol.

chronic diseases: diseases characterized by a slow progression and long duration. Examples include heart disease, cancer, and diabetes.

risk factor: a condition or behavior associated with an elevated frequency of a disease but not proved to be causal. Leading risk factors for chronic diseases include obesity, cigarette smoking, high blood pressure, high blood cholesterol, physical inactivity, and a diet high in saturated fats and low in vegetables, fruits, and whole grains.

TABLE 1-6	Factors Contributing to Deaths in the United States
Factors	**Percentage of Deaths**
Tobacco	20
Poor diet/inactivity	14
Alcohol	6
Microbial agents	4
Pollutants/toxins	3
Firearms	2
Sexual behavior	1
Motor vehicles	1
Illicit drugs	1

SOURCE: Centers for Disease Control, www.cdc.gov.

Physical activity can be both fun and beneficial.

high blood pressure as an older adult, for example. Thus, to minimize the damage, early intervention is most effective.

Risk Factors Cluster Risk factors also tend to cluster. For example, a person who is obese may be physically inactive, have high blood pressure, and have high blood cholesterol—all risk factors associated with heart disease. Intervention that focuses on one risk factor often benefits the others as well. For example, physical activity can help reduce weight. Then both physical activity and weight loss will help to lower blood pressure and blood cholesterol.

Risk Factors in Perspective The most prominent factor contributing to death in the United States is tobacco use,■ followed by diet and activity patterns, and alcohol use (see Table 1-6). Risk factors such as smoking, poor dietary habits, physical inactivity, and alcohol consumption are personal behaviors that can be changed. Decisions to not smoke, to eat a well-balanced diet, to engage in regular physical activity, and to drink alcohol in moderation (if at all) improve the likelihood that a person will enjoy good health. Other risk factors, such as genetics,■ gender, and age, also play important roles in the development of chronic diseases, but they cannot be changed. Health recommendations acknowledge the influence of such factors on the development of disease, but must focus on those that are changeable. For the two out of three Americans who do not smoke or drink alcohol excessively, the one choice that can influence long-term health prospects more than any other is diet.[16]

■ Cigarette smoking is responsible for almost one of every five deaths each year.

■ Chapter 18 considers how findings from the Human Genome Project will influence health care.

IN SUMMARY Within the range set by genetics, a person's choice of diet influences long-term health. Diet has no influence on some diseases, but is linked closely to others. Personal life choices, such as engaging in physical activity and using tobacco or alcohol, also affect health for the better or worse.

Increase the proportion of persons appropriately counseled about health behaviors.

HEALTHY PEOPLE 2010

The next several chapters will provide many more details about nutrients and how they support health. Whenever appropriate, the discussion will show how diet influences each of today's major diseases. Dietary recommendations will appear again and again, as each nutrient's relationships with health are explored. Most people who follow the recommendations will benefit and can enjoy good health into their later years.

Nutrition in Your Life

Your food choices play a key role in keeping you healthy and reducing your risk of chronic diseases.

- What factors most influence your food choices? How often do you think of health and nutrition when choosing foods?

- Do you or members of your family have any of the chronic disease risk factors and conditions (listed on p. 24 in the definition)? If so, which ones?

- What lifestyle changes could you make to improve your chances of enjoying good health?

NUTRITION ON THE NET

 Access these websites for further study of topics covered in this chapter.

- Find updates and quick links to these and other nutrition-related sites at our website: **www.wadsworth.com/nutrition**

- Search for "nutrition" at the U.S. Government health and nutrition information sites: **www.healthfinder.gov** or **www.nutrition.gov**

- Learn more about basic science research from the National Science Foundation and Research!America: **www.nsf.gov** and **researchamerica.org**

- Review the Dietary Reference Intakes: **www.nap.edu**

- Review nutrition recommendations from the Food and Agriculture Organization and the World Health Organization: **www.fao.org** and **www.who.org**

- View Healthy People 2010: **www.healthypeople.gov**

- Visit the Food and Nutrition section of the Healthy Living area in Health Canada: **www.hc-sc.gc.ca**

- Learn about the national nutrition survey: **www.cdc.gov/nchs/nhanes.htm**

- Get information from the Food Surveys Research Group: **www.barc.usda.gov/bhnrc/foodsurvey**

- Visit the food and nutrition center of the Mayo Clinic: **www.mayohealth.org**

- Find reviews of, and links to, nutrition and health websites by Tufts University Nutrition Navigator: **navigator.tufts.edu**

NUTRITION CALCULATIONS

Several chapters end with problems to give you practice in doing simple nutrition-related calculations. Although the situations are hypothetical, the numbers are real, and calculating the answers (check them on p. 29) provides a valuable nutrition lesson. Once you have mastered these examples, you will be prepared to examine your own food choices. Be sure to show your calculations for each problem.

1. Calculate the energy provided by a food's energy-nutrient contents. A cup of fried rice contains 5 grams protein, 30 grams carbohydrate, and 11 grams fat.

   a. How many kcalories does the rice provide from these energy nutrients?

   —————————— = —— kcal protein.
   —————————— = —— kcal carbohydrate.
   —————————— = —— kcal fat.

   Total = —— kcal.

   b. What percentage of the energy in the fried rice comes from each of the energy-yielding nutrients?

   —————————— = —— % kcal from protein.
   —————————— = —— % kcal from carbohydrate.
   —————————— = —— % kcal from fat.

   Total = —— %

Note: The total should add up to 100%; 99% or 101% due to rounding is also acceptable.

   c. Calculate how many of the 146 kcalories provided by a 12-ounce can of beer come from alcohol, if the beer contains 1 gram protein and 13 grams carbohydrate. (Note: The remaining kcalories derive from alcohol.)

   1 g protein = —— kcal protein.
   13 g carbohydrate = —— kcal carbohydrate.
   = —— kcal alcohol.

   How many grams of alcohol does this represent?
   —— g alcohol.

2. Even a little nutrition knowledge can help you identify some bogus claims. Consider an advertisement for a new "super supplement" that claims the product provides 15 grams protein and 10 kcalories per dose. Is this possible? —— Why or why not? —————————— = —— kcal.

STUDY QUESTIONS

These questions will help you review this chapter. You will find the answers in the discussions on the pages provided.

1. Give several reasons (and examples) why people make the food choices that they do. (pp. 4–5)

2. What is a nutrient? Name the six classes of nutrients found in foods. What is an essential nutrient? (pp. 6–7)

3. Which nutrients are inorganic, and which are organic? Discuss the significance of that distinction. (pp. 6, 10–11)

4. Which nutrients yield energy, and how much energy do they yield per gram? How is energy measured? (pp. 7–9)

5. Describe how alcohol resembles nutrients. Why is alcohol not considered a nutrient? (p. 9)

6. What is the science of nutrition? Describe the types of research studies and methods used in acquiring nutrition information. (pp. 11–15)

7. Explain how variables might be correlational but not causal. (p. 15)

8. What are the DRI? Who develops the DRI? To whom do they apply? How are they used? In your description, identify the categories of DRI and indicate how they are related. (pp. 16–19)

9. What judgment factors are involved in setting the energy and nutrient recommendations? (pp. 17–18)

10. What happens when people get either too little or too much energy or nutrients? Define malnutrition, undernutrition, and overnutrition. Describe the four methods used to detect energy and nutrient deficiencies and excesses. (pp. 20–21)

11. What methods are used in nutrition surveys? What kinds of information can these surveys provide? (pp. 22–23)

12. Describe risk factors and their relationships to disease. (pp. 24–25)

These multiple choice questions will help you prepare for an exam. Answers can be found on p. 29.

1. When people eat the foods typical of their families or geographic region, their choices are influenced by:
   a. habit.
   b. nutrition.
   c. personal preference.
   d. ethnic heritage or tradition.

2. Both the human body and many foods are composed mostly of:
   a. fat.
   b. water.
   c. minerals.
   d. proteins.

3. The inorganic nutrients are:
    a. proteins and fats.
    b. vitamins and minerals.
    c. minerals and water.
    d. vitamins and proteins.

4. The energy-yielding nutrients are:
    a. fats, minerals, and water.
    b. minerals, proteins, and vitamins.
    c. carbohydrates, fats, and vitamins.
    d. carbohydrates, fats, and proteins.

5. Studies of populations that reveal correlations between dietary habits and disease incidence are:
    a. clinical trials.
    b. laboratory studies.
    c. case-control studies.
    d. epidemiological studies.

6. An experiment in which neither the researchers nor the subjects know who is receiving the treatment is known as:
    a. double blind.
    b. double control.
    c. blind variable.
    d. placebo control.

7. An RDA represents the:
    a. highest amount of a nutrient that appears safe for most healthy people.
    b. lowest amount of a nutrient that will maintain a specified criterion of adequacy.
    c. average amount of a nutrient considered adequate to meet the known nutrient needs of practically all healthy people.
    d. average amount of a nutrient that will maintain a specific biochemical or physiological function in half the people.

8. Historical information, physical examinations, laboratory tests, and anthropometric measures are:
    a. techniques used in diet planning.
    b. steps used in the scientific method.
    c. approaches used in disease prevention.
    d. methods used in a nutrition assessment.

9. A deficiency caused by an inadequate dietary intake is a(n):
    a. overt deficiency.
    b. covert deficiency.
    c. primary deficiency.
    d. secondary deficiency.

10. Behaviors such as smoking, dietary habits, physical activity, and alcohol consumption that influence the development of disease are known as:
    a. risk factors.
    b. chronic causes.
    c. preventive agents.
    d. disease descriptors.

REFERENCES

1. L. L. Birch, Development of food preferences, *Annual Review of Nutrition* 19 (1999): 41–62; M. B. M. van den Bree, L. J. Eaves, and J. T. Dwyer, Genetic and environmental influences on eating patterns of twins aged ≥ 50 y, *American Journal of Clinical Nutrition* 70 (1999): 456–465.

2. J. E. Tillotson, Our ready-prepared, ready-to-eat nation, *Nutrition Today* 37 (2002): 36–38; F. Katz, "How nutritious?" meets "How convenient?" *Food Technology* 53 (1999): 44–50.

3. L. Canetti, E. Bachar, and E. M. Berry, Food and emotion, *Behavioural Processes* 60 (2002): 157–164.

4. J. D. Fernstrom, Diet, neurochemicals, and mental energy, *Nutrition Reviews* 59 (2001): S22–S24.

5. Position of the American Dietetic Association: Total diet approach to communicating food and nutrition information, *Journal of the American Dietetic Association* 102 (2002): 100–108.

6. D. Shattuck, Nutritional genomics, *Journal of the American Dietetic Association* 103 (2003): 16, 18; P. Trayhurn, Nutritional genomics—"Nutrigenom-ics," *British Journal of Nutrition* 89 (2003): 1–2; F. P. Guengerich, Functional genomics and proteomics applied to the study of nutritional metabolism, *Nutrition Reviews* 59 (2001): 259–263.

7. Committee on Dietary Reference Intakes, *Dietary Reference Intakes for Energy, Carbohydrate, Fiber, Fat, Fatty Acids, Cholesterol, Protein, and Amino Acids* (Washington, D.C.: National Academies Press, 2002); Committee on Dietary Reference Intakes, *Dietary Reference Intakes for Vitamin A, Vitamin K, Arsenic, Boron, Chromium, Copper, Iodine, Iron, Manganese, Molybdenum, Nickel, Silicon, Vanadium, and Zinc* (Washington, D.C.: National Academy Press, 2001); Committee on Dietary Reference Intakes, *Dietary Reference Intakes for Vitamin C, Vitamin E, Selenium, and Carotenoids* (Washington, D.C.: National Academy Press, 2000); Committee on Dietary Reference Intakes, *Dietary Reference Intakes for Thiamin, Riboflavin, Niacin, Vitamin B_6, Folate, Vitamin B_{12}, Pantothenic Acid, Biotin, and Choline* (Washington, D.C.: National Academy Press, 1998); Committee on Dietary Reference Intakes, *Dietary Reference Intakes for Calcium, Phosphorus, Magnesium, Vitamin D, and Fluoride* (Washington, D.C.: National Academy Press, 1997).

8. C. D. Berndanier, Nutrient-gene interactions, *Nutrition Today* 35 (2000): 8–17.

9. S. P. Murphy, S. I. Barr, and M. I. Poos, Using the new Dietary Reference Intakes to assess diets: A map to the maze, *Nutrition Reviews* 60 (2002): 267–275.

10. J. Dwyer and coauthors, Integration of the Continuing Survey of Food Intakes by Individuals and the National Health and Nutrition Examination Survey, *Journal of the American Dietetic Association* 101 (2001): 1142–1143.

11. S. S. Smith, NCHS launches latest National Health and Nutrition Examination Survey, *Public Health Reports* 114 (1999): 190–192.

12. S. J. Crockett and coauthors, Nutrition monitoring application in the food industry, *Nutrition Today* 37 (2002): 130–135.

13. U.S. Department of Health and Human Services, *Healthy People 2010: Understanding and Improving Health*, January 2000.

14. D. S. Satcher, Healthy People at 2000, *Public Health Reports* 114 (1999): 563–564.

15. Joint WHO/FAO expert report on diet, nutrition and the prevention of chronic disease, available at www.who.int/hpr/nutrition/expertconsultationge.htm.

16. *The Surgeon General's Report on Nutrition and Health: Summary and Recommendations*, DHHS (PHS) publication no. 88-50211 (Washington, D.C.: Government Printing Office, 1988).

ANSWERS

Nutrition Calculations

1. a. $5 \text{ g protein} \times 4 \text{ kcal/g} = 20 \text{ kcal protein.}$

   $30 \text{ g carbohydrate} \times 4 \text{ kcal/g} = 120 \text{ kcal carbohydrate.}$

   $11 \text{ g fat} \times 9 \text{ kcal/g} = 99 \text{ kcal fat.}$

   Total = 239 kcal.

   b. $20 \text{ kcal} \div 239 \text{ kcal} \times 100 = 8.4\% \text{ kcal from protein.}$

   $120 \text{ kcal} \div 239 \text{ kcal} \times 100 = 50.2\% \text{ kcal from carbohydrate.}$

   $99 \text{ kcal} \div 239 \text{ kcal} \times 100 = 41.4\% \text{ kcal from fat.}$

   Total = 100%.

   c. $1 \text{ g protein} = 4 \text{ kcal protein.}$

   $13 \text{ g carbohydrate} = 52 \text{ kcal carbohydrate.}$

   $146 \text{ total kcal} - 56 \text{ kcal (protein} + \text{carbohydrate)} = 90$

   kcal alcohol.

   $90 \text{ kcal alcohol} \div 7 \text{ g/kcal} = 12.9 \text{ g alcohol.}$

2. No. 15 g protein $\times$ 4 kcal/g = 60 kcal.

Study Questions (multiple choice)

1. d   2. b   3. c   4. d   5. d   6. a   7. c   8. d

9. c   10. a

HIGHLIGHT

Nutrition Information and Misinformation—On the Net and in the News

© USDA, Agricultural Research Service

People learn about nutrition daily as they watch television, read newspapers, turn the pages of magazines, talk with friends, and search the Internet. They want to know how best to take care of themselves. In some cases, they are seeking miracles: tricks to help them lose weight, foods to forestall aging, and supplements to build muscles. People's heightened interest in nutrition and health translates into billions of dollars spent on services and products sold by both legitimate and fraudulent businesses. Although consumers who obtain legitimate products can improve their health, those enticed by **fraud** may lose their health, their savings, or both.[1] Ironically, nutrition **quackery** prevents people from attaining the health they seek by giving them false hope and delaying the implementation of effective strategies (boldface terms are defined in the glossary on p. 31). Furthermore, the conflicting information that results from a mixture of science and quackery confuses consumers.

Science and quackery may be easy to tell apart at the extremes, but much nutrition information lies between the extremes. How can people distinguish valid nutrition information from misinformation? One excellent approach is to notice *who* is purveying the information. The "who" behind the information is not always evident, though, especially in the world of electronic media. Consumers need to keep in mind that *people* develop CD-ROMs and create websites on the Internet, just as people write books and report the news.

This highlight begins by examining the unique potential and problems of relying on the Internet and the media for nutrition information. It continues with a discussion of how to identify reliable nutrition information that applies to all resources, including the Internet and the news.

or inaccurate information. Simply put: anyone can publish anything.

For experienced users who know which sources are reliable, results are just a mouse click away; not only is access easy, but the information is often more current than that obtainable from other sources. For others, though, answers lie tangled in a web of information overload and questionable reliability.

With hundreds of millions of **websites** on the **World Wide Web,** searching for nutrition information can be an overwhelming experience—much like walking into an enormous bookstore with millions of books, magazines, newspapers, and videos. And like a bookstore, the Internet offers no guarantees of the accuracy of the information found there—and much of it is pure fiction.

When using the Internet, keep in mind that the quality of health-related information available covers a broad range.[2] Just because you find it on the Net doesn't make it true. Websites must be evaluated for their accuracy, just like every other source. The accompanying "How to" provides tips for determining whether a website is reliable.

Increase the proportion of health-related World Wide Web sites that disclose information that can be used to assess the quality of the site.

To help users find reliable nutrition information on the Internet, Tufts University maintains an online rating and review guide called the Nutrition Navigator (**navigator.tufts.edu**). The ratings reflect the opinions of a panel of nutrition experts who have scored selected websites on the basis of their accuracy, depth, and ease of use. In addition to a rating, the Nutrition Navigator provides a review of the website's content and links to recommended sites. The Nutrition Navigator is an excellent site from which to launch your ventures into nutrition **cyberspace.** Similarly, the Health on the Net Foundation (**www.hon.ch**) provides guidance in finding useful and reliable health information online.[3]

Nutrition on the Net

Got a question? The **Internet** has an answer. The Internet offers endless opportunities to obtain high-quality information, but it also delivers an abundance of incomplete, misleading,

Nutrition in the News

Consumers get most of their nutrition information from television news and magazine reports, which have heightened awareness of how diet influences the development of diseases.

Consumers benefit from news coverage of nutrition when they learn to make lifestyle changes that will improve their health. Sometimes, however, magazine articles or television programs reporting on nutrition trends mislead consumers and create confusion. The high-protein diet craze, for example, was featured numerous times in national news magazines and on all major networks. No doubt, it was a hot topic and people wanted to know all about it (see Highlight 8 for coverage of the high-protein weight-loss diets). Unfortunately, many of these reports told a lopsided story based on a few testimonials. They did not present the results of research studies or a balance of expert opinions. Telling the whole story might not have been as entertaining, but it would have been more informative.

Tight deadlines and limited understanding sometimes make it difficult to provide a thorough report. Hungry for the latest news, the media often report scientific findings prematurely—without benefit of careful interpretation, replication, and peer review.[4]

HOW TO Determine Whether a Website Is Reliable

To determine whether a website offers reliable nutrition information, ask the following questions:

- **Who?** Who is responsible for the site? Is it staffed by qualified professionals? Look for the authors' names and credentials. Have experts reviewed the content for accuracy?
- **When?** When was the site last updated? Because nutrition is an ever-changing science, sites need to be dated and updated frequently.
- **Where?** Where is the information coming from? The three letters following the dot in a Web address identify the site's affiliation. Addresses ending in "gov" (government), "edu" (educational institute), and "org" (organization) generally provide reliable information; "com" (commercial) sites represent businesses and, depending on their qualifications and integrity, may or may not offer dependable information.
- **Why?** Why is the site giving you this information? Is the site providing a public service or selling a product? Many commercial sites provide accurate information, but some do not. When money is the prime motivation, be aware that the information may be biased.

If you are satisfied with the answers to all of the above questions, then ask this final question:

- **What?** What is the message, and is it in line with other reliable sources? Information that contradicts common knowledge should be questioned. Many reliable sites provide links to other sites to facilitate your quest for knowledge, but this provision alone does not guarantee a reputable intention. Be aware that any site can link to any other site without permission.

Usually, the reports present findings from a single, recently released study, making the news current and controversial.[5] Consequently, the public receives diet and health news

GLOSSARY

accredited: approved; in the case of medical centers or universities, certified by an agency recognized by the U.S. Department of Education.

American Dietetic Association (ADA): the professional organization of dietitians in the United States. The Canadian equivalent is Dietitians of Canada, which operates similarly.

correspondence schools: schools that offer courses and degrees by mail. Some correspondence schools are accredited; others are not.

cyberspace: a term coined by William Gibson referring to the nonphysical place where all Internet activity occurs.

dietetic technician: a person who has completed a minimum of an associate's degree from an accredited university or college and an approved dietetic technician program that includes a supervised practice experience. See also *dietetic technician, registered (DTR)*.

dietetic technician, registered (DTR): a dietetic technician who has passed a national examination and maintains registration through continuing professional education.

dietitian: a person trained in nutrition, food science, and diet planning. See also *registered dietitian*.

DTR: see *dietetic technician, registered*.

fraud or **quackery:** the promotion, for financial gain, of devices, treatments, services, plans, or products (including diets and supplements) that alter or claim to alter a human condition without proof of safety or effectiveness. (The word *quackery* comes from the term *quacksalver*, meaning a person who quacks loudly about a miracle product— a lotion or a salve.)

Internet (the Net): a worldwide network of millions of computers linked together to share information.

license to practice: permission under state or federal law, granted on meeting specified criteria, to use a certain title (such as dietitian) and offer certain services. **Licensed dietitians** may use the initials **LD** after their names.

misinformation: false or misleading information.

nutritionist: a person who specializes in the study of nutrition. Note that this definition does not specify qualifications and may apply not only to registered dietitians but also to self-described experts whose training is questionable. Most states have licensing laws that define the scope of practice for those calling themselves nutritionists.

public health dietitians: dietitians who specialize in providing nutrition services through organized community efforts.

RD: see *registered dietitian*.

registered dietitian (RD): a person who has completed a minimum of a bachelor's degree from an accredited university or college, has completed approved course work and a supervised practice program, has passed a national examination, and maintains registration through continuing professional education.

registration: listing; with respect to health professionals, listing with a professional organization that requires specific course work, experience, and passing of an examination.

websites: Internet resources composed of text and graphic files, each with a unique URL (Uniform Resource Locator) that names the site (for example, www.usda.gov).

World Wide Web (the Web, commonly abbreviated **www**): a graphical subset of the Internet.

quickly, but not always in perspective. Pressure to write catchy headlines and sensational stories twists inconclusive findings into "meaningful discoveries."

As a result, "surprising new findings" seem to contradict one another, and consumers feel frustrated and betrayed. Occasionally, the reports are downright erroneous, but sometimes the apparent contradictions are simply the normal result of science at work. A single study contributes to the big picture, but when viewed alone, the image is distorted. To be meaningful, its conclusions must be presented cautiously within the context of other research findings.

People who do not understand how science operates may become distrustful as they try to learn nutrition from current news reports: "How am I supposed to know what to eat when the scientists themselves don't know?" General background knowledge about the science of nutrition is the best foundation a person can have for judging the validity of new nutrition information. (Congratulations on your decision to take this course.)

Identifying Nutrition Experts

Regardless of whether the medium is electronic, print, or video, consumers need to ask whether the person behind the information is qualified to speak on nutrition. If the creator of a website on the Internet recommends eating three pineapples a day to lose weight, a trainer at the gym praises a high-protein diet, or a health-store clerk suggests an herbal supplement, should you believe these people? Can you distinguish between accurate news reports and sensational programs on television? Have you noticed that many televised nutrition messages are presented by celebrities, fitness experts, psychologists, food editors, and chefs—that is, almost anyone except a **dietitian**? When you are confused or need sound dietary advice, whom should you ask?

Physicians and Other Health Care Professionals

Many people turn to physicians or other health care professionals for dietary advice, expecting them to know all about health-related matters. But are they the best sources of accurate and current information on nutrition? Only about one-fourth of all medical schools in the United States require students to take even one nutrition course; less than half provide an elective nutrition course.[6] Students attending these classes receive an average of 20 hours of nutrition instruction—an amount they themselves consider inadequate. By comparison, most students reading this text are taking a nutrition class that provides an average of 45 hours of instruction.

The **American Dietetic Association (ADA)** asserts that standardized nutrition education should be included in the curricula for all health care professionals: physicians, nurses, physician's assistants, dental hygienists, physical and occupational therapists, social workers, and all others who provide services directly to clients.[7] When these professionals understand the relevance of nutrition in the treatment and prevention of disease and have command of reliable nutrition information, then all the people they serve will also be better informed.[8]

Most health care professionals appreciate the connections between health and nutrition. Those who have specialized in clinical nutrition are especially well qualified to speak on the subject. Few, however, have the time or experience to develop diet plans and provide detailed diet instructions for clients. Often they wisely refer clients to a qualified nutrition expert—a **registered dietitian (RD)**.

Increase the proportion of physician office visits that provide or order nutrition counseling and educational services.

Registered Dietitians (RD)

A registered dietitian (RD) has the educational background necessary to deliver reliable nutrition advice and care. To become an RD, a person must earn an undergraduate degree requiring some 60 or so semester hours in nutrition, food science, and other related subjects; complete a year's clinical internship or the equivalent; pass a national examination administered by the ADA; and maintain up-to-date knowledge and **registration** by participating in required continuing education activities such as attending seminars, taking courses, or writing professional papers.

Some states allow anyone to use the title dietitian or **nutritionist**, but others allow only RDs or people with certain graduate degrees to call themselves dietitians. Many states provide a further guarantee: a certification or **license to practice.** By requiring dietitians to be licensed, states identify people who have met minimal standards of education and experience.

Dietitians perform a multitude of duties in many settings in most communities. They work in the food industry, pharmaceutical companies, home health agencies, long-term care institutions, private practice, public health departments, research centers, education settings, fitness centers, and hospitals. Depending on their work settings, dietitians can assume a number of different job responsibilities and positions. In hospitals, administrative dietitians manage the foodservice system; clinical dietitians provide client care (see Table H1-1); and nutrition support team dietitians coordinate nutrition care with other health care professionals. In the food industry, dietitians conduct research, develop products, and market services.

Public health dietitians who work in government-funded agencies play a key role in delivering nutrition services to people in the community.[9] Among their many roles, public health dietitians help plan, coordinate, and evaluate food assistance programs; act as consultants to other agencies; manage finances; and much more.

TABLE H1-1	Responsibilities of a Clinical Dietitian

- Assesses clients' nutrition status.
- Determines clients' nutrient requirements.
- Monitors clients' nutrient intakes.
- Develops, implements, and evaluates clients' nutrition care plans.
- Counsels clients to cope with unique diet plans.
- Teaches clients and their families about nutrition needs and diet plans.
- Provides training for other dietitians, nurses, interns, and dietetics students.
- Serves as liaison between clients and the foodservice department.
- Communicates with physicians, nurses, pharmacists, and other health care professionals about clients' progress, needs, and treatments.
- Participates in professional activities to enhance knowledge and skill.

Other Dietary Employees

In some facilities, a **dietetic technician** assists registered dietitians in both administrative and clinical responsibilities. A dietetic technician has been educated and trained to work under the guidance of a registered dietitian; upon passing a national examination, the person's title changes to **dietetic technician, registered (DTR)**.

In addition to the dietetic technician, dietary employees may include clerks, aides, cooks, porters, and other assistants. These dietary employees do not have extensive formal training in nutrition, and their ability to provide accurate information may be limited.

Identifying Fake Credentials

In contrast to registered dietitians, thousands of people possess fake nutrition degrees and claim to be nutrition consultants or doctors of "nutrimedicine." These and other such titles may sound meaningful, but most of these people lack the established credentials and training of an ADA-sanctioned dietitian. If you look closely, you can see signs of their fake expertise.

Consider educational background, for example. The minimal standards of education for a dietitian specify a bachelor of science (BS) degree in food science and human nutrition or related fields from an **accredited** college or university. Such a degree generally requires four to five years of study. In contrast, a fake nutrition expert may display a degree from a six-month correspondence course. Such a degree simply falls short. In some cases, businesses posing as legitimate **correspondence schools** offer even less—they sell certificates to anyone who pays the fees. To obtain these "degrees," a candidate need not attend any classes, read any books, or pass any examinations.

To guard educational quality, an accrediting agency recognized by the U.S. Department of Education (DOE) certifies that certain schools meet criteria established to ensure that an institution provides complete and accurate schooling. Unfor-

tunately, fake nutrition degrees are available from schools "accredited" by more than 30 phony accrediting agencies.

To dramatize the ease with which anyone can obtain a fake nutrition degree, one writer enrolled in a correspondence course for a fee of $82. She made every attempt to fail, intentionally answering all examination questions

Charlie displays his professional credentials.

incorrectly. Even so, she received a "nutritionist" certificate at the end of the course. The "school" explained that it was sure she must have just misread the test.

In a similar stunt, Ms. Sassafras Herbert was named a "professional member" of a professional association. For her efforts, Sassafras has received a wallet card and is listed in a *Who's Who* publication that is distributed at health fairs and trade shows nationwide. Sassafras is a poodle; her master, Victor Herbert, MD, paid $50 to prove that she could be awarded these honors merely by sending in her name. Mr. Charlie Herbert, who is also a professional member of such an organization, is a cat. Admittedly, these examples are dated, but acquiring false credentials is still easy—perhaps even more so, thanks to the Internet.

By knowing what qualifies someone to speak on nutrition, consumers can determine whether that person's advice might be harmful or helpful. Don't be afraid to ask for credentials. Does the personal trainer at the gym have a degree in nutrition from an accredited university? Is the creator of a nutrition website an RD or otherwise qualified to write on nutrition? Have you seen the health-store clerk's license to practice as a dietitian? If not, seek a better-qualified source for your nutrition information. After all, your health depends on it.

Identifying Valid Information

Where do nutrition experts get their information? As Chapter 1 explained, nutrition knowledge derives from scientific research.

Researchers conduct experiments and then record and analyze their results, exercising caution in their interpretation of the findings. For example, in an epidemiological study, scientists may use a specific segment of the population—say, men 18 to 30 years old. When the scientists draw conclusions, they are careful not to generalize the findings to all people. Similarly, scientists performing research studies using animals are cautious in applying their findings to human beings. Conclusions from any one research study are always tentative and take into account findings from studies conducted by other scientists as well. As evidence accumulates, scientists gain confidence about making recommendations that affect people's health and lives. Still, their statements are worded cautiously, as in "A diet high in fruits and vegetables *may* protect against *some* cancers."

Quite often, as they approach an answer to one research question, scientists raise several more questions, so future research projects are never lacking. Further scientific investigation then seeks to answer questions such as "What substance or substances within fruits and vegetables provide protection?" If those substances turn out to be the vitamins found so abundantly in fresh produce, then, "How much is needed to offer protection?" "How do these vitamins protect against cancer?" "Is it their action as antioxidant nutrients?" "If not, might it be another action or even another substance that accounts for the protection fruits and vegetables provide against cancer?" (Highlight 11 explores the answers to these questions and reviews recent research on antioxidant nutrients and disease.)

The findings from a research study are submitted to a board of reviewers composed of other scientists who rigorously evaluate the study to assure that the scientific method was followed—a process known as peer review. The reviewers critique the study's hypothesis, methodology, statistical significance, and conclusions (Table H1-2 describes the parts of a research article). If the reviewers consider the conclusions to be well supported by the evidence, they endorse the work for publication in a scientific journal where others can read it. This raises an important point regarding information found on the Internet: much gets published without the rigorous scrutiny of peer review. Consequently, readers must assume greater responsibility for examining the data and conclusions presented—often without the benefit of journal citations. Until you feel confident in critically evaluating nutrition information, you would be wise to restrict your research to one of

HOW TO   Find Credible Sources of Nutrition Information

Government agencies, volunteer associations, consumer groups, and professional organizations provide consumers with reliable health and nutrition information. Credible sources of nutrition information include:

- Nutrition and food science departments at a university or community college.
- Local agencies such as the health department or County Cooperative Extension Service.
- Government health agencies such as:
  - Department of Agriculture (USDA)   www.usda.gov
  - Department of Health and Human
    Services (DHHS)   www.os.dhhs.gov
  - Food and Drug Administration (FDA)   www.fda.gov
  - Health Canada   www.hc-sc.gc.ca/nutrition
- Volunteer health agencies such as:
  - American Cancer Society   www.cancer.org
  - American Diabetes Association   www.diabetes.org
  - American Heart Association   www.americanheart.org
- Reputable consumer groups such as:
  - American Council on Science and Health   www.acsh.org
  - Federal Citizen Information Center   www.pueblo.gsa.gov
  - International Food Information Council   ific.org
- Professional health organizations such as:
  - American Dietetic Assocation   www.eatright.org
  - American Medical Association   www.ama-assn.org
  - Dietitians of Canada   www.dietitians.ca
- Journals such as:
  - *American Journal of Clinical Nutrition*   www.faseb.org/ajcn
  - *New England Journal of Medicine*   www.nejm.org
  - *Nutrition Reviews*   www.ilsi.org/publications

TABLE H1-2   Parts of a Research Article

- *Abstract.* The abstract provides a brief overview of the article.
- *Introduction.* The introduction clearly states the purpose of the current study by proposing a hypothesis.
- *Review of literature.* A comprehensive review of the literature reveals all that science has uncovered on the subject to date.
- *Methodology.* The methodology section defines key terms and describes the instruments and procedures used in conducting the study.
- *Results.* The results report the findings and may include tables and figures that summarize the information.
- *Conclusions.* The conclusions drawn are those supported by the data and reflect the original purpose as stated in the introduction. Usually, they answer a few questions and raise several more.
- *References.* The references reflect the investigator's knowledge of the subject and should include an extensive list of relevant studies (including key studies several years old as well as current ones).

the many online peer-reviewed journals (see the accompanying "How to" for selected website addresses).

Regardless of whether an article is presented on the Internet, on television, or in print, readers must evaluate the study and assess the findings in light of knowledge gleaned from other studies. Figure H1-1 provides examples of reliable nutrition information.

Even when a new finding is published or released to the media, it is still only preliminary and not very meaningful by itself. Other scientists will need to confirm or disprove the findings through replication. To be accepted into the body of nutrition knowledge, a finding must stand up to rigorous, repeated testing in experiments performed by several different researchers. What we "know" in nutrition results from years of replicating study findings. Communicating the latest finding in its proper context without distorting or oversimplifying the message is a challenge for scientists and journalists alike.

With each report from scientists, the field of nutrition changes a little—each finding contributes another piece to the whole body of knowledge. People who know how science works understand that single findings, like single frames in a movie, are just small parts of a larger story. Over years, the pic-

FIGURE H1-1 Sources of Reliable Nutrition Information

REVIEWS

Articles that examine all the major work on a subject are published in review journals like *Nutrition Reviews*. These articles provide references to all of the original work reviewed.

JOURNALS

Articles that present all the details of the methods, results, and conclusions of a particular study are published in journals like the *American Journal of Clinical Nutrition*.

INDEXES

Indexes provide a listing of research articles on a given subject. Several online indexes are available, but one of the best for nutrition research is PubMed, a service of the National Library of Medicine. For free access, visit www.pubmed.gov

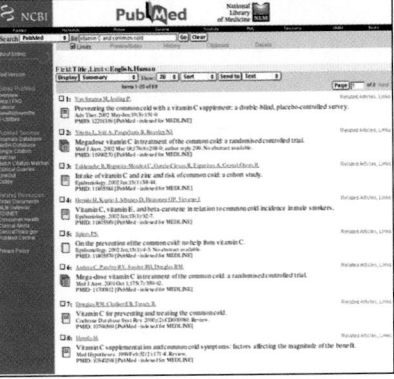

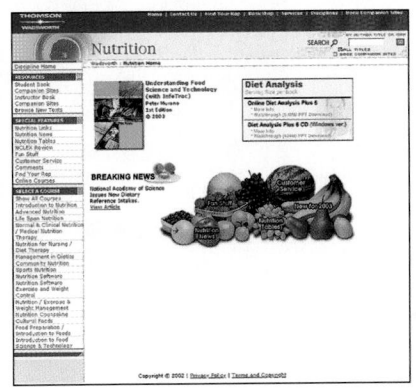

WEBSITES

Websites on the Internet developed by credible sources, such as those listed on p. 34, can provide valuable nutrition information and direct users to other resources. A quick link to many of these nutrition resources is available when you visit www.wadsworth.com/nutrition

---

ture of what is "true" in nutrition gradually changes, and modifications in recommendations then follow. Instead of eating 4 servings of fruits and vegetables as recommended by the old Four Food Group Plan, people are now encouraged to eat 2 to 4 servings of fruits and 3 to 5 servings of vegetables as suggested by the current Daily Food Guide (presented in Chapter 2).

Because science is a step-by-step, information-gathering and testing process, old research still has value. A hypothesis first advanced in 1960 that stands up to decades of validation has real strength. When it comes to scientific information, "new" does not necessarily mean "improved." In fact, any science report based on all new references is suspect, for truly strong research is based on a body of work conducted over many years. This is why, even in books published just this year, you will see references to old reports. Some studies have become classics: they were exciting when they first appeared, and they have stood up to the test of time.

Identifying Misinformation

Did you receive the e-mail warning about Costa Rican bananas causing the disease "necrotizing fasciitis"? If so, you've been scammed by Internet misinformation. Nutrition is a hot topic, and scattered among the valid research findings are thousands of misleading and unfounded claims. How can a person identify nutrition **misinformation** and quackery—on the Net and in the news? Once upon a time, quacks rode into town in wooden wagons hawking snake oil for 50 cents a bottle to "cure what ails you," but those days are gone. Today's purveyors approach consumers in less obvious ways. They deliver their messages via the Internet, on glossy pages of magazines, in televised infomercials, and at social gatherings. The claims may look slick and sound logical, but they lack the research support found in nutrition science. Figure H1-2 presents red flags that alert consumers to nutrition misinformation.[10]

FIGURE H1-2   Red Flags of Nutrition Quackery

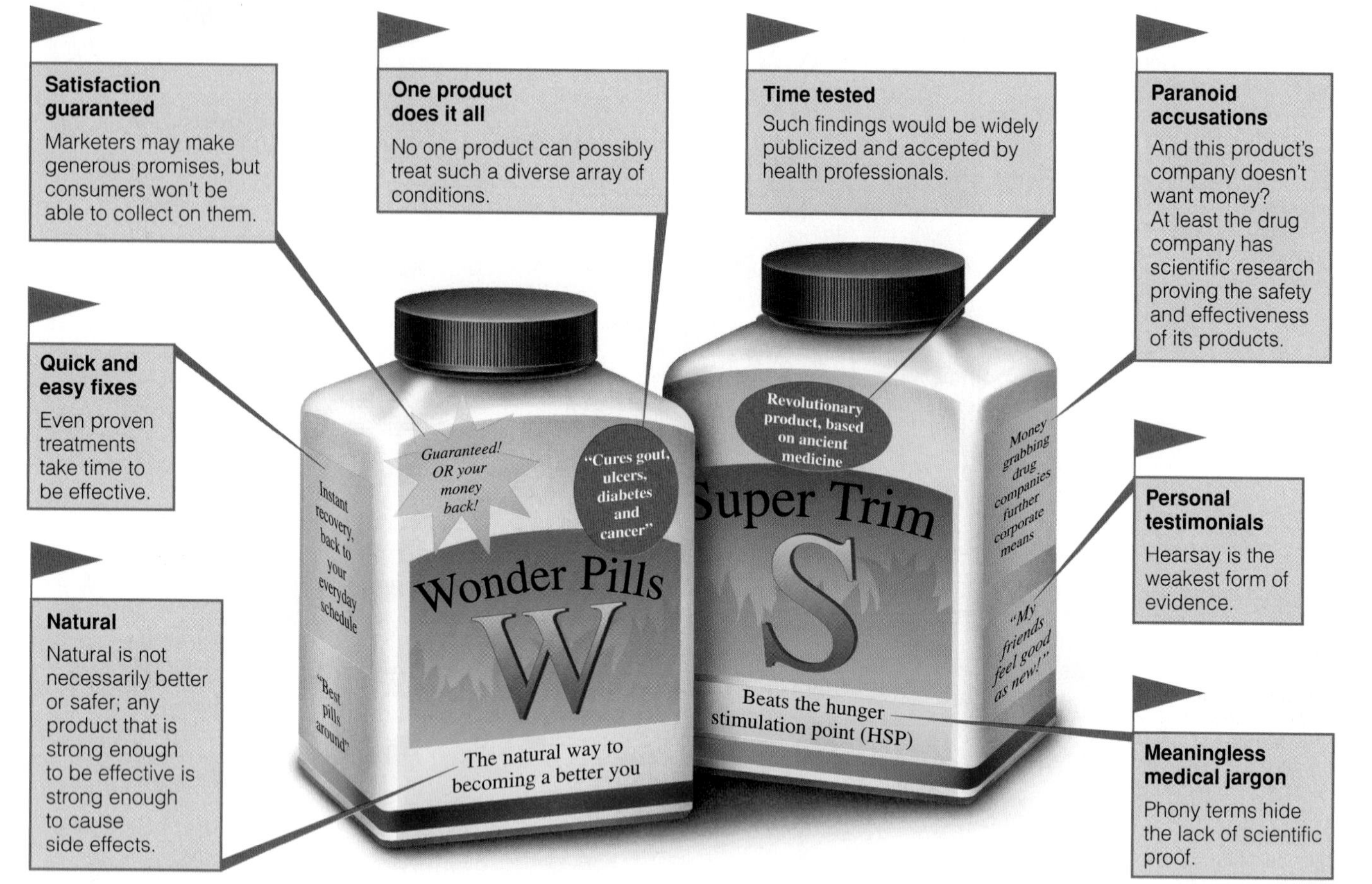

Satisfaction guaranteed
Marketers may make generous promises, but consumers won't be able to collect on them.

One product does it all
No one product can possibly treat such a diverse array of conditions.

Time tested
Such findings would be widely publicized and accepted by health professionals.

Paranoid accusations
And this product's company doesn't want money? At least the drug company has scientific research proving the safety and effectiveness of its products.

Quick and easy fixes
Even proven treatments take time to be effective.

Natural
Natural is not necessarily better or safer; any product that is strong enough to be effective is strong enough to cause side effects.

Personal testimonials
Hearsay is the weakest form of evidence.

Meaningless medical jargon
Phony terms hide the lack of scientific proof.

Guaranteed! OR your money back!

"Cures gout, ulcers, diabetes and cancer"

Instant recovery, back to your everyday schedule

"Best pills around"

Wonder Pills
W
The natural way to becoming a better you

Revolutionary product, based on ancient medicine

Super Trim
S

Money grabbing drug companies further corporate means

"My friends feel good as new!"

Beats the hunger stimulation point (HSP)

Sales of unproven and dangerous products have always been a concern, but the Internet now provides merchants with an easy and inexpensive way to reach millions of customers around the world. Because of the difficulty in regulating the Internet, fraudulent and illegal sales of medical products have hit a bonanza. As is the case with the air, no one owns the Internet, and similarly, no one has control over the pollution. Countries have different laws regarding sales of drugs, dietary supplements, and other health products, but applying these laws to the Internet marketplace is almost impossible. Even if illegal activities could be defined and identified, finding the person responsible for a particular website is not always possible. Websites can open and close in a blink of a cursor. Now, more than ever, consumers must heed the caution "Buyer beware."

In summary, when you hear nutrition news, consider its source. Ask yourself these two questions: Is the person purveying the information qualified to speak on nutrition? Is the information based on valid scientific research? If not, find a better source, for your health is your most precious asset.

NUTRITION ON THE NET

 Access these websites for further study of topics covered in this highlight.

- Find updates and quick links to these and other nutrition-related sites at our website: **www.wadsworth.com/nutrition**

- Visit the National Council Against Health Fraud: **www.ncahf.org**

- Check the ratings and reviews of websites by Tufts University Nutrition Navigator: **navigator.tufts.edu**

- Find a registered dietitian in your area from the American Dietetic Association: **www.eatright.org**

- Find a nutrition professional in Canada from the Dietitians of Canada: **www.dietitians.ca**

- Find out whether a correspondence school is accredited from the Distance Education and Training Council's Accrediting Commission: **www.detc.org**

- Find out whether a school is properly accredited for a dietetics degree from the American Dietetic Association: **www.eatright.org/cade**

- Obtain a listing of accredited institutions, professionally accredited programs, and candidates for accreditation from the American Council on Education: **www.acenet.edu**

- Learn more about quackery from Stephen Barrett's Quackwatch: **www.quackwatch.com**

- Search "quackery" at the U.S. Government health information site: **www.healthfinder.gov**

- Check out health-related hoaxes and rumors: **www.cdc.gov/hoax_rumors.htm** and **www.urbanlegends.com/ulz**

- Find reliable research articles: **www.pubmed.gov**

REFERENCES

1. Position of the American Dietetic Association: Food and nutrition misinformation, *Journal of the American Dietetic Association* 102 (2002): 260–266.
2. G. Eysenbach and coauthors, Empirical studies assessing the quality of health information for consumers on the World Wide Web: A systematic review, *Journal of the American Medical Association* 287 (2002): 2691–2700.
3. S. M. Dorman, Health on the Net Foundation: Advocating for quality health information, *Journal of School Health* 72 (2002): 86.
4. L. M. Schwartz, S. Woloshin, and L. Baczek, Media coverage of scientific meetings: Too much, too soon? *Journal of the American Medical Association* 287 (2002): 2859–2863.

5. N. S. Wellman and coauthors, Do we facilitate the scientific process and the development of dietary guidance when findings from single studies are publicized? An American Society for Nutritional Sciences Controversy Session Report, *American Journal of Clinical Nutrition* 70 (1999): 802–805.
6. J. A. Schulman, Nutrition education in medical schools: Trends and implications for health educators, *Med Ed Online,* **www.med-ed-Online.org/ f0000015.htm** (accessed October 24, 2000).
7. Position of the American Dietetic Association: Nutrition education for health care professionals, *Journal of the American Dietetic Association* 98 (1998): 343–346.

8. Intersociety Professional Nutrition Education Consortium, Bringing nutrition specialists into the mainstream: Rationale for the Intersociety Professional Nutrition Education Consortium, *American Journal of Clinical Nutrition* 68 (1998): 894–898.
9. D. B. Johnson and coauthors, Public health nutrition practice in the United States, *Journal of the American Dietetic Association* 101 (2001): 529–534.
10. Adapted from P. Kurtzweil, How to spot health fraud, *FDA Consumer,* November/-December 1999, pp. 22–26.

Chapter 2

Planning a Healthy Diet

Chapter Outline

Principles and Guidelines: *Diet-Planning Principles* • *Dietary Guidelines for Americans*

Diet-Planning Guides: *Food Group Plans* • *Exchange Lists* • *Putting the Plan into Action* • *From Guidelines to Groceries*

Food Labels: *The Ingredient List* • *Serving Sizes* • *Nutrition Facts* • *The Daily Values* • *Nutrient Claims* • *Health Claims* • *Structure-Function Claims* • *Consumer Education*

Highlight: *A World Tour of Pyramids, Pagodas, and Plates*

Nutrition Explorer CD-ROM Outline

Nutrition Animation: *Diet Planning Using the Healthy Eating Index*

Case Study: *Planning a Healthy Diet*

Student Practice Test

Glossary Terms

Nutrition on the Net

© Brian Hagiwara/FoodPix/Getty Images

Nutrition in Your Life

You make food choices—deciding what to eat and how much to eat—more than 1000 times every year. We eat so frequently that it's easy for us to choose a meal without giving its nutrient contributions or health consequences any thought. Even when we want to make healthy choices, we may not know which foods to select or what quantity to consume. Given a few tools and tips, you can learn to plan a healthy diet.

Chapter 1 explained that the body's many activities are supported by the nutrients delivered by the foods people eat. Food choices made over years influence the body's health, and consistently poor choices increase the risks of developing chronic diseases. This chapter shows how a person can select from the tens of thousands of foods available to create a diet that supports health. Fortunately, most foods provide several nutrients, so one trick for wise diet planning is to select a combination of foods that deliver a full array of nutrients. This chapter begins by introducing the diet-planning principles and dietary guidelines that assist people in selecting foods that will deliver nutrients without excess energy.

Principles and Guidelines

How well you nourish yourself does not depend on the selection of any one food. Instead it depends on the selection of many different foods at numerous meals over days, months, and years. Diet-planning principles and dietary guidelines are key concepts to keep in mind whenever you are selecting foods—whether shopping at the grocery store, choosing from a restaurant menu, or preparing a home-cooked meal.

© Polara Studios Inc.

To ensure an adequate and balanced diet, eat a variety of foods daily, choosing different foods from each group.

■ Diet-planning principles:
- Adequacy.
- Balance.
- kCalorie (energy) control.
- Nutrient Density.
- Moderation.
- Variety.

■ Balance in the diet helps to ensure adequacy.

■ Nutrient density promotes adequacy and kcalorie control.

adequacy (dietary): providing all the essential nutrients, fiber, and energy in amounts sufficient to maintain health.

balance (dietary): providing foods in proportion to each other and in proportion to the body's needs.

kcalorie (energy) control: management of food energy intake.

nutrient density: a measure of the nutrients a food provides relative to the energy it provides. The more nutrients and the fewer kcalories, the higher the nutrient density.

empty-kcalorie foods: a popular term used to denote foods that contribute energy but lack protein, vitamins, and minerals.

Diet-Planning Principles

Diet planners have developed several ways to select foods. Whatever plan or combination of plans they use, though, they keep in mind the six basic diet-planning principles■ listed in the margin.

Adequacy **Adequacy** means that the diet provides sufficient energy and enough of all the nutrients to meet the needs of healthy people. Take the essential nutrient iron, for example. Each day the body loses some iron, so people have to replace it by eating foods that contain iron. A person whose diet fails to provide enough iron-rich foods may develop the symptoms of iron-deficiency anemia: the person may feel weak, tired, and listless; have frequent headaches; and find that even the smallest amount of muscular work brings disabling fatigue. To prevent these deficiency symptoms, a person must include foods that supply adequate iron. The same is true for all the other essential nutrients introduced in Chapter 1.

Balance The art of balancing the diet involves consuming enough—but not too much—of each type of food. The essential minerals calcium and iron, taken together, illustrate the importance of dietary **balance.** Meats, fish, and poultry are rich in iron but poor in calcium. Conversely, milk and milk products are rich in calcium but poor in iron. Use some meat or meat alternates for iron; use some milk and milk products for calcium; and save some space for other foods, too, since a diet consisting of milk and meat alone would not be adequate.■ For the other nutrients, people need grains, vegetables, and fruits.

kCalorie (Energy) Control Designing an adequate diet without overeating requires careful planning. Once again, balance plays a key role. The amount of energy coming into the body from foods should balance with the amount of energy being used by the body to sustain its metabolic and physical activities. Upsetting this balance leads to gains or losses in body weight. The discussion of weight control in Chapter 9 examines this issue in more detail, but the key to **kcalorie control** is to select foods of high **nutrient density.**

Nutrient Density To eat well without overeating, select foods that deliver the most nutrients for the least food energy. Consider foods containing calcium, for example. You can get about 300 milligrams of calcium from either 1½ ounces of cheddar cheese or 1 cup of fat-free milk, but the cheese delivers about twice as much food energy (kcalories) as the milk. The fat-free milk, then, is twice as calcium dense as the cheddar cheese; it offers the same amount of calcium for half the kcalories. Both foods are excellent choices for adequacy's sake alone, but to achieve adequacy while controlling kcalories,■ the fat-free milk is the better choice. (Alternatively, a person could select a low-fat cheddar cheese.) The many bar graphs that appear in Chapters 10 through 13 highlight the most nutrient-dense choices, and the accompanying "How to" describes how to compare foods based on nutrient density.

Just like a person who has to pay for rent, food, clothes, and tuition on a tight budget, a person whose energy allowance is limited has to obtain iron, calcium, and all the other essential nutrients on a tight energy budget. To succeed, the person has to get many nutrients for each kcalorie "dollar." In the cola and grapes example in the margin on p. 41, both provide about the same number of kcalories, but the grapes deliver many more nutrients. A person who makes nutrient-dense choices such as fruit over cola can meet daily nutrient needs on a lower energy budget. Such choices support good health.

Foods that are notably low in nutrient density—such as potato chips, candies, and colas—are sometimes called **empty-kcalorie foods.** The kcalories these foods provide are "empty" in that they deliver only energy (from sugar, fat, or both) with little, or no, protein, vitamins, or minerals.

Moderation Foods rich in fat and sugar provide enjoyment and energy but relatively few nutrients. In addition, they promote weight gain when eaten in excess.

HOW TO Compare Foods Based on Nutrient Density

One way to evaluate foods is simply to notice their nutrient contribution *per serving:* 1 cup of milk provides 301 milligrams of calcium, and ½ cup of fresh, cooked turnip greens provides 99 milligrams. Thus a serving of milk offers three times as much calcium as a serving of turnip greens. To get 300 milligrams of calcium, a person could choose either 1 serving of milk or 3 servings (1½ cups) of turnip greens.

Another valuable way to evaluate foods is to consider their nutrient density—their nutrient contribution *per kcalorie.* Fat-free milk delivers 86 kcalories with its 301 milligrams of calcium. To calculate the nutrient density, divide milligrams by kcalories:

$$\frac{301 \text{ mg calcium}}{86 \text{ kcal}} = 3.5 \text{ mg per kcal.}$$

Do the same for the fresh turnip greens, which provide 15 kcalories with the 99 milligrams of calcium:

$$\frac{99 \text{ mg calcium}}{15 \text{ kcal}} = 6.6 \text{ mg per kcal.}$$

The more milligrams per kcalorie, the greater the nutrient density. Turnip greens are more calcium dense than milk. They provide more calcium *per kcalorie* than milk, but milk offers more calcium *per serving.* Both approaches offer valuable information, especially when combined with a realistic appraisal. What matters most is which are you more likely to consume—1½ cups of turnip greens or 1 cup of milk? You can get 300 milligrams of calcium from either, but the greens will save you about 40 kcalories (the savings would be even greater if you usually use whole milk).

Keep in mind, too, that calcium is only one of the many nutrients that foods provide. Similar calculations for protein, for example, would show that fat-free milk provides more protein both *per kcalorie* and *per serving* than turnip greens—that is, milk is more protein dense. Combining variety with nutrient density helps to ensure the adequacy of all nutrients.

This cola and bunch of grapes illustrate nutrient density. Each provides about 150 kcalories, but the grapes offer a trace of protein, some vitamins, minerals, phytochemicals, and fiber along with the energy; the cola beverage offers only "empty" kcalories from sugar without any other nutrients. Grapes, or any fruit for that matter, are more nutrient dense than cola beverages.

A person practicing **moderation**■ would eat such foods only on occasion and would regularly select foods low in fat and sugar, a practice that automatically improves nutrient density. Returning to the example of cheddar cheese and fat-free milk, the fat-free milk not only offers the same amount of calcium for less energy, but it contains far less fat than the cheese.

■ Moderation contributes to adequacy, balance, and kcalorie control.

Variety A diet may have all of the virtues just described and still lack **variety,** if a person eats the same foods day after day. People should select foods from each of the food groups daily and vary their choices within each food group from day to day for several reasons. First, different foods within the same group contain different arrays of nutrients. Among the fruits, for example, strawberries are especially rich in vitamin C while apricots are rich in vitamin A. Second, no food is guaranteed entirely free of substances that, in excess, could be harmful. The strawberries might contain trace amounts of one contaminant, the cantaloupes another. By alternating fruit choices, a person will ingest very little of either contaminant. (Contamination of foods is discussed in Chapter 19.) Third, as the adage goes, variety is the spice of life. Even if a person eats beans frequently, the person can enjoy pinto beans in Mexican burritos today, garbanzo beans in Greek salad tomorrow, and baked beans with barbecued chicken on the weekend. Eating nutritious meals need never be boring.

Dietary Guidelines for Americans

Figure 2-1 presents the 2000 *Dietary Guidelines for Americans.*[1]■ In general, the *Dietary Guidelines* answer the question, What should an individual eat to stay healthy? The first two guidelines encourage people to aim for fitness by combining sensible eating with regular physical activity to achieve and maintain a healthy weight. The next four guidelines urge people to build a healthy base by using the

■ Look for a revision of the *Dietary Guidelines* in 2005.

moderation (dietary): providing enough but not too much of a substance.

variety (dietary): eating a wide selection of foods within and among the major food groups.

FIGURE 2-1 *Dietary Guidelines for Americans:* The ABC's of Good Health

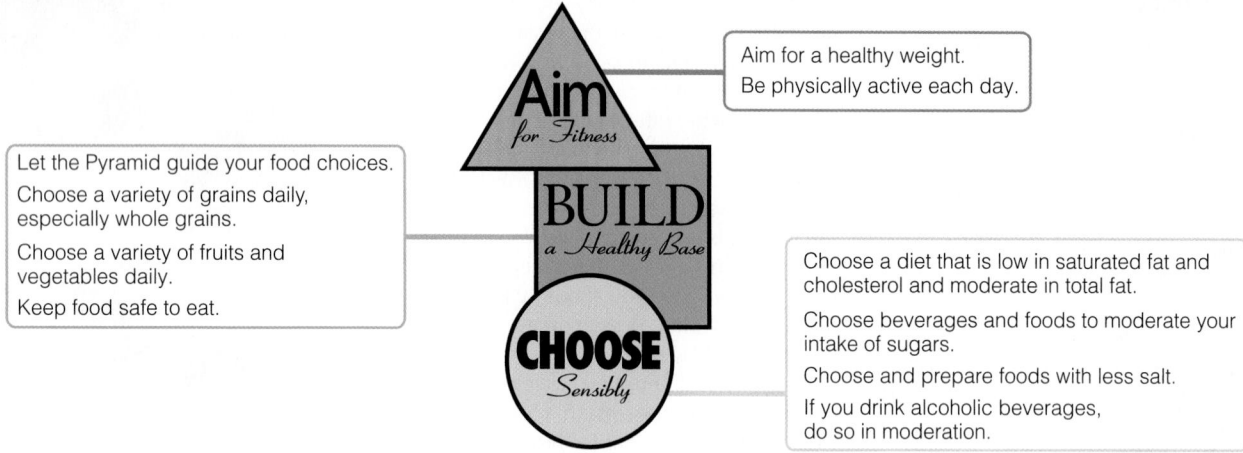

Aim for a healthy weight.
Be physically active each day.

Let the Pyramid guide your food choices.
Choose a variety of grains daily, especially whole grains.
Choose a variety of fruits and vegetables daily.
Keep food safe to eat.

Choose a diet that is low in saturated fat and cholesterol and moderate in total fat.
Choose beverages and foods to moderate your intake of sugars.
Choose and prepare foods with less salt.
If you drink alcoholic beverages, do so in moderation.

NOTE: These guidelines are intended for adults and healthy children ages 2 and older.
SOURCE: U.S. Department of Agriculture and U.S. Department of Health and Human Services, *Nutrition and Your Health: Dietary Guidelines for Americans,* Home and Garden Bulletin no. 232 (Washington, D.C.: 2000).

TABLE 2-1 *Canada's Guidelines for Healthy Eating*

- Enjoy a variety of foods.
- Emphasize cereals, breads, other grain products, vegetables, and fruits.
- Choose lower-fat dairy products, leaner meats, and foods prepared with little or no fat.
- Achieve and maintain a healthy body weight by enjoying regular physical activity and healthy eating.
- Limit salt, alcohol, and caffeine.

SOURCE: These guidelines derive from *Action Towards Healthy Eating—Canada's Guidelines for Healthy Eating and Recommended Strategies for Implementation.*

■ Five food groups:
  • Breads, cereals, and other grain products.
  • Vegetables.
  • Fruits.
  • Meat, poultry, fish, and alternates.
  • Milk, cheese, and yogurt.

food group plans: diet-planning tools that sort foods of similar origin and nutrient content into groups and then specify that people should eat certain numbers of servings from each group.

Food Guide Pyramid in meal planning (introduced on p. 43); choosing a variety of grains, vegetables, and fruits daily; and keeping foods safe. The last four guidelines encourage people to choose sensibly in their use of fats, sugars, salt, and alcoholic beverages for those who partake. Together, these ten guidelines point the way toward better health. Table 2-1 presents *Canada's Guidelines for Healthy Eating,* and Highlight 2 examines dietary guidelines from dozens of countries around the world, as well as those from the United States and Canada.

IN SUMMARY A well-planned diet delivers adequate nutrients, a balanced array of nutrients, and an appropriate amount of energy. It is based on nutrient-dense foods, moderate in substances that can be detrimental to health, and varied in its selections. The *Dietary Guidelines* apply these principles, offering practical advice on how to eat for good health.

Diet-Planning Guides

To plan a diet that achieves all of the dietary ideals just outlined, a person needs tools as well as knowledge. Two of the tools most widely used for diet planning are food group plans and exchange lists.

Food Group Plans

Food group plans build a diet from clusters of foods that are similar in origin and nutrient content. Thus each group represents a set of nutrients that differs from the nutrients supplied by the other groups. Selecting foods from each of the groups eases the task of creating a balanced diet.

Daily Food Guide Figure 2-2 (on pp. 44–45) presents the USDA's Daily Food Guide, a food group plan that assigns foods to five■ major food groups. The figure lists the number of servings recommended, the most notable nutrients of each group, the serving sizes, and the foods within each group categorized by nutrient density. It also includes an illustration of the USDA's Food Guide Pyramid, a pictorial representation of the Daily Food Guide. These guides apply to older children and adults only; young children have their own pyramid, which is presented in Chapter 16. Appendix I presents *Canada's Food Guide to Healthy Eating.*

Notable Nutrients   The beauty of the Daily Food Guide lies in its simplicity and flexibility. For example, a person can substitute cheese for milk because both supply the key nutrients for the milk group. A person following a food group plan receives not only the nutrients for which each group is noted, but small amounts of other nutrients and nonnutrients as well.

Miscellaneous Foods   Some foods do not fit into any of the food groups. Foods that are high in fat, sugar, or alcohol provide energy, but too few nutrients to hold a significant place in the diet. Such foods should be used sparingly and only after basic nutrient needs have been met by the foundation foods. Examples of "miscellaneous" foods include salad dressings, jams, and alcoholic beverages.

Mixtures of Foods   Some foods—such as casseroles, soups, and sandwiches—fall into two or more food groups. With a little practice, users can begin to see the number of servings from each food group. From the Daily Food Guide point of view, a chicken enchilada looks like one serving from each of four different food groups if it is made with a corn tortilla from the bread group; ½ cup chopped onion, pepper, and tomatoes from the vegetable group; 3 ounces of chicken from the meat group; and 1½ ounces of shredded cheese from the milk group.

Nutrient Density   The Daily Food Guide provides a foundation for a healthy diet, but it fails to account for the fat and energy differences between foods within a single group—for example, between fat-free milk and ice cream, baked fish and hot dogs, green beans and french fries, apples and avocados, or bread and biscuits. According to the Daily Food Guide, any of these substitutions would be acceptable. People who have low energy allowances are advised to select the most nutrient-dense foods within each group. Notice that Figure 2-2 provides a key indicating which foods *within each group* are high, moderate, or low in nutrient density.

Recommended Servings   As mentioned earlier, all food groups offer valuable nutrients, and people should make selections from each group daily. The recommended numbers of daily servings are expressed as ranges:

- 6 to 11 servings of breads and cereals.
- 3 to 5 servings of vegetables.
- 2 to 4 servings of fruits.
- 2 to 3 servings of meats and meat alternates.
- 2 servings of milk and milk products. (Older children, teenagers, young adults, women who are pregnant or breastfeeding, and older adults are advised to have 3 servings.)

Within these suggested ranges, the number of servings recommended depends on the person's energy needs, which, as Chapter 8 explains, depend on such personal characteristics as age, gender, and activity level. The lowest number of suggested servings from each group provides about the right amount of food energy (about 1600 kcalories) for young children, sedentary women, and older adults. The middle of each range (about 2200 kcalories) is appropriate for most children, teenage girls, active women, and sedentary men. The upper end of the range (about 2800 kcalories) meets the needs of teenage boys, active men, and very active women. Table 2-2 (on p. 46) shows recommended numbers of servings for each of these three levels. Physical activity increases a person's energy allowance and permits the person to eat more foods, or higher-kcalorie foods, to obtain the needed nutrients without gaining weight.

Serving Sizes   What counts as a serving? The answer differs for each food group and for various foods within a group. For example, a serving of milk is 1 cup

The Pyramid describes *half* of a 2-ounce bagel as one serving, yet most bagels today weigh in at 4 ounces or more—meaning that a person eating one of these large bagels for breakfast is actually getting four or more bread servings, not one.

© Matthew Farruggio

FIGURE 2-2 The Daily Food Guide and Food Guide Pyramid

Key:
- ● Foods generally highest in nutrient density (good first choice).
- ◐ Foods moderate in nutrient density (reasonable second choice).
- ● Foods lowest in nutrient density (limit selections).

BREADS, CEREALS, AND OTHER GRAIN PRODUCTS: 6 TO 11 SERVINGS PER DAY

These foods contribute complex carbohydrates, riboflavin, thiamin, niacin, folate, iron, protein, magnesium, and fiber.

Serving = 1 slice bread; ½ c cooked cereal, rice, or pasta; 1 oz ready-to-eat cereal; ½ bun, bagel, or English muffin; 1 small roll, biscuit, or muffin; 3 to 4 small or 2 large crackers.

- ● Whole grains (wheat, oats, barley, millet, rye, bulgur, couscous, polenta), enriched breads, rolls, tortillas, cereals, bagels, rice, pastas (macaroni, spaghetti), air-popped corn.
- ◐ Pancakes, muffins, cornbread, crackers, cookies, biscuits, presweetened cereals, granola, taco shells, waffles, french toast.
- ● Croissants, fried rice, doughnuts, pastries, cakes, pies.

VEGETABLES: 3 TO 5 SERVINGS PER DAY

(Use dark green, leafy vegetables and legumes several times a week.)
These foods contribute vitamin A, vitamin C, folate, potassium, magnesium, and fiber, and lack fat and cholesterol.

Serving = ½ c cooked or raw vegetables; 1 c leafy raw vegetables; ½ c cooked legumes; ¾ c vegetable juice.

- ● Bamboo shoots, bok choy, bean sprouts, broccoli, brussels sprouts, cabbage, carrots, cauliflower, corn, cucumbers, eggplant, green beans, green peas, leafy greens (spinach, mustard, and collard greens), legumes, lettuce, mushrooms, okra, onions, peppers, potatoes, pumpkin, scallions, seaweed, snow peas, soybeans, sweet potatoes, tomatoes, water chestnuts, winter squash.
- ◐ Candied sweet potatoes.
- ● French fries, tempura vegetables, scalloped potatoes, potato salad.

FRUITS: 2 TO 4 SERVINGS PER DAY

These foods contribute vitamin A, vitamin C, potassium, and fiber, and lack sodium, fat, and cholesterol.

Serving = typical portion (such as 1 medium apple, banana, or orange, ½ grapefruit, 1 melon wedge); ¾ c juice; ½ c berries; ½ c diced, cooked, or canned fruit; ¼ c dried fruit.

- ● Apples, apricots, bananas, blueberries, cantaloupe, grapefruit, guava, kiwi, oranges, papaya, peaches, pears, pineapples, plums, strawberries, watermelon; unsweetened juices.
- ◐ Canned or frozen fruit (in syrup); sweetened juices; dried fruit, coconut, avocados, olives.
- ● Punches, ades, and fruit drinks that contain little juice and lots of added sugars.

MEAT, POULTRY, FISH, AND ALTERNATES: 2 TO 3 SERVINGS PER DAY

Meat, poultry, and fish contribute protein, phosphorus, vitamin B₆, vitamin B₁₂, zinc, iron, niacin, and thiamin; legumes are notable for their protein, fiber, thiamin, folate, vitamin E, potassium, magnesium, iron, and zinc, and for their lack of fat and cholesterol.

Serving = 2 to 3 oz lean, cooked meat, poultry, or fish (total 5 to 7 oz per day); count 1 egg, ½ c cooked legumes, 4 oz tofu, ⅓ c nuts or seeds, or 2 tbs peanut butter as 1 oz meat (or about ⅓ serving).

- ● Poultry (light meat, no skin), fish, shellfish, legumes, egg whites.
- ◐ Lean meat (fat-trimmed beef, lamb, pork); poultry (dark meat, no skin); ham; refried beans; whole eggs, tofu, tempeh, peanut butter, nuts.
- ● Hot dogs, luncheon meats, ground beef, sausage, bacon, fried fish or poultry, duck.

FIGURE 2-2 Daily Food Guide and Food Guide Pyramid—continued

MILK, CHEESE, AND YOGURT: 2 SERVINGS PER DAY

(3 servings per day for older children, teenagers, young adults, pregnant/lactating women, and older adults.)
These foods contribute calcium, riboflavin, protein, vitamin B_{12}, and, when fortified, vitamin D and vitamin A.

Serving = 1 c milk or yogurt; 2 oz process cheese food; $1\frac{1}{2}$ oz cheese.

● Fat-free and 1% low-fat milk (and fat-free products such as buttermilk, cottage cheese, cheese, yogurt); fortified soy milk.

◐ 2% reduced-fat milk (and low-fat products such as yogurt, cheese, cottage cheese); chocolate milk; sherbet; ice milk.

● Whole milk (and whole-milk products such as cheese, yogurt); custard; milk shakes; pudding; ice cream.

> *Note: These serving recommendations were established before the 1997 DRI, which raised the recommended intake for calcium; meeting the calcium recommendation may require an additional serving from the milk, cheese, and yogurt group.*

FATS, SWEETS, AND ALCOHOLIC BEVERAGES: USE SPARINGLY

These foods contribute sugar, fat, alcohol, and food energy (kcalories). They should be used sparingly because they provide food energy while contributing few nutrients. Miscellaneous foods not high in kcalories, such as spices, herbs, coffee, tea, and diet soft drinks, can be used freely.

● Foods high in fat include margarine, salad dressing, oils, lard, mayonnaise, sour cream, cream cheese, butter, gravy, sauces, potato chips, chocolate bars.

● Foods high in sugar include cakes, pies, cookies, doughnuts, sweet rolls, candy, soft drinks, fruit drinks, jelly, syrup, gelatin, desserts, sugar, and honey.

● Alcoholic beverages include wine, beer, and liquor.

© Polara Studios Inc. (all)

Food Guide Pyramid

A Guide to Daily Food Choices
The breadth of the base shows that grains (breads, cereals, rice, and pasta) deserve most emphasis in the diet. The tip is smallest: use fats, oils, and sweets sparingly.

Key:
● Fat (naturally occurring and added)
▼ Sugars (added)
These symbols show fats, oils and added sugars in foods.

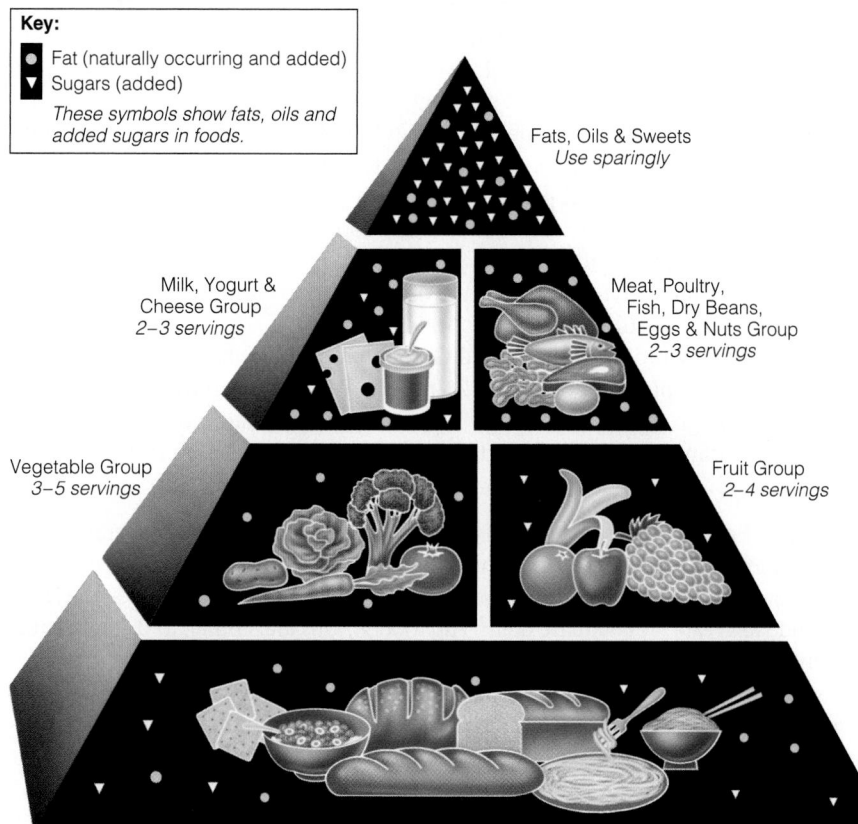

Fats, Oils & Sweets
Use sparingly

Milk, Yogurt & Cheese Group
2–3 servings

Meat, Poultry, Fish, Dry Beans, Eggs & Nuts Group
2–3 servings

Vegetable Group
3–5 servings

Fruit Group
2–4 servings

Bread, Cereal, Rice & Pasta Group
6–11 servings

TABLE 2-2 Recommended Servings for Different Energy Intakes

Food Group	Energy[a] (kcal)		
	1600	2200	2800
Bread, especially whole grain	6	9	11
Vegetable	3	4	5
Fruit	2	3	4
Milk, preferably fat-free or low fat[b]	2–3	2–3	2–3
Meat, preferably lean or low fat	2 (5 oz)	2 (6 oz)	3 (7 oz)

[a]Choose low-fat and lean foods from the five major food groups and use foods from the fats, oils, and sweets group sparingly.
[b]Older children, teenagers, young adults, women who are pregnant or breastfeeding, and older adults need 3 servings. In fact, given the 1997 DRI, which raised calcium recommendations, all individuals may need an additional serving from the milk group.

SOURCE: Adapted from U.S. Department of Agriculture, Center for Nutrition Policy and Promotion, *The Food Guide Pyramid*, Home and Garden Bulletin no. 252, 1996.

■ For quick and easy estimates, visualize each portion as being about the size of a common object:
- 1 c cooked vegetables = a fist.
- 1 medium fruit = a baseball.
- ¼ c dried fruit = a golf ball.
- 3 oz meat = a deck of cards.
- 2 tbs peanut butter = a marshmallow.
- 1½ oz cheese = 6 stacked dice.
- ½ c ice cream = a racquetball.
- 4 small cookies = 4 poker chips.

■ Legumes include a variety of beans and peas:
- Black beans.
- Black-eyed peas.
- Garbanzo beans.
- Great northern beans.
- Kidney beans.
- Lentils.
- Navy beans.
- Peanuts.
- Pinto beans.
- Soybeans.
- Split peas.

legumes (lay-GYOOMS, LEG-yooms): plants of the bean and pea family, rich in protein compared with other plant-derived foods.

whereas a serving of fruit juice is ¾ cup. Be aware that most people do not serve foods in carefully measured portions, nor do the amounts they actually put on their plates reflect the standard serving sizes. Standard serving sizes for the Food Guide Pyramid are generally smaller than people typically eat. Many restaurants, for example, offer 8- to 16-ounce steaks that are equivalent to four or five (2- to 3-ounce) *servings* of meat. Similarly, a bakery may sell muffins or bagels that are two to three times the size of a serving of bread.

Figure 2-2 lists the standard serving sizes for foods within each group. When counting servings, be sure to consider the quantity consumed.■ For example, ½ cup of cooked rice is considered one serving. So, 1 cup of rice counts as 2 of the recommended 6 to 11 daily servings from the bread group. Similarly, ¼ cup counts as ½ serving.

Food Guide Pyramid The Food Guide Pyramid is a graphic depiction of the Daily Food Guide (see Figure 2-2 again). The illustration was designed to depict variety, moderation, and also proportions: the relative size of each section represents the number of daily servings recommended. The broad base at the bottom conveys the message that grains should be abundant and form the foundation of a healthy diet. Fruits and vegetables appear at the next level, showing that they have a slightly less prominent, but still important, place in the diet. Meats and milks appear in a smaller band near the top. Careful selections from the meat and milk groups can contribute valuable nutrients, such as protein, vitamins, and minerals, without too much fat and cholesterol. Fats, oils, and sweets occupy the tiny apex, indicating that they should be used sparingly.

Alcoholic beverages do not appear in the Pyramid, but they too should be limited. Items such as spices, coffee, tea, and diet soft drinks provide few, if any, nutrients, but can add flavor and pleasure to meals when used judiciously.

Tiny dots and triangles are sprinkled over the food groups to represent naturally occurring and added fats and added sugars, respectively. These symbols are meant to remind users that specific foods within the various groups are high in fats, sugars, or both, and so should be eaten in moderation.

The Daily Food Guide plan and Food Guide Pyramid emphasize grains, vegetables, and fruits—all plant foods. Some 75 percent of a day's servings should come from these three groups. This strategy helps all people obtain complex carbohydrates, fibers, vitamins, minerals, and phytochemicals with little fat. It also eases diet planning for vegetarians.

Vegetarian Food Guide Vegetarian diets rely mainly on plant foods: grains, vegetables, **legumes,**■ fruits, seeds, and nuts. Some vegetarian diets include eggs, milk products, or both. People who do not eat meats or milk products can still use the Daily Food Guide to create an adequate diet.[2] The food groups are similar, and the number of servings remains the same. Vegetarians select *meat alternates* from the meat group—foods such as legumes, seeds, nuts, tofu, and, for those who eat them, eggs. Legumes and at least one cup of dark leafy greens help to supply the iron that meats usually provide. Vegetarians who do not drink cow's milk can use soy "milk"—a product made from soybeans that provides similar nutrients if it has been fortified with calcium, vitamin D, and vitamin B_{12}. Highlight 6 presents a Daily Food Guide for Vegetarians, defines vegetarian terms, and provides more information on vegetarian diet planning.

Ethnic Food Choices People can use the Food Guide Pyramid and still enjoy a diverse array of cuisines by sorting ethnic foods into their appropriate food groups.

TABLE 2-3 Ethnic Cuisines and Food Choices

	Grains	Vegetables	Fruits	Meats and Alternates	Milk
Asian	Rice, noodles, millet	Amaranth, baby corn, bamboo shoots, chayote, bok choy, mung bean sprouts, sugar peas, straw mushrooms, water chestnuts, kelp	Carambola, guava, kumquat, lychee, persimmon, melons, mandarin orange	Soybeans, squid, tofu, duck eggs, pork, poultry, fish and other seafood, peanuts, cashews	Soy milk
Mediterranean	Pita pocket bread, pastas, rice, couscous, polenta, bulgur, focaccia, Italian bread	Eggplant, tomatoes, peppers, cucumbers, grape leaves	Olives, grapes, figs	Fish and other seafood, gyros, lamb, chicken, beef, pork, sausage, lentils, fava beans	Ricotta, provolone, parmesan, feta, mozzarella, and goat cheeses; yogurt
Mexican	Tortillas (corn or flour), taco shells, rice	Chayote, corn, jicama, tomato salsa, cactus, cassava, tomatoes, yams, chilies	Guava, mango, papaya, avocado, plantain, bananas, oranges	Refried beans, fish, chicken, chorizo, beef, eggs	Cheese, custard

© Becky Luigart-Stayner/Corbis

© PhotoDisc Inc.

© PhotoDisc Inc.

For example, a person eating Mexican foods would find tortillas in the bread group, jicama in the vegetable group, and guava in the fruit group. Table 2-3 features foods unique to selected cuisines, and Highlight 2 presents food guides from selected countries.

Perceptions and Actual Intakes The Daily Food Guide and Food Guide Pyramid were developed to help people choose a balanced and healthful diet. Are these plans successful? Yes, they can help people select nutrient-rich diets. In fact, one survey reports that only adults who select the recommended number of servings from each of the five food groups meet recommendations for energy, fat, fiber, vitamins, and minerals.[3] Unfortunately, only 1 percent of the more than 8000 people surveyed made such selections. More commonly, they neglected the fruit, milk, and bread groups and, by doing so, raised their percentage of kcalories from fat and lowered their fiber, calcium, and zinc intakes.

Many adults *think* they are selecting foods that reflect the recommendations of the Food Guide Pyramid. In reality they are barely meeting, or falling short of, recommendations for all food groups and are eating too many fats, sweets, and oils.[4] In a sense, our pyramids are top-heavy and "tumbling," as Figure 2-3 shows. They need more support from each of the five food groups to build a balanced diet.

Increase the proportion of persons aged 2 years and older who consume at least two daily servings of fruit; at least three daily servings of vegetables, with at least one-third being dark green or orange vegetables; and at least six daily servings of grain products, with at least three being whole grains.

HEALTHY PEOPLE 2010

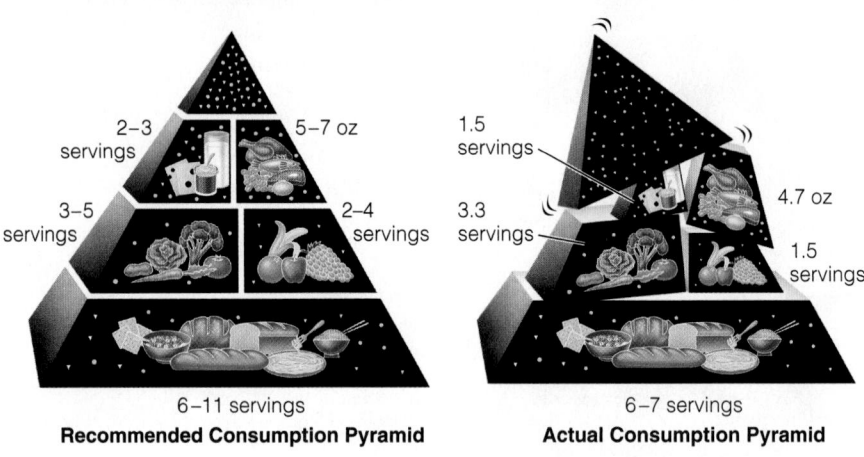

FIGURE 2-3 Recommended Consumption Pyramid and Actual Consumption Pyramid Compared

Recommended Consumption Pyramid

Actual Consumption Pyramid

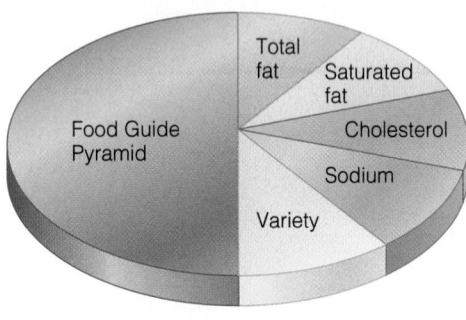

FIGURE 2-4 Healthy Eating Index Components

Nutrition Explorer Use the Healthy Eating Index to compare your diet against the recommendations of the Dietary Guidelines for Americans.

Healthy Eating Index: a measure developed by the USDA for assessing how well a diet conforms to the recommendations of the Food Guide Pyramid and the *Dietary Guidelines for Americans.*

exchange lists: diet-planning tools that organize foods by their proportions of carbohydrate, fat, and protein. Foods on any single list can be used interchangeably.

Healthy Eating Index How can a person know whether his or her diet is doing its share to support good health? To measure how well a diet meets the recommendations of the *Dietary Guidelines* and the Food Guide Pyramid, the USDA developed the **Healthy Eating Index.** Meeting the recommendations of the *Dietary Guidelines* for total fat, saturated fat, cholesterol, sodium, and variety can each provide up to 10 points as can sufficient selections from each of the five food groups of the Food Guide Pyramid—for a possible total of 100 points. Figure 2-4 shows how these dietary components contribute to the overall score, and the accompanying "How to" provides instructions for scoring a diet based on the Healthy Eating Index.

Pyramid Shortcomings The Food Guide Pyramid is not perfect and critics are quick to point out its flaws.[5] Perhaps its greatest weakness is in not distinguishing among the individual foods within each group. Not all fats are created equal, for example, nor are all fats "bad" (as Highlight 5 explains). In fact, as Chapter 5 will explain, some fats are essential to good health. Butter and lard may be damaging to the heart, but the oils of fish, nuts, and vegetables help protect against heart disease. Simply advising consumers to eat fats "sparingly" simplifies and distorts the message that some fats are harmful and others are beneficial. Similarly, gathering all carbohydrate-rich breads, cereals, rice, and pastas together in one food group delivers a mistaken message of equality. Refined products, especially those with added fats and sugars, fall short of supporting health the way whole grains do. The USDA is currently reexamining the Pyramid and considering possible revisions to reflect some of these concerns. Many of the upcoming chapters examine the links between diet and health, and Chapter 18 presents a complete summary, including another look at possible Pyramid revisions.

Exchange Lists

Food group plans are particularly well suited to help a person achieve dietary adequacy, balance, and variety. **Exchange lists** provide additional help in achieving kcalorie control and moderation. Originally developed for people with diabetes, exchange systems have proved useful for general diet planning as well.

Unlike the Daily Food Guide, which sorts foods primarily by their source (such as milks) and their vitamin and mineral contents, the exchange system sorts foods

HOW TO Use the Healthy Eating Index to Assess Your Diet

The Healthy Eating Index appraises a diet according to the Food Guide Pyramid and the *Dietary Guidelines.* Ten different components of the diet are given scores from 0 to 10 each—thus 100 is a perfect score.

The first five scores reflect how well the diet meets the serving recommendations of the Food Guide Pyramid. Each of the five food groups can contribute up to 10 points, if the diet meets at least the recommended minimum number of servings. A diet could receive 10 points for providing six or more grains, 10 points for three or more vegetables, 10 points for two or more fruits, 10 points for two or more milks, and 10 points for two or more meats. If no servings from a particular group were eaten, score 0; if fewer than the recommended number of servings were eaten, give a partial score (such as 5 points for eating one of the two recommended fruits). Extra foods in any group receive no extra points (seven grains receive 10 points, just as six grains do).

Extra foods do not receive penalties either, even though they may contribute to an excessively high energy intake. The Healthy Eating Index accounts for energy intake only in setting the *minimum* recommended number of servings from each food group, not in comparing energy intake with needs. Consequently a person can overeat and still receive an excellent score. Another apparent shortcoming of this index is that a person who scores the maximum number of points in all areas except the milk group, for example, would still end up with an outstanding final score. Yet, if the person's diet consistently lacked foods from the milk group without careful substitutions, then calcium nutrition would suffer. Like many good tools, the Healthy Eating Index accomplishes much of what it sets out to do, but it cannot complete the job alone. Common sense based on nutrition knowledge is needed to round out the evaluation.

The next four dietary components in the Healthy Eating Index reflect nutrients that the *Dietary Guidelines* advise consumers to limit—fat, saturated fat, cholesterol, and sodium. Not exceeding recommended amounts of each of these can contribute up to 10 points. The final component of a healthy diet—variety in foods choices—also contributes up to 10 points if the diet includes at least eight different foods in a day. The accompanying table provides a scoring guide for the *Dietary Guidelines* portion of the Healthy Eating Index.

Review Figure 2-4 to see how these ten dietary components are combined to produce the Healthy Eating Index. Scores above 80 are considered good; those between 50 and 80 need improvement; and those below 50 are considered poor. Higher scores are associated with a greater likelihood of meeting health goals and dietary recommendations for most nutrients.

Scoring Guide
Dietary Guidelines Portion of the Healthy Eating Index

Points	Total Fat (% of kcal)	Saturated Fat (% of kcal)	Cholesterol (mg)	Sodium (mg)	Variety[a]
10	≤30.0	≤10.0	≤300	≤2400	≥8.0
9	31.5	10.5	315	2640	7.5
8	33.0	11.0	330	2880	7.0
7	34.5	11.5	345	3120	6.5
6	36.0	12.0	360	3360	6.0
5	37.5	12.5	375	3600	5.5
4	39.0	13.0	390	3840	5.0
3	40.5	13.5	405	4080	4.5
2	42.0	14.0	420	4320	4.0
1	43.5	14.5	435	4560	3.5
0	≥45.0	≥15.0	≥450	≥4800	≤3.0

[a]Values for variety are based on a one-day period; for a three-day period, the values range from 16 to 6.

according to their energy-nutrient contents. Consequently, foods do not always appear on the exchange list where you might first expect to find them. For example, cheeses are grouped with meats because, like meats, cheeses contribute energy from protein and fat but provide negligible carbohydrate. (In the food group plan presented earlier, cheeses are classed with milk because they are milk products with similar calcium contents.)

TABLE 2-4   Diet Planning Using the Daily Food Guide

This diet plan is one of many possibilities. It follows the minimum number of servings suggested by the Daily Food Guide.

Food Group	Breakfast	Lunch	Dinner
Grains (breads and cereals)— choose 6 to 11 servings	1	2	3
Vegetables—choose 3 to 5 servings			3
Fruit—choose 2 to 4 servings	1	1	
Meat—choose 2 to 3 servings (5 to 7 oz)		1	1
Milk—choose 2 servings	1	1	

For similar reasons, starchy vegetables such as corn, green peas, and potatoes are listed on the starch list in the exchange system, rather than with the vegetables. Likewise, olives are not classed as a "fruit" as a botanist would claim; they are classified as a "fat" because their fat content makes them more similar to butter than to berries. Bacon and nuts are also on the fat list to remind users of their high fat content. These groupings highlight the characteristics of foods that are significant to energy intake. To learn more about this useful diet-planning tool, study Appendix G, which gives complete details of the major exchange system used in the United States, and Appendix I, which provides details of the choice system used in Canada.

Putting the Plan into Action

Table 2-4 and Figure 2-5 show how to use the Daily Food Guide to plan a diet. The Daily Food Guide ensures that a certain number of servings is chosen from each of the five food groups (see the first column of the table). The next step in diet planning is to assign the food groups to meals (and snacks), as in the remaining columns of Table 2-4.

FIGURE 2-5   A Sample Diet Plan and Menu

SAMPLE MENU		Servings	Energy (kcal)
Breakfast	1 oz cereal	1 grain	110
	1 banana	1 fruit	109
	1 c fat-free milk	1 milk	86
Lunch	1 turkey sandwich	2 grains, 1 meat	360
	1 bunch grapes	1 fruit	53
	1 c fat-free milk	1 milk	86
Dinner	1½ c spaghetti with meat sauce	1½ grains, 1 meat, 1 vegetable	554
	1 c tossed salad	1 vegetable	88
	½ c green beans	1 vegetable	22
	1 slice Italian bread	1 grain	54
	2 graham crackers	½ grain	59

Matthew Farruggio (all)

This sample menu provides about 1600 kcalories and meets dietary recommendations to provide 45 to 65 percent of its kcalories from carbohydrate, 20 to 35 percent from fat, and 10 to 35 percent from protein. The mayonnaise in the sandwich and the salad dressing on the salad count as added fat at the tip of the Pyramid.

Next, a person could begin to fill in the plan with real foods to create a menu. For example, the breakfast calls for 1 grain, 1 fruit, and 1 milk. A person might select a bowl of cereal with banana slices and milk:

> 1 ounce cereal = 1 grain.
> 1 banana = 1 fruit.
> 1 cup fat-free milk = 1 milk.

Or ½ bagel and a bowl of cantaloupe pieces topped with yogurt:

> ½ bagel = 1 grain.
> ½ cup melon pieces = 1 fruit.
> 1 cup fat-free plain yogurt = 1 milk.

Then the person could move on to complete the menu for lunch, dinner, and snacks. The final plan might look like the one in Figure 2-5. With the addition of a small amount of fat, this sample diet plan provides about 1600 kcalories.

As you can see, we all make countless food-related decisions daily—whether we have a plan or not. Following a plan, like the Daily Food Guide, that incorporates health recommendations and diet-planning principles helps a person to make wise decisions. Figure 2-6 illustrates how to think "Pyramid" when planning a meal.

From Guidelines to Groceries

Dietary recommendations emphasize nutrient-rich foods such as whole grains, fruits, vegetables, lean meats, fish, poultry, and low-fat milk products. You can design such a diet for yourself, but how do you begin? Start with the foods you enjoy

FIGURE 2-6   Thinking "Pyramid" When Planning Meals

Selecting foods from three or more different food groups helps to add variety and nutrient balance. This meal provides two servings of grains, one serving of vegetables, one serving of meat, and a little fat. Other meals should provide milk, fruit, vegetables, an additional meat (or meat alternate) serving, and whole grains to complete the Pyramid's recommendations.

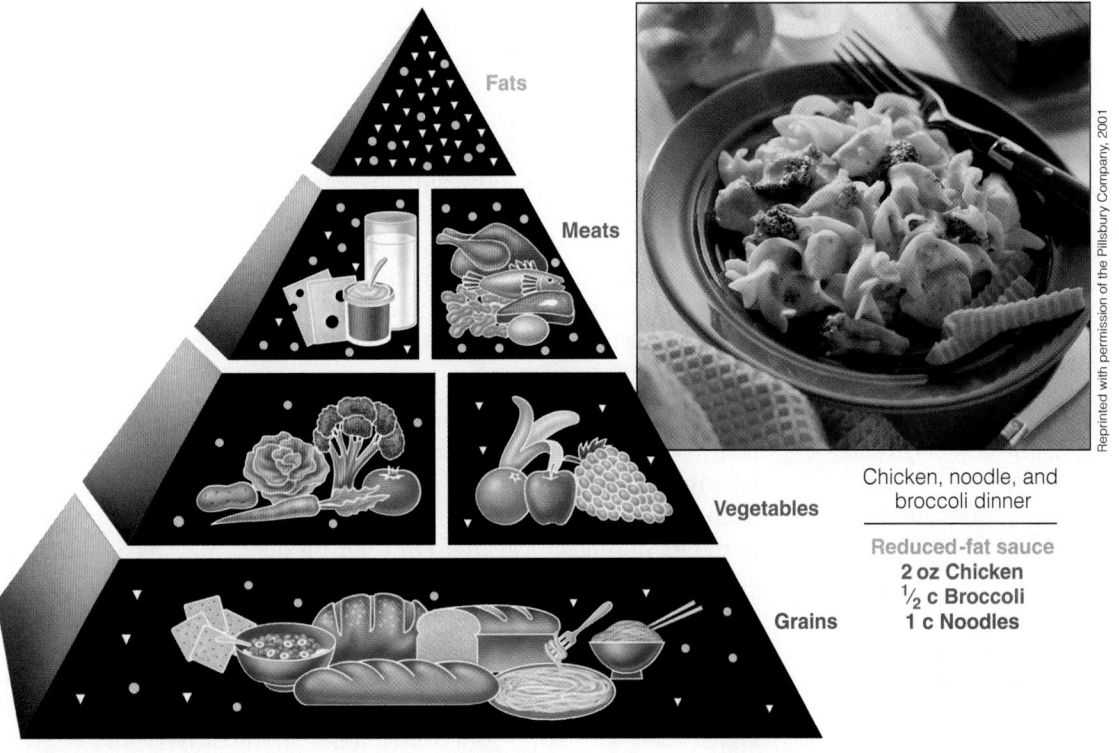

Chicken, noodle, and broccoli dinner

Reduced-fat sauce
2 oz Chicken
½ c Broccoli
1 c Noodles

Reprinted with permission of the Pillsbury Company, 2001

FIGURE 2-7 A Wheat Plant

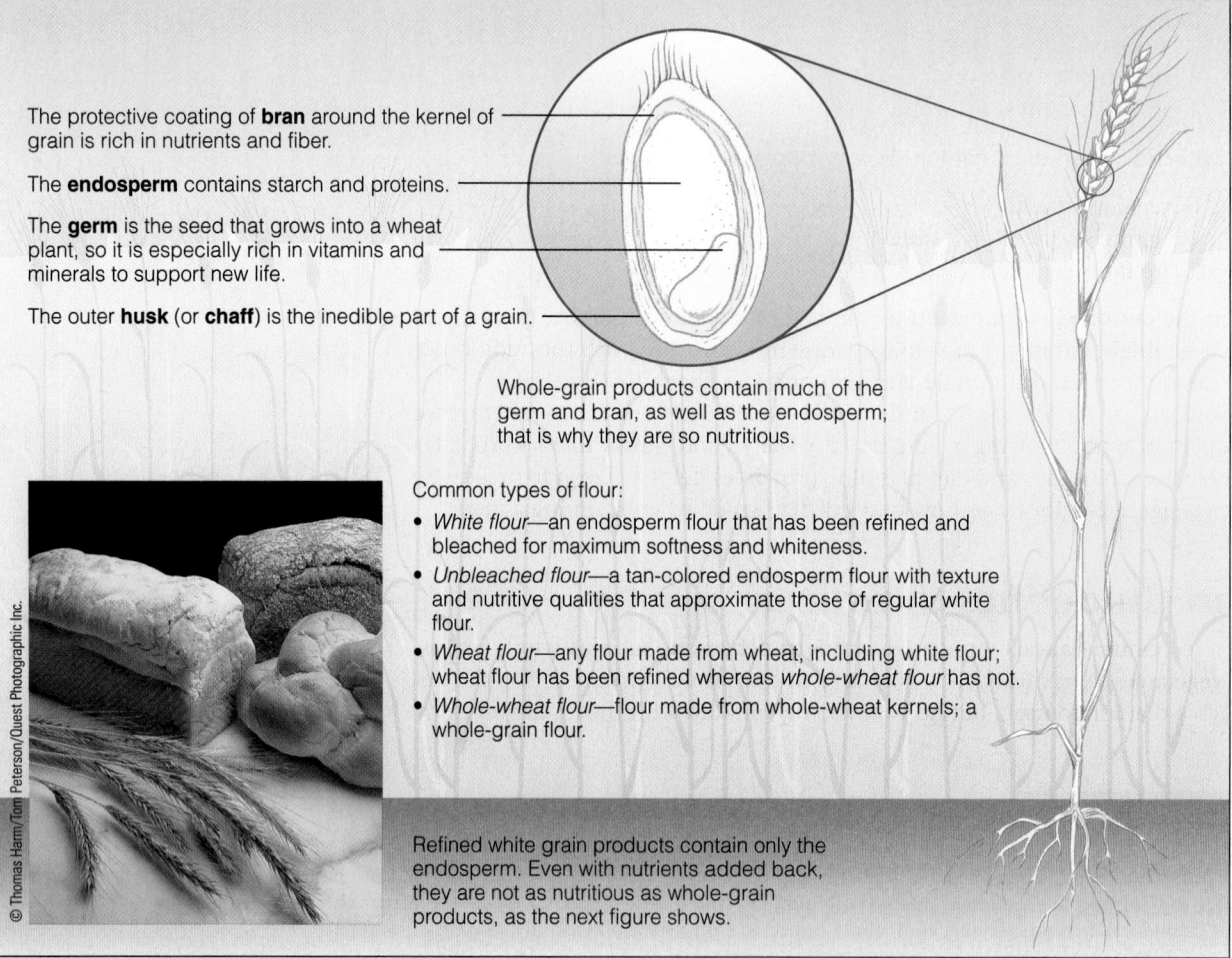

The protective coating of **bran** around the kernel of grain is rich in nutrients and fiber.

The **endosperm** contains starch and proteins.

The **germ** is the seed that grows into a wheat plant, so it is especially rich in vitamins and minerals to support new life.

The outer **husk** (or **chaff**) is the inedible part of a grain.

Whole-grain products contain much of the germ and bran, as well as the endosperm; that is why they are so nutritious.

Common types of flour:

- *White flour*—an endosperm flour that has been refined and bleached for maximum softness and whiteness.
- *Unbleached flour*—a tan-colored endosperm flour with texture and nutritive qualities that approximate those of regular white flour.
- *Wheat flour*—any flour made from wheat, including white flour; wheat flour has been refined whereas *whole-wheat flour* has not.
- *Whole-wheat flour*—flour made from whole-wheat kernels; a whole-grain flour.

Refined white grain products contain only the endosperm. Even with nutrients added back, they are not as nutritious as whole-grain products, as the next figure shows.

© Thomas Harm/Tom Peterson/Quest Photographic Inc.

---

processed foods: foods that have been treated to change their physical, chemical, microbiological, or sensory properties.

fortified: the addition to a food of nutrients that were either not originally present or present in insignificant amounts. Fortification can be used to correct or prevent a widespread nutrient deficiency or to balance the total nutrient profile of a food.

refined: the process by which the coarse parts of a food are removed. When wheat is refined into flour, the bran, germ, and husk are removed, leaving only the endosperm.

enriched: the addition to a food of nutrients that were lost during processing so that the food will meet a specified standard.

whole grain: a grain milled in its entirety (all but the husk), not refined.

eating. Then try to make improvements, little by little. When shopping, think of the food groups, and choose nutrient-dense foods within each group.

Be aware that many of the 50,000 food options available today are **processed foods** that have lost valuable nutrients and gained sugar, fat, and salt as they were transformed from farm-fresh foods to those found in the bags, boxes, and cans that line grocery-store shelves. Their value in the diet depends on the starting food and the type of processing. Sometimes these foods have been **fortified** to improve their nutrient contents.

Breads, Cereals, and Other Grain Products When shopping for grain products, you will find them described as *refined, enriched,* or *whole grain*. These terms refer to the milling process and the making of grain products, and they have different nutrition implications (see Figure 2-7). **Refined** foods may have lost many nutrients during processing; **enriched** products may have had some nutrients added back; and **whole-grain** products may be rich in fiber and all the nutrients found in the original grain. As such, whole-grain products support good health and should be eaten regularly.

When it became a common practice to refine the wheat flour used for bread by milling it and throwing away the bran and the germ, consumers suffered a tragic loss of many nutrients.[6] As a consequence, in the early 1940s Congress passed legislation requiring that all grain products that cross state lines be enriched with iron, thiamin, riboflavin, and niacin. In 1996, this legislation was amended to include folate, a vitamin considered essential in the prevention of some birth defects. Most grain products that have been refined, such as rice, wheat pastas like macaroni and spaghetti, and cereals (both cooked and ready-to-eat types), have subsequently been enriched,■ and their labels say so.

Enrichment doesn't make a slice of bread rich in these added nutrients, but people who eat several slices a day obtain significantly more of these nutrients than they would from unenriched white bread. To a great extent, the enrichment of white flour helps to prevent deficiencies of these nutrients, but it fails to compensate for losses of many other nutrients and fiber. As Figure 2-8 shows, whole-grain items still outshine the enriched ones. Only *whole-grain* flour contains all of the nutritive portions of the grain. Whole-grain products, such as brown rice or oatmeal, not only provide more nutrients and fiber, but do not contain the added salt and sugar of flavored, processed rice or sweetened cereals.

Speaking of cereals, ready-to-eat breakfast cereals are the most highly fortified foods on the market. Like an enriched food, a *fortified* food has had nutrients added during processing, but in a fortified food, the added nutrients may not have been present in the original product. (The terms *fortified* and *enriched* may be used interchangeably.)[7] Some breakfast cereals made from refined flour and fortified with high doses of vitamins and minerals are actually more like supplements

When shopping for bread, look for the descriptive words *whole grain* or *whole wheat* and check the fiber contents on the Nutrition Facts panel of the label—the more fiber, the more likely the bread is a whole-grain product.

■ Grain enrichment nutrients:
- Iron.
- Thiamin.
- Riboflavin.
- Niacin.
- Folate.

FIGURE 2-8   Nutrients in Bread

Whole-grain bread is more nutritious than other breads, even enriched bread. For iron, thiamin, riboflavin, niacin, and folate, enriched bread provides about the same quantities as whole-grain bread and significantly more than unenriched bread. For fiber and the other nutrients (both those shown here and those not shown), enriched bread provides less than whole-grain bread.

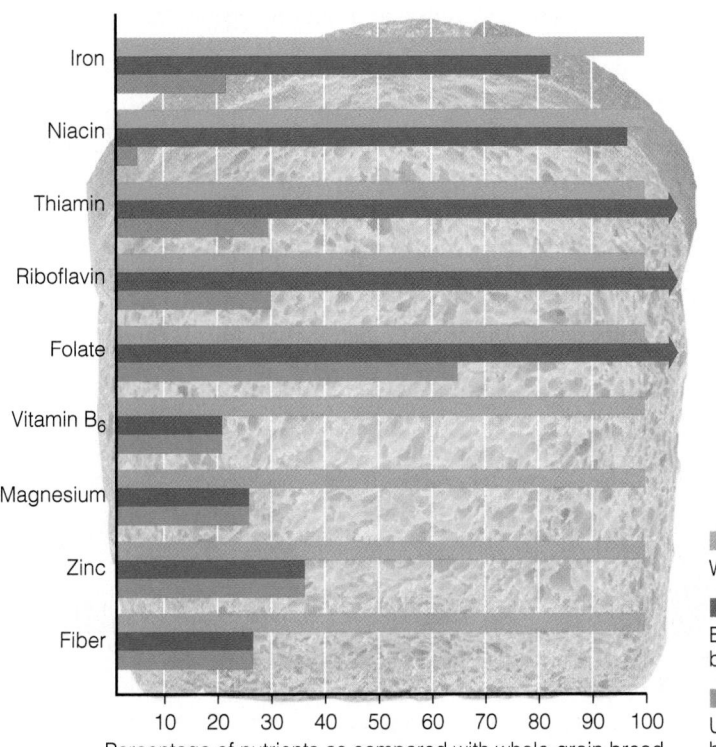

Whole-grain bread

Enriched white bread

Unenriched white bread

Percentage of nutrients as compared with whole-grain bread

Eat 5 to 9 a Day for Better Health

The "5 to 9 a Day" campaign (**www.5aday.gov**) encourages consumers to eat a variety of fruits and vegetables by selecting a serving or two from each of five colors.

disguised as cereals than they are like whole grains. They may be nutritious—with respect to the nutrients added—but they still may fail to convey the full spectrum of nutrients that a whole-grain food or a mixture of such foods might provide. Furthermore, minerals (especially iron) are not as well absorbed from enriched foods as from naturally occurring sources. Still, fortified foods help people meet their vitamin and mineral needs.[8]

Vegetables Posters in the produce section of grocery stores encourage consumers to "eat 5 a day." Such efforts are part of a national educational campaign to increase fruit and vegetable consumption to 5 to 9 servings every day (see Figure 2-9). To help consumers remember to eat a variety of fruits and vegetables, the campaign provides practical tips, such as selecting from each of five colors.

Choose fresh vegetables often, especially dark green leafy and yellow-orange vegetables like spinach, broccoli, and sweet potatoes. Cooked or raw, vegetables are good sources of vitamins, minerals, and fiber. Frozen and canned vegetables without added salt are acceptable alternatives to fresh. To control fat, energy, and sodium intakes, limit butter, salad dressings, and salt on vegetables.

Legumes Choose often from the variety of legumes available. They are an economical, low-fat, nutrient- and fiber-rich food choice.

Fruit Choose fresh fruits often, especially citrus fruits and yellow-orange fruits like cantaloupes and peaches. Frozen, dried, and canned fruits without added sugar are acceptable alternatives to fresh. Fruits supply valuable vitamins, minerals, fibers, and phytochemicals. They add flavors, colors, and textures to meals, and their natural sweetness makes them enjoyable as snacks or desserts.

Fruit juices are healthy beverages, but contain little dietary fiber compared with whole fruits. Whole fruits satisfy the appetite better than juices, thereby helping

Combining legumes with foods from other food groups creates delicious meals.

Add rice to red beans for a hearty meal.

Enjoy a Greek salad topped with garbanzo beans for a little ethnic diversity.

A bit of meat and lots of spices turn kidney beans into chili con carne.

people to limit food energy intakes. For people who need extra food energy, though, juices are a good choice. Be aware that sweetened fruit "drinks" or "ades" contain mostly water, sugar, and a little juice for flavor. Some may have been fortified with vitamin C, but lack any other significant nutritional value.

Meat, Fish, and Poultry Meat, fish, and poultry provide essential minerals, such as iron and zinc, and abundant B vitamins as well as protein. To buy and prepare these foods without excess energy, fat, and sodium takes a little knowledge and planning. When shopping in the meat department, choose fish, poultry, and lean cuts of beef and pork named "round" or "loin" (as in top round or pork tenderloin). As a guide, "prime" and "choice" cuts generally have more fat than "select" cuts. Restaurants usually serve prime cuts. Ground beef, even "lean" ground beef, derives most of its food energy from fat. Have the butcher trim and grind a lean round steak instead. Alternatively, **textured vegetable protein** can be used instead of ground beef in a casserole, spaghetti sauce, or chili, saving fat kcalories.

Weigh meat after it is cooked and the bones and fat are removed. In general, 4 ounces of raw meat is equal to about 3 ounces of cooked meat. Some examples of 3-ounce portions of meat include 1 medium pork chop, ½ chicken breast, or 1 steak or hamburger about the size of a deck of cards. To keep fat intake moderate, bake, roast, broil, grill, or braise meats (but do not fry them in fat); remove the skin from poultry after cooking; trim visible fat before cooking; and drain fat after cooking. Chapter 5 offers many additional strategies for moderating fat intake.

Milk Shoppers will find a variety of fortified foods in the dairy case. Examples are milk, to which vitamins A and D have been added, and soy milk,■ to which calcium, vitamin D, and vitamin B_{12} have been added. In addition, shoppers may find **imitation foods** (such as cheese products), **food substitutes** (such as egg substitutes), and functional foods■ (such as margarine with plant sterols). As food technology advances, many such foods offer alternatives to traditional choices that may help people who want to reduce their fat and cholesterol intakes. Chapter 5 gives other examples.

When shopping, choose fat-free■ or low-fat milk, yogurt, and cheeses. Such selections help consumers moderate their fat intake. Milk products are important sources of calcium, but can provide too much sodium and fat if not selected with care.

■ Be aware that not all soy milks have been fortified. Read labels carefully.

■ Reminder: *Functional foods* contain physiologically active compounds that provide health benefits beyond basic nutrition.

■ • **Fat-free** milk may also be called **nonfat, skim, zero-fat,** or **no-fat.**
   • **Low-fat** milk refers to 1% milk.
   • **Reduced-fat** milk refers to 2% milk; it may also be called **less-fat.**

IN SUMMARY Food group plans select from different families of similar foods to provide adequacy, balance, and variety in the diet. They make it easier to plan a diet that includes abundant grains, vegetables, legumes, and fruits and moderate amounts of meats and milk products. In making any food choice, remember to view the food in the context of your total diet. It is the combination of many different foods that provides the abundance of nutrients so essential to a healthy diet.

Food Labels

Many consumers read food labels to help them select foods with less fat, saturated fat, cholesterol, and sodium and more complex carbohydrates and dietary fiber. Food labels appear on virtually all processed foods, and posters or brochures provide similar nutrition information for fresh meats, fruits, and vegetables (see Figure 2-10). A few foods need not carry nutrition labels: those contributing few nutrients, such as plain coffee, tea, and spices; those produced by small businesses; and those prepared and sold in the same establishment. Producers of some of these items, however, voluntarily use labels. Even markets selling nonpackaged items voluntarily present nutrient information, either in brochures or on signs posted at

textured vegetable protein: processed soybean protein used in vegetarian products such as soy burgers.

imitation foods: foods that substitute for and resemble another food, but are nutritionally inferior to it with respect to vitamin, mineral, or protein content. If the substitute is not inferior to the food it resembles and if its name provides an accurate description of the product, it need not be labeled "imitation."

food substitutes: foods that are designed to replace other foods.

FIGURE 2-10 Example of a Food Label

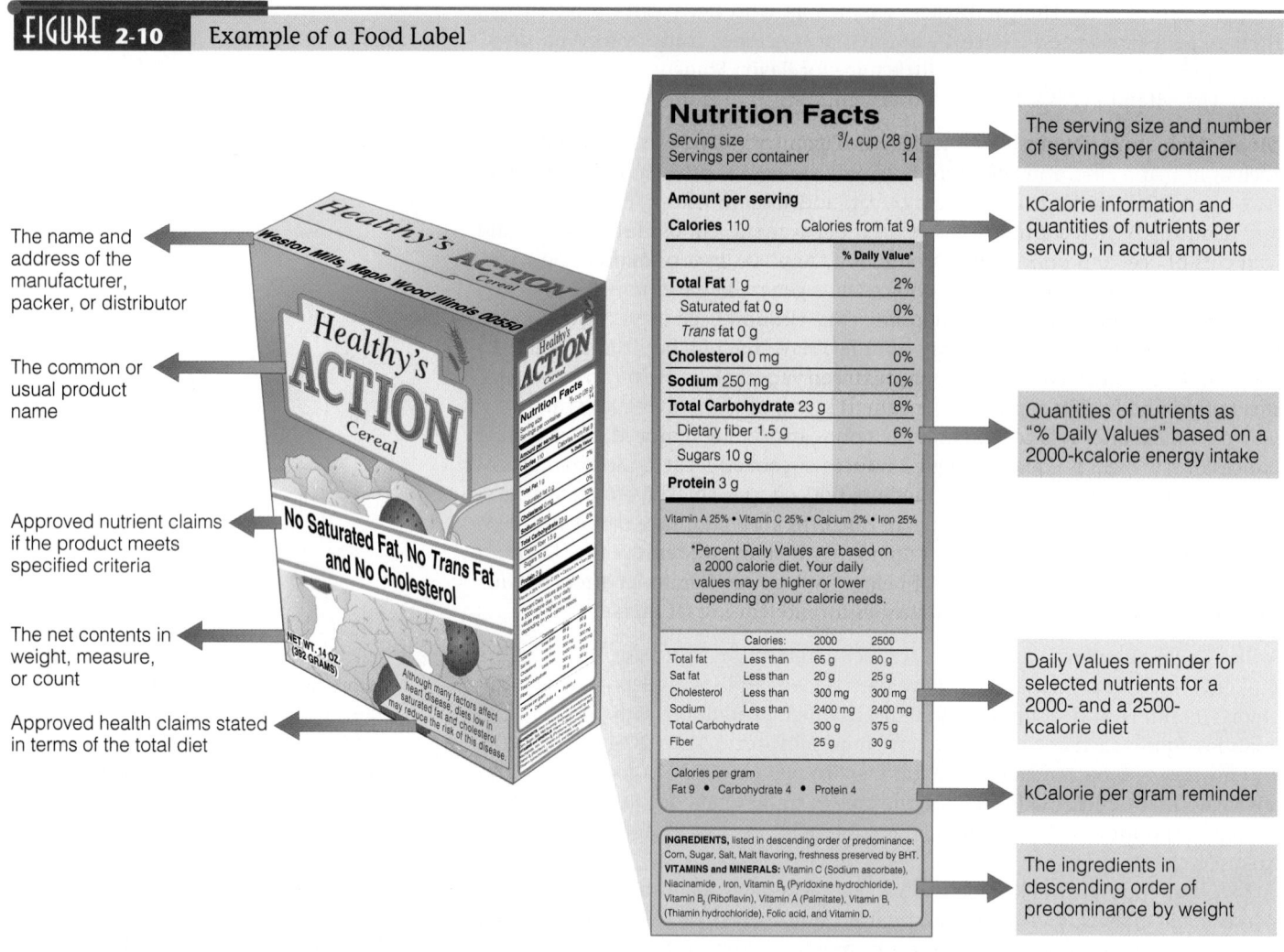

The name and address of the manufacturer, packer, or distributor

The common or usual product name

Approved nutrient claims if the product meets specified criteria

The net contents in weight, measure, or count

Approved health claims stated in terms of the total diet

The serving size and number of servings per container

kCalorie information and quantities of nutrients per serving, in actual amounts

Quantities of nutrients as "% Daily Values" based on a 2000-kcalorie energy intake

Daily Values reminder for selected nutrients for a 2000- and a 2500-kcalorie diet

kCalorie per gram reminder

The ingredients in descending order of predominance by weight

the point of purchase. Restaurants need not supply complete nutrition information for menu items unless claims such as "low fat" or "heart healthy" have been made. When ordering such items, keep in mind that restaurants tend to serve extra-large portions—two to three times standard serving sizes. A "low-fat" ice cream, for example, may have only 3 grams of fat per ½ cup, but you may be served 2 cups for a total of 12 grams of fat and all their accompanying kcalories.

The Ingredient List

All foods must list all ingredients on the label in descending order of predominance by weight. Knowing that the first ingredient predominates by weight, consumers can glean much information. Compare these products, for example:

- A beverage powder that contains "sugar, citric acid, natural flavors . . ." versus a juice that contains "water, tomato concentrate, concentrated juices of carrots, celery. . . ."
- A cereal that contains "puffed milled corn, sugar, corn syrup, molasses, salt . . ." versus one that contains "100 percent rolled oats."
- A canned fruit that contains "sugar, apples, water" versus one that contains simply "apples, water."

In each comparison, consumers can tell that the second product is the more nutrient dense.

Serving Sizes

Because labels present nutrient information per serving, they must identify the size of a serving. The Food and Drug Administration (FDA) has established specific serving sizes for various foods and requires that all labels for a given product use the same serving size. For example, the serving size for all ice creams is ½ cup and for all beverages, 8 fluid ounces. This facilitates comparison shopping. Consumers can see at a glance which brand has more or fewer kcalories or grams of fat, for example. Standard serving sizes are expressed in both common household measures, such as cups, and metric measures, such as milliliters, to accommodate users of both types of measures (see Table 2-5).

When examining the nutrition facts on a food label, consumers need to consider how the serving size compares with the actual quantity eaten. If it is not the same, they will need to adjust the quantities accordingly. For example, if the serving size is four cookies and you only eat two, then you need to cut the nutrient and kcalorie values in half; similarly, if you eat eight cookies, then you need to double the values. Notice, too, that small bags or individually wrapped items, such as chips or candy bars, may contain more than a single serving. The number of servings per container is listed just below the serving size.

Be aware that serving sizes on food labels are not always the same as those of the Pyramid.[9] For example, a serving of rice on a food label is 1 cup, whereas in the Pyramid it is ½ cup. Unfortunately, this discrepancy, coupled with each person's own perception (oftentimes misperception) of standard serving sizes, sometimes creates confusion for consumers trying to follow recommendations.

TABLE 2-5	Household and Metric Measures

- 1 teaspoon (tsp) = 5 milliliters (mL)
- 1 tablespoon (tbs) = 15 mL
- 1 cup (c) = 240 mL
- 1 fluid ounce (fl oz) = 30 mL
- 1 ounce (oz) = 28 grams (g)

NOTE: The Aids to Calculation section at the back of the book provides additional weights and measures.

Nutrition Facts

In addition to the serving size and the servings per container, the FDA requires that the "Nutrition Facts" panel on a label present nutrient information in two ways—in quantities (such as grams) and as percentages of standards called the **Daily Values.** The Nutrition Facts panel must provide the nutrient amount, percent Daily Value, or both for the following:

- Total food energy (kcalories).
- Food energy from fat (kcalories).
- Total fat (grams and percent Daily Value).
- Saturated fat (grams and percent Daily Value).
- *Trans* fat (grams).*
- Cholesterol (milligrams and percent Daily Value).
- Sodium (milligrams and percent Daily Value).
- Total carbohydrate, including starch, sugar, and fiber (grams and percent Daily Value).
- Dietary fiber (grams and percent Daily Value).
- Sugars (grams), including both those naturally present in and those added to the food.
- Protein (grams).

The labels must also present nutrient content information as a percentage of the Daily Values for the following vitamins and minerals:

- Vitamin A.
- Vitamin C.
- Iron.
- Calcium.

Consumers read food labels to learn about the nutrient contents of a food or to compare similar foods.

Daily Values (DV): reference values developed by the FDA specifically for use on food labels.

---

* *Trans* fats are required on food labels by January 1, 2006.

TABLE 2-6 Daily Values for Food Labels

Food labels must present the "% Daily Value" for these nutrients.

Food Component	Daily Value	Calculation Factors
Fat	65 g	30% of kcalories
Saturated fat	20 g	10% of kcalories
Cholesterol	300 mg	—
Carbohydrate (total)	300 g	60% of kcalories
Fiber	25 g	11.5 g per 1000 kcalories
Protein	50 g	10% of kcalories
Sodium	2400 mg	—
Potassium	3500 mg	—
Vitamin C	60 mg	—
Vitamin A	1500 µg	—
Calcium	1000 mg	—
Iron	18 mg	—

NOTE: Daily Values were established for adults and children over 4 years old. The values for energy-yielding nutrients are based on 2000 kcalories a day.

■ % Daily Values:
≥20% = high or excellent source.
10–19% = good source.

The FDA developed the Daily Values for use on food labels because comparing nutrient amounts against a standard helps make them meaningful to consumers. A person might wonder, for example, whether 1 milligram of iron or calcium is a little or a lot. Well, as Table 2-6 shows, the Daily Value for iron is 18 milligrams, so 1 milligram of iron is enough to take notice of: it is over 5 percent. But the Daily Value for calcium on food labels is 1000 milligrams, so 1 milligram of calcium is essentially nothing.

The Daily Values

The Daily Values reflect dietary recommendations for nutrients and dietary components that have important relationships with health. The "% Daily Value" column on a label provides a ballpark estimate of how individual foods contribute to the total diet. It compares key nutrients in a serving of food with the daily goals of a person consuming 2000 kcalories. A 2000-kcalorie diet is considered about right for moderately active women, teenage girls, and sedentary men. Older adults, young children, and sedentary women may need fewer kcalories. Most labels list, at the bottom, Daily Values for both a 2000-kcalorie and a 2500-kcalorie diet, but the "% Daily Value" column on all labels applies only to a 2000-kcalorie diet. A 2500-kcalorie diet is considered about right for many men, teenage boys, and active women. People who are exceptionally active may have still higher energy needs. Labels may also provide a reminder of the kcalories in a gram of carbohydrate, fat, and protein below the Daily Value information (review Figure 2-10).

A person who consumes 2000 kcalories a day can simply add up all the "% Daily Values" for a particular nutrient to see if the day's diet fits with recommendations. If the "% Daily Values" total 100 percent, then recommendations are met. People who require more or less than 2000 kcalories daily must do some calculations to see how foods compare with their personal nutrition goals. They can use the calculation column in Table 2-6 or the suggestions presented in the accompanying "How to" feature.

Daily Values help consumers see easily whether a food contributes "a little" or "a lot" of a nutrient.■ For example, the "% Daily Value" column on a label of macaroni and cheese may say 20 percent for fat. This tells the consumer that each serving of this food contains about 20 percent of the day's allotted 65 grams of fat. A person consuming 2000 kcalories a day could simply keep track of the percentages of Daily Values from foods eaten in a day and try not to exceed 100 percent. Be aware that for some nutrients (such as fat and sodium) you will want to select foods with a low "% Daily Value" and for others (such as calcium and fiber) you will want a high "% Daily Value." To determine whether a particular food is a wise choice, a consumer needs to consider its place in the diet among all the other foods eaten during the day.

Daily Values also make it easy to compare foods. For example, a consumer might discover that frozen macaroni and cheese has a Daily Value for fat of 20 percent, whereas macaroni and cheese prepared from a boxed mix has a Daily Value of 15 percent. By comparing labels, consumers who are concerned about their fat intakes will be able to make informed decisions.

HOW TO Calculate Personal Daily Values

The Daily Values on food labels are designed for a 2000-kcalorie intake, but you can calculate a personal set of Daily Values based on your energy allowance. Consider a person with a 1500-kcalorie intake, for example. To calculate a daily goal for fat, multiply energy intake by 30 percent:

1500 kcal × 0.30 kcal from fat
= 450 kcal from fat.

The "kcalories from fat" are listed on food labels, so a person could then add all the "kcalories from fat" values for a day, using 450 as an upper limit. A person who preferred to count grams of fat could divide this 450 kcalories from fat by 9 kcalories per gram to determine the goal in grams:

450 kcal from fat ÷ 9 kcal/g
= 50 g fat.

Alternatively, a person could calculate that 1500 kcalories is 75 percent of the 2000-kcalorie intake used for Daily Values:

1500 kcal ÷ 2000 kcal = 0.75.
0.75 × 100 = 75%.

Then, instead of trying to achieve 100 percent of the Daily Value, a person consuming 1500 kcalories would aim for 75 percent. Similarly, a person consuming 2800 kcalories would aim for 140 percent:

2800 kcal ÷ 2000 kcal = 1.40 or 140%.

Table 2-6 includes a calculation column that can help you estimate your personal daily value for several nutrients.

Nutrient Claims

Have you noticed phrases such as "good source of fiber" on a box of cereal or "rich in calcium" on a package of cheese? These and other **nutrient claims** may be used on labels as long as they meet FDA definitions, which include the conditions under which each term can be used (see the accompanying glossary for these definitions). For example, in addition to having less than 2 milligrams of cholesterol, a "cholesterol-free" product may not contain more than 2 grams of saturated fat and *trans* fat combined per serving.

Some descriptions *imply* that a food contains, or does not contain, a nutrient. Implied claims are prohibited unless they meet specified criteria. For example, a claim that a product "contains no oil" *implies* that the food contains no fat. If the product is truly fat-free, then it may make the no-oil claim, but if it contains another source of fat, such as butter, it may not.

nutrient claims: statements that characterize the quantity of a nutrient in a food.

GLOSSARY OF TERMS ON FOOD LABELS

GENERAL TERMS

free: "nutritionally trivial" and unlikely to have a physiological consequence; synonyms include "without," "no," and "zero." A food that does not contain a nutrient naturally may make such a claim, but only as it applies to all similar foods (for example, "applesauce, a fat-free food").

good source of: the product provides between 10 and 19% of the Daily Value for a given nutrient per serving.

healthy: a food that is low in fat, saturated fat, cholesterol, and sodium and that contains at least 10% of the Daily Values for vitamin A, vitamin C, iron, calcium, protein, or fiber.

high: 20% or more of the Daily Value for a given nutrient per serving; synonyms include "rich in" or "excellent source."

less: at least 25% less of a given nutrient or kcalories than the comparison food (see individual nutrients); synonyms include "fewer" and "reduced."

light or lite: any use of the term other than as defined must specify what it is referring to (for example, "light in color" or "light in texture").

low: an amount that would allow frequent consumption of a food without exceeding the Daily Value for the nutrient. A food

that is naturally low in a nutrient may make such a claim, but only as it applies to all similar foods (for example, "fresh cauliflower, a low-sodium food"); synonyms include "little," "few," and "low source of."

more: at least 10% more of the Daily Value for a given nutrient than the comparison food; synonyms include "added" and "extra."

organic: on food labels, that at least 95% of the product's ingredients have been grown and processsed according to USDA regulations defining the use of fertilizers, herbicides, insecticides, fungicides, preservatives, and other chemical ingredients (see Chapter 19).

ENERGY

kcalorie-free: fewer than 5 kcal per serving.

light: one-third fewer kcalories than the comparison food.

low kcalorie: 40 kcal or less per serving.

reduced kcalorie: at least 25% fewer kcalories per serving than the comparison food.

FAT AND CHOLESTEROL[a]

percent fat-free: may be used only if the product meets the definition of *low fat* or *fat-free* and must reflect the amount of fat in 100 g (for example, a food that contains 2.5 g of fat

per 50 g can claim to be "95 percent fat free").

fat-free: less than 0.5 g of fat per serving (and no added fat or oil); synonyms include "zero-fat," "no-fat," and "nonfat."

low fat: 3 g or less fat per serving.

less fat: 25% or less fat than the comparison food.

saturated fat-free: less than 0.5 g of saturated fat and 0.5 g of *trans* fat per serving.

low saturated fat: 1 g or less saturated fat and less than 0.5 g of *trans* fat per serving.

less saturated fat: 25% or less saturated fat and *trans* fat combined than the comparison food.

trans fat-free: less than 0.5 g of *trans* fat and less than 0.5 g of saturated fat per serving.

cholesterol-free: less than 2 mg cholesterol per serving and 2 g or less saturated fat and *trans* fat combined per serving.

low cholesterol: 20 mg or less cholesterol per serving and 2 g or less saturated fat and *trans* fat combined per serving.

less cholesterol: 25% or less cholesterol than the comparison food (reflecting a reduction of at least 20 mg per serving), and 2 g or less saturated fat and *trans* fat combined per serving.

extra lean: less than 5 g of fat, 2 g of saturated fat and *trans* fat

combined, and 95 mg of cholesterol per serving and per 100 g of meat, poultry, and seafood.

lean: less than 10 g of fat, 4.5 g of saturated fat and *trans* fat combined, and 95 mg of cholesterol per serving and per 100 g of meat, poultry, and seafood.

light: 50% or less of the fat than in the comparison food (for example, 50% less fat than our regular cookies).

CARBOHYDRATES: FIBER AND SUGAR

high fiber: 5 g or more fiber per serving. A high-fiber claim made on a food that contains more than 3 g fat per serving and per 100 g of food must also declare total fat.

sugar-free: less than 0.5 g of sugar per serving.

SODIUM

sodium-free and **salt-free:** less than 5 mg of sodium per serving.

low sodium: 140 mg or less per serving.

light: a low-kcalorie, low-fat food with a 50% reduction in sodium.

light in sodium: no more than 50% of the sodium of the comparison food.

very low sodium: 35 mg or less per serving.

[a]Foods containing more than 13 grams total fat per serving or per 50 grams of food must indicate those contents immediately after a cholesterol claim. As you can see, all cholesterol claims are prohibited when the food contains more than 2 grams saturated fat and *trans* fat combined per serving.

TABLE 2-7	Food Label Health Claims—The "A" List

- Calcium and reduced risk of osteoporosis
- Sodium and reduced risk of hypertension
- Dietary saturated fat and cholesterol and reduced risk of coronary heart disease
- Dietary fat and reduced risk of cancer
- Fiber-containing grain products, fruits, and vegetables and reduced risk of cancer
- Fruits, vegetables, and grain products that contain fiber, particularly soluble fiber, and reduced risk of coronary heart disease
- Fruits and vegetables and reduced risk of cancer
- Folate and reduced risk of neural tube defects
- Sugar alcohols and reduced risk of tooth decay
- Soluble fiber from whole oats and from psyllium seed husk and reduced risk of heart disease
- Soy protein and reduced risk of heart disease
- Whole grains and reduced risk of heart disease and certain cancers
- Plant sterol and plant stanol esters and heart disease
- Potassium and reduced risk of hypertension and stroke

Health Claims

Until recently, the FDA held manufacturers to the highest standards of scientific evidence before approving **health claims** on food labels. Consumers reading "Diets low in sodium may reduce the risk of high blood pressure," for example, knew that the FDA had examined enough scientific evidence to establish a clear link between diet and health. Such reliable health claims make up the FDA's "A" list (see Table 2-7).

These reliable health claims still appear on some food labels, but finding them may be difficult now that the FDA has created three additional categories of claims based on scientific evidence that is less conclusive (see Table 2-8). These categories were added after a court ruled that "holding only the highest scientific standard for claims interferes with commercial free speech."[10] Food manufacturers had argued that they should be allowed to inform consumers about possible benefits based on less than clear and convincing evidence. The FDA states that the new rules will enable consumers to receive more information about nutrients and foods that show preliminary promise in preventing disease. Consumer groups argue that such information is confusing. Even with required disclaimers for health claims graded "B," "C," or "D," distinguishing "A" claims from others is difficult, as the next section shows. (Health claims on supplement labels are presented in Highlight 10.)

health claims: statements that characterize the relationship between a nutrient or other substance in a food and a disease or health-related condition.

structure-function claims: statements that characterize the relationship between a nutrient or other substance in a food and its role in the body.

Structure-Function Claims

Unlike health claims, which require food manufacturers to collect scientific evidence and petition the FDA, **structure-function claims** can be made without any FDA approval. Products can claim to "slow aging," "improve memory," and "build strong

TABLE 2-8	The FDA's Health Claims Report Card	
Grade	Level of Confidence in Health Claim	Required Label Disclaimers
A	High: Significant scientific agreement	These health claims do not require disclaimers; see Table 2-7 for examples.
B	Moderate: Evidence is supportive, but not conclusive	"[Health claim.] Although there is scientific evidence supporting this claim, the evidence is not conclusive."
C	Low: Evidence is limited and not conclusive	"Some scientific evidence suggests [health claim]. However, FDA has determined that this evidence is limited and not conclusive."
D	Very low: Little scientific evidence supporting this claim	"Very limited and preliminary scientific research suggests [health claim]. FDA concludes that there is little scientific evidence supporting this claim."

bones" without any proof. The only criterion for a structure-function claim is that it must not mention a disease or symptom. Unfortunately, structure-function claims can be deceptively similar to health claims. Consider these statements:

- "May reduce the risk of heart disease."
- "Promotes a healthy heart."

Most consumers would argue that these two claims say the same thing. In fact, the first is a health claim that requires FDA approval and the second is an unproven, but legal, structure-function claim. Table 2-9 lists examples of structure-function claims.

Consumer Education

Labels are valuable only if people know how to use them, so the FDA has designed several programs to educate consumers. Consumers who understand how to read labels will be best able to apply the information to achieve and maintain healthful dietary practices.

Table 2-10 (on p. 62) shows how the messages from the *Dietary Guidelines,* the Food Guide Pyramid, and food labels coordinate with each other. To help consumers understand and coordinate these messages, an alliance of health organizations, the food industry, and government agencies has developed an educational program called "It's All About You." The program is designed to deliver simple messages that will motivate consumers to think positively about making reasonable changes in their eating and physical activity habits.

IN SUMMARY Food labels provide consumers with information they need to select foods that will help them meet their nutrition and health goals. Given labels with relevant information presented in a standardized, easy-to-read format, consumers are well prepared to plan and create healthful diets.

TABLE 2-9 Examples of Structure-Function Claims

- Builds strong bones
- Defends your health
- Promotes relaxation
- Slows aging
- Improves memory
- Guards against colds
- Boosts the immune system
- Lifts your spirits
- Supports heart health

NOTE: Structure-function claims cannot make statements about diseases. See Table 2-7 on p. 60 for examples of health claims.

TABLE 2-10   From Guidelines to Groceries

Dietary Guidelines	Food Guide Pyramid	Food Labels

Aim for a healthy weight.	Build a healthy base by eating vegetables, fruits, and grains Choose foods low in fats and added sugars. Select sensible portion sizes.	Look for foods that describe their kcalorie contents as *free, low, reduced, light,* or *less.*
Be physically active each day.	Increase your physical activity.	
Let the Pyramid guide your food choices.	Choose at least the minimum number of servings from each food group.	Look for foods that describe their vitamin, mineral, or fiber contents as a *good source* or *high.*
Choose a variety of grains daily, especially whole grains.	Choose at least 6 servings of grain products daily.	Look for foods that describe their fiber contents as *high.* Aim for 100% of the Daily Value for fiber from a variety of sources. A heart disease health claim identifies grains with soluble fibers from oats and psyllium seeds. A cancer health claim identifies grains that are good sources of fiber and low in fat.
Choose a variety of fruits and vegetables daily.	Choose at least 2 servings of fruits and 3 servings of vegetables daily.	Look for foods that describe their fiber contents as *high.* Aim for 100% of the Daily Value for fiber from a variety of sources. A cancer health claim identifies fruits and vegetables that are low in fat and good sources of fiber, vitamin A, or vitamin C.
Keep food safe to eat.		Follow the *safe handling instructions* on packages of meat and other safety instructions, such as *keep refrigerated,* on packages of perishable foods.
Choose a diet that is low in saturated fat and cholesterol and moderate in total fat.	Limit foods in the tip of the Pyramid. Choose foods within each group that are low in fat, saturated fat, and cholesterol.	Look for foods that describe their fat, saturated fat, and cholesterol contents as *free, less, low, light, reduced, lean,* or *extra lean.* Keep your intake under 100% of the Daily Value for fat, saturated fat, and cholesterol. A heart disease or a cancer health claim identifies foods low in fat, saturated fat, and cholesterol.
Choose beverages and foods to moderate your intake of sugars.	Limit foods in the tip of the Pyramid. Choose foods within each group that are low in added sugars.	Look for foods that describe their sugar contents as *free* or *reduced.* A tooth decay health claim identifies foods that contain sugar alcohols, but no sugar. A food may be high in sugar if its ingredients list begins with, or contains several of the following: *sugar, sucrose, fructose, malt ose, lactose, honey, syrup, corn syrup, high-fructose corn syrup, molasses,* or *fruit juice concentrate.*
Choose and prepare foods with less salt.	Choose foods within each group that are low in salt and sodium.	Look for foods that describe their salt and sodium contents as *free, low,* or *reduced.* Keep your intake under 100% of the Daily Value for sodium. A high blood pressure health claim identifies foods low in sodium.
If you drink alcoholic beverages, do so in moderation.	Like other foods in the tip, use sparingly (no more than one drink a day for women and two drinks a day for men).	*Light* beverages contain fewer kcalories and less alcohol than regular versions.

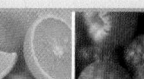

Nutrition in Your Life

The secret to making healthy food choices is learning to incorporate the *Dietary Guidelines* and Food Guide Pyramid into your decisions.

- Do you eat at least the minimum number of servings from each of the five food groups daily?

- Do you try to vary your choices within each food group from day to day? If not, why not?

- What dietary changes could you make to improve your chances of enjoying good health?

NUTRITION ON THE NET

 Access these websites for further study of topics covered in this chapter.

- Find updates and quick links to these and other nutrition-related sites at our website: **www.wadsworth.com/nutrition**

- Search for "diet" and "food labels" at the U.S. Government health information site: **www.healthfinder.gov**

- Learn more about the *Dietary Guidelines for Americans:* **health.gov/dietaryguidelines**

- Find Canadian information on nutrition guidelines and food labels at: **www.hc-sc.gc.ca**

- Visit the Food Guide Pyramid section (including its ethnic/cultural pyramids) of the U.S. Department of Agriculture: **www.nal.usda.gov/fnic**

- Visit the Traditional Diet Pyramids for various ethnic groups at Oldways Preservation and Exchange Trust: **www.oldwayspt.org**

- Search for "exchange lists" at the American Diabetes Association: **www.diabetes.org**

- Learn more about food labeling from the Food and Drug Administration: **www.cfsan.fda.gov**

- Search for "food labels" at the International Food Information Council: **www.ific.org**

- Assess your diet at the CNPP Interactive Healthy Eating Index: **www.usda.gov/cnpp**

- Get healthy eating tips from the "5 a day" programs: **www.5aday.gov** or **www.5aday.org**

NUTRITION CALCULATIONS

These problems will give you practice in doing simple nutrition-related calculations. Although the situations are hypothetical, the numbers are real, and calculating the answers (check them on p. 65) provides a valuable nutrition lesson. Be sure to show your calculations for each problem.

1. *Read a food label.* Look at the cereal label in Figure 2-10 and answer the following questions:
   a. What is the size of a serving of cereal?
   b. How many kcalories are in a serving?
   c. How much fat is in a serving?
   d. How many kcalories does this represent?
   e. What percentage of the kcalories in this product comes from fat?
   f. What does this tell you?

   g. What is the % Daily Value for fat?
   h. What does this tell you?
   i. Does this cereal meet the criteria for a low-fat product (refer to the glossary on p. 59)?
   j. How much fiber is in a serving?
   k. Read the Daily Value chart on the lower section of the label. What is the Daily Value for fiber?
   l. What percentage of the Daily Value for fiber does a serving of the cereal contribute? Show the calculation the label-makers used to come up with the % Daily Value for fiber.
   m. What is the predominant ingredient in the cereal?
   n. Have any nutrients been added to this cereal (is it fortified)?

2. *Calculate a personal Daily Value.* The Daily Values on food labels are for people with a 2000-kcalorie intake.
   a. Suppose a person has a 1600-kcalorie energy allowance. Use the calculation factors listed in Table 2-6 to calculate a set of personal "Daily Values" based on 1600 kcalories. Show your calculations.
   b. Revise the % Daily Value chart of the cereal label in Figure 2-10 based on your "Daily Values" for a 1600-kcalorie diet.

STUDY QUESTIONS

These questions will help you review this chapter. You will find the answers in the discussions on the pages provided.

1. Name the diet-planning principles and briefly describe how each principle helps in diet planning. (pp. 39–41)

2. What recommendations appear in the *Dietary Guidelines for Americans*? (pp. 41–42)

3. Name the five food groups in the Daily Food Guide and identify several foods typical of each group. Explain how such plans group foods and what diet-planning principles the plans best accommodate. How are food group plans used, and what are some of their strengths and weaknesses? (pp. 42–48)

4. Review the *Dietary Guidelines*. What types of grocery selections would you make to achieve those recommendations? (pp. 42, 51–55)

5. What information can you expect to find on a food label? How can this information help you choose between two similar products? (pp. 55–58)

6. What are the Daily Values? How can they help you meet health recommendations? (p. 58)

7. Describe the differences between nutrient claims, health claims, and structure-function claims. (pp. 59–61)

These multiple choice questions will help you prepare for an exam. Answers can be found on p. 65.

1. The diet-planning principle that provides all the essential nutrients in sufficient amounts to support health is:
   a. balance.
   b. variety.
   c. adequacy.
   d. moderation.

2. A person who chooses a chicken leg that provides 0.5 milligram of iron and 95 kcalories instead of two tablespoons of peanut butter that also provide 0.5 milligram of iron but 188 kcalories is using the principle of nutrient:
   a. control.
   b. density.
   c. adequacy.
   d. moderation.

3. Which of the following is consistent with the *Dietary Guidelines for Americans*?
   a. Choose a diet restricted in fat and cholesterol.
   b. Balance the food you eat with physical activity.
   c. Choose a diet with plenty of milk products and meats.
   d. Eat an abundance of foods to ensure nutrient adequacy.

4. According to the Food Guide Pyramid, which food group provides the foundation of a healthy diet?
   a. vegetables
   b. milk, yogurt, and cheese
   c. breads, cereals, rice, and pasta
   d. meat, poultry, fish, dry beans, eggs, and nuts

5. Foods within a given food group of the Pyramid are similar in their contents of:
   a. energy.
   b. proteins and fibers.
   c. vitamins and minerals.
   d. carbohydrates and fats.

6. In the exchange system, each portion of food on any given list provides about the same amount of:
   a. energy.
   b. satiety.
   c. vitamins.
   d. minerals.

7. Enriched grain products are fortified with:
   a. fiber, folate, iron, niacin, and zinc.
   b. thiamin, iron, calcium, zinc, and sodium.
   c. iron, thiamin, riboflavin, niacin, and folate.
   d. folate, magnesium, vitamin B_6, zinc, and fiber.

8. Food labels list ingredients in:
   a. alphabetical order.
   b. ascending order of predominance by weight.
   c. descending order of predominance by weight.
   d. manufacturer's order of preference.

9. "Milk builds strong bones" is an example of a:
   a. health claim.
   b. nutrition fact.
   c. nutrient content claim.
   d. structure-function claim.

10. Daily Values on food labels are based on a:
    a. 1500-kcalorie diet.
    b. 2000-kcalorie diet.
    c. 2500-kcalorie diet.
    d. 3000-kcalorie diet.

REFERENCES

1. U.S. Department of Agriculture and U.S. Department of Health and Human Services, *Nutrition and Your Health: Dietary Guidelines for Americans,* Home and Garden Bulletin no. 232 (Washington, D.C.: 2000).
2. Position of the American Dietetic Association and Dietitians of Canada: Vegetarian diets, *Journal of the American Dietetic Association* 103 (2003): 748–765.
3. S. M. Krebs-Smith and coauthors, Characterizing food intake patterns of American adults, *American Journal of Clinical Nutrition* 65 (1997): S1264–S1268.
4. P. P. Basiotis, M. Lino, and J. M. Dinkins, Consumption of food group servings: People's perceptions vs. reality, *Family Economics and Nutrition Review* 14 (2002): 67–69; A. K. Kant, Consumption of energy-dense, nutrient-poor foods by adult Americans: Nutrition and health implications. The third National Health and Nutrition Examination Survey, 1988–1994, *American Journal of Clinical Nutrition* 72 (2000): 929–936; K. S. Tippett, C. W. Enns, and A. J. Moshfegh, Food consumption surveys in the US Department of Agriculture, *Nutrition Today* 34 (1999): 33–46.
5. W. C. Willett and M. J. Stampfer, Rebuilding the food pyramid, *Scientific American,* January 2003, pp. 64–71.
6. J. R. Backstrand, The history and future of food fortification in the United States: A public health perspective, *Nutrition Reviews* 60 (2002): 15–26; Y. K. Park and coauthors, History of cereal-grain product fortification in the United States, *Nutrition Today* 36 (2001): 124–137.
7. As cited in 21 Code of Federal Regulations—Food and Drugs, Section 104.20, 45 *Federal Register* 6323, January 25, 1980, as amended in 58 *Federal Register* 2228, January 6, 1993.
8. Position of the American Dietetic Association: Food fortification and dietary supplements, *Journal of the American Dietetic Association* 101 (2001): 115–125.
9. D. Herring and coauthors, Serving sizes in the Food Guide Pyramid and on the nutrition facts label: What's different and why? *Family Economics and Nutrition Review* 14 (2002): 71–73; M. B. Hogbin and M. A. Hess, Public confusion over food portions and servings, *Journal of the American Dietetic Association* 99 (1999): 1209–1211.
10. N. Hellmich, FDA to allow qualified health claims on foods, *USA Today,* 11 July 2003, available at www.USATODAY.com.

ANSWERS

Nutrition Calculations

1. a. ¾ cup (28 g).
   b. 110 kcalories.
   c. 1 g fat.
   d. 9 kcalories.
   e. 9 kcal ÷ 110 kcal = 0.08.
      0.08 × 100 = 8%.
   f. This cereal derives 8 percent of its kcalories from fat.
   g. 2%.
   h. A serving of this cereal provides 2 percent of the 65 grams of fat recommended for a 2000-kcalorie diet.
   i. Yes.
   j. 1.5 g fiber.
   k. 25 g.
   l. 1.5 g ÷ 25 g = 0.06.
      0.06 × 100 = 6%.
   m. Corn.
   n. Yes.

2. a. Daily Values for 1600-kcalorie diet:

   Fat: 1600 kcal × 0.30 = 480 kcal from fat.
      480 kcal ÷ 9 kcal/g = 53 g fat.

   Saturated fat: 1600 kcal × 0.10 = 160 kcal from saturated fat.
      160 kcal ÷ 9 kcal/g = 18 g saturated fat.

   Cholesterol: 300 mg.

   Carbohydrate: 1600 kcal × 0.60 = 960 kcal from carbohydrate.
      960 kcal ÷ 4 kcal/g = 240 g carbohydrate.

   Fiber: 1600 kcal ÷ 1000 kcal = 1.6.
      1.6 × 11.5 g = 18.4 g fiber.

   Protein: 1600 kcal × 0.10 = 160 kcal from protein.
      160 kcal ÷ 4 kcal/g = 40 g protein.

   Sodium: 2400 mg.

   Potassium: 3500 mg.

   b.

Total fat	2%	(1 g ÷ 53 g)
Saturated fat	0%	(0 g ÷ 18 g)
Cholesterol	0%	(no calculation needed)
Sodium	10%	(no calculation needed)
Total carbohydrate	10%	(23 g ÷ 240 g)
Dietary fiber	8%	(1.5 g ÷ 18.4 g)

Study Questions (multiple choice)

1. c   2. b   3. b   4. c   5. c   6. a   7. c   8. c
9. d   10. b

HIGHLIGHT

A World Tour of Pyramids, Pagodas, and Plates

© Michael S. Yamashita/CORBIS

In China, a family's meal may include horse meat or fried scorpion; in Great Britain, people traditionally drink a pot of tea and eat a scone between lunch and dinner; and in India, those with a reverence for life called *ahimsa* espouse a vegetarian diet.[1] The food supplies, eating habits, and cultural beliefs of people living in diverse regions of the world are different, but their needs are similar in fundamental ways. People from every corner of the world need foods—whether cheeseburgers or grasshoppers—to provide the nutrients necessary to maintain life and defend against disease.

How should a person select foods to maintain good health? In the United States, the *Dietary Guidelines for Americans* offer direction. Other countries provide similar guidelines, and it can be enlightening to see how U.S. guidelines compare with those from other countries. A tour of dietary guidelines from around the world finds they are similar in many ways and different in a few. This highlight elaborates on the *Dietary Guidelines for Americans* that were introduced in Chapter 2 and compares them with guidelines from around the world.

Dietary Guidelines

Governments develop dietary guidelines based on the nutrition problems, food supplies, eating habits, and cultural beliefs of their populations. These guidelines offer practical suggestions for making food choices that will support good health and reduce the risk of chronic diseases. They also provide the basis for many policy and education decisions. Although each set of guidelines reflects the people and foodways of a specific country, guidelines from around the world display more similarities than differences.

Aim for a Healthy Weight

Healthy body weight plays a key role in reducing the risk of chronic diseases. To that end, the *Dietary Guidelines for Americans* encourage people to combine sensible eating habits with regular exercise that will limit weight gain and support weight loss in those who are overweight. Canada, Germany, Aus-

tralia, and dozens of other developed countries also advise consumers to "watch your weight and stay active."

A notable difference is apparent in developing countries, such as Indonesia, where undernutrition threatens health as much or more than overnutrition does. The focus there is on consuming "foods to provide sufficient energy."

Be Physically Active Each Day

Physical activity not only helps with weight management and disease prevention, but is also improves cardiovascular fitness, strengthens bones and muscles, controls blood pressure, and promotes psychological well-being. Furthermore, a person who is physically active can afford to eat more, which makes it easier to get the nutrients the body needs to stay fit. The *Dietary Guidelines for Americans* recommend making physical activity a regular part of the daily routine and participating in 30 minutes or more of moderate physical activity on most days. This amount of activity offers some health benefits, but it is not enough to maintain a healthy body weight. To prevent weight gain and to accrue additional health benefits, the DRI recommend 60 minutes of moderately intense activity. (Chapter 14 provides many more details, and Chapter 8 includes a table showing energy expenditures for a variety of activities.)

Many of the guidelines from around the world that encourage a healthy body weight mention balancing physical activity with food intake. For example, Japan's guideline says to "match daily energy intake with daily physical activity."

Let the Pyramid Guide Your Food Choices

Each food provides a unique assortment of nutrients. To get all the nutrients in the right amounts, a person needs to eat a variety of foods in sufficient quantities. The *Dietary Guidelines for Americans* depend on the Pyramid to guide daily food choices.

Not all nations use the Pyramid, of course, but most consistently feature "variety" in their message, quite often as the premier guideline. To achieve variety, guidelines in Japan, for example, recommend eating 30 or more different kinds of foods daily.

Figure H2-1 (on pp. 68–69) features shapes of various food guide plans from selected nations.[2] Most food guide plans classify foods into the following groups: grains, vegetables, fruits, meats, and milk products, with a few minor variations. For example, Canada, China, and Portugal cluster fruits and vegetables into one group; Mexico and the Philippines group milk products with meats.

Potatoes and legumes appear in all food guides, but their locations vary. In the United States, potatoes are grouped with vegetables, but in Great Britain, Mexico, and Korea, potatoes are listed with the grains. Sweden separates potatoes and other tubers into their own group, featuring them as a "base food," providing a "foundation for a nutritious and inexpensive diet." As for legumes, the United States includes them in both the meat and the vegetable groups; Sweden, Germany, and Australia put them in the vegetable group; and China places them in the milk group.

Most food guide plans describe the recommended number of servings and serving sizes for each food group. Some, such as Mexico, do not quantify recommendations, but instead offer general guidelines—"muchas verduras y frutas" (many vegetables and fruits) and "pocos alimentos de origen animal" (little food of animal origin). In all cases, the guidelines emphasize abundant grains, vegetables, and fruits and moderate meats and milks.

Choose a Variety of Grains Daily

Grains form the foundation of the Pyramid, highlighting their important contributions of vitamins, minerals, energy, and dietary fiber. Efforts to include at least six servings of grains per day focus on a variety of selections from whole-grain foods such as whole wheat, brown rice, oats, and corn.

Canada, Australia, Germany, and Great Britain also mention the need to "eat plenty of foods rich in starch and fiber," such as bread and cereals. Guidelines from Greece suggest that people "prefer whole grain breads and pastas." Other countries, such as Indonesia, recommend that consumers "obtain about half of total energy from complex carbohydrate–rich foods."

Choose a Variety of Fruits and Vegetables Daily

As Chapter 2 mentioned, the campaign to "eat 5 a day" reflects the dietary guideline to eat at least two servings of fruits and three servings of vegetables each day. Not only do fruits and vegetables supply vitamins and minerals in abundance, but their phytochemicals and fibers help to maintain good health and protect against diseases such as heart disease and cancer. Because different fruits and vegetables deliver different nutrients, it is especially important to select a variety and to include dark green leafy and deep yellow and orange vegetables; citrus fruits, melons, and berries; and legumes often.

Several nations have a guideline that encourages consumers to "eat plenty of fruits and vegetables." Many have adopted a "5 a day" recommendation.

Keep Foods Safe to Eat

Maintaining a healthy weight, exercising daily, and selecting a variety of grains, legumes, fruits, and vegetables protect a person's health over a lifetime. Foods can also affect immediate health. As Chapter 19 will detail, foods that are tainted with harmful bacteria, viruses, parasites, or chemical contaminants can make people sick with flu-like symptoms of nausea and diarrhea—or worse. Keeping foods safe requires washing hands and food preparation surfaces often; separating raw, cooked, and prepared foods while shopping, cooking, and storing; cooking foods to the appropriate temperature; and refrigerating perishable foods promptly.

The prevalence and risks of foodborne illnesses in the developing countries of Southeast Asia are among the highest in the world.[3] Many of these Asian countries include a guideline advising their people to "eat clean and safe food." The message in Indonesia and Thailand is to "consume food which is prepared hygienically." The Philippine guideline goes on to explain that eating clean and safe food "will prevent foodborne diseases."

Choose a Diet Low in Saturated Fat and Cholesterol and Moderate in Total Fat

Some fat in the diet is essential to good health, but certain kinds of fat, most notably saturated fats, can be detrimental to heart health as Highlight 5 explains. For this reason, the *Dietary Guidelines for Americans* encourage people to limit saturated fat intake to less than 10 percent of total kcalories, cholesterol intake to less than 300 milligrams a day, and total fat intake to no more than 30 percent of total kcalories.

Most countries make some statement about not eating "too many foods that contain a lot of fat," but the recommended limitations vary. Canada, for example, agrees with the United States. Korea suggests "keeping fat intake at 20 percent of energy intake," and the Netherlands allows "up to 35 percent." Interestingly, both Korea and the Netherlands have lower rates of heart disease than the United States and Canada.[4] The Germans don't mention "percent kcalories from fat," but instead restrict dietary fat to 70 to 90 grams per day, which is equivalent to 30 to 40 percent of a 2000-kcalorie diet. The German fat message adds a preference for fats of plant origin and cautions that meat, sausages, and eggs should be eaten in moderation. Japan and China—countries with relatively low rates of heart disease—also advise consumers to "eat more vegetable oils than animal fat."

Choose Beverages and Foods to Moderate Sugar Intake

Foods and beverages containing sugar promote tooth decay. In addition, they frequently deliver many kcalories with few, if any, nutrients. As a result, these items may contribute to

FIGURE H2-1   A Comparison of Food Guide Designs from Selected Nations

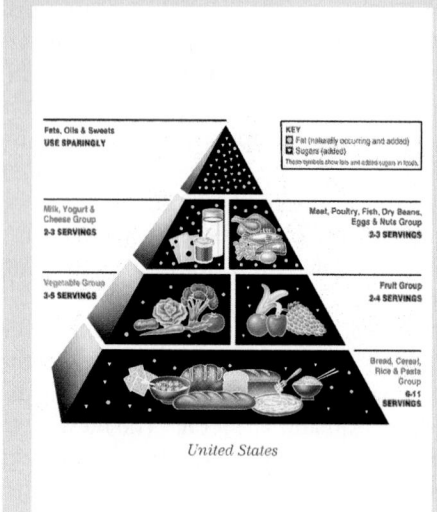

United States

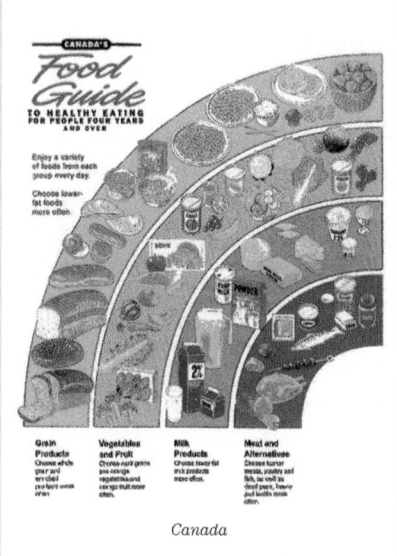

Canada

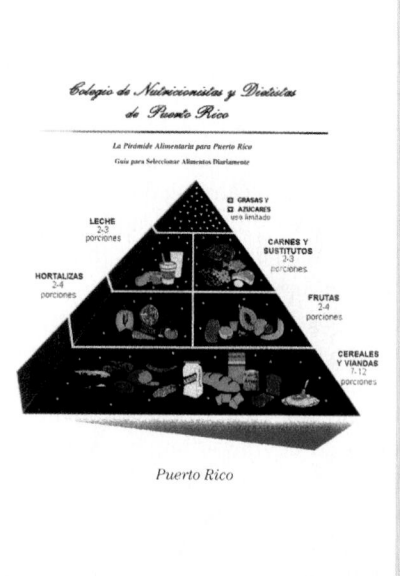

Puerto Rico

The United States uses a pyramid to convey the message of proportionality. (Interestingly, Thailand inverts the pyramid to convey the same message of proportionality, putting the largest amounts at the top and the smaller amounts at the bottom.)

Canada's unique rainbow design also illustrates proportionality.

Puerto Rico (and the Philippines) adopted the pyramid design and then made modifications. Notably, Puerto Rico adds a blue shadow to illustrate water, and the Philippines combines the milk and meat groups into one group of animal foods.

SOURCE: *Journal of The American Dietetic Association.* April 2002, pp. 484–485.

---

weight gain or malnutrition. For these reasons, the *Dietary Guidelines for Americans* encourage people to choose sensibly and to limit their intake of beverages and foods that are high in added sugars. (Chapter 4 presents additional dietary strategies to limit tooth decay.)

Many nations take a similar approach, saying "don't have sugary foods and drinks too often" or "eat only a moderate amount of sugars and foods containing added sugars." Greek guidelines do not address sugar directly, but urge consumers to "prefer fruits and nuts as snacks, instead of sweets or candy bars" and to "prefer water over soft drinks"—suggestions that reflect traditional Greek foodways. Other nations, including Canada, Korea, China, the Philippines, and Japan, do not mention sugars in their dietary guidelines.

Choose and Prepare Foods with Less Salt

Because of the link between salt intake and high blood pressure, the *Dietary Guidelines for Americans* suggest that consumers choose and prepare foods with less salt. This recommendation does not specify a quantity, but food labels use a Daily Value of 2400 milligrams of sodium per day—the amount of sodium in about one teaspoon of salt (6 grams).

Japan is one of the few countries to specify a quantity, and its upper limit on salt is almost twice that of the U.S. guideline—10 grams a day. Most nations address salt intake without specifying quantities. Canada states that "the sodium content of the diet should be reduced"; Australia advises that you "choose low salt foods and use salt sparingly"; and Great Britain suggests that you "minimize salt use." Indonesia and Germany—two countries whose iodine status is less than sufficient—don't limit salt intake, but instead remind consumers to use iodized salt.[5]

Drink Alcoholic Beverages in Moderation (If at All)

Because alcohol consumed in excess is detrimental to health in many ways, the *Dietary Guidelines for Americans* advise adults who drink alcoholic beverages to do so in moderation. Moderation is defined as one drink a day for women and two a day for men—and the size of "a drink" is specified (see Highlight 7).

FIGURE H2-1 A Comparison of Food Guide Designs from Selected Nations—continued

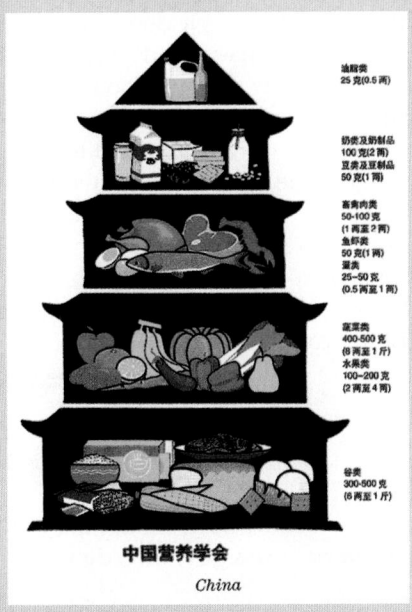

In China (and Korea), a pagoda is used to depict the food guide.

In Germany (and most other European countries, as well as Australia), the food guide is illustrated with a circle.

In Mexico (and Great Britain), a circular food guide is transformed into a sectioned plate of foods.

SOURCE: *Journal of The American Dietetic Association.* April 2002, pp. 484–485.

Most countries agree: "if you drink alcohol, limit your intake." Canada's limit on alcohol is "no more than 5 percent of energy intake as alcohol, or two drinks daily, whichever is less." Greece cautions those who drink alcohol to do so only occasionally and in small quantities—specified as 25 milliliters of pure alcohol (equivalent to a little less than one ounce of pure alcohol, or about two drinks). Great Britain is less specific and simply suggests that "if you drink, keep within sensible limits." Some countries, such as Indonesia, restrict consumption altogether, advising consumers to "avoid drinking alcoholic beverages." In Hungary, "alcohol is forbidden for children and pregnant women." The Netherlands simply acknowledges "that current alcohol consumption is far too high in many cases."

Additional Comments

As you can see, dietary guidelines from around the world look a little different on the surface, but on close examination their messages agree with each other. They all support good health.

Guidelines from other countries sometimes address issues that are not covered in the *Dietary Guidelines for Americans*. For example, Indonesia encourages its people to "eat breakfast." Several countries provide guidelines for iron and calcium. Most of these make general statements, such as "consume iron rich foods." Some countries offer suggestions for specific groups of people, most commonly, pregnant or breastfeeding women. In Indonesia, mothers are advised to "breast feed your baby exclusively for four months," and in Australia, guidelines, "encourage and support breastfeeding."

Water and fluid intake are addressed in several sets of dietary guidelines. People of Indonesia are advised to "drink adequate quantities of fluids that are free from contaminants," and those in Greece are told to drink about 1.5 liters of water daily. Canada's comment on water addresses its fluoride content, rather than intake. In addition, Canada advises consumers to limit caffeine to no more than the equivalent of four cups of regular coffee a day.

Eating Pleasure

The *Dietary Guidelines for Americans* do not specifically acknowledge that "eating is one of life's greatest pleasures," but the guidelines of many nations do. Clearly, eating provides more than just food to the body. Foods bring pleasure through their flavors and promote social interactions, ethnic traditions, and family time together. Great Britain's first guideline says simply, "enjoy your food." The Netherlands also presents a simple message: "food + joy = health." French guidelines also emphasize enjoyment and suggest that you eat "three good meals a day." Greek guidelines advocate that you "eat slowly, preferably at regular times during the day and in a pleasant environment." Similarly, German guidelines want you to "make sure your dishes are prepared gently and taste well" and to "take your time and enjoy eating." Japan's guidelines capture the spirit of enjoying food and family together: "Make all activities pertaining to food and eating pleasurable ones. Use the mealtime as an occasion for family communication and

appreciate home cooking." Vietnam's guideline delivers a similar message—serve "healthy family meals that are delicious, wholesome, economical, and served with affection."

We agree. Eating should be a pleasure. In this fast-paced world of drive-through restaurants and packaged meals, we sometimes forget to take time to enjoy foods. Eating has become another time-consuming task in our busy days—one that is undertaken with little thought of how the foods eaten will nourish the body or how the time spent will nourish the mind. Because eating is so central to our well-being, we should pay attention and do it well. Take time to select fresh foods. Prepare them creatively. Give thanks for the bounty. Share meals with others. Savor the flavors. Enjoy conversations. Experience the pleasure of nourishing yourself well.

NUTRITION ON THE NET

 Access these websites for further study of topics covered in this highlight.

- Find updates and quick links to these and other nutrition-related sites at our website: **www.wadsworth.com/nutrition**

- Link to dietary guidelines from around the world from the USDA: **www.nal.usda.gov/fnic/dga**

REFERENCES

1. P. G. Kittler and K. P. Sucher, *Cultural Foods* (Belmont, Calif.: Wadsworth/ Thomson Learning, 2002).
2. J. Painter, J. Rah, and Y. Lee, Comparison of international food guide pictorial representations, *Journal of the American Dietetic Association* 102 (2002): 483–489.
3. M. D. Miliotis and J. W. Bier, *International Handbook of Foodborne Pathogens* (New York: Marcel Dekker, 2003).
4. World Health Organization, **www.who.int/whosis**, visited June 2003.
5. International Council for the Control of Iodine Deficiency Disorder, www. people.virginia.edu/~jtd/iccidd/mi/regions/americas_map.htm, visited June 2003.

Digestion, Absorption, and Transport

Chapter Outline

Digestion: *Anatomy of the Digestive Tract*
• *The Muscular Action of Digestion* • *The Secretions of Digestion* • *The Final Stage*

Absorption: *Anatomy of the Absorptive System*
• *A Closer Look at the Intestinal Cells*

The Circulatory Systems: *The Vascular System* • *The Lymphatic System*

Regulation of Digestion and Absorption: *Gastrointestinal Hormones and Nerve Pathways* • *The System at Its Best*

Highlight: *Common Digestive Problems*

Nutrition Explorer CD-ROM Outline

Nutrition Animation: *Digestion and Absorption*

Case Study: *Distressed Digestion*

Student Practice Test

Glossary Terms

Nutrition on the Net

Nutrition in Your Life

Have you ever wondered what happens to the food you eat after you swallow it? Or how your body extracts nutrients from food? Have you ever marveled at how it all just seems to happen? Follow foods as they travel through the digestive system. Learn how a healthy digestive system transforms whatever food you give it—whether sirloin steak and potatoes or tofu and brussels sprouts—into the nutrients that will nourish the cells of your body.

This chapter takes you on the journey that transforms the foods you eat into the nutrients featured in the later chapters. Then it follows the nutrients as they travel through the intestinal cells and into the body to do their work. This introduction presents a general overview of the processes common to all nutrients; later chapters discuss the specifics of digesting and absorbing individual nutrients.

Digestion

Digestion is the body's ingenious way of breaking down foods into nutrients in preparation for **absorption.** In the process, it overcomes many obstacles for you without any conscious effort on your part. Consider these obstacles:

1. Human beings breathe, eat, and drink through their mouths. Air taken in through the mouth must go to the lungs; food and liquid must go to the stomach. The throat must be arranged so that swallowing and breathing don't interfere with each other.

2. Below the lungs lies the diaphragm, a dome of muscle that separates the upper half of the major body cavity from the lower half. Food must pass through this wall to reach the stomach.

3. The materials within the tract should be kept moving forward, slowly but steadily, at a pace that permits all reactions to reach completion.

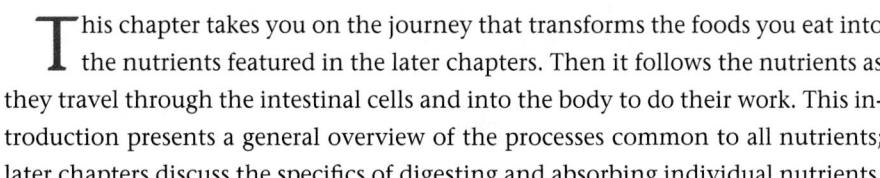

digestion: the process by which food is broken down into absorbable units.
- **digestion** = take apart

absorption: the uptake of nutrients by the cells of the small intestine for transport into either the blood or the lymph.
- **absorb** = suck in

The process of digestion transforms all kinds of *foods* into *nutrients*.

4. To move through the system, food must be lubricated with fluids. Too much would form a liquid that would flow too rapidly; too little would form a paste too dry and compact to move at all. The amount of fluids must be regulated to keep the intestinal contents at the right consistency to move smoothly along.

5. When the digestive enzymes are breaking food down, they need it in finely divided form, suspended in enough liquid so that every particle is accessible. Once digestion is complete and the needed nutrients have been absorbed out of the tract into the body, the system must excrete the residue that remains, but excreting all the water along with the solid residue would be both wasteful and messy. Some water should be withdrawn, leaving a paste just solid enough to be smooth and easy to pass.

6. The enzymes of the digestive tract are designed to digest carbohydrate, fat, and protein. The walls of the tract, composed of living cells, are also made of carbohydrate, fat, and protein. These cells need protection against the action of the powerful digestive juices that they secrete.

7. Once waste matter has reached the end of the tract, it must be excreted, but it would be inconvenient and embarrassing if this function occurred continuously. Provision must be made for periodic, voluntary evacuation.

The following sections show how the body elegantly and efficiently handles these obstacles.

Anatomy of the Digestive Tract

The **gastrointestinal (GI) tract** is a flexible muscular tube from the mouth, through the esophagus, stomach, small intestine, large intestine, and rectum to the anus. Figure 3-1 traces the path followed by food from one end to the other. In a sense, the human body surrounds the GI tract. The inner space within the GI tract, called the **lumen,** is continuous from one end to the other (GI anatomy terms appear in boldface type and are defined in the glossary below). Only when a nutrient or other substance penetrates the GI tract's wall does it enter the body proper; many materials pass through the GI tract without being digested or absorbed.

---

gastrointestinal (GI) tract: the digestive tract. The principal organs are the stomach and intestines.
- **gastro** = stomach
- **intestinalis** = intestine

GLOSSARY OF GI ANATOMY TERMS

These terms are listed in order from start to end of the digestive tract.

mouth: the oral cavity containing the tongue and teeth.

pharynx (FAIR-inks): the passageway leading from the nose and mouth to the larynx and esophagus, respectively.

epiglottis (epp-ee-GLOTT-iss): cartilage in the throat that guards the entrance to the trachea and prevents fluid or food from entering it when a person swallows.
- **epi** = upon (over)
- **glottis** = back of tongue

esophagus (ee-SOFF-ah-gus): the food pipe; the conduit from the mouth to the stomach.

sphincter (SFINK-ter): a circular muscle surrounding, and able to close, a body opening. Sphincters are found at specific points along the GI tract and regulate the flow of food particles.
- **sphincter** = band (binder)

esophageal (ee-SOF-ah-GEE-al) **sphincter:** a sphincter muscle at the upper or lower end of the esophagus. The *lower esophageal sphincter* is also called the *cardiac sphincter.*

stomach: a muscular, elastic, saclike portion of the digestive tract that grinds and churns swallowed food, mixing it with acid and enzymes to form chyme.

pyloric (pie-LORE-ic) **sphincter:** the circular muscle that separates the stomach from the small intestine and regulates the flow of partially digested food into the small intestine; also called *pylorus* or *pyloric valve.*
- **pylorus** = gatekeeper

gallbladder: the organ that stores and concentrates bile. When it receives the signal that fat is present in the duodenum, the gallbladder contracts and squirts bile through the bile duct into the duodenum.

pancreas: a gland that secretes digestive enzymes and juices into the duodenum.

small intestine: a 10-foot length of small-diameter intestine that is the major site of digestion of food and absorption of nutrients. Its segments are the duodenum, jejunum, and ileum.

lumen (LOO-men): the space within a vessel, such as the intestine.

duodenum (doo-oh-DEEN-um, doo-ODD-num): the top portion of the small intestine (about "12 fingers' breadth" long in ancient terminology).
- **duodecim** = twelve

jejunum (je-JOON-um): the first two-fifths of the small intestine beyond the duodenum.

ileum (ILL-ee-um): the last segment of the small intestine.

ileocecal (ill-ee-oh-SEEK-ul) **valve:** the sphincter separating the small and large intestines.

large intestine or **colon** (COAL-un): the lower portion of intestine that completes the digestive process. Its segments are the ascending colon, the transverse colon, the descending colon, and the sigmoid colon.
- **sigmoid** = shaped like the letter S (sigma in Greek)

appendix: a narrow blind sac extending from the beginning of the colon that stores lymph cells.

rectum: the muscular terminal part of the intestine, extending from the sigmoid colon to the anus.

anus (AY-nus): the terminal outlet of the GI tract.

FIGURE 3-1 The Gastrointestinal Tract

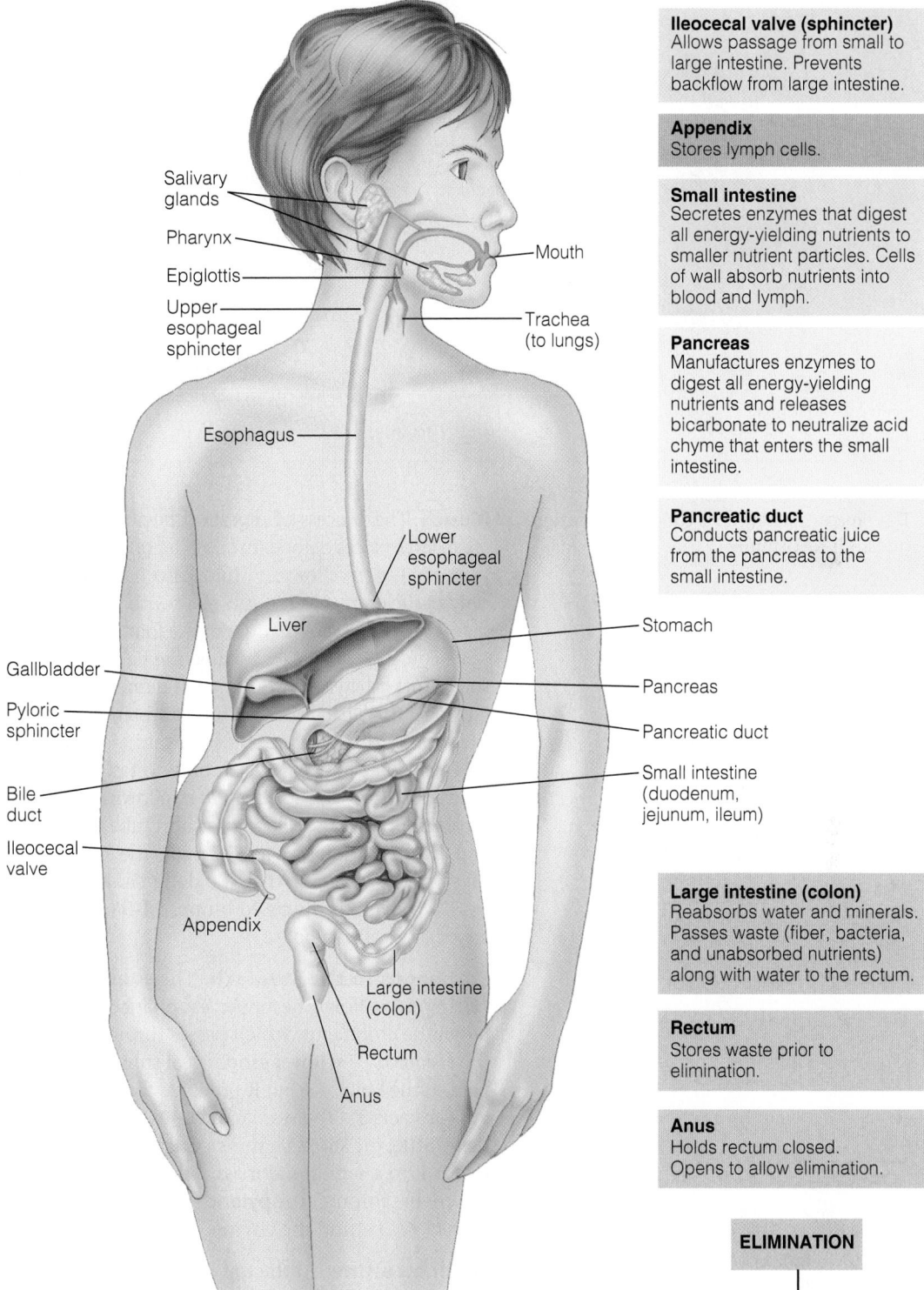

INGESTION

Mouth
Chews and mixes food with saliva.

Pharynx
Directs food from mouth to esophagus.

Salivary glands
Secrete saliva (contains starch-digesting enzymes).

Epiglottis
Protects airway during swallowing.

Trachea
Allows air to pass to and from lungs.

Esophagus
Passes food from the mouth to the stomach.

Esophageal sphincters
Allow passage from mouth to esophagus and from esophagus to stomach. Prevent backflow from stomach to esophagus and from esophagus to mouth.

Stomach
Adds acid, enzymes, and fluid. Churns, mixes, and grinds food to a liquid mass.

Pyloric sphincter
Allows passage from stomach to small intestine. Prevents backflow from small intestine.

Liver
Manufactures bile salts, detergent-like substances, to help digest fats.

Gallbladder
Stores bile until needed.

Bile duct
Conducts bile from the gallbladder to the small intestine.

Ileocecal valve (sphincter)
Allows passage from small to large intestine. Prevents backflow from large intestine.

Appendix
Stores lymph cells.

Small intestine
Secretes enzymes that digest all energy-yielding nutrients to smaller nutrient particles. Cells of wall absorb nutrients into blood and lymph.

Pancreas
Manufactures enzymes to digest all energy-yielding nutrients and releases bicarbonate to neutralize acid chyme that enters the small intestine.

Pancreatic duct
Conducts pancreatic juice from the pancreas to the small intestine.

Large intestine (colon)
Reabsorbs water and minerals. Passes waste (fiber, bacteria, and unabsorbed nutrients) along with water to the rectum.

Rectum
Stores waste prior to elimination.

Anus
Holds rectum closed. Opens to allow elimination.

ELIMINATION

Labels on figure:
Salivary glands
Pharynx
Epiglottis
Upper esophageal sphincter
Mouth
Trachea (to lungs)
Esophagus
Lower esophageal sphincter
Liver
Gallbladder
Pyloric sphincter
Bile duct
Ileocecal valve
Appendix
Stomach
Pancreas
Pancreatic duct
Small intestine (duodenum, jejunum, ileum)
Large intestine (colon)
Rectum
Anus

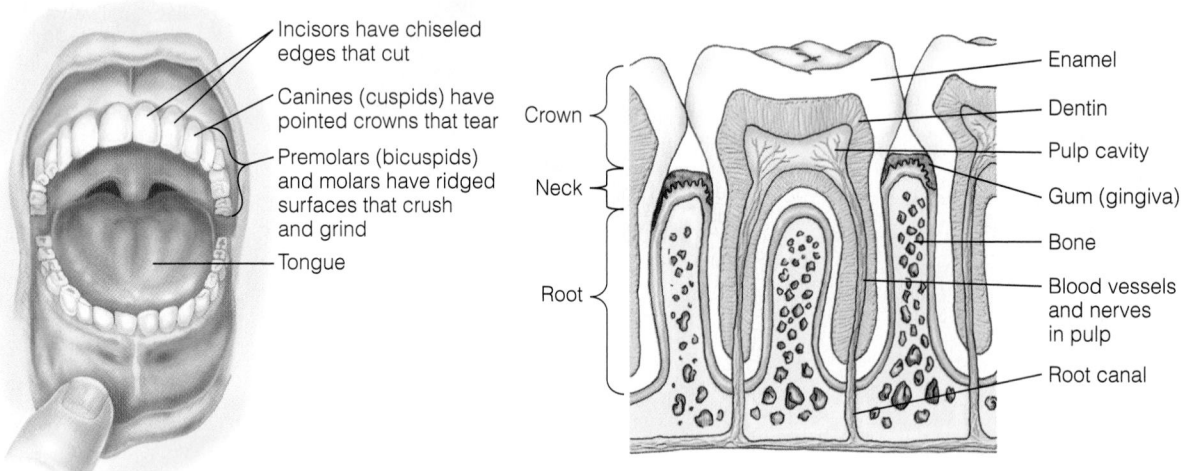

FIGURE 3-2 The Teeth

Incisors have chiseled edges that cut

Canines (cuspids) have pointed crowns that tear

Premolars (bicuspids) and molars have ridged surfaces that crush and grind

Tongue

Crown

Neck

Root

Enamel

Dentin

Pulp cavity

Gum (gingiva)

Bone

Blood vessels and nerves in pulp

Root canal

■ The process of chewing is called **mastication** (mass-tih-KAY-shun).

Mouth The process of digestion begins in the **mouth.** As you chew,■ your teeth crush large pieces of food into smaller ones (see Figure 3-2), and fluids blend with these pieces to ease swallowing. Fluids also help dissolve the food so that you can taste it; only particles in solution can react with taste buds. When stimulated, the taste buds detect one, or a combination, of the four basic taste sensations: sweet, sour, bitter, and salty. Some scientists also include the flavor associated with monosodium glutamate, sometimes called *savory* or its Asian name *umami* (ooh-MOM-ee). In addition to these chemical triggers, aroma, texture, and temperature also affect a food's flavor. In fact, the sense of smell is thousands of times more sensitive than the sense of taste.

The tongue allows you not only to taste food, but also to move food around the mouth, facilitating chewing and swallowing. When you swallow a mouthful of food, it passes through the **pharynx,** a short tube that is shared by both the **digestive system** and the respiration system. To bypass the entrance to your lungs, the **epiglottis** closes off your air passages so that you don't choke when you swallow, thus resolving obstacle 1. (Choking is discussed on pp. 94–95.) After a mouthful of food has been swallowed, it is called a **bolus.**

Esophagus to the Stomach The **esophagus** has a **sphincter** muscle at each end. During a swallow, the upper **esophageal sphincter** opens. The bolus then slides down the esophagus, which passes through a hole in the diaphragm (obstacle 2) to the **stomach.** The lower esophageal sphincter at the entrance to the stomach closes behind the bolus so that it proceeds forward and doesn't slip back into the esophagus (obstacle 3). The stomach retains the bolus for a while in its upper portion. Little by little, the stomach transfers the food to its lower portion, adds juices to it, and grinds it to a semiliquid mass called **chyme.** Then, bit by bit, the stomach releases the chyme through the **pyloric sphincter,** which opens into the **small intestine** and then closes behind the chyme.

digestive system: all the organs and glands associated with the ingestion and digestion of food.

bolus (BOH-lus): a portion; with respect to food, the amount swallowed at one time.
• **bolos** = lump

chyme (KIME): the semiliquid mass of partly digested food expelled by the stomach into the duodenum.
• **chymos** = juice

Small Intestine At the top of the small intestine, the chyme bypasses the opening from the common bile duct, which is dripping fluids (obstacle 4) into the small intestine from two organs outside the GI tract—the **gallbladder** and the **pancreas.** The chyme travels on down the small intestine through its three segments—the **duodenum,** the **jejunum,** and the **ileum**—almost 10 feet of tubing coiled within the abdomen.*

---

*The small intestine is almost two and a half times shorter in living adults than it is at death, when muscles are relaxed and elongated.

Large Intestine (Colon) Having traveled the length of the small intestine, what remains of the chyme arrives at another sphincter (obstacle 3 again): the **ileocecal valve,** at the beginning of the **large intestine (colon)** in the lower right-hand side of the abdomen. As the chyme enters the colon, it passes another opening. Had it slipped into this opening, it would have ended up in the **appendix,** a blind sac about the size of your little finger. The chyme bypasses this opening, however, and travels along the large intestine up the right-hand side of the abdomen, across the front to the left-hand side, down to the lower left-hand side, and finally below the other folds of the intestines to the back side of the body, above the **rectum.**

During the chyme's passage to the rectum, the colon withdraws water from it, leaving semisolid waste (obstacle 5). The strong muscles of the rectum and anal canal hold back this waste until it is time to defecate. Then the rectal muscles relax (obstacle 7), and the two sphincters of the **anus** open to allow passage of the waste.

The Muscular Action of Digestion

The first step in the reduction of food to a liquid takes place in the mouth, where chewing, the addition of saliva, and the action of the tongue reduce the food to a coarse mash. Then you swallow, and thereafter, you are generally unaware of all the activity that follows. As is the case with so much else that happens in the body, the muscles of the digestive tract meet internal needs without your having to exert any conscious effort. They keep things moving■ at just the right pace, slow enough to get the job done and fast enough to make progress.

Peristalsis The entire GI tract is ringed with circular muscles that can squeeze it tightly. Surrounding these rings of muscle are longitudinal muscles. When the rings tighten and the long muscles relax, the tube is constricted. When the rings relax and the long muscles tighten, the tube bulges. This action—called **peristalsis**—occurs continuously and pushes the intestinal contents along (obstacle 3 again). (If you have ever watched a lump of food pass along the body of a snake, you have a good picture of how these muscles work.)

The waves of contraction ripple along the GI tract at varying rates and intensities depending on the part of the GI tract and on whether food is present. For example, waves occur three times per minute in the stomach, but speed up to ten times per minute when chyme reaches the small intestine. When you have just eaten a meal, the waves are slow and continuous; when the GI tract is empty, the intestine is quiet except for periodic bursts of powerful rhythmic waves. Peristalsis, along with the sphincter muscles that surround the digestive tract at key places, keeps things moving along.

Stomach Action The stomach has the thickest walls and strongest muscles of all the GI tract organs. In addition to the circular and longitudinal muscles, it has a third layer of diagonal muscles that also alternately contract and relax (see Figure 3-3). These three sets of muscles work to force the chyme downward, but the pyloric sphincter usually remains tightly closed, preventing the chyme from passing into the duodenum of the small intestine. As a result, the chyme is churned and forced down, hits the pyloric sphincter, and remains in the stomach. Meanwhile, the stomach wall releases gastric juices. When the chyme is completely liquefied, the pyloric sphincter opens briefly, about three times a minute, to allow small portions of chyme through. At this point, the chyme no longer resembles food in the least.

Segmentation The circular muscles of the intestines rhythmically contract and squeeze their contents. These contractions, called **segmentation,** mix the chyme and promote close contact with the digestive juices and the absorbing cells of the intestinal walls before letting the contents move slowly along. Figure 3-4 illustrates peristalsis and segmentation.

■ The ability of the GI tract muscles to move is called their **motility** (moh-TIL-ah-tee).

FIGURE 3-3 | Stomach Muscles

The stomach has three layers of muscles.

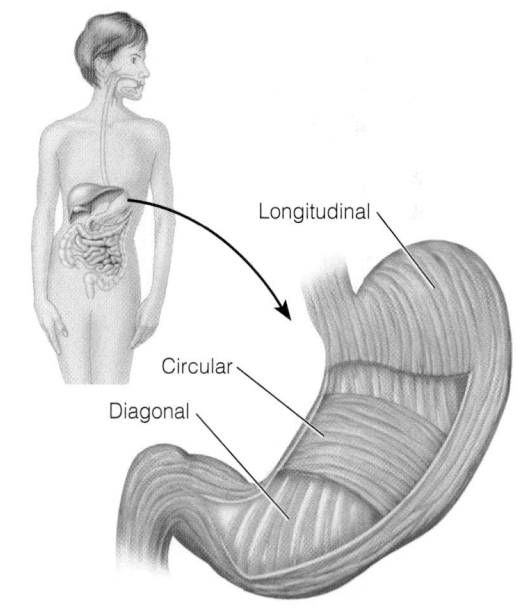

Longitudinal

Circular

Diagonal

peristalsis (per-ih-STALL-sis): wavelike muscular contractions of the GI tract that push its contents along.
• peri = around
• stellein = wrap

segmentation (SEG-men-TAY-shun): a periodic squeezing or partitioning of the intestine at intervals along its length by its circular muscles.

FIGURE 3-4   Peristalsis & Segmentation

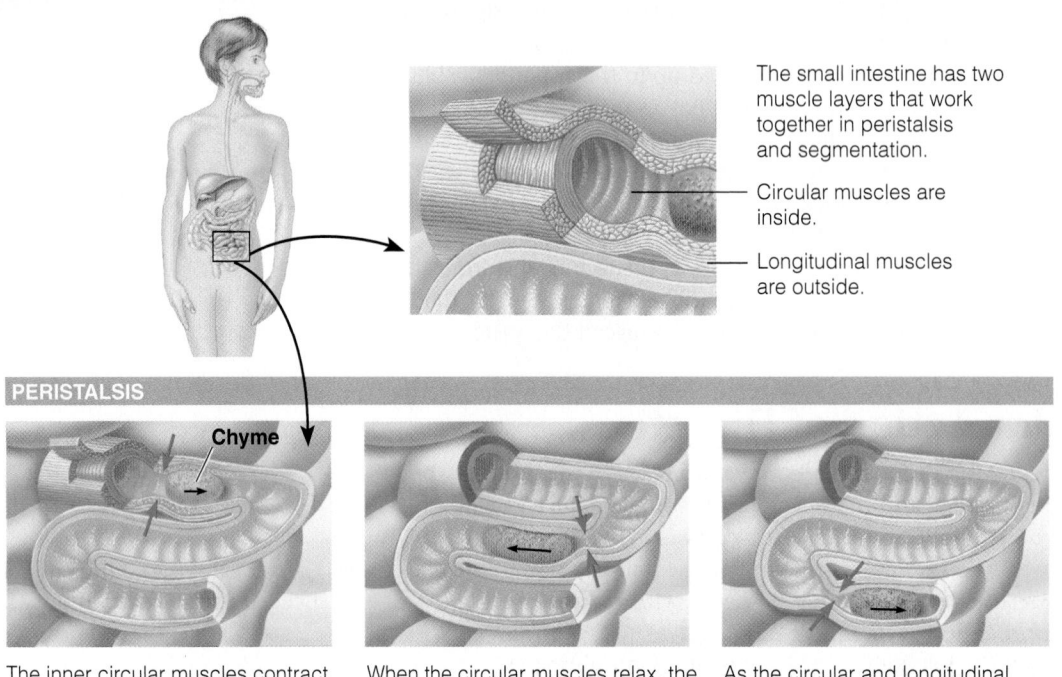

The small intestine has two muscle layers that work together in peristalsis and segmentation.

Circular muscles are inside.

Longitudinal muscles are outside.

PERISTALSIS

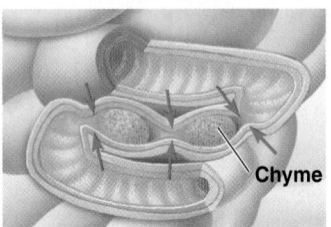

Chyme

The inner circular muscles contract, tightening the tube and pushing the food forward in the intestine.

When the circular muscles relax, the outer longitudinal muscles contract, and the intestinal tube is loose.

As the circular and longitudinal muscles tighten and relax, the chyme moves ahead of the constriction.

SEGMENTATION

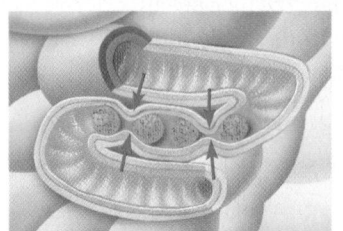

Chyme

Circular muscles contract, creating segments within the intestine.

As each set of circular muscles relaxes and contracts, the chyme is broken up and mixed with digestive juices.

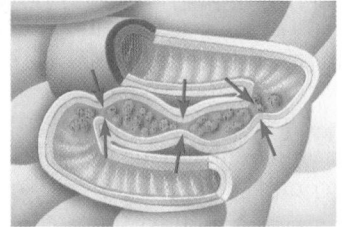

These alternating contractions, occurring 12 to 16 times per minute, continue to mix the chyme and bring the nutrients into contact with the intestinal lining for absorption.

reflux: a backward flow.
- **re** = back
- **flux** = flow

Sphincter Contractions Sphincter muscles periodically open and close, allowing the contents of the GI tract to move along at a controlled pace (obstacle 3 again). At the top of the esophagus, the upper esophageal sphincter opens in response to swallowing. At the bottom of the esophagus, the lower esophageal sphincter (sometimes called the cardiac sphincter because of its proximity to the heart) prevents **reflux** of the stomach contents. At the bottom of the stomach, the pyloric sphincter, which stays closed most of the time, holds the chyme in the stomach long enough for it to be thoroughly mixed with gastric juice and liquefied. The pyloric sphincter also prevents the intestinal contents from backing up into the stomach. At the end of the small intestine, the ileocecal valve performs a similar function, emptying the contents of the small intestine into the large intestine. Finally, the tightness of the rectal muscle is a kind of safety device; together with the two sphincters of the anus, it prevents elimination until you choose to perform

FIGURE 3-5   An Example of a Sphincter Muscle

When the circular muscles of a sphincter contract, the passage closes; when they relax, the passage opens.

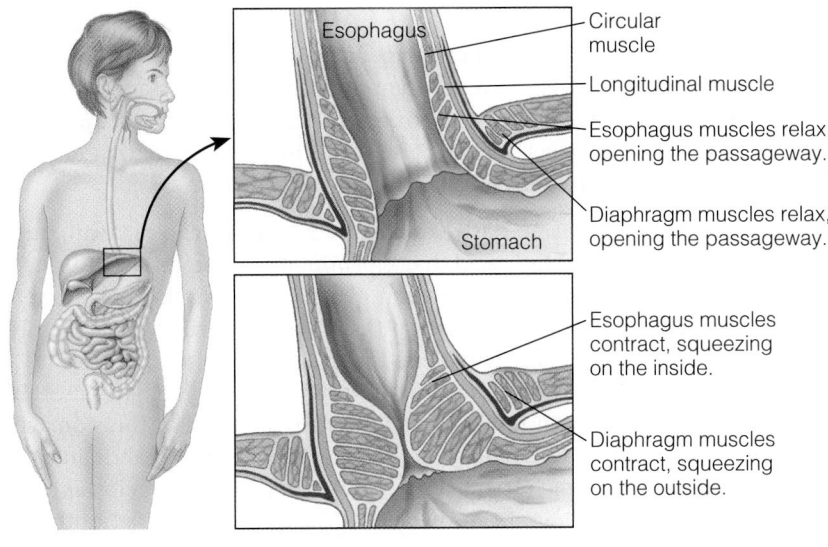

- Esophagus
- Circular muscle
- Longitudinal muscle
- Esophagus muscles relax, opening the passageway.
- Diaphragm muscles relax, opening the passageway.
- Stomach
- Esophagus muscles contract, squeezing on the inside.
- Diaphragm muscles contract, squeezing on the outside.

it voluntarily (obstacle 7). Figure 3-5 illustrates how sphincter muscles contract and relax to close and open passageways.

The Secretions of Digestion

To break down food into small nutrients that the body can absorb, five different organs produce secretions: the salivary glands, the stomach, the pancreas, the liver (via the gallbladder), and the small intestine. These secretions enter the GI tract at various points along the way, bringing an abundance of water (obstacle 3 again) and a variety of enzymes.

Enzymes■ are formally introduced in Chapter 6, but for now a simple definition will suffice. An enzyme is a protein that facilitates a chemical reaction—making a molecule, breaking a molecule, changing the arrangement of a molecule, or exchanging parts of molecules. As a **catalyst,** the enzyme itself remains unchanged. The enzymes involved in digestion facilitate a chemical reaction known as **hydrolysis**—the addition of water *(hydro)* to break *(lysis)* a molecule into smaller pieces. The glossary below identifies some of the common **digestive enzymes** and related terms; later chapters introduce specific enzymes. When learning about enzymes, it

■ All enzymes and some hormones are proteins, but enzymes are not hormones. Enzymes facilitate the making and breaking of bonds in chemical reactions; hormones act as chemical messengers, sometimes regulating enzyme action.

catalyst (CAT-uh-list): a compound that facilitates chemical reactions without itself being changed in the process.

GLOSSARY OF DIGESTIVE ENZYMES

digestive enzymes: proteins found in digestive juices that act on food substances, causing them to break down into simpler compounds.

-ase (ACE): a word ending denoting an enzyme. The word beginning often identifies the

compounds the enzyme works on. Examples include:

- **carbohydrase** (KAR-boe-HIGH-drase), an enzyme that hydrolyzes carbohydrates.
- **lipase** (LYE-pase), an enzyme that hydrolyzes lipids (fats).
- **protease** (PRO-tee-ase), an

enzyme that hydrolyzes proteins.

hydrolysis (high-DROL-ih-sis): a chemical reaction in which a major reactant is split into two products, with the addition of a hydrogen atom (H) to one and a hydroxyl group (OH) to the

other (from water, H_2O). (The noun is **hydrolysis;** the verb is **hydrolyze.**)

- **hydro** = water
- **lysis** = breaking

FIGURE 3-6 — The Salivary Glands

The salivary glands secrete saliva into the mouth and begin the digestive process. Given the short time food is in the mouth, salivary enzymes contribute little to digestion.

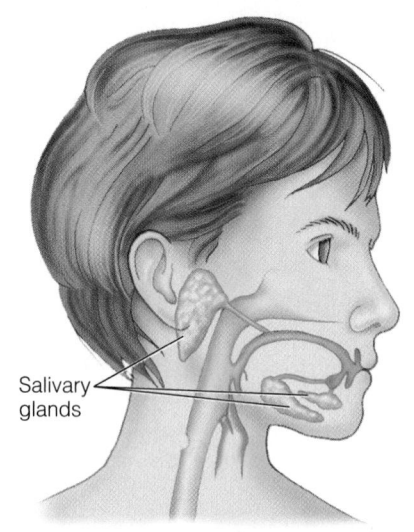

Salivary glands

---

helps to know that the word ending *-ase* denotes an enzyme. Enzymes are often identified by the organ they come from and the compounds they work on; *gastric lipase,* for example, is a stomach enzyme that acts on lipids, whereas *pancreatic lipase* comes from the pancreas (and also works on lipids).

Saliva The **salivary glands,** shown in Figure 3-6, squirt just enough **saliva** to moisten each mouthful of food so that it can pass easily down the esophagus (obstacle 4). (Digestive glands and their secretions are defined in the glossary below.) The saliva contains water, salts, mucus, and enzymes that initiate the digestion of carbohydrates. Saliva also protects the teeth and the linings of the mouth, esophagus, and stomach from attack by substances that might harm them.

Gastric Juice In the stomach, **gastric glands** secrete **gastric juice,** a mixture of water, enzymes, and **hydrochloric acid** that acts primarily in protein digestion. The acid is so strong that it causes the sensation of heartburn if it happens to reflux into the esophagus. Highlight 3, following this chapter, discusses heartburn, ulcers, and other common digestive problems.

The strong acidity of the stomach prevents bacterial growth and kills most bacteria that enter the body with food. It would destroy the cells of the stomach as well, but for their natural defenses. To protect themselves from gastric juice, the cells of the stomach wall secrete **mucus,** a thick, slippery, white substance that coats the cells, protecting them from the acid and enzymes that might otherwise harm them (obstacle 6).

Figure 3-7 shows how the strength of acids is measured—in **pH** units. Note that the acidity of gastric juice registers below "2" on the pH scale—stronger than vinegar. The stomach enzymes work most efficiently in the stomach's strong acid, but the salivary enzymes, which are swallowed with food, do not work in acid this strong. Consequently, the salivary digestion of carbohydrate gradually ceases as the stomach acid penetrates each newly swallowed bolus of food. In fact, salivary enzymes become just other proteins to be digested.

Pancreatic Juice and Intestinal Enzymes By the time food leaves the stomach, digestion of all three energy nutrients (carbohydrates, fats, and proteins) has begun, and the action gains momentum in the small intestine. There the pancreas contributes digestive juices by way of ducts leading into the duodenum. The **pancreatic juice** contains enzymes that act on all three energy nutrients, and the cells of the intestinal wall also possess digestive enzymes on their surfaces.

pH: the unit of measure expressing a substance's acidity or alkalinity (Chapter 12 provides a more detailed definition).

GLOSSARY OF DIGESTIVE GLANDS AND THEIR SECRETIONS

These terms are listed in order from start to end of the digestive tract.

gland: a cell or group of cells that secretes materials for special uses in the body. Glands may be **exocrine (EKS-oh-crin) glands,** secreting their materials "out" (into the digestive tract or onto the surface of the skin), or **endocrine (EN-doe-crin) glands,** secreting their materials "in" (into the blood).
- **exo** = outside
- **endo** = inside
- **krine** = to separate

salivary glands: exocrine glands that secrete saliva into the mouth.

saliva: the secretion of the salivary glands. Its principal enzyme begins carbohydrate digestion.

gastric glands: exocrine glands in the stomach wall that secrete gastric juice into the stomach.
- **gastro** = stomach

gastric juice: the digestive secretion of the gastric glands of the stomach.

hydrochloric acid: an acid composed of hydrogen and chloride atoms (HCl). The gastric glands normally produce this acid.

mucus (MYOO-kus): a slippery substance secreted by cells of the GI lining (and other body linings) that protects the cells from exposure to digestive juices (and other destructive agents). The lining of the GI tract with its coat of mucus is a **mucous membrane.** (The noun is **mucus;** the adjective is **mucous.**)

liver: the organ that manufactures bile. (The liver's many other functions are described in Chapter 7.)

bile: an emulsifier that prepares fats and oils for digestion; an exocrine secretion made by the liver, stored in the gallbladder, and released into the small intestine when needed.

emulsifier (ee-MUL-sih-fire): a substance with both water-soluble and fat-soluble portions that promotes the mixing of oils and fats in a watery solution.

pancreatic (pank-ree-AT-ic) juice: the exocrine secretion of the pancreas, containing enzymes for the digestion of carbohydrate, fat, and protein as well as bicarbonate, a neutralizing agent. The juice flows from the pancreas into the small intestine through the pancreatic duct. (The pancreas also has an endocrine function, the secretion of insulin and other hormones.)

bicarbonate: an alkaline secretion of the pancreas, part of the pancreatic juice. (Bicarbonate also occurs widely in all cell fluids.)

In addition to enzymes, the pancreatic juice contains sodium **bicarbonate,** which is basic or alkaline—the opposite of the stomach's acid (review Figure 3-7). The pancreatic juice thus neutralizes the acid chyme arriving in the small intestine from the stomach. From this point on, the chyme remains at a neutral or slightly alkaline pH. The enzymes of both the intestine and the pancreas work best in this environment.

Bile Bile also flows into the duodenum. The **liver** continuously produces bile, which is then concentrated and stored in the gallbladder. The gallbladder squirts the bile into the duodenum when fat arrives there. Bile is not an enzyme, but an **emulsifier** that brings fats into suspension in water so that enzymes can break them down into their component parts. Thanks to all these secretions, the three energy-yielding nutrients are digested in the small intestine (the summary on p. 82 provides a table of digestive secretions and their actions).

Protective Factors Both the small and the large intestine, being neutral in pH, permit the growth of bacteria (known as the intestinal flora). In fact, a healthy intestinal tract supports a thriving bacterial population that normally does the body no harm and may actually do some good. Bacteria in the GI tract produce several vitamins,■ including a significant amount of vitamin K, although the amount is insufficient to meet the body's total need for that vitamin.

Provided that the normal intestinal flora are thriving, infectious bacteria have a hard time getting established and launching an attack on the system. Diet is one of several factors that influence the bacterial population and its environment. For example, GI bacteria digest some dietary fibers and produce short fragments of fat that the cells of the colon use for energy.[*] In addition to diet, secretions from the GI tract—saliva, mucus, gastric acid, and digestive enzymes—not only help with digestion, but also defend against foreign invaders. The GI tract also maintains several different kinds of defending cells that confer specific immunity against intestinal diseases such as inflammatory bowel disease.

The Final Stage

At this point, the three energy-yielding nutrients—carbohydrate, fat, and protein—have been disassembled and are ready to be absorbed. Most of the other nutrients—vitamins, minerals, and water—need no such disassembly; some vitamins and minerals are altered slightly during digestion, but most are absorbed as they are. Undigested residues, such as some fibers, are not absorbed, but continue through the digestive tract, providing a semisolid mass that helps exercise the muscles and keep them strong enough to perform peristalsis efficiently. Fiber also retains water, accounting for the pasty consistency of **stools,** and carries some bile acids, some minerals, and some additives and contaminants with it out of the body.

By the time the contents of the GI tract reach the end of the small intestine, little remains but water, a few dissolved salts and body secretions, and undigested materials such as fiber. These enter the large intestine (colon).

In the colon, intestinal bacteria ferment some fibers, producing water, gas, and small fragments of fat that provide energy for the cells of the colon. The colon itself retrieves all materials that the body can recycle—water and dissolved salts (see Figure 3-8). The waste that is finally excreted has little or nothing of value left in it. The body has extracted all that it can use from the food. Figure 3-9 summarizes digestion by following a sandwich through the GI tract and into the body.

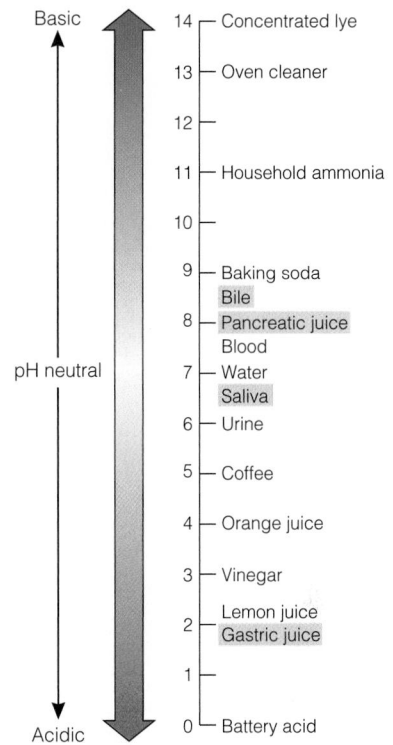

FIGURE 3-7 | The pH Scale

pH's of common substances:

Basic — 14 — Concentrated lye
13 — Oven cleaner
12
11 — Household ammonia
10
9 — Baking soda
— Bile
8 — Pancreatic juice
— Blood
pH neutral — 7 — Water
— Saliva
6 — Urine
5 — Coffee
4 — Orange juice
3 — Vinegar
2 — Lemon juice
— Gastric juice
1
Acidic — 0 — Battery acid

A substance's acidity or alkalinity is measured in pH units. The pH is the negative logarithm of the hydrogen ion concentration. Each increment presents a tenfold increase in concentration of hydrogen particles. For example, a pH of 2 is 1000 times stronger than a pH of 5.

■ Vitamins produced by bacteria include:
- Biotin.
- Folate.
- Vitamin B_6.
- Vitamin B_{12}.
- Vitamin K.

stools: waste matter discharged from the colon; also called **feces** (FEE-seez).

[*]These small fragments of fat are short-chain fatty acids, described in Chapter 5.

FIGURE 3-8   The Colon

The colon begins with the ascending colon rising upward toward the liver. It becomes the transverse colon as it turns and crosses the body toward the spleen. The descending colon turns downward and becomes the sigmoid colon, which extends to the rectum. Along the way, the colon mixes the intestinal contents, absorbs water and salts, and forms stools.

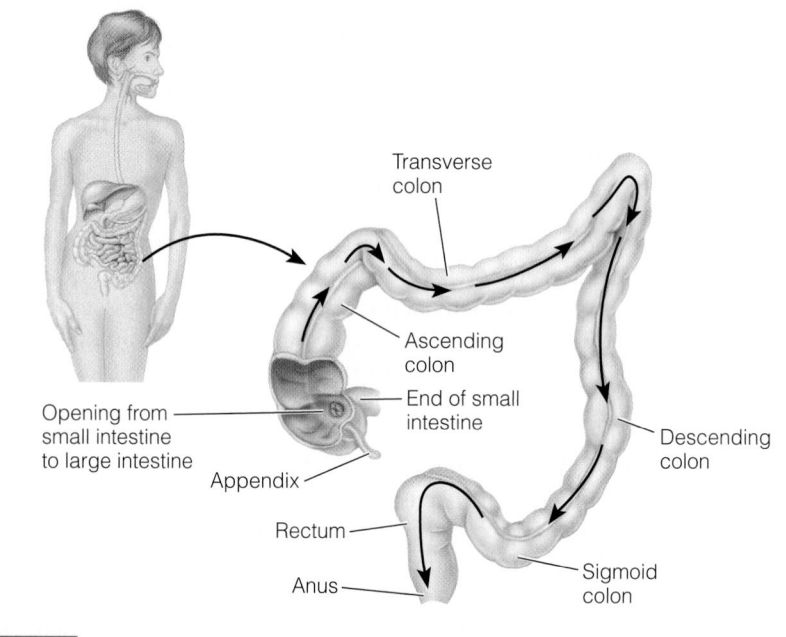

Nutrition Explorer Follow the movement of a bolus of food as it moves through the GI tract. Along the way the functions of the primary organs involved in digestion are illustrated.

IN SUMMARY
As Figure 3-1 shows, food enters the mouth and travels down the esophagus and through the upper and lower esophageal sphincters to the stomach, then through the pyloric sphincter to the small intestine, on through the ileocecal valve to the large intestine, past the appendix to the rectum, ending at the anus. The wavelike contractions of peristalsis and the periodic squeezing of segmentation keep things moving at a reasonable pace. Along the way, secretions from the salivary glands, stomach, pancreas, liver (via the gallbladder), and small intestine deliver fluids and digestive enzymes.

Summary of Digestive Secretions and Their Actions

Organ or Gland	Target Organ	Secretion	Action
Salivary glands	Mouth	Saliva	Fluid eases swallowing; salivary enzyme breaks down **carbohydrate.**
Gastric glands	Stomach	Gastric juice	Fluid mixes with bolus; hydrochloric acid uncoils **proteins**; enzymes break down proteins; mucus protects stomach cells.
Pancreas	Small intestine	Pancreatic juice	Bicarbonate neutralizes acidic gastric juices; pancreatic enzymes break down **carbohydrates, fats,** and **proteins.**
Liver	Gallbladder	Bile	Bile stored until needed.
Gallbladder	Small intestine	Bile	Bile emulsifies **fat** so enzymes can attack.
Intestinal glands	Small intestine	Intestinal juice	Intestinal enzymes break down **carbohydrate, fat,** and **protein** fragments; mucus protects the intestinal wall.

FIGURE 3-9   The Digestive Fate of a Sandwich

To review the digestive processes, follow a peanut butter and banana sandwich on whole-wheat, seasame seed bread through the GI tract.

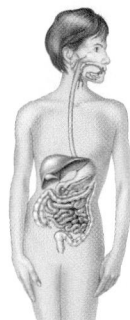

MOUTH: CHEWING AND SWALLOWING, WITH LITTLE DIGESTION

Carbohydrate digestion begins as the salivary enzyme starts to break down the starch from bread and peanut butter.
Fiber covering on the sesame seeds is crushed by the teeth, which exposes the nutrients inside the seeds to the upcoming digestive enzymes.

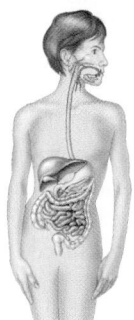

STOMACH: COLLECTING AND CHURNING, WITH SOME DIGESTION

Carbohydrate digestion continues until the mashed sandwich has been mixed with the gastric juices; the stomach acid of the gastric juices inactivates the salivary enzyme, and carbohydrate digestion ceases.
Proteins from the bread, seeds, and peanut butter begin to uncoil when they mix with the gastric acid, making them available to the gastric protease enzymes that begin to digest proteins.
Fat from the peanut butter forms a separate layer on top of the watery mixture.

SMALL INTESTINE: DIGESTING AND ABSORBING

Sugars from the banana require so little digestion that they begin to traverse the intestinal cells immediately on contact.
Starch digestion picks up when the pancreas sends pancreatic enzymes to the small intestine via the pancreatic duct. Enzymes on the surfaces of the small intestinal cells complete the process of breaking down starch into small fragments that can be absorbed through the intestinal cell walls and into the portal vein.
Fat from the peanut butter and seeds is emulsified with the watery digestive fluids by bile. Now the pancreatic and intestinal lipases can begin to break down the fat to smaller fragments that can be absorbed through the cells of the small intestinal wall and into the lymph.
Protein digestion depends on the pancreatic and intestinal proteases. Small fragments of protein are liberated and absorbed through the cells of the small intestinal wall and into the portal vein.
Vitamins and minerals are absorbed.

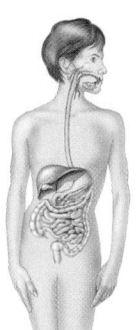

Note: Sugars and starches are members of the carbohydrate family.

LARGE INTESTINE: REABSORBING AND ELIMINATING

Fluids and some minerals are absorbed.
Some fibers from the seeds, whole-wheat bread, peanut butter, and banana are partly digested by the bacteria living there, and some of these products are absorbed.
Most fibers pass through the large intestine and are excreted as feces; some fat, cholesterol, and minerals bind to fiber and are also excreted.

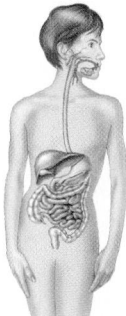

Absorption

Within three or four hours after you have eaten a dinner of beans and rice (or spinach lasagna, or steak and potatoes) with vegetable, salad, beverage, and dessert, your body must find a way to absorb the molecules derived from carbohydrate, protein, and fat digestion—and the vitamin and mineral molecules as well. Most absorption takes place in the small intestine, one of the most elegantly designed organ systems in the

To become part of your body, food must first be digested and absorbed.

■ The problem of food contaminants, which may be absorbed defenselessly by the body, is the subject of Chapter 19.

---

villi (VILL-ee, VILL-eye): fingerlike projections from the folds of the small intestine; singular **villus.**

microvilli (MY-cro-VILL-ee, MY-cro-VILL-eye): tiny, hairlike projections on each cell of every villus that can trap nutrient particles and transport them into the cells; singular **microvillus.**

crypts (KRIPTS): tubular glands that lie between the intestinal villi and secrete intestinal juices into the small intestine.

goblet cells: cells of the GI tract (and lungs) that secrete mucus.

---

body. Within its 10-foot length, which provides a surface area equivalent to a tennis court, the small intestine engulfs and absorbs the nutrient molecules. To remove the molecules rapidly and provide room for more to be absorbed, a rush of circulating blood continuously washes the underside of this surface, carrying the absorbed nutrients away to the liver and other parts of the body. Figure 3-10 describes how nutrients are absorbed by simple diffusion, facilitated diffusion, or active transport. Later chapters provide details on specific nutrients. Before following nutrients through the body, we must look more closely at the anatomy of the absorptive system.

Anatomy of the Absorptive System

The inner surface of the small intestine looks smooth and slippery, but viewed through a microscope, it turns out to be wrinkled into hundreds of folds. Each fold in turn is contoured into thousands of fingerlike projections, as numerous as the hairs on velvet fabric. These small intestinal projections are the **villi.** A single villus, magnified still more, turns out to be composed of hundreds of cells, each covered with its own microscopic hairs, the **microvilli** (see Figure 3-11). In the crevices between the villi lie the **crypts**—tubular glands that secrete the intestinal juices into the small intestine. Near by **goblet cells** secrete mucus.

The villi are in constant motion. Each villus is lined by a thin sheet of muscle, so it can wave, squirm, and wriggle like the tentacles of a sea anemone. Any nutrient molecule small enough to be absorbed is trapped among the microvilli that coat the cells and then drawn into the cells. Some partially digested nutrients are caught in the microvilli, digested further by enzymes there, and then absorbed into the cells.

A Closer Look at the Intestinal Cells

The cells of the villi are among the most amazing in the body, for they recognize and select the nutrients the body needs and regulate their absorption.■ As already described, each cell of a villus is coated with thousands of microvilli, which project from the cell's membrane (review Figure 3-11). In these microvilli and in the membrane lie hundreds of different kinds of enzymes and "pumps," which recognize

---

FIGURE 3-10 Absorption of Nutrients

Absorption of nutrients into intestinal cells typically occurs by simple diffusion or active transport.

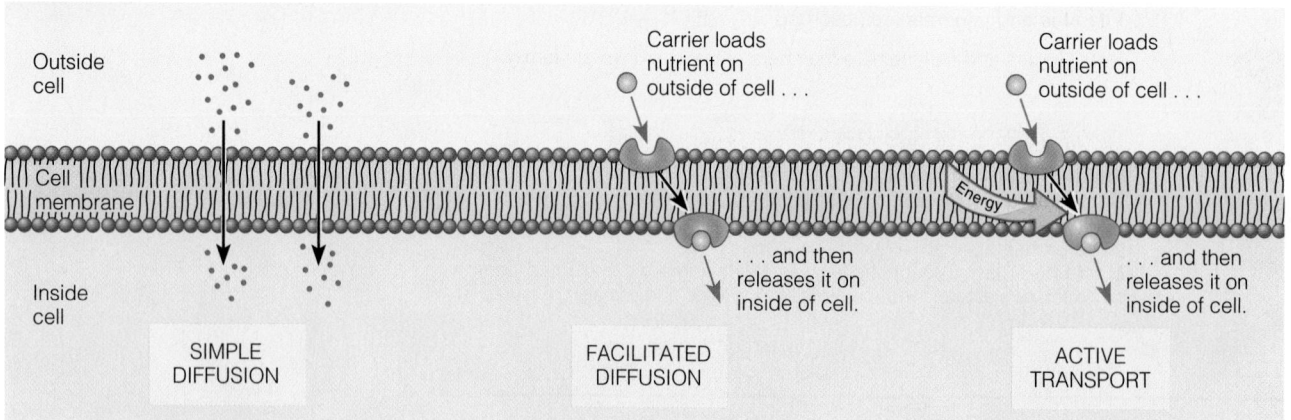

Some nutrients (such as water and small lipids) are absorbed by simple diffusion. They cross into intestinal cells freely.

Some nutrients (such as the water-soluble vitamins) are absorbed by facilitated diffusion. They need a specific carrier to transport them from one side of the cell membrane to the other. (Alternatively, facilitated diffusion may occur when the carrier changes the cell membrane in such a way that the nutrients can pass through.)

Some nutrients (such as glucose and amino acids) must be absorbed actively. These nutrients move against a concentration gradient, which requires energy.

FIGURE 3-11   The Small Intestinal Villi

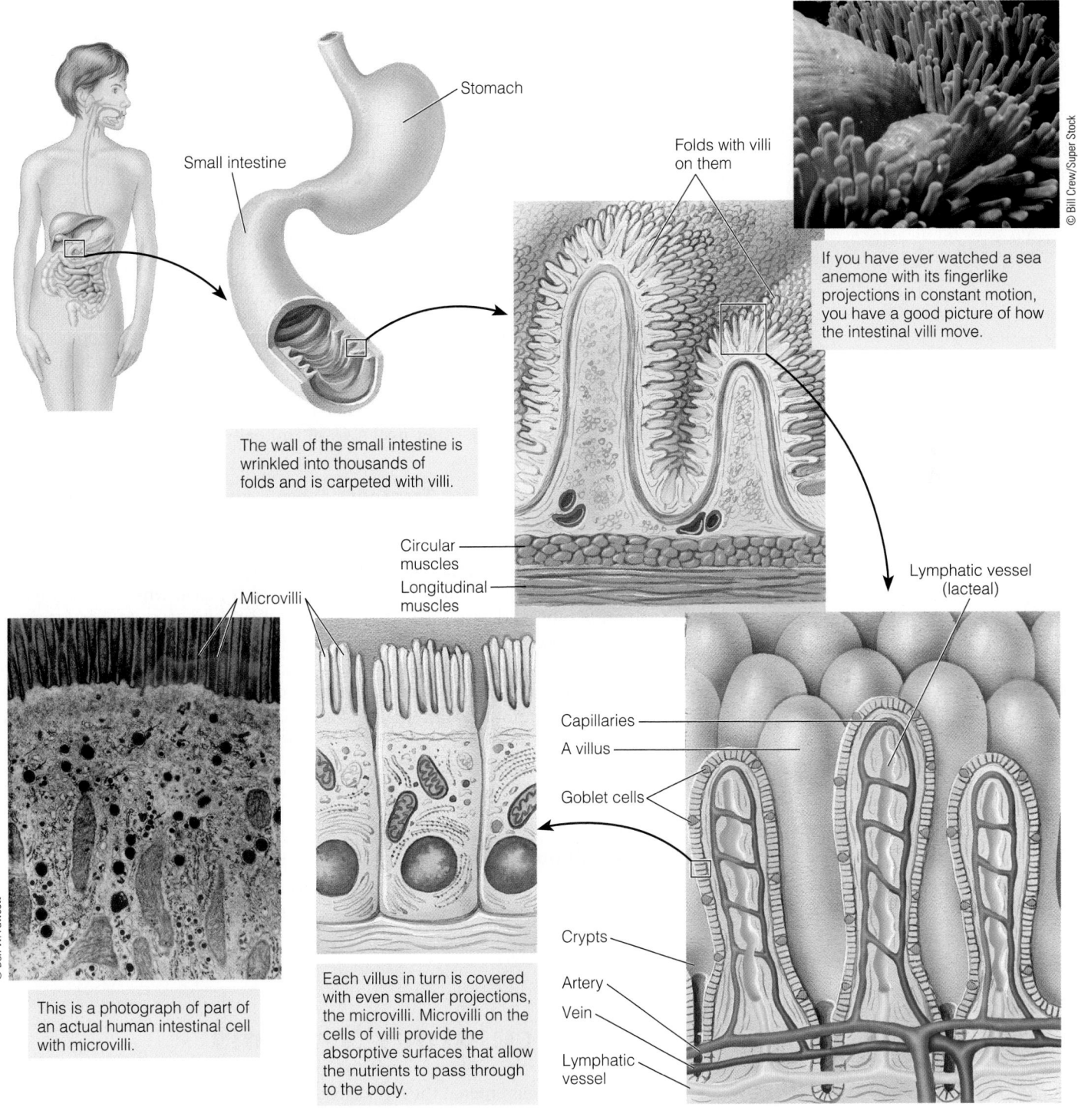

Stomach

Small intestine

Folds with villi on them

If you have ever watched a sea anemone with its fingerlike projections in constant motion, you have a good picture of how the intestinal villi move.

© Bill Crew/Super Stock

The wall of the small intestine is wrinkled into thousands of folds and is carpeted with villi.

Circular muscles

Longitudinal muscles

Microvilli

Lymphatic vessel (lacteal)

Capillaries

A villus

Goblet cells

Crypts

Artery

Vein

Lymphatic vessel

© Don W. Fawcett

This is a photograph of part of an actual human intestinal cell with microvilli.

Each villus in turn is covered with even smaller projections, the microvilli. Microvilli on the cells of villi provide the absorptive surfaces that allow the nutrients to pass through to the body.

and act on different nutrients. Descriptions of specific enzymes and "pumps" for each nutrient are presented in the following chapters where appropriate, but the point here is that the cells are equipped to handle all kinds and combinations of foods and nutrients.

Specialization in the GI Tract A further refinement of the system is that the cells of successive portions of the intestinal tract are specialized to absorb different nutrients. The nutrients that are ready for absorption early are absorbed near the top of the tract; those that take longer to be digested are absorbed farther down. Registered dietitians and medical professionals who treat digestive disorders learn the specialized absorptive functions of different parts of the GI tract so that if one part becomes dysfunctional, the diet can be adjusted accordingly.

The Myth of "Food Combining" The idea that people should not eat certain food combinations (for example, fruit and meat) at the same meal, because the digestive system cannot handle more than one task at a time, is a myth. The art of "food combining" (which actually emphasizes "food separating") is based on this idea, and it represents faulty logic and a gross underestimation of the body's capabilities. In fact, the contrary is often true; foods eaten together can enhance each other's use by the body. For example, vitamin C in a pineapple or other citrus fruit can enhance the absorption of iron from a meal of chicken and rice or other iron-containing foods. Many other instances of mutually beneficial interactions are presented in later chapters.

Preparing Nutrients for Transport When a nutrient molecule has crossed the cell of a villus, it enters either the bloodstream or the lymphatic system. Both transport systems supply vessels to each villus, as shown in Figure 3-11. The water-soluble nutrients and the smaller products of fat digestion are released directly into the bloodstream and guided directly to the liver where their fate and destination will be determined. The larger fats and the fat-soluble vitamins are insoluble in water, however, and blood is mostly water. The intestinal cells assemble many of the products of fat digestion into larger molecules. These larger molecules cluster together with special proteins, forming chylomicrons.■ These chylomicrons cannot pass into the capillaries and are released into the lymphatic system instead; the chylomicrons move through the lymph and later enter the bloodstream at a point near the heart, thus bypassing the liver at first. Details follow.

■ Chylomicrons (kye-lo-MY-cronz) are described in Chapter 5.

> **IN SUMMARY** The many folds and villi of the small intestine dramatically increase its surface area, facilitating nutrient absorption. Nutrients pass through the cells of the villi and enter either the blood (if they are water soluble or small fat fragments) or the lymph (if they are fat soluble).

The Circulatory Systems

Once a nutrient has entered the bloodstream, it may be transported to any of the cells in the body, from the tips of the toes to the roots of the hair. The circulatory systems deliver nutrients wherever they are needed.

The Vascular System

The vascular, or blood circulatory, system is a closed system of vessels through which blood flows continuously, with the heart serving as the pump (see Figure 3-12). As the blood circulates through this system, it picks up and delivers materials as needed.

FIGURE 3-12 The Vascular System

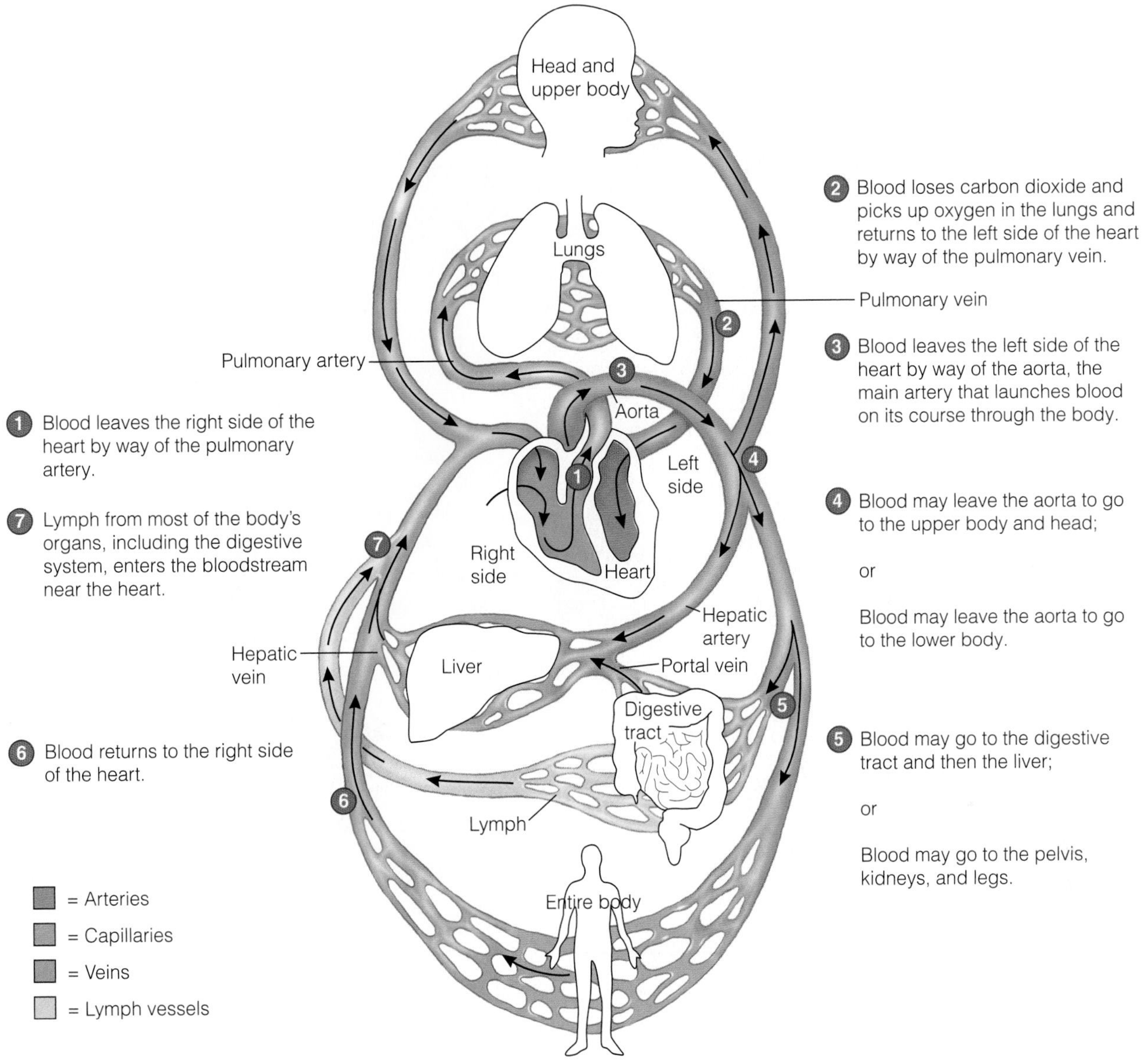

Head and upper body

Lungs

Pulmonary artery

Aorta

Left side

Right side

Heart

Hepatic artery

Hepatic vein

Liver

Portal vein

Digestive tract

Lymph

Entire body

■ = Arteries
■ = Capillaries
■ = Veins
□ = Lymph vessels

1 Blood leaves the right side of the heart by way of the pulmonary artery.

7 Lymph from most of the body's organs, including the digestive system, enters the bloodstream near the heart.

6 Blood returns to the right side of the heart.

2 Blood loses carbon dioxide and picks up oxygen in the lungs and returns to the left side of the heart by way of the pulmonary vein.

Pulmonary vein

3 Blood leaves the left side of the heart by way of the aorta, the main artery that launches blood on its course through the body.

4 Blood may leave the aorta to go to the upper body and head;

or

Blood may leave the aorta to go to the lower body.

5 Blood may go to the digestive tract and then the liver;

or

Blood may go to the pelvis, kidneys, and legs.

All the body tissues derive oxygen and nutrients from the blood and deposit carbon dioxide and other wastes into it. The lungs exchange carbon dioxide (which leaves the blood to be exhaled) and oxygen (which enters the blood to be delivered to all cells). The digestive system supplies the nutrients to be picked up. In the kidneys, wastes other than carbon dioxide are filtered out of the blood to be excreted in the urine.

Blood leaving the right side of the heart circulates through the lungs and then back to the left side of the heart. The left side of the heart then pumps the blood out through **arteries** to all systems of the body. The blood circulates in the **capillaries,** where it exchanges material with the cells, and then collects into **veins,** which return it again to the right side of the heart. In short, blood travels this simple route:

• Heart to arteries to capillaries to veins to heart.

arteries: vessels that carry blood from the heart to the tissues.

capillaries (CAP-ill-aries): small vessels that branch from an artery. Capillaries connect arteries to veins. Exchange of oxygen, nutrients, and waste materials takes place across capillary walls.

veins (VANES): vessels that carry blood to the heart.

FIGURE 3-13   The Liver

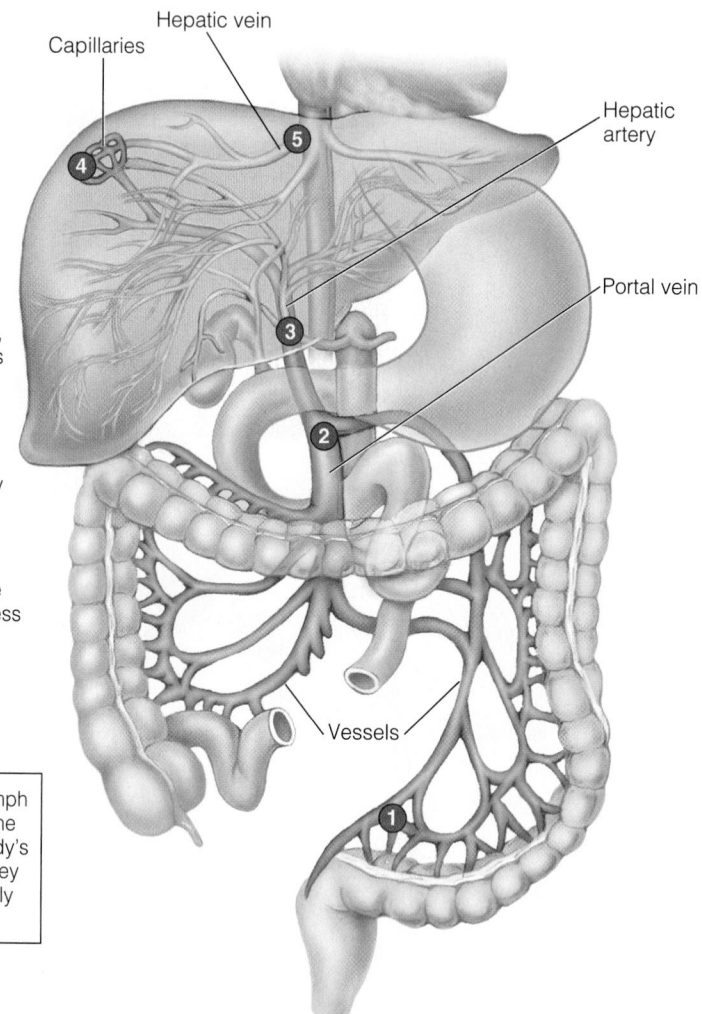

1 Vessels gather up nutrients and reabsorbed water and salts from all over the digestive tract.

Not shown here:
Parallel to these vessels (veins) are other vessels (arteries) that carry oxygen-rich blood from the heart to the intestines.

2 The vessels merge into the portal vein, which conducts all absorbed materials to the liver.

3 The hepatic artery brings a supply of freshly oxygenated blood (not loaded with nutrients) from the lungs to supply oxygen to the liver's own cells.

4 Capillaries branch all over the liver, making nutrients and oxygen available to all its cells and giving the cells access to blood from the digestive system.

5 The hepatic vein gathers up blood in the liver and returns it to the heart.

In contrast, nutrients absorbed into lymph do not go to the liver first. They go to the heart, which pumps them to all the body's cells. The cells remove the nutrients they need, and the liver then has to deal only with the remnants.

Capillaries

Hepatic vein

Hepatic artery

Portal vein

Vessels

The routing of the blood past the digestive system has a special feature. The blood is carried to the digestive system (as to all organs) by way of an artery, which (as in all organs) branches into capillaries to reach every cell. Blood leaving the digestive system, however, goes by way of a vein. The **portal vein** directs blood not back to the heart, but to another organ—the liver. This vein *again* branches into *capillaries* so that every cell of the liver has access to the blood. Blood leaving the liver then *again* collects into a vein, called the **hepatic vein,** which returns blood to the heart.

The route is:

- Heart to arteries to capillaries (in intestines) to vein to capillaries (in liver) to vein to heart.

Figure 3-13 shows the liver's key position in nutrient transport. An anatomist studying this system knows there must be a reason for this special arrangement. The liver's placement ensures that it will be first to receive the materials absorbed from the GI tract. In fact, the liver has many jobs to do in preparing the absorbed nutrients for use by the body. It is the body's major metabolic organ.

You might guess that, in addition, the liver serves as a gatekeeper to defend against substances that might harm the heart or brain. This is why, when people ingest poisons that succeed in passing the first barrier (the intestinal cells), the liver quite often suffers the damage—from viruses such as hepatitis, from drugs

portal vein: the vein that collects blood from the GI tract and conducts it to capillaries in the liver.

- **portal** = gateway

hepatic vein: the vein that collects blood from the liver capillaries and returns it to the heart.

- **hepatic** = liver

such as barbiturates or alcohol, from toxins such as pesticide residues, and from contaminants such as mercury. Perhaps, in fact, you have been undervaluing your liver, not knowing what heroic tasks it quietly performs for you.

The Lymphatic System

The **lymphatic system** provides a one-way route for fluid from the tissue spaces to enter the blood. Unlike the vascular system, the lymphatic system has no pump; instead, **lymph** circulates between the cells of the body and collects into tiny vessels. The fluid moves from one portion of the body to another as muscles contract and create pressure here and there. Ultimately, much of the lymph collects in a large duct behind the heart. This duct terminates in a vein that moves the lymph toward the heart.■ Thus materials from the GI tract that enter lymphatic vessels■ (large fats and fat-soluble vitamins) ultimately enter the bloodstream, circulating through arteries, capillaries, and veins like the other nutrients, with a notable exception—they bypass the liver at first.

Once inside the vascular system, the nutrients can travel freely to any destination and can be taken into cells and used as needed. What becomes of them is described in later chapters.

■ The duct that conveys lymph toward the heart is the **thoracic** (thor-ASS-ic) **duct**. The **subclavian** (sub-KLAY-vee-an) **vein** connects this duct with the right upper chamber of the heart, providing a passageway by which lymph can be returned to the vascular system.

■ The lymphatic vessels of the intestine that take up nutrients and pass them to the lymph circulation are called **lacteals** (LACK-tee-als).

> **IN SUMMARY** Nutrients leaving the digestive system via the blood are routed directly to the liver before being transported to the body's cells. Those leaving via the lymphatic system eventually enter the vascular system, but bypass the liver at first.

Regulation of Digestion and Absorption

There is nothing random about digestion and absorption; they are coordinated in every detail. The ability of the digestive tract to handle its ever-changing contents routinely illustrates an important physiological principle that governs the way all living things function—the principle of **homeostasis.** Simply stated, conditions have to stay about the same for an organism to survive; if they deviate too far from the norm, the organism must "do something" to bring them back to normal. The body's regulation of digestion is one example of homeostatic regulation. The body also regulates its temperature, its blood pressure, and all other aspects of its blood chemistry in similar ways.

The following paragraphs describe the regulation of digestion and absorption in healthy adults, but many factors■ can influence normal GI function. For example, peristalsis and sphincter action are poorly coordinated in newborns, so infants tend to "spit up" during the first several months of life. Older adults often experience constipation, in part because the intestinal wall loses strength and elasticity with age, which slows GI motility. Diseases can also interfere with digestion and absorption and often lead to malnutrition. Lack of nourishment, in general, and lack of certain dietary constituents such as fiber, in particular, alter the structure and function of GI cells. Quite simply, GI tract health depends on adequate nutrition.

■ Factors influencing GI function:
- Physical immaturity.
- Aging.
- Illness.
- Nutrition.

lymphatic (lim-FAT-ic) **system:** a loosely organized system of vessels and ducts that convey fluids toward the heart. The GI part of the lymphatic system carries the products of fat digestion into the bloodstream.

lymph (LIMF): a clear yellowish fluid that is almost identical to blood except that it contains no red blood cells or platelets. Lymph from the GI tract transports fat and fat-soluble vitamins to the bloodstream via lymphatic vessels.

homeostasis (HOME-ee-oh-STAY-sis): the maintenance of constant internal conditions (such as blood chemistry, temperature, and blood pressure) by the body's control systems. A homeostatic system is constantly reacting to external forces so as to maintain limits set by the body's needs.
- **homeo** = the same
- **stasis** = staying

Gastrointestinal Hormones and Nerve Pathways

Two intricate and sensitive systems coordinate all the digestive and absorptive processes: the hormonal (or endocrine) system and the nervous system. Even before the first bite of food is taken, the mere thought, sight, or smell of food can trigger a response from these systems. Then, as food travels through the GI tract, it either stimulates or inhibits digestive secretions by way of messages that are carried from one section of the GI tract to another by both **hormones** and nerve pathways. (Appendix A presents a brief summary of the body's hormonal system and nervous system.)

Notice that the kinds of regulation that will be described are all examples of *feedback* mechanisms. A certain condition demands a response. The response changes that condition, and the change then cuts off the response. Thus the system is self-corrective. Examples follow:

- *The stomach normally maintains a pH between 1.5 and 1.7. How does it stay that way?* Food entering the stomach stimulates cells in the stomach wall to release the hormone **gastrin.** Gastrin, in turn, stimulates the stomach glands to secrete the components of hydrochloric acid. When pH 1.5 is reached, the acid itself turns off the gastrin-producing cells. They stop releasing gastrin, and the glands stop producing hydrochloric acid. Thus the system adjusts itself.

  Nerve receptors in the stomach wall also respond to the presence of food and stimulate both the gastric glands to secrete juices and the muscles to contract. As the stomach empties, the receptors are no longer stimulated, the flow of juices slows, and the stomach quiets down.

- *The pyloric sphincter opens to let out a little chyme, then closes again. How does it know when to open and close?* When the pyloric sphincter relaxes, acidic chyme slips through. The cells of the pyloric muscle on the intestinal side sense the acid, causing the pyloric sphincter to close tightly. Only after the chyme has been neutralized by pancreatic bicarbonate and the juices surrounding the pyloric sphincter have become alkaline can the muscle relax again. This process ensures that the chyme will be released slowly enough to be neutralized as it flows through the small intestine. This is important because the small intestine has less of a mucous coating than the stomach does and so is not as well protected from acid.

- *As the chyme enters the intestine, the pancreas adds bicarbonate to it so that the intestinal contents always remain at a slightly alkaline pH. How does the pancreas know how much to add?* The presence of chyme stimulates the cells of the duodenum wall to release the hormone **secretin** into the blood. When secretin reaches the pancreas, it stimulates the pancreas to release its bicarbonate-rich juices. Thus, whenever the duodenum signals that acidic chyme is present, the pancreas responds by sending bicarbonate to neutralize it. When the need has been met, the cells of the duodenal wall are no longer stimulated to release secretin, the hormone no longer flows through the blood, the pancreas no longer receives the message, and it stops sending pancreatic juice. Nerves also regulate pancreatic secretions.

- *Pancreatic secretions contain a mixture of enzymes to digest carbohydrate, fat, and protein. How does the pancreas know how much of each type of enzyme to provide?* This is one of the most interesting questions physiologists have asked. Clearly, the pancreas does know what its owner has been eating, and it secretes enzyme mixtures tailored to handle the food mixtures that have been arriving lately (over the last several days). Enzyme activity changes proportionately in response to the amounts of carbohydrate, fat, and protein in the diet. If a person has been eating mostly carbohydrates, the pancreas makes and secretes mostly carbohydrases; if the person's diet has been high in fat, the pancreas produces more lipases; and so forth. Presumably, hormones from the GI tract, secreted in response to meals, keep the pancreas informed as to its digestive tasks. The day or two lag between the time a person's diet changes dramatically and the time digestion of the new diet becomes efficient explains why dietary changes can "upset digestion" and should be made gradually.

- *Why don't the digestive enzymes damage the pancreas?* The pancreas protects itself from harm by producing an inactive form of the enzymes.■ Then it releases these proteins into the small intestine where they are activated to

■ The inactive precursor of an enzyme is called a **proenzyme** or **zymogen** (ZYE-mo-jen).
- **pro** = before
- **zym** = concerning enzymes
- **gen** = to produce

hormones: chemical messengers. Hormones are secreted by a variety of glands in response to altered conditions in the body. Each hormone travels to one or more specific target tissues or organs, where it elicits a specific response to maintain homeostasis. In general, any gastrointestinal hormone may be called an **enterogastrone** (EN-ter-oh-GAS-trone), but the term refers specifically to any hormone that slows motility and inhibits gastric secretions.

gastrin: a hormone secreted by cells in the stomach wall. Target organ: the glands of the stomach. Response: secretion of gastric acid.

secretin (see-CREET-in): a hormone produced by cells in the duodenum wall. Target organ: the pancreas. Response: secretion of bicarbonate-rich pancreatic juice.

become enzymes. In pancreatitis, the digestive enzymes somehow become active within the pancreas itself, causing inflammation and damaging the delicate pancreatic tissues.

- *When fat is present in the intestine, the gallbladder contracts to squirt bile into the intestine to emulsify the fat. How does the gallbladder get the message that fat is present?* Fat in the intestine stimulates cells of the intestinal wall to release the hormone **cholecystokinin (CCK).** This hormone, traveling by way of the blood to the gallbladder, stimulates it to contract, releasing bile into the small intestine. Once the fat in the intestine is emulsified and enzymes have begun to work on it, the fat no longer provokes release of the hormone, and the message to contract is canceled. (By the way, fat emulsification can continue even after a diseased gallbladder has been surgically removed because the liver can deliver bile directly to the small intestine.)

- *Fat takes longer to digest than carbohydrate does. When fat is present, intestinal motility slows to allow time for its digestion. How does the intestine know when to slow down?* Cholecystokinin and **gastric-inhibitory peptide** slow GI tract motility. By slowing the digestive process, fat helps to maintain a pace that will allow all reactions to reach completion. Gastric-inhibitory peptide also inhibits gastric acid secretion. Hormonal and nervous mechanisms like these account for much of the body's ability to adapt to changing conditions.

Once a person has started to learn the answers to questions like these, it may be hard to stop. Some people devote their whole lives to the study of physiology. For now, however, these few examples will be enough to illustrate how all the processes throughout the digestive system are precisely and automatically regulated without any conscious effort.

IN SUMMARY Digestion and absorption depend on the coordinated efforts of the hormonal system and the nervous system. Together, they regulate the processes of transforming foods into nutrients.

The System at Its Best

This chapter has described the anatomy of the digestive tract on several levels: the sequence of digestive organs, the cells and structures of the villi, and the selective machinery of the cell membranes. The intricate architecture of the digestive system makes it sensitive and responsive to conditions in its environment. Knowing the optimal conditions will help you to promote the best functioning of the system.

One indispensable condition is good health of the digestive tract itself. This health is affected by such lifestyle factors as sleep, physical activity, and state of mind. Adequate sleep allows for repair and maintenance of tissue and removal of wastes that might impair efficient functioning. Activity promotes healthy muscle tone. Mental state influences the activity of regulatory nerves and hormones; for healthy digestion, you should be relaxed and tranquil at mealtimes.

Another factor is the kind of meals you eat. Among the characteristics of meals that promote optimal absorption of nutrients are those mentioned in Chapter 2: balance, moderation, variety, and adequacy. Balance and moderation require having neither too much nor too little of anything. For example, too much fat can be harmful, but some fat is beneficial in slowing down intestinal motility and providing time for absorption of some of the nutrients that are slow to be absorbed.

Variety is important for many reasons, but one is that some food constituents interfere with nutrient absorption. For example, some compounds common in high-fiber foods such as whole-grain cereals, certain leafy green vegetables, and legumes bind with minerals. To some extent, then, the minerals in those foods may become unavailable for absorption. These high-fiber foods are still valuable, but need to be balanced with a variety of other foods that can provide the minerals.

cholecystokinin (coal-ee-sis-toe-KINE-in), or **CCK:** a hormone produced by cells of the intestinal wall. Target organ: the gallbladder. Response: release of bile and slowing of GI motility.

gastric-inhibitory peptide: a hormone produced by the intestine. Target organ: the stomach. Response: slowing of the secretion of gastric juices and of GI motility.

As for adequacy—in a sense, this entire book is about dietary adequacy. But here, at the end of this chapter, is a good place to underline the interdependence of the nutrients. It could almost be said that every nutrient depends on every other. All the nutrients work together, and all are present in the cells of a healthy digestive tract. To maintain health and promote the functions of the GI tract, you should make balance, moderation, variety, and adequacy features of every day's menus.

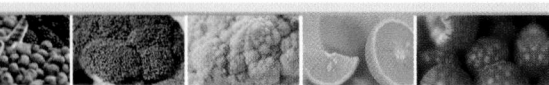

Nutrition in Your Life

A healthy digestive system can adjust to almost any diet and can handle any combination of foods with ease.

- Do you usually enjoy meals without overeating to the point of discomfort?
- Do you experience GI distress regularly?
- What changes can you make in your eating habits to promote GI health?

NUTRITION ON THE NET

 Access these websites for further study of topics covered in this chapter.

- Find updates and quick links to these and other nutrition-related sites at our website: **www.wadsworth.com/nutrition**

- Visit the Center for Digestive Health and Nutrition: **www.gihealth.com**
- Visit the Digest This! section of the American College of Gastroenterology: **www.acg.gi.org**

STUDY QUESTIONS

These questions will help you review this chapter. You will find the answers in the discussions on the pages provided.

1. Describe the obstacles associated with digesting food and the solutions offered by the human body. (pp. 73–80)

2. Describe the path food follows as it travels through the digestive system. Summarize the muscular actions that take place along the way. (pp. 74–79)

3. Name five organs that secrete digestive juices. How do the juices and enzymes facilitate digestion? (pp. 79–81)

4. Describe the problems associated with absorbing nutrients and the solutions offered by the small intestine. (pp. 83–86)

5. How is blood routed through the digestive system? Which nutrients enter the bloodstream directly? Which are first absorbed into the lymph? (pp. 86–89)

6. Describe how the body coordinates and regulates the processes of digestion and absorption. (pp. 89–91)

7. How does the composition of the diet influence the functioning of the GI tract? (pp. 90–91)

8. What steps can you take to help your GI tract function at its best? (pp. 91–92)

These multiple choice questions will help you prepare for an exam. Answers can be found on p. 93.

1. The semiliquid, partially digested food that travels through the intestinal tract is called:
   a. bile.
   b. lymph.
   c. chyme.
   d. secretin.

2. The muscular contractions that move food through the GI tract are called:
   a. hydrolysis.
   b. sphincters.
   c. peristalsis.
   d. bowel movements.

3. The main function of bile is to:
    a. emulsify fats.
    b. catalyze hydrolysis.
    c. slow protein digestion.
    d. neutralize stomach acidity.

4. The pancreas neutralizes stomach acid in the small intestine by secreting:
    a. bile.
    b. mucus.
    c. enzymes.
    d. bicarbonate.

5. Which nutrient passes through the GI tract mostly undigested and unabsorbed?
    a. fat
    b. fiber
    c. protein
    d. carbohydrate

6. Absorption occurs primarily in the:
    a. mouth.
    b. stomach.
    c. small intestine.
    d. large intestine.

7. All blood leaving the GI tract travels first to the:
    a. heart.
    b. liver.
    c. kidneys.
    d. pancreas.

8. Which nutrients leave the GI tract by way of the lymphatic system?
    a. water and minerals
    b. proteins and minerals
    c. all vitamins and minerals
    d. fats and fat-soluble vitamins

9. Digestion and absorption are coordinated by the:
    a. pancreas and kidneys.
    b. liver and gallbladder.
    c. hormonal system and the nervous system.
    d. vascular system and the lymphatic system.

10. Gastrin, secretin, and cholecystokinin are examples of:
    a. crypts.
    b. enzymes.
    c. hormones.
    d. goblet cells.

ANSWERS

Study Questions (multiple choice)
1. c   2. c   3. a   4. d   5. b   6. c   7. b   8. d
9. c   10. c

Common Digestive Problems

© Ronnie Kaufman/CORBIS

The facts of anatomy and physiology presented in Chapter 3 permit easy understanding of some common problems that occasionally arise in the digestive tract. Food may slip into the air passages instead of the esophagus, causing choking. Bowel movements may be loose and watery, as in diarrhea, or painful and hard, as in constipation. Some people complain about belching, while others are bothered by intestinal gas. Sometimes people develop medical problems such as an ulcer. This highlight describes some of the symptoms of these common digestive problems and suggests strategies for preventing them (the glossary on p. 96 defines the relevant terms).

To help a person who is choking, first ask this critical question: "Can you make any sound at all?" If so, relax. You have time to decide what you can do to help. Whatever you do, don't hit him on the back—the particle may become lodged more firmly in his air passage. If the person cannot make a sound, shout for help and perform the **Heimlich maneuver** (described in Figure H3-2). You would do well to take a life-saving course and practice these techniques, for you will have no time for hesitation once you are called on to perform this death-defying act.

Almost any food can cause choking, although some are cited more often than others: chunks of meat, hot dogs, nuts, whole grapes, raw carrots, marshmallows, hard or sticky candies, gum, popcorn, and peanut butter. These foods are particularly difficult for young children to safely chew and swallow. In 2000, more than 17,500 children (under 15 years old) in the United States choked; most of them choked on food, and 160 of them choked to death.[1] Always remain alert to the dangers of choking whenever young children are eating. To prevent choking, cut food into small pieces, chew thoroughly

Choking

A person chokes when a piece of food slips into the **trachea** and becomes lodged so securely that it cuts off breathing (see Figure H3-1). Without oxygen, the person may suffer brain damage or die. For this reason, it is imperative that everyone learn to recognize the international signal for choking (shown in Figure H3-2) and act promptly.

The choking scenario might read like this. A person is dining in a restaurant with friends. A chunk of food, usually meat, becomes lodged in his trachea so firmly that he cannot make a sound. No sound can be made because the **larynx** is in the trachea and makes sounds only when air is pushed across it. Often he chooses to suffer alone rather than "make a scene in public." If he tries to communicate distress to his friends, he must depend on pantomime. The friends are bewildered by his antics and become terribly worried when he "faints" after a few minutes without air. They call for an ambulance, but by the time it arrives, he is dead from suffocation.

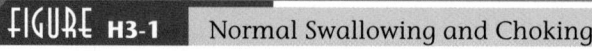

FIGURE H3-1 Normal Swallowing and Choking

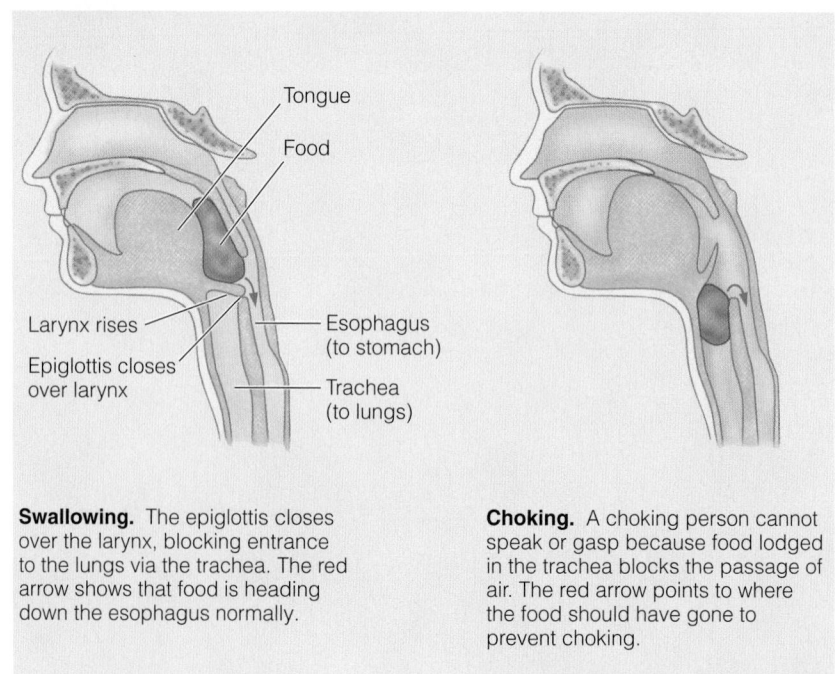

Tongue

Food

Larynx rises

Epiglottis closes over larynx

Esophagus (to stomach)

Trachea (to lungs)

Swallowing. The epiglottis closes over the larynx, blocking entrance to the lungs via the trachea. The red arrow shows that food is heading down the esophagus normally.

Choking. A choking person cannot speak or gasp because food lodged in the trachea blocks the passage of air. The red arrow points to where the food should have gone to prevent choking.

FIGURE H3-2 First Aid for Choking

The strategy most likely to succeed is abdominal thrusts, sometimes called the Heimlich maneuver. Only if all else fails, open the mouth by grasping both the tongue and lower jaw and lifting. Then, and only if you can see the object, use your finger to sweep it out and begin rescue breathing.

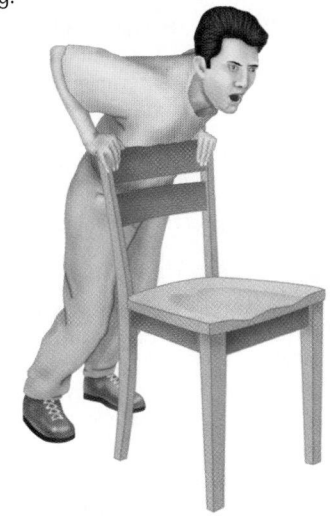

This universal signal for choking alerts others to the need for assistance. Stand behind the person, and wrap your arms around him. Place the thumb side of one fist snugly against his body, slightly above the navel and below the rib cage. Grasp your fist with your other hand and give him a sudden strong hug inward and upward. Repeat thrusts as necessary.

To self-administer first aid, place the thumb side of one fist slightly above the navel and below the rib cage, grasp the fist with your other hand, and then press inward and upward with a quick motion. If this is unsuccessful, quickly press your upper abdomen over any firm surface such as the back of a chair, a countertop, or a railing.

before swallowing, don't talk or laugh with food in your mouth, and don't eat when breathing hard.

Vomiting

Another common digestive mishap is **vomiting.** Vomiting can be a symptom of many different diseases or may arise in situations that upset the body's equilibrium, such as air or sea travel. For whatever reason, the waves of peristalsis reverse direction, and the contents of the stomach are propelled up through the esophagus to the mouth and expelled.

If vomiting continues long enough or is severe enough, the reverse peristalsis will extend beyond the stomach and carry the contents of the duodenum, with its green bile, into the stomach and then up the esophagus. Although certainly unpleasant and wearying for the nauseated person, vomiting such as this is no cause for alarm. Vomiting is one of the body's adaptive mechanisms to rid itself of something irritating. The best advice is to rest and drink small amounts of liquids as tolerated until the nausea subsides.

A physician's care may be needed, however, when large quantities of fluid are lost from the GI tract, causing dehydration.

With massive fluid loss from the GI tract, all of the body's other fluids redistribute themselves so that, eventually, fluid is taken from every cell of the body. Leaving the cells with the fluid are salts that are absolutely essential to the life of the cells, and they must be replaced, which is difficult while the vomiting continues. Intravenous feedings of saline and glucose are frequently necessary while the physician is diagnosing the cause of the vomiting and instituting corrective therapy.

In an infant, vomiting is likely to become serious early in its course, and a physician should be contacted soon after onset. Infants have more fluid between their body cells than adults do, so more fluid can move readily into the digestive tract and be lost from the body. Consequently, the body water of infants becomes depleted and their body salt balance upset faster than in adults.

Self-induced vomiting, such as occurs in bulimia nervosa, also has serious consequences. In addition to fluid and salt imbalances, repeated vomiting can cause irritation and infection of the pharynx, esophagus, and salivary glands; erosion of the teeth and gums; and dental caries. The esophagus may rupture or tear, as may the stomach. Sometimes the eyes become red from pressure during vomiting. Bulimic behavior reflects underlying psychological problems that require intervention. (Bulimia nervosa is discussed fully in Highlight 9.)

Projectile vomiting is also serious. The contents of the stomach are expelled with such force that they leave the mouth in a wide arc like a bullet leaving a gun. This type of vomiting requires immediate medical attention.

Diarrhea

Diarrhea is characterized by frequent, loose, watery stools. Such stools indicate that the intestinal contents have moved too quickly through the intestines for fluid absorption to take place, or that water has been drawn from the cells lining the intestinal tract and added to the food residue. Like vomiting, diarrhea can lead to considerable fluid and salt losses, but the composition of the fluids is different. Stomach fluids lost in vomiting are highly acidic, whereas intestinal fluids lost in diarrhea are nearly neutral. When fluid losses require medical attention, correct replacement is crucial.

Diarrhea is a symptom of a variety of medical conditions and treatments. It may occur abruptly in a healthy person as a result

acid controllers: medications used to prevent or relieve indigestion by suppressing production of acid in the stomach; also called **H2 blockers.** Common brands include Pepcid AC, Tagamet HB, Zantac 75, and Axid AR.

antacids: medications used to relieve indigestion by neutralizing acid in the stomach. Common brands include Alka-Seltzer, Maalox, Rolaids, and Tums.

belching: the expulsion of gas from the stomach through the mouth.

colitis (ko-LYE-tis): inflammation of the colon.

colonic irrigation: the popular, but potentially harmful practice of "washing" the large intestine with a powerful enema machine.

constipation: the condition of having infrequent or difficult bowel movements.

defecate (DEF-uh-cate): to move the bowels and eliminate waste.
• **defaecare** = to remove dregs

diarrhea: the frequent passage of watery bowel movements.

diverticula (dye-ver-TIC-you-la): sacs or pouches that develop in the weakened areas of the intestinal wall (like bulges in an inner tube where the tire wall is weak).
• **divertir** = to turn aside

diverticulitis (DYE-ver-tic-you-LYE-tis): infected or inflamed diverticula.
• **itis** = infection or inflammation

diverticulosis (DYE-ver-tic-you-LOH-sis): the condition of having diverticula. About one in every six people in Western countries develops diverticulosis in middle or later life.
• **osis** = condition

enemas: solutions inserted into the rectum and colon to stimulate a bowel movement and empty the lower large intestine.

gastroesophageal reflux: the backflow of stomach acid into the esophagus, causing damage to the cells of the esophagus and the sensation of heartburn. **Gastroesophageal reflux disease (GERD)** is characterized by symptoms of reflux occurring two or more times a week.

heartburn: a burning sensation in the chest area caused by backflow of stomach acid into the esophagus.

Heimlich (HIME-lick) **maneuver (abdominal thrust maneuver):** a technique for dislodging an object from the trachea of a choking person (see Figure H3-2); named for the physician who developed it.

hemorrhoids (HEM-oh-royds): painful swelling of the veins surrounding the rectum.

hiccups (HICK-ups): repeated cough-like sounds and jerks that are produced when an involuntary spasm of the diaphragm muscle sucks air down the windpipe; also spelled *hiccoughs.*

indigestion: incomplete or uncomfortable digestion, usually accompanied by pain, nausea, vomiting, heartburn, intestinal gas, or belching.
• **in** = not

irritable bowel syndrome: an intestinal disorder of unknown cause. Symptoms include abdominal discomfort and cramping, diarrhea, constipation, or alternating diarrhea and constipation.

larynx: the voice box (see Figure H3-1).

laxatives: substances that loosen the bowels and thereby prevent or treat constipation.

mineral oil: a purified liquid derived from petroleum and used to treat constipation.

peptic ulcer: an erosion in the mucous membrane of either the stomach (a gastric ulcer) or the duodenum (a duodenal ulcer).

trachea (TRAKE-ee-uh): the windpipe; the passageway from the mouth and nose to the lungs.

ulcer: an erosion in the topmost, and sometimes underlying, layers of cells in an area. See also *peptic ulcer.*

vomiting: expulsion of the contents of the stomach up through the esophagus to the mouth.

---

of infections (such as food poisoning) or as a side effect of medications. When used in large quantities, food ingredients such as the sugar alternative sorbitol and the fat alternative olestra may also cause diarrhea in some people. If a food is responsible, then that food must be omitted from the diet, at least temporarily. If medication is responsible, a different medicine, when possible, or a different form (injectable versus oral, for example) may alleviate the problem.

Diarrhea may also occur as a result of disorders of the GI tract, such as irritable bowel syndrome or colitis. **Irritable bowel syndrome** is one of the most common GI disorders and is characterized by a disturbance in the motility of the GI tract.[2] Dietary treatment hinges on identifying and avoiding individual foods that cause intolerance. For most people, a low-fat diet provided in small meals, with a gradual increase in fiber, is helpful. People with **colitis,** an inflammation of the large intestine, may also suffer from severe diarrhea. They often benefit from complete bowel rest and medication. If treatment fails, surgery to remove the colon and rectum may be necessary.

As you can see, treatment for diarrhea depends on its cause and its severity. Mild diarrhea may remit without treatment; simply rest and drink extra liquids to replace fluid losses. If diarrhea persists, though, especially in an infant, young child, or elderly person, call a physician. Severe diarrhea can be life-threatening when it leads to dehydration and electrolyte imbalances. (Chapter 12 provides more information on dehydration and its therapy.)

Constipation

Like diarrhea, **constipation** describes a symptom, not a disease. Each person's GI tract has its own cycle of waste elimination, which depends on its owner's health, the type of food eaten, when it was eaten, and when the person takes time to **defecate.** What's normal for some people may not be normal for others. Some people have bowel movements three times a day; others may have them three times a week. Only when people pass stools that are difficult or painful to expel or when they experience a reduced frequency of bowel movements from their typical pattern are they constipated. Abdominal discomfort, headaches, backaches, and the passing of gas sometimes accompany constipation.

Often a person's lifestyle may cause constipation. Being too busy to respond to the defecation signal is a common complaint. If a person receives the signal to defecate and ignores it, the signal may not return for several hours. In the meantime, water continues to be withdrawn from the fecal matter, so when the person does defecate, the bowel movement is dry and hard. In such a case, a person's daily regimen may need to be revised to allow time to have a bowel movement when the body sends its signal. One possibility is to go to bed earlier in order to rise earlier, allowing ample time for a leisurely breakfast and a movement.

Another cause of constipation is lack of physical activity. Physical activity improves muscle tone, not just of the outer body, but also of the digestive tract. As little as 30 minutes of physical activity a day can help prevent or alleviate constipation.

Although constipation usually reflects lifestyle habits, in some cases it may be a side effect of medication or may reflect a medical problem such as tumors that are obstructing the passage of waste. If discomfort is associated with passing fecal matter, seek medical advice to rule out disease. Once this has been done, dietary or other measures for correction can be considered.

One dietary measure that may be appropriate is to increase dietary fiber. Fibers found in cereal products help to prevent constipation by increasing fecal mass. In the GI tract, fiber attracts water, creating soft, bulky stools that stimulate bowel contractions to push the contents along. These contractions strengthen the intestinal muscles. The improved muscle tone, together with the water content of the stools, eases elimination, reducing the pressure in the rectal veins and helping to prevent **hemorrhoids.** Chapter 4 provides more information on fiber's role in maintaining a healthy colon and reducing the risks of colon cancer and diverticulosis. **Diverticulosis** is a condition in which the intestinal walls develop bulges in weakened areas, most commonly in the colon (see Figure H3-3). These bulging pockets, known as **diverticula,** can worsen constipation, entrap feces, and become painfully infected and inflamed **(diverticulitis).** Treatment may require hospitalization, antibiotics, or surgery.

Drinking plenty of water in conjunction with eating high-fiber foods also helps with constipation. The increased bulk physically stimulates the upper GI tract, promoting peristalsis throughout.

Eating prunes—or "dried plums" as some have renamed them—can also be helpful. Prunes are high in fiber and also contain a laxative substance.[*] If a morning defecation is desired, a person can drink prune juice at bedtime; if the evening is preferred, the person can drink prune juice with breakfast.

Honey can also have a laxative effect due to its incomplete absorption. Although this characteristic may cause problems for people with irritable bowel syndrome, eating honey may

---

[*]This substance is dihydroxyphenyl isatin.

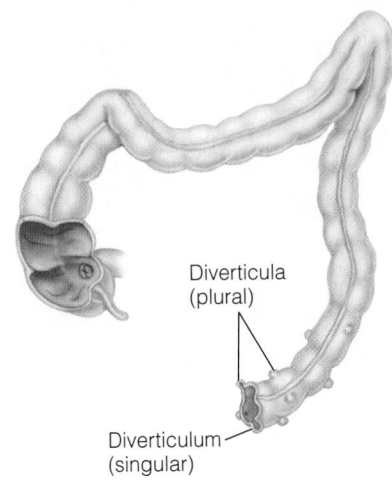

FIGURE H3-3 Diverticula in the Colon

Diverticula may develop anywhere along the GI tract, but are most common in the colon.

Diverticula (plural)

Diverticulum (singular)

be an easy and effective treatment for those who are constipated. Honey should never be fed to infants, however, because of the risk of botulism (as explained in Chapter 16).

Adding fat to the diet can relieve some constipation by stimulating the hormone cholecystokinin, which summons bile into the duodenum. Bile's high salt content draws water from the intestinal wall, which stimulates peristalsis and softens the fecal matter.

These suggested changes in lifestyle or diet should correct chronic constipation without the use of **laxatives, enemas,** or **mineral oil,** although television commercials often try to persuade people otherwise. One of the fallacies often perpetrated by advertisements is that one person's successful use of a product is a good recommendation for others to use that product.

As a matter of fact, diet changes that relieve constipation for one person may increase the constipation of another. For instance, increasing fiber intake stimulates peristalsis and helps the person with a sluggish colon. Some people, though, have a spastic type of constipation, in which peristalsis promotes strong contractions that close off a segment of the colon and prevent passage; for these people, increasing fiber intake would be exactly the wrong thing to do.

A person who seems to need products such as laxatives should seek a physician's advice. Opinions from friends or alternative medicine practitioners may cause more harm than good. One potentially harmful but currently popular practice that is being promoted by some alternative medicine practitioners is **colonic irrigation**—the internal washing of the large intestine with a powerful enema machine. Such an extreme cleansing is not only unnecessary, but can be hazardous, causing illness and death from equipment contamination, electrolyte depletion,

Beans, broccoli, cabbage, and onions produce gas in many people. People troubled by gas need to determine which foods bother them and then eat those foods in moderation.

and intestinal perforation. Less extreme practices can cause problems, too. Frequent use of laxatives and enemas can lead to dependency; upset the body's fluid, salt, and mineral balances; and, in the case of mineral oil, interfere with the absorption of fat-soluble vitamins. (Mineral oil dissolves the vitamins, but is not itself absorbed; instead, it leaves the body, carrying the vitamins with it.)

Belching and Gas

Many people complain of problems that they attribute to excessive gas. For some, **belching** is the complaint. Others blame intestinal gas for abdominal discomforts and embarrassment. Most people believe that the problems occur after they eat certain foods. This may be the case with intestinal gas, but belching results from swallowing air. The best advice for belching seems to be to eat slowly, chew thoroughly, and relax while eating.

Everyone swallows a little bit of air with each mouthful of food, but people who eat too fast may swallow too much air and then have to belch. Ill-fitting dentures, carbonated beverages, and chewing gum can also contribute to the swallowing of air with resultant belching. Occasionally, belching can be a sign of a more serious disorder, such as gallbladder disease or a peptic ulcer.

People who eat or drink too fast may also trigger **hiccups,** the repeated spasms that produce a cough-like sound and jerky movement. Normally, hiccups soon subside and are of no medical significance, but they can be bothersome. The most effective cure is to hold the breath for as long as possible, which helps to relieve the spasms of the diaphragm.

While expelling gas can be a humiliating experience, it is quite normal. (People experiencing painful bloating from malabsorption diseases, however, require medical treatment.) Healthy people expel several hundred milliliters of gas several times a day. Almost all (99 percent) of the gases expelled—nitrogen, oxygen, hydrogen, methane, and carbon dioxide—are odorless. The remaining "volatile" gases are the infamous ones.

Foods that produce gas usually must be determined individually. The most common offenders are foods rich in the carbohydrates—sugars, starches, and fibers. When partially digested carbohydrates reach the large intestine, bacteria digest them, giving off gas as a by-product. People can test foods suspected of forming gas by omitting them individually for a trial period and seeing if there is any improvement.

Heartburn and "Acid Indigestion"

Almost everyone has experienced **heartburn** at one time or another, usually soon after eating a meal. Medically known as **gastroesophageal reflux,** heartburn is the painful sensation a person feels behind the breastbone when the lower esophageal sphincter allows the stomach contents to reflux into the esophagus (see Figure H3-4). This may happen if a person eats or drinks too much (or both). Tight clothing and even changes of position (lying down, bending over) can cause it, too, as can some medications and smoking. A defect of the sphincter muscle itself is a possible, but less common cause.

If the heartburn is not caused by an anatomical defect, treatment is fairly simple. To avoid such misery in the future, the person needs to learn to eat less at a sitting, chew food more thoroughly, and eat it more slowly. Additional strategies are presented in Table H3-1 at the end of this highlight.

As far as "acid indigestion" is concerned, recall from Chapter 3 that the strong acidity of the stomach is a desirable condition—television commercials for **antacids** and **acid controllers** notwithstanding. People who overeat or eat too quickly are likely to suffer from **indigestion.** The muscular reaction of the stomach to unchewed lumps or to being overfilled may be so violent that it causes regurgitation (reverse peristalsis). When this happens, overeaters may taste the stomach acid and feel pain. Responding to advertisements, they may reach for antacids or acid controllers. Both of these drugs were originally designed to treat GI illnesses such as ulcers. As is true of most over-the-counter medicines, antacids and acid controllers should be used only infrequently for occasional heartburn; they may mask or cause problems if used regularly, as the next section explains. Instead of self-

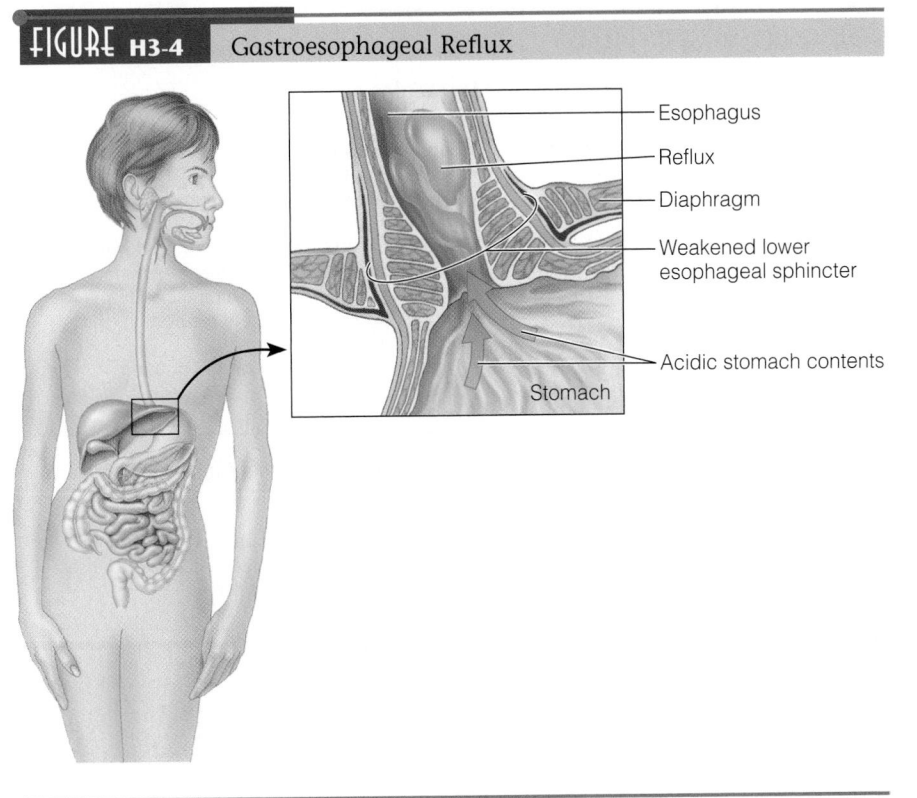

FIGURE H3-4 Gastroesophageal Reflux

- Esophagus
- Reflux
- Diaphragm
- Weakened lower esophageal sphincter
- Acidic stomach contents
- Stomach

medicating, people who suffer from frequent and regular bouts of heartburn and indigestion need to see a physician, who can prescribe specific medication to control gastroesophageal reflux. Without treatment, the repeated splashes of acid can severely damage the cells of the esophagus, creating a condition known as Barrett's esophagus.[3] At that stage, the risk of cancer in the throat or esophagus increases dramatically. To repeat, if symptoms persist, see a doctor—don't self-medicate.

Ulcers

Ulcers of the stomach (gastric ulcers) or the duodenum of the small intestine (duodenal ulcers) are another common digestive problem. (The term **peptic ulcer** includes both types.) An **ulcer** is an erosion of the top layer of cells from an area, such as the wall of the stomach or duodenum. This erosion leaves the underlying layers of cells unprotected and exposed to gastric juices. The erosion may proceed until the gastric juices reach the capillaries that feed the area, leading to bleeding, and reach the nerves, causing pain. If GI bleeding is excessive, iron deficiency may develop. If the erosion penetrates all the way through the GI lining, a life-threatening infection can develop.

Many people naively believe that an ulcer is caused by stress or spicy foods, but this is not the case. The stomach lining in a healthy person is well protected by its mucous coat. What, then, causes ulcers to form?

Three major causes of ulcers have been identified: bacterial infection with *Helicobacter pylori* (commonly abbreviated *H. pylori*); the use of certain anti-inflammatory drugs such as aspirin, ibuprofen, and naproxen; and disorders that cause excessive gastric acid secretion. Most commonly, ulcers develop in response to *H. pylori* infection.[4] The cause of the ulcer dictates the type of medication used in treatment. For example, people with ulcers caused by infection receive antibiotics, whereas those with ulcers caused by medicines discontinue their use. In addition, all treatment plans aim to relieve pain, heal the ulcer, and prevent recurrence.

Diet therapy once played a major role in ulcer treatment, but it no longer does. Current practice is simply to treat for infection, eliminate any food that routinely causes indigestion or pain, and avoid coffee and caffeine- and alcohol-containing beverages. Both regular and decaffeinated coffee stimulate acid secretion and so aggravate *existing* ulcers.

Ulcers and their treatments highlight the importance of not self-medicating when symptoms persist. People with *H. pylori* infection often take over-the-counter acid controllers to relieve the pain of their ulcers when they need physician-prescribed antibiotics instead. Suppressing gastric acidity not only fails to heal the ulcer, but actually worsens inflammation during an *H. pylori* infection. Furthermore, *H. pylori* infection has been linked with stomach cancer as well, making prompt diagnosis and appropriate treatment essential.[5]

Table H3-1 (on p. 100) summarizes strategies to prevent or alleviate common GI problems. Many of these problems reflect hurried lifestyles. For this reason, many of their remedies require that people slow down and take the time to eat leisurely; chew food thoroughly to prevent choking, heartburn, and acid indigestion; rest until vomiting and diarrhea subside; and heed the urge to defecate. In addition, learn how to handle life's day-to-day problems and challenges without overreacting and becoming upset; learn how to relax, to get enough sleep, and to enjoy life. Remember, "what's eating you" may cause more GI distress than what you eat.

TABLE H3-1 Strategies to Prevent or Alleviate Common GI Problems

GI Problem	Strategies	GI Problem	Strategies
Choking	• Take small bites of food. • Chew thoroughly before swallowing. • Don't talk or laugh with food in your mouth. • Don't eat when breathing hard.	Heartburn	• Eat small meals. • Drink liquids between meals. • Sit up while eating; elevate your head when lying down. • Wait 1 hour after eating before lying down. • Wait 2 hours after eating before exercising. • Refrain from wearing tight-fitting clothing. • Avoid foods, beverages, and medications that aggravate your heartburn. • Refrain from smoking cigarettes or using tobacco products. • Lose weight if overweight.
Diarrhea	• Rest. • Drink fluids to replace losses. • Call for medical help if diarrhea persists.		
Constipation	• Eat a high-fiber diet. • Drink plenty of fluids. • Exercise regularly. • Respond promptly to the urge to defecate.		
Belching	• Eat slowly. • Chew thoroughly. • Relax while eating.	Ulcer	• Take medicine as prescribed by your physician. • Avoid coffee and caffeine- and alcohol-containing beverages. • Avoid foods that aggravate your ulcer. • Minimize aspirin, ibuprofen, and naproxen use. • Refrain from smoking cigarettes.
Intestinal gas	• Eat bothersome foods in moderation.		

NUTRITION ON THE NET

 Access these websites for further study of topics covered in this highlight.

• Find updates and quick links to these and other nutrition-related sites at our website: **www.wadsworth.com/nutrition**

• Search for "choking," "vomiting," "diarrhea," "constipation," "heartburn," "indigestion," and "ulcers" at the U.S. Government health information site: **www.healthfinder.gov**

• Visit the Center for Digestive Health and Nutrition: **www.gihealth.com**

• Visit the Digestive Diseases section of the National Institute of Diabetes, Digestive, and Kidney Diseases: **www.niddk.nih.gov/health/health.htm**

• Visit the Digest This! section of the American College of Gastroenterology: **www.acg.gi.org**

• Learn more about *H. pylori* from the Helicobacter Foundation: **www.helico.com**

REFERENCES

1. K. Gotsch, J. L. Annest, and P. Holmgreen, Nonfatal choking-related episodes among children—United States, 2001, *Morbidity and Mortality Weekly Report* 51 (2002): 945–948.
2. B. J. Horwitz and R. S. Fisher, The irritable bowel syndrome, *New England Journal of Medicine* 344 (2001): 1846–1850.
3. N. Shaheen and D. F. Ransohoff, Gastroesophageal reflux, Barrett's esophagus, and esophageal cancer: Scientific review, *Journal of the American Medical Association* 287 (2002): 1972–1981.
4. S. Suerbaum and P. Michetti, *Helicobacter pylori* infection, *New England Journal of Medicine* 347 (2002): 1175–1186.
5. N. Uemura and coauthors, *Helicobacter pylori* infection and the developoment of gastric cancer, *New England Journal of Medicine* 345 (2001): 784–789.

The Carbohydrates: Sugars, Starches, and Fibers

Chapter Outline

The Chemist's View of Carbohydrates

The Simple Carbohydrates: *Monosaccharides • Disaccharides*

The Complex Carbohydrates: *Glycogen • Starches • Fibers*

Digestion and Absorption of Carbohydrates: *Carbohydrate Digestion • Carbohydrate Absorption • Lactose Intolerance*

Glucose in the Body: *A Preview of Carbohydrate Metabolism • The Constancy of Blood Glucose*

Health Effects and Recommended Intakes of Sugars: *Health Effects of Sugars • Accusations against Sugars • Recommended Intakes of Sugars*

Health Effects and Recommended Intakes of Starch and Fibers: *Health Effects of Starch and Fibers • Recommended Intakes of Starch and Fibers • From Guidelines to Groceries*

Highlight: *Alternatives to Sugar*

Nutrition Explorer CD-ROM Outline

Nutrition Animation: *Carbohydrate Digestion*

Case Study: *Simple Sugar and Complex Carbohydrate*

Student Practice Test

Glossary Terms

Nutrition on the Net

© Mary Jan Cardenas/The Image Bank/Getty Images

Nutrition in Your Life

Whether you are cramming for an exam or daydreaming about your next vacation, your brain needs carbohydrate to power its activities. Your muscles need carbohydrate to fuel their work, too, whether you are racing up the stairs to class or moving on the dance floor to your favorite music. Where can you get carbohydrate? And are some foods healthier choices than others? As you will learn from this chapter, whole grains, vegetables, legumes, and fruits naturally deliver ample carbohydrate and fiber with valuable vitamins and minerals and little or no fat. Milk products typically lack fiber, but they also provide carbohydrate along with an assortment of vitamins and minerals.

A student, quietly studying a textbook, is seldom aware that within his brain cells, billions of glucose molecules are splitting to provide the energy that permits him to learn. Yet glucose provides nearly all of the energy the human brain uses daily. Similarly, a marathon runner, bursting across the finish line in an explosion of sweat and triumph, seldom gives thanks to the glycogen fuel her muscles have devoured to help her finish the race. Yet, together, glucose and its storage form glycogen provide about half of all the energy muscles and other body tissues use. The other half of the body's energy comes mostly from fat.

People don't eat glucose and glycogen directly; they eat foods rich in **carbohydrates.** Then their bodies convert the carbohydrates mostly into glucose for immediate energy and into glycogen for reserve energy. All plant foods—whole grains, vegetables, legumes, and fruits—provide ample carbohydrate. Milk also contains carbohydrates.

Many people mistakenly think of carbohydrates as "fattening" and avoid them when trying to lose weight. Such a strategy may be helpful if the carbohydrates are

carbohydrates: compounds composed of carbon, oxygen, and hydrogen arranged as monosaccharides or multiples of monosaccharides. Most, but not all, carbohydrates have a ratio of one carbon molecule to one water molecule: $(CH_2O)_n$.
- **carbo** = carbon (C)
- **hydrate** = with water (H_2O)

the simple sugars of candy bars and cookies, but counterproductive if the carbohydrates are the complex carbohydrates of whole grains, vegetables, and legumes. As the next section explains, not all carbohydrates are created equal.

The Chemist's View of Carbohydrates

The dietary carbohydrate family includes the **simple carbohydrates** (the sugars) and the **complex carbohydrates** (the starches and fibers). The simple carbohydrates are those that chemists describe as:

- Monosaccharides—single sugars.
- Disaccharides—sugars composed of pairs of monosaccharides.

The complex carbohydrates are:

- Polysaccharides—large molecules composed of chains of monosaccharides.

■ Most of the monosaccharides important in nutrition are **hexoses**, simple sugars with six atoms of carbon and the formula $C_6H_{12}O_6$.
- **hex** = six

To understand the structure of carbohydrates, look at the units of which they are made. The monosaccharides most important in nutrition are the 6-carbon hexoses.■ Each contains 6 carbon atoms, 12 hydrogens, and 6 oxygens (written in shorthand as $C_6H_{12}O_6$).

Each atom can form a certain number of chemical bonds with other atoms:

- Carbon atoms can form four bonds.
- Nitrogen atoms, three.
- Oxygen atoms, two.
- Hydrogen atoms, only one.

Chemists represent the bonds as lines between the chemical symbols (such as C, N, O, and H) that stand for the atoms (see Figure 4-1).

Atoms form molecules in ways that satisfy the bonding requirements of each atom. Figure 4-1 includes the structure of ethyl alcohol, the active ingredient of alcoholic beverages, as an example. The two carbons each have four bonds represented by lines; the oxygen has two; and each hydrogen has one bond connecting it to other atoms. Chemical structures obey these bonding rules because the laws of nature demand it.

FIGURE 4-1 Atoms and Their Bonds

The four main types of atoms found in nutrients are hydrogen (H), oxygen (O), nitrogen (N), and carbon (C).

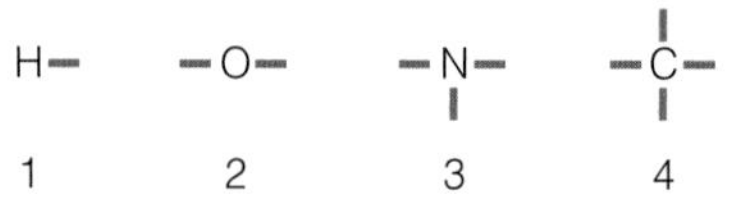

$$H- \quad -O- \quad -N- \quad -\overset{\displaystyle |}{\underset{\displaystyle |}{C}}-$$
$$1 \qquad 2 \qquad 3 \qquad 4$$

Each atom has a characteristic number of bonds it can form with other atoms.

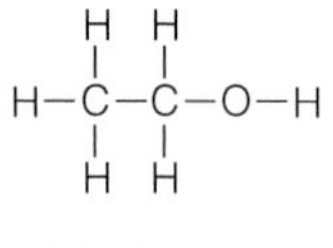

$$H-\overset{\displaystyle H}{\underset{\displaystyle H}{C}}-\overset{\displaystyle H}{\underset{\displaystyle H}{C}}-O-H$$

Notice that in this simple molecule of ethyl alcohol, each H has one bond, O has two, and each C has four.

IN SUMMARY The carbohydrates are made of carbon (C), oxygen (O), and hydrogen (H). Each of these atoms can form a specified number of chemical bonds: carbon forms four, oxygen forms two, and hydrogen forms one.

The Simple Carbohydrates

The following list of the simple carbohydrates most important in nutrition symbolizes them as hexagons and pentagons of different colors.[*] Three are monosaccharides:

- Glucose.
- Fructose.
- Galactose.

simple carbohydrates (sugars): monosaccharides and disaccharides.

complex carbohydrates (starches and fibers): polysaccharides composed of straight or branched chains of monosaccharides.

___

[*]Fructose is shown as a pentagon, but like the other monosaccharides, it has six carbons (as you will see in Figure 4-4).

Three are disaccharides:

- Maltose (glucose + glucose).

- Sucrose (glucose + fructose).

- Lactose (glucose + galactose).

Monosaccharides

The three **monosaccharides** important in nutrition all have the same numbers and kinds of atoms, but in different arrangements. These chemical differences account for the differing sweetness of the monosaccharides. A pinch of purified glucose on the tongue gives only a mild sweet flavor, and galactose hardly tastes sweet at all, but fructose is as intensely sweet as honey and, in fact, is the sugar primarily responsible for honey's sweetness.

Glucose Chemically, **glucose** is a larger and more complicated molecule than the ethyl alcohol shown in Figure 4-1, but it obeys the same rules of chemistry: each carbon atom has four bonds; each oxygen, two bonds; and each hydrogen, one bond. Figure 4-2 illustrates the chemical structure of a glucose molecule.

The diagram of a glucose molecule shows all the relationships between the atoms and proves simple on examination, but chemists have adopted even simpler ways to depict chemical structures. Figure 4-3 shows how the chemical structure of glucose can be simplified by combining or omitting several symbols and still convey the same information.

Commonly known as blood sugar, glucose serves as an essential energy source for all the body's activities. Its significance to nutrition is tremendous. Later sections will explain that glucose is one of the two sugars in every disaccharide and the unit from which the polysaccharides are made almost exclusively. One of these polysaccharides, starch, is the chief food source of energy for the world's people; another, glycogen, is an important storage form of energy in the body. Glucose reappears frequently throughout this chapter and all those that follow.

Fructose Fructose is the sweetest of the sugars. Curiously, fructose has exactly the same chemical *formula* as glucose—$C_6H_{12}O_6$—but its *structure* differs (see Figure 4-4). The arrangement of the atoms in fructose stimulates the taste buds on the tongue to produce the sweet sensation. Fructose occurs naturally in fruits and honey; other sources include products such as soft drinks, ready-to-cereals, and desserts that have been sweetened with high-fructose corn syrup (defined on p. 120).

Galactose The monosaccharide **galactose** rarely occurs naturally as a single sugar. Galactose has the same numbers and kinds of atoms as glucose and fructose in yet another arrangement. Figure 4-5 shows galactose beside a molecule of glucose for comparison.

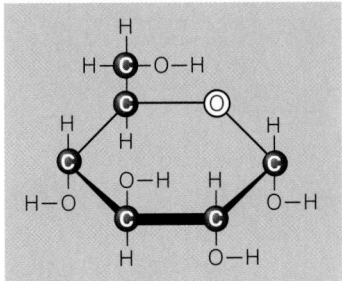

FIGURE 4-2 Chemical Structure of Glucose

On paper, the structure of glucose has to be drawn flat, but in nature the five carbons and oxygen are roughly in a plane. The atoms attached to the ring carbons extend above and below the plane.

monosaccharides (mon-oh-SACK-uh-rides): carbohydrates of the general formula $C_nH_{2n}O_n$ that consist of a single ring. See Appendix C for the chemical structures of the monosaccharides.
- **mono** = one
- **saccharide** = sugar

glucose (GLOO-kose): a monosaccharide; sometimes known as blood sugar or **dextrose**.
- **ose** = carbohydrate
- ⬡ = glucose

fructose (FRUK-tose or FROOK-tose): a monosaccharide. Sometimes known as fruit sugar or **levulose**, fructose is found abundantly in fruits, honey, and saps.
- **fruct** = fruit
- ⬠ = fructose

galactose (ga-LAK-tose): a monosaccharide; part of the disaccharide lactose.
- ⬡ = galactose

FIGURE 4-3 Simplified Diagrams of Glucose

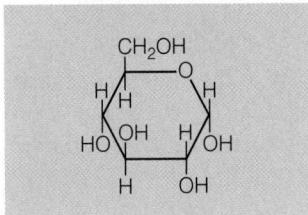

The lines representing some of the bonds and the carbons at the corners are not shown.

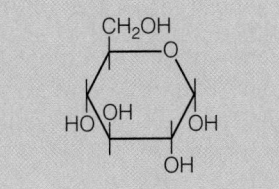

Now the single hydrogens are not shown, but lines still extend upward or downward from the ring to show where they belong.

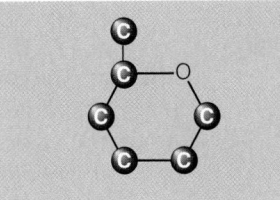

Another way to look at glucose is to notice that its six carbon atoms are all connected.

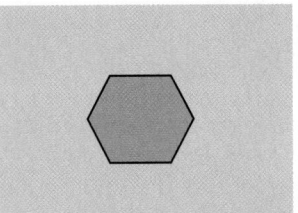

In this and other illustrations throughout this book, glucose is represented as a blue hexagon.

FIGURE 4-4 Two Monosaccharides: Glucose and Fructose

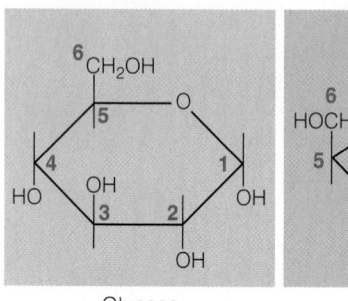
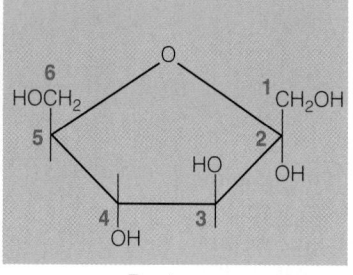

Glucose   Fructose

Can you see the similarities? If you learned the rules in Figure 4-3, you will be able to "see" 6 carbons (numbered), 12 hydrogens (those shown plus one at the end of each single line), and 6 oxygens in both these compounds.

FIGURE 4-5 Two Monosaccharides: Glucose and Galactose

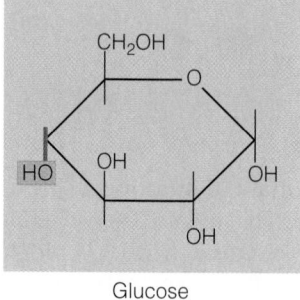
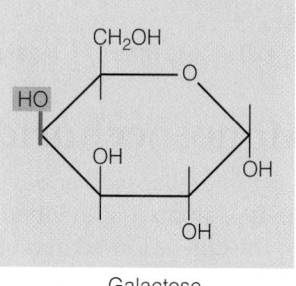

Glucose   Galactose

Notice the similarities and the difference (highlighted in red).

Disaccharides

The **disaccharides** are pairs of the three monosaccharides just described. Glucose occurs in all three; the second member of the pair is either fructose, galactose, or another glucose. These carbohydrates—and all the other energy nutrients—are put together and taken apart by similar chemical reactions: condensation and hydrolysis.

Condensation To make a disaccharide, a chemical reaction known as **condensation** links two monosaccharides together (see Figure 4-6). A hydroxyl (OH) group from one monosaccharide and a hydrogen atom (H) from the other combine to create a molecule of water (H_2O). The two originally separate monosaccharides link together with a single oxygen (O).

Hydrolysis To break a disaccharide in two, a chemical reaction known as hydrolysis■ occurs (see Figure 4-7). A molecule of water splits to provide the H and OH needed to complete the resulting monosaccharides. Hydrolysis reactions commonly occur during digestion.

Maltose The disaccharide **maltose** consists of two glucose units. Maltose is produced whenever starch breaks down—as happens in plants when seeds germinate and in human beings during carbohydrate digestion. It also occurs during the fermentation process that yields alcohol. Maltose is only a minor constituent of a few foods, most notably barley.

Sucrose Fructose and glucose together form **sucrose.** Because the fructose is in a position accessible to the taste receptors, sucrose tastes sweet, accounting for some of the natural sweetness of fruits, vegetables, and grains. To make table sugar, sucrose is refined from the juices of sugarcane and sugar beets, then granulated. Depending on the extent to which it is refined, the product becomes the familiar brown, white, and powdered sugars available at grocery stores.

Lactose The combination of galactose and glucose makes the disaccharide **lactose,** the principal carbohydrate of milk. Known as milk sugar, lactose contributes about 5 percent of milk's weight. Depending on the milk's fat content, lactose contributes 30 to 50 percent of milk's energy.

■ Reminder: A *hydrolysis* reaction splits a molecule into two, with H added to one and OH to the other (from water); Chapter 3 explained that hydrolysis reactions break down molecules during digestion.

disaccharides (dye-SACK-uh-rides): pairs of monosaccharides linked together. See Appendix C for the chemical structures of the disaccharides.
- **di** = two

condensation: a chemical reaction in which two reactants combine to yield a larger product.

maltose (MAWL-tose): a disaccharide composed of two glucose units; sometimes known as malt sugar.
- ◆◇ = maltose

sucrose (SUE-krose): a disaccharide composed of glucose and fructose; commonly known as table sugar, beet sugar, or cane sugar. Sucrose also occurs in many fruits and some vegetables and grains.
- **sucro** = sugar
- ◆⬡ = sucrose

lactose (LAK-tose): a disaccharide composed of glucose and galactose; commonly known as milk sugar.
- **lact** = milk
- ⬡◇ = lactose

IN SUMMARY Six simple carbohydrates, or sugars, are important in nutrition. The three monosaccharides (glucose, fructose, and galactose) all have the same chemical formula ($C_6H_{12}O_6$), but their structures differ. The three disaccharides (maltose, sucrose, and lactose) are pairs of monosaccharides, each

FIGURE 4-6 Condensation of Two Monosaccharides to Form a Disaccharide

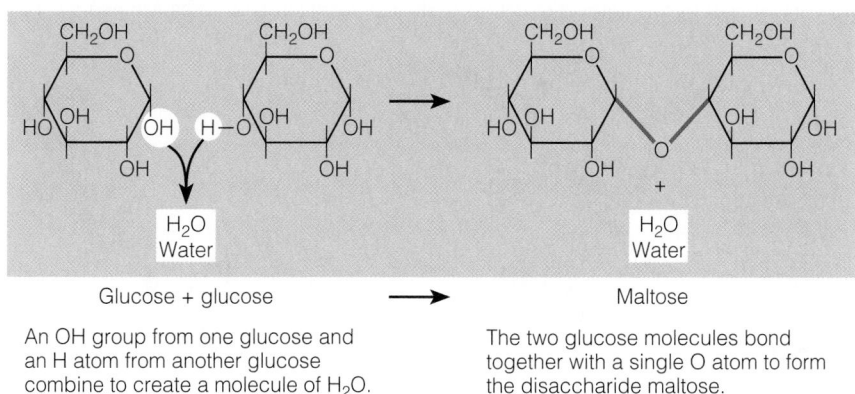

Glucose + glucose ⟶ Maltose

An OH group from one glucose and an H atom from another glucose combine to create a molecule of H_2O.

The two glucose molecules bond together with a single O atom to form the disaccharide maltose.

Fruits package their simple sugars with fibers, vitamins, and minerals, making them a sweet and healthy snack.

containing a glucose paired with one of the three monosaccharides. The sugars derive primarily from plants, except for lactose and its component galactose, which come from milk and milk products. Two monosaccharides can be linked together by a condensation reaction to form a disaccharide and water. A disaccharide, in turn, can be broken into its two monosaccharides by a hydrolysis reaction using water.

The Complex Carbohydrates

The simple carbohydrates are the sugars just mentioned—the monosaccharides glucose, fructose, and galactose and the disaccharides maltose, sucrose, and lactose. In contrast, the complex carbohydrates contain many glucose units and, in some cases, a few other monosaccharides strung together as **polysaccharides.** Three types of polysaccharides are important in nutrition: glycogen, starches, and fibers.

Glycogen is a storage form of energy in the animal body; starches play that role in plants; and fibers provide structure in stems, trunks, roots, leaves, and skins of plants. Both glycogen and starch are built of glucose units; fibers are composed of a variety of monosaccharides and other carbohydrate derivatives.

FIGURE 4-7 Hydrolysis of a Disaccharide

Hydrolysis occurs during digestion.

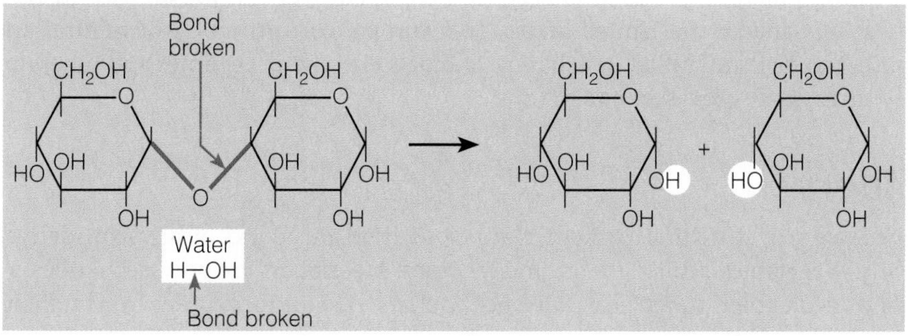

Bond broken

Water
H—OH

Bond broken

Maltose ⟶ Glucose + glucose

The disaccharide maltose splits into two glucose molecules with H added to one and OH to the other (from water).

polysaccharides: compounds composed of many monosaccharides linked together. An intermediate string of three to ten monosaccharides is an **oligosaccharide.**
• **poly** = many
• **oligo** = few

FIGURE 4-8 Glycogen and Starch Molecules Compared (Small Segments)

Notice the more highly branched the structure, the greater the number of ends from which glucose can be released. (These units would have to be magnified millions of times to appear at the size shown in this figure. For details of the chemical structures, see Appendix C.)

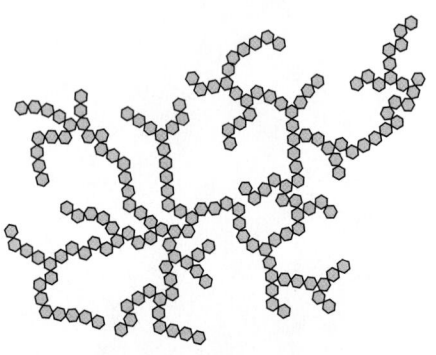

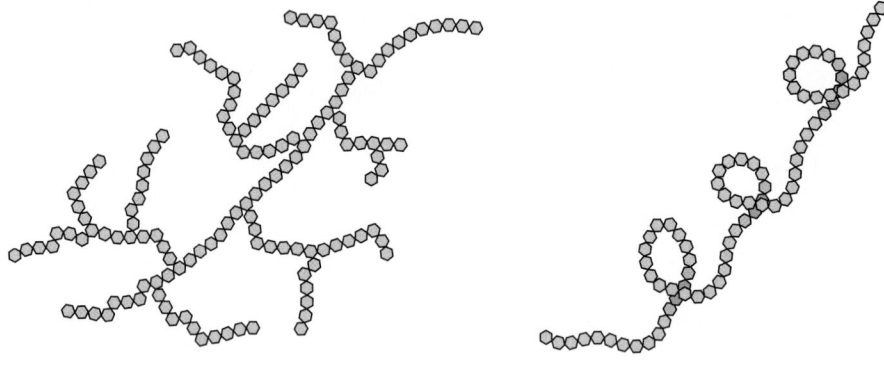

Glycogen

A glycogen molecule contains hundreds of glucose units in long, highly branched chains.

Starch (amylopectin)   Starch (amylose)

A starch molecule contains hundreds of glucose molecules in either occasionally branched chains (amylopectin) or unbranched chains (amylose).

Glycogen

Glycogen is found to only a limited extent in meats and not at all in plants.* For this reason, glycogen is not a significant food source of carbohydrate, but it does perform an important role in the body. The human body stores much of its glucose as glycogen—many glucose molecules linked together in highly branched chains (see the left side of Figure 4-8). This arrangement permits rapid hydrolysis. When the hormonal message "Release energy" arrives at the storage sites in a liver or muscle cell, enzymes respond by attacking all the many branches of each glycogen simultaneously, making a surge of glucose available.†

Starches

Just as the human body stores glucose as glycogen, plant cells store glucose as **starches**—long, branched or unbranched chains of hundreds or thousands of glucose molecules linked together (see the middle and right side of Figure 4-8). These giant starch molecules are packed side by side in grains such as wheat or rice, in root crops and tubers such as yams and potatoes, and in legumes such as peas and beans. When you eat the plant, your body hydrolyzes the starch to glucose and uses the glucose for its own energy purposes.

All starchy foods come from plants. Grains are the richest food source of starch, providing much of the food energy for people all over the world—rice in Asia; wheat in Canada, the United States, and Europe; corn in much of Central and South America; and millet, rye, barley, and oats elsewhere. Legumes and tubers are also important sources of starch.

Fibers

Fibers are the structural parts of plants and thus are found in all plant-derived foods—vegetables, fruits, grains, and legumes. Most fibers are polysaccharides. As mentioned earlier, starches are also polysaccharides, but fibers differ from starches

glycogen (GLY-co-gen): an animal polysaccharide composed of glucose; manufactured and stored in the liver and muscles as a storage form of glucose. Glycogen is not a significant food source of carbohydrate and is not counted as one of the complex carbohydrates in foods.
• **glyco** = glucose
• **gen** = gives rise to

starches: plant polysaccharides composed of glucose.

fibers: in plant foods, the *nonstarch polysaccharides* that are not digested by human digestive enzymes, although some are digested by GI tract bacteria. Fibers include cellulose, hemicelluloses, pectins, gums, and mucilages and the nonpolysaccharides lignins, cutins, and tannins.

---

*Glycogen in animal muscles rapidly hydrolyzes after slaughter.
†Normally, only liver cells can produce glucose from glycogen to be sent *directly* to the blood; muscle cells can also produce glucose from glycogen, but must use it themselves. Muscle cells can restore the blood glucose level *indirectly*, however, as Chapter 7 explains.

in that the bonds between their monosaccharides cannot be broken down by digestive enzymes in the body. Consequently, fibers contribute no monosaccharides, and therefore little or no energy, to the body.

For these reasons, fibers are often described as *nonstarch polysaccharides*. The nonstarch polysaccharide fibers include cellulose, hemicelluloses, pectins, gums, and mucilages. Fibers also include some *nonpolysaccharides* such as lignins, cutins, and tannins. Each of the fibers has a different structure. Most contain monosaccharides, but differ in the types they contain and in the bonds that link the monosaccharides to each other. These differences produce diverse health effects as explained later.

Cellulose Cellulose is the primary constituent of plant cell walls and therefore occurs naturally in all vegetables, fruits, and legumes. Cellulose can also be extracted from wood pulp or cotton and added to foods as an anticaking, thickening, and texturizing agent during processing.

Like starch, cellulose is composed of glucose molecules connected in long chains. Unlike starch, however, the chains do not branch, and the bonds linking the glucose molecules together cannot be broken by human enzymes (see Figure 4-9).

Hemicelluloses The hemicelluloses are the main constituent of cereal fibers. They are composed of various monosaccharide backbones with branching side chains of monosaccharides.[*]

Pectins All pectins consist of a backbone of one type of monosaccharide; some are unbranched, whereas others have side chains of various monosaccharides.[†] Commonly found in vegetables and fruits (especially citrus fruits and apples), pectins may be isolated and used by the food industry to thicken jelly, keep salad dressings from separating, and otherwise control texture and consistency. Pectins can perform these functions because they readily form gels in water.

Gums and Mucilages When cut, a plant secretes gums from the site of the injury. Like the other fibers, gums are composed of various monosaccharides and their derivatives. Gums such as *guar gum* and *gum arabic* are used as additives by the food industry to thicken processed foods. Mucilages are similar to gums in structure; they include *psyllium* and *carrageenan*, which are added to foods as stabilizers.

Lignin This *nonpolysaccharide* fiber has a three-dimensional structure that gives it strength.[‡] Because of its toughness, few of the foods that people eat contain much lignin. It occurs in the woody parts of vegetables such as carrots and the small seeds of fruits such as strawberries.

Resistant Starches A few starches are classified as fibers. Known as **resistant starches,** these starches escape digestion and absorption in the small intestine. Starch may resist digestion for several reasons, including the individual's efficiency in digesting starches and the food's physical properties. Resistant starch is common in whole legumes, raw potatoes, and unripe bananas.

Fiber Characteristics The previous paragraphs described fibers according to their chemistry, but their physical characteristics may better explain their actions in the body. Fibers do not sort neatly into groups, but a few generalizations can be made. Some fibers dissolve in water **(soluble fibers),** form gels **(viscous),** and are easily digested by bacteria in the colon **(fermentable).** Commonly found in legumes and fruits, these fibers are most often associated with protecting against heart disease and diabetes by lowering blood cholesterol and glucose levels, respectively.[1]

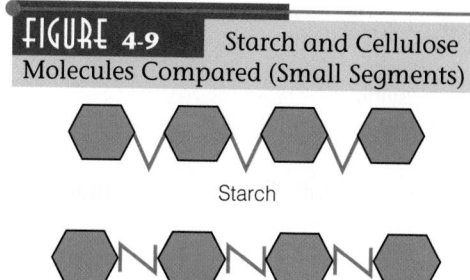

FIGURE 4-9 Starch and Cellulose Molecules Compared (Small Segments)

Starch

Cellulose

The bonds that link the glucose molecules together in cellulose are different from the bonds in starch (and glycogen). Human enzymes cannot digest cellulose. See Appendix C for chemical structures and descriptions of linkages.

© Polara Studios Inc.

Major sources of starch include grains, (such as rice, wheat, millet, rye, barley, and oats), legumes (such as kidney beans, black-eyed peas, pinto beans, navy beans, and garbanzo beans), tubers (such as potatoes), and root crops (such as yams and cassava).

resistant starches: starches that escape digestion and absorption in the small intestine of healthy people.

soluble fibers: indigestible food components that dissolve in water to form a gel. An example is pectin from fruit, which is used to thicken jellies.

viscous: a gel-like consistency.

fermentable: the extent to which bacteria in the GI tract can break down fibers to fragments that the body can use.[*]

[*] Dietary fibers are fermented by bacteria in the colon to short-chain fatty acids, which are absorbed and metabolized by cells in the GI tract and liver (Chapter 5 describes fatty acids).

---

[*] In hemicelluloses, the most common backbone monosaccharides are xylose, mannose, and galactose; the common side chains are arabinose, glucuronic acid, and galactose (see Appendix C for structures).
[†] In pectins, the backbone is usually made of galacturonic acid units.
[‡] Lignins are polymers of several dozen molecules of phenol (an alcohol), with strong internal bonds that make them impervious to digestive enzymes.

■ Viscous (soluble, more fermentable):
  • Gums and mucilages.
  • Pectins.
  • Psyllium.
  • Some hemicelluloses.

■ Nonviscous (insoluble, less fermentable):
  • Cellulose.
  • Lignins.
  • Psyllium.
  • Resistant starch.
  • Many hemicelluloses.

■ *Dietary fibers* occur naturally in intact plants. *Functional fibers* have been extracted from plants or manufactured and have beneficial effects in human beings. *Total fiber* is the sum of dietary fibers and functional fibers.

Other fibers do not dissolve in water **(insoluble fibers),** do not form gels (nonviscous), and are less readily fermented. Found mostly in grains and vegetables, these fibers promote bowel movements and alleviate constipation. These generalizations between viscous■ and nonviscous■ fibers are useful, but exceptions occur. For example, insoluble rice bran also lowers blood cholesterol, and the soluble fiber of the psyllium plant effectively promotes bowel movements.

Recently, the committee on Dietary Reference Intakes (DRI) proposed terms that distinguish fibers not by their chemical or physical properties, but by their source. Fibers that occur naturally in intact plants are called *dietary fibers,* whereas fibers that have been extracted from plants or manufactured and have beneficial health effects are called *functional fibers. Total fiber* refers to the sum of dietary fibers and functional fibers. These definitions■ were created to accommodate the labeling of products that may contain new fiber sources that prove to have beneficial effects.

A compound not classed as a fiber but often found with it in foods is **phytic acid.** Because of this close association, researchers have been unable to determine whether it is the fiber, the phytic acid, or both, that binds with minerals, preventing their absorption. This binding presents a risk of mineral deficiencies, but the risk is minimal when fiber intake is reasonable and mineral intake adequate. The nutrition consequences of such mineral losses are described further in Chapters 12 and 13.

© Banana Stock/SuperStock

When a person eats carbohydrate-rich foods, the body receives a valuable commodity—glucose.

IN SUMMARY The complex carbohydrates are the polysaccharides (chains of monosaccharides): glycogen, starches, and fibers. Both glycogen and starch are storage forms of glucose—glycogen in the body, and starch in plants—and both yield energy for human use. The fibers also contain glucose (and other monosaccharides), but their bonds cannot be broken by human digestive enzymes, so they yield little, if any, energy. The accompanying table summarizes the carbohydrate family of compounds.

The Carbohydrate Family

Simple Carbohydrates (sugars)	Complex Carbohydrates
• Monosaccharides	• Polysaccharides
Glucose	Glycogen[a]
Fructose	Starches
Galactose	Fibers
• Disaccharides	
Maltose	
Sucrose	
Lactose	

[a]Glycogen is a complex carbohydrate (a polysaccharide), but not a *dietary* source of carbohydrate.

insoluble fibers: indigestible food components that do not dissolve in water. Examples include the tough, fibrous structures found in the strings of celery and the skins of corn kernels.

phytic (FYE-tick) **acid:** a nonnutrient component of plant seeds; also called **phytate** (FYE-tate). Phytic acid occurs in the husks of grains, legumes, and seeds and is capable of binding minerals such as zinc, iron, calcium, magnesium, and copper in insoluble complexes in the intestine, which the body excretes unused.

Digestion and Absorption of Carbohydrates

The ultimate goal of digestion and absorption of sugars and starches is to dismantle them into small molecules—chiefly glucose—that the body can absorb and use. The large starch molecules require extensive breakdown; the disaccharides need only be broken once and the monosaccharides not at all. The initial splitting begins in the mouth; the final splitting and absorption occur in the small intestine; and conversion to a common energy currency (glucose) takes place in the liver. The details follow.

Carbohydrate Digestion

Figure 4-10 (on p. 112) traces the digestion of carbohydrates through the GI tract. When a person eats foods containing starch, enzymes hydrolyze the long chains to shorter chains,■ the short chains to disaccharides, and, finally, the disaccharides to monosaccharides. This process begins in the mouth.

In the Mouth In the mouth, thoroughly chewing high-fiber foods slows eating and stimulates the flow of saliva. The salivary enzyme **amylase** starts to work, hydrolyzing starch to shorter polysaccharides and to maltose. In fact, you can taste the change if you hold a piece of starchy food like a cracker in your mouth for a few minutes without swallowing it—the cracker begins tasting sweeter as the enzyme acts on it. Because food is in the mouth for only a short time, very little carbohydrate digestion takes place there. That digestive activity temporarily ceases is of no consequence, however; it picks up again further down the tract.

In the Stomach The swallowed bolus■ mixes with the stomach's acid and protein-digesting enzymes, which inactivate salivary amylase. Thus the role of salivary amylase in starch digestion is relatively minor. To a small extent, the stomach's acid continues breaking down starch, but its juices contain no enzymes to digest carbohydrate. Fibers linger in the stomach and delay gastric emptying, thereby providing a feeling of fullness and **satiety.**

In the Small Intestine The small intestine performs most of the work of carbohydrate digestion. A major carbohydrate-digesting enzyme, pancreatic amylase, enters the intestine via the pancreatic duct and continues breaking down the polysaccharides to shorter glucose chains and disaccharides. The final step takes place on the outer membranes of the intestinal cells. There specific enzymes■ dismantle specific disaccharides:

- **Maltase** breaks maltose into two glucose molecules.
- **Sucrase** breaks sucrose into one glucose and one fructose molecule.
- **Lactase** breaks lactose into one glucose and one galactose molecule.

At this point, all polysaccharides and disaccharides have been broken down to monosaccharides—mostly glucose molecules, with some fructose and galactose molecules as well.

In the Large Intestine Within one to four hours after a meal, all the sugars and most of the starches have been digested.■ Only the fibers remain in the digestive tract. Fibers in the large intestine attract water, which softens the stools for passage without straining. Also, bacteria in the GI tract ferment some fibers. This process generates water, gas, and short-chain fatty acids (described in Chapter 5).* The colon uses these small fat molecules for energy. Metabolism of short-chain fatty acids also occurs in the cells of the liver. Fibers, therefore, can contribute some energy (1.5 to 2.5 kcalories per gram), depending on the extent to which they are broken down by bacteria and the fatty acids are absorbed.

Carbohydrate Absorption

Glucose is unique in that it can be absorbed to some extent through the lining of the mouth, but for the most part, nutrient absorption takes place in the small intestine. Glucose and galactose traverse the cells lining the small intestine by active transport; fructose is absorbed by facilitated diffusion, which slows its entry and produces a smaller rise in blood glucose. Likewise, unbranched chains of starch are digested slowly and produce a smaller rise in blood glucose than branched chains, which have many more places for enzymes to attack and release glucose rapidly.

■ The short chains of glucose units that result from the breakdown of starch are known as **dextrins**. The word sometimes appears on food labels because dextrins can be used as thickening agents in foods.

■ Reminder: A *bolus* is a portion of food swallowed at one time.

■ Reminder: In general, the word ending *–ase* identifies an enzyme, and the word beginning identifies the molecule that the enzyme works on.

■ Starches and sugars are called **available carbohydrates** because human digestive enzymes break them down for the body's use. In contrast, fibers are called **unavailable carbohydrates** because human digestive enzymes cannot break their bonds.

amylase (AM-ih-lace): an enzyme that hydrolyzes amylose (a form of starch). Amylase is a *carbohydrase,* an enzyme that breaks down carbohydrates.

satiety (sah-TIE-eh-tee): the feeling of fullness and satisfaction that food brings (Chapter 8 provides a more detailed description).
- **sate** = to fill

maltase: an enzyme that hydrolyzes maltose.

sucrase: an enzyme that hydrolyzes sucrose.

lactase: an enzyme that hydrolyzes lactose.

---

*The short-chain fatty acids produced by GI bacteria are primarily acetic acid, propionic acid, and butyric acid.

FIGURE 4-10   Carbohydrate Digestion in the GI Tract

STARCH

FIBER

Mouth and salivary glands
The salivary glands secrete saliva into the mouth to moisten the food. The salivary enzyme amylase begins digestion:

Starch $\xrightarrow{\text{amylase}}$ small polysaccharides, maltose

Mouth
The mechanical action of the mouth crushes and tears fiber in food and mixes it with saliva to moisten it for swallowing.

Stomach
Stomach acid inactivates salivary enzymes, halting starch digestion.

Stomach
Fiber is not digested, and it delays gastric emptying.

Small intestine and pancreas
The pancreas produces an amylase that is released through the pancreatic duct into the small intestine:

Starch $\xrightarrow{\text{Pancreatic amylase}}$ Small polysaccharides, disaccharides

Then disaccharidase enzymes on the surface of the small intestinal cells hydrolyze the disaccharides into monosaccharides:

Maltose $\xrightarrow{\text{maltase}}$ glucose + glucose

Sucrose $\xrightarrow{\text{sucrase}}$ fructose + glucose

Lactose $\xrightarrow{\text{lactase}}$ galactose + glucose

Intestinal cells absorb these monosaccharides.

Small intestine
Fiber is not digested, and it delays absorption of other nutrients.

Large intestine
Most fiber passes intact through the digestive tract to the large intestine. Here, bacterial enzymes digest fiber:

Some fiber $\xrightarrow{\text{Bacterial enzymes}}$ Fatty acids, gas

Fiber holds water; regulates bowel activity; and binds substances such as bile, cholesterol, and some minerals, carrying them out of the body.

Salivary glands

Mouth

Stomach

(Liver)

(Gallbladder)

Pancreas

Small intestine

Large intestine

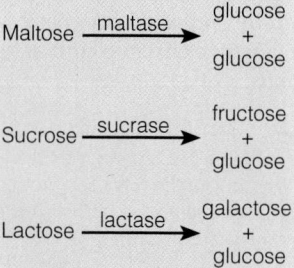

Nutrition Explorer Follow the digestion of starch as it begins in the mouth, and is completed in the small intestine with the release of glucose.

FIGURE 4-11   Absorption of Monosaccharides

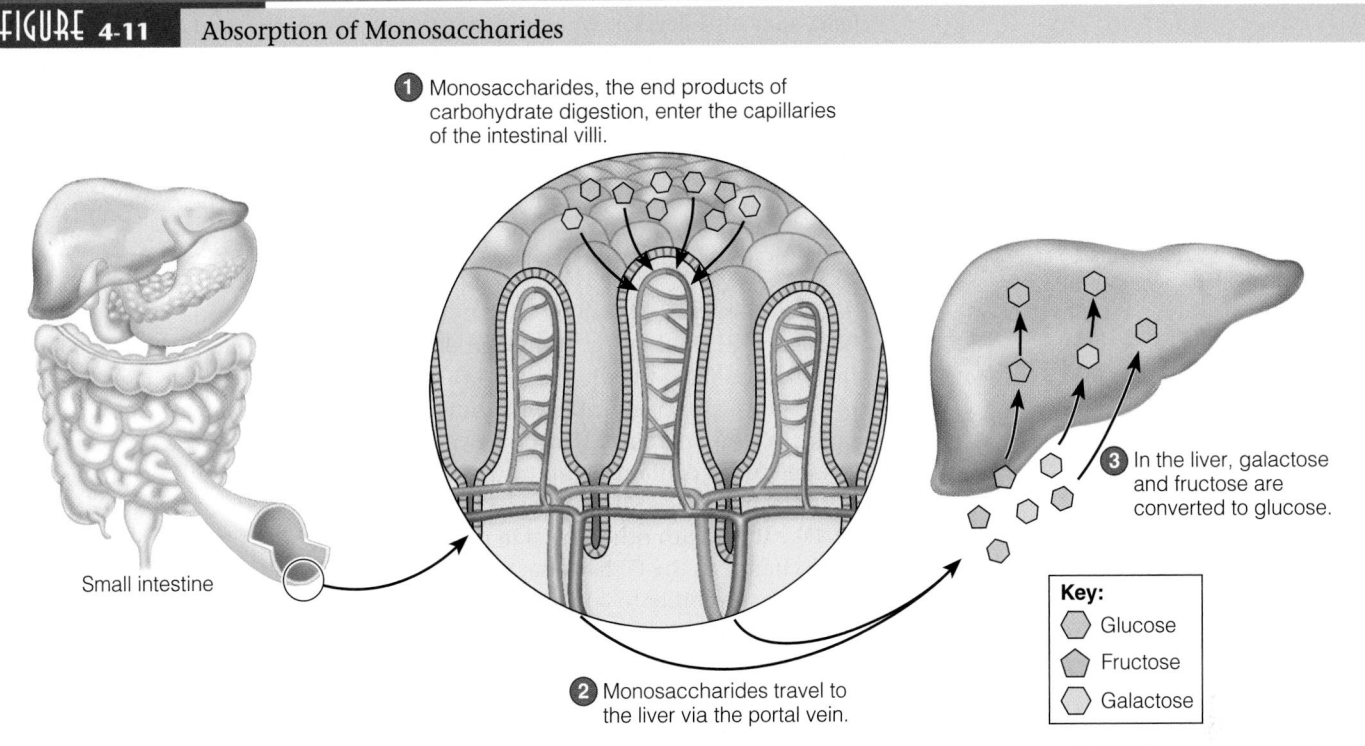

1 Monosaccharides, the end products of carbohydrate digestion, enter the capillaries of the intestinal villi.

Small intestine

2 Monosaccharides travel to the liver via the portal vein.

3 In the liver, galactose and fructose are converted to glucose.

Key:
⬡ Glucose
⬠ Fructose
⬡ Galactose

As the blood from the intestines circulates through the liver, cells there take up fructose and galactose and convert them to other compounds, most often to glucose, as shown in Figure 4-11. Thus all disaccharides provide at least one glucose molecule directly, and they can provide another one indirectly—through the conversion of fructose and galactose to glucose.

IN SUMMARY   In the digestion and absorption of carbohydrates, the body breaks down starches into disaccharides and disaccharides into monosaccharides; it then converts monosaccharides mostly to glucose to provide energy for the cells' work. The fibers help to regulate the passage of food through the GI system and slow the absorption of glucose, but contribute little, if any, energy.

Lactose Intolerance

Normally, the intestinal cells produce enough of the enzyme lactase to ensure that the disaccharide lactose found in milk is both digested and absorbed efficiently. Lactase activity is highest immediately after birth, as befits an infant whose first and only food for a while will be breast milk or infant formula. In the great majority of the world's populations, lactase activity declines dramatically during childhood and adolescence to about 5 to 10 percent of the activity at birth. Only a relatively small percentage (about 30 percent) of the people in the world retain enough lactase to digest and absorb lactose efficiently throughout adult life.

Symptoms When more lactose is consumed than the available lactase can handle, lactose molecules remain in the intestine undigested, attracting water and causing bloating, abdominal discomfort, and diarrhea—the symptoms of **lactose intolerance.** The undigested lactose becomes food for intestinal bacteria, which multiply and produce irritating acid and gas, further contributing to the discomfort and diarrhea.

lactose intolerance: a condition that results from inability to digest the milk sugar lactose; characterized by bloating, gas, abdominal discomfort, and diarrhea. Lactose intolerance differs from milk allergy, which is caused by an immune reaction to the protein in milk.

■ Estimated prevalence of lactose intolerance:
>80% Southeast Asians.
80% Native Americans.
75% African Americans.
70% Mediterranean peoples.
60% Inuits.
50% Hispanics.
20% Caucasians.
<10% Northern Europeans.

■ Lactose in selected foods:

Whole-wheat bread, 1 slice	0.5 g
Dinner roll, 1	0.5 g
Cheese, 1 oz	
Cheddar or American	0.5 g
Parmesan or cream	0.8 g
Doughnut (cake type), 1	1.2 g
Chocolate candy, 1 oz	2.3 g
Sherbet, 1 c	4.0 g
Cottage cheese (low-fat), 1 c	7.5 g
Ice cream, 1 c	9.0 g
Milk, 1 c	12.0 g
Yogurt (low-fat), 1 c	15.0 g

Note: Yogurt is often enriched with nonfat milk solids, which increase its lactose content to a level higher than milk's.

lactase deficiency: a lack of the enzyme required to digest the disaccharide lactose into its component monosaccharides (glucose and galactose).

acidophilus (ASS-ih-DOF-ih-lus) **milk:** a cultured milk created by adding *Lactobacillus acidophilus*, a bacterium that breaks down lactose to glucose and galactose, producing a sweet, lactose-free product.

Causes As mentioned, lactase activity commonly declines with age. **Lactase deficiency** may also develop when the intestinal villi are damaged by disease, certain medicines, prolonged diarrhea, or malnutrition; this can lead to temporary or permanent lactose malabsorption, depending on the extent of the intestinal damage. In extremely rare cases, an infant is simply born with a lactase deficiency.

Prevalence The prevalence■ of lactose intolerance varies widely among ethnic groups, indicating that the trait is genetically determined. The prevalence of lactose intolerance is lowest among Scandinavians and other northern Europeans and highest among native North Americans and Southeast Asians.

Dietary Changes Managing lactose intolerance requires some dietary changes, although total elimination of milk products usually is not necessary. Excluding all milk products from the diet can lead to nutrient deficiencies, for these foods are a major source of several nutrients, notably the mineral calcium, the B vitamin riboflavin, and vitamin D. Fortunately, many people with lactose intolerance can consume foods containing up to 6 grams of lactose (½ cup milk) without symptoms. The most successful strategies are to increase intake of milk products gradually, take them with other foods in meals, and spread their intake throughout the day. A change in the GI bacteria, not the reappearance of the missing enzyme, accounts for the ability to adapt to milk products.

In many cases, lactose-intolerant people can tolerate fermented milk products such as yogurt and **acidophilus milk.**[2] The bacteria in these products digest lactose for their own use, thus reducing the lactose content. Even when the lactose content is equivalent to milk's, yogurt produces fewer symptoms. Hard cheeses and cottage cheese are often well tolerated because most of the lactose is removed with the whey during manufacturing. Lactose continues to diminish as the cheese ages.

Many lactose-intolerant people use commercially prepared milk products that have been treated with an enzyme that breaks down the lactose. Alternatively, they take enzyme tablets with meals or add enzyme drops to their milk. The enzyme hydrolyzes much of the lactose in milk to glucose and galactose, which lactose-intolerant people can absorb without ill effects.

Because people's tolerance to lactose varies widely, lactose-restricted diets must be highly individualized. A completely lactose-free diet can be difficult because lactose appears not only in milk and milk products but also as an ingredient in many nondairy foods■ such as breads, cereals, breakfast drinks, salad dressings, and cake mixes. People on strict lactose-free diets need to read labels and avoid foods that include milk, milk solids, whey (milk liquid), and casein (milk protein, which may contain traces of lactose). They also need to check all medications with the pharmacist because 20 percent of prescription drugs and 5 percent of over-the-counter drugs contain lactose as a filler.

People who consume few or no milk products must take care to meet riboflavin, vitamin D, and calcium needs. Later chapters on the vitamins and minerals offer help with finding good nonmilk sources of these nutrients.

IN SUMMARY Lactose intolerance is a common condition that occurs when there is insufficient lactase to digest the disaccharide lactose found in milk and milk products. Symptoms include GI distress. Because treatment requires limiting milk intake, other sources of riboflavin, vitamin D, and calcium must be included in the diet.

Glucose in the Body

The primary role of the available carbohydrates in human nutrition is to supply the body's cells with glucose to deliver the indispensable commodity, energy. Starch contributes most to the body's glucose supply, but as explained earlier, any of the monosaccharides can also provide glucose.

Scientists have long known that providing energy is glucose's primary role in the body, but only recently uncovered additional roles glucose and other sugars perform in the body.[3]■ Sugar molecules dangle from many of the body's protein and fat molecules, with dramatic consequences. Sugars attached to a protein change the protein's shape and function; when they bind to lipids in a cell's membranes, sugars alter the way cells recognize each other.■ Cancer cells coated with sugar molecules, for example, are able to sneak by the cells of the immune system. Armed with this knowledge, scientists are now trying to use sugar molecules to create an anticancer vaccine. Further advances in knowledge are sure to reveal numerous ways these simple, yet remarkable, sugar molecules influence the health of the body.

■ The study of sugars is known as *glycobiology.*

■ These combination molecules are known as *glycoproteins* and *glycolipids,* respectively.

A Preview of Carbohydrate Metabolism

Glucose plays the central role in carbohydrate metabolism. This brief discussion provides just enough information about carbohydrate metabolism to illustrate that the body needs and uses glucose as a chief energy nutrient. Chapter 7 provides a full description of energy metabolism, and Chapter 10 shows how the B vitamins participate.

Storing Glucose as Glycogen The liver stores about one-third of the body's total glycogen and releases glucose into the bloodstream as needed. After a meal, blood glucose rises, and liver cells link the excess glucose molecules by condensation reactions into long, branching chains of glycogen. When blood glucose falls, the liver cells dismantle the glycogen by hydrolysis reactions into single molecules of glucose and release them into the bloodstream. Thus glucose becomes available to supply energy to the brain and other tissues regardless of whether the person has eaten recently. Muscle cells can also store glucose as glycogen (the other two-thirds), but they hoard most of their supply, using it just for themselves during exercise.

Glycogen holds water and therefore is rather bulky. The body can store only enough glycogen to provide energy for relatively short periods of time—less than a day during rest and a few hours at most during exercise. For its long-term energy reserves, for use over days or weeks of food deprivation, the body uses its abundant, water-free fuel, fat, as Chapter 5 describes.

Using Glucose for Energy Glucose fuels the work of most of the body's cells. Inside a cell, enzymes break glucose in half. These halves can be put back together to make glucose, or they can be further broken down into smaller fragments (never again to be reassembled to form glucose). The small fragments can yield energy when broken down completely to carbon dioxide and water.

As mentioned, the liver's glycogen stores last only for hours, not for days. To keep providing glucose to meet the body's energy needs, a person has to eat dietary carbohydrate frequently. Yet people who do not always attend faithfully to their bodies' carbohydrate needs still survive. How do they manage without glucose from dietary carbohydrate? Do they simply draw energy from the other two energy-yielding nutrients, fat and protein? They do draw energy, but not simply.

Making Glucose from Protein Glucose is the preferred energy source for brain cells, other nerve cells, and developing red blood cells. Body protein can be converted to glucose to some extent, but protein has jobs of its own that no other nutrient can do. Body fat cannot be converted to glucose to any significant extent. Thus, when a person does not replenish depleted glycogen stores by eating carbohydrate, body proteins are dismantled to make glucose to fuel these special cells.

The conversion of protein to glucose is called **gluconeogenesis**—literally, the making of new glucose. Only adequate dietary carbohydrate can prevent this use of protein for energy, and this role of carbohydrate is known as its **protein-sparing action.**

Making Ketone Bodies from Fat Fragments An inadequate supply of carbohydrate can shift the body's energy metabolism in a precarious direction. With less carbohydrate providing glucose to meet the brain's energy needs, fat takes an alternative

The carbohydrates of grains, vegetables, fruits, and legumes supply most of the energy in a healthful diet.

© Brian Leatart/FoodPix/Getty Images

gluconeogenesis (gloo-co-nee-oh-GEN-ih-sis): the making of glucose from a noncarbohydrate source (described in more detail in Chapter 7).
• **gluco** = glucose
• **neo** = new
• **genesis** = making

protein-sparing action: the action of carbohydrate (and fat) in providing energy that allows protein to be used for other purposes.

metabolic pathway; instead of entering the main energy pathway, fat fragments combine with each other, forming **ketone bodies.** Ketone bodies provide an alternate fuel source during starvation, but when their production exceeds their use, they accumulate in the blood, causing **ketosis,** a condition that disturbs the body's normal **acid-base balance,** as Chapter 7 describes. (Highlight 8 explores ketosis and the health consequences of low-carbohydrate diets further.)

To spare body protein and prevent ketosis, the body needs at least 50 to 100 grams of carbohydrate a day. Dietary recommendations urge people to select abundantly from carbohydrate-rich foods to provide for considerably more.

Using Glucose to Make Fat After meeting its energy needs and filling its glycogen stores to capacity, the body must find a way to store any extra glucose. The liver breaks it into smaller molecules and puts them together into the more permanent energy-storage compound—fat. Then the fat travels to the fatty tissues of the body for storage. Unlike the liver cells, which can store only enough glycogen to meet less than a day's worth of energy needs, fat cells can store unlimited quantities of fat.

Even though excess carbohydrate can be converted to fat and stored, this is a relatively minor pathway under normal conditions. Storing carbohydrate as body fat is energetically expensive; the body uses more energy to convert dietary carbohydrate to body fat than it does to convert dietary fat to body fat.

The Constancy of Blood Glucose

Every body cell depends on glucose for its fuel to some extent, and the cells of the brain and the rest of the nervous system depend almost exclusively on glucose for their energy. The activities of these cells never cease, and they do not have the ability to store glucose. Day and night they continually draw on the supply of glucose in the fluid surrounding them. To maintain the supply, a steady stream of blood moves past these cells bringing more glucose from either the intestines (food) or the liver (via glycogen breakdown or glucose synthesis).

Maintaining Glucose Homeostasis To function optimally, the body must maintain blood glucose within limits that permit the cells to nourish themselves. If blood glucose falls below normal,■ the person may become dizzy and weak; if it rises above normal, the person may become fatigued. Left untreated, fluctuations to the extremes—either high or low—can be fatal.

The Regulating Hormones Blood glucose homeostasis■ is regulated primarily by two hormones: insulin, which moves glucose from the blood into the cells, and glucagon, which brings glucose out of storage when necessary. Figure 4-12 depicts these hormonal regulators at work.

After a meal, as blood glucose rises, special cells of the pancreas respond by secreting **insulin** into the blood.* In general, the amount of insulin secreted corresponds with the rise in glucose. As the circulating insulin contacts the receptors on the body's other cells, the receptors respond by ushering glucose from the blood into the cells. Most of the cells take only the glucose they can use for energy right away, but the liver and muscle cells can assemble the small glucose units into long, branching chains of glycogen for storage. The liver cells can also convert glucose to fat for export to other cells. Thus elevated blood glucose returns to normal as excess glucose is stored as glycogen (which can be converted back to glucose) and fat (which cannot be).

When blood glucose falls (as occurs between meals), other special cells of the pancreas respond by secreting **glucagon** into the blood.† Glucagon raises blood glucose by signaling the liver to dismantle its glycogen stores and release glucose into the blood for use by all the other body cells.

■ Normal blood glucose (fasting): 70 to 110 mg/dL.

■ Reminder: *Homeostasis* is the maintenance of constant internal conditions by the body's control systems.

ketone (KEE-tone) **bodies:** the product of the incomplete breakdown of fat when glucose is not available in the cells.

ketosis (kee-TOE-sis): an undesirably high concentration of ketone bodies in the blood and urine.

acid-base balance: the equilibrium in the body between acid and base concentrations (see Chapter 12).

insulin (IN-suh-lin): a hormone secreted by special cells in the pancreas in response to (among other things) increased blood glucose concentration. The primary role of insulin is to control the transport of glucose from the bloodstream into the muscle and fat cells.

glucagon (GLOO-ka-gon): a hormone that is secreted by special cells in the pancreas in response to low blood glucose concentration and elicits release of glucose from liver glycogen stores.

---

*The *beta* (BAY-tuh) *cells,* one of several types of cells in the pancreas, secrete insulin in response to elevated blood glucose concentration.
†The *alpha cells* of the pancreas secrete glucagon in response to low blood glucose.

FIGURE 4-12 Maintaining Blood Glucose Homeostasis

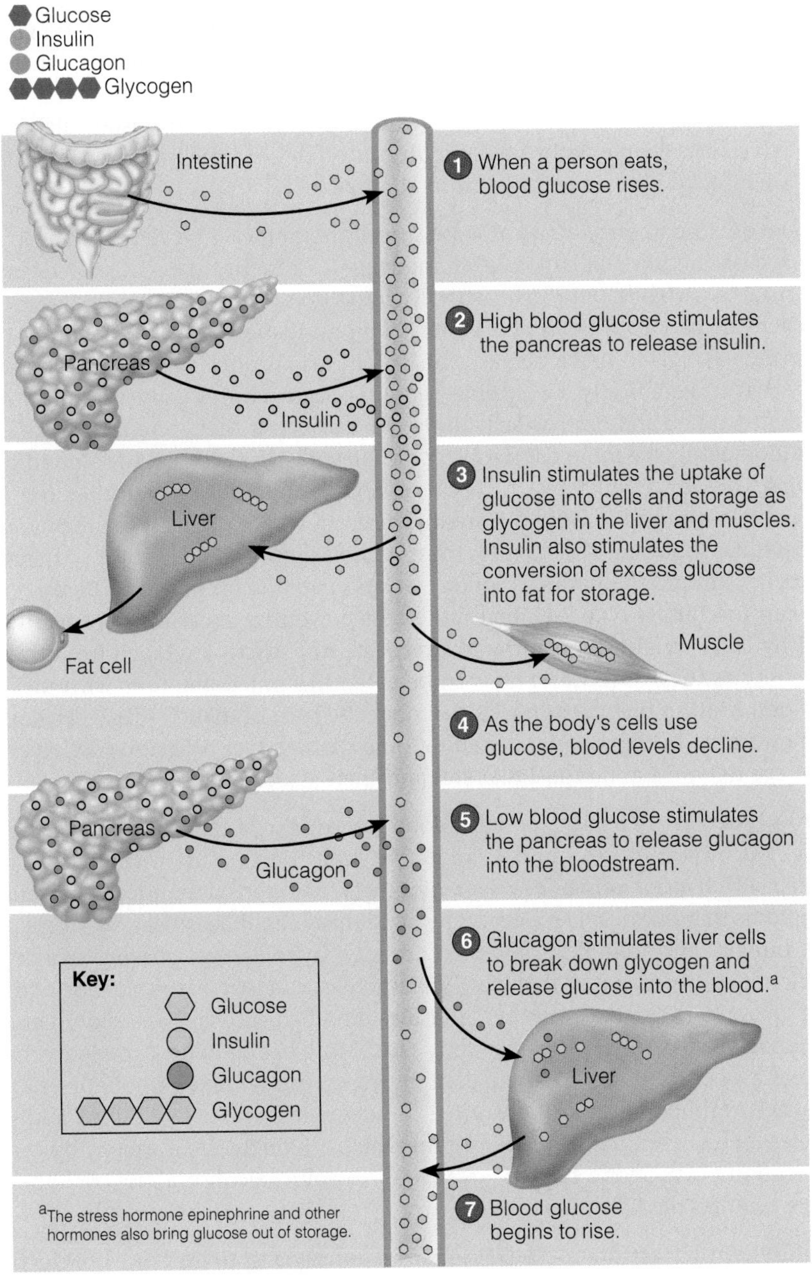

- Glucose
- Insulin
- Glucagon
- Glycogen

Intestine

1. When a person eats, blood glucose rises.

Pancreas

Insulin

2. High blood glucose stimulates the pancreas to release insulin.

Liver

3. Insulin stimulates the uptake of glucose into cells and storage as glycogen in the liver and muscles. Insulin also stimulates the conversion of excess glucose into fat for storage.

Fat cell

Muscle

4. As the body's cells use glucose, blood levels decline.

Pancreas

Glucagon

5. Low blood glucose stimulates the pancreas to release glucagon into the bloodstream.

6. Glucagon stimulates liver cells to break down glycogen and release glucose into the blood.[a]

Liver

7. Blood glucose begins to rise.

Key:
- ⬡ Glucose
- ◯ Insulin
- ⬤ Glucagon
- ⬡⬡⬡⬡ Glycogen

[a] The stress hormone epinephrine and other hormones also bring glucose out of storage.

Another hormone that calls glucose from the liver cells is the "fight-or-flight" hormone, **epinephrine.** When a person experiences stress, epinephrine acts quickly, ensuring that all the body cells have energy fuel in emergencies. Among its many roles in the body, epinephrine works to release glucose from liver glycogen to the blood.

Balancing within the Normal Range The maintenance of normal blood glucose ordinarily depends on two processes. When blood glucose falls below normal, food can replenish it, or in the absence of food, glucagon can signal the liver to break down glycogen stores. When blood glucose rises above normal, insulin can signal the cells to take in glucose for energy. Eating balanced meals at regular intervals helps the body maintain a happy medium between the extremes. Balanced

epinephrine (EP-ih-NEFF-rin): a hormone of the adrenal gland that modulates the stress response; formerly called **adrenaline.**

meals that provide abundant complex carbohydrates, including fibers, and a little fat help to slow down the digestion and absorption of carbohydrate so that glucose enters the blood gradually, providing a steady, ongoing supply.

Falling outside the Normal Range This influence of foods on blood glucose has given rise to the oversimplification that foods *govern* blood glucose concentrations. Foods do not; the body does. In some people, however, blood glucose regulation fails. When this happens, either of two conditions can result: diabetes or hypoglycemia. People with these conditions often plan their diets to help maintain their blood glucose within a normal range.

Diabetes In **diabetes,** blood glucose surges after a meal and remains above normal levels■ because insulin is either inadequate or ineffective. Thus *blood* glucose is central to diabetes, but *dietary* carbohydrates do not cause diabetes.

There are two main types of diabetes. In **type 1 diabetes,** the less common type, the pancreas fails to make insulin; the exact cause is unclear. Some research suggests that in genetically susceptible people, certain viruses activate the immune system to attack and destroy cells in the pancreas as if they were foreign cells. In **type 2 diabetes,** the more common type of diabetes, the cells fail to respond to insulin;■ this condition tends to occur as a consequence of obesity. As the incidence of obesity in the United States has risen in recent decades, the incidence of diabetes has followed. This trend is most notable among children and adolescents, as obesity among the nation's youth reaches epidemic proportions. Because obesity can precipitate type 2 diabetes, the best preventive measure is to maintain a healthy body weight. Concentrated sweets are not strictly excluded from the diabetic diet as they once were, but can be eaten in limited amounts with meals as part of a healthy diet. Chapter 15 describes the type of diabetes that develops in some women during pregnancy (gestational diabetes), and Chapter 18 gives full coverage to type 1 and type 2 diabetes and their associated problems.

Hypoglycemia In healthy people, blood glucose rises after eating and then gradually falls back into the normal range. The transition occurs without notice. In people with **hypoglycemia,** however, blood glucose drops dramatically, producing symptoms that mimic an anxiety attack: weakness, rapid heartbeat, sweating, anxiety, hunger, and trembling. Most commonly, hypoglycemia occurs as a consequence of poorly managed diabetes. Too much insulin, strenuous physical activity, inadequate food intake, or illness can cause blood glucose levels to plummet.

Hypoglycemia in healthy people is rare. Most people who experience hypoglycemia need only adjust their diets by replacing refined carbohydrates with fiber-rich carbohydrates and ensuring an adequate protein intake.[4] In addition, smaller meals eaten more frequently may help. Hypoglycemia caused by certain medications, pancreatic tumors, overuse of insulin, alcohol abuse, or other illnesses requires medical intervention.

The Glycemic Response The **glycemic response** refers to how quickly glucose is absorbed after a person eats, how high blood glucose rises, and how quickly it returns to normal. Slow absorption, a modest rise in blood glucose, and a smooth return to normal are desirable (a low glycemic response); fast absorption, a surge in blood glucose, and an overreaction that plunges glucose below normal are less desirable (a high glycemic response). Different foods have different effects on blood glucose.

The rate of glucose absorption is particularly important to people with diabetes, who may benefit from limiting foods that produce too great a rise, or too sudden a fall, in blood glucose. To aid their choices, such people may be able to use the **glycemic index,** a method of classifying foods according to their potential to raise blood glucose. Figure 4-13 ranks selected foods by their glycemic index.[5] Some studies have shown that selecting foods with a low glycemic index is a practical way to improve glucose control.[6]

Lowering the glycemic index of the *diet* may improve lipid metabolism and prevent heart disease as well.[7] It may also help with weight management.[8] Fibers and

---

■ Blood glucose (fasting):
- Prediabetes: 110 to 125 mg/dL.
- Diabetes: ≥126 mg/dL.

■ The condition of having blood glucose levels higher than normal, but below the diagnosis of diabetes is sometimes called **prediabetes.**

---

diabetes (DYE-ah-BEE-teez): a disorder of carbohydrate metabolism resulting from inadequate or ineffective insulin.

type 1 diabetes: the less common type of diabetes in which the person produces no insulin at all; formerly known as **insulin-dependent diabetes mellitus (IDDM)** or **juvenile-onset diabetes** (because it frequently develops in childhood), although some cases arise in adulthood.

type 2 diabetes: the more common type of diabetes in which the fat cells resist insulin; formerly called **noninsulin-dependent diabetes mellitus (NIDDM)** or **adult-onset diabetes.** Type 2 usually progresses more slowly than type 1.

hypoglycemia (HIGH-po-gligh-SEE-me-ah): an abnormally low blood glucose concentration.

glycemic (gligh-SEEM-ic) **response:** the extent to which a food raises the blood glucose concentration and elicits an insulin response.

glycemic index: a method of classifying foods according to their potential for raising blood glucose.

other slowly digested carbohydrates prolong the presence of foods in the digestive tract, thus providing greater satiety and diminishing the insulin response, which can help with weight control.[9] In contrast, the rapid absorption of glucose from a high-glycemic diet seems to increase the risk of heart disease and promote overeating in some overweight people.[10]

Despite these possible benefits, the usefulness of the glycemic index is surrounded by controversy as researchers debate whether selecting foods based on the glycemic index is practical or offers any real health benefits.[11] Those opposing the use of the glycemic index argue that it is not well enough supported by scientific research.[12] Relatively few foods have had their glycemic index determined, and when the glycemic index has been established, it is based on an average of multiple tests that often result in wide variations. Values vary because of differences in the physical and chemical characteristics of foods, testing methods of laboratories, and digestive processes of individuals.

Furthermore, the practical utility of the glycemic index is limited because this information is nether provided on food labels nor intuitively apparent. Indeed, a food's glycemic index is not always what one might expect. Ice cream, for example, is a high-sugar food, but it produces less of a response than baked potatoes, a high-starch food, most likely because the fat in the ice cream slows GI motility and thus the rate of glucose absorption. Mashed potatoes produce more of a response than honey, probably because the honey's fructose content has little effect on blood glucose. Perhaps most relevant to real life, a food's glycemic effect differs depending on how it is prepared and whether it is eaten alone or with other foods. Most people eat a variety of foods, cooked and raw, that provide different amounts of carbohydrate, fat, and protein—all of which influence the glycemic index of a meal.

Paying attention to the glycemic index may not be necessary because current guidelines already suggest many low glycemic index choices: whole grains, legumes, vegetables, fruits, and milk products.[13] In addition, eating frequent, small meals spreads glucose absorption across the day and thus offers similar metabolic advantages to eating foods with a low glycemic response. People wanting to follow a low-glycemic diet should be careful not to adopt a low-carbohydrate diet.[14] The problems associated with a low-carbohydrate diet are addressed in Highlight 8.

IN SUMMARY Dietary carbohydrates provide glucose that can be used by the cells for energy, stored by the liver and muscles as glycogen, or converted into fat if intakes exceed needs. All of the body's cells depend on glucose; those of the central nervous system are especially dependent on it. Without glucose, the body is forced to break down its protein tissues to make glucose and to alter energy metabolism to make ketone bodies from fats. Blood glucose regulation depends primarily on two pancreatic hormones: insulin to remove glucose from the blood into the cells when levels are high and glucagon to free glucose from glycogen stores and release it into the blood when levels are low. The glycemic index measures how blood glucose responds to foods.

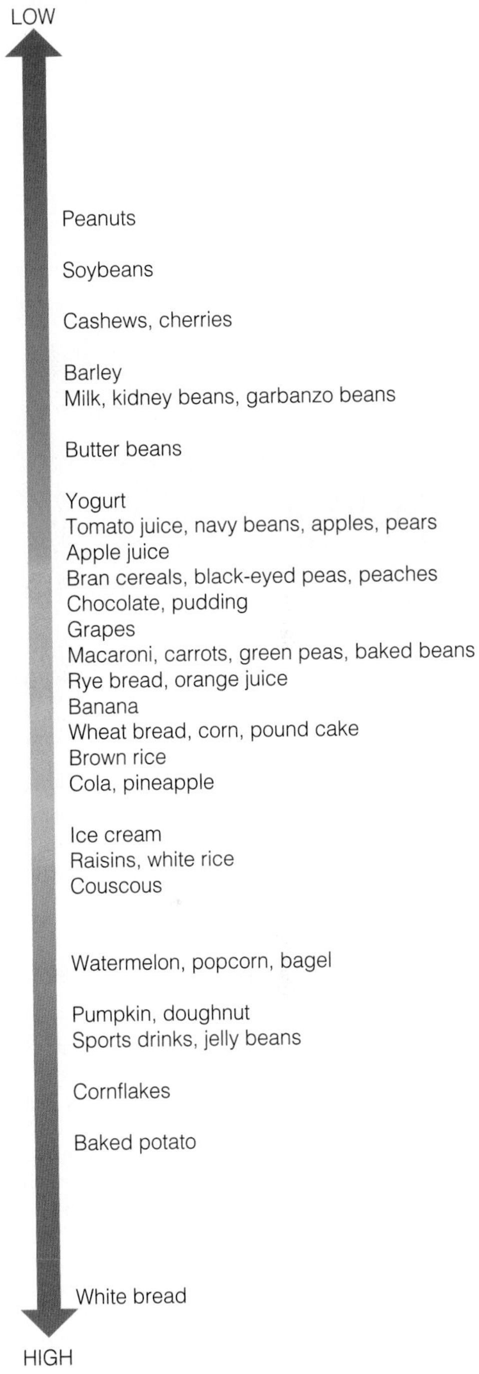

Health Effects and Recommended Intakes of Sugars

Ever since people first discovered honey and dates, they have enjoyed the sweetness of sugars. In the United States, the natural sugars of milk, fruits, vegetables, and grains account for about half of the sugar intake; the other half consists of sugars that have been refined and added to foods for a variety of purposes (see p. 120 margin).■ The use of sweeteners in food manufacturing has risen steadily over the past several decades. These **added sugars** assume various names on food labels: sucrose, invert sugar, corn sugar, corn syrups and solids, high-fructose corn syrup, and honey. A food

FIGURE 4-13 Glycemic Index of Selected Foods

LOW

Peanuts

Soybeans

Cashews, cherries

Barley
Milk, kidney beans, garbanzo beans

Butter beans

Yogurt
Tomato juice, navy beans, apples, pears
Apple juice
Bran cereals, black-eyed peas, peaches
Chocolate, pudding
Grapes
Macaroni, carrots, green peas, baked beans
Rye bread, orange juice
Banana
Wheat bread, corn, pound cake
Brown rice
Cola, pineapple

Ice cream
Raisins, white rice
Couscous

Watermelon, popcorn, bagel

Pumpkin, doughnut
Sports drinks, jelly beans

Cornflakes

Baked potato

White bread

HIGH

added sugars: sugars and syrups used as an ingredient in the processing and preparation of foods such as breads, cakes, beverages, jellies, and ice cream as well as sugars eaten separately or added to foods at the table.

■ As an additive, sugar:
- Enhances flavor.
- Supplies texture and color to baked goods.
- Provides fuel for fermentation, causing bread to rise or producing alcohol.
- Acts as a bulking agent in ice cream and baked goods.
- Acts as a preservative in jams.
- Balances the acidity of tomato- and vinegar-based products.

is likely to be high in added sugars if its ingredient list starts with any of the sugars named in the accompanying glossary or if it includes several of them.

Health Effects of Sugars

In moderate amounts, sugars add pleasure to meals without harming health. In excess, however, they can be detrimental in two ways. One, sugars can contribute to nutrient deficiencies by supplying energy (kcalories) without providing nutrients. Two, sugars contribute to tooth decay.

Nutrient Deficiencies Empty-kcalorie foods that contain lots of added sugar such as cakes, candies, and sodas deliver glucose and energy with few, if any, other nutrients. By comparison, foods such as whole grains, vegetables, legumes, and fruits that contain some natural sugars and lots of starches and fibers deliver protein, vitamins, and minerals along with their glucose and energy.

A person spending 200 kcalories of a day's energy allowance on a 16-ounce soda gets little of value for those kcaloric "dollars." In contrast, a person using 200 kcalories on three slices of whole-wheat bread gets 9 grams of protein, 6 grams of fiber, plus several of the B vitamins with those kcalories. For the person who wants something sweet, perhaps a reasonable compromise would be to have two slices of bread with a teaspoon of jam on each. The amount of sugar a person can afford to eat depends on how many kcalories are available beyond those needed to deliver indispensable vitamins and minerals.

With careful food selections, a person can obtain all the needed nutrients within an allowance of about 1500 kcalories. Some people have more generous energy allowances with which to "purchase" nutrients. For example, an active teenage boy may need as many as 4000 kcalories a day. If he eats mostly nutritious foods, then the "empty kcalories" of cola beverages may be an acceptable addition to his diet. In contrast, an inactive older woman who is limited to fewer than 1500 kcalories a day can afford only the most nutrient-dense foods.

© Polara Studios, Inc.

Over half of the added sugars in our diet come from soft drinks and table sugar, but baked goods, fruit drinks, ice cream, candy, and breakfast cereals also make substantial contributions.

GLOSSARY OF ADDED SUGARS

brown sugar: refined white sugar crystals to which manufacturers have added molasses syrup with natural flavor and color; 91 to 96% pure sucrose.

confectioners' sugar: finely powdered sucrose, 99.9% pure.

corn sweeteners: corn syrup and sugars derived from corn.

corn syrup: a syrup made from cornstarch that has been treated with acid, high temperatures, and enzymes that produce glucose, maltose, and dextrins. See also *high-fructose corn syrup (HFCS)*.

dextrose: an older name for glucose.

granulated sugar: crystalline sucrose; 99.9% pure.

high-fructose corn syrup (HFCS): a syrup made from cornstarch

that has been treated with an enzyme that converts some of the glucose to the sweeter fructose; made especially for use in processed foods and beverages, where it is the predominant sweetener. With a chemical structure similar to sucrose, HFCS has a fructose content of 42, 55, or 90%, with glucose making up the remainder.

honey: sugar (mostly sucrose) formed from nectar gathered by bees. An enzyme splits the sucrose into glucose and fructose. Composition and flavor vary, but honey always contains a mixture of sucrose, fructose, and glucose.

invert sugar: a mixture of glucose and fructose formed by the hydrolysis of sucrose in a chemical process; sold only in

liquid form and sweeter than sucrose. Invert sugar is used as a food additive to help preserve freshness and prevent shrinkage.

levulose: an older name for fructose.

maple sugar: a sugar (mostly sucrose) purified from the concentrated sap of the sugar maple tree.

molasses: the thick brown syrup produced during sugar refining. Molasses retains residual sugar and other by-products and a few minerals; blackstrap molasses contains significant amounts of calcium and iron—the iron comes from the *machinery* used to process the sugar.

raw sugar: the first crop of crystals harvested during sugar processing. Raw sugar cannot

be sold in the United States because it contains too much filth (dirt, insect fragments, and the like). Sugar sold as "raw sugar" domestically has actually gone through over half of the refining steps.

turbinado (ter-bih-NOD-oh) **sugar:** sugar produced using the same refining process as white sugar, but without the bleaching and anti-caking treatment. Traces of molasses give turbinado its sandy color.

white sugar: pure sucrose or "table sugar," produced by dissolving, concentrating, and recrystallizing raw sugar.

| **TABLE 4-1** | Sample Nutrients in Sugar and Other Foods |

The indicated portion of any of these foods provides approximately 100 kcalories. Notice that for a similar number of kcalories and grams of carbohydrate, milk, legumes, fruits, grains, and vegetables offer more of the other nutrients than do the sugars.

	Size of 100 kcal Portion	Carbohydrate (g)	Protein (g)	Calcium (mg)	Iron (mg)	Vitamin A (µg)	Vitamin C (mg)
Foods							
Milk, 1% low-fat	1 c	12	8	300	0.1	144	2
Kidney beans	½ c	20	7	30	1.6	0	2
Apricots	6	24	2	30	1.1	554	22
Bread, whole wheat	1½ slices	20	4	30	1.9	0	0
Broccoli, cooked	2 c	20	12	188	2.2	696	148
Sugars							
Sugar, white	2 tbs	24	0	trace	trace	0	0
Molasses, blackstrap	2½ tbs	28	0	343	12.6	0	0.1
Cola beverage	1 c	26	0	6	trace	0	0
Honey	1½ tbs	26	trace	2	0.2	0	trace

Some people believe that because honey is a natural food, it is nutritious—or, at least, more nutritious than sugar.* A look at their chemical structures reveals the truth. Honey, like table sugar, contains glucose and fructose. The primary difference is that in table sugar the two monosaccharides are bonded together as a disaccharide, whereas in honey some of them are free. Whether a person eats monosaccharides individually, as in honey, or linked together, as in table sugar, they end up the same way in the body: as glucose and fructose.

Honey does contain a few vitamins and minerals, but not many, as Table 4-1 shows. Honey is denser than crystalline sugar, too, so it provides more energy per spoonful.

This is not to say that all sugar sources are alike, for some are more nutritious than others. Consider a fruit, say, an orange. The fruit may give you the same amounts of fructose and glucose and the same number of kcalories as a dose of sugar or honey, but the packaging is more valuable nutritionally. The fruit's sugars arrive in the body diluted in a large volume of water, packaged in fiber, and mixed with valuable minerals, vitamins, and phytochemicals.

As these comparisons illustrate, the significant difference between sugar sources is not between "natural" honey and "purified" sugar but between concentrated sweets and the dilute, naturally occurring sugars that sweeten foods. You can suspect an exaggerated nutrition claim when someone asserts that one product is more nutritious than another because it contains honey.

Sugar can contribute to nutrient deficiencies only by displacing nutrients. For nutrition's sake, the appropriate attitude to take is not that sugar is "bad" and must be avoided, but that nutritious foods must come first. If the nutritious foods end up crowding sugar out of the diet, that is fine—but not the other way around. As always, the goals to seek are balance, variety, and moderation.

Dental Caries Both sugars and starches begin breaking down to sugars in the mouth and so can contribute to tooth decay. Bacteria in the mouth ferment the sugars and in the process produce an acid that dissolves tooth enamel (see Figure 4-14). People can eat sugar without this happening, though, for much depends on how long foods stay in the mouth. Sticky foods stay on the teeth longer and keep yielding acid longer than foods that are readily cleared from the mouth. For that reason, sugar in a juice consumed quickly, for example, is less likely to cause **dental caries** than sugar in a pastry. By the same token, the sugar in sticky foods such as dried fruits is more detrimental than its quantity alone would suggest.

Matthew Farruggio

You receive about the same amount and kinds of sugars from an orange as from a tablespoon of honey, but the packaging makes a big nutrition difference.

dental caries: decay of teeth.
• **caries** = rottenness

---

*Honey should never be fed to infants because of the risk of botulism. Chapters 16 and 19 provide more details.

FIGURE 4-14 Dental Caries

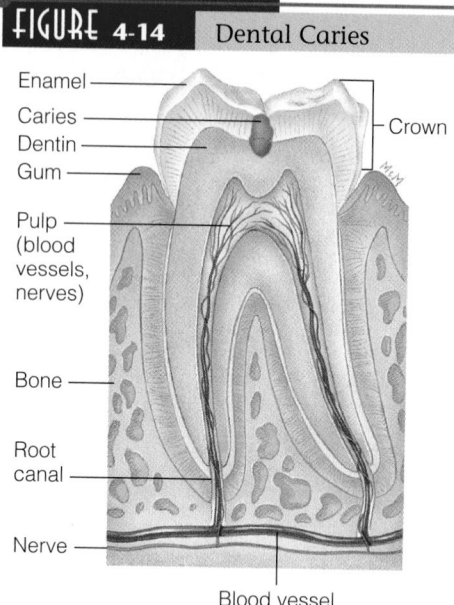

Enamel
Caries
Dentin
Gum
Pulp (blood vessels, nerves)
Bone
Root canal
Nerve
Crown
Blood vessel

Dental caries begins when acid dissolves the enamel that covers the tooth. If not repaired, the decay may penetrate the dentin and spread into the pulp of the tooth, causing inflammation and an abscess.

■ To prevent dental caries:
- Limit between-meal snacks containing sugars and starches.
- Brush and floss teeth regularly.
- If brushing and flossing are not possible, at least rinse with water.

dental plaque: a gummy mass of bacteria that grows on teeth and can lead to dental caries and gum disease.

serotonin (SER-oh-tone-in): a neurotransmitter important in sleep regulation, appetite control, and sensory perception among other roles.

Another concern is how often people eat sugar. Bacteria produce acid for 20 to 30 minutes after each exposure. If a person eats three pieces of candy at one time, the teeth will be exposed to approximately 30 minutes of acid destruction. But, if the person eats three pieces at half-hour intervals, the time of exposure increases to 90 minutes. Likewise, slowly sipping a sugary soft drink may be more harmful than drinking quickly and clearing the mouth of sugar. Nonsugary foods can help remove sugar from tooth surfaces; hence, it is better to eat sugar with meals than between meals. Foods such as milk and cheese may be particularly helpful in minimizing the effects of the acids and in restoring the lost enamel.[15]

The development of caries depends on several factors: the bacteria that reside in **dental plaque,** the saliva that cleanses the mouth, the minerals that form the teeth, and the foods that remain after swallowing. For most people, good oral hygiene will prevent■ dental caries. In fact, regular brushing (twice a day, with a fluoride toothpaste) and flossing may be more effective in preventing dental caries than restricting sugary foods.[16]

Accusations against Sugars

Sugars have been blamed for a variety of other health problems.[17] The following paragraphs evaluate some of these accusations.

Accusation: Sugar Causes Obesity Foods high in added sugars deliver a lot of energy (kcalories). When they are high in fat too, both total energy and fat intakes increase. Exceeding energy needs contributes to weight gain, but sugar is not the sole cause of obesity—and obesity can occur without a high-sugar diet. The notion that eating sweet foods stimulates appetite and promotes overeating has not been supported by research.

Limiting selections of foods and beverages high in added sugars can be an effective weight-loss strategy, however, especially for people whose excess kcalories derive primarily from added sugars. Replacing a can of cola with a glass of water every day, for example, can help a person lose a pound (or at least not gain a pound) in a month. That may not sound like much, but it adds up to over 10 pounds a year, for very little effort.

Accusation: Sugar Causes Heart Disease A diet high in added sugars can alter blood lipids to favor heart disease.[18] (Lipids include fats and cholesterol, as Chapter 5 explains.) This effect is most dramatic in people who respond to sucrose with abnormally high insulin secretions, which promote the making of excess fat.[19] For most people, though, moderate sugar intakes do *not* elevate blood lipids.[20] To keep these findings in perspective, consider that heart disease correlates most closely with factors that have nothing to do with nutrition, such as smoking and genetics. Among dietary risk factors, several—such as saturated fats, *trans* fats, and obesity—have much stronger associations with heart disease than do sugar intakes.

Accusation: Sugar Causes Misbehavior in Children and Criminal Behavior in Adults Sugar has been blamed for the misbehaviors of hyperactive children, delinquent adolescents, and lawbreaking adults. Such speculations have been based on personal stories and have not been confirmed by scientific research. No scientific evidence supports a relationship between sugar and hyperactivity or other misbehaviors. Chapter 16 provides accurate information on diet and children's behavior.

Accusation: Sugar Causes Cravings and Addictions Foods in general, and carbohydrates and sugars more specifically, are not addictive in the biological ways that drugs are. Yet some people describe themselves as having "carbohydrate cravings" or being "sugar addicts." One frequently noted theory is that people seek carbohydrates as a way to increase their levels of the brain neurotransmitter **serotonin,** which elevates mood. Interestingly, when those with self-described carbohydrate cravings indulge, they tend to eat more of everything, but the percentage of energy from carbohydrates remains unchanged.[21] Alcohol also raises serotonin levels, and alcohol-dependent people who crave carbohydrates seem to handle sobriety better when given a high-carbohydrate diet.[22]

One reasonable explanation for the carbohydrate cravings that some people experience involves the self-imposed labeling of a food as both "good" and "bad"—that is, one that is desirable but should be eaten with restraint.[23] Chocolate is a familiar example. Restricting intake heightens the desire further (a "craving"). Then "addiction" is used to explain why resisting the food is so difficult and, sometimes, even impossible. But the "addiction" is not pharmacological; a capsule of the psychoactive substances commonly found in chocolate, for example, does not satisfy the craving.

Recommended Intakes of Sugars

The *Dietary Guidelines*■ urge people to "choose beverages and foods to moderate your intakes of sugars," but they do not define "moderate." The Food Guide Pyramid provides a little more guidance, placing added sugars at the tip and suggesting that consumers use them "sparingly." Sparingly is then defined as shown in the margin.■ Estimates indicate that, on average, each person in the United States consumes about 105 pounds of added sugar per year, or about 30 teaspoons of added sugar a day, an amount that exceeds the Pyramid's recommendations.[24]

Estimating the *added* sugars in a diet is not always easy for consumers. Food labels list the total grams of sugar a food provides, but this total reflects both added sugars and those occurring naturally in foods. To help estimate sugar and energy intakes accurately, the list in the margin■ shows the amounts of concentrated sweets that are equivalent to 1 teaspoon of white sugar. These sugars all provide about 5 grams of carbohydrate and about 20 kcalories per teaspoon. Some are lower (16 kcalories for table sugar), while others are higher (22 kcalories for honey), but a 20-kcalorie average is an acceptable approximation. For a person who uses catsup liberally, it may help to remember that 1 tablespoon of catsup supplies about 1 teaspoon of sugar.

The DRI committee did not set an upper limit for sugar intake, but as mentioned, excessive intakes can interfere with sound nutrition and dental health. Few people can eat lots of sugary treats and still meet all of their nutrient needs without exceeding their kcalorie allowance. Specifically, the DRI suggests that added sugars should account for no more than 25 percent of the day's total energy intake.[25] When added sugars occupy this much of a diet, intakes from the five food groups fall below recommendations.[26] For a person consuming 2000 kcalories a day, 25 percent represents 500 kcalories (that is, 125 grams) from concentrated sugars—and that's a lot of sugar.■ Perhaps an athlete in training whose energy needs are high can afford the added sugars from sports drinks without compromising nutrient intake, but most people would do better using added sugars more "sparingly." The Pyramid recommendations represent roughly 10 percent of the day's total energy intake. A recent report from the World Health Organization (WHO) and the Food and Agriculture Organization (FAO) agrees that people should restrict their consumption of added sugars to less than 10 percent of total energy.

■ *Dietary Guidelines:*
- Limit your intake of beverages and foods that are high in added sugars. Don't let soft drinks or sweets crowd out other foods you need, such as milk or other calcium sources.

■ The Food Guide Pyramid suggests:
- ≤6 tsp for a 1600 kcal diet.
- ≤12 tsp for a 2200 kcal diet.
- ≤18 tsp for a 2800 kcal diet.

■ 1 tsp white sugar =
- 1 tsp brown sugar.
- 1 tsp candy.
- 1 tsp corn sweetener or corn syrup.
- 1 tsp honey.
- 1 tsp jam or jelly.
- 1 tsp maple sugar or maple syrup.
- 1 tsp molasses.
- 1½ oz carbonated soda.
- 1 tbs catsup.

■ For perspective, each of these concentrated sugars provides about 500 kcal:
- 40 oz cola.
- ½ c honey.
- 125 jelly beans.
- 23 marshmallows.
- 30 tsp sugar.

How many kcalories from sugar does your favorite beverage or snack provide?

IN SUMMARY Sugars pose no major health threat except for an increased risk of dental caries. Excessive intakes may displace needed nutrients and fiber and may contribute to obesity when energy intake exceeds needs. A person deciding to limit daily sugar intake should recognize that not all sugars need to be restricted, just concentrated sweets, which are relatively empty of other nutrients and high in kcalories. Sugars that occur naturally in fruits, vegetables, and milk are acceptable.

Foods rich in starch and fiber offer many health benefits.

■ Consuming 5 to 10 g of viscous fiber daily reduces blood cholesterol by 3 to 5%. For perspective, ½ c dry oat bran provides 8 g of fiber, and 1 c cooked barley or ½ c cooked legumes provides about 6 g of fiber.

Health Effects and Recommended Intakes of Starch and Fibers

Carbohydrates and fats are the two major sources of energy in the diet. When one is high, the other is usually low—and vice versa. A diet that provides abundant carbohydrate (45 to 65 percent of energy intake) and some fat (20 to 35 percent of energy intake) within a reasonable energy allowance best supports good health. To increase carbohydrate, focus on whole grains, vegetables, legumes, and fruits—foods noted for their starch, fibers, and naturally occurring sugars.

Health Effects of Starch and Fibers

In addition to starch, fibers, and natural sugars, whole grains, vegetables, legumes, and fruits supply valuable vitamins and minerals and little or no fat. The following paragraphs describe some of the health benefits of diets that include a variety of these foods daily.

Heart Disease High-carbohydrate diets, especially those rich in whole grains, may protect against heart disease and stroke, although sorting out the exact reasons why can be difficult.[27] Such diets are low in animal fat and cholesterol and high in fibers, vegetable proteins, and phytochemicals—all factors associated with a lower risk of heart disease. (The role of animal fat and cholesterol in heart disease is discussed in Chapter 5. The role of vegetable proteins in heart disease is presented in Chapter 6. The benefits of phytochemicals in disease prevention are featured in Highlight 13.)

Foods rich in viscous fibers (such as oat bran, barley, and legumes) lower blood cholesterol■ by binding with bile acids and thereby increasing their excretion.[28] Consequently, the liver must use its cholesterol to make new bile acids. In addition, the bacterial by-products of fiber fermentation in the colon also inhibit cholesterol synthesis in the liver.[29] The net result is lower blood cholesterol.[30]

Several researchers have speculated that fiber may also exert its effect by displacing fats in the diet. While this is certainly helpful, even when dietary fat is low, high intakes of fibers exert a separate and significant cholesterol-lowering effect. In other words, a high-fiber diet helps to prevent heart disease independent of fat intake.

Diabetes High-fiber foods play a key role in reducing the risk of type 2 diabetes.[31] When viscous fibers trap nutrients and delay their transit through the GI tract, glucose absorption is slowed, and this helps to prevent the glucose surge and rebound that seem to be associated with diabetes onset.

GI Health Dietary fibers enhance the health of the large intestine. The healthier the intestinal walls, the better they can block absorption of unwanted constituents. Fibers such as cellulose (as in cereal brans, fruits, and vegetables) increase stool weight, easing passage, and reduce transit time. In this way, the fibers help to alleviate or prevent constipation.

Taken with ample fluids, fibers help to prevent several GI disorders. Large, soft stools ease elimination for the rectal muscles and reduce the pressure in the lower bowel, making it less likely that rectal veins will swell (hemorrhoids). Fiber prevents compaction of the intestinal contents, which could obstruct the appendix and permit bacteria to invade and infect it (appendicitis). In addition, fiber stimulates the GI tract muscles so that they retain their strength and resist bulging out into pouches known as diverticula (illustrated in Figure H3-3 on p. 97).[32]

Cancer Many, but not all, research studies suggest that increasing dietary fiber protects against colon cancer.[33] When the largest study of diet and cancer to date examined the diets of over a half million people in ten countries for four and a half years, the researchers found an inverse association between dietary fiber and colon cancer.[34] People who ate the most dietary fiber (35 grams per day) reduced their risk of colon cancer by 40 percent compared with those who ate the least fiber (15 grams per day). Importantly, the study focused on dietary fiber, not fiber supplements or additives, which lack valuable nutrients and phytochemicals that also help protect against cancer.

Fibers help prevent colon cancer by diluting, binding, and rapidly removing potentially cancer-causing agents from the colon. In addition, some fibers stimulate bacterial fermentation of resistant starch and fiber in the colon, a process that produces short-chain fatty acids that lower the pH.[35] These small fat molecules and the lower pH inhibit cancer growth in the colon.[36]

Discrepancies in research findings may reflect the delay between eating low-fiber diets and developing colon cancer decades later or the differences in effectiveness between various types and sources of fiber. Despite the inconclusive evidence, health care professionals continue to recommend a high-fiber diet that includes at least five servings of vegetables and fruits and generous portions of whole grains and legumes.[37]

Weight Management Foods rich in complex carbohydrates tend to be low in fat and added sugars and can therefore promote weight loss by delivering less energy■ per bite. In addition, as fibers absorb water from the digestive juices, they swell, creating feelings of fullness and delaying hunger.[38]

Many weight-loss products on the market today contain bulk-inducing fibers such as methylcellulose, but buying pure fiber compounds like this is neither necessary nor advisable. To use fiber in a weight-loss plan, select fresh fruits, vegetables, legumes, and whole-grain foods. High-fiber foods not only add bulk to the diet, but are economical and nutritious.

Most experts agree that the health benefits attributed to fiber may come from other constituents of fiber-containing foods, and not from fiber alone.[39] For this reason, consumers should select whole grains, legumes, fruits, and vegetables instead of fiber supplements. Table 4-2 summarizes fibers and their health benefits.

Harmful Effects of Excessive Fiber Intake Despite fiber's benefits to health, a diet high in fiber also has a few drawbacks. A person who has a small capacity and eats mostly high-fiber foods may not be able to take in enough food to meet

■ Reminder:
- Carbohydrate: 4 kcal/g.
- Fat: 9 kcal/g.

TABLE 4-2 Fibers: Their Characteristics, Food Sources, and Health Effects in the Body

Fiber Characteristics	Major Food Sources	Actions in the Body	Health Benefits
Viscous, soluble, more fermentable • Gums and mucilages • Pectins • Psyllium[a] • Some hemicelluloses	Whole-grain products (barley, oats, oat bran, rye), fruits (apples, citrus), legumes, seeds and husks, vegetables; also extracted and used as food additives.	• Lower blood cholesterol by binding bile. • Slow glucose absorption. • Slow transit of food through upper GI tract. • Hold moisture in stools, softening them. • Yield small fat molecules after fermentation that the colon can use for energy.	• Lower risk of heart disease. • Lower risk of diabetes.
Nonviscous, insoluble, less fermentable • Cellulose • Lignins • Psyllium[a] • Resistant starch • Many hemicelluloses	Brown rice, fruits, legumes, seeds, vegetables (cabbage, carrots, brussels sprouts), wheat bran, whole grains; also extracted and used as food additives.	• Increase fecal weight and speed fecal passage through colon. • Provide bulk and feelings of fullness.	• Alleviate constipation. • Lower risks of diverticulosis, hemorrhoids, and appendicitis. • May help with weight management.

[a]Psyllium, a fiber laxative and cereal additive, has both soluble and insoluble properties.

energy or nutrient needs. The malnourished, the elderly, and young children adhering to all-plant (vegan) diets are especially vulnerable to this problem.

Launching suddenly into a high-fiber diet can cause temporary bouts of abdominal discomfort, gas, and diarrhea and, more seriously, can obstruct the GI tract. To prevent such complications, a person adopting a high-fiber diet is advised to:

- Increase fiber intake gradually over several weeks to give the GI tract time to adapt.
- Drink lots of liquids to soften the fiber as it moves through the GI tract.
- Select fiber-rich foods from a variety of sources—fruits, vegetables, legumes, and whole-grain breads and cereals.

Some fibers can limit the absorption of nutrients by speeding the transit of foods through the GI tract and by binding to minerals. When mineral intake is adequate, however, a reasonable intake of high-fiber foods does not seem to compromise mineral balance.

Clearly, fiber is like all the nutrients in that "more" is "better" only up to a point. Again, the key words are balance, moderation, and variety.

IN SUMMARY An adequate intake of fiber:
- Fosters weight management.
- Lowers blood cholesterol.
- May help prevent colon cancer.
- Helps prevent and control diabetes.
- Helps prevent and alleviate hemorrhoids.
- Helps prevent appendicitis.
- Helps prevent diverticulosis.

An excessive intake of fiber:
- Displaces energy- and nutrient-dense foods.
- Causes intestinal discomfort and distention.
- May interfere with mineral absorption.

Recommended Intakes of Starch and Fibers

Dietary recommendations suggest that carbohydrates provide about half (45 to 65 percent) of the energy requirement. A person consuming 2000 kcalories a day should therefore have 900 to 1300 kcalories of carbohydrate, or about 225 to 325 grams.■ This amount is more than adequate to meet the RDA■ for carbohydrate, which is set at 130 grams per day, based on the average minimum amount of glucose used by the brain.[40]

When it established the Daily Values that appear on food labels, the Food and Drug Administration (FDA) used a 60 percent of kcalories guideline in setting the Daily Value■ for carbohydrate at 300 grams per day. For most people, this means increasing total carbohydrate intake. To this end, the *Dietary Guidelines*■ encourage people to choose a variety of whole grains, vegetables, fruits, and legumes daily.

Increase the proportion of people who meet the *Dietary Guidlines* daily goal of at least six servings of grain products and at least five servings of vegetables and fruits.

Recommendations for fiber■ suggest the same foods just mentioned: whole grains, vegetables, fruits, and legumes, which also provide minerals and vitamins. The FDA set the Daily Value■ for fiber at 25 grams or 11.5 grams per 1000-kcalorie

---

Margin notes (left column):

■ • 45% of 2000 kcal:

$$\frac{45}{100} = \frac{x}{2000 \text{ kcal}}.$$

$100x = 90{,}000$ kcal.

$x = 900$ kcal.

900 kcal ÷ 4 kcal/g = 225 g.

• 65% of 2000 kcal:

$$\frac{65}{100} = \frac{x}{2000 \text{ kcal}}.$$

$100x = 130{,}000$ kcal.

$x = 1300$ kcal.

1300 kcal ÷ 4 kcal/g = 325 g.

■ RDA for carbohydrate:
- 130 g/day.
- 45 to 65% of energy intake.

■ Daily Values:
- 300 g carbohydrate (based on 60% of 2000 kcal diet).

■ *Dietary Guidelines:*
- Choose a variety of grains daily, especially whole grains such as whole wheat, brown rice, and oats.
- Choose a variety of fruits and vegetables daily. Eat at least two servings of fruit and at least three servings of vegetables each day.

■ To increase your fiber intake:
- Eat whole-grain cereals that contain ≥5 g fiber per serving for breakfast.
- Eat raw vegetables.
- Eat fruits (such as pears) and vegetables (such as potatoes) with their skins.
- Add legumes to soups, salads, and casseroles.
- Eat fresh and dried fruit for snacks.

■ Daily Values:
- 25 g fiber (based on 11.5 g/1000 kcal).

| **TABLE 4-3** | Fiber in Selected Foods |

Bread, Cereal, Rice, and Pasta Group

Whole-grain products provide about 1 to 2 grams (or more) of fiber per serving:

- 1 slice whole-wheat, pumpernickel, rye bread.
- 1 oz ready-to-eat cereal (100% bran cereals contain 10 grams or more).
- ½ c cooked barley, bulgur, grits, oatmeal.

Vegetable Group

Most vegetables contain about 2 to 3 grams of fiber per serving:

- 1 c raw bean sprouts.
- ½ c cooked broccoli, brussels sprouts, cabbage, carrots, cauliflower, collards, corn, eggplant, green beans, green peas, kale, mushrooms, okra, parsnips, potatoes, pumpkin, spinach, sweet potatoes, swiss chard, winter squash.
- ½ c chopped raw carrots, peppers.

Fruit Group

Fresh, frozen, and dried fruits have about 2 grams of fiber per serving:

- 1 medium apple, banana, kiwi, nectarine, orange, pear.
- ½ c applesauce, blackberries, blueberries, raspberries, strawberries.
- Fruit juices contain very little fiber.

Legumes

Many legumes provide about 6 to 8 grams of fiber per serving:

- ½ c cooked baked beans, black beans, black-eyed peas, kidney beans, navy beans, pinto beans.

Some legumes provide about 5 grams of fiber per serving:

- ½ c cooked garbanzo beans, great northern beans, lentils, lima beans, split peas.

NOTE: Appendix H provides fiber grams for over 2000 foods.

intake. The DRI recommendation■ is slightly higher, at 14 grams per 1000-kcalorie intake. Similarly, the American Dietetic Association suggests 20 to 35 grams of dietary fiber daily, which is about two times higher than the average intake in the United States.[41] An effective way to add fiber while lowering fat is to substitute plant sources of proteins (legumes) for animal sources (meats). Table 4-3 presents a list of fiber sources.

As mentioned earlier, too much fiber is no better than too little. The World Health Organization recommends an upper limit of 40 grams of dietary fiber a day.

From Guidelines to Groceries

A diet following the Food Guide Pyramid, which includes 3 to 5 vegetable servings, 2 to 4 fruit servings, and 6 to 11 bread servings daily, can easily supply the recommended amount of carbohydrates and fiber. In selecting high-fiber foods, keep in mind the principle of variety. The fibers in oats lower cholesterol, whereas those in bran help promote GI tract health. (Review Table 4-2 to see the diverse health effects of various fibers.)

Grains A serving of most foods in the bread and cereal group—a slice of whole-wheat bread, half an English muffin, or a 6-inch tortilla, provides about 15 grams of carbohydrate, mostly as starch. Be aware that some foods in this group, especially snack crackers and baked goods such as biscuits, croissants, and muffins, contain added sugars, added fat, or both. When selecting from the bread and cereal group, be sure to include whole-grain products (see Figure 4-15). The "3 are Key" message may help consumers to remember to choose a whole-grain cereal for breakfast, a whole-grain bread for lunch, and a whole-grain pasta or rice for dinner.[42]

■ Fiber AI:
- 14 g/1000 kcal/day.
- Men:
  19–50 yr: 38 g/day.
  51+ yr: 30 g/day.
- Women:
  19–50 yr: 25 g/day.
  51+ yr: 21 g/day.

Reminder: An *AI (Adequate Intake)* is used as a guide for nutrient intake when an RDA cannot be established (see Chapter 1).

FIGURE 4-15   Bread Labels Compared

Food labels provide the quantities of total carbohydrate, dietary fiber, and sugars. Total carbohydrate and dietary fiber are also stated as "% Daily Values." A close look at these two labels reveals that bread made from whole wheat flour provides almost three times as much fiber as the one made mostly from refined wheat flour. When the words whole wheat or whole grain appear on the label, the bread inside contains all of the nutrients that bread can provide.

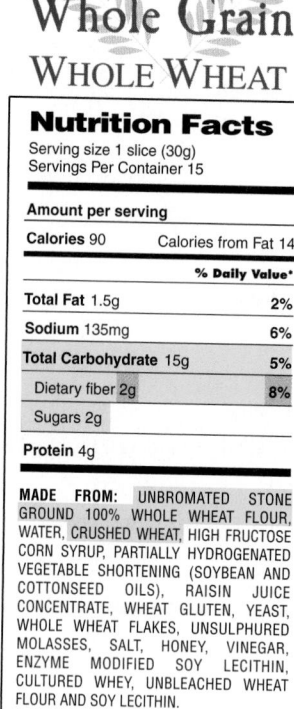

Vegetables The amount of carbohydrate a serving of vegetables provides depends primarily on its starch content. Starchy vegetables—a half-cup of cooked dry beans, corn, peas, plantain, potatoes, or sweet potatoes—provide about 15 grams of carbohydrate per serving. A serving of most other *nonstarchy* vegetables—such as a half-cup of carrots, broccoli, tomatoes, or squash or a cup of salad greens—provides about 5 grams.

Fruits A typical fruit serving—a small banana, apple, or orange, or a half-cup of most canned or fresh fruit—contains an average of about 15 grams of carbohydrate, mostly as sugars, including the fruit sugar fructose. Fruits vary greatly in their water and fiber contents and, therefore, in their sugar concentrations.

Milks and Milk Products A serving (a cup) of milk or yogurt provides about 12 grams of carbohydrate. Cottage cheese provides about 6 grams of carbohydrate per cup, but most other cheeses contain little, if any, carbohydrate.

Meats and Meat Alternates With two exceptions, foods in the meats and meat alternates group deliver almost no carbohydrate to the diet. The exceptions are nuts, which provide a little starch and fiber along with their abundant fat, and legumes, which provide an abundance of both starch and fiber. Just a half-cup serving of legumes provides 15 grams of carbohydrate, about half from fiber.

Read Food Labels Food labels list the amount, in grams, of *total* carbohydrate—including starch, fibers, and sugars—per serving (review Figure 4-15). Fiber grams are also listed separately, as are the grams of sugars. (With this information, you can calculate starch grams■ by subtracting the grams of fibers and sugars from the total carbohydrate.) Sugars reflect both added sugars and those that occur naturally in foods. Total carbohydrate and dietary fiber are also expressed as "% Daily Values" for a person consuming 2000 kcalories; there is no Daily Value for sugars.

■ To calculate starch grams using the first label in Figure 4-15: 15 g total − 4 g (dietary fiber + sugars) = 11 g starch.

> **IN SUMMARY** Clearly, a diet rich in complex carbohydrates—starches and fibers—supports efforts to control body weight and prevent heart disease, cancer, diabetes, and GI disorders. For these reasons, recommendations urge people to eat plenty of whole grains, vegetables, legumes, and fruits—enough to provide 45 to 65 percent of the daily energy intake from carbohydrate.

In today's world, there is one other reason why plant foods rich in complex carbohydrates and natural sugars are a better choice than animal foods or foods high in concentrated sweets: in general, less energy and resources are required to grow and process plant foods than to produce sugar or foods derived from animals. Chapter 20 takes a closer look at the environmental impacts of food production and use.

Nutrition in Your Life

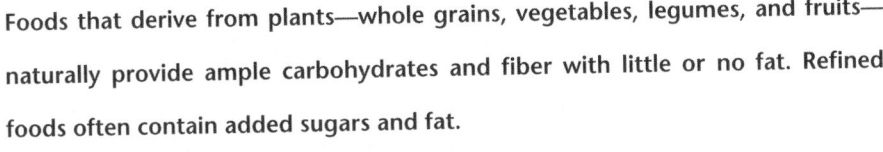

Foods that derive from plants—whole grains, vegetables, legumes, and fruits—naturally provide ample carbohydrates and fiber with little or no fat. Refined foods often contain added sugars and fat.

- Do you eat at least six servings of grain products daily, making sure to include at least three servings of whole-grain foods?

- Do you eat at least five servings of fruits and vegetables daily?

- Do you choose beverages and foods that limit your intake of sugars?

NUTRITION ON THE NET

 Access these websites for further study of topics covered in this chapter.

- Find updates and quick links to these and other nutrition-related sites at our website: **www.wadsworth.com/nutrition**

- Search for "lactose intolerance" at the U.S. Government health information site: **www.healthfinder.gov**

- Search for "sugars" and "fiber" at the International Food Information Council site: **www.ific.org**

- Learn more about dental caries from the American Dental Association and the National Institute of Dental and Craniofacial Research: **www.ada.org** and **www.nidcr.nih.gov**

- Learn more about diabetes from the American Diabetes Association, the Canadian Diabetes Association, and the National Institute of Diabetes and Digestive and Kidney Diseases: **www.diabetes.org**, **www.diabetes.ca**, and **www.niddk.nih.gov**

NUTRITION CALCULATIONS

These problems will give you practice in doing simple nutrition-related calculations. Although the situations are hypothetical, the numbers are real, and calculating the answers (check them on p. 132) provides a valuable lesson. Be sure to show your calculations for each problem.

Health recommendations suggest that 45 to 65 percent of the daily energy intake come from carbohydrates. Stating recommendations in terms of percentage of energy intake is meaningful only if energy intake is known. The following exercises illustrate this concept.

1. Calculate the carbohydrate intake (in grams) for a student who has a high carbohydrate intake (70 percent of energy intake) and a moderate energy intake (2000 kcalories a day).

    How does this carbohydrate intake compare to the Daily Value of 300 grams? To the 45 to 65 percent recommendation?

2. Now consider a professor who eats half as much carbohydrate as the student (in grams) and has the same energy intake. What percentage does carbohydrate contribute to the daily intake?

    How does carbohydrate intake compare to the Daily Value of 300 grams? To the 45 to 65 percent recommendation?

3. Now consider an athlete who eats twice as much carbohydrate (in grams) as the student and has a much higher energy intake (6000 kcalories a day). What percentage does carbohydrate contribute to this person's daily intake?

    How does carbohydrate intake compare to the Daily Value of 300 grams? To the 45 to 65 percent recommendation?

4. One more example. In an attempt to lose weight, a person adopts a diet that provides 150 grams of carbohydrate per day and limits energy intake to 1000 kcalories. What percentage does carbohydrate contribute to this person's daily intake?

    How does this carbohydrate intake compare to the Daily Value of 300 grams? To the 45 to 65 percent recommendation?

These exercises should convince you of the importance of examining actual intake as well the percentage of energy intake.

STUDY QUESTIONS

These questions will help you review this chapter. You will find the answers in the discussions on the pages provided.

1. Which carbohydrates are described as simple, and which are complex? (p. 104)

2. Describe the structure of a monosaccharide and name the three monosaccharides important in nutrition. Name the three disaccharides commonly found in foods and their component monosaccharides. In what foods are these sugars found? (pp. 105–106)

3. What happens in a condensation reaction? In a hydrolysis reaction? (pp. 106–107)

4. Describe the structure of polysaccharides and name the ones important in nutrition. How are starch and glycogen similar, and how do they differ? How do the fibers differ from the other polysaccharides? (pp. 107–110)

5. Describe carbohydrate digestion and absorption. What role does fiber play in the process? (pp. 110–113)

6. What are the possible fates of glucose in the body? What is the protein-sparing action of carbohydrate? (pp. 114–115)

7. How does the body maintain its blood glucose concentration? What happens when the blood glucose concentration rises too high or falls too low? (pp. 116–118)

8. What are the health effects of sugars? What are the dietary recommendations regarding concentrated sugar intakes? (pp. 120–122)

9. What are the health effects of starches and fibers? What are the dietary recommendations regarding these complex carbohydrates? (pp. 124–127)

10. What foods provide starches and fibers? (pp. 127–128)

These multiple choice questions will help you prepare for an exam. Answers can be found on p. 132.

1. Carbohydrates are found in virtually all foods except:
    a. milks.
    b. meats.
    c. breads.
    d. fruits.

2. Disaccharides include:
    a. starch, glycogen, and fiber.
    b. amylose, pectin, and dextrose.
    c. sucrose, maltose, and lactose.
    d. glucose, galactose, and fructose.

3. The making of a disaccharide from two monosaccharides is an example of:
    a. digestion.
    b. hydrolysis.
    c. condensation.
    d. gluconeogenesis.

4. The storage form of glucose in the body is:
    a. insulin.
    b. maltose.
    c. glucagon.
    d. glycogen.

5. The significant difference between starch and cellulose is that:
   a. starch is a polysaccharide, but cellulose is not.
   b. animals can store glucose as starch, but not as cellulose.
   c. hormones can make glucose from cellulose, but not from starch.
   d. digestive enzymes can break the bonds in starch, but not in cellulose.

6. The ultimate goal of carbohydrate digestion and absorption is to yield:
   a. fibers.
   b. glucose.
   c. enzymes.
   d. amylase.

7. The enzyme that breaks a disaccharide into glucose and galactose is:
   a. amylase.
   b. maltase.
   c. sucrase.
   d. lactase.

8. With insufficient glucose in metabolism, fat fragments combine to form:
   a. dextrins.
   b. mucilages.
   c. phytic acids.
   d. ketone bodies.

9. What does the pancreas secrete when blood glucose rises? When blood glucose falls?
   a. insulin; glucagon
   b. glucagon; insulin
   c. insulin; glycogen
   d. glycogen; epinephrine

10. What percentage of the daily energy intake should come from carbohydrates?
    a. 15 to 20
    b. 25 to 30
    c. 45 to 50
    d. 45 to 65

REFERENCES

1. B. M. Davy and C. L. Melby, The effect of fiber-rich carbohydrates on features of Syndrome X, *Journal of the American Dietetic Association* 103 (2003): 86–96.
2. S. W. Rizkalla and coauthors, Chronic consumption of fresh but not heated yogurt improves breath-hydrogen status and short-chain fatty acid profiles: A controlled study in healthy men with or without lactose maldigestion, *American Journal of Clinical Nutrition* 72 (2000): 1474–1479.
3. T. Maeder, Sweet medicines, *Scientific American* 287 (2002): 40–47; J. Travis, The true sweet science—Researchers develop a taster for the study of sugars, *Science News* 161 (2002): 232–233; multiple articles in Carbohydrates and glycobiology—Searching for medicine's sweet spot, *Science* 291 (2001): 2338–2378.
4. G. Pourmotabbed and A. E. Kitabchi, Hypoglycemia, *Obstetrics and Gynecology Clinics of North America* 28 (2001): 383–400.
5. K. Foster-Powell, S. H. A. Holt, and J. C. Brand-Miller, International table of glycemic index and glycemic load values: 2002, *American Journal of Clinical Nutrition* 76 (2002): 5–56.
6. A. E. Buyken and coauthors, Glycemic index in the diet of European outpatients with type 1 diabetes: Relations to glycated hemoglobin and serum lipids, *American Journal of Clinical Nutrition* 73 (2001): 574–581.
7. T. M. S. Wolever, Carbohydrate and the regulation of blood glucose and metabolism, *Nutrition Reviews* 61 (2003): S40–S48; D. J. A. Jenkins and coauthors, Glycemic index: Overview of implications in health and disease, *American Journal of Clinical Nutrition* 76 (2002): 266S–273S; S. Liu and coauthors, Dietary glycemic load assessed by food-frequency questionnaire in relation to plasma high-density lipoprotein cholesterol and fasting plasma triacylglycerols in postmenopausal women, *American Journal of Clinical Nutrition* 73 (2001): 560–566; S. Liu and coauthors, A prospective study of dietary glycemic load, carbohydrate intake, and risk of coronary heart disease in US women, *American Journal of Clinical Nutrition* 71 (2000): 1455–1461; K. L. Morris and M. B. Zemel, Glycemic index, cardiovascular disease, and obesity, *Nutrition Reviews* 57 (1999): 273–276.
8. L. E. Spieth and coauthors, A low-glycemic index diet in the treatment of pediatric obesity, *Archives of Pediatrics and Adolescent Medicine* 154 (2000): 947–951.
9. S. B. Roberts, Glycemic index and satiety, *Nutrition in Clinical Care* 6 (2003): 20–26; D. S. Ludwig and coauthors, Dietary fiber, weight gain, and cardiovascular disease risk factors in young adults, *Journal of the American Medical Association* 282 (1999): 1539–1546.
10. S. Liu and coauthors, Relation between a diet with a high glycemic load and plasma concentrations of high-sensitivity C-reactive protein in middle-aged women, *American Journal of Clinical Nutrition* 75 (2002): 492–498; D. S. Ludwig and coauthors, High glycemic index foods, overeating, and obesity, *Pediatrics* 103 (1999): e26 (www.pediatrics.org).
11. D. S. Ludwig, The glycemic index—Physiological mechanisms relating to obesity, diabetes, and cardiovascular disease, *Journal of the American Medical Association* 287 (2002): 2414–2423.
12. F. X. Pi-Sunyer, Glycemic index and disease, *American Journal of Clinical Nutrition* 76 (2002): 290S–298S.
13. C. Beebe, Diets with a low glycemic index: Not ready for practice yet! *Nutrition Today* 34 (1999): 82–86.
14. E. Saltzman, The low glycemic index diet: Not yet ready for prime time, *Nutrition Reviews* 57 (1999): 297.
15. S. Kashket and D. P. DePaola, Cheese consumption and the development and progression of dental caries, *Nutrition Reviews* 60 (2002): 97–103; Department of Health and Human Services, *Oral Health in America: A Report of the Surgeon General* (Rockville, Md.: National Institutes of Health, 2000), pp. 250–251.
16. S. Gibson and S. Williams, Dental caries in pre-school children: Associations with social class, toothbrushing habit and consumption of sugars and sugar-containing foods. Further analysis of data from the National Diet and Nutrition Survey of children aged 1.5–4.5 years, *Caries Research* 33 (1999): 101–113.
17. J. M. Jones and K. Elam, Sugars and health: Is there an issue? *Journal of the American Dietetic Association* 103 (2003): 1058–1060.
18. B. V. Howard and J. Wylie-Rosett, AHA Scientific Statement: Sugar and cardiovascular disease, *Circulation* 106 (2002): 523.
19. J. M. Schwarz and coauthors, Hepatic de novo lipogenesis in normoinsulinemic and hyperinsulinemic subjects consuming high-fat, low-carbohydrate and low-fat, high-carbohydrate isoenergetic diets, *American Journal of Clinical Nutrition* 77 (2003): 43–50.
20. E. J. Parks and M. K. Hellerstein, Carbohydrate-induced hypertriacylglycerolemia: Historical perspective and review of biological mechanisms, *American Journal of Clinical Nutrition* 71 (2000): 412–433.
21. S. Yanovski, Sugar and fat: Cravings and aversions, *Journal of Nutrition* 133 (2003): 835S–837S.
22. M. Moorhouse and coauthors, Carbohydrate craving by alcohol-dependent men during sobriety: Relationship to nutrition and serotonergic function, *Alcoholism, Clinical and Experimental Research* 24 (2000): 635–643.
23. P. J. Rogers and H. J. Smit, Food craving and food "addiction": A critical review of the evidence from a biopsychosocial perspective, *Pharmacology, Biochemistry, and Behavior* 66 (2000): 3–14.

24. Economic Research Service, Farm Service Agency, and Foreign Agricultural Service, USDA, 2001.
25. Committee on Dietary Reference Intakes, *Dietary Reference Intakes for Energy, Carbohydrate, Fiber, Fat, Fatty Acids, Cholesterol, Protein, and Amino Acids* (Washington, D.C.: National Academies Press, 2002).
26. S. A. Bowman, Diets of individuals based on energy intakes from added sugars, *Family Economics and Nutrition Review* 12 (1999): 31–38.
27. F. B. Hu and W. C. Willett, Optimal diets for prevention of coronary heart disease, *Journal of the American Medical Association* 288 (2002): 2569–2578; N. M. McKeown and coauthors, Whole-grain intake is favorably associated with metabolic risk factors for type 2 diabetes and cardiovascular disease in the Framingham Offspring Study, *American Journal of Clinical Nutrition* 76 (2002): 390–398; S. Liu and coauthors, Whole-grain consumption and risk of coronary heart disease: Results from the Nurses' Health Study, *American Journal of Clinical Nutrition* 70 (1999): 412–419; A. Wolk and coauthors, Long-term intake of dietary fiber and decreased risk of coronary heart disease among women, *Journal of the American Medical Association* 281 (1999): 1998–2004; J. L. Slavin and coauthors, Plausible mechanisms for the protectiveness of whole grains, *American Journal of Clinical Nutrition* 70 (1999): 459S–463S.
28. L. Van Horn and N. Ernst, A summary of the science supporting the new National Cholesterol Education program dietary recommendations: What dietitians should know, *Journal of the American Dietetic Association* 101 (2001): 1148–1154; L. Brown and coauthors, Cholesterol-lowering effects of dietary fiber: A meta-analysis, *American Journal of Clinical Nutrition* 69 (1999): 30–42.
29. M. L. Fernandez, Soluble fiber and nondigestible carbohydrate effects on plasma lipids and cardiovascular risk, *Current Opinion in Lipidology* 12 (2001): 35–40; Brown and coauthors, 1999.
30. B. M. Davy and coauthors, High-fiber oat cereal compared with wheat cereal consumption favorably alters LDL-cholesterol subclass and particle numbers in middle-aged and older men, *American Journal of Clinical Nutrition* 76 (2002): 351–358; D. J. A. Jenkins and coauthors, Soluble fiber intake at a dose approved by the US Food and Drug Administration for a claim of health benefits: Serum lipid risk factors for cardiovascular disease assessed in a randomized controlled crossover trial, *American Journal of Clinical Nutrition* 75 (2002): 834–839; J. W. Anderson and coauthors, Cholesterol-lowering effects of psyllium intake adjunctive to diet therapy in men and women with hypercholesterolemia: A meta-analysis of 8 controlled trials, *American Journal of Clinical Nutrition* 71 (2000): 472–479; Brown and coauthors, 1999.
31. T. T. Fung and coauthors, Whole-grain intake and the risk of type 2 diabetes: A prospective study in men, *American Journal of Clinical Nutrition* 76 (2002): 535–540.
32. W. Aldoori and M. Ryan-Harshman, Preventing diverticular disease: Review of recent evidence on high-fibre diets, *Canadian Family Physician* 48 (2002): 1632–1637.
33. A. Schatzkin and coauthors, Lack of effect of a low-fat, high-fiber diet on the recurrence of colorectal adenomas, *New England Journal of Medicine* 342 (2000): 1149–1155; D. S. Alberts and coauthors, Lack of effect of a high-fiber cereal supplement on the recurrence of colorectal adenomas, *New England Journal of Medicine* 342 (2000): 1156–1162; F. Macrae, Wheat bran fiber and development of adenomatous polyps: Evidence from randomized, controlled clinical trials, *American Journal of Medicine* 106 (1999): 38S–42S; D. Kritchevsky, Protective role of wheat bran fiber: Preclinical data, *American Journal of Medicine* 106 (1999): 28S–31S; C. S. Fuchs and coauthors, Dietary fiber and risk of colorectal cancer and adenoma in women, *New England Journal of Medicine* 340 (1999): 169–176.
34. S. A. Bingham and coauthors, Dietary fibre in food and protection against colorectal cancer in the European Prospective Investigation into Cancer and Nutrition (EPIC): An observational study, *Lancet* 361 (2003): 1496–1501.
35. J. L. Slavin, Mechanisms for the impact of whole grain foods on cancer risk, *Journal of the American College of Nutrition* 19 (2000): 300S–307S.
36. N. J. Emenaker and coauthors, Short-chain fatty acids inhibit invasive human colon cancer by modulating uPA, TIMP-1, TIMP-2, Mutant p53, Bcl-2, Bax, p21, and PCNA protein expression in an in vitro cell culture model, *Journal of Nutrition* 131 (2001): 3041S–3046S.
37. American Gastroenterological Association medical position statement: Impact of dietary fiber on colon cancer occurrence, *Gastroenterology* 118 (2000): 1233–1234.
38. N. C. Howarth, E. Saltzman, and S. B. Roberts, Dietary fiber and weight regulation, *Nutrition Reviews* 59 (2001): 129–139; A. Sparti and coauthors, Effect of diets high or low in unavailable and slowly digestible carbohydrates on the pattern of 24-h substrate oxidation and feelings of hunger in humans, *American Journal of Clinical Nutrition* 72 (2000): 1461–1468.
39. Committee on Dietary Reference Intakes, 2002, p. 7-4.
40. Committee on Dietary Reference Intakes, 2002.
41. Position of the American Dietetic Association: Health implications of dietary fiber, *Journal of the American Dietetic Association* 102 (2002): 993–999.
42. J. L. Slavin and coauthors, The role of whole grains in disease prevention, *Journal of the American Dietetic Association* 101 (2001): 780–785.

ANSWERS

Nutrition Calculations

1. 0.7×2000 total kcal/day = 1400 kcal from carbohydrate/day.

   1400 kcal from carbohydrate ÷ 4 kcal/g = 350 g carbohydrate.

   This carbohydrate intake is higher than the Daily Value and higher than the 45 to 65 percent recommendation.

2. 350 g carbohydrate ÷ 2 = 175 g carbohydrate/day.

   175 g carbohydrate × 4 kcal/g = 700 kcal from carbohydrate.

   700 kcal from carbohydrate ÷ 2000 total kcal/day = 0.35.

   $0.35 \times 100 = 35\%$ kcal from carbohydrate.

   This carbohydrate intake is lower than the Daily Value and lower than the 45 to 65 percent recommendation.

3. 350 g carbohydrate × 2 = 700 g carbohydrate/day.

   700 g carbohydrate × 4 kcal/g = 2800 kcal from carbohydrate.

   2800 kcal from carbohydrate ÷ 6000 total kcal/day = 0.466 (rounded to 0.47)

   $0.47 \times 100 = 47\%$ kcal from carbohydrate.

This carbohydrate intake is higher than the Daily Value and meets the 45 to 65 percent recommendation.

4. 150 g carbohydrate × 4 kcal/g = 600 kcal from carbohydrate.

   600 kcal from carbohydrate ÷ 1000 total kcal/day = 0.60.

   $0.60 \times 100 = 60\%$ kcal from carbohydrate.

This carbohydrate intake is lower than the Daily Value and meets the 45 to 65 percent recommendation.

Study Questions (multiple choice)

1. b   2. c   3. c   4. d   5. d
6. b   7. d   8. d   9. a   10. d

HIGHLIGHT

Alternatives to Sugar

Funnette Division, Hoechst Celanese Corp.

Almost everyone finds sweet tastes pleasing—after all, a preference for sweets is inborn. To a child's taste, the sweeter the food, the better. In adults, the preference for sweets is somewhat diminished, but most still enjoy an occasional sweet food or beverage. Facing the health concerns of overweight and obesity, many consumers turn to alternative sweeteners to help them control kcalories and limit their use of sugar. In doing so, they encounter two sets of alternative sweeteners. One set, the **artificial sweeteners**, provide virtually no energy and are sometimes referred to as nonnutritive sweeteners. The other set, the **sugar replacers**, yield energy and are sometimes referred to as **nutritive sweeteners.**

Artificial Sweeteners

Artificial sweeteners permit people to keep their sugar and energy intakes down, yet still enjoy the delicious sweet tastes of their favorite foods and beverages. The Food and Drug Administration (FDA) has approved the use of several artificial sweeteners—saccharin, aspartame, acesulfame potassium (acesulfame-K), sucralose, and neotame. Two others have petitioned the FDA and are awaiting approval—alitame and cyclamate. Table H4-1 and the glossary below provide general details about each of these sweeteners.

Saccharin, acesulfame-K, and sucralose are not metabolized in the body; in contrast, the body digests aspartame as a protein. In fact, aspartame is *technically* classified as a nutritive sweetener because it yields energy (4 kcalories per gram, as does protein). But because so little is used, its energy contribution is negligible.

Some consumers have challenged the safety of using artificial sweeteners. Considering that all substances are toxic at

GLOSSARY

Acceptable Daily Intake (ADI): the estimated amount of a sweetener that individuals can safely consume each day over the course of a lifetime without adverse effect.

acesulfame (AY-sul-fame) potassium: an artificial sweetener composed of an organic salt that has been approved for use in both the United States and Canada; also known as **acesulfame-K** because K is the chemical symbol for potassium.

alitame (AL-ih-tame): an artificial sweetener composed of two amino acids (alanine and aspartic acid); FDA approval pending.

artificial sweeteners: sugar substitutes that provide negligible, if any, energy; sometimes called **nonnutritive sweeteners.**

aspartame (ah-SPAR-tame or ASS-par-tame): an artificial sweetener composed of two amino acids (phenylalanine and aspartic acid); approved for use in both the United States and Canada.

cyclamate (SIGH-kla-mate): an artificial sweetener that is being considered for approval in the United States and is available in Canada as a tabletop sweetener, but not as an additive.

neotame (NEE-oh-tame): an artificial sweetener composed of two amino acids (phenylalanine and aspartic acid); approved for use in the United States.

nutritive sweeteners: sweeteners that yield energy, including both sugars and sugar replacers.

saccharin (SAK-ah-ren): an artificial sweetener that has

been approved for use in the United States. In Canada, approval for use in foods and beverages is pending; currently available only in pharmacies and only as a tabletop sweetener, not as an additive.

stevia (STEE-vee-ah): a South American shrub whose leaves are used as a sweetener; sold in the United States as a dietary supplement that provides sweetness without kcalories.

sucralose (SUE-kra-lose): an artificial sweetener approved for use in the United States and Canada.

sugar replacers: sugarlike compounds that can be derived from fruits or commercially produced from dextrose; also called **sugar alcohols** or **polyols.** Sugar alcohols are absorbed more slowly than

other sugars and metabolized differently in the human body; they are not readily utilized by ordinary mouth bacteria. Examples are **maltitol, mannitol, sorbitol, xylitol, isomalt,** and **lactitol.**

tagatose (TAG-ah-tose): a monosaccharide structurally similar to fructose that is incompletely absorbed and thus provides only 1.5 kcalories per gram; approved for use as a generally recognized as safe ingredient.

TABLE H4-1   Sweeteners

Sweeteners	Relative Sweetness[a]	Energy (kcal/g)	Acceptable Daily Intake	Average Amount to Replace 1 tsp Sugar	Approved Uses
Approved Sweeteners					
Saccharin	450	0	5 mg/kg body weight	12 mg	Tabletop sweeteners, wide range of foods, beverages, cosmetics, and pharmaceutical products
Aspartame	200	4[b]	50 mg/kg body weight[c] Warning to people with PKU: Contains phenylalanine	18 mg	General purpose sweetener in all foods and beverages
Acesulfame-K	200	0	15 mg/kg body weight[d]	25 mg	Tabletop sweeteners, puddings, gelatins, chewing gum, candies, baked goods, desserts, alcoholic beverages
Sucralose	600	0	5 mg/kg body weight	6 mg	Carbonated beverages, dairy products, baked goods, coffee and tea, fruit spreads, syrups, tabletop sweeteners, chewing gum, frozen desserts, salad dressing
Neotame	8000	0	18 mg/day	0.5μg	Baked goods, nonalcoholic beverages, chewing gum, candies, frostings, frozen desserts, gelatins, puddings, jams and jellies, syrups
Tagatose	0.8	1.5	7.5 g/day	1 tsp	Baked goods, beverages, cereals, chewing gum, confections, dairy products, dietary supplements, health bars, tabletop sweetener
Sweeteners with Approval Pending					**Proposed Uses**
Alitame	2000	4[e]	—		Beverages, baked goods, tabletop sweeteners, frozen desserts
Cyclamate	30	0	—		Tabletop sweeteners, baked goods

[a] Relative sweetness is determined by comparing the approximate sweetness of a sugar substitute with the sweetness of pure sucrose, which has been defined as 1.0. Chemical structure, temperature, acidity, and other flavors of the foods in which the substance occurs all influence relative sweetness.

[b] Aspartame provides 4 kcalories per gram, as does protein, but because so little is used, its energy contribution is negligible. In powdered form it is sometimes mixed with lactose, however, so a 1-gram packet may provide 4 kcalories.

[c] Recommendations from the World Health Organization and in Europe and Canada limit aspartame intake to 40 milligrams per kilogram of body weight.

[d] Recommendations from the World Health Organization limit acesulfame-K intake to 9 milligrams per kilogram of body weight.

[e] Alitame provides 4 kcalories per gram, as does protein, but because so little is used, its energy contribution is negligible.

some dose, it is little surprise that large doses of artificial sweeteners (or their components or metabolic by-products) have toxic effects. The question to ask is whether their ingestion is safe for human beings in quantities people normally use (and potentially abuse). The answer is yes, except in the special case described later for aspartame.

Saccharin

Saccharin, used for over 100 years in the United States, is currently used by some 50 million people—primarily in soft drinks, secondarily as a tabletop sweetener. Saccharin is rapidly excreted in the urine and does not accumulate in the body.

Questions about saccharin's safety surfaced in 1977, when experiments suggested that large doses of saccharin (equivalent to hundreds of cans of diet soda daily for a lifetime) increased the risk of bladder cancer in rats. The FDA proposed banning saccharin as a result. Public outcry in favor of saccharin was so loud, however, that Congress imposed a moratorium

on the ban—a moratorium that was repeatedly extended until 1991, when the FDA withdrew its proposal to ban saccharin. Products containing saccharin were required to carry a warning label: "Use of this product may be hazardous to your health. This product contains saccharin, which has been determined to cause cancer in laboratory animals."

Does saccharin cause cancer? The largest population study to date, involving 9000 men and women, showed that overall saccharin use did not increase the risk of cancer. Among certain small groups of the population, however, such as those who both smoked heavily and used saccharin, the risk of bladder cancer was slightly greater. Other studies involving more than 5000 people with bladder cancer showed no association between bladder cancer and saccharin use.[1] In 2000, saccharin was removed from the list of suspected cancer-causing substances. Warning labels are no longer required.

Common sense dictates that consuming large amounts of any substance is probably not wise, but at current, moderate intake levels, saccharin appears to be safe for most people. It has been approved for use in more than 100 countries.

FIGURE H4-1   Structure of Aspartame

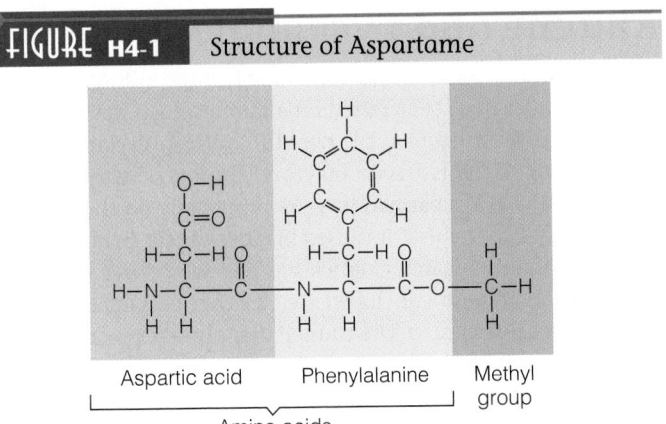

Aspartic acid   Phenylalanine   Methyl group

Amino acids

FIGURE H4-2   Metabolism of Aspartame

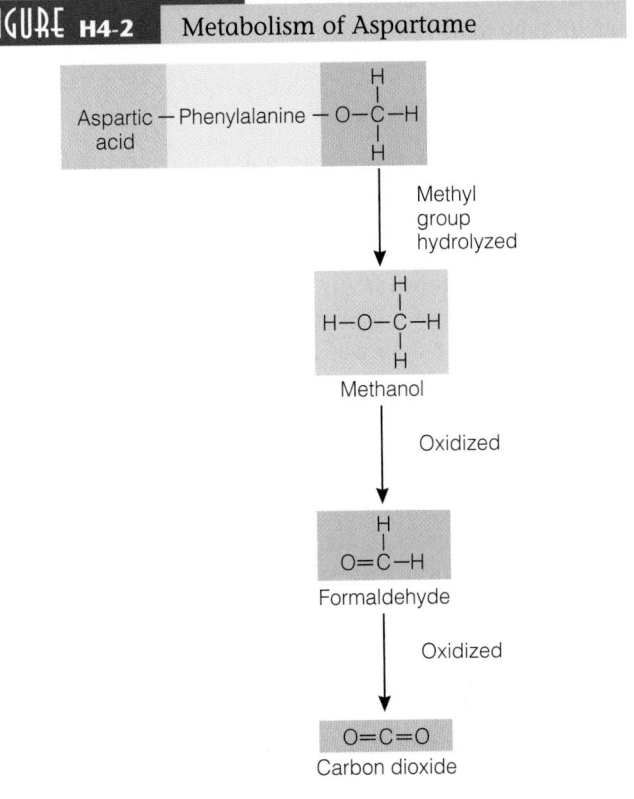

Aspartame

Aspartame is one of the most studied of all food additives; extensive animal and human studies document its safety. Long-term consumption of aspartame is not associated with any adverse health effects.

The nutrients in aspartame may present a problem for certain people, however, and for this reason, aspartame carries a warning on its label. Aspartame is a simple chemical compound made of components common to a many foods: two amino acids (phenylalanine and aspartic acid) and a methyl group (CH_3). Figure H4-1 shows its chemical structure. The flavors of the components give no clue to the combined effect; one of them tastes bitter, and the other is tasteless, but the combination creates a product that is 200 times sweeter than sucrose.

In the digestive tract, enzymes split aspartame into its three component parts. The body absorbs the two amino acids and uses them just as if they had come from food protein, which is made entirely of amino acids including these two.

Because this sweetener contributes phenylalanine, products containing aspartame must bear a warning label for people with the inherited disease phenylketonuria (PKU). People with PKU are unable to dispose of any excess phenylalanine. The accumulation of phenylalanine and its by-products is toxic to the developing nervous system, causing irreversible brain damage. For this reason, all newborns in the United States are screened for PKU. The treatment for PKU is a special diet that must strike a balance, providing enough phenylalanine to support normal growth and health but not enough to cause harm. The question is, does aspartame raise blood phenylalanine high enough to be toxic to people with PKU? Apparently not. The little extra phenylalanine from aspartame poses only a small risk, even in heavy users.

Still, there is a compelling reason why children with PKU need to get all their phenylalanine from foods, and not from an artificial sweetener. The PKU diet excludes such protein- and nutrient-rich foods as milk, meat, fish, poultry, cheese, eggs, nuts, legumes, and many bread products. Only with difficulty can these children obtain the many essential nutrients—such as calcium, iron, and the B vitamins—found along with phenylalanine in these foods. To suggest that children with PKU squander any of their limited phenylalanine allowance on the purified phenylalanine of aspartame, which contributes none of the associated vitamins or minerals essential for good health and normal growth, would open the way for poor nutrition.

Setting aside the special case of PKU, is there any reason to be concerned about the products aspartame yields in the body? During metabolism, the methyl group momentarily becomes methyl alcohol (methanol)—a potentially toxic compound (see Figure H4-2). This breakdown also occurs when aspartame-sweetened beverages are stored at warm temperatures over time. The amount of methanol produced may be safe to consume, but a person may not want to, considering that the beverage has lost its sweetness. In the body, enzymes convert methanol to formaldehyde, another toxic compound. Finally, formaldehyde is broken down to carbon dioxide. Before aspartame could be approved, the quantities of these products generated during metabolism had to be determined; they were found to fall below the threshold at which they would cause harm. In fact, ounce for ounce, tomato juice yields six times as much methanol as a diet soda.

In conclusion, except for people with PKU, aspartame is safe. Some individuals may exhibit vague, but not dangerous, symptoms due to unusual sensitivity to aspartame, but it is generally safe. Like saccharin, aspartame has been approved for use in more than 100 countries.

Acesulfame-K

The FDA approved **acesulfame potassium (acesulfame-K)** in 1988 after reviewing more than 90 safety studies conducted over 15 years. Some consumer groups believe that acesulfame-K causes tumors in rats and should not have been approved. The FDA counters that the tumors were not caused by the sweetener, but were typical of those commonly found in rat studies. Acesulfame-K has been approved for use in more than 60 countries.

Sucralose

Sucralose received FDA approval in 1998 after a review of over 110 safety studies conducted on both animals and human beings. Sucralose is unique among the artificial sweeteners in that it is made from sugar that has had three of its hydroxyl (OH) groups replaced by chlorine atoms. The result is an exceptionally stable molecule that is 600 times sweeter than sugar. Because the body does not recognize sucralose as a carbohydrate, it passes through the GI tract undigested and unabsorbed.

Neotame

Neotame is the most recent artificial sweetener to hit the markets. The FDA approved neotame in 2002 after reviewing more than 110 safety studies conducted on both animals and human beings. Neotame is so intensely sweet—about 8000 times sweeter than sugar—that very little is needed.

Like aspartame, neotame also contains the amino acids phenylalanine and aspartic acid and a methyl group. Unlike aspartame, however, neotame has an additional side group attached. This simple difference makes all the difference to people with PKU because it blocks the digestive enzymes that normally separate phenylalanine and aspartic acid. Consequently, the amino acids are not absorbed and neotame need not carry a warning for people with PKU.

Tagatose

The FDA recently granted the fructose relative **tagatose** the status of generally recognized as safe, making it available as a low-kcalorie sweetener for a variety of foods and beverages. This monosaccharide is naturally found in only a few foods, but it can be derived from lactose. Unlike fructose or lactose, however, 80 percent of tagatose remains unabsorbed until it reaches the large intestine. There, bacteria ferment tagatose, releasing gases and short chain fatty acids that are absorbed. As a result, tagatose provides only 1.5 kcalories per gram. At high doses, tagatose causes flatulence, rumbling, and loose stools; otherwise, no adverse side effects have been noted. Unlike other sugars, tagatose does not promote dental caries and may carry a dental caries health claim.

Alitame and Cyclamate

FDA approval for **alitame** and **cyclamate** is still pending. To date, no safety issues have been raised for alitame, and it has been approved for use in other countries. In contrast, cyclamate has been battling safety issues for 50 years. Approved by the FDA in 1949, cyclamate was banned in 1969 principally on the basis of one study indicating that it caused bladder cancer in rats.

The National Research Council has reviewed dozens of studies on cyclamate and concluded that neither cyclamate nor its metabolites cause cancer. The council did, however, recommend further research to determine if heavy or long-term use poses risks. Although cyclamate does not *initiate* cancer, it may *promote* cancer development once it is started. The FDA currently has no policy on substances that enhance the cancer-causing activities of other substances, but it is unlikely to approve cyclamate soon, if at all. Agencies in more than 50 other countries, including Canada, have approved cyclamate.

Stevia—An Herbal Alternative

The FDA has backed its approval or denial of artificial sweeteners with decades of extensive research. Such research is lacking for the herb **stevia**, a shrub whose leaves have long been used by the people of South America to sweeten their beverages. In the United States, stevia is sold in health-food stores as a dietary supplement. The FDA has reviewed the limited research on the use of stevia as an alternative to artificial sweeteners and found concerns regarding its effect on reproduction, cancer development, and energy metabolism. Used sparingly, stevia may do little harm, but the FDA could not approve its extensive and widespread use in the U.S. market. Canada, the European Union, and the United Nations have reached similar conclusions. That stevia can be sold as a dietary supplement, but not used as a food additive in the United States, highlights key differences in FDA regulations. Food additives must prove their safety and effectiveness before receiving FDA approval, whereas dietary supplements are not required to submit to any testing or receive any approval. (See Highlight 10 for information on dietary supplements and Highlight 18 for more on herbs.)

Acceptable Daily Intake

The amount of artificial sweetener considered safe for daily use is called the **Acceptable Daily Intake (ADI)**. The ADI represents the level of consumption that, if maintained every day throughout a person's life, would still be considered safe by a wide margin.

For example, the ADI for aspartame is 50 milligrams per kilogram of body weight. That is, the FDA approved aspartame based on the assumption that no one would consume more than 50 milligrams per kilogram of body weight in a day. This maximum daily intake is indeed a lot: for a 150-pound adult, it adds up to 97 packets of Equal or 20 cans of soft drinks sweetened only with aspartame. The company that produces aspartame estimates that if all the sugar and saccharin in the U.S. diet were replaced with

TABLE H4-2 Average Aspartame Contents of Selected Foods

Food	Aspartame (mg)
12 oz diet soft drink	170
8 oz powdered drink	100
8 oz sugar-free fruit yogurt	124
4 oz gelatin dessert	80
1 packet sweetener	35

TABLE H4-3 Sugar Replacers

Sugar Alcohols	Relative Sweetness[a]	Energy (kcal/g)	Approved Uses
Isomalt	0.5	2.0	Candies, chewing gum, ice cream, jams and jellies, frostings, beverages, baked goods
Lactitol	0.4	2.0	Candies, chewing gum, frozen dairy desserts, jams and jellies, frostings, baked goods
Maltitol	0.9	2.1	Particularly good for candy coating
Mannitol	0.7	1.6	Bulking agent, chewing gum
Sorbitol	0.5	2.6	Special dietary foods, candies, gums
Xylitol	1.0	2.4	Chewing gum, candies, pharmaceutical and oral health products

[a] Relative sweetness is determined by comparing the approximate sweetness of a sugar replacer with the sweetness of pure sucrose, which has been defined as 1.0. Chemical structure, temperature, acidity, and other flavors of the foods in which the substance occurs all influence relative sweetness.

aspartame, 1 percent of the population would be consuming the FDA maximum. Most people who use aspartame consume less than 5 milligrams per kilogram of body weight per day. But a young child who drinks four glasses of aspartame-sweetened beverages on a hot day and has five servings of other products with aspartame that day (such as pudding, chewing gum, cereal, gelatin, and frozen desserts) takes in the FDA maximum level. Although this presents no proven hazard, it seems wise to offer children other foods so as not to exceed the limit. Table H4-2 lists the average amounts of aspartame in some common foods.

For persons choosing to use artificial sweeteners, the American Dietetic Association wisely advises that they be used in moderation and only as part of a well-balanced nutritious diet.[2] The dietary principles of both moderation and variety help to reduce the possible risks associated with any food.

Artificial Sweeteners and Weight Control

The rate of obesity in the United States has been rising for decades. Foods and beverages sweetened with artificial sweeteners were among the first products developed to help people control their weight. Ironically, a few studies have reported that intense sweeteners, such as aspartame, may stimulate appetite, which could lead to weight gain.[3] Contradicting these reports, most studies find no change in feelings of hunger and no change in food intakes or body weight.[4] Adding to the confusion, some studies report lower energy intakes and greater weight losses when people eat or drink artificially sweetened products.[5]

When studying the effects of artificial sweeteners on food intake and body weight, researchers ask different questions and take different approaches. It matters, for example, whether the people used in a study are of a healthy weight and whether they are following a weight-loss diet. Motivations for using sweeteners differ, too, and this influences a person's actions. For example, one person might drink an artificially sweetened beverage now so as to be able to eat a high-kcalorie food later. This person's energy intake might stay the same or increase. Another person trying to control food energy intake might drink an artificially sweetened beverage now and then choose a low-kcalorie food later. This plan would help reduce the person's energy intake.

In designing experiments on artificial sweeteners, researchers have to distinguish between the effects of sweetness and the effects of a particular substance. If a person is hungry shortly after eating an artificially sweetened snack, is that because the sweet taste (of all sweeteners, including sugars) stimulates appetite? Or is it because the artificial sweetener itself stimulates appetite? Research must also distinguish between the effects of food energy and the effects of the substance. If a person is hungry shortly after eating an artificially sweetened snack, is that because less food energy was available to satisfy hunger? Or is it because the artificial sweetener itself triggers hunger? Furthermore, if appetite is stimulated and a person feels hungry, does that actually lead to increased food intake?

Whether a person compensates for the energy reduction of artificial sweeteners either partially or fully depends on several factors. Using artificial sweeteners will not automatically lower energy intake; to control energy intake successfully, a person needs to make informed diet and activity decisions throughout the day (as Chapter 9 explains).

Sugar Replacers

Some "sugar-free" or reduced-kcalorie products contain sugar replacers.* The term *sugar replacers* describes the sugar alcohols—familiar examples include mannitol, sorbitol, xylitol,

---

*To minimize confusion, the American Diabetes Association prefers the term *sugar replacers* instead of "sugar alcohols" (which connotes alcohol), "bulk sweeteners" (which connotes fiber), or "sugar substitutes" (which connotes aspartame and saccharin).

FIGURE H4-3   Sugar Alternatives on Food Labels

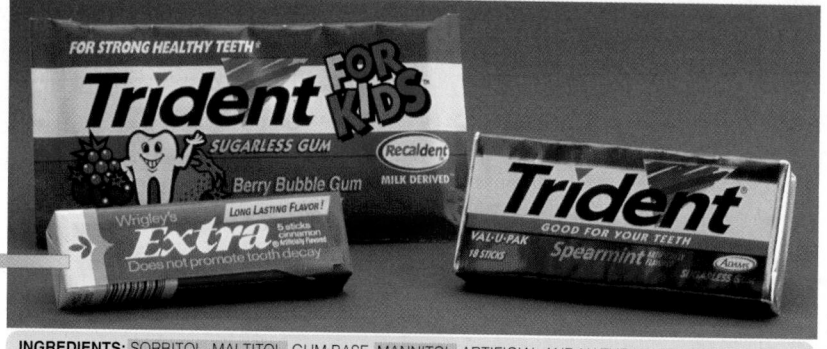

Products containing sugar replacers may claim to "not promote tooth decay" if they meet FDA criteria for dental plaque activity.

Products containing aspartame must carry a warning for people with phenylketonuria.

This ingredient list includes both sugar alcohols and artificial sweetenters.

INGREDIENTS: SORBITOL, MALTITOL, GUM BASE, MANNITOL, ARTIFICIAL AND NATURAL FLAVORING, ACACIA, SOFTENERS, TITANIUM DIOXIDE (COLOR), ASPARTAME, ACESULFAME POTASSIUM AND CANDELILLA WAX.
PHENYLKETONURICS: CONTAINS PHENYLALANINE.

35% FEWER CALORIES THAN SUGARED GUM.

Products containing less than 0.5 g of sugar per serving can claim to be "sugarless" or "sugar-free."

Products that claim to be "reduced kcalories" must provide at least 25% fewer kcalories per serving than the comparison item.

Nutrition Facts

Serving Size 2 pieces (3g)
Servings 6
Calories 5

Amount per serving	% DV*
Total Fat 0g	0%
Sodium 0mg	0%
Total Carb. 2g	1%
Sugars 0g	
Sugar Alcohol 2g	
Protein 0g	

*Percent Daily Values (DV) are based on a 2,000 calorie diet.

Not a significant source of other nutrients.

© Craig Moore

---

maltitol, isomalt, and lactitol—that provide bulk and sweetness in cookies, hard candies, sugarless gums, jams, and jellies. These products claim to be "sugar-free" on their labels, but in this case, "sugar-free" does not mean free of kcalories. Sugar replacers do provide kcalories, but fewer than their carbohydrate cousins, the sugars. Table H4-3 includes their energy values, but a simple estimate can help consumers: divide grams by 2.[6] Sugar alcohols occur naturally in fruits and vegetables; they are also used by manufacturers as a low-energy bulk ingredient in many products.

Sugar alcohols evoke a low glycemic response. The body absorbs sugar alcohols slowly; consequently, they are slower to enter the bloodstream than other sugars. Side effects such as gas, abdominal discomfort, and diarrhea, however, make them less attractive than the artificial sweeteners. For this reason, regulations require food labels to state that "Excess consumption may

have a laxative effect" if reasonable consumption of that food could result in the daily ingestion of 50 grams of a sugar alcohol.

The real benefit of using sugar replacers is that they do not contribute to dental caries. Bacteria in the mouth cannot metabolize sugar alcohols as rapidly as sugar. They are therefore valuable in chewing gums, breath mints, and other products that people keep in their mouths for a while. Figure H4-3 presents labeling information for products using alternatives to sugar.

The sugar replacers, like the artificial sweeteners, can occupy a place in the diet, and provided they are used in moderation, they will do no harm. In fact, they can help, both by providing an alternative to sugar for people with diabetes and by inhibiting caries-causing bacteria. People may find it appropriate to use all three sweeteners at times: artificial sweeteners, sugar replacers, and sugar itself.

NUTRITION ON THE NET

 Access these websites for further study of topics covered in this highlight.

- Find updates and quick links to these and other nutrition-related sites at our website:
  www.wadsworth.com/nutrition

- Search for "artificial sweeteners" at the U.S. Government health information site: **www.healthfinder.gov**

- Search for "sweeteners" at the International Food Information Council site: **www.ific.org**

REFERENCES

1. Position of the American Dietetic Association: Use of nutritive and nonnutritive sweeteners, *Journal of the American Dietetic Association* 98 (1998): 580–587.
2. Position of the American Dietetic Association, 1998.
3. J. E. Blundell and P. J. Rogers, Sweet carbohydrate substitutes (intense sweeteners) and the control of appetite: Scientific issues, in *Appetites and Body Weight Regulation: Sugar, Fat, and Macronutrient Substitutes,* ed. J. D. Fernstrom and G. D. Miller (Boca Raton, Fla,: CRC Press, 1994), pp. 113–124.
4. S. J. Gatenby and coauthors, Extended use of foods modified in fat and sugar content: Nutrition implications in a free-living female population, *American Journal of Clinical Nutrition* 65 (1997): 1867–1873; A. Drewnowski, Intense sweeteners and the control of appetite, *Nutrition Reviews* 53 (1995): 1–7.
5. A. Raben and coauthors, Sucrose compared with artificial sweeteners: Different effects on ad libitum food intake and body weight after 10 wk of supplementation in overweight subjects, *American Journal of Clinical Nutrition* 76 (2002): 721–729; G. L. Blackburn and coauthors, The effect of aspartame as part of a multidisciplinary weight-control program on short- and long-term control of body weight, *American Journal of Clinical Nutrition* 65 (1997): 409–418.
6. K. McNutt, What clients need to know about sugar replacers, *Journal of the American Dietetic Association* 100 (2000): 466–469.

The Lipids: Triglycerides, Phospholipids, and Sterols

Nutrition in Your Life

Most likely, you know what you don't like about body fat, but do you appreciate how it insulates you against the cold or powers your hike around a lake? And what about food fat? You're right to thank fat for providing the delicious flavors and aromas of buttered popcorn and fried chicken—and to curse it for contributing to the weight gain and heart disease so common today. The challenge is to strike a healthy balance of enjoying some fat, but not too much. Learning which kinds of fats are most harmful will also serve you well.

Most people are surprised to learn that fat has some virtues. Only when people consume either too much or too little fat, or too much of some kinds of fat, does ill health follow. It is true, though, that in our society of abundance, people are likely to encounter too much fat.

Fat refers to the class of nutrients known as **lipids**. The lipid family includes triglycerides (**fats** and **oils**), phospholipids, and sterols. The triglycerides■ predominate, both in foods and in the body.

■ Of the lipids in foods, 95% are fats and oils (triglycerides); of the lipids stored in the body, 99% are triglycerides.

The Chemist's View of Fatty Acids and Triglycerides

Like carbohydrates, fatty acids and triglycerides are composed of carbon (C), hydrogen (H), and oxygen (O). These lipids have many more carbons and hydrogens in proportion to their oxygens, however, and so can supply more energy per gram (Chapter 7 provides details).

The many names and relationships in the lipid family can seem overwhelming—like meeting a friend's extended family for the first time. To ease the introductions,

lipids: a family of compounds that includes triglycerides, phospholipids, and sterols. Lipids are characterized by their insolubility in water. (Lipids also include the fat-soluble vitamins, described in Chapter 11.)

fats: lipids that are solid at room temperature (70°F or 25°C).

oils: lipids that are liquid at room temperature (70°F or 25°C).

141

this chapter first presents each of the lipids from a chemist's point of view using both words and diagrams. Then the chapter follows the lipids through digestion and absorption and into the body to examine their roles in health and disease. For people who think more easily in words than in chemical symbols, this *preview* of the upcoming chemistry may be helpful:

1. Every triglyceride contains one molecule of glycerol and three fatty acids (basically, chains of carbon atoms).

2. Fatty acids may be 4 to 24 (even numbers of) carbons long, the 18-carbon ones being the most common in foods and especially noteworthy in nutrition.

3. Fatty acids may be saturated or unsaturated. Unsaturated fatty acids may have one or more points of unsaturation (that is, they may be monounsaturated or polyunsaturated).

4. Of special importance in nutrition are the polyunsaturated fatty acids whose *first* point of unsaturation is next to the third carbon (known as omega-3 fatty acids) or next to the sixth carbon (omega-6).

5. The 18-carbon fatty acids that fit this description are linolenic acid (omega-3) and linoleic acid (omega-6). Each is the primary member of a family of longer-chain fatty acids that help to regulate blood pressure, blood clotting, and other body functions important to health.

The paragraphs, definitions, and diagrams that follow present this information again in much more detail.

Fatty Acids

A **fatty acid** is an organic acid—a chain of carbon atoms with hydrogens attached—that has an acid group (COOH) at one end and a methyl group (CH_3) at the other end. The organic acid shown in Figure 5-1 is acetic acid, the compound that gives vinegar its sour taste. Acetic acid is the shortest such acid, with a "chain" only two carbon atoms long.

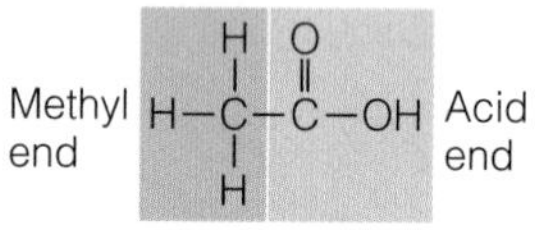

The Length of the Carbon Chain Most naturally occurring fatty acids contain even numbers of carbons in their chains—up to 24 carbons in length. This discussion begins with the 18-carbon fatty acids, which are abundant in our food supply. Stearic acid is the simplest of the 18-carbon fatty acids; the bonds between its carbons are all alike:

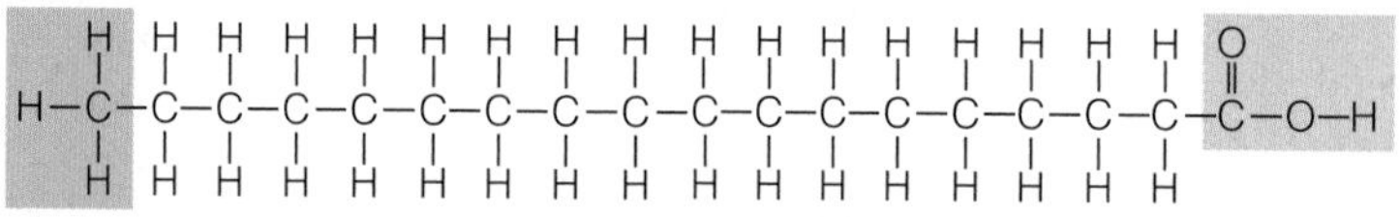

FIGURE 5-1 Acetic Acid

Acetic acid is a two-carbon organic acid.

Methyl end · Acid end

Stearic acid, an 18-carbon saturated fatty acid.

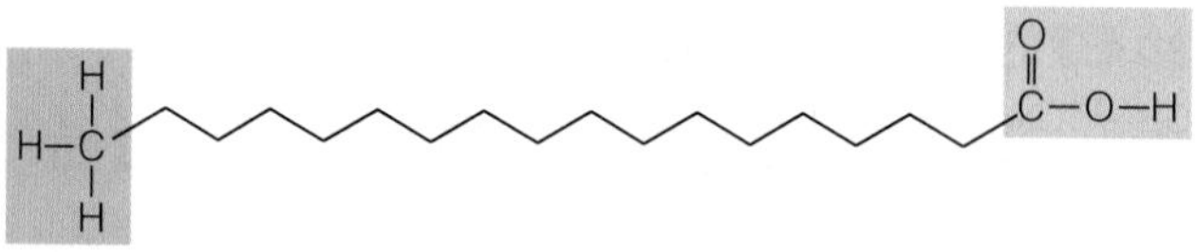

Stearic acid (simplified structure).

As you can see, stearic acid is 18 carbons long, and each atom meets the rules of chemical bonding described in Figure 4-1 on p. 104. The following structure also depicts stearic acid, but in a simpler way, with each "corner" on the zigzag line representing a carbon atom with two attached hydrogens:

As mentioned, the carbon chains of fatty acids vary in length. The long-chain (12 to 24 carbons) fatty acids of meats, fish, and vegetable oils are most common in the diet. Smaller amounts of medium-chain (6 to 10 carbons) and short-chain (fewer than 6 carbons) fatty acids also occur, primarily in dairy products. (Tables C-1 and C-2 in Appendix C provide the names, chain lengths, and sources of fatty acids commonly found in foods.)

fatty acid: an organic compound composed of a carbon chain with hydrogens attached and an acid group (COOH) at one end and a methyl group (CH_3) at the other end.

GLOSSARY OF SATURATION TERMS

These terms are listed in order from the most saturated to the most unsaturated.

saturated fatty acid: a fatty acid carrying the maximum possible number of hydrogen atoms—for example, stearic acid. A **saturated fat** is composed of triglycerides in which most of the fatty acids are saturated.

point of unsaturation: the double bond of a fatty acid,

where hydrogen atoms can easily be added to the structure.

unsaturated fatty acid: a fatty acid that lacks hydrogen atoms and has at least one double bond between carbons (includes monounsaturated and polyunsaturated fatty acids). An **unsaturated fat** is composed of triglycerides in which most of the fatty acids are unsaturated.

monounsaturated fatty acid: a fatty acid that lacks two hydrogen atoms and has one double bond between carbons—for example, oleic acid. A **monounsaturated fat** is composed of triglycerides in which most of the fatty acids are monounsaturated.
- **mono** = one

polyunsaturated fatty acid (PUFA): a fatty acid that lacks

four or more hydrogen atoms and has two or more double bonds between carbons—for example, linoleic acid (two double bonds) and linolenic acid (three double bonds). A **polyunsaturated fat** is composed of triglycerides in which most of the fatty acids are polyunsaturated.
- **poly** = many

The Degree of Unsaturation Stearic acid is a **saturated fatty acid,** (terms that describe the saturation of fatty acids are defined in the glossary above). A saturated fatty acid is fully loaded with hydrogen atoms and contains only single bonds between its carbon atoms. If two hydrogens were missing from the middle of the carbon chain, the remaining structure might be:

An impossible chemical structure.

Such a compound cannot exist, however, because two of the carbons have only three bonds each, and nature requires that every carbon have four bonds. The two carbons therefore form a double bond:

Oleic acid, an 18-carbon monounsaturated fatty acid.

The same structure drawn more simply looks like this:*

Oleic acid (simplified structure).

The double bond is a **point of unsaturation.** Hence, a fatty acid like this—with two hydrogens missing and a double bond—is an **unsaturated fatty acid.** This one is the 18-carbon *mon*ounsaturated fatty acid oleic acid, which is abundant in olive oil and canola oil.

A *poly*unsaturated fatty acid has two or more carbon-to-carbon double bonds. **Linoleic acid,** the 18-carbon fatty acid common in vegetable oils, lacks four hydrogens and has two double bonds:

Linoleic acid, an 18-carbon polyunsaturated fatty acid.

*Remember that each "corner" on the zigzag line represents a carbon atom with two attached hydrogens. In addition, although drawn straight here, the actual shape kinks at the double bonds (as shown in the left side of Figure 5-8).

linoleic (lin-oh-LAY-ick) **acid:** an essential fatty acid with 18 carbons and two double bonds.

TABLE 5-1   18-Carbon Fatty Acids

Name	Number of Carbon Atoms	Number of Double Bonds	Saturation	Common Food Sources
Stearic acid	18	0	Saturated	Most animal fats
Oleic acid	18	1	Monounsaturated	Olive, canola oils
Linoleic acid	18	2	Polyunsaturated	Sunflower, safflower, corn, and soybean oils
Linolenic acid	18	3	Polyunsaturated	Soybean and canola oils, flaxseed, walnuts

Drawn more simply, linoleic acid looks like this (though the actual shape would kink at the double bonds):

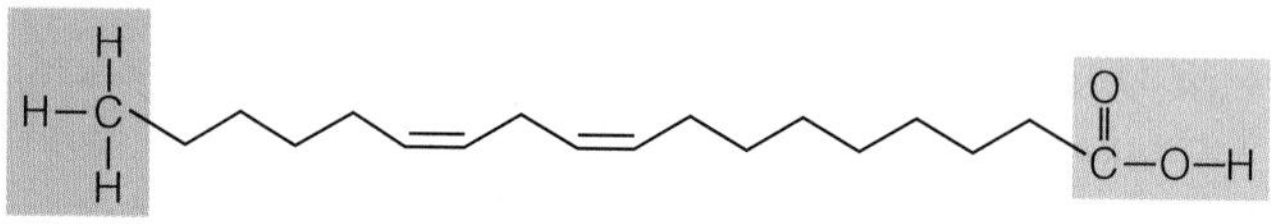

Linoleic acid (simplified structure).

A fourth 18-carbon fatty acid is **linolenic acid,** which has three double bonds. Table 5-1 presents the 18-carbon fatty acids.■

■ Chemists use a shorthand notation to describe fatty acids. The first number indicates the number of carbon atoms; the second, the number of double bonds. For example, the notation for stearic acid is 18:0.

The Location of Double Bonds Fatty acids differ not only in the length of their chains and their degree of saturation, but also in the locations of their double bonds (see Figure 5-2). Chemists identify polyunsaturated fatty acids by the position of the double bond nearest the methyl (CH_3) end of the carbon chain, which is described by an **omega** number. A polyunsaturated fatty acid with its first double bond three carbons away from the methyl end is an **omega-3 fatty acid.** Similarly, an **omega-6 fatty acid** is a polyunsaturated fatty acid with its first double bond six carbons away

FIGURE 5-2   Omega-3 and Omega-6 Fatty Acids Compared

Linolenic acid, an omega-3 fatty acid

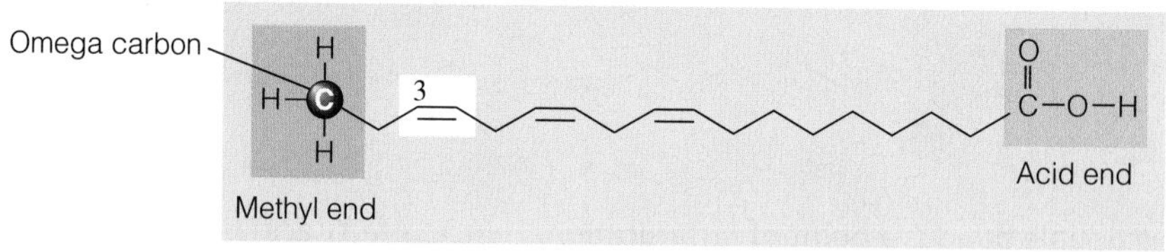

Linoleic acid, an omega-6 fatty acid

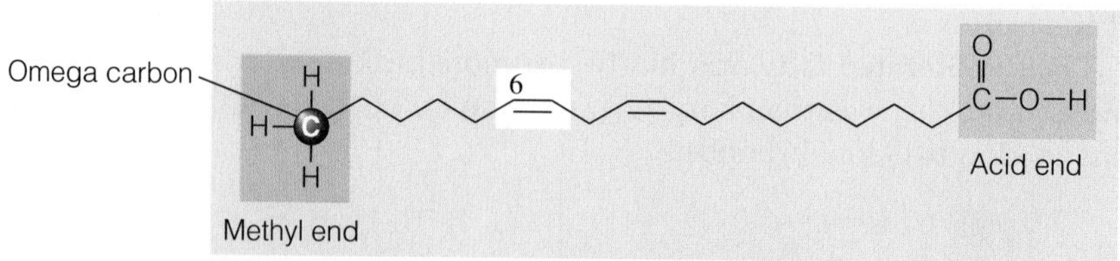

The omega number indicates the position of the first double bond in a fatty acid, counting from the methyl (CH_3) end. Thus an omega-3 fatty acid's first double bond occurs three carbons from the methyl end, and an omega-6 fatty acid's first double bond occurs six carbons from the methyl end. The members of an omega family may have different lengths and different numbers of double bonds, but the first double bond occurs at the same point in all of them. These structures are drawn linearly here to ease counting carbons and locating double bonds, but their shapes actually bend at the double bonds, as shown in Figure 5-8.

linolenic (lin-oh-LEN-ick) **acid:** an essential fatty acid with 18 carbons and three double bonds.

omega: the last letter of the Greek alphabet (ω), used by chemists to refer to the position of the first double bond from the methyl end of a fatty acid.

omega-3 fatty acid: a polyunsaturated fatty acid in which the first double bond is three carbons away from the methyl (CH_3) end of the carbon chain.

omega-6 fatty acid: a polyunsaturated fatty acid in which the first double bond is six carbons from the methyl (CH_3) end of the carbon chain.

from the methyl end. Figure 5-2 compares two 18-carbon fatty acids—linolenic acid (an omega-3 fatty acid) and linoleic acid (an omega-6 fatty acid).

Triglycerides

Few fatty acids occur free in foods or in the body. Most often, they are incorporated into **triglycerides**—lipids composed of three fatty acids attached to a **glycerol**. (Figure 5-3 presents a glycerol molecule.) To make a triglyceride, a series of condensation reactions combine a hydrogen atom (H) from the glycerol and a hydroxyl (OH) group from a fatty acid, forming a molecule of water (H_2O) and leaving a bond between the other two molecules (see Figure 5-4). Most triglycerides contain a mixture of more than one type of fatty acid (see Figure 5-5 on p. 146).

Degree of Unsaturation Revisited

The chemistry of a fatty acid—whether it is short or long, saturated or unsaturated, with its first double bond here or there—influences the characteristics of foods and the health of the body. A later section of this chapter explains how these features affect health; this section describes how the degree of unsaturation influences the fats and oils in foods.

Firmness The degree of unsaturation influences the firmness of fats at room temperature. Generally speaking, the polyunsaturated vegetable oils are liquid at room temperature, and the more saturated animal fats are solid. Not all vegetable oils are polyunsaturated, however. Cocoa butter, palm oil, palm kernel oil, and coconut oil■ are saturated even though they are of vegetable origin; they are firmer than most vegetable oils because of their saturation, but softer than most animal fats because of their short carbon chains (8 to 14 carbons long). Generally, the shorter the carbon chain, the softer the fat is at room temperature. Fatty acid compositions of selected fats and oils are shown in Figure 5-6 (on p. 146), and Appendix H provides the fat and fatty acid contents of many other foods.

Stability Saturation also influences stability. All fats can become rancid when exposed to oxygen. Polyunsaturated fats spoil most readily because their double bonds are unstable; monounsaturated fats are slightly less susceptible. Saturated

FIGURE 5-3 Glycerol

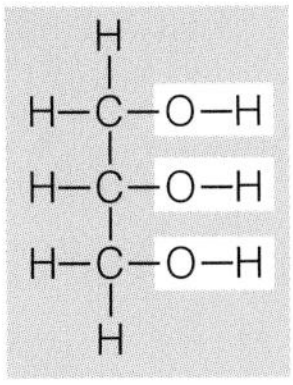

When glycerol is free, an OH group is attached to each carbon. When glycerol is part of a triglyceride, each carbon is attached to a fatty acid by a carbon-oxygen bond.

■ The food industry often refers to these saturated vegetable oils as the "tropical oils."

triglycerides (try-GLISS-er-rides): the chief form of fat in the diet and the major storage form of fat in the body; composed of a molecule of glycerol with three fatty acids attached; also called **triacylglycerols** (try-ay-seel-GLISS-er-ols).*
- **tri** = three
- **glyceride** = of glycerol
- **acyl** = a carbon chain

*Research scientists commonly use the term *triacylglycerols;* this book continues to use the more familiar term *triglycerides,* as do many other health and nutrition books and journals.

glycerol (GLISS-er-ol): an alcohol composed of a three-carbon chain, which can serve as the backbone for a triglyceride.
- **ol** = alcohol

FIGURE 5-4 Condensation of Glycerol and Fatty Acids to Form a Triglyceride

To make a triglyceride, three fatty acids attach to glycerol in condensation reactions:

Glycerol + 3 fatty acids → Triglyceride + 3 water molecules

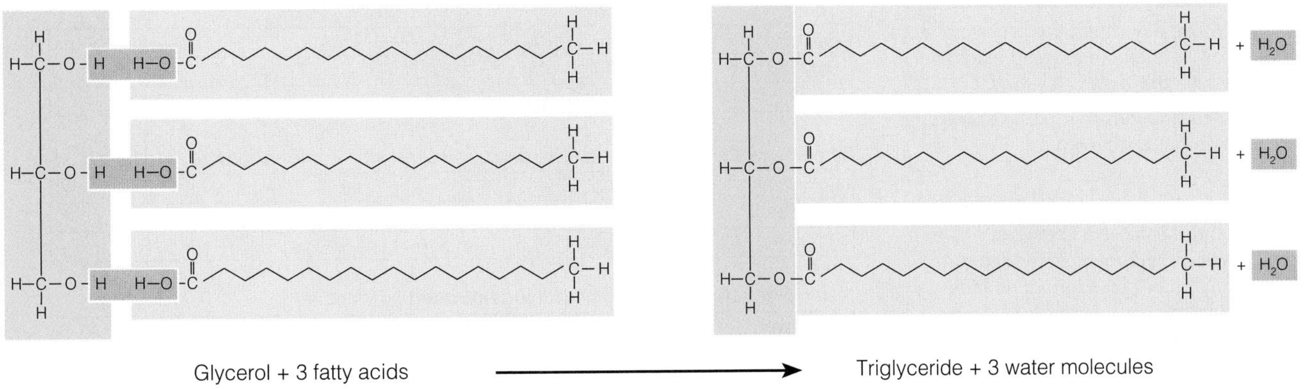

An H atom from glycerol and an OH group from a fatty acid combine to create water, leaving the O on the glycerol and the C at the acid end of each fatty acid to form a bond.

Three fatty acids attached to a glycerol form a triglyceride and yield water. In this example, all three fatty acids are stearic acid, but most often triglycerides contain mixtures of fatty acids (as shown in Figure 5-5).

FIGURE 5-5   A Mixed Triglyceride

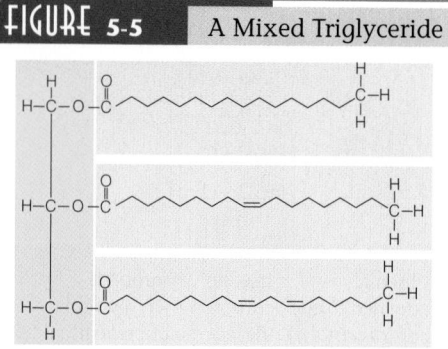

This mixed triglyceride includes a saturated fatty acid, a monounsaturated fatty acid, and a polyunsaturated fatty acid.

fats are most resistant to **oxidation** and thus least likely to become rancid. The oxidation of fats produces a variety of compounds that smell and taste rancid; other types of spoilage can occur due to microbial growth.

Manufacturers can protect fat-containing products against rancidity in three ways—none of them perfect. First, products may be sealed air-tight in nonmetallic containers, protected from light, and refrigerated—an expensive and inconvenient storage system. Second, manufacturers may add **antioxidants** to compete for the oxygen and thus protect the oil (examples are the additives BHA and BHT and vitamin E); the advantages and disadvantages of antioxidants in food processing are presented in Chapter 19. Third, manufacturers may saturate some or all of the points of unsaturation by adding hydrogen molecules—a process known as hydrogenation.

Hydrogenation   **Hydrogenation** offers two advantages. First, it protects against oxidation (thereby prolonging shelf life) by making polyunsaturated fats more saturated (see Figure 5-7). Second, it alters the texture of foods by making liquid vegetable oils more solid (as in margarine and shortening). Hydrogenated fats make margarine spreadable, pie crusts flaky, and puddings creamy.

***Trans*-Fatty Acids**   Figure 5-7 illustrates the total hydrogenation of a polyunsaturated fatty acid to a saturated fatty acid, which rarely occurs during food processing. Most often, a fat is partially hydrogenated, and some of the double bonds that remain after processing change from *cis* to *trans*. In nature, most double bonds are *cis*—meaning that the hydrogens next to the double bonds are on the same side of the carbon chain. Only a few fatty acids (notably those found in milk and meat products) are ***trans*-fatty acids**—meaning that the hydrogens next to the double bonds are on opposite sides of the carbon chain (see Figure 5-8). These arrangements result in different configurations for the fatty acids, and this difference affects function: in the body, *trans*-fatty acids behave more like saturated fats than like unsaturated fats. The relationship between *trans*-fatty acids and heart disease has been the subject of much recent research, as a later section describes.

FIGURE 5-6   Comparison of Dietary Fats

Most fats are a mixture of saturated, monounsaturated, and polyunsaturated fatty acids.

Key:
- ■ Saturated fats
- □ Monounsaturated fats
- ■ Polyunsaturated, omega-6 fats
- ■ Polyunsaturated, omega-3 fats

Animal fats and the tropical oils of coconut and palm are mostly **saturated.**

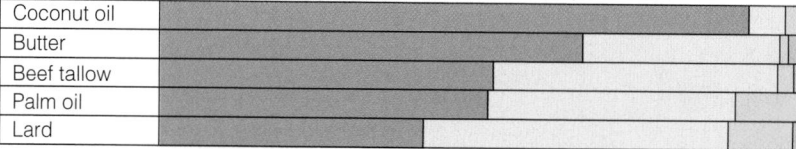

Coconut oil
Butter
Beef tallow
Palm oil
Lard

Some vegetable oils, such as olive and canola, are rich in **monounsaturated** fatty acids.

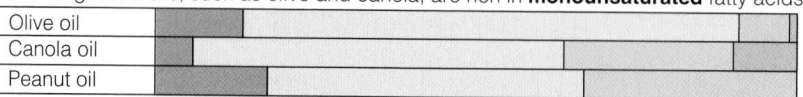

Olive oil
Canola oil
Peanut oil

Many vegetable oils are rich in **polyunsaturated** fatty acids.

Safflower oil
Sunflower oil
Corn oil
Soybean oil
Cottonseed oil

oxidation (OKS-ee-day-shun): the process of a substance combining with oxygen; oxidation reactions involve the loss of electrons.

antioxidants: compounds that protect others from oxidation by being oxidized themselves.

hydrogenation (HIGH-dro-gen-AY-shun or high-DROJ-eh-NAY-shun): a chemical process by which hydrogens are added to monounsaturated or polyunsaturated fatty acids to reduce the number of double bonds, making the fats more saturated (solid) and more resistant to oxidation (protecting against rancidity). Hydrogenation produces *trans*-fatty acids.

***trans*-fatty acids:** fatty acids with hydrogens on opposite sides of the double bond.

FIGURE 5-7 Hydrogenation

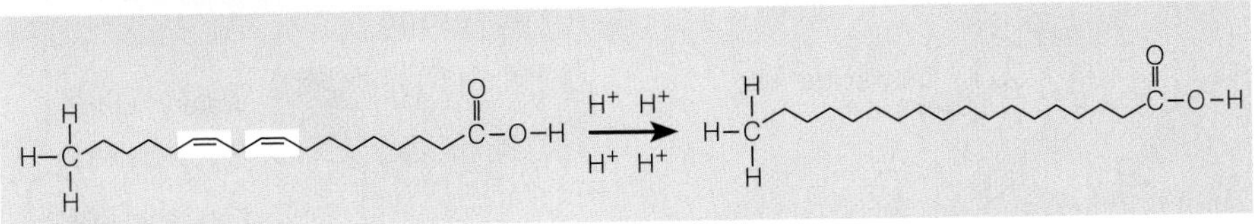

Polyunsaturated fatty acid → Hydrogenated (saturated) fatty acid

Double bonds carry a slightly negative charge and readily accept positively charged hydrogen atoms, creating a saturated fatty acid. Most often, fat is partially hydrogenated, creating a *trans*-fatty acid (shown in Figure 5-8).

At room temperature, saturated fats (such as those found in butter and other animal fats) are solid, whereas unsaturated fats (such as those found in oil) are usually liquid.

IN SUMMARY The predominant lipids both in foods and in the body are triglycerides: glycerol backbones with three fatty acids attached. Fatty acids vary in the length of their carbon chains, their degrees of unsaturation, and the location of their double bond(s). Those that are fully loaded with hydrogens are saturated; those that are missing hydrogens and therefore have double bonds are unsaturated (monounsaturated or polyunsaturated). The vast majority of triglycerides contain more than one type of fatty acid. Fatty acid saturation affects fats' physical characteristics and storage properties. Hydrogenation, which makes polyunsaturated fats more saturated, gives rise to *trans*-fatty acids, altered fatty acids that may have health effects similar to those of saturated fatty acids.

The Chemist's View of Phospholipids and Sterols

The preceding pages have been devoted to one of the three classes of lipids, the triglycerides, and their component parts, the fatty acids. The other two classes of lipids, the phospholipids and sterols, make up only 5 percent of the lipids in the diet.

FIGURE 5-8 Cis- and Trans-Fatty Acids Compared

This example shows the *cis* configuration for an 18-carbon monounsaturated fatty acid (oleic acid) and its corresponding *trans* configuration (elaidic acid).

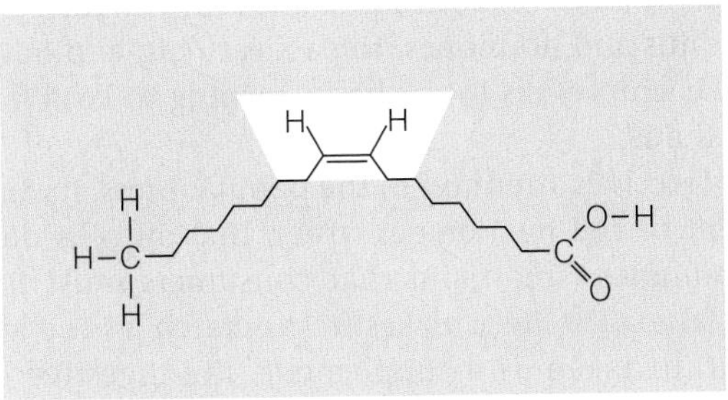

cis-fatty acid

trans-fatty acid

A *cis*-fatty acid has its hydrogens on the same side of the double bond; *cis* molecules fold back into a U-like formation. Most naturally occuring unsaturated fatty acids in foods are *cis*.

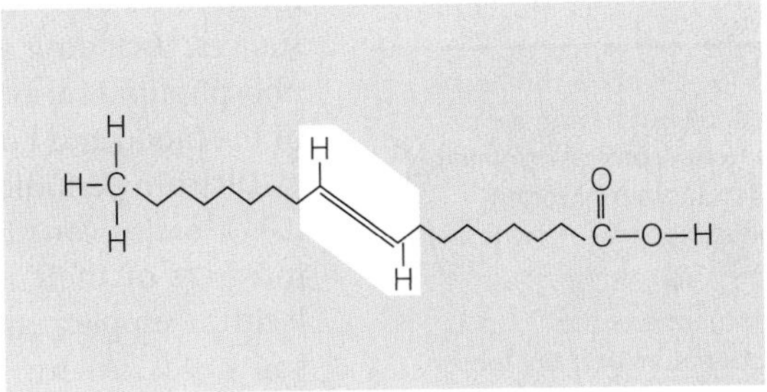

A *trans*-fatty acid has its hydrogens on the opposite sides of the double bond; *trans* molecules are more linear. The *trans* form typically occurs in partially hydrogenated foods when hydrogen atoms shift around some double bonds and change the configuration from *cis* to *trans*.

FIGURE 5-9   Lecithin

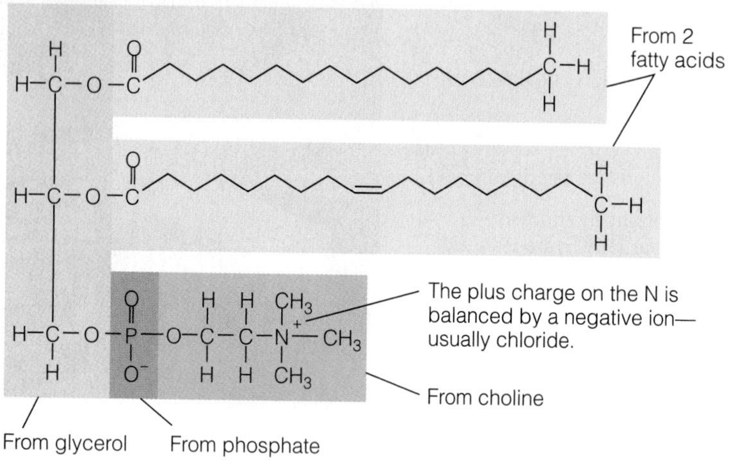

The plus charge on the N is balanced by a negative ion—usually chloride.

From 2 fatty acids

From choline

From glycerol   From phosphate

Lecithin is one of the phospholipids. Other phospholipids have different fatty acids at the upper two positions and different groups attached to phosphate. Notice that a molecule of lecithin is similar to a triglyceride but contains only two fatty acids. The third position is occupied by a phosphate group and a molecule of choline.

FIGURE 5-10   Phospholipids of a Cell Membrane

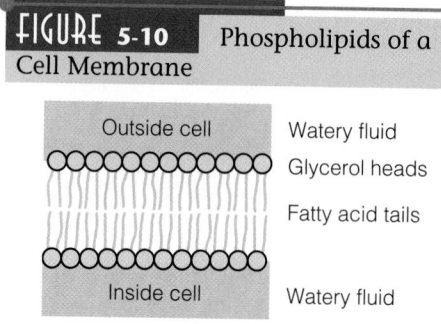

Outside cell

Watery fluid
Glycerol heads
Fatty acid tails
Watery fluid

Inside cell

A cell membrane is made of phospholipids assembled into an orderly formation called a bilayer. The fatty acid "tails" orient themselves away from the watery fluid inside and outside of the cell. The glycerol and phosphate "heads" are attracted to the watery fluid.

Phospholipids

The best-known **phospholipid** is **lecithin**. A diagram of a lecithin molecule is shown in Figure 5-9. Notice that lecithin has a backbone of glycerol with two of its three attachment sites occupied by fatty acids like those in triglycerides. The third site is occupied by a phosphate group and a molecule of **choline**. The fatty acids make phospholipids soluble in fat; the phosphate group allows them to dissolve in water. Such versatility enables the food industry to use phospholipids as emulsifiers■ to mix fats with water in such products as mayonnaise and candy bars.

Phospholipids in Foods In addition to the phospholipids used by the food industry as emulsifiers, phospholipids are also found naturally in foods. The richest food sources of lecithin are eggs, liver, soybeans, wheat germ, and peanuts.

Roles of Phospholipids The lecithins and other phospholipids are important constituents of cell membranes (see Figure 5-10). Because phospholipids are soluble in both water and fat, they can help lipids move back and forth across the cell membranes into the watery fluids on both sides. Thus they enable fat-soluble substances, including vitamins and hormones, to pass easily in and out of cells. The phospholipids also act as emulsifiers in the body, helping to keep fats suspended in the blood and body fluids.

Lecithin periodically receives attention in the popular press. Its fans claim that it is a major constituent of cell membranes (true), that all cells depend on the integrity of their membranes (true), and that consumers must therefore take lecithin supplements (false). The liver makes from scratch all the lecithin a person needs. As for lecithin taken as a supplement, the digestive enzyme lecithinase■ in the intestine hydrolyzes most of it before it passes into the body, so little lecithin reaches the tissues intact. In other words, the lecithins are *not essential nutrients*; they are just another lipid. Like other lipids, they contribute 9 kcalories per gram to the body's energy economy—an unexpected "bonus" many people taking lecithin supplements fail to realize. Furthermore, large doses of lecithin may cause GI distress, sweating, and loss of appetite. Perhaps these symp-

■ Reminder: *Emulsifiers* are substances with both water-soluble and fat-soluble portions that promote the mixing of oils and fats in watery solutions.

■ Reminder: The word ending *-ase* denotes an enzyme. Hence, lecithinase is an enzyme that works on lecithin.

phospholipid (FOS-foe-LIP-id): a compound similar to a triglyceride but having a phosphate group (a phosphorus-containing salt) and choline (or another nitrogen-containing compound) in place of one of the fatty acids.

lecithin (LESS-uh-thin): one of the phospholipids. Both nature and the food industry use lecithin as an emulsifier to combine water-soluble and fat-soluble ingredients that do not ordinarily mix, such as water and oil.

choline (KOH-leen): a nitrogen-containing compound found in foods as part of lecithin and other phospholipids.

toms are beneficial because they may warn people to stop self-dosing with lecithin.

> **IN SUMMARY** Phospholipids, including lecithin, have a unique chemical structure that allows them to be soluble in both water and fat. In the body, phospholipids are part of cell membranes; the food industry uses phospholipids as emulsifiers to mix fats with water.

Sterols

In addition to triglycerides and phospholipids, the lipids include the **sterols,** compounds with a multiple-ring structure.[*] The most famous sterol is **cholesterol;** Figure 5-11 (on p. 150) shows its chemical structure.

Sterols in Foods Foods derived from both plants and animals contain sterols, but only those from animals contain cholesterol: meats, eggs, fish, poultry, and dairy products. Some people, confused about the distinction between dietary and blood cholesterol, have asked which foods contain the "good" cholesterol. "Good" cholesterol is not a type of cholesterol found in foods, but refers to the way the body transports cholesterol in the blood, as explained later (p. 154).

Roles of Sterols Many vitally important body compounds are sterols. Among them are bile acids, the sex hormones (such as testosterone), the adrenal hormones (such as cortisol), and vitamin D, as well as cholesterol itself. Cholesterol in the body can serve as the starting material for the synthesis of these compounds or as a structural component of cell membranes; more than 90 percent of all the body's cholesterol resides in the cells. Despite popular impressions to the contrary, cholesterol is not a villain lurking in some evil foods—it is a compound the body makes and uses.■ Right now, as you read, your liver is manufacturing cholesterol from fragments of carbohydrate, protein, and fat. In fact, the liver makes about 800 to 1500 milligrams of cholesterol per day,■ thus contributing much more to the body's total than does the diet.

Cholesterol's harmful effects in the body occur when it forms deposits in the artery walls. These deposits lead to **atherosclerosis,** a disease that causes heart attacks and strokes (Chapter 18 provides many more details).

> **IN SUMMARY** Sterols have a multiple-ring structure that differs from the structure of other lipids. In the body, sterols include cholesterol, bile, vitamin D, and some hormones. Only animal-derived foods contain cholesterol. To summarize, the members of the lipid family include:
>
> - **Triglycerides** (fats and oils), which are made of:
>   - Glycerol (1 per triglyceride) and
>   - Fatty acids (3 per triglyceride). Depending on the number of double bonds, fatty acids may be:
>   - *Saturated* (no double bonds).
>   - *Monounsaturated* (one double bond).
>   - *Polyunsaturated* (more than one double bond). Depending on the location of the double bonds, polyunsaturated fatty acids may be:
>   - *Omega-3* (first double bond 3 carbons away from methyl end).
>   - *Omega-6* (first double bond 6 carbons away from methyl end).
> - **Phospholipids** (such as lecithin).
> - **Sterols** (such as cholesterol).

Without help from emulsifiers, fats and water don't mix.

Matthew Farruggio

■ The chemical structure is the same, but cholesterol that is made in the body is called **endogenous** (en-DOGDE-eh-nus), whereas cholesterol from outside the body (from foods) is called **exogenous** (eks-ODGE-eh-nus).
- **endo** = within
- **gen** = arising
- **exo** = outside (the body)

■ For perspective, the Daily Value for cholesterol is 300 mg/day.

sterols (STARE-ols or STEER-ols): compounds containing a four-carbon ring structure with any of a variety of side chains attached.

cholesterol (koh-LESS-ter-ol): one of the sterols containing a four-carbon ring structure with a carbon side chain.

atherosclerosis (ath-er-oh-scler-OH-sis): a type of artery disease characterized by accumulations of cholesterol-containing material on the inner walls of the arteries (see Chapter 18).
- **athero** = porridge or soft
- **scleros** = hard
- **osis** = condition

---

[*]The four-ring core structure identifies a steroid; sterols are alcohol derivatives with a steroid ring structure.

FIGURE 5-11 Cholesterol

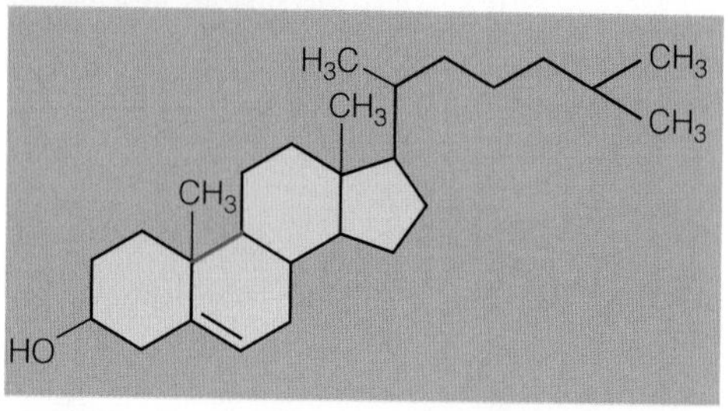

Cholesterol

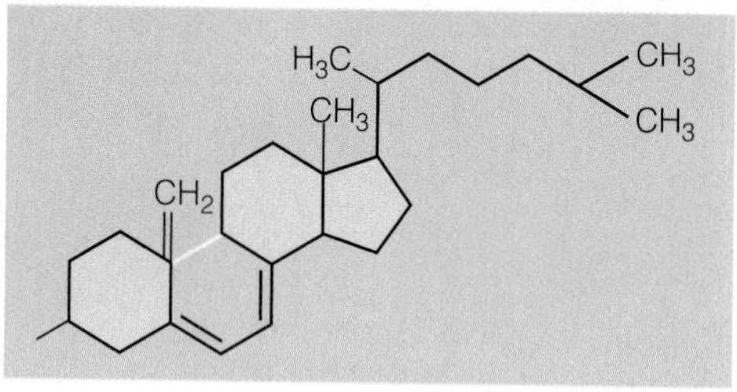

Vitamin D₃

The fat-soluble vitamin D is synthesized from cholesterol; notice the many similarities. The only difference is that cholesterol has a closed ring (highlighted in color), whereas vitamin D's is open, accounting for its vitamin activity. Notice, too, how different cholesterol is from the triglycerides and phospholipids.

■ Reminder: An enzyme that hydrolyzes lipids is called a *lipase; lingual* refers to the tongue.

■ In addition to bile acids and bile salts, bile contains cholesterol, phospholipids (especially lecithin), antibodies, water, electrolytes, and bilirubin and biliverdin (pigments resulting from the breakdown of heme).

hydrophobic (high-dro-FOE-bick): a term referring to water-fearing, or non-water-soluble, substances; also known as **lipophilic** (fat loving).
• hydro = water
• phobia = fear
• lipo = lipid
• phile = love

hydrophilic (high-dro-FIL-ick): a term referring to water-loving, or water-soluble, substances.

monoglycerides: molecules of glycerol with one fatty acid attached. A molecule of glycerol with two fatty acids attached is a **diglyceride.**
• mono = one
• di = two

Digestion, Absorption, and Transport of Lipids

Each day, the GI tract receives, on average, 50 to 100 grams of triglycerides, 4 to 8 grams of phospholipids, and 200 to 350 milligrams of cholesterol. The body faces a challenge in digesting and absorbing these lipids: getting at them. Fats are **hydrophobic**—that is, they tend to separate from the watery fluids of the GI tract—whereas the enzymes for digesting fats are **hydrophilic.** The challenge is keeping the fats mixed in the watery fluids of the GI tract.

Lipid Digestion

The goal of fat digestion is to dismantle triglycerides into small molecules that the body can absorb and use—namely, **monoglycerides,** fatty acids, and glycerol. Figure 5-12 traces the digestion of triglycerides through the GI tract, and the following paragraphs provide the details.

In the Mouth Fat digestion starts off slowly in the mouth, with some hard fats beginning to melt when they reach body temperature. A salivary gland at the base of the tongue releases an enzyme (lingual lipase)■ that plays a minor role in fat digestion in adults and an active role in infants. In infants, this enzyme efficiently digests the short- and medium-chain fatty acids found in milk.

In the Stomach In a quiet stomach, fat would float as a layer above the other components of swallowed food. But the strong muscle contractions of the stomach propel the stomach contents toward the pyloric sphincter. Some chyme passes through the pyloric sphincter periodically, but the remaining partially digested food is propelled back into the body of the stomach. This churning grinds the solid pieces to finer particles, mixes the chyme, and disperses the fat into smaller droplets. These actions help to expose the fat for attack by the gastric lipase enzyme—an enzyme that performs best in the acidic environment of the stomach.[1] Still, little fat digestion takes place in the stomach; most of the action occurs in the small intestine.

In the Small Intestine Fat in the small intestine triggers the release of the hormone cholecystokinin (CCK), which signals the gallbladder to release its stores of bile. (Remember that the liver makes bile, and the gallbladder stores it until it is needed.) Among bile's many ingredients■ are bile acids, which are made in the liver from cholesterol and have a similar structure. In addition, they often pair up with an amino acid (a building block of protein). The amino acid end is attracted to water, and the sterol end is attracted to fat (see Figure 5-13 on p. 152). This structure improves bile's ability to act as an emulsifier, drawing fat molecules into the surrounding watery fluids. There the fats are fully digested as they encounter lipase enzymes from the pancreas and small intestine. The process of emulsification is diagrammed in Figure 5-14 (on p. 152).

Most of the hydrolysis of triglycerides occurs in the small intestine. The major fat-digesting enzymes are pancreatic lipases; some intestinal lipases are also active. These enzymes remove one, then the other, of each triglyceride's outer fatty acids, leaving a monoglyceride. Occasionally, enzymes remove all three fatty acids, leaving a free molecule of glycerol. Hydrolysis of a triglyceride is shown in Figure 5-15 (on p. 153).

Phospholipids are digested similarly—that is, their fatty acids are removed by hydrolysis. The two fatty acids and the remaining phospholipid fragment are then absorbed. Most sterols can be absorbed as is; if any fatty acids are attached, they are first hydrolyzed off.

Bile's Routes After bile has entered the small intestine and emulsified fat, it has two possible destinations, illustrated in Figure 5-16 (on p. 153). Most of the bile is reabsorbed from the intestine and recycled. The other possibility is that some of the bile can be trapped by fibers in the large intestine and carried out of the body

FIGURE 5-12 Fat Digestion in the GI Tract

FAT

Mouth and salivary glands
Some hard fats begin to melt as they reach body temperature. Sublingual salivary gland in the base of the tongue secretes lingual lipase.

Stomach
The acid-stable lingual lipase initiates lipid digestion by hydrolyzing one bond of triglycerides to produce diglycerides and fatty acids. The degree of hydrolysis by lingual lipase is slight for most fats but may be appreciable for milk fats. The stomach's churning action mixes fat with water and acid. A gastric lipase accesses and hydrolyzes (only a very small amount of) fat.

Small intestine
Bile flows in from the gallbladder (via the common bile duct):

Fat $\xrightarrow{\text{bile}}$ emulsified fat

Pancreatic lipase flows in from the pancreas (via the pancreatic duct):

Emulsified fat (triglycerides) $\xrightarrow{\text{Pancreatic (and intestinal) lipase}}$ monoglycerides, glycerol, fatty acids (absorbed)

Large intestine
Some fat and cholesterol trapped in fiber, exit in feces.

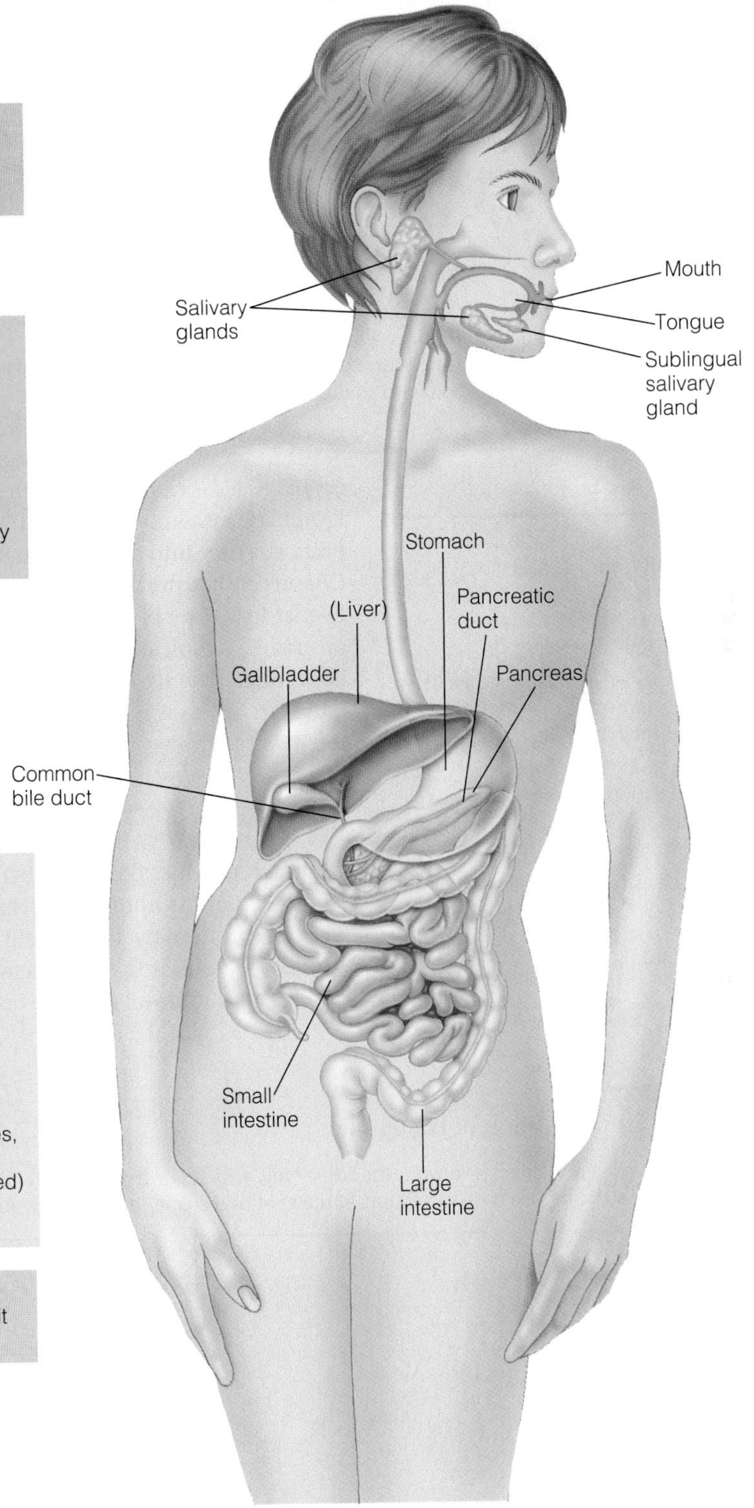

with the feces. Because cholesterol is needed to make bile, the excretion of bile effectively reduces blood cholesterol. As Chapter 4 explains, the fibers most effective at lowering blood cholesterol this way are the pectins and gums commonly found in fruits, oats, and legumes.[2]

FIGURE 5-13 A Bile Acid

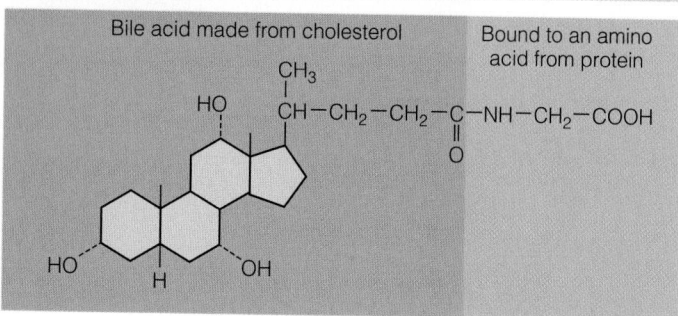

Bile acid made from cholesterol

Bound to an amino acid from protein

This is one of several bile acids the liver makes from cholesterol. It is then bound to an amino acid to improve its ability to form micelles, spherical complexes of emulsified fat. Most bile acids occur as bile salts, usually in association with sodium, but sometimes with potassium or calcium.

micelles (MY-cells): tiny spherical complexes of emulsified fat that arise during digestion; most contain bile salts and the products of lipid digestion, including fatty acids, monoglycerides, and cholesterol.

chylomicrons (kye-lo-MY-cronz): the class of lipoproteins that transport lipids from the intestinal cells to the rest of the body.

lipoproteins (LIP-oh-PRO-teenz): clusters of lipids associated with proteins that serve as transport vehicles for lipids in the lymph and blood.

VLDL (very-low-density lipoprotein): the type of lipoprotein made primarily by liver cells to transport lipids to various tissues in the body; composed primarily of triglycerides.

Lipid Absorption

Figure 5-17 (on p. 154) illustrates the absorption of lipids. Small molecules of digested triglycerides (glycerol and short- and medium-chain fatty acids) can diffuse easily into the intestinal cells; they are absorbed directly into the bloodstream. Larger molecules (the monoglycerides and long-chain fatty acids) merge into spherical complexes, known as **micelles.** Micelles are emulsified fat droplets formed by molecules of bile surrounding monoglycerides and fatty acids. This configuration permits solubility in the watery digestive fluids and transportation to the intestinal cells. Upon arrival, the lipid contents of the micelles diffuse into the intestinal cells. Once inside, the monoglycerides and long-chain fatty acids are reassembled into new triglycerides.

Within the intestinal cells, the newly made triglycerides and other lipids (cholesterol and phospholipids) are packed with protein into transport vehicles known as **chylomicrons.** The intestinal cells then release the chylomicrons into the lymphatic system. The chylomicrons glide through the lymph until they reach a point of entry into the bloodstream at the thoracic duct near the heart. (Recall from Chapter 3 that nutrients from the GI tract that enter the lymph system bypass the liver at first.) The blood carries these lipids to the rest of the body for immediate use or storage. A look at these lipids in the body reveals the kinds of fat the diet has been delivering.[3] The fat stores and muscle cells of people who eat a diet rich in unsaturated fats, for example, contain more unsaturated fats than those of people who select a diet high in saturated fats.

IN SUMMARY The body makes special arrangements to digest and absorb lipids. It provides the emulsifier bile to make them accessible to the fat-digesting lipases that dismantle triglycerides, mostly to monoglycerides and fatty acids, for absorption by the intestinal cells. The intestinal cells assemble freshly absorbed lipids into chylomicrons, lipid packages with protein escorts, for transport so that cells all over the body may select needed lipids from them.

FIGURE 5-14 Emulsification of Fat by Bile

Like bile, detergents are emulsifiers and work the same way, which is why they are effective in removing grease spots from clothes. Molecule by molecule, the grease is dissolved out of the spot and suspended in the water, where it can be rinsed away.

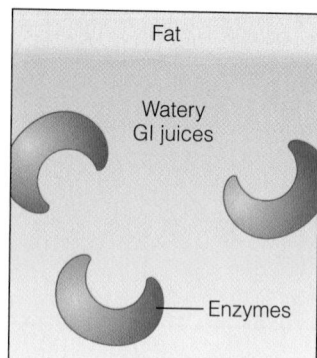

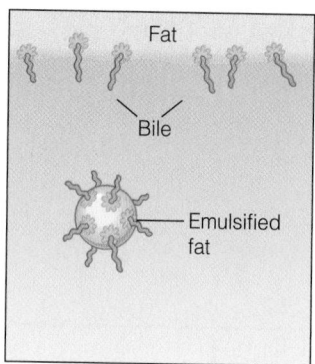

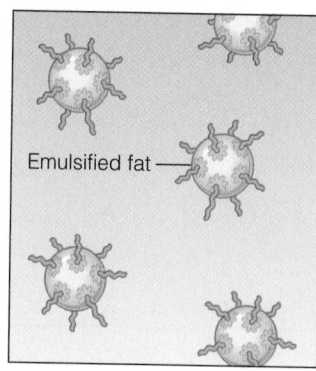

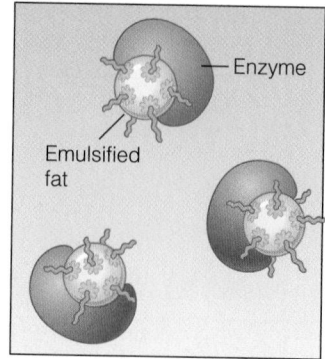

In the stomach, the fat and watery GI juices tend to separate. The enzymes in the GI juices can't get at the fat.

When fat enters the small intestine, the gallbladder secretes bile. Bile has an affinity for both fat and water, so it can bring the fat into the water.

Bile's emulsifying action converts large fat globules into small droplets that repel each other.

After emulsification, more fat is exposed to the enzymes, making fat digestion more efficient.

FIGURE 5-15 Digestion (Hydrolysis) of a Triglyceride

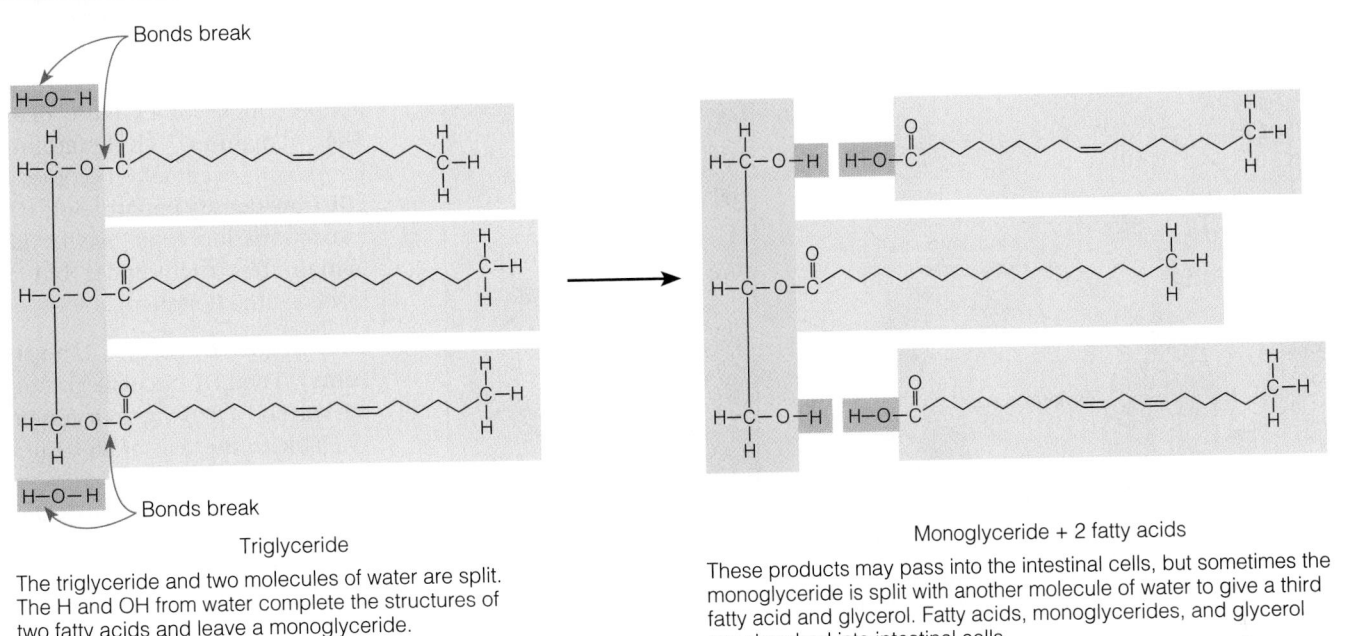

Triglyceride

The triglyceride and two molecules of water are split. The H and OH from water complete the structures of two fatty acids and leave a monoglyceride.

Monoglyceride + 2 fatty acids

These products may pass into the intestinal cells, but sometimes the monoglyceride is split with another molecule of water to give a third fatty acid and glycerol. Fatty acids, monoglycerides, and glycerol are absorbed into intestinal cells.

Lipid Transport

The chylomicrons are only one of several clusters of lipids and proteins that are used as transport vehicles for fats. As a group, these vehicles are known as **lipoproteins,** and they solve the body's problem of transporting fatty materials through the watery bloodstream. The body makes four main types of lipoproteins, distinguished by their size and density.* Each type contains different kinds and amounts of lipids and proteins.■ Figure 5-18 on p. 155 shows the relative compositions and sizes of the lipoproteins.

Chylomicrons The chylomicrons are the largest and least dense of the lipoproteins. They transport *diet*-derived lipids (mostly triglycerides) from the intestine (via the lymph system) to the rest of the body. Cells all over the body remove triglycerides from the chylomicrons as they pass by, so the chylomicrons get smaller and smaller. Within 14 hours after absorption, most of the triglycerides have been depleted, and only a few remnants of protein, cholesterol, and phospholipid remain. Special protein receptors on the membranes of the liver cells recognize and remove these chylomicron remnants from the blood. After collecting the remnants, the liver cells first dismantle them and then either use or recycle the pieces.

VLDL (Very-Low-Density Lipoproteins) Meanwhile, in the liver, the most active site of lipid synthesis, the cells are synthesizing other lipids. The liver cells use fatty acids arriving in the blood to make cholesterol, other fatty acids, and other compounds. At the same time, the liver cells may be making lipids from carbohydrates, proteins, or alcohol. Ultimately, the lipids made in the liver and those collected from chylomicron remnants are packaged with proteins as **VLDL (very-low-density lipoproteins)**■ and shipped to other parts of the body.

■ The more lipids, the lower the density; the more proteins, the higher the density.

■ Chylomicrons and VLDL transport triglycerides.

FIGURE 5-16 Enterohepatic Circulation

Most of the bile released into the small intestine is reabsorbed and sent back to the liver to be reused. This cycle is called the **enterohepatic circulation** of bile. Some bile is excreted.
- **enteron** = intestine
- **hepat** = liver

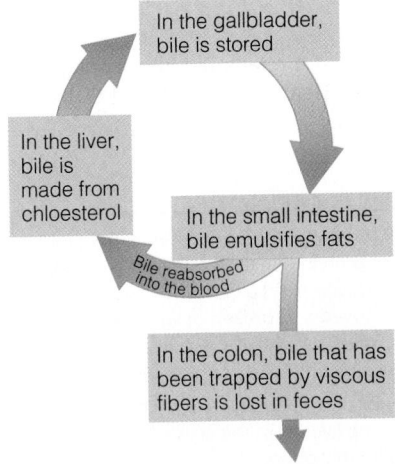

In the gallbladder, bile is stored

In the liver, bile is made from chloesterol

In the small intestine, bile emulsifies fats

Bile reabsorbed into the blood

In the colon, bile that has been trapped by viscous fibers is lost in feces

*Chemists can identify the various lipoproteins by their density by layering a blood sample below a thick fluid in a test tube and spinning the tube in a centrifuge. The most buoyant particles (highest in lipids) rise to the top and have the lowest density; the densest particles (highest in proteins) remain at the bottom and have the highest density. Others distribute themselves in between.

FIGURE 5-17 Absorption of Fat

The end products of fat digestion are mostly monoglycerides, some fatty acids, and very little glycerol. Their absorption differs depending on their size. (In reality, molecules of fatty acid are too small to see without a powerful microscope, while villi are visible to the naked eye.)

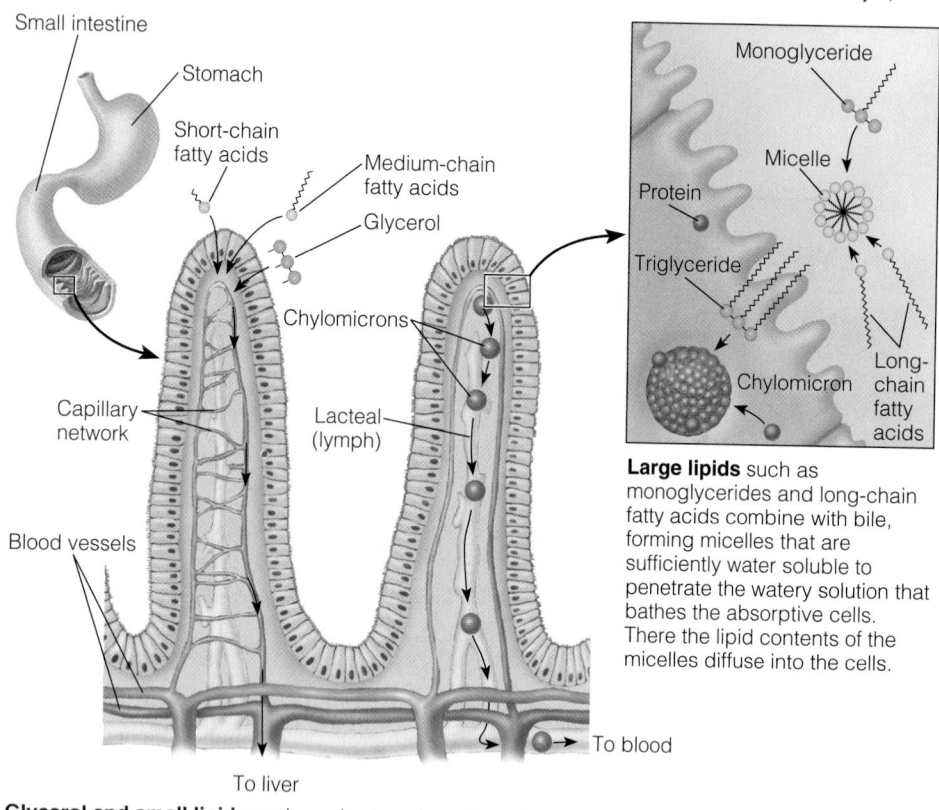

Large lipids such as monoglycerides and long-chain fatty acids combine with bile, forming micelles that are sufficiently water soluble to penetrate the watery solution that bathes the absorptive cells. There the lipid contents of the micelles diffuse into the cells.

Glycerol and small lipids such as short- and medium-chain fatty acids can move directly into the bloodstream.

As the VLDL travel through the body, cells remove triglycerides, causing the VLDL to shrink. As a VLDL loses triglycerides, the proportion of lipids shifts, and the lipoprotein becomes more dense. The remaining cholesterol-rich lipoprotein eventually becomes an **LDL (low-density lipoprotein).*** This trans-formation explains why LDL contain few triglycerides but are loaded with cholesterol.

LDL (Low-Density Lipoproteins) The LDL circulate throughout the body, making their contents available to the cells of all tissues—muscles, including the heart muscle; fat stores; the mammary glands; and others. The cells take triglycerides, cholesterol, and phospholipids to build new membranes, make hormones or other compounds, or store for later use. Special LDL receptors on the liver cells play a crucial role in the control of blood cholesterol concentrations by removing LDL from circulation.

HDL (High-Density Lipoproteins) Fat cells may release glycerol, fatty acids, cholesterol, and phospholipids to the blood. The liver makes **HDL (high-density lipoprotein)** to carry cholesterol■ from the cells back to the liver for recycling or disposal.

■ LDL and HDL transport cholesterol.

■ The transport of cholesterol from the tissues to the liver is sometimes called the *scavenger pathway.*

■ To help you remember, think of elevated HDL as Healthy and elevated LDL as Less healthy.

LDL (low-density lipoprotein): the type of lipoprotein derived from very-low-density lipoproteins (VLDL) as VLDL triglycerides are removed and broken down; composed primarily of cholesterol.

HDL (high-density lipoprotein): the type of lipoprotein that transports cholesterol back to the liver from the cells; composed primarily of protein.

Health Implications The distinction between LDL and HDL has implications for the health of the heart and blood vessels. The blood cholesterol linked to heart disease is LDL cholesterol. HDL also carry cholesterol, but elevated HDL represent cholesterol returning■ from the rest of the body to the liver for breakdown and excretion.[4] High LDL cholesterol is associated with a high risk of heart attack, whereas high HDL cholesterol seems to have a protective effect. This is why some people refer to LDL as "bad," and HDL as "good," cholesterol.■ Keep in mind that the cholesterol itself is the same, and that the differences between LDL and HDL reflect the *proportions* and *types* of lipids and proteins within them—not the type of cholesterol. The margin on p. 155■ lists factors that influence LDL and HDL, and Chapter 18 provides many more details.

Not too surprisingly, numerous genes influence how the body handles the uptake, synthesis, transport, and degradation of the lipoproteins.[5] Much research is currently focusing on how nutrient-gene interactions may direct the progression of heart disease.

*Before becoming LDL, the VLDL are first transformed into intermediate-density lipoproteins (IDL), sometimes called VLDL remnants. Some IDL may be picked up by the liver and rapidly broken down; those IDL that remain in circulation continue to deliver triglycerides to the cells and eventually become LDL. Researchers debate whether IDL are simply transitional particles or a separate class of lipoproteins; normally, IDL do not accumulate in the blood. Measures of blood lipids include IDL and LDL.

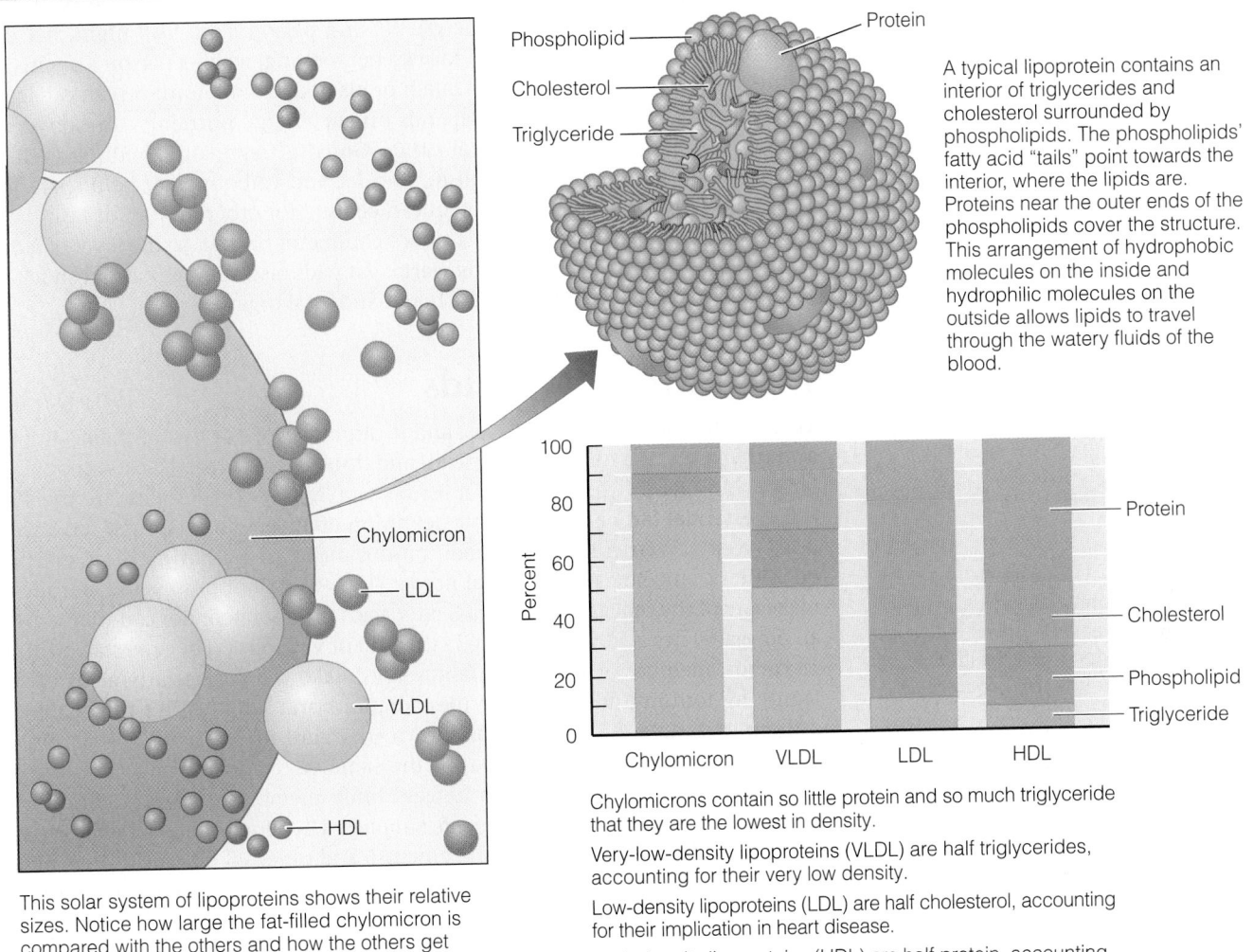

FIGURE 5-18 Size and Compositions of the Lipoproteins

Phospholipid — Protein

Cholesterol

Triglyceride

A typical lipoprotein contains an interior of triglycerides and cholesterol surrounded by phospholipids. The phospholipids' fatty acid "tails" point towards the interior, where the lipids are. Proteins near the outer ends of the phospholipids cover the structure. This arrangement of hydrophobic molecules on the inside and hydrophilic molecules on the outside allows lipids to travel through the watery fluids of the blood.

Chylomicron

LDL

VLDL

HDL

This solar system of lipoproteins shows their relative sizes. Notice how large the fat-filled chylomicron is compared with the others and how the others get progressively smaller as their proportion of fat declines and protein increases.

Protein

Cholesterol

Phospholipid

Triglyceride

Chylomicron   VLDL   LDL   HDL

Chylomicrons contain so little protein and so much triglyceride that they are the lowest in density.

Very-low-density lipoproteins (VLDL) are half triglycerides, accounting for their very low density.

Low-density lipoproteins (LDL) are half cholesterol, accounting for their implication in heart disease.

High-density lipoproteins (HDL) are half protein, accounting for their high density.

Nutrition Explorer View the various types of lipoproteins as they leave their tissue of origin, and follow their metabolism as they deliver their cargo of lipids to specific destinations throughout the body.

IN SUMMARY The liver packages lipids with proteins into lipoproteins for transport around the body. All four types of lipoproteins carry all classes of lipids (triglycerides, phospholipids, and cholesterol), but the chylomicrons are the largest and the highest in triglycerides; VLDL are smaller and are about half triglycerides; LDL are smaller still and are high in cholesterol; and HDL are the smallest and are rich in protein.

■ Factors that lower LDL or raise HDL:
- Weight control.
- Monounsaturated or polyunsaturated, instead of saturated, fat in the diet.
- Soluble, viscous fibers (see Chapter 4).
- Phytochemicals (see Highlight 13).
- *Moderate* alcohol consumption.
- Physical activity.

Lipids in the Body

The blood carries lipids to various sites around the body. Once they arrive at their destinations, the lipids can get to work providing energy, insulating against temperature extremes, protecting against shock, and maintaining cell membranes. This section provides an overview first of the roles of triglycerides and fatty acids and then of the metabolic pathways they can follow within the body's cells.

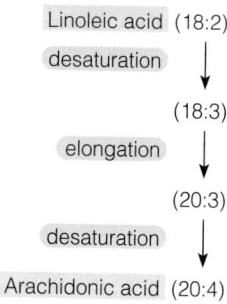

FIGURE 5-19 The Pathway from One Omega-6 Fatty Acid to Another

Linoleic acid (18:2)

desaturation ↓

(18:3)

elongation ↓

(20:3)

desaturation ↓

Arachidonic acid (20:4)

Note: The first number indicates the number of carbons and the second, the number of double bonds. Similar reactions occur when the body makes EPA and DHA from linolenic acid.

■ A nonessential nutrient (such as arachidonic acid) that must be supplied by the diet in special circumstances (as in a linoleic acid deficiency) is considered *conditionally essential*.

essential fatty acids: fatty acids needed by the body, but not made by it in amounts sufficient to meet physiological needs.

arachidonic (a-RACK-ih-DON-ic) **acid:** an omega-6 polyunsaturated fatty acid with 20 carbons and four double bonds; present in small amounts in meat and other animal products and synthesized in the body from linoleic acid.

EPA, or **eicosapentaenoic** (EYE-cossa-PENTA-ee-NO-ick) **acid:** an omega-3 polyunsaturated fatty acid with 20 carbons and five double bonds; present in fish and synthesized in limited amounts in the body from linolenic acid.

DHA, or **docosahexaenoic** (DOE-cossa-HEXA-ee-NO-ick) **acid:** an omega-3 polyunsaturated fatty acid with 22 carbons and six double bonds; present in fish and synthesized in limited amounts in the body from linolenic acid.

eicosanoids (eye-COSS-uh-noyds): derivatives of 20-carbon fatty acids; biologically active compounds that help to regulate blood pressure, blood clotting, and other body functions. They include *prostaglandins* (PROS-tah-GLAND-ins), *thromboxanes* (throm-BOX-ains), and *leukotrienes* (LOO-ko-TRY-eens).

Roles of Triglycerides

First and foremost, the triglycerides—either from food or from the body's fat stores—provide the body with energy. When a person dances all night, her dinner's triglycerides provide the fuel to keep her moving; when a person loses his appetite, his stored triglycerides fuel much of his body's work until he can eat again.

Efficient energy metabolism depends on the energy nutrients—carbohydrate, fat, and protein—supporting each other. Glucose fragments combine with fat fragments during energy metabolism, and fat and carbohydrate help spare protein, providing energy so that protein can be used for other important tasks.

Fat also insulates the body. Fat is a poor conductor of heat, so the layer of fat beneath the skin helps keep the body warm. Fat pads also serve as natural shock absorbers, providing a cushion for the bones and vital organs.

Essential Fatty Acids

The human body needs fatty acids, and it can make all but two of them—linoleic acid (the 18-carbon omega-6 fatty acid) and linolenic acid (the 18-carbon omega-3 fatty acid). These two fatty acids must be supplied by the diet and are therefore called **essential fatty acids**. A simple definition of an essential nutrient has already been given: a nutrient that the body cannot make, or cannot make in sufficient quantities to meet its physiological needs. The cells do not possess the enzymes to make any of the omega-6 or omega-3 fatty acids from scratch; nor can they convert an omega-6 fatty acid to an omega-3 fatty acid or vice versa. They *can* start with the 18-carbon member of an omega family and make the longer fatty acids of that family by forming double bonds (desaturation) and lengthening the chain two carbons at a time (elongation), as shown in Figure 5-19. This is a slow process because the two families compete for the same enzymes. Too much of one can create a deficiency of the other's longer family members, which is critical only when the diet fails to deliver adequate supplies. Therefore, the most effective way to maintain body supplies of all the omega-6 and omega-3 fatty acids is to obtain them directly from foods—most notably, from vegetable oils, seeds, nuts, fish, and other marine foods.

Linoleic Acid and the Omega-6 Family Linoleic acid is the primary member of the omega-6 family. Given linoleic acid, the body can make other members of the omega-6 family—such as the 20-carbon polyunsaturated fatty acid, **arachidonic acid**. Should a linoleic acid deficiency develop, arachidonic acid, and all other fatty acids that derive from linoleic acid, would also become essential and have to be obtained from the diet.■ Normally, vegetable oils and meats supply enough omega-6 fatty acids to meet the body's needs.

Linolenic Acid and the Omega-3 Family Linolenic acid is the primary member of the omega-3 family.* Like linoleic acid, this 18-carbon fatty acid cannot be made in the body and must be supplied by foods. Given dietary linolenic acid, the body can make small amounts of the 20- and 22-carbon members of the omega-3 series, **EPA (eicosapentaenoic acid)** and **DHA (docosahexaenoic acid)**. These omega-3 fatty acids are essential for normal growth and development, especially in the eyes and brain.[6] They may also play an important role in the prevention and treatment of heart disease.

Eicosanoids The body uses arachidonic acid and EPA to make substances known as **eicosanoids**. Eicosanoids are a diverse group of compounds that are sometimes described as "hormonelike," but they differ from hormones in important ways. For one, hormones are secreted in one location and travel to affect cells all over the

---

*This omega-3 linolenic acid is known as alpha-linolenic acid and is the fatty acid referred to in this chapter. Another fatty acid, also with 18 carbons and three double bonds, belongs to the omega-6 family and is known as gamma-linolenic acid.

body, whereas eicosanoids appear to affect only the cells in which they are made or nearby cells in the same localized environment. For another, hormones elicit the same response from all their target cells, whereas eicosanoids often have different effects on different cells.

The actions of various eicosanoids sometimes oppose each other. One causes muscles to relax and blood vessels to dilate, while another causes muscles to contract and blood vessels to constrict, for example. Certain eicosanoids participate in the immune response to injury and infection, producing fever, inflammation, and pain.[7] One of the ways aspirin works to relieve these symptoms is by slowing the synthesis of these eicosanoids.

Eicosanoids that derive from EPA differ slightly from those that derive from arachidonic acid, with those from EPA providing greater health benefits. The EPA eicosanoids help lower blood pressure, prevent blood clot formation, protect against irregular heartbeats, and reduce inflammation. Because the omega-6 and omega-3 fatty acids compete for the same enzymes to make arachidonic acid and EPA and to make the eicosanoids, the body needs these long-chain polyunsaturated fatty acids from the diet to make eicosanoids in sufficient quantities.

Fatty Acid Deficiencies Most diets in the United States and Canada meet the essential fatty acid requirement adequately. Historically, deficiencies have developed only in infants and young children fed fat-free milk and low-fat diets or in hospital clients mistakenly fed formulas that provided no polyunsaturated fatty acids for long periods of time. Classic deficiency symptoms include growth retardation, reproductive failure, skin lesions, kidney and liver disorders, and subtle neurological and visual problems.

Interestingly, a deficiency of omega-3 fatty acids (EPA and DHA) may be associated with depression.[8] Some neurochemical pathways in the brain become more active and others become less active.[9] It is unclear, however, which comes first—whether inadequate intake alters brain activity or depression alters fatty acid metabolism. To find the answers, researchers must untangle a multitude of confounding factors.

Double thanks: The body's fat stores provide energy for a walk, and the heel's fat pads cushion against the hard pavement.

© Jim Cummins/Taxi/Getty Images

IN SUMMARY

In the body, triglycerides:
- Provide an energy reserve when stored in the body's fat tissue.
- Insulate against temperature extremes.
- Protect against shock.
- Help the body use carbohydrate and protein efficiently.

Linoleic acid (18 carbons, omega-6) and linolenic acid (18 carbons, omega-3) are essential nutrients. They serve as structural parts of cell membranes and as precursors to the longer fatty acids that can make eicosanoids—powerful compounds that participate in blood pressure regulation, blood clot formation, and the immune response to injury and infection, among other functions. Because essential fatty acids are common in the diet and stored in the body, deficiencies are unlikely.

A Preview of Lipid Metabolism

The blood delivers triglycerides to the cells for their use. This is a preview of how the cells store and release energy from fat; Chapter 7 provides details.

Storing Fat as Fat The triglycerides, familiar as the fat in foods and as body fat, serve the body primarily as a source of fuel. Fat provides more than twice the energy of carbohydrate and protein,■ making it an extremely efficient storage form of energy. Unlike the liver's glycogen stores, the body's fat stores have virtually unlimited capacity, thanks to the special cells of the **adipose tissue.** Unlike most body cells, which can store only limited amounts of fat, the fat cells of the adipose tissue readily take up and store fat. An adipose cell is depicted in Figure 5-20.

FIGURE 5-20 An Adipose Cell

Newly imported triglycerides first form small droplets at the periphery of the cell, then merge with the large, central globule.

Large central globule of (pure) fat

Cell nucleus

Cytoplasm

As the central globule enlarges, the fat cell membrane expands to accommodate its swollen contents.

■ 1 g fat = 9 kcal.

adipose (ADD-ih-poce) **tissue:** the body's fat tissue; consists of masses of triglyceride-storing cells.

Fat supplies most of the energy during a long-distance run.

© Bob Thomas/Stone/Getty Images

■ Reminder: Gram for gram, fat provides more than twice as much energy as carbohydrate or protein.

■ 1 lb body fat = 3500 kcal.

lipoprotein lipase (LPL): an enzyme that hydrolyzes triglycerides passing by in the bloodstream and directs their parts into the cells, where they can be metabolized for energy or reassembled for storage.

hormone-sensitive lipase: an enzyme inside adipose cells that responds to the body's need for fuel by hydrolyzing triglycerides so that their parts (glycerol and fatty acids) escape into the general circulation and thus become available to other cells for fuel. The signals to which this enzyme responds include epinephrine and glucagon, which oppose insulin (see Chapter 4).

To convert food fats to body fat, the body simply absorbs the parts and puts them (and others) together again in storage. It requires very little energy to do this. An enzyme—**lipoprotein lipase (LPL)**—hydrolyzes triglycerides from lipoproteins, producing glycerol, fatty acids and monoglycerides that enter the adipose cells. Inside the cells, other enzymes reassemble the pieces into triglycerides again for storage. Earlier, Figure 5-4 (on p. 145) showed how the body can make a triglyceride from glycerol and fatty acids. Triglycerides fill the adipose cells, storing a lot of energy in a relatively small space. ■ Adipose cells store fat after meals when a heavy traffic of chylomicrons and VLDL loaded with triglycerides passes by; they release it later whenever the blood needs replenishing.

Using Fat for Energy Fat supplies 60 percent of the body's ongoing energy needs during rest. During prolonged light to moderately intense exercise or extended periods of food deprivation, fat stores may make a slightly greater contribution to energy needs.

When cells demand energy, an enzyme **(hormone-sensitive lipase)** inside the adipose cells responds by dismantling stored triglycerides and releasing the glycerol and fatty acids directly into the blood. Energy-hungry cells anywhere in the body can then capture these compounds and take them through a series of chemical reactions to yield energy, carbon dioxide, and water.

A person who fasts (drinking only water) will rapidly metabolize body fat. A pound of body fat provides 3500 kcalories, ■ so you might think a fasting person who expends 2000 kcalories a day could lose more than half a pound of body fat each day.* Actually, the person has to obtain some energy from lean tissue because the brain, nerves, and red blood cells need glucose. Also, the complete breakdown of fat requires carbohydrate or protein. Even on a total fast, a person cannot lose more than half a pound of pure fat per day. Still, in conditions of enforced starvation—say, during a siege or a famine—a fatter person can survive longer than a thinner person thanks to this energy reserve.

Although fat provides energy during a fast, it can provide very little glucose to give energy to the brain and nerves. Only the small glycerol molecule can be converted to glucose; fatty acids cannot be. (Figure 7-12 on p. 226 illustrates how only 3 of the 50 or so carbon atoms in a molecule of fat can yield glucose.) After prolonged glucose deprivation, brain and nerve cells develop the ability to derive about two-thirds of their minimum energy needs from the ketone bodies that the body makes from fat fragments. Ketone bodies cannot sustain life by themselves, however. As Chapter 7 explains, fasting for too long will cause death, even if the person still has ample body fat.

IN SUMMARY The body can easily store unlimited amounts of fat if given excesses, and this body fat is used for energy when needed. The liver can also convert excess carbohydrate and protein into fat. Fat breakdown requires simultaneous carbohydrate breakdown for maximum efficiency; without carbohydrate, fats break down to ketone bodies.

Health Effects and Recommended Intakes of Lipids

Of all the nutrients, fat is most often linked with heart disease, some types of cancer, and obesity. Fortunately, the same recommendation can help with all of these health problems: choose a diet that is low in saturated fats, *trans* fats, and cholesterol and moderate in total fat.

---

*The reader who knows that 1 pound = 454 grams and that 1 gram of fat = 9 kcalories may wonder why a pound of body fat does not equal 4086 (9 × 454) kcalories. The reason is that body fat contains some cell water and other minerals; it is not quite pure fat.

Health Effects of Lipids

Hearing a physician say, "Your blood lipid profile looks fine," is reassuring. The **blood lipid profile**■ reveals the concentrations of various lipids in the blood, notably triglycerides and cholesterol, and their lipoprotein carriers (VLDL, LDL, and HDL). This information alerts people to their disease risks and their need to change eating habits.

Heart Disease Most people realize that elevated blood cholesterol is a major risk factor for **cardiovascular disease.** Cholesterol accumulates in the arteries, restricting blood flow and raising blood pressure. The consequences are deadly; in fact, heart disease is the nation's number one killer of adults. Blood cholesterol is often used to predict the likelihood of a person's suffering a heart attack or stroke; the higher the cholesterol, the earlier and more likely the tragedy. Efforts to prevent heart disease focus on lowering blood cholesterol.[10]

Commercials advertise products that are low in cholesterol, and magazine articles tell readers how to cut the cholesterol in their favorite recipes. What most people don't realize, though, is that *food* cholesterol does not raise *blood* cholesterol as dramatically as *saturated fat* does.

Risks from Saturated Fats Recall that LDL cholesterol raises the risk of heart disease. Most often implicated in raising LDL cholesterol are the saturated fats. In general, the more saturated fat in the diet, the more LDL cholesterol in the body. Not all saturated fats have the same cholesterol-raising effect, however. Most notable among the saturated fatty acids that raise blood cholesterol are lauric, myristic, and palmitic acids (12, 14, and 16 carbons, respectively). In contrast, stearic acid (18 carbons) does not seem to raise blood cholesterol, but making such distinctions may be impractical in diet planning because these saturated fatty acids typically appear together in the same foods.[11]

Fats from animal sources are the main sources of saturated fats■ in most people's diets (see Figure 5-21). Some vegetable fats (coconut and palm) and hydrogenated fats provide smaller amounts of saturated fats. Selecting poultry or fish and fat-free milk products helps to lower saturated fat intake and heart disease risk.[12] Using nonhydrogenated margarine and unsaturated cooking oil is another simple change that can dramatically lower saturated fat intake.

Risks from *Trans* Fats Research also suggests an association between dietary *trans*-fatty acids and heart disease.[13] In the body, *trans*-fatty acids alter blood cholesterol the same way some saturated fats do: they raise LDL cholesterol.[14] Limiting the intake of *trans*-fatty acids can improve blood cholesterol. *Trans*-fatty acids make up approximately 7 percent of the fat intake (or 2 percent of the energy intake) in the U.S. diet.[15]■

Reports on *trans*-fatty acids have raised consumer doubts about whether margarine is, after all, a better choice than butter for heart health. The American Heart Association has stated that because butter is rich in both saturated fat and cholesterol while margarine is made from vegetable fat with no dietary cholesterol, margarine is still preferable to butter. Be aware that soft margarines (liquid or tub) ■ are less hydrogenated and relatively lower in *trans*-fatty acids; consequently, they do not raise blood cholesterol as the saturated fats of butter or the *trans* fats of hard (stick) margarines do.[16] Some manufacturers are now offering nonhydrogenated margarines that are "*trans* fat free." The last section of this chapter describes how to read food labels and compares butter and margarines; whichever you decide to use, remember to use them sparingly.

Risks from Cholesterol Dietary cholesterol has also been implicated in raising blood cholesterol and increasing the risk of heart disease, although its effect is not as strong as that of saturated fat or *trans* fat. Still, health experts advise limiting cholesterol intake.

Recall that cholesterol is found only in foods derived from animals. Consequently, eating less fat from meats, eggs, and milk products helps lower dietary

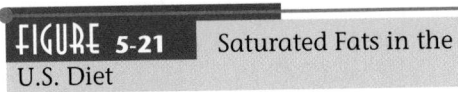

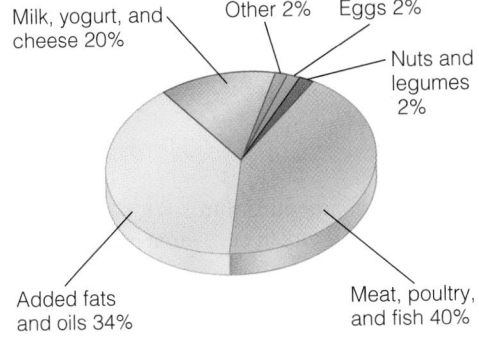

FIGURE 5-21 Saturated Fats in the U.S. Diet

Fruits, grains, and vegetables are insignificant sources, unless saturated fats are intentionally added to them during preparation.

Milk, yogurt, and cheese 20% — Other 2% — Eggs 2% — Nuts and legumes 2% — Meat, poultry, and fish 40% — Added fats and oils 34%

■ Desirable blood lipid profile:
- Total cholesterol: <200 mg/dL.
- LDL cholesterol: <100 mg/dL.
- HDL cholesterol: ≥60 mg/dL.
- Triglycerides: <150 mg/dL.

■ Major sources of saturated fats:
- Whole milk, cream, butter, cheese.
- Fatty cuts of beef and pork.
- Coconut, palm, and palm kernel oils (and products containing them such as candies, pastries, pies, doughnuts, and cookies).

■ Major sources of *trans* fats:
- Deep-fried foods (vegetable shortening).
- Cakes, cookies, doughnuts, pastry, crackers.
- Snack chips.
- Margarine.
- Imitation cheese.
- Meat and dairy products.

■ When selecting margarine, look for:
- Soft (liquid or tub) instead of hard (stick).
- ≤2 g saturated fat.
- Liquid vegetable oil (not hydrogenated or partially hydrogenated) as first ingredient.
- "*Trans* fat free."

blood lipid profile: results of blood tests that reveal a person's total cholesterol, triglycerides, and various lipoproteins.

cardiovascular disease (CVD): a general term for all diseases of the heart and blood vessels. Atherosclerosis is the main cause of CVD. When arteries that carry blood to the heart muscle become blocked, the heart suffers damage known as **coronary heart disease (CHD).**
- **cardio** = heart
- **vascular** = blood vessels

■ Major sources of cholesterol:
- Eggs.
- Milk products.
- Meat, poultry, shellfish.

■ Sources of monounsaturated fats:
- Olive oil, canola oil, peanut oil.
- Avocados.

■ Sources of polyunsaturated fats:
- Vegetable oils (safflower, sesame, soy, corn, sunflower).
- Nuts and seeds.

■ Major sources of omega-3 fats:
- Vegetable oils (canola, soybean, flaxseed).
- Walnuts, flaxseeds.
- Fatty fish (mackerel, salmon, sardines).

cholesterol intake■ (as well as total and saturated fat intakes). Figure 5-22 shows the cholesterol contents of selected foods. Many more foods, with their cholesterol contents, appear in Appendix H. For most people trying to lower blood cholesterol, however, limiting saturated fat is more effective than limiting cholesterol intake.

An egg contains just over 200 milligrams of cholesterol, all of it in the yolk. A person on a strict low-cholesterol diet must curtail the use of egg yolks, but for people with a healthy lipid profile, eating up to one egg a day is not detrimental.[17] Eggs are a valuable part of the diet because they are inexpensive, useful in cooking, and a source of high-quality protein and other nutrients.[18] Food manufacturers have produced several fat-free, no-cholesterol egg substitutes. Alternatively, a person could use the whites of a fresh egg.

Benefits from Monounsaturated Fats and Polyunsaturated Fats Replacing both saturated and *trans* fats with monounsaturated■ and polyunsaturated■ fats may be the most effective dietary strategy in preventing heart disease.[19] The lower rates of heart disease among people in the Mediterranean region of the world are often attributed to their liberal use of olive oil, a rich source of monounsaturated fatty acids.[20] Olive oil also delivers valuable phytochemicals that help to protect against heart disease.[21] Replacing saturated fats with the polyunsaturated fatty acids of other vegetable oils also lowers blood cholesterol. Highlight 5 examines various types of fats and their roles in supporting or harming heart health.

Benefits from Omega-3 Fats Research on the different types of fats has spotlighted the beneficial effects of the omega-3■ polyunsaturated fatty acids in reducing the risks of heart disease.[22] Regular consumption of omega-3 fatty acids helps to prevent blood clots, protect against irregular heartbeats, and lower blood pressure, especially in people with hypertension or atherosclerosis.[23]

Fatty fish are among the best sources of omega-3 fatty acids, and Highlight 5 features their role in supporting heart health. Chapter 18 presents more details on the roles of omega-3 fatty acids in heart disease, and Chapter 19 discusses the adverse consequences of mercury, an environmental contaminant common in some fish that may diminish the health benefits of omega-3 fatty acids.[24]

FIGURE 5-22 Cholesterol in Selected Foods

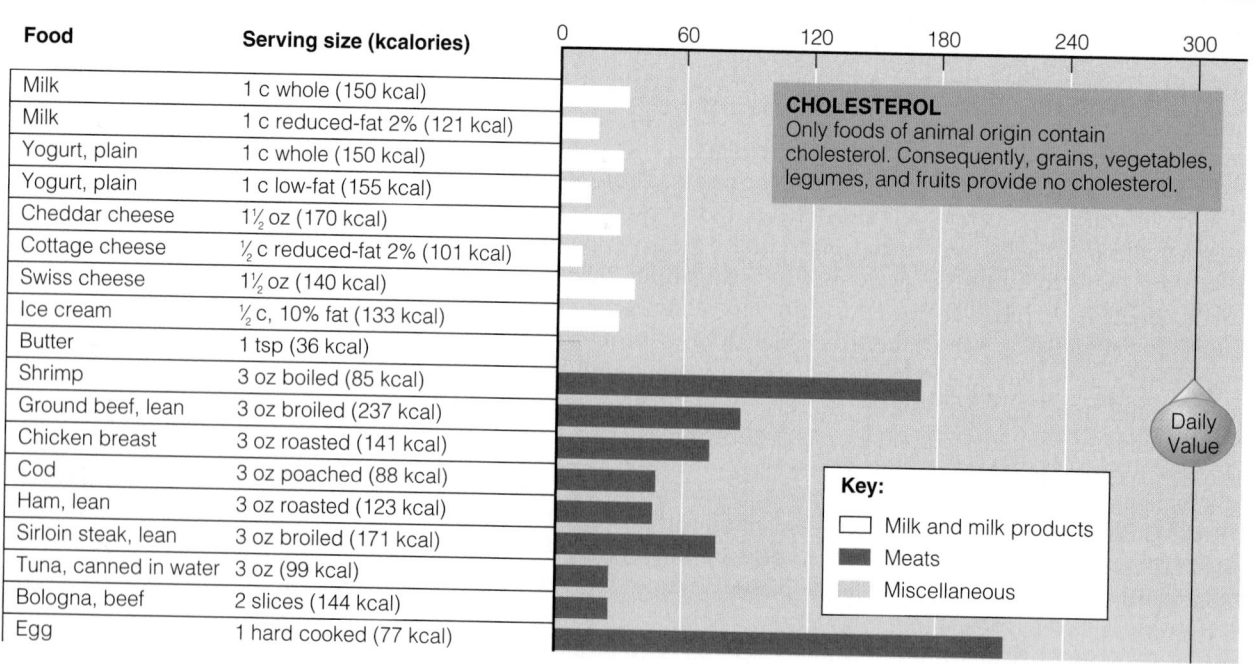

Food	Serving size (kcalories)
Milk	1 c whole (150 kcal)
Milk	1 c reduced-fat 2% (121 kcal)
Yogurt, plain	1 c whole (150 kcal)
Yogurt, plain	1 c low-fat (155 kcal)
Cheddar cheese	1½ oz (170 kcal)
Cottage cheese	½ c reduced-fat 2% (101 kcal)
Swiss cheese	1½ oz (140 kcal)
Ice cream	½ c, 10% fat (133 kcal)
Butter	1 tsp (36 kcal)
Shrimp	3 oz boiled (85 kcal)
Ground beef, lean	3 oz broiled (237 kcal)
Chicken breast	3 oz roasted (141 kcal)
Cod	3 oz poached (88 kcal)
Ham, lean	3 oz roasted (123 kcal)
Sirloin steak, lean	3 oz broiled (171 kcal)
Tuna, canned in water	3 oz (99 kcal)
Bologna, beef	2 slices (144 kcal)
Egg	1 hard cooked (77 kcal)

CHOLESTEROL
Only foods of animal origin contain cholesterol. Consequently, grains, vegetables, legumes, and fruits provide no cholesterol.

Daily Value

Key:
- ☐ Milk and milk products
- ■ Meats
- Miscellaneous

TABLE 5-2   Sources of Omega Fatty Acids

Omega-6

Linoleic acid	Vegetable oils (corn, sunflower, safflower, soybean, cottonseed), poultry fat, nuts, seeds
Arachidonic acid	Meats, poultry, eggs (or can be made from linoleic acid)

Omega-3

Linolenic acid	Oils (flaxseed, canola, walnut, wheat germ, soybean) Nuts and seeds (butternuts, flaxseeds, walnuts, soybean kernels) Vegetables (soybeans)
EPA and DHA	Human milk
	Pacific oysters and fish[a] (mackerel, salmon, bluefish, mullet, sablefish, menhaden, anchovy, herring, lake trout, sardines, tuna) (or can be made from linolenic acid)

[a]All fish contain some EPA and DHA; the amounts vary among species and within a species depending on such factors as diet, season, and environment. The fish listed here except tuna provide at least 1 gram of omega-3 fatty acids in 100 grams of fish (3.5 ounces). Tuna provides fewer omega-3 fatty acids, but because it is commonly consumed, its contribution can be significant.

Balance Omega-6 and Omega-3 Intakes Table 5-2 provides sources of omega-6 and omega-3 fatty acids. To obtain the right balance between omega-6 and omega-3 fatty acids, most people need to eat more fish and less meat. The American Heart Association recommends two 3-ounce servings of fish a week, with an emphasis on fatty fish (salmon, herring, and mackerel, for example).[25] Eating fish instead of meat supports heart health, especially when combined with physical activity. Even one fish meal a month may be enough to make a difference.[26] When preparing fish, grill, bake, or broil, but do not fry. Fried fish from fast-food restaurants and frozen fried fish products are often low in omega-3 fatty acids and high in *trans-* and saturated fatty acids. Fish provides many minerals (except iron) and vitamins and is leaner than most other animal-protein sources. When used in a weight-loss program, eating fish improves blood lipids even more effectively than either measure alone.[27]

In addition to fish, other functional foods■ are being developed to help consumers improve their omega-3 fatty acid intake. For example, hens fed flaxseed produce eggs rich in omega-3 fatty acids. Including even one enriched egg in the diet daily can significantly increase a person's intake of omega-3 fatty acids. Another option may be to select wild game or pasture-fed cattle, which provide more omega-3 fatty acids and less saturated fat than grain-fed cattle.[28]

For most people, fish oil should come from fish, and omega-3 fatty acids should come from foods, not from supplements.[29] Routine supplementation is not recommended for a number of reasons.* Perhaps most importantly, high intakes of omega-3 polyunsaturated fatty acids may increase bleeding time, interfere with wound healing, raise LDL cholesterol, and suppress immune function.[30] Fish oil supplements are made from fish skins and livers, which may contain environmental contaminants. Some fish oils also naturally contain large amounts of the two most potentially toxic vitamins, A and D. Lastly, supplements are expensive; money is better spent on foods that can provide a full array of nutrients. People with heart disease, however, may benefit from doses greater than can be achieved through diet alone; they should consult a physician about including supplements as part of their treatment plan.[31]

Cancer The evidence for links between dietary fats and cancer■ is less convincing than for heart disease, but it does suggest a possible association between fat and some types of cancers.[32] Dietary fat seems not to *initiate* cancer development but to *promote* cancer once it has arisen.

■ Reminder: *Functional foods* contain physiologically active compounds that provide health benefits beyond basic nutrition (see Highlight 13 for a full discussion).

■ Other risk factors for cancer include smoking, alcohol, and environmental contaminants. Chapter 18 provides many more details about these risk factors and the development of cancer.

---

*In Canada, fish oil supplements require a physician's prescription.

The relationship between dietary fat and the risk of cancer differs for various types of cancers. In the case of breast cancer, evidence has been weak and inconclusive.[33] Some studies indicate little or no association between dietary fat and breast cancer; others find that total *energy* intake and obesity are better predictors than percentage of kcalories from fat.[34] In the case of prostate cancer, there appears to be a harmful association with fat, although a specific type of fat has not yet been implicated.[35]

The relationship between dietary fat and the risk of cancer differs for various types of fats as well. The association between cancer and fat appears to be due primarily to saturated fats or dietary fat from meats (which is mostly saturated). Fat from milk or fish has not been implicated in cancer risk. In fact, eating fish rich in omega-3 fatty acids seems to protect against some cancers.[36] Thus health advice to reduce cancer risks parallels that given to reduce heart disease risks: reduce saturated fats and increase omega-3 fatty acids.

Obesity Fat contributes more than twice as many kcalories■ per gram as either carbohydrate or protein. Consequently, people who eat high-fat diets regularly may exceed their energy needs and gain weight, especially if they are inactive.[37] Because fat boosts energy intake, cutting fat can be an effective strategy in cutting kcalories. In some cases, though, choosing a fat-free food offers no kcalorie savings. Fat-free frozen desserts, for example, often have so much sugar added that the kcalorie count can be as high as in the regular-fat product. In that case, cutting fat and adding carbohydrate offers no kcalorie savings or weight-loss advantage. In fact, it may even raise energy intake and exacerbate weight problems. Later chapters revisit the role of dietary fat in the development of obesity.

> ● **IN SUMMARY** High blood LDL cholesterol poses a risk of heart disease, and high intakes of saturated and *trans* fats, specifically, contribute most to high LDL. Cholesterol in foods presents less of a risk. Omega-3 fatty acids appear to be protective.

Recommended Intakes of Fat

Some fat in the diet is essential for good health, but too much fat, especially saturated fat, increases the risks for chronic diseases. Defining the exact amount of fat, saturated fat, or cholesterol that benefits health or begins to harm health, however, is not possible; for this reason, no RDA or upper limit has been set. Instead, recommendations suggest a diet that is low in saturated fat, *trans* fat, and cholesterol and provides 20 to 35 percent of the daily energy intake from fat.■ The top end of this range is slightly higher than previous recommendations. This revision recognizes that diets with up to 35 percent of kcalories from fat can be compatible with good health if energy intake is reasonable and saturated fat intake is low. When total fat exceeds 35 percent, saturated fat increases to unhealthy levels.[38] For a 2000-kcalorie diet, 20 to 35 percent represents 400 to 700 kcalories from fat (roughly 45 to 75 grams). Part of this fat allowance should provide for the essential fatty acids—linoleic acid and linolenic acid. Recommendations suggest that linoleic acid■ provide 5 to 10 percent of the daily energy intake and linolenic acid,■ 0.6 to 1.2 percent.[39]

The *Dietary Guidelines*■ urge people to choose a diet low in saturated fat and cholesterol and moderate in total fat. To help consumers meet that goal, the Food and Drug Administration (FDA) established Daily Values■ on food labels using 30 percent of energy intake as the guideline for fat and 10 percent for saturated fat; the Daily Value for cholesterol is 300 milligrams regardless of energy intake. There is no Daily Value for *trans* fat, but consumers should try to keep intakes within the 10 percent allotted for saturated fat. According to surveys, adults in the United States receive about 35 percent of their total energy from fat, with saturated fat contributing about 12 percent of the total; cholesterol intakes in the United States average 250 milligrams a day for women and 350 for men.[40]

■ Reminder: Fat is a more concentrated energy source than the other energy nutrients: 1 g carbohydrate or protein = 4 kcal, but 1 g fat = 9 kcal.

■ DRI for fat:
- 20 to 35% of energy intake.

■ Linoleic acid AI:
- 5 to 10% of energy intake.
Men:
- 19–50 yr: 17 g/day.
- 51+ yr: 14 g/day.
Women:
- 19–50 yr: 12 g/day.
- 51+ yr: 11 g/day.

■ Linolenic acid AI:
- 0.6 to 1.2% energy intake.
Men: 1.6 g/day.
Women: 1.1 g/day.

■ *Dietary Guidelines:*
- Choose a diet that is low in saturated fat and cholesterol and moderate in total fat.

■ Daily Values:
- 65 g fat (based on 30% of 2000 kcal diet).
- 20 g saturated fat (based on 10% of 2000 kcal diet).
- 300 mg cholesterol.

Increase the proportion of people who consume less than 10 percent of kcalories from saturated fat and no more than 30 percent of kcalories from total fat.

The Food Guide Pyramid places naturally occurring and added fat at the tip and suggests that consumers use them "sparingly" (as defined in the margin).■ To help estimate fat and energy intakes accurately, the list in the margin■ shows fats equivalent to 1 teaspoon of oil. These fats all provide about 5 grams of fat and 45 kcalories.

Although it is very difficult to do, some people actually manage to eat too little fat—to their detriment. Among them are people with eating disorders, described in Highlight 9, and athletes. Athletes following a diet too low in fat (15 percent of total kcalories) fall short on energy, vitamins, minerals, and essential fatty acids as well as on performance.[41] Trained athletes have greater endurance on a high-fat diet (42 to 55 percent) than on a low-fat diet (10 to 15 percent)—even when energy intake is the same.[42] As mentioned earlier, most adults should consume at least 20 percent of their energy intake from fat, but athletes need at least 30 percent.[43] As a practical guideline, it is wise to include the equivalent of at least a teaspoon of fat in every meal—a little peanut butter on toast or mayonnaise on tuna, for example. Dietary recommendations that limit fat were developed for healthy people over age two; Chapter 16 discusses the fat needs of infants and young children.

As the photos in Figure 5-23 (p. 164) show, fat accounts for a lot of the energy in foods, and removing the fat from foods cuts energy and saturated fat intakes dramatically. To reduce dietary fat, eliminate fat as a seasoning and in cooking; remove the fat from high-fat foods; replace high-fat foods with low-fat alternatives; and emphasize grains, fruits, and vegetables. The remainder of the chapter identifies sources of fat in the diet, food group by food group.

From Guidelines to Groceries

Fats accompany protein in foods derived from animals, such as meat, fish, poultry, and eggs, and carbohydrate in foods derived from plants, such as avocados and coconuts. Fats carry with them the four fat-soluble vitamins—A, D, E, and K—together with many of the compounds that give foods their flavor, texture, and palatability. Fat is responsible for the delicious aromas associated with sizzling bacon and hamburgers on the grill, onions being sautéed, or vegetables in a stir-fry. Of course, these wonderful characteristics lure people into eating too much from time to time. With careful selections, a diet following the Food Guide Pyramid can both support good health and meet fat recommendations (see the "How to" feature on p. 165).

Meats and Meat Alternates Many meats and meat alternates■ contain fat, saturated fat, and cholesterol, but also provide high-quality protein and valuable vitamins and minerals. They can be included in a healthy diet if a person makes lean choices and prepares them using the suggestions outlined in the box on p. 165. Another strategy to lower blood cholesterol is to prepare meals using soy protein instead of animal protein.[44]

Milks and Milk Products Like meats, milks and milk products■ should also be selected with an awareness of their fat, saturated fat, and cholesterol contents. Fat-free and low-fat milk products provide as much or more protein, calcium, and other nutrients as their whole-milk versions—but with little or no saturated fat. Preliminary research suggests that selecting fermented milk products, such as yogurt, may also help to lower blood cholesterol.[45] These foods increase the population and activity of bacteria in the colon that ferment fibers. As Chapter 4 explained, this action lowers blood cholesterol by excreting bile and producing short-chain fatty acids that inhibit cholesterol synthesis in the liver.[46]

■ Using fats sparingly means:
- ≤53 g for 1600 kcal diet.
- ≤73 g for 2200 kcal diet.
- ≤93 g for 2800 kcal diet.

■ 1 tsp oil =
- 1 tsp shortening.
- 1 tsp mayonnaise (1 tbs reduced fat).
- 1 tsp butter.
- 1 tsp margarine.
- 1 tbs salad dressing or cream cheese (2 tbs reduced fat).
- 2 tbs sour cream (3 tbs reduced fat).

■ Very lean options:
- Chicken (white meat, no skin); cod, flounder, trout; tuna (canned in water); legumes.

Lean options:
- Beef or pork "round" or "loin" cuts; chicken (dark meat, no skin); herring or salmon; tuna (canned in oil).

Medium-fat options:
- Ground beef; eggs, tofu.

High-fat options:
- Sausage, bacon; luncheon meats; hot dogs; peanut butter, nuts.

■ Fat-free and low-fat options:
- Fat-free or 1% milk or yogurt (plain); fat-free and low-fat cheeses.

Reduced-fat options:
- 2% milk, low-fat yogurt (plain).

High-fat options:
- Whole milk, regular cheeses.

FIGURE 5-23   Cutting Fat Cuts kCalories—and Saturated Fat

Pork chop with fat (340 kcal, 19 g fat, 7 g saturated fat).

Potato with 1 tbs butter and 1 tbs sour cream (350 kcal, 14 g fat, 10 g saturated fat).

Whole milk, 1 c (150 kcal, 8 g fat, 5 g saturated fat).

Pork chop with fat trimmed off (230 kcal, 9 g fat, 3 g saturated fat).

Plain potato (200 kcal, <1 g fat, 0 g saturated fat).

Fat-free milk, 1 c (90 kcal, <1 g fat, <1 g saturated fat).

© Polara Studios, Inc. (all)

Vegetables, Fruits, and Grains Choosing vegetables, fruits, whole grains, and legumes also helps lower the saturated fat, cholesterol, and total fat content of the diet. Most vegetables and fruits naturally contain little or no fat; avocados and olives are exceptions, but most of their fat is unsaturated, which is not harmful to heart health. Most grains contain only trace amounts of fat. Some grain *products* such as fried taco shells, croissants, and biscuits are high in saturated fat, though, so consumers need to read food labels. Similarly, many people add butter, margarine, or cheese sauce to grains and vegetables, which raises their saturated and *trans* fat contents. Because fruits are often eaten without added fat, a diet that includes several servings of fruit daily can help a person meet the dietary recommendations for fat.

A diet rich in vegetables, fruits, whole grains, and legumes offers abundant vitamin C, folate, vitamin A, vitamin E, and dietary fiber—all important in supporting health. Consequently, such a diet protects against disease by both reducing saturated fat, cholesterol, and total fat and by increasing nutrients. It also provides valuable phytochemicals that help defend against heart disease.

Invisible Fat *Visible* fat, such as butter and the fat trimmed from meat, is easy to see. *Invisible* fat is less apparent and can be present in foods in surprising amounts.[47] Invisible fat "marbles" a steak or is hidden in foods like cheese. Any *fried* food contains abundant fat: potato chips, french fries, fried wontons, and fried fish. Many *baked* goods, too, are high in fat: pie crusts, pastries, crackers, biscuits, cornbread, doughnuts, sweet rolls, cookies, and cakes. Most chocolate bars deliver more kcalories from fat than from sugar. Even cream-of-mushroom soup prepared with water derives 66 percent of its energy from fat. Keep invisible fats in mind when making food selections.

Choose Wisely Consumers can find an abundant array of foods that are low in saturated fat, *trans* fat, cholesterol, and total fat. In many cases, they are familiar foods prepared with less fat. For example, fat can be removed by skimming milk or

HOW TO Make Heart-Healthy Choices—by Food Group

Breads and Cereals

- Select breads, cereals, and crackers that are low in saturated and *trans* fat (for example, bagels instead of croissants).
- Prepare pasta with a tomato sauce instead of a cheese or cream sauce.

Vegetables and Fruits

- Enjoy the natural flavor of steamed vegetables for dinner and fruits for dessert.
- Eat at least two vegetables (in addition to a salad) with dinner.
- Snack on raw vegetables or fruits instead of high-fat items like potato chips.
- Buy frozen vegetables without sauce.

Milk and Milk Products

- Switch from whole milk to reduced-fat, from reduced-fat to low-fat, and from low-fat to fat-free (nonfat).
- Use fat-free and low-fat cheeses (such as part-skim ricotta and low-fat mozzarella) instead of regular cheeses.
- Use fat-free or low-fat yogurt or sour cream instead of regular sour cream.
- Use evaporated fat-free milk instead of cream.
- Enjoy fat-free frozen yogurt, sherbet, or ice milk instead of ice cream.

Meat and Meat Alternates

- Fat adds up quickly, even with lean meat; limit intake to about 6 ounces (cooked weight) daily.

- Eat at least two servings of fish per week (particularly fish such as mackerel, lake trout, herring, sardines, and salmon).
- Choose fish, poultry, or lean cuts of pork or beef; look for unmarbled cuts named *round* or *loin* (eye of round, top round, bottom round, round tip, tenderloin, sirloin, center loin, and top loin).
- Choose processed meats such as lunch meats and hot dogs that are low in saturated fat and cholesterol.
- Trim the fat from pork and beef; remove the skin from poultry.
- Grill, roast, broil, bake, stir-fry, stew, or braise meats; don't fry. When possible, place food on a rack so that fat can drain.
- Use lean ground turkey or lean ground beef in recipes; brown ground meats without added fat, then drain off fat.
- Select tuna, sardines, and other canned meats packed in water; rinse oil-packed items with hot water to remove much of the fat.
- Fill kabob skewers with lots of vegetables and slivers of meat; create main dishes and casseroles by combining a little meat, fish, or poultry with a lot of pasta, rice, or vegetables.
- Use legumes often.
- Eat a meatless meal or two daily.
- Use egg substitutes in recipes instead of whole eggs or use two egg whites in place of each whole egg.

Fats and Oils

- Use butter or margarine sparingly; select soft margarines instead of hard margarines.

- Limit use of lard and meat fat.
- Limit use of products made with coconut oil, palm kernel oil, and palm oil (read labels on bakery goods, processed foods, popcorn oils, and nondairy creamers).
- Reduce use of hydrogenated shortenings and margarines and products that contain them (read labels on crackers, cookies, and other commercially prepared baked goods).

Miscellaneous

- Use a nonstick pan or coat the pan lightly with vegetable oil.
- Refrigerate soups and stews; when the fat solidifies, remove it.
- Use wine; lemon, orange, or tomato juice; herbs; spices; fruits; or broth instead of butter or margarine when cooking.
- Stir-fry in a small amount of oil; add moisture and flavor with broth, tomato juice, or wine.
- Use variety to enhance enjoyment of the meal: vary colors, textures, and temperatures—hot cooked versus cool raw foods—and use garnishes to complement food.

SOURCE: Adapted from Third Report of the National Cholesterol Education Program (NCEP) Expert Panel on Detection, Evaluation, and Treatment of High Blood Cholesterol in Adults (Adult Treatment Panel III), NIH publication no. 02-5215 (Bethesda, Md.: National Heart, Lung, and Blood Institute, 2002), pp. V-25–V-27.

---

trimming meats. Manufacturers can dilute fat by adding water or whipping in air. They use fat-free milk in creamy desserts and lean meats in frozen entrées. Sometimes they simply prepare the products differently. For example, fat-free potato chips may be baked instead of fried. Beyond lowering the fat content, manufacturers have developed margarines fortified with phytochemicals that lower blood cholesterol.*[48] (Highlight 13 explores these and other functional foods designed to support health.) Such choices make heart-healthy eating easy.

Fat Replacers Some foods are made with **fat replacers**—ingredients derived from carbohydrate, protein, or fat that replace some or all of the fat in foods. The body may digest and absorb some of these substances, so they may contribute some energy, although significantly less than fat's 9 kcalories per gram.

Some foods are made with **artificial fats**—fat replacers that offer the sensory and cooking qualities of fats, but none of the kcalories. A familiar example of an artificial fat that has been approved for use in snack foods such as potato chips, crackers, and tortilla chips is **olestra**. Olestra's chemical structure is similar to that of a regular fat (a triglyceride) but with important differences. A triglyceride is composed of a glycerol molecule with three fatty acids attached, whereas olestra is

fat replacers: ingredients that replace some or all of the functions of fat and may or may not provide energy.

artificial fats: zero-energy fat replacers that are chemically synthesized to mimic the sensory and cooking qualities of naturally occurring fats, but are totally or partially resistant to digestion.

olestra: a synthetic fat made from sucrose and fatty acids that provides 0 kcalories per gram; also known as **sucrose polyester.**

---

*Margarines that lower blood cholesterol contain plant sterols and are marketed under the brand names Benecol and Take Control.

made of a sucrose molecule with six to eight fatty acids attached. Enzymes in the digestive tract cannot break the bonds of olestra, so unlike sucrose or fatty acids, olestra passes through the system unabsorbed.

The FDA's evaluation of olestra's safety addressed two questions. First, is olestra toxic? Research on both animals and human beings supports the safety of olestra as a partial replacement for dietary fats and oils, with no reports of cancer or birth defects. Second, does olestra affect either nutrient absorption or the health of the digestive tract? When olestra passes through the digestive tract unabsorbed, it binds with some of the fat-soluble vitamins A, D, E, and K and carries them out of the body, robbing the person of these valuable nutrients. To compensate for these losses, the FDA requires the manufacturer to fortify olestra with vitamins A, D, E, and K. Saturating olestra with these vitamins does not make the product a good source of vitamins, but it does block olestra's ability to bind with the vitamins from other foods. An asterisk in the ingredients list informs consumers that these added vitamins are "dietarily insignificant."

Some consumers of olestra experience digestive distress: cramps, gas, bloating, and diarrhea. The FDA initially required a label warning stating that "olestra may cause abdominal cramping and loose stools" and that it "inhibits the absorption of some vitamins and other nutrients," but recently concluded that such a statement is no longer warranted.

Consumers need to keep in mind that low-fat and fat-free foods still deliver kcalories. Decades ago, consumers hailed the arrival of artificial sweeteners as a weight-loss wonder, but in reality, kcalories saved by using artificial sweeteners were readily replaced by kcalories from other foods. Alternatives to fat can help to lower energy intake and support weight loss only when they actually *replace* fat and energy in the diet.

Read Food Labels Labels list total fat, saturated fat, and cholesterol contents of foods in addition to fat kcalories per serving (see Figure 5-24). Because each package provides information for a single serving and serving sizes are standardized, consumers can easily compare similar products. By 2006 labels will provide content information on *trans*-fatty acids as well.[49] In the meantime, keep in mind that a food that lists partially hydrogenated oils among its first three ingredients usually contains substantial amounts of *trans*-fatty acids, as well as some saturated fat. Figure 5-24 includes tips on finding the *trans* fats on food labels.

Total fat, saturated fat, and cholesterol are also expressed as "% Daily Values" for a person consuming 2000 kcalories. People who are consuming more or less than 2000 kcalories daily can calculate their personal Daily Value for fat as described in the "How to" on p. 168 (top). *Trans* fats do not have a Daily Value.

Be aware that the "% Daily Value" for fat is not the same as "% kcalories from fat." This important distinction is explained in the "How to" feature on p. 168 (bottom). Because recommendations apply to average daily intakes and not to individual food items, food labels do not provide "% kcalories from fat." Still, you can get an idea of whether a particular food is high or low in fat.

IN SUMMARY

In foods, triglycerides:

- Deliver fat-soluble vitamins, energy, and essential fatty acids.
- Contribute to the sensory appeal of foods and stimulate appetite.

While some fat in the diet is necessary, health authorities recommend a diet moderate in total fat and low in saturated fat, *trans* fat, and cholesterol. They also recommend replacing saturated fats with monounsaturated and polyunsaturated fats, particularly omega-3 fatty acids from foods such as fish, not from supplements. Many selection and preparation strategies can help bring these goals within reach, and food labels help to identify foods consistent with these guidelines.

© Polara Studios Inc.

Well-balanced, healthy meals provide some fat with an emphasis on monounsaturated and polyunsaturated fats.

FIGURE 5-24   Butter and Margarine Labels Compared

Food labels list the kcalories from fat and the quantities and Daily Values for fat, saturated fat, and cholesterol. Information on polyunsaturated and monounsaturated fats is optional, but if it is provided, you can add the three types of fat together and subtract from the total to calculate *trans* fat. In this example, stick margarine has 3 g *trans* fat, tub margarine has 1 g *trans* fat, and liquid margarine has 0.5 g *trans* fat. Products that contain 0.5 g or less of *trans* fat and 0.5 g or less of saturated fat may claim "no *trans* fat." Similarly, products that contain 2 mg or less of cholesterol and 2 g or less of saturated fat may claim to be "cholesterol-free."

If the list of ingredients includes hydrogenated oils, you know the food contains *trans* fat. Chapter 2 explained that foods list their ingredients in descending order of predominance by weight. As you can see from this example, the closer "partially hydrogenated oils" is to the beginning of the ingredients list, the more *trans* fats the product contains.

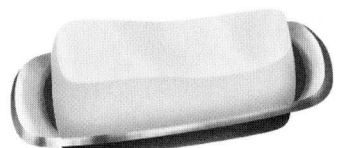

Butter

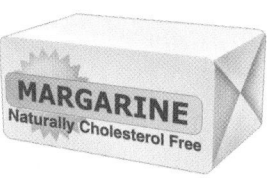

Margarine (stick)

Margarine (tub)

Margarine (liquid)

Butter

Nutrition Facts

Serving size 1 Tbsp (14g)
Servings Per Container 32

Amount per serving

Calories 100 Calories from Fat 100

	%Daily Value*
Total Fat 11g	17%
Saturated Fat 7g	36%
Cholesterol 30mg	10%
Sodium 90mg	4%
Total Carbohydrate 0g	0%
Protein 0g	

Vitamin A 8%

Not a significant source of dietary fiber, sugars, vitamin C, calcium, and iron.

*Percent Daily Values are based on a 2,000 calorie diet.

INGREDIENTS: Cream, salt.

Margarine (stick)

Nutrition Facts

Serving size 1 Tbsp (14g)
Servings Per Container 32

Amount per serving

Calories 90   Calories from Fat 90

	%Daily Value*
Total Fat 10g	15%
Saturated Fat 2g	10%
Polyunsaturated Fat 2g	
Monounsaturated Fat 3g	
Cholesterol 0mg	0%
Sodium 95mg	4%
Total Carbohydrate 0g	0%
Protein 0g	

Vitamin A 10%

Not a significant source of dietary fiber, sugars, vitamin C, calcium, and iron.

*Percent Daily Values are based on a 2,000 calorie diet.

INGREDIENTS: Vegetable oil blend (partially hydrogenated and liquid soybean oils), water, sweet cream buttermilk, salt, vegetable mono- and diglycerides, soy lecithin, citric acid, artificial flavor, vitamin A, colored with beta carotene.

Margarine (tub)

Nutrition Facts

Serving size 1 Tbsp (14g)
Servings Per Container 32

Amount per serving

Calories 90   Calories from Fat 90

	%Daily Value*
Total Fat 10g	15%
Saturated Fat 2g	10%
Polyunsaturated Fat 4.5g	
Monounsaturated Fat 2.5g	
Cholesterol 0mg	0%
Sodium 95mg	4%
Total Carbohydrate 0g	0%
Protein 0g	

Vitamin A 10%

Not a significant source of dietary fiber, sugars, vitamin C, calcium, and iron.

*Percent Daily Values are based on a 2,000 calorie diet.

INGREDIENTS: Water, liquid soybean oil, partially hydrogenated soybean oil, sweet cream, buttermilk, gelatin, salt, vegetable mono- and diglycerides, soy lecithin, lactic acid, artificial flavor, vitamin A, colored with beta carotene.

Margarine (liquid)

Nutrition Facts

Serving size 1 Tbsp (14g)
Servings Per Container 32

Amount per serving

Calories 60   Calories from Fat 60

	%Daily Value*
Total Fat 7g	10%
Saturated Fat 1g	6%
Polyunsaturated Fat 4g	
Monounsaturated Fat 1.5g	
Cholesterol 0mg	0%
Sodium 85mg	3%
Total Carbohydrate 0g	0%
Protein 0g	

Vitamin A 10%

Not a significant source of dietary fiber, sugars, vitamin C, calcium, and iron.

*Percent Daily Values are based on a 2,000 calorie diet.

INGREDIENTS: Liquid soybean oil, water, sweet cream buttermilk, salt, partially hydrogenated cottonseed oil, vegetable mono- and diglycerides, soy lecithin, citric acid, artificial flavor, vitamin A, colored with beta carotene.

If people were to make only one change in their diets, they would be wise to limit their intakes of saturated fat. Sometimes these choices can be difficult, though, because fats make foods taste delicious. To maintain good health, must a person give up all high-fat foods forever—never again to eat marbled steak, hollandaise sauce, or gooey chocolate cake? Not at all. These foods bring pleasure to a meal and can be enjoyed as part of a healthy diet when eaten in small quantities on occasion, but it is true that they are not everyday foods. The key word for fat is not deprivation, but moderation: Appreciate the energy and enjoyment that fat provides, but take care not to exceed your needs.

HOW TO Calculate a Personal Daily Value for Fat

The % Daily Value for fat on food labels is based on 65 grams. To know how your intake compares with this recommendation, you can either count grams until you reach 65, or add the "% Daily Values" until you reach 100 percent—if your energy intake is 2000 kcalories a day. If your energy intake is more or less, you have a couple of options.

You can calculate your personal daily fat allowance in grams. Suppose your energy intake is 1800 kcalories per day and your goal is 30 percent kcalories from fat. Multiply your total energy intake by 30 percent, then divide by 9:

1800 total kcal × 0.30 from fat = 540 fat kcal.
540 fat kcal ÷ 9 kcal/g = 60 g fat.

Another way to calculate your personal fat allowance is to cross out the last digit of your energy intake and divide by 3. For example, 1800 kcalories becomes 180; then you divide by 3:

180 ÷ 3 = 60 g fat/day.

(In familiar measures, 60 grams of fat is about the same as ⅔ stick of butter or ¼ cup of oil.)

The accompanying table shows the numbers of grams of fat allowed per day for various energy intakes. With one of these numbers in mind, you can quickly evaluate the number of fat grams in foods you are considering eating.

Energy (kcal/day)	20% kCalories from Fat	35% kCalories from Fat	Fat (g/day)
1200	240	420	27–47
1400	280	490	31–54
1600	320	560	36–62
1800	360	630	40–70
2000	400	700	44–78
2200	440	770	49–86
2400	480	840	53–93
2600	520	910	58–101
2800	560	980	62–109
3000	600	1050	67–117

HOW TO Understand "% Daily Value" and "% kCalories from Fat"

The "% Daily Value" that is used on food labels to describe the amount of fat in a food is not the same as the "% kcalories from fat" that is used in dietary recommendations to describe the amount of fat in the diet. They may appear similar, but their difference is worth understanding. Consider, for example, a piece of lemon meringue pie that provides 140 kcalories and 12 grams of fat. Because the Daily Value for fat is 65 grams for a 2000-kcalorie intake, 12 grams represent about 18 percent:

12 g ÷ 65 g = 0.18.
0.18 × 100 = 18%.

The pie's "% Daily Value" is 18 percent, or almost one-fifth, of the day's fat allowance.

Uninformed consumers may mistakenly believe that this food meets recommendations to limit fat to "20 to 35 percent kcalories," but it doesn't—for two reasons. First, the pie's 12 grams of fat contribute 108 of the 140 kcalories, for a total of 77 percent kcalories from fat:

12 g fat × 9 kcal/g = 108 kcal.
108 kcal ÷ 140 kcal = 77%.

Second, the "percent kcalories from fat" guideline applies to a day's total intake, not to an individual food. Of course, if every selection throughout the day exceeds 35 percent kcalories from fat, you can be certain that the day's total intake will, too.

© 1998 PhotoDisc, Inc.

Whether a person's energy and fat allowance can afford a piece of lemon meringue pie depends on the other food and activity choices made that day.

Nutrition in Your Life

To maintain good health, eat enough, but not too much, fat and select the right

kinds.

- Do you use polyunsaturated and monounsaturated vegetable oils instead of

  animal fats, hard margarine, and partially hydrogenated shortenings?

- Do you choose fat-free or low-fat milk products and lean meats, fish, and

  poultry?

- Do you eat fish at least twice a week? If not, does your diet include other

  sources of omega-3 fatty acids?

NUTRITION ON THE NET

 Access these websites for further study of topics
covered in this chapter.

- Find updates and quick links to these and other
  nutrition-related sites at our website:
  www.wadsworth.com/nutrition

- Search for "cholesterol" and "dietary fat" at the U.S.
  Government health information site:
  www.healthfinder.gov

- Review the American Dietetic Association's *ABC's of
  Fats, Oils, and Cholesterol:*
  www.eatright.org/nfs2.html

- Search for "fat" at the International Food Information
  Council site: **www.ific.org**

- Find dietary strategies to prevent heart disease at the
  American Heart Association: **www.americanheart.org**

NUTRITION CALCULATIONS

These problems will give you practice in doing simple
nutrition-related calculations (see p. 172 for answers). Show
your calculations for each problem.

1. Be aware of the fats in milks. Following are four
   categories of milk.

	Wt (g)	Fat (g)	Prot (g)	Carb (g)
Milk A (1 c)	244	8	8	12
Milk B (1 c)	244	5	8	12
Milk C (1 c)	244	3	8	12
Milk D (1 c)	244	0	8	12

   a. Based on *weight*, what percentage of each milk is fat
      (round off to a whole number)?

   b. How much energy from fat will a person receive from
      drinking 1 cup of each milk?

   c. How much total energy will the person receive from 1
      cup of each milk?

   d. What percentage of the energy in each milk comes
      from fat?

   e. In the grocery store, how is each milk labeled?

2. Judge foods' fat contents by their labels.

   a. A food label says that one serving of the food contains
      6.5 grams fat. What would the % Daily Value for fat be?
      What does the Daily Value you just calculated mean?

   b. How many kcalories from fat does a serving contain?
      (Round off to the nearest whole number.)

   c. If a *serving* of the food contains 200 kcalories, what
      percentage of the energy is from fat?

This example should show you how easy it is to evaluate
foods' fat contents by reading labels and to see the difference
between the % Daily Value and the percentage of kcalories
from fat.

3. Now consider a piece of carrot cake. Remember that the
   Daily Value suggests 65 grams of fat as acceptable within
   a 2000-kcalorie diet. A serving of carrot cake provides 30
   grams of fat. What percentage of the Daily Value is that?
   What does this mean?

STUDY QUESTIONS

These questions will help you review this chapter. You will find the answers in the discussions on the pages provided.

1. Name three classes of lipids found in the body and in foods. What are some of their functions in the body? What features do fats bring to foods? (pp. 141, 155–157)

2. What features distinguish fatty acids from each other? (pp. 142–145)

3. What does the term *omega* mean with respect to fatty acids? Describe the roles of the omega fatty acids in disease prevention. (pp. 144,160–161)

4. What are the differences between saturated, unsaturated, monounsaturated, and polyunsaturated fatty acids? Describe the structure of a triglyceride. (pp. 142–145)

5. What does hydrogenation do to fats? What are *trans*-fatty acids, and how do they influence heart disease? (pp. 146, 159)

6. How do phospholipids differ from triglycerides in structure? How does cholesterol differ? How do these differences in structure affect function? (pp. 148–149)

7. What roles do phospholipids perform in the body? What roles does cholesterol play in the body? (pp. 148–149)

8. Trace the steps in fat digestion, absorption, and transport. Describe the routes cholesterol takes in the body. (pp. 150–153)

9. What do lipoproteins do? What are the differences among the chylomicrons, VLDL, LDL, and HDL? (pp. 153–154)

10. Which of the fatty acids are essential? Name their chief dietary sources. (pp. 156–157)

11. How does excessive fat intake influence health? What factors influence LDL, HDL, and total blood cholesterol? (pp. 159–162)

12. What are the dietary recommendations regarding fat and cholesterol intake? List ways to reduce intake. (pp. 162–166)

13. What is the Daily Value for fat (for a 2000-kcalorie diet)? What does this number represent? (pp. 162, 168)

These multiple choice questions will help you prepare for an exam. Answers can be found on p. 172.

1. Saturated fatty acids:
   a. are always 18 carbons long.
   b. have at least one double bond.
   c. are fully loaded with hydrogens.
   d. are always liquid at room temperature.

2. A triglyceride consists of:
   a. three glycerols attached to a lipid.
   b. three fatty acids attached to a glucose.
   c. three fatty acids attached to a glycerol.
   d. three phospholipids attached to a cholesterol.

3. The difference between *cis*- and *trans*-fatty acids is:
   a. the number of double bonds.
   b. the length of their carbon chains.
   c. the location of the first double bond.
   d. the configuration around the double bond.

4. Which of the following is *not* true? Lecithin is:
   a. an emulsifier.
   b. a phospholipid.
   c. an essential nutrient.
   d. a constituent of cell membranes.

5. Chylomicrons are produced in the:
   a. liver.
   b. pancreas.
   c. gallbladder.
   d. small intestine.

6. Transport vehicles for lipids are called:
   a. micelles.
   b. lipoproteins.
   c. blood vessels.
   d. monoglycerides.

7. The lipoprotein most associated with a high risk of heart disease is:
   a. CHD.
   b. HDL.
   c. LDL.
   d. LPL.

8. Which of the following is not true? Fats:
   a. contain glucose.
   b. provide energy.
   c. protect against organ shock.
   d. carry vitamins A, D, E, and K.

9. The essential fatty acids include:
   a. stearic acid and oleic acid.
   b. oleic acid and linoleic acid.
   c. palmitic acid and linolenic acid.
   d. linoleic acid and linolenic acid.

10. A person consuming 2200 kcalories a day who wants to meet health recommendations should limit daily fat intake to:
   a. 20 to 35 grams.
   b. 50 to 85 grams.
   c. 75 to 100 grams.
   d. 90 to 130 grams.

REFERENCES

1. C. T. Phan and P. Tso, Intestinal lipid absorption and transport, *Frontiers in Bioscience* 6 (2001): 299–319.
2. L. Brown and coauthors, Cholesterol-lowering effects of dietary fiber: A meta-analysis, *American Journal of Clinical Nutrition* 69 (1999): 30–42.
3. A. Andersson and coauthors, Fatty acid composition of skeletal muscle reflects dietary fat composition in humans, *American Journal of Clinical Nutrition* 76 (2002): 1222–1229; A. Baylin and coauthors, Adipose tissue biomarkers of fatty acid intake, *American Journal of Clinical Nutrition* 76 (2002): 750–757.
4. C. Bruce, R. A. Chouinard, Jr., and A. R. Tall, Plasma lipid transfer proteins, high-density lipoproteins, and reverse cholesterol transport, *Annual Review of Nutrition* 18 (1998): 297–330.
5. C. D. Berdanier, *Advanced Nutrition: Macronutrients* (Boca Raton, Fla.: CRC Press, 2000), pp. 273–280.
6. R. Uauy and P. Mena, Lipids and neuro-development, *Nutrition Reviews* 59 (2001): S34–S48.
7. D. Hwang, Fatty acids and immune responses—A new perspective in searching for clues to mechanism, *Annual Review of Nutrition* 20 (2000): 431–456; O. Morteau, Prostaglandins and inflammation: The cyclooxygenase controversy, *Archivum Immunologiae et Therapiae Experimentalis* 48 (2000): 473–480.
8. J. R. Hibbeln, Seafood consumption, the DHA content of mothers' milk and prevalence rates of postpartum depression: A cross-national, ecological analysis, *Journal of Affective Disorders* 69 (2002): 15–29; K. A. Bruinsma and D. L. Taren, Dieting, essential fatty acid intake, and depression, *Nutrition Reviews* 58 (2000): 98–108; A. L. Stoll and coauthors, Omega 3 fatty acids in bipolar disorder: A preliminary double-blind placebo-controlled trial, *Archives of General Psychiatry* 56 (1999): 407–412.
9. L. Zimmer and coauthors, The dopamine mesocorticolimbic pathway is affected by deficiency in n-3 polyunsaturated fatty acids, *American Journal of Clinical Nutrition* 75 (2002): 662–667.
10. J. Stamler and coauthors, Relationship of baseline serum cholesterol levels in 3 large cohorts of younger men to long-term coronary, cardiovascular, and all-cause mortality and to longevity, *Journal of the American Medical Association* 284 (2000): 311–318; D. Steinberg and A. M. Gotto, Preventing coronary artery disease by lowering cholesterol levels, *Journal of the American Medical Association* 282 (1999): 2043–2050.
11. F. D. Kelly and coauthors, A stearic acid–rich diet improves thrombogenic and atherogenic risk factor profiles in healthy males, *European Journal of Clinical Nutrition* 55 (2001): 88–96; F. B. Hu and coauthors, Dietary saturated fats and their food sources in relation to the risk of coronary heart disease in women, *American Journal of Clinical Nutrition* 70 (1999): 1001–1008; W. E. Connor, Harbingers of coronary heart disease: Dietary saturated fatty acids and cholesterol—Is chocolate benign because of its stearic acid content? *American Journal of Clinical Nutrition* 70 (1999): 951–952.
12. Hu and coauthors, 1999.
13. J. I. Pederson and coauthors, Adipose tissue fatty acids and risk of myocardial infarction, *European Journal of Clinical Nutrition* 54 (2000): 618–625.
14. M. B. Katan, *Trans* fatty acids and plasma lipoproteins, *Nutrition Reviews* 58 (2000): 188–191.
15. D. B. Allison and coauthors, Estimated intakes of *trans* fatty and other fatty acids in the US population, *Journal of the American Dietetic Association* 99 (1999): 166–174.
16. M. A. Denke, B. Adams-Huet, and A. T. Nguyen, Individual cholesterol variation in response to a margarine- or butter-based diet: A study in families, *Journal of the American Medical Association* 284 (2000): 2740–2747; A. H. Lichtenstein and coauthors, Effects of different forms of dietary hydrogenated fats on serum lipoprotein cholesterol levels, *New England Journal of Medicine* 340 (1999): 1933–1940.
17. W. D. Song and J. M. Kerver, Nutritional contribution of eggs to American diets, *Journal of the American College of Nutrition* 19 (2000): 556S–562S; S. B. Kritchevsky and D. Kritchevsky, Egg consumption and coronary heart disease: An epidemiologic review, *Journal of the American College of Nutrition* 19 (2000): 549S–555S; F. B. Hu and coauthors, A prospective study of egg consumption and risk of cardiovascular disease in men and women, *Journal of the American Medical Association* 281 (1999): 1387–1394.
18. Song and Kerver, 2000.
19. Lichtenstein and coauthors, 1999.
20. C. Thomsen and coauthors, Differential effects of saturated and monounsaturated fatty acids on postprandial lipemia and incretin responses in healthy subjects, *American Journal of Clinical Nutrition* 69 (1999): 1135–1143.
21. A. H. Stark and Z. Madar, Olive oil as a functional food: Epidemiology and nutritional approaches, *Nutrition Reviews* 60 (2002): 170–176.
22. F. B. Hu and coauthors, Fish and omega-3 fatty acid intake and risk of coronary heart disease in women, *Journal of the American Medical Association* 287 (2002): 1815–1821; C. M. Albert and coauthors, Blood levels of long-chain n-3 fatty acids and the risk of sudden death, *New England Journal of Medicine* 346 (2002): 1113–1118; C. von Schacky, n-3 Fatty acids and the prevention of coronary atherosclerosis, *American Journal of Clinical Nutrition* 71 (2000): 224S–227S.
23. P. J. H. Jones and V. W. Y. Lau, Effect of n-3 polyunsaturated fatty acids on risk reduction of sudden death, *Nutrition Reviews* 60 (2002): 407–413; P. J. Nestel, Fish oil and cardiovascular disease: Lipids and arterial function, *American Journal of Clinical Nutrition* 71 (2000): 228S–231S.
24. E. Guallar and coauthors, Mercury, fish oils, and the risk of myocardial infarction, *New England Journal of Medicine* 347 (2002): 1747–1754.
25. AHA Dietary Guidelines, published online on October 5, 2000, http://circ.ahajournals.org/cgi/content/full/4304635102.
26. K. He and coauthors, Fish consumption and risk of stroke in men, *Journal of the American Medical Association* 288 (2002): 3130–3136.
27. T. A. Mori and coauthors, Dietary fish as a major component of a weight-loss diet: Effect on serum lipids, glucose, and insulin metabolism in overweight hypertensive subjects, *American Journal of Clinical Nutrition* 70 (1999): 817–825.
28. L. Cordain and coauthors, Fatty acid analysis of wild ruminant tissues: Evolutionary implications for reducing diet-related chronic disease, *European Journal of Clinical Nutrition* 56 (2002): 181–191.
29. T. A. Mori and L. J. Beilin, Long-chain omega 3 fatty acids, blood lipids and cardiovascular risk reduction, *Current Opinion in Lipidology* 12 (2001): 11–17.
30. S. Bechoua and coauthors, Influence of very low dietary intake of marine oil on some functional aspects of immune cells in healthy elderly people, *British Journal of Nutrition* 89 (2003): 523–532; F. Thies and coauthors, Dietary supplementation with eicosapentaenoic acid, but not with other long-chain n-3 or n-6 polyunsaturated fatty acids, decreases natural killer cell activity in healthy subjects aged >55 y, *American Journal of Clinical Nutrition* 73 (2001): 539–548; V. M. Montori and coauthors, Fish oil supplementation in type 2 diabetes: A quantitative systematic review, *Diabetes Care* 23 (2000): 1407–1415.
31. P. M. Kris-Etherton and coauthors, AHA Scientific Statement: Fish consumption, fish oil, omega-3 fatty acids, and cardiovascular disease, *Circulation* 106 (2002): 2747–2757.
32. P. L. Zock, Dietary fats and cancer, *Current Opinion in Lipidology* 12 (2001): 5–10.
33. M. M. Lee and S. S. Lin, Dietary fat and breast cancer, *Annual Review of Nutrition* 20 (2000): 221–248; E. B. Feldman, Breast cancer risk and intake of fat, *Nutrition Reviews* 57 (1999): 353–356.
34. M. D. Holmes and coauthors, Association of dietary intake of fat and fatty acids with risk of breast cancer, *Journal of the American Medical Association* 281 (1999): 914–920.
35. A. R. Kristal and coauthors, Associations of energy, fat, calcium, and vitamin D with prostate cancer risk, *Cancer Epidemiology, Biomarkers and Prevention* 11 (2002): 719–725; L. N. Kolonel, A. M. Nomura, and R. V. Cooney, Dietary fat and prostate cancer: Current status, *Journal of the National Cancer Institute* 91 (1999): 414–428; J. A. Thomas, Diet, micronutrients, and the prostate gland, *Nutrition Reviews* 57 (1999): 95–103.
36. R. F. Gimble and coauthors, The ability of fish oil to suppress tumor necrosis factor α production by peripheral blood mononuclear cells in healthy men is associated with polymorphisms in genes that influence tumor necrosis factor α production, *American Journal of Clinical Nutrition* 76 (2002): 454–459; L. Guangming and coauthors, Omega 3 but not omega 6 fatty acids inhibit AP-1 activity and cell transformation in JB6 cells, *Proceedings of the National Academy of Sciences* 98 (2001): 7510–7515; P. Terry and coauthors, Fatty fish consumption and risk of prostate cancer, *Lancet* 357 (2001): 1764–1766; E. D. Collett and coauthors, n-6 and n-3 polyunsaturated fatty acids differentially modulate oncogenic Ras activation in colonocytes, *American Journal of Physiology: Cell Physiology* 280 (2001): C1066–C1075; E. Fernandez and coauthors, Fish consumption and cancer risk, *American Journal of Clinical Nutrition* 70 (1999): 85–90.
37. Committee on Dietary Reference Intakes, *Dietary Reference Intakes for Energy, Carbohydrate, Fiber, Fat, Fatty Acids, Cholesterol, Protein, and Amino Acids* (Washington, D.C.: National Academies Press, 2002).
38. Committee on Dietary Reference Intakes, 2002.
39. Committee on Dietary Reference Intakes, 2002.
40. N. D. Ernst and coauthors, Consistency between US dietary fat intake and serum

total cholesterol concentrations: The National Health and Nutrition Examination Surveys, *American Journal of Clinical Nutrition* 66 (1997): 965S–972S; National Center for Health Statistics, www.cdc.gov/nchs, site visited on November 6, 2000.

41. P. J. Horvath and coauthors, The effects of varying dietary fat on performance and metabolism in trained male and female runners, *Journal of the American College of Nutrition* 19 (2000): 52–60; P. J. Horvath and coauthors, The effects of varying dietary fat on the nutrient intake in trained male and female runners, *Journal of the American College of Nutrition* 19 (2000): 42–51.

42. D. R. Pendergast, J. J. Leddy, and J. T. Venkatraman, A perspective on fat intake in athletes, *Journal of the American College of Nutrition* 19 (2000): 345–350.

43. Committee on Dietary Reference Intakes, 2002; Pendergast, 2000.

44. S. Tonstad, K. Smerud, and L. Høie, A comparison of the effects of 2 doses of soy protein or casein on serum lipids, serum lipoproteins, and plasma total homocysteine in hyper-cholesterolemic subjects, *American Journal of Clinical Nutrition* 76 (2002): 78–84; S. R. Teixeira and coauthors, Effects of feeding 4 levels of soy protein for 3 and 6 wk on blood lipids and apolipoproteins in moderately hypercholesterolemic men, *American Journal of Clinical Nutrition* 71 (2000): 1077–1084.

45. M. Pfeuffer and J. Schrezenmeir, Bioactive substances in milk with properties decreasing risk of cardiovascular diseases, *British Journal of Nutrition* 84 (2000): S155–S159.

46. B. M. Davy and coauthors, High-fiber oat cereal compared with wheat cereal consumption favorably alters LDL-cholesterol subclass and particle numbers in middle-aged and older men, *American Journal of Clinical Nutrition* 76 (2002): 351–358; D. J. A. Jenkins and coauthors, Soluble fiber intake at a dose approved by the US Food and Drug Administration for a claim of health benefits: Serum lipid risk factors for cardiovascular disease assessed in a randomized controlled crossover trial, *American Journal of Clinical Nutrition* 75 (2002): 834–839; L. Van Horn and N. Ernst, A summary of the science supporting the new National Cholesterol Education program dietary recommendations: What dietitians should know, *Journal of the American Dietetic Association* 101 (2001): 1148–1154; M. L. Fernandez, Soluble fiber and nondigestible carbohydrate effects on plasma lipids and cardiovascular risk, *Current Opinion in Lipidology* 12 (2001): 35–40; J. W. Anderson and coauthors, Cholesterol-lowering effects of psyllium intake adjunctive to diet therapy in men and women with hypercholesterolemia: A meta-analysis of 8 controlled trials, *American Journal of Clinical Nutrition* 71 (2000): 472–479; M. P. St. Onge, E. R. Farnworth, and P. J. H. Jones, Consumption of fermented and nonfermented dairy products: Effects on cholesterol concentrations and metabolism, *American Journal of Clinical Nutrition* 71 (2000): 674–681; Brown and coauthors, 1999.

47. B. M. Popkin and coauthors, Where's the fat? Trends in U.S. diets 1965–1996, *Preventive Medicine* 32 (2001): 245–254.

48. P. J. H. Jones and coauthors, Cholesterol-lowering efficacy of a sitostanol-containing phytosterol mixture with a prudent diet in hyperlipidemic men, *American Journal of Clinical Nutrition* 69 (1999): 1144–1150; M. A. Hallikainen and M. I. J. Uusitupa, Effects of 2 low-fat stanol ester-containing margarines on serum cholesterol concentrations as part of a low-fat diet in hypercholesterolemic subjects, *American Journal of Clinical Nutrition* 69 (1999): 403–410.

49. J. G. Dausch, Trans-fatty acids: A regulatory update, *Journal of the American Dietetic Association* 102 (2002): 18.

ANSWERS

Nutrition Calculations

1. a. Milk A: 8 g fat ÷ 244 g total = 0.03; 0.03 × 100 = 3%.

   Milk B: 5 g fat ÷ 244 g total = 0.02; 0.02 × 100 = 2%.

   Milk C: 3 g fat ÷ 244 g total = 0.01; 0.01 × 100 = 1%.

   Milk D: 0 g fat ÷ 244 g total = 0.00; 0.00 × 100 = 0%.

   b. Milk A: 8 g fat × 9 kcal/g = 72 kcal from fat.

   Milk B: 5 g fat × 9 kcal/g = 45 kcal from fat.

   Milk C: 3 g fat × 9 kcal/g = 27 kcal from fat.

   Milk D: 0 g fat × 9 kcal/g = 0 kcal from fat.

   c. Milk A: (8 g fat × 9 kcal/g) + (8 g prot × 4 kcal/g) + (12 g carb × 4 kcal/g) = 152 kcal.

   Milk B: (5 g fat × 9 kcal/g) + (8 g prot × 4 kcal/g) + (12 g carb × 4 kcal/g) = 125 kcal.

   Milk C: (3 g fat × 9 kcal/g) + (8 g prot × 4 kcal/g) + (12 g carb × 4 kcal/g) = 107 kcal.

   Milk D: (0 g fat × 9 kcal/g) + (8 g prot × 4 kcal/g) + (12 g carb × 4 kcal/g) = 80 kcal.

   d. Milk A: 72 kcal from fat ÷ 152 total kcal = 0.47; 0.47 × 100 = 47%.

   Milk B: 45 kcal from fat ÷ 125 total kcal = 0.36; 0.36 × 100 = 36%.

   Milk C: 27 kcal from fat ÷ 107 total kcal = 0.25; 0.25 × 100 = 25%.

   Milk D: 0 kcal from fat ÷ 80 total kcal = 0.00; 0.00 × 100 = 0%.

   e. Milk A: whole.

   Milk B: reduced-fat, 2%, or less-fat.

   Milk C: low-fat or 1%.

   Milk D: fat-free, nonfat, skim, zero-fat, or no-fat.

2. a. 6.5 g ÷ 65 g = 0.1; 0.1 × 100 = 10%. A Daily Value of 10% means that one serving of this food contributes about ¹⁄₁₀ of the day's fat allotment.

   b. 6.5 g × 9 kcal/g = 58.5, rounded to 59 kcal from fat.

   c. (59 kcal from fat ÷ 200 kcal) × 100 = 30% kcalories from fat.

3. (30 g fat ÷ 65 g fat) × 100 = 46% of the Daily Value for fat; this means that almost half of the day's fat allotment would be used in this one dessert.

Study Questions (multiple choice)

1. c   2. c   3. d   4. c   5. d   6. b   7. c   8. a
9. d   10. b

High-Fat Foods—Friend or Foe?

© Philip Salaverry/FoodPix/Getty Images

Eat less fat. Eat more fatty fish. Give up butter. Use margarine. Give up margarine. Use olive oil. Steer clear of saturated. Seek out omega-3. Stay away from *trans*. Stick with mono- and polyunsaturated. Keep fat intake moderate. Today's fat messages seem to be forever multiplying and changing. No wonder people feel confused about dietary fat. The confusion stems in part from the complexities of fat and in part from the nature of recommendations. As Chapter 5 explained, "dietary fat" refers to several kinds of fats, some fats support health whereas others damage it, and foods typically provide a mixture of fats in varying proportions. It has taken researchers decades to sort through the relationships between the various kinds of fat and their roles in supporting or harming health. Translating these research findings into dietary recommendations is a challenging process. Too little information can mislead consumers, but too much detail can overwhelm them. As research findings accumulate, recommendations slowly evolve and become more refined. That's where we are with fat recommendations today—refining them from the general to the specific. Though they may seem to be "forever multiplying and changing," in fact, they are becoming more meaningful.

This highlight begins with a look at these changing guidelines. It continues by identifying which foods provide which fats and presenting the Mediterranean diet, an example of a food plan that embraces the heart-healthy fats. It closes with strategies to help consumers choose the right amounts of the right kinds of fats for a healthy diet.

Changing Guidelines for Fat Intake

Dietary recommendations for fat have changed in recent years, shifting the emphasis from lowering total fat, in general, to limiting saturated and *trans* fat, specifically. For decades, health experts advised limiting intakes of total fat to 30 percent or less of energy intake. They recognized that saturated fats and *trans* fats were the ones that raise blood cholesterol, but reasoned that by limiting total fat intake, saturated and *trans* fat intake would decline as well. People were simply advised to cut back on all fat so that they would cut back on saturated and *trans* fat. Such advice may have oversimplified the message and unnecessarily restricted total fat.

Low-fat diets have a place in treatment plans for people with elevated blood lipids or heart disease, but researchers question the wisdom of such diets for healthy people as a means of controlling weight and preventing diseases.[1] Several problems accompany low-fat diets. For one, many people find low-fat diets difficult to maintain over time.[2] For another, low-fat diets are not necessarily low-kcalorie diets; if energy intake exceeds energy needs, weight gain follows, and obesity brings a host of health problems, including heart disease. For still another, diets extremely low in fat may exclude fatty fish, nuts, seeds, and vegetable oils—all valuable sources of many essential fatty acids, phytochemicals, vitamins, and minerals. Importantly, the fats from these sources protect against heart disease, as later sections of this highlight explain.

Today, health experts have revised dietary recommendations to acknowledge that not all fats have damaging health consequences. In fact, higher intakes of some kinds of fats (for example, the omega-3 fatty acids) support good health. Instead of urging people to cut back on all fats, current recommendations suggest carefully replacing the "bad" saturated fats with the "good" unsaturated fats and enjoying them in moderation.[3] The goal is to create a diet moderate in kcalories that provides enough of the fats that support good health, but not too much of those that harm health. (Turn to pp. 159–160 for a review of the health consequences of each type of fat.)

With these findings and goals in mind, the DRI committee recently established a healthy range of 20 to 35 percent of energy intake from fat. This range appears to be compatible with low rates of heart disease, diabetes, obesity, and cancer.[4] Heart-healthy recommendations suggest that within this range, consumers should try to minimize their intakes of saturated fat, *trans* fat, and cholesterol and use monounsaturated and polyunsaturated fats instead.[5]

Asking consumers to limit their total fat intake was less than perfect advice, but it was straightforward—find the fat and cut back. Asking consumers to keep their intakes of saturated fats, *trans* fats, and cholesterol low and to use monounsaturated and polyunsaturated fats instead may be more on target with heart health, but it also makes diet planning more complicated. To make appropriate selections, consumers must first learn which foods contain which fats.

High-Fat Foods and Heart Health

Avocados, bacon, walnuts, potato chips, and mackerel are all high-fat foods, yet some of these foods have detrimental effects on heart health when consumed in excess, while others seem neutral or even beneficial. This section presents some of the accumulating evidence that helped to distinguish which high-fat foods belong in a healthy diet and which ones need to be kept to a minimum. As you will see, a little more fat in the diet may be compatible with heart health, but only if the great majority of it is the unsaturated kind.

Cook with Olive Oil

As it turns out, the traditional diets of Greece and other countries in the Mediterranean region offer an excellent example of eating patterns that use "good" fats liberally. Often, these diets are rich in olives and their oil. A classic study of the world's people, the Seven Countries Study, found that death rates from heart disease were strongly associated with diets high in saturated fats, but only weakly linked with total fat.[6] In fact, the two countries with the highest fat intakes, Finland and the Greek island of Crete, had the highest (Finland) and lowest (Crete) rates of heart disease deaths. In both countries, the people consumed 40 percent or more of their kcalories from fat. Clearly, a high-fat diet was not the primary problem, so researchers refocused their attention on the type of fat. They began to notice the benefits of olive oil.

A diet that uses olive oil instead of other cooking fats, especially butter, stick margarine, and meat fats, may offer numerous health benefits.[7] Olive oil helps to protect against heart disease by:

- Lowering total and LDL cholesterol and not lowering HDL cholesterol or raising triglycerides.[8]
- Lowering LDL cholesterol susceptibility to oxidation.[9]
- Lowering blood-clotting factors.[10]
- Providing phytochemicals that act as antioxidants (see Highlight 11).[11]
- Lowering blood pressure.[12]

When compared with other fats, olive oil seems to be a wise choice, but controlled clinical trials are too scarce to support population-wide recommendations to switch to a high-fat diet rich in olive oil.[13] Importantly, olive oil is not a magic potion; drizzling it on foods does not make them healthier. Like other fats, olive oil delivers 9 kcalories per gram, which can contribute to weight gain in people who fail to balance their energy intake with their energy output. Its role in a healthy diet is to *replace* the saturated fats. Other vegetable oils, such as canola or safflower oil, in their liquid unhydrogenated states, are also generally low in saturated fats and high in unsaturated fats. For

Olives and their oil may benefit heart health.

this reason, heart-healthy diets use these unsaturated vegetable oils as substitutes for the more saturated fats of butter, hydrogenated stick margarine, lard, or shortening. (Remember that the tropical oils—coconut, palm, and palm kernel—are too saturated to be included with the heart-healthy vegetable oils.)

Nibble on Nuts

Tree nuts and peanuts are traditionally excluded from low-fat diets, and for good reasons. Nuts provide up to 80 percent of their kcalories from fat, and a quarter cup (about an ounce) of mixed nuts provides over 200 kcalories. In a recent review of the literature, however, researchers found that people who ate a one-ounce serving of nuts on five or more days a week had a reduced risk of heart disease compared with people consuming no nuts.[14] A smaller positive association was noted for any amount greater than one serving of nuts a week. The nuts were those commonly eaten in the United States: almonds, Brazil nuts, cashews, hazelnuts, macadamia nuts, pecans, pistachios, walnuts, and even peanuts. On average, these nuts contain mostly monounsaturated fat (59 percent), some polyunsaturated fat (27 percent), and little saturated fat (14 percent).[15]

Research has shown a benefit from walnuts and almonds in particular. In study after study, walnuts, when substituted for other fats in the diet, produce favorable effects on blood lipids—even in people with elevated total and LDL cholesterol.[16] Results are similar for almonds. In one study, researchers gave men and women one of three kinds of snacks, all of equal kcalories: whole-wheat muffins, almonds (about 2½ ounces), or half muffins and half almonds.[17] At the end of a month, people receiving the full almond snack had the greatest drop in blood LDL cholesterol; those eating the half almond snack had a lesser, but still significant drop in blood lipids; and those eating the muffin-only snack had no change.

Matthew Farruggio

For heart health, snack on nuts instead of potato chips.

Studies on peanuts, macadamia nuts, pecans, and pistachios follow suit, indicating that including nuts may be a wise strategy against heart disease. Nuts may protect against heart disease because they provide:

- Monounsaturated and polyunsaturated fats in abundance, but few saturated fats.
- Fiber, vegetable protein, and other valuable nutrients, including the antioxidant vitamin E (see Highlight 11).
- Phytochemicals that act as antioxidants (see Highlight 13).

Before advising consumers to include nuts in their diets, a caution is in order. As mentioned, most of the energy nuts provide comes from fats. Consequently, they deliver many kcalories per bite. In studies examining the effects of nuts on heart disease, researchers carefully adjust diets to make room for the nuts without increasing the total kcalories—that is, they use nuts *instead of, not in addition to,* other foods (such as meats, potato chips, oils, margarine, and butter). Consumers who do not make similar replacements could end up gaining weight if they simply add nuts on top of their regular diets. Weight gain, in turn, elevates blood lipids and raises the risks of heart disease.

Feast on Fish

Research into the health benefits of the long-chain omega-3 polyunsaturated fatty acids began with a simple observation: The native peoples of Alaska, northern Canada, and Greenland, who eat a diet rich in omega-3 fatty acids, notably EPA and DHA, have a remarkably low rate of heart disease even though their diets are relatively high in fat.[18] These omega-3 fatty acids help to protect against heart disease by:[19]

- Reducing blood triglycerides.
- Preventing blood clots.
- Protecting against irregular heartbeats.
- Lowering blood pressure.
- Defending against inflammation.

- Serving as precursors to eicosanoids.

For people with hypertension or atherosclerosis, these actions can be life saving. (Chapter 18 presents more details on the action of omega-3 fatty acids in preventing heart disease.)

Research studies have provided strong evidence that increasing omega-3 fatty acids in the diet supports heart health and lowers the rate of deaths from heart disease.[20] For this reason, the American Heart Association recommends including fish in a heart-healthy diet. People who eat some fish each week can lower their risks of heart attack and stroke.[21] Table 5-2 on p. 161 lists fish that provide at least 1 gram of omega-3 fatty acids per serving.

Fish is the best source of EPA and DHA in the diet, but it is also a major source of mercury, an environmental contaminant. Most fish contain at least trace amounts of mercury, but tilefish, swordfish, king mackerel, marlin, and shark have especially high levels. For this reason, the FDA advises pregnant and lactating women, women of childbearing age who may become pregnant, and young children to avoid:

- Tilefish, swordfish, king mackeral, marlin, and shark.

And to limit average weekly consumption of:

- Ocean, coastal, and other commercial fish to 12 ounces (cooked or canned) *or* freshwater fish caught by family and friends to 6 ounces (cooked).

Others may want to adopt this advice as well. In addition to the direct toxic effects of mercury, some (but not all) research suggests that mercury may diminish the health benefits of omega-3 fatty acids.[22] Such findings serve as a reminder that our health depends on the health of our planet. The protective effect of fish in the diet is available, provided that the fish and their surrounding waters are not heavily contaminated. (Chapter 19 discusses the adverse consequences of mercury, and Chapter 20 presents the relationships between diet and the environment in more detail.)

In an effort to limit exposure to pollutants, some consumers choose farm-raised fish. Compared with fish caught in the wild, farm-raised fish do tend to be lower in mercury, but they are also lower in

© www.comstock.com

Fish is a good source of the omega-3 fatty acids.

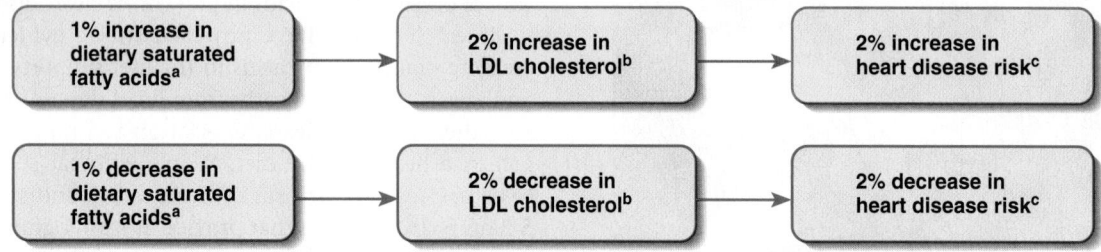

FIGURE H5-1 Potential Relationships among Dietary Saturated Fatty Acids, LDL Cholesterol, and Heart Disease Risk

[a]Percentage of change in total dietary energy from saturated fatty acids.
[b]Percentage of change in blood LDL cholesterol.
[c]Percentage of change in an individual's risk of heart disease; the percentage of change in risk may increase when blood lipid changes are sustained over time.

SOURCE: Third Report of the National Cholesterol Education Program (NCEP) Expert Panel on Detection, Evaluation, and Treatment of High Blood Cholesterol in Adults (Adult Treatment Panel III), NIH publication no. 02-5215 (Bethesda, Md.: National Heart, Lung, and Blood Institute, 2002), p. V-8 and II-4.

omega-3 fatty acids. When selecting fish, keep the diet strategies of variety and moderation in mind. Varying choices and eating moderate amounts helps to limit the intake of contaminants such as mercury.

High-Fat Foods and Heart Disease

The number one dietary determinant of LDL cholesterol is saturated fat. Figure H5-1 shows that each 1 percent increase in energy from saturated fatty acids in the diet may produce a 2 percent jump in heart disease risk by elevating blood LDL cholesterol. Conversely, reducing saturated fat intake by 1 percent can be expected to produce a 2 percent drop in heart disease risk by the same mechanism. Even a 2 percent drop in LDL represents a significant improvement for the health of the heart.[23] Like saturated fats, *trans* fats also raise heart disease risk by elevating LDL cholesterol. A heart-healthy diet limits foods rich in these two types of fat.

Limit Fatty Meats, Whole-Milk Products, and Tropical Oils

The major sources of saturated fats in the U.S. diet are fatty meats, whole-milk products, tropical oils, and products made from any of these foods. To limit saturated fat intake, consumers must choose carefully among these high-fat foods. Over a third of the fat in most meats is saturated. Similarly, over half of the fat is saturated in whole milk and other high-fat dairy products, such as cheese, butter, cream, half-and-half, cream cheese, sour cream, and ice cream. The tropical oils of palm, palm kernel, and coconut are rarely used by consumers in the kitchen, but are used heavily by food manufacturers and

so are commonly found in many commercially prepared foods.

When choosing meats, milk products, and commercially prepared foods, look for those lowest in saturated fat. Labels provide a useful guide for comparing products in this regard, and Appendix H lists the saturated fat in several thousand foods.

Even with careful selections, a nutritionally adequate diet will provide some saturated fat. Zero saturated fat is not possible even when experts design menus with the mission to keep saturated fat as low as possible.[24] Because most saturated fats come from animal foods, vegetarian diets can, and usually do, deliver fewer saturated fats than mixed diets.

Limit Hydrogenated Foods

Chapter 5 explained that solid shortening and margarine are made from vegetable oil that has been hardened through hydrogenation. This process both saturates some of the unsaturated fatty acids and introduces *trans*-fatty acids. Many convenience foods contain *trans* fats, including:

- Fried foods such as french fries, chicken, and other commercially fried foods.
- Commercial baked goods such as cookies, doughnuts, pastries, breads, and crackers.
- Snack foods such as chips.
- Imitation cheeses.

To keep *trans* fat intake low, use these foods sparingly as an occasional taste treat. Chapter 5 describes current labeling regulations for *trans* fats and provides tips for reading food labels.

Table H5-1 summarizes which foods provide which fats. Substituting unsaturated fats for saturated fats at each meal and snack can help protect against heart disease. Table H5-2 provides several examples and shows how such substitutions can lower saturated fat and raise unsaturated fat—even when total fat and kcalories remain unchanged.

TABLE H5-1 Major Sources of Various Fatty Acids

Healthful Fatty Acids

Monounsaturated	Omega-6 Polyunsaturated	Omega-3 Polyunsaturated
Avocado	Margarine (nonyhydrogenated)	Fatty fish (herring, mackerel, salmon, tuna)
Oils (canola, olive, peanut, sesame)	Oils (corn, cottonseed, safflower, soybean)	Flaxseed
Nuts (almonds, cashews, filberts, hazelnuts, macadamia nuts, peanuts, pecans, pistachios)	Nuts (walnuts)	Nuts
Olives	Mayonnaise	
Peanut butter	Salad dressing	
Seeds (sesame)	Seeds (pumpkin, sunflower)	

Harmful Fatty Acids

Saturated	*Trans*
Bacon	Fried foods (hydrogenated shortening)
Butter	Margarine (hydrogenated or partially hydrogenated)
Chocolate	
Coconut	Nondairy creamers
Cream cheese	Many fast foods
Cream, half-and-half	Shortening
Lard	Commercial baked goods (including doughnuts, cakes, cookies)
Meat	
Milk and milk products (whole)	Many snack foods (including microwave popcorn, chips, crackers)
Oils (coconut, palm, palm kernel)	
Shortening	
Sour cream	

NOTE: Keep in mind that foods contain a mixture of fatty acids.

TABLE H5-2 Replacing Saturated Fat with Unsaturated Fat

Examples of ways to replace saturated fats with unsaturated fats include sautéing foods in olive oil instead of butter, garnishing salads with sunflower seeds instead of bacon, snacking on mixed nuts instead of potato chips, using avocado instead of cheese on a sandwich, and eating salmon instead of steak. Portion sizes have been adjusted so that each of these foods provides approximately 100 kcalories. Notice that for a similar number of kcalories and grams of fat, the first choices offer less saturated fat and more unsaturated fat.

	Total Fat (g)	Saturated Fat (g)	Unsaturated Fat (g)
Olive oil vs. butter	11 vs. 11	2 vs. 7	9 vs. 4
Sunflower seeds vs. bacon	8 vs. 9	1 vs. 3	7 vs. 6
Mixed nuts vs. potato chips	9 vs. 7	1 vs. 2	8 vs. 5
Avocado vs. cheese	10 vs. 8	2 vs. 4	8 vs. 4
Salmon vs. steak	4 vs. 5	1 vs. 2	3 vs. 3
Totals	**42 vs. 40**	**7 vs. 18**	**35 vs. 22**

NOTE: Portion sizes that provide approximately 100 kcalories: 1 tbs olive oil, 1 tbs butter, 2 tbs dry roasted sunflower seeds, 2 slices cooked bacon, 2 tbs dry roasted mixed nuts, 10 potato chips, 6 slices avocado, 1 slice cheddar cheese, 2 oz salmon, and 1½ oz steak.

The Mediterranean Diet

The links between good health and traditional Mediterranean diets of the mid-1900s and health were introduced earlier with regard to olive oil.[25] For people who eat these diets, the incidence of heart disease, some cancers, and other chronic diseases is low, and life expectancy is high.[26]

Although each of the many countries that border the Mediterranean Sea has its own culture, traditions, and dietary habits, their similarities are much greater than the use of olive oil alone. In fact, according to a recent study, no one factor alone can be credited with reducing disease risks—the association holds true only when the overall diet pattern is present.[27] Apparently, each of the foods contributes small benefits that harmonize to produce either a substantial cumulative or a synergistic effect.

The Mediterranean people focus their diets on crusty breads, whole grains, potatoes, and pastas; a variety of vegetables (including wild greens) and legumes; feta and mozzarella cheeses and yogurt; nuts; and fruits (especially grapes and figs). They eat some fish, other seafood, poultry, a few eggs, and little meat.

Along with olives and olive oil, their principal sources of fat are nuts and fish; they rarely use butter or encounter hydrogenated fats. Consequently, traditional Mediterranean diets are:

- Low in saturated fat.
- Very low in *trans* fat.
- Rich in unsaturated fat.
- Rich in complex carbohydrate and fiber.
- Rich in nutrients and phytochemicals that support good health.

People following the traditional Mediterranean diet can receive as much as 40 percent of a day's kcalories from fat, but their limited consumption of dairy products and meats provides less than 10 percent from saturated fats. In addition, because the animals graze, the meat, dairy products, and eggs are richer in omega-3 fatty acids than those from animals fed grain. Other foods typical of the Mediterranean, such as wild plants and snails, provide omega-3 fatty acids as well. All in all, the traditional Mediterranean diet has gained a reputation for its health benefits as well as its delicious flavors, but beware of the typical Mediterranean-style cuisine available in U.S. restaurants. It has been adjusted to popular tastes, meaning that it is often much higher in saturated fats and meats, and much lower in the potentially beneficial constituents, than the traditional fare.

Conclusion

Are some fats "good" and others "bad" from the body's point of view? The saturated and *trans* fats indeed seem mostly bad for the health of the heart. Aside from providing energy, which unsaturated fats can do equally well, saturated and *trans* fats bring no indispensable benefits to the body. Furthermore, no harm can come from consuming diets low in them. Still, foods rich in these fats are often delicious, giving them a special place in the diet.

In contrast, the unsaturated fats are mostly good for the health of the heart when consumed in moderation. To date, their one proven fault seems to be that they, like all fats, provide abundant energy to the body and so may promote obesity if they drive kcalorie intakes higher than energy needs.[28] Obesity, in turn, often begets many body ills, as Chapter 8 makes clear.

When judging foods by their fatty acids, keep in mind that the fat in foods is a mixture of "good" and "bad," providing both saturated and unsaturated fatty acids. Even predominantly monounsaturated olive oil delivers some saturated fat. Consequently, even when a person chooses foods with mostly unsaturated fats, saturated fat can still add up if total fat is high. For this reason, fat must be kept below 35 percent of total kcalories if the diet is to be moderate in saturated fat. Even experts run into difficulty when attempting to create nutritious diets from a variety of foods that are low in saturated fats when kcalories from fat exceed 35 percent of the total.[29]

Does this mean that you must forever go without favorite cheeses, ice cream cones, or a grilled steak? The famous French chef Julia Child makes this point about moderation:

> An imaginary shelf labeled INDULGENCES is a good idea. It contains the best butter, jumbo-size eggs, heavy cream, marbled steaks, sausages and pâtés, hollandaise and butter sauces, French butter-cream fillings, gooey chocolate cakes, and all those lovely items that demand disciplined rationing. Thus, with these items high up and almost out of reach, we are ever conscious that they are not everyday foods. They are for special occasions, and when that occasion comes we can enjoy every mouthful.
>
> —Julia Child, *The Way to Cook*, 1989.

Additionally, food manufacturers may come to the assistance of consumers wishing to avoid the health threats from saturated and *trans* fats. A margarine maker has announced that it will no longer offer products containing *trans* fats; a major snack manufacturer will soon reduce the saturated and *trans* fats in some of its products and offer snack foods in single-serving packages. Other companies are likely to follow if consumers respond favorably.

Another idea is to adopt some of the Mediterranean eating habits and simply enjoy a high-fat diet. Including vegetables, fruits, and legumes as part of a balanced daily diet is a good idea, as is *replacing* saturated fats such as butter, shortening, and meat fat with unsaturated fats like olive oil and the oils from nuts and fish. These foods provide vitamins, minerals, and phytochemicals—all valuable in protecting the body's health. The authors of this book would not stop there, however. They would urge you to reduce fats from convenience foods and fast foods; choose small portions of meats, fish, and poultry; and include fresh foods from all the groups each day. Take care to select portion sizes that will best meet your energy needs. Also, exercise daily.

REFERENCES

1. F. B. Hu, J. E. Manson, and W. C. Willett, Types of dietary fat and risk of coronary heart disease: A critical review, *Journal of the American College of Nutrition* 20 (2001): 5–19.

2. M. de Lorgeril and coauthors, Mediterranean diet, traditional risk factors, and the rate of cardiovascular complications after myocardial infarction: Final report of the Lyon Diet Heart Study, *Circulation* 99 (1999): 779–785.

3. Third Report of the National Cholesterol Education Program (NCEP) Expert Panel on Detection, Evaluation, and Treatment of High Blood Cholesterol in Adults (Adult Treatment Panel III), NIH publication no.

02-5215 (Bethesda, Md.: National Heart, Lung, and Blood Institute, 2002); Committee on Dietary Reference Intakes, *Dietary Reference Intakes for Energy, Carbohydrate, Fiber, Fat, Fatty Acids, Cholesterol, Protein, and Amino Acids* (Washington, D.C.: National Academies Press, 2002).

4. Committee on Dietary Reference Intakes, 2002, p. 11–3.

5. Third Report of the National Cholesterol Education Program (NCEP) Expert Panel on Detection, Evaluation, and Treatment of High Blood Cholesterol in Adults (Adult Treatment Panel III), 2002.

6. A. Keys, *Seven Countries: A Multivariate Analysis of Death and Coronary Heart Disease* (Cambridge, Mass.: Harvard University Press, 1980).

7. A. H. Stark and Z. Madar, Olive oil as a functional food: Epidemiology and nutritional approaches, *Nutrition Reviews* 60 (2002): 170–176.

8. P. M. Kris-Etherton and coauthors, High-monounsaturated fatty acid diets lower both plasma cholesterol and triacyglycerol concentrations, *American Journal of Clinical Nutrition* 70 (1999): 1009–1015.

9. R. L. Hargrove and coauthors, Low fat and high monounsaturated fat diets decrease human low density lipoprotein oxidative susceptibility in vitro, *Journal of Nutrition* 131 (2001): 1758–1763.

10. C. M. Williams, Beneficial nutritional properties of olive oil: Implications for postprandial lipoproteins and factor VII, *Nutrition, Metabolism, and Cardiovascular Diseases* 11 (2001): 51–56; J. P. De La Cruz and coauthors, Antithrombotic potential of olive oil administration in rabbits with elevated cholesterol, *Thrombosis Research* 100 (2000): 305–315; L. F. Larsen, J. Jespersen, and P. Marckmann, Are olive oil diets antithrombotic? Diets enriched with olive, rapeseed, or sunflower oil affect postprandial factor VII differently, *American Journal of Clinical Nutrition* 70 (1999): 976–982.

11. F. Visiol and C. Galli, Biological properties of olive oil phytochemicals, *Critical Reviews in Food Science and Nutrition* 42 (2002): 209–221; M. N. Vissers and coauthors, Olive oil phenols are absorbed in humans, *Journal of Nutrition* 132 (2002): 409–417; M. Fito and coauthors, Protective effect of olive oil and its phenolic compounds against low density lipoprotein oxidation, *Lipids* 35 (2000): 633–638; R. W. Owen and coauthors, The antioxidant/anticancer potential of phenolic compounds isolated from olive oil, *European Journal of Cancer* 36 (2000): 1235–1247.

12. L. A. Ferrara and coauthors, Olive oil and reduced need for antihypertensive medications, *Archives of Internal Medicine* 160 (2000): 837–842.

13. L. Van Horn and N. Ernst, A summary of the science supporting the new National Cholesterol Education program dietary recommendations: What dietitians should know, *Journal of the American Dietetic Association* 101 (2001): 1148–1154.

14. P. M. Kris-Etherton and coauthors, The effects of nuts on coronary heart disease risk, *Nutrition Reviews* 59 (2001): 103–111.

15. F. B. Hu and M. J. Stampfer, Nut consumption and risk of coronary heart disease: A review of epidemiologic evidence, *Current Atherosclerosis Reports* 1 (1999): 204–209.

16. E. B. Feldman, The scientific evidence for a beneficial health relationship between walnuts and coronary heart disease, *Journal of Nutrition* 132 (2002): 1062S–1101S; D. Zambón and coauthors, Substituting walnuts for monounsaturated fat improves the serum lipid profile of hypercholesterolemic men and women: A randomized crossover trial, *Annals of Internal Medicine* 132 (2000): 538–546.

17. D. J. Jenkins and coauthors, Dose response of almonds on coronary heart disease risk factors: Blood lipids, oxidized low-density lipoproteins, lipoprotein (a), homocysteine, and pulmonary nitric oxide: A randomized, controlled, crossover trial, *Circulation* 106 (2002): 1327–1332.

18. E. Dewailly and coauthors, Cardiovascular disease risk factors and n-3 fatty acid status in the adult population of James Bay Cree, *American Journal of Clinical Nutrition* 76 (2002): 85–92; E. Dewailly and coauthors, n-3 fatty acids and cardiovascular disease risk factors among the Inuit of Nunavik, *American Journal of Clinical Nutrition* 74 (2001): 464–473.

19. P. J. H. Jones and V. W. Y. Lau, Effect of n-3 polyunsaturated fatty acids on risk reduction of sudden death, *Nutrition Reviews* 60 (2002): 407–413; W. E. Connor, Importance of n-3 fatty acids in health and disease, *American Journal of Clinical Nutrition* 71 (2000): 171S–175S; P. J. Nestel, Fish oil and cardiovascular disease: Lipids and arterial function, *American Journal of Clinical Nutrition* 71 (2000): 228S–231S; C. von Schacky, n-3 fatty acids and the prevention of coronary atherosclerosis, *American Journal of Clinical Nutrition* 71 (2000): 224S–227S.

20. F. B. Hu and coauthors, Fish and omega-3 fatty acid intake and risk of coronary heart disease in women, *Journal of the American Medical Association* 287 (2002): 1815–1821; C. M. Albert and coauthors, Blood levels of long-chain n-3 fatty acids and the risk of sudden death, *New England Journal of Medicine* 346 (2002): 1113–1118; von Schacky, 2000.

21. H. Iso and coauthors, Intake of fish and omega-3 acids and risk of stroke in women, *Journal of the American Medical Association* 285 (2001): 304–312.

22. E. Guallar and coauthors, Mercury, fish oils, and the risk of myocardial infarction, *New England Journal of Medicine* 347 (2002): 1747–1754; K. Yoshizawa and coauthors, Mercury and the risk of coronary heart disease in man, *New England Journal of Medicine* 347 (2002): 1755–1760.

23. Third Report of the National Cholesterol Education Program (NCEP) Expert Panel on Detection, Evaluation, and Treatment of High Blood Cholesterol in Adults (Adult Treatment Panel III), 2002, p.V-8.

24. Committee on Dietary Reference Intakes, 2002, pp. 11–46 and G-1.

25. A. P. Simopoulos, The Mediterranean diets: What is so special about the diet of Greece? The scientific evidence, *Journal of Nutrition* 131 (2001): 3065S–3073S.

26. A. Trichopoulou and coauthors, Cancer and Mediterranean dietary traditions, *Cancer Epidemiology, Biomarkers and Prevention* 9 (2000): 869–873; C. Lasheras, S. Fernandez, and A. M. Patterson, Mediterranean diet and age with respect to overall survival in institutionalized, nonsmoking elderly people, *American Journal of Clinical Nutrition* 71 (2000): 987–992; A. Trichopoulou and E. Vasilopoulou, Mediterranean diet and longevity, *British Journal of Nutrition* 84 (2000): 205–209.

27. A. Trichopoulou and coauthors, Adherence to a Mediterranean diet and survival in a Greek population, *New England Journal of Medicine* 348 (2003): 2599–2608.

28. Committee on Dietary Reference Intakes, 2002, pp. 11–19.

29. Committee on Dietary Reference Intakes, 2002, pp. 11–22.

Protein: Amino Acids

Chapter Outline

Nutrition Explorer CD-ROM Outline

© James Jackson/Stone/Getty Images

Nutrition in Your Life

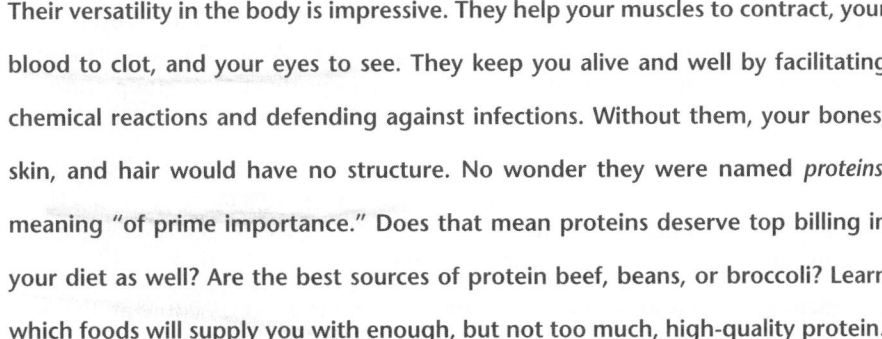

Their versatility in the body is impressive. They help your muscles to contract, your blood to clot, and your eyes to see. They keep you alive and well by facilitating chemical reactions and defending against infections. Without them, your bones, skin, and hair would have no structure. No wonder they were named *proteins,* meaning "of prime importance." Does that mean proteins deserve top billing in your diet as well? Are the best sources of protein beef, beans, or broccoli? Learn which foods will supply you with enough, but not too much, high-quality protein.

People commonly associate protein with strength and meat with protein. Consequently, they eat steak to build their muscles, but their thinking is only partly correct. Protein is a vital structural and working substance in all cells, not just muscle cells. Meat is a good source of protein, but so are milk, eggs, legumes, and many grains and vegetables. People who overvalue protein may overemphasize meat in their diets, sometimes at the expense of other, equally important nutrients and foods. Protein is important, but it is only one of the nutrients needed to maintain the body's health.

The Chemist's View of Proteins

Chemically, **proteins** contain the same atoms as carbohydrates and lipids—carbon (C), hydrogen (H), and oxygen (O)—but proteins also contain nitrogen (N) atoms. These nitrogen atoms give the name *amino* (nitrogen containing) to the amino acids—the links in the chains of proteins.

proteins: compounds composed of carbon, hydrogen, oxygen, and nitrogen atoms, arranged into amino acids linked in a chain. Some amino acids also contain sulfur atoms.

FIGURE 6-1 | Amino Acid Structure

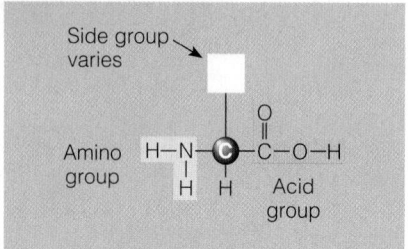

Side group varies

Amino group

Acid group

All amino acids have a carbon (known as the alpha-carbon), with an amino group (NH_2), an acid group (COOH), a hydrogen (H), and a side group attached. The side group is a unique chemical structure that differentiates one amino acid from another.

■ Reminder:
- H forms 1 bond.
- O forms 2 bonds.
- N forms 3 bonds.
- C forms 4 bonds.

■ Some researchers refer to essential amino acids as **indispensable** and to nonessential amino acids as **dispensable**.

amino (a-MEEN-oh) **acids:** building blocks of proteins. Each contains an amino group, an acid group, a hydrogen atom, and a distinctive side group, all attached to a central carbon atom.
- **amino** = containing nitrogen

nonessential amino acids: amino acids that the body can synthesize (see Table 6-1).

essential amino acids: amino acids that the body cannot synthesize in amounts sufficient to meet physiological needs (see Table 6-1).

Amino Acids

All **amino acids** have the same basic structure—a central carbon (C) atom with a hydrogen (H), an amino group (NH_2), and an acid group (COOH) attached to it. Carbon atoms need to form four bonds,■ though, so a fourth attachment is necessary. It is this fourth site that distinguishes each amino acid from the others. Attached to the carbon atom at the fourth bond is a distinct atom, or group of atoms, known as the *side group* or *side chain* (see Figure 6-1).

Unique Side Groups The side groups on amino acids vary from one amino acid to the next, making proteins more complex than either carbohydrates or lipids. A polysaccharide (starch, for example) may be several thousand units long, but every unit is a glucose molecule just like all the others. A protein, on the other hand, is made up of about 20 different amino acids, each with a different side group. Table 6-1 lists the amino acids most common in proteins.*

The simplest amino acid, glycine, has a hydrogen atom as its side group. A slightly more complex amino acid, alanine, has an extra carbon with three hydrogen atoms. Other amino acids have more complex side groups (see Figure 6-2 for examples). Thus, although all amino acids share a common structure, they differ in size, shape, electrical charge, and other characteristics because of differences in these side groups.

Nonessential Amino Acids More than half of the amino acids are **nonessential,** meaning that the body can synthesize them for itself. Proteins in foods usually deliver these amino acids, but it is not essential that they do so. The body can make any nonessential amino acid, given nitrogen to form the amino group and fragments from carbohydrate or fat to form the rest of the structure.

Essential Amino Acids There are nine amino acids that the human body either cannot make at all or cannot make in sufficient quantity to meet its needs. These nine amino acids must be supplied by the diet; they are **essential.**■ The first column in Table 6-1 presents the essential amino acids.

Conditionally Essential Amino Acids Sometimes a nonessential amino acid becomes essential under special circumstances. For example, the body normally

---

* Besides the 20 common amino acids, which can all be components of proteins, others do not occur in proteins, but can be found individually (for example, taurine and ornithine). Some amino acids occur in related forms (for example, proline can acquire an OH group to become hydroxyproline).

TABLE 6-1 | Amino Acids

Proteins are made up of about 20 common amino acids. The first column lists the essential amino acids for human beings (those the body cannot make—that must be provided in the diet). The second column lists the nonessential amino acids. In special cases, some nonessential amino acids may become conditionally essential (see the text). In a newborn, for example, only five amino acids are truly nonessential; the other nonessential amino acids are conditionally essential until the metabolic pathways are developed enough to make those amino acids in adequate amounts.

Essential Amino Acids		Nonessential Amino Acids	
Histidine	(HISS-tuh-deen)	Alanine	(AL-ah-neen)
Isoleucine	(eye-so-LOO-seen)	Arginine	(ARJ-ih-neen)
Leucine	(LOO-seen)	Asparagine	(ah-SPAR-ah-geen)
Lysine	(LYE-seen)	Aspartic acid	(ah-SPAR-tic acid)
Methionine	(meh-THIGH-oh-neen)	Cysteine	(SIS-teh-een)
Phenylalanine	(fen-il-AL-ah-neen)	Glutamic acid	(GLU-tam-ic acid)
Threonine	(THREE-oh-neen)	Glutamine	(GLU-tah-meen)
Tryptophan	(TRIP-toe-fan,	Glycine	(GLY-seen)
	TRIP-toe-fane)	Proline	(PRO-leen)
Valine	(VAY-leen)	Serine	(SEER-een)
		Tyrosine	(TIE-roe-seen)

FIGURE 6-2 Examples of Amino Acids

Note that all amino acids have a common chemical structure but that each has a different side group. Appendix C presents the chemical structures of the 20 amino acids most common in proteins.

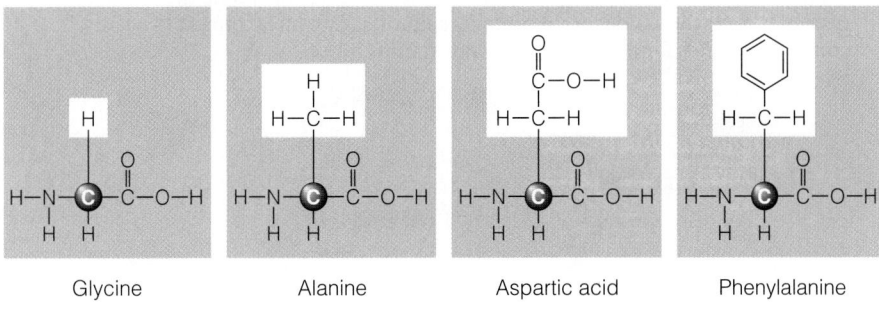

| Glycine | Alanine | Aspartic acid | Phenylalanine |

uses the essential amino acid phenylalanine to make tyrosine (a nonessential amino acid). But if the diet fails to supply enough phenylalanine, or if the body cannot make the conversion for some reason (as happens in the inherited disease phenylketonuria), then tyrosine becomes **conditionally essential.**

Proteins

Cells link amino acids end-to-end in a variety of sequences to form thousands of different proteins. A **peptide bond** unites each amino acid to the next.

Amino Acid Chains Condensation reactions connect amino acids, just as they combine monosaccharides to form disaccharides, and fatty acids with glycerol to form triglycerides. Two amino acids bonded together form a **dipeptide** (see Figure 6-3). By another such reaction, a third amino acid can be added to the chain to form a **tripeptide.** As additional amino acids join the chain, a **polypeptide** is formed. Most proteins are a few dozen to several hundred amino acids long. Figure 6-4 (on p. 184) provides an example—insulin.

Amino Acid Sequences If a person could walk along a carbohydrate molecule like starch, the first stepping stone would be a glucose. The next stepping stone would also be a glucose, and it would be followed by a glucose, and yet another glucose. But if a person were to walk along a polypeptide chain, each stepping stone would be one of 20 different amino acids. The first stepping stone might be the amino acid methionine. The second might be an alanine. The third might be a glycine, and the fourth a tryptophan, and so on. Walking along another

FIGURE 6-3 Condensation of Two Amino Acids to Form a Dipeptide

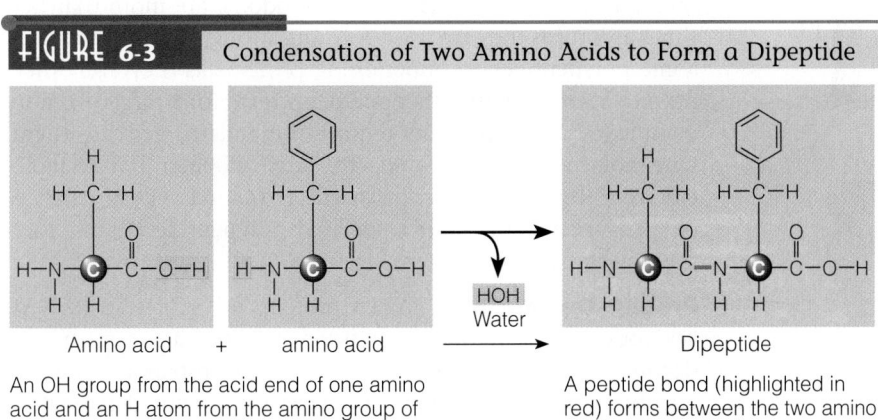

| Amino acid | + | amino acid | → | Dipeptide |

An OH group from the acid end of one amino acid and an H atom from the amino group of another join to form a molecule of water.

A peptide bond (highlighted in red) forms between the two amino acids, creating a dipeptide.

conditionally essential amino acid: an amino acid that is normally nonessential, but must be supplied by the diet in special circumstances when the need for it exceeds the body's ability to produce it.

peptide bond: a bond that connects the acid end of one amino acid with the amino end of another, forming a link in a protein chain.

dipeptide (dye-PEP-tide): two amino acids bonded together.
- **di** = two
- **peptide** = amino acid

tripeptide: three amino acids bonded together.
- **tri** = three

polypeptide: many (ten or more) amino acids bonded together.
- **poly** = many

FIGURE 6-4 Amino Acid Sequence of Human Insulin

Human insulin is a relatively small protein that consists of 51 amino acids in two short polypeptide chains. (For amino acid abbreviations, see Appendix C.) Two bridges link the two chains. A third bridge spans a section within the short chain.

Known as disulfide bridges, these links always involve the amino acid cysteine (Cys), whose side group contains sulfur (S). Cysteines connect to each other when bonds form between these side groups.

polypeptide path, a person might step on a phenylalanine, then a valine, and a glutamine. In other words, amino acid sequences within proteins vary.

The amino acids can act somewhat like the letters in an alphabet. If you had only the letter G, all you could write would be a string of Gs: G–G–G–G–G–G–G. But with 20 different letters available, you could create poems, songs, or novels. Similarly, the 20 amino acids can be linked together in a variety of sequences—even more than are possible for letters in a word or words in a sentence. Thus the variety of possible sequences for polypeptide chains is tremendous.

Protein Shapes Polypeptide chains twist into a variety of complex, tangled shapes, depending on their amino acid sequences. The unique side group of each amino acid gives it characteristics that attract it to, or repel it from, the surrounding fluids and other amino acids. Some amino acid side groups carry electrical charges that are attracted to water molecules (they are hydrophilic). Other side groups are neutral and are repelled by water (they are hydrophobic). As amino acids are strung together to make a polypeptide, the chain folds so that its charged hydrophilic side groups are on the outer surface near water; the neutral hydrophobic groups tuck themselves inside, away from water. The intricate, coiled shape the polypeptide finally assumes gives it maximum stability.

Protein Functions The extraordinary and unique shapes of proteins enable them to perform their various tasks in the body. Some form hollow balls that can carry and store materials within them, and some, such as those of tendons, are more than ten times as long as they are wide, forming strong, rodlike structures. Some polypeptides are functioning proteins as they are; others need to associate with other polypeptides to form larger working complexes. Some proteins require minerals to activate them. One molecule of **hemoglobin**—the large, globular protein molecule that, by the billions, packs the red blood cells and carries oxygen—is made of four associated polypeptide chains, each holding the mineral iron (see Figure 6-5).

Protein Denaturation When proteins are subjected to heat, acid, or other conditions that disturb their stability, they undergo **denaturation**—that is, they uncoil and lose their shapes and, consequently, their ability to function. Past a certain point, denaturation is irreversible. Familiar examples of denaturation include the hardening of an egg when it is cooked, the curdling of milk when acid is added, and the stiffening of egg whites when they are whipped.

hemoglobin (HE-moh-GLOW-bin): the globular protein of the red blood cells that carries oxygen from the lungs to the cells throughout the body.
• **hemo** = blood
• **globin** = globular protein

denaturation (dee-NAY-chur-AY-shun): the change in a protein's shape and consequent loss of its function brought about by heat, agitation, acid, base, alcohol, heavy metals, or other agents.

FIGURE 6-5 The Structure of Hemoglobin

One of the four highly folded polypeptide chains that forms the globular hemoglobin protein

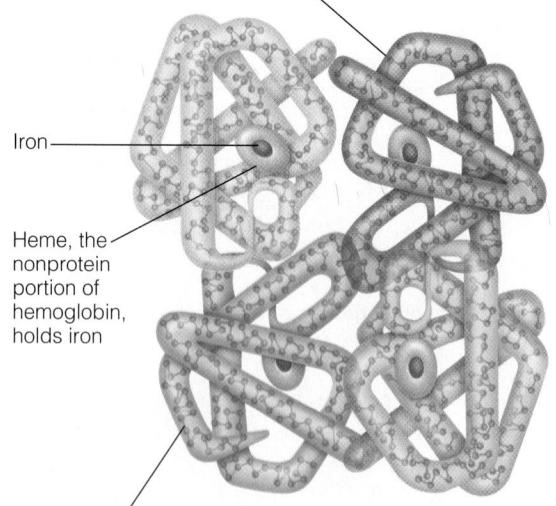

Iron

Heme, the nonprotein portion of hemoglobin, holds iron

The amino acid sequence determines the shape of the polypeptide chain

IN SUMMARY Chemically speaking, proteins are more complex than carbohydrates or lipids, being made of some 20 different amino acids, 9 of which the body cannot make (they are essential). Each amino acid contains an amino group, an acid group, a hydrogen atom, and a distinctive side group. Cells link amino acids together in a series of condensation reactions to create proteins. The distinctive sequence of amino acids in each protein determines its unique shape and function.

Digestion and Absorption of Protein

Proteins in foods do not become body proteins directly. Instead, they supply the amino acids from which the body makes its own proteins. When a person eats foods containing protein, enzymes break the long polypeptide strands into shorter strands, the short strands into tripeptides and dipeptides, and, finally, the tripeptides and dipeptides into amino acids.

Protein Digestion

Figure 6-6 (on p. 186) illustrates the digestion of protein through the GI tract. Proteins are crushed and moistened in the mouth, but the real action begins in the stomach.

In the Stomach The major event in the stomach is the partial breakdown (hydrolysis) of proteins. Hydrochloric acid uncoils (denatures) each protein's tangled strands so that digestive enzymes can attack the peptide bonds. The hydrochloric acid also converts the inactive form■ of the enzyme pepsinogen to its active form, **pepsin**. Pepsin cleaves proteins—large polypeptides—into smaller polypeptides and some amino acids.

In the Small Intestine When polypeptides enter the small intestine, several pancreatic and intestinal **proteases** hydrolyze them further into short peptide chains,■ tripeptides, dipeptides, and amino acids. Then **peptidase** enzymes on the membrane surfaces of the intestinal cells split most of the dipeptides and tripeptides into single amino acids. Only a few peptides escape digestion and enter the blood intact. Figure 6-6 includes names of the digestive enzymes for protein and describes their actions.

Protein Absorption

A number of specific carriers transport amino acids (and some dipeptides and tripeptides) into the intestinal cells. Once inside the intestinal cells, amino acids may be used for energy or to synthesize needed compounds. Those not used by the intestinal cells are transported across the cell membrane into the surrounding fluid where they enter the capillaries on their way to the liver.

Some nutrition faddists fail to realize that most proteins are broken down to amino acids before absorption. They urge consumers to "Eat enzyme A. It will help you digest your food." Or "Don't eat food B. It contains enzyme C, which will digest cells in your body." In reality, though, enzymes in foods are digested, just as all proteins are. Even the digestive enzymes—which function optimally at their specific pH—are denatured and digested when the pH of their environment changes. (For example, the enzyme pepsin, which works best in the low pH of the stomach becomes inactive and is digested when it enters the higher pH of the small intestine.)

Another misconception is that eating predigested proteins (amino acid supplements) saves the body from having to digest proteins and keeps the digestive system

■ The inactive form of an enzyme is called a **proenzyme** or a **zymogen** (ZYE-moh-jen).

■ A string of four to nine amino acids is an **oligopeptide** (OL-ee-go-PEP-tide).
  • **oligo** = few

pepsin: a gastric enzyme that hydrolyzes protein. Pepsin is secreted in an inactive form, **pepsinogen,** which is activated by hydrochloric acid in the stomach.

proteases (PRO-tee-aces): enzymes that hydrolyze protein.

peptidase: a digestive enzyme that hydrolyzes peptide bonds. *Tripeptidases* cleave tripeptides; *dipeptidases* cleave dipeptides. *Endopeptidases* cleave peptide bonds within the chain to create smaller fragments, whereas *exopeptidases* cleave bonds at the ends to release free amino acids.
  • **tri** = three
  • **di** = two
  • **endo** = within
  • **exo** = outside

FIGURE 6-6   Protein Digestion in the GI Tract

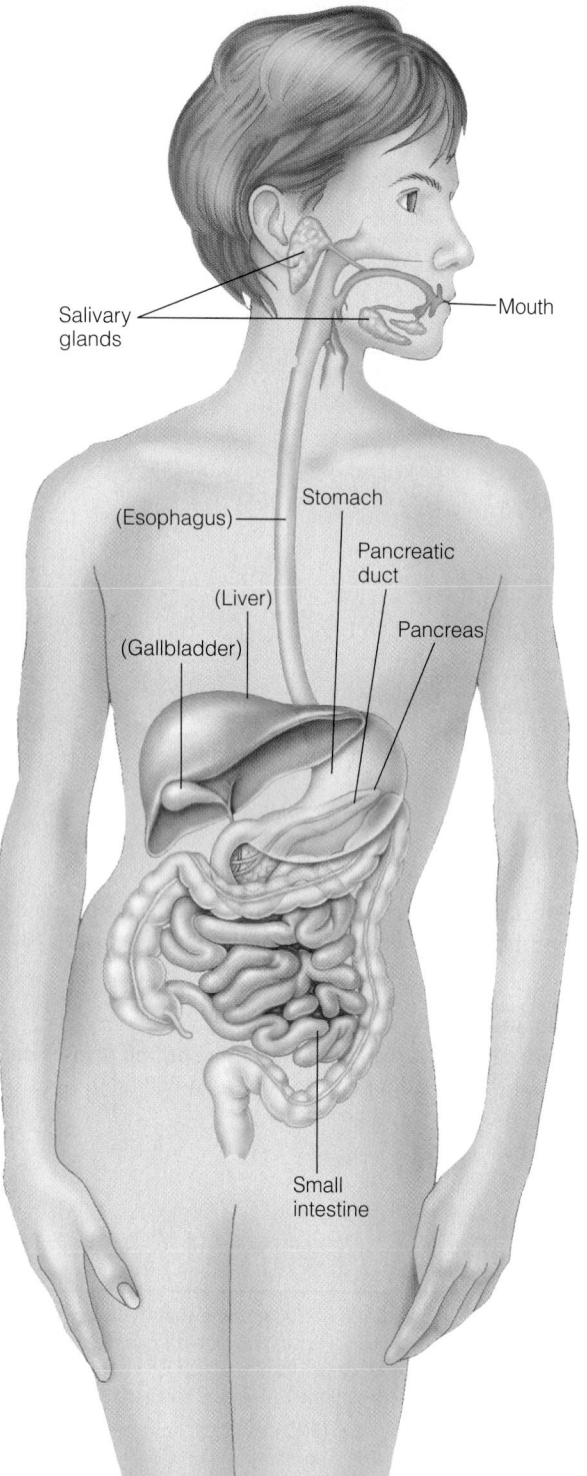

Salivary glands

Mouth

(Esophagus)

Stomach

Pancreatic duct

(Liver)

Pancreas

(Gallbladder)

Small intestine

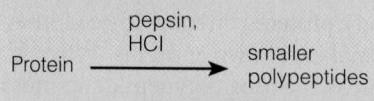

PROTEIN

Mouth and salivary glands

Chewing and crushing moisten protein-rich foods and mix them with saliva to be swallowed.

Stomach

Hydrochloric acid (HCl) uncoils protein strands and activates stomach enzymes:

Protein →[pepsin, HCl]→ smaller polypeptides

In the Stomach:

Hydrochloric acid (HCl)
- Denatures protein structure.
- Activates pepsinogen to pepsin.

Pepsin
- Cleaves proteins to smaller polypeptides and some free amino acids.
- Inhibits pepsinogen synthesis.

Small intestine and pancreas

Pancreatic and small intestinal enzymes split polypeptides further:

Poly-peptides →[pancreatic and intestinal proteases]→ tripeptides, dipeptides, amino acids

Then enzymes on the surface of the small intestinal cells hydrolyze these peptides and the cells absorb them:

Peptides →[intestinal tripeptidases and dipeptidases]→ amino acids (absorbed)

In the Small Intestine:

Enteropeptidase[a]
- Converts pancreatic trypsinogen to trypsin.

Trypsin
- Inhibits trypsinogen synthesis.
- Cleaves peptide bonds next to the amino acids lysine and arginine.
- Converts pancreatic procarboxypeptidases to carboxypeptidases.
- Converts pancreatic chymotrypsinogen to chymotrypsin.

Chymotrypsin
- Cleaves peptide bonds next to the amino acids phenylalanine, tyrosine, tryptophan, methionine, asparagine, and histidine.

Carboxypeptidases
- Cleave amino acids from the acid (carboxyl) ends of polypeptides.

Elastase and collagenase
- Cleave polypeptides into smaller polypeptides and tripeptides.

Intestinal tripeptidases
- Cleave tripeptides to dipeptides and amino acids.

Intestinal dipeptidases
- Cleave dipeptides to amino acids.

Intestinal aminopeptidases
- Cleave amino acids from the amino ends of small polypeptides (oligopeptides).

[a]Enteropeptidase was formerly known as *enterokinase*.

Nutrition Explorer View the digestion of proteins in the stomach and small intestine as they are denatured and their peptide bonds are subsequently hydrolyzed.

from "overworking." Such a belief grossly underestimates the body's abilities. As a matter of fact, the digestive system handles whole proteins *better* than predigested ones because it dismantles and absorbs the amino acids at rates that are optimal for the body's use. (The last section of this chapter discusses amino acid supplements further.)

IN SUMMARY Digestion is facilitated mostly by the stomach's acid and enzymes, which first denature dietary proteins, then cleave them into smaller polypeptides and some amino acids. Pancreatic and intestinal enzymes split these polypeptides further, to oligo-, tri-, and dipeptides, and then split most of these to single amino acids. Then carriers in the membranes of intestinal cells transport the amino acids into the cells, where they are released into the bloodstream.

Proteins in the Body

The human body contains an estimated 10,000 to 50,000 different kinds of proteins. Of these, about 1000 have been studied,■ although with the recent surge in knowledge gained from sequencing the human genome,■ this number is sure to grow rapidly. Only about 10 are described in this chapter—but these should be enough to illustrate the versatility, uniqueness, and importance of proteins. As you will see, each protein has a specific function and that function is determined during protein synthesis.

■ The study of the body's proteins is called **proteomics.**

■ Reminder: The human genome is the full set of chromosomes, including all of the genes and associated DNA.

Protein Synthesis

Each human being is unique because of minute differences in the body's proteins. These differences are determined by the amino acid sequences of proteins, which, in turn, are determined by genes. The following paragraphs describe in words the ways cells synthesize proteins; Figure 6-7 (on p. 188) provides a pictorial description.

The instructions for making every protein in a person's body are transmitted by way of the genetic information received at conception. This body of knowledge, which is filed in the DNA (deoxyribonucleic acid) within the nucleus of every cell, never leaves the nucleus.

Delivering the Instructions To inform a cell of the sequence of amino acids for a needed protein, a stretch of DNA serves as a template for making a strand of RNA (ribonucleic acid) that carries a code. Known as messenger RNA, this molecule escapes through the nuclear membrane. Messenger RNA seeks out and attaches itself to one of the ribosomes (a protein-making machine, which is itself composed of RNA and protein). Thus situated, messenger RNA presents its list, specifying the sequence in which the amino acids are to line up to make a strand of protein.

Lining Up the Amino Acids Other forms of RNA, called transfer RNA, collect amino acids from the cell fluid and bring them to the messenger. Each of the 20 amino acids has a specific transfer RNA. Thousands of transfer RNA, each carrying its amino acid, cluster around the ribosomes, awaiting their turn to unload. When the messenger's list calls for a specific amino acid, the transfer RNA carrying that amino acid moves into position. Then the next loaded transfer RNA moves into place and then the next and the next. In this way, the amino acids line up in the sequence that is called for, and enzymes bind them together. Finally, the completed protein strand is released, the messenger is degraded, and the transfer RNA are freed to return for other loads of amino acids.

Sequencing Errors The sequence of amino acids in each protein determines its shape, which supports a specific function. If a genetic error alters the amino acid sequence of a protein, or if a mistake is made in copying the sequence, an altered protein will result, sometimes with dramatic consequences. The protein hemoglobin offers one example of such a genetic variation. In a person with **sickle-cell anemia,**■ two of hemoglobin's four polypeptide chains (described earlier on p. 184) have the normal sequence of amino acids, but the other two chains do not—they have the amino acid valine in a position that is normally occupied by glutamic acid (see

■ Anemia is not a disease, but a symptom of various diseases. In the case of sickle-cell anemia, a defect in the hemoglobin molecule changes the shape of the red blood cells. Later chapters describe how vitamin and mineral deficiencies affect the size and color of the red blood cells. In all cases, the abnormal blood cells are unable to meet the body's oxygen demands.

sickle-cell anemia: a hereditary form of anemia characterized by abnormal sickle- or crescent-shaped red blood cells. Sickled cells interfere with oxygen transport and blood flow. Symptoms are precipitated by dehydration and insufficient oxygen (as may occur at high altitudes) and include hemolytic anemia (red blood cells burst), fever, and severe pain in the joints and abdomen.

FIGURE 6-7 Protein Synthesis

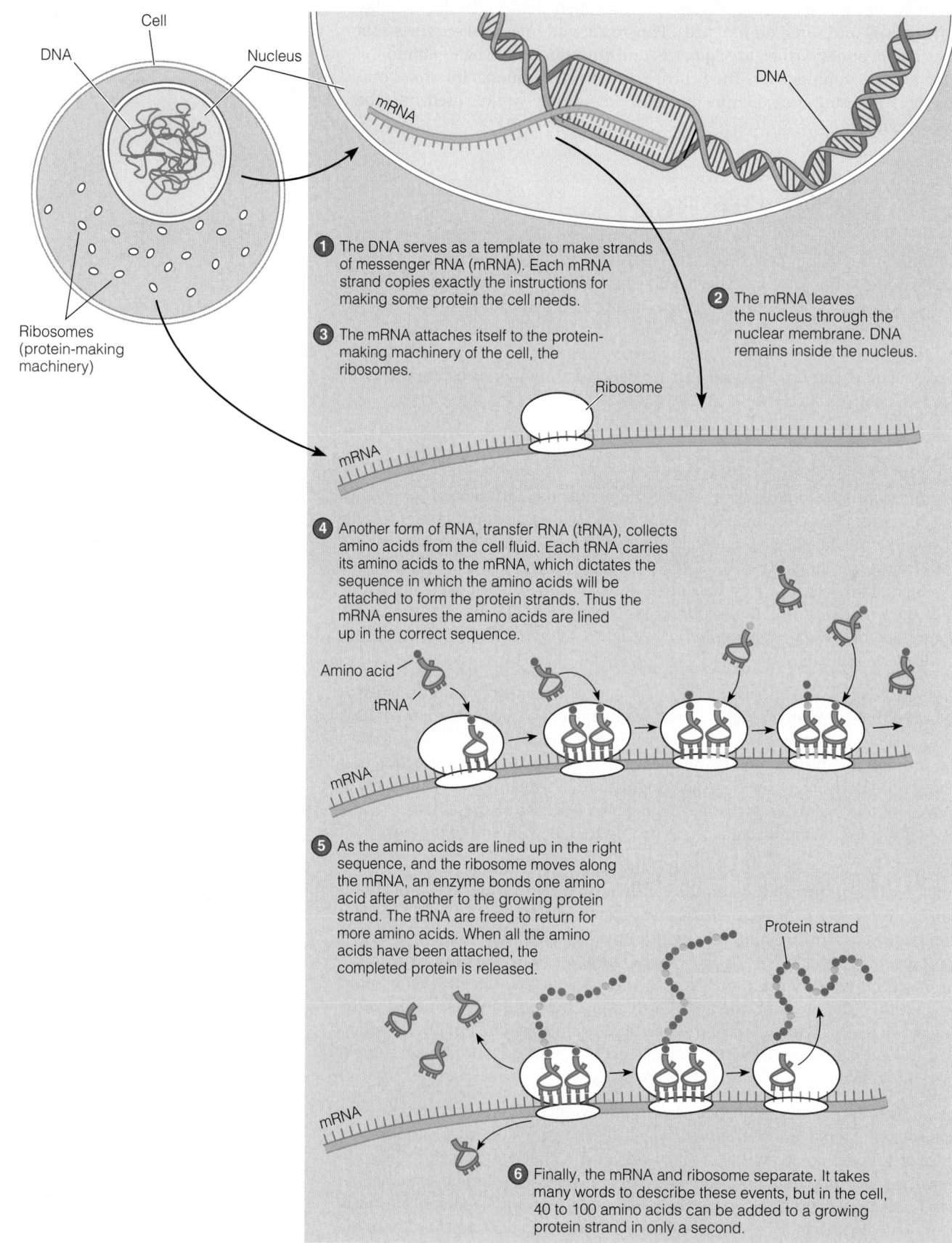

Cell

DNA

Nucleus

DNA

mRNA

Ribosomes (protein-making machinery)

1 The DNA serves as a template to make strands of messenger RNA (mRNA). Each mRNA strand copies exactly the instructions for making some protein the cell needs.

2 The mRNA leaves the nucleus through the nuclear membrane. DNA remains inside the nucleus.

3 The mRNA attaches itself to the protein-making machinery of the cell, the ribosomes.

Ribosome

mRNA

4 Another form of RNA, transfer RNA (tRNA), collects amino acids from the cell fluid. Each tRNA carries its amino acids to the mRNA, which dictates the sequence in which the amino acids will be attached to form the protein strands. Thus the mRNA ensures the amino acids are lined up in the correct sequence.

Amino acid

tRNA

mRNA

5 As the amino acids are lined up in the right sequence, and the ribosome moves along the mRNA, an enzyme bonds one amino acid after another to the growing protein strand. The tRNA are freed to return for more amino acids. When all the amino acids have been attached, the completed protein is released.

Protein strand

mRNA

6 Finally, the mRNA and ribosome separate. It takes many words to describe these events, but in the cell, 40 to 100 amino acids can be added to a growing protein strand in only a second.

Figure 6-8). This single alteration in the amino acid sequence changes the character and shape of hemoglobin so much that it loses its ability to carry oxygen effectively. The red blood cells filled with this abnormal hemoglobin stiffen into elongated sickle, or crescent, shapes instead of maintaining their normal pliable disc shape—hence the name, sickle-cell anemia. Sickle-cell anemia raises energy needs, causes many medical problems, and can be fatal.[1] Caring for children with sickle-cell anemia includes diligent attention to their water needs; dehydration can trigger a crisis.

Nutrients and Gene Expression When a cell makes a protein as described earlier, scientists say that the gene for that protein has been "expressed." Cells can regulate gene expression to make the type of protein, in the amounts and at the rate, they need. Nearly all of the body's cells possess the genes for making all human proteins, but each type of cell makes only the proteins it needs. For example, cells of the pancreas express the gene for insulin; in other cells, that gene is idle. Similarly, the cells of the pancreas do not make the protein hemoglobin, which is needed only by the red blood cells.

Recent research has unveiled some of the fascinating ways nutrients regulate gene expression and protein synthesis.■ These discoveries have begun to explain some of the relationships among nutrients, genes, and disease development. The benefits of polyunsaturated fatty acids in defending against heart disease, for example, are partially explained by their role in influencing gene expression for lipid enzymes. Later chapters provide additional examples of how nutrients influence gene expression.

FIGURE 6-8 Sickle Cells Compared with Normal Red Blood Cells

Normally, red blood cells are disc-shaped; in the inherited disorder sickle-cell anemia, red blood cells are sickle- or crescent-shaped. This alteration in shape occurs because valine replaces glutamic acid in the amino acid sequence of two of hemoglobin's polypeptide chains. As a result of this one alteration, the hemoglobin has a diminished capacity to carry oxygen.

Sickle-shaped blood cells   Normal red blood cells

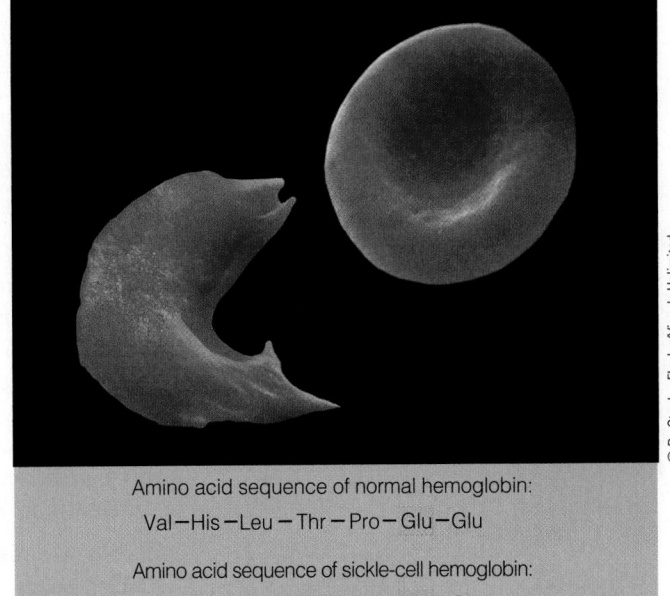

© Dr. Stanley Flegler/Visuals Unlimited

Amino acid sequence of normal hemoglobin:
Val—His—Leu—Thr—Pro—Glu—Glu

Amino acid sequence of sickle-cell hemoglobin:
Val—His—Leu—Thr—Pro—Val—Glu

IN SUMMARY Cells synthesize proteins according to the genetic information provided by the DNA in the nucleus of each cell. This information dictates the order in which amino acids must be linked together to form a given protein. Sequencing errors occasionally occur, sometimes with significant consequences.

■ Nutrients can play key roles in activating or silencing genes. Switching genes on and off, without changing the genetic sequence itself, is known as **epigenetics.**
  • **epi** = among

Roles of Proteins

Whenever the body is growing, repairing, or replacing tissue, proteins are involved. Sometimes their role is to facilitate or to regulate; other times it is to become part of a structure. Versatility is a key feature of proteins.

As Building Materials for Growth and Maintenance From the moment of conception, proteins form the building blocks of muscles, blood, and skin—in fact, of most body structures. For example, to build a bone or a tooth, cells first lay down a **matrix** of the protein **collagen** and then fill it with crystals of calcium, phosphorus, magnesium, fluoride, and other minerals.

Collagen also provides the material of ligaments and tendons and the strengthening glue between the cells of the artery walls that enables the arteries to withstand the pressure of the blood surging through them with each heartbeat. Also made of collagen are scars that knit the separated parts of torn tissues together.

Proteins are also needed for replacement. The life span of a skin cell is only about 30 days. As old skin cells are shed, new cells made largely of protein grow from underneath to compensate. Cells in the deeper skin layers synthesize new proteins to go into hair and fingernails. Muscle cells make new proteins to grow larger and stronger in response to exercise. Cells of the GI tract are replaced every

matrix (MAY-tricks): the basic substance that gives form to a developing structure; in the body, the formative cells from which teeth and bones grow.

collagen (KOL-ah-jen): the protein from which connective tissues such as scars, tendons, ligaments, and the foundations of bones and teeth are made.

FIGURE 6-9 Enzyme Action

Each enzyme facilitates a specific chemical reaction. In this diagram, an enzyme enables two compounds to make a more complex structure, but the enzyme itself remains unchanged.

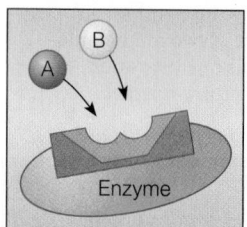

The separate compounds, A and B, are attracted to the enzyme's active site, making a reaction likely.

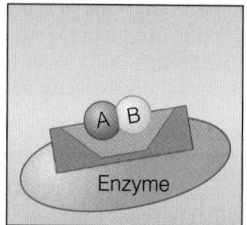

The enzyme forms a complex with A and B.

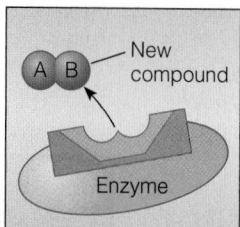

The enzyme is unchanged, but A and B have formed a new compound, AB.

■ Breaking down reactions are **catabolic**, whereas building up reactions are **anabolic** (Chapter 7 provides more details).

TABLE 6-2 Examples of Hormones and Their Actions

Hormones	Actions
Growth hormone	Promotes growth.
Insulin and glucagon	Regulate blood glucose (see Chapter 4).
Thyroxin	Regulates the body's metabolic rate (see Chapter 8).
Calcitonin and parathormone	Regulate blood calcium (see Chapter 12).
Antidiuretic hormone	Regulates fluid and electrolyte balance (see Chapter 12).

NOTE: *Hormones* are chemical messengers that are secreted by endocrine glands in response to altered conditions in the body. Each travels to one or more specific target tissues or organs, where it elicits a specific response. For descriptions of many hormones important in nutrition, see Appendix A.

enzymes: proteins that facilitate chemical reactions without being changed in the process; protein catalysts.

fluid balance: maintenance of the proper types and amounts of fluid in each compartment of the body fluids (see also Chapter 12).

edema (eh-DEEM-uh): the swelling of body tissue caused by excessive amounts of fluid in the interstitial spaces; seen in protein deficiency (among other conditions).

three days. Both inside and outside, then, the body continuously deposits protein into new cells that replace those that have been lost.

As Enzymes Some proteins act as **enzymes.** Digestive enzymes have appeared in every chapter since Chapter 3, but digestion is only one of the many processes enzymes facilitate. Enzymes not only break down substances, they also build substances (such as bone)■ and transform one substance into another (amino acids into glucose, for example). Figure 6-9 diagrams a synthesis reaction.

An analogy may help to clarify the role of enzymes. Enzymes are comparable to the clergy and judges who make and dissolve marriages. When a minister marries two people, they become a couple, with a new bond between them. They are joined together—but the minister remains unchanged. The minister represents enzymes that synthesize large compounds from smaller ones. One minister can perform thousands of marriage ceremonies, just as one enzyme can perform billions of synthetic reactions.

Similarly, a judge who lets married couples separate may decree many divorces before retiring or dying. The judge represents enzymes that hydrolyze larger compounds to smaller ones; for example, the digestive enzymes. The point is that, like the minister and the judge, enzymes themselves are not altered by the reactions they facilitate. They are catalysts, permitting reactions to occur more quickly and efficiently than if substances depended on chance encounters alone.

As Hormones Cells can switch their protein machinery on or off in response to the body's needs. Often hormones do the switching, with marvelous precision. The body's many hormones are messenger molecules, and *some* hormones are proteins. Various endocrine glands in the body release hormones in response to changes that challenge the body. The blood carries the hormones from these glands to their target tissues, where they elicit the appropriate responses to restore normal conditions.

The hormone insulin provides a familiar example. When blood glucose rises, the pancreas releases its insulin. Insulin stimulates the transport proteins of the muscles and adipose tissue to pump glucose into the cells faster than it can leak out. (After acting on the message, the cells destroy the insulin.) Then, as blood glucose falls, the pancreas slows its release of insulin. Many other proteins act as hormones, regulating a variety of actions in the body (see Table 6-2 for examples).

As Regulators of Fluid Balance Proteins help to maintain the body's **fluid balance.** Figure 12-1 in Chapter 12 illustrates a cell and its associated fluids. As the figure explains, the body's fluids are contained inside the cells (intracellular) or outside the cells (extracellular). Extracellular fluids, in turn, can be found either in the spaces between the cells (interstitial) or within the blood vessels (intravascular). The fluid within the intravascular spaces is called plasma (essentially blood without its red blood cells). Fluids can flow freely between these compartments, but being large, proteins cannot. Proteins are trapped primarily within the cells and to a lesser extent in the plasma. Wherever proteins are, they attract water.

The exchange of materials between the blood and the cells takes place across the capillary walls, which allow the passage of fluids and a variety of materials—but usually not plasma proteins. Still some plasma proteins leak out of the capillaries into the interstitial fluid between the cells. These proteins cannot be reabsorbed back into the plasma; they normally reenter circulation via the lymph system. If plasma proteins enter the interstitial spaces faster than they can be cleared, fluid accumulates (because proteins attract water) and causes swelling. Swelling due to an excess of interstitial fluid is known as **edema.** The protein-related causes of edema include:

- Excessive protein losses caused by kidney disease or large wounds (such as extensive burns).

- Inadequate protein synthesis caused by liver disease.
- Inadequate dietary intake of protein.

Whatever the cause of edema, the result is the same: a diminished capacity to deliver nutrients and oxygen to the cells and to remove wastes from them. As a consequence, cells fail to function adequately.

As Acid-Base Regulators Proteins also help to maintain the balance between **acids** and **bases** within the body fluids. Normal body processes continually produce acids and bases, which the blood carries to the kidneys and lungs for excretion. The challenge is to do this without upsetting the blood's acid-base balance.

In an acid solution, hydrogen ions abound; the more hydrogen ions, the more concentrated the acid. Proteins, which have negative charges on their surfaces, attract hydrogen ions, which have positive charges. By accepting and releasing hydrogen ions,■ proteins maintain the acid-base balance of the blood and body fluids.

The blood's acid-base balance is tightly controlled. The extremes of **acidosis** and **alkalosis** lead to coma and death, largely because they denature working proteins. Disturbing a protein's shape renders it useless. To give just one example, denatured hemoglobin loses its capacity to carry oxygen.

As Transporters Some proteins move about in the body fluids, carrying nutrients and other molecules. The protein hemoglobin carries oxygen from the lungs to the cells. The lipoproteins transport lipids around the body. Special transport proteins carry vitamins and minerals.

The transport of the mineral iron provides an especially good illustration of these proteins' specificity and precision. When iron enters an intestinal cell after a meal has been digested and absorbed, it is captured by a protein. Before leaving the intestinal cell, iron is attached to another protein that carries it though the bloodstream to the cells. Once iron enters a cell, it is attached to a storage protein that will hold the iron until it is needed. When it is needed, iron is incorporated into proteins in the red blood cells and muscles that assist in oxygen transport and use. (Chapter 13 provides more details on how these protein carriers transport and store iron.)

Some transport proteins reside in cell membranes and act as "pumps," picking up compounds on one side of the membrane and releasing them on the other as needed. Each transport protein is specific for a certain compound or group of related compounds. Figure 6-10 illustrates how a membrane-bound transport protein

■ Compounds that help keep a solution's acidity or alkalinity constant are called **buffers.**

acids: compounds that release hydrogen ions in a solution.

bases: compounds that accept hydrogen ions in a solution.

acidosis (assi-DOE-sis): above-normal acidity in the blood and body fluids.

alkalosis (alka-LOE-sis): above-normal alkalinity (base) in the blood and body fluids.

FIGURE 6-10 An Example of Protein Transport

This transport protein resides within a cell membrane and acts as a two-door passageway. Molecules enter on one side of the membrane and exit on the other, but the protein doesn't leave the membrane. This example shows how the transport protein moves sodium and potassium in opposite directions across the membrane to maintain a high concentration of potassium and a low concentration of sodium within the cell. This active transport system requires energy.

Key:
- Sodium
- Potassium

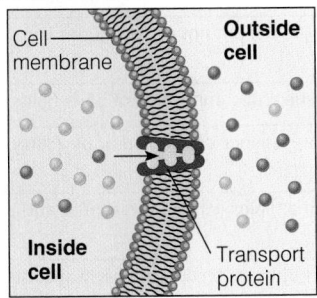

The transport protein picks up sodium from inside the cell.

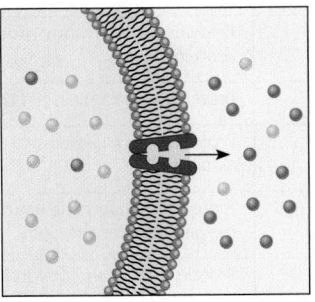

The protein changes shape and releases sodium outside the cell.

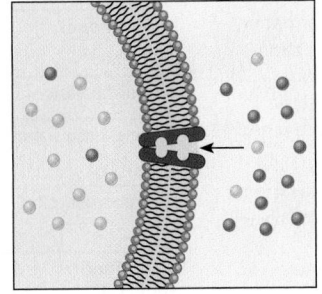

The transport protein picks up potassium from outside the cell.

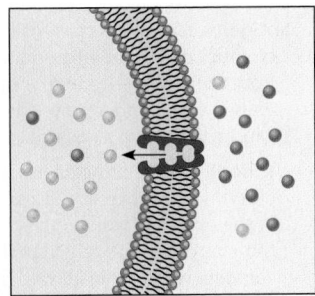

The protein changes shape and releases potassium inside the cell.

Growing children end each day with more bone, blood, muscle, and skin cells than they had at the beginning of the day.

■ Reminder: Protein provides 4 kcal/g. Return to p. 9 for a refresher on how to calculate the protein kcalories from foods.

■ Reminder: The making of glucose from noncarbohydrate sources such as amino acids is *gluconeogenesis*.

helps to maintain the sodium and potassium concentrations in the fluids inside and outside cells. The balance of these two minerals is critical to nerve transmissions and muscle contractions; imbalances can cause irregular heartbeats, muscular weakness, kidney failure, and even death.

As Antibodies Proteins also defend the body against disease. A virus—whether it is one that causes flu, smallpox, measles, or the common cold—enters the cells and multiplies there. One virus may produce 100 replicas of itself within an hour or so. Each replica can then burst out and invade 100 different cells, soon yielding 10,000 virus particles, which invade 10,000 cells. Left free to do their worst, they will soon overwhelm the body with disease.

Fortunately, when the body detects these invading **antigens,** it manufactures **antibodies,** giant protein molecules designed specifically to combat them. The antibodies work so swiftly and efficiently that in a normal, healthy individual, most diseases never have a chance to get started. Without sufficient protein, though, the body cannot maintain its army of antibodies to resist infectious diseases.

Each antibody is designed to destroy just one antigen. Once the body has manufactured antibodies against a particular antigen (such as the measles virus), it remembers how to make them. Consequently, the next time the body encounters that same antigen, it will produce antibodies even more quickly. In other words, the body develops a molecular memory, known as **immunity.** (Chapter 16 describes food allergies—the immune system's response to food antigens.)

As a Source of Energy and Glucose Even though proteins are needed to do the work that only they can perform, they will be sacrificed to provide energy■ and glucose■ if need be. Without energy, cells die; without glucose, the brain and nervous system falter. Chapter 7 provides many more details on energy metabolism.

Other Roles As mentioned earlier, proteins form integral parts of most body structures such as skin, muscles, and bones. They also participate in some of the body's most amazing activities such as blood clotting and vision. When a tissue is injured, a rapid chain of events leads to the production of fibrin, a stringy, insoluble mass of protein fibers that forms a clot from liquid blood. Later, more slowly, the protein collagen forms a scar to replace the clot and permanently heal the wound. The light-sensitive pigments in the cells of the retina are molecules of the protein opsin. Opsin responds to light by changing its shape, thus initiating the nerve impulses that convey the sense of sight to the brain.

IN SUMMARY The protein functions discussed here are summarized in the table below. They are only a few of the many roles proteins play, but they convey some sense of the immense variety of proteins and their importance in the body.

Growth and maintenance	Proteins form integral parts of most body structures such as skin, tendons, membranes, muscles, organs, and bones. As such, they support the growth and repair of body tissues.
Enzymes	Proteins facilitate chemical reactions.
Hormones	Proteins regulate body processes. (Some, but not all, hormones are proteins.)
Fluid balance	Proteins help to maintain the volume and composition of body fluids.
Acid-base balance	Proteins help maintain the acid-base balance of body fluids by acting as buffers.
Transportation	Proteins transport substances, such as lipids, vitamins, minerals, and oxygen, around the body.
Antibodies	Proteins inactivate foreign invaders, thus protecting the body against diseases.
Energy	Proteins provide some fuel for the body's energy needs.

antigens: substances that elicit the formation of antibodies or an inflammation reaction from the immune system. A bacterium, a virus, a toxin, and a protein in food that causes allergy are all examples of antigens.

antibodies: large proteins produced by the immune system in response to the invasion of the body by foreign molecules (usually proteins called antigens). Antibodies combine with and inactivate the foreign invaders, thus protecting the body.

immunity: the body's ability to defend itself against diseases; see Chapter 18.

A Preview of Protein Metabolism

This section previews protein metabolism; Chapter 7 provides a full description. Cells have several metabolic options, depending on their protein and energy needs.

Protein Turnover and the Amino Acid Pool Within each cell, proteins are continually being made and broken down, a process known as **protein turnover.** When proteins break down, they free amino acids to join the general circulation.■ These amino acids mix with amino acids from dietary protein to form an "**amino acid pool**" within the cells and circulating blood. The rate of protein degradation and the amount of protein intake may vary, but the pattern of amino acids within the pool remains fairly constant. Regardless of their source, any of these amino acids can be used to make body proteins or other nitrogen-containing compounds, or they can be stripped of their nitrogen and used for energy (either immediately or stored as fat for later use).

Nitrogen Balance Protein turnover and **nitrogen balance** go hand in hand. In healthy adults, protein synthesis balances with degradation, and protein intake from food balances with nitrogen excretion in the urine, feces, and sweat. When nitrogen intake equals nitrogen output, the person is in nitrogen equilibrium,■ or zero nitrogen balance. Researchers use nitrogen balance studies to estimate protein requirements.[2]

If the body synthesizes more than it degrades and adds protein, nitrogen status becomes positive.■ Nitrogen status is positive in growing infants and children, pregnant women, and people recovering from protein deficiency or illness; their nitrogen intake exceeds their nitrogen output. They are retaining protein in new tissues as they add blood, bone, skin, and muscle cells to their bodies.

If the body degrades more than it synthesizes and loses protein, nitrogen status becomes negative.■ Nitrogen status is negative in people who are starving or suffering other severe stresses such as burns, injuries, infections, and fever; their nitrogen output exceeds their nitrogen intake. During these times, the body loses nitrogen as it breaks down muscle and other body proteins for energy.

Using Amino Acids to Make Proteins or Nonessential Amino Acids As mentioned, cells can assemble amino acids into the proteins they need to do their work. If a particular nonessential amino acid is not readily available, cells can make it from another amino acid. If an essential amino acid is missing, the body may break down some of its own proteins to obtain it.

Using Amino Acids to Make Other Compounds Cells can also use amino acids to make other compounds. For example, the amino acid tyrosine is used to make the **neurotransmitters** norepinephrine and epinephrine, which relay nervous system messages throughout the body. Tyrosine can also be made into the pigment melanin, which is responsible for brown hair, eye, and skin color, or into the hormone thyroxin, which helps to regulate the metabolic rate. For another example, the amino acid tryptophan serves as a precursor for the vitamin niacin and for serotonin, a neurotransmitter important in sleep regulation, appetite control, and sensory perception.[3]

Using Amino Acids for Energy and Glucose As mentioned earlier, when glucose or fatty acids are limited, cells are forced to use amino acids for energy and glucose. The body does not make a specialized storage form of protein as it does for carbohydrate and fat. Glucose is stored as glycogen in the liver and fat as triglycerides in adipose tissue, but protein in the body is available only as the working and structural components of the tissues. When the need arises, the body dismantles its tissue proteins and uses them for energy. Thus, over time, energy deprivation (starvation) always causes wasting of lean body tissue as well as fat loss. An adequate supply of carbohydrates and fats spares amino acids from being used for energy and allows them to perform their unique roles.

■ Amino acids (or proteins) that derive from within the body are **endogenous** (en-DODGE-eh-nus). In contrast, those that derive from foods are **exogenous** (eks-ODGE-eh-nus).
- **endo** = within
- **gen** = arising
- **exo** = outside (the body)

■ Nitrogen equilibrium (zero nitrogen balance): N in = N out.

■ Positive nitrogen: N in > N out.

■ Negative nitrogen: N in < N out.

---

protein turnover: the degradation and synthesis of protein.

amino acid pool: the supply of amino acids derived from either food proteins or body proteins that collect in the cells and circulating blood and stand ready to be incorporated in proteins and other compounds or used for energy.

nitrogen balance: the amount of nitrogen consumed (N in) as compared with the amount of nitrogen excreted (N out) in a given period of time.*

neurotransmitters: chemicals that are released at the end of a nerve cell when a nerve impulse arrives there. They diffuse across the gap to the next cell and alter the membrane of that second cell to either inhibit or excite it.

* The genetic materials DNA and RNA contain nitrogen, but the quantity is insignificant compared with the amount in protein. The average amino acid weighs about 6.25 times as much as the nitrogen it contains, so scientists can estimate the amount of protein in a sample of food, body tissue, or other material by multiplying the weight of the nitrogen in it by 6.25.

Deaminating Amino Acids When amino acids are broken down (as occurs when they are used for energy), they are first deaminated—stripped of their nitrogen-containing amino groups. **Deamination** produces ammonia, which the cells release into the bloodstream. The liver picks up the ammonia, converts it into urea (a less toxic compound), and returns the urea to the blood. (Urea metabolism is described in Chapter 7.) The kidneys filter urea out of the blood; thus the amino nitrogen ends up in the urine. The remaining carbon fragments of the deaminated amino acids may enter a number of metabolic pathways—for example, they may be used for energy or for the production of glucose, ketones, cholesterol, or fat.*

Using Amino Acids to Make Fat Amino acids may be used to make fat when energy and protein intakes exceed needs and carbohydrate intake is adequate. The amino acids are deaminated, the nitrogen is excreted, and the remaining carbon fragments are converted to fat and stored for later use. In this way, protein-rich foods can contribute to weight gain.

> **IN SUMMARY** Proteins are constantly being synthesized and broken down as needed. The body's assimilation of amino acids into proteins and its release of amino acids via protein degradation and excretion can be tracked by measuring nitrogen balance, which should be positive during growth and steady in adulthood. An energy deficit or an inadequate protein intake may force the body to use amino acids as fuel, creating a negative nitrogen balance. Protein eaten in excess of need is degraded and stored as body fat.

Protein in Foods

In the United States and Canada, where nutritious foods are abundant, most people eat protein in such large quantities that they receive all the amino acids they need. In countries where food is scarce and the people eat only marginal amounts of protein-rich foods, however, the *quality* of the protein becomes crucial.

Protein Quality

The protein quality of the diet determines, in large part, how well children grow and how well adults maintain their health. Put simply, **high-quality proteins** provide enough of all the essential amino acids needed to support the body's work, and low-quality proteins don't. Two factors influence protein quality—the protein's digestibility and its amino acid composition.

Digestibility As explained earlier, proteins must be digested before they can provide amino acids. **Protein digestibility** depends on such factors as the protein's source and the other foods eaten with it. The digestibility of most animal proteins is high (90 to 99 percent); plant proteins are less digestible (70 to 90 percent for most, but over 90 percent for soy and legumes).

Amino Acid Composition To make proteins, a cell must have all the needed amino acids available simultaneously. The liver can produce any nonessential amino acid that may be in short supply so that the cells can continue linking amino acids into protein strands. If an essential amino acid is missing, though, a cell must dismantle its own proteins to obtain it. Therefore, to prevent protein breakdown, dietary protein must supply at least the nine essential amino acids

Black beans and rice, a favorite Hispanic combination, together provide a balanced array of amino acids.

deamination (dee-AM-eh-NAY-shun): removal of the amino (NH_2) group from a compound such as an amino acid.

high-quality proteins: dietary proteins containing all the essential amino acids in relatively the same amounts that human beings require. They may also contain nonessential amino acids.

protein digestibility: a measure of the amount of amino acids absorbed from a given protein intake.

---

*Chemists sometimes classify amino acids according to the destinations of their carbon fragments after deamination. If the fragment leads to the production of glucose, the amino acid is called "glucogenic"; if it leads to the formation of ketone bodies, fats, and sterols, the amino acid is called "ketogenic." There is no sharp distinction between glucogenic and ketogenic amino acids, however. A few are both; most are considered glucogenic; only one (leucine) is clearly ketogenic.

plus enough nitrogen-containing amino groups and energy for the synthesis of the others. If the diet supplies too little of any essential amino acid, protein synthesis will be limited. The body makes whole proteins only; if one amino acid is missing, the others cannot form a "partial" protein. An essential amino acid supplied in less than the amount needed to support protein synthesis is called a **limiting amino acid.**

Reference Protein The quality of food proteins is determined based on how they compare with the essential amino acid requirements of preschool-age children. Such a standard is called a **reference protein.**■ The rationale behind using the requirements of this age group is that if a protein will effectively support a young child's growth and development, then it will meet or exceed the requirements of older children and adults.

High-Quality Proteins As mentioned earlier, a high-quality protein contains all the essential amino acids in relatively the same amounts as human beings require; it may or may not contain all the nonessential amino acids. Proteins that are low in an essential amino acid cannot, by themselves, support protein synthesis. Generally, foods derived from animals (meat, fish, poultry, cheese, eggs, yogurt, and milk) provide high-quality proteins, although gelatin is an exception (it lacks tryptophan and cannot support growth and health as a diet's sole protein). Proteins from plants (vegetables, nuts, seeds, grains, and legumes) have more diverse amino acid patterns and tend to be limiting in one or more essential amino acids. Some plant proteins (for example, corn protein) are notoriously low quality. A few others (for example, soy protein) are high quality.

Complementary Proteins In general, plant proteins are of lower quality than animal proteins, and plants also offer less protein (per weight or measure of food). For this reason, many vegetarians improve the quality of proteins in their diets by combining plant-protein foods that have different but complementary amino acid patterns. This strategy yields **complementary proteins** that together contain all the essential amino acids in quantities sufficient to support health. The protein quality of the combination is greater than for either food alone (see Figure 6-11).

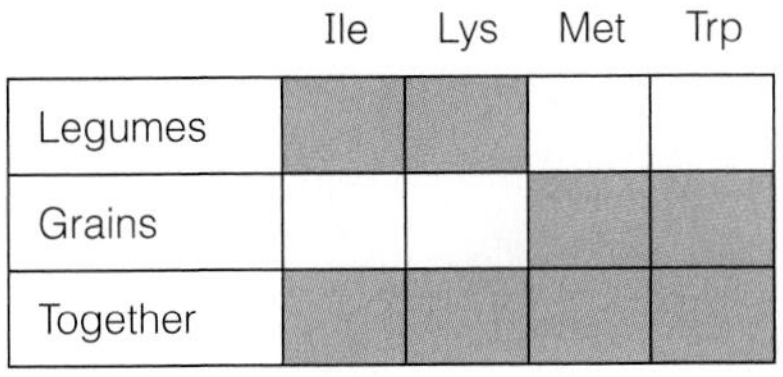

Many people have long believed that combining plant proteins at every meal is critical to protein nutrition. For most healthy vegetarians, though, it is not necessary to balance amino acids at each meal when protein intake is varied and energy intake is sufficient.[4] Vegetarians can receive all the amino acids they need over the course of a day, if they eat a variety of grains, legumes, seeds, nuts, and vegetables. Protein deficiency will develop, however, when fruits and certain vegetables make up the core of the diet, severely limiting both the *quantity* and *quality* of protein. Highlight 6 shows how to plan a nutritious vegetarian diet.

A Measure of Protein Quality—PDCAAS Researchers have developed several methods for evaluating the quality of food proteins and identifying high-quality proteins. The following paragraph briefly describes the measure used by the Committee on Dietary Reference Intakes to evaluate protein quality. Appendix D provides details on other measures.

The **protein digestibility–corrected amino acid score,** or **PDCAAS,** compares the amino acid composition of a protein with human amino acid requirements and corrects for digestibility. First the protein's amino acid composition is determined, and then it is compared against the amino acid requirements of preschool-age children. This comparison reveals the most limiting amino acid—the one that falls shortest compared with the reference. If a food protein's limiting amino acid is 70 percent of the amount found in the reference protein, it receives a score of 70. The amino acid score is multiplied by the food's protein digestibility percentage to determine the PDCAAS. The box on the next page provides an example of how to calculate the PDCAAS, and Table 6-3 lists the PDCAAS values of selected foods.

■ In the past, egg protein was commonly used as the reference protein. Table D-1 in Appendix D presents the amino acid profile of egg. As the reference protein, egg was assigned the value of 100; Table D-2 includes scores of other food proteins for comparison.

FIGURE 6-11 Complementary Proteins

In general, legumes provide plenty of isoleucine (Ile) and lysine (Lys), but fall short in methionine (Met) and tryptophan (Trp). Grains have the opposite strengths and weaknesses, making them a perfect match for legumes.

	Ile	Lys	Met	Trp
Legumes				
Grains				
Together				

limiting amino acid: the essential amino acid found in the shortest supply relative to the amounts needed for protein synthesis in the body. Four amino acids are most likely to be limiting:

• Lysine.
• Methionine.
• Threonine.
• Tryptophan.

reference protein: a standard against which to measure the quality of other proteins.

complementary proteins: two or more dietary proteins whose amino acid assortments complement each other in such a way that the essential amino acids missing from one are supplied by the other.

protein digestibility–corrected amino acid score (PDCAAS): a measure of protein quality assessed by comparing the amino acid score of a food protein with the amino acid requirements of preschool-age children and then correcting for the true digestibility of the protein.

TABLE 6-3 PDCAAS Values of Selected Foods

Casein (milk protein)	1.00
Egg white	1.00
Soybean (isolate)	.99
Beef	.92
Pea flour	.69
Kidney beans (canned)	.68
Chickpeas (canned)	.66
Pinto beans (canned)	.66
Rolled oats	.57
Lentils (canned)	.52
Peanut meal	.52
Whole wheat	.40

NOTE: 1.0 is the maximum PDCAAS a food protein can receive.

Vegetarians obtain their protein from whole grains, legumes, nuts, vegetables, and, in some cases, eggs and milk products.

HOW TO Measure Protein Quality Using PDCAAS

To calculate the PDCAAS (protein digestibility–corrected amino acid score), researchers first determine the amino acid profile of the test protein (in this example, pinto beans). The second column of the table below presents the essential amino acid profile for pinto beans. The third column presents the amino acid reference pattern.

To determine how well the food protein meets human needs, researchers calculate the ratio by dividing the second column by the third column (for example, 30 ÷ 18 = 1.67). The amino acid with the lowest ratio is the most limiting amino acid—in this case, methionine. Its ratio is the amino acid score for the protein—in this case, 0.84.

The amino acid score alone, however, does not account for digestibility. Protein digestibility, as determined by rat studies, yields a value of 79 percent for pinto beans. Together, the amino acid score and the digestibility value determine the PDCAAS:

$$\text{PDCAAS} = \text{protein digestibility} \times \text{amino acid score.}$$
$$\text{PDCAAS for pinto beans} = 0.79 \times 0.84 = 0.66.$$

Thus the PDCAAS for pinto beans is 0.66. Table 6-3 lists the PDCAAS values of selected foods.

The PDCAAS is used to determine the % Daily Value on food labels. To calculate the % Daily Value for protein for canned pinto beans, multiply the number of grams of protein in a standard serving (in the case of pinto beans, 7 grams per ½ cup) by the PDCAAS:

$$7 \text{ g} \times 0.66 = 4.62.$$

This value is then divided by the recommended standard for protein (for children over age four and adults, 50 grams):

$$4.62 \div 50 = 0.09 \text{ (or 9\%)}.$$

The food label for this can of pinto beans would declare that one serving provides 7 grams protein, and if the label included a % Daily Value for protein (which is optional), the value would be 9 percent.

Essential Amino Acids	Amino Acid Profile of Pinto Beans (mg/g protein)	Amino Acid Reference Pattern (mg/g protein)	Amino Acid Score
Histidine	30.0	18	1.67
Isoleucine	42.5	25	1.70
Leucine	80.4	55	1.46
Lysine	69.0	51	1.35
Methionine (+ cystine)	21.1	25	0.84
Phenylalanine (+ tyrosine)	90.5	47	1.93
Threonine	43.7	27	1.62
Tryptophan	8.8	7	1.26
Valine	50.1	32	1.57

IN SUMMARY A diet inadequate in any of the essential amino acids limits protein synthesis. The best guarantee of amino acid adequacy is to eat foods containing high-quality proteins or mixtures of foods containing incomplete but complementary proteins so that each can supply the amino acids missing in the other. Vegetarians can meet their protein needs by eating a variety of whole grains, legumes, seeds, nuts, and vegetables.

Protein Regulations for Food Labels

All food labels must state the *quantity* of protein in grams. The "% Daily Value"■ for protein is not mandatory on all labels, but is required whenever a food makes a protein claim or is intended for consumption by children under four years old.* Whenever the Daily Value percentage is declared, researchers must determine the *quality* of the protein by using the PDCAAS method. Thus, when a % Daily Value is stated for protein, it reflects both quantity and quality.

■ Daily Values:
  • 50 g protein (based on 10% of 2000 kcal diet).

IN SUMMARY The quality of protein is measured by its amino acid content, its digestibility, and its ability to support growth. Such measures are of great importance in dealing with malnutrition worldwide, but in the United States and Canada, where protein deficiency is not common, protein quality scores of individual foods deserve little emphasis.

Health Effects and Recommended Intakes of Protein

As you know by now, protein is indispensable to life. It should come as no surprise that protein deficiency can have devastating effects on people's health. But, like the other nutrients, protein in excess can also be harmful. This section examines the health effects and recommended intakes of protein.

Protein-Energy Malnutrition

When people are deprived of protein, energy, or both, the result is **protein-energy malnutrition (PEM).** Although PEM touches many adult lives, it most often strikes early in childhood. It is one of the most prevalent and devastating forms of malnutrition in the world, afflicting over 500 million children. Most of the 33,000 children who die each day are malnourished.

Inadequate food intake leads to poor growth in children and to weight loss and wasting in adults. Children who are thin for their height may be suffering from **acute PEM** (recent severe food deprivation), whereas children who are short for their age have experienced **chronic PEM** (long-term food deprivation). Poor growth due to PEM is easy to overlook because a small child may look quite normal, but it is the most common sign of malnutrition.

PEM is most prevalent in Africa, Central America, South America, the Middle East, and East and Southeast Asia. In the United States, homeless people and those living in substandard housing in inner cities and rural areas have been diagnosed with PEM. In addition to those living in poverty, elderly people who live alone and adults who are addicted to drugs and alcohol are frequently victims of PEM. PEM develops in young children when parents mistakenly provide "health-food beverages"■ that lack adequate energy or protein instead of milk, most commonly because of nutritional ignorance, perceived milk intolerance, or food faddism.[5] Adult PEM is also seen in people hospitalized with infections such as AIDS or tuberculosis; these infections deplete body proteins, demand extra energy, induce nutrient losses, and alter metabolic pathways. Furthermore, poor nutrient intake during hospitalization worsens malnutrition and impairs recovery, whereas nutrition intervention often improves the body's response to other treatments and the chances of survival.[6] PEM is also common in those suffering from the eating disorder anorexia nervosa

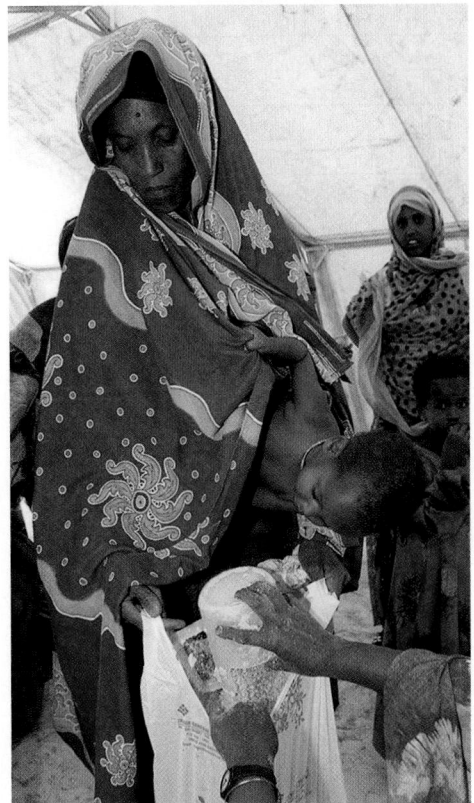

AP/Wide World Photos

Donated food saves some people from starvation, but it is usually insufficient to meet nutrient needs or even to provide a full belly for every person who is hungry.

■ Rice drinks are often sold as milk alternatives, but fail to provide adequate protein, vitamins, and minerals.

protein-energy malnutrition (PEM), also called **protein-kcalorie malnutrition (PCM):** a deficiency of protein, energy, or both, including kwashiorkor, marasmus, and instances in which they overlap (see p. 198).

acute PEM: protein-energy malnutrition caused by recent severe food restriction; characterized in children by thinness for height (wasting).

chronic PEM: protein-energy malnutrition caused by long-term food deprivation; characterized in children by short height for age (stunting).

---

*For labeling purposes, the Daily Values for protein are as follows: for infants, 14 grams; for children under age four, 16 grams; for older children and adults, 50 grams; for pregnant women, 60 grams; and for lactating women, 65 grams.

TABLE 6-4   Features of Marasmus and Kwashiorkor in Children

Separating PEM into two classifications oversimplifies the condition, but at the extremes, marasmus and kwashiorkor exhibit marked differences. Marasmus-kwashiorkor mix presents symptoms common to both marasmus and kwashiorkor. In all cases, children are likely to develop diarrhea, infections, and multiple nutrient deficiencies.

Marasmus	Kwashiorkor
Infancy (less than 2 yr)	Older infants and young children (1 to 3 yr)
Severe deprivation, or impaired absorption, of protein, energy, vitamins, and minerals	Inadequate protein intake or, more commonly, infections
Develops slowly; chronic PEM	Rapid onset; acute PEM
Severe weight loss	Some weight loss
Severe muscle wasting, with no body fat	Some muscle wasting, with retention of some body fat
Growth: <60% weight-for-age	Growth: 60 to 80% weight-for-age
No detectable edema	Edema
No fatty liver	Enlarged fatty liver
Anxiety, apathy	Apathy, misery, irritability, sadness
Good appetite possible	Loss of appetite
Hair is sparse, thin, and dry; easily pulled out	Hair is dry and brittle; easily pulled out; changes color; becomes straight
Skin is dry, thin, and easily wrinkles	Skin develops lesions

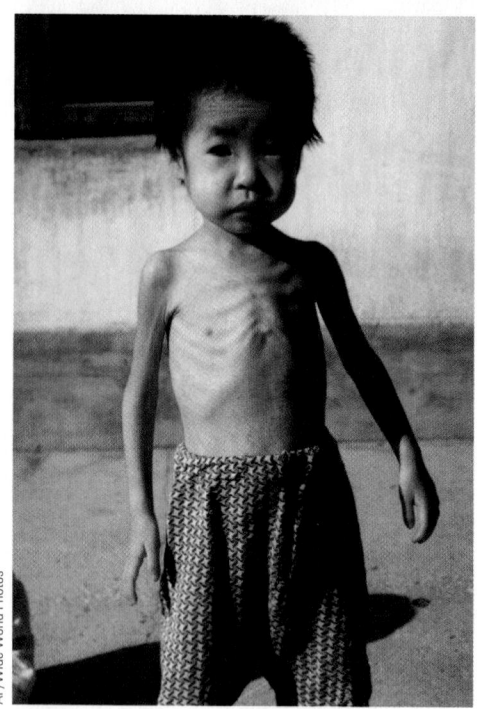

AP/Wide World Photos

The extreme loss of muscle and fat characteristic of marasmus is apparent in this child's "matchstick" arms.

marasmus (ma-RAZ-mus): a form of PEM that results from a severe deprivation, or impaired absorption, of energy, protein, vitamins, and minerals.

kwashiorkor (kwash-ee-OR-core, kwash-ee-or-CORE): a form of PEM that results either from inadequate protein intake or, more commonly, from infections.

(discussed in Highlight 9). Prevention emphasizes frequent, nutrient-dense, energy-dense meals and, equally important, resolution of the underlying causes of PEM—poverty, infections, and illness.

Classifying PEM   PEM occurs in two forms: marasmus and kwashiorkor, which differ in their clinical features (see Table 6-4). The following paragraphs present three clinical syndromes—marasmus, kwashiorkor, and the combination of the two.

Marasmus   Appropriately named from the Greek word meaning "dying away," **marasmus** reflects a severe deprivation of food over a long time (chronic PEM). Put simply, the person is starving and suffering from an inadequate energy *and* protein intake (and inadequate essential fatty acids, vitamins, and minerals as well). Marasmus occurs most commonly in children from 6 to 18 months of age in all the overpopulated urban slums of the world. Children in impoverished nations simply do not have enough to eat and subsist on diluted cereal drinks that supply scant energy and protein of low quality; such food can barely sustain life, much less support growth. Consequently, marasmic children look like little old people—just skin and bones.

Without adequate nutrition, muscles, including the heart, waste and weaken.[7] Because the brain normally grows to almost its full adult size within the first two years of life, marasmus impairs brain development and learning ability. Reduced synthesis of key hormones slows metabolism and lowers body temperature. There is little or no fat under the skin to insulate against cold. Hospital workers find that children with marasmus need to be clothed, covered, and kept warm. Because these children often suffer delays in their mental and behavioral development, they also need loving care, a stimulating environment, and parental attention.

The starving child faces this threat to life by engaging in as little activity as possible—not even crying for food. The body musters all its forces to meet the crisis, so it cuts down on any expenditure of protein not needed for the functioning of the heart, lungs, and brain. Growth ceases; the child is no larger at age four than at age two. Enzymes are in short supply and the GI tract lining deteriorates. Consequently, the child can't digest and absorb what little food is eaten.

Kwashiorkor   Kwashiorkor typically reflects a sudden and recent deprivation of food (acute PEM). Kwashiorkor was originally a Ghanaian word meaning "the evil spirit that infects the first child when the second child is born." When a mother who has been nursing her first child bears a second child, she weans the first child and puts the second one on the breast. The first child, suddenly switched from nutrient-dense, protein-rich breast milk to a starchy, protein-poor cereal, soon begins to sicken and die. Kwashiorkor typically sets in between 18 months and two years.

Kwashiorkor usually develops rapidly as a result of protein deficiency or, more commonly, is precipitated by an illness such as measles or other infection. Other factors may also contribute to the symptoms that accompany kwashiorkor.

The loss of weight and body fat is usually not as severe in kwashiorkor as in marasmus, but there may be some muscle wasting. Proteins and hormones that previously maintained fluid balance diminish, and fluid leaks into the interstitial spaces. The child's limbs and face become swollen with edema, a distinguishing feature of kwashiorkor. The lack of the protein carriers that transport fat out of the liver causes the belly to bulge with a fatty liver. The fatty liver lacks enzymes to clear metabolic toxins from the body, so their harmful effects are prolonged. Inflammation in response to these toxins and to infections further contributes to the edema that accompanies kwashiorkor. Without sufficient tyrosine to make melanin, the child's hair loses its color; inadequate protein synthesis leaves the skin patchy and scaly, often with sores that fail to heal. The lack of proteins to carry or store iron leaves iron free. Unbound iron is common in children with kwashiorkor and may contribute to their illnesses and deaths by promoting bacterial growth and free-radical damage. (Free-radical damage is discussed fully in Highlight 11.)

Marasmus-Kwashiorkor Mix The combination of marasmus and kwashiorkor is characterized by the edema of kwashiorkor with the wasting of marasmus. Most often, the child is suffering the effects of both malnutrition and infections. Some researchers believe that kwashiorkor and marasmus are two stages of the same disease. They point out that kwashiorkor and marasmus often exist side by side in the same community where children consume the same diet. They note that a child who has marasmus can later develop kwashiorkor. Some research indicates that marasmus represents the body's adaptation to starvation and that kwashiorkor develops when adaptation fails.

Infections In PEM, antibodies to fight off invading bacteria are degraded to provide amino acids for other uses, leaving the malnourished child vulnerable to infections. Blood proteins, including hemoglobin, are no longer synthesized, so the child becomes anemic and weak. **Dysentery,** an infection of the digestive tract, causes diarrhea, further depleting the body of nutrients. In the marasmic child, once infection sets in, kwashiorkor often follows and the immune response weakens further.[8]

The combination of infections, fever, fluid imbalances, and anemia often leads to heart failure and occasionally sudden death. Infections combined with malnutrition are responsible for two-thirds of the deaths of young children in developing countries. Measles, which might make a healthy child sick for a week or two, kills a child with PEM within two or three days.

Rehabilitation If caught in time, the life of a starving child may be saved with nutrition intervention. Diarrhea will have incurred dramatic fluid and mineral losses that will require careful correction to help raise the blood pressure and strengthen the heartbeat. After the first 24 to 48 hours, protein and food energy may be given in *small* quantities, with intakes *gradually* increased as tolerated. Severely malnourished people, especially those with edema, recover better with an initial diet that is relatively low in protein (10 percent kcalories from protein).[9]

Experts assure us that we possess the knowledge, technology, and resources to end hunger. Programs that tailor interventions to the local people and involve them in the process of identifying problems and devising solutions have the most success.[10] To win the war on hunger, those who have the food, technology, and resources must make fighting hunger a priority (see Chapter 20 for more on hunger).

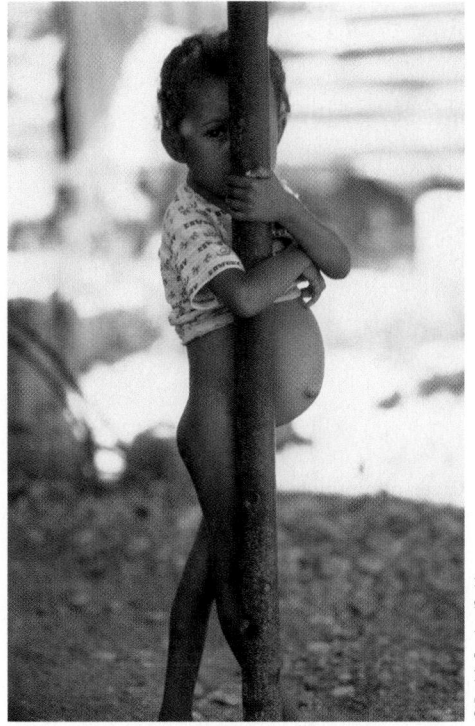

The edema and enlarged liver characteristic of kwashiorkor are apparent in this child's swollen belly. Malnourished children commonly have an enlarged abdomen from parasites as well.

dysentery (DISS-en-terry): an infection of the digestive tract that causes diarrhea.

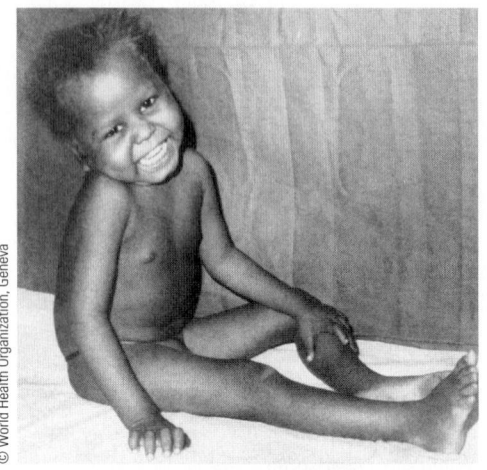

Given appropriate nutrition care, this child has successfully recovered from kwashiorkor.

Health Effects of Protein

While many of the world's people struggle to obtain enough food energy and protein, in developed countries both are so abundant that problems of excess are seen. Overconsumption of protein offers no benefits and may pose health risks. High-protein diets have been implicated in several chronic diseases, including heart disease, cancer, osteoporosis, obesity, and kidney stones, but evidence is insufficient to establish an upper limit.[11]

Researchers attempting to clarify the relationships between excess protein and chronic diseases face several obstacles. Population studies have difficulty determining whether diseases correlate with animal proteins or with their accompanying saturated fats, for example. Studies that rely on data from vegetarians must sort out the many lifestyle factors, other than a "no-meat diet," that might explain relationships between protein and health.

Heart Disease As Chapter 5 mentioned, foods rich in animal protein tend to be rich in saturated fats. Consequently, it is not surprising to find a correlation between animal-protein intake and heart disease, although no independent effect has been demonstrated. On the other hand, substituting soy protein for animal protein lowers blood cholesterol, especially in those with high blood cholesterol.[12]

Research suggests that elevated levels of the amino acid homocysteine may be an independent risk factor for heart disease.[13] Researchers do not yet fully understand the many factors that can raise homocysteine in the blood or whether elevated levels are a cause or an effect of heart disease.[14] Until they can determine the exact role homocysteine plays in heart disease, they are following several leads in pursuit of the answers.[15] Coffee's role in heart disease has been controversial, but research suggests it is among the most influential factors in raising homocysteine, which may explain some of the adverse health effects of heavy consumption.[16] Elevated homocysteine levels are among the many adverse health consequences of smoking cigarettes and drinking alcohol as well.[17] Homocysteine is also elevated with suboptimal intakes of B vitamins and can usually be lowered with supplements of vitamin B_{12}, vitamin B_6, and folate.[18] Such research suggests that a high intake of these vitamins may reduce the risk of heart disease.[19]

In contrast to homocysteine, the amino acid arginine may be a protective factor for heart disease, slowing the progression of atherosclerosis. The exact amount of arginine needed to defend against heart disease has not yet been determined, but it appears to be much greater than a healthy diet or a reasonable quantity of supplements alone can provide. If research confirms the benefits of arginine, look for manufacturers to begin producing functional foods enriched with this amino acid. In the meantime, it would be unwise for consumers to use supplements of arginine, or any other amino acid for that matter (as p. 203 explains).

Cancer As in heart disease, the effects of protein and fats on cancers cannot be easily separated. Population studies suggest a correlation between high intakes of animal proteins and some types of cancer (notably, cancer of the colon, breast, kidneys, pancreas, and prostate).

Adult Bone Loss (Osteoporosis) Chapter 12 presents calcium metabolism, and Highlight 12 elaborates on the main factors that influence osteoporosis. This section briefly describes the relationships between protein intake and bone loss. When protein intake is high, calcium excretion rises. Whether excess protein depletes the bones of their chief mineral may depend upon the ratio of calcium intake to protein intake. After all, bones need both protein and calcium. An ideal ratio has not been determined, but a young woman whose intake meets recommendations for both nutrients has a calcium-to-protein ratio of more than 20 to 1 (milligrams to grams), which probably provides adequate protection for the bones. For most women in the United States, however, average calcium intakes are lower and protein intakes are higher, yielding a 9-to-1 ratio, which may produce calcium losses significant enough to compromise bone health. In other words, the problem may reflect too little calcium, not too much protein. In establishing recommendations, the DRI Committee considered protein's effect on calcium metabolism and bone health, but did not find sufficient evidence to warrant an adjustment for calcium or an upper limit for protein.[20]

Inadequate intakes of protein may also compromise bone health. Osteoporosis is particularly common in elderly women and in adolescents with anorexia nervosa—groups who typically receive less protein than they need. For these people, increasing protein intake may be just what they need to protect their bones.[21]

Weight Control Protein-rich foods are often fat-rich foods that contribute to weight gain with its accompanying health risks. As Highlight 8 explains, weight-loss gimmicks that encourage a high-protein diet may be effective, but only because they are low-kcalorie diets. Diets that provide adequate protein, moderate fat, and sufficient energy from carbohydrates can better support weight loss and good health. Including protein at each meal may help with weight loss by providing satiety; selecting too many protein-rich foods, such as meat and milk, may crowd out fruits, vegetables, and grains, making the diet inadequate in other nutrients.

Kidney Disease Excretion of the end products of protein metabolism depends, in part, on an adequate fluid intake and healthy kidneys. A high protein intake increases the work of the kidneys, but does not appear to cause kidney disease. Restricting dietary protein, however, may help to slow the progression of kidney disease and limit the formation of kidney stones in people who have these conditions.

IN SUMMARY Protein deficiencies arise from both energy-poor and protein-poor diets and lead to the devastating diseases of marasmus and kwashiorkor. Together, these diseases are known as PEM (protein-energy malnutrition), a major form of malnutrition causing death in children worldwide. Excesses of protein offer no advantage; in fact, overconsumption of protein-rich foods may incur health problems as well.

Recommended Intakes of Protein

As mentioned earlier, the body continuously breaks down and loses some protein and cannot store amino acids. To replace protein, the body needs dietary protein for two reasons: first, food protein is the only source of the *essential* amino acids; and second, it is the only practical source of *nitrogen* with which to build the nonessential amino acids and other nitrogen-containing compounds the body needs.

Given recommendations that people's fat intakes should contribute 20 to 35 percent of total food energy, and carbohydrate, 45 to 65 percent, that leaves 10 to 35 percent for protein. In a 2000-kcalorie diet, that represents 200 to 700 kcalories from protein, or 50 to 175 grams. Average intakes in the United States and Canada fall within this range.

Protein RDA The protein RDA■ for adults is 0.8 gram per kilogram of healthy body weight per day. For infants and children, the RDA is slightly higher. The table on the inside front cover lists the RDA for males and females at various ages in two ways—grams per day based on reference body weights and grams per kilogram per day.

The RDA generously covers the needs for replacing worn-out tissue, so it increases for larger people; it also covers the needs for building new tissue during growth, so it increases for infants, children, and pregnant women. The protein RDA is the same for athletes as for others, although some fitness authorities recommend a slightly higher intake, as Chapter 14 explains.[22] The accompanying "How to" shows how to calculate your RDA for protein.

■ RDA for protein:
- 0.8 g/kg/day.
- 10 to 35% of energy intake.

HOW TO Calculate Recommended Protein Intakes

To figure your protein RDA:
- Look up the healthy weight for a person of your height (inside back cover). If your present weight falls within that range, use it for the following calculations. If your present weight falls outside the range, use the midpoint of the healthy weight range as your reference weight.
- Convert pounds to kilograms, if necessary (pounds divided by 2.2 equals kilograms).
- Multiply kilograms by 0.8 to get your RDA in grams per day. (Older teens 14 to 18 years old, multiply by 0.85.) Example:

Weight = 150 lb.

150 lb ÷ 2.2 lb/kg = 68 kg (rounded off).

68 kg × 0.8 g/kg= 54 g protein (rounded off).

Two meat servings of the size depicted here represent the maximum daily meat intake suggested by the Daily Food Guide as health promoting.

© Courtesy National Cattlemen's Beef Association

In setting the RDA, the committee assumes that people are healthy and do not have unusual metabolic needs for protein; that the protein eaten will be of mixed quality (from both high- and low-quality sources); and that the body will use the protein efficiently. In addition, the committee assumes that the protein is consumed along with sufficient carbohydrate and fat to provide adequate energy and that other nutrients in the diet are adequate.

Adequate Energy Note the qualification "adequate energy" in the preceding statement, and consider what happens if energy intake falls short of needs. An intake of 50 grams of protein provides 200 kcalories, which represents 10 percent of the total energy from protein, if the person receives 2000 kcalories a day. But if the person cuts energy intake drastically—to, say, 800 kcalories a day—then an intake of 200 kcalories from protein is suddenly 25 percent of the total; yet it's still the same amount of protein (number of grams). The protein intake is reasonable, but the energy intake is not; the low energy intake will force the body to use the protein to meet energy needs rather than to replace lost body protein. Similarly, if the person's energy intake is high—say, 4000 kcalories—the 50-gram protein intake will represent only 5 percent of the total; yet it *still* is a reasonable protein intake. Again, the energy intake is unreasonable for most people, but in this case, it will permit the protein to be used to meet the body's needs.

Be careful when judging protein (or carbohydrate or fat) intake as a percentage of energy. Always ascertain the number of grams as well, and compare it with the RDA or another standard stated in grams. A recommendation stated as a percentage of energy intake is useful only if the energy intake is within reason.

Protein in Abundance Most people in the United States and Canada receive much more protein than they need. Even athletes in training typically don't need to increase their protein intakes because the additional foods they eat to meet their high energy needs deliver protein as well. (Chapter 14 provides full details on the energy and protein needs of athletes.) That protein intake is high is not surprising considering the abundance of food eaten and the central role meats hold in the North American diet. A single ounce of meat (or ½ cup legumes) delivers about 7 grams of protein, so 8 ounces of meat alone supplies more than the RDA for an average-sized person. Besides meat, well-fed people eat many other nutritious foods, many of which also provide protein. A cup of milk provides 8 grams of protein. Grains and vegetables provide small amounts of protein, but they can add up to significant quantities; fruits and fats provide no protein.

To illustrate how easy it is to overconsume protein, consider the *minimum* recommended servings for the Food Guide Pyramid. Six servings from the bread, cereal, rice, and pasta group provide about 18 grams of protein; three servings of vegetables deliver about 6 grams; two servings of milk offer 16 grams; and two servings (about 5 ounces) of meat contain about 35 grams. This totals 75 grams of protein—higher than recommendations for most people and yet still lower than the average intake of people in the United States.

Just think how much more protein people receive when they eat additional servings. No wonder most people in the United States and Canada get more protein than they need. If they have an adequate *food* intake, they have a more-than-adequate protein intake. The key diet-planning principle to emphasize for protein is moderation. Even though most people receive plenty of protein, some feel compelled to take supplements as well, as the next section describes.

IN SUMMARY Optimally, the diet will be adequate in energy from carbohydrate and fat and will deliver 0.8 gram of protein per kilogram of healthy body weight each day. U.S. and Canadian diets are typically more than adequate in this respect.

Protein and Amino Acid Supplements

Websites, health-food stores, and popular magazine articles advertise a wide variety of protein supplements, and people take these supplements for many different reasons, all of them unfounded. Athletes take protein supplements to build muscle. Dieters take them to spare their bodies' protein while losing weight. Women take them to strengthen their fingernails. People take individual amino acids, too—to cure herpes, to make themselves sleep better, to lose weight, and to relieve pain and depression.* Like many other magic solutions to health problems, protein and amino acid■ supplements don't work these miracles. Furthermore, they may be harmful.[23]

Muscle work builds muscle; protein supplements do not, and athletes do not need them. (Highlight 14 presents more information on protein and other supplements athletes commonly use.) Instead, athletes need a well-balanced diet that provides sufficient dietary protein and adequate food energy. Food energy spares body protein; carbohydrate and fat serve this purpose equally well. Fingernails are not affected by protein supplements, provided the diet is adequate. Furthermore, protein supplements are expensive, less completely digested than protein-rich foods, and, when used as replacements for such foods, often downright dangerous.

Single amino acids do not occur naturally in foods and offer no benefit to the body; in fact, they may be harmful. The body was not designed to handle the high concentrations and unusual combinations of amino acids found in supplements. An excess of one amino acid can create such a demand for a carrier that it limits the absorption of another amino acid, presenting the possibility of a deficiency. Those amino acids winning the competition enter in excess, creating the possibility of a toxicity. Toxicity of single amino acids in animal studies raises concerns about their use in human beings. Anyone considering taking amino acid supplements should check with a registered dietitian or physician first.

In two cases, recommendations for single amino acid supplements have led to widespread public use—lysine to prevent or relieve the infections that cause herpes cold sores on the mouth or genital organs, and tryptophan to relieve pain, depression, and insomnia. In both cases, enthusiastic popular reports preceded careful scientific experiments and health recommendations. A review of the research indicates that lysine may suppress herpes infections in some individuals and appears safe (up to 3 grams per day) when taken in divided doses with meals.[24]

Tryptophan is also effective with respect to pain and sleep, but its use for these purposes is still experimental. More than 1500 people who elected to take tryptophan supplements developed a rare blood disorder known as eosinophilia-myalgia syndrome (EMS). EMS is characterized by severe muscle and joint pain, extremely high fever, and, in over three dozen cases, death. Treatment for EMS usually involves physical therapy and low doses of corticosteroids to relieve symptoms temporarily. The Food and Drug Administration determined that contaminants caused the disease and issued a recall of all products containing manufactured tryptophan.

■ Use of amino acids as dietary supplements is *inappropriate*, especially for:
- All women of childbearing age.
- Pregnant or lactating women.
- Infants, children, and adolescents.
- Elderly people.
- People with inborn errors of metabolism that affect their bodies' handling of amino acids.
- Smokers.
- People on low-protein diets.
- People with chronic or acute mental or physical illnesses who take amino acids without medical supervision.

IN SUMMARY Normal, healthy people never need protein or amino acid supplements. It is safest to obtain lysine, tryptophan, and all other amino acids from protein-rich foods, eaten with abundant carbohydrate and some fat to facilitate their use in the body. With all that we know about science, it is hard to improve on nature.

---

*Canada only allows single amino acid supplements to be sold as drugs or used as food additives.

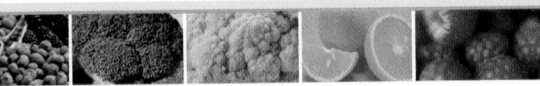

Nutrition in Your Life

Foods that derive from animals—meats, fish, poultry, eggs, and milk products—provide plenty of protein, but are often accompanied by fat. Those that derive from plants—whole grains, vegetables, and legumes—may provide less protein, but also less fat.

- Calculate your daily protein needs. Do you receive enough, but not too much, protein daily?
- What are your dietary sources of proteins? Do you use mostly plant-based or animal-based protein foods in your diet?
- Do you take protein or amino acid supplements?

NUTRITION ON THE NET

 Access these websites for further study of topics covered in this chapter.

- Find updates and quick links to these and other nutrition-related sites at our website: **www.wadsworth.com/nutrition**
- Learn more about sickle-cell anemia from the National Heart, Lung, and Blood Institute or the Sickle Cell Disease Association of America: **www.nhlbi.nih.gov** or **www.sicklecelldisease.org**

- Learn more about protein-energy malnutrition and world hunger from the World Health Organization Nutrition Programme: **www.who.int/nut**
- Chapter 20 offers many more websites on malnutrition and world hunger.

NUTRITION CALCULATIONS

These problems will give you practice in doing simple nutrition-related calculations using hypothetical situations (see p. 207 for answers). Once you have mastered these examples, you will be prepared to examine your own protein needs. Be sure to show your calculations for each problem.

1. Compute recommended protein intakes for people of different sizes. Refer to the "How to" on p. 201 and compute the protein recommendation for the following people. The intake for a woman who weighs 144 pounds is computed for you as an example.

$$144 \text{ lb} \div 2.2 \text{ lb/kg} = 65 \text{ kg.}$$

$$0.8 \text{ g/kg} \times 65 \text{ kg} = 52 \text{ g protein per day.}$$

   a. A woman who weighs 116 pounds.
   b. A man (18 years) who weighs 180 pounds.

2. The chapter warns that recommendations based on percentage of energy intake are not always appropriate. Consider a woman 26 years old who weighs 165 pounds. Her diet provides 1500 kcalories/day with 50 grams carbohydrate and 100 grams fat.
   a. What is this woman's protein intake? Show your calculations.
   b. Is her protein intake appropriate? Justify your answer.
   c. Are her carbohydrate and fat intakes appropriate? Justify your answer.

This exercise should help you develop a perspective on protein recommendations.

STUDY QUESTIONS

These questions will help you review the chapter. You will find the answers in the discussions on the pages provided.

1. How does the chemical structure of proteins differ from the structures of carbohydrates and fats? (pp. 181–182)

2. Describe the structure of amino acids, and explain how their sequence in proteins affects the proteins' shapes. What are essential amino acids? (pp. 182–184)

3. Describe protein digestion and absorption. (pp. 185–186)

4. Describe protein synthesis. (pp. 187–189)

5. Describe some of the roles proteins play in the human body. (pp. 189–192)

6. What are enzymes? What roles do they play in chemical reactions? Describe the differences between enzymes and hormones. (p. 190)

7. How does the body use amino acids? What is deamination? Define nitrogen balance. What conditions are associated with zero, positive, and negative balance? (pp. 193–194)

8. What factors affect the quality of dietary protein? What is a high-quality protein? (pp. 194–195)

9. How can vegetarians meet their protein needs without eating meat? (p. 195)

10. What are the health consequences of ingesting inadequate protein and energy? Describe marasmus and kwashiorkor. How can the two conditions be distinguished, and in what ways do they overlap? (pp. 197–199)

11. How might protein excess, or the type of protein eaten, influence health? (pp. 200–201)

12. What factors are considered in establishing recommended protein intakes? (pp. 201–202)

13. What are the benefits and risks of taking protein and amino acid supplements? (p. 203)

These multiple choice questions will help you prepare for an exam. Answers can be found on p. 207.

1. Which part of its chemical structure differentiates one amino acid from another?
   a. its side group
   b. its acid group
   c. its amino group
   d. its double bonds

2. Isoleucine, leucine, and lysine are:
   a. proteases.
   b. polypeptides.
   c. essential amino acids.
   d. complementary proteins.

3. In the stomach, hydrochloric acid:
   a. denatures proteins and activates pepsin.
   b. hydrolyzes proteins and denatures pepsin.
   c. emulsifies proteins and releases peptidase.
   d. condenses proteins and facilitates digestion.

4. Proteins that facilitate chemical reactions are:
   a. buffers.
   b. enzymes.
   c. hormones.
   d. antigens.

5. If an essential amino acid that is needed to make a protein is unavailable, the cells must:
   a. deaminate another amino acid.
   b. substitute a similar amino acid.
   c. break down proteins to obtain it.
   d. synthesize the amino acid from glucose and nitrogen.

6. Protein turnover describes the amount of protein:
   a. found in foods and the body.
   b. absorbed from the diet.
   c. synthesized and degraded.
   d. used to make glucose.

7. The PDCAAS is used to:
   a. determine protein quality.
   b. assess protein-energy malnutrition.
   c. estimate the weight of nitrogen in a food.
   d. calculate the percentage kcalories from protein.

8. Marasmus develops from:
   a. too much fat clogging the liver.
   b. megadoses of amino acid supplements.
   c. inadequate protein and energy intake.
   d. excessive fluid intake causing edema.

9. The protein RDA for a healthy adult who weighs 180 pounds is:
   a. 50 milligrams/day.
   b. 65 grams/day.
   c. 180 grams/day.
   d. 2000 milligrams/day.

10. Which of these foods has the least protein per serving?
    a. rice
    b. broccoli
    c. pinto beans
    d. orange juice

REFERENCES

1. M. S. Buchowski and coauthors, Equation to estimate resting energy expenditure in adolescents with sickle cell anemia, *American Journal of Clinical Nutrition* 76 (2002): 1335–1344; Committee on Genetics, Health supervision for children with sickle cell disease, *Pediatrics* 109 (2002): 526–535; S. T. Miller and coauthors, Prediction of adverse outcomes in children with sickle cell disease, *New England Journal of Medicine* 342 (2000): 83–89.

2. W. M. Rand, P. L. Pellett, and V. R. Young, Meta-analysis of nitrogen balance studies for estimating protein requirements in healthy adults, *American Journal of Clinical Nutrition* 77 (2003): 109–127.

3. J. Hernández-Rodriguez and G. Manjarrez-Guitiérrez, Macronutrients and neurotransmitter formation during brain development, *Nutrition Reviews* 59 (2001): S49–S59.

4. Position of the American Dietetic Association: Vegetarian diets, *Journal of the American Dietetic Association* 97 (1997): 1317–1321.

5. T. Liu and coauthors, Kwashiorkor in the United States: Fad diets, perceived and true milk allergy, and nutritional ignorance, *Archives of Dermatology* 137 (2001): 630–636; G. Massa, Protein malnutrition due to replacement of milk by rice drink, *European Journal of Pediatrics* 160 (2001): 382–384; N. F. Carvalho and coauthors, Severe nutritional deficiencies in toddlers resulting from health food milk alternatives, *Pediatrics* 107 (2001): e46.

6. G. Akner and T. Cederholm, Treatment of protein-energy malnutrition in chronic nonmalignant disorders, *American Journal of Clinical Nutrition* 74 (2001): 6–24; D. H. Sullivan, S. Sun, and R. C. Walls, Protein-energy undernutrition among elderly hospitalized patients: A prospective study, *Journal of the American Medical Association* 281 (1999): 2013–2019.

7. L. Combaret, D. Taillandier, and D. Attaix, Nutritional and hormonal control of protein breakdown, *American Journal of Kidney Diseases* 37 (2001): S108–S111.

8. M. Reid and coauthors, The acute-phase protein response to infection in edematous and nonedematous protein-energy malnutrition, *American Journal of Clinical Nutrition* 76 (2002): 1409–1415.

9. V. Scherbaum and P. Furst, New concepts on nutritional management of severe malnutrition: The role of protein, *Current Opinion in Clinical Nutrition and Metabolic Care* 3 (2000): 31–38.

10. B. A. Underwood and S. Smitasiri, Micronutrient malnutrition: Policies and programs for control and their implications, *Annual Review of Nutrition* 19 (1999): 303–324; C. G. Victora and coauthors, Potential interventions for the prevention of childhood pneumonia in developing countries: Improving nutrition, *American Journal of Clinical Nutrition* 70 (1999): 309–320.

11. Committee on Dietary Reference Intakes, *Dietary Reference Intakes for Energy, Carbohydrate, Fiber, Fat, Fatty Acids, Cholesterol, Protein, and Amino Acids* (Washington, D.C.: National Academies Press, 2002), p. 10–77.

12. S. Tonstad, K. Smerud, and L. Høie, A comparison of the effects of 2 doses of soy protein or casein on serum lipids, serum lipoproteins, and plasma total homocysteine in hypercholesterolemic subjects, *American Journal of Clinical Nutrition* 76 (2002): 78–84; S. R. Teixeira and coauthors, Effects of feeding 4 levels of soy protein for 3 and 6 wk on blood lipids and apolipoproteins in moderately hypercholesterolemic men, *American Journal of Clinical Nutrition* 71 (2000): 1077–1084.

13. D. S. Wald, M. Law, and J. K. Morris, Homocysteine and cardiovascular disease: Evidence on causality from a meta-analysis, *British Medical Journal* 325 (2002): 1202–1217; The Homocysteine Studies Collaboration, Homocysteine and risk of ischemic heart disease and stroke, *Journal of the American Medical Association* 288 (2002): 2015–2022; P. M. Ridker, Homocysteine and risk of cardiovascular disease among postmenopausal women, *Journal of the American Medical Association* 281 (1999): 1817–1821.

14. L. Brattström and D. E. L. Wilken, Homocysteine and cardiovascular disease: Cause or effect? *American Journal of Clinical Nutrition* 72 (2000): 315–323; R. Meleady and I. Graham, Plasma homocysteine as a cardiovascular risk factor: Causal, consequential, or of no consequence? *Nutrition Reviews* 57 (1999): 299–305.

15. J. Selhub, Homocysteine metabolism, *Annual Review of Nutrition* 19 (1999): 217–246.

16. P. Verhoef and coauthors, Contribution of caffeine to the homocysteine-raising effect of coffee: A randomized controlled trial in humans, *American Journal of Clinical Nutrition* 76 (2002): 1244–1248; M. J. Grubben and coauthors, Unfiltered coffee increases plasma homocysteine concentrations in healthy volunteers: A randomized trial, *American Journal of Clinical Nutrition* 71 (2000): 480–484.

17. L. I. Mennen and coauthors, Homocysteine, cardiovascular disease risk factors, and habitual diet in the French Supplementation and Antioxidant Vitamins and Minerals Study, *American Journal of Clinical Nutrition* 76 (2002): 1279–1289; A. DeBree and coauthors, Lifestyle factors and plasma homocysteine concentrations in a general population sample, *American Journal of Epidemiology* 153 (2001): 150–154.

18. G. Schnyder and coauthors, Decreased rate of coronary restenosis after lowering of plasma homocysteine levels, *New England Journal of Medicine* 345 (2001): 1539–1600; P. F. Jacques and coauthors, The effect of folic acid fortification on plasma folate and total homocysteine concentrations, *New England Journal of Medicine* 340 (1999): 1449–1454; A. Chait and coauthors, Increased dietary micronutrients decrease serum homocysteine concentrations in patients at high risk of cardiovascular disease, *American Journal of Clinical Nutrition* 70 (1999): 881–887; I. A. Brouwer and coauthors, Low-dose folic acid supplementation decreases plasma homocysteine concentrations: A randomized trial, *American Journal of Clinical Nutrition* 69 (1999): 99–104.

19. G. Schnyder and coauthors, Effect of homocysteine-lowering therapy with folic acid, vitamin B_{12}, and vitamin B_6 on clinical outcome after percutaneous coronary intervention—The Swiss Heart Study: A randomized controlled trial, *Journal of the American Medical Association* 288 (2002): 973–979; B. J. Venn and coauthors, Dietary counseling to increase natural folate intake: A randomized, placebo-controlled trial in free-living subjects to assess effects on serum folate and plasma total homocysteine, *American Journal of Clinical Nutrition* 76 (2002): 758–765; E. B. Rimm and coauthors, Folate and vitamin B_6 from diet and supplements in relation to risk of coronary heart disease among women, *Journal of the American Medical Association* 279 (1998): 359–364.

20. Committee on Dietary Reference Intakes, 2002, pp. 11-50–11-51; Committee on Dietary Reference Intakes, *Dietary Reference Intakes for Calcium, Phosphorus, Magnesium, Vitamin D, and Fluoride* (Washington, D.C.: National Academy Press, 1997), pp. 75–76.

21. J. Bell and S. J. Whiting, Elderly women need dietary protein to maintain bone mass, *Nutrition Reviews* 60 (2002): 337–341; M. T. Munoz and J. Argente, Anorexia nervosa in female adolescents: Endocrine and bone mineral density disturbances, *European Journal of Endocrinology* 147 (2002): 275–286.

22. Position of the American Dietetic Association, Dietitians of Canada, and the American College of Sports Nutrition, Nutrition and athletic performance, *Journal of the American Dietetic Association* 100 (2000): 1543–1556.

23. P. J. Garlick, Assessment of the safety of glutamine and other amino acids, *Journal of Nutrition* 131 (2001): 2556S–2561S.

24. N. W. Flodin, The metabolic roles, pharmacology, and toxicology of lysine, *Journal of the American College of Nutrition* 16 (1997): 7–21.

ANSWERS

Nutrition Calculations

1. a. 116 lb ÷ 2.2 lb/kg = 53 kg.

   0.8 g/kg × 53 kg = 42 g protein per day.

   b. 180 lb ÷ 2.2 lb/kg = 82 kg.

   He is 18 years old, so use 0.85 g/kg. 0.85 g/kg × 82 kg = 70 g protein per day.

2. a. 50 g carbohydrate × 4 kcal/g = 200 kcal from carbohydrate.

   100 g fat × 9 kcal/g = 900 kcal from fat.

   1500 kcal − (200 + 900 kcal) = 400 kcal from protein.

   400 kcal ÷ 4 kcal/g = 100 g protein.

   b. Using the RDA guideline of 0.8 g/kg, an appropriate protein intake for this woman would be 60 g protein/day (165 lb ÷ 2.2 lb/kg = 75 kg; 0.8 g/kg × 75 = 60 g/day). Her intake is higher than her RDA. Using the guideline that protein should contribute 10 to 35% of energy intake, her intake of 100 g protein on a 1500 kcal diet falls within the suggested range (400 kcal protein ÷ 1500 total kcal = 27%).

   c. Using the guideline that carbohydrate should contribute 45 to 65% and fat should contribute 20 to 35% of energy intake, her intake of 50 g carbohydrate is low (200 kcal carbohydrate ÷ 1500 total kcal = 13%), and her intake of 100 g fat is high (900 kcal fat ÷ 1500 total kcal = 60%).

Study Questions (multiple choice)

1. a   2. c   3. a   4. b   5. c   6. c   7. a

8. c   9. b   10. d

HIGHLIGHT

Vegetarian Diets

© Polara Studios, Inc.

The waiter presents this evening's specials: a fresh spinach salad topped with mandarin oranges, raisins, and sunflower seeds, served with a bowl of pasta smothered in a mushroom and tomato sauce and topped with grated parmesan cheese. Then this one: a salad made of chopped parsley, scallions, celery, and tomatoes mixed with bulgur wheat and dressed with olive oil and lemon juice, served with a spinach and feta cheese pie. Do these meals sound good to you? Or is something missing . . . a pork chop or ribeye, perhaps?

Would vegetarian fare be acceptable to you some of the time? Most of the time? Ever? Perhaps it is helpful to recognize that dietary choices fall along a continuum—from one end, where people eat no meat or foods of animal origin, to the other end, where they eat generous quantities daily. Meat's place in the diet has been the subject of much research and controversy, as this highlight will reveal. One of the missions of this highlight, in fact, is to identify the *range* of meat intakes most compatible with health.

People who choose to exclude meat and other animal-derived foods from their diets today do so for many of the same reasons the Greek philosopher Pythagoras cited in the sixth century B.C.: physical health, ecological responsibility, and philosophical concerns. They might also cite world hunger issues, economic reasons, ethical concerns, or religious beliefs as motivating factors. Whatever their reasons—and even if they don't have a particular reason—people who exclude meat will be better prepared to plan well-balanced meals if they understand the nutrition and health implications of vegetarian diets.

Vegetarians generally are categorized, not by their motivations, but by the foods they choose to exclude (see the glossary below). Some exclude red meat only; some also exclude chicken or fish; others also exclude eggs; and still others exclude milk and milk products as well. In fact, finding agreement on the definition of the term *vegetarian* is a challenge.[1]

As you will see, though, the foods a person *excludes* are not nearly as important as the foods a person *includes* in the diet. Vegetarian diets that include a variety of whole grains, vegetables, legumes, seeds, nuts, and fruits offer abundant complex carbohydrates and fibers, an assortment of vitamins and minerals, and little fat—characteristics that reflect current dietary recommendations aimed at promoting health and reducing obesity. This highlight examines the health benefits and potential problems of vegetarian diets and shows how to plan a well-balanced vegetarian diet.

Health Benefits of Vegetarian Diets

Research on the health impacts of vegetarianism would be relatively easy if vegetarians differed from other people only in not eating meat. Many vegetarians, however, have adopted lifestyles

GLOSSARY

lactovegetarians: people who include milk and milk products, but exclude meat, poultry, fish, seafood, and eggs from their diets.
• **lacto** = milk

lacto-ovo-vegetarians: people who include milk, milk products, and eggs, but exclude meat, poultry, fish, and seafood from their diets.
• **ovo** = egg

macrobiotic diets: extremely restrictive diets limited to a few grains and vegetables; based on metaphysical beliefs and not on nutrition.

meat replacements: products formulated to look and taste like meat, fish, or poultry; usually made of textured vegetable protein.

omnivores: people who have no formal restriction on the eating of any foods.
• **omni** = all
• **vores** = to eat

tempeh (TEM-pay): a fermented soybean food, rich in protein and fiber.

textured vegetable protein: processed soybean protein used in vegetarian products such as soy burgers; see also *meat replacements*.

tofu (TOE-foo): a curd made from soybeans, rich in protein and often fortified with calcium; used in many Asian and vegetarian dishes in place of meat.

vegans (VEE-guns, VAY-guns, or VEJ-ans): people who exclude all animal-derived foods (including meat, poultry, fish, eggs, and dairy products) from their diets; also called **pure vegetarians, strict vegetarians,** or **total vegetarians.**

vegetarians: a general term used to describe people who exclude meat, poultry, fish, or other animal-derived foods from their diets.

that differ from others: they typically maintain a healthy weight, use no tobacco or illicit drugs, use little (if any) alcohol, and are physically active. Researchers must account for these lifestyle differences before they can determine which aspects of health correlate just with diet. Even then, *correlations* merely reveal what health factors *go with* the vegetarian diet, not what health effects may be *caused by* the diet. Without more evidence, conclusions remain tentative. Still, with all these qualifications, research findings suggest that well-planned vegetarian diets offer sound nutrition and health benefits to adults.[2]

In general, vegetarians maintain a healthier body weight than nonvegetarians.[3] Studies report higher weights among people eating a mixed diet compared with vegetarians and that body weight increases as frequency of meat consumption increases.[4] Vegetarians' lower body weights correlate with their high intakes of fiber and low intakes of fat. Since obesity impairs health in a number of ways, this gives vegetarians a health advantage.

Vegetarians tend to have lower blood pressure and lower rates of hypertension than nonvegetarians. Appropriate body weight helps to maintain a healthy blood pressure, as does a diet low in total fat and saturated fat and high in fiber, fruits, and vegetables. Lifestyle factors also seem to influence blood pressure: smoking and alcohol intake raise blood pressure, and physical activity lowers it.

The incidence of heart disease and related deaths is much lower for vegetarians than for meat eaters. The dietary factor most directly related to heart disease is saturated animal fat, and in general, vegetarian diets are lower in total fat, saturated fat, and cholesterol than typical meat-based diets. The fats common in plant-based diets—the monounsaturated fats of olives, seeds, and nuts and the polyunsaturated fats of vegetable oils—are associated with a decreased risk of heart disease.[5] Furthermore, vegetarian diets are generally higher in dietary fiber, another factor that helps control blood lipids and protect against heart disease.

Many vegetarians include soy products such as **tofu** in their diets, and these foods offer additional benefits. Even when their intakes of energy, protein, carbohydrate, total fat, saturated fat, unsaturated fat, alcohol, and fiber are the same, people eating meals based on tofu have lower blood cholesterol and triglyceride levels than those eating meat.[6] Soy products, such as tofu, contain phytochemicals that may be responsible for their ability to lower blood cholesterol (as Highlight 13 explains in greater detail).[7]

Vegetarians have a significantly lower rate of cancer than the general population. Their low cancer rates may be due to their high intakes of fruits and vegetables.

Some scientific findings indicate that vegetarian diets are associated not only with lower cancer mortality in general, but with lower incidence of cancer at specific sites as well, most notably, colon cancer. People with colon cancer seem to eat more meat, more saturated fat, and fewer vegetables than others without cancer. High-protein, high-fat, low-fiber diets create an environment in the colon that promotes the devel-opment of cancer in some people. A high-meat diet has been associated with stomach cancer as well.[8]

Vegetarian Diet Planning

The vegetarian has the same meal-planning task as any other person—using a variety of foods that will deliver all the needed nutrients within an energy allowance that maintains a healthy body weight (as discussed in Chapter 2). An added challenge is to do so with fewer options.

Vegetarians who include milk products and eggs can meet recommendations for most nutrients about as easily as nonvegetarians. Such diets provide enough energy, protein, and other nutrients to support the health of adults and the growth of children and adolescents.

Vegetarians are often advised to follow the Daily Food Guide presented in Chapter 2 with a few modifications. Those who include milk products and eggs can follow the regular plan, using legumes, nuts, and seeds and products made from them, such as peanut butter, **tempeh,** and tofu, in place of meat. Those who do not use milk can use soy milk and tofu fortified with calcium, vitamin D, and vitamin B_{12}. Dark green vegetables and legumes help meet iron and zinc needs.

Several food guides have been developed specifically for vegetarian diets.[9] They all address the particular nutrition concerns of vegetarians, but differ slightly. Figure H6-1 presents one version and the USDA Food Guide Pyramid for comparison.[10] At first glance, it looks like the familiar Food Guide Pyramid, but it actually has notable differences. The vegetable and fruit groups have subgroups that emphasize particularly good sources of calcium and iron, respectively. Green leafy vegetables are featured in the vegetable group because they provide almost five times as much calcium per serving as other vegetables. Similarly, dried fruits receive special notice in the fruit group because they deliver six times as much iron as other fruits. A separate group for nuts and seeds provides additional sources of protein, iron, zinc, and essential fatty acids. A group for oils at the tip encourages the use of vegetable oils rich in unsaturated fats and omega-3 fatty acids. The meat and meat alternate group has been revised and renamed "beans and protein foods," and soy milks are found in the dairy group. To ensure adequate intakes of vitamin B_{12}, vitamin D, and calcium, vegetarians need to select fortified foods or use supplements daily. This design is flexible enough that a variety of people can use it: people who have adopted various vegetarian diets, those who want to make the transition to a vegetarian diet, and those who simply want to include more plant-based meals in their diets. This vegetarian food guide also includes other lifestyle factors that contribute to good health: physical activity and water intake.

Most vegetarians easily obtain large quantities of the nutrients that are abundant in plant foods: thiamin, folate, and vitamins B_6, C, A, and E. Vegetarian food guides help to ensure adequate intakes of the main nutrients vegetarian diets might otherwise lack: iron, zinc, calcium, vitamin B_{12}, and vitamin D.

FIGURE H6-1   Vegetarian and USDA Food Pyramids Compared

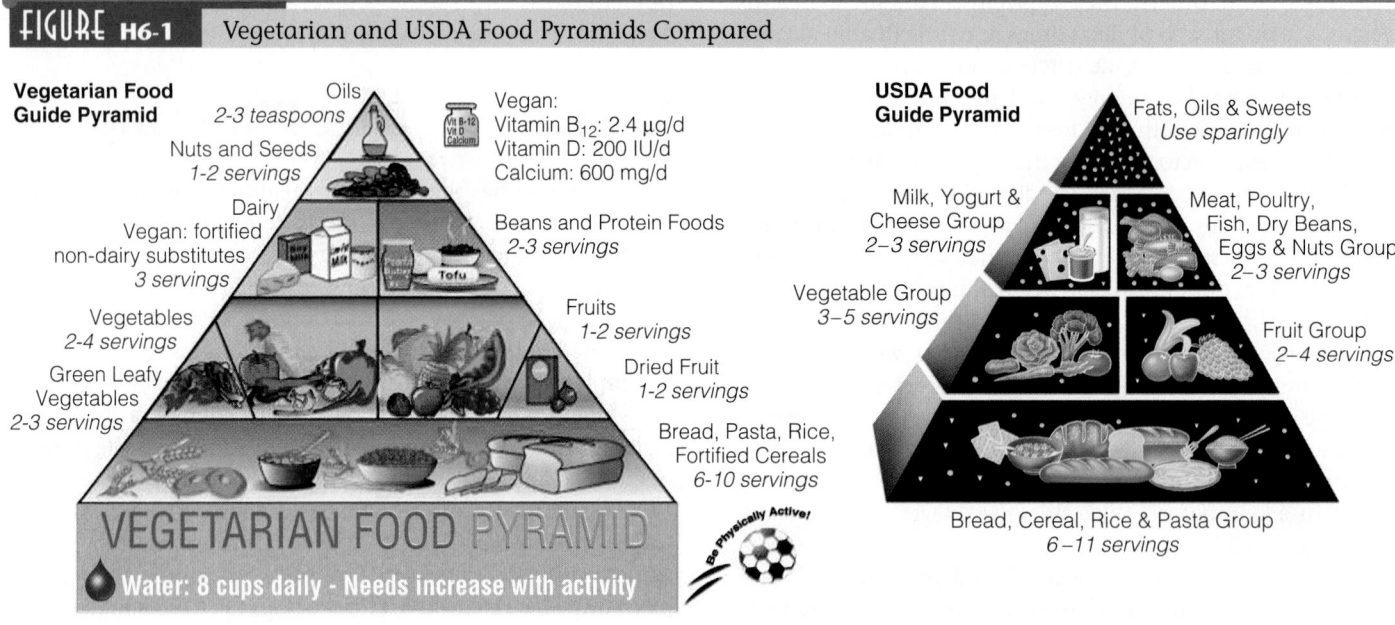

SOURCE: C. A. Venti and C. S. Johnston, Modifiied food guide pyramid for lactovegetarians and vegans *Journal of Nutrition* 132 (2002): 1050–1054.

Protein

The protein RDA for vegetarians is the same as for others, although some have suggested that it should be higher because of the lower digestibility of plant proteins.[11] **Lacto-ovo-vegetarians** who use animal-derived foods such as milk and eggs receive high-quality proteins and are likely to meet their protein needs. Even those who adopt only plant-based diets are likely to meet protein needs provided that energy intakes are adequate and the protein sources varied.[12] The proteins of whole grains, legumes, seeds, nuts, and vegetables can provide adequate amounts of all the amino acids. An advantage of many vegetarian protein foods is that they are generally lower in saturated fat than meats and are often higher in fiber and richer in some vitamins and minerals.

To ease meal preparation, vegetarians sometimes use **meat replacements** made of **textured vegetable protein** (soy protein). These foods are formulated to look and taste like meat, fish, or poultry. Many of these products are fortified to provide the known nutrient contents of animal-protein foods, but sometimes they fall short. A wise vegetarian learns to use a variety of whole, unrefined foods often and commercially prepared foods less frequently. Vegetarians may also use soybeans in the form of tofu to bolster protein intake.

Vitamins and Minerals

Getting enough iron can be a problem even for meat eaters, and those who eat no meat must pay special attention to their iron intake. The iron in plant foods such as legumes, dark green leafy vegetables, iron-fortified cereals, and whole-grain breads and cereals is poorly absorbed.[13] Because the bioavailability of iron from a vegetarian diet is low, the iron RDA for vegetarians is higher than for others (see Chapter 13 for more details).

Fortunately, the body seems to adapt to a vegetarian diet by absorbing iron more efficiently.[14] Furthermore, iron absorption is enhanced by vitamin C, and vegetarians typically eat many vitamin C–rich fruits and vegetables. Consequently, vegetarians suffer no more iron deficiency than other people do.[15]

Zinc is similar to iron in that meat is its richest food source and zinc from plant sources is not well absorbed.[16] In addition, soy, which is commonly used as a meat alternate in vegetarian meals, interferes with zinc absorption. Nevertheless, most vegetarian adults are not zinc deficient. Perhaps the best advice to vegetarians regarding zinc is to eat a variety of nutrient-dense foods; include whole grains, nuts, and legumes such as black-eyed peas, pinto beans, and kidney beans; and maintain an adequate energy intake. For those who include seafood in their diets, oysters, crabmeat, and shrimp are rich in zinc.

The calcium intakes of **lactovegetarians** are similar to those of the general population, but people who use no milk products risk deficiency. Careful planners select calcium-rich foods, such as calcium-fortified juices, soy milk, and breakfast cereals, in ample quantities regularly. This is especially important for children and adolescents. Soy formulas for infants are fortified with calcium and can be used in cooking, even for adults. Other good calcium sources include figs, some legumes, some green vegetables such as broccoli and turnip greens, some nuts such as almonds, certain seeds such as sesame seeds, and calcium-set tofu.* The choices should be

---

*Calcium salts are often added during processing to coagulate the tofu.

varied because calcium absorption from some plant foods may be limited (as Chapter 12 explains).

The requirement for vitamin B_{12} is small, but this vitamin is found only in animal-derived foods. Fermented soy products such as tempeh may contain some vitamin B_{12} from the bacteria that did the fermenting, but unfortunately, much of the vitamin B_{12} found in these products may be an inactive form. Seaweeds such as nori and chlorella supply some vitamin B_{12}, but not much, and excessive intakes of these foods can lead to iodine toxicity. To defend against vitamin B_{12} deficiency, **vegans** must rely on vitamin B_{12}–fortified sources (such as soy milk or breakfast cereals) or supplements. Without vitamin B_{12}, the nerves suffer damage, leading to such health consequences as loss of vision.[17]

People who do not use vitamin D–fortified foods and do not receive enough exposure to sunlight to synthesize adequate vitamin D may need supplements to defend against bone loss.[18] This is particularly important for infants, children, and older adults. In northern climates during winter months, young children on vegan diets can readily develop rickets, the vitamin D–deficiency disease.

Omega-3 Fatty Acids

Vegetarian diets typically provide enough omega-6 fatty acids, but lack omega-3 fatty acids. This imbalance slows production of EPA and DHA in the body, and without fish, eggs, or sea vegetables in the diet, intake of EPA and DHA falls short as well. To compensate for this inadequacy, vegetarians need to include good sources of linolenic acid, such as flaxseed, walnuts, soybeans, and their oils, in their diets daily.

Vegetarian Diets through the Life Span

Vegetarians who plan their diets carefully easily obtain all the nutrients they need to support good health. Achieving adequate energy and nutrient intakes may be difficult, however, for the vegan who excludes all animal products, and particularly for growing children and pregnant and lactating women. Foods of plant origin generally offer much less energy per bite than foods of animal origin. While a diet that delivers a lot of food with relatively little energy may be advantageous for overweight adults wanting to lose weight, it can be detrimental during stages of the life span involving growth. Diet planning during pregnancy, lactation, infancy, childhood, and adolescence must provide for the increases in energy and nutrients needed during those times—when the consequences of poor nutrition can be great.

Pregnancy and Lactation

In general, a vegetarian diet favors a healthy pregnancy and successful lactation if it provides adequate energy; includes milk and milk products; and contains a wide variety of legumes, cereals, fruits, and vegetables.[19] Many vegetarian women are well nourished, with nutrient intakes from diet alone exceeding the RDA for all vitamins and minerals except iron, which is low for most women. In contrast, vegan women who restrict themselves to an exclusively plant-based diet generally have low food energy intakes and are thin; for pregnant women, this can be a problem. Women with low prepregnancy weights and small weight gains during pregnancy jeopardize a healthy pregnancy.

Vegan diets, which exclude all foods of animal origin, may require supplementation with vitamin B_{12}, calcium, and vitamin D, or the addition of foods fortified with these nutrients. Infants of vegan parents may suffer spinal cord damage and develop severe psychomotor retardation due to a lack of vitamin B_{12} in the mother's diet during pregnancy. Breastfed infants of vegan mothers have been reported to develop vitamin B_{12} deficiency and severe movement disorders. Giving the infants vitamin B_{12} supplements corrects the blood and neurological symptoms of deficiency, as well as the structural abnormalities, but cognitive and language development delays may persist. A vegan mother needs a regular source of vitamin B_{12}–fortified foods or a supplement that provides 2.6 micrograms daily.

A pregnant woman who cannot meet her calcium needs through diet alone may need 600 milligrams of supplemental calcium daily, taken with meals. Pregnant women who do not receive sufficient dietary vitamin D or enough exposure to sunlight may need a supplement that provides 10 micrograms daily.

Infancy

The newborn infant is a lactovegetarian. As long as the infant has access to sunlight as a source of vitamin D and to sufficient quantities of either infant formulas or breast milk from a mother who eats an adequate diet, the infant will thrive during the early months. "Health-food beverages," such as rice milk, are inappropriate choices because they lack the protein, vitamins, and minerals infants and toddlers need; in fact, their use can lead to severe nutritional deficiencies.[20]

Infants beyond about four months of age present a greater challenge in terms of meeting nutrient needs by way of vegetarian and, especially, vegan diets. Continued breastfeeding or formula feeding is recommended, but supplementary feedings are necessary to ensure adequate energy and iron intakes. Infants and young children in vegetarian families should be given iron-fortified infant cereals well into the second year of life. Mashed or pureed legumes, tofu, and cooked eggs can be added to their diets in place of meat.

The risks of poor nutrition status in infants increase with weaning and reliance on table foods. Infants who receive a well-balanced vegetarian diet that includes milk products and a variety of other foods can easily meet their nutritional requirements for growth. This is not always true for vegan infants. Restrictive vegan diets pose a threat to infants' health. Parents or caregivers who choose to feed their infants vegan diets should consult with their pediatrician and a registered dietitian to ensure a nutritionally adequate diet that will support growth.[21]

The growth of vegan infants slows significantly around the time of transition from breast milk to solid foods. Protein-energy malnutrition and deficiencies of vitamin D, vitamin B_{12}, iron, and calcium have been reported in infants fed vegan diets. Vegan diets that are high in fiber, other complex carbohydrates, and water fill an infant's stomach before meeting energy needs. This problem can be partially alleviated by providing more energy-dense foods: nut butters, legumes, dried fruit spreads, and mashed avocado. Using soy formulas (or milk) fortified with calcium, vitamin B_{12}, and vitamin D and including vitamin C–containing foods at meals to enhance iron absorption will help prevent other nutrient deficiencies in vegan diets.

Childhood and Adolescence

Well-planned vegetarian diets, especially those that include eggs, milk, and milk products, can easily provide adequate nutrient intakes for growing children. The growth of vegetarian children is similar to that of their peers.[22]

Vegan diets, on the other hand, can fail to provide sufficient energy to support the growth of a child within a quantity of food small enough for the child to eat. A child's small stomach can hold only so much food, and a vegan child may feel full before eating enough to meet nutrient and energy needs. A vegan child's diet should emphasize cereals, legumes, and nuts to meet protein and energy needs in a small volume. Meat, which contains abundant protein, iron, and food energy in less bulk, supports the growth of children more efficiently. Compared with meat-eating children, vegan children tend to be shorter in height and lighter in weight; their low energy intakes can impair growth.

When vegan children get their protein only from plant foods, they may need protein intakes higher than the RDA for normal growth and health. The standard protein recommendations may be inadequate to support the growth of vegan children, but specific recommendations have not been established.

Other nutritional concerns for vegans include vitamin B_{12}, calcium, and vitamin D. Children who were raised on vegan diets and then switched to more liberal diets have difficulty achieving an adequate vitamin B_{12} status even with a moderate consumption of animal products.[23] Adolescents following a vegan diet low in calcium and vitamin D have a reduced bone density, which may have implications for bone health later in life. Numerous fortified animal-free foods are now available to help meet the nutrient needs of growing children on a vegan diet.[24]

For many adolescents, vegetarian diets offer the advantages of more fruits and vegetables and fewer sweets, fast foods, and salty snacks. As a result, these teens meet many of the Healthy People 2010 objectives and enjoy good growth and health.[25] Some teens, however, use vegetarian diets as a camouflage for eating disorders.[26] By hiding behind a vegetarian diet, these teens can greatly limit their food choices while distracting their parents from an eating disorder that threatens growth and health (see Highlight 9 for more details).

Healthy Food Choices

In general, adults who eat vegetarian diets have lowered their risks of mortality and several chronic diseases, including obesity, high blood pressure, heart disease, and cancer.[26] But there is nothing mysterious or magical about the vegetarian diet; vegetarianism is not a religion like Buddhism or Hinduism, but merely an eating plan that selects plant foods to deliver needed nutrients. The quality of the diet depends not on whether it includes meat, but on whether the other food choices are nutritionally sound. A diet that includes ample fruits, vegetables, whole grains, legumes, nuts, and seeds is higher in fiber, antioxidant vitamins, and phytochemicals, and lower in saturated fats than meat-based diets. Variety is key to nutritional adequacy in a vegetarian diet. Restrictive plans, such as **macrobiotic diets,** that limit selections to a few grains and vegetables cannot possibly deliver a full array of nutrients.

Having learned some of the relationships between diet and health, many people may discover that their strategies for planning meals need to change. In the past, they decided what cut of beef, ham, pork, lamb, poultry, or fish to prepare and then filled in the menu with an accompanying "starch" (potato, rice, or noodles), salad or other vegetable, and bread. Now they fill their dinner plates with legumes, whole grains, vegetables, and fruits. Then they may add small quantities of milk products, eggs, lean meat, fish, or poultry. They have decreased their use of animal products and increased their consumption of plant foods without any intention of "becoming a vegetarian." Such a plan offers many of the same health advantages of a vegetarian diet if it limits meat intake to the recommended 5 to 7 ounces daily and includes lean cuts, as well as abundant whole grains, fruits, and vegetables.

For the most part, it seems that nonmeat and low-meat diets can both support good health. Conversely, both plant-based and meat-based diets can be detrimental to health when overloaded with fat. Vegetarians who dine on cheddar cheese, butter sauces, sour cream, and deep-fried vegetables invite the same health hazards as **omnivores** who overeat high-fat meats. And both diets, if not properly balanced, can lack nutrients. Poorly planned vegetarian diets typically lack iron, zinc, calcium, vitamin B_{12}, and vitamin D; without planning, the meat eater's diet may lack vitamin A, vitamin C, folate, and fiber, among others. Quite simply, the negative health aspects of any diet, including vegetarian diets, reflect poor diet planning. Careful attention to energy intake and specific problem nutrients can ensure adequacy.

Keep in mind, too, that diet is only one factor influencing health. Whatever a diet consists of, its context is also important: no smoking, alcohol consumption in moderation (if at all), regular physical activity, adequate rest, and medical attention when needed all contribute to a healthy life. Establishing these healthy habits early in life seems to be the most important step one can take to reduce the risks of later diseases (as Highlight 16 explains).

NUTRITION ON THE NET

 Access these websites for further study of topics covered in this highlight.

- Find updates and quick links to these and other nutrition-related sites at our website: **www.wadsworth.com/nutrition**

- Search for "vegetarian" at the Food and Drug Administration's site: **www.fda.gov**

- Visit the Vegetarian Resource Group: **www.vrg.org**

- Review another vegetarian diet pyramid developed by Oldways Preservation & Exchange Trust: **www.oldwayspt.org**

REFERENCES

1. S. I. Barr and G. E. Chapman, Perceptions and practices of self-defined current vegetarian, former vegetarian, and nonvegetarian women, *Journal of the American Dietetic Association* 102 (2002): 354–360; R. Weinsier, Use of the term vegetarian, *American Journal of Clinical Nutrition* 71 (2000): 1211–1212; P. K. Johnston and J. Sabate, Reply to R. Weinsier, *American Journal of Clinical Nutrition* 71 (2000): 1212–1213.
2. Position of the American Dietetic Association and Dietitians of Canada: Vegetarian diets, *Journal of the American Dietetic Association* 103 (2003): 748–765; T. J. Key, G. K. Davey, and P. N. Appleby, Health benefits of a vegetarian diet, *The Proceedings of the Nutrition Society* 58 (1999): 271–273; T. J. Key and coauthors, Mortality in vegetarians and nonvegetarians: Detailed findings from a collaborative analysis of 5 prospective studies, *American Journal of Clinical Nutrition* 70 (1999): 516S–524S.
3. Key, Davey, and Appleby, 1999.
4. G. E. Fraser, Associations between diet and cancer, ischemic heart disease, and all-cause mortality in non-Hispanic white California Seventh-day Adventists, *American Journal of Clinical Nutrition* 70 (1999): 532S–538S; P. N. Appleby and coauthors, The Oxford vegetarian study: An overview, *American Journal of Clinical Nutrition* 70 (1999): 525S–531S.
5. Third Report of the National Cholesterol Education Program (NCEP) Expert Panel on Detection, Evaluation, and Treatment of High Blood Cholesterol in Adults (Adult Treatment Panel III), NIH publication no. 02-5215 (Bethesda, Md.: National Heart, Lung, and Blood Institute, 2002); A. M. Coulston, The role of dietary fats in plant-based diets, *American Journal of Clinical Nutrition* 70 (1999): 512S–515S.
6. E. L. Ashton, F. S. Dalais, and M. J. Ball, Effect of meat replacement by tofu on CHD risk factors including copper induced LDL oxidation, *Journal of the American College of Nutrition* 19 (2000): 761–767.
7. C. D. Gardner and coauthors, The effect of soy protein with or without isoflavones relative to milk protein on plasma lipids in hypercholesterolemic postmenopausal women, *American Journal of Clinical Nutrition* 73 (2001): 667–668.
8. H. Chen and coauthors. Dietary patterns and adenocarcinoma of the esophagus and distal stomach, *American Journal of Clinical Nutrition* 75 (2002): 137–144.
9. M. Virginia, V. Melina, and A. R. Mangels, A new food guide for North American vegetarians, *Journal of the American Dietetic Association* 103 (2003): 771–775; C. A. Venti and C. S. Johnston, Modified food guide pyramid for lactovegetarians and vegans, *Journal of Nutrition* 132 (2002): 1050–1054; E. H. Haddad, J. Sabaté, and C. G. Whitten, Vegetarian food guide pyramid: A conceptual framework, *American Journal of Clinical Nutrition* 70 (1999): 615S–619S.
10. Venti and Johnston, 2002.
11. Venti and Johnston, 2002; V. Messina and A. R. Mangels, Considerations in planning vegan diets: Children, *Journal of the American Dietetic Association* 101 (2001): 661–669.
12. Position of the American Dietetic Association and Dietitians of Canada, 2003.
13. J. R. Hunt, Moving toward a plant-based diet: Are iron and zinc at risk? *Nutrition Reviews* 60 (2002): 127–134.
14. J. R. Hunt and Z. K. Roughead, Nonheme-iron absorption, fecal ferritin excretion, and blood indexes of iron status in women consuming controlled lactoovovegetarian diets for 8 wk, *American Journal of Clinical Nutrition* 69 (1999): 944–952.
15. C. L. Larsson and G. K. Johansson, Dietary intake and nutritional status of young vegans and omnivores in Sweden, *American Journal of Clinical Nutrition* 76 (2002): 100–106; M. J. Ball and M. A. Bartlett, Dietary intake and iron status of Australian vegetarian women, *American Journal of Clinical Nutrition* 70 (1999): 353–358.
16. Hunt, 2002.
17. D. Milea, N. Cassoux, and P. LeHoang, Blindness in a strict vegan, *New England Journal of Medicine* 342 (2000): 897–898.
18. T. A. Outila and coauthors, Dietary intake of vitamin D in premenopausal, healthy vegans was insufficient to maintain concentrations of serum 25-hydroxyvitamin D and intact parathyroid hormone within normal ranges during the winter in Finland, *Journal of the American Dietetic Association* 100 (2000): 434–441.
19. Position of the American Dietetic Association and Dietitians of Canada, 2003.
20. T. Liu and coauthors, Kwashiorkor in the United States: Fad diets, perceived and true milk allergy, and nutritional ignorance, *Archives of Dermatology* 137 (2001): 630–636; G. Massa, Protein malnutrition due to replacement of milk by rice drink, *European Journal of Pediatrics* 160 (2001): 382–384; N. F. Carvalho and coauthors, Severe nutritional deficiencies in toddlers resulting from health food milk alternatives, *Pediatrics* 107 (2001): e46.
21. A. R. Mangels and V. Messina, Considerations in planning vegan diets: Infants, *Journal of the American Dietetic Association* 101 (2001): 670–677.
22. M. Hebbelinck, P. Clarys, and A. DeMalsche, Growth, development, and physical fitness of Flemish vegetarian children, adolescents, young adults, *American Journal of Clinical Nutrition* 70 (1999): 579S–585S.
23. M. van Dusseldorp and coauthors, Risk of persistent cobalamin deficiency in adolescents fed a macrobiotic diet in early life, *American Journal of Clinical Nutrition* 69 (1999): 664–671.
24. Messina and Mangels, 2001.
25. C. L. Perry and coauthors, Adolescent vegetarians: How well do their dietary patterns meet the Healthy People 2010 objectives? *Archives of Pediatrics and Adolescent Medicine* 156 (2002): 431–437.
26. Y. Martins, P. Pliner, and R. O'Connor, Restrained eating among vegetarians: Does a vegetarian eating style mask concerns about weight? *Appetite* 32 (1999): 145–154.
27. Fraser, 1999.

Chapter 7

Metabolism: Transformations and Interactions

© Mark Thomas/FoodPix/Getty Images

Chapter Outline

Chemical Reactions in the Body

Breaking Down Nutrients for Energy: *Glucose • Glycerol and Fatty Acids • Amino Acids • Breaking Down Nutrients for Energy—In Summary • The Final Steps of Catabolism*

The Body's Energy Budget: *The Economics of Feasting • The Transition from Feasting to Fasting • The Economics of Fasting*

Highlight: *Alcohol and Nutrition*

Nutrition Explorer CD-ROM Outline

Nutrition Animations:

1. *Metabolism: Glucose to Pyruvate/Lactate and on to Acetyl CoA*
2. *Metabolism: Fatty Acids to Acetyl CoA*
3. *Metabolism: The TCA Cycle*
4. *Metabolism: The Electron Transport Chain and ATP Synthesis*

Case Study: *Feasting and Fasting*

Student Practice Test

Glossary Terms

Nutrition on the Net

Nutrition in Your Life

You eat breakfast and hustle off to class. After lunch, you study for tomorrow's exam. Dinner is followed by an evening of dancing. Do you ever think about how the food you eat powers the activities of your life? What happens when you don't eat—or when you eat too much? Learn how the cells of your body transform carbohydrates, fats, and proteins into energy—and what happens when you give your cells too much or too little of any of these nutrients. Discover the metabolic pathways that lead to body fat and those that support physical activity. It's really quite fascinating.

Energy enables people to breathe, ride bicycles, compose music, and do everything else they do. All the energy that sustains human life initially comes from the sun. During **photosynthesis,** plants make simple sugars from carbon dioxide and capture the sun's light energy in the chemical bonds of those sugars. Then human beings eat either the plants or animals that have eaten the plants. These foods provide energy, but how does the body obtain that energy from foods? This chapter presents the nutrients that provide the body with **fuel** and follows them through a series of reactions that release energy from their chemical bonds. As the bonds break, they release energy in a controlled version of the process by which wood burns in a fire. Both wood and food have the potential to provide energy. When wood burns in the presence of oxygen, it generates heat and light (energy), steam (water), and some carbon dioxide and ash (waste). Similarly, during the body's **metabolism,** energy, water, and carbon dioxide are released.

By studying metabolism, you will understand how the body uses foods to meet its needs and why some foods meet those needs better than others. Readers who are interested in weight control will discover which foods contribute most to body fat and which to select when trying to gain or lose weight safely. Physically active readers

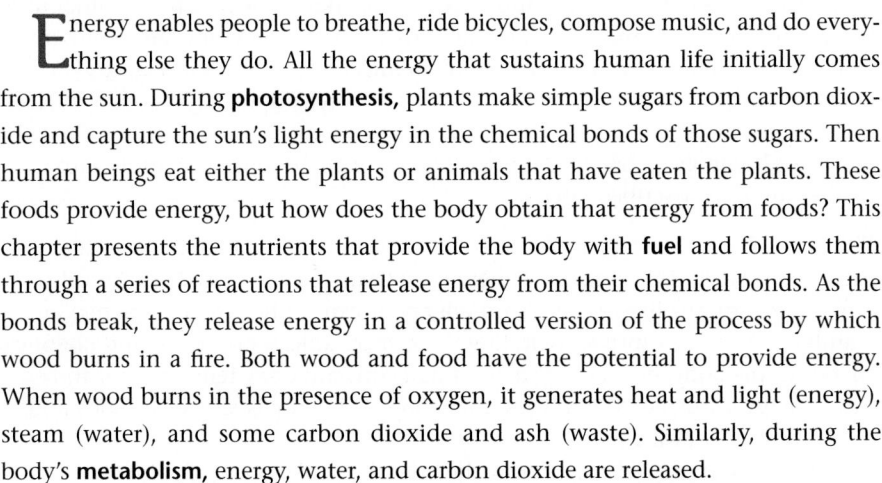

photosynthesis: the process by which green plants use the sun's energy to make carbohydrates from carbon dioxide and water.
- **photo** = light
- **synthesis** = put together (making)

fuel: compounds that cells can use for energy. The major fuels include glucose, fatty acids, and amino acids; other fuels include ketone bodies, lactic acid, glycerol, and alcohol.

metabolism: the sum total of all the chemical reactions that go on in living cells. Energy metabolism includes all the reactions by which the body obtains and spends the energy from food.
- **metaballein** = change

will discover which foods best support endurance activities and which to select when trying to build lean body mass.

Chemical Reactions in the Body

Earlier chapters introduced some of the body's chemical reactions: the making and breaking of the bonds in carbohydrates, lipids, and proteins. Metabolism is the sum of these and all the other chemical reactions that go on in living cells; *energy metabolism* includes all the ways the body obtains and uses energy from food.

Chapters 4, 5, and 6 laid the groundwork for the study of metabolism; a brief review may be helpful. During digestion, the body breaks down the three energy-yielding nutrients—carbohydrates, lipids, and proteins—into four basic units that can be absorbed into the blood:

- From carbohydrates—glucose (and other monosaccharides).
- From fats (triglycerides)—glycerol and fatty acids.
- From proteins—amino acids.

The body uses carbohydrates and fats for most of its energy needs; amino acids are used primarily as building blocks for proteins, but they also enter energy pathways, contributing about 10 to 15 percent of the day's energy use.

Look for these four basic units to appear again and again in the metabolic reactions described in this chapter. Alcohol also enters many of the metabolic pathways; Highlight 7 focuses on how alcohol disrupts metabolism and how the body handles it.

Building Reactions—Anabolism Earlier chapters described how condensation reactions combine the basic units of energy-yielding nutrients to build body compounds. Glucose molecules may be joined together to make glycogen chains. Glycerol and fatty acids may be assembled into triglycerides. Amino acids may be linked together to make proteins. Each of these reactions starts with small, simple compounds and uses them as building blocks to form larger, more complex structures. Such reactions involve doing work and so require energy. The building up of body compounds is known as **anabolism;** this book represents anabolic reactions, wherever possible, with "up" arrows in chemical diagrams (such as those shown in Figure 7-1).

Breakdown Reactions—Catabolism The breaking down of body compounds is known as **catabolism;** catabolic reactions release energy and are represented, wherever possible, by "down" arrows in chemical diagrams (as in Figure 7-1). Earlier chapters described how hydrolysis reactions break down glycogen to glucose, triglycerides to fatty acids and glycerol, and proteins to amino acids. When the body needs energy, it breaks down any or all of these four basic units into even smaller units, as described later.

The Transfer of Energy in Reactions As Chapter 1 explained, *energy* is the capacity to do work. The concept of energy can be difficult to grasp because although every aspect of our lives depends on energy, it cannot be seen or touched and it manifests in various forms, including heat, mechanical, electrical, and chemical energy. In the body, heat energy maintains a constant body temperature, and electrical energy sends nerve impulses, for example. Energy is stored in foods and in the body as chemical energy.

Some of the energy released during the breakdown of glucose, glycerol, fatty acids, and amino acids from foods is captured by "high-energy storage compounds" in the body. One such compound is **ATP (adenosine triphosphate).** ATP, as its name indicates, contains three phosphate groups (see Figure 7-2).■ The bonds connecting the phosphate groups are often described as "high-energy" bonds, referring to the bonds'

■ ATP = A-P~P~P.
(Each ~ denotes a "high-energy" bond.)

anabolism (an-ABB-o-lism): reactions in which small molecules are put together to build larger ones. Anabolic reactions require energy.
- **ana** = up

catabolism (ca-TAB-o-lism): reactions in which large molecules are broken down to smaller ones. Catabolic reactions release energy.
- **kata** = down

ATP or **adenosine** (ah-DEN-oh-seen) **triphosphate** (try-FOS-fate): a common high-energy compound composed of a purine (adenine), a sugar (ribose), and three phosphate groups.

FIGURE 7-1 Anabolic and Catabolic Reactions Compared

NOTE: You need not memorize a color code to understand the figures in this chapter, but you may find it helpful to know that blue is used for carbohydrates, yellow for fats, and red for proteins.

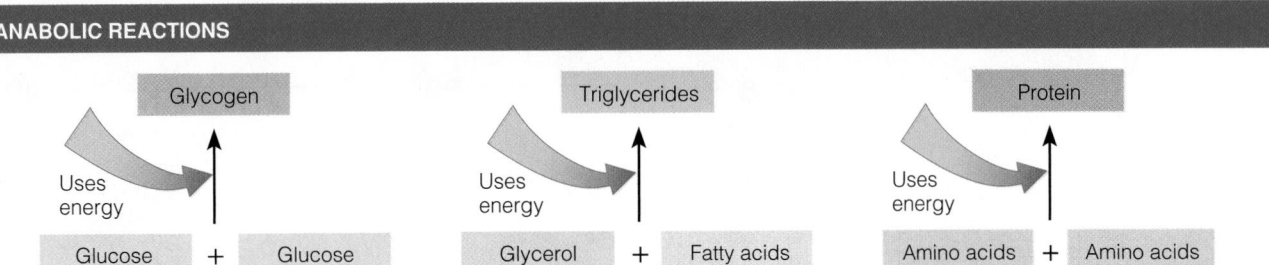

ANABOLIC REACTIONS

Anabolic reactions include the making of glycogen, triglycerides, and protein; these reactions require differing amounts of energy.

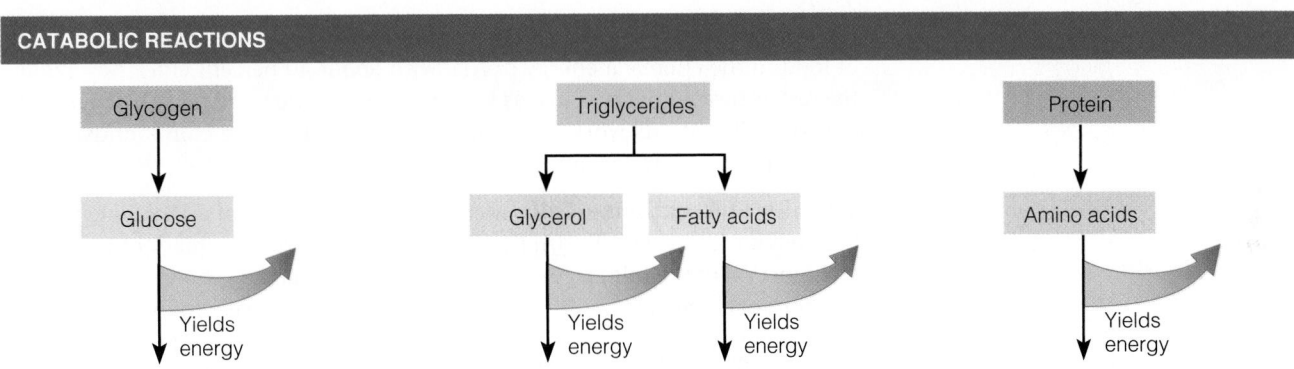

CATABOLIC REACTIONS

Catabolic reactions include the breakdown of glycogen, triglycerides, and protein; the further catabolism of glucose, glycerol, fatty acids, and amino acids releases differing amounts of energy. Much of the energy released is captured in the bonds of adenosine triphosphate (ATP), introduced on p. 216.

readiness to release their energy. The negative charges on the phosphate groups make ATP vulnerable to hydrolysis. Whenever cells do any work that requires energy, hydrolysis reactions readily break these high-energy bonds of ATP, splitting off one or two phosphate groups and releasing their energy. Quite often, the hydrolysis of ATP occurs simultaneously with reactions that will use that energy—a metabolic duet known as **coupled reactions.**

Figure 7-3 illustrates how the body captures and releases energy in the bonds of ATP. In essence, the body uses ATP to transfer the energy released during catabolic

FIGURE 7-2 ATP (Adenosine Triphosphate)

ATP is one of the body's quick-energy molecules. Notice that the bonds connecting the three phosphate groups have been drawn as wavy lines, indicating a high-energy bond. When these bonds are broken, energy is released.

Adenosine + 3 phosphate groups

coupled reactions: pairs of chemical reactions in which some of the energy released from the breakdown of one compound is used to create a bond in the formation of another compound.

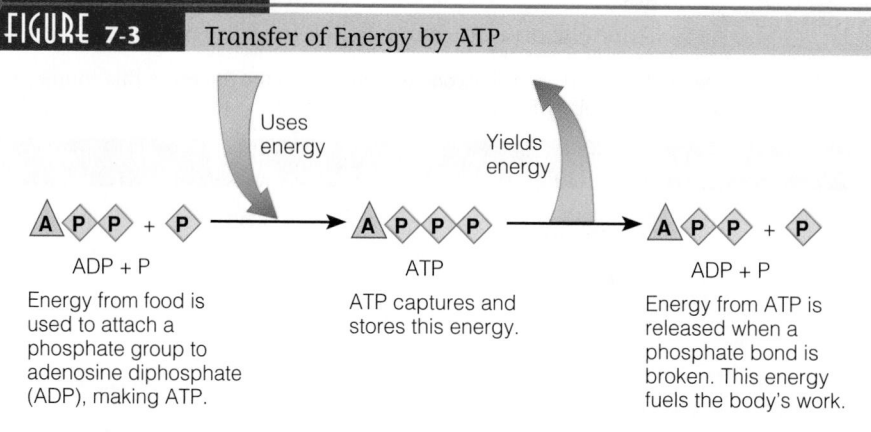

FIGURE 7-3   Transfer of Energy by ATP

Uses energy

Yields energy

ADP + P

Energy from food is used to attach a phosphate group to adenosine diphosphate (ADP), making ATP.

ATP

ATP captures and stores this energy.

ADP + P

Energy from ATP is released when a phosphate bond is broken. This energy fuels the body's work.

reactions to power its anabolic reactions. The body converts the chemical energy of food to the chemical energy of ATP with about 40 percent efficiency, radiating the rest as heat.[1] Energy is lost as heat again when the body uses the chemical energy of ATP to do its work—moving muscles, synthesizing compounds, or transporting nutrients, for example.

The Site of Reactions—Cells   Metabolic work is going on all the time within all the body's trillions of cells. (Appendix A presents a brief summary of the structure and function of the cell.) Figure 7-4 depicts a typical cell and shows where the major reactions of energy metabolism take place. The type and extent of metabolic activity

FIGURE 7-4   A Typical Cell (Simplified Diagram)

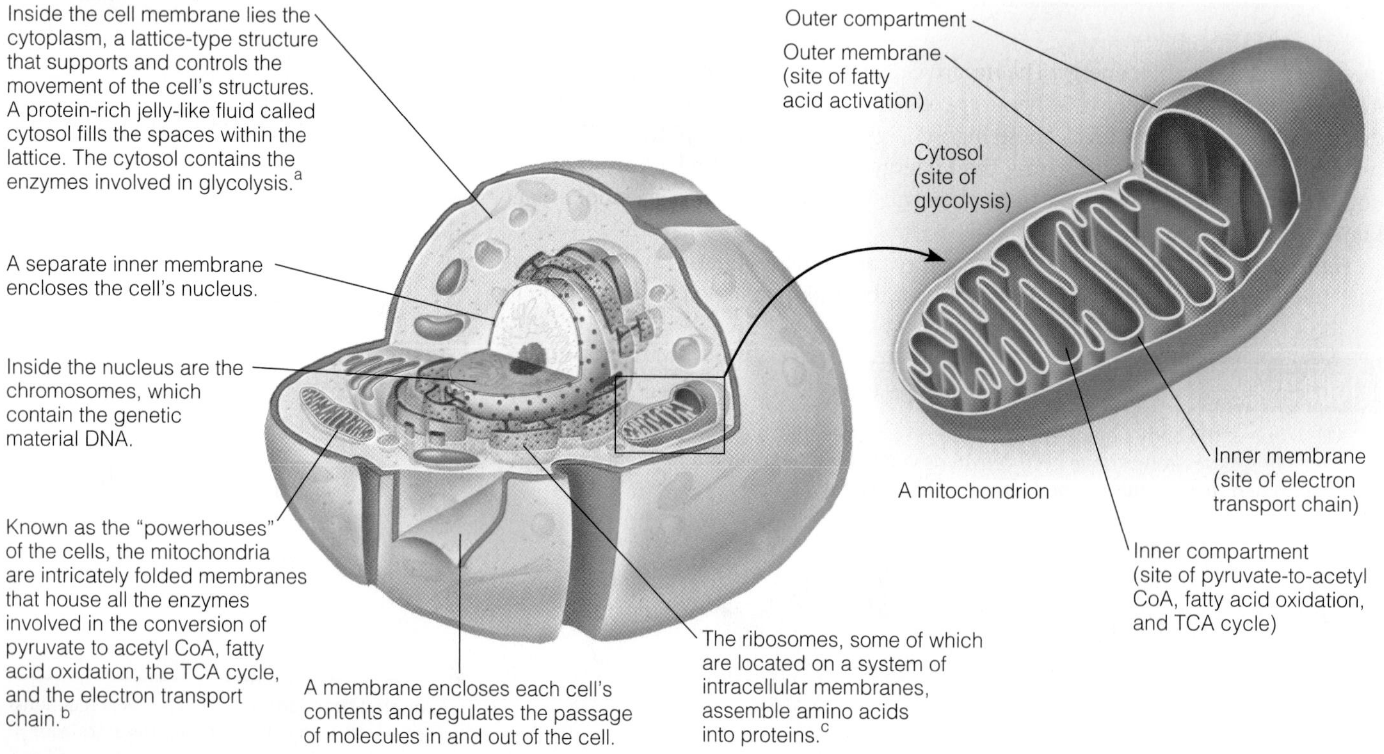

Inside the cell membrane lies the cytoplasm, a lattice-type structure that supports and controls the movement of the cell's structures. A protein-rich jelly-like fluid called cytosol fills the spaces within the lattice. The cytosol contains the enzymes involved in glycolysis.[a]

A separate inner membrane encloses the cell's nucleus.

Inside the nucleus are the chromosomes, which contain the genetic material DNA.

Known as the "powerhouses" of the cells, the mitochondria are intricately folded membranes that house all the enzymes involved in the conversion of pyruvate to acetyl CoA, fatty acid oxidation, the TCA cycle, and the electron transport chain.[b]

A membrane encloses each cell's contents and regulates the passage of molecules in and out of the cell.

The ribosomes, some of which are located on a system of intracellular membranes, assemble amino acids into proteins.[c]

Outer compartment

Outer membrane (site of fatty acid activation)

Cytosol (site of glycolysis)

A mitochondrion

Inner membrane (site of electron transport chain)

Inner compartment (site of pyruvate-to-acetyl CoA, fatty acid oxidation, and TCA cycle)

[a] Glycolysis is described on p. 221.
[b] The conversion of pyruvate to acetyl CoA, fatty acid oxidation, the TCA cycle, and the electron transport chain are described on pp. 223, 224–225, 229, 230–231.
[c] Figure 6-7 on p. 188 describes protein synthesis.

TABLE 7-1 Metabolic Work of the Liver

The liver is the most active processing center in the body. When nutrients enter the body, the liver receives them first; then it metabolizes, packages, stores, or ships them out for use by other organs. When alcohol, drugs, or poisons enter the body, they are also sent directly to the liver; here they are detoxified and their by-products shipped out for excretion. An enthusiastic anatomy and physiology professor once remarked that given the many vital activities of the liver, we should express our feelings for others by saying, "I love you with all my liver," instead of with all my heart. Granted, this declaration lacks romance, but it makes a valid point. Here are just some of the many jobs performed by the liver.

Carbohydrates:

- Converts fructose and galactose to glucose.
- Makes and stores glycogen.
- Breaks down glycogen and releases glucose.
- Breaks down glucose for energy when needed.
- Makes glucose from some amino acids and glycerol when needed.
- Converts excess glucose to fatty acids.

Lipids:

- Builds and breaks down triglycerides, phospholipids, and cholesterol as needed.
- Breaks down fatty acids for energy when needed.
- Packages extra lipids in lipoproteins for transport to other body organs.
- Manufactures bile to send to the gallbladder for use in fat digestion.
- Makes ketone bodies when necessary.

Proteins:

- Manufactures nonessential amino acids that are in short supply.
- Removes from circulation amino acids that are present in excess of need and deaminates them or converts them to other amino acids.
- Removes ammonia from the blood and converts it to urea to be sent to the kidneys for excretion.
- Makes other nitrogen-containing compounds the body needs (such as bases used in DNA and RNA).
- Makes plasma proteins such as clotting factors.

Other:

- Detoxifies alcohol, other drugs, and poisons; prepares waste products for excretion.
- Helps dismantle old red blood cells and captures the iron for recycling.
- Stores most vitamins and many minerals.

To renew your appreciation for this remarkable organ, you might want to review Figure 3-13 on p. 88.

---

vary depending on the type of cell, but of all the body's cells, the liver cells are the most versatile and metabolically active. Table 7-1 offers insights into the liver's work.

The Helpers in Reactions—Enzymes and Coenzymes Metabolic reactions almost always require enzymes■ to facilitate their action. In many cases, the enzymes need assistants to help them. Enzyme helpers are called **coenzymes.**■

Coenzymes are complex organic molecules that associate closely with most enzymes, but are not proteins themselves. The relationships between various coenzymes and their respective enzymes may differ in detail, but one thing is true of all: without its coenzyme, an enzyme cannot function. Some of the B vitamins serve as coenzymes that participate in the energy metabolism of glucose, glycerol, fatty acids, and amino acids (Chapter 10 provides more details).

■ Reminder: *Enzymes* are protein catalysts—proteins that facilitate chemical reactions without being changed in the process.

■ The general term for substances that facilitate enzyme action is **cofactors;** they include both organic coenzymes such as vitamins and inorganic substances such as minerals.

IN SUMMARY During digestion the energy-yielding nutrients—carbohydrates, lipids, and proteins—are broken down to glucose (and other monosaccharides), glycerol, fatty acids, and amino acids. Aided by enzymes and coenzymes, the cells use these products of digestion to build more complex compounds (anabolism) or break them down further to release energy (catabolism). The energy released during catabolism may be captured by high-energy compounds such as ATP.

coenzymes: complex organic molecules that work with enzymes to facilitate the enzymes' activity. Many coenzymes have B vitamins as part of their structures (Figure 10-1 in Chapter 10 illustrates coenzyme action).
- **co** = with

All the energy used to keep the heart beating, the brain thinking, and the legs running comes from the carbohydrates, fats, and proteins in foods.

pyruvate (PIE-roo-vate): a 3-carbon compound that plays a key role in energy metabolism.

acetyl CoA (ASS-eh-teel, or ah-SEET-il, coh-AY): a 2-carbon compound (acetate, or acetic acid, shown in Figure 5-1 on p. 142) to which a molecule of CoA is attached.

CoA (coh-AY): coenzyme A; the coenzyme derived from the B vitamin pantothenic acid and central to energy metabolism.

TCA cycle or tricarboxylic (try-car-box-ILL-ick) acid cycle: a series of metabolic reactions that break down molecules of acetyl CoA to carbon dioxide and hydrogen atoms; also called the Kreb's cycle after the biochemist who elucidated its reactions.

electron transport chain: the final pathway in energy metabolism that transports electrons from hydrogen to oxygen and captures the energy released in the bonds of ATP.

Breaking Down Nutrients for Energy

Glucose, glycerol, fatty acids, and amino acids are the basic units derived from food, but a molecule of each of these compounds is made of still smaller units, the atoms—carbons, nitrogens, oxygens, and hydrogens. During catabolism, the body separates these atoms from one another. To follow this action, recall how many carbons are in the "backbones" of these compounds:

- Glucose has 6 carbons:

- Glycerol has 3 carbons:

- A fatty acid usually has an even number of carbons, commonly 16 or 18 carbons:[*]

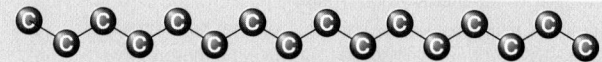

- An amino acid has 2, 3, or more carbons with a nitrogen attached:[†]

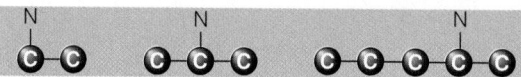

Full chemical structures and reactions appear both in the earlier chapters and in Appendix C; this chapter diagrams the reactions using just the compounds' carbon and nitrogen backbones.

As you will see, each of the compounds—glucose, glycerol, fatty acids, and amino acids—starts down a different path. Along the way, two new names appear—**pyruvate** (a 3-carbon structure) and **acetyl CoA** (a 2-carbon structure with a coenzyme, **CoA**, attached)—and the rest of the story falls into place around them.[§] Two major points to notice in the following discussion:

- Pyruvate can be used to make glucose.
- Acetyl CoA cannot be used to make glucose.

Eventually, all of the energy-yielding nutrients can enter the common pathways of the **TCA cycle** and the **electron transport chain**. (Similarly, people from three different locations can all enter an interstate highway and travel to the same destination.) The TCA cycle and electron transport chain have central roles in energy metabolism and receive full attention later in the chapter, but first the text describes how each of the energy-yielding nutrients is broken down to acetyl CoA and other compounds in preparation for their entrance into these final energy pathways.

Glucose

What happens to glucose, glycerol, fatty acids, and amino acids during energy metabolism can best be understood by starting with glucose. This discussion features glucose because of its central role in carbohydrate metabolism and because liver cells convert the other monosaccharides (fructose and galactose) to compounds that enter the same energy pathways.

---

[*]The figures in this chapter show 16- or 18-carbon fatty acids. Fatty acids may have 4 to 20 or more carbons, with chain lengths of 16 and 18 carbons most prevalent.
[†]The figures in this chapter usually show amino acids as compounds of 2, 3, or 5 carbons arranged in a straight line, but in reality amino acids may contain other numbers of carbons and assume other structural shapes (see Appendix C).
[§]The term *pyruvate* means a salt of *pyruvic acid*. (Throughout this book, the ending -ate is used interchangeably with -ic acid; for our purposes they mean the same thing.)

FIGURE 7-5 — Glycolysis: Glucose-to-Pyruvate

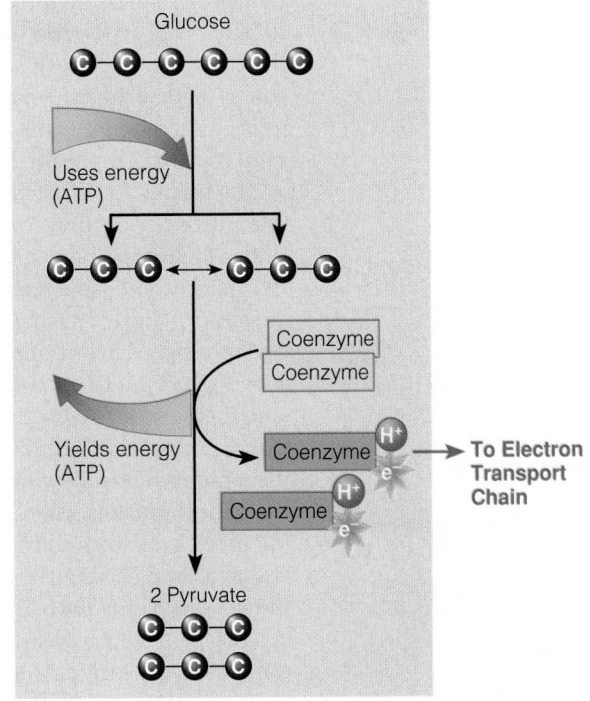

A little ATP is used to start the splitting of the 6-carbon compound glucose into two interchangeable 3-carbon compounds. These compounds are converted to pyruvate in a series of reactions.

Any of the monosaccharides can enter the glycolysis pathway at various points.

A little ATP is produced, and coenzymes carry the hydrogens and their electrons to the electron transport chain.

Glycolysis of one molecule of glucose produces two molecules of pyruvate.

NOTE: These arrows point down indicating the breakdown of glucose to pyruvate during energy metabolism. Alternatively, the arrows could point up indicating the making of glucose from pyruvate, but that is not the focus of this discussion.

Nutrition Explorer — Watch an animated explanation of the first pathway glucose takes on its way to yield energy.

Glucose-to-Pyruvate The first pathway glucose takes on its way to yield energy is called **glycolysis** (glucose splitting).* Figure 7-5 shows a simplified drawing of glycolysis. (This pathway actually involves several steps and several enzymes, which are shown in Appendix C.) In glycolysis, the 6-carbon glucose is split in half, forming two 3-carbon compounds. These 3-carbon compounds continue along the pathway until they are converted to pyruvate. Thus the net yield of one glucose molecule is two pyruvate molecules. The net yield of energy at this point is small; to start glycolysis, the cell uses a little energy and then produces only a little more than it had to invest initially.† In addition, as glucose breaks down to pyruvate, hydrogen atoms with their electrons are released and carried to the electron transport chain by coenzymes made from the B vitamin niacin. A later section of the chapter explains how oxygen accepts the electrons and combines with the hydrogens to form water and how the process captures energy in the bonds of ATP.

This discussion focuses primarily on the breakdown of glucose for energy, but if needed, cells in the liver (and to some extent, the kidneys) can make glucose again from pyruvate in a process similar to the reversal of glycolysis. Making glucose requires energy, however, and a few different enzymes. Still, glucose can be made from pyruvate, so the arrows between glucose and pyruvate could point up as well as down.■

Pyruvate's Options Pyruvate may enter either an anaerobic or an aerobic energy pathway. When the body needs energy quickly—as occurs when you run a quarter mile as fast as you can—pyruvate is converted to lactic acid in an **anaerobic** pathway. When energy expenditure proceeds at a slower pace—as occurs when you ride a bike for an hour—pyruvate breaks down to acetyl CoA in an **aerobic** pathway. The following paragraphs explain these pathways.

Glucose ⇕ Pyruvate

■ Glucose may go "down" to make pyruvate, or pyruvate may go "up" to make glucose, depending on the cell's needs.

glycolysis (gligh-COLL-ih-sis): the metabolic breakdown of glucose to pyruvate. Glycolysis does not require oxygen (anaerobic).
• **glyco** = glucose
• **lysis** = breakdown

anaerobic (AN-air-ROE-bic): not requiring oxygen.
• **an** = not

aerobic (air-ROE-bic): requiring oxygen.

*Glycolysis takes place in the cytosol of the cell (see Figure 7-4).
†The cell uses 2 ATP to begin the breakdown of glucose to pyruvate, but then gains 4 ATP for a net gain of 2 ATP.

Pyruvate-to-Lactic Acid As mentioned earlier, coenzymes carry the hydrogens from glucose breakdown to the electron transport chain. If the electron transport chain is unable to accept these hydrogens, as may occur when cells lack sufficient **mitochondria** (review Figure 7-4) or in the absence of sufficient oxygen, pyruvate can accept the hydrogens. By accepting the hydrogens, pyruvate becomes **lactic acid,** and the coenzymes are freed to return to glycolysis to pick up more hydrogens. In this way, glucose can continue providing energy anaerobically for a while (see the left side of Figure 7-6).

The production of lactic acid occurs to a limited extent even at rest. During high-intensity exercise, however, the concentration of lactic acid increases dramatically. Under these conditions, the muscles rely heavily on anaerobic glycolysis to produce ATP quickly. The rapid rate of glycolysis produces abundant pyruvate and releases hydrogen-carrying coenzymes more rapidly than the mitochondria can handle them. To enable exercise to continue at this intensity, pyruvate is converted to lactic acid, which allows glycolysis to continue (as mentioned earlier). The accumulation of lactic acid (and the subsequent drop in pH) in the muscles produces the burning pain and fatigue that are commonly associated with intense exercise. In contrast, a person performing the same exercise following endurance training experiences less discomfort in part because the number of mitochondria in the muscle cells have increased. This adaptation improves the mitochondria's ability to keep pace with the rapid rate of glycolysis, and less lactic acid accumulates. One possible fate of lactic acid is to be transported from the muscles to the liver. There the liver can convert the lactic acid produced in muscles to glucose in a recycling process called the **Cori cycle** (see Figure 7-6). (Muscle cells cannot recycle lactic acid to glucose because they lack a necessary enzyme.)

Whenever carbohydrates, fats, or proteins are broken down to provide energy, oxygen is always ultimately involved in the process. The role of oxygen in metabolism is worth noticing, for it helps our understanding of physiology and metabolic reactions. Chapter 14 will describe the body's use of the energy nutrients to fuel physical activity, but the facts just presented offer a sneak preview. The breakdown of glucose-to-pyruvate-to-lactic acid proceeds without oxygen (it is anaerobic). This anaerobic

mitochondria (my-toh-KON-dree-uh): the cellular organelles responsible for producing ATP aerobically; made of membranes (lipid and protein) with enzymes mounted on them.
- **mitos** = thread (referring to their slender shape)
- **chondros** = cartilage (referring to their external appearance)

lactic acid: a 3-carbon compound produced from pyruvate during anaerobic metabolism.

$$CH_3$$
$$|$$
$$C-OH$$
$$|$$
$$COOH$$

Cori cycle: the path from muscle lactic acid (which travels to the liver) to glucose (which can travel back to the muscle); named after the scientist who elucidated this pathway.

FIGURE 7-6 Pyruvate-to-Lactic Acid (Anaerobic)

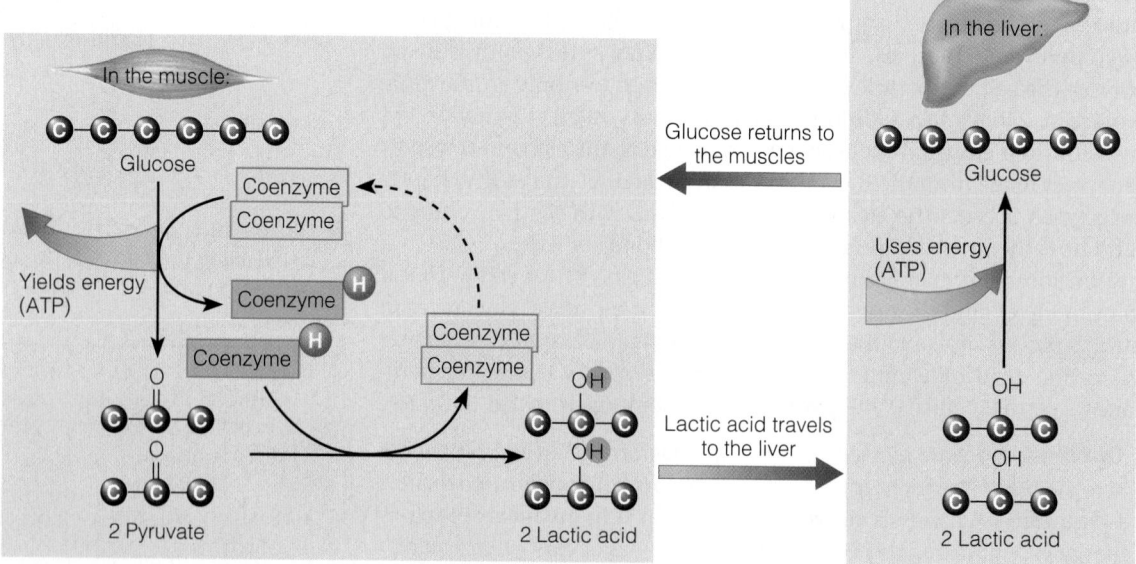

Working muscles break down most of their glucose molecules anaerobically to pyruvate. If the cells lack sufficient mitochondria or in the absence of sufficient oxygen, pyruvate can accept the hydrogens from glucose breakdown and become lactic acid. This conversion frees the coenzymes so that glycolysis can continue.

Liver enzymes can convert lactic acid to glucose, but this reaction requires energy. The recycling of glucose from lactic acid is known as the Cori cycle.

pathway yields energy quickly, but it cannot be sustained for long—a couple of minutes at most. Conversely, the aerobic pathways produce energy more slowly, but because they can be sustained for a long time, their total energy yield is greater.

Pyruvate-to-Acetyl CoA If the cell needs energy and oxygen is available, pyruvate molecules enter the mitochondria of the cell where they will be converted to acetyl CoA. A carbon group (COOH) from the 3-carbon pyruvate is removed to produce a 2-carbon compound that bonds with a molecule of CoA, becoming acetyl CoA. The carbon group from pyruvate becomes carbon dioxide, which is released into the blood, circulated to the lungs, and breathed out. Figure 7-7 diagrams the pyruvate-to-acetyl CoA reaction.

The step from pyruvate to acetyl CoA is metabolically irreversible: a cell cannot retrieve the shed carbons from carbon dioxide to remake pyruvate and then glucose. It is a one-way step and is therefore shown with only a "down" arrow in Figure 7-8.

Acetyl CoA's Options Acetyl CoA has two main options—it may be used to synthesize fats or to generate ATP. When ATP is abundant, acetyl CoA makes fat, the most efficient way to store energy for later use when energy may be needed. Thus any molecule that can make acetyl CoA—including glucose, glycerol, fatty acids, and amino acids—can make fat. In reviewing Figure 7-8, notice that acetyl CoA can be used as a building block for fatty acids, but it cannot be used to make glucose or amino acids.

When ATP is low and the cell needs energy, acetyl CoA may proceed through the TCA cycle, releasing hydrogens with their electrons to the electron transport chain. The story of acetyl CoA continues on p. 229 after a discussion of how fat and protein arrive at the same crossroads. For now, know that when acetyl CoA from the breakdown of glucose enters the aerobic pathways of the TCA cycle and electron transport chain, much more ATP is produced than during glycolysis. The role of glycolysis is to provide energy for short bursts of activity and to prepare glucose for later energy pathways.

FIGURE 7-7 Pyruvate-to-Acetyl CoA (Aerobic)

Each pyruvate loses a carbon as carbon dioxide and picks up a molecule of CoA, becoming acetyl CoA. The arrow goes only one way (down), because the step is not reversible. Result from 1 glucose: 2 carbon dioxide and 2 acetyl CoA.

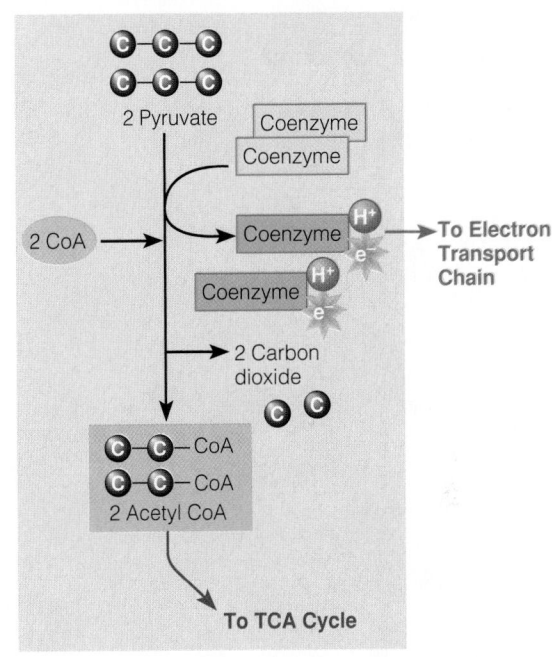

IN SUMMARY The breakdown of glucose to energy begins with glycolysis, a pathway that produces pyruvate. Keep in mind that glucose can be synthesized only from pyruvate or compounds earlier in the pathway. Pyruvate may be converted to lactic acid anaerobically or to acetyl CoA aerobically. Once the commitment to acetyl CoA is made, glucose is not retrievable; acetyl CoA cannot go back to glucose. Figure 7-9 summarizes the breakdown of glucose.

FIGURE 7-8 The Paths of Pyruvate and Acetyl CoA

Pyruvate may follow several reversible paths, but the path from pyruvate to acetyl CoA is irreversible.

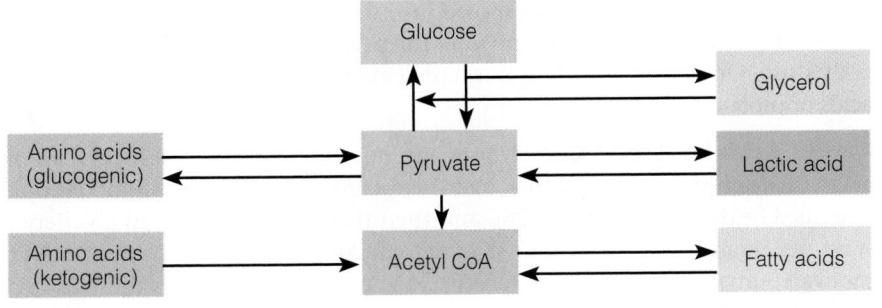

NOTE: Amino acids that can be used to make glucose are called *glucogenic*; amino acids that are converted to acetyl CoA are called *ketogenic*.

The anaerobic breakdown of glucose-to-pyruvate-to-lactic acid is the major source of energy for short, intense exercise.

FIGURE 7-9 Glucose Enters the
Energy Pathway

This figure combines Figure 7-5 and Figure 7-7 to
show the breakdown of glucose-to-pyruvate-to-acetyl
CoA. Details of the TCA cycle and the electron
transport chain are given later and in Appendix C.

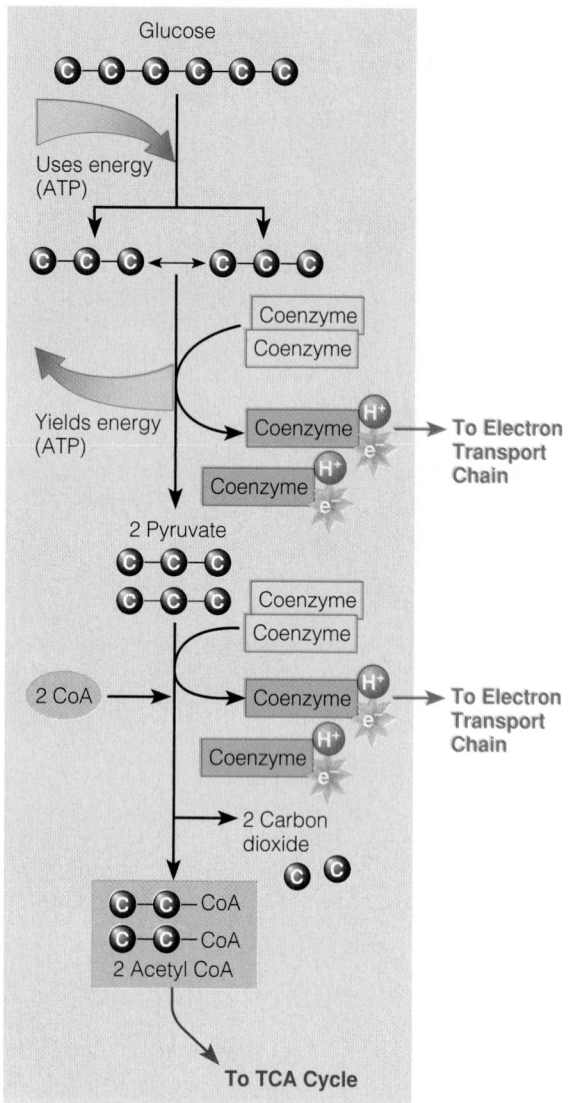

Glycerol and Fatty Acids

Once glucose breakdown is understood, fat and protein breakdown are
easily learned, for all three eventually enter the same metabolic pathways. Recall that triglycerides can break down to glycerol and fatty acids.

Glycerol-to-Pyruvate Glycerol (a 3-carbon compound like pyruvate,
but with a different arrangement of H and OH on the C) is easily converted to another 3-carbon compound. This compound may go either
"up" the pathway to form glucose or "down" to form pyruvate and then
acetyl CoA (review Figure 7-8 on p. 223).

Fatty Acids-to-Acetyl CoA Fatty acids are taken apart 2 carbons at a
time in a series of reactions known as **fatty acid oxidation.***■ Figure 7-10
illustrates fatty acid oxidation and shows that in the process, each 2-
carbon fragment splits off and combines with a molecule of CoA to
make acetyl CoA. As each 2-carbon fragment breaks off from a fatty acid
during oxidation, hydrogens and their electrons are released and carried
to the electron transport chain by coenzymes made from the B vitamins
riboflavin and niacin. Figure 7-11 (on p. 226) summarizes the breakdown of fats.

Fatty Acids Cannot Be Used to Synthesize Glucose When carbohydrate is unavailable, the liver cells can make glucose from pyruvate and
other 3-carbon compounds, such as glycerol,■ but they cannot make glucose from the 2-carbon fragments of fatty acids. In chemical diagrams, the
arrow between pyruvate and acetyl CoA always points only one way—
down—and fatty acid fragments enter the metabolic path below this arrow (review Figure 7-8). Thus fatty acids cannot be used to make glucose.

The significance of this is that red blood cells and the brain and
nervous system depend primarily on glucose as fuel. Remember that
almost all dietary fats are triglycerides and that triglycerides contain
only one small molecule of glycerol with three fatty acids. The glycerol
can yield glucose, but that represents only 3 of the 50 or so carbon
atoms in a triglyceride—about 5 percent of its weight (see Figure 7-12
on p. 226). The other 95 percent cannot be converted to glucose.

IN SUMMARY The body can convert the small glycerol portion of a triglyceride to either pyruvate (and then glucose) or acetyl
CoA. The fatty acids of a triglyceride, on the other hand, cannot
make glucose, but they can provide abundant acetyl CoA. Acetyl
CoA from either source may then enter the TCA cycle to release energy or combine with other molecules of acetyl CoA to make body fat.

Amino Acids

The preceding two sections have shown how the breakdown of carbohydrate and
fat produces acetyl CoA, which can enter the pathways that provide energy for the
body's use. One energy-yielding nutrient remains: protein or, rather, the amino
acids of protein.

Amino Acids-to-Acetyl CoA Before entering the metabolic pathways, amino
acids are deaminated (that is, they lose their nitrogen-containing amino group as
described in the section on p. 226), and then they are catabolized in a variety of
ways. As Figure 7-13 (on p. 227) shows, some amino acids can be converted to
pyruvate; others are converted to acetyl CoA; and still others enter the TCA cycle
directly as compounds other than acetyl CoA.

■ Note: The *oxidation* of energy nutrients
refers to the metabolic reactions that lead
to the release of energy.

■ Reminder: The making of glucose from
noncarbohydrate sources is called *gluconeogenesis*. The glycerol portion of a triglyceride
and most amino acids can be used to make
glucose (review Figure 7-8). The liver is the
major site of gluconeogenesis, but the kidneys become increasingly involved under
certain circumstances, such as starvation.

fatty acid oxidation: the metabolic
breakdown of fatty acids to acetyl CoA; also
called **beta oxidation.**

*Oxidation of fatty acids occurs in the mitochondria of the cells (see Figure 7-4).

FIGURE 7-10   Fatty Acid to Acetyl CoA

Fatty acids are taken apart to 2-carbon fragments that combine with CoA to make acetyl CoA.

The fatty acid is first activated by coenzyme A.

As each carbon-carbon bond is cleaved, hydrogens and their electrons are released and coenzymes pick them up.

Another CoA joins the chain, and the bond at the second carbon (the beta-carbon) weakens. Acetyl CoA splits off, leaving a fatty acid that is two carbons shorter.

The shorter fatty acid enters the pathway and the cycle repeats, releasing more hydrogens with their electrons and more acetyl CoA. The molecules of acetyl CoA enter the TCA cycle, and the coenzymes carry the hydrogens and their electrons to the electron transport chain.

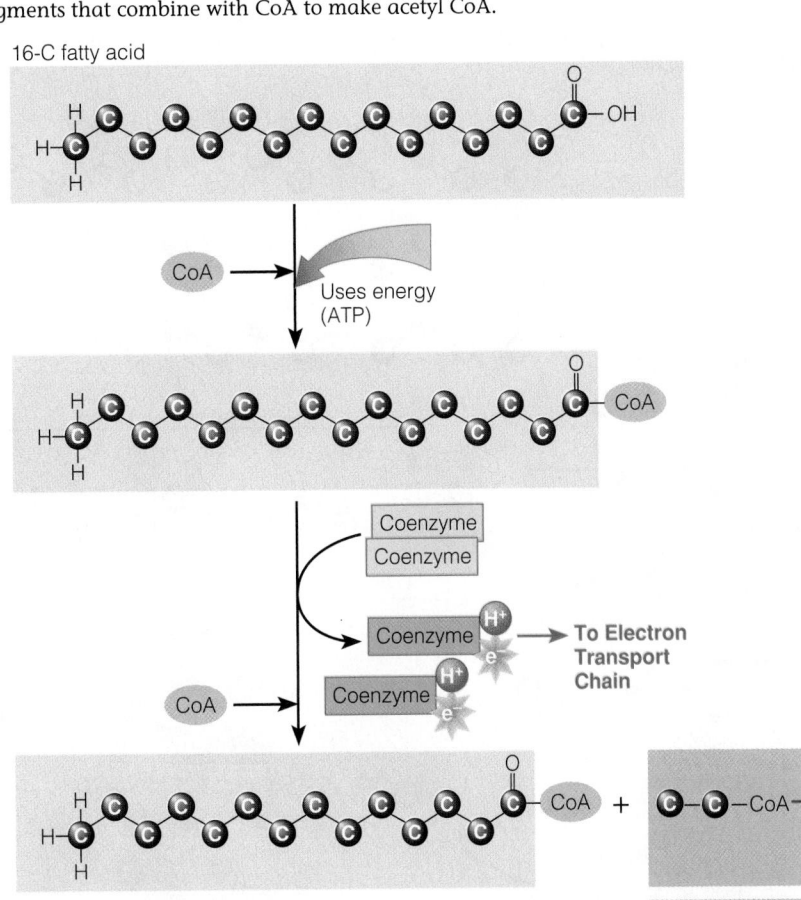

16-C fatty acid

CoA → Uses energy (ATP)

Coenzyme
Coenzyme
Coenzyme   H+ e- → To Electron Transport Chain
CoA →   Coenzyme   H+ e-

CoA + C-C—CoA → To TCA Cycle

Net result from a 16-C fatty acid:	14-C fatty acid CoA	+	1 acetyl CoA
Cycle repeats, leaving:	12-C fatty acid CoA	+	2 acetyl CoA
Cycle repeats, leaving:	10-C fatty acid CoA	+	3 acetyl CoA
Cycle repeats, leaving:	8-C fatty acid CoA	+	4 acetyl CoA
Cycle repeats, leaving:	6-C fatty acid CoA	+	5 acetyl CoA
Cycle repeats, leaving:	4-C fatty acid CoA	+	6 acetyl CoA
Cycle repeats, leaving:	2-C fatty acid CoA*	+	7 acetyl CoA

*Notice that 2-C fatty acid CoA = acetyl CoA, so that the final yield from a 16-C fatty acid is 8 acetyl CoA.

Nutrition Explorer Watch as fatty acids are taken apart two carbons at a time in a series of aerobic reactions known as fatty acid oxidation.

Amino Acids-to-Glucose As you might expect, amino acids that are used to make pyruvate can provide glucose, whereas those used to make acetyl CoA can provide additional energy or make body fat but cannot make glucose.■ Amino acids entering the TCA cycle directly can continue in the cycle and generate energy; alternatively, they can generate glucose.[2] Thus protein, unlike fat, is a fairly good source of glucose when carbohydrate is not available.

A key to understanding these metabolic pathways is learning which fuels can be converted to glucose and which cannot. The parts of protein and fat that can be converted to pyruvate *can* provide glucose for the body, whereas the parts that are converted to acetyl CoA *cannot* provide glucose, but can readily provide fat. The body must have glucose to fuel the activities of the central nervous system and red blood cells. Without glucose from food, the body will devour its own lean (protein-containing) tissue to provide the amino acids to make glucose. Therefore, to keep this from happening, the body needs foods that can provide glucose—primarily carbohydrate. Giving the body only fat, which delivers mostly acetyl CoA, puts it in the position of having to break down protein tissue to make glucose. Giving the

■ Amino acids that can make glucose via either pyruvate or TCA cycle intermediates are *glucogenic;* amino acids that are degraded to acetyl CoA are *ketogenic.*

FIGURE 7-11 Fats Enter the Energy Pathway

Glycerol enters the glycolysis pathway about midway between glucose and pyruvate and can be converted to either; fatty acids are broken down into 2-carbon fragments that combine with CoA to form acetyl CoA (shown in Figure 7-10). Net from a 16-carbon fatty acid: 8 acetyl CoA molecules.

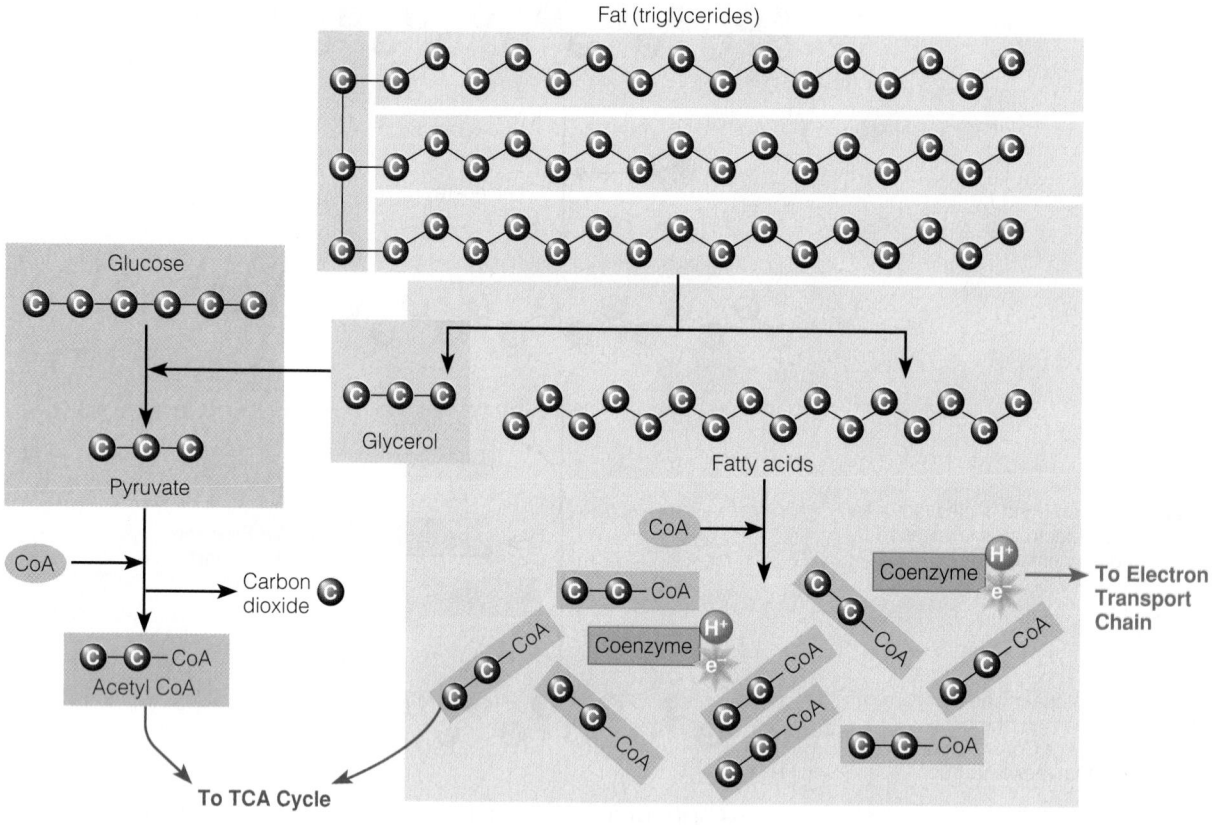

- A healthy diet provides:
  - 45–65% kcalories from carbohydrate.
  - 10–35% kcalories from protein.
  - 20–35% kcalories from fat.

body only protein puts it in the position of having to convert protein to glucose. Clearly, the best diet■ supplies ample carbohydrate, adequate protein, and some fat.

Deamination When amino acids are metabolized for energy or used to make fat, they must be deaminated first. Two products result from deamination. One is the carbon structure without its amino group—often a **keto acid** (see Figure 7-14). The other

FIGURE 7-12 The Carbons of a Typical Triglyceride

A typical triglyceride contains only one small molecule of glycerol (3 C), but has three fatty acids (each commonly 16 C or 18 C, or about 48 C to 54 C in total). Only the glycerol portion of a triglyceride can yield glucose

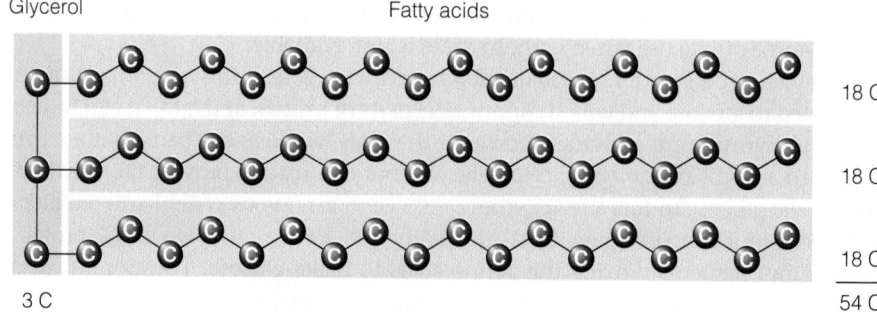

keto (KEY-toe) **acid:** an organic acid that contains a carbonyl group (C=O).

FIGURE 7-13   Amino Acids Enter the Energy Pathway

NOTE: The arrows from pyruvate and the TCA cycle to amino acids are possible only for *nonessential* amino acids; remember, the body cannot make essential amino acids.

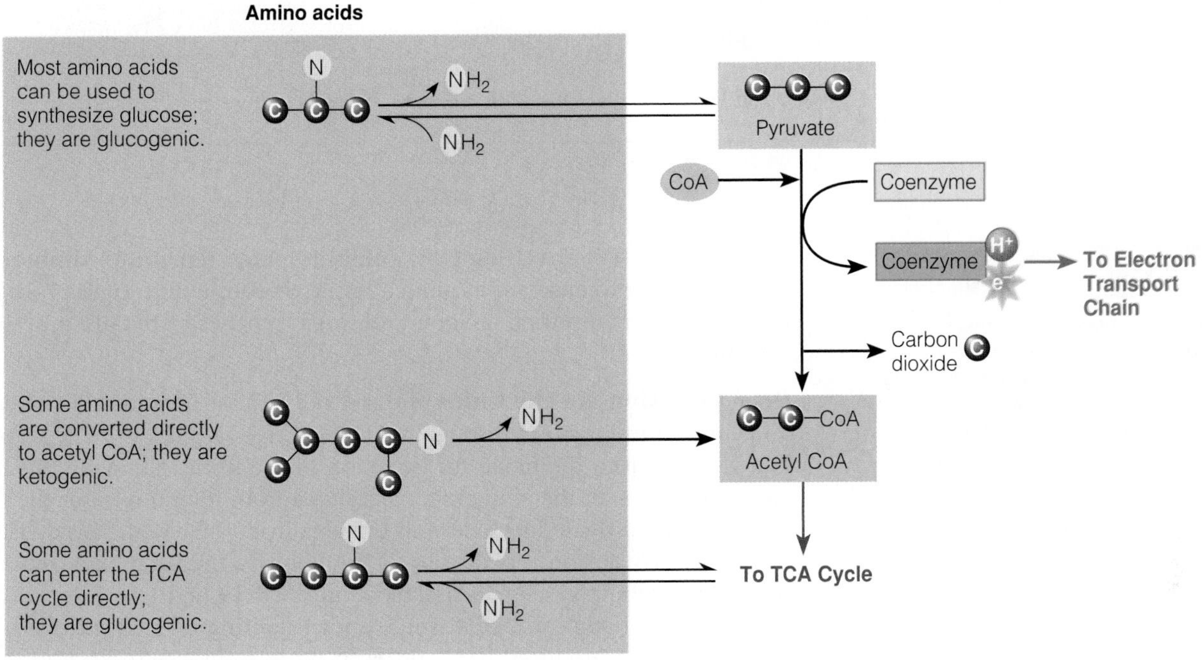

Amino acids

Most amino acids can be used to synthesize glucose; they are glucogenic.

Pyruvate

CoA

Coenzyme

Coenzyme H^+ e^- → To Electron Transport Chain

Carbon dioxide

Some amino acids are converted directly to acetyl CoA; they are ketogenic.

Acetyl CoA

Some amino acids can enter the TCA cycle directly; they are glucogenic.

To TCA Cycle

---

product is **ammonia** (NH_3), a toxic compound chemically identical to the strong-smelling ammonia in bottled cleaning solutions. Ammonia is a base, and if the body produces larger quantities than it can handle, the blood's critical acid-base balance becomes upset.

Transamination As the discussion of protein in Chapter 6 pointed out, only some amino acids are essential; others can be made in the body, given a source of nitrogen. By transferring an amino group from one amino acid to its corresponding keto acid, cells can make a new amino acid and a new keto acid, as shown in Figure 7-15 (on p. 228). Through many such **transamination** reactions, involving many different keto acids, the liver cells can synthesize the nonessential amino acids.

Ammonia-to-Urea in the Liver The liver continuously produces small amounts of ammonia in deamination reactions. Some of this ammonia provides the nitrogen needed for the synthesis of nonessential amino acids (review

FIGURE 7-14   Deamination and Synthesis of a Nonessential Amino Acid

The deamination of an amino acid produces ammonia (NH_3) and a keto acid.

Given a source of NH_3, the body can make nonessential amino acids from keto acids.

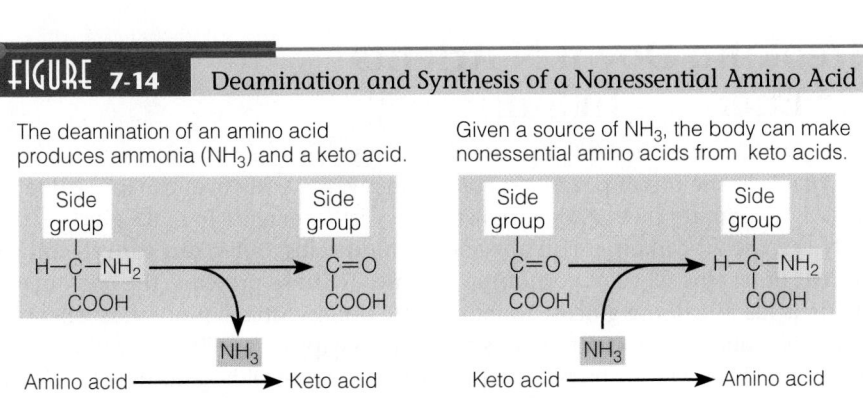

ammonia: a compound with the chemical formula NH_3; produced during the deamination of amino acids.

transamination (TRANS-am-ih-NAY-shun): the transfer of an amino group from one amino acid to a keto acid, producing a new nonessential amino acid and a new keto acid.

FIGURE 7-15 Transamination and Synthesis of a Nonessential Amino Acid

The body can transfer amino groups (NH_2) from an amino acid to a keto acid, forming a new *nonessential* amino acid and a new keto acid.

Transamination reactions require the vitamin B_6 coenzyme.

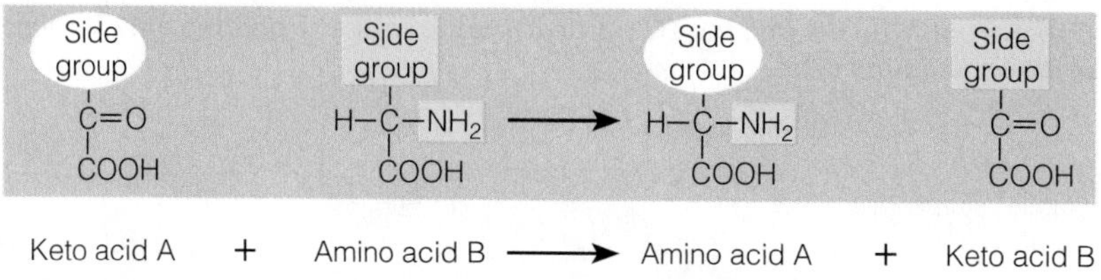

Side group	Side group	Side group	Side group
C=O	H–C–NH_2	H–C–NH_2	C=O
COOH	COOH	COOH	COOH

Keto acid A + Amino acid B ⟶ Amino acid A + Keto acid B

FIGURE 7-16 Urea Synthesis

When amino nitrogen is stripped from amino acids, ammonia is produced. The liver detoxifies ammonia before releasing it into the bloodstream by combining it with another waste product, carbon dioxide, to produce urea. See Appendix C for details.

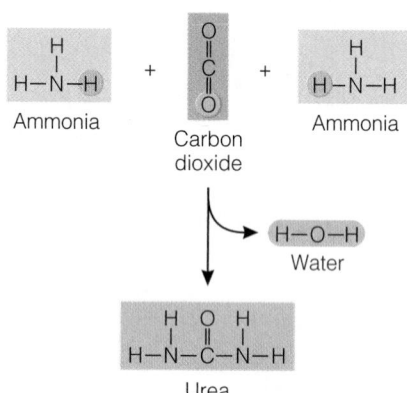

Ammonia + Carbon dioxide + Ammonia

→ Water (H–O–H)

Urea

Figure 7-14). The liver quickly combines any remaining ammonia with carbon dioxide to make **urea**, a much less toxic compound. Figure 7-16 provides a greatly oversimplified diagram of urea synthesis; details are shown in Appendix C.

Urea Excretion via the Kidneys Liver cells release urea into the blood, where it circulates until it passes through the kidneys (see Figure 7-17). The kidneys then remove urea from the blood for excretion in the urine. Normally, the liver efficiently captures all the ammonia, makes urea from it, and releases the urea into the blood; then the kidneys clear all the urea from the blood. This division of labor allows easy diagnosis of diseases of both organs. In liver disease, blood ammonia will be high; in kidney disease, blood urea will be high.

Urea is the body's principal vehicle for excreting unused nitrogen, and the amount of urea produced increases with protein intake. To keep urea in solution, the body needs water. For this reason, a person who regularly consumes a high-protein diet (say, 100 grams a day or more) must drink plenty of water to dilute and excrete urea from the body. Without extra water, a person on a high-protein diet risks dehydration because the body uses its water to rid itself of urea. This explains some of the water loss that accompanies high-protein diets. Such losses may make high-protein diets *appear* to be effective, but water loss, of course, is of no value to the person who wants to lose body fat (as Highlight 8 explains).

IN SUMMARY The body can use some amino acids to produce glucose, while others can be used either to generate energy or to make fat. Before an amino acid enters one of these metabolic pathways, its nitrogen-containing amino group must be removed through deamination. Some of the nitrogen may be used to make nonessential amino acids and other nitrogen-containing compounds; the rest is cleared from the body via urea synthesis in the liver and excretion in the kidneys.

Breaking Down Nutrients for Energy—In Summary

To review the ways the body can use the energy-yielding nutrients, see the summary table at the top of p. 229. To obtain energy, the body uses glucose and fatty acids as its primary fuels, and amino acids to a lesser extent. To make glucose, the body can use all carbohydrates and most amino acids, but it can convert only 5 percent of fat (the glycerol portion) to glucose. To make proteins, the body needs amino acids. It can use glucose to make some nonessential amino acids when nitrogen is available; it cannot use fats to make body proteins. Finally, when energy is consumed beyond the body's needs, all three energy-yielding nutrients can contribute to fat storage.

urea (you-REE-uh): the principal nitrogen-excretion product of protein metabolism. Two ammonia fragments are combined with carbon dioxide to form urea.

IN SUMMARY

Nutrient	Yields Energy?	Yields Glucose?	Yields Amino Acids and Body Proteins?	Yields Fat Stores?[a]
Carbohydrates (glucose)	Yes	Yes	Yes—when nitrogen is available, can yield *nonessential* amino acids	Yes
Lipids (fatty acids)	Yes	No	No	Yes
Lipids (glycerol)	Yes	Yes—when carbohydrate is unavailable	Yes—when nitrogen is available, can yield *nonessential* amino acids	Yes
Proteins (amino acids)	Yes	Yes—when carbohydrate is unavailable	Yes	Yes

[a]When energy intake exceeds needs, any of the energy-yielding nutrients can contribute to body fat stores.

The Final Steps of Catabolism

Thus far the discussion has followed each of the energy-yielding nutrients down three different pathways. All lead to the point where acetyl CoA enters the TCA cycle. The TCA cycle reactions take place in the inner compartment of the mitochondria (see Figure 7-4 on p. 218). The significance of this location will become evident as details unfold.

The TCA Cycle Acetyl CoA enters the TCA cycle, a busy metabolic traffic center. The TCA cycle is called a cycle, but that doesn't mean it regenerates acetyl CoA. Acetyl CoA goes one way only—down to two carbon dioxide molecules and a coenzyme (CoA). The TCA cycle is a circular path, though, in the sense that a 4-carbon compound known as **oxaloacetate** is needed in the first step and synthesized in the last step.

Oxaloacetate's role in replenishing the TCA cycle is critical. When oxaloacetate is insufficient, the TCA cycle slows down, and the cells face an energy crisis. Oxaloacetate is made primarily from pyruvate, although it can be made from certain amino acids. Importantly, oxaloacetate cannot be made from fat. That oxaloacetate must be available for acetyl CoA to enter the TCA cycle underscores the importance of carbohydrates in the diet. A diet that provides ample carbohydrate ensures an adequate supply of oxaloacetate (because glucose produces pyruvate during glycolysis). (Highlight 8 presents more information on the consequences of low-carbohydrate diets.)

As Figure 7-18 shows, oxaloacetate is the first 4-carbon compound to enter the TCA cycle. Oxaloacetate picks up acetyl CoA (a 2-carbon compound), drops off one carbon (as carbon dioxide), then another carbon (as carbon dioxide), and returns to pick up another acetyl CoA. As for the acetyl CoA, its carbons go only one way—to carbon dioxide (see Appendix C for additional details).[*]

As acetyl CoA molecules break down to carbon dioxide, hydrogen atoms with their electrons are removed from the compounds in the cycle. Each turn of the TCA cycle loses a total of eight electrons. Coenzymes made from the B vitamins niacin and riboflavin receive the hydrogens and their electrons from the TCA cycle

---

[*]Actually, the carbons that enter the cycle in acetyl CoA may not be the exact ones that are given off as carbon dioxide. In one of the steps of the cycle, a 6-carbon compound of the cycle becomes symmetrical, both ends being identical. Thereafter it loses carbons to carbon dioxide at one end or the other. Thus only half of the carbons from acetyl CoA are given off as carbon dioxide in any one turn of the cycle; the other half become part of the compound that returns to pick up another acetyl CoA. It is true to say, though, that for each acetyl CoA that enters the TCA cycle, 2 carbons are given off as carbon dioxide. It is also true that with each turn of the cycle the energy equivalent of one acetyl CoA is released.

FIGURE 7-17 Urea Excretion

The liver and kidneys both play a role in disposing of excess nitrogen. Can you see why the person with liver disease has high blood ammonia, while the person with kidney disease has high blood urea? (Figure 12-2 provides details of how the kidneys work.)

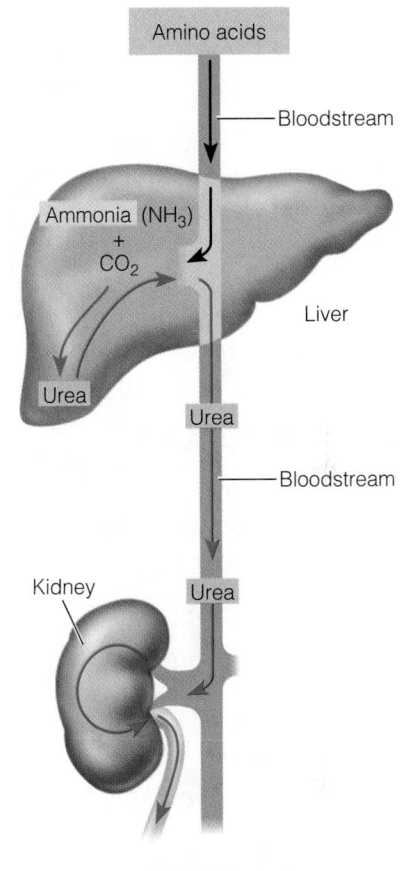

To bladder and out of body

---

oxaloacetate (OKS-ah-low-AS-eh-tate): a carbohydrate intermediate of the TCA cycle.

FIGURE 7-18 The TCA Cycle

Oxaloacetate, a compound made primarily from pyruvate, starts the TCA cycle. (Knowing that glucose produces pyruvate during glycolysis and that oxaloacetate must be available to start the TCA cycle, you can understand why the complete oxidation of fat requires carbohydrate.) The 4-carbon oxaloacetate joins with the 2-carbon acetyl CoA. The new 6-carbon compound releases carbons as carbon dioxide, becoming a 5- and then a 4-carbon compound. Each reaction changes the structure slightly until finally the original 4-carbon oxaloacetate forms again and picks up another acetyl CoA—from the breakdown of glucose, glycerol, fatty acids, and amino acids—and starts the cycle over again. The breakdown of acetyl CoA releases hydrogens with their electrons, which are carried by coenzymes made from the B vitamins niacin and riboflavin to the electron transport chain. (For more details, see Appendix C.)

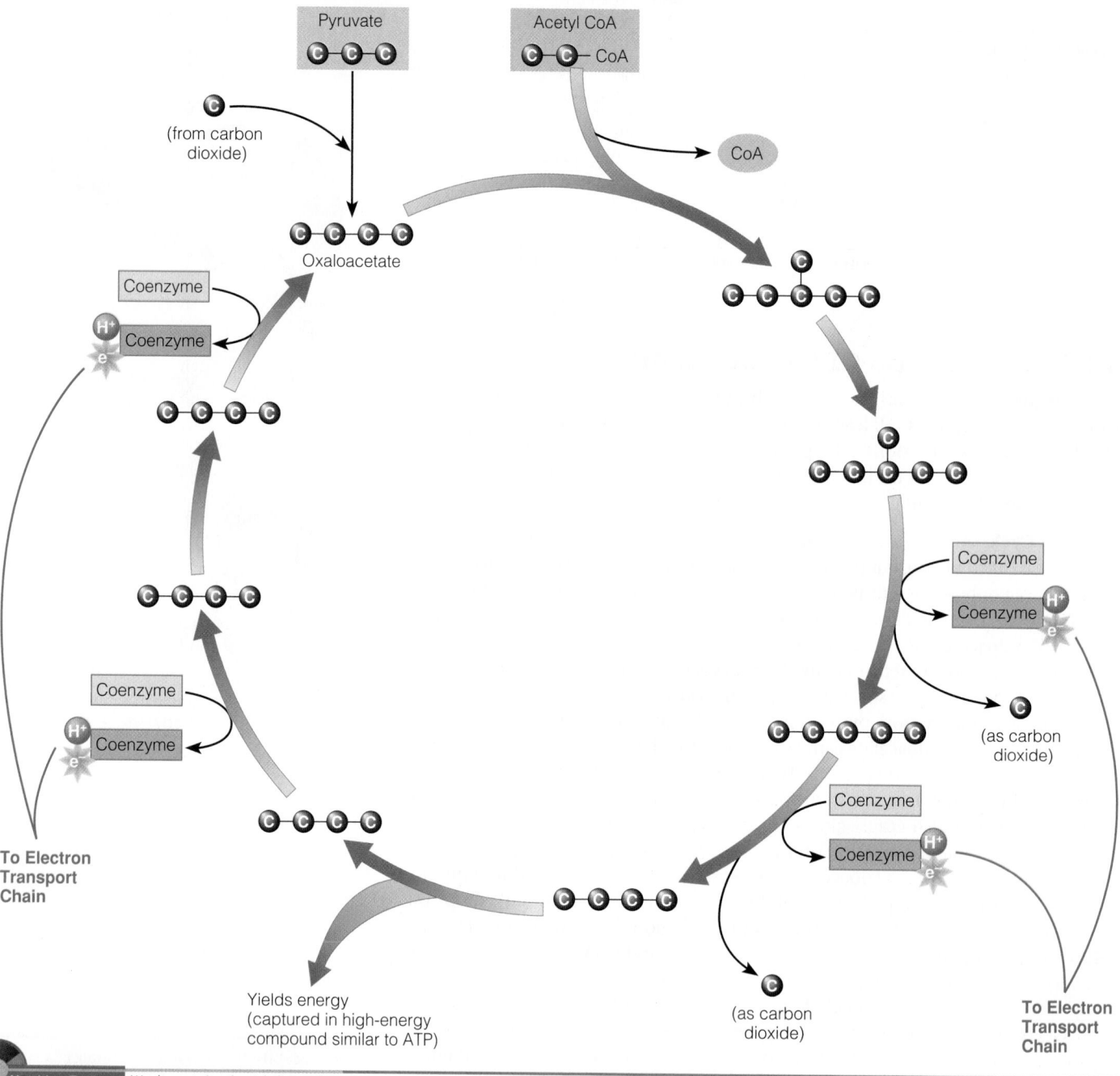

Nutrition Explorer Watch as a series of metabolic reactions break down molecules of acetyl CoA to carbon dioxide and hydrogen atoms.

and transfer them to the electron transport chain—much like a taxi cab that picks up passengers in one location and drops them off in another.

The Electron Transport Chain In the final pathway, the electron transport chain, energy is captured in the high-energy bonds of ATP. The electron transport chain consists of a series of proteins that serve as electron "carriers." These carriers are mounted in sequence on the inner membrane of the mitochondria (review Figure 7-4 on p. 218). As the coenzymes deliver their electrons from the TCA cycle, glycolysis, and fatty acid oxidation to the electron transport chain, each carrier receives the electrons and then passes them on to the next carrier. These electron carriers continue passing the electrons down until they reach oxygen at the end of the chain. Oxygen (O) accepts the electrons and combines with hydrogen atoms (H) to form water (H_2O).■ That oxygen must be available for energy metabolism explains why it is essential to life.

As electrons are passed from carrier to carrier, enough energy is released to pump hydrogen ions across the membrane to the outer compartment of the mitochondria. The rush of hydrogen ions back into the inner compartment powers the synthesis of ATP. In this way, energy is captured in the bonds of ATP. The ATP leaves the mitochondria and enters the cytoplasm, where it can be used for energy. Figure 7-19 provides a simple diagram of the electron transport chain; see Appendix C for details.

The kCalories-per-Gram Secret Revealed Of the three energy-yielding nutrients, fat provides the most energy per gram.■ The reason may be apparent from Figure 7-20, which compares a fatty acid with a glucose molecule. Notice that nearly all the bonds in the fatty acid are between carbons and hydrogens. Oxygen can be added to all of them (forming carbon dioxide with the carbons, and water with the hydrogens). As this happens, hydrogens are released to coenzymes heading for the electron transport chain. In glucose, on the other hand, an oxygen is already bonded to each carbon; thus there is less potential for oxidation, and fewer hydrogens released when the remaining bonds are broken.

■ The results of the electron transport chain:
- O_2 consumed.
- H_2O and CO_2 produced.
- Energy captured in ATP.

■ Fat = 9 kcal/g.
Carbohydrate = 4 kcal/g.
Protein = 4 kcal/g.

 **Nutrition Explorer** Watch an animated explanation of the electron transport chain, the final pathway in energy metabolism where electrons from hydrogen are passed to oxygen and the energy released is trapped in the bonds of ATP.

FIGURE 7-19 Electron Transport Chain and ATP Synthesis

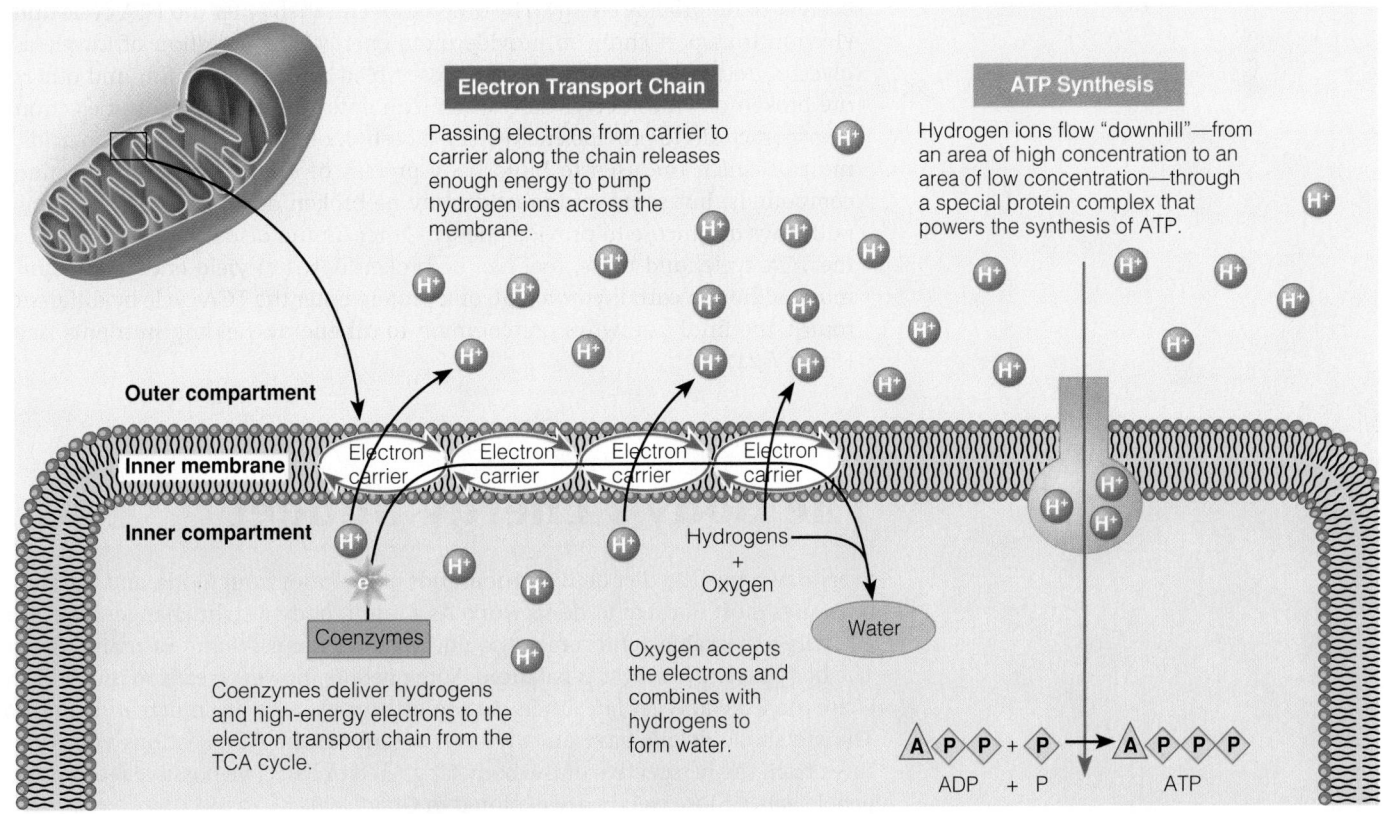

FIGURE 7-20   Chemical Structures of a Fatty Acid and Glucose Compared

To ease comparison, the structure shown here for glucose is not the ring structure shown in Chapter 4, but an alternative way of drawing its chemical structure.

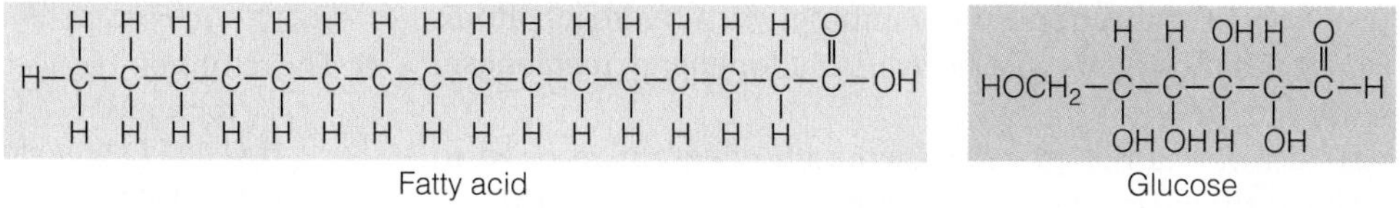

Because fat contains many carbon-hydrogen bonds that can be readily oxidized, it sends numerous coenzymes with their hydrogens and electrons to the electron transport chain where that energy can be captured in the bonds of ATP. This explains why fat yields more kcalories per gram than carbohydrate or protein. (Remember that each ATP holds energy and that kcalories measure energy; thus the more ATP generated, the more kcalories have been collected.) For example, one glucose molecule will yield 36 to 38 ATP when completely oxidized. In comparison, one 16-carbon fatty acid molecule will yield 129 ATP when completely oxidized. Fat is a more efficient fuel source. Gram for gram, fat can provide much more energy than either of the other two energy-yielding nutrients, making it the body's preferred form of energy storage. (Similarly, you might prefer to fill your car with a fuel that provides 130 miles per gallon versus one that provides 35 miles per gallon.)

IN SUMMARY   After a balanced meal, the body handles the nutrients as follows. The digestion of carbohydrate yields glucose (and other monosaccharides); some is stored as glycogen, and some is broken down to pyruvate and acetyl CoA to provide energy. The acetyl CoA can then enter the TCA cycle and electron transport chain to provide more energy. The digestion of fat yields glycerol and fatty acids; some are reassembled and stored as fat, and others are broken down to acetyl CoA, which can enter the TCA cycle and electron transport chain to provide energy. The digestion of protein yields amino acids, most of which are used to build body protein or other nitrogen-containing compounds, but some amino acids may be broken down through the same pathways as glucose to provide energy. Other amino acids enter directly into the TCA cycle, and these, too, can be broken down to yield energy. In summary, although carbohydrate, fat, and protein enter the TCA cycle by different routes, the final pathways are common to all energy-yielding nutrients (see Figure 7-21).

The Body's Energy Budget

Every day, a healthy diet delivers thousands of kcalories from foods, and the active body uses most of them to do its work. As a result, body weight changes little, if at all. This remarkable achievement could be called the economy of maintenance. The body's energy budget is balanced. Some people, however, eat too much or exercise too little and get fat; others eat too little or exercise too much and get thin. The metabolic details have already been described; the next sections will review them from the perspective of the body fat gained or lost. The possible reasons why people gain or lose weight are explored in Chapter 8.

FIGURE 7-21 The Central Pathways of Energy Metabolism

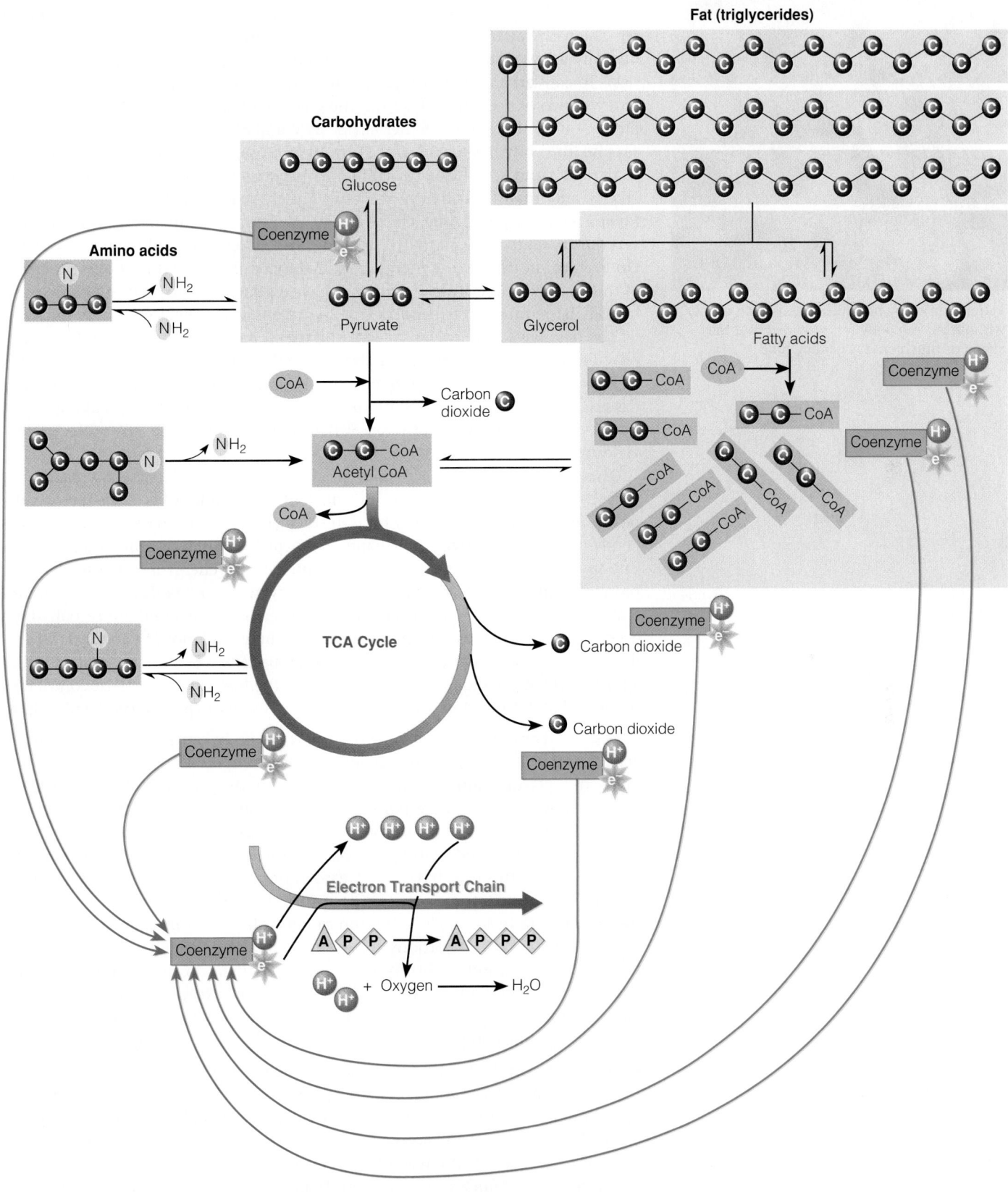

People can enjoy bountiful meals such as this without storing body fat, provided that they spend as much energy as they take in.

The Economics of Feasting

When a person eats too much, metabolism favors fat formation. Fat cells enlarge regardless of whether the excess in kcalories derives from protein, carbohydrate, or fat. The pathway from dietary fat to body fat, however, is the most direct (requiring only a few metabolic steps) and the most efficient (costing only a few kcalories). To convert a dietary triglyceride to a triglyceride in adipose tissue, the body removes two of the fatty acids from the glycerol backbone, absorbs the parts, and puts them (and others) together again. By comparison, to convert a molecule of sucrose, the body has to split glucose from fructose, absorb them, dismantle them to pyruvate and acetyl CoA, assemble many acetyl CoA molecules into fatty acid chains, and finally attach fatty acids to a glycerol backbone to make a triglyceride for storage in adipose tissue. Quite simply, the body uses much less energy to convert dietary fat to body fat than it does to convert dietary carbohydrate to body fat. On average, storing excess energy from dietary fat in body fat uses only 5 percent of the ingested energy intake, but storing excess energy from dietary carbohydrate in body fat requires an expenditure of 25 percent of the ingested energy intake.

The pathways from excess protein and excess carbohydrate to body fat are not only indirect and inefficient, but also less preferred (having other priorities). Before entering fat storage, protein must first tend to its many roles in the body's lean tissues, and carbohydrate must fill the glycogen stores. Simply put, making fat is a low priority for these two nutrients. Still, if eaten in abundance, any of the energy-yielding nutrients can make fat.

This chapter has described each of the energy-yielding nutrients individually, but cells use a mixture of these fuels. How much of which nutrient is in the fuel mix depends, in part, on its availability from the diet. (The proportion of each fuel also depends on physical activity, as Chapter 14 explains.) Dietary protein and dietary carbohydrate influence the mixture of fuel used during energy metabolism. Usually, protein's contribution to the fuel mix is relatively minor and fairly constant, but protein oxidation does increase when protein is eaten in excess. Similarly, carbohydrate eaten in excess significantly enhances carbohydrate oxidation. In contrast, fat oxidation does *not* respond to dietary fat intake, especially when dietary changes occur abruptly. The more protein or carbohydrate in the fuel mix, the less fat contributes to the fuel mix. Instead of being oxidized, fat accumulates in storage. Details follow.

Excess Protein Recall from Chapter 6 that the body cannot store excess amino acids as such; it has to convert them to other compounds. Contrary to popular opinion, a person cannot grow muscle simply by overeating protein. Lean tissue such as muscle develops in response to a stimulus such as hormones or physical activity. When a person overeats protein, the body uses the surplus first by replacing normal daily losses and then by increasing protein oxidation. The body achieves protein balance this way, but any increase in protein oxidation displaces fat in the fuel mix. Any additional protein is then deaminated and the remaining carbons used to make fatty acids. Thus a person can grow fat by eating too much protein.

People who eat huge portions of meat and other protein-rich foods may wonder why they have weight problems. Not only does the fat in those foods lead to fat storage, but the protein can, too, when energy intake exceeds energy needs. Many fad weight-loss diets encourage high protein intakes based on the false assumption that protein builds only muscle, not fat (see Highlight 8 for more details).

Excess Carbohydrate Compared with protein, the proportion of carbohydrate in the fuel mix changes more dramatically when a person overeats. The body handles abundant carbohydrate by first storing it as glycogen, but glycogen storage areas are limited and fill quickly. Because maintaining glucose balance is critical, the body uses glucose frugally when the diet provides only small amounts and freely when stores are abundant. In other words, glucose oxidation rapidly adjusts to the dietary intake of carbohydrate.

Excess glucose can be converted to fat directly, but this is a minor pathway.[3] As mentioned earlier, converting glucose to fat is energetically expensive and does not occur until after glycogen stores have been filled. Even then, only a little, if any, new fat is made from carbohydrate.[4]

Nevertheless, excess dietary carbohydrate can lead to weight gain when extra carbohydrate displaces fat in the fuel mix. When this occurs, carbohydrate spares both dietary fat and body fat from oxidation—an effect that may be more pronounced in overweight people than in lean people.[5] The net result: excess carbohydrate contributes to obesity or at least to the maintenance of an overweight body.

Excess Fat Unlike excess protein and carbohydrate, which both enhance their own oxidation, eating too much fat does not promote fat oxidation. Instead, excess dietary fat moves efficiently into the body's fat stores; almost all of the excess is stored.

IN SUMMARY If energy intake exceeds the body's energy needs, the result will be weight gain—regardless of whether the excess intake is from protein, carbohydrate, or fat. The difference is that the body is much more efficient at storing energy when the excess derives from dietary fat.

The Transition from Feasting to Fasting

Figure 7-22 shows the metabolic pathways operating in the body as it shifts from feasting (part A) to fasting (parts B and C). After a meal, glucose, glycerol, and fatty acids from foods are used as needed and then stored. Later, as the body shifts from

FIGURE 7-22 Feasting and Fasting

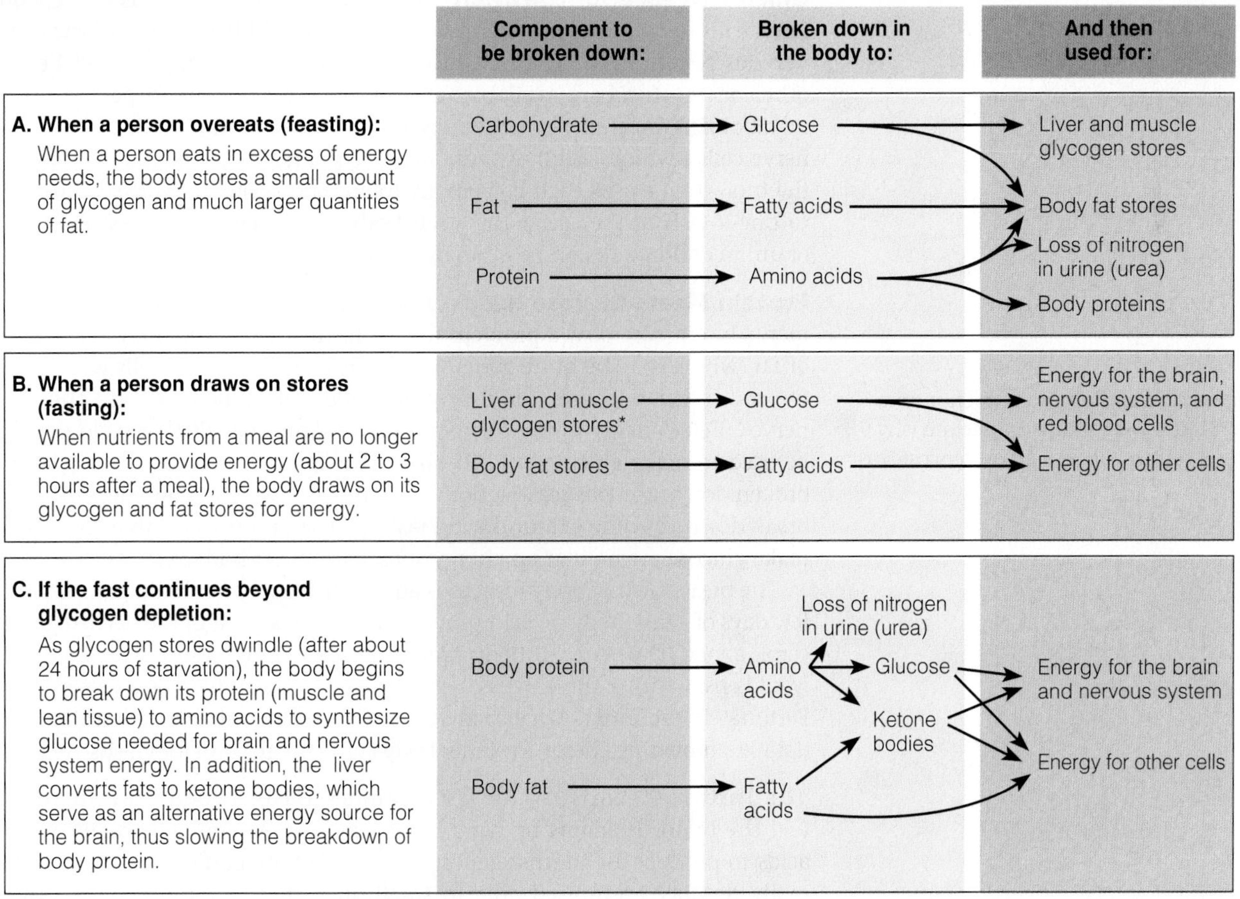

	Component to be broken down:	Broken down in the body to:	And then used for:
A. When a person overeats (feasting): When a person eats in excess of energy needs, the body stores a small amount of glycogen and much larger quantities of fat.	Carbohydrate → Fat → Protein →	Glucose Fatty acids Amino acids	Liver and muscle glycogen stores Body fat stores Loss of nitrogen in urine (urea) Body proteins
B. When a person draws on stores (fasting): When nutrients from a meal are no longer available to provide energy (about 2 to 3 hours after a meal), the body draws on its glycogen and fat stores for energy.	Liver and muscle glycogen stores* Body fat stores →	Glucose Fatty acids	Energy for the brain, nervous system, and red blood cells Energy for other cells
C. If the fast continues beyond glycogen depletion: As glycogen stores dwindle (after about 24 hours of starvation), the body begins to break down its protein (muscle and lean tissue) to amino acids to synthesize glucose needed for brain and nervous system energy. In addition, the liver converts fats to ketone bodies, which serve as an alternative energy source for the brain, thus slowing the breakdown of body protein.	Body protein → Body fat →	Amino acids ↔ Glucose Loss of nitrogen in urine (urea) Ketone bodies Fatty acids	Energy for the brain and nervous system Energy for other cells

*The muscles' stored glycogen provides glucose only for the muscle in which the glycogen is stored.

■ The cells' work that maintains all life processes refers to the body's *basal metabolism*, which is described in Chapter 8.

a fed state to a fasting one, it begins drawing on these stores. Glycogen and fat are released from storage to provide more glucose, glycerol, and fatty acids for energy.

Energy is needed all the time. Even when a person is asleep and totally relaxed, the cells of many organs are hard at work. In fact, this work—the cells' work that maintains all life processes■ without any conscious effort—represents about two-thirds to three-fourths of the total energy a person spends in a day. The small remainder is the work that a person's muscles perform voluntarily during waking hours.

The body's top priority is to meet the cells' needs for energy, and it normally does this by periodic refueling—that is, by eating several times a day. When food is not available, the body turns to its own tissues for other fuel sources. If people choose not to eat, we say they are fasting; if they have no choice, we say they are starving. The body makes no such distinction. In either case, the body is forced to switch to a wasting metabolism, drawing on its reserves of carbohydrate and fat and, within a day or so, on its vital protein tissues as well.

The Economics of Fasting

During fasting, carbohydrate, fat, and protein are all eventually used for energy—fuel must be delivered to every cell. As the fast begins, glucose from the liver's stored glycogen and fatty acids from the adipose tissue's stored fat are both flowing into cells, then breaking down to yield acetyl CoA, and delivering energy to power the cells' work. Several hours later, however, most of the glucose is used up—liver glycogen is exhausted and blood glucose begins to fall. Low blood glucose serves as a signal that promotes further fat breakdown and release of amino acids from muscles.

Glucose Needed for the Brain At this point, most of the cells are depending on fatty acids to continue providing their fuel. But red blood cells and the cells of the nervous system need glucose. Glucose is their major energy fuel, and even when other energy fuels are available, glucose must be present to permit the energy-metabolizing machinery of the nervous system to work. Normally, the brain and nerve cells—which weigh only about three pounds—consume about two-thirds of the total *glucose* used each day (about 400 to 600 kcalories' worth). About one-fifth to one-fourth of the *energy* the adult body uses when it is at rest is spent by the brain; in children, it can be up to one-half.

■ Red blood cells contain no mitochondria. Review Figure 7-4 to fully appreciate why red blood cells must depend on glucose for energy.

Protein Meets Glucose Needs The red blood cells' and brain's special requirements for glucose pose a problem for the fasting body. The body can use its stores of fat, which may be quite generous, to furnish most of its cells with energy, but the red blood cells are completely dependent on glucose,■ and the brain and nerves prefer energy in the form of glucose. Amino acids that yield pyruvate can be used to make glucose; and to obtain the amino acids, body proteins must be broken down. For this reason, body protein tissues such as muscle and liver always break down to some extent during fasting. The amino acids that can't be used to make glucose are used as an energy source for other body cells.

The breakdown of body protein is an expensive way to obtain glucose. In the first few days of a fast, body protein provides about 90 percent of the needed glucose; glycerol, about 10 percent. If body protein losses were to continue at this rate, death would ensue within three weeks, regardless of the quantity of fat a person had stored. Fortunately, fat breakdown also increases with fasting—in fact, fat breakdown almost doubles, providing energy for other body cells and glycerol for glucose production.

■ Reminder: *Ketone bodies* are compounds produced during the incomplete breakdown of fat when glucose is not available.

The Shift to Ketosis As the fast continues, the body finds a way to use its fat to fuel the brain. It adapts by combining acetyl CoA fragments derived from fatty acids to produce an alternate energy source, ketone bodies (see Figure 7-23). Normally produced and used only in small quantities, ketone bodies■ can provide fuel for some brain cells. Ketone body production rises until, after about ten days of fasting, it is meeting much of the nervous system's energy needs. Still, many ar-

FIGURE 7-23 Ketone Body Formation

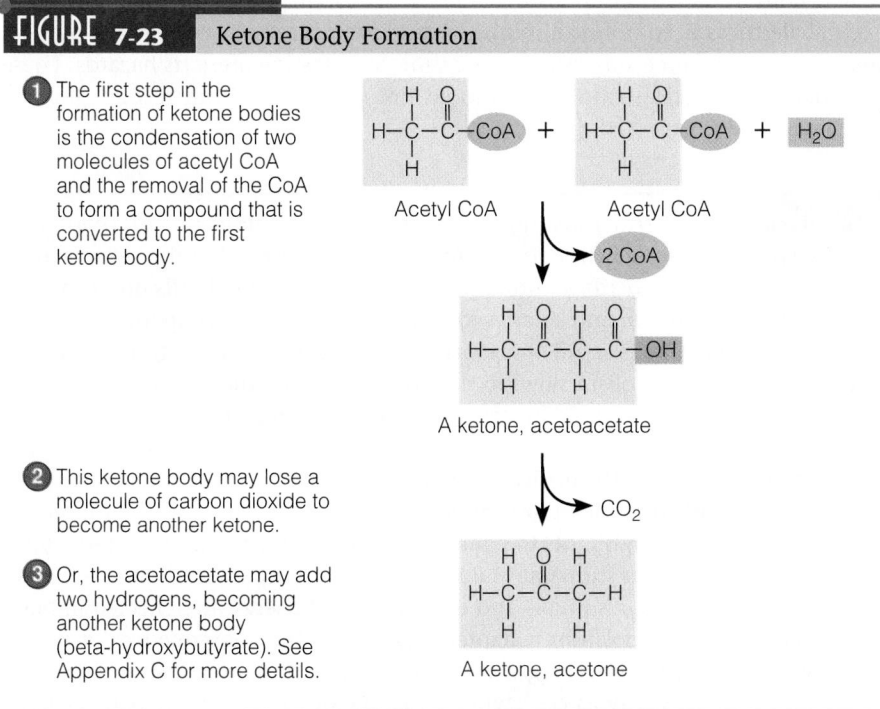

1 The first step in the formation of ketone bodies is the condensation of two molecules of acetyl CoA and the removal of the CoA to form a compound that is converted to the first ketone body.

Acetyl CoA   Acetyl CoA

2 CoA

A ketone, acetoacetate

2 This ketone body may lose a molecule of carbon dioxide to become another ketone.

CO_2

3 Or, the acetoacetate may add two hydrogens, becoming another ketone body (beta-hydroxybutyrate). See Appendix C for more details.

A ketone, acetone

eas of the brain rely exclusively on glucose, and to produce it, the body continues to sacrifice protein—albeit at a slower rate than in the early days of fasting.

When ketone bodies contain an acid group (COOH), they are called keto acids. Small amounts of keto acids are a normal part of the blood chemistry, but when their concentration rises, the pH of the blood drops. This is ketosis, a sign that the body's chemistry is going awry. Elevated blood ketones (ketonemia) are excreted in the urine (ketonuria). A fruity odor on the breath (known as acetone breath) develops, reflecting the presence of the ketone acetone.

Suppression of Appetite Ketosis also induces a loss of appetite. As starvation continues, this loss of appetite becomes an advantage to a person without access to food, because the search for food would be a waste of energy. When the person finds food and eats again, the body shifts out of ketosis, the hunger center gets the message that food is again available, and the appetite returns. Highlight 8 includes a discussion of the risks of ketosis-producing diets in its review of popular weight-loss diets.

Slowing of Metabolism In an effort to conserve body tissues for as long as possible, the hormones of fasting slow metabolism. As the body shifts to the use of ketone bodies, it simultaneously reduces its energy output and conserves both its fat and its lean tissue. Still the lean (protein-containing) organ tissues shrink in mass and perform less metabolic work, reducing energy expenditures. As the muscles waste, they can do less work and so demand less energy, reducing expenditures further. Because of the slowed metabolism, the loss of fat falls to a bare minimum—less, in fact, than the fat that would be lost on a low-kcalorie diet. Thus, although *weight* loss during fasting may be quite dramatic, *fat* loss may be less than when at least some food is eaten.

Symptoms of Starvation The adaptations just described—slowing of energy output and reduction in fat loss—occur in the starving child, the hungry homeless adult, the fasting religious person, the adolescent with anorexia nervosa, and the malnourished hospital patient. Such adaptations help to prolong their lives and explain the physical symptoms of energy deprivation: wasting, slowed metabolism, lowered body temperature, and reduced resistance to disease.

The body's adaptations to fasting are sufficient to maintain life for a long time. Mental alertness need not be diminished, and even some physical energy may remain unimpaired for a surprisingly long time. Still, fasting presents hazards. These remarkable adaptations, however, should not prevent us from recognizing the very real hazards that fasting presents.

> **IN SUMMARY** When fasting, the body makes a number of adaptations: increasing the breakdown of fat to provide energy for most of the cells, using glycerol and amino acids to make glucose for the red blood cells and central nervous system, producing ketones to fuel the brain, suppressing the appetite, and slowing metabolism. All of these measures conserve energy and minimize losses. In fact, metabolism slows to such an extent that the loss of fat eventually slows to less than would be achieved with a low-kcalorie diet.

This chapter has probed the intricate details of metabolism at the level of the cells, exploring the transformations of nutrients to energy and to storage compounds. Several chapters and highlights to come build on this information. The highlight that follows this chapter shows how alcohol disrupts normal metabolism. Chapter 8 describes how a person's intake and expenditure of energy are reflected in body weight and body composition. Chapter 9 examines the consequences of unbalanced energy budgets—overweight and underweight. Chapter 10 shows the vital roles the B vitamins play as coenzymes assisting all the metabolic pathways described here. And Chapter 14 revisits metabolism to show how it supports the work of physically active people and how athletes can best apply that information in their choices of foods to eat.

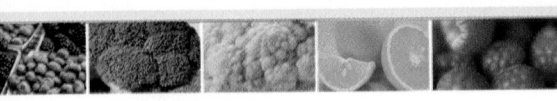

Nutrition in Your Life

All day, every day, your cells dismantle carbohydrates, fats, and proteins, with the help of vitamins, minerals, and water, releasing energy to meet your body's immediate needs or storing it as fat for later use.

- What types of foods best support aerobic and anaerobic activities?

- Do you eat more protein, carbohydrate, or fat than your body needs?

- Do you follow a low-carbohydrate diet that forces your body into ketosis?

STUDY QUESTIONS

These questions will help you review the chapter. You will find the answers in the discussions on the pages provided.

1. Define metabolism, anabolism, and catabolism; give an example of each. (pp. 215–216)

2. Name one of the body's high-energy molecules, and describe how it is used. (pp. 216–218)

3. What are coenzymes, and what service do they provide in metabolism? (p. 219)

4. Name the four basic units, derived from foods, that are used by the body in metabolic transformations.

How many carbons are in the "backbones" of each? (p. 220)

5. Define aerobic and anaerobic metabolism. How does insufficient oxygen influence metabolism? (pp. 221–223)

6. How does the body dispose of excess nitrogen? (pp. 227–229)

7. Summarize the main steps in the metabolism of glucose, glycerol, fatty acids, and amino acids. (pp. 228–229)

8. Describe how a surplus of the three energy nutrients contributes to body fat stores. (pp. 234–235)

9. What adaptations does the body make during a fast? What are ketone bodies? Define ketosis. (pp. 236–237)

10. Distinguish between a loss of *fat* and a loss of *weight,* and describe how each might happen. (pp. 237–238)

These multiple choice questions will help you prepare for an exam. Answers can be found below.

1. Hydrolysis is an example of a(n):
   a. coupled reaction.
   b. anabolic reaction.
   c. catabolic reaction.
   d. synthesis reaction.

2. During metabolism, released energy is captured and transferred by:
   a. enzymes.
   b. pyruvate.
   c. acetyl CoA.
   d. adenosine triphosphate.

3. Glycolysis:
   a. requires oxygen.
   b. generates abundant energy.
   c. converts glucose to pyruvate.
   d. produces ammonia as a by-product.

4. The pathway from pyruvate to acetyl CoA:
   a. produces lactic acid.
   b. is known as gluconeogenesis.
   c. is metabolically irreversible.
   d. requires more energy than it produces.

5. For complete oxidation, acetyl CoA enters:
   a. glycolysis.
   b. the TCA cycle.
   c. the Cori cycle.
   d. the electron transport chain.

6. Deamination of an amino acid produces:
   a. vitamin B_6 and energy.
   b. pyruvate and acetyl CoA.
   c. ammonia and a keto acid.
   d. carbon dioxide and water.

7. Before entering the TCA cycle, each of the energy-yielding nutrients is broken down to:
   a. ammonia.
   b. pyruvate.
   c. electrons.
   d. acetyl CoA.

8. The body stores energy for future use in:
   a. proteins.
   b. acetyl CoA.
   c. triglycerides.
   d. ketone bodies.

9. During a fast, when glycogen stores have been depleted, the body begins to synthesize glucose from:
   a. acetyl CoA.
   b. amino acids.
   c. fatty acids.
   d. ketone bodies.

10. During a fast, the body produces ketone bodies by:
   a. hydrolyzing glycogen.
   b. condensing acetyl CoA.
   c. transaminating keto acids.
   d. converting ammonia to urea.

REFERENCES

1. J. H. Wilmore and D. L. Costill, Physical energy: Fuel metabolism, *Nutrition Reviews* 59 (2001): S13–S16; A. D. Kriketos, J. C. Peters, and J. O. Hill, Cellular and whole-animal energetics, in *Biochemical and Physiological Aspects of Human Nutrition,* ed. M. H. Stipanuk (Philadelphia: W. B. Saunders, 2000), pp. 411–424.
2. J. L. Groff and S. S. Gropper, *Advanced Nutrition and Human Metabolism* (Belmont, Calif.: Wadsworth/Thomson Learning, 2000), p. 188.
3. M. K. Hellerstein, De novo lipogenesis in humans: Metabolic and regulatory aspects, *European Journal of Clinical Nutrition* 53 (1999): S53–S65.
4. R. M. Devitt and coauthors, De novo lipogenesis during controlled overfeeding with sucrose or glucose in lean and obese women, *American Journal of Clinical Nutrition* 74 (2001): 707–708.
5. I. Marques-Lopes and coauthors, Postprandial de novo lipogenesis and metabolic changes induced by a high-carbohydrate, low-fat meal in lean and overweight men, *American Journal of Clinical Nutrition* 73 (2001): 253–261.

ANSWERS

Study Questions (multiple choice)

1. c   2. d   3. c   4. c   5. b   6. c   7. d   8. c
9. b   10. b

HIGHLIGHT

Alcohol and Nutrition

With the understanding of metabolism gained from Chapter 7, you are in a position to understand how the body handles alcohol, how alcohol interferes with metabolism, and how alcohol impairs health and nutrition. The potential health benefits of drinking alcohol in *moderation* are presented in Chapter 18.

Alcohol in Beverages

To the chemist, **alcohol** refers to a class of organic compounds containing hydroxyl (OH) groups (the accompanying glossary defines alcohol and related terms). The glycerol to which fatty acids are attached in triglycerides is an example of an alcohol to a chemist. To most people, though, *alcohol* refers to the intoxicating ingredient in **beer, wine,** and **distilled liquor (hard liquor).** The chemist's name for this particular alcohol is *ethyl alcohol,* or **ethanol.** Glycerol has 3 carbons with 3 hydroxyl groups attached; ethanol has only 2 carbons and 1 hydroxyl group (see Figure H7-1). The remainder of this highlight talks about the particular alcohol, ethanol, but refers to it simply as *alcohol.*

Alcohols affect living things profoundly, partly because they act as lipid solvents. Their ability to dissolve lipids out of cell membranes allows alcohols to penetrate rapidly into cells, destroying cell structures and thereby killing the cells. For this reason, most alcohols are toxic in relatively small amounts; by the same token, because they kill microbial cells, they are useful as disinfectants.

Ethanol is less toxic than the other alcohols. Sufficiently diluted and taken in small enough doses, its action in the brain produces an effect that people seek—not with zero risk, but with a low enough risk (if the doses are low enough) to be tolerable. Used in this way, alcohol is a **drug**—that is, a substance that modifies body functions. Like all drugs, alcohol both offers benefits and poses hazards. "If you drink alcoholic beverages, do so in moderation" are the words of caution from the *Dietary Guidelines.*

The term **moderation** is important in describing alcohol use. How many drinks constitute moderate use, and how much is "a drink"? First, a **drink** is any alcoholic beverage that delivers ½ ounce of *pure ethanol:*

FIGURE H7-1 — Two Alcohols: Glycerol and Ethanol

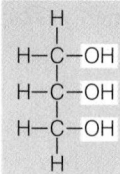

Glycerol is the alcohol used to make triglycerides.

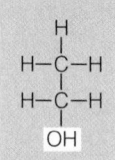

Ethanol is the alcohol in beer, wine, and distilled liquor.

- 5 ounces of wine.
- 10 ounces of wine cooler.
- 12 ounces of beer.
- 1½ ounces of distilled liquor (80 proof whiskey, scotch, rum, or vodka).

Beer, wine, and liquor deliver different amounts of alcohol. The amount of alcohol in distilled liquor is stated as **proof:** 100 proof liquor is 50 percent alcohol, 80 proof is 40 percent alcohol, and so forth. Wine and beer have less alcohol than distilled liquor, although some fortified wines and beers have more alcohol than the regular varieties (see photo caption on p. 241).

Second, because people have different tolerances for alcohol, it is impossible to name an exact daily amount of alcohol that is appropriate for everyone. Authorities have attempted to identify amounts that are acceptable for most healthy people. An accepted definition of moderation is not more than two drinks a day for the average-sized man and not more than one drink a day for the average-sized woman. (Pregnant women are advised to abstain from alcohol, as Highlight 15 explains.) Notice that this advice is stated as a maximum, not as an average; seven

12 oz beer

10 oz wine cooler

1½ oz hard liquor (80 proof whiskey, gin, brandy, rum, vodka)

5 oz wine

Each of these servings equals one drink.

GLOSSARY

acetaldehyde (ass-et-AL-duh-hide): an intermediate in alcohol metabolism.

alcohol: a class of organic compounds containing hydroxyl (OH) groups.

alcohol abuse: a pattern of drinking that includes failure to fulfill work, school, or home responsibilities; drinking in situations that are physically dangerous (as in driving while intoxicated); recurring alcohol-related legal problems (as in aggravated assault charges); or continued drinking despite ongoing social problems that are caused by or worsened by alcohol.

alcohol dehydrogenase (dee-high-DROJ-eh-nayz): an enzyme active in the stomach and the liver that converts ethanol to acetaldehyde.

alcoholism: a pattern of drinking that includes a strong craving for alcohol, a loss of control and an inability to stop drinking once begun, withdrawal symptoms (nausea, sweating, shakiness, and anxiety) after heavy drinking, and the need for increasing amounts of alcohol in order to feel "high."

antidiuretic hormone (ADH): a hormone produced by the pituitary gland in response to dehydration (or a high sodium concentration in the blood). It stimulates the kidneys to reabsorb more water and therefore to excrete less. This ADH should not be confused with the enzyme alcohol dehydrogenase, which is also sometimes abbreviated ADH.

beer: an alcoholic beverage brewed by fermenting malt and hops.

cirrhosis (seer-OH-sis): advanced liver disease in which liver cells turn orange, die, and harden, permanently losing their function; often associated with alcoholism.

• **cirrhos** = an orange

distilled liquor or **hard liquor:** an alcoholic beverage made by fermenting and distilling grains; sometimes called *distilled spirits.*

drink: a dose of any alcoholic beverage that delivers ½ oz of pure ethanol:
• 5 oz of wine.
• 10 oz of wine cooler.
• 12 oz of beer.
• 1½ oz of hard liquor (80 proof whiskey, scotch, rum, or vodka).

drug: a substance that can modify one or more of the body's functions.

ethanol: a particular type of alcohol found in beer, wine, and distilled liquor; also called *ethyl alcohol* (see Figure H7-1). Ethanol is the most widely used—and abused—drug in our society. It is also the only legal, nonprescription drug that produces euphoria.

fatty liver: an early stage of liver deterioration seen in several diseases, including kwashiorkor and alcoholic liver disease. Fatty liver is characterized by an accumulation of fat in the liver cells.

fibrosis (fye-BROH-sis): an intermediate stage of liver deterioration seen in several diseases, including viral hepatitis and alcoholic liver disease. In fibrosis, the liver cells lose their function and assume the characteristics of connective tissue cells (fibers).

MEOS or **microsomal** (my-krow-SO-mal) **ethanol-oxidizing system:** a system of enzymes in the liver that oxidize not only alcohol, but also several classes of drugs.

moderation: in relation to alcohol consumption, not more than two drinks a day for the average-sized man and not more than one drink a day for the average-sized woman.

NAD (nicotinamide adenine dinucleotide): the main coenzyme form of the vitamin niacin. Its reduced form is NADH.

narcotic (nar-KOT-ic): a drug that dulls the senses, induces sleep, and becomes addictive with prolonged use.

proof: a way of stating the percentage of alcohol in distilled liquor. Liquor that is 100 proof is 50% alcohol; 90 proof is 45%, and so forth.

Wernicke-Korsakoff (VER-nee-key KORE-sah-kof) **syndrome:** a neurological disorder typically associated with chronic alcoholism and caused by a deficiency of the B vitamin thiamin; also called *alcohol-related dementia.*

wine: an alcoholic beverage made by fermenting grape juice.

Matthew Farruggio

Wines contain 7 to 24 percent alcohol by volume; those containing 14 percent or more must state their alcohol content on the label, whereas those with less than 14 percent may simply state "table wine" or "light wine." Beers typically contain less than 5 percent alcohol by volume and malt liquors, 5 to 8 percent; regulations vary, with some states requiring beer labels to show the alcohol content and others prohibiting such statements.

drinks one night a week would not be considered moderate, even though one a day would be. Doubtless some people could consume slightly more; others could not handle nearly so much without risk. The amount a person can drink safely is highly individual, depending on genetics, health, gender, body composition, age, and family history.

Reduce average annual alcohol consumption.

HEALTHY
PEOPLE
2010

Alcohol in the Body

From the moment an alcoholic beverage enters the body, alcohol is treated as if it has special privileges. Unlike foods, which require time for digestion, alcohol needs no digestion and is quickly absorbed. About 20 percent is absorbed directly across the walls of an empty stomach and can reach the brain within a minute. Consequently, a person can immediately feel euphoric when drinking, especially on an empty stomach.

When the stomach is full of food, alcohol has less chance of touching the walls and diffusing through, so its influence on the brain is slightly delayed. This information leads to a practical tip: eat snacks when drinking alcoholic beverages. Carbohydrate snacks slow alcohol absorption and high-fat snacks slow peristalsis, keeping the alcohol in the stomach longer. Salty snacks make a person thirsty; to quench thirst, drink water instead of more alcohol.

The stomach begins to break down alcohol with its **alcohol dehydrogenase** enzyme. This action can reduce the amount of alcohol entering the blood by about 20 percent. Women produce less of this stomach enzyme than men; consequently, more alcohol reaches the intestine for absorption into the bloodstream. As a result, women absorb about one-third more alcohol than men of the same size who drink the same amount of alcohol. Consequently, they are more likely to become more intoxicated on less alcohol than men. These differences between men and women help explain why women have a lower alcohol tolerance and a lower recommendation for moderate intake.

In the small intestine, alcohol is rapidly absorbed. From this point on, alcohol receives priority treatment: it gets absorbed and metabolized before most nutrients. Alcohol's priority status helps to ensure a speedy disposal and reflects two facts: alcohol cannot be stored in the body, and it is potentially toxic.

Alcohol Arrives in the Liver

The capillaries of the digestive tract merge into veins that carry the alcohol-laden blood to the liver. These veins branch and rebranch into capillaries that touch every liver cell. Liver cells are the only other cells in the body that can make enough of the alcohol dehydrogenase enzyme to oxidize alcohol at an appreciable rate. The routing of blood through the liver cells gives them the chance to dispose of some alcohol before it moves on.

Alcohol affects every organ of the body, but the most dramatic evidence of its disruptive behavior appears in the liver. If liver cells could talk, they would describe alcohol as demanding, egocentric, and disruptive of the liver's efficient way of running its business. For example, liver cells normally prefer fatty acids as their fuel, and they like to package excess fatty acids into triglycerides and ship them out to other tissues. When alcohol is present, however, the liver cells are forced to metabolize alcohol and let the fatty acids accumulate, sometimes in huge stockpiles. Alcohol metabolism can also permanently change liver cell structure, impairing the liver's ability to metabolize fats. This explains why heavy drinkers develop fatty livers.

The liver can process about ½ ounce *ethanol* per hour (the amount in a typical drink), depending on the person's body size, previous drinking experience, food intake, and general health. This maximum rate of alcohol breakdown is set by the amount of alcohol dehydrogenase available. If more alcohol arrives at the liver than the enzymes can handle, the extra alcohol travels to all parts of the body, circulating again and again until liver enzymes are finally available to process it. Another practical tip derives from this information: drink slowly enough to allow the liver to keep up—no more than one drink per hour.

The amount of alcohol dehydrogenase enzyme present in the liver varies with individuals, depending on the genes they have inherited and on how recently they have eaten. Fasting for as little as a day forces the body to degrade its proteins, including the alcohol-processing enzymes, and this can slow the rate of alcohol metabolism by half. Drinking after not eating all day thus causes the drinker to feel the effects more promptly for two reasons: rapid absorption and slowed breakdown. By maintaining higher blood alcohol concentrations for longer times, alcohol can anesthetize the brain more completely (as described later in this highlight).

The alcohol dehydrogenase enzyme breaks down alcohol by removing hydrogens in two steps. (Figure H7-2 provides a simplified diagram of alcohol metabolism; Appendix C provides the chemical details.) In the first step, alcohol dehydrogenase oxidizes alcohol to **acetaldehyde**. High concentrations of acetaldehyde in the brain and other tissues are responsible for many of the damaging effects of **alcohol abuse.**

In the second step, a related enzyme, acetaldehyde dehydrogenase, converts acetaldehyde to acetate, which is then converted to acetyl CoA—the "crossroads" compound introduced in Chapter 7 that can enter the TCA cycle to generate energy. These

FIGURE H7-2   Alcohol Metabolism

The conversion of alcohol to acetyl CoA requires the B vitamin niacin in its role as the coenzyme NAD. When the enzymes oxidize alcohol, they remove H atoms and attach them to NAD. Thus NAD is used up, and NADH accumulates. (Note: More accurately, NAD+ is converted to NADH + H+.)

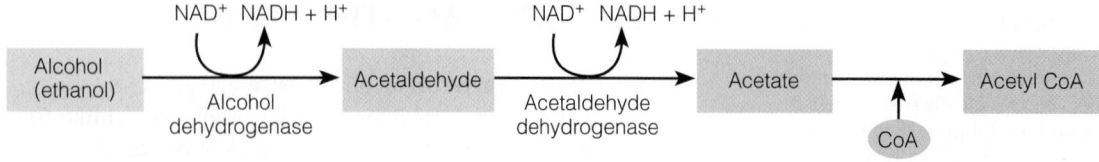

FIGURE H7-3 Alternate Route for Acetyl CoA: To Fat

Acetyl CoA molecules are blocked from getting into the TCA cycle by the high level of NADH. Instead of being used for energy, the acetyl CoA molecules become building blocks for fatty acids.

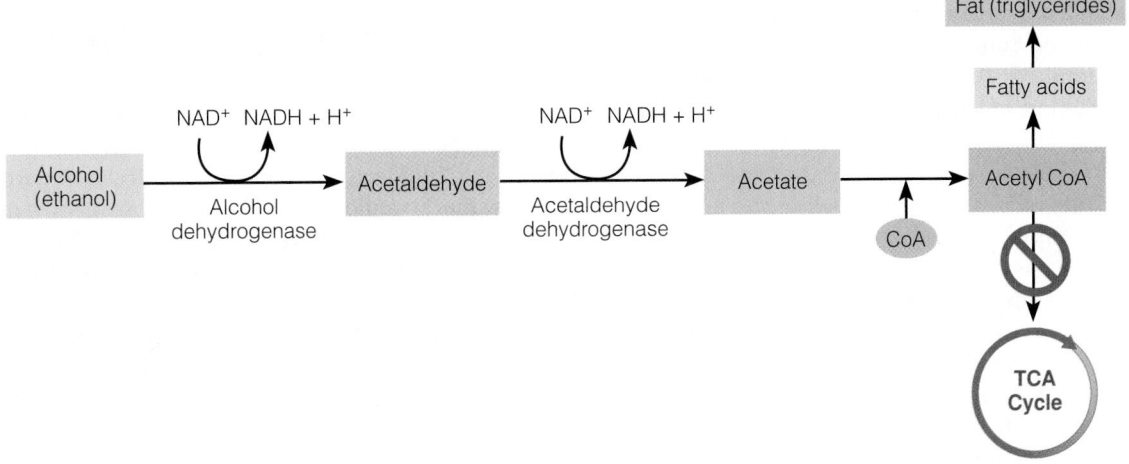

Alcohol Disrupts the Liver

During alcohol metabolism, the multitude of other metabolic processes for which NAD is required, including glycolysis, the TCA cycle, and the electron transport chain, falter. Its presence is sorely missed in these energy pathways because it is the chief carrier of the hydrogens that travel with their electrons along the electron transport chain. Without adequate NAD, these energy pathways cannot function. Traffic either backs up, or an alternate route is taken. Such changes in the normal flow of energy pathways have striking physical consequences.

For one, the accumulation of hydrogen ions during alcohol metabolism shifts the body's acid-base balance toward acid. For another, the accumulation of NADH slows the TCA cycle, so pyruvate and acetyl CoA build up. Excess acetyl CoA then takes the route to fatty acid synthesis (as Figure H7-3 illustrates), and fat clogs the liver.

As you might expect, a liver overburdened with fat cannot function properly. Liver cells become less efficient at performing a number of tasks. Much of this inefficiency impairs a person's nutritional health in ways that cannot be corrected by diet alone. For example, the liver has difficulty activating vitamin D, as well as producing and releasing bile. To overcome such problems, a person needs to stop drinking alcohol.

reactions produce hydrogen ions (H+). The B vitamin niacin, in its role as the coenzyme **NAD** (nicotinamide adenine dinucleotide), helpfully picks up these hydrogen ions (becoming NADH). Thus, whenever the body breaks down alcohol, NAD diminishes and NADH accumulates. (Chapter 10 presents information on NAD and the other coenzyme roles of the B vitamins.)

The synthesis of fatty acids accelerates with exposure to alcohol. Fat accumulation can be seen in the liver after a single night of heavy drinking. **Fatty liver,** the first stage of liver deterioration seen in heavy drinkers, interferes with the distribution of nutrients and oxygen to the liver cells. Fatty liver is reversible with abstinence from alcohol. If fatty liver lasts long enough, however, the liver cells will die and form fibrous scar tissue. This second stage of liver deterioration is called **fibrosis.** Some liver cells can regenerate with good nutrition and abstinence from alcohol, but in the most advanced stage, **cirrhosis,** damage is the least reversible.

The fatty liver has difficulty generating glucose from protein. The lack of glucose together with the overabundance of acetyl CoA sets the stage for ketosis. The body uses the acetyl CoA to make ketone bodies; their acidity pushes the acid-base balance further toward acid and suppresses nervous system activity.

Excess NADH also promotes the making of lactic acid from pyruvate. The conversion of pyruvate to lactic acid uses the hydrogens from NADH and restores some NAD, but a lactic acid buildup has serious consequences of its own—it adds still further to the body's acid burden and interferes with the excretion of another acid, uric acid, causing inflammation of the joints.

Alcohol alters both amino acid and protein metabolism. Synthesis of proteins important in the immune system slows down, weakening the body's defenses against infection. Protein deficiency can develop, both from a diminished synthesis of protein and from a poor diet. Normally, the cells would at least use the amino acids from the protein foods a person eats, but the drinker's liver deaminates the amino acids and uses the carbon fragments primarily to make fat or ketones. Eating well does not protect the drinker from protein depletion; a person has to stop drinking alcohol.

The liver's priority treatment of alcohol affects its handling of drugs as well as nutrients. In addition to the dehydrogenase enzyme already described, the liver possesses an enzyme system that metabolizes *both* alcohol and several other types of drugs. Called the **MEOS (microsomal ethanol-oxidizing system)**, this system handles about one-fifth of the total alcohol a person consumes. At high blood concentrations or with repeated exposures, alcohol stimulates the synthesis of enzymes in the MEOS. The result is a more efficient metabolism of alcohol and tolerance to its effects.

As a person's blood alcohol rises, alcohol competes with—and wins out over—other drugs whose metabolism also relies on the MEOS. If a person drinks and uses another drug at the same time, the MEOS will dispose of the alcohol first and metabolize the drug more slowly. While the drug waits to be handled later, the dose may build up so that its effects are greatly amplified—sometimes to the point of being fatal.

In contrast, once a heavy drinker stops drinking and alcohol is no longer competing with other drugs, the enhanced MEOS metabolizes drugs much faster than before. As a result, determining the correct dosages of medications can be challenging.

This discussion has emphasized the major way that the blood is cleared of alcohol—metabolism by the liver—but there is another way. About 10 percent of the alcohol leaves the body through the breath and in the urine. This is the basis for the breath and urine tests for drunkenness. The amounts of alcohol in the breath and in the urine are in proportion to the amount still in the bloodstream and brain. In nearly all states, legal drunkenness is set at 0.10 percent or less, reflecting the relationship between alcohol use and traffic and other accidents.

Alcohol Arrives in the Brain

Alcohol is a **narcotic.** People used it for centuries as an anesthetic because it can deaden pain. But alcohol was a poor anesthetic because one could never be sure how much a person would need and how much would be a fatal dose. Consequently, new, more predictable anesthetics have replaced alcohol. Nonetheless, alcohol continues to be used today as a kind of social anesthetic to help people relax or to relieve anxiety. People think that alcohol is a stimulant because it seems to relieve inhibitions. Actually, though, it accomplishes this by sedating *inhibitory* nerves, which are more numerous than excitatory nerves. Ultimately, alcohol acts as a depressant and affects all the nerve cells. Figure H7-4 describes alcohol's effects on the brain.

It is lucky that the brain centers respond to a rising blood alcohol concentration in the order described in Figure H7-4 because a person usually passes out before managing to drink a lethal dose. It is possible, though, to drink so fast that the effects of alcohol continue to accelerate after the person has passed out. Occasionally, a person dies from drinking enough to stop the heart before passing out. Table H7-1 shows the

---

FIGURE H7-4   Alcohol's Effects on the Brain

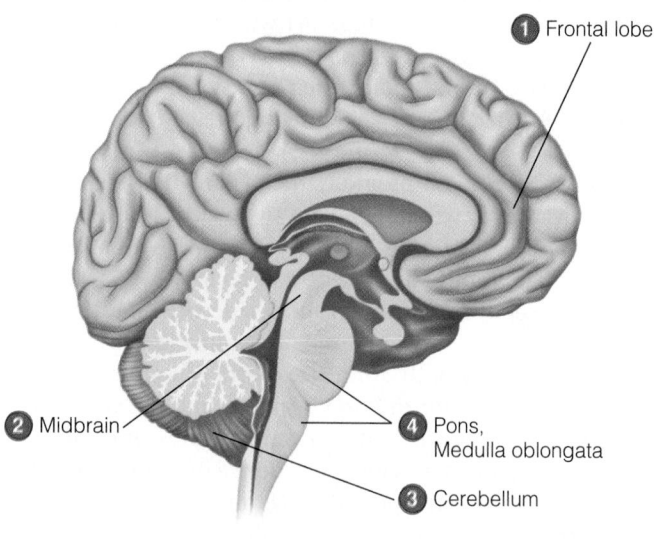

1 Frontal lobe

2 Midbrain

4 Pons, Medulla oblongata

3 Cerebellum

① Judgment and reasoning centers are most sensitive to alcohol. When alcohol flows to the brain, it first sedates the frontal lobe, the center of all conscious activity. As the alcohol molecules diffuse into the cells of these lobes, they interfere with reasoning and judgment.

② Speech and vision centers in the midbrain are affected next. If the drinker drinks faster than the rate at which the liver can oxidize the alcohol, blood alcohol concentrations rise: the speech and vision centers of the brain become sedated.

③ Voluntary muscular control is then affected. At still higher concentrations, the cells in the cerebellum responsible for coordination of voluntary muscles are affected, including those used in speech, eye-hand coordination, and limb movements. At this point people under the influence stagger or weave when they try to walk, or they may slur their speech.

④ Respiration and heart action are the last to be affected. Finally, the conscious brain is completely subdued, and the person passes out. Now the person can drink no more; this is fortunate because higher doses would anesthetize the deepest brain centers that control breathing and heartbeat, causing death.

TABLE H7-1 · Alcohol Doses and Blood Levels

Number of Drinks[a]	Percentage of Blood Alcohol by Body Weight				
	100 lb	120 lb	150 lb	180 lb	200 lb
2	0.08	0.06	0.05	0.04	0.04
4	0.15	0.13	0.10	0.08	0.08
6	0.23	0.19	0.15	0.13	0.11
8	0.30	0.25	0.20	0.17	0.15
12	0.45	0.36	0.30	0.25	0.23
14	0.52	0.42	0.35	0.34	0.27

NOTE: In some states driving under the influence is proved when an adult's blood contains 0.08 percent alcohol, and in others, 0.10. Many states have adopted a "zero-tolerance" policy for drivers under age 21, using 0.02 percent as the limit.

[a] Taken within an hour or so; each drink equivalent to ½ ounce pure ethanol.

TABLE H7-2 · Alcohol Blood Levels and Brain Responses

Blood Alcohol Concentration	Effect on Brain
0.05	Impaired judgment, relaxed inhibitions, altered mood, increased heart rate
0.10	Impaired coordination, delayed reaction time, exaggerated emotions, impaired peripheral vision, impaired ability to operate a vehicle
0.15	Slurred speech, blurred vision, staggered walk, seriously impaired coordination and judgment
0.20	Double vision, inability to walk
0.30	Uninhibited behavior, stupor, confusion, inability to comprehend
0.40 to 0.60	Unconsciousness, shock, coma, death (cardiac or respiratory failure)

NOTE: Blood alcohol concentration depends on a number of factors, including alcohol in the beverage, the rate of consumption, the person's gender, and body weight. For example, a 100-pound female can become legally drunk (0.10 concentration) by drinking three beers in an hour, whereas a 220-pound male consuming that amount at the same rate would have a 0.05 blood alcohol concentration.

blood alcohol levels that correspond to progressively greater intoxication, and Table H7-2 shows the brain responses that occur at these blood levels.

Like liver cells, brain cells die with excessive exposure to alcohol. Liver cells may be replaced, but not all brain cells can regenerate. Thus some heavy drinkers suffer permanent brain damage. Whether alcohol impairs cognition in moderate drinkers is unclear.[1]

People who drink alcoholic beverages may notice that they urinate more, but they may be unaware of the vicious cycle that results. Alcohol depresses production of **antidiuretic hormone (ADH)**, a hormone produced by the pituitary gland that retains water. Loss of body water leads to thirst, and thirst leads to more drinking. Water will relieve dehydration, but the thirsty drinker may drink alcohol instead, which only worsens the problem. Such information provides another practical tip: drink water when thirsty and before each alcoholic drink. Drink an extra glass or two before going to bed. This strategy will help lessen the effects of a hangover.

Water loss is accompanied by the loss of important minerals. As Chapters 12 and 13 will explain, these minerals are vital to the body's fluid balance and to many chemical reactions in the cells, including muscle action. Detoxification treatment includes restoration of mineral balance as quickly as possible.

Alcohol and Malnutrition

For many moderate drinkers, alcohol does not suppress food intake and may actually stimulate appetite.[2] Moderate drinkers usually consume alcohol as *added* energy—on top of their normal food intake. In addition, alcohol in moderate doses is efficiently metabolized. Consequently, alcohol can contribute to body fat and weight gain.[3] Metabolically, alcohol is almost as efficient as fat in promoting obesity; each

ounce of alcohol represents about a half-ounce of fat. Alcohol's contribution to body fat is most evident in the central obesity that commonly accompanies alcohol consumption, popularly—and appropriately—known as the "beer belly." Alcohol in heavy doses, though, is not efficiently metabolized, generating more heat than fat. Heavy drinkers usually consume alcohol as *substituted* energy—instead of their normal food intake. They tend to eat poorly and suffer malnutrition.

Alcohol is rich in energy (7 kcalories per gram), but as with pure sugar or fat, the kcalories are empty of nutrients. The more alcohol people drink, the less likely that they will eat enough food to obtain adequate nutrients. The more kcalories spent on alcohol, the fewer kcalories available to spend on nutritious foods. Table H7-3 shows the kcalorie amounts of typical alcoholic beverages.

Chronic alcohol abuse not only displaces nutrients from the diet, but also interferes with the body's metabolism of nutrients.[4] Most dramatic is alcohol's effect on the B vitamin folate. The liver loses its ability to retain folate, and the kidneys increase their excretion of it. Alcohol abuse creates a folate deficiency that devastates digestive system function. The intestine normally releases and retrieves folate continuously, but it becomes damaged by folate deficiency and alcohol toxicity, so it fails to retrieve its own folate and misses any that may trickle in from food as well. Alcohol also interferes with the action of folate in converting homocysteine to methionine. The result is an excess of homocysteine, which has been linked to heart disease, and an inadequate supply of methionine, which slows the production of new cells, especially the rapidly dividing cells of the intestine and the blood. The combination of poor folate status and alcohol consumption has also been implicated in promoting colorectal cancer.

The inadequate food intake and impaired nutrient absorption that accompany chronic alcohol abuse frequently lead to

TABLE H7-3 kCalories in Alcoholic Beverages and Mixers

Beverage	Amount (oz)	Energy (kcal)
Beer		
Regular	12	150
Light	12	78–131
Nonalcoholic	12	32–82
Distilled liquor (gin, rum, vodka, whiskey)		
80 proof	1½	100
86 proof	1½	105
90 proof	1½	110
Liqueurs		
Coffee liqueur, 53 proof	1½	175
Coffee and cream liqueur, 34 proof	1½	155
Crème de menthe, 72 proof	1½	185
Mixers		
Club soda	12	0
Cola	12	150
Cranberry juice cocktail	8	145
Diet drinks	12	2
Ginger ale or tonic	12	125
Grapefruit juice	8	95
Orange juice	8	110
Tomato or vegetable juice	8	45
Wine		
Dessert	3½	110–135
Nonalcoholic	8	14
Red or rosé	3½	75
White	3½	70
Wine cooler	12	170

a deficiency of another B vitamin—thiamin. In fact, the cluster of thiamin-deficiency symptoms commonly seen in chronic **alcoholism** has its own name—the **Wernicke-Korsakoff syndrome.** This syndrome is characterized by paralysis of the eye muscles, poor muscle coordination, impaired memory, and damaged nerves; it and other alcohol-related memory problems may respond to thiamin supplements.[5]

Acetaldehyde, an intermediate in alcohol metabolism (review Figure H7-2 on p. 242), interferes with nutrient use, too. For example, acetaldehyde dislodges vitamin B_6 from its protective binding protein so that it is destroyed, causing a vitamin B_6 deficiency and, thereby, lowered production of red blood cells.

Malnutrition occurs not only because of lack of intake and altered metabolism, but because of direct toxic effects as well. Alcohol causes stomach cells to oversecrete both gastric acid and histamine, an immune system agent that produces inflammation. Beer in particular stimulates gastric acid secretion, irritating the linings of the stomach and esophagus and making them vulnerable to ulcer formation.

Overall, nutrient deficiencies are virtually inevitable in alcohol abuse, not only because alcohol displaces food but also because alcohol directly interferes with the body's use of nutrients, making them ineffective even if they are present. Intestinal cells fail to absorb B vitamins, notably, thiamin,

folate, and vitamin B_{12}. Liver cells lose efficiency in activating vitamin D. Cells in the retina of the eye, which normally process the alcohol form of vitamin A (retinol) to its aldehyde form needed in vision (retinal), find themselves processing ethanol to acetaldehyde instead. Likewise, the liver cannot convert the aldehyde form of vitamin A to its acid form (retinoic acid), which is needed to support the growth of its (and all) cells.[6]

Regardless of dietary intake, excessive drinking over a lifetime creates deficits of all the nutrients mentioned in this discussion and more. No diet can compensate for the damage caused by heavy alcohol consumption.

Alcohol's Short-Term Effects

The effects of abusing alcohol may be apparent immediately, or they may not become evident for years to come. Among the immediate consequences, all of the following involve alcohol use:[7]

- One-quarter of all emergency-room admissions.
- One-third of all suicides.
- One-half of all homicides.
- One-half of all domestic violence incidents.
- One-half of all traffic fatalities.
- One-half of all fire victim fatalities.

These statistics are sobering. The consequences of heavy drinking touch all elements of society—men and women, black and white, young and old, rich and poor. One group particularly hard hit by heavy drinking is college students—not because they are prone to alcoholism, but because they are living in an environment and during a developmental stage of life in which heavy drinking is considered acceptable.[8]

Heavy or binge drinking (defined as at least four drinks in a row for women and five drinks in a row for men) is widespread on college campuses and poses serious health and social consequences to drinkers and nondrinkers alike.*[9] In fact, binge drinking can kill: the respiratory center of the brain becomes anesthetized, and breathing stops. Acute alcohol intoxication can cause coronary artery spasms, leading to heart attacks.

Binge drinking is especially common among college students who live in a fraternity or sorority house, attend parties frequently, engage in other risky behaviors, and have a history of binge drinking in high school.[10] Compared with nondrinkers or moderate drinkers, people who frequently binge drink (at least three times within two weeks) are more likely to

HEALTHY PEOPLE 2010 Reduce the proportion of persons engaging in binge drinking of alcoholic beverages.

*This definition of binge drinking, without specification of time elapsed, is consistent with standard practice in alcohol research.

TABLE H7-4 Signs of Alcoholism

- Tolerance—the person needs higher and higher intakes of alcohol to achieve intoxication.
- Withdrawal—the person who stops drinking experiences anxiety, agitation, increased blood pressure, or seizures, or seeks alcohol to relieve these symptoms.
- Impaired control—the person intends to have 1 or 2 drinks, but has 9 or 10 instead, or the person tries to control or quit drinking, but fails.
- Disinterest—the person neglects important social, family, job, or school activities because of drinking.
- Time—the person spends a great deal of time obtaining and drinking alcohol or recovering from excessive drinking.
- Impaired ability—the person's intoxication or withdrawal symptoms interfere with work, school, or home.
- Problems—the person continues drinking despite physical hazards or medical, legal, psychological, family, employment, or school problems.

The presence of three or more of these conditions is required to make a diagnosis.

SOURCE: Adapted from *Diagnostic and Statistical Manual of Mental Disorders*, 4th ed. (Washington, D.C.: American Psychiatric Association, 1994).

engage in unprotected sex, have multiple sex partners, damage property, and assault others.[11] On average, *every day* alcohol is involved in the:[12]

- Death of 4 college students.
- Sexual assault of 192 college students.
- Injury of 1370 college students.
- Assault of 1644 college students.

Binge drinkers skew the statistics on college students' alcohol use. The median number of drinks consumed by college students is 1.5 per week, but for binge drinkers, it is 14.5. Nationally, only 20 percent of all students are frequent binge drinkers; yet they account for two-thirds of all the alcohol students report consuming and most of the alcohol-related problems.[13]

Binge drinking is not limited to college campuses, of course, but that environment seems most accepting of such behavior despite its problems. Social acceptance may make it difficult for binge drinkers to recognize themselves as problem drinkers. For this reason, interventions must focus both on educating individuals and on changing the campus social environment.[14] The damage alcohol causes only becomes worse if the pattern is not broken. Alcohol abuse sets in much more quickly in young people than in adults. Those who start drinking at an early age more often suffer from alcoholism than people who start later on. Table H7-4 lists the key signs of alcoholism.

Alcohol's Long-Term Effects

The most devastating long-term effect of alcohol is the damage done to a child whose mother abused alcohol during pregnancy. The effects of alcohol on the unborn, and the message that pregnant women should not drink alcohol, are presented in Highlight 15.

For nonpregnant adults, a drink or two sets in motion many destructive processes in the body, but the next day's ab-

stinence reverses them. As long as the doses are moderate, the time between them is ample, and nutrition is adequate, recovery is probably complete.

If the doses of alcohol are heavy and the time between them short, complete recovery cannot take place. Repeated onslaughts of alcohol gradually take a toll on all parts of the body (see Table H7-5). Compared with nondrinkers and moderate drinkers, heavy drinkers—especially those under age 35—have significantly greater risks of dying from all causes.[15]

Personal Strategies

One obvious option available to people attending social gatherings is to enjoy the conversation, eat the food, and drink nonalcoholic beverages. Several nonalcoholic beverages are available that mimic the look and taste of their alcoholic counterparts. For those who enjoy champagne or beer, sparkling ciders and beers without alcohol are available. Instead of drinking a cocktail, a person can sip tomato juice with a slice of lime and a stalk of celery or just a plain cola beverage. Any of these drinks can ease conversation.

The person who chooses to drink alcohol should sip each drink slowly with food. The alcohol should arrive at the liver cells slowly enough that the enzymes can handle the load. It is best to space drinks, too, allowing about an hour or so to metabolize each drink.

If you want to help sober up a friend who has had too much to drink, don't bother walking arm in arm around the block. Walking muscles have to work harder, but muscle cells can't metabolize alcohol; only liver cells can. Remember that each person has a limited amount of the alcohol dehydrogenase enzyme that clears the blood at a steady rate. Time alone will do the job.

Nor will it help to give your friend a cup of coffee. Caffeine is a stimulant, but it won't speed up alcohol metabolism. The police say ruefully, "If you give a drunk a cup of coffee, you'll just have a wide-awake drunk on your hands." Table H7-6 presents other alcohol myths.

TABLE H7-5   Health Effects of Heavy Alcohol Consumption

Health Problem	Effects of Alcohol
Arthritis	Increases the risk of inflamed joints.
Cancer	Increases the risk of cancer of the liver, pancreas, rectum, and breast; increases the risk of cancer of the mouth, pharynx, larynx, and esophagus, where alcohol interacts synergistically with tobacco.
Fetal alcohol syndrome	Causes physical and behavioral abnormalities in the fetus (see Highlight 15).
Heart disease	In heavy drinkers, raises blood pressure, blood lipids, and the risk of stroke and heart disease; when compared with those who abstain, heart disease risk is generally lower in light-to-moderate drinkers (see Chapter 18).
Hyperglycemia	Raises blood glucose.
Hypoglycemia	Lowers blood glucose, especially in people with diabetes.
Infertility	Increases the risks of menstrual disorders and spontaneous abortions (in women); suppresses luteinizing hormone (in women) and testosterone (in men).
Kidney disease	Enlarges the kidneys, alters hormone functions, and increases the risk of kidney failure.
Liver disease	Causes fatty liver, alcoholic hepatitis, and cirrhosis.
Malnutrition	Increases the risk of protein-energy malnutrition; low intakes of protein, calcium, iron, vitamin A, vitamin C, thiamin, vitamin B_6, and riboflavin; and impaired absorption of calcium, phosphorus, vitamin D, and zinc.
Nervous disorders	Causes neuropathy and dementia; impairs balance and memory.
Obesity	Increases energy intake, but is not a primary cause of obesity.
Psychological disturbances	Causes depression, anxiety, and insomnia.

NOTE: This list is by no means all-inclusive. Alcohol has direct toxic effects on all body systems.

TABLE H7-6   Myths and Truths concerning Alcohol

Myth: Hard liquors such as rum, vodka, and tequila are more harmful than wine and beer.
Truth: The damage caused by alcohol depends largely on the *amount* consumed. Compared with hard liquor, beer and wine have relatively low percentages of alcohol, but they are often consumed in larger quantities.

Myth: Consuming alcohol with raw seafood diminishes the likelihood of getting hepatitis.
Truth: People have eaten contaminated oysters while drinking alcoholic beverages and not gotten as sick as those who were not drinking. But do not be misled: hepatitis is too serious an illness for anyone to depend on alcohol for protection.

Myth: Alcohol stimulates the appetite.
Truth: For some people, alcohol may stimulate appetite, but it seems to have the opposite effect in heavy drinkers. Heavy drinkers tend to eat poorly and suffer malnutrition.

Myth: Drinking alcohol is healthy.
Truth: Moderate alcohol consumption is associated with a lower risk for heart disease (see Chapter 18 for more details). Higher intakes, however, raise the risks for high blood pressure, stroke, heart disease, some cancers, accidents, violence, suicide, birth defects, and deaths in general. Furthermore, excessive alcohol consumption damages the liver, pancreas, brain, and heart. No authority recommends that nondrinkers begin drinking alcoholic beverages to obtain health benefits.

Myth: Wine increases the body's absorption of minerals.
Truth: Wine may increase the body's absorption of potassium, calcium, phosphorus, magnesium, and zinc, but the alcohol in wine also promotes the body's excretion of these minerals, so no benefit is gained.

Myth: Alcohol is legal and, therefore, not a drug.
Truth: Alcohol is legal for adults 21 years old and older, but it is also a drug—a substance that alters one or more of the body's functions.

Myth: A shot of alcohol warms you up.
Truth: Alcohol diverts blood flow to the skin making you *feel* warmer, but it actually cools the body.

Myth: Wine and beer are mild; they do not lead to alcoholism.
Truth: Alcoholism is not related to the kind of beverage, but rather to the quantity and frequency of consumption.

Myth: Mixing different types of drinks gives you a hangover.
Truth: Too much alcohol in any form produces a hangover.

Myth: Alcohol is a stimulant.
Truth: People think alcohol is a stimulant because it seems to relieve inhibitions, but it does so by depressing the activity of the brain. Alcohol is medically defined as a depressant drug.

Myth: Beer is a great source of carbohydrate, vitamins, minerals, and fluids.
Truth: Beer does provide some carbohydrate, but most of its kcalories come from alcohol. The few vitamins and minerals in beer cannot compete with rich food sources. And the diuretic effect of alcohol causes the body to lose more fluid in urine than is provided by the beer.

People who have passed out from drinking need 24 hours to sober up completely. Let them sleep, but watch over them. Encourage them to lie on their sides, instead of their backs. That way, if they vomit, they won't choke.

Don't drive too soon after drinking. The lack of glucose for the brain's function and the length of time needed to clear the blood of alcohol make alcohol's adverse effects linger long after its blood concentration has fallen. Driving coordination is still impaired the morning *after* a night of drinking, even if the drinking was moderate. Responsible aircraft pilots know that they must allow 24 hours for their bodies to clear alcohol completely, and they refuse to fly any sooner. The Federal Aviation Administration and major airlines enforce this rule.

Reduce deaths and injuries caused by alcohol- and drug-related motor vehicle crashes.

HEALTHY PEOPLE 2010

Look again at the drawing of the brain in Figure H7-4 and note that when someone drinks, judgment fails first. Judgment might tell a person to limit alcohol consumption to two drinks at a party, but if the first drink takes judgment away, many more drinks may follow. The failure to stop drinking as planned, on repeated occasions, is a danger sign warning that the person should not drink at all. The accompanying Nutrition on the Net provides websites for organizations that offer information about alcohol and alcohol abuse.

Ethanol interferes with a multitude of chemical and hormonal reactions in the body—many more than have been enumerated here. With heavy alcohol consumption, the potential for harm is great. The best way to escape the harmful effects of alcohol is, of course, to refuse alcohol altogether. If you do drink alcoholic beverages, do so with care, and in moderation.

NUTRITION ON THE NET

 Access these websites for further study of topics covered in this highlight.

- Find updates and quick links to these and other nutrition-related sites at our website: **www.wadsworth.com/nutrition**

- Search for "alcohol" at the U.S. Government health site: **www.healthfinder.gov**

- Gather information on alcohol and drug abuse from the National Clearinghouse for Alcohol and Drug Information (NCADI): **www.health.org**

- Learn more about alcoholism and drug dependence from the National Council on Alcoholism and Drug Dependence (NCADD): **www.ncadd.org**

- Find help for a family alcohol problem from Alateen and Al-Anon Family support groups: **www.al-anon.alateen.org**

- Find help for an alcohol or drug problem from Alcoholics Anonymous (AA) or Narcotics Anonymous: **www.aa.org** or **www.wsoinc.com**

- Search for "party" to find tips for hosting a safe party from Mothers Against Drunk Driving (MADD): **www.madd.org**

REFERENCES

1. D. Krahn and coauthors, Alcohol use and cognition at mid-life: The importance of adjusting for baseline cognitive ability and educational attainment, *Alcoholism: Clinical and Experimental Research* 27 (2003): 1162–1166.

2. M. S. Westerterp-Plantenga and C. R. Verwegen, The appetizing effect of an aperitif in overweight and normal-weight humans, *American Journal of Clinical Nutrition* 69 (1999): 205–212.

3. S. G. Wannamethee and A. G. Shaper, Alcohol, body weight, and weight gain in middle-aged men, *American Journal of Clinical Nutrition* 77 (2003): 1312–1317.

4. C. S. Lieber, Alcohol: Its metabolism and interaction with nutrients, *Annual Review of Nutrition* 20 (2000): 395–430.

5. M. L. Ambrose, S. C. Bowden, and G. Whelan, Thiamin treatment and working memory function of alcohol-dependent people: Preliminary findings, *Alcoholism: Clinical and Experimental Research* 25 (2001): 112–116.

6. X.-D. Wang, Chronic alcohol intake interferes with retinoid metabolism and signaling, *Nutrition Reviews* 57 (1999): 51–59.

7. Position paper on drug policy: Physician Leadership on National Drug Policy (PLNDP), Brown University Center for Alcohol and Addiction Studies, 2000.

8. A. M. Brower, Are college students alcoholics? *Journal of American College Health* 50 (2002): 253–255.

9. H. Wechsler and coauthors, Trends in college binge drinking during a period of increased prevention efforts—Findings from Harvard School of Public Health College Alcohol Study Surveys: 1993–2001, *Journal of American College Health* 50 (2002): 203–217.

10. P. W. Meilman, J. S. Leichliter, and C. A. Presley, Greeks and athletes: Who drinks more? *Journal of American College Health* 47 (1999): 187–190; B. E. Borsari and K. B. Carey, Understanding fraternity drinking: Five recurring themes in the literature, 1980–1998, *Journal of American College Health* 48 (1999): 30–37.

11. Wechsler and coauthors, 2002.

12. R. W. Hingson and coauthors, Magnitude of alcohol-related mortality and morbidity among U.S. college students ages 18–24, *Journal of Studies on Alcohol* 63 (2002): 136–144; National Institute of Alcohol Abuse and Alcoholism, *A Call to Action: Changing the Culture of Drinking at U.S. Colleges*, 2002, available from **www. collegedrinkingprevention.gov**.

13. H. Wechsler and coauthors, College alcohol use: A full or empty glass? *Journal of American College Health* 47 (1999): 247–252.

14. A. Ziemelis, R. B. Bucknam, and A. M. Elfessi, Prevention efforts underlying decreases in binge drinking at institutions of higher learning, *Journal of American College Health* 50 (2002): 238–252.

15. I. R. White, D. R. Altmann, and K. Nanchahal, Alcohol consumption and mortality: Modeling risks for men and women at different ages, *British Medical Journal* 325 (2002): 191–197.

Energy Balance and Body Composition

Chapter Outline

Energy Balance

Energy In: The kCalories Foods Provide: *Food Composition • Food Intake*

Energy Out: The kCalories the Body Expends: *Components of Energy Expenditure • Estimating Energy Requirements*

Body Weight, Body Composition, and Health: *Defining Healthy Body Weight • Body Fat and Its Distribution • Health Risks Associated with Body Weight and Body Fat*

Highlight: *The Latest and Greatest Weight-Loss Diet—Again*

Nutrition Explorer CD-ROM Outline

Nutrition Animation: *Energy Balance and Body Composition*

Case Study: *Satiation and Appetite*

Student Practice Test

Glossary Terms

Nutrition on the Net

© Charles Thatcher/Stone/Getty Images

Nutrition in Your Life

It's a simple mathematical equation: energy in + energy out = energy balance. The reality, of course, is much more complex. One day you may devour a dozen doughnuts at midnight and sleep through your morning workout—tipping the scales toward weight gain. Another day you may snack on veggies and train for this weekend's 10K race—shifting the balance toward weight loss. Your body weight—especially as it relates to your body fat—and your level of fitness have consequences for your health. So, how are you doing? Are you ready to see how your "energy in" and "energy out" balance and whether your body weight and fat measures are consistent with good health?

The body's remarkable machinery can cope with many extremes of diet. As Chapter 7 explained, both excess carbohydrate (glucose) and excess protein (amino acids) can contribute to body fat. To some extent, amino acids can be used to make glucose. To a very limited extent, even fat (the glycerol portion) can be used to make glucose. But a grossly unbalanced diet imposes hardships on the body. If energy intake is too low or if too little carbohydrate or protein is supplied, the body must degrade its own lean tissue to meet its glucose and protein needs. If energy intake is too high, the body stores fat.

Both excessive and deficient body fat result from unbalanced energy budgets. The simple picture is as follows. People who have consumed more food energy than they have spent bank the surplus as body fat. To reduce body fat, they need to expend more energy than they take in from food. In contrast, people who have consumed too little food energy to support their bodies' activities have relied on their bodies' fat stores and possibly some of their lean tissues as well. To gain

When energy in balances with energy out, a person's body weight is stable.

251

weight, these people need to take in more food energy than they expend. As you will see, though, the details of the body's weight regulation are quite complex. This chapter describes energy balance and body composition and examines the health problems associated with having too much or too little body fat; the next chapter presents strategies toward resolving these problems.

Energy Balance

People expend energy continuously and eat periodically to refuel. Ideally, their energy intakes cover their energy expenditures without too much excess. Excess energy is stored as fat, and stored fat is used for energy between meals. The amount of body fat a person deposits in, or withdraws from, "storage" on any given day depends on the energy balance for that day—the amount consumed (energy in) versus the amount expended (energy out). When a person is maintaining weight, energy in equals energy out. When the balance shifts, weight changes. For each 3500 kcalories■ eaten in excess, a pound of body fat is stored; similarly, a pound of fat is lost for each 3500 kcalories expended beyond those consumed. The fat stores of even a healthy-weight adult represent an ample reserve of energy—50,000 to 200,000 kcalories.

Quick changes in body weight are not simple changes in fat stores. Weight gained or lost rapidly includes some fat, large amounts of fluid, and some lean tissues such as muscles and bone minerals. (Because water constitutes about 60 percent of an adult's body weight, retention or loss of water influences body weight significantly.) Even over the long term, the composition of weight gained or lost is normally about 75 percent fat and 25 percent lean. During starvation, losses of fat and lean are about equal. (Recall from Chapter 7 that without adequate carbohydrate, protein-rich lean tissues break down to provide glucose.) Invariably, though, *fat* gains and losses are gradual. The next two sections examine the two sides of the energy-balance equation: energy in and energy out.

■ 1 lb body fat = 3500 kcal.
Body fat, or adipose tissue, is composed of a mixture of mostly fat, some protein, and water. A pound of body fat (454 g) is approximately 87% fat, or (454 × 0.87) 395 g, and 395 g × 9 kcal/g = 3555 kcal.

> **IN SUMMARY** When the energy consumed equals the energy expended, a person is in energy balance and body weight is stable. If more energy is taken in than is expended, a person gains weight. If more energy is spent than is taken in, a person loses weight.

Energy In: The kCalories Foods Provide

Foods and beverages provide the "energy in" part of the energy-balance equation. How much energy a person receives depends on the composition of the foods and beverages and on the amount the person eats and drinks.

Food Composition

To find out how many kcalories a food provides, a scientist can burn the food in a **bomb calorimeter** (see Figure 8-1). When the food burns, the chemical bonds between the carbon and hydrogen atoms break, releasing energy in the form of heat. The amount of heat given off provides a *direct* measure of the food's energy value (remember that kcalories are units of heat energy). In addition to releasing heat, these reactions generate carbon dioxide and water—just as the body's cells do when they metabolize the energy-yielding nutrients. When the food burns and

FIGURE 8-1 Bomb Calorimeter

When food is burned, the chemical bonds between the carbons and hydrogens are broken, and energy is released in the form of heat. The amount of heat generated provides a direct measure of the amount of energy stored in the food's chemical bonds.

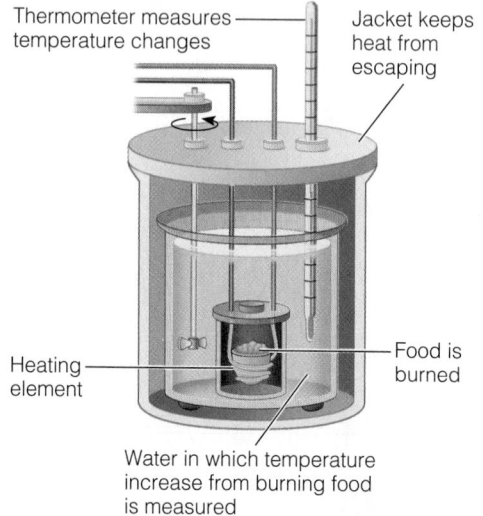

Thermometer measures temperature changes

Jacket keeps heat from escaping

Heating element

Food is burned

Water in which temperature increase from burning food is measured

---

bomb calorimeter (KAL-oh-RIM-eh-ter): an instrument that measures the heat energy released when foods are burned, thus providing an estimate of the potential energy of the foods.
• **calor** = heat
• **metron** = measure

the chemical bonds break, the carbons (C) and hydrogens (H) combine with oxygen (O) to form carbon dioxide (CO_2) and water (H_2O). The amount of oxygen consumed gives an *indirect* measure■ of the amount of energy released.

A bomb calorimeter measures the available energy in foods, but overstates the amount of energy that the human body■ derives from foods. The body is less efficient than a calorimeter and cannot metabolize all of a food's energy-yielding nutrients all the way to carbon dioxide and water. Researchers can correct for this discrepancy mathematically to create useful tables of the energy values of foods (such as Appendix H). These values provide reasonable estimates, but do not reflect the *precise* amount of energy a person will derive from the foods consumed.

The energy values of foods can also be computed from the amounts of carbohydrate, fat, and protein (and alcohol, if present) in the foods.[*] For example, a food■ containing 12 grams of carbohydrate, 5 grams of fat, and 8 grams of protein would provide 48 carbohydrate kcalories, 45 fat kcalories, and 32 protein kcalories, for a total of 125 kcalories. (To review how to calculate the energy available from foods, turn to p. 9.)

Food Intake

To achieve energy balance, the body must meet its needs without taking in too much or too little energy. Somehow the body decides how much and how often to eat—when to start eating and when to stop. As you will see, many signals initiate or delay eating.

Hunger People eat for a variety of reasons, most obviously (although not necessarily most commonly) because they are hungry. Most people recognize **hunger** as an irritating feeling that prompts thoughts of food and motivates them to start eating. In the body, hunger is the physiological response to a need for food triggered by chemical messengers originating and acting in the brain, primarily in the **hypothalamus.**[1] Hunger can be influenced by the presence or absence of nutrients in the bloodstream, the size and composition of the preceding meal, customary eating patterns, climate (heat reduces food intake; cold increases it), exercise, hormones, and physical and mental illnesses.

The stomach is ideally designed to handle periodic batches of food, and people typically eat meals at roughly four-hour intervals. Four hours after a meal, most, if not all, of the food has left the stomach. Most people do not feel like eating again until the stomach is either empty or almost so. Even then, a person may not feel hungry for quite a while.

Appetite Hunger is only one of the signals determining whether a person will eat. **Appetite** also initiates eating. A person may experience appetite without hunger, for example, when presented with a hot piece of apple pie after having eaten a large dinner. The sight and smell of the pie, not hunger, trigger appetite. In contrast, a person may feel hungry but have no appetite for food when faced with unfamiliar foods, a stressful situation, or illness; in such circumstances, eating becomes a chore.

Satiation During the course of a meal, as food enters the GI tract and hunger diminishes, **satiation** develops. Receptors in the stomach stretch, and the person begins to feel full. The response: satiation occurs and the person stops eating.

Satiety After a meal, the feeling of **satiety** continues to suppress hunger and allows a person to not eat again for a while. Whereas *satiation* tells us to "stop eating," *satiety* reminds us to "not start eating again." Figure 8-2 summarizes the relationships among hunger, satiation, and satiety. Of course, people can override these signals, especially when presented with stressful situations or favorite foods.

■ Food energy values can be determined by:
  - **Direct calorimetry,** which measures the amount of heat released.
  - **Indirect calorimetry,** which measures the amount of oxygen consumed.

■ The number of kcalories that the body derives from a food, as contrasted with the number of kcalories determined by calorimetry, is the **physiological fuel value.**

■ Reminder:
  - 1 g carbohydrate = 4 kcal.
  - 1 g fat = 9 kcal.
  - 1 g protein = 4 kcal.
  - 1 g alcohol = 7 kcal.

As Chapter 1 mentioned, many scientists measure food energy in kilojoules instead. Conversion factors for these and other measures are in the Aids to Calculation section on the last two pages of the book.

hunger: the painful sensation caused by a lack of food that initiates food-seeking behavior.

hypothalamus (high-po-THAL-ah-mus): a brain center that controls activities such as maintenance of water balance, regulation of body temperature, and control of appetite.

appetite: the integrated response to the sight, smell, thought, or taste of food that initiates or delays eating.

satiation (say-she-AY-shun): the feeling of satisfaction and fullness that occurs during a meal and halts eating. Satiation determines how much food is consumed during a meal.

satiety (sah-TIE-eh-tee): the feeling of satisfaction that occurs after a meal and inhibits eating until the next meal. Satiety determines how much time passes between meals.

---

[*]Some of the food energy values in the table of food composition in Appendix H were derived by bomb calorimetry, and many were calculated from their energy-yielding nutrient contents.

FIGURE 8-2 Hunger, Satiation, and Satiety

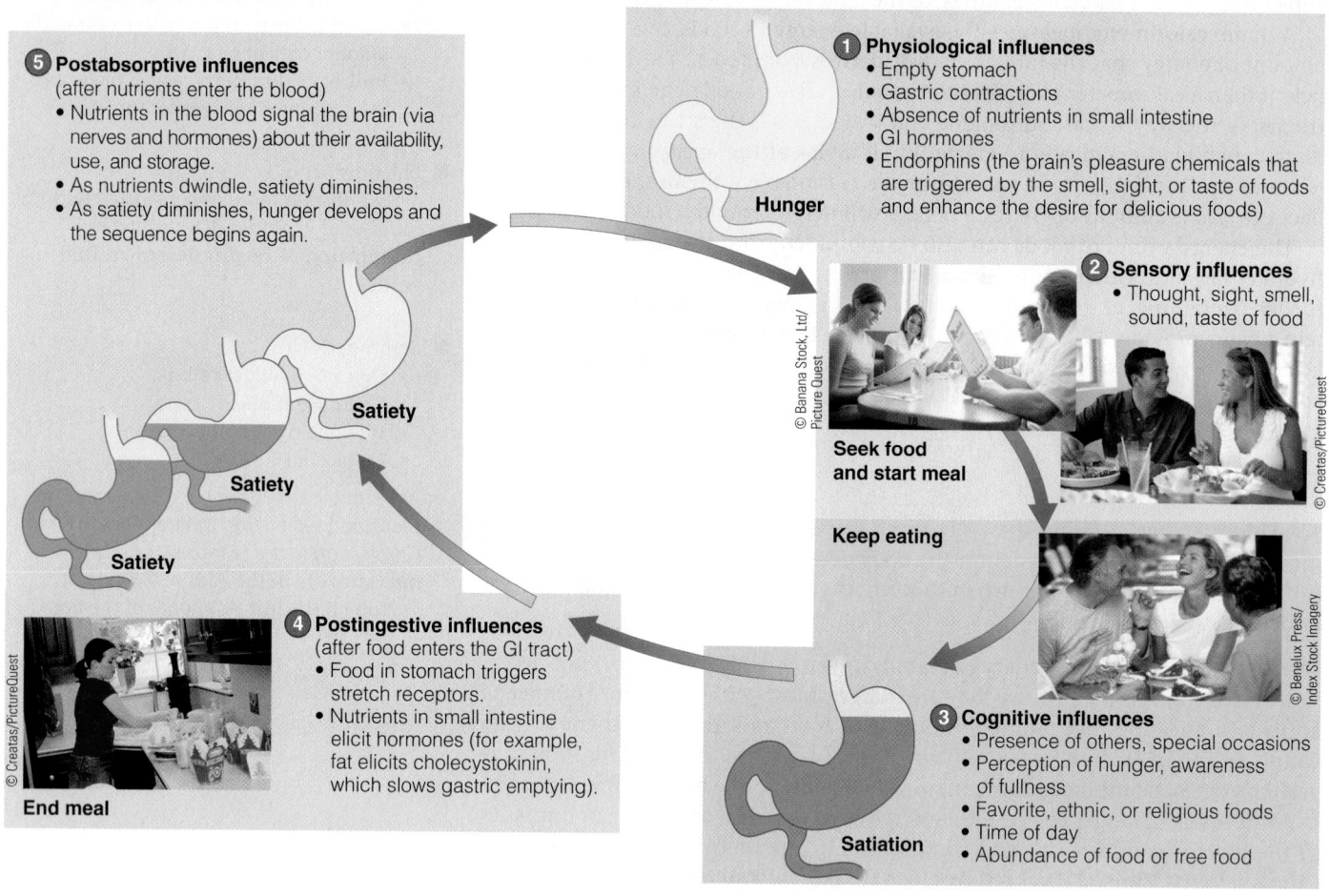

⑤ Postabsorptive influences
(after nutrients enter the blood)
• Nutrients in the blood signal the brain (via nerves and hormones) about their availability, use, and storage.
• As nutrients dwindle, satiety diminishes.
• As satiety diminishes, hunger develops and the sequence begins again.

Satiety
Satiety
Satiety

④ Postingestive influences
(after food enters the GI tract)
• Food in stomach triggers stretch receptors.
• Nutrients in small intestine elicit hormones (for example, fat elicits cholecystokinin, which slows gastric emptying).

End meal

① Physiological influences
• Empty stomach
• Gastric contractions
• Absence of nutrients in small intestine
• GI hormones
• Endorphins (the brain's pleasure chemicals that are triggered by the smell, sight, or taste of foods and enhance the desire for delicious foods)

Hunger

② Sensory influences
• Thought, sight, smell, sound, taste of food

Seek food and start meal

Keep eating

③ Cognitive influences
• Presence of others, special occasions
• Perception of hunger, awareness of fullness
• Favorite, ethnic, or religious foods
• Time of day
• Abundance of food or free food

Satiation

■ Eating in response to arousal is called **stress eating**.

■ Cognitive influences include perceptions, memories, intellect, and social interactions.

satiating: having the power to suppress hunger and inhibit eating.

Overriding Hunger and Satiety Not surprisingly, eating can be triggered by signals other than hunger, even when the body does not need food. Some people experience food cravings when they are bored or anxious. In fact, they may eat in response to any kind of stress,■ negative or positive. ("What do I do when I'm grieving? Eat. What do I do when I'm celebrating? Eat!") Some people respond to external cues such as the time of day ("It's time to eat") or the availability, sight, and taste of food ("I'd love a piece of chocolate even though I'm stuffed"). Being presented with large portion sizes, favorite foods, or an abundance or variety of foods also stimulates eating and increases energy intake.[2] These cognitive influences■ can easily lead to weight gain.

Eating can also be suppressed by signals other than satiety, even when a person is hungry. People with the eating disorder anorexia nervosa, for example, use tremendous discipline to ignore the pangs of hunger. Some people simply cannot eat during times of stress, negative or positive. ("I'm too sad to eat. I'm too excited to eat!") Why some people overeat in response to stress and others cannot eat at all remains a bit of a mystery. Factors that appear to be involved include how the person perceives the stress and whether usual eating behaviors are restrained. (Highlight 9 features anorexia nervosa and other eating disorders.)

Sustaining Satiation and Satiety The extent to which foods produce satiation and sustain satiety depends in part on the nutrient composition of a meal.[3] Of the three energy-yielding nutrients, protein is the most **satiating**. Foods rich in complex carbohydrates and fibers also effectively provide satiation by filling the stomach and delaying the absorption of nutrients. In contrast, fat has a weak effect on satiation; consequently, eating high-fat foods may lead to passive overconsumption. High-fat foods are flavorful, which stimulates the appetite and entices people to eat more. High-fat foods are also energy dense; consequently, they deliver more kcalories per bite. (Chapter 1 introduced the concept of energy density, and Chapter 9 describes how considering a food's energy density can help with weight management.) Although fat provides little satiation during a meal, it produces strong satiety signals once it enters the intestine. Fat in the intestine triggers the release of cholecystokinin—a hormone that signals satiety and inhibits food intake.[4]

Eating high-fat foods while trying to limit energy intake requires small portion sizes, which can leave a person feeling unsatisfied. By comparison, simple, whole foods such as potatoes, apples, oranges, whole-grain pastas, fish, and steak are highly satiating—and they provide a rich array of nutrients. Instead of feeling deprived eating small portions of high-fat foods, a person can feel satisfied eating large portions of high-protein and high-fiber foods. Portion size correlates directly with a food's satiety.[5] Figure 8-3 illustrates how fat influences portion size.

Message Central—The Hypothalamus As you can see, eating is a complex behavior controlled by a variety of psychological, social, metabolic, and physiological factors. The hypothalamus appears to be the control center, integrating messages about energy intake, expenditure, and storage from other parts of the brain and from the mouth, GI tract, and liver. Some of these messages influence satiation, controlling the size of a meal; others influence satiety, determining the frequency of meals.

Dozens of chemicals in the brain participate in appetite control and energy balance.[6] By understanding the action of these brain chemicals, researchers may one day be able to control appetite. The greatest challenge now is in sorting out the many actions of these brain chemicals. For example, one of these chemicals, **neuropeptide Y,** causes carbohydrate cravings, initiates eating, decreases energy expenditure, and increases fat storage—all factors favoring a positive energy balance and weight gain.

Regardless of hunger, people typically overeat when offered the abundance and variety of an "all you can eat" buffet.

neuropeptide Y: a chemical produced in the brain that stimulates appetite, diminishes energy expenditure, and increases fat storage.

FIGURE 8-3   How Fat Influences Portion Sizes

837 kcal
71 g fat

55 kcal
3 g fat

For the same size portion, peanuts deliver more than 15 times the kcalories and 20 times the fat of popcorn.

100 kcal
9 g fat

100 kcal
5 g fat

For the same number of kcalories, a person can have a few high-fat peanuts or almost 2 cups of high-fiber popcorn. (This comparison used oil-based popcorn; using air-popped popcorn would double the amount of popcorn in this example.)

■ Energy expenditure, like food energy, can be determined by:
- **Direct calorimetry**, which measures the amount of heat released.
- **Indirect calorimetry**, which measures the amount of oxygen consumed and carbon dioxide expelled.

FIGURE 8-4 Components of Energy Expenditure

The amount of energy spent in a day differs for each individual, but in general, basal metabolism is the largest component of energy expenditure (60 to 65 percent), and the thermic effect of food is the smallest (only 10 percent). The amount spent in voluntary physical activities has the greatest variability, depending on a person's activity patterns. For a sedentary person, physical activities may account for less than half as much energy as basal metabolism, whereas an extremely active person may expend as much on activity as for basal metabolism.

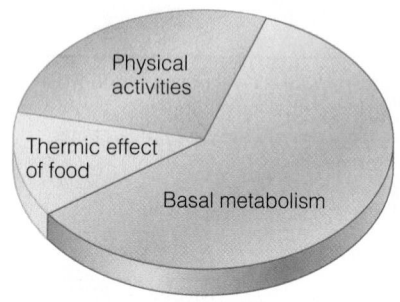

■ Quick and easy estimates for basal energy needs:
- Men: Slightly >1 kcal/min (1.1 to 1.3 kcal/min) or 24 kcal/kg/day.
- Women: Slightly <1 kcal/min (0.8 to 1.0 kcal/min) or 23 kcal/kg/day.

For perspective, a burning candle or a 75-watt light bulb releases about 1 kcal/min.

thermogenesis: the generation of heat; used in physiology and nutrition studies as an index of how much energy the body is expending.

basal metabolism: the energy needed to maintain life when a body is at complete digestive, physical, and emotional rest.

IN SUMMARY A mixture of signals governs a person's eating behaviors. Hunger and appetite initiate eating, whereas satiation and satiety stop and delay eating. Each responds to messages from the nervous and hormonal systems. Superimposed on these are complex factors involving emotions, habits, and other aspects of human behavior.

Energy Out: The kCalories the Body Expends

Chapter 7 explained that heat is released whenever the body breaks down carbohydrate, fat, or protein for energy and again when that energy is used to do work. The work itself, as it is done, generates heat as well. The body's generation of heat is known as **thermogenesis,** and it can be measured to determine the amount of energy expended.■ The total energy a body expends reflects three main categories of thermogenesis:
- Energy expended for basal metabolism.
- Energy expended for physical activity.
- Energy expended for food consumption.

A fourth category is sometimes involved:
- Energy expended for adaptation.

Components of Energy Expenditure

People expend energy when they are physically active, of course, but they also expend energy when they are resting quietly. In fact, quiet metabolic activities account for the lion's share of most people's energy expenditures, as Figure 8-4 shows.

Basal Metabolism About two-thirds of the energy the average person expends in a day supports the body's **basal metabolism.** Metabolic activities maintain the body temperature, keep the lungs inhaling and exhaling air, the bone marrow making new red blood cells, the heart beating 100,000 times a day, and the kidneys filtering wastes—in short, they support all the basic processes of life.

The **basal metabolic rate (BMR)** is the rate at which the body expends energy for these maintenance activities.■ The rate may vary dramatically from person to person and may vary for the same individual with a change in circumstance or physical condition. The rate is slowest when a person is sleeping undisturbed, but it is usually measured in a room with a comfortable temperature when the person is awake, but lying still, after a restful sleep and an overnight (12- to 14-hour) fast. A similar measure of energy output—called the **resting metabolic rate (RMR)**—is slightly higher than the BMR because its criteria for recent food intake and physical activity are not as strict.

In general, the more a person weighs, the more *total* energy is expended on basal metabolism, but the amount of energy *per pound* of body weight may be lower. For example, an adult's BMR might be 1500 kcalories per day and an infant's only 500, but compared to body weight, the infant's BMR is more than twice as fast. Similarly, a normal-weight adult may have a metabolic rate one and a half times that of an obese adult when compared to body weight because lean tissue is metabolically more active than body fat.

Table 8-1 summarizes the factors that raise and lower the BMR. For the most part, the BMR is highest in people who are growing (children and pregnant women) and in those with considerable **lean body mass** (physically fit people and males). One way to increase the BMR then is to participate in endurance and strength-training activities regularly to maximize lean body mass.[7] The BMR is also high in people with fever or under stress and in people with highly active thyroid glands. The BMR slows down with a loss of lean body mass and during fasting and malnutrition.

Physical Activity The second component of a person's energy output is physical activity: voluntary movement of the skeletal muscles and support systems. Physical activity is the most variable—and the most changeable—component of energy expenditure. Consequently, its influence on both weight gain and weight loss can be significant.

During physical activity, the muscles need extra energy to move, and the heart and lungs need extra energy to deliver nutrients and oxygen and dispose of wastes. The amount of energy needed for any activity, whether playing tennis or studying for an exam, depends on three factors: muscle mass, body weight, and activity. The larger the muscle mass required and the heavier the weight of the body part being moved, the more energy is spent. Table 8-2 (on p. 258) gives average energy expenditures for people of different body weights engaged in various activities and shows that a heavy person usually uses more energy per minute to perform a task than a light person does. The activity's duration, frequency, and intensity also influence energy expenditure: the longer, the more frequent, and the more intense the activity, the more kcalories spent. (Chapter 14 describes how an activity's duration, frequency, and intensity also influence the body's use of the energy-yielding nutrients.)

Thermic Effect of Food When a person eats, the GI tract muscles speed up their rhythmic contractions, the cells that manufacture and secrete digestive juices begin their tasks, and some nutrients are absorbed by active transport. This acceleration

It feels like work and it may make you tired, but studying requires only a kcalorie or two per minute.

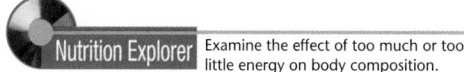

Examine the effect of too much or too little energy on body composition.

TABLE 8-1	Factors That Affect the BMR
Factor	**Effect on BMR**
Age	Lean body mass diminishes with age, slowing the BMR.[a]
Height	In tall, thin people, the BMR is higher.[b]
Growth	In children and pregnant women, the BMR is higher.
Body composition (gender)	The more lean tissue, the higher the BMR (which is why males usually have a higher BMR than females). The more fat tissue, the lower the BMR.
Fever	Fever raises the BMR.[c]
Stresses	Stresses (including many diseases and certain drugs) raise the BMR.
Environmental temperature	Both heat and cold raise the BMR.
Fasting/starvation	Fasting/starvation lowers the BMR.[d]
Malnutrition	Malnutrition lowers the BMR.
Hormones (gender)	The thyroid hormone thyroxin, for example, can speed up or slow down the BMR.[e] Premenstrual hormones slightly raise the BMR.
Smoking	Nicotine increases energy expenditure.
Caffeine	Caffeine increases energy expenditure.
Sleep	BMR is lowest when sleeping.

[a]The BMR begins to decrease in early adulthood (after growth and development cease) at a rate of about 2 percent/decade. A reduction in voluntary activity as well brings the total decline in energy expenditure to 5 percent/decade.
[b]If two people weigh the same, the taller, thinner person will have the faster metabolic rate, reflecting the greater skin surface, through which heat is lost by radiation, in proportion to the body's volume (see the margin drawing on p. 259).
[c]Fever raises the BMR by 7 percent for each degree Fahrenheit.
[d]Prolonged starvation reduces the total amount of metabolically active lean tissue in the body, although the decline occurs sooner and to a greater extent than body losses alone can explain. More likely, the neural and hormonal changes that accompany fasting are responsible for changes in the BMR.
[e]The thyroid gland releases hormones that travel to the cells and influence cellular metabolism. Thyroid hormone activity can speed up or slow down the rate of metabolism by as much as 50 percent.

basal metabolic rate (BMR): the rate of energy use for metabolism under specified conditions: after a 12-hour fast and restful sleep, without any physical activity or emotional excitement, and in a comfortable setting. It is usually expressed as kcalories per kilogram body weight per hour.

resting metabolic rate (RMR): similar to the BMR, a measure of the energy use of a person at rest in a comfortable setting, but with less stringent criteria for recent food intake and physical activity. Consequently, the RMR is slightly higher than the BMR.

lean body mass: the weight of the body minus the fat content.

TABLE 8-2   Energy Spent on Various Activities

The values listed in this table reflect both the energy spent in physical activity *and* the amount used for BMR.

Activity	kCal/lb/min[a]	kCalories per Minute at Different Body Weights				
		110 lb	125 lb	150 lb	175 lb	200 lb
Aerobic dance (vigorous)	.062	6.8	7.8	9.3	10.9	12.4
Basketball (vigorous, full court)	.097	10.7	12.1	14.6	17.0	19.4
Bicycling						
13 mph	.045	5.0	5.6	6.8	7.9	9.0
15 mph	.049	5.4	6.1	7.4	8.6	9.8
17 mph	.057	6.3	7.1	8.6	10.0	11.4
19 mph	.076	8.4	9.5	11.4	13.3	15.2
21 mph	.090	9.9	11.3	13.5	15.8	18.0
23 mph	.109	12.0	13.6	16.4	19.0	21.8
25 mph	.139	15.3	17.4	20.9	24.3	27.8
Canoeing, flat water, moderate pace	.045	5.0	5.6	6.8	7.9	9.0
Cross-country skiing						
8 mph	.104	11.4	13.0	15.6	18.2	20.8
Golf (carrying clubs)	.045	5.0	5.6	6.8	7.9	9.0
Handball	.078	8.6	9.8	11.7	13.7	15.6
Horseback riding (trot)	.052	5.7	6.5	7.8	9.1	10.4
Rowing (vigorous)	.097	10.7	12.1	14.6	17.0	19.4
Running						
5 mph	.061	6.7	7.6	9.2	10.7	12.2
6 mph	.074	8.1	9.2	11.1	13.0	14.8
7.5 mph	.094	10.3	11.8	14.1	16.4	18.8
9 mph	.103	11.3	12.9	15.5	18.0	20.6
10 mph	.114	12.5	14.3	17.1	20.0	22.9
11 mph	.131	14.4	16.4	19.7	22.9	26.2
Soccer (vigorous)	.097	10.7	12.1	14.6	17.0	19.4
Studying	.011	1.2	1.4	1.7	1.9	2.2
Swimming						
20 yd/min	.032	3.5	4.0	4.8	5.6	6.4
45 yd/min	.058	6.4	7.3	8.7	10.2	11.6
50 yd/min	.070	7.7	8.8	10.5	12.3	14.0
Table tennis (skilled)	.045	5.0	5.6	6.8	7.9	9.0
Tennis (beginner)	.032	3.5	4.0	4.8	5.6	6.4
Walking (brisk pace)						
3.5 mph	.035	3.9	4.4	5.2	6.1	7.0
4.5 mph	.048	5.3	6.0	7.2	8.4	9.6
Weight lifting						
light-to-moderate effort	.024	2.6	3.0	3.6	4.2	4.8
vigorous effort	.048	5.2	6.0	7.2	8.4	9.6
Wheelchair basketball	.084	9.2	10.5	12.6	14.7	16.8
Wheeling self in wheelchair	.030	3.3	3.8	4.5	5.3	6.0

[a]To calculate kcalories spent per minute of activity for your own body weight, multiply kcal/lb/min by your exact weight and then multiply that number by the number of minutes spent in the activity. For example, if you weigh 142 pounds, and you want to know how many kcalories you spent doing 30 minutes of vigorous aerobic dance: 0.062 × 142 = 8.8 kcalories per minute; 8.8 × 30 (minutes) = 264 total kcalories spent.

of activity requires energy and produces heat; it is known as the **thermic effect of food (TEF).**

The thermic effect of food is proportional to the food energy taken in and is usually estimated at 10 percent of energy intake. Thus a person who ingests 2000 kcalories probably expends about 200 kcalories on the thermic effect of food. The proportions vary for different foods, however, and are also influenced by factors such as meal size and frequency. In general, the thermic effect of food is greater for high-protein foods than for high-fat foods■ and for a meal eaten all at once rather than spread out over a couple of hours. Some research suggests that the thermic effect of food is reduced in obese people and may contribute to their efficient storage of fat.[8] For most purposes, however, the thermic effect of food can be ignored when estimating energy expenditure because its contribution to total energy output is smaller than the probable errors involved in estimating overall energy intake and output.

Adaptive Thermogenesis Some additional energy is spent when a person must adapt to dramatically changed circumstances **(adaptive thermogenesis).** When the body has to adapt to physical conditioning, extreme cold, overfeeding, starvation, trauma, or other types of stress, it has extra work to do, building the tissues and producing the enzymes and hormones necessary to cope with the demand. In some circumstances, this energy makes a considerable difference in the total energy spent. Because this component of energy expenditure is so variable and specific to individuals, it is not included when calculating energy requirements.

Estimating Energy Requirements

In estimating energy requirements, the DRI Committee developed equations that consider how the following factors influence energy expenditure:■

- *Gender.* In general, women have a lower BMR than men, in large part because men typically have more lean body mass. In addition, menstrual hormones influence the BMR in women, raising it just prior to menstruation. Two sets of energy equations—one for men and one for women—were developed to accommodate the influence of gender on energy expenditure.

- *Growth.* The BMR is high in people who are growing. For this reason, pregnant women and children have their own sets of energy equations.

- *Age.* The BMR declines during adulthood as lean body mass diminishes.[9] This change in body composition occurs, in part, because some hormones that influence metabolism become more, or less, active with age. Physical activities tend to decline as well, bringing the average reduction in energy expenditure to about 5 percent per decade. The decline in the BMR that occurs when a person becomes less active reflects the loss of lean body mass and may be prevented with ongoing physical activity. Because age influences energy expenditure, it is also factored into the energy equations.

- *Physical activity.* Using individual values for various physical activities (as in Table 8-2) is time-consuming and impractical for estimating the energy needs of a population. Instead, various activities are clustered according to the typical intensity of a day's efforts. Energy equations include a physical activity factor for various levels of intensity for each gender.

- *Body composition and body size.* The BMR is high in people who are tall and so have a large surface area.■ Similarly, the more a person weighs, the more energy is expended on basal metabolism. For these reasons, the energy equations include a factor for both height and weight.

As just explained, energy needs vary between individuals depending on such factors as gender, growth, age, physical activity, and body composition. Even when two people are similarly matched, however, their energy needs will still differ because of genetic differences. Perhaps one day genetic research will reveal how to estimate

■ Thermic effect of foods:
- Carbohydrate: 5–10%.
- Fat: 0–5%.
- Protein: 20–30%.
- Alcohol: 20%.

The percentages are calculated by dividing the energy expended during digestion and absorption (above basal) by the energy content of the food.

■ Note that Table 8-1 (p. 257) listed these factors among those that influence BMR and consequently energy expenditure.

■ Each of these structures is made of 8 blocks. They weigh the same, but they are arranged differently. The short, wide structure has 24 sides and the tall, thin one has 34. Because the tall, thin structure has a greater surface area, it will lose more heat (expend more energy) than the short, wide one. Similarly, two people of different heights might weigh the same, but the taller, thin one will have a higher BMR (expending more energy) because of the greater skin surface.

thermic effect of food (TEF): an estimation of the energy required to process food (digest, absorb, transport, metabolize, and store ingested nutrients); also called the **specific dynamic effect (SDE)** of food or the **specific dynamic activity (SDA)** of food. The sum of the TEF and any increase in the metabolic rate due to overeating is known as **diet-induced thermogenesis (DIT).**

adaptive thermogenesis: adjustments in energy expenditure related to changes in environment such as extreme cold and to physiological events such as overfeeding, trauma, and changes in hormone status.

HOW TO Estimate Energy Requirements

To determine your estimated energy requirements (EER), use the appropriate equation, inserting your age in years, weight (wt) in kilograms, height (ht) in meters, and physical activity (PA) factor from the accompanying table. (To convert pounds to kilograms, divide by 2.2; to convert inches to meters, divide by 39.37.)

- For men 19 years and older:
  $$EER = 662 - 9.53 \times age + PA \times [(15.91 \times wt) + (539.6 \times ht)].$$
- For women 19 years and older:
  $$EER = 354 - 6.91 \times age + PA \times [(9.36 \times wt) + (726 \times ht)].$$

For example, consider an active 30-year-old male who is 5 feet 11 inches tall and weighs 178 pounds. First, he converts his weight from pounds to kilograms and his height from inches to meters, if necessary:

$$178\ lb \div 2.2 = 80.9\ kg.$$
$$71\ in \div 39.37 = 1.8\ m.$$

Next, he considers his level of daily physical activity and selects the appropriate PA factor from the accompanying table (in this example, 1.25 for an active male). Then, he inserts his age, PA factor, weight, and height into the appropriate equation:

$$EER = 662 - 9.53 \times 30 + 1.25 \times [(15.91 \times 80.9) + (539.6 \times 1.8)].$$

(A reminder: do calculations within the parentheses first, and multiplication before addition and subtraction.) He calculates:

$$EER = 662 - 9.53 \times 30 + 1.25 \times (1287 + 971).$$
$$EER = 662 - 9.53 \times 30 + 1.25 \times 2258.$$
$$EER = 662 - 286 + 2823.$$
$$EER = 3199.$$

The estimated energy requirement for an active 30-year-old male who is 5 feet 11 inches tall and weighs 178 pounds is about 3200 kcalories/day. His actual requirement probably falls within a range■ of 200 kcalories above and below this estimate.

Physical Activity (PA) Factors for EER Equations

	Men	Women	Physical Activity
Sedentary	1.0	1.0	Only those physical activities required for normal independent living
			For an average-weight person, activities equivalent to walking at a pace of 2–4 mph for the following distances:
Low active	1.11	1.12	1.5 to 3.0 miles/day
Active	1.25	1.27	3 to 10 miles/day
Very active	1.48	1.45	10 or more miles/day

■ For *most* people, the actual energy requirement falls within these ranges:
- For men, EER ± 200 kcal.
- For women, EER ± 160 kcal.

For *almost all* people, the actual energy requirement falls within these ranges:
- For men, EER ± 400 kcal.
- For women, EER ± 320 kcal.

■ Appendix F presents tables that provide a shortcut to estimating total energy expenditure and instructions to help you determine the appropriate physical activity factor to use in the equation.

requirements for each individual. For now, the accompanying "How to" provides instructions on calculating your estimated energy requirements using the DRI equations and physical activity factors.■

IN SUMMARY A person in energy balance takes in energy from food and expends most of it on basal metabolic activities, some of it on physical activities, and a little on the thermic effect of food. Because energy requirements vary from person to person, such factors as gender, age, weight, and height as well as the intensity and duration of physical activity must be considered when estimating energy requirements.

Body Weight, Body Composition, and Health

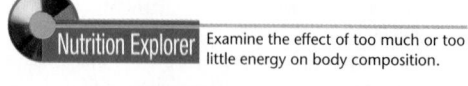

Nutrition Explorer Examine the effect of too much or too little energy on body composition.

A person 5 feet 10 inches tall who weighs 150 pounds may carry only about 30 of those pounds as fat. The rest is mostly water and lean tissues—muscles, organs such as the heart and liver, and the bones of the skeleton. Direct measures of

body composition are impossible in living human beings; instead, researchers assess body composition indirectly based on the following assumption:

Body weight = fat + lean tissue (including water).

Weight gains and losses tell us nothing about how the body's composition may have changed, yet weight is the measure most people use to judge their "fatness." For many people, overweight means overfat, but this is not always the case. Athletes with dense bones and well-developed muscles may be overweight by some standards, but have little body fat. Conversely, inactive people may seem to have acceptable weights, when, in fact, they may have too much body fat.

Defining Healthy Body Weight

How much should a person weigh? How can a person know if her weight is appropriate for her height? How can a person know if his weight is jeopardizing his health? Such questions seem so simple, yet the answers can be complex—and quite different depending on whom you ask.

The Criterion of Fashion In asking what is ideal, people often mistakenly turn to fashion for the answer. No doubt our society sets unrealistic ideals for body weight, especially for women. Miss America, our nation's icon of beauty, has never been overweight, and she has grown progressively thinner over the years (see Figure 8-5).[10] Magazines, movies, and television all convey the message that to be thin is to be beautiful and happy. As a result, the media have a great influence on the weight concerns and dieting patterns of people of all ages, but most tragically on young, impressionable children and adolescents.[11] Even five-year-olds are concerned about their body weight.[12] One-half of preteen girls and one-third of preteen boys are dissatisfied with their body weight and shape.[13]

Importantly, perceived body image has little to do with actual body weight or size. People of all shapes, sizes, and ages—including extremely thin fashion models with anorexia nervosa and fitness instructors with ideal body composition—have learned to be unhappy with their "overweight" bodies. Such dissatisfaction can lead to damaging behaviors, such as starvation diets, diet pill abuse, and health care avoidance.[14] The first step toward making healthy changes may be self-acceptance.

At 6 feet 3 inches tall and 245 pounds, Mike O'Hearn would be considered over*weight* by most standards. Yet he is clearly not over*fat*.

---

body composition: the proportions of muscle, bone, fat, and other tissue that make up a person's total body weight.

FIGURE 8-5   The Declining Weight of Miss America

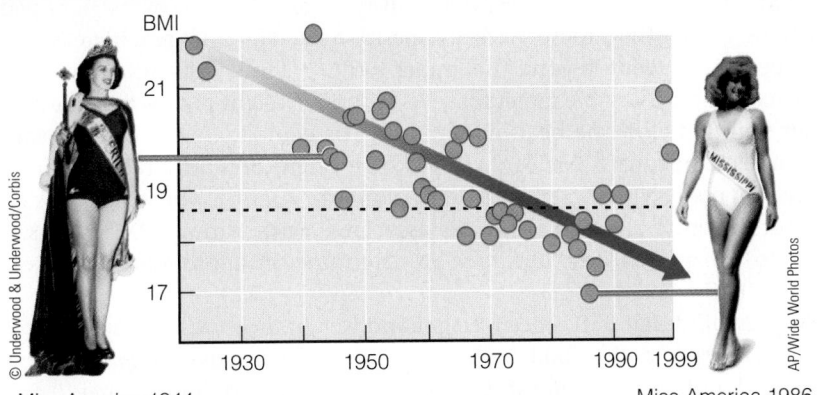

Miss America 1944

Miss America 1986

As explained on p. 262, the body mass index (BMI) describes relative weight for height. Over the years, the BMI of Miss America has declined steadily. Since the mid-1960s, most have fallen below 18.5, the cutoff point indicating underweight with its associated health problems.

SOURCE: S. Rubenstein and B. Caballero, Is Miss America an undernourished role model? *Journal of the American Medical Association* 283 (2000): 1569. Used with permission.

TABLE 8-3	Tips for Accepting a Healthy Body Weight

- Value yourself and others for human attributes other than body weight. Realize that prejudging people by weight is as harmful as prejudging them by race, religion, or gender.
- Use positive, nonjudgmental descriptions of your body.
- Accept positive comments from others.
- Focus on your whole self including your intelligence, social grace, and professional and scholastic achievements.
- Accept that no magic diet exists.
- Stop dieting to lose weight. Adopt a lifestyle of healthy eating and physical activity permanently.
- Follow the Food Guide Pyramid. Never restrict food intake below the minimum levels that meet nutrient needs.
- Become physically active, not because it will help you get thin but because it will make you feel good and enhance your health.
- Seek support from loved ones. Tell them of your plan for a healthy life in the body you have been given.
- Seek professional counseling, *not* from a weight-loss counselor, but from someone who can help you make gains in self-esteem without weight as a factor.
- Join with others to fight weight discrimination and fashion stereotypes. (Search your local paper, or see p. 304 for names of support groups.)

© Lori Adamski Peek/Stone/Getty Images

A healthy body contains enough lean tissue to support health and the right amount of fat to meet body needs.

■ To convert pounds to kilograms:
lb ÷ 2.2 lb/kg = kg.
To convert inches to meters:
in ÷ 39.37 in/m = m.

body mass index (BMI): an index of a person's weight in relation to height; determined by dividing the weight (in kilograms) by the square of the height (in meters).

underweight: body weight below some standard of acceptable weight that is usually defined in relation to height (such as BMI).

overweight: body weight above some standard of acceptable weight that is usually defined in relation to height (such as BMI).

Keep in mind that fashion is fickle; body shapes that our society values change with time. Furthermore, body shapes that our society values differ from those of other societies. The standards defining "ideal" are subjective and frequently have little in common with health. Table 8-3 offers some tips for adopting health as an ideal, rather than society's misconceived image of beauty.

The Criterion of Health Even if our society were to accept fat as beautiful, obesity would still be a major risk factor for several life-threatening diseases. For this reason, the most important criterion for determining how much a person should weigh and how much body fat a person needs is not appearance but good health and longevity. Ideally, a person has enough fat to meet basic needs but not so much as to incur health risks. This range of healthy body weights has been identified using a common measure of weight and height—the body mass index.[15]

Body Mass Index The **body mass index (BMI)** describes relative weight for height:■

$$\text{BMI} = \frac{\text{weight (kg)}}{\text{height (m)}^2} \quad \text{or} \quad \frac{\text{weight (lb)} \times 703.}{\text{height (in)}^2}$$

Weight classifications based on BMI are presented in Figure 8-6. Notice that healthy weight falls between a BMI of 18.5 and 24.9, with **underweight** below 18.5, **overweight** above 25, and obese above 30. Well over half of adults in the United States have a BMI greater than 25, as Figure 8-7 shows.[16]

A BMI of 25 for adults represents a healthy target either for overweight people to achieve or for others to not exceed. Obesity-related diseases and increased mortality become evident beyond this upper limit.[17] The lower end of the healthy range may be a reasonable target for severely underweight people to achieve. BMI values slightly below the healthy range may be compatible with good health if food intake is adequate, but below a BMI of 17, signs of illness, reduced work capacity, and poor reproductive function become apparent.[18] The inside back cover pre-sents weights and visual images associated with various BMI values. The "How to" on p. 264 describes how to determine an appropriate body weight based on BMI.

Keep in mind that BMI reflects height and weight measures and not body composition. Consequently, a bodybuilder may be classified as over*weight* by BMI standards and not be over*fat*. At the peak of his bodybuilding career, Arnold Schwarzenegger won the Mr. Olympia competition with a BMI of 31; the model on p. 261 also has a BMI greater than 30. Yet neither would be considered obese. Striking differences in body composition are also apparent among people of various ethnic and racial groups. For example, blacks tend to have a greater bone density and protein content than whites; consequently, using BMI as the standard may overestimate the prevalence of obesity among blacks.[19]

FIGURE 8-6   BMI Values Used to Assess Weight

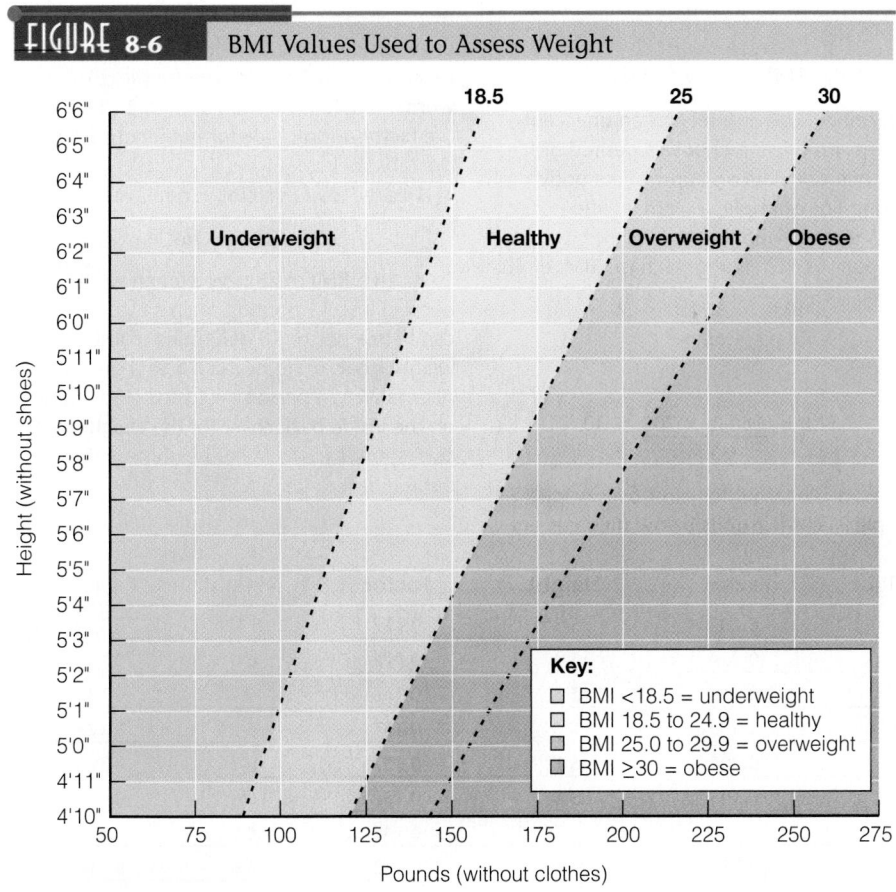

NOTE: Chapter 16 presents BMI values for children and adolescents age 2 to 20.
SOURCE: U.S. Department of Agriculture and U.S. Department of Health and Human Services, *Nutrition and Your Health: Dietary Guidelines for Americans* (Washington, D.C.: 2000), p. 7.

FIGURE 8-7   Distribution of Body Weights in U.S. Adults

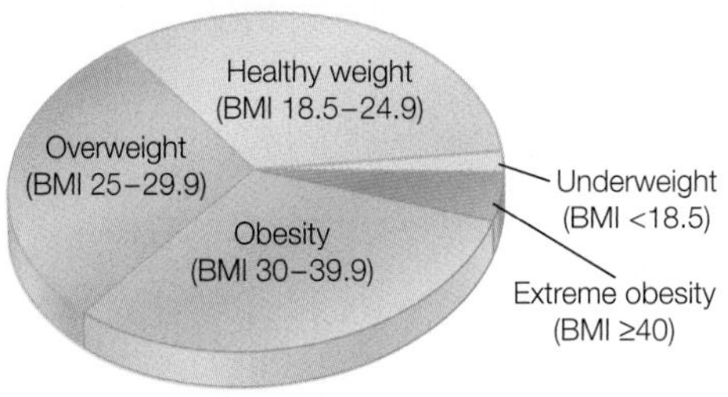

Healthy weight
(BMI 18.5–24.9)

Overweight
(BMI 25–29.9)

Obesity
(BMI 30–39.9)

Underweight
(BMI <18.5)

Extreme obesity
(BMI ≥40)

IN SUMMARY   Current standards for body weight are based on a person's weight in relation to height, called the body mass index (BMI), and reflect disease risks. To its disadvantage, BMI does not reflect body fat, and it may misclassify very muscular people as overweight.

Body Fat and Its Distribution

Although weight measures are inexpensive, easy to take, and highly accurate, they fail to reveal two valuable pieces of information in assessing disease risk: how much of the weight is fat and where the fat is located. The ideal amount of body fat depends partly on the person. A normal-weight man may have from 13 to 21 percent body fat; a woman, because of her greater quantity of essential fat, 23 to 31 percent. In general, health problems typically develop when body fat exceeds 22 percent in young men, 25 percent in men over age 40, 32 percent in young women, and 35 percent in women over age 40. Body fat may contribute as much as 70 percent in excessively obese adults. Figure 8-8 (on p. 265) compares the body composition of healthy weight men and women.

Some People Need Less Body Fat   For many athletes, a lower percentage of body fat may be ideal—just enough fat to provide fuel, insulate and protect the body, assist in nerve impulse transmissions, and support normal hormone activity, but not so much as to burden the body with excess bulk. For some athletes, then, ideal body fat might be 5 to 10 percent for men and 15 to 20 percent for women.

HOW TO Determine Body Weight Based on BMI

A person whose BMI reflects an unacceptable health risk can choose a desired BMI and then calculate an appropriate body weight. For example, a woman who is 5 feet 5 inches (1.65 meters) tall and weighs 180 pounds (82 kilograms) has a BMI of 30:

$$BMI = \frac{82 \text{ kg}}{1.65 \text{ m}^2} = 30$$

or

$$BMI = \frac{180 \text{ lb} \times 703}{65 \text{ in}^2} = 30.$$

A reasonable target for most overweight people is a BMI 2 units below their current one. To determine a desired goal weight based on a BMI of 28, for example, the woman could divide the desired BMI by the factor appropriate for her height from the table below:

Desired BMI ÷ factor = goal weight.

$$28 \div 0.166 = 169 \text{ lb}.$$

To reach a BMI of 28, this woman would need to lose 11 pounds. Such a calculation can help a person to determine realistic weight goals using health risk as a guide. Alternatively, a person could search the table on the inside back cover for the weight that corresponds to his or her height and the desired BMI.

Height	Factor	Height	Factor	Height	Factor
4'7"	0.232	5'3"	0.177	5'11"	0.139
4'8"	0.224	5'4"	0.172	6'0"	0.136
4'9"	0.216	5'5"	0.166	6'1"	0.132
4'10"	0.209	5'6"	0.161	6'2"	0.128
4'11"	0.202	5'7"	0.157	6'3"	0.125
5'0"	0.195	5'8"	0.152	6'4"	0.122
5'1"	0.189	5'9"	0.148	6'5"	0.119
5'2"	0.183	5'10"	0.143	6'6"	0.116

SOURCE: R. P. Abernathy, Body mass index: Determination and use. Copyright the American Dietetic Association. Reprinted by permission from *Journal of the American Dietetic Association* 91 (1991): 843.

(You may want to review the photo on p. 261 to appreciate what 8 percent body fat looks like.)

Some People Need More Body Fat For an Alaska fisherman, a higher percentage of body fat is probably beneficial because fat provides an insulating blanket to prevent excessive loss of body heat in cold climates. A woman starting a pregnancy needs sufficient body fat to support conception and fetal growth. Below a certain threshold for body fat, hormone synthesis falters, and individuals may become infertile, develop depression, experience abnormal hunger regulation, or become unable to keep warm. These thresholds differ for each function and for each individual; much remains to be learned about them.

Fat Distribution The distribution of fat on the body may be more critical than the total amount of fat alone. **Intra-abdominal fat** that is stored around the organs of the abdomen is referred to as **central obesity** or upper-body fat and, independently of total body fat, is associated with increased risks of heart disease, stroke, diabetes, hypertension, and some types of cancer.[20]

Abdominal fat is most common in men and to a lesser extent in women past menopause. Even when total body fat is similar, men have more abdominal fat than women. Regardless of gender, the risks of cardiovascular disease, diabetes, and mortality are increased for those with excessive abdominal fat.

Fat around the hips and thighs, sometimes referred to as lower-body fat, is most common in women during their reproductive years and seems relatively harmless. In fact, overweight people who do not have abdominal fat are less susceptible to health problems than overweight people with abdominal fat. Figure 8-9 compares the body shapes of people with upper-body fat and lower-body fat.

intra-abdominal fat: fat stored within the abdominal cavity in association with the internal abdominal organs, as opposed to the fat stored directly under the skin (subcutaneous fat).

central obesity: excess fat around the trunk of the body; also called **abdominal fat** or **upper-body fat.**

FIGURE 8-8 Male and Female Body Compositions Compared

The differences between male and female body compositions become apparent during adolescence. Lean body mass (primarily muscle) increases more in males than in females. Fat assumes a larger percentage of female body composition as essential body fat is deposited in the mammary glands and pelvic region in preparation for childbearing. Both men and women have essential fat associated with the bone marrow, the central nervous system, and the internal organs.

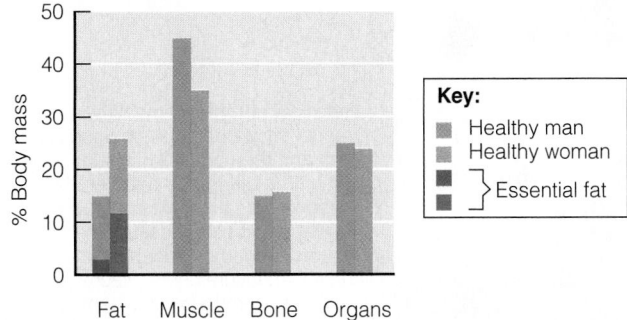

Key:
- Healthy man
- Healthy woman
- } Essential fat

SOURCE: R. E. C. Wildman and D. M. Medeiros, *Advanced Human Nutrition* (Boca Raton, Fla.: CRC Press, 2000), pp. 321–323. Used with permission.

FIGURE 8-9 "Apple" and "Pear" Body Shapes Compared

Upper-body fat is more common in men than in women and is closely associated with heart disease, stroke, diabetes, hypertension, and some types of cancer. In contrast, lower-body fat is more common in women than in men and is not usually associated with chronic diseases. Popular articles sometimes call bodies with upper-body fat "apples" and those with lower-body fat, "pears." Researchers sometimes refer to upper-body fat as "android" (manlike) obesity and to lower-body fat as "gynoid" (womanlike) obesity.

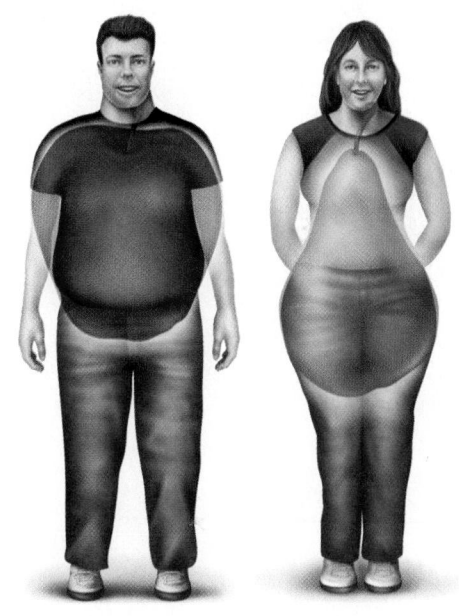

Waist Circumference A person's **waist circumference** is the most practical indicator of fat distribution and abdominal fat.*[21] In general, women with a waist circumference of greater than 35 inches (88 centimeters) and men with a waist circumference of greater than 40 inches (102 centimeters) have a high risk of central obesity–related health problems, such as diabetes and cardiovascular disease.[22] Appendix E includes instructions for measuring waist circumference and assessing abdominal fat.

Other Measures of Body Composition Health care professionals commonly use BMI and waist circumference to access a person's body weight and to monitor changes over time because these measures are relatively easy to determine and inexpensive.[23] Researchers needing more precise measures of body composition may choose any of several other techniques to estimate body fat and its distribution (see Figure 8-10 on p. 266). Mastering these techniques requires proper instruction and practice to ensure reliability. Appendix E provides additional details and includes many of the tables and charts routinely used in assessment procedures.†

IN SUMMARY The ideal amount of body fat varies from person to person, but researchers have found that body fat in excess of 22 percent for young men and 32 percent for young women (the levels rise slightly with age) poses health risks. Central obesity in which excess fat is distributed around the trunk of the body presents greater health risks than excess fat distributed on the lower body.

Health Risks Associated with Body Weight and Body Fat

Body weight and fat distribution correlate with disease risks and life expectancy.[24] These risks indicate a greater *likelihood* of developing a chronic disease and shortening life expectancy. Not all overweight and underweight people will get sick and

waist circumference: an anthropometric measurement used to assess a person's abdominal fat.

---

*The National Heart, Lung, and Blood Institute has replaced waist-hip ratio with waist circumference for assessment of obesity health risks.
†In addition to the methods shown in Figure 8-10, researchers sometimes estimate body composition using these methods: total body water, radioactive potassium count, near-infrared spectrophotometry, ultrasound, computed tomography, and magnetic resonance imaging. Each has advantages and disadvantages with respect to cost, technical difficulty, and precision of estimating body fat (see Appendix E for a comparison).

FIGURE 8-10   Methods Used to Assess Body Fat

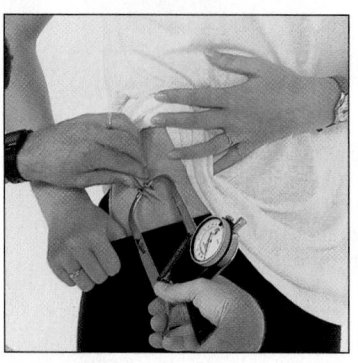

Fatfold measures estimate body fat by using a caliper to gauge the thickness of a fold of skin on the back of the arm (over the triceps), below the shoulder blade (subscapular), and in other places (including lower-body sites) and then comparing these measurements with standards.

© Fitness & Wellness, Boise, Idaho

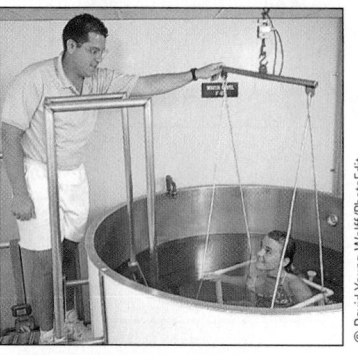

Hydrodensitometry measures body density by weighing the person first on land and then again while submerged in water. The difference between the person's actual weight and underwater weight provides a measure of the body's volume. A mathematical equation using the two measurements (volume and actual weight) determines body density, from which the percentage of body fat can be estimated.

© David Young-Wolff/PhotoEdit

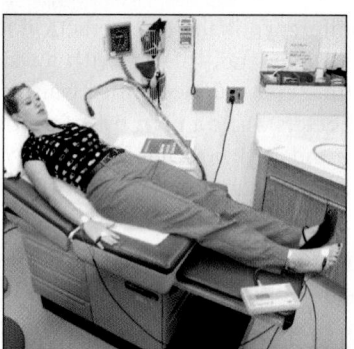

Bioelectrical impedance measures body fat by using a low-intensity electrical current. Because electrolyte-containing fluids, which readily conduct an electrical current, are found primarily in lean body tissues, the leaner the person, the less resistance to the current. The measurement of electrical resistance is then used in a mathematical equation to estimate the percentage of body fat.

© Geri Engberg

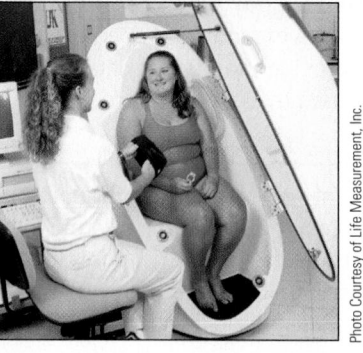

Air displacement plethysmography estimates body composition by having a person sit inside a chamber while computerized sensors determine the amount of air displaced by the person's body.

Photo Courtesy of Life Measurement, Inc.

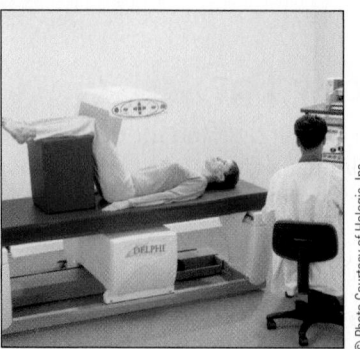

Dual energy X-ray absorptiometry (DEXA) uses two low-dose X-rays that differentiate among fat-free soft tissue (lean body mass), fat tissue, and bone tissue, providing a precise measurement of total fat and its distribution in all but extremely obese subjects.

© Photo Courtesy of Hologic, Inc.

die before their time nor will all normal-weight people live long healthy lives. These are *correlations,* not *causes.* For the most part, people with a BMI between 18.5 and 24.9 have relatively few health risks; risks increase as BMI falls below or rises above this range, indicating that both too little and too much body fat impair health.[25] Epidemiological data show a J- or U-shaped relationship between body weights and mortality (see Figure 8-11).[26] People who are extremely underweight or extremely obese carry higher risks of early deaths than those whose weights fall within the acceptable range. These mortality risks decline with age.[27]

Independently of BMI, factors such as smoking habits raise health risks, and physical fitness lowers them.[28] A man with a BMI of 22 who smokes two packs of cigarettes a day is jeopardizing his health, whereas a woman with a BMI of 32 who walks briskly for an hour a day is improving her health.

Health Risks of Underweight Some underweight people enjoy an active, healthy life, but others are underweight because of malnutrition, smoking habits, substance abuse, or illnesses. Weight and fat measures alone would not reveal these underlying causes, but a complete assessment that includes a diet and medical history, physical examination, and biochemical analysis would.

An underweight person, especially an older adult, may be unable to preserve lean tissue during the fight against a wasting disease such as cancer or a digestive disorder, especially when the disease is accompanied by malnutrition. Without adequate nutrient and energy reserves, an underweight person will have a particularly tough battle against such medical stresses. In fact, many people with cancer die, not from the cancer itself, but from malnutrition. Underweight women develop menstrual irregularities and become infertile. Exactly how infertility develops is unclear, but contributing factors include not only body weight but also restricted energy and fat intake and depleted body fat stores. Those who do conceive may give birth to unhealthy infants. An underweight woman can improve her chances of having a healthy infant by gaining weight prior to conception, during pregnancy, or both. Underweight and significant weight loss are also associated with osteoporosis and bone fractures.[29] For all these reasons, underweight people may benefit from enough of a weight gain to provide an energy reserve and protective amounts of all the nutrients that can be stored.

Health Risks of Overweight As for excessive body fat, the health risks are so many that it has been designated a disease: obesity. Among the health risks associated with obesity are diabetes, hypertension, cardiovascular disease, sleep apnea (abnormal ceasing of breathing during sleep), osteoarthritis, some cancers, gallbladder disease, respiratory problems (including Pickwickian syndrome, a breathing blockage linked with sudden death), and complications in pregnancy and surgery. Each year, these obesity-related illnesses cost our nation billions of dollars—in fact, as much as the medical costs of smoking.[30]

The cost in terms of lives is also great: an estimated 300,000 people die each year from obesity-related diseases.[31] In fact, obesity is second only to tobacco in causing preventable illnesses and premature deaths. Mortality increases as excess weight increases; people with a BMI greater than 35 are twice as likely to die prematurely as others. The risks associated with a high BMI appear to be greater for whites than for blacks.[32] In fact, the health risks associated with obesity do not become apparent in black women until a BMI of 37.[33]

Equally important, both central obesity and weight gains of more than 20 pounds between early and middle adulthood correlate with increased mortality. Fluctuations in body weight, as typically occur with "yo-yo" dieting, also increase the risks of chronic diseases and premature death.[34] In contrast, sustained weight loss improves physical well-being, reduces disease risks, and increases life expectancy.[35]

Cardiovascular Disease The relationship between obesity and cardiovascular disease risk is strong, with links to both elevated blood cholesterol and hypertension. Central obesity may raise the risk of heart attack and stroke as much as the

FIGURE 8-11 BMI and Mortality

This J-shaped curve describes the relationship between body mass index (BMI) and mortality and shows that both underweight and overweight present risks of a premature death.

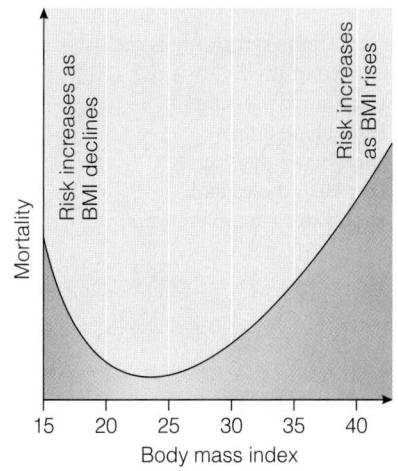

■ Cardiovascular disease risk factors associated with obesity:
- High LDL cholesterol.
- Low HDL cholesterol.
- High blood pressure (hypertension).
- Diabetes.

Chapter 18 provides many more details.

■ The metabolic syndrome is a cluster of at least three of the following:
- High blood pressure.
- High blood glucose.
- High blood triglycerides.
- Low HDL cholesterol.
- High waist circumference.

Being active—even if overweight—is healthier than being sedentary. With a BMI of 36, aerobics instructor Jennifer Portnick is considered obese, but her daily workout routine helps to keep her in good health.

three leading risk factors (high LDL cholesterol, hypertension, and smoking) do.■ In addition to body fat and its distribution, weight gain also increases the risk of cardiovascular disease. Weight loss, on the other hand, can effectively lower both blood cholesterol and blood pressure in obese people. Of course, lean and normal-weight people may also have high blood cholesterol and blood pressure, and these factors are just as dangerous in lean people as in obese people.

Diabetes Diabetes (type 2) is three times more likely to develop in an obese person than in a nonobese person. Furthermore, the person with type 2 diabetes often has central obesity. Central-body fat cells appear to be larger and more insulin-resistant than lower-body fat cells. The association between **insulin resistance** and obesity is strong.[36] Both are major risk factors for the development of type 2 diabetes. Chapter 18 describes the central role obesity plays, together with other risk factors■—collectively known as the metabolic syndrome or Syndrome X—in raising health risks dramatically.[37]

Diabetes appears to be influenced by weight gains as well as by body weight. A weight gain of more than 10 pounds since age 18 doubles the risk of developing diabetes, even in women of average weight. In contrast, weight loss is effective in improving glucose tolerance and insulin resistance.[38]

Cancer The risk of some cancers increases with both body weight and weight gain, but researchers do not fully understand the relationships.[39] One possible explanation may be that obese people have elevated levels of hormones that could influence cancer development.[40] For example, adipose tissue is the major site of estrogen synthesis in women, obese women have elevated levels of estrogen, and estrogen has been implicated in the development of cancers of the female reproductive system—cancers that account for half of all cancers in women.

Fit and Fat versus Sedentary and Slim Importantly, BMI and weight gains and losses do not tell all of the story. Cardiorespiratory fitness also plays a major role in health and longevity, independently of BMI.[41] Normal-weight people who are fit have a lower risk of mortality than normal-weight people who are unfit. Furthermore, overweight but fit people have lower risks than normal-weight, unfit ones. Clearly, a healthy body weight is good, but it may not be good enough. Fitness, in and of itself, offers many health benefits, as Chapter 14 confirms. The next chapter explores weight management and the benefits of achieving and maintaining a healthy weight.

IN SUMMARY The weight appropriate for an individual depends largely on factors specific to that individual, including body fat distribution, family health history, and current health status. At the extremes, both overweight and underweight carry clear risks to health.

Nutrition in Your Life

When combined with fitness, a healthy body weight will help you to defend against chronic diseases.

- Do you balance your food intake with physical activity?
- Is your BMI between 18.5 and 24.9?
- Is your waist circumference less than 35 inches for a woman or 40 inches for a man?

insulin resistance: the reduced ability of insulin to regulate glucose metabolism.

NUTRITION ON THE NET

 Access these websites for further study of topics covered in this chapter.

- Find updates and quick links to these and other nutrition-related sites at our website: **www.wadsworth.com/nutrition**

- Obtain food composition data from the USDA Nutrient Data Laboratory: **www.nal.usda.gov/fnic/foodcomp**

- Learn about the 10,000 Steps Program at Shape Up America: **www.shapeup.org**

- Visit the special web pages and interactive applications for Healthy Weight: **www.nhlbi.nih.gov/subsites/index.htm**

NUTRITION CALCULATIONS

These problems give you practice in estimating energy needs. Once you have mastered these examples, you will be prepared to examine your own energy intakes and energy expenditures. Be sure to show your calculations for each problem and check p. 271 for answers.

1. Compare the energy a person might spend on various physical activities. Refer to Table 8-2 on p. 258, and compute how much energy a person who weighs 142 pounds would spend doing each of the following. You may want to compare various activities based on your weight.

   30 min vigorous aerobic dance:

   $$0.062 \text{ kcal/lb/min} \times 142 \text{ lb} = 8.8 \text{ kcal/min.}$$

   $$8.8 \text{ kcal/min} \times 30 \text{ min} = 264 \text{ kcal.}$$

   a. 2 hours golf, carrying clubs.
   b. 20 minutes running at 9 mph.
   c. 45 minutes swimming at 20 yd/min.
   d. 1 hour walking at 3.5 mph.

2. Consider the effect of age on BMR. An infant who weighs 20 pounds has a BMR of 500 kcalories/day; an adult who weighs 170 pounds has a BMR of about 1500. Based on body weight, who has the faster BMR?

3. Compute daily energy needs for a woman, age 20, who is 5 feet 6 inches tall, weighs 130 pounds, and is lightly active.

4. Discover what weight is needed to achieve a desired BMI. Refer to the table on p. 264 and consider a person who is 5 feet 4 inches tall. Suppose this person wants to have a BMI of 21. What should this person weigh? Does this agree with the table on the inside back cover?

STUDY QUESTIONS

These questions will help you review the chapter. You will find the answers in the discussions on the pages provided.

1. What are the consequences of an unbalanced energy budget? (p. 252)

2. Define hunger, appetite, satiation, and satiety and describe how each influences food intake. (pp. 253–255)

3. Describe each component of energy expenditure. What factors influence each? How can energy expenditure be estimated? (pp. 256–260)

4. Distinguish between body weight and body composition. What assessment techniques are used to measure each? (pp. 260–266)

5. What problems are involved in defining "ideal" body weight? (pp. 261–262)

6. What is central obesity, and what is its relationship to disease? (pp. 264–265)

7. What risks are associated with excess body weight and excess body fat? (pp. 265–268)

These multiple choice questions will help you prepare for an exam. Answers can be found on p. 271.

1. A person who consistently consumes 1700 kcalories a day and spends 2200 kcalories a day for a month would be expected to:
   a. lose ½ to 1 pound.
   b. gain ½ to 1 pound.
   c. lose 4 to 5 pounds.
   d. gain 4 to 5 pounds.

2. A bomb calorimeter measures:
   a. physiological fuel.
   b. energy available from foods.
   c. kcalories a person derives from foods.
   d. heat a person releases in basal metabolism.

3. The psychological desire to eat that accompanies the sight, smell, or thought of food is known as:
   a. hunger.
   b. satiety.
   c. appetite.
   d. palatability.

4. A person watching television after dinner reaches for a snack during a commercial in response to:
   a. external cues.
   b. hunger signals.
   c. stress arousal.
   d. satiety factors.

5. The largest component of energy expenditure is:
   a. basal metabolism.
   b. physical activity.
   c. indirect calorimetry.
   d. thermic effect of food.

6. A major factor influencing BMR is:
   a. hunger.
   b. food intake.
   c. body composition.
   d. physical activity.

7. The thermic effect of an 800-kcalorie meal is about:
   a. 8 kcalories.
   b. 80 kcalories.
   c. 160 kcalories.
   d. 200 kcalories.

8. For health's sake, a person with a BMI of 21 might want to:
   a. lose weight.
   b. maintain weight.
   c. gain weight.

9. Which of the following reflects height and weight?
   a. body mass index
   b. central obesity
   c. waist circumference
   d. body composition

10. Which of the following increases disease risks?
   a. BMI 19–21
   b. BMI 22–25
   c. lower-body fat
   d. central obesity

REFERENCES

1. A. Del Parigi and coauthors, Sex differences in the human brain's response to hunger and satiation, *American Journal of Clinical Nutrition* 75 (2002): 1017–1022.
2. B. J. Rolls, E. L. Morris, and L. S. Roe, Portion size of food affects energy intake in normal-weight and overweight men and women, *American Journal of Clinical Nutrition* 76 (2002): 1207–1213; R. J. Stubbs and coauthors, Effect of altering the variety of sensorially distinct foods, of the same macronutrient content, on food intake and body weight in men, *European Journal of Clinical Nutrition* 55 (2001): 19–28.
3. C. Marmonier, D. Chapelot, and J. Louis-Sylvestre, Effects of macronutrient content and energy density of snacks consumed in a satiety state on the onset of the next meal, *Appetite* 34 (2000): 161–168; M. S. Westerterp-Plantenga and coauthors, Satiety related to 24 h diet-induced thermogenesis during high protein/carbohydrate vs high fat diets measured in a respiration chamber, *European Journal of Clinical Nutrition* 53 (1999): 495–502.
4. B. Burton-Freeman, P. A. Davis, and B. O. Schneeman, Plasma cholecystokinin is associated with subjective measures of satiety in women, *American Journal of Clinical Nutrition* 76 (2002): 659–667; C. Feinle, D. O'Donovan, and M. Horowitz, Carbohydrate and satiety, *Nutrition Reviews* 60 (2002): 155–169; G. A. Bray, Afferent signals regulating food intake, *Proceedings of the Nutrition Society* 59 (2000): 373–384; T. H. Moran, Cholecystokinin and satiety: Current perspectives, *Nutrition* 16 (2000): 858–865.
5. S. H. A. Holt, J. C. Brand-Miller, and P. A. Stitt, The effects of equal-energy portions of different breads on blood glucose levels, feelings of fullness and subsequent food intake, *Journal of the American Dietetic Association* 101 (2001): 767–773.
6. M. W. Schwartz and coauthors, Model for the regulation of energy balance and adiposity, *American Journal of Clinical Nutrition* 69 (1999): 584–596.
7. J. T. Lemmer and coauthors, Effect of strength training on resting metabolic rate and physical activity: Age and gender comparisons, *Medicine and Science in Sports and Exercise* 33 (2001): 532–541.
8. G. P. Granata and L. J. Brandon, The thermic effect of food and obesity: Discrepant results and methodological variations, *Nutrition Reviews* 60 (2002): 223–233; L. Jonge and G. A. Bray, The thermic effect of food is reduced in obesity, *Nutrition Reviews* 60 (2002): 295–297.
9. F. X. Pi-Sunyer, Overnutrition and undernutrition as modifiers of metabolic processes in disease states, *American Journal of Clinical Nutrition* 72 (2000): 533S–537S.
10. S. Rubinstein and B. Caballero, Is Miss America an undernourished role model? *Journal of the American Medical Association* 283 (2000): 1569.
11. J. Wardle, J. Waller, and E. Fox, Age of onset and body dissatisfaction in obesity, *Addictive Behaviors* 27 (2002): 561–573; A. E. Field and coauthors, Peer, parent, and media influences on the development of weight concerns and frequent dieting among preadolescent and adolescent girls and boys, *Pediatrics* 107 (2001): 54–60.
12. K. K. Davison and L. L. Birch, Weight status, parent reaction, and self-concept in five-year-old girls, *Pediatrics* 107 (2001): 46–53.
13. H. Truby and S. J. Paxton, Development of the Children's Body Image Scale, *British Journal of Clinical Psychology* 41 (2002): 185–203; H. A. Hausenblas and coauthors, Body image in middle school children, *Eating and Weight Disorders* 7 (2002): 244–248.
14. C. A. Drury and M. Louis, Exploring the association between body weight, stigma of obesity, and health care avoidance, *Journal of the American Academy of Nurse Practitioners* 14 (2002): 554–561.
15. R. J. Kuczmarski and K. M. Flegal, Criteria for definition of overweight in transition: Background and recommendations for the United States, *American Journal of Clinical Nutrition* 72 (2000): 1074–1081.
16. K. M. Flegal and coauthors, Prevalence and trends in obesity among US adults, *Journal of the American Medical Association* 288 (2002): 1723–1727.
17. J. Stevens and coauthors, Evaluation of WHO and NHANES II standards for overweight using mortality rates, *Journal of the American Dietetic Association* 100 (2000): 825–827; A. Must and coauthors, The disease burden associated with overweight and obesity, *Journal of the American Medical Association* 282 (1999): 1523–1529.
18. L. J. Hoffer, Metabolic consequences of starvation, in M. E. Shils and coeditors, *Modern Nutrition in Health and Disease* (Baltimore: Williams & Wilkins, 1999), pp. 645–665.
19. D. R. Wagner and V. H. Heyward, Measures of body composition in blacks and whites: A comparative review, *American Journal of Clinical Nutrition* 71 (2000): 1392–1402.
20. T. B. Nguyen-Duy and coauthors, Visceral fat and liver fat are independent predictors of metabolic risk factors in men, *American Journal of Physiology. Endocrinology and Metabolism* (2003); G. Davì and coauthors, Platelet activation in obese women—Role of inflammation and oxidant stress, *Journal of the American Medical Association* 288 (2002): 2008–2014; J. M. Oppert and coauthors, Anthropometric estimates of muscle and fat mass in relation to cardiac and cancer mortality in men: The Paris Prospective Study, *American Journal of Clinical Nutrition* 75 (2002): 1107–1113.
21. I. Janssen and coauthors, Body mass index and waist circumference independently contribute to the prediction of nonabdominal, abdominal subcutaneous, and visceral fat, *American Journal of Clinical Nutrition* 75 (2002): 683–688.
22. S. K. Zhu and coauthors, Waist circumference and obesity-associated risk factors among whites in the third National Health and Nutrition Examination Survey: Clinical action thresholds, *American Journal of Clinical Nutrition* 76 (2002): 743–749.

23. J. C. Seidell and coauthors, Report from a Centers for Disease Control and Prevention workshop on use of adult anthropometry for public health and primary health care, *American Journal of Clinical Nutrition* 73 (2001): 123–126.

24. A. H. Mokdad and coauthors, Prevalence of obesity, diabetes, and obesity-related health risk factors, 2001, *Journal of the American Medical Association* 289 (2003): 76–79; K. R. Fontaine and coauthors, Years of life lost due to obesity, *Journal of the American Medical Association* 289 (2003): 187–193; Must and coauthors, 1999.

25. A. Thorogood and coauthors, Relation between body mass index and mortality in an unusually slim cohort, *Journal of Epidemiology and Community Health* 57 (2003) 130–133; P. T. Katzmarzyk, C. L. Craig, and C. Bouchard, Original article underweight, overweight and obesity: Relationships with mortality in the 13-year follow-up of the Canada Fitness Study, *Journal of Clinical Epidemiology* 54 (2001): 916–920; R. Bender and coauthors, Effect of age on excess mortality in obesity, *Journal of the American Medical Association* 281 (1999): 1498–1504.

26. R. G. Rogers, R. A. Hummer, and P. M. Krueger, The effect of obesity on overall, circulatory disease- and diabetes-specific mortality, *Journal of Biosocial Science* 35 (2003): 107–129; D. B. Allison and coauthors, Differential associations of body mass index and adiposity with all-cause mortality among men in the first and second National Health and Nutrition Examination Surveys (NHANES I and NHANES II) follow-up studies, *International Journal of Obesity and Related Metabolic Disorders* 26 (2002): 410–416; H. E. Meyer and coauthors, Body mass index and mortality: The influence of physical activity and smoking, *Medicine and Science in Sports and Exercise* 34 (2002): 1065–1070; P. N. Singh, K. D. Lindsted, and G. E. Fraser, Body weight and mortality among adults who never smoked, *American Journal of Epidemiology* 150 (1999): 1152–1164; E. E. Calle and coauthors, Body-mass index and mortality in a prospective cohort of U.S. adults, *New England Journal of Medicine* 341 (1999): 1097–1105.

27. I. Baik and coauthors, Adiposity and mortality in men, *American Journal of Epidemiology* 152 (2000): 264–271; J. Stevens, Impact of age on associations between weight and mortality, *Nutrition Reviews* 58 (2000): 129–137; Bender and coauthors, 1999.

28. A. Peeters and coauthors, Obesity in adulthood and its consequences for life expectancy: A life-table analysis, *Annals of Internal Medicine* 138 (2003): 24–32; Meyer and coauthors, 2002; M. Wei and coauthors, Relationship between low cardiorespiratory fitness and mortality in normal-weight, overweight, and obese men, *Journal of the American Medical Association* 282 (1999): 1547–1553.

29. L. M. Salamone and coauthors, Effect of a lifestyle intervention on bone mineral density in premenopausal women: A randomized trial, *American Journal of Clinical Nutrition* 70 (1999): 97–103.

30. E. A. Finkelstein, I. C. Fiebelkorn, and G. Wang, National medical expenditures attributable to overweight and obesity: How much, and who's paying? 2003, available at **www.healthaffairs.org/WebExclusives/Finkelstein_Web_Excl_051403.htm**.

31. D. B. Allison and coauthors, Annual deaths attributable to obesity in the United States, *Journal of the American Medical Association* 282 (1999): 1530–1538.

32. J. Stevens and coauthors, The effect of decision rules on the choice of a body mass index cutoff for obesity: Examples from African American and white women, *American Journal of Clinical Nutrition* 75 (2002): 986–992; J. Stevens, Obesity and mortality in African-Americans, *Nutrition Reviews* 58 (2000): 346–353; Calle and coauthors, 1999.

33. J. E. Manson and S. S. Bassuk, Obesity in the United States: A fresh look at its high toll, *Journal of the American Medical Association* 289 (2003): 229–230.

34. K. A. Petersmark and coauthors, The effect of weight cycling on blood lipids and blood pressure in the Multiple Risk Factor Intervention Trial special intervention group, *International Journal of Obesity and Related Metabolic Disorders* 23 (1999): 1246–1255.

35. Davì, 2002; D. F. Williamson and coauthors, Intentional weight loss and mortality among overweight individuals with diabetes, *Diabetes Care* 23 (2000): 1499–1504; J. T. Fine and coauthors, A prospective study of weight change and health-related quality of life in women, *Journal of the American Medical Association* 282 (1999): 2136–2142; K. Karason and coauthors, Weight loss and progression of early atherosclerosis in the carotid artery: A four-year controlled study of obese subjects, *International Journal of Obesity and Related Metabolic Disorders* 23 (1999): 948–956; G. Oster and coauthors, Lifetime health and economic benefits of weight loss among obese persons, *American Journal of Public Health* 89 (1999): 1536–1542.

36. J. L. Sievenpiper and coauthors, Simple skinfold-thickness measurements complement conventional anthropometric assessments in predicting glucose tolerance, *American Journal of Clinical Nutrition* 73 (2001): 567–573; D. H. Bessesen, Obesity as a factor, *Nutrition Reviews* 58 (2000): S12–S15; M. Rosenbaum and coauthors, Effects of changes in body weight on carbohydrate metabolism, catecholamine excretion, and thyroid function, *American Journal of Clinical Nutrition* 71 (2000): 1421–1432.

37. P. Maison and coauthors, Do different dimensions of the metabolic syndrome change together over time? Evidence supporting obesity as the central feature, *Diabetes Care* 24 (2001): 1758–1763.

38. B. A. Gower and coauthors, Effects of weight loss on changes in insulin sensitivity and lipid concentrations in premenopausal African American and white women, *American Journal of Clinical Nutrition* 76 (2002): 923–927.

39. D. S. Michaud and coauthors, Physical activity, obesity, height, and the risk of pancreatic cancer, *Journal of the American Medical Association* 286 (2001): 921–929; G. H. Rauscher, S. T. Mayne, and D. T. Janerich, Relation between body mass index and lung cancer risk in men and women never and former smokers, *American Journal of Epidemiology* 152 (2000): 506–513; W. H. Chow and coauthors, Obesity, hypertension, and the risk of kidney cancer in men, *New England Journal of Medicine* 343 (2000): 1305–1311; S. D. Li and S. Mobarhan, Association between body mass index and adenocarcinoma of the esophagus and gastric cardia, *Nutrition Reviews* 58 (2000): 54–56; J. Kermström and E. Barrett-Connor, Obesity, weight change, fasting insulin, proinsulin, C-peptide, and insulin-like growth factor-1 levels in women with and without breast cancer: The Rancho Bernardo Study, *Journal of Women's Health and Gender-based Medicine* 8 (1999): 1265–1272.

40. G. A. Bray, The underlying basis for obesity: Relationship to cancer, *Journal of Nutrition* 132 (2002): 3451S–3455S.

41. S. W. Farrell and coauthors, The relation of body mass index, cardiorespiratory fitness, and all-cause mortality in women, *Obesity Research* 10 (2002): 417–423; C. D. Lee and S. N. Blair, Cardiorespiratory fitness and smoking-related and total cancer mortality in men, *Medicine and Science in Sports and Exercise* 34 (2002): 735–739; C. D. Lee and S. N. Blair, Cardiorespiratory fitness and stroke mortality in men, *Medicine and Science in Sports and Exercise* 34 (2002): 592–595; M. Wei and coauthors, Relationship between low cardiorespiratory fitness and mortality in normal-weight, overweight, and obese men, *Journal of the American Medical Association* 282 (1999): 1547–1553.

ANSWERS

Nutrition Calculations

1. a. 0.045 kcal/lb/min × 142 lb = 6.4 kcal/min.

   6.4 kcal/min × 120 min = 768 kcal.

   b. 0.103 kcal/lb/min × 142 lb = 14.6 kcal/min.

   14.6 kcal/min × 20 min = 292 kcal.

   c. 0.032 kcal/lb/min × 142 lb = 4.5 kcal/min.

   4.5 kcal/min × 45 min = 202.5 kcal.

   d. 0.035 kcal/lb/min × 142 lb = 5 kcal/min.

   5 kcal/min × 60 min = 300 kcal.

2. The infant has the faster BMR (500 kcal/day ÷ 20 lb = 25 kcal/lb/day and 1500 kcal/day ÷ 170 lb = 8.8 kcal/lb/day). Because the infant has a BMR of 25 kcal/lb, whereas the adult has a BMR of 8.8 kcal/lb, the infant's BMR is almost 3 times faster than the adult's based on body weight.

3. EER = 354 − 6.91 × 20 + 1.12 × [(9.36 × 59) + (726 × 1.68)].

   EER = 354 − 138.2 + 1.12 (552.24 + 1219.68).

   EER = 354 − 138.2 + 1984.6 = 2200 kcal/day.

4. 21 ÷ 0.172 = 122 lb. Yes.

Study Questions (multiple choice)

1. c   2. b   3. c   4. a   5. a   6. c   7. b   8. b
9. a   10. d

HIGHLIGHT

The Latest and Greatest Weight-Loss Diet—Again

© Geri Engberg

To paraphrase William Shakespeare, "a fad diet by any other name would still be a fad diet." And the names are legion: the Atkins New Diet Revolution, the Calories Don't Count diet, the Protein Power diet, the Carbohydrate Addict's diet, the Lo-Carbo diet, the Healthy for Life diet, the Zone diet.* Year after year, "new and improved" diets appear on bookstore shelves and circulate among friends. People of all sizes eagerly try the best diet on the market ever, hoping that this one will really work. Sometimes fad diets seem to work for a while, but more often than not, their success is short-lived. Then another fad diet takes the spotlight. Here's how Dr. K. Brownell, an obesity researcher at Yale University, describes this phenomenon: "When I get calls about the latest diet fad, I imagine a trick birthday cake candle that keeps lighting up and we have to keep blowing it out."

Realizing that fad diets do not offer a safe and effective plan for weight loss, health professionals speak out, but they never get the candle blown out permanently. New fad diets can keep making outrageous claims because no one requires their advocates to prove what they say. Fad diet gurus do not have to conduct credible research on the benefits or dangers of their diets. They can simply make recommendations and then later, if questioned, search for bits and pieces of research that support the conclusions they have already reached. That's backwards. Diet and health recommendations should *follow* years of sound research that has been reviewed by panels of scientists *before* being offered to the public.

Because anyone can publish anything—in books or on the Internet—peddlers of fad diets can make unsubstantiated statements that fall far short of the truth, but sound impressive to the uninformed. They often offer distorted bits of legitimate research. They may start with one or more actual facts, but then leap from one erroneous conclusion to the next. Anyone who wants to believe them is forced to wonder how the thousands of scientists working on obesity research over the past century could possibly have missed such obvious connections. Table H8-1 presents some of the claims and truths of fad diets.

TABLE H8-1	The Claims and Truths of Fad Diets
The Claim:	You can lose weight with "exceptionally easy rules."
The Truth:	Most fad diet plans have complicated rules that require you to calculate protein requirements, count carbohydrate grams, combine certain foods, time meal intervals, purchase special products, plan daily menus, and measure serving sizes.
The Claim:	You can lose weight by eating a specific ratio of carbohydrates, protein, and fat.
The Truth:	Weight loss depends on spending more energy than you take in.
The Claim:	This "revolutionary diet" can "reset your genetic code."
The Truth:	You inherited your genes and cannot alter your genetic code.
The Claim:	High-protein diets are popular, selling more than 20 million books, because they work.
The Truth:	Weight-loss books are popular because people grasp for quick fixes and simple solutions to their weight problems. If book sales were an indication of weight-loss success, we would be a lean nation—but they're not, and neither are we.
The Claim:	People gain weight on low-fat diets.
The Truth:	People can gain weight on low-fat diets if they overindulge in carbohydrates and proteins while cutting fat; low-fat diets are not necessarily low-kcalorie diets. But people can also lose weight on low-fat diets if they cut kcalories as well as fat.
The Claim:	High-protein diets energize the brain.
The Truth:	The brain depends on glucose for its energy; the primary dietary source of glucose is carbohydrate, not protein.
The Claim:	Thousands of people have been successful with this plan.
The Truth:	Authors of fad diets have not published their research findings in scientific journals. Success stories are anecdotal and failures are not reported.
The Claim:	Carbohydrates raise blood glucose levels, triggering insulin production and fat storage.
The Truth:	Insulin promotes fat storage when energy intake exceeds energy needs. Furthermore, insulin is only one hormone involved in the complex processes of maintaining the body's energy balance and health.
The Claim:	Eat protein and lose weight.
The Truth:	For every complicated problem, there is a simple—and wrong—solution.

*The following sources offer comparisons and evaluations of various fad diets for your review: S. T. St. Jeor and coauthors, Dietary protein and weight reduction: A statement for healthcare professionals from the nutrition committee of the Council on Nutrition, Physical Activity, and Metabolism of the American Heart Association, *Circulation* 104 (2001): 1869–1874; G. L. Blackburn and V. H. He, The changing nature of obesity in the U.S.: How serious is the problem? *Nutrition & the M.D.*, June 1999, pp. 3–7.

Matthew Farruggio

High-protein, low-carbohydrate meals overemphasize meat, fish, poultry, eggs, and cheeses, and shun breads, pastas, fruits, and vegetables.

No matter what their names are, most fad diets espouse essentially the same high-protein, low-carbohydrate diet. After all, diets may come in all flavors, but only in three proportions: high fat, high carbohydrate, or high protein. Few consumers would believe that high-fat diets could lead to weight loss; contrary to such a claim, dietary fat does not promote fat oxidation. Consumers already hear from many free sources that high-carbohydrate diets support good health, so peddling that idea would not be a profitable venture. That leaves high-protein diets, and they surface regularly in various guises as the best way to lose weight. High-protein diets are by design relatively low in carbohydrate. This highlight examines some of the science and the science fiction behind high-protein, low-carbohydrate fad diets.

The Diet's Appeal

Perhaps the greatest appeal of a high-protein, low-carbohydrate diet is that it turns current diet recommendations upside down. Foods such as meats and milk products that need to be selected carefully to limit saturated fat can now be eaten with abandon. Grains, legumes, vegetables, and fruits that we are told to eat in abundance can now be ignored. For some people, this is a dream come true: steaks without the potatoes, ribs without the coleslaw, and meatballs without the pasta. Who can resist the promise of weight loss while eating freely from a list of favorite foods?

To lure dieters in, proponents of high-protein diets often blame the currently recommended high-carbohydrate, low-fat diet for our obesity troubles. They claim that the incidence of obesity is rising because we are eating less fat. Such a claim may impress the naive, but it sends skeptical people running for the facts. True, the incidence of obesity has risen dramatically over the past two decades.[1] True, our intake of fat has dropped from 35 to 33 percent of daily energy intake.[2] Such facts might seem to imply that lowering fat intake leads to obesity, but this is an erroneous conclusion. The *percentage* declined only because average energy intakes increased by almost 200 kcalories a day (from 1878 kcalories a day to 2056). Actual fat intake *increased* by 3 grams a day (from 73 grams to 76). Furthermore, fewer than half of us engage in regular physical activity.[3] Obesity experts blame our high energy intakes and low energy outputs for the increase in obesity. Weight loss, after all, depends on a negative energy balance. To their credit, some of these diet plans recommend exercise—and regular physical activity is an integral component of successful weight loss.[4]

Dieters are also lured into fad diets by sophisticated—yet often erroneous—explanations of the metabolic consequences of eating certain foods. Terms such as *eicosanoids* and *de novo lipogenesis* are scattered about, intimidating readers into believing that the authors must be right given their brilliance in understanding the body.

One common misconception currently circulating amongst fad diets focuses on insulin. High-protein diet proponents claim that carbohydrates are bad. Some go so far as to equate carbohydrates with toxic poisons or addictive drugs. Starches and sugars are considered evil because they are absorbed easily and raise blood glucose. The pancreas then responds by secreting insulin—and insulin is touted as the real villain responsible for our nation's epidemic of obesity.

What does insulin do? Among its roles, insulin facilitates the transport of glucose into the cells, the storage of fatty acids as fat, and the synthesis of cholesterol. It is an anabolic hormone that builds and stores. True—but not the whole truth and nothing but the truth. Insulin is only one of many factors involved in the body's metabolism of nutrients and regulation of body weight.[5] Furthermore, as Chapter 4's discussion of the glycemic index pointed out, blood glucose and insulin do not always respond to foods as might be expected. Many carbohydrates—fruits, vegetables, legumes, and whole grains—are rich in fibers that slow glucose absorption and moderate insulin response. Most importantly, insulin is critical to maintaining health, as any person with type 1 diabetes can attest. Insulin causes problems only when a person develops insulin resistance—that is, when the body's cells do not respond to the large quantities of insulin that the pancreas continues to pump out in an effort to get a response. Insulin resistance is a major health problem—but it is not caused by carbohydrate, or by protein, or by fat. It results from being overweight. When a person loses weight, insulin response improves.

Another crazy distortion of the facts is the claim that high-protein foods expend more energy. As Chapter 8 mentioned, the thermic effect of food for protein is higher than for carbohydrate or fat, but the increase is still insignificant—perhaps the equivalent of two pounds per year.

If high-protein diets were as successful as some people claim, then consumers who tried them would lose lots of weight, and our obesity problems would be solved. Obviously, this is not the case. Similarly, if high-protein diets were as worthless as others claim, then consumers would eventually stop pursuing them. Clearly, this is not happening either. These diets have enough going for them that they work for some people at least for a short time, but they fail to produce long-lasting results for most people. Studies report that people following high-protein, low-carbohydrate diets do lose weight.[6] In fact, they lose more than people following conventional high-carbohydrate, low-fat diets—but only for the first six months. Their later gains make up the difference, so total weight loss is no different after one year.[7] The following sections examine some of the apparent achievements and shortcomings of high-protein diets.[8]

The Diet's Achievements

With over half of our nation's adults overweight and many more concerned about their weight, the market for a weight-loss book, product, or program is huge (no pun intended). Americans spend an estimated $33 billion a year on weight-loss books and products. Even a plan that offers only minimal weight-loss success easily attracts a following. High-protein diet plans offer a little success to some people for a short time. Here's why.

Don't Count kCalories

Who wants to count kcalories? Even experienced dieters find counting kcalories burdensome, not to mention timeworn. They want a new, easy way to lose weight, and high-protein diet plans seem to offer this boon. But while these diets often claim to disregard kcalories, their design typically ensures a low energy intake. They advise dieters to stop counting kcalories, but then recommend three meals "not to exceed 500 kcalories each and two snacks of less than 100 kcalories each." Most of the sample menu plans provided by these diets are designed to deliver 800 to 1200 kcalories a day.

Even when counting kcalories is truly not necessary, the total tends to be low simply because food intake is so limited. Without its refried beans, tortilla wrapping, and chopped vegetables, a burrito is reduced to a pile of ground beef. Without the baked potato, there's no need for butter and sour cream. Weight loss occurs because of the low energy intake—not the proportion of energy nutrients.[9] Success, then, depends on the restricted intake, not on protein's magical powers or carbohydrate's evil forces. This is an important point. Any diet can produce weight loss, at least temporarily, if intake is restricted. The real value of a diet is determined by its ability to maintain weight loss and support good health over the long term. The goal is not simply weight loss, but health gains—and whether high-protein, low-carbohydrate diets can support optimal health over time remains unknown.

Satisfy Hunger

As Chapter 8 mentioned, of the three energy-yielding nutrients, protein is the most satiating. Consequently, high-protein meals may suppress hunger and delay the start of the next meal. Furthermore, studies have reported that people tend to eat less after a high-protein meal than after a low-protein one.[10] In real-life situations, though, there is a strong association between a person's protein intake and BMI—the higher the intake, the higher the BMI.[11] This association remains apparent even after adjusting for energy intake and physical activity. All meals—whether designed for weight loss or not—should include enough protein to satisfy hunger, but not so much as to contribute to weight gain.

Follow a Plan

Most people need specific instructions and examples to make dietary changes. Fad diets offer dieters a plan. The user doesn't have to decide what foods to eat, how to prepare them, or how much to eat. Unfortunately, these instructions serve short-term weight-loss needs only. They do not provide for long-term changes in lifestyle that will support weight maintenance or health goals.

The success of any weight-loss diet depends on the person adopting the plan and sticking with it. People who prefer the high-protein, low-carbohydrate diet over the high-carbohydrate, low-fat diet may have more success at sticking with it. Again, weight loss occurs because of the duration of a low-kcalorie plan—not the proportion of energy nutrients.[12]

Limit Choices

Diets that omit hundreds of foods and several food groups limit a person's options and lack variety. Chapter 2 praised variety as a valuable way to ensure an adequate intake of nutrients, but variety also entices people to eat more food and gain more weight.[13] Without variety, some people lose interest in eating, which further reduces energy intake. Even if the allowed foods are favorites, eating the same foods week after week can become monotonous.

The Diet's Shortcomings

People who have followed high-protein diet plans for several months have lost weight. But can these diets also be harmful?

Too Much Fat

Some fad diets focus so intently on promoting protein and curbing carbohydrate that they fail to account for the fat that accompanies many high-protein foods. A breakfast of bacon and eggs, lunch of ham and cheese, and dinner of barbecued short ribs would provide 100 grams of protein—and 121 grams of fat! Yet this day's meals, even with a snack of peanuts, provide only 1600 kcalories. Without careful selection, protein-rich diets can be extraordinarily high in saturated fat and cholesterol—dietary factors that raise LDL cholesterol and the risks for heart disease.

Overall, studies report that people following high-protein, low-carbohydrate diets have little or no change in blood pressure or blood lipids—risk factors for heart disease.[14] Some researchers speculate that the weight loss that occurs on these diets offsets the adverse effects of a diet high in saturated fat and low in fruits and vegetables.[15]

Too Much Protein

Moderation has been a recurring theme throughout this text, with recommendations to get enough, but not too much, and cautions that too much can be as harmful as too little. The DRI Committee did not establish an upper limit for protein, but it does recognize that high-protein diets have been implicated in chronic diseases such as osteoporosis, kidney stones and kidney disease, some cancers, heart disease, and obesity.[16] One four-month study reports no adverse effects on bone metabolism.[17] Health recommendations typically advise a protein intake of 50 to 100 grams per day and within the range of 10 to 35 percent of energy intake.[18] Popular high-protein diets suggest a protein intake of 70 to 160 grams per day, representing 25 to 65 percent of energy intake.[19]

Too Little Everything Else

The quality of the diet suffers when carbohydrates are restricted.[20] Without fruits, vegetables, and whole grains, high-protein diets lack not only carbohydrate, but fiber, vitamins, minerals, and phytochemicals as well—all dietary factors protective against disease. To help shore up some of these inadequacies, fad diets often recommend a daily supplement. Conveniently, many of the companies selling fad diets also peddle these supplements. But as Highlights 10 and 11 explain, foods offer many more health benefits than any supplement can provide. Quite simply, if the diet is inadequate, it needs to be improved, not supplemented.

The Body's Perspective

When a person consumes a low-carbohydrate diet, a metabolism similar to that of fasting prevails (see Chapter 7 for a review of fasting). With little dietary carbohydrate coming in, the body uses its glycogen stores to provide glucose for the cells of the brain, nerves, and blood. Once the body depletes its glycogen reserves, it begins making glucose from the amino acids of protein (gluconeogenesis). A low-carbohydrate diet may provide abundant protein from food, but the body still uses some protein from body tissues.

Dieters can know glycogen depletion has occurred and gluconeogenesis has begun by monitoring their urine. Whenever glycogen or protein is broken down, water is released and urine production increases. Low-carbohydrate diets also induce keto-sis, and ketones can be detected in the urine. Ketones form whenever glucose is lacking and fat breakdown is incomplete.

Many fad diets regard ketosis as the key to losing weight, but a study comparing weight-loss diets found no relation between ketosis and weight loss.[21] People in ketosis may experience a loss of appetite and a dramatic weight loss within the first few days. They would be disillusioned if they were aware that much of this weight loss reflects the loss of glycogen and protein together with large quantities of body fluids and important minerals.[22] They need to learn to appreciate the difference between loss of *fat* and loss of *weight*. Fat losses on ketogenic diets are no greater than on other diets providing the same number of kcalories. Once the dieter returns to well-balanced meals that provide adequate energy, carbohydrate, fat, protein, vitamins, and minerals, the body avidly retains these needed nutrients. The weight will return, quite often to a level higher than the starting point. Table H8-2 lists other consequences of a ketogenic diet.

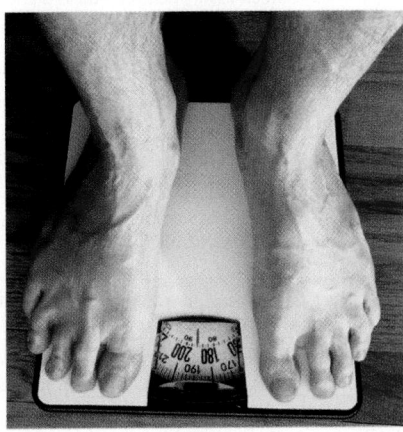

The wise consumer distinguishes between loss of fat and loss of weight.

1998 Photo Disc Inc.

Table H8-3 offers guidelines for identifying fad diets and other weight-loss scams; it includes the hallmarks of a reasonable weight-loss program as well. Diets that overemphasize protein and fall short on carbohydrate may not harm healthy people if used for only a little while, but they cannot support optimal health for long. Chapter 9 includes reasonable approaches to weight management and concludes that the ideal diet is one you can live with for the rest of your life. Keep that criterion in mind when you evaluate the next "latest and greatest weight-loss diet" that comes along.

TABLE H8-2 Adverse Side Effects of Low-Carbohydrate, Ketogenic Diets

- Nausea
- Fatigue (especially if physically active)
- Constipation
- Low blood pressure
- Elevated uric acid (which may exacerbate kidney disease and cause inflammation of the joints in those predisposed to gout)
- Stale, foul taste in the mouth (bad breath)
- In pregnant women, fetal harm and stillbirth

TABLE H8-3   Guidelines for Identifying Fad Diets and Other Weight-Loss Scams

1. They promise dramatic, rapid weight loss. Weight loss should be gradual and not exceed 2 pounds per week.

2. They promote diets that are nutritionally unbalanced or extremely low in kcalories. Diets should provide:

   • A reasonable number of kcalories (not fewer than 1200 kcalories per day for women and 1500 kcalories per day for men).

   • Enough, but not too much, protein (between the RDA and twice the RDA).

   • Enough, but not too much fat (between 20 and 35 percent of daily energy intake from fat).

   • Enough carbohydrate to spare protein and prevent ketosis (at least 100 grams per day) and 20 to 30 grams of fiber from food sources.

   • A balanced assortment of vitamins and minerals from a variety of foods from each of the food groups.

   • At least 1 liter (about 1 quart) of water daily or 1 milliliter per kcalorie daily—whichever is more.

3. They use liquid formulas rather than foods. Foods should accommodate a person's ethnic background, taste preferences, and financial means.

4. They attempt to make clients dependent upon special foods or devices. Programs should teach clients how to make good choices from the conventional food supply.

5. They fail to encourage permanent, realistic lifestyle changes. Programs should provide physical activity plans that involve spending at least 300 kcalories a day and behavior-modification strategies that help to correct poor eating habits.

6. They misrepresent salespeople as "counselors" supposedly qualified to give guidance in nutrition and/or general health. Even if adequately trained, such "counselors" would still be objectionable because of the obvious conflict of interest that exists when providers profit directly from products they recommend and sell.

7. They collect large sums of money at the start or require that clients sign contracts for expensive, long-term programs. Programs should be reasonably priced and run on a pay-as-you-go basis.

8. They fail to inform clients of the risks associated with weight loss in general or the specific program being promoted. They should provide information about dropout rates, the long-term success of their clients, and possible side effects.

9. They promote unproven or spurious weight-loss aids such as human chorionic gonadotropin hormone (HCG), starch blockers, diuretics, sauna belts, body wraps, passive exercise, ear stapling, acupuncture, electric muscle-stimulating (EMS) devices, spirulina, amino acid supplements (e.g., arginine, ornithine), glucomannan, methylcellulose (a "bulking agent"), "unique" ingredients, and so forth.

10. They fail to provide for weight maintenance after the program ends.

SOURCES: Adapted from American College of Sports Medicine, *ACSM's Guidelines for Exercise Testing and Prescription* (Baltimore: Williams & Wilkins, 1995), pp. 218–219; J. T. Dwyer, Treatment of obesity: Conventional programs and fad diets, in *Obesity,* ed. P. Björntorp and B. N. Brodoff (Philadelphia: J. B. Lippincott, 1992), p. 668; *National Council Against Health Fraud Newsletter,* March/April 1987, National Council Against Health Fraud, Inc.

NUTRITION ON THE NET

 Access these websites for further study of topics covered in this highlight.

• Find updates and quick links to these and other nutrition-related sites at our website: **www.wadsworth.com/nutrition**

• Search for the Great Nutrition Debate at the USDA's site: **www.usda.gov**

REFERENCES

1. A. H. Mokdad and coauthors, The spread of the obesity epidemic in the United States, 1991–1998, *Journal of the American Medical Association* 282 (1999): 1519–1522.

2. P. Chanmugam and coauthors, Did fat intake in the United States really decline between 1989–1991 and 1994–1996? *Journal of the American Dietetic Association* 103 (2003): 867–872.

3. P. M. Barnes and C. A. Schoenborn, *Physical Activity among Adults: United States, 2000,* 2003, available at **www.cdc.gov/nchs/about/major/nhis/released200306.htm#7**.

4. M. Jakicic and coauthors, Appropriate intervention strategies for weight loss and prevention of weight regain for adults, *Medicine and Science in Sports and Exercise* 33 (2001): 2145–2156.

5. J. C. Brüning and coauthors, Role of brain insulin receptor in control of body weight and reproduction, *Science* 289 (2000): 2122–2125.

6. E. C. Westman and coauthors, Effect of 6-month adherence to a very low carbohydrate diet program, *American Journal of Medicine* 113 (2002): 30–36.

7. G. D. Foster and coauthors, A randomized trial of a low-carbohydrate diet for obesity, *New England Journal of Medicine* 348 (2003): 2082–2090.

8. J. Eisenstein and coauthors, High-protein weight-loss diets: Are they safe and do they work? A review of the experimental and epidemiologic data, *Nutrition Reviews* 60 (2002): 189–200.

9. D. K. Layman and coauthors, A reduced ratio of dietary carbohydrate to protein improves body composition and blood lipid profiles during weight loss in adult women, *Journal of Nutrition* 133 (2003): 411–417; D. M. Bravata and coauthors, Efficacy and safety of low-carbohydrate diets, *Journal of the American Medical Association* 289 (2003): 1837–1850; M. R. Freedman, J. King, and E. Kennedy, Popular

diets: A scientific review, *Obesity Research* 9 (2001): 1S–5S; A. Golay and coauthors, Similar weight loss with low-energy food combining or balanced diets, *International Journal of Obesity and Related Metabolic Disorders* 24 (2000): 492–496; N. H. Baba and coauthors, High protein vs high carbohydrate hypoenergetic diet for the treatment of obese hyperinsulinemic subjects, *International Journal of Obesity and Related Metabolic Disorders* 23 (1999): 1202–1206.

10. A. R. Skov and coauthors, Randomized trial on protein vs carbohydrate in ad libitum fat reduced diet for the treatment of obesity, *International Journal of Obesity and Related Metabolic Disorders* 23 (1999): 528–536.

11. A. Trichopoulou and coauthors, Lipid, protein and carbohydrate intake in relation to body mass index, *European Journal of Clinical Nutrition* 56 (2002): 37-43.

12. Bravata and coauthors, 2003.

13. M. A. McCrory and coauthors, Dietary variety within food groups: Association with energy and body fatness in men and women, *American Journal of Clinical Nutrition* 69 (1999): 440–447.

14. Bravata and coauthors, 2003.

15. Foster and coauthors, 2003.

16. Committee on Dietary Reference Intakes, *Dietary Reference Intakes for Energy, Carbohydrate, Fiber, Fat, Fatty Acids, Cholesterol, Protein, and Amino Acids* (Washington, D.C.: National Academies Press, 2002).

17. E. Farnsworth and coauthors, Effect of a high-protein, energy-restricted diet on body composition, glycemic control, and lipid concentrations in overweight and obese hyperinsulinemic men and women, *American Journal of Clinical Nutrition* 78 (2003): 31–39.

18. Committee on Dietary Reference Intakes, 2002; S. T. St. Jeor and coauthors, Dietary protein and weight reduction: A statement for healthcare professionals from the nutrition committee of the Council on Nutrition, Physical Activity, and Metabolism of the American Heart Association, *Circulation* 104 (2001): 1869–1874.

19. St. Jeor and coauthors, 2001.

20. E. T. Kennedy and coauthors, Popular diets: Correlation to health, nutrition, and obesity, *Journal of the American Dietetic Association* 101 (2001): 411–420.

21. Foster and coauthors, 2003.

22. St. Jeor and coauthors, 2001.

Weight Management: Overweight and Underweight

Chapter Outline

Overweight: *Fat Cell Development* • *Fat Cell Metabolism* • *Set-Point Theory*

Causes of Obesity: *Genetics* • *Environment*

Problems of Obesity: *Health Risks* • *Perceptions and Prejudices* • *Dangerous Interventions*

Aggressive Treatments of Obesity: *Drugs* • *Surgery*

Weight-Loss Strategies: *Eating Plans* • *Physical Activity* • *Behavior and Attitude* • *Weight Maintenance* • *Prevention* • *Public Health Programs*

Underweight: *Problems of Underweight* • *Weight-Gain Strategies*

Highlight: *Eating Disorders*

Nutrition Explorer CD-ROM Outline

Nutrition Animation: *Balancing Energy In and Energy Out*

Case Study: *Diet Strategies for Overweight and Obese Individuals*

Student Practice Test

Glossary Terms

Nutrition on the Net

© Ken Scott/Stone/Getty Images

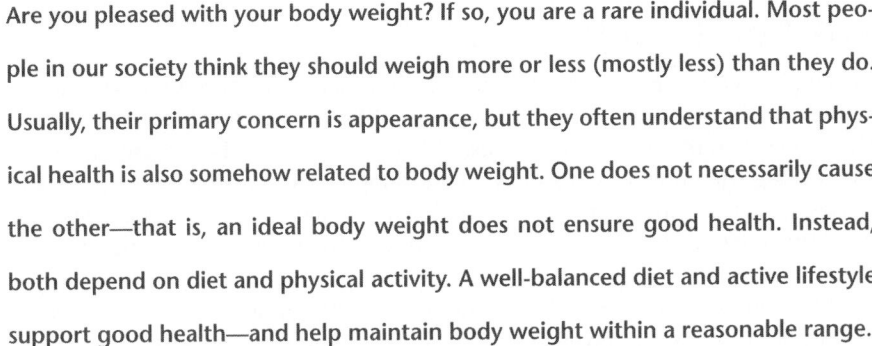

Nutrition in Your Life

Are you pleased with your body weight? If so, you are a rare individual. Most people in our society think they should weigh more or less (mostly less) than they do. Usually, their primary concern is appearance, but they often understand that physical health is also somehow related to body weight. One does not necessarily cause the other—that is, an ideal body weight does not ensure good health. Instead, both depend on diet and physical activity. A well-balanced diet and active lifestyle support good health—and help maintain body weight within a reasonable range.

The previous chapter described how body weight is stable when energy in equals energy out. Weight gains occur when energy intake exceeds energy expended, and conversely, weight losses occur when energy expended exceeds energy intake. At the extremes, both overweight and underweight present health risks. This chapter emphasizes overweight, partly because it has been more intensively studied and partly because it is a widespread health problem in developed countries and a growing concern in developing countries. Information on underweight is presented wherever appropriate. The highlight that follows this chapter delves into the eating disorders anorexia nervosa and bulimia nervosa.

Overweight

Despite our preoccupation with body image and weight loss, the prevalence of overweight and obesity in the United States continues to rise dramatically.[1] In the past decade, obesity increased in every state, in both genders, and across all ages, races, and educational levels (see Figure 9-1). Over half of the adults in the United States are now considered overweight or obese, as defined by a BMI of 25 or

■ BMI:
- Healthy weight: 18.5–24.9.
- Overweight: 25.0–29.9.
- Obese: ≥30.

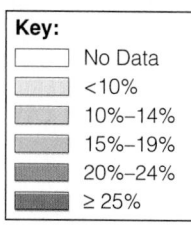

FIGURE 9-1 Increasing Prevalence of Obesity (BMI ≥30) among U.S. Adults

Key:
	No Data
	<10%
	10%–14%
	15%–19%
	20%–24%
	≥ 25%

1991: Only four states had obesity rates greater than 15 percent.

1996: Over half of the states had obesity rates greater than 15 percent.

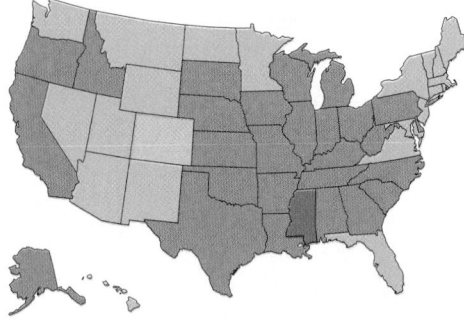

2001: Only one state had an obesity rate below 15 percent, most had obesity rates greater than 20 percent, and one had an obesity rate greater than 25 percent.

SOURCE: Mokdad A. H., et al. *Journal of the American Medical Association* 1999, 282:16; 2001, 286:10.

greater.■ The prevalence of overweight is especially high among women, the poor, blacks, and Hispanics.

The prevalence of overweight among children in the United States is also rising at an alarming rate. An estimated 15 percent of children and adolescents ages 6 to 19 years are overweight. Chapter 16 presents information on overweight during childhood and adolescence.

Obesity is so widespread and its prevalence is rising so rapidly that many refer to it as an **epidemic.**[2] According to the World Health Organization, this epidemic of obesity has spread worldwide, affecting over 300 million adults. Contrary to popular opinion, obesity is not limited to industrialized nations; over 115 million people in developing countries suffer from obesity-related problems. Before examining the suspected causes of obesity and the various strategies used to treat it, it may be helpful to understand the development and metabolism of body fat.

HEALTHY PEOPLE 2010

Increase the proportion of adults who are at healthy weight. Reduce the proportion of adults who are obese.

Fat Cell Development

When more energy is consumed than is spent, much of the excess energy is stored in the fat cells of adipose tissue. The amount of fat in a person's body reflects both the *number* and the *size* of the fat cells. The number of fat cells increases most rapidly during the growing years of late childhood and early puberty. After growth ceases, fat cell number may continue to increase whenever energy balance is positive. Obese people have more fat cells than healthy-weight people; their fat cells are also larger.

When energy intake exceeds expenditure, the fat cells accumulate triglycerides and expand in size (review Figure 5-20 on p. 157). When the cells enlarge, they stimulate cell proliferation so that their numbers increase again.[3] Thus obesity develops■ when a person's fat cells increase in number, in size, or quite often both. Figure 9-2 illustrates fat cell development.

When energy out exceeds energy in, the size of fat cells dwindles, but not their number. People with extra fat cells tend to regain lost weight rapidly; with weight gain, their many fat cells readily fill. In contrast, people with an average number of enlarged fat cells may be more successful in maintaining weight losses; when their cells shrink, both cell size and number are normal. Prevention of obesity is most critical, then, during the growing years when fat cells increase in number.

As mentioned, excess fat is typically stored in adipose tissue. This stored fat may be well tolerated, but fat accumulation in organs such as the heart or liver clearly plays a key role in the development of diseases such as heart failure or fatty liver.[4]■

Fat Cell Metabolism

The enzyme lipoprotein lipase (LPL)■ promotes fat storage in both adipose and muscle cells. Obese people generally have much more LPL activity in their fat cells than lean people do (their muscle cell LPL activity is similar, though). This high LPL activity makes fat storage especially efficient. Consequently, even modest excesses in energy intake have a more dramatic impact on obese people than on lean people.

The activity of LPL is partially regulated by gender-specific hormones—estrogen in women and testosterone in men. In women, fat cells in the breasts, hips, and thighs produce abundant LPL, putting fat away in those body sites; in men, fat

FIGURE 9-2 Fat Cell Development

Fat cells are capable of increasing their size by 20-fold and their number by several thousandfold.

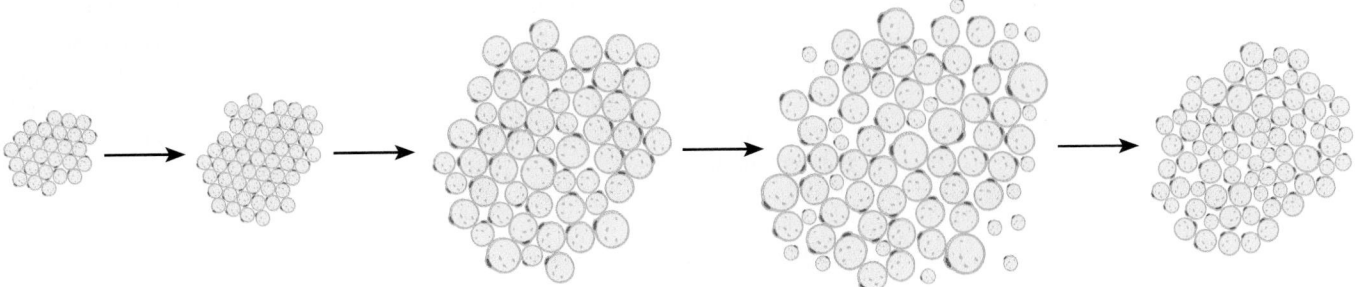

During growth, fat cells increase in number.

When energy intake exceeds expenditure, fat cells increase in size.

When fat cells have enlarged and energy intake continues to exceed energy expenditure, fat cells increase in number again.

With fat loss, the size of the fat cells shrinks, but not the number.

cells in the abdomen produce abundant LPL. This enzyme activity explains why men tend to develop central obesity around the abdomen whereas women more readily develop lower-body fat around the hips and thighs.

Gender differences are also apparent in the activity of the enzymes controlling the release and breakdown of fat in various parts of the body.[5] The release of lower-body fat is less active in women than in men, whereas the release of upper-body fat is similar. Furthermore, basal fat oxidation is lower in women than in men. Consequently, women may have a more difficult time losing fat in general, and from the hips and thighs in particular.

Enzyme activity may also explain why some people who lose weight regain it so easily. After weight loss, LPL activity increases, and it does so most dramatically in people who were fattest prior to weight loss. Apparently, weight loss serves as a signal to the gene that produces the LPL enzyme, saying "Make more enzyme to store fat." People easily regain weight after having lost it because they are battling against enzymes that want to store fat. The activities of these and other proteins provide an explanation for the observation that some inner mechanism seems to set a person's weight or body composition at a fixed point; the body will adjust to restore that **set point** if the person tries to change it.

Set-Point Theory

Many internal physiological variables, such as blood glucose, blood pH, and body temperature, remain fairly stable under a variety of conditions. The hypothalamus and other regulatory centers constantly monitor and delicately adjust conditions so as to maintain homeostasis.[6] The stability of such complex systems may depend on set-point regulators that maintain variables within specified limits.

Researchers have confirmed that after weight gains or losses, the body adjusts its metabolism so as to restore the original weight. Energy expenditure increases after weight gain and decreases after weight loss. These changes in energy expenditure differ from those expected based on body composition and help to explain why it is so difficult for an underweight person to maintain weight gains and an overweight person to maintain weight losses.

- Obesity due to an increase in the *number* of fat cells is **hyperplastic obesity**. Obesity due to an increase in the *size* of fat cells is **hypertrophic obesity**.

- The adverse effects of fat in nonadipose tissues are known as **lipotoxicity**.

- Reminder: *Lipoprotein lipase (LPL)* is an enzyme that hydrolyzes triglycerides passing by in the bloodstream and directs their parts into the cells, where they can be metabolized or reassembled for storage.

epidemic (EP-ee-DEM-ick): the appearance of a disease (usually infectious) or condition that attacks many people at the same time in the same region.
- **epi** = upon
- **demos** = people

set point: the point at which controls are set (for example, on a thermostat). The set-point theory that relates to body weight proposes that the body tends to maintain a certain weight by means of its own internal controls.

IN SUMMARY Fat cells develop by increasing in number and size. Prevention of excess weight gain depends on maintaining a reasonable number of fat cells. With gains or losses, the body adjusts to return to its previous status.

Causes of Obesity

Why do people accumulate excess body fat? The obvious answer is that they take in more food energy than they spend. But that answer falls short of explaining why they do this. Is it genetic? Environmental? Cultural? Behavioral? Socioeconomic? Psychological? Metabolic? All of these? Most likely, obesity has many interrelated causes. Why an imbalance between energy intake and energy expenditure occurs remains a bit of a mystery; the next sections summarize possible explanations.

Genetics

Genetics plays a true causative role in relatively few cases of obesity, for example, in Prader-Willi syndrome—a genetic disorder characterized by excessive appetite, massive obesity, short stature, and often mental retardation. Most cases of obesity, however, do not stem from a genetic mutation, yet genetic influences do seem to be involved.

Researchers have found that adopted children tend to be similar in weight to their biological parents, not to their adoptive parents. Studies of twins yield similar findings: identical twins are twice as likely to weigh the same as fraternal twins—even when reared apart. These findings suggest an important role for genetics in determining a person's *susceptibility* to obesity.[7] In other words, even if genes do not *cause* obesity, genetic factors may influence the food intake and activity patterns that lead to it and the metabolic pathways that maintain it.[8]

Clearly, something genetic makes a person more or less likely to gain or lose weight when overeating or undereating. Some people gain more weight than others on comparable energy intakes. Given an extra 1000 kcalories a day for 100 days, some pairs of identical twins gain less than 10 pounds while others gain up to 30 pounds. Within each pair, the amounts of weight gained, percentages of body fat, and locations of fat deposits are similar. Similarly, some people lose more weight than others following comparable exercise routines.

Researchers have been examining several genes in search of answers to obesity questions. As Chapter 6's section on protein synthesis described, each cell expresses only the genes for the proteins it needs, and each protein performs a unique function. The following paragraphs describe some recent research involving proteins that might help explain energy regulation and obesity development.[9]

Leptin Researchers have identified an obesity gene, called *ob,* that is expressed in the fat cells and codes for the protein **leptin.** Leptin acts as a hormone, primarily in the hypothalamus. Research suggests that leptin signals sufficient energy stores and promotes a negative energy balance by suppressing appetite and increasing energy expenditure.[10] Changes in energy expenditure primarily reflect changes in basal metabolism, but may also include changes in physical activity patterns.

Mice with a defective *ob* gene do not produce leptin and can weigh up to three times as much as normal mice and have five times as much body fat (see Figure 9-3). When injected with a synthetic form of leptin, the mice rapidly lose body fat. (Because leptin is a protein, it would be destroyed during digestion if given orally; consequently, it must be given by injection.) The fat cells not only lose fat, but they self-destruct (reducing cell number), which may explain why weight gains are delayed when the mice are fed again.

Although extremely rare, a genetic deficiency of leptin has been identified in human beings as well. An error in the gene that codes for leptin was discovered in three extremely obese children with barely detectable blood levels of leptin. Without leptin, the children have little appetite control; they are constantly hungry and eat considerably more than their siblings or peers. Given daily injections of leptin, these children lost a substantial amount of weight, confirming leptin's role in regulating appetite and body weight.[11]

leptin: a protein produced by fat cells under direction of the *ob* gene that decreases appetite and increases energy expenditure; sometimes called the *ob* **protein.**

• **leptos** = thin

FIGURE 9-3   Mice with and without Leptin Compared

Both of these mice have a defective *ob* gene. Consequently, they do not produce leptin. They both became obese, but the one on the right received daily injections of leptin, which suppressed food intake and increased energy expenditure, resulting in weight loss.

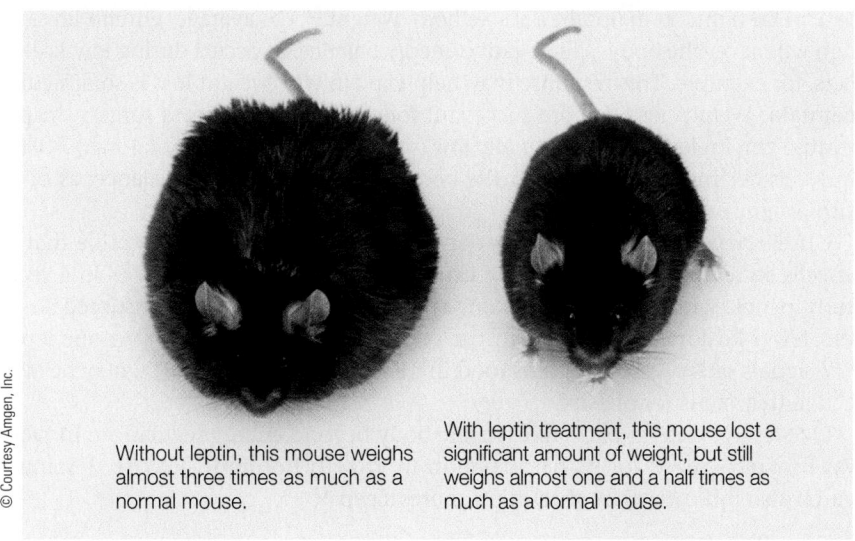

Without leptin, this mouse weighs almost three times as much as a normal mouse.

With leptin treatment, this mouse lost a significant amount of weight, but still weighs almost one and a half times as much as a normal mouse.

© Courtesy Amgen, Inc.

Not too surprisingly, leptin injections are only effective in suppressing appetite and supporting weight loss when overeating and obesity are the result of a leptin deficiency. Very few obese people have a leptin deficiency, however. In fact, blood levels of leptin usually correlate directly with body fat: the more body fat, the more leptin.[12] Obese people generally have high leptin levels, and weight gain increases leptin concentrations.[13] Researchers speculate that in obesity, leptin rises in an effort to overcome an insensitivity or resistance to leptin.[14]

Some researchers have reexamined the evidence on leptin from another point of view—one of undernutrition.[15] Instead of focusing on leptin's role as a satiety signal that might help prevent obesity by regulating food intake, they view leptin as a starvation hormone that signals energy deficits.[16] When energy intake is low, as occurs during starvation or undernutrition, leptin levels decline, and metabolism slows in an effort to reduce energy demands. Clearly, leptin plays a major role in energy regulation, but additional research is needed to clarify its actions when intake is either excessive or deficient.

In addition to serving as a satiety signal, leptin plays several other roles in the body.[17] For example, leptin may inform the female reproductive system about body fat reserves; stimulate growth of new blood vessels, especially in the cornea of the eye; enhance the maturation of bone marrow cells; promote formation of red blood cells; and help support a normal immune response.[18]

Ghrelin Researchers have recently discovered another protein that also acts as a hormone primarily in the hypothalamus, but it works in the opposite direction of leptin. Known as **ghrelin,** this protein is secreted primarily by the stomach cells and promotes a positive energy balance by stimulating appetite and promoting efficient energy storage.[19] The role ghrelin plays in regulating food intake and body weight is the subject of much intense research.[20]

Ghrelin triggers the desire to eat. Blood levels of ghrelin typically rise before and fall rapidly after a meal—reflecting the hunger and satiety that precede and follow eating.[21] In general, fasting blood levels correlate inversely with body weight: lean people have high ghrelin levels and obese people have low levels.[22] Interestingly, while ghrelin levels are high in underweight people, they are exceptionally high in anorexia nervosa and return to normal with nutrition intervention—indicating

ghrelin (GRELL-in): a protein produced by the stomach cells that enhances appetite and decreases energy expenditure.
• **ghre** = growth

that both body weight and nutrition status influence ghrelin levels.[23] Also noteworthy, ghrelin levels in Prader-Willi syndrome are markedly high and remain elevated even after a meal, which helps to explain the excessive appetite commonly seen in this disorder.[24] Similarly, ghrelin levels do not seem to decline after a meal in obese people, as they do for lean people.[25]

Ghrelin fights to maintain a stable body weight.[26] On average, ghrelin levels are high whenever the body is in negative energy balance, as occurs during low-kcalorie diets, for example. This response may help explain why weight loss is so difficult to maintain. Weight loss is more successful following gastric bypass surgery, in part, because ghrelin levels are abnormally low (why this is so remains unknown).[27] Ghrelin levels decline again whenever the body is in positive energy balance, as occurs with weight gains.[28]

Ghrelin levels also decline in response to high levels of PYY, a peptide that the GI cells secrete after a meal in proportion to the kcalories ingested.[29] In a recent study, people who were given PYY and then offered buffet meals consumed 30 percent fewer kcalories in a day than the control group.[30] Like the hormone leptin, PYY signals satiety and decreases food intake, but unlike leptin, PYY may be an effective treatment for obesity.

Like leptin, ghrelin plays roles in the body beyond energy regulation. In fact, it was first recognized for its participation in growth hormone activity.[31] Some research also indicates that ghrelin promotes sleep.[32]

Uncoupling Proteins Other genes code for proteins involved in energy metabolism. These proteins may influence the storing or expending of energy with different efficiencies or in different types of fat. The body has two types of fat: white and **brown adipose tissue.** White adipose tissue stores fat for other cells to use for energy; brown adipose tissue releases stored energy as heat. Recall from Chapter 7 that when fat is oxidized, some of the energy is released in heat and some is captured in ATP. In brown adipose tissue, oxidation may be uncoupled■ from ATP formation; it produces heat only. Radiating energy away as heat enables the body to spend, rather than store, energy. Heat production is particularly important in newborns, in adults who live in extremely cold climates, and in animals that hibernate; they have plenty of brown adipose tissue. In contrast, most human adults have small amounts of brown fat, and its role in body weight regulation, though probably minimal, is only just beginning to be understood.

Uncoupling proteins are active not only in brown fat, but in white fat and many other tissues as well.[33] Their actions seem to influence the basal metabolic rate (BMR) and oppose the development of obesity.[34] Animals with abundant amounts of these uncoupling proteins resist weight gain, whereas those with minimal amounts gain weight easily. Similarly, people with a genetic variant of an uncoupling protein have lower metabolic rates and are more overweight than others.[35] Whether the body dissipates the energy from an ice cream sundae as heat or stores it in body fat has major consequences for a person's body weight.

Environment

Although genetic studies indicate that body weight may be at least partially heritable, they do not fully explain obesity. In contrast to the studies mentioned earlier that found similar weights between identical twins, some identical twins have dramatically different body weights.[36] With obesity rates rising over the past three decades and the **gene pool** remaining relatively unchanged, environment must play a role as well. The environment includes all of the circumstances that we encounter daily that push us toward fatness or thinness. Keep in mind that genetic and environmental factors are not mutually exclusive; genes can influence eating behaviors, for example.

Overeating One explanation for obesity is that overweight people overeat, although diet histories may not always reflect high intakes. Diet histories are not always accurate records of actual intakes; both normal-weight and obese people commonly

■ Reminder: In *coupled reactions,* the energy released from the breakdown of one compound is used to create a bond in the formation of another compound. In *uncoupled reactions,* the energy is released as heat.

brown adipose tissue: masses of specialized fat cells packed with pigmented mitochondria that produce heat instead of ATP.

gene pool: all the genetic information of a population at a given time.

underreport their dietary intakes.[37] Most importantly, current dietary intakes may not reflect the eating habits that lead to obesity. Obese people who had a positive energy balance for years and accumulated excess body fat may not currently have a positive energy balance. This reality highlights an important point: the energy-balance equation must consider time. Both present *and* past eating and activity patterns influence current body weight.

We live in an environment that exposes us to an abundance of high-kcalorie, high-fat foods that are readily available, relatively inexpensive, heavily advertised, ■ and reasonably tasty.[38] Food is available everywhere, all the time—thanks largely to fast food. Our highways are lined with fast-food restaurants, and convenience stores and service stations offer fast food as well. Fast food is available in our schools, malls, and airports. It's convenient and it's available morning, noon, and night—and all times in between. Most alarming are the extraordinarily large serving sizes and ready-to-go meals that offer supersize■ combinations. People buy the large sizes and combinations, perceiving them to be a good value, but then they eat more than they need—a bad deal. Simply put, large portion sizes deliver more kcalories.[39] And portion sizes of virtually all foods and beverages have increased markedly in the past several decades, most notably at fast-food restaurants.[40] Not only have portion sizes increased over time, but they are now two to eight times larger than standard serving sizes.[41] The trend toward large portion sizes parallels the prevalence of overweight and obesity in the United States, beginning in the 1970s, increasing sharply in the 1980s, and continuing today.[42]

Restaurant food, especially fast food, is a major player in the development of obesity.[43] Fast food is often high in fat. Fat's 9 kcalories per gram quickly add up, amplifying people's energy intakes and enlarging their body fat stores. The combination of large portions and energy-dense foods is a double whammy. Reducing portion sizes is somewhat helpful, but the real kcalorie savings come from lowering the energy density.[44] After all, large portions of foods with low energy density such as fruits and vegetables can help with weight loss.

Physical Inactivity Our environment fosters physical inactivity as well. Life requires little exertion—escalators carry us up stairs, automobiles take us across town, buttons roll down windows, and remote controls change television channels. Modern technology has replaced physical activity at home, at work, and in transportation. Inactivity contributes to weight gain and poor health.[45] In turn, watching television, playing video games, and using the computer may contribute most to physical inactivity. The more time people spend in these sedentary activities, the more likely they are to be overweight.[46]

These sedentary activities contribute to weight gain in several ways. First, they require little energy beyond the resting metabolic rate. Second, they replace time spent in more vigorous activities. Watching television also influences food purchases and correlates with between-meal snacking on the high-kcalorie, high-fat foods most heavily advertised.

People may be obese, then, not because they eat too much, but because they move too little. Some obese people are so extraordinarily inactive that even when they eat less than lean people, they still have an energy surplus. Reducing their food intake further would jeopardize health and incur nutrient deficiencies. Physical activity is a necessary component of nutritional health. People must be physically active if they are to eat enough food to deliver all the nutrients they need without unhealthy weight gain. In fact, to prevent weight gain, the DRI■ suggests an accumulation of 60 minutes of moderately intense physical activities every day in addition to less intense activities of daily living.

■ The food industry spends $30 billion a year on advertising. The message? "Eat more."

■ Want fries with that? A supersize portion delivers over 600 kcalories.

■ DRI for physical activity: 60 min/day (moderate intensity).

IN SUMMARY Obesity has many causes and different combinations of causes in different people. Some causes, such as overeating and physical inactivity, may be within a person's control, and some, such as genetics, may be beyond it.

Problems of Obesity

An estimated 35 to 45 percent of all U.S. women (and 20 to 30 percent of U.S. men) are trying to lose weight at any given time, spending up to $40 billion each year to do so.[47] Some of these people do not even need to lose weight. Others may benefit from weight loss, but they are not successful; relatively few succeed, and even fewer succeed permanently. Whether an overweight person needs to lose weight is a question of health.

Health Risks

Chapter 8 described some of the health problems that commonly accompany obesity. In evaluating the risks to health from obesity, health care professionals use three indicators:[48]

- Body mass index■ (BMI, as described in Chapter 8).
- Waist circumference■ (also described in Chapter 8).
- Disease risk profile, taking into account family history, life-threatening diseases, and common risk factors for chronic diseases.[49]

The higher the BMI, the greater the waist circumference, and the more risk factors, the greater the urgency to treat obesity.

People can best decide whether weight loss might be beneficial by considering their health status and motivation. People who are overweight by BMI standards, but otherwise in good health, might not benefit from losing weight; they might focus on preventing further weight gains instead. In contrast, those who are obese and suffering from a life-threatening disease such as diabetes might improve their health substantially by adopting a diet and exercise plan that supports weight loss. Regarding motivation, a person needs to be ready and willing to make lifestyle changes.

Overweight in Good Health Often a person's motivations for weight loss have nothing to do with health. A healthy young woman with a BMI of 26■ might want to lose a few pounds for spring break, but doing so might not improve her health. In fact, if she opts for a starvation diet or diet pills, she would be healthier *not* trying to lose weight.

Obese or Overweight with Risk Factors Weight loss is recommended for people who are obese and those who are overweight (or who have a high waist circumference) with two or more risk factors for chronic diseases.■ A 50-year-old man with a BMI of 28■ who has high blood pressure and a family history of heart disease can improve his health by adopting a diet low in saturated fat and a regular exercise plan.

Obese or Overweight with Life-Threatening Condition Weight loss is also recommended for a person who is either overweight or obese and suffering from a life-threatening condition such as heart disease, diabetes, or sleep apnea.■ The health benefits of weight loss are clear. For example, a 30-year-old man with a BMI of 40■ might be able to prevent or control the diabetes that runs in his family by losing 75 pounds. The effort required to do so may be great, but it is far less than the effort and consequences of living with diabetes.

Perceptions and Prejudices

Many people assume that every obese person can achieve slenderness and should pursue that goal. First consider that most obese people cannot—for whatever reason—successfully lose weight and maintain their losses.[50] Then consider the prejudice involved in that assumption. People come with varying weight tendencies, just as they come with varying potentials for height and degrees of health, yet we

---

■ BMI 25.0–29.9 = overweight.
BMI ≥30 = obese.

■ Men: >102 cm (>40 in).
Women: >88 cm (>35 in).

■ For reference, a woman with a BMI of 26 might be:
- 5 ft 3 in, 146 lb.
- 5 ft 5 in, 156 lb.
- 5 ft 7 in, 166 lb.

■ Obese people and overweight people with two or more of these risk factors require aggressive treatment:
- Hypertension.
- Cigarette smoking.
- High LDL.
- Low HDL.
- Impaired glucose tolerance.
- Family history of heart disease.
- Men ≥45 yr; women ≥55 yr.

■ For reference, a man with a BMI of 28 might be:
- 5 ft 8 in, 184 lb.
- 5 ft 10 in, 195 lb.
- 6 ft, 206 lb.

■ Obese people and overweight people with any of these diseases require aggressive treatment:
- Heart disease.
- Diabetes (type 2).
- Sleep apnea, a disturbance of breathing during sleep, including temporarily stopping.

■ For reference, a man with a BMI of 40 might be:
- 5 ft 8 in, 265 lb.
- 5 ft 10 in, 280 lb.
- 6 ft, 295 lb.

do not expect tall people to shrink or healthy people to get sick in an effort to become "normal."

Social Consequences Large segments of our society place such enormous value on thinness that obese people face prejudice and discrimination on the job, at school, and in social situations: they are judged on their appearance more than on their character.[51] Socially, obese people are stereotyped as lazy and lacking in self-control. Such a critical view of overweight is not prevalent in many other cultures, including segments of our society. Instead, overweight is simply accepted or even embraced as a sign of robust health and beauty. Many overweight people today are tired of our obsession with weight control and simply want to be accepted as they are. To free our society of its obsession with body weight and prejudice against obesity, we must first learn to judge others for who they are and not for what they weigh.

Psychological Problems Psychologically, obese people may suffer embarrassment when others treat them with hostility and contempt, and some have even come to view their own bodies as grotesque and loathsome. Parents and friends may scold them for lacking the discipline to resolve their weight problems. Health care professionals, including dietitians, are among the chief offenders. Criticism from others hurts self-esteem. Feelings of rejection, shame, or depression are common among obese people.

Most weight-loss programs assume that the problem can be solved simply by applying willpower and hard work. If determination were the only factor involved, though, the success rate would be far greater than it is. Overweight people may readily assume the blame for their failures to lose weight and maintain the losses when, in fact, the programs have failed. Ineffective treatment and its associated sense of failure add to a person's psychological burden. Figure 9-4 illustrates how the devastating psychological effects of obesity and dieting perpetuate themselves.

Dangerous Interventions

People attach so many dreams of happiness to weight loss that they willingly risk huge sums of money for the slightest chance of success. As a result, weight-loss schemes flourish. Of the tens of thousands of claims, treatments, and theories for losing weight, few are effective—and many are downright dangerous. The negative effects must be carefully considered before embarking on any weight-loss program. Some interventions■ entail greater dangers than the risk of being overweight. Physical problems may arise from fad diets, "yo-yo" dieting, and drug use, and psychological problems may emerge from repeated "failures."

Some of the nation's most popular diet books and weight-loss programs have misled consumers with unsubstantiated claims and deceptive testimonials. Furthermore, they fail to provide an assessment of the short- and long-term results of their treatment plans, even though such evaluations are possible and would permit consumers to make informed decisions. Of course, some weight-loss programs are better than others in terms of cost, approach, and customer satisfaction, but few are particularly successful in helping people keep lost weight off. Clients can expect reputable programs to abide by a consumer bill of rights that explains the risks associated with weight-loss programs and provides honest predictions of success (see Table 9-1).

Fad Diets **Fad diets** often sound good, but typically fall short of delivering on their promises. They espouse exaggerated or false theories of weight loss and advise consumers to follow inadequate diets. Some fad diets are hazardous to health. Adverse reactions can be as minor as headaches, nausea, and dizziness or as serious as death. Table H8-3 in Highlight 8 (on p. 276) offers guidelines for identifying unsound weight-loss schemes and diets.

Over-the-Counter Drugs Millions of people in the United States use nonprescription weight-loss products. Most of them are women, especially young obese women, but almost 10 percent are of normal weight.[52] Only one over-the-counter

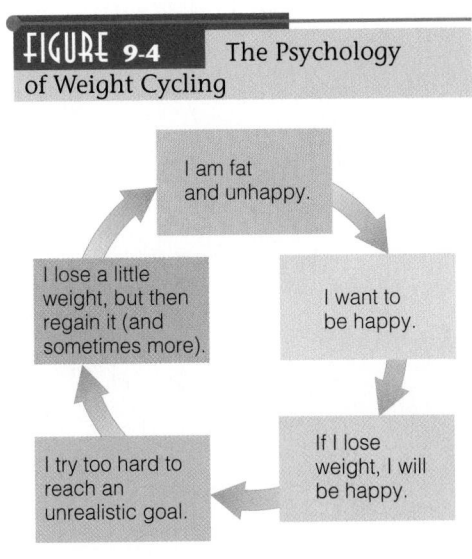

FIGURE 9-4 The Psychology of Weight Cycling

I am fat and unhappy.

I want to be happy.

If I lose weight, I will be happy.

I try too hard to reach an unrealistic goal.

I lose a little weight, but then regain it (and sometimes more).

■ Scrutinize fad diets, magic potions, and wonder gizmos with a healthy dose of skepticism.

fad diets: popular eating plans that promise quick weight loss. Most fad diets severely limit certain foods or overemphasize others (for example, never eat potatoes or pasta or eat cabbage soup daily).

TABLE 9-1	**Weight-Loss Consumer Bill of Rights (An Example)**

1. *WARNING:* Rapid weight loss may cause serious health problems. Rapid weight loss is weight loss of more than 1½ to 2 pounds per week or weight loss of more than 1 percent of body weight per week after the second week of participation in a weight-loss program.

2. Consult your personal physician before starting any weight-loss program.

3. Only permanent lifestyle changes, such as making healthful food choices and increasing physical activity, promote long-term weight loss and successful maintenance.

4. Qualifications of this provider are available upon request.

5. *YOU HAVE A RIGHT TO:*

   - Ask questions about the potential health risks of this program and its nutritional content, psychological support, and educational components.

   - Receive an itemized statement of the actual or estimated price of the weight-loss program, including extra products, services, supplements, examinations, and laboratory tests.

   - Know the actual or estimated duration of the program.

   - Know the name, address, and qualifications of the dietitian or nutritionist who has reviewed and approved the weight-loss program.

■ **Benzocaine** is marketed under the trade names:
- Diet Ayds (candy).
- Slim Mint (gum).

■ Read labels of over-the-counter products to determine if they contain **phenylpropanolamine** (fen-ill-pro-pa-NOLE-a-mean).

■ Ephedrine is an amphetamine-like substance extracted from the Chinese ephedra herb *ma huang.*

So many promises, so little success.

serotonin (SER-oh-tone-in): a neurotransmitter important in sleep regulation, appetite control, and sensory perception, among other roles. Serotonin is synthesized in the body from the amino acid tryptophan with the help of vitamin B_6.

medication to help with weight loss has been approved by the Food and Drug Administration (FDA). It contains benzocaine■ (in a candy or gum form), which anesthetizes the tongue, reducing taste sensations.

In 2000, the FDA recommended that manufacturers voluntarily discontinue marketing over-the-counter products containing phenylpropanolamine,■ an ingredient commonly used in products to suppress appetite.* Reported side effects include dry mouth, rapid pulse, nervousness, sleeplessness, hypertension, irregular heartbeats, kidney failure, seizures, and strokes.

Herbal Products and Dietary Supplements In their search for weight-loss magic, some consumers turn to "natural" herbal products and dietary supplements, even though few have proved to be effective.[53] St. John's wort, for example, contains substances that inhibit the uptake of **serotonin** and thus suppress appetite. In addition to the many cautions that accompany the use of any herbal remedies, consumers should be aware that St. John's wort is often prepared in combination with the herbal stimulant ephedrine.■ Ephedrine-containing supplements promote modest short-term weight loss (about 2 pounds a month), but the associated risks are high.[54] These supplements have been implicated in several cases of heart attacks and seizures and have been linked to about 100 deaths. For this reason, the FDA has banned the sale of dietary supplements containing ephedra.† Table 9-2 presents the claims and the dangers behind ephedrine and several other weight-loss supplements.[55]

Herbal laxatives containing senna, aloe, rhubarb root, cascara, castor oil, and buckthorn (or various combinations) are commonly sold as "dieter's tea." Such concoctions commonly cause nausea, vomiting, diarrhea, cramping, and fainting and may have contributed to the deaths of four women who had drastically reduced their food intakes. Consumers mistakenly believe that laxatives will diminish nutrient absorption and reduce kcalorie intake, but remember that absorption occurs primarily in the upper small intestine and these laxatives act on the lower large intestine. Highlight 18 explores the possible benefits and potential dangers of herbal products and other alternative therapies. As it explains, current laws do not require manufacturers of dietary supplements to test the safety or effectiveness of any product. Consumers cannot assume that an herb or supplement of any kind is safe or effective just because it is available on the market. Supplements may contain contaminants and may not contain the amounts of active ingredients listed on the labels.[56] Anyone using dietary supplements for weight loss should first consult with a physician.

---

*Phenylpropanolamine is not commercially available in Canada.
†Ma huang (ephedrine) is illegal in Canada.

TABLE 9-2 Selected Herbal and Other Dietary Supplements Marketed for Weight Loss

Product	Manufacturers' Claims	Research Findings	Adverse Effects
Chitosan[a] (pronounced KITE-oh-san; derived from chitin, the substance that forms the hard shells of lobsters, crabs, and other crustaceans)	Binds to dietary fat, preventing digestion and absorption	Ineffective	Impaired absorption of fat-soluble vitamins
Chromium (trace mineral)	Eliminates body fat	Ineffective; weight gain reported when not accompanied by exercise	Headaches, sleep disturbances, and mood swings; hexavalent form is toxic and carcinogenic
Conjugated linoleic acid (CLA; a group of fatty acids related to linoleic acid, but with different cis- and trans-configurations)	Reduces body fat and suppresses appetite	Some evidence in animal studies, but ineffective in human studies	None known
Ephedrine[b] (amphetamine-like substance derived from the Chinese ephedra herb ma huang)	Speeds body's metabolism	Weight loss and dangerous side effects	Insomnia, tremors, heart attacks, strokes, and death; FDA has banned the sale of these products
Hydroxycitric acid[c] (active ingredient derived from the rind of the tropical fruit garcinia cambogia)	Inhibits the enzyme that converts citric acid to fat; suppresses appetite	Ineffective	Toxicity symptoms reported in animal studies
Pyruvate[d] (3-carbon compound produced during glycolysis)	Speeds body's metabolism	Modest weight loss with high doses	GI distress
Triiodothyroacetic acid[e] (TRIAC, a potent thyroid hormone)	Speeds up body's metaboiosm	Weight loss and dangerous side effects	Diarrhea, fatigue, drowsiness, insomnia, nervousness, sweating, heart attacks, and strokes; FDA warning issued
Yohimbine (derived from the bark of a West African tree)	Promotes weight loss	Ineffective	Nervousness, insomnia, anxiety, dizziness, tremors, headaches, nausea, vomiting, hypertension

NOTE: The FDA has not approved the use of any of these products; most products are used in conjunction with a 1000- to 1800-kcalorie diet.
[a]Marketed under the trade names Chitorich, Exofat, Fat Breaker, Fat Blocker, Fat Magnet, Fat Trapper, and Fatsorb.
[b]Marketed under the trade names Diet Fuel, Metabolife, and Nature's Nutrition Formula One.
[c]Marketed under the trade names Ultra Burn, Citralean, CitriMax, Citrin, Slim Life, Brindleslim, Medislim, and Beer Belly Busters.
[d]Marketed under the trade names Exercise in a Bottle, Pyruvate Punch, Pyruvate-c, and Provate.
[e]Marketed under the trade name Triax Metabolic Accelerator.

Other Gimmicks Other gimmicks don't help with weight loss either. Hot baths do not speed up metabolism so that pounds can be lost in hours. Steam and sauna baths do not melt the fat off the body, although they may dehydrate people so that they lose water weight. Brushes, sponges, wraps, creams, and massages intended to move, burn, or break up "**cellulite**" do nothing of the kind, because there is no such thing as cellulite.

IN SUMMARY The question whether a person should lose weight depends on many factors: the extent of overweight, age, health, and genetic makeup among them. Not all obesity will cause disease or shorten life expectancy. Just as there are unhealthy, normal-weight people, there are healthy, obese people. Some people may risk more in the process of losing weight than in remaining overweight. Weight-loss diets and supplements can be physically and psychologically damaging.

■ The field of medicine that specializes in treating obesity is called **bariatrics**.
• **bar** = weight

cellulite (SELL-you-light or SELL-you-leet): supposedly, a lumpy form of fat; actually, a fraud. Fatty areas of the body may appear lumpy when the strands of connective tissue that attach the skin to underlying muscles pull tight where the fat is thick. The fat itself is the same as fat anywhere else in the body. If the fat in these areas is lost, the lumpy appearance disappears.

clinically severe obesity: a BMI of 40 or greater or a BMI of 35 or greater with additional risk factors. A less preferred term used to describe the same condition is morbid obesity.

Aggressive Treatments of Obesity

The degree of obesity and the risk of disease guide the selection of appropriate strategies for weight reduction. An overweight person in good health may need to improve eating habits and increase physical activity, but someone with **clinically severe obesity** may need more aggressive treatment■ options—drugs or surgery.[57]

Drugs

Based on new understandings of obesity's genetic basis and its classification as a chronic disease, much research effort has focused on drug treatments for obesity. Experts reason that if obesity is a chronic disease, it should be treated as such—and the treatment of most chronic diseases includes drugs. The challenge, then, is to develop an effective drug that can be used over time without adverse side effects or the potential for abuse. No such drug currently exists.[58]

Several drugs for weight loss have been tried over the years. When used as part of a long-term, comprehensive weight-loss program, drugs can help obese people to lose up to 10 percent of their initial weight and maintain that loss for at least a year.[59] Because weight regain commonly occurs with the discontinuation of drug therapy, treatment must be long term. Yet the long-term use of drugs poses risks. We don't yet know whether a person would benefit more from maintaining a 20-pound excess or from taking a drug for a decade to keep the 20 pounds off. Physicians must prescribe drugs appropriately, inform consumers of the potential risks, and monitor side effects carefully. Two prescription drugs are currently on the market: sibutramine and orlistat. One reduces food intake; the other reduces nutrient absorption.[60]

Sibutramine Sibutramine suppresses appetite.[*] The drug is most effective when used in combination with a reduced-kcalorie diet and increased physical activity. Side effects include dry mouth, headache, constipation, rapid heart rate, and high blood pressure. The FDA advises those with high blood pressure not to use sibutramine and others to monitor their blood pressure.

Orlistat Orlistat takes a different approach to weight control.[†] It inhibits pancreatic lipase activity in the GI tract, thus blocking dietary fat digestion and absorption by about 30 percent. The drug is taken with meals and is most effective when accompanied by a reduced-kcalorie, low-fat diet. Side effects include gas, frequent bowel movements, and reduced absorption of fat-soluble vitamins.

Other Drugs Several other drugs are currently under study, including combinations that both decrease appetite and increase metabolism. Another approach being tested uses anticancer drugs to deprive the body's fat stores of needed blood vessels.[61]

Surgery

Surgery as an approach to weight loss is justified in some specific cases of clinically severe obesity. Surgical procedures effectively limit food intake by reducing the capacity of the stomach and suppress hunger by reducing production of the hormone ghrelin.[62] They reduce the size of the outlet as well, so they delay the passage of food from the stomach into the intestine for digestion and absorption (see Figure 9-5). The results are dramatic: most people achieve a lasting weight loss of more than 50 percent of their excess body weight.[63]

The long-term safety and effectiveness of gastric surgery depend, in large part, on compliance with dietary instructions. Common immediate postsurgical complications include infections, nausea, vomiting, and dehydration; in the long term, vitamin and mineral deficiencies and psychological problems are common. Lifelong medical supervision is necessary for those who choose the surgical route, but in suitable candidates, the health benefits of weight loss may prove worth the risks.[64]

Another surgical procedure is used, not to treat obesity, but to remove the evidence. Plastic surgeons can extract some fat deposits by suction lipectomy, or "liposuction." This cosmetic procedure has little effect on body weight, but can alter body shape slightly in specific areas. Liposuction is a popular procedure in part be-

sibutramine (sigh-BYOO-tra-mean): a drug used in the treatment of obesity that slows the reabsorption of serotonin in the brain, thus suppressing appetite and creating a feeling of fullness.

orlistat (OR-leh-stat): a drug used in the treatment of obesity that inhibits the absorption of fat in the GI tract, thus limiting kcaloric intake.

---

[*]Sibutramine is marketed under the trade name Meridia.
[†]Orlistat is marketed under the trade name Xenical.

FIGURE 9-5 Surgical Procedures Used in the Treatment of Severe Obesity

Both of these surgical procedures reduce the size of the stomach. Notice that the first procedure maintains a relatively normal flow, whereas the second one bypasses most of the stomach, all of the duodenum, and some of the jejunum. The dark pink areas highlight the flow of food through the GI tract. The pale pink areas indicate the sections that have been bypassed.

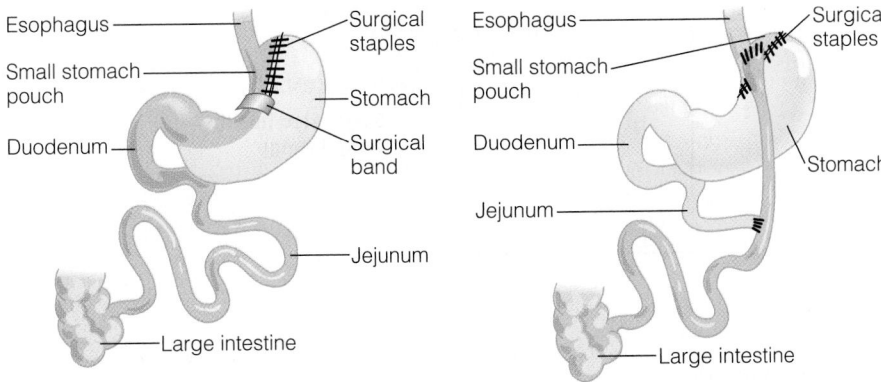

Esophagus — Surgical staples
Small stomach pouch
Duodenum
Stomach
Surgical band
Jejunum
Large intestine

Esophagus — Surgical staples
Small stomach pouch
Duodenum
Jejunum
Stomach
Large intestine

In vertical banded gastroplasty, the surgeon constructs a small stomach pouch and restricts the outlet from the stomach to the intestine.

In gastric bypass, the surgeon constructs a small stomach pouch and creates an outlet directly to the jejunum.

cause of its perceived safety, but, in fact, there can be serious complications that occasionally result in death.[65]

IN SUMMARY Obese people with high risks of medical problems may need aggressive treatment, including drugs or surgery. Others may benefit most from improving eating and exercise habits.

Weight-Loss Strategies

Successful weight-loss strategies embrace small changes, moderate losses, and reasonable goals. A 200-pound woman who loses 10 to 20 pounds in a year is much more likely to maintain losses and reap health benefits than if she were to drop down to 130 pounds in that same time. In keeping with this philosophy, the *Dietary Guidelines* suggest that for good health, a person should "aim for a healthy weight." The focus is not on weight loss per se, but on health gains. In fact, the *Guidelines* go on to say, "If you are already overweight, first aim to prevent further weight gain, and then lose weight to improve your health."

Modest weight loss, even when a person is still overweight, can improve control of diabetes and reduce the risks of heart disease by lowering blood pressure and blood cholesterol, especially for those with central obesity. Improvements in physical capabilities and bodily pain become evident with even a 5-pound weight loss.[66] For these reasons, parameters such as blood pressure, blood cholesterol, or even vitality are more useful than body weight in marking success. People less concerned with disease risks may prefer to set goals for personal fitness, such as being able to play with children or climb stairs without becoming short of breath. Importantly, they can enjoy living a healthy life instead of focusing on the elusive goal of losing weight.

Whether the goal is health or fitness, expectations need to be reasonable. Unreachable targets ensure frustration and failure. If goals are achieved or exceeded, there will be rewards instead of disappointments.

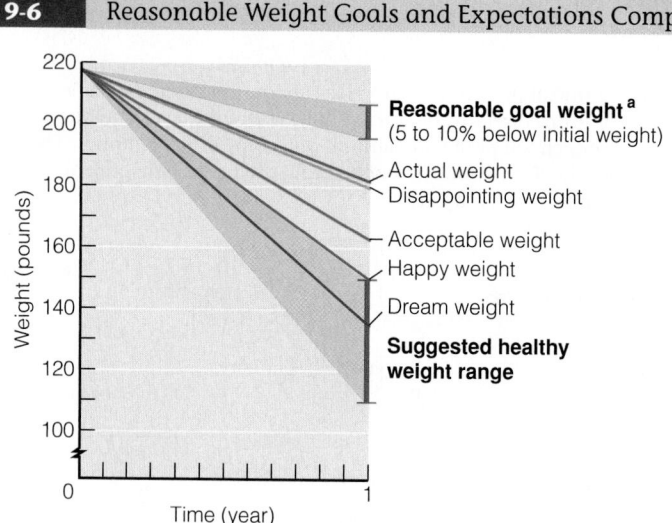

FIGURE 9-6 Reasonable Weight Goals and Expectations Compared

[a]Reasonable goal weights reflect pounds lost over time. Given more time, reasonable goals may eventually fall within the suggested healthy-weight range.

SOURCE: Adapted from G. D. Foster and coauthors, What is a reasonable weight loss? Patients' expectations and evaluations of obesity treatment outcomes, *Journal of Consulting and Clinical Psychology* 65 (1997): 79–85.

Research findings highlight the great disparity between lofty expectations and reasonable success.[67] Before beginning a weight-loss program, obese women identified the weights they would describe as "dream," "happy," "acceptable," and "disappointing" (see Figure 9-6). All of these weights were below their starting weight. Their goal weights far exceeded the 5 to 10 percent recommended by experts, or even the 15 percent reported by the most successful weight-loss studies. Even their "disappointing" weights exceeded recommended goals. Close to a year later, and after an average loss of 35 pounds, almost half of the women did not achieve even their "disappointing" weights. They did, however, experience more physical, social, and psychological benefits than they had predicted for that weight. Still, in a culture that overvalues thinness, these women were not satisfied with a 16 percent reduction in weight—not because their efforts were unsuccessful, but because their expectations were unrealistic.

Depending on initial body weight, a reasonable rate of weight loss for overweight people is ½ to 2 pounds a week,■ or 10 percent of body weight over six months.[68] For a person weighing 250 pounds, a 10 percent loss is 25 pounds, or about 1 pound a week for six months. Such gradual weight losses are more likely to be maintained than rapid losses. Keep in mind that pursuing good health is a lifelong journey. Most adults are keenly aware of their body weights and shapes and realize that what they eat and what they do can make a difference to some extent. Those who are most successful at weight management seem to have fully incorporated healthful eating and physical activity into their daily lives.[69] Such advice—to reduce kcalorie intake and increase physical activity—would hardly surprise anyone, yet only one in five people trying to control their weight follows these recommendations.[70]

Eating Plans

Contrary to the claims of fad diets, no one food plan is magical, and no specific food must be included or avoided in a weight-management program. In designing a plan, people need only consider foods that they like or can learn to like, that are available, and that are within their means.

■ Safe rate for weight loss:
- ½ to 2 lb/week.
- 10% body weight/6 mo.

TABLE 9-3 Recommendations for a Weight-Loss Diet

Nutrient	Recommended Intake
kCalories	
For people with BMI ≥ 35	Approximately 500 to 1000 kcalories per day reduction from usual intake
For people with BMI between 27 and 35	Approximately 300 to 500 kcalories per day reduction from usual intake
Total fat	30% or less of total kcalories
Saturated fatty acids[a]	8 to 10% of total kcalories
Monounsaturated fatty acids	Up to 15% of total kcalories
Polyunsaturated fatty acids	Up to 10% of total kcalories
Cholesterol[a]	300 mg or less per day
Protein[b]	Approximately 15% of total kcalories
Carbohydrate[c]	55% or more of total kcalories
Sodium chloride	No more than 2400 mg of sodium or approximately 6 g of sodium chloride (salt) per day
Calcium	1000 to 1500 mg per day
Fiber[c]	20 to 30 g per day

[a]People with high blood cholesterol should aim for less than 7 percent kcalories from saturated fat and 200 milligrams of cholesterol per day.
[b]Protein should be derived from plant sources and lean sources of animal protein.
[c]Carbohydrates and fiber should be derived from vegetables, fruits, and whole grains.
SOURCE: National Institutes of Health Obesity Education Initiative, *The Practical Guide: Identification, Evaluation, and Treatment of Overweight and Obesity in Adults* (Washington, D.C.: U.S. Department of Health and Human Services, 2000), p. 27.

Be Realistic about Energy Intake The main characteristic of a weight-loss diet is that it provides less energy than the person needs to maintain present body weight. If food energy is restricted too severely, dieters may not receive sufficient nutrients and may lose lean tissue. Rapid weight loss usually means excessive loss of lean tissue, a lower BMR, and a rapid weight gain to follow. In addition, restrictive eating may set in motion the unhealthy behaviors of eating disorders (described in Highlight 9).

Table 9-3 outlines the recommendations of a weight-loss diet. Energy intake should provide nutritional adequacy without excess—that is, somewhere between deprivation and complete freedom to eat whatever, whenever. A reasonable suggestion is that an adult needs to increase activity and reduce food intake enough to create a deficit of 500 kcalories per day.[71] Such a deficit produces a weight loss of about 1 pound per week—a rate that supports the loss of fat efficiently while retaining lean tissue. In general, weight-loss diets provide 1200 to 1600 kcalories a day.[72]

Emphasize Nutritional Adequacy Nutritional adequacy is difficult to achieve on fewer than 1200 kcalories a day, and most healthy adults need never consume any less than that. A plan that provides an adequate intake supports a healthier and more successful weight loss than a restrictive plan that creates feelings of starvation and deprivation, which can lead to an irresistible urge to binge.

Following the Food Guide Pyramid's■ minimum suggested servings supplies about 1600 kcalories, with 30 percent of kcalories as fat. Such an intake would allow most people to lose weight and still meet their nutrient needs with careful, nutrient-dense food selections. (Women might need iron or calcium supplements.) Keep in mind, too, that well-balanced diets that emphasize fruits, vegetables, whole grains, lean meats or meat alternates, and low-fat milk products offer many health rewards even when they don't result in weight loss.[73]

Eat Small Portions As mentioned earlier, portion sizes at markets, at restaurants, and even at home have increased dramatically over the years.[74] We have come to expect large portions, and we have learned to clean our plates. Many of us pay more attention to these outside cues defining how much to eat than to our internal cues

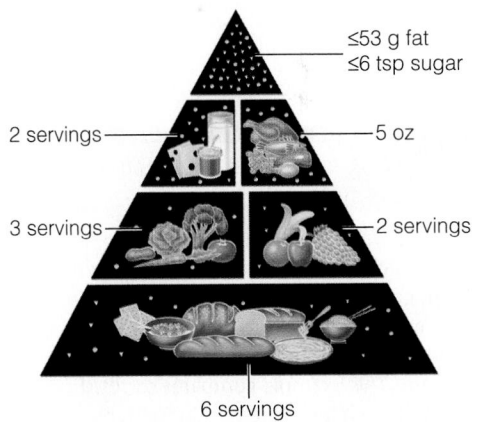

≤53 g fat
≤6 tsp sugar
2 servings — — 5 oz
3 servings — — 2 servings
6 servings

■ For an adequate 1600-kcalorie diet, select the minimum number of suggested servings. To reduce kcalories further, choose low-kcalorie and low-fat options from within each group and restrict foods from the tip.

FIGURE 9-7 Energy Density

Decreasing the energy density (kcal/g) of foods allows a person to eat satisfying portions while still reducing energy intake. To lower energy density, select foods high in water or fiber and low in fat.

Selecting grapes with their high water content instead of raisins increases the volume and cuts the energy intake in half.

Even at the same weight and similar serving sizes, the fiber-rich broccoli delivers twice the fiber of the potatoes for about one-fourth the energy.

By selecting the water-packed tuna (on the right), a person can enjoy the same amount for fewer kcalories.

of hunger and satiety. For health's sake, we may need to learn to eat less food at each meal—one piece of chicken for dinner instead of two, a teaspoon of butter on vegetables instead of a tablespoon, and one cookie for dessert instead of six. The goal is to eat enough food for energy, nutrients, and pleasure, but not more. This amount should leave a person feeling satisfied—not stuffed.

Keep in mind that even fat-free and low-fat foods can deliver a lot of kcalories when a person eats large quantities. A low-fat cookie or two can be a sweet treat even on a weight-loss diet, but larger portions defeat the savings.

Lower Energy Density Most people take their cues about how much to eat based on portion sizes and typically choose similar amounts from day to day.[75] To lower energy intake, a person can either reduce the portion size or reduce the energy density.[76] Figure 9-7 illustrates how water, fiber, and fat influence energy density, and the accompanying "How to" feature compares foods based on their energy density. Foods containing water, those rich in fiber, and those low in fat help to lower energy density, providing more satiety for fewer kcalories.

Remember Water Water helps with weight management in several ways. For one, foods with a high water content (such as broth-based soups) increase fullness, reduce hunger, and consequently reduce energy intake.[77] For another, drinking water fills the stomach between meals and satisfies the thirst that was formerly met by eating extra food (remember that foods provide water). Water also helps the GI tract adapt to a high-fiber diet.

Focus on Complex Carbohydrates Healthy meals and snacks center on complex carbohydrate foods. Fresh

HOW TO Compare Foods Based on Energy Density

Chapter 2 described how to evaluate foods based on their nutrient density—their nutrient contribution per kcalorie. Another way to evaluate foods is to consider their energy density—their energy contribution per gram. This example compares carrot sticks with french fries. The conclusion is no surprise, but understanding the mathematics may offer valuable insight into the concept of energy density. A carrot weighing 72 grams delivers 31 kcalories. To calculate the energy density, divide kcalories by grams:

$$\frac{31 \text{ kcal}}{72 \text{ g}} = 0.43 \text{ kcal/g.}$$

Do the same for french fries weighing 50 grams and contributing 167 kcalories:

$$\frac{167 \text{ kcal}}{50 \text{ g}} = 3.34 \text{ kcal/g.}$$

The more kcalories per gram, the greater the energy density. French fries are more energy dense than carrots. They provide more energy per gram—and per bite. Considering a food's energy density is especially useful in planning diets for weight management. Foods with a high energy density help with weight gain, whereas foods with a low energy density help with weight loss.

fruits, vegetables, legumes, and whole grains offer abundant vitamins, minerals, and fiber but little fat. Consequently, high-carbohydrate diets tend to be relatively low in energy and high in nutrients.[78]

High-fiber foods also require effort to eat—an added bonus. People who eat these foods in abundance spontaneously eat for longer times and take in fewer kcalories than when eating foods of high energy density. The satiety signal indicating fullness is sent after a 20-minute lag, so a person who slows down and savors each bite eats less before the signal reaches the brain. Of course, much depends on whether the person pays attention to internal satiety signals and stops eating or responds to external cognitive influences and continues.

Choose Fats Sensibly Ideally, a weight-loss diet will be both high in fiber and low in fat.[79] Lowering the fat content of a food lowers its energy density—for example, selecting fat-free milk instead of whole milk. That way, a person can consume the usual amount (say, a cup of milk) at a lower energy intake (85 instead of 150 kcalories).

Fat has a weak satiating effect, and satiation plays a key role in determining food intake during a meal. Consequently, a person eating a high-fat meal raises energy intake by adding both more food and more fat kcalories. For these reasons, measure fat with extra caution. Less fat in the diet means less fat in the body (review p. 165 for strategies to lower fat in the diet). Be careful not to take this advice to extremes, however; too little fat in the diet or in the body carries health risks as well, as Chapter 5 explained.

Whether a low-fat diet is the best option for weight loss is the subject of some controversy and much debate. An important point to notice in any discussion on weight-loss diets is total energy intake. *Low fat* simply means the energy derived from fat is relatively low compared with the total energy intake; it does not mean total energy intake is low. And reducing energy intake to less than expended is essential for weight loss. One way to lower energy intake is to lower fat intake. In these cases, adopting a low-fat diet can help with weight loss.[80]

Another currently popular way to lower energy intake is to lower carbohydrate intake. Highlight 8 discusses these diets fully, but findings from a recent study are worth mentioning here as well.[81] In this study, people were randomly assigned to one of two diets—either a low-carbohydrate diet or a low-fat diet. They were given descriptions of the diets and then fed themselves, as would be typical of many dieters. Both groups lost weight, but those on the low-carbohydrate diet lost more weight during the first six months; their diets produced a greater energy deficit. Interestingly, the differences in weight loss between the two groups disappeared by the end of one year. Between six months and one year, weight remained fairly stable in the low-fat group, but regains were evident in the low-carbohydrate group. These findings highlight an important point: weight loss requires a commitment to long-term changes in food choices. They also confirm another critical point: weight loss depends on a low energy intake—not the proportion of energy nutrients.[82]

Watch for Other Empty kCalories A person trying to achieve or maintain a healthy weight needs to pay attention not only to fat, but to sugar and alcohol, too.[83] Using them for pleasure on occasion is compatible with health as long as most daily choices are of nutrient-dense foods. Not only does alcohol add kcalories, but accompanying mixers can also add both kcalories and fat, especially in creamy drinks such as piña coladas (review Table H7-3 on p. 246). Furthermore, drinking alcohol reduces a person's inhibitions, which can sabotage weight-control efforts—at least temporarily.

IN SUMMARY A person who adopts a lifelong "eating plan for good health" rather than a "diet for weight loss" will be more likely to keep the lost weight off. Table 9-4 provides several tips for successful weight management.

TABLE 9-4	Weight-Management Strategies

In General

- Focus on healthy eating and activity habits, not on weight losses or gains.
- Adopt reasonable expectations about health and fitness goals and about how long it will take to achieve them.
- Make nutritional adequacy a high priority.
- Learn, practice, and follow a healthful eating plan for the rest of your life.
- Participate in some form of physical activity regularly.
- Adopt permanent lifestyle changes to achieve and maintain a healthy weight.

For Weight Loss

- Energy out should exceed energy in by about 500 kcalories/day. Increase your physical activity enough to spend more energy than you consume from foods.
- Emphasize foods with a low energy density and a high nutrient density.
- Eat small portions. Share a restaurant meal with a friend or take home half for lunch tomorrow.
- Eat slowly.
- Limit high-fat foods. Make legumes, whole grains, vegetables, and fruits central to your diet plan.
- Limit low-fat treats to the serving size on the label.
- Limit concentrated sweets and alcoholic beverages.
- Drink a glass of water before you begin to eat and another while you eat. Drink plenty of water throughout the day (8 glasses or more a day).
- Keep a record of diet and exercise habits; it reveals problem areas, the first step toward improving behaviors.
- Learn alternative ways to deal with emotions and stresses.
- Attend support groups regularly or develop supportive relationships with others.

For Weight Gain

- Energy in should exceed energy out by at least 500 kcalories/day. Increase your food intake enough to store more energy than you spend in exercise. Exercise and eat to build muscles.
- Expect weight gain to take time (1 pound per month would be reasonable).
- Emphasize energy-dense foods.
- Eat at least three meals a day.
- Eat large portions of foods and expect to feel full.
- Eat snacks between meals.
- Drink plenty of juice and milk.

The key to good health is to combine sensible eating with regular exercise.

Physical Activity

The best approach to weight management combines diet and physical activity. To prevent weight gains and support weight losses, current recommendations advise 60 minutes of moderately intense physical activity a day in addition to activities of daily life.[84] People who combine diet and exercise typically lose more fat, retain more muscle, and regain less weight than those who only diet. Even when people who include physical activity in their weight-management program do not lose more weight, they seem to follow their diet plans more closely and maintain their losses better than those who do not exercise.[85] Consequently, they benefit both from taking in a little less energy and from expending a little more energy in physical activity. Importantly, those who exercise reduce abdominal obesity and improve their blood pressure, insulin resistance, and cardiorespiratory fitness, regardless of weight loss.[86] Chapter 14 presents the many health benefits of physical activity; the focus here is on its role in weight management.

Activity and Energy Expenditure Table 8-2 (on p. 258) shows how much energy each of several activities uses. The number of kcalories spent in an activity depends on body weight, intensity, and duration. For example, a person who weighs 150 pounds and walks 3½ miles in 60 minutes expends about 315 kcalories. That same person running 3 miles in 30 minutes uses a similar amount. By comparison, a 200-pound person running 3 miles in 30 minutes expends an additional 100 kcalories or so. The goal is to expend as much energy as your time allows. The greater the energy deficit created by exercise, the greater the fat loss. And be careful not to compensate for the energy spent in exercise by eating more food. Otherwise, energy balance won't shift and fat loss will be less significant.

Activity and Metabolism Activity also contributes to energy expenditure in an indirect way—by speeding up metabolism. It does this both immediately and over the long term. On any given day, metabolism remains slightly elevated for several hours after intense and prolonged exercise.■ Over the long term, a person who engages in daily vigorous activity gradually develops more lean tissue. Metabolic rate rises accordingly, and this supports continued weight loss or maintenance.

Activity and Body Composition Physically active people have less body fat than sedentary people do—even if they have the same BMI.[87] Physical activity, even without weight loss, changes body composition: body fat decreases and lean body mass increases.[88] Furthermore, exercise specifically decreases abdominal fat.[89]

Activity and Appetite Control Physical activity also helps to control appetite. Many people think that exercising will make them eat more, but this is not entirely true. Active people do have healthy appetites, but immediately after an intense workout, most people do not feel like eating. They may be thirsty and want to shower, but they are not hungry. The reason is that the body has released fuels from storage to support the exercise, so glucose and fatty acids are abundant in the blood. At the same time, the body has suppressed its digestive functions. Hard physical work and eating are not compatible. A person must calm down, put energy fuels back in storage, and relax before eating. Thus exercise may actually help curb appetite, especially the inappropriate appetite that accompanies boredom, anxiety, or depression. Weight-management programs encourage people who feel the urge to eat when not hungry to go out and exercise instead. The activity passes time, relieves anxiety, and prevents inappropriate eating.

Activity and Psychological Benefits Activity also helps reduce stress. Since stress itself cues inappropriate eating for many people, activity can help here, too. In addition, the fit person looks and feels healthy and, as a result, gains self-esteem. High self-esteem motivates a person to persist in seeking good health and fitness, which keeps the beneficial■ cycle going.

Choosing Activities Clearly, physical activity■ is a plus in a weight-management program. What kind of physical activity is best? People should choose activities that they enjoy and are willing to do regularly. What schedule of physical activity is best? It doesn't matter; whether a person chooses several short bouts of exercise or one continuous workout, the fitness and weight-loss benefits are the same—and any activity is better than being sedentary.[90]

Health care professionals frequently advise people to engage in activities of low-to-moderate intensity for a long duration, such as an hour-long fast-paced walk. The reasoning behind such advice is that people exercising at low-to-moderate intensity are more likely to stick with their activity for longer times and are less likely to injure themselves. A person who stays with an activity routine long enough to enjoy the rewards will be less inclined to give it up and will, over the long term, reap many health benefits. Activity of low-to-moderate intensity■ that expends at least 2000 kcalories per week is especially helpful for weight management.[91]

In addition to exercise, a person can incorporate hundreds of energy-spending activities into daily routines: take the stairs instead of the elevator, walk to the neighbor's apartment instead of making a phone call, and rake the leaves instead of using a blower. Remember that sitting uses more kcalories than lying down, standing uses more kcalories than sitting, and moving uses more kcalories than standing. A 175-pound person who replaces a 30-minute television program with

Nutrition Explorer Examine how much activity is needed to expend the energy contained in some commonly eaten foods.

■ This postexercise effect raises the energy expenditure of exercise by about 15 percent.

■ Benefits of physical activity in a weight-management program:
- Short-term increase in energy expenditure (from exercise and from a slight rise in metabolism).
- Long-term increase in BMR (from an increase in lean tissue).
- Improved body composition.
- Appetite control.
- Stress reduction and control of stress eating.
- Physical, and therefore psychological, well-being.
- Improved self-esteem.

Chapter 14 presents additional benefits of physical activity.

■ For an active life, limit sedentary activities, engage in strength and flexibility activities and enjoy leisure activities often, engage in vigorous activities regularly, and be as active as possible every day (see the activity pyramid in Chapter 14).

■ Estimated energy expended when walking at a moderate pace = 1 kcal/mi/kg body wt.

a 2-mile walk a day can spend enough energy to lose (or at least not gain) 18 pounds in a year. One program recommends an activity goal of 10,000 steps a day. By wearing a pedometer, a person can easily track a day's activities without measuring miles or watching the clock. The point is: be active. Walk. Run. Swim. Dance. Cycle. Climb. Skip. Do whatever you enjoy doing—and do it often.

Spot Reducing People sometimes ask about "spot reducing." Unfortunately, muscles do not "own" the fat that surrounds them. Fat cells all over the body release fat in response to the demand of physical activity for use by whatever muscles are active. No exercise can remove the fat from any one particular area.

Exercise can help with trouble spots in another way, though. The "trouble spot" for most men is the abdomen, their primary site of fat storage. During aerobic exercise, abdominal fat readily releases its stores, providing fuel to the physically active body. With regular exercise and weight loss, men will deplete these abdominal fat stores before those in the lower body. Women may also deplete abdominal fat with exercise, but their "trouble spots" are more likely to be their hips and thighs.

In addition to aerobic activity, strength training can help to improve the tone of muscles in a trouble area, and stretching to gain flexibility can help with associated posture problems. A combination of aerobic, strength, and flexibility workouts best improves fitness and physical appearance.

IN SUMMARY Physical activity should be an integral part of a weight-control program. Physical activity can increase energy expenditure, improve body composition, help control appetite, reduce stress and stress eating, and enhance physical and psychological well-being.

Behavior and Attitude

Behavior modification once held a key position in weight-loss programs, but its status has diminished in recent years. Still, behavior and attitude play an important role in supporting efforts to achieve and maintain appropriate body weight and composition. Changing the hundreds of small behaviors of overeating and underexercising that lead to, and perpetuate, obesity requires time and effort. A person must commit to take action.

Adopting a positive, matter-of-fact attitude helps to ensure success. Healthy eating and activity choices are an essential part of healthy living and should simply be incorporated into the day—much like brushing one's teeth or wearing a safety belt.

Become Aware of Behaviors To solve a problem, a person must first identify all the behaviors that created the problem. Keeping a record will help to identify eating and exercise behaviors that may need changing (see Figure 9-8). It will also establish a baseline against which to measure future progress.

Change Behaviors Strategies■ focus on learning desired eating and exercise behaviors and eliminating unwanted behaviors. With so many possible behavior changes, a person can choose where to begin. Start simply and don't try to master them all at once. Attempting too many changes at one time can be overwhelming. Pick one trouble area that is manageable and start there. Practice a desired behavior until it becomes routine. Then select another trouble area to work on, and so on. Another bit of advice along the same lines: don't try to tackle major changes during a particularly stressful time of life.

Personal Attitude For many people, overeating and being overweight have become an integral part of their identity. Those who fully understand their personal relationships with food are best prepared to make healthful changes in eating and exercise behaviors.

Sometimes habitual behaviors that are hazardous to health, such as smoking or drinking alcohol, contribute positively by helping people adapt to stressful situa-

---

■ Examples of behavioral strategies to support weight change:
- Do not grocery shop when hungry.
- Eat slowly (pause during meals, chew thoroughly, put down utensils between bites).
- Exercise when watching television.

---

behavior modification: the changing of behavior by the manipulation of antecedents (cues or environmental factors that trigger behavior), the behavior itself, and consequences (the penalties or rewards attached to behavior).

FIGURE 9-8   Food Record

The entries in a food record should include the times and places of meals and snacks, the types and amounts of foods eaten, and a description of the individual's feelings when eating. The diary should also record physical activities: the kind, the intensity level, the duration, and the person's feelings about them.

Time	Place	Activity or food eaten	People present	Mood
10:30–10:40	School vending machine	6 peanut butter crackers and 12 oz. cola	by myself	Starved
12:15–12:30	Restaurant	Sub sandwich and 12 oz. cola	friends	relaxed & friendly
3:00–3:45	Gym	Weight training	work out partner	tired
4:00–4:10	Snack bar	Small frozen yogurt	by myself	OK

tions. Similarly, many people overeat to cope with the stresses of life. To break out of that pattern, they must first identify the particular stressors that trigger the urge to overeat. Then, when faced with these situations, they must learn and practice problem-solving skills that will help them to respond appropriately.[92]

All this is not to imply that psychological therapy holds the magic answer to a weight problem. Still, efforts to improve one's general well-being may result in healthy eating and activity habits even when weight loss is not the primary goal. When the problems that trigger the urge to overeat are resolved in alternative ways, people may find they eat less. They may begin to respond appropriately to internal cues of hunger rather than inappropriately to external cues of stress. Sound emotional health supports a person's ability to take care of physical health in all ways—including nutrition, weight management, and fitness.

Support Groups Group support can prove helpful when making life changes. Some people find it useful to join a group such as Take Off Pounds Sensibly (TOPS), Weight Watchers (WW), Overeaters Anonymous (OA), or others. Some dieters prefer to form their own self-help groups or find support online. The Internet offers numerous opportunities for weight-loss education and counseling that may be effective alternatives to face-to-face programs.[93] As always, consumers need to choose wisely and avoid rip-offs.

Increase the proportion of worksites that offer nutrition or weight management classes or counseling.

HEALTHY PEOPLE 2010

IN SUMMARY   A surefire remedy for obesity has yet to be found, although many people find a combination of the approaches just described to be most effective. Diet and exercise shift energy balance so that more energy is being spent than is taken in. Physical activity increases energy expenditure, builds lean tissue, and improves health. Energy intake should be reduced by 500 to 1000 kcalories per day. Behavior modification retrains habits to support a healthy eating and exercise plan. This treatment package requires time, individualization, and sometimes the assistance of a registered dietitian.

Weight Maintenance

People who are successful often experience much of their weight loss within half a year and then reach a plateau. This slowdown can be disappointing, but should be recognized as an opportunity for the body to adjust to its new weight. Reaching a plateau provides a little relief from the distraction of weight-loss dieting. An appropriate goal at this point is to continue eating and activity behaviors that will maintain weight. Attempting to lose additional weight at this point would require heroic efforts and would almost certainly meet with failure.

The prevalence of **successful weight-loss maintenance** is difficult to determine, in part because researchers have used different criteria.[94] Some look at success after one year and others after five years; some quantify success as 10 or more pounds lost and others as 5 or 10 percent of initial body weight lost. Furthermore, most research studies examine the success of one episode of weight loss in a structured program, but this scenario does not necessarily reflect the experiences of the general population. In reality, most people have lost weight several times in their lifetimes and did so on their own, not in a formal program. One survey reports that almost 50 percent of the overweight people who intentionally lost at least 10 percent of their initial body weight maintained the loss for at least a year and 25 percent maintained it for at least five years.[95] Using the same criteria, a review of research studies suggests a success rate of approximately 20 percent.[96]

Those who are successful in maintaining their weight loss have established vigorous exercise regimens and careful eating patterns, taking in less energy and a lower percentage of kcalories from fat than the national average.[97] Because these people are more efficient at storing fat, they do not have the same flexibility in their food and activity habits as their friends who have never been overweight.[98] With weight loss, metabolism shifts downward so that formerly overweight people require less energy than might be expected given their current body weight and body composition.[99] Consequently, to keep weight off, they must either eat less or exercise more than people the same size who have never been obese.

Physical activity plays a key role in maintaining weight.[100] Those who exercise vigorously are far more successful than those who are inactive.[101] On average, weight maintenance requires a person to expend about 2000 kcalories in physical activity per week.[102] To accomplish this, a person might exercise either moderately (such as brisk walking) for 60 minutes a day or vigorously (such as fast bicycling) for 35 minutes a day, for example.

In addition to limiting energy intake and exercising regularly, one other strategy may help with weight maintenance: frequent self-monitoring.[103] People who weigh themselves periodically and monitor their eating and exercise habits regularly can detect weight gains in the early stages and promptly initiate changes to prevent relapse.

Losing weight and maintaining the loss may not be as easy as gaining the weight in the first place, but it is possible. Those who have been successful find that it gets easier with time—the changes in diet and activity patterns become permanent.

Prevention

Given the information presented up to this point in the chapter, the adage "An ounce of prevention is worth a pound of cure" seems particularly apropos. Obesity is a major risk factor for numerous diseases, and losing weight is challenging and

successful weight-loss maintenance: achieving a weight loss of at least 10 percent of initial body weight and maintaining the loss for at least one year.

TABLE 9-5	Suggested Public Health Strategies	
Strategies	**Examples of Suggested Nutritional Strategies**	**Examples of Successful Nonnutritional Strategies**
Impose safety standards to reduce the potential for harm.	• Regulate the kcalorie or fat density of foods. • Regulate the size of packages of high-fat foods.	• Mandate safety glass in automobiles. • Regulate the lead content of paint.
Control commercial advertising to limit the influence of harmful products.	• Improve nutrition labeling and product packaging. • Restrict the promotion of high-fat foods (especially when directed at children).	• Restrict cigarette advertising (especially when directed at children). • Add health warnings to alcoholic beverages.
Control the conditions under which products are sold to limit exposure to hazardous substances.	• Remove high-fat, low–nutrient density foods from school vending machines. • Restrict the number of vendors licensed to sell high-fat foods.	• Mandate minimum-age laws for the use of tobacco, alcohol, and automobiles. • Restrict the number of vendors licensed to sell alcohol.
Control prices to reduce consumption.	• Tax soft drinks and other foods high in kcalories, fat, or sugar.	• Tax alcohol and tobacco.

SOURCES: Adapted from M. Nestle and M. F. Jacobson, Halting the obesity epidemic: A public health policy approach, *Public Health Reports* 115 (2000): 12–24; R. W. Jeffery, Public health approaches to the management of obesity, in K. B. Brownell and C. G. Fairburn, eds., *Eating Disorder and Obesity—A Comprehensive Handbook* (New York: Guilford Press, 1995), pp. 558–563.

often temporary. Strategies for preventing weight gain■ are very similar to those for losing weight, with one exception: they begin early and continue throughout life. Over the years, they become an integral part of a person's life. It is much easier for a person to resist doughnuts for breakfast if he rarely eats them. Similarly, a person will have little trouble walking each morning if she has always been active.

Public Health Programs

Is there anyone in the United States who hasn't heard the message that obesity raises the risks of chronic diseases and that overweight people should aim for a healthy weight by eating sensibly and becoming physically active? Not likely. Yet implementing such advice is difficult in an environment of abundant food and physical inactivity. To successfully treat obesity, we may have to change the environment in which we live.[104] Table 9-5 provides examples of public health strategies that have been suggested to improve our nation's nutrition environment. Some of these strategies may seem radical, but dramatic measures may be needed if we are to curb the obesity epidemic that is sweeping across the nation.[105]

IN SUMMARY Preventing weight gains and maintaining weight losses require vigilant attention to diet and physical activity. Taking care of oneself is a lifelong responsibility.

Underweight

Underweight■ is a far less prevalent problem than overweight, affecting no more than 5 percent of U.S. adults (review Figure 8-7 on p. 263). Whether the underweight person needs to gain weight is a question of health and, like weight loss, a highly individual matter. People who are healthy at their present weights may stay there; there are no compelling reasons to try to gain weight. Those who are thin because of malnourishment or illness, however, might benefit from a diet that supports weight gain. Medical advice can help make the distinction.

Thin people may find gaining weight difficult.[106] Those who wish to gain weight for appearance's sake or to improve their athletic performance need to be aware that healthful weight gains can be achieved only by physical conditioning

■ To prevent excessive weight gain:
- Eat regular meals and limit snacking.
- Drink water instead of high-kcalorie beverages.
- Select sensible portion sizes and limit daily energy intake to no more than energy expended.
- Become physically active and limit sedentary activities.

■ Reminder: *Underweight* is a body weight so low as to have adverse health effects; it is generally defined as BMI <18.5.

combined with high energy intakes. On a high-kcalorie diet alone, a person may gain weight, but it will be mostly fat. Even if the gain improves appearance, it can be detrimental to health and might impair athletic performance. Therefore, in weight gain, as in weight loss, physical activity and energy intake are essential components of a sound plan.

Problems of Underweight

The causes of underweight may be as diverse as those of overweight—hunger, appetite, and satiety irregularities; psychological traits; metabolic factors; and hereditary tendencies. Habits learned early in childhood, especially food aversions, may perpetuate themselves.

The demand for energy to support physical activity and growth often contributes to underweight. An active, growing boy may need more than 4000 kcalories a day to maintain his weight and may be too busy to take time to eat. Underweight people find it hard to gain weight due, in part, to their expenditure of energy in adaptive thermogenesis. So much energy may be spent adapting to a higher food intake that at first as many as 750 to 800 extra kcalories a day may be needed to gain a pound a week. Like those who want to lose weight, people who want to gain must learn new habits and learn to like new foods. They are also similarly vulnerable to potentially harmful schemes and would be wise to review the consumer bill of rights on p. 288, using "weight gain" instead of "weight loss" where appropriate.

An underweight condition known as anorexia nervosa sometimes develops in people who employ self-denial to control their weight. They go to such extremes that they become severely undernourished, achieving final body weights of 70 pounds or even less. The distinguishing feature of a person with anorexia nervosa, as opposed to other underweight people, is that the starvation is intentional. Anorexia nervosa and other eating disorders are the subject of the highlight that follows this chapter.

Weight-Gain Strategies

Weight-gain strategies center on eating foods that provide many kcalories in a small volume and exercising to build muscle. Following the Food Guide Pyramid's■ maximum suggested servings provides a diet pattern for 2800 kcalories a day and incorporates many of the principles for planning a healthy diet.

Energy-Dense Foods Energy-dense foods (the very ones eliminated from a successful weight-loss diet) hold the key to weight gain. Pick the highest-kcalorie items from each food group—that is, milk shakes instead of fat-free milk, salmon instead of snapper, avocados instead of cucumbers, a cup of grape juice instead of a small apple, and whole-wheat muffins instead of whole-wheat bread. Because fat provides more than twice as many kcalories per teaspoon as sugar does, fat adds kcalories without adding much bulk.

Be aware that health experts routinely recommend a low-fat diet because the biggest health problems in the United States involve obesity and heart disease. Eating high-kcalorie, high-fat foods is not healthy for most people, but may be essential for an underweight individual who needs to gain weight. An underweight person who is physically active and eating a nutritionally adequate diet can afford a few extra kcalories from fat. For health's sake, it would be wise to select foods with monounsaturated and polyunsaturated fats instead of those with saturated or *trans* fats: for example, sautéing vegetables in olive oil instead of butter or hydrogenated margarine.

Regular Meals Daily People who are underweight need to make meals a priority and take the time to plan, prepare, and eat each meal. They should eat at least

93 g fat
18 tsp sugar

3 servings — 7 oz

5 servings — 4 servings

11 servings

■ For an adequate 2800-kcalorie diet, select the maximum number of suggested servings. To increase kcalories further, use larger portion sizes and choose high-kcalorie options from within each group.

three healthy meals every day and learn to eat more food within the first 20 minutes of a meal. Another suggestion is to eat meaty appetizers or the main course first and leave the soup or salad until later.

Large Portions Underweight people need to learn to eat more food at each meal. Put extra slices of ham and cheese on the sandwich for lunch, drink milk from a larger glass, and eat cereal from a larger bowl.

The person should expect to feel full. Most underweight individuals are accustomed to small quantities of food. When they begin eating significantly more, they feel uncomfortable. This is normal and passes over time.

Extra Snacks Since a substantially higher energy intake is needed each day, in addition to eating more food at each meal, it is necessary to eat more frequently. Between-meal snacks do not interfere with later meals; they can readily lead to weight gains.[107] For example, a student might make three sandwiches in the morning and eat them between classes in addition to the day's three regular meals. Snacking on dried fruit, nuts, and seeds is also an easy way to add kcalories.

Juice and Milk Beverages provide an easy way to increase energy intake. Consider that 6 cups of cranberry juice add almost 1000 kcalories to the day's intake. kCalories can be added to milk by mixing in powdered milk or packets of instant breakfast.

For people who are underweight due to illness, concentrated liquid formulas are often recommended because a weak person can swallow them easily. A physician or registered dietitian can recommend high-protein, high-kcalorie formulas to help an underweight person maintain or gain. Used in addition to regular meals, these can help considerably.

Exercising to Build Muscles To gain weight, use strength training primarily and increase energy intake to support that exercise. Eating extra food will then support a gain of both muscle and fat. About 700 to 1000 kcalories a day above normal energy needs is enough to support both the exercise and the building of muscle.

IN SUMMARY Both the incidence of underweight and the health problems associated with it are less prevalent than overweight and its associated problems. To gain weight, a person must train physically and increase energy intake by selecting energy-dense foods, eating regular meals, taking larger portions, and consuming extra snacks and beverages. Table 9-4 (on p. 296) includes a summary of weight-gain strategies.

Nutrition in Your Life

To enjoy good health and maintain a reasonable body weight, combine sensible eating habits and regular physical activity.

- Do you try to lose or gain weight even though your BMI falls between 18.5 and 24.9?
- Does your weight fluctuate up and down dramatically over time?
- Do you follow fad diets or take over-the-counter drugs or herbal supplements?

NUTRITION ON THE NET

 Access these websites for further study of topics covered in this chapter.

- Find updates and quick links to these and other nutrition-related sites at our website: **www.wadsworth.com/nutrition**

- Search for "obesity" and "weight control" at the U.S. Government health information site: **www.healthfinder.gov**

- Review the Clinical Guidelines on the Identification, Evaluation, and Treatment of Overweight and Obesity in Adults: **www.nhlbi.nih.gov/guidelines/obesity/ob_home.htm**

- Learn about the drugs used for weight loss from the Center for Drug Evaluation and Research: **www.fda.gov/cder**

- Learn about weight control and the WIN program from the Weight-control Information Network: **www.niddk.nih.gov/health/nutrit/win.htm**

- Visit weight-loss support groups, such as Take Off Pounds Sensibly (TOPS), Overeaters Anonymous (OA), and Weight Watchers: **www.tops.org**, **www.oa.org**, and **www.weightwatchers.com**

- See what the obesity professionals think at the North American Association for the Study of Obesity and the American Society for Bariatric Surgery: **www.naaso.org** and **www.asbs.org**

- Consider the nondietary approaches of HUGS International: **www.hugs.com**

- Learn about the 10,000 Step Program from Shape Up America!: **www.shapeup.org/10000steps.html**

- Find helpful information on achieving and maintaining a healthy weight from the Calorie Control Council: **www.caloriecontrol.org**

- Learn how to end size discrimination and improve the quality of life for fat people from the National Association to Advance Fat Acceptance: **www.naafa.org**

- Find good advice on starting a weight-loss program from the Partnership for Healthy Weight Management: **www.consumer.gov/weightloss**

- Consider ways to live a healthy life at any weight: **www.bodypositive.com**

NUTRITION CALCULATIONS

These problems give you practice in doing simple energy-balance calculations (see p. 309 for answers). Once you have mastered these examples, you will be prepared to examine your own food choices. Be sure to show your calculations for each problem.

1. Critique a commercial weight-loss plan. Consumers spend billions of dollars a year on weight-loss programs such as Slim-Fast, Sweet Success, Weight Watchers, Nutri/System, Jenny Craig, Optifast, Medifast, and Formula One. One such plan calls for a milk shake in the morning, at noon, and as an afternoon snack and "a sensible, balanced, low-fat dinner" in the evening. One shake mixed in 8 ounces of vitamin A– and D–fortified fat-free milk offers 190 kcalories; 32 grams of carbohydrate, 13 grams of protein, and 1 gram of fat; at least one-third of the Daily Value for all vitamins and minerals; plus 2 grams of fiber.

   a. Calculate the kcalories and grams of carbohydrate, protein, and fat that three shakes provide.

   b. How do these values compare with the criteria listed in item 2 in Table H8-3 on p. 276?

   c. Plan "a sensible, balanced, low-fat dinner" that will help make this weight-loss plan adequate and balanced. Now, how do the day's totals compare with the criteria in item 2 in Table H8-3 on p. 276?

   d. Critique this plan using the other criteria described in Table H8-3 on p. 276 as a guide.

2. Evaluate a weight-gain attempt. People attempting to gain weight sometimes have a hard time because they choose low-kcalorie, high-bulk foods that make it hard to consume enough energy. Consider the following lunch: a chef's salad consisting of 2 cups iceberg lettuce, 1 whole tomato, 1 ounce swiss cheese, 1 ounce roasted ham (lean and fat), 1 hard-boiled egg, ½ cucumber, and ¼ cup mayonnaise-type salad dressing. If you weighed these foods, you'd find that they totaled 552 grams. This is a pretty filling meal.

   a. How much does this meal weigh in pounds?

   b. The meal provides 541 kcalories. What is the energy density of this meal, expressed in kcalories per gram?

   c. To gain weight, this person is advised to eat an additional 500 kcalories at this meal. Using foods with this same energy density, how much more chef's salad will this person have to eat?

   d. Suppose a person simply can't do this. Try to reduce the bulk of this meal by replacing some of the lettuce with more energy-dense foods. Delete 1 cup lettuce from the salad and add 1 ounce roast beef and 1 ounce cheddar cheese. Show how these changes influence the weight and kcalories of this meal. (Use Appendix H.)

Item No./Food	Weight (g)	Energy (kcal)
Original totals:	552	541
Minus:		
#867 Lettuce, 1 c	−_____	−_____
Plus:		
#603 Roast beef, 1 oz	+_____	+_____
#37 Cheddar cheese, 1 oz	+_____	+_____
Totals:	_____	_____

e. How many kcalories did the changes add?

f. How much more *weight* of food did these changes add?

This exercise should reveal why people attempting to gain weight are advised to add high-fat items, within reason, to their daily meals.

STUDY QUESTIONS

These questions will help you review the chapter. You will find the answers in the discussions on the pages provided.

1. Describe how body fat develops, and suggest some reasons why it is difficult for an obese person to maintain weight loss. (pp. 280–281)

2. What factors contribute to obesity? (pp. 282–285)

3. List several aggressive ways to treat obesity, and explain why such methods are not recommended for every overweight person. (pp. 289–291)

4. Discuss reasonable dietary strategies for achieving and maintaining a healthy body weight. (pp. 291–295)

5. What are the benefits of increased physical activity in a weight-loss program? (pp. 296–298)

6. Describe the behavioral strategies for changing an individual's dietary habits. What role does personal attitude play? (pp. 298–299)

7. Describe strategies for successful weight gain. (pp. 302–303)

These multiple choice questions will help you prepare for an exam. Answers can be found on p. 309.

1. With weight loss, fat cells:
   a. decrease in size only.
   b. decrease in number only.
   c. decrease in both number and size.
   d. decrease in number, but increase in size.

2. Obesity is caused by:
   a. overeating.
   b. inactivity.
   c. defective genes.
   d. multiple factors.

3. The protein produced by the fat cells under the direction of the *ob* gene is called:
   a. leptin.
   b. serotonin.
   c. sibutramine.
   d. phentermine.

4. The biggest problem associated with the use of drugs in the treatment of obesity is:
   a. cost.
   b. chronic dosage.
   c. ineffectiveness.
   d. adverse side effects.

5. A realistic goal for weight loss is to reduce body weight:
   a. down to the weight a person was at age 25.
   b. down to the ideal weight in the weight-for-height tables.
   c. by 10 percent over six months.
   d. by 15 percent over three months.

6. A nutritionally sound weight-loss diet might restrict daily energy intake to create a:
   a. 1000-kcalorie-per-month deficit.
   b. 500-kcalorie-per-month deficit.
   c. 500-kcalorie-per-day deficit.
   d. 1000-kcalorie-per-day deficit.

7. Successful weight loss depends on:
   a. avoiding fats and limiting water.
   b. taking supplements and drinking water.
   c. increasing proteins and restricting carbohydrates.
   d. reducing energy intake and increasing physical activity.

8. Physical activity does not help a person to:
   a. lose weight.
   b. retain muscle.
   c. maintain weight loss.
   d. lose fat in trouble spots.

9. Which strategy would *not* help an overweight person to lose weight?
   a. Exercise.
   b. Eat slowly.
   c. Limit high-fat foods.
   d. Eat energy-dense foods regularly.

10. Which strategy would *not* help an underweight person to gain weight?
    a. Exercise.
    b. Drink plenty of water.
    c. Eat snacks between meals.
    d. Eat large portions of foods.

REFERENCES

1. K. M. Flegal and coauthors, Prevalence and trends in obesity among US adults, *Journal of the American Medical Association* 288 (2002): 1723–1727; C. E. Lewis, Weight gain continues in the 1990s: 10-year trends in weight and overweight from the CARDIA study, *American Journal of Epidemiology* 151 (2000): 1172–1181; A. H. Mokdad and coauthors, The spread of the obesity epidemic in the United States, 1991–1998, *Journal of the American Medical Association* 282 (1999): 1519–1522.

2. T. E. Kottke, L. A. Wu, and R. S. Hoffman, Economic and psychological implications of the obesity epidemic, *Mayo Clinic Proceedings* 78 (2003): 92–94; M. Kohn and M. Booth, The worldwide epidemic of obesity in adolescents, *Adolescent Medicine* 14 (2003): 1–9; G. du Toit and M. T. van der Merwe, The epidemic of childhood obesity, *South African Medical Journal* 93 (2003): 49–50; C. J. Schrodt, The obesity epidemic and physician responsibility, *Journal of the Kentucky Medical Association* 101 (2003): 27–28; W. H. Dietz and coauthors, Policy tools for the childhood obesity epidemic, *Journal of Law, Medicine, and Ethics* 30 (2002): 83–87; M. Chopra, S. Galbraith, and I. Darnton-Hill, A global response to a global problem: The epidemic of overnutrition, *Bulletin of the World Health Organization* 80 (2002): 952–958; J. P. Koplan and W. H. Dietz, Caloric imbalance and public health policy, *Journal of the American Medical Association* 282 (1999): 1579; Mokdad and coauthors, 1999.

3. E. D. Rosen, The molecular control of adipogenesis with special reference to lymphatic pathology, *Annals of the New York Academy of Sciences* 979 (2002): 143–158; D. B. Hausman and coauthors, The biology of white adipocyte proliferation, *Obesity Reviews* 2 (2001): 239–254.

4. J. E. Schaffer, Lipotoxicity: When tissues overeat, *Current Opinion in Lipidology* 14 (2003): 281–287.

5. E. Blaak, Gender differences in fat metabolism, *Current Opinion in Clinical Nutrition and Metabolic Care* 4 (2001): 499–502.

6. J. Webber and I. A. Macdonald, Signalling in body-weight homeostasis: Neuroendocrine efferent signals, *Proceedings of the Nutrition Society* 59 (2000): 397–404.

7. L. Pérusse and C. Bouchard, Gene-diet interactions in obesity, *American Journal of Clinical Nutrition* 72 (2000): 1285S–1290S.

8. P. Froguel and P. Boutin, Genetics of pathways regulating body weight in the development of obesity in humans, *Experimental Biology and Medicine* 226 (2001): 991–996.

9. D. E. Cummings and M. W. Schwartz, Genetics and pathophysiology of human obesity, *Annual Reviews of Medicine* 54 (2003): 453–471; J. Altman, Weight in the balance, *Neuroendocrinology* 76 (2002): 131–136; B. M. Spiegelman and J. S. Flier, Obesity and the regulation of energy balance, *Cell* 104 (2001): 531–543; M. W. Schwartz and coauthors, Central nervous system control of food intake, *Nature* 404 (2000): 661–671.

10. R. B. Ceddia, W. N. William Jr., and R. Curi, The response of skeletal muscle to leptin, *Frontiers in Bioscience* 6 (2001): D90–D97; C A. Baile, M. A. Della-Fera, and R. J. Martin, Regulation of metabolism and body fat mass by leptin, *Annual Review of Nutrition* 20 (2000): 105–127.

11. I. S. Farooqi and coauthors, Effects of recombinant leptin therapy in a child with congenital leptin deficiency, *New England Journal of Medicine* 341 (1999): 879–884.

12. C. E. Ruhl and J. E. Everhart, Leptin concentrations in the United States: Relations with demographic and anthropometric measures, *American Journal of Clinical Nutrition* 74 (2001): 295–301; B. Lönnerdal and P. J. Havel, Serum leptin concentrations in infants: Effects of diet, sex, and adiposity, *American Journal of Clinical Nutrition* 72 (2000): 484–489; H. Fors and coauthors, Serum leptin levels correlate with growth hormone secretion and body fat in children, *Journal of Clinical Endocrinology and Metabolism* 84 (1999): 3586–3590.

13. A. Polito and coauthors, Basal metabolic rate in anorexia nervosa: Relation to body composition and leptin concentrations, *American Journal of Clinical Nutrition* 71 (2000): 1495–1502.

14. J. Proietto and A. W. Thorburn, The therapeutic potential of leptin, *Expert Opinion on Investigational Drugs* 12 (2003): 373–378; W. A. Banks, Leptin transport across the blood-brain barrier: Implications for the cause and treatment of obesity, *Current Pharmaceutical Design* 7 (2001): 125–133; N. F. Chu and coauthors, Plasma leptin concentrations and four-year weight gain among US men, *International Journal of Obesity and Related Metabolic Disorders* 25 (2001): 346–353; Ceddia, William, and Curi, 2001.

15. R. H. Unger, Leptin physiology: A second look, *Regulatory Peptides* 92 (2000): 87–95.

16. A. M. Prentice and coauthors, Leptin and undernutrition, *Nutrition Reviews* 60 (2002): S56–S67.

17. J. Harvey and M. L. Ashford, *Neuropharmacology* 44 (2003): 845–854.

18. S. Takeda, F. Elefteriou, and G. Karsenty, Common endocrine control of body weight, reproduction, and bone mass, *Annual Review of Nutrition* 23 (2003): 403–411; Leptin: A key regulator in nutrition, *Nutrition Reviews* 60 (2002): entire issue; S. Moschos, J. L. Chan, and C. S. Mantzoros, Leptin and reproduction: A review, *Fertility and Sterility* 77 (2002): 433–444; R. B. S. Harris, Leptin—Much more than a satiety signal, *Annual Review of Nutrition* 20 (2000): 45–75; J. C. Fleet, Leptin and bone: Does the brain control bone biology? *Nutrition Reviews* 58 (2000): 209–211.

19. M. Kojima and K. Kangawa, Ghrelin, an orexigenic signaling molecule from the gastrointestinal tract, *Current Opinion in Pharmacology* 2 (2002): 665–668.

20. J. Eisenstein and A. Greenberg, Ghrelin: Update 2003, *Nutrition Reviews* 61 (2003): 101–104; O. Ukkola and S. Poykko, Ghrelin, growth and obesity, *Annals of Medicine* 34 (2002): 102–108.

21. G. Schaller and coauthors, Plasma ghrelin concentrations are not regulated by glucose or insulin: A double-blind, placebo-controlled crossover clamp study, *Diabetes* 52 (2003): 16–20; G. Iniguez and coauthors, Fasting and post-glucose ghrelin levels in SGA infants: Relationships with size and weight gain at one year of age, *Journal of Clinical Endocrinology and Metabolism* 87 (2002): 5830–5833.

22. M. Tanaka and coauthors, Habitual binge/purge behavior influences circulating ghrelin levels in eating disorders, *Journal of Psychiatric Research* 37 (2003): 17–22; J. H. Lindeman and coauthors, Ghrelin and the hyposomatotropism of obesity, *Obesity Research* 10 (2002): 1161–1166.

23. V. Tolle and coauthors, Balance in ghrelin and leptin plasma levels in anorexia nervosa patients and constitutionally thin women, *Journal of Clinical Endocrinology and Metabolism* 88 (2003): 109–116; M. F. Saad and coauthors, Insulin regulates plasma ghrelin concentration, *Journal of Clinical Endocrinology and Metabolism* 87 (2002): 3997–4000.

24. A. M. Haqq and coauthors, Serum ghrelin levels are inversely correlated with body mass index, age, and insulin concentrations in normal children and are markedly increased in Prader-Willi syndrome, *Journal of Clinical Endocrinology and Metabolism* 88 (2003): 174–178; A. DelParigi and coauthors, High circulating ghrelin: A potential cause for hyperphagia and obesity in Prader-Willi syndrome, *Journal of Clinical Endocrinology and Metabolism* 87 (2002): 5461–5464.

25. P. J. English and coauthors, Food fails to suppress ghrelin levels in obese humans, *Journal of Clinical Endocrinology and Metabolism* 87 (2002): 2984.

26. D. E. Cummings and coauthors, Plasma ghrelin levels after diet-induced weight loss or gastric bypass surgery, *New England Journal of Medicine* 346 (2002): 1623–1630.

27. Cummings and coauthors, 2002.

28. Iniguez and coauthors, 2002.

29. J. Korner and R. L. Leibel, To eat or not to eat—How the gut talks to the brain, *New England Journal of Medicine* 349 (2003): 926–930.

30. R. L. Batterham and coauthors, Inhibition of food intake in obese subjects by peptide YY_{3-36}, *New England Journal of Medicine* 349 (2003): 941–948.

31. F. Broglio and coauthors, Ghrelin: Endocrine and non-endocrine actions, *Journal of Pediatric Endocrinology and Metabolism* 15 (2002): 1219–1227.

32. J. C. Weikel and coauthors, Ghrelin promotes slow-wave sleep in humans, *American Journal of Physiology: Endocrinology and Metabolism* 284 (2003): E407–E415.

33. G. Wolf, The uncoupling proteins UCP2 and UCP3 in skeletal muscle, *Nutrition Reviews* 59 (2001): 56–57.

34. L. P. Kozak and M. E. Harper, Mitochondrial uncoupling proteins in energy expenditure, *Annual Review of Nutrition* 20 (2000): 339–363.

35. S. Y. S. Kimm and coauthors, Racial differences in the relation between uncoupling protein genes and resting energy expenditure, *American Journal of Clinical Nutrition* 75 (2002): 714–719; J. A. Yanovski and coauthors, Associations between uncoupling protein 2, body composition, and resting energy expenditure in lean and obese African American, white, and Asian children, *American Journal of Clinical Nutrition* 71 (2000): 1405–1412.

36. P. Hakala and coauthors, Environmental factors in the development of obesity in identical twins, *American Journal of Obesity and Related Metabolic Disorders* 23 (1999): 746–753.

37. A. H. C. Goris, M. S. Westerterp-Plantenga, and K. R. Westerterp, Undereating and underrecording of habitual food intake in obese men: Selective underreporting of fat intake, *American Journal of Clinical Nutrition* 71 (2000): 130–134.

38. J. C. Peters, The challenge of managing body weight in the modern world, *Asia Pacific Journal of Clinical Nutrition* 11 (2002): S714–S717.

39. B. J. Rolls, E. L. Morris, and L. S. Roe, Portion size of food affects energy intake

in normal-weight and overweight men and women, *American Journal of Clinical Nutrition* 76 (2002): 1207–1213.

40. S. J. Nielsen and B. M. Popkin, Patterns and trends in food portion sizes, 1977–1998, *Journal of the American Medical Association* 289 (2003): 450–453; H. Smiciklas-Wright and coauthors, Foods commonly eaten in the United States, 1989–1991 and 1994–1996: Are portion sizes changing? *Journal of the American Dietetic Association* 103 (2003): 41–47.

41. L. R. Young and M. Nestle, Expanding portion sizes in the US marketplace: Implications for nutrition counseling, *Journal of the American Dietetic Association* 103 (2003): 231–234.

42. L. R. Young and M. Nestle, The contribution of expanding portion sizes to the US obesity epidemic, *American Journal of Public Health* 92 (2002): 246–249.

43. J. K. Binkley, J. Eales, and M. Jekanowski, The relation between dietary change and rising US obesity, *International Journal of Obesity and Related Metabolic Disorders* 24 (2000): 1032–1039.

44. B. J. Rolls, The supersizing of America: Portion size and the obesity epidemic, *Nutrition Today* 38 (2003): 42–53.

45. M. Lahti-Koski and coauthors, Associations of body mass index and obesity with physical activity, food choices, alcohol intake, and smoking in the 1982–1997 FINRISK Studies, *American Journal of Clinical Nutrition* 75 (2002): 809–817; M. Wei and coauthors, The association between cardiorespiratory fitness and impaired glucose and type 2 diabetes mellitus in men, *Annals of Internal Medicine* 130 (1999): 89–96; U.S. Department of Health and Human Services, *Physical Activity and Health—A Report of the Surgeon General Executive Summary*, 1996.

46. F. B. Hu and coauthors, Television watching and other sedentary behaviors in relation to risk of obesity and type 2 diabetes mellitus in women, *Journal of the American Medical Association* 289 (2003): 1785–1791; J. Salmon and coauthors, The association between television viewing and overweight among Australian adults participating in varying levels of leisure-time physical activity, *International Journal of Obesity and Related Metabolic Disorders* 24 (2000): 600–606.

47. *U.S. News and World Report,* June 16, 2003, p. 36; www.niddk.nih.gov/healthnutrit/pubs/statobes.htm.

48. National Institutes of Health Obesity Education Initiative, *The Practical Guide: Identification, Evaluation, and Treatment of Overweight and Obesity in Adults* (Washington, D.C.: U.S. Department of Health and Human Services, 2000).

49. National Institutes of Health Obesity Education Initiative, 2000.

50. S. Sarlio-Lähteenkorva, A. Rissanen, and J. Kaprio, A descriptive study of weight loss maintenance: 6 and 15 year follow-up of initially overweight adults, *International Journal of Obesity and Related Metabolic Disorders* 24 (2000): 116–125.

51. N. S. Wellman and B. Friedberg, Causes and consequences of adult obesity: Health, social and economic impacts in the United States, *Asia Pacific Journal of Clinical Nutrition* 11 (2002): S705–S709.

52. H. M. Blanck, L. K. Khan, and M. K. Serdula, Use of nonprescription weight loss products: Results from a multistate survey, *Journal of the American Medical Association* 286 (2001): 930–935.

53. D. J. Dyck, Dietary fat intake, supplements, and weight loss, *Canadian Journal of Applied Physiology* 25 (2000): 495–523.

54. P. G. Shekelle and coauthors, Efficacy and safety of ephedra and ephedrine for

weight loss and athletic performance: A meta-analysis, *Journal of the American Medical Association* 289 (2003): 1537–1545.

55. G. Egger, D. Cameron-Smith, and R. Stanton, The effectiveness of popular, non-prescription weight loss supplements, *Medical Journal of Australia* 171 (1999): 604–608; A. Sarubin, *The Health Professional's Guide to Popular Dietary Supplements* (Chicago: The American Dietetic Association, 1999); S. Foster and V. E. Tyler, *Tyler's Honest Herbal—A Sensible Guide to the Use of Herbs and Related Remedies* (New York: Haworth Herbal Press, 1999).

56. S. P. Dolan and coauthors, Analysis of dietary supplements for arsenic, cadmium, mercury, and lead using inductively coupled plasma mass spectrometry, *Journal of Agricultural and Food Chemistry* 51 (2003): 1307–1312; A. H. Feifer, N. E. Fleshner, and L. Klotz, Analytical accuracy and reliability of commonly used nutritional supplements in prostate disease, *Journal of Urology* 168 (2002): 150–154.

57. S. Z. Yanovski and J. A. Yanovski, Obesity, *New England Journal of Medicine* 346 (2002): 591–602.

58. C. H. Halsted, Is blockade of pancreatic lipase the answer? *American Journal of Clinical Nutrition* 69 (1999): 1059–1060.

59. G. Glazer, Long-term pharmacotherapy of obesity 2000: A review of efficacy and safety, *Archives of Internal Medicine* 161 (2001): 1814–1824.

60. S. Schurgin and R. D. Siegel, Pharmacotherapy of obesity: An update, *Nutrition in Clinical Care* 6 (2003): 27–37.

61. M. A. Rupnick and coauthors, Adipose tissue mass can be regulated through the vasculature, *Proceedings of the National Academy of Science* 99 (2002): 10730–10735.

62. Cummings and coauthors, 2002.

63. E. C. Mun, G. L. Blackburn, and J. B. Matthews, Current status of medical and surgical therapy for obesity, *Gastroenterology* 120 (2001): 669–681; H. J. Sugarman, The epidemic of severe obesity: The value of surgical treatment, *Mayo Clinic Proceedings* 75 (2000): 669–672.

64. R. E. Brolin, Bariatric surgery and long-term control of morbid obesity, *Journal of the American Medical Association* 288 (2002): 2793–2796.

65. J. G. Bruner and R. H. de Jong, Lipoplasty claims experience of U.S. insurance companies, *Plastic and Reconstructive Surgery* 107 (2001): 1285–1291; R. B. Rao, S. F. Ely, and R. S. Hoffman, Deaths related to liposuction, *New England Journal of Medicine* 340 (1999): 1471–1475.

66. J. T. Fine and coauthors, A prospective study of weight change and health-related quality of life in women, *Journal of the American Medical Association* 282 (1999): 2136–2142.

67. G. D. Foster and coauthors, Obese patients' perceptions of treatment outcomes and the factors that influence them, *Archives of Internal Medicine* 161 (2001): 2133–2139.

68. National Institutes of Health Obesity Education Initiative, 2000, p. 2.

69. Position of the American Dietetic Association: Weight management, *Journal of the American Dietetic Association* 102 (2002): 1145–1155.

70. M. K. Serdula and coauthors, Prevalence of attempting weight loss and strategies for controlling weight, *Journal of the American Medical Association* 282 (1999): 1353–1358.

71. National Institutes of Health Obesity Education Initiative, *Clinical Guidelines on the Identification, Evaluation, and Treatment of Overweight and Obesity in Adults* (Washington, D.C.: U.S. Department of Health and Human Services, 1998).

72. National Institutes of Health Obesity Education Initiative, 2000, pp. 26–27.

73. A. K. Kant and coauthors, A prospective study of diet quality and mortality in women, *Journal of the American Medical Association* 283 (2000): 2109–2115.

74. Nielsen and Popkin, 2003; Smiciklas-Wright and coauthors, 2003; Young and Nestle, 2002.

75. E. A. Bell and B. J. Rolls, Energy density of foods affects energy intake across multiple levels of fat content in lean and obese women, *American Journal of Clinical Nutrition* 73 (2001): 1010–1018.

76. T. V. Kral, L. S. Roe, and B. J. Rolls, Does nutrition information about the energy density of meals affect food intake in normal-weight women? *Appetite* 39 (2002): 137–145; B. J. Rolls and E. A. Bell, Dietary approaches to the treatment of obesity, *Medical Clinics of North America* 84 (2000): 401–418, vi.

77. B. J. Rolls, E. A. Bell, and M. L. Thorwart, Water incorporated into a food but not served with a food decreases energy intake in lean women, *American Journal of Clinical Nutrition* 70 (1999): 448–455.

78. S. A. Bowman and J. T. Spence, A comparison of low-carbohydrate vs. high-carbohydrate diets: Energy restriction, nutrient quality and correlation to body mass index, *Journal of the American College of Nutrition* 21 (2002): 268–274.

79. M. Yao and S. B. Roberts, Dietary energy density and weight regulation, *Nutrition Reviews* 59 (2001): 247–258.

80. A. Astrup and coauthors, Low-fat diets and energy balance: How does the evidence stand in 2002? *Proceedings of the Nutrition Society* 61 (2002): 299–309; S. D. Poppitt and coauthors, Long-term effects of ad libitum low-fat, high-carbohydrate diets on weight and serum lipids in overweight subjects with metabolic syndrome, *American Journal of Clinical Nutrition* 75 (2002): 11–20; S. E. Kasim-Karakas and coauthors, Changes in plasma lipoproteins during low-fat, high-carbohydrate diets: Effects of energy intake, *American Journal of Clinical Nutrition* 71 (2000): 1439–1447.

81. G. D. Foster and coauthors, A randomized trial of a low-carbohydrate diet for obesity, *New England Journal of Medicine* 348 (2003): 2082–2090.

82. D. K. Layman and coauthors, A reduced ratio of dietary carbohydrate to protein improves body composition and blood lipid profiles during weight loss in adult women, *Journal of Nutrition* 133 (2003): 411–417; D. M. Bravata and coauthors, Efficacy and safety of low-carbohydrate diets, *Journal of the American Medical Association* 289 (2003): 1837–1850; S. Pirozzo and coauthors, Advice on low-fat diets for obesity, *Cochrane Database of Systematic Review* (2002), available at **www.update-software.com/abstracts/ab003640.htm**; M. R. Freedman, J. King, and E. Kennedy, Popular diets: A scientific review, *Obesity Research* 9 (2001): 1S–5S; A. Golay and coauthors, Similar weight loss with low-energy food combining or balanced diets, *International Journal of Obesity and Related Metabolic Disorders* 24 (2000): 492–496; N. H. Baba and coauthors, High protein vs high carbohydrate hypoenergetic diet for the treatment of obese hyperinsulinemic subjects, *International Journal of Obesity and Related Metabolic Disorders* 23 (1999): 1202–1206.

83. Astrup and coauthors, 2002.

84. Committee on Dietary Reference Intakes, *Dietary Reference Intakes for Energy, Carbohydrate, Fiber, Fat, Fatty Acids, Cholesterol, Protein, and Amino Acids* (Washington, D.C.: National Academies Press, 2002).

85. J. W. Anderson and coauthors, Long-term weight-loss maintenance: A meta-analysis

of US studies, *American Journal of Clinical Nutrition* 74 (2001): 579–584.

86. J. F. Carroll and C. K. Kyser, Exercise training in obesity lowers blood pressure independent of weight change, *Medicine and Science in Sports and Exercise* 34 (2002): 596–601; B. Gutin and coauthors, Effects of exercise intensity on cardiovascular fitness, total body composition, and visceral adiposity of obese adolescents, *American Journal of Clinical Nutrition* 75 (2002): 818–826; R. Ross and coauathors, Reduction in obesity and related comorbid conditions after diet-induced weight loss and exercise induced weight loss in men, *Annals of Internal Medicine* 133 (2000): 92–103; A. L. Dunn and coauthors, Comparison of lifestyle and structured interventions to increase physical activity and cardiorespiratory fitness: A randomized trial, *Journal of the American Medical Association* 281 (1999): 327–334; J. M. Jakicic and coauthors, Effects of intermittent exercise and use of home exercise equipment on adherence, weight loss, and fitness in overweight women—A randomized trial, *Journal of the American Medical Association* 282 (1999): 1554–1560.

87. U. G. Kyle and coauthors, Physical activity and fat-free and fat mass by bioelectrical impedance in 3853 adults, *Medicine and Science in Sports and Exercise* 33 (2001): 576–584.

88. Gutin and coauthors, 2002; R. Ross and coauthors, Reduction in obesity and related comorbid conditions after diet-induced weight loss or exercise-induced weight loss in men: A randomized, controlled trial, *Annals of Internal Medicine* 133 (2000): 92–103; G. Benedetti and coauthors, Body composition and energy expenditure after weight loss following bariatric surgery, *Journal of the American College of Nutrition* 19 (2000): 270–274.

89. Gutin and coauthors, 2002; Ross and coauthors, 2000.

90. W. D. Schmidt, C. J. Biwer, and L. K. Kalscheuer, Effects of long *versus* short bout exercise on fitness and weight loss in overweight females, *Journal of the American College of Nutrition* 20 (2001): 494–501.

91. American College of Sports Medicine, Position stand: Appropriate intervention strategies for weight loss and prevention of weight regain for adults, *Medicine and Science in Sports and Exercise* 33 (2001): 2145–2156.

92. S. M. Byrne, Psychological aspects of weight maintenance and relapse in obesity, *Journal of Psychosomatic Research* 53 (2002): 1029–1036.

93. D. F. Tate, E. H. Jackvony, and R. R. Wing, Effects of Internet behavioral counseling on weight loss in adults at risk for type 2 diabetes: A randomized trial, *Journal of the American Medical Association* 289 (2003): 1833–1836.

94. R. R. Wing and J. O. Hill, Successful weight loss maintenance, *Annual Review of Nutrition* 21 (2001): 323–341.

95. M. T. Mcguire, R. R. Wing, and J. O. Hill, The prevalence of weight loss maintenance among American adults, *International Journal of Obesity and Related Metabolic Disorders* 23 (1999): 1314–1319.

96. Wing and Hill, 2001; M. R. Lowe, K. Miller-Kovach, and S. Phelan, Weight-loss maintenance in overweight individuals one to five years following successful completion of a commercial weight loss program, *International Journal of Obesity and Related Metabolic Disorders* 25 (2001): 325–331.

97. M. S. Leser, S. Z. Yanovski, and J. A. Yanovski, A low-fat intake and greater activity level are associated with lower weight regain 3 years after completing a very-low-calorie diet, *Journal of the American Dietetic Association* 102 (2002): 1252–1256.

98. A. Raben and coauthors, Diurnal metabolic profiles after 14d of an ad libitum high-starch, high-sucrose, or high-fat diet in normal-weight, never-obese and postobese women, *American Journal of Clinical Nutrition* 73 (2001): 177–189.

99. A. Astrup and coauthors, Meta-analysis of resting metabolic rate in formerly obese subjects, *American Journal of Clinical Nutrition* 69 (1999): 1117–1122.

100. R. L. Weinsier and coauthors, Free-living activity energy expenditure in women successful and unsuccessful at maintaining a normal body weight, *American Journal of Clinical Nutrition* 75 (2002): 499–504.

101. D. A. Schoeller, K. Shay, and R. F. Kushner, How much physical activity is needed to minimize weight gain in previously obese women? *American Journal of Clinical Nutrition* 66 (1997): 551–556.

102. American College of Sports Medicine, 2001.

103. Wing and Hill, 2001.

104. R. E. Killingsworth, Health promoting community design: A new paradigm to promote healthy and active communities, *American Journal of Health Promotion* 17 (2003): 169–170; M. Nestle and M. F. Jacobson, Halting the obesity epidemic: A public health policy approach, *Public Health Reports* 115 (2000): 12–24.

105. S. L. Mercer and coauthors, Possible lessons from the tobacco experience for obesity control, *American Journal of Clinical Nutrition* 77 (2003): 1073S–1082S.

106. T. B. VanItallie, Resistance to weight gain during overfeeding: A NEAT explanation, *Nutrition Reviews* 59 (2001): 48–51.

107. C. Marmonier and coauthors, Snacks consumed in a nonhungry state have poor satiating efficiency: Influence of snack composition on substrate utilization and hunger, *American Journal of Clinical Nutrition* 76 (2002): 518–528.

ANSWERS

Nutrition Calculations

1. a. Three milk shakes provide: 3×190 kcal = 570 kcal; 3×32 g carbohydrate = 96 g carbohydrate; 3×13 g protein = 39 g protein; and 3×1 g fat = 3 g fat.

   b. To meet this criteria, the plan needs *at least* an additional 630 kcalories (1200 kcal − 570 kcal = 630 kcal); an additional 5 to 24 grams of protein, depending on the person's RDA based on gender and age (63 g − 39 g = 24 g and 44 g − 39 g = 5 g); an additional 4 grams of carbohydrate (100 g − 96 g = 4 g); and some additional fat.

   c. Of course, there are many possible dinners that you could plan. One might be:

   Salad made with 1 c lettuce, 1 c chopped tomatoes and onions, ¼ c garbanzo beans, and 2 tbs low-fat dressing

   4 oz grilled chicken

   1 medium baked potato

   1 c summer squash and zucchini

   1 c melon cubes

   This meal brings the day's totals to 1215 kcalories, 90 g of protein, 192 g of carbohydrate, and 13 g of fat, which meets the goals for kcalories, protein, and carbohydrate. Because the milk shake has been fortified, all vitamin and mineral needs are covered as well. The only possible dietary shortcoming is that the day's percent kcalories from fat is low (only 10%), but because energy and nutrient recommendations have been met and the goal is weight loss, this may be acceptable.

   d. This weight-loss plan uses a liquid formula rather than foods, making clients dependent on a special device (the formula) rather than teaching them how to make good choices from the conventional food supply. It provides no information about dropout rates, the long-term success of clients, or weight maintenance after the program ends.

2. a. More than a pound (552 g ÷ 454 g/lb = 1.2 lb).

   b. 541 kcal ÷ 552 g = 0.98 kcal/g.

   c. More than another whole pound (0.98 kcal/g $\times$ 500 kcal = 490 g; 490 g ÷ 454 g/lb = 1.1 lb).

   d.

Item No./Food	Weight (g)	Energy (kcal)
Original totals:	552	541
Minus:		
#867 Lettuce, 1 c	−55	−7
Plus:		
#603 Roast beef, 1 oz	+28	+68
#37 Cheddar cheese, 1 oz	+28	+113
Totals:	553 g	715 kcal

   e. 715 kcal − 541 kcal = 174 kcal added.

   f. 553 g − 552 g = 1 g added.

Study Questions (multiple choice)

1. a   2. d   3. a   4. d   5. c   6. c   7. d   8. d
9. d   10. b

Eating Disorders

© Steve Niedorf Photography/The Image Bank/Getty Images

For some people, dieting to lose weight progresses to a dangerous and obsessive point. An estimated 5 million people in the United States, primarily girls and young women, suffer from the **eating disorders** anorexia nervosa and bulimia nervosa (the accompanying glossary defines these and related terms).[1] Many more suffer from binge-eating disorders or other unspecified conditions that do not meet the strict criteria for anorexia nervosa or bulimia nervosa, but still imperil a person's well-being.

Why do so many people in our society suffer from eating disorders? Most experts agree that the causes are multifactorial: sociocultural, psychological, and perhaps neurochemical. Excessive pressure to be thin is at least partly to blame. When low body weight becomes an important goal, people begin to view normal healthy body weight as being too fat, and they take unhealthy actions to lose weight.

Young people who attempt extreme weight loss may have learned to identify discomforts such as anger, jealousy, or disappointment with "feeling fat." They may also be depressed or suffer social anxiety. As weight loss becomes more of a focus, psychological problems worsen, and the likelihood of developing eating disorders intensifies. Athletes are among those most likely to develop eating disorders.[2]

The Female Athlete Triad

At age 14, Suzanne was a top contender for a spot on the state gymnastics team. Each day her coach reminded team members that they must weigh no more than their assigned weights in order to qualify for competition. The coach chastised gymnasts who gained weight, and Suzanne was terrified of being singled out. Convinced that the less she weighed the better she would perform, Suzanne weighed herself several times a day to confirm that she had not exceeded her 80-pound limit. Driven to excel in her sport, Suzanne kept her weight down by eating very little and training very hard. Unlike many of her friends, Suzanne never began to menstruate. A few months before her fifteenth birthday, Suzanne's coach dropped her back to the second-level team. Suzanne blamed her poor performance on a slow-healing stress fracture. Mentally stressed and physically exhausted, she quit gymnastics and began overeating between periods of self-starvation. Suzanne had developed the dangerous combination of problems that characterize the **female athlete triad**—disordered eating, amenorrhea, and osteoporosis (see Figure H9-1).

Disordered Eating

Part of the reason many athletes engage in **disordered eating** behaviors may be that they and their coaches have embraced unsuitable weight standards. An athlete's body must be heavier for a given height than a nonathlete's body because the athlete's body is dense, containing more healthy bone and muscle and less fat. When athletes rely on scales, they may mistakenly be-

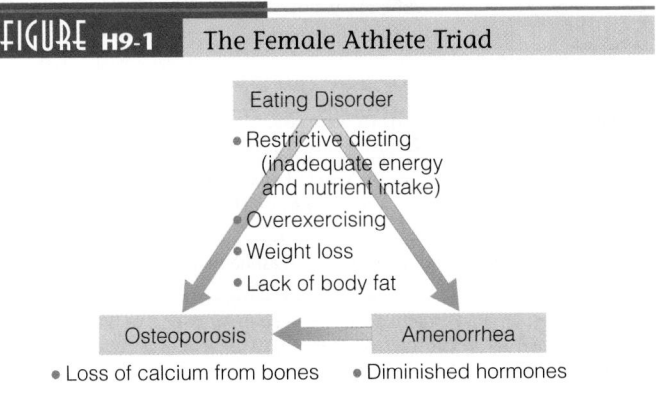

FIGURE H9-1 The Female Athlete Triad

Eating Disorder
- Restrictive dieting (inadequate energy and nutrient intake)
- Overexercising
- Weight loss
- Lack of body fat

Osteoporosis
- Loss of calcium from bones

Amenorrhea
- Diminished hormones

lieve they are too fat because weight standards, such as the BMI, do not provide adequate information about body composition.

Many young athletes severely restrict energy intakes to improve performance, enhance the aesthetic appeal of their performance, or meet the weight guidelines of their specific sports. They fail to realize that the loss of lean tissue that accompanies energy restriction actually impairs their physical performance. The increasing incidence of abnormal eating habits among athletes is cause for concern. Male athletes, especially wrestlers and gymnasts, are affected by these disorders as well, but females are most vulnerable. Risk factors for eating disorders among athletes include:

- Young age (adolescence).
- Pressure to excel at a chosen sport.
- Focus on achieving or maintaining an "ideal" body weight or body fat percentage.

A few years earlier, this Olympic gold medalist would have been too weak and malnourished from anorexia nervosa to have set a world record in the cycling road race.

- Participation in endurance sports or competitions that emphasize a lean appearance or judge performance on aesthetic appeal such as gymnastics, wrestling, figure skating, or dance.
- Weight-loss dieting at an early age.
- Unsupervised dieting.

Amenorrhea

The prevalence of **amenorrhea** among premenopausal women in the United States is about 2 to 5 percent overall, but among female athletes, it may be as high as 66 percent. Contrary to previous notions, amenorrhea is *not* a normal adaptation to strenuous physical training: it is a symptom of something going wrong.[4] Amenorrhea is characterized by low blood estrogen, infertility, and often bone mineral losses. Excessive training, depleted body fat, low body weight, and inadequate nutrition all contribute to amenorrhea.[5] However amenorrhea develops, it threatens the integrity of the bones. Bone losses remain significant even after recovery.[6] (Women with bulimia frequently have menstrual irregularities, but they rarely cease menstruating, and so may be spared this loss of bone integrity.[7])

Osteoporosis

For most people, weight-bearing physical activity, dietary calcium, and the hormone estrogen protect against the bone loss of osteoporosis. For young women with disordered eating and amenorrhea, strenuous activity can impair bone health. Vigorous training combined with inadequate food intake raises stress hormones.[8] These stress hormones compromise bone health, greatly increasing the risks of **stress fractures** today and of osteoporosis in later life. Stress fractures, a serious form of bone injury, commonly occur among dancers and other athletes with amenorrhea, low calcium intakes, and disordered eating.[9] Many underweight young athletes have bones like those of postmenopausal women, and they may never recover their lost bone even after diagnosis and treatment—which makes prevention critical.[10] Athletes should be encouraged to consume at least 1300 milligrams of calcium each day, to eat nutrient-dense foods, and to obtain enough energy to support both weight gain and the energy expended in physical activity.

Other Dangerous Practices of Athletes

Only females face the threats of the female athlete triad, of course, but many male athletes face pressure to achieve a certain body weight and may develop eating disorders. Each week throughout the season, David drastically restricts his food and fluid intake before a wrestling match in an effort to "make weight." For at least three wrestlers in 1997, the consequences were deadly. Wrestlers and their coaches believe that competing in a lower weight class will give them a competitive advantage over smaller opponents. To that end, David practices in rubber suits, sits in saunas, and takes diuretics to lose 4 to 6 pounds.[11] He hopes to replenish the lost fluids, glycogen, and lean tissue during the hours between his weigh-in and competition, but the body needs days to correct this metabolic mayhem. Reestablishing fluid and electrolyte balances may take a day or two, replenishing glycogen stores may take two to three days, and replacing lean tissue may take even longer.

Ironically, the combination of food deprivation and dehydration impairs physical performance by reducing muscle strength, decreasing anaerobic power, reducing endurance capacity, and lowering oxygen consumption. For optimal performance, wrestlers need to first achieve their competitive weight during the off-season and then eat well-balanced meals and drink plenty of fluids during the competitive season.

Some athletes go to extreme measures to bulk up and *gain* weight. People afflicted with **muscle dysmorphia** eat high-protein diets, take dietary supplements, weight train for hours at a time, and often abuse steroids in an attempt to bulk up. Their bodies are large and muscular, yet they see themselves as puny 90-pound weaklings. They are preoccupied with the idea that their bodies are too small or inadequately muscular. Like others with distorted body images, people with muscle dysmorphia weigh themselves frequently and center their lives on diet and exercise. Paying attention to diet and pumping iron for fitness is admirable, but obsessing over it can cause serious social, occupational, and physical problems.

Preventing Eating Disorders in Athletes

To prevent eating disorders in athletes and dancers, the performers, their coaches, and their parents must learn about inappropriate body weight ideals, improper weight-loss techniques, eating disorder development, proper nutrition, and safe weight-control methods. Young people naturally search for identity and will often follow the advice of a person in authority without question. Therefore, coaches and dance instructors should never encourage unhealthy weight loss to qualify for competition or to conform with distorted artistic ideals. Athletes who truly need to lose weight should try to do so during the off-season and under the supervision of a health care professional. Frequent weighings can push young people who are striving to lose weight into a cycle of starving to confront the scale, then bingeing uncontrollably afterward. The erosion of self-esteem that accompanies these events can interfere with normal psychological development and set the stage for serious problems later on.

Table H9-1 includes suggestions to help athletes and dancers protect themselves against developing eating disor-

TABLE H9-1	Tips for Combating Eating Disorders

General Guidelines

- Never restrict food servings to below the numbers suggested for adequacy by the Food Guide Pyramid.
- Eat frequently. Include healthy snacks between meals. The person who eats frequently never gets so hungry as to allow hunger to dictate food choices.
- If not at a healthy weight, establish a reasonable weight goal based on a healthy body composition.
- Allow a reasonable time to achieve the goal. A reasonable loss of excess fat can be achieved at the rate of about 10 percent of body weight in six months.
- Establish a weight-maintenance support group with people who share interests.

Specific Guidelines for Athletes and Dancers

- Replace weight-based goals with performance-based goals.
- Restrict weight-loss activities to the off-season.
- Remember that eating disorders impair physical performance. Seek confidential help in obtaining treatment if needed.
- Focus on proper nutrition as an important facet of your training, as important as proper technique.

ders. The remaining sections describe eating disorders that anyone, athlete or nonathlete, may experience.

Anorexia Nervosa

Julie is 18 years old and a superachiever in school. She watches her diet with great care, and she exercises daily, maintaining a rigorous schedule of self-discipline. She is thin, but she is determined to lose more weight. She is 5 feet 6 inches tall and weighs 85 pounds. She has **anorexia nervosa.**

Characteristics of Anorexia Nervosa

Julie is unaware that she is undernourished, and she sees no need to obtain treatment. She developed amenorrhea several months ago and has become moody and chronically depressed. She insists that she is too fat, although her eyes are sunk in deep hollows in her face. Julie denies that she is ever tired, although she is close to physical exhaustion and no longer sleeps easily. Her family is concerned, and though reluctant to push her, they have finally insisted that she see a psychiatrist. Julie's psychiatrist has diagnosed anorexia nervosa (see Table H9-2) and prescribed group therapy as a start. If she does not begin to gain weight soon, she may need to be hospitalized.

As mentioned in the introduction, most anorexia nervosa victims are females; males account for only about 1 in 20 reported cases. Central to the diagnosis of anorexia nervosa is a

TABLE H9-2 Criteria for Diagnosis of Anorexia Nervosa

A person with anorexia nervosa demonstrates the following:

A. Refusal to maintain body weight at or above a minimal normal weight for age and height (e.g., weight loss leading to maintenance of body weight less than 85 percent of that expected; or failure to make expected weight gain during period of growth, leading to body weight less than 85 percent of that expected).

B. Intense fear of gaining weight or becoming fat, even though underweight.

C. Disturbance in the way in which one's body weight or shape is experienced, undue influence of body weight or shape on self-evaluation, or denial of the seriousness of the current low body weight.

D. In females past puberty, amenorrhea, i.e., the absence of at least three consecutive menstrual cycles. (A woman is considered to have amenorrhea if her periods occur only following hormone, e.g., estrogen, administration.)

Two types:

Restricting type: During the episode of anorexia nervosa, the person does not regularly engage in binge eating or purging behavior (i.e., self-induced vomiting or the misuse of laxatives, diuretics, or enemas).

Binge eating/purging type: During the episode of anorexia nervosa, the person regularly engages in binge eating or purging behavior (i.e., self-induced vomiting or the misuse of laxatives, diuretics, or enemas).

SOURCE: Reprinted with permission from American Psychiatric Association, *Diagnostic and Statistical Manual of Mental Disorders,* 4th ed. Text Revision. (Washington, D.C.: American Psychiatric Association, 2000).

distorted body image that overestimates personal body fatness. When Julie looks at herself in the mirror, she sees a "fat" 85-pound body. The more Julie overestimates her body size, the more resistant she is to treatment, and the more unwilling to examine her faulty values and misconceptions. Malnutrition is known to affect brain functioning and judgment in this way, causing lethargy, confusion, and delirium.

Anorexia nervosa cannot be self-diagnosed. Nearly everyone in our society is engaged in the pursuit of thinness, and denial runs high among people with anorexia nervosa. Some women have all the attitudes and behaviors associated with the condition, but without the dramatic weight loss.

Self-Starvation How can a person as thin as Julie continue to starve herself? Julie uses tremendous discipline against her hunger to strictly limit her portions of low-kcalorie foods. She will deny her hunger, and having adapted to so little food, she feels full after eating only a half-dozen carrot sticks. She knows the kcalorie contents of dozens of foods and the kcalorie costs of as many exercises. If she feels that she has gained an ounce of weight, she runs or jumps rope until she is sure she has exercised it off. If she fears that the food she has eaten outweighs the exercise, she may take laxatives to hasten the passage of food from her system. She drinks water incessantly to fill her stomach, risking dangerous mineral imbalances. She is desperately hungry. In fact, she is starving, but she doesn't eat because her need for self-control dominates.

Many people, on learning of this disorder, say they wish they had "a touch" of it to get thin. They mistakenly think that people with anorexia nervosa feel no hunger. They also fail to recognize the pain of the associated psychological and physical trauma.

Physical Consequences The starvation of anorexia nervosa damages the body just as the starvation of war and poverty does. In fact, after a few months, most people with anorexia nervosa have protein-energy malnutrition (PEM) that is similar to marasmus (described in Chapter 6). Their bodies have been depleted of both body fat and protein.[13] Victims are dying to be thin—quite literally. In young people, growth ceases and normal development falters. They lose so much lean tissue that basal metabolic rate slows. In addition, the heart pumps inefficiently and irregularly, the heart muscle becomes weak and thin, the chambers diminish in size, and the blood pressure falls.[14] Minerals that help to regulate heartbeat become unbalanced. Many deaths occur due to multiple organ system failure: the heart, kidneys, and liver cease to function.

Starvation brings other physical consequences as well: loss of brain tissue, impaired immune response, anemia, and a loss of digestive functions that worsens malnutrition. Peristalsis becomes sluggish, the stomach empties slowly, and the lining of the intestinal tract atrophies. The deteriorated GI tract fails to provide sufficient digestive enzymes and absorptive surfaces for handling any food that is eaten. The pancreas slows its production of digestive enzymes. The person may suffer from diarrhea, further worsening malnutrition.

Other effects of starvation include altered blood lipids, high blood vitamin A and vitamin E, low blood proteins, dry thin skin, abnormal nerve functioning, reduced bone density, low body temperature, low blood pressure, and the development of fine body hair (the body's attempt to keep warm). The electrical activity of the brain becomes abnormal, and insomnia is common. Both women and men lose their sex drives.

Women with anorexia nervosa develop amenorrhea (it is one of the diagnostic criteria). In young girls, the onset of menstruation is delayed. Menstrual periods typically resume with recovery, although some women never restart even after they have gained weight. Should an underweight woman with anorexia nervosa become pregnant, she is likely to give birth to an underweight baby—and low-birthweight babies face many health problems (as Chapter 15 explains). Mothers with anorexia nervosa may underfeed their children who then fail to grow and suffer the other consequences of starvation.

Treatment of Anorexia Nervosa

Treatment of anorexia nervosa requires a multidisciplinary approach.[15] Teams of physicians, nurses, psychiatrists, family therapists, and dietitians work together to resolve two sets of issues and behaviors: those relating to food and weight, and those involving relationships with oneself and others. The first dietary objective is to stop weight loss while establishing regular eating patterns. Appropriate diet is crucial to recovery and must be tailored individually to each client's needs.[16] Because body weight is low and fear of weight gain is high, initial food intake may be small. As eating becomes more comfortable, energy intake should increase gradually. Initially, clients may be unwilling to eat for themselves; they may need to be fed by tube. Those who will eat have a good chance of recovering without other interventions. Even after recovery, energy intakes may not fully return to normal.[17]

Because anorexia nervosa is like starvation physically, health care professionals classify clients based on indicators of PEM.* Low-risk clients need nutrition counseling. Intermediate-risk clients may need supplements such as high-kcalorie, high-protein formulas in addition to regular meals. High-risk clients may require hospitalization and may need to be fed by tube at first to prevent death. This step may cause psychological trauma. Drugs are commonly prescribed, but play a limited role in treatment.

Denial runs high among those with anorexia nervosa. Few seek treatment on their own. About half of the women who are treated can maintain their body weight at 85 percent or more of a healthy weight; at that weight, many of them begin menstruating again.[18] The other half have poor to fair treatment outcomes, relapse into abnormal eating behaviors, or die. Anorexia nervosa has one of the highest mortality rates among psychiatric disorders.[19] An estimated 1000 women die each year of anorexia nervosa—most commonly from cardiac complications due to malnutrition or suicide.[20]

Before drawing conclusions about someone who is extremely thin or who eats very little, remember that diagnosis requires professional assessment. Several national organizations offer information for people who are seeking help with anorexia nervosa, either for themselves or for others.†

Bulimia Nervosa

Kelly is a charming, intelligent, 30-year-old flight attendant of normal weight who thinks constantly about food. She alternatively starves herself and secretly binges; when she has eaten too much, she makes herself vomit. Most readers recognize these symptoms as those of **bulimia nervosa.**

Characteristics of Bulimia Nervosa

Bulimia nervosa is distinct from anorexia nervosa and is more prevalent, although the true incidence is difficult to establish because bulimia nervosa is not as physically apparent. More

---

*Indicators of protein-energy malnutrition: a low percentage of body fat, low serum albumin, low serum transferrin, and impaired immune reactions.
†Internet sites are listed at the end of this highlight.

TABLE H9-3	**Criteria for Diagnosis of Bulimia Nervosa**

A person with bulimia nervosa demonstrates the following:

A. Recurrent episodes of binge eating. An episode of binge eating is characterized by both of the following:

   1. Eating, in a discrete period of time (e.g., within any two-hour period), an amount of food that is definitely larger than most people would eat during a similar period of time and under similar circumstances.

   2. A sense of lack of control over eating during the episode (e.g., a feeling that one cannot stop eating or control what or how much one is eating).

B. Recurrent inappropriate compensatory behavior in order to prevent weight gain, such as self-induced vomiting; misuse of laxatives, diuretics, enemas, or other medications; fasting; or excessive exercise.

C. Binge eating and inappropriate compensatory behaviors both occur, on average, at least twice a week for three months.

D. Self-evaluation unduly influenced by body shape and weight.

E. The disturbance does not occur exclusively during episodes of anorexia nervosa.

Two types:

   Purging type: The person regularly engages in self-induced vomiting or the misuse of laxatives, diuretics, or enemas.

   Nonpurging type: The person uses other inappropriate compensatory behaviors, such as fasting or excessive exercise, but does not regularly engage in self-induced vomiting or the misuse of laxatives, diuretics, or enemas.

SOURCE: Reprinted with permission from American Psychiatric Association, *Diagnostic and Statistical Manual of Mental Disorders,* 4th ed. Text Revision. (Washington, D.C.: American Psychiatric Association, 2000).

---

men suffer from bulimia nervosa than from anorexia nervosa, but bulimia nervosa is still more common in women than in men. The secretive nature of bulimic behaviors makes recognition of the problem difficult, but once it is recognized, diagnosis is based on the criteria listed in Table H9-3.

Like the typical person with bulimia nervosa, Kelly is single, female, and white. She is well educated and close to her ideal body weight, although her weight fluctuates over a range of 10 pounds or so every few weeks. She prefers to weigh less than the weight that her body maintains naturally.

Kelly seldom lets her eating disorder interfere with work or other activities, although a third of all bulimics do. From early childhood she has been a high achiever and emotionally dependent on her parents. As a young teen, Kelly frequently followed severely restricted diets, but could never maintain the weight loss.[21] Kelly feels anxious at social events and cannot easily establish close personal relationships. She is usually depressed, is often impulsive, and has low self-esteem. When crisis hits, Kelly responds by replaying events, worrying excessively, and blaming herself but never asking for help— behaviors that interfere with effective coping.

Binge Eating Like the person with anorexia nervosa, the person with bulimia nervosa spends much time thinking

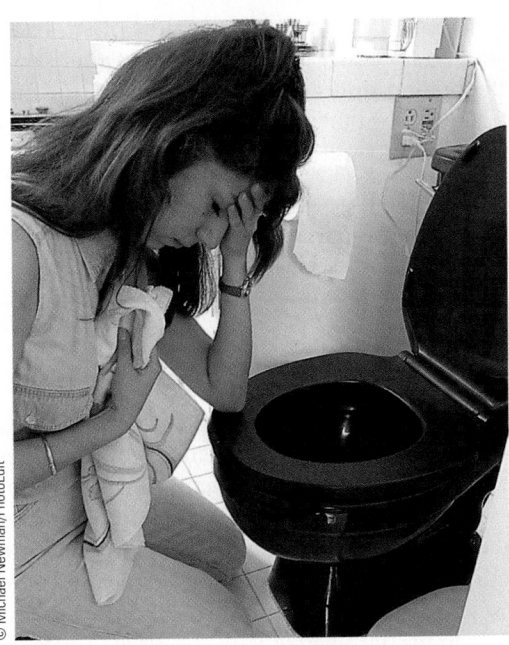

Bulimic binges are often followed by self-induced vomiting and feelings of shame or disgust.

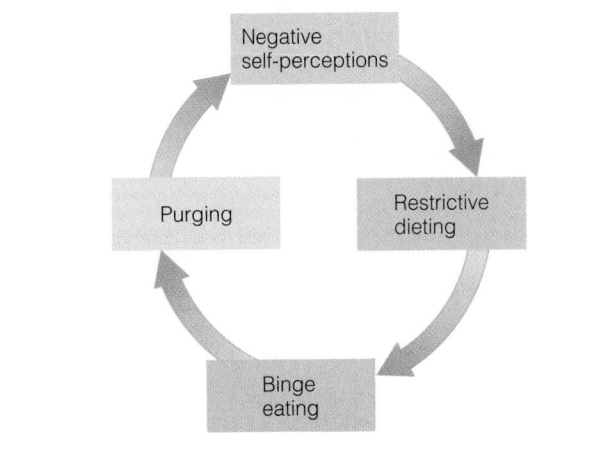

FIGURE H9-2 The Vicious Cycle of Restrictive Dieting and Binge Eating

Negative self-perceptions → Restrictive dieting → Binge eating → Purging → Negative self-perceptions

about her body weight and food. Her preoccupation with food manifests itself in secret binge-eating episodes, which usually progress through several emotional stages: anticipation and planning, anxiety, urgency to begin, rapid and uncontrollable consumption of food, relief and relaxation, disappointment, and finally shame or disgust.

A bulimic binge is characterized by a sense of lacking control over eating. During a binge, the person consumes food for its emotional comfort and cannot stop eating or control what or how much is eaten. A typical binge occurs periodically, in secret, usually at night, and lasts an hour or more. Because a binge frequently follows a period of rigid dieting, eating is accelerated by intense hunger. Energy restriction followed by bingeing can set in motion a pattern of weight cycling, which may make weight loss and maintenance more difficult over time.

During a binge, Kelly consumes thousands of kcalories of easy-to-eat, low-fiber, high-fat, and, especially, high-carbohydrate foods. Typically, she chooses cookies, cakes, and ice cream—and she eats the entire bag of cookies, the whole cake, and every last spoonful in a carton of ice cream. After the binge, Kelly pays the price with swollen hands and feet, bloating, fatigue, headache, nausea, and pain.

Purging To purge the food from her body, Kelly may use a **cathartic**—a strong laxative that can injure the lower intestinal tract. Or she may induce vomiting, with or without the use of an **emetic**—a drug intended as first aid for poisoning. These purging behaviors are often accompanied by feelings of shame or guilt. Hence a vicious cycle develops: negative self-perceptions followed by dieting, bingeing, and purging, which in turn lead to negative self-perceptions (see Figure H9-2).

On first glance, purging seems to offer a quick and easy solution to the problems of unwanted kcalories and body weight. Many people perceive such behavior as neutral or even positive, when, in fact, binge eating and purging have serious physical consequences. Signs of subclinical malnutrition are evident in a compromised immune system. Fluid and mineral imbalances caused by vomiting or diarrhea can lead to abnormal heart rhythms and injury to the kidneys. Urinary tract infections can lead to kidney failure. Vomiting causes irritation and infection of the pharynx, esophagus, and salivary glands; erosion of the teeth; and dental caries. The esophagus may rupture or tear, as may the stomach. Sometimes the eyes become red from pressure during vomiting. The hands may be calloused or cut by the teeth while inducing vomiting. Overuse of emetics depletes potassium concentrations and can lead to death by heart failure.

Unlike Julie, Kelly is aware that her behavior is abnormal, and she is deeply ashamed of it. She wants to recover, and this makes recovery more likely for her than for Julie, who clings to denial. Feeling inadequate ("I can't even control my eating"), Kelly tends to be passive and to look to others for confirmation of her sense of worth. When she experiences rejection, either in reality or in her imagination, her bulimia nervosa becomes worse. If Kelly's depression deepens, she may seek solace in drug or alcohol abuse or other addictive behaviors. Clinical depression is common in people with bulimia nervosa, and the rates of substance abuse are high.

Treatment of Bulimia Nervosa

Kelly needs to establish regular eating patterns. She may also benefit from a regular exercise program.[22] Weight maintenance, rather than cyclic weight gains and losses, is the treatment goal. Major steps toward recovery include discontinuing

TABLE H9-4 Diet Strategies for Combating Bulimia Nervosa

Planning Principles

- Plan meals and snacks; record plans in a food diary prior to eating.
- Plan meals and snacks that require eating at the table and using utensils.
- Refrain from finger foods.
- Refrain from "dieting" or skipping meals.

Nutrition Principles

- Eat a well-balanced diet and regularly timed meals consisting of a variety of foods.
- Include raw vegetables, salad, or raw fruit at meals to prolong eating times.
- Choose whole-grain, high-fiber breads, pasta, rice, and cereals to increase bulk.
- Consume adequate fluid, particularly water.

Other Tips

- Choose foods that provide protein and fat for satiety and bulky, fiber-rich carbohydrates for immediate feelings of fullness.
- Try including soups and other water-rich foods for satiety.
- Choose portions that meet the definition of "a serving" according to the Daily Food Guide (pp. 44–45).
- For convenience (and to reduce temptation) select foods that naturally divide into portions. Select one potato, rather than rice or pasta that can be overloaded onto the plate; purchase yogurt and cottage cheese in individual containers; look for small packages of precut steak or chicken; choose frozen dinners with measured portions.
- Include 30 minutes of physical activity every day—exercise may be an important tool in defeating bulimia.

purging and restrictive dieting habits and learning to eat three meals a day plus snacks.[23] Initially, energy intake should provide enough food to satisfy hunger and maintain body weight. Table H9-4 offers diet strategies to correct the eating problems of bulimia nervosa. About half of the women diagnosed with bulimia nervosa recover completely after five to ten years, with or without treatment, but treatment probably speeds the recovery process.

A mental health professional should be on the treatment team to help clients with their depression and addictive behaviors. Some physicians prescribe the antidepressant drug fluoxetine in the treatment of bulimia nervosa.* Another drug that may be useful in the management of bulimia nervosa is naloxone, an opiate antagonist that suppresses the consumption of sweet and high-fat foods in binge-eaters.

Anorexia nervosa and bulimia nervosa are distinct eating disorders, yet they sometimes overlap in important ways. Anorexia victims may purge, and victims of both disorders share an overconcern with body weight and a tendency to drastically undereat. Many perceive foods as "forbidden" and "give in" to an eating binge. The two disorders can also appear

*Fluoxetine is marketed under the trade name Prozac.

in the same person, or one can lead to the other. Treatment is challenging and relapses are not unusual. Other people have **unspecified eating disorders** that fall short of the criteria for anorexia nervosa or bulimia nervosa, but share some of their features. One such condition is binge-eating disorder.

HEALTHY PEOPLE 2010 Reduce the relapse rates for persons with eating disorders including anorexia nervosa and bulimia nervosa.

Binge-Eating Disorder

Charlie is a 40-year-old schoolteacher who has been overweight all his life. His friends and family are forever encouraging him to lose weight, and he has come to believe that if he only had more willpower, dieting would work. He periodically gives dieting his best shot—restricting energy intake for a day or two only to succumb to uncontrollable cravings, especially for high-fat foods. Like Charlie, up to half of the obese people who try to lose weight periodically binge; unlike people with bulimia nervosa, however, they typically do not purge. Such an eating disorder does not meet the criteria for either anorexia nervosa or bulimia nervosa—yet such compulsive overeating is a problem and occurs in people of normal weight as well as those who are severely overweight. Table H9-5 lists criteria for unspecified eating disorders, including binge eating. Obesity alone is not an eating disorder.

Clinicians note differences between people with bulimia nervosa and those with binge-eating disorder.[24] People with **binge-eating disorder** consume less during a binge, rarely purge, and exert less restraint during times of dieting. Similarities also exist, including feeling out of control, disgusted, depressed, embarrassed, guilty, or distressed because of their self-perceived gluttony.[25]

There are also differences between obese binge-eaters and obese people who do not binge. Those with the binge-eating disorder report higher rates of self-loathing, disgust about body size, depression, and anxiety. Their eating habits differ as well. Obese binge-eaters tend to consume more kcalories and more dessert and snack-type foods during regular meals and binges than obese people who do not binge.

Binge eating is a behavioral disorder that can be resolved with treatment. Resolving such behavior may not bring weight loss, but it may make participation in weight-control programs easier. It also improves physical health, mental health, and the chances of success in breaking the cycle of rapid weight losses and gains.

Eating Disorders in Society

Proof that society plays a role in eating disorders is found in their demographic distribution—they are known only in developed nations, and they become more prevalent as wealth increases

TABLE H9-5 Unspecified Eating Disorders, including Binge-Eating Disorder

Criteria for Diagnosis of Unspecified Eating Disorders, in General

Many people have eating disorders but do not meet all the criteria to be classified as having anorexia nervosa or bulimia nervosa. Some examples include those who:

A. Meet all of the criteria for anorexia nervosa, except irregular menses.

B. Meet all of the criteria for anorexia nervosa, except that their current weights fall within the normal ranges.

C. Meet all of the criteria for bulimia nervosa, except that binges occur less frequently than stated in the criteria.

D. Are of normal body weight and who compensate inappropriately for eating small amounts of food (example: self-induced vomiting after eating two cookies).

E. Repeatedly chew food, but spit it out without swallowing.

F. Have recurrent episodes of binge eating but who do not compensate as do those with bulimia nervosa.

Criteria for Diagnosis of Binge-Eating Disorder, Specifically

A person with a binge-eating disorder demonstrates the following:

A. Recurrent episodes of binge eating. An episode of binge eating is characterized by both of the following:

  1. Eating, in a discrete period of time (e.g., within any two-hour period) an amount of food that is definitely larger than most people would eat in a similar period of time under similar circumstances.

  2. A sense of lack of control over eating during the episode (e.g., a feeling that one cannot stop eating or control what or how much one is eating).

B. Binge-eating episodes are associated with at least three of the following:

  1. Eating much more rapidly than normal.

  2. Eating until feeling uncomfortably full.

  3. Eating large amounts of food when not feeling physically hungry.

  4. Eating alone because of being embarrassed by how much one is eating.

  5. Feeling disgusted with oneself, depressed, or very guilty after overeating.

C. The binge eating causes marked distress.

D. The binge eating occurs, on average, at least twice a week for six months.

E. The binge eating is not associated with the regular use of inappropriate compensatory behaviors (e.g., purging, fasting, excessive exercise) and does not occur exclusively during the course of anorexia nervosa or bulimia nervosa.

SOURCE: Reprinted with permission from American Psychiatric Association, *Diagnostic and Statistical Manual of Mental Disorders,* 4th ed. Text Revision. (Washington, D.C.: American Psychiatric Association, 2000).

and food becomes plentiful. Some people point to the vomitoriums of ancient times and claim that bulimia nervosa is not new, but the two are actually distinct. Ancient people were eating for pleasure, without guilt, and in the company of others; they vomited so that they could rejoin the feast. Bulimia nervosa is a disorder of isolation and is often accompanied by low self-esteem.

Chapter 8 described how our society sets unrealistic ideals for body weight, especially in women, and devalues those who do not conform to them. Anorexia nervosa and bulimia nervosa are not a form of rebellion against these unreasonable expectations, but rather an exaggerated acceptance of them. In fact, body dissatisfaction is a primary factor in the development of eating disorders.[26] Not that everyone who is dissatisfied will develop an eating disorder, but everyone with an eating disorder is dissatisfied. Characteristics of disordered eating such as restrained eating, fasting, binge eating, purging, fear of fatness, and distortion of body image are extraordinarily common among young girls. Most are "on diets," and many are poorly nourished.[27] Some eat too little food to support normal growth; thus they miss out on their adolescent growth spurts and may never catch up. Many eat so little that hunger propels them into binge-purge cycles.

Perhaps a person's best defense against these disorders is to learn to appreciate his or her own uniqueness. When people discover and honor the body's real physical needs, they become unwilling to sacrifice health for conformity. To respect and value oneself may be lifesaving.

NUTRITION ON THE NET

 Access these websites for further study of topics covered in this highlight.

- Find updates and quick links to these and other nutrition-related sites at our website: **www.wadsworth.com/nutrition**

- Search for "anorexia," "bulimia," and "eating disorders" at the U.S. Government health information site: **www.healthfinder.gov**

- Learn more about anorexia nervosa and related eating disorders from Anorexia Nervosa and Related Eating Disorders: **www.anred.com**

- Find out about help-lines, referral networks, support groups, and prevention programs from the American Anorexia Bulimia Association: **www.aabainc.org**

- Get facts about eating disorders from the National Institute of Mental Health: **www.nimh.nih.gov/publicat/eatingdisorder.cfm**

REFERENCES

1. Position of the American Dietetic Association: Nutrition intervention in the treatment of anorexia nervosa, bulimia nervosa, and eating disorders not otherwise specified (EDNOS), *Journal of the American Dietetic Association* 101 (2001): 810–819.

2. Position of the American Dietetic Association, 2001.

3. K. Kazis and E. Iglesias, The female athlete triad, *Adolescent Medicine* 14 (2003): 87–95; S. Sabatini, The female athlete triad, *American Journal of the Medical Sciences* 322 (2001): 193–195; Committee on Sports Medicine and Fitness, Medical concerns in the female athlete, *Pediatrics* 106 (2000): 610–613.

4. N. H. Golden, A review of the female athlete triad (amenorrhea, osteoporosis and disordered eating), *International Journal of Adolescent Medicine and Health* 14 (2002): 9–17.

5. M. Bass, L. Turner, and S. Hunt, Counseling female athletes: Application of the stages of change model to avoid disordered eating, amenorrhea, and osteoporosis, *Psychological Reports* 88 (2001): 1153–1160; M. P. Warren and A. L. Stiehl, Exercise and female adolescents: Effects on the reproductive and skeletal systems, *Journal of the American Medical Women's Association* 54 (1999): 115–120, 138.

6. D. Hartman and coauthors, Bone density of women who have recovered from anorexia nervosa, *International Journal of Eating Disorders* 28 (2000): 107–112.

7. S. J. Crow and coauthors, Long-term menstrual and reproductive function in patients with bulimia nervosa, *American Journal of Psychiatry* 159 (2002): 1048–1050; J. R. Newton and coauthors, Osteoporosis and normal weight bulimia nervosa—Which patients are at risk? *Journal of Psychosomatic Research* 37 (1993): 239–247.

8. J. A. McLean, S. I. Barr, and J. C. Prior, Cognitive dietary restraint is associated with higher urinary cortisol excretion in healthy premenopausal women, *American Journal of Clinical Nutrition* 73 (2001): 7–12.

9. A. I. Zeni and coauthors, Stress injury to the bone among women athletes, *Physical Medicine and Rehabilitation Clinics of North America* 11 (2000): 929–947; A. Nattiv, Stress fractures and bone health in track and field athletes, *Journal of Science and Medicine in Sport* 3 (2000): 268–279.

10. J. A. Hobart and D. R. Smucker, The female athlete triad, *American Family Physician* 61 (2000): 3357–3364, 3367.

11. R. B. Kiningham and D. W. Gorenflo, Weight loss methods of high school wrestlers, *Medicine and Science in Sports and Exercise* 33 (2001): 810–813.

12. D. Neumark-Sztainer and coauthors, Disordered eating among adolescents: Associations with sexual/physical abuse and other familial/psychosocial factors, *International Journal of Eating Disorders* 28 (2000): 249–258.

13. K. P. Kerruish and coauthors, Body composition in adolescents with anorexia nervosa, *American Journal of Clinical Nutrition* 75 (2002): 31–37.

14. C. Romano and coauthors, Reduced hemodynamic load and cardiac hypotrophy in patients with anorexia nervosa, *American Journal of Clinical Nutrition* 77 (2003): 308–312; C. Panagiotopoulos and coauthors, Electrocardiographic findings in adolescents with eating disorders, *Pediatrics* 105 (2000): 1100–1105.

15. Committee on Adolescence, Identifying and treating eating disorders, *Pediatrics* 111 (2003): 204–211.

16. A. E. Becker and coauthors, Eating disorders, *New England Journal of Medicine* 340 (1999): 1092–1098.

17. B. R. Carruth and J. D. Skinner, Dietary and physical activity patterns of young females with histories of eating disorders, *Topics in Clinical Nutrition* 16 (2000): 13–23.

18. H. C. Steinhausen, The outcome of anorexia nervosa in the 20th century, *American Journal of Psychiatry* 159 (2002): 1284–1293; B. Lowe and coauthors, Long-term outcome of anorexia nervosa in a prospective 21-year follow-up study, *Psychological Medicine* 31 (2001): 881–890.

19. P. K. Keel and coauthors, Predictors of mortality in eating disorders, *Archives of General Psychiatry* 60 (2003): 179–183.

20. M. B. Tamburrino and R. A. McGinnis, Anorexia nervosa: A review, *Panminerva Medica* 44 (2002): 301–311.

21. G. C. Patton and coauthors, Onset of adolescent eating disorders: Population based cohort study over 3 years, *British Medical Journal* 318 (1999): 765–768.

22. J. Sundgot-Borgen and coauthors, The effect of exercise, cognitive therapy, and nutritional counseling in treating bulimia nervosa, *Medicine and Science in Sports and Exercise* 34 (2002): 190–195.

23. Position of the American Dietetic Association, 2001.

24. A. E. Dingemans, M. J. Bruna, and E. F. van Furth, Binge eating disorder: A review, *International Journal of Obesity and Related Metabolic Disorders* 26 (2002): 299–307.

25. D. M. Ackard and coauthors, Overeating among adolescents: Prevalence and associations with weight-related characteristics and psychological health, *Pediatrics* 111 (2003): 67–74.

26. J. Polivy and C. P. Herman, Causes of eating disorders, *Annual Review of Psychology* 53 (2002): 187–213.

27. Federal Interagency Forum on Child and Family Statistics, *America's Children: Key National Indicators of Well-Being*, 1999, a report from the National Institutes of Health, available from National Maternal and Child Health Clearinghouse, 2070 Chain Bridge Road, Suite 450, Vienna, VA 22182 or on the Internet at **http://childstats.gov**.

Chapter 10

The Water-Soluble Vitamins: B Vitamins and Vitamin C

© Stephen Wilkes/The Image Bank/Getty Images

Chapter Outline

The Vitamins—An Overview

The B Vitamins—As Individuals:
*Thiamin • Riboflavin • Niacin • Biotin
• Pantothenic Acid • Vitamin B_6 • Folate
• Vitamin B_{12} • Non-B Vitamins*

The B Vitamins—In Concert: *B Vitamin
Roles • B Vitamin Deficiencies • B Vitamin
Toxicities • B Vitamin Food Sources*

Vitamin C: *Vitamin C Roles • Vitamin C
Recommendations • Vitamin C Deficiency
• Vitamin C Toxicity • Vitamin C Food Sources*

Highlight: *Vitamin and Mineral Supplements*

Nutrition Explorer CD-ROM Outline

Nutrition Animation: *Protecting the
Vitamins in Foods*

Case Study: *Absorption of Water
Soluble Vitamins*

Student Practice Test

Glossary Terms

Nutrition on the Net

Nutrition in Your Life

If you were playing a word game and your partner said "vitamins," how would you respond? If "pills" and "supplements" immediately come to mind, you may be missing the main message of the vitamin story—that hundreds of foods deliver over a dozen vitamins that participate in thousands of activities throughout your body. Quite simply, foods supply vitamins to support all that you are and all that you do—and supplements of any one of them, or even a combination of them, can't compete with foods in keeping you healthy.

Earlier chapters focused on the energy-yielding nutrients, which play leading roles in the body. The vitamins and minerals are their supporting cast. This chapter begins with an overview of the vitamins and then examines each of the water-soluble vitamins and a nonvitamin relative named choline; the next chapter features the fat-soluble vitamins. Chapters 12 and 13 present the minerals.

The Vitamins—An Overview

Researchers first recognized that there were substances in foods that were "vital to life" in the early 1900s. Since then, the world of vitamins has opened up dramatically. The vitamins■ are powerful substances, as their *absence* attests. Vitamin A deficiency can cause blindness; a lack of the B vitamin niacin can cause dementia; and a lack of vitamin D can retard bone growth. The consequences of deficiencies are so dire, and the effects of restoring the needed vitamins so dramatic, that people spend billions of dollars every year in the belief that vitamin pills will cure a host of ailments (see Highlight 10). Vitamins certainly support sound nutritional health, but they do not cure all ills. Furthermore, vitamin supplements do not offer the many benefits that come from vitamin-rich foods.

■ Reminder: The *vitamins* are organic, essential nutrients required in tiny amounts to perform specific functions that promote growth, reproduction, or the maintenance of health and life.
- **vita** = life
- **amine** = containing nitrogen (the first vitamins discovered contained nitrogen)

The *presence* of the vitamins also attests to their power. Vitamin C not only prevents the deficiency disease scurvy, but also seems to protect against certain types of cancer. Similarly, vitamin E seems to help protect against some facets of cardiovascular disease. The B vitamin folate helps to prevent birth defects. As you will see, the vitamins' roles in supporting optimal health extend far beyond preventing deficiency diseases. In fact, some of the credit given to low-fat diets in preventing disease actually belongs to the vitamins that diets rich in vegetables, fruits, and whole grains deliver (see Highlight 11 for more on vitamins in disease prevention). Chapter 18 highlights the roles of vitamins in supporting a strong immune system.

The vitamins differ from carbohydrates, fats, and proteins in the following ways:

- *Structure.* Vitamins are individual units; they are not linked together (as are molecules of glucose or amino acids). Appendix C presents the chemical structure for each of the vitamins.

- *Function.* Vitamins do not yield usable energy when broken down; they assist the enzymes that release energy from carbohydrates, fats, and proteins.

- *Food contents.* The amounts of vitamins people ingest daily from foods and the amounts they require are measured in *micrograms* (µg) or *milligrams* (mg), rather than grams (g). ■

■ 1 g = 1000 mg.
 1 mg = 1000 µg.
 For perspective, a dollar bill weighs about 1 g.

The vitamins are similar to the energy-yielding nutrients, though, in that they are vital to life, organic, and available from foods.

Bioavailability The amount of vitamins available from foods depends not only on the quantity provided by a food but also on the amount absorbed and used by the body—referred to as the vitamins' **bioavailability.** The quantity of vitamins in a food can be determined in a rather straightforward manner. Researchers analyze foods to determine their vitamin contents and publish the results in tables of food composition such as Appendix H. Determining the bioavailability of a vitamin is a more complex task because it depends on many factors, including:

- Efficiency of digestion and time of transit through the GI tract.
- Previous nutrient intake and nutrition status.
- Other foods consumed at the same time. (Chapters 10–13 describe factors that inhibit or enhance the absorption of individual vitamins and minerals.)
- Method of food preparation (raw, cooked, or processed).
- Source of the nutrient (synthetic, fortified, or naturally occurring).

Experts consider these factors when estimating recommended intakes.

Precursors Some of the vitamins are available from foods in inactive forms known as **precursors,** or provitamins. Once inside the body, the precursor is converted to an active form of the vitamin. Thus, in measuring a person's vitamin intake, it is important to count both the amount of the active vitamin and the potential amount available from its precursors. The summary tables throughout this chapter and the next indicate which vitamins have precursors.

Organic Nature Being organic, vitamins can be destroyed and left unable to perform their duties. Therefore, they must be handled with care during storage and in cooking. Prolonged heating may destroy much of the thiamin in food. Because riboflavin can be destroyed by the ultraviolet rays of the sun or by fluorescent light, foods stored in transparent glass containers are most likely to lose riboflavin. Oxygen destroys vitamin C, so losses occur when foods are cut or broken and thereby exposed to air. Table 10-1 summarizes ways to minimize nutrient losses in the kitchen, and Chapter 19 provides more details.

Solubility As you may recall, carbohydrates and proteins are hydrophilic and lipids are hydrophobic. The vitamins divide along the same lines—the hy-

To minimize vitamin losses, wrap cut fruits and vegetables or store them in airtight containers.

Polara Studios, Inc.

bioavailability: the rate at and the extent to which a nutrient is abosrbed and used.

precursors: substances that precede others; with regard to vitamins, compounds that can be converted into active vitamins; also known as **provitamins.**

TABLE 10-1 | Minimizing Nutrient Losses

- To slow the degradation of vitamins, refrigerate (most) fruits and vegetables.
- To minimize the oxidation of vitamins, store fruits and vegetables that have been cut in airtight wrappers and juices that have been opened in closed containers (and refrigerate them).
- To prevent losses during washing, rinse fruits and vegetables before cutting.
- To minimize losses during cooking, use a microwave oven or steam vegetables in a small amount of water. Add vegetables after water has come to a boil. Use the cooking water in mixed dishes such as casseroles and soups. Avoid high temperatures and long cooking times.

drophilic, water-soluble ones are the eight B vitamins and vitamin C; the hydrophobic, fat-soluble ones are vitamins A, D, E, and K. As each vitamin was discovered, it was given a name and sometimes a letter and number as well. Many of the water-soluble vitamins have multiple names, which has led to some confusion. The margin ■ lists the standard names; summary tables throughout this chapter provide the common alternative names.

Solubility is apparent in the food sources of the different vitamins, and it affects their absorption, transport, storage, and excretion by the body. The water-soluble vitamins are found in the watery compartments of foods; the fat-soluble vitamins usually occur together in the fats and oils of foods. On being absorbed, the water-soluble vitamins move directly into the blood; like fats, the fat-soluble vitamins must first enter the lymph, then the blood. Once in the blood, many of the water-soluble vitamins travel freely; many of the fat-soluble vitamins require protein carriers for transport. Upon reaching the cells, water-soluble vitamins freely circulate in the water-filled compartments of the body; fat-soluble vitamins are held in fatty tissues and the liver until needed. The kidneys, monitoring the blood that flows through them, detect and remove small excesses of water-soluble vitamins (large excesses, however, may overwhelm the system, creating adverse effects); fat-soluble vitamins tend to remain in fat-storage sites in the body rather than being excreted, and so are more likely to reach toxic levels when consumed in excess.

Because the body stores fat-soluble vitamins, they can be eaten in large amounts once in a while and still meet the body's needs over time. Water-soluble vitamins are retained for varying periods in the body; a single day's omission from the diet does not bring on a deficiency, but still, the water-soluble vitamins must be eaten more regularly than the fat-soluble vitamins.

Toxicity Knowledge about some of the amazing roles of vitamins has prompted many people to begin taking supplements, assuming that more is better. But just as an inadequate intake can cause harm, so can an excessive intake. As mentioned, even some of the water-soluble vitamins have adverse effects when taken in large doses.

That a vitamin can be both essential and harmful may seem surprising, but the same is true of most nutrients. The effects of every substance depend on its dose, and this is one reason consumers should not self-prescribe supplements for their ailments. See the "How to" on the next page for a perspective on doses.

The Committee on Dietary Reference Intakes (DRI) addresses the possibility of adverse effects from high doses of nutrients by establishing Tolerable Upper Intake Levels. An Upper Level defines the highest amount of a nutrient that is likely not to cause harm for most healthy people when consumed daily. The risk of harm increases as intakes rise above the Upper Level. Of the nutrients discussed in this chapter, niacin, vitamin B_6, folate, choline, and vitamin C have had Upper Levels set, and these values are presented in their respective summary tables. Data are lacking to establish Upper Levels for the remaining B vitamins, but this does not mean that excessively high intakes would be without risk. (The inside front cover presents Upper Levels for the vitamins and minerals.)

■ **Water-soluble vitamins:**
- B vitamins:
  - Thiamin.
  - Riboflavin.
  - Niacin.
  - Biotin.
  - Pantothenic acid.
  - Vitamin B_6.
  - Folate.
  - Vitamin B_{12}.
- Vitamin C.

Fat-soluble vitamins:
- Vitamin A.
- Vitamin D.
- Vitamin E.
- Vitamin K.

HOW TO Understand Dose Levels and Effects

A substance may have a beneficial or harmful effect, but a critical thinker would not conclude that the substance itself was beneficial or harmful without first asking what dose was used. The accompanying figure shows three possible relationships between dose levels and effects. The third diagram represents the situation with nutrients—more is better up to a point, but beyond that point, still more is harmful.

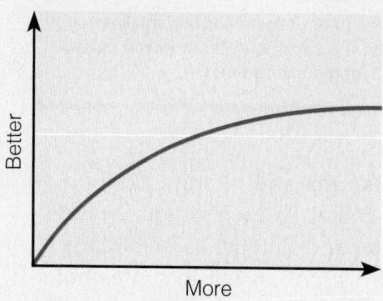

As you progress in the direction of more, the effect gets better and better, with no end in sight (real life is seldom, if ever, like this).

As you progress in the direction of more, the effect reaches a maximum and then a plateau, becoming no better with higher doses.

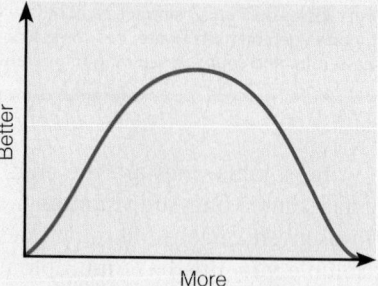

As you progress in the direction of more, the effect reaches an optimum at some intermediate dose and then declines, showing that more is better up to a point and then harmful. That too much is as harmful as too little represents the situation with nutrients.

IN SUMMARY The vitamins are essential nutrients needed in tiny amounts in the diet both to prevent deficiency diseases and to support optimal health. The water-soluble vitamins are the B vitamins and vitamin C; the fat-soluble vitamins are vitamins A, D, E, and K. The accompanying table summarizes the differences between the water-soluble and fat-soluble vitamins.

	Water-Soluble Vitamins: B Vitamins and Vitamin C	Fat-Soluble Vitamins: Vitamins A, D, E, and K
Absorption	Directly into the blood.	First into the lymph, then the blood.
Transport	Travel freely.	Many require protein carriers.
Storage	Circulate freely in water-filled parts of the body.	Stored in the cells associated with fat.
Excretion	Kidneys detect and remove excess in urine.	Less readily excreted; tend to remain in fat-storage sites.
Toxicity	Possible to reach toxic levels when consumed from supplements.	Likely to reach toxic levels when consumed from supplements.
Requirements	Needed in frequent doses (perhaps 1 to 3 days).	Needed in periodic doses (perhaps weeks or even months).

NOTE: Exceptions occur, but these differences between the water-soluble and fat-soluble vitamins are valid generalizations.

The discussion of B vitamins that follows begins with a brief description of each of them, then offers a look at the ways they work together. Thus a preview of the individuals is followed by a survey of them all together, in concert.

The B Vitamins—As Individuals

Despite supplement advertisements that claim otherwise, the vitamins do not provide the body with fuel for energy. It is true, though, that without B vitamins the body would lack energy. The energy-yielding nutrients—carbohydrate, fat, and

FIGURE 10-1   Coenzyme Action

Some vitamins form part of the coenzymes that enable enzymes either to synthesize compounds (as illustrated by the lower enzymes in this figure) or to dismantle compounds (as illustrated by the upper enzymes).

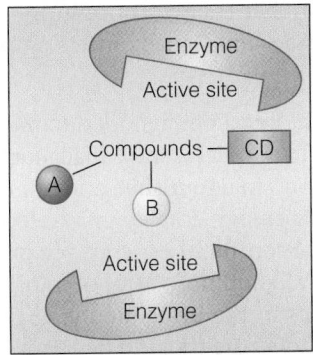

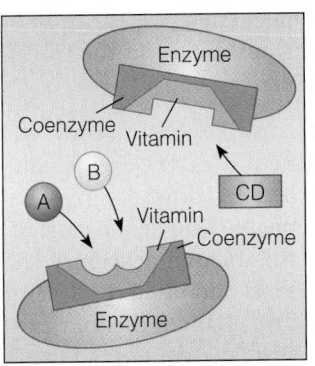

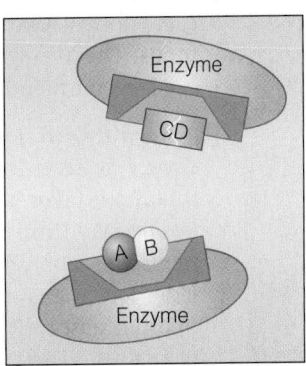

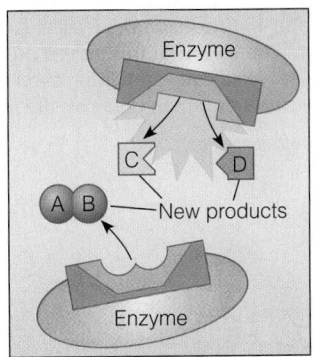

Without coenzymes, compounds A, B, and CD don't respond to their enzymes.

With the coenzymes in place, compounds are attracted to their sites on the enzymes . . .

. . . and the reactions proceed instantaneously. The coenzymes often donate or accept electrons, atoms, or groups of atoms.

The reactions are completed with either the formation of a new product, AB, or the breaking apart of a compound into two new products, C and D, and the release of energy.

protein—are used for fuel; the B vitamins help the body to use that fuel. Several of the B vitamins—thiamin, riboflavin, niacin, pantothenic acid, and biotin—form part of the coenzymes■ that assist certain enzymes in the release of energy from carbohydrate, fat, and protein. Other B vitamins play other indispensable roles in metabolism. Vitamin B_6 assists enzymes that metabolize amino acids; folate and vitamin B_{12} help cells to multiply. Among these cells are the red blood cells and the cells lining the GI tract—cells that deliver energy to all the others.

The vitamin portion of a coenzyme allows a chemical reaction to occur; the remaining portion of the coenzyme binds to the enzyme. Without its coenzyme, an enzyme cannot function. Thus symptoms of B vitamin deficiencies directly reflect the disturbances of metabolism incurred by a lack of coenzymes. Figure 10-1 illustrates coenzyme action.

The following sections describe individual B vitamins and note many coenzymes and metabolic pathways. Keep in mind that a later section will assemble these pieces of information into a whole picture.

The following sections also present the recommendations, deficiency and toxicity symptoms, and food sources for each vitamin. The recommendations for the B vitamins and vitamin C reflect the 1998 and 2000 DRI, respectively.[1] For thiamin, riboflavin, niacin, vitamin B_6, folate, vitamin B_{12}, and vitamin C, sufficient data were available to establish an RDA; for biotin, pantothenic acid, and choline, an Adequate Intake (AI) was set; only niacin, vitamin B_6, folate, choline, and vitamin C have Tolerable Upper Intake Levels. These values appear in the summary tables and figures that follow, as well as on the inside front covers.

■ Reminder: A *coenzyme* is a small organic molecule that associates closely with certain enzymes; many B vitamins form an integral part of coenzymes.

Thiamin

Thiamin is the vitamin part of the coenzyme TPP (thiamin pyrophosphate), which assists in energy metabolism. The TPP coenzyme participates in the conversion of pyruvate to acetyl CoA (described in Chapter 7). The reaction removes one carbon from the 3-carbon pyruvate to make the 2-carbon acetyl CoA and carbon dioxide (CO_2). Later TPP participates in a similar step in the TCA cycle where it helps convert a 5-carbon compound to a 4-carbon compound. Besides playing these pivotal roles in the energy metabolism of all cells, thiamin occupies a special site on the

thiamin (THIGH-ah-min): a B vitamin. The coenzyme form is **TPP (thiamin pyrophosphate)**.

FIGURE 10-2 Thiamin-Deficiency Symptom—The Edema of Beriberi

Beriberi may be characterized as "wet" (referring to edema) or "dry" (with muscle wasting, but no edema). Physical examination confirms that this person has wet beriberi. Notice how the impression of the physician's thumb remains on the leg.

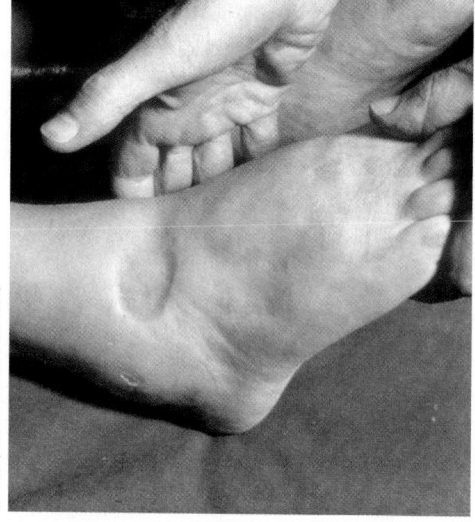

© NMSB/Custom Medical Stock Photo

■ Severe thiamin deficiency in alcohol abusers is called the **Wernicke-Korsakoff** (VER-nee-key KORE-sah-kof) **syndrome.** Symptoms include disorientation, loss of short-term memory, jerky eye movements, and staggering gait.

© Polara Studios Inc.

Pork is the richest source of thiamin, but enriched or whole-grain products typically make the greatest contribution to a day's intake because of the quantities eaten. Legumes such as split peas are also valuable sources of thiamin.

beriberi: the thiamin-deficiency disease.
- **beri** = weakness
- **beriberi** = "I can't, I can't"

membranes of nerve cells. Consequently, processes in nerves and in their responding tissues, the muscles, depend heavily on thiamin.

Thiamin Recommendations Dietary recommendations are based primarily on thiamin's role in enzyme activity. Generally, thiamin needs will be met if a person eats enough food to meet energy needs and obtains that energy from nutritious foods. The average thiamin intake in the United States and Canada meets or exceeds recommendations.

Thiamin Deficiency and Toxicity People who fail to eat enough food to meet energy needs risk nutrient deficiencies, including thiamin deficiency. Inadequate thiamin intakes have been reported among the nation's malnourished and homeless people. Similarly, people who derive most of their energy from empty-kcalorie items, like alcohol,■ risk thiamin deficiency. Alcohol contributes energy, but provides few, if any, nutrients and often displaces food. In addition, alcohol impairs thiamin absorption and enhances thiamin excretion in the urine, doubling the risk of deficiency. An estimated four out of five alcoholics are thiamin deficient.

Prolonged thiamin deficiency can result in the disease **beriberi,** which was first observed in Indonesia when the custom of polishing rice became widespread.[2] Rice provided 80 percent of the energy intake of the people of that area, and the germ and bran of the rice grain was their principal source of thiamin. When the germ and bran were removed in the preparation of white rice, beriberi spread like wildfire. The symptoms of beriberi include damage to the nervous system as well as to the heart and other muscles. Figure 10-2 presents one of the symptoms of beriberi. No adverse effects have been associated with excesses of thiamin; no Upper Level has been determined.

Thiamin Food Sources Before examining Figure 10-3, you may want to read the accompanying "How to," which describes the many features found in this and similar figures in this chapter and the next three chapters. When you look at Figure 10-3, notice that thiamin occurs in small quantities in many nutritious foods; highly refined foods contain almost no thiamin. The long red bar near the bottom of the graph shows that meats in the pork family are exceptionally rich in thiamin.

As mentioned earlier, prolonged cooking can destroy thiamin. Also, like other water-soluble vitamins, thiamin leaches into water when foods are boiled or blanched. Cooking methods that require little or no water such as steaming and microwave heating conserve thiamin and other water-soluble vitamins. The accompanying table summarizes thiamin's main functions, food sources, and deficiency symptoms.

IN SUMMARY Thiamin

Other Names	Deficiency Disease
Vitamin B₁	Beriberi (wet, with edema; dry, with muscle wasting)

1998 RDA

Men: 1.2 mg/day

Women: 1.1 mg/day

Deficiency Symptoms*

Enlarged heart, cardiac failure; muscular weakness; apathy, poor short-term memory, confusion, irritability; anorexia, weight loss

Chief Functions in the Body

Part of coenzyme TPP (thiamin pyrophosphate) used in energy metabolism

Toxicity Symptoms

None reported

Significant Sources

Whole-grain, fortified, or enriched grain products; moderate amounts in all nutritious food; pork

Easily destroyed by heat

*Severe thiamin deficiency is often related to heavy alcohol consumption.

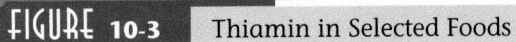

FIGURE 10-3 Thiamin in Selected Foods

Food	Serving size (kcalories)
Bread, whole wheat	1 oz slice (70 kcal)
Cornflakes, fortified	1 oz (110 kcal)
Spaghetti pasta	½ c cooked (99 kcal)
Tortilla, flour	1 10"-round (234 kcal)
Broccoli	½ c cooked (22 kcal)
Carrots	½ c shredded raw (24 kcal)
Potato	1 medium baked w/skin (133 kcal)
Tomato juice	¾ c (31 kcal)
Banana	1 medium raw (109 kcal)
Orange	1 medium raw (62 kcal)
Strawberries	½ c fresh (22 kcal)
Watermelon	1 slice (92 kcal)
Milk	1 c reduced-fat 2% (121 kcal)
Yogurt, plain	1 c low-fat (155 kcal)
Cheddar cheese	1½ oz (171 kcal)
Cottage cheese	½ c low-fat 2% (101 kcal)
Pinto beans	½ c cooked (117 kcal)
Peanut butter	2 tbs (188 kcal)
Sunflower seeds	1 oz dry (165 kcal)
Tofu (soybean curd)	½ c (76 kcal)
Ground beef, lean	3 oz broiled (244 kcal)
Chicken breast	3 oz roasted (140 kcal)
Tuna, canned in water	3 oz (99 kcal)
Egg	1 hard cooked (78 kcal)
Excellent, and sometimes unusual, sources:	
Pork chop, lean	3 oz broiled (169 kcal)
Soy milk	1 c (81 kcal)
Squash, acorn	½ c baked (69 kcal)

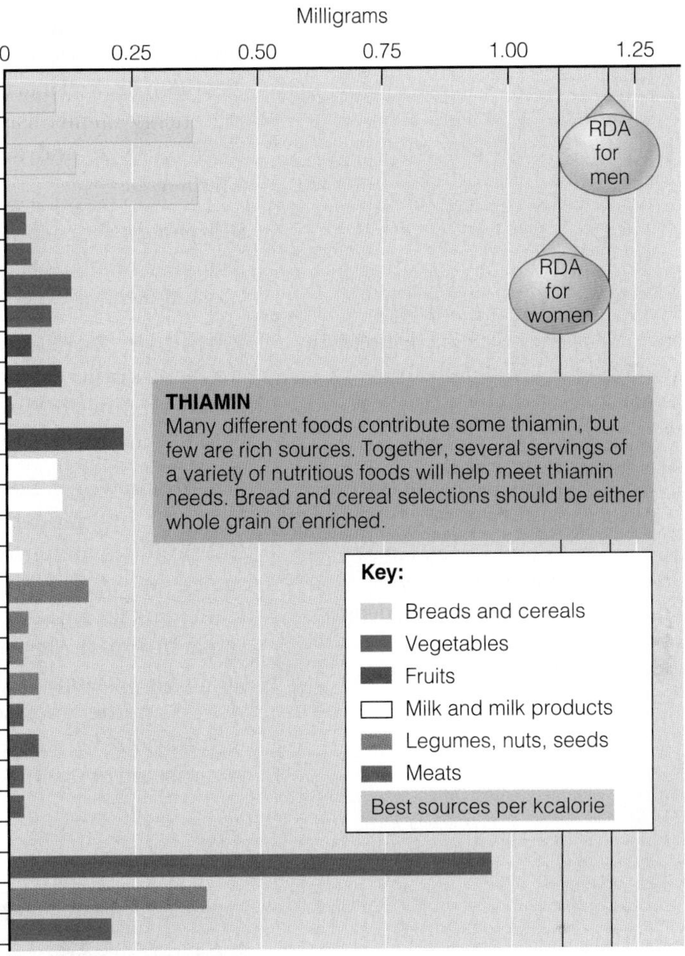

Milligrams scale: 0, 0.25, 0.50, 0.75, 1.00, 1.25

RDA for men

RDA for women

THIAMIN
Many different foods contribute some thiamin, but few are rich sources. Together, several servings of a variety of nutritious foods will help meet thiamin needs. Bread and cereal selections should be either whole grain or enriched.

Key:
- Breads and cereals
- Vegetables
- Fruits
- Milk and milk products
- Legumes, nuts, seeds
- Meats
- Best sources per kcalorie

Note: See below for more information on using this figure.

HOW TO Evaluate Foods for Their Nutrient Contributions

Figure 10-3 is the first of a series of figures in this and the next three chapters that present the vitamins and minerals in foods. Each figure presents the same 24 foods, which were selected to ensure a variety of choices representative of each of the food groups as suggested by the Food Guide Pyramid. From its base, for example, a bread, a cereal, and a pasta were chosen. The suggestion to include a variety of vegetables was also considered: dark green, leafy vegetables (broccoli); deep yellow vegetables (carrots); starchy vegetables (potatoes); legumes (pinto beans); and other vegetables (tomato juice). The selection of fruits followed suggestions to use whole fruits (bananas); citrus fruits (oranges); melons (watermelon); and berries (strawberries). Items were selected from the milk and meat groups in a similar way. In addition to the 24 foods that appear in all of the figures, three

different foods were selected for each of the nutrients to add variety and often reflect excellent, and sometimes unusual, sources.

Notice that the figures list the food, the serving size, and the food energy (kcalories) on the left and graph the amount of the nutrient per serving on the right along with the RDA (or AI) for adults, so you can see how many servings would be needed to meet recommendations. Serving sizes reflect those used by the Food Guide Pyramid. In some cases, the Pyramid specifies ambiguous serving sizes, recommending "1 medium potato" and "1 slice melon." For these foods, serving sizes reflect those presented in Appendix H.

The colored bars show at a glance which food groups best provide a nutrient: yellow for breads and cereals; green for vegetables; purple for fruits; white for milk and milk products; brown for legumes; and red for meat, fish, and poultry. Because the Food Guide Pyramid mentions legumes with both the meat group and the vegetable group

and because legumes are especially rich in many vitamins and minerals, they have been given their own color to highlight their nutrient contributions.

Notice how the bar graphs shift in the various figures. Careful study of all of the figures taken together will confirm that variety is the key to nutrient adequacy.

Another way to evaluate foods for their nutrient contributions is to consider their nutrient density (their thiamin *per 100 kcalories*, for example). Quite often, vegetables rank higher on a nutrient-per-kcalorie list than they do on a nutrient-per-serving list (see p. 41 to review how to evaluate foods based on nutrient density). The left column in the figure highlights five or so foods that offer the best deal for your energy "dollar" (the kcalorie). Notice how many of them are vegetables.

Realistically, people cannot eat for single nutrients. Fortunately, most foods deliver more than one nutrient, allowing people to combine foods into nourishing meals.

Riboflavin

Like thiamin, **riboflavin** serves as a coenzyme in many reactions, most notably in the release of energy from nutrients in all body cells. The coenzyme forms of riboflavin are FMN (flavin mononucleotide) and FAD (flavin adenine dinucleotide); both can accept and then donate two hydrogens (see Figure 10-4). During energy metabolism, FAD picks up two hydrogens (with their electrons) from the TCA cycle and delivers them to the electron transport chain (described in Chapter 7).

Riboflavin Recommendations Like thiamin's RDA, riboflavin's RDA is based primarily on its role in enzyme activity. Most people in the United States and Canada meet or exceed riboflavin recommendations.

Riboflavin Deficiency and Toxicity Riboflavin deficiency■ most often accompanies other nutrient deficiencies. Lack of the vitamin causes inflammation of the membranes of the mouth, skin, eyes, and GI tract. Excesses of riboflavin appear to cause no harm; no Upper Level has been established.

Riboflavin Food Sources The greatest contributions of riboflavin come from milk and milk products (see Figure 10-5). Whole-grain or enriched bread and cereal products are also valuable sources because of the quantities typically consumed. When riboflavin sources are ranked by nutrient density (per kcalorie),■ many dark green, leafy vegetables (such as broccoli, turnip greens, asparagus, and spinach) appear high on the list. Vegans and others who don't use milk must rely on ample servings of dark greens and enriched grains for riboflavin. Nutritional yeast is another good source.

Ultraviolet light and irradiation destroy riboflavin. For these reasons, milk is sold in cardboard or opaque plastic containers, and precautions are taken when vitamin D

■ Riboflavin deficiency is called **ariboflavinosis** (ay-RYE-boh-FLAY-vin-oh-sis).
  • **a** = not
  • **osis** = condition

■ Turn to p. 41 for a review of how to evaluate foods based on nutrient density (per kcalorie).

FIGURE 10-4 Riboflavin Coenzyme, Accepting and Donating Hydrogens

This figure shows the chemical structure of the riboflavin portion of the coenzyme only; the remainder of the coenzyme structure is represented by dotted lines (see Appendix C for the complete chemical structures of FAD and FMN). The reactive sites that accept and donate hydrogens are highlighted in white.

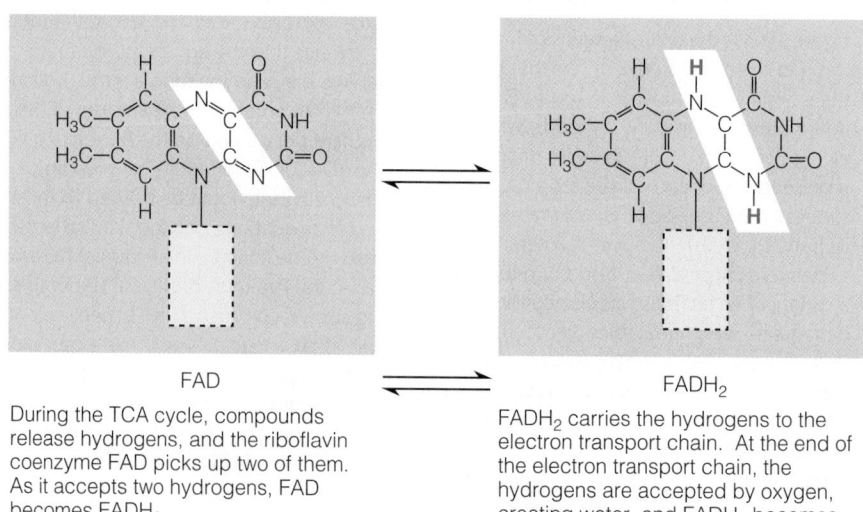

FAD

FADH₂

During the TCA cycle, compounds release hydrogens, and the riboflavin coenzyme FAD picks up two of them. As it accepts two hydrogens, FAD becomes FADH₂.

FADH₂ carries the hydrogens to the electron transport chain. At the end of the electron transport chain, the hydrogens are accepted by oxygen, creating water, and FADH₂ becomes FAD again. For every FADH₂ that passes through the electron transport chain, 2 ATP are generated.

riboflavin (RYE-boh-flay-vin): a B vitamin. The coenzyme forms are **FMN (flavin mononucleotide)** and **FAD (flavin adenine dinucleotide)**.

FIGURE 10-5 Riboflavin in Selected Foods

Food	Serving size (kcalories)	Milligrams
Bread, whole wheat	1 oz slice (70 kcal)	
Cornflakes, fortified	1 oz (110 kcal)	
Spaghetti pasta	½ c cooked (99 kcal)	
Tortilla, flour	1 10"-round (234 kcal)	
Broccoli	½ c cooked (22 kcal)	
Carrots	½ c shredded raw (24 kcal)	
Potato	1 medium baked w/skin (133 kcal)	
Tomato juice	¾ c (31 kcal)	
Banana	1 medium raw (109 kcal)	
Orange	1 medium raw (62 kcal)	
Strawberries	½ c fresh (22 kcal)	
Watermelon	1 slice (92 kcal)	
Milk	1 c reduced-fat 2% (121 kcal)	
Yogurt, plain	1 c low-fat (155 kcal)	
Cheddar cheese	1½ oz (171 kcal)	
Cottage cheese	½ c low-fat 2% (101 kcal)	
Pinto beans	½ c cooked (117 kcal)	
Peanut butter	2 tbs (188 kcal)	
Sunflower seeds	1 oz dry (165 kcal)	
Tofu (soybean curd)	½ c (76 kcal)	
Ground beef, lean	3 oz broiled (244 kcal)	
Chicken breast	3 oz roasted (140 kcal)	
Tuna, canned in water	3 oz (99 kcal)	
Egg	1 hard cooked (78 kcal)	
Excellent, and sometimes unusual, sources:		
Liver	3 oz fried (184 kcal)	
Clams, canned	3 oz (126 kcal)	
Mushrooms	½ c cooked (21 kcal)	

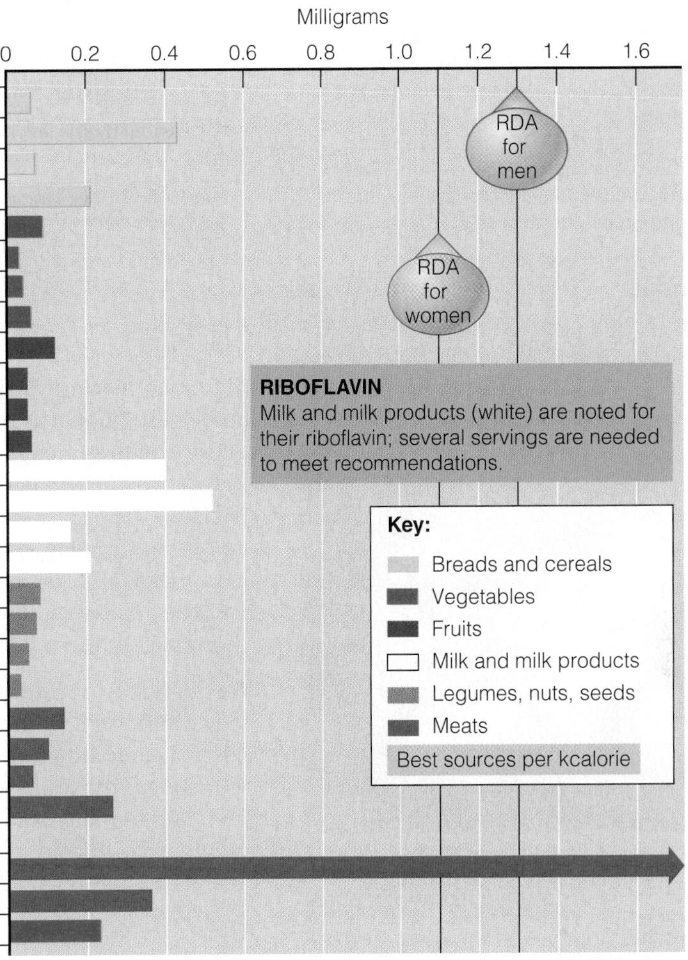

Milligram scale: 0 0.2 0.4 0.6 0.8 1.0 1.2 1.4 1.6

RDA for men

RDA for women

RIBOFLAVIN
Milk and milk products (white) are noted for their riboflavin; several servings are needed to meet recommendations.

Key:
- Breads and cereals
- Vegetables
- Fruits
- Milk and milk products
- Legumes, nuts, seeds
- Meats
- Best sources per kcalorie

Note: See p. 327 for more information on using this figure.

is added to milk by irradiation.* In contrast, riboflavin is stable to heat, so cooking does not destroy it. The following summary table lists riboflavin's chief functions, food sources, and deficiency symptoms.

IN SUMMARY Riboflavin

Other Names

Vitamin B₂

1998 RDA

Men: 1.3 mg/day

Women: 1.1 mg/day

Chief Functions in the Body

Part of coenzymes FMN (flavin mononucleotide) and FAD (flavin adenine dinucleotide) used in energy metabolism

Significant Sources

Milk products (yogurt, cheese); enriched or whole grains; liver

Easily destroyed by ultraviolet light and irradiation

Deficiency Disease

Ariboflavinosis (ay-RYE-boh-FLAY-vin-oh-sis)

(continued)

All of these foods are rich in riboflavin, but milk and milk products provide much of the riboflavin in the diets of most people.

© Polara Studios Inc.

*Vitamin D can be added to milk by feeding cows irradiated yeast or by irradiating the milk itself.

FIGURE 10-6 Niacin-Deficiency Symptom—The Dermatitis of Pellagra

In the dermatitis of pellagra, the skin darkens and flakes away as if it were sunburned. The protein-deficiency disease kwashiorkor also produces a "flaky paint" dermatitis, but the two are easily distinguished. The dermatitis of pellagra is bilateral and symmetrical and occurs only on those parts of the body exposed to the sun.

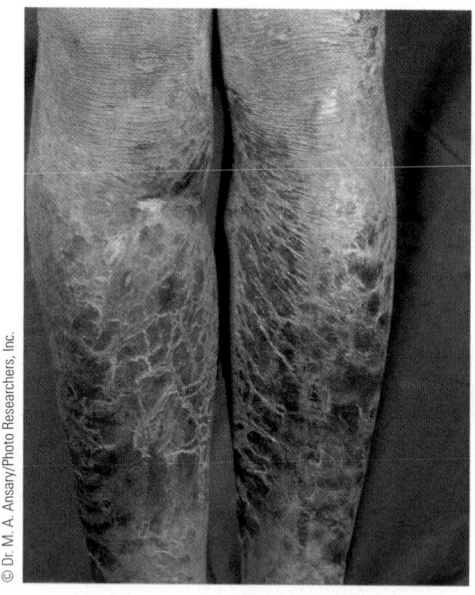

© Dr. M. A. Ansary/Photo Researchers, Inc.

■ 1 NE = 1 mg niacin or 60 mg tryptophan.

■ When a normal dose of a nutrient (levels commonly found in foods) provides a normal blood concentration, the nutrient is having a *physiological* effect. When a large dose (levels commonly available only from supplements) overwhelms some body system and acts like a drug, the nutrient is having a *pharmacological* effect.
  • **physio** = natural
  • **pharma** = drug

niacin (NIGH-a-sin): a B vitamin. The coenzyme forms are **NAD (nicotinamide adenine dinucleotide)** and **NADP (the phosphate form of NAD)**. Niacin can be eaten preformed or made in the body from its precursor, tryptophan, one of the amino acids.

niacin equivalents (NE): the amount of niacin present in food, including the niacin that can theoretically be made from its precursor, tryptophan, present in the food.

pellagra (pell-AY-gra): the niacin-deficiency disease.
  • **pellis** = skin
  • **agra** = rough

niacin flush: a temporary burning, tingling, and itching sensation that occurs when a person takes a large dose of nicotinic acid; often accompanied by a headache and reddened face, arms, and chest.

Riboflavin (continued)

Deficiency Symptoms	Toxicity Symptoms
Inflamed eyelids and sensitivity to light,[a] reddening of cornea; sore throat; cracks and redness at corners of mouth;[b] painful, smooth, purplish red tongue;[c] inflammation characterized by skin lesions covered with greasy scales	None reported

[a]Hypersensitivity to light is *photophobia* (FOE-toe-FOE-bee-ah).
[b]Cracks at the corners of the mouth are termed *cheilosis* (kye-LOH-sis or kee-LOH-sis).
[c]Smoothness of the tongue is caused by loss of its surface structures and is termed *glossitis* (gloss-EYE-tis).

Niacin

The name **niacin** describes two chemical structures: nicotinic acid and nicotinamide (also known as niacinamide). The body can easily convert nicotinic acid to nicotinamide, which is the major form of niacin in the blood.

The two coenzyme forms of niacin, NAD (nicotinamide adenine dinucleotide) and NADP (the phosphate form), participate in numerous metabolic reactions. They are central in energy-transfer reactions, especially the metabolism of glucose, fat, and alcohol. NAD is similar to the riboflavin coenzymes in that it carries hydrogens (and their electrons) during metabolic reactions, including the pathway from the TCA cycle to the electron transport chain.

Niacin Recommendations Niacin is unique among the B vitamins in that the body can make it from the amino acid tryptophan. To make 1 milligram of niacin requires approximately 60 milligrams of dietary tryptophan. For this reason, recommended intakes are stated in **niacin equivalents (NE).**■ A food containing 1 milligram of niacin and 60 milligrams of tryptophan provides the equivalent of 2 milligrams of niacin, or 2 niacin equivalents. The RDA for niacin allows for this conversion and is stated in niacin equivalents.

Niacin Deficiency The niacin-deficiency disease, **pellagra,** produces the symptoms of diarrhea, dermatitis, dementia, and eventually death (often called "the four Ds"). In the early 1900s, pellagra caused widespread misery and some 87,000 deaths in the U.S. South, where many people subsisted on a low-protein diet centered on corn. This diet supplied neither enough niacin nor enough tryptophan. At least 70 percent of the niacin in corn is bound to complex carbohydrates and small peptides, making it unavailable for absorption. Furthermore, corn is high in the amino acid leucine, which interferes with the tryptophan-to-niacin conversion, thus further contributing to the development of pellagra. Figure 10-6 illustrates the dermatitis of pellagra.

Pellagra was first believed to be caused by an infection. Medical researchers spent many years and much effort searching for infectious microbes until they realized that the problem was not what was present in the food, but what was *absent* from it. That a disease such as pellagra could be caused by diet—and not by germs—was a groundbreaking discovery. It contradicted commonly held medical opinions that diseases were caused only by infectious agents and advanced the science of nutrition dramatically.[*]

Niacin Toxicity Naturally occurring niacin from foods■ causes no harm, but large doses from supplements or drugs produce a variety of adverse effects, most notably **"niacin flush."** Niacin flush occurs when nicotinic acid is taken in doses only three to four times the RDA. It dilates the capillaries and causes a tingling sensation that can be painful. The nicotinamide form does not produce this effect— nor does it lower blood cholesterol.

---

[*]Dr. Joseph Goldberger, a physician for the U.S. government, headed the investigations that determined that pellagra was a dietary disorder, not an infectious disease. He died several years before Conrad Elevjhem discovered that a deficiency of niacin caused pellagra.

Large doses of nicotinic acid have been used to help lower blood cholesterol and prevent heart disease.[3] Such therapy must be closely monitored because of its adverse side effects (causes liver damage and aggravates peptic ulcers, among others). People with the following conditions may be particularly susceptible to the toxic effects of niacin: liver disease, diabetes, peptic ulcers, gout, irregular heartbeats, inflammatory bowel disease, migraine headaches, and alcoholism.

Niacin Food Sources Tables of food composition typically list preformed niacin only, but as mentioned, niacin can also be made in the body from the amino acid tryptophan. Hence diets that are high in protein are never lacking niacin. The "How to" on p. 332 shows how to estimate the total amount of niacin available from the diet. Dietary tryptophan could meet about half the daily need for most people, but the average diet easily supplies enough preformed niacin.

The predominance of the red bars in Figure 10-7 (on p. 332) explains why meat, poultry, and fish contribute about half the niacin most people need. An additional fourth of most people's niacin comes from enriched and whole grains. Mushrooms, asparagus, and leafy green vegetables are among the richest vegetable sources (per kcalorie) and can provide abundant niacin when eaten in generous amounts.

Niacin is less vulnerable to losses during food preparation and storage than other water-soluble vitamins. Being fairly heat-resistant, niacin can withstand reasonable cooking times, but like other water-soluble vitamins, it will leach into cooking water. The summary table includes food sources as well as niacin's various names, functions, and deficiency and toxicity symptoms.

Protein-rich foods such as meat, fish, poultry, and peanut butter contribute much of the niacin in people's diets. Enriched breads and cereals and a few vegetables are also rich in niacin.

IN SUMMARY Niacin

Other Names

Nicotinic acid, nicotinamide, niacinamide, vitamin B_3; precursor is dietary tryptophan (an amino acid)

1998 RDA

Men: 16 mg NE/day

Women: 14 mg NE/day

Upper Level

Adults: 35 mg/day

Chief Functions in the Body

Part of coenzymes NAD (nicotinamide adenine dinucleotide) and NADP (its phosphate form) used in energy metabolism

Significant Sources

Milk, eggs, meat, poultry, fish, whole-grain and enriched breads and cereals, nuts, and all protein-containing foods

Deficiency Disease

Pellagra

Deficiency Symptoms

Diarrhea, abdominal pain, vomiting; inflamed, swollen, smooth, bright red tongue;[a] depression, apathy, fatigue, loss of memory, headache; bilateral symmetrical rash on areas exposed to sunlight

Toxicity Symptoms

Painful flush, hives, and rash ("niacin flush"); excessive sweating; blurred vision; liver damage, impaired glucose tolerance

[a]Smoothness of the tongue is caused by loss of its surface structures and is termed *glossitis* (gloss-EYE-tis).

Biotin

Biotin plays an important role in metabolism as a coenzyme that carries activated carbon dioxide. This role is critical in the TCA cycle: biotin delivers a carbon to 3-carbon pyruvate, thus replenishing oxaloacetate, the 4-carbon compound needed to combine with acetyl CoA to keep the TCA cycle turning. The biotin coenzyme also serves crucial roles in gluconeogenesis,■ fatty acid synthesis, and the breakdown of certain fatty acids and amino acids.

■ Reminder: The synthesis of glucose from noncarbohydrate sources such as amino acids or glycerol is called *gluconeogenesis*.

biotin (BY-oh-tin): a B vitamin that functions as a coenzyme in metabolism.

HOW TO | Estimate Niacin Equivalents

To obtain a rough approximation of niacin equivalents:

- Calculate total protein consumed (grams).
- Assuming that the RDA amount of protein will be used first to make body protein, subtract the RDA to obtain "leftover" protein available to make niacin (grams). (Actually, the RDA provides a generous protein allowance, so "leftover" protein may be even greater than this.)
- About 1 gram of every 100 grams of high-quality protein is tryptophan, so divide by 100 to obtain the tryptophan in this leftover protein (grams).
- Multiply by 1000 to express this amount of tryptophan in milligrams.

- Divide by 60 to get niacin equivalents (milligrams).
- Finally, add the amount of preformed niacin obtained in the diet (milligrams).

For example, suppose that a 19-year-old woman who weighs 130 pounds consumes 75 grams of protein in a day. To calculate her protein RDA, first convert pounds to kilograms if necessary, and then multiply by 0.8 g/kg:

$$130 \text{ lb} \div 2.2 \text{ lb/kg} = 59 \text{ kg}$$
$$59 \text{ kg} \times 0.8 \text{ g/kg} = 47 \text{ g.}$$

Then determine her leftover protein by subtracting her RDA from her intake:

$$75 \text{ g protein intake} - 47 \text{ g protein RDA} = 28 \text{ g protein leftover.}$$

Next calculate the amount of tryptophan in this leftover protein:

$$28 \text{ g protein} \div 100 = 0.28 \text{ g tryptophan.}$$

$$0.28 \text{ g tryptophan} \times 1000 = 280 \text{ mg tryptophan.}$$

Then convert milligrams of tryptophan to niacin equivalents:

$$280 \text{ mg tryptophan} \div 60 = 4.7 \text{ mg NE.}$$

To determine the total amount of niacin available from the diet, add the amount available from tryptophan (4.7 mg NE) to the amount of preformed niacin obtained from the diet.

FIGURE 10-7 | Niacin in Selected Foods

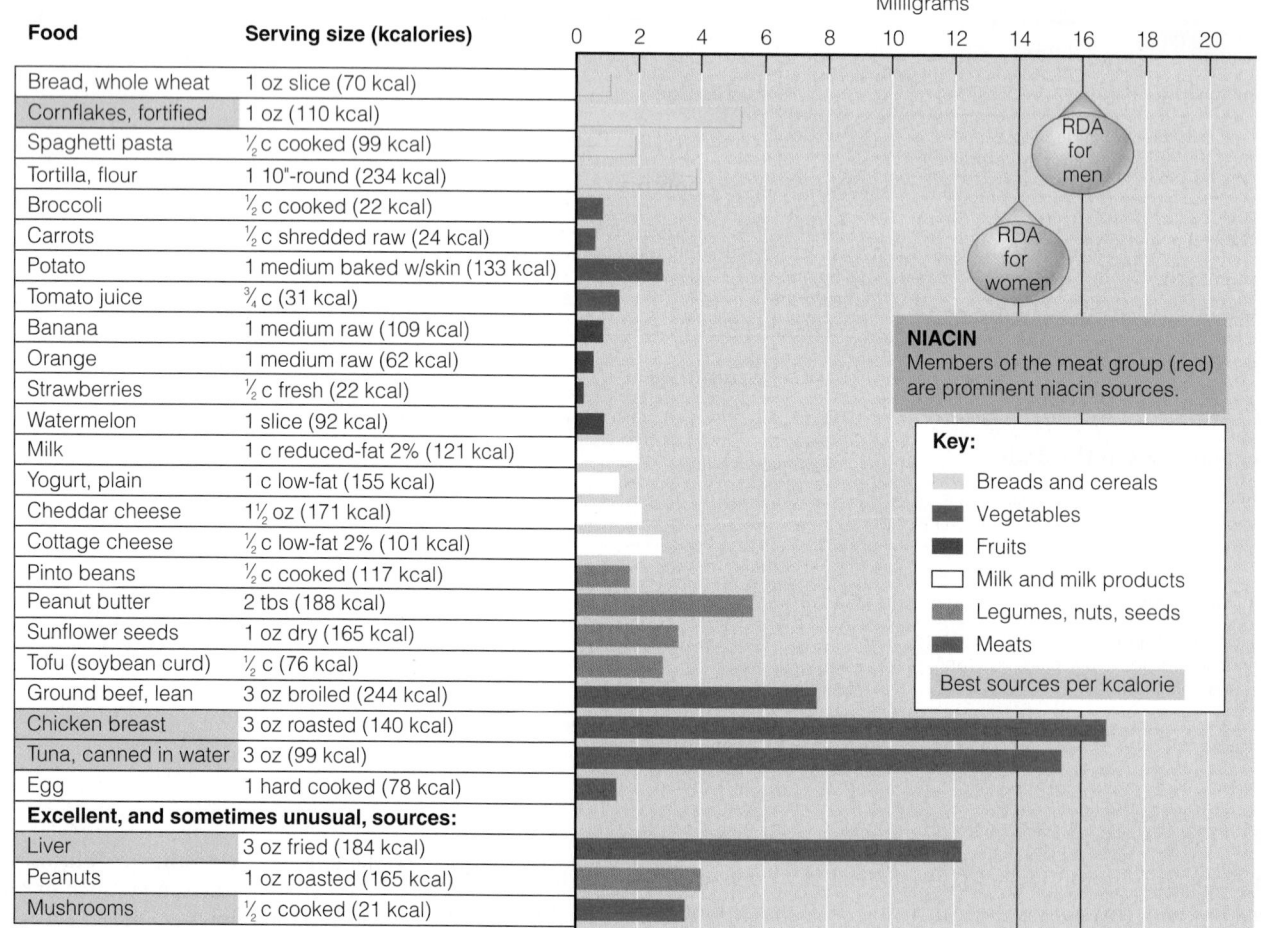

Food	Serving size (kcalories)
Bread, whole wheat	1 oz slice (70 kcal)
Cornflakes, fortified	1 oz (110 kcal)
Spaghetti pasta	½ c cooked (99 kcal)
Tortilla, flour	1 10"-round (234 kcal)
Broccoli	½ c cooked (22 kcal)
Carrots	½ c shredded raw (24 kcal)
Potato	1 medium baked w/skin (133 kcal)
Tomato juice	¾ c (31 kcal)
Banana	1 medium raw (109 kcal)
Orange	1 medium raw (62 kcal)
Strawberries	½ c fresh (22 kcal)
Watermelon	1 slice (92 kcal)
Milk	1 c reduced-fat 2% (121 kcal)
Yogurt, plain	1 c low-fat (155 kcal)
Cheddar cheese	1½ oz (171 kcal)
Cottage cheese	½ c low-fat 2% (101 kcal)
Pinto beans	½ c cooked (117 kcal)
Peanut butter	2 tbs (188 kcal)
Sunflower seeds	1 oz dry (165 kcal)
Tofu (soybean curd)	½ c (76 kcal)
Ground beef, lean	3 oz broiled (244 kcal)
Chicken breast	3 oz roasted (140 kcal)
Tuna, canned in water	3 oz (99 kcal)
Egg	1 hard cooked (78 kcal)
Excellent, and sometimes unusual, sources:	
Liver	3 oz fried (184 kcal)
Peanuts	1 oz roasted (165 kcal)
Mushrooms	½ c cooked (21 kcal)

NIACIN Members of the meat group (red) are prominent niacin sources.

Key:
- Breads and cereals
- Vegetables
- Fruits
- Milk and milk products
- Legumes, nuts, seeds
- Meats
- Best sources per kcalorie

Note: See p. 327 for more information on using this figure.

Biotin Recommendations Biotin is needed in very small amounts. Instead of an RDA, an Adequate Intake (AI) has been determined.

Biotin Deficiency and Toxicity Biotin deficiencies rarely occur. Researchers can induce a biotin deficiency in animals or human beings by feeding them raw egg whites, which contain a protein■ that binds biotin and thus prevents its absorption. Biotin-deficiency symptoms include skin rash, hair loss, and neurological impairment. More than two dozen egg whites must be consumed daily for several months to produce these effects, however, and the eggs have to be raw; cooking denatures the binding protein. No adverse effects from high biotin intakes have been reported; it does not have an Upper Level.

■ The protein **avidin** (AV-eh-din) in egg whites binds biotin.
  • **avid** = greedy

Biotin Food Sources Biotin is widespread in foods (including egg yolks), so eating a variety of foods protects against deficiencies. Some biotin is also synthesized by GI tract bacteria, but this amount may not contribute much to the biotin absorbed. A review of biotin facts is provided in the summary table.

IN SUMMARY Biotin

1998 Adequate Intake (AI)

Adults: 30 µg/day

Chief Functions in the Body

Part of a coenzyme used in energy metabolism, fat synthesis, amino acid metabolism, and glycogen synthesis

Significant Sources

Widespread in foods; organ meats, egg yolks, soybeans, fish, whole grains; also produced by GI bacteria

Deficiency Symptoms

Depression, lethargy, hallucinations, numb or tingling sensation in the arms and legs; red, scaly rash around the eyes, nose, and mouth; hair loss

Toxicity Symptoms

None reported

Pantothenic Acid

Pantothenic acid is involved in more than 100 different steps in the synthesis of lipids, neurotransmitters, steroid hormones, and hemoglobin as part of the chemical structure of coenzyme A—the same CoA that forms acetyl CoA, the "crossroads" compound in several metabolic pathways, including the TCA cycle. (Appendix C presents the chemical structures of these two molecules and shows that coenzyme A is made up in part of pantothenic acid.)

Pantothenic Acid Recommendations An Adequate Intake (AI) for pantothenic acid has been set. It reflects the amount needed to replace daily losses.

Pantothenic Acid Deficiency and Toxicity Pantothenic acid deficiency is rare. Its symptoms involve a general failure of all the body's systems and include fatigue, GI distress, and neurological disturbances. The "burning feet" syndrome that affected prisoners of war in Asia during World War II is thought to have been caused by pantothenic acid deficiency. No toxic effects have been reported, and no Upper Level has been established.

Pantothenic Acid Food Sources Pantothenic acid is widespread in foods, and typical diets seem to provide adequate intakes. Beef, poultry, whole grains, potatoes, tomatoes, and broccoli are particularly good sources. Losses of pantothenic acid during food production can be substantial because it is readily destroyed by the freezing, canning, and refining processes. The following summary table presents pantothenic acid facts.

pantothenic (PAN-toe-THEN-ick) **acid:** a B vitamin. The principal active form is part of coenzyme A, called "CoA" throughout Chapter 7.
  • **pantos** = everywhere

> **IN SUMMARY** Pantothenic Acid
>
> **1998 Adequate Intake (AI)**
>
> Adults: 5 mg/day
>
> **Chief Functions in the Body**
>
> Part of coenzyme A, used in energy metabolism
>
> **Significant Sources**
>
> Widespread in foods; organ meats, mushrooms, avocados, broccoli, whole grains
>
> Easily destroyed by food processing
>
> **Deficiency Symptoms**
>
> Vomiting, nausea, stomach cramps; insomnia, fatigue, depression, irritability, restlessness, apathy; hypoglycemia, increased sensitivity to insulin
>
> **Toxicity Symptoms**
>
> None reported

Vitamin B$_6$

Vitamin B$_6$ occurs in three forms—pyridoxal, pyridoxine, and pyridoxamine. All three can be converted to the coenzyme PLP (pyridoxal phosphate), which is active in amino acid metabolism. Because PLP can transfer amino groups (NH_2) from an amino acid to a keto acid, the body can make nonessential amino acids (review Figure 7-15 on p. 228). The ability to add and remove amino groups makes PLP valuable in protein and urea metabolism as well. The conversions of the amino acid tryptophan to niacin or to the neurotransmitter serotonin■ also depend on PLP as does the synthesis of heme (the nonprotein portion of hemoglobin), nucleic acids (such as DNA and RNA), and lecithin.

A surge of research in the last decade has revealed that vitamin B$_6$ influences cognitive performance, immune function, and steroid hormone activity. Unlike other water-soluble vitamins, vitamin B$_6$ is stored extensively in muscle tissue.

■ Reminder: *Serotonin* is a neurotransmitter important in appetite control, sleep regulation, and sensory perception, among other roles; it is synthesized in the body from the amino acid tryptophan with the help of vitamin B$_6$.

Vitamin B$_6$ Recommendations Because the vitamin B$_6$ coenzymes play many roles in amino acid metabolism, previous RDA were expressed in terms of protein intakes; the current RDA for vitamin B$_6$, however, is not. Research does not support claims that large doses of vitamin B$_6$ enhance muscle strength or physical endurance. As Highlight 14 explains, vitamin supplements cannot compete with a nutritious diet and physical training.

Vitamin B$_6$ Deficiency Without adequate vitamin B$_6$, synthesis of key neurotransmitters diminishes, and abnormal compounds produced during tryptophan metabolism accumulate in the brain. Early symptoms of vitamin B$_6$ deficiency include depression and confusion; advanced symptoms include abnormal brain wave patterns and convulsions.

Alcohol contributes to the destruction and loss of vitamin B$_6$ from the body. As Highlight 7 described, when the body breaks down alcohol, it produces acetaldehyde. If allowed to accumulate, acetaldehyde dislodges the PLP coenzyme from its enzymes; once loose, PLP breaks down and is excreted.

Another drug that acts as a vitamin B$_6$ **antagonist** is INH, a medication that inhibits the growth of the tuberculosis bacterium.* This drug has saved countless lives, but as a vitamin B$_6$ antagonist, INH binds and inactivates the vitamin, inducing a deficiency. Whenever INH is used to treat tuberculosis, vitamin B$_6$ supplements must be given to protect against deficiency.

Oral contraceptives have raised concerns, but they may be unwarranted. Early studies reported signs of vitamin B$_6$ deficiency in oral contraceptive users, but that was when the pills contained estrogen at three to five times the quantities used today. Estrogen creates a shortage of vitamin B$_6$ by stimulating the breakdown of tryptophan, a process that requires the vitamin.

Vitamin B$_6$ Toxicity The first major report of vitamin B$_6$ toxicity appeared in 1983. Until that time, everyone (including researchers and dietitians) believed

vitamin B$_6$: a family of compounds—pyridoxal, pyridoxine, and pyridoxamine. The primary active coenzyme form is **PLP (pyridoxal phosphate)**.

antagonist: a competing factor that counteracts the action of another factor. When a drug displaces a vitamin from its site of action, the drug renders the vitamin ineffective and thus acts as a vitamin antagonist.

---

*INH stands for isonicotinic acid hydrazide.

that, like the other water-soluble vitamins, vitamin B_6 could not reach toxic concentrations in the body. The report described neurological damage in people who had been taking more than 2 grams of vitamin B_6 daily (20 times the current Upper Level) for two months or more.

Some women use vitamin B_6 supplements in an attempt to treat premenstrual syndrome (PMS), the cluster of physical, emotional, and psychological symptoms that some women experience seven to ten days prior to menstruation. The cause of PMS remains undefined, although researchers generally agree that the hormonal changes of the menstrual cycle must be responsible. Without a full understanding of PMS causes, medical treatments flounder, and quack treatments abound. Among nutritional approaches, the taking of vitamin B_6 has received much attention, but seems to have done more harm than good.

Some people have taken vitamin B_6 supplements in an attempt to cure **carpal tunnel syndrome** and sleep disorders even though such treatment seems to be ineffective.[4] Self-prescribing is ill-advised because large doses of vitamin B_6 taken for months or years may cause irreversible nerve degeneration.

Vitamin B_6 Food Sources As you can see from the colored bars in Figure 10-8 (on p. 336), meats, fish, and poultry (red), potatoes and a few other vegetables (green), and fruits (purple) offer vitamin B_6. As is true of most of the other vitamins, fruits and vegetables would rank considerably higher if foods were ranked by nutrient density (vitamin B_6 per kcalorie). Several servings of vitamin B_6–rich foods are needed to meet recommended intakes.

Foods lose vitamin B_6 when heated. Information is limited, but vitamin B_6 bioavailability from plant-derived foods seems to be lower than from animal-derived foods; fiber does not appear to interfere with absorption. The summary table lists food sources of vitamin B_6 as well as its chief functions in the body and common symptoms of both deficiency and toxicity.

Most protein-rich foods such as meat, fish, and poultry provide ample vitamin B_6; some vegetables and fruits are good sources, too.

IN SUMMARY   Vitamin B_6

Other Names

Pyridoxine, pyridoxal, pyridoxamine

1998 RDA

Adults (19–50 yr): 1.3 mg/day

Upper Level

Adults: 100 mg/day

Chief Functions in the Body

Part of coenzymes PLP (pyridoxal phosphate) and PMP (pyridoxamine phosphate) used in amino acid and fatty acid metabolism; helps to convert tryptophan to niacin and to serotonin; helps to make red blood cells

Significant Sources

Meats, fish, poultry, potatoes, legumes, noncitrus fruits, fortified cereals, liver, soy products

Easily destroyed by heat

Deficiency Symptoms

Scaly dermatitis; anemia (small-cell type);[a] depression, confusion, abnormal brain wave pattern, convulsions

Toxicity Symptoms

Depression, fatigue, irritability, headaches, nerve damage causing numbness and muscle weakness leading to an inability to walk and convulsions; skin lesions

[a]Small-cell–type anemia is *microcytic anemia.*

Folate

Folate, also known as folacin or folic acid, has a chemical name that would fit a flying dinosaur: pteroylglutamic acid (PGA for short). Its primary coenzyme form, THF (tetrahydrofolate), serves as part of an enzyme complex that transfers one-carbon compounds that arise during metabolism. This action helps convert vitamin B_{12} to one of its coenzyme forms and helps synthesize the DNA required for all rapidly growing cells.

carpal tunnel syndrome: a pinched nerve at the wrist, causing pain or numbness in the hand. It is often caused by repetitive motion of the wrist.

folate (FOLE-ate): a B vitamin; also known as folic acid, folacin, or pteroylglutamic (tare-o-EEL-glue-TAM-ick) acid (PGA). The coenzyme forms are **DHF (dihydrofolate)** and **THF (tetrahydrofolate).**

FIGURE 10-8 Vitamin B₆ in Selected Foods

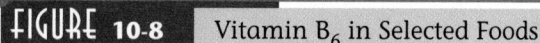

Food	Serving size (kcalories)
Bread, whole wheat	1 oz slice (70 kcal)
Cornflakes, fortified	1 oz (110 kcal)
Spaghetti pasta	½ c cooked (99 kcal)
Tortilla, flour	1 10"-round (234 kcal)
Broccoli	½ c cooked (22 kcal)
Carrots	½ c shredded raw (24 kcal)
Potato	1 medium baked w/skin (133 kcal)
Tomato juice	¾ c (31 kcal)
Banana	1 medium raw (109 kcal)
Orange	1 medium raw (62 kcal)
Strawberries	½ c fresh (22 kcal)
Watermelon	1 slice (92 kcal)
Milk	1 c reduced-fat 2% (121 kcal)
Yogurt, plain	1 c low-fat (155 kcal)
Cheddar cheese	1½ oz (171 kcal)
Cottage cheese	½ c low-fat 2% (101 kcal)
Pinto beans	½ c cooked (117 kcal)
Peanut butter	2 tbs (188 kcal)
Sunflower seeds	1 oz dry (165 kcal)
Tofu (soybean curd)	½ c (76 kcal)
Ground beef, lean	3 oz broiled (244 kcal)
Chicken breast	3 oz roasted (140 kcal)
Tuna, canned in water	3 oz (99 kcal)
Egg	1 hard cooked (78 kcal)
Excellent, and sometimes unusual, sources:	
Prune juice	¾ c (137 kcal)
Bluefish	3 oz baked (135 kcal)
Squash, acorn	½ c baked (69 kcal)

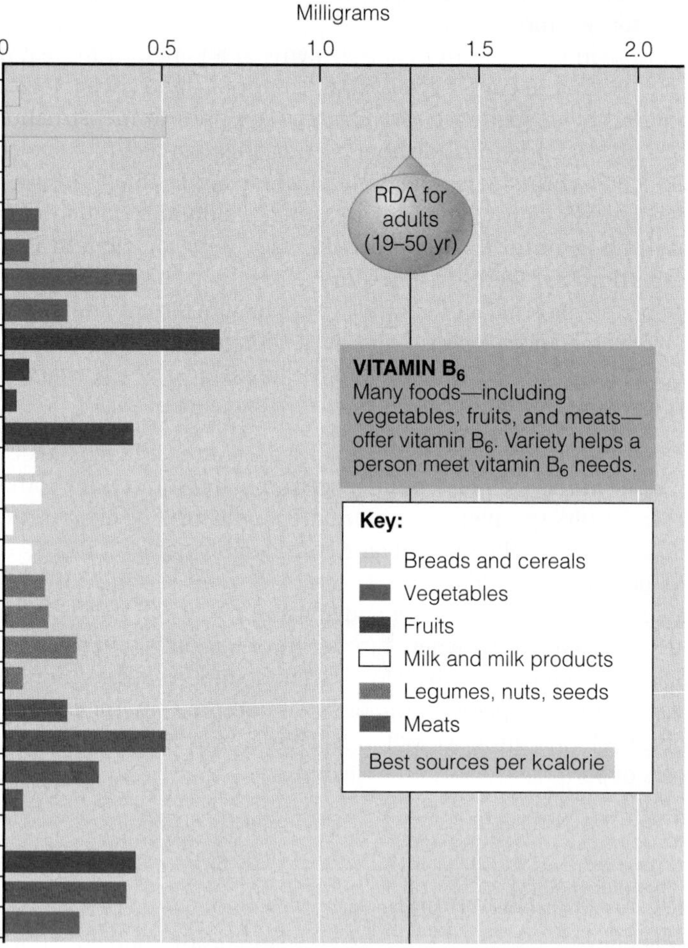

VITAMIN B₆
Many foods—including vegetables, fruits, and meats—offer vitamin B₆. Variety helps a person meet vitamin B₆ needs.

Key:
- Breads and cereals
- Vegetables
- Fruits
- Milk and milk products
- Legumes, nuts, seeds
- Meats

Best sources per kcalorie

Note: See p. 327 for more information on using this figure.

Foods deliver folate mostly in the "bound" form—that is, combined with a string of amino acids (glutamate), known as polyglutamate (see Appendix C for the chemical structure). The intestine prefers to absorb the "free" folate form—folate with only one glutamate attached (the monoglutamate form). Enzymes on the intestinal cell surfaces hydrolyze the polyglutamate to monoglutamate and several glutamates. Then the monoglutamate is attached to a methyl group (CH_3). Special transport systems deliver the monoglutamate with its methyl group to the liver and other body cells.

In order for the folate coenzyme to function, the methyl group must be removed by an enzyme that requires the help of vitamin B_{12}. Without that help, folate becomes trapped inside cells in its methyl form, unavailable to support DNA synthesis and cell growth. Figure 10-9 summarizes folate's absorption and activation.

To dispose of excess folate, the liver secretes most of it into bile and ships it to the gallbladder. Thus folate returns to the intestine in an enterohepatic circulation route like that of bile itself (review Figure 5-16 on p. 153).

This complicated system for handling folate is vulnerable to GI tract injuries. Since folate is actively secreted back into the GI tract with bile, it has to be reabsorbed repeatedly. If the GI tract cells are damaged, then folate is rapidly lost from the body. Such is the case in alcohol abuse; folate deficiency rapidly develops and, ironically, damages the GI tract further. The folate coenzymes, remember, are active in cell multiplication—and the cells lining the GI tract are among the most rapidly renewed cells in the body. Unable to make new cells, the GI tract deteriorates and not only loses folate, but also fails to absorb other nutrients.

dietary folate equivalents (DFE): the amount of folate available to the body from naturally occurring sources, fortified foods, and supplements, accounting for differences in the bioavailability from each source.

neural tube defects: malformations of the brain, spinal cord, or both during embryonic development. The two main types of neural tube defects are **spina bifida** (literally, "split spine") and **anencephaly** ("no brain").

FIGURE 10-9 Folate's Absorption and Activation

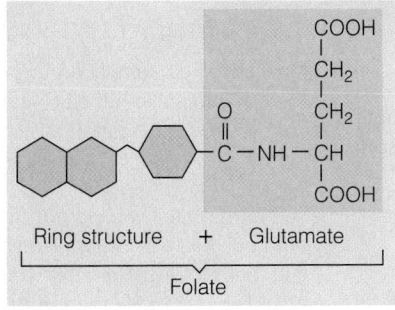

Ring structure + Glutamate

Folate

In foods, folate naturally occurs as polyglutamate. (Folate occurs as mono-glutamate in fortified foods and supplements.)

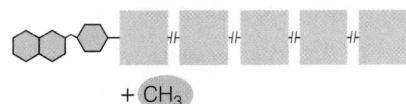

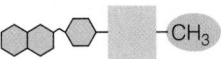

In the intestine, digestion breaks glutamates off . . . and adds a methyl group. Folate is absorbed and delivered to cells.

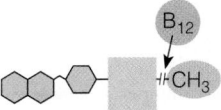

In the cells, folate is trapped in its inactive form.

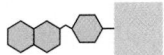 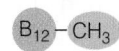

To activate folate, vitamin B_{12} removes and keeps the methyl group, which activates vitamin B_{12}.

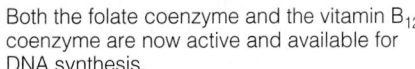

Both the folate coenzyme and the vitamin B_{12} coenzyme are now active and available for DNA synthesis.

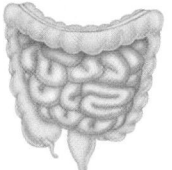

Spinach

Intestine

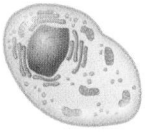

Cell

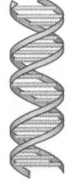

DNA

Folate Recommendations The bioavailability of folate ranges from 50 percent for foods to 100 percent for supplements taken on an empty stomach. These differences in bioavailability were considered in establishing the folate RDA. Naturally occurring folate from foods is given full credit. Synthetic folate from fortified foods and supplements is given extra credit because, on average, it is 1.7 times more available than naturally occurring food folate. Thus a person consuming 100 micrograms of folate from foods and 100 micrograms from a supplement receives 270 **dietary folate equivalents (DFE).**■ (The "How to" on p. 338 describes how to estimate dietary folate equivalents.) The need for folate rises considerably during pregnancy and whenever cells are multiplying, so the recommendations for pregnant women are considerably higher than for other adults.

Folate and Neural Tube Defects Several research studies have confirmed the importance of folate in reducing the risks of **neural tube defects.**[5] The brain and spinal cord develop from the neural tube, and defects in its orderly formation during the early weeks of pregnancy may result in various central nervous system disorders and death. (Chapter 15 provides photos of neural tube development and a figure showing a neural tube defect.)

Folate supplements taken one month before conception and continued throughout the first trimester of pregnancy can help prevent neural tube defects.[6] For this reason, all women of childbearing age■ who are capable of becoming pregnant should consume 0.4 milligram (400 micrograms) of folate daily,■ although only one-third of them actually do.[7] This recommendation can be met through a diet that includes at least five servings of fruits and vegetables daily, but many

■ To calculate DFE:
DFE = µg food folate + (1.7 × µg synthetic folate).
Using the example in the text:
100 µg food
+ 170 µg supplement (1.7 × 100 µg)
270 µg DFE

■ Women of childbearing age (15 to 45 yr) should:
• Eat folate-rich foods.
• Eat folate-fortified foods.
• Take a multivitamin daily (most provide 400 µg folate).

■ Reminder: A milligram (mg) is one-thousandth of a gram. A microgram (µg) is one-thousandth of a milligram (or one-millionth of a gram).
• 0.4 mg = 400 µg.

HOW TO Estimate Dietary Folate Equivalents

Folate is expressed in terms of DFE (dietary folate equivalents) because synthetic folate from supplements and fortified foods is absorbed at almost twice (1.7 times) the rate of naturally occurring folate from other foods. Use the following equation to calculate:

$$DFE = \mu g \text{ food folate} + (1.7 \times \mu g \text{ synthetic folate}).$$

Consider, for example, a pregnant woman who takes a supplement and eats a bowl of fortified cornflakes, 2 slices of fortified bread, and a cup of fortified pasta. From the supplement and fortified foods, she obtains synthetic folate:

Supplement	100 μg folate
Fortified cornflakes	100 μg folate
Fortified bread	40 μg folate
Fortified pasta	60 μg folate
	300 μg folate

To calculate the DFE, multiply the amount of synthetic folate by 1.7:

$$300 \ \mu g \times 1.7 = 510 \ \mu g \text{ DFE}.$$

Now add the naturally occurring folate from the other foods in her diet—in this example, another 90 μg of folate.

$$510 \ \mu g \text{ DFE} + 90 \ \mu g = 600 \ \mu g \text{ DFE}.$$

Notice that if we had not converted synthetic folate from supplements and fortified foods to DFE, then this woman's intake would appear to fall short of the 600 μg recommendation for pregnancy (300 μg + 90 μg = 390 μg). But as our example shows, her intake does meet the recommendation. At this time, supplement and fortified food labels list folate in μg only, not μg DFE, making such calculations necessary.

women typically fail to do so and receive only half this amount from foods. Furthermore, because of the enhanced bioavailability of synthetic folate, supplementation or fortification improves folate status significantly. Women who have given birth to infants with neural tube defects previously should consume 4 milligrams of folate daily before conception and throughout the first trimester of pregnancy.

Because half of the pregnancies each year are unplanned and because neural tube defects occur early in development before most women realize they are pregnant, the Food and Drug Administration (FDA) has mandated that grain products be fortified to deliver folate to the U.S. population.* Labels on fortified products may claim that "adequate intake of folate has been shown to reduce the risk of neural tube defects." Fortification has improved folate status in women of childbearing age and lowered the number of neural tube defects that occur each year, as Figure 10-10 shows.[8]

Folate fortification raises safety concerns as well, especially since folate intakes from fortified foods are more than twice as high as originally predicted.[9] Because high intakes of folate complicate the diagnosis of a vitamin B_{12} deficiency, folate consumption should not exceed 1 milligram daily without close medical supervision.[10]

Recent research has uncovered relationships between folate deficiency and non–neural tube birth defects such as Down syndrome.[11] Folate's exact role in preventing these birth defects, however, remains unclear. Some women whose infants develop these defects are *not* deficient in folate, and others with severe folate deficiencies do *not* give birth to infants with birth defects. Researchers continue to look for other factors that must also be involved.

Folate and Heart Disease The FDA's decision to fortify grain products with folate was strengthened by research indicating an important role for folate in defending against heart disease. As Chapter 6 mentioned, research indicates that high levels of the amino acid homocysteine and low levels of folate increase the risk of fatal heart disease.[12] One of folate's key roles in the body is to break down homocysteine. Without folate, homocysteine accumulates, which seems to en-

---

*Bread products, flour, corn grits, cornmeal, farina, rice, macaroni, and noodles must be fortified with 140 micrograms of folate per 100 grams of grain. For perspective, 100 grams is roughly 3 slices of bread; 1 cup of flour, ½ cup of corn grits, cornmeal, farina, or rice; or ¾ cup of macaroni or noodles.

FIGURE 10-10 | Decreasing Spina Bifida Rates since Folate Fortification

Neural tube defects have declined since folate fortification began in 1996.

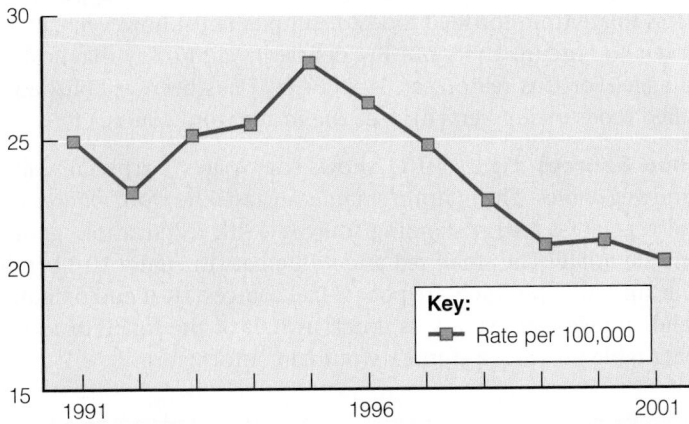

SOURCE: National Vital Statistics System, National Center for Health Statistics, Centers for Disease Control.

hance blood clot formation and arterial wall deterioration. Fortified foods and folate supplements raise blood folate and reduce blood homocysteine levels to an extent that may help to prevent heart disease.[13]

Folate and Cancer Folate may also play a role in preventing cancer.[14] Notably, folate may be most effective in protecting those most likely to develop cancers: men who smoke (against pancreatic cancer) and women who drink alcohol (against breast cancer).[15]

Folate Deficiency Folate deficiency impairs cell division and protein synthesis—processes critical to growing tissues. In a folate deficiency, the replacement of red blood cells and GI tract cells falters. Not surprisingly, then, two of the first symptoms of a folate deficiency are **anemia** and GI tract deterioration.

The anemia of folate deficiency is characterized by large,■ immature red blood cells. Without folate, DNA synthesis slows and the cells lose their ability to divide. The nucleus of the cell is not released as normally occurs during development. As a result, the immature blood cells are enlarged and oval-shaped. They cannot carry oxygen or travel through the capillaries as efficiently as normal red blood cells.

Folate deficiencies may develop from inadequate intake and have been reported in infants fed goat's milk, which is notoriously low in folate. Folate deficiency may also result from impaired absorption or an unusual metabolic need for the vitamin. Metabolic needs increase wherever cell multiplication must speed up: in pregnancies involving twins and triplets; in cancer; in skin-destroying diseases such as chicken pox and measles; and in burns, blood loss, GI tract damage, and the like.

Of all the vitamins, folate appears to be most vulnerable to interactions with drugs, which can lead to a secondary deficiency. Some medications, notably anticancer drugs, have a chemical structure similar to folate's structure and can displace the vitamin from enzymes and interfere with normal metabolism. Cancer cells, like all cells, need the real vitamin to multiply; without it, they die. Unfortunately, these drugs affect both cancerous cells and healthy cells and create a folate deficiency for all cells. (Highlight 17 discusses nutrient-drug interactions and includes a figure illustrating the similarities between the vitamin folate and the anticancer drug methotrexate.)

Aspirin and antacids also interfere with the body's handling of folate. Healthy adults who use these drugs to relieve an occasional headache or upset stomach need not be concerned, but people who rely heavily on aspirin or antacids should

■ Large-cell anemia is known as **macrocytic** or **megaloblastic anemia.**
- **macro** = large
- **cyte** = cell
- **mega** = large

anemia (ah-NEE-me-ah): literally, "too little blood." Anemia is any condition in which too few red blood cells are present, or the red blood cells are immature (and therefore large) or too small or contain too little hemoglobin to carry the normal amount of oxygen to the tissues. It is not a disease itself but can be a symptom of many different disease conditions, including many nutrient deficiencies, bleeding, excessive red blood cell destruction, and defective red blood cell formation.
- **an** = without
- **emia** = blood

Leafy dark green vegetables (such as spinach and broccoli), legumes (such as black beans, kidney beans, and black-eyed peas), liver, and some fruits (notably citrus fruits and juices) are naturally rich in folate.

be aware of the nutrition consequences. Oral contraceptives may also impair folate status, as may smoking.

Folate Toxicity Naturally occurring folate from foods alone appears to cause no harm. Excess folate from fortified foods or supplements, however, can reach high enough levels to obscure a vitamin B_{12} deficiency and delay diagnosis of neurological damage. For this reason, an Upper Level has been established for folate from fortified foods or supplements (see the inside front cover).

Folate Food Sources Figure 10-11 shows that folate is especially abundant in legumes and vegetables. The vitamin's name suggests the word *foliage*, and indeed, leafy green vegetables are outstanding sources. With fortification, grain products also contribute folate. The small red and white bars in Figure 10-11 indicate that meats, milk, and milk products are poor folate sources. Heat and oxidation during cooking and storage can destroy as much as half of the folate in foods. The accompanying table provides a summary of folate information.

IN SUMMARY   Folate

Other Names	Significant Sources
Folic acid, folacin, pteroylglutamic acid (PGA)	Fortified grains, leafy green vegetables, legumes, seeds, liver
1998 RDA	Easily destroyed by heat and oxygen
Adults: 400 µg/day	**Deficiency Symptoms**
Upper Level	Anemia (large-cell type);[a] smooth, red tongue;[b] mental confusion, weakness, fatigue, irritability, headache
Adults: 1000 µg/day	
Chief Functions in the Body	**Toxicity Symptoms**
Part of coenzymes THF (tetrahydrofolate) and DHF (dihydrofolate) used in DNA synthesis and therefore important in new cell formation	Masks vitamin B_{12}–deficiency symptoms

[a]Large-cell–type anemia is known as either *macrocytic* or *megaloblastic anemia.*
[b]Smoothness of the tongue is caused by loss of its surface structures and is termed *glossitis* (gloss-EYE-tis).

Vitamin B_{12}

Vitamin B_{12} and folate are closely related: each depends on the other for activation. Recall that vitamin B_{12} removes a methyl group to activate the folate coenzyme; when folate gives up its methyl group, the vitamin B_{12} coenzyme becomes activated (review Figure 10-9 on p. 337).

The regeneration of the amino acid methionine and the synthesis of DNA and RNA depend on both folate and vitamin B_{12}.[*] In addition, without any help from folate, vitamin B_{12} maintains the sheath that surrounds and protects nerve fibers and promotes their normal growth. Bone cell activity and metabolism also depend on vitamin B_{12}.

In the stomach, hydrochloric acid and the digestive enzyme pepsin release vitamin B_{12} from the proteins to which it is attached in foods. As the vitamin passes to the small intestine, it binds with a molecule called **intrinsic factor.** Bound together, intrinsic factor and vitamin B_{12} travel to the end of the small intestine, where receptors recognize the complex. Importantly, the receptors do not recognize vitamin B_{12} alone without intrinsic factor. There the intrinsic factor is degraded, and the vitamin is gradually absorbed into the bloodstream. Transport of vitamin B_{12} in the blood depends on specific binding proteins.

vitamin B_{12}: a B vitamin characterized by the presence of cobalt (see Figure 13-12 in Chapter 13). The active forms of coenzyme B_{12} are **methylcobalamin** and **deoxyadenosylcobalamin.**

intrinsic factor: a glycoprotein (a protein with short polysaccharide chains attached) manufactured in the stomach that aids in the absorption of vitamin B_{12}.
• **intrinsic** = on the inside

[*]In the body, methionine serves as a methyl (CH_3) donor. In doing so, methionine can be converted to other amino acids. Some of these amino acids can regenerate methionine, but methionine is still considered an essential amino acid that is needed in the diet.

FIGURE 10-11 Folate in Selected Foods

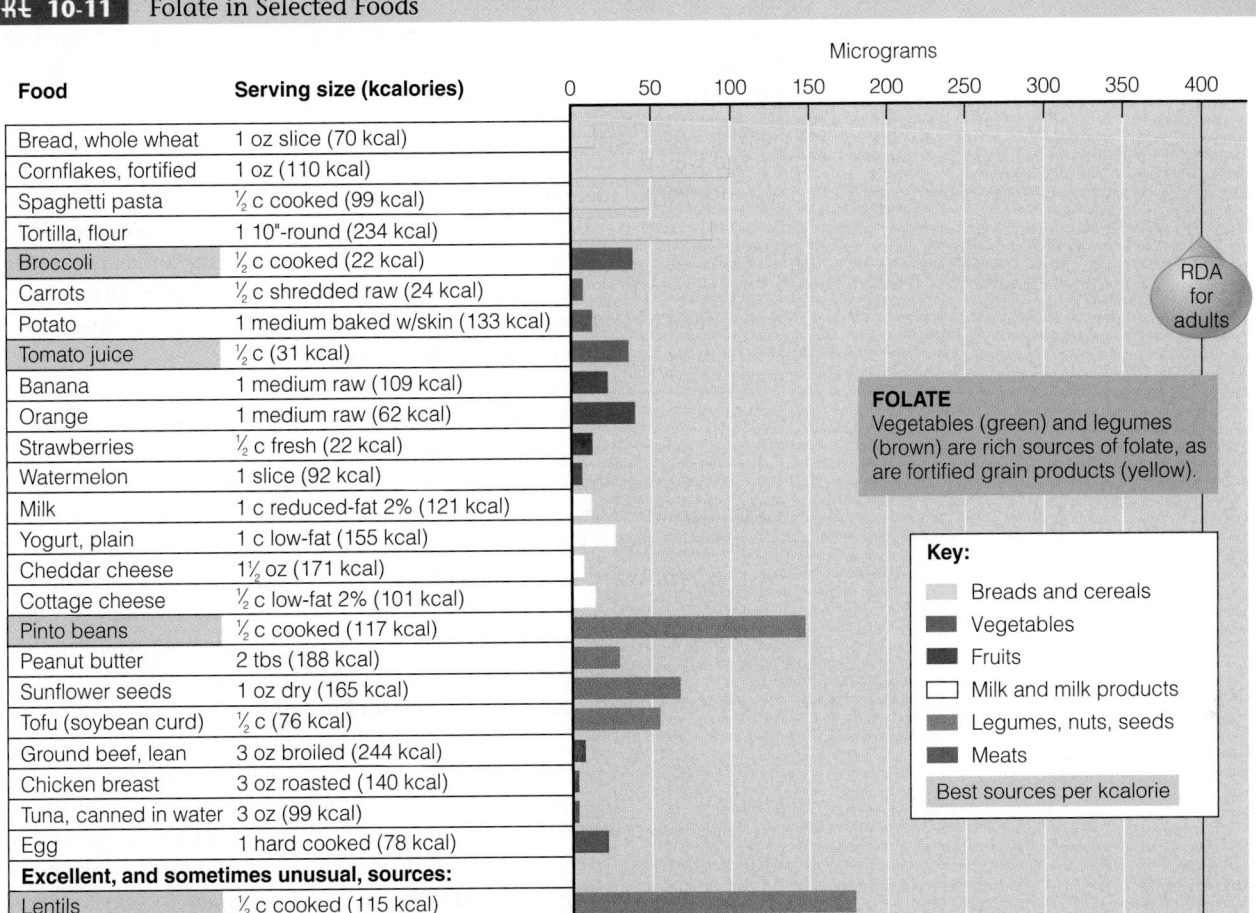

Food	Serving size (kcalories)
Bread, whole wheat	1 oz slice (70 kcal)
Cornflakes, fortified	1 oz (110 kcal)
Spaghetti pasta	½ c cooked (99 kcal)
Tortilla, flour	1 10"-round (234 kcal)
Broccoli	½ c cooked (22 kcal)
Carrots	½ c shredded raw (24 kcal)
Potato	1 medium baked w/skin (133 kcal)
Tomato juice	½ c (31 kcal)
Banana	1 medium raw (109 kcal)
Orange	1 medium raw (62 kcal)
Strawberries	½ c fresh (22 kcal)
Watermelon	1 slice (92 kcal)
Milk	1 c reduced-fat 2% (121 kcal)
Yogurt, plain	1 c low-fat (155 kcal)
Cheddar cheese	1½ oz (171 kcal)
Cottage cheese	½ c low-fat 2% (101 kcal)
Pinto beans	½ c cooked (117 kcal)
Peanut butter	2 tbs (188 kcal)
Sunflower seeds	1 oz dry (165 kcal)
Tofu (soybean curd)	½ c (76 kcal)
Ground beef, lean	3 oz broiled (244 kcal)
Chicken breast	3 oz roasted (140 kcal)
Tuna, canned in water	3 oz (99 kcal)
Egg	1 hard cooked (78 kcal)
Excellent, and sometimes unusual, sources:	
Lentils	½ c cooked (115 kcal)
Asparagus	½ c cooked (22 kcal)
Orange juice	¾ c fresh (84 kcal)

Microgram axis: 0, 50, 100, 150, 200, 250, 300, 350, 400

RDA for adults

FOLATE
Vegetables (green) and legumes (brown) are rich sources of folate, as are fortified grain products (yellow).

Key:
- Breads and cereals
- Vegetables
- Fruits
- Milk and milk products
- Legumes, nuts, seeds
- Meats

Best sources per kcalorie

Note: See p. 327 for more information on using this figure.

Like folate, vitamin B_{12} follows the enterohepatic circulation route. It is continually secreted into bile and delivered to the intestine, where it is reabsorbed. Because most vitamin B_{12} is reabsorbed, healthy people rarely develop a deficiency even when their intake is minimal.

Vitamin B_{12} Recommendations The RDA for adults is only 2.4 micrograms of vitamin B_{12} a day—just over two-millionths of a gram. The ink in the period at the end of this sentence may weigh about 2.4 micrograms. But tiny though this amount appears to the human eye, it contains billions of molecules of vitamin B_{12}, enough to provide coenzymes for all the enzymes that need its help.

Vitamin B_{12} Deficiency and Toxicity Most vitamin B_{12} deficiencies reflect inadequate absorption, not poor intake. Inadequate absorption typically occurs for one of two reasons: a lack of hydrochloric acid or a lack of intrinsic factor. Without hydrochloric acid, the vitamin is not released from the dietary proteins and so is not available for binding with the intrinsic factor. Without the intrinsic factor, the vitamin cannot be absorbed.

Many people, especially those over 50, develop **atrophic gastritis,** a common condition in older people that damages the cells of the stomach. Atrophic gastritis may also develop in response to iron deficiency or infection with *Helicobacter pylori,* the bacterium implicated in ulcer formation. Without healthy stomach cells, production of hydrochloric acid and intrinsic factor diminishes. Even with an adequate intake

atrophic (a-TRO-fik) **gastritis** (gas-TRY-tis): chronic inflammation of the stomach accompanied by a diminished size and functioning of the mucous membrane and glands.
- **atrophy** = wasting
- **gastro** = stomach
- **itis** = inflammation

■ Vitamin B_{12} is found primarily in foods derived from animals.

from foods, vitamin B_{12} status suffers. The vitamin B_{12} deficiency caused by atrophic gastritis and a lack of intrinsic factor is known as **pernicious anemia.**

Some people inherit a defective gene for the intrinsic factor. In such cases, or when the stomach has been injured and cannot produce enough of the intrinsic factor, vitamin B_{12} must be injected to bypass the need for intestinal absorption. Alternatively, the vitamin may be delivered by nasal spray; absorption is rapid, high, and well tolerated.

A prolonged inadequate intake, as can occur with a vegetarian diet,■ may also create a vitamin B_{12} deficiency.[16] People who stop eating foods containing vitamin B_{12} may take several years to develop deficiency symptoms because the body recycles much of its vitamin B_{12}, reabsorbing it over and over again. Even when the body fails to absorb vitamin B_{12}, deficiency may take up to three years to develop because the body conserves its supply.

Because vitamin B_{12} is required to convert folate to its active form, one of the most obvious vitamin B_{12}–deficiency symptoms is the anemia of folate deficiency. This anemia is characterized by large, immature red blood cells, which are indicative of slow DNA synthesis and an inability to divide (see Figure 10-12). When folate is trapped in its inactive (methyl folate) form due to vitamin B_{12} deficiency, or is unavailable due to folate deficiency itself, DNA synthesis slows.

First to be affected in a vitamin B_{12} or folate deficiency are the rapidly growing blood cells. Either vitamin B_{12} or folate will clear up the anemia, but if folate is given when vitamin B_{12} is needed, the result is disastrous: devastating neurological symptoms. Remember that vitamin B_{12}, but not folate, maintains the sheath that surrounds and protects nerve fibers and promotes their normal growth. Folate "cures" the *blood* symptoms of a vitamin B_{12} deficiency, but cannot stop the *nerve* symptoms from progressing. By doing so, folate "masks" a vitamin B_{12} deficiency. Marginal vitamin B_{12} deficiency impairs performance on tests measuring intelligence, spatial ability, and short-term memory.[17] Advanced neurological symptoms include a creeping paralysis that begins at the extremities and works inward and up the spine. Early detection and correction are necessary to prevent permanent nerve damage and paralysis. With sufficient folate in the diet, the neurological symptoms of vitamin B_{12} deficiency can develop without evidence of anemia. Such interactions between folate and vitamin B_{12} highlight some of the safety is-

pernicious (per-NISH-us) **anemia:** a blood disorder that reflects a vitamin B_{12} deficiency caused by lack of intrinsic factor and characterized by abnormally large and immature red blood cells. Other symptoms include muscle weakness and irreversible neurological damage.
• pernicious = destructive

FIGURE 10-12 Normal and Anemic Blood Cells

The anemia of folate deficiency is indistinguishable from that of vitamin B_{12} deficiency. Appendix E describes the biochemical tests used to differentiate the two conditions.

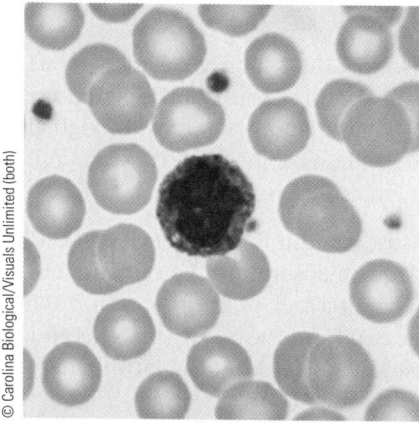

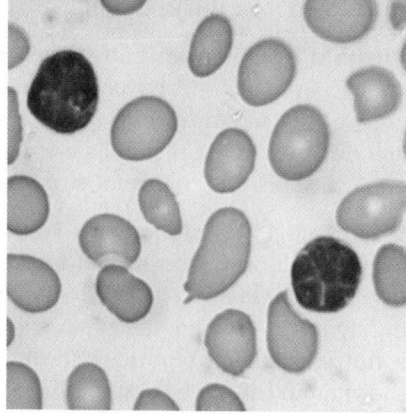

© Carolina Biological/Visuals Unlimited (both)

Normal blood cells. The size, shape, and color of the red blood cells show that they are normal.

Blood cells in pernicious anemia (megaloblastic). Megaloblastic blood cells are slightly larger than normal red blood cells, and their shapes are irregular.

sues surrounding the use of supplements and the fortification of foods. No adverse effects have been reported for excess vitamin B_{12}, and no Upper Level has been set.

Vitamin B_{12} Food Sources Vitamin B_{12} is unique among the vitamins in being found almost exclusively in foods derived from animals. Anyone who eats reasonable amounts of meat is guaranteed an adequate intake, and vegetarians who use milk products or eggs are also protected from deficiency. Vegans, who restrict all foods derived from animals, need a reliable source, such as vitamin B_{12}–fortified soy milk or vitamin B_{12} supplements. Yeast grown on a vitamin B_{12}–enriched medium and mixed with that medium provides some vitamin B_{12}, but yeast itself does not contain active vitamin B_{12}. Fermented soy products such as miso (a soybean paste) and sea algae such as spirulina also do *not* provide active vitamin B_{12}. Extensive research shows that the amounts listed on the labels of these plant products are inaccurate and misleading because the vitamin B_{12} is in an inactive, unavailable form.

As mentioned earlier, the water-soluble vitamins are particularly vulnerable to losses in cooking. For most of these nutrients, microwave heating minimizes losses as well as, or better than, traditional cooking methods. Such is not the case for vitamin B_{12}, however. Microwave heating inactivates vitamin B_{12}. To preserve this vitamin, use the oven or stovetop instead of a microwave to cook meats and milk products (major sources of vitamin B_{12}). The accompanying table provides a summary of information about vitamin B_{12}.

IN SUMMARY Vitamin B_{12}

Other Names

Cobalamin (and related forms)

1998 RDA

Adults: 2.4 µg/day

Chief Functions in the Body

Part of coenzymes methylcobalamin and deoxyadenosylcobalamin used in new cell synthesis; helps to maintain nerve cells; reforms folate coenzyme; helps to break down some fatty acids and amino acids

Significant Sources

Animal products (meat, fish, poultry, shellfish, milk, cheese, eggs), fortified cereals

Easily destroyed by microwave cooking

Deficiency Disease

Pernicious anemia[a]

Deficiency Symptoms

Anemia (large-cell type);[b] fatigue, degeneration of peripheral nerves progressing to paralysis

Toxicity Symptoms

None reported

[a]The name *pernicious anemia* refers to the vitamin B_{12} deficiency caused by atrophic gastritis and a lack of intrinsic factor, but not to that caused by inadequate dietary intake.
[b]Large-cell–type anemia is known as either *macrocytic* or *megaloblastic anemia.*

Non-B Vitamins

Nutrition scientists debate whether other dietary compounds might also be considered vitamins. In some cases, the compounds may be conditionally essential—that is, needed by the body from foods when synthesis becomes insufficient to support normal growth and metabolism. In other cases, the compounds may be vitamin impostors—not needed under any circumstances.

Choline The essentiality of **choline** has been blurry for decades, in part because the body can make choline from the amino acid methionine. Furthermore, choline is commonly found in many foods as part of the lecithin molecule (review Figure 5-9 on p. 148). Consequently, choline deficiencies are rare. Without any

choline (KOH-leen): a nitrogen-containing compound found in foods and made in the body from the amino acid methionine. Choline is used to make the phospholipid lecithin and the neurotransmitter acetylcholine.

dietary choline, however, synthesis alone appears to be insufficient to meet the body's needs, making choline a conditionally essential nutrient. For this reason, the 1998 DRI report established an Adequate Intake (AI) for choline. The body uses choline to make the neurotransmitter acetylcholine and the phospholipid lecithin. The accompanying table summarizes key choline facts.

IN SUMMARY Choline

1998 Adequate Intake (AI)

Men: 550 mg/day

Women: 425 mg/day

Upper Level

Adults: 3500 mg/day

Chief Functions in the Body

Needed for the synthesis of the neurotransmitter acetylcholine and the phospholipid lecithin

Deficiency Symptoms

Liver damage

Toxicity Symptoms

Body odor, sweating, salivation, reduced growth rate, low blood pressure, liver damage

Significant Sources

Milk, liver, eggs, peanuts

Inositol and Carnitine **Inositol** is a part of cell membrane structures, and **carnitine** transports long-chain fatty acids from the cytosol to the mitochondria for oxidation. Like choline, these two substances can be made by the body, but unlike choline, no recommendations have been established. Researchers continue to explore the possibility that these substances may be essential. Even if they are essential, though, supplements are unnecessary because these compounds are widespread in foods.

Some vitamin companies include choline, inositol, and carnitine in their formulations to make their vitamin pills look more "complete" than others, but this strategy offers no real advantage. For a rational way to compare vitamin-mineral supplements, read Highlight 10.

Vitamin Impostors Other substances have been mistaken for essential nutrients for human beings because they are needed for growth by bacteria or other forms of life. Among them are PABA (para-aminobenzoic acid, a component of folate's ring structure), the bioflavonoids (vitamin P or hesperidin), pyrroloquinoline quinone (methoxatin), orotic acid, lipoic acid, and ubiquinone (coenzyme Q_{10}). Other names erroneously associated with vitamins are "vitamin O" (oxygenated salt water), "vitamin B_5" (another name for pantothenic acid), "vitamin B_{15}" (also called "pangamic acid," a hoax), and "vitamin B_{17}" (laetrile, an alleged "cancer cure" and not a vitamin or a cure by any stretch of the imagination—in fact, laetrile is a potentially dangerous substance).

IN SUMMARY The B vitamins serve as coenzymes that facilitate the work of every cell. They are active in carbohydrate, fat, and protein metabolism and in the making of DNA and thus new cells. Historically famous B vitamin–deficiency diseases are beriberi (thiamin), pellagra (niacin), and pernicious anemia (vitamin B_{12}). Pellagra can be prevented by adequate protein because the amino acid tryptophan can be converted to niacin in the body. A high intake of folate can mask the blood symptom of a vitamin B_{12} deficiency, but it will not prevent the associated nerve damage. Vitamin B_6 participates in amino acid metabolism and can be harmful in excess. Biotin and pantothenic acid serve important roles in energy metabolism and are common in a variety of foods. Many substances that people claim as B vitamins are not.

inositol (in-OSS-ih-tall): a nonessential nutrient that can be made in the body from glucose. Inositol is a part of cell membrane structures.

carnitine (CAR-neh-teen): a nonessential nutrient made in the body from the amino acid lysine. Carnitine transports long-chain fatty acids from the cytosol to the mitochondria for oxidation.

The B Vitamins—In Concert

This chapter has described some of the impressive ways that vitamins work individually, as if their many actions in the body could easily be disentangled. In fact, oftentimes it is difficult to tell which vitamin is truly responsible for a given effect because the nutrients are interdependent; the presence or absence of one affects another's absorption, metabolism, and excretion. You have already seen this interdependence with folate and vitamin B_{12}.

Riboflavin and vitamin B_6 provide another example. One of the riboflavin coenzymes, FMN, assists the enzyme that converts vitamin B_6 to its coenzyme form PLP. Consequently, a severe riboflavin deficiency can impair vitamin B_6 activity.[18] Thus a deficiency of one nutrient may alter the action of another. Furthermore, a deficiency of one nutrient may create a deficiency of another. For example, both riboflavin and vitamin B_6 (as well as iron) are required for the conversion of tryptophan to niacin. Consequently, an inadequate intake of either riboflavin or vitamin B_6 can diminish the body's niacin supply. These interdependent relationships are evident in many of the roles B vitamins play in the body.

B Vitamin Roles

Figure 10-13 (on p. 346) is intended to convey an *impression* of the many ways B vitamins busily work in metabolic pathways all over the body. Metabolism is the body's work, and the B vitamin coenzymes are indispensable to every step. In scanning the pathways of metabolism depicted in the figure, note the many abbreviations for the coenzymes that keep the processes going.

Look at the first step in the now-familiar pathway of glucose breakdown. To break down glucose to pyruvate, the cells must have certain enzymes. For the enzymes to work, they must have the niacin coenzyme NAD. To make NAD, the cells must be supplied with niacin (or enough of the amino acid tryptophan to make niacin). They can make the rest of the coenzyme without dietary help.

The next step is the breakdown of pyruvate to acetyl CoA. The enzymes involved in this step require both NAD and the thiamin and riboflavin coenzymes TPP and FAD, respectively. The cells can manufacture the enzymes they need from the vitamins, if the vitamins are in the diet.

Another coenzyme needed for this step is CoA. Predictably, the cells can make CoA except for an essential part that must be obtained in the diet—pantothenic acid. Another coenzyme requiring biotin serves the enzyme complex involved in converting pyruvate to oxaloacetate, the compound that combines with acetyl CoA to start the TCA cycle.

These and other coenzymes participate throughout all the metabolic pathways. When the diet provides riboflavin, the body synthesizes FAD—a needed coenzyme in the TCA cycle. Vitamin B_6 is an indispensable part of PLP—a coenzyme required for many amino acid conversions, for a crucial step in the making of the iron-containing portion of hemoglobin for red blood cells, and for many other reactions. Folate becomes THF—the coenzyme required for the synthesis of new genetic material and therefore new cells. The vitamin B_{12} coenzyme, in turn, regenerates THF to its active form; thus vitamin B_{12} is also necessary for the formation of new cells.

Thus each of the B vitamin coenzymes is involved, directly or indirectly, in energy metabolism. Some facilitate the energy-releasing reactions themselves; others help build new cells to deliver the oxygen and nutrients that allow the energy reactions to occur.

B Vitamin Deficiencies

Now suppose the body's cells lack one of these B vitamins—niacin, for example. Without niacin, the cells cannot make NAD. Without NAD, the enzymes involved in every step of the glucose-to-energy pathway cannot function. Then, because all

FIGURE 10-13 Metabolic Pathways Involving B Vitamins

These metabolic pathways were introduced in Chapter 7 and are presented here to highlight the many coenzymes that facilitate the reactions. These coenzymes depend on the following vitamins:

- NAD and NADP: niacin.
- TPP: thiamin.
- CoA: pantothenic acid.
- B_{12}: vitamin B_{12}.
- FMN and FAD: riboflavin.
- THF: folate.
- PLP: vitamin B_6.
- Biotin.

Pathways leading toward acetyl CoA and the TCA cycle are catabolic, and those leading toward amino acids, glycogen, and fat are anabolic. For further details, see Appendix C.

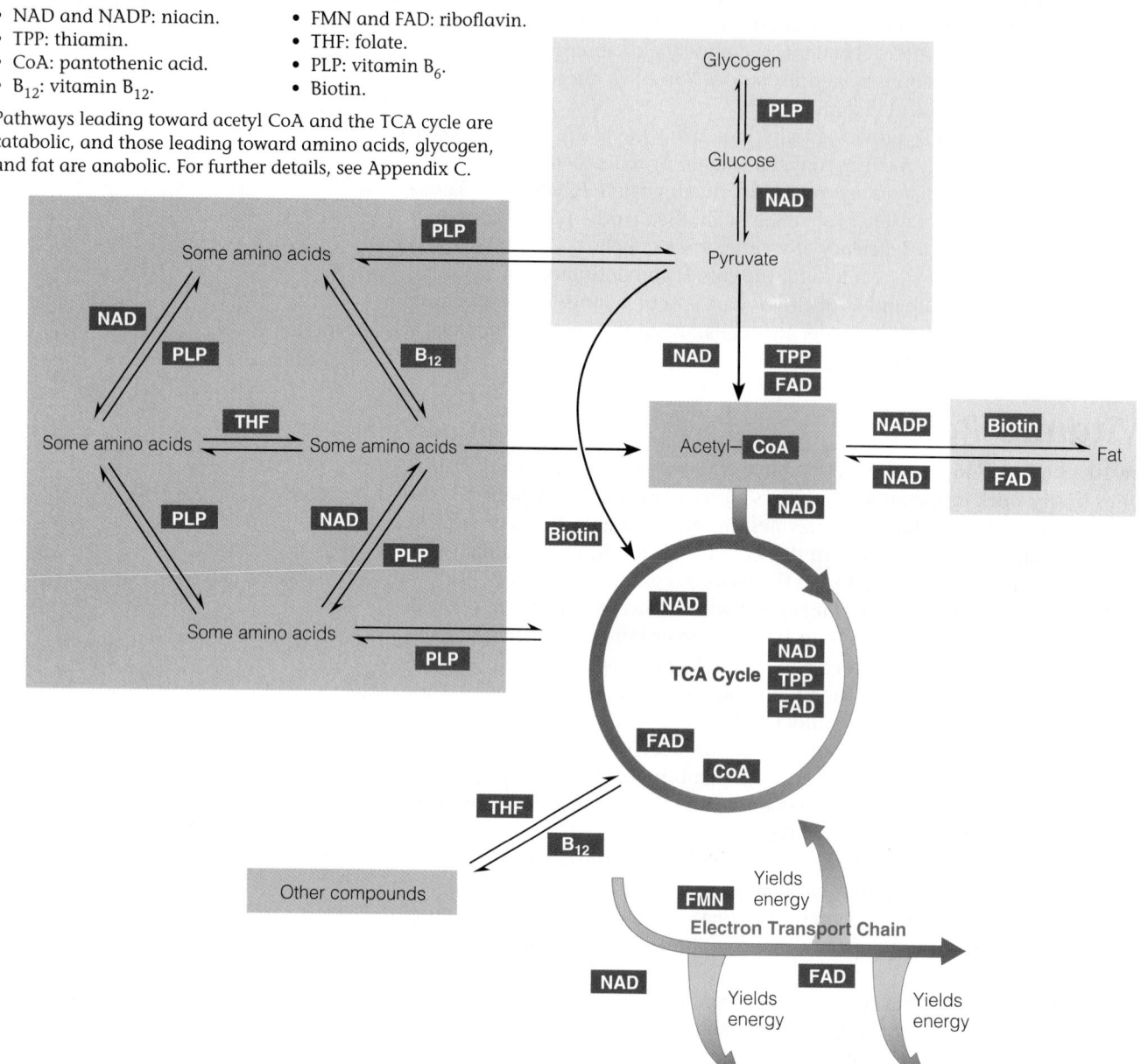

the body's activities require energy, literally everything begins to grind to a halt. This is no exaggeration. The deadly disease pellagra, caused by niacin deficiency, produces the "devastating four Ds": dermatitis, which reflects a failure of the skin; dementia, a failure of the nervous system; diarrhea, a failure of digestion and absorption; and eventually, as would be the case for any severe nutrient deficiency, death. These symptoms are the obvious ones, but a niacin deficiency affects all other organs, too, because all are dependent on the energy pathways. In short, niacin is like the horseshoe nail■ for want of which a war was lost.

All the vitamins are like horseshoe nails. With any B vitamin deficiency, many body systems become deranged, and similar symptoms may appear. A lack of "horseshoe nails" can have disastrous and far-reaching effects.

■ For want of a nail, a horseshoe was lost.
For want of a horseshoe, a horse was lost.
For want of a horse, a soldier was lost.
For want of a soldier, a battle was lost.
For want of a battle, the war was lost,
And all for the want of a horseshoe nail!

—Mother Goose

Deficiencies of single B vitamins seldom show up in isolation, however. After all, people do not eat nutrients singly; they eat foods, which contain mixtures of nutrients. Only in two cases described earlier—beriberi and pellagra—have dietary deficiencies associated with single B vitamins been observed on a large scale in human populations. Even in these cases, the deficiencies were not pure. Both diseases were attributed to deficiencies of single vitamins, but both were deficiencies of several vitamins in which one vitamin stood out above the rest. When foods containing the vitamin known to be needed were provided, the other vitamins that were in short supply came as part of the package.

Major deficiency diseases of epidemic proportions such as pellagra and beriberi are no longer seen in the United States and Canada, but lesser deficiencies of nutrients, including the B vitamins, sometimes occur in people whose food choices are poor because of poverty, ignorance, illness, or poor health habits like alcohol abuse. (Review Highlight 7 to fully appreciate how alcohol induces vitamin deficiencies and interferes with energy metabolism.) Remember from Chapter 1 that deficiencies can arise not only from deficient intakes (primary causes), but also for other (secondary) reasons.

In identifying nutrient deficiencies, it is important to realize that a particular symptom may not always have the same cause. The skin and the tongue (shown in Figure 10-14) appear to be especially sensitive to B vitamin deficiencies, but isolating these body parts in the summary tables earlier in this chapter gives them undue emphasis. Both the skin and the tongue■ are readily visible in a physical examination. The physician sees and reports the deficiency's outward symptoms, but the full impact of a vitamin deficiency occurs inside the cells of the body. If the skin develops a rash or lesions, other tissues beneath it may be degenerating, too. Similarly, the mouth and tongue are the visible part of the digestive system; if they are abnormal, most likely the rest of the GI tract is, too. The "How to" on the next page offers other insights into symptoms and their causes.

B Vitamin Toxicities

Toxicities of the B vitamins from foods alone are unknown, but they can occur when people overuse supplements. With supplements, the quantities can quickly overwhelm the cells. Consider that one small capsule can easily deliver 2 milligrams of vitamin B_6, but it would take more than 3000 bananas, 6600 cups of rice, or 3600 chicken breasts to supply an equivalent amount. When the cells become oversaturated with a vitamin, they must work to eliminate the excess. The cells dispatch water-soluble vitamins to the urine for excretion, but sometimes they cannot keep pace with the onslaught. Homeostasis becomes disturbed and symptoms of toxicity develop.

B Vitamin Food Sources

Significantly, the deficiency diseases of beriberi and pellagra were eliminated by supplying foods—not pills. Vitamin pill advertisements make much of the fact that vitamins are indispensable to life, but human beings obtained their nourishment from foods for centuries before vitamin pills existed. If the diet lacks a vitamin, the first solution is to adjust food intake to obtain that vitamin.

Manufacturers of so-called *natural* vitamins boast that their pills are purified from real foods rather than synthesized in a laboratory. Think back on the course of human evolution; it is not *natural* to take any kind of pill. In reality, the finest, most natural vitamin "supplements" available are whole grains, vegetables, fruits, meat, fish, poultry, eggs, legumes, nuts, and milk and milk products.

The food figures presented in this chapter, taken together, sing the praises of a balanced diet. The cereal and bread group delivers thiamin, riboflavin, niacin, and folate. The fruit and vegetable groups excel in folate. The meat group serves thiamin, niacin, vitamin B_6, and vitamin B_{12} well. The milk group stands out for riboflavin and vitamin B_{12}. A diet that offers a variety of foods from each group, prepared with reasonable care, serves up ample B vitamins.

FIGURE 10-14 B Vitamin–Deficiency Symptom—The Smooth Tongue of Glossitis

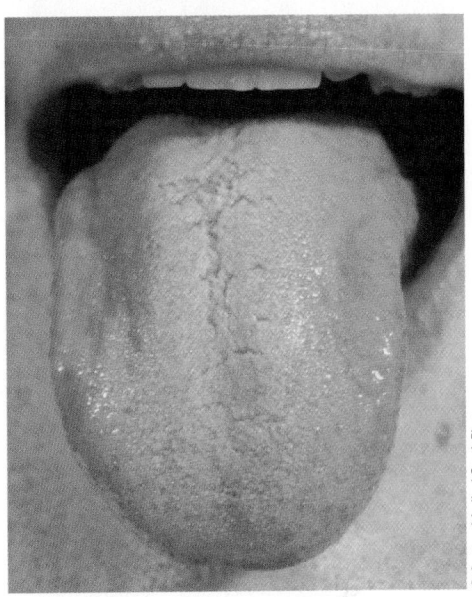

A healthy tongue has a rough and somewhat bumpy surface.

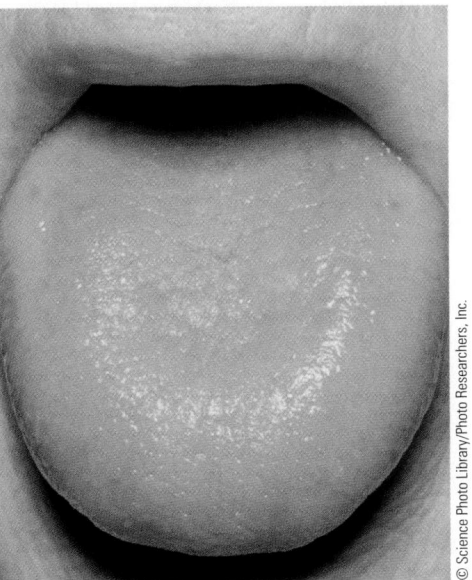

In a B vitamin deficiency, the tongue becomes smooth and swollen due to atrophy of the tissue (glossitis).

■ Two symptoms commonly seen in B vitamin deficiencies are **glossitis** (gloss-EYE-tis), an inflammation of the tongue, and **cheilosis** (kye-LOH-sis or kee-LOH-sis), a condition of reddened lips with cracks at the corners of the mouth.
- **glossa** = tongue
- **cheilos** = lip

The cause of a symptom is not always apparent. The summary tables in this chapter show that deficiencies of riboflavin, niacin, biotin, and vitamin B_6 can all cause skin rashes. But so can a deficiency of protein, linoleic acid, or vitamin A. Because skin is on the outside and easy to see, it is a useful indicator of things-going-wrong-inside-cells. But, by itself, a skin symptom says nothing about its possible cause.

The same is true of anemia. Anemia is often caused by iron deficiency, but it can also be caused by a folate or vitamin B_{12} deficiency; by digestive tract failure to absorb any of these nutrients; or by such nonnutritional causes as infections, parasites, cancer, or loss of blood. No one specific nutrient will always cure a given symptom.

A person who feels chronically tired may be tempted to self-diagnose iron-deficiency anemia and self-prescribe an iron supplement. But this will relieve tiredness only if the cause is indeed iron-deficiency anemia. If the cause is a folate deficiency, taking iron will only prolong the fatigue. A person who is better informed may decide to take a vitamin supplement with iron, covering the possibility of a vitamin deficiency. But the symptom may have a nonnutritional cause. If the cause of the tiredness is actually hidden blood loss due to cancer, the postponement of a diagnosis may be fatal. When fatigue is caused by a lack of sleep, of course, no nutrient or combination of nutrients can replace a good night's rest. A person who is chronically tired should see a physician rather than self-prescribe. If the condition is nutrition related, a registered dietitian should be consulted as well.

IN SUMMARY The B vitamin coenzymes work together in energy metabolism. Some facilitate the energy-releasing reactions themselves; others help build cells to deliver the oxygen and nutrients that permit the energy pathways to run. These vitamins depend on each other to function optimally; a deficiency of any of them creates multiple problems. Fortunately, a variety of foods from each of the food groups will provide an adequate supply of all of the B vitamins.

Vitamin C

Two hundred and fifty years ago, any man who joined the crew of a seagoing ship knew he had at best a 50–50 chance of returning alive—not because he might be slain by pirates or die in a storm, but because he might contract the dread disease **scurvy.** As many as two-thirds of a ship's crew might die of scurvy on a long voyage. Only men on short voyages, especially around the Mediterranean Sea, were free of scurvy. No one knew the reason: that on long ocean voyages, the ship's cook used up the fresh fruits and vegetables early and then served cereals and meats until the return to port.

The first nutrition experiment ever performed on human beings was devised in the mid-1700s to find a cure for scurvy. James Lind, a British physician, divided 12 sailors with scurvy into six pairs. Each pair received a different supplemental ration: cider, vinegar, sulfuric acid, seawater, oranges and lemons, or a strong laxative mixed with spices. Those receiving the citrus fruits quickly recovered, but sadly, it was 50 years before the British navy required all vessels to provide every sailor■ with lime juice daily.

The antiscurvy "something" in limes and other foods was dubbed the **antiscorbutic factor.** Nearly 200 years later, the factor was isolated and found to be a six-carbon compound similar to glucose; it was named **ascorbic acid.** Shortly thereafter, it was synthesized, and today hundreds of millions of vitamin C pills are produced in pharmaceutical laboratories each year.

■ The tradition of providing British sailors with citrus juice daily to prevent scurvy gave them the nickname "limeys."

scurvy: the vitamin C–deficiency disease.

antiscorbutic (AN-tee-skor-BUE-tik) **factor:** the original name for vitamin C.
• anti = against
• scorbutic = causing scurvy

ascorbic acid: one of the two active forms of vitamin C (see Figure 10-15). Many people refer to vitamin C by this name.
• a = without
• scorbic = having scurvy

FIGURE 10-15 | Active Forms of Vitamin C

The two hydrogens highlighted in yellow give vitamin C its acidity and its ability to act as an antioxidant.

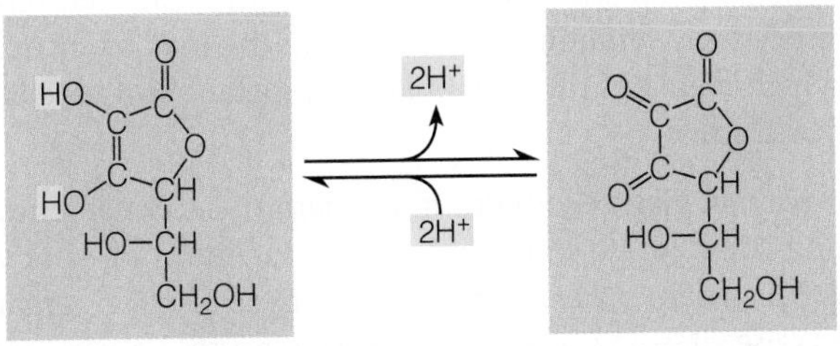

Ascorbic acid protects against oxidative damage by donating its two hydrogens with their electrons to free radicals (molecules with unpaired electrons). In doing so, ascorbic acid becomes dehydroascorbic acid.

Dehydroascorbic acid can readily accept hydrogens to become ascorbic acid. The reversibility of this reaction is key to vitamin C's role as an antioxidant.

Vitamin C Roles

Vitamin C parts company with the B vitamins in its mode of action. In some settings, vitamin C serves as a cofactor■ helping a specific enzyme perform its job, but in others, it acts as an antioxidant participating in more general ways.

As an Antioxidant Vitamin C loses electrons easily, a characteristic that allows it to perform as an **antioxidant.** In the body, antioxidants defend against **free radicals.** Free radicals are discussed in Highlight 11, but for now, a simple definition will suffice. A free radical is a molecule with one or more unpaired electrons, making it unstable and highly reactive. By donating an electron or two, antioxidants neutralize free radicals and protect other substances from their damage. Figure 10-15 illustrates how vitamin C can give up electrons to stop free-radical damage and then receive them again to become reactivated. This recycling of vitamin C is key to limiting losses and maintaining a reserve of antioxidants in the body.

Vitamin C is like a bodyguard for water-soluble substances; it stands ready to sacrifice its own life to save theirs. In the cells and body fluids, vitamin C protects tissues from **oxidative stress** and thus may play an important role in preventing diseases. In the intestines, vitamin C enhances iron absorption by protecting iron from oxidation. (Chapter 13 provides more details on the relationship between vitamin C and iron.)

As a Cofactor in Collagen Formation Vitamin C helps to form the fibrous structural protein of connective tissues known as collagen.■ Collagen serves as the matrix on which bones and teeth are formed. When a person is wounded, collagen glues the separated tissues together, forming scars. Cells are held together largely by collagen; this is especially important in the artery walls, which must expand and contract with each beat of the heart, and in the thin capillary walls, which must withstand a pulse of blood every second or so without giving way.

Chapter 6 described how the body makes proteins by stringing together chains of amino acids. During the synthesis of collagen, each time a proline or lysine is added to the growing protein chain, an enzyme hydroxylates it (adds an OH group to it), making the amino acid hydroxyproline or hydroxylysine, respectively. These two special amino acids facilitate the binding together of collagen fibers to make strong, ropelike structures. The conversion of proline to hydroxyproline requires both vitamin C and iron. Iron works as a cofactor in the reaction, and vitamin C protects iron from oxidation, thereby allowing iron to perform its duty. Without vitamin C and iron, the hydroxylation step does not occur.

■ Reminder: A *cofactor* is a small, inorganic or organic substance that facilitates the action of an enzyme.

■ Reminder: *Collagen* is the structural protein from which connective tissues such as scars, tendons, ligaments, and the foundations of bones and teeth are made.

antioxidant: a substance in foods that significantly decreases the adverse effects of free radicals on normal physiological functions in the human body.

free radicals: unstable molecules with one or more unpaired electrons.

oxidative stress: an imbalance between the production of free radicals and the body's ability to handle them and prevent damage.

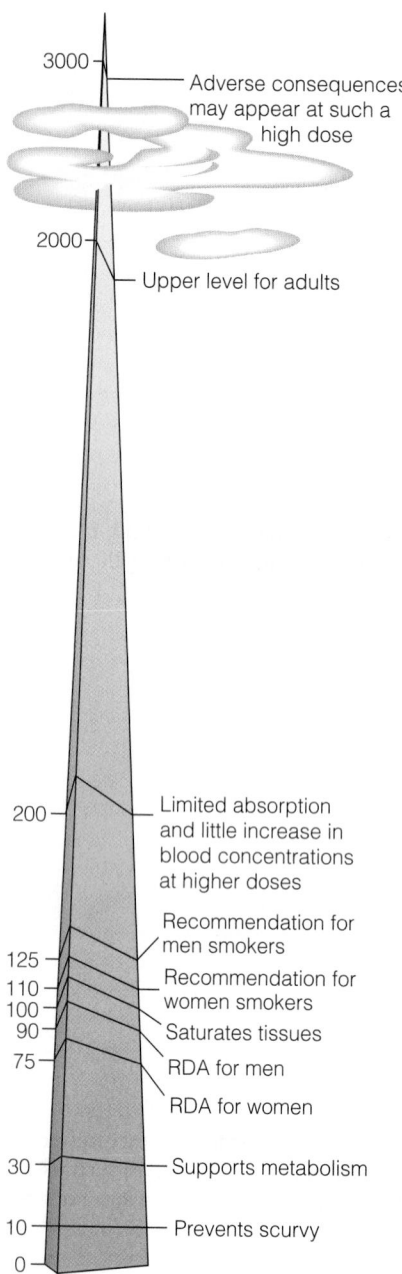

FIGURE 10-16 Vitamin C Intake (mg/day)

Recommendations are generously above the minimum requirement and below the toxicity level.

3000 — Adverse consequences may appear at such a high dose

2000 — Upper level for adults

200 — Limited absorption and little increase in blood concentrations at higher doses

125 — Recommendation for men smokers
110 — Recommendation for women smokers
100 —
90 — Saturates tissues
75 — RDA for men

RDA for women

30 — Supports metabolism

10 — Prevents scurvy

0 —

■ For perspective, 1 c orange juice provides >100 mg vitamin C.

histamine (HISS-tah-mean or HISS-tah-men): a substance produced by cells of the immune system as part of a local immune reaction to an antigen; participates in causing inflammation.

As a Cofactor in Other Reactions Vitamin C also serves as a cofactor in the synthesis of several other compounds. As in collagen formation, vitamin C helps in the hydroxylation of carnitine, a compound that transports long-chain fatty acids into the mitochondria of a cell for energy metabolism. It participates in the conversions of the amino acids tryptophan and tyrosine to the neurotransmitters serotonin and norepinephrine, respectively. Vitamin C also assists in the making of hormones, including thyroxin, which regulates the metabolic rate; metabolism speeds up under times of extreme physical stress.

In Stress The adrenal glands contain more vitamin C than any other organ in the body, and during stress, these glands release the vitamin, together with hormones, into the blood. The vitamin's exact role in the stress reaction remains unclear, but physical stresses raise vitamin C needs. Among the stresses known to increase vitamin C needs are infections; burns; extremely high or low temperatures; intakes of toxic heavy metals such as lead, mercury, and cadmium; the chronic use of certain medications, including aspirin, barbiturates, and oral contraceptives; and cigarette smoking. When immune system cells are called into action, they use a lot of oxygen and produce free radicals. In this case, free radicals are helpful. They act as ammunition in an "oxidative burst" that demolishes the offending viruses and bacteria and destroys the damaged cells. Vitamin C steps in as an antioxidant to control this oxidative activity.

As a Cure for the Common Cold Newspaper headlines touting vitamin C as a cure for colds have appeared frequently over the years, but research supporting such claims has been conflicting and controversial. Some studies find no relationship between vitamin C and the occurrence of the common cold, whereas others report fewer colds, fewer days, and shorter duration of severe symptoms.[19] A review of the research on vitamin C in the treatment and prevention of the common cold reveals a modest benefit—a significant difference in duration of less than a day per cold in favor of those taking a daily dose of at least 1 gram of vitamin C.[20] The term *significant* means that *statistical* analysis suggests that the findings probably didn't arise by chance, but from the experimental treatment being tested. Is a day enough savings to warrant routine daily supplementation? Supplement users seem to think so.

Interestingly, those who received the placebo *but thought they were receiving vitamin C* had fewer colds than the group who received vitamin C *but thought they were receiving the placebo.* (Never underestimate the healing power of faith!)

Discoveries of the ways vitamin C works in the body provide possible links between the vitamin and the common cold. Anyone who has ever had a cold knows the discomfort of a runny or stuffed-up nose. Nasal congestion develops in response to elevated blood **histamine,** and people commonly take antihistamines for relief. Like an antihistamine, vitamin C comes to the rescue and deactivates histamine.

In Disease Prevention Whether vitamin C may help in preventing or treating cancer, heart disease, cataracts, and other diseases is still being studied, and findings are presented in Highlight 11. Conducting research in the United States and Canada can be difficult, however, because diets typically contribute enough vitamin C to provide optimal health benefits.

Vitamin C Recommendations

How much vitamin C does a person need? As Figure 10-16 illustrates, recommendations are set generously above the minimum requirement to prevent scurvy and well below the toxicity level.[21] Current recommendations are higher than the previous RDA, but not as high as some experts had proposed.[22]

The requirement—the amount needed to prevent the overt symptoms of scurvy—is only 10 milligrams daily. However, 10 milligrams a day does not saturate all the body tissues; higher intakes will increase the body's total vitamin C. At about 100 milligrams■ per day, 95 percent of the population probably reaches tis-

FIGURE 10-17 Vitamin C–Deficiency Symptoms—Scorbutic Gums and Pinpoint Hemmorhages

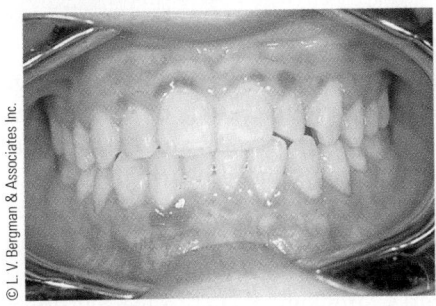

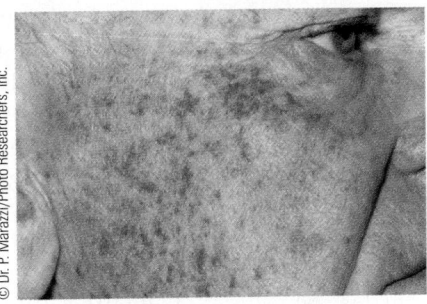

Scorbutic gums. Unlike other lesions of the mouth, scurvy presents a symmetrical appearance without infection.

Pinpoint hemorrhages. Small red spots appear in the skin, indicating spontaneous bleeding internally.

sue saturation. At about 200 milligrams, absorption reaches a maximum, and there is little, if any, increase in blood concentrations at higher doses. Excess vitamin C is readily excreted.

As mentioned earlier, cigarette smoking increases the need for vitamin C. Cigarette smoke contains oxidants, which greedily deplete this potent antioxidant. Exposure to cigarette smoke, especially when accompanied by low intakes of vitamin C, depletes the body's pool in both active and passive smokers; similarly, people who chew tobacco have low levels of vitamin C as well.[23] Because people who smoke cigarettes regularly suffer significant oxidative stress, their requirement for vitamin C is increased an additional 35 milligrams; nonsmokers regularly exposed to cigarette smoke should be sure to meet their RDA for vitamin C.

After oral surgery, dentists may prescribe supplemental vitamin C to hasten healing. After major operations or extensive burns, when scar tissue is forming, a physician may prescribe 1000 milligrams (1 gram) a day or even more. Self-medication is not recommended.

Vitamin C Deficiency

Two of the most notable signs of a vitamin C deficiency reflect its role in maintaining the integrity of blood vessels. The gums bleed easily around the teeth, and capillaries under the skin break spontaneously, producing pinpoint hemorrhages (see Figure 10-17).

When the vitamin C pool falls to about a fifth of its optimal size (this may take more than a month on a diet lacking vitamin C), scurvy symptoms begin to appear. Inadequate collagen synthesis causes further hemorrhaging. Muscles, including the heart muscle, degenerate. The skin becomes rough, brown, scaly, and dry. Wounds fail to heal because scar tissue will not form. Bone rebuilding falters; the ends of the long bones become softened, malformed, and painful, and fractures develop. The teeth become loose as the cartilage around them weakens. Anemia and infections are common. There are also characteristic psychological signs, including hysteria and depression. Sudden death is likely, caused by massive internal bleeding.

Once diagnosed, scurvy is readily resolved by vitamin C.[24] Moderate doses in the neighborhood of 100 milligrams per day are sufficient, curing the scurvy within about five days. Such an intake is easily achieved by including vitamin C–rich foods in the diet.

When dietitians say "vitamin C," people think "citrus fruits."

But these foods are also rich in vitamin C.

■ Reminder: *Gout* is a metabolic disease in which uric acid crystals precipitate in the joints.

false positive: a test result indicating that a condition is present (positive) when in fact it is not (therefore false).

false negative: a test result indicating that a condition is not present (negative) when in fact it is present (therefore false).

Vitamin C Toxicity

The easy availability of vitamin C supplements and the publication of books recommending vitamin C to prevent colds and cancer have led thousands of people to take large doses of vitamin C. Not surprisingly, toxic effects such as nausea, abdominal cramps, and diarrhea are often reported.

Several instances of interference with medical regimens are also known. Large amounts of vitamin C excreted in the urine obscure the results of tests used to detect diabetes, giving a **false positive** result in some instances and a **false negative** in others. People taking anticlotting medications may unwittingly counteract the effect if they also take massive doses of vitamin C.* Those with kidney disease, a tendency toward gout,■ or a genetic abnormality that alters vitamin C's breakdown to its excretion products are prone to forming kidney stones if they take large doses of vitamin C.† Vitamin C supplements may adversely affect people with iron overload. (Chapter 13 describes the damaging effects of too much iron.) Vitamin C enhances iron absorption and releases iron from body stores; free iron causes the kind of cellular damage typical of free radicals. These events illustrate how vitamin C can act as a *pro*oxidant when quantities exceed the body's needs.

The estimated average intake from both diet and supplements is 187 milligrams of vitamin C a day. Few instances warrant consuming more than 200 milligrams a day. For adults who dose themselves with up to 2 grams a day (and relatively few do), the risks may not be great; those taking more should be aware of the distinct possibility of adverse effects.

Vitamin C Food Sources

Fruits and vegetables can easily provide a generous amount of vitamin C. A cup of orange juice at breakfast, a salad for lunch, and a stalk of broccoli and a potato for dinner alone provide more than 300 milligrams. Clearly, a person making such food choices needs no vitamin C pills.

Figure 10-18 shows the amounts of vitamin C in various common foods. The overwhelming abundance of purple and green bars reveals not only that the citrus fruits are justly famous for being rich in vitamin C, but that other fruits and vegetables are in the same league. A single serving of broccoli, bell pepper, or strawberries provides more than 50 milligrams of the vitamin (and an array of other nutrients). Because vitamin C is vulnerable to heat, raw fruits and vegetables usually have a higher nutrient density than their cooked counterparts. Similarly, because vitamin C is readily destroyed by oxygen, foods and juices should be stored properly and consumed within a week of opening.[25]

The potato is an important source of vitamin C, not because one potato by itself meets the daily need, but because potatoes are such a common staple that they make significant contributions. In fact, scurvy was unknown in Ireland until the potato blight of the mid-1840s when some two million people died of malnutrition and infection.

The lack of yellow, white, brown, and red bars in Figure 10-18 confirms that grains, milk (except breast milk), legumes, and meats are notoriously poor sources of vitamin C. Organ meats (liver, kidneys, and others) and raw meats contain some vitamin C, but most people don't eat large quantities of these. Raw meats and fish contribute enough vitamin C to be significant in parts of Alaska, Canada, and Japan, but elsewhere fruits and vegetables are necessary to supply sufficient vitamin C.

---

*Vitamin C interferes with such anticoagulant drugs as warfarin, dicumarol, heparin, and coumadin. It is unclear whether vitamin C inhibits the absorption or the action of these drugs.
†Vitamin C is inactivated and degraded by several routes, and sometimes oxalate, which can form kidney stones, is produced along the way. People may also develop oxalate crystals in their kidneys regardless of vitamin C status.

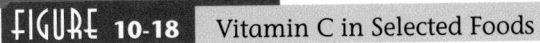

FIGURE 10-18 Vitamin C in Selected Foods

Food	Serving size (kcalories)
Bread, whole wheat	1 oz slice (70 kcal)
Cornflakes, fortified	1 oz (110 kcal)
Spaghetti pasta	½ c cooked (99 kcal)
Tortilla, flour	1 10"-round (234 kcal)
Broccoli	½ c cooked (22 kcal)
Carrots	½ c shredded raw (24 kcal)
Potato	1 medium baked w/skin (133 kcal)
Tomato juice	¾ c (31 kcal)
Banana	1 medium raw (109 kcal)
Orange	1 medium raw (62 kcal)
Strawberries	½ c fresh (22 kcal)
Watermelon	1 slice (92 kcal)
Milk	1 c reduced-fat 2% (121 kcal)
Yogurt, plain	1 c low-fat (155 kcal)
Cheddar cheese	1½ oz (171 kcal)
Cottage cheese	½ c low-fat 2% (101 kcal)
Pinto beans	½ c cooked (117 kcal)
Peanut butter	2 tbs (188 kcal)
Sunflower seeds	1 oz dry (165 kcal)
Tofu (soybean curd)	½ c (76 kcal)
Ground beef, lean	3 oz broiled (244 kcal)
Chicken breast	3 oz roasted (140 kcal)
Tuna, canned in water	3 oz (99 kcal)
Egg	1 hard cooked (78 kcal)
Excellent, and sometimes unusual, sources:	
Red bell pepper	½ c raw chopped (20 kcal)
Kiwi	1 (46 kcal)
Brussels sprouts	½ c cooked (30 kcal)

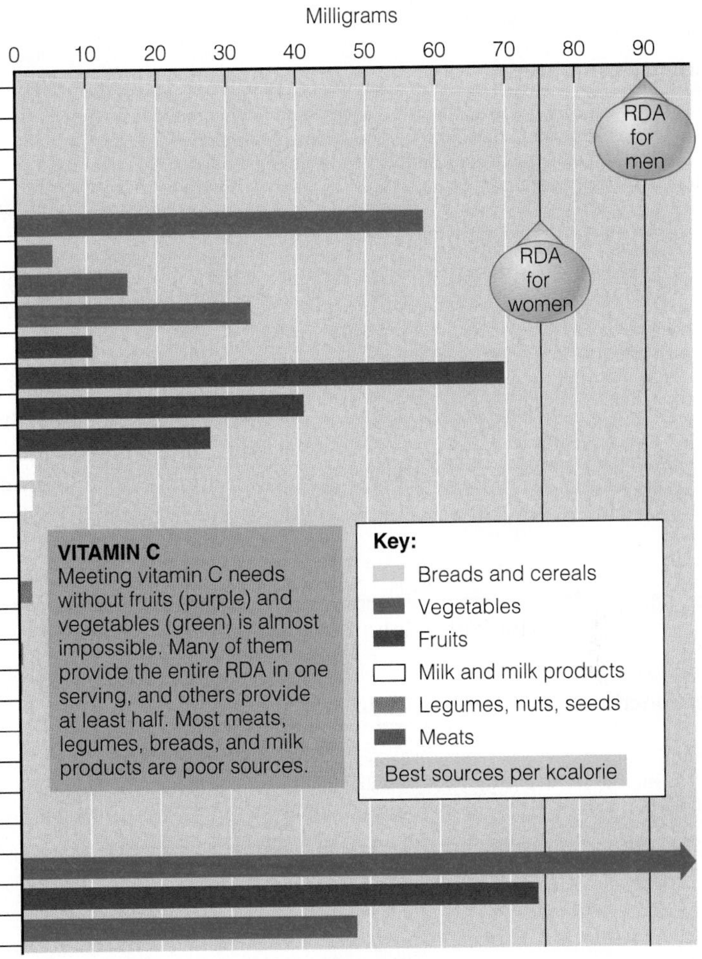

Milligrams: 0 10 20 30 40 50 60 70 80 90

RDA for men

RDA for women

VITAMIN C
Meeting vitamin C needs without fruits (purple) and vegetables (green) is almost impossible. Many of them provide the entire RDA in one serving, and others provide at least half. Most meats, legumes, breads, and milk products are poor sources.

Key:
- Breads and cereals
- Vegetables
- Fruits
- Milk and milk products
- Legumes, nuts, seeds
- Meats

Best sources per kcalorie

Note: See p. 327 for more information on using this figure.

Because of vitamin C's antioxidant property, food manufacturers sometimes add a variation of vitamin C to some beverages and most cured meats, such as luncheon meats, to prevent oxidation and spoilage. This compound safely preserves these foods, but it does not have vitamin C activity in the body. Simply put, "ham and bacon cannot replace fruits and vegetables." See the accompanying table for a summary of vitamin C.

IN SUMMARY Vitamin C

Other Names

Ascorbic acid

2000 RDA

Men: 90 mg/day

Women: 75 mg/day

Smokers: +35 mg/day

Upper Level

Adults: 2000 mg/day

Chief Functions in the Body

Collagen synthesis (strengthens blood vessel walls, forms scar tissue, provides matrix for bone growth), antioxidant, thyroxin synthesis, amino acid metabolism, strengthens resistance to infection, helps in absorption of iron

Significant Sources

Citrus fruits, cabbage-type vegetables, dark green vegetables (such as bell peppers and broccoli), cantaloupe, strawberries, lettuce, tomatoes, potatoes, papayas, mangoes

Easily destroyed by heat and oxygen

(continued)

Vitamin C (continued)

Deficiency Disease

Scurvy

Deficiency Symptoms

Anemia (small-cell type),[a] atherosclerotic plaques, pinpoint hemorrhages; bone fragility, joint pain; poor wound healing, frequent infections; bleeding gums, loosened teeth; muscle degeneration and pain, hysteria, depression; rough skin, blotchy bruises

Toxicity Symptoms

Nausea, abdominal cramps, diarrhea; headache, fatigue, insomnia; hot flashes, rashes; interference with medical tests, aggravation of gout symptoms, urinary tract problems, kidney stones[b]

---

[a] Small-cell–type anemia is *microcytic anemia.*

[b] People with kidney disease, a tendency toward gout, or a genetic abnormality that alters the breakdown of vitamin C are prone to forming kidney stones. Vitamin C is inactivated and degraded by several routes, sometimes producing oxalate, which can form stones in the kidneys.

Vita means life. After this discourse on the vitamins, who could dispute that they deserve their name? Their regulation of metabolic processes makes them vital to the normal growth, development, and maintenance of the body. The accompanying summary table condenses the information provided in this chapter for a quick review. The remarkable roles of the vitamins continue in the next chapter.

IN SUMMARY   The Water-Soluble Vitamins

Vitamin and Chief Functions	Deficiency Symptoms	Toxicity Symptoms	Food Sources
Thiamin Part of coenzyme TPP in energy metabolism	Beriberi (edema or muscle wasting), anorexia and weight loss, neurological disturbances, muscular weakness, heart enlargement and failure	None reported	Enriched, fortified, or whole-grain products; pork
Riboflavin Part of coenzymes FAD and FMN in energy metabolism	Inflammation of the mouth, skin, and eyelids; sensitivity to light	None reported	Milk products; enriched, fortified, or whole-grain products; liver
Niacin Part of coenzymes NAD and NADP in energy metabolism	Pellagra (diarrhea, dermatitis, and dementia)	Niacin flush, liver damage, impaired glucose tolerance	Protein-rich foods
Biotin Part of coenzyme in energy metabolism	Skin rash, hair loss, neurological disturbances	None reported	Widespread in foods; GI bacteria synthesis
Pantothenic acid Part of coenzyme A in energy metabolism	Digestive and neurological disturbances	None reported	Widespread in foods
Vitamin B$_6$ Part of coenzymes used in amino acid and fatty acid metabolism	Scaly dermatitis, depression, confusion, convulsions, anemia	Nerve degeneration, skin lesions	Protein-rich foods
Folate Activates vitamin B$_{12}$; helps synthesize DNA for new cell growth	Anemia, glossitis, neurological disturbances, elevated homocysteine	Masks vitamin B$_{12}$ deficiency	Legumes, vegetables, fortified grain products
Vitamin B$_{12}$ Activates folate; helps synthesize DNA for new cell growth; protects nerve cells	Anemia; nerve damage and paralysis	None reported	Foods derived from animals
Vitamin C Synthesis of collagen, carnitine, hormones, neurotransmitters; antioxidant	Scurvy (bleeding gums, pinpoint hemorrhages, abnormal bone growth, and joint pain)	Diarrhea, GI distress	Fruits and vegetables

Nutrition in Your Life

To obtain all the vitamins you need each day, be sure to select from a variety of foods.

- Do you often choose whole or enriched grains, dark green leafy vegetables, citrus fruits, and legumes?

- If you are a woman of childbearing age, do you eat folate-rich foods or take supplements regularly?

- Do you take supplements that provide more than the upper limit of the vitamins?

 Practice planning several lunch menus with varying energy needs while simultaneously meeting the recommended intakes of several vitamins.

NUTRITION ON THE NET

 Access these websites for further study of topics covered in this chapter. Be aware that many websites on the Internet are peddling vitamin supplements, not accurate information.

- Find updates and quick links to these and other nutrition-related sites at our website: **www.wadsworth.com/nutrition**

- Search for "vitamins" at the American Dietetic Association: **www.eatright.org**

- Review the Dietary Reference Intakes for the water-soluble vitamins: **www.nap.edu/readingroom**

- Visit the World Health Organization to learn about "vitamin deficiencies" around the world: **www.who.int**

- Learn more about neural tube defects from the Spina Bifida Association of America: **www.sbaa.org**

- Read about Dr. Joseph Goldberger and his groundbreaking discovery linking pellagra to diet by searching for his name at: **www.nih.gov** or **www.pbs.org**

- Learn how fruits and vegetables support a healthy diet rich in vitamins from the National Cancer Institute or the 5 A Day for Better Health program: **www.5aday.gov** or **5aday.org**

NUTRITION CALCULATIONS

These problems give you practice in doing simple vitamin-related calculations (answers are provided on p. 358). Be sure to show your calculations for each problem.

1. Review the units in which vitamins are measured (a spot check).
   a. For each of these vitamins, note the unit of measure:

   Thiamin   Folate
   Riboflavin   Vitamin B_{12}
   Niacin   Vitamin C
   Vitamin B_6

   b. Recall from the chapter's description of people's self-dosing with vitamin B_6 that people who suffer toxicity symptoms may be taking more than 2 grams a day, whereas the RDA is less than 2 *mil-*

   ligrams. How much higher than 2 milligrams is 2 grams?
   c. Vitamin B_{12} is measured in micrograms. How many micrograms are in a gram? How many grams are in a teaspoon of a granular powder? How many micrograms does that represent? What is your RDA for vitamin B_{12}?

This exercise should convince you that the amount of vitamins a person needs is indeed quite small—yet still essential.

2. Be aware of how niacin intakes are affected by dietary protein availability.
   a. Refer to the "How to" on p. 332, and calculate how much niacin a woman receives from a diet that

delivers 90 grams protein and 9 milligrams niacin. (Assume her RDA for protein is 46 grams/day.)

   b. Is this woman getting her RDA of niacin (14 milligrams NE)?

This exercise should demonstrate that protein helps meet niacin needs.

STUDY QUESTIONS

These questions will help you review the chapter. You will find the answers in the discussions on the pages provided.

1. How do the vitamins differ from the energy nutrients? (p. 322)

2. Describe some general differences between fat-soluble and water-soluble vitamins. (pp. 322–324)

3. Which B vitamins are involved in energy metabolism? Protein metabolism? Cell division? (pp. 324–325)

4. For thiamin, riboflavin, niacin, biotin, pantothenic acid, vitamin B_6, folate, vitamin B_{12}, and vitamin C, state:
   - Its chief function in the body.
   - Its characteristic deficiency symptoms.
   - Its significant food sources. (See respective summary tables.)

5. What is the relationship of tryptophan to niacin? (p. 330)

6. Describe the relationship between folate and vitamin B_{12}. (pp. 336, 337, 340, 342)

7. What risks are associated with high doses of niacin? Vitamin B_6? Vitamin C? (pp. 330, 334–335, 352)

These questions will help you prepare for an exam. Answers can be found on p. 358.

1. Vitamins:
   a. are inorganic compounds.
   b. yield energy when broken down.
   c. are soluble in either water or fat.
   d. perform best when linked in long chains.

2. The rate at and the extent to which a vitamin is absorbed and used in the body is known as its:
   a. bioavailability.
   b. intrinsic factor.
   c. physiological effect.
   d. pharmacological effect.

3. Many of the B vitamins serve as:
   a. coenzymes.
   b. antagonists.
   c. antioxidants.
   d. serotonin precursors.

4. With respect to thiamin, which of the following is the most nutrient dense?
   a. 1 slice whole-wheat bread (69 kcalories and 0.1 milligram thiamin)
   b. 1 cup yogurt (144 kcalories and 0.1 milligram thiamin)
   c. 1 cup snow peas (69 kcalories and 0.22 milligram thiamin)
   d. 1 chicken breast (141 kcalories and 0.06 milligram thiamin)

5. The body can make niacin from:
   a. tyrosine.
   b. serotonin.
   c. carnitine.
   d. tryptophan.

6. The vitamin that protects against neural tube defects is:
   a. niacin.
   b. folate.
   c. riboflavin.
   d. vitamin B_{12}.

7. A lack of intrinsic factor may lead to:
   a. beriberi.
   b. pellagra.
   c. pernicious anemia.
   d. atrophic gastritis.

8. Which of the following is a B vitamin?
   a. inositol
   b. carnitine
   c. vitamin B_{15}
   d. pantothenic acid

9. Vitamin C serves as a(n):
   a. coenzyme.
   b. antagonist.
   c. antioxidant.
   d. intrinsic factor.

10. The requirement for vitamin C is highest for:
   a. smokers.
   b. athletes.
   c. alcoholics.
   d. the elderly.

REFERENCES

1. Committee on Dietary Reference Intakes, *Dietary Reference Intakes for Vitamin C, Vitamin E, Selenium, and Carotenoids* (Washington, D.C.: National Academy Press, 2000); Committee on Dietary Reference Intakes, *Dietary Reference Intakes for Thiamin, Riboflavin, Niacin, Vitamin B$_6$, Folate, Vitamin B$_{12}$, Pantothenic Acid, Biotin, and Choline* (Washington, D.C.: National Academy Press, 1998).

2. K. J. Carpenter, *Beriberi, White Rice, and Vitamin B: A Disease, a Cause, and a Cure* (Berkeley: University of California Press, 2000).

3. B. G. Brown and coauthors, Simvastatin and niacin, antioxidant vitamins, or the combination for the prevention of coronary disease, *New England Journal of Medicine* 345 (2001): 1583–1592; T. A. Jacobson, Combination lipid-altering therapy: An emerging treatment paradigm for the 21st century, *Current Atherosclerosis Reports* 3 (2001): 373–382.

4. A. A. Gerritsen and coauthors, Conservative treatment options for carpal tunnel syndrome: A systematic review of randomized controlled trials, *Journal of Neurology* 249 (2002): 272–280; R. Luboshitzky and coauthors, The effect of pyridoxine administration on melatonin secretion in normal men, *Neuroendocrinology Letters* 23 (2002): 213–217.

5. A. Fleming, The role of folate in the prevention of neural tube defects: Human and animal studies, *Nutrition Reviews* 59 (2001): S13–S20; S. M. Gross and coauthors, Inadequate folic acid intakes are prevalent among young women with neural tube defects, *Journal of the American Dietetic Association* 101 (2001): 342–345; Committee on Genetics, Folic acid for the prevention of neural tube defects, *Pediatrics* 104 (1999): 325–327; L. D. Botto and coauthors, Neural-tube defects, *New England Journal of Medicine* 341 (1999): 1509–1519.

6. R. J. Berry and coauthors, Prevention of neural-tube defects with folic acid in China, *New England Journal of Medicine* 341 (1999): 1485–1490; Committee on Genetics, 1999.

7. Knowledge and use of folic acid by women of childbearing age—United States, 1995 and 1998, *Morbidity and Mortality Weekly Report* 48 (1999): 327–328; C. J. Lewis and coauthors, Estimated folate intakes: Data updated to reflect food fortification, increased bioavailability, and dietary supplement use, *American Journal of Clinical Nutrition* 70 (1999): 198–207.

8. J. Erickson, Folic acid and prevention of spina bifida and anencephaly, *Morbidity and Mortality Weekly Report* 51 (2002): 1–3; M. A. Honein and coauthors, Impact of folic acid fortification of the US food supply on the occurrence of neural tube defects, *Journal of the American Medical Association* 285 (2001): 2981–2986; R. E. Stevenson and coauthors, Decline in prevalence of neural tube defects in a high-risk region of the United States, *Pediatrics* 106 (2000): 677–683.

9. E. P. Quinlivan and J. F. Gregory III, Effect of food fortification on folic acid intake in the United States, *American Journal of Clinical Nutrition* 77 (2003): 221–225; G. J. Cuskelly, H. McNulty, and J. M. Scott, Fortification with low amounts of folic acid makes a significant difference in folate status in young women: Implications for the prevention of neural tube defects, *American Journal of Clinical Nutrition* 70 (1999): 234–239; P. F. Jacques and coauthors, The effect of folic acid fortification on plasma folate and total homocysteine concentrations, *New England Journal of Medicine* 340 (1999): 1449–1454.

10. Committee on Dietary Reference Intakes, 1998.

11. S. Moyers and L. B. Bailey, Fetal malformations and folate metabolism: Review of recent evidence, *Nutrition Reviews* 59 (2001): 215–224; S. J. James and coauthors, Abnormal folate metabolism and mutation in the methylenetetrahydrofolate reductase gene may be maternal risk factors for Down syndrome, *American Journal of Clinical Nutrition* 70 (1999): 495–501; D. S. Rosenblatt, Folate and homocysteine metabolism and gene polymorphisms in the etiology of Down syndrome, *American Journal of Clinical Nutrition* 70 (1999): 429–430.

12. D. S. Wald, M. Law, and J. K. Morris, Homocysteine and cardiovascular disease: Evidence on causality from a meta-analysis, *British Medical Journal* 325 (2002): 1202; M. L. Bots and coauthors, Homocysteine and short-term risk of myocardial infarction and stroke in the elderly: The Rotterdam Study, *Archives of Internal Medicine* 159 (1999): 38–44.

13. F. V. van Oort and coauthors, Folic acid and reduction of plasma homocysteine concentrations in older adults: A dose-response study, *American Journal of Clinical Nutrition* 77 (2003): 1318–1323; B. J. Venn and coauthors, Dietary counseling to increase natural folate intake: A randomized placebo-controlled trial in free-living subjects to assess effects on serum folate and plasma total homocysteine, *American Journal of Clinical Nutrition* 76 (2002): 758–765; G. Schnyder and coauthors, Decreased rate of coronary restenosis after lowering of plasma homocysteine levels, *New England Journal of Medicine* 345 (2001): 1539–1600; L. J. Riddell and coauthors, Dietary strategies for lowering homocysteine concentrations, *American Journal of Clinical Nutrition* 71 (2000): 448–454; Jacques and coauthors, 1999; I. A. Brouwer and coauthors, Low-dose folic acid supplementation decreases plasma homocysteine concentrations: A randomized trial, *American Journal of Clinical Nutrition* 69 (1999): 99–104.

14. G. C. Rampersaud, L. B. Bailey, and G. P. A. Kauwell, Relationship of folate to colorectal and cervical cancer: Review and recommendations for practitioners, *Journal of the American Dietetic Association* 102 (2002): 1273–1282; S. W. Choi and J. B. Mason, Folate and carcinogenesis: An integrated scheme, *Journal of Nutrition* 130 (2000): 129–132; Y. I. Kim, Methylenetetrahydrofolate reductase polymorphisms, folate, and cancer risk: A paradigm of gene-nutrient interactions in carcinogenesis, *Nutrition Reviews* 58 (2000): 205–209.

15. Y. Kim, Folate and cancer prevention: A new medical application of folate beyond hyperhomocysteinemia and neural tube defects, *Nutrition Reviews* 57 (1999): 314–321; S. Zhang and coauthors, A prospective study of folate intake and the risk of breast cancer, *Journal of the American Medical Association* 281 (1999): 1632–1637.

16. B. D. Hokin and T. Butler, Cyanocobalamin (vitamin B-12) status in Seventh-day Adventist ministers in Australia, *American Journal of Clinical Nutrition* 70 (1999): 576S–578S.

17. M. W. J. Louwman and coauthors, Signs of impaired cognitive function in adolescents with marginal cobalamin status, *American Journal of Clinical Nutrition* 72 (2000): 762–769.

18. H. J. Powers, Riboflavin (vitamin B-2) and health, *American Journal of Clinical Nutrition* 77 (2003): 1352–1360.

19. H. Hemilä and coauthors, Vitamin C, vitamin E, and beta-carotene in relation to common cold incidence in male smokers, *Epidemiology* 13 (2002): 32–37; B. Takkouche and coauthors, Intake of vitamin C and zinc and risk of common cold: A cohort study, *Epidemiology* 13 (2002): 38–44; M. van Straten and P. Josling, Preventing the common cold with a vitamin C supplement: A double-blind, placebo-controlled survey, *Advances in Therapy* 19 (2002): 151–159; H. C. Gorton and K. Jarvis, The effectiveness of vitamin C in preventing and relieving the symptoms of virus-induced respiratory infections, *Journal of Manipulative and Physiological Therapeutics* 22 (1999): 530–533.

20. R. M. Douglas, E. B. Chalker, and B. Treacy, Vitamin C for preventing and treating the common cold (Cochrane Review), *Cochrane Database of Systematic Reviews* 2 (2000): CD000980; H. Hemilä and Z. S. Herman, Vitamin C and the common cold: A retrospective analysis of Chalmer's review, *Journal of the American College of Nutrition* 14 (1995): 116–123.

21. Committee on Dietary Reference Intakes, 2000.

22. Researchers proposed 120 mg/day. M. Levine and coauthors, Criteria and recommendations for vitamin C intake, *Journal of the American Medical Association* 281 (1999): 1415–1423; A. C. Carr and B. Frei, Toward a new recommended dietary allowance for vitamin C based on antioxidant and health effects in humans, *American Journal of Clinical Nutrition* 69 (1999): 1086–1107.

23. A. M. Preston and coauthors, Influence of environmental tobacco smoke on vitamin C status in children, *American Journal of Clinical Nutrition* 77 (2003): 167–172; R. S. Strauss, Environmental tobacco smoke and serum vitamin C levels in children, *Pediatrics* 107 (2001): 540–542; J. Lykkesfeldt and coauthors, Ascorbate is depleted by smoking and repleted by moderate supplementation: A study in male smokers and nonsmokers with matched dietary antioxidant intakes, *American Journal of Clinical Nutrition* 71 (2000): 530–536.

24. M. Weinstein, P. Babyn, and S. Zlotkin, An orange a day keeps the doctor away: Scurvy in the year 2000, **http://www.pediatrics.org/cgi/content/full/108/3/e55**.

25. C. S. Johnston and D. L. Bowling, Stability of ascorbic acid in commercially available orange juices, *Journal of the American Dietetic Association* 102 (2002): 525–529.

ANSWERS

Nutrition Calculations

1. a. Thiamin: mg.   Folate: μg DFE.

   Riboflavin: mg.   Vitamin B_{12}: μg.

   Niacin: mg NE.   Vitamin C: mg.

   Vitamin B_6: mg.

   b. A thousand times higher (2 g × 1000 mg/g = 2000 mg; 2000 mg ÷ 2 mg = 1000).

   c. 1 g = 1000 mg; 1 mg = 1000 μg (1000 × 1000 = 1,000,000); 1 million μg = 1 g.

   1 tsp = 5 g.

   5 × 1,000,000 μg = 5,000,000 μg/tsp.

   See inside front cover for your RDA based on age and gender.

2. a. She eats 90 g protein. Assume she uses 46 g as protein. This leaves 90 g − 46 g = 44 g protein "leftover."

   44 g protein ÷ 100 = 0.44 g tryptophan.

   0.44 g tryptophan × 1000 = 440 mg tryptophan.

   440 mg tryptophan ÷ 60 = 7.3 mg NE.

   7.3 mg NE + 9 mg niacin = 16.3 mg NE.

   b. Yes.

Study Questions (multiple choice)

1. c   2. a   3. a   4. c   5. d   6. b   7. c   8. d

9. c   10. a

HIGHLIGHT

Vitamin and Mineral Supplements

© J. Share/Stone/Getty Images

Almost half of the population in the United States takes vitamin and mineral supplements regularly, spending billions of dollars on them each year.[1] Many people take supplements as dietary insurance—in case they are not meeting their nutrient needs from foods alone. Others take supplements as health insurance—to protect against certain diseases.

One out of every five people takes multinutrient pills daily. Others take large doses of single nutrients, most commonly, vitamin C, vitamin E, beta-carotene, iron, and calcium. In many cases, taking supplements is a costly but harmless practice; sometimes, it is both costly and harmful to health.

For the most part, people self-prescribe supplements, taking them on the advice of friends, television, websites, or books that may or may not be reliable. Sometimes, they take supplements on the recommendation of a physician. When such advice follows a valid nutrition assessment, supplementation may be warranted, but even then the preferred course of action is to improve food choices and eating habits.[2] Without an assessment, the advice to take supplements may be inappropriate. A registered dietitian can help with the decision.[3]

When people think of supplements, they often think of vitamins, but minerals are important, too, of course. People whose diets lack vitamins, for whatever reason, probably lack several minerals as well. This highlight asks several questions related to vitamin-mineral **supplements** (the accompanying glossary defines supplements and related terms). What are the arguments *for* taking supplements? What are the arguments *against* taking them? Finally, if people do take supplements, how can they choose the appropriate ones? (In addition to vitamins and minerals, supplements may also provide amino acids or herbs, which are discussed in Chapter 6 and Highlight 18, respectively.)

Arguments for Supplements

Vitamin-mineral supplements may be appropriate in some circumstances. In some cases, they can correct deficiencies; in others, they can reduce the risk of diseases.

Correct Overt Deficiencies

In the United States and Canada, adults rarely suffer nutrient deficiency diseases such as scurvy, pellagra, and beriberi, but they do still occur. To correct an overt deficiency disease, a physician may prescribe therapeutic doses two to ten times the RDA (or AI) of a nutrient. At such high doses, the supplement is acting as a drug.

Improve Nutrition Status

In contrast to the classical deficiencies, which present a multitude of symptoms and are relatively easy to recognize, subclinical deficiencies are subtle and easy to overlook—and they are also more likely to occur. People who do not eat enough food to deliver the needed amounts of nutrients, such as habitual dieters and the elderly, risk developing subclinical deficiencies. Similarly, vegetarians who restrict their use of entire food groups without appropriate substitutions may fail to fully meet their nutrient needs. If there is no way for these people to eat enough nutritious foods to meet their needs, then vitamin-mineral supplements may be appropriate to help prevent nutrient deficiencies.

GLOSSARY

FDA (Food and Drug Administration): a federal agency that is responsible for, among other things, supplement safety, manufacturing, and information, including product labeling, package inserts, and accompanying literature. Another federal agency, the **FTC (Federal Trade Commission)**, is responsible for, among other things, supplement advertising.

high potency: 100% or more of the Daily Value for the nutrient in a single supplement and for at least two-thirds of the nutrients in a multinutrient supplement.

supplements: pills, capsules, tablets, liquids, or powders that contain vitamins, minerals, herbs, or amino acids; intended to increase dietary intake of these substances.

Reduce Disease Risks

Few people consume the optimal amounts of all the vitamins and minerals by diet alone. Inadequate intakes have been linked to chronic diseases such as heart disease, some cancers, and osteoporosis.[4] For this reason, some physicians recommend that all adults take vitamin-mineral supplements.[5] Others recognize the lack of conclusive evidence and the potential harm of supplementation and advise against such a recommendation.[6]

Highlight 11 reviews the relationships between supplement use and disease prevention. It describes some of the accumulating evidence suggesting that intakes of certain nutrients at levels much higher than can be attained from foods alone may be beneficial in reducing disease risks. It also presents research confirming the associated risks. Clearly, consumers must be cautious in taking supplements to prevent disease.

Many people, especially postmenopausal women and those who are intolerant to lactose or allergic to milk, may not receive enough calcium to forestall the bone degeneration of old age, osteoporosis. For them, nonmilk calcium-rich foods are especially valuable, but calcium supplements may also be appropriate (Highlight 12 provides more details).

Support Increased Nutrient Needs

As Chapters 15–17 explain, nutrient needs increase during certain stages of life, making it difficult to meet some of those needs without supplementation. For example, women who lose a lot of blood and therefore a lot of iron during menstruation each month may need an iron supplement. Women of childbearing age need folate supplements to reduce the risks of neural tube defects. Similarly, pregnant women and women who are breastfeeding their infants have exceptionally high nutrient needs and so usually need special supplements. Newborns routinely receive a single dose of vitamin K at birth to prevent abnormal bleeding. Infants may need other supplements as well, depending on whether they are breastfed or receiving formula, and on whether their water contains fluoride.

Improve the Body's Defenses

Health care professionals may provide special supplementation to people being treated for addictions to alcohol or other drugs and to people with prolonged illnesses, extensive injuries, or other severe stresses such as surgery. Illnesses that interfere with appetite, eating, or nutrient absorption limit nutrient intakes, yet nutrient needs are often heightened by diseases or medications. In all these cases, supplements are appropriate.

Who Needs Supplements?

In summary, the following list acknowledges that in these specific conditions, these people may need to take supplements:

- People with nutrient deficiencies.
- People with low food energy intakes (fewer than 1200 kcalories per day) need a multivitamin and mineral supplement.
- People who eat all-plant diets (vegans) and those with atrophic gastritis need vitamin B_{12}.
- Women who bleed excessively during menstruation need iron.
- People with lactose intolerance or milk allergies, or who otherwise do not consume enough dairy products to forestall extensive bone loss, need calcium.
- People in certain stages of the life cycle who have increased nutrient needs (for example, infants need iron and fluoride, women of childbearing age need folate, pregnant women need iron, and the elderly need vitamins B_{12} and D).
- People with limited milk intake and sun exposure need vitamin D.
- People who have diseases, infections, or injuries or who have undergone surgery that interferes with the intake, absorption, metabolism, or excretion of nutrients.
- People taking medications that interfere with the body's use of specific nutrients.

Except for people in these circumstances, most adults can normally get all the nutrients they need by eating a varied diet of nutrient-dense foods. Even athletes can meet their nutrient needs without the help of supplements, as Chapter 14 explains.

Arguments against Supplements

Foods rarely cause nutrient imbalances or toxicities, but supplements can. The higher the dose, the greater the risk of harm. People's tolerances for high doses of nutrients vary, just as their risks of deficiencies do. Amounts that some can tolerate may be harmful for others, and no one knows who falls where along the spectrum. It is difficult to determine just how much of a nutrient is enough—or too much. The Tolerable Upper Intake Levels of the DRI answer the question how much is too much by defining the highest amount that appears safe for most healthy people. Table H10-1 presents these suggested Upper Levels and Daily Values for selected vitamins and minerals and the quantities typically found in supplements.

Toxicity

The extent and severity of supplement toxicity remain unclear. Only a few alert health care professionals can recognize toxicity, even when it is acute. When it is chronic, with the effects developing subtly and progressing slowly, it often goes unrecognized. In view of the potential hazards, some authori-

TABLE H10-1 Vitamin and Mineral Intakes for Adults

Nutrient	Tolerable Upper Intake Levels[a]	Daily Values	Typical Multivitamin-Mineral Supplement	Average Single-Nutrient Supplement
Vitamins				
Vitamin A	3000 µg (10,000 IU)	5000 IU	5000 IU	8000 to 10,000 IU
Vitamin D	50 µg (2000 IU)	400 IU	400 IU	400 IU
Vitamin E	1000 mg (1500 to 2200 IU)[b]	30 IU	30 IU	100 to 1000 IU
Vitamin K	—[c]	80 µg	40 µg	—[e]
Thiamin	—[c]	1.5 mg	1.5 mg	50 mg
Riboflavin	—[c]	1.7 mg	1.7 mg	25 mg
Niacin (as niacinamide)	35 mg[b]	20 mg	20 mg	100 to 500 mg
Vitamin B6	100 mg	2 mg	2 mg	100 to 200 mg
Folate	1000 µg[b]	400 µg	400 µg	400 µg
Vitamin B12	—[c]	6 µg	6 µg	100 to 1000 µg
Pantothenic acid	—[c]	10 mg	10 mg	100 to 500 mg
Biotin	—[c]	300 µg	30 µg	300 to 600 µg
Vitamin C	2000 mg	60 mg	10 mg	500 to 2000 mg
Choline	3500 mg	—	10 mg	250 mg
Minerals				
Calcium	2500 mg	1000 mg	160 mg	250 to 600 mg
Phosphorus	4000 mg	1000 mg	110 mg	—[e]
Magnesium	350 mg[d]	400 mg	100 mg	250 mg
Iron	45 mg	18 mg	18 mg	18 to 30 mg
Zinc	40 mg	15 mg	15 mg	10 to 100 mg
Iodine	1100 µg	150 µg	150 µg	—[e]
Selenium	400 µg	70 µg	10 µg	50 to 200 µg
Fluoride	10 mg	—	—	—[e]
Copper	10 mg	2 mg	0.5 mg	—[e]
Manganese	11 mg	2 mg	5 mg	—[e]
Chromium	—[c]	120 µg	25 µg	200 to 400 µg
Molybdenum	2000 µg	75 µg	25 µg	—[e]

[a]Unless otherwise noted, Upper Levels represent total intakes from food, water, and supplements.
[b]Upper Levels represent intakes from supplements, fortified foods, or both.
[c]These nutrients have been evaluated by the DRI Committee for Tolerable Upper Intake Levels, but none were established because of insufficient data. No adverse effects have been reported with intakes of these nutrients at levels typical of supplements, but caution is still advised, given the potential for harm that accompanies excessive intakes.
[d]Upper Levels represent intakes from supplements only.
[e]Available as a single supplement by prescription.

ties believe supplements should bear warning labels, advising consumers that large doses may be toxic.

Toxic overdoses of vitamins and minerals in children are more readily recognized and, unfortunately, fairly common. Fruit-flavored, chewable vitamins shaped like cartoon characters entice young children to eat them like candy in amounts that can cause poisoning. High-potency iron supplements (30 milligrams of iron or more per tablet) are especially toxic and are the leading cause of accidental ingestion fatalities among children. Even mild overdoses cause GI distress, nausea, and black diarrhea that reflects gastric bleeding. Severe overdoses result in bloody diarrhea, shock, liver damage, coma, and death.

Life-Threatening Misinformation

Another problem arises when people who are ill come to believe that high doses of vitamins or minerals can be therapeutic. Not only can high doses be toxic, but the person may take them instead of seeking medical help. Furthermore, there are no guarantees that the supplements will be effective.

Marketing materials for supplements often make health statements that are required to be "truthful and not misleading," but often fall far short of both. Highlight 18 revisits this topic and includes a discussion of herbal preparations and other alternative therapies.

Unknown Needs

Another argument against the use of supplements is that no one knows exactly how to formulate the "ideal" supplement. What nutrients should be included? Which, if any, of the phytochemicals should be included? How much of each? On whose needs should the choices be based? Surveys have repeatedly shown little relationship between the supplements people take and the nutrients they actually need.

False Sense of Security

Another argument against supplement use is that it may lull people into a false sense of security. A person might eat irresponsibly, thinking, "My supplement will cover my needs." Or, experiencing a warning symptom of a disease, a person might postpone seeking a diagnosis, thinking, "I probably just need a supplement to make this go away." Such self-diagnosis is potentially dangerous.

Other Invalid Reasons

Other invalid reasons why people might take supplements include:

- The belief that the food supply or soil contains inadequate nutrients.
- The belief that supplements can provide energy.
- The belief that supplements can enhance athletic performance or build lean body tissues without physical work or faster than work alone (see Highlight 14).
- The belief that supplements will help a person cope with stress.
- The belief that supplements can prevent, treat, or cure conditions ranging from the common cold to cancer.

Ironically, people with health problems are more likely to take supplements than other people, yet today's health problems are more likely to be due to overnutrition and poor lifestyle choices than to nutrient deficiencies. The truth—that most people would benefit from improving their eating and exercise habits—is harder to swallow than a supplement pill.

Bioavailability and Antagonistic Actions

In general, the body absorbs nutrients best from foods in which the nutrients are diluted and dispersed among other substances that may facilitate their absorption. Taken in pure,

concentrated form, nutrients are likely to interfere with one another's absorption or with the absorption of nutrients in foods eaten at the same time. Documentation of these effects is particularly extensive for minerals: zinc hinders copper and calcium absorption, iron hinders zinc absorption, calcium hinders magnesium and iron absorption, and magnesium hinders the absorption of calcium and iron. Similarly, binding agents in supplements limit mineral absorption.

Although minerals provide the most familiar and best-documented examples, interference among vitamins is now being seen as supplement use increases. The vitamin A precursor beta-carotene, long thought to be nontoxic, interferes with vitamin E metabolism when taken over the long term as a dietary supplement. Vitamin E, on the other hand, antagonizes vitamin K activity and so should not be used by people being treated for blood-clotting disorders. Consumers who want the benefits of optimal absorption of nutrients should eat ordinary foods, selected for nutrient density and variety.

Whenever the diet is inadequate, the person should first attempt to improve it so as to obtain the needed nutrients from foods. If that is truly impossible, then the person needs a multivitamin-mineral supplement that supplies between 50 and 150 percent of the Daily Value for each of the nutrients. These amounts reflect the ranges commonly found in foods and therefore are compatible with the body's normal handling of nutrients (its physiologic tolerance). The next section provides some pointers to assist in the selection of an appropriate supplement.

Selection of Supplements

Whenever a physician or registered dietitian recommends a supplement, follow the directions carefully. When selecting a supplement yourself, look for a single, balanced vitamin-mineral supplement.

If you decide to take a vitamin-mineral supplement, ignore the eye-catching art and meaningless claims. Pay attention to the form the supplements are in, the list of ingredients, and the price. Here's where the truth lies, and from it you can make a rational decision based on facts. You have two basic questions to answer.

Form

The first question: What form do you want—chewable, liquid, or pills? If you'd rather drink your supplements than chew them, fine. (If you choose a chewable form, though, be aware that chewable vitamin C can dissolve tooth enamel.) If you choose pills, look for statements about the disintegration time. The U.S. Pharmacopeia (USP) suggests that supplements should completely disintegrate within 30 to 45 minutes.[*] Obviously, supplements that don't dissolve have little chance of

---

[*]The USP establishes standards for quality, strength, and purity of supplements.

entering the bloodstream, so look for a brand that claims to meet USP disintegration standards.

Contents

The second question: What vitamins and minerals do *you* need? Generally, an appropriate supplement provides vitamins and minerals in amounts that do not exceed recommended intakes.[7] Avoid supplements that, in a daily dose, provide more than the Tolerable Upper Intake Level for *any* nutrient. Avoid preparations with more than 10 milligrams of iron per dose, except as prescribed by a physician. Iron is hard to get rid of once it's in the body, and an excess of iron can cause problems, just as a deficiency can (see Chapter 13).

Misleading Claims

Ignore "organic" or "natural" claims. Such supplements are no better than others and often cost more. The word *synthetic* may sound like "fake," but to synthesize just means to put together. Whether vitamins are synthesized in a laboratory or synthesized by plants and animals, your body uses them similarly. Only your wallet can tell the difference.

Avoid products that make **"high potency"** claims. More is not better (review the "How to" on p. 324). Remember that foods are also providing these nutrients. Nutrients can build up and cause unexpected problems. For example, a man who takes vitamins and begins to lose his hair may think his hair loss means he needs *more* vitamins, when in fact it may be the early sign of a vitamin A overdose. (Of course, it may be completely unrelated to nutrition as well.)

Be wise to fake vitamins and preparations that contain items not needed in human nutrition, such as carnitine and inositol. Such ingredients reveal a marketing strategy aimed at your pocket, not at your health. The manufacturer wants you to believe that its pills contain the latest "new" nutrient that other brands omit, but in reality, these substances are not known to be needed by human beings.

Realize that the claim that supplements "relieve stress" is another marketing ploy. If you give even passing thought to what people mean by "stress," you'll realize manufacturers could never design a supplement to meet everyone's needs. Is it stressful to take an exam? Well, yes. Is it stressful to survive a major car wreck with third-degree burns and multiple bone fractures? Definitely yes. The body's responses to these stresses are different. The body does use vitamins and minerals in mounting a stress response, but a body fed a well-balanced diet can meet the needs of most minor stresses. As for the major ones, medical intervention is needed. In any case, taking a vitamin supplement won't make life any less stressful.

Other marketing tricks to sidestep are "green" pills that contain dehydrated, crushed parsley, alfalfa, and other fruit and vegetable extracts. The nutrients and phytochemicals advertised can be obtained from a serving of vegetables more easily and for less money. Such pills may also provide enzymes, but these are inactivated in the stomach during protein digestion.

Be aware that some geriatric "tonics" are low in vitamins and minerals, yet so high in alcohol as to threaten inebriation. The liquids designed for infants are more complete.

Recognize the latest nutrition buzzwords. Manufacturers were marketing "antioxidant" supplements before the print had time to dry on the first scientific reports of antioxidant vitamins' action in preventing cancer and cardiovascular disease. Remember, too, that high doses can alter a nutrient's action in the body. An antioxidant in physiological quantities may be beneficial, but in pharmacological quantities, it may act as a prooxidant and produce harmful by-products. Highlight 11 explores antioxidants and supplement use in more detail.

Finally, be aware that advertising on the Internet is cheap and not closely regulated. Promotional e-mails can be sent to millions of people in an instant. Internet messages can easily cite references and provide links to other sites, implying an endorsement when in fact none has been given.[8] Be cautious when examining unsolicited information and search for a balanced perspective.

Cost

When shopping for supplements, remember that local or store brands may be just as good as nationally advertised brands. If they are less expensive, it may be because the price does not have to cover the cost of national advertising.

Regulation of Supplements

The Dietary Supplement Health and Education Act of 1994 was intended to enable consumers to make informed choices about nutrient supplements. The act subjects supplements to the same general labeling requirements that apply to foods. Specifically:

- Nutrition labeling for dietary supplements is required.
- Labels may make nutrient claims (as "high" or "low") according to specific criteria (for example, "an excellent source of vitamin C").
- Labels may claim that the lack of a nutrient can cause a deficiency disease, but if they do, they must also include the prevalence of that deficiency disease in the United States.
- Labels may make health claims that are supported by significant scientific agreement and are not brand specific (for example, "folate protects against neural tube defects"). To date, the following health claims have been approved for supplements: folate and neural tube defects, calcium and osteoporosis, soluble fiber from

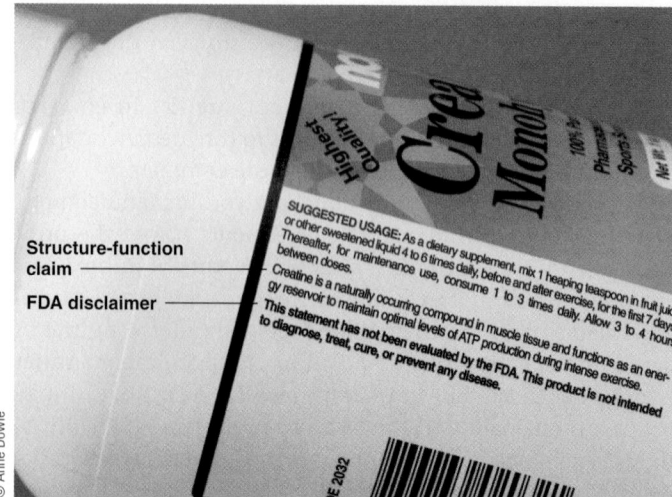

Structure-function claim

FDA disclaimer

Structure-function claims do not need FDA authorization, but they must be accompanied by a disclaimer.

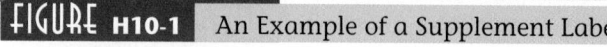

FIGURE H10-1   An Example of a Supplement Label

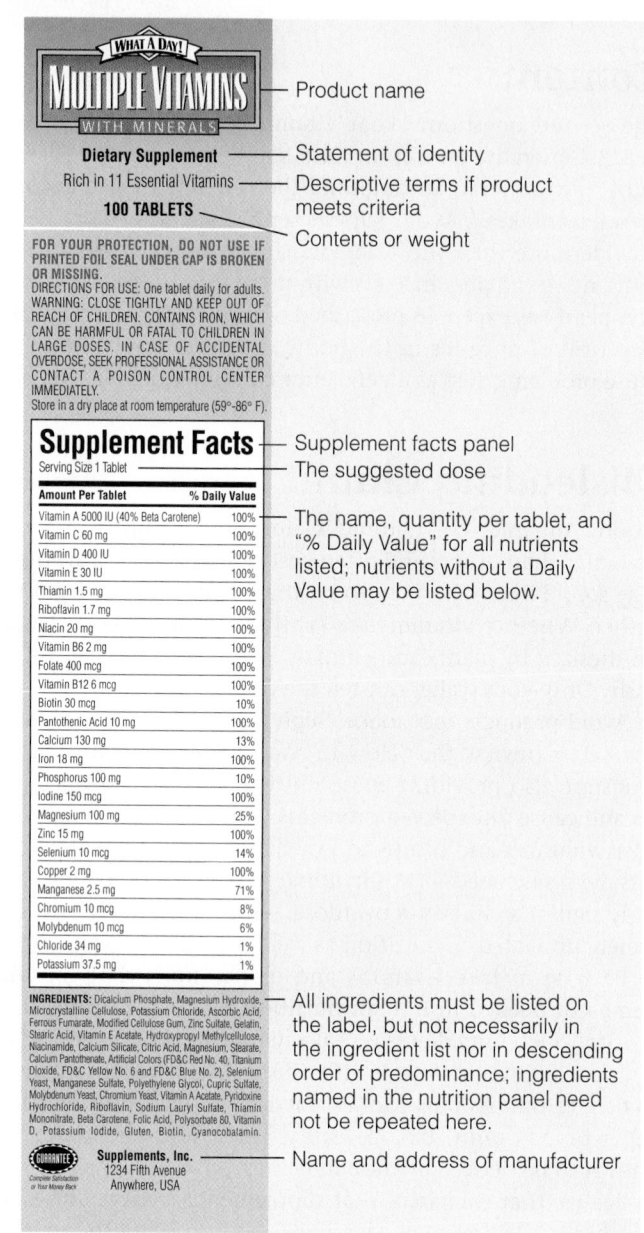

Product name

Statement of identity

Descriptive terms if product meets criteria

Contents or weight

Supplement facts panel

The suggested dose

The name, quantity per tablet, and "% Daily Value" for all nutrients listed; nutrients without a Daily Value may be listed below.

All ingredients must be listed on the label, but not necessarily in the ingredient list nor in descending order of predominance; ingredients named in the nutrition panel need not be repeated here.

Name and address of manufacturer

---

oat bran and from psyllium husks and cardiovascular disease, and omega-3 fatty acids and cardiovascular disease.

- Labels may claim to diagnose, treat, cure, or relieve common complaints such as menstrual cramps or memory loss, but may *not* make claims about specific diseases (except as noted above).

- Labels may make structure-function claims about the role a nutrient plays in the body, how the nutrient performs its function, and how consuming the nutrient is associated with general well-being. These claims must be accompanied by an **FDA** disclaimer statement: "This statement has not been evaluated by the Food and Drug Administration. This product is not intended to diagnose, treat, cure or prevent any disease." Figure H10-1 provides an example of a supplement label that complies with the requirements.

The multibillion-dollar-a-year supplement industry spends much money and effort influencing these regulations. The net effect of the Dietary Supplement Health and Education Act was a deregulation of the supplement industry. Unlike food additives or drugs, supplements do not need to be proved safe and effective, nor do they need the FDA's approval before being marketed. Furthermore, there are no standards for potency or dosage. Should a problem arise, the burden falls to the FDA to prove that the supplement poses an unreasonable risk and should be removed from the market. When asked, most Americans express support for greater regulation of dietary supplements.[9] Health professionals agree.[10]

If all the nutrients we need can come from food, why not just eat food? Foods have so much more to offer than supplements do. Nutrients in foods come in an infinite variety of combinations with a multitude of different carriers and absorption enhancers. They come with water, fiber, and an array of beneficial phytochemicals. Foods stimulate the GI tract to keep it healthy. They provide energy, and since you need energy each day, why not have nutritious foods deliver it? Foods offer pleasure, satiety, and opportunities for socializing while eating. In no way can nutrient supplements hold a candle to foods as a means of meeting human health needs. For further proof, read Highlight 11.

NUTRITION ON THE NET

 Access these websites for further study of topics covered in this highlight.

- Find updates and quick links to these and other nutrition-related sites at our website:
  www.wadsworth.com/nutrition

- Gather information from the Office of Dietary Supplements or Health Canada:
  dietary-supplements.info.nih.gov
  or **www.hc-sc.gc.ca**

- Report adverse reactions associated with dietary supplements to the FDA's MedWatch program:
  www.fda.gov/medwatch

- Search for "supplements" at the American Dietetic Association: **www.eatright.org**

- Learn more about supplements from the FDA Center for Food Safety and Applied Nutrition:
  www.cfsan.fda.gov/~dms/supplmnt.html

- Obtain consumer information on dietary supplements from the U.S. Pharmacopeia: **www.usp.org**

- Review the Federal Trade Commission policies for dietary supplement advertising: **www.ftc.gov/bcp/conline/pubs/buspubs/dietsupp.htm**

REFERENCES

1. L. S. Balluz and coauthors, Vitamin and mineral supplement use in the United States: Results from the third National Health and Nutrition Examination Survey, *Archives of Family Medicine* 9 (2000): 258–262.
2. Position of the American Dietetic Association: Food fortification and dietary supplements, *Journal of the American Dietetic Association* 101 (2001): 115–125.
3. J. R. Hunt, Tailoring advice on dietary supplements: An opportunity for dietetics professionals, *Journal of the American Dietetic Association* 102 (2002): 1754–1755; C. Thomson and coauthors, Guidelines regarding the recommendation and sale of dietary supplements, *Journal of the American Dietetic Association* 102 (2002): 1158–1164.

4. K. M. Fairfield and R. H. Fletcher, Vitamins for chronic disease prevention in adults: Scientific review, *Journal of the American Medical Association* 287 (2002): 3116–3126.
5. R. H. Fletcher and K. M. Fairfield, Vitamins for chronic disease prevention in adults: Clinical applications, *Journal of the American Medical Association* 287 (2002): 3127–3129.
6. C. D. Morris and S. Carson, Routine vitamin supplementation to prevent cardiovascular disease: A summary of the evidence for the U.S. Preventive Services Task Force, *Annals of Internal Medicine* 139 (2003): 56–70; B. Hasanain and A. D. Mooradian, Antioxidant vitamins and their influence in diabetes mellitus, *Current Diabetes Reports* 2 (2002): 448–456.

7. W. C. Willett and M. J. Stampfer, What vitamins should I be taking, doctor? *New England Journal of Medicine* 345 (2001): 1819–1824.
8. J. M. Drazen, Inappropriate advertising of dietary supplements, *New England Journal of Medicine* 348 (2003): 777–778.
9. R. J. Blendon and coauthors, Americans' views on the use and regulation of dietary supplements, *Archives of Internal Medicine* 161 (2001): 805–810.
10. P. B. Fontanarosa, D. Rennie, and C. D. DeAngelis, The need for regulation of dietary supplements—Lessons from ephedra, *Journal of the American Medical Association* 289 (2003): 1568–1570.

The Fat-Soluble Vitamins: A, D, E, and K

Nutrition in Your Life

Realizing that vitamin A from vegetables participates in vision, a mom encourages her children to "eat your carrots" because "they're good for your eyes." A dad takes his children outside to "enjoy the fresh air and sunshine" because they need the vitamin D that is made with the help of the sun. A physician recommends that a patient use vitamin E to slow the progression of heart disease. Another physician gives a newborn a dose of vitamin K to protect against life-threatening blood loss. These common daily occurrences highlight some of the heroic work of the fat-soluble vitamins.

The fat-soluble vitamins A, D, E, and K■ differ from the water-soluble vitamins in several significant ways (review the table on p. 324). Being insoluble in water, the fat-soluble vitamins require bile for their absorption. Upon absorption, fat-soluble vitamins travel through the lymphatic system within chylomicrons before entering the bloodstream, where many of them require protein carriers for transport. The fat-soluble vitamins participate in numerous activities all over the body, but excesses are stored primarily in the liver and adipose tissue. The body maintains blood concentrations by retrieving these vitamins from storage as needed; thus people can eat less than their daily need for days, weeks, or even months or years without ill effects. They need only ensure that over time *average* daily intakes approximate recommendations. By the same token, because fat-soluble vitamins are not readily excreted, the risk of toxicity is greater than it is for the water-soluble vitamins.

■ The fat-soluble vitamins:
- • Vitamin A.
- • Vitamin D.
- • Vitamin E.
- • Vitamin K.

Vitamin A and Beta-Carotene

Vitamin A was the first fat-soluble vitamin to be recognized. Almost a century later, vitamin A and its precursor, **beta-carotene,** continue to intrigue researchers with their diverse roles and profound effects on health.

vitamin A: all naturally occurring compounds with the biological activity of retinol (RET-ih-nol), the alcohol form of vitamin A.

beta-carotene (BAY-tah KARE-oh-teen): one of the carotenoids; an orange pigment and vitamin A precursor found in plants. A **precursor** is a compound that can be converted into an active vitamin.

FIGURE 11-1　Forms of Vitamin A

In this diagram, corners represent carbon atoms, as in all previous diagrams in this book. A further simplification here is that methyl groups (CH₃) are understood to be at the ends of the lines extending from corners. (See Appendix C for complete structures.)

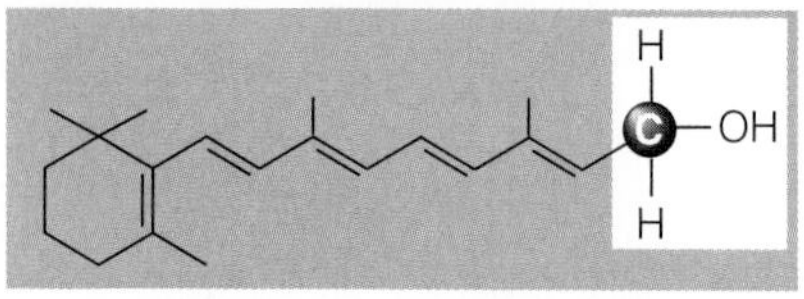

Retinol, the alcohol form

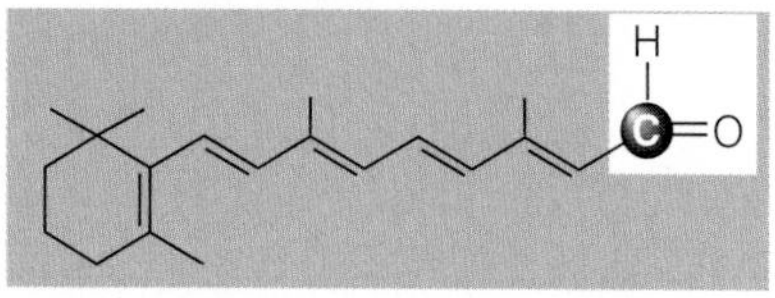

Retinal, the aldehyde form

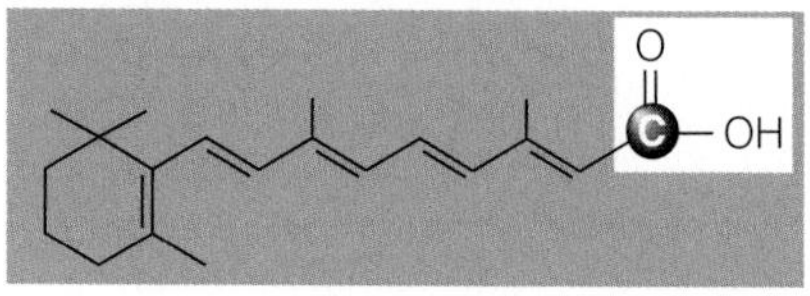

Retinoic acid, the acid form

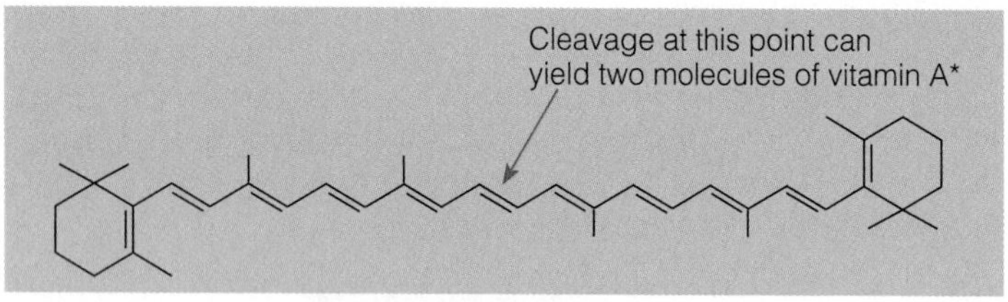

Cleavage at this point can yield two molecules of vitamin A*

Beta-carotene, a precursor

*Sometimes cleavage occurs at other points as well, so that one molecule of beta-carotene may yield only one molecule of vitamin A. Furthermore, not all beta-carotene is converted to vitamin A, and absorption of beta-carotene is not as efficient as that of vitamin A. For these reasons, 12 µg of beta-carotene are equivalent to 1 µg of vitamin A. Conversion of other carotenoids to vitamin A is even less efficient.

■ Carotenoids are among the best-known phytochemicals.

Three different forms of vitamin A are active in the body: retinol, retinal, and retinoic acid. Collectively, these compounds are known as **retinoids.** Foods derived from animals provide compounds (retinyl esters) that are easily converted to retinol in the intestine. Foods derived from plants provide **carotenoids,**■ some of which have **vitamin A activity.*** The most studied of the carotenoids is beta-carotene, which can be split to form retinol in the intestine and liver. Beta-carotene's absorption and conversion are significantly less efficient than those of the retinoids.[1] Figure 11-1 illustrates the structural similarities and differences of these vitamin A compounds and the cleavage of beta-carotene.[2]

The cells can convert retinol and retinal to the other active forms of vitamin A as needed. The conversion of retinol to retinal is reversible, but the further conversion of retinal to retinoic acid is irreversible (see Figure 11-2). This irreversibility is significant because each form of vitamin A performs a function that the others cannot.

A special transport protein, **retinol-binding protein (RBP),** picks up vitamin A from the liver, where it is stored, and carries it in the blood. Cells that use vitamin A have special protein receptors for it, as if the vitamin were fragile and had to be passed carefully from hand to hand without being dropped.[3] Each form of vitamin A has its own receptor protein (retinol has several) within the cells.

---

*Carotenoids with vitamin A activity include alpha-carotene, beta-carotene, and beta-cryptoxanthin; carotenoids with no vitamin A activity include lycopene, lutein, and zeaxanthin.

retinoids (RET-ih-noyds): chemically related compounds with biological activity similar to that of retinol; metabolites of retinol.

carotenoids (kah-ROT-eh-noyds): pigments commonly found in plants and animals, some of which have vitamin A activity. The carotenoid with the greatest vitamin A activity is beta-carotene.

vitamin A activity: a term referring to both the active forms of vitamin A and the precursor forms in foods without distinguishing between them.

retinol-binding protein (RBP): the specific protein responsible for transporting retinol.

FIGURE 11-2　Conversion of Vitamin A Compounds

Notice that the conversion from retinol to retinal is reversible, whereas the pathway from retinal to retinoic acid is not.

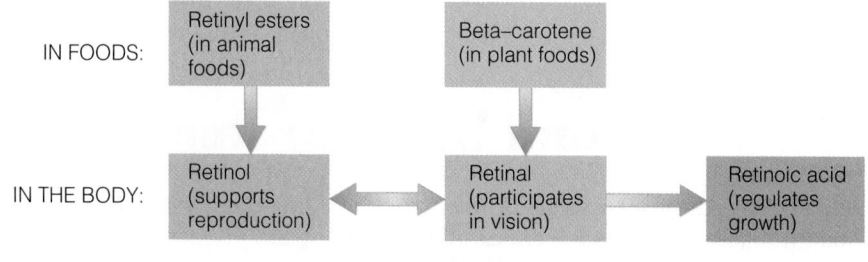

IN FOODS:

| Retinyl esters (in animal foods) | Beta–carotene (in plant foods) |

IN THE BODY:

| Retinol (supports reproduction) | Retinal (participates in vision) | Retinoic acid (regulates growth) |

Roles in the Body

Vitamin A is a versatile vitamin. Its major roles include:

- Promoting vision.
- Participating in protein synthesis and cell differentiation (and thereby maintaining the health of epithelial tissues and skin).
- Supporting reproduction and growth.

As mentioned, each form of vitamin A performs specific tasks. Retinol supports reproduction and is the major transport and storage form of the vitamin. Retinal is active in vision and is also an intermediate in the conversion of retinol to retinoic acid (review Figure 11-2). Retinoic acid acts like a hormone, regulating cell differentiation, growth, and embryonic development.[4] Animals raised on retinoic acid as their sole source of vitamin A can grow normally, but they become blind because retinoic acid cannot be converted to retinal (review Figure 11-2).

Vitamin A in Vision Vitamin A plays two indispensable roles in the eye: it helps maintain a crystal-clear outer window, the **cornea,** and it participates in the conversion of light energy into nerve impulses at the **retina** (see Figure 11-3 for details). The cells of the retina contain **pigment** molecules called **rhodopsin;** each rhodopsin molecule is composed of a protein called **opsin** bonded to a molecule of retinal.■ When light passes through the cornea of the eye and strikes the cells of the retina, rhodopsin responds by changing shape and becoming bleached. As it does, the retinal shifts from a *cis* to a *trans* configuration, just as fatty acids do during hydrogenation (see pp. 146–147). The *trans*-retinal cannot remain bonded to opsin. When retinal is released, opsin changes shape, thereby disturbing the membrane of the cell and generating an electrical impulse that travels along the cell's length. At the other end of the cell, the impulse is transmitted to a nerve cell, which conveys the message to the brain. Much of the retinal is then converted back to its active *cis* form and combined with the opsin protein to regenerate the pigment rhodopsin. Some retinal, however, may be oxidized to retinoic acid, a biochemical dead end for the visual process.

Visual activity leads to repeated small losses of retinal, necessitating its constant replenishment either directly from foods or indirectly from retinol stores. Ultimately, foods supply all the retinal in the pigments of the eye.

Vitamin A in Protein Synthesis and Cell Differentiation Despite its important role in vision, only one-thousandth of the body's vitamin A is in the retina. Much more is in the cells lining the body's surfaces. There the vitamin participates

■ Over 100 million cells reside in the retina, and each contains about 30 million molecules of vitamin A–containing visual pigments.

---

cornea (KOR-nee-uh): the transparent membrane covering the outside of the eye.

retina (RET-in-uh): the layer of light-sensitive nerve cells lining the back of the inside of the eye; consists of rods and cones.

pigment: a molecule capable of absorbing certain wavelengths of light so that it reflects only those that we perceive as a certain color.

rhodopsin (ro-DOP-sin): a light-sensitive pigment of the retina; contains the retinal form of vitamin A and the protein opsin.
- **rhod** = red (pigment)
- **opsin** = visual protein

opsin (OP-sin): the protein portion of the visual pigment molecule.

FIGURE 11-3 Vitamin A's Role in Vision

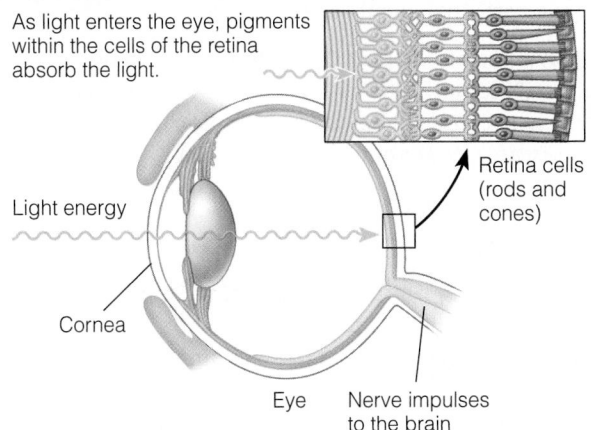

As light enters the eye, pigments within the cells of the retina absorb the light.

Light energy

Cornea

Eye   Nerve impulses to the brain

Retina cells (rods and cones)

The cells of the retina contain rhodopsin, a molecule composed of opsin (a protein) and *cis*-retinal (vitamin A).

cis-Retinal   *trans*-Retinal

As rhodopsin absorbs light, retinal changes from *cis* to *trans*, which triggers a nerve impulse that carries visual information to the brain.

FIGURE 11-4 Mucous Membrane Integrity

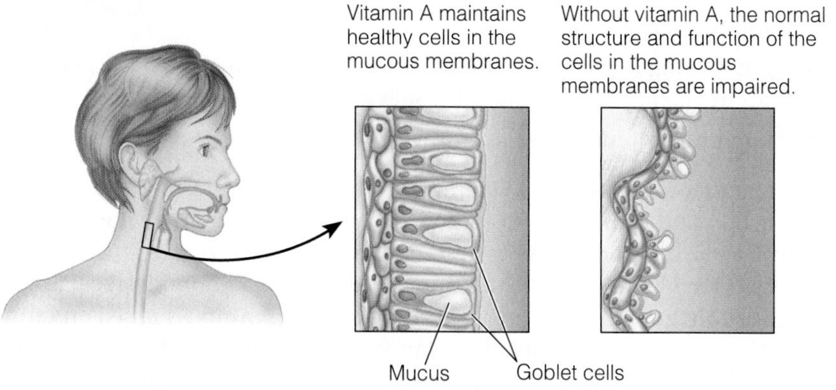

Vitamin A maintains healthy cells in the mucous membranes.

Without vitamin A, the normal structure and function of the cells in the mucous membranes are impaired.

Mucus   Goblet cells

in protein synthesis and **cell differentiation,** a process by which each type of cell develops to perform a specific function.

All body surfaces, both inside and out, are covered by layers of cells known as **epithelial cells.** The **epithelial tissue** on the outside of the body is, of course, the skin. The epithelial tissues that line the inside of the body are the **mucous membranes:** the linings of the mouth, stomach, and intestines; the linings of the lungs and the passages leading to them; the linings of the urinary bladder and urethra; the linings of the uterus and vagina; and the linings of the eyelids and sinus passageways. Within the body, the mucous membranes of the GI tract alone line an area larger than a quarter of a football field, and vitamin A helps to maintain their integrity (see Figure 11-4).

Vitamin A promotes differentiation of both epithelial cells and goblet cells, one-celled glands that synthesize and secrete mucus. Mucus coats and protects the epithelial cells from invasive microorganisms and other harmful substances, such as gastric juices.

Vitamin A in Reproduction and Growth As mentioned, vitamin A also supports reproduction and growth. In men, retinol participates in sperm development, and in women, vitamin A supports normal fetal development during pregnancy. Children lacking vitamin A fail to grow. When given vitamin A supplements, these children gain weight and grow taller.[5]

The growth of bones illustrates that growth is a complex phenomenon of **remodeling.** To convert a small bone into a large bone, the bone-remodeling cells must "undo" some parts of the bone as they go,■ and vitamin A participates in the dismantling. The cells that break down bone contain sacs of degradative enzymes.■ With the help of vitamin A, these enzymes eat away at selected sites in the bone, removing the parts that are not needed.

Beta-Carotene as an Antioxidant In the body, beta-carotene serves primarily as a vitamin A precursor.[6] Not all dietary beta-carotene is converted to active vitamin A, however. Some beta-carotene may act as an antioxidant capable of protecting the body against disease (see Highlight 11 for details).

Vitamin A Deficiency

Vitamin A status depends mostly on the adequacy of vitamin A stores, 90 percent of which are in the liver. Vitamin A status also depends on a person's protein status because retinol-binding proteins serve as the vitamin's transport carriers inside the body.

If a person were to stop eating vitamin A–rich foods, deficiency symptoms would not begin to appear until after stores were depleted—one to two years for a

---

■ The cells that destroy bone during growth are **osteoclasts;** those that build bone are **osteoblasts.**
- **osteo** = bone
- **clast** = break
- **blast** = build

■ The sacs of degradative enzymes are **lysosomes** (LYE-so-zomes).

cell differentiation (DIF-er-EN-she-AY-shun): the process by which immature cells develop specific functions different from those of the original that are characteristic of their mature cell type.

epithelial (ep-i-THEE-lee-ul) **cells:** cells on the surface of the skin and mucous membranes.

epithelial tissue: the layer of the body that serves as a selective barrier between the body's interior and the environment (examples are the cornea, the skin, the respiratory lining, and the lining of the digestive tract).

mucous (MYOO-kus) **membranes:** the membranes, composed of mucus-secreting cells, that line the surfaces of body tissues.

remodeling: the dismantling and re-formation of a structure, in this case, bone.

healthy adult but much sooner for a growing child. Then the consequences would be profound and severe. Vitamin A deficiency is uncommon in the United States, but it is one of the developing world's major nutrition problems.[7] More than 100 million children worldwide have some degree of vitamin A deficiency, and so are vulnerable to infectious diseases and blindness.

Infectious Diseases In developing countries around the world, measles is a devastating infectious disease, killing as many as 2 million children each year. The severity of the illness often correlates with the degree of vitamin A deficiency; deaths are usually due to related infections such as pneumonia and severe diarrhea.[8] Providing large doses of vitamin A reduces the risk of dying from these infections.

The World Health Organization (WHO) and UNICEF (the United Nations International Children's Emergency Fund) have made the control of vitamin A deficiency a major goal in their quest to improve child health and survival throughout the developing world. They recommend routine vitamin A supplementation for all children with measles in areas where vitamin A deficiency is a problem or where the measles death rate is high. In the United States, the American Academy of Pediatrics recommends vitamin A supplementation for certain groups of measles-infected infants and children. Vitamin A supplementation also protects against the complications of other life-threatening infections, including malaria, lung diseases, and HIV (human immunodeficiency virus, the virus that causes AIDS).[9]

Night Blindness **Night blindness** is one of the first detectable signs of vitamin A deficiency and permits early diagnosis.[10] In night blindness, the retina does not receive enough retinal to regenerate the visual pigments bleached by light. The person loses the ability to recover promptly from the temporary blinding that follows a flash of bright light at night or to see after the lights go out. In many parts of the world, after the sun goes down, vitamin A–deficient people become night-blind: children cannot find their shoes or toys, and women cannot fetch water or wash dishes.[11] They often cling to others or sit still, afraid that they may trip and fall or lose their way if they try to walk alone. In many developing countries, night blindness due to vitamin A deficiency is so common that the people have special words to describe it. In Indonesia, the term is *buta ayam,* which means "chicken eyes" or "chicken blindness." (Chickens do not have the cells of the retina that respond to dim light and therefore cannot see at night.) Figure 11-5 shows the eyes' slow recovery in response to a flash of bright light in night blindness.

Blindness (Xerophthalmia) Beyond night blindness is total blindness—failure to see at all. Night blindness is caused by a lack of vitamin A at the back of the eye, the retina; total blindness is caused by a lack at the front of the eye, the cornea.

> **night blindness:** slow recovery of vision after flashes of bright light at night or an inability to see in dim light; an early symptom of vitamin A deficiency.

FIGURE 11-5 Vitamin A–Deficiency Symptom—Night Blindness

These photographs illustrate the eyes' slow recovery in response to a flash of bright light at night. In animal research studies, the response rate is measured with electrodes.

 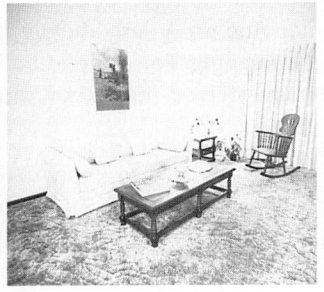  

© David Farr/Image Smythe (all)

In dim light, you can make out the details in this room. You are using your rods for vision.

A flash of bright light momentarily blinds you as the pigment in the rods is bleached.

You quickly recover and can see the details again in a few seconds.

With inadequate vitamin A, you do not recover but remain blinded for many seconds.

FIGURE 11-6 Vitamin A–Deficiency Symptom—The Rough Skin of Keratinization

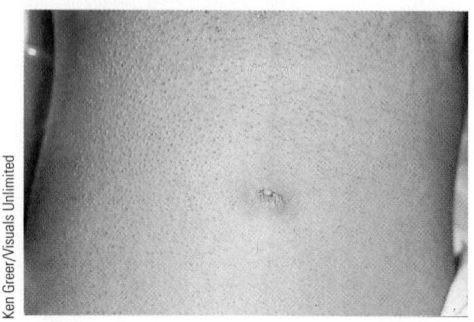

In vitamin A deficiency, the epithelial cells secrete the protein keratin in a process known as *keratinization.* (Keratinization doesn't occur in the GI tract, but mucus-producing cells dwindle, and mucus production declines.) The progression of this condition to the extreme is *hyperkeratinization* or *hyperkeratosis.* When keratin accumulates around each hair follicle, the condition is known as *follicular hyperkeratosis.*

■ Multivitamin supplements typically provide:
  • 750 µg (2500 IU).
  • 1500 µg (5000 IU).
For perspective, the RDA for vitamin A is 700 µg for women and 900 µg for men.

■ For perspective, 10,000 IU ≈ 3000 µg vitamin A, roughly four times the RDA for women.

xerophthalmia (zer-off-THAL-mee-uh): progressive blindness caused by severe vitamin A deficiency.
• **xero** = dry
• **ophthalm** = eye

xerosis (zee-ROW-sis): abnormal drying of the skin and mucous membranes; a sign of vitamin A deficiency.

keratomalacia (KARE-ah-toe-ma-LAY-shuh): softening of the cornea that leads to irreversible blindness; seen in severe vitamin A deficiency.

keratin (KARE-uh-tin): a water-insoluble protein; the normal protein of hair and nails. Keratin-producing cells may replace mucus-producing cells in vitamin A deficiency.

keratinization: accumulation of keratin in a tissue; a sign of vitamin A deficiency.

preformed vitamin A: dietary vitamin A in its active form.

teratogenic (ter-AT-oh-jen-ik): causing abnormal fetal development and birth defects.
• **terato** = monster
• **genic** = to produce

Severe vitamin A deficiency is the major cause of childhood blindness in the world, causing more than half a million preschool children to lose their sight each year.[12] Blindness due to vitamin A deficiency, known as **xerophthalmia**, develops in stages. First, the cornea becomes dry and hard, a condition known as **xerosis.** Corneal xerosis can quickly progress to **keratomalacia,** the softening of the cornea that leads to irreversible blindness.

Keratinization Elsewhere in the body, vitamin A deficiency affects other surfaces. Without vitamin A, the goblet cells in the GI tract diminish in number and activity, limiting the secretion of mucus. With less mucus, normal digestion and absorption of nutrients falter, and this, in turn, worsens malnutrition by limiting the absorption of whatever nutrients the diet may deliver. Similar changes in the cells of other epithelial tissues weaken defenses, making infections of the respiratory tract, the GI tract, the urinary tract, the vagina, and possibly the inner ear likely. On the body's outer surface, the epithelial cells change shape and begin to secrete the protein **keratin**—the hard, inflexible protein of hair and nails. As Figure 11-6 shows, the skin becomes dry, rough, and scaly as lumps of keratin accumulate **(keratinization).**

Vitamin A Toxicity

Just as a deficiency of vitamin A affects all body systems, so does a toxicity. Symptoms of toxicity begin to develop when all the binding proteins are swamped, and free vitamin A damages the cells. Such effects are unlikely when a person depends on a balanced diet for nutrients, but with concentrated amounts of **preformed vitamin A** from foods derived from animals, fortified foods, or supplements, toxicity is a real possibility. Children are most vulnerable to toxicity because they need less and are more sensitive to overdoses. An Upper Level has been set for preformed vitamin A (see inside front cover).

Beta-carotene, which is found in a wide variety of fruits and vegetables, is not converted efficiently enough in the body to cause vitamin A toxicity; instead, it is stored in the fat just under the skin. Overconsumption of beta-carotene from foods may turn the skin yellow, but this is not harmful (see Figure 11-7).[13] In contrast, overconsumption of beta-carotene from supplements may be quite harmful. In excess, this antioxidant may act as a prooxidant, promoting cell division and destroying vitamin A.[14] Furthermore, the adverse effects of beta-carotene supplements are most evident in people who drink alcohol and smoke cigarettes.[15]

Bone Defects Excessive amounts of vitamin A over the years may weaken the bones and contribute to osteoporosis.[16] People consuming large amounts of vitamin A either from supplements or from foods containing retinol have a significantly greater risk of hip fractures.[17] Such findings suggest that most people should not take vitamin A supplements. Even multivitamin supplements■ provide more vitamin A than most people need.[18]

Birth Defects Excessive vitamin A poses a **teratogenic** risk. High intakes (10,000 IU■ of supplemental vitamin A daily) before the seventh week of pregnancy appear to be the most damaging. For this reason, vitamin A is not given as a supplement in the first trimester of pregnancy unless there is specific evidence of deficiency, which is rare.

Not for Acne Adolescents need to know that massive doses of vitamin A have no beneficial effect on **acne.** The prescription medicine Accutane is made from vitamin A but is chemically different. Taken orally, Accutane is effective against the deep lesions of cystic acne. It is highly toxic, however, especially during growth, and has caused birth defects in infants when women have taken it during their pregnancies. For this reason, women taking Accutane must begin using an effective form of contraception at least one month before taking the drug and continue using contraception at least one month after discontinuing its use.

Another vitamin A relative, Retin-A, fights acne, the wrinkles of aging, and other skin disorders. Applied topically, this ointment smooths and softens skin; it also lightens skin that has become darkly pigmented after inflammation. During treatment, the skin becomes red and tender and peels.

Vitamin A Recommendations

Because the body can derive vitamin A from various retinoids and carotenoids, its contents in foods and its recommendations are expressed as **retinol activity equivalents (RAE)**. A microgram of retinol counts as 1 RAE,■ as does 12 micrograms of dietary beta-carotene. Most food and supplement labels report their vitamin A contents using international units (IU),■ an old measure of vitamin activity used before direct chemical analysis was possible.

Vitamin A in Foods

The richest sources of the retinoids are foods derived from animals—liver, fish liver oils, milk and milk products, butter, and eggs. Since vitamin A is fat soluble, it is lost when milk is skimmed. To compensate, reduced-fat and fat-free milks are often fortified so as to supply 6 to 10 percent of the Daily Value per cup.* Margarine is usually fortified so as to provide the same amount of vitamin A as butter.

Plants contain no retinoids, but many vegetables and some fruits contain vitamin A precursors—the carotenoids, red and yellow pigments of plants. Only a few carotenoids have vitamin A activity; the carotenoid with the greatest vitamin A activity is beta-carotene.

The Colors of Vitamin A Foods The dark leafy greens (like spinach—not celery or cabbage) and the rich yellow or deep orange vegetables and fruits (such as winter squash, cantaloupe, carrots, and sweet potatoes—not corn or bananas) help people meet their vitamin A needs (see Figure 11-8 on p. 374). A diet including several servings of such carotene-rich sources helps to ensure a sufficient intake.

An attractive meal that includes foods of different colors most likely supplies vitamin A as well. Most foods with vitamin A activity are brightly colored—green, yellow, orange, and red. Any plant food with significant vitamin A activity must have some color, since beta-carotene is a rich, deep yellow, almost orange compound. The beta-carotene in dark green, leafy vegetables is abundant but masked by large amounts of the green pigment **chlorophyll.**

Bright color is not always a sign of vitamin A activity, however. Beets and corn, for example, derive their colors from the red and yellow **xanthophylls,** which have no vitamin A activity. As for white plant foods such as potatoes, cauliflower, pasta, and rice, they also offer little or no vitamin A.

Vitamin A–Poor Fast Foods Fast foods often lack vitamin A. Anyone who dines frequently on hamburgers, french fries, and colas would be wise to emphasize colorful vegetables and fruits at other meals.

Vitamin A–Rich Liver People sometimes wonder if eating liver too frequently can cause vitamin A toxicity. Liver is a rich source because vitamin A is stored there in animals, just as in humans.† Arctic explorers who have eaten large quantities of polar bear liver have become ill with symptoms suggesting vitamin A toxicity, as have young children who regularly ate a chicken liver spread that provided three times their daily recommended intake. Liver offers many nutrients, and eating it periodically may improve a person's nutrition status, but caution is warranted not to eat too

FIGURE 11-7 | Symptom of Beta-Carotene Excess—Discoloration of the Skin

The hand on the right shows the skin discoloration that occurs when blood levels of beta-carotene rise in response to a low-kcalorie diet that features carrots, pumpkins, and orange juice. (The hand on the left belongs to someone else and is shown for comparison.)

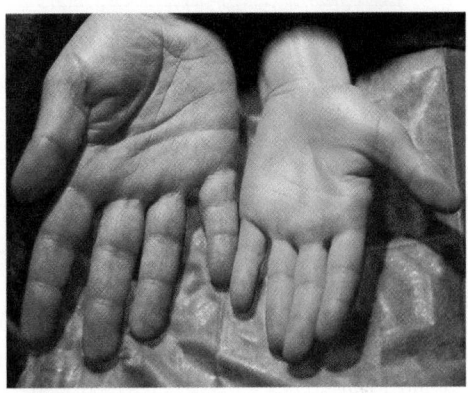

© 2002 Massachusetts Medical Society

■ 1 µg RAE = 1 µg retinol.
  = 2 µg beta-carotene (supplement).
  = 12 µg beta-carotene (dietary).
  = 24 µg of other vitamin A precursor carotenoids.

■ 1 IU retinol = 0.3 µg retinol or 0.3 µg RAE.
  1 IU beta-carotene (supplement) = 0.5 IU retinol or 0.15 µg RAE.
  1 IU beta-carotene (dietary) = 0.165 IU retinol or 0.05 µg RAE.
  1 IU other vitamin A precursor carotenoids = 0.025 µg RAE.

acne: a chronic inflammation of the skin's follicles and oil-producing glands, which leads to an accumulation of oils inside the ducts that surround hairs; usually associated with the maturation of young adults.

retinol activity equivalents (RAE): a measure of vitamin A activity; the amount of retinol that the body will derive from a food containing preformed retinol or its precursor beta-carotene.

chlorophyll (KLO-row-fil): the green pigment of plants, which absorbs light and transfers the energy to other molecules, thereby initiating photosynthesis.

xanthophylls (ZAN-tho-fills): pigments found in plants; responsible for the color changes seen in autumn leaves.

FIGURE 11-8   Vitamin A in Selected Foods

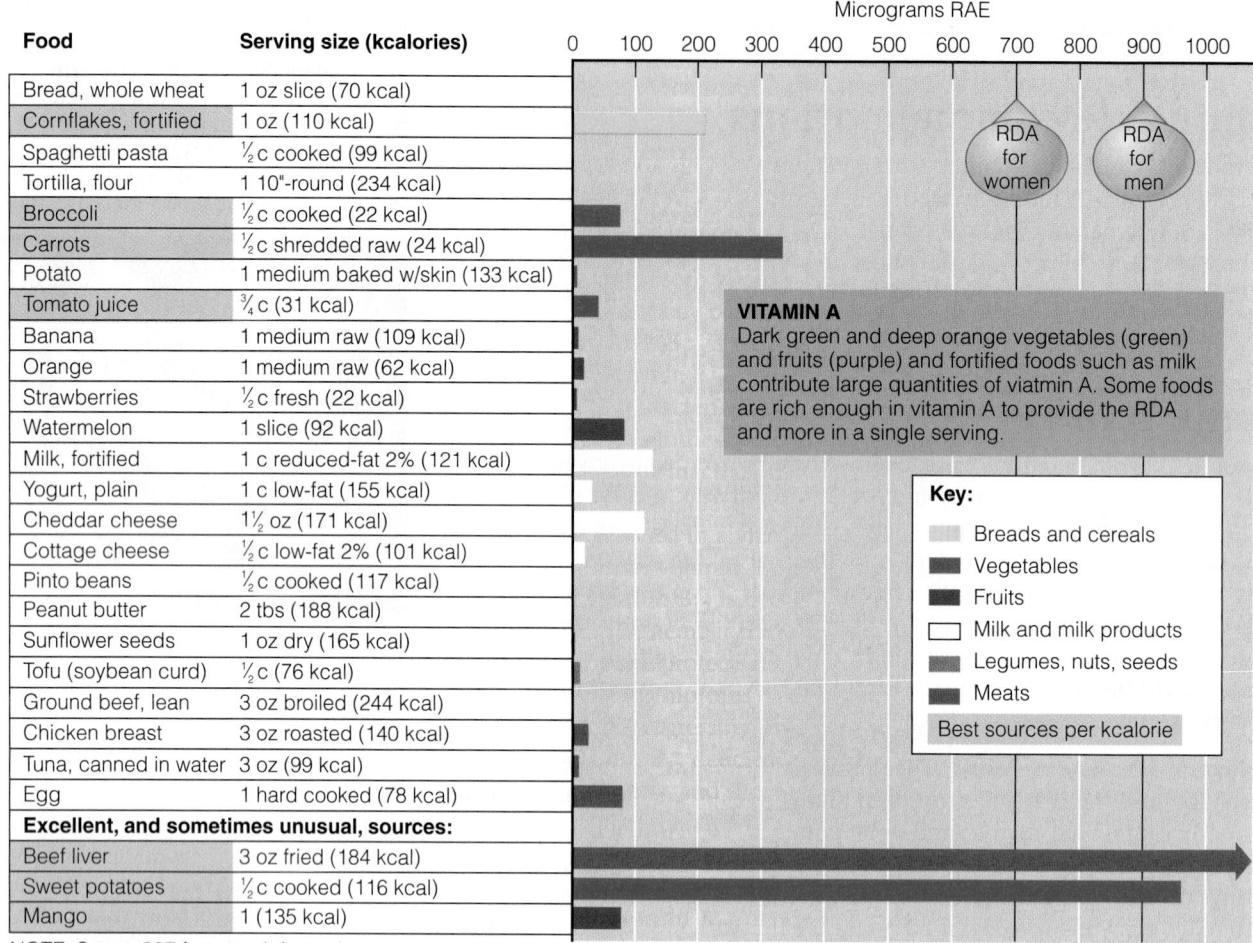

Food	Serving size (kcalories)
Bread, whole wheat	1 oz slice (70 kcal)
Cornflakes, fortified	1 oz (110 kcal)
Spaghetti pasta	½ c cooked (99 kcal)
Tortilla, flour	1 10"-round (234 kcal)
Broccoli	½ c cooked (22 kcal)
Carrots	½ c shredded raw (24 kcal)
Potato	1 medium baked w/skin (133 kcal)
Tomato juice	¾ c (31 kcal)
Banana	1 medium raw (109 kcal)
Orange	1 medium raw (62 kcal)
Strawberries	½ c fresh (22 kcal)
Watermelon	1 slice (92 kcal)
Milk, fortified	1 c reduced-fat 2% (121 kcal)
Yogurt, plain	1 c low-fat (155 kcal)
Cheddar cheese	1½ oz (171 kcal)
Cottage cheese	½ c low-fat 2% (101 kcal)
Pinto beans	½ c cooked (117 kcal)
Peanut butter	2 tbs (188 kcal)
Sunflower seeds	1 oz dry (165 kcal)
Tofu (soybean curd)	½ c (76 kcal)
Ground beef, lean	3 oz broiled (244 kcal)
Chicken breast	3 oz roasted (140 kcal)
Tuna, canned in water	3 oz (99 kcal)
Egg	1 hard cooked (78 kcal)
Excellent, and sometimes unusual, sources:	
Beef liver	3 oz fried (184 kcal)
Sweet potatoes	½ c cooked (116 kcal)
Mango	1 (135 kcal)

Micrograms RAE

RDA for women

RDA for men

VITAMIN A
Dark green and deep orange vegetables (green) and fruits (purple) and fortified foods such as milk contribute large quantities of viatmin A. Some foods are rich enough in vitamin A to provide the RDA and more in a single serving.

Key:
- Breads and cereals
- Vegetables
- Fruits
- Milk and milk products
- Legumes, nuts, seeds
- Meats
- Best sources per kcalorie

NOTE: See p. 327 for more information on using this figure.

much too often, especially for pregnant women. With one ounce of beef liver providing more than three times the RDA for vitamin A, intakes can rise quickly.

IN SUMMARY   Vitamin A is found in the body in three forms: retinol, retinal, and retinoic acid. Together, they are essential to vision, healthy epithelial tissues, and growth. Vitamin A deficiency is a major health problem worldwide, leading to infections, blindness, and keratinization. Toxicity can also cause problems and is most often associated with supplement abuse. Animal-derived foods such as liver and milk provide retinoids, whereas brightly colored plant-derived foods such as spinach, carrots, and pumpkins provide beta-carotene and other carotenoids. In addition to serving as a precursor for vitamin A, beta-carotene may act as an antioxidant in the body. The accompanying table summarizes vitamin A's functions in the body, deficiency symptoms, toxicity symptoms, and food sources.

Vitamin A

Other Names

Retinol, retinal, retinoic acid; precursors are carotenoids such as beta-carotene

2001 RDA

Men: 900 µg RAE/day

Women: 700 µg RAE/day

(continued)

Vitamin A (continued)

Upper Level

Adults: 3000 µg/day

Chief Functions in the Body

Vision; maintenance of cornea, epithelial cells, mucous membranes, skin; bone and tooth growth; reproduction; immunity

Significant Sources

Retinol: fortified milk, cheese, cream, butter, fortified margarine, eggs, liver

Beta-carotene: spinach and other dark leafy greens; broccoli, deep orange fruits (apricots, cantaloupe) and vegetables (squash, carrots, sweet potatoes, pumpkin)

Deficiency Disease

Hypovitaminosis A

Deficiency Symptoms

Night blindness, corneal drying (xerosis), triangular gray spots on eye (Bitot's spots), softening of the cornea (keratomalacia), and corneal degeneration and blindness (xerophthalmia); impaired immunity (infectious diseases); plugging of hair follicles with keratin, forming white lumps (hyperkeratosis)

Toxicity Disease

Hypervitaminosis A[a]

Chronic Toxicity Symptoms

Increased activity of osteoclasts[b] causing reduced bone density; liver abnormalities; birth defects

Acute Toxicity Symptoms

Blurred vision, nausea, vomiting, vertigo; increase of pressure inside skull, mimicking brain tumor; headaches

The carotenoids in foods bring colors to meals; the retinoids in our eyes allow us to see them.

© Polara Studios Inc.

[a] A related condition, *hypercarotenemia*, is caused by the accumulation of too much of the vitamin A precursor beta-carotene in the blood, which turns the skin noticeably yellow. Hypercarotenemia is not, strictly speaking, a toxicity symptom.
[b] *Osteoclasts* are the cells that destroy bone during its growth. Those that build bone are *osteoblasts*.

Vitamin D

Vitamin D (calciferol)■ is different from all the other nutrients in that the body can synthesize it, with the help of sunlight, from a precursor that the body makes from cholesterol. Therefore, vitamin D is not an essential nutrient: given enough time in the sun, people need no vitamin D from foods.[19]

Figure 11-9 (on p. 376) diagrams the pathway for making and activating vitamin D. Ultraviolet rays from the sun hit the precursor in the skin and convert it to previtamin D_3. This compound works its way into the body and slowly, over the next 36 hours, is converted to its active form with the help of the body's heat. The biological activity of the active vitamin is 500- to 1000-fold greater than that of its precursor.

Regardless of whether the body manufactures vitamin D_3 or obtains it directly from foods, two hydroxylation reactions must occur before the vitamin becomes fully active. First, the liver adds an OH group, and then the kidneys add another OH group to produce the active vitamin. A review of Figure 11-9 reveals how diseases affecting either the liver or the kidneys can interfere with the activation of vitamin D and produce symptoms of deficiency.

■ Vitamin D comes in many forms, the two most important being a plant version called **vitamin D_2** or **ergocalciferol** (ER-go-kal-SIF-er-ol) and an animal version called **vitamin D_3** or **cholecalciferol** (KO-lee-kal-SIF-er-ol).

Roles in the Body

Though called a vitamin, vitamin D is actually a hormone—a compound manufactured by one part of the body that causes another part to respond. Like vitamin A, vitamin D has a binding protein that carries it to the target organs—most notably, the intestines, the kidneys, and the bones. All respond to vitamin D by making the bone minerals available.

Vitamin D in Bone Growth Vitamin D is a member of a large and cooperative bone-making and maintenance team composed of nutrients and other compounds, including vitamins A, C, and K; the hormones parathormone and calcitonin; the protein collagen; and the minerals calcium, phosphorus, magnesium,

FIGURE 11-9   Vitamin D Synthesis and Activation

The precursor of vitamin D is made in the liver from cholesterol (see Figure 5-11 on p. 150 and Appendix C). The activation of vitamin D is a closely regulated process. The final product, active vitamin D, is also known as 1,25-dihydroxycholecalciferol (or calcitriol).

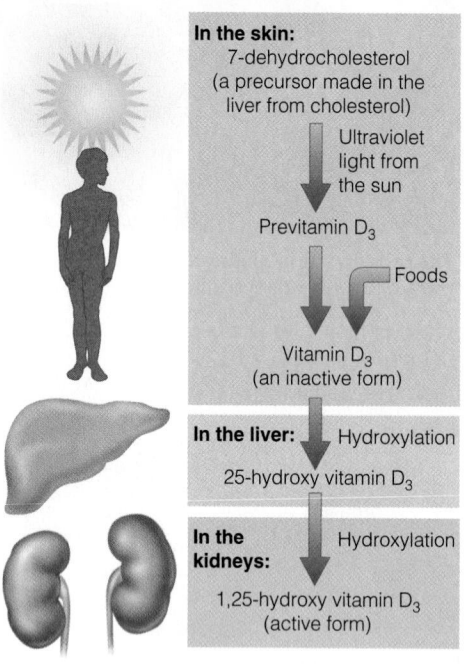

In the skin:
7-dehydrocholesterol (a precursor made in the liver from cholesterol)

↓ Ultraviolet light from the sun

Previtamin D_3

↓ Foods

Vitamin D_3 (an inactive form)

In the liver: Hydroxylation

25-hydroxy vitamin D_3

In the kidneys: Hydroxylation

1,25-hydroxy vitamin D_3 (active form)

■ Because the poorly formed rib attachments resemble rosary beads, this symptom is commonly known as **rachitic** (ra-KIT-it) **rosary** ("the rosary of rickets").

rickets: the vitamin D–deficiency disease in children characterized by inadequate mineralization of bone (manifested in bowed legs or knock-knees, outward-bowed chest, and knobs on ribs). A rare type of rickets, not caused by vitamin D deficiency, is known as *vitamin D–refractory rickets.*

osteomalacia (OS-tee-oh-ma-LAY-shuh): a bone disease characterized by softening of the bones. Symptoms include bending of the spine and bowing of the legs. The disease occurs most often in adult women.
- **osteo** = bone
- **malacia** = softening

and fluoride. Vitamin D's special role in bone growth is to maintain blood concentrations of calcium and phosphorus. The bones grow denser and stronger as they absorb and deposit these minerals.

Vitamin D raises blood concentrations of these minerals in three ways. It enhances their absorption from the GI tract, their reabsorption by the kidneys, and their mobilization from the bones into the blood. The vitamin may work alone, as it does in the GI tract, or in combination with parathormone, as it does in the bones and kidneys. Vitamin D is the director, but the star of the show is calcium. Details of calcium balance appear in Chapter 12.

Vitamin D in Other Roles Scientists have discovered many other vitamin D target tissues, including cells of the immune system, brain and nervous system, pancreas, skin, muscles and cartilage, and reproductive organs. Because vitamin D has numerous functions, it may be valuable in treating a number of disorders. Recent evidence suggests that vitamin D may protect against multiple sclerosis.[20]

Vitamin D Deficiency

Factors that contribute to vitamin D deficiency include dark skin, breastfeeding without supplementation, lack of sunlight, and use of nonfortified milk.[21] In vitamin D deficiency, production of the protein that binds calcium in the intestinal cells slows. Thus, even when calcium in the diet is adequate, it passes through the GI tract unabsorbed, leaving the bones undersupplied. Consequently, a vitamin D deficiency creates a calcium deficiency. Adolescents may not reach their peak bone mass.[22]

Rickets Worldwide, the vitamin D–deficiency disease **rickets** still afflicts many children.[23] The bones fail to calcify normally, causing growth retardation and skeletal abnormalities. The bones become so weak that they bend when they have to support the body's weight (see Figure 11-10). A child with rickets who is old enough to walk characteristically develops bowed legs, often the most obvious sign of the disease. Another sign is the beaded ribs■ that result from the poorly formed attachments of the bones to the cartilage.

Osteomalacia The adult form of rickets, **osteomalacia,** occurs most often in women who have low calcium intakes and little exposure to sun and who go through repeated pregnancies and periods of lactation. Given this combination of risk factors, the leg bones may soften to such an extent that a young woman who is tall and straight at 20 may become bent, bowlegged, and stooped before she is 30.

Osteoporosis Any failure to synthesize adequate vitamin D or obtain enough from foods sets the stage for a loss of calcium from the bones, which can result in fractures. In a group of women with osteoporosis hospitalized for hip fractures, half had an undetected vitamin D deficiency.[24] Highlight 12 describes the many factors that lead to osteoporosis, a condition of reduced bone density.

The Elderly Vitamin D deficiency is especially likely in older adults for several reasons. For one, the skin, liver, and kidneys lose their capacity to make and activate vitamin D with advancing age. For another, older adults typically drink little or no milk—the main dietary source of vitamin D. And finally, older adults typically spend much of the day indoors, and when they do venture outside, many of them cautiously wear protective clothing or apply sunscreen to all sun-exposed areas of their skin. Dark-skinned people living in northern regions are particularly vulnerable.[25] All of these factors increase the likelihood of vitamin D deficiency and its consequences: bone losses and fractures.

Vitamin D Toxicity

Vitamin D clearly illustrates how nutrients in optimal amounts support health, but both inadequacies and excesses cause trouble. Vitamin D is the most likely of the vitamins to have toxic effects when consumed in excessive amounts. The

FIGURE 11-10 | Vitamin D–Deficiency Symptoms—Bowed Legs and Beaded Ribs of Rickets

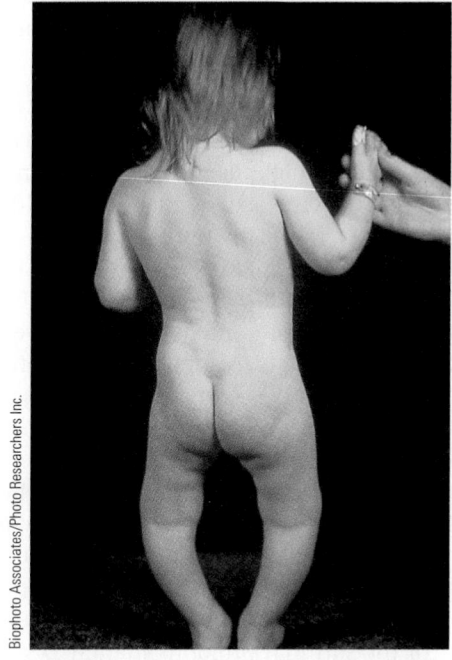

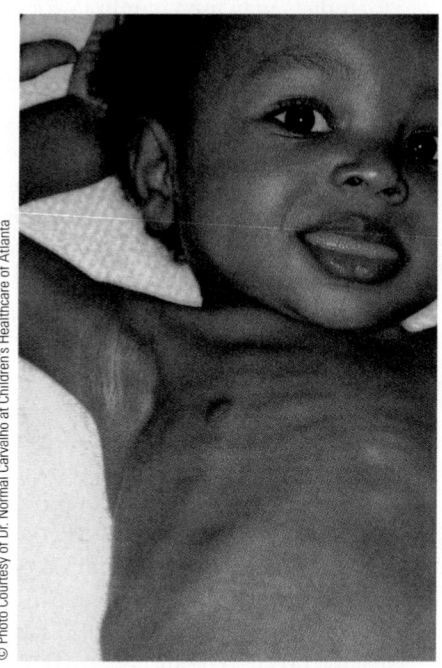

Bowed legs. In rickets, the poorly formed long bones of the legs bend outward as weight-bearing activities such as walking begin.

Beaded ribs. In rickets, a series of "beads" develop where the cartilages and bones attach.

amounts of vitamin D made by the skin and found in foods are well within the safe limits set by the Upper Level, but supplements containing the vitamin in concentrated form should be kept out of the reach of children and used cautiously, if at all, by adults.

An excess of vitamin D raises the concentration of blood calcium.■ Excess blood calcium tends to precipitate in the soft tissue, forming stones, especially in the kidneys where calcium is concentrated in the effort to excrete it. Calcification may also harden the blood vessels and is especially dangerous in the major arteries of the heart and lungs, where it can cause death.

■ High blood calcium is known as **hypercalcemia** and may develop from a variety of disorders, including vitamin D toxicity. It does *not* develop from a high calcium intake.

Vitamin D Recommendations and Sources

Only a few foods contain vitamin D naturally. Fortunately, the body can make all the vitamin D it needs with the help of a little sunshine. In setting dietary recommendations, however, the DRI Committee assumed that no vitamin D was available from this source.

Vitamin D in Foods Most adults, especially in sunny regions, need not make special efforts to obtain vitamin D from food. People who are not outdoors much or who live in northern or predominantly cloudy or smoggy areas are advised to drink at least 2 cups of vitamin D–fortified milk a day. The fortification of milk with vitamin D is the best guarantee that people will meet their needs and underscores the importance of milk in a well-balanced diet.* For those who use margarine in place of butter, fortified margarine is also a significant source. A plant

A cool glass of milk refreshes as it replenishes vitamin D and other bone-building nutrients.

*Vitamin D fortification of milk in the United States is 10 micrograms cholecalciferol (400 IU) per quart; in Canada, 9.6 micrograms (385 IU) per liter.

version of vitamin D may yield an active compound on irradiation, but its contribution is minor. Without adequate sunshine, fortification, or supplementation, a vegan diet cannot meet vitamin D needs. Vegetarians who do not include milk in their diets may use vitamin D–fortified soy milk. Importantly, feeding infants and young children nonfortified "health beverages" instead of milk or infant formula can create severe nutrient deficiencies, including rickets.[26]

Vitamin D from the Sun Most of the world's population relies on natural exposure to sunlight to maintain adequate vitamin D nutrition. The sun imposes no risk of vitamin D toxicity; prolonged exposure to sunlight degrades the vitamin D precursor in the skin, preventing its conversion to the active vitamin. Even lifeguards on southern beaches are safe from vitamin D toxicity from the sun.

Prolonged exposure to sunlight does, however, prematurely wrinkle the skin and present the risk of skin cancer. Sunscreens help reduce these risks, but unfortunately, sunscreens with sun protection factors (SPF) of 8 and higher also prevent vitamin D synthesis. A strategy to avoid this dilemma is to apply sunscreen after enough time has elapsed to provide sufficient vitamin D synthesis. For most people, exposing hands, face, and arms on a clear summer day for 10 to 15 minutes a few times a week should be sufficient to maintain vitamin D nutrition.

The pigments of dark skin provide some protection from the sun's damage, but they also reduce vitamin D synthesis. Dark-skinned people require longer sunlight exposure than light-skinned people: heavily pigmented skin achieves the same amount of vitamin D synthesis in three hours as fair skin in 30 minutes. Latitude, season, and time of day■ also have dramatic effects on vitamin D synthesis (see Figure 11-11). The ultraviolet (UV) rays of the sun that promote vitamin D synthesis are blocked by heavy clouds, smoke, or smog. Differences in skin pigmentation, latitude, and smog may account for the finding that African American people, especially those in northern, smoggy cities, are most likely to develop rickets.[27] For these people, and for those who are unable to go outdoors frequently, dietary vitamin D is essential. Vitamin D stores from summer synthesis alone are insufficient to meet winter needs.[28]

■ Factors that may limit sun exposure and, therefore, vitamin D synthesis:
  • Geographic location.
  • Season of the year.
  • Time of day.
  • Air pollution.
  • Clothing.
  • Tall buildings.
  • Indoor living.
  • Sunscreens.

The sunshine vitamin: vitamin D.

FIGURE 11-11 Vitamin D Synthesis and Latitude

Above 40° north latitude (and below 40° south latitude in the southern hemisphere), vitamin D synthesis essentially ceases for the four months of winter. Synthesis increases as spring approaches, peaks in summer, and declines again in the fall. People living in regions of extreme northern (or extreme southern) latitudes may miss as much as six months of vitamin D production.

Depending on the radiation used, the UV rays from tanning lamps and tanning booths may also stimulate vitamin D synthesis, but the hazards outweigh any possible benefits.* The Food and Drug Administration (FDA) warns that if the lamps are not properly filtered, people using tanning booths risk burns, damage to the eyes and blood vessels, and skin cancer.

IN SUMMARY Vitamin D can be synthesized in the body with the help of sunlight or obtained from foods derived from animals. It sends signals to three primary target sites: the GI tract to absorb more calcium and phosphorus, the bones to release more, and the kidneys to retain more. These actions maintain blood calcium concentrations and support bone formation. A deficiency causes rickets in childhood and osteomalacia in later life. Fortified milk is an important food source. The accompanying table summarizes vitamin D facts.

Vitamin D

Other Names

Calciferol (kal-SIF-er-ol), 1,25-dihydroxy vitamin D (calcitriol); the animal version is vitamin D_3 or cholecalciferol; the plant version is vitamin D_2 or ergocalciferol; precursor is the body's own cholesterol

1997 Adequate Intake (AI)■

Adults: 5 µg/day (19–50 yr)

10 µg/day (51 –70 yr)

15 µg/day (>70 yr)

Upper Level

Adults: 50 µg/day

Chief Functions in the Body

Mineralization of bones (raises blood calcium and phosphorus by increasing absorption from digestive tract, withdrawing calcium from bones, stimulating retention by kidneys)

Significant Sources

Synthesized in the body with the help of sunlight; fortified milk, margarine, butter, cereals, and chocolate mixes; veal, beef, egg yolks, liver, fatty fish (herring, salmon, sardines) and their oils

Deficiency Diseases

Rickets, osteomalacia

Deficiency Symptoms

Rickets in Children

Inadequate calcification, resulting in misshapen bones (bowing of legs); enlargement of ends of long bones (knees, wrists); deformities of ribs (bowed, with beads or knobs);[a] delayed closing of fontanel, resulting in rapid enlargement of head (see figure below); lax muscles resulting in protrusion of abdomen; muscle spasms

Osteomalacia in Adults

Loss of calcium, resulting in soft, flexible, brittle, and deformed bones; progressive weakness; pain in pelvis, lower back, and legs

Toxicity Disease

Hypervitaminosis D

Toxicity Symptoms

Elevated blood calcium; calcification of soft tissues (blood vessels, kidneys, heart, lungs, tisues around joints), frequent urination

■ Vitamin D activity was previously expressed in international units (IU), but is now expressed in micrograms of cholecalciferol. To convert, use the following factor:
  1 IU = 0.025 µg cholecalciferol.

For example:
• 100 IU = 2.5 µg (100 IU × 0.025 µg).
• 400 (IU) = 10 µg (400 IU × 0.025 µg).

Fontanel
A fontanel is an open space in the top of a baby's skull before the bones have grown together. In rickets, closing of the fontanel is delayed.

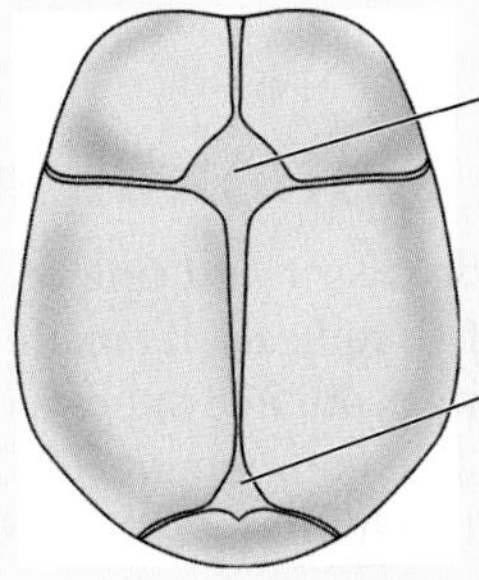

Anterior fontanel normally closes by the end of the second year

Posterior fontanel normally closes by the end of the first year

[a]Bowing of the ribs causes the symptoms known as *pigeon breast*. The beads that form on the ribs resemble rosary beads; thus this symptom is known as *rachitic* (ra-KIT-ik) *rosary* ("the rosary of rickets").

*The best wavelengths for vitamin D synthesis are UV-B rays between 290 and 310 nanometers. Some tanning parlors advertise "UV-A rays only, for a tan without the burn," but in fact, UV-A rays can damage the skin.

FIGURE 11-12 Free-Radical Formation and Antioxidant Protection

Free-radical formation and damage

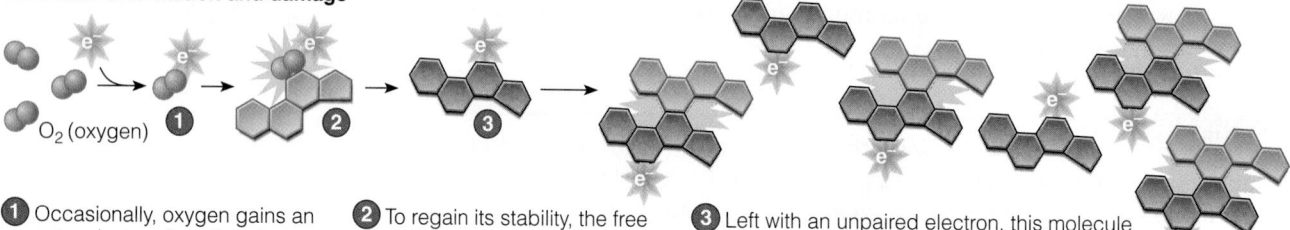

O₂ (oxygen)

① Occasionally, oxygen gains an extra electron from the electron transport chain, thereby generating a free radical.

② To regain its stability, the free radical attacks a nearby molecule (such as a lipid or protein) and steals an electron.

③ Left with an unpaired electron, this molecule becomes a free radical itself and attacks another nearby molecule. The chain reaction continues, causing widespread damage.

Antioxidant protection

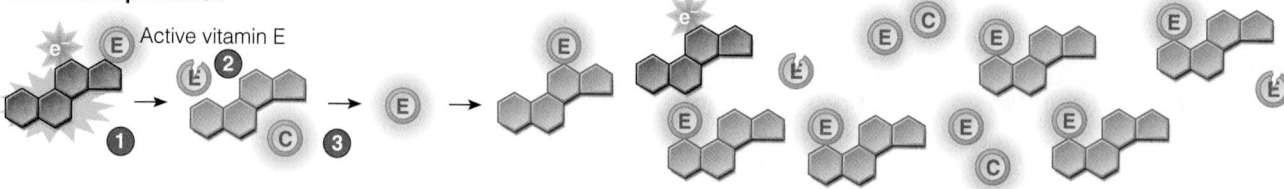

Active vitamin E

① Antioxidants, such as vitamin E, neutralize free radicals by donating one of their own electrons.

② The destructive chain reaction is stopped, but vitamin E is no longer active.

③ Like vitamin E, vitamin C acts as an antioxidant; it also restores vitamin E to its active form. An abundance of dietary antioxidants minimizes free-radical damage.

Nutrition Explorer Watch how free radicals damage macromolecules such as proteins, lipids, and DNA in a self-perpetuating chain reaction. Observe how antioxidants stabilize these highly reactive compounds and interrupt this destructive cycle.

Vitamin E

Researchers discovered a component of vegetable oils necessary for reproduction in rats and named this antisterility factor **tocopherol,** which means "to bring forth offspring." When chemists isolated four different tocopherol compounds, they designated them by the first four letters of the Greek alphabet: alpha, beta, gamma, and delta. The tocopherols consist of a complex ring structure and a long saturated side chain (Appendix C provides the chemical structures). The positions of methyl groups (CH₃) on the side chain and their chemical rotations distinguish one tocopherol from another. **Alpha-tocopherol** is the only one with vitamin E activity in the human body.[29] The other tocopherols are not readily converted to alpha-tocopherol in the body, nor do they perform the same roles. Whether these other tocopherols might be beneficial in other ways is the subject of current research.[30]

Vitamin E as an Antioxidant

Vitamin E is a fat-soluble antioxidant and one of the body's primary defenders against the adverse effects of free radicals. Its main action is to stop the chain reaction of free radicals producing more free radicals (see Figure 11-12). In doing so, vitamin E protects the vulnerable components of the cells and their membranes from destruction. Most notably, vitamin E prevents the oxidation of the polyunsaturated fatty acids, but it protects other lipids and related compounds (for example, vitamin A) as well.

Accumulating evidence suggests that vitamin E may reduce the risk of heart disease by protecting low-density lipoproteins (LDL) against oxidation. The oxidation of LDL has been implicated as a key factor in the development of heart disease. Highlight 11 provides many more details on how vitamin E and other antioxidants protect against chronic diseases, such as heart disease and cancer.

tocopherol (tuh-KOFF-er-ol): a general term for several chemically related compounds, one of which has vitamin E activity (see Appendix C for chemical structures).

alpha-tocopherol: the active vitamin E compound.

While research continues to reveal possible roles for vitamin E, it also has clearly discredited claims that vitamin E improves physical performance, enhances sexual performance, or cures sexual dysfunction in males. Vitamin E does not slow or prevent the processes of aging such as hair turning gray or skin wrinkling. Nor does it slow the progression of Parkinson's disease.

Fat-soluble vitamin E is found predominantly in vegetable oils, seeds, and nuts.

Vitamin E Deficiency

In human beings, a primary deficiency of vitamin E (from a poor intake) is rare; deficiency is usually associated with diseases of fat malabsorption such as cystic fibrosis. Without vitamin E, the red blood cells break open and spill their contents, probably due to oxidation of the polyunsaturated fatty acids in their membranes. This classic sign of vitamin E deficiency, known as **erythrocyte hemolysis,** is seen in premature infants, born before the transfer of vitamin E from the mother to the infant that takes place in the last weeks of pregnancy. Vitamin E treatment corrects **hemolytic anemia.**

Prolonged vitamin E deficiency also causes neuromuscular dysfunction involving the spinal cord and retina of the eye. Common symptoms include loss of muscle coordination and reflexes and impaired vision and speech. Vitamin E treatment corrects these neurological symptoms of vitamin E deficiency, but it does *not* prevent or cure the hereditary **muscular dystrophy** that afflicts children. Children with this condition do not benefit from vitamin E treatment and usually die at an early age when their respiratory muscles deteriorate.

Two other conditions seem to respond to vitamin E treatment, although results are inconsistent. One is a nonmalignant breast disease **(fibrocystic breast disease),** and the other is an abnormality of blood flow that causes cramping in the legs **(intermittent claudication).**

Vitamin E Toxicity

Vitamin E supplement use has risen in recent years as its protective actions against chronic diseases have been recognized. Still, toxicity is rare, and its effects are not as detrimental as with vitamins A and D. The Upper Level for vitamin E (1000 milligrams) is more than 65 times greater than the recommended intake for adults (15 milligrams). Extremely high doses of vitamin E may interfere with the blood-clotting action of vitamin K and enhance the effects of drugs used to oppose blood clotting, causing hemorrhage.

Vitamin E Recommendations

The current RDA for vitamin E differs from previous recommendations in being based on the alpha-tocopherol form only. As mentioned earlier, the other tocopherols cannot be converted to alpha-tocopherol, nor can they perform the same metabolic roles in the body. A person who consumes large quantities of polyunsaturated fatty acids needs more vitamin E. Fortunately, vitamin E and polyunsaturated fatty acids tend to occur together in the same foods.

Vitamin E in Foods

Vitamin E is widespread in foods. Much of the vitamin E in the diet comes from vegetable oils and products made from them, such as margarine and salad dressings. Wheat germ oil is especially rich in vitamin E.

Vitamin E is readily destroyed by heat processing (such as deep-fat frying) and oxidation, so fresh or lightly processed foods are preferable sources. Most processed and convenience foods do not contribute enough vitamin E to ensure an adequate intake.

erythrocyte (eh-RITH-ro-cite) **hemolysis** (he-MOLL-uh-sis): the breaking open of red blood cells (erythrocytes); a symptom of vitamin E–deficiency disease in human beings.
• **erythro** = red
• **cyte** = cell
• **hemo** = blood
• **lysis** = breaking

hemolytic (HE-moh-LIT-ick) **anemia:** the condition of having too few red blood cells as a result of erythrocyte hemolysis.

muscular dystrophy (DIS-tro-fee): a hereditary disease in which the muscles gradually weaken. Its most debilitating effects arise in the lungs.

fibrocystic (FYE-bro-SIS-tik) **breast disease:** a harmless condition in which the breasts develop lumps, sometimes associated with caffeine consumption. In some, it responds to abstinence from caffeine; in others, it can be treated with vitamin E.
• **fibro** = fibrous tissue
• **cyst** = closed sac

intermittent claudication (klaw-dih-KAY-shun): severe calf pain caused by inadequate blood supply. It occurs when walking and subsides during rest.
• **intermittent** = at intervals
• **claudicare** = to limp

■ Appendix H accurately presents vitamin E data in milligrams of alpha-tocopherol.

Prior to 2000, values of the vitamin E in foods reflected all of the various tocopherols and were expressed in "milligrams of tocopherol equivalents."■ These measures overestimated the amount of alpha-tocopherol. To estimate the alpha-tocopherol content of foods stated in tocopherol equivalents, multiply by 0.8.[31]

IN SUMMARY

Vitamin E acts as an antioxidant, defending lipids and other components of the cells against oxidative damage. Deficiencies are rare, but do occur in premature infants, the primary symptom being erythrocyte hemolysis. Vitamin E is found predominantly in vegetable oils and appears to be one of the least toxic of the fat-soluble vitamins. The summary table reviews vitamin E's functions, deficiency symptoms, toxicity symptoms, and food sources.

Vitamin E

Other Names

Alpha-tocopherol

2000 RDA

Adults: 15 mg/day

Upper Level

Adults: 1000 mg/day

Chief Functions in the Body

Antioxidant (stabilization of cell membranes, regulation of oxidation reactions, protection of polyunsaturated fatty acids [PUFA] and vitamin A)

Significant Sources

Polyunsaturated plant oils (margarine, salad dressings, shortenings), leafy green vegetables, wheat germ, whole grains, liver, egg yolks, nuts, seeds

Easily destroyed by heat and oxygen

Deficiency Symptoms

Red blood cell breakage,[a] nerve damage

Toxicity Symptoms

Augments the effects of anticlotting medication

[a]The breaking of red blood cells is called *erythrocyte hemolysis*.

Vitamin K

■ K stands for the Danish word *koagulation* ("coagulation" or "clotting").

Like vitamin D, vitamin K can be obtained from a nonfood source. Bacteria in the GI tract synthesize vitamin K that the body can absorb. Vitamin K■ acts primarily in blood clotting, where its presence can make the difference between life and death. Blood has a remarkable ability to remain a liquid, but can turn solid within seconds when the integrity of that system is disturbed. (If blood did not clot, a single pinprick could drain the entire body of all its blood, just as a tiny hole in a bucket makes the bucket forever useless for holding water.)

Roles in the Body

More than a dozen different proteins and the mineral calcium are involved in making a blood clot. Vitamin K is essential for the activation of several of these proteins, among them prothrombin, made by the liver as a precursor of the protein thrombin (see Figure 11-13). When any of the blood-clotting factors is lacking, **hemorrhagic disease** results. If an artery or vein is cut or broken, bleeding goes unchecked. (Of course, this is not to say that hemorrhaging is always caused by vitamin K deficiency. Another cause is the hereditary disorder **hemophilia**, which is not curable with vitamin K.)

Vitamin K also participates in the synthesis of bone proteins. Without vitamin K, the bones produce an abnormal protein that cannot bind to the minerals that normally form bones; bone density is low.[32] An adequate intake of vitamin K helps to make the bone protein correctly and protect against hip fractures.[33]

hemorrhagic (hem-oh-RAJ-ik) **disease:** a disease characterized by excessive bleeding.

hemophilia (HE-moh-FEEL-ee-ah): a hereditary disease that is caused by a genetic defect and has no relation to vitamin K. The blood is unable to clot because it lacks the ability to synthesize certain clotting factors.

FIGURE 11-13 Blood-Clotting Process

When blood is exposed to air, foreign substances, or secretions from injured tissues, platelets (small, cell-like structures in the blood) release a phospholipid known as thromboplastin. Thromboplastin catalyzes the conversion of the inactive protein prothrombin to the active enzyme thrombin. Thrombin then catalyzes the conversion of the precursor protein fibrinogen to the active protein fibrin that forms the clot.

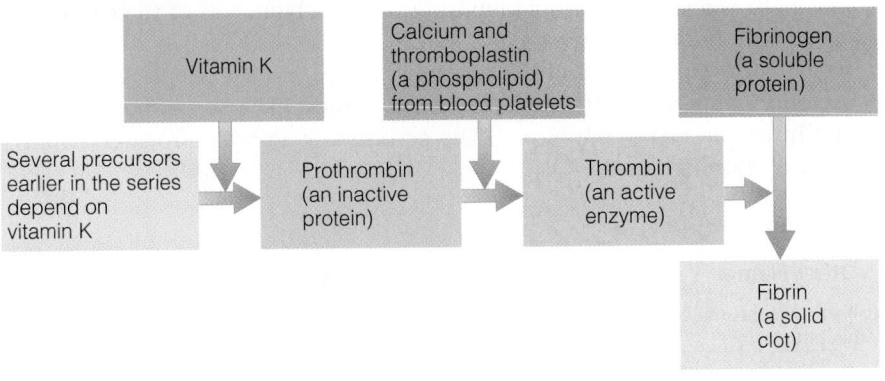

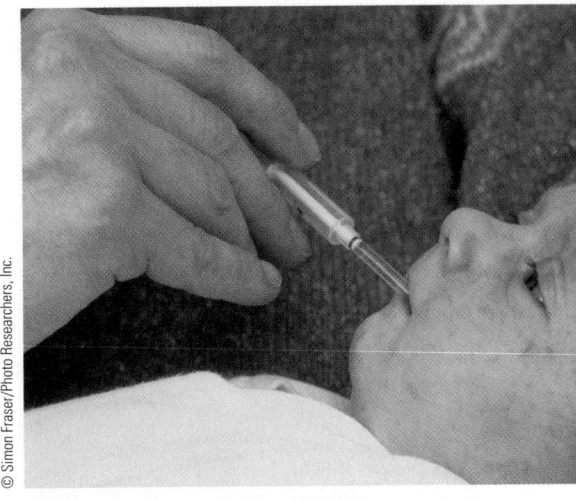

Soon after birth, newborn infants receive a dose of vitamin K to prevent hemorrhagic disease.

Vitamin K is historically known for its role in blood clotting, and more recently for its participation in bone building, but researchers continue to discover proteins needing vitamin K's assistance. These proteins have been identified in the plaques of atherosclerosis, the kidneys, and the nervous system.

Vitamin K Deficiency

A primary deficiency■ of vitamin K is rare, but a secondary deficiency may occur in two circumstances. First, whenever fat absorption falters, as occurs when bile production fails, vitamin K absorption diminishes. Second, some drugs disrupt vitamin K's synthesis and action in the body: antibiotics kill the vitamin K–producing bacteria in the intestine, and anticoagulant drugs interfere with vitamin K metabolism and activity. When vitamin K deficiency does occur, it can be fatal.

Newborn infants present a unique case of vitamin K nutrition because they are born with a **sterile** intestinal tract, and the vitamin K–producing bacteria take weeks to establish themselves. At the same time, plasma prothrombin concentrations are low (this reduces the likelihood of fatal blood clotting during the stress of birth). To prevent hemorrhagic disease in the newborn, a single dose of vitamin K■ (usually as the naturally occurring form, phylloquinone) is given at birth either orally or by intramuscular injection. Concerns that vitamin K given at birth raises the risks of childhood cancer are unproved and unlikely.

Vitamin K Toxicity

Toxicity is not common, and no adverse effects have been reported with high intakes of vitamin K. Therefore, an Upper Level has not been established. High doses of vitamin K can reduce the effectiveness of anticoagulant drugs used to prevent blood clotting. People taking these drugs should eat vitamin K–rich foods in moderation and keep their intakes consistent from day to day.

Vitamin K Recommendations and Sources

As mentioned earlier, vitamin K is made in the GI tract by the billions of bacteria that normally reside there. Once synthesized, vitamin K is absorbed and stored in the liver. This source provides only about half of a person's needs. Vitamin K–rich

■ Reminder: A *primary deficiency* develops in response to an inadequate dietary intake whereas a *secondary deficiency* occurs for other reasons.

■ The natural form of vitamin K is **phylloquinone** (FILL-oh-KWIN-own); the synthetic form is **menadione** (men-uh-DYE-own). See Appendix C for the chemistry of these structures.

sterile: free of microorganisms, such as bacteria.

Notable food sources of vitamin K include milk, eggs, brussels sprouts, collards, liver, cabbage, spinach, and broccoli.

foods such as liver, leafy green vegetables, and members of the cabbage family can easily supply the rest. Milk, meats, eggs, cereals, fruits, and vegetables provide smaller, but still significant, amounts.

IN SUMMARY Vitamin K helps with blood clotting, and its deficiency causes hemorrhagic disease (uncontrolled bleeding). Bacteria in the GI tract can make the vitamin; people typically receive about half of their requirements from bacterial synthesis and half from foods such as liver, leafy green vegetables, and members of the cabbage family. Because people depend on bacterial synthesis for vitamin K, deficiency is most likely in newborn infants and in people taking antibiotics. The accompanying table provides a summary of vitamin K facts.

Vitamin K

Other Names

Phylloquinone, menaquinone, menadione, naphthoquinone

2001 AI

Men: 120 µg/day

Women: 90 µg/day

Chief Functions in the Body

Synthesis of blood-clotting proteins and bone proteins

Significant Sources

Bacterial synthesis in the digestive tract;[a] liver; leafy green vegetables, cabbage-type vegetables; milk

Deficiency Symptoms

Hemorrhaging

Toxicity Symptoms

None known

[a]Vitamin K needs cannot be met from bacterial synthesis alone; however, it is a potentially important source in the small intestine, where absorption efficiency ranges from 40 to 70 percent.

The Fat-Soluble Vitamins—In Summary

The four fat-soluble vitamins play many specific roles in the growth and maintenance of the body. Their presence affects the health and function of the eyes, skin, GI tract, lungs, bones, teeth, nervous system, and blood; their deficiencies become apparent in these same areas. Toxicities of the fat-soluble vitamins are possible, especially when people use supplements, because the body stores excesses.

As with the water-soluble vitamins, the function of one fat-soluble vitamin often depends on the presence of another. Recall that vitamin E protects vitamin A from oxidation. In vitamin E deficiency, vitamin A absorption and storage are impaired. Three of the four fat-soluble vitamins—A, D, and K—play important roles in bone growth and remodeling. As mentioned, vitamin K helps synthesize a specific bone protein, and vitamin D regulates that synthesis. Vitamin A, in turn, may control which bone-building genes respond to vitamin D.

Fat-soluble vitamins also interact with minerals: vitamin D and calcium cooperate in bone formation; and zinc is required for the synthesis of vitamin A's transport protein, retinol-binding protein. Zinc also assists the enzyme that regenerates retinal from retinol in the eye.

The roles of the fat-soluble vitamins differ from those of the water-soluble vitamins, and they appear in different foods, yet they are just as essential to life. The need for them underlines the importance of eating a wide variety of nourishing foods daily. The following table condenses the information on fat-soluble vitamins into a short summary.

IN SUMMARY   The Fat-Soluble Vitamins

Vitamin and Chief Functions	Deficiency Symptoms	Toxicity Symptoms	Significant Sources
Vitamin A Vision; maintenance of cornea, epithelial cells, mucous membranes, skin; bone and tooth growth; reproduction; immunity	Infectious diseases, night blindness, blindness (xerophthalmia), keratinization	Reduced bone mineral density, liver abnormalities, birth defects	Retinol: milk and milk products Beta-carotene: dark green leafy and deep yellow/orange vegetables
Vitamin D Mineralization of bones (raises blood calcium and phosphorus by increasing absorption from digestive tract, withdrawing calcium from bones, stimulating retention by kidneys)	Rickets, osteomalacia	Calcium imbalance (calcification of soft tisues and formation of stones)	Synthesized in the body with the help of sunshine; fortified milk
Vitamin E Antioxidant (stabilization of cell membranes, regulation of oxidation reactions, protection of polyunsaturated fatty acids [PUFA] and vitamin A)	Erythrocyte hemolysis, nerve damage	Hemorrhagic effects	Vegetable oils
Vitamin K Synthesis of blood-clotting proteins and bone proteins	Hemorrhage	None known	Synthesized in the body by GI bacteria; green leafy vegetables

Nutrition in Your Life

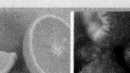

For the fat-soluble vitamins, select colorful fruits and vegetables, fortified milk or

soy products, and vegetable oils; use supplements with caution, if at all.

- Do you eat dark green, leafy or deep yellow vegetables daily?

- Do you drink vitamin D–fortified milk or go outside in the sunshine regularly?

- Do you use vegetable oils when you cook?

NUTRITION ON THE NET

 Access these websites for further study of topics covered in this chapter. Be aware that many websites on the Internet are peddling vitamin supplements, not accurate information.

- Find updates and quick links to these and other nutrition-related sites at our website: **www.wadsworth.com/nutrition**

- Search for "vitamins" at the American Dietetic Association: **www.eatright.org**

- Review the Dietary Reference Intakes for vitamins A, D, E, and K and the carotenoids by searching for "DRI": **www.nap.edu**

- Visit the World Health Organization to learn about "vitamin deficiencies" around the world: **www.who.int**

- Search for "vitamins" at the U.S. Government health information site: **www.healthfinder.gov**

- Learn how fruits and vegetables support a healthy diet rich in vitamins from the 5 A Day for Better Health program: **www.5aday.com** or **www.5aday.gov**

NUTRITION CALCULATIONS

These exercises will help you learn the best food sources for the vitamins and prepare you to examine your own food choices. See p. 388 for answers.

1. Review the units in which vitamins are measured (a spot check). For each of these vitamins, note the unit of measure:

   Vitamin A   Vitamin D

   Vitamin E   Vitamin K

2. Analyze the vitamin contents of foods. Review the figures, photos, and food sources sections in Chapters 10 and 11 and list the food group(s) that contributed the most of each vitamin. Which food groups offer the most thiamin? The most riboflavin? The most niacin? The most vitamin B_6? The most folate? The most vitamin B_{12}? The most vitamin C? The most vitamin A? The most vitamin D? The most vitamin E?

   List the groups that provided "the most" and compare them with the Food Guide Pyramid in Chapter 2.

This exercise should convince you that each of the food groups provides some, but not all, of the vitamins needed daily. For a full array, a person needs to eat a variety of foods from each of the food groups regularly.

STUDY QUESTIONS

These questions will help you review the chapter. You will find the answers in the discussions on the pages provided.

1. List the fat-soluble vitamins. What characteristics do they have in common? How do they differ from the water-soluble vitamins? (p. 367)

2. Summarize the roles of vitamin A and the symptoms of its deficiency. (pp. 369–372)

3. What is meant by vitamin precursors? Name the precursors of vitamin A, and tell in what classes of foods they are located. Give examples of foods with high vitamin A activity. (pp. 367, 373–374)

4. How is vitamin D unique among the vitamins? What is its chief function? What are the richest sources of this vitamin? (pp. 375–376, 377–379)

5. Describe vitamin E's role as an antioxidant. What are the chief symptoms of vitamin E deficiency? (pp. 380–381)

6. What is vitamin K's primary role in the body? What conditions may lead to vitamin K deficiency? (pp. 382–383)

These multiple choice questions will help you prepare for an exam. Answers can be found on p. 388.

1. Fat-soluble vitamins:
   a. are easily excreted.
   b. seldom reach toxic levels.
   c. require bile for absorption.
   d. are not stored in the body's tissues.

2. The form of vitamin A active in vision is:
   a. retinal.
   b. retinol.
   c. rhodopsin.
   d. retinoic acid.

3. Vitamin A–deficiency symptoms include:
   a. rickets and osteomalacia.
   b. hemorrhaging and jaundice.
   c. night blindness and keratomalacia.
   d. fibrocystic breast disease and erythrocyte hemolysis.

4. Good sources of vitamin A include:
   a. oatmeal, pinto beans, and ham.
   b. apricots, turnip greens, and liver.
   c. whole-wheat bread, green peas, and tuna.
   d. corn, grapefruit juice, and sunflower seeds.

5. To keep minerals available in the blood, vitamin D targets:
   a. the skin, the muscles, and the bones.
   b. the kidneys, the liver, and the bones.
   c. the intestines, the kidneys, and the bones.
   d. the intestines, the pancreas, and the liver.

6. Vitamin D can be synthesized from a precursor that the body makes from:
   a. bilirubin.
   b. tocopherol.
   c. cholesterol.
   d. beta-carotene.

7. Vitamin E's most notable role is to:
   a. protect lipids against oxidation.
   b. activate blood-clotting proteins.
   c. support protein and DNA synthesis.
   d. enhance calcium deposits in the bones.

8. The classic sign of vitamin E deficiency is:
   a. rickets.
   b. xeropthalmia.
   c. muscular dystrophy.
   d. erythrocyte hemolysis.

9. Without vitamin K:
   a. muscles atrophy.
   b. bones become soft.
   c. skin rashes develop.
   d. blood fails to clot.

10. A significant amount of vitamin K comes from:
    a. vegetable oils.
    b. sunlight exposure.
    c. bacterial synthesis.
    d. fortified grain products.

REFERENCES

1. S. J. Hickenbottom and coauthors, Variability in conversion of ß-carotene to vitamin A in men as measured by using a double-tracer study design, *American Journal of Clinical Nutrition* 75 (2002): 900–907; K. J. Yeum and R. M. Russell, Carotenoid bioavailability and bioconversion, *Annual Review of Nutrition* 22 (2002): 483–504.

2. G. Wolf, The enzymatic cleavage of ß-carotene: End of a controversy, *Nutrition Reviews* 59 (2001): 116–118.

3. J. L. Napoli, A gene knockout corroborates the integral function of cellular retinol-binding protein in retinoic metabolism, *Nutrition Reviews* 58 (2000): 230–236.

4. M. Clagett-Dame and H. F. DeLuca, The role of vitamin A in mammalian reproduction and embryonic development, *Annual Review of Nutrition* 22 (2002): 347–381.

5. H. Hadi and coauthors, Vitamin A supplementation selectively improves the linear growth of Indonesian preschool children: Results from a randomized controlled trial, *American Journal of Clinical Nutrition* 71 (2000): 507–513.

6. Committee on Dietary Reference Intakes, *Dietary Reference Intakes for Vitamin C, Vitamin E, Selenium, and Carotenoids* (Washington, D.C.: National Academy Press, 2000).

7. C. Ballew and coauthors, Serum retinol distributions in residents of the United States: Third National Health and Nutrition Examination Survey, 1988–1994, *American Journal of Clinical Nutrition* 73 (2001): 586–593.

8. C. E. West, Vitamin A and measles, *Nutrition Reviews* 58 (2000): S46–S54.

9. E. Villamor and coauthors, Vitamin A supplements ameliorate the adverse effect of HIV-1, malaria, and diarrheal infections on child growth, *Pediatrics* 109 (2002): e6; C. Duggan and W. Fawzi, Micronutrients and child health: Studies in international nutrition and HIV infection, *Nutrition Reviews* 59 (2001): 358–369; A. H. Shankar and coauthors, Effect of vitamin A supplementation on morbidity due to *Plasmodium falciparum* in young children in Papua New Guinea randomised trial, *Lancet* 354 (1999): 203–209; J. E. Tyson and coauthors, Vitamin A supplementation for extremely-low-birthweight infants, *New England Journal of Medicine* 340 (1999): 1962–1968; F. Sempértegui and coauthors, The beneficial effects of weekly low-dose vitamin A supplementation on acute lower respiratory infections and diarrhea in Ecuadorian children, *Pediatrics* 104 (1999): e6.

10. R. M. Russell, The vitamin A spectrum: From deficiency to toxicity, *American Journal of Clinical Nutrition* 71 (2000): 878–884.

11. P. Christian and coauthors, Working after the sun goes down: Exploring how night blindness impairs women's work activities in rural Nepal, *European Journal of Clinical Nutrition* 52 (1998): 519–524.

12. A. Sommer, Xerophthalmia and vitamin A status, *Progress in Retinal and Eye Research* 17 (1998): 9–31.

13. A. Mazzone and A. dal Canton, Images in clinical medicine—Hypercarotenemia, *New England Journal of Medicine* 346 (2002): 821.

14. X.-D. Wang and coauthors, Retinoid signaling and activator protein-1 expression in ferrets given ß-carotene supplements and exposed to tobacco smoke, *Journal of the National Cancer Institute* 91 (1999): 60–66.

15. M. A. Leo and C. S. Lieber, Alcohol, vitamin A, and ß-carotene: Adverse interactions, including hepatotoxicity and carcinogenicity, *American Journal of Clinical Nutrition* 69 (1999): 1071–1085.

16. S. Johansson and coauthors, Subclinical hypervitaminosis A causes fragile bones in rats, *Bone* 31 (2002): 685–689; N. Binkley and D. Krueger, Hypervitaminosis A and bone, *Nutrition Reviews* 58 (2000): 138–144.

17. K. Michaelsson and coauthors, Serum retinol levels and the risk of fractures, *New England Journal of Medicine* 348 (2003): 287–294; D. Feskanich and coauthors, Vitamin A intake and hip fractures among postmenopausal women, *Journal of the American Medical Association* 287 (2002): 47–54; S. J. Whiting and B. Lemke, Excess retinol intake may explain the high incidence of osteoporosis in northern Europe, *Nutrition Reviews* 57 (1999): 192–195.

18. L. M. Voyles and coauthors, High levels of retinol intake during the first trimester of pregnancy result from use of over-the-counter vitamin/mineral supplements, *Journal of the American Dietetic Association* 100 (2000): 1068–1070.

19. R. P. Heaney, Lessons for nutritional science from vitamin D, *American Journal of Clinical Nutrition* 69 (1999): 825–826.

20. I. A. van der Mei and coauthors, Past exposure to sun, skin phenotype, and risk of multiple sclerosis: Case-control study, *British Medical Journal* 327 (2003): 316–321.

21. I. N. Sills, Nutritional rickets: A preventable disease, *Topics in Clinical Nutrition* 17 (2001): 36–43.

22. M. K. M. Lehtonen-Veromaa and coauthors, Vitamin D and attainment of peak bone mass among peripubertal Finnish girls: A 3-y prospective study, *American Journal of Clinical Nutrition* 76 (2002): 1446–1453; T. A. Outila, M. U. M. Kärkkäinen, and C. J. E. Lamberg-Allardt, Vitamin D status affects serum parathyroid hormone concentrations during winter in female adolescents: Associations with forearm bone mineral density, *American Journal of Clinical Nutrition* 74 (2001): 206–210.

23. S. A. Abrams, Nutritional rickets: An old disease returns, *Nutrition Reviews* 60 (2002): 111–115.

24. M. S. LeBoff and coauthors, Occult vitamin D deficiency in postmenopausal US women with acute hip fracture, *Journal of the American Medical Association* 281 (1999): 1505–1511.

25. M. S. Calvo and S. J. Whiting, Prevalence of vitamin D insufficiency in Canada and the United States: Importance to health status and efficacy of current food fortification and dietary supplement use, *Nutrition Reviews* 61 (2003): 107–113.

26. N. F. Carvalho and coauthors, Severe nutritional deficiencies in toddlers resulting from health food milk alternatives, *Pediatrics* 107 (2001): e46.

27. T. A. Sentongo and coauthors, Vitamin D status in children, adolescents, and young adults with Crohn disease, *American Journal of Clinical Nutrition* 76 (2002): 1077–1081; S. Nesby-O'Dell and coauthors, Hypovitaminosis D prevalence and determinants among African American and white women of reproductive age: Third National Health and Nutrition Examination Survey, 1988–1994, *American Journal of Clinical Nutrition* 76 (2002): 187–192; S. R. Kreiter and coauthors, Nutritional rickets in African American breast-fed infants, *Journal of Pediatrics* 137 (2000): 153–157.

28. R. P. Heaney and coauthors, Human serum 25-hydroxycholecalciferol response to extended oral dosing with cholecalciferol, *American Journal of Clinical Nutrition* 77 (2003): 204–210.

29. Committee on Dietary Reference Intakes, 2000.

30. A. M. Papas, Beyond α-tocopherol: The role of the other tocopherols and tocotrienols, in M. S. Meskin and coeditors, *Phytochemicals in Nutrition and Health* (Boca Raton, Fla.: CRC Press, 2002), pp. 61–77; Q. Jiang and coauthors, γTocopherol, the major form of vitamin E in the US diet, deserves more attention, *American Journal of Clinical Nutrition* 74 (2001): 714–722.

31. Committee on Dietary Reference Intakes, 2000.

32. S. L. Booth and coauthors, Vitamin K intake and bone mineral density in women and men, *American Journal of Clinical Nutrition* 77 (2003): 512–516.

33. N. C. Binkley and coauthors, A high phylloquinone intake is required to achieve maximal osteocalcin γ-carboxylation, *American Journal of Clinical Nutrition* 76 (2002): 1055–1060; N. C. Binkley and coauthors, Vitamin K supplementation reduces serum concentrations of under-γ-carboxylated osteocalcin in healthy young and elderly adults, *American Journal of Clinical Nutrition* 72 (2000): 1523–1528; S. L. Booth and coauthors, Dietary vitamin K intakes are associated with hip fracture but not with bone mineral density in elderly men and women, *American Journal of Clinical Nutrition* 71 (2000): 1201–1208; D. Feskanich and coauthors, Vitamin K intake and hip fractures in women: A prospective study, *American Journal of Clinical Nutrition* 69 (1999): 74–79.

ANSWERS

Nutrition Calculations

1. Vitamin A: μg RAE.   Vitamin D: μg.
   Vitamin E: mg.   Vitamin K: μg.

2. Thiamin: Legumes and grains
   Riboflavin: Milks, grains, and meats
   Niacin: Meats and grains
   Vitamin B$_6$: Meats
   Folate: Legumes and vegetables
   Vitamin B$_{12}$: Meats and milks
   Vitamin C: Vegetables and fruits

   Vitamin A: Vegetables, fruits, and milks
   Vitamin D: Milks
   Vitamin E: Legumes and oils

Taken together, "the most" groups form the Pyramid—grains, vegetables, legumes, fruits, milks, meats, and oils.

Study Questions (multiple choice)

1. c   2. a   3. c   4. b   5. c   6. c   7. a   8. d
9. d   10. c

HIGHLIGHT

Antioxidant Nutrients in Disease Prevention

© Nick Clements/Taxi/Getty Images

Count on supplement manufacturers to exploit the day's hot topics in nutrition. The moment bits of research news surface, new supplements appear—and terms like "antioxidants" and "lycopene" become household words. Friendly faces in TV commercials try to persuade us that these supplements hold the magic in the fight against aging and disease. New supplements hit the market and cash registers ring. Vitamin C, for years the leading single nutrient supplement, gains new popularity, and sales of lutein, beta-carotene, and vitamin E supplements soar as well.

In the meantime, scientists and medical experts around the world continue their work to clarify and confirm the roles of antioxidants in preventing chronic diseases.[1] This highlight summarizes some of the accumulating evidence. It also revisits the advantages of foods over supplements. But first it is important to introduce the troublemakers—the **free radicals** (the accompanying glossary defines free radicals and related terms).

known as free radicals. In addition to normal body processes, environmental factors such as ultraviolet radiation, air pollution, and tobacco smoke generate free radicals.

A free radical is a molecule with one or more unpaired electrons.[*] An electron without a partner is unstable and highly reactive. To regain its stability, the free radical quickly finds a stable but vulnerable compound from which to steal an electron (see Figure H11-1 on p. 390).

With the loss of an electron, the formerly stable molecule becomes a free radical itself and steals an electron from another nearby molecule. Thus an electron-snatching chain reaction is under way with free radicals producing more free radicals. Antioxidants neutralize free radicals by donating one of their own electrons, thus ending the chain reaction. When they lose electrons, antioxidants do not become free radicals because they are stable in either form. (Review Figure 10-15 on p. 349 to see how ascorbic acid can give up two hydrogens with their electrons and become dehydroascorbic acid.)

Once formed, free radicals attack. Occasionally, these free-radical attacks are helpful. For example, cells of the immune system use free radicals as ammunition in an "oxidative burst" that demolishes disease-causing viruses and bacteria. Most often, however, free-radical attacks cause widespread damage. They commonly damage the polyunsaturated fatty acids in lipoproteins and in cell membranes, disrupting the transport of substances into and out of cells. Free radicals also damage

Free Radicals and Disease

Chapter 7 described how the body's cells use oxygen in metabolic reactions. In the process, oxygen sometimes reacts with body compounds and produces highly unstable molecules

---

[*]Many free radicals exist, but the oxygen-derived ones are most common in the human body. Examples of oxygen-derived free radicals include superoxide radical ($O_2^{\cdot-}$), hydroxyl radical ($OH\cdot$), and nitric oxide ($NO\cdot$). (The dots in the symbols represent the unpaired electrons.) Technically, hydrogen peroxide (H_2O_2) and singlet oxygen are not free radicals because they contain paired electrons, but the unstable conformation of their electrons makes radical-producing reactions likely. Scientists sometimes use the term *reactive oxygen species (ROS)* to describe all of these compounds.

GLOSSARY

free radicals: unstable and highly reactive atoms or molecules that have one or more unpaired electrons in the outer orbital (see Appendix B for a review of basic chemistry concepts).

oxidants (OK-see-dants): compounds (such as oxygen itself) that oxidize other

compounds. Compounds that prevent oxidation are called *anti*oxidants, whereas those that promote it are called *pro*oxidants.

• anti = against
• pro = for

oxidative stress: a condition in which the production of oxidants and free radicals

exceeds the body's ability to defend itself.

prooxidants: substances that significantly induce oxidative stress.

Reminder: *Dietary antioxidants* are substances typically found in foods that significantly decrease the adverse effects of free radicals

on normal functions in the body. *Nonnutrients* are compounds in foods that do not fit into the six classes of nutrients. *Phytochemicals* are nonnutrient compounds found in plant-derived foods that have biological activity in the body.

FIGURE H11-1   The Actions of Free Radicals and Antioxidants

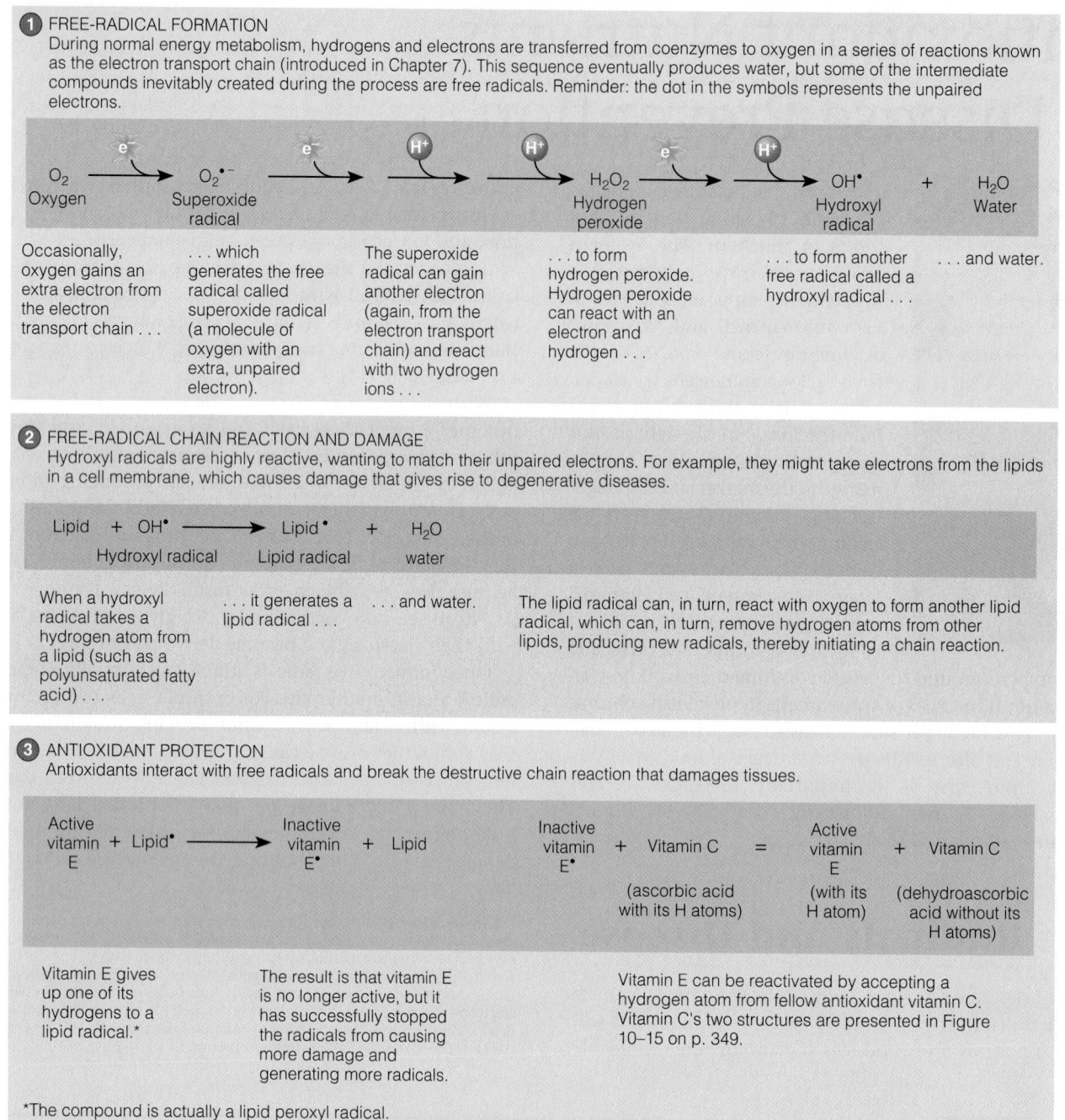

1 FREE-RADICAL FORMATION

During normal energy metabolism, hydrogens and electrons are transferred from coenzymes to oxygen in a series of reactions known as the electron transport chain (introduced in Chapter 7). This sequence eventually produces water, but some of the intermediate compounds inevitably created during the process are free radicals. Reminder: the dot in the symbols represents the unpaired electrons.

$$O_2 \xrightarrow{e^-} O_2^{\bullet -} \xrightarrow{e^-} \xrightarrow{H^+} \xrightarrow{H^+} H_2O_2 \xrightarrow{e^-} \xrightarrow{H^+} OH^\bullet + H_2O$$

Oxygen / Superoxide radical / Hydrogen peroxide / Hydroxyl radical / Water

| Occasionally, oxygen gains an extra electron from the electron transport chain . . . | . . . which generates the free radical called superoxide radical (a molecule of oxygen with an extra, unpaired electron). | The superoxide radical can gain another electron (again, from the electron transport chain) and react with two hydrogen ions . . . | . . . to form hydrogen peroxide. Hydrogen peroxide can react with an electron and hydrogen . . . | . . . to form another free radical called a hydroxyl radical . . . | . . . and water. |

2 FREE-RADICAL CHAIN REACTION AND DAMAGE

Hydroxyl radicals are highly reactive, wanting to match their unpaired electrons. For example, they might take electrons from the lipids in a cell membrane, which causes damage that gives rise to degenerative diseases.

$$Lipid + OH^\bullet \longrightarrow Lipid^\bullet + H_2O$$

Hydroxyl radical / Lipid radical / water

| When a hydroxyl radical takes a hydrogen atom from a lipid (such as a polyunsaturated fatty acid) . . . | . . . it generates a lipid radical . . . | . . . and water. | The lipid radical can, in turn, react with oxygen to form another lipid radical, which can, in turn, remove hydrogen atoms from other lipids, producing new radicals, thereby initiating a chain reaction. |

3 ANTIOXIDANT PROTECTION

Antioxidants interact with free radicals and break the destructive chain reaction that damages tissues.

$$\text{Active vitamin E} + Lipid^\bullet \longrightarrow \text{Inactive vitamin E}^\bullet + Lipid$$

$$\text{Inactive vitamin E}^\bullet + \text{Vitamin C} = \text{Active vitamin E} + \text{Vitamin C}$$

(ascorbic acid with its H atoms) / (with its H atom) / (dehydroascorbic acid without its H atoms)

| Vitamin E gives up one of its hydrogens to a lipid radical.* | The result is that vitamin E is no longer active, but it has successfully stopped the radicals from causing more damage and generating more radicals. | Vitamin E can be reactivated by accepting a hydrogen atom from fellow antioxidant vitamin C. Vitamin C's two structures are presented in Figure 10–15 on p. 349. |

*The compound is actually a lipid peroxyl radical.

cell proteins (altering their functions) and DNA (creating mutations).

The body's natural defenses and repair systems try to control the destruction caused by free radicals, but these systems are not 100 percent effective. In fact, they become less effective with age, and the unrepaired damage accumulates. To some extent, dietary antioxidants defend the body against **oxidative stress,** but if antioxidants are unavailable, or if free-radical pro-duction becomes excessive, health problems may develop. Oxygen-derived free radicals may cause diseases not only by indiscriminately destroying the valuable components of cells, but also by serving as signals for specific activities within the cells. Scientists have identified oxidative stress as a causative factor and antioxidants as a protective factor in cognitive performance and the aging process and in the development of diseases such as cancer, arthritis, cataracts, and heart disease.[2]

Defending against Free Radicals

The body maintains a couple lines of defense against free-radical damage. A system of enzymes disarms the most harmful **oxidants.**[*] The action of these enzymes depends on the minerals selenium, copper, manganese, and zinc. If the diet fails to provide adequate supplies of these minerals, this line of defense weakens. The body also uses the antioxidant vitamins: vitamin E and vitamin C. Vitamin E defends the body's lipids (cell membranes and lipoproteins, for example) by efficiently stopping the free-radical chain reaction. Vitamin C protects the body's watery components, such as the fluid of the blood, against free-radical attacks. Vitamin C seems especially adept at neutralizing free radicals from polluted air and cigarette smoke; it may also restore oxidized vitamin E to its active state.

Dietary antioxidants may also include *non*nutrients—some of the phytochemicals (featured in Highlight 13). Together, nutrients and phytochemicals with antioxidant activity minimize damage by:

- Limiting free-radical formation.
- Destroying free radicals or their precursors.
- Stimulating antioxidant enzyme activity.
- Repairing oxidative damage.
- Stimulating repair enzyme activity.

These actions play key roles in defending the body against cancer and heart disease.

Defending against Cancer

Cancers arise when cellular DNA is damaged—sometimes by free-radical attacks. Antioxidants may reduce cancer risks by protecting DNA from this damage. Many researchers have reported low rates of cancer in people whose diets include abundant vegetables and fruits, rich in antioxidants.[3] Preliminary reports suggest an inverse relationship between DNA damage and vegetable intake and a positive relationship with beef and pork intake. Laboratory studies with animals and with cells in tissue culture also seem to support such findings.

Foods rich in vitamin C seem to protect against certain types of cancers, especially those of the mouth, larynx, esophagus, and stomach. Such a correlation may reflect the benefits of a diet rich in fruits and vegetables and low in fat; it does not necessarily support taking vitamin C supplements to treat or prevent cancer.

Evidence that vitamin E helps guard against cancer is less consistent than for vitamin C. Still, people with low blood levels of vitamin E have high rates of some cancers. Several studies report a cancer-preventing benefit of vegetables and fruits rich in beta-carotene and the other carotenoids as well.[4]

Defending against Heart Disease

High blood cholesterol carried in LDL is a major risk factor for cardiovascular disease, but how do LDL exert their damage? One scenario is that free radicals within the arterial walls oxidize LDL, changing their structure and function. The oxidized LDL then accelerate the formation of artery-clogging plaques. These free radicals also oxidize the polyunsaturated fatty acids of the cell membranes, sparking additional changes in the arterial walls, which impede the flow of blood. Susceptibility to such oxidative damage within the arterial walls is heightened by a diet high in saturated fat or cigarette smoke. In contrast, diets that include plenty of fruits and vegetables, especially when combined with little saturated fat, strengthen antioxidant defenses against LDL oxidation.[5] Antioxidant nutrients taken as supplements also seem to slow the early progression of atherosclerosis.[6]

Antioxidants, especially vitamin E, may protect against cardiovascular disease.[7] Epidemiological studies suggest that people who eat foods rich in vitamin E have low rates of death from heart disease.[8] Similarly, large doses of vitamin E supplements may slow the progression of heart disease. Among its many protective roles, vitamin E defends against LDL oxidation, inflammation, arterial injuries, and blood clotting.[9] Less clear is whether vitamin E supplements benefit people who already have heart disease or multiple risk factors for it.[10] Antioxidant supplements may not be beneficial and, in fact, may even be harmful for these people.[11]

Some studies suggest that vitamin C protects against LDL oxidation, raises HDL, lowers total cholesterol, and improves blood pressure. Vitamin C may also minimize the free-radical action within the arterial wall that typically follows a high-fat meal.[12] In fact, blood flow through the arteries is similar to that seen after a low-fat meal.

Foods, Supplements, or Both?

In the process of scavenging and quenching free radicals, antioxidants themselves become oxidized. To some extent, they can be regenerated, but still, losses occur and free radicals attack continuously. To maintain defenses, a person must replenish dietary antioxidants regularly. But should antioxidants be replenished from foods or from supplements?

---

[*]These enzymes include glutathione peroxidase, thioredoxin reductase, superoxide dismutase, and catalase.

Foods—especially fruits and vegetables—offer not only antioxidants, but an array of other valuable vitamins and minerals as well. Importantly, deficiencies of these nutrients can damage DNA as readily as free radicals can.[13] Eating fruits and vegetables in abundance protects against both deficiencies and diseases. A major review of the evidence gathered from metabolic studies, epidemiologic studies, and dietary intervention trials identified three dietary strategies most effective in preventing heart disease:[14]

- Use unsaturated fats (that have not been hydrogenated) instead of saturated or *trans* fats (see Highlight 5).

- Select foods rich in omega-3 fatty acids (see Chapter 5).

- Consume a diet high in fruits, vegetables, nuts, and whole grains and low in refined grain products.

Such a diet combined with exercise, weight control, and not smoking serves as the best prescription for health. Notably, taking supplements is not among these disease-prevention recommendations.

Some research suggests a protective effect from as little as a daily glass of orange juice or carrot juice (rich sources of vitamin C and beta-carotene, respectively). Other intervention studies, however, have used levels of nutrients that far exceed current recommendations and can be achieved only by taking supplements. In making their recommendations for the antioxidant nutrients, members of the DRI Committee considered whether these studies support substantially higher intakes to help protect against chronic diseases. They did raise the recommendations for vitamins C and E, but do not support taking vitamin pills over eating a healthy diet.

While awaiting additional research, should people anticipate the go-ahead and start taking antioxidant supplements now? Most scientists agree that it is too early to make such a recommendation. Though fruits and vegetables containing many antioxidant nutrients and phytochemicals have been associated with a diminished risk of many cancers, supplements have not always proved beneficial. In fact, sometimes the benefits are more apparent when the vitamins come from foods rather than from supplements. Without data to confirm the benefits of supplements, we cannot accept the potential risks. And the risks are real.

Consider the findings from a study to determine whether daily supplements of vitamin E, beta-carotene, or both would reduce the incidence of lung cancer among smokers. After five to eight years of supplementation, there was no reduction in the incidence of lung cancer; in fact, the researchers found a *higher* incidence of lung cancer among smokers receiving the beta-carotene. Another group of researchers reported similar findings: smokers and asbestos workers receiving beta-carotene and vitamin A supplements for four years had a higher incidence of lung cancer and risk of death than those taking a placebo. These findings brought the study to an end much earlier than planned. Given the association between high intakes of *foods* rich in beta-carotene and low rates of lung cancer reported in earlier epidemiological studies, findings of increased risk were surprising, to say the least.[15] As is often true of most nutrients, the action of beta-carotene differs dramatically at various levels of intake.[16] Amounts commonly found in foods may be beneficial, but high doses trigger the production of keratin in the delicate cells of the lungs, and this effect is magnified with smoking.[17]

Clearly, remedies to life-threatening diseases such as lung cancer are not as simple as taking supplements. Smokers are much wiser to stop smoking than to rely on pills to protect them from lung cancer.

Even if research clearly proves that a particular nutrient is the ultimate protective ingredient in foods, supplements would not be the answer because their contents are limited. Vitamin E supplements, for example, usually contain alpha-tocopherol, but foods provide an assortment of tocopherols among other nutrients, many of which provide valuable protection against free-radical damage. Supplements shortchange users.

Furthermore, much more research is needed to define optimal and dangerous levels of intake. This much we know: antioxidants behave differently under various conditions. At physiological levels typical of a healthy diet, they act as antioxidants, but at pharmacological doses typical of supplements, they may act as **prooxidants,** stimulating the production of free radicals.[18] This is especially likely in the presence of other antioxidants or minerals such as iron. Until the optimum intake of these nutrients can be determined, the risks of supplement use remain unclear. The best way to add antioxidants to the diet is to eat generous servings of fruits and vegetables daily.

It should be clear by now that we cannot know the identity and action of every chemical in every food. Even if we did, why create a supplement to replicate a food? Why not eat foods and enjoy the pleasure, nourishment, and health benefits they provide? The beneficial constituents in foods are widespread among plants. Among the fruits, pomegranates,

Many cancer-fighting products are available now at your local produce counter.

© Jeffry Myers/Stock, Boston/PictureQuest

berries, and citrus rank high in antioxidants; top antioxidant vegetables include kale, spinach, and brussels sprouts; millet and oats contain the most antioxidants among the grains; pinto beans and soybeans are the outstanding legumes; and walnuts outshine the other nuts.[19] But don't try to single out one particular food for its magic nutrient, antioxidant, or phytochemical. Instead, eat a wide variety of fruits and vegetables in generous quantities every day—and get *all* the magic compounds these foods have to offer.

REFERENCES

1. P. Møller and S. Loft, Oxidative DNA damage in human white blood cells in dietary antioxidant intervention studies, *American Journal of Clinical Nutrition* 76 (2002): 303–310; B. Halliwell, Why and how should we measure oxidative DNA damage in nutritional studies? How far have we come? *American Journal of Clinical Nutrition* 72 (2000): 1082–1087.

2. F. Grodstein, J. Chen, and W. C. Willett, High-dose antioxidant supplements and cognitive function in community-dwelling elderly women, *American Journal of Clinical Nutrition* 77 (2003): 975–984; M. J. Engelhart and coauthors, Dietary intake of antioxidants and risk of Alzheimer disease, *Journal of the American Medical Association* 287 (2002): 3223–3229; H. L. Hu and coauthors, Mechanisms of ageing and development, *Science Direct* 121 (2001): 217–230; M. Meydani, Antioxidants and cognitive function, *Nutrition Reviews* 59 (2001): S75–S80; J. W. Miller, Vitamin E and memory: Is it vascular protection? *Nutrition Reviews* 58 (2000): 109–111.

3. A. Martin and coauthors, Roles of vitamins E and C on neurodegenerative diseases and cognitive performance, *Nutrition Reviews* 60 (2002): 308–326; H. Chen and coauthors, Dietary patterns and adenocarcinoma of the esophagus and distal stomach, *American Journal of Clinical Nutrition* 75 (2002): 137–144; B. Halliwell, Establishing the significance and optimal intake of dietary antioxidants: The biomarker concept, *Nutrition Reviews* 57 (1999): 104–113.

4. E. R. Berton and coauthors, A population-based case-control study of carotenoid and vitamin A intake and ovarian cancer (United States), *Cancer Causes Control* 12 (2001): 83–90; D. S. Michaud and coauthors, Intake of specific carotenoids and risk of lung cancer in 2 prospective US cohorts, *American Journal of Clinical Nutrition* 72 (2000): 990–997; M. L. Slattery and coauthors, Carotenoids and colon cancer, *American Journal of Clinical Nutrition* 71 (2000): 575–582; N. McKeown and coauthors, Antioxidants and breast cancer, *Nutrition Reviews* 57 (1999): 321–324; D. A. Cooper, A. L. Eldridge, and J. C. Peters, Dietary carotenoids and lung cancer: A review of recent research, *Nutrition Reviews* 57 (1999): 133–134; P. Riso and coauthors, Does tomato consumption effectively increase the resistance of lymphocyte DNA to oxidative damage? *American Journal of Clinical Nutrition* 69 (1999): 712–718; E. Giovannucci, Tomatoes, tomato-based products, lycopene, and cancer: Review of the epidemiologic literature, *Journal of the National Cancer Institute* 91 (1999): 317–331.

5. R. A. Jacob, Evidence that diet modification reduces in vivo oxidant damage, *Nutrition Reviews* 57 (1999): 255–258.

6. L. Liu and M. Meydani, Combined vitamin C and E supplementation retards early progression of arteriosclerosis in heart transplant patients, *Nutrition Reviews* 60 (2002): 368–371; H. Y. Huang and coauthors, Effects of vitamin C and vitamin E on in vivo lipid peroxidation: Results of a randomized controlled trial, *American Journal of Clinical Nutrition* 76 (2002): 549–555; C. R. Gale, H. E. Ashurst, and H. J. Powers, Antioxidant vitamin status and carotid atherosclerosis in the elderly, *American Journal of Clinical Nutrition* 74 (2001): 402–408; F. Nappo, Impairment of endothelial functions by acute hyperhomocysteinemia and reversal by antioxidant vitamins, *Journal of the American Medical Association* 281 (1999): 2113–2118.

7. A. Iannuzzi and coauthors, Dietary and circulating antioxidant vitamins in relation to carotid plaques in middle-aged women, *American Journal of Clinical Nutrition* 76 (2002): 582–587; M. Meydani, Effect of functional food ingredients: Vitamin E modulation of cardiovascular diseases and immune status in the elderly, *American Journal of Clinical Nutrition* 71 (2000): 1665S–1668S.

8. L. A. Yochum, A. R. Folsom, and L. H. Kushi, Intake of antioxidant vitamins and risk of death from stroke in postmenopausal women, *American Journal of Clinical Nutrition* 72 (2000): 476–483.

9. S. Devaraj, A. Harris, and I. Jialal, Modulation of monocyte-macrophage function with α-tocopherol: Implications for atherosclerosis, *Nutrition Reviews* 60 (2002): 8–14; L. J. van Tits and coauthors, α-Tocopherol supplementation decreases production of superoxide and cytokines by leukocytes ex- vivo in both normolipidemic and hyper-triglyceridemic individuals, *American Journal of Clinical Nutrition* 71 (2000): 458–464; M. Meydani, Vitamin E and prevention of heart disease in high-risk patients, *Nutrition Reviews* 58 (2000): 278–281.

10. B. G. Brown and coauthors, Simvastatin and niacin, antioxidant vitamins, or the combination for the prevention of coronary disease, *New England Journal of Medicine* 345 (2001): 1583–1592.

11. D. D. Waters and coauthors, Effects of hormone replacement therapy and antioxidant vitamin supplements on coronary atherosclerosis in postmenopausal women: A randomized controlled trial, *Journal of the American Medical Association* 288 (2002): 2432–2440; Collaborative Group of the Primary Prevention Project (PPP), Low-dose aspirin and vitamin E in people at cardiovascular risk: A randomized trial in general practice, *Lancet* 357 (2001): 89–95.

12. L. Liu and coauthors, Vitamin C preserves endothelial function in patients with coronary heart disease after a high-fat meal, *Clinical Cardiology* 25 (2002): 219–224.

13. B. N. Ames, Micronutrient deficiencies: A major cause of DNA damage, *Annals of the New York Academy of Sciences* 889 (1999): 87–106.

14. F. B. Hu and W. C. Willett, Optimal diets for prevention of coronary heart disease, *Journal of the American Medical Association* 288 (2002): 2569–2578.

15. W. A. Pryor, W. Stahl, and C. L. Rock, Beta carotene: From biochemistry to clinical trials, *Nutrition Reviews* 58 (2000): 39–53.

16. J. S. Bertram, Carotenoids and gene regulation, *Nutrition Reviews* 57 (1999): 182–191.

17. G. Wolf, The effect of low and high doses of ß-carotene and exposure to cigarette smoke on the lungs of ferrets, *Nutrition Reviews* 60 (2002): 88–90.

18. X. D. Wang and R. M. Russell, Procarcinogenic and anticarcinogenic effects of ß-carotene, *Nutrition Reviews* 57 (1999): 263–272.

19. B. L. Halvorsen and coauthors, A systematic screening of total antioxidants in dietary plants, *Journal of Nutrition* 132 (2002): 461–471.

Water and the Major Minerals

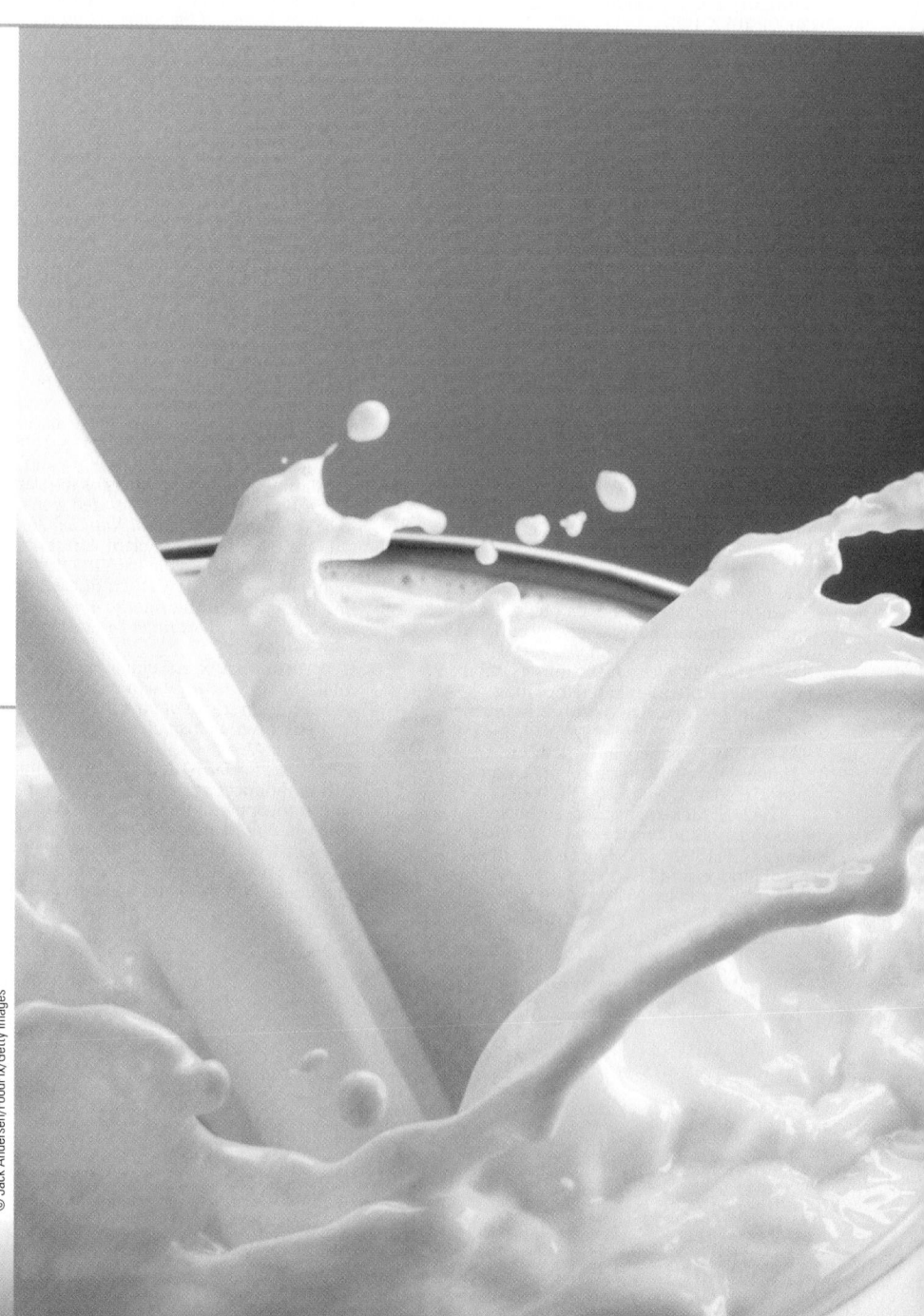

Nutrition in Your Life

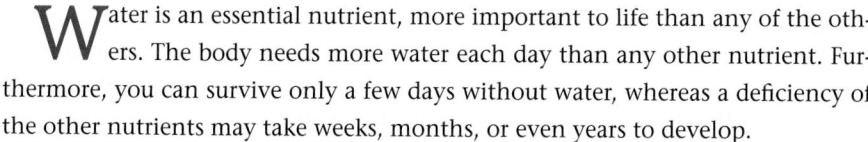

What's your beverage of choice? If you said water, then congratulate yourself for recognizing its importance in maintaining your body's fluid balance. If you answered milk, then pat yourself on the back for taking good care of your bones. Faced with a lack of water, you would realize within days how vital it is to your very survival. The consequences of a lack of milk (or other calcium-rich foods) are also dramatic, but may not become apparent for decades. Water, calcium, and all the other major minerals support fluid balance and bone health. Before getting too comfortable reading this chapter, you might want to get yourself a glass of water or milk. Your body will thank you.

Water is an essential nutrient, more important to life than any of the others. The body needs more water each day than any other nutrient. Furthermore, you can survive only a few days without water, whereas a deficiency of the other nutrients may take weeks, months, or even years to develop.

This chapter begins with a look at water and the body's fluids. The body maintains an appropriate balance and distribution of water with the help of another class of nutrients—the minerals. In addition to introducing the minerals that help regulate body fluids, the chapter describes many of the other important functions minerals perform in the body. Chapter 19 revisits water as a beverage and addresses consumer concerns about its safety.

Water and the Body Fluids

Water constitutes about 60 percent of an adult's body weight and a higher percentage of a child's. Because water makes up about three-fourths of the weight of lean tissue and less than one-fourth of the weight of fat, a person's body composition

Water is the most indispensable nutrient.

influences how much of the body's weight is water. The proportion of water is generally smaller in females, obese people, and the elderly because of their smaller proportion of lean tissue.

In the body, water becomes the fluid in which all life processes occur. The water in the body fluids:

- Carries nutrients and waste products throughout the body.
- Maintains the structure of large molecules such as proteins and glycogen.
- Participates in metabolic reactions.
- Serves as the solvent for minerals, vitamins, amino acids, glucose, and many other small molecules so that they can participate in metabolic activities.
- Acts as a lubricant and cushion around joints and inside the eyes, the spinal cord, and, in pregnancy, the amniotic sac surrounding the fetus in the womb.
- Aids in the regulation of normal body temperature. As Chapter 14 explains, evaporation of sweat from the skin removes excess heat from the body.
- Maintains blood volume.

To support these and other vital functions, the body actively maintains an appropriate **water balance.**■

Water Balance and Recommended Intakes

Every cell contains fluid of the exact composition that is best for that cell **(intracellular fluid)** and is bathed externally in another such fluid **(interstitial fluid).** Interstitial fluid is the largest component of **extracellular fluid.** Figure 12-1 illustrates a cell and its associated fluids. These fluids continually lose and replace their components, yet the composition in each compartment remains remarkably constant under normal conditions. Because imbalances can be devastating, the body quickly responds by adjusting both water intake and excretion as needed. Consequently, the entire system of cells and fluids remains in a delicate but controlled state of homeostasis.

Water Intake **Thirst** and satiety influence water intake, apparently in response to changes sensed by the mouth, hypothalamus,■ and nerves. When the blood becomes concentrated (having lost water, but not the dissolved substances within it), the mouth becomes dry, and the hypothalamus initiates drinking behavior. Stretch receptors in the stomach send signals to stop drinking as do receptors in the heart that monitor blood volume.

Thirst drives a person to seek water, but it lags behind the body's need. When too much water is lost from the body and not replaced, **dehydration** develops. A first sign of dehydration is thirst, the signal that the body has already lost some of its fluid. If a person is unable to obtain fluid or, as in many elderly people, fails to perceive the thirst message, the symptoms of dehydration may progress rapidly from thirst to weakness, exhaustion, and delirium and end in death if not corrected (see Table 12-1). Dehydration may easily develop with either water deprivation or excessive water losses. (Chapter 14 revisits dehydration and the fluid needs of athletes.)

Water intoxication, on the other hand, is rare but can occur with excessive water ingestion and kidney disorders that reduce urine production. The symptoms may include confusion, convulsions, and even death in extreme cases. Excessive water ingestion contributing to the dangerous condition known as hyponatremia in endurance athletes is discussed in Chapter 14.

Water Sources The obvious dietary sources of water are water itself and other beverages, but nearly all foods also contain water. Most fruits and vegetables contain up to 90 percent water; many meats and cheeses contain at least 50 percent (see Table 12-2 for selected foods and Appendix H for many more). Water is also generated during metabolism. Recall that when the energy-yielding nutrients

---

■ Water balance: intake = output.

Nutrition Explorer Watch how water balance is maintained in the body through equalization of water sources and water losses.

■ Reminder: The *hypothalamus* is a brain center that controls activities such as maintenance of water balance, regulation of body temperature, and control of appetite.

water balance: the balance between water intake and output (losses).

intracellular fluid: fluid within the cells, usually high in potassium and phosphate. Intracellular fluid accounts for approximately two-thirds of the body's water.
- **intra** = within

interstitial (IN-ter-STISH-al) **fluid:** fluid between the cells (intercellular), usually high in sodium and chloride. Interstitial fluid is a large component of extracellular fluid.
- **inter** = in the midst, between

extracellular fluid: fluid outside the cells. Extracellular fluid includes two main components—the interstitial fluid and plasma. Extracellular fluid accounts for approximately one-third of the body's water.
- **extra** = outside

thirst: a conscious desire to drink.

dehydration: the condition in which body water output exceeds water input. Symptoms include thirst, dry skin and mucous membranes, rapid heartbeat, low blood pressure, and weakness.

water intoxication: the rare condition in which body water contents are too high in all body fluid compartments.

TABLE 12-1 Signs of Dehydration

Body Weight Lost (%)	Symptoms
1–2	Thirst, fatigue, weakness, vague discomfort, loss of appetite
3–4	Impaired physical performance, dry mouth, reduction in urine, flushed skin, impatience, apathy
5–6	Difficulty in concentrating, headache, irritability, sleepiness, impaired temperature regulation, increased respiratory rate
7–10	Dizziness, spastic muscles, loss of balance, delirium, exhaustion, collapse

NOTE: The onset and severity of symptoms at various percentages of body weight lost depend on the activity, fitness level, degree of acclimation, temperature, and humidity. If not corrected, dehydration can lead to death.

break down, their carbons and hydrogens combine with oxygen to yield carbon dioxide (CO_2)—and water (H_2O). As Table 12-3 (on p. 398) shows, the water derived daily from these three sources averages about 2½ liters (roughly 2½ quarts).

Water Losses The body must excrete a minimum of about 500 milliliters of water each day■ as urine—enough to carry away the waste products generated by a day's metabolic activities. Above this amount, excretion adjusts to balance intake. If a person drinks more water, the kidneys excrete more urine, and the urine becomes more dilute. In addition to urine, water is lost from the lungs as vapor and from the skin as sweat; some is also lost in feces.* The amount of fluid lost from each source varies, depending on the environment (such as heat or humidity) and physical conditions (such as exercise or fever). On average, daily losses total about 2½ liters. Table 12-3 (on p. 398) shows how water excretion balances intake; maintaining this balance requires healthy kidneys and an adequate intake of fluids.

Water Recommendations Because water needs vary depending on diet, activity, environmental temperature, and humidity, a general water requirement is difficult to establish. In the past, recommendations■ were expressed in proportion to the amount of energy expended under average environmental conditions. The recommended water intake for a person who expends 2000 kcalories a day, for example, would be about 2 to 3 liters of water (about 7 to 11 cups). This recommendation is in line with the Adequate Intake (AI) for *total* water set by the DRI Committee.■ Total water includes not only drinking water, but water in other beverages and in foods as well.[1]

*Water lost from the lungs and skin accounts for almost one-half of the daily losses even when a person is not visibly perspiring; these losses are commonly referred to as *insensible water losses*.

TABLE 12-2 Percentage of Water in Selected Foods

100%	Water, diet sodas
90–99%	Fat-free milk, strawberries, watermelon, lettuce, cabbage, celery, spinach, broccoli
80–89%	Fruit juice, yogurt, apples, grapes, oranges, carrots
70–79%	Shrimp, bananas, corn, potatoes, avocados, cottage cheese, ricotta cheese
60–69%	Pasta, legumes, salmon, ice cream, chicken breast
50–59%	Ground beef, hot dogs, feta cheese
40–49%	Pizza
30–39%	Cheddar cheese, bagels, bread
20–29%	Pepperoni sausage, cake, biscuits
10–19%	Butter, margarine, raisins
1–9%	Crackers, cereals, pretzels, taco shells, peanut butter, nuts
0%	Oils

FIGURE 12-1 One Cell and Its Associated Fluids

Fluids are found within the cells (intracellular) or outside the cells (extracellular). Extracellular fluids include plasma (the fluid portion of blood in the intravascular spaces of blood vessels) and interstitial fluids (the tissue fluid that fills the intercellular spaces between the cells).

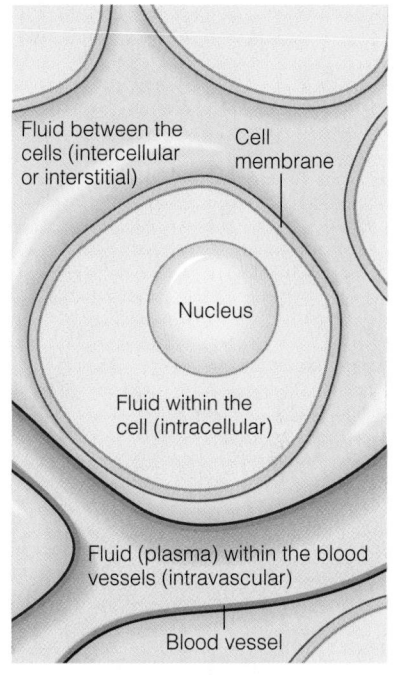

Fluid between the cells (intercellular or interstitial)

Cell membrane

Nucleus

Fluid within the cell (intracellular)

Fluid (plasma) within the blood vessels (intravascular)

Blood vessel

■ The amount of water the body has to excrete each day to dispose of its wastes is the **obligatory** (ah-BLIG-ah-TORE-ee) **water excretion**—about 500 mL (about 2 c, or a pint).

■ Water recommendation:
  • 1.0 to 1.5 mL/kcal expended (adults).[†]
  • 1.5 mL/kcal expended (infants and athletes).
  Conversion factors:
  • 1 mL = 0.03 fluid ounce.
  • 125 mL ≈ ½ c.
  Easy estimation: ½ c per 100 kcal expended.

[†]For those using kilojoules: 4.2 to 6.3 mL/kJ expended.

■ AI for *total* water:
  • Men: 3.7 L/day.
  • Women: 2.7 L/day.
  Conversion factors:
  • 1 L = 33.8 fluid oz.
  • 1 L = 1.06 qt.
  • 1 c = 8 fluid oz.

TABLE 12-3 Water Balance

Water Sources	Amount (mL)	Water Losses	Amount (mL)
Liquids	550 to 1500	Kidneys (urine)	500 to 1400
Foods	700 to 1000	Skin (sweat)	450 to 900
Metabolic water	200 to 300	Lungs (breath)	350
		GI tract (feces)	150
Total	1450 to 2800	Total	1450 to 2800

Because a wide range of water intakes will prevent dehydration and its harmful consequences, the AI is based on average intakes. People who are physically active or who live in hot environments may need more. Survey data show that people are drinking more water, juice, soft drinks, coffee, tea, and alcoholic beverages than previously.[2] Regardless of which beverages people drink, they lose some fluid within a few hours, which is why they have to replenish daily.

Some research indicates that people who drink caffeinated beverages lose a little more fluid than when drinking water because caffeine acts as a diuretic.[3] The DRI Committee considered such findings in their recommendations for water intake and concluded that "caffeinated beverages contribute to the daily total water intake similar to that contributed by non-caffeinated beverages."[4] In other words, it doesn't seem to matter whether people rely on caffeine-containing beverages or other beverages to meet their fluid needs.

As Highlight 7 explains, alcohol acts as a diuretic, and it has many adverse effects on health and nutrition status. Alcohol should not be used to meet fluid needs.

Health Effects of Water In addition to meeting the body's fluid needs, drinking plenty of water may protect the bladder against cancer by diluting the urine and reducing its holding time. The risk of bladder cancer in men decreases when fluid intake is high.[5] An adequate water intake may also protect against kidney stones, prostate cancer, and breast cancer.[6]

The kind of water you drink may also make a difference to health. Water is usually either hard or soft. **Hard water** has high concentrations of calcium and magnesium; sodium or potassium is the principal mineral of **soft water** (see the accompanying glossary for these and other common terms used to describe water). In practical terms, soft water makes more bubbles with less soap; hard water leaves

GLOSSARY OF WATER TERMS

artesian water: water drawn from a well that taps a confined aquifer in which the water is under pressure.

bottled water: drinking water sold in bottles.

carbonated water: water that contains carbon dioxide gas, either naturally occurring or added, that causes bubbles to form in it; also called *bubbling* or *sparkling water*. Seltzer, soda, and tonic waters are legally soft drinks and are not regulated as water.

distilled water: water that has been vaporized and recondensed, leaving it free of dissolved minerals.

filtered water: water treated by filtration, usually through *activated carbon filters* that reduce the lead in tap water, or by *reverse osmosis* units that force pressurized water across a membrane removing lead, arsenic, and some microorganisms from tap water.

hard water: water with a high calcium and magnesium content.

mineral water: water from a spring or well that typically contains 250 to 500 parts per million (ppm) of minerals. Minerals give water a distinctive flavor. Many mineral waters are high in sodium.

natural water: water obtained from a spring or well that is certified to be safe and sanitary. The mineral content may not be changed, but the water may be treated in other ways such as with ozone or by filtration.

public water: water from a municipal or county water system that has been treated and disinfected.

purified water: water that has been treated by distillation or other physical or chemical processes that remove dissolved solids. Because purified water contains no minerals or contaminants, it is useful for medical and research purposes.

soft water: water with a high sodium or potassium content.

spring water: water originating from an underground spring or well. It may be bubbly (carbonated), or "flat" or "still," meaning not carbonated. Brand names such as "Spring Pure" do not necessarily mean that the water comes from a spring.

well water: water drawn from ground water by tapping into an aquifer.

a ring on the tub, a crust of rocklike crystals in the teakettle, and a gray residue in the laundry.

Soft water may seem more desirable around the house, and some homeowners purchase water softeners that replace magnesium and calcium with sodium. In the body, however, soft water with sodium may aggravate hypertension and heart disease.[7] In contrast, hard water may benefit these conditions by virtue of its calcium content.[8]

Soft water also more easily dissolves certain contaminant minerals, such as cadmium and lead, from old plumbing pipes. As Chapter 13 explains, these contaminant minerals harm the body by displacing the nutrient minerals from their normal sites of action. People who live in old buildings should run the cold water tap a minute to flush out harmful minerals whenever the water faucet has been off for more than six hours.

Many people turn to **bottled water** as an alternative to tap water, believing it to be safer than tap water and therefore worth its substantial cost. Chapter 19 offers a discussion of bottled water safety and regulations.

IN SUMMARY Water makes up about 60 percent of the adult body's weight. It assists with the transport of nutrients and waste products throughout the body, participates in chemical reactions, acts as a solvent, serves as a shock absorber, and regulates body temperature. To maintain water balance, intake from liquids, foods, and metabolism must equal losses from the kidneys, skin, lungs, and GI tract. The amount and type of water a person drinks may have positive or negative health effects.

Blood Volume and Blood Pressure

Fluids maintain the blood volume, which in turn influences blood pressure. Central to the regulation of blood volume and blood pressure are the kidneys. All day, every day, the kidneys reabsorb needed substances and water and excrete wastes with some water in the urine (see Figure 12-2 on p. 400). The kidneys meticulously adjust the volume and the concentration of the urine to accommodate changes in the body, including variations in the day's food and beverage intakes. Instructions on whether to retain or release substances or water come from ADH, renin, angiotensin, and aldosterone.

ADH and Water Retention Whenever blood volume or blood pressure falls too low, or whenever the extracellular fluid becomes too concentrated, the hypothalamus signals the pituitary gland to release **antidiuretic hormone (ADH).** ADH is a water-conserving hormone■ that stimulates the kidneys to reabsorb water. Consequently, the more water you need, the less your kidneys excrete. These events also trigger thirst. Drinking water and retaining fluids raise the blood volume and dilute the concentrated fluids, thus helping to restore homeostasis.

Renin and Sodium Retention Cells in the kidneys respond to low blood pressure by releasing an enzyme called **renin.** Through a complex series of events, renin causes the kidneys to reabsorb sodium. Sodium reabsorption, in turn, is always accompanied by water retention, which helps to restore blood volume and blood pressure.

Angiotensin and Blood Vessel Constriction Renin also activates the blood protein angiotensinogen to **angiotensin.** Angiotensin is a powerful **vasoconstrictor**: it narrows the diameters of blood vessels, thereby raising the blood pressure.

Aldosterone and Sodium Retention Angiotensin also mediates the release of the hormone **aldosterone** from the **adrenal glands.** Aldosterone signals the kidneys to retain more sodium and therefore water because when sodium moves, fluids follow. Again, the effect is that when more water is needed, less is excreted.

■ Recall from Highlight 7 that alcohol depresses ADH activity, thus promoting fluid losses and dehydration. In addition to its antidiuretic effect, ADH elevates blood pressure and so is also called **vasopressin** (VAS-oh-PRES-in).
- **vaso** = vessel
- **press** = pressure

antidiuretic hormone (ADH): a hormone released by the pituitary gland in response to highly concentrated blood. The kidneys respond by reabsorbing water, thus preventing water loss.
- **anti** = against
- **dia** = through
- **ure** = urine

renin (REN-in): an enzyme from the kidneys that activates angiotensin.

angiotensin (AN-gee-oh-TEN-sin): a hormone involved in blood pressure regulation. Its precursor protein is called **angiotensinogen.**

vasoconstrictor (VAS-oh-kon-STRIK-tor): a substance that constricts or narrows the blood vessels.

aldosterone (al-DOS-ter-own): a hormone secreted by the adrenal glands that stimulates the reabsorption of sodium by the kidneys. Aldosterone also regulates chloride and potassium concentrations.

adrenal glands: glands adjacent to, and just above, each kidney.

FIGURE 12-2   A Nephron, One of the Kidney's Many Functioning Units

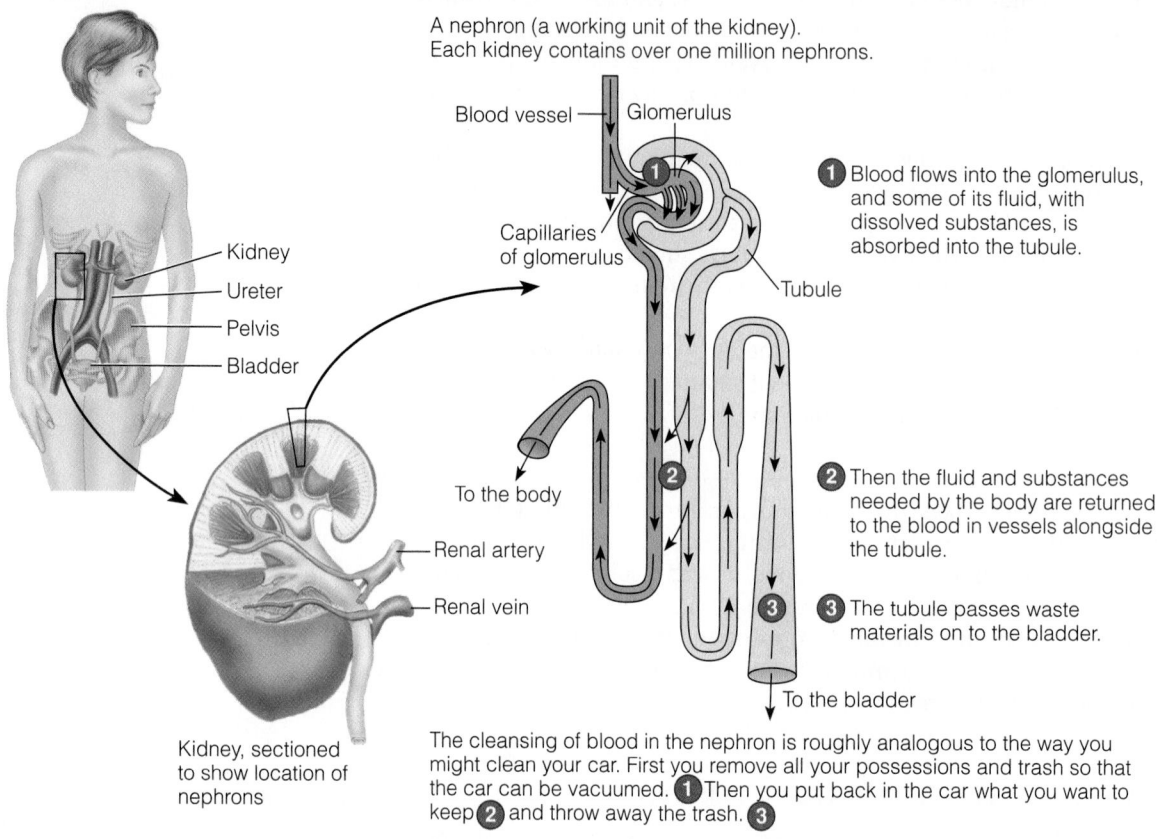

A nephron (a working unit of the kidney).
Each kidney contains over one million nephrons.

Blood vessel — Glomerulus

Capillaries of glomerulus

Tubule

Kidney
Ureter
Pelvis
Bladder

To the body

Renal artery

Renal vein

Kidney, sectioned to show location of nephrons

1 Blood flows into the glomerulus, and some of its fluid, with dissolved substances, is absorbed into the tubule.

2 Then the fluid and substances needed by the body are returned to the blood in vessels alongside the tubule.

3 The tubule passes waste materials on to the bladder.

To the bladder

The cleansing of blood in the nephron is roughly analogous to the way you might clean your car. First you remove all your possessions and trash so that the car can be vacuumed. **1** Then you put back in the car what you want to keep **2** and throw away the trash. **3**

All of these actions are presented in Figure 12-3 and help to explain why high-sodium diets aggravate conditions such as hypertension or edema. Too much sodium causes water retention and an accompanying rise in blood pressure or swelling in the interstitial spaces. Chapter 18 discusses hypertension in detail.

IN SUMMARY   In response to low blood volume, low blood pressure, or highly concentrated body fluids, these actions combine to effectively restore homeostasis:

- ADH retains water.
- Renin retains sodium.
- Angiotensin constricts blood vessels.
- Aldosterone retains sodium.

These actions can maintain water balance only if a person drinks enough water.

Fluid and Electrolyte Balance

Maintaining a balance of about two-thirds of the body fluids inside the cells and one-third outside is vital to the life of the cells. If too much water were to enter the cells, it might rupture them; if too much water were to leave, they would collapse. To control the movement of water, the cells direct the movement of the major minerals.■

Dissociation of Salt in Water   When a mineral **salt** such as sodium chloride (NaCl) dissolves in water, it separates (**dissociates**) into **ions**—positively and neg-

---

■ The major minerals:
- Sodium.
- Chloride.
- Potassium.
- Calcium.
- Phosphorus.
- Magnesium.
- Sulfur.

salt: a compound composed of a positive ion other than H^+ and a negative ion other than OH^-. An example is sodium chloride ($Na^+ Cl^-$).
- **Na** = sodium.
- **Cl** = chloride.

dissociates (dis-SO-see-ates): physically separates.

ions (EYE-uns): atoms or molecules that have gained or lost electrons and therefore have electrical charges. Examples include the positively charged sodium ion (Na^+) and the negatively charged chloride ion (Cl^-). For a closer look at ions, see Appendix B.

FIGURE 12-3   How the Body Regulates Blood Volume

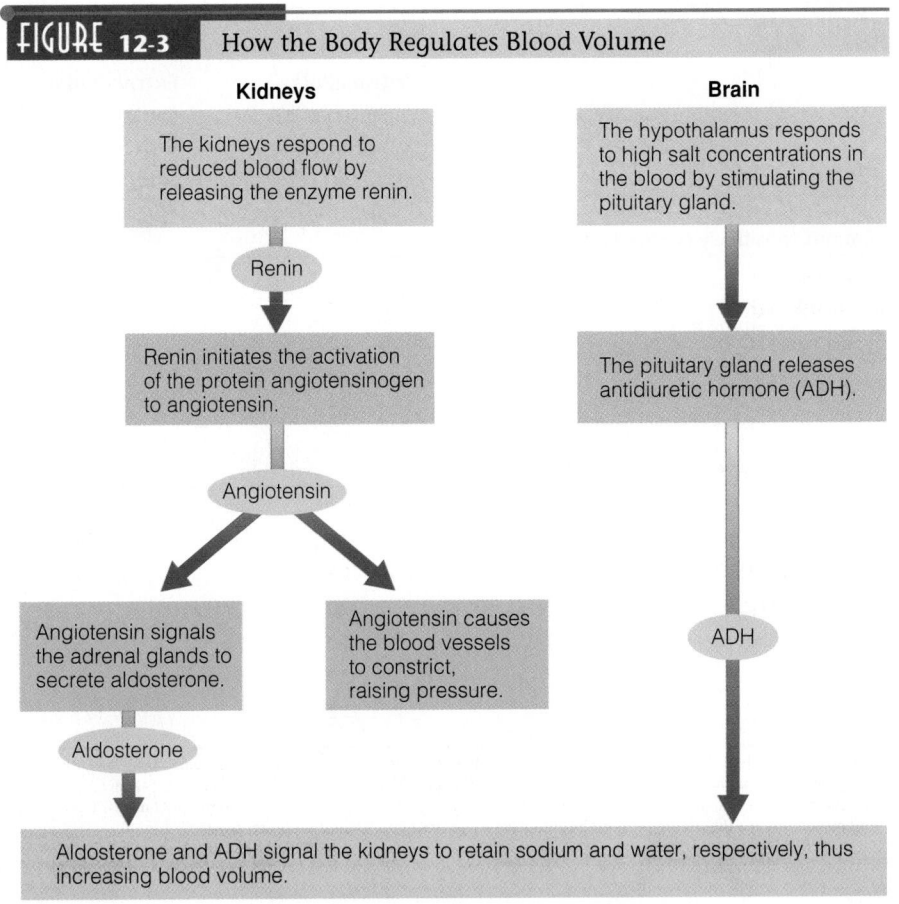

Kidneys

The kidneys respond to reduced blood flow by releasing the enzyme renin.

Renin

Renin initiates the activation of the protein angiotensinogen to angiotensin.

Angiotensin

Angiotensin signals the adrenal glands to secrete aldosterone.

Angiotensin causes the blood vessels to constrict, raising pressure.

Aldosterone

Brain

The hypothalamus responds to high salt concentrations in the blood by stimulating the pituitary gland.

The pituitary gland releases antidiuretic hormone (ADH).

ADH

Aldosterone and ADH signal the kidneys to retain sodium and water, respectively, thus increasing blood volume.

---

atively charged particles (Na^+ and Cl^-). The positive ions are **cations;** the negative ones are **anions.** ■ Unlike pure water, which conducts electricity poorly, ions dissolved in water carry electrical current. For this reason, salts that dissociate into ions are called **electrolytes,** and fluids that contain them are **electrolyte solutions.**

In all electrolyte solutions, anion and cation concentrations balance (the number of negative and positive charges are equal). If a fluid contains 1000 negative charges, it must contain 1000 positive charges, too. If an anion enters the fluid, a cation must accompany it or another anion must leave so that electrical neutrality will be maintained. Whenever Na^+ ions leave a cell, other positive ions enter: potassium (K^+) ions, for example. In fact, it's a good bet that whenever Na^+ and K^+ ions are moving, they are going in opposite directions.

Table 12-4 (on p. 402) shows that, indeed, the positive and negative charges inside and outside cells are perfectly balanced even though the numbers of each kind of ion differ over a wide range. Inside the cells, the positive charges total 202 and the negative charges balance these perfectly. Outside the cells, the amounts and proportions of the ions differ from those inside, but again the positive and negative charges balance. (Scientists count these charges in **milliequivalents, mEq.**)

Electrolytes Attract Water Electrolytes attract water. Each water molecule has a net charge of zero,■ but the oxygen side of the molecule is slightly negatively charged, and the hydrogens are slightly positively charged. Figure 12-4 shows the result in an electrolyte solution: both positive and negative ions attract clusters of water molecules around them. This attraction dissolves salts in water and enables the body to move fluids into appropriate compartments.

■ To help you remember the difference, think of the "t" in cations as a "plus" (+) sign and the "n" in anions as "negative."

■ A neutral molecule, such as water, that has opposite charges spatially separated within the molecule is **polar;** see Appendix B for more details.

cations (CAT-eye-uns): positively charged ions.

anions (AN-eye-uns): negatively charged ions.

electrolytes: salts that dissolve in water and dissociate into charged particles called ions.

electrolyte solutions: solutions that can conduct electricity.

milliequivalents (mEq): the concentration of electrolytes in a volume of solution. Milliequivalents are a useful measure when considering ions because the number of charges reveals characteristics about the solution that are not evident when the concentration is expressed in terms of weight.

TABLE 12-4   Important Body Electrolytes

Electrolytes	Intracellular (inside cells) Concentration (mEq/L)	Extracellular (outside cells) Concentration (mEq/L)
Cations (positively charged ions)		
Sodium (Na$^+$)	10	142
Potassium (K$^+$)	150	5
Calcium (Ca^{++})	2	5
Magnesium (Mg^{++})	40	3
	202	155
Anions (negatively charged ions)		
Chloride (Cl$^-$)	2	103
Bicarbonate (HCO$_3^-$)	10	27
Phosphate (HPO$_4^=$)	103	2
Sulfate (SO$_4^=$)	20	1
Organic acids (lactate, pyruvate)	10	6
Proteins	57	16
	202	155

NOTE: The numbers of positive and negative charges in a given fluid are the same. For example, in extracellular fluid, the cations and anions both equal 155 milliequivalents per liter (mEq/L). Of the cations, sodium ions make up 142 mEq/L; and potassium, calcium, and magnesium ions make up the remainder. Of the anions, chloride ions number 103 mEq/L; bicarbonate ions number 27; and the rest are provided by phosphate ions, sulfate ions, organic acids, and protein.

■ The word ending *-ate* denotes a salt of the mineral. Thus phosphate is the salt form of the mineral phosphorus, and sulfate is the salt form of sulfur.

Water Follows Electrolytes Some electrolytes reside primarily outside the cells (notably, sodium and chloride), while others are predominantly inside the cells (notably, potassium, magnesium, phosphate,■ and sulfate). Cell membranes are selectively permeable, meaning that they allow the passage of some molecules, but not of others. Whenever electrolytes move across the membrane, water follows.

FIGURE 12-4   Water Dissolves Salts and Follows Electrolytes

The structural arrangement of the two hydrogen atoms and one oxygen atom enables water to dissolve salts. Water's role as a solvent is one of its most valuable characteristics.

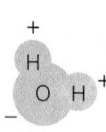

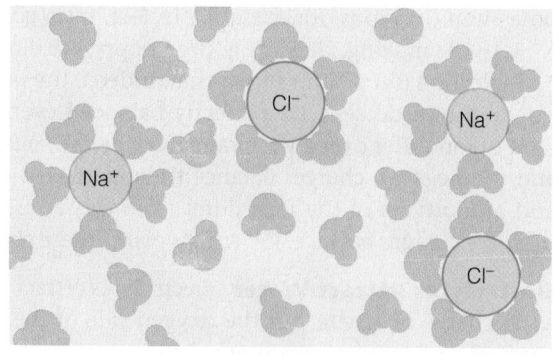

The negatively charged electrons that bond the hydrogens to the oxygen spend most of their time near the oxygen atom. As a result, the oxygen is slightly negative, and the hydrogens are slightly positive (see Appendix B).

In an electrolyte solution, water molecules are attracted to both anions and cations. Notice that the negative oxygen atoms of the water molecules are drawn to the sodium cation (Na$^+$), while the positive hydrogen atoms of the water molecules are drawn to the chloride ions (Cl$^-$).

FIGURE 12-5 | Osmosis

Water flows in the direction of the more highly concentrated solution.

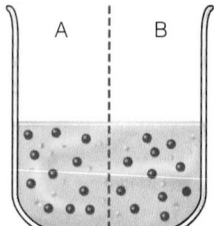
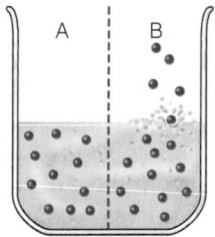
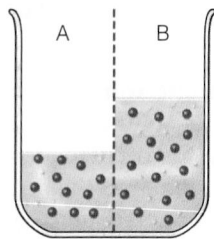

1 With equal numbers of solute particles on both sides, the concentrations are equal, and the tendency of water to move in either direction is about the same.

2 Now additional solute is added to side B. Solute cannot flow across the divider (in the case of a cell, its membrane).

3 Water can flow both ways across the divider, but has a greater tendency to move from side A to side B, where there is a greater concentration of solute. The volume of water becomes greater on side B, and the concentrations on side A and B become equal.

When sprinkled with salt, vegetables "sweat" because water moves toward the higher concentration of salt outside the eggplant.

When immersed in water, raisins get plump because water moves toward the higher concentration of sugar inside the raisins.

The movement of water across a membrane toward the more concentrated **solutes** is called **osmosis**. The amount of pressure needed to prevent the movement of water across a membrane is called the **osmotic pressure**. Figure 12-5 presents osmosis and the photos of salted eggplant and rehydrated raisins provide familiar examples.

Proteins Regulate Flow of Fluids and Ions Chapter 6 described how proteins attract water and help to regulate fluid movement. In addition, transport proteins in the cell membranes regulate the passage of positive ions and other substances from one side of the membrane to the other. Negative ions follow positive ions, and water flows toward the more concentrated solution.

A well-understood protein that regulates the flow of fluids and ions in and out of cells is the sodium-potassium pump. The pump actively exchanges sodium for potassium across the cell membrane, using ATP as an energy source. Figure 6-10 on p. 191 illustrates this action.

Regulation of Fluid and Electrolyte Balance The amounts of various minerals in the body must remain nearly constant. Regulation occurs chiefly at two sites: the GI tract and the kidneys.

The digestive juices of the GI tract contain minerals. These minerals and those from foods are reabsorbed in the large intestine as needed. Each day, 8 liters of fluids and associated minerals are recycled this way, providing ample opportunity for the regulation of electrolyte balance.

The kidneys' control of the body's *water* content by way of the hormone ADH has already been described (see p. 399). To regulate the *electrolyte* contents, the kidneys depend on the adrenal glands, which send out messages by way of the hormone aldosterone (also explained on p. 399). If the body's sodium is low, aldosterone stimulates sodium reabsorption from the kidneys. As sodium is reabsorbed, potassium (another positive ion) is excreted in accordance with the rule that total positive charges must remain in balance with total negative charges.

Fluid and Electrolyte Imbalance

Normally, the body defends itself successfully against fluid and electrolyte imbalances. Certain situations and some medications, however, may overwhelm the body's ability to compensate. Severe and prolonged vomiting and diarrhea and heavy sweating, burns, and traumatic wounds may incur such great fluid and electrolyte losses as to precipitate a medical emergency.

solutes (SOLL-yutes): the substances that are dissolved in a solution. The number of molecules in a given volume of fluid is the **solute concentration**.

osmosis: the movement of water across a membrane *toward* the side where the solutes are more concentrated.

osmotic pressure: the amount of pressure needed to prevent the movement of water across a membrane.

Physically active people must remember to replace their body fluids.

■ Health care workers use **oral rehydration therapy (ORT)**—a simple solution of sugar, salt, and water, taken by mouth—to treat dehydration caused by diarrhea. A simple ORT recipe (cool before giving):
- ½ L boiling water.
- A small handful of sugar (4 tsp).
- 3 pinches of salt (½ tsp).

pH: a measure of the concentration of H^+ ions (see Appendix B). The lower the pH, the higher the H^+ ion concentration and the stronger the acid. A pH above 7 is alkaline, or base (a solution in which OH^- ions predominate).

bicarbonate: a compound with the formula HCO_3 that results from the dissociation of carbonic acid; of particular importance in maintaining the body's acid-base balance. (Bicarbonate is also an alkaline secretion of the pancreas, part of the pancreatic juice.)

carbonic acid: a compound with the formula H_2CO_3 that results from the combination of carbon dioxide (CO_2) and water (H_2O); of particular importance in maintaining the body's acid-base balance.

Sodium and Chloride Most Easily Lost Because sodium and chloride are the body's principal extracellular cation and anion, they are first to be lost when fluid is lost by sweating, bleeding, or excretion. It is no coincidence that after sweating excessively or losing fluid in other ways, a person craves salty foods and refreshing drinks. (Chapter 14 presents a discussion of sport drinks.)

Different Solutes Lost by Different Routes If fluid is lost by vomiting or diarrhea, sodium is lost indiscriminately. If the adrenal glands oversecrete aldosterone, as occurs when a tumor develops, the kidneys may excrete too much potassium. And the person with uncontrolled diabetes may lose a solute not normally excreted: glucose, and with it, large amounts of fluid. All three situations bring on dehydration, but drinking water alone cannot restore electrolyte balance. In each case, medical intervention is required.

Replacing Lost Fluids and Electrolytes In many cases, people can replace the fluids and minerals lost in sweat or in a temporary bout of diarrhea by drinking plain cool water and eating regular foods. Some cases, however, demand rapid replacement of fluids and electrolytes—for example, when diarrhea threatens the life of a malnourished child. Caregivers around the world have learned to use simple formulas■ to treat mild-to-moderate cases of diarrhea. These lifesaving formulas do not require hospitalization and can be prepared from ingredients available locally. Caregivers need only learn to measure ingredients carefully and use sanitary water. Once rehydrated, a person can begin eating foods.

Acid-Base Balance

The body uses its ions not only to help maintain fluid and electrolyte balance, but also to regulate the acidity (**pH**) of its fluids. The pH scale of Chapter 3 is repeated here, in Figure 12-6, with the normal and abnormal pH ranges of the blood added. As you can see, the body must maintain the pH within a narrow range to avoid life-threatening consequences. Slight deviations in either direction can damage proteins, causing metabolic mayhem. Enzymes couldn't catalyze reactions and hemoglobin couldn't carry oxygen—to name just two examples.

The acidity of the body's fluids is determined by the concentration of hydrogen ions (H^+). A high concentration of hydrogen ions would be very acidic. Normal energy metabolism generates hydrogen ions, as well as many other acids, that must be neutralized. Three systems defend the body against fluctuations in pH—buffers in the blood, respiration in the lungs, and excretion in the kidneys.

Regulation by the Buffers Bicarbonate (a base) and **carbonic acid** (an acid) in the body fluids, as well as some proteins, protect the body against changes in acidity by acting as buffers—substances that can neutralize acids or bases. These buffer systems serve as a first line of defense against changes in the fluids' acid-base balance.

Regulation by the Lungs Respiration provides another defense. Carbon dioxide, which is formed all the time during cellular metabolism, forms carbonic acid in the blood, which then dissociates to form hydrogen ions and bicarbonate ions. The appropriate balance between carbonic acid and bicarbonate is essential to maintaining optimal blood pH. If too much carbonic acid builds up, the respiration rate speeds up; this hyperventilation increases the amount of carbon dioxide exhaled, thereby lowering the carbonic acid concentration and restoring homeostasis. Conversely, if bicarbonate builds up, the respiration rate slows; carbon dioxide is retained and forms more carbonic acid. Again, homeostasis is restored.

Regulation by the Kidneys The kidneys play the primary role in maintaining long-term control of acid-base balance. By selecting which ions to retain and which to excrete, the kidneys adjust the body's acid-base balance. Their work is complex, but its net effect is easy to sum up. The *body's* total acid burden remains nearly constant; *urine's* acidity fluctuates to accommodate that balance.

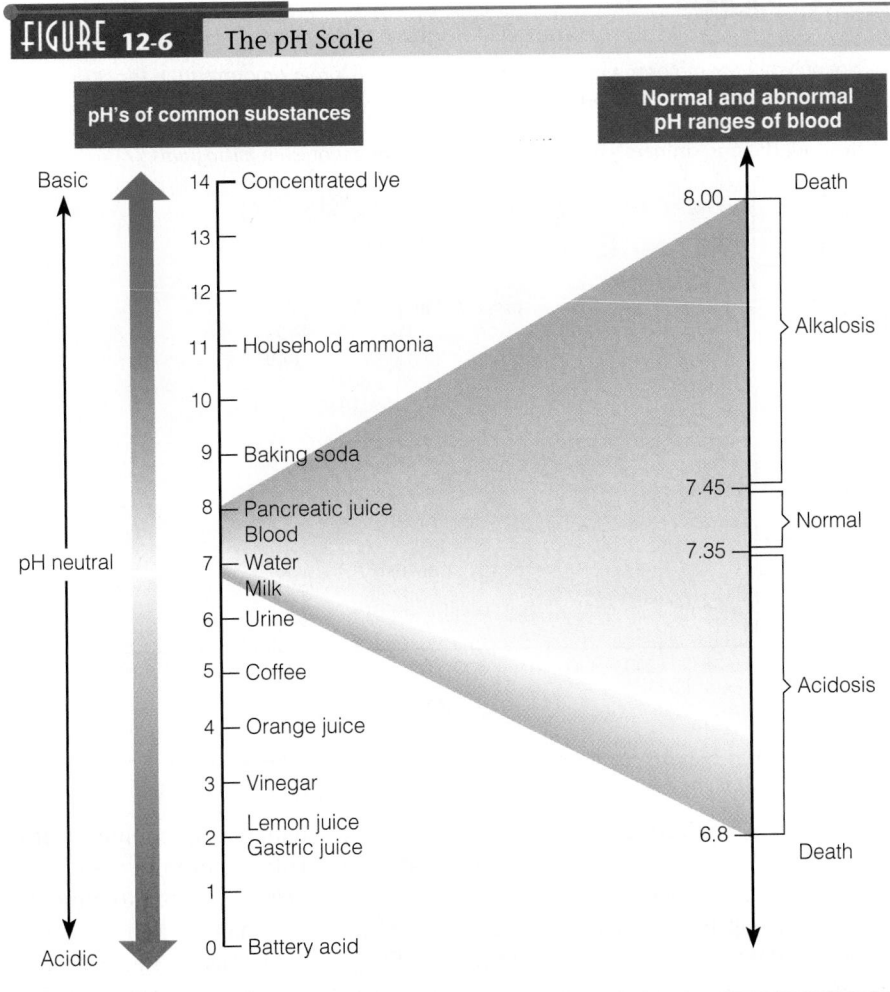

FIGURE 12-6 The pH Scale

pH's of common substances

Basic

14 — Concentrated lye
13 —
12 —
11 — Household ammonia
10 —
9 — Baking soda
8 — Pancreatic juice
    Blood
pH neutral   7 — Water
    Milk
6 — Urine
5 — Coffee
4 — Orange juice
3 — Vinegar
2 — Lemon juice
    Gastric juice
1 —
0 — Battery acid

Acidic

Normal and abnormal pH ranges of blood

8.00 — Death

Alkalosis

7.45 —
      Normal
7.35 —

Acidosis

6.8 — Death

NOTE: Each step is ten times as concentrated in base ($\frac{1}{10}$ as much acid, or H^+) as the one below it.

IN SUMMARY Electrolytes (charged minerals) in the fluids help distribute the fluids inside and outside the cells, thus ensuring the appropriate water balance and acid-base balance to support all life processes. Excessive losses of fluids and electrolytes upset these balances; the kidneys play a key role in restoring homeostasis.

The Minerals—An Overview

Figure 12-7 shows the amounts of the **major minerals** found in the body and, for comparison, some of the trace minerals. The distinction between the major and trace minerals does not mean that one group is more important than the other—all minerals are vital. The major minerals are so named because they are present, and needed, in larger amounts in the body. They are shown at the top of the figure and are discussed in this chapter. The trace minerals (shown at the bottom) are discussed in Chapter 13. A few generalizations pertain to all of the minerals and distinguish them from the vitamins. Especially notable is their chemical nature.

Inorganic Elements Unlike the organic vitamins, which are easily destroyed, minerals are inorganic elements that always retain their chemical identity. Once minerals enter the body proper, they remain there until excreted; they cannot be

major minerals: essential mineral nutrients found in the human body in amounts larger than 5 g; sometimes called **macrominerals.**

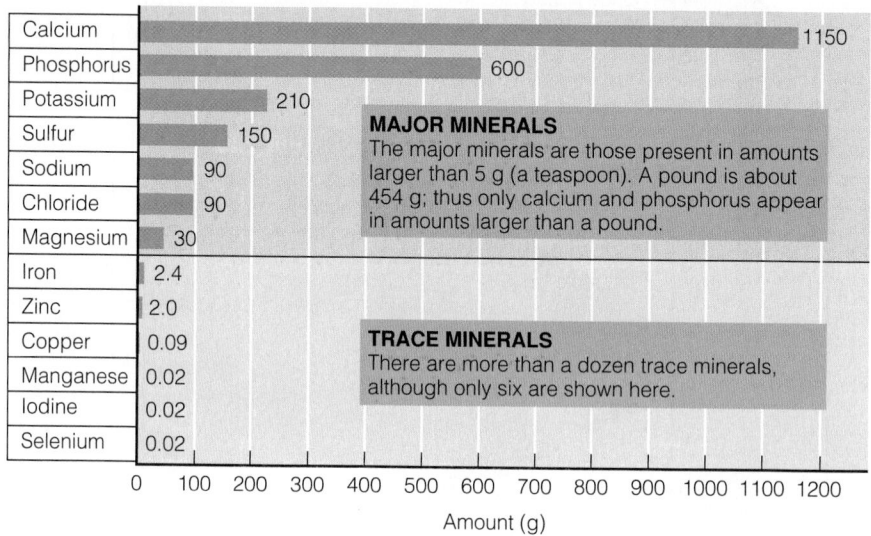

FIGURE 12-7 Minerals in a 60-kilogram (132-pound) Human Body

Not only are the major minerals present in the body in larger amounts than the trace minerals, but they are also needed by the body in larger amounts. Recommended intakes for the major minerals are stated in *hundreds of milligrams* or *grams*, whereas those for the trace minerals are listed in *tens of milligrams* or even *micrograms*.

changed into anything else. Iron, for example, may temporarily combine with other charged elements in salts, but it is always iron. Neither can minerals be destroyed by heat, air, acid, or mixing; consequently, little care is needed to preserve minerals during food preparation. In fact, the ash that remains when a food is burned contains all the minerals that were in the food originally. Minerals can be lost from food only when they leach into water that is then poured down the drain.

The Body's Handling of Minerals The minerals also differ from the vitamins in the amounts the body can absorb and in the extent to which they must be specially handled. Some minerals, such as potassium, are easily absorbed into the blood, transported freely, and readily excreted by the kidneys, much like the water-soluble vitamins. Other minerals, such as calcium, are more like fat-soluble vitamins in that they must have carriers to be absorbed and transported. And, like some of the fat-soluble vitamins, minerals taken in excess can be toxic.

Variable Bioavailability The bioavailability■ of minerals varies. Some foods contain **binders** that combine chemically with minerals, preventing their absorption and carrying them out of the body with other wastes. Examples of binders include phytates, which are found primarily in legumes and grains, and oxalates, which are present in rhubarb and spinach, among other foods. These foods contain more minerals than the body actually receives for use.

Nutrient Interactions Chapter 10 described how the presence or absence of one vitamin can affect another's absorption, metabolism, and excretion. The same is true of the minerals. The interactions between sodium and calcium, for example, cause both to be excreted when sodium intakes are high. Phosphorus binds with magnesium in the GI tract, so magnesium absorption is limited when phosphorus intakes are high. These are just two examples of the interactions involving minerals featured in this chapter. Discussions in both this chapter and the next point out additional problems that arise from such interactions. Notice how often they reflect an excess of one mineral creating an inadequacy of another and how supplements—not foods—are most often to blame.

■ Reminder: *Bioavailability* refers to the rate at and the extent to which a nutrient is absorbed and used.

binders: chemical compounds in foods that combine with nutrients (especially minerals) to form complexes the body cannot absorb. Examples include **phytates** (FYE-tates) and **oxalates** (OCK-sa-lates).

Varied Roles While all the major minerals help to maintain the body's fluid balance described earlier, sodium, chloride, and potassium are most noted for that role. For this reason, these three minerals are discussed first here. Later sections describe the minerals most noted for their roles in bone growth and health—calcium, phosphorus, and magnesium.

IN SUMMARY The major minerals are found in larger quantities in the body, whereas the trace minerals occur in smaller amounts. Minerals are inorganic elements that retain their chemical identities; they usually receive special handling and regulation in the body; and they may bind with other substances or interact with other minerals, thus limiting their absorption.

Sodium

People have held salt (sodium chloride) in high regard throughout recorded history. We say "you are the salt of the earth" to someone we admire and "you are not worth your salt" to someone we consider worthless. Even the word *salary* comes from the Latin word for salt.

Cultures vary in their use of salt, but most people find its taste innately appealing. Salt brings its own tangy taste and enhances other flavors, most likely by suppressing the bitter flavors. You can taste this effect for yourself: tonic water with its bitter quinine tastes sweeter with a little salt added.

Sodium Roles in the Body **Sodium** is the principal cation of the extracellular fluid and the primary regulator of its volume. Sodium also helps maintain acid-base balance and is essential to nerve impulse transmission and muscle contraction.[*]

Foods usually provide more sodium than the body needs. Sodium is readily absorbed by the intestinal tract and travels freely in the blood until it reaches the kidneys, which filter all the sodium out of the blood; then, with great precision, they return to the bloodstream the exact amount the body needs. Normally, the amount excreted is approximately equal to the amount ingested on a given day. When blood sodium rises, as when a person eats salted foods, thirst signals the person to drink until the appropriate sodium-to-water ratio is restored. Then the kidneys excrete both the excess water and the excess sodium together.

Sodium Recommendations Diets rarely lack sodium and even when intakes are low, the body adapts by reducing sodium losses in urine and sweat, thus making deficiencies unlikely. Sodium recommendations■ are set low enough to protect against high blood pressure, but high enough to allow an adequate intake of other nutrients. Because high sodium intakes correlate with high blood pressure, the Upper Level for adults is set at 2300 milligrams per day, slightly lower than the Daily Value used on food labels (2400 milligrams).

■ AI for sodium:
- 1500 mg/day (19–50 yr).
- 1300 mg/day (51–70 yr).
- 1200 mg/day (>70 yr).

Increase the proportion of persons aged 2 years and older who consume 2400 mg or less of sodium daily.

Sodium and Hypertension For years, a high *sodium* intake was considered the primary factor responsible for high blood pressure. Then research pointed to *salt* (sodium chloride) as the dietary culprit. Salt has a greater effect on blood pressure than either sodium or chloride alone or in combination with other ions.

sodium: the principal cation in the extracellular fluids of the body; critical to the maintenance of fluid balance, nerve impulse transmissions, and muscle contractions.

---

[*]One of the ways the kidneys regulate acid-base balance is by excreting hydrogen ions (H^+) in exchange for sodium ions (Na^+).

Fresh herbs add flavor to a recipe without adding salt.

■ Salt (sodium chloride) is about 40% sodium.
1 g salt contributes 400 mg sodium.
5 g salt = 1 tsp.
1 tsp salt contributes 2000 mg sodium.

salt sensitivity: a characteristic of individuals who respond to a high salt intake with an increase in blood pressure or to a low salt intake with a decrease in bloood pressure.

Some individuals respond sensitively to excesses in salt intake and experience high blood pressure. People most likely to have a **salt sensitivity** include those whose parents had high blood pressure; those with chronic kidney disease or diabetes; African Americans; and people over 50 years of age.* Overweight people also appear to be particularly sensitive to the effect of salt on blood pressure. For them, a high salt intake correlates strongly with heart disease, and salt restriction helps to lower their blood pressure.[9]

In fact, a salt-restricted diet lowers blood pressure in people without hypertension as well.[10] Because reducing salt intake causes no harm and diminishes the risk of hypertension and heart disease, the *Dietary Guidelines* advise limiting daily salt intake to less than 6 grams■ (the equivalent of 2.4 grams or 2400 milligrams of *sodium*). Higher intakes seem to be well tolerated in most healthy people, however. The accompanying "How to" offers strategies for cutting salt (and therefore sodium) intake.

One diet plan, known as the DASH (Dietary Approaches to Stop Hypertension) diet, also lowers blood pressure.[11] The DASH approach emphasizes fruits, vegetables, and low-fat dairy products; includes whole grains, nuts, poultry, and fish; and calls for reduced intakes of red meat, butter, and other high-fat foods. The DASH diet in combination with a reduced sodium intake is even more effective in lowering blood pressure than either strategy alone. Chapter 18 offers a complete discussion of hypertension and the dietary recommendations for its prevention and treatment.

Sodium and Osteoporosis A high sodium intake is also associated with calcium excretion, but whether it influences bone loss is less clear.[12] One review of the research concludes that evidence is insufficient to recommend reducing sodium intakes to prevent osteoporosis.[13] Other researchers disagree, acknowledging that while no long-term evidence shows that a reduced sodium intake prevents osteoporosis, less convincing evidence, together with the absence of harm, supports such a recommendation.[14] In other words, reducing sodium intake can't hurt and it may help. Some research shows that potassium may counteract the effects of sodium on calcium excretion.[15] Dietary advice to prevent osteoporosis might therefore suggest selecting foods that are high in calcium and potassium and low in sodium.

Sodium in Foods In general, processed foods have the most sodium, while unprocessed foods such as fresh fruits, vegetables, milk, and meats have the least. In fact, as much as 75 percent of the sodium in people's diets comes from salt added to foods by manufacturers; about 15 percent comes from salt added during cooking and at the table; and only 10 percent comes from the natural content in foods.

Because processed foods may contain sodium without chloride, as in additives such as sodium bicarbonate or sodium saccharin, they do not always taste salty. Most people are surprised to learn that 1 ounce of cornflakes contains more sodium than 1 ounce of salted peanuts— and that ½ cup of instant chocolate pudding contains still more. (The peanuts taste saltier because the salt is all on the surface, where the tongue's sensors immediately pick it up.)

HOW TO   Cut Salt Intake

Most people eat more salt (and therefore sodium) than they need, and some people can lower their blood pressure by avoiding highly salted foods and removing the salt-shaker from the table. Foods eaten without salt may seem less tasty at first, but with repetition, people can learn to enjoy the natural flavors of many unsalted foods. Strategies to cut salt intake include:

- Cook with little or no added salt.
- Prepare foods with sodium-free spices such as basil, bay leaves, curry, garlic, ginger, mint, oregano, pepper, rosemary, and thyme; lemon juice; vinegar; or wine.
- Add little or no salt at the table; taste foods before adding salt.
- Read labels with an eye open for salt. (See the glossary on p. 59 for terms used to describe the sodium contents of foods on labels.)

- Select low-salt or salt-free products when available.

Use these foods sparingly:

- Foods prepared in brine, such as pickles, olives, and sauerkraut.
- Salty or smoked meats, such as bologna, corned or chipped beef, bacon, frankfurters, ham, lunch meats, salt pork, sausage, and smoked tongue.
- Salty or smoked fish, such as anchovies, caviar, salted and dried cod, herring, sardines, and smoked salmon.
- Snack items such as potato chips, pretzels, salted popcorn, salted nuts, and crackers.
- Condiments such as bouillon cubes; seasoned salts; MSG; soy, teriyaki, Worcestershire, and barbeque sauces; prepared horseradish, catsup, and mustard.
- Cheeses, especially processed types.
- Canned and instant soups.

---

*Compared with others, salt-sensitive individuals have elevated concentrations of renin in their blood.

FIGURE 12-8 What Processing Does to the Sodium and Potassium Contents of Foods

People who eat foods high in salt often happen to be eating fewer potassium-containing foods at the same time. Note how potassium is lost and sodium is gained as foods become more processed, causing the potassium-to-sodium ratio to fall dramatically. Even when potassium isn't lost, the addition of sodium still lowers the potassium-to-sodium ratio. Limiting sodium intake may help in two ways, then—by lowering blood pressure in salt-sensitive individuals and by indirectly raising potassium intakes in all individuals.

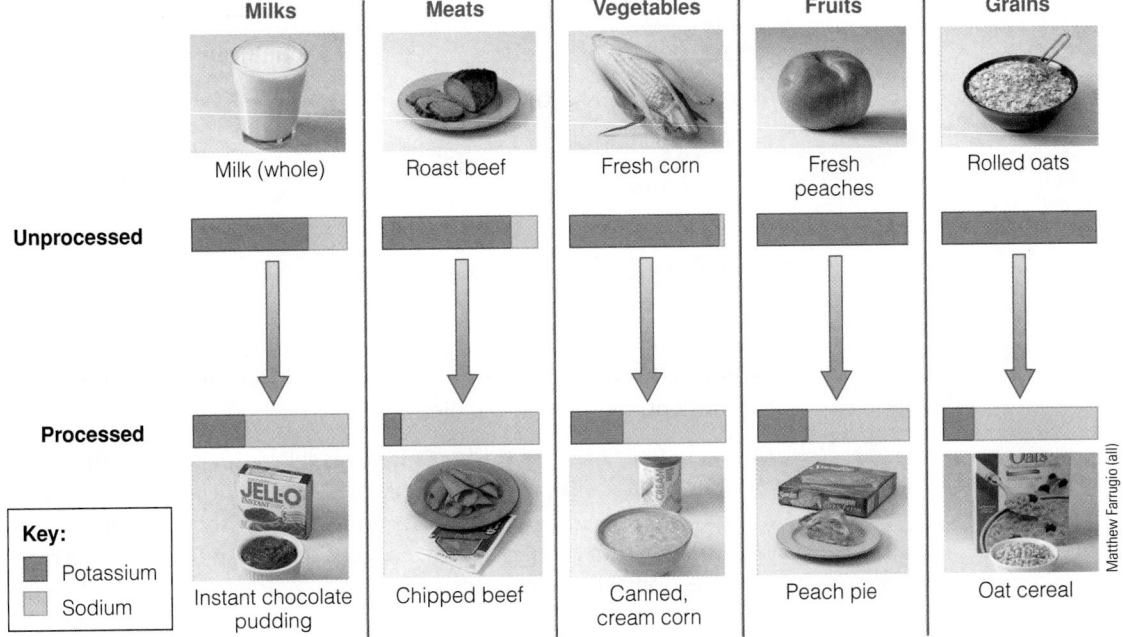

Figure 12-8 shows that processed foods not only contain more sodium than their less processed counterparts, but also have less potassium. Low potassium may be as significant as high sodium when it comes to blood pressure regulation, so processed foods have two strikes against them.

Sodium Deficiency If blood sodium drops, as may occur with vomiting, diarrhea, or heavy sweating, both sodium and water must be replenished. Under normal conditions of sweating due to exercise, salt losses can easily be replaced later in the day with ordinary foods. Salt tablets are not recommended because too much salt, especially if taken with too little water, can induce dehydration. During intense activities, such as ultra-endurance events, athletes can lose so much sodium and drink so much water that they develop hyponatremia—too little sodium in the blood. Chapter 14 offers details about hyponatremia and guidelines for ultra-endurance athletes.

Sodium Toxicity and Excessive Intakes The immediate symptoms of acute sodium toxicity are edema and hypertension, but such toxicity poses no problem as long as water needs are met. Prolonged excessive sodium intake■ may contribute to hypertension in some people, as explained earlier.

■ UL for sodium: 2300 mg/day.

IN SUMMARY
Sodium is the main cation outside cells and one of the primary electrolytes responsible for maintaining fluid balance. Dietary deficiency is rare, and excesses may aggravate hypertension in some people. For this reason, health professionals advise a diet moderate in salt and sodium. The table on the next page summarizes information about sodium.

Sodium

2004 Adequate Intake (AI)

Adults: 1500 mg/day (19–50 yr)
1300 mg/day (51–70 yr)
1200 mg/day (>70 yr)

Upper Level

Adults: 2300 mg/day

Chief Functions in the Body

Maintains normal fluid and electrolyte balance; assists in nerve impulse transmission and muscle contraction

Deficiency Symptoms

Muscle cramps, mental apathy, loss of appetite

Toxicity Symptoms

Edema, acute hypertension

Significant Sources

Table salt, soy sauce; moderate amounts in meats, milks, breads, and vegetables; large amounts in processed foods

Chloride

The element *chlorine* (Cl$_2$) is a poisonous gas. When chlorine reacts with sodium or hydrogen, however, it forms the negative chloride ion (Cl$^-$). *Chloride* is an essential nutrient, required in the diet.

Chloride Roles in the Body   Chloride is the major anion of the extracellular fluids (outside the cells), where it occurs mostly in association with sodium. Chloride can move freely across membranes and so also associates with potassium inside cells. Like sodium and potassium, chloride maintains fluid and electrolyte balance.

In the stomach, the chloride ion is part of hydrochloric acid, which maintains the strong acidity of the gastric juice. One of the most serious consequences of vomiting is the loss of this acid■ from the stomach, which upsets the acid-base balance.* Such imbalances are commonly seen in bulimia nervosa, as Highlight 9 describes.

Chloride Recommendations and Intakes   Chloride is abundant in foods (especially processed foods) as part of sodium chloride and other salts. Because the proportion of chloride in salt is greater than sodium,■ chloride recommendations are slightly higher than, but still equivalent to, those of sodium. In other words, ¾ teaspoon of salt will deliver some sodium, more chloride, and still meet the AI for both.

Chloride Deficiency and Toxicity   Diets rarely lack chloride. Chloride losses may occur in conditions such as heavy sweating, chronic diarrhea, and vomiting. The only known cause of high blood chloride concentrations is dehydration due to water deficiency. In both cases, consuming ordinary foods and beverages can restore chloride balance.

■ Reminder: The loss of acid can lead to *alkalosis,* an above-normal alkalinity in the blood and body fluids.

■ Salt (sodium chloride) is about 60% chloride.
1 g salt contributes 600 mg chloride.
5 g salt = 1 tsp.
1 tsp salt contributes 3000 mg chloride.

IN SUMMARY   Chloride is the major anion outside cells, and it associates closely with sodium. In addition to its role in fluid balance, chloride is part of the stomach's hydrochloric acid. The accompanying table summarizes information on chloride.

Chloride

2004 Adequate Intake (AI)

Adults: 2300 mg/day (19–50 yr)
2000 mg/day (51—70 yr)
1800 mg/day (>70 yr)

Upper Level

Adults: 3600 mg/day

(continued)

chloride (KLO-ride): the major anion in the extracellular fluids of the body. Chloride is the ionic form of chlorine, Cl$^-$; see Appendix B for a description of the chlorine-to-chloride conversion.

*Hydrochloric acid secretion into the stomach involves the addition of bicarbonate ions (base) to the plasma. These bicarbonate ions (HCO$_3^-$) are neutralized by hydrogen ions (H$^+$) from the gastric secretions that are reabsorbed into the plasma. When hydrochloric acid is lost during vomiting, these hydrogen ions are no longer available for reabsorption, and so, in effect, the concentrations of bicarbonate ions in the plasma are increased. In this way, excessive vomiting of acidic gastric juices leads to *metabolic alkalosis.*

Chloride (continued)

Chief Functions in the Body

Maintains normal fluid and electrolyte balance; part of hydrochloric acid found in the stomach, necessary for proper digestion

Deficiency Symptoms

Do not occur under normal circumstances

Toxicity Symptoms

Vomiting

Significant Sources

Table salt, soy sauce; moderate amounts in meats, milks, eggs; large amounts in processed foods

Potassium

Like sodium, **potassium** is a positively charged ion. In contrast to sodium, potassium is the body's principal cation *inside* the body cells.

Potassium Roles in the Body Potassium plays a major role in maintaining fluid and electrolyte balance and cell integrity. During nerve impulse transmission and muscle contraction, potassium and sodium briefly trade places across the cell membrane. The cell then quickly pumps them back into place. Controlling potassium distribution is a high priority for the body because it affects many aspects of homeostasis, including a steady heartbeat.

Potassium Recommendations and Intakes Potassium is abundant in all living cells, both plant and animal. Because cells remain intact unless foods are processed, the richest sources of potassium are *fresh* foods of all kinds—as Figure 12-9 (on p. 412) shows. In contrast, most processed foods such as canned vegetables, ready-to-eat cereals, and luncheon meats contain less potassium—and more sodium (recall Figure 12-8). To meet the AI for potassium, most people need to increase their intake of fruits and vegetables to five to nine servings daily.

Potassium and Hypertension Diets low in potassium seem to play an important role in the development of high blood pressure. Low potassium intakes raise blood pressure, whereas high potassium intakes appear to both prevent and correct hypertension.[16]■ Potassium-rich fruits and vegetables also appear to reduce the risk of stroke—more so than can be explained by the reduction in blood pressure alone.

Potassium Deficiency Potassium deficiency is the most common electrolyte imbalance. It is more often caused by excessive losses than by deficient intakes. Conditions such as diabetic acidosis, dehydration, or prolonged vomiting or diarrhea can create a potassium deficiency, as can the regular use of certain drugs, including diuretics, steroids, and strong laxatives.[*] For this reason, many physicians prescribe potassium supplements along with these potassium-wasting drugs. One of the earliest symptoms of deficiency is muscle weakness.

Potassium Toxicity Potassium toxicity does not result from overeating foods high in potassium; therefore an Upper Level was not set. It can result from overconsumption of potassium salts or supplements (including some "energy fitness shakes") and from certain diseases or treatments.[17] Given more potassium than the body needs, the kidneys accelerate their excretion. If the GI tract is bypassed, however, and potassium is injected directly into a vein, it can stop the heart.

Fresh foods, especially fruits and vegetables, provide potassium in abundance.

■ Reminder: The DASH diet, used to lower blood pressure, emphasizes potassium-rich foods such as fruits and vegetables.

IN SUMMARY Potassium, like sodium and chloride, is an electrolyte that plays an important role in maintaining fluid balance. Potassium is the primary cation inside cells; fresh foods, notably fruits and vegetables, are its best sources. The table on the next page summarizes facts about potassium.

potassium: the principal cation within the body's cells; critical to the maintenance of fluid balance, nerve impulse transmissions, and muscle contractions.

---

[*]People using diuretics to control hypertension should know that some cause potassium excretion and can induce a deficiency. Those using these drugs must be particularly careful to include rich sources of potassium in their daily diets. (Some diuretics are designed to spare potassium.)

FIGURE 12-9 Potassium in Selected Foods

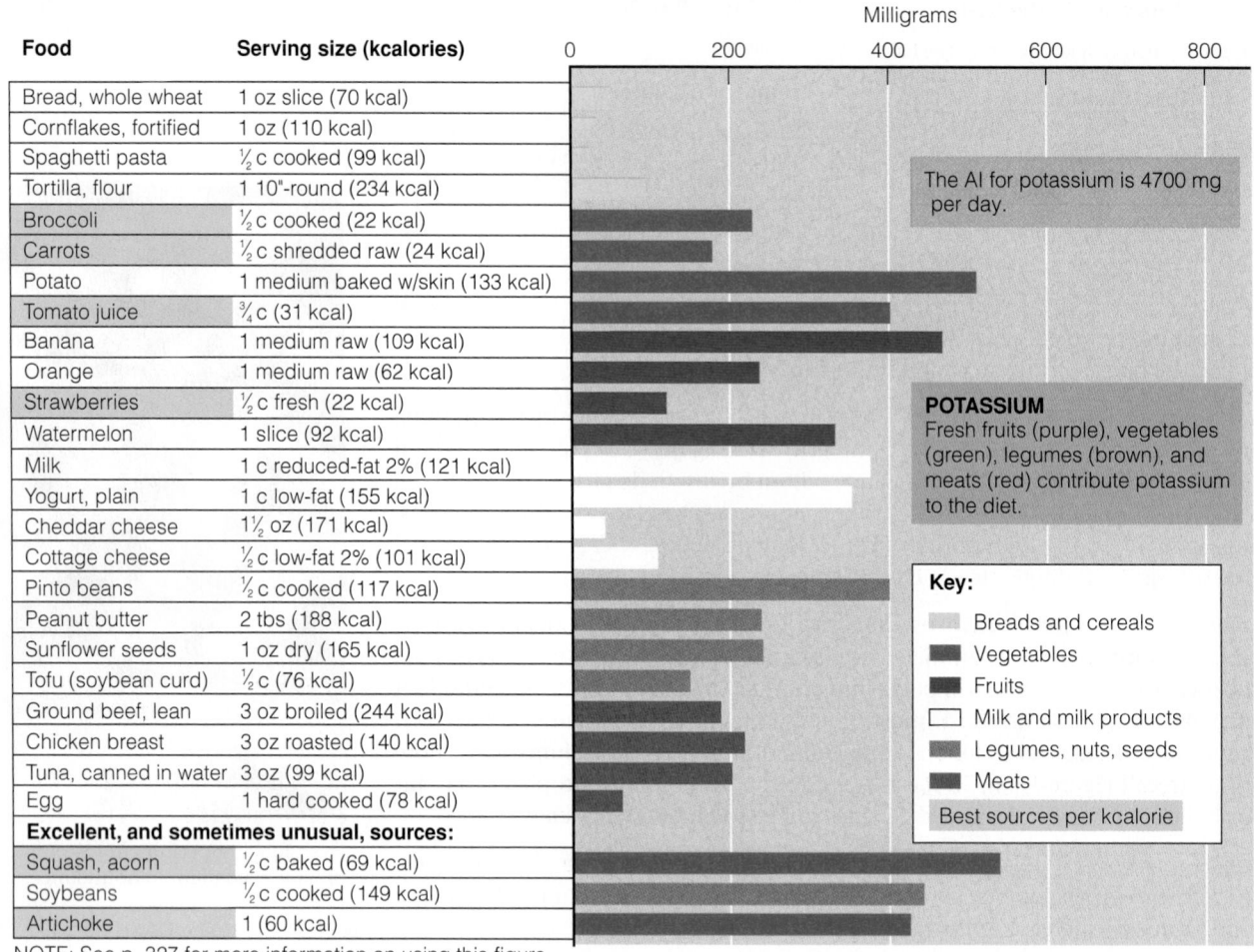

Food	Serving size (kcalories)	Milligrams
Bread, whole wheat	1 oz slice (70 kcal)	
Cornflakes, fortified	1 oz (110 kcal)	
Spaghetti pasta	½ c cooked (99 kcal)	
Tortilla, flour	1 10"-round (234 kcal)	
Broccoli	½ c cooked (22 kcal)	
Carrots	½ c shredded raw (24 kcal)	
Potato	1 medium baked w/skin (133 kcal)	
Tomato juice	¾ c (31 kcal)	
Banana	1 medium raw (109 kcal)	
Orange	1 medium raw (62 kcal)	
Strawberries	½ c fresh (22 kcal)	
Watermelon	1 slice (92 kcal)	
Milk	1 c reduced-fat 2% (121 kcal)	
Yogurt, plain	1 c low-fat (155 kcal)	
Cheddar cheese	1½ oz (171 kcal)	
Cottage cheese	½ c low-fat 2% (101 kcal)	
Pinto beans	½ c cooked (117 kcal)	
Peanut butter	2 tbs (188 kcal)	
Sunflower seeds	1 oz dry (165 kcal)	
Tofu (soybean curd)	½ c (76 kcal)	
Ground beef, lean	3 oz broiled (244 kcal)	
Chicken breast	3 oz roasted (140 kcal)	
Tuna, canned in water	3 oz (99 kcal)	
Egg	1 hard cooked (78 kcal)	
Excellent, and sometimes unusual, sources:		
Squash, acorn	½ c baked (69 kcal)	
Soybeans	½ c cooked (149 kcal)	
Artichoke	1 (60 kcal)	

The AI for potassium is 4700 mg per day.

POTASSIUM
Fresh fruits (purple), vegetables (green), legumes (brown), and meats (red) contribute potassium to the diet.

Key:
- Breads and cereals
- Vegetables
- Fruits
- Milk and milk products
- Legumes, nuts, seeds
- Meats

Best sources per kcalorie

NOTE: See p. 327 for more information on using this figure.

Potassium

2004 Adequate Intake (AI)

Adults: 4700 mg/day

Chief Functions in the Body

Maintains normal fluid and electrolyte balance; facilitates many reactions; supports cell integrity; assists in nerve impulse transmission and muscle contractions

Deficiency Symptoms[a]

Muscular weakness, paralysis, confusion

Toxicity Symptoms

Muscular weakness; vomiting; if given into a vein, can stop the heart

Significant Sources

All whole foods: meats, milks, fruits, vegetables, grains, legumes

[a]Deficiency accompanies dehydration.

Calcium

calcium: the most abundant mineral in the body; found primarily in the body's bones and teeth.

Calcium is the most abundant mineral in the body. It receives much emphasis in this chapter and in the highlight that follows because an adequate intake helps grow a healthy skeleton in early life and minimize bone loss in later life.

Calcium Roles in the Body

Ninety-nine percent of the body's calcium is in the bones (and teeth), where it plays two roles. First, it is an integral part of bone structure, providing a rigid frame that holds the body upright and serves as attachment points for muscles, making motion possible. Second, it serves as a calcium bank, offering a readily available source of the mineral to the body fluids should a drop in blood calcium occur.

Calcium in Bones As bones begin to form, calcium salts form crystals, called **hydroxyapatite,** on a matrix of the protein collagen. During **mineralization,** as the crystals become denser, they give strength and rigidity to the maturing bones. As a result, the long leg bones of children can support their weight by the time they have learned to walk.

Many people have the idea that once a bone is built, it is inert like a rock. Actually, the bones are gaining and losing minerals continuously in an ongoing process of remodeling. Growing children gain more bone than they lose, and healthy adults maintain a reasonable balance. When withdrawals substantially exceed deposits, problems such as osteoporosis develop (as described in Highlight 12).

The formation of teeth follows a pattern similar to that of bones. The turnover of minerals in teeth is not as rapid as in bone, however; fluoride hardens and stabilizes the crystals of teeth, opposing the withdrawal of minerals from them.

Calcium in Body Fluids The 1 percent of the body's calcium that circulates in the fluids as ionized calcium is vital to life. The calcium ion participates in the regulation of muscle contractions, the clotting of blood, the transmission of nerve impulses, the secretion of hormones, and the activation of some enzyme reactions.

Calcium also activates a protein called **calmodulin.** This protein relays messages from the cell surface to the inside of the cell. Several of these messages help to maintain normal blood pressure.

Calcium and Disease Prevention Calcium may protect against hypertension.[18] For this reason, restricting sodium to treat hypertension is narrow advice, especially considering the success of the DASH diet in lowering blood pressure. The DASH diet is not particularly low in sodium, but it is rich in calcium, as well as in magnesium and potassium. As mentioned earlier, the DASH diet, together with a reduced sodium intake, is more effective in lowering blood pressure than either strategy alone. Some research also suggests protective relationships between dietary calcium and blood cholesterol, diabetes, and colon cancer.[19] Highlight 12 explores calcium's role in preventing osteoporosis.

Calcium and Obesity Calcium may also play a role in maintaining a healthy body weight.[20] Analyses of national survey data as well as small clinical studies show an inverse relationship between calcium intake and body fatness: the higher the calcium intake, the lower the body fatness.[21] In particular, calcium from dairy foods seems to exert greater effects on body weight than calcium in supplement form.[22] Notably, calcium intake does not simply reflect a healthy lifestyle; the relationship remains strong even when factors such as energy intake and exercise are considered. An adequate dietary calcium intake may help prevent excessive fat accumulation by stimulating hormonal action that targets the breakdown of stored fat.[23] Large, well-designed clinical studies are needed to confirm the effects of dietary calcium intake on body weight.

Calcium Balance Calcium homeostasis is one of the body's highest priorities and involves a system of hormones and vitamin D. Whenever blood calcium falls too low or rises too high, three organ systems respond: the intestines, bones, and kidneys. Figure 12-10 illustrates how vitamin D and the hormones **parathormone** and **calcitonin** return blood calcium to normal.

The calcium in bone provides a nearly inexhaustible bank of calcium for the blood. The blood borrows and returns calcium as needed so that even with a dietary deficiency, *blood* calcium remains normal—even as *bone* calcium diminishes

---

hydroxyapatite (high-drox-ee-APP-ah-tite): crystals made of calcium and phosphorus.

mineralization: the process in which calcium, phosphorus, and other minerals crystallize on the collagen matrix of a growing bone, hardening the bone.

calmodulin (cal-MOD-you-lin): an inactive protein that becomes active when bound to calcium. Once activated, it becomes a messenger that tells other proteins what to do. The system serves as an interpreter for hormone- and nerve-mediated messages arriving at cells.

parathormone (PAIR-ah-THOR-moan): a hormone from the parathyroid glands that regulates blood calcium by raising it when levels fall too low; also known as **parathyroid hormone.**

calcitonin (KAL-see-TOE-nin): a hormone from the thyroid gland that regulates blood calcium by lowering it when levels rise too high.

FIGURE 12-10 Calcium Balance

Blood calcium is regulated in part by vitamin D and two hormones—calcitonin and parathormone. Bone serves as a reservoir when blood calcium is high and as a source of calcium when blood calcium is low. Osteoclasts break down bone and release calcium into the blood; osteoblasts build new bone using calcium from the blood.

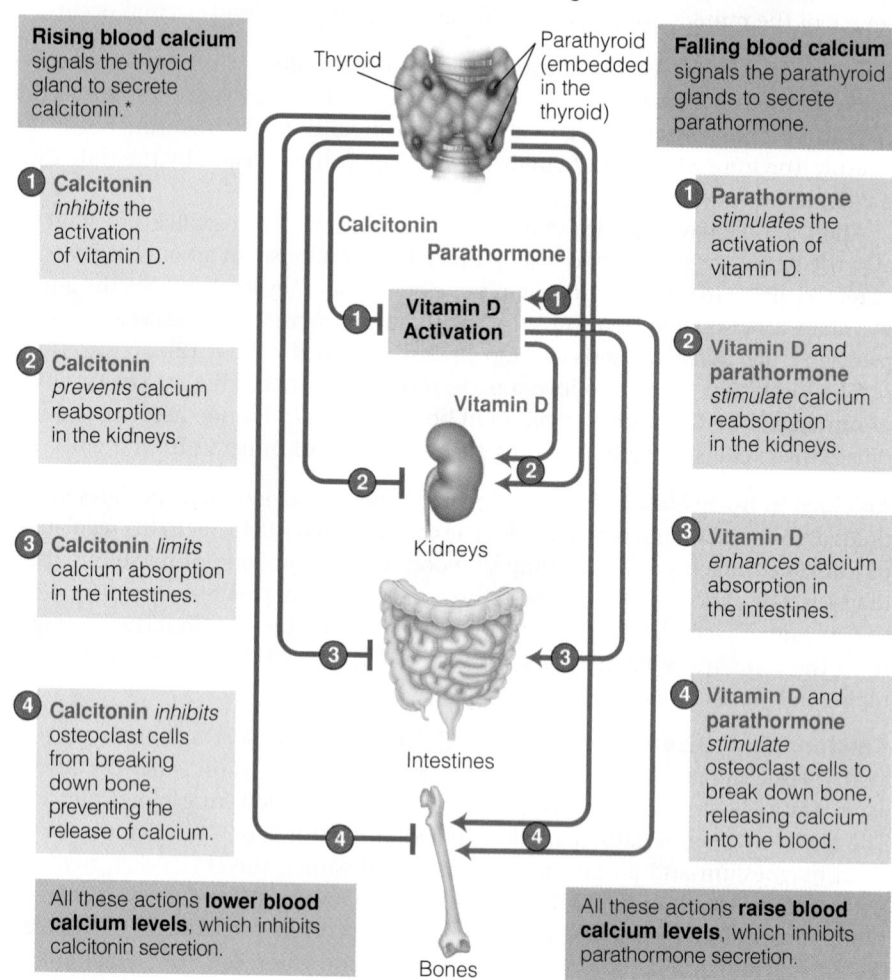

Rising blood calcium signals the thyroid gland to secrete calcitonin.*

1 Calcitonin *inhibits* the activation of vitamin D.

2 Calcitonin *prevents* calcium reabsorption in the kidneys.

3 Calcitonin *limits* calcium absorption in the intestines.

4 Calcitonin *inhibits* osteoclast cells from breaking down bone, preventing the release of calcium.

All these actions **lower blood calcium levels**, which inhibits calcitonin secretion.

Falling blood calcium signals the parathyroid glands to secrete parathormone.

1 Parathormone *stimulates* the activation of vitamin D.

2 Vitamin D and **parathormone** *stimulate* calcium reabsorption in the kidneys.

3 Vitamin D *enhances* calcium absorption in the intestines.

4 Vitamin D and **parathormone** *stimulate* osteoclast cells to break down bone, releasing calcium into the blood.

All these actions **raise blood calcium levels**, which inhibits parathormone secretion.

Thyroid · Parathyroid (embedded in the thyroid) · Calcitonin · Parathormone · **Vitamin D Activation** · Vitamin D · Kidneys · Intestines · Bones

*Calcitonin plays a major role in defending infants and young children against the dangers of rising blood calcium that can occur when regular feedings of milk deliver large quanities of calcium to a small body. In contrast, calcitonin plays a relatively minor role in adults because their absorption of calcium is less efficient and their bodies are larger, making elevated blood calcium unlikely.

(see Figure 12-11). Blood calcium changes only in response to abnormal regulatory control, not to diet. A person can have an inadequate calcium intake for years and suffer no noticeable symptoms. Only late in life does it become apparent that the integrity of the bones has been compromised.

Blood calcium above normal results in **calcium rigor:** the muscles contract and cannot relax. Similarly, blood calcium below normal causes **calcium tetany**—also characterized by uncontrolled muscle contraction. These conditions do *not* reflect a *dietary* excess or lack of calcium; they are caused by a lack of vitamin D or by abnormal secretion of the regulatory hormones. A chronic *dietary* deficiency of calcium, or a chronic deficiency due to poor absorption over the years, depletes the savings account in the bones. Again: the *bones*, not the blood, are robbed by a calcium deficiency.

Calcium Absorption Many factors affect calcium absorption, but on average, adults absorb about 25 percent of the calcium they ingest. The stomach's acidity helps to keep calcium soluble, and vitamin D helps to make the **calcium-binding protein** needed for absorption. (This explains why calcium-rich milk is the best food for vitamin D fortification.)

calcium rigor: hardness or stiffness of the muscles caused by high blood calcium concentrations.

calcium tetany (TET-ah-nee): intermittent spasm of the extremities due to nervous and muscular excitability caused by low blood calcium concentrations.

calcium-binding protein: a protein in the intestinal cells, made with the help of vitamin D, that facilitates calcium absorption.

Whenever calcium is needed, the body increases its production of the calcium-binding protein to improve calcium absorption. The result is obvious in the case of a pregnant woman, who absorbs 50 percent of the calcium from the milk she drinks. Similarly, growing children absorb 50 to 60 percent of the calcium they consume. Then, when bone growth slows or stops, absorption falls to the adult level of about 25 percent. In addition, absorption becomes more efficient during times of inadequate intakes.[24]

Many of the conditions that enhance calcium absorption inhibit its absorption when they are absent. For example, sufficient vitamin D supports absorption, while a deficiency impairs it. In addition, fiber, in general, and the binders phytate and oxalate, in particular, interfere with calcium absorption, but their effects are relatively minor at intakes typical of U.S. diets. Vegetables with oxalates and whole grains with phytates are nutritious foods, of course, but they are not useful calcium sources. The margin■ presents factors that influence calcium balance.

Calcium Recommendations and Sources

Calcium is unlike most other nutrients, in that hormones maintain its *blood* concentration regardless of dietary intake. As Figure 12-11 shows, when intake is high, the *bones* benefit; when intake is low, the *bones* suffer. Calcium recommendations are therefore based on the amount needed to retain the most calcium in bones. By retaining the most calcium possible, the bones can develop to their fullest potential in size and density—their **peak bone mass**—within genetic limits.

Calcium Recommendations Because obtaining enough calcium during growth helps to ensure that the skeleton will be strong and dense, recommendations have been set high at 1300 milligrams daily for adolescents up to the age of 18 years. Between the ages of 19 and 50, recommendations are lowered to 1000 milligrams a day; for later life, recommendations are raised again to 1200 milligrams a day to minimize the bone loss that tends to occur later in life. Some authorities advocate as much as 1500 milligrams a day for women over 50. Many people in the United States and Canada, particularly women, have calcium intakes far below current recommendations. High intakes of calcium from supplements may have adverse effects such as kidney stone formation; for this reason, an Upper Level has been established (see inside front cover).

High intakes of both dietary protein and sodium increase calcium losses, but whether these losses impair bone development remains unclear.[25] In establishing an Adequate Intake (AI) for calcium, the DRI Committee considered these nutrient interactions, but did not adjust dietary recommendations based on this information.

Calcium in Milk Products Figure 12-12 on p. 416 shows that calcium is found most abundantly in a single class of foods—milk■ and milk products. Unfortunately, many people, for a variety of reasons, cannot or do not drink milk. For example, some people are lactose intolerant, while others simply do not enjoy the taste of milk.

The person who doesn't like to drink milk may prefer to eat cheese or yogurt. Alternatively, milk and milk products can be concealed in foods. Powdered fat-free milk can be added to casseroles, soups, and other mixed dishes during preparation; 5 heaping tablespoons offer the equivalent of a cup of milk. This simple step is an excellent way for older women to obtain not only extra calcium, but more protein, vitamins, and minerals as well.

It is especially difficult for children who don't drink milk to meet their calcium needs. Children who don't drink milk have lower calcium intakes and poorer bone

FIGURE 12-11 Maintaining Blood Calcium from the Diet and from the Bones

With an adequate intake of calcium-rich food, blood calcium remains normal . . .

With a dietary deficiency, blood calcium still remains normal . . .

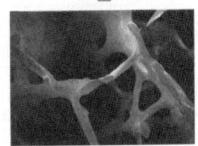

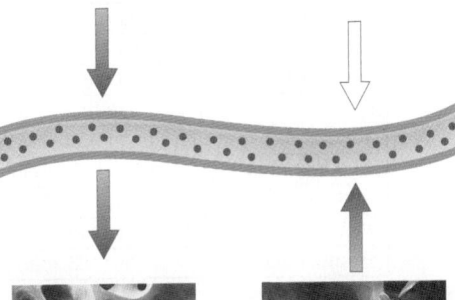

. . . and bones deposit calcium. The result is strong, dense bones.

. . . because bones give up calcium to the blood. The result is weak, osteoporotic bones.

© David Dempster from J Bone Miner Res,1986 (both)

■ Factors that *enhance* calcium absorption:
- Stomach acid.
- Vitamin D.
- Lactose.
- Growth hormones.

Factors that *inhibit* calcium absorption:
- Lack of stomach acid.
- Vitamin D deficiency.
- High phosphorus intake.
- High-fiber diet.
- Phytates (in seeds, nuts, grains).
- Oxalates (in beet greens, rhubarb, spinach).

■ Suggested minimum daily milk servings:
- Young children (2 to 8 yr): 2 c.
- Adults: 2 c.
- Older children, teenagers, and young adults (9 to 24 yr): 3 c.
- Older adults (over 50 yr): 3 c.
- Pregnant or lactating women: 3 c.

peak bone mass: the highest attainable bone size and density for an individual, developed during the first three decades of life.

FIGURE 12-12 Calcium in Selected Foods

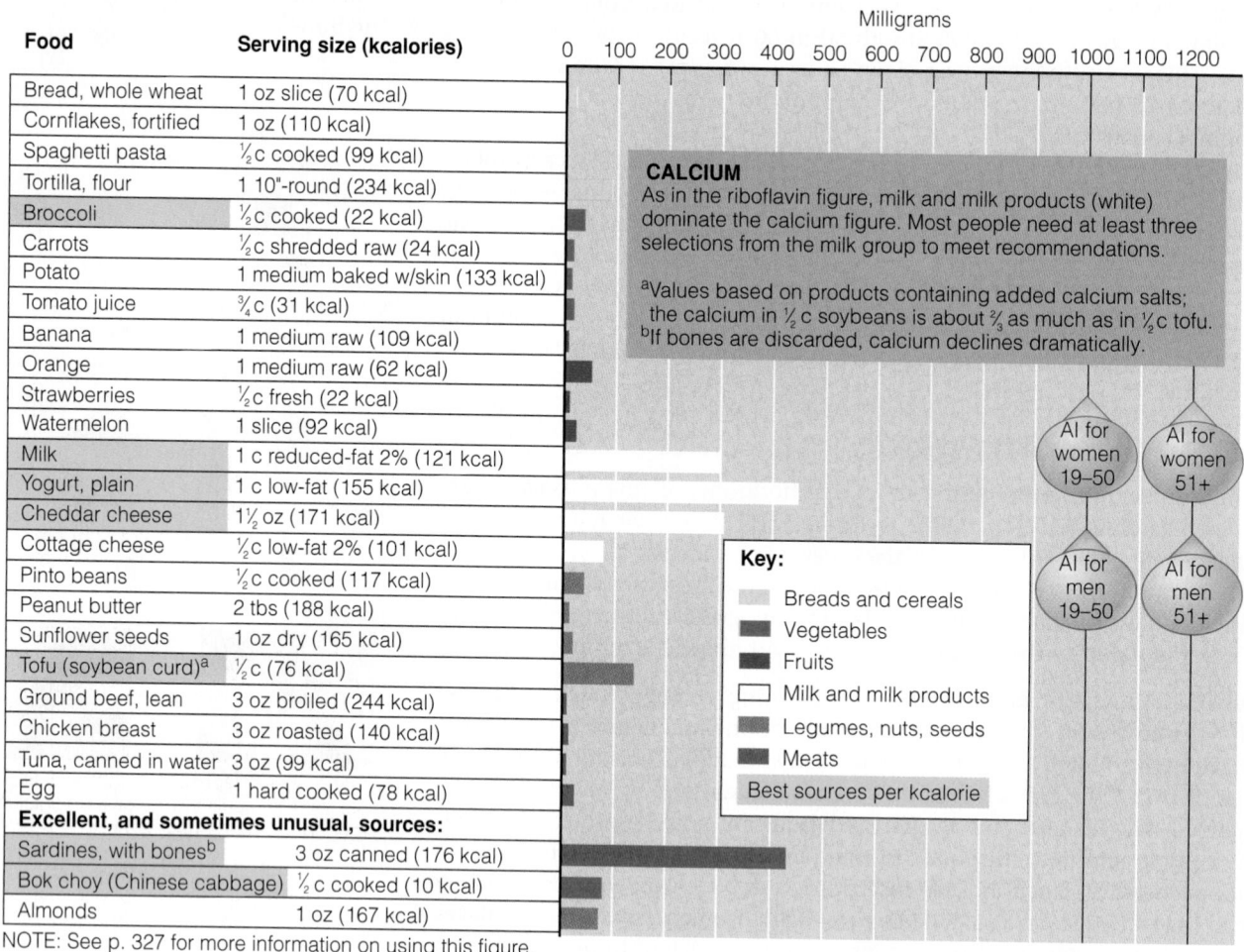

Food	Serving size (kcalories)
Bread, whole wheat	1 oz slice (70 kcal)
Cornflakes, fortified	1 oz (110 kcal)
Spaghetti pasta	½ c cooked (99 kcal)
Tortilla, flour	1 10"-round (234 kcal)
Broccoli	½ c cooked (22 kcal)
Carrots	½ c shredded raw (24 kcal)
Potato	1 medium baked w/skin (133 kcal)
Tomato juice	¾ c (31 kcal)
Banana	1 medium raw (109 kcal)
Orange	1 medium raw (62 kcal)
Strawberries	½ c fresh (22 kcal)
Watermelon	1 slice (92 kcal)
Milk	1 c reduced-fat 2% (121 kcal)
Yogurt, plain	1 c low-fat (155 kcal)
Cheddar cheese	1½ oz (171 kcal)
Cottage cheese	½ c low-fat 2% (101 kcal)
Pinto beans	½ c cooked (117 kcal)
Peanut butter	2 tbs (188 kcal)
Sunflower seeds	1 oz dry (165 kcal)
Tofu (soybean curd)[a]	½ c (76 kcal)
Ground beef, lean	3 oz broiled (244 kcal)
Chicken breast	3 oz roasted (140 kcal)
Tuna, canned in water	3 oz (99 kcal)
Egg	1 hard cooked (78 kcal)
Excellent, and sometimes unusual, sources:	
Sardines, with bones[b]	3 oz canned (176 kcal)
Bok choy (Chinese cabbage)	½ c cooked (10 kcal)
Almonds	1 oz (167 kcal)

CALCIUM
As in the riboflavin figure, milk and milk products (white) dominate the calcium figure. Most people need at least three selections from the milk group to meet recommendations.

[a]Values based on products containing added calcium salts; the calcium in ½ c soybeans is about ⅔ as much as in ½ c tofu.
[b]If bones are discarded, calcium declines dramatically.

AI for women 19–50
AI for women 51+
AI for men 19–50
AI for men 51+

Key:
- Breads and cereals
- Vegetables
- Fruits
- Milk and milk products
- Legumes, nuts, seeds
- Meats

Best sources per kcalorie

NOTE: See p. 327 for more information on using this figure.

■ People with lactose intolerance may be able to consume small quantities of milk, as Chapter 4 explains.

health than those who drink milk regularly, and are shorter as well.[26] The consequences of drinking too little milk during childhood and adolescence persist into adulthood. Women who seldom drank milk as children or teenagers have lower bone density and greater risk of fractures than those who drank milk regularly.[27] It is possible for people who do not drink milk to obtain adequate calcium, but only if they carefully select other calcium-rich foods.

Calcium in Other Foods Some cultures do not use milk in their cuisines; some vegetarians exclude milk as well as meat; and some people are allergic to milk protein or are lactose intolerant.■ These people need to find nonmilk sources of calcium to help meet their calcium needs. Some brands of tofu, corn tortillas, some nuts (such as almonds), and some seeds (such as sesame seeds) can supply calcium for the person who doesn't use milk products. A slice of most breads contains only about 5 to 10 percent of the calcium found in milk, but can be a major source for people who eat many slices because the calcium is well absorbed.

Among the vegetables, mustard and turnip greens, bok choy, kale, parsley, watercress, and broccoli are good sources of available calcium. So are some seaweeds such as the nori popular in Japanese cooking. Some dark green, leafy vegetables—notably spinach and Swiss chard—appear to be calcium-rich but actually provide little, if any, calcium to the body because of the binders they contain. It would take 8 cups of spinach—containing six times as much calcium as 1 cup of milk—to deliver the equivalent in *absorbable* calcium.[28]

With the exception of foods such as spinach that contain calcium binders, however, the calcium content of foods is usually more important than bioavailability.[29] Consequently, recognizing that people eat a variety of foods containing calcium, the DRI Committee did not consider calcium bioavailability when setting recommendations. The margin drawing■ ranks selected foods according to their calcium bioavailability.

Oysters are also a rich source of calcium, as are small fish eaten with their bones, such as canned sardines. Many Asians prepare a stock from bones that helps account for their adequate calcium intake without the use of milk. They soak the cracked bones from chicken, turkey, pork, or fish in vinegar and then slowly boil the bones until they become soft. The bones release calcium into the acidic broth, and most of the vinegar boils off. Cooks then use the stock, which contains more than 100 milligrams of calcium per tablespoon, in place of water to prepare soups, vegetables, and rice. Similarly, cooks in the Navajo tribe use an ash prepared from the branches and needles of the juniper tree in their recipes. One teaspoon of juniper ash provides about as much calcium as a cup of milk.

Some mineral waters provide as much as 500 milligrams of calcium per liter, offering a convenient way to meet both calcium and water needs.[30] Similarly, calcium-fortified orange juice and other fruit and vegetable juices allow a person to meet both calcium and vitamin needs easily. Other examples of calcium-fortified foods include high-calcium milk (milk with extra calcium added) and calcium-fortified cereals. The "How to" on p. 418 describes a shortcut method for estimating your calcium intake. Highlight 12 discusses calcium supplements.

A generalization that has been gaining strength throughout this book is supported by the information given here about calcium. A balanced diet that supplies a variety of foods is the best plan to ensure adequacy for all essential nutrients. All food groups should be included, and none should be overemphasized. In our culture, calcium intake is usually inadequate wherever milk is lacking in the diet—whether through ignorance, poverty, simple dislike, fad dieting, lactose intolerance, or allergy. By contrast, iron is usually lacking whenever milk is overemphasized, as Chapter 13 explains.

Calcium Deficiency

A low calcium intake during the growing years limits the bones' ability to reach their optimal mass and density. Most people achieve a peak bone mass by their late 20s, and dense bones best protect against age-related bone loss and fractures (see Figure 12-13). All adults lose bone as they grow older, beginning between the ages of 30 and 40. Should bone losses reach the point of causing fractures under common, everyday stresses, the condition is known as **osteoporosis.** Osteoporosis afflicts more than 25 million people in the United States, mostly older women.

Milk and milk products are rightly famous for their calcium contents.

© Thomas Harm, Tom Peterson/Quest Photographic Inc.

■ Bioavailability of Calcium from Selected Foods

≥50% absorbed	Cauliflower, watercress, brussels sprouts, rutabaga, kale, mustard greens, bok choy, broccoli, turnip greens
≈30% absorbed	Milk, calcium-fortified soy milk, calcium-set tofu, cheese, yogurt, calcium-fortified foods and beverages
≈20% absorbed	Almonds, sesame seeds, pinto beans, sweet potatoes
≤5% absorbed	Spinach, rhubarb, Swiss chard

FIGURE 12-13 Phases of Bone Development throughout Life

The active growth phase occurs from birth to approximately age 20. The next phase of peak bone mass development occurs between the ages of 12 and 30. The final phase, when bone resorption exceeds formation, begins between the ages of 30 and 40 and continues through the remainder of life.

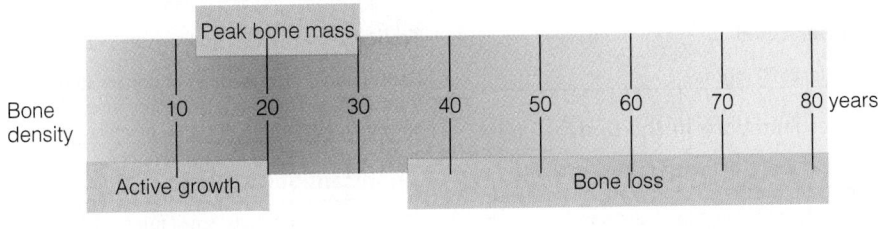

Bone density

Peak bone mass

10   20   30   40   50   60   70   80 years

Active growth

Bone loss

osteoporosis (OS-tee-oh-pore-OH-sis): a disease in which the bones become porous and fragile due to a loss of minerals; also called **adult bone loss.**
• osteo = bone
• porosis = porous

HOW TO Estimate Your Calcium Intake

Most dietitians have developed useful shortcuts to help them estimate nutrient intakes and "see" inadequacies in the diet. They can tell at a glance whether a day's meals fall short of calcium recommendations, for example.

To estimate calcium intakes, keep two bits of information in mind:

- A cup of milk provides about 300 milligrams of calcium.
- Adults need between 1000 and 1200 milligrams of calcium per day, which represents 3 to 4 cups of milk—or the equivalent:

$$1000 \text{ mg} \div 300 \text{ mg/c} = 3\frac{1}{3} \text{ c.}$$
$$1200 \text{ mg} \div 300 \text{ mg/c} = 4 \text{ c.}$$

If a person drinks 3 to 4 cups of milk a day, it's easy to see that calcium needs are being met. If not, it takes some detective work to identify the other sources and estimate total calcium intake.

To estimate a person's daily calcium intake, use this shortcut, which compares the calcium in calcium-rich foods to the calcium content of milk. The calcium in a cup of milk is assigned 1 point, and the goal is to attain 3 to 4 points per day. Foods are given points as follows:

- 1 c milk, yogurt, or fortified soy milk or 1½ oz cheese = 1 point.
- 4 oz canned fish with bones (sardines) = 1 point.
- 1 c ice cream, cottage cheese, or calcium-rich vegetable (see the text) = ½ point.

Then, because other foods also contribute small amounts of calcium, together they are given a point.

- Well-balanced diet containing a variety of foods = 1 point.

Now consider a day's meals with calcium in mind. Cereal with 1 cup of milk for breakfast (1 point for milk), a ham and cheese sub sandwich for lunch (1 point for cheese), and a cup of broccoli and lasagna for dinner (½ point for calcium-rich vegetable and 1 point for cheese in lasagna)—plus 1 point for all other foods eaten that day—adds up to 4½ points. This shortcut estimate indicates that calcium recommendations have been met, and a diet analysis of these few foods reveals a calcium intake of over 1000 milligrams. By knowing the best sources of each nutrient, you can learn to scan the day's meals and quickly see if you are meeting your daily goals.

HEALTHY PEOPLE 2010

Reduce the prevalence of osteoporosis, among people aged 50 and over.

Unlike many diseases that make themselves known through symptoms such as pain, shortness of breath, skin lesions, tiredness, and the like, osteoporosis is silent. The body sends no signals saying bones are losing their calcium and, as a result, their integrity. Blood samples offer no clues because blood calcium remains normal regardless of bone content, and measures of bone density are not routinely taken. Highlight 12 suggests strategies to protect against bone loss, of which eating calcium-rich foods is only one. Even during adulthood, however, high calcium intakes may promote bone strength, prevent further deterioration, and reverse bone loss.[31]

IN SUMMARY

Most of the body's calcium is in the bones where it provides a rigid structure and a reservoir of calcium for the blood. Blood calcium participates in muscle contraction, blood clotting, and nerve impulses and is closely regulated by a system of hormones and vitamin D. Calcium is found predominantly in milk and milk products, but some other foods including certain vegetables and tofu also provide calcium. Even when calcium intake is inadequate, blood calcium remains normal, but at the expense of bone loss, which can lead to osteoporosis. Calcium's roles, deficiency symptoms, and food sources are summarized below.

Calcium

1997 Adequate Intake (AI)

Adults: 1000 mg/day (19–50 yr)
1200 mg/day (>51 yr)

Upper Level

Adults: 2500 mg/day

Chief Functions in the Body

Mineralization of bones and teeth; also involved in muscle contraction and relaxation, nerve functioning, blood clotting, blood pressure, and immune defenses

Deficiency Symptoms

Stunted growth in children; bone loss (osteoporosis) in adults

Toxicity Symptoms

Constipation; increased risk of urinary stone formation and kidney dysfunction; interference with absorption of other minerals

Significant Sources

Milk and milk products, small fish (with bones), tofu (bean curd), greens (broccoli, chard), legumes

Phosphorus

Phosphorus is the second most abundant mineral in the body. About 85 percent of it is found combined with calcium in the hydroxyapatite crystals of bones and teeth.

Phosphorus Roles in the Body Phosphorus salts (phosphates) are found not only in bones and teeth, but in all body cells as part of a major buffer system (phosphoric acid and its salts). Phosphorus is also part of DNA and RNA and is therefore necessary for all growth.

Phosphorus assists in energy metabolism. Many enzymes and the B vitamins become active only when a phosphate group is attached. ATP itself, the energy currency of the cells, uses three phosphate groups to do its work.

Lipids containing phosphorus as part of their structures (phospholipids) help to transport other lipids in the blood. Phospholipids are also the major structural components of cell membranes, where they control the transport of nutrients into and out of the cells. Some proteins, such as the casein in milk, contain phosphorus as part of their structures (phosphoproteins).

Phosphorus Recommendations and Intakes Diets that provide adequate energy and protein also supply adequate phosphorus. Dietary deficiencies of phosphorus are unknown. As Figure 12-14 shows, foods rich in proteins are the best sources of phosphorus. In addition to legumes and foods from the milk and meat groups, processed foods (including soft drinks) are usually high in phosphorus (from the additives).

phosphorus: a major mineral found mostly in the body's bones and teeth.

FIGURE 12-14 Phosphorus in Selected Foods

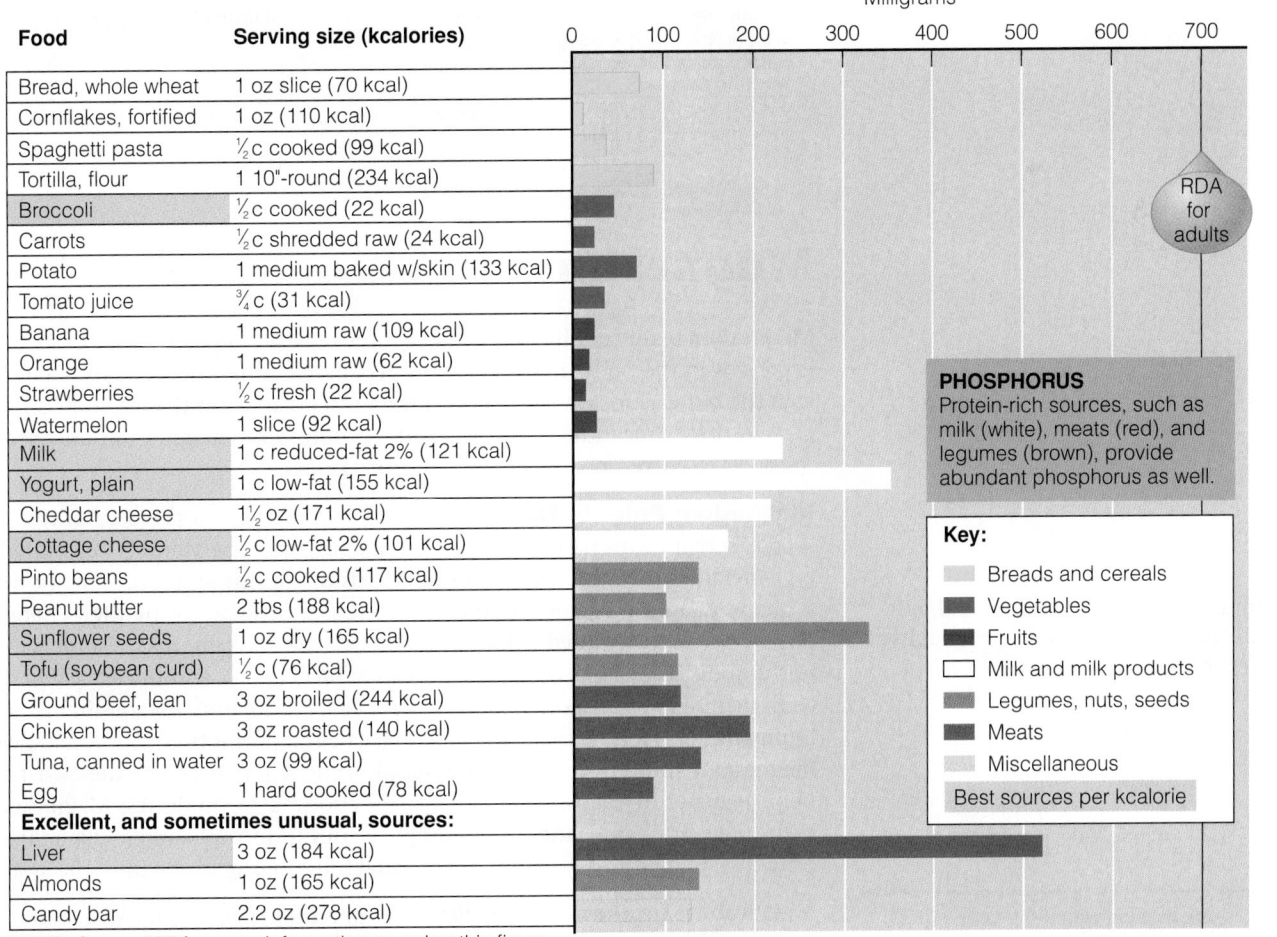

PHOSPHORUS
Protein-rich sources, such as milk (white), meats (red), and legumes (brown), provide abundant phosphorus as well.

Key:
- Breads and cereals
- Vegetables
- Fruits
- Milk and milk products
- Legumes, nuts, seeds
- Meats
- Miscellaneous

Best sources per kcalorie

Food	Serving size (kcalories)
Bread, whole wheat	1 oz slice (70 kcal)
Cornflakes, fortified	1 oz (110 kcal)
Spaghetti pasta	½ c cooked (99 kcal)
Tortilla, flour	1 10"-round (234 kcal)
Broccoli	½ c cooked (22 kcal)
Carrots	½ c shredded raw (24 kcal)
Potato	1 medium baked w/skin (133 kcal)
Tomato juice	¾ c (31 kcal)
Banana	1 medium raw (109 kcal)
Orange	1 medium raw (62 kcal)
Strawberries	½ c fresh (22 kcal)
Watermelon	1 slice (92 kcal)
Milk	1 c reduced-fat 2% (121 kcal)
Yogurt, plain	1 c low-fat (155 kcal)
Cheddar cheese	1½ oz (171 kcal)
Cottage cheese	½ c low-fat 2% (101 kcal)
Pinto beans	½ c cooked (117 kcal)
Peanut butter	2 tbs (188 kcal)
Sunflower seeds	1 oz dry (165 kcal)
Tofu (soybean curd)	½ c (76 kcal)
Ground beef, lean	3 oz broiled (244 kcal)
Chicken breast	3 oz roasted (140 kcal)
Tuna, canned in water	3 oz (99 kcal)
Egg	1 hard cooked (78 kcal)
Excellent, and sometimes unusual, sources:	
Liver	3 oz (184 kcal)
Almonds	1 oz (165 kcal)
Candy bar	2.2 oz (278 kcal)

RDA for adults

NOTE: See p. 327 for more information on using this figure.

In the past, researchers emphasized the importance of an ideal calcium-to-phosphorus ratio from the diet to support calcium metabolism, but there is little or no evidence to support this concept.[32] The quantities of calcium and phosphorus in the diet are far more important than their ratio to each other. A high phosphorus intake has been blamed for bone loss when, in fact, a low calcium intake—not a phosphorus toxicity or an improper ratio—is responsible.[33] Research shows that the displacement of milk in the diet by cola drinks, not the phosphoric acid content of the beverages, has adverse effects on bone.[34] No adverse effects of high dietary phosphorus intakes have been reported; still, an Upper Level has been established (see inside front cover).

IN SUMMARY Phosphorus accompanies calcium both in the crystals of bone and in many foods such as milk. Phosphorus is also important in energy metabolism, as part of phospholipids, and as part of the genetic materials DNA and RNA. The summary table below lists functions of, and other information about, phosphorus.

Phosphorus

1997 RDA	**Deficiency Symptoms**
Adults: 700 mg/day	Muscular weakness, bone pain[a]

Upper Level	**Toxicity Symptoms**
Adults (19–70 yr): 4000 mg/day	Calcification of nonskeletal tissues, particularly the kidneys

Chief Functions in the Body	**Significant Sources**
Mineralization of bones and teeth; part of every cell; important in genetic material, part of phospholipids, used in energy transfer and in buffer systems that maintain acid-base balance	All animal tissues (meat, fish, poultry, eggs, milk)

[a]Dietary deficiency rarely occurs, but some drugs can bind with phosphorus making it unavailable and resulting in bone loss that is characterized by weakness and pain.

Magnesium

Magnesium barely qualifies as a major mineral: only about 1 ounce of magnesium is present in the body of a 130-pound person. Over half of the body's magnesium is in the bones. Most of the rest is in the muscles and soft tissues, with only 1 percent in the extracellular fluid. As with calcium, bone magnesium may serve as a reservoir to ensure normal blood concentrations.

Magnesium Roles in the Body Magnesium acts in all the cells of the soft tissues, where it forms part of the protein-making machinery and is necessary for energy metabolism. It participates in hundreds of enzyme systems. A major role is as a catalyst■ in the reaction that adds the last phosphate to the high-energy compound ATP. As a required component for ATP metabolism, magnesium is essential to the body's use of glucose; the synthesis of protein, fat, and nucleic acids; and the cells' membrane transport systems. Together with calcium, magnesium is involved in muscle contraction and blood clotting: calcium promotes the processes, whereas magnesium inhibits them. This dynamic interaction between the two minerals helps regulate blood pressure and the functioning of the lungs. Magnesium also helps prevent dental caries by holding calcium in tooth enamel. Like many other nutrients, magnesium supports the normal functioning of the immune system.

Magnesium Intakes Average dietary magnesium estimates for U.S. adults fall below recommendations. Dietary intake data, however, do not include the contribution

■ Reminder: A *catalyst* is a compound that facilitates chemical reactions without itself being changed in the process.

magnesium: a cation within the body's cells, active in many enzyme systems.

made by water. In areas with hard water, the water contributes both calcium and magnesium to daily intakes. Mineral waters noted earlier for their calcium content may also be magnesium-rich and can be important sources of this mineral for those who drink them.[35] Bioavailability of magnesium from mineral water is about 50 percent, but improves when the water is consumed with a meal.[36]

The brown bars in Figure 12-15 indicate that legumes, seeds, and nuts make significant magnesium contributions. Magnesium is part of the chlorophyll molecule, so leafy green vegetables are also good sources.

Magnesium Deficiency Even with average magnesium intakes below recommendations, deficiency symptoms rarely appear except with diseases. Magnesium deficiency may develop in cases of alcohol abuse, protein malnutrition, kidney disorders, and prolonged vomiting or diarrhea. People using diuretics may also show symptoms. A severe magnesium deficiency causes a tetany similar to the calcium tetany described earlier. Magnesium deficiencies also impair central nervous system activity and may be responsible for the hallucinations experienced during alcohol withdrawal.

Magnesium and Hypertension Magnesium is critical to heart function and seems to protect against hypertension and heart disease. Interestingly, people living in areas of the country with hard water, which contains high concentrations of calcium and magnesium, tend to have low rates of heart disease. With magnesium deficiency, the walls of the arteries and capillaries tend to constrict, a possible explanation for the hypertensive effect.

FIGURE 12-15 Magnesium in Selected Foods

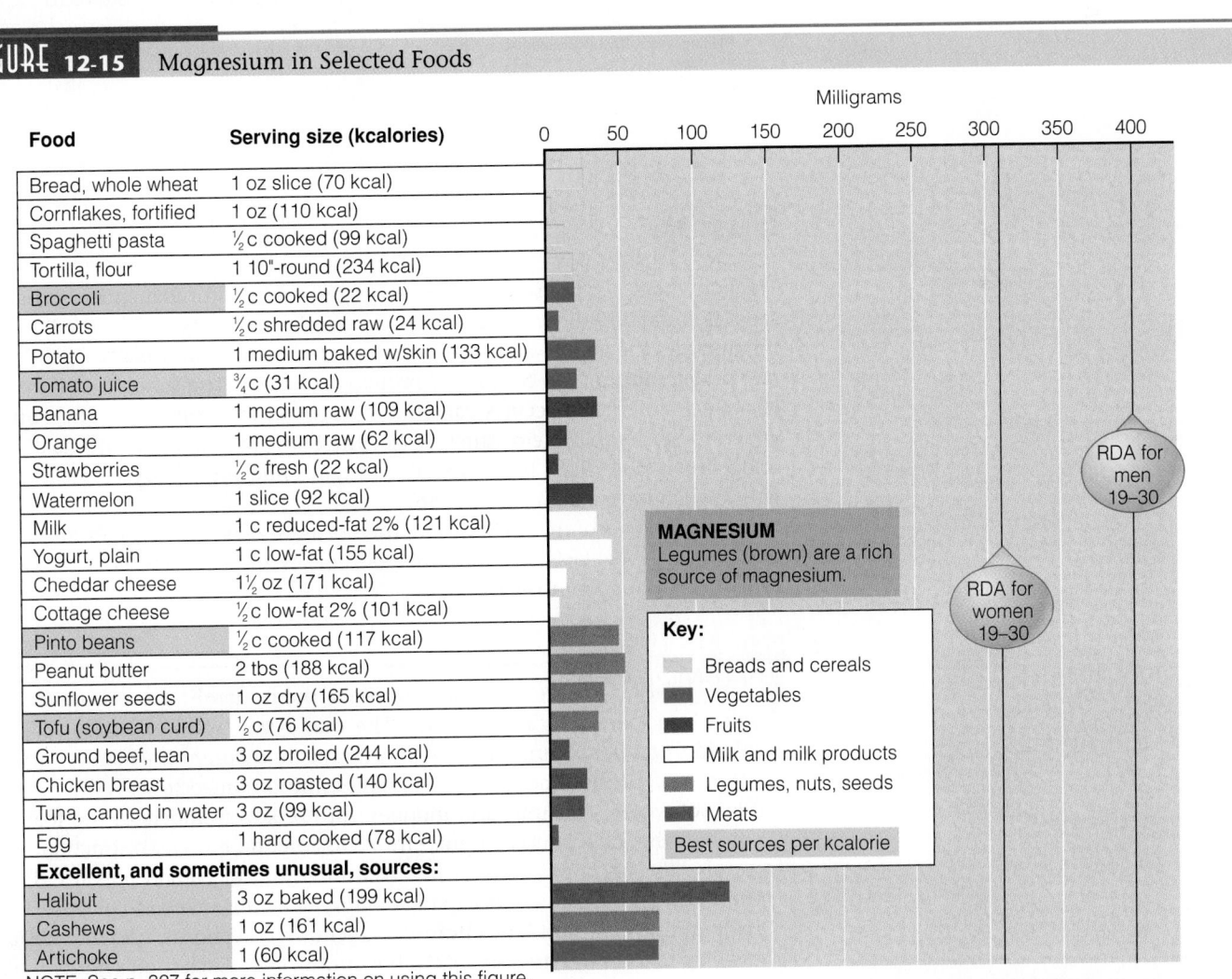

NOTE: See p. 327 for more information on using this figure.

Magnesium Toxicity Magnesium toxicity is rare, but it can be fatal.[37] The Upper Level for magnesium applies only to nonfood sources such as supplements or magnesium salts.

IN SUMMARY Like calcium and phosphorus, magnesium supports bone mineralization. Magnesium is also involved in numerous enzyme systems and in heart function. It is found abundantly in legumes and leafy green vegetables and, in some areas, in water. The table below offers a summary.

Magnesium

1997 RDA	**Deficiency Symptoms**
Men (19–30 yr): 400 mg/day	Weakness; confusion; if extreme, convulsions, bizarre muscle movements (especially of eye and face muscles), hallucinations, and difficulty in swallowing; in children, growth failure[a]
Women (19–30 yr): 310 mg/day	
Upper Level	
Adults: 350 mg nonfood magnesium/day	**Toxicity Symptoms**
	From nonfood sources only; diarrhea, alkalosis, dehydration
Chief Functions in the Body	
Bone mineralization, building of protein, enzyme action, normal muscle contraction, nerve impulse transmission, maintenance of teeth, and functioning of immune system	**Significant Sources**
	Nuts, legumes, whole grains, dark green vegetables, seafood, chocolate, cocoa

[a]A still more severe deficiency causes tetany, an extreme, prolonged contraction of the muscles similar to that caused by low blood calcium.

Sulfur

The body does not use **sulfur** by itself as a nutrient. Sulfur is mentioned here because it is a major mineral that occurs in essential nutrients such as the B vitamin thiamin and the amino acids methionine and cysteine. Sulfur plays a well-known role in determining the contour of protein molecules. The sulfur-containing side chains in cysteine molecules can link to each other, forming disulfide bridges, which stabilize the protein structure (see the drawing of insulin with its disulfide bridges on p. 184). Skin, hair, and nails contain some of the body's more rigid proteins, which have a high sulfur content.

There is no recommended intake for sulfur, and no deficiencies are known. Only when people lack protein to the point of severe deficiency will they lack the sulfur-containing amino acids.

IN SUMMARY Like the other nutrients, the minerals' actions are coordinated to get the body's work done. The major minerals, especially sodium, chloride, and potassium, influence the body's fluid balance; whenever an anion moves, a cation moves—always maintaining homeostasis. Sodium, chloride, potassium, calcium, and magnesium are key members of the team of nutrients that direct nerve impulse transmission and muscle contraction; they are also the primary nutrients involved in regulating blood pressure. Phosphorus and magnesium participate in many reactions involving glucose, fatty acids, amino acids, and the vitamins. Calcium, phosphorus, and magnesium combine to form the structure of the bones and teeth. Each major mineral also plays other specific roles in the body. (See the summary table on p. 423.)

sulfur: a mineral present in the body as part of some proteins.

The Major Minerals

Mineral and Chief Functions	Deficiency Symptoms	Toxicity Symptoms	Significant Sources
Sodium Maintains normal fluid and electrolyte balance; assists in nerve impulse transmission and muscle contraction	Muscle cramps, mental apathy, loss of appetite	Edema, acute hypertension	Table salt, soy sauce; moderate amounts in meats, milks, breads, and vegetables; large amounts in processed foods
Chloride Maintains normal fluid and electrolyte balance; part of hydrochloric acid found in the stomach, necessary for proper digestion	Do not occur under normal circumstances	Vomiting	Table salt, soy sauce; moderate amounts in meats, milks, eggs; large amounts in processed foods
Potassium Maintains normal fluid and electrolyte balance; facilitates many reactions; supports cell integrity; assists in nerve impulse transmission and muscle contractions	Muscular weakness, paralysis, confusion	Muscular weakness; vomiting; if given into a vein, can stop the heart	All whole foods; meats, milks, fruits, vegetables, grains, legumes
Calcium Mineralization of bones and teeth; also involved in muscle contraction and relaxation, nerve functioning, blood clotting, blood pressure, and immune defenses	Stunted growth in children; bone loss (osteoporosis) in adults	Constipation; increased risk of urinary stone formation and kidney dysfunction; interference with absorption of other minerals	Milk and milk products, small fish (with bones), tofu, greens (broccoli, chard), legumes
Phosphorus Mineralization of bones and teeth; part of every cell; important in genetic material, part of phospholipids, used in energy transfer and in buffer systems that maintain acid-base balance	Muscular weakness, bone pain[a]	Calcification of nonskeletal tissues, particularly the kidneys	All animal tissues (meat, fish, poultry, eggs, milk)
Magnesium Bone mineralization, building of protein, enzyme action, normal muscle contraction, nerve impulse transmission, maintenance of teeth, and functioning of immune system	Weakness; confusion; if extreme, convulsions, bizarre muscle movements (especially of eye and face muscles), hallucinations, and difficulty in swallowing; in children, growth failure[b]	From nonfood sources only; diarrhea, alkalosis, dehydration	Nuts, legumes, whole grains, dark green vegetables, seafood, chocolate, cocoa
Sulfur As part of proteins, stabilizes their shape by forming disulfide bridges; part of the vitamins biotin and thiamin and the hormone insulin	None known; protein deficiency would occur first	Toxicity would occur only if sulfur-containing amino acids were eaten in excess; this (in animals) depresses growth	All protein-containing foods (meats, fish, poultry, eggs, milk, legumes, nuts)

[a]Dietary deficiency rarely occurs, but some drugs can bind with phosphorus making it unavailable and resulting in bone loss that is characterized by weakness and pain.
[b]A still more severe deficiency causes tetany, an extreme, prolonged contraction of the muscles similar to that caused by low blood calcium.

With all of the tasks these minerals perform, they are of great importance to life. Consuming enough of each of them every day is easy, given a variety of foods from each of the food groups. Whole-grain breads supply magnesium; fruits, vegetables, and legumes also provide magnesium and potassium, too; milks offer calcium and phosphorus; meats also offer phosphorus and sulfur as well; all foods provide sodium and chloride, excesses being more problematic than inadequacies. The message is quite simple and has been repeated throughout this text: for an adequate intake of all the nutrients, including the major minerals, choose different foods from each of the five food groups. And drink plenty of water.

Nutrition in Your Life

Many people may miss the mark when it comes to drinking enough water to keep their bodies well hydrated or obtaining enough calcium to promote strong bones; in contrast, sodium intakes often exceed those recommended for health.

- Do you drink plenty of water—about 8 glasses—every day?
- Do you select and prepare foods with less salt?
- Do you drink at least 3 glasses of milk—or get the equivalent in calcium—every day?

NUTRITION ON THE NET

 Access these websites for further study of topics covered in this chapter.

- Find updates and quick links to these and other nutrition-related sites at our website: **www.wadsworth.com/nutrition**
- Search for "minerals" at the American Dietetic Association site: **www.eatright.org**

- Learn about sodium in foods and on food labels from the Food and Drug Administration: **www.fda.gov/fdac/foodlabel/sodium.html**
- Find tips and recipes for including more milk in the diet: **www.whymilk.com**
- Learn about the benefits of calcium from the National Dairy Council: **www.nationaldairycouncil.org**

NUTRITION CALCULATIONS

These problems give you an appreciation for the minerals in foods. Be sure to show your calculations (see p. 427 for answers).

1. For each of these minerals, note the unit of measure:
   Calcium   Magnesium   Phosphorus
   Potassium   Sodium

2. Learn to appreciate calcium-dense foods. The foods in the accompanying table are ranked in order of their calcium contents per serving.
   a. Which foods offer the most calcium per kcalorie? To calculate calcium density, divide calcium (mg) by energy (kcal). Record your answer in the table (round your answers); the first one is done for you.
   b. The top five items ranked in order of calcium contents per serving are sardines > milk > cheese > salmon > broccoli. What are the top five items in order of calcium contents per kcalorie?

Food	Calcium (mg)	Energy (kcal)	Calcium Density (mg/kcal)
Sardines, 3 oz canned	325	176	1.85
Milk, fat-free, 1 c	301	85	
Cheddar cheese, 1 oz	204	114	
Salmon, 3 oz canned	182	118	
Broccoli, cooked from fresh, chopped, ½ c	36	22	
Sweet potato, baked in skin, 1 ea	32	140	
Cantaloupe melon, ½	29	93	
Whole-wheat bread, 1 slice	21	64	
Apple, 1 medium	15	125	
Sirloin steak, lean, 3 oz	9	171	

This information should convince you that milk, milk products, fish eaten with their bones, and dark green vegetables are the best choices for calcium.

3.   a. Consider how the rate of absorption influences the amount of calcium available for the body's use. Use the drawing on p. 417 to determine how much calcium the body actually receives from the foods listed in the accompanying table by multiplying the milligrams of calcium in the food by the percentage absorbed. The first one is done for you.

   b. To appreciate how the absorption rate influences the amount of calcium available to the body, compare broccoli with almonds. Which provides more calcium in foods and to the body?

   c. To appreciate how the calcium content of foods influences the amount of calcium available to the body, compare cauliflower with milk. How much cauliflower would a person have to eat to receive an equivalent amount of calcium as from 1 cup of milk? How does your answer change when you account for differences in their absorption rates?

Food	Calcium in the Food (mg)	Absorption Rate (%)	Calcium in the Body (mg)
Cauliflower, ½ c cooked, fresh	10	≥50	≥5
Broccoli, ½ c cooked, fresh	36		
Milk, 1 c 1% low-fat	300		
Almonds, 1 oz	75		
Spinach, 1 c raw	55		

STUDY QUESTIONS

These questions will help you review the chapter. You will find the answers in the discussions on the pages provided.

1. List the roles of water in the body. (p. 396)

2. List the sources of water intake and routes of water excretion. (pp. 396–397)

3. What is ADH? Where does it exert its action? What is aldosterone? How does it work? (p. 399)

4. How does the body use electrolytes to regulate fluid balance? (pp. 400–403)

5. What do the terms *major* and *trace* mean when describing the minerals in the body? (p. 405)

6. Describe some characteristics of minerals that distinguish them from vitamins. (pp. 405–407)

7. What is the major function of sodium in the body? Describe how the kidneys regulate blood sodium. Is a dietary deficiency of sodium likely? Why or why not? (pp. 407–409)

8. List calcium's roles in the body. How does the body keep blood calcium constant regardless of intake? (pp. 413–414)

9. Name significant food sources of calcium. What are the consequences of inadequate intakes? (pp. 415–418)

10. List the roles of phosphorus in the body. Discuss the relationships between calcium and phosphorus. Is a dietary deficiency of phosphorus likely? Why or why not? (pp. 419–420)

11. State the major functions of chloride, potassium, magnesium, and sulfur in the body. Are deficiencies of these nutrients likely to occur in your own diet? Why or why not? (pp. 410, 411, 420–421, 422)

These multiple choice questions will help you prepare for an exam. Answers can be found on p. 427.

1. The body generates water during the:
   a. buffering of acids.
   b. dismantling of bone.
   c. metabolism of minerals.
   d. breakdown of energy nutrients.

2. Regulation of fluid and electrolyte balance and acid-base balance depends primarily on the:
   a. kidneys.
   b. intestines.
   c. sweat glands.
   d. specialized tear ducts.

3. The distinction between the major and trace minerals reflects the:
   a. ability of their ions to form salts.
   b. amounts of their contents in the body.
   c. importance of their functions in the body.
   d. capacity to retain their identity after absorption.

4. The principal cation in extracellular fluids is:
   a. sodium.
   b. chloride.
   c. potassium.
   d. phosphorus.

5. The role of chloride in the stomach is to help:
   a. support nerve impulses.
   b. convey hormonal messages.
   c. maintain a strong acidity.
   d. assist in muscular contractions.

6. Which would provide the most potassium?
   a. bologna
   b. potatoes
   c. pickles
   d. whole-wheat bread

7. Calcium homeostasis depends on:
   a. vitamin K, aldosterone, and renin.
   b. vitamin K, parathormone, and renin.
   c. vitamin D, aldosterone, and calcitonin.
   d. vitamin D, calcitonin, and parathormone.

8. Calcium absorption is hindered by:
   a. lactose.
   b. oxalates.

   c. vitamin D.
   d. stomach acid.

9. Phosphorus assists in many activities in the body, but *not*:
   a. energy metabolism.
   b. the clotting of blood.
   c. the transport of lipids.
   d. bone and teeth formation.

10. Most of the body's magnesium can be found in the:
    a. bones.
    b. nerves.
    c. muscles.
    d. extracellular fluids.

REFERENCES

1. Committee on Dietary Reference Intakes, *Dietary Reference Intakes for Water, Potassium, Sodium, Chloride, and Sulfate* (Washington, D.C.: National Academies Press, 2004), p. 67.
2. U.S. Department of Agriculture, *1994–1996, 1998 Continuing Survey of Food Intakes by Individuals (CSFII) 1994–1996, and Diet and Health Knowledge Survey, 2000.* (Available from the National Technical Information Service, Springfield, VA: tel 1-800-553-6847; CD-ROM accession number PB2000-500027).
3. M. Neuhäuser-Berthold and coauthors, Coffee consumption and total body water homeostasis as measured by fluid balance and bioelectrical impedance analysis, *Annals of Nutrition and Metabolism* 41 (1997): 29–36.
4. Committee on Dietary Reference Intakes, 2004, pp. 120–121.
5. D. S. Michaud and coauthors, Fluid intake and the risk of bladder cancer in men, *New England Journal of Medicine* 340 (1999): 1390–1397.
6. S. M. Kleiner, Water: An essential but overlooked nutrient, *Journal of the American Dietetic Association* 99 (1999): 200–206.
7. M. P. Sauvant and D. Pepin, Geographic variation of the mortality from cardiovascular disease and drinking water in a French small area (Puy de Dome), *Environmental Research* 84 (2000): 219–227.
8. H. Bohmer, H. Muller, and K. L. Resch, Calcium supplementation with calcium-rich mineral waters: A systematic review and meta-analysis of its bioavailability, *Osteoporosis International* 11 (2000): 938–943; R. Maheswaran and coauthors, Magnesium in drinking water supplies and mortality from acute myocardial infarction in north west England, *Heart* 82 (1999): 455–460.
9. J. He and coauthors, Dietary sodium intake and subsequent risk of cardiovascular disease in overweight adults, *Journal of the American Medical Association* 282 (1999): 2027–2034.
10. F. M. Sacks and coauthors, Effects on blood pressure of reduced dietary sodium and the Dietary Approaches to Stop Hypertension (DASH) diet, *New England Journal of Medicine* 344 (2001): 3–10.
11. Sacks and coauthors, 2001.
12. M. Harrington and K. D. Cashman, High salt intake appears to increase bone resorption in postmenopausal women but high potassium intake ameliorates this adverse effect, *Nutrition Reviews* 61 (2003): 179–183.
13. A. J. Cohen and F. J. Roe, Review of risk factors for osteoporosis with particular reference to a possible aetiological role of dietary salt, *Food and Chemical Toxicology* 38 (2000): 237–253.
14. F. P. Cappuccio and coauthors, Unravelling the links between calcium excretion, salt intake, hypertension, kidney stones and bone metabolism, *Journal of Nephrology* 13 (2000): 169–177.
15. D. E. Sellmeyer, M. Schloetter, and A. Sebastin, Potassium citrate prevents increased urine calcium excretion and bone resorption induced by a high sodium chloride diet, *Journal of Clinical Endocrinology and Metabolism* 87 (2002): 2008–2012.
16. F. J. He and G. A. MacGregor, Beneficial effects of potassium, *British Medical Journal* 323 (2001): 497–501.
17. K. Kathleen, Hyperkalemia, *American Journal of Nursing* 100 (2000): 55–56.
18. R. Jorde and K. H. Bønaa, Calcium from dairy products, vitamin D intake, and blood pressure: The Tromsø study, *American Journal of Clinical Nutrition* 71 (2000): 1530–1535.
19. M. Jacqmain and coauthors, Calcium intake, body composition, and lipoprotein-lipid concentrations, *American Journal of Clinical Nutrition* 77 (2003): 1448–1452; E. Kampman and coauthors, Calcium, vitamin D, sunshine exposure, dairy products and colon cancer risk (United States), *Cancer Causes and Control* 5 (2000): 459–466; E. Kallay and coauthors, Dietary calcium and growth modulation of human colon cancer cells: Role of the extracellular calcium-sensing receptor, *Cancer Detection and Prevention* 24 (2000): 127–136.
20. S. J. Parikh and J. A. Yanovski, Calcium and adiposity, *American Journal of Clinical Nutrition* 77 (2003): 281–287; D. Teegarden, Calcium intake and reduction in weight or fat mass, *Journal of Nutrition* 133 (2003): 249S–251S; R. P. Heaney, K. M. Davies, and M. J. Barger-Lux, Calcium and weight: Clinical studies, *Journal of the American College of Nutrition* 21 (2002): 152–155.
21. Jacqmain and coauthors, 2003; Teegarden, 2003; Heaney, Davies, and Barger-Lux,
2002; M. B. Zemel and coauthors, Regulation of adiposity by dietary calcium, *FASEB Journal* 14 (2000): 1132–1138.
22. M. B. Zemel and coauthors, Dietary calcium and dairy products accelerate weight and fat loss during energy restriction in obese adults, *American Journal of Clinical Nutrition* 75 (2002): 342S.
23. M. B. Zemel, Mechanisms of dairy modulation of adiposity, *Journal of Nutrition* 133 (2003): 252S–256S.
24. R. L. Wolf and coauthors, Factors associated with calcium absorption efficiency in pre- and perimenopausal women, *American Journal of Clinical Nutrition* 72 (2000): 466–471; Committee on Dietary Reference Intakes, *Dietary Reference Intakes for Calcium, Phosphorus, Magnesium, Vitamin D, and Fluoride* (Washington, D.C.: National Academy Press, 1997), pp. 72–73.
25. Committee on Dietary Reference Intakes, 1997, pp. 75–76.
26. R. E. Black and coauthors, Children who avoid drinking cow milk have low dietary calcium intakes and poor bone health, *American Journal of Clinical Nutrition* 76 (2002): 675–680.
27. H. J. Kalkwarf, J. C. Khoury, and B. P. Lanphear, Milk intake during childhood and adolescence, adult bone density, and osteoporotic fractures in US women, *American Journal of Clinical Nutrition* 77 (2003): 257–265.
28. C. M. Weaver, W. R. Proulx, and R. Heaney, Choices for achieving adequate dietary calcium with a vegetarian diet, *American Journal of Clinical Nutrition* 70 (1999): 543S–548S.
29. Committee on Dietary Reference Intakes, 1997, pp. 73–74.
30. P. Galan and coauthors, Contribution of mineral waters to dietary calcium and magnesium intake in a French adult population, *Journal of the American Dietetic Association* 102 (2002): 1658–1662; J. Guillemant and coauthors, Mineral water as a source of dietary calcium: Acute effects on parathyroid function and bone resorption in young men, *American Journal of Clinical Nutrition* 71 (2000): 999–1002.
31. R. P. Heaney and coauthors, Dietary changes favorably affect bone remodeling in older adults, *Journal of the American Dietetic Association* 99 (1999): 1228–1233.

32. Committee on Dietary Reference Intakes, 1997, pp. 152–154.
33. Committee on Dietary Reference Intakes, 1997, p. 182.
34. R. P. Heaney and K. Rafferty, Carbonated beverages and urinary calcium excretion, *American Journal of Clinical Nutrition* 74 (2001): 343–347.

35. Galan and coauthors, 2002.
36. M. Sabatier and coauthors, Meal effect on magnesium bioavailability from mineral water in healthy women, *American Journal of Clinical Nutrition* 75 (2002): 65–71.

37. J. K. McGuire, M. S. Kulkarni, and H. P. Baden, Fatal hypermagnesemia in a child treated with megavitamin/megamineral therapy, *Pediatrics* 105 (2000): e18.

ANSWERS

Nutrition Calculations

1. Calcium: mg.   Magnesium: mg.   Phosphorus: mg.
   Potassium: mg.   Sodium: mg.

2. a.

Food	Calcium Density (mg/kcal)
Sardines, 3 oz canned	325 mg ÷ 176 kcal = 1.85 mg/kcal
Milk, fat-free, 1 c	301 mg ÷ 85 kcal = 3.54 mg/kcal
Cheddar cheese, 1 oz	204 mg ÷ 114 kcal = 1.79 mg/kcal
Salmon, 3 oz canned	182 mg ÷ 118 kcal = 1.54 mg/kcal
Broccoli, cooked from fresh, chopped, ½ c	36 mg ÷ 22 kcal = 1.64 mg/kcal
Sweet potato, baked in skin, 1 ea	32 mg ÷ 140 kcal = 0.23 mg/kcal
Cantaloupe melon, ½	29 mg ÷ 93 kcal = 0.31 mg/kcal
Whole-wheat bread, 1 slice	21 mg ÷ 64 kcal = 0.33 mg/kcal
Apple, 1 medium	15 mg ÷ 125 kcal = 0.12 mg/kcal
Sirloin steak, lean, 3 oz	9 mg ÷ 171 kcal = 0.05 mg/kcal

   b. Milk > sardines > cheese > broccoli > salmon.

3. a.

Food	Calcium in Food (mg) ✖ Absorption rate (%) = Calcium in the Body (mg)
Cauliflower, ½ c cooked, fresh	10 mg × 0.50 = 5 mg (or more)
Broccoli, ½ c cooked, fresh	36 mg × 0.50 = 18 mg (or more)
Milk, 1 c 1% low-fat	300 mg × 0.30 = 90 mg
Almonds, 1 oz	75 mg × 0.20 = 15 mg
Spinach, 1 c raw	55 mg × 0.05 = 3 mg (or less)

   b. The almonds offer more than twice as much calcium per serving, but an equivalent amount after absorption.

   c. To equal the 300 milligrams provided by milk, a person would need to eat 15 cups of cauliflower (300 mg/c milk ÷ 10 mg/½ c cauliflower = 30 ½ c or 15 c). After considering the better absorption rate of cauliflower, a person would need to eat 9 cups of cauliflower (5 mg/½ c or 10 mg/c; 90 mg ÷ 10 mg/c = 9 c) to match the 90 milligrams available to the body from milk after absorption. The better absorption rate reduced the quantity of cauliflower significantly, but that's still a lot of cauliflower.

Study Questions (multiple choice)

1. d   2. a   3. b   4. a   5. c   6. b   7. d   8. b
9. b   10. a

HIGHLIGHT

Osteoporosis and Calcium

© Photo Disc Inc.

Osteoporosis becomes apparent during the later years, but it develops much earlier—and without warning. Few people are aware that their bones are being robbed of their strength. The problem often first becomes evident when someone's hip suddenly gives way. People say, "She fell and broke her hip," but in fact the hip may have been so fragile that it broke *before* she fell. Even bumping into a table may be enough to shatter a porous bone into fragments so numerous and scattered that they cannot be reassembled. Removing them and replacing them with an artificial joint requires major surgery. An estimated 300,000 people in the United States are hospitalized each year because of hip fractures related to osteoporosis. About a fourth die of complications within a year. A fourth of those who survive will never walk or live independently again. Their quality of life slips downward.

This highlight examines osteoporosis, one of the most prevalent diseases of aging, affecting more than 44 million people in the United States—most of them women over 50. It reviews the many factors that contribute to the 1.5 million breaks in the bones of the hips, vertebrae, wrists, arms, and ankles each year. And it presents strategies to reduce the risks, paying special attention to the role of dietary calcium.

Bone Development and Disintegration

Bone has two compartments: the outer, hard shell of **cortical bone,** and the inner, lacy matrix of **trabecular bone.** (The glossary defines these and other bone-related terms.) Both can lose minerals, but in different ways and at different rates. The photograph on p. 429 shows a human leg bone sliced lengthwise, exposing the lacy, calcium-containing crystals of trabecular bone. These crystals give up calcium to the blood when the diet runs short, and they take up calcium again when the supply is plentiful (review Figure 12-11 on p. 415). For people who have eaten calcium-rich foods throughout the bone-forming years of their youth, these deposits make bones dense and provide a rich reservoir of calcium.

Surrounding and protecting the trabecular bone is a dense, ivorylike exterior shell—the cortical bone. Cortical bone composes the shafts of the long bones, and a thin cortical shell caps the end of the bone, too. Both compartments confer strength on bone: cortical bone provides the sturdy outer wall, while trabecular bone provides support along the lines of stress.

The two types of bone play different roles in calcium balance and osteoporosis. Supplied with blood vessels and metabolically active, trabecular bone is sensitive to hormones that govern day-to-day deposits and withdrawals of calcium. It readily gives up minerals whenever blood calcium needs replenishing.

GLOSSARY

antacids: acid-buffering agents used to counter excess acidity in the stomach. Calcium-containing preparations (such as Tums) contain available calcium. Antacids with aluminum or magnesium hydroxides (such as Rolaids) can accelerate calcium losses.

bone meal or **powdered bone:** crushed or ground bone preparations intended to supply calcium to the diet. Calcium from bone is not well absorbed and is often contaminated with

toxic minerals such as arsenic, mercury, lead, and cadmium.

bone density: a measure of bone strength. When minerals fill the bone matrix (making it dense), they give it strength.

cortical bone: the very dense bone tissue that forms the outer shell surrounding trabecular bone and comprises the shaft of a long bone.

dolomite: a compound of minerals (calcium magnesium carbonate) found in limestone and marble. Dolomite is

powdered and is sold as a calcium-magnesium supplement, but may be contaminated with toxic minerals, is not well absorbed, and interacts adversely with absorption of other esssential minerals.

oyster shell: a product made from the powdered shells of oysters that is sold as a calcium supplement, but is not well absorbed by the digestive system.

trabecular (tra-BECK-you-lar) **bone:** the lacy inner structure of

calcium crystals that supports the bone's structure and provides a calcium storage bank.

type I osteoporosis: osteoporosis characterized by rapid bone losses, primarily of trabecular bone.

type II osteoporosis: osteoporosis characterized by gradual losses of both trabecular and cortical bone.

Reminder: *Osteoporosis* is a disease characterized by porous and fragile bones.

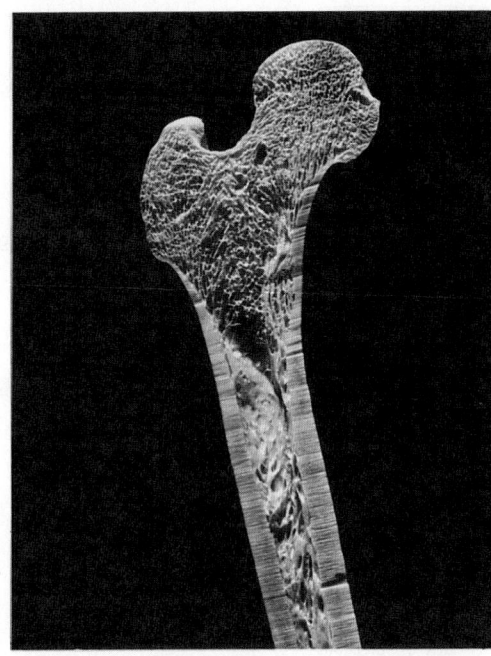

Trabecular bone is the lacy network of calcium-containing crystals that fills the interior. Cortical bone is the dense, ivorylike bone that forms the exterior shell.

Losses of trabecular bone start becoming significant for men and women in their 30s, although losses can occur whenever calcium withdrawals exceed deposits.

Cortical bone also gives up calcium, but slowly and at a steady pace. Cortical bone losses typically begin at about age 40 and continue slowly but surely thereafter.

FIGURE H12-1 Healthy and Osteoporotic Trabecular Bones

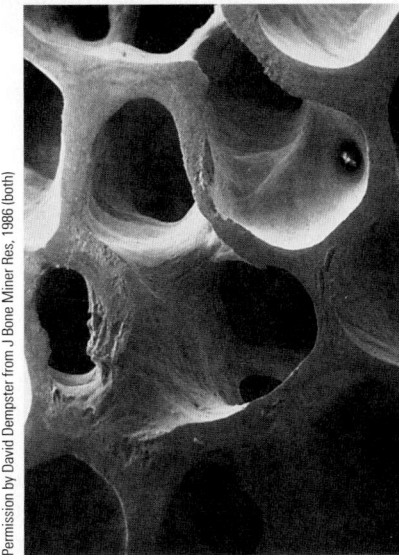

Electron micrograph of healthy trabecular bone.

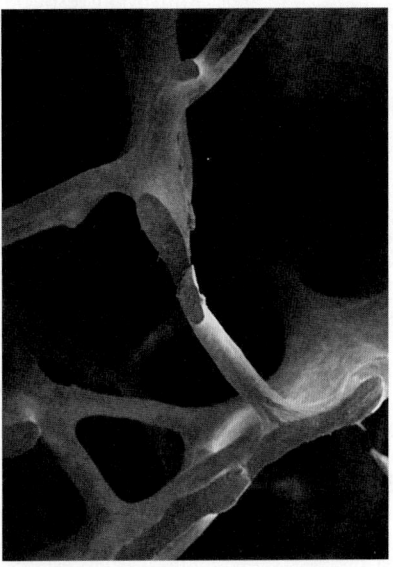

Electron micrograph of trabecular bone affected by osteoporosis.

Losses of trabecular and cortical bone reflect two types of osteoporosis, which cause two types of bone breaks. **Type I osteoporosis** involves losses of trabecular bone (see Figure H12-1). These losses sometimes exceed three times the expected rate, and bone breaks may occur suddenly. Trabecular bone becomes so fragile that even the body's own weight can overburden the spine—vertebrae may suddenly disintegrate and crush down, painfully pinching major nerves. Wrists may break as bone ends weaken, and teeth may loosen or fall out as the trabecular bone of the jaw recedes. Women are most often the victims of this type of osteoporosis, outnumbering men six to one.

In **type II osteoporosis,** the calcium of both cortical and trabecular bone is drawn out of storage, but slowly over the years. As old age approaches, the vertebrae may compress into wedge shapes, forming what is often called a "dowager's hump," the posture many older people assume as they "grow shorter." Figure H12-2 (on p. 430) shows the effect of compressed spinal bone on a woman's height and posture. Because both the cortical shell and the trabecular interior weaken, breaks most often occur in the hip, as mentioned in the introductory paragraph. A woman is twice as likely as a man to suffer type II osteoporosis.

Table H12-1 summarizes the differences between the two types of osteoporosis. Physicians can diagnose osteoporosis and assess the risk of bone fractures by measuring **bone density** using dual-energy X-ray absorptiometry (DEXA scan) or ultrasound. They also consider risk factors that predict bone fractures, including age, personal and family history of fracture, heritage, BMI, and physical inactivity.[1] Table H12-2 summarizes the major risk factors and protective factors for osteoporosis. The more risk factors that apply to a person, the greater the chances of bone loss. Notice that several risk factors that are influential in the development of osteoporosis—such as age, gender, and heritage—cannot be changed. Other risk factors—such as diet, physical activity, body weight, smoking, and alcohol use—are personal behaviors that can be changed. By eating a well-balanced diet rich in calcium, being physically active, abstaining from smoking, and drinking alcohol in moderation (if at all), people can defend themselves against osteoporosis. These decisions are particularly important for those with other risk factors that cannot be changed.

Whether a person develops osteoporosis seems to depend on the interactions of several factors, including nutrition. The strongest predictor of bone density is age: osteoporosis is responsible for 90 percent of the hip fractures in women and 80 percent in men over the age of 65.

FIGURE H12-2   Loss of Height in a Woman Caused by Osteoporosis

The woman on the left is about 50 years old. On the right, she is 80 years old. Her legs have not grown shorter: only her back has lost length, due to collapse of her spinal bones (vertebrae). Collapsed vertebrae cannot protect the spinal nerves from pressure that causes excruciating pain.

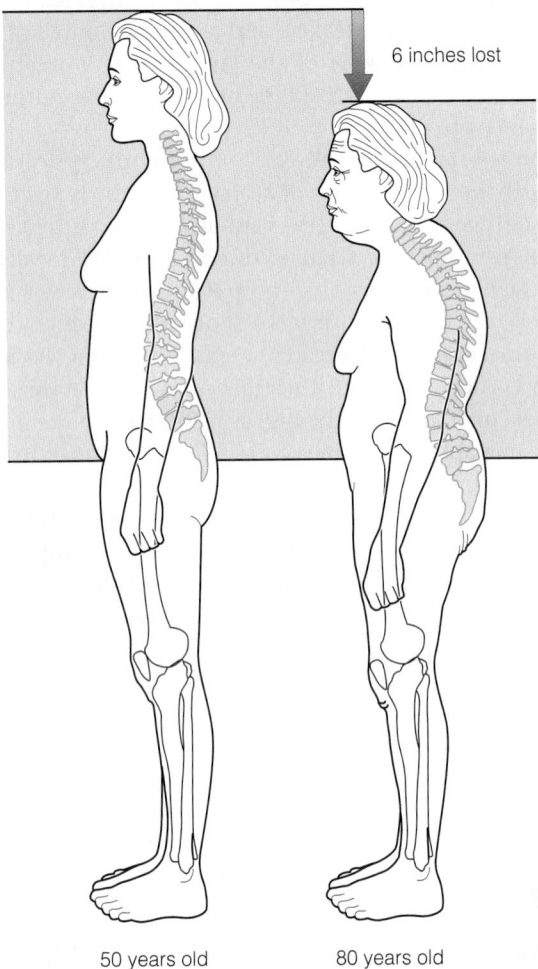

6 inches lost

50 years old   80 years old

TABLE H12-1   Types of Osteoporosis Compared

	Type I	Type II
Other name	Postmenopausal osteoporosis	Senile osteoporosis
Age of onset	50 to 70 years old	70 years and older
Bone loss	Trabecular bone	Both trabecular and cortical bone
Fracture sites	Wrist and spine	Hip
Gender incidence	6 women to 1 man	2 women to 1 man
Primary causes	Rapid loss of estrogen in women following menopause; loss of testosterone in men with advancing age	Reduced calcium absorption, increased bone mineral loss, increased propensity to fall

TABLE H12-2   Risk Factors and Protective Factors for Osteoporosis

Risk Factors	Protective Factors
• Older age	• Younger age
• Low BMI	• High BMI
• Caucasian, Asian, or Hispanic heritage	• African American heritage
• Cigarette smoking	• No smoking
• Alcohol consumption in excess	• Alcohol consumption in moderation
• Sedentary lifestyle	• Regular weight-bearing exercise
• Use of glucocorticoids or anticonvulsants	• Use of diuretics
• Female gender	• Male gender
• Maternal history of osteoporosis fracture or personal history of fracture	• Bone density assessment and treatment (if necessary)
• Estrogen deficiency in women (amenorrhea or menopause, especially early or surgically induced); testosterone deficiency in men	• Use of estrogen therapy
• Lifetime diet inadequate in calcium and vitamin D	• Lifetime diet rich in calcium and vitamin D

Age and Bone Calcium

Two major stages of life are critical in the development of osteoporosis. The first is the bone-acquiring stage of childhood and adolescence. The second is the bone-losing decades of late adulthood (especially in women after menopause). The bones gain strength and density all through the growing years and into young adulthood. As people age, the cells that build bone gradually become less active, but those that dismantle bone continue working. The result is that bone loss exceeds bone formation. Some bone loss is inevitable, but losses can be curtailed by maximizing bone mass.

Maximizing Bone Mass

To maximize bone mass, the diet must deliver an adequate supply of calcium during the first three decades of life. Children and teens who get enough calcium and vitamin D have denser bones than those with inadequate intakes. With little or no calcium from the diet, the body must depend on bone to supply calcium to the blood; bone mass diminishes, and bones lose their density and strength. When people reach the bone-losing years of middle age, those who formed dense bones during their youth have the advantage. They simply have more bone starting out and can lose more before suffering ill effects. Figure H12-3 demonstrates this effect.

FIGURE H12-3 Bone Losses over Time Compared

Peak bone mass is achieved by age 30. Women gradually lose bone mass until menopause, when losses accelerate dramatically and then gradually taper off.

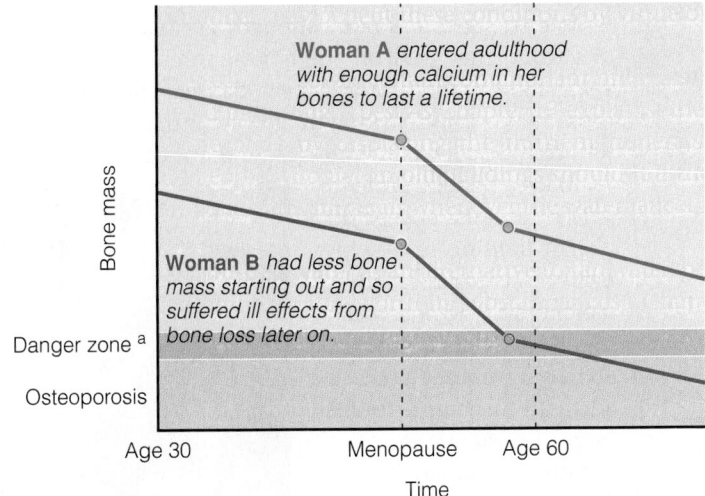

Woman A entered adulthood with enough calcium in her bones to last a lifetime.

Woman B had less bone mass starting out and so suffered ill effects from bone loss later on.

Bone mass

Danger zone ª

Osteoporosis

Age 30 Menopause Age 60

Time

ªPeople with a moderate degree of bone mass reduction are said to have *osteopenia* and are at increased risk of fractures.
SOURCE: Data from Committee on Dietary Reference Intakes, *Dietary Reference Intakes for Calcium, Phosphorus, Magnesium, Vitamin D, and Fluoride* (Washington, D.C.: National Academy Press, 1997), pp. 71–145.

Minimizing Bone Loss

Not only does dietary calcium build strong bones in youth, but it remains important in protecting against losses in the later years.[2] Unfortunately, calcium intakes of older adults are typically low, and calcium absorption declines after about the age of 65 years. The kidneys do not activate vitamin D as well as they did earlier (recall that active vitamin D enhances calcium absorption). Also, sunlight is needed to form vitamin D, and many older people spend little or no time outdoors in the sunshine. For these reasons, and because intakes of vitamin D are typically low anyway, blood vitamin D declines.

Some of the hormones that regulate bone and calcium metabolism also change with age and accelerate bone mineral withdrawal.[*] Together, these age-related factors contribute to bone loss: inefficient bone remodeling, reduced calcium intakes, impaired calcium absorption, poor vitamin D status, and hormonal changes that favor bone mineral withdrawal.

Gender and Hormones

After age, gender is the next strongest predictor of osteoporosis: men have greater bone density than women at maturity, and women have greater losses than men in later life. Conse-

quently, women account for four out of five cases of osteoporosis. Menopause imperils women's bones. Bone dwindles rapidly when the hormone estrogen diminishes and menstruation ceases. Women may lose up to 20 percent of their bone mass during the six to eight years following menopause. Eventually, losses taper off so that women again lose bone at the same rate as men their age. Losses of bone minerals continue throughout the remainder of a woman's lifetime, but not at the free-fall pace of the menopause years (review Figure H12-3).

Rapid bone losses also occur when *young* women's ovaries fail to produce enough estrogen, causing menstruation to cease. In some, diseased ovaries are to blame and must be removed; in others, the ovaries fail to produce sufficient estrogen because the women suffer from anorexia nervosa and have unreasonably restricted their body weight (see Highlight 9). The amenorrhea and low body weights explain much of the bone loss seen in these young women, even years after diagnosis and treatment.[3] Estrogen therapy can help nonmenstruating women prevent further bone loss and reduce the incidence of fractures.[4] Because estrogen therapy may increase the risks for heart disease and breast cancer, women must carefully weigh any potential benefits against the possible dangers.[5] Other drugs used to prevent or treat osteoporosis include raloxifene, alendronate, risedronate, and calcitonin.[†] A combination of hormone replacement and a drug may be an option for some women.[6]

Some women who choose not to use estrogen therapy turn to soy as an alternative treatment. Interestingly, the phytochemicals commonly found in soybeans mimic the actions of estrogen in the body. When natural estrogen is lacking, as after menopause, these phytochemicals may step in to stimulate estrogen-sensitive tissues. By way of this action, soy and its phytochemicals may help to prevent the rapid bone losses of the menopause years.[7] Research is far from conclusive, but some evidence suggests that soy may indeed offer some protection.[8]

If estrogen deficiency is a major cause of osteoporosis in women, what is the cause of bone loss in men? The male sex hormone testosterone appears to play a role. Men with low levels of testosterone, as occurs after removal of diseased testes or when testes lose function with aging, suffer more fractures. Treatment for men with osteoporosis includes

---

[*]Among the hormones suggested as influential are parathormone, calcitonin, and estrogen.

[†]Raloxifene (rah-LOX-ih-feen) is a selective estrogen-receptor modulator (SERM), marketed as Evista; alendronate (a-LEN-droe-nate) is a bisphosphonate, marketed as Fosamax; risedronate (rih-SEH-droe-nate) is a bisphosphonate, marketed as Actonel; and calcitonin is a hormone, marketed as Calcimar and Miacalcin.

testosterone replacement therapy. Thus both male and female sex hormones participate in the development and treatment of osteoporosis.

Most hormone and drug treatments for osteoporosis work by inhibiting the activities of the bone-dismantling cells, thus allowing the bone-building cells to slowly shore up bone tissue with new calcium deposits.* Research is also under way to develop drugs that will stimulate the bone-building cells to accelerate fracture healing and restore bone strength.[9] Leading contenders include parathormone, cholesterol-lowering drugs (statins), and leptin.

Genetics and Ethnicity

Osteoporosis may, in part, be hereditary; family history of osteoporosis or fracture is a risk factor. The exact role of genetics is unclear, but most likely it influences both the peak bone mass achieved during growth and the bone loss incurred during the later years. The extent to which a given genetic potential is realized, however, depends on many outside factors. Diet and physical activity, for example, can maximize peak bone density during growth, whereas alcohol and tobacco abuse can accelerate bone losses later in life.

Risks of osteoporosis appear to run along racial lines and reflect genetic differences in bone development. African Americans, for example, seem to use and conserve calcium more efficiently than Caucasians. Consequently, even though their calcium intakes are typically lower, black people have denser bones than white people do. Greater bone density expresses itself in a lower rate of osteoporosis among blacks. Fractures, for example, are about twice as likely in white women age 65 or older as in black women.

Other ethnic groups have a high risk of osteoporosis. Asians from China and Japan, Mexican Americans, Hispanic people from Central and South America, and Inuit people from St. Lawrence Island typically have lower bone density than Caucasians. One might expect that these groups would suffer more bone fractures, but this is not always the case. Again, genetic differences may explain why. Asians, for example, generally have small, compact hips, which makes them less susceptible to fractures.

Findings from around the world demonstrate that although a person's genes may lay the groundwork, environmental factors influence the genes' ultimate expression. Diet in general, and calcium in particular, are among those environmental factors. Others include physical activity, body weight, smoking, and alcohol. Importantly, all of these factors are within a person's control.

Physical Activity and Body Weight

Muscle strength and bone strength go together. When muscles work, they pull on the bones, stimulating them to develop more trabeculae and grow denser. The hormones that promote new muscle growth also favor the building of bone. As a result, active bones are denser than sedentary bones.[10]

Strength training helps to build strong bones.

To keep bones healthy, a person should engage in weight-training or weight-bearing activities (such as dancing and vigorous walking) daily. Regular physical activity combined with an adequate calcium intake helps to maximize bone density in adolescence.[11] Adults can also maximize and maintain bone density with a regular program of weight training.[12] Even past menopause when most women are losing bone, weight training improves bone density.[13]

Heavier body weights and weight gains place a similar stress on the bones and promote their density. In fact, underweight and weight losses are significant and consistent predictors of bone density losses and risk of fractures.[14] As mentioned in Highlight 9, the combination of underweight, severely restricted energy intake, extreme daily exercise, and amenorrhea reliably predicts bone loss. Interestingly, some evidence suggests that the bone density associated with overweight may be due not to body weight alone but to the lack of, or inability to respond to, leptin.[15]

Cells respond, with the help of the necessary regulators, to the demands put upon them. Then they select the nutrients they need from what is offered. To increase bone density, put a demand on the bones, make them work, and then provide the raw materials from which they can grow strong: calcium and all the other nutrients in the right balance.

Smoking and Alcohol

Add bone damage to the list of ill consequences associated with smoking. Bones of smokers are less dense than those of nonsmokers—even after controlling for differences in age, body weight, and physical activity habits.[16] Blood levels of vitamin D and bone-related hormones in smokers favor decreased calcium absorption and increased bone resorption.[17]

---

*The generic name of the drug is aldendronate, marketed as Fosamax. This drug belongs to a group of nonhormonal medicines called bisphosphonates.

Fortunately, these damaging effects can be reversed with smoking cessation. Blood indicators of beneficial bone activity are apparent six weeks after a person stops smoking.[18] In time, bone density is similar for former smokers and nonsmokers.

People who abuse alcohol often suffer from osteoporosis and experience more bone breaks than others. Several factors appear to be involved: alcohol enhances fluid excretion, leading to excessive calcium losses in the urine; upsets the hormonal balance required for healthy bones; slows bone formation, leading to lower bone density; stimulates bone breakdown; and increases the risk of falling. Alcohol in moderate amounts, however, may protect bone density by decreasing remodeling activity.[19]

Dietary Calcium Is the Key to Prevention

Bone strength later in life depends most on how well the bones were built during childhood and adolescence. Adequate calcium nutrition during the growing years is essential to achieving optimal peak bone mass.[20] Simply put, growing children who do not get enough calcium do not have strong bones.[21] Neither do adults who did not get enough calcium during their childhood and adolescence.[22] To that end, the DRI Committee recommends 1300 milligrams of calcium per day for everyone 9 through 18 years of age. Unfortunately, few girls meet the recommendations for calcium during these bone-forming years. (Boys generally obtain intakes close to those recommended because they eat more food.) Consequently, most girls start their adult years with less-than-optimal bone density. As for adults, women rarely meet their recommended intakes of 1000 to 1200 milligrams from food. Some authorities suggest 1500 milligrams of calcium for postmenopausal women who are not receiving estrogen, but warn that intakes exceeding 2500 milligrams a day could cause health problems.

Other Nutrients Play Supporting Roles

Much research has focused on calcium, but other nutrients support bone health, too. Adequate protein protects bones and reduces the likelihood of hip fractures.[23] As mentioned earlier, vitamin D is needed for optimal bone health.[24] Supplementation with vitamin D reduces bone loss and the risk of fractures.[25] Vitamin K protects against hip fractures.[26] The minerals magnesium and potassium also help to maintain bone mineral density.[27] Vitamin A is needed in the bone-remodeling process, but too much may be associated with osteoporosis.[28] Additional research points to the bone benefits not of a specific nutrient, but of a diet rich in fruits and vegetables.[29] In contrast, diets containing too much salt, candy, or caffeine are associated with bone losses.[30] Clearly, a well-balanced diet that depends on all the food groups to supply a full array of nutrients is central to bone health.

A Perspective on Supplements

Bone health improves when people increase their intake of calcium-rich foods.[31] People who do not consume milk products or other calcium-rich foods in amounts that provide even half the recommended calcium should consider consulting a registered dietitian who can assess the diet and suggest food choices to correct any inadequacies. For those who are unable to consume enough calcium-rich foods to forestall osteoporosis, taking calcium supplements may be appropriate.

Selecting a calcium supplement requires a little investigative work to sort through the many options. Before examining calcium supplements, recognize that multivitamin-mineral pills contain little or no calcium. The label may list a few milligrams of calcium, but remember that the recommended intake is a gram or more for adults.

Calcium supplements are typically sold as compounds of calcium carbonate (common in **antacids** and fortified chocolate candies), citrate, gluconate, lactate, malate, or phosphate. These supplements often include magnesium, vitamin D, or both. In addition, there are calcium supplements made from **bone meal, oyster shell,** or **dolomite** (limestone). Many calcium supplements, especially those derived from these natural products, contain lead—which impairs health in numerous ways, as Chapter 13 points out.[32] Fortunately, calcium interferes with the absorption and action of lead in the body.

The first question to ask is how much calcium the supplement provides. Most calcium supplements provide between 250 and 1000 milligrams of calcium. To be safe, total calcium intake from both foods and supplements should not exceed 2500 milligrams a day. Read the label to find out how much a dose supplies. Be aware that a 1000-milligram tablet of calcium carbonate contains only 400 milligrams of calcium. Unless the label states otherwise, supplements of calcium carbonate are 40 percent calcium; those of calcium citrate are 21 percent; lactate, 13 percent; and gluconate, 9 percent. Select a low-dose supplement and take it several times a day rather than taking a large-dose supplement all at once. Taking supplements in doses of 500 milligrams or less improves absorption. Small doses also help ease the GI distress (constipation, intestinal bloating, and excessive gas) that sometimes accompanies calcium supplement use.

The next question to ask is how well the body absorbs and uses the calcium from various supplements. Most healthy people absorb calcium equally well—and as well as from milk—from any of these supplements: calcium carbonate, citrate, or phosphate. More important than supplement solubility is tablet disintegration. When manufacturers compress large quantities of calcium into small pills, the stomach acid has difficulty penetrating the pill. To test a supplement's ability to dissolve, drop it into a 6-ounce cup of vinegar, and stir occasionally. A high-quality formulation will dissolve within half an hour.

Finally, having chosen a supplement, a person must take it regularly, but when should you take it? To circumvent adverse nutrient interactions, take calcium supplements between, not with, meals. (Importantly, do not take calcium supplements with iron supplements or iron-rich meals; calcium inhibits iron absorption.) To enhance calcium absorption, take supplements with meals. If such contradictory advice drives you crazy, reconsider the benefits of food sources of calcium. Most experts agree that foods are the best source of calcium.

First, ensure an optimal peak bone mass during childhood and adolescence by eating a balanced diet rich in calcium and engaging in regular physical activity. Then maintain that bone mass by continuing those healthy diet and activity habits, abstaining from cigarette smoking, and using alcohol moderately, if at all. Finally, minimize bone loss by maintaining an adequate nutrition and activity regimen and, for women, consult a physician about calcium supplements or other drug therapies that may be effective both in preventing bone loss and in restoring lost bone. The reward is the best possible chance of preserving bone health throughout life.

Some Closing Thoughts

Unfortunately, many of the strongest risk factors for osteoporosis are beyond people's control: age, gender, and genetics. But there are still several effective strategies for prevention.[33]

NUTRITION ON THE NET

 Access these websites for further study of topics covered in this highlight.

- Find updates and quick links to these and other nutrition-related sites at our website: **www.wadsworth.com/nutrition**

- Search for "falls and fractures" at the National Institute on Aging: **www.nih.gov/nia**

- Visit the National Institutes of Health Osteoporosis and Related Bone Diseases' National Resource Center: **www.osteo.org**

- Obtain additional information from the National Osteoporosis Foundation: **www.nof.org**

REFERENCES

1. E. S. Siris and coauthors, Identification and fracture outcomes of undiagnosed low bone mineral density in postmenopausal women: Results from the National Osteoporosis Risk Assessment, *Journal of the American Medical Association* 286 (2001): 2815–2822; D. J. van der Voort and coauthors, Screening for osteoporosis using easily obtainable biometrical data: Diagnostic accuracy of measured, self-reported and recalled BMI, and related costs of bone mineral density measurements, *Osteoporosis International* 11 (2000): 233–239; L. W. Turner, P. A. Faile, and R. Tomlinson, Jr., Osteoporosis diagnosis and fracture, *Orthopaedic Nursing*, September/October 1999, pp. 21–27.

2. B. A. Peterson and coauthors, The effects of an educational intervention on calcium intake and bone mineral content in young women with low calcium intake, *American Journal of Health Promotion* 14 (2000): 149–156.

3. D. Hartman and coauthors, Bone density of women who have recovered from anorexia nervosa, *International Journal of Eating Disorders* 28 (2000): 107–112.

4. H. J. Kloosterboer and A. G. Ederveen, Pros and cons of existing treatment modalities in osteoporosis: A comparison between tibolone, SERMs and estrogen (+/− progestogen) treatments, *Journal of Steroid Biochemistry and Molecular Biology* 83 (2002): 157–165; R. A. Sayegh and P. G. Stubblefield, Bone metabolism and the perimenopause overview, risk factors, screening, and osteoporosis preventive measures, *Obstetrics and Gynecology Clinics of North America* 29 (2002): 495–510.

5. R. T. Chlebowski and coauthors, Influence of estrogen plus progestin on breast cancer and mammography in healthy post-menopausal women: The Women's Health Initiative Randomized Trial, *Journal of the American Medical Association* 289 (2003): 3243–3253;

C. G. Solomon and R. G. Dluhy, Rethinking postmenopausal hormone therapy, *New England Journal of Medicine* 348 (2003): 579–580; Writing Group for the Women's Health Initiative Investigators, Risks and benefits of estrogen plus progestin in healthy postmenopausal women: Principal results from the Women's Health Initiative Randomized Controlled Trial, *Journal of the American Medical Association* 288 (2002): 321–333; O. Ylikorkala and M. Metsa-heikkila, Hormone replacement therapy in women with a history of breast cancer, *Gynecological Endocrinology* 16 (2002): 469–478.

6. S. L. Greenspan, N. M. Resnick, and R. A. Parker, Combination therapy with hormone replacement and alendronate for prevention of bone loss in elderly women: A randomized controlled trial, *Journal of the American Medical Association* 289 (2003): 2525–2533.

7. R. Brynin, Soy and its isoflavones: A review of their effects on bone density, *Alternative Medicine Review* 7 (2002): 317–327.

8. B. H. Arjmandi and coauthors, Soy protein has a greater effect on bone in postmenopausal women not on hormone replacement therapy, as evidenced by reducing bone resorption and urinary calcium excretion, *Journal of Clinical Endocrinology and Metabolism* 88 (2003): 1048–1054; T. Uesugi, Y. Fukui, and Y. Yamori, Beneficial effects of soybean isoflavone supplementation on bone metabolism and serum lipids in postmenopausal Japanese women: A four-week study, *Journal of the American College of Nutrition* 21 (2002): 97–102.

9. J. F. Whitfield, How to grow bone to treat osteoporosis and mend fractures, *Current Rheumatology Reports* 5 (2003): 45–56.

10. K. Delvaux and coauthors, Bone mass and lifetime physical activity in Flemish males: A 17-year follow-up study, *Medicine and Science in Sports and Exercise* 33 (2001): 1868–1875; L. Metcalfe and coauthors, Postmenopausal women and exercise for prevention of osteoporosis: The Bone, Estrogen, Strength Training (BEST) Study, *ACSM'S Health and Fitness Journal,* May/June 2001, pp. 6–14.

11. M. C. Wang and coauthors, Diet in midpuberty and sedentary activity in prepuberty predict peak bone mass, *American Journal of Clinical Nutrition* 77 (2003): 495–503; S. J. Stear and coauthors, Effect of a calcium and exercise intervention on the bone mineral status of 16-18-y-old adolescent girls, *American Journal of Clinical Nutrition* 77 (2003): 985–992; J. J. B. Anderson, Calcium requirements during adolescence to maximize bone health, *Journal of the American College of Nutrition* 20 (2001): 186S–191S.

12. J. E. Layne and M. E. Nelson, The effects of progressive resistance training on bone density: A review, *Medicine and Science in Sports and Exercise* 31 (1999): 25–30.

13. E. C. Cussler and coauthors, Weight lifted in strength training predicts bone change in postmenopausal women, *Medicine and Science in Sports and Exercise* 35 (2003): 10–17; Metcalfe and coauthors, 2001.

14. T. A. Ricci and coauthors, Moderate energy restriction increases bone resorption in obese postmenopausal women, *American Journal of Clinical Nutrition* 73 (2001): 347–352; D. Chao and coauthors, Effect of voluntary weight loss on bone mineral density in older overweight women, *Journal of the American Geriatrics Society* 48 (2000): 753–759; L. M. Salamone and coauthors, Effect of a lifestyle intervention on bone mineral density in premenopausal women: A randomized trial, *American Journal of Clinical Nutrition* 70 (1999): 97–103.

15. J. C. Fleet, Leptin and bone: Does the brain control bone biology? *Nutrition Reviews* 58 (2000): 209–211.

16. P. Gerdhem and K. J. Obrant, Effects of cigarette-smoking on bone mass as assessed by dual-energy X-ray absorptiometry and ultrasound, *Osteoporosis International* 13 (2002): 932–936; K. D. Ward and R. C. Klesges, A meta-analysis of the effects of cigarette smoking on bone mineral density,

17. P. B. Rapuri and coauthors, Smoking and bone metabolism in elderly women, *Bone* 27 (2000): 429–436; A. P. Hermann and coauthors, Premenopausal smoking and bone density in 2015 perimenopausal women, *Journal of Bone and Mineral Research* 15 (2000): 780–787.

18. C. Oncken and coauthors, Effects of smoking cessation or reduction on hormone profiles and bone turnover in postmenopausal women, *Nicotine and Tobacco Research* 4 (2002): 451–458.

19. P. B. Rapuri and coauthors, Alcohol intake and bone metabolism in elderly women, *American Journal of Clinical Nutrition* 72 (2000): 1206–1213; O. Ganry, C. Baudoin, and P. Fardellone, for the EPIDOS, Effect of alcohol intake on bone mineral density in elderly women: The EPIDOS Study, *American Journal of Epidemiology* 151 (2000): 773–780.

20. D. Teegarden and coauthors, Previous milk consumption is associated with greater bone density in young women, *American Journal of Clinical Nutrition* 69 (1999): 1014–1017.

21. R. E. Black and coauthors, Children who avoid drinking cow milk have low dietary calcium intakes and poor bone health, *American Journal of Clinical Nutrition* 76 (2002): 675–680.

22. H. J. Kalkwarf, J. C. Khoury, and B. P. Lanphear, Milk intake during childhood and adolescence, adult bone density, and osteoporotic fractures in US women, *American Journal of Clinical Nutrition* 77 (2003): 257–265.

23. J. Bell, Elderly women need dietary protein to maintain bone mass, *Nutrition Reviews* 60 (2002): 337–341; B. Dawson-Hughes and S. S. Harris, Calcium intake influences the association of protein intake with rates of bone loss in elderly men and women, *American Journal of Clinical Nutrition* 75 (2002): 773–779; J. H. E. Promislow and coauthors, Protein consumption and bone mineral density in the elderly: The Rancho Bernardo Study, *American Journal of Epidemiology* 155 (2002): 636–644; R. G. Munger, J. R. Cerhan, and B. C. Chiu, Prospective study of dietary protein intake and risk of hip fracture in postmenopausal women, *American Journal of Clinical Nutrition* 69 (1999): 147–152.

24. A. G. Need and coauthors, Vitamin D status: Effects on parathyroid hormone and 1,25-dihydroxyvitamin D in postmenopausal women, *American Journal of Clinical Nutrition* 71 (2000): 1577–1581.

25. D. Feskanich, W. C. Willett, and G. A. Colditz, Calcium, vitamin D, milk consumption, and hip fractures: A prospective study among postmenopausal women, *American Journal of Clinical Nutrition* 77 (2003): 504–511.

26. N. C. Binkley and coauthors, A high phylloquinone intake is required to achieve maximal osteocalcin γ-carboxylation, *American Journal of Clinical Nutrition* 76 (2002): 1055–1060; S. L. Booth and coauthors, Dietary vitamin K intakes are associated

with hip fracture but not with bone mineral density in elderly men and women, *American Journal of Clinical Nutrition* 71 (2000): 1201–1208; D. Feskanich and coauthors, Vitamin K intake and hip fractures in women: A prospective study, *American Journal of Clinical Nutrition* 69 (1999): 74–79.

27. K. L. Tucker and coauthors, Potassium, magnesium, and fruit and vegetable intakes are associated with greater bone mineral density in elderly men and women, *American Journal of Clinical Nutrition* 69 (1999): 727–736.

28. K. Michaelsson and coauthors, Serum retinol levels and the risk of fractures, *New England Journal of Medicine* 348 (2003): 287–294; D. Feskanich and coauthors, Vitamin A intake and hip fractures among postmenopausal women, *Journal of the American Medical Association* 287 (2002): 47–54; S. Johansson and coauthors, Subclinical hypervitaminosis A causes fragile bones in rats, *Bone* 31 (2002): 685–689; N. Binkley and D. Krueger, Hypervitaminosis A and bone, *Nutrition Reviews* 58 (2000): 138–144; S. J. Whiting and B. Lemke, Excess retinol intake may explain the high incidence of osteoporosis in northern Europe, *Nutrition Reviews* 57 (1999): 192–195.

29. K. L. Tucker and coauthors, Bone mineral density and dietary patterns in older adults: The Framingham Osteoporosis Study, *American Journal of Clinical Nutrition* 76 (2002): 245–252; D. M. Hegsted, Fractures, calcium, and the modern diet, *American Journal of Clinical Nutrition* 74 (2001): 571–573; S. A. New and coauthors, Dietary influences on bone mass and bone metabolism: Further evidence of a positive link between fruit and vegetable consumption and bone health? *American Journal of Clinical Nutrition* 71 (2000): 142–151; J. J. B. Anderson, Plant-based diets and bone health: Nutritional implications, *American Journal of Clinical Nutrition* 70 (1999): 539S–542S; Tucker and coauthors, 1999.

30. M. Harrington and K. D. Cashman, High salt intake appears to increase bone resorption in postmenopausal women but high potassium intake ameliorates this adverse effect, *Nutrition Reveiws* 61 (2003): 179–183; Tucker and coauthors, 2002; P. B. Rapuri and coauthors, Caffeine intake increases the rate of bone loss in elderly women and interacts with vitamin D receptor genotypes, *American Journal of Clinical Nutrition* 74 (2001): 694–700.

31. R. P. Heaney and coauthors, Dietary changes favorably affect bone remodeling in older adults, *Journal of the American Dietetic Association* 99 (1999): 1228–1233.

32. E. A. Ross, N. J. Szabo, and I. R. Tebbett, Lead content of calcium supplements, *Journal of the American Medical Association* 284 (2000): 1425–1429.

33. NIH Consensus Development Panel on Osteoporosis Prevention, Diagnosis, and Therapy, Osteoporosis prevention, diagnosis, and therapy, *Journal of the American Medical Association* 285 (2001): 785–795.

The Trace Minerals

© Brian Hagiwara/FoodPix/Getty Images

Chapter Outline

Nutrition Explorer CD-ROM Outline

Nutrition in Your Life

Trace—barely a perceptible amount. But the trace minerals tackle big jobs. Your blood can't carry oxygen without iron, and insulin can't deliver glucose without chromium. Teeth become decayed without fluoride, and thyroid glands develop goiter without iodine. Together, the trace minerals—iron, zinc, iodine, selenium, copper, manganese, fluoride, chromium, and molybdenum—keep you healthy and strong. Where can you get these amazing minerals? A variety of foods, especially those from the meat and meat alternate group, sprinkled with a little iodized salt and complemented by a glass of fluoridated water will do the trick. It's remarkable what your body can do with only a few milligrams—or even micrograms—of the trace minerals.

Figure 12-7 in the last chapter (p. 406) showed the tiny quantities of **trace minerals** in the human body. The trace minerals are so named because they are present, and needed, in relatively small amounts in the body. All together, they would produce only a bit of dust, hardly enough to fill a teaspoon. Yet they are no less important than the major minerals or any of the other nutrients. Each of the trace minerals performs a vital role. A deficiency of any of them may be fatal, and an excess of many is equally deadly. Remarkably, people's diets normally supply just enough of these minerals to maintain health.

The Trace Minerals—An Overview

The body requires the trace minerals in minuscule quantities. They participate in diverse tasks all over the body, each having special duties that only it can perform.

Food Sources The trace mineral contents of foods depend on soil and water composition and on how foods are processed. Furthermore, many factors in the

trace minerals: essential mineral nutrients found in the human body in amounts smaller than 5 g; sometimes called **microminerals.**

■ Reminder: *Bioavailability* refers to the rate at and the extent to which a nutrient is absorbed and used.

diet and within the body affect the minerals' bioavailability.■ Still, outstanding food sources for each of the trace minerals, just like those for the other nutrients, include a wide variety of foods, especially unprocessed, whole foods.

Deficiencies Severe deficiencies of the better-known minerals are easy to recognize. Deficiencies of the others may be harder to diagnose, and for all minerals, mild deficiencies are easy to overlook. Because the minerals are active in all the body systems—the GI tract, cardiovascular system, blood, muscles, bones, and central nervous system—deficiencies can have wide-reaching effects and can affect people of all ages. The most common result of a deficiency in children is failure to grow and thrive.

Toxicities Some of the trace minerals are toxic at intakes not far above the estimated requirements. Thus it is important not to habitually exceed the Upper Level of recommended intakes. Many vitamin-mineral supplements contain trace minerals, making it easy for users to exceed their needs. The Food and Drug Administration (FDA) has no authority to limit the amounts of trace minerals in supplements; consumers have demanded the freedom to choose their own doses of nutrients.* Individuals who take supplements must therefore be aware of the possible dangers and select supplements that contain no more than 100 percent of the Daily Value. It would be easier and safer to meet nutrient needs by selecting a variety of foods than by combining an assortment of supplements (see Highlight 10).

Interactions Interactions among the trace minerals are common and often well coordinated to meet the body's needs. For example, several of the trace minerals support insulin's work, influencing its synthesis, storage, release, and action.[1]

At other times, interactions lead to nutrient imbalances. An excess of one may cause a deficiency of another. (A slight manganese overload, for example, may aggravate an iron deficiency.) A deficiency of one may interfere with the work of another. (A selenium deficiency halts the activation of the iodine-containing thyroid hormones.) A deficiency of a trace mineral may even open the way for a contaminant mineral to cause a toxic reaction. (Iron deficiency, for example, makes the body vulnerable to lead poisoning.) These examples reinforce the need to balance intakes and to use supplements wisely, if at all. A good food source of one nutrient may be a poor food source of another; and factors that enhance the action of some trace minerals may hinder others. (Meats are a good source of iron, but a poor source of calcium; vitamin C enhances the absorption of iron but hinders that of copper.) Research on the trace minerals is active, suggesting that we have much more to learn about them.

IN SUMMARY Although the body uses only tiny amounts of the trace minerals, they are vital to health. Because so little is required, the trace minerals can be toxic at levels not far above estimated requirements—a consideration for supplement users. Like the other nutrients, the trace minerals are best obtained by eating a variety of whole foods.

Iron

Iron is an essential nutrient, vital to many of the cells' activities, but it poses a problem for millions of people: some people simply don't eat enough iron-containing foods to support their health optimally, while others absorb so much iron that it threatens their health. Iron exemplifies the principle that both too little and too much of a nutrient in the body can be harmful.

---

*Canada limits the amounts of trace minerals in supplements.

Iron Roles in the Body

Iron has the knack of switching back and forth between two ionic states.■ In the reduced state, iron has lost two electrons and therefore has a net positive charge of two; it is known as *ferrous iron*. In the oxidized state, iron has lost a third electron, has a net positive charge of three, and is known as *ferric iron*. Because ferrous iron can be oxidized to ferric iron and ferric iron can be reduced to ferrous iron, iron can serve as a cofactor■ to enzymes involved in oxidation-reduction reactions—reactions so widespread in metabolism that they occur in all cells. Iron is also required by enzymes involved in the making of amino acids, collagen, hormones, and neurotransmitters. (For details about ions, oxidation, and reduction, see Appendix B.)

Iron forms a part of the electron carriers that participate in the electron transport chain (discussed in Chapter 7).* In this pathway, these carriers transfer hydrogens and electrons to oxygen, forming water, and in the process make ATP for the cells' energy use.

Most of the body's iron is found in two proteins: hemoglobin■ in the red blood cells and **myoglobin** in the muscle cells. In both, iron helps accept, carry, and then release oxygen.

Iron Absorption and Metabolism

The body conserves iron. Because it is difficult to excrete iron once it is in the body, balance is maintained primarily through absorption: more iron is absorbed when stores are empty, and less is absorbed when stores are full.[2]

Iron Absorption Special proteins help the body absorb iron from food (see Figure 13-1). One protein, called mucosal **ferritin,** receives iron from food and stores it in the mucosal cells■ of the small intestine. When the body needs iron, mucosal ferritin

■ Iron's two ionic states:
  • Ferrous iron (reduced): Fe^{++}.
  • Ferric iron (oxidized): Fe^{+++}.

■ Reminder: A *cofactor* is a substance that works with an enzyme to facilitate a chemical reaction.

■ Reminder: *Hemoglobin* is the oxygen-carrying protein of the red blood cells that transports oxygen from the lungs to tissues throughout the body; hemoglobin accounts for 80% of the body's iron.

■ A mucous membrane such as the one that lines the GI tract is sometimes called the **mucosa** (mu-KO-sa). The adjective of mucosa is **mucosal** (mu-KO-sal).

myoglobin: the oxygen-holding protein of the muscle cells.
• **myo** = muscle

ferritin (FAIR-ih-tin): the iron storage protein.

---

*The iron-containing electron carriers of the electron transport chain are known as *cytochromes*. See Appendix C for details of this pathway.

FIGURE 13-1   Iron Absorption

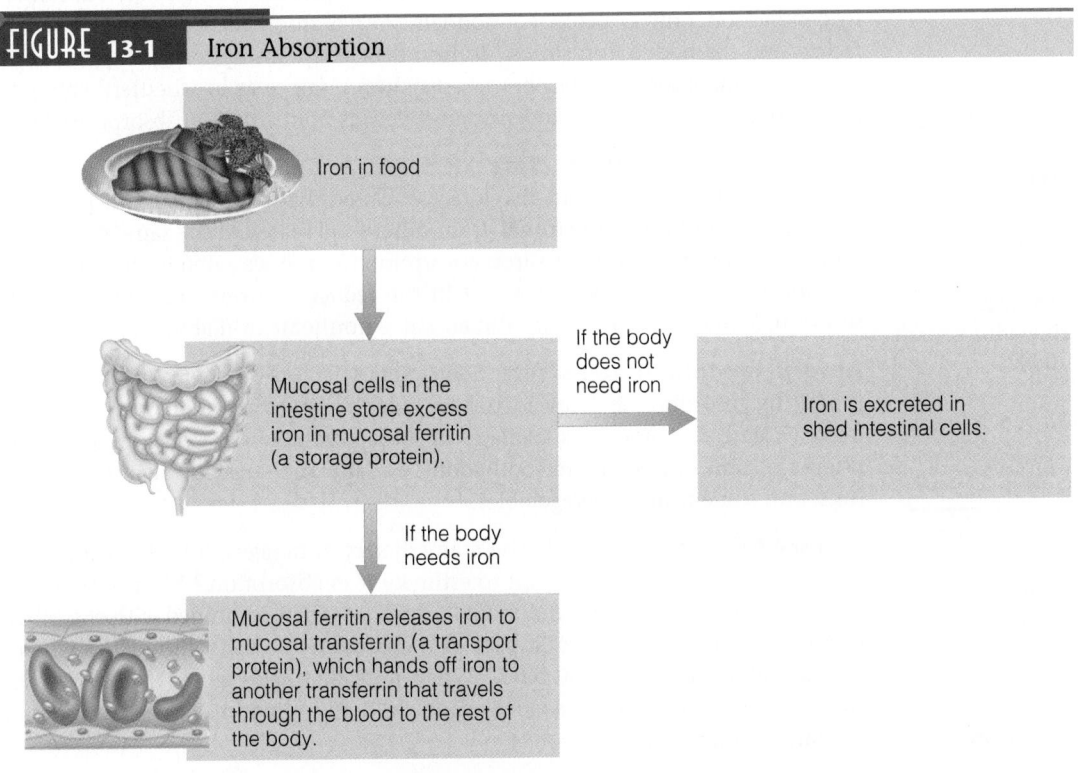

Iron in food

Mucosal cells in the intestine store excess iron in mucosal ferritin (a storage protein).

If the body does not need iron

Iron is excreted in shed intestinal cells.

If the body needs iron

Mucosal ferritin releases iron to mucosal transferrin (a transport protein), which hands off iron to another transferrin that travels through the blood to the rest of the body.

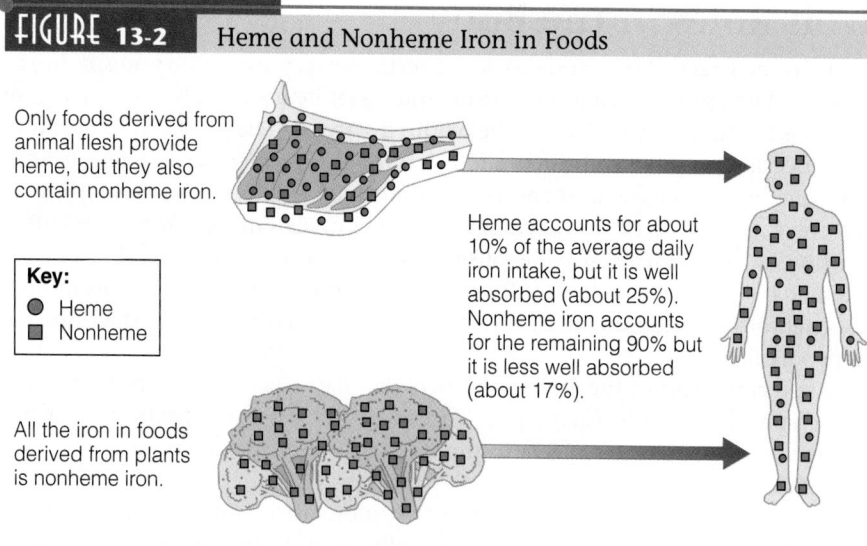

FIGURE 13-2 Heme and Nonheme Iron in Foods

Only foods derived from animal flesh provide heme, but they also contain nonheme iron.

Key:
● Heme
■ Nonheme

All the iron in foods derived from plants is nonheme iron.

Heme accounts for about 10% of the average daily iron intake, but it is well absorbed (about 25%). Nonheme iron accounts for the remaining 90% but it is less well absorbed (about 17%).

releases some iron to another protein, called mucosal **transferrin.** Mucosal transferrin transfers the iron to another protein, *blood transferrin,* which transports the iron to the rest of the body. If the body does not need iron, it is carried out when the intestinal cells are shed and excreted in the feces; intestinal cells are replaced about every three days. By holding iron temporarily, these cells can either deliver iron when the day's intake falls short or dispose of it when intakes exceed needs.

Heme and Nonheme Iron Iron absorption depends in part on its source. Iron occurs in two forms in foods: as **heme** iron, which is found only in foods derived from the flesh of animals, such as meats, poultry, and fish; and as nonheme iron, which is found in both plant-derived and animal-derived foods (see Figure 13-2). On average, heme iron represents about 10 percent of the iron a person consumes in a day. Even though heme iron accounts for only a small proportion of the intake, it is so well absorbed that it contributes significant iron: about 25 percent of heme iron is absorbed. By comparison, only 17 percent of nonheme iron is absorbed, depending on dietary factors and the body's iron stores.[3] In iron deficiency, absorption increases, and in iron overload, absorption declines.[4] Researchers disagree as to whether heme iron absorption responds to iron stores as sensitively as nonheme iron absorption does.

Absorption-Enhancing Factors Meat, fish, and poultry contain not only the well-absorbed heme iron, but also a factor (called the **MFP factor**) that promotes the absorption of nonheme iron■ from other foods eaten at the same meal. Vitamin C also enhances nonheme iron absorption from foods eaten in the same meal by capturing the iron and keeping it in the reduced ferrous form, ready for absorption. Some acids and sugars also enhance nonheme iron absorption.

Absorption-Inhibiting Factors Some dietary factors bind with nonheme iron, inhibiting absorption.■ These factors include the phytates and fibers in soy products, whole grains, and nuts; oxalates in some vegetables; the calcium and phosphorus in milk; the EDTA in food additives;* and the tannic acid in tea, coffee, nuts, and some fruits and vegetables.

Dietary Factors Combined The many dietary enhancers, inhibitors, and their combined effects make it difficult to estimate iron absorption.[5] Most of these factors exert a strong influence individually, but not when combined with the others in a meal. Furthermore, the impact of the combined effects diminishes when a diet is evaluated over several days. When multiple meals are analyzed together, three factors appear to be most relevant: MFP and vitamin C as enhancers and phytates as inhibitors.[6]

■ Factors that *enhance* nonheme iron absorption:
- MFP factor.
- Vitamin C (ascorbic acid).
- Citric acid and lactic acid from foods and HCl acid from the stomach.
- Sugars (including the sugars in wine).

■ Factors that *inhibit* nonheme iron absorption:
- Phytates and fibers (grains and vegetables).
- Oxalates (spinach, beets, rhubarb).
- Calcium and phosphorus (milk).
- EDTA (food additives).
- Tannic acid (and other polyphenols in tea and coffee).

transferrin (trans-FAIR-in): the iron transport protein.

heme (HEEM): the iron-holding part of the hemoglobin and myoglobin proteins. About 40% of the iron in meat, fish, and poultry is bound into heme; the other 60% is **nonheme** iron.

MFP factor: a factor associated with the digestion of **meat, fish, and poultry** that enhances nonheme iron absorption.

*EDTA is ethylenediamine tetra acetate, a chelating agent that is used in food processing to retard crystal formation and promote color retention.

Individual Variation Overall, about 18 percent of dietary iron is absorbed from mixed diets and only about 10 percent from vegetarian diets.[7] As you might expect, vegetarian diets do not have the benefit of easy-to-absorb heme iron or the help of MFP in enhancing absorption, but even the absorption of nonheme iron is low.[8] In addition to dietary influences, iron absorption also depends on an individual's health, stage in the life cycle, and iron status. Absorption can be as low as 2 percent in a person with GI disease or as high as 35 percent in a rapidly growing, healthy child. The body adapts to absorb more iron when a person's iron stores fall short, or when the need increases for any reason (such as pregnancy). The body makes more mucosal transferrin to absorb more iron from the intestines and more blood transferrin to carry more iron around the body. Similarly, when iron stores are sufficient, the body adapts to absorb less iron.[9]

Iron Transport and Storage Blood transferrin delivers iron to the bone marrow and other tissues. The bone marrow uses large quantities to make new red blood cells, whereas other tissues use less. Surplus iron is stored in the protein ferritin, primarily in the liver, but also in the bone marrow and spleen. When dietary iron has been plentiful, ferritin is constantly and rapidly made and broken down, providing an ever-ready supply of iron. When iron concentrations become abnormally high, the liver converts some ferritin into another storage protein called **hemosiderin.** Hemosiderin releases iron more slowly than ferritin does. By storing excess iron, the body protects itself: free iron acts as a free radical, attacking cell lipids, DNA, and protein.[10] (See Highlight 11 for more information on free radicals and the damage they can cause.)

Iron Recycling The average red blood cell lives about four months; then the spleen and liver cells remove it from the blood, take it apart, and prepare the degradation products for excretion or recycling. The iron is salvaged: the liver attaches it to blood transferrin, which transports it back to the bone marrow to be reused in making new red blood cells. Thus, although red blood cells live for only about four months, the iron recycles through each new generation of cells (see Figure 13-3).

This chili dinner provides several factors that may enhance iron absorption: heme and non-heme iron and MFP from meat, nonheme iron from legumes, and vitamin C from tomatoes.

hemosiderin (heem-oh-SID-er-in): an iron storage protein primarily made in times of iron overload.

FIGURE 13-3 Iron Recycled in the Body

Once iron enters the body, most of it is recycled. Some is lost with body tissues and must be replaced by eating iron-containing food.

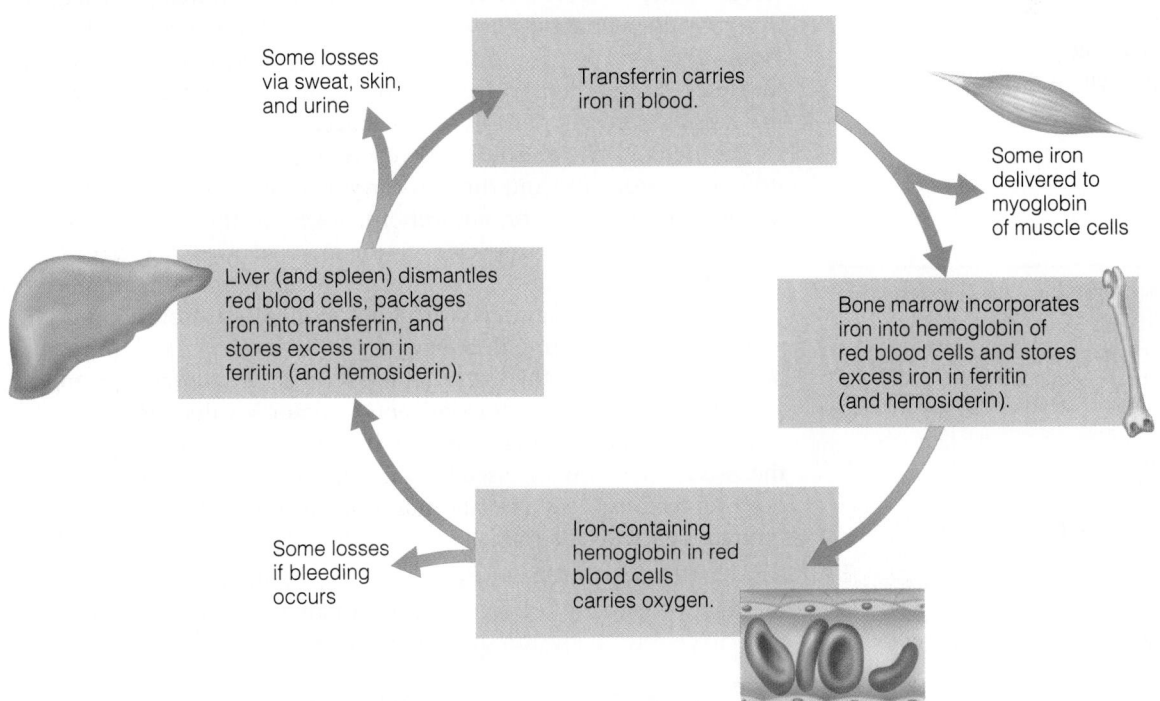

Some losses via sweat, skin, and urine

Transferrin carries iron in blood.

Some iron delivered to myoglobin of muscle cells

Liver (and spleen) dismantles red blood cells, packages iron into transferrin, and stores excess iron in ferritin (and hemosiderin).

Bone marrow incorporates iron into hemoglobin of red blood cells and stores excess iron in ferritin (and hemosiderin).

Some losses if bleeding occurs

Iron-containing hemoglobin in red blood cells carries oxygen.

The body loses some iron daily via the GI tract and, if bleeding occurs, in blood; only tiny amounts of iron are lost in urine, sweat, and shed skin.*

Iron Deficiency

Worldwide, **iron deficiency** is the most common nutrient deficiency, affecting more than 1.2 billion people.[11] In developing countries, almost half of the preschool children and pregnant women suffer from **iron-deficiency anemia**.[12] In the United States, iron deficiency is less prevalent, but still affects 10 percent of toddlers, adolescent girls, and women of childbearing age; preventing and correcting iron deficiency are high priorities.[13]

Vulnerable Stages of Life Some stages of life■ both demand more iron and provide less, making deficiency likely.[14] Women in their reproductive years are especially prone to iron deficiency because of repeated blood losses during menstruation. Pregnancy demands additional iron to support the added blood volume, growth of the fetus, and blood loss during childbirth. Infants and young children receive little iron from their high-milk diets, yet need extra iron to support their rapid growth. The rapid growth of adolescence, especially for males, and the menstrual losses of females also demand extra iron that a typical teen diet may not provide. An adequate iron intake is especially important during these stages of life.

- ■ High risk for iron deficiency:
  - Women in their reproductive years.
  - Pregnant women.
  - Infants and young children.
  - Teenagers.

Reduce iron deficiency among young children, females of childbearing age, and pregnant females.

■ The iron content of blood is about 0.5 mg/100 mL blood. A person donating a pint of blood (approximately 500 mL) loses about 2.5 mg of iron.

Blood Losses Bleeding■ from any site incurs iron losses. In some cases, as in an active ulcer, the bleeding may not be obvious, but even small chronic blood losses significantly deplete iron reserves; treating the ulcer resolves the iron deficiency.[15] In developing countries, blood loss is often brought on by malaria and parasitic infections of the GI tract. People who donate blood regularly also incur losses and may benefit from iron supplements. As mentioned, menstrual losses can be considerable as they tap women's iron stores regularly.

■ Stages of iron deficiency:
- Iron stores diminish.
- Transport iron decreases.
- Hemoglobin production declines.

Assessment of Iron Deficiency Iron deficiency develops in stages.■ This section provides a brief overview of how to detect these stages, and Appendix E provides more details. In the first stage of iron deficiency, iron stores diminish. Measures of serum ferritin (in the blood) reflect iron stores and are most valuable in assessing iron status at this earliest stage.

The second stage of iron deficiency is characterized by a decrease in transport iron: serum iron falls, and the iron-carrying protein transferrin *increases* (an adaptation that enhances iron absorption). Together, these two measures can determine the severity of the deficiency—the more transferrin and the less iron in the blood, the more advanced the deficiency is. Transferrin saturation—the percentage of transferrin that is saturated with iron—decreases as iron stores decline.

The third stage of iron deficiency occurs when the lack of iron limits hemoglobin production. Now the hemoglobin precursor, **erythrocyte protoporphyrin**, begins to accumulate as hemoglobin and **hematocrit** values decline.

Hemoglobin and hematocrit tests are easy, quick, and inexpensive, so they are the tests most commonly used in evaluating iron status; their usefulness is limited, however, because they are late indicators of iron deficiency. Furthermore, other nutrient deficiencies and medical conditions can influence their values.

Iron Deficiency and Anemia Iron deficiency and iron-deficiency anemia are not the same: people may be iron deficient without being anemic. The term *iron deficiency* refers to depleted body iron stores without regard to the degree of depletion or to the

iron deficiency: the state of having depleted iron stores.

iron-deficiency anemia: severe depletion of iron stores that results in low hemoglobin and small, pale red blood cells. Anemias that impair hemoglobin synthesis are **microcytic**.
- **micro** = small
- **cytic** = cell

erythrocyte protoporphyrin (PRO-toe-PORE-fe-rin): a precursor to hemoglobin.

hematocrit (hee-MAT-oh-krit): measurement of the volume of the red blood cells packed by centrifuge in a given volume of blood.

---

*Adults lose about 1.0 milligram of iron per day. Women lose additional iron in menses. Menstrual losses vary considerably, but over a month, they average about 0.5 milligram per day.

FIGURE 13-4   Normal and Anemic Blood Cells

Both size and color are normal in these blood cells.

Blood cells in iron-deficiency anemia are small (microcytic) and pale (hypochromic) because they contain less hemoglobin.

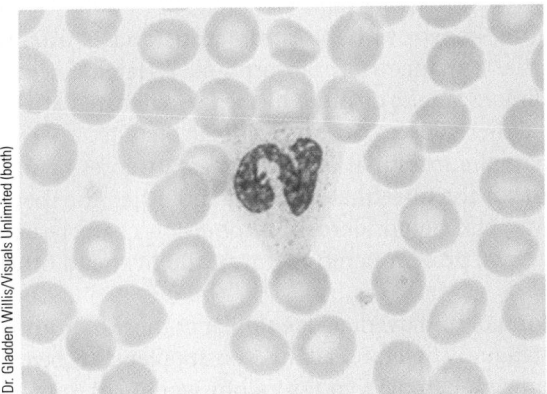
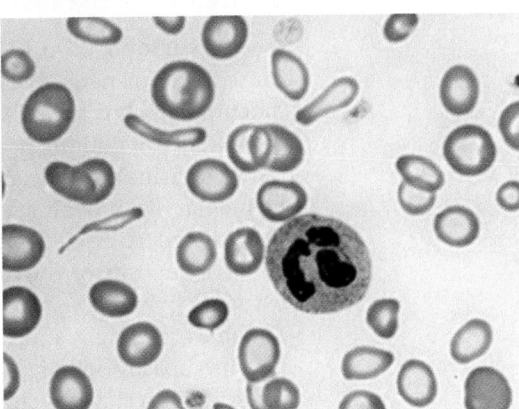

© Dr. Gladden Willis/Visuals Unlimited (both)

presence of anemia. The term *iron-deficiency anemia* refers to the severe depletion of iron stores that results in a low hemoglobin concentration. In iron-deficiency anemia, red blood cells are pale and small■ (see Figure 13-4). They can't carry enough oxygen from the lungs to the tissues. Without adequate iron, energy metabolism in the cells falters. The result is fatigue, weakness, headaches, apathy, pallor, and poor resistance to cold temperatures. Since hemoglobin is the bright red pigment of the blood, the skin of a fair person who is anemic may become noticeably pale. In a dark-skinned person, the tongue and eye lining, normally pink, will be very pale.

The fatigue that accompanies iron-deficiency anemia differs from the tiredness a person experiences from a simple lack of sleep. People with anemia feel fatigue only when they exert themselves. Iron supplementation can relieve the fatigue and improve the body's response to physical activity.[16] (The iron needs of physically active people and the special iron deficiency known as sports anemia are discussed in Chapter 14.)

Iron Deficiency and Behavior Long before the red blood cells are affected and anemia is diagnosed, a developing iron deficiency affects behavior. Even at slightly lowered iron levels, energy metabolism is impaired and neurotransmitter synthesis is altered, reducing physical work capacity and mental productivity.[17] Without the physical energy and mental alertness to work, plan, think, play, sing, or learn, people simply do these things less. They have no obvious deficiency symptoms; they just appear unmotivated, apathetic, and less physically fit. Work productivity and voluntary activities decline.[18]

Many of the symptoms associated with iron deficiency are easily mistaken for behavioral or motivational problems. A restless child who fails to pay attention in class might be thought contrary. An apathetic homemaker who has let housework pile up might be thought lazy. No responsible dietitian would ever claim that all behavioral problems are caused by nutrient deficiencies, but poor nutrition is always a possible contributor to problems like these. When investigating a behavioral problem, check the adequacy of the diet and seek a routine physical examination before undertaking more expensive, and possibly harmful, treatment options. (The effects of iron deficiency on children's behavior are discussed further in Chapter 16.)

Iron Deficiency and Pica A curious behavior seen in some iron-deficient people, especially in women and children of low-income groups, is **pica**—an appetite for ice, clay, paste, and other nonfood substances. These substances contain no iron and cannot remedy a deficiency; in fact, clay actually inhibits iron absorption, which may explain the iron deficiency that accompanies such behavior.

■ Iron-deficiency anemia is a **microcytic** (my-cro-SIT-ic) **hypochromic** (high-po-KROME-ic) **anemia.**
- **micro** = small
- **cytic** = cell
- **hypo** = too little
- **chrom** = color

pica (PIE-ka): a craving for nonfood substances. Also known as **geophagia** (gee-oh-FAY-gee-uh) when referring to clay eating and **pagophagia** (pag-oh-FAY-gee-uh) when referring to ice craving.

Iron Toxicity

In general, even a diet that includes fortified foods poses no special risk for iron toxicity.[19] The body normally absorbs less iron when its stores are full, but some individuals are poorly defended against excess iron. Once considered rare, **iron overload** has emerged as an important disorder of iron metabolism and regulation.

Iron Overload Iron overload is known as **hemochromatosis** and is usually caused by a genetic disorder that enhances iron absorption.[20] Hereditary hemochromatosis is the most common genetic disorder in the United States, affecting some 1.5 million people. Other causes of iron overload include repeated blood transfusions (which bypass the intestinal defense), massive doses of supplementary iron (which overwhelm the intestinal defense), and other rare metabolic disorders. Excess iron may cause **hemosiderosis,** a condition characterized by large deposits of the iron storage protein hemosiderin in the liver and other tissues.

Some of the signs and symptoms of iron overload are similar to those of iron deficiency: apathy, lethargy, and fatigue. Therefore, taking iron supplements before assessing iron status is clearly unwise; hemoglobin tests alone would fail to make the distinction because excess iron accumulates in storage. Iron overload assessment tests measure transferrin saturation and serum ferritin.

Iron overload is characterized by tissue damage, especially in iron-storing organs such as the liver. Infections are likely because bacteria thrive on iron-rich blood. Symptoms are most severe in alcohol abusers because alcohol damages the intestine, further impairing its defenses against absorbing excess iron. Untreated hemochromatosis aggravates the risks of diabetes, liver cancer, heart disease, and arthritis.

Iron overload is more common in men than in women and is twice as prevalent among men as iron deficiency. The widespread fortification of foods with iron makes it difficult for people with hemochromatosis to follow a low-iron diet, and greater dangers lie in the indiscriminate use of iron and vitamin C supplements. Vitamin C not only enhances iron absorption, but releases iron from ferritin, allowing free iron to wreak the damage typical of free radicals.[21] This example shows how vitamin C acts as a *pro*oxidant when taken in high doses. (See Highlight 11 for a discussion of free radicals and their effects on disease development.)

Iron and Heart Disease Some research suggests a link between heart disease and elevated iron stores, but the evidence is inconsistent and unconvincing.[22] As mentioned, free radicals can attack ferritin, causing it to release iron from storage. Free iron, in turn, acts as an oxidant that can generate more free radicals. Whether iron's oxidation of LDL plays a role in the development of heart disease has not been proved.[23]

Iron and Cancer There may be an association between iron and some cancers. Explanations for how iron might be involved in causing cancer focus on its free-radical activity, which can damage DNA (see Highlight 11). One of the benefits of a high-fiber diet may be that its phytates bind iron, making it less available for such reactions.

Iron Poisoning Large doses of iron supplements cause GI distress, including constipation, nausea, vomiting, and diarrhea. These effects may not be as serious as other consequences of iron toxicity, but they are consistent enough to establish an Upper Level of 45 milligrams per day for adults.

Ingestion of iron-containing supplements remains a leading cause of accidental poisoning in small children.[24] Symptoms of intoxication include nausea, vomiting, diarrhea, a rapid heartbeat, a weak pulse, dizziness, shock, and confusion. As few as five iron tablets containing as little as 200 milligrams of iron have caused the deaths of dozens of young children. The exact cause of death is uncertain, but excessive free-radical damage is thought to play a role in heart failure and respiratory distress; autopsy reports reveal iron deposits and cell death in the stomach, small intestine, liver, and blood vessels (which can cause internal bleeding).[25]

iron overload: toxicity from excess iron.

hemochromatosis (HE-moh-KRO-ma-toe-sis): a hereditary defect in iron absorption characterized by deposits of iron-containing pigment in many tissues, with tissue damage.

hemosiderosis (HE-moh-sid-er-OH-sis): a condition characterized by the deposition of hemosiderin in the liver and other tissues.

To calculate the recommended daily iron intake, the DRI Committee considers a number of factors. For example, for a woman of childbearing age (19 to 50):

- Losses from feces, urine, sweat, and shed skin: 1.0 milligram.
- Losses through menstruation (about 14 milligrams total averaged over 28 days): 0.5 milligram.

These losses reflect an average daily need (total) of 1.5 milligrams of *absorbed* iron.

An estimated average requirement is determined based on the daily need and the assumption that an average of 18 percent of ingested iron is absorbed:

1.5 mg iron (needed) ÷ 0.18 (percent iron absorbed) = 8 mg iron (estimated average requirement).

Then a margin of safety is added to cover the needs of essentially all women of childbearing age, and the RDA is set at 18 milligrams.

Keep iron-containing tablets out of the reach of children. If you suspect iron poisoning, call the nearest poison control center or a physician immediately.

Iron Recommendations and Sources

To obtain enough iron, people must first select iron-rich foods and then eat so as to maximize iron absorption. This discussion begins by identifying iron-rich foods, then reviews factors affecting absorption.

Recommended Iron Intakes The usual diet in the United States provides about 6 to 7 milligrams of iron for every 1000 kcalories. The recommended daily intake for men is 8 milligrams, and most men eat more than 2000 kcalories a day, so they can meet their iron needs with little effort. Women in their reproductive years, however, need 18 milligrams a day. (The accompanying "How to" explains how to calculate the recommended intake.) Vegetarians need 1.8 times as much iron to make up for the low bioavailability typical of their diets.[26]■

Because women have higher iron needs and lower energy needs, they sometimes have trouble obtaining enough iron. On average, women receive only 12 to 13 milligrams of iron per day, not enough until after menopause. To meet their iron needs from foods, premenopausal women need to select iron-rich foods at every meal.

Iron in Foods Figure 13-5 (on p. 446) shows the amounts of iron in selected foods. Meats, fish, and poultry contribute the most iron; other protein-rich foods such as legumes and eggs are also good sources. Although an indispensable part of the diet, foods in the milk group are notoriously poor in iron. Grain foods vary, with whole-grain, enriched, and fortified breads and cereals providing the most iron. Finally, dark greens (such as broccoli) and dried fruits (such as raisins) contribute some iron.

Iron-Enriched Foods Iron is one of the enrichment nutrients for grain products. One serving of enriched bread or cereal provides only a little iron, but because people eat many servings of these foods, the contribution can be significant. Iron added to foods is not absorbed as well as naturally occurring iron, but when eaten with absorption-enhancing foods, enrichment iron can make a difference. In cases of iron overload, enrichment may exacerbate the problem.[27]

Maximizing Iron Absorption In general, the bioavailability of iron in meats, fish, and poultry is high; in grains and legumes, intermediate; and in most vegetables, especially those high in oxalates such as spinach, low. As mentioned earlier, the amount of iron ultimately absorbed from a meal depends on the combined effects of several enhancing and inhibiting factors. For maximum absorption of nonheme iron, eat meat for MFP and fruits or vegetables for vitamin C. The iron of baked beans, for example, will be enhanced by the MFP in a piece of ham served with them; the iron of bread will be enhanced by vitamin C in a slice of tomato on a sandwich.

■ To calculate the RDA for vegetarians, multiply by 1.8:
- 8 mg × 1.8 = 14 mg/day (vegetarian men).
- 18 mg × 1.8 = 32 mg/day (vegetarian women, 19 to 50 yr).

When the label on a grain product says "enriched," it means iron and several B vitamins have been added.

FIGURE 13-5 Iron in Selected Foods

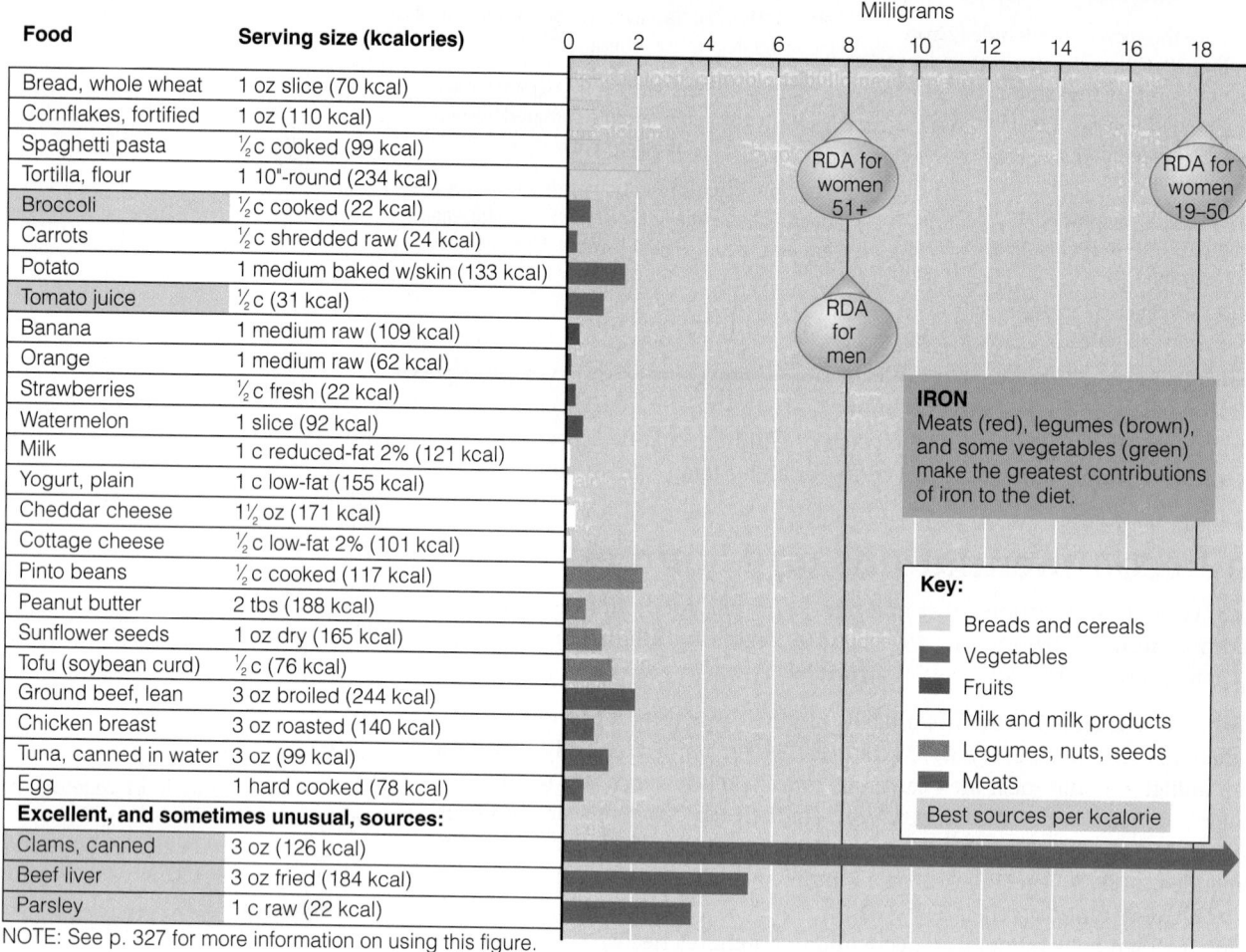

Food	Serving size (kcalories)
Bread, whole wheat	1 oz slice (70 kcal)
Cornflakes, fortified	1 oz (110 kcal)
Spaghetti pasta	½ c cooked (99 kcal)
Tortilla, flour	1 10"-round (234 kcal)
Broccoli	½ c cooked (22 kcal)
Carrots	½ c shredded raw (24 kcal)
Potato	1 medium baked w/skin (133 kcal)
Tomato juice	½ c (31 kcal)
Banana	1 medium raw (109 kcal)
Orange	1 medium raw (62 kcal)
Strawberries	½ c fresh (22 kcal)
Watermelon	1 slice (92 kcal)
Milk	1 c reduced-fat 2% (121 kcal)
Yogurt, plain	1 c low-fat (155 kcal)
Cheddar cheese	1½ oz (171 kcal)
Cottage cheese	½ c low-fat 2% (101 kcal)
Pinto beans	½ c cooked (117 kcal)
Peanut butter	2 tbs (188 kcal)
Sunflower seeds	1 oz dry (165 kcal)
Tofu (soybean curd)	½ c (76 kcal)
Ground beef, lean	3 oz broiled (244 kcal)
Chicken breast	3 oz roasted (140 kcal)
Tuna, canned in water	3 oz (99 kcal)
Egg	1 hard cooked (78 kcal)
Excellent, and sometimes unusual, sources:	
Clams, canned	3 oz (126 kcal)
Beef liver	3 oz fried (184 kcal)
Parsley	1 c raw (22 kcal)

NOTE: See p. 327 for more information on using this figure.

IRON
Meats (red), legumes (brown), and some vegetables (green) make the greatest contributions of iron to the diet.

Key:
- Breads and cereals
- Vegetables
- Fruits
- Milk and milk products
- Legumes, nuts, seeds
- Meats
- Best sources per kcalorie

RDA for women 51+
RDA for women 19–50
RDA for men

Iron Contamination and Supplementation

In addition to the iron from foods, **contamination iron** from nonfood sources of inorganic iron salts can contribute to the day's intakes. People can also get iron from supplements.

Contamination Iron Foods cooked in iron cookware take up iron salts. The more acidic the food, and the longer it is cooked in iron cookware, the higher the iron content. The iron content of eggs can triple in the time it takes to scramble them in an iron pan. Admittedly, the absorption of this iron may be poor (perhaps only 1 to 2 percent), but every little bit helps a person who is trying to increase iron intake.

Iron Supplements People who are iron deficient may need supplements as well as an iron-rich, absorption-enhancing diet. Many physicians routinely recommend iron supplements to pregnant women, infants, and young children. Iron from supplements is less well absorbed than that from food, so the doses have to be high. The absorption of iron taken as ferrous sulfate or as an iron **chelate** is better than that from other iron supplements. Absorption also improves when supplements are taken between meals or at bedtime on an empty stomach, and with liquids other than milk, tea, or coffee, which inhibit absorption. Taking iron supplements in a single dose instead of several doses per day is equally effective and may improve a person's willingness to take it regularly.[28]

There is no benefit to taking iron supplements with orange juice because vitamin C does not enhance absorption from supplements as it does from foods. (Vitamin C

contamination iron: iron found in foods as the result of contamination by inorganic iron salts from iron cookware, iron-containing soils, and the like.

chelate (KEY-late): a substance that can grasp the positive ions of a metal.
- **chele** = claw

enhances iron absorption by converting insoluble ferric iron in foods to the more soluble ferrous iron, and supplemental iron is already in the ferrous form.) Constipation is a common side effect of iron supplementation; drinking plenty of water may help to relieve this problem.

IN SUMMARY Most of the body's iron is in hemoglobin and myoglobin where it carries oxygen for use in energy metabolism; some iron is also required for enzymes involved in a variety of reactions. Special proteins assist with iron absorption, transport, and storage—all helping to maintain an appropriate balance, because both too little and too much iron can be damaging. Iron deficiency is most common among infants and young children, teenagers, women of childbearing age, and pregnant women; symptoms include fatigue and anemia. Iron overload is most common in men. Heme iron, which is found only in meat, fish, and poultry, is better absorbed than nonheme iron, which occurs in most foods. Nonheme iron absorption is improved by eating iron-containing foods with foods containing the MFP factor and vitamin C; absorption is limited by phytates and oxalates. The summary table presents a few iron facts.

An old-fashioned iron skillet adds iron to foods.

Iron

2001 RDA

Men: 8 mg/day

Women: 18 mg/day (19–50 yr)

8 mg/day (51+)

Upper Level

Adults: 45 mg/day

Chief Functions in the Body

Part of the protein hemoglobin, which carries oxygen in the blood; part of the protein myoglobin in muscles, which makes oxygen available for muscle contraction; necessary for the utilization of energy as part of the cells' metabolic machinery

Significant Sources

Red meats, fish, poultry, shellfish, eggs, legumes, dried fruits

Deficiency Symptoms

Anemia: weakness, fatigue, headaches; impaired work performance and cognitive function; impaired immunity; pale skin, nailbeds, mucous membranes, and palm creases; concave nails; inability to regulate body temperature; pica

Toxicity Symptoms

GI distress
Iron overload: infections, fatigue, joint pain, skin pigmentation, organ damage

Zinc

Zinc is a versatile trace element required as a cofactor■ by more than 100 enzymes. Virtually all cells contain zinc, but the highest concentrations are in muscle and bone.

Zinc Roles in the Body

Zinc supports the work of numerous proteins in the body, including the **metalloenzymes,**■ which are involved in a variety of metabolic processes.* In addition, zinc stabilizes cell membranes, helping to strengthen their defense against free-radical attacks. Zinc also assists in immune function and in growth and development. Zinc participates in the synthesis, storage, and release of the hormone insulin in the pancreas, although it does not appear to play a direct role in insulin's action. Zinc interacts with platelets in blood clotting, affects thyroid hormone function, and influences behavior and learning performance. It is needed to produce the active form of vitamin A (retinal) in visual pigments and the retinol-binding protein that transports vitamin A. It is essential to normal taste perception, wound healing, the making of sperm, and

■ Reminder: A *cofactor* is a substance that works with an enzyme to facilitate a chemical reaction.

■ Metalloenzymes that require zinc:
- Help make parts of the genetic materials DNA and RNA.
- Manufacture heme for hemoglobin.
- Participate in essential fatty acid metabolism.
- Release vitamin A from liver stores.
- Metabolize carbohydrates.
- Synthesize proteins.
- Metabolize alcohol in the liver.
- Dispose of damaging free radicals.

metalloenzymes (meh-TAL-oh-EN-zimes): enzymes that contain one or more minerals as part of their structures.

---

*Among the metalloenzymes requiring zinc are carbonic anhydrase, deoxythymidine kinase, DNA and RNA polymerase, and alkaline phosphatase.

fetal development. A zinc deficiency impairs all these and other functions, underlining the vast importance of zinc in supporting the body's proteins.

Zinc Absorption and Metabolism

The body's handling of zinc resembles that of iron in some ways and differs in others. A key difference is the circular passage of zinc from the intestine to the body and back again.

Zinc Absorption The rate of zinc absorption varies from about 15 to 40 percent, depending on a person's zinc status: if more is needed, more is absorbed. Also, dietary factors influence zinc absorption. For example, fiber and phytates bind zinc, thus limiting its bioavailability.[29]

Upon absorption into an intestinal cell, zinc has several options. It may become involved in the metabolic functions of the cell itself. Alternatively, it may be retained within the cell by **metallothionein,** a special binding protein similar to the iron storage protein, mucosal ferritin.

Metallothionein in the intestinal cells helps to regulate zinc absorption by holding it in reserve until the body needs zinc. Then metallothionein releases it into the blood where it can be transported around the body. Metallothionein in the liver performs a similar role, binding zinc until other body tissues signal a need for it.

Zinc Recycling Some zinc eventually reaches the pancreas, where it is incorporated into many of the digestive enzymes that the pancreas releases into the intestine at mealtimes. The intestine thus receives two doses of zinc with each meal—one from foods and the other from the zinc-rich pancreatic secretions. The recycling of zinc in the body from the pancreas to the intestine and back to the pancreas is referred to as the **enteropancreatic circulation** of zinc. As this zinc circulates through the intestine, it may be refused entry by the intestinal cells or retained in them on any of its times around (see Figure 13-6). The body loses zinc

---

metallothionein (meh-TAL-oh-THIGH-oh-neen): a sulfur-rich protein that avidly binds with and transports metals such as zinc.
- **metallo** = containing a metal
- **thio** = containing sulfur
- **ein** = a protein

enteropancreatic (EN-ter-oh-PAN-kree-AT-ik) **circulation:** the circulatory route from the pancreas to the intestine and back to the pancreas.

FIGURE 13-6   Enteropancreatic Circulation of Zinc

Some zinc from food is absorbed by the small intestine and sent to the pancreas to be incorporated into digestive enzymes that return to the small intestine. This cycle is called the enteropancreatic circulation of zinc.

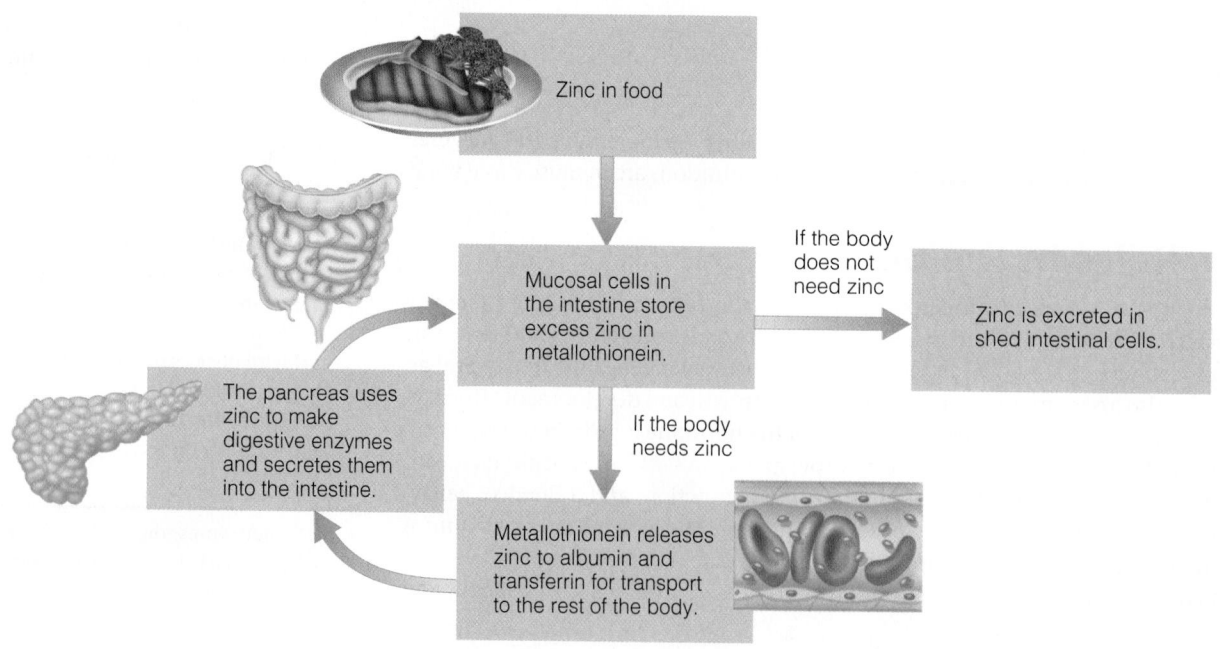

primarily in feces. Smaller losses occur in urine, shed skin, hair, sweat, menstrual fluids, and semen.

Zinc Transport Zinc's main transport vehicle in the blood is the protein albumin. Some zinc also binds to transferrin—the same transferrin that carries iron in the blood. In healthy individuals, transferrin is usually less than 50 percent saturated with iron, but in iron overload, it is more saturated. Diets that deliver more than twice as much iron as zinc leave too few transferrin sites available for zinc. The result: poor zinc absorption. The converse is also true: large doses of zinc inhibit iron absorption.

Large doses of zinc create a similar problem with another essential mineral, copper. These nutrient interactions highlight one of the many reasons why people should use supplements conservatively, if at all: supplementation can easily create imbalances.

Zinc Deficiency

Severe zinc deficiencies are not widespread in developed countries, but they do occur in vulnerable groups—pregnant women, young children, the elderly, and the poor. Human zinc deficiency was first reported in the 1960s in children and adolescent boys in Egypt, Iran, and Turkey. Children have especially high zinc needs because they are growing rapidly and synthesizing many zinc-containing proteins; the native diets among those populations were not meeting these needs. Middle Eastern diets are typically low in the richest zinc source, meats, and the staple foods are legumes, unleavened breads, and other whole-grain foods—all high in fiber and phytates, which inhibit zinc absorption.[*]

Figure 13-7 shows the severe growth retardation and mentions the arrested sexual maturation characteristic of zinc deficiency. In addition, zinc deficiency hinders digestion and absorption, causing diarrhea, which worsens malnutrition not only for zinc, but for all nutrients. It impairs the immune response, making infections likely—among them, infections of the GI tract, which worsen malnutrition, including zinc malnutrition (a classic downward spiral of events). Chronic zinc deficiency damages the central nervous system and brain and may lead to poor motor development and cognitive performance. Because zinc deficiency directly impairs vitamin A metabolism, vitamin A–deficiency symptoms often appear. Zinc deficiency also disturbs thyroid function and the metabolic rate. It alters taste, causes loss of appetite, and slows wound healing—in fact, its symptoms are so all-pervasive that generalized malnutrition and sickness are more likely to be the diagnosis than simple zinc deficiency.

Zinc Toxicity

High doses (50 to 450 milligrams) of zinc may cause vomiting, diarrhea, headaches, exhaustion, and other symptoms. An Upper Level for adults was set at 40 milligrams based on zinc's interference in copper metabolism—an effect that, in animals, leads to degeneration of the heart muscle.

Zinc Recommendations and Sources

Figure 13-8 shows zinc amounts in foods per serving. Zinc is highest in protein-rich foods such as shellfish (especially oysters), meats, poultry, and liver. Legumes and whole-grain products are good sources of zinc if eaten in large quantities; in typical U.S. diets, phytate intake from grains is not high enough to impair zinc absorption. Vegetables vary in zinc content depending on the soil in which they are grown. Average intakes in the United States are slightly higher than recommendations.

---

[*]Unleavened bread contains no yeast, which normally breaks down phytates during fermentation.

FIGURE 13-7 Zinc-Deficiency Symptoms—The Stunted Growth of Dwarfism

The Egyptian man on the right is an adult of average height. The Egyptian boy on the left is 17 years old but is only 4 feet tall, like a 7-year-old in the United States. His genitalia are like those of a 6-year-old. The growth retardation, known as dwarfism, is rightly ascribed to zinc deficiency because it is partially reversible when zinc is restored to the diet.

© H. Sanstead, University of Texas at Galveston

Zinc is highest in protein-rich foods such as oysters, beef, poultry, legumes, and nuts.

© Polara Studios Inc.

FIGURE 13-8 Zinc in Selected Foods

Food	Serving size (kcalories)
Bread, whole wheat	1 oz slice (70 kcal)
Cornflakes, fortified	1 oz (110 kcal)
Spaghetti pasta	½ c cooked (99 kcal)
Tortilla, flour	1 10"-round (234 kcal)
Broccoli	½ c cooked (22 kcal)
Carrots	½ c shredded raw (24 kcal)
Potato	1 medium baked w/skin (133 kcal)
Tomato juice	¾ c (31 kcal)
Banana	1 medium raw (109 kcal)
Orange	1 medium raw (62 kcal)
Strawberries	½ c fresh (22 kcal)
Watermelon	1 slice (92 kcal)
Milk	1 c reduced-fat 2% (121 kcal)
Yogurt, plain	1 c low-fat (155 kcal)
Cheddar cheese	1½ oz (171 kcal)
Cottage cheese	½ c low-fat 2% (101 kcal)
Pinto beans	½ c cooked (117 kcal)
Peanut butter	2 tbs (188 kcal)
Sunflower seeds	1 oz dry (165 kcal)
Tofu (soybean curd)	½ c (76 kcal)
Ground beef, lean	3 oz broiled (244 kcal)
Chicken breast	3 oz roasted (140 kcal)
Tuna, canned in water	3 oz (99 kcal)
Egg	1 hard cooked (78 kcal)
Excellent, and sometimes unusual, sources:	
Oysters	3 oz cooked (139 kcal)
Sirloin steak, lean	3 oz broiled (172 kcal)
Crab	3 oz cooked (94 kcal)

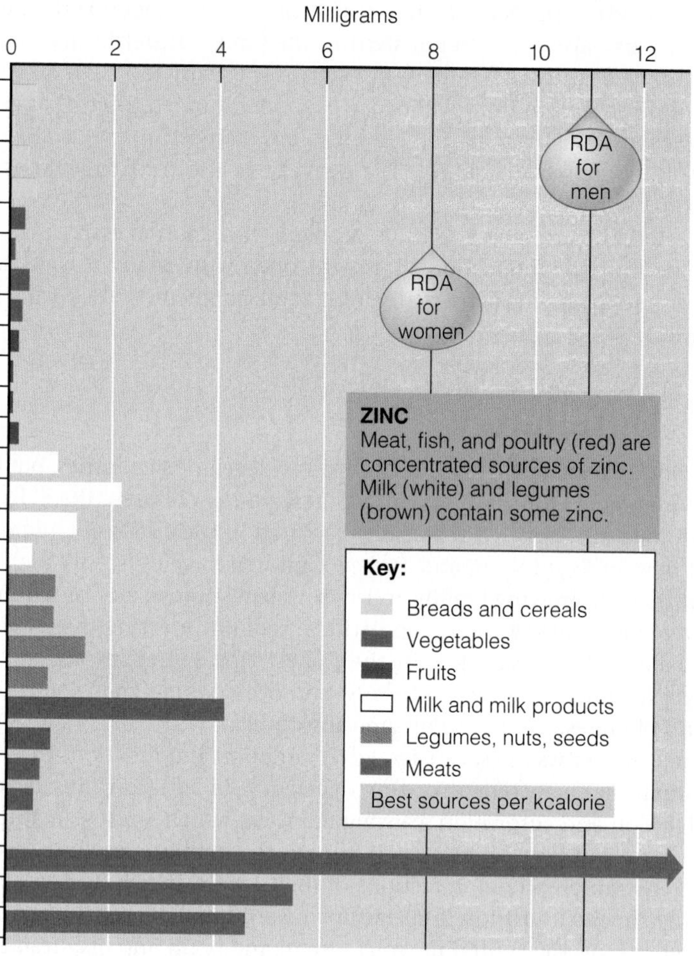

Milligrams: 0 2 4 6 8 10 12

RDA for men
RDA for women

ZINC
Meat, fish, and poultry (red) are concentrated sources of zinc. Milk (white) and legumes (brown) contain some zinc.

Key:
- Breads and cereals
- Vegetables
- Fruits
- Milk and milk products
- Legumes, nuts, seeds
- Meats
- Best sources per kcalorie

NOTE: See p. 327 for more information on using this figure.

Zinc Supplementation

In developed countries, most people can get enough zinc from the diet without resorting to supplements. In developing countries, zinc supplements play a major role in the treatment of childhood infectious diseases. Zinc supplements effectively reduce the incidence of disease and death associated with diarrhea.[30]

The use of zinc lozenges to treat the common cold has been controversial and inconclusive, with some studies finding them effective and others not.[31] The different study results may reflect the effectiveness of various zinc compounds. Some studies using zinc gluconate report shorter duration of cold symptoms, whereas most studies using other combinations of zinc report no effect. Common side effects of zinc lozenges include nausea and bad taste reactions.

IN SUMMARY Zinc-requiring enzymes participate in a multitude of reactions affecting growth, vitamin A activity, and pancreatic digestive enzyme synthesis, among others. Both dietary zinc and zinc-rich pancreatic secretions (via enteropancreatic circulation) are available for absorption. Absorption is monitored by a special binding protein (metallothionein) in the intestine. Protein-rich foods derived from animals are the best sources of bioavailable zinc. Fiber and phytates in cereals bind zinc, limiting absorption. Growth retardation and sexual immaturity are hallmark symptoms of zinc deficiency. These facts and others are included in the following table.

Zinc

2001 RDA

Men: 11 mg/day

Women: 8 mg/day

Upper Level

Adults: 40 mg/day

Chief Functions in the Body

Part of many enzymes; associated with the hormone insulin; involved in making genetic material and proteins, immune reactions, transport of vitamin A, taste perception, wound healing, the making of sperm, and the normal development of the fetus

Significant Sources

Protein-containing foods: red meats, shellfish, whole grains

Deficiency Symptoms[a]

Growth retardation, delayed sexual maturation, impaired immune function, hair loss, eye and skin lesions, loss of appetite

Toxicity Symptoms

Loss of appetite, impaired immunity, low HDL, copper and iron deficiencies

[a]A rare inherited disease of zinc malabsorption, *acrodermatitis* (AK-roh-der-ma-TIE-tis) *enteropathica* (EN-ter-oh-PATH-ick-ah), causes additional and more severe symptoms.

Iodine

Traces of the iodine ion (called iodide)■ are indispensable to life. In the GI tract, iodine from foods becomes iodide; this chapter uses *iodine* when referring to the nutrient in foods and *iodide* when referring to it in the body. Iodide occurs in the body in minuscule amounts, but its principal role in the body and its requirement are well established.

Iodide Roles in the Body Iodide is an integral part of the thyroid hormones■ that regulate body temperature, metabolic rate, reproduction, growth, blood cell production, nerve and muscle function, and more. By controlling the rate at which the cells use oxygen, these hormones influence the amount of energy released during basal metabolism.

Iodine Deficiency The hypothalamus regulates thyroid hormone production by controlling the release of the pituitary's thyroid-stimulating hormone (TSH).■ With iodine deficiency, thyroid hormone production declines, and the body responds by secreting more TSH in a futile attempt to accelerate iodide uptake by the thyroid gland. If a deficiency persists, the cells of the thyroid gland enlarge, so as to trap as much iodide as possible. Sometimes the gland enlarges until it makes a visible lump in the neck, a simple **goiter** (shown in Figure 13-9 on p. 452).

Goiter afflicts about 200 million people the world over, many of them in South America, Asia, and Africa. In all but 4 percent of these cases, the cause is iodine deficiency. As for the 4 percent (8 million), most have goiter because they regularly eat excessive amounts of foods■ that contain an antithyroid substance **(goitrogen)** whose effect is not counteracted by dietary iodine. The goitrogens present in plants remind us that even natural components of foods can cause harm when eaten in excess.

Goiter may be the earliest and most obvious sign of iodine deficiency, but the most tragic and prevalent damage occurs in the brain. Children with even a mild iodine deficiency typically have goiters and perform poorly in school; with treatment, mental performance in the classroom improves.[32]

A severe iodine deficiency during pregnancy causes the extreme and irreversible mental and physical retardation known as **cretinism.**■ Cretinism affects approximately 6 million people worldwide and can be averted by the early diagnosis and treatment of maternal iodine deficiency. A worldwide effort to provide iodized salt to people living in iodine-deficient areas has been dramatically successful.

■ The ion form of *iodine* is called *iodide.*

■ The thyroid gland releases tetraiodothyronine (T_4), commonly known as **thyroxine** (thigh-ROCKS-in), to its target tissues. Upon reaching the cells, T_4 is deiodinated to triiodothyronine (T_3), which is the active form of the hormone.

■ Thyroid-stimulating hormone is also called *thyrotropin.*

■ Examples of goitrogen-containing foods:
- Cabbage, spinach, radishes, rutabagas.
- Soybeans, peanuts.
- Peaches, strawberries.

■ The underactivity of the thyroid gland is known as *hypothyroidism* and may be caused by iodine deficiency or any number of other causes. Without treatment, an infant with *congenital hypothyroidism* will develop the physical and mental retardation of *cretinism.*

goiter (GOY-ter): an enlargement of the thyroid gland due to an iodine deficiency, malfunction of the gland, or overconsumption of a goitrogen. Goiter caused by iodine deficiency is **simple goiter.**

goitrogen (GOY-troh-jen): a substance that enlarges the thyroid gland and causes **toxic goiter.** Goitrogens occur naturally in such foods as cabbage, kale, brussels sprouts, cauliflower, broccoli, and kohlrabi.

cretinism (CREE-tin-ism): a congenital disease characterized by mental and physical retardation and commonly caused by maternal iodine deficiency during pregnancy.

FIGURE 13-9 Iodine-Deficiency Symptom—The Enlarged Thyroid of Goiter

In iodine deficiency, the thyroid gland enlarges—a condition known as simple goiter.

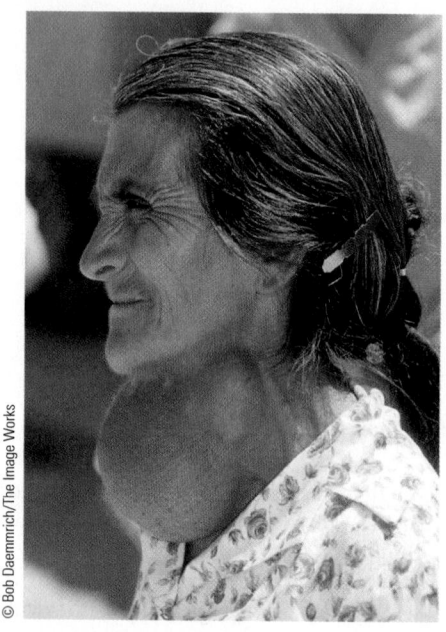

© Bob Daemmrich/The Image Works

■ Iodized salt contains about 60 µg iodine per gram salt.

■ On average, ½ tsp iodized salt provides the RDA for iodine.

© Craig M. Moore

Only "iodized salt" has had iodine added.

Iodine Toxicity Excessive intakes of iodine can enlarge the thyroid gland, just as deficiency can. During pregnancy, exposure to excessive iodine from foods, prenatal supplements, or medications is especially damaging to the developing infant. An infant exposed to toxic amounts of iodine during gestation may develop a goiter so severe as to block the airways and cause suffocation. The Upper Level is over 1000 micrograms per day for an adult—several times higher than average intakes.

Iodine Recommendations and Sources The ocean is the world's major source of iodine. In coastal areas, seafood, water, and even iodine-containing sea mist are dependable iodine sources. Further inland, the amount of iodine in foods is variable and generally reflects the amount present in the soil in which plants are grown or on which animals graze. Landmasses that were once under the ocean have soils rich in iodine; those in flood-prone areas where water leaches iodine from the soil are poor in iodine. In the United States and Canada, the iodization of salt■ has eliminated the widespread misery caused by iodine deficiency during the 1930s, but iodized salt is not available in many parts of the world. Some countries add iodine to bread, fish paste, or drinking water instead.

Average consumption of iodine in the United States exceeds recommendations, but falls below toxic levels as well. Some of the excess iodine in the U.S. diet stems from fast foods, which use iodized salt liberally. Some iodine comes from bakery products and from milk. The baking industry uses iodates (iodine salts) as dough conditioners, and most dairies feed cows iodine-containing medications and use iodine to disinfect milking equipment. Now that these sources have been identified, food industries have reduced their use of these compounds, but the sudden emergence of this problem points to a need for continued surveillance of the food supply. Processed foods in the United States do not use iodized salt.

The recommended intake of iodine for adults is a minuscule amount. The need for iodine is easily met by consuming seafood, vegetables grown in iodine-rich soil, and iodized salt.■ In the United States, labels indicate whether salt is iodized; in Canada, all table salt is iodized.

IN SUMMARY

Iodide, the ion of the mineral iodine, is an essential component of the thyroid hormone. An iodine deficiency can lead to simple goiter—enlargement of the thyroid gland—and can impair fetal development, causing cretinism. Iodization of salt has largely eliminated iodine deficiency in the United States and Canada. The table provides a summary of iodine.

Iodine

2001 RDA	**Deficiency Disease**
Adults: 150 µg/day	Simple goiter, cretinism
Upper Level	**Deficiency Symptoms**
1100 µg/day	Underactive thyroid gland, goiter, mental and physical retardation in infants (cretinism)
Chief Functions in the Body	
A component of two thyroid hormones that help to regulate growth, development, and metabolic rate	**Toxicity Symptoms**
	Underactive thyroid gland, elevated TSH, goiter
Significant Sources	
Iodized salt, seafood, bread, dairy products, plants grown in iodine-rich soil and animals fed those plants	

Selenium

The essential mineral **selenium** shares some of the chemical characteristics of the mineral sulfur. This similarity allows selenium to substitute for sulfur in the amino acids methionine, cysteine, and cystine.[33]

Selenium Roles in the Body Selenium is one of the body's antioxidant nutrients, working primarily as a part of the enzyme glutathione peroxidase. Glutathione peroxidase and vitamin E work in concert. Glutathione peroxidase prevents free-radical formation, thus blocking the chain reaction before it begins; if free radicals do form and a chain reaction starts, vitamin E stops it. (Highlight 11 describes free-radical formation, chain reactions, and antioxidant action in detail.) Another enzyme that converts the thyroid hormone to its active form also contains selenium.

Selenium Deficiency Selenium deficiency is associated with a heart disease■ that is prevalent in regions of China where the soil and foods lack selenium. The primary cause of this heart disease is probably a virus, but selenium deficiency appears to predispose people to it, and adequate selenium seems to prevent it.

Selenium and Cancer Some research suggests that selenium may protect against some types of cancers. Given the potential for harm and the lack of conclusive evidence, however, recommendations to take selenium supplements would be premature—and perhaps ineffective as well. Selenium from foods is far more effective in inhibiting cancer growth than selenium from supplements.[34] Such a finding reinforces a theme that has been repeated throughout this text—foods offer many more health benefits than supplements.

Selenium Recommendations and Sources The soil in many regions of the United States and Canada contains selenium. People living in regions with selenium-poor soil may still get enough selenium, partly because they eat vegetables and grains transported from other regions and partly because they eat meats and other animal products, which are reliable sources of selenium. Average intakes in the United States and Canada are above the RDA, which is based on the amount needed to maximize glutathione peroxidase activity.

Selenium Toxicity Because high doses of selenium are toxic, an Upper Level has been set. Selenium toxicity causes loss and brittleness of hair and nails, garlic breath odor, and nervous system abnormalities.

■ The heart disease associated with selenium deficiency is named **Keshan** (KESH-an or ka-SHAWN) **disease** for one of the provinces of China where it was studied. Keshan disease is characterized by heart enlargement and insufficiency; fibrous tissue replaces the muscle tissue that normally composes the middle layer of the walls of the heart.

IN SUMMARY Selenium is an antioxidant nutrient that works closely with the glutathione peroxidase enzyme and vitamin E. Selenium is found in association with protein in foods. Deficiencies are associated with a predisposition to a type of heart disease and possibly with some kinds of cancer. See the table below for a summary of selenium.

Selenium

2000 RDA

Adults: 55 µg/day

Upper Level

Adults: 400 µg/day

Chief Functions in the Body

Defends against oxidation; regulates thyroid hormone

Significant Sources

Seafood, meat, whole grains, vegetables (depending on soil content)

Deficiency Symptoms

Predisposition to heart disease characterized by cardiac tissue becoming fibrous (Keshan disease)

Toxicity Symptoms

Loss and brittleness of hair and nails; skin rash, fatigue, irritability, and nervous system disorders; garlic breath odor

selenium (se-LEEN-ee-um): a trace element.

Copper

The body contains about 100 milligrams of copper. It is found in a variety of cells and tissues.

Copper Roles in the Body Copper serves as a constituent of several enzymes. The copper-containing enzymes have diverse metabolic roles with one common characteristic: all involve reactions that consume oxygen or oxygen radicals. For example, copper-containing enzymes catalyze the oxidation of ferrous iron to ferric iron.*[35] Copper's role in iron metabolism makes it a key factor in hemoglobin synthesis. Two copper- and zinc-containing enzymes participate in the body's natural defense against free radicals.† Still another copper enzyme helps to manufacture collagen and heal wounds.‡ Copper, like iron, is needed in many of the metabolic reactions related to the release of energy.§

Copper Deficiency and Toxicity Copper deficiency is rare. In animals, copper deficiency raises blood cholesterol and damages blood vessels, raising questions about whether low dietary copper might contribute to cardiovascular disease in humans. Typical U.S. diets provide adequate amounts. Some genetic disorders create a copper toxicity, but excessive intakes from foods are unlikely. Excessive intakes from supplements may cause liver damage, and therefore an Upper Level has been set.

Two rare genetic disorders affect copper status in opposite directions. In Menkes disease, the intestinal cells absorb copper, but cannot release it into circulation, causing a life-threatening deficiency. In Wilson's disease, copper accumulates in the liver and brain, creating a life-threatening toxicity. Wilson's disease can be controlled by reducing copper intake, using chelating agents such as penicillamine, and taking zinc supplements, which interfere with copper absorption. (The use of chelation in health care is mentioned in Highlight 18's discussion of alternative therapies.)

Copper Recommendations and Sources The richest food sources of copper are legumes, whole grains, nuts, shellfish, and seeds. Over half of the copper from foods is absorbed, and the major route of elimination appears to be bile. Water may also provide copper, depending on the type of plumbing pipe and the hardness of the water.

IN SUMMARY Copper is a component of several enzymes, all of which are involved in some way with oxygen or oxidation. Some act as antioxidants; others are essential to iron metabolism. Legumes, whole grains, and shellfish are good sources of copper. See the table for a summary of copper facts.

Copper

2001 RDA	**Significant Sources**
Adults: 900 µg/day	Seafood, nuts, whole grains, seeds, legumes
Upper Level	**Deficiency Symptoms**
Adults: 10,000 µg/day (10 mg/day)	Anemia, bone abnormalities
Chief Functions in the Body	**Toxicity Symptoms**
Necessary for the absorption and use of iron in the formation of hemoglobin; part of several enzymes	Liver damage

---

*The copper-containing enzyme *ceruloplasmin* participates in the oxidation of ferrous iron to ferric iron.
†Two copper-containing *superoxide dismutase* enzymes defend against free radicals.
‡The copper-containing enzyme *lysyl oxidase* helps synthesize connective tissues.
§The copper-containing enzyme *cytochrome C oxidase* participates in the electron transport chain.

Manganese

The human body contains a tiny 20 milligrams of manganese. Most of it can be found in the bones and metabolically active organs such as the liver, kidneys, and pancreas.

Manganese Roles in the Body Manganese acts as a cofactor for many enzymes that facilitate the metabolism of carbohydrate, lipids, and amino acids. In addition, manganese-containing metalloenzymes assist in bone formation and the conversion of pyruvate to a TCA cycle compound.

Manganese Deficiency and Toxicity Manganese requirements are low, and many plant foods contain significant amounts of this trace mineral, so deficiencies are rare. As is true of other trace minerals, however, dietary factors such as phytates inhibit its absorption. In addition, high intakes of iron and calcium limit manganese absorption, so people who use supplements of these minerals regularly may impair their manganese status.

Toxicity is more likely to occur from an environment contaminated with manganese than from dietary intake. Miners who inhale large quantities of manganese dust on the job over prolonged periods show symptoms of a brain disease, along with abnormalities in appearance and behavior. Still, an Upper Level has been established based on intakes from food, water, and supplements.

Manganese Recommendations and Sources Grain products make the greatest contribution of manganese to the diet. With insufficient information to establish an RDA, an AI was set based on average intakes.

IN SUMMARY Manganese-dependent enzymes are involved in bone formation and various metabolic processes. Manganese is widespread in plant foods, so deficiencies are rare, although regular use of calcium and iron supplements may limit manganese absorption. A summary of manganese appears in the table below.

Manganese

2001 AI	Significant Sources
Men: 2.3 mg/day	Nuts, whole grains, leafy vegetables, tea
Women: 1.8 mg/day	**Deficiency Symptoms**
Upper Level	Rare
Adults: 11 mg/day	**Toxicity Symptoms**
Chief Functions in the Body	Nervous system disorders
Cofactor for several enzymes	

Fluoride

Fluoride is present in virtually all soils, water supplies, plants, and animals. Only a trace of fluoride occurs in the human body, but with this amount, the crystalline deposits in bones and teeth are larger and more perfectly formed.

Fluoride Roles in the Body During the mineralization of bones and teeth, calcium and phosphorus form crystals called hydroxyapatite. Then fluoride replaces the hydroxyl (OH) portions of the hydroxyapatite crystal, forming **fluorapatite**, which makes the bones stronger and the teeth more resistant to decay.

Fluoride and Dental Caries Dental caries ranks as the nation's most widespread health problem: an estimated 95 percent of the population have decayed,

fluorapatite (floor-APP-uh-tite): the stabilized form of bone and tooth crystal, in which fluoride has replaced the hydroxyl groups of hydroxyapatite.

FIGURE 13-10 U.S. Population with Access to Fluoridated Water through Public Water Systems

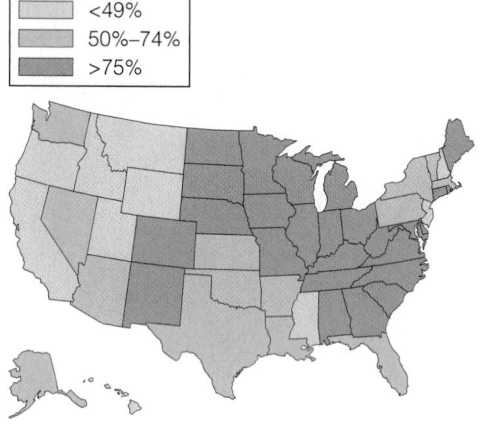

Key:
- ▢ <49%
- ▢ 50%–74%
- ▢ >75%

■ For perspective, 1 part per million (1 ppm) is approximately 1 mg per liter.

■ To prevent fluorosis:
- Monitor the fluoride content of the local water supply.
- Supervise toddlers when they brush their teeth and use only a little toothpaste (pea-size amount).
- Use fluoride supplements only as prescribed by a physician.

FIGURE 13-11 Fluoride-Toxicity Symptom—The Mottled Teeth of Fluorosis

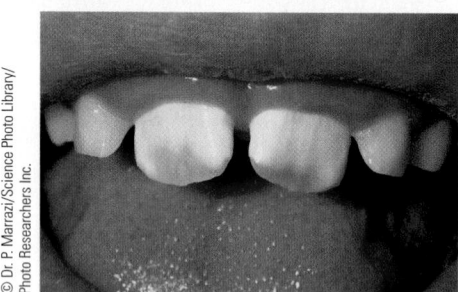

© Dr. P. Marrazi/Science Photo Library/ Photo Researchers Inc.

■ Small organic compounds that enhance insulin's action are called **glucose tolerance factors (GTF).** Some glucose tolerance factors contain chromium.

fluorosis (floor-OH-sis): discoloration and pitting of tooth enamel caused by excess fluoride during tooth development.

missing, or filled teeth. By interfering with a person's ability to chew and eat a wide variety of foods, these dental problems can quickly lead to a multitude of nutrition problems. Where fluoride is lacking, dental decay is common.

Drinking water is usually the best source of fluoride; over 65 percent of the U.S. population receives fluoride through the public water system (see Figure 13-10).[36] (Most bottled waters lack fluoride.) Fluoridation of drinking water to raise the concentration to 1 part fluoride per 1 million■ parts water offers the greatest protection against dental caries at virtually no risk of toxicity.[37] By fluoridating the drinking water, a community offers its residents, particularly the children, a safe, economical, practical, and effective way to defend against dental caries.[38]

Fluoride Toxicity Too much fluoride can damage the teeth, causing **fluorosis.** For this reason, an Upper Level has been established. In mild cases, the teeth develop small white specks; in severe cases, the enamel becomes pitted and permanently stained (as shown in Figure 13-11). Fluorosis occurs only during tooth development and cannot be reversed, making its prevention■ a high priority. To limit fluoride ingestion, take care not to swallow fluoride-containing dental products such as toothpaste and mouthwash.

Fluoride Recommendations and Sources As mentioned earlier, much of the U.S. population has access to water with an optimal fluoride concentration, which typically delivers about 1 milligram per person per day.[39] Fish and most teas contain appreciable amounts of natural fluoride.

Increase the proportion of the U.S. population served by community water systems with optimally fluoridated water.

IN SUMMARY Fluoride makes bones stronger and teeth more resistant to decay. Fluoridation of public water supplies can significantly reduce the incidence of dental caries, but an excess of fluoride during tooth development can cause fluorosis—discolored and pitted tooth enamel. The table below summarizes fluoride information.

Fluoride

1997 AI	Significant Sources
Men: 3.8 mg/day	Drinking water (if fluoride containing or fluoridated), tea, seafood
Women: 3.1 mg/day	

Upper Level	Deficiency Symptoms
Adults: 10 mg/day	Susceptibility to tooth decay

Chief Functions in the Body	Toxicity Symptoms
Involved in the formation of bones and teeth; helps to make teeth resistant to decay	Fluorosis (pitting and discoloration of teeth)

Chromium

Chromium is an essential mineral that participates in carbohydrate and lipid metabolism. Like iron, chromium assumes different charges. In the case of chromium, the Cr^{+++} ion is the most stable and most commonly found in foods.

Chromium Roles in the Body Chromium helps maintain glucose homeostasis by enhancing the activity of the hormone insulin.■ When chromium is lacking, a diabeteslike condition may develop with elevated blood glucose and impaired glucose tolerance, insulin response, and glucagon response. In spite of these relationships, research findings suggest that chromium supplements do not effectively improve glucose or insulin responses in diabetes.[40]

Chromium Recommendations and Sources Chromium is present in a variety of foods. The best sources are unrefined foods, particularly liver, brewer's yeast, and whole grains. The more refined foods people eat, the less chromium they ingest.

Chromium Supplements Supplement advertisements have succeeded in convincing consumers that they can lose fat and build muscle by taking chromium picolinate. Whether chromium—picolinate or plain—supplements reduce body fat or improve muscle strength remains controversial. (Highlight 14 revisits chromium picolinate and other supplements athletes use in the hopes of improving their performance.)

IN SUMMARY Chromium enhances insulin's action. A deficiency can result in a diabeteslike condition. Chromium is widely available in unrefined foods including brewer's yeast, whole grains, and liver. The table below provides a summary of chromium.

Chromium

2001 AI	Deficiency Symptoms
Men: 35 µg/day	Diabeteslike condition
Women: 25 µg/day	
	Toxicity Symptoms
Chief Functions in the Body	None reported
Enhances insulin action	
Significant Sources	
Meats (especially liver), whole grains, brewer's yeast	

Molybdenum

Molybdenum acts as a working part of several metalloenzymes. Dietary deficiencies of molybdenum are unknown because the amounts needed are minuscule—as little as 0.1 part per million parts of body tissue. Legumes, breads and other grain products, leafy green vegetables, milk, and liver are molybdenum-rich foods. Average daily intakes fall within the suggested range of intakes.

Molybdenum toxicity is rare, but has been reported in animal studies, and an Upper Level has been established. Characteristics include kidney damage and reproductive abnormalities. For a summary of molybdenum facts, see the accompanying table.

Molybdenum

2001 RDA	Significant Sources
Adults: 45 µg/day	Legumes, cereals, organ meats
Upper Level	**Deficiency Symptoms**
Adults: 2 mg/day	Unknown
Chief Functions in the Body	**Toxicity Symptoms**
Cofactor for several enzymes	None reported; reproductive effects in animals

molybdenum (mo-LIB-duh-num): a trace element.

FIGURE 13-12 Cobalt with Vitamin B₁₂

The intricate vitamin B_{12} molecule contains one atom of the mineral cobalt. The alternative name for vitamin B_{12}, cobalamin, reflects the presence of cobalt in its structure.

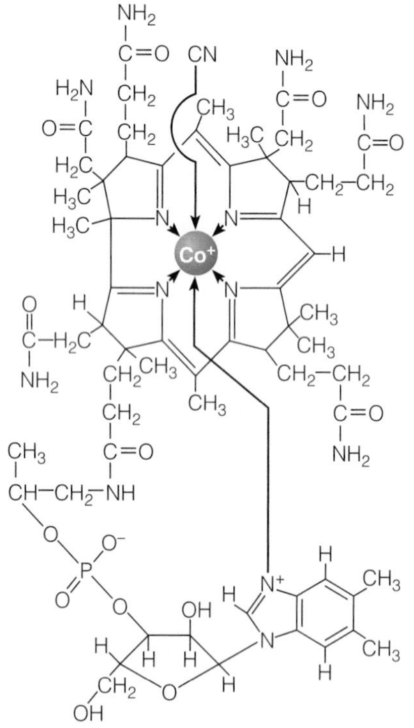

TABLE 13-1 Symptoms of Lead Toxicity

In Children

- Learning disabilities (reduced short-term memory; impaired concentration)
- Low IQ
- Behavior problems
- Slow growth
- Iron-deficiency anemia
- Dental caries
- Sleep disturbances (night waking, restlessness, head banging)
- Nervous system disorders; seizures
- Slow reaction time; poor coordination
- Impaired hearing

In Adults

- Hypertension
- Reproductive complications
- Kidney failure

Other Trace Minerals

Research to determine whether other trace minerals are essential is difficult, both because their quantities in the body are so small and because human deficiencies are unknown. Guessing their functions in the body can be particularly problematic. Much of the available knowledge comes from research using animals.

Nickel may serve as a cofactor for certain enzymes. Silicon is involved in the formation of bones and collagen. Vanadium, too, is necessary for growth and bone development and also for normal reproduction. Cobalt is a key mineral in the large vitamin B_{12} molecule (see Figure 13-12), but it is not an essential nutrient and no recommendation has been established. Boron may play a key role in brain activities; in animals, boron strengthens bones.[41]

In the future many other trace minerals may turn out to play key nutritional roles. Even arsenic—famous as a poison used by murderers and known to be a carcinogen—may turn out to be essential for human beings in tiny quantities; it has already proved useful in the treatment of some types of leukemia.

Contaminant Minerals

Chapter 12 and this chapter have told of the many ways minerals serve the body—maintaining fluid and electrolyte balance, providing structural support to the bones, transporting oxygen, and assisting enzymes. In contrast to those minerals that the body requires, contaminant minerals impair the body's growth, work capacity, and general health. Contaminant minerals include the **heavy metals** lead, mercury, and cadmium that enter the food supply by way of soil, water, and air pollution. This section focuses on lead poisoning because it is the most serious environmental threat to young children, but all contaminant minerals disrupt body processes and impair nutrition status similarly.

Like other minerals, lead is indestructible; the body cannot change its chemistry. Chemically similar to nutrient minerals like iron, calcium, and zinc (cations with two positive charges), lead displaces them from some of the metabolic sites they normally occupy, but is then unable to perform their roles. For example, lead competes with iron in heme, but then cannot carry oxygen; similarly, lead competes with calcium in the brain, but then cannot signal messages from nerve cells. Excess lead in the blood also deranges the structure of red blood cell membranes, making them leaky and fragile. Lead interacts with white blood cells, too, impairing their ability to fight infection, and it binds to antibodies, thwarting their effort to resist disease.

In addition to its effects on the blood, lead damages many body systems, particularly the vulnerable nervous system, kidneys, and bone marrow. It impairs such normal activities as growth by interfering with hormone activity.[42] It interferes with tooth development and may contribute to dental caries as well.[43] Even at low levels, blood lead concentrations correlate with poor IQ scores.[44] In short, lead's interactions in the body have profound adverse effects. The greater the exposure, the more damaging the effects. The American Academy of Pediatrics recommends testing children who have been identified as having a high risk for lead poisoning. Those with high blood lead levels are treated with drugs that bind to lead and carry it out in the urine. Table 13-1 lists symptoms of lead toxicity.

Lead typifies the ways all heavy metals behave in the body: they interfere with nutrients that are trying to do their jobs. The "good guy" nutrients are shoved aside by the "bad guy" contaminants. Then the contaminants cannot perform the roles of the nutrients, and health diminishes. To safeguard our health, we must defend ourselves against contamination by eating nutrient-rich foods and preserving a clean environment.

Closing Thoughts on the Nutrients

This chapter completes the introductory lessons on the nutrients. Each nutrient from the amino acids to zinc has been described rather thoroughly—its chemistry, roles in the body, sources in the diet, symptoms of deficiency and toxicity, and influences on health and disease. Such a detailed examination is informative, but it can also be misleading. It is important to step back from the myopic study of the individual nutrients to look at them as a whole. After all, people eat foods, not nutrients, and most foods deliver dozens of nutrients. Furthermore, nutrients work cooperatively with each other in the body; their actions are most often *interactions*. This chapter alone mentioned how iron depends on vitamin C to keep it in its active form and copper to incorporate it into hemoglobin; how zinc is needed to activate and transport vitamin A; and how both iodine and selenium are needed for the synthesis of thyroid hormone. The accompanying table condenses the information on the trace minerals for your review.

heavy metals: any of a number of mineral ions such as mercury and lead, so called because they are of relatively high atomic weight. Many heavy metals are poisonous.

IN SUMMARY The Trace Minerals

Mineral and Chief Functions	Deficiency Symptoms	Toxicity Symptoms[a]	Significant Sources
Iron Part of the protein hemoglobin, which carries oxygen in the blood; part of the protein myoglobin in muscles, which makes oxygen available for muscle contraction; necessary for energy metabolism	Anemia: weakness, fatigue, headaches; impaired work performance; impaired immunity; pale skin, nail beds, mucous membranes, and palm creases; concave nails; inability to regulate body temperature; pica	GI distress; iron overload: infections, fatigue, joint pain, skin pigmentation, organ damage	Red meats, fish, poultry, shellfish, eggs, legumes, dried fruits
Zinc Part of insulin and many enzymes; involved in making genetic material and proteins, immune reactions, transport of vitamin A, taste perception, wound healing, the making of sperm, and normal fetal development	Growth retardation, delayed sexual maturation, impaired immune function, hair loss, eye and skin lesions, loss of appetite.	Loss of appetite, impaired immunity, low HDL, copper and iron deficiencies	Protein-containing foods: red meats, fish, shellfish, poultry, whole grains
Iodine A component of the thyroid hormones that help to regulate growth, development, and metabolic rate	Underactive thyroid gland, goiter, mental and physical retardation (cretinism)	Underactive thyroid gland, elevated TSH, goiter	Iodized salt; seafood; plants grown in iodine-rich soil and animals fed those plants
Selenium Part of an enzyme that defends against oxidation; regulates thyroid hormone	Associated with Keshan disease	Nail and hair brittleness and loss; fatigue, irritability, and nervous system disorders, skin rash, garlic breath odor	Seafoods, organ meats; other meats, whole grains, and vegetables (depending on soil content)
Copper Helps form hemoglobin; part of several enzymes	Anemia, bone abnormalities	Liver damage	Seafood, nuts, legumes, whole grains, seeds

[a]Acute toxicities of many minerals cause abdominal pain, nausea, vomiting, and diarrhea.

continued

The Trace Minerals—*continued*

Mineral and Chief Functions	Deficiency Symptoms	Toxicity Symptoms[a]	Significant Sources
Manganese Cofactor for several enzymes	Rare	Nervous symptom disorders	Nuts, whole grains, leafy vegetables, tea
Fluoride Helps form bones and teeth; confers decay resistance on teeth	Susceptibility to tooth decay	Fluorosis (pitting and discoloration) of teeth,	Drinking water if fluoride containing or fluoridated, tea, seafood
Chromium Enhances insulin action	Diabeteslike condition	None reported	Meats (liver), whole grains, brewer's yeast
Molybdenum Cofactor for several enzymes	Unknown	None reported	Legumes, cereals, organ meats

Estimates of how much of each particular nutrient the body needs fall between intakes that are inadequate and cause illness and intakes that are excessive and cause illness. Between deficiency and toxicity lies a wide range of intakes that support health—to varying degrees. In the past, nutrient needs were determined by how much was needed to prevent deficiency symptoms. If lack of a nutrient caused illness, it was defined as essential. Today, nutrient needs are based on how much is needed to support optimal health. The amount of vitamin C needed to prevent scurvy is much less than the amount correlated with reducing the risk of cancer, for example. Furthermore, nutrients are being examined within the context of the whole diet. Health benefits are not credited to vitamin C alone, but to the vitamin C–rich fruits and vegetables that also provide many other nutrients—and nonnutrients (phytochemicals)—important to health.

People can also improve their health with physical activity. Energy expenditure is unlike money expenditure: it is desirable to *spend* energy, not to save it (within reason, of course). The more energy people spend, the more food they can afford to eat—food that delivers both nutrients and pleasure. The next chapter presents details on nutrition and physical activity.

Nutrition in Your Life

Trace minerals from a variety of foods, especially those in the meat and meat alternate group, support many of your body's activities.

- Do you eat a variety of foods, including some meats, seafood, poultry, or legumes, daily?
- Do you use iodized salt?
- Do you drink fluoridated water?

Nutrition Explorer Practice planning meals with varying energy needs to meet recommended intakes of several minerals.

NUTRITION ON THE NET

 Access these websites for further study of topics covered in this chapter.

- Find updates and quick links to these and other nutrition-related sites at our website: **www.wadsworth.com/nutrition**

- Search for "minerals" at the American Dietetic Association: **www.eatright.org**

- Search for the individual minerals by name at the U.S. Government health information site: **www.healthfinder.gov**

- Learn more about iron overload from the Iron Overload Diseases Association: **www.ironoverload.org**

- Learn more about iodine and thyroid disease from the American Thyroid Association: **www.thyroid.org**

NUTRITION CALCULATIONS

Once you have mastered these examples, you will understand minerals a little better and be prepared to examine your own food choices. Be sure to show your calculations for each problem. (see p. 464 for answers.)

1. For each of these minerals, note the unit of measure for recommendations:

Iron	Manganese
Zinc	Fluoride
Iodine	Chromium
Selenium	Molybdenum
Copper	

2. Appreciate foods for their iron density. Following is a list of foods with the energy amount and the iron content per serving.
   a. Rank these foods by iron per serving.
   b. Calculate the iron density (divide milligrams by kcalories) for these foods and rank them by their iron per kcalorie.
   c. Name three foods that are higher on the second list than they were on the first list.
   d. What do these foods have in common?

Food	Iron (mg)	Energy (kcal)	Iron Density (mg/kcal)
Milk, fat-free, 1 c	0.10	85	
Cheddar cheese, 1 oz	0.19	114	
Broccoli, cooked from fresh, chopped, 1 c	1.31	44	
Sweet potato, baked in skin, 1 ea	0.51	117	
Cantaloupe melon, ½	0.56	93	
Carrots, from fresh, ½ c	0.48	35	
Whole-wheat bread, 1 slice	0.87	64	
Green peas, cooked from frozen, ½ c	1.26	62	
Apple, medium	0.38	125	
Sirloin steak, lean, 4 oz	3.81	228	
Pork chop, lean, broiled, 1 ea	0.66	166	

STUDY QUESTIONS

These questions will help you review the chapter. You will find the answers in the discussions on the pages provided.

1. Distinguish between heme and nonheme iron. Discuss the factors that enhance iron absorption. (pp. 440–441)

2. Distinguish between iron deficiency and iron-deficiency anemia. What are the symptoms of iron-deficiency anemia? (pp. 442–443)

3. What causes iron overload? What are its symptoms? (p. 444)

4. Describe the similarities and differences in the absorption and regulation of iron and zinc. (pp. 440–441, 448–449)

5. Discuss possible reasons for a low intake of zinc. What factors affect the bioavailability of zinc? (p. 449)

6. Describe the principal functions of iodide, selenium, copper, manganese, fluoride, chromium, and molybdenum in the body. (pp. 451, 453, 454, 455, 456, 457)

7. What public health measure has been used in preventing simple goiter? What measure has been recommended for protection against tooth decay? (pp. 451, 456)

8. Discuss the importance of balanced and varied diets in obtaining the essential minerals and avoiding toxicities. (pp. 458–459)

9. Describe some of the ways trace minerals interact with each other and with other nutrients. (p. 459)

These multiple choice questions will help you prepare for an exam. Answers can be found on p. 464.

1. Iron absorption is impaired by:
   a. heme.
   b. phytates.

c. vitamin C.

d. MFP factor.

2. Which of these people is *least* likely to develop an iron deficiency?

   a. 3-year-old boy

   b. 52-year-old man

   c. 17-year-old girl

   d. 24-year-old woman

3. Which of the following would *not* describe the blood cells of a severe iron deficiency?

   a. anemic

   b. microcytic

   c. pernicious

   d. hypochromic

4. Which provides the most absorbable iron?

   a. 1 apple

   b. 1 c milk

   c. 3 oz steak

   d. ½ c spinach

5. The intestinal protein that helps to regulate zinc absorption is:

   a. albumin.

   b. ferritin.

   c. hemosiderin.

   d. metallothionein.

6. A classic sign of zinc deficiency is:

   a. anemia.

   b. goiter.

   c. mottled teeth.

   d. growth retardation.

7. Cretinism is caused by a deficiency of:

   a. iron.

   b. zinc.

   c. iodine.

   d. selenium.

8. The mineral best known for its role as an antioxidant is:

   a. copper.

   b. selenium.

   c. manganese.

   d. molybdenum.

9. Fluorosis occurs when fluoride:

   a. is excessive.

   b. is inadequate.

   c. binds with phosphorus.

   d. interacts with calcium.

10. Which mineral enhances insulin activity?

    a. zinc

    b. iodine

    c. chromium

    d. manganese

REFERENCES

1. R. A. Anderson, Role of dietary factors: Micronutrients, *Nutrition Reviews* 58 (2000): S10–S11.
2. M. Wessling-Resnick, Iron transport, *Annual Review of Nutrition* 20 (2000): 129–151; N. C. Andrews, Disorders of iron metabolism, *New England Journal of Medicine* 341 (1999): 1986–1995.
3. Committee on Dietary Reference Intakes, *Dietary Reference Intakes for Vitamin A, Vitamin K, Arsenic, Boron, Chromium, Copper, Iodine, Iron, Manganese, Molybdenum, Nickel, Silicon, Vanadium, and Zinc* (Washington, D.C.: National Academy Press, 2001), p. 315.
4. S. Miret, R. J. Simpson, and A. T. McKie, Physiology and molecular biology of dietary iron absorption, *Annual Review of Nutrition* 23 (2003): 283–301.
5. L. Hallberg and L. Hulthén, Prediction of dietary iron absorption: An algorithm for calculating absorption and bioavailability of dietary iron, *American Journal of Clinical Nutrition* 71 (2000): 1147–1160.
6. M. B. Reddy, R. F. Hurrell, and J. D. Cook, Estimation of nonheme-iron bioavailability from meal composition, *American Journal of Clinical Nutrition* 71 (2000): 937–943.
7. Committee on Dietary Reference Intakes, 2001, p. 351.
8. J. R. Hunt and Z. K. Roughead, Nonheme-iron absorption, fecal ferritin excretion, and blood indexes of iron status in women consuming controlled lactoovovegetarian diets for 8 wk, *American Journal of Clinical Nutrition* 69 (1999): 944–952.

9. J. R. Hunt and Z. K. Roughead, Adaptation of iron absorption in men consuming diets with high or low iron bioavailability, *American Journal of Clinical Nutrition* 71 (2000): 94–102.
10. R. S. Eisenstein, Iron regulatory proteins and the molecular control of mammalian iron metabolism, *Annual Review of Nutrition* 20 (2000): 627–662.
11. J. L. Beard and J. R. Connor, Iron status and neural functioning, *Annual Review of Nutrition* 23 (2003): 41–58.
12. World Health Organization, http://www.who.int/nut/ida.htm.
13. Iron deficiency—United States, 1999–2000, *Morbidity and Mortality Weekly Report* 51 (2002): 897–899.
14. L. Hallberg, Perspectives on nutritional iron deficiency, *Annual Review of Nutrition* 21 (2001): 1–21.
15. B. Annibale and coauthors, Reversal of iron deficiency anemia after *Helicobacter pylori* eradication in patients with asymptomatic gastritis, *Annals of Internal Medicine* 131 (1999): 668–672.
16. T. Brownlie and coauthors, Marginal iron deficiency without anemia impairs aerobic adaptation among previously untrained women, *American Journal of Clinical Nutrition* 75 (2002): 734–742.
17. J. Beard, Iron deficiency alters brain development and functioning, *Journal of Nutrition* 133 (2003): 1468S–1472S; E. M. Ross, Evaluation and treatment of iron deficiency in adults, *Nutrition in Clinical Care* 5 (2002): 220–224.

18. J. D. Haas and T. Brownlie, Iron deficiency and reduced work capacity: A critical review of the research to determine a causal relationship, *Journal of Nutrition* 131 (2001): 676S–690S.
19. A. L. M. Heath and S. J. Fairweather-Tait, Health implications of iron overload: The role of diet and genotype, *Nutrition Reviews* 61 (2003): 45–62.
20. R. E. Fleming and W. S. Sly, Mechanisms of iron accumulation in hereditary hemochromatosis, *Annual Review of Physiology* 64 (2002): 663–680; R. J. Wood, The "anemic" enterocyte in hereditary hemochromatosis: Molecular insights into the control of intestinal iron absorption, *Nutrition Reviews* 60 (2002): 144–148; M. J. Nowicki and B. R. Bacon, Hereditary hemochromatosis in siblings: Diagnosis by genotyping, *Pediatrics* 105 (2000): 426–429; A. S. Tavill, Clinical implications of the hemochromatosis gene, *New England Journal of Medicine* 341 (1999): 755–757.
21. B. Lachili and coauthors, Increased lipid peroxidation in pregnant women after iron and vitamin C supplementation, *Biological Trace Element Research* 83 (2001): 103–110; V. Herbert, S. Shaw, and E. Jayatilleke, Vitamin C–driven free radical generation from iron, *Journal of Nutrition* 126 (1996): 1213S–1220S.
22. U. Ramakrishnan, E. Kuklina, and A. D. Stein, Iron stores and cardiovascular disease risk factors in women of reproductive age in the United States, *American Journal of Clinical Nutrition* 76 (2002): 1256–1260;

C. T. Stempos and coauthors, Serum ferritin and death from all causes and cardiovascular disease: The NHANES II Mortality Study, National Health and Nutrition Examination Study, *Annals of Epidemiology* 10 (2000): 441–448; J. Danesh and P. Appleby, Coronary heart disease and iron status: Meta-analyses of prospective studies, *Circulation* 99 (1999): 852–854; B. de Valk and J. J. Marx, Iron, atherosclerosis, and ischemic heart disease, *Archives of Internal Medicine* 159 (1999): 1542–1548; K. Klipstein-Grobusch and coauthors, Serum ferritin and risk of myocardial infarction in the elderly: The Rotterdam Study, *American Journal of Clinical Nutrition* 69 (1999): 1231–1236.

23. J. L. Derstine and coauthors, Iron status in association with cardiovascular disease risk in 3 controlled feeding studies, *American Journal of Clinical Nutrition* 77 (2003): 56–62; K. Klipstein-Grobusch and coauthors, Dietary iron and risk of myocardial infarction in the Rotterdam Study, *American Journal of Epidemiology* 149 (1999): 421–428.

24. M. Shannon, Ingestion of toxic substances by children, *New England Journal of Medicine* 342 (2000): 186–191; C. C. Morris, Pediatric iron poisonings in the United States, *Southern Medicine Journal* 93 (2000): 352–358.

25. W. J. Bartfay and coauthors, Cytotoxic aldehyde generation in heart following acute iron-loading, *Journal of Trace Elements in Medicine and Biology* 14 (2000): 14–20; A. S. Ioannides and J. M. Panisello, Acute respiratory distress syndrome in children with acute iron poisoning: The role of intravenous desferrioxamine, *European Journal of Pediatrics* 159 (2000): 158–159; J. P. Pestaner and coauthors, Ferrous sulfate toxicity: A review of autopsy findings, *Biological Trace Element Research* 69 (1999): 191–198.

26. Committee on Dietary Reference Intakes, 2001, p. 351.

27. J. R. Backstrand, The history and future of food fortification in the United States: A public health perspective, *Nutrition Reviews* 60 (2002): 15–26.

28. S. Zlotkin and coauthors, Randomized, controlled trial of single versus 3-times-daily ferrous sulfate drops for treatment of anemia, *Pediatrics* 108 (2001): 613–616.

29. C. L. Adams and coauthors, Zinc absorption from a low-phytic acid maize, *American Journal of Clinical Nutrition* 76 (2002): 556–559.

30. T. A. Strand and coauthors, Effectiveness and efficacy of zinc for the treatment of acute diarrhea in young children, *Pediatrics* 109 (2002): 898–903; N. Bhandari and coauthors, Substantial reduction in severe diarrheal morbidity by daily zinc supplementation in young North Indian children, *Pediatrics* 109 (2002): e86; C. Duggan and W. Fawzi, Micronutrients and child health: Studies in international nutrition and HIV infection, *Nutrition Reviews* 59 (2001): 358–369; The Zinc Investigators' Collaborative Group, Therapeutic effects of oral zinc in acute and persistent diarrhea in children in developing countries: Pooled analysis of randomized controlled trials, *American Journal of Clinical Nutrition* 72 (2000): 1516–1522; R. B. Costello and J. Grumstrup-Scott, Zinc: What role might supplements play? *Journal of the American Dietetic Association* 100 (2000): 371–375.

31. B. H. McElroy and S. P. Miller, Effectiveness of zinc gluconate glycine lozenges (Cold-Eeze) against the common cold in school-aged subjects: A retrospective chart review, *American Journal of Therapeutics* 9 (2002): 472–475; I. Marshall, Zinc for the common cold, *Cochrane Database of Systematic Reviews* 2 (2000): CD001364; J. L. Jackson, E. Lesho, and C. Peterson, Zinc and the common cold: A meta-analysis revisited, *Journal of Nutrition* 130 (2000): 1512S–1515S; A. S. Prasad and coauthors, Duration of symptoms and plasma cytokine levels in patients with the common cold treated with zinc acetate: A randomized, double-blind, placebo-controlled trial, *Annals of Internal Medicine* 133 (2000): 245–252; R. B. Turner and W. E. Cetnarowski, Effect of treatment with zinc gluconate or zinc acetate on experimental and natural colds, *Clinical Infectious Diseases* 31 (2000): 1202–1208.

32. T. van den Briel and coauthors, Improved iodine status is associated with improved mental performance of schoolchildren in Benin, *American Journal of Clinical Nutrition* 72 (2000): 1179–1185.

33. D. M. Driscoll and P. R. Copeland, Mechanism and regulation of selenoprotein synthesis, *Annual Review of Nutrition* 23 (2003): 17–40.

34. J. W. Finley and C. D. Davis, Selenium (Se) from high-selenium broccoli is utilized differently than selenite, selanate and selenomethionine, but is more effective in inhibiting colon carcinogenesis, *Biofactors* 14 (2001): 191–196.

35. N. E. Hellman and J. D. Gitlin, Ceruloplasmin metabolism and function, *Annual Review of Nutrition* 22 (2002): 439–458.

36. Populations receiving optimally fluoridated public drinking water—United States, 2000, *Morbidity and Mortality Weekly Report* 51 (2002): 144–147.

37. Position of the American Dietetic Association: The impact of fluoride on health, *Journal of the American Dietetic Association* 101 (2001): 126–132.

38. Recommendations for using fluoride to prevent and control dental caries in the United States, *Morbidity and Mortality Weekly Report* 50 (2001): entire supplement.

39. Populations receiving optimally fluoridated public drinking water—United States, 2000, 2002.

40. M. D. Althuis and coauthors, Glucose and insulin responses to dietary chromium supplements: A meta-analysis, *American Journal of Clinical Nutrition* 76 (2002): 148–155; L. G. Trow and coauthors, Lack of effect of dietary chromium supplementation on glucose tolerance, plasma insulin and lipoprotein levels in patients with type 2 diabetes, *International Journal of Vitamin and Nutrition Research* 70 (2000): 14–18.

41. T. A. Devirian and S. L. Volpe, The physiological effects of dietary boron, *Critical Reviews in Food and Science Nutrition* 43 (2003): 219–231.

42. S. G. Selevan and coauthors, Blood lead concentration and delayed puberty in girls, *New England Journal of Medicine* 348 (2003): 1527–1536.

43. M. E. Moss, B. P. Lanphear, and P. A. Auinger, Association of dental caries and blood lead levels, *Journal of the American Medical Association* 281 (1999): 2294–2298.

44. R. L. Canfield and coauthors, Intellectual impairment in children with blood lead concentrations below 10 µg per deciliter, *New England Journal of Medicine* 348 (2003): 1517–1526.

ANSWERS

Nutrition Calculations

1. Iron: mg.   Selenium: µg.   Fluoride: mg.

   Zinc: mg.   Copper: µg.   Chromium: µg.

   Iodine: µg.   Manganese: mg.   Molybdenum: µg.

2. a. Sirloin steak > broccoli > green peas > bread > pork chop > cantaloupe > sweet potato > carrots > apple > cheese > milk.

   b.

Food	Iron Density (mg/kcal)
Milk, fat-free, 1 c	0.10 mg ÷ 85 kcal = 0.0012 mg/kcal
Cheddar cheese, 1 oz	0.19 mg ÷ 114 kcal = 0.0017 mg/kcal
Broccoli, cooked from fresh, chopped, 1 c	1.31 mg ÷ 44 kcal = 0.0298 mg/kcal
Sweet potato, baked in skin, 1 ea	0.51 mg ÷ 117 kcal = 0.0044 mg/kcal
Cantaloupe melon, ½	0.56 mg ÷ 93 kcal = 0.0060 mg/kcal
Carrots, from fresh, ½ c	0.48 mg ÷ 35 kcal = 0.0137 mg/kcal
Whole-wheat bread, 1 slice	0.87 mg ÷ 64 kcal = 0.0136 mg/kcal
Green peas, cooked from frozen, ½ c	1.26 mg ÷ 62 kcal = 0.0203 mg/kcal
Apple, medium	0.38 mg ÷ 125 kcal = 0.0030 mg/kcal
Sirloin steak, lean, 4 oz	3.81 mg ÷ 228 kcal = 0.0167 mg/kcal
Pork chop, lean broiled, 1 ea	0.66 mg ÷ 166 kcal = 0.0040 mg/kcal

Broccoli > green peas > sirloin steak > carrots > bread > cantaloupe > sweet potato > pork chop > apple > cheese > milk.

c. Broccoli, green peas, and carrots are all higher on the per-kcalorie list.

d. They are all vegetables.

Study Questions (multiple choice)

1. b   2. b   3. c   4. c   5. d   6. d   7. c   8. b

9. a   10. c

HIGHLIGHT

Phytochemicals and Functional Foods

© John E. Kelly/FoodPix/Getty Images

Chapter 13 completes the introductory discussions on the six classes of nutrients—carbohydrates, lipids, proteins, vitamins, minerals, and water. In addition to these nutrients, foods contain thousands of nonnutrient compounds, including the phytochemicals. Chapter 1 introduced the phytochemicals as compounds found in plant-derived foods (*phyto* means plant) that have biological activity in the body. Research on phytochemicals is unfolding daily, adding to our knowledge of their roles in human health, but there are still many questions and only tentative answers. Just a few of the tens of thousands of phytochemicals have been researched at all, and only a sampling are mentioned in this highlight—enough to illustrate their wide variety of food sources and roles in supporting health.

The concept that foods provide health benefits beyond those of the nutrients emerged from numerous epidemiological studies showing the protective effects of plant-based diets on cancer and heart disease. People have been using foods to maintain health and prevent disease for years, but now these foods have been given a name—they are called **functional foods** (the accompanying glossary defines this and related terms). Much of this text touts the benefits of nature's func-

tional foods—grains rich in dietary fibers, fish rich in omega-3 fatty acids, and fruits rich in phytochemicals, for example. This highlight begins with a look at some of these familiar functional foods, the phytochemicals they contain, and their roles in disease prevention. Then the discussion turns to examine the most controversial of functional foods—novel foods to which phytochemicals have been added to promote health.[1] How these foods fit into a healthy diet is still unclear.[2]

The Phytochemicals

In foods, phytochemicals impart tastes, aromas, colors, and other characteristics. They give hot peppers their burning sensation, garlic its pungent flavor, and tomatoes their dark red color. In the body, phytochemicals can have profound physiological effects, acting as antioxidants, mimicking hormones, and suppressing the development of diseases.[3] Table H13-1 (on p. 466) presents the names, possible effects, and food sources of some of the better-known phytochemicals.

Defending against Cancer

A variety of phytochemicals from a variety of foods appear to protect against DNA damage and defend the body against cancer.[4] A few examples follow.

GLOSSARY

flavonoids (FLAY-von-oyds): yellow pigments in foods; phytochemicals that may exert physiological effects on the body.

flaxseed: the small brown seed of the flax plant; used in baking, cereals, or other foods and valued by industry as a source of linseed oil and fiber.

functional foods: foods that contain physiologically active compounds that provide health benefits beyond basic nutrition; sometimes called *designer foods* or *nutraceuticals.*

lignans: phytochemicals present in flaxseed, but not in flax oil, that are converted to phytosterols by intestinal bacteria and are under study as possible anticancer agents.

lutein (LOO-teen): a plant pigment of yellow hue; a phytochemical believed to play roles in eye functioning and health.

lycopene (LYE-koh-peen): a pigment responsible for the red color of tomatoes and other red-hued vegetables; a

phytochemical that may act as an antioxidant in the body.

phytoestrogens: plant-derived compounds that have structural and functional similarities to human estrogen. Phytoestrogens include genistein, daidzein, and glycitein.

phytosterols: plant-derived compounds that have structural similarities to cholesterol and lower blood cholesterol by competing with cholesterol for absorption. Phytosterols include sterol esters and stanol esters.

probiotics: microbial food ingredients that are beneficial to health. Nondigestible food ingredients that encourage the growth of favorable bacteria are called **prebiotics.**

- **pro** = for
- **bios** = life
- **pre** = before

yogurt: milk fermented by specific bacterial cultures.

Reminder: *Phytochemicals* are nonnutrient compounds found in plant-derived foods that have biological activity in the body.

TABLE H13-1   Phytochemicals—Their Food Sources and Actions

Name	Possible Effects	Food Sources
Capsaicin	Modulates blood clotting, possibly reducing the risk of fatal clots in heart and artery disease.	Hot peppers
Carotenoids (include beta-carotene, lycopene, lutein, and hundreds of related compounds)[a]	Act as antioxidants, possibly reducing risks of cancer and other diseases.	Deeply pigmented fruits and vegetables (apricots, broccoli, cantaloupe, carrots, pumpkin, spinach, sweet potatoes, tomatoes)
Curcumin	May inhibit enzymes that activate carcinogens.	Tumeric, a yellow-colored spice
Flavonoids (include flavones, flavonols, isoflavones, catechins, and others)[b,c]	Act as antioxidants; scavenge carcinogens; bind to nitrates in the stomach, preventing conversion to nitrosamines; inhibit cell proliferation.	Berries, black tea, celery, citrus fruits, green tea, olives, onions, oregano, purple grapes, purple grape juice, soybeans and soy products, vegetables, whole wheat, wine
Indoles[d]	May trigger production of enzymes that block DNA damage from carcinogens; may inhibit estrogen action.	Broccoli and other cruciferous vegetables (brussels sprouts, cabbage, cauliflower), horseradish, mustard greens
Isothiocyanates (including sulforaphane)	Inhibit enzymes that activate carcinogens; trigger production of enzymes that detoxify carcinogens.	Broccoli and other cruciferous vegetables (brussels sprouts, cabbage, cauliflower), horseradish, mustard greens
Lignans[e]	Block estrogen activity in cells, possibly reducing the risk of cancer of the breast, colon, ovaries, and prostate.	Flaxseed and its oil, whole grains
Monoterpenes (include limonene)	May trigger enzyme production to detoxify carcinogens; inhibit cancer promotion and cell proliferation.	Citrus fruit peels and oils
Organosulfur compounds	May speed production of carcinogen-destroying enzymes; slow production of carcinogen-activating enzymes.	Chives, garlic, leeks, onions
Phenolic acids[c]	May trigger enzyme production to make carcinogens water soluble, facilitating excretion.	Coffee beans, fruits (apples, blueberries, cherries, grapes, oranges, pears, prunes), oats, potatoes, soybeans
Phytic acid	Binds to minerals, preventing free-radical formation, possibly reducing cancer risk.	Whole grains
Phytoestrogens (genistein and daidzein)	Estrogen inhibition may produce these actions: inhibit cell replication in GI tract; reduce risk of breast, colon, ovarian, prostate, and other estrogen-sensitive cancers; reduce cancer cell survival. Estrogen mimicking may reduce risk of osteoporosis.	Soybeans, soy flour, soy milk, tofu, textured vegetable protein, other legume products
Protease inhibitors	May suppress enzyme production in cancer cells, slowing tumor growth; inhibit hormone binding; inhibit malignant changes in cells.	Broccoli sprouts, potatoes, soybeans and other legumes, soy products
Resveratrol	Offsets artery-damaging effects of high-fat diets.	Red wine, peanuts
Saponins	May interfere with DNA replication, preventing cancer cells from multiplying; stimulate immune response.	Alfalfa sprouts, other sprouts, green vegetables, potatoes, tomatoes
Tannins[c]	May inhibit carcinogen activation and cancer promotion; act as antioxidants.	Black-eyed peas, grapes, lentils, red and white wine, tea

[a]Other carotenoids include alpha-carotene, beta-cryptoxanthin, and zeaxanthin.
[b]Other flavonoids of interest include ellagic acid and ferulic acid; see also *phytoestrogens*.
[c]A subset of the larger group *phenolic phytochemicals*.
[d]Indoles include dithiothiones, isothiocyantes, and others.
[e]Lignans act as phytosterols, but their food sources are limited.

Soybeans and products made from them correlate with low rates of cancer, especially cancers of the breast and prostate.[5] Soybeans—as well as **flaxseed** oil, whole grains, fruits, and vegetables—are a rich source of an array of phytochemicals, among them the **phytoestrogens.** Phytoestrogens are plant compounds that weakly mimic or modulate the effects of the steroid hormone estrogen in the body.[6] These phytoestrogens have antioxidant activity and appear to slow the growth of breast and prostate cancers.[7]

Tomatoes seem to offer protection against cancers of the esophagus, lung, prostate, and stomach. Among the phyto-chemicals responsible for this effect is **lycopene,** one of beta-carotene's many carotenoid relatives. Lycopene is the pigment that gives apricots, guava, papaya, pink grapefruits, and watermelon their red color—and it is especially abundant in tomatoes and cooked tomato products. Lycopene is a powerful antioxidant that seems to inhibit the growth of cancer cells.[8] Importantly, these benefits are seen when people eat *foods* containing lycopene-rich tomato products.[9]

Soybeans and tomatoes are only two of the many fruits and vegetables credited with providing anticancer activity. Researchers speculate that people might cut their risks of cancers in half simply by meeting current recommendations to eat five or more servings of fruits and vegetables a day.

Defending against Heart Disease

Diets based primarily on unprocessed foods appear to support heart health better than those founded on highly refined foods—perhaps because of the abundance of nutrients, fiber, or phytochemicals such as the **flavonoids**.[10] Flavonoids, a large group of phytochemicals known for their health-promoting qualities, are found in whole grains, legumes, soy, vegetables, fruits, herbs, spices, teas, chocolate, nuts, olive oil, and red wines.[11] Flavonoids are powerful antioxidants that may help to protect LDL cholesterol against oxidation and reduce blood platelet stickiness, making blood clots less likely.[12] An abundance of flavonoid-containing *foods* in the diet lowers the risks of chronic diseases.[13] Importantly, no claims can be made for flavonoids themselves as the protective factor, particularly when they are extracted from foods and sold as supplements.[14]

In addition to flavonoids, fruits and vegetables are rich in carotenoids. Studies suggest that a diet rich in carotenoids is also associated with a lower risk of heart disease.[15] Notable among the carotenoids that may defend against heart disease are **lutein** and lycopene.[16]

The **phytosterols** of soybeans and other vegetables may also protect against heart disease.[17] These cholesterol-like molecules are naturally found in all plants and inhibit cholesterol absorption in the body.[18] As a result, blood cholesterol levels decline. The phytoestrogens of soy may also protect against heart disease by acting as antioxidants and lowering blood pressure.[19]

The Phytochemicals in Perspective

Because foods deliver thousands of phytochemicals in addition to dozens of nutrients, researchers must be careful in giving credit for particular health benefits to any one compound. Diets rich in whole grains, legumes, vegetables, fruits, and nuts seem to be protective against heart disease and cancer, but identifying *the* specific foods or components of foods that are responsible is difficult.[20] Each food possesses a unique array of phytochemicals—citrus fruits provide monoterpenes; grapes, resveratrol; and flaxseed, **lignans.** Broccoli may contain as many as 10,000 different phytochemicals—each with the potential to influence some action in the body. Beverages such as wine, spices such as oregano, and oils such as olive oil contain phytochemicals that may explain, in part, why people who live in the Mediterranean region have reduced risks of heart disease.[21] Even identifying all of the phytochemicals and their effects doesn't answer all the questions because the actions of phytochemicals may be complementary or overlapping—which reinforces the principle of variety in diet planning.[22] For an

Nature offers a variety of functional foods that provide us with many health benefits.

appreciation of the array of phytochemicals offered by a variety of fruits and vegetables, see Figure H13-1 (on p. 468).

Functional Foods

Because foods naturally contain thousands of phytochemicals that are biologically active in the body, virtually all of them have some special value in supporting health. In other words, even simple, whole foods, in reality, are functional foods. Cranberries may help protect against urinary tract infections; garlic may lower blood cholesterol; and tomatoes may protect against some cancers, just to name a few examples.[23] But that hasn't stopped food manufacturers from trying to create functional foods as well. The creation of more functional foods has become the fastest-growing trend and the greatest influence transforming the American food supply.[24]

Many processed foods become functional foods when they are fortified with nutrients or enhanced with phytochemicals or herbs (calcium-fortified orange juice, for example). Less frequently, an entirely new food is created, as in the case of a meat substitute made of mycoprotein—a protein derived from a fungus.*[25] This functional food not only provides dietary fiber, polyunsaturated fats, and high-quality protein, but it lowers LDL cholesterol, raises HDL cholesterol, improves glucose response, and prolongs satiety after a meal. Such a novel functional food raises the question—is it a food or a drug?

Foods as Pharmacy

Not too long ago, most of us could agree on what was a food and what was a drug. Today, functional foods blur the distinctions. They have characteristics similar to both foods and drugs, but do not fit neatly into either category.

---

*This mycoprotein product is marketed under the trade name Quorn (pronounced KWORN).

FIGURE H13-1   An Array of Phytochemicals in a Variety of Fruits and Vegetables

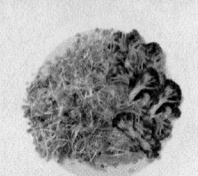

Broccoli and broccoli sprouts contain an abundance of the cancer-fighting phytochemical sulforaphane.

An apple a day—rich in flavonoids—may protect against lung cancer.

The phytoestrogens of soybeans seem to starve cancer cells and inhibit tumor growth; the phytosterols may lower blood cholesterol and protect cardiac arteries.

Garlic, with its abundant organosulfur compounds, may lower blood cholesterol and protect against stomach cancer.

The phytochemical resveratrol found in grapes (and nuts) protects against cancer by inhibiting cell growth and against heart disease by limiting clot formation and inflammation.

The ellagic acid of strawberries may inhibit certain types of cancer.

Tomatoes, with their abundant lycopene, may defend against cancer by protecting DNA from oxidative damage.

The monoterpenes of citrus fruits (and cherries) may inhibit cancer growth.

The flavonoids in black tea may protect against heart disease, whereas those in green tea may defend against cancer.

The flavonoids in cocoa and chocolate defend against oxidation and reduce the tendency of blood to clot.

Spinach and other colorful vegetables contain the carotenoids lutein and zeaxanthin, which help protect the eyes against macular degeneration.

Flaxseed, the richest source of lignans, may prevent the spread of cancer.

Blueberries, a rich source of flavonoids, improve memory in animals.

Consider the healing powers of **yogurt,** for example. Yogurt contains *Lactobacillus* and other living bacteria that ferment milk into yogurt. These microorganisms, called **probiotics,** change the population of microbes in the GI tract, which improves defenses against GI disorders.[26] Research is under way to determine whether probiotics may help to alleviate diarrhea, inflammatory bowel disease, and lactose intolerance; enhance immune function; protect against GI cancer; and lower blood cholesterol.[27] As information on the potential benefits of probiotics unfolds, food manufacturers may begin to include these microorganisms in a variety of other foods. Food companies have already developed products with added phytochemicals. Consider margarine, for example.

Eating nonhydrogenated margarine sparingly instead of butter generously may lower blood cholesterol slightly over several months and clearly falls into the food category. Taking the drug Lipitor, on the other hand, lowers blood cholesterol significantly within weeks and clearly falls into the drug category. But margarine enhanced with a phytosterol that lowers blood cholesterol is in a gray area between the two. The margarine looks and tastes like a food, but it acts like a drug.

The use of functional foods as drugs creates a whole new set of diet-planning problems. Not only must foods provide an adequate intake of all the nutrients to support good health, but they must also deliver druglike ingredients to protect against disease. Like drugs used to treat chronic diseases, functional foods may need to be eaten several times a day for several months or years to have a beneficial effect. Sporadic users may be disappointed in the results. When used four times a day for four weeks, margarine enriched with phytosterols reduces cholesterol by 8 percent, much more than regular margarine does, but not nearly as much as the 32 percent reduction seen with cholesterol-lowering drugs.[28] For this reason, functional foods may be more useful for prevention and mild cases of disease than for intervention and more severe cases.

Foods and drugs differ dramatically in cost as well. Functional foods such as fruits and vegetables incur no added costs, of course, but foods that have been manufactured with added phytochemicals can be expensive, costing up to six times as much as their conventional counterparts. The price of functional foods typically falls between that of traditional foods and medicines.

Unanswered Questions

To achieve a desired health effect, which is the better choice: to eat a food designed to affect some body function or simply to adjust the diet? Does it make more sense to use a margarine enhanced with a phytosterol that lowers blood cholesterol or simply to limit the amount of butter eaten?[*][29] Is it smarter to eat eggs enriched with omega-3 fatty acids or to restrict egg consumption?[30] Might functional foods offer a sensible solution for improving our nation's health—if done correctly? Perhaps so—but there is a problem with functional foods: the food industry is moving too fast for either scientists or the Food and Drug Administration to keep up. Consumers were able to buy soup with St. John's wort that claimed to enhance mood and fruit juice with echinacea that was supposed to fight colds while scientists were still conducting their studies on these ingredients. Research to determine the safety and effectiveness of these substances is still in progress. Until this work is complete, consumers are on their own in finding the answers to the following questions:[31]

- *Does it work?* Research is generally lacking and findings are often inconclusive.
- *How much does it contain?* Food labels are not required to list the quantities of added phytochemicals. Even if they were, consumers have no standard for comparison and cannot deduce whether the amounts listed are a little or a lot. Most importantly, until research is complete, food manufacturers do not know what amounts (if any) are most effective—or most toxic.

Functional foods currently on the market promise to "enhance mood," "promote relaxation and good karma," "increase alertness," and "improve memory," among other claims.

- *Is it safe?* Functional foods can act like drugs. They contain ingredients that can alter body functions and cause allergies, drug interactions, drowsiness, and other side effects. Yet, unlike drug labels, food labels do not provide instructions for the dosage, frequency, or duration of treatment.
- *Is it healthy?* Adding phytochemicals to a food does not magically make it a healthy choice. A candy bar may be fortified with phytochemicals, but it is still made mostly of sugar and fat.

Critics suggest that the designation "functional foods" may be nothing more than a marketing tool. After all, even the most experienced researchers cannot yet identify the perfect combination of nutrients and phytochemicals to support optimal health. Yet manufacturers are freely experimenting with various concoctions as if they possessed that knowledge. Is it okay for them to sprinkle phytochemicals on fried snack foods and label them "functional," thus implying health benefits? Do we want our children receiving their nourishment from fortified caramel candies and chocolate cakes?

Future Foods

Nature has elegantly designed foods to provide us with a complex array of dozens of nutrients and thousands of additional compounds that may benefit health—most of which we have yet to identify or understand. Over the years, we have taken those foods and first deconstructed them and then reconstructed them in an effort to "improve" them. With new scientific understandings of how nutrients—and the myriad of other compounds in foods—interact with genes, we may someday be able to design foods to meet the *exact* health needs of *each* individual.[32] Indeed, our knowledge of the human

---

*Margarine products that lower blood cholesterol contain either sterol esters from vegetable oils, soybeans, and corn or stanol esters from wood pulp.

genome and of human nutrition may well merge to allow for specific recommendations for individuals based on their predisposition to diet-related diseases.[33]

If the present trend continues, then someday physicians may be able to prescribe the perfect foods to enhance your health, and farmers will be able to grow them. As Highlight 19 explains, scientists have developed gene technology to alter the composition of food crops. They can grow rice enriched with vitamin A and tomatoes containing a hepatitis vaccine. It seems quite likely that foods can be created to meet every possible human need. But then, in a sense, that was largely true 100 years ago when we relied on the bounty of nature.

NUTRITION ON THE NET

 Access these websites for further study of topics covered in this highlight.

- Find updates and quick links to these and other nutrition-related sites at our website: **www.wadsworth.com/nutrition**

- Search for "functional foods" at the International Food Information Council: **www.ificinfo.org**

- Search for "functional foods" at the Center for Science in the Public Interest: **www.cspinet.org**

- Find out if warnings have been issued for any food ingredients at the FDA website: **www.fda.gov**

REFERENCES

1. J. A. Milner, Functional foods: The US perspective, *American Journal of Clinical Nutrition* 71 (2000): 1654S–1659S.
2. C. H. Halsted, Dietary supplements and functional foods: 2 sides of a coin? *American Journal of Clinical Nutrition* 77 (2003): 1001S–1007S.
3. P. M. Kris-Etherton and coauthors, Bioactive compounds in foods: Their role in the prevention of cardiovascular disease and cancer, *American Journal of Medicine* 113 (2002): 71S–88S.
4. C. S. Yang and coauthors, Inhibition of carcinogenesis by dietary polyphenolic compounds, *Annual Review of Nutrition* 21 (2001): 381–406; M. Abdulla and P. Gruber, Role of diet modification in cancer prevention, *Biofactors* 12 (2000): 45–51.
5. C. A. Lamartiniere, Protection against breast cancer with genistein: A component of soy, *American Journal of Clinical Nutrition* 71 (2000): 1705S–1707S.
6. I. C. Munro and coauthors, Soy isoflavones: A safety review, *Nutrition Reviews* 61 (2003): 1–33.
7. C. A. Lamartiniere and coauthors, Genistein chemoprevention: Timing and mechanisms of action in murine mammary and prostate, *Journal of Nutrition* 132 (2002): 552S–558S.
8. D. Heber and Q. Y. Lu, Overview of mechanisms of action of lycopene, *Experimental Biology and Medicine* 227 (2002): 920–923; T. M. Vogt and coauthors, Serum lycopene, other serum carotenoids, and risk of prostate cancer in US blacks and whites, *American Journal of Epidemiology* 155 (2002): 1023–1032; Q. Y. Lu and coauthors, Inverse associations between plasma lycopene and other carotenoids and prostate cancer, *Cancer Epidemiology, Biomarkers and Prevention* 10 (2001): 749–756.
9. E. Giovannucci and coauthors, A prospective study of tomato products, lycopene, and prostate cancer risk, *Journal of the National Cancer Institute* 94 (2002): 391–398; L. Chen and coauthors, Oxidative DNA damage in prostate cancer patients consuming tomato sauce–based entrees as a whole-food intervention, *Journal of the National Cancer Institute* 93 (2001): 1872–1879.
10. J. A. Ross and C. M. Kasum, Dietary flavonoids: Bioavailability, metabolic effects, and safety, *Annual Review of Nutrition* 22 (2002): 19–34.
11. F. M. Steinberg, M. M. Bearden, and C. L. Keen, Cocoa and chocolate flavonoids: Implications for cardiovascular health, *Journal of the American Dietetic Association* 103 (2003): 215–223; F. Visioli, and C. Galli, Biological properties of olive oil phytochemicals, *Critical Reviews in Food Science and Nutrition* 42 (2002): 209–221; Y. J. Surh, Anti-tumor promoting potential of selected spice ingredients with antioxidative and anti-inflammatory activities: A short review, *Food and Chemical Toxicology* 40 (2002): 1091–1097; J. M. Geleijnse and coauthors, Inverse association of tea and flavonoid intakes with incident myocardial infarction: The Rotterdam Study, *American Journal of Clinical Nutrition* 75 (2002): 880–886, C. L. Keen, Chocolate: Food as medicine/medicine as food, *Journal of the American College of Nutrition* 20 (2001): 436S–439S.
12. B. Fuhrman and M. Aviram, Flavonoids protect LDL from oxidation and attenuate atherosclerosis, *Current Opinion in Lipidology* 12 (2001): 41–48.
13. M. Messina, C. Gardner, and S. Barnes, Gaining insight into the health effects of soy but a long way still to go: Commentary on the Fourth International Symposium on the Role of Soy in Preventing and Treating Chronic Disease, *Journal of Nutrition* 132 (2002): 547S–551S; P. Knekt and coauthors, Flavonoid intake and risk of chronic diseases, *American Journal of Clinical Nutrition* 76 (2002): 560–568.
14. Ross and Kasum, 2002.
15. S. K. Osganian and coauthors, Dietary carotenoids and risk of coronary artery disease in women, *American Journal of Clinical Nutrition* 77 (2003): 1390–1399; S. Liu and coauthors, Intake of vegetables rich in carotenoids and risk of coronary heart disease in men: The Physicians' Heart Study, *International Journal of Epidemiology* 30 (2001): 130–135; S. B. Kritchevsky, beta-Carotene, carotenoids and the prevention of coronary heart disease, *Journal of Nutrition* 129 (1999): 5–8.
16. T. H. Rissanen and coauthors, Serum lycopene concentrations and carotid atherosclerosis: The Kuopio Ischaemic Heart Disease Risk Factor Study, *American Journal of Clinical Nutrition* 77 (2003): 133–138; Heber and Lu, 2002; J. H. Dwyer and coauthors, Oxygenated carotenoid lutein and progression of early atherosclerosis: The Los Angeles atherosclerosis study, *Circulation* 103 (2001): 2922–2927; L. Arab and S. Steck, Lycopene and cardiovascular disease, *American Journal of Clinical Nutrition* 71 (2000): 1691S–1695S.
17. R. E. Ostlund, Jr., Phytosterols in human nutrition, *Annual Review of Nutrition* 22 (2002): 533–549.
18. C. A. Vanstone and coauthors, Unesterified plant sterols and stanols lower LDL-cholesterol concentrations equivalently in hypercholesterolemic persons, *American Journal of Clinical Nutrition* 76 (2002): 1272–1278.
19. M. Rivas and coauthors, Soy milk lowers blood pressure in men and women with mild to moderate essential hypertension, *Journal of Nutrition* 132 (2002): 1900–1902.
20. Kris-Etherton and coauthors, 2002; C. M. Steinmaus, S. Nunez, and A. H. Smith, Diet and bladder cancer: A meta-analysis of six dietary variables, *American Journal of Epidemiology* 151 (2000): 693–702.
21. F. Visioli, A. Poli, and C. Gall, Antioxidant and other biological activities of phenols from olives and olive oil, *Medicinal Research Reviews* 22 (2002): 65–75; A. Trichopoulou, E. Vasilopoulou, and A. Lagiou, Mediterranean diet and coronary heart disease: Are antioxidants critical? *Nutrition Reviews* 57 (1999): 253–255.
22. J. W. Lampe, Health effects of vegetables and fruit: Assessing mechanism of action in human experimental studies, *American Journal of Clinical Nutrition* 70 (1999): 475S–490S.

23. A. B. Howell and B. Foxman, Cranberry juice and adhesion of antibiotic resistant uropathogens, *Journal of the American Medical Association* 287 (2002): 3082–3083; C. W. Hadley and coauthors, Tomatoes, lycopene, and prostate cancer: Progress and promise, *Experimental Biology and Medicine* 227 (2002): 869–880; R. T. Ackermann and coauthors, Garlic shows promise for improving some cardiovascular risk factors, *Archives of Internal Medicine* 161 (2001): 813–824.

24. Position of the American Dietetic Association: Functional foods, *Journal of the American Dietetic Association* 99 (1999): 1278–1285.

25. T. Peregrin, Mycoprotein: Is America ready for a meat substitute derived from a fungus? *Journal of the American Dietetic Association* 102 (2002): 628.

26. M. E. Sanders, Probiotics: Considerations for human health, *Nutrition Reviews* 61 (2003): 91–99; M. H. Floch and J. Hong-Curtiss, Probiotics and functional foods in gastrointestinal disorders, *Current Gastroenterology Reports* 3 (2001): 343–350; Probiotics and prebiotics, *American Journal of Clinical Nutrition* (supplement) 73 (2001): entire issue.

27. J. M. Saavedra and A. Tschernia, Human studies with probiotics and prebiotics: Clinical implications, *British Journal of Nutrition* 87 (2002): S241–S246; P. Marteau and M. C. Boutron-Ruault, Nutritional advantages of probiotics and prebiotics, *British Journal of Nutrition* 87 (2002): S153–S157; G. T. Macfarlane and J. H. Cummings, Probiotics, infection and immunity, *Current Opinion in Infectious Diseases* 15 (2002): 501–506; L. Kopp-Hoolihan, Prophylactic and therapeutic uses of probiotics: A review, *Journal of the American Dietetic Association* 101 (2001): 229–238; M. B. Roberfroid, Prebiotics and probiotics: Are they functional foods? *American Journal of Clinical Nutrition* 71 (2000): 1682S–1687S.

28. L. A. Simons, Additive effect of plant sterol-ester margarine and cerivastatin in lowering low-density lipoprotein cholesterol in primary hypercholesterolemia, *American Journal of Cardiology* 90 (2002): 737–740.

29. P. Nestel and coauthors, Cholesterol-lowering effects of plant sterol esters and non-esterified stanols in margarine, butter and low-fat foods, *European Journal of Clinical Nutrition* 55 (2001): 1084–1090.

30. D. J. Farrell, Enrichment of hen eggs with n-3 long-chain fatty acids and evaluation of enriched eggs in humans, *American Journal of Clinical Nutrition* 68 (1998): 538–544.

31. C. Hasler and coauthors, How to evaluate the safety, efficacy, and quality of functional foods and their ingredients, *Journal of the American Dietetic Association* 101 (2001): 733–736; B. Brophy and D. Schardt, Functional foods, *Nutrition Action Healthletter,* April 1999, pp. 3–7.

32. J. A. Milner, Functional foods and health: A US perspective, *British Journal of Nutrition* 88 (2002): S151–S158.

33. C. M. Hasler, The changing face of functional foods, *Journal of the American College of Nutrition* 19 (2000): 499S–506S; I. H. Rosenberg, *What Is a Nutrient? Defining the Food-Drug Continuum* (Washington, D.C.: Center for Food and Nutrition Policy, 1999).

Chapter

14

Fitness: Physical Activity, Nutrients, and Body Adaptations

Chapter Outline

Fitness: *Benefits of Fitness • Developing Fitness • Cardiorespiratory Endurance*

Energy Systems, Fuels, and Nutrients to Support Activity: *The Energy Systems of Physical Activity—ATP and CP • Glucose Use during Physical Activity • Fat Use during Physical Activity • Protein Use during Physical Activity—and between Times • Vitamins and Minerals to Support Activity • Fluids and Electrolytes to Support Activity • Poor Beverage Choices: Caffeine and Alcohol*

Diets for Physically Active People: *Choosing a Diet to Support Fitness • Meals before and after Competition*

Highlight: *Supplements as Ergogenic Aids*

Nutrition Explorer CD-ROM Outline

Nutrition Animation: *Exercise Intensity and Fuel Use*

Case Study: *Can Physical Fitness Come in a Bottle?*

Student Practice Test

Glossary Terms

Nutrition on the Net

Nutrition in Your Life

You choose to be physically active or inactive, and your choice can make a **huge** difference in how well you feel and how long you live. Today's world makes it easy to be inactive—too easy in fact—but the many health rewards of being physically active make it well worth the effort. You may even discover how much fun it is to be active, and with a little perseverance, you may become physically fit as well. As you become more active, you will find that the foods you eat can make a difference in how fast you run, how far you swim, or how much weight you lift. It's up to you. The choice is yours.

A re you physically fit? If so, the following description applies to you. Your joints are flexible, your muscles are strong, and your body is lean with enough, but not too much, fat. You have the endurance to engage in daily physical activities with enough reserve energy to handle added challenges. Carrying heavy suitcases, opening a stuck window, or climbing four flights of stairs, which might strain an unfit person, is easy for you. What's more, you are prepared to meet mental and emotional challenges, too. All these characteristics of **fitness** describe the same wonderful condition of a healthy body.

Or perhaps you are leading a **sedentary** life. Today's world encourages inactivity. As people go through life exerting minimal physical effort, they become weak and unfit and may begin to feel unwell. In fact, a sedentary lifestyle fosters the development of several chronic diseases.

Regardless of your level of fitness, this chapter is written for "you," whoever you are and whatever your goals—whether you want to improve your health, lose weight, hone your athletic skills, ensure your position on a sports team, or simply adopt an active lifestyle. This chapter begins by discussing fitness and its benefits

fitness: the characteristics that enable the body to perform physical activity; more broadly, the ability to meet routine physical demands with enough reserve energy to rise to a physical challenge; or the body's ability to withstand stress of all kinds.

sedentary: physically inactive (literally, "sitting down a lot").

Physical activity, or its lack, exerts a significant and pervasive influence on everyone's nutrition and overall health.

■ Each comparison influences the risks associated with chronic disease and death similarly:
- Vigorous exercise vs. minimal exercise.
- Healthy weight vs. 20% overweight.
- Nonsmoking vs. smoking (one pack a day).

physical activity: bodily movement produced by muscle contractions that substantially increase energy expenditure.

exercise: planned, structured, and repetitive bodily movement that promotes or maintains physical fitness.

and then goes on to explain how the body uses energy nutrients to fuel physical activity. Finally, it describes diets to support fitness.

Fitness

Fitness depends on a certain minimum amount of **physical activity** or **exercise**. Physical activity and exercise both involve bodily movement, muscle contraction, and enhanced energy expenditure, but by definition, a distinction is made between the two terms. Exercise is often considered to be a vigorous, structured, and planned type of physical activity. This chapter focuses on the active body's use of energy nutrients—whether that body is pedaling a bike across campus or pedaling a stationary bike in a gym. Thus, for our purposes here, the terms *physical activity* and *exercise* will be used interchangeably.

Benefits of Fitness

Extensive evidence confirms that regular physical activity promotes health■ and reduces the risk of developing a number of diseases.[1] Still, despite an increasing awareness of the health benefits that physical activity confers, more than 60 percent of adults in the United States are not regularly active, and 25 percent are completely inactive.[2] Physical inactivity is linked to the major degenerative diseases—heart disease, cancer, stroke, diabetes, and hypertension—that are the primary killers of adults in developed countries.[3] Every year an estimated $24 billion is spent on health care costs attributed to physical inactivity in the United States.[4]

As a person becomes physically fit, the health of the entire body improves. In general, physically fit people enjoy:

- *Restful sleep.* Rest and sleep occur naturally after periods of physical activity. During rest, the body repairs injuries, disposes of wastes generated during activity, and builds new physical structures.

- *Nutritional health.* Physical activity expends energy and thus allows people to eat more food. If they choose wisely, active people will consume more nutrients and be less likely to develop nutrient deficiencies.

- *Optimal body composition.* A balanced program of physical activity limits body fat and increases or maintains lean tissue. Thus physically active people have relatively less body fat than sedentary people at the same body weight.[5]

- *Optimal bone density.* Weight-bearing physical activity builds bone strength and protects against osteoporosis.[6]

- *Resistance to colds and other infectious diseases.* Fitness enhances immunity.*[7]

- *Low risks of some types of cancers.* Lifelong physical activity may help to protect against colon cancer, breast cancer, and some other cancers.[8]

- *Strong circulation and lung function.* Physical activity that challenges the heart and lungs strengthens the circulatory system.

- *Low risk of cardiovascular disease.* Physical activity lowers blood pressure, slows resting pulse rate, and lowers blood cholesterol, thus reducing the risks of heart attacks and strokes.[9] Some research suggests that physical activity may reduce the risk of cardiovascular disease in another way as well—by reducing intra-abdominal fat stores.[10]

---

*Moderate physical activity can stimulate immune function. Intense, vigorous, prolonged activity such as marathon running, however, may compromise immune function.

- *Low risk of type 2 diabetes.* Physical activity normalizes glucose tolerance.[11] Regular physical activity reduces the risk of developing type 2 diabetes and benefits those who already have the condition.

- *Reduced risk of gallbladder disease in women.* Regular physical activity reduces women's risk of gallbladder disease—perhaps by facilitating weight control and lowering blood lipid levels.[12]

- *Low incidence and severity of anxiety and depression.* Physical activity may improve mood and enhance the quality of life by reducing depression and anxiety.[13]

- *Strong self-image.* The sense of achievement that comes from meeting physical challenges promotes self-confidence.

- *Long life and high quality of life in the later years.* Active people have a lower mortality rate than sedentary people.[14] Even a two-mile walk daily can add years to a person's life. In addition to extending longevity, physical activity supports independence and mobility in later life by reducing the risk of falls and minimizing the risk of injury should a fall occur.[15]

Physical activity helps you look good, feel good, and have fun, and it brings many long-term health benefits as well.

What does a person have to do to reap the health rewards of physical activity? For *health's* sake, the American College of Sports Medicine (ACSM) and the *Dietary Guidelines for Americans* specify that people need to spend an accumulated minimum of 30 minutes in some sort of physical activity on most days of each week.[16] Eight minutes spent climbing up stairs, another 10 spent pulling weeds, and 12 more spent walking the dog all contribute to the day's total (see Figure 14-1). The DRI Committee, however, advises that 30 minutes of physical activity each day is

---

FIGURE 14-1 Physical Activity Pyramid

DO SPARINGLY—
Limit sedentary activities.
- Watch TV, videos, or movies
- Play computer games

2–3 DAYS/WEEK—
Engage in strength and flexibility activities and enjoy leisure activities often.
- Sit-ups, push-ups
- Strength training such as weight lifting
- Stretching exercises such as yoga
- Leisure activities such as canoeing, dancing, golfing, horseback riding, bowling

3–5 DAYS/WEEK—
Engage in vigorous activities regularly.
- Aerobic activities such as running, biking, swimming, roller-blading, rowing, cross-country skiing, kickboxing, power walking, dancing, jumping rope
- Sports activities such as basketball, soccer, volleyball, tennis, football, racquetball, softball

EVERY DAY—
Be as active as possible.
- Use the stairs
- Walk or bike to class, work, or shops
- Scrub floors, wash windows
- Walk your dog
- Mow grass, rake leaves, turn compost, shovel snow, tend garden
- Wash and wax your car
- Play with children

TABLE 14-1 Guidelines for Physical Fitness

	Cardiorespiratory	Strength	Flexibility
Type of Activity	Aerobic activity that uses large-muscle groups and can be maintained continuously	Resistance activity that is performed at a controlled speed and through a full range of motion	Stretching activity that uses the major muscle groups
Frequency	3 to 5 days per week	2 to 3 days per week	2 to 3 days per week
Intensity	55 to 90% of maximum heart rate	Enough to enhance muscle strength and improve body composition	Enough to develop and maintain a full range of motion
Duration	20 to 60 minutes	8 to 12 repetitions of 8 to 10 different exercises (minimum)	4 repetitions of 10 to 30 seconds per muscle group (minimum)

SOURCE: Adapted from American College of Sports Medicine, Position stand: The recommended quantity and quality of exercise for developing and maintaining cardiorespiratory and muscular fitness, and flexibility in healthy adults, *Medicine and Science in Sports and Exercise* 30 (1998): 975–991.

not enough for adults to maintain a healthy body weight (BMI of 18.5 to 24.9) and recommends at least 60 minutes of moderately intense activity such as walking or jogging each day.[17] The hour or more of activity can be split into shorter sessions throughout the day—two 30-minute sessions, or four 15-minute sessions, for example.[18]

Increase the proportion of adults who engage regularly, preferably daily, in moderate physical activity for at least 30 minutes per day.

To develop and maintain *fitness,* the ACSM recommends the types and amounts of physical activities presented in Table 14-1.[19] These guidelines help adults develop programs to improve their cardiorespiratory endurance, body composition, strength, and flexibility. At this level of fitness, a person can reap even greater health benefits (further reduction of cardiovascular disease risk, for example).[20]

■ Reminder: *Body composition* refers to the proportions of muscle, bone, fat, and other tissue that make up a person's total body weight.

The bottom line is that any physical activity, even moderate activity, provides some health benefits, and these benefits follow a dose-response relationship. Therefore, some activity is better than none, and more activity is better still—up to a point.[21] (Pursued in excess, intense physical activity, especially when combined with poor eating habits, can undermine health, as Highlight 9 explains.)

Reduce the proportion of adults who engage in no leisure-time physical activity.

flexibility: the capacity of the joints to move through a full range of motion; the ability to bend and recover without injury.

muscle strength: the ability of muscles to work against resistance.

muscle endurance: the ability of a muscle to contract repeatedly without becoming exhausted.

Developing Fitness

To be physically fit, a person needs to develop enough flexibility, muscle strength and endurance, and cardiorespiratory endurance to meet the everyday demands of life with some to spare and to achieve a reasonable body weight and body composition. **Flexibility** allows the joints to move freely, reducing the risk of injury. **Muscle strength** and **muscle endurance** enable muscles to work harder and longer without fatigue. **Cardiorespiratory endurance** supports the ongoing action of the heart and lungs.

Physical activity supports lean body tissues and reduces excess body fat. A person who practices a physical activity *adapts* by becoming better able to perform it after each session—with more flexibility, more strength, and more endurance.

Increase the proportion of adults who perform physical activities that enhance and maintain flexibility.

The principles of **conditioning** apply to each component of fitness—flexibility, strength, and endurance. During conditioning, the body adapts microscopically to perform the work asked of it. The way to achieve conditioning is by **training**, primarily by applying the **progressive overload principle**—that is, by asking a little more of the body in each training session.

The Overload Principle You can apply the progressive overload principle in several different ways. You can perform the activity more often—that is, increase its **frequency.** You can perform it more strenuously—that is, increase its **intensity.** Or you can do it for longer times—that is, increase its **duration.** All three strategies, individually or in combination, work well. The rate of progression depends on individual characteristics such as fitness level, health status, age, and preference. If you enjoy your workout, do it more often. If you do not have much time, increase intensity. If you dislike hard work, take it easy and go longer. If you want continuous improvements, remember to overload progressively as you reach higher levels of fitness.

When increasing the frequency, intensity, or duration of a workout, however, exercise to a point that only *slightly* exceeds the comfortable capacity to work. It is better to progress slowly than to risk injury by overexertion.

The Body's Response to Physical Activity Fitness develops in response to demand and wanes when demand ceases. Muscles gain size and strength after being made to work repeatedly, a response called **hypertrophy.** Conversely, without activity, muscles diminish in size and lose strength, a response called **atrophy.**

Hypertrophy and atrophy are adaptive responses to the muscles' greater and lesser work demands, respectively. Thus cyclists often have strong, well-developed legs but less arm or chest strength; a tennis player may have one superbly strong arm, while the other is just average. A variety of physical activities produces the best overall fitness, and to this end, people need to work different muscle groups from day to day. This strategy provides a day or two of rest for different muscle groups, giving them time to replenish nutrients and to repair any minor damage incurred by the activity.

cardiorespiratory endurance: the ability to perform large-muscle, dynamic exercise of moderate-to-high intensity for prolonged periods.

conditioning: the physical effect of training; improved flexibility, strength, and endurance.

training: practicing an activity regularly, which leads to conditioning. (Training is what you do; conditioning is what you get.)

progressive overload principle: the training principle that a body system, in order to improve, must be worked at frequencies, durations, or intensities that gradually increase physical demands.

frequency: the number of occurrences per unit of time (for example, the number of activity sessions per week).

intensity: the degree of exertion while exercising (for example, the amount of weight lifted or the speed of running).

duration: length of time (for example, the time spent in each activity session).

hypertrophy (high-PER-tro-fee): growing larger; with regard to muscles, an increase in size (and strength) in response to use.

atrophy (AT-ro-fee): becoming smaller; with regard to muscles, a decrease in size (and strength) because of disuse, undernutrition, or wasting diseases.

People's bodies are shaped by the activities they perform.

Other tips for building fitness and minimizing the risk of overuse injuries are:

- Be active all week, not just on the weekends.
- Use proper equipment and attire.
- Perform exercises using proper form.
- Include **warm-up** and **cool-down** activities in each session. Warming up helps to prepare muscles, ligaments, and tendons for the upcoming activity and mobilizes fuels to support strength and endurance activities; cooling down reduces muscle cramping and allows the heart rate to slow gradually.
- Train hard enough to challenge your strength or endurance a few times each week, not every time you work out. Between challenges, do moderate workouts and include at least one day of rest each week.
- Pay attention to body signals. Symptoms such as abnormal heartbeats, dizziness, lightheadedness, cold sweat, confusion, or pain or pressure in the middle of the chest, teeth, jaw, neck, or arm demand immediate medical attention.

Work out wisely. Do not start with activities so demanding that pain stops you within two days. Learn to enjoy small steps toward improvement. Fitness builds slowly.

Cautions on Starting   Before beginning a fitness program, make sure it is safe for you to do so. Most apparently healthy people can begin a **moderate exercise** program such as walking or increasing daily activities without a medical examination, but people with any of the risk factors listed in the margin■ may need medical advice.[22]

Weight Training   **Weight training** has been recognized as a means to build lean body mass and develop and maintain muscle strength and endurance for many years. Additional benefits of weight training, however, have emerged only recently. Progressive weight training not only increases muscle strength and endurance, but it prevents and manages several chronic diseases, including cardiovascular disease, and enhances psychological well-being.[23] Weight training can also help to maximize and maintain bone mass.[24] Even in women past menopause (when most women are losing bone), a one-year program of weight training improved bone density; in fact, the more weight lifted, the greater the improvement.[25]

Weight training enhances performance in other sports, too. Swimmers can develop a more efficient stroke and tennis players, a more powerful serve, when they train with weights.

Depending on the technique, weight training can emphasize either muscle strength or muscle endurance. To emphasize muscle strength, combine high resistance (heavy weight) with a low number (8 to 10) of repetitions. To emphasize muscle endurance, combine less resistance (lighter weight) with more (12 to 15) repetitions.

HEALTHY PEOPLE 2010   Increase the proportion of adults who perform physical activities that enhance and maintain muscular strength and endurance.

Cardiorespiratory Endurance

The length of time a person can remain active with an elevated heart rate—that is, the ability of the heart, lungs, and blood to sustain a given demand—defines a person's cardiorespiratory endurance. Cardiorespiratory endurance training improves a person's ability to sustain vigorous activities such as running, brisk walking, or swimming. Such training enhances the capacity of the heart, lungs, and blood to deliver oxygen to, and remove waste from, the body's cells. Cardiorespiratory endurance training, therefore, is *aerobic.*■ As the cardiorespiratory system gradually

---

■ Major coronary risk factors:
- Family history of heart disease.
- Cigarette smoking.
- Hypertension.
- Serum cholesterol >200 mg/dL or HDL <35 mg/dL, or on lipid-lowering medication.
- Diabetes.
- Sedentary lifestyle.
- Obesity (BMI ≥30).

■ Reminder: Recall from Chapter 7 that *aerobic* means requiring oxygen.

warm-up: 5 to 10 minutes of light activity, such as easy jogging or cycling, prior to a workout to prepare the body for more vigorous activity.

cool-down: 5 to 10 minutes of light activity, such as walking or stretching, following a vigorous workout to return the body's core gradually to near-normal temperature.

moderate exercise: activity equivalent to the rate of exertion reached when walking at a speed of 4 miles per hour (15 minutes to walk one mile).

weight training (also called **resistance training**): the use of free weights or weight machines to provide resistance for developing muscle strength and endurance. A person's own body weight may also be used to provide resistance as when a person does push-ups, pull-ups, or abdominal crunches.

adapts to the demands of aerobic activity, the body delivers oxygen more efficiently. In fact, the accepted measure of a person's cardiorespiratory fitness is maximal oxygen uptake (**VO₂max**). The benefits of cardiorespiratory training are not just physical, though, because all of the body's cells, including the brain cells, require oxygen to function. When the cells receive more oxygen more readily, both the body and the mind benefit.

Cardiorespiratory Conditioning **Cardiorespiratory conditioning**■ occurs as aerobic workouts improve heart and lung activities. **Cardiac output** increases, thus enhancing oxygen delivery. The heart becomes stronger, and each beat pumps more blood. Because the heart pumps more blood with each beat, fewer beats are necessary, and the resting heart rate slows down. The average resting pulse rate for adults is around 70 beats per minute, but people who achieve cardiorespiratory conditioning may have resting pulse rates of 50 or even lower. The muscles that work the lungs become stronger, too, so breathing becomes more efficient. Circulation through the arteries and veins improves. Blood moves easily, and blood pressure falls.

Cardiorespiratory endurance reflects the health of the heart and circulatory system, on which all other body systems depend. Figure 14-2 (on p. 480) shows the major relationships among the heart, circulatory system, and lungs.

To improve your cardiorespiratory endurance, the activity you choose must be sustained for 20 minutes or longer and use most of the large-muscle groups of the body (legs, buttocks, and abdomen). You must also train at an intensity that elevates your heart rate.

■ Cardiorespiratory conditioning:
- Increases cardiac output and oxygen delivery.
- Increases stroke volume.
- Slows resting pulse.
- Increases breathing efficiency.
- Improves circulation.
- Reduces blood pressure.

Increase the proportion of adults who engage in vigorous physical activity that promotes the development and maintenance of cardiorespiratory fitness 3 or more days per week for 20 or more minutes per occasion.

A person's own perceived effort is usually a reliable indicator of the intensity of an activity. In general, when you're working out, do so at an intensity that raises your heart rate, but still leaves you able to talk comfortably. If you are more competitive and want to work to your limit on some days, a treadmill test can reveal your maximum heart rate. You can work out safely at up to 90 percent of that rate. The ACSM guidelines for developing and maintaining cardiorespiratory fitness were given in Table 14-1 on p. 476.

Muscle Conditioning A fringe benefit of cardiorespiratory training is that fit muscles use oxygen efficiently, reducing the heart's workload. An added bonus is that muscles that use oxygen efficiently can burn fat longer—a plus for body composition and weight control.

A Balanced Fitness Program The intensity and type of physical activities that are best for one person may not be good for another. The intensity■ to choose depends on your present fitness: work so as to breathe hard, but not so hard as to incur an oxygen debt. If you have been sedentary, the level of intensity that you initially perform will differ dramatically from the intensity of a fit person.

The type of physical activity that is best for you depends, too, on what you want to achieve■ and what you enjoy doing. Some people love walking, while others prefer to dance or ride a bike. If you want to be stronger and firmer, lift weights. And remember, muscle is more metabolically active than body fat, so the more muscle you have, the more energy you'll burn.

In a balanced fitness program, aerobic activity improves cardiorespiratory fitness, stretching enhances flexibility, and weight training develops muscle strength and endurance. Table 14-2 provides an example of a balanced fitness program.

■ Fitness tip: Select an activity and an intensity level that are challenging, but not overwhelming.

■ Fitness tip: Select an activity that will help you meet your goals.

VO₂max: the maximum rate of oxygen consumption by an individual at sea level.

cardiorespiratory conditioning: improvements in heart and lung function and increased blood volume, brought about by aerobic training.

cardiac output: the volume of blood discharged by the heart each minute; determined by multiplying the stroke volume by the heart rate. The stroke volume is the amount of oxygenated blood the heart ejects toward the tissues at each beat. Cardiac output (volume/minute) = stroke volume (volume/beat) × heart rate (beats/minute).

The key to regular physical activity is finding an activity that you enjoy.

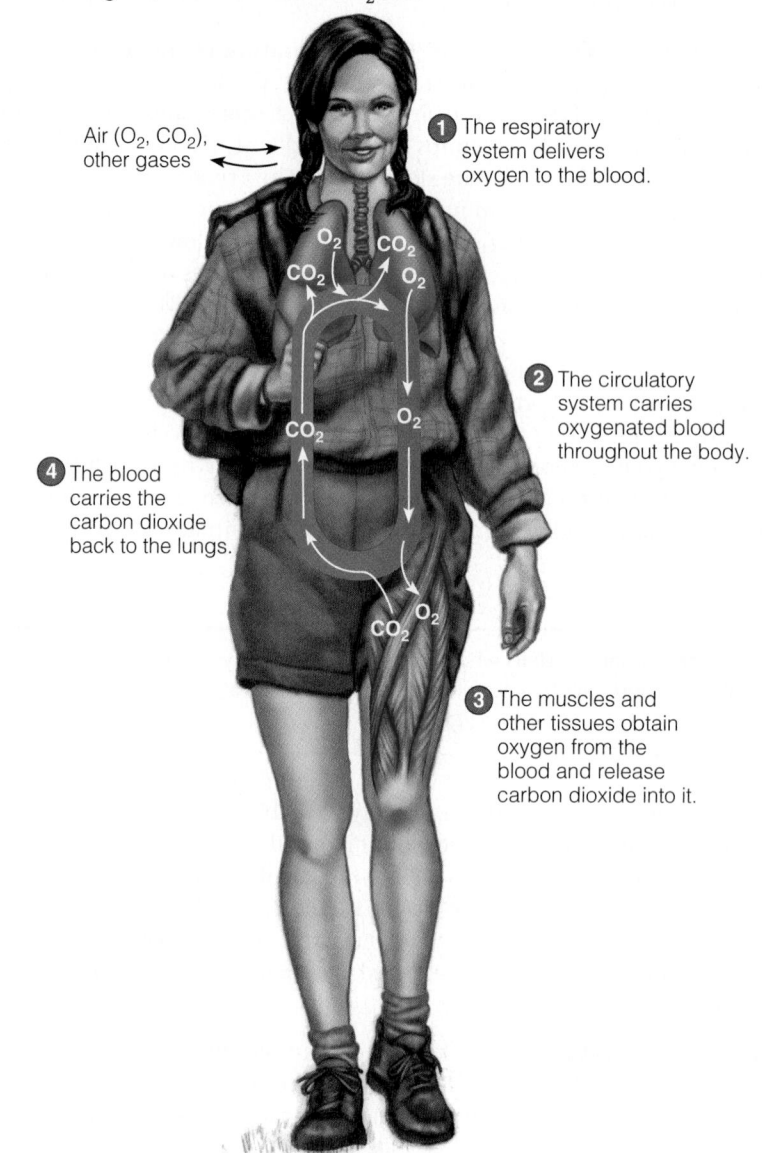

FIGURE 14-2 Delivery of Oxygen by the Heart and Lungs to the Muscles

The cardiorespiratory system responds to the muscles' demand for oxygen by building up its capacity to deliver oxygen. Researchers can measure cardiorespiratory fitness by measuring the maximum amount of oxygen a person consumes per minute while working out, a measure called VO_2max.

Air (O_2, CO_2), other gases

1 The respiratory system delivers oxygen to the blood.

2 The circulatory system carries oxygenated blood throughout the body.

4 The blood carries the carbon dioxide back to the lungs.

3 The muscles and other tissues obtain oxygen from the blood and release carbon dioxide into it.

TABLE 14-2 A Sample Balanced Fitness Program (45 minutes a day)

Monday, Wednesday, Friday:
- 5 minutes of warm-up activity
- 30 minutes of aerobic activity
- 10 minutes of cool-down activity and stretching

Tuesday, Thursday:
- 5 minutes of warm-up activity
- 30 minutes of weight training
- 10 minutes of cool-down activity and stretching

Saturday or Sunday:
- Sports, walking, hiking, biking, or swimming

IN SUMMARY Physical activity brings positive rewards: good health and long life. To develop fitness—whose components are flexibility, muscle strength and endurance, and cardiorespiratory endurance—a person must condition the body, through training, to adapt to the activity performed.

Energy Systems, Fuels, and Nutrients to Support Activity

Nutrition and physical activity go hand in hand. Activity demands carbohydrate and fat as fuel, protein to build and maintain lean tissues, vitamins and minerals to support both energy metabolism and tissue building, and water to help distribute the fuels and to dissipate the resulting heat and wastes. This section describes how nutrition supports a person who decides to get up and go.

The Energy Systems of Physical Activity—ATP and CP

Muscles contract fast. When called upon, they respond quickly without taking time to metabolize fat or carbohydrate for energy. In the first fractions of a second, muscles starting to move depend on their supplies of quick-energy compounds to power their movements. Exercise physiologists know these compounds by their abbreviations, ATP and CP.

ATP As Chapter 7 described, all of the energy-yielding nutrients—carbohydrate, fat, and protein—can enter metabolic pathways that make the high-energy compound ATP (adenosine triphosphate). ATP is present in small amounts in all body tissues all the time, and it can deliver energy instantaneously. In the muscles, ATP provides the chemical driving force for contraction. When ATP is split, its energy is released, and the muscle cells channel some of that energy into mechanical movement and most of it into heat—heat the exerciser can feel building up. A tiny but essential pool of ATP is always ready to meet the cells' sudden demands for movement.

CP Immediately after the onset of a demand, before muscle ATP pools dwindle, a muscle enzyme begins to break down another high-energy compound that is stored in the muscle, **CP,** or **creatine phosphate.** CP is made from creatine, a compound commonly found in muscles, with a phosphate group attached, and it can split (anaerobically)■ to release phosphate and replenish ATP supplies. Supplies of CP in a muscle last for only about 10 seconds, producing enough quick energy without oxygen for a 100-meter dash.

When activity ceases and the muscles are resting, ATP feeds energy back to CP by giving up one of its phosphate groups to creatine. Thus CP is produced during rest by reversing the process that occurs during muscular activity.■ (Highlight 14 includes creatine supplements in its discussion of substances commonly used in the pursuit of fitness.)

The Energy-Yielding Nutrients To meet the more prolonged demands of sustained activity, the muscles generate ATP from the more abundant fuels: carbohydrate, fat, and protein. The breakdown of these nutrients generates ATP all day every day, and so maintains the supply. Muscles always use a mixture of fuels—never just one.

During rest, the body derives more than half of its ATP from fatty acids and most of the rest from glucose, along with a small percentage from amino acids. During physical activity, the body adjusts its mixture of fuels. How much of which fuel■ the muscles use during physical activity depends on an interplay among the fuels available from the diet, the intensity and duration of the activity, and the degree to which the body is conditioned to perform that activity. The next sections explain these relationships by examining each of the energy-yielding nutrients individually, but keep in mind that while one fuel may predominate at a given time, the other two will still be involved. Table 14-3 (on p. 482) shows how fuel use changes according to the intensity and duration of the activity.

As you read about each of the energy-yielding nutrients, notice how its contribution to the fuel mixture shifts depending on whether the activity is anaerobic or aerobic. Anaerobic activities are associated with strength, agility, and split-second surges of power. The jump of the basketball player, the slam of the tennis serve, the heave of the weight lifter at the barbells, and the blast of the fullback through the opposing line involve anaerobic work. Such high-intensity, short-duration activities depend mostly on glucose as the chief energy fuel.

Endurance activities of low-to-moderate intensity and long duration depend more on fat to provide energy aerobically. The ability to continue swimming to

Split-second surges of power as in the heave of a barbell or jump of a basketball player involve *anaerobic* work.

© Simon Watson/The Image Bank/Getty Images

■ Reminder: Recall from Chapter 7 that *anaerobic* means not requiring oxygen.

■ During rest: ATP + creatine → CP.
During activity: CP → ATP + creatine.

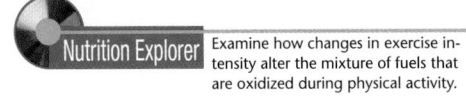

Nutrition Explorer Examine how changes in exercise intensity alter the mixture of fuels that are oxidized during physical activity.

■ Fuel mixture during activity depends on:
- Diet.
- Intensity and duration of activity.
- Training.

CP, creatine phosphate (also called **phosphocreatine**): a high-energy compound in muscle cells that acts as a reservoir of energy that can maintain a steady supply of ATP. CP provides the energy for short bursts of activity.

TABLE 14-3 Fuels Used for Activities of Different Intensities and Durations

Activity Intensity	Activity Duration	Preferred Fuel Source	Oxygen Needed?	Activity Example
Extreme[a]	8 to 10 sec	ATP-CP (immediate availability)	No	100-yard dash, shot put
Very high	20 sec to 3 min	ATP from carbohydrate (lactic acid)	No (anaerobic)	¼-mile run at maximal speed
High	3 min to 20 min	ATP from carbohydrate	Yes (aerobic)	Cycling, swimming, or running
Moderate	More than 20 min	ATP from fat	Yes (aerobic)	Hiking

[a]All levels of activity intensity use the ATP-CP system initially; extremely intense short-term activities rely solely on the ATP-CP system.

Sustained muscular efforts as in a long-distance rowing event or cross-country run involve *aerobic* work.

■ Fitness tip: To fill glycogen stores, eat plenty of carbohydrate-rich foods.

■ Reminder: *Lactic acid* is the product of anaerobic glycolysis.

the shore, to keep on hiking to the top of the mountain, or to continue pedaling all the way home reflects aerobic capacity. As mentioned earlier, aerobic capacity is also crucial to maintaining the health of the heart and circulatory system. The relationships among fuels and physical activity bear heavily on what foods best support your chosen activities.

Glucose Use during Physical Activity

Glucose, stored in the liver and muscles as glycogen, is vital to physical activity. During exertion, the liver breaks down its glycogen and releases the glucose into the bloodstream. The muscles use both this glucose and their own private glycogen stores to fuel their work. Glycogen supplies can easily support everyday activities, but are limited to less than 2000 kcalories of energy, enough for about 20 miles of running.[26] The more glycogen the muscles store, the longer the stores will last during physical activity, which in turn influences performance. When glycogen is depleted, the muscles become fatigued.

Diet Affects Glycogen Storage and Use How much carbohydrate a person eats influences how much glycogen is stored.■ A classic study compared fuel use during activity among three groups of runners on different diets.[27] For several days before testing, one group consumed a normal mixed diet, a second group consumed a high-carbohydrate diet, and the third group consumed a no-carbohydrate diet (fat and protein diet). As Figure 14-3 shows, the high-carbohydrate diet allowed the runners to keep going longer before exhaustion. This study and many others that followed have confirmed that high-carbohydrate diets enhance endurance by enlarging glycogen stores.

Intensity of Activity Affects Glycogen Use How long an exercising person's glycogen will last depends not only on diet, but also on the intensity of the activity. Moderate activities, such as jogging, during which breathing is steady and easy, use glycogen slowly. The lungs and circulatory system have no trouble keeping up with the muscles' need for oxygen. The individual breathes easily, and the heart beats steadily—the activity is aerobic. The muscles derive their energy from both glucose and fatty acids. By depending partly on fatty acids, moderate aerobic activity conserves glycogen.

Intense activities—the kind that make it difficult "to catch your breath," such as a quarter-mile race—use glycogen quickly. The muscles break down glucose to pyruvate anaerobically, producing ATP quickly.

Lactic Acid When the rate of glycolysis exceeds the capacity of the mitochondria to accept hydrogens with their electrons for the electron transport chain, the accumulating pyruvate molecules are converted to lactic acid.■ At low intensities, lactic acid is readily cleared from the blood, but at higher intensities, lactic acid accumulates. When the rate of lactic acid production exceeds the rate of clearance, intense activity can be maintained for only 1 to 3 minutes (as in a 400- to 800-meter race or a boxing match).

FIGURE 14-3   The Effect of Diet on Physical Endurance

A high-carbohydrate diet can increase an athlete's endurance. In this study, the fat and protein diet provided 94 percent of kcalories from fat and 6 percent from protein; the normal mixed diet provided 55 percent of kcalories from carbohydrate; and the high-carbohydrate diet provided 83 percent of kcalories from carbohydrate.

Maximum endurance time:

Fat and protein diet — 57 min

Normal mixed diet — 114 min

High-carbohydrate diet — 167 min

© Bonnie Kamin/PhotoEdit, Inc.

When production of lactic acid exceeds the ability of the muscles to use it, they release it, and it travels in the blood to the liver. There, liver enzymes convert the lactic acid back into glucose. Glucose can then return to the muscles to fuel additional activity. (The recycling process that regenerates glucose from lactic acid, known as the *Cori cycle,* is shown in Figure 7-6 on p. 222.)

Duration of Activity Affects Glycogen Use Glycogen use depends not only on the intensity of an activity, but also on its duration. Within the first 20 minutes or so of moderate activity, a person uses mostly glycogen for fuel—about one-fifth of the available glycogen. As the muscles devour their own glycogen, they become ravenous for more glucose, and the liver responds by emptying out its glycogen stores.

After 20 minutes, a person who continues exercising moderately (mostly aerobically) begins to use less and less glycogen and more and more fat for fuel (review Table 14-3 on p. 482). Still, glycogen use continues, and if the activity lasts long enough and is intense enough, muscle and liver glycogen stores will be depleted. Physical activity can continue for a short time thereafter only because the liver scrambles to produce, from lactic acid and certain amino acids, the minimum amount of glucose needed to briefly forestall total depletion.

Glucose Depletion After a couple of hours of strenuous activity, glucose stores are depleted. When depletion occurs, it brings nervous system function to a near halt, making continued exertion almost impossible. Marathon runners refer to this point of glucose exhaustion as "hitting the wall."

To avoid such debilitation, endurance athletes try to maintain their blood glucose for as long as they can. To maximize glucose supply, endurance athletes:

- Eat a high-carbohydrate diet (approximately 8 grams of carbohydrate per kilogram of body weight or about 70 percent of energy intake) regularly.*

Moderate- to high-intensity aerobic exercises that can be sustained for only a short time (less than 20 minutes) use some fat, but more glucose for fuel.

© David Leahy/Taxi/Getty Images

---

*Percentage of energy intake is meaningful only when total energy intake is known. Consider that at high energy intakes (say, 5000 kcalories/day), even a moderate carbohydrate diet (40 percent of energy intake) supplies 500 grams of carbohydrate—enough for a 137-pound athlete in heavy training. By comparison, at a moderate energy intake (2000 kcalories/day), a high carbohydrate intake (70 percent of energy intake) supplies 350 grams—plenty of carbohydrate for most people, but not enough for athletes in heavy training.

HOW TO Maximize Glycogen Stores: Carbohydrate Loading

Some athletes use a technique called carbohydrate loading to trick their muscles into storing extra glycogen before a competition. Carbohydrate loading can nearly double muscle glycogen concentrations. In general, the athlete tapers training during the week before the competition and then eats a high-carbohydrate diet during the three days just prior to the event.[a] Specifically, the athlete would follow the plan in the accompanying table. In this carbohydrate loading plan, glycogen storage occurs slowly, and athletes must alter their training for several days before the event.

In contrast, a group of researchers have designed a quick method of carbohydrate

Before the Event	Training Intensity	Training Duration	Dietary Carbohydrate
6 days	Moderate (70% VO$_2$max)	90 min	Normal (5 g/kg body weight)
4–5 days	Moderate (70% VO$_2$max)	40 min	Normal (5 g/kg body weight)
2–3 days	Moderate (70% VO$_2$max)	20 min	High-carbohydrate (10 g/kg body weight)
1 day	Rest	—	High-carbohydrate (10 g/kg body weight)

loading that has produced promising preliminary results. The researchers found that athletes could attain above-normal concentrations of muscle glycogen by eating a high-carbohydrate diet (10 g/kg body weight) after a short (3 minutes) but very intense bout of exercise.[b] More studies are needed to confirm these findings and to determine whether an exercise session of

less intensity and shorter duration would accomplish the same results.

Extra glycogen gained through carbohydrate loading can benefit an athlete who must keep going for 90 minutes or longer. Those who exercise for shorter times simply need a regular high-carbohydrate diet. In a hot climate, extra glycogen confers an additional advantage: as glycogen breaks down, it releases water, which helps to meet the athlete's fluid needs.

[a]E. Coleman, Carbohydrate and exercise, in *Sports Nutrition: A Guide for the Professional Working with Active People*, 3rd ed., ed. C. A. Rosenbloom (Chicago: The American Dietetic Association, 2000), pp. 13–31.

[b]T. J. Fairchild and coauthors, Rapid carbohydrate loading after a short bout of near maximal-intensity exercise, *Medicine and Science in Sports and Exercise* 34 (2001): 980–986.

■ For perspective, snack ideas providing 60 g carbohydrate:
- 16 oz sports drink and a small bagel.
- 16 oz milk and 4 oatmeal cookies.
- 8 oz pineapple juice and a granola bar.

■ Fitness tip: To help delay exhaustion during long competitive events, eat or drink a light carbohydrate-rich snack during the event.

carbohydrate loading: a regimen of moderate exercise followed by the consumption of a high-carbohydrate diet that enables muscles to store glycogen beyond their normal capacities; also called **glycogen loading** or **glycogen super compensation.**

- Take glucose (usually in sports drinks) periodically during activities that last for 45 minutes or more.
- Eat carbohydrate-rich foods (approximately 60 grams of carbohydrate)■ immediately following activity.
- Train the muscles to store as much glycogen as possible.

The last section of this chapter, "Diets for Physically Active People," discusses how to design a high-carbohydrate diet for performance, and the accompanying "How to" describes **carbohydrate loading**—a technique used to maximize glycogen stores for long endurance competitions.

Glucose during Activity Muscles can obtain the glucose they need not only from glycogen stores, but also from foods and beverages consumed during activity. Consuming sugar is especially useful during exhausting endurance activities (lasting more than 45 minutes) and during games such as soccer or hockey, which last for hours and demand repeated bursts of intense activity.[28] Endurance athletes often run short of glucose by the end of competitive events, and they are wise to take light carbohydrate snacks or drinks (under 200 kcalories) periodically■ during activity.[29] During the last stages of an endurance competition, when glycogen is running low, glucose consumed during the event can slowly make its way from the digestive tract to the muscles and augment the body's supply of glucose enough to forestall exhaustion.

Glucose after Activity Eating high-carbohydrate foods *after* physical activity also enlarges glycogen stores. A high-carbohydrate meal eaten within 15 minutes after physical activity accelerates the rate of glycogen storage by 300 percent. After two hours, the rate of glycogen storage declines by almost half. Despite this slower rate of glycogen restoration, muscles continue to accumulate glycogen as long as athletes eat carbohydrate-rich foods within two hours following activity.[30] This is particularly important to athletes who train hard more than once a day.

Chapter 4 introduced the glycemic effect and discussed the possible health benefits of eating a *low*-glycemic diet. For athletes wishing to maximize muscle glycogen synthesis after strenuous training, however, eating foods with a *high* glycemic index may be more beneficial (see Figure 4-13 on p. 119).[31]

Training Affects Glycogen Use Training,■ too, affects how much glycogen muscles will store. Muscle cells that repeatedly deplete their glycogen through hard work adapt to store greater amounts of glycogen to support that work.

Conditioned muscles also rely less on glycogen and more on fat for energy, so glycogen breakdown and glucose use occur more slowly in trained than in untrained individuals at a given work intensity.[32] A person attempting an activity for the first time uses much more glucose than an athlete who is trained to perform it. Oxygen delivery to the muscles by the heart and lungs plays a role, but equally importantly, trained muscles are better equipped to use the oxygen because their cells contain more mitochondria.■ Untrained muscles depend more heavily on anaerobic glucose breakdown, even when physical activity is just moderate.

■ Fitness tip: To make glycogen, muscles need carbohydrate, but they also need rest, so vary daily exercise routines to work different muscles on different days.

■ Reminder: The *mitochondria* are the structures within a cell responsible for producing ATP (see Figure 7-19 on p. 231).

Fat Use during Physical Activity

As Figure 14-3 showed, researchers have long recognized the importance of a high-carbohydrate diet for endurance performance. When endurance athletes "fat load" by consuming high-fat, low-carbohydrate diets for one to three days, their performance is impaired because their small glycogen stores are depleted quickly.[33] Endurance athletes who adhere to a high-fat, low-carbohydrate diet for more than a week, however, adapt by relying more on fat to fuel activity. Even with fat adaptation, however, performance benefits are not consistently evident.[34] In some cases, athletes on high-fat diets experience greater fatigue and perceive the activity to be more strenuous than athletes on high-carbohydrate diets.[35]

Of course, high-fat diets carry risks of heart disease and cannot be recommended without careful consideration. Most nutrition experts agree that the potential for adverse health effects of prolonged high-fat diets continues to outweigh any possible benefit to performance.

Sports nutrition experts recommend that endurance athletes consume 20 to 30 percent of their energy from fat in order to meet nutrient and energy needs.[36] Athletes who restrict fat below 20 percent of total energy intake may fail to consume adequate energy and nutrients.

In contrast to *dietary* fat, *body* fat stores are of tremendous importance during physical activity, as long as the activity is not too intense. Unlike glycogen stores, the body's fat stores can usually provide more than 70,000 kcalories and fuel hours of activity without running out.[37]

The fat used in physical activity is liberated as fatty acids from the internal fat stores and from the fat under the skin. Areas that have the most fat to spare donate the greatest amounts to the blood (although they may not be the areas that appear fattiest). This is why "spot reducing" doesn't work—muscles do not own the fat that surrounds them. Fat cells release fatty acids into the blood, not into the underlying muscles. Then the blood gives to each muscle the amount of fat that it needs. Proof of this is found in a tennis player's arms—the fatfolds measure the same in both arms, even though the muscles of one arm work much harder and may be larger than those of the other. A balanced fitness program that includes strength training, however, will tighten muscles underneath the fat, improving the overall appearance. Keep in mind that some body fat is essential to good health.

Duration of Activity Affects Fat Use Early in an activity, as the muscles draw on fatty acids, blood levels fall. If the activity continues for more than a few minutes, the hormone epinephrine signals the fat cells to begin breaking down their stored triglycerides and liberating fatty acids into the blood. After about 20 minutes■ of physical activity, the blood fatty acid concentration surpasses the normal resting concentration. Thereafter, sustained, moderate activity uses body fat stores as its major fuel.

Intensity of Activity Affects Fat Use The intensity of physical activity also affects fat use. As the intensity of activity increases, fat makes less and less of a contribution to the fuel mixture. Remember that fat can be broken down for energy

■ Fitness tip: To burn fat, exercise for 20 minutes or more.

Abundant energy from the breakdown of fat can come only from aerobic metabolism.

© 2001 PhotoDisc, Inc.

■ Fitness tip: To burn more fat, train aerobically.

Low- to moderate-intensity aerobic exercises that can be sustained for a long time (more than 20 minutes) use some glucose, but more fat for fuel.

only by aerobic metabolism. For fat to fuel activity, then, oxygen must be abundantly available. If a person is breathing easily during activity, the muscles are getting all the oxygen they need and are able to use more fat in the fuel mixture.

Training Affects Fat Use Training—repeated aerobic activity—produces the adaptations that permit the body to draw more heavily on fat■ for fuel. Training stimulates the muscle cells to manufacture more and larger mitochondria, the "powerhouse" structures of the cells that produce ATP for energy. Another adaptation: the heart and lungs become stronger and better able to deliver oxygen to muscles at high activity intensities. Still another: hormones in the body of a trained person slow glucose release from the liver and speed up the use of fat instead. These adaptations reward not only trained athletes but all active people; a person who trains by way of aerobic activities such as distance running or cycling becomes well suited to the activity.

Protein Use during Physical Activity—and between Times

Table 14-3 on p. 482 summarized the fuel uses discussed so far, but did not include the third energy-yielding nutrient, protein, because protein is not a major fuel for physical activity. Nevertheless, physically active people use protein just as other people do—to build muscle and other lean tissues and, to some extent, to fuel activity. The body does, however, handle protein differently during activity than during rest.

Protein Used in Muscle Building Synthesis of body proteins is suppressed during activity. In the hours of recovery following activity, though, protein synthesis accelerates beyond normal resting levels. As noted earlier, eating high-carbohydrate foods immediately after exercise accelerates muscle glycogen storage. Similarly, research shows that eating carbohydrate, together with protein, enhances muscle protein synthesis.[38] Remember that the body adapts and builds the molecules, cells, and tissues it needs for the next period of activity. Whenever the body remodels a part of itself, it also tears down old structures to make way for new ones. Repeated activity, with just a slight overload, triggers the protein-dismantling and protein-synthesizing equipment of each muscle cell to make needed changes—that is, to adapt.

The physical work of each muscle cell acts as a signal to its DNA and RNA to begin producing the kinds of proteins that will best support that work. Take jogging, for example. In the first difficult sessions, the body is not yet equipped to perform aerobic work easily, but with each session, the cells' genetic material gets the message that an overhaul is needed. In the hours that follow the session, the genes send molecular messages to the protein-building equipment that tell it what old structures to break down and what new structures to build, and within the limits of its genetic potential, it responds. An athlete may add between ¼ ounce and 1 ounce (between 7 and 28 grams) of body protein to existing muscle mass each day during active muscle-building phases of training. Also among the new structures are more mitochondria to facilitate efficient aerobic metabolism. Over a few weeks' time, remodeling occurs and jogging becomes easier.

Protein Used as Fuel Not only do athletes retain more protein in their muscles, but they also use more protein as fuel: muscles speed up their use of amino acids for energy during physical activity, just as they speed up their use of fat and carbohydrate. Still, protein contributes at most about 10 percent of the total fuel used, both during activity and during rest. The most active people of all, endurance athletes, use up large amounts of all energy fuels, including protein, during performance, but such athletes also eat more food and therefore usually consume enough protein.

Diet Affects Protein Use during Activity The factors that affect how much protein is used during activity seem to be the same three that influence the use of fat and carbohydrate—for one, diet. People who consume diets adequate in energy and rich in *carbohydrate*■ use less protein than those who eat protein- and fat-rich diets. Recall that carbohydrates spare proteins from being broken down to make glucose when needed. Since physical activity requires glucose, a diet lacking in carbohydrate necessitates the conversion of amino acids to glucose. So does a diet high in fat, because fatty acids can never provide glucose.

Intensity and Duration of Activity Affect Protein Use during Activity A second factor, the intensity and duration of activity, also modifies protein use. Endurance athletes who train for over an hour a day, engaging in aerobic activity of moderate intensity and long duration, may deplete their glycogen stores by the end of their workouts and become somewhat more dependent on body protein for energy.

In contrast, anaerobic strength training does not use more protein for energy, but does demand more protein to build muscle. Thus the protein needs of both endurance and strength athletes are higher than those of sedentary people, but certainly not as high as the protein intakes many athletes consume.

Training Affects Protein Use A third factor that influences a person's use of protein during physical activity is the extent of training. Particularly in strength athletes such as bodybuilders, the higher the degree of training, the less protein a person uses during an activity.

Protein Recommendations for Active People As mentioned, all active people, and especially athletes in training, probably need more protein than sedentary people do. Endurance athletes, such as long-distance runners and cyclists, use more protein for fuel than strength or power athletes do, and they retain some, especially in the muscles used for their sport. Strength athletes, such as bodybuilders, and power athletes, such as football players, use less protein for fuel, but they still use some and retain much more. Therefore, *all* athletes in training should attend to protein needs, but should first meet their energy needs with adequate carbohydrate intakes. Otherwise, they will burn off as fuel the very protein that they wish to retain in muscle.

How much protein, then, should an active person consume? Although the DRI Committee does not recommend greater than normal protein intakes for athletes, other authorities do.[39] These recommendations specify different protein intakes for athletes pursuing different activities (see Table 14-4).[40] A later section translates protein recommendations into a diet plan and shows that no one needs protein supplements, or even large servings of meat, to obtain the highest recommended protein intakes. (Chapter 6 concluded that most people receive more than enough protein without supplements and reviewed the potential dangers of using protein and amino acid supplements.)

■ Fitness tip: To conserve protein, eat a diet adequate in energy and rich in carbohydrate.

TABLE 14-4 Recommended Protein Intakes for Athletes

	Recommendations (g/kg/day)	Protein Intakes (g/day)	
		Males	Females
RDA for adults	0.8	56	44
Recommended intake for power (strength or speed) athletes	1.6–1.7	112–119	88–94
Recommended intake for endurance athletes	1.2–1.6	84–112	66–88
U.S. average intake		95	65

NOTE: Daily protein intakes are based on a 70-kilogram (154-pound) man and 55-kilogram (121-pound) woman.
SOURCES: Committee on Dietary Reference Intakes, *Dietary Reference Intakes for Energy, Carbohydrate, Fiber, Fat, Fatty Acids, Cholesterol, Protein and Amino Acids* (Washington, D.C.: National Academies Press, 2002), pp. 10-52–10-53; Position of the American Dietetic Association, Dietitians of Canada, and the American College of Sports Medicine: Nutrition and athletic performance, *Journal of the American Dietetic Association* 100 (2000): 1543–1556.

IN SUMMARY The mixture of fuels the muscles use during physical activity depends on diet, the intensity and duration of the activity, and training. During intense activity, the fuel mix is mostly glucose, whereas during less intense, moderate activity, fat makes a greater contribution. With endurance training, muscle cells adapt to store more glycogen and to rely less on glucose and more on fat for energy. Athletes in training may need more protein than sedentary people do, but they typically eat more food as well and therefore obtain enough protein.

Vitamins and Minerals to Support Activity

Many of the vitamins and minerals assist in releasing energy from fuels and in transporting oxygen. This knowledge has led many people to believe, mistakenly, that vitamin and mineral *supplements* offer physically active people both health benefits and athletic advantages. (Highlight 10 focuses on vitamin and mineral supplements, and Highlight 14 explores supplements and other products people use in the hope of enhancing athletic performance.)

Supplements Nutrient supplements do not enhance the performance of well-nourished people. Deficiencies of vitamins and minerals, however, do impede performance. In general, active people who eat enough nutrient-dense foods to meet energy needs also meet their vitamin and mineral needs. After all, active people eat more food; it stands to reason that with the right choices, they'll get more nutrients.

Athletes who lose weight to meet low body-weight requirements, however, may eat so little food that they fail to obtain all the nutrients they need.[41] The practice of "making weight" is opposed by many health and fitness organizations, but for athletes who choose this course of action, a single daily multivitamin-mineral supplement that provides no more than the DRI recommendations for nutrients may be beneficial.

Some athletes believe that taking vitamin or mineral supplements directly before competition will enhance performance. These beliefs are contrary to scientific reality. Most vitamins and minerals function as small parts of larger working units. After entering the blood, they have to wait for the cells to combine them with their appropriate other parts so that they can do their work. This takes time—hours or days. Vitamins or minerals taken right before an event are useless for improving performance, even if the person is actually suffering deficiencies of them.

In general, then, active people who eat well-balanced meals need no vitamins or minerals in supplement form. Two nutrients, vitamin E and iron, do merit special attention here, however, each for a different reason. Vitamin E is discussed because so many athletes take supplements of it. Iron is discussed because some athletes may be unaware that they need supplements.

Vitamin E Vitamin E and other antioxidant nutrients may be especially effective for athletes exercising in extreme environments, such as heat, cold, and high altitudes. During prolonged, high-intensity physical activity, the muscles' consumption of oxygen increases tenfold or more, enhancing the production of damaging free radicals in the body.[42] Vitamin E is a potent antioxidant that vigorously defends cell membranes against oxidative damage. Some athletes and active people take megadoses of vitamin E in hopes of preventing such oxidative damage to muscles.■ Supplementation with vitamin E does seem to protect against exercise-induced oxidative stress, but there is little evidence that vitamin E supplements can improve performance.[43] Clearly, more research is needed, but in the meantime, active people can benefit by using vegetable oils and eating generous servings of antioxidant-rich fruits and vegetables regularly.

■ Reminder: The Tolerable Upper Intake Level (UL) for vitamin E is 1000 mg per day.

Iron Physically active young women, especially those who engage in endurance activities such as distance running, are prone to iron deficiency.[44] Physical activity can affect iron status in several ways. For one thing, iron losses in sweat can contribute to iron deficiency. For another, physical activity may cause small blood losses through the digestive tract, at least in some athletes. Perhaps more significant than these losses are deficits caused by poor iron absorption in some athletes and the high demands of muscles for the iron-containing electron carriers of the mitochondria and the muscle protein myoglobin.

In contrast, a type of red blood cell destruction associated with physical activity does not impair iron status. Blood cells are squashed when body tissues (such as the soles of the feet) make high-impact contact with an unyielding surface (such as the ground). The iron released from the bursting, or hemolysis, of these red blood cells is recycled, however, not lost in urine. Thus *exertional hemolysis,* as it is called, rarely, if ever, contributes to anemia in athletes.[45]

For perfect functioning, every nutrient is needed.

Iron Deficiency Iron deficiency affects more young women than men, and physically active people are no exception. Habitually low intakes of iron-rich foods and high iron losses through menstruation, as well as through the other routes mentioned, can cause iron deficiency in physically active young women.

Adolescent female athletes who eat vegetarian diets may be particularly vulnerable to iron deficiency.[46] As Chapter 13 explained, the bioavailability of iron is often poor in vegetarian diets. To protect against iron deficiency, vegetarian athletes need to select good dietary sources of iron (fortified cereals, legumes, nuts, and seeds) and include vitamin C–rich foods with each meal.[47] As long as vegetarian athletes, like all athletes, consume enough nutrient-dense foods, they can perform as well as anyone.

Iron-Deficiency Anemia Iron-deficiency anemia impairs physical performance because the hemoglobin in red blood cells is needed to deliver oxygen to the cells for energy metabolism. Whether iron deficiency without clinical signs of anemia impairs physical performance is less clear.[48] Central to the issue is how to define anemia in athletes. Should anemia be defined based on hemoglobin values alone? If so, is there one standard "cutoff point"? If not, what other factors need to be considered? One leading sports medicine expert suggests that anemia is relative and is best defined as a hemoglobin (and ferritin) that is too low for the *individual*.[49] Iron values considered normal for one athlete may be anemic for another. Without adequate oxygen, an active person cannot perform aerobic activities and tires easily.

Sports Anemia Early in training, athletes may develop low blood hemoglobin for a while. This condition, sometimes called **"sports anemia,"** is not a true iron-deficiency condition. Strenuous aerobic activity promotes destruction of the more fragile, older red blood cells, and the resulting cleanup work reduces the blood's iron content temporarily. Strenuous activity also expands the blood's plasma volume, thereby reducing the red blood cell count per unit of blood, but the red blood cells do not diminish in size or number as in anemia, so oxygen-carrying capacity is not hindered. Most researchers view sports anemia as an *adaptive*, temporary response to endurance training. Iron-deficiency anemia requires iron supplementation, but sports anemia does not.

Iron Recommendations for Athletes The best strategy concerning iron depends on the individual. Menstruating women may border on iron deficiency even without the iron losses incurred by physical activity. Active teens of both genders have high iron needs because they are growing. Especially for women and teens, then, prescribed supplements may be needed to correct deficiencies of iron, but blood tests should guide this decision. (Chapter 13 provides many more details about iron and the tests used in assessing its status.)

IN SUMMARY With the possible exception of iron, well-nourished active people and athletes do not need nutrient supplements. Female athletes need to pay special attention to their iron needs.

sports anemia: a transient condition of low hemoglobin in the blood, associated with the early stages of sports training or other strenuous activity.

To prevent dehydration and the fatigue that accompanies it, drink plenty of liquids before, during, and after physical activity.

■ Note: 10 degrees on the Celsius scale is about 18 degrees on the Fahrenheit scale.

■ Fitness tip: To prevent heat stroke, drink fluids, rest when tired, and wear appropriate clothing.

■ Symptoms of dehydration and heat stroke:
- Headache.
- Nausea.
- Dizziness.
- Clumsiness.
- Stumbling.
- Sudden cessation of sweating (hot, dry skin).
- Confusion or other mental changes.

■ Fitness tip: To prevent hypothermia, drink fluids and wear appropriate clothing.

hyperthermia: an above-normal body temperature.

heat stroke: a dangerous accumulation of body heat with accompanying loss of body fluid.

hypothermia: a below-normal body temperature.

Fluids and Electrolytes to Support Activity

The need for water far surpasses the need for any other nutrient. The body relies on watery fluids as the medium for all of its life-supporting activities, and if it loses too much water, its well-being becomes compromised.

Obviously, the body loses water via sweat. Breathing uses water, too, exhaled as vapor. During physical activity, water losses from both routes are significant, and dehydration becomes a threat. Dehydration's first symptom is fatigue: a water loss of even 1 to 2 percent of body weight can reduce a person's capacity to do muscular work.[50] With a water loss of about 7 percent, a person is likely to collapse.

Fluid Losses via Sweat Recall from Chapter 7 that working muscles produce heat as a by-product of energy metabolism. During intense activity, muscle heat production can be 15 to 20 times greater than at rest. The body cools itself by sweating. Each liter of sweat dissipates almost 600 kcalories of heat, preventing a rise in body temperature of almost 10 degrees■ on the Celsius scale. The body routes its blood supply through the capillaries just under the skin, and the skin secretes sweat to evaporate and cool the skin and the underlying blood. The blood then flows back to cool the deeper body chambers.

Hyperthermia In hot, humid weather, sweat doesn't evaporate well because the surrounding air is already laden with water. In **hyperthermia,** body heat builds up and triggers maximum sweating, but without sweat evaporation, little cooling takes place. In such conditions, active people must take precautions to prevent **heat stroke.** To reduce the risk of heat stroke,■ drink enough fluid before and during the activity, rest in the shade when tired, and wear lightweight clothing that allows sweat to evaporate.[51] (Hence the danger of rubber or heavy suits that supposedly promote weight loss during physical activity—they promote profuse sweating, prevent sweat evaporation, and invite heat stroke.) If you ever experience any of the symptoms of heat stroke listed in the margin,■ stop your activity, sip fluids, seek shade, and ask for help. Heat stroke can be fatal, young people often die of it, and these symptoms demand attention.

Hypothermia In cold weather, **hypothermia,** or low body temperature, can pose as serious a threat as heat stroke in hot weather. Inexperienced, slow runners participating in long races on cold or wet, chilly days are especially vulnerable to hypothermia. Slow runners who produce little heat can become too cold if clothing is inadequate. Early symptoms of hypothermia include shivering and euphoria. As body temperature continues to fall, shivering may stop, and weakness, disorientation, and apathy may occur. Each of these symptoms can impair a person's ability to act against a further drop in body temperature. Even in cold weather, however, the active body still sweats and still needs fluids.■ The fluids should be warm or at room temperature to help protect against hypothermia.

Fluid Replacement via Hydration Endurance athletes can easily lose 1.5 liters or more of fluid during *each hour* of activity. To prepare for fluid losses, a person must hydrate before activity. To replace fluid losses,■ the person must rehydrate during and after activity. (Table 14-5 presents one schedule of hydration for physical activity.) Even then, in hot weather, the GI tract may not be able to absorb enough water fast enough to keep up with sweat losses, and some degree of dehydration may be inevitable.

Athletes who are preparing for competition are often advised to drink extra fluids in the *days* immediately before the event, especially if they are still training. The extra water is not stored in the body, but drinking extra water ensures maximum hydration at the start of the event. Full hydration is imperative for every athlete both in training and in competition. The athlete who arrives at an event even slightly dehydrated arrives with a disadvantage.

What is the best fluid for an exercising body? For noncompetitive, everyday active people, plain, cool water is recommended, especially in warm weather, for two reasons: it rapidly leaves the digestive tract to enter the tissues where it is needed,

TABLE 14-5	Hydration Schedule for Physical Activity
When to Drink	**Amount of Fluid**
2 hr before activity	2 to 3 c
15 min before activity	1 to 2 c
Every 15 min during activity	1 to 1½ c
After activity	2 to 3 c for each pound of body weight lost

SOURCE: R. Murray, Fluid and electrolytes, in *Sports Nutrition: A Guide for the Professional Working with Active People,* 3rd ed., ed. C. A. Rosenbloom (Chicago: The American Dietetic Association, 2000), pp. 95–106; National Athletic Trainer's Association Position Statement: Fluid replacement for athletes, *Journal of Athletic Training* 35 (2000): 212–224.

Although water is the best fluid for most physically active people, many good-tasting sports drinks are available.

and it cools the body from the inside out. For endurance athletes, carbohydrate-containing beverages may be appropriate. Fluid ingestion during the event has the dual purposes of replenishing water lost through sweating and providing a source of carbohydrate to supplement the body's limited glycogen stores. Carbohydrate depletion brings on fatigue in the athlete, but as already mentioned, fluid loss and the accompanying buildup of body heat can be life-threatening. Thus the first priority for endurance athletes should be to replace fluids. Many good-tasting drinks are marketed for active people; the "How to" on p. 492 compares them with water.

Electrolyte Losses and Replacement When a person sweats, small amounts of electrolytes—the electrically charged minerals sodium, potassium, chloride, and magnesium—are lost from the body along with water. Losses are greatest in beginners; training improves electrolyte retention.

To replenish lost electrolytes, a person ordinarily needs only to eat a regular diet that meets energy and nutrient needs. In events lasting more than one hour, sports drinks may be needed to replace fluids and electrolytes. Salt tablets can worsen dehydration and impair performance; they increase potassium losses, irritate the stomach, and cause vomiting.

Hyponatremia When athletes compete in endurance sports lasting longer than three hours, replenishing electrolytes becomes crucial. If athletes sweat profusely over a long period of time and do not replace lost sodium, a dangerous condition known as **hyponatremia** may result. Recent research shows that some athletes who sweat profusely may also lose more sodium in their sweat than others—and are prone to debilitating heat cramps.[52] These athletes lose twice as much sodium in sweat as athletes who don't cramp. Depending on individual variation, exercise intensity, and changes in ambient temperature and humidity, sweat rates can exceed 3 liters per hour.[53]

Hyponatremia may also occur when endurance athletes drink such large amounts of water over the course of a long event that they overhydrate, diluting the body's fluids to such an extent that the sodium concentration becomes extremely low. During long competitions, when athletes lose sodium through heavy sweating and consume excessive amounts of liquids, especially water, hyponatremia becomes likely.

Some athletes may still be vulnerable to hyponatremia even when they drink sports drinks during an event.[54] Sports drinks do contain sodium, but as the "How to" points out, the sodium content of sports drinks is low and, in some cases, too low to replace sweat losses. Still, sports drinks do offer more sodium than plain water.

To prevent hyponatremia, athletes need to replace sodium during prolonged events. They should favor sports drinks over water and eat pretzels in the last half of a long race.[55] Some may need beverages with higher sodium concentrations than commercial sports drinks. In the days before the event, especially an event in the heat, athletes should not restrict salt in their diets. The symptoms of hyponatremia are similar to, but not the same as, those of dehydration (see the margin). ■

■ Symptoms of hyponatremia:
- Severe headache.
- Vomiting.
- Bloating, puffiness from water retention (shoes tight, rings tight).
- Confusion.
- Seizure.

hyponatremia (HIGH-poe-na-TREE-mee-ah): a decreased concentration of sodium in the blood.
- **hypo** = below
- **natrium** = sodium (Na)
- **emia** = blood

HOW TO Evaluate Sports Drinks

Hydration is critical to optimal performance. Water best meets the fluid needs of most people, yet manufacturers market many good-tasting sports drinks for active people. More than 20 "power beverages" compete for their share of the $1 billion market. What do sports drinks have to offer?

- *Fluid.* Sports drinks offer fluids to help offset the loss of fluids during physical activity, but plain water can do this, too. Alternatively, diluted fruit juices or flavored water can be used, if preferred to plain water.

- *Glucose.* Sports drinks offer simple sugars or **glucose polymers** that help maintain hydration and blood glucose and enhance performance as effectively as, or maybe even better than, water. Such measures are especially beneficial for strenuous endurance activities lasting longer than 45 minutes, during intense activities, or during prolonged competitive games that demand repeated intermittent activity.[a] Sports drinks are also suitable for events

lasting less than 45 minutes although plain water is appropriate as well.[b]

Fluid transport to the tissues from beverages containing up to 8 percent glucose is rapid. Most sports drinks contain about 7 percent carbohydrate (about half the sugar of ordinary soft drinks, or about 5 teaspoons in each 12 ounces). Less than 6 percent may not enhance performance, and more than 8 percent may cause abdominal cramps, nausea, and diarrhea.

Although glucose does enhance endurance performance in strenuous competitive events, for the moderate exerciser, it can be counterproductive if weight loss is the goal. Glucose is sugar, and like candy, it provides only empty kcalories—no vitamins or minerals. Most sports drinks provide between 50 and 100 kcalories per cup.

- *Sodium and other electrolytes.* Sports drinks offer sodium and other electrolytes to help replace those lost during physical activity. Sodium in sports drinks also helps to increase the rate of fluid absorption from the

GI tract and maintain plasma volume during activity and recovery.

Most physically active people do not need to replace the minerals lost in sweat immediately; a meal eaten within hours of competition replaces these minerals soon enough. Most sports drinks are relatively low in sodium, however, so those who choose to use these beverages run little risk of excessive intake.

- *Good taste.* Manufacturers reason that if a drink tastes good, people will drink more, thereby ensuring adequate hydration. For athletes who prefer the flavors of sports drinks over water, it may be worth paying for good taste to replace lost fluids.

- *Psychological edge.* Sports drinks provide a psychological edge for some people who associate the drinks with athletes and sports. The need to belong is valid. If the drinks boost morale and are used with care, they may do no harm.

For athletes who exercise for 45 minutes or more, sports drinks provide an advantage over water. For most physically active people, though, water is the best fluid to replenish lost fluids. The most important thing to do is drink—even if you don't feel thirsty.

[a] E. Coleman, Fluid replacement for athletes, *Sports Medicine Digest* 25 (2003): 76–77; Inter-association Task Force on Exertional Heat Illness Consensus Statement, *NATA News*, June 2003.

[b] Position of the American Dietetic Association, Dietitians of Canada, and the American College of Sports Medicine: Nutrition and athletic performance, *Journal of the American Dietetic Association* 100 (2000): 1543–1556.

■ Beer facts:
- *Beer is not carbohydrate-rich.* Beer is kcalorie-rich, but only ⅓ of its kcalories are from carbohydrates. The other ⅔ are from alcohol.
- *Beer is mineral-poor.* Beer contains a few minerals, but to replace those lost in sweat, athletes need good sources such as fruit juices.
- *Beer is vitamin-poor.* Beer contains traces of some B vitamins, but it cannot compete with food sources.
- *Beer causes fluid losses.* Beer is a fluid, but alcohol is a diuretic and causes the body to lose valuable fluid.

glucose polymers: compounds that supply glucose, not as single molecules, but linked in chains somewhat like starch. The objective is to attract less water from the body into the digestive tract (osmotic attraction depends on the number, not the size, of particles).

Poor Beverage Choices: Caffeine and Alcohol

Athletes, like others, sometimes drink beverages that contain caffeine or alcohol. Each of these substances can influence physical performance.

Caffeine Caffeine is a stimulant, and athletes sometimes use it to enhance performance as Highlight 14 explains. Carbonated soft drinks, whether they contain caffeine or not, may not be a wise choice for athletes: bubbles make a person feel full quickly and so limit fluid intake.

Alcohol Some athletes mistakenly believe that they can replace fluids and load up on carbohydrates by drinking beer.■ A 12-ounce beer provides 13 grams of carbohydrate—one-third the amount of carbohydrate in a glass of orange juice the same size. In addition to carbohydrate, beer also contains alcohol, of course. Energy from alcohol breakdown generates heat, but does not fuel muscle work because alcohol is metabolized in the liver.

It is difficult to overstate alcohol's detrimental effects on physical activity. Alcohol's diuretic effect impairs the body's fluid balance, making dehydration likely; after physical activity, a person needs to replace fluids, not lose them by drinking beer. Alcohol also impairs the body's ability to regulate its temperature, increasing the likelihood of hypothermia or heat stroke.

Alcohol also alters perceptions; slows reaction time; reduces strength, power, and endurance; and hinders accuracy, balance, eye-hand coordination, and coordination in general—all opposing optimal athletic performance. In addition, it deprives people of their judgment, thereby compromising their safety in sports; many sports-related fatalities and injuries involve alcohol or other drugs.

Clearly, alcohol impairs performance, but physically active people do drink on occasion. A word of caution: do not drink alcohol before exercising, and drink plenty of water after exercising before drinking alcohol.

IN SUMMARY Active people need to drink plenty of water; endurance athletes need to drink both water and carbohydrate-containing beverages, especially during training and competition. During events lasting longer than three hours, athletes need to pay special attention to replacing sodium losses to prevent hyponatremia.

Diets for Physically Active People

No one diet best supports physical performance. Active people who choose foods within the framework of the diet-planning principles presented in Chapter 2 can design many excellent diets.

Choosing a Diet to Support Fitness

First, remember that water is depleted more rapidly than any other nutrient. A diet to support fitness must provide water, energy, and all the other nutrients.

Water Even casual exercisers must attend conscientiously to their fluid needs. Physical activity blunts the thirst mechanism, especially in cold weather. During activity, thirst signals come too late, so don't wait to feel thirsty before drinking. To find out how much water is needed to replenish activity losses, weigh yourself before and after the activity—the difference is almost all water. One pound equals roughly 2 cups (500 milliliters) of fluid.

Nutrient Density A healthful diet is based on nutrient-dense foods—foods that supply adequate vitamins and minerals for the energy they provide. Active people need to eat both for nutrient adequacy and for energy. A diet that is high in carbohydrate (60 to 70 percent of total kcalories), moderate in fat (20 to 30 percent), and adequate in protein (10 to 20 percent) ensures full glycogen and other nutrient stores.

Carbohydrate On two occasions, the active person's regular high-carbohydrate, fiber-rich diet may require temporary adjustment. Both of these exceptions involve training for competition rather than fitness. One special occasion is the pregame meal, when fiber-rich, bulky foods are best avoided. The pregame meal is discussed in a later section.

The other occasion is during intensive training, when energy needs may be so high as to outstrip the person's capacity to eat enough food to meet them. At that point, added sugar and fat may be needed. The person can add concentrated carbohydrate foods, such as dried fruits, sweet potatoes, and nectars, and even high-fat foods, such as avocados, nuts, cookies, and ice cream. Still, a nutrient-rich diet remains central for adequacy's sake. Though vital, energy alone is not enough to support performance.

Some athletes use commercial high-carbohydrate liquid supplements to obtain the carbohydrate and energy needed for heavy training and top performance. These supplements do not *replace* regular food; they are meant to be used in *addition* to it. Unlike the sports beverages discussed in the "How to" on p. 492, these high-carbohydrate supplements are too concentrated in carbohydrate to be used for fluid replacement.

■ Carbohydrate recommendation for athletes in heavy training: 8 to 10 g/kg body weight.

A variety of foods is the best source of nutrients for athletes.

Protein In addition to carbohydrate and some fat (and the energy they provide), physically active people need protein. How much of what kinds of foods supply enough protein to meet their needs? Meats and milk products are rich protein sources, but to recommend that active people emphasize these foods would be narrow advice. As mentioned repeatedly, active people need diets rich in carbohydrate, and of course, meats have none to offer. Legumes, whole grains, and vegetables provide protein with abundant carbohydrate. Table 14-4 (on p. 487) showed recommended protein intakes for active people.

A Performance Diet Example A person who engages in vigorous physical activity on a daily basis could easily require more than 3000 kcalories per day. To meet this need, the person can choose a variety of nutrient-dense foods. Figure 14-4 shows one example of meals that provide 3300 kcalories. These meals supply about 125 grams of protein, equivalent to the highest recommended intake for an athlete weighing 160 pounds. Obviously, the more energy a person requires, the more protein that person will receive, assuming the foods chosen are nutrient dense. This relationship between energy and protein intakes breaks down only when people meet their energy needs with high-fat, high-sugar confections. The meals shown in Figure 14-4 provide almost 520 grams of carbohydrate, or over 60 percent of total kcalories. Athletes who train exhaustively for endurance events may want to aim for somewhat higher carbohydrate intakes. Beyond these specific concerns of total energy, protein, and carbohydrate, the diet most beneficial to athletic performance is remarkably similar to the diet recommended for most people.[56]

Meals before and after Competition

No single food improves speed, strength, or skill in competitive events, although some *kinds* of foods do support performance better than others as already explained. Still, a competitor may eat a particular food before or after an event for psychological reasons. One eats a steak the night before wrestling, while another

---

FIGURE 14-4 An Athlete's Meal Selections

Breakfast
1 c shredded wheat with
  low-fat milk and banana.
2 slices whole-wheat toast
  with jelly.
1½ c orange juice.

Lunch
2 turkey sandwiches.
1½ c low-fat milk.
Large bunch of grapes.

Snack
3 c plain popcorn.
A smoothie made from:
  1½ c apple juice.
  1½ frozen banana.

Dinner
Salad: 1 c spinach, carrots,
  and mushrooms with
  ½ c garbanzo beans,
  1 tbs sunflower seeds, and
  1 tbs ranch salad dressing.
1 c spaghetti with meat sauce.
1 c green beans.
1 corn on the cob.
2 slices Italian bread.
4 tsp butter.
1 piece angel food cake with
  fresh strawberries and
  whipping cream
1 c low-fat milk.

Total kcal: 3000

63% kcal from carbohydrate
22% kcal from fat
15% kcal from protein

All vitamin and mineral intakes exceed
the RDA for both men and women.

takes some honey five minutes after diving. As long as these practices remain harmless, they should be respected.

Pregame Meals Science indicates that the pregame meal or snack should include plenty of fluids and be light and easy to digest. It should provide between 300 and 800 kcalories, primarily from carbohydrate-rich foods that are familiar and well tolerated by the athlete. The meal should end three to four hours before competition to allow time for the stomach to empty before exertion.

Breads, potatoes, pasta, and fruit juices—that is, carbohydrate-rich foods low in fat and fiber—form the basis of the best pregame meal (see Figure 14-5 for some examples). Bulky, fiber-rich foods such as raw vegetables or high-bran cereals, although usually desirable, are best avoided just before competition. Fiber in the digestive tract attracts water and can cause stomach discomfort during performance. Liquid meals■ are easy to digest, and many such meals are commercially available. Alternatively, athletes can mix fat-free milk or juice, frozen fruits, and flavorings in a blender.

Postgame Meals As mentioned earlier, eating high-carbohydrate foods *after* physical activity enhances glycogen storage. Since people are usually not hungry immediately following physical activity, carbohydrate-containing beverages such as sports drinks or fruit juices may be preferred. If an active person does feel hungry after an event, then foods high in carbohydrate and low in protein, fat, and fiber are the ones to choose—the same ones recommended prior to competition.

■ High-carbohydrate, liquid pregame meal ideas:
- Apple juice, frozen banana, and cinnamon.
- Papaya juice, frozen strawberries, and mint.
- Fat-free milk, frozen banana, and vanilla.

IN SUMMARY The person who wants to excel physically will apply accurate nutrition knowledge along with dedication to rigorous training. A diet that provides ample fluid and includes a variety of nutrient-dense foods in quantities to meet energy needs will enhance not only athletic performance, but overall health as well. Carbohydrate-rich foods that are light and easy-to-digest are recommended for both the pregame and the postgame meal. Training and genetics being equal, who will win a competition—the athlete who habitually consumes inadequate amounts of needed nutrients or the competitor who arrives at the event with a long history of full nutrient stores and well-met metabolic needs?

FIGURE 14-5 | Examples of High-Carbohydrate Pregame Meals

Pregame meals should be eaten three to four hours before the event and provide 300 to 800 kcalories, primarily from carbohydrate-rich foods. Each of these sample meals provides at least 65% of total kcalories from carbohydrate.

Matthew Farruggio (all)

300-kcalorie meal
1 large apple
4 saltine crackers
1½ tbs reduced-fat peanut butter

500-kcalorie meal
1 large whole-wheat bagel
2 tbs jelly
1½ c low-fat milk

750-kcalorie meal
1 large baked potato
2 tsp margarine
1 c steamed broccoli
1 c mixed carrots and green peas
5 vanilla wafers
1½ c apple or pineapple juice

Some athletes learn that nutrition can support physical performance and turn to pills and powders instead of foods. In case you need further convincing that a healthful diet surpasses such potions, the following highlight addresses this issue.

Nutrition in Your Life

The foods and beverages you eat and drink provide fuel and other nutrients to support your physical activity.

- Are you physically active for at least 30 minutes, and preferably 60 minutes, a day on most or all days of the week?
- Do you eat a carbohydrate-rich diet regularly?
- Do you drink fluids, especially water, before, during, and after physical activity?

NUTRITION ON THE NET

 Access these websites for further study of topics covered in this chapter.

- Find updates and quick links to these and other nutrition-related sites at our website: **www.wadsworth.com/nutrition**
- Visit the U.S. Government site: **www.fitness.gov**
- Search for "physical fitness" at the American College of Sports Medicine information site: **www.acsm.org**
- Review the Surgeon General's Report on Physical Activity: **www.cdc.gov/nccdphp/sgr/sgr.htm**
- Review resources offered on the Nutrition and Physical Activity site from the Centers for Disease Control and Prevention: **www.cdc.gov/nccdphp/dnpa**

- Learn about the President's Council on Physical Fitness and Sports: **www.presidentschallenge.com**
- Visit Shape Up America: **www.shapeup.org**
- Visit the American Council on Exercise (ACE): **www.acefitness.org**
- Visit the Sport Medicine and Science Council of Canada: **www.smscc.ca**
- Find fitness information at the Cooper Institute for Aerobics Research: **www.cooperinst.org**
- Find information on sports drinks and other nutrition and fitness topics at the Gatorade Sports Science Institute site: **www.gssiweb.com**

STUDY QUESTIONS

These questions will help you review the chapter. You will find the answers in the discussions on the pages provided.

1. Define fitness, and list its benefits. (pp. 473–476)
2. Explain the overload principle. (p. 477)
3. Define cardiorespiratory conditioning and list some of its benefits. (p. 479)
4. What types of activity are anaerobic? Which are aerobic? (pp. 481–482)
5. Describe the relationships among energy expenditure, type of activity, and oxygen use. (pp. 481–482)
6. What factors influence the body's use of glucose during physical activity? How? (pp. 482–485)
7. What factors influence the body's use of fat during physical activity? How? (pp. 485–486)

8. What factors influence the body's use of protein during physical activity? How? (pp. 486–487)
9. Why are some athletes likely to develop iron-deficiency anemia? Compare iron-deficiency anemia and sports anemia, explaining the differences. (pp. 488–489)
10. Discuss the importance of hydration during training, and list recommendations to maintain fluid balance. (pp. 490–492)
11. Describe the components of a healthy diet for athletic performance. (pp. 493–494)

These multiple choice questions will help you prepare for an exam. Answers can be found on p. 498.

1. Physical inactivity is linked to all of the following diseases except:

a. cancer.

b. diabetes.

c. emphysema.

d. hypertension.

2. The progressive overload principle can be applied by performing:

a. an activity less often.

b. an activity with more intensity.

c. an activity in a different setting.

d. a different activity each day of the week.

3. The process that regenerates glucose from lactic acid is known as the:

a. Cori cycle.

b. ATP-CP cycle.

c. adaptation cycle.

d. cardiac output cycle.

4. "Hitting the wall" is a term runners sometimes use to describe:

a. dehydration.

b. competition.

c. indigestion.

d. glucose depletion.

5. The technique endurance athletes use to maximize glycogen stores is called:

a. aerobic training.

b. muscle conditioning.

c. carbohydrate loading.

d. progressive overloading.

6. Conditioned muscles rely less on ____ and more on ____ for energy.

a. protein; fat

b. fat; protein

c. glycogen; fat

d. fat; glycogen

7. Vitamin or mineral supplements taken right before an event are useless for improving performance because the:

a. athlete sweats the nutrients out during the event.

b. stomach can't digest supplements during physical activity.

c. nutrients are diluted by all the fluids the athlete drinks.

d. body needs hours or days for the nutrients to do their work.

8. Physically active young women, especially those who are endurance athletes, are prone to:

a. energy excess.

b. iron deficiency.

c. protein overload.

d. vitamin A toxicity.

9. The body's need for ____ far surpasses its need for any other nutrient.

a. water

b. protein

c. vitamins

d. carbohydrate

10. A recommended pregame meal includes plenty of fluids and provides between:

a. 300 and 800 kcalories, mostly from fat-rich foods.

b. 50 and 100 kcalories, mostly from fiber-rich foods.

c. 1000 and 2000 kcalories, mostly from protein-rich foods.

d. 300 and 800 kcalories, mostly from carbohydrate-rich foods.

REFERENCES

1. J. Myers and coauthors, Exercise capacity and mortality among men referred for exercise testing, *New England Journal of Medicine* 346 (2002): 793–801; I. M. Lee and R. S. Paffenbarger, Associations of light, moderate, and vigorous intensity physical activity with longevity: The Harvard Alumni Health Study, *American Journal of Epidemiology* 151 (2000): 293–299; F. B. Hu and coauthors, Walking compared with vigorous physical activity and risk of type 2 diabetes in women, *Journal of the American Medical Association* 282 (1999): 1433–1439.

2. P. M. Barnes and C. A. Schoenborn, *Physical Activity among Adults: United States, 2000,* Advance Data from Vital and Health Statistics document no. 333 (2003), available from www.cdc.gov/nchs/Default.htm; American College of Sports Medicine, *ACSM's Guidelines for Exercise Testing and Prescription,* 6th ed. (Philadelphia: Williams & Wilkins, 2000), pp. vii-ix.

3. I. Lee and coauthors, Physical activity and coronary heart disease in women: Is "no pain, no gain" passé? *Journal of the American Medical Association* 285 (2001): 1447–1454; K. Moreau and coauthors, Increasing daily walking lowers blood pressure in postmenopausal women, *Medi-*

cine and Science in Sports and Exercise 33 (2001): 1825–1831; F. W. Booth and coauthors, Waging war on modern chronic diseases: Primary prevention through exercise biology, *Journal of Applied Physiology* 88 (2000): 774–787.

4. G. A. Colditz, Economic costs of obesity and inactivity, *Medicine and Science in Sports and Exercise* 31 (1999): S663–S667.

5. U. G. Kyle and coauthors, Physical activity and fat-free and fat mass by bioelectrical impedance in 3853 adults, *Medicine and Science in Sports and Exercise* 33 (2001): 576–584.

6. L. Metcalfe and coauthors, Postmenopausal women and exercise for prevention of osteoporosis: The Bone, Estrogen, Strength Training (BEST) Study, *ACSM's Health and Fitness Journal,* May/June 2001, pp. 6–14; K. Delvaux and coauthors, Bone mass and lifetime physical activity in Flemish males: A 27-year follow-up study, *Medicine and Science in Sports and Exercise* 33 (2001): 1868–1875.

7. C. E. Matthews and coauthors, Moderate to vigorous physical activity and risk of upper-respiratory tract infection, *Medicine and Science in Sports and Exercise* 34 (2002): 1242–1248.

8. C. M. Friedenreich, Physical activity and cancer: Lessons learned from nutritional epidemiology, *Nutrition Reviews* 59 (2001): 349–357; M. E. Martinez and coauthors, Physical activity, body mass index, and prostaglandin E_2 levels in rectal mucosa, *Journal of the National Cancer Institute* 91 (1999): 950–953.

9. J. E. Manson and coauthors, Walking compared with vigorous exercise for the prevention of cardiovascular events in women, *New England Journal of Medicine* 347 (2002): 716–725; C. D. Lee and S. N. Blair, Cardiorespiratory fitness and stroke mortality in men, *Medicine and Science in Sports and Exercise* 34 (2002): 592–595; Y.-Z. Kimberly and coauthors, Physical activity and cardiovascular risk factors in a developing population, *Medicine and Science in Sports and Exercise* 33 (2001): 1598–1604; F. B. Hu and coauthors, Physical activity and risk of stroke in women, *Journal of the American Medical Association* 283 (2000): 2961–2967.

10. A. Trichopoulou and coauthors, Physical activity and energy intake selectively predict the waist-to-hip ratio in men but not in women, *American Journal of Clinical Nutrition* 74 (2001): 574–578.

11. R. M. van Dam and coauthors, Physical activity and glucose tolerance in elderly men: The Zutphen Elderly Study, *Medicine and Science in Sports and Exercise* 34 (2002): 1132–1136; K. J. Stewart, Exercise training and the cardiovascular consequences of type 2 diabetes and hypertension: Plausible mechanisms for improving cardiovascular health, *Journal of the American Medical Association* 288 (2002): 1622–1631; J. Tuomilehto and coauthors, Prevention of type 2 diabetes mellitus by changes in lifestyle among subjects with impaired glucose tolerance, *New England Journal of Medicine* 344 (2001): 1343–1350; American College of Sports Medicine, Position stand: Exercise and type 2 diabetes, *Medicine and Science in Sports and Exercise* 32 (2000): 1345–1360.

12. G. Misciagna and coauthors, Diet, physical activity, and gallstones—a population-based, case-control study in southern Italy, *American Journal of Clinical Nutrition* 69 (1999): 120–126; M. F. Leitzmann and coauthors, Recreational physical activity and the risk of cholecystectomy in women, *New England Journal of Medicine* 341 (1999): 777–784.

13. W. J. Strawbridge and coauthors, Physical activity reduces the risk of subsequent depression for older adults, *American Journal of Epidemiology* 156 (2002): 328–334.

14. Myers and coauthors, 2002; Lee and Paffenbarger, 2000.

15. T. Rantanen and coauthors, Midlife hand grip strength as a predictor of old age disability, *Journal of the American Medical Association* 281 (1999): 558–560; American College of Sports Medicine, Position stand: Exercise and physical activity for older adults, *Medicine and Science in Sports and Exercise* 30 (1998): 992–1008.

16. American College of Sports Medicine, *ACSM's Guidelines*, 2000, pp. 137–164; U.S. Department of Agriculture, U.S. Department of Health and Human Services, *Dietary Guidelines for Americans* (Washington, D.C.: Government Printing Office, 2000).

17. Committee on Dietary Reference Intakes, *Dietary Reference Intakes for Energy, Carbohydrate, Fiber, Fat, Fatty Acids, Cholesterol, Protein and Amino Acids* (Washington, D.C.: National Academies Press, 2002), pp. 12-1–12-39.

18. W. D. Schmidt, C. J. Biwer, and L. K. Kalscheuer, Effects of long *versus* short bout exercise on fitness and weight loss in overweight females, *Journal of the American College of Nutrition* 20 (2001): 494–501.

19. American College of Sports Medicine, Position stand: The recommended quantity and quality of exercise for developing and maintaining cardiorespiratory and muscular fitness, and flexibility in healthy adults, *Medicine and Science in Sports and Exercise* 30 (1998): 975–991.

20. Manson and coauthors, 2002; Lee and Blair, 2002; S. Carroll, C. B. Cooke, and R. J. Butterly, Metabolic clustering, physical activity and fitness in nonsmoking, middle-aged men, *Medicine and Science in Sports and Exercise* 32 (2000): 2079–2086.

21. American College of Sports Medicine, *ACSM's Guidelines*, 2000, pp. 3–21.

22. American College of Sports Medicine, *ACSM's Guidelines*, 2000, pp. 22–32.

23. American College of Sports Medicine, Position stand: Progression models in resistance training for healthy adults, *Medicine and Science in Sports and Exercise* 34 (2002): 364–380; M. L. Pollock and coauthors, AHA Science Advisory: Resistance exercise in individuals with and without cardiovascular disease: Benefits, rationale, safety, and prescription, *Circulation* 101 (2000): 828–833.

24. J. E. Layne and M. E. Nelson, The effects of progressive resistance training on bone density: A review, *Medicine and Science in Sports and Exercise* 31 (1999): 25–30.

25. E. C. Cussler and coauthors, Weight lifted in strength training predicts bone change in postmenopausal women, *Medicine and Science in Sports and Exercise* 35 (2003): 10–17; Metcalfe and coauthors, 2001.

26. J. H. Wilmore and D. L. Costill, Physical energy: Fuel metabolism, *Nutrition Reviews* 59 (2001): S13–S16.

27. J. Bergstrom and coauthors, Diet, muscle glycogen and physical performance, *Acta Physiologica Scandanavica* 71 (1967): 140–150.

28. R. S. Welsh and coauthors, Carbohydrates and physical/mental performance during intermittent exercise to fatigue, *Medicine and Science in Sports and Exercise* 34 (2002): 723–731.

29. Position of the American Dietetic Association, Dietitians of Canada, and the American College of Sports Medicine: Nutrition and athletic performance, *Journal of the American Dietetic Association* 100 (2000): 1543–1556; American College of Sports Medicine, Position stand, Exercise and fluid replacement, *Medicine and Science in Sports and Exercise* 28 (1996): i–vii.

30. Position of the American Dietetic Association, Dietitians of Canada, and the American College of Sports Medicine, 2000.

31. E. Coleman, Carbohydrate and exercise, in *Sports Nutrition: A Guide for the Professional Working with Active People*, 3rd ed., ed. C. A. Rosenbloom (Chicago: The American Dietetic Association, 2000), pp. 13–31.

32. J. Manetta and coauthors, Fuel oxidation during exercise in middle-aged men: Role of training and glucose disposal, *Medicine and Science in Sports and Exercise* 34 (2002): 423–429.

33. L. M. Burke and J. A. Hawley, Effects of short-term fat adaptation on metabolism and performance of prolonged exercise, *Medicine and Science in Sports and Exercise* 34 (2002): 1492–1498.

34. Burke and Hawley, 2002; L. M. Burke and coauthors, Adaptations to short-term high-fat diet persist during exercise despite high carbohydrate availability, *Medicine and Science in Sports and Exercise* 34 (2002): 83–91; A. L. Staudacher and coauthors, Effects of fat adaptation and carbohydrate restoration on prolonged endurance exercise, *Journal of Applied Physiology* 91 (2001): 115–122.

35. J. W. Helge, Long-term fat adaptation, effects on performance, training capacity, and fat utilization, *Medicine and Science in Sports and Exercise* 34 (2002): 1499–1504; N. D. Stepto and coauthors, Effect of short-term fat adaptation on high-intensity training, *Medicine and Science in Sports and Exercise* 34 (2002): 449–455.

36. E. Coleman, Does a low-fat diet impair nutrition and performance? *Sports Medicine Digest* 22 (2000): 41; Position of the American Dietetic Association, Dietitians of Canada, and the American College of Sports Medicine, 2000.

37. Wilmore and Costill, 2001.

38. M. Suzuki, Glycemic carbohydrates consumed with amino acids or protein right after exercise enhance muscle formation, *Nutrition Reviews* 61 (2003): S88–S94.

39. Committee on Dietary Reference Intakes, 2002; Position of the American Dietetic Association, Dietitians of Canada, and the American College of Sports Medicine, 2000.

40. Position of the American Dietetic Association, Dietitians of Canada, and the American College of Sports Medicine, 2000.

41. Position of the American Dietetic Association, Dietitians of Canada, and the American College of Sports Medicine, 2000.

42. S. K. Powers, L. L. Ji, and C. Leeuwenburgh, Exercise training-induced alterations in skeletal muscle antioxidant capacity: A brief review, *Medicine and Science in Sports and Exercise* 31 (1999): 987–997.

43. M. L. Urso and P. M. Clarkson, Oxidative stress, exercise, and antioxidant supplementation. *Toxicology* 189 (2003): 41–54; W. J. Evans, Vitamin E, vitamin C and exercise, *American Journal of Clinical Nutrition* 72 (2000): 647S–652S.

44. J. Beard and B. Tobin, Iron status and exercise, *American Journal of Clinical Nutrition* 72 (2000): 594S–597S.

45. E. R. Eichner, Non-anemias in athletes—Sports anemia and footstrike hemolysis: Friends not foes? *Sports Medicine Digest* 23 (2001): 53.

46. Position of the American Dietetic Association, Dietitians of Canada, and the American College of Sports Medicine, 2000; Beard and Tobin, 2000.

47. D. C. Nieman, Physical fitness and vegetarian diets: Is there a relation? *American Journal of Clinical Nutrition* 70 (1999): 570S–575S.

48. T. Brownlie and coauthors, Marginal iron deficiency without anemia impairs aerobic adaptation among previously untrained women, *American Journal of Clinical Nutrition* 75 (2002): 734–742; E. R. Eichner, Anemia in female athletes, *Sports Medicine Digest* 22 (2000): 42–43.

49. Eichner, 2000.

50. M. N. Sawka and S. J. Montain, Fluid and electrolyte supplementation for exercise heat stress, *American Journal of Clinical Nutrition* 72 (2000): 564S–572S.

51. American College of Sports Medicine, Position stand: Heat and cold illnesses during distance running, *Medicine and Science in Sports and Exercise* 28 (1996): i–x.

52. J. R. Stofan and coauthors, Sweat and sodium losses in NCAA Division 1 football players with a history of whole-body muscle cramping, presented at the annual meeting of the American College of Sports Medicine, 2003, unpublished.

53. N. J. Rehrer, Fluid and electrolyte balance in ultra-endurance sport, *Sports Medicine* 31 (2001): 701–715.

54. M. Hsieh and coauthors, Hyponatremia in runners requiring on-site medical treatment at a single marathon, *Medicine and Science in Sports and Exercise* 34 (2002): 185–189.

55. E. R. Eichner, Exertional hyponatremia: Why so many women? *Sports Medicine Digest* 24 (2002): 54, 56.

56. Position of the American Dietetic Association, Dietitians of Canada, and the American College of Sports Medicine, 2000.

ANSWERS

Study Questions (multiple choice)

1. c   2. b   3. a   4. d   5. c   6. c   7. d   8. b   9. a   10. d

Supplements as Ergogenic Aids

© Ellen Stagg/Stone/Getty Images

Athletes gravitate to promises that they can enhance their performance by taking pills, powders, or potions. Unfortunately, they often hear such promises from their coaches and peers, who advise them to use nutrient supplements, take drugs, or follow procedures that claim to deliver results without effort. When such aids are harmless, they are only a waste of money; when they impair performance or harm health, they waste athletic potential and cost lives. This highlight looks at some promises of magic to improve physical performance.

Ergogenic Aids

Many substances or treatments claim to be *ergogenic,* meaning work enhancing. The glossary below defines several of the commonly used **ergogenic aids** discussed in this highlight.

The glossary on p. 500 presents additional substances promoted as ergogenic aids. For the large majority of these substances, research findings do not support those claims.[1] Athletes who hear that a product is ergogenic should ask who is making the claim and who will profit from the sale.

Sometimes it is difficult to distinguish valid claims from bogus ones. Fitness magazines and Internet websites are particularly troublesome because many of them present both valid and invalid nutrition information alongside slick advertisements for nutrition products. Advertisements often feature colorful anatomical figures, graphs, and tables that appear scientific. Some ads even include references, citing or linking to such credible sources as the *American Journal of Clinical Nutrition* and the *Journal of the American Medical Association*. These ads create the illusion of endorsement and credibility to gain readers' trust. Keep in mind, however, that the ads are created not to teach, but to sell. A careful reading of the cited research might reveal that the ads have presented the research findings out of context.[2] In one such case, an ad cited a research article to support the invalid conclusion that its human growth hormone supplement "increases lean body mass and bone mineral." Researchers reporting in the cited article had reached another conclusion: "its general use now or in the immediate

GLOSSARY OF SUBSTANCES PROMOTED AS ERGOGENIC AIDS

arginine: a nonessential amino acid falsely promoted as enhancing the secretion of human growth hormone, the breakdown of fat, and the development of muscle.

bee pollen: a product consisting of bee saliva, plant nectar, and pollen that supposedly aids in weight loss and boosts athletic performance. It does neither and may cause an allergic reaction in individuals sensitive to it.

boron: a nonessential mineral that is promoted as a "natural" steroid replacement.

brewer's yeast: a preparation of yeast cells, containing a concentrated amount of B vitamins and some minerals; falsely promoted as an energy booster.

cell salts: a preparation of minerals supposedly harvested from living cells, sold as a health-promoting supplement.

coenzyme Q10: a lipid found in cells (mitochondria) shown to improve exercise performance in heart disease patients, but not effective in improving the performance of healthy athletes.

desiccated liver: dehydrated liver powder that supposedly contains all the nutrients found in liver in concentrated form; possibly not dangerous, but has no particular nutritional merit and is considerably more expensive than fresh liver.

DNA (deoxyribonucleic acid): the genetic material of cells necessary in protein synthesis; falsely promoted as an energy booster.

epoetin (eh-poy-EE-tin): a drug derived from the human hormone erythropoietin and marketed under the trade name Epogen; illegally used to increase oxygen capacity.

gelatin: a soluble form of the protein collagen, used to thicken foods; sometimes falsely promoted as a strength enhancer.

ginseng: a plant whose extract supposedly boosts energy. Side effects of chronic use include nervousness, confusion, and depression.

glycine: a nonessential amino acid, promoted as an ergogenic aid because it is a precursor of the high-energy compound creatine phosphate. Other amino acids commonly packaged for athletes that are equally useless include tryptophan, ornithine, arginine, lysine, and the branched-chain amino acids.

growth hormone releasers: herbs or pills that supposedly regulate hormones; falsely promoted as enhancing athletic performance.

guarana: a reddish berry found in Brazil's Amazon valley that is used as an ingredient in carbonated sodas and taken in powder or tablet form. Guarana is marketed as an ergogenic aid to enhance speed and endurance, an aphrodisiac, a "cardiac tonic," an "intestinal disinfectant," and a smart drug that supposedly improves memory and concentration and wards off senility. Because guarana contains seven times as much caffeine as its relative the coffee bean, there are concerns that high doses can stress the heart and cause panic attacks.

herbal steroids or **plant sterols:** curious mixtures of herbs, "adaptogens," and "aphrodisiacs" that supposedly enhance hormone activity. Products marketed as herbal steroids include astragalus, damiana, dong quai, fo ti teng, ginseng root, licorice root, palmetto berries, sarsaparilla, schizardra, unicorn root, yohimbe bark, and yucca.

HMB (beta-hydroxy-beta-methylbutyrate): a metabolite of the branched-chain amino acid leucine. Claims that HMB increases muscle mass and strength are based on the results of two studies from the lab that developed HMB as a supplement.

inosine: an organic chemical that is falsely said to "activate cells, produce energy, and facilitate exercise," but actually has been shown to reduce the endurance of runners.

ma huang: an evergreen plant derivative that supposedly boosts energy and helps with weight control. Ma huang contains ephedrine, a cardiac stimulant, and has been associated with high blood pressure, rapid heart rate, nerve damage, muscle injury, psychosis, stroke, and memory loss.

niacin: a B vitamin that when taken in excess rushes blood to the skin, producing vascularity and a red tint—physical attributes bodybuilders strive to attain prior to performance. These attributes do not enhance performance, and excess niacin can cause headaches and nausea.

octacosanol: an alcohol isolated from wheat germ; often falsely promoted as enhancing athletic performance.

ornithine: a nonessential amino acid falsely promoted as enhancing the secretion of human growth hormone, the breakdown of fat, and the development of muscle.

oryzanol: a plant sterol that supposedly provides the same physical responses as anabolic steroids without the adverse side effects; also known as *ferulic acid, ferulate,* or *FRAC.*

pangamic acid: also called vitamin B_{15} (but not a vitamin, nor even a specific compound—it can be anything with that label); falsely claimed to speed oxygen delivery.

phosphate pills: a product demonstrated to increase the levels of a metabolically important phosphate compound (diphosphoglycerate) in red blood cells and the potential of the cells to deliver oxygen to the body's muscle cells. However, it does not extend endurance or increase efficiency of aerobic metabolism and may cause calcium losses from the bones if taken in excess.

pyruvate: a 3-carbon compound derived during the metabolism of glucose, certain amino acids, and glycerol. Claims that pyruvate burns fat and enhances endurance are based on two studies of untrained individuals by the same author. Common side effects include intestinal gas and diarrhea.

ribose: a 5-carbon sugar falsely promoted as improving the regeneration of ATP and thereby the speed of recovery after high-power exercise.

RNA (ribonucleic acid): the genetic material of cells necessary for protein synthesis; falsely promoted as enhancing athletic performance.

royal jelly: the substance produced by worker bees and fed to the queen bee; falsely promoted as increasing strength and enhancing performance.

sodium bicarbonate: baking soda; an alkaline salt believed to neutralize blood lactic acid and thereby to reduce pain and enhance possible workload. "Soda loading" may cause intestinal bloating and diarrhea.

spirulina: a kind of alga ("blue-green manna") that supposedly contains large amounts of protein and vitamin B_{12}, suppresses appetite, and improves athletic performance. It does none of these things and is potentially toxic.

succinate: a compound synthesized in the body and involved in the TCA cycle; falsely promoted as a metabolic enhancer.

superoxide dismutase (SOD): an enzyme that protects cells from oxidation. When it is taken orally, the body digests and inactivates this protein; it is useless to athletes.

wheat germ oil: the oil from the wheat kernel; often falsely promoted as an energy aid.

future is not justified." Scientific facts had been exaggerated and twisted to promote sales. Highlight 1 describes ways to recognize misinformation and quackery.

Dietary Supplements

A variety of supplements make claims based on misunderstood nutrition principles. The claims may sound good, but for the most part, they have little or no factual basis.

Protein Powders

Because the body builds muscle protein from amino acids, many athletes take protein powders with the false hope of stimulating muscle growth. Protein powders can supply amino acids to the body, but nature's protein sources—lean meat, milk, and legumes—supply all these amino acids and more.

Whey protein appears to be particularly popular among athletes hoping to achieve greater muscle gains. A waste product of cheese manufacturing, whey protein is a common ingredient in many low-cost protein powders. Athletes and active people who want bigger muscles should know that whey protein does not increase muscle mass. To build bigger muscles, they need to eat food with adequate energy and protein to support the weight-training work that does increase muscle mass. Those who still think they need more whey should pour a glass of milk; one cup provides 1.5 grams of whey.

Purified protein preparations contain none of the other nutrients needed to support the building of muscle, and the protein they supply is not needed by athletes who eat food. It is excess protein, and the body dismantles it and uses it for energy or stores it as body fat. The deamination of excess amino acids places an extra burden on the kidneys to excrete unused nitrogen.

Amino Acid Supplements

Chapter 6 (p. 203) describes how the competition among amino acids for carriers can cause the absorption of some amino acids to be limited, creating a deficiency, and the absorption of others to be enhanced, creating the possibility of toxicity. Most healthy athletes eating well-balanced diets do not need amino acid supplements.

Advertisers point to research that identifies the **branched-chain amino acids** as the main ones used as fuel by exercising muscles. What the ads leave out is that compared to glucose and fatty acids, branched-chain amino acids provide very little fuel and that ordinary foods provide them in abundance anyway. Large doses of branched-chain amino acids can raise plasma ammonia concentrations, which can be toxic to the brain. Branched-chain amino acid supplements are neither effective nor safe and are not recommended.

Carnitine

Carnitine, a nonessential nutrient, is often promoted as a "fat burner." Some athletes use it, hoping carnitine will help them burn more fat, thereby sparing glycogen during endurance events.

In the body, carnitine facilitates the transfer of fatty acids across the mitochondrial membrane. Supplement manufacturers suggest that with more carnitine available, fat oxidation will be enhanced, but this does not seem to be the case. Carnitine supplementation neither raises muscle carnitine concentrations nor enhances exercise performance.[3] It does, however, produce diarrhea in about half of the people who use it. Milk and meat products are good sources of carnitine, and supplements are not needed.

Chromium Picolinate

Chapter 13 introduced chromium as an essential trace mineral involved in carbohydrate and lipid metabolism. Advertisements in bodybuilding magazines claim that **chromium picolinate,** which is supposed to be more easily absorbed than chromium alone, builds muscle, enhances energy, and burns fat. Such claims derive from one or two initial studies reporting that men who weight trained while taking chromium picolinate supplements increased lean body mass and reduced body fat. Most subsequent studies, however, show no effects of chromium picolinate supplementation on strength, lean body mass, or body fat. Whether high doses of chromium picolinate affect iron stores remains unclear, with limited research finding either no effect or a slight decrease.[4]

Complete Nutrition Supplements

Several drinks and candy bars appeal to athletes by claiming to provide "complete" nutrition. These products usually taste good and provide extra food energy, but fall short of providing "complete" nutrition. They can be useful as a pregame meal or a between-meal snack, but they should not replace regular meals.

A nutritionally "complete" drink may help a nervous athlete who cannot tolerate solid food on the day of an event. A liquid meal two or three hours before competition can supply some of the fluid and carbohydrate needed in a pregame meal, but a shake of fat-free milk or juice (such as apple or papaya) and ice milk or frozen fruit (such as strawberries or bananas) can do the same thing less expensively.

Creatine

Interest in—and use of—**creatine** supplements to enhance performance during intense activity has grown dramatically in the last few years. Power athletes such as weight lifters use creatine supplements to enhance stores of the high-energy compound creatine phosphate (CP) in muscles. Theoretically, the more creatine phosphate in muscles, the higher the intensity at which an athlete can train. High-intensity training stimulates the muscles to adapt, which, in turn, improves performance.

The results of some studies suggest that creatine supplementation does enhance performance of high-intensity activity such as weight lifting or sprinting.[5] Creatine may improve performance by increasing muscle strength and size, cell hydration, or glycogen loading capacity.[6]

The question of whether creatine supplements are safe cannot yet be answered with certainty.[7] Even short-term (5 to 7 days) creatine supplementation may pose risks to athletes with kidney disease or other conditions, and studies on the safety of long-term creatine supplementation are not available. One side effect of creatine supplementation that no one disputes is weight gain. For some athletes, weight gain, especially muscle gain, is beneficial, but for others, it is not.

Some medical and fitness experts voice concern that, like many performance enhancement supplements before it, creatine is being taken in huge doses (5 to 30 grams per day) before evidence of its value has been ascertained. Even people who eat red meat, which is a creatine-rich food, do not consume near the amount supplements provide. (Creatine content varies, but on average, pork, chicken, and beef provide 65 to 180 milligrams per ounce.) Despite the uncertainties, creatine supplements are not illegal in international competition. The American Academy of Pediatrics discourages creatine supplementation in adolescents under 18 years old.[8]

Caffeine

Some research supports the use of **caffeine** to enhance endurance.[9] Caffeine supposedly enhances endurance by stimulating fatty acid release, thereby slowing glycogen use, and by attenuating the perception of fatigue. At least one study reports that this is not the case.[10] Light activity before a workout stimulates fat release also, but in addition, the activity warms the muscles and connective tissues, making them flexible and resistant to injury. Caffeine does not offer these added benefits.

Caffeine is a stimulant that elicits a number of physiological and psychological effects in the body. (The table at the start of Appendix H provides a list of common caffeine-containing items and the doses they deliver.) The possible benefits of caffeine use must be weighed against its adverse effects—stomach upset, nervousness, irritability, headaches, and diarrhea. Caffeine-containing beverages should be used in moderation, if at all, and *in addition* to other fluids, not as a substitute for them. College, national, and international athletic competitions prohibit the use of caffeine in amounts greater than the equivalent of 5 or 6 cups of coffee consumed in a two-hour period prior to competition. Urine tests that detect more caffeine than this disqualify athletes from competition.

Oxygenated Water

Oxygenated water—water infused with oxygen—claims to "improve athletic performance, increase endurance, and sharpen concentration" by delivering extra oxygen to the muscles. What's wrong with this claim? First is the assumption that oxygen can enter the bloodstream by way of the GI tract—which it can't. Second is the assumption that the body can use more oxygen than it receives from the lungs—which it can't. The only time athletes might benefit from oxygen (inhaled, not swallowed) might be when exercising at elevations higher than they are accustomed to. Any benefits of oxygenated water come from the water, not the oxygen.

Hormonal Supplements

The dietary supplements discussed this far may or may not help athletic performance, but in the doses commonly taken, they seem to cause little harm. The remaining discussion features supplements that are clearly damaging.

Anabolic Steroids

Among the most dangerous and illegal ergogenic practices is the taking of **anabolic steroids.** These drugs are derived from the male sex hormone testosterone, which promotes the development of male characteristics and lean body mass. Athletes take steroids to stimulate muscle bulking.

To athletes struggling to excel, the promise of bigger, stronger muscles than training alone can produce has been tempting. Athletes who lack superstar genetic material and who normally would not be able to break into the elite ranks can, with the help of steroids, suddenly compete with true champions. Especially in professional circles, where monetary rewards for excellence are sky-high, steroid use is common despite its illegality and side effects.

The American Academy of Pediatrics and the American College of Sports Medicine condemn athletes' use of anabolic steroids, and the International Olympic Committee bans their use. These authorities cite the known toxic side effects and maintain that taking these drugs is a form of cheating. Other athletes are put in the difficult position of either conceding an unfair advantage to competitors who use steroids or taking them and accepting the risk of harmful side effects (see Table H14-1). Young athletes should not be forced to make such a choice.

The price for the potential competitive edge that steroids confer is high—sometimes it is life itself. Steroids are not simple pills that build bigger muscles, but complex chemicals to which the body reacts in many ways, particularly when bodybuilders and other athletes take large amounts. The safest, most effective way to build muscle has always been through

TABLE H14-1 Anabolic Steroids: Side Effects and Adverse Reactions

Mind

- Extreme aggression with hostility ("steroid rage"); mood swings; anxiety; dizziness; drowsiness; unpredictability; insomnia; psychotic depression; personality changes, suicidal thoughts

Face and Hair

- Swollen appearance; greasy skin; severe, scarring acne; mouth and tongue soreness; yellowing of whites of eyes (jaundice)
- In females, male-pattern hair loss and increased growth of face and body hair

Voice

- In females, irreversible deepening of voice

Chest

- In males, breathing difficulty, breast development
- In females, breast atrophy

Heart

- Heart disease; elevated or reduced heart rate; heart attack; stroke; hypertension; increased LDL; reduced HDL

Abdominal Organs

- Nausea; vomiting; bloody diarrhea; pain; edema; liver tumors (possibly cancerous); liver damage, disease, or rupture leading to fatal liver failure; kidney stones and damage; gallstones; frequent urination; possible rupture of aneurysm or hemorrhage

Blood

- Blood clots; high risk of blood poisoning; those who share needles risk contracting HIV (the AIDS virus) or other disease-causing organisms; septic shock (from injections)

Reproductive System

- In males, permanent shrinkage of testes; prostate enlargement with increased risk of cancer; sexual dysfunction; loss of fertility; excessive and painful erections
- In females, loss of menstruation and fertility; permanent enlargement of external genitalia; fetal damage, if pregnant

Muscles, Bones, and Connective Tissues

- Increased susceptibility to injury with delayed recovery times; cramps; tremors; seizurelike movements; injury at injection site
- In adolescents, failure to grow to normal height

Other

- Fatigue; increased risk of cancer

---

hard training and a sound diet, and—despite popular misconceptions—it still is.

Some manufacturers peddle specific herbs as legal substitutes for steroid drugs. They falsely claim that these herbs contain hormones, enhance the body's hormonal activity, or both. In some cases, an herb may contain plant sterols, such as oryzanol, but these compounds are poorly absorbed. Even if absorption occurs, the body cannot convert herbal compounds to anabolic steroids. None of these products has any proven anabolic steroid activity, none enhances muscle strength, and some contain natural toxins. In short, "natural" does not mean "harmless."

DHEA and Androstenedione

Some athletes use **DHEA** and **androstenedione** as alternatives to anabolic steroids. Androstenedione made headlines in the late 1990s when the media reported that baseball great Mark McGwire had been using it. DHEA (dehydroepiandrosterone) and androstenedione are hormones made in the adrenal glands that serve as precursors to the male hormone testosterone. Advertisements claim the hormones "burn fat," "build muscle," and "slow aging," but evidence to support such claims is lacking.[11]

Short-term side effects of DHEA and androstenedione may include oily skin, acne, body hair growth, liver enlargement, testicular shrinkage, and aggressive behavior.[12] Long-term effects of these hormones remain to be seen and may take years to become evident. The potential for harm from DHEA and androstenedione supplements is great, and athletes, as well as others, should avoid them. DHEA and androstenedione are banned by the International Olympic Committee and the National Collegiate Athletic Association.

Human Growth Hormone

Some short or average-sized athletes sometimes use **hGH (human growth hormone)** to build lean tissue and increase their height if they are still in their growing years. Athletes in power sports such as weight lifting and judo are most likely to experiment with hGH, believing the injectable hormone will provide the benefits of anabolic steroids without the dangerous side effects.

Abuse of hGH is not as extensive as abuse of steroids or other such drugs, in part because of the cost. A dose of hGH that will produce the effect sought might cost $2000 a week on the black market. As with other drugs sold on the black market, athletes often do not get what they think they are buying.

Taken in large quantities, hGH causes the disease acromegaly, in which the body becomes huge and the organs and bones overenlarge. Other effects include diabetes, thyroid disorder, heart disease, menstrual irregularities, diminished sexual desire, and shortened life span. The U.S. Olympic Committee bans hGH use, but tests cannot definitively distinguish between naturally occurring hGH and hGH used as a drug.[13] The committee maintains that the use of hGH is a form of cheating that undermines the quest for physical excellence and that its use is coercive to other athletes.

The search for a single food, nutrient, drug, or technique that will safely and effectively enhance athletic performance will no doubt continue as long as people strive to achieve excellence in sports. So far, when athletic performance does improve after use of an ergogenic aid, the improvement can often be attributed to the placebo effect, which is strongly at work in athletes. Even if a reliable source reports a performance boost from a newly tried product, give the effect time to fade away. Chances are excellent that it simply reflects the power of the mind over the body.

The overwhelming majority of potions sold for athletes are frauds. Wishful thinking will not substitute for talent, hard training, adequate diet, and mental preparedness in competition. But don't discount the power of mind over body for a minute—it is formidable, and sports psychologists dedicate their work to harnessing it. You can use it by imagining yourself a winner and visualizing yourself excelling in your sport. You don't have to buy magic to obtain a winning edge; you already possess it—your physically fit mind and body.

NUTRITION ON THE NET

 Access these websites for further study of topics covered in this highlight.

- Find updates and quick links to these and other nutrition-related sites at our website: **www.wadsworth.com/ nutrition**

- Find information on sports drinks and other nutrition and fitness topics at the Gatorade Sports Science Institute site: **www.gssiweb.com**

REFERENCES

1. F. Brouns and coauthors, Functional foods and food supplements for athletes: From myths to benefit claims substantiation through the study of selected biomarkers, *British Journal of Nutrition* 88 (2002): S177–S186.
2. J. M. Drazen, Inappropriate advertising of dietary supplements, *New England Journal of Medicine* 348 (2003): 777–778.
3. E. P. Brass, Supplemental carnitine and exercise, *American Journal of Clinical Nutrition* 72 (2000): 628S–623S.
4. Committee on Dietary Reference Intakes, *Dietary Reference Intakes for Vitamin A, Vitamin K, Arsenic, Boron, Chromium, Copper, Iodine, Iron, Manganese, Molybdenum, Nickel, Silicon, Vanadium, and Zinc* (Washington, D.C.: National Academy Press, 2001), p. 200.
5. J. S. Volek and coauthors, Physiological responses to short-term exercise in the heat after creatine loading, *Medicine and Science in Sports and Medicine* 33 (2001): 1101–1108; D. Preen and coauthors, Effect of creatine loading on long-term sprint exercise performance and metabolism, *Medicine and Science in Sports and Medicine* 33 (2001): 814–821; A. Casey and P. L. Greenhaff, Does

dietary creatine supplementation play a role in skeletal muscle metabolism and performance? *American Journal of Clinical Nutrition* 72 (2000): 607S–617S; K. Vandenberghe and coauthors, Phosphocreatine resynthesis is not affected by creatine loading, *Medicine and Science in Sports and Medicine* 31 (1999): 236–242.
6. A. G. Nelson, Muscle glycogen supercompensation is enhanced by prior creatine supplementation, *Medicine and Science in Sports and Medicine* 33 (2001): 1096–1100; M. G. Bemben and coauthors, Creatine supplementation during resistance training in college football athletes, *Medicine and Science in Sports and Medicine* 33 (2001): 1667–1673; D. Willoughby and J. Rosene, Effects of oral creatine and resistance training on myosin heavy chain expression, *Medicine and Science in Sports and Medicine* 33 (2001): 1674–1681.
7. E. B. Feldman, Creatine: A dietary supplement and ergogenic aid, *Nutrition Reviews* 57 (1999): 45–50.
8. J. D. Metzl and coauthors, Creatine use among young athletes, *Pediatrics* 108 (2001): 421–425.

9. C. R. Bruce and coauthors, Enhancement of 2000-m rowing performance after caffeine ingestion, *Medicine and Science in Sports and Medicine* 32 (2000): 1958–1963.
10. D. Laurent and coauthors, Effects of caffeine on muscle glycogen utilization and the neuroendocrine axis during exercise, *Journal of Clinical Endocrinology and Metabolism* 85 (2000): 2167–2169.
11. B. B. Rasmussen and coauthors, Androstenedione does not stimulate muscle protein anabolism in young healthy men, *Journal of Clinical Endocrinology and Metabolism* 85 (2000): 55–59.
12. R. Skinner, E. Coleman, and C. A. Rosenbloom, Ergogenic aids, in *Sports Nutrition: A Guide for the Professional Working with Active People,* 3rd ed., ed. C. A. Rosenbloom (Chicago: The American Dietetic Association, 2000), pp. 107–146.
13. L. di Luigi and L. Guidetti, IGF-1, IGFBP-2, and -3: Do they have a role in detecting rhGH abuse in trained men? *Medicine and Science in Sports and Medicine* 34 (2002): 1270–1278.

Life Cycle Nutrition: Pregnancy and Lactation

© Wally Eberhart/Botanica/Getty Images

Chapter Outline

Nutrition prior to Pregnancy

Growth and Development during Pregnancy: *Placental Development • Fetal Growth and Development • Critical Periods*

Maternal Weight: *Weight prior to Conception • Weight Gain during Pregnancy • Exercise during Pregnancy*

Nutrition during Pregnancy: *Energy and Nutrient Needs during Pregnancy • Common Nutrition-Related Concerns of Pregnancy*

High-Risk Pregnancies: *The Infant's Birthweight • Malnutrition and Pregnancy • Food Assistance Programs • Maternal Health • The Mother's Age • Practices Incompatible with Pregnancy*

Nutrition during Lactation: *Lactation: A Physiological Process • Breastfeeding: A Learned Behavior • The Mother's Nutrient Needs • Practices Incompatible with Lactation • Maternal Health*

Highlight: *Fetal Alcohol Syndrome*

Nutrition Explorer CD-ROM Outline

Nutrition Animation: *Nutrient Needs of Women and Infants*

Case Study: *Preparing for Pregnancy*

Student Practice Test

Glossary Terms

Nutrition on the Net

Nutrition in Your Life

Food choices have consequences. Sometimes they happen immediately, as when you get heartburn after eating a pepperoni and jalapeño pizza. Other times they sneak up on you, as when you gain weight after indulging in double hot fudge sundaes. Quite often, they are temporary and easily resolved, as when hunger pangs strike after you drink only a diet cola for lunch. During pregnancy, however, the consequences of a woman's food choices are dramatic. They affect not only her health, but also the growth and development of another human being—not just for today, but for years to come. Making smart food choices is a huge responsibility, but fortunately, it's fairly simple.

All people—pregnant and lactating women, infants, children, adolescents, and adults—need the same nutrients, but the amounts they need vary depending on their stage of life. This chapter focuses on nutrition in preparation for, and support of, pregnancy and lactation. The next two chapters address the needs of infants, children, adolescents, and older adults.

Nutrition prior to Pregnancy

A section on nutrition prior to pregnancy must, by its nature, focus mainly on women. A man's nutrition may affect his **fertility** and possibly the genetic contributions he makes to his children, but nutrition exerts its primary influence through the woman. Her body provides the environment for the growth and development of a new human being. Prior to pregnancy, a woman has a unique opportunity to prepare herself physically, mentally, and emotionally for the many

fertility: the capacity of a woman to produce a normal ovum periodically and of a man to produce normal sperm; the ability to reproduce.

changes to come. In preparation for a healthy pregnancy, a woman can establish the following habits:

- *Achieve and maintain a healthy body weight.* Both underweight and overweight women, and their newborns, face increased risks of complications.
- *Choose an adequate and balanced diet.* Malnutrition reduces fertility and impairs the early development of an infant should a woman become pregnant.
- *Be physically active.* A women who wants to be physically active *when* she is pregnant needs to become physically active *beforehand*.
- *Avoid harmful influences.* Both maternal and paternal ingestion of harmful substances (such as cigarettes, alcohol, drugs, or environmental contaminants) can alter genes or their expression, interfering with fertility and causing abnormalities.

Young adults who nourish and protect their bodies do so not only for their own sakes, but also for future generations.

Growth and Development during Pregnancy

A whole new life begins at **conception**. Organ systems develop rapidly, and nutrition plays many supportive roles. This section describes placental development and fetal growth, paying close attention to times of intense developmental activity.

Placental Development

In the early days of pregnancy, a spongy structure known as the **placenta** develops in the **uterus.** Two associated structures also form (see Figure 15-1). One is the **amniotic sac,** a fluid-filled balloonlike structure that houses the developing fetus. The other is the **umbilical cord,** a ropelike structure containing fetal blood vessels that extends through the fetus's "belly button" (the umbilicus) to the placenta. These three structures play crucial roles during pregnancy and then are expelled from the uterus during childbirth.

The placenta develops as an interweaving of fetal and maternal blood vessels embedded in the uterine wall. The maternal blood transfers oxygen and nutrients to the fetus's blood and picks up fetal waste products. By exchanging oxygen, nutrients, and waste products, the placenta performs the respiratory, absorptive, and excretory functions that the fetus's lungs, digestive system, and kidneys will provide after birth.

The placenta is a versatile, metabolically active organ. Like all body tissues, the placenta uses energy and nutrients to support its work. Like a gland, it produces an array of hormones that maintain pregnancy and prepare the mother's breasts for lactation (making milk). A healthy placenta is essential for the developing fetus to attain its full potential.

Fetal Growth and Development

Fetal development begins with the fertilization of an **ovum** by a **sperm.** Three stages follow: the zygote, the embryo, and the fetus (see Figure 15-2).

The Zygote The newly fertilized ovum, or **zygote,** begins as a single cell and divides to become many cells during the days after fertilization. Within two weeks, the zygote embeds itself in the uterine wall—a process known as **implantation.** Cell division continues—each set of cells divides into many other cells. As development proceeds, the zygote becomes an embryo.

conception: the union of the male sperm and the female ovum; fertilization.

placenta (plah-SEN-tuh): the organ that develops inside the uterus early in pregnancy, through which the fetus receives nutrients and oxygen and returns carbon dioxide and other waste products to be excreted.

uterus (YOU-ter-us): the muscular organ within which the infant develops before birth.

amniotic (am-nee-OTT-ic) **sac:** the "bag of waters" in the uterus, in which the fetus floats.

umbilical (um-BILL-ih-cul) **cord:** the ropelike structure through which the fetus's veins and arteries reach the placenta; the route of nourishment and oxygen to the fetus and the route of waste disposal from the fetus. The scar in the middle of the abdomen that marks the former attachment of the umbilical cord is the **umbilicus** (um-BILL-ih-cus), commonly known as the "belly button."

ovum (OH-vum): the female reproductive cell, capable of developing into a new organism upon fertilization; commonly referred to as an egg.

sperm: the male reproductive cell, capable of fertilizing an ovum.

zygote (ZY-goat): the product of the union of ovum and sperm; so-called for the first two weeks after fertilization.

implantation: the stage of development in which the zygote embeds itself in the wall of the uterus and begins to develop; occurs during the first two weeks after conception.

FIGURE 15-1   The Placenta and Associated Structures

To understand how placental villi absorb nutrients without maternal and fetal blood interacting directly, think of how the intestinal villi work. The GI side of the intestinal villi is bathed in a nutrient-rich fluid (chyme). The intestinal villi absorb the nutrient molecules and release them into the body via capillaries. Similarly, the maternal side of the placental villi is bathed in nutrient-rich maternal blood. The placental villi absorb the nutrient molecules and release them to the fetus via fetal capillaries.

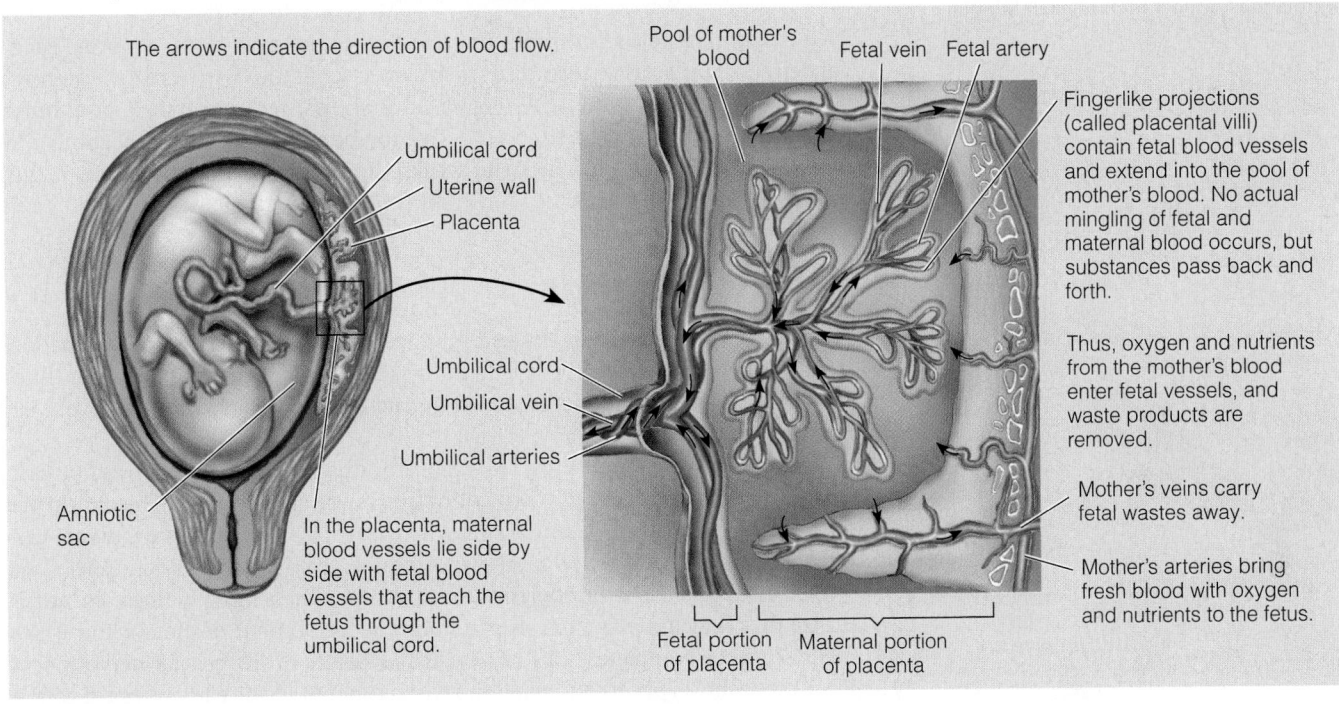

The arrows indicate the direction of blood flow.

Pool of mother's blood   Fetal vein   Fetal artery

Umbilical cord
Uterine wall
Placenta

Umbilical cord
Umbilical vein
Umbilical arteries

Amniotic sac

In the placenta, maternal blood vessels lie side by side with fetal blood vessels that reach the fetus through the umbilical cord.

Fingerlike projections (called placental villi) contain fetal blood vessels and extend into the pool of mother's blood. No actual mingling of fetal and maternal blood occurs, but substances pass back and forth.

Thus, oxygen and nutrients from the mother's blood enter fetal vessels, and waste products are removed.

Mother's veins carry fetal wastes away.

Mother's arteries bring fresh blood with oxygen and nutrients to the fetus.

Fetal portion of placenta   Maternal portion of placenta

FIGURE 15-2   Stages of Embryonic and Fetal Development

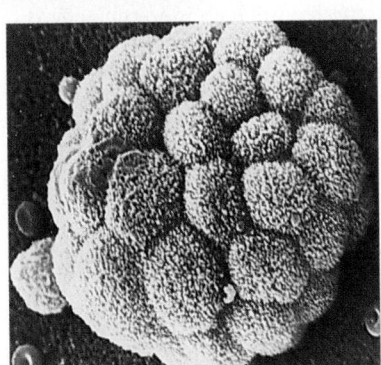

1 A newly fertilized ovum is about the size of a period at the end of this sentence. This **zygote** at less than one week after fertilization is not much bigger and is ready for implantation.

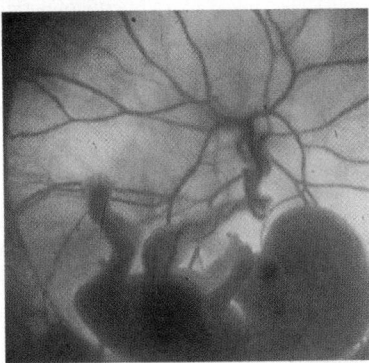

3 A **fetus** after 11 weeks of development is just over an inch long. Notice the umbilical cord and blood vessels connecting the fetus with the placenta.

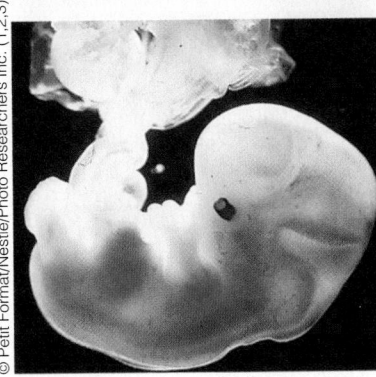

2 After implantation, the placenta develops and begins to provide nourishment to the developing embryo. An **embryo** five weeks after fertilization is about ¹/₂ inch long.

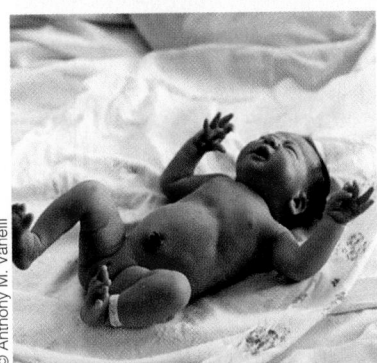

4 A **newborn infant** after nine months of development measures close to 20 inches in length. From eight weeks to term, this infant grew 20 times longer and 50 times heavier.

© Petit Format/Nestle/Photo Researchers Inc. (1,2,3)

© Anthony M. Vanelli

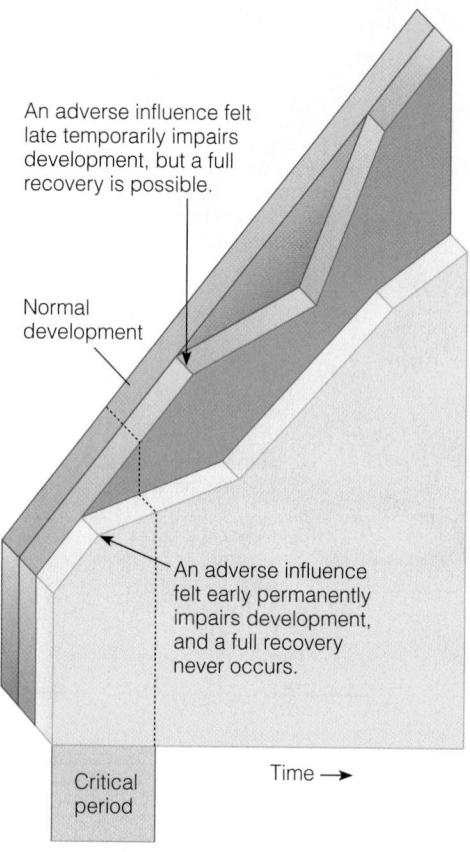

FIGURE 15-3 The Concept of Critical Periods

Critical periods occur early in development. An adverse influence felt early can have a much more severe and prolonged impact than one felt later on.

An adverse influence felt late temporarily impairs development, but a full recovery is possible.

Normal development

An adverse influence felt early permanently impairs development, and a full recovery never occurs.

Critical period

Time →

■ Reminder: The *neural tube* is the structure that eventually becomes the brain and spinal cord.

embryo (EM-bree-oh): the developing infant from two to eight weeks after conception.

fetus (FEET-us): the developing infant from eight weeks after conception until term.

critical periods: finite periods during development in which certain events occur that will have irreversible effects on later developmental stages; usually a period of rapid cell division.

gestation (jes-TAY-shun): the period from conception to birth. For human beings, the average length of a healthy gestation is 40 weeks. Pregnancy is often divided into thirds, called **trimesters.**

The Embryo The **embryo** develops at an amazing rate. At first, the number of cells in the embryo doubles approximately every 24 hours; later the rate slows, and only one doubling occurs during the final ten weeks of pregnancy. The embryo's size changes very little, but at eight weeks, the 1¼-inch embryo has a complete central nervous system, a beating heart, a digestive system, well-defined fingers and toes, and the beginnings of facial features.

The Fetus The **fetus** continues to grow during the next seven months. Each organ grows to maturity according to its own schedule, with greater intensity at some times than at others. As Figure 15-2 shows, fetal growth is phenomenal: weight increases from less than an ounce to about 7½ pounds (3500 grams). Most successful pregnancies last 39 to 41 weeks and produce a healthy infant weighing between 6½ and 9 pounds.

Critical Periods

Times of intense development and rapid cell division are called **critical periods**—critical in the sense that those cellular activities can occur only at those times. If cell division and number are limited during a critical period, full recovery is not possible (see Figure 15-3).

Each organ and tissue is most vulnerable to adverse influences (such as nutrient deficiencies or toxins) during its own critical period (see Figure 15-4). The critical period for neural tube■ development, for example, is from 17 to 30 days **gestation.** Consequently, neural tube development is most vulnerable to nutrient deficiencies, nutrient excesses, or toxins during this critical time—when most women do not even realize that they are pregnant. Any abnormal development of the neural tube or its failure to close completely can cause a major defect in the central nervous system. Figure 15-5 shows photos of neural tube development in the early weeks of gestation.

FIGURE 15-4 Critical Periods of Development

During embryonic development (from 2 to 8 weeks), many of the tissues are in their critical periods (purple area of the bars); events occur that will have irreversible effects on the development of those tissues. In the later stages of development (green area of the bars), the tissues continue to grow and change, but the events are less critical in that they are relatively minor or reversible.

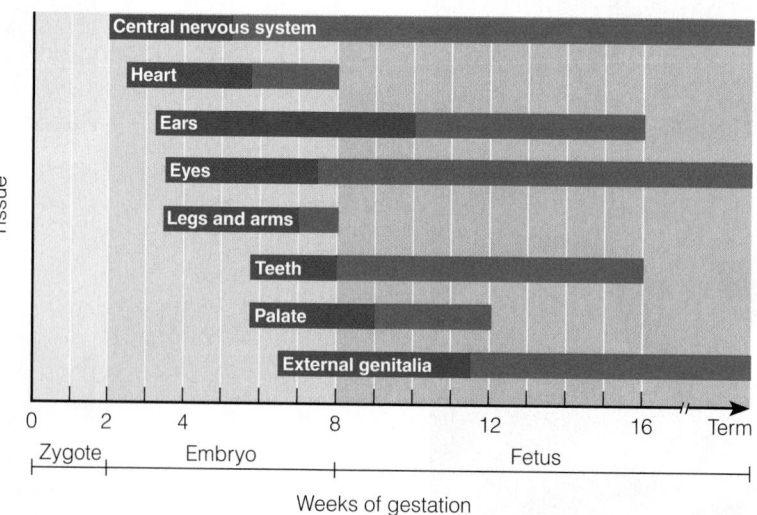

Key:
■ Critical development
■ Continued development

Central nervous system
Heart
Ears
Eyes
Legs and arms
Teeth
Palate
External genitalia

Tissue

0   2   4   8   12   16   Term
Zygote | Embryo | Fetus

Weeks of gestation

SOURCE: Adapted with permission from *Before We Were Born* by K. L. Moore. W. B. Saunders.

FIGURE 15-5 Neural Tube Development

The neural tube is the beginning structure of the brain and spinal cord. Any failure of the neural tube to close or to develop normally results in central nervous system disorders such as spina bifida and anencephaly. Successful development of the neural tube depends, in part, on the vitamin folate.

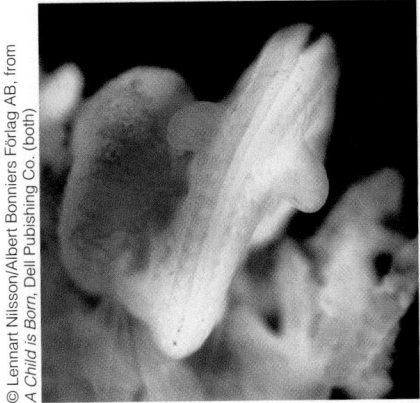 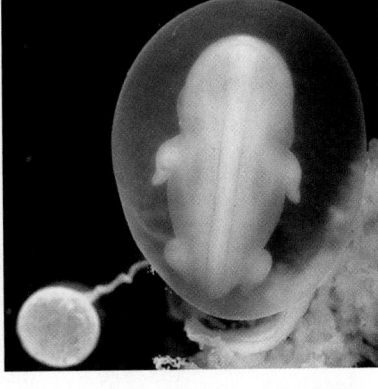

© Lennart Nilsson/Albert Bonniers Förlag AB, from *A Child is Born,* Dell Publishing Co. (both)

At four weeks, the neural tube has yet to close (notice the gap at the top).

At six weeks, the neural tube (outlined by the delicate red vertebral arteries) has successfully closed.

Neural Tube Defects In the United States, approximately 30 of every 100,000 infants are born with a **neural tube defect;** some 1000 or so infants are affected each year.* Many other pregnancies with neural tube defects end in abortions or stillbirths.

The two most common types of neural tube defects are anencephaly and spina bifida. In **anencephaly,** the upper end of the neural tube fails to close. Consequently, the brain is either missing or fails to develop. Pregnancies affected by anencephaly often end in miscarriage; infants born with anencephaly die shortly after birth.

Spina bifida is characterized by incomplete closure of the spinal cord and its bony encasement (see Figure 15-6 on p. 512). The meninges membranes covering the spinal cord often protrude as a sac, which may rupture and lead to meningitis, a life-threatening infection. Spina bifida is accompanied by varying degrees of paralysis, depending on the extent of the spinal cord damage. Mild cases may not even be noticed, but severe cases lead to death. Common problems include clubfoot, dislocated hip, kidney disorders, curvature of the spine, muscle weakness, mental handicaps, and motor and sensory losses.

A pregnancy affected by a neural tube defect can occur in any woman, but these factors make it more likely:

- A previous pregnancy affected by a neural tube defect.
- Maternal diabetes (type 1).
- Maternal use of antiseizure medications.
- Maternal obesity.
- Exposure to high temperatures early in pregnancy (prolonged fever or hot-tub use).
- Race/ethnicity (neural tube defects are more common among whites and Hispanics than among others).
- Low socioeconomic status.

Folate supplementation reduces the risk.

neural tube defect: a serious central nervous system birth defect that often results in lifelong disability or death.

anencephaly (AN-en-SEF-a-lee): an uncommon and always fatal type of neural tube defect; characterized by the absence of a brain.

- **an** = not (without)
- **encephalus** = brain

spina (SPY-nah) **bifida** (BIFF-ih-dah): one of the most common types of neural tube defects; characterized by the incomplete closure of the spinal cord and its bony encasement.

- **spina** = spine
- **bifida** = split

---

*Worldwide, some 300,000 to 400,000 infants are born with neural tube defects each year.

FIGURE 15-6 Spina Bifida

Spina bifida, a common neural tube defect, occurs when the vertebrae of the spine fail to close around the spinal cord, leaving it unprotected. The B vitamin folate helps prevent spina bifida and other neural tube defects.

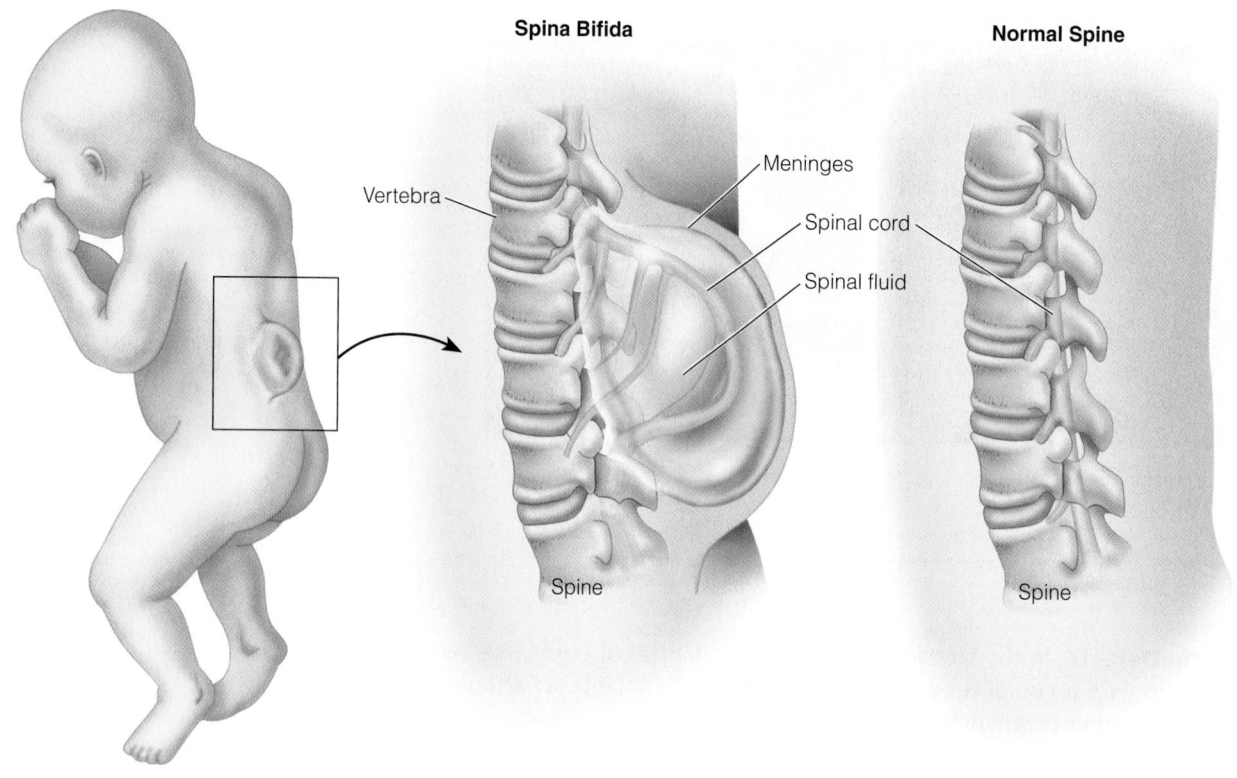

SOURCE: From the *Journal of the American Medical Association*, June 20, 2001, Vol. 285, No. 23, p. 3050. Reprinted with permission of the American Medical Association.

■ Folate RDA:
  • For women: 400 μg (0.4 mg)/day.
  • During pregnancy: 600 μg (0.6 mg)/day.

Folate Supplementation Chapter 10 described how folate supplements taken one month before conception and continued throughout the first trimester can help prevent neural tube defects. For this reason, all women of childbearing age■ who are capable of becoming pregnant should consume 400 micrograms (0.4 milligram) of folate daily. Most over-the-counter multivitamin supplements contain 400 micrograms of folate; prenatal supplements usually contain at least 800 micrograms. A woman who has previously had an infant with a neural tube defect may be advised by her physician to take folate supplements in doses ten times larger—4 milligrams daily. Because high doses of folate can mask the pernicious anemia of a vitamin B_{12} deficiency, quantities of 1 milligram or more require a prescription.

Because half of the pregnancies each year are unplanned and because neural tube defects occur early in development before most women realize they are pregnant, grain products in the United States are fortified with folate to ensure an adequate intake. Labels on fortified products may claim that an "adequate intake of folate has been shown to reduce the risk of neural tube defects." Fortification has improved folate status in women of childbearing age and lowered the number of neural tube defects that occur each year, as Figure 10-10 on p. 339 shows.[1]

HEALTHY PEOPLE 2010

Reduce the occurrence of spina bifida and other neural tube defects. Increase the proportion of pregnancies begun with an optimum folate level.

Chronic Diseases Some research suggests that adverse influences at critical times set the stage for chronic diseases in adult life.[2] Poor maternal diet during critical periods may permanently alter body functions such as blood pressure, glucose tolerance, and immune functions that influence disease development.[3] For example, maternal malnutrition may alter blood vessel growth and program lipid metabolism and lean body mass development in such a way that the infant will develop risk factors for cardiovascular disease as an adult.[4]

Malnutrition during the critical period of pancreatic cell growth provides an example of how type 2 diabetes may develop in adulthood.[5] The pancreatic cells responsible for producing insulin (the beta cells) normally increase more than 130-fold between 12 weeks gestation and five months after birth. Nutrition is a primary determinant of beta cell growth, and infants who have suffered prenatal malnutrition have significantly fewer beta cells than well-nourished infants. They are also more likely to be low-birthweight infants—and low birthweight correlates with a high risk of type 2 diabetes during adulthood.[6] One hypothesis suggests that diabetes may develop from the interaction of inadequate nutrition early in life with abundant nutrition later in life: the small mass of beta cells developed in lean times during fetal development may be insufficient in times of overnutrition during adulthood when the body needs more insulin.

Hypertension may develop from a similar scenario of inadequate growth during placental and gestational development followed by accelerated growth during early childhood: the small mass of kidney cells developed in lean times may be insufficient to handle the excessive demands of later life.[7] Low-birthweight infants who gain weight rapidly as young children are likely to develop hypertension and heart disease as adults.[8]

Both of these mice have the gene that tends to produce fat, yellow pups, but their mothers had different diets. The mother of the mouse on the right received a dietary supplement, which silenced the gene, resulting in brown pups with normal appetites.

Fetal Programming Recent genetic research may help to explain this phenomenon of substances such as nutrients influencing the development of diseases later on in adulthood—a process known as **fetal programming.** Researchers know that simply having a certain gene does not ensure that its associated trait will be expressed; the gene has to be activated.[9] (Similarly, owning lamps does not ensure you will have light in your home unless you turn them on.) Nutrients play key roles in activating or silencing genes. Switching genes on and off does not change the genetic sequence itself,■ but it can have dramatic consequences for a person's health. In the case of pregnancy, the mother's nutrition can permanently change gene expression in her fetus as well.[10]

Whether silencing or activating a gene is good or bad depends on what the gene does. Silencing a gene that stimulates cancer growth, for example, would be good, but silencing a gene that suppresses cancer growth would be bad. Similarly, activating a gene that defends against obesity would be good, but activating a gene that promotes obesity would be bad. Much research is under way to determine which nutrients activate or silence which genes.

■ The study of heritable changes in gene function that occur without a change in the DNA sequence is called **epigenetics.**

IN SUMMARY Maternal nutrition before and during pregnancy affects both the mother's health and the infant's growth. As the infant develops through its three stages—the zygote, embryo, and fetus—its organs and tissues grow, each on its own schedule. Times of intense development are critical periods that depend on nutrients to proceed smoothly. Without folate, for example, the neural tube fails to develop completely during the first month of pregnancy, prompting recommendations that all women of childbearing age take folate daily.

Because critical periods occur throughout pregnancy, a woman should continuously take good care of her health. That care should include achieving and maintaining a healthy body weight prior to pregnancy and gaining sufficient weight during pregnancy to support a healthy infant.

fetal programming: the influence of substances during fetal growth on the development of diseases in later life.

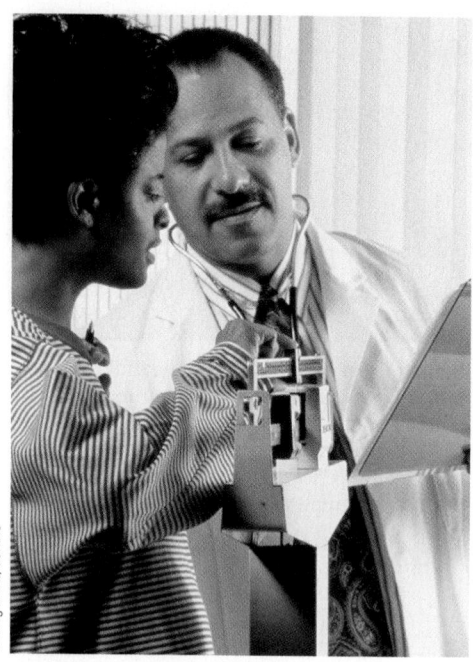

Fetal growth and maternal health depend on a sufficient weight gain during pregnancy.

■ The term **macrosomia** (mak-roh-SO-me-ah) describes high-birthweight infants (roughly 9 lb, or 4000 g, or more); macrosomia results from prepregnancy obesity, excessive weight gain during pregnancy, or uncontrolled diabetes.
- **macro** = large
- **soma** = body

preterm (infant): an infant born prior to the 38th week of pregnancy; also called a **premature infant.** A **term** infant is born between the 38th and 42nd week of pregnancy.

post term (infant): an infant born after the 42nd week of pregnancy.

cesarean section: a surgically assisted birth involving removal of the fetus by an incision into the uterus, usually by way of the abdominal wall.

Maternal Weight

Birthweight is the most reliable indicator of an infant's health. As a later section of this chapter explains, an underweight infant is more likely to have physical and mental defects, become ill, and die than a normal-weight infant. In general, higher birthweights present fewer risks for infants. Two characteristics of the mother's weight influence an infant's birthweight: her weight *prior* to conception and her weight gain *during* pregnancy.

Weight prior to Conception

A woman's weight prior to conception influences fetal growth. Even with the same weight gain during pregnancy, underweight women tend to have smaller babies than heavier women.

Underweight An underweight woman has a high risk of having a low-birthweight infant, especially if she is unable to gain sufficient weight during pregnancy. In addition, the rates of **preterm** births and infant deaths are higher for underweight women. An underweight woman improves her chances of having a healthy infant by gaining sufficient weight prior to conception or by gaining extra pounds during pregnancy. To gain weight, an underweight woman can follow the dietary recommendations for pregnant women (described in Figure 15-11 on p. 520).

Overweight and Obesity Overweight and obesity also create problems related to pregnancy and childbirth.[11] Obese women have an especially high risk of medical complications such as hypertension, gestational diabetes, and postpartum infections.[12] Compared with other women, obese women are also more likely to have other complications of labor and delivery.[13]

Overweight women have the lowest rate of low-birthweight infants. In fact, infants of overweight women are more likely to be born **post term** and to weigh more than 9 pounds.■ Large newborns increase the likelihood of a difficult labor and delivery, birth trauma, and **cesarean section.** Consequently, these infants have a greater risk of poor health and death than infants of normal weight.

Of greater concern than infant birthweight is the poor development of infants born to obese mothers. Obesity may double the risk for neural tube defects. In addition, both overweight and obese women have a greater risk of giving birth to infants with heart defects and other abnormalities.[14]

Weight-loss dieting during pregnancy is never advisable. Overweight women should try to achieve a healthy body weight before becoming pregnant, avoid excessive weight gain during pregnancy, and postpone weight loss until after childbirth. Weight loss is best achieved by eating moderate amounts of nutrient-dense foods and exercising to lose body fat.

Weight Gain during Pregnancy

All pregnant women must gain weight—fetal growth and maternal health depend on it. Maternal weight gain during pregnancy correlates closely with infant birthweight, which is a strong predictor of the health and subsequent development of the infant.

Recommended Weight Gains Table 15-1 presents recommended weight gains for various prepregnancy weights. The recommended gain for a woman who begins pregnancy at a healthy weight and is carrying a single fetus is 25 to 35 pounds. An underweight woman needs to gain between 28 and 40 pounds; and an overweight woman, between 15 and 25 pounds. Some women should strive for gains at the upper end of the target range, notably, adolescents who are still growing themselves and black women whose infants tend to be smaller than white infants even with the same maternal weight gain. Short women (5 feet 2 inches and under) should strive for gains at the lower end of the target range. Women who are

TABLE 15-1	Recommended Weight Gains Based on Prepregnancy Weight
Prepregnancy Weight	**Recommended Weight Gain**
Underweight (BMI <18.5)	28 to 40 lb (12.5 to 18.0 kg)
Healthy weight (BMI 18.5 to 24.9)	25 to 35 lb (11.5 to 16.0 kg)
Overweight (BMI 25.0 to 29.9)	15 to 25 lb (7.0 to 11.5 kg)
Obese (BMI ≥30)	15 lb minimum (6.8 kg minimum)

NOTE: These classifications for BMI are slightly different from those developed in 1990 by the Committee on Nutritional Status during Pregnancy and Lactation for the publication *Nutrition during Pregnancy* (Washington, D.C.: National Academy Press). That committee acknowledged that because such classifications had not been validated by research on pregnancy outcome, "any cut off points will be arbitrary for women of reproductive age." For these reasons, it seems appropriate to use the values developed for adults in 1998 by the National Institutes of Health (see Chapter 8).

carrying twins should aim for a weight gain of 35 to 45 pounds. If a woman gains more than is recommended early in pregnancy, she should not restrict her energy intake later in order to lose weight. A large weight gain over a short time, however, indicates excessive fluid retention and may be the first sign of the serious medical complication preeclampsia, discussed later.

Increase the proportion of mothers who achieve a recommended weight gain during their pregnancies.

HEALTHY PEOPLE 2010

Weight-Gain Patterns For the normal-weight woman, weight gain ideally follows a pattern of 3½ pounds during the first trimester and 1 pound per week thereafter. Health care professionals monitor weight gain using a prenatal weight-gain grid (see Figure 15-7).

Components of Weight Gain Women often express concern about the weight gain that accompanies a healthy pregnancy. They may find comfort in a reminder that most of the gain supports the growth and development of the placenta,

FIGURE 15-7	Recommended Prenatal Weight Gain Based on Prepregnancy Weight

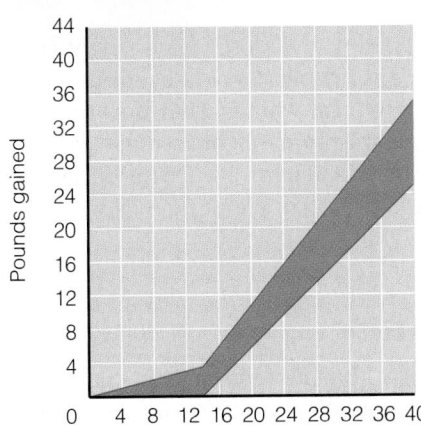

Normal-weight women should gain about 3½ pounds in the first trimester and just under 1 pound/week thereafter, achieving a total gain of 25 to 35 pounds by term.

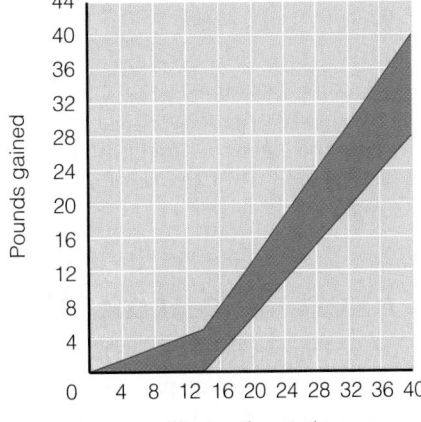

Underweight women should gain about 5 pounds in the first trimester and just over 1 pound/week thereafter, achieving a total gain of 28 to 40 pounds by term.

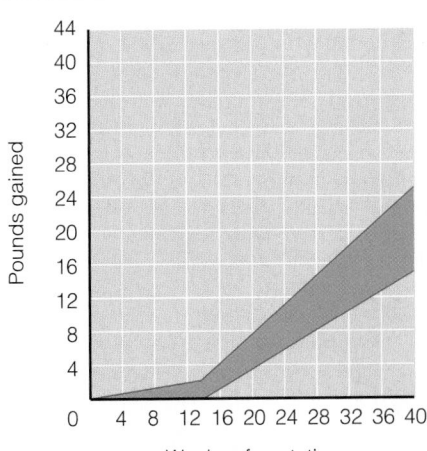

Overweight women should gain about 2 pounds in the first trimester and ⅔ pound/week thereafter, achieving a total gain of 15 to 25 pounds.

FIGURE 15-8 Components of Weight Gain during Pregnancy

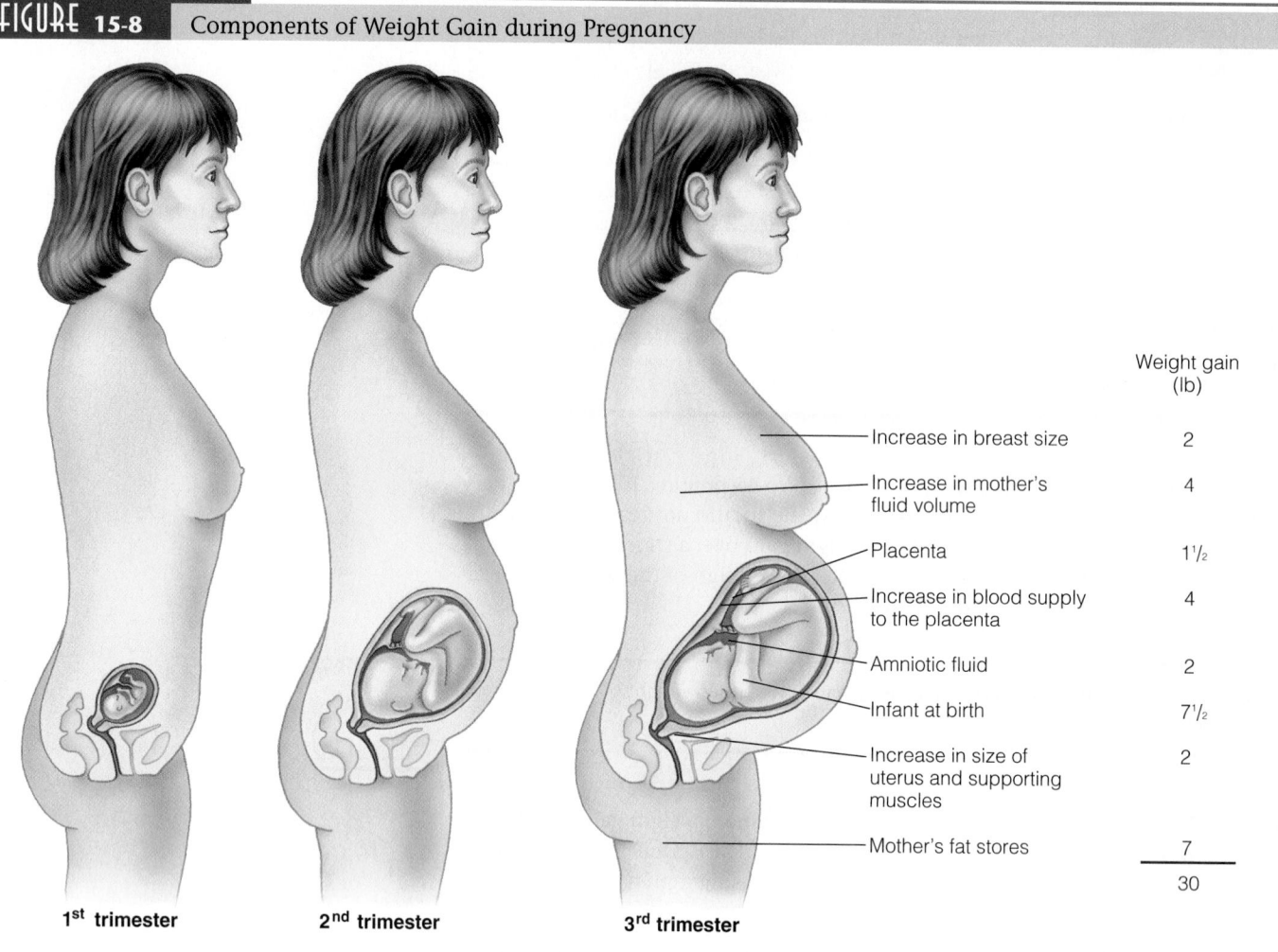

	Weight gain (lb)
Increase in breast size	2
Increase in mother's fluid volume	4
Placenta	1½
Increase in blood supply to the placenta	4
Amniotic fluid	2
Infant at birth	7½
Increase in size of uterus and supporting muscles	2
Mother's fat stores	7
	30

1st trimester   2nd trimester   3rd trimester

uterus, blood, and breasts, as well as an optimally healthy 7½-pound infant. A small amount goes into maternal fat stores, and even that fat is there for a special purpose: to provide energy for labor and lactation. Figure 15-8 shows the components of a typical 30-pound weight gain.

Weight Loss after Pregnancy The pregnant woman loses some weight at delivery. In the following weeks, she loses more as her blood volume returns to normal and she sheds accumulated fluids. The typical woman does not, however, return to her prepregnancy weight. In general, the more weight a woman gains beyond the needs of pregnancy, the more she will retain. Even with an average weight gain, though, most women tend to retain a couple of pounds with each pregnancy.

Exercise during Pregnancy

An active, physically fit woman experiencing a normal pregnancy can continue to exercise throughout pregnancy, adjusting the duration and intensity as the pregnancy progresses. Staying active can improve fitness, prevent or manage gestational diabetes, facilitate labor, and reduce stress. Women who exercise during pregnancy report fewer discomforts throughout their pregnancies. Regular exercise develops the strength and endurance a woman needs to carry the extra weight through pregnancy and to labor through an intense delivery. It also maintains the habits that help a woman lose excess weight and get back into shape after the birth.

FIGURE 15-9   Exercise Guidelines during Pregnancy

DO

Do exercise regularly (at least three times a week).

Do warm up with 5 to 10 minutes of light activity.

Do exercise for 20 to 30 minutes at your target heart rate.

Do cool down with 5 to 10 minutes of slow activity and gentle stretching.

Do drink water before, after, and during exercise.

Do eat enough to support the additional needs of pregnancy plus exercise.

Pregnant women can enjoy the benefits of exercise.

DON'T

Don't exercise vigorously after long periods of inactivity.

Don't exercise in hot, humid weather.

Don't exercise when sick with fever.

Don't exercise while lying on your back after the first trimester of pregnancy or stand motionless for prolonged periods.

Don't exercise if you experience any pain or discomfort.

Don't participate in activities that may harm the abdomen or involve jerky, bouncy movements.

A pregnant woman should participate in "low-impact" activities and avoid sports in which she might fall or be hit by other people or objects. For example, playing singles tennis with one person on each side of the net is safer than a fast-moving game of racquetball in which the two competitors can collide. Swimming and water aerobics are particularly beneficial because they allow the body to remain cool and move freely with the water's support, thus reducing back pain.[15] Figure 15-9 provides some guidelines for exercise during pregnancy.[16] Several of the guidelines are aimed at preventing excessively high internal body temperature and dehydration, both of which can harm fetal development. To this end, pregnant women should also stay out of saunas, steam rooms, and hot whirlpools.

IN SUMMARY   A healthy pregnancy depends on a sufficient weight gain. Women who begin their pregnancies at a healthy weight need to gain about 30 pounds, which covers the growth and development of the placenta, uterus, blood, breasts, and infant. By remaining active throughout pregnancy, a woman can develop the strength she needs to carry the extra weight and maintain habits that will help her lose it after the birth.

Nutrition during Pregnancy

A woman's body changes dramatically during pregnancy. Her uterus and its supporting muscles increase in size and strength; her blood volume increases by half to carry the additional nutrients and other materials; her joints become more flexible in preparation for childbirth; her feet swell in response to high concentrations of the hormone estrogen, which promotes water retention and helps to ready the uterus for delivery; and her breasts enlarge in preparation for lactation. The hormones that mediate all these changes may influence her mood. She can best prepare to handle these changes given a nutritious diet, regular physical activity, plenty of rest, and caring companions. This section highlights the role of nutrition.

A pregnant woman's food choices support both her health and her infant's growth and development.

Energy and Nutrient Needs during Pregnancy

From conception to birth, all parts of the infant—bones, muscles, organs, blood cells, skin, and other tissues—are made from nutrients in the foods the mother eats. For most women, nutrient needs during pregnancy and lactation■ are higher than at any other time (see Figure 15-10). To meet the high nutrient demands of pregnancy, a woman will need to make careful food choices, but her body will also help by maximizing absorption and minimizing losses.[17]

Energy The energy needs of pregnant women are greater than those of nonpregnant women—an additional 340 kcalories during the second trimester and an extra 450 kcalories during the third.■ Underweight women and physically active women may require more. A woman can easily get these added kcalories with an extra serving from each of the five food groups—a slice of bread, a serving of vegetables, a couple ounces of lean meat, a piece of fruit, and a cup of low-fat milk (see Figure 15-11 on p. 520). Alternatively, a woman who has been neglecting her calcium needs may want to use her additional kcalories for milk and milk products. A variety of strategies are appropriate in meeting the energy demands of pregnancy.[18]

For a 2000-kcalorie daily intake, these added kcalories represent about 15 to 20 percent more food energy than before pregnancy. The increase in nutrient needs is often greater than this, so nutrient-dense foods should supply the extra kcalories: foods such as whole-grain breads and cereals, legumes, dark green vegetables, citrus fruits, low-fat milk and milk products, and lean meats, fish, poultry, and eggs. Ample carbohydrate (ideally, 175 grams or more per day and certainly no less than 135 grams) is necessary to fuel the fetal brain and spare the protein needed for growth.

Protein The protein RDA■ for pregnancy is an additional 25 grams per day higher than for nonpregnant women. Pregnant women can easily meet their protein needs by selecting meats, milk products, and protein-containing plant foods such as legumes, whole grains, nuts, and seeds. Use of high-protein supplements during pregnancy may be harmful and is discouraged.

Essential Fatty Acids The high nutrient requirements of pregnancy leave little room in the diet for excess fat, but the essential long-chain polyunsaturated fatty acids are particularly important to the growth and development of the fetus.[19] The brain is largely made of lipid material, and it depends heavily on the long-chain omega-3 and omega-6 fatty acids for its growth, function, and structure. (See Table 5-2 on p. 161 for a list of good food sources of the omega fatty acids.)

Nutrients for Blood Production and Cell Growth New cells are laid down at a tremendous pace as the fetus grows and develops. At the same time, the mother's red blood cell mass expands. All nutrients are important in these processes, but for folate, vitamin B_{12}, iron, and zinc, the needs are especially great due to their key roles in the synthesis of DNA and new cells.

The requirement for folate increases dramatically during pregnancy.■ It is best to obtain sufficient folate from a combination of supplements, fortified foods, and a diet that includes fruits, juices, green vegetables, and whole grains.[20] The "How to" feature in Chapter 10 on p. 338 describes how folate from each of these sources contributes to a day's intake.

The pregnant woman also has a slightly greater need for the B vitamin that activates the folate enzyme—vitamin B_{12}.■ Generally, even modest amounts of meat, fish, eggs, or milk products together with body stores easily meet the need for vitamin B_{12}. Vegans who exclude all foods of animal origin, however, need daily supplements of vitamin B_{12} or vitamin B_{12}–fortified foods to prevent the neurological complications of a deficiency.

Pregnant women need iron■ to support their enlarged blood volume and to provide for placental and fetal needs. The developing fetus draws on maternal iron stores to create stores of its own to last through the first four to six months after birth

---

■ The table on the inside front cover provides separate listings for women during pregnancy and lactation, reflecting their heightened nutrient needs.

■ Energy requirement during pregnancy:
- 2nd trimester: +340 kcal/day.
- 3rd trimester: +450 kcal/day.

■ Protein RDA during pregnancy:
- +25 g/day.

■ Folate RDA during pregnancy:
- 600 µg/day.

■ Vitamin B_{12} RDA during pregnancy:
- 2.6 µg/day.

■ Iron RDA during pregnancy:
- 27 mg/day.

FIGURE 15-10   Comparison of Nutrient Recommendations for Nonpregnant, Pregnant, and Lactating Women

For actual values, turn to the table on the inside front cover.

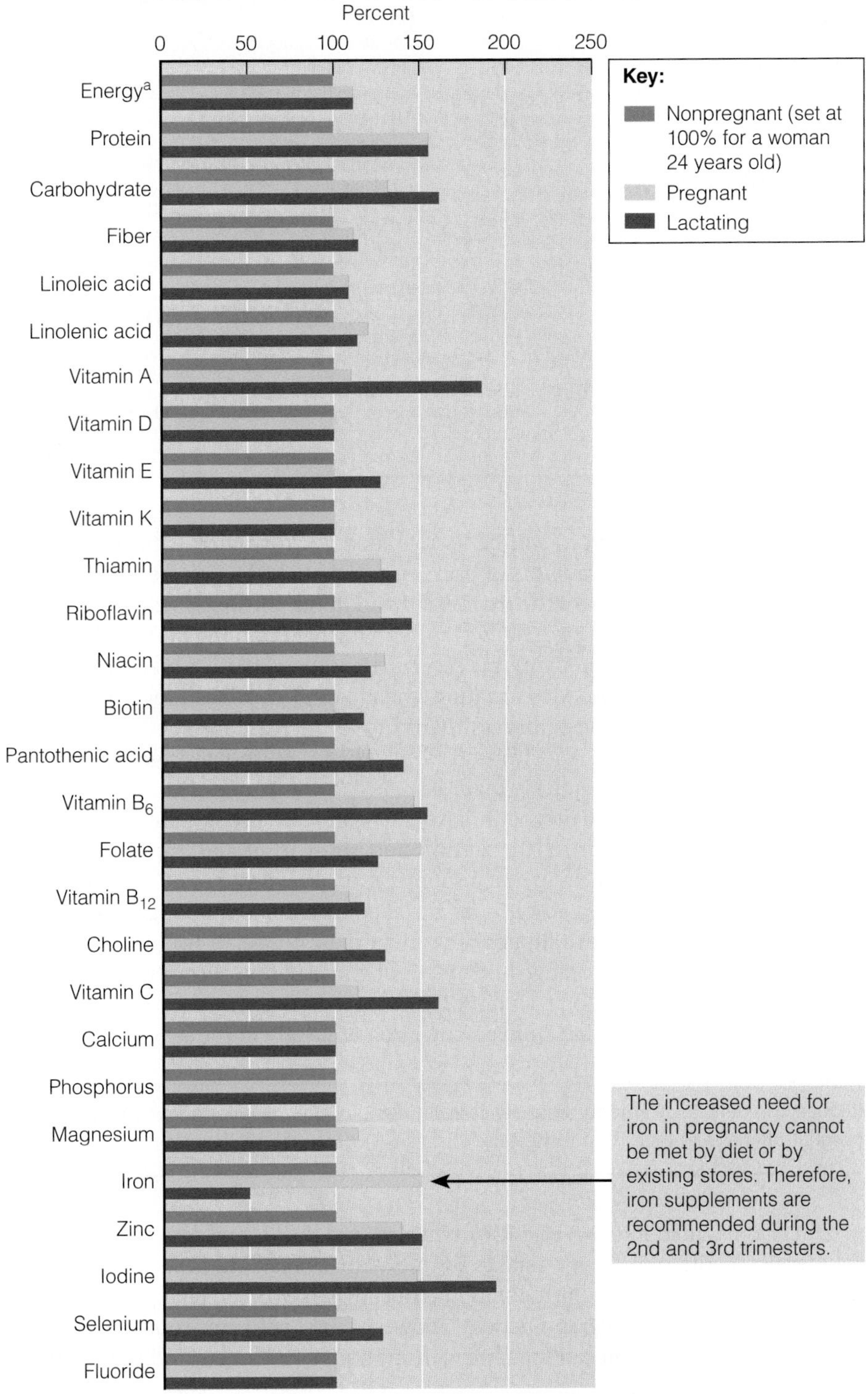

Key:
- Nonpregnant (set at 100% for a woman 24 years old)
- Pregnant
- Lactating

Percent

Energy[a]
Protein
Carbohydrate
Fiber
Linoleic acid
Linolenic acid
Vitamin A
Vitamin D
Vitamin E
Vitamin K
Thiamin
Riboflavin
Niacin
Biotin
Pantothenic acid
Vitamin B_6
Folate
Vitamin B_{12}
Choline
Vitamin C
Calcium
Phosphorus
Magnesium
Iron
Zinc
Iodine
Selenium
Fluoride

The increased need for iron in pregnancy cannot be met by diet or by existing stores. Therefore, iron supplements are recommended during the 2nd and 3rd trimesters.

[a]Energy allowance during pregnancy is for 2nd trimester; energy allowance during the 3rd trimester is slightly higher; no additional allowance is provided during the 1st trimester. Energy allowance during lactation is for the first 6 months; energy allowance during the second 6 months is slightly higher.

Nutrition Explorer   Examine how pregnancy and lactation alter the nutrient needs of women, and examine the nutrient needs of infants.

FIGURE 15-11 Daily Food Choices for Pregnant and Lactating Women

Pregnant or Lactating Women

3 to 4

3

4 to 5

3 to 4

7 to 11

SAMPLE MENU

Breakfast
1 English muffin
2 tbs peanut butter
1 c low-fat vanilla yogurt
½ c fresh strawberries
1 c orange juice

Midmorning snack
1 c cranberry juice
1 oz pretzels

Lunch
Sandwich (tuna salad on whole-wheat bread)
½ carrot (sticks)
1 c low-fat milk

Dinner
Chicken cacciatore
  4 oz chicken
  ¾ c stewed tomatoes
1 c rice
¾ c summer squash
1½ c salad (spinach, mushrooms, onions)
1 tbs salad dressing
2 slices Italian bread
2 tsp butter or margarine
1 c low-fat milk

Evening snack
1 c low fat-milk
3 oatmeal cookies

NOTE: This sample meal plan follows the Daily Food Guide for pregnant and lactating women and provides about 2500 kcalories (55% from carbohydrate, 20% from protein, and 25% from fat). Figure 2-2 in Chapter 2 provides a detailed summary of the Daily Food Guide.

when milk, which lacks iron, will be its sole food. Even women with inadequate stores transfer significant amounts of iron, suggesting that the iron needs of the fetus have priority over those of the mother.[21] In addition, the blood losses inevitable at birth, especially during a cesarean section, can further drain the mother's supply.*

During pregnancy, the body makes several adaptations to help meet the exceptionally high need for iron. Menstruation, the major route of iron loss in women, ceases, and iron absorption improves thanks to an increase in blood transferrin, the body's iron-absorbing and iron-carrying protein. Without sufficient intake, though, iron stores would quickly dwindle.

Few women enter pregnancy with adequate iron stores, so a daily iron supplement is recommended during the second and third trimesters for all pregnant women. For this reason, most prenatal supplements provide 30 to 60 milligrams of iron a day. To enhance iron absorption, the supplement should be taken between meals or at bedtime and with liquids other than milk, coffee, or tea, which inhibit iron absorption. (Drinking orange juice does not enhance iron absorption from supplements as it does from foods; vitamin C enhances iron absorption by converting iron from ferric to ferrous, but supplemental iron is already in the ferrous form.)

HEALTHY PEOPLE 2010 Reduce iron deficiency among pregnant females. Reduce anemia among low-income pregnant females in their third trimester.

■ Zinc RDA during pregnancy:
  • 12 mg/day (≤18 yr).
  • 11 mg/day (19–50 yr).

Zinc■ is required for DNA and RNA synthesis and thus for protein synthesis and cell development. Typical zinc intakes for pregnant women are lower than recommendations, but routine supplementation is not advised. Women taking iron supplements (more than 30 milligrams per day), however, may need zinc supplementation because large doses of iron can interfere with the body's absorption and use of zinc.

Nutrients for Bone Development Vitamin D and the bone-building minerals calcium, phosphorus, magnesium, and fluoride are in great demand during pregnancy. Insufficient intakes may produce abnormal fetal bones and teeth.

■ The AI for Vitamin D does not increase during pregnancy.

Vitamin D■ plays a vital role in calcium absorption and utilization. Consequently, severe maternal vitamin D deficiency interferes with normal calcium metabolism, resulting in rickets in the fetus and osteomalacia in the mother. Regular

---

*On average almost twice as much blood is lost during a cesarean delivery as during the average vaginal delivery of a single fetus.

exposure to sunlight and consumption of vitamin D–fortified milk are usually sufficient to provide the recommended amount of vitamin D during pregnancy. Routine supplementation is not recommended because of the toxicity risk. Vegans who avoid milk, eggs, and fish may receive enough vitamin D from regular exposure to sunlight and from fortified soy milk.

Calcium absorption more than doubles early in pregnancy, helping the mother to meet the calcium needs of pregnancy.■ During the last trimester, as the fetal bones begin to calcify, over 300 milligrams a day are transferred to the fetus. Recommendations to ensure an adequate calcium intake during pregnancy are aimed at conserving maternal bone while supplying fetal needs.

Calcium intakes for pregnant women■ typically fall below recommendations. Because bones are still actively depositing minerals until about age 25 or so, adequate calcium is especially important for young women. Pregnant women under age 25 who receive less than 600 milligrams of dietary calcium daily need to increase their intake of milk, cheese, yogurt, and other calcium-rich foods. Alternatively, and less preferably, they may need a daily supplement of 600 milligrams of calcium.

Other Nutrients The nutrients mentioned here are those most intensely involved in blood production, cell growth, and bone growth. Of course, other nutrients are also needed during pregnancy to support the growth and health of both fetus and mother. Even with adequate nutrition, repeated pregnancies less than a year apart deplete nutrient reserves: fetal growth may be compromised and maternal health may decline. The optimal interval between pregnancies is 18 to 23 months.[22]

Nutrient Supplements Physicians routinely recommend daily multivitamin-mineral supplements for pregnant women. Prenatal supplements typically contain greater amounts of folate, iron, and calcium than regular vitamin-mineral supplements. These supplements are particularly beneficial for women who do not eat adequately and for those in high-risk groups: women carrying multiple fetuses, cigarette smokers, and alcohol and drug abusers. The use of prenatal supplements may help reduce the risks of preterm delivery, low infant birthweights, and birth defects.[23] Figure 15-12 presents a label from a standard prenatal supplement.

■ The AI for calcium does not increase during pregnancy.

■ The Food Guide Pyramid suggests 3 servings of milk for women who are pregnant or breastfeeding.

FIGURE 15-12 Example of a Prenatal Supplement

Supplement Facts
Serving Size 1 Tablet

Amount Per Tablet	% Daily Value for Pregnant/ Lactating Women
Vitamin A 4000 IU	50%
Vitamin C 100 mg	167%
Vitamin D 400 IU	100%
Vitamin E 11 IU	37%
Thiamin 1.84 mg	108%
Riboflavin 1.7 mg	85%
Niacin 18 mg	90%
Vitamin B6 2.6 mg	104%
Folate 800 mcg	100%
Vitamin B12 4 mcg	50%
Calcium 200 mg	15%
Iron 27 mg	150%
Zinc 25 mg	167%

INGREDIENTS: calcium carbonate, microcrystalline cellulose, dicalcium phosphate, ascorbic acid, ferrous fumarate, zinc oxide, acacia, sucrose ester, niacinamide, modified cellulose gum, di-alpha tocopheryl acetate, hydroxypropyl methylcellulose, hydroxypropyl cellulose, artificial colors (FD&C blue no. 1 lake, FD&C red no. 40 lake, FD&C yellow no. 6 lake, titanium dioxide), polyethylene glycol, starch, pyridoxine hydrochloride, vitamin A acetate, riboflavin, thiamin mononitrate, folic acid, beta carotene, cholecalciferol, maltodextrin, gluten, cyanocobalamin, sodium bisulfite.

TABLE 15-2 Strategies to Alleviate Maternal Discomforts

To Alleviate the Nausea of Pregnancy	To Prevent or Alleviate Constipation	To Prevent or Relieve Heartburn
• On waking, arise slowly. • Eat dry toast or crackers. • Chew gum or suck hard candies. • Eat small, frequent meals. • Avoid foods with offensive odors. • When nauseated, do not drink citrus juice, water, milk, coffee, or tea.	• Eat foods high in fiber (fruits, vegetables, and whole-grain cereals). • Exercise regularly. • Drink at least eight glasses of liquids a day. • Respond promptly to the urge to defecate. • Use laxatives only as prescribed by a physician; do not use mineral oil, because it interferes with absorption of fat-soluble vitamins.	• Relax and eat slowly. • Chew food thoroughly. • Eat small, frequent meals. • Drink liquids between meals. • Avoid spicy or greasy foods. • Sit up while eating; elevate the head while sleeping. • Wait an hour after eating before lying down. • Wait two hours after eating before exercising.

Common Nutrition-Related Concerns of Pregnancy

Nausea, constipation, heartburn, and food sensitivities are common nutrition-related concerns during pregnancy. A few simple strategies can help alleviate maternal discomforts (see Table 15-2).

Nausea Not all women have queasy stomachs in the early months of pregnancy, but many do. The nausea of "morning" (actually, anytime) sickness ranges from mild queasiness to debilitating nausea and vomiting. Severe and continued vomiting may require hospitalization if it results in acidosis, dehydration, or excessive weight loss. The hormonal changes of early pregnancy seem to be responsible for a woman's sensitivities to the appearance, texture, or smell of foods. Traditional strategies for quelling nausea are listed in Table 15-2, but some women benefit most from simply eating the foods they want when they feel like eating. They may also find comfort in a cleaner, quieter, and more temperate environment.

Constipation and Hemorrhoids As the hormones of pregnancy alter muscle tone and the growing fetus crowds intestinal organs, an expectant mother may experience constipation. She may also develop hemorrhoids (swollen veins of the rectum). These can be painful, and straining during bowel movements may cause bleeding. She can gain relief by following the strategies listed in Table 15-2.

Heartburn Heartburn is another common complaint during pregnancy. The hormones of pregnancy relax the digestive muscles, and the growing fetus puts increasing pressure on the mother's stomach. This combination allows stomach acid to back up into the lower esophagus and create a burning sensation near the heart. Tips to help relieve heartburn are included in Table 15-2.

Food Cravings and Aversions Some women develop cravings for, or aversions to, particular foods and beverages during pregnancy. These **food cravings** and **food aversions** are fairly common, but do not seem to reflect real physiological needs. In other words, a woman who craves pickles does not necessarily need salt. Similarly, cravings for ice cream are common in pregnancy, but do not signify a calcium deficiency. Cravings and aversions that arise during pregnancy are most likely due to hormone-induced changes in sensitivity to taste and smell.

Nonfood Cravings Some pregnant women develop cravings for nonfood items■ such as laundry starch, clay, soil, or ice—a practice known as pica.[24] Pica is a cultural phenomenon that reflects a society's folklore; it is especially common among African American women. Pica is often associated with iron-deficiency anemia, but whether iron deficiency leads to pica or pica leads to iron deficiency is unclear. Eating clay or soil may interfere with iron absorption and displace iron-rich foods from the diet.

■ Reminder: The general term for eating nonfood items is *pica*. The specific craving for nonfood items that come from the earth, such as clay or dirt, is known as *geophagia*.

food cravings: strong desires to eat particular foods.

food aversions: strong desires to avoid particular foods.

IN SUMMARY Energy and nutrient needs are high during pregnancy. A balanced diet that includes an extra serving from each of the five food groups can usually meet these needs, with the exception of iron and folate (supplements are recommended). The nausea, constipation, and heartburn that sometimes accompany pregnancy can usually be alleviated with a few simple strategies. Food cravings do not typically reflect physiological needs.

High-Risk Pregnancies

Some pregnancies jeopardize the life and health of the mother and infant. Table 15-3 identifies several characteristics of a **high-risk pregnancy.** A woman with none of these risk factors is said to have a **low-risk pregnancy.** The more factors that apply, the higher the risk. All pregnant women, especially those in high-risk categories, need prenatal care, including dietary■ advice.

■ Nutrition advice in prenatal care:
- Eat well-balanced meals.
- Gain enough weight to support fetal growth.
- Take prenatal supplements as prescribed.
- Stop drinking alcohol.

Increase the proportion of pregnant women who receive early and adequate prenatal care.

The Infant's Birthweight

A high-risk pregnancy is likely to produce an infant with **low birthweight.** Low-birthweight infants, defined as infants who weigh 5½ pounds or less, are classified according to their gestational age. Preterm infants are born before they are fully developed; they are often underweight and have trouble breathing because their lungs are immature. Preterm infants may be small, but if their size and weight are appropriate for their age,■ they can catch up in growth given adequate nutrition support. In contrast, small-for-gestational-age infants have

■ Some preterm infants are of a weight **appropriate for gestational age (AGA);** others are **small for gestational age (SGA),** often reflecting malnutrition.

TABLE 15-3	High-Risk Pregnancy Factors
Factor	**Condition That Raises Risk**
Maternal weight	
Prior to pregnancy	Prepregnancy BMI either <18.5 or >25
During pregnancy	Insufficient or excessive pregnancy weight gain
Maternal nutrition	Nutrient deficiencies or toxicities; eating disorders
Socioeconomic status	Poverty, lack of family support, low level of education, limited food available
Lifestyle habits	Smoking, alcohol or other drug use
Age	Teens, especially 15 years or younger; women 35 years or older
Previous pregnancies	
Number	Many previous pregnancies (3 or more to mothers under age 20; 4 or more to mothers age 20 or older)
Interval	Short intervals between pregnancies (<18 months)
Outcomes	Previous history of problems
Multiple births	Twins or triplets
Birthweight	Low- or high-birthweight infants
Maternal health	
High blood pressure	Development of pregnancy-related hypertension
Diabetes	Development of gestational diabetes
Chronic diseases	Diabetes; heart, respiratory, and kidney disease; certain genetic disorders; special diets and medications

high-risk pregnancy: a pregnancy characterized by indicators that make it likely the birth will be surrounded by problems such as premature delivery, difficult birth, retarded growth, birth defects, and early infant death.

low-risk pregnancy: a pregnancy characterized by indicators that make a normal outcome likely.

low birthweight (LBW): a birthweight of 5½ lb (2500 g) or less; indicates probable poor health in the newborn and poor nutrition status in the mother during pregnancy, before pregnancy, or both. Normal birthweight for a full-term baby is 6½ to 8¾ lb (about 3000 to 4000 g).

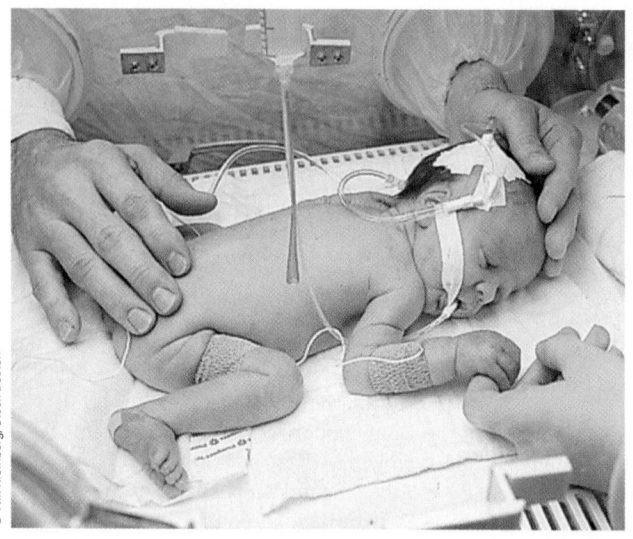

Low-birthweight babies need special care and nourishment.

suffered growth failure in the uterus and do not catch up as well. For the most part, survival improves with increased gestational age and birthweight.

Low-birthweight infants are more likely to experience complications during delivery than normal-weight babies. They also have a statistically greater chance of having physical and mental birth defects, contracting diseases, and dying early in life. Of infants who die before their first birthdays, about two-thirds were low-birthweight newborns. Very-low-birthweight infants (3½ pounds or less) struggle not only for their immediate physical health and survival, but for their future cognitive development and abilities as well.[25]

A strong relationship is evident between socioeconomic disadvantage and low birthweight. Low socioeconomic status impairs fetal development by causing stress and by limiting access to medical care and to nutritious foods. Low socioeconomic status often accompanies teen pregnancies, smoking, and alcohol and drug abuse—all predictors of low birthweight.

HEALTHY PEOPLE 2010

Reduce low birthweight (LBW) and very low birthweight (VLBW).

Malnutrition and Pregnancy

Good nutrition clearly supports a pregnancy. In contrast, malnutrition interferes with the ability to conceive, the likelihood of implantation, and the subsequent development of a fetus should conception and implantation occur.

Malnutrition and Fertility The nutrition habits and lifestyle choices people make can influence the course of a pregnancy they are not even planning at the time. Severe malnutrition and food deprivation can reduce fertility: women may develop amenorrhea,■ and men may lose their ability to produce viable sperm. Furthermore, both men and women lose sexual interest during times of starvation. Starvation arises predictably during famines, wars, and droughts, but can also occur amidst peace and plenty. Many young women who diet excessively are starving and suffering from malnutrition (see Highlight 9).

Malnutrition and Early Pregnancy If a malnourished woman does become pregnant, she faces the challenge of supporting both the growth of a baby and her own health with inadequate nutrient stores. Malnutrition prior to and around conception prevents the placenta from developing fully. A poorly developed placenta cannot deliver optimum nourishment to the fetus, and the infant will be born small and possibly with physical and cognitive abnormalities. If this small infant is a female, she may develop poorly and have an elevated risk of developing a chronic condition that could impair her ability to give birth to a healthy infant. Thus a woman's malnutrition can adversely affect not only her children but her *grandchildren*.

Malnutrition and Fetal Development Without adequate nutrition during pregnancy, fetal growth and infant health are compromised. In general, consequences of malnutrition during pregnancy include fetal growth retardation, congenital malformations (birth defects), spontaneous abortion and stillbirth, premature birth, and low infant birthweight. Malnutrition, coupled with low birthweight, is a factor in more than half of all deaths of children under four years of age worldwide.

Food Assistance Programs

Women in high-risk pregnancies can find assistance from the WIC program—a high-quality, cost-effective health care and nutrition services program for women, infants, and children in the United States.[26] Formally known as the Special Sup-

■ Reminder: *Amenorrhea* is the temporary or permanent absence of menstrual periods. Amenorrhea is normal before puberty, after menopause, during pregnancy, and during lactation; otherwise it is abnormal.

plemental Nutrition Program for Women, Infants, and Children, WIC provides nutrition education and nutritious foods to infants, children up to age five, and pregnant and breastfeeding women who qualify financially and have a high risk of medical or nutritional problems. The program is both remedial and preventive: services include health care referrals, nutrition education, and food packages or vouchers for specific foods. These foods supply nutrients known to be lacking in the diets of the target population: most notably, protein, calcium, iron, vitamin A, and vitamin C. WIC-sponsored foods include tuna fish, carrots, eggs, milk, iron-fortified cereal, vitamin C–rich juice, cheese, legumes, peanut butter, and infant formula.

Over 7 million people—most of them young children—receive WIC benefits each month. Prenatal WIC participation can effectively reduce infant mortality, low birthweight, and maternal and newborn medical costs. In 2002, Congress appropriated over $4 billion for WIC. For every dollar spent on WIC, an estimated three dollars in medical costs are saved in the first two months after birth.

Maternal Health

Medical disorders can threaten the life and health of both mother and fetus. If diagnosed and treated early, many diseases can be managed to ensure a healthy outcome—another strong argument for early prenatal care.

Preexisting Diabetes Whether diabetes presents risks depends on how well it is controlled before and during pregnancy. Without proper management of maternal diabetes, women face high infertility rates, and those who do conceive may experience episodes of severe hypoglycemia or hyperglycemia, spontaneous abortions, and pregnancy-related hypertension. Infants may be large, suffer physical and mental abnormalities, and experience other complications such as severe hypoglycemia or respiratory distress, both of which can be fatal. Ideally, a woman with diabetes will receive the prenatal care needed to achieve glucose control before conception and continued glucose control throughout pregnancy.

Gestational Diabetes Approximately 1 in 14 women who does not have diabetes develops a condition known as **gestational diabetes** during pregnancy. Gestational diabetes usually develops during the second half of pregnancy, with subsequent return to normal after childbirth. Some women with gestational diabetes, however, develop diabetes (usually type 2) after pregnancy, especially if they are overweight. For this reason, health care professionals advise against excessive weight gain.

The most common consequences of gestational diabetes are complications during labor and delivery and a high infant birthweight.[27] Birth defects associated with gestational diabetes include heart damage, limb deformities, and neural tube defects. To ensure that the problems of gestational diabetes are dealt with promptly, physicians screen for the risk factors■ listed in the margin and test high-risk women for glucose intolerance immediately and average-risk women between 24 and 28 weeks gestation.[28] Dietary recommendations should meet the needs of pregnancy and maternal blood glucose goals.[29] To maintain normal blood glucose levels, carbohydrates should be restricted to 35 to 40 percent of energy intake. To limit excessive weight gain, obese women should limit energy intake to about 25 kcalories per kilogram body weight. Diet and moderate exercise may control gestational diabetes, but if blood glucose fails to normalize, insulin or other drugs may be required.[30]

Preexisting Hypertension Hypertension complicates pregnancy and affects its outcome in different ways, depending on when the hypertension first develops and on how severe it becomes. In addition to the threats hypertension always carries (such as heart attack and stroke), high blood pressure increases the risks of a low-birthweight infant or the separation of the placenta from the wall of the uterus before the birth, resulting in stillbirth. Ideally, before a woman with hypertension becomes pregnant, her blood pressure will be under control.

■ Risk factors for gestational diabetes:
- Age 35 or older.
- BMI >25 or excessive weight gain.
- Complications in previous pregnancies, including high-birthweight infant.
- Symptoms of diabetes.
- Family history of diabetes.
- Hispanic, black, Native American, South or East Asian, Pacific Islander, or indigenous Australian.

gestational diabetes: abnormal glucose tolerance during pregnancy.

■ The hypertensive diseases of pregnancy are sometimes called **toxemia**.

■ The normal edema of pregnancy responds to gravity; fluid pools in the ankles. The edema of preeclampsia is a generalized edema. The differences between these two types of edema help with the diagnosis of preeclampsia.

■ Warning signs of preeclampsia:
  • Hypertension.
  • Protein in the urine.
  • Upper abdominal pain.
  • Severe and constant headaches.
  • Swelling, especially of the face.
  • Dizziness.
  • Blurred vision.
  • Sudden weight gain (1 lb/day).
  • Fetal growth retardation.

transient hypertension of pregnancy: high blood pressure that develops in the second half of pregnancy and resolves after childbirth, usually without affecting the outcome of the pregnancy.

preeclampsia (PRE-ee-KLAMP-see-ah): a condition characterized by hypertension, fluid retention, and protein in the urine; formerly known as *pregnancy-induced hypertension.**

eclampsia (eh-KLAMP-see-ah): a severe stage of preeclampsia characterized by convulsions.

---

*The Working Group on High Blood Pressure in Pregnancy, convened by the National High Blood Pressure Education Program of the National Heart, Lung, and Blood Institute, suggested abandoning the term *pregnancy-induced hypertension* because it failed to differentiate between the mild, transient hypertension of pregnancy and the life-threatening hypertension of preeclampsia.

Transient Hypertension of Pregnancy Some women develop hypertension during the second half of pregnancy.* Most often, the rise in blood pressure is mild and does not affect the pregnancy adversely. Blood pressure usually returns to normal during the first few weeks after childbirth. This **transient hypertension of pregnancy** differs from the life-threatening hypertensive diseases■ of pregnancy—preeclampsia and eclampsia.

Preeclampsia and Eclampsia Hypertension may signal the onset of **preeclampsia,** a condition characterized not only by high blood pressure but also by protein in the urine and fluid retention (edema). The edema■ of preeclampsia is a whole-body edema, distinct from the localized fluid retention women normally experience late in pregnancy.

Preeclampsia usually occurs with first pregnancies■ and most often after 20 weeks gestation. Symptoms typically regress within two days of delivery. Both men and women who were born of pregnancies complicated by preeclampsia are more likely to have a child born of a pregnancy complicated by preeclampsia, suggesting a genetic predisposition.[31] Black women have a much greater risk of preeclampsia than white women.

Preeclampsia affects almost all of the mother's organs—the circulatory system, liver, kidneys, and brain. Blood flow through the vessels that supply oxygen and nutrients to the placenta diminishes. For this reason, preeclampsia often retards fetal growth. In some cases, the placenta separates from the uterus, resulting in premature birth or stillbirth.

Preeclampsia can progress rapidly to **eclampsia**—a condition characterized by convulsive seizures and coma. Maternal death during pregnancy and childbirth is extremely rare in developed countries, but when it does occur, eclampsia is a common cause. The rate of death for black women with eclampsia is over four times the rate for white women.[32]

Preeclampsia demands prompt medical attention. Treatment focuses on controlling blood pressure and preventing convulsions. If preeclampsia develops early and is severe, induced labor or cesarean section may be necessary, regardless of gestational age. The infant will be preterm, with all of the associated problems, including poor lung development and special care needs. Several dietary factors have been studied, but none have proved conclusive in preventing preeclampsia.[33] Calcium supplementation may be effective for some women.[34]

The Mother's Age

Maternal age also influences the course of a pregnancy. Compared with women of the physically ideal childbearing age of 20 to 25, both younger and older women face more complications of pregnancy.

Pregnancy in Adolescents Many adolescents become sexually active before age 19, and over 800,000 adolescent girls face pregnancies each year in the United States; over half of them give birth.[35] Put another way, about 1 out of every 20 babies is born to a teenager. Nourishing a growing fetus adds to a teenage girl's nutrition burden, especially if her growth is still incomplete. Simply being young increases the risks of pregnancy complications independently of important socioeconomic factors.

Common complications among adolescent mothers include iron-deficiency anemia (which may reflect poor diet and inadequate prenatal care) and prolonged labor (which reflects the mother's physical immaturity). On a positive note, maternal death is lowest for mothers under age 20.

---

*Blood pressure of 140/90 millimeters mercury during the second half of pregnancy in a woman who has not previously exhibited hypertension indicates high blood pressure. So does a rise in systolic blood pressure of 30 millimeters or in diastolic blood pressure of 15 millimeters on at least two occasions more than six hours apart. By this rule, an apparently "normal" blood pressure of 120/85 would be high for a woman whose normal value was 90/70.

Pregnant teenagers have higher rates of stillbirths, preterm births, and low-birthweight infants than do adult women. Many of these infants suffer physical problems, require intensive care, and die within the first year. The care of infants born to teenagers costs our society an estimated $1 billion annually. Because teenagers have few financial resources, they cannot pay these costs. Furthermore, their low economic status contributes significantly to the complications surrounding their pregnancies. At a time when prenatal care is most important, it is less accessible. And the pattern of teenage pregnancies continues from generation to generation, with almost 40 percent of the daughters born to teenage mothers becoming teenage mothers themselves. Clearly, teenage pregnancy is a major public health problem.

Reduce pregnancies among adolescent females.

To support the needs of both mother and fetus, young teenagers (13 to 16 years old) are encouraged to strive for the highest weight gains recommended for pregnancy. For a teen who enters pregnancy at a healthy body weight, a weight gain of approximately 35 pounds is recommended; this amount minimizes the risk of delivering a low-birthweight infant. Gaining less weight may limit fetal growth. Pregnant and lactating teenagers can use the food guide presented in Figure 15-11 (on p. 520), making sure to select at least 4 servings of milk or milk products daily.

Without the appropriate economic, social, and physical support, a young mother will not be able to care for herself during her pregnancy and for her child after the birth. To improve her chances for a successful pregnancy and a healthy infant, she must seek prenatal care. WIC helps pregnant teenagers obtain adequate food for themselves and their infants (WIC was introduced on p. 524).

Pregnancy in Older Women In the last several decades, many women have delayed childbearing while they pursue education and careers. As a result, the number of first births to women 35 and older has increased dramatically. Most of these women, even those over age 50, have healthy pregnancies.[36]

The few complications associated with later childbearing often reflect chronic conditions such as hypertension and diabetes, which can complicate an otherwise healthy pregnancy. These complications may result in a cesarean section, which is twice as common in women over 35 as among younger women. For all these reasons, maternal death rates are higher in women over 35 than in younger women.

The babies of older mothers face problems of their own including higher rates of premature births and low birthweight.[37] Their rates of birth defects are also high. Because 1 out of 50 pregnancies in older women produces an infant with genetic abnormalities, obstetricians routinely screen women older than 35. For a 40-year-old mother, the risk of having a child with **Down syndrome,** for example, is about 1 in 100 compared with 1 in 300 for a 35-year-old and 1 in 10,000 for a 20-year-old. In addition, fetal death is twice as high for women 35 years and older than for younger women. Why this is so remains a bit of a mystery. One possibility is that the uterine blood vessels of older women may not fully adapt to the increased demands of pregnancy.

Practices Incompatible with Pregnancy

Besides malnutrition, a variety of lifestyle factors can have adverse effects on pregnancy; and some may be teratogenic.■ People who are planning to have children can make the choice to practice healthy behaviors.

Alcohol One out of eight pregnant women drinks alcohol at some time during her pregnancy; 1 out of 30 drinks frequently.[38] Alcohol consumption during pregnancy can cause irreversible mental and physical retardation of the fetus—fetal alcohol

■ Reminder: The word *teratogenic* describes a factor that causes abnormal fetal development and birth defects.

Down syndrome: a genetic abnormality that causes mental retardation, short stature, and flattened facial features.

syndrome (FAS). Of the leading causes of mental retardation, FAS is the only one that is totally *preventable*. To that end, the surgeon general urges all pregnant women to refrain from drinking alcohol. Fetal alcohol syndrome is the topic of Highlight 15, which includes mention of how alcohol consumption by men may also affect fertility and fetal development.

HEALTHY PEOPLE 2010 — Increase abstinence from alcohol among pregnant women. Reduce the occurrence of fetal alcohol syndrome (FAS).

Medicinal Drugs Drugs other than alcohol can also cause complications during pregnancy, problems in labor, and serious birth defects. For these reasons, pregnant women should not take any medicines without consulting their physicians, who must weigh the benefits against the risks.[39]

Herbal Supplements Similarly, pregnant women should seek a physician's advice before using herbal supplements as well. Women sometimes seek herbal preparations during their pregnancies to induce labor, aid digestion, promote water loss, support restful sleep, and fight depression. As Highlight 18 explains, some herbs may be safe, but many others are definitely harmful.

Illicit Drugs The recommendation to avoid drugs during pregnancy also includes illicit drugs, of course. Unfortunately, use of illicit drugs, such as cocaine and marijuana, is common among some pregnant women.*

Drugs of abuse, such as cocaine, easily cross the placenta and impair fetal growth and development.[40] Furthermore, they are responsible for preterm births, low-birthweight infants, perinatal deaths,■ and sudden infant deaths. If these newborns survive, central nervous system damage is evident: their cries, sleep, and behaviors early in life are abnormal, and their cognitive development later in life is impaired.[41] They may be hypersensitive or underaroused; those who test positive for drugs suffer the greatest effects of toxicity and withdrawal.[42]

Smoking and Chewing Tobacco Smoking cigarettes or chewing tobacco at any time exerts harmful effects, and pregnancy dramatically magnifies the hazards of these practices. Smoking restricts the blood supply to the growing fetus and so limits oxygen and nutrient delivery and waste removal. A mother who smokes is more likely to have a complicated birth and a low-birthweight infant.[43] Indeed, of all preventable causes of low birthweight in the United States, smoking has the greatest impact. Although most infants born to cigarette smokers are low birthweight, some are not, suggesting that the effect of smoking on birthweight also depends, in part, on genes involved in the metabolism of smoking toxins.[44] Smokers also tend to eat less nutritious foods during their pregnancies than do nonsmokers, further impairing fetal development. Unfortunately, an estimated one out of eight pregnant women smokes, and rates are even higher for unmarried women and those who have not graduated from high school.[45]

In addition to contributing to low birthweight, smoking can cause death in an otherwise healthy fetus or newborn. A positive relationship exists between **sudden infant death syndrome (SIDS)** and both cigarette smoking during pregnancy and postnatal exposure to passive smoke. Smoking during pregnancy may even harm the intellectual and behavioral development of the child later in life. The margin■ lists other complications of smoking during pregnancy.

■ The word *perinatal* refers to the time between the 28th week of gestation and one month after birth.

■ Complications associated with smoking during pregnancy:
- Fetal growth retardation.
- Low birthweight.
- Complications at birth (prolonged final stage of labor).
- Mislocation of the placenta.
- Premature separation of the placenta.
- Vaginal bleeding.
- Spontaneous abortion.
- Fetal death.
- Sudden infant death syndrome (SIDS).
- Middle ear diseases.
- Cardiac and respiratory diseases.

sudden infant death syndrome (SIDS): the unexpected and unexplained death of an apparently well infant; the most common cause of death of infants between the second week and the end of the first year of life; also called *crib death*.

HEALTHY PEOPLE 2010 — Increase smoking cessation during pregnancy. Increase abstinence from cigarettes among pregnant women.

---

*It is estimated that 17 percent of pregnant women use marijuana and at least 6 percent use cocaine.

Infants of mothers who chew tobacco also have low birthweights and high rates of fetal deaths. Any woman who smokes cigarettes or chews tobacco and is considering pregnancy or who is already pregnant should try to quit.

Environmental Contaminants Infants and young children of pregnant women exposed to environmental contaminants such as lead show signs of delayed mental and psychomotor development. During pregnancy, lead readily moves across the placenta, inflicting severe damage on the developing fetal nervous system.[46] In addition, infants exposed to even low levels of lead during gestation weigh less at birth and consequently struggle to survive. For these reasons, it is particularly important that pregnant women receive foods and beverages grown and prepared in environments free of contamination. A diet high in calcium will also help to defend against lead contamination.[47] Breastfeeding the infant may also help to counterbalance the developmental damage incurred from contamination during pregnancy.[48]

Among the contaminants of concern is mercury. As Chapter 5 mentioned, fatty fish are a good source of omega-3 fatty acids, but some fish contain large amounts of the pollutant mercury, which can harm the developing brain and nervous system.[49] For this reason, pregnant (and lactating) women should:[50]

- Avoid shark, swordfish, king mackerel, and tilefish.
- Limit average weekly consumption to 12 ounces (cooked or canned) of ocean, coastal, and other commercial fish *or* to 6 ounces (cooked) of freshwater fish caught by family and friends.

Supplements of fish oil are not recommended both because they may contain concentrated toxins and because their effects on pregnancy remain unknown.

Foodborne Illness As Chapter 19 explains, foodborne illnesses arise when people eat foods that contain infectious microbes or microbes that produce toxins. At best, the vomiting and diarrhea associated with these illnesses can leave a pregnant woman exhausted and dehydrated; at worse, foodborne illnesses■ can cause meningitis, pneumonia, or even fetal death. Chapter 19 presents precautions to minimize the risks of foodborne illness.

Vitamin-Mineral Megadoses The pregnant woman who is trying to eat well may mistakenly assume that more is better when it comes to vitamin-mineral supplements. This is simply not true; many vitamins and minerals are toxic when taken in excess. Excessive vitamin A is particularly infamous for its role in malformations of the cranial nervous system. Intakes before the seventh week appeared to be the most damaging. (Review Figure 15-4 on p. 510 to see how many tissues are in their critical periods prior to the seventh week.) For this reason, vitamin A is not given as a supplement in the first trimester of pregnancy unless there is specific evidence of deficiency, which is rare. A pregnant woman can obtain all the vitamin A and most of the other vitamins and minerals she needs by making wise food choices. She should take supplements only on the advice of a registered dietitian or physician.

Caffeine Caffeine crosses the placenta, and the developing fetus has a limited ability to metabolize it. Research studies have not proved that caffeine (even in high doses) causes birth defects in human infants (as it does in animals), but some evidence suggests that moderate-to-heavy use may increase the risk of spontaneous abortion.[51](Heavy caffeine use was defined as the equivalent of 3 to 6 cups of coffee a day.) All things considered, it might be most sensible to limit caffeine consumption to the equivalent of a cup of coffee or two 12-ounce cola beverages a day. (The caffeine contents of selected beverages, foods, and drugs are listed at the beginning of Appendix H.)

Weight-Loss Dieting Weight-loss dieting, even for short periods, is hazardous during pregnancy. Low-carbohydrate diets or fasts that cause ketosis deprive the fetal brain of needed glucose and may impair cognitive development. Such diets are also likely to lack other nutrients vital to fetal growth. Regardless of prepregnancy weight, pregnant women should never intentionally lose weight.

Young adults can prepare for a healthy pregnancy by taking care of themselves today.

© Jose I. Pelaez, Inc./CORBIS

■ Pregnant women are about 20 times more likely than other healthy adults to get the foodborne illness *listeriosis*. To prevent listeriosis:
- Use only pasteurized dairy products; avoid Mexican soft cheeses, feta cheese, brie, Camembert, and blue-veined cheeses such as Roquefort.
- Thoroughly cook meat, poultry, and seafood.
- Thoroughly reheat hot dogs, luncheon meats, and deli meats, including cured meats such as salami.
- Wash all fruits and vegetables.
- Avoid refrigerated pâté, meat spreads, smoked seafood such as salmon or trout, and any fish labeled "nova," "lox," or "kippered," unless prepared in a cooked dish.

Sugar Substitutes Artificial sweeteners have been extensively investigated and found to be acceptable during pregnancy if used within the Food and Drug Administration's guidelines (presented in Highlight 4).[52] Still, it would be prudent for pregnant women to use sweeteners in moderation and within an otherwise nutritious and well-balanced diet. Women with phenylketonuria should not use aspartame, as Highlight 4 explains.

> **IN SUMMARY** High-risk pregnancies, especially for teenagers, threaten the life and health of both mother and infant. Proper nutrition and abstinence from smoking, alcohol, and other drugs improve the outcome. In addition, prenatal care includes monitoring pregnant women for gestational diabetes and preeclampsia.

In general, most women can enjoy a healthy pregnancy if they:[53]

- Get prenatal care.
- Eat a balanced diet, safely prepared.
- Take prenatal supplements as prescribed.
- Gain a healthy amount of weight.
- Refrain from cigarettes, alcohol, and drugs (including herbs) unless prescribed by a physician.

Childbirth marks the end of pregnancy and the beginning of a new set of parental responsibilities—including feeding the newborn.

Nutrition during Lactation

■ To learn about breastfeeding, a pregnant woman can read at least one of the many books available. Nutrition on the Net at the end of this chapter provides a list of other nutrition resources, including LaLeche League International.

Before the end of her pregnancy, a woman will need to consider whether to feed her infant breast milk,■ infant formula, or both. These options are the only recommended foods for an infant during the first four to six months of life. The rate of breastfeeding is close to the Healthy People 2010 goal of 75 percent at birth, but falls far short of goals at six months and a year.[54] This section focuses on how the mother's nutrition supports the making of breast milk; the next chapter describes how the infant benefits from drinking breast milk.

Increase the proportion of mothers who breastfeed their babies.

In many countries around the world, a woman breastfeeds her newborn without considering the alternatives or consciously making a decision. In other parts of the world, a woman feeds her newborn formula simply because she knows so little about breastfeeding. She may have misconceptions or feel uncomfortable about a process she has never seen or experienced. Breastfeeding offers many health benefits to both mother and infant, and every pregnant woman should seriously consider it (see Table 15-4).[55] Still, there are valid reasons for not breastfeeding, and formula-fed infants grow and develop into healthy children.

Lactation: A Physiological Process

lactation: production and secretion of breast milk for the purpose of nourishing an infant.

mammary glands: glands of the female breast that secrete milk.

Lactation naturally follows pregnancy, as the mother's body continues to nourish the infant. The **mammary glands** secrete milk for this purpose. The mammary glands develop during puberty, but remain fairly inactive until pregnancy. During pregnancy, hormones promote the growth and branching of a duct system in the breasts and the development of the milk-producing cells.

TABLE 15-4	Benefits of Breastfeeding

For Infants:

- Provides the appropriate composition and balance of nutrients with high bioavailability.
- Provides hormones that promote physiological development.
- Improves cognitive development.
- Protects against a variety of infections.
- May protect against some chronic diseases, such as diabetes (type 1) and hypertension, later in life.
- Protects against food allergies.

For Mothers:

- Contracts the uterus.
- Delays the return of regular ovulation, thus lengthening birth intervals. (It is not, however, a dependable method of contraception.)
- Conserves iron stores (by prolonging amenorrhea).
- May protect against breast and ovarian cancer.

Other:

- Cost savings from not needing medical treatment for childhood illnesses or time off work to care for them.
- Cost savings from not needing to purchase formula (even after adjusting for added foods in the diet of a lactating mother).[a]
- Environmental savings to society from not needing to manufacture, package, and ship formula and dispose of the packaging.

[a] A nursing mother produces over 35 gallons of milk during the first six months, saving her roughly $450 in formula costs.

A woman who decides to breastfeed offers her infant a full array of nutrients and protective factors to support optimal health and development.

The hormones **prolactin** and **oxytocin** finely coordinate lactation. The infant's demand for milk stimulates the release of these hormones, which signal the mammary glands to supply milk. Prolactin is responsible for milk production. Prolactin concentrations remain high and milk production continues as long as the infant is nursing.

The hormone oxytocin causes the mammary glands to eject milk into the ducts, a response known as the **let-down reflex.** The mother feels this reflex as a contraction of the breast, followed by the flow of milk and the release of pressure. By relaxing and eating well, the nursing mother promotes easy let-down of milk and greatly enhances her chances of successful lactation.

Breastfeeding: A Learned Behavior

Lactation is an automatic physiological process that virtually all mothers are capable of doing. Breastfeeding, on the other hand, is a learned behavior that not all mothers decide to do. Of women who do breastfeed, those who receive early and repeated information and support breastfeed their infants longer than others. Health care professionals■ play an important role in providing encouragement and accurate information on breastfeeding.[56] Women who have been successful breastfeeding can offer advice and dispel misperceptions about lifestyle issues.[57] Table 15-5 (on p. 532) lists ten steps maternity facilities and health care professionals can take to promote successful breastfeeding among new mothers.[58]

The mother's partner also plays an important role in encouraging breastfeeding.[59] When partners support the decision, mothers are more likely to start and continue breastfeeding. Clearly, educating those closest to the mother could change attitudes and promote breastfeeding.

Most healthy women who want to breastfeed can do so with a little preparation. Physical obstacles to breastfeeding are rare, although most nursing mothers quit before the recommended six months because of perceived difficulties.[60] Successful

■ Some hospitals employ *certified lactation consultants* who specialize in helping new mothers establish a healthy breastfeeding relationship with their newborn. These consultants are often registered nurses with specialized training in breast and infant anatomy and physiology.

prolactin (pro-LAK-tin): a hormone secreted from the anterior pituitary gland that acts on the mammary glands to initiate and sustain milk production.
- **pro** = promote
- **lacto** = milk

oxytocin (OCK-see-TOH-sin): a hormone that stimulates the mammary glands to eject milk during lactation and the uterus to contract during childbirth.

let-down reflex: the reflex that forces milk to the front of the breast when the infant begins to nurse.

TABLE 15-5   Ten Steps to Successful Breastfeeding

To promote breastfeeding, every maternity facility should:

- Develop a written breastfeeding policy that is routinely communicated to all health care staff.
- Train all health care staff in the skills necessary to implement the breastfeeding policy.
- Inform all pregnant women about the benefits and management of breastfeeding.
- Help mothers initiate breastfeeding within ½ hour of birth.
- Show mothers how to breastfeed and how to maintain lactation, even if they need to be separated from their infants.
- Give newborn infants no food or drink other than breast milk, unless medically indicated.
- Practice rooming-in, allowing mothers and infants to remain together 24 hours a day.
- Encourage breastfeeding on demand.
- Give no artificial nipples or pacifiers to breastfeeding infants.[a]
- Foster the establishment of breastfeeding support groups and refer mothers to them at discharge from the facility.

[a]Compared with nonusers, infants who use pacifiers breastfeed less frequently and stop breastfeeding at a younger age. C. G. Victora and coauthors, Pacifier use and short breastfeeding duration: Cause, consequence, or coincidence? *Pediatrics* 99 (1997): 445–453.

SOURCE: United Nations Children's Fund and World Health Organization, *Protecting, Promoting and Supporting Breastfeeding: The Special Role of Maternity Services.*

breastfeeding requires adequate nutrition and rest. This, plus the support of all who care, will help to enhance the well-being of mother and infant.

The Mother's Nutrient Needs

Ideally, the mother who chooses to breastfeed her infant will continue to eat nutrient-dense foods throughout lactation. An adequate diet is needed to support the stamina, patience, and self-confidence that nursing an infant demands.

Energy Intake and Exercise   A nursing mother produces about 25 ounces of milk per day, with considerable variation from woman to woman and in the same woman from time to time, depending primarily on the infant's demand for milk. To produce an adequate supply of milk, a woman needs extra energy—almost 500 kcalories a day above her regular need during the first six months of lactation. To meet this energy need,■ she can eat an extra 330 kcalories of food each day and let the fat reserves she accumulated during pregnancy provide the rest. Most women need at least 1800 kcalories a day to receive all the nutrients required for successful lactation. Severe energy restriction may hinder milk production.

After the birth of the infant, many women are in a hurry to lose the extra body fat they accumulated during pregnancy. Opinions differ as to whether breastfeeding helps with postpartum weight loss. In general, most women lose 1 to 2 pounds a month during the first four to six months of lactation; some may lose more, and others may maintain or even gain weight. Neither the quality nor the quantity of breast milk is adversely affected by moderate weight loss, and infants grow normally.[61]

Women often exercise to lose weight and improve fitness, and this is compatible with breastfeeding and infant growth.[62] Intense physical activity can raise the lactic acid concentration of breast milk, which may influence the milk's taste. Some infants may prefer milk produced prior to exercise (which has a lower lactic acid content). In these cases, mothers can either breastfeed before exercise or express their milk before exercise for use afterward.

Energy Nutrients   Recommendations for protein and fatty acids remain about the same during lactation as during pregnancy, but increase for carbohydrates and fibers. Nursing mothers need additional carbohydrate to replace the glucose used to make the lactose in breast milk. The fiber recommendation is 1 gram higher simply because it is based on kcalorie intake, which increases during lactation.

■ Energy requirement during lactation:
- 1st 6 mo: +330 kcal/day.
- 2nd 6 mo: +400 kcal/day.

A jog through the park provides an opportunity for physical activity and fresh air.

© Ariel Skelley/CORBIS

Vitamins and Minerals A question often raised is whether a mother's milk may lack a nutrient if she fails to get enough in her diet. The answer differs from one nutrient to the next, but in general, nutritional inadequacies reduce the *quantity,* not the *quality,* of breast milk. Women can produce milk with adequate protein, carbohydrate, fat, and most minerals, even when their own supplies are limited. For these nutrients and for the vitamin folate as well, milk quality is maintained at the expense of maternal stores. This is most evident in the case of calcium: dietary calcium has no effect on the calcium concentration of breast milk, but maternal bones lose some of their density during lactation. Bone density increases again when lactation ends; breastfeeding has no long-term harmful effects on bones.[63] Nutrients in breast milk most likely to decline in response to prolonged inadequate intakes are the vitamins—especially vitamins B_6, B_{12}, A, and D. Review Figure 15-10 (on p. 519) to compare a lactating woman's nutrient needs with those of pregnant and nonpregnant women.

Nutritious foods support successful lactation.

Water Despite misconceptions, a mother who drinks more fluid does not produce more breast milk. To protect herself from dehydration, however, a lactating woman needs to drink plenty of fluids.■ A sensible guideline is to drink a glass of milk, juice, or water at each meal and each time the infant nurses.

■ AI for *total* water (including drinking water, other beverages, and foods) during lactation: 3 L/day.

Nutrient Supplements Most lactating women can obtain all the nutrients they need from a well-balanced diet without taking vitamin-mineral supplements. Nevertheless, some may need iron supplements, not to enhance the iron in their breast milk, but to refill their depleted iron stores. Maternal iron stores dwindle during pregnancy when the fetus takes iron to meet its own needs during the first four to six months after birth. In addition, childbirth may have incurred blood losses. Thus a woman may need iron supplements during lactation, even though, until menstruation resumes, her iron requirement is about half that of other nonpregnant women her age.

Particular Foods Foods with strong or spicy flavors (such as garlic) may alter the flavor of breast milk. A sudden change in the taste of the milk may annoy some infants. Familiar flavors may enhance enjoyment.[64]

Infants who develop symptoms of food allergy may be more comfortable if the mother's diet excludes the most common offenders—cow's milk, eggs, fish, peanuts, and tree nuts.[65] Generally, infants with a strong family history of food allergies benefit from breastfeeding.

A nursing mother can usually eat whatever nutritious foods she chooses. If she suspects a particular food is causing the infant discomfort, her physician may recommend a dietary challenge: eliminate the food from the diet to see if the infant's reactions subside; then return the food to the diet, and again monitor the infant's reactions. If a food must be eliminated for an extended time, appropriate substitutions must be made to ensure nutrient adequacy.

Practices Incompatible with Lactation

Some substances impair milk production or enter breast milk and interfere with infant development. This section discusses practices that a breastfeeding mother should avoid.

Alcohol Alcohol easily enters breast milk, and its concentration peaks within an hour of ingestion. Infants drink less breast milk when their mothers have consumed even small amounts of alcohol (equivalent to a can of beer). Three possible reasons, acting separately or together, may explain why. For one, the alcohol may have altered the flavor of the breast milk and thereby the infants' acceptance of it. For another, because infants metabolize alcohol inefficiently, even low doses may be potent enough to suppress their feeding and cause sleepiness. Third, the alcohol may have interfered with lactation by inhibiting the hormone oxytocin.

In the past, alcohol has been recommended to mothers to facilitate lactation despite a lack of scientific evidence that it does so. The research summarized here suggests that alcohol actually hinders breastfeeding. An occasional glass of wine or beer is considered within safe limits, but in general, lactating women should consume little or no alcohol.

Medicinal Drugs Most medicines are compatible with breastfeeding, but some are contraindicated, either because they suppress lactation or because they are secreted into breast milk and can harm the infant.[66] As a precaution, a nursing mother should consult with her physician prior to taking any drug, including herbal supplements.

Illicit Drugs Illicit drugs, of course, are harmful to the physical and emotional health of both the mother and the nursing infant. Breast milk can deliver such high doses of illicit drugs as to cause irritability, tremors, hallucinations, and even death in infants. Women whose infants have overdosed on illicit drugs contained in breast milk have been convicted of murder.

Smoking Cigarette smoking reduces milk volume, so smokers may produce too little milk to meet their infants' energy needs. The milk they do produce contains nicotine, which alters its smell and flavor. Consequently, infants of breastfeeding mothers who smoke gain less weight than infants of those who do not smoke. Furthermore, infant exposure to passive smoke negates the protective effect breastfeeding offers against SIDS and increases the risks dramatically.

Environmental Contaminants Chapter 19 discusses environmental contaminants in the food supply. Some of these environmental contaminants, such as DDT, PCBs, and dioxin, can find their way into breast milk. Inuit mothers living in Arctic Québec who eat seal and beluga whale blubber have high concentrations of DDT and PCBs in their breast milk, but the impact on infant development is unclear. Preliminary studies indicate that the children of these Inuit mothers are developing normally. Researchers speculate that the abundant omega-3 fatty acids of the Inuit diet may protect against damage to the central nervous system. Breast milk tainted with dioxins interferes with tooth development during early infancy, producing soft, mottled teeth that are vulnerable to dental caries.[67]

Caffeine Caffeine enters breast milk and may make an infant irritable and wakeful. As during pregnancy, caffeine consumption should be moderate—the equivalent of 1 to 2 cups of coffee a day. Larger doses of caffeine may interfere with the bioavailability of iron from breast milk and impair the infant's iron status.

Maternal Health

If a woman has an ordinary cold, she can continue nursing without worry. If susceptible, the infant will catch it from her anyway. (Thanks to the immunological protection of breast milk, the baby may be less susceptible than a formula-fed baby would be.) With appropriate treatment, a woman who has an infectious disease such as tuberculosis or hepatitis can breastfeed; transmission is rare.[68] Women with HIV (human immunodeficiency virus) infections, however, should consider other options.

HIV Infection and AIDS Mothers with HIV infections can transmit the virus (which causes AIDS) to their infants through breast milk, especially during the early months of breastfeeding.[69] Where safe alternatives are available, HIV-positive women should *not* breastfeed their infants. In developing countries, where the feeding of inappropriate or contaminated formulas causes 1.5 million infant deaths each year, the decision is less obvious.[70] To prevent the mother-to-child transmission of HIV, WHO and UNICEF urge mothers in developing countries *not* to breastfeed, but stress the importance of finding suitable feeding alternatives to prevent the malnutrition, disease, and death that commonly occur when women in these countries do not breastfeed.

Diabetes Women with diabetes (type 1) may need careful monitoring and counseling to ensure successful lactation. These women need to adjust their energy intakes and insulin doses to meet the heightened needs of lactation. Maintaining good glucose control helps to initiate lactation and support milk production.

Postpartum Amenorrhea Women who breastfeed experience prolonged **postpartum amenorrhea.** Absent menstrual periods, however, do not protect a woman from pregnancy. To prevent pregnancy, a couple must use some form of contraception—but not oral contraceptive agents. Standard oral contraceptives contain estrogen, which reduces milk volume and the protein content of breast milk.

Breast Health Some women fear that breastfeeding will cause their breasts to sag. The breasts do swell and become heavy and large immediately after the birth, but even when they are producing enough milk to nourish a thriving infant, they eventually shrink back to their prepregnant size. Given proper support, diet, and exercise, breasts often return to their former shape and size when lactation ends. Breasts change their shape as the body ages, but breastfeeding does not accelerate this process.

Whether the physical and hormonal events of lactation protect women from later breast cancer is an area of active research. Some research suggests no association between breastfeeding and breast cancer, whereas other research suggests a protective effect.[71] The reduction in breast cancer risk is most apparent for premenopausal women who were young when they breastfed and who breastfed for a long time.

> **IN SUMMARY** The lactating woman needs extra fluid and enough energy and nutrients to produce about 25 ounces of milk a day. Alcohol, other drugs, smoking, and contaminants may reduce milk production or enter breast milk and impair infant development.

This chapter has focused on the nutrition needs of the mother during pregnancy and lactation. The next chapter explores the dietary needs of infants, children, and adolescents.

> **postpartum amenorrhea:** the normal temporary absence of menstrual periods immediately following childbirth.

Nutrition in Your Life

The choices a woman makes in preparation for, and in support of, pregnancy and lactation can influence both her health and her infant's development—today and for decades to come.

- For women of childbearing age, do you consume at least 400 micrograms of folate daily?

- For women who are pregnant, are you paying attention to your nutrition needs and gaining the amount of weight recommended?

- For women who are about to give birth, have you carefully considered all the advantages of breastfeeding your infant and received the advice and support you need to be successful?

NUTRITION ON THE NET

 Access these websites for further study of topics covered in this chapter.

- Find updates and quick links to these and other nutrition-related sites at our website: **www.wadsworth.com/nutrition**

- Visit the pregnancy and child health center of the Mayo Clinic: **www.mayohealth.org**

- Learn more about having a healthy baby and about birth defects from the March of Dimes and the National Center on Birth Defects and Developmental Disabilities: **www.modimes.org** and **www.cdc.gov/ncbddd**

- Learn more about neural tube defects from the Spina Bifida Association of America: **www.sbaa.org**

- Search for "birth defects," "pregnancy," "adolescent pregnancy," "maternal and infant health," and

"breastfeeding" at the U.S. Government health information site: **www.healthfinder.gov**

- Search for "pregnancy" at the American Dietetic Association site: **www.eatright.org**

- Learn more about the WIC program: **www.fns.usda.gov/fns**

- Visit the American College of Obstetricians and Gynecologists: **www.acog.org**

- Learn more about gestational diabetes from the American Diabetes Association: **www.diabetes.org**

- Learn more about breastfeeding from LaLeche League International: **www.lalecheleague.org**

- Obtain prenatal nutrition guidelines from Health Canada: **www.hc-sc.gc.ca**

STUDY QUESTIONS

These questions will help you review the chapter. You will find the answers in the discussions on the pages provided.

1. Describe the placenta and its function. (pp. 508–509)

2. Describe the normal events of fetal development. How does malnutrition impair fetal development? (pp. 508–510, 524)

3. Define the term *critical period*. How do adverse influences during critical periods affect later health? (pp. 510–513)

4. Explain why women of childbearing age need folate in their diets. How much is recommended, and how can women ensure that these needs are met? (pp. 511–512, 518)

5. What is the recommended pattern of weight gain during pregnancy for a woman at a healthy weight? For an underweight woman? For an overweight woman? (pp. 514–515)

6. What does a pregnant woman need to know about exercise? (pp. 516–517)

7. Which nutrients are needed in the greatest amounts during pregnancy? Why are they so important? Describe wise food choices for the pregnant woman. (pp. 518–521)

8. Define low-risk and high-risk pregnancies. What is the significance of infant birthweight in terms of the child's future health? (pp. 523–524)

9. Describe some of the special problems of the pregnant adolescent. Which nutrients are needed in increased amounts? (pp. 526–527)

10. What practices should be avoided during pregnancy? Why? (pp. 527–530)

11. How do nutrient needs during lactation differ from nutrient needs during pregnancy? (pp. 532–533)

These multiple choice questions will help you prepare for an exam. Answers can be found on (p. 538).

1. The spongy structure that delivers nutrients to the fetus and returns waste products to the mother is called the:
   a. embryo.
   b. uterus.
   c. placenta.
   d. amniotic sac.

2. Which of these strategies is *not* a healthy option for an overweight woman?
   a. Limit weight gain during pregnancy.
   b. Postpone weight loss until after pregnancy.
   c. Follow a weight-loss diet during pregnancy.
   d. Try to achieve a healthy weight before becoming pregnant.

3. A reasonable weight gain during pregnancy for a normal-weight woman is about:
   a. 10 pounds.
   b. 20 pounds.
   c. 30 pounds.
   d. 40 pounds.

4. Energy needs during pregnancy increase by about:
   a. 100 kcalories/day.
   b. 300 kcalories/day.
   c. 500 kcalories/day.
   d. 700 kcalories/day.

5. To help prevent neural tube defects, grain products are now fortified with:
   a. iron.
   b. folate.
   c. protein.
   d. vitamin C.

6. Pregnant women should *not* take supplements of:
   a. iron.
   b. folate.
   c. vitamin A.
   d. vitamin C.

7. The combination of high blood pressure, protein in the urine, and edema signals:
   a. jaundice.
   b. preeclampsia.
   c. gestational diabetes.
   d. gestational hypertension.

8. To facilitate lactation, a mother needs:
   a. about 5000 kcalories a day.
   b. adequate nutrition and rest.

   c. vitamin and mineral supplements.
   d. a glass of wine or beer before each feeding.

9. A breastfeeding woman should drink plenty of water to:
   a. produce more milk.
   b. suppress lactation.
   c. prevent dehydration.
   d. dilute nutrient concentration.

10. A woman may need iron supplements during lactation:
   a. to enhance the iron in her breast milk.
   b. to provide iron for the infant's growth.
   c. to replace the iron in her body's stores.
   d. to support the increase in her blood volume.

REFERENCES

1. J. Erickson, Folic acid and prevention of spina bifida and anencephaly, *Morbidity and Mortality Weekly Report* 51 (2002): 1–3; M. A. Honein and coauthors, Impact of folic acid fortification of the US food supply on the occurrence of neural tube defects, *Journal of the American Medical Association* 285 (2001): 2981–2986.

2. C. N. Hales and S. E. Ozanne, For debate: Fetal and early postnatal growth restriction lead to diabetes, the metabolic syndrome and renal failure, *Diabetologia* 46 (2003): 1013–1019; K. M. Rasmussen, The "fetal origins" hypothesis: Challenges and opportunities for maternal and child nutrition, *Annual Review of Nutrition* 21 (2001): 73–95; K. M. Godfrey and D. J. P. Barker, Fetal nutrition and adult disease, *American Journal of Clinical Nutrition* 71 (2000): 1344S–1352S; A. Lucas, M. S. Fewtrell, and J. Cole, Fetal origins of adult disease—The hypothesis revisited, *British Medical Journal* 319 (1999): 245–249.

3. B. E. Birgisdottir and coauthors, Size at birth and glucose intolerance in a relatively genetically homogeneous, high-birth weight population, *American Journal of Clinical Nutrition* 76 (2002): 399–403; T. W. McDade and coauthors, Prenatal undernutrition, postnatal environments, and antibody response to vaccination in adolescence, *American Journal of Clinical Nutrition* 74 (2001): 543–548.

4. P. Szitányi, J. Janda, and R. Poledne, In-trauterine undernutrition and programming as a new risk of cardiovascular disease in later life, *Physiological Research* 52 (2003): 389–395; A. Singhal and coauthors, Programming of lean body mass: A link between birth weight, obesity, and cardiovascular disease? *American Journal of Clinical Nutrition* 77 (2003): 726–730; T. J. Roseboom and coauthors, Plasma lipid profiles in adults after prenatal exposure to the Dutch famine, *American Journal of Clinical Nutrition* 72 (2000): 1101–1106.

5. G. Wolf, Adult type 2 diabetes induced by intrauterine growth retardation, *Nutrition Reviews* 61 (2003): 176–179.

6. J. W. Rich-Edwards and coauthors, Birth-weight and the risk for type 2 diabetes mellitus in adult women, *Annals of Internal Medicine* 130 (1999): 278–284.

7. K. M. Moritz, M. Dodic, and E. M. Wintour, Kidney development and the fetal programming of adult disease, *Bioessays* 25 (2003): 212–220; M. Symonds and coauthors, Maternal nutrient restriction during placental growth, programming of fetal adiposity and juvenile blood pressure control, *Archives of*

Physiology and Biochemistry 111 (2003): 45–52; J. Eriksson and coauthors, Fetal and child-hood growth and hypertension in adult life, *Hypertension* 36 (2000): 790–794.

8. C. M. Law and coauthors, Fetal, infant, and childhood growth and adult blood pressure: A longitudinal study from birth to 22 years of age, *Circulation* 105 (2002): 1088–1092; J. G. Eriksson and coauthors, Early growth and coronary heart disease in later life: Longitudinal study, *British Medical Journal* 322 (2001): 949–953.

9. E. Pennisi, Behind the scenes of gene expression, *Science* 293 (2001): 1064–1067.

10. R. A. Waterland and R. L. Jirtle, Transposable elements: Targets for early nutritional effects on epigenetic gene regulation, *Molecular and Cellular Biology* 23 (2003): 5293–5300.

11. J. M. Baeten, E. A. Bukusi, and M. Lambe, Pregnancy complications and outcomes among overweight and obese nulliparous women, *American Journal of Public Health* 91 (2001): 436–440.

12. F. Galtier-Dereure, C. Boegner, and J. Bringer, Obesity and pregnancy: Complications and cost, *American Journal of Clinical Nutrition* 71 (2000): 1242S–1248S.

13. T. K. Young and B. Woodmansee, Factors that are associated with cesarean delivery in a large private practice: The importance of pregnancy body mass index and weight gain, *American Journal of Obstetrics and Gynecology* 187 (2002): 312–318.

14. M. L. Watkins and coauthors, Maternal obesity and risk for birth defects, *Pediatrics* 111 (2003): 1152–1158.

15. M. Kihlstrand and coauthors, Water gymnastics reduced the intensity of back/low back pain in pregnant women, *Acta Obstetrica Gynecologica Scandinavica* 78 (1999): 180–185.

16. R. Artal and M. O'Toole, Guidelines of the American College of Obstetricians and Gynecologists for exercise during pregnancy and the postpartum period, *British Journal of Sports Medicine* 37 (2003): 6–12; Committee on Obstetric Practice, Exercise during pregnancy and the postpartum period, *Obstetrics and Gynecology* 99 (2002): 171–173.

17. J. C. King, Physiology of pregnancy and nutrient metabolism, *American Journal of Clinical Nutrition* 71 (2000): 1218S–1225S.

18. D. L. Dufour, J. C. Reina, and G. B. Spurr, Energy intake and expenditure of free-living, pregnant Colombian women in an urban setting, *American Journal of Clinical Nutrition* 70 (1999): 269–276; L. E. Kopp-Hoolihan and coauthors, Longitudinal assessment of energy balance in well-nourished pregnant women, *American Journal of Clinical Nutrition* 69 (1999): 697–704.

19. M. Makrides and R. A. Gibson, Long-chain polyunsaturated fatty acid requirements during pregnancy and lactation, *American Journal of Clinical Nutrition* 71 (2000): 307S–311S; M. D. M. Al, A. C. van Houwelin-gen, and G. Hornstra, Long-chain polyunsaturated fatty acids, pregnancy, and pregnancy outcome, *American Journal of Clinical Nutrition* 71 (2000): 285S–291S; A. K. Dutta-Roy, Transport mechanisms for long-chain polyunsaturated fatty acids in the human placenta, *American Journal of Clinical Nutrition* 71 (2000): 315S–322S; S. M. Innis, Maternal diet, length of gestation, and long-chain polyunsaturated fatty acid status of infants at birth, *American Journal of Clinical Nutrition* 70 (1999): 181–182.

20. Committee on Dietary Reference Intakes, *Dietary Reference Intakes for Thiamin, Riboflavin, Niacin, Vitamin B_6, Folate, Vitamin B_{12}, Pantothenic Acid, Biotin, and Choline* (Washington, D.C.: National Academy Press, 1998), pp. 196–305.

21. K. O. O'Brien and coauthors, Maternal iron status influences iron transfer to the fetus during the third trimester of pregnancy, *American Journal of Clinical Nutrition* 77 (2003): 924–930.

22. B. P. Zhu and coauthors, Effect of the interval between pregnancies on perinatal outcomes, *New England Journal of Medicine* 340 (1999): 589–594.

23. M. M. Werler and coauthors, Multivitamin supplementation and risk of birth defects, *American Journal of Epidemiology* 150 (1999): 675–682.

24. E. A. Rose, J. H. Porcerelli, and A. V. Neale, Pica: Common but commonly missed, *Journal of the American Board of Family Practice* 13 (2000): 353–358.

25. S. Saigal and coauthors, School difficulties at adolescence in a regional cohort of children who were extremely low birth weight, *Pediatrics* 105 (2000): 325–331.

26. American Academy of Pediatrics, WIC program, *Pediatrics* 108 (2001): 1216–1217.

27. W. van Wootten and R. E. Turner, Macrosomia in neonates of mothers with gestational diabetes is associated with body mass index and previous gestational diabetes, *Journal of the American Dietetic Association* 102 (2002): 241–243.

28. Report of the Expert Committee on the Diagnosis and Classification of Diabetes Mellitus, *Diabetes Care* 26 (2003): S5–S20.

29. Position statement from the American Diabetes Association: Gestational diabetes mellitus, *Diabetes Care* 26 (2003): S103–S105.

30. O. Langer and coauthors, A comparison of glyburide and insulin women with

gestational diabetes mellitus, *New England Journal of Medicine* 343 (2000): 1134–1138.

31. M. S. Esplin and coauthors, Paternal and maternal components of the predisposition to preeclampsia, *New England Journal of Medicine* 344 (2001): 867–872.

32. *National Vital Statistics Reports* 47 (1999): 14.

33. D. Maine, Role of nutrition in the prevention of toxemia, *American Journal of Clinical Nutrition* 72 (2000): 298S–300S.

34. J. Villar and J. M. Belizan, Same nutrient, different hypotheses: Disparities in trials of calcium supplementation during pregnancy, *American Journal of Clinical Nutrition* 71 (2000): 1375S–1379S.

35. National Center for Health Statistics, Centers for Disease Control, www.cdc.gov/nchs, visited April 25, 2003; National and state-specific pregnancy rates among adolescents —United States, 1995–1997, *Morbidity and Mortality Weekly Report* 49 (2000): 605–611.

36. R. J. Paulson and coauthors, Pregnancy in the sixth decade of life—Obstetric outcomes in women of advanced reproductive age, *Journal of the American Medical Association* 288 (2002): 2320–2323.

37. S. C. Tough and coauthors, Delayed childbearing and its impact on population rate changes in lower birth weight, multiple birth, and preterm delivery, *Pediatrics* 109 (2002): 399–403.

38. Alcohol use among women of childbearing age—United States, 1991–1999, *Morbidity and Mortality Weekly Report* 51 (2002): 273–276.

39. Committee on Drugs, American Academy of Pediatrics, Use of psychoactive medication during pregnancy and possible effects on the fetus and newborn, *Pediatrics* 105 (2000): 880–887.

40. E. S. Bandstra and coauthors, Intrauterine growth of full-term infants: Impact of prenatal cocaine exposure, *Pediatrics* 108 (2001): 1309–1319; D. A. Frank and coauthors, Level of in utero cocaine exposure and neonatal ultrasound findings, *Pediatrics* 104 (1999): 1101–1105; G. A. Richardson and coauthors, Growth of infants prenatally exposed to cocaine/crack: Comparison of a prenatal care and a no prenatal care sample, *Pediatrics* 104 (1999): e8.

41. L. T. Singer and coauthors, Cognitive and motor outcomes of cocaine-exposed infants, *Journal of the American Medical Association* 287 (2002): 1952–1960; M. S. Scher, G. A. Richardson, and N. L. Day, Effects of prenatal cocaine/crack and other drug exposure on electroencephalographic sleep studies at birth and one year, *Pediatrics* 105 (2000): 39–48.

42. Committee on Drugs, American Academy of Pediatrics, Neonatal drug withdrawal, *Pediatrics* 101 (1998): 1079–1088.

43. J. M. Lightwood, C. S. Phibbs, and S. A. Glantz, Short-term health and economic benefits of smoking cessation: Low birth weight, *Pediatrics* 104 (1999): 1312–1320; S. Cnattingius and coauthors, The influence of gestational age and smoking habits on the risk of subsequent preterm deliveries, *New England Journal of Medicine* 341 (1999): 943–948.

44. X. Wang and coauthors, Maternal cigarette smoking, metabolic gene polymorphism, and infant birth weight, *Journal of the American Medical Association* 287 (2002): 195–202.

45. S. J. Ventura and coauthors, Trends and variations in smoking during pregnancy and low birth weight. Evidence from the birth certificate, 1990–2000, *Pediatrics* 111 (2003): 1176–1180; S. H. Ebrahim and coauthors, Trends in pregnancy-related smoking rates in the United States, 1987–1996, *Journal of the American Medical Association* 283 (2000): 361–366.

46. A. Gomaa and coauthors, Maternal bone lead as an independent risk factor for fetal neurotoxicity: A prospective study, *Pediatrics* 110 (2002): 110–118.

47. M. A. Johnson, High calcium intake blunts pregnancy-induced increases in maternal blood lead, *Nutrition Reviews* 59 (2001): 152–156.

48. N. Ribas-Fitó and coauthors, Breastfeeding, exposure to organochlorine compounds, and neurodevelopment in infants, *Pediatrics* 111 (2003): e580–e585.

49. S. E. Schober and coauthors, Blood mercury levels in US children and women of child-bearing age, 1999–2000, *Journal of the American Medical Association* 289 (2003): 1667–1674.

50. Fish advisories, www.epa.gov/ost/fish visited April 23, 2003.

51. S. Cnattingius and coauthors, Caffeine intake and the risk of first-trimester spontaneous abortion, *New England Journal of Medicine* 343 (2000): 1839–1845; M. A. Klebanoff and coauthors, Maternal serum paraxanthine, a caffeine metabolite, and the risk of spontaneous abortion, *New England Journal of Medicine* 341 (1999): 1639–1644.

52. Position of the American Dietetic Association: Use of nutritive and nonnutritive sweeteners, *Journal of the American Dietetic Association* 98 (1998): 580–587.

53. Position of the American Dietetic Association: Nutrition and lifestyle for a healthy pregnancy outcome, *Journal of the American Dietetic Association* 102 (2002): 1479–1490.

54. R. Li and coauthors, Prevalence of breastfeeding in the United States: The 2001 National Immunization Survey, *Pediatrics* 111 (2003): 1198–1201; A. S. Ryan, Z. Wenjun, and A. Acosta, Breastfeeding continues to increase into the new millennium, *Pediatrics* 110 (2002): 1103–1109.

55. Position of the American Dietetic Association: Breaking the barriers to breastfeeding, *Journal of the American Dietetic Association* 101 (2001): 1213–1220; Department of Health and Human Services, *Breastfeeding—HHS Blueprint for Action on Breastfeeding*, 2000.

56. Physicians and breastfeeding promotion in the United States: A call to action, *Pediatrics* 107 (2001): 584–588.

57. D. R. Zimmerman and N. Guttman, "Breast is best": Knowledge among low-income mothers is not enough, *Journal of Human Lactation* 17 (2001): 14–19.

58. B. L. Philipp and coauthors, Baby-friendly hospital initiative improves breastfeeding initiation rates in a US hospital setting, *Pediatrics* 108 (2001): 677–681.

59. C. L. Dennis, Breastfeeding initiation and duration: A 1990–2000 literature review, *Journal of Obstetric, Gynecologic and Neonatal Nursing* 31 (2002): 12–32; S. Arora and coauthors, Major factors influencing breastfeeding rates: Mother's perception of father's attitude and milk supply, *Pediatrics* 106 (2000): e67.

60. Dennis, 2002.

61. M. A. McCrory, Does dieting during lactation put infant growth at risk? *Nutrition Reviews* 59 (2001): 18–21.

62. C. A. Lovelady and coauthors, The effect of weight loss in overweight, lactating women on the growth of their infants, *New England Journal of Medicine* 342 (2000): 449–453; M. A. McCrory and coauthors, Randomized trial of the short-term effects of dieting compared with dieting plus aerobic exercise on lactation performance, *American Journal of Clinical Nutrition* 69 (1999): 959–967.

63. L. M. Paton and coauthors, Pregnancy and lactation have no long-term deleterious effect on measures of bone mineral in healthy women: A twin study, *American Journal of Clinical Nutrition* 77 (2003): 707–714.

64. J. A. Mennella, C. P. Jagnow, and G. K. Beauchamp, Prenatal and postnatal flavor learning by human infants, *Pediatrics* 107 (2001): e88.

65. Committee on Nutrition, American Academy of Pediatrics, Hypoallergenic infant formulas, *Pediatrics* 106 (2000): 346–349.

66. S. Ito and A. Lee, Drug excretion into breast milk—Overview, *Advanced Drug Delivery Reviews* 55 (2003): 617–627; Committee on Drugs, American Academy of Pediatrics, The transfer of drugs and other chemicals into human milk, *Pediatrics* 108 (2001): 776–789.

67. S. Alaluusua and coauthors, Developing teeth as biomarker of dioxin exposure, *Lancet* 353 (1999): 206.

68. J. S. Wang, Q. R. Zhu, and X. H. Wang, Breastfeeding does not pose any additional risk of immunoprophylaxis failure on infants of HBV carrier mothers, *International Journal of Clinical Practice* 57 (2003): 100–102; J. B. Hill and coauthors, Risk of hepatitis B transmission in breast-fed infants of chronic hepatitis B carriers, *Obstetrics and Gynecology* 99 (2002): 1049–1052; M. L. Newell and L. Pembrey, Mother-to-child transmission of hepatitis C virus infection, *Drugs of Today* 38 (2002): 321–337; E. R. Stiehm and M. A. Keller, Breast milk transmission of viral disease, *Advances in Nutritional Research* 10 (2001): 105–122.

69. J. S. Read and the Committee on Pediatric AIDS, human milk, breastfeeding, and transmission of human immunodeficiency virus type 1 in the United States, *Pediatrics* 112 (2003): 1196–1205; R. Nduati and coauthors, Effect of breastfeeding and formula feeding on transmission of HIV-1: A randomized clinical trial, *Journal of the American Medical Association* 283 (2000): 1167–1174.

70. J. Humphrey and P. Iliff, Is breast not best? Feeding babies born to HIV-positive mothers: Bringing balance to a complex issue, *Nutrition Reviews* 59 (2001): 119–127; E. Hormann, Breast-feeding and HIV: What choices does a mother really have? *Nutrition Today* 34 (1999): 189–196; M. G. Fowler, J. Bertolli, and P. Nieburg, When is breastfeeding not best? The dilemma facing HIV-infected women in resource poor settings, *Journal of the American Medical Association* 282 (1999): 781–783.

71. L. Lipworth, L. R. Bailey, and D. Trichopoulon, History of breast-feeding in relation to breast cancer risk: A review of the epidemiologic literature, *Journal of the National Cancer Institute* 92 (2000): 302–312.

ANSWERS

Study Questions (multiple choice)

1. c   2. c   3. c   4. b   5. b   6. c   7. b   8. b   9. c   10. c

Fetal Alcohol Syndrome

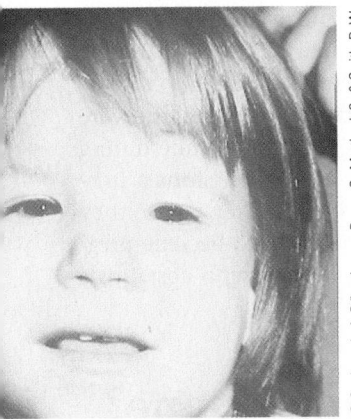

© Streissguth, A. P./Landesman-Dwyer, S. Martin, J. C. & Smith, D. W.

As Chapter 15 mentioned, drinking alcohol during pregnancy endangers the fetus. Alcohol crosses the placenta freely and deprives the developing fetus of both nutrients and oxygen. The result may be **fetal alcohol syndrome (FAS;** see the glossary on p. 540), a cluster of physical, mental, and neurobehavioral symptoms that includes:

- Prenatal and postnatal growth retardation.

- Impairment of the brain and central nervous system, with consequent mental retardation, poor motor skills and coordination, and hyperactivity.

- Abnormalities of the face and skull (see Figure H15-1).

- Increased frequency of major birth defects: cleft palate, heart defects, and defects in ears, eyes, genitals, and urinary system.

Tragically, the damage evident at birth persists: children with FAS never fully recover.[1]

Each year, as many as 12,000 infants are born with FAS because their mothers drank too much alcohol during pregnancy. In addition, three times as many infants are born with less serious, yet still significant, damage because their mothers drank alcohol—just not as much. These abnormalities fall short of FAS and were formerly known as **fetal alcohol effects (FAE).** This catchall term has been replaced by two terms that describe the mental and physical symptoms of **prenatal alcohol exposure.**[2] The cluster of mental problems associated with prenatal alcohol exposure is known as **alcohol-related neurodevelopmental disorder (ARND),** and the physical malformations are referred to as **alcohol-related birth defects (ARBD).** Some children with ARBD and ARND have no outward signs; others may be short or have only minor facial abnormali-

ties. They often go undiagnosed even when they develop learning difficulties in the early school years. Mood disorders and problem behaviors, such as aggression, are common.[3]

The surgeon general states that pregnant women should drink absolutely no alcohol. Abstinence from alcohol is the best policy for pregnant women both because alcohol consumption during pregnancy has such severe consequences and because FAS can only be prevented—it cannot be treated. Further, because the most severe damage occurs around the time of conception—*before a woman may even realize that she is pregnant*—even a woman planning to conceive should abstain.

Drinking during Pregnancy

As mentioned in Chapter 15, one out of eight pregnant women drinks alcohol at some time during her pregnancy; 1 out of 30 uses alcohol frequently, and 1 out of 40 admits to binge drinking.[4] When a woman drinks during pregnancy, she causes damage in two ways: directly, by intoxication, and indirectly, by malnutrition. Prior to the complete formation of the placenta (approximately 12 weeks), alcohol diffuses

FIGURE H15-1 Typical Facial Characteristics of FAS

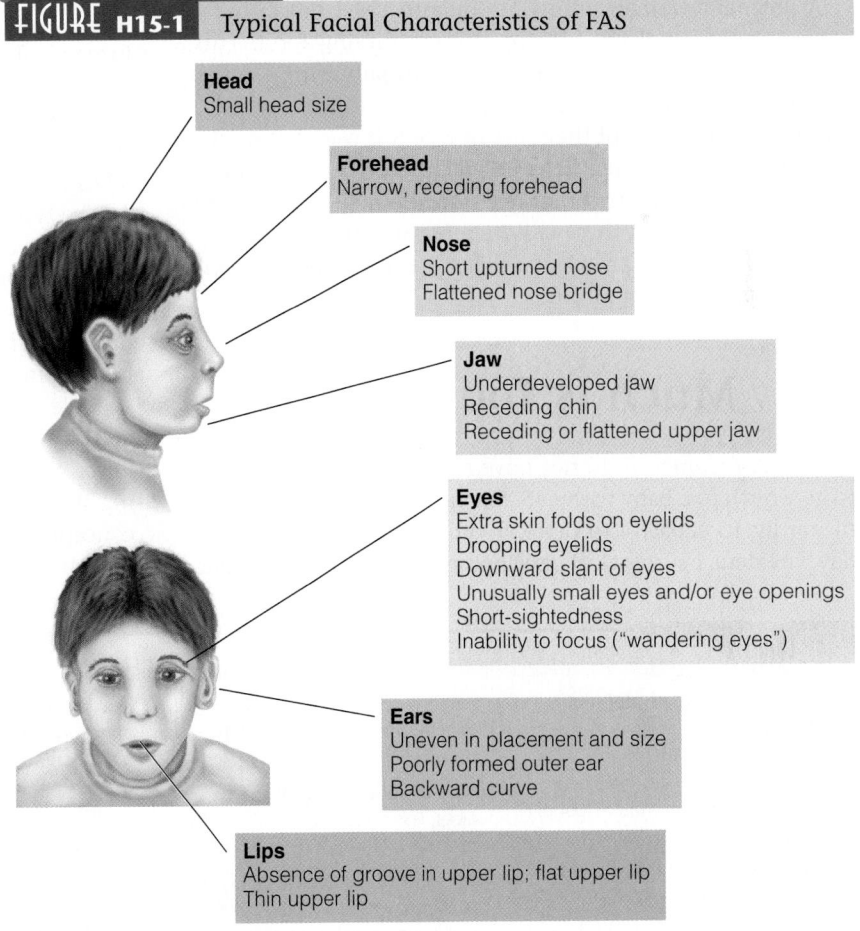

Head
Small head size

Forehead
Narrow, receding forehead

Nose
Short upturned nose
Flattened nose bridge

Jaw
Underdeveloped jaw
Receding chin
Receding or flattened upper jaw

Eyes
Extra skin folds on eyelids
Drooping eyelids
Downward slant of eyes
Unusually small eyes and/or eye openings
Short-sightedness
Inability to focus ("wandering eyes")

Ears
Uneven in placement and size
Poorly formed outer ear
Backward curve

Lips
Absence of groove in upper lip; flat upper lip
Thin upper lip

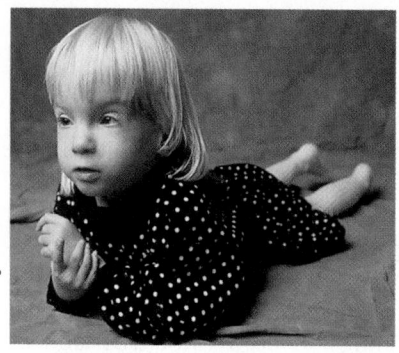
© 1995 George Steinmetz

Characteristic facial features may diminish with time, but children with FAS typically continue to be short and underweight for their age.

directly into the tissues of the developing embryo, causing incredible damage. (Review Figure 15-4 on p. 510 and note that the critical periods for most tissues occur during embryonic development.) Alcohol interferes with the orderly development of tissues during their critical periods, reducing the number of cells and damaging those that are produced. The damage of alcohol toxicity during brain development is apparent in its reduced size and impaired function.[5]

When alcohol crosses the placenta, fetal blood alcohol rises until it reaches an equilibrium with maternal blood alcohol. The mother may not even appear drunk, but the fetus may be poisoned. The fetus's body is small, its detoxification system is immature, and alcohol remains in fetal blood long after it has disappeared from maternal blood.

A pregnant woman harms her unborn child not only by consuming alcohol but also by not consuming food. This combination enhances the likelihood of malnutrition and a poorly developed infant. It is important to realize, however, that malnutrition is not the cause of FAS. It is true that mothers of FAS children often have unbalanced diets and nutrient deficiencies. It is also true that malnutrition may augment the clinical signs seen in these children, but it is the *alcohol* that causes the damage. An adequate diet alone will not prevent FAS if alcohol abuse continues.

How Much Is Too Much?

A pregnant woman need not have an alcohol-abuse problem to give birth to a baby with FAS. She need only drink in excess of her liver's capacity to detoxify alcohol. Even one drink a day threatens neurological development and behaviors.[6] Four drinks a day dramatically worsens the risk of having an infant with physical malformations.

In addition to total alcohol intake, drinking patterns play an important role. Most FAS studies report their findings in terms of average intake per day, but people usually drink more heavily on some days than on others. For example, a woman who drinks an *average* of 1 ounce of alcohol (2 drinks) a day may not drink at all during the week, but then have 10 drinks on Saturday night, exposing the fetus to extremely toxic quantities of alcohol. Whether various drinking patterns incur damage depends on the frequency of consumption, the quantity consumed, and the stage of fetal development at the time of each drinking episode.

An occasional drink may be innocuous, but researchers are unable to say how much alcohol is safe to consume during pregnancy. For this reason, health care professionals urge women to stop drinking alcohol as soon as they realize they are pregnant or better, as soon as they *plan* to become pregnant. Why take any risk? Only the woman who abstains is sure of protecting her infant from FAS.

When Is the Damage Done?

The first month or two of pregnancy is a critical period of fetal development. Because pregnancy usually cannot be confirmed before five to six weeks, a woman may not even realize she is pregnant during that critical time. Therefore, it is advisable for women who are trying to conceive, or who suspect they might be pregnant, to abstain or curtail their alcohol intakes to ensure a healthy start.

The type of abnormality observed in an FAS infant depends on the developmental events occurring at the times of alcohol exposure. During the first trimester, developing organs such as the brain, heart, and kidneys may be malformed. During the second trimester, the risk of spontaneous abortion increases. During the third trimester, body and brain growth may be retarded.

Male alcohol ingestion may also affect fertility and fetal development.[7] Animal studies have found smaller litter sizes, lower birthweights, reduced survival rates, and impaired learning ability in the offspring of males consuming alcohol prior to conception. An association between paternal alcohol intake one month prior to conception and low infant birthweight is also

GLOSSARY

alcohol-related birth defects (ARBD): malformations in the skeletal and organ systems (heart, kidneys, eyes, ears) associated with prenatal alcohol exposure.

alcohol-related neurodevelopmental disorder (ARND): abnormalities in the central nervous system and cognitive development associated with prenatal alcohol exposure.

fetal alcohol effects (FAE): an older, less preferred, term used to describe both ARBD and ARND.

fetal alcohol syndrome (FAS): a cluster of physical, behavioral, and cognitive abnormalities associated with prenatal alcohol exposure, including facial malformations, growth retardation, and central nervous system disorders.

prenatal alcohol exposure: subjecting a fetus to a pattern of excessive alcohol intake characterized by substantial regular use or heavy episodic drinking.

NOTE: See Highlight 7 for other alcohol-related terms and information.

© 1995 George Steinmetz

Children born with FAS must live with the long-term consequences of prenatal brain damage.

Matthew Farruggio

All containers of beer, wine, and liquor warn women not to drink alcoholic beverages during pregnancy because of the risk of birth defects.

apparent in human beings. (Paternal alcohol intake was defined as an average of 2 or more drinks daily or at least 5 drinks on one occasion.) This relationship was independent of either parent's smoking and of the mother's use of alcohol, caffeine, or other drugs.

In view of the damage caused by FAS, prevention efforts focus on educating women not to drink during pregnancy.[8] Everyone should know of the potential dangers. Women who drink alcohol and who are sexually active may benefit from counseling and effective contraception to prevent pregnancy.[9] Almost half of all pregnancies are unintended, with many conceived during a binge drinking episode.[10]

Public service announcements and alcohol beverage warning labels help to raise awareness. Everyone should hear the message loud and clear: Don't drink alcohol prior to conception or during pregnancy.

NUTRITION ON THE NET

Access these websites for further study of topics covered in this highlight.

- Find updates and quick links to these and other nutrition-related sites at our website: **www.wadsworth.com/nutrition**
- Visit the National Organization on Fetal Alcohol Syndrome: **www.nofas.org**
- Search for "fetal alcohol syndrome" at the U.S. Government health information site: **www.healthfinder.gov**

- Request information on fetal alcohol syndrome from the National Clearinghouse for Alcohol and Drug Information: **www.health.org**
- Request information on drinking during pregnancy from the National Institute on Alcohol Abuse and Alcoholism: **www.niaaa.nih.gov**
- Gather facts on fetal alcohol syndrome from the March of Dimes: **www.modimes.org**

REFERENCES

1. N. L. Day and coauthors, Prenatal alcohol exposure predicts continued deficits in offspring size at 14 years of age, *Alcoholism: Clinical and Experimental Research* 26 (2002): 1584–1591; M. D. Cornelius and coauthors, Alcohol, tobacco and marijuana use among pregnant teenagers: 6-year follow-up of offspring growth effects, *Neurotoxicology and Teratology* 24 (2002): 703–710.
2. Committee on Substance Abuse and Committee on Children with Disabilities, American Academy of Pediatrics, Fetal alcohol syndrome and alcohol-related neurodevelopmental disorders, *Pediatrics* 106 (2000): 358–361.
3. M. J. O'Connor and coauthors, Psychiatric illness in a clinical sample of children with prenatal alcohol exposure, *American Journal of Drug and Alcohol Abuse* 28 (2002):

743–754; B. Sood and coauthors, Prenatal alcohol exposure and childhood behavior at age 6 to 7 years: Dose-response effect, *Pediatrics* 108 (2001): e34.
4. Alcohol use among women of childbearing age—United States 1991–1999, *Morbidity and Mortality Weekly Report* 51 (2002): 273–276.
5. J. W. Olney and coauthors, The enigma of fetal alcohol neurotoxicity, *Annals of Medicine* 34 (2002): 109–119.
6. S. W. Jacobson and coauthors, Validity of maternal report of prenatal alcohol, cocaine, and smoking in relation to neurobehavioral outcome, *Pediatrics* 109 (2002): 815–825.
7. H. Klonoff-Cohen, P. Lam-Kruglick, and C. Gonzalez, Effects of maternal and paternal alcohol consumption on the success

rates of in vitro fertilization and gamete intrafallopian transfer, *Fertility and Sterility* 79 (2003): 330–339.
8. J. R. Hankin, Fetal alcohol syndrome prevention research, *Alcohol Research and Health* 26 (2002): 58–65.
9. The Project CHOICES Intervention Research Group, Reducing the risk of alcohol-exposed pregnancies: A study of a motivational intervention in community settings, *Pediatrics* 111 (2003): 1131–1135.
10. T. S. Naimi and coauthors, Binge drinking in the preconception period and the risk of unintended pregnancy: Implications for women and their children, *Pediatrics* 111 (2003): 1136–1141.

Life Cycle Nutrition: Infancy, Childhood, and Adolescence

Chapter Outline

Nutrition during Infancy: *Energy and Nutrient Needs • Breast Milk • Infant Formula • Special Needs of Preterm Infants • Introducing Cow's Milk • Introducing Solid Foods • Mealtimes with Toddlers*

Nutrition during Childhood: *Energy and Nutrient Needs • Hunger and Malnutrition in Children • The Malnutrition-Lead Connection • Hyperactivity and "Hyper" Behavior • Food Allergy and Intolerance • Childhood Obesity • Mealtimes at Home • Nutrition at School*

Nutrition during Adolescence: *Growth and Development • Energy and Nutrient Needs • Food Choices and Health Habits • Problems Adolescents Face*

Highlight: *Childhood Obesity and the Early Development of Chronic Diseases*

Nutrition Explorer CD-ROM Outline

Nutrition Animation: *Nutrition in Childhood*

Case Study: *Food Choices Differ among Age Groups*

Student Practice Test

Glossary Terms

Nutrition on the Net

© Simon Watson/FoodPix/Getty Images

Nutrition in Your Life

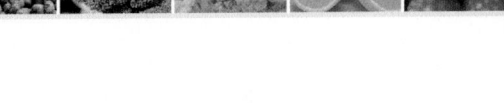

Much of this book has focused on you—your food choices and how they might affect your health. This chapter shifts the focus from you the recipient to you the caregiver. One day (if not already), children may depend on you to feed them well and teach them wisely. The responsibility of nourishing children can seem overwhelming at times, but the job is fairly simple: offer children a variety of nutritious foods to support their growth and teach them how to make healthy food and activity choices. Presenting foods in a relaxed and supportive environment nourishes both physical and emotional well-being.

The first year of life is a time of phenomenal growth and development. After the first year, a child continues to grow and change, but more slowly. Still, the cumulative effects over the next decade are remarkable. Then, as the child enters the teen years, the pace toward adulthood accelerates dramatically. This chapter examines the special nutrient needs of infants, children, and adolescents.

Nutrition during Infancy

Initially, the infant drinks only breast milk or formula, but later begins to eat some foods, as appropriate. Common sense in the selection of infant foods and a nurturing, relaxed environment support an infant's health and well-being.

FIGURE 16-1 Weight Gain of Infants in Their First Five Years of Life

In the first year, an infant's birthweight may triple, but over the following several years, the rate of weight gain gradually diminishes.

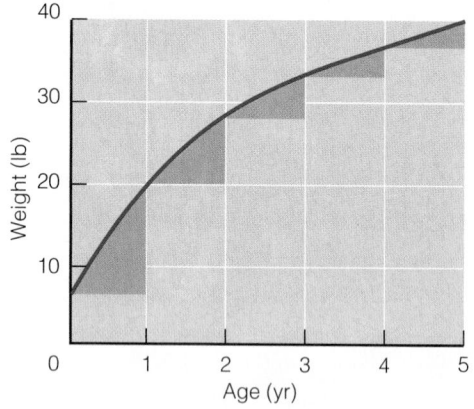

	Infants	Adults
Heart rate (beats/minute)	120 to 140	70 to 80
Respiration rate (breaths/minute)	20 to 40	15 to 20
Energy needs (kcal/body weight)	45/lb (100/kg)	<18/lb (<40/kg)

After six months, energy saved by slower growth is spent in increased activity.

Energy and Nutrient Needs

An infant grows fast during the first year, as Figure 16-1 shows. Growth directly reflects nutrient intake and is an important parameter in assessing the nutrition status of infants and children. Health care professionals measure the heights and weights of infants and children at intervals and compare measures both with standard growth curves for gender and age and with previous measures of each child (see the "How to" on p. 546).[1]

Energy Intake and Activity A healthy infant's birthweight doubles by about five months of age and triples by one year, typically reaching 20 to 25 pounds. The infant's length changes more slowly than weight, increasing about 10 inches from birth to one year. By the end of the first year, infant growth slows considerably; an infant typically gains less than 10 pounds during the second year and grows about 5 inches in height.

Not only do infants grow rapidly, but their basal metabolic rate is remarkably high—about twice that of an adult, based on body weight. A newborn baby requires about 450 kcalories per day, whereas most adults require about 2000 kcalories per day. In terms of body weight, the difference is remarkable. Infants require about 100 kcalories per kilogram of body weight per day, whereas most adults need fewer than 40.■ If an infant's energy needs were superimposed on an adult, a 170-pound adult would require over 7000 kcalories a day. After six months, metabolic needs decline as the growth rate slows, but some of the energy saved by slower growth is spent in increased activity.

Energy Nutrients Recommendations for the energy nutrients—carbohydrate, fat, and protein—during the first six months of life are based on the average intakes of healthy, full-term infants fed breast milk.[2] During the second six months of life, recommendations reflect typical intakes from solid foods as well as breast milk.

As discussed in Chapter 4, carbohydrates provide energy to all the cells of the body, especially those in the brain, which depend primarily on glucose to fuel activities. Relative to the size of the body, an infant's brain is larger than an adult's and uses relatively more glucose—about 60 percent of the day's total energy intake.[3]

Fat provides most of the energy in breast milk and standard infant formula. Its high energy density supports the rapid growth of early infancy.

No single nutrient is more essential to growth than protein. All of the body's cells and most of its fluids contain protein; it is the basic building material of the body's tissues. Chapter 6 detailed the problems inadequate protein can cause. Excess dietary protein can cause problems, too, especially in a small infant. Too much protein stresses the liver and kidneys, which have to metabolize and excrete the excess nitrogen. Signs of protein overload include acidosis, dehydration, diarrhea, elevated blood ammonia, elevated blood urea, and fever. Such problems are not common, but have been observed in infants fed inappropriate foods, such as fat-free milk or concentrated formula.

Vitamins and Minerals Like the recommendations for the energy nutrients, those for the vitamins and minerals are based on the average amount of nutrients consumed by thriving infants breastfed by well-nourished mothers. An infant's needs for most of these nutrients, in proportion to body weight, are more than double those of an adult. Figure 16-2 illustrates this by comparing a five-month-old infant's needs per unit of body weight with those of an adult man. Some of the differences are extraordinary.

Water One of the most essential nutrients for infants, as for everyone, is water. The younger the infant, the greater the percentage of body weight is water. During early infancy, breast milk or infant formula normally provides enough water to replace fluid losses in a healthy infant. Even in hot, dry climates, neither breastfed nor bottle-fed infants need supplemental water.[4] Because much of the fluid in an infant's body is located *outside* the cells—between the cells and in the blood ves-

FIGURE 16-2 Recommended Intakes of an Infant and an Adult Compared on the Basis of Body Weight

Because infants are small, they need smaller total amounts of the nutrients than adults do, but when comparisons are based on body weight, infants need over twice as much of many nutrients. Infants use large amounts of energy and nutrients, in proportion to their body size, to keep all their metabolic processes going.

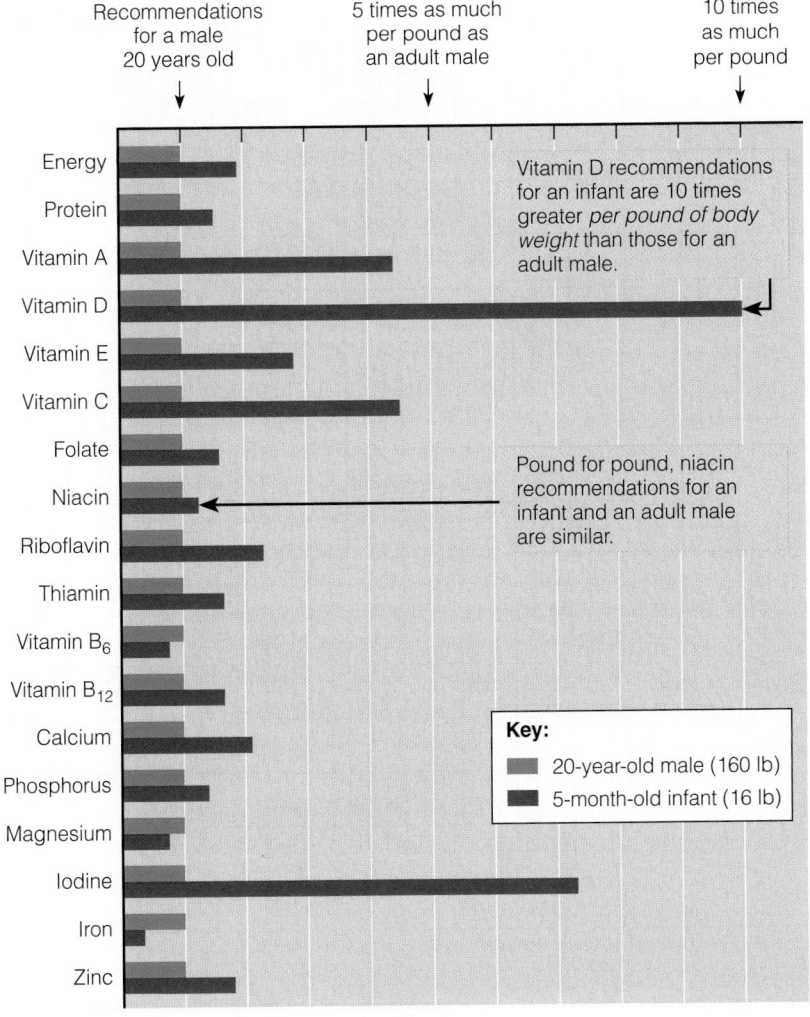

Recommendations for a male 20 years old

5 times as much per pound as an adult male

10 times as much per pound

Vitamin D recommendations for an infant are 10 times greater *per pound of body weight* than those for an adult male.

Pound for pound, niacin recommendations for an infant and an adult male are similar.

Key:
- 20-year-old male (160 lb)
- 5-month-old infant (16 lb)

Energy, Protein, Vitamin A, Vitamin D, Vitamin E, Vitamin C, Folate, Niacin, Riboflavin, Thiamin, Vitamin B$_6$, Vitamin B$_{12}$, Calcium, Phosphorus, Magnesium, Iodine, Iron, Zinc

sels—rapid fluid losses and the resulting dehydration can be life-threatening. Conditions that cause rapid fluid loss, such as diarrhea or vomiting, require an electrolyte solution designed for infants.

Breast Milk

In the United States and Canada, the two dietary practices that have the most effect on an infant's nutrition are the milk the infant receives and the age at which solid foods are introduced. A later section discusses the introduction of solid foods, but as to the milk, both the American Academy of Pediatrics (AAP) and the Canadian Paediatric Society strongly recommend breastfeeding for full-term infants, except where specific contraindications exist. The American Dietetic Association (ADA) also advocates breastfeeding for the nutritional health it confers on the infant as well as for the many other benefits it provides both infant and mother (review Table 15-4 on p. 531).[5]

HOW TO Plot Measures on a Growth Chart

You can assess the growth of infants and children by plotting their measurements on a percentile graph. Percentile graphs divide the measures of a population into 100 equal divisions so that half of the population falls at or above the 50th percentile, and half falls below. Using percentiles allows for comparisons among people of the same age and gender.

To plot measures on a growth chart, follow these steps:

- Select the appropriate chart based on age and gender. For this example, use the accompanying chart, which gives percentiles for weight for girls from birth to 36 months. (Appendix E provides other growth charts for both boys and girls of various ages.)
- Locate the infant's age along the horizontal axis at the bottom of the chart (in this example, 6 months).
- Locate the infant's weight in pounds or kilograms along the vertical axis of the chart (in this example, 17 pounds or 7.7 kilograms).
- Mark the chart where the age and weight lines intersect (shown here with a red dot), and read off the percentile.

This six-month-old infant is at the 75th percentile. Her pediatrician will weigh her again over the next few months and expect the growth curve to follow the same percentile throughout the first year. In general, dramatic changes or measures much above the 80th percentile or much below the 10th percentile may be cause for concern.

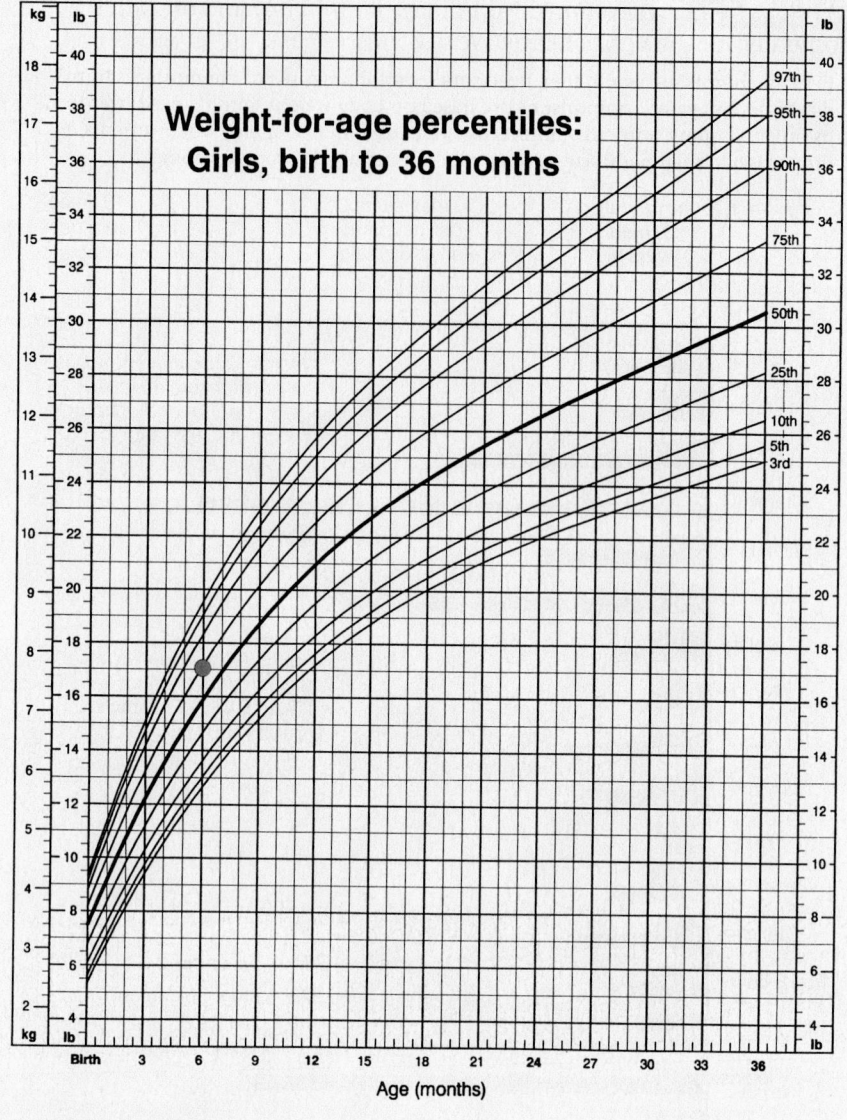

Weight-for-age percentiles: Girls, birth to 36 months

SOURCE: Developed by the National Center for Health Statistics in collaboration with the National Center for Chronic Disease Prevention and Health Promotion (2000).

■ Reminder: Chapter 15 discusses breastfeeding, breastfeeding support, reasons why some women choose not to breastfeed, and contraindications to breastfeeding.

Breast milk excels as a source of nutrients for infants. Its unique nutrient composition and protective factors promote optimal infant health and development throughout the first year of life. Both the AAP and the ADA recognize exclusive breastfeeding for 6 months, and breastfeeding with complementary foods for at least 12 months, as the best feeding pattern for infants.[6] Experts add, though, that iron-fortified formula, which imitates the nutrient composition of breast milk, is an acceptable alternative. After all, the primary goal is to provide the infant nourishment in a relaxed and loving environment.■

Frequency and Duration of Breastfeeding Breast milk is more easily and completely digested than formula, so breastfed infants usually need to eat more frequently than formula-fed infants do. During the first few weeks, approximately 8 to 12 feedings a day, on demand, or whenever the infant cries with hunger, promote optimal milk production and infant growth. An infant who nurses every two to three hours and sleeps contentedly between feedings is adequately nourished.

As the infant gets older, stomach capacity enlarges and the mother's milk production increases, allowing for longer feeding intervals.

Even though the infant obtains about half the milk from the breast during the first two or three minutes of sucking, breastfeeding is encouraged for about 10 to 15 minutes on each breast. The infant's sucking, as well as the complete removal of milk from the breast, stimulates lactation.

Energy Nutrients The energy-nutrient composition of breast milk differs dramatically from that recommended for adult diets (see Figure 16-3). Yet for infants, breast milk is nature's most nearly perfect food, providing the clear lesson that people at different stages of life have different nutrient needs.

The carbohydrate in breast milk (and infant formula) is the disaccharide lactose. In addition to being easily digested, lactose enhances calcium absorption.

The amount of protein in breast milk is less than in cow's milk, but this quantity is actually beneficial because it places less stress on the infant's immature kidneys to excrete the major end product of protein metabolism, urea. Much of the protein in breast milk is **alpha-lactalbumin,** which is efficiently digested and absorbed.

As for the lipids, breast milk contains a generous proportion of the essential fatty acids linoleic acid and linolenic acid, as well as their longer-chain derivatives arachidonic acid and DHA (docosahexaenoic acid). Until recently, infant formula provided only linoleic acid and linolenic acid. Formula with arachidonic acid and DHA added is now commercially available.[7] Infants can produce arachidonic acid and DHA from linoleic and linolenic acid, respectively, but researchers are trying to determine whether some infants need more than they can make.

Arachidonic acid and DHA are found abundantly in both the retina of the eye and the brain, and research has focused on the visual and mental development of breastfed infants and infants fed standard formula without DHA and arachidonic acid added.[8] Breastfed infants generally score higher on tests of mental development than formula-fed infants do, but whether this difference can be attributed to DHA and arachidonic acid is difficult to ascertain.[9] In one study, researchers found no developmental or visual differences between infants fed standard formula and those fed formula with added DHA and arachidonic acid.[10] In another study, infants were breastfed for the first six weeks of life and then were given either standard formula or formula with added DHA and arachidonic acid. The infants fed the formula fortified with DHA and arachidonic acid had sharper vision at one year of age than those who were fed standard formula.[11] In a study of young children, those who had been breastfed as infants had sharper vision than those who were fed formulas.[12] The enhanced visual development was attributed to the DHA in breast milk based on reports from the mothers on their consumption of DHA-rich fish. A large study of preterm infants also found improved visual acuity in those given formulas supplemented with DHA and arachidonic acid compared with controls.[13]

Vitamins With the exception of vitamin D, the vitamins in breast milk are ample to support infant growth. The vitamin D in breast milk is low, and vitamin D deficiency impairs bone mineralization. Vitamin D deficiency is most likely in infants who are not exposed to sunlight daily, have darkly pigmented skin, and receive breast milk without vitamin D supplementation.[14] Reports of infants in the United States developing the vitamin D–deficiency disease rickets and recommendations by the AAP to keep infants under six months of age out of direct sunlight have prompted new vitamin D guidelines. The AAP now recommends a vitamin D supplement for all infants who are breastfed exclusively, and for any infants who do not receive at least 500 milliliters (15 ounces) per day of vitamin D–fortified formula.[15]

Minerals The calcium content of breast milk is ideal for infant bone growth, and the calcium is well absorbed. Breast milk contains relatively small amounts of iron, but the iron has a high bioavailability. Zinc also has a high bioavailability, thanks to the presence of a zinc-binding protein. Breast milk is low in sodium, another benefit for immature kidneys. Fluoride promotes the development of strong teeth, but breast milk is not a good source.

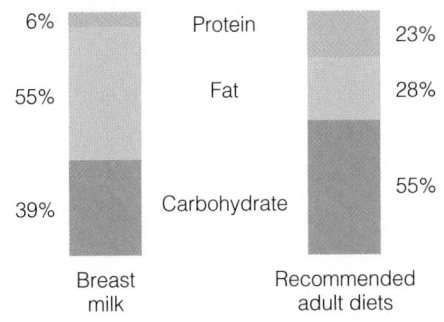

FIGURE 16-3 Percentages of Energy-Yielding Nutrients in Breast Milk and in Recommended Adult Diets

The proportions of energy-yielding nutrients in human breast milk differ from those recommended for adults.[a]

	Breast milk	Recommended adult diets
Protein	6%	23%
Fat	55%	28%
Carbohydrate	39%	55%

[a]The values listed for adults represent the midpoints of the acceptable ranges for protein (10 to 35 percent), fat (20 to 35 percent), and carbohydrate (45 to 65 percent).

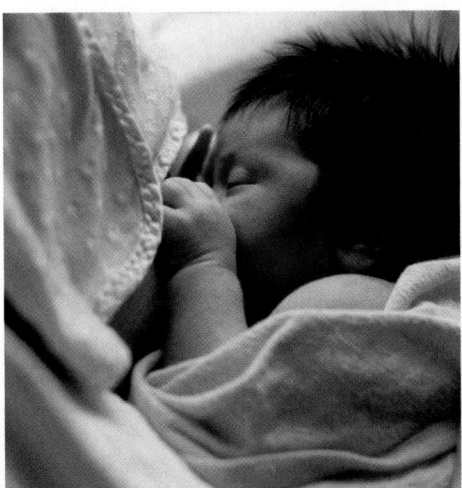

Women are encouraged to breastfeed whenever possible because breast milk offers infants many nutrient and health advantages.

alpha-lactalbumin (lact-AL-byoo-min): a major protein in human breast milk, as opposed to **casein** (CAY-seen), a major protein in cow's milk.

TABLE 16-1 Supplements for Full-Term Infants

	Vitamin D[a]	Iron[b]	Fluoride[c]
Breastfed infants:			
Birth to six months of age	✔		
Six months to one year	✔	✔	✔
Formula-fed infants:			
Birth to six months of age			
Six months to one year		✔	✔

[a]Vitamin D supplements are recommended for all infants who are exclusively breastfed, and for any infants who do not receive at least 500 milliliters (15 ounces) of vitamin D–fortified formula.

[b]Infants four to six months of age need additional iron, preferably in the form of iron-fortified cereal for both breastfed and formula-fed infants and iron-fortified infant formula for formula-fed infants.

[c]At six months of age, breastfed infants and formula-fed infants who receive ready-to-use formulas (these are prepared with water low in fluoride) or formula mixed with water that contains little or no fluoride (less than 0.3 ppm) need supplements.
SOURCE: Adapted from Committee on Nutrition, American Academy of Pediatrics, *Pediatric Nutrition Handbook*, 5th ed., ed. R. E. Kleinman (Elk Grove Village, Ill.: American Academy of Pediatrics, 2004).

Supplements Pediatricians may routinely prescribe liquid supplements containing vitamin D, iron, and fluoride. Table 16-1 offers a schedule of supplements during infancy. In addition, the AAP recommends that a single dose of vitamin K be given to infants at birth to protect them from bleeding to death (see Chapter 11 for a description of vitamin K's role in blood clotting).

Immunological Protection In addition to nutritional benefits, breast milk also offers immunological protection. Not only is breast milk sterile, but it actively fights disease and protects infants from illnesses.[16] Such protection is most valuable during the first year, when the infant's immune system is not fully prepared to mount a response against infection.

During the first two or three days after delivery, the breasts produce **colostrum,** a premilk substance containing mostly serum with antibodies and white blood cells. Colostrum (like breast milk) helps protect the newborn from infections against which the mother has developed immunity. The maternal antibodies swallowed with the milk inactivate disease-causing bacteria within the digestive tract before they can start infections. This explains, in part, why breastfed infants have fewer intestinal infections than formula-fed infants.

In addition to antibodies, colostrum and breast milk provide other powerful agents■ that help to fight against bacterial infection. Among them are **bifidus factors,** which favor the growth of the "friendly" bacterium *Lactobacillus bifidus* in the infant's digestive tract, so that other, harmful bacteria cannot gain a foothold there. An iron-binding protein in breast milk, **lactoferrin,** keeps bacteria from getting the iron they need to grow, helps absorb iron into the infant's bloodstream, and kills some bacteria directly.[17] The protein **lactadherin** in breast milk fights off the virus that causes most infant diarrhea. Also present is a growth factor that stimulates the development and maintenance of the infant's digestive tract and its protective factors. Several breast milk enzymes such as lipase also help protect the infant against infection. Clearly, breast milk is a very special substance.

Allergy and Disease Protection In addition to protection against infection, breast milk may offer protection against the development of allergies and diseases as well. Compared with formula-fed infants, breastfed infants have a lower incidence of allergic reactions, such as asthma, recurrent wheezing, and skin rash.[18] This protection is especially noticeable among infants with a family history of allergies.[19] Similarly, breast milk may offer protection against the development of cardiovascular disease. Compared with formula-fed infants, breastfed infants have lower blood cholesterol as adults.[20]

■ Protective factors in breast milk:
- Antibodies.
- Bifidus factors.
- Lactoferrin.
- Lactadherin.
- Growth factor.
- Lipase enzyme.

colostrum (ko-LAHS-trum): a milklike secretion from the breast, present during the first day or so after delivery before milk appears; rich in protective factors.

bifidus (BIFF-id-us, by-FEED-us) **factors:** factors in colostrum and breast milk that favor the growth of the "friendly" bacterium *Lactobacillus* (lack-toh-ba-SILL-us) *bifidus* in the infant's intestinal tract, so that other, less desirable intestinal inhabitants will not flourish.

lactoferrin (lack-toh-FERR-in): a protein in breast milk that binds iron and keeps it from supporting the growth of the infant's intestinal bacteria.

lactadherin (lack-tad-HAIR-in): a protein in breast milk that attacks diarrhea-causing viruses.

Other Potential Benefits Breastfeeding may also help protect against excessive weight gain later on. A well-controlled survey of more than 15,000 adolescents and their mothers indicates that those who were mostly breastfed for the first six months of life were less likely to become overweight than those who were fed formula.[21] Another survey of more than 2000 children (ages nine to ten) had similar findings.[22] A study of much younger children (three to five years of age), however, found no clear evidence that breastfeeding influences body weight.[23] These researchers noted that other factors, especially the mother's weight, strongly predict overweight in children.

Breastfeeding may have a positive effect on later intelligence.[24] In one study, young adults who had been breastfed as long as nine months scored higher on two different intelligence tests than those who had been breastfed less than one month. Many other studies suggest a beneficial effect of breastfeeding on intelligence, but when subjected to strict standards of methodology (for example, large sample size and appropriate intelligence testing), the evidence is less convincing.[25] Nevertheless, the possibility that breastfeeding may positively affect later intelligence is intriguing. It may be that some specific component of breast milk, such as DHA, stimulates brain development or that certain factors associated with the feeding process itself promote intellect; most likely, a combination of factors contributes to the positive association. More large, well-controlled studies are needed to confirm the effects, if any, of breastfeeding on later intelligence.

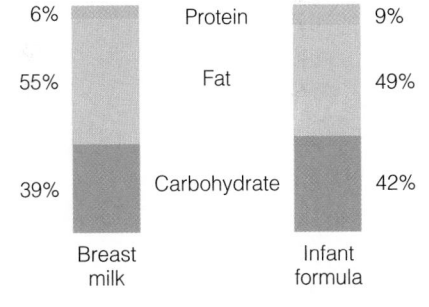

The infant thrives on infant formula offered with affection.

Infant Formula

A woman who breastfeeds for a year can **wean** her infant to cow's milk, bypassing the need for infant formula. However, a woman who decides to feed her infant formula from birth, to wean to formula after less than a year of breastfeeding, or to substitute formula for breastfeeding on occasion must select an appropriate infant formula and learn to prepare it.

Infant Formula Composition Formula manufacturers attempt to copy the nutrient composition of breast milk as closely as possible. Figure 16-4 illustrates the energy-nutrient balance of both. The AAP recommends that all formula-fed infants receive iron-fortified infant formulas. The increasing use of iron-fortified formulas during the past few decades is a major reason for the decline in iron-deficiency anemia among U.S. infants.

Risks of Formula Feeding Infant formulas contain no protective antibodies for infants, but in general, vaccinations, purified water, and clean environments in developed countries help protect infants from infections. Formulas can be prepared safely by following the rules of proper food handling and using water that is free of contamination. Of particular concern is lead-contaminated water, a major source of lead poisoning in infants. Because the first water drawn from the tap each day is highest in lead, a person living in a house with old, lead-soldered plumbing should let the water run a few minutes before drinking or using it to prepare formula or food.

In developing countries and in poor areas of the United States, formula may be unavailable, prepared with contaminated water, or overdiluted in an attempt to save money. More than 1.2 billion people in developing countries do not have access to safe drinking water. Contaminated formulas often cause infections, leading to diarrhea, dehydration, and malabsorption. Without sterilization and refrigeration, bottles of formula are an ideal breeding ground for bacteria. Whenever such risks are present, breastfeeding can be a life-saving option: breast milk is sterile, and its antibodies enhance an infant's resistance to infections.

Infant Formula Standards National and international standards have been set for the nutrient contents of infant formulas. In the United States, the standard developed by the AAP reflects "human milk taken from well-nourished mothers during the first or second month of lactation, when the infant's growth rate is high."

FIGURE 16-4 Percentages of Energy-Yielding Nutrients in Breast Milk and in Infant Formula

The average proportions of energy-yielding nutrients in human breast milk and formula differ slightly. In contrast, cow's milk provides too much protein and too little carbohydrate.

	Breast milk	Infant formula
Protein	6%	9%
Fat	55%	49%
Carbohydrate	39%	42%

wean: gradually replacing breast milk with infant formula or other foods appropriate to an infant's diet.

FIGURE 16-5 Nursing Bottle Tooth Decay

This child was frequently put to bed sucking on a bottle filled with apple juice, so the teeth were bathed in carbohydrate for long periods of time—a perfect medium for bacterial growth. The upper teeth show signs of decay.

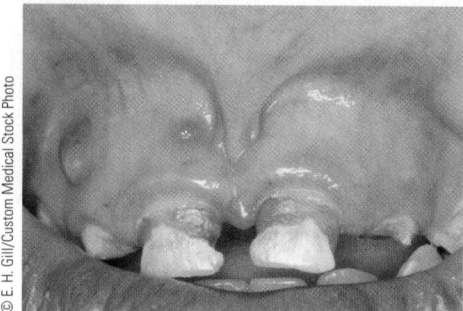

© E. H. Gill/Custom Medical Stock Photo

The Food and Drug Administration (FDA) mandates the safety and nutritional quality of infant formulas. Formulas meeting these standards have similar nutrient compositions; small differences are sometimes confusing, but usually unimportant.

Special Formulas Standard formulas are inappropriate for some infants. Special formulas have been designed to meet the dietary needs of infants with specific conditions such as prematurity or inherited diseases. Infants allergic to milk protein can drink special **hypoallergenic formulas** or formulas based on soy protein.[26] Soy formulas also use cornstarch and sucrose instead of lactose and so are recommended for infants with lactose intolerance as well. They are also useful as an alternative to milk-based formulas for vegan families. Despite these limited uses, soy formulas account for one-fourth of the infant formulas sold today. While soy formulas support the normal growth and development of infants, for infants who don't need them, they offer no advantage over milk formulas.

Inappropriate Formulas Caregivers must use only products designed for infants; soy *beverages,* for example, are nutritionally incomplete and inappropriate for infants. Goat's milk is also inappropriate for infants in part because of its low folate content. An infant receiving goat's milk is likely to develop "goat's milk anemia," an anemia characteristic of folate deficiency.

Nursing Bottle Tooth Decay An infant cannot be allowed to sleep with a bottle because of the potential damage to developing teeth. Salivary flow, which normally cleanses the mouth, diminishes as the infant falls asleep. Prolonged sucking on a bottle of formula, milk, or juice bathes the upper teeth in a carbohydrate-rich fluid that nourishes decay-producing bacteria. (The tongue covers and protects most of the lower teeth, but they, too, may be affected.) The result is extensive and rapid tooth decay (see Figure 16-5). To prevent **nursing bottle tooth decay,** no infant should be put to bed with a bottle of nourishing fluid.

Special Needs of Preterm Infants

An estimated one out of nine pregnancies in the United States results in a preterm birth.[27] The terms *preterm* and *premature* imply incomplete fetal development, or immaturity, of many body systems. As might be expected, preterm birth is a leading cause of infant deaths. Preterm infants face physical independence before some of their organs and body tissues are ready. The rate of weight gain in the fetus is greater during the last trimester of gestation than at any other time. Therefore, a preterm infant is most often a low-birthweight infant as well. With a premature birth, the infant is deprived of the nutritional support of the placenta during a time of maximal growth.

The last trimester of gestation is also a time of building nutrient stores. Being born with limited nutrient stores intensifies the precarious situation for the infant. Further compromising the nutrition status of preterm infants is their physical and metabolic immaturity. Nutrient absorption, especially of fat and calcium, from an immature GI tract is limited. Consequently, preterm, low-birthweight infants are candidates for nutrient imbalances. Deficiencies of the fat-soluble vitamins, calcium, iron, and zinc are common.

Preterm infants may miss out on the transfer of the long-chain fatty acids arachidonic acid and DHA, so critical to the healthy growth and development of the blood vessels and brain.[28] Supplementing breast milk or enriching infant formulas with these fatty acids may be beneficial for preterm infants.[29]

Preterm breast milk is well suited to meet a preterm infant's needs. During early lactation, preterm milk contains higher concentrations of protein and is lower in volume than term milk. The low milk volume is advantageous because preterm infants consume small quantities of milk per feeding, and the higher protein concentration allows for better growth. In many instances, supplements of nutrients specifically designed for preterm infants are added to the mother's expressed breast

hypoallergenic formulas: clinically tested infant formulas that do not provoke reactions in 90% of infants or children with confirmed cow's milk allergy. Like all infant formulas, hypoallergenic formulas must demonstrate nutritional suitability to support infant growth and development. Extensively hydrolyzed and free amino acid–based formulas are examples.

nursing bottle tooth decay: extensive tooth decay due to prolonged tooth contact with formula, milk, fruit juice, or other carbohydrate-rich liquid offered to an infant in a bottle.

milk and fed to the infant from a bottle. When fortified with a preterm supplement, preterm breast milk supports growth at a rate that approximates the growth rate that would have occurred within the uterus.

Introducing Cow's Milk

The age at which whole cow's milk should be introduced to the infant's diet has long been a source of controversy. The AAP advises that whole cow's milk is not appropriate during the first year.[30] Children one to two years of age should not be given reduced-fat, low-fat, or fat-free milk routinely; they need the fat of whole milk. Between the ages of two and five years, a gradual transition from whole milk to the lower-fat milks can take place, but care should be taken to avoid excessive restriction of dietary fat.

In some infants, particularly those younger than six months of age, whole cow's milk causes intestinal bleeding, which can lead to iron deficiency. Cow's milk is also a poor source of iron. Consequently, it both causes iron loss and fails to replace iron. Furthermore, the bioavailability of iron from infant cereal and other foods is reduced when cow's milk replaces breast milk or iron-fortified formula during the first year. Compared to breast milk or iron-fortified formula, cow's milk is higher in calcium and lower in vitamin C, characteristics that inhibit iron absorption. Furthermore, the higher protein concentration of cow's milk can stress the infant's kidneys. In short, cow's milk is a poor choice during the first year of life; infants need breast milk or iron-fortified infant formula.

Research examining the relationships between early exposure to cow's milk (or formula using cow's milk) and the development of type 1 diabetes (insulin-dependent) has been inconclusive and contradictory.[31] Families with a strong history of type 1 diabetes may want to breastfeed and avoid products containing cow's milk protein during the first year.

Introducing Solid Foods

The high nutrient needs of infancy are met first by breast milk or formula only and then by a limited diet to which foods■ are gradually added. Infants gradually develop the ability to chew, swallow, and digest the wide variety of foods available to adults. The caregiver's selection of appropriate foods at the appropriate stages of development is prerequisite to the infant's optimal growth and health.

When to Begin In addition to breast milk or formula, an infant can begin eating solid foods between four and six months. The main purpose of introducing solid foods is to provide nutrients that are no longer supplied adequately by breast milk or formula alone. The foods chosen must be foods that the infant is developmentally capable of handling both physically and metabolically.■ The exact timing depends on the individual infant's needs and developmental readiness (see Table 16-2), which vary from infant to infant because of differences in growth rates, activities, and environmental conditions.

Food Allergies To prevent allergy and to facilitate its prompt identification should it occur, experts recommend introducing single-ingredient foods, one at a time, in small portions, and waiting four to five days before introducing the next new food. For example, rice cereal is usually the first cereal introduced because it is least allergenic. When it is clear that rice cereal is not causing an allergy, another grain, perhaps barley or oats, is introduced. Wheat cereal is offered last because it is the most common offender. If a cereal causes an allergic reaction such as a skin rash, digestive upset, or respiratory discomfort, its use should be discontinued before introducing the next food. A later section in this chapter offers more on food allergies.

Choice of Infant Foods Infant foods should be selected to provide variety, balance, and moderation. Commercial baby foods offer a wide variety of palatable, nutritious

■ The German word **beikost** (BYE-cost) describes any nonmilk foods given to an infant.

■ Digestive secretions gradually increase throughout the first year of life, making the digestion of solid foods more efficient.

TABLE 16-2 · Infant Development and Recommended Foods

NOTE: Because each stage of development builds on the previous stage, the foods from an earlier stage continue to be included in all later stages.

Age (mo)	Feeding Skill	Appropriate Foods Added to the Diet
0–4	Turns head toward any object that brushes cheek. Initially swallows using back of tongue; gradually begins to swallow using front of tongue as well. Strong reflex (extrusion) to push food out during first 2 to 3 months.	Feed breast milk or infant formula.
4–6	Extrusion reflex diminishes, and the ability to swallow nonliquid foods develops. Indicates desire for food by opening mouth and leaning forward. Indicates satiety or disinterest by turning away and leaning back. Sits erect with support at 6 months. Begins chewing action. Brings hand to mouth. Grasps objects with palm of hand.	Begin iron-fortified cereal mixed with breast milk, formula, or water. Begin pureed vegetables and fruits.
6–8	Able to feed self finger foods. Develops pincer (finger to thumb) grasp. Begins to drink from cup.	Begin textured vegetables and fruits. Begin unsweetened, diluted fruit juices from cup.
8–10	Begins to hold own bottle. Reaches for and grabs food and spoon. Sits unsupported.	Begin breads and cereals from table. Begin yogurt. Begin pieces of soft, cooked vegetables and fruit from table. Gradually begin finely cut meats, fish, casseroles, cheese, eggs, and mashed legumes.
10–12	Begins to master spoon, but still spills some.	Include at least 4 servings of breads and cereals from table, in addition to infant cereal.[a] Include at least 2 servings of fruits and 3 servings of vegetables.[a] Include 2 servings of meat, fish, poultry, eggs, or legumes.[a]

[a] Serving sizes for infants and young children are smaller than those for an adult. For example, a serving might be ½ slice of bread instead of 1 slice, or ¼ cup rice instead of ½ cup.

SOURCE: Adapted in part from Committee on Nutrition, American Academy of Pediatrics, *Pediatric Nutrition Handbook,* 5th ed., ed. R. E. Kleinman (Elk Grove Village, Ill.: American Academy of Pediatrics, 2004), pp. 103–115.

foods in a safe and convenient form. Homemade infant foods can be as nutritious as commercially prepared ones, as long as the cook minimizes nutrient losses during preparation. Ingredients for homemade foods should be fresh, whole foods without added salt, sugar, or seasonings. Pureed food can be frozen in ice cube trays, providing convenient-sized blocks of food that can be thawed, warmed, and fed to the infant. To guard against foodborne illnesses, hands and equipment must be kept clean.

Because recommendations to restrict fat do not apply to children under age two, labels on foods for children under two (such as infant meats and cereals) cannot carry information about fat. Fat information is omitted from infant food labels to prevent parents from restricting fat in infants' diets. Fearing that their infant will become overweight, parents may unintentionally malnourish the infant by limiting fat. In fact, infants and young children, because of their rapid growth, need more fat than older children and adults.

Foods to Provide Iron Rapid growth demands iron. At about four to six months, the infant begins to need more iron than stores plus breast milk or iron-fortified formula can provide. In addition to breast milk or iron-fortified formula, infants can receive iron from iron-fortified cereals and, once they readily accept solid foods, from meat or meat alternates such as legumes.[32] Iron-fortified cereals contribute a significant amount of iron to an infant's diet, but the iron's bioavailability is poor.[33] Caregivers can enhance iron absorption from iron-fortified cereals by serving vitamin C–rich foods and juices with meals.

Foods to Provide Vitamin C The best sources of vitamin C are fruits and vegetables. Some authorities suggest that an infant who is introduced to fruits before vegetables may develop a preference for sweets and find the vegetables less palatable. To prevent this, introduce vegetables first, fruits later.

Foods such as iron-fortified cereals and formulas, mashed legumes, and strained meats provide iron.

© Polara Studios Inc.

Fruit juice may be a good source of vitamin C, but infants and young children may fail to grow and thrive when they drink so much juice each day that other, more energy- and nutrient-dense foods are displaced from their diets.[34] AAP recommendations set limits on juice consumption for infants and young children (one to six years of age): 4 to 6 ounces per day.[35] Fruit juices should be diluted and served in a cup, not a bottle, once the infant is six months of age or older.

Foods to Omit Concentrated sweets, including baby food "desserts," have no place in an infant's diet. They convey no nutrients to support growth, and the extra food energy can promote obesity. Products containing sugar alcohols such as sorbitol should also be limited, as they may cause diarrhea. Canned vegetables are also inappropriate for infants, as they often contain too much sodium. Honey and corn syrup should never be fed to infants because of the risk of **botulism.***

Infants and even young children cannot safely chew and swallow any of the foods listed in the margin;■ they can easily choke on these foods, a risk not worth taking. Nonfood items may present even greater choking hazards to infants and young children.[36] Parents and caregivers must pay careful attention to eliminate choking hazards in children's environments.

Foods at One Year At one year of age, whole cow's milk can become a primary source of most of the nutrients an infant needs; 2 to 3½ cups a day meets those needs sufficiently. More milk than this displaces iron-rich foods and can lead to **milk anemia.** If powdered milk is used, it should contain fat.

Other foods—meats, iron-fortified cereals, enriched or whole-grain breads, fruits, and vegetables—should be supplied in variety and in amounts sufficient to round out total energy needs. Ideally, a one-year-old will sit at the table, eat many of the same foods everyone else eats, and drink liquids from a cup, not a bottle. Figure 16-6 (on p. 554) shows a meal plan that meets a one-year-old's requirements.

Mealtimes with Toddlers

The nurturing of a young child involves more than nutrition. Those who care for young children are responsible for providing not only nutritious milk, foods, and water, but also a safe, loving, secure environment in which the children may grow and develop. In light of toddlers' developmental and nutrient needs and their often contrary and willful behavior, a few feeding guidelines may be helpful:

- Discourage unacceptable behavior, such as standing at the table or throwing food, by removing the young child from the table to wait until later to eat. Be consistent and firm, not punitive. The child will soon learn to sit and eat.

- Let toddlers explore and enjoy food, even if this means eating with fingers for a while. Use of the spoon will come in time.

- Don't force food on children. Rejecting new foods is normal; acceptance is more likely as children become familiar with new foods through repeated opportunities to taste them.

- Provide nutritious foods, and let children choose which ones, and how much, they will eat. Gradually, they will acquire a taste for different foods.

- Limit sweets. Infants and young children have little room for empty-kcalorie foods in their daily energy allowance. Do not use sweets as a reward for eating meals.

- Don't turn the dining table into a battleground. Make mealtimes enjoyable. Teach healthy food choices and eating habits in a pleasant environment.

Toddlers need vitamin A– and vitamin D–fortified whole milk.

■ To prevent choking, do not give infants or young children:
- Raw carrots.
- Cherries.
- Gum.
- Hard or gel-type candies.
- Hot dog slices.
- Marshmallows.
- Nuts.
- Peanut butter.
- Popcorn.
- Raw celery.
- Whole beans.
- Whole grapes.

Keep these nonfood items out of their reach:
- Balloons.
- Coins.
- Pen tops.
- Small balls.

botulism (BOT-chew-lism): an often fatal foodborne illness caused by the ingestion of foods containing a toxin produced by bacteria that grow without oxygen (see Chapter 19 for details).

milk anemia: iron-deficiency anemia that develops when an excessive milk intake displaces iron-rich foods from the diet.

---

*In infants, but not older individuals, ingestion of *Clostridium botulinum* spores can cause illness when the spores germinate in the intestine and produce toxin, which is absorbed. Symptoms include poor feeding, constipation, loss of tension in the arteries and muscles, weakness, and respiratory compromise. Infant botulism has been implicated in 5 percent of cases of sudden infant death syndrome (SIDS).

FIGURE 16-6 Sample Meal Plan for a One-Year-Old

✿ SAMPLE MENU ✿

Breakfast	½ c iron-fortified breakfast cereal
	¼ c whole milk (with cereal)
	½ c orange juice
Morning snack	1 to 2 oz cheese cubes
	Teething crackers
	½ c vitamin C–fortified fruit juice
Lunch	½ sandwich: 1 slice bread with 2 tbs tuna salad or egg salad
	½ c vegetables[a] (steamed carrots)
	1 c whole milk
Afternoon snack	1 slice toast
	1 to 2 tbs apple butter
	½ c whole milk
Dinner	2 to 3 oz chopped meat or well-cooked mashed legumes
	¼ c potato, rice, or pasta
	¼ c vegetables[a] (chopped broccoli)
	¼ c fruit[b] (sliced strawberries)
	1 c whole milk

[a]Include dark green, leafy and deep yellow vegetables.
[b]Include citrus fruits, melons, and berries.

IN SUMMARY The primary food for infants during the first 12 months is either breast milk or iron-fortified formula. In addition to nutrients, breast milk also offers immunological protection. At about four to six months, infants should gradually begin eating solid foods. By one year, they are drinking from a cup and eating many of the same foods as the rest of the family.

Nutrition during Childhood

Each year from age one to adolescence, a child typically grows taller by 2 to 3 inches and heavier by 5 to 6 pounds. Growth charts provide valuable clues to a child's health. Weight gains out of proportion to height gains may reflect overeating and inactivity, whereas measures significantly below the standard suggest malnutrition.

Increases in height and weight are only two of the many changes growing children experience (see Figure 16-7). At age one, children can stand alone and are beginning to toddle; by two, they can walk and are learning to run; by three, they can jump and are climbing with confidence. Bones and muscles increase in mass and density to make these accomplishments possible. Thereafter, further lengthening of the long bones and increases in musculature proceed unevenly and more slowly until adolescence.

Energy and Nutrient Needs

Children's appetites begin to diminish around one year, consistent with the slowing growth. Thereafter, children spontaneously vary their food intakes to coincide with their growth patterns; they demand more food during periods of rapid growth than during slow growth. Sometimes they seem insatiable; other times they seem to live on air and water.

Children's energy intakes also vary widely from meal to meal. Even so, their total daily intakes remain remarkably constant. If children eat less at one meal, they typically eat more at the next, and vice versa. Overweight children are an excep-

Ideally, a one-year-old eats many of the same foods as the rest of the family.

FIGURE 16-7 Body Shape of One-Year-Old and Two-Year-Old Compared

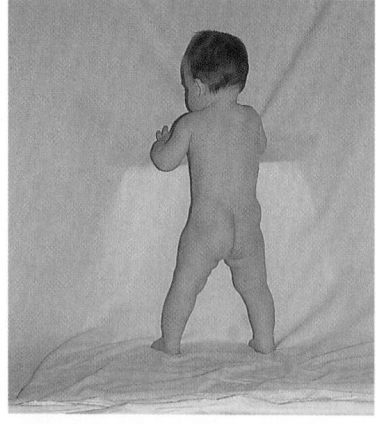
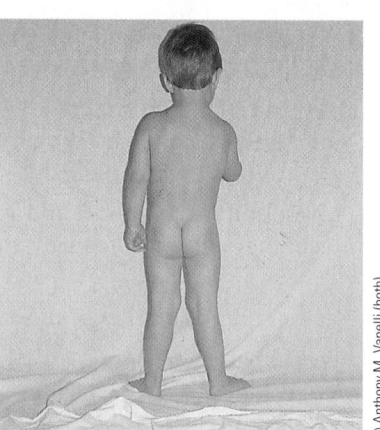

The body shape of a one-year-old (left) changes dramatically by age two (right). The two-year-old has lost much of the baby fat; the muscles (especially in the back, buttocks, and legs) have firmed and strengthened; and the leg bones have lengthened.

tion: they do not always adjust their energy intakes appropriately and may eat in response to external cues, disregarding hunger and satiety signals.

Energy Intake and Activity Individual children's energy needs vary widely, depending on their growth and physical activity. A one-year-old child needs about 800 kcalories a day; an active six-year-old needs twice as many kcalories a day. By age ten, an active child needs about 2000 kcalories a day. Total energy needs increase slightly with age, but energy needs per kilogram of body weight actually decline gradually.

Inactive children can become obese even when they eat less food than the average. Unfortunately, our nation's children are becoming less and less active, with young girls showing a marked reduction in their physical activity. Schools would serve our children well by offering activities to promote physical fitness.[37] Children who learn to enjoy physical play and exercise, both at home and at school, are best prepared to maintain active lifestyles as adults.

Some children, notably those adhering to a vegan diet, may have difficulty meeting their energy needs. Grains, vegetables, and fruits provide plenty of fiber, adding bulk, but may provide too few kcalories to support growth. Soy products, other legumes, and nut or seed butters offer more concentrated sources of energy to support optimal growth and development.[38]

Carbohydrate and Fiber Carbohydrate recommendations are based on glucose use by the brain. After a year of age, brain glucose use remains fairly constant and is within the adult range. Carbohydrate recommendations for children from the age of one year on are therefore the same as for adults (see inside front cover).[39]

Fiber recommendations derive from adult intakes shown to reduce the risk of coronary heart disease and are based on energy intakes. Consequently, children who have low energy intakes need less fiber than those with high intakes.[40] At a minimum, children's fiber intakes should at least equal their "age plus 5 grams."[41]

Fat and Fatty Acids As long as children's energy intakes are adequate, fat intakes below 30 percent of total energy do not impair growth.[42] Children who eat low-fat diets, however, have low intakes of some vitamins and minerals. The energy of dietary fat is important for young children who eat less food than older children and adults. No RDA for total fat has been established, but the DRI Committee recommends a fat intake of 30 to 40 percent of energy for children 1 to 3 years of age and 25 to 35 percent for children 4 to 18 years of age.[43] Recommended intakes of the essential fatty acids are based on average intakes (see inside front cover).

Protein Like energy needs, total protein needs increase slightly with age, but when the child's body weight is considered, the protein requirement actually declines slightly (see inside front cover). The estimation of protein needs considers the requirements for maintaining nitrogen balance, the quality of protein consumed, and the added needs of growth.

Vitamins and Minerals The vitamin and mineral needs of children increase with age (see inside front cover). A balanced diet of nutritious foods can meet children's needs for these nutrients, with the notable exception of iron. Iron-deficiency anemia is a major problem worldwide, as well as being the most prevalent nutrient deficiency among U.S. and Canadian children, especially toddlers one to two years of age.[44] During the second year of life, toddlers progress from a diet of iron-rich infant foods such as breast milk, iron-fortified formula, and iron-fortified infant cereal to a diet of adult foods and iron-poor cow's milk. In addition, their appetites often fluctuate; some become finicky about the foods they eat, and others prefer milk and juice to solid foods.[45] All of these situations can interfere with children eating iron-rich foods at a critical time for brain growth and development.

Reduce iron deficiency among young children.

HEALTHY PEOPLE 2010

To prevent iron deficiency, children's foods must deliver approximately 10 milligrams of iron per day. To achieve this goal, snacks and meals should include iron-rich foods, and milk intake should be reasonable so that it will not displace lean meats, fish, poultry, eggs, legumes, and whole-grain or enriched products. (Chapter 13 describes iron-rich foods and ways to maximize iron absorption.)

Supplements With the exception of specific recommendations for fluoride, iron, and vitamin D during infancy and childhood, the AAP and other professional groups agree that well-nourished children do not need vitamin and mineral supplements.[46] Despite this, many children and adolescents take supplements.[46] Ironically, children with poor nutrient intakes do not receive supplements, and those who do take supplements receive extra nutrients they do not need. Furthermore, researchers are still studying the safety of supplement use by children.[47] In the meantime, parents and other caregivers need to rely on foods instead of supplements to nourish children.

The Federal Trade Commission has issued a consumer guidance warning parents about giving supplements advertised to prevent or cure childhood illnesses such as colds, ear infections, or asthma. The term *supplement* today includes many herbal products that have not been tested for safety and effectiveness in children.[48]

Planning Children's Meals To provide all the needed nutrients, children's meals should include a variety of foods from each food group—in amounts suited to their appetites and needs. Figure 16-8 presents the Food Guide Pyramid for young children. Notice that the number of servings from each group is the same as the lower number for adults. For two- to three-year-olds, serving sizes are smaller, about two-thirds the portion for a child over four years old.

Children whose diets follow the Food Guide Pyramid pattern meet their nutrient needs fully, but few children eat according to these recommendations. Based on an analysis of the most recent national food intake data, the USDA found that most (81 percent) children between two and nine years of age have diets that need substantial improvement.[49] In another study, only 5 percent of the children consumed the suggested number of servings from even four of the five food groups. Consequently, intakes of several nutrients, notably calcium, iron, and zinc, typically fall below recommendations.[50]

Hunger and Malnutrition in Children

Most children in the United States and Canada have access to regular meals, but hunger and malnutrition do appear in certain circumstances. Low-income children, for example, may be hungry and malnourished. An estimated 12 million U.S. children are hungry at least some of the time and living in poverty.[51] Chapter 20 examines the causes and consequences of hunger in the United States and around the world.

When hunger is chronic, children become malnourished and suffer growth retardation. Worldwide, malnutrition takes a devastating toll on children, contributing to nearly half of the deaths of children under four years old. Vitamin A deficiency afflicts 3 to 10 million children worldwide, inducing blindness, stunted growth, and infections.[52] Zinc deficiency also retards growth and typically accompanies protein-energy malnutrition and vitamin A deficiency.

HEALTHY PEOPLE 2010 | Reduce growth retardation among low-income children under age 5 years.

The United Nations Children's Fund, known as UNICEF, helps children living in poverty in developing countries get the nutrition and health care they need. UNICEF works with more than 160 countries through national governments, private-sector partners, and other international agencies to protect children and their rights, and to reduce childhood death and illness.

Nutrition Explorer Compare the nutrient density of several child-friendly meals and snacks, and evaluate how well they meet the dietary needs of children of several different ages.

FIGURE 16-8   Food Guide Pyramid for Young Children

FOOD Guide PYRAMID

for Young Children

A Daily Guide for
2- to 6-Year-Olds

Fats & Sweets
Eat less

Milk Group
2 servings

Meat Group
2 servings

Vegetable Group
3 servings

Fruit Group
2 servings

Grain Group
6 servings

EAT a variety of FOODS AND ENJOY!

What counts as one serving?

GRAIN GROUP
1 slice bread
$^1/_2$ c cooked rice or pasta
$^1/_2$ c cooked cereal
1 oz ready-to-eat cereal

VEGETABLE GROUP
$^1/_2$ c chopped raw or cooked vegetables
1 c raw leafy vegetables

FRUIT GROUP
1 piece of fruit or melon wedge
$^3/_4$ c juice
$^1/_2$ c canned fruit
$^3/_4$ c dried fruit

MILK GROUP
1 c milk or yogurt
2 oz cheese

MEAT GROUP
2 to 3 oz cooked lean meat, poultry, or fish
$^1/_2$ c cooked dry beans, or 1 egg counts
as 1 oz lean meat; 2 tbs of peanut butter
count as 1 oz meat

FATS AND SWEETS
Limit kcalories from these.

Four- to six-year-olds can eat these
serving sizes. Offer two- to three-year-olds
less, except for milk. Two- to six-year-old
children need a total of 2 servings from
the milk group each day.

SOURCE: USDA Center for Nutrition and Policy Promotion, March 1999, Program AID 1649.

Hunger and Behavior Even when hunger is temporary, as when a child misses one meal, behavior and academic performance are affected. Children who eat nutritious breakfasts improve their school performance and are tardy or absent significantly less often than their peers who do not. Without breakfast, children perform poorly in tasks requiring concentration, their attention spans are shorter, and they even score lower on intelligence tests than their well-fed peers; malnourished children are particularly vulnerable. Unfortunately, an estimated 4 out of 30 students miss breakfast each day. Common sense dictates that it is unreasonable to expect anyone to learn and perform without fuel. For the child who hasn't had breakfast, the morning's lessons may be lost altogether. Even if a child has eaten breakfast, discomfort from hunger may become distracting by late morning.

The problem children face when attempting morning schoolwork on an empty stomach appears to be at least partly due to low blood glucose. The average child up to age ten or so needs to eat about every four hours to maintain a blood glucose concentration high enough to support the activity of the brain and the rest of the nervous system. A child's brain is as big as an adult's, and the brain is the body's chief glucose consumer, using about three times as much glucose per day as the rest of the body. A child's liver is much smaller than an adult's, however, and the liver is responsible for storing glucose as glycogen and releasing it into the blood

Healthy, well-nourished children are alert in the classroom and energetic at play.

as needed. A child's liver can store only about four hours' worth of glycogen—hence the need to eat fairly often. Teachers aware of the late-morning slump in their classrooms wisely request that midmorning snacks be provided; snacks improve classroom performance all the way to lunchtime.

Eating breakfast also helps children to meet their nutrient needs each day. Children who skip breakfast typically do not make up the deficits at later meals—they simply have lower intakes of energy, vitamins, and minerals than those who eat breakfast.

Iron Deficiency and Behavior Iron deficiency has well-known and widespread effects on children's behavior and intellectual performance.[53] In addition to carrying oxygen in the blood, iron transports oxygen within cells, which use it in energy metabolism. Iron is also used to make neurotransmitters—most notably, those that regulate the ability to pay attention, which is crucial to learning. Consequently, iron deficiency not only causes an energy crisis, but also directly affects attention span and learning ability.

Iron deficiency is often diagnosed by a quick, easy, inexpensive hemoglobin or hematocrit test that detects a deficit of iron in the *blood*. A child's *brain*, however, is sensitive to low iron concentrations long before the blood effects appear. Iron deficiency lowers the "motivation to persist in intellectually challenging tasks" and impairs overall intellectual performance. Anemic children perform less well on tests and are more disruptive than their nonanemic classmates; iron supplementation improves learning and memory. When combined with other nutrient deficiencies, iron-deficiency anemia has synergistic effects that are especially detrimental to learning. Furthermore, children who had iron-deficiency anemia *as infants* continue to perform poorly as they grow older, even if their iron status improves.[54] The long-term damaging effects on mental development make prevention of iron deficiency during infancy and early childhood a high priority.

Other Nutrient Deficiencies and Behavior A child with any of several nutrient deficiencies may be irritable, aggressive, disagreeable, or sad and withdrawn. Such a child may be labeled "hyperactive," "depressed," or "unlikable," when in fact these traits may arise from simple, even marginal, malnutrition. Parents and medical practitioners often overlook the possibility that malnutrition may account for abnormalities of appearance and behavior. Any departure from normal healthy appearance and behavior is a sign of possible poor nutrition (see Table 16-3). In any such case, inspection of the child's diet by a registered dietitian or other qualified health care professional is in order. Any suspicion of dietary inadequacies, no matter what other causes may be implicated, should prompt steps to correct those inadequacies immediately.

The Malnutrition-Lead Connection

Children who are malnourished are vulnerable to lead poisoning. They absorb more lead if their stomachs are empty; if they have low intakes of calcium, zinc, vitamin C, or vitamin D; and, of greatest concern because it is so common, if they have iron deficiencies.[55] Iron deficiency weakens the body's defenses against lead absorption, and lead poisoning can cause iron deficiency. Common to both iron deficiency and lead poisoning are a low socioeconomic background and a lack of immunizations against infectious diseases. Another common factor is pica—a craving for nonfood items. Many children with lead poisoning eat dirt or chips of old paint, two common sources of lead.

The anemia brought on by lead poisoning may be mistaken for a simple iron deficiency and therefore may be incorrectly treated. Like iron deficiency, mild lead toxicity has nonspecific symptoms, including diarrhea, irritability, and fatigue. The symptoms are not reversed by adding iron to the diet; exposure to lead must stop. With further exposure, the signs become more pronounced: children develop learning disabilities and behavioral problems. Still more severe lead toxicity can cause irreversible nerve damage, paralysis, mental retardation, and death.

Old, lead-based paint threatens the health of an exploring child.

TABLE 16-3 Physical Signs of Malnutrition in Children

	Well-Nourished	Malnourished	Possible Nutrient Deficiencies
Hair	Shiny, firm in the scalp	Dull, brittle, dry, loose; falls out	PEM
Eyes	Bright, clear pink membranes; adjust easily to light	Pale membranes; spots; redness; adjust slowly to darkness	Vitamin A, the B vitamins, zinc, and iron
Teeth and gums	No pain or caries, gums firm, teeth bright	Missing, discolored, decayed teeth; gums bleed easily and are swollen and spongy	Minerals and vitamin C
Face	Clear complexion without dryness or scaliness	Off-color, scaly, flaky, cracked skin	PEM, vitamin A, and iron
Glands	No lumps	Swollen at front of neck, cheeks	PEM and iodine
Tongue	Red, bumpy, rough	Sore, smooth, purplish, swollen	B vitamins
Skin	Smooth, firm, good color	Dry, rough, spotty; "sandpaper" feel or sores; lack of fat under skin	PEM, essential fatty acids, vitamin A, B vitamins, and vitamin C
Nails	Firm, pink	Spoon-shaped, brittle, ridged	Iron
Internal systems	Regular heart rhythm, heart rate, and blood pressure; no impairment of digestive function, reflexes, or mental status	Abnormal heart rate, heart rhythm, or blood pressure; enlarged liver, spleen; abnormal digestion; burning, tingling of hands, feet; loss of balance, coordination; mental confusion, irritability, fatigue	PEM and minerals
Muscles and bones	Muscle tone; posture, long bone development appropriate for age	"Wasted" appearance of muscles; swollen bumps on skull or ends of bones; small bumps on ribs; bowed legs or knock-knees	PEM, minerals, and vitamin D

More than 400,000 children—most of them under age six—have blood lead concentrations high enough to cause mental, behavioral, and other health problems.[56] Lead intoxication in young children comes from their own behaviors and activities—putting their hands in their mouths, playing in dirt and dust, and chewing on nonfood items. Unfortunately, the body readily absorbs lead during times of rapid growth and hoards it possessively thereafter. Lead is not easily excreted and accumulates mainly in the bones, but also in the brain, teeth, and kidneys. Tragically, a child's neuromuscular system is also maturing during these first few years of life. No wonder children with elevated lead levels experience impairment of balance, motor development, and the relaying of nerve messages to and from the brain. Unfortunately, deficits in intellectual development are only partially reversed when lead levels decline.[57]

Eliminate elevated blood lead levels in children. Increase the proportion of persons living in pre-1950s housing that has been tested for the presence of lead-based paint.

Federal laws mandating reductions in leaded gasolines, lead-based solder, and other products over the past three decades have helped to reduce the amounts of lead in food and in the environment in the United States. As a consequence, the prevalence of lead toxicity in children has declined dramatically for most of the United States, but lead exposure is still a threat in certain communities.[58] The "How to" on the next page presents strategies for defending children against lead toxicity.

Hyperactivity and "Hyper" Behavior

All children are naturally active, and many of them become overly active on occasion—for example, in anticipation of a birthday party. Such behavior is markedly different from true **hyperactivity.**

hyperactivity: inattentive and impulsive behavior that is more frequent and severe than is typical of others a similar age; professionally called **attention-deficit/hyperactivity disorder (ADHD).**

HOW TO Protect against Lead Toxicity

Researchers simultaneously made three major discoveries about lead toxicity: lead poisoning has *subtle* effects, the effects are *permanent,* and they occur at *low levels of exposure.* The amount of lead recognized to cause harm is only 10 micrograms per 100 milliliters of blood. Some research shows that blood lead concentrations *below* this amount may adversely affect children's scores on intelligence tests.[a] Consequently, consumers should take ultraconservative measures to protect themselves, and especially their infants and young children, from lead poisoning. The American Academy of Pediatrics and the Centers for Disease Control recommend screening in communities with a substantial number of houses built before 1950 and in those with a substantial number of children with elevated lead levels. In addition to screening children most likely to be exposed, pediatricians should alert all parents to the possible dangers of lead exposure and explain prevention strategies.

Preventive strategies include:

- In contaminated environments, keep small children from putting dirty or old painted objects in their mouths, and make sure children wash their hands before eating.

Similarly, keep small children from eating any nonfood items. Lead poisoning has been reported in young children who have eaten crayons or pool cue chalk.

- Wet-mop floors and damp-sponge walls regularly. Children's blood lead levels decline when the homes they live in are cleaned regularly.

- Be aware that other countries do not have the same regulations protecting consumers against lead. Children have been poisoned by eating crayons made in China and drinking fruit juice canned in Mexico.

- Do not use lead-contaminated water to make infant formula.

- Once you have opened canned food, store it in a lead-free container to prevent lead migration into the food.

- Do not store acidic foods or beverages (such as vinegar or orange juice) in ceramic dishware or alcoholic beverages in pewter or crystal decanters.

- Many manufacturers are now making lead-safe products.[b] Old, handmade, or imported ceramic cups and bowls may contain lead and should not be used to heat coffee or tea or acidic foods such as tomato soup.

- U.S. wineries have stopped using lead in their foil seals, but older bottles may still be around and other countries may still use

lead; to be safe, wipe the foil-sealed rim of a wine bottle with a clean wet cloth before removing the cork.

- Feed children nutritious meals regularly.

- Before using your newspaper to wrap food, mulch garden plants, or add to your compost, confirm with the publisher that the paper uses no lead in its ink.

The Environmental Protection Agency (EPA) also publishes a booklet, *Lead and Your Drinking Water,* in which the following cautions appear:

- Have the water in your home tested by a competent laboratory.

- Use only cold water for drinking, cooking, and making formula (cold water absorbs less lead).

- When water has been standing in pipes for more than two hours, flush the cold-water pipes by running water through them for 30 seconds before using it for drinking, cooking, or mixing formulas.

- If lead contamination of your water supply seems probable, obtain additional information and advice from the EPA and your local public health agency.

By taking these steps, parents can protect themselves and their children from this preventable danger.[c]

---

[a]R. L. Canfield and coauthors, Intellectual impairment in children with blood lead concentrations below 10 µg per deciliter, *New England Journal of Medicine* 348 (2003): 1517–1526.

[b]*A Shopper's Guide to Low-Lead China* is available from the Environmental Defense Fund, 257 Park Avenue South, New York, NY 10010; telephone (800) 284-3322.

[c]The National Lead Information Center provides two hotlines; call (800) LEAD-FYI (532-3394) for general information or (800) 424-LEAD (424-5323) with specific questions.

Hyperactivity Hyperactive children have trouble sleeping, cannot sit still for more than a few minutes at a time, act impulsively, and have difficulty paying attention. These behaviors interfere with social development and academic progress. The cause of hyperactivity remains unknown, but it affects 3 to 5 percent of young school-age children. To resolve the problems surrounding hyperactivity, physicians often recommend specific behavioral strategies, special educational programs, and psychological counseling; in many cases, they prescribe medication.[59]

Parents of hyperactive children sometimes seek help from alternative therapies, including special diets. They mistakenly believe a solution may lie in manipulating the diet—most commonly, by excluding sugar or food additives. Adding carrots or eliminating candy is such a simple solution that many parents eagerly give dietary advice a try. These dietary changes will not solve the problem of true hyperactivity. Studies have consistently found no convincing evidence that sugar causes hyperactivity or worsens behavior; in fact, sugar may actually have a sedative effect. Recommendations to restrict sugar in children's diets to prevent or treat behavioral problems are groundless.

Misbehaving Even a child who is not truly hyperactive can be difficult to manage at times. Michael may act unruly out of a desire for attention, Jessica may be cranky because of a lack of sleep, Christopher may react violently after watching too much television, and Sheila may be unable to sit still in class due to a lack of exercise. All of these children may benefit from more consistent care—regular hours of sleep, regular mealtimes, and regular outdoor activity.

Food Allergy and Intolerance

Food allergy is frequently blamed for physical and behavioral abnormalities in children, but only about 3 to 5 percent of children are diagnosed with true food allergies.[60] Food allergies diminish with age, until in adulthood they affect about 1 or 2 percent of the population.[61]

A true food allergy occurs when fractions of a food protein or other large molecule are absorbed into the blood and elicit an immunologic response. (Recall that proteins are normally dismantled in the digestive tract to amino acids that are absorbed without such a reaction.) The body's immune system reacts to these large food molecules as it does to other antigens—by producing antibodies, histamines, or other defensive agents.

Detecting Food Allergy Allergies may have one or two components. They always involve antibodies; they may or may not involve symptoms.■ This means that allergies can be diagnosed only by testing for antibodies. Even symptoms exactly like those of an allergy may not be caused by one. Once a food allergy has been diagnosed, therapy requires strict elimination of the offending food. Children with allergies, like all children, need all their nutrients, so it is important to include other foods that offer the same nutrients as the omitted foods.[62]

Allergic reactions to food may be immediate or delayed. In both cases, the antigen interacts immediately with the immune system, but the timing of symptoms varies from minutes to 24 hours after consumption of the antigen. Identifying the food that causes an immediate allergic reaction is fairly easy because the symptoms appear shortly after the food is eaten. Identifying the food that causes a delayed reaction is more difficult because the symptoms may not appear until much later. By this time, many other foods may have been eaten, complicating the picture.

Anaphylactic Shock The life-threatening food allergy reaction of **anaphylactic shock** is most often caused by peanuts, tree nuts, milk, eggs, wheat, soybeans, fish, or shellfish.[63] Among these foods, eggs, milk, soy, and peanuts most often cause problems in children. Children are more likely to outgrow allergies to eggs, milk, and soy than allergies to peanuts. Peanuts cause more life-threatening reactions than do all other food allergies combined. Research is currently under way to help those with peanut allergies tolerate small doses, thus saving lives and minimizing reactions.[64] One possible solution depends on finding a natural, hypoallergenic peanut among the 14,000 varieties of peanuts that grow. Families of children with a life-threatening food allergy and school personnel who supervise them must guard them against any exposure to the allergen.[65] The child must learn to identify which foods pose a problem and then learn and use refusal skills for all foods that may contain the allergen.

Reduce deaths from anaphylaxis caused by food allergies.

Parents of allergic children can pack safe foods for lunches and snacks and ask school officials to strictly enforce a "no swapping" policy in the lunchroom. The child must be able to recognize the symptoms of impending anaphylactic shock,■ such as a tingling of the tongue, throat, or skin, or difficulty breathing. Any person with food allergies severe enough to cause anaphylactic shock should wear a medical alert bracelet or necklace. Finally, the responsible child and the school staff should be prepared with injections of **epinephrine,** which prevents anaphylaxis after exposure to the allergen.[66] Many preventable deaths occur each year when people with food allergies accidentally ingest the allergen but have no epinephrine.[67]

These normally wholesome foods may cause life-threatening symptoms in people with allergies.

■ A person who produces antibodies *without* having any symptoms has an **asymptomatic allergy;** a person who produces antibodies *and* has symptoms has a **symptomatic allergy.**

■ Symptoms of impending anaphylactic shock:
- Tingling sensation in mouth.
- Swelling of the tongue and throat.
- Irritated, reddened eyes.
- Difficulty breathing, asthma.
- Hives, swelling, rashes.
- Vomiting, abdominal cramps, diarrhea.
- Drop in blood pressure.
- Loss of consciousness.
- Death.

food allergy: an adverse reaction to food that involves an immune response; also called **food-hypersensitivity reaction.**

anaphylactic (an-AFF-ill-LAC-tic) **shock:** a life-threatening whole-body allergic reaction.

epinephrine (EP-ih-NEFF-rin): a hormone of the adrenal gland administered by injection to counteract anaphylactic shock by opening the airways and maintaining heartbeat and blood pressure.

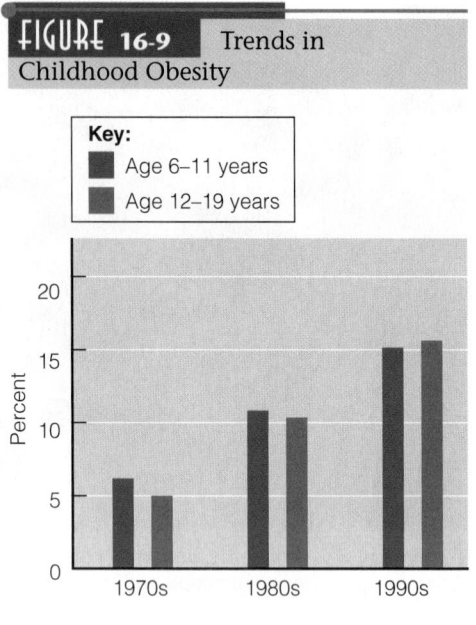

FIGURE 16-9 Trends in Childhood Obesity

Allergens often sneak into foods in unexpected ways. To protect people with allergies, regulations require food manufacturers to declare common allergens, including food additives, on food labels.[68] Manufacturers must also prevent cross-contamination during production. Eliminating peanuts per se may not be too difficult, but avoiding peanut dust in a chocolate cookie can be a challenge. Equipment used for mixing peanut butter cookies must be disassembled and cleaned thoroughly before being used to make other products. When cross-contamination is likely, food labels must state that the product may contain an allergy-producing food. The FDA is currently working with the food industry to develop more clearly worded labels to alert consumers to the presence of food allergens.[69]

Food Intolerances Not all **adverse reactions** to foods are food allergies, although even physicians may describe them as such. Signs of adverse reactions to foods include stomachaches, headaches, rapid pulse rate, nausea, wheezing, hives, bronchial irritation, coughs, and other such discomforts. Among the causes may be reactions to chemicals in foods, such as the flavor enhancer monosodium glutamate (MSG), the natural laxative in prunes, or the mineral sulfur; digestive diseases, such as obstructions or injuries; enzyme deficiencies, such as lactose intolerance; and even psychological aversions. These reactions involve symptoms but no antibody production. Therefore, they are **food intolerances,** not allergies.

Pesticides on produce may also cause adverse reactions. Pesticides may linger on the foods to which they were applied in the field. Health risks from pesticide exposure are probably small for healthy adults, but children may be vulnerable. Chapter 19 revisits the issues surrounding the use of pesticides on food crops.

Hunger, lead poisoning, hyperactivity, and allergic reactions can all adversely affect a child's nutrition status and health. Fortunately, each of these problems has solutions. They may not be easy solutions, but at least we have a reasonably good understanding of the problems and ways to correct them. Such is not the case with the most pervasive health problem for children in the United States—obesity.

Childhood Obesity

The number of overweight children has increased dramatically over the past three decades (see Figure 16-9). Like their parents, children in the United States are becoming fatter. An estimated 15 percent of U.S. children and adolescents 6 to 19 years of age are overweight.[70] Based on data from the BMI-for-age growth charts, children and adolescents are categorized as *at risk of overweight* above the 85th percentile and as *overweight* at the 95th percentile and above. Prevalence data reflect only children and adolescents in the overweight category; if those at risk of overweight were also included, the estimated 15 percent would likely double. Figure 16-10 presents the BMI for children and adolescents, indicating cutoff points for overweight and at risk of overweight.

HEALTHY PEOPLE 2010

Reduce the proportion of children and adolescents who are overweight or at risk of overweight.

The problem is especially troubling because overweight children have the potential of becoming obese adults with all the social, economic, and medical ramifications that often accompany obesity. They have additional problems, too, arising from differences in their growth, physical health, and psychological development. In trying to explain the rise in childhood obesity, researchers point to both genetic and environmental factors.

Genetic and Environmental Factors Parental obesity predicts an early increase in a young child's BMI, and it more than doubles the chances that a young

adverse reactions: unusual responses to food (including intolerances and allergies).

food intolerances: adverse reactions to foods that do not involve the immune system.

FIGURE 16-10 Body Mass Index-for-Age Percentiles: Boys and Girls, Age 2 to 20

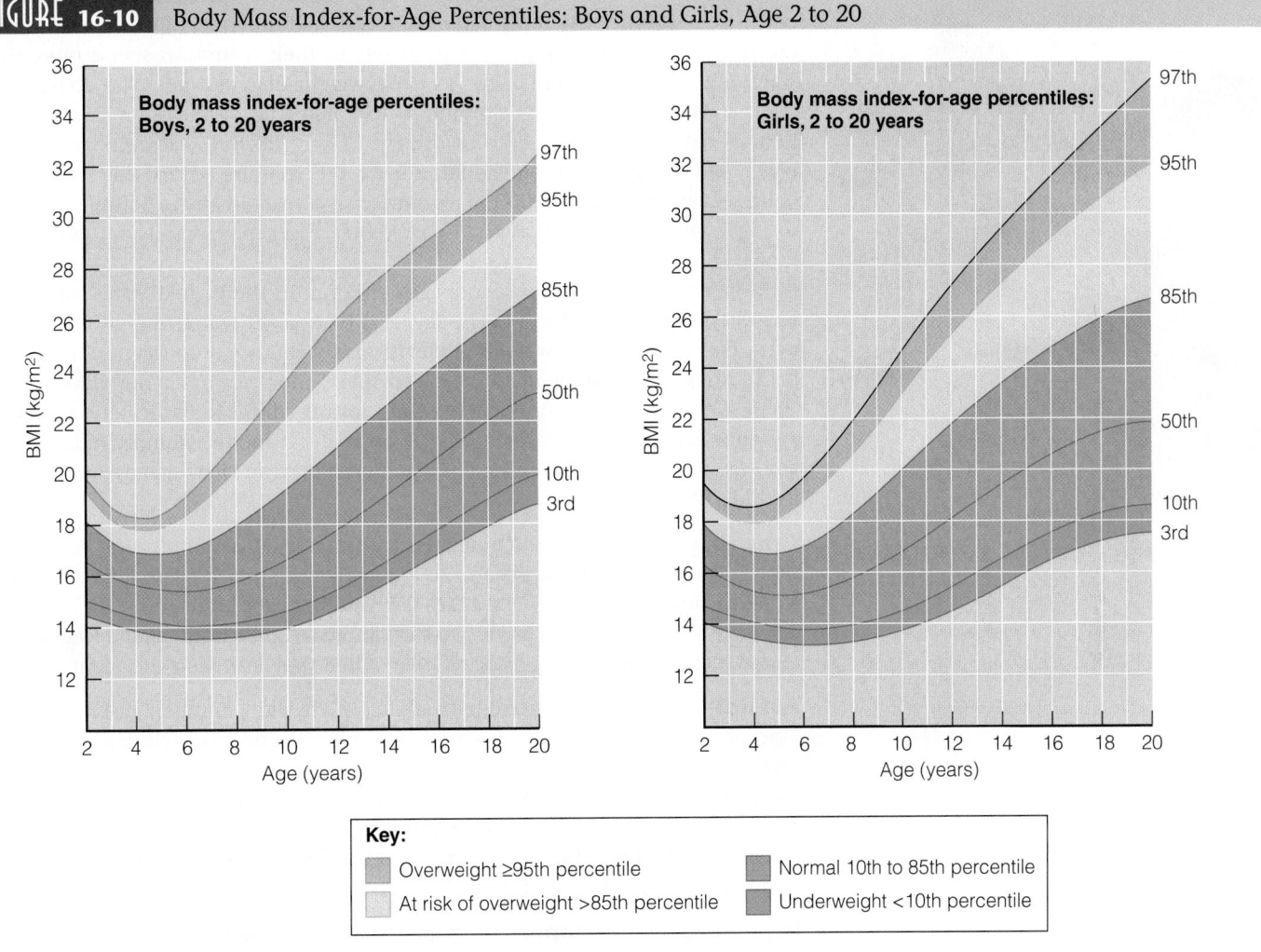

child will become an obese adult.[71] Nonobese children with neither parent obese have a less than 10 percent chance of becoming obese in adulthood, whereas obese teens with at least one obese parent have a greater than 80 percent chance of being obese adults. As children grow older, their own obesity also becomes an important factor in determining their obesity as adults.[72] The link between parental and child obesity reflects both genetic and environmental factors.

Diet and physical inactivity must also play a role in explaining why children are heavier today than they were 30 or so years ago. Children's dietary fat intakes vary, of course, but children who prefer high-fat foods tend to be more overweight than their peers.[73] Particularly noteworthy is that they also tend to have overweight parents. Such findings confirm the significant roles parents play in teaching children about healthy food choices, providing children with nutrient-dense foods, and serving as role models. When parents eat fruits and vegetables frequently, their children do, too.[74] The more fruits and vegetables children eat, the more vitamins and minerals and the less fat in their diets.

Most likely, children have grown more overweight because of their lack of physical activity.[75] An inactive child can become obese even while eating less food than an active child. Today's children are more sedentary and less physically fit than children were even 20 years ago.

Television watching■ may contribute most to physical inactivity. One study compared children who were experimentally set up with a stationary bicycle that they had to pedal in order to activate the television with children who did not

■ TV fosters obesity because it:
- Requires no energy beyond basal metabolism.
- Replaces vigorous activities.
- Encourages snacking.
- Promotes a sedentary lifestyle.

Playing video games influences children's activity patterns similarly.

Television watching influences children's eating habits and activity patterns.

have to pedal to watch TV, but could if they chose to.[76] The children who had to pedal were significantly more active, and watched much less TV, than those who did not. Children who have television sets in their bedrooms spend more time watching TV and are more likely to be overweight than children who do not have televisions in their rooms.[77] Children who watch a lot of television (four or more hours a day) are most likely to be obese and least likely to eat fruits and vegetables.[78] They often snack on the fattening foods that are advertised. The average child sees an estimated 30,000 TV commercials a year—many peddling foods high in sugar, fat, and salt such as sugar-coated breakfast cereals, candy bars, chips, fast foods, and carbonated beverages.

More than 25 percent of school-aged children in the United States watch four or more hours of television every day, and 67 percent watch two or more hours every day.[79] Time spent watching television is second only to sleep in time spent being physically inactive. Children also spend more time playing video games. These activities use no more energy than resting, displace participation in more vigorous activities, and foster snacking on high-fat foods.[80] Simply reducing the amount of time spent watching television (and playing video games) can improve a child's BMI.

Growth Overweight children develop a characteristic set of physical traits. They typically begin puberty earlier and so grow taller than their peers at first, but then stop growing at a shorter height. They develop greater bone and muscle mass in response to the demand of having to carry more weight—both fat and lean weight. Consequently, they appear "stocky" even when they lose their excess fat.

Physical Health Like overweight adults, overweight children display a blood lipid profile indicative that atherosclerosis is beginning to develop: high levels of total cholesterol, triglycerides, and LDL cholesterol. Overweight children also tend to have high blood pressure; in fact, obesity is the leading cause of pediatric hypertension. Their risks for developing type 2 diabetes and respiratory diseases (such as asthma) are also exceptionally high.[81] These relationships between childhood obesity and chronic diseases are discussed fully in Highlight 16.

Psychological Development In addition to the physical consequences, childhood obesity brings a host of emotional and social problems.[82] Because people frequently judge others on appearance more than on character, overweight children are often victims of prejudice. Many suffer discrimination by adults and rejection by their peers. They may have poor self-images, a sense of failure, and a passive approach to life. Television shows, which are a major influence in children's lives, often portray the fat person as the bumbling misfit. Overweight children themselves may come to accept this negative stereotype. Researchers investigating children's reactions to various body types find that both normal-weight and underweight children respond unfavorably to overweight bodies.

Prevention and Treatment of Obesity Medical science has worked wonders in preventing or curing many of even the most serious childhood diseases, but obesity remains a challenge.[83] Once excess fat has been stored, it is stubbornly difficult to remove. In light of all this, parents are encouraged to make major efforts to prevent childhood obesity or to begin treatment early—before adolescence.[84] Treatment must consider the many aspects of the problem and possible solutions. An integrated approach is recommended, involving diet, physical activity, psychological support, and behavioral changes.

Diet The initial goal for overweight children is to reduce the rate of weight gain; that is, to maintain weight while the child grows taller. Continued growth will then accomplish the desired change in weight for height. Weight loss is usually not recommended because diet restriction can interfere with growth and development.[85]

Whether the goal is to treat or prevent obesity, these strategies may be helpful:

- Serve family meals that reflect kcalorie control both in the foods offered and in the ways foods have been prepared.

- Involve children in shopping for, and preparing, meals.

- Encourage children to eat only when they are hungry, to eat slowly, to pause and enjoy their table companions, and to stop eating when they are full.

- Teach them how to select nutrient-dense foods that will meet their nutrient needs within their energy allowances and to serve themselves appropriate portions at meals; the amount of food offered influences the amount of food eaten.[86]

- Limit, but don't overly restrict, high-fat, high-sugar foods, including sugar-sweetened soft drinks.

- Never force children to clean their plates.

- Plan for snack times and provide a variety of nutritious snacks (see Table 16-5 later in this chapter).

- Discourage eating while watching TV.

Physical Activity   The many benefits of physical activity are well known, but often are not incentive enough to motivate overweight people, especially children. Yet regular vigorous activity can improve a child's weight, body composition, and physical fitness.[87] Ideally, parents will limit sedentary activities and encourage daily physical activity to promote strong skeletal, muscular, and cardiovascular development and instill in their children the desire to be physically active throughout life. Most importantly, parents need to set a good example. Physical activity is a natural and lifelong behavior of healthy living. It can be as simple as riding a bike, playing tag, jumping rope, or doing chores. It need not be an organized sport; it just needs to be some activity on a regular basis. The AAP supports the efforts of schools to include more physical activity in the curriculum and encourages parents to support their children's participation.[88]

Increase the proportion of the nation's public and private schools that require daily school physical education for all students. Increase the proportion of adolescents who engage in moderate physical activity for at least 30 minutes on 5 or more days per week. Increase the proportion of adolescents who engage in vigorous physical activity that promotes cardiorespiratory fitness 3 or more days per week for 20 or more minutes per occasion. Increase the proportion of adolescents who spend physical education class time being physically active.

Psychological Support   Weight-loss programs that involve parents and other caregivers in treatment report greater success than those without parental involvement. Because obesity in parents and their children tends to be positively correlated, both benefit from a weight-loss program. Parental attitudes about food greatly influence children's eating behavior, so it is important that the influence be positive. Otherwise, eating problems may become exacerbated. Unaware that they are teaching their children, parents pass on lessons at the dinner table, on television trays, and in drive-through restaurants. Overweight parents may model for their children the behaviors that have led to their weight gains—eating too much, dieting inappropriately, exercising too little. This pattern is especially evident between mothers and daughters.[89]

Behavioral Changes   In contrast to traditional weight-loss programs that focus on *what* to eat, behavioral programs focus on *how* to eat. These techniques involve changing learned habits that lead a child to eat excessively.

Obesity is prevalent in our society. Its far-reaching effects lend urgency to the need to find a remedy. Because treatment of obesity is frequently unsuccessful, it is most important to prevent its onset. Above all, be sensible in teaching children how to

Eating is more fun for children when friends are there.

maintain appropriate body weight. Children can easily get the impression that their worth is tied to their body weight. Parents and the media are most influential in shaping self-concept, weight concerns, and dieting practices.[90] Some parents fail to realize that society's ideal of slimness can be perilously close to starvation and that a child encouraged to "diet" cannot obtain the energy and nutrients required for normal growth and development. Even healthy children without diagnosable eating disorders have been observed to limit their growth through "dieting." Weight gain in truly overweight children can be managed without compromising growth, but should be overseen by a health care professional.

Mealtimes at Home

Traditionally, parents served as **gatekeepers**, determining what foods and activities were available in their children's lives. Then the children made their own selections. Gatekeepers who wanted to promote nutritious choices and healthful habits provided access to nutrient-dense, delicious foods and opportunities for active play at home.

In today's consumer-oriented society, children have greater influence over family decisions concerning food—the fast-food restaurant the family chooses when eating out, the type of food the family eats at home, and the specific brands the family purchases at the grocery store. Parental guidance in food choices is still necessary, but equally important is teaching children consumer skills to help them make informed choices.

Honoring Children's Preferences Researchers attempting to explain children's food preferences encounter contradictions. Children say they like colorful foods, yet most often reject green and yellow vegetables while favoring brown peanut butter and white potatoes, apple wedges, and bread. They seem to like raw vegetables better than cooked ones, so it is wise to offer vegetables that are raw or slightly undercooked, served separately, and easy to eat. Foods should be warm, not hot, because a child's mouth is much more sensitive than an adult's. The flavor should be mild because a child has more taste buds, and smooth foods such as mashed potatoes or split-pea soup should contain no lumps (a child wonders, with some disgust, what the lumps might be). Children prefer foods that are familiar, so offer various foods regularly.

Make mealtimes fun for children. Young children like to eat at little tables and to be served small portions of food. They like sandwiches cut in different geometric shapes and common foods called silly names. They also like to eat with other children, and they tend to eat more when in the company of their friends. Children are also more likely to give up their prejudices against foods when they see their peers eating them.

Learning through Participation Allowing children to help plan and prepare the family's meals provides enjoyable learning experiences and encourages children to eat the foods they have prepared. Vegetables are pretty, especially when fresh, and provide opportunities for children to learn about color, growing vegetables and their seeds, and shapes and textures—all of which are fascinating to young children. Measuring, stirring, washing, and arranging foods are skills that even a young child can practice with enjoyment and pride (see Table 16-4).

Avoiding Power Struggles Problems over food often arise during the second or third year, when children begin asserting their independence. Many of these problems stem from the conflict between children's developmental stages and capabilities and parents who, in attempting to do what they think is best for their children, try to control every aspect of eating. Such conflicts can disrupt children's abilities to regulate their own food intakes or to determine their own likes and dis-

gatekeepers: with respect to nutrition, key people who control other people's access to foods and thereby exert profound impacts on their nutrition. Examples are the spouse who buys and cooks the food, the parent who feeds the children, and the caregiver in a day-care center.

likes. For example, many people share the misconception that children must be persuaded or coerced to try new foods. In fact, the opposite is true. When children are forced to try new foods, even by way of rewards, they are less likely to try those foods again than are children who are left to decide for themselves. Similarly, when children are restricted from eating their favorite foods, they are more likely to want those foods.[91] The parent is responsible for providing healthful foods, but the child is responsible for *how much* and even *whether* to eat.

When introducing new foods at the table, offer them one at a time and only in small amounts at first. The more often a food is presented to a young child, the more likely the child will accept that food. Offer the new food at the beginning of the meal, when the child is hungry, and allow the child to make the decision to accept or reject it. Never make an issue of food acceptance. A power struggle almost invariably sets a firm pattern of resistance and permanently closes the child's mind.

Choking Prevention Parents must always be alert to the dangers of choking. A choking child is silent, so an adult should be present whenever a child is eating. Make sure the child sits when eating; choking is more likely when a child is running or falling. (See p. 553 for a list of foods and nonfood items most likely to cause choking.)

Playing First Children may be more relaxed and attentive at mealtime if outdoor play or other fun activities are scheduled before, rather than immediately after, mealtime. Otherwise children "hurry up and eat" so that they can go play.

Snacking Parents may find that their children snack so much that they aren't hungry at mealtimes. Instead of teaching children *not* to snack, parents might be wise to teach them *how* to snack. Provide snacks that are as nutritious as the foods served at mealtime. Snacks can even be mealtime foods served individually over time, instead of all at once on one plate. When providing snacks to children, think of the five food groups and offer such snacks as pieces of cheese, tangerine slices, and peanut butter on whole-wheat crackers (see Table 16-5 on p. 568). Snacks that are easy to prepare should be readily available to children, especially if they arrive home from school before their parents.

To ensure that children have healthy appetites and plenty of room for nutritious foods when they are hungry, parents and teachers must limit access to candy, cola, and other concentrated sweets. Limiting access includes limiting the amount of pocket money children have to buy such foods themselves.[92] If these foods are permitted in large quantities, the only possible outcomes are nutrient deficiencies, obesity, or both. The preference for sweets is innate; most children do not naturally select nutritious foods on the basis of taste. When children are allowed to create meals freely from a variety of foods, they typically select foods that provide a lot of sugar. When their parents are watching, or even when they think their parents are watching, children improve their selections.[93]

Sweets need not be banned altogether. Children who are exceptionally active can enjoy high-kcalorie foods such as ice cream or pudding from the milk group or pancakes from the bread group. As for sedentary children, they need to become more active, so they can also enjoy some of these foods without unhealthy weight gain.

Preventing Dental Caries Children frequently snack on sticky, sugary foods that stay on the teeth and provide an ideal environment for the growth of bacteria that cause dental caries. Teach children to brush and floss after meals, to brush or rinse after eating snacks, to avoid sticky foods, and to select crisp or fibrous foods frequently.

Serving as Role Models In an effort to practice these many tips, parents may overlook perhaps the single most important influence on their children's food habits—themselves. Parents who don't eat carrots shouldn't be surprised when their children refuse to eat carrots. Likewise, parents who comment negatively on the smell of brussels sprouts may not be able to persuade children to try them. Children learn much through imitation. It is not surprising that children prefer the foods other family

TABLE 16-4 Food Skills of Preschool Children[a]

Age 1–2 years, when large muscles develop, the child:

- Uses short-shanked spoon.
- Helps feed self.
- Lifts and drinks from cup.
- Helps scrub, tear, break, or dip foods.

Age 3 years, when medium hand muscles develop, the child:

- Spears food with fork.
- Feeds self independently.
- Helps wrap, pour, mix, shake, or spread foods.
- Helps crack nuts with supervision.

Age 4 years, when small finger muscles develop, the child:

- Uses all utensils and napkin.
- Helps roll, juice, mash, or peel foods.
- Cracks egg shells.

Age 5 years, when fine coordination of fingers and hands develops, the child:

- Helps measure, grind, grate, and cut (soft foods with dull knife).
- Uses hand-cranked egg beater with supervision.

[a]These ages are approximate. Healthy, normal children develop at their own pace.

Children enjoy eating the foods they help to prepare.

TABLE 16-5 Healthful Snack Ideas—Think Food Groups, Alone and in Combination

Selecting two or more foods from different food groups adds variety and nutrient balance to snacks. The combinations are endless, so be creative.

Grain Products

Grain products are filling snacks, especially when combined with other foods:

- Cereal with fruit and milk
- Crackers and cheese
- Whole-grain toast with peanut butter
- Popcorn with grated cheese
- Oatmeal raisin cookies with milk

Vegetables

Cut-up fresh, raw vegetables make great snacks alone or in combination with foods from other food groups:

- Celery with peanut butter
- Broccoli, cauliflower, and carrot sticks with a flavored cottage cheese dip

Fruits

Fruits are delicious snacks and can be eaten alone—fresh, dried, or juiced—or combined with other foods:

- Apples and cheese
- Bananas and peanut butter
- Peaches with yogurt
- Raisins mixed with sunflower seeds or nuts

Meats and Meat Alternates

Meat and meat alternates add protein to snacks:

- Refried beans with nachos and cheese
- Tuna on crackers
- Luncheon meat on whole-grain bread

Milk and Milk Products

Milk can be used as a beverage with any snack, and many other milk products, such as yogurt and cheese, can be eaten alone or with other foods as listed above.

---

members enjoy and dislike foods that are never offered to them.[94] Parents, older siblings, and other caregivers set an irresistible example by sitting with younger children, eating the same foods, and having pleasant conversations during mealtimes.

While serving and enjoying food, caregivers can promote both physical and emotional growth at every stage of a child's life. They can help their children to develop both a positive self-concept and a positive attitude toward food. If the beginnings are right, children will grow without the conflicts and confusions over food that can lead to nutrition and health problems.

Nutrition at School

While parents are doing what they can to establish good eating habits in their children at home, others are preparing and serving foods to their children at day-care centers and schools. In addition, children begin to learn about food and nutrition in the classroom. Meeting the nutrition and education needs of children is critical to supporting their healthy growth and development.[95]

Meals at School The U.S. government assists schools financially so that every student can receive nutritious meals at school. Both the School Breakfast Program and the National School Lunch Program provide meals at a reasonable cost to children from families with the financial means to pay. Meals are available free or at reduced cost to children from low-income families. In addition, schools can obtain food commodities. Nationally, the U.S. Department of Agriculture (USDA) administers the programs; on the state level, state departments of education operate them.* The programs usually cost local school districts little, but the educational rewards are great. Several studies have reported that children who participate in school food programs show improvements in learning.[96]

Increase the proportion of children and adolescents aged 6 to 19 years whose intakes of meals and snacks at school contribute to good overall dietary quality.

Approximately 27 million children receive lunches through the National School Lunch Program—half of them free or at a reduced price.[97] School lunches offer a variety of food choices and help children meet at least one-third of their recommended intakes for energy, protein, vitamin A, vitamin C, iron, and calcium. Table 16-6 shows school lunch patterns for children of different ages and specifies the numbers of servings of milk, protein-rich foods (meat, poultry, fish, cheese, eggs, legumes, or peanut butter), vegetables, fruits, and breads or other grain foods. Over a week's menus, these lunches are also required to meet the *Dietary Guidelines.* Schools making special efforts to lower fat in school lunches typically have trouble providing enough energy and nutrients, especially iron, to meet specifications. The American Dietetic Association (ADA) advocates the development of dietary guidelines specifically for children to ensure that school lunches will both provide adequate energy and nutrients and support health.[98] Other health professionals agree that there is a need for separate guidelines that address children's unique needs.[99]

*School lunches in Canada are administered locally and therefore vary from area to area.

			Grade School through High School (Grade)		
TABLE 16-6 **School Lunch Patterns for Different Ages**[a]					
Food Group	**Preschool (Age)**				
	1 to 2	3 to 4	K to 3	4 to 6	7 to 12
Meat or meat alternate 1 serving:					
Lean meat, poultry, or fish	1 oz	1½ oz	1½ oz	2 oz	3 oz
Cheese	1 oz	1½ oz	1½ oz	2 oz	3 oz
Large egg(s)	½	¾	¾	1	1½
Cooked dry beans or peas	¼ c	⅜ c	⅜ c	½ c	¾ c
Peanut butter	2 tbs	3 tbs	3 tbs	4 tbs	6 tbs
Yogurt	½ c	¾ c	¾ c	1 c	1½ c
Peanuts, soynuts, tree nuts, or seeds[b]	½ oz	¾ oz	¾ oz	1 oz	1½ oz
Vegetable and/or fruit 2 or more servings, both to total	½ c	½ c	½ c	¾ c	¾ c
Bread or bread alternate[c] Servings	5 per week	8 per week	8 per week	8 per week	10 per week
Milk 1 serving of fluid milk	¾ c	¾ c	1 c	1 c	1 c

[a]The quantities listed represent per-lunch minimums for each age and grade except those for the oldest group, which are recommendations. Schools unable to serve the recommended quantities for grades 7 to 12 must provide at least the amount shown for grades 4 to 6.
[b]These meat alternates may be used to meet no more than half of the meat or meat alternate requirement; therefore, they must be used in a meal with another meat or meat alternate.
[c]Schools must serve daily at least ½ serving of bread or bread alternate to the youngest age group and at least 1 serving to older children.
SOURCE: U.S. Department of Agriculture, National School Lunch Program Regulations, revised January 1, 1998.

School lunches provide children with nourishment at little or no charge.

The School Breakfast Program is available in slightly more than half of the nation's schools, and about 7 million children participate in it.[100] The school breakfast must provide at least a fourth of the RDA for each of many nutrients and contain at least one serving of milk; one serving of fruit, juice, or vegetable; and either two servings of bread (or bread alternates), two servings of meat (or meat alternates), or one serving of each.

Another federal program, the Child and Adult Care Food Program (CACFP), operates similarly and provides funds to organized child-care programs. All eligible children, centers, and family day-care homes may participate. Sponsors are reimbursed for most meal costs and may also receive USDA commodity foods.

Competing Influences at School Serving healthful lunches is only half the battle; students need to eat them, too. Short lunch periods and long waiting lines prevent some students from eating a school lunch and leave others with too little time to complete their meals.[101] Nutrition efforts at schools are also undermined when students can buy meals from fast-food restaurants or a la carte foods such as pizza or snack foods and carbonated beverages from snack bars, school stores, and vending machines.[102] These items compete with nutritious school lunches and are often high in fat and sugar.[103] Some states restrict the sale of competing foods and have higher rates of participation in school meal programs than the national average. Nutrition professionals advocate prohibiting sales of food and beverages from vending machines or school stores in middle and high schools until 30 minutes after the end of the last meal unless they are part of the school foodservice and meet *Dietary Guidelines* standards.[104]

Nutrition Education at School Coincident with the school breakfast and lunch programs is a program of nutrition education and training (NET) in all public schools. This program is minimally funded, but program administrators are ingenious and creative in accomplishing its highest-priority objectives. School health clinics offer another opportunity to provide nutrition education and intervention. Children need to be fed well *and* to learn enough about nutrition to make healthful food choices when the choices become theirs to make. Effective nutrition education programs involve children's families and focus on changing specific behaviors rather than teaching general nutrition facts.[105]

HEALTHY PEOPLE 2010

Increase the proportion of middle, junior high, and senior high schools that provide school health education to prevent health problems in several topics (including unhealthy dietary patterns and inadequate physical activity).

IN SUMMARY Children's appetites and nutrient needs reflect their stage of growth. Those who are chronically hungry and malnourished suffer growth retardation; when hunger is temporary and nutrient deficiencies are mild, the problems are usually more subtle—such as poor academic performance. Iron deficiency is widespread and has many physical and behavioral consequences. "Hyper" behavior is not caused by poor nutrition; misbehavior may reflect inconsistent care. Childhood obesity has become a major health problem. Adults at home and at school need to provide children with nutrient-dense foods and teach them how to make healthful diet and activity choices.

Nutrition during Adolescence

Teenagers make many more choices for themselves than they did as children. They are not fed, they eat; they are not sent out to play, they choose to go. At the same time, social pressures thrust choices at them: whether to drink alcoholic beverages and whether to develop their bodies to meet extreme ideals of slimness or athletic prowess. Their interest in nutrition—both valid information and misinformation—derives from personal, immediate experiences. They are concerned with how diet can improve their lives now—they engage in fad dieting in order to fit into a new bathing suit, avoid greasy foods in an effort to clear acne, or eat a pile of spaghetti to prepare for a big sporting event. In presenting information on the nutrition and health of adolescents, this section includes many topics of interest to teens.

Growth and Development

With the onset of **adolescence,** the steady growth of childhood speeds up abruptly and dramatically, and the growth patterns of female and male become distinct. Hormones direct the intensity of the adolescent growth spurt, profoundly affecting every organ of the body, including the brain. After two to three years of intense growth and a few more at a slower pace, physically mature adults emerge.

In general, the adolescent growth spurt begins at age 10 or 11 for females and at 12 or 13 for males. It lasts about two and a half years. Before **puberty,** male and female body compositions differ only slightly, but during the adolescent spurt, differences between the genders become apparent in the skeletal system, lean body mass, and fat stores. In females, fat assumes a larger percentage of the total body weight, and in males, the lean body mass—principally muscle and bone—increases much more than in females (review Figure 8-8 on p. 265). On average, males grow 8 inches taller, and females, 6 inches taller. Males gain approximately 45 pounds, and females, about 35 pounds.

Nutritious snacks contribute valuable nutrients to an active teen's diet.

Energy and Nutrient Needs

Energy and nutrient needs are greater during adolescence than at any other time of life, except pregnancy and lactation. In general, nutrient needs rise throughout childhood, peak in adolescence, and then level off or even diminish as the teen becomes an adult.

Energy Intake and Activity The energy needs of adolescents vary greatly, depending on their current rate of growth, gender, body composition, and physical activity.[106] Boys' energy needs may be especially high; they typically grow faster than girls and, as mentioned, develop a greater proportion of lean body mass. An exceptionally active boy of 15 may need 3500 kcalories or more a day just to maintain his weight. Girls start growing earlier than boys and attain shorter heights and lower weights, so their energy needs peak sooner and decline earlier than those of their male peers. A sedentary girl of 15 whose growth is nearly at a standstill may need only 1700 kcalories a day if she is to avoid excessive weight gain. Thus adolescent girls need to pay special attention to being physically active and selecting foods of high nutrient density so as to meet their nutrient needs without exceeding their energy needs.

The insidious problem of obesity becomes ever more apparent in adolescence and often continues into adulthood. The problem is most evident in females, especially those of African American descent. Without intervention, overweight adolescents will face numerous physical and socioeconomic consequences for years to come. The consequences of obesity are so dramatic and our society's attitude toward obese people is so negative that even teens of normal or below-normal weight may perceive a need to lose weight. When taken to extremes, restrictive diets bring dramatic physical consequences of their own, as Highlight 9 explains.

Vitamins The RDA (or AI) for most vitamins increase during the adolescent years (see the table on the inside front cover). Several of the vitamin recommendations for adolescents are similar to those for adults, including the recommendation for vitamin D. During puberty, both the activation of vitamin D and the absorption of

adolescence: the period from the beginning of puberty until maturity.

puberty: the period in life in which a person becomes physically capable of reproduction.

calcium are enhanced, thus supporting the intense skeletal growth of the adolescent years without additional vitamin D.

Iron The need for iron increases during adolescence for both females and males, but for different reasons. Iron needs increase for females as they start to menstruate, and for males, as their lean body mass develops. Hence, the RDA increases at age 14 for both males and females. For females, the RDA remains high into late adulthood. For males, the RDA returns to preadolescent values in early adulthood.

In addition, iron needs increase when the adolescent growth spurt begins, whether that occurs before or after age 14. Therefore, boys in a growth spurt need an additional 2.9 milligrams of iron per day above the RDA for their age; girls need an additional 1.1 milligrams per day.[107]

Furthermore, iron recommendations for girls before age 14 do not reflect the iron losses of menstruation. The average age of menarche (first menstruation) in the United States is 12.5 years, however.[108] Therefore, for girls under the age of 14 who have started to menstruate, an additional 2.5 milligrams of iron per day is recommended.[109] Thus the RDA for iron depends not only on age and gender but also on whether the individual is in a growth spurt or has begun to menstruate, as listed in the margin.■

Iron intakes often fail to keep pace with increasing needs, especially for females, who typically consume less iron-rich foods such as meat and fewer total kcalories than males. Not surprisingly, iron deficiency is most prevalent among adolescent girls. Iron-deficient children and teens score lower on standardized tests than those who are not iron deficient.[110]

Calcium Adolescence is a crucial time for bone development, and the requirement for calcium reaches its peak during these years. Unfortunately, many adolescents have calcium intakes below current recommendations. Low calcium intakes during times of active growth, especially if paired with physical inactivity, may compromise the development of peak bone mass, which is considered the best protection against adolescent fractures and adulthood osteoporosis.[111] Increasing milk products■ in the diet to meet calcium recommendations greatly increases bone density.[112] Once again, however, teenage girls are most vulnerable, for their milk—and therefore their calcium—intakes begin to decline at the time when their calcium needs are greatest.[113] Furthermore, women have much greater bone losses than men in later life. In addition to dietary calcium, sports activities during adolescence build strong bones.

■ Iron RDA for males:
- 9–13 yr (8 mg/day).
- 9–13 yr in growth spurt (10.9 mg/day).
- 14–18 yr (11 mg/day).
- 14–18 yr in growth spurt (13.9 mg/day).

Iron RDA for females:
- 9–13 yr (8 mg/day).
- 9–13 yr in menarche (10.5 mg/day).
- 9–13 yr in menarche and growth spurt (11.6 mg/day).
- 14–18 yr (15 mg/day).
- 14–18 yr in growth spurt (16.1 mg/day).

■ Teenagers need to select at least 4 servings from the milk group daily to meet their calcium goal of 1300 mg/day. Chapter 12 presents other calcium-rich food choices.

Food Choices and Health Habits

Teenagers like the freedom to come and go as they choose. They eat what they want if it is convenient and if they have the time.[114] With a multitude of after-school, social, and job activities, they almost inevitably fall into irregular eating habits. At any given time on any given day, a teenager may be skipping a meal, eating a snack, preparing a meal, or consuming food prepared by a parent or restaurant. Adolescents who frequently eat meals with their families, however, eat more fruits, vegetables, grains, and calcium-rich foods, and drink fewer soft drinks, than those who seldom eat with their families.[115] Furthermore, they are less likely to smoke, drink alcohol, or use illegal drugs than adolescents who seldom eat meals with their families.[116]

Snacks Snacks typically provide at least a fourth of the average teenager's daily food energy intake. Most often, favorite snacks are high in fat and sodium and low in calcium, iron, vitamin A, vitamin C, folate, and fiber. Most adolescents need to eat a greater variety of foods to obtain these nutrients. Table 16-5 on p. 568 shows how to combine foods from different food groups to create healthy snacks. Vending machines rarely offer nutrient-dense options, and nutrition information alone does not convince people to make healthy choices.

Because their lunches rarely include fruits, vegetables, or milk, many teens fail to get all the vitamins and minerals they need each day.

Beverages Most frequently, adolescents drink soft drinks instead of fruit juice or milk with lunch, supper, and snacks. About the only time they select fruit juices is at breakfast. When they drink milk, they are more likely to consume it with a meal (especially breakfast) than as a snack. Because of their greater food intakes, boys are more likely than girls to drink enough milk to meet their calcium needs.

Over the past three decades, teens (especially girls) have been drinking more soft drinks and less milk, as Figure 16-11 shows.[117] Adolescents who drink soft drinks regularly have a higher energy intake and a lower calcium intake than those who do not; they are also more likely to be overweight.[118] The National Institutes of Health calls the low calcium intakes of America's teens "a crisis with long-term health effects." About 85 percent of girls and 64 percent of boys ages 12 to 19 years do not get enough calcium and are at serious risk for developing osteoporosis and other bone diseases in later life.[119]

For adolescents who can afford the kcalories and are meeting their calcium needs, soft drinks are an acceptable part of the diet. Soft drinks may present a different problem, however, when caffeine■ intake becomes excessive. Caffeine seems to be relatively harmless when used in moderate doses (the equivalent of fewer than four 12-ounce cola beverages a day). In greater amounts, it can cause the symptoms associated with anxiety—sweating, tenseness, and inability to concentrate.

Eating Away from Home Adolescents eat about one-third of their meals away from home, and their nutritional welfare is enhanced or hindered by the choices they make. A lunch of a hamburger, a chocolate shake, and french fries supplies substantial quantities of many nutrients at a kcalorie cost of about 800, an energy intake some adolescents can afford. When they eat this sort of lunch, teens can adjust their breakfast and dinner choices to include fruits and vegetables for vitamin A, vitamin C, folate, and fiber and lean meats and legumes for iron and zinc. (See Appendix H for the nutrient contents of fast foods.) The more adolescents eat at fast-food restaurants, the fewer servings of fruits, vegetables, and milk they eat or drink.[120]

Peer Influence Many of the food and health choices adolescents make reflect the opinions and actions of their peers. When others perceive milk as "babyish," a teen will choose soft drinks instead; when others skip lunch and hang out in the parking lot, a teen may join in for the camaraderie, regardless of hunger. Adults need to remember that adolescents have the right to make their own decisions—even if they are contrary to the adults' views. Gatekeepers can set up the environment so that nutritious foods are available and can stand by with reliable nutrition information and advice, but the rest is up to the adolescents. Ultimately, they make the choices. (Highlight 9 examines the influence of social pressures on the development of eating disorders.)

Problems Adolescents Face

Physical maturity and growing independence present adolescents with new choices to make. The consequences of those choices will influence their nutritional health both today and throughout life. Some teenagers begin using drugs, alcohol, and tobacco; others wisely refrain. Information about the use of these substances is presented here because most people are first exposed to them during adolescence, but it actually applies to people of all ages.

Marijuana Almost half of the high school students in the United States report having at least tried marijuana.[121] Marijuana is unique among drugs in that it seems to enhance the enjoyment of eating, especially of sweets, a phenomenon commonly known as "the munchies." The active ingredient in marijuana is similar to chemicals that occur naturally in the brain. Known as endogenous cannabinoids, or endocannabinoids, these chemicals have receptors throughout the body and brain and may play roles in regulating appetite, pain, and memory. Research on how these chemicals work may shed light on why marijuana induces "the

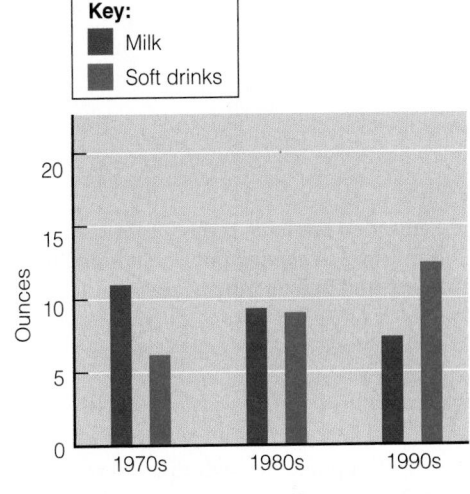

FIGURE 16-11 Average Daily Intakes of Milk and Soft Drinks Compared

Over the years, adolescent milk intakes have decreased as soft drink intakes have increased.

Key:
■ Milk
■ Soft drinks

■ For perspective, caffeine-containing soft drinks typically deliver between 30 and 55 mg of caffeine per 12-ounce can. A pharmacologically active dose of caffeine is defined as 200 mg. Appendix H starts with a table listing the caffeine contents of selected foods, beverages, and drugs.

munchies." Whatever the mechanism, prolonged use of marijuana does not seem to bring about a weight gain.

Cocaine One in 11 high school seniors reports having used cocaine at least once.[122] Cocaine stimulates the nervous system and elicits the stress response—constricted blood vessels, raised blood pressure, widened pupils of the eyes, and increased body temperature. It also drives away feelings of fatigue. Cocaine occasionally causes immediate death—usually by heart attack, stroke, or seizure in an already damaged body system.

Weight loss is common, and cocaine abusers often develop eating disorders. Notably, the craving for cocaine replaces hunger; rats given unlimited cocaine will choose it over food until they starve to death. Thus, unlike marijuana use, cocaine use has major nutritional consequences.

Ecstasy The designer drug ecstasy has also lured 1 in 12 high school seniors at least once. Ecstasy signals the nerve cells to dump all their stored serotonin ■ at once and then prevents its reabsorption. The rush of serotonin flooding the gap between the nerve cells (synapse) alters a person's mood, but may also damage nerve cells and impair memory. Because serotonin helps to regulate body temperature, overheating is a common and potentially dangerous side effect. People who use ecstasy regularly tend to lose weight.

Drug Abuse, in General The nutrition problems associated with other drugs vary in degree, but drug abusers in general face multiple nutrition problems.■ During withdrawal from drugs, an important part of treatment is to identify and correct nutrient deficiencies.

Alcohol Abuse Sooner or later all teenagers face the decision of whether to drink alcohol. The law forbids the sale of alcohol to people under 21, but most adolescents who want it can get it. Four out of five high school students have had at least one alcoholic beverage; about half drink regularly; and one in three students drinks heavily (defined as five or more drinks on at least one occasion in the previous month).[123]

Highlight 7 describes how alcohol affects nutrition status. To sum it up, alcohol provides energy but no nutrients, and it can displace nutritious foods from the diet. Alcohol alters nutrient absorption and metabolism, so imbalances develop. People who cannot keep their alcohol use moderate must abstain to maintain their health. Highlight 7 lists resources for people with alcohol-related problems.

Smoking Slightly less than 30 percent of U.S. high school students report smoking a cigarette in the previous month.[124] This is the lowest rate of smoking among high schoolers since 1991. Cigarette smoking is a pervasive health problem causing thousands of people to suffer from cancer and diseases of the cardiovascular, digestive, and respiratory systems. These effects are beyond the scope of nutrition, but smoking cigarettes does influence hunger, body weight, and nutrient status.[125]

■ Reminder: *Serotonin* is a neurotransmitter important in the regulation of appetite, sleep, and body temperature.

■ Nutrition problems of drug abusers:
- They buy drugs with money that could be spent on food.
- They lose interest in food during "highs."
- They use drugs that depress appetite.
- Their lifestyle fails to promote good eating habits.
- If they use intravenous (IV) drugs, they may contract AIDS, hepatitis, or other infectious diseases, which increase their nutrient needs. Hepatitis also causes taste changes and loss of appetite.
- Medicines used to treat drug abuse may alter nutrition status.

Reduce tobacco use by adolescents. Increase tobacco use cessation attempts by adolescent smokers.

Smoking a cigarette eases feelings of hunger. When smokers receive a hunger signal, they can quiet it with cigarettes instead of food. Such behavior ignores body signals and postpones energy and nutrient intake. Indeed, smokers tend to weigh less than nonsmokers and to gain weight when they stop smoking. People contemplating giving up cigarettes should know that the average weight gain is about 10 pounds in the first year. Smokers wanting to quit should prepare for the possibility of weight gain and adjust their diet and activity habits so as to maintain weight during and after quitting. Smoking cessation programs need to include strategies for weight management.[126]

Nutrient intakes of smokers and nonsmokers differ. Smokers tend to have lower intakes of dietary fiber, vitamin A, beta-carotene, folate, and vitamin C. The association between smoking and low intakes of fruits and vegetables rich in these nutrients may be noteworthy, considering their protective effect against lung cancer (see Highlight 11).

Compared to nonsmokers, smokers require more vitamin C■ to maintain steady body pools. Oxidants in cigarette smoke accelerate vitamin C metabolism and deplete smokers' body stores of this antioxidant.[127] This depletion is even evident to some degree in nonsmokers who are exposed to passive smoke.[128]

■ The vitamin C requirement for people who regularly smoke cigarettes is an additional 35 mg/day.

Beta-carotene enhances the immune response and protects against some cancer activity. Specifically, the risk of lung cancer is greatest for smokers who have the lowest intakes. Of course, such evidence should not be misinterpreted. It does not mean that as long as people eat their carrots, they can safely use tobacco. Nor does it mean that beta-carotene *supplements* would be beneficial; smokers taking beta-carotene supplements actually had a higher incidence of lung cancer and risk of death than those taking a placebo (see Highlight 11 for more details). Smokers are ten times more likely to get lung cancer than nonsmokers. Both smokers and nonsmokers, however, can reduce their cancer risks by eating fruits and vegetables rich in antioxidants (see Highlight 11 for details on antioxidant nutrients and disease prevention).

Smokeless Tobacco Nationwide, 1 in 15 high school students reports having used smokeless tobacco products.[129] Like cigarettes, smokeless tobacco use is linked to many health problems, from minor mouth sores to tumors in the nasal cavities, cheeks, gums, and throat. The risk of mouth and throat cancers is even greater than for smoking tobacco. Other drawbacks to tobacco chewing and snuff dipping include bad breath, stained teeth, and blunted senses of smell and taste. Tobacco chewing also damages the gums, tooth surfaces, and jawbones, making teeth loss later in life likely.

IN SUMMARY Nutrient needs rise dramatically as children enter the rapid growth spurt of the teen years. The busy lifestyles of adolescents add to the challenge of meeting their nutrient needs—especially for iron and calcium. In addition to making wise food choices, adolescents need to refrain from using substances that will impair their health—including illicit drugs, alcohol, and tobacco.

The nutrition and lifestyle choices people make as children and adolescents have long-term, as well as immediate, effects on their health. Highlight 16 describes how sound choices and good habits during childhood and adolescence can help prevent chronic diseases later in life.

Nutrition in Your Life

Encouraging children to eat nutritious foods today helps them learn how to make healthy food choices tomorrow.

- If there are children in your life, do they receive enough food for healthy growth, but not so much as to lead to obesity?
- Are they physically active at home and at school?
- Do they get enough calcium and iron?

NUTRITION ON THE NET

 Access these websites for further study of topics covered in this chapter.

- Find updates and quick links to these and other nutrition-related sites at our website: **www.wadsworth.com/nutrition**

- Search for "infants," "baby bottle tooth decay," "premature birth," "hyperactivity," "food allergies," and "adolescent health," at the U.S. Government health information site: **www.healthfinder.gov**

- Learn how to care for infants, children, and adolescents from the American Academy of Pediatrics and the Canadian Paediatric Society: **www.aap.org** and **www.cps.ca**

- Download the current growth charts and learn about their most recent revision: **www.cdc.gov/ growthcharts**

- Get information on the Food Guide Pyramid for young children from the USDA: **www.usda.gov/cnpp**

- Get tips for feeding children from the American Dietetic Association and the Kids Food Cyber Club: **www.eatright.org** and **www.kidfood.org**

- Get tips for keeping children healthy from the Nemours Foundation: **www.kidshealth.org**

- Visit the National Center for Education in Maternal & Child Health and the National Institute of Child Health and Human Development: **www.ncemch.org** and **www.nichd.nih.gov**

- Learn about the Child Nutrition Programs: **www.fns.usda.gov/fns**

- Learn how UNICEF works to protect children: **www.unicef.org**

- Learn how to reduce lead exposure in your home from the U.S. Department of Housing and Urban Development Office of Lead Hazard Control: **www.hud.gov/lead**

- Learn more about food allergies from the American Academy of Allergy, Asthma, and Immunology; the Food Allergy Network; and the International Food Information Council: **www.aaaai.org**, **www.foodallergy.org**, and **www.ific.org**

- Learn more about hyperactivity from Children and Adults with Attention Deficit Disorders: **www.chadd.org**

- Visit the Milk Matters section of the National Institute of Child Health and Human Development (NICHD): **www.nichd.nih.gov**

- Learn more about caffeine from the International Food Information Council: **www.ific.org**

- To learn about healthy foods and to find recipes and ideas for physical activities, visit: **www.kidnetic.com**

- Get weight-loss tips for children and adolescents: **www.shapedown.com**

- Learn about nondietary approaches to weight loss from HUGS International: **www.hugs.com**

- Read the message for parents and teens on the risks of tobacco use from the American Academy of Pediatrics: **www.aap.org**

- Get help quitting smoking at QuitNet: **www.quitnet.com**

- Visit the Tobacco Information and Prevention Source (TIPS) of the Centers for Disease Control and Prevention: **www.cdc.gov/tobacco/sgr/sgr_2000**

STUDY QUESTIONS

These questions will help you review the chapter. You will find the answers in the discussions on the pages provided.

1. Describe some of the nutrient and immunological attributes of breast milk. (pp. 547–548)

2. What are the appropriate uses of formula feeding? What criteria would you use in selecting an infant formula? (pp. 549–550)

3. Why are solid foods not recommended for an infant during the first few months of life? When is an infant ready to start eating solid food? (pp. 551–553)

4. Identify foods that are inappropriate for infants and explain why they are inappropriate. (p. 553)

5. What nutrition problems are most common in children? What strategies can help prevent these problems? (pp. 555–556)

6. Describe the relationships between nutrition and behavior. How does television influence nutrition? (pp. 557–558, 563–564)

7. Describe a true food allergy. Which foods most often cause allergic reactions? How do food allergies influence nutrition status? (pp. 561–562)

8. Describe the problems associated with childhood obesity and the strategies for prevention and treatment. (pp. 562–566)

9. List strategies for introducing nutritious foods to children. (pp. 566–568)

10. What impact do school meal programs have on the nutrition status of children? (pp. 568–570)

11. Describe the changes in nutrient needs from childhood to adolescence. Why is an adolescent girl more likely to develop an iron deficiency than is a boy? (pp. 571–572)

12. How do adolescents' eating habits influence their nutrient intakes? (pp. 572–573)

13. How does the use of illicit drugs influence nutrition status? (pp. 573–574)

14. How do the nutrient intakes of smokers differ from those of nonsmokers? What impacts can those differences exert on health? (pp. 574–575)

These multiple choice questions will help you prepare for an exam. Answers can be found on p. 580.

1. A reasonable weight for a healthy five-month-old infant who weighed 8 pounds at birth might be:
    a. 12 pounds.
    b. 16 pounds.
    c. 20 pounds.
    d. 24 pounds.

2. Dehydration can develop quickly in infants because:
    a. much of their body water is extracellular.
    b. they lose a lot of water through urination and tears.
    c. only a small percentage of their body weight is water.
    d. they drink lots of breast milk or formula, but little water.

3. An infant should begin eating solid foods between:
    a. 2 and 4 weeks.
    b. 1 and 3 months.
    c. 4 and 6 months.
    d. 8 and 10 months.

4. Among U.S. and Canadian children, the most prevalent nutrient deficiency is of:
    a. iron.
    b. folate.
    c. protein.
    d. vitamin D.

5. A true food allergy always:
    a. elicits an immune response.
    b. causes an immediate reaction.
    c. creates an aversion to the offending food.
    d. involves symptoms such as headaches or hives.

6. Which of the following strategies is *not* effective?
    a. Play first, eat later.
    b. Provide small portions.
    c. Encourage children to help prepare meals.
    d. Use dessert as a reward for eating vegetables.

7. To help teenagers consume a balanced diet, parents can:
    a. monitor the teens' food intake.
    b. give up—parents can't influence teenagers.
    c. keep the pantry and refrigerator well stocked.
    d. forbid snacking and insist on regular, well-balanced meals.

8. During adolescence, energy and nutrient needs:
    a. reach a peak.
    b. fall dramatically.
    c. rise, but do not peak until adulthood.
    d. fluctuate so much that generalizations can't be made.

9. The nutrients most likely to fall short in the adolescent diet are:
    a. sodium and fat.
    b. folate and zinc.
    c. iron and calcium.
    d. protein and vitamin A.

10. To balance the day's intake, an adolescent who eats a hamburger, fries, and cola at lunch might benefit most from a dinner of:
    a. fried chicken, rice, and banana.
    b. ribeye steak, baked potato, and salad.
    c. pork chop, mashed potatoes, and apple juice.
    d. spaghetti with meat sauce, broccoli, and milk.

REFERENCES

1. R. J. Kuczmarski and coauthors, CDC Growth charts: United States, *Advanced Data* 314 (2000): 1–28.
2. Committee on Dietary Reference Intakes, *Dietary Reference Intakes for Energy, Carbohydrate, Fiber, Fat, Fatty Acids, Cholesterol, Protein, and Amino Acids* (Washington, D.C.: National Academies Press, 2002).
3. Committee on Dietary Reference Intakes, 2002, pp. 6-12–6-13.
4. Committee on Nutrition, American Academy of Pediatrics, *Pediatric Nutrition Handbook*, 5th ed., ed. R. E. Kleinman (Elk Grove Village, Ill.: American Academy of Pediatrics, 2004), pp. 103–115.
5. Position of the American Dietetic Association: Breaking the barriers to breastfeeding, *Journal of the American Dietetic Association* 101 (2001): 1213–1220.
6. Committee on Nutrition, American Academy of Pediatrics, 2004, pp. 55–85; Position of the American Dietetic Association, 2001.
7. J. D. Carver, Advances in nutritional modifications of infant formulas, *American Journal of Clinical Nutrition* 77 (2003): 1550S–1554S.
8. N. Auestad and coauthors, Growth and development in term infants fed long-chain polyunsaturate fatty acids: A double-masked, randomized, parallel, prospective, multivariate study, *Pediatrics* 108 (2001): 372–381; J. W. Anderson, B. M. Johnstone, and D. T. Remley, Breast-feeding and cognitive development: A meta-analysis, *American Journal of Clinical Nutrition* 70 (1999): 525–535.
9. Anderson, Johnstone, and Remley, 1999.
10. Auestad and coauthors, 2001.
11. E. E. Birch and coauthors, A randomized controlled trial of long-chain polyunsaturated fatty acid supplementation of formula in term infants after weaning at 6 wk of age, *American Journal of Clinical Nutrition* 75 (2002): 570–580.
12. C. Williams and coauthors, Stereoacuity at age 3.5 in children born fullterm is associated with prenatal and postnatal dietary factors: A report from a population-based cohort study, *American Journal of Clinical Nutrition* 73 (2001): 316–322.
13. D. L. O'Connor and coauthors, Growth and development in preterm infants fed long-chain polyunsaturated fatty acids: A prospective, randomized, controlled trial, *Pediatrics* 108 (2001): 359–371.
14. L. M. Gartner, F. R. Greer, and the Section on Breastfeeding and Committee on Nutrition, Prevention of rickets and vitamin D deficiency: New guidelines for vitamin D intake, *Pediatrics* 111 (2003): 908–910; S. Fitzpatrick and coauthors, Vitamin D–deficient rickets: A multifactorial disease, *Nutrition Reviews* 58 (2000): 218–222.
15. Gartner, Greer, and the Section on Breastfeeding and Committee on Nutrition, 2003.
16. Position of the American Dietetic Association, 2001: S. Arifeen and coauthors, Exclusive breastfeeding reduces acute respiratory infection and diarrhea deaths among infants in Dhaka slums, *Pediatrics* 108 (2001): e67.
17. B. Lönnerdal, Nutritional and physiologic significance of human milk proteins, *American Journal of Clinical Nutrition* 77 (2003): 1537S–1543S.
18. W. H. Oddy and coauthors, Association between breastfeeding and asthma in 6

year old children: Findings of a prospective birth cohort study, *British Medical Journal* 319 (1999): 815–819.

19. M. Gdalevich, D. Mimouni, and M. Mimouni, Breastfeeding and the risk of bronchial asthma in childhood: A systematic review with meta-analysis of prospective studies, *Journal of Pediatrics* 139 (2001): 261–266.

20. C. G. Owen and coauthors, Infant feeding and blood cholesterol: A study in adolescents and systematic review, *Pediatrics* 110 (2002): 597–608.

21. M. W. Gillman and coauthors, Risk of overweight among adolescents who were breastfed as infants, *Journal of the American Medical Association* 285 (2001): 2461–2467.

22. A. D. Liese and coauthors, Inverse association of overweight and breastfeeding in 9 to 10 year-old-children in Germany, *International Journal of Obesity and Related Metabolic Disorders* 25 (2001): 1644–1650.

23. M. L. Hediger and coauthors, Association between infant breastfeeding and overweight in young children, *Journal of the American Medical Association* 285 (2001): 2453–2460.

24. E. L. Mortensen and coauthors, The association between duration of breastfeeding and adult intelligence, *Journal of the American Medical Association* 287 (2002): 2365–2371.

25. A. Jain, J. Concato, and J. M. Leventhal, How good is the evidence linking breastfeeding and intelligence? *Pediatrics* 109 (2002): 1044–1053.

26. Committee on Nutrition, American Academy of Pediatrics, Hypoallergenic infant formulas, *Pediatrics* 106 (2000): 346–349.

27. M. F. MacDorman and coauthors, Annual summary of vital statistics—2001, *Pediatrics* 110 (2002): 1037–1052.

28. M. A. Crawford, Placental delivery of arachidonic and docosahexaenoic acids: Implications for the lipid nutrition of preterm infants, *American Journal of Clinical Nutrition* 71 (2000): 275S–284S.

29. O'Connor and coauthors, 2001; R. Uauy and D. R. Hoffman, Essential fat requirements of preterm infants, *American Journal of Clinical Nutrition* 71 (2000): 245S–250S.

30. Committee on Nutrition, American Academy of Pediatrics, 2004, p. 111.

31. L. Monetini and coauthors, Bovine beta-casein antibodies in breast- and bottle-fed infants: Their relevance in type 1 diabetes, *Diabetes/Metabolism Research and Reviews* 17 (2001): 51–54; C. D. Bernadier, Diabetes mellitus: Is there a connection with infant feeding practices? *Nutrition Today* 36 (2001): 241–248; M. Hummel and coauthors, No major association of breast-feeding, vaccinations, and child viral diseases with early islet autoimmunity in the German BABY-DIAB study, *Diabetes Care* 23 (2000): 969–974.

32. S. J. Fomon, Feeding normal infants: Rationale for recommendations, *Journal of the American Dietetic Association* 101 (2001): 1002–1005.

33. L. Hallberg and coauthors, The role of meat to improve the critical iron balance during weaning, *Pediatrics* 111 (2003): 864–870; Fomon, 2001; L. Davidsson and coauthors, Iron bioavailability in infants from an infant cereal fortified with ferric pyrophosphate or ferrous fumarate, *American Journal of Clinical Nutrition* 71 (2000): 1597–1602.

34. B. A. Dennison and coauthors, Children's growth parameters vary by type of fruit juice consumed, *Journal of the American College of Nutrition* 18 (1999): 346–352.

35. Committee on Nutrition, American Academy of Pediatrics, The use and misuse of fruit juice in pediatrics, *Pediatrics* 107 (2001): 1210–1213.

36. Centers for Disease Control and Prevention, Nonfatal choking—related episodes among children—United States, 2001, *Morbidity and Mortality Weekly Report* 51 (2002): 945–948.

37. Committee on Sports Medicine and Fitness and Committee on School Health, Physical fitness and activity in schools, *Pediatrics* 105 (2000): 1156–1157.

38. V. Messina and A. R. Mangels, Considerations in planning vegan diets: Children, *Journal of the American Dietetic Association* 101 (2001): 661–669.

39. Committee on Dietary Reference Intakes, 2002, Chapter 6.

40. Committee on Dietary Reference Intakes, 2002, Chapter 7.

41. Position of the American Dietetic Association: Health implications of dietary fiber, *Journal of the American Dietetic Association* 102 (2002): 993–1000.

42. Committee on Dietary Reference Intakes, 2002, Chapter 8.

43. Committee on Dietary Reference Intakes, 2002, Chapter 11.

44. Centers for Disease Control and Prevention, Iron deficiency—United States, 1999–2000, *Morbidity and Mortality Weekly Report* 51 (2002): 897–899; M. F. Picciano and coauthors, Nutritional guidance is needed during dietary transition in early childhood, *Pediatrics* 106 (2000): 109–114.

45. S. L. Johnson, Children's food acceptance patterns: The interface of ontogeny and nutrition needs, *Nutrition Reviews* 60 (2002): S91–S94; Eden, 2001.

46. R. E. Kleinman, Current approaches to standards of care for children: How does the pediatric community currently approach this issue? *Nutrition Today* 37 (2002): 177–178.

47. D. J. Raiten, M. F. Picciano, and P. Coates, Dietary supplement use in children: Who, what, why, and where do we go from here: Executive summary, *Nutrition Today* 37 (2002): 167–169.

48. Federal Trade Commission, Consumer Features, Promotions for kids' dietary supplements leaves sour taste, **www.ftc.gov/bcp/conline/features/kidsupp.htm.** May 2000.

49. M. Lino, and coauthors, U.S. Department of Agriculture, Center for Nutrition Policy and Promotion, The quality of young children's diets, *Family Economics and Nutrition Review* 14 (2002): 52–59.

50. S. B. Roberts and M. B. Heyman, Micronutrient shortfalls in young children's diets: Common, and owing to inadequate intakes both at home and at child care centers, *Nutrition Reviews* 58 (2000): 27–29.

51. Federal Interagency Forum on Child and Family Statistics, *America's Children: Key National Indicators of Well-Being 2002,* available at **www.childstats.gov.**

52. Committee on Dietary Reference Intakes, *Dietary Reference Intakes for Vitamin A, Vitamin K, Arsenic, Boron, Chromium, Copper, Iodine, Iron, Manganese, Molybdenum, Nickel, Silicon, Vanadium, and Zinc* (Washington, D.C.: National Academy Press, 2001), pp. 82–161.

53. Committee on Dietary Reference Intakes, 2001, pp. 290–393; H. Saloojee and J. M. Pettifor, Iron deficiency and impaired child development, *British Medical Journal* 323 (2001): 1377–1378; J. S. Halterman and coauthors, Iron deficiency and cognitive achievement among school-aged children and adolescents in the United States, *Pediatrics* 107 (2001): 1381–1386; R. J. Stoltzfus, Iron-deficiency anemia: Reexamining the nature and magnitude of the public health problem. Summary:

Implications for research and programs, *Journal of Nutrition* 131 (2001): 697S–700S.

54. B. Lozoff and coauthors, Poorer behavioral and developmental outcome more than 10 years after treatment for iron deficiency in infancy, *Pediatrics* 105 (2000): e51; E. K. Hurtado, A. H. Claussen, and K. G. Scott, Early childhood anemia and mild or moderate mental retardation, *American Journal of Clinical Nutrition* 69 (1999): 115–119.

55. J. A. Simon and E. S. Hudes, Relationship of ascorbic acid to blood lead levels, *Journal of the American Medical Association* 281 (1999): 2289–2293.

56. Centers for Disease Control and Prevention, Childhood lead poisoning, **www.cdc.gov/nceh/lead/factsheets/childhoodlead.htm**, site visited April 22, 2003.

57. X. Liu and coauthors, Do children with falling blood lead levels have improved cognition? *Pediatrics* 110 (2002): 787–791; W. J. Rogan and coauthors, The effect of chelation therapy with succimer on neuropsychological development in children exposed to lead, *New England Journal of Medicine* 344 (2001): 1421–1426; J. F. Rosen and P. Mushak, Primary prevention of childhood lead poisoning—The only solution, *New England Journal of Medicine* 344 (2001): 1470–1471.

58. Blood lead levels in young children—United States and selected states, 1996–1999, *Morbidity and Mortality Weekly Report* 49 (2000): 1133–1137.

59. S. Parmet, C. Lynm, and R. M. Glass, Attention-deficit/hyperactivity disorder, *Journal of the American Medical Association* 288 (2002): 1804; Subcommittee on Attention-Deficit/Hyperactivity Disorder, American Academy of Pediatrics, Clinical practice guideline: Treatment of the school-aged child with attention-deficit/hyperactivity disorder, *Pediatrics* 108 (2001): 1033–1044; The MTA Cooperative Group, A 14-month randomized clinical trial of treatment strategies for attention-deficit/hyperactivity disorder: Multimodal treatment study of children with ADHD, *Archives of General Psychiatry* 56 (1999): 1073–1086.

60. Food Allergy and Intolerances, National Institutes of Health Fact Sheet, **www.niaid.nih.gov/factsheets/food.htm**, site visited on April 23, 2003; R. Formanek, Food allergies: When food becomes the enemy, *FDA Consumer,* July/August 2001, pp. 10–16.

61. Formanek, 2001.

62. L. Christie and coauthors, Food allergies in children affect nutrient intake and growth, *Journal of the American Dietetic Association* 102 (2002): 1648–1651.

63. K. J. Falci, K. L. Gombas, and E. L. Elliot, Food allergen awareness: An FDA priority, *Food Safety Magazine,* February/March 2001, available at **www.cfsan.fda.gov/~dms/.**

64. H. Metzger, Two approaches to peanut allergy, *New England Journal of Medicine* 348 (2003): 1046–1048.

65. B. Wuthrich, Lethal or life-threatening allergic reactions to food, *Journal of Investigational Allergology and Clinical Immunology* 10 (2000): 59–65.

66. G. S. Rhim and M. S. McMorris, School readiness for children with food allergies, *Annals of Allergy, Asthma and Immunology* 86 (2001): 172–176.

67. S. A. Bock, A. Munoz-Furlong, and H. A. Sampson, Fatalities due to anaphylactic reactions to foods, *Journal of Allergy and Clinical Immunology* 107 (2001): 191–193.

68. J. M. Yeung, R. S. Applebaum, and R. Hildwine, Criteria to determine food

allergen priority, *Journal of Food Protection* 63 (2000): 982–986.

69. Formanek, 2001.

70. C. L. Ogden and coauthors, Prevalence and trends in overweight among US children and adolescents, 1999–2000, *Journal of the American Medical Association* 288 (2002): 1728–1732.

71. A. R. Dorosty and coauthors, Factors associated with early adiposity rebound, *Pediatrics* 105 (2000): 1115–1118.

72. A. Must, Does overweight in childhood have an impact on adult health? *Nutrition Reviews* 61 (2003): 139–142; S. S. Guo and coauthors, Predicting overweight and obesity in adulthood from body mass index values in childhood and adolescence, *American Journal of Clinical Nutrition* 76 (2002): 653–658; A. D. Salbe and coauthors, Assessing risk factors for obesity between childhood and adolescence: I. Birth weight, childhood adiposity, parental obesity, insulin, and leptin, *Pediatrics* 110 (2002): 299–306; M. Mijailovic, V. Mijailovic, and D. Micic, Childhood onset of obesity: Does an obese child become an obese adult? *Journal of Pediatric Endocrinology* 14 (2001): 1335S–1365S.

73. S. M. Robertson and coauthors, Factors related to adiposity among children aged 3 to 7 years, *Journal of the American Dietetic Association* 99 (1999): 938–943.

74. J. O. Fisher and coauthors, Parental influences on young girls' fruit and vegetable, micronutrient, and fat intakes, *Journal of the American Dietetic Association* 102 (2002): 58–64.

75. Centers for Disease Control and Prevention, Physical activity levels among children aged 9–13 years—United States, 2002, *Morbidity and Mortality Weekly Report* 52 (2003): 785–788; Committee on Nutrition, American Academy of Pediatrics, Prevention of pediatric overweight and obesity, *Pediatrics* 112 (2003): 424–430; R. Chatrath and coauthors, Physical fitness of urban American children, *Pediatric Cardiology* 23 (2002): 608–612; A. D. Salbe and coauthors, Assessing risk factors for obesity between childhood and adolescence: II. Energy metabolism and physical activity, *Pediatrics* 110 (2002): 307–314.

76. M. S. Faith and coauthors, Effects of contingent television on physical activity and television viewing in obese children, *Pediatrics* 107 (2001): 1043–1048.

77. B. A. Dennison, T. A. Erb, and P. L. Jenkins, Television viewing and television in bedroom associated with overweight risk among low-income preschool children, *Pediatrics* 109 (2002): 1028–1035.

78. C. J. Crespo and coauthors, Television watching, energy intake, and obesity in US children: Results from the third National Health and Nutrition Examination Survey, 1988–1994, *Archives of Pediatric and Adolescent Medicine* 155 (2001): 360–365; K. A. Coon and coauthors, Relationships between use of television during meals and children's food consumption patterns, *Pediatrics* 107 (2001): e71.

79. R. E. Andersen and coauthors, Relationship of physical activity and television watching with body weight and level of fatness among children, *Journal of the American Medical Association* 279 (1998): 938–942.

80. J. Utter and coauthors, Couch potatoes or french fries: Are sedentary behaviors associated with body mass index, physical activity, and dietary behaviors among adolescents? *Journal of the American Dietetic Association* 103 (2003): 1298–1305.

81. A. Must and S. E. Anderson, Effects of obesity on morbidity in children and adolescents, *Nutrition in Clinical Care* 6 (2003): 4–12; R. Sinha and coauthors, Prevalence of impaired glucose tolerance among children and adolescents with marked obesity, *New England Journal of Medicine* 346 (2002): 802–810.

82. J. B. Schwimmer, T. M. Burwinkle, and J. W. Varni, Health-related quality of life of severely obese children and adolescents, *Journal of the American Medical Association* 289 (2003): 1813–1819.

83. M. I. Goran, Metabolic precursors and effects of obesity in children: A decade of progress, 1990–1999, *American Journal of Clinical Nutrition* 73 (2001): 158–171.

84. American Heart Association's statement for health professionals on cardiovascular health in children, posted on July 1, 2002, www.americanheart.org.

85. Helping Your Overweight Child, www.niddk.nih.gov/health/nutrit/pubs/helpchld.htm, site visited on April 24, 2003.

86. B. J. Rolls, D. Engell, and L. L. Birch, Serving portion size influences 5-year-old but not 3-year-old children's food intakes, *Journal of the American Dietetic Association* 100 (2000): 232–234.

87. B. Gutin and coauthors, Effects of exercise intensity on cardiovascular fitness, total body composition, and visceral adiposity of obese adolescents, *American Journal of Clinical Nutrition* 75 (2002): 818–826; L. M. LeMura and M. T. Maziekas, Factors that alter body fat, body mass, and fat-free mass in pediatric obesity, *Medicine and Science in Sports and Exercise* 34 (2001): 487–496.

88. Committee on Sports Medicine and Fitness and Committee on School Health, American Academy of Pediatrics, Physical fitness and activity in schools, *Pediatrics* 105 (2000): 1156–1157.

89. T. M. Cutting and coauthors, Like mother, like daughter: Familiar patterns of overweight are mediated by mothers' dietary disinhibition, *American Journal of Clinical Nutrition* 69 (1999): 608–613.

90. D. Spruijt-Metz and coauthors, Relation between mothers' child-feeding practices and childrens' adiposity, *American Journal of Clinical Nutrition* 75 (2002): 581–586; A. E. Field and coauthors, Peer, parent, and media influences on the development of weight concerns and frequent dieting among preadolescent and adolescent girls and boys, *Pediatrics* 107 (2001): 54–60; K. K. Davison and L. L. Birch, Weight status, parent reaction, and self-concept in five-year-old girls, *Pediatrics* 107 (2001): 46–53.

91. J. O. Fisher and L. L. Birch, Restricting access to palatable foods affects children's behavioral response, food selection, and intake, *American Journal of Clinical Nutrition* 69 (1999): 1264–1272.

92. B. P. Roberts, A. S. Blinkhorn, and J. T. Duxbury, The power of children over adults when obtaining sweet snacks, *International Journal of Paediatric Dentistry* 13 (2003): 76–84.

93. R. E. Klesges and coauthors, Parental influence on food selection in young children and its relationships to childhood obesity, *American Journal of Clinical Nutrition* 53 (1991): 859–864.

94. J. D. Skinner and coauthors, Children's food preferences: A longitudinal analysis, *Journal of the American Dietetic Association* 102 (2002): 1638–1647.

95. Position of the American Dietetic Association, Society of Nutrition Education, and American School Food Service Association—Nutrition services: An essential component of comprehensive school health programs, *Journal of the American Dietetic Association* 103 (2003): 505–514; Position of the American Dietetic Association: Nutrition standards for child-care programs, *Journal of the American Dietetic Association* 99 (1999): 981–988.

96. Position of the American Dietetic Association, Society of Nutrition Education, and American School Food Service Association, 2003.

97. Position of the American Dietetic Association: Child and adolescent food and nutrition programs, *Journal of the American Dietetic Association* 103 (2003): 887–893.

98. Position of the American Dietetic Association: Dietary guidance for healthy children aged 2 to 11 years, *Journal of the American Dietetic Association* 99 (1999): 93–101.

99. M. F. Picciano, L. D. McBean, and V. A. Stallings, How to grow a healthy child: A conference report, *Nutrition Today* 34 (1999): 6–14.

100. Position of the American Dietetic Association, 2003.

101. E. A. Bergman and coauthors, Time spent by schoolchildren to eat lunch, *Journal of the American Dietetic Association* 100 (2000): 696–698.

102. J. L. Kramer-Atwood and coauthors, Fostering healthy food consumption in schools: Focusing on the challenges of competitive foods, *Journal of the American Dietetic Association* 102 (2002): 1228–1233; Position of the American Dietetic Association: Local support for nutrition integrity in schools, *Journal of the American Dietetic Association* 100 (2000): 108–111; K. W. Cullen and coauthors, Effect of a la carte and snack bar foods at school on children's lunchtime intake of fruits and vegetables, *Journal of the American Dietetic Association* 100 (2000): 1482–1486.

103. M. B. Wildey and coauthors, Fat and sugar levels are high in snacks purchased from student stores in middle schools, *Journal of the American Dietetic Association* 100 (2000): 319–322; L. Harnack and coauthors, Availability of a la carte food items in junior and senior high schools: A needs assessment, *Journal of the American Dietetic Association* 100 (2000): 701–703.

104. Position of the American Dietetic Association, Society for Nutrition Education, and American School Food Service Association, 2003.

105. Position of the American Dietetic Association, Society for Nutrition Education, and American School Food Service Association, 2003.

106. Committee on Dietary Reference Intakes, 2002, Chapter 5.

107. Committee on Dietary Reference Intakes, 2001, pp. 290–393.

108. W. C. Chumlea and coauthors, Age at menarche and racial comparisons in US girls, *Pediatrics* 111 (2003): 110–113.

109. Committee on Dietary Reference Intakes, 2001, pp. 290–393.

110. Halterman and coauthors, 2001.

111. Committee on Nutrition, American Academy of Pediatrics, Calcium requirements of infants, children, and adolescents, *Pediatrics* 104 (1999): 1152–1157.

112. H. J. Kalkwarf, J. C. Khoury, and B. P. Lanphear, Milk intake during childhood and adolescence, adult bone density, and osteoporotic fractures in US women, *American Journal of Clinical Nutrition* 77 (2003): 257–265.

113. S. A. Bowman, Beverage choices of young females: Changes and impact on nutrient intakes, *Journal of the American Dietetic Association* 102 (2002): 1234–1239.

114. M. Story, D. Neumark-Sztainer, and S. French, Individual and environmental influences on adolescent eating behaviors, *Journal of the American Dietetic Association* 102 (2002): S40–S51.

115. D. Neumark-Sztainer and coauthors, Family meal patterns: Associations with sociodemographic characteristics and improved dietary intake among adolescents, *Journal of the American Dietetic Association* 103 (2003): 317–322.

116. The National Center on Addiction and Substance Abuse of Columbia University (CASA), **www.casacolumbia.org**.

117. S. A. French, B. H. Lin, and J. F. Guthrie, National trends in soft drink consumption among children and adolescents age 6 to 17 years: Prevalence, amounts, and sources, 1977/1978 to 1994/1998, *Journal of the American Dietetic Association* 103 (2003): 1326–1331; Bowman, 2002.

118. D. S. Ludwig, K. E. Peterson, and S. L. Gortmaker, Relation between consumption of sugar-sweetened drinks and childhood obesity: A prospective, observational analysis, *Lancet* 357 (2001): 505–508; L. Harnack, J. Stang, and M. Story, Soft drink consumption among US children and adolescents: Nutritional consequences, *Journal of the American Dietetic Association* 99 (1999): 436–441.

119. Federal Update, Milk matters, *Journal of the American Dietetic Association* 102 (2002): 469.

120. S. A. French and coauthors, Fast food restaurant use among adolescents: Associations with nutrient intake, food choices, and behavioral and psychosocial variables, *International Journal of Obesity and Related Metabolic Disorders* 25 (2001): 1823–1833.

121. L. Kann and coauthors, Youth risk behavior surveillance—United States, 1999, *Morbidity and Mortality Weekly Report* 49 (2000): entire supplement.

122. Kann and coauthors, 2000.

123. Kann and coauthors, 2000.

124. Centers for Disease Control and Prevention, Trends in cigarette smoking among high school students—United States, 1991–2001, *Morbidity and Mortality Weekly Report* 51 (2002): 409–412.

125. J. S. Hampl and N. M. Betts, Cigarette use during adolescence: Effects on nutritional status, *Nutrition Reviews* 57 (1999): 215–221.

126. L. M. Varner, Impact of combined weight-control and smoking-cessation interventions on body weight: Review of the literature, *Journal of the American Dietetic Association* 99 (1999): 1272–1275.

127. J. Lykkesfeldt and coauthors, Ascorbate is depleted by smoking and repleted by moderate supplementation: A study in male smokers and nonsmokers with matched dietary antioxidant intakes, *American Journal of Clinical Nutrition* 71 (2000): 530–536.

128. A. M. Preston and coauthors, Influence of environmental tobacco smoke on vitamin C status in children, *American Journal of Clinical Nutrition* 77 (2003): 167–172.

129. Centers for Disease Control and Prevention, Youth tobacco surveillance—United States, 2000, *Morbidity and Mortality Weekly Report* (supplement) 50 (2001): 5–10.

ANSWERS

Study Questions (multiple choice)

1. b   2. a   3. c   4. a   5. a   6. d   7. c   8. a   9. c   10. d

Childhood Obesity and the Early Development of Chronic Diseases

When people think of the health problems of children and adolescents, they typically think of measles and acne, not heart disease or diabetes. They think of heart disease as the number one killer of adults in the United States and Canada, but it begins in childhood. Similarly, diabetes (type 2) has historically been called "adult onset," but it is now an epidemic among children and adolescents.[1] Risk factors for these two diseases develop early and are most evident in overweight youngsters.[2]

This highlight focuses on efforts to prevent childhood obesity and the development of heart disease and type 2 diabetes, but the benefits extend to other obesity-related diseases as well. The years of childhood (ages 2 to 18 years) are emphasized here, for the earlier in life health-promoting habits become established, the better they will stick. Chapter 18 fills in the rest of the story of nutrition's role in reducing chronic disease risk.

Invariably, questions arise as to what extent genetics is involved in disease development. For heart disease and type 2 diabetes, genetics does not appear to play a *determining* role; that is, a person is not simply destined at birth to develop these diseases. Instead, genetics appears to play a *permissive* role—the potential is inherited and then will develop, if given a push by poor health choices such as excessive weight gain, poor diet, sedentary lifestyle, and cigarette smoking.

Many experts agree that preventing or treating obesity in childhood will reduce the rate of chronic diseases in adulthood. Without intervention, most overweight children become overweight adolescents who become overweight adults, and being overweight exacerbates every chronic disease that adults face.[3]

Early Development of Type 2 Diabetes

Type 2 diabetes, a chronic disease closely linked with obesity, has been on the rise among children and adolescents as the prevalence of obesity in U.S. youth has increased in recent years.[4] An estimated 85 percent of the children diagnosed with type 2 diabetes are obese. Most are diagnosed during puberty, but as children become more obese and less active, the trend is shifting to younger children. Type 2 diabetes is most likely to occur in those who are obese and sedentary and have a family history of diabetes.

In type 2 diabetes, the cells become insulin-resistant—that is, insulin can no longer escort glucose from the blood into the cells. The combination of obesity and insulin resistance produces a cluster of symptoms, including high blood cholesterol and high blood pressure, which, in turn, promotes the development of atherosclerosis and the early development of heart disease.[5] Other common problems evident by early adulthood include kidney disease, blindness, and miscarriages. The complications of diabetes, especially when encountered at a young age, can shorten life expectancy.

Prevention and treatment of type 2 diabetes depend on weight management, which can be particularly difficult in a youngster's world of video games and candy bars. The activity and dietary suggestions to help defend against heart disease later in this highlight apply to type 2 diabetes as well.

Early Development of Heart Disease

Most people consider heart disease to be an adult disease because its incidence rises with advancing age, and symptoms rarely appear before age 30. In actuality, the disease process begins much earlier.

Atherosclerosis

Most **cardiovascular disease** involves **atherosclerosis** (see the glossary on p. 582 for these and related terms). Atherosclerosis develops when regions of an artery's walls become progressively thickened with **plaque**—an accumulation of fatty deposits, smooth muscle cells, and fibrous connective tissue. If it progresses, atherosclerosis may eventually block the flow of blood to the heart and cause a heart attack or cut off blood flow to the brain and cause a stroke. Infants are born with healthy, smooth, clear arteries, but within the first decade of life, **fatty streaks** may begin to appear (see Figure H16-1). During adolescence, these fatty streaks may begin to accumulate fibrous connective tissue. By early adulthood, the fibrous plaques may begin to calcify and become raised lesions, especially in boys and young men. As the lesions grow more numerous and enlarge, the heart disease rate begins to rise, most dramatically at about age 45 in men and 55 in women. From this point on, arterial damage and blockage progress rapidly, and heart attacks and strokes threaten life. In short, the consequences of atherosclerosis, which become apparent only in adulthood, have their beginnings in the first decades of life.[6]

Atherosclerosis is not inevitable; people can grow old with relatively clear arteries. Early lesions may either progress or regress, depending on several factors, many of which reflect lifestyle behaviors. Smoking, for example, is strongly associated with the prevalence of fatty streaks and raised lesions, even in young adults.

Blood Cholesterol

As blood cholesterol rises, atherosclerotic lesion coverage increases. Cholesterol values at birth are similar in all populations; differences emerge in early childhood. Standard values for cholesterol in children and adolescents (ages 2 to 18 years) are listed in Table H16-1.[7]

In general, blood cholesterol tends to rise as dietary saturated fat increases. In recent years, for example, Japanese children have adopted a diet more like that of the United States—higher in saturated fat—and as expected, their blood cholesterol has increased. Although their saturated fat intakes are still lower than intakes typical of U.S. children, their blood cholesterol has responded more dramatically, exceeding the levels seen in U.S. children.[8] Such findings suggest a genetic influence on the cholesterol response to dietary fat.[9]

Blood cholesterol also correlates with childhood obesity, especially central obesity.[10] LDL cholesterol rises with obesity, and HDL declines. These relationships are apparent throughout childhood, and their magnitude increases with age.

Children who are both overweight and have high blood cholesterol are quite likely to have parents who develop heart disease early.[11] For this reason, selective screening is recommended for children and adolescents whose parents (or grandparents) have heart disease; those whose parents have elevated blood cholesterol; and those

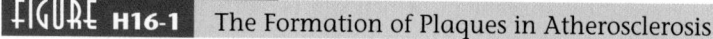

FIGURE H16-1   The Formation of Plaques in Atherosclerosis

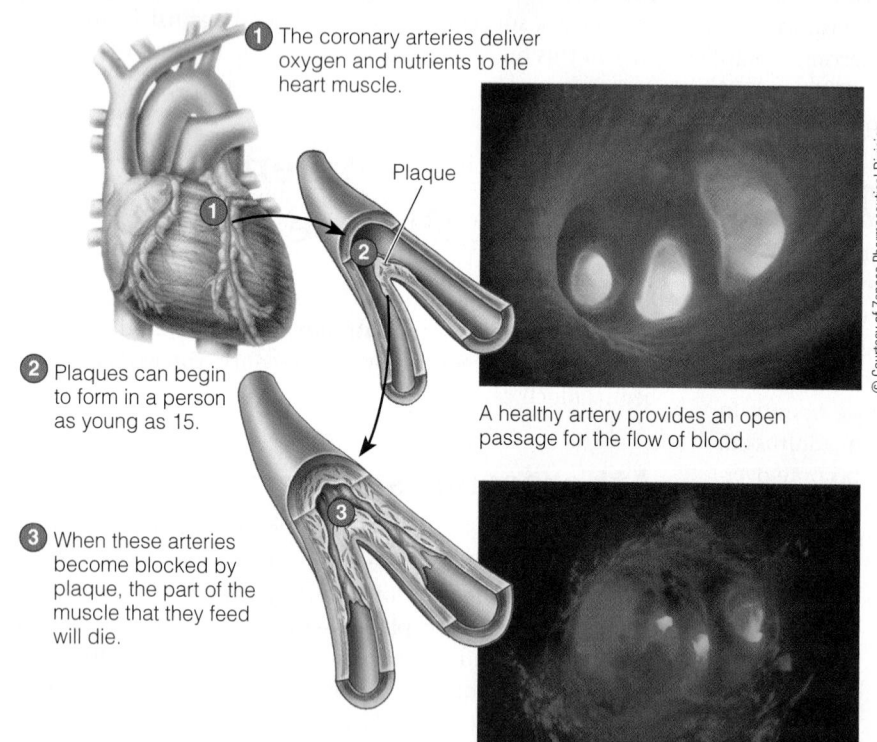

1. The coronary arteries deliver oxygen and nutrients to the heart muscle.

2. Plaques can begin to form in a person as young as 15.

3. When these arteries become blocked by plaque, the part of the muscle that they feed will die.

Plaque

© Courtesy of Zeneca Pharmaceutical Division, Cheshire, England (both)

A healthy artery provides an open passage for the flow of blood.

Plaques form along the artery's inner wall, reducing blood flow. Clots can form, aggravating the problem.

TABLE H16-1 Cholesterol Values for Children and Adolescents

Disease Risk	Total Cholesterol (mg/dL)	LDL Cholesterol (mg/dL)
Acceptable	<170	<110
Borderline	170–199	110–129
High	≥200	≥130

NOTE: Adult values appear in Table 18-4 on p. 622.

TABLE H16-2 Hypertension Standards for Children and Adolescents

	Systolic over Diastolic Pressure (mm Hg)			
	6 to 9 yr	10 to 12 yr	13 to 15 yr	16 to 18 yr
Mild hypertension	111–121 over 70–77	117–125 over 75–81	124–135 over 77–85	127–141 over 80–91
Moderate hypertension	122–129 over 70–85	126–133 over 82–89	136–143 over 86–91	142–149 over 92–97
Severe hypertension	>129 over >85	>133 over >89	>143 over >91	>149 over >97

NOTE: Adult values appear in Table 18-4 on p. 622.

whose family history is unavailable, especially if other risk factors are evident.[12] Because blood cholesterol in children is a good predictor of adult values, some experts recommend universal screening to identify all children with high blood cholesterol. They note that many children who have high blood cholesterol would be missed under current screening criteria.

Among those children who may have high blood cholesterol, but may not meet screening criteria are those who are overweight. The incidence of high blood cholesterol in obese children with no other criteria is similar to that of nonobese children with family histories of heart disease. In addition to overweight, health care professionals should consider whether children smoke or consume a diet high in saturated fat.[13]

Early—but not advanced—atherosclerotic lesions are reversible, making screening and education a high priority. Both those with family histories of heart disease and those with multiple risk factors need intervention. Children with the highest risks of developing heart disease are sedentary and obese, with high blood pressure and high blood cholesterol.[14] In contrast, children with the lowest risks of heart disease are physically active and of normal weight, with low blood pressure and favorable lipid profiles. Routine pediatric care should identify these known risk factors and provide intervention when needed.

Blood Pressure

Pediatricians routinely monitor blood pressure in children and adolescents. High blood pressure may signal an underlying disease or the early onset of hypertension. Hypertension accelerates the development of atherosclerosis. Standard values for hypertension in children and adolescents are given in Table H16-2.

Like atherosclerosis and high blood cholesterol, hypertension may develop in the first decades of life, especially among obese children.[15] Children can control their hypertension by participating in regular aerobic activity and by losing weight or maintaining their weight as they grow taller. No evidence suggests that restricting sodium lowers blood pressure in children and adolescents.

Physical Activity

Research has also confirmed an association between blood lipids and physical activity in children, similar to that seen in adults. Active children have a better lipid profile than physically inactive children.

Just as blood cholesterol and obesity track over the years, so does a youngster's level of physical activity. Those who are inactive now are likely to still be inactive years later. Similarly, those who are physically active now tend to remain so. Compared with inactive teens, those who are physically active weigh less, smoke less, eat a diet lower in saturated fats, and have better blood lipid profiles. Both obesity and blood cholesterol also correlate with the inactive pastime of watching television. The message is clear: physical activity offers numerous health benefits, and children who are active today are most likely to be active for years to come.

Dietary Recommendations for Children

Regardless of family history, experts agree that all children should eat a variety of foods and maintain desirable weight. There is less agreement, however, as to whether it is wise to restrict fat in children's diets.[16] Still, health experts recommend that children over age two receive at least 25 percent and no more than 35 percent of total energy from fat. Such a diet appears to improve blood lipids without compromising nutrient adequacy, physical growth, or neurological development.[17]

Moderation, Not Deprivation

Healthy children over age two can begin the transition to eating according to recommendations by eating fewer high-fat foods, replacing some high-fat foods with low-fat choices, and selecting more fruits and vegetables. All high-fat foods need not be eliminated, though. Healthy meals can still include moderate amounts of a child's favorite foods, even if they are high-fat selections such as french fries and ice cream. Without such additions, diets may be too low in fat, not to mention unappetizing and boring.

Parents and caregivers play a key role in helping children establish healthy eating habits.[18] Balanced meals need to provide lean meat, poultry, fish, and legumes; fruits and vegetables; whole grains; and low-fat milk products. Such meals can provide enough energy and nutrients to support growth and maintain blood cholesterol within a healthy range.

Pediatricians warn parents to avoid extremes; they caution that while intentions may be good, excessive food restriction may create nutrient deficiencies and impair growth. Furthermore, parental control over eating may instigate battles and foster attitudes about foods that can lead to inappropriate eating behaviors.

Diet First, Drugs Later

Experts agree that children with high blood cholesterol should first be treated with diet. If blood cholesterol remains high in children ten years and older after 6 to 12 months of dietary intervention, then drugs may be necessary to lower blood cholesterol. Drugs can effectively lower blood cholesterol without interfering with adolescent growth or development.[19]

Smoking

Even though the focus of this text is nutrition, another risk factor for heart disease that starts in childhood and carries over into adulthood must also be addressed—cigarette smoking. Each day 3000 children light up for the first time—typically in grade school. Among high school students, almost two out of three have tried smoking, and one in seven smokes

Cigarette smoking is the number one preventable cause of deaths.

regularly.[20] Approximately 80 percent of all adult smokers began smoking before the age of 18.[21]

Of those teenagers who continue smoking, half will eventually die of smoking-related causes. Efforts to teach children about the dangers of smoking need to be aggressive. Children are not likely to consider the long-term health consequences of tobacco use. They are more likely to be struck by the immediate health consequences, such as shortness of breath when playing sports, or social consequences, such as having bad breath. Whatever the context, the message to all children and teens should be clear: don't start smoking. If you've already started, quit.

In conclusion, *adult* heart disease is a major *pediatric* problem. Without intervention, some 60 million children are destined to suffer its consequences within the next 30 years. Optimal prevention efforts focus on children, especially on those who are overweight.[22]

Just as young children receive vaccinations against infectious diseases, they need screening for, and education about, chronic diseases. Many health education programs have been implemented in schools around the country. These programs are most effective when they include education in the classroom, heart-healthy meals in the lunchroom, fitness activities on the playground, and parental involvement at home.

NUTRITION ON THE NET

 Access these websites for further study of topics covered in this highlight.

- Find updates and quick links to these and other nutrition-related sites at our website: **www.wadsworth.com/nutrition**
- Get weight-loss tips for children and adolescents: **www.shapedown.com**

- Learn about nondietary approaches to weight loss from HUGS International: **www.hugs.com**
- Visit the Nemours Foundation: **www.kidshealth.org**
- Find information on diabetes in children at the American Diabetes Association and Juvenile Diabetes Research Foundation: **www.diabetes.org** and **www.jdrf.org**

REFERENCES

1. T. Aye and L. L. Levitsky, Type 2 diabetes: An epidemic disease in childhood, *Current Opinion in Pediatrics* 15 (2003): 411–415.

2. G. D. Ball and L. J. McCargar, Childhood obesity in Canada: A review of prevalence estimates and risk factors for cardiovascular diseases and type 2 diabetes, *Canadian Journal of Applied Physiology* 28 (2003): 117–140.

3. A. Must, Does overweight in childhood have an impact on adult health? *Nutrition Reviews* 61 (2003): 139–142; D. S. Freedman, Clustering of coronary heart disease risk factors among obese children, *Journal of Pediatric Endocrinology and Metabolism* 15 (2002): 1099–1108.

4. D. S. Ludwig and C. B. Ebbeling, Type 2 diabetes mellitus in children: Primary care and public health considerations, *Journal of the American Medical Association* 286 (2001): 1427–1430; American Diabetes Association, Type 2 diabetes in children and adolescents, *Pediatrics* 105 (2000): 671–680.

5. R. Kohen-Avramoglu, A. Theriault, and K. Adeli, Emergence of the metabolic syndrome in childhood: An epidemiological overview and mechanistic link to dyslipidemia, *Clinical Biochemistry* 36 (2003): 413–420.

6. S. Li and coauthors, Childhood cardiovascular risk factors and carotid vascular changes in adulthood: The Bogalusa Heart Study, *Journal of the American Medical Association* 290 (2003): 2271–2276; K. B. Keller and L. Lemberg, Obesity and the metabolic syndrome, *American Journal of Clinical Care* 12 (2003): 167–170; H. C. McGill Jr. and coauthors, Origin of atherosclerosis in childhood and adolescence, *American Journal of Clinical Nutrition* 72 (2000): 1307S–1315S.

7. Committee on Nutrition, American Academy of Pediatrics, Cholesterol in childhood, *Pediatrics* 101 (1998): 141–147.

8. S. C. Couch and coauthors, Rapid westernization of children's blood cholesterol in 3 countries: Evidence for nutrient-gene interactions? *American Journal of Clinical Nutrition* 72 (2000): 1266S–1274S.

9. S. Q. Ye and P. O. Kwiterovich Jr., Influence of genetic polymorphisms on responsiveness to dietary fat and cholesterol, *American Journal of Clinical Nutrition* 72 (2000): 1275S–1284S.

10. O. Fiedland and coauthors, Obesity and lipid profiles in children and adolescents, *Journal of Pediatric Endocrinology and Metabolism* 15 (2002): 1011–1016; T. Dwyer and coauthors, Syndrome X in 8-y-old Australian children: Stronger associations with current body fatness than with infant size or growth, *International Journal of Obesity and Related Metabolic Disorders* 26 (2002): 1301–1309.

11. B. Glowinska, M. Urban, and A. Koput, Cardiovascular risk factors in children with obesity, hypertension and diabetes: Lipoprotein (a) levels and body mass index correlate with family history of cardiovascular disease, *European Journal of Pediatrics* 161 (2002): 511–518.

12. A. Wiegman and coauthors, Family history and cardiovascular risk in familial hypercholesterolemia: Data in more than 1000 children, *Circulation* 107 (2003): 1473–1478; Committee on Nutrition, American Academy of Pediatrics, 1998.

13. Committee on Nutrition, American Academy of Pediatrics, 1998.

14. V. N. Muratova and coauthors, The relation of obesity to cardiovascular risk factors among children: The CARDIAC project, *West Virginia Medical Journal* 98 (2002): 263–267.

15. Dwyer and coauthors, 2002.

16. R. E. Olson, Is it wise to restrict fat in the diets of children? *Journal of the American Dietetic Association* 100 (2000): 28–32; E. Satter, A moderate view on fat restriction, *Journal of the American Dietetic Association* 100 (2000): 32–36; L. A. Lytle, In defense of a low-fat diet for healthy children, *Journal of the American Dietetic Association* 100 (2000): 39–41.

17. E. Obarzanek and coauthors, Long-term safety and efficacy of a cholesterol-lowering diet in children with elevated low-density lipoprotein cholesterol: Seven-year results of the Dietary Intervention Study in Children (DISC), *Pediatrics* 107 (2001): 256–264; L. Rask-Nissilä and coauthors, Neurological development of 5-year-old children receiving a low-saturated fat, low-cholesterol diet since infancy: A randomized controlled study, *Journal of the American Medical Association* 284 (2000): 993–1000; R. M. Lauer and coauthors, Efficacy and safety of lowering dietary intake of total fat, saturated fat, and cholesterol in children with elevated LDL cholesterol: The Dietary Intervention Study in Children, *American Journal of Clinical Nutrition* 72 (2000): 1332S–1342S; N. F. Butte, Fat intake of children in relation to energy requirements, *American Journal of Clinical Nutrition* 72 (2000): 1246S–1252S.

18. T. A. Nicklas and coauthors, Family and child-care provider influences on preschool children's fruit, juice, and vegetable consumption, *Nutrition Reviews* 59 (2001): 224–235.

19. S. de Jongh and coauthors, Efficacy and safety of statin therapy in children with familial hypercholesterolemia: A randomized, double-blind, placebo-controlled trial with simvastatin, *Circulation* 106 (2002): 2231–2237.

20. Trends in cigarette smoking among high school students—United States, 1991–2001, *Morbidity and Mortality Weekly Report* 51 (2002): 409–412.

21. Youth tobacco surveillance—United States, 2000, *Morbidity and Mortality Weekly Report* 50 (2001): entire supplement.

22. Committee on Nutrition, American Academy of Pediatrics, Prevention of pediatric overweight and obesity, *Pediatrics* 112 (2003): 424–430.

Life Cycle Nutrition: Adulthood and the Later Years

Chapter Outline

Nutrition and Longevity: *Observation of Older Adults • Manipulation of Diet*

The Aging Process: *Physiological Changes • Other Changes*

Energy and Nutrient Needs of Older Adults: *Water • Energy and Energy Nutrients • Vitamins and Minerals • Nutrient Supplements*

Nutrition-Related Concerns of Older Adults: *Cataracts and Macular Degeneration • Arthritis • The Aging Brain*

Food Choices and Eating Habits of Older Adults: *Food Assistance Programs • Meals for Singles*

Highlight: *Nutrient-Drug Interactions*

Nutrition Explorer CD-ROM Outline

Nutrition Animation: *Nutrient Needs in Later Years*

Case Study: *Drug and Nutrient Interactions in the Elderly*

Student Practice Test

Glossary Terms

Nutrition on the Net

© Judd Pilossof/FoodPix/Getty Images

Nutrition in Your Life

Take a moment to envision yourself 20, 40, or even 60 years from now. Are you physically fit and healthy? Can you see yourself walking on the beach with friends or tossing a ball with children? Are you able to climb stairs and carry your own groceries? Importantly, are you enjoying life? If you're lucky, you will grow old with good health, but much of that depends on your actions today—and every day from now until then. Making nutritious foods and physical activities a priority in your life can help bring rewards of continued health and enjoyment in later life.

Wise food choices, made throughout adulthood, can support a person's ability to meet physical, emotional, and mental challenges and to enjoy freedom from disease. Two goals motivate adults to pay attention to their diets: promoting health and slowing aging. Much of this text has focused on nutrition to support health, and Chapter 18 features prevention of chronic diseases such as cancer and heart disease; this chapter focuses on aging and the nutrition needs of older adults.

The U.S. population is growing older. The majority is now middle-aged, and the ratio of old people to young is increasing, as Figure 17-1 (on p. 588) shows. In 1900, only 1 out of 25 people was 65 or older. In 2000, one out of eight had reached age 65. Projections for 2030 are one out of five.

Our society uses the arbitrary age of 65 years to define the transition point between middle age and old age, but growing "old" happens day by day, with changes occurring gradually over time. Since 1950 the population of those over 65 has almost tripled. Remarkably, the fastest-growing age group has been people over 85 years; since 1950 their numbers have increased sevenfold. The number of people in the United States age 100 or older doubled in the last decade. Similar trends are occurring in populations worldwide.[1]

FIGURE 17-1 The Aging of the U.S. Population

In general the percentage of older people in the population has increased over the decades while the percentage of younger people has decreased.

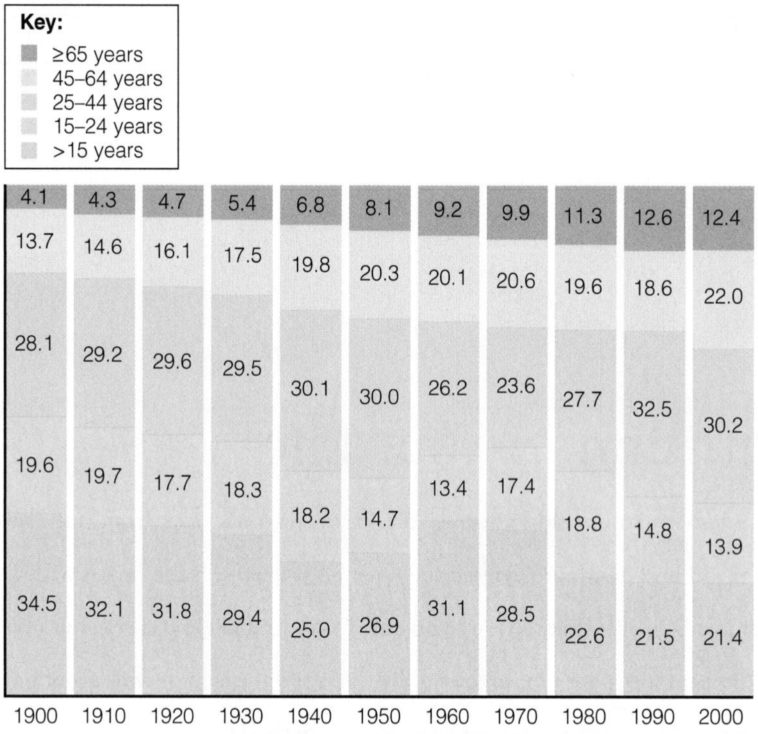

Key:
- ≥65 years
- 45–64 years
- 25–44 years
- 15–24 years
- >15 years

	1900	1910	1920	1930	1940	1950	1960	1970	1980	1990	2000
≥65	4.1	4.3	4.7	5.4	6.8	8.1	9.2	9.9	11.3	12.6	12.4
45–64	13.7	14.6	16.1	17.5	19.8	20.3	20.1	20.6	19.6	18.6	22.0
25–44	28.1	29.2	29.6	29.5	30.1	30.0	26.2	23.6	27.7	32.5	30.2
15–24	19.6	19.7	17.7	18.3	18.2	14.7	13.4	17.4	18.8	14.8	13.9
>15	34.5	32.1	31.8	29.4	25.0	26.9	31.1	28.5	22.6	21.5	21.4

SOURCE: U.S. Census Bureau, Decennial census of population, 1900 to 2000.

Life expectancy in the United States for white women is 80 years and for black women, 75 years; for white men, it is 75 years and for black men, 68 years—all record highs and much higher than the average life expectancy of 47 years in 1900.[2] Women who live to 80 can expect to survive an additional nine years, on average; men, an additional seven. Advances in medical science—antibiotics and other treatments—are largely responsible for almost doubling the life expectancy in the twentieth century. Improved nutrition and an abundant supply of food have also contributed to lengthening life expectancy. The **life span** has not lengthened as dramatically; human **longevity** appears to have an upper limit. The potential human life span is currently 130 years. With recent advances in medical technology and genetic knowledge, however, researchers may one day be able to extend the life span even further by slowing, or perhaps preventing, aging and its accompanying diseases.

Nutrition and Longevity

Research in the field of aging is active—and difficult. Researchers are challenged by the diversity of older adults. When older adults experience health problems, it is hard to know whether to attribute these problems to genetics, aging, or other environmental factors such as nutrition. The idea that nutrition can influence the aging process is particularly appealing, because people can control and change their eating habits. The questions being asked include:

- To what extent is aging inevitable, and can it be slowed through changes in lifestyle and environment?

life expectancy: the average number of years lived by people in a given society.

life span: the maximum number of years of life attainable by a member of a species.

longevity: long duration of life.

- What role does nutrition play in the aging process, and what role can it play in slowing aging?

With respect to the first question, it seems that aging is an inevitable, natural process, programmed into the genes at conception. People can, however, slow the process within genetic limits by adopting healthy lifestyle habits such as eating nutritious food and engaging in physical activity. In fact, an estimated 70 to 80 percent of the average person's life expectancy may depend on individual health-related behaviors; genes determine the remaining 20 to 30 percent.[3]

With respect to the second question, good nutrition helps to maintain a healthy body and can therefore ease the aging process in many significant ways. Clearly, nutrition can improve the **quality of life** in the later years.

Observation of Older Adults

The strategies adults use to meet the two goals mentioned at the start of this chapter—promoting health and slowing aging—are actually very much the same. What to eat, when to sleep, how physically active to be, and other lifestyle choices greatly influence both physical health and the aging process.

Healthy Habits A person's **physiological age** reflects his or her health status and may or may not reflect the person's **chronological age.** Quite simply, some people seem younger, and others older, than their years. Six lifestyle behaviors seem to have the greatest influence on people's health and therefore on their physiological age:

- Sleeping regularly and adequately.
- Eating well-balanced meals, including breakfast, regularly.
- Engaging in physical activity regularly.
- Not smoking.
- Not using alcohol, or using it in moderation.
- Maintaining a healthy body weight.

Over the years, the effects of these lifestyle choices accumulate—that is, people who follow most of these practices live longer and have fewer disabilities as they age.[4] They are in better health, even when older in chronological age, than people who do not adopt these behaviors. Even though people cannot change their birth dates, they may be able to add years to, and enhance the quality of, their lives. Physical activity seems to be most influential in preventing or slowing the many changes that define a stereotypical "old" person. After all, many of the physical limitations that accompany aging occur because people become inactive, not because they become older.

Physical Activity The many remarkable benefits of regular physical activity outlined in Chapter 14 are not limited to the young. Compared with those who are inactive, older adults who are active weigh less; have greater flexibility, more endurance, better balance, and better health; and live longer.[5] They reap additional benefits from various activities as well: aerobic activities improve cardiorespiratory endurance, blood pressure, and blood lipid concentrations; moderate endurance activities improve the quality of sleep; and strength training improves posture and mobility. In fact, regular physical activity is the most powerful predictor of a person's mobility in the later years. Physical activity also increases blood flow to the brain, thereby preserving mental ability, alleviating depression, and supporting independence.[6]

Muscle mass and muscle strength tend to decline with aging, making older people vulnerable to falls and immobility. Falls are a major cause of fear, injury, disability, and even death among older adults. Many lose their independence as a result of falls. Regular physical activity tones, firms, and strengthens muscles, helping to improve confidence, reduce the risk of falling, and lessen the risk of injury should a fall occur.

quality of life: a person's perceived physical and mental well-being.

physiological age: a person's age as estimated from her or his body's health and probable life expectancy.

chronological age: a person's age in years from his or her date of birth.

Regular physical activity promotes a healthy, independent lifestyle.

Even without a fall, older adults may become so weak that they can no longer perform life's daily tasks, such as climbing stairs, carrying packages, and opening jars. By improving muscle strength, which allows a person to perform these tasks, strength training helps to maintain independence. Even in frail, elderly people over 85 years of age, strength training not only improves balance, muscle strength, and mobility, but also increases energy expenditure and energy intake, thereby enhancing nutrient intakes. This finding highlights another reason to be physically active: a person spending energy can afford to eat more food and thus receives more nutrients as a result. People who are committed to an ongoing fitness program can benefit from higher energy and nutrient intakes and still maintain their body weights.

Ideally, physical activity should be part of each day's schedule and should be intense enough to prevent muscle atrophy and to speed up the heartbeat and respiration rate. Although aging reduces both speed and endurance to some degree, older adults can still train and achieve exceptional performances. Healthy older adults who have not been active can ease into a suitable routine. They can start by walking short distances until they are walking at least 10 minutes continuously, and then gradually increase their distance to a 30- to 45-minute workout at least five days a week. Table 17-1 provides exercise guidelines for seniors; people with medical conditions should check with a physician before beginning an exercise routine, as should sedentary men over 40 and women over 50 who want to participate in a vigorous program.

Manipulation of Diet

In their efforts to understand longevity, researchers have not only observed people, but have also manipulated influencing factors, such as diet, in animals. This research has given rise to some interesting and suggestive findings.

Energy Restriction in Animals Animals live longer and have fewer age-related diseases when their energy intakes are restricted.[7] These life-prolonging benefits become evident when the diet provides enough food to prevent malnutrition and an energy intake of about 70 percent of normal. Exactly how energy restriction

TABLE 17-1	Exercise Guidelines for Older Adults			
	Endurance	**Strength**	**Balance**	**Flexibility**
Examples				
Start easy	Be active 5 minutes on most or all days.	Using 0- to 2-pound weights, do 1 set of 8 repetitions twice a week.	Hold onto table or chair with one hand, then with one finger.	Hold stretch 10 seconds; do each stretch 3 times.
Progress gradually to goal	Be active 30 minutes (minimum) on most or all days.	Increase weight as able; do 2 sets of 8–15 repetitions twice a week.	Do not hold onto table or chair; then close eyes.	Hold stretch 30 seconds; do each stretch 5 times.
Cautions and comments	Stop if you are breathing so hard you can't talk or if you feel dizziness or chest pain.	Breathe out as you contract and in as you relax (do not hold breath); use smooth, steady movements.	Incorporate balance techniques with strength exercises as you progress.	Stretch after strength and endurance exercises for 20 minutes, 3 times a week; use slow, steady movements; bend joints slightly.

SOURCE: *Exercise: A Guide from the National Institute on Aging,* **www.nia.nih.gov**, accessed May 2003.

prolongs life remains largely unexplained, although gene activity appears to play a key role.[8] The genetic activity of old mice differs from that of young mice, with some genes becoming more active with age and others less active. With an energy-restricted diet, many of the genetic activities of older mice revert to those of younger mice. These "slow-aging" genetic changes are apparent in as little as one month on a restricted diet.[9]

The consequences of energy restriction include a delay in the onset, or prevention, of diseases such as atherosclerosis; prolonged growth and development; and improved blood glucose, insulin sensitivity, and blood lipids.[10] In addition, energy metabolism slows and body temperature drops—indications of a reduced rate of oxygen consumption.[11] As Highlight 11 explained, the use of oxygen during energy metabolism produces free radicals, which have been implicated in the aging process. Restricting energy intake not only produces fewer free radicals, but also increases antioxidant activity and enhances DNA repair. Reducing oxidative stress may at least partially explain how restricting energy intake lengthens life expectancy.[12]

Interestingly, longevity appears to depend on restricting energy intake and not on the amount of body fat. Genetically obese rats live longer when given a restricted diet even though their body fat is similar to that of other rats allowed to eat freely.

Energy Restriction in Human Beings Research on a variety of animals■ confirms the relationship between energy restriction and longevity. Applying the results of animal studies to human beings is problematic, however, and conducting studies on human beings raises numerous questions—beginning with how to define energy restriction.[13] Does it mean eating less or just weighing less? Is it less than you want or less than the average? Does eating less have to result in weight loss? Does it matter whether weight loss results from more exercise or from less food? Or whether weight loss is intentional or unintentional? Answers await research.

Extreme starvation to extend life, like any extreme, is rarely, if ever, worth the price. Moderation, on the other hand, may be valuable. Many of the physiological responses to energy restriction seen in animals also occur in people whose intakes are *moderately* restricted. When people cut back on their usual energy intake by 10 to 20 percent,■ body weight, body fat, and blood pressure drop, and blood lipids and insulin response improve—favorable changes for preventing chronic diseases. The reduction in oxidative damage that occurs with energy restriction in animals also occurs in people whose diets include antioxidant nutrients and phytochemicals. As Highlights 11 and 13 explain, diets, such as the Mediterranean diet, that include an abundance of fruits, vegetables, olive oil, and red wine—with their array of antioxidants and phytochemicals—support good health and long life.[14]

■ kCalorie-restricted research has been conducted on various species, including mice, rats, rhesus monkeys, cynomolgus monkeys, spiders, and fish.

■ For perspective, a person with a usual energy intake of 2000 kcalories might cut back to 1600 to 1800 kcalories.

IN SUMMARY Life expectancy in the United States increased dramatically in the twentieth century. Factors that enhance longevity include limited or no alcohol use, regular balanced meals, weight control, adequate sleep, abstinence from smoking, and regular physical activity. Energy restriction in animals seems to lengthen their lives. Whether such dietary intervention in human beings is beneficial remains unknown. At the very least, nutrition—especially when combined with regular physical activity—can influence aging and longevity in human beings by supporting good health and preventing disease.

The Aging Process

As people get older, each person becomes less and less like anyone else. The older people are, the more time has elapsed for such factors as nutrition, genetics, physical activity, and everyday **stress** to influence physical and psychological aging.

stress: any threat to a person's well-being; a demand placed on the body to adapt.

Both physical **stressors** (such as alcohol abuse, other drug abuse, smoking, pain, and illness) and psychological stressors (such as exams, divorce, moving, and the death of a loved one) elicit the body's **stress response.** The body responds to such stressors with an elaborate series of physiological steps, as the nervous and hormonal systems bring about defensive readiness in every body part. These effects favor physical action—the classic fight-or-flight response. Stress that is prolonged or severe can drain the body of its reserves and leave it weakened, aged, and vulnerable to illness, especially if physical action is not taken. As people age, they lose their ability to adapt to both external and internal disturbances. When disease strikes, the reduced ability to adapt makes the aging individual more vulnerable to death than a younger person.

Highlight 11 described the oxidative stresses that occur when free radicals exceed the body's ability to defend itself. Increased free-radical activity and decreased antioxidant protection are common, but not inevitable, features of aging—and antioxidants seem to help slow the aging process.[15] Healthy people over 100 years old who had higher intakes of vegetables showed less evidence of oxidative stress than people 70 to 99 years old. Such findings seem to suggest that the fountain of youth may actually be a cornucopia of fruits and vegetables rich in antioxidants. (Return to Highlight 11 for more details on the antioxidant action of fruits and vegetables in defending against oxidative stress.)

Physiological Changes

As aging progresses, inevitable changes in each of the body's organs contribute to the body's declining function. These physiological changes influence nutrition status, just as growth and development do in the earlier stages of the life cycle.

Body Weight Over half of the adults in the United States are now considered overweight or obese. Chapter 8 presented the many health problems that accompany obesity and the BMI guidelines for a healthy body weight (18.5 to 24.9). These guidelines apply to all adults, regardless of age, but they may be too restrictive for older adults. The importance of body weight in defending against chronic diseases differs for older adults. Being moderately overweight may not be harmful. For adults over 65, health risks do not become apparent until BMI reaches at least 27—and the relationship tends to diminish with age until it disappears by age 75.[16] Not all older adults are overweight, of course. In fact, the prevalence of overweight decreases with increasing age after age 55.[17]

For older adults, a low body weight may be more detrimental than a high one. Low body weight often reflects malnutrition and the trauma associated with a fall. Many older adults experience unintentional weight loss, in large part because of an inadequate food intake.[18] Without adequate nutrient reserves, an underweight person may be unprepared to fight against diseases. Even a slight weight loss (5 percent) increases the likelihood of disease and premature death, making every meal a life-saving event.[19]

Body Composition In general, older people tend to lose bone and muscle and gain body fat. Many of these changes occur because some hormones that regulate appetite and metabolism become less active with age, while others become more active.*

Loss of muscle, known as **sarcopenia,** can be significant in the later years, and its consequences can be quite dramatic (see Figure 17-2).[20] As muscles diminish and weaken, people lose their ability to move and maintain balance, making falls likely. The limitations that accompany the loss of muscle mass and strength play a key role in the diminishing health that often accompanies aging. Optimal nutri-

---

stressors: environmental elements, physical or psychological, that cause stress.

stress response: the body's response to stress, mediated by both nerves and hormones.

sarcopenia (SAR-koh-PEE-nee-ah): loss of skeletal muscle mass, strength, and quality.
• **sarco** = flesh
• **penia** = loss or lack

---

*Examples of hormones that change with age include growth hormone and androgens, which decline with advancing age, thus contributing to the decrease in lean body mass, and prolactin, which increases with age, helping to maintain body fat. Insulin sensitivity also diminishes as people grow older, most likely because of increases in body fat and decreases in physical activity.

FIGURE 17-2 Sarcopenia

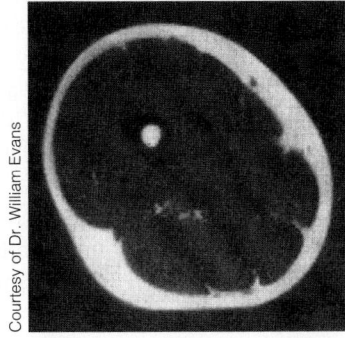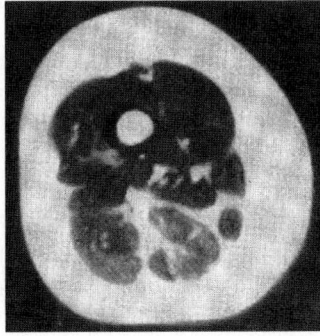

Courtesy of Dr. William Evans

These cross sections of two women's thighs may appear to be about the same size from the outside, but the 20-year-old woman's thigh (left) is dense with muscle tissue. The 64-year-old woman's thigh (right) has lost muscle and gained fat, changes that may be largely preventable with strength-building physical activities.

tion and regular physical activity can help maintain muscle mass and strength and minimize the changes in body composition associated with aging.[21]

Immune System Changes in the immune system also bring declining function with age. In addition, the immune system is compromised by nutrient deficiencies. Thus the combination of old age and malnutrition makes older people vulnerable to infectious diseases.[22] Adding insult to injury, antibiotics often are not effective against infections in people with compromised immune systems. Consequently, infectious diseases are a major cause of death in older adults. Older adults may improve their immune system responses by exercising regularly.[23]

GI Tract In the GI tract, the intestinal wall loses strength and elasticity with age, and GI hormone secretions change.[24] All of these actions slow motility. Constipation is much more common in the elderly than in the young.

Atrophic gastritis, a condition that affects almost one-third of those over 60, is characterized■ by an inflamed stomach, bacterial overgrowth, and a lack of hydrochloric acid and intrinsic factor. All of these can impair the digestion and absorption of nutrients, most notably, vitamin B_{12}, but also biotin, folate, calcium, iron, and zinc.

Difficulty in swallowing, medically known as **dysphagia,** occurs in all age groups, but especially in the elderly. Being unable to swallow a mouthful of food can be scary, painful, and dangerous. Even swallowing liquids can be a problem for some people. Consequently, the person may eat less food and drink fewer beverages, resulting in weight loss, malnutrition, and dehydration. Dietary intervention for dysphagia is highly individualized based on the person's abilities and tolerances. The diet typically provides moist, soft-textured, tender-cooked, or pureed foods and thickened liquids.

Tooth Loss Regular dental care over a lifetime protects against tooth loss and gum disease, which are common in old age. These conditions make chewing difficult or painful. Dentures, even when they fit properly, are less effective than natural teeth, and inefficient chewing can cause choking. People with tooth loss,■ gum disease, and ill-fitting dentures tend to limit their food selections to soft foods. If foods such as corn on the cob, apples, and hard rolls are replaced by creamed corn, applesauce, and rice, then nutrition status may not be greatly affected, but when food groups are eliminated and variety is limited, poor nutrition follows. People without teeth typically eat fewer fruits and vegetables and have less variety in their diets.[25] Consequently, they have low intakes of fiber and vitamins. To determine whether a visit to the dentist is needed, an older adult can check the conditions listed in the margin.■

■ Consequences of atrophic gastritis:
  • Inflamed stomach.
  • Increased bacterial growth.
  • Reduced hydrochloric acid.
  • Reduced intrinsic factor.
  • Increased risk of nutrient deficiencies, notably of vitamin B_{12}.

■ The medical term for lack of teeth is **edentulous** (ee-DENT-you-lus).
  • **e** = without
  • **dens** = teeth

■ Conditions requiring dental care:
  Dry mouth.
  Eating difficulty.
  No dental care within 2 years.
  Tooth or mouth pain.
  Altered food selections.
  Lesions, sores, or lumps in mouth.

dysphagia (dis-FAY-gee-ah): difficulty in swallowing.

Shared meals can brighten the day and enhance the appetite.

Sensory Losses and Other Physical Problems Sensory losses and other physical problems can also interfere with an older person's ability to obtain adequate nourishment. Failing eyesight, for example, can make driving to the grocery store impossible and shopping for food a frustrating experience. It may become so difficult to read food labels and count money that the person doesn't buy needed foods. Carrying bags of groceries may be an unmanageable task. Similarly, a person with limited mobility may find cooking and cleaning up too hard to do. Not too surprisingly, the prevalence of undernutrition is high among those who are homebound.

Sensory losses can also interfere with a person's ability or willingness to eat. Taste and smell sensitivities tend to diminish with age and may make eating less enjoyable. If a person eats less, then weight loss and nutrient deficiencies may follow. Loss of vision and hearing may contribute to social isolation, and eating alone may lead to poor intake.

Other Changes

In addition to the physiological changes that accompany aging, adults are changing in many other ways that influence their nutrition status.[26] Psychological, economic, and social factors play big roles in a person's ability and willingness to eat.

Psychological Changes Although not an inevitable component of aging, depression is common among older adults.[27] Depressed people, even those without disabilities, lose their ability to perform simple physical tasks. They frequently lose their appetite and the motivation to cook or even to eat. An overwhelming sense of grief and sadness at the death of a spouse, friend, or family member may leave a person, especially an elderly person, feeling powerless to overcome depression. When a person is suffering the heartache and loneliness of bereavement, cooking meals may not seem worthwhile. The support and companionship of family and friends, especially at mealtimes, can help overcome depression and enhance appetite.

Economic Changes Overall, older adults today have higher incomes than their cohorts of previous generations.[28] Still, poverty is a major problem for about 20 percent of the people over age 65. Factors such as living arrangements and income make significant differences in the food choices, eating habits, and nutrition status of older adults, especially those over age 80. People of low socioeconomic means are likely to have inadequate food and nutrient intakes. Only about one-third of the needy elderly receive assistance from federal programs.

Social Changes Malnutrition among older adults is most common in hospitals and nursing homes.[29] In the community, malnutrition is most likely to occur among those living alone, especially men; those with the least education; those living in federally funded housing (an indicator of low income); and those who have recently experienced a change in lifestyle. Adults who live alone do not necessarily make poor food choices, but they often consume too little food: loneliness is directly related to nutritional inadequacies, especially of energy intake.

IN SUMMARY Many changes that accompany aging can impair nutrition status. Among physiological changes, hormone activity alters body composition, immune system changes raise the risk of infections, atrophic gastritis interferes with digestion and absorption, and tooth loss limits food choices. Psychological changes such as depression, economic changes such as loss of income, and social changes such as loneliness contribute to poor food intake.

Energy and Nutrient Needs of Older Adults

Knowledge about the nutrient needs and nutrition status of older adults has grown considerably in recent years. The Dietary Reference Intakes (DRI) cluster people over 50 into two age categories—one group of 51 to 70 years and one of 71 and older. Increasingly, research is showing that the nutrition needs of people 50 to 70 years old differ from those of people over 70.

Setting standards for older people is difficult because individual differences become more pronounced as people grow older. People start out with different genetic predispositions and ways of handling nutrients, and the effects of these differences become magnified with years of unique dietary habits. For example, one person may tend to omit fruits and vegetables from his diet, and by the time he is old, he may have a set of nutrition problems associated with a lack of fiber and antioxidants. Another person may have omitted milk and milk products all her life—her nutrition problems may be related to a lack of calcium. Also, as people age, they suffer different chronic diseases and take various medicines—both of which will affect nutrient needs. For all of these reasons, researchers have difficulty even defining "healthy aging," a prerequisite to developing recommendations to meet the "needs of practically all healthy persons." The following discussion gives special attention to the nutrients of greatest concern.

Growing old can be enjoyable for people who take care of their health and live each day fully.

Water

Despite real fluid needs, many older people do not seem to feel thirsty or notice mouth dryness. Many nursing home employees say it is hard to persuade their elderly clients to drink enough water and fruit juices. Older adults may find it difficult and bothersome to get a drink or to get to a bathroom. Those who have lost bladder control may be afraid to drink too much water.

Dehydration is a risk for older adults. Total body water decreases as people age, so even mild stresses such as fever or hot weather can precipitate rapid dehydration in older adults. Dehydrated older adults seem to be more susceptible to urinary tract infections, pneumonia, **pressure ulcers,** and confusion and disorientation. To prevent dehydration, older adults need to drink at least 6 glasses of water a day.[30]

Energy and Energy Nutrients

On average, energy needs decline an estimated 5 percent per decade. One reason is that people usually reduce their physical activity as they age, although they need not do so. Another reason is that basal metabolic rate declines 1 to 2 percent per decade as lean body mass diminishes.

The lower energy expenditure of older adults means that they need to eat less food to maintain their weights. Accordingly, the estimated energy requirements■ for adults decrease steadily after age 19, as the "How to" on p. 596 explains.

On limited energy allowances, people must select mostly nutrient-dense foods. There is little leeway for sugars, fats, oils, or alcohol. The Daily Food Guide (on pp. 44–45) offers a dietary framework for adults of all ages. Most older adults would do well to select the lower number of recommended servings from each food group. Those who need additional food energy can choose extra servings.

Protein Because energy needs decrease, protein must be obtained from low-kcalorie sources of high-quality protein, such as lean meats, poultry, fish, and eggs; fat-free and low-fat milk products; and legumes. Protein is especially important for the elderly to support a healthy immune system and to prevent muscle wasting.

Examine the various challenges elderly individuals face in meeting their nutrient needs.

■ Estimated energy requirements:
- Men: Subtract 10 kcal/day for each year of age above 19.
- Women: Subtract 7 kcal/day for each year of age above 19.

pressure ulcers: damage to the skin and underlying tissues as a result of compression and poor circulation; commonly seen in people who are bedridden or chairbound.

HOW TO Estimate Energy Requirements for Older Adults

The "How to" on p. 260 described how to estimate the energy requirements for adults using an equation that accounts for age, physical activity, weight, and height. Alternatively, energy requirements for older adults can be "guesstimated" by using the values listed in the tables in Appendix F for adults 30 years of age and subtracting 7 kcalories for women and 10 kcalories for men per day for each year over 30.

For example, Table F-4 lists 2556 kcalories per day for a woman who is 5 feet 5 inches tall, weighs 150 pounds, and has a low activity level. To estimate the energy requirements of a similar 50-year-old woman, subtract 7 kcalories per day for each year over 30:

$$50 - 30 = 20 \text{ yr.}$$
$$20 \text{ yr} \times 7 \text{ kcal/day} = 140 \text{ kcal/day.}$$
$$2556 \text{ kcal/day (at age 30)} - 140 \text{ kcal/day}$$
$$= 2416 \text{ kcal/day (at age 50).}$$

Similarly, using Table F-5 to estimate the energy requirements of a sedentary 65-year-old man who is 5 feet 11 inches tall and weighs 250 pounds, subtract 10 kcalories per day for each year over 30:

$$65 - 30 = 35 \text{ yr.}$$
$$35 \text{ yr} \times 10 \text{ kcal/day} = 350 \text{ kcal/day.}$$
$$3088 \text{ kcal/day (at age 30)} - 350 \text{ kcal/day}$$
$$= 2738 \text{ kcal/day (at age 65).}$$

(Adults between the ages of 19 and 30 can also use the values listed in the tables in Appendix F by adding 7 kcalories for women and 10 kcalories for men per day for each year below 30.)

Underweight or malnourished older adults need protein- and energy-dense snacks such as hard-boiled eggs, tuna fish and crackers, peanut butter on graham crackers, and hearty soups. Drinking liquid nutritional formulas between meals can also boost energy and nutrient intakes.[31] Importantly, the diet should provide enjoyment as well as nutrients.[32]

Carbohydrate and Fiber As always, abundant carbohydrate is needed to protect protein from being used as an energy source. Sources of complex carbohydrates such as legumes, vegetables, whole grains, and fruits are also rich in fiber and essential vitamins and minerals. Average fiber intakes among older adults are lower than current recommendations (14 grams per 1000 kcalories).[33] Eating high-fiber foods and drinking water can alleviate constipation—a condition common among older adults, especially nursing home residents. Physical inactivity and medications also contribute to the high incidence of constipation.

Fat As is true for people of all ages, fat intake needs to be moderate in the diets of most older adults—enough to enhance flavors and provide valuable nutrients, but not so much as to raise the risks of cancer, atherosclerosis, and other degenerative diseases. This recommendation should not be taken too far; limiting fat too severely may lead to nutrient deficiencies and weight loss—two problems that carry greater health risks in the elderly than overweight.

Vitamins and Minerals

Most people can achieve adequate vitamin and mineral intakes simply by including foods from all food groups in their diets, but older adults often omit fruits and vegetables. Similarly, few older adults consume the recommended amounts of milk or milk products.

Vitamin B$_{12}$ An estimated 10 to 30 percent of adults over 50 have atrophic gastritis.■ As Chapter 10 explained, people with atrophic gastritis are particularly vulnerable to vitamin B$_{12}$ deficiency: the bacterial overgrowth that accompanies this condition uses up the vitamin, and without hydrochloric acid and intrinsic factor, digestion and absorption of vitamin B$_{12}$ are inefficient. Given the poor cognition, anemia, and devastating neurological effects associated with a vitamin B$_{12}$ deficiency, an adequate intake is imperative.[34] The RDA for older adults is the same as for younger adults, but with the added suggestion to obtain most of a day's intake from vitamin B$_{12}$–fortified foods and supplements.[35] The bioavailability of vitamin B$_{12}$ from these sources is better than from foods.

Vitamin D Vitamin D deficiency is a problem among older adults. Only vitamin D–fortified milk provides significant vitamin D, and many older adults drink little or

■ Reminder: *Atrophic gastritis* is a chronic inflammation of the stomach characterized by inadequate hydrochloric acid and intrinsic factor—two key players in vitamin B$_{12}$ absorption.

no milk. Further compromising the vitamin D status of many older people, especially those in nursing homes, is their limited exposure to sunlight. Finally, aging reduces the skin's capacity to make vitamin D and the kidneys' ability to convert it to its active form. Not only are older adults not getting enough vitamin D, but they may actually need more. To prevent bone loss and to maintain vitamin D status, especially in those who engage in minimal outdoor activity, adults 51 to 70 years old need 10 micrograms daily and those over 70 need 15 micrograms.[36]

Calcium Chapter and Highlight 12 emphasized the importance of abundant dietary calcium throughout life, and especially for women after menopause, to protect against osteoporosis. The DRI Committee recommends 1200 milligrams of calcium daily, but the calcium intakes of older people in the United States are well below recommendations.[37] Some older adults avoid milk and milk products because they dislike these foods or associate them with stomach discomfort. Simple solutions include using calcium-fortified juices, adding powdered milk to recipes, and taking supplements; Chapter 12 offers many other strategies for including nonmilk sources of calcium for those who do not drink milk.

Iron The iron needs of men remain unchanged throughout adulthood. For women, iron needs decrease substantially when blood loss through menstruation ceases. Consequently, iron-deficiency anemia is less common in older adults than in younger people. In fact, elevated iron stores are more likely than deficiency in older people, especially those who take iron supplements, eat red meat regularly, and include vitamin C–rich fruits in their daily diet.[38]

Nevertheless, iron deficiency may develop in older adults, especially when their food energy intakes are low. Aside from diet, two other factors may lead to iron deficiency in older people: chronic blood loss from diseases and medicines, and poor iron absorption due to reduced stomach acid secretion and antacid use. For older people with infectious diseases, the consequences of iron-deficiency anemia can be life-threatening.[39] Anyone concerned with older people's nutrition should keep these possibilities in mind.

Nutrient Supplements

People judge for themselves how to manage their nutrition, and some turn to supplements. Advertisers target older people with appeals to take supplements and eat "health" foods, claiming that these products prevent disease and promote longevity. About half of all women over 65 take some type of dietary supplement, while about one-fifth of older men do. When recommended by a physician or registered dietitian, vitamin D and calcium supplements for osteoporosis or vitamin B_{12} for pernicious anemia may be beneficial. Many health care professionals recommend a daily multivitamin-mineral supplement that provides 100 percent or less of the Daily Value for the listed nutrients.[40] They reason that such a supplement is more likely to be beneficial than to cause harm.

People with small energy allowances would do well to become more active so they can afford to eat more food. Food is the best source of nutrients for everybody. Supplements are just that—supplements to foods, not substitutes for them. For anyone who is motivated to obtain the best possible health, it is never too late to learn to eat well, drink water, exercise regularly, and adopt other lifestyle habits such as quitting smoking and moderating alcohol use.

IN SUMMARY The following table summarizes the nutrient concerns of aging. Although some nutrients need special attention in the diet, supplements are not routinely recommended. The ever-growing number of older people creates an urgent need to learn more about how their nutrient requirements differ from those of others and how such knowledge can enhance their health.

Nutrient	Effect of Aging	Comments
Water	Lack of thirst and decreased total body water make dehydration likely.	Mild dehydration is a common cause of confusion. Difficulty obtaining water or getting to the bathroom may compound the problem.
Energy	Need decreases as muscle mass decreases (sarcopenia).	Physical activity moderates the decline.
Fiber	Likelihood of constipation increases with low intakes and changes in the GI tract.	Inadequate water intakes and lack of physical activity, along with some medications, compound the problem.
Protein	Needs may stay the same or increase slightly.	Low-fat, high-fiber legumes and grains meet both protein and other nutrient needs.
Vitamin B_{12}	Atrophic gastritis is common.	Deficiency causes neurological damage; supplements may be needed.
Vitamin D	Increased likelihood of inadequate intake; skin synthesis declines.	Daily sunlight exposure in moderation or supplements may be beneficial.
Calcium	Intakes may be low; osteoporosis is common.	Stomach discomfort commonly limits milk intake; calcium substitutes or supplements may be needed.
Iron	In women, status improves after menopause; deficiencies are linked to chronic blood losses and low stomach acid output.	Adequate stomach acid is required for absorption; antacid or other medicine use may aggravate iron deficiency; vitamin C and meat increase absorption.

Nutrition-Related Concerns of Older Adults

Nutrition may play a greater role than has been realized in preventing many changes once thought to be inevitable consequences of growing older. The following discussions of cataracts and macular degeneration, arthritis, and the aging brain show that nutrition may provide at least some protection against some of the conditions associated with aging.

Cataracts and Macular Degeneration

Cataracts are age-related thickenings in the lenses of the eyes that impair vision. If not surgically removed, they ultimately lead to blindness. Cataracts occur even in well-nourished individuals as a result of ultraviolet light exposure, oxidative stress, injury, viral infections, toxic substances, and genetic disorders. Many cataracts, however, are vaguely called senile cataracts—meaning "caused by aging." In the United States, more than half of all adults 65 and older have a cataract.

Oxidative stress appears to play a significant role in the development of cataracts, and the antioxidant nutrients may help minimize the damage. Studies have reported an inverse relationship between cataracts and dietary intakes of vitamin C, vitamin E, and carotenoids; taking supplements of these antioxidant nutrients seems to slow the progression or reduce the risk of developing age-related cataracts.[41]

One other diet-related factor may play a role in the development of cataracts: overweight.[42] Overweight appears to be associated with cataracts, but its role has not been identified. Risk factors that typically accompany overweight, such as inactivity, diabetes, or hypertension, do not explain the association.

Another common cause of visual loss among older people is **macular degeneration,** a deterioration of the macular region of the retina.[43] Similarly to cataracts, risk factors for age-related macular degeneration include oxidative stress from sunlight, and preventive factors include supplements of antioxidant vitamins plus zinc and of the carotenoids lutein and zeaxanthin.[44] Total dietary fat may also be a risk factor for macular degeneration, but the omega-3 fatty acids of fish may be protective.[45]

Arthritis

Over 40 million people in the United States have some form of **arthritis.**[46] As the population ages, it is expected that the prevalence will increase to 60 million by 2020.

cataracts (KAT-ah-rakts): thickenings of the eye lenses that impair vision and can lead to blindness.

macular (MACK-you-lar) **degeneration:** deterioration of the macular area of the eye that can lead to loss of central vision and eventual blindness. The **macula** is a small, oval, yellowish region in the center of the retina that provides the sharp, straight-ahead vision so critical to reading and driving.

arthritis: inflammation of a joint, usually accompanied by pain, swelling, and structural changes.

Osteoarthritis The most common type of arthritis that disables older people is **osteoarthritis,** a painful deterioration of the cartilage in the joints. During movement, the ends of bones are normally protected from wear by cartilage and by small sacs of fluid that act as a lubricant. With age, bones sometimes disintegrate, and the joints become malformed and painful to move.

One known connection between osteoarthritis■ and nutrition is overweight. Weight loss may relieve some of the pain for overweight persons with osteoarthritis, partly because the joints affected are often weight-bearing joints that are stressed and irritated by having to carry excess poundage. Interestingly, though, weight loss often relieves the worst of the pain of arthritis in the hands as well, even though they are not weight-bearing joints. Jogging and other weight-bearing exercises do not worsen arthritis. In fact, both aerobic activity and strength training offer modest improvements in physical performance and pain relief.

Rheumatoid Arthritis Another type of arthritis known as **rheumatoid arthritis** has possible links to diet through the immune system. In rheumatoid arthritis, the immune system mistakenly attacks the bone coverings as if they were made of foreign tissue. In some individuals, certain foods, notably vegetables and olive oil, may moderate the inflammatory responses and provide some relief.[47]

The omega-3 fatty acids commonly found in fish oil reduce joint tenderness and improve mobility in some people with rheumatoid arthritis.[48] The same diet recommended for heart health—one low in saturated fat from meats and milk products and high in omega-3 fats from fish—helps prevent or reduce the inflammation in the joints that makes arthritis so painful.

Another possible link between nutrition and rheumatoid arthritis involves the oxidative damage to the membranes within joints that causes inflammation and swelling. The antioxidant vitamins C and E defend against oxidation, and supplements of these nutrients may help prevent or relieve the pain of rheumatoid arthritis.[49]

Treatment Treatment for arthritis—dietary or otherwise—may help relieve discomfort and improve mobility, but it does not cure the condition. Traditional medical intervention for arthritis includes medication and surgery. Alternative therapies to treat arthritis abound, but none have proved safe and effective in scientific studies. Two currently popular supplements—glucosamine and chondroitin—may relieve pain and improve mobility as well as over-the-counter pain relievers, but stronger research studies are needed to confirm reports.[50] Drugs and supplements used to relieve arthritis can impose nutrition risks; many affect appetite and alter the body's use of nutrients, as Highlight 17 explains.

The Aging Brain

The brain, like all of the body's organs, responds to both genetic and environmental factors that can enhance or diminish its amazing capacities. One of the challenges researchers face when studying the human brain is to distinguish among normal age-related physiological changes, changes caused by diseases, and changes that result from cumulative, environmental factors such as diet.

The brain normally changes in some characteristic ways as it ages. For one thing, its blood supply decreases. For another, the number of **neurons,** the brain cells that specialize in transmitting information, diminishes as people age. When the number of nerve cells in one part of the cerebral cortex diminishes, hearing and speech are affected. Losses of neurons in other parts of the cortex can impair memory and cognitive function. When the number of neurons in the hindbrain diminishes, balance and posture are affected. Losses of neurons in other parts of the brain affect still other functions. Some of the cognitive loss and forgetfulness generally attributed to aging may be due in part to environmental, and therefore controllable, factors—including nutrient deficiencies.

Nutrient Deficiencies and Brain Function Nutrients influence the development and activities of the brain.[51] The ability of neurons to synthesize specific

■ Risk factors for osteoarthritis:
- Age.
- Smoking.
- BMI at age 40.
- Lack of hormone therapy (in women).

osteoarthritis: a painful, degenerative disease of the joints that occurs when the cushioning cartilage in a joint deteriorates; joint structure is damaged, with loss of function; also called **degenerative arthritis.**

rheumatoid (ROO-ma-toyd) **arthritis:** a disease of the immune system involving painful inflammation of the joints and related structures.

neurons: nerve cells; the structural and functional units of the nervous system. Neurons initiate and conduct nerve transmissions.

TABLE 17-2 Summary of Nutrient-Brain Relationships

Brain Function	Depends on an Adequate Intake of:
Short-term memory	Vitamin B_{12}, vitamin C, vitamin E
Performance in problem-solving tests	Riboflavin, folate, vitamin B_{12}, vitamin C
Mental health	Thiamin, niacin, zinc, folate
Cognition	Folate, vitamin B_6, vitamin B_{12}, iron, vitamin E
Vision	Essential fatty acids, vitamin A
Neurotransmitter synthesis	Tyrosine, tryptophan, choline

neurotransmitters depends in part on the availability of precursor nutrients that are obtained from the diet.[52] The neurotransmitter serotonin, for example, derives from the amino acid tryptophan. To function properly, the enzymes involved in neurotransmitter synthesis require vitamins and minerals. Thus nutrient deficiencies may contribute to the loss of memory and cognition that some older adults experience.[53] Such losses may be preventable or at least diminished or delayed through diet. Table 17-2 summarizes some of the better-known connections between brain function and nutrients.

In some instances, the degree of cognitive loss is extensive. Such **senile dementia** may be attributable to a specific disorder such as a brain tumor or Alzheimer's disease.

Alzheimer's Disease Much attention has focused on the *abnormal* deterioration of the brain called **Alzheimer's disease,** which affects 10 percent of U.S. adults by age 65 and 30 percent of those over 85. Diagnosis of Alzheimer's disease depends on its characteristic symptoms: the victim gradually loses memory and reasoning, the ability to communicate, physical capabilities, and eventually life itself.[54] Nerve cells in the brain die, and communication between the cells breaks down.

Researchers are closing in on the exact cause of Alzheimer's disease.* Clearly, genetic factors are involved. Free radicals may also be involved.[55] Nerve cells in the brains of people with Alzheimer's disease show evidence of free-radical attack—damage to DNA, cell membranes, and proteins. They also show evidence of the minerals that trigger free-radical attacks—iron, copper, zinc, and aluminum.

In Alzheimer's disease, the brain is littered with clumps of a protein fragment called beta-amyloid. Free radicals and beta-amyloid have a sinister relationship: free radicals accelerate the clumping of beta-amyloid, and beta-amyloid produces more free radicals. Scientists believe beta-amyloid clogs the brain and damages or kills certain nerve cells, causing memory loss. Some research suggests that the antioxidant nutrients can limit free-radical damage and delay or prevent Alzheimer's disease.[56] Drug research is focusing on developing an immunization for beta-amyloid or an enzyme to block its production.[57]

Late in the course of the disease there is a decline in the activity of the enzyme that assists in the production of the neurotransmitter acetylcholine from choline and acetyl CoA.[58] Acetylcholine is essential to memory, but supplements of choline (or of lecithin, which contains choline) have no effect on memory or on the progression of the disease. Drugs that inhibit the breakdown of acetylcholine, on the other hand, have proved beneficial.

Research suggests that cardiovascular disease risk factors such as high blood pressure, diabetes, and elevated levels of homocysteine may be related to the development of Alzheimer's disease.[59] Diets designed to support a healthy heart may benefit a healthy brain as well.

Treatment for Alzheimer's disease involves providing care to clients and support to their families. Drugs are used to improve or at least to slow the loss of short-term memory and cognition, but they do not treat the disease.[60] Other drugs may be used to control depression, anxiety, and behavior problems.

Maintaining appropriate body weight may be the most important nutrition concern for the person with Alzheimer's disease.[61] Depression and forgetfulness can lead to changes in eating behaviors and poor food intake. Furthermore, changes in the body's weight-regulation system may contribute to weight loss. Perhaps the best that a caregiver can do nutritionally for a person with Alzheimer's disease is to supervise food planning and mealtimes. Providing well-liked and well-balanced meals and snacks in a cheerful atmosphere encourages food consumption. To minimize confusion, offer a few ready-to-eat foods, in bite-size pieces, with seasonings and sauces. To avoid mealtime disruptions, control distractions such as music, television, children, and the telephone.

senile dementia: the loss of brain function beyond the normal loss of physical adeptness and memory that occurs with aging.

Alzheimer's disease: a degenerative disease of the brain involving memory loss and major structural changes in neuron networks; also known as *senile dementia of the Alzheimer's type (SDAT), primary degenerative dementia of senile onset,* or *chronic brain syndrome.*

---

*A report on the genetic and other aspects of Alzheimer's is available from Alzheimer's Disease Education and Referral Center, P.O. Box 8250, Silver Springs, MD 20907-8250.

IN SUMMARY Senile dementia and other losses of brain function afflict millions of older adults, while others face loss of vision due to cataracts or macular degeneration, or cope with the pain of arthritis. As the number of people over age 65 continues to grow, the need for solutions to these problems becomes urgent. Some problems may be inevitable, but others are preventable and good nutrition may play a key role.

Food Choices and Eating Habits of Older Adults

Older people are an incredibly diverse group, and for the most part they are independent, socially sophisticated, mentally lucid, fully participating members of society who report themselves to be happy and healthy. In fact, the quality of life among the elderly has improved, and their chronic disabilities have declined dramatically in recent years.[62] By practicing stress-management skills, maintaining physical fitness, participating in activities of interest, and cultivating spiritual health, as well as obtaining adequate nourishment, people can support a high quality of life into old age (see Table 17-3 for some strategies).

Older people spend more money per person on foods to eat at home than other age groups and less money on foods away from home. Manufacturers would be wise to cater to the preferences of older adults by providing good-tasting, nutritious foods in easy-to-open, single-serving packages with labels that are easy to read. Such services enable older adults to maintain their independence and to feel a sense of control and involvement in their own lives. Another way older adults can take care of themselves is by remaining or becoming physically active. As mentioned earlier, physical activity helps preserve one's ability to perform daily tasks and so promotes independence.

Familiarity, taste, and health beliefs are most influential on older people's food choices. Eating foods that are familiar, especially ethnic foods that recall family meals and pleasant times, can be comforting. People 65 and over are less likely to

Both foods and mental challenges nourish the brain.

TABLE 17-3	Strategies for Growing Old Healthfully

- Choose nutrient-dense foods.
- Be physically active. Walk, run, dance, swim, bike, or row for aerobic activity. Lift weights, do calisthenics, or pursue some other activity to tone, firm, and strengthen muscles. Practice balancing on one foot or doing simple movements with your eyes closed. Modify activities to suit changing abilities and tastes.
- Maintain appropriate body weight.
- Reduce stress (cultivate self-esteem, maintain a positive attitude, manage time wisely, know your limits, practice assertiveness, release tension, and take action).
- For women, discuss with a physician the risks and benefits of estrogen replacement therapy.
- For people who smoke, discuss with a physician strategies and programs to help you quit.
- Expect to enjoy sex, and learn new ways of enhancing it.
- Use alcohol only moderately, if at all; use drugs only as prescribed.
- Take care to prevent accidents.
- Expect good vision and hearing throughout life; obtain glasses and hearing aids if necessary.
- Take care of your teeth; obtain dentures if necessary.

- Be alert to confusion as a disease symptom, and seek diagnosis.
- Take medications as prescribed; see a physician before self-prescribing medicines or herbal remedies and a registered dietitian before self-prescribing supplements.
- Control depression through activities and friendships; seek professional help if necessary.
- Drink 6 to 8 glasses of water every day.
- Practice mental skills. Keep on solving math problems and crossword puzzles, playing cards or other games, reading, writing, imagining, and creating.
- Make financial plans early to ensure security.
- Accept change. Work at recovering from losses; make new friends.
- Cultivate spiritual health. Cherish personal values. Make life meaningful.
- Go outside for sunshine and fresh air as often as possible.
- Be socially active—play bridge, join an exercise or dance group, take a class, teach a class, eat with friends, volunteer time to help others.
- Stay interested in life—pursue a hobby, spend time with grandchildren, take a trip, read, grow a garden, or go to the movies.
- Enjoy life.

Social interactions at a congregate meal site can be as nourishing as the foods served.

diet to lose weight than younger people are, but are more likely to diet in pursuit of medical goals such as controlling blood glucose and cholesterol.

Food Assistance Programs

The Nutrition Screening Initiative is part of a national effort to identify and treat nutrition problems in older persons; it uses a screening checklist. To *determine* the risk of malnutrition in older clients, health care professionals can keep in mind the characteristics and questions listed in Table 17-4.

The U.S. government funds the federal Elderly Nutrition Program to improve older people's nutrition status and enable them to avoid medical problems, continue living in communities of their own choice, and stay out of institutions. Its specific goals are to provide low-cost, nutritious meals; opportunities for social interaction; homemaker education and shopping assistance; counseling and referral to social services; and transportation. The program's mission has always been to provide "more than a meal."

The Elderly Nutrition Program provides for **congregate meals** at group settings such as community centers. Administrators try to select sites for congregate meals where as many eligible people as possible can participate. Volunteers may also deliver meals to those who are homebound either permanently or temporarily; these home-delivered meals are known as **Meals on Wheels.** The home-delivery program ensures nutrition, but its recipients miss out on the social benefits of the congregate meals; every effort is made to persuade older people to come to the shared meals, if they can. All persons aged 60 years and older and their spouses are eligible to receive meals from these programs, regardless of their income. Priority is given to those who are economically and socially needy. An estimated 3 million of our nation's older adults benefit from these meals.

These programs provide at least one meal a day that meets a third of the RDA for this age group; they must operate five or more days a week. Many programs voluntarily offer additional services designed to appeal to older adults: provisions for special diets (to meet medical needs or religious preferences), food pantries, ethnic meals, and delivery of meals to the homeless.

HEALTHY
PEOPLE
2010

Increase the receipt of home foodservices by people aged 65 and older who have difficulty in preparing their own meals or are otherwise in need of home-delivered meals.

congregate meals: nutrition programs that provide food for the elderly in conveniently located settings such as community centers.

Meals on Wheels: a nutrition program that delivers food for the elderly to their homes.

TABLE 17-4  Risk Factors for Malnutrition in Older Adults

	These questions help *determine* the risk of malnutrition in older adults:
Disease	• Do you have an illness or condition that changes the types or amounts of foods you eat?
Eating poorly	• Do you eat fewer than two meals a day? Do you eat fruits, vegetables, and milk products daily?
Tooth loss or mouth pain	• Is it difficult or painful to eat?
Economic hardship	• Do you have enough money to buy the food you need?
Reduced social contact	• Do you eat alone most of the time?
Multiple medications	• Do you take three or more different prescribed or over-the-counter medications daily?
Involuntary weight loss or gain	• Have you lost or gained 10 pounds or more in the last six months?
Needs assistance	• Are you physically able to shop, cook, and feed yourself?
Elderly person	• Are you older than 80?

Older adults can learn about the available programs in their communities by looking in the Yellow Pages of the telephone book under "Social Services" or "Senior Citizens' Organizations."[*] In addition, the local senior center and hospital can usually direct people to programs providing nutrition and other health-related services.

Meals for Singles

Many older adults live alone, and singles of all ages face challenges in purchasing, storing, and preparing food. Large packages of meat and vegetables are often intended for families of four or more, and even a head of lettuce can spoil before one person can use it all. Many singles live in small dwellings and have little storage space for foods. A limited income presents additional obstacles. This section offers suggestions that can help to solve some of the problems singles face.

Spend Wisely People who have the means to shop and cook for themselves can cut their food bills just by being wise shoppers. Large supermarkets are usually less expensive than convenience stores. A grocery list helps reduce impulse buying, and specials and coupons can save money when the items featured are those that the shopper needs and uses.

Buying the right amount so as not to waste any food is a challenge for people eating alone. They can buy fresh milk in the size best suited for personal needs. Pint-size and even cup-size boxes■ of milk are available and can be stored unopened on a shelf for up to three months without refrigeration.

Many foods that offer a variety of nutrients for practically pennies have a long shelf life; staples such as rice, pastas, dry powdered milk, and dried legumes can be purchased in bulk and stored for months at room temperature. Other foods that are usually a good buy include whole pieces of cheese rather than sliced or shredded cheese, fresh produce in season, variety meats such as chicken livers, and cereals that require cooking instead of ready-to-serve cereals.

A person who has ample freezing space can buy large packages of meat, such as pork chops, ground beef, or chicken, when they are on sale. Then the meat can be immediately wrapped into individual servings for the freezer. All the individual servings can be put in a bag marked appropriately with the contents and the date.

Frozen vegetables are more economical in large bags than in small boxes. The amount needed can be taken out, and the bag closed tightly with a twist tie or rubber band. If the package is returned quickly to the freezer each time, the vegetables will stay fresh for a long time.

Finally, breads and cereals usually must be purchased in larger quantities. Again the amount needed for a few days can be taken out and the rest stored in the freezer.

Grocers will break open a package of wrapped meat and rewrap the portion needed. Similarly, eggs can be purchased by the half-dozen. Eggs do keep for long periods, though, if stored properly in the refrigerator.

Fresh fruits and vegetables can be purchased individually. A person can buy fresh fruit at various stages of ripeness: a ripe one to eat right away, a semiripe one to eat soon after, and a green one to ripen on the windowsill. If vegetables are packaged in large quantities, the grocer can break open the package so that a smaller amount can be purchased. Small cans of fruits and vegetables, even though they are more expensive per unit, are a reasonable alternative, considering that it is expensive to buy a regular-size can and let the unused portion spoil.

Be Creative Creative chefs think of various ways to use foods when only large amounts are available. For example, a head of cauliflower can be divided into thirds. Then one-third is cooked and eaten hot. Another third is put into a vinegar and oil marinade for use in a salad. And the last third can be used in a casserole or stew.

A variety of vegetables and meats can be enjoyed stir-fried; inexpensive vegetables such as cabbage, celery, and onion are delicious when crisp cooked in a little

Taking time to nourish your body well is a gift you give yourself.

■ Boxes of milk that can be stored at room temperature have been exposed to temperatures above those of pasteurization just long enough to sterilize the milk—a process called **ultrahigh temperature (UHT).**

Buy only what you will use.

*To find a local provider, call Eldercare Locator at (800) 677-1116.

Invite guests to share a meal.

oil with herbs or lemon added. Interesting frozen vegetable mixtures are available in larger grocery stores. Cooked, leftover vegetables can be dropped in at the last minute. A bonus of a stir-fried meal is that there is only one pan to wash. Similarly, a microwave oven allows a chef to use fewer pots and pans. Meals and leftovers can also be frozen or refrigerated in microwavable containers to reheat as needed.

Many frozen dinners offer nutritious options. Adding a fresh salad, a whole-wheat roll, and a glass of milk can make a nutritionally balanced meal.

Also, single people shouldn't hesitate to invite someone to share meals with them whenever there is enough food. It's likely that the person will return the invitation, and both parties will get to enjoy companionship and a meal prepared by others.

IN SUMMARY Older people can benefit from both the nutrients provided and the social interaction available at congregate meals. Other government programs deliver meals to those who are homebound. With creativity and careful shopping, those living alone can prepare nutritious, inexpensive meals. Physical activity, mental challenges, stress management, and social activities can also help people grow old comfortably.

Nutrition in Your Life

By eating a balanced diet, maintaining a healthy body weight, and engaging in a variety of physical, social, and mental activities, you can enjoy good health in later life.

- If there are older adults in your life, do they have the financial means, physical ability, and social support they need to eat adequately?
- Have they experienced an unintentional loss of weight recently?
- Are they active physically, socially, and mentally?

NUTRITION ON THE NET

 Access these websites for further study of topics covered in this chapter.

- Find updates and quick links to these and other nutrition-related sites at our website: **www.wadsworth.com/nutrition**
- Search for "aging," "arthritis," and "Alzheimer's" on the U.S. Government health information site: **www.healthfinder.gov**
- Visit the National Aging Information Center of the Administration on Aging: **www.aoa.gov**
- Visit the American Geriatrics Society: **www.americangeriatrics.org**
- Visit the National Institute on Aging: **www.nia.nih.gov**
- Visit the American Association of Retired Persons: **www.aarp.org**
- Get nutrition tips for growing older in good health from the American Dietetic Association: **www.eatright.org**

- Learn more about cataracts and macular degeneration from the National Eye Institute, the Macular Degeneration Partnership, and the American Society of Cataract and Refractive Surgery: **www.nei.nih.gov**, **www.macd.net**, and **www.ascrs.org**
- Learn more about arthritis from the Arthritis Society, the Arthritis Foundation, and the National Institute of Arthritis and Musculoskeletal and Skin Diseases: **www.arthritis.ca**, **www.arthritis.org**, and **www.niams.nih.gov**
- Learn more about Alzheimer's disease from the NIA Alzheimer's Disease Education and Referral Center and the Alzheimer's Association: **www.alzheimers.org** and **www.alz.org**
- Find out about federal government programs designed to help senior citizens maintain good health: **www.seniors.gov**

STUDY QUESTIONS

These questions will help you review the chapter. You will find the answers in the discussions on the pages provided.

1. What roles does nutrition play in aging, and what roles can it play in retarding aging? (pp. 588–591)

2. What are some of the physiological changes that occur in the body's systems with aging? To what extent can aging be prevented? (pp. 592–594)

3. Why does the risk of dehydration increase as people age? (p. 595)

4. Why do energy needs usually decline with advancing age? (p. 595)

5. Which vitamins and minerals need special consideration for the elderly? Explain why. Identify some factors that complicate the task of setting nutrient standards for older adults. (pp. 596–597)

6. Discuss the relationships between nutrition and cataracts and between nutrition and arthritis. (pp. 598–599)

7. What characteristics contribute to malnutrition in older people? (pp. 601–602)

These multiple choice questions will help you prepare for an exam. Answers can be found on p. 607.

1. Life expectancy in the United States is:
   a. 48 to 60 years.
   b. 58 to 70 years.
   c. 68 to 80 years.
   d. 78 to 90 years.

2. The human life span is about:
   a. 85 years.
   b. 100 years.
   c. 115 years.
   d. 130 years.

3. A 72-year-old person whose physical health is similar to that of people 10 years younger has a(n):
   a. chronological age of 62.
   b. physiological age of 72.
   c. physiological age of 62.
   d. absolute age of minus 10.

4. Rats live longest when given diets that:
   a. eliminate all fat.
   b. provide lots of protein.
   c. allow them to eat freely.
   d. restrict their energy intakes.

5. Which characteristic is *not* commonly associated with atrophic gastritis?
   a. inflamed stomach
   b. vitamin B_{12} toxicity
   c. bacterial overgrowth
   d. lack of intrinsic factor

6. On average, adult energy needs:
   a. decline 5 percent per year.
   b. decline 5 percent per decade.
   c. remain stable throughout life.
   d. rise gradually throughout life.

7. Which nutrients seem to protect against cataract development?
   a. minerals
   b. lecithins
   c. antioxidants
   d. amino acids

8. The best dietary advice for a person with osteoarthritis might be to:
   a. avoid milk products.
   b. take fish oil supplements.
   c. take vitamin E supplements.
   d. lose weight, if overweight.

9. Congregate meal programs are preferable to Meals on Wheels because they provide:
   a. nutritious meals.
   b. referral services.
   c. social interactions.
   d. financial assistance.

10. The Elderly Nutrition Program is available to:
    a. all people 65 years and older.
    b. all people 60 years and older.
    c. homebound people only, 60 years and older.
    d. low-income people only, 60 years and older.

REFERENCES

1. Trends in aging—United States and worldwide, *Morbidity and Mortality Weekly Report* 52 (2003): 101–106.

2. National vital statistics report, **www.cdc.gov/nchs**, site visited May 1, 2003.

3. T. Perls, Genetic and environmental influences on exceptional longevity and the AGE nomogram, *Annals of the New York Academy of Sciences* 959 (2002): 1–13.

4. G. E. Fraser and D. J. Shavlik, Ten years of life: Is it a matter of choice? *Archives of Internal Medicine* 161 (2001): 1645–1652.

5. E. W. Gregg and coauthors, Relationship of changes in physical activity and mortality among older women, *Journal of the Ameri-*

can Medical Association 289 (2003): 2379–2386; American College of Sports Medicine, Position stand: Exercise and physical activity for older adults, *Medicine and Science in Sports and Exercise* 30 (1998): 992–1008.

6. W. J. Strawbridge and coauthors, Physical activity reduces the risk of subsequent depression for older adults, *American Journal of Epidemiology* 156 (2002): 328–334; A. J. Schuit and coauthors, Physical activity and cognitive decline, the role of the apolipoprotein e4 allele, *Medicine and Science in Sports and Exercise* 33 (2001): 772–777.

7. I. M. Lee and coauthors, Epidemiologic data on the relationships of caloric intake, energy

balance, and weight gain over the life span with longevity and morbidity, *Journals of Gerontology: Series A, Biological Sciences and Medical Sciences* 56 (2001): 7–19.

8. C. K. Lee and coauthors, Gene expression profile of aging and its retardation by calorie restriction, *Science* 285 (1999): 1390–1393.

9. S. X. Cao and coauthors, Genomic profiling of short- and long-term caloric restriction effects in the liver of aging mice, *Proceeding of the National Academy of Sciences of the United States of America* 98 (2001): 10630–10635.

10. J. M. Dhahbi and coauthors, Caloric restriction alters the feeding response of key metabolic enzyme genes, *Mechanisms of*

Ageing and Development 122 (2001): 1033–1048; J. J. Ramsey and coauthors, Dietary restriction and aging in rhesus monkeys: The University of Wisconsin study, *Experimental Gerontology* 35 (2000): 1131–1149; A. C. Gazdag and coauthors, Effect of long-term calorie restriction on GLUT4, phosphatidylinositol-3 kinase p85 subunit, and insulin receptor substrate-1 protein levels in rhesus monkey skeletal muscle, *Journals of Gerontology: Series A, Biological Sciences and Medical Sciences* 55 (2000): B44–B46; W. T. Cefalu and coauthors, Influence of caloric restriction on the development of atherosclerosis in nonhuman primates: Progress to date, *Toxicological Sciences* 52 (1999): 49–55.

11. J. P. DeLany and coauthors, Long-term calorie restriction reduces energy expenditure in aging monkeys, *Journals of Gerontology: Series A, Biological Sciences and Medical Sciences* 54 (1999): B5–B11.

12. G. Barja, Endogenous oxidative stress: Relationship to aging, longevity and caloric restriction, *Ageing Research and Reviews* 1 (2002): 397–411; B. J. Merry, Molecular mechanisms linking calorie restriction and longevity, *International Journal of Biochemistry and Cell Biology* 34 (2002): 1340–1354; J. Wanagat, D. B. Allison, and R. Weindruch, Calorie intake and aging: Mechanisms in rodents and a study in nonhuman primates, *Toxicological Sciences* 52 (1999): 35–40.

13. L. K. Heilbronn and E. Ravussin, Calorie restriction and aging: Review of the literature and implications for studies in humans, *American Journal of Clinical Nutrition* 78 (2003): 361–369; Lee and coauthors, 2001.

14. A. Trichopoulou and E. Vasilopoulou, Mediterranean diet and longevity, *British Journal of Nutrition* 84 (2000): S205–S209.

15. H. Hu and coauthors, Antioxidants may contribute in the fight against ageing: An in vitro model, *Mechanisms of Ageing and Development* 121 (2001): 217–230.

16. A. Heiat, V. Vaccarino, and H. M. Krumholz, An evidence-based assessment of federal guidelines for overweight and obesity as they apply to elderly persons, *Archives of Internal Medicine* 161 (2001): 1194–1203.

17. Surveillance for selected public health indicators affecting older adults—United States, *Morbidity and Mortality Weekly Report* 48 (1999): 94–95.

18. S. B. Roberts, Energy regulation and aging: Recent findings and their implications, *Nutrition Reviews* 58 (2000): 91–97; M. J. Toth and E. T. Poehlman, Energetic adaptation to chronic disease in the elderly, *Nutrition Reviews* 58 (2000): 61–66.

19. A. B. Newman and coauthors, Weight change in old age and its association with mortality, *Journal of the American Geriatrics Society* 49 (2001): 1309–1318; F. Landi and coauthors, Body mass index and mortality among older people living in the community, *Journal of the American Geriatrics Society* 47 (1999): 1072–1076.

20. H. K. Kamel, Sarcopenia and aging, *Nutrition Reviews* 61 (2003): 157–167; C. W. Bales and C. S. Ritchie, Sarcopenia, weight loss, and nutritional frailty in the elderly, *Annual Review of Nutrition* 22 (2002): 309–323; R. Roubenoff and C. Castaneda, Sarcopenia—Understanding the dynamics of aging muscle, *Journal of the American Medical Association* 286 (2001): 1230–1231.

21. R. D. Hansen and B. J. Allen, Habitual physical activity, anabolic hormones, and potassium content of fat-free mass in postmenopausal women, *American Journal of Clinical Nutrition* 75 (2002): 314–320; R. N. Baumgartner and coauthors, Predictors of skeletal muscle mass in elderly men

and women, *Mechanisms of Ageing and Development* 107 (1999): 123–136.

22. G. Ravaglia and coauthors, Effect of micronutrient status on natural killer cell immune function in healthy free-living subjects aged ≥ 90 y, *American Journal of Clinical Nutrition* 71 (2000): 590–598.

23. M. J. M. Chin a Paw and coauthors, Immunity in frail elderly: A randomized controlled trial of exercise and enriched foods, *Medicine and Science in Sports and Exercise* 32 (2000): 2005–2011.

24. C. G. MacIntosh and coauthors, Effects of age on concentrations of plasma cholecystokinin, glucagon-like peptide 1, and peptide YY and their relation to appetite and pyloric motility, *American Journal of Clinical Nutrition* 69 (1999): 999–1006.

25. N. R. Sahyoun, C. L. Lin, and E. Krall, Nutritional status of the older adult is associated with dentition status, *Journal of the American Dietetic Association* 103 (2003): 61–66.

26. Position of the American Dietetic Association: Nutrition, aging, and the continuum of care, *Journal of the American Dietetic Association* 100 (2000): 580–595.

27. D. G. Blazer, Depression in late life: Review and commentary, *Journals of Gerontology: Series A, Biological Sciences and Medical Sciences* 58 (2003): 249–265; J. Unutzer, M. L. Bruce, and NIMH Affective Disorders Workgroup, The elderly, *Mental Health Services Research* 4 (2002): 245–247.

28. T. Hungerford and coauthors, Trends in the economic status of the elderly, 1976–2000, *Social Security Bulletin* 64 (2001): 12–22.

29. N. L. Crogan and A. Pasvogel, The influence of protein-calorie malnutrition on quality of life in nursing homes, *Journals of Gerontology: Series A, Biological Sciences and Medical Sciences* 58 (2003): 159–164; Y. Guigoz, S. Lauque, and B. J. Vellas, Identifying the elderly at risk for malnutrition: The Mini Nutritional Assessment, *Clinics in Geriatric Medicine* 18 (2002): 737–757; W. O. Seiler, Clinical pictures of malnutrition in ill elderly subjects, *Nutrition* 17 (2001): 496–498.

30. D. H. Holben and coauthors, Fluid intake compared with established standards and symptoms of dehydration among elderly residents of a long-term-care facility, *Journal of the American Dietetic Association* 99 (1999): 1447–1450.

31. M. M. G. Wilson, R. Purushothaman, and J. E. Morley, Effect of liquid dietary supplements on energy intake in the elderly, *American Journal of Clinical Nutrition* 75 (2002): 944–947.

32. Position of the American Dietetic Association: Liberalized diets for older adults in long-term care, *Journal of the American Dietetic Association* 102 (2002): 1316–1323.

33. Committee on Dietary Reference Intakes, *Dietary Reference Intakes for Energy, Carbohydrate, Fiber, Fat, Fatty Acids, Cholesterol, Protein, and Amino Acids* (Washington, D.C.: National Academies Press, 2002).

34. M. A. Johnson and coauthors, Hyperhomocysteinemia and vitamin B-12 deficiency in elderly using Title IIIc nutrition services, *American Journal of Clinical Nutrition* 77 (2003): 211–220; C. Ho, G. P. A. Kauwell, and L. B. Bailey, Practitioners' guide to meeting the vitamin B-12 Recommended Dietary Allowance for people aged 51 years and older, *Journal of the American Dietetic Association* 99 (1999): 725–727; H. W. Baik and R. M. Russell, Vitamin B12 deficiency in the elderly, *Annual Review of Nutrition* 19 (1999): 357–377.

35. Committee on Dietary Reference Intakes, *Dietary Reference Intakes for Thiamin, Riboflavin, Niacin, Vitamin B6, Folate, Vitamin*

B12, *Pantothenic Acid, Biotin, and Choline* (Washington, D.C.: National Academy Press, 2000), p. 338.

36. Committee on Dietary Reference Intakes, *Dietary Reference Intakes for Calcium, Phosphorus, Magnesium, Vitamin D, and Fluoride* (Washington, D.C.: National Academy Press, 1997).

37. Committee on Dietary Reference Intakes, 1997.

38. D. J. Fleming and coauthors, Dietary factors associated with the risk of high iron stores in the elderly Framingham Heart Study cohort, *American Journal of Clinical Nutrition* 76 (2002): 1375–1384; D. J. Fleming and coauthors, Iron status of the free-living, elderly Framingham Heart Study cohort: An iron-replete population with a high prevalence of elevated iron stores, *American Journal of Clinical Nutrition* 73 (2001): 638–646.

39. G. J. Izaks, R. G. J. Westendorp, and D. L. Knook, The definition of anemia in older persons, *Journal of the American Medical Association* 281 (1999): 1714–1717.

40. R. H. Fletcher and K. M. Fairfield, Vitamins for chronic disease prevention in adults, *Journal of the American Medical Association* 287 (2002): 3127–3129; W. C. Willett and M. J. Stampfer, What vitamins should I be taking, doctor? *New England Journal of Medicine* 345 (2001): 1819–1824.

41. The REACT Group, The Roche European American Cataract Trial (REACT): A randomized clinical trial to investigate the efficacy of an oral antioxidant micronutrient mixture to slow progression of age-related cataract, *Ophthalmic Epidemiology* 9 (2002): 49–80; A. Taylor and coauthors, Long-term intake of vitamins and carotenoids and odds of early age-related cortical and posterior subcapsular lens opacities, *American Journal of Clinical Nutrition* 75 (2002): 540–549; L. Brown and coauthors, A prospective study of carotenoid intake and risk of cataract extraction in US men, *American Journal of Clinical Nutrition* 70 (1999): 517–524; L. Chason-Taber and coauthors, A prospective study of carotenoid and vitamin A intake and risk of cataract extraction in US women, *American Journal of Clinical Nutrition* 70 (1999): 509–516.

42. D. A. Schaumberg and coauthors, Relations of body fat distribution and height with cataract in men, *American Journal of Clinical Nutrition* 73 (2000): 1495–1502.

43. J. L. Gottlieb, Age-related macular degeneration, *Journal of the American Medical Association* 288 (2002): 2233–2236.

44. N. I. Krinsky, J. T. Landrum, and R. A. Bone, Biologic mechanisms of the protective role of lutein and zeaxanthin in the eye, *Annual Review of Nutrition* 23 (2003): 171–201; Age-Related Eye Disease Study Research Group, Antioxidants and zinc to prevent progression of age-related macular degeneration, *Journal of the American Medical Association* 286 (2001): 2466–2468.

45. E. Cho and coauthors, Prospective study of dietary fat and the risk of age-related macular degeneration, *American Journal of Clinical Nutrition* 73 (2001): 209–218.

46. Prevalence of arthritis—United States, 1997, *Morbidity and Mortality Weekly Report* 50 (2001): 334–336.

47. L. Skoldstam, L. Hagfors, and G. Johansson, An experimental study of a Mediterranean diet intervention for patients with rheumatoid arthritis, *Annals of the Rheumatic Diseases* 62 (2003): 208–214; A. Linos and coauthors, Dietary factors in relation to rheumatoid arthritis: A role for olive oil and cooked vegetables, *American Journal of Clinical Nutrition* 70 (1999): 1077–1082;

J. Kjeldsen-Kragh, Rheumatoid arthritis treated with vegetarian diets, *American Journal of Clinical Nutrition* 70 (1999): 594S–600S.

48. L. Cleland, M. James, and S. Proudman, The role of fish oils in the treatment of rheumatoid arthritis, *Drugs* 63 (2003): 845–853; J. M. Kremer, n-3 Fatty acid supplements in rheumatoid arthritis, *American Journal of Clinical Nutrition* 71 (2000): 349S–351S.

49. J. R. Cerhan and coauthors, Antioxidant micronutrients and risk of rheumatoid arthritis in a cohort of older women, *American Journal of Epidemiology* 157 (2003): 345–354; S. Tidow-Kebritchi and S. Mobarhan, Effects of diets containing fish oil and vitamin E on rheumatoid arthritis, *Nutrition Reviews* 59 (2001): 335–341.

50. T. E. McAlindon and coauthors, Glucosamine and chondroitin for treatment of osteoarthritis: A systematic quality assessment and meta-analysis, *Journal of the American Medical Association* 283 (2000): 1469–1475.

51. R. J. Kaplan and coauthors, Dietary protein, carbohydrate, and fat enhance memory performance in the healthy elderly, *American Journal of Clinical Nutrition* 74 (2001): 687–693; L. Dye, A. Lluch, and J. E. Blundell, Macronutrients and mental performance, *Nutrition* 16 (2000): 1021–1034; J. D. Fernstrom, Can nutrient supplements modify brain function? *American Journal of Clinical Nutrition* 71 (2000): 1669S–1673S.

52. R. J. Wurtman and coauthors, Effects of normal meals rich in carbohydrates or proteins on plasma tryptophan and tyrosine ratios, *American Journal of Clinical Nutrition* 77 (2003): 128–132.

53. S. J. Duthie and coauthors, Homocysteine, B vitamin status, and cognitive function in the elderly, *American Journal of Clinical*

Nutrition 75 (2002): 908–913; J. Selhub and coauthors, B vitamins, homocysteine, and neurocognitive function in the elderly, *American Journal of Clinical Nutrition* 71 (2000): 614S–620S.

54. J. L. Cummings and G. Cole, Alzheimer disease, *Journal of the American Medical Association* 287 (2002): 2335–2338.

55. Y. Christen, Oxidative stress and Alzheimer disease, *American Journal of Clinical Nutrition* 71 (2000): 621S–629S.

56. M. J. Engelhart and coauthors, Dietary intake of antioxidants and risk of Alzheimer disease, *Journal of the American Medical Association* 287 (2002): 3223–3229; M. C. Morris, Dietary intake of antioxidant nutrients and the risk of incident Alzheimer disease in a biracial community study, *Journal of the American Medical Association* 287 (2002): 3230–3237; M. Grundman, Vitamin E and Alzheimer disease: The basis for additional clinical trials, *American Journal of Clinical Nutrition* 71 (2000): 630S–636S.

57. D. Schenk and coauthors, Immunization with amyloid-beta attenuates Alzheimer-disease-like pathology in the PDAPP mouse, *Nature* 400 (1999): 173–177; R. Vassar and coauthors, Beta-secretase cleavage of Alzheimer's amyloid precursor protein by the transmembrane aspartic protease BACE, *Science* 286 (1999): 735; I. Hussain and coauthors, Identification of novel aspartic protease (Asp 2) as beta-secretase, *Molecular and Cellular Neuroscience* 14 (1999): 419–427.

58. K. L. Davis and coauthors, Cholinergic markers in elderly patients with early signs of Alzheimer disease, *Journal of the American Medical Association* 281 (1999): 1401–1406.

59. S. Seshadri and coauthors, Plasma homocysteine as a risk factor for dementia and Alzheimer's disease, *New England Journal of*

Medicine 346 (2002): 476–483; D. Snowdon and coauthors, Serum folate and the severity of atrophy of the neocortex in Alzheimer disease: Findings from the Nun Study, *American Journal of Clinical Nutrition* 71 (2000): 993–998; D. G. Weir and A. M. Molloy, Microvascular disease and dementia in the elderly: Are they related to hyperhomocysteinemia? *American Journal of Clinical Nutrition* 71 (2000): 859–860; J. W. Miller, Homocysteine and Alzheimer's disease, *Nutrition Reviews* 57 (1999): 126–129.

60. R. Mayeux and M. Sand, Treatment of Alzheimer's disease, *New England Journal of Medicine* 341 (1999): 1670–1679.

61. S. Gillette-Guyonnet and coauthors, Weight loss in Alzheimer disease, *American Journal of Clinical Nutrition* 71 (2000): 637S–642S; E. T. Poehlman and R. V. Dvorak, Energy expenditure, energy intake, and weight loss in Alzheimer disease, *American Journal of Clinical Nutrition* 71 (2000): 650S–655S; S. Rivière and coauthors, Nutrition and Alzheimer's disease, *Nutrition Reviews* 57 (1999): 363–367.

62. V. A. Freedman, L. G. Martin, and R. F. Schoeni, Recent trends in disability and functioning among older adults in the United States: A systematic review, *Journal of the American Medical Association* 288 (2002): 3137–3146; Y. Liao and coauthors, Quality of the last year of life of older adults: 1986 vs 1993, *Journal of the American Medical Association* 283 (2000): 512–518.

ANSWERS

Study Questions (multiple choice)

1. c   2. d   3. c   4. d   5. b   6. b   7. c   8. d   9. c   10. b

Nutrient-Drug Interactions

People over the age of 65 take about one-third of all the over-the-counter and prescription drugs sold in the United States. They receive an average of 13 prescriptions a year and may take as many as six drugs at a time. Most often, they take drugs for heart disease, but also to treat arthritis, respiratory problems, and gastrointestinal disorders. They often go to different doctors for each of these different conditions and receive different prescriptions from each. To avoid harmful drug interactions, they need to inform all of their physicians and pharmacists of all the medicines being taken. These medicines enable people of all ages to enjoy better health, but they also bring side effects and risks.

This highlight focuses on some of the nutrition-related consequences of medical drugs, both prescription drugs and nonprescription (over-the-counter) drugs. Highlight 7 described the relationships between nutrition and the drug alcohol, and Highlight 18 presents information on herbal supplements and other alternative therapies.

The Actions of Drugs

Most people think of drugs either as medicines that help them recover from illnesses or as illegal substances that lead to bodily harm and addiction. Actually, both uses of the term *drug* are correct because any substance that modifies one or more of the body's functions is, technically, a drug. Even medical drugs set in motion both desirable and undesirable events within the body.

Consider aspirin. One action of aspirin is to limit the production of certain prostaglandins. Some prostaglandins help to produce fevers, some sensitize pain receptors, some cause contractions of the uterus, some stimulate digestive tract motility, some control nerve impulses, some regulate blood pressure, some promote blood clotting, and some cause inflammation. By interfering with prostaglandin actions, aspirin reduces fever and inflammation, relieves pain, and slows blood clotting, among other things.

A person cannot use aspirin to produce one of its effects without producing all of its other effects. Someone who is prone to strokes and heart attacks might take aspirin to prevent blood clotting, but it would also dull that person's sense of pain. Another person who took aspirin only for pain would also experience slow blood clotting. The anticlotting effect might be dangerous if it caused abnormal bleeding. A single two-tablet dose of aspirin doubles the bleeding time of wounds, an effect that lasts from four to seven hours. For this reason, physicians caution clients to refrain from taking aspirin before surgery.

The Interactions between Drugs and Nutrients

Hundreds of drugs and nutrients interact, and these interactions can lead to nutrient imbalances or interfere with drug effectiveness.[1] Adverse nutrient-drug interactions are most likely if drugs are taken over long periods, if several drugs are taken, or if nutrition status is poor or deteriorating. Understandably, then, elderly people with chronic diseases are most vulnerable. Studies of institutionalized elderly people confirm that multiple drug use significantly impairs nutrition status in this population.

Nutrients and medications may interact in many ways:

- Drugs can alter food intake and the absorption, metabolism, and excretion of nutrients.
- Foods and nutrients can alter the absorption, metabolism, and excretion of drugs.

The following paragraphs describe these interactions, and Table H17-1 (on p. 609) summarizes this information and provides specific examples.

Altered Food Intake

Many medicines can lead to malnutrition by interfering with food intake. Drugs can influence appetite, alter taste or smell, cause sores or irritation in the mouth, reduce the flow of saliva, or induce nausea or vomiting. Amphetamines used to treat depression provide an example. They suppress appetite, alter taste perceptions, dry the mouth, and cause nausea. Conversely, some medicines stimulate the appetite and lead to undesirable weight gain. An example is astemizole (Hismanal), an antihistamine used to relieve allergy symptoms.

TABLE H17-1 Mechanisms and Examples of Nutrient-Drug Interactions

Drugs Can Alter Food Intake by:

- Altering the appetite (amphetamines suppress the appetite).
- Interfering with taste or smell (methotrexate changes taste sensations).
- Inducing nausea or vomiting (digitalis can do both).[a]
- Changing the oral environment (phenobarbital can cause dry mouth).[b]
- Irritating the GI tract (cyclophosphamide induces mucosal ulcers).[c]
- Causing sores or inflammation of the mouth (methotrexate can cause painful mouth ulcers).

Drugs Can Alter Nutrition Absorption by:

- Changing the acidity of the digestive tract (antacids can interfere with iron absorption).
- Altering digestive juices (cimetidine can improve fat absorption).[d]
- Altering motility of the digestive tract (laxatives speed motility, causing the malabsorption of many nutrients).
- Inactivating enzyme systems (neomycin may reduce lipase activity).[e]
- Damaging mucosal cells (chemotherapy can damage mucosal cells).
- Binding to nutrients (antacids bind phosphorus).

Foods Can Alter Drug Absorption by:

- Changing the acidity of the digestive tract (candy can change the acidity, thus dissolving slow-acting asthma medication too quickly).
- Stimulating secretion of digestive juices (griseofulvin is absorbed better when taken with foods that stimulate the release of digestive enzymes).
- Altering rate of absorption (aspirin is absorbed more slowly when taken with food).
- Binding to drugs (calcium binds to tetracycline, limiting drug absorption).
- Competing for absorption sites in the intestines (dietary amino acids interfere with levodopa absorption this way).[f]

Drugs and Nutrients Can Interact and Alter Metabolism by:

- Acting as structural analogs (as anticoagulants and vitamin K do).
- Competing with each other for metabolic enzyme systems (as phenobarbital and folate do).[b]
- Altering enzyme activity and contributing pharmacologically active substances (as monoamine oxidase inhibitors and tyramine do).

Drugs Can Alter Nutrient Excretion by:

- Altering reabsorption in the kidneys (some diuretics increase the excretion of sodium and potassium).
- Displacing nutrients from their plasma protein carriers (aspirin displaces folate).

Foods Can Alter Drug Excretion by:

- Changing the acidity of the urine (vitamin C can alter urinary pH and limit the excretion of aspirin).

NOTE: Most of these drugs are mentioned in the text; others are introduced in their respective footnotes.
[a]Digitalis is used in the treatment of congestive heart failure.
[b]Phenobarbital is an anticonvulsant used in the treatment of epilepsy.
[c]Cyclophosphamide is an immunosuppressant used in the treatment of organ transplants.
[d]Cimetidine is an H2 blocker used in the treatment of ulcers.
[e]Neomycin is an antibiotic used in the treatment of infections.
[f]Levodopa is used in the treatment of Parkinson's disease.

Altered Nutrient Absorption

Laxatives provide an example of how drugs can interfere with nutrient absorption. Laxatives cause foods to move so rapidly through the intestine that many vitamins do not have enough time to be absorbed. The use of mineral oil as a laxative robs the person of the fat-soluble vitamins, most notably, vitamin D. The vitamins from foods dissolve in the indigestible oil and are excreted; calcium, too, is lost this way. A person who uses laxatives daily for a long time may find that the intestines can no longer function without them. Such dependence can lead to malnutrition.

Altered Drug Absorption

A classic example of how foods can interfere with drug absorption is the interactions between the antibiotic tetracycline and the minerals calcium and iron. When calcium and tetracycline, or iron and tetracycline, are taken at the same time, they bind to each other, thus reducing the absorption of both. People are therefore advised not to take tetracycline with milk, milk products, or calcium-containing antacids, such as Tums. Similarly, iron supplements should be taken two hours apart from tetracycline doses.

Another example of how foods can interfere with the absorption of a drug is the interaction between acidic foods and the nicotine gum that people sometimes use to help quit smoking cigarettes. Certain acid-containing foods and beverages interfere with the absorption of nicotine through the lining of the mouth into the blood (see Table H17-2). For maximum effectiveness, people should refrain from ingesting foods and beverages for 15 minutes before, and while, chewing the gum. When a food or beverage blocks nicotine's absorption from the mouth, the person swallows the nicotine, and this may cause nausea and hiccups as well as interfere with the drug's effectiveness.

Some medications are absorbed better with foods than without them. For this reason, the antifungal drug griseofulvin is always given with meals. Similarly, a glass of grapefruit juice significantly enhances the absorption of several common medications; because it can potentiate the effects of these drugs, grapefruit juice may be restricted when these drugs are taken. In many cases, though, foods delay the rate at which drugs are absorbed. In some instances this, too, can be helpful. An aspirin

TABLE H17-2 Foods and Beverages That Limit the Effectiveness of Nicotine Gum

• Apple Juice	• Lemon-lime soda
• Beer	• Mustard
• Catsup	• Orange juice
• Coffee	• Pineapple juice
• Colas	• Soy sauce
• Grape Juice	• Tomato juice

Methotrexate (a drug used in the treatment of cancer and rheumatoid arthritis) is structurally similar to the B vitamin folate. When this medication is used, it competes for the enzyme that normally activates folate, creating a secondary deficiency of folate. Notice the similarities in their chemical structures.

Folate

Methotrexate

taken on an empty stomach works faster than when it is given with food, but because aspirin can irritate the GI tract, taking it with food can reduce nausea and prevent bleeding.

Altered Metabolism

To appreciate how nutrient-drug interactions can affect metabolism, consider medicines that resemble vitamins in structure. Vitamin K and the anticlotting medication warfarin (Coumadin) provide an example. Warfarin opposes clotting by interfering with vitamin K's action. To be effective, the warfarin dose must be large enough to counteract vitamin K from the diet. If a person's vitamin K intake increases, as occurs when green leafy vegetables are abundant in the diet, then the physician must increase the drug dose. To avoid potential problems, people taking warfarin should try to consume the same amount of vitamin K every day, and that amount should meet current dietary recommendations.[2]

Another example is methotrexate, used to treat cancer and rheumatoid arthritis. Structurally similar to the B vitamin folate, methotrexate displaces folate and causes a folate deficiency (see Figure H17-1). Because of this, the risks of neural tube defects and cardiovascular disease increase whenever people take drugs that act as folate antagonists.[*][3] Aspirin can also alter folate metabolism, but in a different way. Aspirin competes with folate for its protein carrier, thus interfering with the body's use of the vitamin. When aspirin is used over long periods of time, health care professionals should ensure that either the diet or supplements supply sufficient folate to meet the added demands.

[*]Other folate antagonists include aminopterin, sulfasalazine, pyrimethamine, trimethoprim, triamterene, carbamazepine, phenytoin, phenobarbital, and primidone.

The effects of tyramine provide another example of a substance in foods that alters a drug's action. Tyramine is a substance found in some foods, and it interacts with monoamine oxidase inhibitors (MAO inhibitors), which are prescribed to treat certain forms of severe depression. MAO inhibitors block the action of the enzyme in the brain that normally inactivates tyramine. When people take the drug, the enzyme fails to act. Thus tyramine remains active and stimulates the release of the neurotransmitter norepinephrine. This action can lead to severe hypertension and headaches. If blood pressure rises high enough, it can be fatal. For this reason, people taking MAO inhibitors must restrict their intakes of foods rich in tyramine (see Table H17-3).

Sometimes the combination of a specific food and a drug improves the drug's action in the body. Citrus fruits, for example, enhance the effectiveness of the anticancer drug tamoxifen. Oranges, grapefruits, and tangerines contain flavonoids that assist the drug in halting the growth of cancer cells.

Altered Drug Excretion

The acidity of the urine affects the reabsorption of drugs back into the blood by the kidneys. An acidic urine limits the excretion of acidic drugs like aspirin. Some nutrients, such as

TABLE H17-3   Foods Restricted in a Tyramine-Controlled Diet

Beverages:	Red wines including chianti, sherry[a]
Cheeses:	Aged cheeses, American, camembert, cheddar, gouda, gruyère, mozzarella, parmesan, provolone, romano, roquefort, stilton[b]
Meats:	Liver; dried, salted, smoked, or pickled fish; sausage, pepperoni; dried meats
Vegetables:	Fava beans; Italian broad beans; sauerkraut; fermented pickles and olives
Other:	Brewer's yeast;[c] all aged and fermented products; soy sauce in large amounts; cheese-filled breads, crackers, and desserts; salad dressings containing cheese

NOTE: The tyramine contents of foods vary from product to product depending on the methods used to prepare, process, and store the food. In some cases, as little as 1 ounce of cheese can cause a severe hypertensive reaction in people taking monoamine oxidase inhibitors. In general, the following foods contain small enough amounts of tyramine that they can be consumed in small quantities: ripe avocado, banana, yogurt, sour cream, acidophilus milk, buttermilk, raspberries, and peanuts.
[a]Most wine and domestic beer can be consumed in small quantities.
[b]Unfermented cheeses, such as ricotta, cottage cheese, and cream cheese, are allowed.
[c]Products made with baker's yeast are allowed.

vitamin C, can lower the pH of urine, making it more acidic. Therefore, large doses of vitamin C given with aspirin increase the urine's acidity, keeping aspirin in the blood longer.

Altered Nutrient Excretion

Medicines can also alter urinary excretion of nutrients. For example, some diuretics accelerate the excretion of the minerals calcium, potassium, magnesium, and zinc.

Other Ingredients in Drugs

Besides the active ingredients, medicines may contain other substances such as sugar, sorbitol, lactose, and sodium. For most people who use medicines on occasion and in small amounts, such ingredients pose no problem. When medicines are taken regularly or in large doses, however, people on special diets may need to be aware of these additional ingredients and their effects.

Many liquid preparations contain sugar or sorbitol to make them taste better. For people who must regulate their intakes of carbohydrates, such as people with diabetes, the amount of sugar in these medicines may need to be considered. Large doses of liquids containing sorbitol may cause diarrhea. The lactose added as filler to some medications may cause problems for people who are lactose intolerant.

Antibiotics and antacids often contain sodium. People who take Alka Seltzer, for example, may not realize that a single two-tablet dose may exceed their recommended sodium intake for a whole day. In addition, antacids neutralize stomach acid, and many nutrients depend on acid for their digestion. Taking any antacid regularly will reduce the absorption of many nutrients.

The Health Professional and Nutrient-Drug Interactions

Hundreds of nutrient-drug interactions have been identified, and information continues to accumulate. It would be difficult, if not impossible, to remember all the potential effects of these interactions on nutrition status. Instead, health care professionals would serve their clients well to:

- Keep in mind that nutrient-drug interactions can and do occur, especially when medicine use is long term.
- Record and review drug and diet histories of clients with potential interactions in mind.
- Be aware of groups of people who are likely to develop drug-related nutrient deficiencies; be prepared to look up the nutrition effects of medications that these clients are taking.
- Reassess nutrition status frequently for high-risk clients.[4]
- Become familiar with the nutrient interactions of drugs commonly used to treat the disorders of their clients.
- Provide accurate information about nutrient-drug interactions and clarify common misconceptions.[5]

Nutrient interactions and risks are not unique to prescription drugs. People who buy over-the-counter drugs also need to protect themselves. The increasing availability of over-the-counter drugs allows people to treat themselves for many ailments from arthritis to yeast infections. Consumers need to ask their physicians about potential interactions and check with their pharmacists for instructions on taking drugs with foods. Should problems arise, they should seek professional care without delay.

REFERENCES

1. J. M. Sorensen, Herb-drug, food-drug, nutrient-drug, and drug-drug interactions: Mechanisms involved and their medical implications, *Journal of Alternative and Complementary Medicine* 8 (2002): 293–308.
2. S. L. Booth and M. A. Centurelli, Vitamin K: A practical guide to the dietary management of patients on warfarin, *Nutrition Reviews* 57 (1999): 288–296.
3. S. Hernández-Díaz and coauthors, Folic acid antagonists during pregnancy and the risk of birth defects, *New England Journal of Medicine* 343 (2000): 1608–1614.
4. M. G. Sanford and coauthors, Protocols for identifying drug-nutrient interactions in patients: The role of the dietitian, *Journal of the American Dietetic Association* 102 (2002): 729–731.
5. K. E. Anderson and D. J. Greenblatt, Assessing and managing drug-nutrient interactions, *Journal of the American Pharmaceutical Association* 42 (2002): S28–S29.

Chapter | 18

Diet and Health

Chapter Outline

Nutrition Explorer CD-ROM Outline

© Tif Hunter/Stone/Getty Images

Nutrition in Your Life

You've heard it all before. Eat more veggies. Eat more fiber. Eat more fish. Put down the saltshaker. Limit the fat. Be active. Don't smoke. And don't drink too much alcohol. What's the deal? If you follow this advice, will it really make a difference in how well or how long you live? In a word, yes. You can bet your life on it. If you could grow old in good health without having a heart attack or stroke, or getting diabetes, hypertension, or cancer, wouldn't you be willing to do just about anything—including improving your diet and activity habits? Of course, you would. And you can start today.

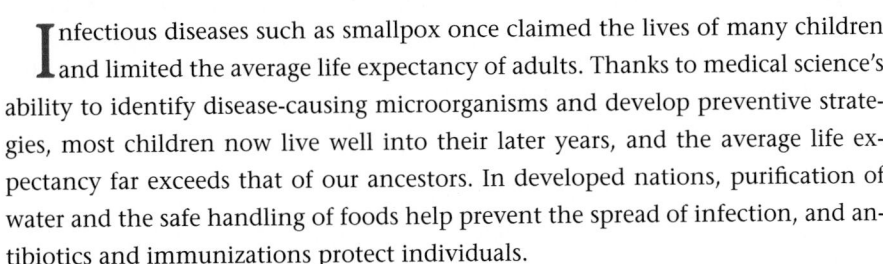

Infectious diseases such as smallpox once claimed the lives of many children and limited the average life expectancy of adults. Thanks to medical science's ability to identify disease-causing microorganisms and develop preventive strategies, most children now live well into their later years, and the average life expectancy far exceeds that of our ancestors. In developed nations, purification of water and the safe handling of foods help prevent the spread of infection, and antibiotics and immunizations protect individuals.

Despite these advances, some infectious diseases still endanger many lives today. Growing threats around the globe include **bioterrorism,** the emergence of new diseases such as SARS (sudden acute respiratory syndrome), and disease strains such as tuberculosis and some foodborne infections that have become resistant to antibiotics.[1] Although government security and public health measures such as emergency preparedness, safe food and water supplies, and medical care do much to contain infectious diseases, people are exposed to millions of microbes each day. Nutrition cannot directly prevent or cure infectious diseases, but good nutrition can strengthen, and malnutrition can weaken, the body's defenses against them.

bioterrorism: the intentional spreading of disease-causing microorganisms or toxins.

■ Other lifestyle factors that contribute to the development of chronic diseases:
- Physical inactivity.
- Overweight.
- Tobacco use.
- Alcohol and drug abuse.

This chapter begins with a description of the immune system and the relationships between nutrition and infectious diseases, but the bulk of the chapter focuses on the chronic diseases that pose the greatest threat to the lives of most people in developed countries. These chronic diseases develop over a lifetime as a result of metabolic abnormalities induced by such factors as genetics, age, gender, and lifestyle. Diet is among the many lifestyle factors■ that influence the development of chronic diseases.[2]

Nutrition and Infectious Diseases

It is difficult to know exactly where infectious diseases fall among the leading causes of death. Compared with chronic diseases, infectious diseases pose a much greater challenge for public health officials tracking prevalence. One physician might classify an ear infection as an infectious disease, while another calls it a disease of the ear; similarly, loss of hearing due to an infection, though not an infectious condition itself, may be assigned to the infectious disease category. Trends change quickly as well. A disease, such as AIDS, that did not even exist until the early 1980s may suddenly appear and become one of the leading causes of death. A preventive strategy, such as food irradiation, may just as quickly eliminate hundreds of thousands of cases of foodborne infections each year. Public health strategies help the entire country defend against the spread of infection; each individual's immune system provides a personal line of defense.

The Immune System

The **immune system** defends the body so diligently and silently that people do not even notice the thousands of enemy attacks mounted against them every day (the accompanying glossary defines immune system terms). If the immune system fails, though, the body suddenly becomes vulnerable to every wayward disease-causing agent that comes its way; infectious disease invariably follows.

The body's first lines of defense against foreign substances—the skin, mucous membranes, and GI tract—normally deter invaders. If an invader penetrates these barriers and gains entry into the body, then the organs■ and cells of the immune system race into action. Foreign substances that elicit such a response are called **antigens**. Examples include bacteria, viruses, toxins, and food proteins that cause allergies.

Of the 100 trillion cells that make up the human body, one in every hundred is a white blood cell. Two types of white blood cells,■ the phagocytes and lymphocytes, defend the body against infectious diseases.

■ Organs of the immune system:
- Spleen.
- Lymph nodes.
- Thymus.

■ Cells of the immune system:
- Phagocytes.
- Lymphocytes:
  - B-cells.
  - T-cells.

GLOSSARY OF IMMUNE SYSTEM TERMS

B-cells: lymphocytes that produce antibodies. *B* stands for bursa, an organ in the chicken associated with the first identification of the B-cells.

cytokines (SIGH-toe-kines): special proteins that direct immune and inflammatory responses.

immune system: the body's natural defense system against foreign materials that have penetrated the skin or mucous membranes.

immunoglobulins (IM-you-noh-GLOB-you-linz): proteins capable of acting as antibodies.

lymphocytes (LIM-foh-sites): white blood cells that participate in acquired immunity; B-cells and T-cells.

phagocytes (FAG-oh-sites): white blood cells (neutrophils and macrophages) that have the ability to ingest and destroy foreign substances.
- **phagein** = to eat

phagocytosis (FAG-oh-sigh-TOH-sis): the process by which phagocytes engulf and destroy foreign materials.

T-cells: lymphocytes that attack antigens. *T* stands for the thymus gland, where the T-cells are stored for a while.

Reminder: *Antibodies* are large proteins of the blood and body fluids, produced by the immune system in response to the invasion of the body by foreign molecules (usually proteins called *antigens*). Antibodies combine with and inactivate the foreign invaders, thus protecting the body.

Antigens are substances that elicit the formation of antibodies or an inflammation reaction from the immune system.

Phagocytes **Phagocytes,** the scavengers of the immune system, are the first to arrive at the scene if an invader gains entry. Upon recognizing the foreign invader, the phagocyte engulfs and digests it, if possible, in a process called **phagocytosis.**■ Phagocytes also secrete special proteins called **cytokines** that activate the metabolic and immune responses to infection.

Lymphocytes: B-cells The **lymphocytes** are of two distinct types: B-cells and T-cells. **B-cells** respond to infection by rapidly dividing and producing large proteins known as **antibodies.** Antibodies travel in the bloodstream to the site of the infection. There they stick to the surfaces of the foreign particles and kill or otherwise inactivate them, making the foreign particles easy for the phagocytes to ingest.

The antibodies are members of a class of proteins known as **immunoglobulins**—literally, large globular proteins that produce immunity. Antibodies react selectively to a specific foreign organism, and the B-cells retain a memory of how to make them. The next time the immune system encounters the same foreign organism, it can respond with greater speed than it did the first time. B-cells play a major role in resistance to infection.

Lymphocytes: T-cells The **T-cells** travel directly to the invasion site to battle the invaders. T-cells recognize the antigens displayed on the surfaces of their partner phagocyte cells and multiply in response. Then they release powerful chemicals to destroy all the foreign particles that have this antigen on their surfaces. As the T-cells begin to win the battle against infection, they release signals to slow down the immune response.

Unlike the phagocytes, which are capable of inactivating many different types of invaders, T-cells are highly specific. Each T-cell can attack only one type of antigen. This specificity is remarkable, for nature creates millions of antigens. After making enough T-cells to destroy a particular antigen, some lymphocytes retain the necessary information to serve as memory cells so that the immune system can rapidly produce the same type of T-cells again should the identical infection recur.

T-cells actively defend the body against fungi, viruses, parasites, and a few types of bacteria; they can also destroy cancer cells. T-cells participate in the rejection of newly transplanted tissues, which is why physicians prescribe immunosuppressive drugs following transplantation surgery. It is the T-cells that are inactivated by the virus that causes AIDS, as explained later.

Nutrition and Immunity

Of all the body's systems, the immune system responds most sensitively to subtle changes in nutrition status. Malnutrition compromises immunity.[3] Impaired immunity then opens the way for infectious diseases, which typically raise nutrient needs and lower food intake. Consequently, nutrition status suffers further.[4] Thus disease and malnutrition create a **synergistic** downward spiral that must be broken for recovery to occur (see Figure 18-1).

Impaired immunity is a hallmark of protein-energy malnutrition (PEM). Table 18-1 (on p. 616) presents the effects of PEM on the body's defenses. Deficiencies of vitamins and minerals also diminish the immune response, as may excesses.[5] Likewise, interactions between nutrients may enhance or impair immunity. Quite simply, optimal immunity depends on optimal nutrition—enough, but not too much, of each of the nutrients.■ People with depressed immune systems, such as the elderly, may benefit from supplements of selected nutrients.

HIV and AIDS

Perhaps the most infamous infectious disease today is **AIDS (acquired immune deficiency syndrome).** AIDS develops from infection by **HIV (human immunodeficiency virus),** which attacks the immune system and disables the body's defenses against other diseases. Then these diseases, which would produce only mild, if any, illness in people with healthy immune systems, destroy health and life. The glossary on the next page defines AIDS-related terms.

■ Two types of immune system cells ingest and destroy foreign antigens by phagocytosis: neutrophils and macrophages.

FIGURE 18-1 Nutrition and Immunity

Regardless of where a person enters the spiral, malnutrition, illness, and weakened immunity interact to compromise recovery and worsen malnutrition.

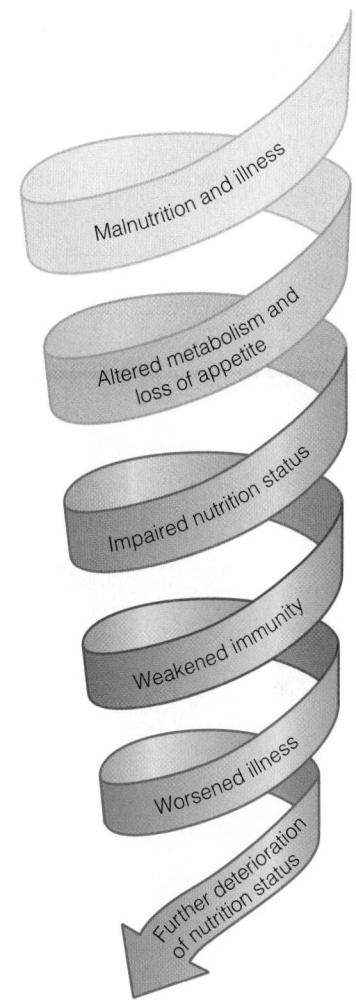

Malnutrition and illness
Altered metabolism and loss of appetite
Impaired nutrition status
Weakened immunity
Worsened illness
Further deterioration of nutrition status

■ Nutrients known to affect immunity:
- Protein.
- Fatty acids.
- Vitamin A.
- Vitamin E.
- Vitamin B_6.
- Folate.
- Vitamin C.
- Iron.
- Zinc.
- Selenium.

synergistic (SIN-er-JIST-ick): multiple factors operating together in such a way that their combined effects are greater than the sum of their individual effects.

TABLE 18-1 Effects of Protein-Energy Malnutrition (PEM) on the Body's Defense Systems

Body's Defense System	Effects of PEM
Skin	Thinned, with less connective tissue to serve as a barrier to protect underlying tissues; delayed skin sensitivity reaction to antigens.
Digestive tract and other body linings	Antibody secretions and immune cell number reduced.
Lymph tissues[a]	Immune system organs reduced in size; cells of immune defense depleted.
General response	Invader kill time prolonged; circulating immune cells reduced; antibody response impaired.

[a]Lymph tissues include the thymus gland, lymph nodes, and spleen.

TABLE 18-2 HIV and AIDS Epidemic at a Glance, 2002

	World	United States
Living with HIV or AIDS	42,000,000	850,000
Newly infected with HIV	5,000,000	40,000
AIDS deaths	3,100,000	14,500

The HIV/AIDS epidemic continues to sweep across countries, especially in sub-Saharan Africa. Table 18-2 shows its impact worldwide and in the United States. For many years, the devastating effects of HIV infection seemed unstoppable, but in the mid-to-late 1990s, the death rate in the United States from AIDS began to decline, and the progression from HIV to AIDS slowed dramatically. The disease still has no cure, but remarkable progress has been made in understanding and treating HIV infection. Without a cure, the best course is prevention. HIV is transmitted by direct contact with contaminated body fluids, including semen, vaginal secretions, and blood (but not saliva), or by passage of the infection from a mother to her infant during pregnancy, birth, or breastfeeding.

Once a person has been infected with HIV, laboratory tests can detect antibodies within three months, typically in three to four weeks. Because people remain symptom-free in the early stages of infection, however, they may not even consider being tested for HIV for several years following infection. Thus early detection to prevent the spread of HIV infection and to ensure early treatment for the person infected are important health goals.

HEALTHY PEOPLE 2010

Reduce AIDS and the number of cases of HIV infection among adolescents and adults.

How AIDS Develops

HIV infection attacks the immune system and leaves its victims defenseless against **opportunistic infections** and disorders from which most people are protected. The disorder begins with infection by the virus and progresses in stages. The virus gradually destroys cells with a specific protein called CD4+ on their surfaces. Among the cells most affected are **CD4+ T-lymphocytes,** essential components of the immune system. At first, CD4+ T-lymphocytes decline gradually, and the HIV-infected individual remains symptom-free. As the infection progresses, though, depletion of CD4+ T-lymphocytes greatly impairs immune function. Symptoms may include fatigue, skin rashes, fevers, diarrhea, muscle pain, night sweats, weight loss, oral lesions and infections, and other opportunistic infections that are not life-threatening.

GLOSSARY OF AIDS TERMS

AIDS (acquired immune deficiency syndrome): the end stage of HIV infection, in which severe complications are manifested. The cluster of mild symptoms that sometimes occurs early in the course of AIDS is called **AIDS-related complex (ARC).**

CD4+ T-lymphocytes: circulating white blood cells that contain the CD4+ protein on their surfaces and are a necessary component of the immune system.

HIV (human immunodeficiency virus): the virus that causes AIDS. The infection progresses to become an immune system disorder that leaves its victims defenseless against numerous infections.

opportunistic infections: infections from microorganisms that normally do not cause disease in the general population but can cause great harm in people once their immune systems are compromised (as in HIV infection).

wasting syndrome: an involuntary loss of more than 10% of body weight, common in AIDS and cancer.

Later, frequent and often fatal complications arise, such as severe weight loss; tuberculosis; recurrent bacterial pneumonia; serious infections of the central nervous system, GI tract, and skin; cancers; and severe diarrhea. About half of the people with an HIV infection develop AIDS within ten years, although this time varies greatly from person to person depending on such factors as nutrition and health status and medical interventions. Clinicians monitor the progression of the disease by measuring the concentrations of CD4+ T-lymphocytes and the circulating virus (called the viral load).

The HIV Wasting Syndrome

People with AIDS frequently experience malnutrition and wasting. The **wasting syndrome** often begins early in the disease and becomes progressively worse. The degree of wasting in people with AIDS, especially in the last few months before death, is similar to that seen in people who die from starvation.

The causes of malnutrition and wasting in HIV infection are related to the disease itself, its complications, and its treatments, all of which can result in inadequate nutrient intakes, malabsorption, excessive nutrient losses, and an accelerated metabolism (people with cancer often experience wasting for similar reasons).

Even when other complications ultimately cause death, malnutrition appears to be an important contributing factor. For people with AIDS, the severity of wasting may determine the duration of survival, suggesting that nutrition intervention can help extend lives by minimizing weight loss, preserving lean body mass, and strengthening the immune system.[6]

Nutrition Support for People with HIV Infections

In an era of improved treatments and prolonged survival for people with HIV infections, measures that improve the quality of life assume great importance. Attention to nutrition cannot change the ultimate outcome of an HIV infection, but it can prevent and reverse malnutrition, which may improve the quality of life and slow disease progression. Zinc deficiency, for example, is the most prevalent vitamin or mineral deficiency seen in people infected with HIV. When patients' plasma zinc concentrations are normalized, the rate of opportunistic infections declines, and disease progression slows.[7] Good nutrition status may improve a person's response to drug therapy, reduce duration of hospital stays, and promote physical independence. At a minimum, meeting nutrient needs eliminates the additional stresses imposed by malnutrition.

A specific dietary strategy for the treatment of AIDS has not been devised. Instead, practitioners rely on clinical experience to make nutrient recommendations based on the complications that arise in each case. Malnutrition, tissue wasting, fat accumulation in the belly and breast areas, and risk of additional chronic diseases are all potential problems. The primary goal is to enhance health and well-being and restore appropriate amounts of lean and fat tissue.

People who are unable to eat enough food to obtain an adequate nutrient intake might benefit from a multivitamin-mineral supplement, but prescribed medicines must be considered when recommending types and amounts of supplements. Individuals who are unable to eat enough food to prevent nutrition complications and unintentional weight loss may need more aggressive nutrition support.

People with weakened immune systems, such as those with HIV infection, are vulnerable to foodborne infections. These infections further weaken the immune system, making a simple case of food poisoning a life-threatening event. People with HIV infections must carefully follow the guidelines for buying, preparing,

The countless lives touched by AIDS serve as a potent reminder of the need to continue the search for a cure.

and storing foods safely (presented in Chapter 19). In addition, the Food and Drug Administration (FDA) warns people with HIV infections to avoid raw or undercooked seafood.[8]

With no present cure available, people with HIV infection and other life-threatening diseases often seek alternative therapies. The highlight that follows addresses the role of alternative therapies in medical care today.

> **IN SUMMARY** Everyone should adopt an effective personal strategy to prevent HIV infection. (Call the AIDS hotline for the information you need.[*]) Should a person contract HIV, nutrition intervention can help prevent malnutrition and minimize the wasting that accompanies the progression of AIDS.

Nutrition and Chronic Diseases

Figure 18-2 shows the ten leading causes of death in the United States.[9] Four of these causes, including the top three, have some relationship with diet. Taken together, these four conditions account for about two-thirds of the nation's more than 2 million deaths each year. Worldwide, statistics are similar, with developing nations sharing many of the same chronic diseases as developed nations.[10]

This chapter explains how the major chronic diseases develop and summarizes their major links with nutrition. Earlier chapters described individual nutrients' connections with diseases and may have left the mistaken impression of "one disease–one nutrient" relationships. Indeed, valid links do exist between saturated fat and heart disease, calcium and osteoporosis, and antioxidant vitamins and cancer, but focusing only on these links oversimplifies the story. In reality, each nutrient may have connections with several diseases because its role in the body is not specific to a disease, but to a body function. Furthermore, each of the chronic diseases develops in response to multiple risk factors,■ including many nondietary factors

■ Some risk factors, such as diet and physical activity, are *modifiable,* meaning that they can be changed; others, such as genetics, age, and gender, cannot be changed.

---

[*]AIDS hotline: (800) 342-AIDS.

FIGURE 18-2   The Ten Leading Causes of Death in the United States[a]

Many deaths have multiple causes, but diet influences the development of several chronic diseases—notably, heart disease, some types of cancer, stroke, and diabetes.

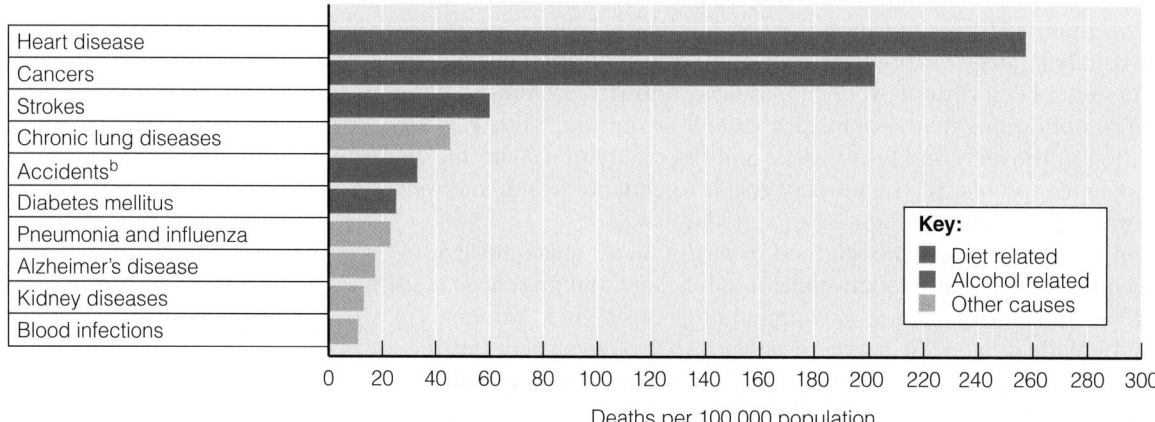

[a]Rates are age adjusted to allow relative comparisons of mortality among groups and over time.
[b]Motor vehicle and other accidents are the leading cause of death among people aged 15–24, followed by homicide, suicide, cancer, and heart disease. Alcohol contributes to about half of all accident fatalities.
SOURCE: Data from National Center for Health Statistics, 2002.

such as genetics, physical inactivity, and smoking. An integrated and balanced approach to disease prevention, therefore, includes attention to all of the many factors involved. Figure 18-3 illustrates some of the relationships between risk factors and chronic diseases.

Notice how many of the diseases have a genetic component. A family history of a certain disease is a powerful indicator of a person's tendency to contract that disease. Still, lifestyle factors are often pivotal in determining whether that tendency will be expressed. Genetics and lifestyle often work synergistically; for instance, cigarette smoking is especially likely to bring on heart disease in people who are genetically predisposed to develop it. Not smoking would benefit everyone's health, of course, regardless of genetic predisposition, but some recommendations to prevent chronic diseases best meet an individual's needs when family history is considered. For example, women with a family history of breast cancer might reduce their risks if they abstain from alcohol, whereas those with a family history of heart disease might benefit from one or two glasses of wine a week.

Vegetables rich in fiber, phytochemicals, and the antioxidant nutrients (beta-carotene, vitamin C, and vitamin E) help to protect against chronic diseases.

IN SUMMARY Heart disease, cancers, and strokes are the three leading causes of death in the United States, and diabetes also ranks among the top ten. All four of these chronic diseases have significant links with nutrition, although other lifestyle risk factors and genetics are also important.

FIGURE 18-3 Risk Factors and Chronic Diseases

	Diet Risk Factors						Other Risk Factors					
Chronic Diseases	Diet high in fat, saturated fat, and/or trans fat	Excessive alcohol intake	Low complex carbohydrate/fiber intake	Low vitamin and/or mineral intake	High sugar intake	High intake of salty or pickled foods	Genetics	Age	Sedentary lifestyle	Smoking and tobacco use	Stress	Environmental contaminants
Cancers	✔	✔	✔	✔		✔	✔	✔	✔	✔		✔
Hypertension	✔	✔		✔		✔	✔	✔	✔	✔	✔	✔
Diabetes (type 2)	✔		✔				✔	✔	✔			
Osteoporosis		✔		✔			✔	✔	✔	✔		
Atherosclerosis	✔	✔	✔	✔			✔	✔	✔	✔	✔	✔
Obesity	✔	✔	✔		✔		✔		✔			
Stroke	✔	✔	✔				✔	✔	✔	✔	✔	✔
Diverticulosis	✔		✔	✔					✔	✔		
Dental and oral disease				✔	✔		✔			✔		

This chart shows that the same risk factor can affect many chronic diseases. Notice, for example, how many diseases have been linked to a sedentary lifestyle. The chart also shows that a particular disease, such as atherosclerosis, may have several risk factors.

This flow chart shows that many of these conditions are themselves risk factors for other chronic diseases. For example, a person with diabetes is likely to develop atherosclerosis and hypertension. These two conditions, in turn, worsen each other and may cause a stroke or heart attack. Notice how all of these chronic diseases are linked to obesity.

Cardiovascular Disease

The major causes of death around the world today are diseases of the heart and blood vessels, collectively known as **cardiovascular disease (CVD).** (The accompanying glossary defines this and other heart disease terms.) In the United States, cardiovascular disease claims the lives of nearly 1 million people each year.[11]

Coronary heart disease (CHD) is the most common form of cardiovascular disease and is usually caused by **atherosclerosis** in the **coronary arteries** that supply blood to the heart muscle. Atherosclerosis is the accumulation of lipids and other materials in the arteries.

HEALTHY PEOPLE 2010

Reduce coronary heart disease deaths. Reduce stroke deaths.

How Atherosclerosis Develops

Nutrition Explorer Heart disease is the most common cardiovascular disease and usually involves atherosclerosis and hypertension. Watch an animated explanation of how these diseases develop and how each makes the other worse.

As Highlight 16 pointed out, no one is free of the fatty streaks that may one day become the **plaques** of atherosclerosis. For most adults, the question is not whether you have plaques, but how far advanced they are and what you can do to slow or reverse their progression.

Atherosclerosis or "hardening of the arteries" usually begins with the accumulation of soft fatty streaks along the inner arterial walls, especially at branch points (see Figure H16-1 on p. 582). These fatty streaks gradually enlarge and harden as they fill with lipids and minerals and become encased in fibrous connective tissue, forming plaques. Plaques stiffen the arteries and narrow the passages through them. Most people have well-developed plaques by the age of 30. As Chapter 5 pointed out, a diet high in saturated fat is a major contributor to the development of plaques and the progression of atherosclerosis.[12] But atherosclerosis is much

---

GLOSSARY OF HEART DISEASE TERMS

angina (an-JYE-nah or AN-ji-nah): a painful feeling of tightness or pressure in and around the heart, often radiating to the back, neck, and arms; caused by a lack of oxygen to an area of heart muscle.

CHD risk equivalents: disorders that raise the risk of heart attacks, strokes, and other complications associated with cardiovascular disease to the same degree as existing CHD. These disorders include symptomatic carotid artery disease, peripheral arterial disease, abdominal aortic aneurysm, and diabetes mellitus.

coronary heart disease (CHD): the damage that occurs when the blood vessels carrying blood to the heart (the **coronary arteries**) become narrow and occluded.

embolism (EM-boh-lizm): the obstruction of a blood vessel by an **embolus** (EM-boh-luss), or traveling clot, causing sudden tissue death.
- **embol** = to insert, plug

heart attack: sudden tissue death caused by blockages of vessels that feed the heart muscle; also called **myocardial** (my-oh-KAR-dee-al) **infarction** (in-FARK-shun) or **cardiac arrest.**
- **myo** = muscle
- **cardial** = heart
- **infarct** = tissue death

hypertension: higher-than-normal blood pressure. Hypertension that develops without an identifiable cause is known as **essential** or **primary hypertension**; hypertension that is caused by a specific disorder

such as kidney disease is known as **secondary hypertension.**

prehypertension: slightly higher-than-normal blood pressure, but not as high as hypertension.

stroke: an event in which the blood flow to a part of the brain is cut off; also called **cerebrovascular accident (CVA).**
- **cerebro** = brain
- **vascular** = blood vessels

thrombosis (throm-BOH-sis): the formation of a **thrombus** (THROM-bus), or a blood clot, that may obstruct a blood vessel, causing gradual tissue death.
- **thrombo** = clot

transient ischemic (is-KEY-mik) **attack (TIA):** a temporary reduction in blood flow to the brain, which causes temporary symptoms that vary depending

on the part of the brain affected. Common symptoms include light-headedness, visual disturbances, paralysis, staggering, numbness, and inability to swallow.

Reminder: *Atherosclerosis* is a type of artery disease characterized by plaques along the inner walls of the arteries.

Cardiovascular disease (CVD) is a general term for all diseases of the heart and blood vessels.

Plaques are mounds of lipid material, mixed with smooth muscle cells and calcium, that develop in the artery walls in atherosclerosis. Plaque associated with atherosclerosis is known as **atheromatous** (ATH-er-OH-ma-tus) **plaque.**

more than the simple accumulation of lipids within the artery wall—it is a complex inflammatory response to tissue damage.

Causes of Atherosclerosis The cells lining the blood vessels may incur damage from high LDL cholesterol, hypertension, toxins from cigarette smoking, elevated homocysteine, or some viral and bacterial infections.[13] Such damage increases the permeability of the blood vessel walls and elicits an inflammatory response. The immune system sends in macrophages,■ and the smooth muscle cells of the artery wall try to repair the damage. Particles of LDL cholesterol become trapped in the blood vessel walls. Free radicals produced during inflammatory responses oxidize the LDL cholesterol, and the macrophages engulf it. The macrophages swell with large quantities of oxidized LDL cholesterol and eventually become the cells of plaque. Arterial damage and the inflammatory response also favor the formation of blood clots and allow minerals to harden plaque and form the fibrous connective tissue that encapsulates it.

The recognition that atherosclerosis involves an inflammatory response that weakens the walls of the arteries has led researchers to look for signs or markers of inflammation in the blood vessel walls. The most promising of these markers is a protein known as **C-reactive protein (CRP).** In a study of 28,000 women, high levels of CRP were even more predictive of future heart attack than high LDL cholesterol, which has a strong relationship with atherosclerosis, as a later section explains.[14]

Blood Clots and Atherosclerosis Once plaques have formed, a sudden spasm or surge in blood pressure in an artery can tear away part of the fibrous coat covering a plaque. When this happens, the body responds to the damage as it would to other tissue injuries. **Platelets,** tiny disc-shaped bodies, cover the damaged area, and together with other factors, they form a clot.

The action of platelets is under the control of certain eicosanoids, known as prostaglandins and thromboxanes, which are made from the 20-carbon omega-6 and omega-3 fatty acids (introduced in Chapter 5). Each eicosanoid plays a specific role in helping to regulate■ many of the body's activities. Sometimes their actions oppose each other. For example, one eicosanoid prevents, and another promotes, clot formation; similarly, one dilates, and another constricts, the blood vessels. When omega-3 fatty acids are abundant in the diet, they make more of the kinds of eicosanoids that favor heart health.[15]

Abnormal blood clotting can trigger life-threatening events. A blood clot may stick to a plaque in an artery and gradually grow large enough to restrict or close off a blood vessel **(thrombosis).** A coronary thrombosis blocks blood flow through an artery that feeds the heart muscle. A cerebral thrombosis blocks blood flow through an artery that feeds the brain. A clot may also break free from the artery wall and travel through the circulatory system until it lodges in a small artery and suddenly shuts off flow to the tissues fed by this artery **(embolism).**

Blood Pressure and Atherosclerosis The heart must create enough pressure to push blood through the circulatory system. When arteries are narrowed by plaques, clots, or both, blood flow is restricted, and the heart must then generate more pressure to deliver blood to the tissues. This higher blood pressure further damages the artery walls, and plaques and clots are especially likely to form at damage points. Thus the development of atherosclerosis is a self-accelerating process. (A later section describes additional consequences of high blood pressure.)

The Result: Heart Attacks and Strokes When atherosclerosis in the coronary arteries becomes severe enough to restrict blood flow and deprive the heart muscle of oxygen, CHD develops. The person with CHD often experiences pain and pressure in the area around the heart **(angina).** If blood flow to the heart is cut off and that area of the heart muscle dies, a **heart attack** results. Restricted blood flow to the brain causes a **transient ischemic attack** or **stroke.** Coronary heart disease and strokes are the first and third leading causes of death, respectively, for adults in the Unites States.

■ Reminder: *Macrophages* are large, phagocytic cells of the immune system.
• **macro** = large
• **phagein** = to eat

■ Eicosanoids help to regulate:
• Blood pressure.
• Blood clot formation.
• Blood vessel contractions.
• Immune response.
• Nerve impulse transmissions.

C-reactive protein (CRP): a protein produced during the acute phase of infection or inflammation that enhances immunity by promoting phagocytosis and activating platelets. Its presence may be used to assess a person's risk of an impending heart attack or stroke.

platelets: tiny, disc-shaped bodies in the blood, important in blood clot formation.

TABLE 18-3 Risk Factors for CHD

Major Risk Factors for CHD (not modifiable)

- Increasing age.
- Male gender.
- Family history of premature heart disease.

Major Risk Factors for CHD (modifiable)

- High blood LDL cholesterol.
- Low blood HDL cholesterol.
- High blood pressure (hypertension).
- Diabetes.
- Obesity (especially abdominal obesity).
- Physical inactivity.
- Cigarette smoking.
- An "atherogenic" diet (high in saturated fats and low in vegetables, fruits, and whole grains).

NOTE: Risk factors highlighted in color have relationships with diet.
SOURCE: Expert Panel on Detection, Evaluation, and Treatment of High Blood Cholesterol in Adults (Adult Treatment Panel III), *Third Report of the National Cholesterol Education Program (NCEP)*, NIH publication no. 02-5215 (Bethesda, MD.: National Heart, Lung, and Blood Institute, 2002), pp. II-15–II-20.

Risk Factors for Coronary Heart Disease

Although atherosclerosis can develop in any blood vessel, the coronary arteries are most often affected, leading to CHD. Table 18-3 lists the major risk factors for CHD. The criteria for defining blood lipids, blood pressure, and obesity in relation to CHD risk are shown in Table 18-4; Tables H16-1 and H16-2 on p. 583 present standards for children and adolescents.

By middle age, most adults have at least one risk factor for CHD, and many have more than one.[16] Public health officials in both the United States and Canada recommend screening to identify risk factors in individuals and offer preventive advice for the population. Such public health programs are proving successful: since 1960, both blood cholesterol levels and deaths from cardiovascular disease among U.S. adults have shown a continuous and substantial downward trend.[17] These trends reflect behavior changes in individuals. As adults grow older, many of them stop smoking, limit alcohol consumption, and become mindful that their food choices can improve their cardiovascular health.

With respect to **hypertension,** a major national effort to identify and treat hypertension has resulted in the publication of new guidelines.[18] As Table 18-4 notes, values only slightly higher than desirable are now called **prehypertension.** This new classification recognizes the continuous, consistent, and independent relationship between rising blood pressure levels and increasing risk of heart attack or stroke.

Age, Gender, and Family History A review of Table 18-3 shows that three of the major risk factors for CHD cannot be modified by diet or otherwise: age, gender, and family history. As men and women grow older, the risk of CHD rises. The increasing risk of CHD with advancing age reflects the steady progression of atherosclerosis.[19] On average, older people have more atherosclerosis than younger people do.

In men, aging becomes a significant risk factor at age 45 or older. CHD occurs about 10 to 15 years later in women than in men. Women younger than 45 tend to have lower LDL cholesterol than men of the same age, but a woman's blood cholesterol typically begins to rise between ages 45 and 55. Thus aging becomes a significant risk factor for women who are 55 or older. The gender difference has been attributed to a protective effect of estrogen in women, but CHD rates do not suddenly accelerate at menopause as naturally occurring estrogen levels taper off.[20] Rather, as in men, heart disease rates increase linearly with age. And, as in men, all of the major risk factors raise the risk of CHD in women. Ultimately, CHD kills as many women as men—and kills more women in the United States than any other disease.

TABLE 18-4 Standards for CHD Risk Factors

Risk Factors	Desirable	Borderline	High
Total blood cholesterol (mg/dL)	<200	200–239	≥240
LDL cholesterol (mg/dL)	<100[a]	130–159	160–189[b]
HDL cholesterol (mg/dL)	≥60	59–40	<40
Triglycerides, fasting (mg/dL)	<150	150–199	200–499[c]
Body mass index (BMI)[d]	18.5–24.9	25–29.9	≥30
Blood pressure (systolic and/or diastolic pressure)	<120/<80	120–139/80–89[e]	≥140/≥90[f]

[a]100–129 mg/dL LDL indicates a near or above optimal level.
[b]≥190 mg/dL LDL indicates a very high risk.
[c]≥500 md/dL triglycerides indicates a very high risk.
[d]Body mass index (BMI) was defined in Chapter 8; BMI standards are found on the inside back cover.
[e]These values indicate prehypertension.
[f]These values indicate stage one hypertension; ≥160/≥100 indicates stage two hypertension. Physicians use these classifications to determine medical treatment.

Nonetheless, at every age men have a greater risk of CHD than women do. The reasons for this gender difference are not completely understood, but can be partly explained by the earlier onset of risk factors such as elevated LDL cholesterol and blood pressure in men.

A history of early CHD in immediate family members is an independent risk factor even when other risk factors are considered. The more family members affected and the earlier the age of onset, the greater the risk.[21]

High LDL and Low HDL Cholesterol In population studies, the relationship between total blood cholesterol and atherosclerosis is strong—and most of the total cholesterol is made up of LDL cholesterol. If LDL cholesterol remains in the blood after the body's cells take up the amount they need, then the excess becomes available for oxidation. The higher the LDL cholesterol, the greater the risk of CHD.

LDL are clearly the most atherogenic lipoproteins. As explained in Chapter 5, HDL also carry cholesterol, but raised HDL represent cholesterol returning from the cells to the liver and thus indicate a *reduced* risk of atherosclerosis and heart attack. High LDL and low HDL correlate *directly* with heart disease,■ whereas low LDL and high HDL correlate *inversely* with risk.

■ Cholesterol is carried in several lipoproteins, chief among them LDL and HDL (see Chapter 5 for details). Remember them this way:
  • LDL = Low-density lipoproteins = Less healthy.
  • HDL = High-density lipoproteins = Healthy.

Reduce the mean total blood cholesterol levels among adults. Reduce the proportion of adults with high total blood cholesterol levels.

HEALTHY PEOPLE 2010

How elevated LDL increase the risk of CHD remains unclear, but as mentioned earlier, LDL cholesterol plays a role in the development of fibrous plaques. When the plaques weaken and become unstable, they can rupture, causing a heart attack. Evidence shows that elevated LDL contribute to plaque instability. The goal of LDL-lowering treatment in people with advanced atherosclerosis is to stabilize plaques. In the earlier stages, the goal of treatment is to slow the development of plaque. In clinical trials, lowering LDL significantly reduces the incidence of CHD.[22]

High Blood Pressure (Hypertension) Chronic high blood pressure (hypertension) frequently accompanies atherosclerosis, diabetes, and obesity. The higher the blood pressure above normal, the greater the risk of heart disease. The relationship between hypertension and heart disease risk holds for men and women, young and old. High blood pressure injures the artery walls and accelerates plaque formation, thus initiating or worsening the progression of atherosclerosis. Then the plaques and reduced blood flow raise blood pressure further, and hypertension and atherosclerosis become mutually aggravating conditions.

Diabetes Diabetes—a major independent risk factor for all forms of cardiovascular disease—substantially increases the risk of death from CHD.[23] In diabetes, blood vessels often become blocked and circulation diminishes. Atherosclerosis progresses rapidly. For many people with diabetes, the risk of CHD is similar to that of people with established CHD.[24] In fact, physicians describe diabetes and other disorders that have risks similar to CHD as **CHD risk equivalents.** Treatment to lower LDL cholesterol in diabetes follows the same recommendations as in CHD.

Obesity and Physical Inactivity Obesity, especially abdominal obesity, and physical inactivity significantly modify several of the risk factors for CHD, contributing to high LDL cholesterol, low HDL cholesterol, hypertension, and diabetes.[25] Conversely, weight loss and physical activity protect against CHD by lowering LDL, raising HDL, improving insulin sensitivity, and lowering blood pressure. Regular physical activity also increases energy expenditure and builds lean body mass, thereby improving body composition and physical fitness.

Cigarette Smoking Cigarette smoking is a powerful risk factor for CHD and other forms of cardiovascular disease. The risk increases the more a person smokes and is the same for men and women. Smoking damages the heart directly by increasing blood pressure and the heart's workload. It deprives the heart of oxygen and damages platelets, making blood clot formation likely. Toxins in cigarette smoke damage blood vessels, setting the stage for atherosclerosis. When people quit smoking, their risk of CHD declines within a few months.[26]

Atherogenic Diet Diet influences the risk of CHD. An "atherogenic diet"—high in saturated fats and cholesterol and low in fruits and vegetables—elevates LDL cholesterol. Conversely, diets rich in fruits, vegetables, and whole grains seem to lower the risk of CHD even more than might be expected based on risk factors such as LDL cholesterol alone. The specific nutrients responsible for this benefit remain to be defined, but some of the likely contenders include the antioxidant nutrients and omega-3 fatty acids. Dietary strategies to reduce the risk of CHD are discussed in a later section.

Other Risk Factors The major risk factors for CHD listed in Table 18-3 and discussed in the previous sections have solid associations with the development of CHD. Nevertheless, other factors also seem to influence a person's risk of CHD. These factors, known as **emerging risk factors,** may be helpful in assessing an individual's risk of CHD. For example, some people with CHD, especially those with diabetes and those who are overweight, have elevated triglycerides. Whether elevated blood triglycerides represent an independent risk factor for CHD remains debatable. In the latest report by the National Cholesterol Education Program Expert Panel, elevated blood triglycerides are considered a marker for other risk factors (high LDL, low HDL, overweight, and diabetes, for example), but are not designated as a major risk factor.

Metabolic Syndrome It befits a nutrition book to focus on dietary strategies to prevent heart disease. As Table 18-3 shows, most of the modifiable risk factors are directly related to diet. Several of these diet-related risk factors—low HDL, high blood pressure, **insulin resistance,** and abdominal obesity—along with high blood triglycerides comprise a cluster of health risks known as the **metabolic syndrome.**■ Each of these factors increases the likelihood of developing CHD independently, but when they occur together, they elevate the risk synergistically.[27] Overeating and physical inactivity play a major role in the development of the metabolic syndrome. About 47 million people in the United States have the metabolic syndrome.[28]

Recommendations for Reducing Coronary Heart Disease Risk

Recommendations to reduce cardiovascular disease risk include both screening and intervention. The accompanying "How to" provides a tool to assess a person's ten-year heart disease risk. Notice that total cholesterol and HDL cholesterol are among the risk factors, but LDL cholesterol is not. LDL cholesterol is routinely estimated from measures of total cholesterol and HDL cholesterol and thus would not add information to this assessment.[29] Once a person's risks have been identified, treatment focuses on lowering LDL cholesterol. The higher a person's risk, the lower the LDL target and the more aggressive the treatment as the risk categories in the "How to" explain. Treatment plans may include major lifestyle changes in diet, physical activity, and smoking cessation; medications; or both. The LDL cholesterol goals and treatment plans are specific to individuals, so they are best prescribed by a qualified health care provider.

Cholesterol Screening To determine an individual's risk of CHD, health care professionals review the person's health history and measure several blood lipids including total cholesterol, LDL cholesterol, HDL cholesterol, and triglycerides. Ideally, at least two measurements are taken at least one week apart and then com-

---

■ The metabolic syndrome includes any three of the following:
- Abdominal obesity: waist circumference >40 in (for men) or >35 in (for women).
- Triglycerides: ≥150 mg/dL.
- HDL: <40 mg/dL (in men) or <50 mg/dL (in women).
- Blood pressure: ≥130/85 mm Hg.
- Fasting glucose: ≥110 mg/dL.

emerging risk factors: recently identified factors that enhance the ability to predict disease risk in an individual.

insulin resistance: the condition in which a normal amount of insulin produces a subnormal effect, resulting in an elevated fasting glucose; a metabolic consequence of obesity that precedes type 2 diabetes.

metabolic syndrome: a combination of risk factors—insulin resistance, hypertension, abnormal blood lipids, and abdominal obesity—that greatly increase a person's risk of developing coronary heart disease; also called **Syndrome X, insulin resistance syndrome,** or **dysmetabolic syndrome.**

HOW TO Assess Your Risk of Heart Disease

Do you know your heart disease risk score? This assessment estimates your ten-year risk for CHD using charts from the Framingham Heart Study.* Be aware that a high score does not mean that you *will* develop heart disease, but it should warn you of the possibility and prompt you to consult a physician about your health. You will need to know your blood cholesterol (ideally, the average of at least two recent measurements) and blood pressure (ideally, the average of several recent measurements). With this information in hand, find yourself in the five tables below and add the points for each risk factor.

Age (years)

	Men	Women
20–34	−9	−7
35–39	−4	−3
40–44	0	0
45–49	3	3
50–54	6	6
55–59	8	8
60–64	10	10
65–69	11	12
70–74	12	14
75–79	13	16

HDL (mg/dL)

	Men	Women
≥60	−1	−1
50–59	0	0
40–49	1	1
<40	2	2

Systolic Blood Pressure (mm Hg)

	Untreated		Treated	
	Men	Women	Men	Women
<120	0	0	0	0
120–129	0	1	1	3
130–139	1	2	2	4
140–159	1	3	2	5
≥160	2	4	3	6

Total Cholesterol (mg/dL)

	Age 20–39		Age 40–49		Age 50–59		Age 60–69		Age 70–79	
	Men	Women	Men	Women	Men	Women	Men	Women	Men	Women
<160	0	0	0	0	0	0	0	0	0	0
160–199	4	4	3	3	2	2	1	1	0	1
200–239	7	8	5	6	3	4	1	2	0	1
240–279	9	11	6	8	4	5	2	3	1	2
≥280	11	13	8	10	5	7	3	4	1	2

Smoking (any cigarette smoking in the past month)

	Men	Women	Men	Women	Men	Women	Men	Women	Men	Women
Smoker	8	9	5	7	3	4	1	2	1	1
Nonsmoker	0	0	0	0	0	0	0	0	0	0

Scoring Your Heart Disease Risk

Add up your total points: _____ . Now find your total in the first column for your gender in the table at the right and then look to the next column for your approximate risk of developing heart disease within the next ten years. Depending on your risk category, the following strategies can help reduce your risk:

- *>20% = High risk (CHD risk equivalent).* Try to lower LDL using all lifestyle changes and, most likely, lipid-lowering medications as well.

- *10–20% = Moderate risk.* Try to lower LDL using all lifestyle changes and, possibly, lipid-lowering medications.

- *<10% = Low risk.* Maintain or initiate lifestyle choices that help prevent elevation of LDL to prevent future heart disease.

Men		Women	
Total	Risk	Total	Risk
<0	<1%	<9	<1%
0–4	1%	9–12	1%
5–6	2%	13–14	2%
7	3%	15	3%
8	4%	16	4%
9	5%	17	5%
10	6%	18	6%
11	8%	19	8%
12	10%	20	11%
13	12%	21	14%
14	16%	22	17%
15	20%	23	22%
16	25%	24	27%
≥17	≥30%	≥25	≥30%

*An electronic version of this assessment is available on the ATP III page of the National Heart, Lung, and Blood Institute's website (www.nhlbi.nih.gov/guidelines/cholesterol). Another risk inventory is available from the American Heart Association (www.americanheart.org).
SOURCE: Adapted from Expert Panel on Detection, Evaluation, and Treatment of High Blood Cholesterol in Adults (Adult Treatment Panel III), *Third Report of the National Cholesterol Education Program (NCEP)*, NIH publication no. 02-5216 (Bethesda, MD.: National Heart, Lung, and Blood Institute, 2002), section III.

pared to standards (shown earlier in Table 18-4 on p. 622). Single measurements may fail to identify those at risk or may misclassify them because blood cholesterol and other lipid concentrations vary significantly from day to day.

Lifestyle Changes Recommendations to reduce the risk of CHD focus on lifestyle changes. To that end, people are encouraged to increase physical activity, lose weight (if necessary), implement dietary changes, and reduce exposure to tobacco smoke

TABLE 18-5 **Summary of AHA Dietary Guidelines for the General Population**

A Healthy Eating Pattern

- Consume a variety of fruits, vegetables, and whole-grain products.
- Include fat-free and low-fat milk products, fish, legumes, poultry, and lean meats.

A Healthy Body Weight

- Balance energy intake with energy needs.
- Achieve a level of physical activity that either balances with energy needs (for weight maintenance) or exceeds energy needs (for weight reduction).

A Desirable Blood Cholesterol and Lipoprotein Profile

- Limit foods with a high content of saturated fatty acids and *trans*-fatty acids (<10 percent of total energy intake) and cholesterol (<300 milligrams).[a]
- Replace saturated fats with unsaturated fats (both long-chain omega-3 polyunsaturated and monounsaturated fatty acids) from vegetables, fish, and nuts.

A Desirable Blood Pressure

- Limit the intake of salt (sodium chloride) to <6 grams per day.
- Limit alcohol consumption (no more than 1 drink per day for women and 2 drinks per day for men).
- Maintain a healthy body weight and follow a diet that emphasizes vegetables, fruits, and low-fat or fat-free milk products.

[a] For individuals with elevated LDL cholesterol or cardiovascular disease, the saturated and *trans* fat target should be much lower (<7 percent of total energy intake). For these same individuals, as well as for those with diabetes, the cholesterol target should also be much lower (<200 milligrams per day).

either by quitting smoking or by avoiding secondhand smoke. Treatment plans for people with existing CHD or conditions that place them at high risk for heart attacks and strokes (CHD risk equivalents) also focus on lifestyle changes first, but their target LDL is lower. If lifestyle changes fail to lower LDL or blood pressure to acceptable levels, then medications are prescribed. Estimates suggest that a staggering 65 million people in the United States need to implement lifestyle changes to prevent CHD and that another 36 million need medications as well.

Diet strategies to both prevent and treat CHD focus on four main goals:[30]

- A healthy eating pattern.
- A healthy body weight.
- A desirable blood cholesterol and lipoprotein profile.
- A desirable blood pressure.

The specific strategies for achieving a desirable blood cholesterol and lipoprotein profile vary somewhat, depending on whether the objective is to prevent CHD in a person who has a healthy profile or to improve the blood lipid profile in high-risk groups—people with elevated LDL cholesterol, preexisting CHD, insulin resistance, or diabetes mellitus. Table 18-5 summarizes diet strategies recommended by the American Heart Association (AHA) for preventing CHD and reducing elevated LDL cholesterol.

A Healthy Eating Pattern In the past, the AHA's dietary recommendations for preventing and treating CHD centered specifically on the total energy and the amount and type of fat in the diet. Although guidelines for energy and fat remain pivotal, the current recommendations have a broader focus on eating patterns that foster both general and cardiovascular health.

In addition to limiting fats (described later), the heart-healthy diet encourages consumption of complex carbohydrate–rich foods. The viscous (soluble) fiber found in oats, barley, and pectin-rich fruits and vegetables, for example, helps to improve blood lipids.[31] The minerals of carbohydrate-rich foods help control blood pressure (described later); the antioxidant nutrients help protect against LDL oxidation (discussed in Highlight 11); and the B vitamins help lower blood homocysteine levels. (Research shows a positive association between elevated blood homocysteine and the risk of CHD.[32] Although an adequate supply of folate, vitamin B_6, and vitamin B_{12} lowers homocysteine levels, whether it also reduces CHD risk remains to be answered.)

A Healthy Body Weight Not too surprisingly, the AHA recommends that people achieve and maintain a desirable body weight. When overweight people lose weight (even 10 pounds), CHD risk factors improve: blood pressure, blood cholesterol, and blood triglycerides decline.

A Desirable Blood Cholesterol and Lipoprotein Profile High intakes of saturated fatty acids, *trans*-fatty acids, and, to a lesser extent, dietary cholesterol raise blood LDL cholesterol. To achieve a healthy blood cholesterol and lipoprotein profile, the AHA recommends limiting saturated fatty acids and *trans*-fatty acids to less than 10 percent of the total energy intake.■ Dietary cholesterol should be less than 300 milligrams a day. In most cases, foods high in cholesterol are also high in saturated fat. Exceptions are eggs, which are low in saturated fats and high in cholesterol, and coconut, palm, and palm kernel oils, which are high in saturated fats and contain no cholesterol.

Polyunsaturated fatty acids and monounsaturated fatty acids lower LDL cholesterol. Either unsaturated fat or carbohydrate can replace saturated fats in a heart-healthy diet. High-carbohydrate diets, however, can elevate blood triglycerides and reduce HDL cholesterol, which may be a problem for people with type 2 diabetes.■ In these cases, replacing saturated fats with unsaturated fats instead of carbohydrate may help improve the lipoprotein profile.

Fish oils, rich in omega-3 polyunsaturated fatty acids, lower triglycerides, prevent blood clots, and may reduce the risk of sudden death associated with CHD.[33] For these reasons, AHA guidelines recommend at least two servings of fish per week.■ Plant sources of omega-3 fatty acids, which include flaxseed and flaxseed oil, canola oil, soybean oil, and nuts, may also confer benefits.[34] There is not, however, enough evidence to recommend the use of fish oil supplements, which carry their own risks (see p. 161).

A Desirable Blood Pressure Diet strategies to lower blood pressure have traditionally included recommendations to control weight, reduce salt intake, increase potassium intake, and limit alcohol. Also beneficial are diets that provide calcium and magnesium and limit saturated fats and cholesterol. Dietary approaches to control hypertension are presented later in the chapter (see p. 631).

Other Dietary Strategies In general, a nutritionally balanced diet improves many measures of CHD risk. In addition, specific dietary strategies also help to protect against CHD, as Table 18-6 (on p. 628) summarizes; earlier chapters and highlights provided details.

Other Dietary Strategies—Viscous (Soluble) Fiber Viscous (soluble) fibers can bind cholesterol and bile in the intestinal tract and reduce their absorption. An extra 5 to 10 grams of viscous (soluble) fiber daily lowers LDL cholesterol levels by about 5 percent.[35] Dietary sources of viscous (soluble) fibers include oats, barley, legumes, and fruit (see Table 18-6).■

Other Dietary Strategies—Phytosterols Margarines enriched with phytosterols are being marketed as functional foods that lower cholesterol. Phytosterols resemble cholesterol in structure and thus compete with it for absorption; by limiting cholesterol absorption, they lower blood cholesterol. Phytosterols are found most abundantly in vegetable oils, but it may be difficult to consume enough from nonfortified foods to lower blood cholesterol. Enriched margarines deliver up to ten times as much as corn oil, one of the richest sources of phytosterols. Because even a couple of tablespoons of margarine provides some fat, fat-free foods enriched with phytosterols may be a beneficial option.[36] Research suggests these products are safe as currently used.[37] Many health professionals recommend that people with elevated LDL cholesterol or preexisting CHD or CHD risk equivalents use margarines made with phytosterols. Improvements appear to be greater for

■ Highlight 5 provides suggestions for lowering saturated fat intake (p. 176) and lists major food sources of saturated fats and *trans*-fatty acids (p. 177).

■ People with elevated triglycerides should avoid simple sugars, which often cause triglycerides to rise.

■ Table 5-2 on p. 161 lists fish that are particularly rich in omega-3 fatty acids, as well as other sources.

■ Table 4-2 on p. 125 lists fibers, their characteristics, actions in the body, and health benefits.

TABLE 18-6 Dietary Factors Protecting against CHD

In addition to the dietary strategies mentioned in Table 18-5, these dietary factors may also lower the risk of heart attack, especially in people with CHD.

Dietary Factor	Protection against CHD
Viscous (soluble) fiber (apples and other fruits, oats, barley, legumes) and phytosterols	• Lowers blood cholesterol, especially in those with high cholesterol
Omega-3 fatty acids (fish oils)	• Prevent clot formation • Prevent irregular heartbeats • Lower triglycerides • Defend against inflammation • Lower blood pressure
Alcohol (in moderation)	• Raises HDL cholesterol • Prevents clot formation
Folate, vitamin B$_6$, vitamin B$_{12}$	• Reduce homocysteine
Vitamin E (vegetable oils and margarines, some nuts, wheat germ)	• Slows progression of plaque formation • Limits LDL oxidation
Soy (protein and isoflavones)	• Lowers blood cholesterol, LDL cholesterol, and triglycerides

those with higher blood cholesterol and higher intakes of fat, saturated fat, cholesterol, and energy.[38]

Other Dietary Strategies—Soy Protein and Isoflavones When soy proteins are substituted for animal proteins, blood levels of LDL, cholesterol, and triglycerides fall, and HDL levels do not. The cholesterol-lowering effect may be due to soy isoflavones—phytochemicals that have an estrogen-like effect.[39] For people who have elevated cholesterol levels and are following low-saturated fat diets, adding soy protein daily can significantly lower LDL cholesterol.[40] The amount of soy protein needed for significant benefit is estimated to be at least 25 grams (about four servings) daily. The FDA allows foods that contain 6.25 grams of soy protein per serving to carry a health claim for reduced risk of heart disease.

Other Dietary Strategies—Moderate Alcohol Intake Perhaps most controversial are findings that *moderate* alcohol consumption (no more than one drink daily for women and two for men) may reduce overall mortality in general and the risk of heart disease in particular by raising HDL cholesterol, preventing blood clot formation, and improving insulin sensitivity.[41] These benefits are most apparent in people over age 50, those with one or more risk factors, and those with high LDL cholesterol. Benefits are not always apparent. At least one study found that abstainers and moderate drinkers shared similar risks of dying from heart disease.[42] Genetics may play a key role in determining the beneficial effects, if any, of alcohol.[43]

As Highlight 7 described, alcohol has many negative effects on body systems, and a later section in this chapter describes its link with cancer. Any benefits that alcohol may confer on cardiovascular health must be weighed against the risks of incurring these negative health effects, as well as the possibility of alcohol abuse.

Physical Activity Physical activity deserves attention in any program to reduce CHD risk. Frequent and sustained aerobic activity may be most effective in lowering LDL and raising HDL, but weight training (an anaerobic exercise) can also raise HDL if undertaken regularly. Furthermore, as Chapter 14 explained, aerobic, endurance-type activities, such as brisk walking, undertaken faithfully for 30 minutes or more as a daily or every-other-day routine can strengthen the heart and blood vessels; improve body composition; expand the volume of oxygen the heart can deliver to the tissues at each beat and so reduce the heart's workload; change the hormonal climate in which the body does its work in such a way as to lower blood pressure; and bring about a redistribution of body water that eases the transit of blood through the peripheral arteries. Regular physical activity also has favorable effects on the metabolic syndrome. These changes are so beneficial that some experts believe that physical activity should be *the* primary focus of efforts to prevent cardiovascular disease.[44]

If heart and artery disease has already set in, a monitored program of physical activity may actually help to reverse it. Activity may stimulate development of new arteries to feed the heart muscle, which may account for the excellent recovery seen in some heart attack victims who exercise regularly.

Some researchers wonder if physical activity itself raises HDL or if the weight loss that often accompanies exercise is the real protective factor. For women, weight loss through diet alone appears to *lower* HDL, but when diet is combined with moderate aerobic activity, HDL do not decline. In fact, HDL increase substantially in women who exercise regularly. In men, diet raises HDL, and the combination of activity and diet results in a significantly greater rise in HDL than diet alone. Diet helps a little, physical activity helps a little, and the combination is better still.

Drug Therapy Because lipid-lowering drugs■ carry potential risks and are costly, physicians generally do not prescribe drugs until after a three-month trial of intensive diet therapy and physical activity has proved unsuccessful in lowering blood lipids. For people with very high LDL cholesterol (190 milligrams or greater per deciliter), a shorter diet trial may be considered. In addition to lipid-lowering drugs, aspirin and anticoagulants may be used to prevent clot formation and anti-hypertensives to reduce blood pressure. All of these drugs incur potential risks, including nutrition-related side effects, a problem compounded because treatment often entails multiple drugs and continues for many years or even for life. (See Highlight 17 for more on nutrient-drug interactions.)

Regular aerobic exercise can help to defend against heart disease by strengthening the heart muscle, promoting weight loss, and improving blood lipid and blood glucose levels.

■ As Chapter 10 explained, large doses of niacin can effectively lower blood cholesterol, but they have adverse side effects as well. Self-medication is never advisable.

IN SUMMARY Plaques in atherosclerosis raise blood pressure and, when they rupture, trigger abnormal blood clotting, which can cause heart attacks and strokes. Dietary recommendations to lower the risks of cardiovascular disease are summarized in Tables 18-5 and 18-6. Quitting smoking and engaging in regular physical activity also improve heart health.

Hypertension

Anyone concerned about atherosclerosis and the risk it presents must also be concerned about hypertension. The two together are a life-threatening combination. The higher the blood pressure is above normal, the greater the risk. (Low blood pressure, on the other hand, is generally a sign of long life expectancy and low heart disease risk.) Hypertension is believed to affect close to 60 million people in the United States, or about a third of the adult population.[45] It contributes to over a million heart attacks and half a million strokes each year. In fact, hypertension is the most consistent and powerful predictor of stroke.[46] People cannot feel the physical effects of high blood pressure, but it can impair life's quality and end life prematurely.

Reduce the proportion of adults with high blood pressure. Increase the proportion of adults with high blood pressure who are taking action (for example, losing weight, increasing physical activity, or reducing sodium intake) to help control their blood pressure.

HEALTHY PEOPLE 2010

How Hypertension Develops

What triggers chronic hypertension remains for the most part unknown, although one of the mechanisms involving the kidneys has been defined. When blood flow to the kidneys is reduced (as occurs in atherosclerosis), the kidneys respond by setting in motion actions that raise blood pressure by expanding blood volume and constricting peripheral blood vessels (review Figure 12-3 on p. 401). Unfortunately, the pressure increases not only in the kidneys, but all over the body. High blood pressure stresses the heart, which has to pump extra hard to push the blood against resistant arteries. Hypertension worsens atherosclerosis, as described earlier, by mechanically injuring the artery linings and accelerating plaque formation. Then the plaques induce a further rise in blood pressure, intensifying the problem.

Strain on the heart's pump, the left ventricle, can enlarge and weaken it, until it gradually fails (heart failure). Constant elevated pressure in an artery may cause it to gradually balloon out and eventually burst (aneurysm). Aneurysms that go undetected can lead to massive bleeding and death, particularly if a large vessel such as the aorta is affected. In the small arteries of the brain, an aneurysm may lead to stroke, and in the eye, it may lead to blindness. Similarly, the kidneys can be damaged (kidney disease) when the heart is unable to adequately pump enough blood through them.

Risk Factors for Hypertension

Several major risk factors predicting the development of hypertension have been identified, including:

- *Smoking.* Smoking increases the heart's workload, raising blood pressure.
- *Alcohol.* Alcohol, especially if consumed regularly in amounts greater than two drinks per day, may raise blood pressure. Furthermore, alcohol may interfere with drug therapy, and it is associated with strokes independently of hypertension.
- *High blood lipids.* High blood lipids contribute to both atherosclerosis and hypertension.
- *Diabetes.* Elevated insulin signals the kidneys to retain sodium and raise blood pressure. Hypertension is two to three times more common in people with type 2 diabetes (a disease characterized by overweight and insulin resistance) than in the general population.
- *Gender.* In general, blood pressure is higher in men than in women and in women past menopause than in other women.
- *Age.* Arteries lose their elasticity and blood pressure increases with age; most people who develop hypertension do so after age 60.
- *Heredity.* A family history of hypertension and heart disease in women under 65 and in men under 55 significantly raises the risk of developing hypertension.
- *Obesity.* The added adipose tissue that obesity incurs means miles of extra capillaries through which the blood must be pumped. The combination of hypertension, atherosclerosis, and obesity puts a severe strain on the heart and arteries, which intensifies cardiovascular complications.
- *Race.* The prevalence of hypertension differs among racial and ethnic groups; for African Americans, it is among the highest in the world.

Again, diet and physical inactivity interact with many of these risk factors. Notice that a high salt intake is not a risk factor for the development of hypertension, although a low salt intake may lower blood pressure, as a later section explains.

Recommendations for Reducing Hypertension Risk

The single most effective step people can take against hypertension is to find out whether they have it. At checkup time, a health care professional can provide an accurate resting blood pressure reading.*■ Under normal conditions, blood pres-

■ The optimal resting blood pressure for adults is <120 over <80 mm Hg. For adults 40 to 70 years of age, each increase of 20 mm Hg in systolic, or 10 mm Hg in diastolic, blood pressure doubles the risk of cardiovascular disease.

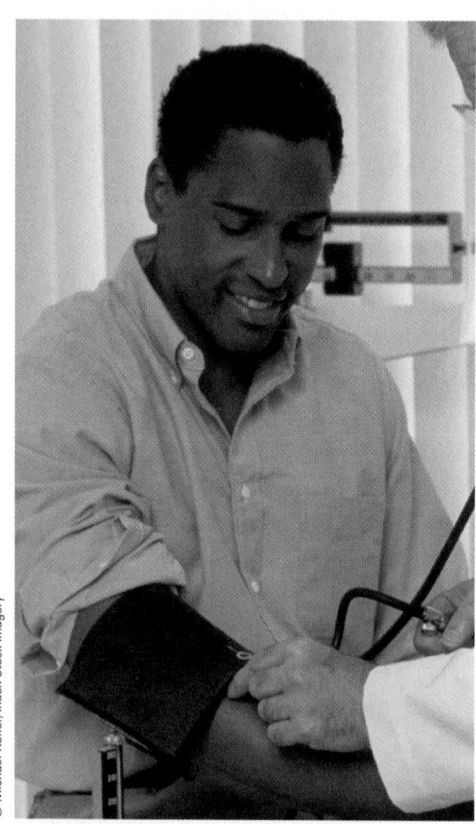

© Michael Keller/Index Stock Imagery

To guard against hypertension, have your blood pressure checked regularly.

---

*Blood pressure is measured in millimeters of mercury (mm Hg). The first number is the systolic and refers to the usual rhythmic contraction of the heart; the second number is the diastolic and refers to the usual rhythmic expansion of the heart.

sure fluctuates continuously in response to a variety of factors including such actions as talking or shifting position. Some people react emotionally to the procedure, which raises the blood pressure reading. For these reasons, if the resting blood pressure is above normal, the reading should be repeated before confirming the diagnosis of hypertension. Thereafter, the blood pressure should be checked regularly.

Weight Control Efforts to reduce high blood pressure focus on weight control. Weight loss alone is one of the most effective nondrug treatments for hypertension. Those who are using drugs to control their blood pressure can often reduce or discontinue the drugs if they lose weight. Even a modest weight loss of 10 pounds can lower blood pressure significantly.

Physical Activity The higher the blood pressure and the less active a person is to begin with, the greater the effect physical activity has in reducing blood pressure. Physical activity helps with weight control, of course, but moderate aerobic activity, such as 30 to 60 minutes of brisk walking most days, also helps to lower blood pressure directly. Those who engage in regular aerobic activity may not need medication for mild hypertension.

Alcohol Those who drink alcohol should do so in moderation—no more than one to two drinks a day. Such amounts appear safe from a blood pressure point of view.

The DASH Diet After decades of research, results of the Dietary Approaches to Stop Hypertension (DASH) trial show that a diet rich in fruits, vegetables, and low-fat milk■ products and low in total fat and saturated fat can significantly lower blood pressure. As Table 18-7 shows, the DASH eating plan emphasizes fruits and vegetables—even more than the Food Guide Pyramid. When the DASH diet is combined with a limited intake of sodium, the effects on blood pressure are greater still.[47] In addition to lowering blood pressure, the DASH diet lowers total cholesterol and LDL cholesterol as well.[48]■ Thus the heart-healthy dietary guidelines embrace these strategies in an overall diet to prevent and treat CHD.

For many years, controversy surrounded recommendations to restrict sodium or salt, but strong evidence supports the important role this strategy plays in preventing and reducing hypertension. Lowering sodium intakes reduces blood pressure regardless of gender or race, presence or absence of preexisting hypertension, or whether people follow the DASH diet or a typical American diet. Furthermore, the lower the sodium intake,■ the greater the drop in blood pressure. (The box in Chapter 12 on p. 408 includes suggestions for limiting sodium.)

Drug Therapy When diet and physical activity fail to reduce blood pressure, diuretics and antihypertensive agents may be prescribed. Diuretics lower blood pressure by increasing fluid loss. Some diuretics can lead to a potassium deficiency. People taking these diuretics need to include rich sources of potassium or supplements daily and watch for signs of potassium imbalances such as weakness (particularly of the legs), unexplained numbness or tingling sensation, cramps, irregular heartbeats, and excessive thirst and urination. Blood potassium should be monitored regularly.

Although some diuretics can lead to a potassium deficiency, others spare potassium. A combination of these two types of diuretics may be prescribed to prevent potassium deficiency. In such cases, excessive potassium intakes and potassium supplements should be avoided; too much potassium is also life-threatening.

IN SUMMARY The most effective dietary strategy for preventing hypertension is weight control. Also beneficial are diets rich in fruits, vegetables, and low-fat milk products and low in fat, saturated fat, and sodium.

TABLE 18-7 The DASH Eating Plan and the Food Guide Pyramid Compared

Recommended Number of Daily Servings

Food Group	DASH	Pyramid
Grains	7–8	6–11
Vegetables	4–5	3–5
Fruits	4–5	2–4
Milk (fat-free/low-fat)	2–3	2–3
Meat (lean)[a]	2 or less	2–3
kCalories	2000	1600–2800

NOTE: The DASH eating plan, like the Food Guide Pyramid, recommends that fats, oils, and sweets be used sparingly.
[a]The DASH eating plan also includes recommended servings for nuts, seeds, and dry beans (4 to 5 per week), whereas the Food Guide Pyramid includes these foods with the meat group.

■ The DASH diet is rich in potassium and calcium.

■ Like other low-fat diets, the DASH diet also lowers HDL—a seemingly undesirable outcome. Whether a lowered HDL raises the risk of CHD is unknown, although some studies suggest that people with both low LDL and low HDL do not have an increased risk of CHD.

■ The DASH diet trial studied the effects of three levels of sodium restriction: 3300 mg, 2400 mg, and 1500 mg.

The richest sources of potassium are *fresh* foods of all kinds.

© Stefan Hallberg/Index Stock Imagery/PictureQuest

Diabetes Mellitus

The incidence of diabetes among children and adults has risen dramatically in the last decade and is expected to double in the next 50 years (see Figure 18-4). Diabetes mellitus ranks sixth among the leading causes of death (review Figure 18-2 on p. 618). In addition, diabetes underlies, or contributes to, several other major diseases, including heart disease and stroke. In fact, people with diabetes are twice as likely to develop these cardiovascular problems as those without diabetes.

Prevent diabetes. Reduce diabetes-related deaths among persons with diabetes.

How Diabetes Develops

Diabetes mellitus describes a group of metabolic disorders characterized by high blood glucose caused by insufficient insulin, ineffective insulin, or a combination of the two. The accompanying glossary defines diabetes, and Table 18-8 shows the distinguishing features of its two main forms, type 1 diabetes and type 2 diabetes. As described in the next section, the development of type 1 and type 2 diabetes differs, but some of the complications are similar.

To appreciate the problems presented by an absolute or relative lack of insulin, consider insulin's normal action. After a meal, insulin signals the body's cells to receive the energy nutrients from the blood—amino acids, glucose, and fatty acids. Insulin helps to maintain blood glucose within normal limits and stimulates protein synthesis, glycogen synthesis in liver and muscle, and fat synthesis. Without insulin, glucose regulation falters, and metabolism of the energy-yielding nutrients changes.

Type 1 Diabetes In **type 1 diabetes,** the less common type of diabetes (about 5 to 10 percent of all diagnosed cases), the pancreas loses its ability to synthesize the hormone insulin.[49] Type 1 diabetes is an **autoimmune disorder.**[49] In most cases, the individual inherits a defect in which immune cells mistakenly attack and destroy the insulin-producing beta cells of the pancreas. The rate of beta cell destruction in type 1 diabetes varies. In some people (mainly infants and children), destruction is rapid; in others (mainly adults), it is slow. Type 1 diabetes commonly occurs in childhood and adolescence, but it can occur at any age, even late in life.[50]

Without insulin, the body's energy metabolism changes, with such severe consequences as to threaten survival. The cells must have insulin to take up the

FIGURE 18-4 Prevalence of Diabetes among Adults in the United States

Key:

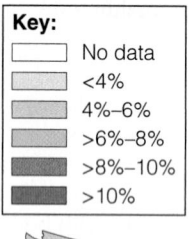

	No data
	<4%
	4%–6%
	>6%–8%
	>8%–10%
	>10%

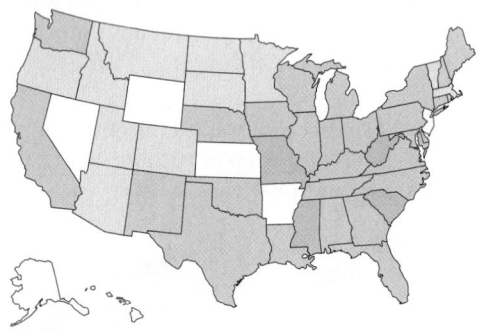

1990

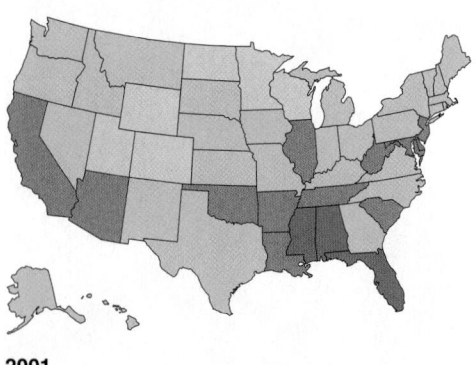

2001

SOURCE: Centers for Disease Control and Prevention, www.cdc.gov/nccdphp/aag/aag_ddt.htm.

TABLE 18-8 Features of Type 1 and Type 2 Diabetes Compared

	Type 1 Diabetes	Type 2 Diabetes
Other names	Insulin-dependent diabetes[a] mellitus (IDDM)	Noninsulin-dependent diabetes[a] mellitus (NIDDM)
Average age of onset	<20 (mean age, 12)	10–19; >40
Associated conditions	Viral infection, heredity	Obesity, heredity, aging
Insulin required?	Yes	Sometimes
Cell response to insulin	Normal	Resistant
Symptoms	Relatively severe	Relatively moderate
Prevalence in diabetic population	5 to 10%	90 to 95%

[a]The terms *insulin-dependent diabetes (IDDM)* and *noninsulin-dependent diabetes (NIDDM)* have been eliminated from use by the Expert Committee on the Diagnosis and Classification of Diabetes Mellitus. The committee states that these terms have been confusing and often result in classifying patients based on treatment rather than the cause of the disorder.

needed fuels from the blood. People with type 1 diabetes must inject insulin or use external pumps; it cannot be taken orally because insulin is a protein and the enzymes of the GI tract would digest it.

Type 2 Diabetes **Type 2 diabetes** is the predominant form of diabetes (90 to 95 percent of all cases). Unfortunately, as many as 50 percent of people with type 2 diabetes—8 million people in the United States—are undiagnosed.[51] Undiagnosed, and therefore untreated, type 2 diabetes is a serious problem that significantly increases a person's risk of CHD, stroke, and peripheral vascular disease.

Type 2 diabetes usually develops in people over 40 years old, as the insulin-producing cells of the pancreas progressively lose their function with age. But type 2 diabetes is also seen in overweight children because it is closely associated with obesity and physical inactivity; almost everyone with type 2 diabetes is overweight.[52]

One of the many metabolic consequences of obesity is insulin resistance. Insulin is available, but it is not effective in moving glucose into the cells. Compared with normal-weight people, obese people require much more insulin to maintain normal blood glucose. In an effort to lower blood glucose, the pancreas produces more insulin, but as body fat increases, insulin receptors diminish in number or in function. Consequently, the cells respond less sensitively to insulin; that is, they become insulin resistant. Blood glucose rises. This prediabetic condition of having a slightly elevated blood glucose level is known as **impaired glucose tolerance.** At some point, the pancreas cannot produce enough insulin to keep up, glucose rises even higher, and type 2 diabetes develops. Age and obesity alone do not predict the onset of type 2 diabetes; genetics also plays a role.[53]

Complications of Diabetes

In both types of diabetes, glucose fails to gain entry into the cells and consequently accumulates in the blood. These two problems lead to both acute and chronic complications. Figure 18-5 summarizes the metabolic changes and acute complications that can arise in uncontrolled diabetes. Notice that when some glucose enters the cells, as in type 2 diabetes, many of the symptoms of type 1 do not occur.

Over the long term, the person with diabetes suffers not only from the acute complications shown in Figure 18-5, but also from its chronic effects. Chronically elevated blood glucose alters glucose metabolism in virtually every cell of the body. Some cells begin to convert excess glucose to sugar alcohols, for example, causing toxicity and cell distention; distended cells in the lenses of the eyes cause blurry vision. Some cells produce glycoproteins by attaching excess glucose to an amino acid in a protein; these proteins cannot function normally, which leads to a host of other problems. The structures of the blood vessels and nerves become damaged, leading to loss of circulation and nerve function. Infections occur due to poor circulation coupled with glucose-rich blood and urine. People with diabetes must pay special attention to hygiene and keep alert for early signs of infection. Early, aggressive treatment to control blood glucose significantly reduces the risk of long-term diabetes-related complications.[54]

autoimmune disorder: a condition in which the body develops antibodies to its own proteins and then proceeds to destroy cells containing these proteins. In type 1 diabetes, the body develops antibodies to its insulin and destroys the pancreatic cells that produce the insulin, creating an insulin deficiency.

FIGURE 18-5 Metabolic Consequences of Untreated Diabetes

The metabolic consequences of type 1 diabetes are more immediate and severe than those of type 2. In type 1, no insulin is available to allow any glucose to enter the cells. When glucose cannot enter the cells, a cascade of metabolic changes follows. In type 2 diabetes, some glucose enters the cells. Because the cells are not "starved" for glucose, the body does not shift into the metabolism of fasting (losing weight and producing ketones).

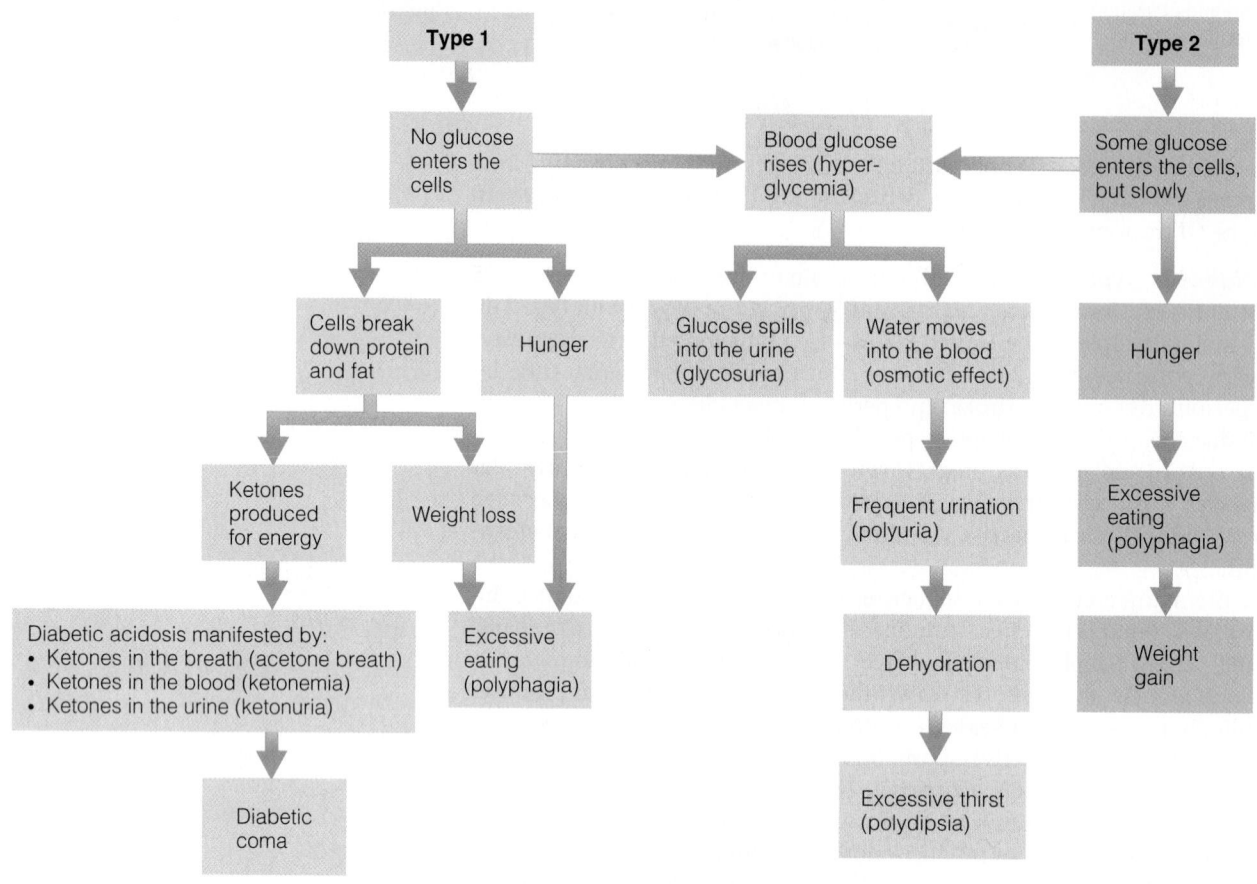

Diseases of the Large Blood Vessels As mentioned, atherosclerosis tends to develop early, progress rapidly, and be more severe in people with diabetes. The interrelationships among insulin resistance, obesity, hypertension, and atherosclerosis help explain why more than 80 percent of people with diabetes die as a consequence of cardiovascular diseases, especially heart attacks. If nerve function is also impaired, the person may have a heart attack and not even realize it. Research shows that long-term, intensive intervention, targeting multiple risk factors (blood glucose levels, blood pressure, blood lipids) can reduce the risk of cardiovascular disease among those with type 2 diabetes.[55]

Diseases of the Small Blood Vessels Disorders of the small blood vessels (capillaries)■ may also develop and lead to loss of kidney function and retinal degeneration with accompanying loss of vision. About 85 percent of people with diabetes have impaired kidney function, loss of vision, or both. Consequently, diabetes is a leading cause of both kidney failure and blindness.

Diseases of the Nerves Nerve tissues may also deteriorate, expressed at first as a painful prickling sensation, often in the arms and legs. Later, the person loses sensation in the hands and feet. Injuries to these areas may go unnoticed, and infections can progress rapidly. With loss of both circulation and nerve function, undetected injury and infection may lead to death of tissue (gangrene),■ necessitating amputation of the limbs (most often the legs or feet). People with diabetes are advised to take conscientious care of their feet and visit a podiatrist regularly.

■ Disorders of the small blood vessels are called **microangiopathies.**
  • **micro** = small
  • **angeion** = vessel
  • **pathos** = disease

■ The death of tissue, usually due to deficient blood supply, is **gangrene** (GANG-green).

Nerve damage can also delay gastric emptying. When the stomach empties slowly after a meal, the person may experience a premature feeling of fullness, bloating, nausea, vomiting, weight loss, and poor blood glucose control due to irregular nutrient absorption.

Recommendations for Diabetes

Diet is an important component of diabetes treatment. To maintain near-normal blood glucose levels, the diet is designed to deliver the same amount of carbohydrate each day, spaced evenly throughout the day. Several approaches can be used to plan such diets, but many people with diabetes learn to count carbohydrates using the exchange system that is presented in Appendixes G and I.

Providing a consistent carbohydrate intake spaced throughout the day helps people with diabetes maintain appropriate blood glucose levels and maximizes the effectiveness of drug therapy. Eating too much carbohydrate at one time can raise blood glucose too high, stressing the already-compromised insulin-producing cells. Eating too little carbohydrate can lead to hypoglycemia. The *amount* of carbohydrate affects blood glucose levels more than the source of the carbohydrate.[56] Put another way, people with diabetes must pay attention to their carbohydrate intake, regardless of whether they eat bread or cake. They can eat sweets and sugar as part of a healthy diet if they can afford the kcalories; they just need to remember to count the carbohydrates.

People with diabetes who have elevated blood lipids may need to watch their fat intake as well as their carbohydrate intake. When they lower their fat intake, the percentage of kcalories from carbohydrates increases. When they increase their carbohydrate intake, however, they may have difficulty controlling both their blood glucose and their blood lipid levels because high-carbohydrate diets raise triglycerides and lower HDL. In addition, people accustomed to a high-fat diet sometimes have difficulty complying with a low-fat diet. Mounting evidence suggests that people can control their blood glucose and improve their blood lipids with a fat intake up to 35 percent of total energy and a carbohydrate intake of 50 percent of total energy, provided that fat kcalories come primarily from unsaturated fats. Diets high in monounsaturated fat may reduce the susceptibility of LDL to oxidation and thereby help reduce the risk of cardiovascular disease. Such diets, however, may also increase energy intake and thereby promote weight gain.[57] Results such as these remind practitioners that diet plans must be individualized to be successful.

Recommendations for Type 1 Diabetes Normally, the body secretes a constant, baseline amount of insulin at all times and secretes more as blood glucose rises following meals. People with type 1 diabetes however, produce little or no insulin. They must learn to adjust their insulin doses and schedule of administration to accommodate meals, physical activity, and health status. To maintain blood glucose within a fairly normal range requires a lifelong commitment to a carefully coordinated program of diet, physical activity, and insulin.

Nutrition therapy focuses on maintaining optimal nutrition status, controlling blood glucose, achieving a desirable blood lipid profile, controlling blood pressure, and preventing and treating the complications of diabetes. In addition to meeting basic nutrient requirements, the diet must provide a fairly consistent carbohydrate intake from day to day and at each meal and snack to help minimize fluctuations in blood glucose. Further alterations in diet may be necessary for the person with chronic complications such as cardiovascular or kidney disease.

Participation in all levels of physical activity is possible for people with type 1 diabetes who have good blood glucose control and no complications, but they should check with their physician first. One potential problem is hypoglycemia, which can occur during, immediately after, or many hours after physical activity.[58] To avoid hypoglycemia, the person must monitor blood glucose before and

For a person with type 1 diabetes, good health depends on coordinating the timing of meals, activities, and insulin.

after activity to identify when changes in insulin or food intake are needed. Carbohydrate-rich foods should be readily available during and after activity.

Recommendations for Type 2 Diabetes In people with type 2 diabetes, even moderate weight loss (10 to 20 pounds) can help improve insulin resistance, blood lipids, and blood pressure. Together with diet, a regular routine of moderate physical activity not only supports weight loss, but also improves blood glucose control, blood lipid profiles, and blood pressure.[59] Thus the benefits of regular, long-term physical activity for the treatment and prevention of type 2 diabetes are substantial.[60]

> **IN SUMMARY** Diabetes is characterized by high blood glucose and either insufficient insulin, ineffective insulin, or a combination of the two. People with type 1 diabetes coordinate diet, insulin injections, and physical activity to help control their blood glucose. Those with type 2 benefit most from a diet and physical activity program that controls glucose fluctuations and promotes weight loss.

Cancer

The thought of cancer often strikes fear in people. Indeed, cancer ranks just below cardiovascular disease as a cause of death for the entire population, and it is the number one cause of death for those between the ages of 45 and 64. Consequently, many people have personal experiences with cancer. As with cardiovascular disease, the prognosis for cancer today is far brighter than in the past. Identification of risk factors, new detection techniques, and innovative therapies offer hope and encouragement.

Cancer is not a single disorder. There are many **cancers,** that is, many different kinds of malignancies (see the accompanying glossary of cancer terms). They have different characteristics, occur in different locations in the body, take different courses, and require different treatments.

HEALTHY
PEOPLE
2010

Reduce the overall cancer death rate.

GLOSSARY OF CANCER TERMS

antipromoters: factors that oppose the development of cancer.

cancers: diseases that result from the unchecked growth of malignant tumors.

carcinogens (car-SIN-oh-jenz): substances or agents that are capable of causing cancer.
- **carcin** = cancer
- **gen** = gives rise to

initiators: factors that cause mutations that give rise to cancer.

metastasize (me-TAS-tah-size): to spread from one part of the body to another.

promoters: factors that favor the development of cancer once it has started.

tumor: a new growth of tissue forming an abnormal mass with no function; also called a **neoplasm** (NEE-oh-plazm). Tumors that multiply out of control, threaten health, and require treatment are **malignant** (ma-LIG-nant). Tumors that stop growing

without intervention or can be removed surgically and pose no threat to health are **benign** (bee-NINE).
- **benign** = mild
- **malignus** = of bad kind

Cancers are classified by the tissues or cells from which they develop:
- **adenomas** (ADD-eh-NOH-mahz): cancers that arise from glandular tissues.
- **carcinomas** (KAR-see-NOH-mahz): cancers that arise from epithelial tissues.

- **gliomas** (gly-OH-mahz): cancers that arise from glial cells of the central nervous system.
- **leukemias** (loo-KEE-mee-ahz): cancers that arise from the white blood cells.
- **lymphomas** (lim-FOH-mahz): cancers that arise from lymph tissue.
- **melanomas** (MEL-ah-NOH-mahz): cancers that arise from pigmented skin cells.
- **sarcomas** (sar-KOH-mahz): cancers that arise from muscle, bone, or connective tissues.

How Cancer Develops

The genes in a healthy body work together regulating cell division to ensure that each new cell is a replica of the parent cell. In this way, the healthy body grows, replacing dead cells and repairing damaged ones. Cancers develop from mutations in the genes that regulate cell division. The mutations silence the genes that ordinarily monitor replicating DNA for chemical errors. The affected cells seemingly have no built-in brakes to halt cell division. As the abnormal mass of cells, called a malignant **tumor,** grows, blood vessels form to supply the tumor with the nutrients it needs to support its growth. Eventually, the tumor invades healthy tissue and may **metastasize.** Clinicians describe cancers by their location, size, and extent, specifically noting if the tumor has spread to surrounding lymph nodes or to distant sites in the body. Figure 18-6 illustrates tumor formation.

Genetic Factors All cancers have a genetic component in that a mutation causes abnormal cell growth, but some cancers have a genetically inherited component. A person with a family history of breast cancer, for example, has a greater risk of developing breast cancer than a person without such a genetic predisposition. This does not mean, however, that the person *will* develop cancer, only that the risk is greater.

Immune Factors A healthy immune system recognizes foreign cells and destroys them. Researchers theorize that an ineffective immune system may not recognize tumor cells as foreign, thus allowing unchecked growth. Aging affects immune function, and the incidence of cancer increases with age. Drugs that suppress the immune system and viral infections (including HIV infection) and other disorders that severely tax the immune system may also increase the risk of cancer.

FIGURE 18-6 Tumor Formation

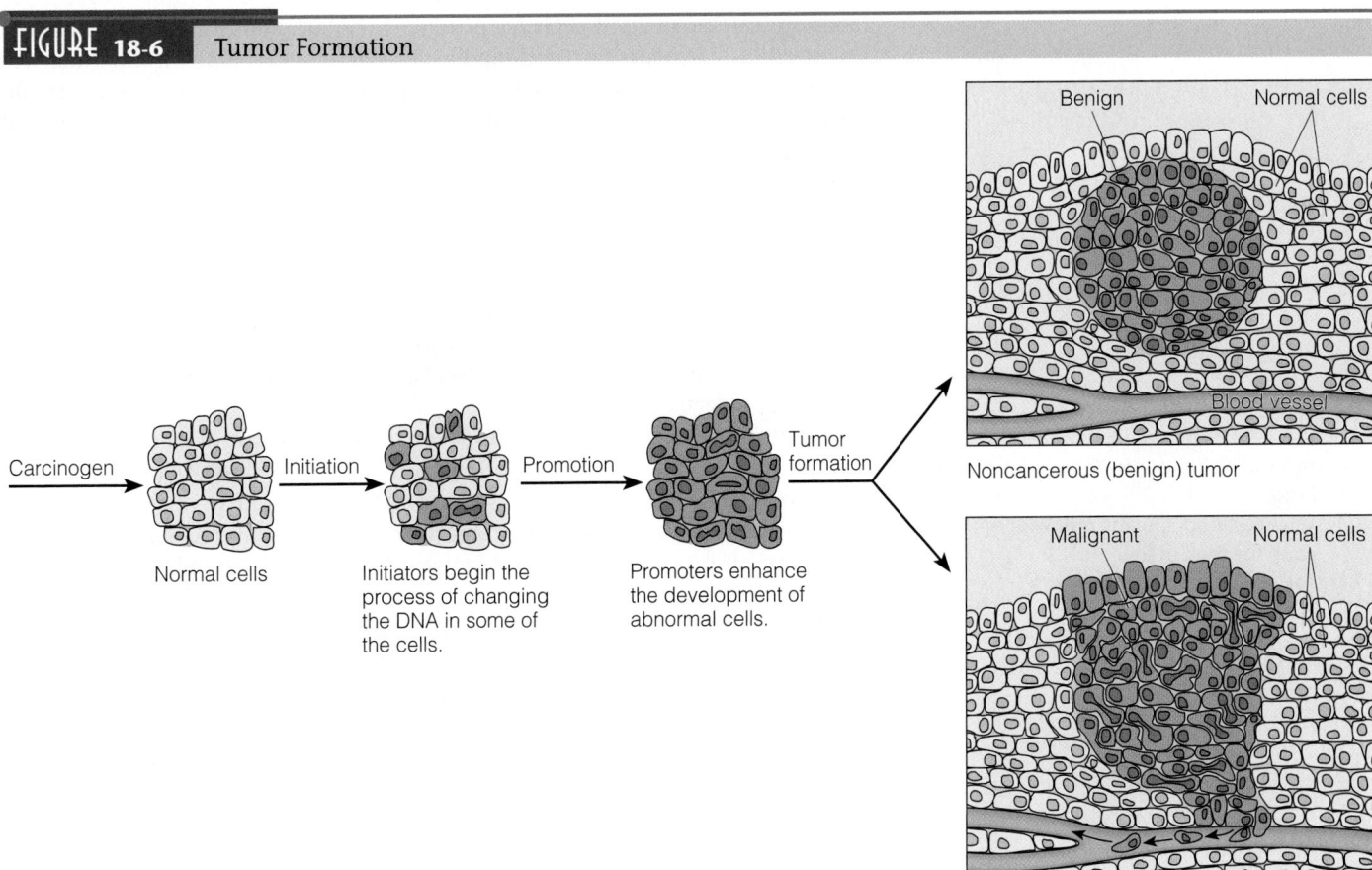

Carcinogen → Initiation → Promotion → Tumor formation

Normal cells

Initiators begin the process of changing the DNA in some of the cells.

Promoters enhance the development of abnormal cells.

Benign — Normal cells

Blood vessel

Noncancerous (benign) tumor

Malignant — Normal cells

Cancerous (malignant) tumor releases cells into the bloodstream (metastasis)

People with cancer take comfort from the support of others and from the knowledge that medical science is waging an unrelenting battle in their defense.

Environmental Factors Among environmental factors, exposure to radiation and sun, water and air pollution, and smoking are known to cause cancer. Lack of physical activity may also play a role in the development of some types of cancer.[61] Men and women whose lifestyles include regular, vigorous physical activity have the lowest risk of colon cancer.[62] Physical activity may also protect against breast cancer by reducing body weight and by other mechanisms not related to body weight.[63]

Obesity itself is clearly a risk factor for certain types of cancer (such as colon, breast in postmenopausal women, endometrial, kidney, and esophageal) and possibly for other types (such as ovarian and prostate) as well.[64] Different cancers have various causes, so the exact ways in which obesity influences the development of a cancer depends on the site as well as other factors, including hormonal interactions. In the case of breast cancer in postmenopausal women, for example, the hormone estrogen is implicated. Obese postmenopausal women have much higher levels of estrogen than lean women do because fat tissue produces estrogen. Researchers believe that obese women's extended exposure to estrogen is linked to an increased risk of breast cancer after menopause.[65] The relationships between excessive body weight and certain cancers provide yet another reason to adopt a lifestyle that embraces physical activity and sound nutrition.

As Table 18-9 shows, dietary constituents are also associated with an increased risk of certain cancers. Some dietary factors may initiate cancer development (**initiators**), others may promote cancer development once it has started (**promoters**), and still others may protect against the development of cancer (**antipromoters**).

Dietary Factors—Cancer Initiators We do not know to what extent diet contributes to cancer development, although some experts estimate that diet may be linked to up to a third of all cases. Consequently, many people think that certain foods are carcinogenic, especially those that contain additives or pesticides. As Chapter 19 will explain, our food supply is one of the safest in the world. Additives that have been approved for use in foods are not **carcinogens.** Some pesticides are carcinogenic at high doses, but not at the concentrations allowed on fruits and vegetables. The benefits of eating fruits and vegetables are far greater than any potential risk.[66]

Cancers of the head and neck correlate strongly with the combination of alcohol and tobacco use and with low intakes of green and yellow fruits and vegetables. Alcohol intake alone is associated with cancers of the mouth, throat, and breast, and alcoholism often damages the liver and precedes the development of liver cancer. These findings illustrate clearly why any potential benefit of moderate alcohol consumption on cardiovascular disease must be weighed against the potential dangers.

Grilling meat, fish, or other foods over a direct flame causes fat and added oils to splash on the fire and then vaporize, creating carcinogens■ that rise and stick to the food.* Eating grilled food introduces these carcinogens to the digestive system, where they may damage the stomach and intestinal lining. Once these compounds are absorbed into the blood, however, they are detoxified by the liver.

Evidence from population studies spanning the globe for over 20 years supports the theory that diets high in meat, and especially red meat, are related to a greater risk of developing colon cancer.[67] In particular, processed meats and meats cooked to the crispy well-done stage may be at fault. Remember, however, that even strong correlation is not causation—certain foods may be implicated, but they have not been found guilty of actually causing cancer. Nevertheless, replacing most servings of red meats with poultry, fish, or legumes, and choosing only occasional servings of grilled, fried, highly browned, or smoked foods is in the best interests of health.

Another reason to moderate consumption of fried foods such as french fries and potato chips is the presence of a substance called acrylamide, which is a po-

■ To minimize carcinogen formation during cooking:
- When grilling, line the grill with foil, or wrap the food in foil.
- Take care not to burn foods.
- Marinate meats beforehand.

---
*The carcinogens of greatest concern are some of those called *polycyclic aromatic hydrocarbons.*

TABLE 18-9 Factors Associated with Cancer at Specific Sites

Cancer Sites	Associated with:	Probable Protective Effect from:
Bladder cancer	Cigarette smoking and alcohol; weak association with coffee and chlorinated drinking water	Fruits and vegetables (especially fruits); adequate fluid intake
Breast cancer	High intakes of food energy, alcohol intake; low vitamin A intake; obesity, sedentary lifestyle, probably not associated with dietary fat	Monounsaturated fats; physical activity
Cervical cancer	Folate deficiency; viral infection; possibly, cigarette smoking	Adequate folate intake; possibly, fruits and vegetables
Colorectal cancer	High intakes of fat (particularly saturated fat), red meat, alcohol, and supplemental iron; low intakes of fiber, folate, vitamin D, and vegetables; inactivity; cigarette smoking	Vegetables, especially cruciferous (cabbage-type) vegetables; calcium, vitamin D, and dairy intake; possibly, whole wheat, wheat bran; high levels of physical activity
Kidney cancer	Possibly, high intakes of red meat (especially fried, sautéed, charred, burned, or cooked well-done); cigarette smoking; obesity	Fruit and vegetables, especially orange-colored and dark green ones
Mouth, throat, and esophagus cancers	Heavy use of alcohol, tobacco, and especially combined use; heavy use of preserved foods (such as pickles); low intakes of vitamins and minerals; obesity (esophageal)	Fruits and vegetables
Liver cancer	Infection with hepatitis virus; high intakes of alcohol; iron overload; toxins of a mold (aflatoxin) or other toxicity	Vegetables, especially yellow and green ones
Lung cancer	Smoking; low vitamin A; supplements of beta-carotene (in smokers)	Fruits and vegetables
Ovarian cancer	Possibly, high lactose intake from milk products; inversely correlated with oral contraceptive use	Vegetables, especially green leafy ones
Pancreatic and lung cancer	Possibly, high intakes of red meat (pancreatic cancer); correlated with cigarette smoking and air pollution	Fruits and vegetables, especially green and yellow ones
Prostate cancer	High intakes of fats, especially saturated fats from red meats and possibly milk products	Possibly, cooked tomatoes, soybeans, soy products, and flaxseed; adequate selenium intake
Stomach cancer	High intakes of smoke- or salt-preserved foods (such as dried, salted fish); cigarette smoking; possibly, refined flour or starch; infection with ulcer-causing bacteria	Fresh fruits and vegetables, especially tomatoes

SOURCES: National Cancer Policy Board, Institute of Medicine, S. J. Curry, T. Byers, and M. Hewitt, eds., *Fulfilling the Potential of Cancer Prevention and Early Detection* (Washington, D.C.: National Academies Press, 2003), pp. 66–86; S. E. McCann and coauthors, Risk of human ovarian cancer is related to dietary intake of selected nutrients, phytochemicals and food groups, *Journal of Nutrition* 133 (2003): 1937–1942; S. A. Smith-Warner and coauthors, Intake of fruits and vegetables and risk of breast cancer, *Journal of the American Medical Association* 285 (2001): 769–776; R. L. Nelson, Iron and colorectal risk: Human studies, *Nutrition Reviews* 59 (2001): 140–148; M. C. Jansen and coauthors, Dietary fiber and plant foods in relation to colorectal cancer mortality: The Seven Countries Study, *International Journal of Cancer* 81 (1999): 174–179; B. S. Reddy, Role of dietary fiber in colon cancer: An overview, *American Journal of Medicine* 106 (1999): S16–S19.

tential carcinogen. Acrylamide is produced when certain starches such as potatoes are fried or baked at high temperatures. Chapter 19 offers a discussion of acrylamide in foods.

Dietary Factors—Cancer Promoters Unlike carcinogens, which initiate cancers, some dietary components promote cancers. That is, once the initiating step has taken place, these components may accelerate tumor development.

Studies of animals suggest that high-fat diets may promote cancer. In studies of human beings, however, evidence is mixed as to whether a diet high in fat promotes cancer.[68] Comparisons of populations around the world reveal that high-fat diets often, but not always, correlate with high cancer rates. Within a single population, however, cancer rates do not reliably reflect fat intakes.[69] Studies on whether dietary fat promotes breast cancer seem to indicate that it does not.[70] For the risk of prostate cancer, evidence seems to implicate meat fats but not vegetable fats, while consuming fatty fish may be protective.[71]

One attribute of dietary fat is energy density—gram for gram, fat provides more kcalories than either carbohydrate or protein. Diets high in *kcalories* do seem to promote cancer, especially in laboratory settings, so researchers still must untangle the effects of fat alone from those of total energy.

The type of fat in the diet may influence cancer promotion or prevention. Some evidence implicates saturated and *trans*-fatty acids in cancer promotion, while suggesting that omega-3 fatty acids from fish may protect against some cancers.[72]

Cruciferous vegetables, such as cauliflower, broccoli, and brussels sprouts, contain nutrients and phytochemicals that may inhibit cancer development.

Thus the same dietary fat advice applies to cancer protection as to heart disease: reduce saturated and *trans*-fatty acids and increase omega-3 fatty acids.

Dietary Factors—Antipromoters It seems apparent that foods may also contain antipromoters. Almost without exception, epidemiological studies find a link between eating plenty of fruits and vegetables and a low incidence of cancers. The fiber in fruits and vegetables may help to protect against some cancers by speeding up the transit time of all materials through the colon so that the colon walls are not exposed to cancer-causing substances for long. Many studies have found that fiber-rich diets protect against some forms of cancer, including colon cancer.[73] A few studies dispute such findings, but the majority of evidence supports the premise that fiber-rich diets protect against colon cancer.[74] People who choose generous servings of high-fiber, nutrient-rich fruits, vegetables, and whole grains may reduce their risk of some cancers as well as heart disease, diabetes, and other chronic diseases.[75]

In addition to fiber, fruits and vegetables contain both nutrients and nonnutrients (phytochemicals) that protect against cancer. By acting as scavengers of oxygen-derived free radicals, the antioxidant nutrients beta-carotene, vitamin C, and vitamin E may help to prevent cell and tissue damage that can give rise to cancer (see Highlight 11). Phytochemicals common to many vegetables, especially cruciferous vegetables, can activate enzymes that destroy carcinogens (see Highlight 13).

Recommendations for Reducing Cancer Risk

On the basis of current knowledge and available evidence, the following guidelines are recommended for cancer prevention:[76]

- *Eat a variety of healthful foods, with an emphasis on plant sources.* Eat five or more servings of a variety of vegetables and fruits each day. Choose whole grains in preference to processed (refined) grains and sugars. Limit consumption of red meats, especially those high in fat and processed. Choose foods that maintain a healthful weight.

- *Adopt a physically active lifestyle.* Adults: engage in at least moderate activity for 30 minutes or more on five or more days of the week; 45 minutes or more of moderate-to-vigorous activity on five or more days per week may further reduce the risk of breast and colon cancer. Children and adolescents: engage in at least 60 minutes of moderate-to-vigorous physical activity on at least five days per week.

- *Maintain a healthful weight throughout life.* Balance kcaloric intake with physical activity. Lose weight if currently overweight or obese.

- *If you drink alcoholic beverages, limit consumption.*

And most importantly:

- *Do not smoke or use tobacco in any form.*

Dietary supplements are "unnecessary and may be unhelpful" for those who follow these recommendations.

One additional recommendation is in order: vary food choices. This last suggestion is based on an important concept that applies specifically to the prevention of cancer initiation—dilution. Eating a variety of foods dilutes the negative qualities of any one food.

IN SUMMARY Some dietary factors, such as alcohol and heavily smoked foods, may initiate cancer development; others, such as saturated fat or *trans*-fatty acids, may promote cancer once it has gotten started; and still others, such as fiber, antioxidant nutrients, and phytochemicals, may serve as antipromoters that protect against the development of cancer. By eating many fruits, vegetables, legumes, and whole grains and reducing saturated and *trans* fat intake, people obtain the best possible nutrition at the lowest possible risk.

Recommendations for Chronic Diseases

This chapter's discussion of chronic diseases began with the major cardiovascular diseases, described diabetes, and then went on to cancer—three different conditions with distinct sets of causes. Yet dietary excesses, particularly excess food energy and fat intakes, increase the likelihood of all three diseases.[77] Similarly, all are responsive to diet, and in most cases, the beneficial foods are similar.

Not all diet recommendations apply equally to all of the diseases or to all people with a particular disease, but fortunately for the consumer, dietary recommendations■ do not contradict one another. In fact, they support each other. Most people can gain some disease-prevention benefits by making dietary changes. To that end, the recommendations of the American Heart Association (Table 18-5 on p. 626), the DASH diet (p. 631), and the recommendations of the American Cancer Society (p. 640) describe the kinds of foods people should include or limit. As this book goes to press, plans are being made to revise and improve the Food Guide Pyramid for the benefit of health. Figure 18-7 presents recommendations for disease prevention overlaid on the familiar Pyramid.[78]

Several recommendations are aimed at weight control. Obesity is common in the United States, and it is linked with most of the chronic diseases that threaten life (review Figure 18-3 on p. 619). The problems of overweight people multiply when medical conditions develop. For example, overweight people readily develop diabetes, which is often accompanied by high blood pressure and high blood cholesterol. Such a combination of problems may require only one treatment: adopting a healthful diet and regular exercise program.

■ A summary of the *Diet, Nutrition, and Prevention of Chronic Diseases* report from the World Health Organization (WHO) is presented in Appendix I.

FIGURE 18-7   Food Guide Pyramid for Disease Prevention

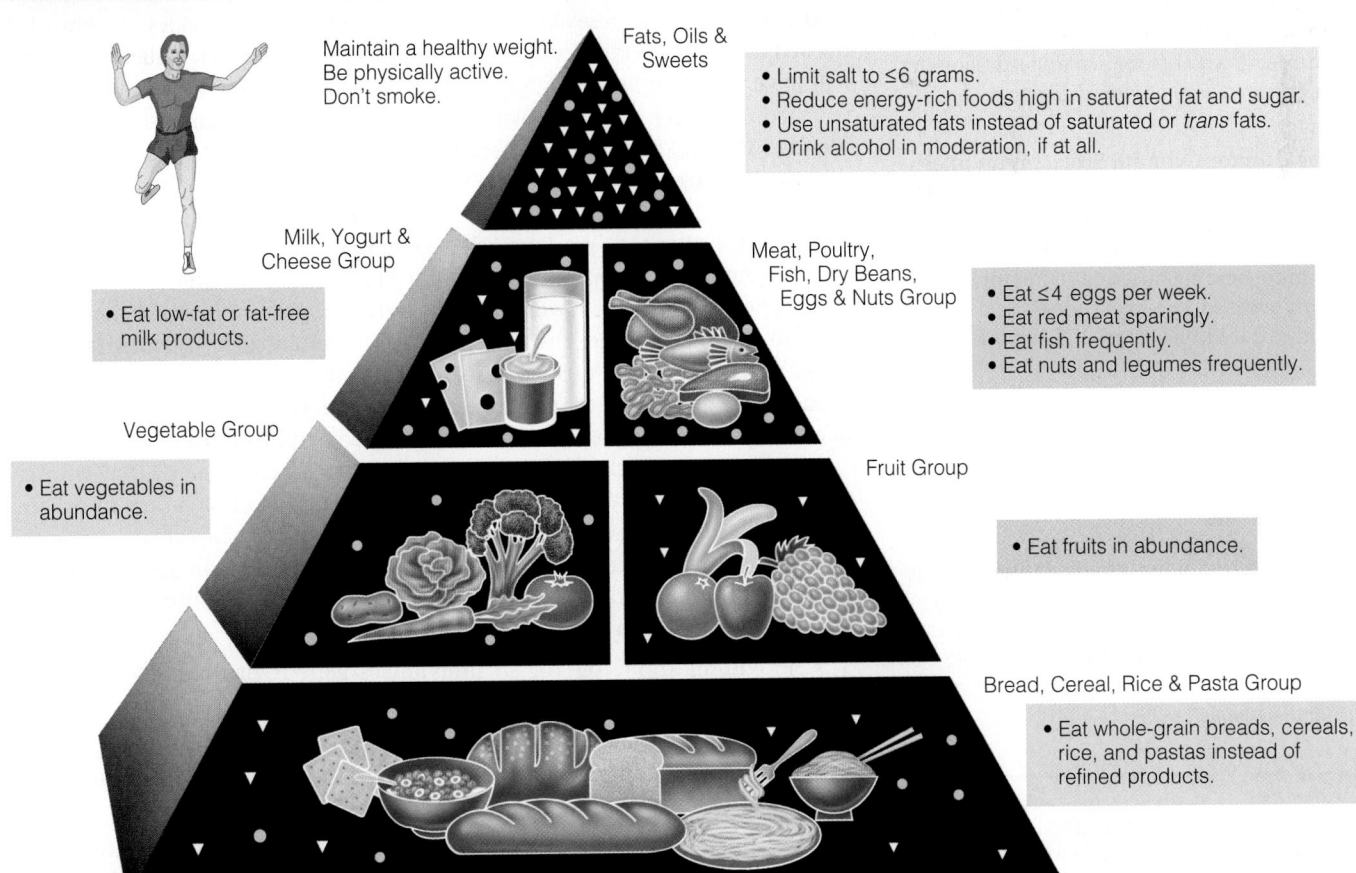

Maintain a healthy weight. Be physically active. Don't smoke.

Fats, Oils & Sweets
- Limit salt to ≤6 grams.
- Reduce energy-rich foods high in saturated fat and sugar.
- Use unsaturated fats instead of saturated or *trans* fats.
- Drink alcohol in moderation, if at all.

Milk, Yogurt & Cheese Group
- Eat low-fat or fat-free milk products.

Meat, Poultry, Fish, Dry Beans, Eggs & Nuts Group
- Eat ≤4 eggs per week.
- Eat red meat sparingly.
- Eat fish frequently.
- Eat nuts and legumes frequently.

Vegetable Group
- Eat vegetables in abundance.

Fruit Group
- Eat fruits in abundance.

Bread, Cereal, Rice & Pasta Group
- Eat whole-grain breads, cereals, rice, and pastas instead of refined products.

Physical activity and a moderate weight loss of even 10 to 20 pounds can help improve blood glucose, blood lipids, and blood pressure.

Recommendations for the Population The recommendations to prevent chronic diseases address the general population■ in the hope that all people at all levels of risk may benefit. Such a strategy is similar to national efforts to vaccinate to prevent measles, fluoridate water to prevent dental caries, and fortify grains with folate to prevent neural tube defects.

Recommendations for Individuals People's hereditary susceptibility to diseases and their responsiveness to dietary measures vary. Unlike nutrient-deficiency diseases, which develop when nutrients are lacking and disappear when the nutrients are provided, chronic diseases are neither caused nor prevented by diet alone. Many people have followed dietary advice and developed heart disease or cancer anyway; others have ignored all advice and lived long and healthy lives. For many people, though, diet does influence the time of onset and course of some chronic diseases, and many health care professionals urge dietary measures as part of a disease-prevention strategy.

To determine whether dietary recommendations are important to you personally, look at your family history to see which diseases are common to your relatives. In addition, examine your personal history, taking note of your blood pressure, blood lipid profile, and lifestyle habits such as smoking and physical activity.

Increase the proportion of physician office visits made by patients with a diagnosis of cardiovascular disease, diabetes, or hyperlipidemia that include counseling or education related to diet and nutrition.

■ Recommendations that urge all people to make dietary changes believed to forestall or prevent diseases are taking a *preventive* or *population approach*. Alternatively, recommendations that urge dietary changes only for people who are known to need them are taking a *medical* or *individual approach*.

■ The **Human Genome Project** is an international project whose purpose is to determine the sequence of the human genome, develop genetic and physical maps of the human genome, locate and identify human genes, and explore the ethical, legal, and social implications of this work.

■ Nutrients and nonnutrients influence gene expression, which in turn affects growth and metabolism. Genes also influence the activities of nutrients in the body. The study of these relationships is called *nutritional genomics*.

genome (GEE-nome): the full complement of genetic material (DNA) in the chromosomes of a cell. In human beings, the genome consists of 46 chromosomes. The study of genomes is called **genomics**.

Recommendations for Each Individual Even when recommendations are made "for individuals," they apply to large groups of people—those with hypertension or those with diabetes, for example. But that's expected to change in the next decade or so as research on the human **genome**■ provides the knowledge needed to create *specific* recommendations for *each* individual.[79] Such information will have major implications for society in general, and for health care in particular.[80] For example, one eventual goal is to provide personalized dietary recommendations based on a person's predicted response to nutrients derived from the genetic profiling of the individual.[81]

In the past, our understanding of the relationship between genetics and health was relatively limited. We knew that genetic diseases were caused by an extra chromosome (as in Down syndrome) or a missing chromosome or part of a chromosome or by a specific mutation in a single gene (as in phenylketonuria, PKU). The cause and effect were clear—those with the genetic defect got the disease and those without it, didn't. Such diseases are genetically predetermined, and although they greatly affect those touched by them, they are relatively rare. Consequently, genetics has played a relatively small role in health care.

In contrast, most chronic diseases have a genetic component that *predisposes* the prevention or development of a disease, depending on a variety of other factors (such as diet and physical activity). Because these diseases are quite common, affecting virtually everyone, an understanding of the human genome will have widespread ramifications for health care. This new understanding of the human genome is expected to change health care by:

• Providing knowledge of an individual's genetic predisposition to specific diseases.

• Allowing physicians to develop "designer" therapies—prescribing the most effective schedule of screening, behavior changes (including diet),■ and medical interventions based on each individual's genetic profile.

• Enabling manufacturers to create new medications for each genetic variation so that physicians can prescribe the best medicine in the exact dose and frequency to enhance effectiveness and eliminate side effects.

- Providing a better understanding of the nongenetic factors that influence disease development.

Using a cotton swab to gather a person's genetic profile from the inside of the cheek and then transcribing that information onto a wallet-size identification card may sound like a scene from a science fiction movie. Yet such a scenario is expected to be a reality in our lifetimes.

IN SUMMARY Clearly, optimal nutrition plays a key role in keeping people healthy and reducing the risk of chronic diseases.[82] To have the greatest impact possible, dietary recommendations are aimed at the entire population, and not just at the individuals who might benefit most. Recommendations focus on weight control and urge people to limit fat, increase complex carbohydrates, and balance food intake with activity. A person can do no better than to incorporate those suggestions into his or her daily life.

Nutrition in Your Life

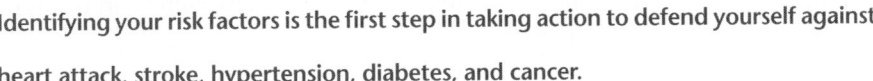

Identifying your risk factors is the first step in taking action to defend yourself against

heart attack, stroke, hypertension, diabetes, and cancer.

- Do you have a personal or family history of heart disease, hypertension,

  diabetes, or cancer?

- Are you sedentary or overweight? Do you smoke cigarettes?

- Do you have high blood cholesterol or high blood pressure?

NUTRITION ON THE NET

 Access these websites for further study of topics covered in this chapter.

- Find updates and quick links to these and other nutrition-related sites at our website: **www.wadsworth.com/nutrition**

- Find AIDS information at the Office of AIDS Research: **www.nih.gov/od/oar**

- Learn about HIV infections: **www.hivpositive.com**

- Review resources offered by the National Center for Chronic Disease Prevention and Health Promotion: **www.cdc.gov/nccdphp**

- Find information on health at the NIH Consumer Health Information site: **www.health.nih.gov**

- Visit the National Health Information Center site: **www.health.gov**

- Find information about health statistics at the National Center for Health Statistics site: **www.cdc.gov/nchswww**

- Search for "chronic diseases," "disease prevention," "men's health," "women's health," "heart disease," "stroke," "high blood pressure," "cancer," and "diabetes" at the U.S. Government site: **www.healthfinder.gov**

- Learn about women's health from the National Women's Health Information Center site: **www.4women.org**

- Review the Surgeon General's Reports on Physical Activity and Health and Reducing Tobacco Use: **www.surgeongeneral.gov/library**

- Assess your heart disease risk at the American Heart Association site: **www.americanheart.org**

- Visit the National Stroke Association: **www.stroke.org**

- Find information on the DASH diet: **dash.bwh.harvard.edu**

- Visit the National Heart, Lung, and Blood Institute site and click on "clinical practice guidelines" for cholesterol and hypertension guidelines: **www.nhlbi.nih.gov/index.htm**

- Learn about diabetes at the American Diabetes Association site: **www.diabetes.org**

- Learn about diabetes at the Canadian Diabetes Association site: **www.diabetes.ca**

- Visit the National Institute of Diabetes & Digestive & Kidney Diseases site: **www.niddk.nih.gov**

- Visit the National Cancer Institute site: **www.nci.nih.gov** or **www.cancer.gov**

- Visit the American Cancer Society site: **www.cancer.org**

- Learn about cancer research from the American Institute for Cancer Research site: **www.aicr.org**

- Find resources to help you quit smoking: **www.lungusa.org**

- Visit the CDC National Prevention Information Network: **www.cdcnpin.org**

STUDY QUESTIONS

These questions will help you review the chapter. You will find the answers in the discussions on the pages provided.

1. How do the major diseases of today as a group differ from those of several decades ago as a group? Why is nutrition considered so important in connection with today's major diseases? (pp. 613–614)

2. What is HIV infection? What are the consequences of HIV infection? What is the HIV wasting syndrome? (pp. 615–617)

3. In what ways might good nutrition status alter the course of HIV infection? (pp. 617–618)

4. Identify the major diet-related risk factors for atherosclerosis, hypertension, diabetes, and cancer. (p. 619)

5. Describe some ways in which people can alter their diets to lower their blood cholesterol levels. (pp. 625–628)

6. Describe some steps that people with hypertension can take to lower their blood pressure. (pp. 630–631)

7. Name the two major types of diabetes and describe some differences between them. How do dietary recommendations for diabetes compare with the healthy diet recommended for all people? (pp. 632–636)

8. Differentiate between cancer initiators, promoters, and antipromoters. Which nutrients or foods fit into each of these categories? (pp. 638–640)

9. Describe the characteristics of a diet that might offer the best protection against the onset of cancer. (p. 640)

These multiple choice questions will help you prepare for an exam. Answers can be found on p. 646.

1. The immune cells most seriously damaged by HIV are:
   a. B-cells.
   b. T-cells.
   c. antigens.
   d. immunoglobulins.

2. People with HIV infections are most susceptible to:
   a. diabetes.
   b. hypertension.
   c. heart attacks.
   d. food poisoning.

3. The leading cause of death in the United States is:
   a. AIDS.
   b. cancer.
   c. diabetes.
   d. heart disease.

4. Plaques in the arteries contribute to the development of:
   a. cancer.
   b. diabetes.
   c. atherosclerosis.
   d. infectious diseases.

5. Which blood lipid correlates directly with heart disease?
   a. HDL
   b. LDL
   c. VLDL
   d. triglycerides

6. Moderate amounts of alcohol may protect against heart disease by:
   a. promoting LDL oxidation.
   b. preventing clot formation.
   c. raising LDL and lowering HDL.
   d. accelerating plaque formation.

7. What is the most effective strategy for most people to lower their blood pressure?
   a. lose weight
   b. restrict salt
   c. monitor glucose
   d. supplement protein

8. In diabetes, when glucose fails to enter the cells and metabolism shifts to break down protein and fat for energy:
   a. weight gain occurs.
   b. hypertension worsens.
   c. atherosclerosis develops.
   d. ketone production increases.

9. The most important dietary strategy in diabetes is to:
   a. provide for a consistent carbohydrate intake.
   b. restrict fat to 30 percent of daily kcalories.
   c. limit carbohydrate intake to 300 milligrams a day.
   d. take multiple vitamin and mineral supplements daily.

10. Which of the following help(s) to protect against cancer?
    a. alcohol
    b. pickled foods
    c. phytochemicals
    d. omega-6 fatty acids

REFERENCES

1. Centers for Disease Control and Prevention, World TB Day—March 24, 2002, *Morbidity and Mortality Weekly Reports* 51 (2002): 229; H. Gillespie, Antibiotic resistance in the absence of selective pressure, *International Journal of Antimicrobial Agents* 17 (2001): 171–176; D. K. Warren and V. J. Fraser, Infection control measures to limit antimicrobial resistance, *Critical Care Medicine* 29 (2001): N128–N134; S. Binder, A. M. Levitt, and J. M. Hughes, Preventing emerging infectious diseases as we enter the 21st century: CDC's strategy, *Public Health Reports* 114 (1999): 130–134.

2. Committee on Dietary Reference Intakes, *Dietary Reference Intakes for Energy, Carbohydrate, Fiber, Fat, Fatty Acids, Cholesterol, Protein, and Amino Acids* (Washington, D.C.: National Academies Press, 2002), Chapter 11; F. B. Hu and W. C. Willett, Optimal diets for prevention of coronary heart disease, *Journal of the American Medical Association* 288 (2002): 2569–2578; Position of the American Dietetic Association: The role of dietetics professionals in health promotion and disease prevention, *Journal of the American Dietetic Association* 102 (2002): 1680–1687; K. P. Moritsugu, A report from the office of the Surgeon General, *Journal of the American Dietetic Association* 100 (2000): 1013–1014.

3. G. T. Keusch, The history of nutrition: Malnutrition, infection and immunity, *Journal of Nutrition* 133 (2003): 336S–340S; C. J. Field, I. R. Johnson, and P. D. Schley, Nutrients and their role in host resistance to infection, *Journal of Leukocyte Biology* 71 (2002): 16–32.

4. P. Bhaskaram, Micronutrient malnutrition, infection, and immunity: An overview, *Nutrition Reviews* 60 (2002): S40–S45.

5. Bhaskaram, 2002; Position of the American Dietetic Association and Dietitians of Canada: Nutrition intervention in the care of persons with human immunodeficiency virus infection, *Journal of the American Dietetic Association* 100 (2000): 708–717.

6. J. D. Bogden and coauthors, Status of selected nutrients and progression of human immunodeficiency virus type 1 infection, *American Journal of Clinical Nutrition* 72 (2000): 809–815; Position of the American Dietetic Association and Dietitians of Canada, 2000.

7. M. K. Baum and coauthors, Zinc status in human immunodeficiency virus type 1 infection and illicit drug use, *Clinical Infectious Diseases* 37 (2003): S117–S123.

8. Position of the American Dietetic Association and Dietitians of Canada, 2000.

9. A. M. Miniño and coauthors, Deaths: Final data for 2000, *National Vital Statistics Report*, September 16, 2002.

10. Executive Summary: Joint WHO/FAO Expert Report on Diet, Nutrition, and the Prevention of Chronic Disease, www.who.int/hpr/nutrition/expertconsultationge.htm.

11. American Heart Association, *Heart Disease and Strokes Statistics—2003 Update*, www.americanheart.org/statistics/cvd.html, site visited September 18, 2003.

12. Hu and Willett, 2002; R. P. Lauber and N. F. Sheard, The American Heart Association Dietary Guidelines for 2000: A summary report, *Nutrition Reviews* 59 (2001): 298–306.

13. F. Pellegatta and coauthors, Different short- and long-term effects of resveratrol on nuclear factor-kB phosphorylation and nuclear appearance in human endothelial cells, *American Journal of Clinical Nutrition* 77 (2003): 1220–1228; R. Ross, Atherosclerosis—An inflammatory disease, *New England Journal of Medicine* 340 (1999): 115–126.

14. P. M. Ridker and coauthors, Comparison of C-reactive protein and low-density lipoprotein cholesterol levels in the prediction of first cardiovascular events, *New England Journal of Medicine* 347 (2002): 1557–1565.

15. M. Laidlaw and B. J. Holub, Effects of supplementation with fish oil-derived n-3 fatty acids and γ-linolenic acid on circulating plasma lipids and fatty acid profiles in women, *American Journal of Clinical Nutrition* 77 (2003): 37–42; P. J. H. Jones, Effect of n-3 polyunsaturated fatty acids on risk reduction of sudden death, *Nutrition Reviews* 60 (2002): 407–413; W. E. Connor, Importance of n-3 fatty acids in health and disease, *American Journal of Clinical Nutrition* 71 (2000): 171S–175S.

16. E. S. Ford, W. H. Giles, and W. H. Dietz, Prevalence of the metabolic syndrome among US adults: Findings from the Third National Health and Nutrition Examination Survey, *Journal of the American Medical Association* 287 (2002): 356–359; G. S. Berenson and coauthors, Association between multiple cardiovascular risk factors and atherosclerosis in children and young adults, *New England Journal of Medicine* 338 (1998): 1650–1656.

17. M. Szklo and coauthors, Trends in plasma cholesterol levels in the Atherosclerosis Risk in Communities (ARIC) study, *Preventive Medicine* 30 (2000): 252–259.

18. Joint National Committee, *Prevention, Detection, Evaluation, and Treatment of High Blood Pressure, Seventh Report*, NIH publication no. 03-5233 (Bethesda, Md.: National Heart, Lung, and Blood Institute, 2003), pp. 1–3.

19. Expert Panel on Detection, Evaluation, and Treatment of High Blood Cholesterol in Adults (Adult Treatment Panel III), *Third Report of the National Cholesterol Education Program (NCEP)*, NIH publication no. 02-5215 (Bethesda, Md.: National Heart, Lung, and Blood Institute, 2002), p. II-18.

20. Expert Panel on Detection, Evaluation, and Treatment of High Blood Cholesterol in Adults (Adult Treatment Panel III), 2002, p. VIII-2.

21. Expert Panel on Detection, Evaluation, and Treatment of High Blood Cholesterol in Adults (Adult Treatment Panel III), 2002, p. II-19.

22. Expert Panel on Detection, Evaluation, and Treatment of High Blood Cholesterol in Adults (Adult Treatment Panel III), 2002, pp. II-1–II-4.

23. J. A. Beckman, M. A. Creager, and P. Libby, Diabetes and atherosclerosis: Epidemiology, pathophysiology, and management, *Journal of the American Medical Association* 287 (2002): 2570–2581.

24. Expert Panel on Detection, Evaluation, and Treatment of High Blood Cholesterol in Adults (Adult Treatment Panel III), 2002, pp. II-16, 11-50–11-53.

25. S. Bioletto and coauthors, Acute hyperinsulinemia and very-low-density and low-density lipoprotein subfractions in obese subjects, *American Journal of Clinical Nutrition* 71 (2000): 443–449; A. Must and coauthors, The disease burden associated with overweight and obesity, *Journal of the American Medical Association* 282 (1999): 1523–1529.

26. Expert Panel on Detection, Evaluation, and Treatment of High Blood Cholesterol in Adults (Adult Treatment Panel III), 2002, p. II-16.

27. H. M. Lakka and coauthors, The metabolic syndrome and total and cardiovascular disease mortality in middle-aged men, *Journal of the American Medical Association* 288 (2002): 2709–2716.

28. Ford, Giles, and Dietz, 2002.

29. Expert Panel on Detection, Evaluation, and Treatment of High Blood Cholesterol in Adults (Adult Treatment Panel III), 2002, p. III-6.

30. R. M. Krauss and coauthors, AHA dietary guidelines revision 2000: A statement for healthcare professionals from the Nutrition Committee of the American Heart Association, *Circulation* 102 (2000): 2284–2299.

31. S. Liu and coauthors, Is intake of breakfast cereals related to total and cause-specific mortality in men? *American Journal of Clinical Nutrition* 77 (2003): 594–599; L. A. Bazzano and coauthors, Fruit and vegetable intake and risk of cardiovascular disease in US adults: The first National Health and Nutrition Examination Survey Epidemiologic Follow-up Study, *American Journal of Clinical Nutrition* 76 (2002): 93–99.

32. D. S. Wald, Homocysteine and cardiovascular disease: Evidence on causality from a meta-analysis, *British Medical Journal* 325 (2002): 1202.

33. R. N. Lemaitre and coauthors, n-3 polyunsaturated fatty acids, fatal ischemic heart disease, and nonfatal myocardial infarction in older adults: The Cardiovascular Health Study, *American Journal of Clinical Nutrition* 77 (2003): 319–325; C. M. Albert and coauthors, Blood levels of long chain n-3 polyunsaturated fatty acids and the risk of sudden death, *New England Journal of Medicine* 346 (2002): 1113–1118; P. Nestel and coauthors, The n-3 fatty acids eicosapentaenoic acid and docosahexaenoic acid increase systemic arterial compliance in humans, *American Journal of Clinical Nutrition* 76 (2002): 326–330; W. S. Harris, Cardioprotective effects of σ-3 fatty acids, *Nutrition in Clinical Practice* 16 (2001): 6–12; C. R. Harper and T. A. Jacobson, The role of omega-3 fatty acids in the prevention of coronary heart disease, *Archives of Internal Medicine* 161 (2001): 2185–2192.

34. P. M. Kris-Etherton, W. S. Harris, and L. J. Appel, Fish consumption, fish oil, omega-3 fatty acids, and cardiovascular disease, *Circulation* 106 (2002): 2747–2757; C. Von Shacky and coauthors, The effect of dietary omega-3 fatty acids on coronary atherosclerosis: A randomized, double-blind, placebo-controlled trial, *Annals of Internal Medicine* 130 (1999): 554–562.

35. Expert Panel on Detection, Evaluation, and Treatment of High Blood Cholesterol in Adults (Adult Treatment Panel III), 2002, p. V-13.

36. C. A. Spilburg and coauthors, Fat-free foods supplemented with soy stanol-lecithin powder reduce cholesterol absorption and LDL cholesterol, *Journal of the American Dietetic Association* 103 (2003): 577–581.

37. R. E. Ostlund, Phytosterols in human nutrition, *Annual Review of Nutrition* 22 (2002): 533–549.

38. M. J. Mussner and coauthors, Effects of phytosterol ester-enriched margarine on plasma lipoproteins in mild to moderate hypercholesterolemia are related to basal cholesterol and fat intake, *Metabolism* 51 (2002): 189–194.

39. J. R. Crouse and coauthors, A randomized trial comparing the effect of casein with that of soy protein containing varying amounts of isoflavones on plasma concentrations of lipids and lipoproteins, *Archives of Internal Medicine* 159 (1999): 2070–2072.

40. S. R. Teixeira and coauthors, Effects of feeding 4 levels of soy protein for 3 and 6 weeks on blood lipids and apolipoproteins in moderately hypercholesterolemic men, *American Journal of Clinical Nutrition* 71 (2000): 1077–1084.

41. K. J. Mukamal and coauthors, Roles of drinking pattern and type of alcohol consumed in coronary heart disease in men, *New England Journal of Medicine* 348 (2003): 109–118; M. J. Davies and coauthors, Effects of moderate alcohol intake on fasting insulin and glucose concentrations and insulin sensitivity in postmenopausal women: A randomized controlled trial, *Journal of the American Medical Association* 287 (2002): 2559–2562; D. J. Baer and coauthors, Moderate alcohol consumption lowers risk factors for cardiovascular disease in postmenopausal women fed a controlled diet, *American Journal of Clinical Nutrition* 75 (2002): 593–599.

42. C. L. Hart and coauthors, Alcohol consumption and mortality from all causes, coronary heart disease, and stroke: Results from a prospective cohort study of Scottish men with 21 years of follow up, *British Medical Journal* 318 (1999): 1725–1729.

43. L. M. Hines and coauthors, Genetic variation in alcohol dehydrogenase and the beneficial effect of moderate alcohol consumption on myocardial infarction, *New England Journal of Medicine* 344 (2001): 549–555.

44. F. W. Booth and coauthors, Waging war on modern chronic diseases: Primary prevention through exercise biology, *Journal of Applied Physiology* 88 (2000): 774–787; F. B. Hu and coauthors, Physical activity and risk of stroke in women, *Journal of the American Medical Association* 283 (2000): 2961–2967.

45. I. Hajjar and T. A. Kotchen, Trends in prevalence, awareness, treatment, and control of hypertension in the United States, 1988–2000, *Journal of the American Medical Association* 290 (2003): 199–206; Joint National Committee, 2003, pp. 1–3.

46. J. A. Staessen, T. Kuznetsova, and K. Stolarz, Hypertension prevalence and stroke mortality across populations, *Journal of the American Medical Association* 289 (2003): 2420–2422.

47. F. M. Sacks and coauthors, Effects on blood pressure of reduced sodium and the Dietary Approaches to Stop Hypertension (DASH) diet: DASH-Sodium Collaborative Research Group, *New England Journal of Medicine* 344 (2001): 3–10.

48. E. Obarzanek and coauthors, Effects on blood lipids of a blood pressure-lowering diet: The Dietary Approaches to Stop Hypertension (DASH) Trial, *American Journal of Clinical Nutrition* 74 (2001): 80–89.

49. The Expert Committee on the Diagnosis and Classification of Diabetes Mellitus, Report of the Expert Committee on the Diagnosis and Classification of Diabetes Mellitus, *Diabetes Care* 26 (2003): S5–S20.

50. The Expert Committee on the Diagnosis and Classification of Diabetes Mellitus, 2003.

51. The Expert Committee on the Diagnosis and Classification of Diabetes Mellitus, 2003.

52. D. H. Amschler, The alarming increase of type 2 diabetes in children, *Journal of School Health* 72 (2002): 39–41.

53. Centers for Disease Control and Prevention, Chronic Disease Prevention, Diabetes: Disabling, deadly, and on the rise, *At A Glance,* 2003, available from www.cdc.gov/nccdphp/aag/aag_ddt.htm; J. V. Neel, The "thrifty genotype" in 1998, *Nutrition Reviews* 57 (1999): S2–S9; C. D. Berdanier, Diabetes mellitus: A genetic disease, *Nutrition Today* 34 (1999): 89–98.

54. C. M. Clark and coauthors, Promoting early diagnosis and treatment of type 2 diabetes, *Journal of the American Medical Association* 284 (2000): 363–364.

55. P. Gæde and coauthors, Multifactorial intervention and cardiovascular disease in patients with type 2 diabetes, *New England Journal of Medicine* 348 (2003): 383–393.

56. American Diabetes Association, Position statement: Evidence-based nutrition principles and recommendations for the treatment and prevention of diabetes and related complications, *Diabetes Care* 25 (2002): S50–S60.

57. American Diabetes Association, 2002.

58. American Diabetes Association, Position statement: Physical activity/exercise and diabetes mellitus, *Diabetes Care* 26 (2003): S73–S77.

59. American College of Sports Medicine, Position stand: Exercise and type 2 diabetes, *Medicine and Science in Sports and Exercise* 32 (2000): 1345–1360.

60. American Diabetes Association, 2003.

61. Y. Mao and coauthors, Physical inactivity, energy intake, obesity and the risk of rectal cancer in Canada, *International Journal of Cancer* 105 (2003): 831–837; A. S. Furberg and I. Thune, Metabolic abnormalities (hypertension, hyperglycemia and overweight) lifestyle (high energy intake and physical inactivity) and endometrial cancer risk in a Norwegian cohort, *International Journal of Cancer* 104 (2003): 669–676; E. Giovannucci, Diet, body weight, and colorectal cancer: A summary of the epidemiologic evidence, *Journal of Women's Health* 12 (2003): 173–182; H. Vainio, R. Kaaks, and F. Bianchini, Weight control and physical activity in cancer prevention: International evaluation of the evidence, *European Journal of Cancer Prevention* 2 (2002): S94–S100.

62. National Cancer Policy Board, Institute of Medicine, S. J. Curry, T. Byers, and M. Hewitt, eds., *Fulfilling the Potential of Cancer Prevention and Early Detection* (Washington, D.C.: National Academies Press, 2003), pp. 58–61; M. L. Slattery and coauthors, Lifestyle and colon cancer: An assessment of factors associated with risk, *American Journal of Epidemiology* 150 (1999): 869–877.

63. National Cancer Policy Board, 2003, pp. 59–60; J. B. Barnett, The relationship between obesity and breast cancer risk and mortality, *Nutrition Reviews* 61 (2003): 73–76.

64. E. Calle and coauthors, Overweight, obesity, and mortality from cancer in a prospectively studied cohort of U.S. adults, *New England Journal of Medicine* 348 (2003): 1625–1638; National Cancer Policy Board, 2003, pp. 61–66.

65. Barnett, 2003.

66. Food and Nutrition Science Alliance (FANSA) Statement on diet and cancer prevention in the United States, November 1999.

67. T. Norat and E. Riboli, Meat consumption and colorectal cancer: A review of epidemiologic evidence, *Nutrition Reviews* 59 (2001): 37–47.

68. P. L. Zock, Dietary fats and cancer, *Current Opinions in Lipidology* 12 (2001): 5–10.

69. Zock, 2001.

70. Zock, 2001; M. D. Holmes and coauthors, Association of dietary intake of fat and fatty acids with the risk of breast cancer, *Journal of the American Medical Association* 281 (1999): 914–920.

71. P. Terry and coauthors, Fatty fish consumption and risk of prostate cancer, *Lancet* 357 (2001): 1764–1766; Zock, 2001; W. C. Willet, Diet and cancer, *Oncologist* 5 (2000): 393–404.

72. National Cancer Policy Board, 2003, p. 77; P. D. Terry, T. E. Rohan, and A. Wolk, Intakes of fish and marine fatty acids and the risks of cancers of the breast and prostate and of other hormone-related cancers: A review of the epidemiologic evidence, *American Journal of Clinical Nutrition* 77 (2003): 532–543.

73. B. S. Reddy, Role of dietary fiber in colon cancer: An overview, *American Journal of Medicine* 106 (1999): S16–S19; D. Kritchevsky, Protective role of wheat bran fiber: Preclinical data, *American Journal of Medicine* 106 (1999): S28–S31; M. C. Jansen and coauthors, Dietary fiber and plant foods in relation to colorectal cancer mortality: The Seven Countries Study, *International Journal of Cancer* 81 (1999): 174–179.

74. A. Flood and coauthors, Fruit and vegetable intakes and the risk of colorectal cancer in the Breast Cancer Detection Demonstration Project follow-up cohort, *American Journal of Clinical Nutrition* 75 (2002): 936–943; A. Schatzkin and coauthors, Lack of effect of a low-fat, high-fiber diet on the recurrence of colorectal adenomas, *New England Journal of Medicine* 342 (2000): 1149–1155; C. S. Fuchs and coauthors, Dietary fiber and the risk of colorectal cancer and adenoma in women, *New England Journal of Medicine* 340 (1999): 169–176.

75. Bazzano and coauthors, 2002.

76. American Cancer Society 2001 Nutrition and Physical Activity Guidelines Advisory Committee, *American Cancer Society Guidelines on Nutrition and Physical Activity for Cancer,* available from www.cancer.org or upon request from the American Cancer Society (800) ACS-2345.

77. R. J. Deckelbaum and coauthors, AHA Conference Proceedings—Summary of a scientific conference on preventive nutrition: Pediatrics to geriatrics, *Circulation* 100 (1999): 450–456.

78. Hu and Willett, 2002.

79. F. S. Collins and V. A. McKusick, Implications of the Human Genome Project for medical science, *Journal of the American Medical Association* 285 (2001): 540–544.

80. N. Fogg-Johnson and J. Kaput, Nutrigenomics: An emerging scientific discipline, *Food Technology* 57 (2003): 60–67; R. Weinshilboum, Inheritance and drug response, *New England Journal of Medicine* 348 (2003): 529–537; A. E. Guttmacher and F. S. Collins, Genomic medicine—A primer, *New England Journal of Medicine* 347 (2002): 1512–1520; A. Guttmacher of the National Human Genome Research Institute, speaking at the annual meeting of the American Dietetic Association, October 2000; M. J. Friedrich, Relating genomic research to patient care, *Journal of the American Medical Association* 284 (2000): 2581–2582; M. J. Friedrich, Genetic screening to offset adult disease, *Journal of the American Medical Association* 284 (2000): 2308.

81. P. Trayhurn, Nutritional genomics—"Nutrigenomics," *British Journal of Nutrition* 89 (2003): 1–2.

82. Position of the American Dietetic Association: The role of nutrition in health promotion and disease prevention programs, *Journal of the American Dietetic Association* 98 (1998): 205–208.

ANSWERS

Study Questions (multiple choice)

1. b 2. d 3. d 4. c 5. b 6. b 7. a 8. d 9. a 10. c

HIGHLIGHT

Complementary and Alternative Medicine

© Art Montes De Oca/Taxi/Getty Images

If you suffered from migraine headaches or severe joint pain, where would you turn for relief? Would you visit a physician? Or are you more likely to go to an herbalist or an acupuncturist? Most physicians diagnose and treat medical conditions in ways that are accepted by the established medical community; herbalists and acupuncturists, among others, offer alternatives to standard medical practice. Instead of taking two aspirin, for example, you might be advised to chew two fresh leaves of the herb feverfew or to swallow a tincture of white willow bark. Or you might receive a massage and several acupuncture needles.

Complementary and alternative medicine has become increasingly popular in recent decades (see the accompanying glossary for this and related terms).[1] People use these therapies for a variety of reasons. Some want to take more responsibility for both maintaining their own health and finding cures for their own diseases, especially when traditional medical therapies prove ineffective. Others have become distrustful of, and feel overwhelmed by, the high-tech diagnostic tests and costly treatments that **conventional medicine** offers. This highlight explores alternative therapies in search of their possible benefits and with an awareness of their potential harms.

Defining Complementary and Alternative Medicine

By definition, complementary and alternative medicine is not conventional medicine. It includes a variety of approaches, philosophies, and treatments, some of which are defined in the glossary of alternative therapies on the next page. When these therapies are used instead of conventional medicine, they are called *alternative;* when used together with conventional medicine, they are called *complementary.*

A growing number of health care professionals are learning about alternative therapies; half of U.S. medical schools now offer elective courses in alternative medicine, and even more include discussions of these therapies in their required courses.[2] By incorporating some of the beneficial alternative therapies into their practices, an approach called **integrative medicine,** health care professionals take advantage of the best of both kinds of medicine.[3] To best serve their clients, these health care professionals provide balanced advice, guard against bias, and maintain trusting relationships.[4]

For some alternative therapies, preliminary and limited scientific evidence suggests some effectiveness; but for most, well-designed scientific studies have yet to determine safety and effectiveness.[5] If proved safe and effective, an alternative therapy may be adopted by conventional medicine. Cancer radiation therapy, for example, was once considered an unconventional therapy, but it proved its clinical value and became part of accepted medical practice. In some cases, a therapy that is accepted by conventional medicine for a specific ailment is used for a different purpose in an alternative therapy. For example, chelation therapy, the preferred medical treatment for lead poisoning, is a common alternative therapy for cardiovascular disease.

Sound Research, Loud Controversy

Much information on alternative therapies comes from folklore, tradition, and testimonial accounts. Relatively few clinical trials have been conducted. Consequently, scientific evidence proving

GLOSSARY OF ALTERNATIVE THERAPIES

acupuncture (AK-you-PUNK-cher): a technique that involves piercing the skin with long thin needles at specific anatomical points to relieve pain or illness. Acupuncture sometimes uses heat, pressure, friction, suction, or electromagnetic energy to stimulate the points.

aroma therapy: a technique that uses oil extracts from plants and flowers (usually applied by massage or baths) to enhance physical, psychological, and spiritual health.

ayurveda (AH-your-VAY-dah): a traditional Hindu system of improving health by using herbs, diet, meditation, massage, and yoga to stimulate the body, mind, and spirit to prevent and treat disease.

bioelectromagnetic medical applications: the use of electrical energy, magnetic energy, or both to stimulate bone repair, wound healing, and tissue regeneration.

biofeedback: the use of special devices to convey information about heart rate, blood pressure, skin temperature, muscle relaxation, and the like to enable a person to learn how to consciously control these medically important functions.

biofield therapeutics: a manual healing method that directs a healing force from an outside source (commonly God or another supernatural being) through the practitioner and into the client's body; commonly known as "laying on of hands."

cartilage therapy: the use of cleaned and powdered connective tissue, such as collagen, to improve health.

chelation (kee-LAY-shun) **therapy:** the use of ethylene diamine tetraacetic acid (EDTA) to bind with metallic ions, thus healing the body by removing toxic metals.

chiropractic (KYE-roh-PRAK-tik): a manual healing method of manipulating the spine to restore health.

faith healing: healing by invoking divine intervention without the use of medical, surgical, or other traditional therapy.

herbal (ERB-al) **medicine:** the use of plants to treat disease or improve health; also known as *botanical medicine* or *phytotherapy.*

homeopathy (hoh-me-OP-ah-thee): a practice based on the theory that "like cures like," that is, that substances that cause symptoms in healthy people can cure those symptoms when given in very dilute amounts.
- homeo = like
- pathos = suffering

hydrotherapy: the use of water (in whirlpools, as douches, or packed as ice, for example) to promote relaxation and healing.

hypnotherapy: a technique that uses hypnosis and the power of suggestion to improve health behaviors, relieve pain, and heal.

imagery: a technique that guides clients to achieve a desired physical, emotional, or spiritual state by visualizing themselves in that state.

iridology: the study of changes in the iris of the eye and their relationships to disease.

macrobiotic diets: extremely restrictive diets limited to a few grains and vegetables; based on metaphysical beliefs and not nutrition. A macrobiotic diet might consist of brown rice, miso soup, and sea vegetables, for example.

massage therapy: a healing method in which the therapist manually kneads muscles to reduce tension, increase blood circulation, improve joint mobility, and promote healing of injuries.

meditation: a self-directed technique of relaxing the body and calming the mind.

naturopathic (nay-chur-oh-PATH-ick) **medicine:** a system that taps the natural healing forces within the body by integrating several practices, including traditional medicine, herbal medicine, clinical nutrition, homeopathy, acupuncture, East Asian medicine, hydrotherapy, and manipulative therapy.

orthomolecular medicine: the use of large doses of vitamins to treat chronic disease.

ozone therapy: the use of ozone gas to enhance the body's immune system.

qi gong (chée GUNG): a Chinese system that combines movement, meditation, and breathing techniques to enhance the flow of qi (vital energy) in the body.

---

the safety and effectiveness of many alternative therapies is lacking. Some say that alternative therapies simply do not work; others suggest these therapies have not been given a fair trial. In an effort to "explore complementary and alternative healing practices through vigorous science," the National Center for Complementary and Alternative Medicine supports clinical trials of these therapies. Articles reporting the results of these clinical trials are available in a subset of PubMed created specifically for scientifically based, peer-reviewed journals on complementary and alternative therapies.

Sound research would answer two important questions. First, does the treatment offer better results than either doing nothing or giving a placebo? Second, do the benefits clearly outweigh the risks? Each of these points is worthy of elaboration.

Placebo Effect

Stories abound that credit alternative therapies with miraculous cures. Without scientific research to determine effectiveness, however, one is left to wonder whether it is the therapies or the placebo effect that produces the cure. Recall from Chapter 1 that giving a placebo often brings about a healing effect in people who believe they are receiving the treatment. Traditional medicine tends to neglect this powerful remedy, whereas many alternative therapies embrace it.

Risks versus Benefits

Ideally, a therapy provides benefits with little or no risk. Figure H18-1 presents several examples of herbal remedies that appear to be generally safe and reasonably effective in treating various conditions.[6] Such findings, if replicated, hold promise that these alternative therapies may one day be integrated into conventional medicine.

Some alternative therapies are innocuous, providing little or no benefit for little or no risk. Sipping a cup of warm tea with a pleasant aroma, for example, won't cure heart disease, but it may improve one's mood and help relieve tension. Given no physical hazard and little financial risk, such therapies are acceptable.

In contrast, other products and procedures are downright dangerous, posing great risks while providing no benefits. One example is the folk practice of geophagia (eating earth or clay), which

FIGURE H18-1 Examples of Herbal Remedies

Echinacea may help reduce the duration and severity of upper respiratory infections.

Ginkgo may slow the loss of cognitive function associated with age.

St. John's wort may be effective in treating mild depression.

American ginseng may improve glucose control in people with type 2 diabetes.

Saw palmetto may improve the symptoms associated with an enlarged prostate.

The gel of an aloe vera plant soothes a minor burn.

can cause GI impaction and impair iron absorption. Another is the taking of laetrile to treat cancer, which can cause cyanide poisoning. Clearly, such therapies are too harmful to be used.

Perhaps most controversial are alternative therapies that may provide benefits, but also carry significant, unknown, or debatable risks. Smoking marijuana is an example of such an alternative therapy.[7] The compounds in marijuana seem to provide relief from symptoms such as nausea, vomiting, and pain that commonly accompany cancer, AIDS, and other diseases, but also pose risks that some people, including many physicians, consider acceptable and others, mainly politicians, deem intolerable. Physicians have focused on individuals and recognize that marijuana stimulates the appetite in their nauseated clients;

politicians and others have focused on society and realize that marijuana is one of many drugs that can be abused. Figure H18-2 (on p. 650) summarizes the relationships between risks and benefits.

Nutrition-Related Alternative Therapies

Most alternative therapies fall outside the field of nutrition, but nutrition itself can be an alternative therapy. Furthermore, many alternative therapies prescribe specific dietary regimens

FIGURE H18-2   Risk-Benefit Relationships

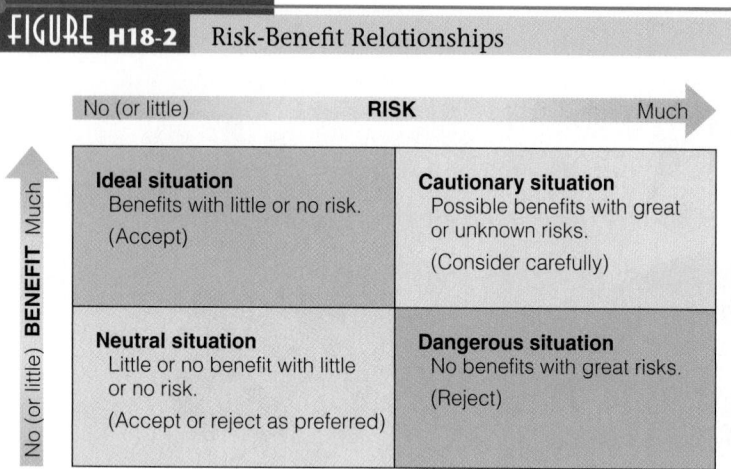

RISK — No (or little) → Much

BENEFIT — No (or little) ↑ Much

Ideal situation Benefits with little or no risk. (Accept)	**Cautionary situation** Possible benefits with great or unknown risks. (Consider carefully)
Neutral situation Little or no benefit with little or no risk. (Accept or reject as preferred)	**Dangerous situation** No benefits with great risks. (Reject)

even though most practitioners are not registered dietitians (see Highlight 1). Nutrition-related alternative therapies include the use of foods, vitamin and mineral supplements, and herbs to prevent and treat illnesses.

Foods

The many dietary recommendations presented throughout this text are based on scientific evidence and do *not* fall into the alternative therapies category; strategies that are still experimental, however, do. For example, alternative therapists may recommend macrobiotic diets to help prevent chronic diseases, whereas most registered dietitians would advise people to eat a balanced diet that includes at least five servings of fresh vegetables and fruits daily. Similarly, enough scientific evidence is available to recommend including soy protein in the diet to protect against heart disease, but not to determine whether the phytoestrogens of soy are safe or beneficial in managing the symptoms of menopause.

Highlight 13 explored the potential health benefits of soy and many other functional foods and concluded that no one food is magical. As part of a balanced diet, these foods can support good health and protect against disease. Importantly, the benefits derive from a variety of *foods*. More research is needed to determine the safety and effectiveness of taking supplements of the phytochemicals found in these foods.

Vitamin and Mineral Supplements

Like foods, vitamin and mineral supplements may fall into either the conventional or the alternative realm of medicine. For example, conventional advice recommends consuming 400 micrograms of folate to prevent neural tube defects, but not the taking of 1000 milligrams of vitamin C to prevent the common cold. Highlight 10 examined the appropriate use of supplements and potential dangers of excessive intakes.

As research on nutrition and chronic diseases has revealed many of the roles played by the vitamins and minerals in supporting health, conventional medicine has warmed up to the possibility that vitamin and mineral supplements might be an appropriate preventive therapy.[8] Some vitamin and mineral supplements appear to be in transition from alternative medicine to conventional medicine; that is, they have begun to prove their safety and effectiveness. Table H18-1 includes several nutrition-related therapies among those recognized to slow the progression of cancer and treat related symptoms. Herbal remedies, however, still remain clearly in the realm of complementary and alternative medicine.

Herbal Remedies

From earliest times, people have used myriad herbs and other plants to cure aches and ills with varying degrees of success (review Figure H18-1).[9] Upon scientific study, dozens of these folk remedies reveal their secrets.[10] For example, myrrh, a plant resin used as a painkiller in ancient times, does indeed have an analgesic effect. The herb valerian, which has long been used as a tranquilizer, contains oils that have a sedative effect. Senna leaves, brewed as a laxative tea, produce compounds that act as a potent cathartic drug. Green tea, brewed from the dried leaves of *Camellia sinensis,* contains phytochemicals that induce cancer cells to self-destruct. Naturally occurring salicylates provide the same protective effects as low doses of aspirin. Salicylates are found in spices such as curry, paprika, and thyme; fruits; vegetables; teas; and candies flavored with wintergreen (methylsalicylate).

Beneficial compounds from wild species contribute to about half of our modern medicines. By analyzing these compounds, pharmaceutical labs can synthesize pure forms of the drugs. Unlike herbs and wild species, which vary from batch to batch, synthesized medicines deliver exact dosages. By synthesizing drugs, we are also able to conserve endangered species. Consider that it took all of the bark from one 40-foot-tall, 100-

Digoxin, the most commonly prescribed heart medication, derives from the leaves of the foxglove plant *(Digitalis purpurea).*

TABLE H18-1 Advice and Precautions on Alternative Therapies for Cancer and Related Conditions	
Therapy	**Precautions**
Accept/Consider Recommending—Evidence supports effectiveness and safety.	
Vitamin E (for prostate cancer)	Not appropriate for people with a low platelet count; those taking anticoagulant medications; or those undergoing radiation, chemotherapy, or surgery
Acupuncture (for nausea and vomiting)	Not appropriate for people with a low platelet count or those taking anticoagulant medications
Massage (for anxiety, nausea, and lymph drainage)	Not appropriate directly over tumors, stents, or prosthetic devices and in areas damaged by surgery or radiation; or in people with bleeding abnormalities
Accept—Evidence supports safety, but inconclusive on effectiveness.	
Low-fat diet (for breast and prostate cancer)	Not appropriate for people with poor nutrition status
Macrobiotic diet[a]	Not appropriate for people with poor nutrition status or those who have breast or endometrial cancer
Vitamin E (for some cancers)	Not appropriate for people with a low platelet count; those taking anticoagulant medications; or those undergoing radiation, chemotherapy, or surgery
Soy (for prostate cancer)	Not appropriate for people with a low platelet count or those taking anticoagulant medications or undergoing surgery
Shark cartilage	Not appropriate for people with hypercalcemia
Mind-body therapies	Not appropriate for people who do not have reasonable expectations
Acupuncture (for chronic pain)	Not appropriate for people with a low platelet count or those taking anticoagulant medications
Massage (for pain)	Not appropriate directly over tumors, stents, or prosthetic devices and in areas damaged by surgery or radiation; or in people with bleeding abnormalities
Discourage—Evidence indicates either ineffectiveness or serious risk.	
Vitamin A supplements (both retinols and carotenoid precursors)	May increase the incidence of cancer in high-risk populations
Vitamin C supplements	May have anticoagulant effects
Soy (for breast or endometrial cancer)	May stimulate tumor growth and inhibit platelet aggregation

NOTE: Alternative therapies may be appropriate as an adjunct to, not a replacement of, conventional treatment; physicians need to monitor progress and revise recommendations as needed.
[a]When carefully planned, macrobiotic diets can provide adequate nutrition, little fat, and abundant phytoestrogens from soy. Restrictive macrobiotic diets, however, can cause malnutrition.

SOURCE: Adapted from W. A. Weiger and coauthors, Advising patients who seek complementary and alternative medical therapies for cancer, *Annals of Internal Medicine* 137 (2002): 889–903.

year-old Pacific yew tree to produce one 300-milligram dose of the anticancer drug paclitaxel (Taxol), until scientists learned how to synthesize it. Many yet undiscovered cures may be forever lost as wild species are destroyed, long before their secrets are revealed to medicine.

Herbal Precautions

Simply because plants are "natural" does not mean that they are beneficial or even safe. Nothing could be more natural—and deadly—than the poisonous herb hemlock. Several herbal remedies have toxic effects. The popular Chinese herbal potion jin bu huan, which is used as a pain and insomnia remedy, has been linked with several cases of acute hepatitis. Germanium, a nonessential mineral commonly found in many herbal products, has been associated with chronic kidney failure. Paraguay tea produces symptoms of agitation, confusion, flushed skin, and fever. Kombucha tea, commonly used in the hopes of preventing cancer, relieving arthritis, curing insomnia, and stimulating hair regrowth, can cause severe metabolic acidosis. Table H18-2 (on p. 652) lists selected herbs, their common uses, and risks.[11]

Some people use herbs to treat or prevent disease, but herbs are not drugs; they are dietary supplements. The Food and Drug Administration (FDA) does not evaluate dietary supplements for safety or effectiveness, nor does it monitor their contents.[12] Under the Dietary Supplement Health and Education Act, rather than the herb manufacturers having to prove the safety of their products, the FDA has the burden of proving that a product is not safe. Consequently, consumers may lack information about or find discrepancies regarding:

- *True identification of herbs.* Most mint teas are safe, for instance, but some varieties contain the highly toxic pennyroyal oil. Mistakenly used to soothe a colicky baby, mint tea laden with pennyroyal has been blamed for the liver and neurological injuries of at least two infants, one of whom died.

- *Purity of herbal preparations.* A young child diagnosed with lead poisoning had taken an herbal vitamin that contained large quantities of lead and mercury for four years.[13] Potentially toxic quantities of lead have been detected in 11 different dietary supplements.[14]

TABLE H18-2   Selected Herbs, Their Common Uses, and Risks

Common Name	Scientific Source Name	Claims and Uses	Risks[a]
Aloe (gel)	*Aloe vera*	Promote wound healing	Generally considered safe
Black cohosh	*Actaea racemos* (formerly *Cimicifuga racemosa*)	Ease menopause symptoms	May cause clotting in blood vessels of the eye, change the curvature of the cornea
Chamomile (flowers)	*Matricaria chamomilla*	Relieve indigestion	Generally considered safe
Chaparral (leaves and twigs)	*Larrea tridentata*	Slow aging, "cleanse" blood, heal wounds, cure cancer, treat acne	Acute, toxic hepatitis; liver damage
Comfrey (leafy plant)	*Symphytum officinale, S. asperum, S. x uplandicum*	Soothe nerves	Liver damage
Echinacea (roots)	*Enchinacea angustifolia, E. pallida, E. purpurea*	Alleviate symptoms of colds, flus, and infections; promote wound healing; boost immunity	Generally considered safe
Ephedra (stems)	*Ephedra sinica*	Promote weight loss	Rapid heart rate, tremors, seizures, insomnia, headaches, hypertension
Feverfew (leaves)	*Tanacetum parthenium*	Prevent migraine headaches	Generally considered safe; may cause mouth irritation, swelling, ulcers, and GI distress
Garlic (bulbs)	*Allium sativum*	Lower blood lipids and blood pressure	Generally considered safe; may cause garlic breath, body odor, gas, and GI distress; inhibits blood clotting
Ginger	*Zingiber officinale*	Prevent motion sickness, nausea	Generally considered safe
Ginkgo (tree leaves)	*Ginkgo biloba*	Improve memory, relieve vertigo	Generally considered safe; may cause headache, GI distress, dizziness; may inhibit blood clotting
Ginseng (roots)	*Panax ginseng* (Asian), *P. quinquefolius* (American)	Boost immunity, increase endurance	Generally considered safe; may cause insomnia and high blood pressure
Goldenseal (roots)	*Hydrastis canadensis*	Relieve indigestion, treat urinary infections	Generally considered safe
Kava	*Piper methysticum*	Relieves anxiety, promotes relaxation	Liver failure
Saw palmetto (ripe fruits)	*Serenoa repens*	Relieve symptoms of enlarged prostate; diuretic; enhance sexual vigor; enlarge mammary glands	Generally considered safe
St. John's wort (leaves and tops)	*Hypericum perforatum*	Relieve depression and anxiety	Generally considered safe; may cause fatigue and GI distress
Valerian (roots)	*Valeriana officinalis*	Calm nerves, improve sleep	Generally considered safe
Yohimbe (tree bark)	*Pausinystalia yohimbe*	Enhance "male performance"	Kidney failure, seizures

[a]Allergies are always a possible risk; see Table H18-3 for drug interactions.

- *Appropriate uses and contraindications of herbs.* Herbal remedies alone may be appropriate for minor ailments—a cup of chamomile tea to ease gastric discomfort or the gel of an aloe vera plant to soothe a sunburn, for example—but not for major health problems such as cancer or AIDS.

- *Effectiveness of herbs.* Herbal remedies may claim to work wonders without having to prove effectiveness. Research studies often report conflicting findings, with some suggesting a benefit and others indicating no effectiveness.[15]

- *Variability of herbs.* Not all species are created equal. The various species of coneflower provide an example. *Echinacea purpurea,* for example, may help fight respiratory infections, but *Echinacea augustifolia* may not. Similarly, not all parts of a plant provide the same compounds. Leaves, roots, and oils contain different compounds and extracts,

and the temperatures used during manufacturing may affect their potency. Consumers are not always aware of such differences, and manufacturers do not always make such distinctions when preparing and labeling supplements.

- *Accuracy of labels.* Supplements may contain none of an herb or mixed species, and labels are often inaccurate. More often than not, supplements do not contain the species or the quantities of active ingredients stated on their labels.[16] In at least two cases, supplements did not even contain herbs, but prescription medicines instead.[17] Such discrepancies in the contents of supplements interfere with scientific research and make it difficult to interpret the findings.

- *Safe dosages of herbs.* Herbs may contain active ingredients—compounds that affect the body. Each of these ac-

TABLE H18-3 Herb and Drug Interactions

Herb	Drug	Interaction
American ginseng	Estrogens, corticosteroids	Enhances hormonal response.
American ginseng	Breast cancer therapeutic agent	Synergistically inhibits cancer cell growth.
American ginseng, karela	Blood glucose regulators	Affect blood glucose levels.
Echinacea (possible immunostimulant)	Cyclosporine and corticosteroids (immunosuppressants)	May reduce drug effectiveness.
Evening primrose oil, borage	Anticonvulsants	Lower seizure threshold.
Feverfew	Aspirin, ibuprofen, and other nonsteroidal anti-inflammatory drugs	Negates the effect of the herb in treating migraine headaches.
Feverfew, garlic, ginkgo, ginger, and Asian ginseng	Warfarin, coumarin (anticlotting drugs, "blood thinners")	Prolong bleeding time; increase likelihood of hemorrhage.
Garlic	Protease inhibitor (HIV drug)	May reduce drug effectiveness.
Kava, valerian	Anesthetics	May enhance drug action.
Kelp (iodine source)	Synthroid or other thyroid hormone replacers	Interferes with drug action.
Kyushin, licorice, plantain, uzara root, hawthorn, Asian ginseng	Digoxin (cardiac antiarrhythmic drug derived from the herb foxglove)	Interfere with drug action and monitoring.
St. John's wort, saw palmetto, black tea	Iron	Tannins in herbs inhibit iron absorption.
Valerian	Barbiturates	Causes excessive sedation.

tive ingredients has a different potency, time of onset, duration of activity, and consequent effects, making the plant itself too unpredictable to be useful. Foxglove leaves, for example, contain dozens of compounds that have an effect on the heart; digoxin, a drug derived from foxglove, offers a standard dosage that allows for a more predictable cardiac response. Even when herbs are manufactured into capsules or liquids, their concentrations of active ingredients differ dramatically from batch to batch and from the quantities stated on the labels.[18]

- *Interactions of herbs with medicines and other herbs.* Like drugs, herbs may interfere with, or potentiate, the effects of other herbs and drugs (see Table H18-3). A person taking cardiac medication who takes the herb foxglove may be headed for disaster from the combined effect on the heart. Similarly, taking St. John's wort with medicines used to treat heart disease, depression, seizures, and certain cancers might diminish or exaggerate the intended effects.[19] Because *Ginkgo biloba* impairs blood clotting, it can cause bleeding problems for people taking aspirin or other blood-thinning medicines regularly.

- *Adverse reactions and toxicity levels of herbs.* Herbs may produce undesirable reactions. The herb ephedra, commonly known as ma huang and used to promote weight loss, acts as a strong central nervous system stimulant, causing rapid heart rate, headaches, insomnia, tremors, seizures, and even death.[20] The herbal root kava, commonly used to treat anxiety and insomnia, may have such a sedating effect as to impair driving. Chinese herbal treatments containing *Aristolochia fangchi* are notorious for causing kidney damage and cancers.[21] Table H18-2 (on p. 651) includes risks associated with commonly used herbs.

To ensure the safety and standardization of herbal remedies, Congress needs to establish new regulations.[22]

Because herbal medicines are sold as dietary supplements, their labels cannot claim to cure a disease, but they can make various other claims. Not surprisingly, when a label claims that an herbal product may strengthen immunity, improve memory, support eyesight, or maintain heart health, consumers believe that taking the product will provide those benefits. Beware. Manufacturers need not prove effectiveness; they only need to state on the product label that this claim "Has not been evaluated by the FDA." Consumers who decide to use herbs need to become informed of the possible risks.

Internet Precautions

As Highlight 1 pointed out, just because something appears on the Internet, "it ain't necessarily so." Keep in mind that the thousands of websites touting the benefits of herbal medicines and other dietary supplements are marketing their products. Most claim to prevent or treat specific diseases, but few include the FDA disclaimer statement.[23] Many of the websites promote products by quoting researchers or physicians. Such quotations lend an air of authority to advertisements, but be aware that these sources may not even exist—and if they do, their comments may have been taken out of context. When asked, they may not agree at all with the claims attributed to them by the manufacturer.

Other deceits and dangers lurk in cyberspace as well. Potentially toxic substances, illegal and unavailable in many countries, are now easy to obtain via the Internet. Electronic access to products such as absinthe and oil of wormwood could be deadly. When the FDA discovers websites selling unapproved drugs, such as laetrile, it can order the business to shut down.[24] But consumers need to remain vigilant because other similar businesses pop up quickly.

The Consumer's Perspective

Some health care professionals may dismiss alternative therapies as ineffective and perhaps even dangerous, but consumers think otherwise. In a survey of over 2000 people, two-thirds had used at least one alternative therapy for a variety of medical complaints ranging from anxiety and headaches to cancer and tumors.[25] Interestingly, those who seek alternative therapies seem to do so not so much because they are dissatisfied with conventional medicine as because they find these alternatives more in line with their beliefs about health and life.

Most often, people use alternative therapies in addition to, rather than in place of, conventional therapies.[26] Few consult an alternative therapist without also seeing a physician. In fact, most people seek alternative therapies for nonserious medical conditions or for health promotion. They simply want to feel better and access is easy. Sometimes their symptoms are chronic and subjective, such as pain and fatigue, and difficult to treat. In these cases, the chances of finding relief are often as good with a placebo, standard medical intervention, or even nonintervention.

Consumers spend billions of dollars on alternative health services and related products such as herbs, crystals, and aromas. As Highlight 1 pointed out, selecting a reliable practitioner depends on finding out about training, qualifications, and licenses. (To review how a person can identify health fraud and quackery, turn to pp. 35–36. For a list of credible sources of nutrition information, see p. 34.)

In addition, consumers should inform their physicians about the use of any alternative therapies so that a comprehensive treatment plan can be developed and potential problems can be averted. As mentioned, herb-drug interactions can create problems, and one in six clients who takes prescription drugs also uses herbal products.[27] When considering herbal products, remember to include supplements, teas, and garden plants.[28] Some herbal products may need to be discontinued, especially before surgery when interactions with anesthesia or normal blood clotting can be life-threatening.[29]

Alternative therapies come in a variety of shapes and sizes. Both their benefits and their risks may be small, none, or great. Wise consumers and health care professionals accept the beneficial, or even neutral, practices with an open mind and reject those practices known to cause harm. Making healthful choices requires understanding all the choices.

NUTRITION ON THE NET

 Access these websites for further study of topics covered in this highlight.

- Find updates and quick links to these and other nutrition-related sites at our website: **www.wadsworth.com/nutrition**

- Search for "alternative medicine," "herbs," "holistic health," "homeopathy," and "preventive medicine" at the U.S. Government health information site: **www.healthfinder.gov**

- Learn about complementary and alternative medicine from the National Institutes of Health's National Center for Complementary and Alternative Medicine: **ww.nccam.nih.gov**

- Search CAM on PubMed for a literature search of the complementary and alternative subset of PubMed: **www.nlm.nih.gov/nccam/camonpubmed.html**

- Find out more about herbs from the American Botanical Council: **www.herbalgram.org**

- Report adverse effects associated with herbal remedies to the FDA MedWatch: **www.fda.gov/medwatch**

- Obtain information on herbal medications from HerbMed or from the Integrative Medicine Service at Memorial Sloan-Kettering Cancer Center: **www.herbmed.org** or **www.mskcc.org/aboutherbs**

- Get dietary supplement information from the National Institutes of Health's Office of Dietary Supplements: **dietary-supplements.info.nih.gov**

- Review the backgrounds and practices of many popular practitioners of alternative treatments: **www.quackwatch.com**

REFERENCES

1. R. C. Kessler and coauthors, Long-term trends in the use of complementary and alternative medical therapies in the United States, *Annals of Internal Medicine* 135 (2001): 262–268.

2. B. Barzansky, H. S. Jonas, and S. I. Etzel, Educational programs in US medical schools, 1999–2000, *Journal of the American Medical Association* 284 (2000): 1114–1120.

3. R. Touger-Decker and C. A. Thomson, Complementary and alternative medicine: Competencies for dietetics professionals, *Journal of the American Dietetic Association* 103 (2003): 1465–1469; M. A. Frenkel and J. M. Borkan, An approach for integrating complementary-alternative medicine into primary care, *Family Practice* 20 (2003): 324–332.

4. Committee on Children with Disabilities, American Academy of Pediatrics, Counseling families who choose complementary and alternative medicine for their child with chronic illness or disability, *Pediatrics* 107 (2001): 598–601.

5. nccam.nih.gov/news/2001, posted February 2001 and visited September 2003.

6. G. Y. Yeh and coauthors, Systematic review of herbs and dietary supplements for glycemic control in diabetes, *Diabetes Care* 26 (2003): 1277–1294; S. S. Percival, Use of echinacea in medicine, *Biochemical Pharmacology* 60 (2000): 155–158; V. Vuksan and coauthors, American ginseng (*Panax quinquefolius* L.) reduces postprandial glycemia in nondiabetic subjects and subjects with type 2 diabetes mellitus, *Archives of Internal Medicine* 160 (2000): 1009–1013; K. A. Wesnes and coauthors, The memory enhancing effects of a Ginkgo biloba/Panax ginseng combination in healthy middle-aged volunteers, *Psychopharmacology* 152 (2000): 353–361.

7. M. Mitka, Therapeutic marijuana use supported while thorough proposed study done, *Journal of the American Medical Association* 281 (1999): 1473–1474.

8. C. D. Morris and S. Carson, Summary of evidence: Routine vitamin supplementation to prevent cardiovascular disease, *Annals of Internal Medicine* 139 (2003): 56–70; C. Ritenbaugh, K. Streit, and M. Helfand, Summary of evidence from randomized controlled trials: Routine vitamin supplementation to prevent cancer, available from the Agency for Healthcare Research and Quality, **www.preventiveservices.ahrq.gov**.

9. A. T. Borchers and coauthors, Inflammation and Native American medicine: The role of botanicals, *American Journal of Clinical Nutrition* 72 (2000): 339–347.

10. W. J. Craig, Health-promoting properties of common herbs, *American Journal of Clinical Nutrition* 70 (1999): 491S–499S.

11. E. Ernst, The risk-benefit profile of commonly used herbal therapies: Ginkgo, St. John's wort, ginseng, echinacea, saw palmetto, and kava, *Annals of Internal Medicine* 136 (2002): 42–53; Hepatic toxicity possibly associated with kava-containing products—United States, Germany, and Switzerland, 1999–2002, *Morbidity and Mortality Weekly Report* 51 (2002): 1065–1067; S. Foster and V. E. Tyler, *Tyler's Honest Herbal: A Sensible Guide to the Use of Herbs and Related Remedies* (New York: Haworth Press, 1999).

12. J. Hankin, Keeping up with the increasing popularity of nonvitamin, nonmineral supplements, *Journal of the American Dietetic Association* 100 (2000): 419–420.

13. C. Moore and R. Adler, Herbal vitamins: Lead toxicity and developmental delay, *Pediatrics* 106 (2000): 600–602.

14. S. P. Dolan and coauthors, Analysis of dietary supplements for arsenic, cadmium, mercury, and lead using inductively coupled plasma mass spectrometry, *Journal of Agricultural and Food Chemistry* 51 (2003): 1307–1312.

15. Hypericum Depression Trial Study Group, Effect of *Hypericum perforatum* (St John's wort) in major depressive disorder: A randomized controlled trial, *Journal of the American Medical Association* 287 (2002): 1807–1814; P. R. Solomon and coauthors, Ginkgo for memory enhancement: A randomized controlled trial, *Journal of the American Medical Association* 288 (2002): 835–840; Wesnes and coauthors, 2000).

16. C. M. Gilroy and coauthors, Echinacea and truth in labeling, *Archives of Internal Medicine* 163 (2003): 699–704; A. H. Feifer, N. E. Fleshner, and L. Klotz, Analytical accuracy and reliability of commonly used nutritional supplements in prostate disease, *Journal of Urology* 168 (2002): 150–154.

17. Dietary-supplement recall, *Consumer Reports on Health*, April 2002, p. 3.

18. M. R. Karkey and co-authors, Variability in commercial ginseng products: An analysis of 25 preparations, *American Journal of Clinical Nutrition* 73 (2001): 1101–1106.

19. J. S. Markowitz and coauthors, Effect of St. John's wort on drug metabolism by induction of cytochrome P450 3A4 enzyme, *Journal of the American Medical Association* 290 (2003): 1500–1504; Risk of drug interactions with St. John's wort, *Journal of the American Medical Association* 283 (2000): 1679.

20. C. A. Haller and N. L. Benowitz, Adverse cardiovascular and central nervous system events associated with dietary supplements containing ephedra alkaloids, *New England Journal of Medicine* 343 (2000): 1833–1838.

21. J. L. Nortier and coauthors, Urothelial carcinoma associated with the use of a Chinese herb (*Aristolochia fangchi*), *New England Journal of Medicine* 342 (2000): 1686–1692.

22. C. D. DeAngelis and P. B. Fontanarosa, Drugs alias dietary supplements, *Journal of the American Medical Association* 290 (2003): 1519–1520; D. M. Marcus and A. P. Grollman, Botanical medicines—The need for new regulations, *New England Journal of Medicine* 347 (2002): 2073–2076.

23. C. A. Morris and J. Avorn, Internet marketing of herbal products, *Journal of the American Medical Association* 290 (2003): 1505–1509.

24. C. Lewis, Online laetrile vendor ordered to shut down, *FDA Consumer*, March/April 2001, pp. 37–38.

25. Kessler and coauthors, 2001.

26. B. G. Druss and R. A. Rosenheck, Association between use of unconventional therapies and conventional medical services, *Journal of the American Medical Association* 282 (1999): 651–656.

27. D. W. Kaufman and coauthors, Recent patterns of medication use in the ambulatory adult population of the United States: The Slone Survey, *Journal of the American Medical Association* 287 (2002): 337–344.

28. M. A. Kuhn, Herbal remedies: Drug-herb interactions, *Critical Care Nurse* 22 (2002): 22–28.

29. M. K. Ang-Lee, J. Moss, and C. S. Yuan, Herbal medicines and perioperative care, *Journal of the American Medical Association* 286 (2001): 208–216.

Consumer Concerns about Foods and Water

© Leigh Beisch/FoodPix/Getty Images

Chapter Outline

Foodborne Illnesses: *Foodborne Infections and Food Intoxications* • *Food Safety in the Marketplace* • *Food Safety in the Kitchen* • *Food Safety While Traveling* • *Advances in Food Safety*

Nutritional Adequacy of Foods and Diets: *Obtaining Nutrient Information* • *Minimizing Nutrient Losses*

Environmental Contaminants: *Harmfulness of Environmental Contaminants* • *Guidelines for Consumers*

Natural Toxicants in Foods

Pesticides: *Hazards and Regulation of Pesticides* • *Monitoring Pesticides* • *Consumer Concerns*

Food Additives: *Regulations Governing Additives* • *Intentional Food Additives* • *Indirect Food Additives*

Consumer Concerns about Water: *Sources of Drinking Water* • *Water Systems and Regulations*

Highlight: *Food Biotechnology*

Nutrition Explorer CD-ROM Outline

Nutrition Animation: *Consumer Concerns about Food Safety*

Case Study: *Food Safety on the Go*

Student Practice Test

Glossary Terms

Nutrition on the Net

Nutrition in Your Life

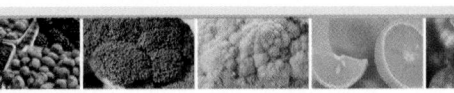

Do you know what causes food poisoning and how to protect yourself against it? Were you alarmed to learn that french fries contain acrylamide or that fish contain mercury? Are you concerned about the pesticides that might linger on fruits and vegetables—or the hormones and antibiotics that remain in beef and chicken? Do you wonder whether foods contain enough nutrients—or too many additives? Making informed decisions and practicing a few food safety tips will allow you to enjoy a variety of foods while limiting your risks of experiencing food-related illnesses.

With the benefits of a safe and abundant food supply comes the responsibility to select, prepare, and store foods safely.

Take a moment to consider the task of supplying food to over 280 million people in the United States (and millions more in all corners of the world). To feed this nation, farmers grow and harvest crops; dairy producers supply milk products; ranchers raise livestock; shippers deliver foods to manufacturers by land, sea, and air; manufacturers prepare, process, preserve, and package products for refrigerated food cases and grocery-store shelves; and grocers store the food and supply it to consumers. After much time, much labor, and extensive transport, an abundant supply of a large variety of safe foods finally reaches consumers at reasonable market prices.

The **FDA** and other government and international agencies monitor this huge system using a network of people and sophisticated equipment (the glossary on p. 658 identifies the various food regulatory agencies by their abbreviations). These agencies focus on the potential **hazard** of foods, which differs from the **toxicity** of a substance—a distinction worth understanding. Anything can be toxic. Toxicity simply means that a substance *can* cause harm *if* enough is consumed. We consume many substances that are toxic, without **risk,** because the amounts are so small. The term *hazard,* on the other hand, is more relevant to our daily lives because it refers to the

hazard: a source of danger; used to refer to circumstances in which harm is possible under normal conditions of use.

toxicity: the ability of a substance to harm living organisms. All substances are toxic if high enough concentrations are used.

risk: a measure of the probability and severity of harm.

657

harm that is *likely* under real-life conditions. Consumers rely on these monitoring agencies to set **safety** standards and can learn to protect themselves from food hazards by taking a few preventive steps.

After the events of September 11, 2001, the threat of deliberate microbial contamination of the U.S. food supply became a pressing issue.[1] To tighten security around the nation's food supply, the FDA has established guidelines for firms that produce, process, transport, or otherwise handle food.* The **USDA** has also created a Food Biosecurity Action Team to protect agriculture and other aspects of the food supply. Other agencies are also taking action, but details of the war against domestic bioterrorism are beyond the scope of this discussion.

This chapter focuses on actions of individuals to promote food safety. It addresses the following food safety concerns:

- Foodborne illnesses.
- Nutritional adequacy of foods.
- Environmental contaminants.
- Naturally occurring toxicants.
- Pesticides.
- Food additives.
- Water safety.

The chapter begins with the FDA's highest priority—the serious and prevalent threat of foodborne illness. The highlight that follows looks at genetically engineered foods.

■ Get medical help when these symptoms occur:
- Bloody stools.
- Diarrhea lasting more than 3 days.
- Difficulty breathing.
- Difficulty swallowing.
- Double vision.
- Fever lasting more than 24 hours.
- Headache accompanied by muscle stiffness and fever.
- Numbness, muscle weakness, and tingling sensations in the skin.
- Rapid heart rate, fainting, and dizziness.

safety: the condition of being free from harm or danger.

foodborne illness: illness transmitted to human beings through food and water, caused by either an infectious agent (foodborne infection) or a poisonous substance (food intoxication); commonly known as **food poisoning.**

Foodborne Illnesses

The FDA lists **foodborne illness** as the leading food safety concern because episodes of food poisoning far outnumber episodes of any other kind of food contamination. The **CDC** estimates that 76 million people experience foodborne illness each year in the United States. For some 5000 people each year, the symptoms■ can be

_____
*FDA guidance documents on biosecurity are available at www.cfsan.fda.gov/dms/guidance.html.

GLOSSARY OF AGENCIES THAT MONITOR THE FOOD SUPPLY

CDC (Centers for Disease Control): a branch of the Department of Health and Human Services that is responsible for, among other things, monitoring foodborne diseases.
www.cdc.gov

EPA (Environmental Protection Agency): a federal agency that is responsible for, among other things, regulating pesticides and establishing water quality standards.
www.epa.gov

FAO (Food and Agriculture Organization): an international agency (part of the United Nations) that has adopted standards to regulate pesticide use among other responsibilities.
www.fao.org

FDA (Food and Drug Administration): a part of the Department of Health and Human Services' Public Health Service that is responsible for ensuring the safety and wholesomeness of all foods processed and sold in interstate commerce except meat, poultry, and eggs (which are under the jurisdiction of the USDA); inspecting food plants and imported foods; and setting standards for food composition.
www.fda.gov

USDA (U.S. Department of Agriculture): the federal agency responsible for enforcing standards for the wholesomeness and quality of meat, poultry, and eggs produced in the United States; conducting nutrition research; and educating the public about nutrition.
www.usda.gov

WHO (World Health Organization): an international agency that has adopted standards to regulate pesticide use among other responsibilities.
www.who.ch

so severe as to cause death. Most vulnerable are pregnant women; very young, very old, sick, or malnourished people; and those with a weakened immune system (as in AIDS). By taking the proper precautions, people can minimize their chances of contracting foodborne illnesses.

Foodborne Infections and Food Intoxications

Foodborne illness can be caused by either an infection or an intoxication. Table 19-1 (on p. 660) summarizes the most common or severe foodborne illnesses, their food sources, general symptoms, and prevention methods.

Foodborne Infections Foodborne infections are caused by eating foods contaminated by infectious microbes. Two of the most common foodborne **pathogens** are *Campylobacter jejuni* and *Salmonella*,■ which enter the GI tract in contaminated foods such as undercooked poultry and unpasteurized milk. Symptoms generally include abdominal cramps, fever, and diarrhea. If a person experiences these symptoms as the major or only symptoms of a bout of "flu," chances are excellent that what the person really has is a foodborne infection.*

■ One out of three deaths caused by foodborne infections can be attributed to *Salmonella.*

Reduce infections caused by key foodborne pathogens.

HEALTHY PEOPLE 2010

Food Intoxications Food intoxications are caused by eating foods containing natural toxins or, more likely, microbes that produce toxins. The most common food toxin is produced by *Staphylococcus aureus;* it affects more than one million people each year. Less common, but more infamous, is *Clostridium botulinum,* an organism that produces a deadly toxin in anaerobic conditions such as improperly canned (especially home-canned) foods and homemade garlic or herb-flavored oils stored at room temperature. Botulism paralyzes muscles; consequently, seeing, speaking, swallowing, and breathing become difficult.[2] Because death can occur within 24 hours of onset, botulism demands immediate medical attention. Even then, survivors may suffer the effects for months or years.

Food Safety in the Marketplace

Transmission of foodborne illness has changed as our food supply and lifestyles have changed.[3] In the past, foodborne illness was caused by one person's error in a small setting, such as improperly refrigerated egg salad at a family picnic, and affected only a few victims. Today, we are eating more foods prepared and packaged by others. Consequently, when a food manufacturer or restaurant chef makes an error, foodborne illness can become epidemic. An estimated 80 percent of reported foodborne illnesses are caused by errors in a commercial setting, such as the improper **pasteurization** of milk at a large dairy.

In the mid-1990s, when a fast-food restaurant served undercooked burgers tainted with an infectious strain of *Escherichia coli,* hundreds of patrons became ill, and at least three people died. In the late-1990s, a national meat-processing plant had to recall 15 million pounds of hot dogs and lunch meats after *Listeria* poisoning killed 15 people and made over 100 others sick. These incidents and others since have focused the national spotlight on two important safety issues: disease-causing organisms are commonly found in raw foods, and thorough cooking kills most of these foodborne pathogens. This heightened awareness sparked a much needed overhaul of national food safety programs.

To prevent food intoxication from homemade flavored oils, wash and dry the herbs before adding them to the oil and keep the oil refrigerated.

pathogens (PATH-oh-jens): microorganisms capable of producing disease.

pasteurization: heat processing of food that inactivates some, but not all, microorganisms in the food; not a sterilization process. Bacteria that cause spoilage are still present.

---

*Some viruses do cause intestinal distress, and those that do are usually transmitted via food; true influenza viruses cause symptoms primarily in the upper respiratory tract.

TABLE 19-1 Foodborne Illnesses

Disease and Organism That Causes It	Most Frequent Food Sources	Onset and General Symptoms	Prevention Methods[a]
Foodborne Infections			
Campylobacteriosis (KAM-pee-loh-BAK-ter-ee-OH-sis) *Campylobacter jejuni* bacterium	Raw poultry, beef, lamb, unpasteurized milk (foods of animal origin eaten raw or undercooked or recontaminated after cooking).	Onset: 2 to 5 days. Diarrhea, nausea, vomiting, abdominal cramps, fever; sometimes bloody stools; lasts 7 to 10 days.	Cook foods thoroughly; use pasteurized milk; use sanitary food-handling methods.
Giardiasis (JYE-are-DYE-ah-sis) *Giardia lamblia* protozoon	Contaminated water; uncooked foods.	Onset: 5 to 25 days. Diarrhea (but occasionally constipation), abdominal pain, gas, abdominal distention, digestive disturbances, anorexia, nausea, and vomiting.	Use sanitary food-handling methods; avoid raw fruits and vegetables where protozoa are endemic; dispose of sewage properly.
Hepatitis (HEP-ah-TIE-tis) Hepatitis A virus	Undercooked or raw shellfish.	Onset: 15 to 50 days (28 to 30 days average). Inflammation of the liver with tiredness; nausea, vomiting, or indigestion; jaundice (yellowed skin and eyes from buildup of wastes); muscle pain.	Cook foods thoroughly.
Listeriosis (lis-TER-ee-OH-sis) *Listeria monocytogenes* bacterium	Raw meat, poultry, and seafood; raw milk; soft cheeses; leafy vegetables.	Onset: 1 to 21 days. Mimics flu; blood poisoning, complications in pregnancy, and meningitis (stiff neck, severe headache, and fever).	Use sanitary food-handling methods; cook foods thoroughly; use pasteurized milk.
Perfringens (per-FRINGE-enz) **food poisoning** *Clostridium perfringens* bacterium	Meats and meat products stored at between 120° and 130°F.	Onset: 8 to 24 hr. Abdominal pain, diarrhea, nausea, and vomiting; symptoms last a day or less and are usually mild; can be serious in old or weak people.	Use sanitary food-handling methods; cook foods thoroughly; refrigerate foods promptly and properly.
Salmonellosis (sal-moh-neh-LOH-sis) *Salmonella* bacteria (>2300 types)	Raw or undercooked eggs, meats, poultry, raw milk and other dairy products, shrimp, frog legs, yeast, coconut, pasta, and chocolate.	Onset: 8 to 72 hr. Nausea, fever, chills, vomiting, abdominal cramps, diarrhea, and headache; lasts 1 to 2 days; can be fatal.	Use sanitary food-handling methods; use pasteurized milk; cook foods thoroughly; refrigerate foods promptly and properly.
E. coli infection *Escherichia coli*[b] bacterium	Undercooked ground beef, unpasteurized milk and milk products, contaminated water, and person-to-person contact.	Onset: 2 to 5 days. Severe bloody diarrhea, abdominal cramps, acute kidney failure; lasts 8 days; can be fatal.	Cook ground beef thoroughly; avoid raw milk and milk products; use sanitary food-handling methods; use treated, boiled, or bottled water.
Shigellosis (shi-gel-LOH-sis) *Shigella* bacteria (>30 types)	Person-to-person contact, raw foods, salads, dairy products, and contaminated water.	Onset: 12 to 50 hr. Bloody diarrhea, chills, vomiting, cramps, fever; lasts 2 to 14 days	Use sanitary food-handling methods; cook foods thoroughly; proper refrigeration.
Vibrio (VIB-ree-oh) **bacteria** *Vibrio vulnificus*[c] bacterium	Raw seafood and contaminated water.	Onset: 1 to 7 days. Diarrhea, abdominal cramps, fever, chills; can be fatal.	Use sanitary food-handling methods; cook foods thoroughly.
Food Intoxications			
Botulism (BOT-chew-lizm) Botulinum toxin [produced by *Clostridium botulinum* bacterium, which grows without oxygen, in low-acid foods, and at temperatures between 40° and 120°F; the **botulinum** (BOT-chew-line-um) **toxin** responsible for botulism is called **botulin** (BOT-chew-lin)]	Anaerobic environment of low acidity (canned corn, peppers, green beans, soups, beets, asparagus, mushrooms, ripe olives, spinach, tuna, chicken, chicken liver, liver pâté, luncheon meats, ham, sausage, stuffed eggplant, lobster, and smoked and salted fish).	Onset: 4 to 36 hr. Nervous system symptoms, including double vision, inability to swallow, speech difficulty, and progressive paralysis of the respiratory system; often fatal; leaves prolonged symptoms in survivors.	Use proper canning methods for low-acid foods; refrigerate homemade garlic and herb oils; avoid commercially prepared foods with leaky seals or with bent, bulging, or broken cans.
Staphylococcal (STAF-il-oh-KOK-al) **food poisoning** Staphylococcal toxin (produced by *Staphylococcus aureus* bacterium)	Toxin produced in meats, poultry, egg products, tuna, potato and macaroni salads, and cream-filled pastries.	Onset: 1 to 6 hr. Diarrhea, nausea, vomiting, abdominal cramps, and fatigue; mimics flu; lasts 2 to 3 days; rarely fatal.	Use sanitary food-handling methods; cook food thoroughly; refrigerate foods promptly and properly; use proper home-canning methods.

NOTE: Travelers' diarrhea is most commonly caused by *E. coli, Campylobacter jejuni, Shigella,* and *Salmonella.*

[a] The "How to" on pp. 664–665 provides more details on the proper handling, cooking, and refrigeration of foods.

[b] The most serious strain is *E. coli* O157:H7.

[c] Most cases of *Vibrio vulnificus* occur in persons with underlying illness, particularly those with liver disorders, diabetes, cancer, and AIDS, and those who require long-term steroid use. The fatality rate is 50 percent for this population.

Improve food employee behaviors and food preparation practices that directly relate to foodborne illnesses in retail food establishments.

Industry Controls To make our food supply safe for consumers, the USDA, the FDA, and the food-processing industries have developed and implemented programs to control foodborne illness.* The **Hazard Analysis Critical Control Points (HACCP)** system requires food manufacturers to identify points of contamination and implement controls to prevent foodborne disease. For example, after tracing two large outbreaks of salmonellosis to imported cantaloupe, producers began using chlorinated water to wash the melons and to make ice for packing and shipping. Safety procedures such as this prevent hundreds of thousands of foodborne illnesses each year.

This example raises another issue regarding the safety of imported foods. FDA inspectors cannot keep pace with the increasing numbers of imported foods; they inspect fewer than 2 percent of the almost 3 million shipments of fruits, vegetables, and seafood coming into more than 300 ports in the United States each year. The FDA is working with other countries to adopt the safe food-handling practices used in the United States.

Consumer Awareness Canned and packaged foods sold in grocery stores are easily controlled, but rare accidents do happen. Batch numbering makes it possible to recall contaminated foods through public announcements via newspapers, television, and radio. In the grocery store, consumers can inspect the safety seals and wrappers of packages. A broken seal, bulging can lid, or mangled package fails to protect the consumer against microbes, insects, spoilage, or even vandalism.

State and local health regulations provide guidelines on the cleanliness of facilities and the safe preparation of foods for restaurants, cafeterias, and fast-food establishments. Even so, consumers can also take these actions to help prevent foodborne illnesses when dining out:

- Wash hands with hot, soapy water before meals.
- Expect clean tabletops, dinnerware, utensils, and food preparation areas.
- Expect cooked foods to be served piping hot and salads to be fresh and cold.
- Refrigerate doggy bags within two hours.

Improper handling of foods can occur anywhere along the line from commercial manufacturers to large supermarkets to small restaurants to private homes. Maintaining a safe food supply requires everyone's efforts (see Figure 19-1 on p. 662).

Food Safety in the Kitchen

Whether microbes multiply and cause illness depends, in part, on what happens in the kitchen—whether the kitchen is in your home, a school cafeteria, a gourmet restaurant, or a canning manufacturer. For the most part, foodborne illness can be prevented by doing four simple things:

- *Keep a clean, safe kitchen.* Wash countertops, cutting boards, hands, sponges, and utensils in hot, soapy water before and after each step of food preparation.
- *Avoid cross-contamination.* Keep raw eggs, meat, poultry, and seafood separate from other foods. Wash all utensils and surfaces (such as cutting boards or platters) that have been in contact with these foods with hot, soapy water before using them again. Bacteria inevitably left on the surfaces from the raw meat can recontaminate the cooked meat or other foods—a problem known as **cross-contamination.**
- *Keep hot foods hot.* Cook foods long enough to reach internal temperatures that will kill microbes, and maintain adequate temperatures to prevent bacterial growth until the foods are served.

 Evaluate the relative health risks posed by food toxins, additives, and contaminants that are a part of our daily diet.

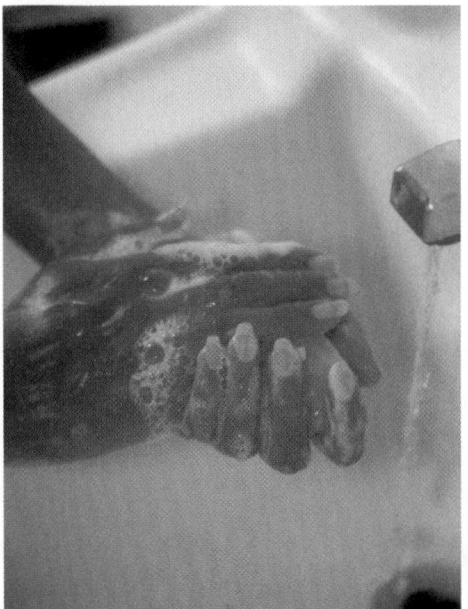

© 2002 PhotoDisc/Getty Images

Wash your hands with warm water and soap before preparing or eating food to reduce the chance of microbial contamination.

Hazard Analysis Critical Control Points (HACCP): a systematic plan to identify and correct potential microbial hazards in the manufacturing, distribution, and commercial use of food products; commonly referred to as "HASS-ip."

cross-contamination: the contamination of food by bacteria that occurs when the food comes into contact with surfaces previously touched by raw meat, poultry, or seafood.

---

*In addition to HACCP, these programs include the Emerging Infections Program (EIP), the Foodborne Diseases Active Surveillance Network (FoodNet), and the Food Safety Inspection Service (FSIS).

FIGURE 19-1 Food Safety from Farms to Consumers

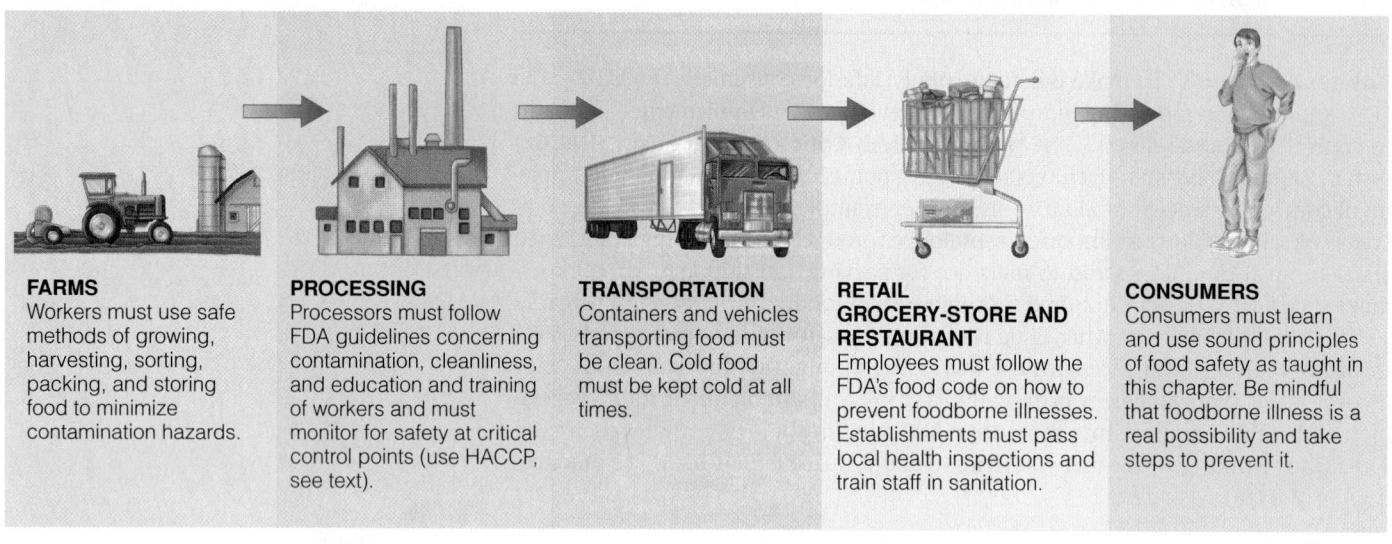

FARMS
Workers must use safe methods of growing, harvesting, sorting, packing, and storing food to minimize contamination hazards.

PROCESSING
Processors must follow FDA guidelines concerning contamination, cleanliness, and education and training of workers and must monitor for safety at critical control points (use HACCP, see text).

TRANSPORTATION
Containers and vehicles transporting food must be clean. Cold food must be kept cold at all times.

RETAIL GROCERY-STORE AND RESTAURANT
Employees must follow the FDA's food code on how to prevent foodborne illnesses. Establishments must pass local health inspections and train staff in sanitation.

CONSUMERS
Consumers must learn and use sound principles of food safety as taught in this chapter. Be mindful that foodborne illness is a real possibility and take steps to prevent it.

- *Keep cold foods cold.* Go directly home upon leaving the grocery store and immediately unpack foods into the refrigerator or freezer upon arrival. After a meal, refrigerate any leftovers immediately.

See the "How to" on pp. 664–665 for additional food safety tips.

HEALTHY PEOPLE 2010

Increase the proportion of consumers who follow key food safety practices.

Safe Handling of Meats and Poultry Figure 19-2 presents label instructions for the safe handling of meat and poultry and two types of USDA seals. Meats and poultry contain bacteria and provide a moist, nutrient-rich environment that favors microbial growth. Ground meat is especially susceptible because it receives more handling than other kinds of meat and has more surface exposed to bacterial contamination. Consumers cannot detect the harmful bacteria in or on meat. For safety's sake, cook meat thoroughly, using a thermometer to test the internal temperature (see Figure 19-3).

Other Meat Concerns Reports on mad cow disease from dozens of countries, including Canada and the United States, have sparked consumer concerns.[4] Mad cow disease is a fatal condition that affects the central nervous system of cattle.* A similar disease develops in people who have eaten contaminated beef from infected cows (milk products appear to be safe).† Over 150 cases have been reported worldwide, almost all in the United Kingdom. The USDA has taken numerous steps to prevent mad cow disease from entering the U.S. food supply. Because the infectious agents occur in the intestines, central nervous system, and other organs, but not in muscle meat, concerned consumers may want to select whole cuts of meat instead of ground beef or sausage. A few recent reports of hunters developing fatal neurological disorders have raised concerns about a similar disease in wild game. Hunters and consumers who regularly eat elk, deer, or antelope should check the advisories of their state department of agriculture.

Safe Handling of Seafood Most seafood available in the United States and Canada is safe, but eating it undercooked or raw can cause severe illnesses—hepatitis, worms, parasites, viral intestinal disorders, and other diseases.‡ Rumor has it

© EyeWire Inc.

Cook hamburgers to 160°F; color alone cannot determine doneness. Some burgers will turn brown before reaching 160°F, while others may retain some pink color, even when cooked to 175°F.

---

*Mad cow disease is technically known as bovine spongiform encephalopathy (BSE).
†The human form of BSE is called new variant Creutzfeldt-Jakob Disease (nvCJD).
‡Diseases caused by toxins from the sea include ciguatera poisoning, scombroid poisoning, and paralytic and neurotoxic shellfish poisoning.

FIGURE 19-2   Meat and Poultry Safety, Grading, and Inspection Seals

Inspection is mandatory; grading is voluntary. Neither guarantees that the product will not cause foodborne illnesses, but consumers can help to prevent foodborne illnesses by following the safe handling instructions.

The voluntary "Graded by USDA" seal indicates that the product has been graded for tenderness, juiciness, and flavor. Beef is graded Prime (abundant marbling of the meat muscle), Choice (less marbling), and Select (lean). Similarly, poultry is graded A, B, and C.

The mandatory "Inspected and Passed by the USDA" seal ensures that meat and poultry products are safe, wholesome, and correctly labeled. Inspection does not guarantee that the meat is free of potentially harmful bacteria.

Safe Handling Instructions

THIS PRODUCT WAS PREPARED FROM INSPECTED AND PASSED MEAT AND/OR POULTRY. SOME FOOD PRODUCTS MAY CONTAIN BACTERIA THAT CAN CAUSE ILLNESS IF THE PRODUCT IS MISHANDLED OR COOKED IMPROPERLY. FOR YOUR PROTECTION, FOLLOW THESE SAFE HANDLING INSTRUCTIONS.

KEEP REFRIGERATED OR FROZEN. THAW IN REFRIGERATOR OR MICROWAVE.

KEEP RAW MEAT AND POULTRY SEPARATE FROM OTHER FOODS. WASH WORKING SURFACES (INCLUDING CUTTING BOARDS), UTENSILS, AND HANDS AFTER TOUCHING RAW MEAT OR POULTRY.

COOK THOROUGHLY.

KEEP HOT FOODS HOT. REFRIGERATE LEFTOVERS IMMEDIATELY OR DISCARD.

The USDA requires that safe handling instructions appear on all packages of meat and poultry.

Eating raw seafood is a risky proposition.

that freezing fish will make it safe to eat raw, but this is only partly true. Commercial freezing will kill mature parasitic worms, but only cooking can kill all worm eggs and other microorganisms that can cause illness. For safety's sake, all seafood should be cooked until it is opaque. Even **sushi** can be safe to eat when chefs combine cooked seafood and other ingredients into delicacies.

FIGURE 19-3   Recommended Safe Temperatures (Fahrenheit)

Bacteria multiply rapidly at temperatures between 40° and 140°F. Cook foods to the temperatures shown on this thermometer and hold them at 140°F or higher.

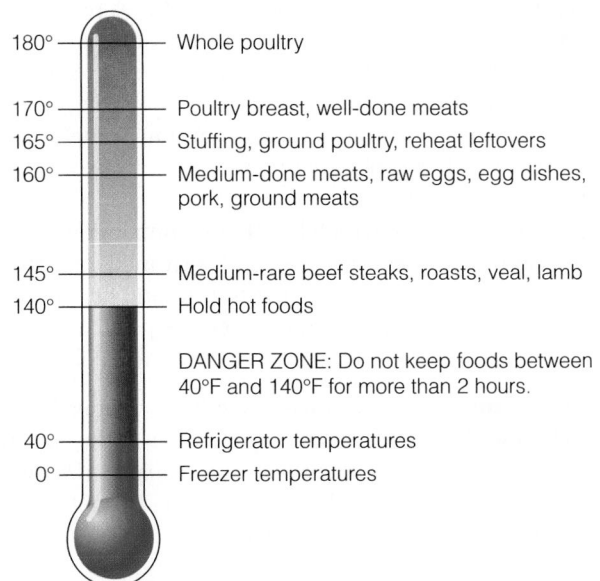

180° — Whole poultry
170° — Poultry breast, well-done meats
165° — Stuffing, ground poultry, reheat leftovers
160° — Medium-done meats, raw eggs, egg dishes, pork, ground meats
145° — Medium-rare beef steaks, roasts, veal, lamb
140° — Hold hot foods
DANGER ZONE: Do not keep foods between 40°F and 140°F for more than 2 hours.
40° — Refrigerator temperatures
0° — Freezer temperatures

sushi: vinegar-flavored rice and seafood, typically wrapped in seaweed and stuffed with colorful vegetables. Some sushi is stuffed with raw fish; other varieties contain cooked seafood.

 **HOW TO** Prevent Foodborne Illnesses

Most foodborne illnesses can be prevented by following four simple rules: keep a clean kitchen, avoid cross-contamination, keep hot foods hot, and keep cold foods cold.

Keep a Clean Kitchen

- Wash fruits and vegetables in a clean sink with a scrub brush and warm water; store washed and unwashed produce separately.
- Use hot, soapy water to wash hands, utensils, dishes, nonporous cutting boards, and countertops before handling food and between tasks when working with different foods. Use a bleach solution on cutting boards (one capful per gallon of water).
- Cover cuts with clean bandages before food preparation; dirty bandages carry harmful microorganisms.
- Mix foods with utensils, not hands; keep hands and utensils away from mouth, nose, and hair.
- Anyone may be a carrier of bacteria and should avoid coughing or sneezing over food. A person with a skin infection or infectious disease should not prepare food.
- Wash or replace sponges and towels regularly.

- Clean up food spills and crumb-filled crevices.

Avoid Cross-Contamination

- Wash all surfaces that have been in contact with raw meats, poultry, eggs, fish, and shellfish before reusing.
- Serve cooked foods on a clean plate. Separate raw foods from those that have been cooked.
- Don't use marinade that was in contact with raw meat for basting or sauces.

Keep Hot Foods Hot

- When cooking meats or poultry, use a thermometer to test the internal temperature. Insert the thermometer between the thigh and the body of a turkey or into the thickest part of other meats, making sure the tip of the thermometer is not in contact with bone or the pan. Cook to the temperature indicated for that particular meat (see Figure 19-3 on p. 663); cook hamburgers to at least medium well-done. If you have safety questions, call the USDA Meat and Poultry Hotline: (800) 535-4555.
- Cook stuffing separately, or stuff poultry just prior to cooking.
- Do not cook large cuts of meat or turkey in a microwave oven; it leaves some parts undercooked while overcooking others.

- Cook eggs before eating them (soft-boiled for at least 3½ minutes; scrambled until set, not runny; fried for at least 3 minutes on one side and 1 minute on the other).
- Cook seafood thoroughly. If you have safety questions about seafood call the FDA hotline: (800) FDA-4010.
- When serving foods, maintain temperatures at 140°F or higher.
- Heat leftovers thoroughly to at least 165°F.

Keep Cold Foods Cold

- When running errands, stop at the grocery store last. When you get home, refrigerate the perishable groceries (such as meats and dairy products) immediately. Do not leave perishables in the car any longer than it takes for ice cream to melt.
- Put packages of raw meat, fish, or poultry on a plate before refrigerating to prevent juices from dripping on food stored below.
- Buy only foods that are solidly frozen in store freezers.
- Keep cold foods at 40°F or less; keep frozen foods at 0°F or less (keep a thermometer in the refrigerator).
- Marinate meats in the refrigerator, not on the counter.

TABLE 19-2 Safe Refrigerator Storage Times (≤40°F)

1 to 2 Days

Raw ground meats, breakfast or other raw sausages, raw fish or poultry; gravies

3 to 5 Days

Raw steaks, roasts, or chops; cooked meats, poultry, vegetables, and mixed dishes; lunch meats (packages opened); mayonnaise salads (chicken, egg, pasta, tuna)

1 Week

Hard-cooked eggs, bacon or hot dogs (opened packages); smoked sausages or seafood

2 to 4 Weeks

Raw eggs (in shells); lunch meats, bacon, or hot dogs (packages unopened); dry sausages (pepperoni, hard salami); most aged and processed cheeses (Swiss, brick)

2 Months

Mayonnaise (opened jar); most dry cheese (parmesan, romano)

Eating raw oysters can be dangerous for anyone, but people with liver disease and weakened immune systems are most vulnerable. At least ten species of bacteria found in raw oysters can cause serious illness and even death.[*] Raw oysters may also carry the hepatitis A virus, which can cause liver disease. Some hot sauces can kill many of these bacteria, but not the virus; alcohol may also protect some people against some oyster-borne illnesses, but not enough to guarantee protection (or to recommend drinking alcohol). Pasteurization of raw oysters—holding them at a specified temperature for a specified time—holds promise for killing bacteria without cooking the oyster or altering its texture or flavor.

As population density increases along the shores of seafood-harvesting waters, pollution inevitably invades the sea life there. Preventing seafood-borne illness is in large part a task of controlling water pollution. To help ensure a safe seafood market, the FDA requires processors to adopt food safety practices based on the HACCP system mentioned earlier.

Chemical pollution and microbial contamination lurk not only in the water, but in the boats and warehouses where seafood is cleaned, prepared, and refrigerated. Seafood is one of the most perishable foods: time and temperature are critical to its freshness and flavor. To keep seafood as fresh as possible, people in the industry "keep it cold, keep it clean, and keep it moving." Wise consumers eat it cooked.

Other Precautions and Procedures Fresh food generally smells fresh. Not all types of food poisoning are detectable by odor, but some bacterial wastes produce "off" odors. If an abnormal odor exists, the food is spoiled. Throw it out or, if it was recently purchased, return it to the grocery store. Do not taste it. Table 19-2 lists safe refrigerator storage times for selected foods.

---

[*]Raw oysters can carry the bacterium *Vibrio vulnificus;* see Table 19-1 for details.

HOW TO Prevent Foodborne Illnesses—continued

- Refrigerate leftovers promptly; use shallow containers to cool foods faster; use leftovers within 3 to 4 days.
- Thaw meats or poultry in the refrigerator, not at room temperature. If you must hasten thawing, use cool water (changed every 30 minutes) or a microwave oven.
- Freeze meat, fish, or poultry immediately if not planning to use within a few days.

In General

- Do not reuse disposable containers; use nondisposable containers or recycle instead.
- Do not taste food that is suspect. "If in doubt, throw it out."
- Throw out foods with danger-signaling odors. Be aware, though, that most food-poisoning bacteria are odorless, colorless, and tasteless.
- Do not buy or use items that have broken seals or mangled packaging; such containers cannot protect against microbes, insects, spoilage, or even vandalism. Check safety seals, buttons, and expiration dates.
- Follow label instructions for storing and preparing packaged and frozen foods; throw out foods that have been thawed or refrozen.

- Discard foods that are discolored, moldy, or decayed or that have been contaminated by insects or rodents.

For Specific Food Items

- *Canned goods.* Carefully discard food from cans that leak or bulge so that other people and animals will not accidentally ingest it; before canning, seek professional advice from the USDA Extension Service (check your phone book under U.S. government listings, or ask directory assistance).
- *Milk and cheeses.* Use only pasteurized milk and milk products. Aged cheeses, such as cheddar and swiss, do well for an hour or two without refrigeration, but should be refrigerated or stored in an ice chest for longer periods.
- *Eggs.* Use clean eggs with intact shells. Do not eat eggs, even pasteurized eggs, raw; raw eggs are commonly found in Caesar salad dressing, eggnog, cookie dough, hollandaise sauce, and key lime pie. Cook eggs until whites are firmly set and yolks begin to thicken.
- *Honey.* Honey may contain dormant bacterial spores, which can awaken in the human body to produce botulism. In adults, this poses little hazard, but infants under one year of age should never be fed honey.

Honey can accumulate enough toxin to kill an infant; it has been implicated in several cases of sudden infant death. (Honey can also be contaminated with environmental pollutants picked up by the bees.)
- *Mayonnaise.* Commercial mayonnaise may actually help a food to resist spoilage because of the acid content. Still, keep it cold after opening.
- *Mixed salads.* Mixed salads of chopped ingredients spoil easily because they have extensive surface area for bacteria to invade, and they have been in contact with cutting boards, hands, and kitchen utensils that easily transmit bacteria to food (regardless of their mayonnaise content). Chill them well before, during, and after serving.
- *Picnic foods.* Choose foods that last without refrigeration such as fresh fruits and vegetables, breads and crackers, and canned spreads and cheeses that can be opened and used immediately. Pack foods cold, layer ice between foods, and keep foods out of water.
- *Seafood.* Buy only fresh seafood that has been properly refrigerated or iced. Cooked seafood should be stored separately from raw seafood to avoid cross-contamination.

Local health departments and the USDA Extension Service can provide additional information about food safety. Should precautions fail and mild foodborne illness develop, drink clear liquids to replace fluids lost through vomiting and diarrhea. If serious foodborne illness is suspected, first call a physician. Then wrap the remainder of the suspected food and label the container so that the food cannot be mistakenly eaten, place it in the refrigerator, and hold it for possible inspection by health authorities. The margin■ identifies foods most commonly implicated in foodborne illnesses.

Food Safety While Traveling

People who travel to other countries have a 50–50 chance of contracting an illness, most often **travelers' diarrhea**.[5] Like many other foodborne illnesses, travelers' diarrhea is a sometimes serious, always annoying bacterial infection of the digestive tract. The risk is high because, for one thing, some countries' cleanliness standards for food and water may be lower than those in the United States and Canada. For another, every region's microbes are different, and while people are immune to those in their own neighborhoods, they have had no chance to develop immunity to the pathogens in places they are visiting for the first time. The "How to" on the next page offers tips for food safety while traveling.[6]

Advances in Food Safety

Advances in technology have dramatically improved the quality and safety of foods available on the market. From pasteurization in the early 1900s■ to irradiation in the early 2000s, these advances offer numerous benefits, but they also raise consumer concerns.[7]

■ Frequently unsafe:
  - Raw milk and milk products.
  - Raw or undercooked seafood, meat, poultry, or eggs.
  - Raw sprouts and scallions.
Occasionally unsafe:
  - Soft cheeses (Mexican style, feta, brie, camembert, blue-veined).
  - Salad bar items.
  - Unwashed berries and grapes.
  - Sandwiches.
  - Hamburgers.
Rarely unsafe:
  - Peeled fruit.
  - High-sugar foods.
  - Steaming-hot foods.

■ During the last century, pasteurization of milk helped to control typhoid fever, tuberculosis, scarlet fever, diphtheria, and other infectious diseases.

travelers' diarrhea: nausea, vomiting, and diarrhea caused by consuming food or water contaminated by any of several organisms, most commonly, *E. coli*, *Shigella*, *Campylobacter jejuni,* and *Salmonella*.

Foodborne illnesses contracted while traveling are colloquially known as travelers' diarrhea. A bout of this ailment can ruin the most enthusiastic tourist's trip. To avoid foodborne illness, follow the food safety tips outlined on pp. 664–665. In addition, while traveling:

- Wash your hands often with soap and hot water, especially before handling food or eating. Use antiseptic gel or hand wipes.
- Eat only well-cooked and hot or canned foods. Eat raw fruits or vegetables only if you have washed them in purified water and peeled them yourself. Skip salads and raw fish and shellfish.
- Be aware that water, and ice made from it, may be unsafe. Use safe, bottled water for drinking, making ice cubes, and brushing teeth. Alternatively, take along disinfecting tablets or a device to boil water. Do not use ice unless it was made from purified or bottled water.
- Drink no beverages made with tap water. Drink only treated, boiled, canned, or bottled beverages, and drink them without ice, even if they are not chilled to your liking.
- Refuse dairy products unless they have been properly pasteurized and refrigerated.
- Do not buy food and drinks from street vendors.
- Before you leave on the trip, ask your physician to recommend an antimotility agent and an antibiotic to take with you in case your efforts to avoid illness fail.

To sum up these recommendations, "Boil it, cook it, peel it, or forget it." Chances are excellent that if you follow these rules, you will remain well.

Irradiation The use of low-dose **irradiation** protects consumers from foodborne illnesses by:[8]

- Controlling mold in grains.
- Sterilizing spices and teas for storage at room temperature.
- Controlling insects and extending shelf life in fresh fruits and vegetables (inhibits the growth of sprouts on potatoes and onions and delays ripening in some fruits such as strawberries and mangoes).
- Destroying harmful bacteria in fresh and frozen beef, poultry, lamb, and pork.

Some foods are not candidates for the treatment. For example, when irradiated, high-fat meats develop off-odors, egg whites turn milky, grapefruits become mushy, and milk products change flavor. (Incidentally, the milk in those boxes kept at room temperature on grocery-store shelves is not irradiated, but sterilized with an **ultrahigh temperature treatment.**)

The use of food irradiation has been extensively evaluated over the past 50 years; approved for use in more than 40 countries; and supported by numerous health agencies, including the **FAO, WHO,** and the American Medical Association. Irradiation does not noticeably change the taste, texture, or appearance of approved foods,■ nor does it make foods radioactive. Vitamin loss is minimal and comparable to amounts lost in other food-processing methods such as canning. Because irradiation kills bacteria without the use of heat, it is sometimes called "cold pasteurization."

Consumer Concerns about Irradiation Many consumers, associating radiation with cancer, birth defects, and mutations, have strong negative emotions about the use of irradiation on foods. Some confuse it with food contamination by radioactive particles, such as occurs in the aftermath of a nuclear accident. Some balk at the idea of irradiating, and thus sterilizing, contaminated foods and prefer instead the elimination of unsanitary slaughtering and food preparation conditions. Food producers, on the other hand, are eager to use irradiation, but hesitate to do so until consumers are ready to accept it and willing to pay for it. Once consumers understand the benefits of irradiation, about half are willing to use irradiated foods, but only a fourth are willing to pay more.[9]

Regulation of Irradiation The FDA has established regulations governing the specific uses of irradiation and allowed doses. Each food that has been treated with irradiation must say so on its label.■ Labels can be misleading, however. Products that use irradiated foods as ingredients are not required to say so on the label. Furthermore, consumers may interpret the *absence* of the irradiation symbol to mean that the food was produced without any kind of treatment. This is not true; it is

■ Foods approved for irradiation:
- Eggs.
- Raw beef, lamb, poultry, pork.
- Spices, tea.
- Wheat.
- Vegetables (potatoes, tomatoes, onions).
- Fresh fruit (strawberries, citrus, papaya).

■ This international symbol, called the *radura*, identifies retail foods that have been irradiated. The words "Treated by irradiation" or "Treated with irradiation" must accompany the symbol. The irradiation label is not required on commercially prepared foods that contain irradiated ingredients, such as spices.

irradiation: sterilizing a food by exposure to energy waves, similar to ultraviolet light and microwaves.

ultrahigh temperature (UHT) treatment: sterilizing a food by brief exposure to temperatures above those normally used.

just that the FDA does not require label statements for other treatments used for the same purpose, such as postharvest fumigation with pesticides. If all treatment methods were declared, consumers could make fully informed choices.

Other Pasteurizing Systems Other technologies using high-intensity pulsed light or electron beams have also been approved by the FDA. Like irradiation, these technologies kill microorganisms and extend the shelf life of foods without diminishing their nutrient content.

> **IN SUMMARY** Millions of people suffer mild to life-threatening symptoms caused by foodborne illnesses (review Table 19-1). As the "How to" on pp. 664–665 describes, most of these illnesses can be prevented by storing and cooking foods at their proper temperatures and by preparing them in sanitary conditions.

Nutritional Adequacy of Foods and Diets

In years past, when most foods were whole and farm fresh, the task of meeting nutrient needs primarily involved balancing servings from the various food groups. Today, however, foods have changed. Many "new" foods are available to appeal to consumers' demands for convenience and flavor, but not necessarily to deliver a balanced assortment of needed nutrients.

Obtaining Nutrient Information

To help consumers find their way among these foods and combine them into healthful diets, the FDA has developed extensive nutrition labeling regulations, as Chapter 2 described. In addition, the USDA's *Dietary Guidelines* help consumers "eat to stay healthy," and the Food Guide Pyramid helps them to put those recommendations into practice (see Chapter 2).

Minimizing Nutrient Losses

In addition to selecting nutritious foods and preparing them safely, consumers can improve their nutritional health by learning to store and cook foods in ways that minimize nutrient losses. Water-soluble vitamins are the most vulnerable of the nutrients, but both vitamins and minerals can be lost when they dissolve in water that is then discarded.

Fruits and vegetables contain enzymes that both synthesize and degrade vitamins. After a fruit or vegetable has been picked, vitamin synthesis stops, but degradation continues. To slow the degradation of vitamins, most fruits and vegetables should be kept refrigerated until used (degradative enzymes are most active at warmer temperatures).

Some vitamins are easily destroyed by oxygen. To minimize the destruction of vitamins, store fruits and vegetables that have been cut and juice that has been opened in airtight containers and refrigerate them.

Water-soluble vitamins readily dissolve in water. To prevent losses during washing, wash fruits and vegetables before cutting. To minimize losses during cooking, steam or microwave vegetables. Alternatively, use the cooking water when preparing meals such as casseroles and soups.

Finally, keep in mind that most vitamin losses are not catastrophic and that a law of diminishing returns operates. Do not fret over small losses or waste time that may be valuable in improving your health in other ways. Be assured that if you start with plenty of fruits and vegetables and are reasonably careful in their storage and preparation, you will receive a sufficient supply of all the nutrients they provide.

> **IN SUMMARY** In the marketplace, food labels, the *Dietary Guidelines,* and the Food Guide Pyramid all help consumers learn about nutrition and how to plan healthy diets. At home, consumers can minimize nutrient losses from fruits and vegetables by refrigerating them, washing them before cutting them, storing them in airtight containers, and cooking them for short times and in minimal water.

Environmental Contaminants

Concern about environmental contamination of foods is growing as the world becomes more populated and more industrialized. Industrial processes pollute the air, water, and soil. Plants absorb the **contaminants,** and people consume the plants (grains, vegetables, legumes, and fruits) or the meat and milk products from livestock that have eaten the plants. Similarly, polluted water contaminates the fish and other seafood that people eat.[10]

Harmfulness of Environmental Contaminants

The potential harmfulness of a contaminant depends in part on its **persistence**—the extent to which it lingers in the environment or in the body. Some contaminants in the environment are short-lived because microorganisms or agents such as sunlight or oxygen can break them down. Some contaminants in the body may linger for only a short time because the body rapidly excretes them or metabolizes them to harmless compounds. These contaminants present little cause for concern. Some contaminants, however, resist breakdown and can accumulate. Each level of the **food chain,** then, has a greater concentration than the one below **(bioaccumulation).** Figure 19-4 shows how bioaccumulation leads to high concentrations of toxins in people at the top of the food chain.

Contaminants enter the environment in various ways. Accidental spills are rare, but can have devastating effects. More commonly, small amounts are released over long periods. The following paragraphs describe how three contaminants found their way into the food supply in the past. The first example involves a **heavy metal;** the others involve **organic halogens.**

Methylmercury A classic example of acute contamination occurred in 1953 when a number of people in Minamata, Japan, became ill with a disease no one had seen before. By 1960, 121 cases had been reported, including 23 in infants. Mortality was high; 46 died, and the survivors suffered blindness, deafness, lack of coordination, and intellectual deterioration. The cause was ultimately revealed to be methylmercury contamination of fish from the bay where these people lived. The infants who contracted the disease had not eaten any fish, but their mothers had, and even though the mothers exhibited no symptoms during their pregnancies, the poison had affected their unborn babies. Manufacturing plants in the region were discharging mercury into the waters of the bay, the mercury was turning to methylmercury on leaving the factories, and the fish in the bay were accumulating this poison in their bodies. Some of the affected families had been eating fish from the bay every day.

PBB and PCB In 1973, half a ton of **PBB (polybrominated biphenyl),** a toxic organic compound, was accidentally mixed into some livestock feed that was distributed throughout the state of Michigan. The chemical found its way into

---

contaminants: substances that make a food impure and unsuitable for ingestion.

persistence: stubborn or enduring continuance; with respect to food contaminants, the quality of persisting, rather than breaking down, in the bodies of animals and human beings.

food chain: the sequence in which living things depend on other living things for food.

bioaccumulation: the accumulation of contaminants in the flesh of animals high on the food chain.

heavy metal: any of a number of mineral ions such as mercury and lead, so called because they are of relatively high atomic weight. Many heavy metals are poisonous.

organic halogens: organic compounds containing one or more atoms of a halogen—fluorine, chlorine, iodine, or bromine.

PBB (polybrominated biphenyl) and **PCB (polychlorinated biphenyl):** toxic organic compounds used in pesticides, paints, and flame retardants.

FIGURE 19-4   Bioaccumulation of Toxins in the Food Chain

This example features fish as the food for human consumption, but bioaccumulation of toxins occurs on land as well when cows, pigs, and chickens eat or drink contaminated foods or water.

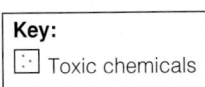

Key:

☐ Toxic chemicals

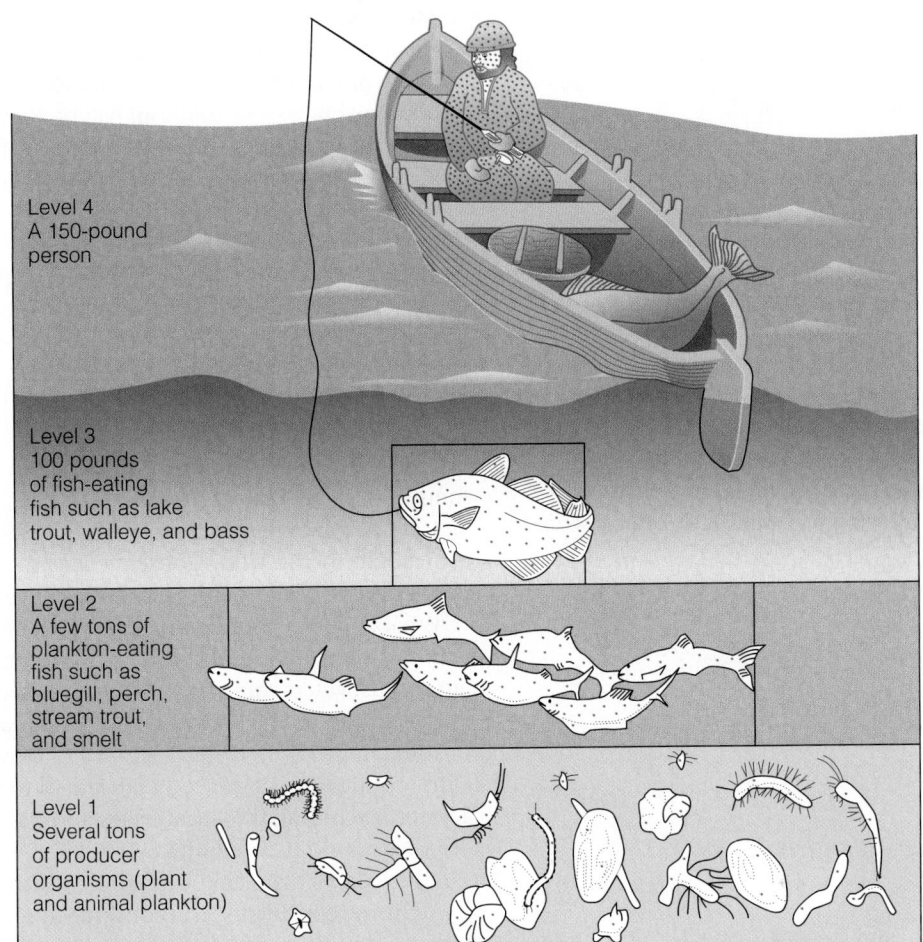

4 If none of the chemicals are lost along the way, people ultimately receive all of the toxic chemicals that were present in the original plants and plankton.

Level 4
A 150-pound person

3 Contaminants become further concentrated in larger fish that eat the small fish from the lower part of the food chain.

Level 3
100 pounds of fish-eating fish such as lake trout, walleye, and bass

2 Contaminants become more concentrated in small fish that eat the plants and plankton.

Level 2
A few tons of plankton-eating fish such as bluegill, perch, stream trout, and smelt

1 Plants and plankton at the bottom of the food chain become contaminated with toxic chemicals, such as methylmercury (shown as red dots).

Level 1
Several tons of producer organisms (plant and animal plankton)

---

millions of animals and then into people who ate the meat. The seriousness of the accident began to come to light when dairy farmers reported their cows were going dry, aborting their calves, and developing abnormal growths on their hooves. Although more than 30,000 cattle, sheep, and swine and more than a million chickens were destroyed, an estimated 97 percent of Michigan's residents had been exposed to PBB. Some of the exposed farm residents suffered nervous system aberrations and liver disorders.

A similar accident occurred in 1979 when **PCB (polychlorinated biphenyl)** contaminated rice oil in Taiwan. Women who had eaten the tainted rice oil gave birth to children with developmental problems. Decades later, young men who were exposed to PCB during gestation have reduced fertility.[11]

Guidelines for Consumers

How much of a threat do environmental contaminants pose to the food supply? For the most part, the hazards appear to be small. The FDA regulates the presence of contaminants in foods and requires foods with unsafe amounts to be removed from the market. Similarly, health agencies may issue advisories informing consumers about the potential dangers of eating contaminated foods.

Reduce exposure of the population to heavy metals and other toxic chemicals.

■ Pregnant and lactating women and young children should avoid:
- Tilefish, swordfish, king mackerel, shark.

And limit average weekly consumption of:
- Ocean, coastal, and other commercial fish to 12 oz (cooked or canned) *or*
- Freshwater fish caught by family and friends to 6 oz (cooked).

■ For perspective, 1 ppm (part per million) is equivalent to about 1 minute in 2 years or 1 cent in $10,000.

■ Fish relatively high in mercury:
- Tilefish, swordfish, king mackerel, shark.

Fish relatively low in mercury:
- Cod, haddock, pollock, salmon, sole, tilapia.
- Most shellfish.

■ Fish relatively high in omega-3 fatty acids:
- Anchovy, bluefish, herring, lake trout, mackerel, menhaden, mullet, sablefish, salmon, sardines.

Most recently, mercury poisoning has aroused concerns—even at levels one-tenth of those in the Minamata catastrophe. Most vulnerable are pregnant and lactating women and young children■ because mercury toxicity damages the developing brain.[12] Virtually all fish have at least trace amounts of mercury (on average, 0.12 parts per million).■ Mercury, PCB, chlordane, dioxins, and DDT are the toxins most responsible for fish contamination, but mercury leads the list by threefold.[13]

Review Figure 19-4 (on p. 669) and notice how toxins such as mercury become more concentrated in animals and in people high in the food chain. Because of bioaccumulation, large gamefish at the top of the aquatic food chain generally have the highest concentrations of mercury (ten times the average).■ Consumers who enjoy eating these fish should select the smaller, younger ones (within legal limits). Also because of bioaccumulation, the concentrations in fish may be a million times higher than the concentrations in the water itself.

The **EPA** regulates commercial fishing to help ensure that fish destined for consumption in the United States meet safety standards for mercury and other contaminants. Farm-raised fish usually have lower concentrations of mercury than fish caught in the wild. Consequently, most consumers in the United States are not in danger of receiving harmful levels of mercury from fish.

What about the noncommercial fish a person catches from a local lake, river, or ocean? After all, it's almost impossible to tell whether water is contaminated without sophisticated equipment. Each state monitors its waters and issues advisories to inform the public if chemical contaminants have been found in the local fish. In 2001, in the United States advisories were issued for 71 percent of the coastline of the contiguous 48 states, 14 percent of the river miles, and 28 percent of the lake acreage (excluding the Great Lakes and their connecting waters, which are all under advisory).[14] To find out whether a fish advisory has been posted in your region, call the local or state environmental health department.

All things considered, fish continue to support a healthy diet, providing valuable protein, omega-3 fatty acids,■ and minerals. Mercury may increase the risk of heart disease, however; researchers are still working to determine whether high concentrations of mercury in fish diminish the heart-protecting effects of the omega-3 fatty acids.[15]

Reduce the potential human exposure to persistent chemicals by decreasing fish contaminant levels.

IN SUMMARY Environmental contamination of foods is a concern, but so far, the hazards appear relatively small. In all cases, two principles apply. First, remain alert to the possibility of contamination of foods, and keep an ear open for public health announcements and advice. Second, eat a variety of foods. Varying food choices is an effective defensive strategy against the accumulation of toxins in the body. Each food eaten dilutes contaminants that may be present in other components of the diet.

Natural Toxicants in Foods

Consumers concerned about food contamination may think that they can eliminate all poisons from their diets by eating only "natural" foods. On the contrary, nature has provided plants with an abundant array of toxicants. A few examples will show how even "natural" foods may contain potentially harmful substances. They also show that while the *potential* for harm exists, *actual* harm rarely occurs.

Poisonous mushrooms are a familiar example of plants that can be harmful when eaten. Few people know, though, that other commonly eaten foods contain substances that can cause illnesses. Cabbage, turnips, mustard greens, kale, brussels sprouts, cauliflower, broccoli, kohlrabi, and radishes contain small quantities of goitrogens—compounds that can enlarge the thyroid gland. Eating exceptionally large amounts of goitrogen-containing vegetables can aggravate a preexisting thyroid problem, but usually does not initiate one.

Lima beans and fruit seeds such as apricot pits contain cyanogens—inactive compounds that produce the deadly poison cyanide upon activation by a specific plant enzyme. For this reason, many countries restrict commercially grown lima beans to those varieties with the lowest cyanogen contents. As for fruit seeds, they are seldom deliberately eaten. An occasional swallowed seed or two presents no danger, but a couple of dozen seeds can be fatal to a small child. Perhaps the most infamous cyanogen in seeds is laetrile—a compound erroneously represented as a cancer cure. True, laetrile kills cancer, but only at doses that kill the person, too. Research over the past hundred years has never proved laetrile to be an effective cancer treatment. In fact, laetrile is more dangerous than no treatment at all. The combination of cyanide poisoning and lack of medical attention is life-threatening.

The humble potato contains many natural poisons including **solanine,** a powerful narcotic-like substance. The small amounts of solanine normally found in potatoes are harmless, but solanine is toxic and presents a hazard when consumed in large quantities. Physical symptoms of solanine poisoning include headache, vomiting, abdominal pain, diarrhea, and fever; neurological symptoms include apathy, restlessness, drowsiness, confusion, stupor, hallucinations, and visual disturbances. Solanine production increases when potatoes are improperly stored in the light and in either very cold or fairly warm places. Cooking does not destroy solanine, but because most of a potato's solanine is in the green layer that develops just beneath the skin, it can be peeled off, making the potato safe to eat.

IN SUMMARY Natural toxicants include the goitrogens in cabbage, cyanogens in lima beans, and solanine in potatoes. These examples of naturally occurring toxicants illustrate two familiar principles. First, any substance can be toxic when consumed in excess. Second, poisons are poisons, whether made by people or by nature. Remember: it is not the source of a chemical that makes it hazardous, but its chemical structure and the quantity consumed.

Pesticides

The use of **pesticides** is controversial. They help to ensure the survival of crops, but they leave **residues** in the environment and on some of the foods we eat.

Hazards and Regulation of Pesticides

Ideally, a pesticide would destroy the pest and quickly degenerate to nontoxic products without accumulating in the food chain. Then, by the time consumers ate the food, no harmful residues would remain. Unfortunately, no such perfect pesticide exists. As new pesticides are developed, government agencies assess their risks and benefits and vigilantly monitor their use.

Hazards of Pesticides Pesticides may linger on the foods to which they were applied in the field. Health risks from pesticide exposure are probably small for healthy adults, but children, the elderly, and people with weakened immune systems may be vulnerable to some types of pesticide poisoning. To protect infants

solanine (SOH-lah-neen): a poisonous narcotic-like substance present in potato peels and sprouts.

pesticides: chemicals used to control insects, weeds, fungi, and other pests on plants, vegetables, fruits, and animals. Used broadly, the term includes herbicides (to kill weeds), insecticides (to kill insects), and fungicides (to kill fungi).

residues: whatever remains. In the case of pesticides, those amounts that remain on or in foods when people buy and use them.

and children, government agencies set a **tolerance level** for each pesticide by first identifying foods that children commonly eat in large amounts and then considering the effects of pesticide exposure during each developmental stage.

Reduce pesticide exposures.

Regulation of Pesticides Consumers depend on the EPA and the FDA to keep pesticide use within safe limits. These agencies evaluate the risks and benefits of a pesticide's use by asking such questions as, How dangerous is it? How much residue is left on the crop? How much harm does the pesticide do to the environment? How necessary is it? What are the alternatives to its use?

If the pesticide is approved, the EPA establishes a tolerance level for its presence in foods, well below the level at which it could cause any conceivable harm. Tolerance regulations also state the specific crops to which each pesticide can be applied. If a pesticide is misused, growers risk fines, lawsuits, and destruction of their crops.

Once tolerances are set, the FDA enforces them by monitoring foods and livestock feeds for the presence of pesticides. Over the past several decades of testing, the FDA has seldom found residues above tolerance levels, so it appears that pesticides are generally used according to regulations. Minimal pesticide use means lower costs for growers. In addition to costs, many farmers are also concerned about the environment, the quality of their farmland, and a safe food supply. Where violations are found, they are usually due to unusual weather conditions, use of unapproved pesticides, or misuse—for example, application of a particular pesticide to a crop for which it has not been approved.

As many as 400 varieties of fruits and vegetables are imported from other countries.

© Angelo Cavalli/The Image Bank/Getty Images

Pesticides from Other Countries Because other countries may not have the same pesticide regulations as the United States and Canada, imported foods may contain both pesticides that have been banned and permitted pesticides at concentrations higher than are allowed in domestic foods. A loophole in federal regulations allows U.S. companies to manufacture and sell, to other countries, pesticides that are banned in this country. The banned pesticides then return to the United States on imported foods—a circuitous route that concerned consumers have called the "circle of poison." Federal inspectors sample imported foods and refuse entry if they are found to contain illegal pesticide residues. The United States, Mexico, and Canada work together to establish a pesticide policy for all of North America.

Monitoring Pesticides

The FDA collects and analyzes samples of both domestic and imported foods. If the agency finds samples in violation of regulations, it can seize the products or order them destroyed. The FDA may also invoke a **certification** requirement that forces manufacturers, at their own expense, to have their foods periodically inspected and certified safe by an independent testing agency. Individual states also scan for pesticides (as well as for industrial chemicals) and provide information to the FDA.

Food in the Fields In addition to its ongoing surveillance, the FDA also conducts selective surveys to determine the presence of particular pesticides in specific crops. For example, one year the agency searched for aldicarb in potatoes, captan in cherries, and diaminozide (the chemical name for Alar) in apples, among others. Actions taken that year required several certifications. Thus one shipper in Australia had to certify apples; one in Canada, peppers; one in Costa Rica, chayotes. All grapes from Mexico had to be certified and so did all mangoes from anywhere. This shows, incidentally, how many foods come from abroad—not only those already named, but hundreds more—and that the FDA monitors them as carefully as it does the domestic food supply.

tolerance level: the maximum amount of a residue permitted in a food when a pesticide is used according to label directions.

certification: the process in which a private laboratory inspects shipments of a product for selected chemicals and then, if the product is free of violative levels of those chemicals, issues a guarantee to that effect.

Food on the Plate In addition to monitoring foods in the field for pesticides, the FDA also monitors people's actual intakes. The agency conducts the Total Diet Study (sometimes called the "Market Basket Survey") to estimate the dietary intakes of pesticide residues by eight age and gender groups from infants to senior citizens. Four times a year, FDA surveyors buy over 200 foods from U.S. grocery stores, each time in several cities. They prepare the foods table ready and then analyze them not only for pesticides, but for essential minerals, industrial chemicals, heavy metals, and radioactive materials. In all, the survey reports on over 10,000 samples a year, and recently more than half have been imported foods. Most heavily sampled are fresh vegetables, fruits, and dairy products.

The Total Diet Study provides a direct estimate of the amounts of pesticide residues that remain in foods as they are usually eaten—after they have been washed, peeled, and cooked. The FDA finds the intake of almost all pesticides to be less than 1 percent of the amount considered acceptable. The amount considered acceptable is "the daily intake of a chemical which, if ingested over a lifetime, appears to be without appreciable risk."■ All in all, these findings corroborate the safety of the U.S. food supply with respect to pesticide residues.

Consumer Concerns

Despite these reassuring reports, consumers still worry that the monitoring of foods may not be adequate. For one thing, manufacturers develop new pesticides all the time. For another, as described earlier, other countries use pesticides that are illegal for use here. For still another, although the regulations may protect U.S. foods adequately, they may not necessarily protect the environment or the people who work in the fields. Concerns over poisoning of soil, waterways, wildlife, and workers may well be valid.

The FDA does not sample *all* food shipments or test for *all* pesticides in each sample. The FDA is a *monitoring* agency, and as such, it cannot, nor can it be expected to, guarantee 100 percent safety in the food supply. Instead, it sets standards so that substances do not become a hazard, checks enough samples to adequately assess average food safety, and acts promptly when problems or suspicions arise.

Minimizing Risks Whether consumers are ingesting pesticide residues depends on a number of factors. How much of a given food is the consumer eating? What pesticide was used on it? How much was used? How long ago was the food last sprayed? Did environmental conditions promote pest growth or pesticide breakdown? How well was the produce washed? Was it peeled or cooked? With so many factors, consumers cannot know for sure whether pesticide residues remain on foods, but they can minimize their risks by following the guidelines offered in the "How to" feature on p. 674. The food supply is protected well enough that consumers who take these precautions can feel secure that the foods they eat are safe.

Alternatives to Pesticides To feed a nation while using fewer pesticides requires creative farming methods. Highlight 19 describes how scientists have genetically altered plants to produce their own pesticides, and Highlight 20 presents alternative, or sustainable, agriculture methods. These methods include such practices as rotating crops and using plants that produce natural pesticides. Among natural pesticides are the nicotine in tobacco and psoralens in celery. Natural pesticides are less damaging to other living things and less persistent in the environment than most human-made ones. Other alternatives to heavy pesticide use include releasing organisms into fields to destroy pests and planting nonfood crops nearby to kill pests or attract them away from the food crops. For example, releasing sterile male fruit flies into orchards helps to curb the population growth of these pests; some flowers, such as marigolds, release natural insecticides and are often planted near crops such as tomatoes. Such alternative farming methods are more labor-intensive and may produce smaller yields than conventional methods, at least initially. Over time, though, by eliminating expensive pesticides, fertilizers, and fuels, these alternatives may actually cut costs more than they cut yields.

Washing fresh fruits and vegetables removes most, if not all, of the pesticide residues that might have been present.

■ *Without appreciable risk* means "practical certainty that injury will not result even after a lifetime of exposure."

People can grow pesticide-free crops when their gardens or farms are relatively small.

■ Organic foods that have met USDA standards may use this seal on their labels.

organic: in agriculture, crops grown and processed according to USDA regulations defining the use of fertilizers, herbicides, insecticides, fungicides, preservatives, and other chemical ingredients.

Organically Grown Crops Alternative methods may be especially useful if farmers want to produce and market **organic** crops grown and processed according to USDA regulations defining the use of synthetic fertilizers, herbicides, insecticides, fungicides, preservatives, and other chemical ingredients.■ Similarly, meat, poultry, eggs, and dairy products may be called organic if the livestock has been raised according to USDA regulations defining the grazing conditions and the use of organic feed, hormones, and antibiotics. In addition, producers may *not* claim products are organic if they have been irradiated, genetically engineered, or grown with fertilizer made from sewer sludge. Figure 19-5 shows examples of food labels for products using organic ingredients.

Implied in the marketing of *organic foods* is that organic products are safer or healthier for consumers than those grown using other methods, which may not be

FIGURE 19-5 Food Labels for Organic Products

United States Department of Agriculture

Foods made with 100 percent organic ingredients may claim "100% organic" and use the seal.

Foods made with a least 95 percent organic ingredients may claim "organic" and use the seal.

Foods made with at least 70 percent organic ingredients may list up to three of those ingredients on the front panel.

Foods made with less than 70 percent organic ingredients may list them on the side panel, but cannot make any claims on the front.

the case. Using unprocessed animal manure as an organic fertilizer, for example, may transmit bacteria, such as *E. coli,* to human beings. Both organic and conventional methods may have advantages and disadvantages, and consumers must remain informed.

> **IN SUMMARY** Pesticides can safely improve crop yields when used according to regulations, but can also be hazardous when used inappropriately. The FDA tests both domestic and imported foods for pesticide residues in the fields and in market basket surveys of foods prepared table ready. Consumers can minimize their ingestion of pesticide residues on foods by following the suggestions in the "How to" on p. 674. Alternative farming methods may allow farmers to grow crops with few or no pesticides.

Many consumers are willing to pay a little more for pesticide-free produce.

Food Additives

Additives confer many benefits on foods. Some reduce the risk of foodborne illness (for example, nitrites used in curing meat prevent poisoning from the botulinum toxin). Others enhance nutrient quality (as in vitamin D–fortified milk). Most additives are **preservatives** that help prevent spoilage during the time it takes to deliver foods long distances to grocery stores and then to kitchens. Some additives simply make foods look and taste good.

Intentional additives are put into foods on purpose, while indirect additives may get in unintentionally before or during processing. This discussion begins with the regulations that govern additives, then presents intentional additives class by class, and finally goes on to say a word about the indirect additives.

Regulations Governing Additives

The FDA's concern with additives hinges primarily on their safety. To receive permission to use a new additive in food products, a manufacturer must satisfy the FDA that the additive is:

- Effective (it does what it is supposed to do).
- Detectable and measurable in the final food product.
- Safe (when fed in large doses to animals under strictly controlled conditions, it causes no cancer, birth defects, or other injury).

On approving an additive's use, the FDA writes a regulation stating in what amounts and in what foods the additive may be used. No additive receives permanent approval; all must undergo periodic review.

The GRAS List Many familiar substances are exempted from complying with the FDA's approval procedure because they are **generally recognized as safe (GRAS)**, based either on their extensive, long-term use in foods or on current scientific evidence. Several hundred substances are on the GRAS list, including such items as salt, sugar, caffeine, and many spices. Whenever substantial scientific evidence or public outcry has questioned the safety of any substance on the GRAS list, it has been reevaluated. If a legitimate question has been raised about a substance, it has been removed or reclassified. Meanwhile, the entire GRAS list is subjected to ongoing review.

The Delaney Clause One risk that the U.S. law on additives refuses to tolerate at any level is the risk of cancer. To remain on the GRAS list, an additive must not have been found to be a carcinogen in any test on animals or human beings. The **Delaney Clause** (the part of the law that states this criterion) is uncompromising in addressing carcinogens in foods and drugs; in fact, it has been under fire for many years for being too strict and inflexible.

additives: substances not normally consumed as foods but added to food either intentionally or by accident.

preservatives: antimicrobial agents, antioxidants, and other additives that retard spoilage or maintain desired qualities, such as softness in baked goods.

generally recognized as safe (GRAS): food additives that have long been in use and are believed safe. First established by the FDA in 1958, the GRAS list is subject to revision as new facts become known.

Delaney Clause: a clause in the Food Additive Amendment to the Food, Drug, and Cosmetic Act that states that no substance that is known to cause cancer in animals or human beings at any dose level shall be added to foods.

Without additives, bread would quickly get moldy, and salad dressing would go rancid.

■ For perspective, one part per trillion is equivalent to about one grain of sugar in an Olympic-size swimming pool; or 1 second in 32,000 years; or one hair on 10 million heads, assuming none are bald.

■ The *de minimis* rule defines risk as a cancer rate of less than one cancer per million people exposed to a contaminant over a 70-year lifetime.

The Delaney Clause states that "no additive shall be deemed to be safe if it is found to induce cancer when ingested [at any level] by man or animal." That sounds clear enough, yet some products that fail to meet that standard still remain on the market. The artificial sweetener saccharin was the first exception to the rule. In the 1970s, when the FDA tried to ban saccharin because tests had revealed that it caused cancer in animals, consumers raised an outcry asking that it still be allowed in foods. In an attempt to balance the Delaney Clause with current food safety and cancer knowledge, Congress created a special exception that allowed saccharin to remain on the market as long as products containing it carried a warning.

The Delaney Clause is best understood as a product of a different historical era. It was adopted almost 50 years ago at a time when scientists knew less about the relationships between carcinogens and cancer development. At that time, most substances were detectable in foods only in relatively large amounts, such as parts per thousand. Today, scientific understanding of cancer has progressed, and technology has advanced so that carcinogens in foods can be detected even when they are present only in parts per billion or even per trillion.■ Earlier, "zero risk" may have seemed attainable, but today we know it is not: all substances, no matter how pure, can be shown to be contaminated at some level with one carcinogen or another. For these reasons, the FDA prefers to deem additives (and pesticides and other contaminants) safe if lifetime use presents no more than a one-in-a-million risk of cancer to human beings. Thus, instead of the "zero-risk" policy of the Delaney Clause, the FDA uses a "negligible-risk" standard, sometimes referred to as a *de minimis rule.*■

Margin of Safety Whatever risk level is permitted, actual risks must be determined by experiments. To determine risks posed by an additive, researchers feed test animals the additive at several concentrations throughout their lives. The additive is then permitted in foods in amounts 100 times *below* the lowest level that is found to cause any harmful effect, that is, at a 1/100 **margin of safety.** In many foods, *naturally* occurring substances occur with narrower margins of safety. Even nutrients pose risks at dose levels above those recommended and normally consumed: for young adults, the recommendation for vitamin D is only 1/10 of the Upper Level. People consume common table salt daily in amounts only three to five times less than those that pose a hazard.

Risks versus Benefits Of course, additives would not be added to foods if they only presented risks. Additives are in foods because they offer benefits that outweigh the risks they present, or make the risks worth taking. In the case of color additives that only enhance the appearance of foods but do not improve their health value or safety, no amount of risk may be deemed worth taking. In contrast, the FDA finds that it is worth taking the small risks associated with the use of nitrites on meat products, for example, because nitrites inhibit the formation of the deadly botulinum toxin. The choice involves a compromise between the risks of using additives and the risks of doing without them.

It is the manufacturers' responsibility to use only the amounts of additives that are necessary to get the needed effect, and no more. The FDA also requires that additives *not* be used:

• To disguise faulty or inferior products.

• To deceive the consumer.

• Where they significantly destroy nutrients.

• Where their effects can be achieved by economical, sound manufacturing processes.

Intentional Food Additives

Intentional food additives are added to foods to give them some desirable characteristic: resistance to spoilage, color, flavor, texture, stability, or nutritional value. The accompanying glossary defines the categories of additives, and the next sections describe additives people most often ask about.

margin of safety: when speaking of food additives, a zone between the concentration normally used and that at which a hazard exists. For common table salt, for example, the margin of safety is 1/5 (five times the amount normally used would be hazardous).

intentional food additives: additives intentionally added to foods, such as nutrients, colors, and preservatives.

GLOSSARY OF INTENTIONAL FOOD ADDITIVES

antimicrobial agents: preservatives that prevent microorganisms from growing.

antioxidants: preservatives that delay or prevent rancidity of fats in foods and other damage to food caused by oxygen.

artificial colors: certified food colors added to enhance appearance. (*Certified* means approved by the FDA.)
Bleaching agents may be used to whiten foods such as flour and cheese.

artificial flavors, flavor enhancers: chemicals that mimic natural flavors and those that enhance flavor.

nutrient additives: vitamins and minerals added to improve nutritive value.

emulsifiers and **gums:** thickening and stabilizing agents that maintain emulsions, foams, or suspensions or lend a desirable thick consistency to foods.

Antimicrobial Agents Foods can go bad in two ways. One way is relatively harmless: by losing their flavor and attractiveness. (Additives to prevent this kind of spoilage include antioxidants, discussed later.) The other way is by becoming contaminated with microbes that cause foodborne illnesses, a hazard that justifies the use of **antimicrobial agents.**

The most widely used antimicrobial agents■ are ordinary salt and sugar. Salt has been used throughout history to preserve meat and fish; sugar serves the same purpose in canned and frozen fruits and in jams and jellies. Both exert their protective effect primarily by capturing water and making it unavailable to microbes. Other additives, such as potassium sorbate and sodium propionate, are used to extend the shelf life of baked goods, cheeses, beverages, mayonnaise, margarine, and other products.

Other antimicrobial agents, the **nitrites,** are added to foods for three main purposes: to preserve color, especially the pink color of hot dogs and other cured meats; to enhance flavor by inhibiting rancidity, especially in cured meats and poultry; and to protect against bacterial growth. In amounts smaller than those needed to confer color, nitrites prevent the growth of the bacteria that produce the deadly botulinum toxin.

Nitrites clearly serve a useful purpose, but their use has been controversial. In the human body, nitrites can be converted to **nitrosamines.** At nitrite levels higher than those used in food products, nitrosamine formation causes cancer in animals. The food industry uses the minimal amount of nitrites necessary to achieve results, and nitrosamine formation has not been shown to cause cancer in human beings.

Detectable amounts of nitrosamine-related compounds are found in malt beverages (beer) and cured meats (primarily bacon). Yet even the quantities found in beer and bacon hardly make a difference in a person's overall exposure to nitrosamine-related compounds. An average cigarette smoker inhales 100 times the nitrosamines that the average bacon eater ingests. A beer drinker ingests twice as much as the bacon eater, but even so, nitrosamine exposure from new car interiors and cosmetics is higher than this.

Antioxidants Another way food can go bad is by exposure to oxygen (oxidation). Oftentimes, these changes involve no hazard to health, but they damage the food's appearance, flavor, and nutritional quality. Oxidation is easy to detect when sliced apples or potatoes turn brown or when oil goes rancid. **Antioxidants** prevent these reactions. Among the antioxidants■ approved for use in foods are vitamin C (ascorbate) and vitamin E (tocopherol).

Another group of antioxidants, the **sulfites,**■ cost less than the vitamins. Sulfites prevent oxidation in many processed foods, alcoholic beverages (especially wine), and drugs. Because some people experience adverse reactions, the FDA prohibits sulfite use on foods intended to be consumed raw, with the exception of grapes, and requires foods and drugs that contain sulfite additives to declare it on their labels. For most people, sulfites pose no hazard in the amounts used in products, but there is one more consideration: sulfites destroy the B vitamin thiamin. For this reason, the FDA prohibits their use in foods that are important sources of the vitamin, such as enriched grain products.

Both salt and sugar act as preservatives by withdrawing water from food; microbes cannot grow without water.

■ Common antimicrobial additives:
- Salt.
- Sugar.
- Nitrites and nitrates (such as sodium nitrate).

■ Common antioxidant additives:
- Vitamin C (erythorbic acid, sodium ascorbate).
- Vitamin E (tocopherol).
- Sulfites.
- BHA and BHT.

■ Sulfites appear on food labels as:
- Sulfur dioxide.
- Sodium sulfite.
- Sodium bisulfite.
- Potassium bisulfite.
- Sodium metabisulfite.
- Potassium metabisulfite.

nitrites (NYE-trites): salts added to food to prevent botulism. One example is sodium nitrite, which is used to preserve meats.

nitrosamines (nye-TROHS-uh-meens): derivatives of nitrites that may be formed in the stomach when nitrites combine with amines. Nitrosamines are carcinogenic in animals.

sulfites: salts containing sulfur that are added to foods to prevent spoilage.

■ Common artificial color additives:
- Blue #1 and #2 (brilliant blue and indigotine).
- Green #3 (fast green).
- Red #40 and #3 (allura red and erythrosine).
- Yellow #5 and #6 (tartrazine and sunset yellow).

■ Common natural color additives:
- Annatto (yellow).
- Caramel (yellowish brown).
- Carotenoids (yellowish orange).
- Dehydrated beets (reddish brown).
- Grape skins (red, green).

■ Common flavor additives:
- Salt.
- Sugar.
- Spices.
- Artificial sweeteners.
- MSG.

■ Common emulsifying additives:
- Lecithin.
- Alginates.
- Mono- and diglycerides.

■ Common gum additives:
- Agar, alginates, carrageenan.
- Guar, locust bean, psyllium.
- Pectin.
- Xanthan gum.
- Gum arabic.
- Cellulose derivatives.

■ Common nutrient additives:
- Thiamin, niacin, riboflavin, folate, and iron in grain products.
- Iodine in salt.
- Vitamins A and D in milk.
- Vitamin C and calcium in fruit drinks.
- Vitamin B$_{12}$ in vegetarian foods.

BHA and BHT: preservatives commonly used to slow the development of off-flavors, odors, and color changes caused by oxidation.

monosodium glutamate (MSG): a sodium salt of the amino acid glutamic acid commonly used as a flavor enhancer. The FDA classifies MSG as a "generally recognized as safe" ingredient.

MSG symptom complex: an acute, temporary intolerance reaction that may occur after the ingestion of the additive MSG (monosodium glutamate). Symptoms include burning sensations, chest and facial flushing and pain, and throbbing headaches.

Two other antioxidants in wide use are **BHA** and **BHT,** which prevent rancidity in baked goods and snack foods.* Several tests have shown that animals fed large amounts of BHT develop *less* cancer when exposed to carcinogens and live *longer* than controls. Apparently, BHT protects against cancer through its antioxidant effect, which is similar to that of the antioxidant nutrients. The amount of BHT ingested daily from the U.S. diet, however, contributes little to the body's antioxidant defense system. A caution: at intakes higher than those that protect against cancer, BHT has *produced* cancer. Vitamins E and C remain the most important dietary antioxidants to strengthen defenses against cancer. (See Highlight 11 for a full discussion.)

Colors Only a few **artificial colors**■ remain on the FDA's list of additives approved for use in foods—a highly select group that has survived considerable testing. Colors derived from the natural pigments of plants must also meet standards of purity and safety. Examples of natural pigments■ commonly used by the food industry are the caramel that tints cola beverages and baked goods and the carotenoids that color margarine, cheeses, and pastas.

Artificial Flavors and Flavor Enhancers Natural flavors, **artificial flavors,** and **flavor enhancers**■ are the largest single group of food additives. Many foods taste wonderful because manufacturers have added the natural flavors of spices, herbs, essential oils, fruits, and fruit juices. Some spices, notably those used in Mediterranean cooking, provide antioxidant protection as well as flavors.[16] Oftentimes, natural flavors are used in combination with artificial flavors. The sugar alternatives discussed in Highlight 4 are among the most widely used artificial flavor additives.

One of the best-known flavor enhancers is **monosodium glutamate,** or **MSG**—a sodium salt of the amino acid glutamic acid. MSG is used widely in a number of foods, especially Asian foods, canned vegetables, soups, and processed meats. Besides enhancing the well-known sweet, salty, bitter, and sour tastes, MSG itself may possess a pleasant flavor. Adverse reactions to MSG—known as the **MSG symptom complex**—may occur in people with asthma and in sensitive individuals who consume large amounts of MSG, especially on an empty stomach. Otherwise, MSG is considered safe for adults. It is not allowed in foods designed for infants, however. Food labels require ingredient lists to itemize all additives, including MSG.

Texture and Stability Some additives help to maintain a desirable consistency in foods. **Emulsifiers**■ keep mayonnaise stable, control crystallization in syrups, keep spices dispersed in salad dressings, and allow powdered coffee creamer to dissolve easily. **Gums**■ are added to thicken foods and help form gels. Yeast may be added to provide leavening, and bicarbonates and acids may be used to control acidity.

Nutrient Additives As mentioned earlier, manufacturers sometimes add nutrients to fortify or maintain the nutritional quality of foods. Included among **nutrient additives**■ are the five nutrients added to refined grains, the iodine added to salt, the vitamins A and D added to milk, and the nutrients added to fortified breakfast cereals. A nutrient-poor food with nutrients added may appear to be nutrient-rich, but it is rich only in those nutrients chosen for addition. Appropriate uses of nutrient additives are to:

- Correct dietary deficiencies known to result in diseases.
- Restore nutrients to levels found in the food before storage, handling, and processing.
- Balance the vitamin, mineral, and protein contents of a food in proportion to the energy content.
- Correct nutritional inferiority in a food that replaces a more nutritious traditional food.

---

*BHA is butylated hydroxyanisole; BHT is butylated hydroxytoluene.

As mentioned earlier, nutrients are sometimes also added for other purposes. Vitamins C and E are used for their antioxidant properties, and beta-carotene and other carotenoids are sometimes used for color.

Indirect Food Additives

Indirect or **incidental additives** find their way into foods during harvesting, production, processing, storage, or packaging. Incidental additives may include tiny bits of plastic, glass, paper, tin, and other substances from packages as well as chemicals from processing, such as the solvent used to decaffeinate coffee. The following paragraphs discuss six different types of indirect additives that sometimes make headline news.

Acrylamide Raw potatoes don't have it, but french fries do—acrylamide, a compound that forms when carbohydrate-rich foods are cooked at high temperatures. Apparently, acrylamide has been in foods ever since we started baking, frying, and roasting, but it only recently came to our attention. And now scientists are trying to determine whether its presence poses a problem. At high doses, acrylamide causes cancer and nerve damage in animals. As such, scientists classify it as both a carcinogen and a genotoxicant,■ but quantities commonly found in foods appear to be well below the amounts that cause such damage. The FDA is currently investigating how acrylamide is formed in foods, how its formation can be limited, and whether its presence is harmful.

Microwave Packaging Some microwave products are sold in "active packaging" that helps to cook the food; for example, pizzas are often heated on a metalized film laminated to paperboard. This film absorbs the microwave energy in the oven and reaches temperatures as high as 500°F. At such temperatures, packaging components migrate into the food. For this reason, manufacturers must perform specific tests to determine whether materials are migrating into foods; if they are, their safety must be confirmed by strict procedures similar to those governing intentional additives.

Most microwave products are sold in "passive packaging" that is transparent to microwaves and simply holds the food as it cooks. These containers don't get much hotter than the foods, but materials still migrate at high temperatures. Consumers should not reuse these containers. Instead they should use only glass or ceramic containers■ designed for use in microwave ovens and avoid using disposable styrofoam or plastic containers such those used for carryout or margarine.

Dioxins Coffee filters, milk cartons, paper plates, and frozen food packages, if made from bleached paper, can contaminate foods with minute quantities of **dioxins**—compounds formed during chlorine treatment of wood pulp during paper manufacture. Dioxin contamination of foods from such products appears only in trace quantities—in the parts-per-trillion range (recall, for perspective, that one part per trillion is equal to 1 second in 32,000 years). Such levels appear to present no health risks to people, but scientists recognize that dioxins are extremely toxic and are known to cause cancer in animals. Accordingly, the paper industry has reduced its use of chlorine to cut dioxin exposure; in the meantime, the FDA has concluded that drinking milk from bleached-paper cartons presents no health hazard.

Decaffeinated Coffee Many consumers have tried to eliminate caffeine from their diets by selecting decaffeinated coffee. To remove caffeine from coffee beans, manufacturers often use methylene chloride in a process that leaves traces of the chemical in the final product. The FDA estimates that the average cup of coffee decaffeinated this way contains about 0.1 part per million of methylene chloride, which seems to pose no significant threat. A person drinking decaffeinated coffee containing 100 times as much methylene chloride every day for a lifetime has a one-in-a-million chance of developing cancer from it. People are exposed to much more methylene chloride from other sources such as hair sprays and paint stripping solutions. Still, some consumers prefer either to return to caffeine or to select coffee

© 1998 PhotoDisc Inc.

Color additives not only make foods attractive, but identify flavors as well. Everyone agrees that yellow jellybeans should taste lemony and black ones like licorice.

■ A *carcinogen* is a substance that causes cancer, and a *genotoxicant* is a substance that mutates or damages genetic material.

■ Quick test for using glass or ceramic containers in a microwave oven: Microwave the empty container for 1 min.
- If it's warm, it's unsafe for the microwave.
- If it's lukewarm, it's safe for short-term reheating in the microwave.
- If it's cool, it's safe for long-term cooking in the microwave.

indirect or **incidental additives:** substances that can get into food as a result of contact with foods during growing, processing, packaging, storing, cooking, or some other stage before the foods are consumed; sometimes called **accidental additives.**

dioxins (dye-OCK-sins): a class of chemical pollutants created as by-products of chemical manufacturing, incineration, chlorine bleaching of paper pulp, and other industrial processes. Dioxins persist in the environment and accumulate in the food chain.

decaffeinated in another way, perhaps by steam. Unfortunately, manufacturers are not required to state on their labels the type of decaffeination process used in their products. Many labels provide consumer-information telephone numbers for those who have such questions.

Hormones Hormones are a unique type of incidental additive in that their use is intentional, but their presence in the final food product is not. The FDA has approved about a dozen hormones for use in food-producing animals, and the USDA has established limits for residues allowed in meat products.

Some ranchers in the United States treat young calves with **bovine growth hormone (BGH).** Hormone-treated animals produce leaner meats, and dairy cows produce more milk. All cows make BGH naturally. Scientists can also genetically alter bacteria to produce BGH, which allows laboratories to harvest huge quantities of the hormone and sell it to farmers as a drug. Genetic engineering practices such as this have aroused some consumer concerns (see Highlight 19).

Indeed, traces of BGH do remain in the meat and milk of both hormone-treated and untreated cows. BGH residues have not been tested for safety in human beings because residues of the natural hormone have always been present in milk and meat and the amount found in treated cows is within the range that can occur naturally. Furthermore, BGH, being a peptide hormone, is denatured by the heat used in processing milk and cooking meat and is also digested by enzymes in the GI tract. If any BGH were to enter the bloodstream, it would have no effect because the chemical structures of animal growth hormones differ from those in human beings; therefore, BGH does not stimulate receptors for *human* growth hormone. According to the National Institutes of Health, "As currently used in the United States, meat and milk from [hormone] treated cows are as safe as those from untreated cows." Whether hormones that have passed through the animals into feces and then contaminated the soil and water interfere with plants or animals in the environment remains controversial.[17]

Antibiotics Like hormones, antibiotics are also intentionally given to livestock, and residues may remain in the meats and milks. Consequently, people consuming these foods receive tiny doses of antibiotics regularly, and those with sensitivity to antibiotics may suffer allergic reactions[18] To minimize drug residues in foods, the FDA requires a specified time between the time of medication and the time of slaughter to allow for drug metabolism and excretion.

Of greater concern to the public's health is the development of antibiotic resistance. Resistance develops when antibiotics are overused. Physicians and veterinarians use an estimated 5 million pounds of antibiotics to treat infections in people and animals, but farmers add five times as much to livestock feed to enhance growth. Not surprisingly, meat from these animals contains resistant bacteria. In one study, 20 percent of the meat samples contained *Salmonella;* most of the *Salmonella* were resistant to at least one antibiotic, and over half were resistant to at least three antibiotics.[19] Such indiscriminate use of antibiotics can be catastrophic to the treatment of disease in human beings.[20] Antibiotics are less effective in treating people who are infected with resistant bacteria. The FDA continues to monitor the use of antibiotics in the food industry with the goal of ensuring that antibiotics remain effective in treating human disease.

HEALTHY
PEOPLE
2010

Prevent an increase in the proportion of pathogens of the *Salmonella* species from humans and from animals at slaughter that are resistant to antimicrobial drugs.

bovine growth hormone (BGH): a hormone produced naturally in the pituitary gland of a cow that promotes growth and milk production; now produced for agricultural use by bacteria.

• **bovine** = of cattle

IN SUMMARY On the whole, the benefits of food additives seem to justify the risks associated with their use. The FDA regulates the use of the following intentional additives: antimicrobial agents (such as nitrites) to prevent microbial spoilage; antioxidants (such as vitamins C and E, sulfites, and BHA and BHT) to prevent oxidative changes; colors (such as tartrazine) and flavor enhancers (such as MSG) to appeal to senses; and nutrients (such as iodine in salt) to enrich or fortify foods. Incidental additives sometimes get into foods during processing, but rarely present a hazard, although some processes such as treating livestock with hormones and antibiotics raise consumer concerns.

Consumer Concerns about Water

Foods are not alone in transmitting foodborne diseases; water is guilty, too. In fact, *Cryptosporidium* and *Cyclospora,* commonly found in fresh fruits and vegetables, and *Vibrio vulnificus,* found in raw oysters, are commonly transmitted through contaminated water. In addition to microorganisms, water may contain many of the same impurities that foods do: environmental contaminants, pesticides, and additives such as chlorine used to kill pathogenic microorganisms and fluoride used to protect against dental caries. A glass of "water" is more than just H_2O. This discussion examines the sources of drinking water,■ harmful contaminants, and ways to ensure water safety.

■ Water that is suitable for drinking is called **potable** (POT-ah-bul). Only 1% of all the earth's water is potable.

Sources of Drinking Water

Drinking water comes from two sources—surface water and groundwater. Each source supplies water for about half of the population.

Most major cities obtain their drinking water from surface water—the water in lakes, rivers, and reservoirs. Surface water is readily contaminated because it is directly exposed to acid rain, runoff from highways and urban areas, pesticide runoff from agricultural areas, and industrial wastes that are dumped directly into it. Surface water contamination is reversible, however, because fresh rain constantly replaces the water. It is also cleansed to some degree by aeration, sunlight, and plants and microorganisms that live in it.

Groundwater is the water in underground aquifers—rock formations that are saturated with and yield usable water. People who live in rural areas rely mostly on groundwater pumped up from private wells. Groundwater is contaminated more slowly than surface water, but also more permanently. Contaminants deposited on the ground migrate slowly through the soil before reaching groundwater. Once there, the contaminants break down less rapidly than in surface water due to the lack of aeration, sunlight, and aerobic microorganisms. The slow replacement of groundwater also helps contaminants remain for a long time. Groundwater is especially susceptible to contamination from hazardous waste sites, dumps and landfills, underground tanks storing gasoline and other chemicals, and improperly discarded household chemicals and solvents.

Water Systems and Regulations

Public water systems treat water to remove contaminants that have been detected above acceptable levels. During treatment, a disinfectant (usually, chlorine) is added to kill bacteria. The addition of chlorine to public water is an important public health measure that appears to offer great benefits and small risks. On the one hand, chlorinated water has eliminated such water-borne diseases as typhoid fever, which once ravaged communities, killing thousands of people. On the other hand, it has been associated with a slight increase in bladder and rectal cancers and with contamination of the environment with the toxic by-product dioxin.

Clean rivers represent irreplaceable water resources.

©1998 PhotoDisc Inc.

The EPA is responsible for ensuring that public water systems meet minimum standards for protecting the public health.*

> Increase the proportion of persons served by community water systems who receive a supply of drinking water that meets the regulations of the Safe Drinking Water Act.

Even safe water may have characteristics that some consumers find unpleasant. Most of these problems reflect the mineral content of the water. For example, manganese and copper give water a metallic taste, and sulfur produces a "rotten egg" odor. Iron leaves a rusty brown stain on plumbing fixtures and laundry. Calcium and magnesium (commonly found in "hard water") build up in coffeemakers and hot water heaters; similarly, soap is not easily rinsed away in hard water, leaving bathtubs and laundry looking dingy. For these and other reasons, some consumers have adopted alternatives to the public water system.

Home Water Treatments To ease concerns about the quality of drinking water, some people purchase home water-treatment systems. Because the EPA does not certify or endorse these water-treatment systems, consumers must shop carefully. Manufacturers offer a variety of units for removing contaminants from drinking water. None of them removes all contaminants, and each has its own advantages and disadvantages. Choosing the right treatment unit depends on the kinds of contaminants in the water. For example, activated carbon filters are particularly effective in removing chlorine, heavy metals such as mercury, and organic contaminants from sediment; reverse osmosis, which forces pressurized water across a membrane, flushes out sodium, arsenic, and some microorganisms such as *Giardia;* and distillation systems, which boil water and condense the steam to water, remove contaminants such as lead and kill microorganisms in the process. Therefore, before purchasing a home water-treatment unit, a consumer must first determine the quality of the water. In some cases, a state or county health department will test water samples or can refer the consumer to a certified laboratory.

Bottled Water Many people turn to bottled water as an alternative to tap water. The average consumer drinks more than 20 gallons of bottled water a year. Bottled water is classified as a food, so it is regulated nationwide by the FDA and locally by state health and environmental agencies. The FDA has established quality and safety standards for bottled drinking waters compatible with those set by the EPA for public water systems. In addition, all bottled waters must be processed, packaged, and labeled in accordance with FDA regulations. Bottled water is expensive compared to public water, costing upwards of 200 times as much. Its quality varies among brands because of variations in the source water used and company practices.

Labels on bottled water must identify the water's source. Approximately 75 percent of bottled waters derive from protected groundwater (from springs or wells) that has been disinfected with ozone rather than chlorine. Ozone kills microorganisms, then disintegrates spontaneously into water and oxygen, leaving behind no toxic by-products. Other bottled waters derive from municipal tap water that has been treated by carbon filtration to remove chlorine and inorganic compounds. Bottled waters may also be treated by reverse osmosis or ion exchange to remove inorganic compounds. Alternatively, the water may be distilled or deionized to remove dissolved solids. Most bottled waters do not contain fluoride; consequently, they do not provide the tooth protection of fluoridated water from community public water systems.

Despite government regulations, some contamination has been detected in some bottled waters. While the amounts of most contaminants found in bottled waters are probably insignificant, consumers should be aware that bottled water is not always purer than the water from their taps. As a safeguard, the FDA recommends that bottled water be handled like other foods and be refrigerated after opening.

---

*The EPA's safe drinking water hotline: (800) 426–4791.

HOW TO Disinfect Water

In an extreme emergency, when safe water is unavailable, the EPA advises disinfecting water for drinking, cooking, and brushing teeth. Well water is safest, but lake or stream water may be used. Clear water is most easily treated; filter cloudy or discolored water through several layers of clean cloth or a coffee filter before disinfecting.

The preferred method of disinfecting water is to boil it vigorously for at least one minute to kill *all* disease-causing organisms. If boiling is not possible, *most* disease-causing microorganisms can be killed using chlorine or iodine disinfecting tablets available from drugstores and sporting good stores. Follow label directions.

Alternatively, use chlorine-containing laundry bleach and follow the directions on the bottle. If there are no directions, mix 5 drops of regular (not concentrated, scented, or color-safe) bleach with each quart of clear water. If water is cloudy or colored, double the amount. Let water stand for at least 30 minutes before using it. Properly treated water smells slightly of chlorine; if no chlorine odor is present, repeat the dosage. To remove the odor, pour water back and forth between clean containers to aerate it. Iodine tincture, a common first aid antiseptic for wounds, kills *some* disease-causing organisms, but it is less effective than chlorine. Add 5 drops of 2 percent iodine tincture to each quart of water (add 10 drops if water is cloudy). Let water stand for at least 30 minutes before using it.

Protection of drinking water is the subject of an ongoing battle between environmentalists and industry. It may soon become a source of conflict between the world's nations as the population continues to grow and the renewable water supply remains constant. Estimates are that within the next 50 years, half of the world's people will not have enough clean water to meet their needs. To avert this potential calamity, we must take active steps to conserve water, clean polluted water, desalinate seawater, and curb population growth. The accompanying box describes how to disinfect bacterially contaminated water.

IN SUMMARY Like foods, water may contain infectious microorganisms, environmental contaminants, pesticide residues, and additives. The EPA monitors the safety of the public water system, but many consumers choose home water-treatment systems or bottled water instead of tap water.

As this chapter said at the start, supplying food safely to hundreds of millions of people is an incredible challenge—one that gets met, for the most part, with incredible efficiency. The following chapter describes a contrasting situation—that of the food supply not reaching the people.

Nutrition in Your Life

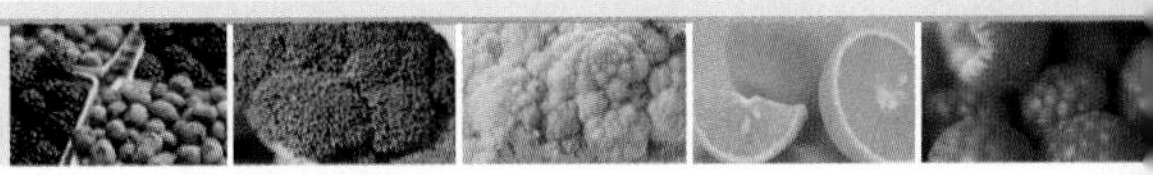

Practicing food safety allows you to eat a variety of foods, with little risk of food-related illnesses.

- Do you wash your hands, utensils, and kitchen surfaces with hot, soapy water when preparing foods?

- Do you separate raw and cooked foods while storing and preparing them?

- Do you cook foods to a safe temperature and refrigerate perishable foods promptly?

NUTRITION ON THE NET

 Access these websites for further study of topics covered in this chapter.

- Find updates and quick links to these and other nutrition-related sites at our website: **www.wadsworth.com/nutrition**

- Get food safety tips from the Government Food Safety Information site or from the Fight BAC! Campaign of the Partnership for Food Safety Education: **www.foodsafety.gov** or **www.fightbac.org**

- Learn more about foodborne illnesses from the National Center for Infectious Diseases at the Centers for Disease Control and Prevention: **www.cdc.gov/ncidod**

- Learn about the various types of food thermometers and how and when to use them from the USDA Thermy Campaign: **www.fsis.usda.gov/thermy**

- Find commonsense health tips for travelers at the Centers for Disease Control and Prevention site: **www.cdc.gov/travel**

- Learn more about food irradiation from the International Consultative Group on Food Irradiation and the International Food Information Council: **www.iaea.org/icgfi** and **www.ific.org**

- Report adverse reactions to the FDA's MedWatch program at (800) 332-1088 or: **www.fda.gov/medwatch**

- Get fish advisories from the Environmental Protection Agency: **www.epa.gov/ost/fish**

- Review tips from the Environmental Protection Agency on methods of food buying and preparation that will help minimize pesticide exposure: **www.epa.gov/pesticides/food**

- Visit the Canadian Food Inspection Agency (CFIA): **www.inspection.gc.ca**

- Learn more about food safety in the marketplace from the Food Safety and Inspection Service: **www.usda.gov/fsis**

- Learn more about organic foods and national organic food standards from the National Organic Program: **www.ams.usda.gov/nop**

- Learn more about safe drinking water from the Environmental Protection Agency: **www.epa.gov/safewater**

- Enjoy the humor and music of food toxicologist Carl Winter at: **foodsafe.ucdavis.edu/music.html**

STUDY QUESTIONS

These questions will help you review the chapter. You will find the answers in the discussions on the pages provided.

1. To what extent does food poisoning present a real hazard to consumers eating U.S. foods? How often does it occur? (p. 658)

2. Distinguish between the two types of foodborne illnesses and provide an example of each. Describe measures that help prevent foodborne illnesses. (pp. 659, 664–665)

3. What special precautions apply to meats? To seafood? (pp. 662–664)

4. What is meant by a "persistent" contaminant of foods? Describe how contaminants get into foods and build up in the food chain. (pp. 668–669)

5. What dangers do natural toxicants present? (p. 671)

6. How do pesticides become a hazard to the food supply, and how are they monitored? In what ways can people reduce the concentrations of pesticides in and on foods that they prepare? (pp. 671–675)

7. What is the difference between a GRAS substance and a regulated food additive? Give examples of each. Name and describe the different classes of additives. (pp. 675–680)

These multiple choice questions will help you prepare for an exam. Answers can be found on p. 685.

1. Eating a contaminated food such as undercooked poultry or unpasteurized milk might cause a:
   a. food allergy.
   b. food infection.
   c. food intoxication.
   d. botulinum reaction.

2. The temperature danger zone for foods ranges from:
   a. −20°F to 120°F.
   b. 0°F to 100°F.
   c. 20°F to 120°F.
   d. 40°F to 140°F.

3. Examples of foods that frequently cause foodborne illness are:
   a. canned foods.
   b. steaming-hot foods.
   c. fresh fruits and vegetables.
   d. raw milk, seafood, meat, and eggs.

4. Irradiation can help improve our food supply by:
   a. cooking foods quickly.
   b. killing microorganisms.
   c. minimizing the use of preservatives.
   d. improving the nutrient content of foods.

5. Solanine is an example of a(n):
   a. heavy metal.
   b. artificial color.
   c. natural toxicant.
   d. animal hormone.

6. The standard that deems additives safe if lifetime use presents no more than a one-in-a-million risk of cancer is known as the:
   a. Delaney Clause.
   b. zero-risk policy.
   c. GRAS list of standards.
   d. negligible-risk policy.

7. Common antimicrobial additives include:
   a. salt and nitrites.
   b. carrageenan and MSG.
   c. dioxins and sulfites.
   d. vitamin C and vitamin E.

8. Common antioxidants include:
   a. BHA and BHT.
   b. tartrazine and MSG.
   c. sugar and vitamin E.
   d. nitrosamines and salt.

9. Incidental additives that may enter foods during processing include:
   a. dioxins and BGH.
   b. dioxins and folate.
   c. beta-carotene and agar.
   d. nitrites and irradiation.

10. Chlorine is added to water to:
    a. protect against dental caries.
    b. destroy harmful minerals such as lead and mercury.
    c. kill pathogenic microorganisms.
    d. remove the sulfur that produces a "rotten egg" odor.

REFERENCES

1. B. Bruemmer, Food biosecurity, *Journal of the American Dietetic Association* 103 (2003): 687–691; T. Peregrin, Bioterrorism and food safety: What nutrition professionals need to know to educate the American public, *Journal of the American Dietetic Association* 102 (2002): 14, 16; Food and Drug Administration, Food security guidance: Availability, *Federal Register* 67 (2002): 1224–1225; J. Sobel, A. S. Khan, and D. L. Swerdlow, Threat of a biological terrorist attack on the US food supply: The CDC perspective, *Lancet* 359 (2002): 874–880.

2. E. A. Coleman and M. E. Yergler, Botulism, *American Journal of Nursing* 102 (2002): 44–47.

3. Position of the American Dietetic Association: Food and water safety, *Journal of the American Dietetic Association* 103 (2003): 1203–1218.

4. U.S. Food and Drug Administration, Commonly asked questions about BSE in products regulated by FDA's Center for Food Safety and Applied Nutrition (CFSAN), **www.cfsan.fda.gov/~comm/bsefaq.html**, site updated January 30, 2004 and visited February 6, 2004; U.S. Department of Agriculture, Bovine spongiform encephalopathy (BSE) Q & A's, **www.aphis.usda.gov/lpa/issues/bse/bse_q&a.html**, site updated January 21, 2004 and visited February 6, 2004.

5. E. T. Ryan, M. E. Wilson, and K. C. Kain, Illness after international travel, *New England Journal of Medicine* 347 (2002): 505–516.

6. E. T. Ryan and K. C. Kain, Health advice and immunizations for travelers, *New England Journal of Medicine* 342 (2000): 1716–1725.

7. K. M. Shea and the Committee on Environmental Health, Technical report: Irradiation of food, *Pediatrics* 106 (2000): 1505–1510.

8. Position of the American Dietetic Association: Food irradiation, *Journal of the American Dietetic Association* 100 (2000): 246–253.

9. P. Frenzen and coauthors, Consumer acceptance of irradiated meat and poultry products, **www.cdc.gov/foodnet/pub/publications**.

10. C. P. Dougherty and coauthors, Dietary exposures to food contaminants across the United States, *Environmental Research* 84 (2000): 170–185.

11. Y. L. Guo and coauthors, Semen quality after prenatal exposure to polychlorinated biphenyls and dibenzofurans, *Lancet* 356 (2000): 1240–1241.

12. National Report on Human Exposure to Environmental Chemicals, March 2001, www.cdc.gov/nceh/dls/report; L. R. Goldman, M. W. Shannon, and the Committee on Environmental health, Technical report: Mercury in the environment—Implications for pediatricians, *Pediatrics* 108 (2001): 197–205; Blood and hair mercury levels in young children and women of childbearing age—United States, 1999, *Morbidity and Mortality Weekly Report* 50 (2001): 140–143.

13. EPA Fact Sheet, Update: National listing of fish and wildlife advisories, May 2002, available online at **www.epa.gov/ost/fish**.

14. EPA Fact Sheet, 2002.

15. E. Guallar and coauthors, Mercury, fish oils, and the risk of myocardial infarction, *New England Journal of Medicine* 347 (2002): 1747–1754; K. Yoshizawa and coauthors, Mercury and the risk of coronary heart disease in men, *New England Journal of Medicine* 347 (2002): 1755–1760.

16. M. Martínez-Tomé and coauthors, Antioxidant properties of Mediterranean spices compared with common food additives, *Journal of Food Protection* 64 (2001): 1412–1419.

17. J. Raloff, Hormones: Here's the beef, *Science News* 161 (2002): 10–12.

18. D. H. Hammer and C. J. Gill, From the farm to the kitchen table: The negative impact of antimicrobial use in animals on humans, *Nutrition Reviews* 60 (2002): 261–264.

19. D. G. White and coauthors, The isolation of antibiotic-resistant salmonella from retail ground meats, *New England Journal of Medicine* 345 (2001): 1147–1154.

20. Hammer and Gill, 2002; S. L. Gorbach, Antimicrobial use in animal feed—Time to stop, *New England Journal of Medicine* 345 (2001): 1202–1203; B. R. Berends and coauthors, Human health hazards associated with the administration of antimicrobials to slaughter animals. Part II. An assessment of the risks of resistant bacteria in pigs and pork, *Veterinary Quarterly* 23 (2001): 10–21; P. D. Fey and coauthors, Ceftriaxone-resistant salmonella infection acquired by a child from cattle, *New England Journal of Medicine* 342 (2000): 1242–1249.

ANSWERS

Study Questions (multiple choice)

1. b   2. d   3. d   4. b   5. c   6. d   7. a   8. a   9. a   10. c

HIGHLIGHT

Food Biotechnology

Advances in food **biotechnology** promise just about everything from the frivolous (a tear-free onion) to the profound (a hunger-free world). Already biotechnology has produced leaner meats, longer shelf lives, better nutrient composition, and greater crop yields grown with fewer pesticides. Overall, biotechnology offers numerous opportunities to overcome food shortages, improve the environment, and eliminate disease.[1] But it also raises concerns about possible risks to the environment and human health.[2] Critics assert that biotechnology will exacerbate world hunger, destroy the environment, and endanger health. This highlight presents some of the many issues surrounding genetically engineered foods, and the accompanying glossary defines the terms used.

The Promises of Genetic Engineering

For centuries farmers have been selectively breeding plants and animals to shape the characteristics of their crops and livestock. They have created prettier flowers, hardier vegetables, and leaner animals. Consider the success of selectively breeding corn. Early farmers in Mexico began with a wild, native plant called teosinte (tea-oh-SIN-tay) that bears only five or six kernels on each small spike. Many years of patient selective breeding have produced large ears filled with hundreds of plump kernels aligned in perfect formation, row after row.

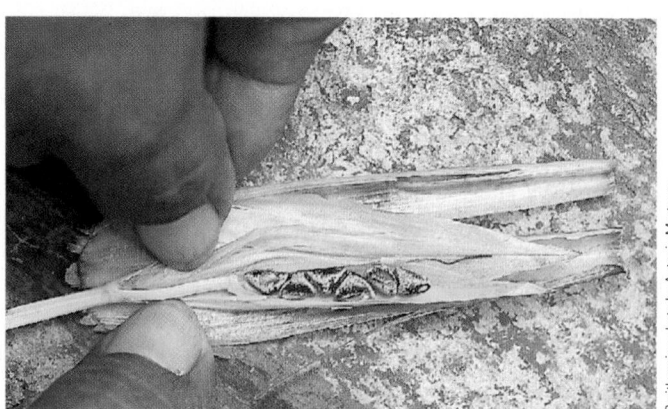

This wild predecessor of corn, with its sparse five or six kernels, bears little resemblance to today's large, full, sweet ears.

Such genetic improvements, together with the use of irrigation, fertilizers, and pesticides, were responsible for over half of the increases in U.S. crop yields in the twentieth century. Farmers still use selective breeding, but now, in the twenty-first century, advances in **genetic engineering** have brought rapid and dramatic changes to agriculture and food production.

Although selective breeding works, it is slow and imprecise because it involves mixing thousands of genes from two plants and hoping for the best. With genetic engineering, scientists can improve crops (or livestock) by introducing a copy of the specific gene needed to produce the desired trait. Figure H19-1 illustrates the difference. Once introduced, the selected gene acts like any other gene—it provides instructions for making a protein. The protein then determines a characteristic in the genetically modified plant or animal. In short, the process has been speeded up and refined. Farmers no longer need to wait patiently for breeding to yield improved crops and animals, nor must they even respect natural lines of re-

GLOSSARY

biotechnology: the use of biological systems or organisms to create or modify products. Examples include the use of bacteria to make yogurt, yeast to make beer, and cross-breeding to enhance crop production.

genetic engineering: the use of biotechnology to modify the genetic material of living cells so that they will produce new substances or perform new functions. Foods produced via this technology are called **genetically modified (GM)** or **genetically engineered (GE) foods.**

plant-pesticides: pesticides made by the plants themselves.

rennin: an enzyme that coagulates milk; found in the gastric juice of cows, but not human beings.

FIGURE **H19-1** Selective Breeding and Genetic Engineering Compared

Traditional Selective Breeding

Traditional selective breeding combines many genes from two varieties of the same species to produce one with the desired characteristics.

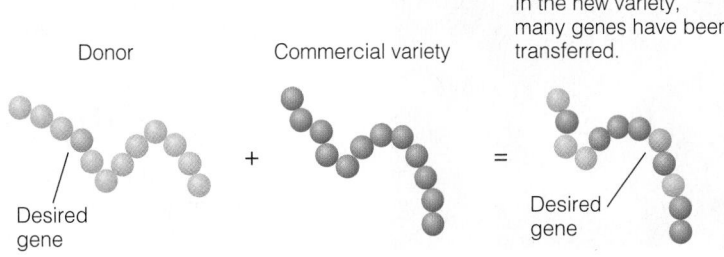

Donor Commercial variety In the new variety, many genes have been transferred.

Desired gene + = Desired gene

Genetic Engineering

Through genetic engineering, a single gene (or several) are transferred from the same or different species to produce one with the desired characteristics.

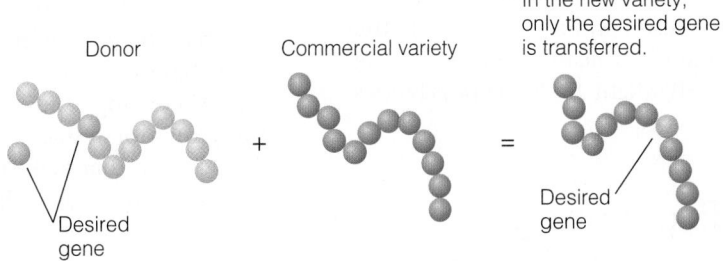

Donor Commercial variety In the new variety, only the desired gene is transferred.

Desired gene + = Desired gene

SOURCE: © 1995 Monsanto Company.

production among species. Laboratory scientists can now copy genes from any organism and insert them into almost any other organism—plant, animal, or microbe. Their work is changing not only the way farmers plant, fertilize, and harvest their crops, but also the ways the food industry processes food and consumers receive nutrients, phytochemicals, and drugs.

Extended Shelf Life

Among the first products of genetic engineering to hit the market were tomatoes that stay firm and ripe longer than regular tomatoes that are typically harvested green and ripened in the stores. These genetically modified tomatoes promise less waste and higher profits. Normally, tomatoes produce a protein that softens them after they have been picked. Scientists can now introduce into a tomato plant a gene that is a mirror image of the one that codes for the "softening" enzyme. This gene fastens itself to the RNA of the native gene and blocks synthesis of the softening protein. Without this protein, the genetically altered tomato softens more slowly

than a regular tomato, allowing growers to harvest it at its most flavorful and nutritious vine-ripe stage.

Improved Nutrient Composition

Genetic engineering can also improve the nutrient composition of foods.[3] Instead of manufacturers adding nutrients to foods during processing, plants can be genetically altered to do their own fortification work—a strategy called "biofortification."[4] Soybeans may be implanted with a gene that upgrades soy protein to a quality approaching that of milk. Corn may be modified to contain lysine and tryptophan, its two limiting amino acids. Soybean and canola plants can be genetically modified to alter the composition of their oils, making them richer in the heart-healthy monounsaturated fatty acids.[5] "Golden rice," which has received genes from a daffodil and a bacterium that enable it to make beta-carotene, offers some promise in helping to correct vitamin A deficiency worldwide. (Chapter 11 described how vitamin A deficiency contributes to the deaths of 2 million children and the blindness of a half million each year.)

Genetically modified cauliflower is orange, reflecting a change in a single gene that increases its production of beta-carotene 100-fold.

As you might predict, enhancing the chemical composition of plants is not limited to the essential nutrients. Genetically modified crops can also produce more of the phytochemicals that help maintain health and reduce the risks of chronic diseases (see Highlight 13).[6] The possibilities seem endless.

Efficient Food Processing

Genetic engineering also helps to process foods more efficiently, which saves money. For example, the protein **rennin**, which is used to coagulate milk in the production of cheese, has traditionally been harvested from the stomachs of calves, a costly process. Now scientists can insert a copy of the rennin gene into bacteria and then use bacterial cultures to mass-produce rennin—saving time, money, space, and animals.

Genetic engineering can also bypass costly food-processing steps. At present, people who are lactose intolerant can buy milk that has been treated with the lactase enzyme. Wouldn't it be more convenient, and less expensive, if scientists could induce cows to make lactose-free milk directly? They're working on it. They have already successfully inserted into mice the genetic material needed to make lactase in their mammary glands, thereby producing a low-lactose milk. Decaffeinated coffee beans are another real possibility.

Efficient Drug Delivery

Genetic research today has progressed well beyond tweaking a gene here and there to produce a desired trait. Scientists can now clone animals. By cloning animals, scientists have the ability to produce both needed food and pharmaceutical products. Using animals and other organisms in the development of pharmaceuticals is whimsically called "biopharming." For example, a cow with the genetic equipment to make a vaccine in its milk could provide both nourishment and immunization to a whole village of people now left unprotected because they lack food and medical help. Similarly, researchers have figured out how to induce bananas and potatoes to make hepatitis vaccines and tobacco leaves to make AIDS drugs. They can also harvest vaccines by genetically altering hydroponically grown tomato plants to secrete a protein through their root systems into the water.

Genetically Assisted Agriculture

Genetic engineering has helped farmers to grow crops that resist herbicides. About half of the soybean crops in the United States have been genetically engineered to withstand a potent herbicide. As a result, farmers can spray whole fields with this herbicide and kill the weeds without harming the soybeans.

Similarly, farmers can grow crops that produce their own pesticides—substances known as **plant-pesticides.** Corn, broccoli, and potatoes have received a gene from a bacterium that produces a protein that is toxic to leaf-chewing caterpillars (but not to humans). Yellow squash has been given two viral genes that confer resistance to the most common viral diseases. Potatoes can now produce a beetle-killing toxin in their leaves. These crops and many others like them are currently being grown or tested in fields around the United States. Growing crops that make their own pesticides allows farmers to save time, increase yields, and use fewer, or less harmful, pesticides.[7]

Other Possibilities

Many other biotechnology possibilities are envisioned for the near future. Shrimp may be empowered to fight diseases with genetic ammunition borrowed from sea urchins. Plants may be given special molecules to help them grow in polluted soil. With these and other advances, farmers may reliably produce bumper crops of food every year on far fewer acres of land, with less loss of water and topsoil, and far less use of toxic pesticides and herbicides. Supporters of biotechnology predict that these efforts will enhance food production and help meet the challenge of feeding an ever-increasing world population. They contend that genetically modified crops have the potential to eliminate hunger and starvation. Others suggest that the problems of world hunger are more complex than biotechnology alone can resolve and that the potential risks of genetic engineering may outweigh the potential benefits.[8]

The projects mentioned in this highlight are already in progress. Close on their heels are many more ingenious ideas. What if salt tolerance could be transplanted from a coastal marsh plant into crop plants? Could crops then be irrigated with seawater, thus conserving dwindling fresh water supplies? Would the world food supply increase if rice farmers could grow plants that were immune to disease? What if consumers could dictate which traits scientists insert into food plants? Would they choose to add phytochemicals to fight cancer or reduce the risk of heart disease? These and other possibilities seem unlimited, and though they may sound incredible, many such products have already been developed and are awaiting approval from the FDA, EPA, and USDA.

The Potential Problems and Concerns

Although many scientists hail biotechnology with confidence, others have reservations. Some consumers also have concerns about what they call "Frankenfoods." Those who oppose biotechnology fear for the safety of a world where genetic tampering produces effects that are not yet fully understood. They suspect that the food industry may be driven by potential profits, without ethical considerations or laws to harness its effects.[9] They point out that even the scientists who developed the techniques cannot predict the ultimate outcomes of their discoveries. These consumers don't want to eat a scientific experiment or interfere with natural systems. Genetic decisions, they say, are best left to the powers of nature.

If science and the marketplace are allowed to drive biotechnology without restraint, critics fear that these problems may result:

- *Disruption of natural ecosystems.* New, genetically unusual organisms that have no natural place in the food chain or evolutionary biological systems could escape into the environment and reproduce.

- *Introduction of diseases.* Newly created viruses may mutate to cause deadly diseases that may attack plants, animals, or human beings. Genetically modified bacteria may develop resistance to antibiotics, making the drugs useless in fighting infections.

- *Introduction of allergens and toxins.* Genetically modified crops may contain new substances that cause allergies or concentrate toxins that cause poisoning.

- *Creation of biological weapons.* Fatal bacterial and viral diseases may be developed for use as weapons.

- *Ethical dilemmas.* Critics pose the question, "How many human genes does an organism have to contain before it is considered human? For instance, how many human genes would a green pepper have to contain before one would have qualms about eating it?"[10]

Proponents of biotechnology respond that evidence to date does not justify these concerns.[11] Such opposing views illustrate the tension between the forward thrust of science and the hesitancy of consumers. Table H19-1 (on p. 690) summarizes the issues.

From another perspective, some argue that the concerns expressed by those protesting genetically engineered foods re-

Some consumers believe that food biotechnology will cause more harm than good.

flect prejudices acquired in an elitist world of fertile land and abundant food.[12] Those living in poverty-stricken areas of the world do not have the luxury of determining how to grow crops and process foods. They cannot afford the delays created when protesters destroy test crops and disrupt scientific meetings. They need solutions now. People are starving, and genetic engineering holds great promise for providing them with food.

At a minimum, critics of biotechnology have made a strong case for rigorous safety testing and labeling of new products. They contend, for example, that when a new gene has been introduced into a food, tests should ensure that other, unwanted genes have not accompanied it. If a disease-producing microorganism has donated genetic material, scientists must prove that no dangerous characteristic from the microorganism has also entered the food. If the inserted genetic material comes from a source to which some people develop allergies, such as nuts, then the new product should be labeled to alert them. Furthermore, if the newly altered genetic material creates proteins that have never before been encountered by the human body, their effects should be studied to ensure that people can eat them safely.

FDA Regulations

The FDA has taken the position that foods produced through biotechnology are not substantially different from others and require no special testing, regulations, or labeling. After all, most foods available today have already been genetically altered by years of selective breeding. The new vegetable broccoflower, a product of sophisticated cross-breeding of broccoli with cauliflower, met no testing or approval barriers on its way to the dinner plate. When the vegetable became available on the market, scientists studied its nutrient contents (see Appendix H), but they did not question its safety.

TABLE H19-1 Food Biotechnology: Point, Counterpoint

Arguments in Opposition to Genetic Engineering	Arguments in Support of Genetic Engineering
1. **Ethical and moral issues.** It's immoral to "play God" by mixing genes from organisms unable to do so naturally. Religious and vegetarian groups object to genes from prohibited species occurring in their allowable foods.	1. **Ethical and moral issues.** Scientists throughout history have been persecuted and even put to death by fearful people who accuse them of playing God. Yet, today many of the world's citizens enjoy a long and healthy life of comfort and convenience due to once-feared scientific advances put to practical use.
2. **Imperfect technology.** The technology is young and imperfect—genes rarely function in just one way, their placement is imprecise ("shotgun"), and all of their potential effects are impossible to predict. Toxins are as likely to be produced as the desired trait. Over 95 percent of DNA is called "junk" because scientists have not yet determined its function.	2. **Advanced technology.** Recombinant DNA technology is precise and reliable. Many of the most exciting recent advances in medicine, agriculture, and technology were made possible by the application of this technology.
3. **Environmental concerns.** Environmental side effects are unknown. The power of a genetically modified organism to change the world's environments is unknown until such changes actually occur—then the "genie is out of the bottle." Once out, the genie cannot be put back in the bottle because insects, birds, and the wind distribute genetically altered seed and pollen to points unknown.	3. **Environmental protection.** Genetic engineering may be the only hope of saving rain forest and other habitats from destruction. Through genetic engineering, farmers can make use of previously unproductive lands such as salt-rich soils and arid areas.
4. **"Genetic pollution."** Other kinds of pollution can often be cleaned up with money, time, and effort. Once genes are spliced into living things, those genes forever bear the imprint of human tampering.	4. **Genetic improvements.** Genetic side effects are more likely to benefit the environment than to harm it.
5. **Crop vulnerability.** Pests and disease can quickly adapt to overtake genetically identical plants or animals around the world. Diversity is key to defense.	5. **Improved crop resistance.** Pests and diseases can be specifically fought on a case-by-case basis. Biotechnology is the key to defense.
6. **Loss of gene pool.** Loss of genetic diversity threatens to deplete valuable gene banks from which scientists can develop new agricultural crops.	6. **Gene pool preserved.** Thanks to advances in genetics, laboratories around the world are able to stockpile the genetic material of millions of species that, without such advances, would have been lost forever.
7. **Profit motive.** Genetic engineering will profit industry more than the world's poor and hungry.	7. **Everyone profits.** Industries benefit from genetic engineering, and a thriving food industry benefits the nation and its people, as witnessed by countries lacking such industries. Genetic engineering promises to provide adequate nutritious food for millions who lack such food today. Developed nations gain cheaper, more attractive, more delicious foods with greater variety and availability year round.
8. **Unproven safety for people.** Human safety testing of genetically altered products is generally lacking. The population is an unwitting experimental group in a nationwide laboratory study for the benefit of industry.	8. **Safe for people.** Human safety testing of genetically altered products is unneeded because the products are essentially the same as the original foodstuffs.
9. **Increased allergens.** Allergens can unwittingly be transferred into foods.	9. **Control of allergens.** A few allergens can be transferred into foods, but these are known, and foods likely to contain them are clearly labeled to warn consumers.
10. **Decreased nutrients.** A fresh-looking tomato or other produce held for several weeks may have lost substantial nutrients.	10. **Increased nutrients.** Genetic modifications can easily enhance the nutrients in foods.
11. **No product tracking.** Without labeling, the food industry cannot track problems to the source.	11. **Excellent product tracking.** The identity and location of genetically altered foodstuffs are known, and they can be tracked should problems arise.
12. **Overuse of herbicides.** Farmers, knowing that their crops resist herbicide effects, will use them liberally.	12. **Conservative use of herbicides.** Farmers will not waste expensive herbicides in second or third applications when the prescribed amount gets the job done the first time.
13. **Increased consumption of pesticides.** When a pesticide is produced by the flesh of produce, consumers cannot wash it off the skin of the produce with running water as they can with ordinary sprays.	13. **Reduced pesticides on foods.** Pesticides produced by produce in tiny amounts known to be safe for consumption are more predictable than applications by agricultural workers who make mistakes. Because other genetic manipulations will eliminate the need for postharvest spraying, fewer pesticides will reach the dinner table.
14. **Lack of oversight.** Government oversight is run by industry people for the benefit of industry—no one is watching out for the consumer.	14. **Sufficient regulation and rapid response.** Government agencies are efficient in identifying and correcting problems as they occur in the industry.

In most cases, the new genetically modified food differs from the old conventional one only by a gene or two.[13] The rennin produced by bacteria is structurally and functionally the same as the rennin produced by calves, for example. For that reason, the FDA considers it "generally recognized as safe (GRAS)."

A product such as the tomato described earlier need not be tested because its new genes *prevent* synthesis of a protein and add nothing but a tiny fragment of genetic material. Nor does this tomato require special labeling because it is not significantly different from the many other varieties of tomatoes on the market. On the other hand, any substances introduced into a food (such as a hormone or protein) by way of bioengineering must meet the same safety standards applied to all additives. A

tomato plant with a gene that, for example, produces a pesticide cannot be marketed until tests prove it safe for consumption. The FDA assures consumers that all bioengineered foods on the market today are as safe as their traditional counterparts.

Foods produced through biotechnology that are substantially different from others must be labeled to identify that difference. For example, if the nutrient composition of the new product differs from its traditional counterpart, as in the soybean and canola oils mentioned earlier, then labeling is required. Similarly, if an allergy-causing protein has been introduced to a non-allergenic food, then labeling must warn consumers.

Most consumers want all genetically altered products clearly labeled. Consumer advocacy groups claim that by not requiring such labeling, the FDA forces millions of consumers to be guinea pigs, unwittingly testing genetically modified foods. Additionally, they say, people who have religious objections to consuming foods to which genes of prohibited organisms have been added have no way of identifying those foods. For example, someone keeping a kosher kitchen may unknowingly use a food containing genes from a pig. Currently labeling is voluntary. Manufacturers may state that a product has been "genetically engineered." Those who do would be wise to explain its purpose and benefit. When consumers recognize a personal health benefit, most tend to accept genetically engineered foods.[14]

Speaking in defense of the FDA's position are the FDA itself, recognized as the nation's leading expert and advocate for food

safety, and the American Dietetic Association, which represents current scientific thinking in nutrition. Many other scientific organizations agree, contending that biotechnology can deliver an improved food supply if we give it a fair chance to do so.

Will our impressive new technologies provide foods to meet the needs of the future? Optimists would say yes. Biotechnology holds a world of promise, and with proper safeguards and controls, it may yield products that meet the needs of consumers almost perfectly.

NUTRITION ON THE NET

 Access these websites for further study of topics covered in this highlight.

- Find updates and quick links to these and other nutrition-related sites at our website: **www.wadsworth.com/nutrition**

- Visit the USDA Biotechnology Information Resource site: **www.nal.usda.gov/bic**

- Get a "pro" biotechnology perspective from the Council for Biotechnology Information: **www.whybiotech.com**

- For another "for" view, search for "biotechnology" at the International Food Information Council: **www.ific.org**

- Get a "con" biotechnology perspective from the Genetic Engineering section of Greenpeace, USA: **www.greenpeaceusa.org**

- Another "against" view is available from the Union of Concerned Scientists: **www.ucsusa.org**

- Find discussions on major issues influencing the quality of life on planet earth at the Turning Point Project: **www.turnpoint.org**

REFERENCES

1. P. W. Phillips, Biotechnology in the global agri-food system, *Trends in Biotechnology* 20 (2002): 376–381.
2. J. Robinson, Ethics and transgenic crops: A review, *Electronic Journal of Biotechnology* 2 (1999): **www.ejb.org/content/vol2/issue2/full/3/index.html** (ISSN 0717 3458).
3. L. Yan and P. S. Kerr, Genetically engineered crops: Their potential use for improvement of human nutrition, *Nutrition Reviews* 60 (2002): 135–141.
4. H. E. Bouis, Plant breeding: A new tool for fighting micronutrient malnutrition, *Journal of Nutrition* 132 (2002): 491S–494S.
5. P. Broun, S. Gettner, and C. Somerville, Genetic engineering of plant lipids, *Annual Review of Nutrition* 19 (1999): 197–216.
6. M. A. Grusak, Phytochemicals in plants: Genomics-assisted plant improvement for nutritional and health benefits, *Current Opinion in Biotechnology* 13 (2002): 508–511; L. Kochian and D. F. Garvin,

Agricultural approaches to improving phytonutrient content in plants: An overview, *Nutrition Reviews* 57 (1999): S13–S18; M. W. Farnham, P. W. Simon, and J. R. Stommel, Improved phytonutrient content through plant genetic improvement, *Nutrition Reviews* 57 (1999): S19–S26.
7. J. Huang, C. Pray, and S. Rozelle, Enhancing the crops to feed the poor, *Nature* 418 (2002): 678–684.
8. A. Bakshi, Potential adverse health effects of genetically modified crops, *Journal of Toxicology and Environmental Health: Part B, Critical Reviews* 6 (2003): 211–215; B. C. Babcock and C. A. Francis, Solving global nutrition challenges requires more than new biotechnologies, *Journal of the American Dietetic Association* 100 (2000): 1308–1311.
9. J. Walsh, Brave new farm, *Time*, January 11, 1999, pp. 86–88.
10. R. Epstein, Redesigning the world: Ethical questions about genetic engineering, an

essay available at **online.sfsu.edu/~rone/GE%20Essays/Redesigning.htm**.
11. J. A. Thomas, Safety of foods derived from genetically modified plants, *Texas Medicine* 99 (2003): 66–69.
12. H. Adamu, Nigeria's minister of agriculture and rural development, "Hungry Africans want biotech crops, so get out of the way," *International Herald Tribune*, September 12, 2000, p. 8.
13. K. T. Atherton, Safety assessment of genetically modified crops, *Toxicology* 181 (2002): 421–426.
14. J. L. Brown and Y. Ping, Consumer perception of risk associated with eating genetically engineered soybeans is less in the presence of a perceived consumer benefit, *Journal of the American Dietetic Association* 103 (2003): 208–214.

Hunger and the Global Environment

© Nino Mascardi/The Image Bank/Getty Images

Nutrition in Your Life

Imagine living with hunger from the moment you wake up until the time you thankfully fall asleep—and all through your dreams as well. Meal after meal, day after day, you have little or no food to eat. You know you need food, but you have no money. Would you beg on the street corner or go "dumpster diving" at the nearest fast-food restaurant? And then where would you find your next meal? How will you ever get enough to eat as long as you live in poverty? Resolving the hunger problem—whether in your community or across the globe—depends on alleviating poverty and using resources wisely.

One person in every eight worldwide experiences persistent hunger—not the healthy appetite triggered by anticipation of a hearty meal, but the painful sensation caused by a lack of food. The physical feelings are the same, but in this chapter, hunger takes on greater meaning because the lack of food is recurrent and involuntary. Hunger deprives a person of the physical and mental energy needed to enjoy a full life and often leads to severe malnutrition and death. Tens of thousands of people die of starvation each day—one child every seven seconds.

The enormity of the hunger problem reflects not only huge numbers, but major challenges. As people populate and pollute the earth, resources become depleted, making food less available. Hunger and poverty, population growth, and environmental degradation are linked together; thus they tend to worsen each other. Because their causes overlap, so do their solutions: any initiative a person takes to help solve one problem will help solve many others. Eliminating hunger requires a balance among the distribution of food, the numbers of people, and the care of the environment.

■ "Never doubt that a small group of thoughtful, committed people can change the world. Indeed, it is the only thing that ever has."—Margaret Mead

■ An estimated one out of six children lives in poverty.

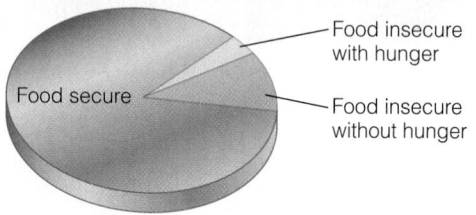

FIGURE 20-1 Prevalence of Food Insecurity and Hunger in U.S. Households, 2001

SOURCE: Economic Research Service, U.S. Department of Agriculture, **www.ers.usda.gov/publications/fanrr29**, posted October 2002 and visited on June 3, 2003.

Increase food security among U.S. households and in so doing reduce hunger.

food security: certain access to enough food for all people at all times to sustain a healthy and active life.

food insecurity: limited or uncertain access to foods of sufficient quality or quantity to sustain a healthy and active life.

food insufficiency: an inadequate amount of food due to a lack of resources.

food poverty: hunger resulting from inadequate access to available food for various reasons, including inadequate resources, political obstacles, social disruptions, poor weather conditions, and lack of transportation.

Resolving the hunger problem may seem at first beyond the influence of the ordinary person. Can one person's choice to limit family size or to recycle a bottle or to volunteer at a food recovery program make a difference?■ In truth, such choices produce several benefits. For one, a person's action may influence many other people over time. For another, an action repeated becomes a habit, with compounded benefits. For still another, making choices with an awareness of the consequences gives a person a sense of personal control, hope, and effectiveness. The daily actions of many concerned people can help solve the problems of hunger in their own neighborhoods or on the other side of the world.

Hunger in the United States

Ideally, all people at all times would have access to enough food to support an active, healthy life; in other words, they would experience **food security**. Unfortunately, almost 34 million people in the United States, including 13 million children,■ live in poverty and cannot afford to buy enough food to maintain good health.[1] Said another way, one out of ten households experiences hunger or the threat of hunger. Given the agricultural bounty and enormous wealth in this country, do these numbers surprise you? The limited or uncertain availability of nutritionally adequate and safe foods is known as **food insecurity** and is a major social problem in our nation today. Inadequate diets lead to poor health in adults and impaired physical, psychological, and cognitive development in children.

The accompanying "How to" presents the questions used in national surveys to identify food insecurity in the United States, and Figure 20-1 shows the most recent findings. Responses to these questions provide crude, but necessary, data to estimate the degree of hunger in this country.[2]

Defining Hunger in the United States

At its most extreme, people experience hunger because they have absolutely no food. More often, they have too little food **(food insufficiency)** and try to stretch their limited resources by eating small meals or skipping meals—often for days at a time. Sometimes hungry people obtain enough food to satisfy their hunger, perhaps by seeking food assistance or finding food through socially unacceptable ways—begging from strangers, stealing from markets, or scavenging through garbage cans, for example. Sometimes obtaining food raises concerns for food safety—for example, when rot, slime, mold, or insects have damaged foods or when people eat others' leftovers or meat from roadkill.[3]

Hunger has many causes, but in developed countries, the primary cause is **food poverty.** People are hungry not because there is no food nearby to purchase, but because they lack money. An estimated one out of nine people in the United States lives in poverty. Even those above the poverty line may not have food security. Physical and mental illnesses and disabilities, unemployment, low-paying jobs, unexpected or ongoing medical expenses, and high living expenses threaten their financial stability. When money is tight, people are forced to choose between food and life's other necessities—utilities, housing, and medical care.[4] Food costs are more variable and flexible; a person can purchase fewer groceries to lower the monthly food bill, but usually can't pay only a portion of the bills for electricity, rent, or medication. Further contributing to food poverty are other problems such

HOW TO Identify Food Insecurity in a U.S. Household

To determine the extent of food insecurity in a household, surveys ask the following questions. These questions reflect stages people in a household experience as food insecurity becomes progressively worse. Positive responses to the first several questions identify those who have concerns about their food supplies and food budgets and make adjustments to meet their basic needs. As food insecurity worsens, the middle questions identify adults who experience hunger and eat less. Most often, adults tend to protect their children from this experience. In the most severe cases, children also suffer from hunger and eat less.

- We worried whether our food would run out before we got money to buy more.
- The food that we bought just didn't last, and we didn't have money to get more.
- We couldn't afford to eat balanced meals.
- We relied on only a few kinds of low-cost food to feed our children because we were running out of money to buy food.
- We couldn't feed our children a balanced meal because we couldn't afford that.
- Our children were not eating enough because we just couldn't afford enough food.

- In the last 12 months, did you or other adults in your household ever cut the size of your meals or skip meals because there wasn't enough money for food?
- How often did this happen? (Considered a positive response if it occurred in three or more months during the previous year.)
- In the last 12 months, did you ever eat less than you felt you should because there wasn't enough money for food?
- In the last 12 months, were you ever hungry but didn't eat because you couldn't afford enough food?
- In the last 12 months, did you lose weight because you didn't have enough money for food?
- In the last 12 months, did you or other adults in your household ever not eat for a whole day because there wasn't enough money for food?
- How often did this happen? (Considered a positive response if it occurred in three or more months during the previous year.)
- In the last 12 months, did you ever cut the size of your children's meals because there wasn't enough money for food?
- In the last 12 months, were the children ever hungry but you just couldn't afford more food?

- In the last 12 months, did your children ever skip a meal because there wasn't enough money for food?
- How often did this happen? (Considered a positive response if it occurred in three or more months during the previous year.)
- In the last 12 months, did your children ever not eat for a whole day because there wasn't enough money for food?

The more positive responses, the greater the food insecurity. Households with children answer all of the questions and are categorized as follows:

$\leq$2 positive responses = food secure.

3–7 positive responses = food insecure without hunger.

$\geq$8 positive responses = food insecure with hunger.

Households without children answer fewer questions and are categorized as follows:

$\leq$2 positive responses = food secure.

3–5 positive responses = food insecure without hunger.

$\geq$6 positive responses = food insecure with hunger.

Figure 20-1 (on p. 694) shows the results of the 2001 surveys.

---

as abuse of alcohol and other drugs; lack of awareness of available food assistance programs; and the reluctance of people, particularly the elderly, to accept what they perceive as "welfare" or "charity." Lack of resources remains the major cause of food poverty, and solving this problem would do a lot to relieve hunger.

In the United States, poverty and hunger reach across various segments of society, touching single mothers living in households with their children; Hispanics, Native Americans, and African Americans; and those living in the inner cities more than others. People living in poverty are simply unable to buy sufficient amounts of nourishing foods, even if they are wise shoppers. For many of the children in these families, school lunch (and breakfast, where available) may be the only nourishment for the day. Otherwise they go hungry, waiting for an adult to find money for food. Not surprisingly, these children perform poorly in school and in social situations.[5]

Relieving Hunger in the United States

The American Dietetic Association (ADA) calls for aggressive action to bring an end to domestic food insecurity and hunger and to achieve food and nutrition security for everybody living in the United States.[6] Many federal and local programs aim to prevent or relieve malnutrition and hunger in the United States.

Federal Food Assistance Programs Adequate nutrition and food security are essential in supporting good health and achieving the public health goals of the United States.[7] To that end, an extensive network of federal assistance programs provides life-giving food daily to millions of U.S. citizens. One out of every six Americans receives food assistance of some kind, at a total cost of almost $40 billion per

Feeding the hungry—in the United States.

Food banks depend on the helping hands of caring volunteers.

year. Even so, the programs are not fully successful in preventing hunger, but they do seem to improve the nutrient intakes of those who participate. Programs described in earlier chapters include the WIC program for low-income pregnant women, breastfeeding mothers, and their young children (Chapter 15); the school lunch, breakfast, and child-care food programs for children (Chapter 16); and the food assistance programs for older adults such as congregate meals and Meals on Wheels (Chapter 17).

The Food Stamp Program, administered by the U.S. Department of Agriculture (USDA), is the largest of the federal food assistance programs, both in amount of money spent and in number of people served. It provides assistance to almost 20 million people at a cost of over $20 billion per year; over half of the recipients are children.[8] The USDA issues food stamp coupons or debit cards through state agencies to households—people who buy and prepare food together. The amount a household receives depends on its size, resources, and income. The average monthly benefit is about $80 per person.[9] Recipients may use the coupons or cards like cash to purchase food and food-bearing plants and seeds, but not to buy tobacco, cleaning items, alcohol, or other nonfood items. The accompanying "How to" offers shopping tips for those on a limited budget.

The Food Stamp Program improves nutrient intakes significantly, but hunger continues to plague the United States. Of the estimated 2 million homeless people in the United States who are eligible for food assistance, only 15 percent of single adults and 50 percent of families receive food stamps.

National Food Recovery Programs Efforts to resolve the problem of hunger in the United States do not depend solely on federal assistance programs. National **food recovery** programs have made a dramatic difference. The largest program, Second Harvest, coordinates the efforts of more than 250 **food pantries, emergency kitchens,** and homeless shelters in providing more than 1 billion pounds of food to 45,000 local agencies that feed over 23 million people a year.

Each year, an estimated one-fifth of our food supply is wasted in fields, commercial kitchens, grocery stores, and restaurants—that's enough food to feed 49 million people. Food recovery programs collect and distribute good food that would otherwise go to waste. Volunteers might pick corn left in an already harvested field, a grocer might deliver ripe bananas to a local **food bank,** and a caterer might take leftover chicken salad to a community shelter, for example. All of these efforts help to feed the hungry in the United States.

Community Efforts Food recovery programs depend on volunteers. Concerned citizens work through local agencies and churches to feed the hungry. Community-based food pantries provide groceries, and soup kitchens serve prepared meals. Meals often deliver adequate nourishment, but most homeless people receive fewer than one and a half meals a day, so many are still inadequately nourished.

IN SUMMARY Food insecurity and hunger are widespread in the United States among those living in poverty. Government assistance programs help to relieve poverty and hunger, but equally important are food recovery programs and other community efforts.

World Hunger

As distressing as hunger is in the United States, its prevalence is greater and its consequences more severe in developing countries. Although the hunger in these countries has diverse causes, once again the primary cause is poverty, and the poverty is more extreme than in the United States. Most people cannot grasp the severity of poverty in the developing world. One-fifth of the world's 6 billion (plus) people have no land and no possessions *at all*. They are the "poorest poor."

food recovery: collecting wholesome food for distribution to low-income people who are hungry. Four common methods of food recovery are:
- *Field gleaning:* collecting crops from fields that either have already been harvested or are not profitable to harvest.
- *Perishable food rescue or salvage:* collecting perishable produce from wholesalers and markets.
- *Prepared food rescue:* collecting prepared foods from commercial kitchens.
- *Nonperishable food collection:* collecting processed foods from wholesalers and markets.

food pantries: programs that provide groceries to be prepared and eaten at home.

emergency kitchens: programs that provide prepared meals to be eaten on site; often called *soup kitchens.*

food bank: a facility that collects and distributes food donations to authorized organizations feeding the hungry.

HOW TO Plan Healthy, Thrifty Meals

Chapter 2 introduced the Food Guide Pyramid and principles for planning a healthy diet. Meeting that goal on a limited budget adds to the challenge. To save money and spend wisely, plan and shop for healthy meals with the following tips in mind:

Planning

- Make a grocery list before going to the store to avoid expensive "impulse" items. Do not shop when hungry.
- Use leftovers.
- Center meals on rice, noodles, and other grains.
- Use small quantities of meat, poultry, fish, or eggs.
- Use legumes instead of meat, poultry, fish, or eggs several times a week.
- Use cooked cereals such as oatmeal instead of ready-to-eat breakfast cereals.
- Cook large quantities when time and money allow.
- Check for sales and clip coupons for products you need; plan meals to take advantage of sale items.

Shopping

- Buy day-old bread and other products from the bakery outlet.
- Select whole foods instead of convenience foods (potatoes instead of instant mashed potatoes, for example).
- Try store brands.
- Buy fresh produce that is in season; buy canned or frozen items at other times.
- Buy only the amount of fresh foods that you will eat before it spoils. Buy large bags of frozen items or dry goods; when cooking, take out the amount needed and store the remainder.
- Buy fat-free dry milk; mix and refrigerate quantities needed for a day or two. Buy fresh milk by the gallon or half-gallon.
- Buy less expensive cuts of meat. Chuck and bottom round roast are usually inexpensive; cover during cooking and cook long enough to make meat tender. Buy whole chickens instead of pieces.
- Compare the unit price (cost per ounce, for example) of similar foods so that you can select the least expensive brand or size.
- Buy nonfood items such as toilet paper and laundry detergent at discount stores instead of grocery stores.

For daily menus and recipes for healthy, thrifty meals, visit the USDA Center for Nutrition Policy and Promotion: **www.usda.gov/cnpp**

They survive on less than $1 a day each, they lack clean drinking water, and they cannot read or write. The average U.S. housecat receives twice as much protein every day as one of these people, and the cost of keeping that cat is greater than such a person's annual income.

The "poorest poor" are usually female. Many societies around the world under-value females, providing girls with poorer diets and fewer opportunities than boys. Malnourished girls become malnourished mothers who give birth to low-birthweight infants—and the cycle of hunger, malnutrition, and poverty continues.

Not only does poverty cause hunger, but tragically, hunger worsens poverty by robbing a person of the good health and the physical and mental energy needed to be active and productive. Hungry people simply cannot work hard enough to get themselves out of poverty. Malnourished people with a low BMI earn less money performing manual labor; an increase in BMI of 1 percent correlates with an increase in wages of 2 percent. Economists calculate that cutting world hunger and malnutrition in half by 2015 would generate a value of more than $120 billion in longer, healthier, and more productive lives.[10]

Food Shortages

World hunger brings to mind victims of **famine,** a severe food shortage in an area that causes widespread starvation and death. In recent years, the natural causes of famine—drought, flood, and pests—have become less important than the political causes created by people. Figure 20-2 (on p. 698) shows the hunger hotspots in the world and their primary causes.

Political Turbulence A sudden increase in food prices, a drop in workers' incomes, or a change in government policy can quickly leave millions hungry. Between 15 million and 30 million people died during the Chinese famine of 1959 through 1961, the worst famine of the twentieth century. The main cause was government policies associated with the Great Leap Forward; this government initiative was intended to transform China's economy, but the poorly planned communal farm system and the widespread waste of resources devastated the Chinese agricultural system.

Armed Conflicts In the past decade, armed conflict and political unrest were the dominant cause of famine worldwide. Farmers become warriors, their agricultural fields become battlegrounds, the citizens go hungry, and the warring factions

famine: extreme scarcity of food in an area that causes starvation and death in a large portion of the population.

Feeding the hungry—in Calcutta, India.

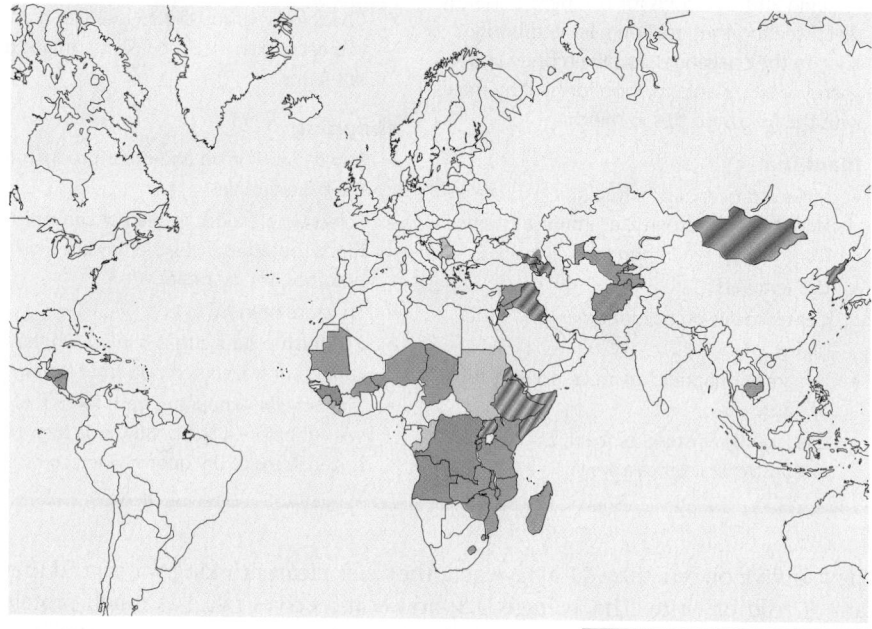

FIGURE 20-2   Hunger Hotspots

Natural disasters (primarily drought) and conflict (war and civil strife) are the most common causes of food shortages. Africa is currently the most affected region.

Note: Areas with stripes of color have multiple causes of hunger.

Key:
- No food emergencies
- Economic problems
- Natural disasters
- Past conflict
- Conflict

often block famine relief. The world continues to struggle to find a middle ground between respecting the sovereignty of nations and insisting that all nations allow humanitarian assistance to reach the people.

Natural Disasters   Sometimes drought and other poor weather conditions create food shortages. During natural disasters without war, food aid from countries around the world has provided a safety net for countries whose crops fail. But food aid now does more than just offset poor harvests; it also delivers food relief to countries, such as Ethiopia, that are chronically short of food because of ongoing drought and without resources to buy food.

Malnutrition

Although we usually associate world hunger with famine, the numbers affected by famine are relatively small compared with those suffering from persistent hunger and malnutrition. Over 800 million people, mostly women and children, are malnourished, with another billion perilously close.[11] The nutrients most likely to be lacking are iron, iodine, and vitamin A.[12] The prevalence and consequences of these deficiencies stagger the mind. Over 30 percent of the world's population have iron-deficiency anemia, a leading cause of maternal deaths, premature births, low birthweights, infections, and premature deaths. Iodine deficiency affects one out of seven, resulting in stillbirths and irreversible mental retardation (cretinism). Almost 80 million young children (under age five) suffer from symptoms of vitamin A deficiency—blindness, growth retardation, and poor resistance to common childhood infections such as measles.■ The deficiency symptoms of these nutri-

■ To help prevent blindness and reduce measles mortality, health care workers distribute vitamin A supplements to millions of children worldwide.

ents and those of the other vitamins and minerals were presented in Chapters 10 through 13; Chapter 6 described protein-energy malnutrition; and Chapters 15 through 17 examined the effects of malnutrition during various stages of the life cycle. The consequences of nutrient deficiencies are felt not only by individuals, but by entire nations. When people suffer from mental retardation, growth failure, blindness, infections, and other consequences of malnutrition, the economy of their country declines as productivity decreases and health care costs increase.[13] The dramatic signs of malnutrition are most evident at each end of the life span in a nation's high infant mortality rates and low life expectancy.

In addition to specific nutrient deficiencies, one child in six worldwide is born underweight, and almost two in five children are underweight by the age of five. These underweight children are malnourished and readily develop the diseases of poverty: parasitic and infectious diseases that cause diarrhea (dysentery and cholera), acute respiratory illnesses (pneumonia and whooping cough), measles, and malaria. The synergistic combination of infectious disease and malnutrition dramatically increases the likelihood of early death.[14] Compared with adequately nourished children, the risk of death is 2.5 times greater for children with mild malnutrition, 4.6 times greater for children with moderate malnutrition, and 8.4 times greater for children with severe malnutrition. Each year, 6 million children die as a result of hunger and malnutrition. Most of them do not starve to death—they die from the diarrhea and dehydration that accompany infections. In fact, their BMI alone would not provide a clue as to how close to death they actually are.[15] Health care workers around the world save millions of lives■ each year by effectively reversing dehydration and correcting the diarrhea with **oral rehydration therapy (ORT).**

Diminishing Food Supply

Most disturbingly, such misery and starvation exist side by side with ample food supplies. The demand for food is great, but technological advances in farming have increased crop yields, and prices of many foods have fallen in response. Wheat and corn, for example, the staple foods of many nations, are abundantly available and are now priced at less than half of their cost of 40 years ago.

At the present rate of growth, however, the world's population may soon outstrip the rate of food production. Environmental degradation and dwindling water supplies may limit further growth in the world's food production in many agricultural areas. No part of the world is safely insulated against future food shortages. Developed countries may be the last to feel the effects, but they will ultimately go as the world goes.

> **IN SUMMARY** Natural causes such as drought, flood, and pests and political causes such as armed conflicts and government policies all contribute to the extreme hunger and poverty seen in the developing countries. To meet future demands for food, technology must improve food production and nations must control overpopulation.

Poverty and Overpopulation

The world's population is rising at an alarming rate, as Figure 20-3 (on p. 700) shows. Skyrocketing human numbers threaten the earth's capacity■ to provide safe water and adequate food for its inhabitants.

The sheer magnitude of the world's annual population increase of over 70 million people is difficult to comprehend. Every half-second, the population increases by another person. Every 40 days, the world adds the equivalent of another New York City. During the six months of the terrible 1992 famine in Somalia, an estimated

Photo by Chris Hondros/Getty Images

International efforts help to relieve hunger and poverty in Afghanistan and around the world.

■ To prevent death from diarrheal disease, provide:
- Adequate sanitation.
- Safe water.
- Oral rehydration therapy.

■ The maximum number of people the earth can support over time is its **human carrying capacity.**

oral rehydration therapy (ORT): the administration of a simple solution of sugar, salt, and water, taken by mouth, to treat dehydration caused by diarrhea. A simple ORT recipe:
- 1 c boiling water.
- 2 tsp sugar.
- A pinch of salt.

FIGURE 20-3 World Population Growth

World Population Growth chart: Billion (y-axis, 0–10) vs. Mid-decade totals and projections (1950–2050)

Mid-decade totals and projections

300,000 people starved to death. Yet it took the world only 29 *hours* to replace their numbers!

As the world's population continues to grow, much of the increase is occurring in developing countries where hunger and malnutrition are already widespread. More people sharing the little food available can only worsen the problem. Stabilizing the population may be the only way the world's food production will be able to keep up with demands. Without population stabilization, the world can neither support the lives of people already born nor halt environmental deterioration around the globe. And before the population problem can be resolved, it may be necessary to remedy the poverty problem. In countries around the world, economic growth has been accompanied by slowed population growth.

Population growth is a central factor contributing to poverty and hunger. The reverse is also true: poverty and hunger contribute to population growth.

Population Growth Leads to Hunger and Poverty The first of these cause-and-effect relationships is easy to understand. As a population grows larger, more mouths must be fed, and the worse poverty and hunger become.

Population growth also contributes to hunger indirectly by preempting good agricultural land for growing cities and industry and forcing people onto marginal land, where they cannot produce sufficient food for themselves. The world's poorest people live in the world's most damaged and inhospitable environments.

Hunger and Poverty Lead to Population Growth How does poverty lead to overpopulation? Poverty and its consequences—inadequate food and shelter—leave women vulnerable to physical abuse, forced marriages, and prostitution. Furthermore, they lack access to reproductive health care and family planning counseling. Also, in some regions of the world, families depend on children to farm the land, haul water, and care for adults in their old age. Children are an economic asset for these families. Poverty claims many young children, who are among the most likely to die from malnutrition and disease. If a family faces ongoing poverty, the parents may choose to have many children to ensure that some will survive to adulthood. People are willing to risk having fewer children only if they are sure that their children will live and that the family can develop other economic assets (skills, businesses, land).

Breaking the Cycle Relieving poverty and hunger, then, may be a necessary first step in curbing population growth. When people attain better access to health care, education, and family planning, the death rate falls. At first, births outnumber deaths, but as the standard of living continues to improve, families become

Families in developing countries depend on their children to help provide for daily needs.

© Alexandra Avakian/Contact Press/PictureQuest

willing to risk having fewer children. Then the birth rate falls. Thus improvements in living standards help stabilize the population.

The link between improved economic status and slowed population growth has been demonstrated in several countries. Central to achieving this success is sustainable development that includes not only economic growth, but a sharing of resources among all groups. Where this has happened, population growth has slowed the most: in parts of Sri Lanka, Taiwan, Malaysia, and Costa Rica, for example. Where economic growth has occurred but only the rich have grown richer, population growth has remained high; examples include Brazil, the Philippines, and Thailand, where large families continue to be a major economic asset for the poor.

As a society gains economic footing, education also becomes a higher priority. A society that educates its children, both males and females, experiences a drop in birth rates. Education, particularly for girls and women, brings improvements in family life, including improved nutrition, better sanitation, effective birth control, and elevated status. With improved conditions, more infants live to adulthood, making smaller families feasible.

IN SUMMARY More people means more mouths to feed, which worsens the problems of poverty and hunger. Poverty and hunger, in turn, encourage parents to have more children. Breaking this cycle requires improving the economic status of the people and providing them with health care, education, and family planning.

Environmental Degradation and Hunger

Hunger, poverty, and overpopulation interact with another force: environmental degradation. The environment suffers as more and more people must share fewer and fewer resources. In developing countries, people living in poverty sell everything they own to obtain money for food, even the seeds that would have provided next year's crops. They cut trees for firewood or timber to sell, then lose the soil to erosion. Without these resources, they become poorer still. Thus poverty contributes to environmental ruin, and the ruin leads to hunger. Figure 20-4 shows the vicious cycle of poverty, population growth, and environmental degradation leading to hunger. Environmental degradation threatens the world's ability to produce enough food to feed its many people.

FIGURE 20-4 Poverty, Overpopulation, and Environmental Degradation

The interactions of poverty, overpopulation, and environmental degradation worsen hunger.

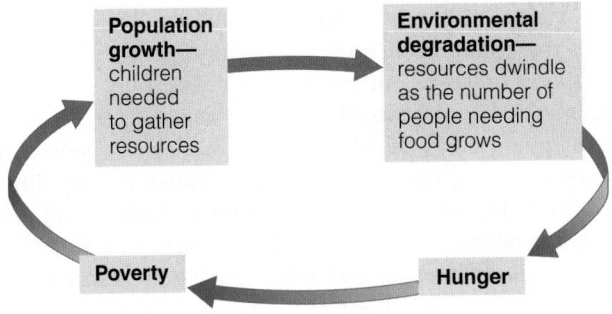

Without water, croplands become deserts.

Environmental Limitations in Food Production

Environmental problems that are slowing food production include:

- *Soil erosion, compaction, and salinization* due to overtillage and overirrigation, which result in extensive loss of productive croplands.

- *Deforestation and desertification* due to overgrazing, which leads to soil erosion.

- *Air pollution* produced from the burning of **fossil fuels,** which damages crops and depletes the ozone.

- *Ozone depletion,* which allows harmful radiation from the sun to damage crops, especially radiation-sensitive crops such as soybeans.

- *Climate changes* caused by destruction of forests and concentration of heat-trapping carbon dioxide produced by fossil fuels. A rise in global temperature may reduce soil moisture, impair pollination of major food crops such as rice and corn, slow growth, weaken disease resistance, and disrupt many other factors affecting crop yields.

- *Water pollution* from agricultural sediments, salts, fertilizers, pesticides, and manure, which limits agricultural yields, drinking water, and fishery production.

- *Water scarcity* due to overuse of surface and ground water for irrigation, which may limit human population growth even before food scarcity does. In many areas, the supplies of fresh water are already inadequate to fully support the survival of crops, livestock, and people.

- *Extensive overgrazing,* which is causing rangelands to deteriorate. In nearly all developing countries, the feed needs of livestock now exceed the capacity of their rangelands.

- *Overfishing and water pollution,* which are destroying fisheries and diminishing the supply of seafood.

All in all, environmental problems are reducing the world's ability to feed its people.

Other Limitations in Food Production

With crop fields, rangelands, and fish yields diminishing, can advances in agriculture compensate for the losses caused by environmental degradation? Historically, agriculture has improved yields by making greater investments in irrigation, fertilizer, and improved genetic strains. Today, however, the contributions these measures can make are reaching their limits, in part because they have also created environmental problems. Irrigation can no longer compensate by improving crop yields because almost all the land that can benefit from irrigation is already receiving it. In fact, rising concentrations of salt in the soil—a by-product of irrigation—are *lowering* yields on close to a quarter of the world's irrigated cropland. Nor can fertilizer use enhance agricultural production much. Much of the fertilizing that can be done is being done—and with great effect; fertilizer use supports some 40 percent of the world's total crop yields. Adding more fertilizer, however, brings no further rise in yield and adds to the pollution of nearby waterways. As for the development of high-yielding strains of crops, recent advances have been dramatic, but even they may be inadequate to change the overall trends. Furthermore, the raw materials necessary for developing new crops have become less available as genetic variation for many plant species is lost. Of the 5000 food plants grown throughout the world a few centuries ago, only 150 are cultivated in commercial agriculture today. Most of the world's population relies on only five cereals, three legumes, and three root crops to meet their energy needs. Even among these, valuable strains are vanishing.

Since 2000, the world's grain harvest has fallen short of projected consumption needs. Meanwhile, the world's population continues to rise at the rate of at least 1 percent a year. Many authorities in many fields are calling for a reduction in the

fossil fuels: coal, oil, and natural gas. These fuels are nonrenewable resources that pollute. (Renewable or alternative fuels, such as solar and wind energy, pollute less or not at all.)

growth rate of the world's population as the only way to enable the world's food production to keep pace with its growing numbers.

The world still produces enough food to feed all its people,■ and the problem of hunger today remains a problem of unequal distribution of land to grow crops or income to purchase foods. If present trends continue, however, the time is fast approaching when there will be an absolute deficit of food. This conclusion seems inescapable. The world's increasing population threatens the world's capacity to produce adequate food. Until the nations of the world resolve the population problem, they can neither support the lives of people already born nor remedy global trends toward environmental deterioration. And to resolve the population problem, a necessary first step is to remedy the poverty problems, for reasons already discussed. Of the 70 million people being added to the population each year, 95 percent are born in the most poverty-stricken areas of the world.

■ World agriculture produces enough food to provide each person with 2720 kcal/day.

IN SUMMARY Increasing environmental degradation reduces our ability to produce enough food to feed the world's people. Exacerbating the situation is the rapid increase in the world's population.

Solutions

Slowly but surely, improvements are evident in developing nations. Most nations have seen a rise in their gross domestic product, a key measure of economic well-being. Adult literacy rates and the proportion of children being sent to school have risen. The proportion of undernourished people has declined. Optimism abounds, though problems remain.

The keys to solving the world's hunger, poverty, and environmental problems are in the hands of both the poor and the rich nations, but require different efforts from them. The poor nations need to provide contraceptive technology and family planning information to their citizens, develop better programs to assist the poor, and slow and reverse the destruction of environmental resources. The rich nations need to stem their wasteful and polluting uses of resources and energy, which are contributing to global environmental degradation. They also must become willing to ease the debt burden that many poor nations face.

Sustainable Development Worldwide

Many nations now recognize that improving all nations' economies is a prerequisite to meeting the world's other urgent needs: hunger relief, population stabilization, environmental preservation, and **sustainable** resources. Over 100 nations have agreed to a set of principles of sustainable development—development that would equitably meet both the economic and the environmental needs of present and future generations. They recognize that relieving poverty will help relieve environmental degradation and hunger.

To rephrase a well-known adage: If you give a man a fish, he will eat for a day. If you teach him to fish and enable him to buy and maintain his own gear and bait, he will eat for a lifetime and help to feed others. Unlike food giveaways and money doles, which are only stop-gap measures, social programs that permanently improve the lives of the poor can permanently solve the hunger problem.

Activism and Simpler Lifestyles at Home

Every segment of our society can join in the fight against hunger, poverty, and environmental degradation. The federal government, the states, local communities, big business and small companies, educators, and all individuals, including dietitians and foodservice managers, have many opportunities to resolve these problems.

sustainable: able to continue indefinitely. Here, the term means using resources at such a rate that the earth can keep on replacing them and producing pollutants at a rate with which the environment and human cleanup efforts can keep pace, so that no net accumulation of pollution occurs.

Each person's choice to get involved and be heard can help lead to needed change.

■ A popular adage urges us to "Think globally, act locally."

Government Action Government policies can change to promote sustainability. For example, the government can use tax dollars and other resources to develop energy conservation services and crop protection.

Business Involvement Businesses can take the initiative to help; some already have. Several large corporations are major supporters of antihunger programs. Many grocery stores and restaurants participate in food recovery programs by giving their leftover foods to community distribution centers.

Education Educators, including nutrition educators, can teach others about the underlying social and political causes of poverty, the root cause of hunger. At the college level, they can teach the relationships between hunger and population, hunger and environmental degradation, hunger and the status of women, and hunger and global economics. They can advocate legislation to address these problems. They can teach the poor to develop and run nutrition programs in their own communities and to fight on their own behalf for antipoverty, antihunger legislation.

Foodservice Efforts Dietitians and foodservice managers have a special role to play, and their efforts can make an impressive difference. Their professional organization, the ADA, urges members to conserve resources and minimize waste in both their professional and their personal lives.[16] In addition, the ADA urges its members to educate themselves and others on hunger, its consequences, and programs to fight it; to conduct research on the effectiveness and benefits of programs; and to serve as advocates on the local, state, and national levels to help end hunger in the United States.[17] Globally, the ADA supports programs that combat malnutrition, provide food security, promote self-sufficiency, respect local cultures, protect the environment, and sustain the economy.[18]

Individual Choices Individuals can assist the global community in solving its poverty and hunger problems by joining and working for hunger-relief organizations (see Table 20-1). They can also support organizations that lobby for the needed changes in economic policies toward developing countries.

Most importantly, all individuals can try to make lifestyle choices■ that consider the environmental consequences. Many small decisions each day have major consequences for the environment. The accompanying "How to" describes how consumers can conserve resources and minimize waste when making food-related choices.

TABLE 20-1	Hunger-Relief Organizations

Action without Borders
79 Fifth Ave., 17th Floor
New York, NY 10118
(212) 843-3973
www.idealist.org

Bread for the World
50 F St. NW, Suite 500
Washington, DC 20001
(800) 82-BREAD or (800) 822-7323
(202) 639-9400; fax (202) 639-9401
www.bread.org

Center on Hunger and Poverty
Brandeis University
Mailstop 077
Waltham, MA 02454
(781) 736-8885
www.centeronhunger.org

Community Food Security Coalition
P.O. Box 909
Venice, CA 90294
(310) 822-5410
www.foodsecurity.org

Congressional Hunger Center
229½ Pennsylvania Ave.
Washington, DC 20003
(202) 547-7022
www.hungercenter.org

Oxfam America
26 West St.
Boston, MA 02111-1206
(800) 77-OXFAM or
(800) 776-9326
www.oxfamamerica.org

Pan American Health Organization
525 23rd St. NW
Washington, DC 20037
(202) 974-3000
www.paho.org

Second Harvest
35 E. Wacker Dr., #2000
Chicago, IL 60601
(800) 771-2303
www.secondharvest.org

Society of St. Andrew
3383 Sweet Hollow Rd.
Big Island, VA 24526
(800) 333-4597
www.endhunger.org

Food and Agriculture
   Organization (FAO) of the
   United Nations
2175 K St. NW, Suite 300
Washington, DC 20437
(202) 653-2400
www.fao.org

United Nations Children's Fund
   (UNICEF)
3 United Nations Plaza
New York, NY 10017-4414
(212) 326-7035
www.unicef.org

World Food Program
Via Vittorio Emanuele Orlando, 83
Rome, Italy 00148
www.wfp.org

World Health Organization (WHO)
525 23rd St. NW
Washington, DC 20037
(202) 974-3000
www.who.org

World Hunger Program
Brown University
Box 1831
Providence, RI 02912
(401) 863-1000
**www.brown.edu/Departments/
World_Hunger_Program/
hungerweb/WHP/overview.html**

World Hunger Year (WHY)
505 Eighth Ave., 21st Floor
New York, NY 10018-6582
(800) GleanIt
www.worldhungeryear.org

HOW TO Make Environmentally Friendly Food-Related Choices

Food production taxes environmental resources and causes pollution. Consumers can make environmentally friendly choices at every step from food shopping to cooking and use of kitchen appliances to serving, cleanup, and waste disposal.

Food Shopping

Transportation:

- Whenever possible, walk or ride a bicycle; use car pools and mass transit.
- Shop only once a week, share trips, or take turns shopping for each other.
- When buying a car, choose an energy-efficient one.

Food choices:

- Choose foods low on the food chain; that is, eat more plants and fewer animals that eat plants (this suggestion complements the *Dietary Guidelines* for eating for good health).
- Eat small portions of meat; select range-fed beef, buffalo, poultry, and fish.
- Select local foods; they require less transportation, packaging, and refrigeration.

Food packages:

- Whenever possible, select foods with no packages; next best are minimal, reusable, or recyclable ones.
- Buy juices and sodas in large glass or recyclable plastic bottles (not small individual cans or cartons); grains in bulk (not separate little packages); and eggs in pressed fiber cartons (not foam, unless it is recycled locally).
- Carry reusable string or cloth shopping bags; alternatively, ask for plastic bags if they are recyclable.

Gardening

- Grow some of your own food, even if it is only herbs planted in pots on your kitchen windowsill.
- Compost all vegetable scraps, fruit peelings, and leftover plant foods.

Cooking Food

- Cook foods quickly in a stir-fry, pressure cooker, or microwave oven.
- When using the oven, bake a lot of food at one time and keep the door closed tightly.
- Use nondisposable utensils, dishes, and pans.
- Use pumps instead of spray products.

Kitchen Appliances

- Use fewer small electrical appliances; open cans, mix batters, sharpen knives, and chop vegetables by hand.
- When buying a large appliance, choose an energy-efficient one.
- Consider solar power to meet home electrical needs.
- Set the water heater at 130°F (54°C), no hotter; put it on a timer; wrap it and the hot-water pipes in insulation; install water-saving faucets.

Food Serving, Dish Washing, and Waste Disposal

- Use "real" plates, cups, and glasses instead of disposable ones.
- Use cloth towels and napkins, reusable storage containers with lids, reusable pans, and dishcloths instead of paper towels, plastic wrap, plastic storage bags, aluminum foil, and sponges.
- Run the dishwasher only when it is full.
- Recycle all glass, plastic, and aluminum.

These suggested lifestyle changes can easily be extended from food to other areas.

The personal rewards of the behaviors presented in the "How to" above are many, from saving money to the satisfaction of knowing that you are treading lightly on the earth. But do they really help? They do, if enough people join in. Because we number more than 6 billion, individual actions can add up to exert an immense impact.

"Be part of the solution, not part of the problem," an adage says. In other words, don't waste time or energy moaning and groaning about how bad things are: do something to improve them. This adage is as applicable to today's global environmental problems as it is to an unwashed dish in the kitchen sink. They are our problems: human beings created them, and human beings must solve them.

IN SUMMARY The global environment, which supports all life, is deteriorating, largely because of our irresponsible use of resources and energy. Governments, businesses, and all individuals have many opportunities to make environmentally conscious choices, which may help solve the hunger problem, improve the quality of life, and generate jobs. Personal choices, made by many people, can have a great impact.

Good planets are hard to find.

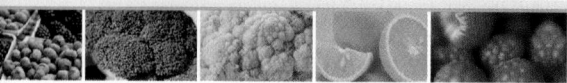

Nutrition in Your Life

Your choice to get involved in the fight against hunger—whether in your community

or across the globe—can make a big difference in the health and survival of others.

- Do you volunteer in local hunger-relief programs?

- Do you write to your legislators and voice your opinions on issues such as food

  assistance programs, environmental degradation, and international debt relief?

- Have you adopted environmentally friendly behaviors when making food-

  related choices?

NUTRITION ON THE NET

 Access these websites for further study of topics
covered in this chapter.

- Find updates and quick links to these and other
  nutrition-related sites at our website:
  www.wadsworth.com/nutrition

- Learn about constructive, community-based solutions
  to the problems of poverty and hunger within and
  between the public and private sectors from the
  National Hunger Clearinghouse:
  www.worldhungeryear.org/nhc

- Visit the USDA Food Stamp Program:
  www.fns.usda.gov/fsp

- Download 40 recipes, two sample weekly menus, and
  numerous tips for planning, shopping for, and cooking

healthy meals on a tight budget from the USDA cook-
book entitled "Recipes and Tips for Healthy, Thrifty
Meals": **www.usda.gov/cnpp**

- Review the Best Practices Manual for Food Recovery and
  Gleaning at the USDA Food and Nutrition Service site:
  www.fns.usda.gov/fdd/gleaning/gleanintro.htm

- Find information on feeding the hungry from the
  Emergency Food and Shelter Program:
  www.efsp.unitedway.org

- Donate free food at The Hunger Site:
  www.thehungersite.com

- See Table 20-1 (on p. 704) for additional websites.

STUDY QUESTIONS

These questions will help you review the chapter. You will
find the answers in the discussions on the pages provided.

1. Identify some reasons why hunger is present in a country as wealthy as the United States. (pp. 694–695)

2. Identify some reasons why hunger is present in the
   developing countries of the world. (pp. 696–698)

3. Explain why relieving environmental problems will also
   help to alleviate hunger and poverty. (pp. 701–703)

4. Discuss the different paths by which rich and poor
   countries can attack the problems of world hunger and
   the environment. (p. 703)

5. Describe some strategies that consumers can use to minimize negative environmental impacts when shopping
   for food, preparing meals, and disposing of garbage.
   (p. 705)

These multiple choice questions will help you prepare for an
exam. Answers can be found on p. 707.

1. Food insecurity refers to the:
   a. uncertainty of foods' safety.
   b. fear of eating too much food.
   c. limited availability of foods.
   d. reliability of food production.

2. The most common cause of hunger in the United States is:
   a. poverty.
   b. alcohol abuse.
   c. mental illness.
   d. lack of education.

3. Food stamp coupons cannot be used to purchase:
   a. tomato plants.
   b. birthday cakes.

c. cola beverages.

d. laundry detergent.

4. Which action is not typical of a food recovery program?

    a. gathering potatoes from a harvested field

    b. collecting overripe tomatoes from a wholesaler

    c. offering food stamp coupons to low-income people

    d. delivering restaurant leftovers to a community shelter

5. The primary cause of the worst famine in the twentieth century was:

    a. armed conflicts.

    b. natural disasters.

    c. food contaminations.

    d. government policies.

6. The most likely cause of death in malnourished children is:

    a. growth failure.

    b. diarrheal disease.

    c. simple starvation.

    d. vitamin A deficiency.

7. Which of the following is most critical in providing food to all the world's people?

    a. decreasing air pollution

    b. increasing water supplies

c. decreasing population growth

d. increasing agricultural land

8. Which of these items is the most environmentally benign choice?

    a. sponges

    b. plastic bags

    c. aluminum foil

    d. cotton towels

9. Which of these methods uses the most fuel?

    a. baking

    b. stir-frying

    c. microwaving

    d. pressure cooking

10. Which of these purchases is the best choice, for environmental reasons?

    a. fresh fish from a local merchant

    b. frozen fish from a developing country

    c. canned fish from a nationally known food manufacturer

    d. packaged fish from the freezer section of a local supermarket

REFERENCES

1. United States Department of Agriculture, *Household Food Security in the United States, 2001*, ERS Food Assistance and Nutrition Research Report no. FANRR-29, October 2002, available from **www.ers.usda.gov/publications/fanrr29**.
2. J. S. Hampl and R. Hill, Dietetic approaches to US hunger and food insecurity, *Journal of the American Dietetic Association* 102 (2002): 919–923.
3. K. M. Kempson and coauthors, Food management practices used by people with limited resources to maintain food sufficiency as reported by nutrition educators, *Journal of the American Dietetic Association* 102 (2002): 1795–1799.
4. M. Kim, J. Ohls, and R. Cohen, *Hunger in America, 2001—National Report Prepared for America's Second Harvest* (Princeton, N.J.: Mathematica Policy Research, Inc., 2001).
5. K. Alaimo, C. M. Olson, and E. A. Frongillo, Jr., Food insufficiency and American school-aged children's cognitive, academic, and psychosocial development, *Pediatrics* 108 (2001): 44–53.

6. Position of the American Dietetic Association: Domestic food and nutrition security, *Journal of the American Dietetic Association* 102 (2002): 1840–1847.
7. Position of the American Dietetic Association, 2002.
8. USDA Food and Nutrition Service, **www.fns.usda.gov/**, site visited June 3, 2003.
9. USDA Food and Nutrition Service, 2002 data, **www.fns.usda.gov/pd/fsavgben.htm**.
10. Food and Agriculture Organization, www.fao.org, site visited June 3, 2003.
11. U. Ramakrishnan, Prevalence of micronutrient malnutrition worldwide, *Nutrition Reviews* 60 (2002): S46–S52; *State of Food Insecurity in the World 2002*, Food and Agriculture Organization of the United Nations.
12. U. Kapil and A. Bhavna, Adverse effects of poor micronutrient status during childhood and adolescence, *Nutrition Reviews* 60 (2002): S84–S90.
13. R. Martorell, The role of nutrition in economic development, *Nutrition Reviews* 54 (1996): S66–S71.

14. M. Peña and J. Bacallao, Malnutrition and poverty, *Annual Review of Nutrition* 22 (2002): 241–253.
15. S. Collins and M. Myatt, Short-term prognosis in severe adult and adolescent malnutrition during famine: Use of a sample prognostic model based on counting clinical signs, *Journal of the American Medical Association* 284 (2000): 621–626.
16. Position of the American Dietetic Association: Dietetic professionals can implement practices to conserve natural resources and protect the environment, *Journal of the American Dietetic Association* 101 (2001): 1221–1227.
17. Position of the American Dietetic Association, 2002.
18. Position of the American Dietetic Association: Addressing world hunger, malnutrition, and food insecurity, *Journal of the American Dietetic Association* 103 (2003): 1046–1057.

ANSWERS

Study Questions (multiple choice)

1. c   2. a   3. d   4. c   5. d   6. b   7. c   8. d   9. a   10. a

Progress toward Sustainable Food Production

While some individuals are attempting to make their own personal lifestyles more environmentally benign, as suggested in Chapter 20, others are seeking ways to improve whole sectors of human enterprise, among them, agriculture. To date, large agricultural enterprises have been among the world's biggest resource users and polluters. Is it possible for agriculture to become sustainable? Do new technologies hold promise for advancing sustainability? And if so, can the change be made without hurting farmers? These questions are addressed in this highlight; the accompanying glossary presents terms important to these concepts.

Costs of Producing Food Unsustainably

The current environmental and social costs of agriculture and the food industry take many forms. Among them are resource waste and pollution (including energy overuse) and disruption of farm communities.

Resource Waste and Pollution

Producing food costs the earth dearly. First of all, to grow food, we clear land—prairie, wetland, and forest—losing native ecosystems and wildlife. Then we plant crops or graze animals on the land.

Planting Crops The soil loses nutrients as each crop is taken from it, so fertilizer is applied. Some fertilizer runs off, polluting the waterways. Some plowed soil runs off, which clouds the waterways and interferes with the growth of aquatic plants and animals.

To protect crops against weeds and pests, we apply herbicides and pesticides. These chemicals also pollute the water and, wherever the wind carries them, the air. Most herbicides and pesticides kill not only weeds and pests, but also native plants, native insects, and animals that eat those plants and insects. Ironically, widespread use of pesticides and herbicides causes resistant pests and weeds to evolve. Consequently, farmers must use still more pesticides and herbicides. These chemicals pose hazards for farm workers who handle them, and their residues may create health concerns for consumers as well (see Chapter 19).

Finally, we irrigate, a practice that causes salts to accumulate on the soil surface. The water evaporates, but the salts do not. As the surface soil becomes increasingly salty, plant growth suffers. Irrigation can also deplete the water supply over time because it pulls water from surface waters or from underground; then, the water evaporates or runs off. This process, carried to the extreme, can dry up rivers and lakes and lower the water table of a whole region. A vicious cycle develops: the drier the region becomes, the more the farmers must irrigate, and the more they irrigate, the drier the region becomes and the deeper they dig their wells.

Raising Livestock Raising livestock also takes a toll. Like plant crops, herds of livestock occupy land that once maintained itself in a natural state. The land pays a price in losses of native plants and animals, soil erosion, water depletion, and desert formation. Alternatively, if animals are raised in large concentrated areas such as cattle feedlots or giant hog "farms," a high price is paid when huge masses of animal wastes produced in overcrowded, factory-style farms leach into local soils and water supplies, polluting them. To prevent contamination of drinking water, the Environmental Protection Agency suggests several strategies for managing livestock, poultry, and

GLOSSARY

agribusiness: agriculture practiced on a massive scale by large corporations owning vast acreages and employing intensive technological, fuel, and chemical inputs.

integrated pest management (IPM): management of pests using a combination of natural and biological controls and few or no pesticides.

nonpoint water pollution: water pollution caused by runoff from all over an area rather than from discrete "point" sources. An example is the pollution caused by runoff from agricultural fields.

sustainable agriculture: agricultural practices that use individualized approaches appropriate to local conditions so as to minimize technological, fuel, and chemical inputs.

SOURCE: Idea and data from T. R. Reid, Feeding the planet, *National Geographic,* October 1998, pp. 58–74.

horse waste.[1] In addition to manure, cows produce large quantities of methane—a potent gas that may contribute to global warming. In addition to the waste problems, animals have to be fed; grain is grown for them on other land. That land may require fertilizers, herbicides, pesticides, and irrigation, too. In the United States, more cropland is used to produce grains for livestock than to produce grains for people. Figure H20-1 compares the grain required to produce various foods.

Fishing Fishing also incurs environmental costs. Fishing easily becomes overfishing and depletes stocks of the very fish that people need to eat. Some fishing methods (such as nets and filament line) kill aquatic animals other than the ones sought and deplete large populations of nonfood animals, such as dolphins. Also, fishing is energy-intensive, requiring fuel for boats, refrigeration, processing, packing, and transport. Water pollution incurs health risks when people eat contaminated fish. Bioaccumulation of toxins in fish is a serious problem in some areas; in others it rules out fish consumption altogether (see Chapter 19 for more details).

Energy Overuse The entire food industry, whether based on growing crops, raising livestock, or fishing, requires energy, which entails burning fossil fuels. Massive fossil fuel use threatens our planet by causing air and water pollution, global warming, ozone depletion, and other environmental ills.

In the United States, the food industry consumes about 20 percent of all the energy the nation uses. Each year, we use 1500 liters of oil (over 350 gallons) *per person* to produce, process, distribute, and prepare our food. Most of this energy is used to run farm machinery and to produce fertilizers and pesticides. Energy is also used to prepare, package, transport, refrigerate, store, cook, and wash our foods.

The Cumulative Effects Many national and international agencies are concerned about the environmental ramifications of agriculture. The prestigious National Academy of Sciences has reported that agriculture is the largest single source of **nonpoint water pollution** in the nation. Pollution from "point sources," such as sewage plants or factories, is relatively easy to control, but runoff from fields and pastures enters waterways from all over broad regions and is nearly impossible to control.

Agriculture is destroying its own foundation. In just the last 40 years, agricultural activities have ruined more than 10 percent of the earth's most fertile land, an area the size of China and India combined. Over 20 million acres have been so damaged that they will be impossible to reclaim.

Agriculture is also weakening its own underpinnings by failing to conserve species diversity. By the year 2050, some 40,000 more plant species may become extinct. The United Nations' Food and Agriculture Organization attributes many of the losses, which are already occurring daily, to modern farming practices, as well as to population growth. The increasing uniformity of global eating habits is also having an effect. Wheat, rice, and maize provide more than half of the food energy around the world; only another two dozen crops provide the remainder. As people everywhere eat the same limited array of foods, local regions' native, genetically diverse plants no longer seem worth preserving. Yet, in the future, as the climate warms and the earth changes, those may be the very plants that people will need for food sources. A wild species of corn that grows in a dry climate, for example, might contain the genetic information necessary to help make domestic corn resistant to drought. Highlight 19 offered several examples of how biotechnology is being used to improve food crops.

The culprits that attend the growing of crops—land clearing; irrigation; fertilizer, pesticide, and herbicide overuse; and loss of genetic diversity—have taken a tremendous toll on the earth. In short, our ways of producing foods are, for the most part, not sustainable.

From Family Farms to Agribusiness

In recent decades, U.S. agriculture has encountered serious economic problems: declining markets for U.S. farm products abroad as other countries have increased their agricultural production and exports, reduced economic ability of many countries to purchase imported grain, and increases in the cost of energy and other supplies needed to produce food domestically.[2] Many U.S. farmers, particularly those who specialize in export crops, have suffered heavy financial losses.

Some have been unable to pay their debts and have had to leave farming. From 1987 to 1997, more than 155,000 family farms were lost.[3] Others remain frustrated and in debt, threatened with poverty.[4]

Small family farms face competition not only from foreign producers, but from huge corporation-owned farms and ranches as well. Huge farms and ranches, collectively part of the massive food-producing enterprise called **agribusiness,** are more likely to engage in technology-intensive practices that use large machinery and large land areas. They tend to use little local labor, and the profits earned tend not to stay in local communities. U.S. agribusinesses located in foreign countries tend to hire local workers willing to work for much less than workers in the United States.

Agribusinesses also tend to place a higher priority on producing abundant, inexpensive food than on protecting soil, water, and biodiversity. As a consequence, they may overuse fertilizers and pesticides, overuse land at the cost of soil erosion, and use irrigation water wastefully. Their impacts are enormous. In an effort to compete with agribusinesses, small farmers, too, may be driven to adopt similar unsustainable practices.

Because of economies of scale, agribusinesses can price their products so low that consumers buy more from them than from smaller, local farms. Consequently, grocery stores can offer broccoli from Mexico, tomatoes from Florida, carrots from California, bananas from Central America, and pineapples from Hawaii at prices no local farmers could beat, even if they could grow those products in their climates. Roadside stands and farmers' markets may offer baskets of local fruits and vegetables, but sometimes less conveniently and at higher prices than many shoppers are willing to pay.

If food prices had to include the costs for pollution cleanup, water protection, and land restoration, the prices of the products produced unsustainably would be higher. If they included a living wage, education, and benefits for the migrant farm workers, they would be higher still.

Proposed Solutions

For each of the problems described above, solutions are being devised, and indeed, some are being put into practice. To fully employ **sustainable agriculture** techniques across the country will require some new learning. Sustainable agriculture is not one system but a set of practices that can be adapted to meet the particular needs of a local area. The first of these ideas, *alternative,* or *low-input, agriculture,* emphasizes careful use of natural processes wherever possible, rather than chemically intensive methods.

Low-Input Agriculture

One form of low-input agriculture is **integrated pest management.** Farmers using this system employ many techniques, such as crop rotation and natural predators, to control pests

rather than depending on heavy use of pesticides alone. Not all crops can grow reliably without pesticides, but many can. Table H20-1 contrasts low-input, sustainable agriculture methods with high-input, unsustainable methods. Many sustainable techniques are not really new, incidentally; they would be familiar to our great grandparents. Farmers today are rediscovering the benefits of old techniques as they adapt and experiment with them in the search for sustainable methods.

Low-input agriculture has some apparent disadvantages, but advantages offset them. For example, as chemical use falls, yields per acre also fall somewhat, but costs per acre also fall, so the return per acre may be the same as or greater than before. More money goes to farmers and less to the fuels, fertilizers, pesticides, and irrigation. The end result of such farming is to make both farmers and consumers better off financially and environmentally.

Low-input agriculture works. As the world's population grows, and its fertile land and clean water dwindle, the need to adopt sustainable agriculture and development around the globe grows urgent. More than 30,000 U.S. farmers are successfully using sustainable techniques such as those described in Table H20-1. They see it as a food production system that can indefinitely sustain a healthy food supply, restore soil and water resources, and revitalize farming communities, while reducing reliance on fossil fuels.

Precision Agriculture

An exciting development in agriculture is the application of computer technologies to food production. Through techniques collectively known as *precision agriculture,* farmers can adjust their management of soil and crops to meet the exact needs of each area of the farm. For example, if one section of a field needs more nitrogen while another needs more potassium, the farmer can program the computer to apply fertilizer in the needed proportions and in the appropriate amounts for each area. Likewise, pesticide application can be programmed to prescribed amounts. A system can be programmed to turn off the flow of pesticides in designated safety zones, avoiding areas too close to streams and other water sources, for example. Such controls offer environmental benefits.

Precision agriculture depends on a *global positioning satellite (GPS)* system. In a GPS system, satellites beam information about land positions and elevations of an area, such as a field, to receivers on a farm. The information is shown on a grid map that pinpoints various locations on the farm. Farmers can use the grid to target, within a meter's accuracy, areas that need specific treatments. They can then program computerized farm equipment to apply chemicals in varying amounts or to till the soil to different depths, for example. This system of precision farming allows workers to till deeply enough to prepare seedbeds properly and to apply enough chemicals to control weeds and pests, but not so much as to waste fuel, worsen erosion, or damage the environment. Finally, at harvest, a GPS system produces an ac-

TABLE H20-1 Agricultural Methods Compared

Unsustainable Methods	Sustainable Methods
• Grow the same crop repeatedly on the same patch of land. This takes more and more nutrients out of the soil, making fertilizer use necessary; favors soil erosion; and invites weeds and pests to become established, making pesticide use necessary.	• Rotate crops. This increases nitrogen in the soil so there is less need to use fertilizers. If used with appropriate plowing methods, rotation reduces soil erosion. Rotation also reduces problems caused by weeds and pests.
• Use fertilizers generously. Excess fertilizer pollutes ground and surface water and costs both farmers' household money and consumers' tax money.	• Reduce the use of fertilizers, and use livestock manure more effectively. Store manure during the nongrowing season and apply it during the growing season. • Alternate nutrient-devouring crops with nutrient-restoring crops, such as legumes. • Compost on a large scale, including all plant residues not harvested. Plow the compost into the soil to improve its water-holding capacity.
• Feed livestock in feedlots where their manure produces a major water pollution probem. Piled in heaps, it also releases methane, a global-warming gas.	• Feed livestock or buffalo on the open range where their manure will fertilize the ground on which plants grow and will release no methane. Alternatively, at least collect feedlot animals' manure and use it for fertilizer or, at the very least, treat it before release.
• Spray herbicides and pesticides over large areas to wipe out weeds and pests.	• Apply ingenuity in weed and pest control. Use precision techniques if affordable or rotary hoes twice instead of herbicides once. Spot treat weeds by hand. • Rotate crops to foil pests that lay their eggs in the soil where last year's crop was grown. • Use resistant crops. • Use biological controls such as predators that destroy the pests.
• Plow the same way everywhere, allowing unsustainable water runoff and erosion.	• Plow in ways tailored to different areas. Conserve both soil and water by using cover crops, crop rotation, no-till planting, and contour plowing.
• Inject animals with antibiotics to prevent disease in livestock.	• Maintain animals' health so that they can resist disease.
• Irrigate on a large scale. Irrigation depletes water supplies and concentrates salts in the soil.	• Irrigate only during dry spells and apply only spot irrigation.

curate accounting of crop yields, acre by acre, so that spot adjustments may be made in the next planting season.

The future of precision agriculture seems bright, and the potential savings to farmers in terms of water, fertilizers, and pesticides are enormous. The accompanying reduction in polluting chemicals introduced into the environment means that everyone benefits.

Agricultural Biotechnology

Although not every farmer worldwide may be in a position to reap benefits from the technologies of precision agriculture, the advances of biotechnology may one day reach around the world. Biotechnology promises economic, environmental, and agricultural benefits by shrinking the acreage needed for crops, reducing soil losses, minimizing use of chemical pesticides, and improving crop protection. Salt-resistant and drought-resistant crops that can grow under stressful conditions may one day thrive in what is now barren land. Genetically modified cereals and legumes may deliver complete proteins to populations who now lack quality protein in their diets. Genetically engineered microbes can continuously re-

new soil structure by fixing nitrogen and releasing other nutrients into the soil, thereby reducing the need for chemical fertilizers and easing the environmental burden.[5] Similarly, microbes may be designed to recycle agricultural, industrial, and household wastes into fertilizers, an obvious boon to the environment. Highlight 19 provides a full discussion of food biotechnology and describes both its potential contributions and possible calamities.

Energy-Efficient Agriculture

Table H20-2 (on p. 712) lists several ways to save energy in the production of food. The last item in the table suggests that consumers should eat low on the food chain, which means choosing plants over meats most of the time, as described next.

Consumer Choices

Consumers can reduce the amount of energy used in food production by centering their diets on foods that require little energy to produce. For the most part, that means eating more

TABLE H20-2 Sustainable Energy-Saving Agricultural Techniques

- Use machinery scaled to the job at hand and operate it at efficient speeds.
- Combine operations. Harrow, plant, and fertilize in the same operation.
- Use diesel fuel. Use solar and wind energy on farms. Use methane from manure. Be open-minded to alternative energy sources.
- Save on technological and chemical inputs and spend some of the savings paying people to do manual jobs. Increasing labor has been considered inefficient. Reverse this thinking: creating more jobs is preferable to using more machinery and fuel.
- Partially return to the techniques of using animal manure and crop rotation. This will save energy because chemical fertilizers require much energy to produce.
- Choose crops that require low energy (fertilizer, pesticides, irrigation).
- Educate people to cook food efficiently and to eat low on the food chain.

foods derived from plants and fewer foods derived from animals. It also means eating more foods grown locally and fewer foods produced elsewhere.[6]

Plant versus Animal Some foods require more energy for their production than others. The least energy is needed for grains: it takes about one-third kcalorie of fuel to produce each kcalorie of grain. Fruits and vegetables are intermediate, and most animal-derived foods require from 10 to 90 kcalories of fuel per kcalorie of edible food. In general, meat-based diets require much more energy, as well as more land and water, than do plant-based diets.[7] An exception is livestock raised on the open range; these animals require about as much energy as most plant foods. We raise so much more grain-fed, than range-fed, livestock, however, that the average energy requirement for meat production is high. Figure H20-2 shows how much less fuel vegetarian diets require than meat-based diets and shows that vegan diets require the least fuel of all.

To support our meat intake, we maintain several billion livestock, about four times our own weight in animals. Livestock consume ten times as much grain each day as we do. We could use much of that grain to make grain products for ourselves and for others around the world. The shift could free up enough grain to feed 400 million people while using less fuel, water, and land.

Part of the solution to the livestock problem may be to cease feeding grain to animals and return to grazing them on the open range, which can be a sustainable practice. Ranchers have to manage the grazing carefully to hold the cattle's numbers to what the land can support without environmental degradation. To accomplish this, the economic benefits of traditional livestock and feed-growing operations would have to end. If producers were to pay the true costs of the environmental damage incurred by irrigation water, fertilizers, pesticides, and fuels, the prices of meats might double or triple. According to classic economic theory, people would then buy less meat (reducing demand), and producers would respond by producing less meat (reducing supply). Meat production would then fall to a sustainable level.

The United Nations describes a nation's impact on the environment as its "ecological footprint"—a measure of the resources used to support a nation's consumption of food, materials, and energy. This measure takes into account the two most challenging aspects of sustainability—per capita resource consumption and population growth. As Figure H20-3 shows, the people of North America are the world's greatest consumers on a per capita basis. Some have estimated that it would take four more planet Earths to accommodate every person in the world using resources at the level used in the United States.[8]

FIGURE H20-2 Amounts of Fuel Required to Feed People Eating at Different Points on the Food Chain

Three people who eat differently are compared here. Each has the same energy intake: 3300 kcalories a day. The fossil fuel amounts necessary to produce these different diets are calculated based on U.S. conditions.

The meat eater consumes a typical U.S. diet of meat, other animal products, and plant foods:

Meat and animal products 2000 kcal — Plant foods 1300 kcal

OIL — 33,900 kcal

Fuel required to produce this food

The lacto-ovo-vegetarian eats a diet that excludes meats, but includes milk products and eggs:

Animal products 1000 kcal — Plant foods 2300 kcal

OIL — 18,900 kcal

Fuel required to produce this food

The vegan eats a diet of plant foods only:

Plant foods 3300 kcal

9,900 kcal

Fuel required to produce this food

SOURCE: Adapted from D. Pimentel, *Food, Energy and the Future of Society* (Boulder, Colo.: Associated University Press, 1980), Figure 5, p. 27.

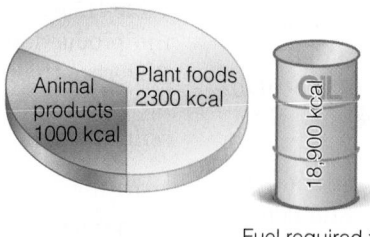
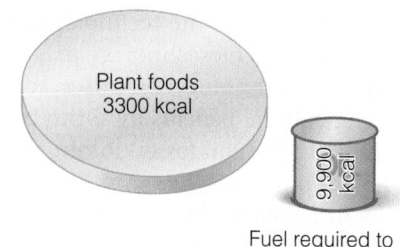

FIGURE H20-3 Ecological Footprints

The width of a bar represents the region's population, and the height represents per capita consumption (in terms of area of productive land or sea required to produce the natural resources consumed). Thus the footprint of the bar represents the region's total consumption. For example, Asia's population is over ten times greater than North America's, but its consumption is only one-sixth as large; thus their footprints are similar in size.

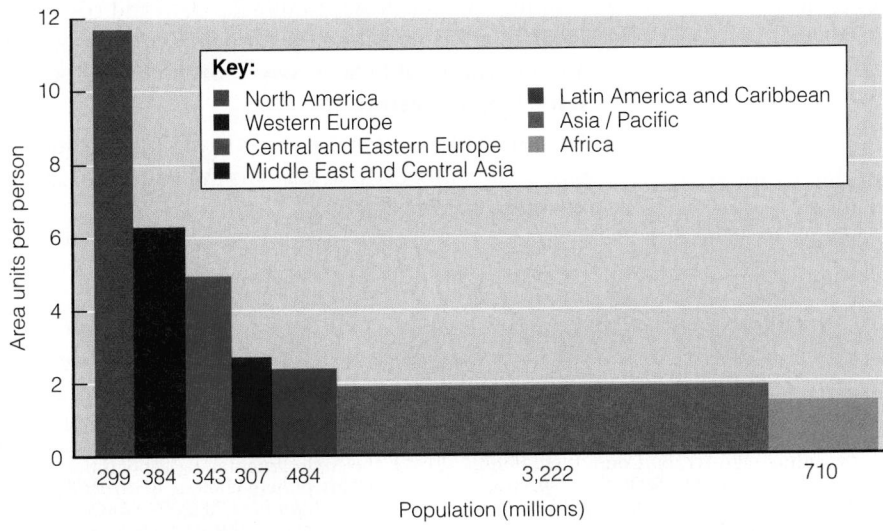

Key:
- North America
- Western Europe
- Central and Eastern Europe
- Middle East and Central Asia
- Latin America and Caribbean
- Asia / Pacific
- Africa

Y-axis: Area units per person (0–12)
X-axis: Population (millions) — 299, 384, 343, 307, 484, 3,222, 710

SOURCE: United Nations Population Fund, **www.unfpa.org/swp/2001/english/ch03.html**, visited June 2003.

Some consumers are taking action to do their part to solve some of these problems. Some are choosing smaller portions of meat or selecting range-fed beef or buffalo only. Livestock on the range eat grass, which people cannot eat. "Range-burger" buffalo also offers nutrient advantages over grain-fed beef because it is lower in fat, and the fat has more polyunsaturated fatty acids, including the omega-3 type.

Some consumers are opting for vegetarian, and even vegan, diets—at least occasionally. Vegetarian diets have less of an environmental impact than meat-based diets.[9] Shifting to a fish diet does not appear to be a practical alternative at present, although fish farming shows promise of providing nutritious food at a price both people and the environment can afford.

Local versus Global Plant-centered diets have an environmental advantage over meat-based diets, but some would argue that they don't go far enough.[10] The most ecologically responsible diets are also based on locally grown products. That "our foods now travel more than we do" has several costly ramifications. Buying *globally* is:

- *Energetically costly.* Foods must be refrigerated and transported thousands of miles to provide a full array of all produce all year round.
- *Socially unjust.* Farmers in impoverished countries, where the people are malnourished, are paid meager wages to grow food for wealthy nations.

- *Economically unwise.* It supports agribusinesses that buy land and labor cheaply in foreign countries instead of supporting local farmers raising crops in our communities.
- *Biologically risky.* Highly perishable foods are shipped from countries with unsafe drinking water and sanitation practices.

For all these reasons, consumers can best improve the global environment by buying locally. Adopting a local diet presents a bit of a challenge at first, especially when local fruits and vegetables are "out of season." But a nutritionally balanced diet of delicious foods is quite possible with a little creative planning.

Chapter 20 and this highlight have presented many problems and have suggested that, although many of the problems are global in scope, the solutions depend on the actions of individual people at the local level. On learning of this, concerned people may take a perfectionist attitude, believing that they "should" be doing more than they realistically can, and so feel defeated. Yet, striving for perfection, even while falling short, is a way to achieve progress

Locally grown foods offer benefits to both the local economy and the global evironment.

© Paul Barton/CORBIS

well worth celebrating. A positive attitude can bring about improvement, and sometimes improvement is enough. Celebrate the changes that are possible today by making them a permanent part of your life; do the same with changes that become possible tomorrow and every day thereafter. The results may surprise you.

NUTRITION ON THE NET

Access these websites for further study of topics covered in this highlight.

- Find updates and quick links to these and other nutrition-related sites at our website: **www.wadsworth.com/nutrition**

- Visit the USDA Alternative Farming Systems Information Center: **www.nal.usda.gov/afsic**

- Visit the Sustainable Agriculture Research and Education Program at UC Davis and the Leopold Center for Sustainable Agriculture at IA State: **www.sarep.ucdavis.edu** and **www.leopold.iastate.edu**

- Find discussions on major issues involving the quality of life on planet Earth at the Turning Point Project site: **www.turnpoint.org**

REFERENCES

1. Source Water Protection Practices Bulletin: Managing livestock, poultry, and horse waste to prevent contamination of drinking water, www.epa.gov/OGWDW/protect/pdfs/livestock.pdf, posted July 2001, visited June 2003.
2. USDA Interagency Agricultural Projections Committee, *USDA Agricultural Baseline Projections to 2008*, 1999, available at usda.mannlib.cornell.edu.
3. Sustainable agriculture: Definitions and terms, www.nalusda.gov/afsic/AFSIC_pubs/srb9902.htm, posted September 1999, visited June 2003.

4. USDA National Agricultural Statistics Service, *Farms and Land in Farms, 1998* (Washington, D.C.: USDA Interagency Agricultural Projections Committee, 1999).
5. R. P. Tengerdy and G. Szakacs, Perspectives in agrobiotechnology, *Journal of Biotechnology* 66 (1998): 91–99.
6. J. D. Gussow, Is local vs. global the next environmental imperative? *Nutrition Today* 35 (2000): 29–35.
7. D. Pimentel and M. Pimentel, Sustainability of meat-based and plant-based diets and the environment, *American Journal of Clinical Nutrition* 78 (2003): 660S–663S.

8. E. O. Wilson, The bottleneck, *Scientific American,* February 2002, pp. 82–91.
9. L. Reijnders and S. Soret, Quantification of the environmental impact of different dietary protein choices, *American Journal of Clinical Nutrition* 78 (2003): 664S–668S; C. Leitzmann, Nutrition ecology: The contribution of vegetarian diets, *American Journal of Clinical Nutrition* 78 (2003): 657S–659S.
10. Gussow, 2000.

Appendixes

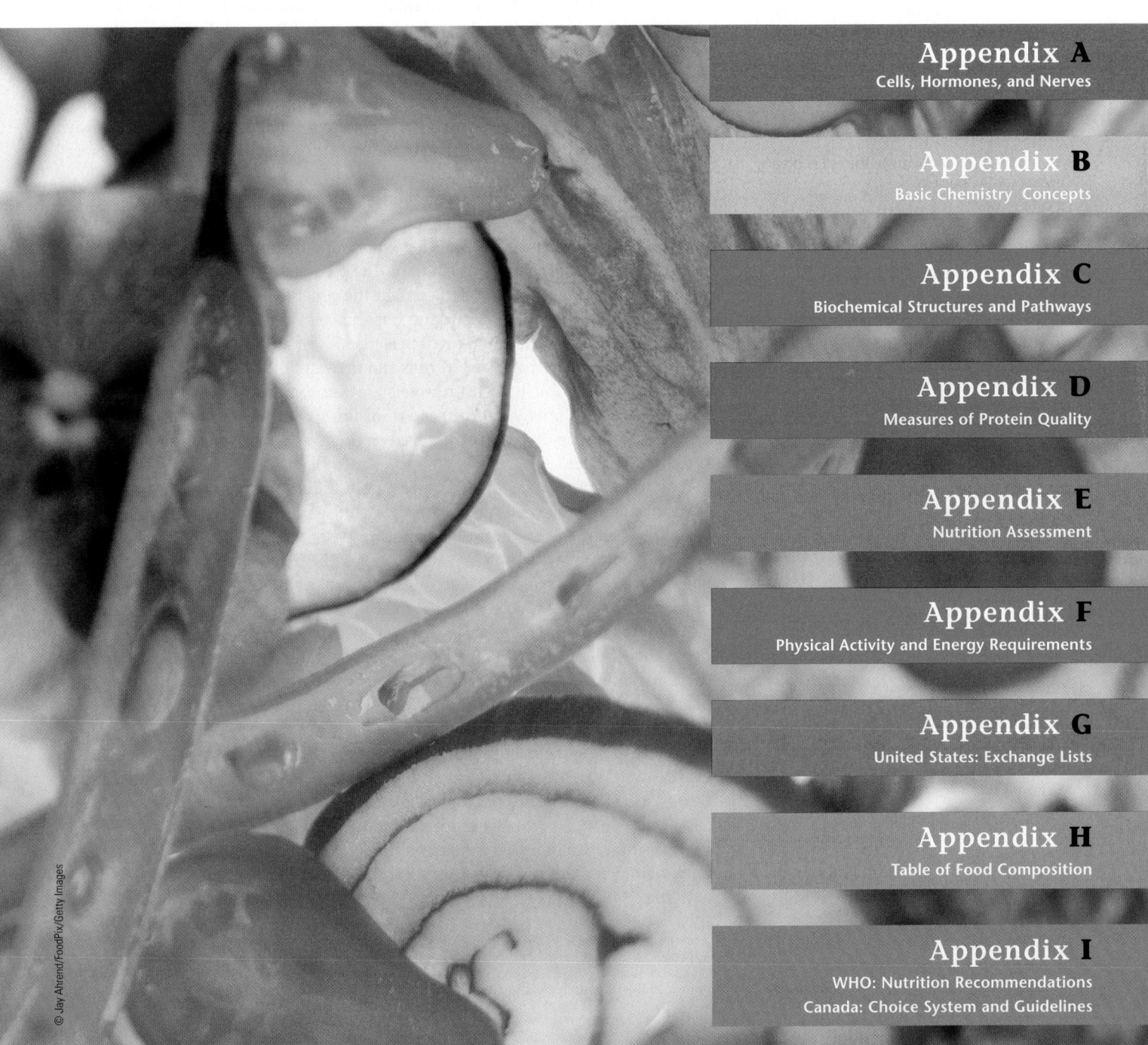

© Jay Ahrend/FoodPix/Getty Images

Appendix A

GLOSSARY OF CELL STRUCTURES

cell: the basic structural unit of all living things.

cell membrane: the thin layer of tissue that surrounds the cell and encloses its contents; made primarily of lipid and protein.

chromosomes: a set of structures within the nucleus of every cell that contains the cell's genetic material, DNA, associated with other materials (primarily proteins).

cytoplasm (SIGH-toh-plazm): the cell contents, except for the nucleus.
- **cyto** = cell
- **plasm** = a form

cytosol: the fluid of cytoplasm; contains water, ions, nutrients, and enzymes.

endoplasmic reticulum (en-doh-PLAZ-mic reh-TIC-you-lum): a complex network of intracellular membranes. The **rough endoplasmic reticulum** is dotted with ribosomes, where protein synthesis takes place. The **smooth endoplasmic reticulum** bears no ribosomes.
- **endo** = inside
- **plasm** = the cytoplasm

Golgi (GOAL-gee) **apparatus:** a set of membranes within the cell where secretory materials are packaged for export.

lysosomes (LYE-so-zomes): cellular organelles; membrane-enclosed sacs of degradative enzymes.
- **lysis** = dissolution

mitochondria (my-toh-KON-dree-uh); singular **mitochondrion:** the cellular organelles responsible for producing ATP aerobically; made of membranes (lipid and protein) with enzymes mounted on them.
- **mitos** = thread (referring to their slender shape)
- **chondros** = cartilage (referring to their external appearance)

nucleus: a major membrane-enclosed body within every cell, which contains the cell's genetic material, DNA, embedded in chromosomes.
- **nucleus** = a kernel

organelles: subcellular structures such as ribosomes, mitochondria, and lysosomes.
- **organelle** = little organ

Cells, Hormones, and Nerves

This appendix is offered as an optional chapter for readers who want to enhance their understanding of how the body coordinates its activities. The text presents a brief summary of the structure and function of the body's basic working unit (the cell) and of the body's two major regulatory systems (the hormonal system and the nervous system).

The Cell

The body's organs are made up of millions of cells and of materials produced by them. Each **cell** is specialized to perform its organ's functions, but all cells have common structures (see the accompanying glossary and Figure A-1). Every cell is contained within a **cell membrane.** The cell membrane assists in moving materials into and out of the cell, and some of its special proteins act as "pumps" (described in Chapter 6). Some features of cell membranes, such as microvilli (Chapter 3), permit cells to interact with other cells and with their environments in highly specific ways.

Inside the membrane lies the **cytoplasm,** which is filled with **cytosol,** or cell "fluid." The cytoplasm contains much more than just fluid, though. It is a highly organized system of fibers, tubes, membranes, particles, and subcellular **organelles** as complex as a city. These parts intercommunicate, manufacture and exchange materials, package and prepare materials for export, and maintain and repair themselves.

Within each cell is another membrane-enclosed body, the **nucleus.** Inside the nucleus are the **chromosomes,** which contain the genetic material, DNA. The DNA encodes all the instructions for carrying out the cell's activities. The role of DNA in coding for cell proteins is summarized in Figure 6-7 on p. 188. Chapter 6 also describes the variety of proteins produced by cells and the ways they perform the body's work.

Among the organelles within a cell are ribosomes, mitochondria, and lysosomes. Figure 6-7 briefly refers to the **ribosomes;** they assemble amino acids into proteins, following directions conveyed to them by RNA copies from the DNA in the chromosomes.

The **mitochondria** are made of intricately folded membranes that bear thousands of highly organized sets of enzymes on their inner and outer surfaces. Mitochondria are crucial to energy metabolism, described in Chapter 7, and muscles conditioned to work aerobically are packed with them. Their presence is implied whenever the TCA cycle and electron transport chain are mentioned because the mitochondria house the needed enzymes.[*]

The **lysosomes** are membranes that enclose degradative enzymes. When a cell needs to self-destruct or to digest materials in its surroundings, its lysosomes free their enzymes. Lysosomes are active when tissue repair or remodeling is taking place—for example, in cleaning up infections, healing wounds, shaping embryonic organs, and remodeling bones.

Besides these and other cellular organelles, the cell's cytoplasm contains a highly organized system of membranes, the **endoplasmic reticulum.** The ribosomes may either float free in the cytoplasm or be mounted on these membranes. A membranous surface dotted with ribosomes looks speckled under the microscope and is called "rough" endoplasmic reticulum; such a surface without ribosomes is called "smooth." Some intracellular membranes are organized into tubules that collect cellular materials, merge with the cell membrane, and discharge their contents to the outside of the cell; these

---

[*]For the reactions of glycolysis, the TCA cycle, and the electron transport chain, see Chapter 7 and Appendix C. The reactions of glycolysis take place in the cytoplasm; the conversion of pyruvate to acetyl CoA takes place in the mitochondria, as do the TCA cycle and electron transport chain reactions. The mitochondria then release carbon dioxide, water, and ATP as their end products.

FIGURE A-1 The Structure of a Typical Cell

The cell shown might be one in a gland (such as the pancreas) that produces secretory products (enzymes) for export (to the intestine). The rough endoplasmic reticulum with its ribosomes produces the enzymes; the smooth reticulum conducts them to the Golgi region; the Golgi membranes merge with the cell membrane, where the enzymes can be released into the extracellular fluid.

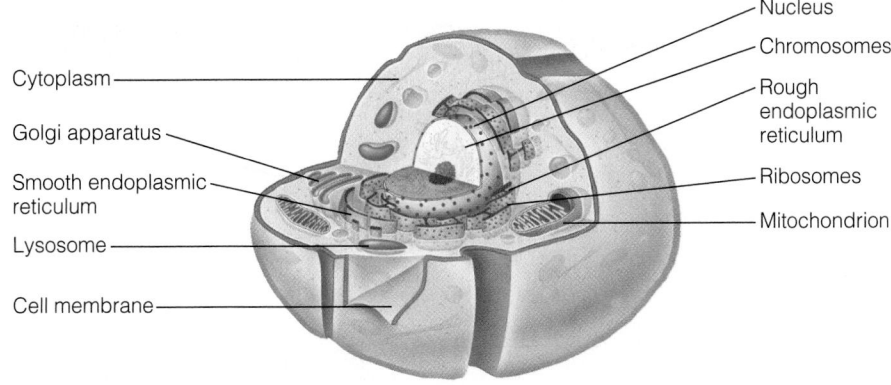

Cytoplasm

Golgi apparatus

Smooth endoplasmic reticulum

Lysosome

Cell membrane

Nucleus

Chromosomes

Rough endoplasmic reticulum

Ribosomes

Mitochondrion

- The study of hormones and their effects is **endocrinology**.

- The **pituitary gland** in the brain has two parts—the **anterior** (front) and the **posterior** (hind) parts.

ribosomes (RYE-boh-zomes): protein-making organelles in cells; composed of RNA and protein.
  - **ribo** = containing the sugar ribose (in RNA)
  - **some** = body

membrane systems are named the **Golgi apparatus,** after the scientist who first described them. The rough and smooth endoplasmic reticula and the Golgi apparatus are continuous with one another, so secretions produced deep in the interior of the cell can be efficiently transported to the outside and released. These and other cell structures enable cells to perform the multitudes of functions for which they are specialized.

The actions of cells are coordinated by both hormones and nerves, as the next sections show. Among the types of cellular organelles are receptors for the hormones delivering instructions that originate elsewhere in the body. Some hormones penetrate the cell and its nucleus and attach to receptors on chromosomes, where they activate certain genes to initiate, stop, speed up, or slow down synthesis of certain proteins as needed. Other hormones attach to receptors on the cell surface and transmit their messages from there. The hormones■ are described in the next section; the nerves, in the one following.

The Hormones

A chemical compound—a **hormone**—originates in a gland and travels in the bloodstream. The hormone flows everywhere in the body, but only its target organs respond to it, because only they possess the receptors to receive it.

The hormones, the glands they originate in, and their target organs and effects are described in this section. Many of the hormones you might be interested in are included, but only a few are discussed in detail. Figure A-2 identifies the glands that produce the hormones, and the accompanying glossary defines the hormones discussed in this section.

Hormones of the Pituitary Gland and Hypothalamus

The anterior pituitary gland■ produces the following hormones, each of which acts on one or more target organs and elicits a characteristic response:

- **Adrenocorticotropin (ACTH)** acts on the adrenal cortex, promoting the production and release of its hormones.

- **Thyroid-stimulating hormone (TSH)** acts on the thyroid gland, promoting the production and release of thyroid hormones.

- **Growth hormone (GH)** acts on all tissues, promoting growth, fat breakdown, and the formation of antibodies.

GLOSSARY OF HORMONES

adrenocorticotropin (ad-REE-noh-KORE-tee-koh-TROP-in) **ACTH:** a hormone, so named because it stimulates (trope) the adrenal cortex. The adrenal gland, like the pituitary, has two parts, in this case an outer portion (cortex) and an inner core (medulla). The realease of ACTH is mediated by **corticotropin-releasing hormone (CRH).**

aldosterone: a hormone from the adrenal gland involved in blood pressure regulation.
  - **aldo** = aldehyde

angiotensin: a hormone involved in blood pressure regulation that is activated by **renin** (REN-in), an enzyme from the kidneys.
  - **angio** = blood vessels
  - **tensin** = pressure
  - **ren** = kidneys

antidiuretic hormone (ADH): the hormone that prevents water loss in urine (also called **vasopressin**).
  - **anti** = against
  - **di** = through
  - **ure** = urine
  - **vaso** = blood vessels
  - **pressin** = pressure

calcitonin (KAL-see-TOH-nin): a hormone secreted by the thyroid gland that regulates (tones) calcium metabolism.

erythropoietin (eh-REE-throh-POY-eh-tin): a hormone that stimulates red blood cell production.
  - **erythro** = red (blood cell)
  - **poiesis** = creating (like poetry)

estrogens: hormones responsible for the menstrual cycle and other female characteristics.
  - **oestrus** = the egg-making cycle
  - **gen** = gives rise to

FIGURE A-2 The Endocrine System

These organs and glands release hormones that regulate body processes. An *endocrine gland* secretes its product directly into *(endo)* the blood; for example, the pancreas cells that produce insulin. An *exocrine gland* secretes its product(s) out *(exo)* to an epithelial surface either directly or through a duct; the sweat glands of the skin and the enzyme-producing glands of the pancreas are both examples. The pancreas is therefore both an endocrine and an exocrine gland.

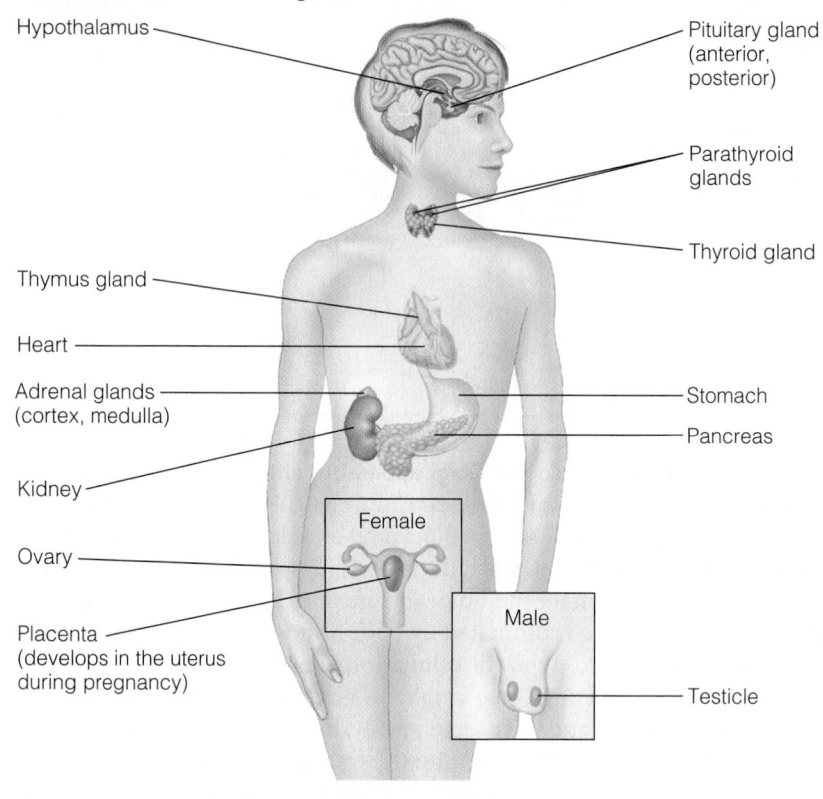

■ Hormones that are turned off by their own effects are said to be regulated by **negative feedback.**

follicle-stimulating hormone (FSH): a hormone that stimulates maturation of the ovarian follicles in females and the production of sperm in males. (The ovarian follicles are part of the female reproductive system where the eggs are produced.) The release of FSH is mediated by **follicle-stimulating hormone releasing hormone (FSH–RH).**

glucocorticoids: hormones from the adrenal cortex that affect the body's management of glucose.
- **gluco** = glucose
- **corticoid** = from the cortex

growth hormone (GH): a hormone secreted by the pituitary that regulates the cell division and protein synthesis needed for normal growth. The release of GH is mediated by **GH-releasing hormone (GHRH).**

hormone: a chemical messenger. Hormones are secreted by a variety of endocrine glands in response to altered conditions in the body. Each hormone travels to one or more specific target tissues or organs, where it elicits a specific response to maintain homeostasis.

luteinizing (LOO-tee-in-EYE-zing) hormone (LH): a hormone that stimulates ovulation and the development of the corpus luteum (the small tissue that develops from a ruptured ovarian follicle and secretes hormones); so called because the follicle turns yellow as it matures. In men, LH stimulates testosterone secretion. The release of LH is mediated by **luteinizing hormone–releasing hormone (LH–RH).**
- **lutein** = a yellow pigment

- **Follicle-stimulating hormone (FSH)** acts on the ovaries in the female, promoting their maturation, and on the testicles in the male, promoting sperm formation.

- **Luteinizing hormone (LH)** also acts on the ovaries, stimulating their maturation, the production and release of progesterone and estrogens, and ovulation; and on the testicles, promoting the production and release of testosterone.

- **Prolactin,** secreted in the female during pregnancy and lactation, acts on the mammary glands to stimulate their growth and the production of milk.

Each of these hormones has one or more signals that turn it on and another (or others) that turns it off.■ Among the controlling signals are several hormones from the hypothalamus:

- **Corticotropin-releasing hormone (CRH),** which promotes release of ACTH, is turned on by stress and turned off by ACTH when enough has been released.

- **TSH-releasing hormone (TRH),** which promotes release of TSH, is turned on by large meals or low body temperature.

- **GH-releasing hormone (GHRH),** which stimulates the release of growth hormone, is turned on by insulin.

- **GH-inhibiting hormone (GHIH** or **somatostatin),** which inhibits the release of GH and interferes with the release of TSH, is turned on by hypoglycemia and/or physical activity and is rapidly destroyed by body tissues so that it does not accumulate.

- **FSH/LH–releasing hormone (FSH/LH–RH)** is turned on in the female by nerve messages or low estrogen and in the male by low testosterone.
- **Prolactin-inhibiting hormone (PIH)** is turned on by high prolactin levels and off by estrogen, testosterone, and suckling (by way of nerve messages).

Let's examine some of these controls. PIH, for example, responds to high prolactin levels (remember, prolactin promotes milk production). High prolactin levels ensure that milk is made and—by calling forth PIH—ensure that prolactin levels don't get too high. But when the infant is suckling—and creating a demand for milk—PIH is not allowed to work (suckling turns off PIH). The consequence: prolactin remains high, and milk production continues. Demand from the infant thus directly adjusts the supply of milk. The need is met through the interaction of the nerves and hormones.

As another example, consider CRH. Stress, perceived in the brain and relayed to the hypothalamus, switches on CRH. On arriving at the pituitary, CRH switches on ACTH. Then ACTH acts on its target organ, the adrenal cortex, which responds by producing and releasing stress hormones. The stress hormones trigger a cascade of events involving every body cell and many other hormones.

The numerous steps required to set the stress response in motion make it possible for the body to fine-tune the response; control can be exerted at each step. These two examples illustrate what the body can do in response to two different stimuli—producing milk in response to an infant's need and gearing up for action in an emergency.

The posterior pituitary gland produces two hormones, each of which acts on one or more target cells and elicits a characteristic response:

- **Antidiuretic hormone (ADH),** or **vasopressin,** acts on the arteries, promoting their contraction, and on the kidneys, preventing water excretion. ADH is turned on whenever the blood volume is low, the blood pressure is low, or the salt concentration of the blood is high (see Chapter 12). It is turned off by the return of these conditions to normal.
- **Oxytocin** acts on the uterus, inducing contractions, and on the mammary glands, causing milk ejection. Oxytocin is produced in response to reduced progesterone levels, suckling, or the stretching of the cervix.

Hormones That Regulate Energy Metabolism

Hormones produced by a number of different glands have effects on energy metabolism:

- Insulin from the pancreas beta cells is turned on by many stimuli, including raised blood glucose. It acts on cells to increase glucose and amino acid uptake into them and to promote the secretion of GHRH.
- Glucagon from the pancreas alpha cells responds to low blood glucose and acts on the liver to promote the breakdown of glycogen to glucose, the conversion of amino acids to glucose, and the release of glucose.
- Thyroxin from the thyroid gland responds to TSH and acts on many cells to increase their metabolic rate, growth, and heat production.
- Norepinephrine and epinephrine■ from the adrenal medulla respond to stimulation by sympathetic nerves and produce reactions in many cells that facilitate the body's readiness for fight or flight: increased heart activity, blood vessel constriction, breakdown of glycogen and glucose, raised blood glucose levels, and fat breakdown. Norepinephrine and epinephrine also influence the secretion of the many hormones from the hypothalamus that exert control on the body's other systems.
- Growth hormone (GH) from the anterior pituitary (already mentioned).
- **Glucocorticoids** from the adrenal cortex become active during times of stress and carbohydrate metabolism.

■ Norepinephrine and epinephrine were formerly called **noradrenalin** and **adrenalin,** respectively.

oxytocin (OCK-see-TOH-sin): a hormone that stimulates the mammary glands to eject milk during lactation and the uterus to contract during childbirth.
- **oxy** = quick
- **tocin** = childbirth

progesterone: the hormone of gestation (pregnancy).
- **pro** = promoting
- **gest** = gestation (pregnancy)
- **sterone** = a steroid hormone

prolactin (proh-LAK-tin): a hormone so named because it promotes *(pro)* the production of milk *(lacto)*. The release of prolactin is mediated by **prolactin-inhibiting hormone (PIH).**

relaxin: the hormone of late pregnancy.

somatostatin (GHIH): a hormone that inhibits the release of growth hormone; the opposite of **somatotropin (GH).**
- **somato** = body
- **stat** = keep the same
- **tropin** = make more

testosterone: a steroid hormone from the testicles, or testes. The steroids, as explained in Chapter 5, are chemically related to, and some are derived from, the lipid cholesterol.
- **sterone** = a steroid hormone

thyroid-stimulating hormone (TSH): a hormone secreted by the pituitary that stimulates the thyroid gland to secrete its hormones—thyroxine and triiodothyronine. The release of TSH is mediated by **TSH-releasing hormone (TRH).**

Every body part is affected by these hormones. Each different hormone has unique effects; and hormones that oppose each other are produced in carefully regulated amounts, so each can respond to the exact degree that is appropriate to the condition.

Hormones That Adjust Other Body Balances

Hormones are involved in moving calcium into and out of the body's storage deposits in the bones:

- **Calcitonin** from the thyroid gland acts on the bones, which respond by storing calcium from the bloodstream whenever blood calcium rises above the normal range. It also acts on the kidneys to increase excretion of both calcium and phosphorus in the urine. Calcitonin plays a major role in infants and young children, but is less active in adults.

- Parathormone (parathyroid hormone or PTH) from the parathyroid gland responds to the opposite condition—lowered blood calcium—and acts on three targets: the bones, which release stored calcium into the blood; the kidneys, which slow the excretion of calcium; and the intestine, which increases calcium absorption.

- Vitamin D from the skin and activated in the kidneys acts with parathormone and is essential for the absorption of calcium in the intestine.

Figure 12-10 on p. 414 diagrams the ways vitamin D and the hormones calcitonin and parathormone regulate calcium homeostasis.

Another hormone has effects on blood-making activity:

- **Erythropoietin** from the kidneys is responsive to oxygen depletion of the blood and to anemia. It acts on the bone marrow to stimulate the making of red blood cells.

Another hormone is special for pregnancy:

- **Relaxin** from the ovaries is secreted in response to the raised progesterone and estrogen levels of late pregnancy. This hormone acts on the cervix and pelvic ligaments to allow them to stretch so that they can accommodate the birth process without strain.

Other agents help regulate blood pressure:

- **Renin** (an enzyme), from the kidneys, in cooperation with **angiotensin** in the blood responds to a reduced blood supply experienced by the kidneys and acts in several ways to increase blood pressure. Renin and angiotensin also stimulate the adrenal cortex to secrete the hormone aldosterone.

- **Aldosterone,** a hormone from the adrenal cortex, targets the kidneys, which respond by reabsorbing sodium. The effect is to retain more water in the bloodstream—thus, again, raising the blood pressure. Figure 12-3 (on p. 401) in Chapter 12 provides more details.

The Gastrointestinal Hormones

Several hormones are produced in the stomach and intestines in response to the presence of food or the components of food:

- Gastrin from the stomach and duodenum stimulates the production and release of gastric acid and other digestive juices and the movement of the GI contents through the system.

- Cholecystokinin from the duodenum signals the gallbladder and pancreas to release their contents into the intestine to aid in digestion.

- Secretin from the duodenum calls forth acid-neutralizing bicarbonate from the pancreas into the intestine and slows the action of the stomach and its secretion of acid and digestive juices.

- Gastric-inhibitory peptide from the duodenum and jejunum inhibits the secretion of gastric acid and slows the process of digestion.

These hormones are defined and presented in more detail in Chapter 3.

The Sex Hormones

There are three major sex hormones:

- **Testosterone** from the testicles is released in response to LH (described earlier) and acts on all the tissues that are involved in male sexuality, promoting their development and maintenance.

- **Estrogens** from the ovary are released in response to both FSH and LH and act similarly in females.

- **Progesterone** from the ovary's corpus luteum and from the placenta acts on the uterus and mammary glands, preparing them for pregnancy and lactation.

This brief description of the hormones and their functions should suffice to provide an awareness of the enormous impact these compounds have on body processes. The other overall regulating agency is the nervous system.

The Nervous System

The nervous system has a central control system—a sort of computer—that can evaluate information about conditions within and outside the body, and a vast system of wiring that receives information and sends instructions. The control unit is the brain and spinal cord, called the **central nervous system;** and the vast complex of wiring between the center and the parts is the **peripheral nervous system.** The smooth functioning that results from the system's adjustments to changing conditions is homeostasis.

The nervous system has two general functions: it controls voluntary muscles in response to sensory stimuli from them, and it controls involuntary, internal muscles and glands in response to nerve-borne and chemical signals about their status. In fact, the nervous system is best understood as two systems that use the same or similar pathways to receive and transmit their messages. The **somatic nervous system** controls the voluntary muscles; the **autonomic nervous system** controls the internal organs.

When scientists were first studying the autonomic nervous system, they noticed that when something hurt one organ of the body, some of the other organs reacted as if in sympathy for the afflicted one. They therefore named the nerve network they were studying the sympathetic nervous system. The term is still used today to refer to that branch of the autonomic nervous system that responds to pain and stress. The other branch is called the parasympathetic nervous system. (Think of the sympathetic branch as the responder when homeostasis needs restoring and the parasympathetic branch as the commander of function during normal times.) Both systems transmit their messages through the brain and spinal cord. Nerves of the two branches travel side by side along the same pathways to transmit their messages, but they oppose each other's actions (see Figure A-3).

An example will show how the sympathetic and parasympathetic nervous systems work to maintain homeostasis. When you go outside in cold weather, your skin's temperature receptors send "cold" messages to the spinal cord and brain. Your conscious mind may intervene at this point to tell you to zip your jacket, but let's say you have no jacket. Your sympathetic nervous system reacts to the external stressor, the cold. It signals your skin-surface capillaries to shut down so that your blood will circulate deeper in your tissues, where it will conserve heat. Your sympathetic nervous system also signals involuntary contractions of the small muscles just under the skin surface. The product of these muscle contractions is heat, and the visible result is goose bumps. If these measures do not raise your body temperature enough, then the sympathetic nerves signal your large muscle groups to shiver; the contractions of these large muscles produce still more heat. All of this activity adds up to a set of adjustments that maintain your homeostasis (with respect

A Appendix

GLOSSARY OF THE NERVOUS SYSTEM

autonomic nervous system: the division of the nervous system that controls the body's automatic responses. Its two branches are the **sympathetic** branch, which helps the body respond to stressors from the outside environment, and the **parasympathetic** branch, which regulates normal body activities between stressful times.

- **autonomos** = self-governing

central nervous system: the central part of the nervous system; the brain and spinal cord.

peripheral (puh-RIFF-er-ul) **nervous system:** the peripheral (outermost) part of the nervous system; the vast complex of wiring that extends from the central nervous system to the body's outermost areas. It contains both somatic and autonomic components.

somatic (so-MAT-ick) **nervous system:** the division of the nervous system that controls the voluntary muscles, as distinguished from the autonomic nervous system, which controls involuntary functions.

- **soma** = body

FIGURE A-3 The Organization of the Nervous System

The brain and spinal cord evaluate information about conditions within and outside the body, and the peripheral nerves receive information and send instructions.

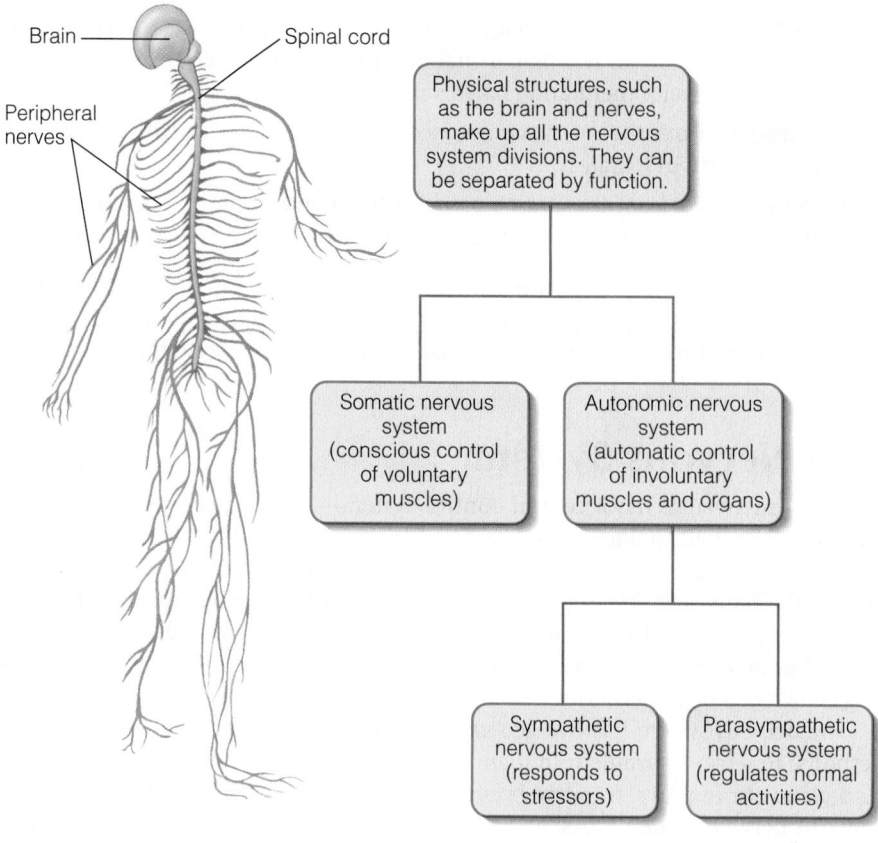

Brain — Spinal cord

Peripheral nerves

Physical structures, such as the brain and nerves, make up all the nervous system divisions. They can be separated by function.

Somatic nervous system (conscious control of voluntary muscles)

Autonomic nervous system (automatic control of involuntary muscles and organs)

Sympathetic nervous system (responds to stressors)

Parasympathetic nervous system (regulates normal activities)

to temperature) under conditions of external extremes (cold) that would throw it off balance. The cold was a stressor; the body's response was resistance.

Now let's say you come in and sit by a fire and drink hot cocoa. You are warm and no longer need all that sympathetic activity. At this point, your parasympathetic nerves take over; they signal your skin-surface capillaries to dilate again, your goose bumps to subside, and your muscles to relax. Your body is back to normal. This is recovery.

Putting It Together

The hormonal and nervous systems coordinate body functions by transmitting and receiving messages. The point-to-point messages of the nervous system travel through a central switchboard (the spinal cord and brain), whereas the messages of the hormonal system are broadcast over the airways (the bloodstream), and any organ with the appropriate receptors can pick them up. Nerve impulses travel faster than hormonal messages do—although both are remarkably swift. Whereas your brain's command to wiggle your toes reaches the toes within a fraction of a second and stops as quickly, a gland's message to alter a body condition may take several seconds or minutes to get started and may fade away equally slowly.

Together, the two systems possess every characteristic a superb communication network needs: varied speeds of transmission, along with private communication lines or public broadcasting systems, depending on the needs of the moment. The hormonal system, together with the nervous system, integrates the whole body's functioning so that all parts act smoothly together.

Basic Chemistry Concepts

This appendix is intended to provide the background in basic chemistry you need to understand the nutrition concepts presented in this book. Chemistry is the branch of natural science that is concerned with the description and classification of **matter**, the changes that matter undergoes, and the **energy** associated with these changes. The accompanying glossary defines matter, energy, and other related terms.

Matter: The Properties of Atoms

Every substance has physical and chemical properties that distinguish it from all other substances and thus give it a unique identity. The physical properties include such characteristics as color, taste, texture, and odor, as well as the temperatures at which a substance changes its state (from a solid to a liquid or from a liquid to a gas) and the weight of a unit volume (its density). The chemical properties of a substance have to do with how it reacts with other substances or responds to a change in its environment so that new substances with different sets of properties are produced.

A physical change does not change a substance's chemical composition. The three physical states—ice, water, and steam—all consist of two hydrogen atoms and one oxygen atom bound together. In contrast, a chemical change occurs when an electric current passes through water. The water disappears, and two different substances are formed: hydrogen gas, which is flammable, and oxygen gas, which supports life.

Substances: Elements and Compounds

The smallest part of a substance that can exist separately without losing its physical and chemical properties is a **molecule.** If a molecule is composed of **atoms** that are alike, the substance is an **element** (for example, O_2). If a molecule is composed of two or more different kinds of atoms, the substance is a **compound** (for example, H_2O).

Just over 100 elements are known, and these are listed in Table B-1. A familiar example is hydrogen, whose molecules are composed only of hydrogen atoms linked together in pairs (H_2). On the other hand, over a million compounds are known. An example is the sugar glucose. Each of its molecules is composed of 6 carbon, 6 oxygen, and 12 hydrogen atoms linked together in a specific arrangement (as described in Chapter 4).

The Nature of Atoms

Atoms themselves are made of smaller particles. Within the atomic nucleus are protons (positively charged particles), and surrounding the nucleus are electrons (negatively charged particles). The number of protons (+) in the nucleus of an

GLOSSARY

atoms: the smallest components of an element that have all of the properties of the element.

compound: a substance composed of two or more different atoms—for example, water (H_2O).

element: a substance composed of atoms that are alike—for example, iron (Fe).

energy: the capacity to do work.

matter: anything that takes up space and has mass.

molecule: two or more atoms of the same or different elements joined by chemical bonds. Examples are molecules of the element oxygen, composed of two oxygen atoms (O_2), and molecules of the compound water, composed of two hydrogen atoms and one oxygen atom (H_2O).

atom determines the number of electrons (−) around it. The positive charge on a proton is equal to the negative charge on an electron, so the charges cancel each other out and leave the atom neutral to its surroundings.

The nucleus may also include neutrons, subatomic particles that have no charge. Protons and neutrons are of equal mass, and together they give an atom its weight. Electrons bond atoms together to make molecules, and they are involved in chemical reactions.

Each type of atom has a characteristic number of protons in its nucleus. The hydrogen atom is the simplest of all. It possesses a single proton, with a single electron associated with it:

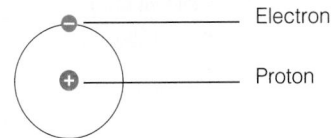

Hydrogen atom (H), atomic number 1.

Just as hydrogen always has one proton, helium always has two, lithium three, and so on. The atomic number of each element is the number of protons in the nucleus of that atom, and this never changes in a chemical reaction; it gives the atom its identity. The atomic numbers for the known elements are listed in Table B-1.

TABLE B-1   Chemical Symbols for the Elements

Number of Protons (Atomic Number)	Element	Number of Electrons in Outer Shell	Number of Protons (Atomic Number)	Element	Number of Electrons in Outer Shell
1	Hydrogen (H)	1	52	Tellurium (Te)	6
2	Helium (He)	2	53	Iodine (I)	7
3	Lithium (Li)	1	54	Xenon (Xe)	8
4	Beryllium (Be)	2	55	Cesium (Cs)	1
5	Boron (B)	3	56	Barium (Ba)	2
6	Carbon (C)	4	57	Lanthanum (La)	2
7	Nitrogen (N)	5	58	Cerium (Ce)	2
8	Oxygen (O)	6	59	Praseodymium (Pr)	2
9	Fluorine (F)	7	60	Neodymium (Nd)	2
10	Neon (Ne)	8	61	Promethium (Pm)	2
11	Sodium (Na)	1	62	Samarium (Sm)	2
12	Magnesium (Mg)	2	63	Europium (Eu)	2
13	Aluminum (Al)	3	64	Gadolinium (Gd)	2
14	Silicon (Si)	4	65	Terbium (Tb)	2
15	Phosphorus (P)	5	66	Dysprosium (Dy)	2
16	Sulfur (S)	6	67	Holmium (Ho)	2
17	Chlorine (Cl)	7	68	Erbium (Er)	2
18	Argon (Ar)	8	69	Thulium (Tm)	2
19	Potassium (K)	1	70	Ytterbium (Yb)	2
20	Calcium (Ca)	2	71	Lutetium (Lu)	2
21	Scandium (Sc)	2	72	Hafnium (Hf)	2
22	Titanium (Ti)	2	73	Tantalum (Ta)	2
23	Vanadium (V)	2	74	Tungsten (W)	2
24	Chromium (Cr)	1	75	Rhenium (Re)	2
25	Manganese (Mn)	2	76	Osmium (Os)	2
26	Iron (Fe)	2	77	Iridium (Ir)	2
27	Cobalt (Co)	2	78	Platinum (Pt)	1
28	Nickel (Ni)	2	79	Gold (Au)	1
29	Copper (Cu)	1	80	Mercury (Hg)	2
30	Zinc (Zn)	2	81	Thallium (Tl)	3
31	Gallium (Ga)	3	82	Lead (Pb)	4
32	Germanium (Ge)	4	83	Bismuth (Bi)	5
33	Arsenic (As)	5	84	Polonium (Po)	6
34	Selenium (Se)	6	85	Astatine (At)	7
35	Bromine (Br)	7	86	Radon (Rn)	8
36	Krypton (Kr)	8	87	Francium (Fr)	1
37	Rubidium (Rb)	1	88	Radium (Ra)	2
38	Strontium (Sr)	2	89	Actinium (Ac)	2
39	Yttrium (Y)	2	90	Thorium (Th)	2
40	Zirconium (Zr)	2	91	Protactinium (Pa)	2
41	Niobium (Nb)	1	92	Uranium (U)	2
42	Molybdenum (Mo)	1	93	Neptunium (Np)	2
43	Technetium (Tc)	1	94	Plutonium (Pu)	2
44	Ruthenium (Ru)	1	95	Americium (Am)	2
45	Rhodium (Rh)	1	96	Curium (Cm)	2
46	Palladium (Pd)	—	97	Berkelium (Bk)	2
47	Silver (Ag)	1	98	Californium (Cf)	2
48	Cadmium (Cd)	2	99	Einsteinium (Es)	2
49	Indium (In)	3	100	Fermium (Fm)	2
50	Tin (Sn)	4	101	Mendelevium (Md)	2
51	Antimony (Sb)	5	102	Nobelium (No)	2
			103	Lawrencium (Lr)	2

Key:

Elements found in energy-yielding nutrients, vitamins, and water.

Major minerals.

Trace minerals.

Besides hydrogen, the atoms most common in living things are carbon (C), nitrogen (N), and oxygen (O), whose atomic numbers are 6, 7, and 8, respectively. Their structures are more complicated than that of hydrogen, but each of them possesses the same number of electrons as there are protons in the nucleus. These electrons are found in orbits, or shells (shown below).

Carbon atom (C), atomic number 6. | Nitrogen atom (N), atomic number 7. | Oxygen atom (O), atomic number 8.

In these and all diagrams of atoms that follow, only the protons and electrons are shown. The neutrons, which contribute only to atomic weight, not to charge, are omitted.

The most important structural feature of an atom for determining its chemical behavior is the number of electrons in its outermost shell. The first, or innermost, shell is full when it is occupied by two electrons; so an atom with two or more electrons has a filled first shell. When the first shell is full, electrons begin to fill the second shell.

The second shell is completely full when it has eight electrons. A substance that has a full outer shell tends not to enter into chemical reactions. Atomic number 10, neon, is a chemically inert substance because its outer shell is complete. Fluorine, atomic number 9, has a great tendency to draw an electron from other substances to complete its outer shell, and thus it is highly reactive. Carbon has a half-full outer shell, which helps explain its great versatility; it can combine with other elements in a variety of ways to form a large number of compounds.

Atoms seek to reach a state of maximum stability or of lowest energy in the same way that a ball will roll down a hill until it reaches the lowest place. An atom achieves a state of maximum stability:

- By gaining or losing electrons to either fill or empty its outer shell.
- By sharing its electrons through bonding together with other atoms and thereby completing its outer shell.

The number of electrons determines how the atom will chemically react with other atoms. The atomic number, not the weight, is what gives an atom its chemical nature.

Chemical Bonding

Atoms often complete their outer shells by sharing electrons with other atoms. In order to complete its outer shell, a carbon atom requires four electrons. A hydrogen atom requires one. Thus, when a carbon atom shares electrons with four hydrogen atoms, each completes its outer shell (as shown in the next column). Electron sharing binds the atoms together and satisfies the conditions of maximum stability for the molecule. The outer shell of each atom is complete, since hydrogen effectively has the required two electrons in its first (outer)

shell, and carbon has eight electrons in its second (outer) shell; and the molecule is electrically neutral, with a total of ten protons and ten electrons.

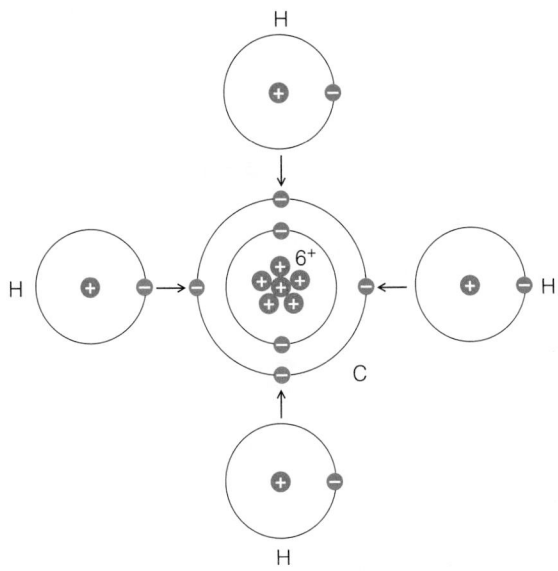

When a carbon atom shares electrons with four hydrogen atoms, a methane molecule is made.

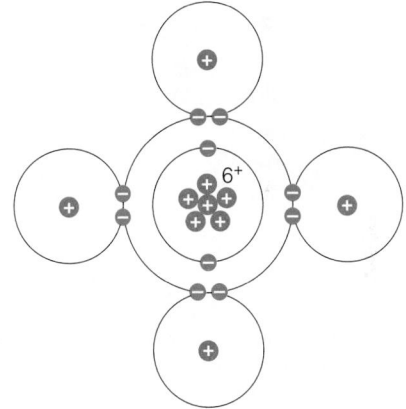

The chemical formula for methane is CH_4. Note that by sharing electrons, every atom achieves a filled outer shell.

Bonds that involve the sharing of electrons, like the bonds between carbon and the four hydrogens, are the most stable kind of association that atoms can form with one another. These bonds are called covalent bonds, and the resulting combination of atoms are called molecules. A single pair of shared electrons forms a single bond. A simplified way to represent a single bond is with a single line. Thus the structure of methane (CH_4) could be represented like this:

$$\begin{array}{c} \text{H} \\ | \\ \text{H} - \text{C} - \text{H} \\ | \\ \text{H} \end{array}$$

Methane (CH_4).

Similarly, one nitrogen atom and three hydrogen atoms can share electrons to form one molecule of ammonia (NH_3):

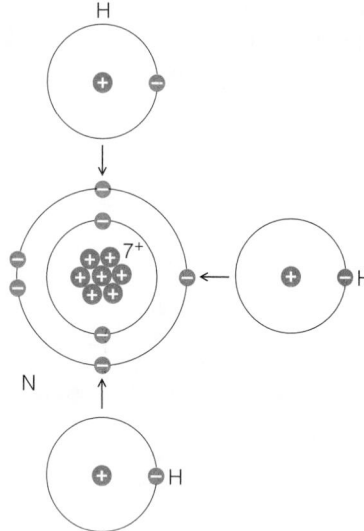

When a nitrogen atom shares electrons with three hydrogen atoms, an ammonia molecule is made.

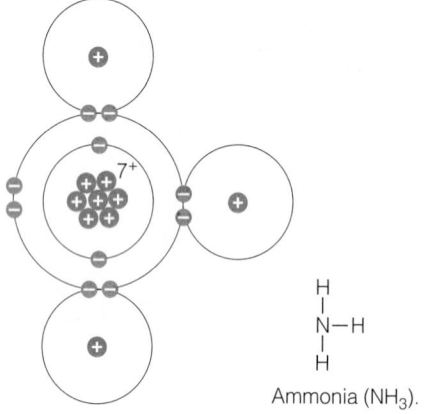

H
|
N—H
|
H

Ammonia (NH₃).

The chemical formula for ammonia is NH₃. Count the electrons in each atom's outer shell to confirm that it is filled.

One oxygen atom may be bonded to two hydrogen atoms to form one molecule of water (H₂O):

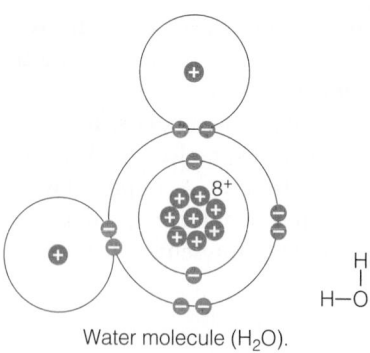

H
|
H—O

Water molecule (H₂O).

When two oxygen atoms form a molecule of oxygen, they must share two pairs of electrons. This double bond may be represented as two single lines:

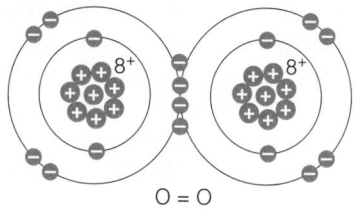

O = O

Oxygen molecule (O₂).

Small atoms form the tightest, most stable bonds. H, O, N, and C are the smallest atoms capable of forming one, two, three, and four electron-pair bonds (respectively). This is the basis for the statement in Chapter 4 that in drawings of compounds containing these atoms, hydrogen must always have one, oxygen two, nitrogen three, and carbon four bonds radiating to other atoms:

$$H— \qquad —O— \qquad -\overset{|}{\underset{}{N}}- \qquad -\overset{|}{\underset{|}{C}}-$$

The stability of the associations between these small atoms and the versatility with which they can combine make them very common in living things. Interestingly, all cells, whether they come from animals, plants, or bacteria, contain the same elements in very nearly the same proportions. The elements commonly found in living things are shown in Table B-2.

TABLE B-2	Elemental Composition of the Human Body	
Element	**Chemical Symbol**	**By Weight (%)**
Oxygen	O	65
Carbon	C	18
Hydrogen	H	10
Nitrogen	N	3
Calcium	Ca	1.5
Phosphorus	P	1.0
Potassium	K	0.4
Sulfur	S	0.3
Sodium	Na	0.2
Chloride	Cl	0.1
Magnesium	Mg	0.1
Total		99.6[a]

[a] The remaining 0.40 percent by weight is contributed by the trace elements: chromium (Cr), copper (Cu), zinc (Zn), selenium (Se), molybdenum (Mo), fluorine (F), iodine (I), manganese (Mn), and iron (Fe). Cells may also contain variable traces of some of the following: boron (B), cobalt (Co), lithium (Li), strontium (Sr), aluminum (Al), silicon (Si), lead (Pb), vanadium (V), arsenic (As), bromine (Br), and others.

Formation of Ions

An atom such as sodium (Na, atomic number 11) cannot easily fill its outer shell by sharing. Sodium possesses a filled first shell of two electrons and a filled second shell of eight; there is only one electron in its outermost shell:

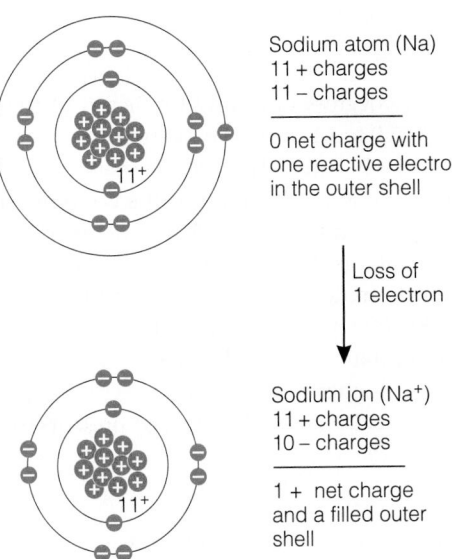

Sodium atom (Na)
11 + charges
11 – charges
_____
0 net charge with
one reactive electron
in the outer shell

Loss of
1 electron

Sodium ion (Na⁺)
11 + charges
10 – charges
_____
1 + net charge
and a filled outer
shell

If sodium loses this electron, it satisfies one condition for stability: a filled outer shell (now its second shell counts as the outer shell). However, it is not electrically neutral. It has 11 protons (positive) and only 10 electrons (negative). It therefore has a net positive charge. An atom or molecule that has lost or gained one or more electrons and so is electrically charged is called an ion.

An atom such as chlorine (Cl, atomic number 17), with seven electrons in its outermost shell, can share electrons to fill its outer shell, or it can gain one electron to complete its outer shell and thus give it a negative charge:

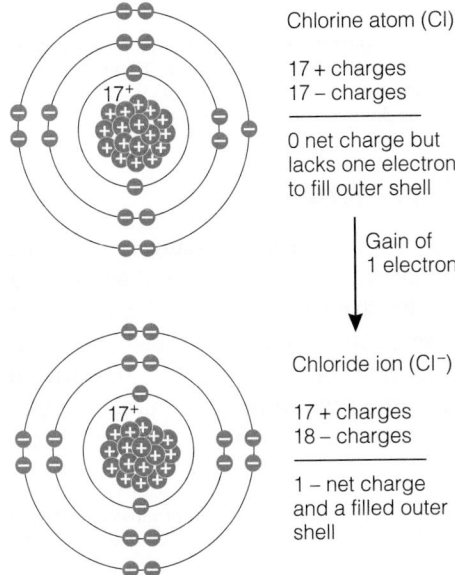

Chlorine atom (Cl)

17 + charges
17 – charges
_____
0 net charge but
lacks one electron
to fill outer shell

Gain of
1 electron

Chloride ion (Cl⁻)

17 + charges
18 – charges
_____
1 – net charge
and a filled outer
shell

A positively charged ion such as sodium ion (Na⁺) is called a cation; a negatively charged ion such as a chloride ion (Cl⁻) is called an anion. Cations and anions attract one another to form salts:

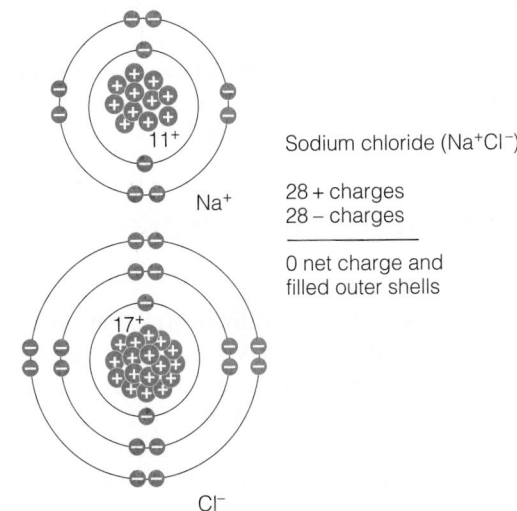

Na⁺

Sodium chloride (Na⁺Cl⁻)

28 + charges
28 – charges
_____
0 net charge and
filled outer shells

Cl⁻

With all its electrons, sodium is a shiny, highly reactive metal; chlorine is the poisonous greenish yellow gas that was used in World War I. But after sodium and chlorine have transferred electrons, they form the stable white salt familiar to you as table salt, or sodium chloride (Na⁺Cl⁻). The dramatic difference illustrates how profoundly the electron arrangement can influence the nature of a substance. The wide distribution of salt in nature attests to the stability of the union between the ions. Each meets the other's needs (a good marriage).

When dry, salt exists as crystals; its ions are stacked very regularly into a lattice, with positive and negative ions alternating in a three-dimensional checkerboard structure. In water, however, the salt quickly dissolves, and its ions separate from one another, forming an electrolyte solution in which they move about freely. Covalently bonded molecules rarely dissociate like this in a water solution. The most common exception is when they behave like acids and release H^+ ions, as discussed in the next section.

An ion can also be a group of atoms bound together in such a way that the group has a net charge and enters into reactions as a single unit. Many such groups are active in the fluids of the body. The bicarbonate ion is composed of five atoms—one H, one C, and three Os—and has a net charge of -1 (HCO_3^-). Another important ion of this type is a phosphate ion with one H, one P, and four O, and a net charge of -2 (HPO_4^{-2}).

Whereas many elements have only one configuration in the outer shell and thus only one way to bond with other elements, some elements have the possibility of varied configurations. Iron is such an element. Under some conditions iron loses two electrons, and under other circumstances it loses three. If iron loses two electrons, it then has a net charge of $+2$, and we call it ferrous iron (Fe^{++}). If it donates three electrons to another atom, it becomes the $+3$ ion, or ferric iron (Fe^{+++}).

B
Appendix

Ferrous iron (Fe^{++})
(had 2 outer-shell electrons
but has lost them)

| 26 + charges |
| 24 − charges |
| 2 + net charge |

Ferric iron (Fe^{+++})
(had 3 outer-shell electrons
but has lost them)

| 26 + charges |
| 23 − charges |
| 3 + net charge |

Remember that a positive charge on an ion means that negative charges—electrons—have been lost and not that positive charges have been added to the nucleus.

Water, Acids, and Bases

Water

The water molecule is electrically neutral, having equal numbers of protons and electrons. When a hydrogen atom shares its electron with oxygen, however, that electron will spend most of its time closer to the positively charged oxygen nucleus. This leaves the positive proton (nucleus of the hydrogen atom) exposed on the outer part of the water molecule. We know, too, that the two hydrogens both bond toward the same side of the oxygen. These two facts explain why water molecules are polar: they have regions of more positive and more negative charge.

Polar molecules like water are drawn to one another by the attractive forces between the positive polar areas of one and the negative poles of another. These attractive forces, sometimes known as polar bonds or hydrogen bonds, occur among many molecules and also within the different parts of single large molecules. Although very weak in comparison with covalent bonds, polar bonds may occur in such abundance that they become exceedingly important in determining the structure of such large molecules as proteins and DNA.

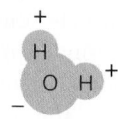

This diagram of the polar water molecule shows displacement of electrons toward the O nucleus; thus the negative region is near the O and the positive regions are near the Hs.

Water molecules have a slight tendency to ionize, separating into positive (H$^+$) and negative (OH$^-$) ions. In pure water, a small but constant number of these ions is present, and the number of positive ions exactly equals the number of negative ions.

Acid

An acid is a substance that releases H$^+$ ions (protons) in a water solution. Hydrochloric acid (HCl$^-$) is such a substance because it dissociates in a water solution into H$^+$ and Cl$^-$ ions. Acetic acid is also an acid because it dissociates in water to acetate ions and free H$^+$:

$$H-\overset{\overset{\displaystyle H}{|}}{\underset{\underset{\displaystyle H}{|}}{C}}-\overset{\overset{\displaystyle O}{\|}}{C}-O-H \longrightarrow H-\overset{\overset{\displaystyle H}{|}}{\underset{\underset{\displaystyle H}{|}}{C}}-\overset{\overset{\displaystyle O}{\|}}{C}-O^- + H^+$$

Acetic acid dissociates into an acetate ion and a hydrogen ion.

The more H$^+$ ions released, the stronger the acid.

pH

Chemists define degrees of acidity by means of the pH scale, which runs from 0 to 14. The pH expresses the concentration of H$^+$ ions: a pH of 1 is extremely acidic, 7 is neutral, and 13 is very basic. There is a tenfold difference in the concentration of H$^+$ ions between points on this scale. A solution with pH 3, for example, has *ten times* as many H$^+$ ions as a solution with pH 4. At pH 7, the concentrations of free H$^+$ and OH$^-$ are exactly the same—1/10,000,000 moles per liter (10^{-7} moles per liter).* At pH 4, the concentration of free H$^+$ ions is 1/10,000 (10^{-4}) moles per liter. This is a higher concentration of H$^+$ ions, and the solution is therefore acidic. Figure 3-7 on p. 81 presents the pH scale.

Bases

A base is a substance that can combine with H$^+$ ions, thus reducing the acidity of a solution. The compound ammonia is such a substance. The ammonia molecule has two electrons that are not shared with any other atom; a hydrogen ion (H$^+$) is just a naked proton with no shell of electrons at all. The proton readily combines with the ammonia molecule to form an ammonium ion; thus a free proton is withdrawn from the solution and no longer contributes to its acidity. Many compounds containing nitrogen are important bases in living systems. Acids and bases neutralize each other to produce substances that are neither acid nor base.

$$:\overset{\overset{\displaystyle H}{|}}{\underset{\underset{\displaystyle H}{|}}{N}}-H + H^+ \longrightarrow H-\overset{\overset{\displaystyle H}{|}}{\underset{\underset{\displaystyle H}{|}}{N}}{}^+-H$$

Ammonia captures a hydrogen ion from water. The two dots here represent the two electrons not shared with another atom. These dots are ordinarily not shown in chemical structure drawings. Compare this drawing with the earlier diagram of an ammonia molecule (p. B-4).

Chemical Reactions

A chemical reaction, or chemical change, results in the breakdown of substances and the formation of new ones. Almost all such reactions involve a change in the bonding of atoms. Old bonds are broken, and new ones are formed. The nuclei of atoms are never involved in chemical reactions—only their outer-shell electrons take part. At the end of a chemical reaction, the number of atoms of each type is always the same as at the beginning.

---

*A mole is a certain number (about 6×10^{23}) of molecules. The pH of a solution is defined as the negative logarithm of the hydrogen ion concentration of the solution. Thus, if the concentration is 10^{-2} (moles per liter), the pH is 2; if 10^{-8}, the pH is 8; and so on.

Diagrams:

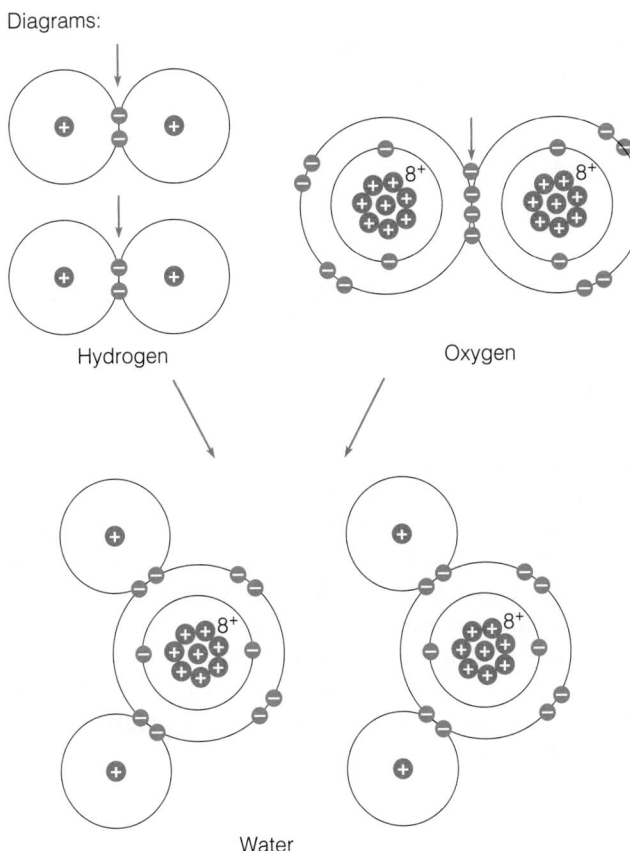

Hydrogen

Oxygen

Water

Structures:

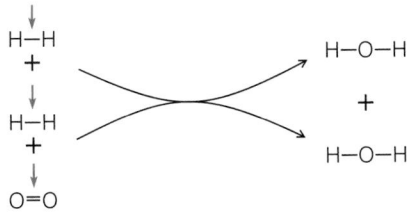

Formulas:

$$2H_2 + O_2 \longrightarrow 2H_2O$$

Hydrogen and oxygen react to form water.

cepted by another. The addition of an atom of oxygen is also oxidation because oxygen (with six electrons in the outer shell) accepts two electrons in becoming bonded. Oxidation, then, is loss of electrons, gain of protons, or addition of oxygen (with six electrons); reduction is the opposite—gain of electrons, loss of protons, or loss of oxygen. The addition of hydrogen atoms to oxygen to form water can thus be described as the reduction of oxygen *or* the oxidation of hydrogen.

If a reaction results in a net increase in the energy of a compound, it is called an endergonic, or "uphill," reaction (energy, *erg,* is added into, *endo,* the compound). An example is the chief result of photosynthesis, the making of sugar in a plant from carbon dioxide and water using the energy of sunlight. Conversely, the oxidation of sugar to carbon dioxide and water is an exergonic, or "downhill," reaction because the end products have less energy than the starting products. Oftentimes, but not always, reduction reactions are endergonic, resulting in an increase in the energy of the products. Oxidation reactions often, but not always, are exergonic.

Chemical reactions tend to occur spontaneously if the end products are in a lower energy state and therefore are more stable than the reacting compounds. These reactions often give off energy in the form of heat as they occur. The generation of heat by wood burning in a fireplace and the maintenance of human body warmth both depend on energy-yielding chemical reactions. These downhill reactions occur easily, although they may require some activation energy to get them started, just as a ball requires a push to start rolling downhill.

Uphill reactions, in which the products contain more energy than the reacting compounds started with, do not occur until an energy source is provided. An example of such an energy source is the sunlight used in photosynthesis, where carbon dioxide and water (low-energy compounds) are combined to form the sugar glucose (a higher-energy compound). Another example is the use of the energy in glucose to combine two low-energy compounds in the body into the high-energy compound ATP (see Chapter 7). The energy in ATP may be used to power many other energy-requiring, uphill reactions. Clearly, any of many different molecules can be used as a temporary storage place for energy.

For example, two hydrogen molecules ($2H_2$) can react with one oxygen molecule (O_2) to form two water molecules ($2H_2O$). In this reaction two substances (hydrogen and oxygen) disappear, and a new one (water) is formed, but at the end of the reaction there are still four H atoms and two O atoms, just as there were at the beginning. Because the atoms are now linked in a different way, their characteristics or properties have changed.

In many instances chemical reactions involve not the relinking of molecules but the exchanging of electrons or protons among them. In such reactions the molecule that gains one or more electrons (or loses one or more hydrogen ions) is said to be reduced; the molecule that loses electrons (or gains protons) is oxidized. A hydrogen ion is equivalent to a proton. Oxidation and reduction reactions take place simultaneously because an electron or proton that is lost by one molecule is ac-

Energy change as reaction occurs

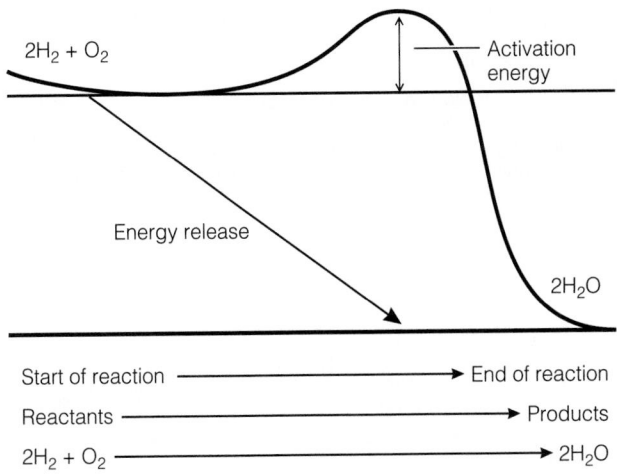

Neither downhill nor uphill reactions occur until something sets them off (activation) or until a path is provided for them to follow. The body uses enzymes as a means of providing paths and controlling chemical reactions (see Chapter 6). By controlling the availability and the action of its enzymes, the cells can "decide" which chemical reactions to prevent and which to promote.

Formation of Free Radicals

Normally, when a chemical reaction takes place, bonds break and re-form with some redistribution of atoms and rearrangement of bonds to form new, stable compounds. Normally, bonds don't split in such a way as to leave a molecule with an odd, unpaired electron. When they do, free radicals are formed. Free radicals are highly unstable and quickly react with other compounds, forming more free radicals in a chain reaction. A cascade may ensue in which many highly reactive radicals are generated, resulting finally in the disruption of a living structure such as a cell membrane.

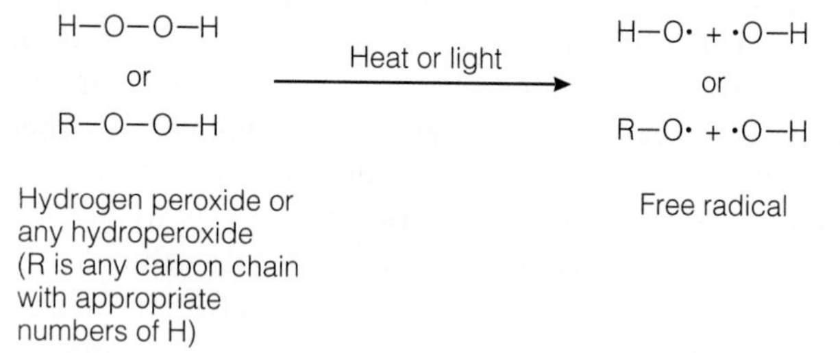

Hydrogen peroxide or any hydroperoxide (R is any carbon chain with appropriate numbers of H)

Free radical

Free radicals are formed. The dots represent single electrons that are available for sharing (the atom needs another electron to fill its outer shell).

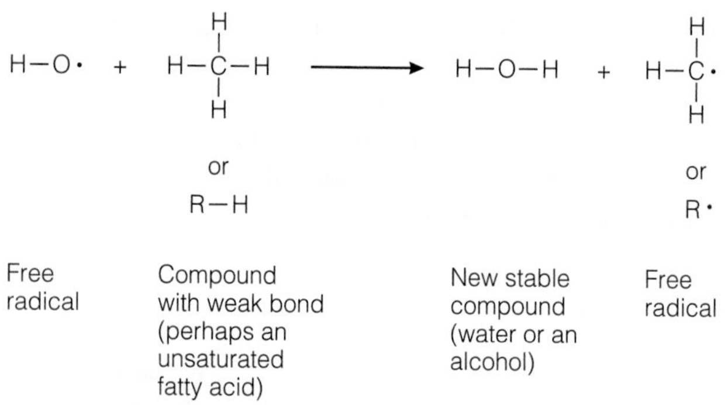

| Free radical | Compound with weak bond (perhaps an unsaturated fatty acid) | New stable compound (water or an alcohol) | Free radical |

Destruction of biological compounds by free radicals. The free radical attacks a weak bond in a biological compound, disrupting it and forming a new stable molecule and another free radical. This free radical can attack another biological compound, and so on.

Oxidation of some compounds can be induced by air at room temperature in the presence of light. Such reactions are thought to take place through the formation of compounds called peroxides:

Peroxides:

H—O—O—H   Hydrogen peroxide

R—O—O—H   Hydroperoxides (R is any carbon chain with appropriate numbers of H)

R—O—O—R   Peroxide

Some peroxides readily disintegrate into free radicals, initiating chain reactions like those just described.

Free radicals are of special interest in nutrition because the antioxidant properties of vitamins C and E as well as beta-carotene and the mineral selenium are thought to protect against the destructive effects of these free radicals (see Highlight 11). For example, vitamin E on the surface of the lungs reacts with, and is destroyed by, free radicals, thus preventing the radicals from reaching underlying cells and oxidizing the lipids in their membranes.

Biochemical Structures and Pathways

The diagrams of nutrients presented here are meant to enhance your understanding of the most important organic molecules in the human diet. Following the diagrams of nutrients are sections on the major metabolic pathways mentioned in Chapter 7—glycolysis, fatty acid oxidation, amino acid degradation, the TCA cycle, and the electron transport chain—and a description of how alcohol interferes with these pathways. Discussions of the urea cycle and the formation of ketone bodies complete the appendix.

C Appendix

Carbohydrates

Monosaccharides

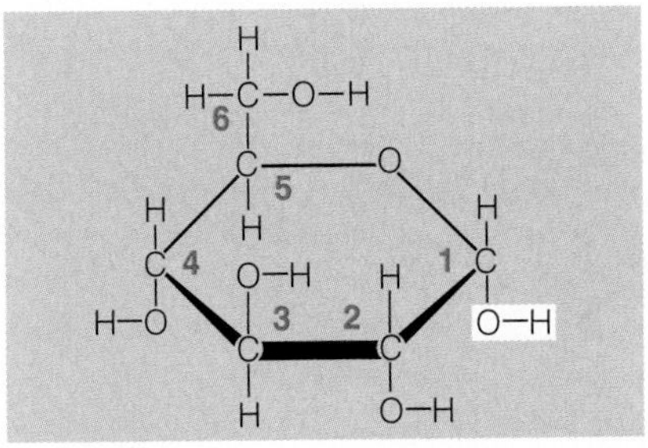

Glucose (alpha form). The ring would be at right angles to the plane of the paper. The bonds directed upward are above the plane; those directed downward are below the plane. This molecule is considered an alpha form because the OH on carbon 1 points downward.

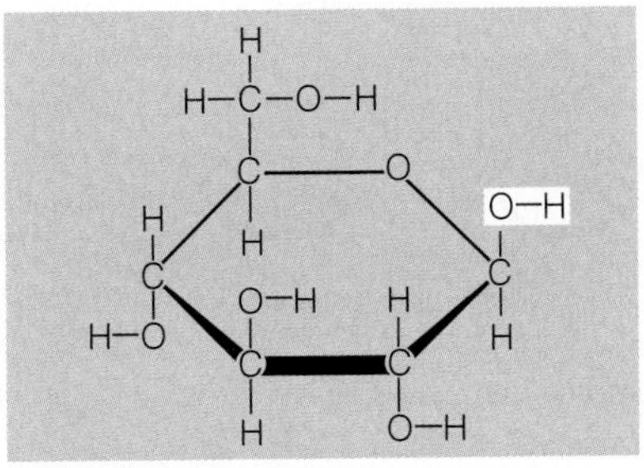

Glucose (beta form). The OH on carbon 1 points upward.
Fructose, galactose: see Chapter 4.

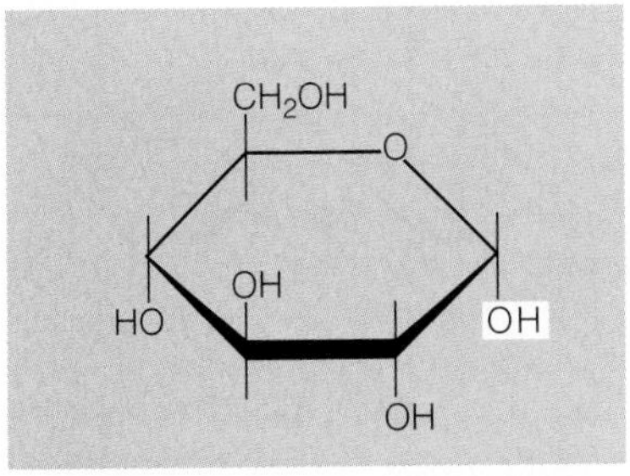

Glucose (alpha form) shorthand notation. This notation, in which the carbons in the ring and single hydrogens have been eliminated, will be used throughout this appendix.

Disaccharides

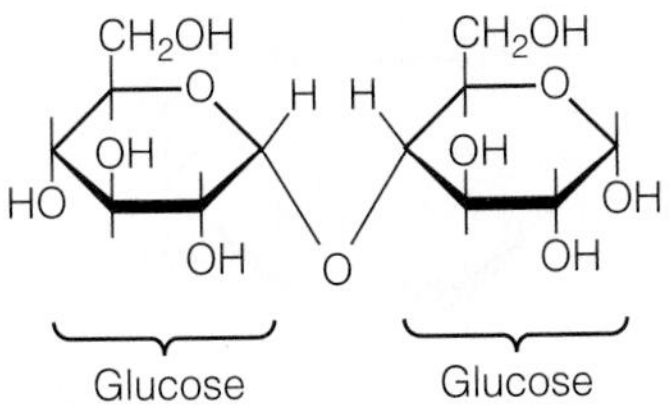

Glucose   Glucose

Maltose.

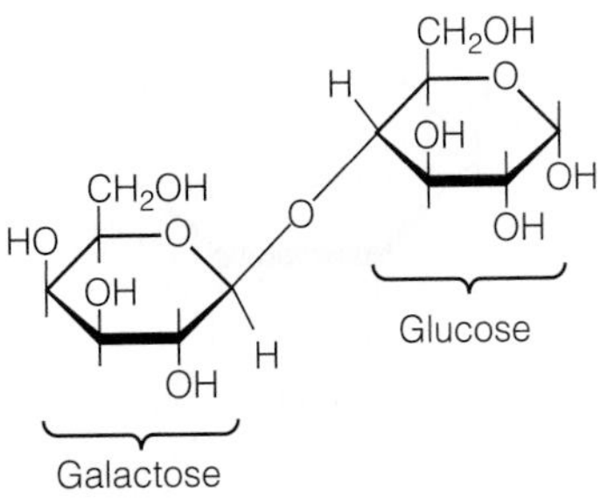

Galactose

Glucose

Lactose (alpha form).

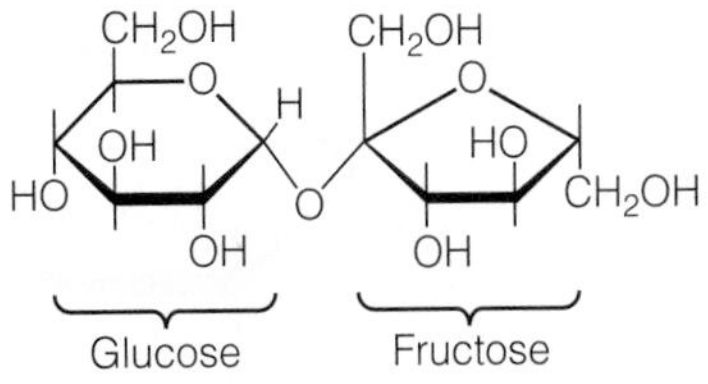

Glucose   Fructose

Sucrose.

Polysaccharides

As described in Chapter 4, starch, glycogen, and cellulose are all long chains of glucose molecules covalently linked together.

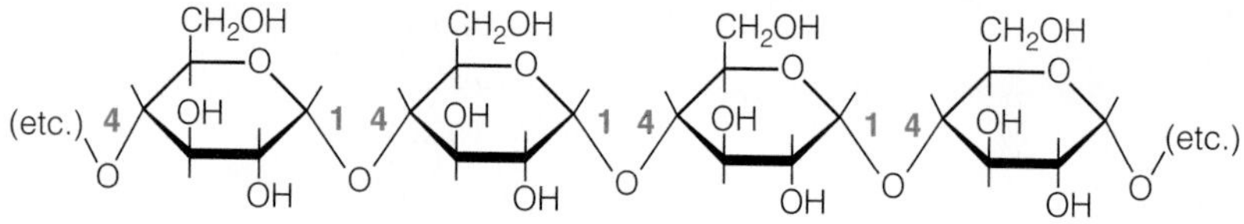

Amylose (unbranched starch)

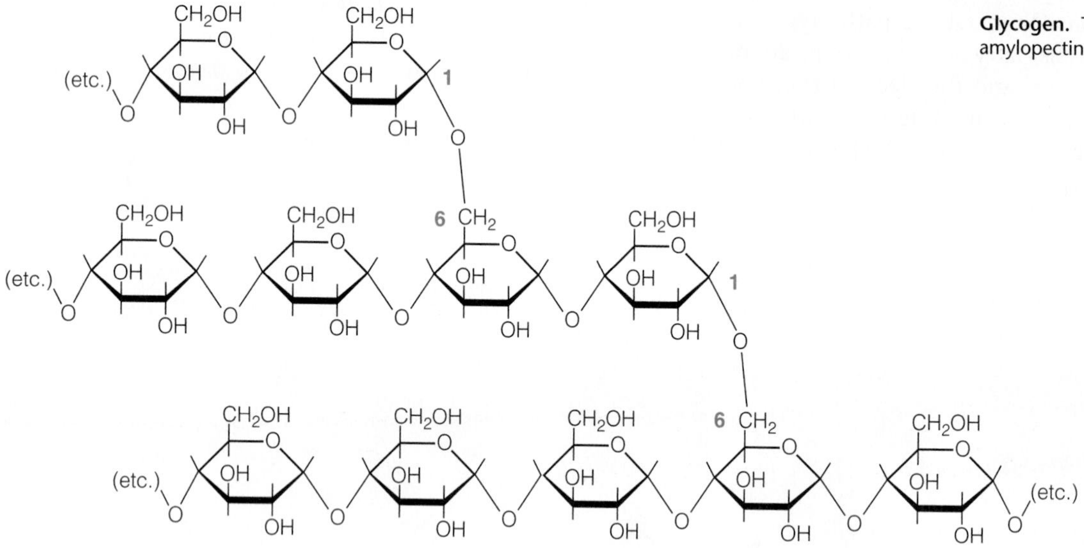

Starch. Two kinds of covalent linkages occur between glucose molecules in starch, giving rise to two kinds of chains. Amylose is composed of straight chains, with carbon 1 of one glucose linked to carbon 4 of the next (α-1,4 linkage). Amylopectin is made up of straight chains like amylose but has occasional branches arising where the carbon 6 of a glucose is also linked to the carbon 1 of another glucose (α-1,6 linkage).

Glycogen. The structure of glycogen is like amylopectin but with many more branches.

Amylopectin (branched starch)

Cellulose. Like starch and glycogen, cellulose is also made of chains of glucose units, but there is an important difference: in cellulose, the OH on carbon 1 is in the beta position (see p. C-1). When carbon 1 of one glucose is linked to carbon 4 of the next, it forms a β-1,4 linkage, which cannot be broken by digestive enzymes in the human GI tract.

Fibers, such as hemicelluloses, consist of long chains of various monosaccharides.

Monosaccharides common in the backbone chain of hemicelluloses:

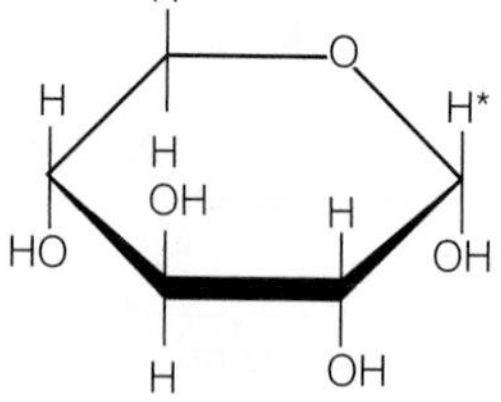

Xylose

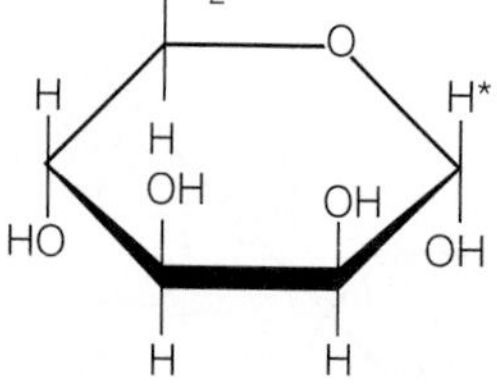

Mannose

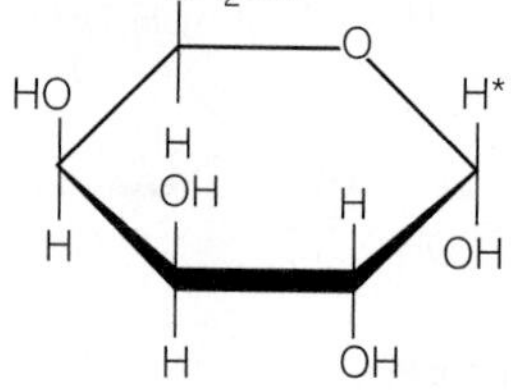

Galactose

---

*These structures are shown in the alpha form with the H on the carbon pointing upward and the OH pointing downward, but they may also appear in the beta form with the H pointing downward and the OH upward.

Monosaccharides common in the side chains of hemicelluloses:

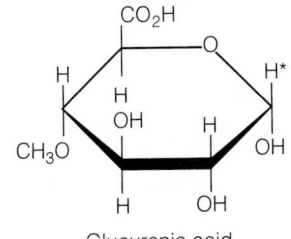

Arabinose　　　　　Glucuronic acid　　　　　Galactose

Hemicelluloses. The most common hemicelluloses are composed of a backbone chain of xylose, mannose, and galactose, with branching side chains of arabinose, glucuronic acid, and galactose.

Lipids

TABLE C-1 Saturated Fatty Acids Found in Natural Fats

Saturated Fatty Acids	Chemical Formulas	Number of Carbons	Major Food Sources
Butyric	C_3H_7COOH	4	Butterfat
Caproic	$C_5H_{11}COOH$	6	Butterfat
Caprylic	$C_7H_{15}COOH$	8	Coconut oil
Capric	$C_9H_{19}COOH$	10	Palm oil
Lauric	$C_{11}H_{23}COOH$	12	Coconut oil, palm oil
Myristic[a]	$C_{13}H_{27}COOH$	14	Coconut oil, palm oil
Palmitic[a]	$C_{15}H_{31}COOH$	16	Palm oil
Stearic[a]	$C_{17}H_{35}COOH$	18	Most animal fats
Arachidic	$C_{19}H_{39}COOH$	20	Peanut oil
Behenic	$C_{21}H_{43}COOH$	22	Seeds
Lignoceric	$C_{23}H_{47}COOH$	24	Peanut oil

[a]Most common saturated fatty acids.

TABLE C-2 Unsaturated Fatty Acids Found in Natural Fats

Unsaturated Fatty Acids	Chemical Formulas	Number of Carbons	Number of Double Bonds	Standard Notation[a]	Omega Notation[b]	Food Sources
Palmitoleic	$C_{15}H_{29}COOH$	16	1	16:1;9	16:1ω7	Seafood, beef
Oleic	$C_{17}H_{33}COOH$	18	1	18:1;9	18:1ω9	Olive oil, canola oil
Linoleic	$C_{17}H_{31}COOH$	18	2	18:2;9,12	18:2ω6	Sunflower oil, safflower oil
Linolenic	$C_{17}H_{29}COOH$	18	3	18:3;9,12,15	18:3ω3	Soybean oil, canola oil
Arachidonic	$C_{19}H_{31}COOH$	20	4	20:4;5,8,11,14	20:4ω6	Eggs, most animal fats
Eicosapentaenoic	$C_{19}H_{29}COOH$	20	5	20:5;5,8,11,14,17	20:5ω3	Seafood
Docosahexaenoic	$C_{21}H_{31}COOH$	22	6	22:6;4,7,10,13,16,19	22:6ω3	Seafood

NOTE: A fatty acid has two ends; designated the methyl (CH_3) end and the carboxyl, or acid (COOH), end.
[a]Standard chemistry notation begins counting carbons at the acid end. The number of carbons the fatty acid contains comes first, followed by a colon and another number that indicates the number of double bonds; next comes a semicolon followed by a number or numbers indicating the positions of the double bonds. Thus the notation for linoleic acid, an 18-carbon fatty acid with two double bonds between carbons 9 and 10 and between carbons 12 and 13, is 18:2;9,12.
[b]Because fatty acid chains are lengthened by adding carbons at the acid end of the chain, chemists use the omega system of notation to ease the task of identifying them. The omega system begins counting carbons at the methyl end. The number of carbons the fatty acid contains comes first, followed by a colon and the number of double bonds; next come the omega symbol (ω) and a number indicating the position of the double bond nearest the methyl end. Thus linoleic acid with its first double bond at the sixth carbon from the methyl end would be noted 18:2ω6 in the omega system.

Protein: Amino Acids

The common amino acids may be classified into the seven groups listed on the next page. Amino acids marked with an asterisk (*) are essential.

1. Amino acids with aliphatic side chains, which consist of hydrogen and carbon atoms (hydrocarbons):

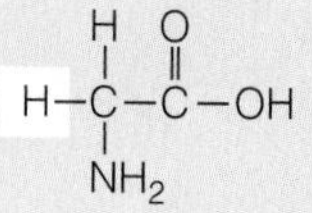

Glycine (Gly)

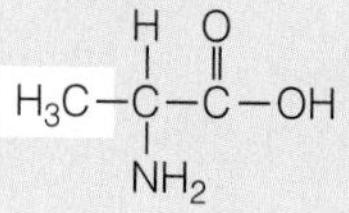

Alanine (Ala)

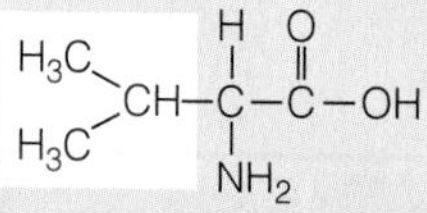

Valine* (Val)

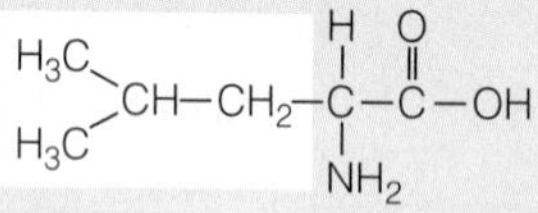

Leucine* (Leu)

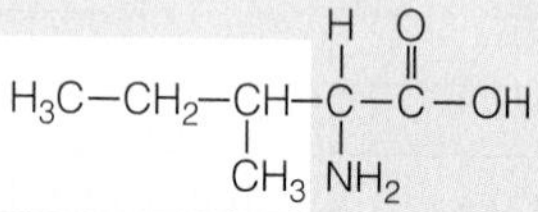

Isoleucine* (Ile)

2. Amino acids with hydroxyl (OH) side chains:

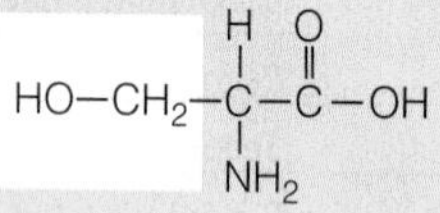

Serine (Ser)

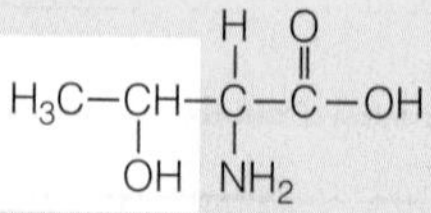

Threonine* (Thr)

3. Amino acids with side chains containing acidic groups or their amides, which contain the group NH_2:

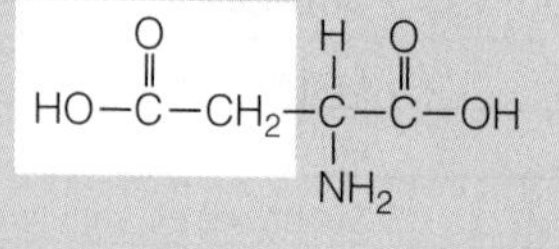

Aspartic acid (Asp)

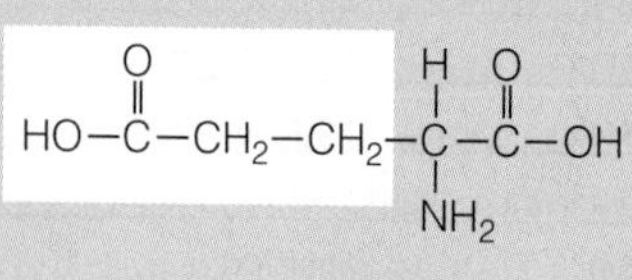

Glutamic acid (Glu)

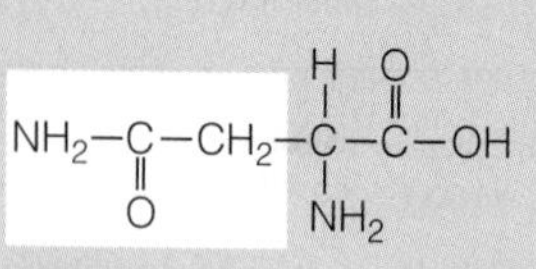

Asparagine (Asn)

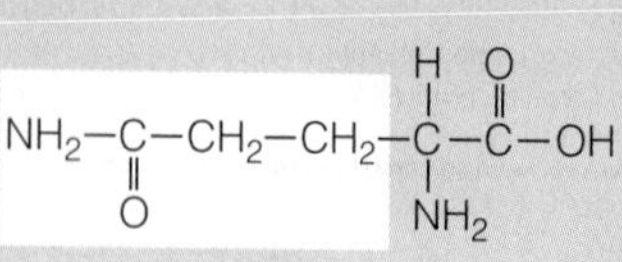

Glutamine (Gln)

4. Amino acids with basic side chains:

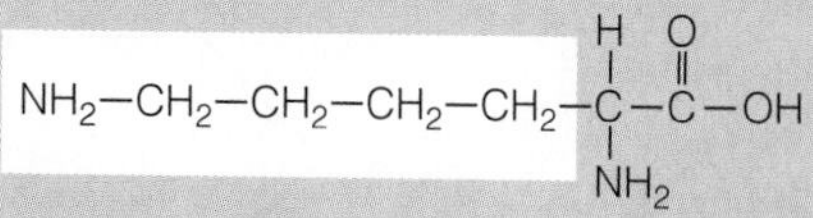

Lysine* (Lys)

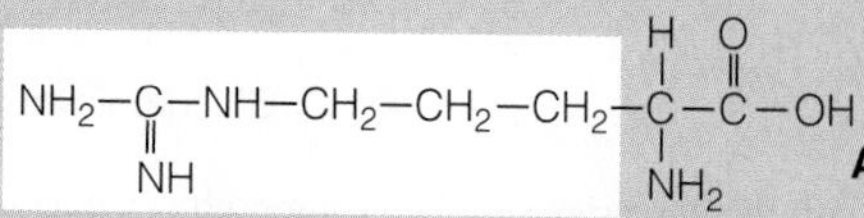

Arginine (Arg)

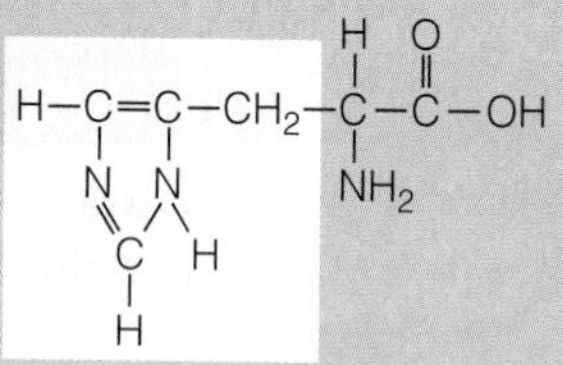

Histidine* (His)

5. Amino acids with aromatic side chains, which are characterized by the presence of at least one ring structure:

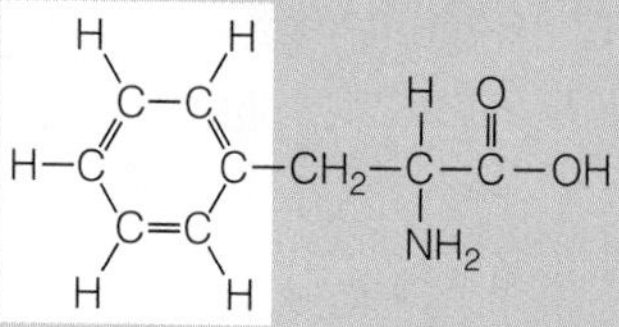

Phenylalanine* (Phe)

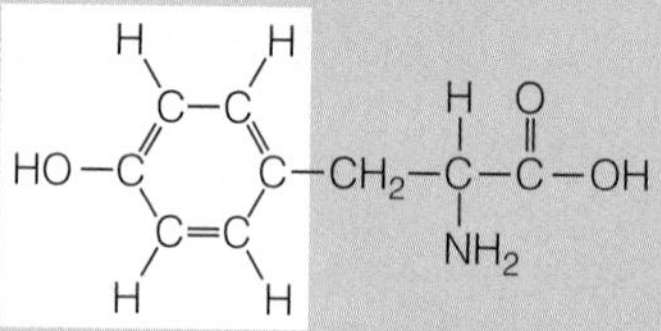

Tyrosine (Tyr)

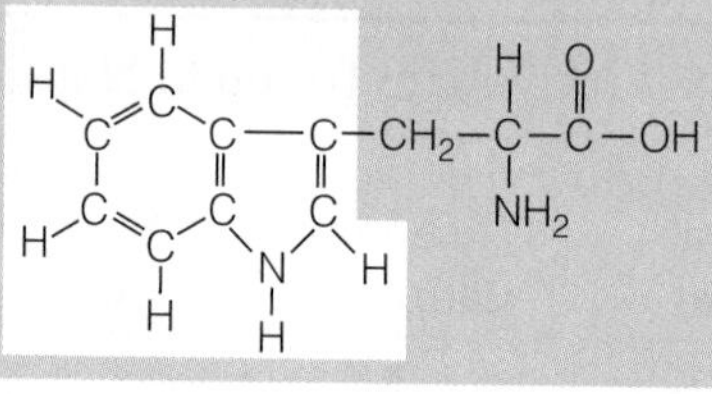

Tryptophan* (Trp)

6. Amino acids with side chains containing sulfur atoms:

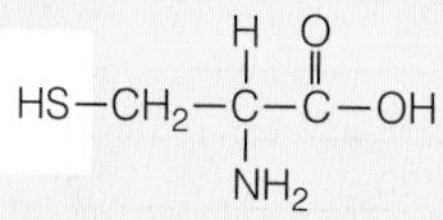

Cysteine (Cys)

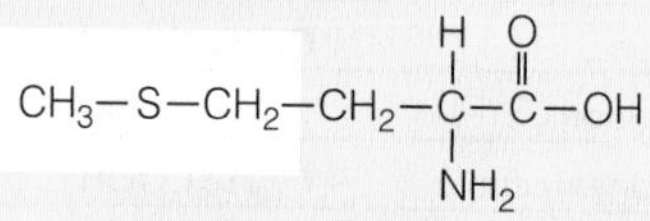

Methionine* (Met)

7. Imino acid:

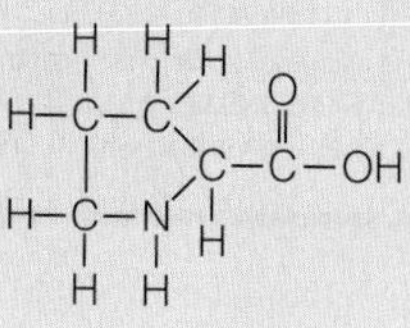

Proline (Pro)

Proline has the same chemical structure as the other amino acids, but its amino group has given up a hydrogen to form a ring.

Vitamins and Coenzymes

Vitamin A: retinol. This molecule is the alcohol form of vitamin A.

Vitamin A: retinal. This molecule is the aldehyde form of vitamin A.

Vitamin A: retinoic acid. This molecule is the acid form of vitamin A.

Vitamin A precursor: beta-carotene. This molecule is the carotenoid with the most vitamin A activity.

w573 UNC-6a

Thiamin. This molecule is part of the coenzyme thiamin pyrophosphate (TPP).

Thiamin pyrophosphate (TPP). TPP is a coenzyme that includes the thiamin molecule as part of its structure.

Riboflavin. This molecule is a part of two coenzymes—flavin mononucleotide (FMN) and flavin adenine dinucleotide (FAD).

Flavin mononucleotide (FMN). FMN is a coenzyme that includes the riboflavin molecule as part of its structure.

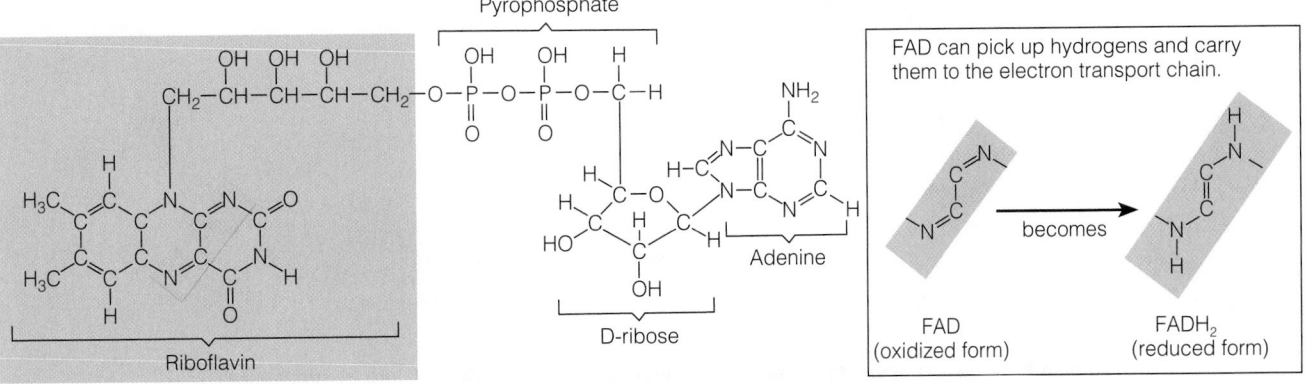

Pyrophosphate

Riboflavin

D-ribose

Adenine

FAD can pick up hydrogens and carry them to the electron transport chain.

becomes

FAD (oxidized form)

FADH₂ (reduced form)

Flavin adenine dinucleotide (FAD). FAD is a coenzyme that includes the riboflavin molecule as part of its structure.

Appendix **C**

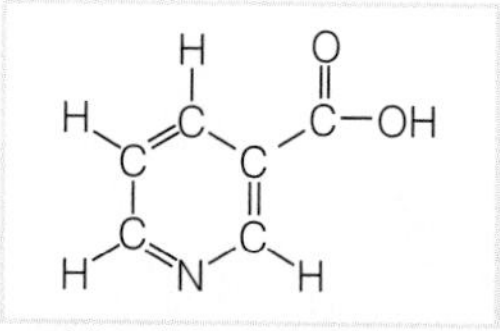

Nicotinic acid

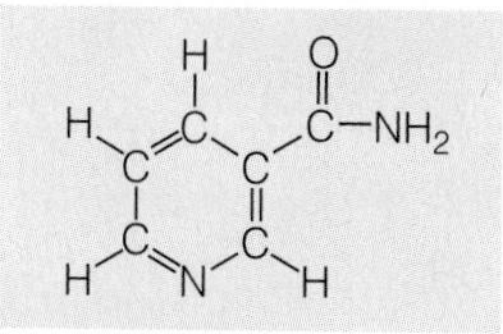

Nicotinamide

Niacin (nicotinic acid and nicotinamide). These molecules are a part of two coenzymes—nicotinamide adenine dinucleotide (NAD+) and nicotinamide adenine dinucleotide phosphate (NADP+).

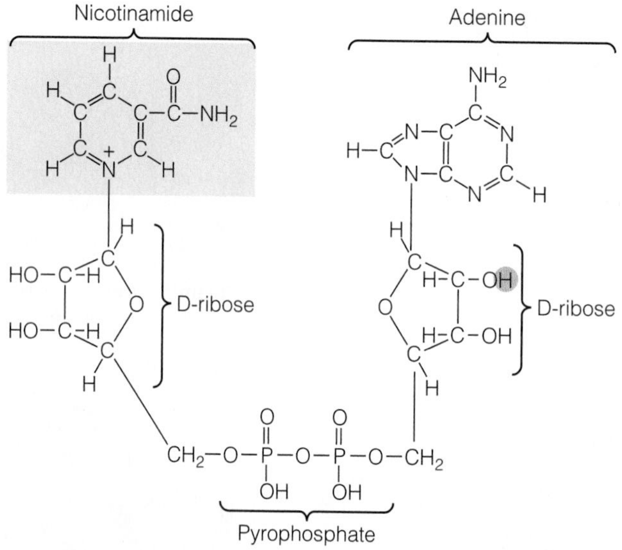

Nicotinamide

Adenine

D-ribose

D-ribose

Pyrophosphate

Nicotinamide adenine dinucleotide (NAD+) and nicotinamide adenine dinucleotide phosphate (NADP+). NADP has the same structure as NAD but with a phosphate group attached to the O instead of the H.

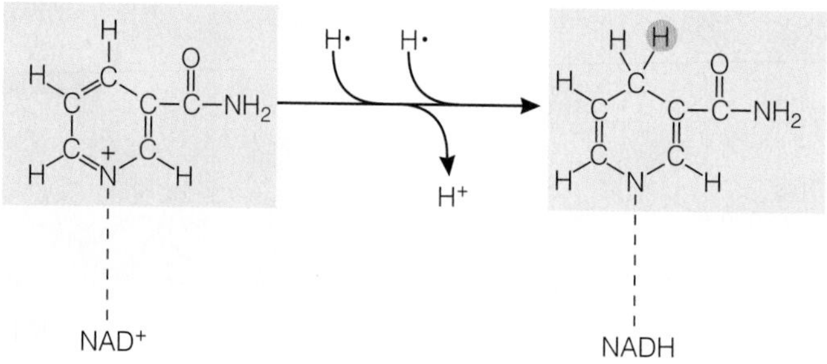

NAD+

NADH

Reduced NAD+ (NADH). When NAD+ is reduced by the addition of H+ and two electrons, it becomes the coenzyme NADH. (The dots on the H entering this reaction represent electrons—see Appendix B.)

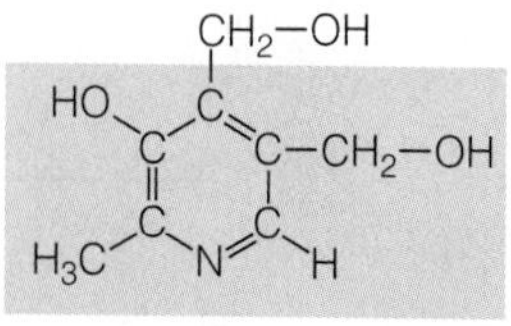

Pyridoxine

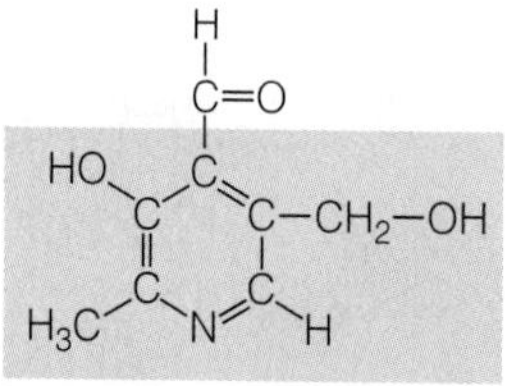

Pyridoxal

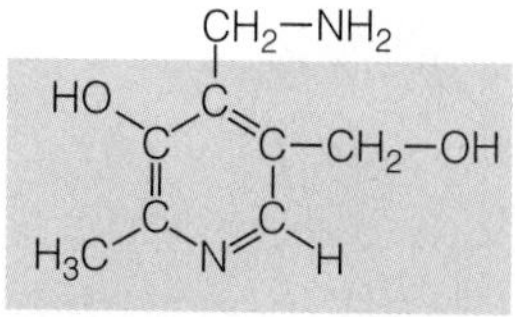

Pyridoxamine

Vitamin B₆ (a general name for three compounds—pyridoxine, pyridoxal, and pyridoxamine). These molecules are a part of two coenzymes—pyridoxal phosphate and pyridoxamine phosphate.

Pyridoxal phosphate (PLP) and pyridoxamine phosphate. These coenzymes include vitamin B_6 as part of their structures.

Vitamin B_{12} (cyanocobalamin). The arrows in this diagram indicate that the spare electron pairs on the nitrogens attract them to the cobalt.

Folate (folacin or folic acid). This molecule consists of a double ring combined with a single ring and at least one glutamate (a nonessential amino acid marked in the box). Folate's biologically active form is tetrahydrofolic acid.

Tetrahydrofolic acid. This active coenzyme form of folate has four added hydrogens. An intermediate form, dihydrofolate, has two added hydrogens.

Pantothenic acid. This molecule is part of coenzyme A (CoA).

Coenzyme A (CoA). Coenzyme A is a coenzyme that includes pantothenic acid as part of its structure.

Biotin.

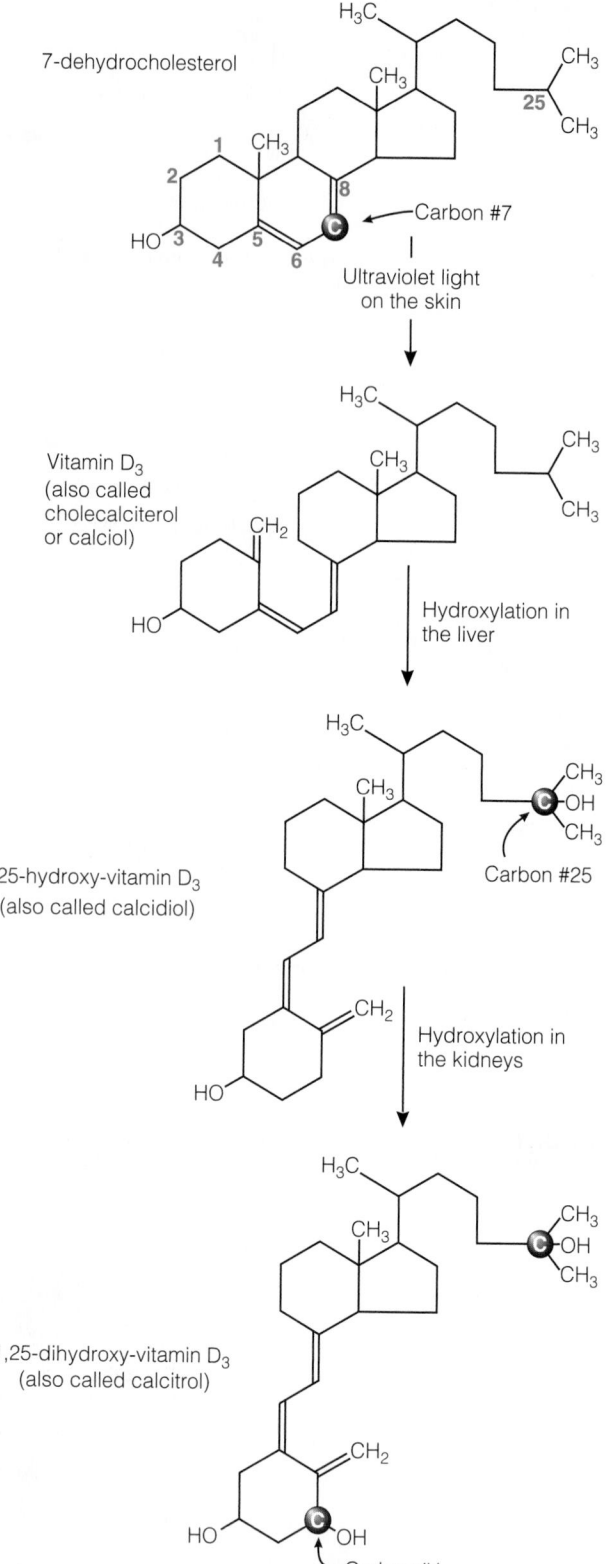

Vitamin C. Two hydrogen atoms with their electrons are lost when ascorbic acid is oxidized and gained when it is reduced again.

Ascorbic acid (reduced form)

Dehydroascorbic acid (oxidized form)

2H⁺

7-dehydrocholesterol

Carbon #7

Ultraviolet light on the skin

Vitamin D₃ (also called cholecalciterol or calciol)

Hydroxylation in the liver

25-hydroxy-vitamin D₃ (also called calcidiol)

Carbon #25

Hydroxylation in the kidneys

1,25-dihydroxy-vitamin D₃ (also called calcitrol)

Carbon #1

Vitamin D. The synthesis of active vitamin D begins with 7-dehydrocholesterol. (The carbon atoms at which changes occur are numbered.)

Vitamin E (alpha-tocopherol). The number and position of the methyl groups (CH_3) bonded to the ring structure differentiate among the tocopherols.

Tocotrienols contain double bonds here.

Vitamin K. Naturally occurring compounds with vitamin K activity include phylloquinones (from plants) and menaquinones (from bacteria).

Menadione. This synthetic compound has the same activity as natural vitamin K.

Adenosine triphosphate (ATP), the energy carrier. The cleavage point marks the bond that is broken when ATP splits to become ADP + P.

Adenosine diphosphate (ADP).

Glycolysis

Figure C-1 depicts the events of glycolysis. The following text describes key steps as numbered on the figure.

FIGURE C-1 Glycolysis

Notice that galactose and fructose enter at different places but all continue on the same pathway.

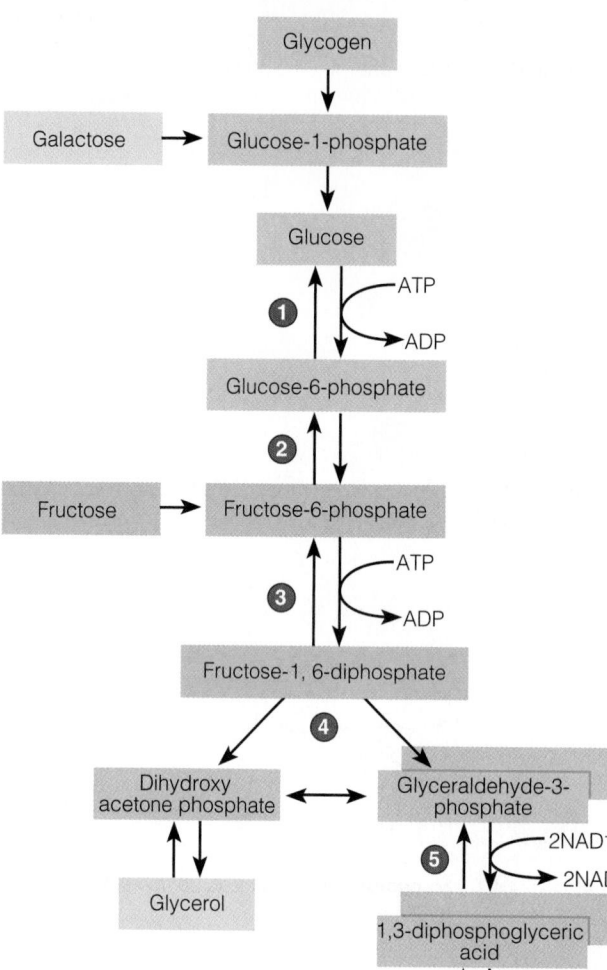
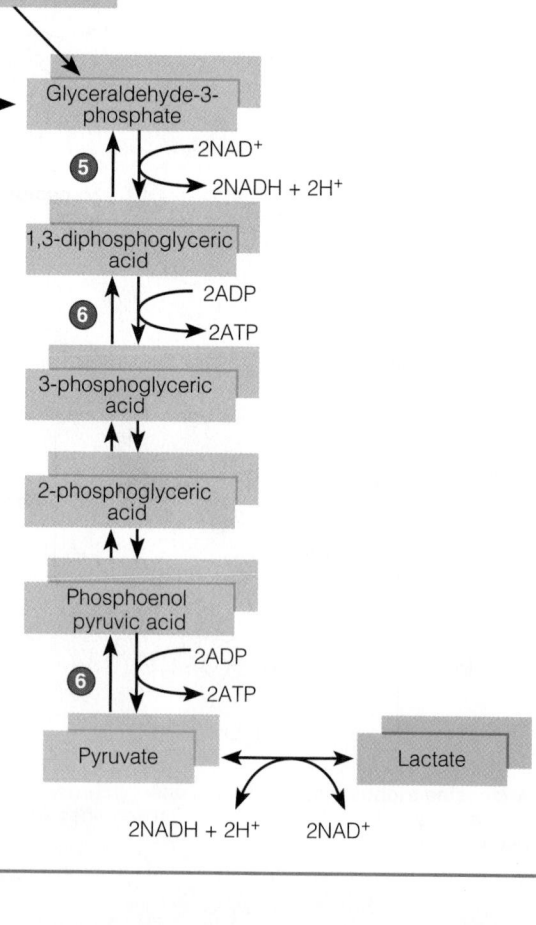

1. A phosphate is attached to glucose at the carbon that chemists call number 6 (review the first diagram of glucose on p. C-1 to see how chemists number the carbons in a glucose molecule). The product is called, logically enough, glucose-6-phosphate. One ATP molecule is used to accomplish this.

2. Glucose-6-phosphate is rearranged by an enzyme.

3. A phosphate is added in another reaction that uses another molecule of ATP. The product this time is fructose-1,6-diphosphate. At this point the six-carbon sugar has a phosphate group on its first and sixth carbons and is ready to break apart.

4. When fructose-1,6-diphosphate breaks in half, the two three-carbon compounds are not identical. Each has a phosphate group attached, but only glyceraldehyde-3-phosphate converts directly to pyruvate. The other compound, however, converts easily to glyceraldehyde-3-phosphate.

5. In the next step, enough energy is released to convert NAD^+ to $NADH + H^+$.

6. In two of the following steps ATP is regenerated.

Remember that in effect two molecules of glyceraldehyde-3-phosphate are produced from glucose; therefore, four ATP molecules are generated from each glucose molecule. Two ATP were needed to get the sequence started, so the net gain at this point is two ATP and two molecules of $NADH + H^+$. As you will see later, each $NADH + H^+$ moves to the electron transport chain to unload its hydrogens onto oxygen, producing two ATP. Thus the total yield from glucose to pyruvate is eight ATP.

Fatty Acid Oxidation

Figure C-2 presents fatty acid oxidation. The sequence is as follows.

1. The fatty acid is activated by combining with coenzyme A (CoA). In this reaction, ATP loses two phosphorus atoms (PP, or pyrophosphate) and becomes AMP (adenosine monophosphate)—the equivalent of a loss of two ATP.

2. In the next reaction, two H with their electrons are removed and transferred to FAD, forming $FADH_2$.

3. In a later reaction, two H are removed and go to NAD^+ (forming $NADH + H^+$).

4. The fatty acid is cleaved at the "beta" carbon, the second carbon from the carboxyl (COOH) end. This break results in a fatty acid that is two carbons shorter than the previous one and a two-carbon molecule of acetyl CoA. At the same time, another CoA is attached to the fatty acid, thus activating it for its turn through the series of reactions.

5. The sequence is repeated with each cycle producing an acetyl CoA and a shorter fatty acid until only a 2-carbon fatty acid remains—acetyl CoA.

In the example shown in Figure C-2, palmitic acid (a 16-carbon fatty acid) will go through this series of reactions seven times, using the equivalent of two ATP for the initial activation and generating seven $FADH_2$, seven $NADH + H^+$, and eight acetyl CoA. As you will see later, each of the seven $FADH_2$ will enter the electron transport chain to unload its hydrogens onto oxygen, yielding two ATP (for a total of 14). Similarly, each $NADH + H^+$ will enter the electron transport chain to unload its hydrogens onto oxygen, yielding three ATP (for a total of 21). Thus the oxidation of a 16-carbon fatty acid uses 2 ATP and generates 35 ATP. When the eight acetyl CoA enter the TCA cycle, even more ATP will be generated, as a later section describes.

Amino Acid Degradation

The first step in amino acid degradation is the removal of the nitrogen-containing amino group through either deamination (Figure 7-14 on p. 227) or transamination (Figure 7-15 on p. 228) reactions. Then the remaining carbon skeletons may enter the metabolic pathways at different places, as shown in Figure C-3.

The TCA Cycle

The tricarboxylic acid, or TCA, cycle is the set of reactions that break down acetyl CoA to carbon dioxide and hydrogens. To link glycolysis to the TCA cycle, pyruvate enters the mitochondrion, loses a carbon group, and bonds with a molecule of CoA to become acetyl CoA. The TCA cycle uses any substance that can be converted to acetyl CoA directly or indirectly through pyruvate.

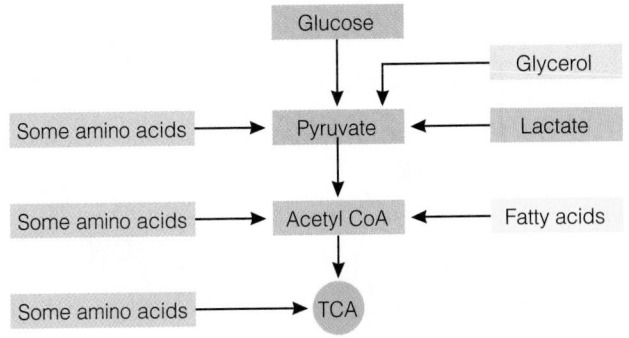

The step from pyruvate to acetyl CoA is complex. We have included only those substances that will help you understand

---

FIGURE C-2 Fatty Acid Oxidation

Palmitic acid (16C)

CoA + ATP →① → AMP + PP

Activated palmitic acid

FAD →② → $FADH_2$

H_2O

NAD^+ →③ → $NADH + H^+$

CoA →④

Activated myristic acid (14C) + Acetyl CoA (2C)

⑤

FIGURE C-3 Amino Acid Degradation

After losing their amino groups, carbon skeletons can be converted to one of seven molecules that can enter the TCA cycle (presented in Figure C-4).

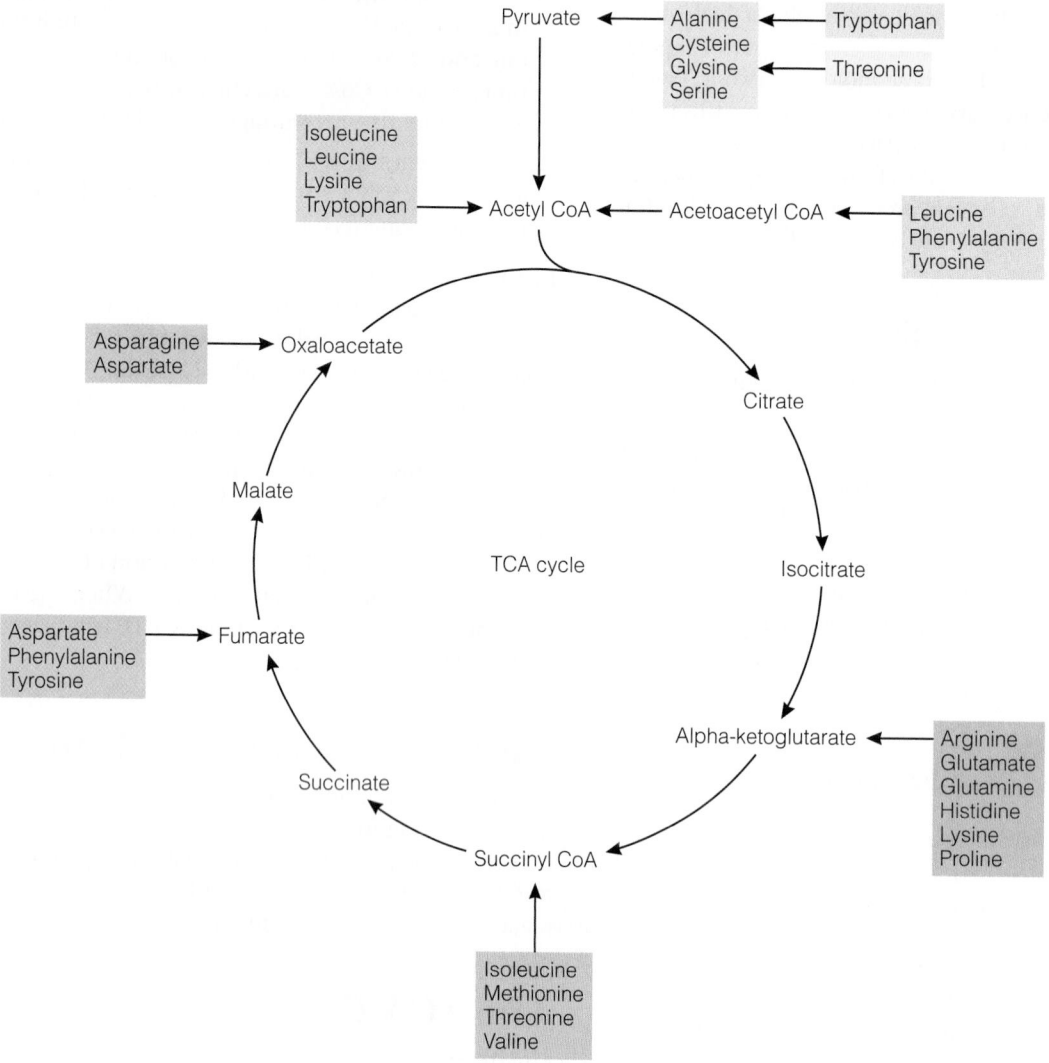

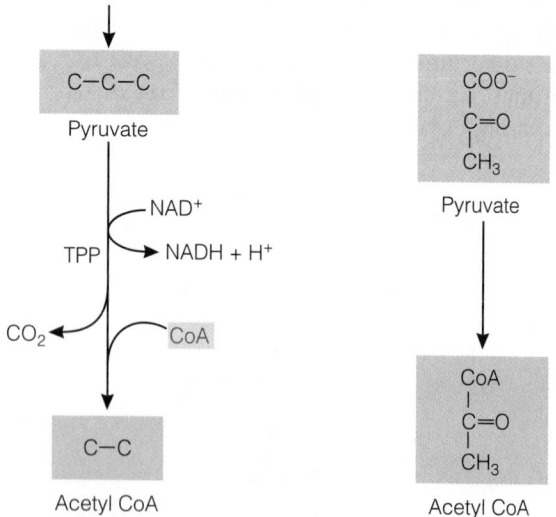

The step from pyruvate to acetyl CoA. (TPP and NAD are coenzymes containing the B vitamins thiamin and niacin, respectively.)

the transfer of energy from the nutrients. Pyruvate loses a carbon to carbon dioxide and is attached to a molecule of CoA. In the process, NAD$^+$ picks up two hydrogens with their associated electrons, becoming NADH + H$^+$.

Let's follow the steps of the TCA cycle (see the corresponding numbers in Figure C-4).

1. The two-carbon acetyl CoA combines with a four-carbon compound, oxaloacetate. The CoA comes off, and the product is a six-carbon compound, citrate.

2. The atoms of citrate are rearranged to form isocitrate.

3. Now two H (with their two electrons) are removed from the isocitrate. One H becomes attached to the NAD$^+$ with the two electrons; the other H is released as H$^+$. Thus NAD$^+$ becomes NADH + H$^+$. (Remember this NADH + H$^+$, but let's follow the carbons first.) A carbon is combined with two oxygens, forming carbon dioxide (which diffuses away into the blood and is exhaled). What is left is the five-carbon compound alpha-ketoglutarate.

FIGURE C-4   The TCA Cycle

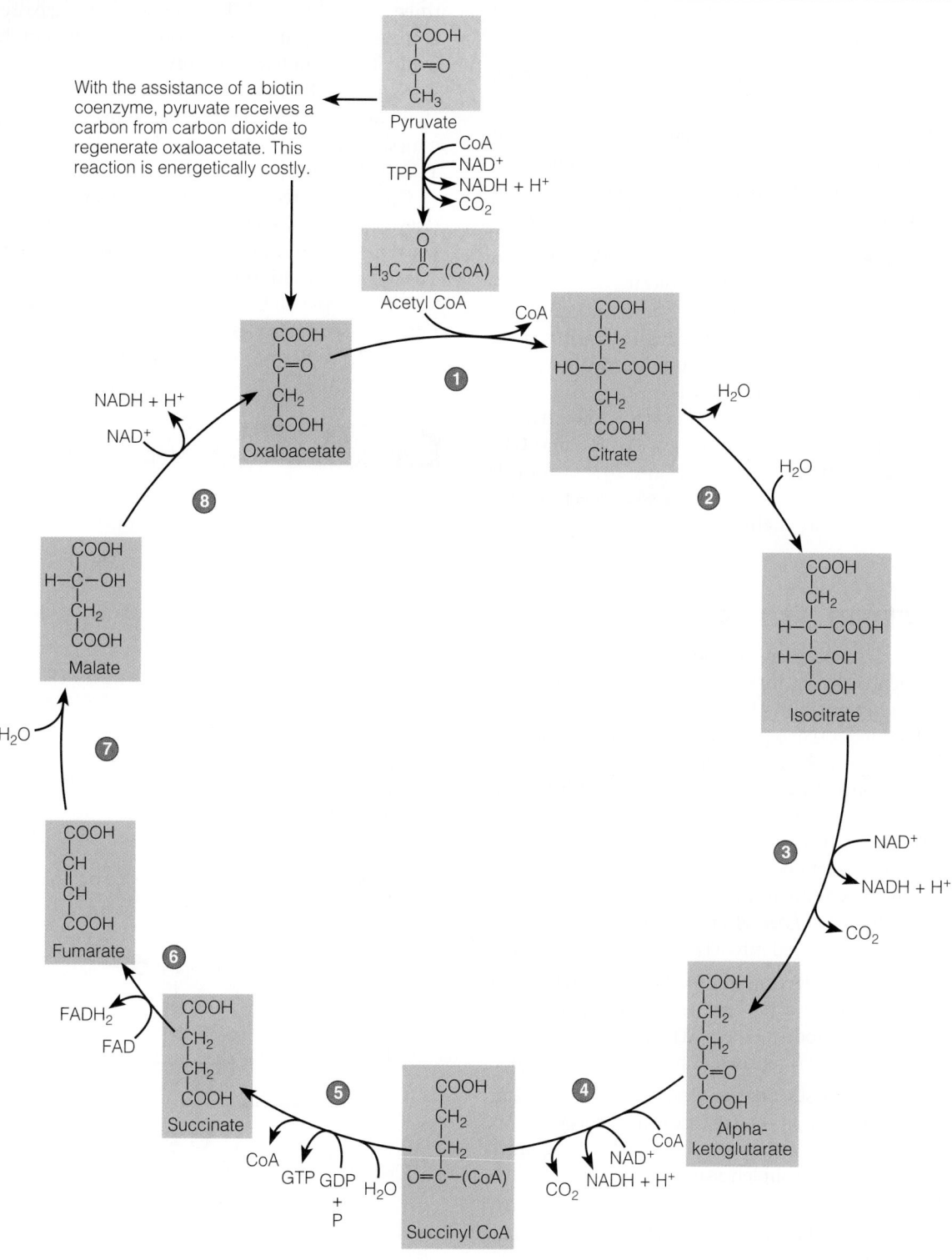

4. Now two compounds interact with alpha-ketoglutarate —a molecule of CoA and a molecule of NAD^+. In this complex reaction, a carbon and two oxygens are removed (forming carbon dioxide); two hydrogens are removed and go to NAD^+ (forming $NADH + H^+$); and the remaining four-carbon compound is attached to the CoA, forming succinyl

CoA. (Remember this $NADH + H^+$ also. You will see later what happens to it.)

5. Now two molecules react with succinyl CoA—a molecule called GDP and one of phosphate (P). The CoA comes off, the GDP and P combine to form the high-energy compound GTP (similar to ATP), and succinate remains. (Remember this GTP.)

6. In the next reaction, two H with their electrons are removed from succinate and are transferred to a molecule of FAD (a coenzyme like NAD^+) to form $FADH_2$. The product that remains is fumarate. (Remember this $FADH_2$.)

7. Next a molecule of water is added to fumarate, forming malate.

8. A molecule of NAD^+ reacts with the malate; two H with their associated electrons are removed from the malate and form $NADH + H^+$. The product that remains is the four-carbon compound oxaloacetate. (Remember this $NADH + H^+$.)

We are back where we started. The oxaloacetate formed in this process can combine with another molecule of acetyl CoA (step 1), and the cycle can begin again, as shown in Figure C-4.

So far, we have seen two carbons brought in with acetyl CoA and two carbons ending up in carbon dioxide. But where are the energy and the ATP we promised?

A review of the eight steps of the TCA cycle shows that the compounds $NADH + H^+$ (three molecules), $FADH_2$, and GTP capture energy originally found in acetyl CoA. To see how this energy ends up in ATP, we must follow the electrons further—into the electron transport chain.

The Electron Transport Chain

The six reactions described here are those of the electron transport chain, which is shown in Figure C-5. Since oxygen is required for these reactions, and ADP and P are combined to form ATP in several of them (ADP is phosphorylated), these reactions are also called oxidative phosphorylation.

An important concept to remember at this point is that an electron is not a fixed amount of energy. The electrons that bond the H to NAD^+ in NADH have a relatively large amount of energy. In the series of reactions that follow, they release this energy in small amounts, until at the end they are attached (with H) to oxygen (O) to make water (H_2O). In some of the steps, the energy they release is captured into ATP in coupled reactions.

1. In the first step of the electron transport chain, NADH reacts with a molecule called a flavoprotein, losing its electrons (and their H). The products are NAD^+ and reduced flavoprotein. A little energy is released as heat in this reaction.

2. The flavoprotein passes on the electrons to a molecule called coenzyme Q. Again they release some energy as heat, but ADP and P bond together and form ATP, storing much of the energy. This is a coupled reaction: ADP + P → ATP.

3. Coenzyme Q passes the electrons to cytochrome b. Again the electrons release energy.

4. Cytochrome b passes the electrons to cytochrome c in a coupled reaction in which ATP is formed: ADP + P → ATP.

5. Cytochrome c passes the electrons to cytochrome a.

6. Cytochrome a passes them (with their H) to an atom of oxygen (O), forming water (H_2O). This is a coupled reaction in which ATP is formed: ADP + P → ATP.

As Figure C-5 shows, each time NADH is oxidized (loses its electrons) by this means, the energy it releases is captured into three ATP molecules. When the electrons are passed on to water at the end, they are much lower in energy than they were originally. This completes the story of the electrons from NADH.

As for $FADH_2$, its electrons enter the electron transport chain at coenzyme Q. From coenzyme Q to water, ATP is generated in only two steps. Therefore, $FADH_2$ coming out of the TCA cycle yields just two ATP molecules.

One energy-receiving compound of the TCA cycle (GTP) does not enter the electron transport chain but gives its energy directly to ADP in a simple phosphorylation reaction. This reaction yields one ATP.

It is now possible to draw up a balance sheet of glucose metabolism (see Table C-3). Glycolysis has yielded $4 NADH + H^+$ and 4 ATP molecules and has spent 2 ATP. The 2 acetyl CoA

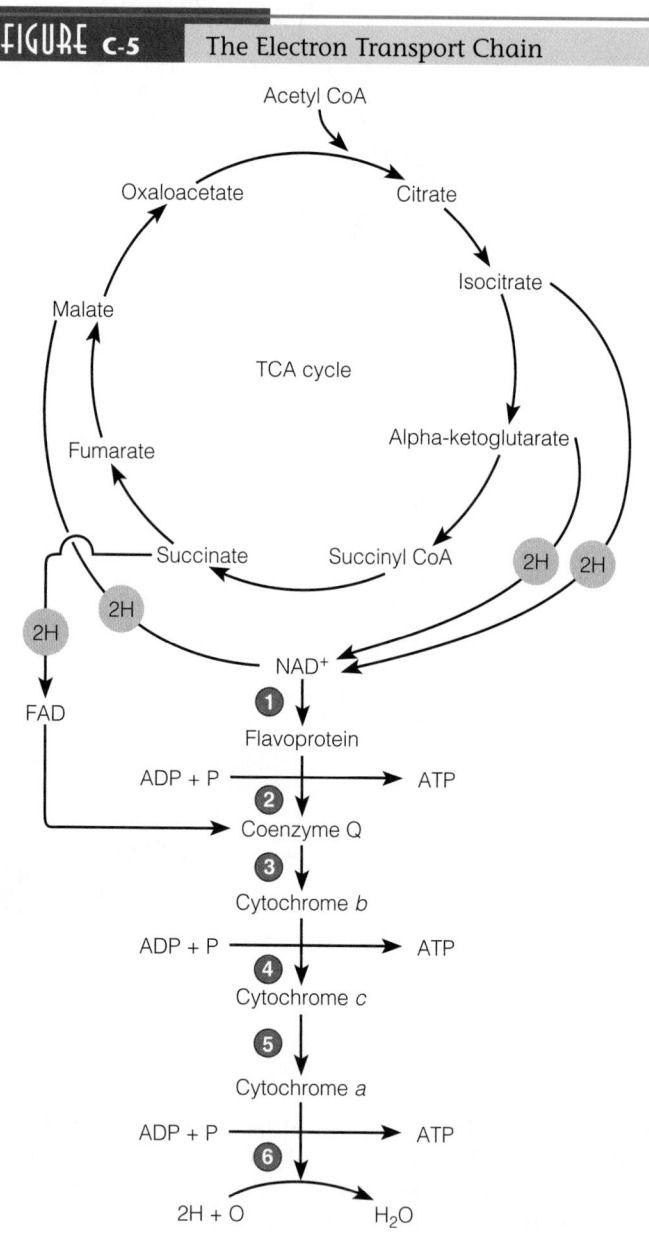

FIGURE C-5 The Electron Transport Chain

going through the TCA cycle have yielded 6 NADH + H$^+$, 2 FADH$_2$, and 2 GTP molecules. After the NADH + H$^+$ and FADH$_2$ have gone through the electron transport chain, there are 34 ATP. Added to these are the 4 ATP from glycolysis and the 2 ATP from GTP, making the total 40 ATP generated from one molecule of glucose. After the expense of 2 ATP is subtracted, there is a net gain of 38 ATP.*

A similar balance sheet from the complete breakdown of one 16-carbon fatty acid would show a net gain of 129 ATP. As mentioned earlier, 35 ATP were generated from the seven FADH$_2$ and seven NADH + H$^+$ produced during fatty acid oxidation. The eight acetyl CoA produced will each generate 12 ATP as they go through the TCA cycle and the electron transport chain, for a total of 96 more ATP. After subtracting the 2 ATP needed to activate the fatty acid initially, the net yield from one 16-carbon fatty acid: 35 + 96 − 2 = 129 ATP.

These calculations help explain why fat yields more energy (measured as kcalories) per gram than carbohydrate or protein. The more hydrogen atoms a fuel contains, the more ATP will be generated during oxidation. The 16-carbon fatty acid molecule, with its 32 hydrogen atoms, generates 129 ATP, whereas glucose, with its 12 hydrogen atoms, yields only 38 ATP.

The TCA cycle and the electron transport chain are the body's major means of capturing the energy from nutrients in ATP molecules. Other means, such as anaerobic glycolysis, contribute energy quickly, but the aerobic processes are the most efficient. Biologists and chemists understand much more about these processes than has been presented here.

Alcohol's Interference with Energy Metabolism

Highlight 7 provides an overview of how alcohol interferes with energy metabolism. With an understanding of the TCA cycle, a few more details may be appreciated. During alcohol metabolism, the enzyme alcohol dehydrogenase oxidizes alcohol to acetaldehyde while it simultaneously reduces a molecule of NAD$^+$ to NADH + H$^+$. The related enzyme acetaldehyde dehydrogenase reduces another NAD$^+$ to NADH + H$^+$ while it oxidizes acetaldehyde to acetyl CoA, the compound that enters the TCA cycle to generate energy. Thus, whenever alcohol is being metabolized in the body, NAD$^+$ diminishes, and NADH + H$^+$ accumulates. Chemists say that the body's "redox state" is altered, because NAD$^+$ can oxidize, and NADH + H$^+$ can reduce, many other body compounds. During alcohol metabolism, NAD$^+$ becomes unavailable for the multitude of reactions for which it is required.

As the previous sections just explained, for glucose to be completely metabolized, the TCA cycle must be operating, and NAD$^+$ must be present. If these conditions are not met (and when alcohol is present, they may not be), the pathway will be blocked, and traffic will back up—or an alternate route will be taken. Think about this as you follow the pathway shown in Figure C-6.

In each step of alcohol metabolism in which NAD$^+$ is converted to NADH + H$^+$, hydrogen ions accumulate, resulting in a dangerous shift of the acid-base balance toward acid (Chapter 12 explains acid-base balance). The accumulation of NADH + H$^+$ depresses TCA cycle activity, so pyruvate and acetyl CoA build up. This condition favors the conversion of pyruvate to lactic acid, which serves as a temporary storage place for hydrogens from NADH + H$^+$. The conversion of pyruvate to lactic acid restores some NAD$^+$, but a lactic acid buildup has serious consequences of its own. It adds to the body's acid burden and interferes with the excretion of uric acid, causing goutlike symptoms. Molecules of acetyl CoA become building blocks for fatty acids or ketone bodies. The making of ketone bodies consumes acetyl CoA and generates NAD$^+$; but some ketone bodies are acids, so they push the acid-base balance further toward acid.

Thus alcohol cascades through the metabolic pathways, wreaking havoc along the way. These consequences have physical effects, which Highlight 7 describes.

The Urea Cycle

Chapter 7 sums up the process by which waste nitrogen is eliminated from the body by stating that ammonia molecules combine with carbon dioxide to produce urea. This is true, but it is not the whole story. Urea is produced in a multistep process within the cells of the liver.

TABLE C-3	Balance Sheet for Glucose Metabolism	
	Expenditures	Income
Glycolysis:		
1 glucose	2 ATP	4 ATP
1 fructose-1,6-diphosphate		2 NADH + H$^+$
2 pyruvate		2 NADH + H$^+$
TCA cycle:		
2 isocitrate		2 NADH + H$^+$
2 alpha-ketoglutarate		2 NADH + H$^+$
2 succinyl CoA		2 GTP
2 succinate		2 FADH$_2$
2 malate		2 NADH + H$^+$
Total ATP collected:		
From glycolysis	2 ATP	4 ATP
From 2 NADH + H$^+$		4–6 ATP[a]
From 8 NADH + H$^+$		24 ATP
From 2 GTP		2 ATP
From 2 FADH$_2$		4 ATP
Totals:	2 ATP	38–40 ATP
Balance on hand from 1 molecule of glucose:		36–38 ATP

[a]Each NADH + H$^+$ from glycolysis can yield 2 or 3 ATP. See the accompanying text.

---

*The total may sometimes be 36 or 37, rather than 38, ATP. The NADH + H$^+$ generated in the cytoplasm during glycolysis pass their electrons on to shuttle molecules, which move them into the mitochondria. One shuttle, malate, contributes its electrons to the electron transport chain before the first site of ATP synthesis, yielding 3 ATP. Another, glycerol phosphate, adds its electrons into the chain beyond that first site, yielding 2 ATP. Thus sometimes 3, and sometimes only 2, ATP result from the NADH + H$^+$ that arise from glycolysis. The amount depends on the cell.

FIGURE C-6 Ethanol Enters the Metabolic Path

This is a simplified version of the glucose-to-energy pathway showing the entry of ethanol. The coenzyme NAD (which is the active form of the B vitamin niacin) is the only one shown here; however, many others are involved.

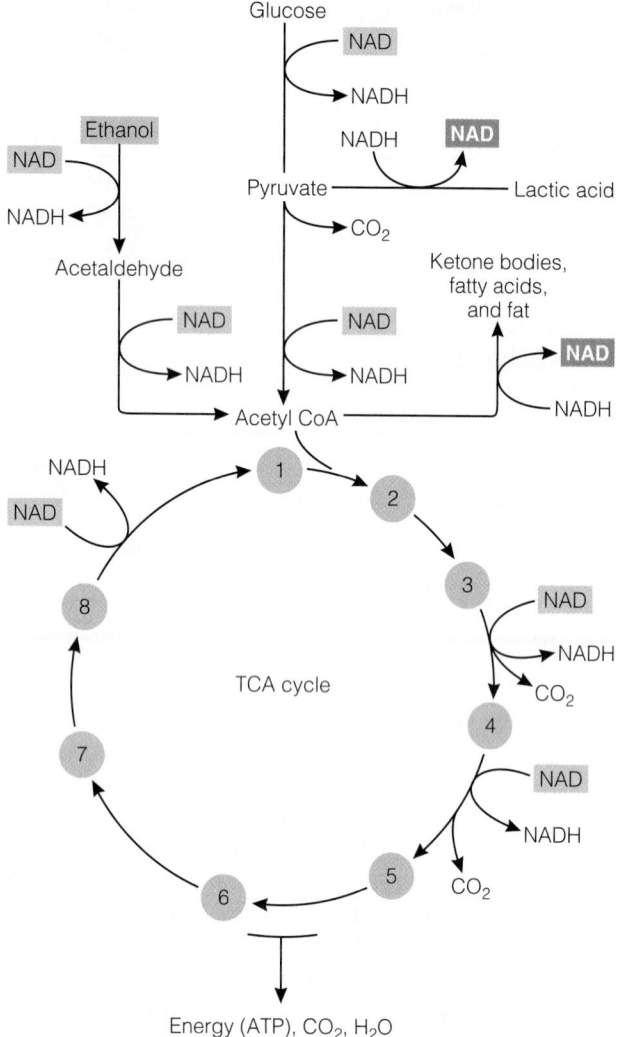

Ammonia, freed from an amino acid or other compound during metabolism anywhere in the body, arrives at the liver by way of the bloodstream and is taken into a liver cell. There, it is first combined with carbon dioxide and a phosphate group from ATP to form carbamyl phosphate:

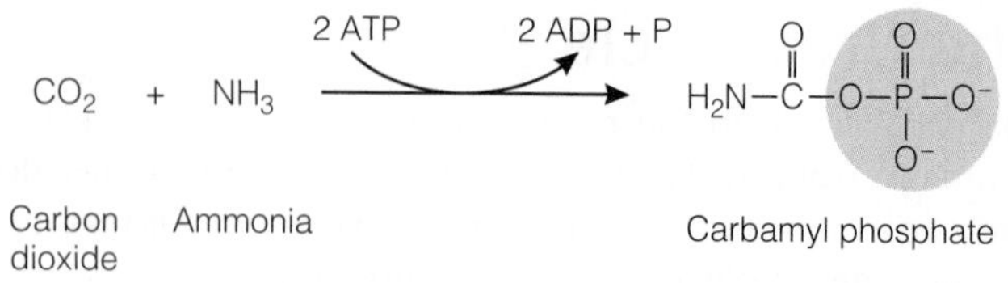

$$CO_2 \; + \; NH_3 \xrightarrow[\text{2 ADP + P}]{\text{2 ATP}} H_2N-\overset{\displaystyle O}{\underset{}{C}}-O-\overset{\displaystyle O}{\underset{\displaystyle O^-}{P}}-O^-$$

Carbon dioxide   Ammonia   Carbamyl phosphate

FIGURE C-7 The Urea Cycle

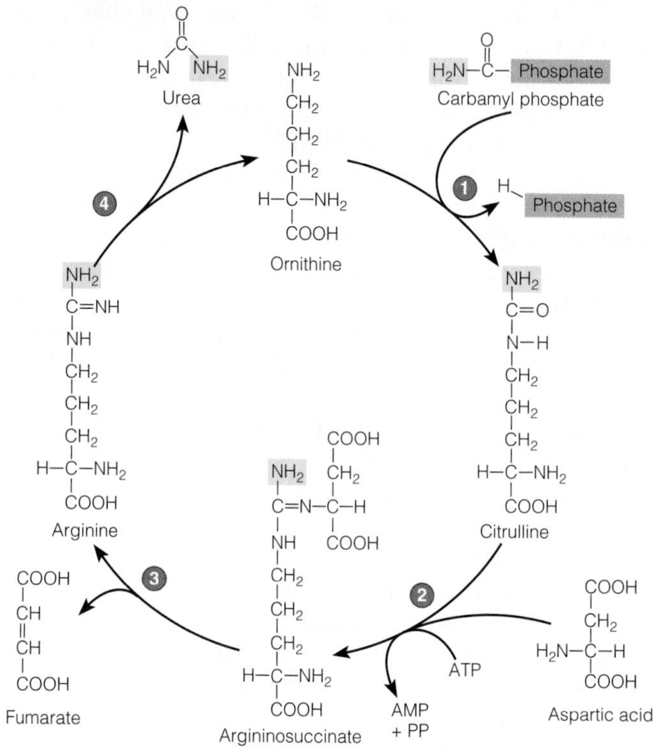

Figure C-7 shows the cycle of four reactions that follow.

1. Carbamyl phosphate combines with the amino acid ornithine, losing its phosphate group. The compound formed is citrulline.

2. Citrulline combines with the amino acid aspartic acid, to form argininosuccinate. The reaction requires energy from ATP. (ATP was shown earlier losing one phosphorus atom in a phosphate group, P, to become ADP. In this reaction, it loses two phosphorus atoms joined together, PP, and becomes adenosine monophosphate, AMP.)

3. Argininosuccinate is split, forming another acid, fumarate, and the amino acid arginine.

4. Arginine loses its terminal carbon with two attached amino groups and picks up an oxygen from water. The end product is urea, which the kidneys excrete in the urine. The compound that remains is ornithine, identical to the ornithine with which this series of reactions began, and ready to react with another molecule of carbamyl phosphate and turn the cycle again.

Formation of Ketone Bodies

Normally, fatty acid oxidation proceeds all the way to carbon dioxide and water. However, in ketosis (discussed in Chapter 7), an intermediate is formed from the condensation of two molecules of acetyl CoA: acetoacetyl CoA. Figure C-8 shows the formation of ketone bodies from that intermediate.

FIGURE C-8   The Formation of Ketone Bodies

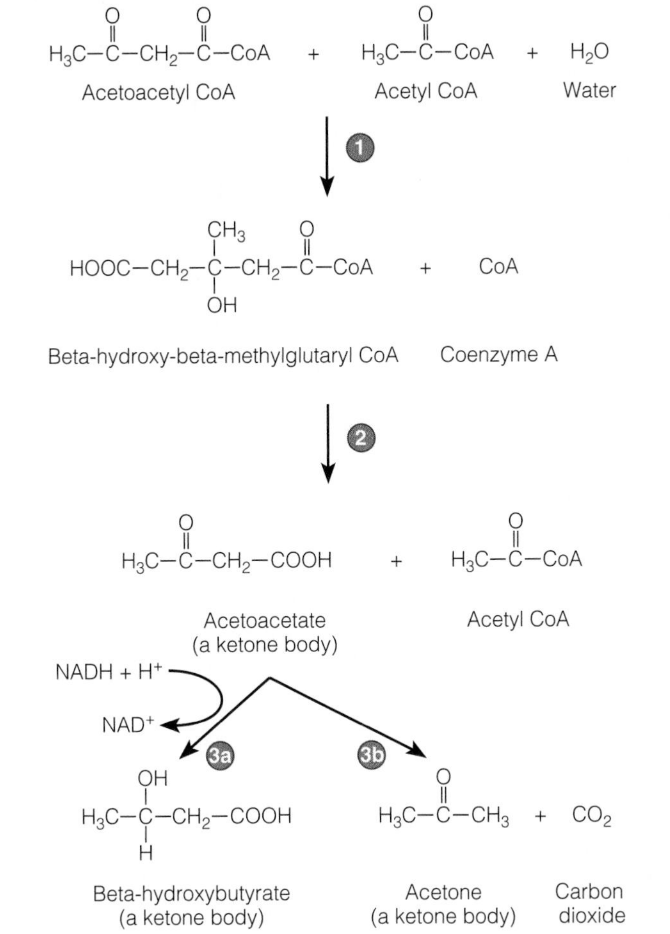

1. Acetoacetyl CoA condenses with acetyl CoA to form a six-carbon intermediate, beta-hydroxy-betamethylglutaryl CoA.

2. This intermediate is cleaved to acetyl CoA and acetoacetate.

3. Acetoactate can be metabolized either to beta-hydroxybutyrate acid (step 3a) or to acetone (3b).

Acetoacetate, beta-hydroxybutyrate, and acetone are the so-called ketone bodies of ketosis. Two are real ketones (they have a C=O group between two carbons); the other is an alcohol that has been produced during ketone formation—hence the term *ketone bodies,* rather than ketones, to describe the three of them. There are many other ketones in nature; these three are characteristic of ketosis in the body.

Measures of Protein Quality

In a world where food is scarce and many people's diets contain marginal or inadequate amounts of protein, it is important to know which foods contain the highest-quality protein. Chapter 6 describes protein quality, and this appendix presents different measures researchers use to assess the quality of a food protein. The accompanying glossary defines related terms.

Amino Acid Scoring

Amino acid scoring evaluates a protein's quality by determining its amino acid composition and comparing it with that of a reference protein. The advantages of amino acid scoring are that it is simple and inexpensive, it easily identifies the limiting amino acid, and it can be used to score mixtures of different proportions of two or more proteins mathematically without having to make up a mixture and test it. Its chief weaknesses are that it fails to estimate the digestibility of a protein, which may strongly affect the protein's quality; it relies on a chemical procedure in which certain amino acids may be destroyed, making the pattern that is analyzed inaccurate; and it is blind to other features of the protein (such as the presence of substances that may inhibit the digestion or utilization of the protein) that would only be revealed by a test in living animals.

Table D-1 shows the reference pattern for the nine essential amino acids. To interpret the table, read, "For every 3210 units of essential amino acids, 145 must be histidine, 340 must be isoleucine, 540 must be leucine," and so on. To compare a test

GLOSSARY

amino acid scoring: a measure of protein quality assessed by comparing a protein's amino acid pattern with that of a reference protein; sometimes called **chemical scoring.**

biological value (BV): a measure of protein quality assessed by measuring the amount of protein nitrogen that is retained from a given amount of protein nitrogen absorbed.

net protein utilization (NPU): a measure of protein quality assessed by measuring the amount of protein nitrogen that is retained from a given amount of protein nitrogen eaten.

PDCAAS (protein digestibility–corrected amino acid score): a measure of protein quality assessed by comparing the amino acid score of a food protein with the amino acid requirements of preschool-age children and then correcting for the true digestibility of the protein; recommended by the FAO/WHO and used to establish protein quality of foods for Daily Value percentages on food labels.

protein efficiency ratio (PER): a measure of protein quality assessed by determining how well a given protein supports weight gain in growing rats; used to establish the protein quality for infant formulas and baby foods.

protein with the reference protein, the experimenter first obtains a chemical analysis of the test protein's amino acids. Then, taking 3210 units of the amino acids, the experimenter compares the amount of each amino acid to the amount found in 3210 units of essential amino acids in egg protein. For example, suppose the test protein contained (per 3210 units) 360 units of isoleucine; 500 units of leucine; 350 of lysine; and for each of the other amino acids, more units than egg protein contains. The two amino acids that are low are leucine (500 as compared with 540 in egg) and lysine (350 versus 440 in egg). The ratio, amino acid in the test protein divided by amino acid in egg, is 500/540 (or about 0.93) for leucine and 350/440 (or about 0.80) for lysine. Lysine is the limiting amino acid (the one that falls shortest compared with egg). If the protein's limiting amino acid is 80 percent of the amount found in the reference protein, it receives a score of 80.

PDCAAS

PDCAAS (protein digestibility–corrected amino acid score) takes the amino acid scoring method a step further by correcting for the digestibility of the protein. Chapter 6 presents PDCAAS in detail.

TABLE D-1 A Reference Pattern for Amino Acid Scoring of Proteins

Essential Amino Acids	Reference Protein—Whole Egg (mg amino acid/g nitrogen)
Histidine	145
Isoleucine	340
Leucine	540
Lysine	440
Methionine + cystine[a]	355
Phenylalanine + tyrosine[b]	580
Threonine	294
Tryptophan	106
Valine	410
Total	3210

[a]Methionine is essential and is also used to make cystine. Thus the methionine requirement is lower if cystine is supplied.
[b]Phenylalanine is essential and is also used to make tyrosine if not enough of the latter is available. Thus the phenylalanine requirement is lower if tyrosine is also supplied.

D Appendix

TABLE D-2	Biological Values (BV) of Selected Foods
Egg	100
Milk	93
Beef	75
Fish	75
Corn	72

NOTE: 100 is the maximum BV a food protein can receive.

Biological Value

The **biological value (BV)** of a protein measures its efficiency in supporting the body's needs. In a test of biological value, two nitrogen balance studies are done. In the first, no protein is fed, and nitrogen (N) excretions in the urine and feces are measured. It is assumed that under these conditions, N lost in the urine is the amount the body always necessarily loses by filtration into the urine each day, regardless of what protein is fed (endogenous N). The N lost in the feces (called metabolic N) is the amount the body invariably loses into the intestine each day, whether or not food protein is fed. (To help you remember the terms: endogenous N is "urinary N on a zero-protein diet"; metabolic N is "fecal N on a zero-protein diet.")

In the second study, an amount of protein slightly below the requirement is fed. Intake and losses are measured; then the BV is derived using this formula:

$$BV = \frac{N \text{ retained}}{N \text{ absorbed}} \times 100.$$

The denominator of this equation expresses the amount of nitrogen *absorbed:* food N minus fecal N (excluding the metabolic N the body would lose in the feces anyway, even without food). The numerator expresses the amount of N *retained* from the N absorbed: absorbed N (as in the denominator) minus the N excreted in the urine (excluding the endogenous N the body would lose in the urine anyway, even without food). The more nitrogen retained, the higher the protein quality. (Recall that when an essential amino acid is missing, protein synthesis stops, and the remaining amino acids are deaminated and the nitrogen excreted.)

Egg protein has a BV of 100, indicating that 100 percent of the nitrogen absorbed is retained. Supplied in adequate quantity, a protein with a BV of 70 or greater can support human growth as long as energy intake is adequate. Table D-2 presents the BV for selected foods.

This method has the advantages of being based on experiments with human beings (it can be done with animals, too, of course) and of measuring actual nitrogen retention. But it is also cumbersome, expensive, and often impractical, and it is based on several assumptions that may not be valid. For example, the physiology, normal environment, or typical food intake of the subjects used for testing may not be similar to those for whom the test protein may ultimately be used. For another example, the retention of protein in the body does

not necessarily mean that it is being well utilized. Considerable exchange of protein among tissues (protein turnover) occurs, but is hidden from view when only N intake and output are measured. The test of biological value wouldn't detect if one tissue were shorted.

Net Protein Utilization

Like BV, **net protein utilization (NPU)** measures how efficiently a protein is used by the body and involves two balance studies. The difference is that NPU measures retention of food nitrogen rather than food nitrogen absorbed (as in BV). The formula for NPU is:

$$NPU = \frac{N \text{ retained}}{N \text{ intake}} \times 100.$$

The numerator is the same as for BV, but the denominator represents food N intake only—not N absorbed.

This method offers advantages similar to those of BV determinations and is used more frequently, with animals as the test subjects. A drawback is that if a low NPU is obtained, the test results offer no help in distinguishing between two possible causes: a poor amino acid composition of the test protein or poor digestibility. There is also a limit to the extent to which animal test results can be assumed to be applicable to human beings.

Protein Efficiency Ratio

The **protein efficiency ratio (PER)** measures the weight gain of a growing animal and compares it to the animal's protein intake. Until recently, the PER was generally accepted in the United States and Canada as the official method for assessing protein quality, and it is still used to evaluate proteins for infants.

Young rats are fed a measured amount of protein and weighed periodically as they grow. The PER is expressed as:

$$PER = \frac{\text{weight gain (g)}}{\text{protein intake (g)}}.$$

This method has the virtues of economy and simplicity, but it also has many drawbacks. The experiments are time-consuming; the amino acid needs of rats are not the same as those of human beings; and the amino acid needs for growth are not the same as for the maintenance of adult animals (growing animals need more lysine, for example). Table D-3 presents PER values for selected foods.

TABLE D-3	Protein Efficiency Ratio (PER) Values of Selected Proteins
Casein (milk)	2.8
Soy	2.4
Glutein (wheat)	0.4

Nutrition Assessment

Nutrition assessment evaluates a person's health from a nutrition perspective. Many factors influence or reflect nutrition status. Consequently, the assessor, usually a registered dietitian assisted by other qualified health care professionals, gathers information from many sources, including:

- Historical information.
- Anthropometric measurements.
- Physical examinations.
- Biochemical analyses (laboratory tests).

Each of these methods involves collecting data in a variety of ways and interpreting each finding in relation to the others to create a total picture.

The accurate gathering of this information and its careful interpretation are the basis for a meaningful evaluation. The more information gathered about a person, the more accurate the assessment will be. Gathering information is a time-consuming process, however, and time is often a rare commodity in the health care setting. Nutrition care is only one part of total care. It may not be practical or essential to collect detailed information on each person.

A strategic compromise is to screen clients by collecting preliminary data. Data such as height-weight and hematocrit are easy to obtain and can alert health care workers to potential problems. **Nutrition screening** identifies clients who will require additional nutrition assessment. This appendix provides a sample of the procedures, standards, and charts commonly used in nutrition assessment.

Historical Information

Clues about present nutrition status become evident with a careful review of a person's historical data (see Table E-1). Even when the data are subjective, they reveal important facts about a person. A thorough history identifies risk factors associated with poor nutrition status (see Table E-2) and provides a sense of the whole person. As you can see, many aspects of a person's life influence nutrition status and provide clues to possible problems.

An adept history taker uses the interview both to gather facts and to establish a rapport with the client. This section briefly reviews the major areas of nutrition concern in a person's history: health, socioeconomic factors, drugs, and diet.

nutrition screening: the use of preliminary nutrition assessment techniques to identify people who are malnourished or are at risk for malnutrition.

E *Appendix*

TABLE E-1	Historical Data Used in Nutrition Assessments
Type of History	**What It Identifies**
Health history	Current and previous health problems and family health history that affect nutrient needs, nutrition status, or the need for intervention to prevent health problems
Socioeconomic history	Personal, cultural, financial, and environmental influences on food intake, nutrient needs, and diet therapy options
Drug history	Medications (prescription and over-the-counter), illicit drugs, dietary supplements, and alternative therapies that affect nutrition status
Diet history	Nutrient intake excesses or deficiencies and reasons for imbalances

TABLE E-2   Risk Factors for Poor Nutrition Status

Health History

- Acquired immune deficiency syndrome (AIDS)
- Alcoholism
- Anorexia (lack of appetite)
- Anorexia nervosa
- Bulimia nervosa
- Burns
- Cancer
- Chewing or swallowing difficulties (including poorly fitted dentures, dental caries, missing teeth, and mouth ulcers)
- Chronic obstructive pulmonary disease
- Circulatory problems
- Constipation
- Crohn's disease
- Cystic fibrosis
- Decubitus ulcers (pressure sores)
- Dementia
- Depleted blood proteins
- Depression
- Diabetes mellitus

- Diarrhea, prolonged or severe
- Drug addiction
- Dysphagia
- Failure to thrive
- Feeding disabilities
- Fever
- GI tract disorders or surgery
- Heart disease
- HIV infection
- Hormonal imbalance
- Hyperlipidemia
- Hypertension
- Infections
- Kidney disease
- Liver disease
- Lung disease
- Malabsorption
- Mental illness
- Mental retardation
- Multiple pregnancies

- Nausea
- Neurologic disorders
- Organ failure
- Overweight
- Pancreatic insufficiency
- Paralysis
- Physical disability
- Pneumonia
- Pregnancy
- Radiation therapy
- Recent major illness
- Recent major surgery
- Recent weight loss or gain
- Tobacco use
- Trauma
- Ulcerative colitis
- Ulcers
- Underweight
- Vomiting, prolonged or severe

Socioeconomic History

- Access to groceries
- Activities
- Age
- Education

- Ethnic identity
- Income
- Kitchen facilities
- Number of people in household

- Occupation
- Religious affiliation

Drug History

- Amphetamines
- Analgesics
- Antacids
- Antibiotics
- Anticonvulsant agents
- Antidepressant agents
- Antidiabetic agents

- Antidiarrheals
- Antifungal agents
- Antihyperlipemics
- Antihypertensives
- Antineoplastics
- Antiulcer agents
- Antiviral agents

- Catabolic steroids
- Diuretics
- Hormonal agents
- Immunosuppressive agents
- Laxatives
- Oral contraceptives
- Vitamin and other dietary supplements

Diet History

- Deficient or excessive food intakes
- Frequently eating out
- Intravenous fluids (other than total parenteral nutrition) for 7 or more days

- Monotonous diet (lacking variety)
- No intake for 7 or more days
- Poor appetite
- Restricted or fad diets

- Unbalanced diet (omitting any food group)
- Recent weight gains or losses

Health History

The assessor can obtain a **health history** from records completed by the attending physician, nurse, or other health care professional. In addition, conversations with the client can uncover valuable information previously overlooked because no one thought to ask or because the client was not thinking clearly when asked.

An accurate, complete health history can reveal conditions that increase a client's risk for malnutrition (review Table E-2). Diseases and their therapies can have either immediate or long-term effects on nutrition status by interfering with ingestion, digestion, absorption, metabolism, or excretion of nutrients.

health history: an account of a client's current and past health status and disease risks.

Socioeconomic History

A **socioeconomic history** reveals factors that profoundly affect nutrition status. The ethnic background and educational level of both the client and the other members of the household influence food availability and food choices. An understanding of the community environment is also important in assessing nutrition status. For example, the interviewer should be familiar with the food habits of the major ethnic groups within the locale, regional food preferences, and nutrition resources and programs available in the community. Local health departments and social agencies often can provide such information.

Level of income also influences the diet. In general, the quality of the diet declines as income falls. At some point, the ability to purchase the foods required to meet nutrient needs is lost; an inadequate income puts an adequate diet out of reach. Agencies use poverty indexes to identify people at risk for poor nutrition and to qualify people for government food assistance programs.

Low income affects not only the power to purchase foods but also the ability to shop for, store, and cook them. A skilled assessor will note whether a person has transportation to a grocery store that sells a sufficient variety of low-cost foods, and whether the person has access to a refrigerator and stove.

Drug History

The many interactions of foods and drugs require that health care professionals take a **drug history** and pay special attention to any client who takes drugs routinely. If a person is taking any drug, the assessor records the name of the drug; the dose, frequency, and duration of intake; the reason for taking the drug; and signs of any adverse effects.

The interactions of drugs and nutrients may take many forms:

- Drugs can alter food intake and the absorption, metabolism, and excretion of nutrients.
- Foods and nutrients can alter the absorption, metabolism, and excretion of drugs.

Highlight 17 discusses nutrient-drug interactions in more detail, and Table H17-1 (on p. 609) summarizes the mechanisms by which these interactions occur and provides specific examples.

Diet History

A **diet history** provides a record of a person's eating habits and food intake and can help identify possible nutrient imbalances. Food choices are an important part of lifestyle and often reflect a person's philosophy. The assessor who asks nonjudgmental questions about eating habits and food intake encourages trust and enhances the likelihood of obtaining accurate information.

Assessors evaluate food intake using various tools such as the 24-hour recall, the usual intake record, the food record, and the food frequency questionnaire. Food models or photos and measuring devices can help clients identify the types of foods and quantities consumed. The assessor also needs to know how the foods are prepared and when they are eaten. In addition to asking about foods, assessors will ask about beverage consumption, including beverages containing alcohol or caffeine.

Besides identifying possible nutrient imbalances, diet histories provide valuable clues about how a person will accept diet changes should they be necessary. Information about what and how a person eats provides the background for realistic and attainable nutrition goals.

24-Hour Recall The **24-hour recall** provides data for one day only and is commonly used in nutrition surveys to obtain estimates of the typical food intakes for a population. The assessor asks the client to recount everything eaten or drunk in the past 24 hours or for the previous day.

socioeconomic history: a record of a person's social and economic background, including such factors as education, income, and ethnic identity.

drug history: a record of all the drugs, over-the-counter and prescribed, that a person takes routinely.

diet history: a record of eating behaviors and the foods a person eats.

24-hour recall: a record of foods eaten by a person for one 24-hour period.

E Appendix

An advantage of the 24-hour recall is that it is easy to obtain. It is also more likely to provide accurate data, at least about the past 24 hours, than estimates of average intakes over long periods. It does not, however, provide enough information to allow accurate generalizations about an individual's usual food intake. The previous day's intake may not be typical, for example, or the person may be unable to report portion sizes accurately or may conceal or forget information about foods eaten. This limitation is partially overcome when 24-hour recalls are collected on several nonconsecutive days.

Usual Intake To obtain data about a person's usual intake, an inquiry might begin with "What is the first thing you usually eat or drink during the day?" Similar questions follow until a typical daily intake pattern emerges. This method can be useful, especially in verifying food intake when the past 24 hours have been atypical. It also helps the assessor verify food habits. For example, one person may always eat an afternoon snack; another may never eat breakfast. A person whose intake varies widely from day to day, however, may find it difficult to answer such general questions, and in that case, another food intake tool should be used to estimate nutrient intake.

Food Record Another tool for history taking is the **food record,** in which the person records food eaten, including the quantity and method of preparation. Chapter 9 (p. 299) provides an example. A food record can help both the assessor and the client to determine factors associated with eating that may affect dietary balance and adequacy.

Food records work especially well with cooperative people but require considerable time and effort on their part. A prime advantage is that the record keeper assumes an active role and may for the first time become aware of personal food habits and assume responsibility for them. It also provides the assessor with an accurate picture of the person's lifestyle and factors that affect food intake. For these reasons, a food record can be particularly useful in outpatient counseling for such nutrition problems as overweight, underweight, or food allergy. The major disadvantages stem from poor compliance in recording the data and conscious or unconscious changes in eating habits that may occur while the person is keeping the record.

Food Frequency Questionnaire An assessor uses a **food frequency questionnaire** to compare a client's food intake with the Daily Food Guide. Clients may be asked how many servings of each of the following they eat in a typical day: breads, cereals, or grain products; vegetables; fruits; meat, poultry, fish, and alternatives; milk, cheese, and yogurt; and fats, oils, and sweets. This information helps pinpoint food groups, and therefore nutrients, that may be excessive or deficient in the diet. That a person ate no vegetables yesterday may not seem particularly significant, but never eating vegetables is a warning of possible nutrient deficiencies. When used with the usual intake or 24-hour recall approach, the food frequency questionnaire enables the assessor to double-check the accuracy of the information obtained.

Analysis of Food Intake Data After collecting food intake data, the assessor estimates nutrient intakes, either informally by using food guides or formally by using food composition tables. The assessor compares these intakes with standards, usually nutrient recommendations or dietary guidelines, to determine how closely the person's diet meets the standards. Are the types and amounts of proteins, carbohydrates (including fiber), and fats (including cholesterol) appropriate? Are all food groups included in appropriate amounts? Is caffeine or alcohol consumption excessive? Are intakes of any vitamins or minerals (including sodium and iron) excessive or deficient? An informal evaluation is possible only if the assessor has enough prior experience with formal calculations to "see" nutrient amounts in reported food intakes without calculations. Even then, such an informal analysis is best followed by a spot check for key nutrients by actual calculation.

food record: an extensive, accurate log of all foods eaten over a period of several days or weeks. A food record that includes associated information such as when, where, and with whom each food is eaten is sometimes called a **food diary.**

food frequency questionnaire: a checklist of foods on which a person can record the frequency with which he or she eats each food.

Formal calculations can be performed either manually (by looking up each food in a table of food composition, recording its nutrients, and adding them up) or by using a computer diet analysis program. The assessor then compares the intakes with standards such as the RDA.

Limitations of Food Intake Analysis Diet histories can be superbly informative, but the skillful assessor also keeps their limitations in mind. For example, a computer diet analysis tends to imply greater accuracy than is possible to obtain from data as uncertain as the starting information. Nutrient contents of foods listed in tables of food composition or stored in computer databases are averages and, for some nutrients, incomplete. In addition, the available data on nutrient contents of foods do not reflect the amounts of nutrients a person actually absorbs. Iron is a case in point: its availability from a given meal may vary depending on the person's iron status; the relative amounts of heme iron, nonheme iron, vitamin C, meat, fish, and poultry eaten at the meal; and the presence of inhibitors of iron absorption such as tea, coffee, and nuts. Chapter 13 describes the many factors that influence iron absorption from a meal.

Furthermore, reported portion sizes may not be correct. The person who reports eating "a serving" of greens may not distinguish between ¼ cup and 2 whole cups; only trained individuals can accurately report serving sizes. Children tend to remember the serving sizes of foods they like as being larger than serving sizes of foods they dislike.

An estimate of nutrient intakes from a diet history, combined with other sources of information, allows the assessor to confirm or eliminate the possibility of suspected food intake problems. The assessor must constantly remember that nutrient intakes in adequate amounts do not guarantee adequate nutrient status for an individual. Likewise, insufficient intakes do not always indicate deficiencies, but instead alert the assessor to possible problems. Each person digests, absorbs, metabolizes, and excretes nutrients in a unique way; individual needs vary. Intakes of nutrients identified by diet histories are only pieces of a puzzle that must be put together with other indicators of nutrition status in order to extract meaning.

Anthropometric Measurements

Anthropometrics are physical measurements that reflect body composition and development (see Table E-3). They serve three main purposes: first, to evaluate the progress of growth in pregnant women, infants, children, and adolescents; second, to detect undernutrition and overnutrition in all age groups; and third, to measure changes in body composition over time.

Health care professionals compare anthropometric measurements taken on an individual with population standards specific for gender and age or with previous measures of the individual. Measurements taken periodically and compared with previous measurements reveal changes in an individual's status.

anthropometrics: measurements of the physical characteristics of the body, such as height and weight.

- **anthropos** = human
- **metric** = measuring

E Appendix

TABLE E-3 Anthropometric Measurements Used in Nutrition Assessments	
Type of Measurement	**What It Reflects**
Abdominal girth measurement	Abdominal fluid retention and abdominal organ size
Height-weight	Overnutrition and undernutrition; growth in children
Head circumference	Brain growth and development in infants and children under age two
Fatfold	Subcutaneous and total body fat
Waist circumference	Body fat distribution

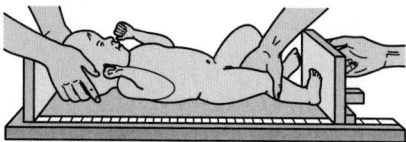

FIGURE E-1 Length Measurement of an Infant

An infant is measured lying down on a measuring board with a fixed headboard and a movable footboard. Note that two people are needed to measure the infant's length.

Mastering the techniques for taking anthropometric measurements requires proper instruction and practice to ensure reliability. Once the correct techniques are learned, taking measurements is easy and requires minimal equipment.

Height and weight are well-recognized anthropometrics; other anthropometrics include fatfold measurements and various measures of lean tissue. Other measures are useful in specific situations. For example, a head circumference measurement may help to assess brain development in an infant, and an abdominal girth measurement supplies information about abdominal fluid retention in individuals with liver disease.

Measures of Growth and Development

Height and weight are among the most common and useful anthropometric measurements. Length measurements for infants and children up to age three and height measurements for children over three are particularly valuable in assessing growth and therefore nutrition status. For adults, height measurements alone are not critical, but help to estimate healthy weight and to interpret other assessment data. Once adult height has been reached, changes in body weight provide useful information in assessing overnutrition and undernutrition.

Height For infants and children younger than three, health care professionals may use special equipment to measure length. The assessor lays the barefoot infant on a measuring board that has a fixed headboard and movable footboard attached at right angles to the surface (see Figure E-1). Often two people are needed to obtain an accurate measurement: one to hold the infant's head against the headboard, and the other to keep the legs straight and do the measuring. This method provides the most accurate measure possible, but many health care professionals use a less exacting method. They may simply hold the infant straight with its head against the headboard or other vertical support, mark the blanket with a chalk or pen at the infant's heel, and then measure the distance from the headboard to the mark. Even more informally and less accurately, they may lay the infant on a flat surface and extend a nonstretchable measuring tape along the side of the infant from the top of the head to the heel of the foot.

The procedure for measuring a child who can stand erect and cooperate is the same as for an adult. The best way to measure standing height is with the person's back against a flat wall to which a nonstretchable measuring tape or stick has been fixed (see Figure E-2). The person stands erect, without shoes, with heels together. The person's line of sight should be horizontal, with the heels, buttocks, shoulders, and head touching the wall. The assessor places a ruler, book, or other inflexible object on top of the head at a right angle to the wall; carefully checks the height measurement; and records it immediately in either inches or centimeters so that the correct measurement will not be forgotten.

The measuring rod of a scale is commonly used, but is less accurate because it bends easily. The assessor follows the same general procedure, asking the person to face away from the scale and to take extra care to stand erect.

Unfortunately, many health care professionals merely ask clients how tall they are rather than measuring their height. Self-reported height is often inaccurate and should be used only as a last resort when measurement is impractical (in the case of an uncooperative client, an emergency admission, or the like).

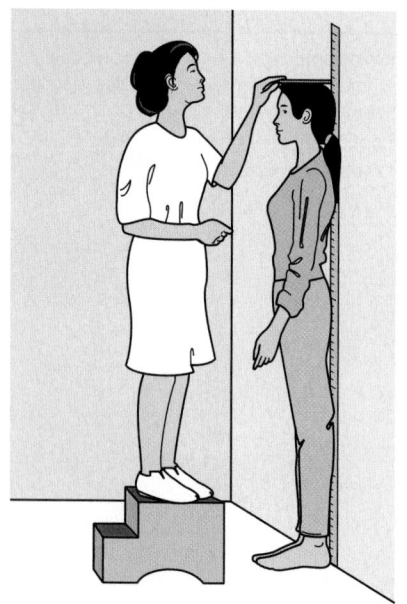

FIGURE E-2 Height Measurement of an Older Child or Adult

Height is measured most accurately when the person stands against a flat wall to which a measuring tape has been affixed.

Weight Valid weight measurements require scales that have been carefully maintained, calibrated, and checked for accuracy at regular intervals. Beam balance and electronic scales are the most accurate types of scales. To measure infants' weight, assessors use special scales that allow infants to lie or sit (see Figure E-3). Weighing infants naked, without diapers, is standard procedure. Children who can stand are weighed in the same way as adults (see Figure E-4). To make repeated measures useful, standardized conditions are necessary. Each weighing should take place at the same time of day (preferably before breakfast), in the same amount of clothing (without shoes), after the person has voided, and on the same scale. Special scales and hospital beds with built-in scales are available for weighing people who are bedridden. Bathroom scales are inaccurate and inappropriate in a professional setting. As with all measurements, the assessor records the observed weight immediately in either pounds or kilograms.

Head Circumference Assessors may also measure head circumference to confirm that infant growth is proceeding normally or to help detect protein-energy malnutrition (PEM) and evaluate the extent of its impact on brain size. To measure head circumference, the assessor places a nonstretchable tape so that it encircles the largest part of the infant's or child's head: just above the eyebrow ridges, just above the point where the ears attach, and around the occipital prominence at the back of the head. To ensure accurate recording, the assessor immediately notes the measure in either inches or centimeters.

Analysis of Measures in Infants and Children Growth retardation is a sign of poor nutrition status. Obesity is also a sign that dietary intervention may be needed.

Health professionals generally evaluate physical development by monitoring the growth rate of a child and comparing this rate with standard charts. Standard charts compare weight to age, height to age, and weight to height; ideally, height and weight are in roughly the same percentile. Although individual growth patterns may vary, a child's growth curve will generally stay at about the same percentile throughout childhood. In children whose growth has been retarded, nutrition rehabilitation will ideally induce height and weight to increase to higher percentiles. In overweight children, the goal is for weight to remain stable as height increases, until weight becomes appropriate for height.

To evaluate growth in infants, an assessor uses charts such as those in Figures E-5 (A and B) through E-10 (A and B).■ The assessor follows these steps to plot a weight measurement on a percentile graph:

- Select the appropriate chart based on age and gender.
- Locate the child's age along the horizontal axis on the bottom of the chart.
- Locate the child's weight in pounds or kilograms along the vertical axis.
- Mark the chart where the age and weight lines intersect, and read off the percentile.

To assess length, height, or head circumference, the assessor follows the same procedure, using the appropriate chart. (When length is measured, use the chart for birth to 36 months; when height is measured, use the chart for 2 to 20 years.) Head circumference percentile should be similar to the child's height and weight percentiles. With height, weight, and head circumference measures plotted on growth percentile charts, a skilled clinician can begin to interpret the data.

Percentile charts divide the measures of a population into 100 equal divisions. Thus half of the population falls above the 50th percentile, and half falls below. The use of percentile measures allows for comparisons among people of the same age and gender. For example, a six-month-old female infant whose weight is at the 75 percentile weighs more than 75 percent of the female infants her age.

Head circumference is generally measured in children under two years of age. Since the brain grows rapidly before birth and during early infancy, extreme and chronic malnutrition during these times can impair brain development, curtailing the number of brain cells and the size of head circumference. Nonnutritional factors, such as certain disorders and genetic variation, can also influence head circumference.

Analysis of Measures in Adults For adults, health care professionals typically compare weights with weight-for-height standards. One such standard is the body mass index (BMI),■ described in Chapter 8 (pp. 262–263), which is useful for estimating the risk to health associated with overnutrition. The back cover shows BMI for various heights and weights.

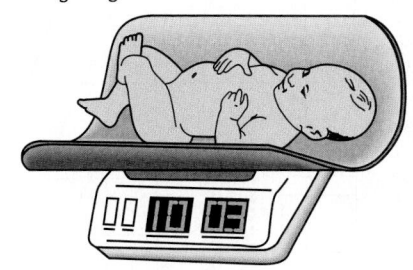

FIGURE E-3 Weight Measurement of an Infant

Infants sit or lie down on scales that are designed to hold them while they are being weighed.

■ Chapter 16 presents BMI charts for children and adolescents.

■ Reminder: The *body mass index (BMI)* is an index of a person's weight in relation to height, determined by dividing the weight in kilograms by the square of the height in meters:

$$BMI = \frac{Weight\ (kg)}{Height\ (m)^2}.$$

FIGURE E-4 Weight Measurement of an Older Child or Adult

Whenever possible, children and adults are measured on beam balance or electronic scales to ensure accuracy.

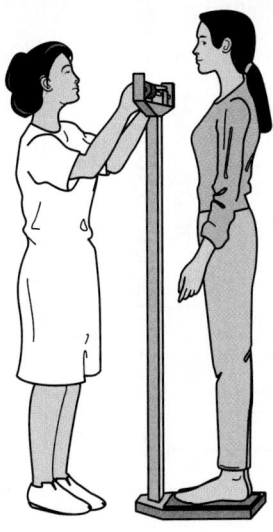

E Appendix

FIGURE E-5A Weight-for-Age Percentiles: Boys, Birth to 36 Months

FIGURE E-5B Weight-for-Age Percentiles: Girls, Birth to 36 Months

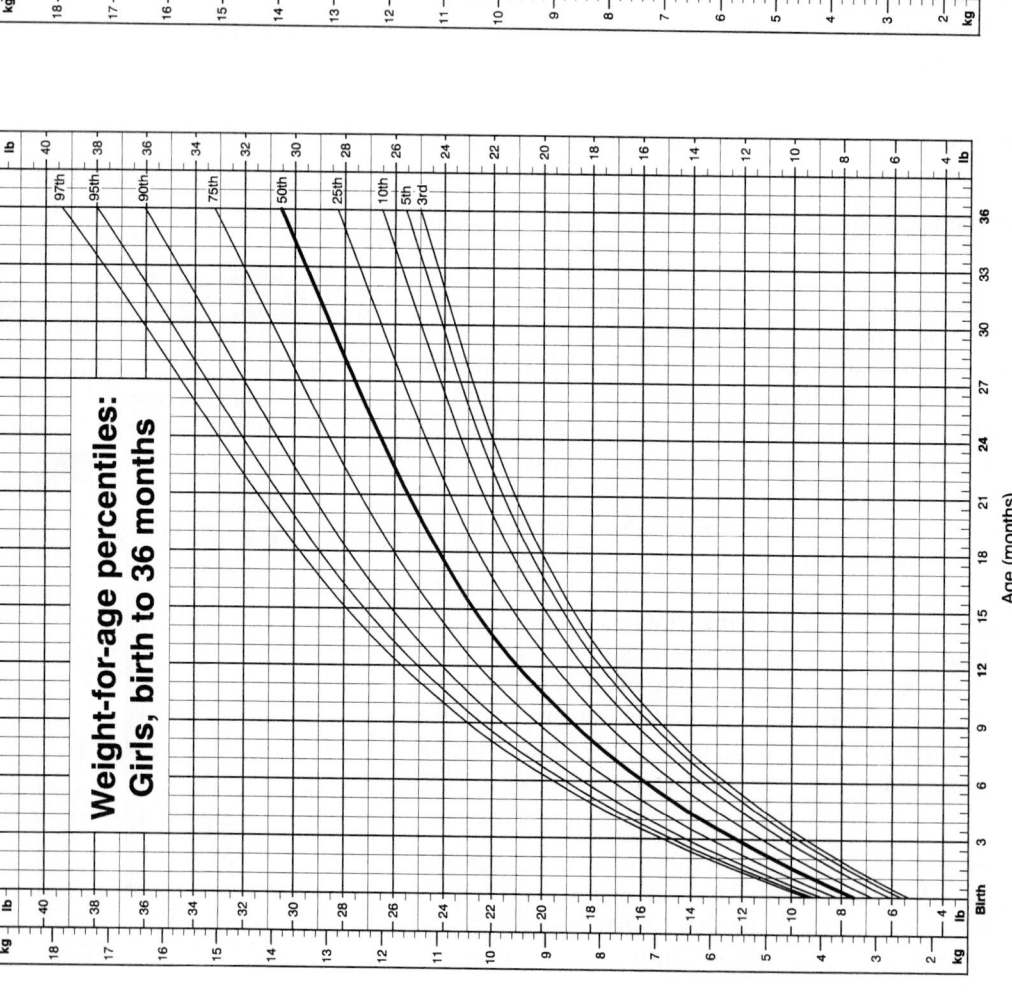

Weight-for-age percentiles: Boys, birth to 36 months

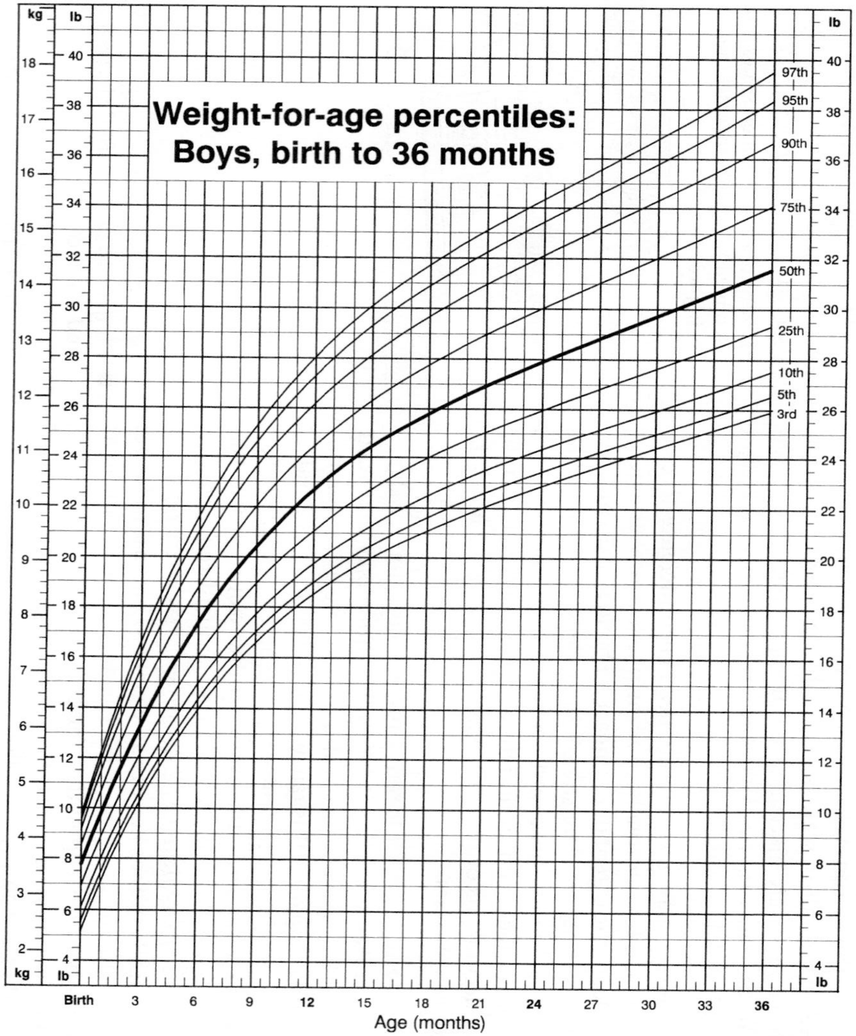

SOURCE: Developed by the National Center for Health Statistics in collaboration with the National Center for Chronic Disease Prevention and Health Promotion (2000).

Figure 1. Weight-for-age percentiles, boys, birth to 36 months, CDC growth charts: United States

Weight-for-age percentiles: Girls, birth to 36 months

SOURCE: Developed by the National Center for Health Statistics in collaboration with the National Center for Chronic Disease Prevention and Health Promotion (2000).

Figure 2. Weight-for-age percentiles, girls, birth to 36 months, CDC growth charts: United States

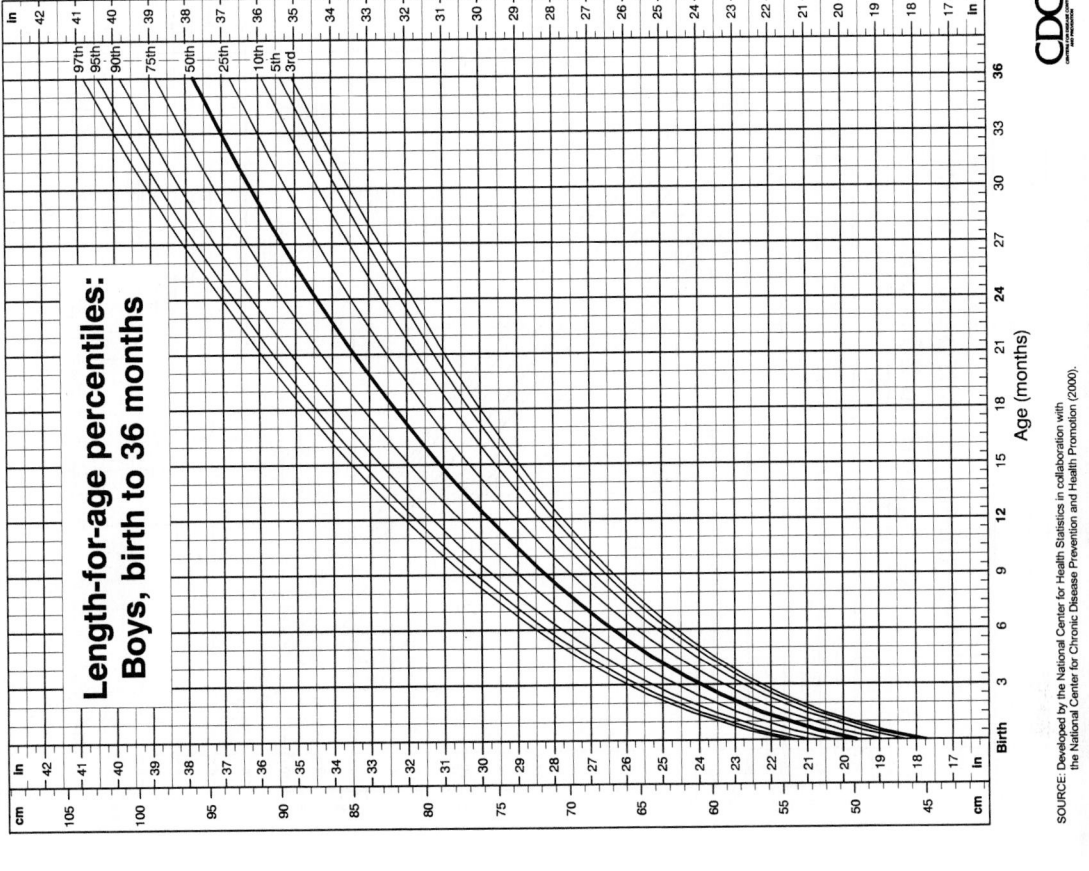

FIGURE E-6A Length-for-Age Percentiles: Boys, Birth to 36 Months

SOURCE: Developed by the National Center for Health Statistics in collaboration with the National Center for Chronic Disease Prevention and Health Promotion (2000).

Figure 3. Length-for-age percentiles, boys, birth to 36 months, CDC growth charts: United States

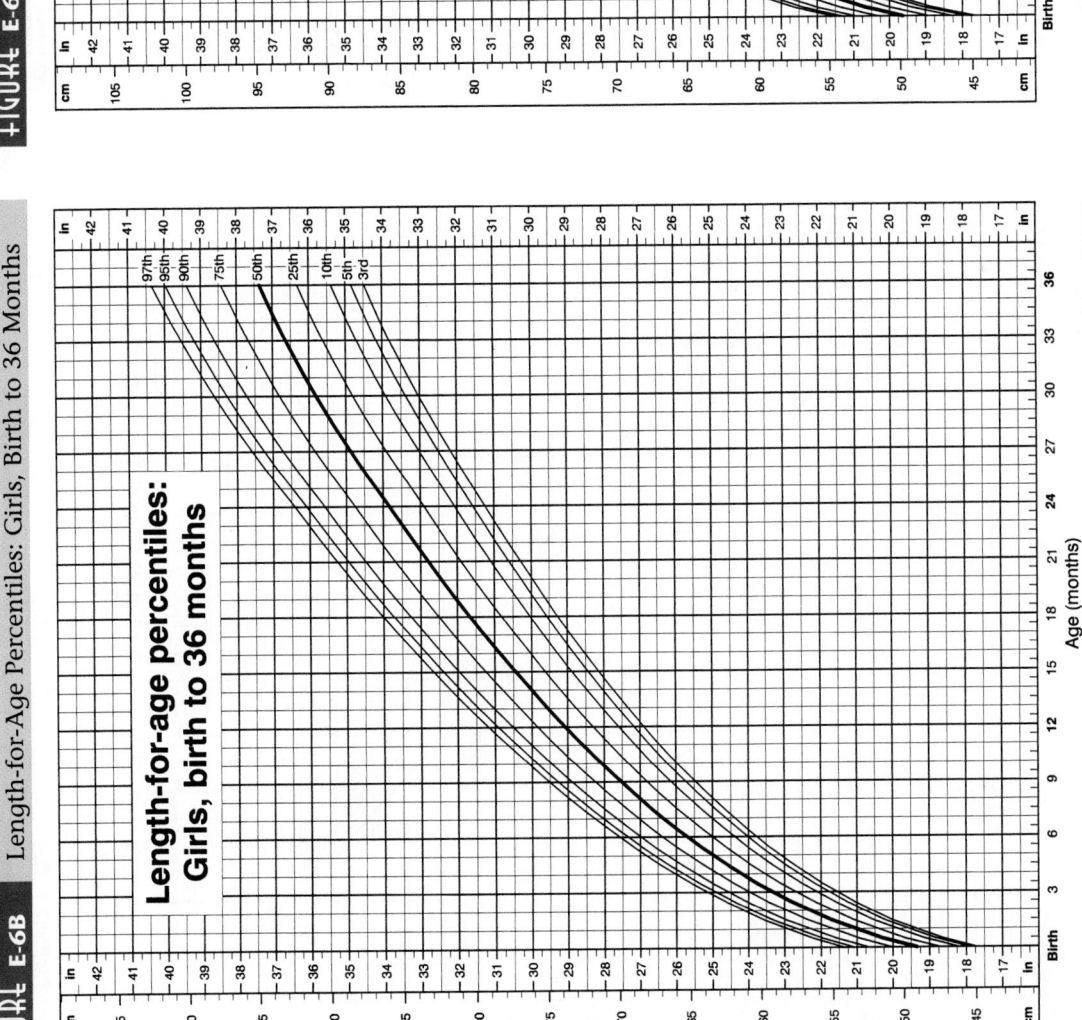

FIGURE E-6B Length-for-Age Percentiles: Girls, Birth to 36 Months

SOURCE: Developed by the National Center for Health Statistics in collaboration with the National Center for Chronic Disease Prevention and Health Promotion (2000).

Figure 4. Length-for-age percentiles, girls, birth to 36 months, CDC growth charts: United States

E Appendix

Appendix E

Weight-for-length percentiles:
Boys, birth to 36 months

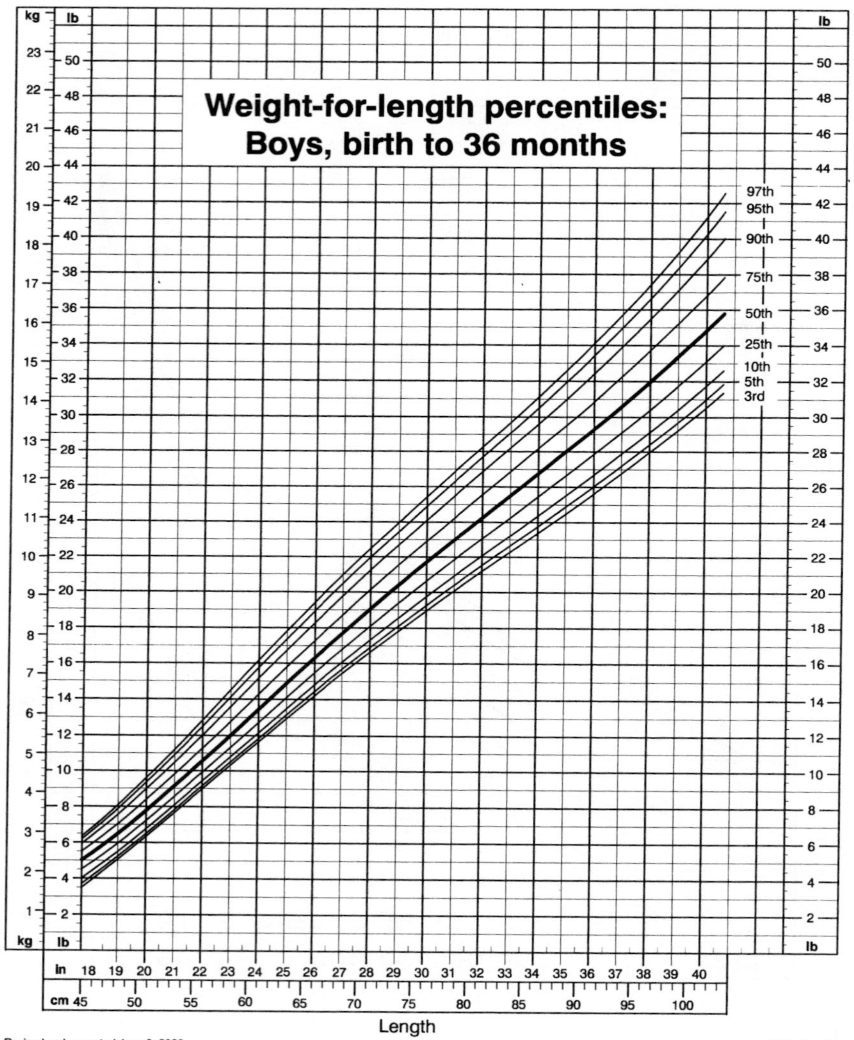

Length

Revised and corrected June 8, 2000.
SOURCE: Developed by the National Center for Health Statistics in collaboration with
the National Center for Chronic Disease Prevention and Health Promotion (2000).

Figure 5. Weight-for-length percentiles, boys, birth to 36 months, CDC growth charts: United States

Weight-for-length percentiles:
Girls, birth to 36 months

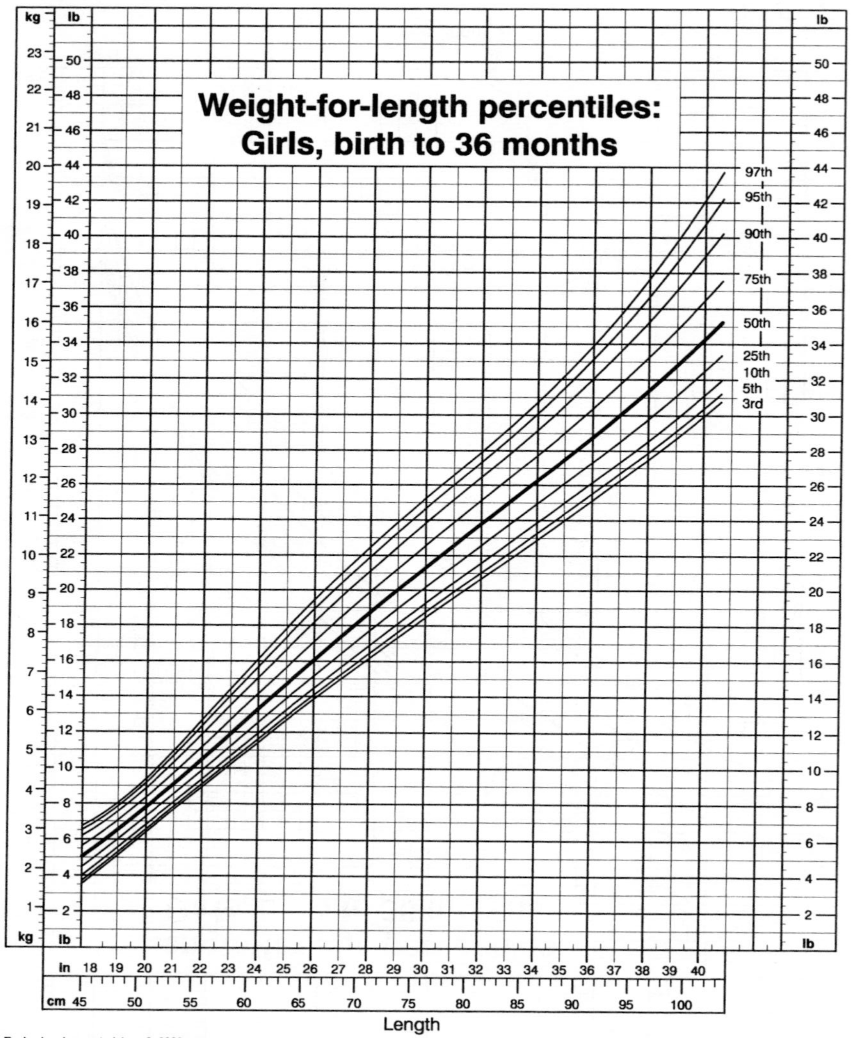

Length

Revised and corrected June 8, 2000.
SOURCE: Developed by the National Center for Health Statistics in collaboration with
the National Center for Chronic Disease Prevention and Health Promotion (2000).

Figure 6. Weight-for-length percentiles, girls, birth to 36 months, CDC growth charts: United States

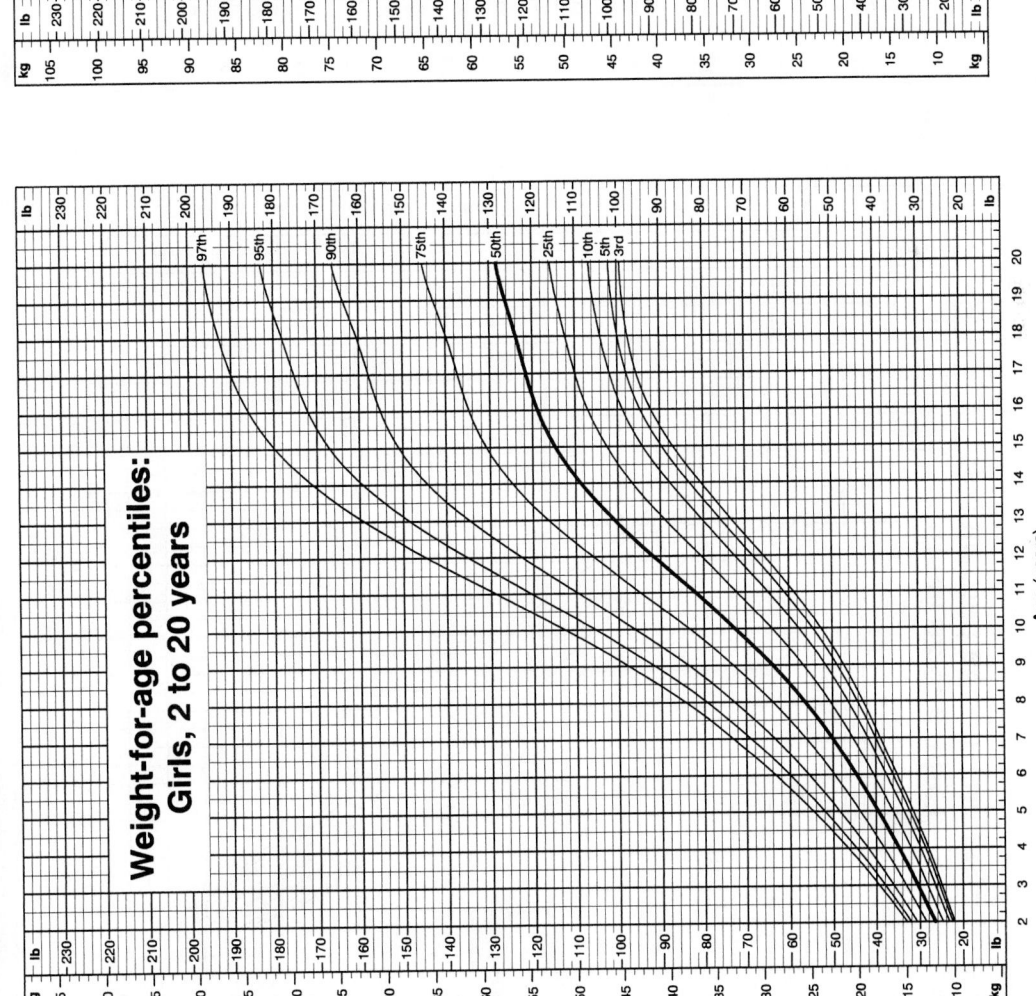

FIGURE E-8A Weight-for-Age Percentiles: Boys, 2 to 20 Years

Weight-for-age percentiles: Boys, 2 to 20 years

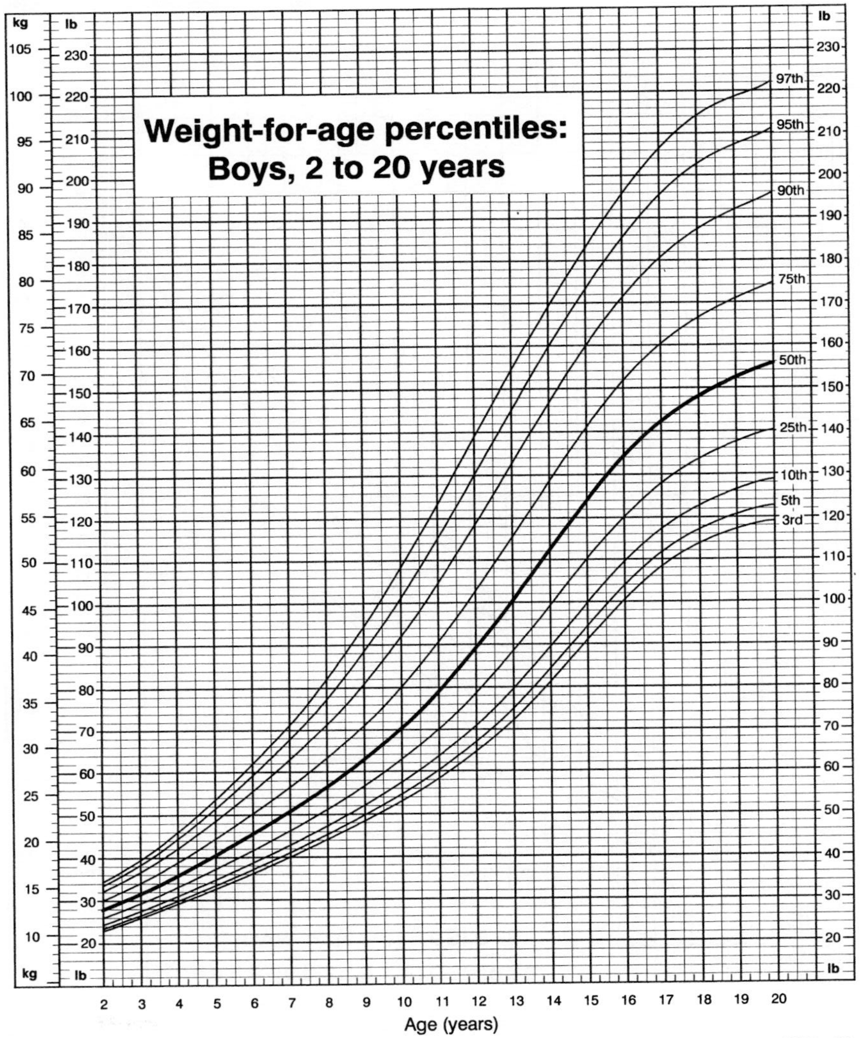

SOURCE: Developed by the National Center for Health Statistics in collaboration with the National Center for Chronic Disease Prevention and Health Promotion (2000).

Figure 9. Weight-for-age percentiles, boys, 2 to 20 years, CDC growth charts: United States

FIGURE E-8B Weight-for-Age Percentiles: Girls, 2 to 20 Years

Weight-for-age percentiles: Girls, 2 to 20 years

SOURCE: Developed by the National Center for Health Statistics in collaboration with the National Center for Chronic Disease Prevention and Health Promotion (2000).

Figure 10. Weight-for-age percentiles, girls, 2 to 20 years, CDC growth charts: United States

Appendix **E**

FIGURE E-9A Stature-for-Age Percentiles: Boys, 2 to 20 Years

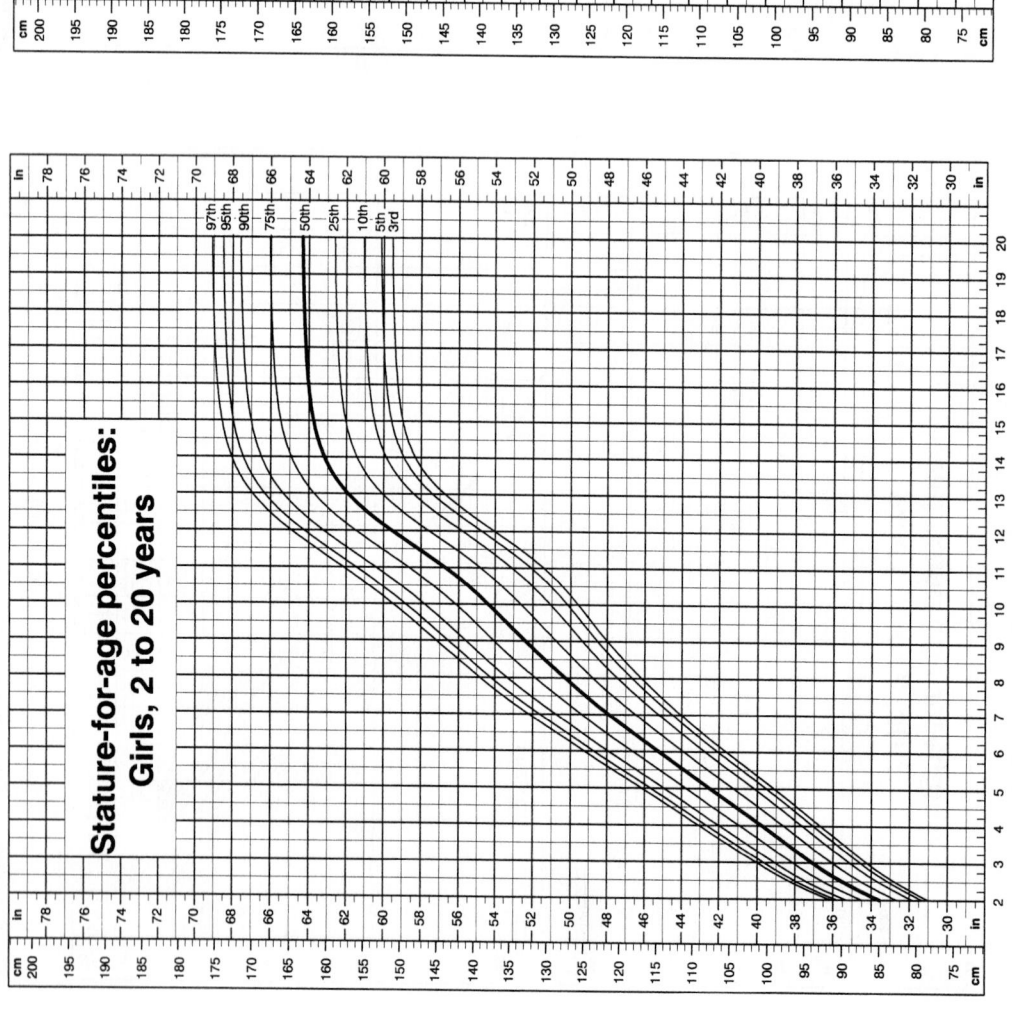

Stature-for-age percentiles:
Boys, 2 to 20 years

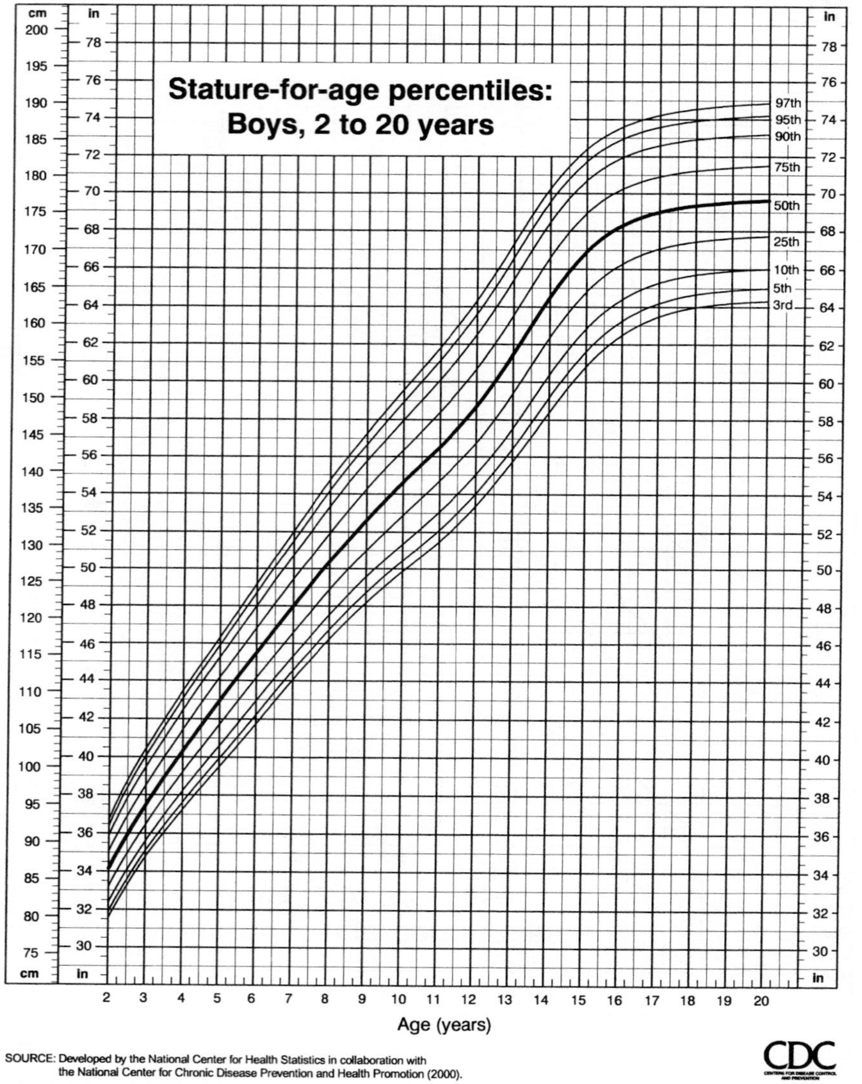

SOURCE: Developed by the National Center for Health Statistics in collaboration with
the National Center for Chronic Disease Prevention and Health Promotion (2000).

Figure 11. Stature-for-age percentiles, boys, 2 to 20 years, CDC growth charts: United States

FIGURE E-9B Stature-for-Age Percentiles: Girls, 2 to 20 Years

Stature-for-age percentiles:
Girls, 2 to 20 years

SOURCE: Developed by the National Center for Health Statistics in collaboration with
the National Center for Chronic Disease Prevention and Health Promotion (2000).

Figure 12. Stature-for-age percentiles, girls, 2 to 20 years, CDC growth charts: United States

FIGURE E-10A Weight-for-Stature Percentiles: Boys, 2 to 20 Years

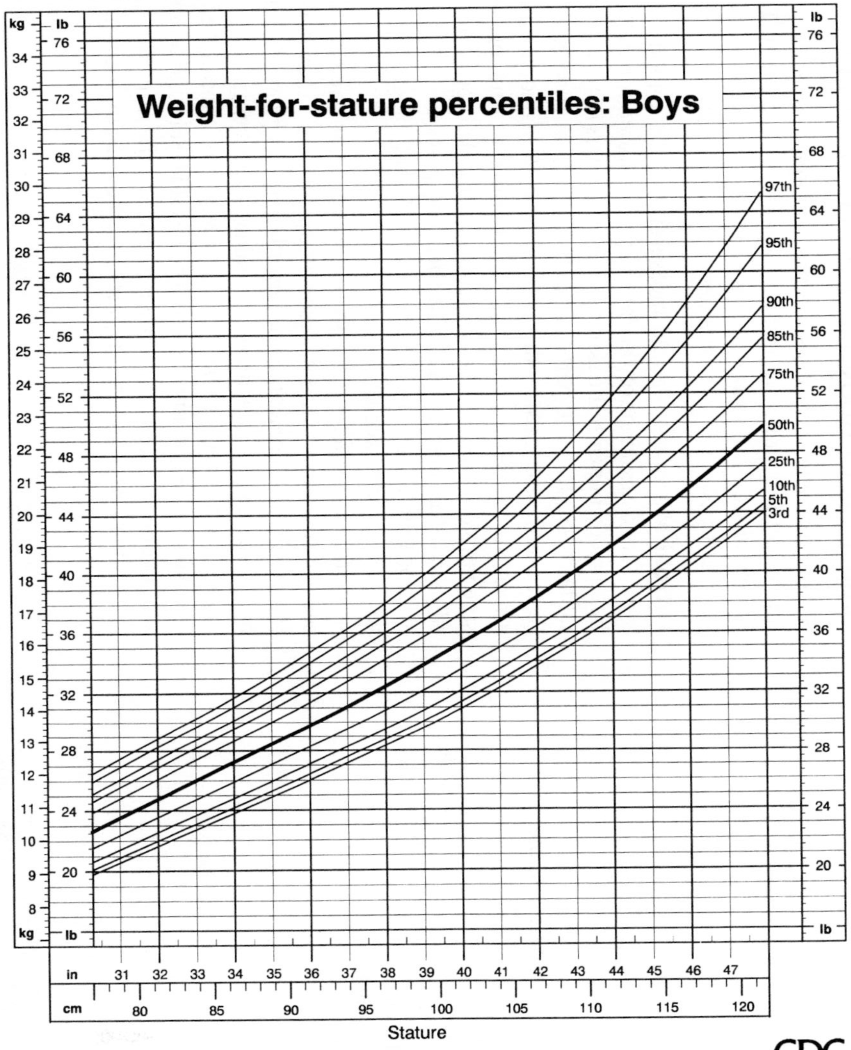

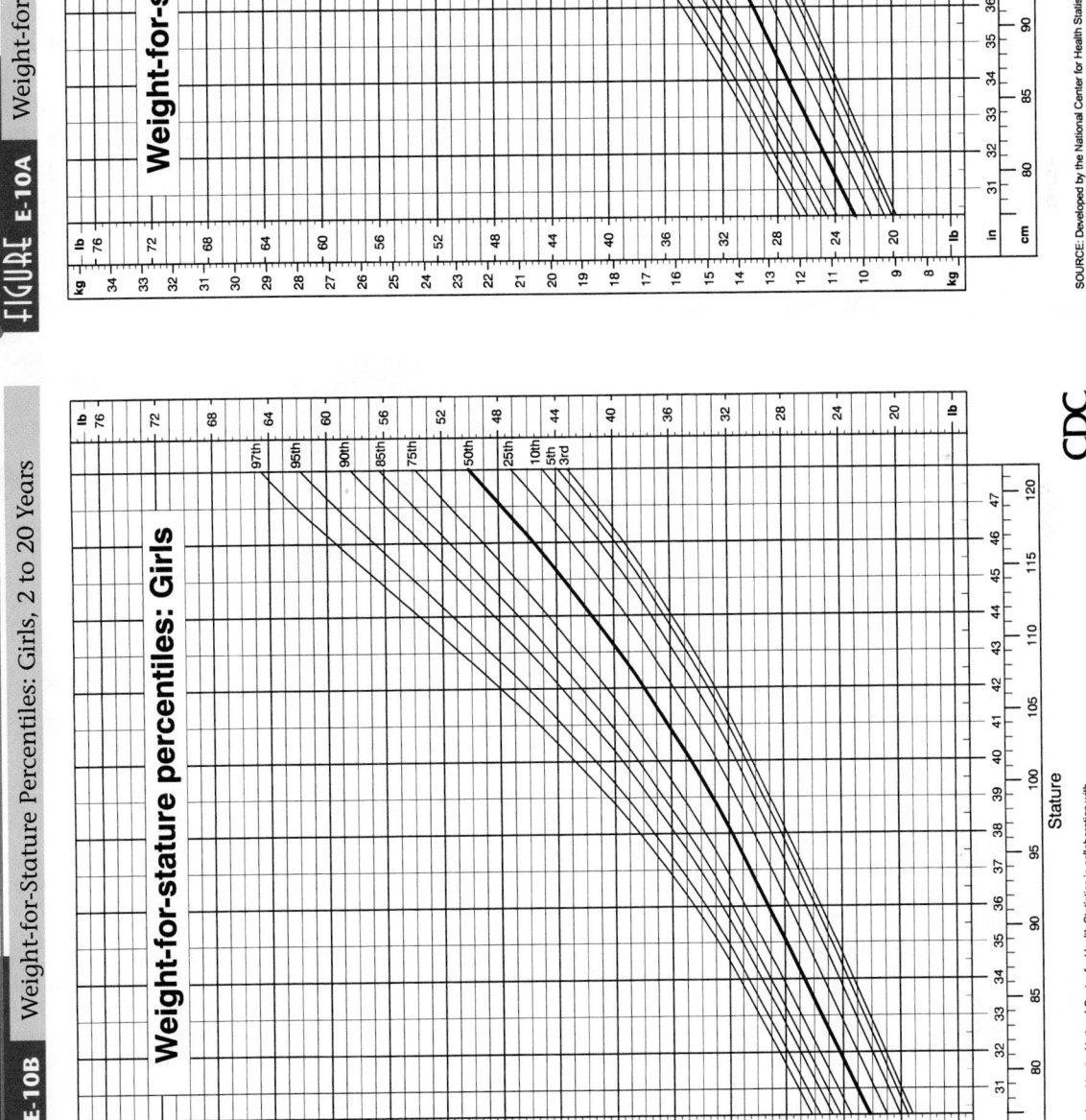

Weight-for-stature percentiles: Boys

SOURCE: Developed by the National Center for Health Statistics in collaboration with the National Center for Chronic Disease Prevention and Health Promotion (2000).

Figure 13. Weight-for-stature percentiles, boys, CDC growth charts: United States

FIGURE E-10B Weight-for-Stature Percentiles: Girls, 2 to 20 Years

Weight-for-stature percentiles: Girls

SOURCE: Developed by the National Center for Health Statistics in collaboration with the National Center for Chronic Disease Prevention and Health Promotion (2000).

Figure 14. Weight-for-stature percentiles, girls, CDC growth charts: United States

Measures of Body Fat and Lean Tissue

Significant weight changes in both children and adults can reflect overnutrition and undernutrition with respect to energy and protein. To estimate the degree to which fat stores or lean tissues are affected by overnutrition or malnutrition, several anthropometric measurements are useful (review Table E-3 on p. E-5).

Fatfold Measures Fatfold measures provide a good estimate of total body fat and a fair assessment of the fat's location. Approximately half the fat in the body lies directly beneath the skin, and the thickness of this subcutaneous fat reflects total body fat. In some parts of the body, such as the back and the back of the arm over the triceps muscle, this fat is loosely attached;■ a person can pull it up between the thumb and forefinger to obtain a measure of fatfold thickness. To measure fatfold, a skilled assessor follows a standard procedure using reliable calipers (illustrated in Figure E-11) and then compares the measurement with standards. Triceps fatfold measures greater than 15 millimeters in men or 25 millimeters in women suggest excessive body fat.

Fatfold measurements correlate directly with the risk of heart disease. They assess central obesity and its associated risks better than do weight measures alone. If a person gains body fat, the fatfold increases proportionately; if the person loses fat, it decreases. Measurements taken from central-body sites (around the abdomen) better reflect changes in fatness than those taken from upper sites (arm and back). A major limitation of the fatfold test is that fat may be thicker under the skin in one area than in another. A pinch at the side of the waistline may not yield the same measurement as a pinch on the back of the arm. This limitation can be

■ Common sites for fatfold measures:
- Triceps.
- Biceps.
- Subscapular (below shoulder blade).
- Suprailiac (above hip bone).
- Abdomen.
- Upper thigh.

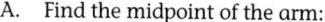

FIGURE E-11   How to Measure the Triceps Fatfold

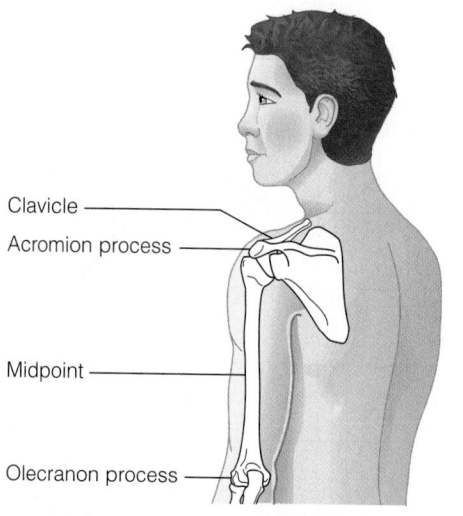
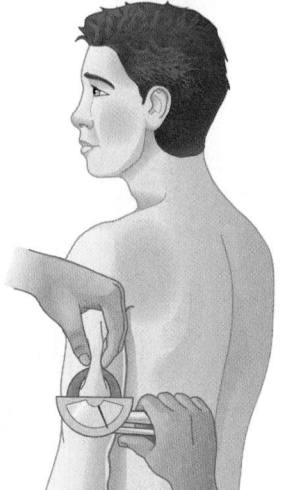

Clavicle
Acromion process
Midpoint
Olecranon process

A. Find the midpoint of the arm:
1. Ask the subject to bend his or her arm at the elbow and lay the hand across the stomach. (If he or she is right-handed, measure the left arm, and vice versa.)
2. Feel the shoulder to locate the acromion process. It helps to slide your fingers along the clavicle to find the acromion process. The olecranon process is the tip of the elbow.
3. Place a measuring tape from the acromion process to the tip of the elbow. Divide this measurement by 2, and mark the midpoint of the arm with a pen.

B. Measure the fatfold:
1. Ask the subject to let his or her arm hang loosely to the side.
2. Grasp a fold of skin and subcutaneous fat between the thumb and forefinger slightly above the midpoint mark. Gently pull the skin away from the underlying muscle. (This step takes a lot of practice. If you want to be sure you don't have muscle as well as fat, ask the subject to contract and relax the muscle. You should be able to feel if you are pinching muscle.)

3. Place the calipers over the fatfold at the midpoint mark, and read the measurement to the nearest 1.0 millimeter in two to three seconds. (If using plastic calipers, align pressure lines, and read the measurement to the nearest 1.0 millimeter in two to three seconds.)
4. Repeat steps 2 and 3 twice more. Add the three readings, and then divide by 3 to find the average.

FIGURE E-12   How to Measure Waist Circumference

Place the measuring tape around the waist just above the bony crest of the hip. The tape runs parallel to the floor and is snug (but does not compress the skin). The measurement is taken at the end of normal expiration.

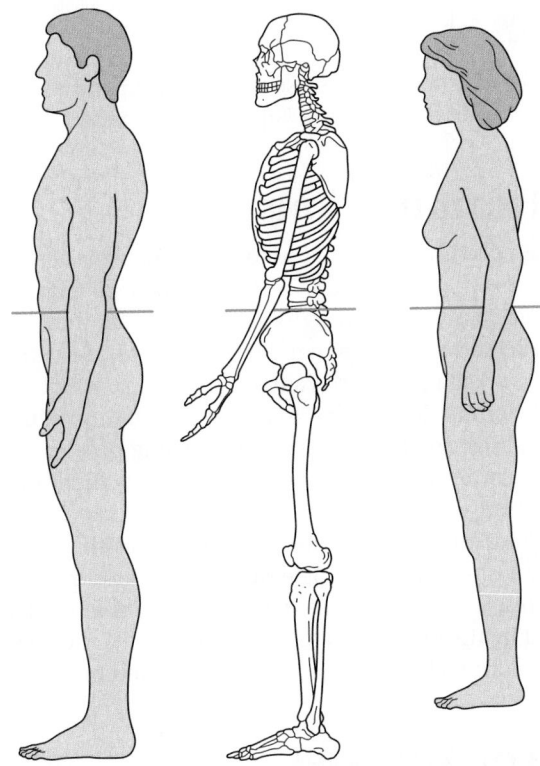

SOURCE: National Institutes of Health Obesity Education Initiative, *Clinical Guidelines on the Identification, Evaluation, and Treatment of Overweight and Obesity in Adults* (Washington, D.C.: U.S. Department of Health and Human Services, 1998), p. 59.

overcome by taking fatfold measurements at several (often three) different places on the body (including upper-, central-, and lower-body sites) and comparing each measurement with standards for that site. Multiple measures are not always practical in clinical settings, however, and most often, the triceps fatfold measurement alone is used because it is easily accessible.

Waist Circumference Chapter 8 described how fat distribution correlates with health risks and mentioned that the waist circumference is a valuable indicator of fat distribution. To measure waist circumference, the assessor places a nonstretchable tape around the person's body, crossing just above the upper hip bones and making sure that the tape remains on a level horizontal plane on all sides (see Figure E-12). The tape is tightened slightly, but without compressing the skin.

Waist-to Hip Ratio Alternatively, some clinicians measure both the waist and the hips. The waist-to-hip ratio■ also assesses abdominal obesity, but provides no more information than using the waist circumference alone. In general, women with a waist-to-hip ratio of 0.80 or greater and men with a waist-to-hip ratio of 0.90 or greater have a high risk of health problems.

Hydrodensitometry To estimate body density using hydrodensitometry, the person is weighed twice—first on land and then again when submerged under water. Underwater weighing usually generates a good estimate of body fat and is useful in research, although the technique has drawbacks: it requires bulky, expensive, and nonportable equipment. Furthermore, submerging some people (especially those who are very young, very old, ill, or fearful) under water is not always practical.

■ To calculate the waist-to-hip ratio, divide the waistline measurement by the hip measurement. For example, a woman with a 28-inch waist and 38-inch hips would have a ratio of 28 ÷ 38 = 0.74.

E
Appendix

Bioelectric Impedance To measure body fat using the bioelectric impedance technique, a very-low-intensity electrical current is briefly sent through the body by way of electrodes placed on the wrist and ankle. As is true of other anthropometric techniques, bioelectrical impedance requires standardized procedures and calibrated instruments to provide reliable results. Recent food intake and hydration status, for example, influence results.

Clinicians use many other methods to estimate body fat and its distribution. Each has its advantages and disadvantages as Table E-4 summarizes.

Physical Examinations

An assessor can use a physical examination to search for signs of nutrient deficiency or toxicity. Like the other assessment methods, such an examination requires knowledge and skill. Many physical signs are nonspecific; they can reflect any of several nutrient deficiencies as well as conditions not related to nutrition (see Table E-5). For example, cracked lips may be caused by sunburn, windburn, dehydration, or any of several B vitamin deficiencies, to name just a few possible causes. For this reason, physical findings are most valuable in revealing problems for other assessment techniques to confirm or for confirming other assessment measures.

With this limitation understood, physical symptoms can be most informative and communicate much information about nutrition health. Many tissues and organs can reflect signs of malnutrition. The signs appear most rapidly in parts of the body where cell replacement occurs at a high rate, such as in the hair, skin, and digestive tract (including the mouth and tongue). The summary tables in Chapters 10, 11, 12, and 13 list additional physical signs of vitamin and mineral malnutrition.

Biochemical Analyses

All of the approaches to nutrition assessment discussed so far are external approaches. Biochemical analyses or laboratory tests help to determine what is happening to the body internally. Common tests are based on analysis of blood and

TABLE E-4   Methods of Estimating Body Fat and Its Distribution

Method	Cost	Ease of Use	Accuracy	Measures Fat Distribution
Height and weight	Low	Easy	High	No
Fatfolds	Low	Easy	Low	Yes
Circumferences	Low	Easy	Moderate	Yes
Ultrasound	Moderate	Moderate	Moderate	Yes
Hydrodensitometry	Low	Moderate	High	No
Heavy water tritiated	Moderate	Moderate	High	No
Deuterium oxide, or heavy oxygen	High	Moderate	High	No
Potassium isotope (^{40}K)	Very high	Difficult	High	No
Total body electrical conductivity (TOBEC)	High	Moderate	High	No
Bioelectric impedance (BIA)	Moderate	Easy	High	No
Dual energy X-ray absorptiometry (DEXA)	High	Easy	High	No
Computed tomography (CT)	Very high	Difficult	High	Yes
Magnetic resonance imaging (MRI)	Very high	Difficult	High	Yes

SOURCE: Adapted with permisssion from G. A. Bray, a handout presented at the North American Association for the Study of Obesity and Emory University School of Medicine Conference on Obesity Update: Pathophysiology, Clinical Consequences, and Therapeutic Options, Atlanta, Georgia, August 31–September 2, 1992.

TABLE E-5 Physical Findings Used in Nutrition Assessments

Body System	Acceptable Findings	Malnutrition Findings	What the Findings Reflect
Hair	Shiny, firm in the scalp	Dull, brittle, dry, loose; falls out	PEM
Eyes	Bright, clear pink membranes; adjust easily to light	Pale membranes; spots; redness; adjust slowly to darkness	Vitamin A, B vitamin, zinc, and iron status
Teeth and gums	No pain or caries, gums firm, teeth bright	Missing, discolored, decayed teeth; gums bleed easily and are swollen and spongy	Mineral and vitamin C status
Face	Clear complexion without dryness or scaliness	Off-color, scaly, flaky, cracked skin	PEM, vitamin A, and iron status
Glands	No lumps	Swollen at front of neck	PEM and iodine status
Tongue	Red, bumpy, rough	Sore, smooth, purplish, swollen	B vitamin status
Skin	Smooth, firm, good color	Dry, rough, spotty; "sandpaper" feel or sores; lack of fat under skin	PEM, essential fatty acid, vitamin A, B vitamin, and vitamin C status
Nails	Firm, pink	Spoon-shaped, brittle, ridged, pale	Iron status
Internal systems	Regular heart rhythm, heart rate, and blood pressure; no impairment of digestive function, reflexes, or mental status	Abnormal heart rate, heart rhythm, or blood pressure; enlarged liver, spleen; abnormal digestion; burning, tingling of hands, feet; loss of balance, coordination; mental confusion, irritability, fatigue	PEM and mineral status
Muscles and bones	Muscle tone; posture, long bone development appropriate for age	"Wasted" appearance of muscles; swollen bumps on skull or ends of bones; small bumps on ribs; bowed legs or knock-knees	PEM, mineral, and vitamin D status

urine samples, which contain nutrients, enzymes, and metabolites that reflect nutrition status. Other tests, such as serum■ glucose, help pinpoint disease-related problems with nutrition implications. Tests that define fluid and electrolyte balance, acid-base balance, and organ function also have nutrition implications. Table E-6 lists biochemical tests most useful for assessing vitamin and mineral status.

The interpretation of biochemical data requires skill. Long metabolic sequences lead to the production of the end products and metabolites seen in blood and urine. No single test can reveal nutrition status because many factors influence test results. The low blood concentration of a nutrient may reflect a primary deficiency of that nutrient, but it may also be secondary to the deficiency of one or several other nutrients or to a disease. Taken together with other assessment data, however, laboratory test results help to create a picture that becomes clear with careful interpretation. They are especially useful in helping to detect subclinical malnutrition by uncovering early signs of malnutrition before the clinical signs of a classic deficiency disease appear.

Laboratory tests used to assess vitamin and mineral status (review Table E-6) are particularly useful when combined with diet histories and physical findings. Vitamin and mineral levels present in the blood and urine sometimes reflect recent rather than long-term intakes. This makes detecting subclinical deficiencies■ difficult. Furthermore, many nutrients interact; therefore, the amounts of other nutrients in the body can affect a lab value for a particular nutrient. It is also important to remember that nonnutrient conditions such as diseases influence biochemical measures.

It is beyond the scope of this text to describe all lab tests and their relations to nutrition status. Instead, the emphasis is on lab tests used to detect protein-energy malnutrition (PEM) and nutritional anemias.

■ The **serum** is the watery portion of the blood that remains after removal of the cells and clot-forming material; **plasma** is the fluid that remains when unclotted blood is centrifuged. In most cases, serum and plasma concentrations are similar, but plasma samples are more likely to clog mechanical blood analyzers, so serum samples are preferred.

■ Reminder: A *subclinical deficiency* is a nutrient deficiency in the early stages before the outward signs have appeared.

Protein-Energy Malnutrition (PEM)

No single biochemical analysis can adequately evaluate PEM. Numerous procedures have been used over the years. This discussion focuses on the measures commonly used today—transthyretin, retinol-binding protein, serum transferrin,

TABLE E-6 Biochemical Tests Useful for Assessing Vitamin and Mineral Status

Nutrient	Assessment Tests
Vitamins	
Vitamin A	Serum retinol, retinol-binding protein
Thiamin[a]	Erythrocyte (red blood cell) transketolase activity, erythrocyte thiamin pyrophosphate
Riboflavin[a]	Erythrocyte glutathione reductase activity
Vitamin B_6[a]	Urinary xanthurenic acid excretion after tryptophan load test, erythrocyte transaminase activity, plasma pyridoxal 5'-phosphate (PLP)
Niacin	Plasma or urinary metabolites NMN (N-methyl nicotinamide) or 2-pyridone, or preferably both expressed as a ratio
Folate[b]	Serum folate, erythrocyte folate (reflects liver stores)
Vitamin B_{12}[b]	Serum vitamin B_{12}, serum and urinary methylmalonic acid, Schilling test
Biotin	Urinary biotin, urinary 3-hydroxyisovaleric acid
Vitamin C	Plasma vitamin C[c], leukocyte vitamin C
Vitamin D	Serum vitamin D
Vitamin E	Serum α-tocopherol, erythrocyte hemolysis
Vitamin K	Serum vitamin K, plasma prothrombin; blood-clotting time (prothrombin time) is not an adequate indicator
Minerals	
Phosphorus	Serum phosphate
Sodium	Serum sodium
Chloride	Serum chloride
Potassium	Serum potassium
Magnesium	Serum magnesium, urinary magnesium
Iron	Hemoglobin, hematocrit, serum ferritin, total iron-binding capacity (TIBC), erythrocyte protoporphyrin, serum iron, transferrin saturation
Iodine	Serum thyroxine or thyroid-stimulating hormone (TSH), urinary iodine
Zinc	Plasma zinc, hair zinc
Copper	Erythrocyte superoxide dismutase, serum copper, serum ceruloplasmin
Selenium	Erythrocyte selenium, glutathione peroxidase activity

[a]Urinary measurements for these vitamins are common, but may be of limited use. Urinary measurements reflect recent dietary intakes and may not provide reliable information concerning the severity of a deficiency.
[b]Folate assessments should always be conducted in conjunction with vitamin B_{12} assessments (and vice versa) to help distinguish the cause of common deficiency symptoms.
[c]Vitamin C shifts between the plasma and the white blood cells known as leukocytes; thus a plasma determination may not accurately reflect the body's pool. A measurement of leukocyte vitamin C can provide information about the body's stores of vitamin C. A combination of both tests may be more reliable than either one alone.
SOURCE: Adapted from H. E. Sauberlich, *Laboratory Tests for the Assessment of Nutritional Status* (Boca Raton, Fla.: CRC Press, 1999).

and IGF-1 (insulin-like growth factor 1). Table E-7 provides standards for these indicators. Although serum albumin is easily and routinely measured, it lacks the sensitivity to assess PEM because of its long turnover rate.[*]

Transthyretin and Retinol-Binding Protein Transthyretin■ and retinol-binding protein occur as a complex in the plasma. They have a rapid turnover and thus respond quickly to dietary protein inadequacy and therapy.[†] Conditions other than malnutrition that lower transthyretin include metabolic stress, hemodialysis, and hypothyroidism; those that raise transthyretin include kidney disease and corticosteroid use. Conditions other than protein malnutrition that lower retinol-binding protein include vitamin A deficiency, metabolic stress,

■ Transthyretin is also known as *prealbumin* or *thyroxine-binding prealbumin*.

---

[*]The half-life of albumin is 18 days, an indication of a slow degradation rate.
[†]The half-lives of transthyretin and retinol-binding protein are 2 days and 12 hours, respectively.

TABLE E-7 Normal Values for Serum Proteins

Indicator	Normal
Albumin (g/dL)[a]	3.5–5.4
Transferrin (mg/dL)[b]	200–400
Transthyretin (mg/dL)	23–43
Retinol-binding protein (mg/dL)	3–7
IGF-1 (μg/L)	300

NOTE: Levels less than normal suggest compromised protein status.
[a] A deciliter (dL) is one-tenth of a liter or 100 milliliters. To convert albumin (g/100 mL or g/dL) to standard international units (g/L), multiply by 100.
[b] To convert transferrin (mg/100 mL) to standard international units (g/L), multiply by 0.01.

hyperthyroidism, liver disease, and cystic fibrosis; kidney disease raises retinol-binding protein levels.

Serum Transferrin Serum transferrin transports iron; consequently, its concentrations reflect both protein and iron status. Using transferrin as an indicator of protein status is complicated when an iron deficiency is present. Transferrin rises as iron deficiency grows worse and falls as iron status improves. Markedly reduced transferrin levels indicate severe PEM; in mild-to-moderate PEM, transferrin levels may vary, limiting their usefulness. Conditions other than protein malnutrition that lower transferrin include liver disease, kidney disease, and metabolic stress; those that raise transferrin include pregnancy, iron deficiency, hepatitis, blood loss, and oral contraceptive use. Although transferrin breaks down in the body more quickly than albumin, it is still relatively slow to respond to changes in protein intake and is not a sensitive indicator of the response to therapy.[*]

IGF-1 (Insulin-like Growth Factor 1) IGF-1 (insulin-like growth factor 1) declines in PEM. IGF-1 has a relatively short half-life and responds specifically to dietary protein rather than energy.[†] For these reasons, it is a sensitive indicator of protein status and response to therapy. Conditions that decrease IGF-1 include anorexia nervosa, inflammatory bowel disease, celiac disease, HIV infection, and fasting.

Nutritional Anemias

Anemia, a symptom of a wide variety of nutrition- and nonnutrition-related disorders, is characterized by a reduced number of red blood cells. Iron, folate, and vitamin B_{12} deficiencies caused by inadequate intake, poor absorption, or abnormal metabolism of these nutrients are the most common nutritional anemias. Some nonnutrition-related causes of anemia include massive blood loss, infections, hereditary blood disorders such as sickle-cell anemia, and chronic liver or kidney disease.

Assessment of Iron-Deficiency Anemia

Iron deficiency, a common mineral deficiency, develops in stages.■ Chapter 13 describes iron deficiency in detail. This section describes tests used to uncover iron deficiency as it progresses. Table E-8 shows which laboratory tests detect various nutrition-related anemias, and Table E-9 provides values used for assessing iron status. Although other tests are more specific in detecting early deficiencies, hemoglobin and hematocrit are the commonly available tests.

■ Stages of iron deficiency:
1. Iron stores diminish.
2. Transport iron decreases.
3. Hemoglobin production falls.

E Appendix

---

[*] The half-life of transferrin is 8 days.
[†] The half-life of IGF-1 is 12 to 15 hours.

TABLE E-8 Laboratory Tests Useful in Evaluating Nutrition-Related Anemias

Test or Test Result	What It Reflects
For Anemia (general)	
Hemoglobin (Hg)	Total amount of hemoglobin in the red blood cells (RBC)
Hematocrit (Hct)	Percentage of RBC in the total blood volume
Red blood cell (RBC) count	Number of RBC
Mean corpuscular volume (MCV)	RBC size; helps to determine if anemia is microcytic (iron deficiency) or macrocytic (folate or vitamin B_{12} deficiency)
Mean corpuscular hemoglobin concentration (MCHC)	Hemoglobin concentration within the average RBC; helps to determine if anemia is hypochromic (iron deficiency) or normochromic (folate or vitamin B_{12} deficiency)
Bone marrow aspiration	The manufacture of blood cells in different developmental states
For Iron-Deficiency Anemia	
↓ Serum ferritin	Early deficiency state with depleted iron stores
↓ Transferrin saturation	Progressing deficiency state with diminished transport iron
↑ Erythrocyte protoporphyrin	Later deficiency state with limited hemoglobin production
For Folate-Deficiency Anemia	
↓ Serum folate	Progressing deficiency state
↓ RBC folate	Later deficiency state
For Vitamin B_{12}–Deficiency Anemia	
↓ Serum vitamin B_{12}	Progressing deficiency state
Schilling test	Absorption of vitamin B_{12}

Hemoglobin Iron forms an integral part of the hemoglobin molecule that transports oxygen to the cells. In iron deficiency, the body cannot synthesize hemoglobin. Low hemoglobin values signal depleted iron stores. Table E-9 provides hemoglobin values used in nutrition assessment. Hemoglobin's usefulness in evaluating iron status is limited, however, because hemoglobin concentrations drop fairly late in the development of iron deficiency, and other nutrient deficiencies and medical conditions can also alter hemoglobin concentrations.

Hematocrit Hematocrit is commonly used to diagnose iron deficiency, even though it is an inconclusive measure of iron status. To measure the hematocrit, a clinician spins a volume of blood in a centrifuge to separate the red blood cells from the plasma. The hematocrit is the percentage of red blood cells in the total blood volume. Table E-9 includes values used to assess hematocrit status. Low values indicate incomplete hemoglobin formation, which is manifested by microcytic (abnormally small-celled), hypochromic (abnormally lacking in color) red blood cells.

Low hemoglobin and hematocrit values alert the assessor to the possibility of iron deficiency. However, many nutrients and other conditions can affect hemoglobin and hematocrit. The other tests of iron status help pinpoint true iron deficiency.

Serum Ferritin In the first stage of iron deficiency, iron stores diminish. Serum ferritin measures provide a noninvasive estimate of iron stores. Such information is most valuable to iron assessment. Table E-9 shows serum ferritin cutoff values that indicate iron store depletion in children and adults. Serum ferritin is not reliable for diagnosing iron deficiency in infants, since normal serum ferritin values are often present in conjunction with iron-responsive anemia.

A decrease in transport iron characterizes the second stage of iron deficiency. This is revealed by an increase in the iron-binding capacity of the protein trans-

TABLE E-9 Criteria for Assessing Iron Status

Test	Age (yr)	Gender	Deficiency Value
Hemoglobin (g/dL)	0.5–10	M–F	<11
	11–15	M	<12
		F	<11.5
	>15	M	<13
		F	<12
	Pregnancy		<11
Hematocrit (%)	0.5–4	M–F	<32
	5–10	M–F	<33
	11–15	M	<35
		F	<34
	>15	M	<40
		F	<36
Serum ferritin (μg/L)	0.5–15	M–F	<10
	>15	M–F	<12
Total iron-binding capacity (μg/dL)	>15	M–F	>400
Serum iron (μg/dL)	>15	M–F	<60
Transferrin saturation (%)	0.5–4	M–F	<12
	5–10	M–F	<14
	>10	M–F	<16
Erythrocyte protoporphyrin (μg/dL RBC)	0.5–4	M–F	>80
	>4	M–F	>70

ferrin and a decrease in serum iron. These changes are reflected by the transferrin saturation, which is calculated from the ratio of the other two values as described in the following paragraphs.

Total Iron-Binding Capacity (TIBC) Iron travels through the blood bound to the protein transferrin. TIBC is a measure of the total amount of iron that transferrin can carry. Lab technicians measure iron-binding capacity directly. Table E-9 includes the cutoff for TIBC.

Serum Iron Lab technicians can also measure serum iron directly. Elevated values indicate iron overload; reduced values indicate iron deficiency. Table E-9 shows the deficient value for serum iron.

Transferrin Saturation The percentage of transferrin that is saturated with iron is an indirect measure that is derived from the serum iron and total iron-binding capacity measures as follows:

$$\%\text{Transferrin} = \frac{\text{serum iron}}{\text{total iron-binding capacity}} \times 100.$$

Table E-9 shows deficient transferrin saturation values for various age groups.

The third stage of iron deficiency occurs when the supply of transport iron diminishes to the point that it limits hemoglobin production. It is characterized by increases in erythrocyte protoporphyrin, a decrease in mean corpuscular volume, and decreased hemoglobin and hematocrit.

Erythrocyte Protoporphyrin The iron-containing portion of the hemoglobin molecule is heme. Heme is a combination of iron and protoporphyrin. Protoporphyrin accumulates in the blood when iron supplies are inadequate for the formation of heme. Lab technicians can measure erythrocyte protoporphyrin directly in a blood sample. The cutoffs for abnormal values of erythrocyte protoporphyrin are shown in Table E-9.

E Appendix

Mean Corpuscular Volume (MCV) A direct or calculated measure of the mean corpuscular volume (MCV) determines the average size of a red blood cell. Such a measure helps to classify the type of nutrient anemia. In iron deficiency, the red blood cells are smaller than average.

Assessment of Folate and Vitamin B_{12} Anemias

Folate deficiency and vitamin B_{12} deficiency present a similar clinical picture—an anemia characterized by abnormally large red blood cell precursors (megaloblasts) in the bone marrow and abnormally large, mature red blood cells (macrocytic cells) in the blood. Distinguishing between these two deficiencies is particularly important because their treatments differ. Giving folate to a person with vitamin B_{12} deficiency improves many of the lab test results indicative of vitamin B_{12} deficiency, but this is a dangerous error because vitamin B_{12} deficiency causes nerve damage that folate cannot correct. Thus inappropriate folate administration masks vitamin B_{12}–deficiency anemia, and nerve damage worsens. For this reason, it is critical to determine whether the anemia results from a folate deficiency or from a vitamin B_{12} deficiency. The following biochemical assessment techniques help to make this distinction.

Mean Corpuscular Volume (MCV) As previously mentioned, the MCV is a measure of red blood cell size. In folate and vitamin $B1_2$ deficiencies, the red blood cells are larger than average (macrocytic). Additional tests must be performed to differentiate folate from vitamin B_{12} deficiency.

Folate Levels Serum folate levels fluctuate with changes in folate intake and metabolism. Thus serum folate concentrations reflect current status, but provide little information about folate stores. As folate deficiency progresses and low serum levels persist, folate stores decline, resulting in folate depletion. Folate depletion is characterized by a fall in the folate concentrations of red blood cells (erythrocytes). As erythrocyte folate levels diminish, folate-deficiency anemia develops. Because low erythrocyte folate concentrations also occur with vitamin B_{12} deficiency, serum vitamin B_{12} concentrations must also be measured. Table E-10 shows standards for folate assessment.

Vitamin B_{12} Levels Serum and urinary methylmalonic acid are elevated in vitamin B_{12} deficiency, but not in folate deficiency. Thus this measure is useful in distinguishing between the two. Vitamin B_{12} deficiency usually arises from malabsorption. To determine whether malabsorption is the cause, a small oral dose of vitamin B_{12} is given, and urinary excretion is measured. This procedure measures vitamin B_{12} absorption and is called a Schilling test.

Early stages of vitamin B_{12} deficiency can be detected by a low percentage saturation of its transport protein, a measure similar to iron's transferrin saturation. As the deficiency progresses, serum vitamin B_{12} concentrations fall. Table E-10 shows standards for vitamin B_{12} assessment.

TABLE E-10   Criteria for Assessing Folate and Vitamin B_{12}

	Deficient	Borderline	Acceptable
Serum folate (ng/mL)[a]	<3.0	3.0–5.9	>6.0
Erythrocyte folate (ng/mL)[a]	<140	140–159	>160
Serum vitamin B_{12} (pg/mL)	<150	150–200	≥201
Serum methylmalonic acid (nmol/L)	<376	—	—

NOTE: A nanogram (ng) is one-billionth of a gram; a picogram (pg) is one-trillionth of a gram.
[a] To convert folate values (ng/mL) to international standard units (nmol/L), multiply by 2.266.

Cautions about Nutrition Assessment

To give all the details of nutrition assessment procedures would entail writing another textbook. Nevertheless, any student of nutrition should know the basics of a proper nutrition assessment procedure for two reasons.

First, competent medical care includes attention to nutrition. Physicians should either employ a person skilled in nutrition assessment techniques or refer all clients to such a person to ensure the sound nutrition health of their clients. Health care facilities should make nutrition assessment a routine part of the initial workup on every client so that nutritional handicaps will not hinder the response to medical treatment and the recovery from illness.

Second, because nutrition is such a popular subject today, fraudulent practices are even more abundant than they have been in the past (and they have always been rampant). The knowledgeable consumer needs to know what procedures to expect in a nutrition assessment and what kinds of information they yield. This appendix has presented the basics of nutrition assessment for these reasons.

This caution is added: the tests outlined here yield information that becomes meaningful only when integrated into a whole picture by a skilled, experienced, and educated interpreter. Potential sources of error are many, from the taking of the initial data to their reporting and analysis. Each assessment method and measure is useful only as a part of the whole to confirm or eliminate the possibility of suspected nutrition problems. For example, the assessor must constantly remember that a sufficient intake of a nutrient does not guarantee adequate nutrient status for an individual. Conversely, the apparent inadequate intake of a nutrient does not, by itself, establish that a deficiency exists.

Similarly, many uncertainties, such as the calibration of the equipment, the skills of the measurer, and the perspective of the interpreter, limit the accuracy and value of anthropometric measures. This is also true of the results of the physical examination. Physical signs suggestive of malnutrition are nonspecific: they can reflect nutrient deficiencies or may be totally unrelated to nutrition. Assessors must interpret physical findings in light of other assessment findings. Finally, the usefulness of biochemical tests is also limited; the assessor must use caution in interpreting results. Vitamin and mineral blood concentrations may reflect disease processes, abnormal hormone levels, or other aberrations rather than dietary intake. Even if concentrations do reflect dietary intake, they may reflect what the person has been eating recently and not give a true picture of the person's nutrient status. Such complications sometimes make it difficult to detect a subclinical deficiency. Furthermore, many nutrients interact. The assessor has to keep in mind that an abnormal lab value for one nutrient may reflect abnormal status of other nutrients. The final diagnosis is therefore appropriately tentative, and its confirmation comes only after careful remedial steps successfully alleviate the observed problems.

Physical Activity and Energy Requirements

Chapter 8 described how to calculate your estimated energy requirements by using an equation that accounts for your gender, age, weight, height, and physical activity level. This appendix first helps you determine the correct physical activity factor to use in the equation, either by calculating your physical activity level or by guesstimating it. Then the appendix presents tables that provide a shortcut to estimating total energy expenditure.*

Calculating Your Physical Activity Level

To calculate your physical activity level, record all of your activities for a typical 24-hour day, noting the type of activity, the level of intensity, and the duration. Then, using a copy of Table F-1, find your activity in the first column (or an activity that is reasonably similar) and multiply the number of minutes spent on that activity by the factor in the third column. Put your answer in the last column and total the accumulated values for the day. Now add the subtotal of the last column to 1.1 (to account for basal energy and the thermic effect of food) as shown. This score indicates your level of physical activity. Using Table F-2, find the PA (physical activity) factor for your gender that correlates with your physical activity level score and use it in the energy equation presented on p. 260 of Chapter 8.

Guesstimating Your Physical Activity Level

As an alternative to recording your activities for a day, you can use the first two columns of Table F-3 to decide if your daily activity is sedentary, low active, active, or very active. Find the PA factor for your gender that correlates with your typical physical activity level and use it in the energy equation presented on p. 260.

Using a Shortcut to Estimate Total Energy Expenditure

The DRI Committee has developed estimates of total energy expenditure based on the equations presented in Chapter 8. These estimates are presented in Table F-4 for women and Table F-5 for men. You can use these tables to estimate your energy requirement—that is, the number of kcalories needed to maintain your current body weight. On the table appropriate for your gender, find your height in meters (or inches) in the left-hand column. Then follow the row across to find your weight in kilograms (or pounds). (If you can't find your exact height and weight, choose a value between the two closest ones.) Look down the column to find the number of kcalories that corresponds to your activity level.

Importantly, the values given in the tables are for 30-year-old people. Women 19 to 29 should add 7 kcalories per day for each year below age 30; older women should subtract 7 kcalories per day for each year above age 30. Similarly, men 19 to 29 should add 10 kcalories per day for each year below age 30; older men should subtract 10 kcalories per day for each year above age 30.

*This appendix, including the tables, is adapted from Committee on Dietary Reference Intakes, *Dietary Reference Intakes for Energy, Carbohydrate, Fiber, Fat, Fatty Acids, Cholesterol, Protein, and Amino Acids* (Washington, D.C.: National Academies Press, 2002).

F Appendix

TABLE F-1 — Physical Activities and Their Scores

If your activity was equivalent to this...	Then list the number of minutes here and ...	Multiply by this factor ...	Add this column to get your physical activity level score:
Activities of Daily Living			
Gardening (no lifting)		0.0032	
Household tasks (moderate effort)		0.0024	
Lifting items continuously		0.0029	
Loading/unloading car		0.0019	
Lying quietly		0.0000	
Mopping		0.0024	
Mowing lawn (power mower)		0.0033	
Raking lawn		0.0029	
Riding in a vehicle		0.0000	
Sitting (idle)		0.0000	
Sitting (doing light activity)		0.0005	
Taking out trash		0.0019	
Vacuuming		0.0024	
Walking the dog		0.0019	
Walking from house to car or bus		0.0014	
Watering plants		0.0014	
Additional Activities			
Billiards		0.0013	
Calisthenics (no weight)		0.0029	
Canoeing (leisurely)		0.0014	
Chopping wood		0.0037	
Climbing hills (carrying 11 lb load)		0.0061	
Climbing hills (no load)		0.0056	
Cycling (leisurely)		0.0024	
Cycling (moderately)		0.0045	
Dancing (aerobic or ballet)		0.0048	
Dancing (ballroom, leisurely)		0.0018	
Dancing (fast ballroom or square)		0.0043	
Golf (with cart)		0.0014	
Golf (without cart)		0.0032	
Horseback riding (walking)		0.0012	
Horseback riding (trotting)		0.0053	
Jogging (6 mph)		0.0088	
Music (playing accordion)		0.0008	
Music (playing cello)		0.0012	
Music (playing flute)		0.0010	
Music (playing piano)		0.0012	
Music (playing violin)		0.0014	
Rope skipping		0.0105	
Skating (ice)		0.0043	
Skating (roller)		0.0052	
Skiing (water or downhill)		0.0055	
Squash		0.0106	
Surfing		0.0048	
Swimming (slow)		0.0033	
Swimming (fast)		0.0057	
Tennis (doubles)		0.0038	
Tennis (singles)		0.0057	
Volleyball (noncompetitive)		0.0018	
Walking (2 mph)		0.0014	
Walking (3 mph)		0.0022	
Walking (4 mph)		0.0033	
Walking (5 mph)		0.0067	
Subtotal			
Factor for basal energy and the thermic effect of food			1.1
Your physical activity level score			

TABLE F-2 — Physical Activity Level Scores and Their PA Factors

Physical Activity Level Score	Description	Men: PA Factor	Women: PA Factor
1.0 to 1.39	Sedentary	1.0	1.0
1.4 to 1.59	Low active	1.11	1.12
1.6 to 1.89	Active	1.25	1.27
1.9 and above	Very active	1.48	1.45

TABLE F-3 — Physical Activity Equivalents and Their PA Factors

Description	Physical Activity Equivalents	Men: PA Factor	Women: PA Factor
Sedentary	Only those physical activities required for normal independent living	1.0	1.0
	Activities equivalent to walking at a pace of 2–4 mph for the following distances:		
Low active	1.5 to 3.0 miles/day	1.11	1.12
Active	3 to 10 miles/day	1.25	1.27
Very active	10 or more miles/day	1.48	1.45

TABLE F-4 — Total Energy Expenditure (TEE in kCalories per Day) for Women 30 Years of Age[a] at Various Levels of Activity and Various Heights and Weights

Heights m (in)	Physical Activity Level	Weight[b] kg (lb)					
1.45 (57)		38.9 (86)	45.2 (100)	52.6 (116)	63.1 (139)	73.6 (162)	84.1 (185)
		kCalories					
	Sedentary	1564	1623	1698	1813	1927	2042
	Low active	1734	1800	1912	2043	2174	2304
	Active	1946	2021	2112	2257	2403	2548
	Very active	2201	2287	2387	2553	2719	2886
1.50 (59)		41.6 (92)	48.4 (107)	56.3 (124)	67.5 (149)	78.8 (174)	90.0 (198)
		kCalories					
	Sedentary	1625	1689	1771	1894	2017	2139
	Low active	1803	1874	1996	2136	2276	2415
	Active	2025	2105	2205	2360	2516	2672
	Very active	2291	2382	2493	2671	2849	3027
1.55 (61)		44.4 (98)	51.7 (114)	60.1 (132)	72.1 (159)	84.1 (185)	96.1 (212)
		kCalories					
	Sedentary	1688	1756	1846	1977	2108	2239
	Low active	1873	1949	2081	2230	2380	2529
	Active	2104	2190	2299	2466	2632	2798
	Very active	2382	2480	2601	2791	2981	3171
1.60 (63)		47.4 (104)	55.0 (121)	64.0 (141)	76.8 (169)	89.6 (197)	102.4 (226)
		kCalories					
	Sedentary	1752	1824	1922	2061	2201	2340
	Low active	1944	2025	2168	2327	2486	2645
	Active	2185	2276	2396	2573	2750	2927
	Very active	2474	2578	2712	2914	3116	3318

continued

[a]For each year below 30, add 7 kcalories/day to TEE. For each year above 30, subtract 7 kcalories/day from TEE.
[b]These columns represent a BMI of 18.5, 22.5, 25, 30, 35, and 40, respectively.

TABLE F-4 Total Energy Expenditure (TEE in kCalories per Day) for Women 30 Years of Age[a] at Various Levels of Activity and Various Heights and Weights—continued

Heights m (in)	Physical Activity Level	Weight[b] kg (lb)					
1.65 (65)		50.4 (111)	58.5 (129)	68.1 (150)	81.7 (180)	95.3 (210)	108.9 (240)
		kCalories					
	Sedentary	1816	1893	1999	2148	2296	2444
	Low active	2016	2102	2556	2425	2594	2763
	Active	2267	2364	2494	2682	2871	3059
	Very active	2567	2678	2824	3039	3254	3469
1.70 (67)		53.5 (118)	62.1 (137)	72.3 (159)	86.7 (191)	101.2 (223)	115.6 (255)
		kCalories					
	Sedentary	1881	1963	2078	2235	2393	2550
	Low active	2090	2180	2345	2525	2705	2884
	Active	2350	2453	2594	2794	2994	3194
	Very active	2662	2780	2938	3166	3395	3623
1.75 (69)		56.7 (125)	65.8 (145)	76.6 (169)	91.9 (202)	107.2 (236)	122.5 (270)
		kCalories					
	Sedentary	1948	2034	2158	2325	2492	2659
	Low active	2164	2260	2437	2627	2817	3007
	Active	2434	2543	2695	2907	3119	3331
	Very active	2758	2883	3054	3296	3538	3780
1.80 (71)		59.9 (132)	69.7 (154)	81.0 (178)	97.2 (214)	113.4 (250)	129.6 (285)
		kCalories					
	Sedentary	2015	2106	2239	2416	2593	2769
	Low active	2239	2341	2529	2731	2932	3133
	Active	2519	2634	2799	3023	3247	3472
	Very active	2855	2987	3172	3428	3684	3940
1.85 (73)		63.3 (139)	73.6 (162)	85.6 (189)	102.7 (226)	119.8 (264)	136.9 (302)
		kCalories					
	Sedentary	2083	2179	2322	2509	2695	2882
	Low active	2315	2422	2624	2836	3049	3262
	Active	2605	2727	2904	3141	3378	3615
	Very active	2954	3093	3292	3562	3833	4103
1.90 (75)		66.8 (147)	77.6 (171)	90.3 (199)	108.3 (239)	126.4 (278)	144.4 (318)
		kCalories					
	Sedentary	2151	2253	2406	2603	2800	2996
	Low active	2392	2505	2720	2944	3168	3393
	Active	2693	2821	3011	3261	3511	3760
	Very active	3053	3200	3414	3699	3984	4270
1.95 (77)		70.3 (155)	81.8 (180)	95.1 (209)	114.1 (251)	133.1 (293)	152.1 (335)
		kCalories					
	Sedentary	2221	2328	2492	2699	2906	3113
	Low active	2470	2589	2817	3053	3290	3526
	Active	2781	2917	3119	3383	3646	3909
	Very active	3154	3309	3538	3838	4139	4439

[a] For each year below 30, add 7 kcalories/day to TEE. For each year above 30, subtract 7 kcalories/day from TEE.
[b] These columns represent a BMI of 18.5, 22.5, 25, 30, 35, and 40, respectively.

TABLE F-5 Total Energy Expenditure (TEE in kCalories per Day) for Men 30 Years of Age[a] at Various Levels of Activity and Various Heights and Weights

Heights m (in)	Physical Activity Level	Weight[b] kg (lb)					
1.45 (57)		38.9 (86)	47.3 (100)	52.6 (116)	63.1 (139)	73.6 (163)	84.1 (185)
		kCalories					
	Sedentary	1777	1911	2048	2198	2347	2496
	Low active	1931	2080	2225	2393	2560	2727
	Active	2127	2295	2447	2636	2826	3015
	Very active	2450	2648	2845	3075	3305	3535
1.50 (59)		41.6 (92)	50.6 (107)	56.3 (124)	67.5 (149)	78.8 (174)	90.0 (198)
		kCalories					
	Sedentary	1848	1991	2126	2286	2445	2605
	Low active	2009	2168	2312	2491	2670	2849
	Active	2215	2394	2545	2748	2951	3154
	Very active	2554	2766	2965	3211	3457	3703
1.55 (61)		44.4 (98)	54.1 (114)	60.1 (132)	72.1 (159)	84.1 (185)	96.1 (212)
		kCalories					
	Sedentary	1919	2072	2205	2376	2546	2717
	Low active	2089	2259	2401	2592	2783	2974
	Active	2305	2496	2646	2862	3079	3296
	Very active	2660	2887	3087	3349	3612	3875
1.60 (63)		47.4 (104)	57.6 (121)	64.0 (141)	76.8 (169)	89.6 (197)	102.4 (226)
		kCalories					
	Sedentary	1993	2156	2286	2468	2650	2831
	Low active	2171	2351	2492	2695	2899	3102
	Active	2397	2601	2749	2980	3210	3441
	Very active	2769	3010	3211	3491	3771	4051
1.65 (65)		50.4 (111)	61.3 (129)	68.1 (150)	81.7 (180)	95.3 (210)	108.9 (240)
		kCalories					
	Sedentary	2068	2241	2369	2562	2756	2949
	Low active	2254	2446	2585	2801	3017	3234
	Active	2490	2707	2854	3099	3345	3590
	Very active	2880	3136	3339	3637	3934	4232
1.70 (67)		53.5 (118)	65.0 (137)	72.3 (159)	86.7 (191)	101.2 (223)	115.6 (255)
		kCalories					
	Sedentary	2144	2328	2454	2659	2864	3069
	Low active	2338	2542	2679	2909	3139	3369
	Active	2586	2816	2961	3222	3483	3743
	Very active	2992	3265	3469	3785	4101	4417
1.75 (69)		56.7 (125)	68.9 (145)	76.6 (169)	91.9 (202)	107.2 (236)	122.5 (270)
		kCalories					
	Sedentary	2222	2416	2540	2757	2975	3192
	Low active	2425	2641	2776	3020	3263	3507
	Active	2683	2927	3071	3347	3623	3900
	Very active	3108	3396	3602	3937	4272	4607
1.80 (71)		59.9 (132)	72.9 (154)	81.0 (178)	97.2 (214)	113.4 (250)	129.6 (285)
		kCalories					
	Sedentary	2301	2507	2628	2858	3088	3318
	Low active	2513	2741	2875	3132	3390	3648
	Active	2782	3040	3183	3475	3767	4060
	Very active	3225	3530	3738	4092	4447	4801

continued

[a]For each year below 30, add 10 kcalories/day to TEE. For each year above 30, subtract 10 kcalories/day from TEE.
[b]These columns represent a BMI of 18.5, 22.5, 25, 30, 35, and 40, respectively.

TABLE F-5 Total Energy Expenditure (TEE in kCalories per Day) for Men 30 Years of Age[a] at Various Levels of Activity and Various Heights and Weights—continued

Heights m (in)	Physical Activity Level	Weight[b] kg (lb)					
1.85 (73)		63.3 (139)	77.0 (162)	85.6 (189)	102.7 (226)	119.8 (264)	136.9 (302)
		kCalories					
	Sedentary	2382	2599	2718	2961	3204	3447
	Low active	2602	2844	2976	3248	3520	3792
	Active	2883	3155	3297	3606	3915	4223
	Very active	3344	3667	3877	4251	4625	4999
1.90 (75)		66.8 (147)	81.2 (171)	90.3 (199)	108.3 (239)	126.4 (278)	144.4 (318)
		kCalories					
	Sedentary	2464	2693	2810	3066	3322	3579
	Low active	2693	2948	3078	3365	3652	3939
	Active	2986	3273	3414	3739	4065	4390
	Very active	3466	3806	4018	4413	4807	5202
1.95 (77)		70.3 (155)	85.6 (180)	95.1 (209)	114.1 (251)	133.1 (293)	152.1 (335)
		kCalories					
	Sedentary	2547	2789	2903	3173	3443	3713
	Low active	2786	3055	3183	3485	3788	4090
	Active	3090	3393	3533	3875	4218	4561
	Very active	3590	3948	4162	4578	4993	5409

[a] For each year below 30, add 10 kcalories/day to TEE. For each year above 30, subtract 10 kcalories/day from TEE.
[b] These columns represent a BMI of 18.5, 22.5, 25, 30, 35, and 40, respectively.

United States: Exchange Lists

Chapter 2 introduced the exchange system, and this appendix provides details from the 2003 edition. Appendix I presents Canada's choice system for meal planning.

The Exchange Groups and Lists

The exchange system sorts foods into three main groups by their proportions of carbohydrate, fat, and protein. These three groups—the carbohydrate group, the fat group, and the meat and meat substitutes group (protein)—organize foods into several exchange lists (see Table G-1). Then any food on a list can be "exchanged" for any other on that same list. The carbohydrate group covers these exchange lists:

- Starch (cereals, grains, pasta, breads, crackers, snacks, starchy vegetables, and dried beans, peas, and lentils).
- Fruit.
- Milk (fat-free, reduced fat, and whole).
- Other carbohydrates (desserts and snacks with added sugars and fats).
- Vegetables.

The fat group covers this exchange list:

- Fats.

The meat and meat substitutes group (protein) covers these exchange lists:

- Meat and meat substitutes (very lean, lean, medium-fat, and high-fat).

TABLE G-1 The Exchange Groups and Lists

Group/Lists	Typical Item/Portion Size	Carbohydrate (g)	Protein (g)	Fat (g)	Energy[a] (kcal)
Carbohydrate Group					
Starch[b]	1 slice bread	15	3	0–1	80
Fruit	1 small apple	15	—	—	60
Milk					
Fat-free, low-fat	1 c fat-free milk	12	8	0–3	90
Reduced-fat	1 c reduced-fat milk	12	8	5	120
Whole	1 c whole milk	12	8	8	150
Other carbohydrates[c]	2 small cookies	15	varies	varies	varies
Vegetable (nonstarchy)	½ c cooked carrots	5	2	—	25
Meat and Meat Substitute Group [d]					
Meat					
Very lean	1 oz chicken (white meat, no skin)	—	7	0–1	35
Lean	1 oz lean beef	—	7	3	55
Medium-fat	1 oz ground beef	—	7	5	75
High-fat	1 oz pork sausage	—	7	8	100
Fat Group					
Fat	1 tsp butter	—	—	5	45

[a]The energy value for each exchange list represents an approximate average for the group and does not reflect the precise number of grams of carbohydrate, protein, and fat. For example, a slice of bread contains 15 grams of carbohydrate (that's 60 kcalories), 3 grams protein (that's another 12 kcalories), and a little fat—rounded to 80 kcalories for ease in calculating. A half-cup of vegetables (not including starchy vegetables) contains 5 grams carbohydrate (20 kcalories) and 2 grams protein (8 more), which has been rounded down to 25 kcalories.
[b]The starch list includes cereals, grains, breads, crackers, snacks, starchy vegetables (such as corn, peas, and potatoes), and legumes (dried beans, peas, and lentils).
[c]The other carbohydrates list includes foods that contain added sugars and fats such as cakes, cookies, doughnuts, ice cream, potato chips, pudding, syrup, and frozen yogurt.
[d]The meat and meat substitutes list includes legumes, cheeses, and peanut butter.

FIGURE G-1 Seeing Exchanges on a Food Label

HOME▲TASTE

Lasagna Dinner
WITH MEAT SAUCE

Nutrition Facts

Serving size 10¹/₂ oz (298 g)
Servings per Package 1

Amount per serving

Calories 361	Calories from Fat 117

	% Daily Value
Total Fat 13 g	20%
Saturated Fat 8 g	40%
Cholesterol 87 mg	29%
Sodium 860 mg	36%
Total Carbohydrate 37 g	12%
Dietary fiber 0 g	
Sugars 8 g	
Protein 26 g	

Can you "see" these exchanges in the label above?

Exchange	Carbohydrate	Protein	Fat
2 starches	30 g	6 g	—
1 vegetable	5 g	2 g	—
3 medium-fat meats	—	21 g	15 g
Exchange totals	**35**	**29**	**15**
Label totals	**37**	**26**	**13**

Knowing that foods on the starch list provide 15 grams of carbohydrate and those on the vegetable list provide 5, you can count a lasagna dinner that provides 37 grams of carbohydrate as "2 starches and 1 vegetable"; knowing that foods on the meat list provide 7 grams of protein, you might count it as "3 meats"; the grams of fat suggest that the meat (and cheese) is probably medium-fat.

Portion Sizes The exchange system helps people control their energy intakes by paying close attention to portion sizes. The portion sizes have been carefully adjusted and defined so that a portion of any food on a given list provides roughly the same amount of carbohydrate, fat, and protein and, therefore, total kcalories. Any food on a list can then be exchanged, or traded, for any other food on that same list without significantly affecting the diet's balance or total kcalories. For example, a person may select either 17 small grapes or ½ large grapefruit as one fruit portion, and either choice would provide roughly 60 kcalories. A whole grapefruit, however, would count as 2 portions.

A *portion* in the exchange system is not always the same as a *serving* in the Daily Food Guide, especially when it comes to meats. The exchange system lists meats and most cheeses in single ounces; that is, one *portion* (or *exchange*) of meat is 1 ounce, whereas one *serving* in the Daily Food Guide is 2 to 3 ounces. Calculating meat by the ounce encourages a person to keep close track of the exact amounts eaten. This in turn helps control energy and fat intakes. Be aware, too, that most people do not serve foods in carefully measured portions, nor do the amounts reflect the exchange system or Daily Food Guide serving sizes. Many restaurants, for example, offer 8- to 16-ounce steaks that are the equivalent to four or five (2- to 3-ounce) *servings* of meat. Similarly, a bakery may sell muffins or bagels that are two to three times the size of a typical bread serving.

To apply the system successfully, users must become familiar with portion sizes. A convenient way to remember the portion sizes and energy values is to keep in mind a typical item from each list (review Table G-1).

The Foods on the Lists Foods do not always appear on the exchange list where you might first expect to find them. They are grouped according to their energy-nutrient contents rather than by their source (such as milks), their outward appearance, or their vitamin and mineral contents. Notice, for example, that cheeses are grouped with meats (not milk) because, like meats, cheeses contribute energy from protein and fat but provide negligible carbohydrate. Similarly, starchy vegetables such as potatoes are found on the starch list with breads and cereals, not with the vegetables, and bacon is with the fats and oils, not with the meats.

Users of the exchange lists learn to view mixtures of foods, such as casseroles and soups, as combinations of foods from different exchange lists. They also learn to interpret food labels with the exchange system in mind (see Figure G-1).

Controlling Energy and Fat By assigning items like bacon to the fat list, the exchange system alerts consumers to foods that are unexpectedly high in fat. Even the starch list specifies which grain products contain added fat (such as biscuits, muffins, and waffles). In addition, the exchange system encourages users to think of fat-free milk as milk and of whole milk as milk with added fat, and to think of very lean meats as meats and of lean, medium-fat, and high-fat meats as meats with added fat. To that end, foods on the milk and meat lists are separated into categories based on their fat contents. The milk group is classed as fat-free, reduced-fat, and whole; the meat group as very lean, lean, medium-fat, and high-fat.

Control of food energy and fat intake can be highly successful with the exchange system. Exchange plans do not, however, guarantee adequate intakes of vitamins and minerals. Food group plans work better from that standpoint because the food groupings are based on similarities in vitamin-mineral content. In the exchange system, for example, meats are grouped with cheeses, yet the meats are iron-rich and calcium-poor, whereas the cheeses are iron-poor and calcium-rich. To take advantage of the strengths of both food group plans and exchange patterns, and to compensate for their weaknesses, diet planners often combine these two diet-planning tools.

TABLE G-2 Diet Planning with the Exchange System Using the Daily Food Guide Pattern

Patterns from Daily Food Guide Plan	Selections Made Using the Exchange System	Energy Cost (kcal)
Grains (breads and cereals)—6 to 11 servings	Starch list—select 9 exchanges	720
Vegetables—3 to 5 servings	Vegetable list—select 4 exchanges	100
Fruits—2 to 4 servings	Fruit list—select 3 exchanges	180
Meat—2 to 3 servings[a]	Meat list—select 6 lean exchanges	330
Milk—2 servings	Milk list—select 2 fat-free exchanges	180
	Fat list—select 5 exchanges	225
Total		1735

[a]In the food group plan, 1 serving is 2 to 3 ounces; in the exchange system, 1 exchange is 1 ounce. The Daily Food Guide suggests that amounts should total 5 to 7 ounces of meat daily.

Combining Food Group Plans and Exchange Lists

A person may find that using a food group plan together with the exchange lists eases the task of choosing foods that provide all the nutrients. The food group plan ensures the all classes of nutritious foods are included, thus promoting adequacy, balance, and variety. The exchange system classifies the food selections by their energy-yielding nutrients, thus controlling energy and fat intakes.

Table G-2 shows how to use the Daily Food Guide plan together with the exchange lists to plan a diet. The Daily Food Guide ensures that a certain number of servings is chosen from each of the five food groups (see the first column of the table). The second column translates the number of servings (using the midpoint) into exchanges. With the addition of a small amount of fat, this sample diet plan provides abut 1750 kcalories. Most people can meet their needs for all the nutrients within this reasonable energy allowance. The next step in diet planning is to assign the exchanges to meals and snacks. The final plan might look like the one in Table G-3.

Next, a person could begin to fill in the plan with real foods to create a menu (use Tables G-4 through G-12). For example, the breakfast plan calls for 2 starch exchanges, 1 fruit exchange, and 1 fat-free milk exchange. A person might select a bowl of shredded wheat with banana slices and milk:

1 cup shredded wheat = 2 starch exchanges.

1 small banana = 1 fruit exchange.

1 cup fat-free milk = 1 milk exchange.

TABLE G-3 A Sample Diet Plan and Menu

This diet plan is one of many possibilities. It follows the number of servings suggested by the Daily Food Guide and meets dietary recommendations to provide 45 to 65 percent of its kcalories from carbohydrate, 10 to 35 percent from protein, and 20 to 35 percent from fat.

Exchange	Breakfast	Lunch	Snack	Dinner	Snack
9 starch	2	2	1	3	1
4 vegetables				4	
3 fruit	1	1	1		
6 lean meat		2		4	
2 fat-free milk	1				1
5 fat		1		4	

SAMPLE MENU

Breakfast:	Cereal with banana and milk
Lunch:	Turkey sandwich and a small bunch of grapes
Snack:	Popcorn and apple juice
Dinner:	Spaghetti with meat sauce; salad with sunflower seeds and dressing; green beans; corn on the cob
Snack:	Graham crackers and milk

G Appendix

Or half a bagel and a bowl of cantaloupe pieces topped with yogurt:

½ bagel = 2 starch exchanges.

⅓ cantaloupe melon = 1 fruit exchange.

⅔ cup fat-free plain yogurt = 1 milk exchange.

TABLE G-4   U.S. Exchange System: Starch List

1 starch exchange = 15 g carbohydrate, 3 g protein, 0–1 g fat, and 80 kcal
NOTE: In general, one starch exchange is ½ c cooked cereal, grain, or starchy vegetable; ⅓ c cooked rice or pasta; 1 oz of bread; ¾ to 1 oz snack food.

Serving Size	Food
Bread	
¼ (1 oz)	Bagel, 4 oz
2 slices (1½ oz)	Bread, reduced-kcalorie
1 slice (1 oz)	Bread, white (including French and Italian), whole-wheat, pumpernickel, rye
4 (⅔ oz)	Bread sticks, crisp, 4″ x ½″
½	English muffin
½ (1 oz)	Hot dog or hamburger bun
¼	Naan, 8″ x 2″
1	Pancake, 4″ across, ¼″ thick
½	Pita, 6″ across
1 (1 oz)	Plain roll, small
1 slice (1 oz)	Raisin bread, unfrosted
1	Tortilla, corn, 6″ across
1	Tortilla, flour, 6″ across
⅓	Tortilla, flour, 10″ across
1	Waffle, 4″ square or across, reduced-fat
Cereals and Grains	
½ c	Bran cereals
½ c	Bulgur, cooked
½ c	Cereals, cooked
¾ c	Cereals, unsweetened, ready-to-eat
3 tbs	Cornmeal (dry)
⅓ c	Couscous
3 tbs	Flour (dry)
¼ c	Granola, low-fat
¼ c	Grape nuts
½ c	Grits, cooked
½ c	Kasha
⅓ c	Millet
¼ c	Muesli
½ c	Oats
⅓ c	Pasta, cooked
1½ c	Puffed cereals
⅓ c	Rice, white or brown, cooked
½ c	Shredded wheat
½ c	Sugar-frosted cereal
3 tbs	Wheat germ
Starchy Vegetables	
⅓ c	Baked beans
½ c	Corn
½ cob (5 oz)	Corn on cob, large
1 c	Mixed vegetables with corn, peas, or pasta

Serving Size	Food
½ c	Peas, green
½ c	Plantains
½ medium (3 oz) or ½ c	Potato, boiled
¼ large (3 oz)	Potato, baked with skin
½ c	Potatoes, mashed
1 c	Squash, winter (acorn, butternut, pumpkin)
½ c	Yams, sweet potatoes, plain
Crackers and Snacks	
8	Animal crackers
3	Graham crackers, 2½″ square
¾ oz	Matzoh
4 slices	Melba toast
24	Oyster crackers
3 c	Popcorn (popped, no fat added or low-fat microwave)
¾ oz	Pretzels
2	Rice cakes, 4″ across
6	Saltine-type crackers
15–20 (¾ oz)	Snack chips, fat-free or baked (tortilla, potato)
2–5 (¾ oz)	Whole-wheat crackers, no fat added
Beans, Peas, and Lentils (count as 1 starch + 1 very lean meat)	
½ c	Beans and peas, cooked (garbanzo, lentils, pinto, kidney, white, split, black-eyed)
⅔ c	Lima beans
3 tbs	Miso 🖉
Starchy Foods Prepared with Fat (count as 1 starch + 1 fat)	
1	Biscuit, 2½″ across
½ c	Chow mein noodles
1 (2 oz)	Cornbread, 2″ cube
6	Crackers, round butter type
1 c	Croutons
1 c (2 oz)	French-fried potatoes (oven baked)
¼ c	Granola
⅓ c	Hummus
⅙ (1 oz)	Muffin, 5 oz
3 c	Popcorn, microwave
3	Sandwich crackers, cheese or peanut butter filling
9–13 (¾ oz)	Snack chips (potato, tortilla)
⅓ c	Stuffing, bread (prepared)
2	Taco shells, 6″ across
1	Waffle, 4½″ square or across
4–6 (1 oz)	Whole-wheat crackers, fat added

🖉 = 400 mg or more of sodium per serving.

Then the person could move on to complete the menu for lunch, dinner, and snacks. (Table G-3 includes a sample menu.) As you can see, we all make countless food-related decisions daily—whether we have a plan or not. Following a plan, like the Daily Food Guide, that incorporates health recommendations and diet-planning principles helps a person to make wise decisions.

TABLE G-5 U.S. Exchange System: Fruit List

1 fruit exchange = 15 g carbohydrate and 60 kcal
NOTE: In general, one fruit exchange is 1 small fresh fruit; ½ c canned or fresh fruit or unsweetened fruit juice; ¼ c dried fruit.

Serving Size	Food	Serving Size	Food
1 (4 oz)	Apple, unpeeled, small	½ (8 oz) or 1 c cubes	Papaya
½ c	Applesauce, unsweetened	1 (4 oz)	Peach, medium, fresh
4 rings	Apples, dried	½ c	Peaches, canned
4 whole (5½ oz)	Apricots, fresh	½ (4 oz)	Pear, large, fresh
8 halves	Apricots, dried	½ c	Pears, canned
½ c	Apricots, canned	¾ c	Pineapple, fresh
1 (4 oz)	Banana, small	½ c	Pineapple, canned
¾ c	Blackberries	2 (5 oz)	Plums, small
¾ c	Blueberries	½ c	Plums, canned
⅓ melon (11 oz) or 1 c cubes	Cantaloupe, small	3	Plums, dried (prunes)
12 (3 oz)	Cherries, sweet, fresh	2 tbs	Raisins
½ c	Cherries, sweet, canned	1 c	Raspberries
3	Dates	1¼ c whole berries	Strawberries
1½ large or 2 medium (3½ oz)	Figs, fresh	2 (8 oz)	Tangerines, small
1½	Figs, dried	1 slice (13½ oz) or 1¼ c cubes	Watermelon
½ c	Fruit cocktail		
½ (11 oz)	Grapefruit, large	**Fruit Juice, unsweetened**	
¾ c	Grapefruit sections, canned	½ c	Apple juice/cider
17 (3 oz)	Grapes, small	⅓ c	Cranberry juice cocktail
1 slice (10 oz) or 1 c cubes	Honeydew melon	1 c	Cranberry juice cocktail, reduced-kcalorie
1 (3½ oz)	Kiwi	⅓ c	Fruit juice blends, 100% juice
¾ c	Mandarin oranges, canned	⅓ c	Grape juice
½ (5½ oz) or ½ c	Mango, small	½ c	Grapefruit juice
1 (5 oz)	Nectarine, small	½ c	Orange juice
1 (6½ oz)	Orange, small	½ c	Pineapple juice
		⅓ c	Prune juice

TABLE G-6 U.S. Exchange System: Milk List

NOTE: In general, one milk exchange is 1 c milk or yogurt.

Serving Size	Food	Serving Size	Food
Fat-Free and Low-Fat Milk		**Reduced-Fat Milk**	
1 fat-free/low-fat milk exchange = 12 g carbohydrate, 8 g protein, 0–3 g fat, 90 kcal		1 reduced-fat milk exchange = 12 g carbohydrate, 8 g protein, 5 g fat, 120 kcal	
1 c	Fat-free milk	1 c	2% milk
1 c	½% milk	1 c	Soy milk
1 c	1% milk	1 c	Sweet acidophilus milk
1 c	Fat-free or low-fat buttermilk	¾ c	Yogurt, plain low-fat
½ c	Evaporated fat-free milk		
⅓ c dry	Fat-free dry milk	**Whole Milk**	
1 c	Soy milk, low-fat or fat-free	1 whole milk exchange = 12 g carbohydrate, 8 g protein, 8 g fat, 150 kcal	
⅔ c (6 oz)	Yogurt, fat-free or low-fat, flavored, sweetened with nonnutritive sweetener and fructose	1 c	Whole milk
		½ c	Evaporated whole milk
⅔ c (6 oz)	Yogurt, plain fat-free	1 c	Goat's milk
		1 c	Kefir
		¾ c	Yogurt, plain (made from whole milk)

G
Appendix

TABLE G-7   U.S. Exchange System: Sweets, Desserts, and Other Carbohydrates List

1 other carbohydrate exchange = 15 g carbohydrate, or 1 starch, or 1 fruit, or 1 milk exchange

Food	Serving Size	Exchanges per Serving
Angel food cake, unfrosted	1/12 cake (2 oz)	2 carbohydrates
Brownies, small, unfrosted	2" square (1 oz)	1 carbohydrate, 1 fat
Cake, unfrosted	2" square (1 oz)	1 carbohydrate, 1 fat
Cake, frosted	2" square (2 oz)	2 carbohydrates, 1 fat
Cookies or sandwich cookies with creme filling	2 small (⅔ oz)	1 carbohydrate, 1 fat
Cookies, sugar-free	3 small or 1 large (¾–1 oz)	1 carbohydrate, 1–2 fats
Cranberry sauce, jellied	¼ c	1½ carbohydrates
Cupcake, frosted	1 small (2 oz)	2 carbohydrates, 1 fat
Doughnut, plain cake	1 medium (1½ oz)	1½ carbohydrates, 2 fats
Doughnut, glazed	3¾" across (2 oz)	2 carbohydrates, 2 fats
Energy, sport, or breakfast bar	1 bar (1⅓ oz)	1½ carbohydrates, 0–1 fat
Energy, sport, or breakfast bar	1 bar (2 oz)	2 carbohydrates, 1 fat
Fruit cobbler	½ c (3½ oz)	3 carbohydrates, 1 fat
Fruit juice bar, frozen, 100% juice	1 bar (3 oz)	1 carbohydrate
Fruit snacks, chewy (pureed fruit concentrate)	1 roll (¾ oz)	1 carbohydrate
Fruit spreads, 100% fruit	1½ tbs	1 carbohydrate
Gelatin, regular	½ c	1 carbohydrate
Gingersnaps	3	1 carbohydrate
Granola or snack bar, regular or low-fat	1 bar (1 oz)	1½ carbohydrates
Honey	1 tbs	1 carbohydrate
Ice cream	½ c	1 carbohydrate, 2 fats
Ice cream, light	½ c	1 carbohydrate, 1 fat
Ice cream, low-fat	½ c	1½ carbohydrates
Ice cream, fat-free, no sugar added	½ c	1 carbohydrate
Jam or jelly, regular	1 tbs	1 carbohydrate
Milk, chocolate, whole	1 c	2 carbohydrates, 1 fat
Pie, fruit, 2 crusts	⅙ of 8" commercially prepared pie	3 carbohydrates, 2 fats
Pie, pumpkin or custard	⅛ of 8" commercially prepared pie	2 carbohydrates, 2 fats
Pudding, regular (made with reduced-fat milk)	½ c	2 carbohydrates
Pudding, sugar-free (made with fat-free milk)	½ c	1 carbohydrate
Reduced-calorie meal replacement (shake)	1 can (10–11 oz)	1½ carbohydrates, 0–1 fats
Rice milk, low-fat or fat-free, plain	1 c	1 carbohydrate
Rice milk, low-fat, flavored	1 c	1½ carbohydrates
Salad dressing, fat-free 	¼ c	1 carbohydrate
Sherbet, sorbet	½ c	2 carbohydrates
Spaghetti or pasta sauce, canned 	½ c	1 carbohydrate, 1 fat
Sports drinks	8 oz (1 c)	1 carbohydrate
Sugar	1 tbs	1 carbohydrate
Sweet roll or danish	1 (2½ oz)	2½ carbohydrates, 2 fats
Syrup, light	2 tbs	1 carbohydrate
Syrup, regular	1 tbs	1 carbohydrate
Syrup, regular	¼ c	4 carbohydrates
Vanilla wafers	5	1 carbohydrate, 1 fat
Yogurt, frozen	½ c	1 carbohydrate, 0–1 fat
Yogurt, frozen, fat-free	⅓ c	1 carbohydrate
Yogurt, low-fat with fruit	1 c	3 carbohydrates, 0–1 fat

 = 400 mg or more sodium per exchange.

TABLE G-8   U.S. Exchange System: Nonstarchy Vegetable List

1 vegetable exchange = 5 g carbohydrate, 2 g protein, 0 g fat, and 25 kcal
NOTE: In general, one vegetable exchange is ½ c cooked vegetables or vegetable juice; 1 c raw vegetables. Starchy vegetables such as corn, peas, and potatoes are on the starch list (Table G-4).

Artichokes	Mushrooms
Artichoke hearts	Okra
Asparagus	Onions
Beans (green, wax, Italian)	Pea pods
Bean sprouts	Peppers (all varieties)
Beets	Radishes
Broccoli	Salad greens (endive, escarole, lettuce, romaine, spinach)
Brussels sprouts	
Cabbage	Sauerkraut
Carrots	Spinach
Cauliflower	Summer squash (crookneck)
Celery	Tomatoes
Cucumbers	Tomatoes, canned
Eggplant	Tomato sauce
Green onions or scallions	Tomato/vegetable juice
Greens (collard, kale, mustard, turnip)	Turnips
Kohlrabi	Water chestnuts
Leeks	Watercress
Mixed vegetables (without corn, peas, or pasta)	Zucchini

 = 400 mg or more sodium per exchange.

TABLE G-9   U.S. Exchange System: Meat and Meat Substitutes List

NOTE: In general, a meat exchange is 1 oz meat, poultry, or cheese; ½ c dried beans (weigh meat and poultry and measure beans after cooking).

Serving Size	Food
Very Lean Meat and Substitutes	
1 very lean meat exchange = 7 g protein, 0–1 g fat, 35 kcal	
1 oz	Poultry: Chicken or turkey (white meat, no skin), Cornish hen (no skin)
1 oz	Fish: Fresh or frozen cod, flounder, haddock, halibut, trout, lox (smoked salmon) ✎ , tuna, fresh or canned in water
1 oz	Shellfish: Clams, crab, lobster, scallops, shrimp, imitation shellfish
1 oz	Game: Duck or pheasant (no skin), venison, buffalo, ostrich
	Cheese with ≤1g fat/oz:
¼ c	Fat-free or low-fat cottage cheese
1 oz	Fat-free cheese
1 oz	Processed sandwich meats with ≤1 g fat/oz (such as deli thin, shaved meats, chipped beef ✎ , turkey ham)
2	Egg whites
¼ c	Egg substitutes, plain
1 oz	Hot dogs with ≤1 g fat/oz ✎
1 oz	Kidney (high in cholesterol)
1 oz	Sausage with ≤1 g fat/oz
Count as 1 very lean meat + 1 starch exchange:	
½ c	Beans, peas, lentils (cooked)
Lean Meat and Substitutes	
1 lean meat exchange = 7 g protein, 3 g fat, 55 kcal	
1 oz	Beef: USDA Select or Choice grades of lean beef trimmed of fat (round, sirloin, and flank steak); tenderloin; roast (rib, chuck, rump); steak (T-bone, porterhouse, cubed), ground round
1 oz	Pork: Lean pork (fresh ham); canned, cured, or boiled ham; Canadian bacon ✎ ; tenderloin, center loin chop
1 oz	Lamb: Roast, chop, leg
1 oz	Veal: Lean chop, roast
1 oz	Poultry: Chicken, turkey (dark meat, no skin), chicken (white meat, with skin), domestic duck or goose (well drained of fat, no skin)
	Fish:
1 oz	Herring (uncreamed or smoked)
6 medium	Oysters
1 oz	Salmon (fresh or canned), catfish
2 medium	Sardines (canned)
1 oz	Tuna (canned in oil, drained)
1 oz	Game: Goose (no skin), rabbit
	Cheese:
¼ c	4.5%-fat cottage cheese

Serving Size	Food
2 tbs	Grated Parmesan
1 oz	Cheeses with ≤3 g fat/oz
1½ oz	Hot dogs with ≤3 g fat/oz ✎
1 oz	Processed sandwich meat with ≤3 g fat/oz (turkey pastrami or kielbasa)
1 oz	Liver, heart (high in cholesterol)
Medium-Fat Meat and Substitutes	
1 medium-fat meat exchange = 7 g protein, 5 g fat, and 75 kcal	
1 oz	Beef: Most beef products (ground beef, meatloaf, corned beef, short ribs, Prime grades of meat trimmed of fat, such as prime rib)
1 oz	Pork: Top loin, chop, Boston butt, cutlet
1 oz	Lamb: Rib roast, ground
1 oz	Veal: Cutlet (ground or cubed, unbreaded)
1 oz	Poultry: Chicken (dark meat, with skin), ground turkey or ground chicken, fried chicken (with skin)
1 oz	Fish: Any fried fish product
	Cheese with ≤5 g fat/oz:
1 oz	Feta
1 oz	Mozzarella
¼ c (2 oz)	Ricotta
1	Egg (high in cholesterol, limit to 3/week)
1 oz	Sausage with ≤5 g fat/oz
1 c	Soy milk
¼ c	Tempeh
4 oz or ½ c	Tofu
High-Fat Meat and Substitutes	
1 high-fat meat exchange = 7 g protein, 8 g fat, 100 kcal	
1 oz	Pork: Spareribs, ground pork, pork sausage
1 oz	Cheese: All regular cheeses (American ✎ , cheddar, Monterey Jack, swiss)
1 oz	Processed sandwich meats with ≤8 g fat/oz (bologna, pimento loaf, salami)
1 oz	Sausage (bratwurst, Italian, knockwurst, Polish, smoked)
1 (10/lb)	Hot dog (turkey or chicken) ✎
3 slices (20 slices/lb)	Bacon
1 tbs	Peanut butter (contains unsaturated fat)
Count as 1 high-fat meat + 1 fat exchange:	
1 (10/lb)	Hot dog (beef, pork, or combination) ✎

✎ = 400 mg or more of sodium per serving.

TABLE G-10   U.S. Exchange System: Fat List

1 fat exchange = 5 g fat and 45 kcal
NOTE: In general, one fat exchange is 1 tsp regular butter, margarine, or vegetable oil; 1 tbs regular salad dressing. Many fat-free and reduced fat foods are on the Free Foods List (Table G-11).

Serving Size	Food
Monounsaturated Fats	
2 tbs (1 oz)	Avocado
1 tsp	Oil (canola, olive, peanut)
8 large	Olives, ripe (black)
10 large	Olives, green, stuffed ✐
6 nuts	Almonds, cashews
6 nuts	Mixed nuts (50% peanuts)
10 nuts	Peanuts
4 halves	Pecans
½ tbs	Peanut butter, smooth or crunchy
1 tbs	Sesame seeds
2 tsp	Tahini or sesame paste
Polyunsaturated Fats	
4 halves	English walnuts
1 tsp	Margarine, stick, tub, or squeeze
1 tbs	Margarine, lower-fat spread (30% to 50% vegetable oil)
1 tsp	Mayonnaise, regular
1 tbs	Mayonnaise, reduced-fat
1 tsp	Oil (corn, safflower, soybean)
1 tbs	Salad dressing, regular ✐
2 tbs	Salad dressing, reduced-fat
2 tsp	Mayonnaise type salad dressing, regular
1 tbs	Mayonnaise type salad dressing, reduced-fat
1 tbs	Seeds (pumpkin, sunflower)
Saturated Fats*	
1 slice (20 slices/lb)	Bacon, cooked
1 tsp	Bacon, grease
1 tsp	Butter, stick
2 tsp	Butter, whipped
1 tbs	Butter, reduced-fat
2 tbs (½ oz)	Chitterlings, boiled
2 tbs	Coconut, sweetened, shredded
1 tbs	Coconut milk
2 tbs	Cream, half and half
1 tbs (½ oz)	Cream cheese, regular
1½ tbs (¾ oz)	Cream cheese, reduced-fat
	Fatback or salt pork† ✐
1 tsp	Shortening or lard
2 tbs	Sour cream, regular
3 tbs	Sour cream, reduced-fat

✐ = 400 mg or more sodium per exchange
*Saturated fats can raise blood cholesterol levels.
†Use a piece 1″ × 1″ × ¼″ if you plan to eat the fatback cooked with vegetables. Use a piece 2″ × 1″ × ½″ when eating only the vegetables with the fatback removed.

TABLE G-11 U.S. Exchange System: Free Foods List

NOTE: A serving of free food contains less than 20 kcalories or no more than 5 grams of carbohydrate; those with serving sizes should be limited to 3 servings a day whereas those without serving sizes can be eaten freely.

Serving Size	Food
Fat-Free or Reduced-Fat Foods	
1 tbs (½ oz)	Cream cheese, fat-free
1 tbs	Creamers, nondairy, liquid
2 tsp	Creamers, nondairy, powdered
4 tbs	Margarine spread, fat-free
1 tsp	Margarine spread, reduced-fat
1 tbs	Mayonnaise, fat-free
1 tsp	Mayonnaise, reduced-fat
1 tbs	Mayonnaise type salad dressing, fat-free
1 tsp	Mayonnaise type salad dressing, reduced-fat
	Nonstick cooking spray
1 tbs	Salad dressing, fat-free or low-fat
2 tbs	Salad dressing, fat-free, Italian
1 tbs	Sour cream, fat-free, reduced-fat
1 tbs	Whipped topping, regular
2 tbs	Whipped topping, light or fat-free
Sugar-Free Foods	
1 piece	Candy, hard, sugar-free
	Gelatin dessert, sugar-free
	Gelatin, unflavored
	Gum, sugar-free
2 tsp	Jam or jelly, light
	Sugar substitutes
2 tbs	Syrup, sugar-free
Drinks	
	Bouillon, broth, consommé 🖋
	Bouillon or broth, low-sodium
	Carbonated or mineral water
	Club soda
1 tbs	Cocoa powder, unsweetened

Serving Size	Food
	Coffee
	Diet soft drinks, sugar-free
	Drink mixes, sugar-free
	Tea
	Tonic water, sugar-free
Condiments	
1 tbs	Catsup
	Horseradish
	Lemon juice
	Lime juice
	Mustard
1 tbs	Pickle relish
1½ medium	Pickles, dill 🖋
2 slices	Pickles, sweet (bread and butter)
¾ oz	Pickles, sweet (gherkin)
¼ c	Salsa
1 tbs	Soy sauce, regular or light 🖋
1 tbs	Taco sauce
	Vinegar
2 tbs	Yogurt
Seasonings	
	Flavoring extracts
	Garlic
	Herbs, fresh or dried
	Hot pepper sauces
	Pimento
	Spices
	Wine, used in cooking
	Worcestershire sauce

🖋 = 400 mg or more of sodium per serving.

TABLE G-12   U.S. Exchange System: Combination Foods List

Food	Serving Size	Exchanges per Serving
Entrées		
Tuna noodle casserole, lasagna, spaghetti with meatballs, chili with beans, macaroni and cheese 🥄	1 c (8 oz)	2 carbohydrates, 2 medium-fat meats
Chow mein (without noodles or rice)	2 c (16 oz)	1 carbohydrate, 2 lean meats
Tuna or chicken salad	½ c (3½ oz)	½ carbohydrate, 2 lean meats, 1 fat
Frozen Entrées and Meals		
Dinner-type meal 🥄	Generally 14–17 oz	3 carbohydrates, 3 medium-fat meats, 3 fats
Entrée or meal with <340 kcal 🥄	About 8–11 oz	2–3 carbohydrates, 1–2 lean meats
Meatless burger, soy based	3 oz	½ carbohydrate, 2 lean meats
Meatless burger, vegetable and starch based	3 oz	1 carbohydrate, 1 lean meat
Pizza, cheese, thin crust 🥄	¼ of 12″ (6 oz)	2 carbohydrates, 2 medium-fat meats, 1 fat
Pizza, meat topping, thin crust 🥄	¼ of 12″ (6 oz)	2 carbohydrates, 2 medium-fat meats, 2 fats
Pot pie 🥄	1 (7 oz)	2½ carbohydrates, 1 medium-fat meat, 3 fats
Soups		
Bean 🥄	1 c	1 carbohydrate, 1 very lean meat
Cream (made with water) 🥄	1 c (8 oz)	1 carbohydrate, 1 fat
Instant 🥄	6 oz prepared	1 carbohydrate
Instant with beans/lentils 🥄	8 oz prepared	2½ carbohydrates, 1 very lean meat
Split pea (made with water) 🥄	½ c (4 oz)	1 carbohydrate
Tomato (made with water)	1 c (8 oz)	1 carbohydrate
Vegetable beef, chicken noodle, or other broth-type 🥄	1 c (8 oz)	1 carbohydrate
Fast Foods		
Burrito with beef 🥄	1 (5–7 oz)	3 carbohydrates, 1 medium-fat meat, 1 fat
Chicken nuggets 🥄	6	1 carbohydrate, 2 medium-fat meats, 1 fat
Chicken breast and wing, breaded and fried 🥄	1 each	1 carbohydrate, 4 medium-fat meats, 2 fats
Chicken sandwich, grilled 🥄	1	2 carbohydrates, 3 very lean meats
Chicken wings, hot 🥄	6 (5 oz)	1 carbohydrate, 3 medium-fat meats, 4 fats
Fish sandwich/tartar sauce 🥄	1	3 carbohydrates, 1 medium-fat meat, 3 fats
French fries 🥄	1 medium serving (5 oz)	4 carbohydrates, 4 fats
Hamburger, regular	1	2 carbohydrates, 2 medium-fat meats
Hamburger, large 🥄	1	2 carbohydrates, 3 medium-fat meats, 1 fat
Hot dog with bun 🥄	1	1 carbohydrate, 1 high-fat meat, 1 fat
Individual pan pizza 🥄	1	5 carbohydrates, 3 medium-fat meats, 3 fats
Pizza, cheese, thin crust 🥄	¼ of 12″ (about 6 oz)	2½ carbohydrates, 2 medium-fat meats
Pizza, meat, thin crust 🥄	¼ of 12″ (about 6 oz)	2½ carbohydrates, 2 medium-fat meats, 1 fat
Soft serve cone	1 small (5 oz)	2½ carbohydrates, 1 fat
Submarine sandwich 🥄	1 sub (6″)	3 carbohydrates, 1 vegetable, 2 medium-fat meats, 1 fat
Submarine sandwich (<6 g fat) 🥄	1 sub (6″)	2½ carbohydrates, 2 lean meats
Taco, hard or soft shell	1 (3–3½ oz)	1 carbohydrate, 1 medium-fat meat, 1 fat

🥄 = 400 mg or more sodium per exchange.

Table of Food Composition

This edition of the table of food composition includes a wide variety of foods from all food groups. It is updated with each edition to reflect nutrient changes for current foods, remove outdated foods, and add foods that are new to the marketplace.*

The nutrient database for this appendix is compiled from a variety of sources, including the USDA Standard Reference database (Release 16), literature sources, and manufacturers' data. The USDA database provides data for a wider variety of foods and nutrients than other sources. Because laboratory analysis for each nutrient can be quite costly, manufacturers tend to provide data only for those nutrients mandated on food labels. Consequently, data for their foods are often incomplete; any missing information is designated in this table as a blank space. Keep in mind that a blank space means only that the information is unknown and should not be interpreted as a zero.

Whenever using nutrient data, remember that many factors influence the nutrient contents of foods, including the mineral content of the soil, the diet of the animal or the fertilizer of the plant, the season of harvest, the method of processing, the length and method of storage, the method of cooking, the method of analysis, and the moisture content of the sample analyzed. With so many factors involved, users must view nutrient data as a close approximation of the actual amount.

For updates, corrections, and a list of 6000 foods and codes found in the diet analysis software that accompanies this text, visit **www.wadsworth.com/nutrition** and click on *Diet Analysis.*

- *Fats* Total fats, as well as the breakdown of total fats to saturated, monounsaturated, and polyunsaturated fats, are listed in the table. The fatty acids seldom add up to the total due to rounding and to other fatty acid components that are not included in these basic categories, such as *trans*-fatty acids and glycerol. *Trans*-fatty acids can comprise a large share of the total fat in margarine and shortening (hydrogenated oils) and in any foods that include them as ingredients.

- *Vitamin A and Vitamin E* In keeping with the 2001 RDA for vitamin A, which established a new measure of vitamin A activity—retinol activity equivalents (RAE)—this appendix presents data for vitamin A in micrograms (μg) RAE. Similarly, because the 2000 RDA for vitamin E is based only on the alpha-tocopherol form of vitamin E, this appendix reports vitamin E data in milligrams (mg) alpha-tocopherol (listed in the table as mg α).

- *Bioavailability* Keep in mind that the availability of nutrients from foods depends not only on the quantity provided by a food, but also on the amount absorbed and used by the body—the bioavailability. The bioavailability of folate from fortified foods, for example, is greater than from naturally occurring sources. Similarly, the body can make niacin from the amino acid tryptophan, but niacin values in this table (and most databases) report preformed niacin only. Chapter 10 provides conversion factors and additional details.

- *Using the Table* The items in this table have been organized into several categories, which are listed at the head of each right-hand page. Page numbers have been provided, and each group has been color-coded to make it easier to find individual items.

  In an effort to conserve space, the following abbreviations have been used in the food descriptions and nutrient breakdowns:

  - diam = diameter
  - ea = each

- enr = enriched
- f/ = from
- frzn = frozen
- g = grams
- liq = liquid
- pce = piece
- pkg = package
- w/ = with
- w/o = without
- t = trace
- 0 = zero (no nutrient value)

- blank space = information not available

- *Caffeine Sources* Caffeine occurs in several plants, including the familiar coffee bean, the tea leaf, and the cocoa bean from which chocolate is made. Most human societies use caffeine regularly, most often in beverages, for its stimulant effect and flavor. Caffeine contents of beverages vary depending on the plants they are made from, the climates and soils where the plants are grown, the grind or cut size, the method and duration of brewing, and the amounts served. The accompanying table shows that in general, a cup of coffee contains the most caffeine; a cup of tea, less than half as much; and cocoa or chocolate, less still. As for cola beverages, they are made from kola nuts, which contain caffeine, but most of their caffeine is added, using the purified compound obtained from decaffeinated coffee beans.

    The FDA lists caffeine as a multipurpose GRAS substance■ that may be added to foods and beverages. Drug manufacturers use caffeine in many kinds of drugs: stimulants, pain relievers, cold remedies, diuretics, and weight-loss aids.

■ Reminder: A GRAS substance is one that is "generally recognized as safe."

TABLE Caffeine Content of Beverages, Foods, and Over-the-Counter Drugs

Beverages and Foods	Average (mg)	Range (mg)	Drugs[a]	Average (mg)
Coffee (5-oz cup)			Cold remedies (standard dose)	
Brewed, drip method	130	110–150	Dristan	0
Brewed, percolator	94	64–124	Coryban-D, Triaminicin	30
Instant	74	40–108	Diuretics (standard dose)	
Decaffeinated, brewed or instant	3	1–5	Aqua-ban, Permathene H_2Off	200
Tea (5-oz cup)			Pre-Mens Forte	100
Brewed, major U.S. brand	40	20–90	Pain relievers (standard dose)	
Brewed, imported brands	60	25–110	Excedrin	130
Instant	30	25–50	Midol, Anacin	65
Iced (12-oz can)	70	67–76	Aspirin, plain (any brand)	0
Soft drinks (12-oz can)			Stimulants	
Dr. Pepper	40		Caffedrin, NoDoz, Vivarin	200
Colas and cherry cola			Weight-control aids (daily dose)	
Regular		30–46	Prolamine	280
Diet		2–58	Dexatrim, Dietac	200
Caffeine-free		0–trace		
Jolt	72			
Mountain Dew, Mello Yello	52			
Fresca, Hires Root Beer, 7-Up, Sprite, Squirt, Sunkist Orange	0			
Cocoa beverage (5-oz cup)	4	2–20		
Chocolate milk beverage (8 oz)	5	2–7		
Milk chocolate candy (1 oz)	6	1–15		
Dark chocolate, semisweet (1 oz)	20	5–35		
Baker's chocolate (1 oz)	26			
Chocolate flavored syrup (1 oz)	4			

NOTE: A pharmacologically active dose of caffeine is defined as 200 milligrams.

[a]Because products change, contact the manufacturer for an update on products you use regularly.

Table H–1

Food Composition   (Computer code number is for West Diet Analysis program)   (For purposes of calculations, use "0" for t, <1, <.1, <.01, etc.)

Computer Code Number	Food Description	Measure	Wt (g)	H₂O (%)	Ener (kcal)	Prot (g)	Carb (g)	Dietary Fiber (g)	Fat (g)	Fat Breakdown (g) Sat	Mono	Poly
	BEVERAGES											
	Alcoholic:											
	Beer:											
22500	Regular (12 fl oz)	1½ c	356	94	117	1	6	<1	<1	0	0	0
22512	Light (12 fl oz)	1½ c	354	95	99	1	5	0	0	0	0	0
22679	Budweiser (12 fl oz)	1½ c	357		145	1	11		0	0	0	0
22616	Miller (12 fl oz)	1½ c	356	96	143	1	13		0	0	0	0
22617	Miller Light (12 fl oz)	1½ c	356		110	1	7		0	0	0	0
	Gin, rum, vodka, whiskey:											
22670	80 proof	1½ fl oz	42	67	97	0	0	0	0	0	0	0
22654	86 proof	1½ fl oz	42	64	105	0	<1	0	0	0	0	0
22661	90 proof	1½ fl oz	42	62	110	0	0	0	0	0	0	0
	Liqueur:											
22519	Coffee liqueur, 53 proof	1½ fl oz	52	31	175	<1	24	0	<1	t	t	t
22520	Coffee & cream liqueur, 34 proof	1½ fl oz	47	46	154	1	10	0	7	4.5	2.1	0.3
22521	Crème de menthe, 72 proof	1½ fl oz	50	28	186	0	21	0	<1	t	t	t
22551	Kahlua	1½ fl oz	30	30	106	<1	13	0	<1	t	t	t
	Wine, 4 fl oz:											
22673	Dessert, sweet	½ c	118	71	189	<1	16	0	0	0	0	0
22501	Red	½ c	118	88	85	<1	2	0	0	0	0	0
22502	Rosé	½ c	118	89	84	<1	2	0	0	0	0	0
22504	White medium	½ c	118	90	80	<1	1	0	0	0	0	0
20077	Nonalcoholic light	1 c	232	98	14	1	3	0	0	0	0	0
22681	Wine cooler	1 c	227	90	113	<1	13	<1	<1	t	t	t
	Carbonated:											
20006	Club soda (12 fl oz)	1½ c	355	100	0	0	0	0	0	0	0	0
20005	Cola beverage (12 fl oz)	1½ c	372	89	156	<1	40	0	0	0	0	0
20030	Diet cola w/aspartame (12 fl oz)	1½ c	355	100	4	<1	<1	0	0	0	0	0
20007	Diet soda pop w/saccharin (12 fl oz)	1½ c	355	100	0	0	<1	0	0	0	0	0
20008	Ginger ale (12 fl oz)	1½ c	366	91	124	0	32	0	0	0	0	0
20031	Grape soda (12 fl oz)	1½ c	372	89	160	0	42	0	0	0	0	0
20032	Lemon-lime (12 fl oz)	1½ c	368	90	147	0	38	0	0	0	0	0
20027	Pepper-type soda (12 fl oz)	1½ c	368	89	151	0	38	0	<1	0.3	0	0
20009	Root beer (12 fl oz)	1½ c	370	89	152	0	39	0	0	0	0	0
20149	Cherry Coke (12 fl oz)	1½ c	375		156	0	42	0	0	0	0	0
20148	Coca Cola Classic (12 fl oz)	1½ c	373		145	0	40	0	0	0	0	0
20150	Diet Coke (12 fl oz)	1½ c	359		1	0	<1	0	0	0	0	0
20207	Diet 7 UP (12 fl oz)	1½ c	360		0	0	0	0	0	0	0	0
20167	Diet Pepsi (12 fl oz)	1½ c	360		0	0	0	0	0	0	0	0
20166	Pepsi cola (12 fl oz)	1½ c	360	88	150	0	41	0	0	0	0	0
20163	Sprite (12 fl oz)	1½ c	373		144	0	39	0	0	0	0	0
	Coffees:											
20012	Brewed	1 c	237	99	9	<1	0	0	2	0	0	0
20592	Cappuccino w/lowfat milk	1½ c	244		110	8	11	0	4	2.5		
20639	Cappuccino w/whole milk	1½ c	244		140	7	11	0	7	4.5		
20668	Latte w/lowfat milk	1½ c	366		170	12	17	0	6	4		
20023	Prepared from instant	1 c	238	99	5	<1	1	0	0			
	Fruit drinks:											
20024	Fruit punch drink, canned	1 c	248	88	117	0	30	<1	0	0	0	0
20142	Gatorade	1 c	241	94	60	0	15	0	0	0	0	0
20026	Grape drink, canned	1 c	250	87	125	<1	32	0	0	0	0	0
20016	Koolade sweetened with sugar	1 c	262	90	97	0	25	0	<1	t	t	t
20017	Koolade sweetened with nutrasweet	1 c	240	95	43	0	11	0	0	0	0	0
20001	Lemonade, frzn concentrate (6-oz can)	¾ c	219	52	396	1	103	<1	<1	t	t	0.1
20000	Lemonade, from concentrate	1 c	248	86	131	<1	34	<1	<1	t	t	t
20003	Limeade, frzn concentrate (6-oz can)	¾ c	218	50	427	<1	106	<1	<1	0	0	0
20002	Limeade, from concentrate	1 c	247	89	104	<1	26	<1	<1	0	0	0
20059	Pineapple grapefruit, canned	1 c	250	88	118	<1	29	0	<1	t	t	t
20025	Pineapple orange, canned	1 c	250	87	125	3	30	<1	0	0	0	0
20559	Powerade	1 c	247		72	0	19	0	0	0	0	0
20737	Snapple, fruit punch	1 c	252	88	110	0	29		0	0	0	0
20761	Snapple, tropical	1 c	252	89	110	0	27		0	0	0	0

PAGE KEY: H–2 = Beverages H–4 = Dairy H–8 = Eggs H–10 = Fat/Oil H–12 = Fruit H–18 = Bakery
H–24 = Grain H–30 = Fish H–30 = Meats H–34 = Poultry H–36 = Sausage H–38 = Mixed/Fast H–42 = Nuts/Seeds H–44 = Sweets
H–46 = Vegetables/Legumes H–56 = Vegetarian H–60 = Misc H–60 = Soups/Sauces H–62 = Fast H–76 = Convenience H–78 = Baby foods

Chol (mg)	Calc (mg)	Iron (mg)	Magn (mg)	Pota (mg)	Sodi (mg)	Zinc (mg)	VT-A (RAE)	Thia (mg)	VT-E (mg α)	Ribo (mg)	Niac (mg)	V-B$_6$ (mg)	Fola (µg)	VT-C (mg)
0	18	0.07	21	89	14	0.04	0	0.02	0	0.09	1.61	0.18	21	0
0	18	0.14	18	64	11	0.11	0	0.03	0	0.11	1.39	0.12	14	0
0	16	<.01	22	111	9	<.01		<.01		0.14	1.4	0.16		0
0					7									
0					6									
0	0	0.02	0	1	<1	0.02	0	<.01	0	<.01	<.01	<.01	0	0
0	0	0.02	0	1	<1	0.02	0	<.01	0	<.01	<.01	<.01	0	0
0	0	0.02	0	1	<1	0.02	0	<.01	0	<.01	<.01	<.01	0	0
0	1	0.03	2	16	4	0.02	0	<.01	0	<.01	0.07	0	0	0
27	8	0.06	1	15	43	0.08	81	<.01	0.21	0.03	0.04	<.01	1	0
0	0	0.04	0	0	2	0.02	0	0	0	0	<.01	0	0	0
0	0	0.02	0	4	2	0.01	0	<.01	0	<.01	0.02	0	0	0
0	9	0.28	11	109	11	0.08	0	0.02	0	0.02	0.25	0	0	0
0	9	0.51	15	132	6	0.11	0	<.01	0	0.03	0.1	0.04	2	0
0	9	0.45	12	117	6	0.07	0	<.01	0	0.02	0.09	0.03	1	0
0	11	0.38	12	94	6	0.08	0	<.01	0	<.01	0.08	0.02	0	0
0	21	0.93	23	204	16	0.19	0	0	0	0.02	0.23	0.05	2	0
0	13	0.62	12	102	19	0.13	<1	0.01	0.01	0.02	0.1	0.03	3	4
0	18	0.04	4	7	75	0.36	0	0	0	0	0	0	0	0
0	11	0.07	4	4	15	0.04	0	0	0	0	0	0	0	0
0	11	0.11	4	21	18	0	0	0.02	0	0.08	0	0	0	0
0	14	0.07	4	14	57	0.11	0	0	0	0	0	0	0	0
0	11	0.66	4	4	26	0.18	0	0	0	0	0	0	0	0
0	11	0.3	4	4	56	0.26	0	0	0	0	0	0	0	0
0	7	0.26	4	4	40	0.18	0	0	0	0	0.06	0	0	0
0	11	0.15	0	4	37	0.15	0	0	0	0	0	0	0	0
0	18	0.18	4	4	48	0.26	0	0	0	0	0	0	0	0
0				0	6		0		0					0
0				0	13		0							0
0			18	6		0			0					0
0					52		0	0		0			0	0
0	0	0		8	35		0							0
0	0	0			35		0							0
0				0	34		0							0
0	2	0.02	5	114	2	0.02	0	0	0.05	0.12	0	<.01	5	0
15	250	0			110									2
30	250	0			105									2
25	400	0			170									4
0	10	0.1	7	71	5	0.02	0	0	0	<.01	0.56	0	0	0
0	20	0.22	7	62	94	0.02	5	0.05	0.05	0.06	0.05	0.03	10	73
0	0	0.12	2	27	96	0.05	0	0.01	0	<.01	0	0	0	0
0	8	0.25	8	82	2	0.05	<1	0.02	0	0.03	0.17	0.04	2	40
0	42	0.13	3	3	37	0.08	0	0	0	<.01	<.01	0	0	31
0	17	0.65	5	50	50	0.26	1	0.02	0	0.05	0.05	0	5	78
0	15	1.58	11	147	9	0.18	1	0.06	0.09	0.21	0.16	0.05	11	39
0	10	0.52	5	50	7	0.07	<1	0.02	0.02	0.07	0.05	0.02	2	13
0	28	0.07	11	94	20	0.09	<1	0.02	0	0.03	0.08	0.04	7	25
0	7	0.02	2	22	5	0.02	<1	<.01	0	<.01	0.02	<.01	2	6
0	18	0.78	15	152	35	0.15	<1	0.08	0.02	0.04	0.67	0.1	22	115
0	12	0.68	15	115	8	0.15	2	0.08	0.08	0.05	0.52	0.12	22	56
0	0	0		32	28	0								0
0					10									0
0					10									

Table H–1

Food Composition

(Computer code number is for West Diet Analysis program)   (For purposes of calculations, use "0" for t, <1, <.1, <.01, etc.)

Computer Code Number	Food Description	Measure	Wt (g)	H₂O (%)	Ener (kcal)	Prot (g)	Carb (g)	Dietary Fiber (g)	Fat (g)	Fat Breakdown (g) Sat	Mono	Poly
	BEVERAGES—Continued											
	Fruit and vegetable juices: see Fruit and Vegetable sections											
	Ultra Slim Fast, ready to drink, can:											
	Chocolate Royale	1 ea	350	83	220	10	40	5	3	1	1.5	0.5
	French Vanilla	1 ea	350	84	220	10	40	5	3	0.5	1.5	0.5
	Strawberries n' cream	1 ea	350	84	220	10	40	5	2	0.5	1.5	0.5
20041	Water, municiple	1 c	237	100	0	0	0	0	0	0	0	0
792	La Croix	1 c	236	100	0	0	0	0	0	0	0	0
20010	Tonic water	1½ c	366	91	124	0	32	0	0	0	0	0
	Tea:											
20014	Brewed, regular	1 c	237	100	2	0	1	0	0	0	0	0
20036	Brewed, herbal	1 c	237	100	2	0	<1	0	0			
20022	From instant, sweetened	1 c	259	91	88	<1	22	0	<1	t	t	t
20020	From instant, unsweetened	1 c	237	100	2	<1	<1	0	0	0	0	0
20443	Green tea bag	1 ea	2		0	0	0	0	0	0	0	0
	DAIRY											
	Butter: see Fats and Oils											
	Cheese:											
	Natural:											
1003	Blue	1 oz	28	42	99	6	1	0	8	5.2	2.2	0.2
1037	Brick	1 oz	28	41	104	7	1	0	8	5.3	2.4	0.2
1004	Brie	1 oz	28	48	94	6	<1	0	8	4.9	2.2	0.2
1006	Camembert	1 oz	28	52	84	6	<1	0	7	4.3	2	0.2
1007	Cheddar:	1 oz	28	37	113	7	<1	0	9	5.9	2.6	0.3
1008	Shredded	1 c	113	37	455	28	1	0	37	23.8	10.6	1.1
1088	Shredded, low fat, sodium	1 oz	28	65	48	7	1	0	2	1.2	0.6	t
1050	Edam	1 oz	28	42	100	7	<1	0	8	4.9	2.3	0.2
1016	Feta	1 oz	28	55	74	4	1	0	6	4.2	1.3	0.2
1054	Gouda	1 oz	28	41	100	7	1	0	8	4.9	2.2	0.2
1073	Gruyere	1 oz	28	33	116	8	<1	0	9	5.3	2.8	0.5
1538	Gorgonzola	1 oz	28	42	97	6	1	0	8	5		
1055	Limburger	1 oz	28	48	92	6	<1	0	8	4.7	2.4	0.1
1017	Monterey Jack	1 oz	28	41	104	7	<1	0	8	5.3	2.5	0.3
1056	Mozzarella whole milk	1 oz	28	50	84	6	1	0	6	3.7	1.8	0.2
1019	Part-skim, low moisture	1 oz	28	46	85	7	1	0	6	3.5	1.6	0.2
1021	Muenster	1 oz	28	42	103	7	<1	0	8	5.4	2.4	0.2
1060	Neufchatel	1 oz	28	62	73	3	1	0	7	4.1	1.9	0.2
1075	Parmesan, grated:	1 oz	28	21	121	11	1	0	8	4.8	2.3	0.3
57	Cup	1 c	100	21	431	38	4	0	29	17.3	8.4	1.2
1023	Provolone	1 oz	28	41	98	7	1	0	7	4.8	2.1	0.2
1064	Ricotta, whole milk	1 c	246	72	428	28	7	0	32	20.4	8.9	0.9
1024	Part-skim milk	1 c	246	74	339	28	13	0	19	12.1	5.7	0.6
1027	Swiss	1 oz	28	37	106	8	2	0	8	5	2	0.3
	Substitute:											
47947	Vegan cheese substitute, slice	1 ea	21		20	2	3		0	0	0	0
47946	Vegetarian cheese substitute topping	2 tsp	5		15	2	<1		<1	0		
	Cottage:											
1099	Low sodium, low fat	1 c	225	84	162	28	6	0	2	1.4	0.6	t
1013	Creamed, large curd	1 c	225	79	232	28	6	0	10	6.4	2.9	0.3
1012	Creamed, small curd	1 c	210	79	216	26	6	0	9	6	2.7	0.3
1049	With fruit	1 c	226	80	219	24	10	<1	9	5.2	2.3	0.3
1014	Low fat 2%	1 c	226	79	203	31	8	0	4	2.8	1.2	0.1
1047	Low fat 1%	1 c	226	82	163	28	6	0	2	1.5	0.7	t
	Cream:											
1015	Regular	1 tbs	15	54	52	1	<1	0	5	3.3	1.5	0.2
47949	Substitute	1 oz	28		49	3	3		3	0		
	Pasteurized processed:											
1456	American	1 oz	28	39	105	6	<1	0	9	5.5	2.5	0.3
1458	Swiss	1 oz	28	42	94	7	1	0	7	4.5	2	0.2
1437	American cheese food, jar	½ c	57	43	188	10	4	0	14	8.5	4.1	0.6

PAGE KEY: H–2 = Beverages   H–4 = Dairy   H–8 = Eggs   H–10 = Fat/Oil   H–12 = Fruit   H–18 = Bakery
H–24 = Grain   H–30 = Fish   H–30 = Meats   H–34 = Poultry   H–36 = Sausage   H–38 = Mixed/Fast   H–42 = Nuts/Seeds   H–44 = Sweets
H–46 = Vegetables/Legumes   H–56 = Vegetarian   H–60 = Misc   H–60 = Soups/Sauces   H–62 = Fast   H–76 = Convenience   H–78 = Baby foods

Chol (mg)	Calc (mg)	Iron (mg)	Magn (mg)	Pota (mg)	Sodi (mg)	Zinc (mg)	VT-A (RAE)	Thia (mg)	VT-E (mg α)	Ribo (mg)	Niac (mg)	V-B$_6$ (mg)	Fola (µg)	VT-C (mg)
5	400	2.7	140	600	220	2.25		0.52	13.64	0.6	7	0.7	120	60
5	400	2.7	140	600	220	2.25		0.52	13.64	0.6	7	0.7	120	60
5	400	2.7	140	600	220	2.25		0.52	13.64	0.6	7	0.7	120	60
0	5	0	2	0	5	0	0	0	0	0	0	0	0	0
0					5									
0	4	0.04	0	0	15	0.37	0	0	0	0	0	0	0	0
0	0	0.05	7	88	7	0.05	0	0	0	0.03	0	0	12	0
0	5	0.19	2	21	2	0.09	0	0.02	0	<.01	0	0	2	0
0	5	0.05	5	49	8	0.03	0	0	0	0.04	0.09	<.01	0	0
0	7	0.05	5	47	7	0.02	0	0	0	<.01	0.09	<.01	0	0
0	0	0			0		0							0
21	148	0.09	6	72	391	0.74	55	<.01	0.07	0.11	0.28	0.05	10	0
26	189	0.12	7	38	157	0.73	82	<.01	0.07	0.1	0.03	0.02	6	0
28	52	0.14	6	43	176	0.67	49	0.02	0.07	0.15	0.11	0.07	18	0
20	109	0.09	6	52	236	0.67	67	<.01	0.06	0.14	0.18	0.06	17	0
29	202	0.19	8	27	174	0.87	74	<.01	0.08	0.1	0.02	0.02	5	0
119	815	0.77	32	111	702	3.51	299	0.03	0.33	0.42	0.09	0.08	20	0
6	197	0.2	8	31	6	0.87		<.01	0.05	<.01	0.03	0.02	5	0
25	205	0.12	8	53	270	1.05	68	0.01	0.07	0.11	0.02	0.02	4	0
25	138	0.18	5	17	312	0.81	35	0.04	0.05	0.24	0.28	0.12	9	0
32	196	0.07	8	34	229	1.09	46	<.01	0.07	0.09	0.02	0.02	6	0
31	283	0.05	10	23	94	1.09	76	0.02	0.08	0.08	0.03	0.02	3	0
30	170	0.18			280									0
25	139	0.04	6	36	224	0.59	95	0.02	0.06	0.14	0.04	0.02	16	0
25	209	0.2	8	23	150	0.84	55	<.01	0.07	0.11	0.03	0.02	5	0
22	141	0.12	6	21	176	0.82	50	<.01	0.05	0.08	0.03	0.01	2	0
15	205	0.07	7	27	148	0.88	38	0.03	0.1	0.09	0.03	0.02	3	0
27	201	0.11	8	38	176	0.79	83	<.01	0.07	0.09	0.03	0.02	3	0
21	21	0.08	2	32	112	0.15	83	<.01	0.26	0.05	0.04	0.01	3	0
25	311	0.25	11	35	428	1.08	34	<.01	0.07	0.14	0.03	0.01	3	0
88	1109	0.9	38	125	1529	3.87	120	0.03	0.26	0.49	0.11	V-.05	10	0
19	212	0.15	8	39	245	0.9	66	<.01	0.06	0.09	0.04	0.02	3	0
125	509	0.93	27	258	207	2.85	295	0.03	0.27	0.48	0.26	0.11	30	0
76	669	1.08	37	308	308	3.3	263	0.05	0.17	0.46	0.19	0.05	32	0
26	221	0.06	11	22	54	1.22	62	0.02	0.11	0.08	0.03	0.02	2	0
0	100			10	220									
0	60			50	80									
9	137	0.32	11	194	29	0.86	25	0.04	0.25	0.36	0.29	0.16	27	0
34	135	0.32	11	189	911	0.83	99	0.05	0.09	0.37	0.28	0.15	27	0
32	126	0.29	10	176	850	0.78	92	0.04	0.08	0.34	0.26	0.14	25	0
29	120	0.36	16	203	777	0.75	86	0.07	0.09	0.32	0.34	0.15	25	3
18	156	0.36	14	217	918	0.95	47	0.05	0.05	0.42	0.33	0.17	29	0
9	138	0.32	11	194	918	0.86	25	0.05	0.02	0.37	0.29	0.15	27	0
16	12	0.18	1	18	44	0.08	55	<.01	0.04	0.03	0.02	<.01	2	0
0	99			15	89									
26	155	0.05	8	47	417	0.8	71	<.01	0.08	0.1	0.02	0.02	2	0
24	216	0.17	8	60	384	1.01	55	<.01	0.1	0.08	0.01	0.01	2	0
46	325	0.32	18	166	721	1.82	115	0.04	0.13	0.29	0.1	0.04	4	0

Table H–1

Food Composition

(Computer code number is for West Diet Analysis program)   (For purposes of calculations, use "0" for t, <1, <.1, <.01, etc.)

Computer Code Number	Food Description	Measure	Wt (g)	H₂O (%)	Ener (kcal)	Prot (g)	Carb (g)	Dietary Fiber (g)	Fat (g)	Fat Breakdown (g) Sat	Mono	Poly
	DAIRY—Continued											
1002	American cheese spread	1 tbs	15	48	44	2	1	0	3	2	0.9	t
1081	Nonfat cheese (Kraft Singles)	1 oz	28	61	44	6	4	0	0	0	0	0
1094	Velveeta cheese spread, low fat, low sodium, slice	1 pce	34	62	61	8	1	0	2	1.5	0.7	t
	Cream:											
	Sweet:											
69	Half & half (cream & milk)	1 c	242	81	315	7	10	0	28	17.3	8	1
500	Tablespoon	1 tbs	15	81	20	<1	1	0	2	1.1	0.5	t
501	Light, coffee or table	1 tbs	15	74	29	<1	1	0	3	1.8	0.8	0.1
511	Light, whipping cream, liquid	1 tbs	15	64	44	<1	<1	0	5	2.9	1.4	0.1
502	Heavy whipping cream, liquid	1 tbs	15	58	52	<1	<1	0	6	3.5	1.6	0.2
510	Whipped cream, pressurized	1 tbs	4	61	10	<1	<1	0	1	0.6	0.3	t
	Sour, cultured:											
79	Regular	1 c	230	71	492	7	10	0	48	30	13.9	1.8
504	Tablespoon	1 tbs	14	71	30	<1	1	0	3	1.8	0.8	0.1
577	Fat free	1 tbs	15	79	12	1	2	0	0	0	0	0
	Imitation and part-dairy:											
507	Coffee whitener, frozen or liquid	1 tbs	15	77	20	<1	2	0	1	1.4	t	t
506	Coffee whitener, powdered	1 tsp	2	2	11	<1	1	0	1	0.7	t	t
508	Dessert topping, frozen, nondairy	1 tbs	5	50	16	<1	1	0	1	1.1	t	t
509	Dessert topping, mix with whole milk	1 tbs	5	67	9	<1	1	0	1	0.5	t	t
514	Dessert topping, pressurized	1 c	70	60	185	1	11	0	16	13.2	1.3	0.2
505	Sour cream, imitation	1 tbs	14	71	29	<1	1	0	3	2.5	t	t
516	Sour dressing, part dairy	1 tbs	15	75	27	<1	1	0	2	2	0.3	t
	Milk: Fluid											
1	Whole milk	1 c	244	88	146	8	11	0	8	4.6	2	0.5
2	2% lowfat milk	1 c	244	89	122	8	11		5	2.3	2	0.2
147	2% milk solids added	1 c	245	89	125	9	11	0	5	2.9	1.4	0.2
4	1% lowfat milk	1 c	244	90	102	8	12		2	1.5	0.7	t
148	1% milk solids added	1 c	245	90	105	9	12	0	2	1.5	0.7	t
6	Nonfat milk, vitamin A added	1 c	245	91	83	8	12	0	<1	0.1	t	t
149	Nonfat milk solids added	1 c	245	90	91	9	12	0	1	0.4	0.2	t
7	Buttermilk, skim	1 c	245	90	98	8	12	0	2	1.3	0.6	t
	Canned:											
11	Sweetened condensed	1 c	306	27	982	24	166	0	27	16.8	7.4	1
10	Evaporated, nonfat	1 c	256	79	200	19	29	0	1	0.3	0.2	t
	Dried:											
32	Buttermilk, sweet	1 c	120	3	464	41	59	0	7	4.3	2	0.3
9	Instant, nonfat, vit A added (makes 1 qt)	1 ea	91	4	326	32	47	0	1	0.4	0.2	t
8	Instant nonfat, vit A added, cup	1 c	68	4	243	24	35	0	<1	0.3	0.1	t
23	Goat milk	1 c	244	87	168	9	11	0	10	6.5	2.7	0.4
51	Kefir	1 c	233	88	149	8	11	0	8			
	Milk beverages and powdered mixes:											
	Chocolate:											
20	Whole	1 c	250	82	208	8	26	2	8	5.3	2.5	0.3
18	2% fat	1 c	250	84	180	8	26	1	5	3.1	1.5	0.2
19	1% fat	1 c	250	84	158	8	26	1	2	1.5	0.8	t
	Chocolate-flavored beverages:											
12	Powder containing nonfat dry milk:	1 oz	28	2	111	2	24	1	1	0.7	0.4	t
48	Prepared with water	1 c	275	86	151	2	32	1	2	0.9	0.5	t
14	Powder without nonfat dry milk:	1 oz	28	1	98	1	25	1	1	0.5	0.3	t
39	Prepared with whole milk	1 c	266	81	226	9	32	1	9	4.9	2.2	0.5
	Eggnog:											
17	Commercial	1 c	254	74	343	10	34	0	19	11.3	5.7	0.9
98	2% low-fat	1 c	254	85	191	12	17	0	8	3.7	2.7	0.7
	Instant breakfast:											
24	Envelope, pwd only	1 ea	37	7	131	7	24		1	0.2	0.1	0.1
25	Prepared with whole milk	1 c	281	78	253	15	36	<1	5	3.1		
26	Prepared with 2% milk	1 c	281	77	280	15	36	<1	9	5.3		

PAGE KEY: H–2 = Beverages H–4 = Dairy H–8 = Eggs H–10 = Fat/Oil H–12 = Fruit H–18 = Bakery
H–24 = Grain H–30 = Fish H–30 = Meats H–34 = Poultry H–36 = Sausage H–38 = Mixed/Fast H–42 = Nuts/Seeds H–44 = Sweets
H–46 = Vegetables/Legumes H–56 = Vegetarian H–60 = Misc H–60 = Soups/Sauces H–62 = Fast H–76 = Convenience H–78 = Baby foods

Chol (mg)	Calc (mg)	Iron (mg)	Magn (mg)	Pota (mg)	Sodi (mg)	Zinc (mg)	VT-A (RAE)	Thia (mg)	VT-E (mg α)	Ribo (mg)	Niac (mg)	V-B$_6$ (mg)	Fola (µg)	VT-C (mg)
8	84	0.05	4	36	202	0.39	26	<.01	0.03	0.06	0.02	0.02	1	0
7	221	0		88	398	0.88				0.15				0
12	233	0.15	8	61	2	1.13		0.01	0.17	0.13	0.03	0.03	3	0
90	254	0.17	24	315	99	1.23	235	0.08	0.8	0.36	0.19	0.09	7	2
6	16	0.01	2	20	6	0.08	15	<.01	0.05	0.02	0.01	<.01	<1	0
10	14	<.01	1	18	6	0.04	27	<.01	0.08	0.02	<.01	<.01	<1	0
17	10	<.01	1	15	5	0.04	42	<.01	0.13	0.02	<.01	<.01	1	0
21	10	<.01	1	11	6	0.03	62	<.01	0.16	0.02	<.01	<.01	1	0
3	4	<.01	0	6	5	0.01	8	<.01	0.03	<.01	<.01	<.01	<1	0
101	267	0.14	25	331	122	0.62	407	0.08	1.38	0.34	0.15	0.04	25	2
6	16	<.01	2	20	7	0.04	25	<.01	0.08	0.02	<.01	<.01	2	0
2	28	0			23									0
0	1	<.01	0	29	12	<.01	1	0	0.24	0	0	0	0	0
0	0	0.02	0	16	4	0.01	<1	0	0.01	<.01	0	0	0	0
0	0	<.01	0	1	1	<.01	<1	0	0.05	0	0	0	0	0
0	4	<.01	0	8	3	0.01	1	<.01	0.02	<.01	<.01	<.01	<1	0
0	4	0.01	1	13	43	<.01	3	0	0.6	0	0	0	0	0
0	0	0.05	1	23	14	0.17	0	0	0.1	0	0	0	0	0
1	17	<.01	2	24	7	0.06	<1	<.01	0.2	0.02	0.01	<.01	2	0
24	246	0.07	24	325	105	0.93	68	0.11	0.15	0.45	0.26	0.09	12	0
20	271	0.24	27	342	115	1.17	134	0.1	0.07	0.45	0.22	0.09	12	0
20	314	0.12	34	397	127	0.98	137	0.1	0.1	0.42	0.22	0.11	12	2
12	264	0.85	27	290	122	2.12	142	0.05	0.02	0.45	0.23	0.09	12	0
10	314	0.12	34	397	127	0.98	145	0.1	0.1	0.42	0.22	0.11	12	2
5	223	1.23	22	238	108	2.08	149	0.11	0.02	0.45	0.23	0.09	12	0
5	316	0.12	37	419	130	1	149	0.1	0	0.43	0.22	0.11	12	2
10	284	0.12	27	370	257	1.03	17	0.08	0.12	0.38	0.14	0.08	12	2
104	869	0.58	80	1135	389	2.88	226	0.28	0.49	1.27	0.64	0.16	34	8
10	742	0.74	69	850	294	2.3	302	0.12	0	0.79	0.45	0.14	23	3
83	1421	0.36	132	1910	620	4.82	59	0.47	0.12	1.89	1.05	0.41	56	7
16	1120	0.28	106	1552	500	4.01	645	0.38	<.01	1.59	0.81	0.31	46	5
12	837	0.21	80	1159	373	3	482	0.28	<.01	1.19	0.61	0.23	34	4
27	327	0.12	34	498	122	0.73	139	0.12	0.17	0.34	0.68	0.11	2	3
		0.3	33	373	107									
30	280	0.6	32	418	150	1.02	65	0.09	0.15	0.4	0.31	0.1	12	2
18	285	0.6	32	422	150	1.02	138	0.09	0.1	0.41	0.32	0.1	12	2
8	288	0.6	32	425	152	1.02	145	0.1	0.05	0.42	0.32	0.1	12	2
2	39	0.33	23	199	141	0.41	<1	0.03	0.15	0.16	0.16	0.03	0	1
3	60	0.47	33	270	195	0.58	<1	0.04	0.19	0.21	0.22	0.04	0	1
0	10	0.88	27	165	59	0.43	<1	<.01	0.01	0.04	0.14	<.01	2	0
24	253	0.8	48	458	154	1.28	69	0.11	0.16	0.48	0.38	0.09	13	0
150	330	0.51	48	419	137	1.17	114	0.09	0.51	0.48	0.27	0.13	3	4
194	270	0.71	32	369	155	1.26		0.11	0.58	0.55	0.21	0.15	30	2
4	105	4.74	84	350	142	3.16		0.31	5.31	0.07	5.27	0.42	105	28
24	403	4.87	119	726	264	4.12		0.41	5.48	0.48	5.48	0.53	118	31
38	396	4.87	117	721	262	4.09	0	0.41	5.55	0.47	5.47	0.52	118	31

Table H–1

Food Composition   (Computer code number is for West Diet Analysis program)   (For purposes of calculations, use "0" for t, <1, <.1, <.01, etc.)

Computer Code Number	Food Description	Measure	Wt (g)	H₂O (%)	Ener (kcal)	Prot (g)	Carb (g)	Dietary Fiber (g)	Fat (g)	Fat Breakdown (g) Sat	Mono	Poly
	DAIRY—Continued											
101	Prepared with 1% milk	1 c	281	79	233	15	36	<1	3	1.8		
27	Prepared with nonfat milk	1 c	282	80	216	16	36		1	0.7		
	Malted Milk:											
30	Chocolate Powder	3 tsp	21	1	79	1	18	1	1	0.5	0.2	t
34	Prepared with whole milk	1 c	265	81	225	9	30	1	9	5	2.2	0.6
38	Ovaltine with whole milk	1 c	265	81	223	9	29	1	9	5	2.2	0.5
28	Natural Powder:	3 tsp	21	2	87	2	16	<1	2	0.9	0.4	0.3
29	Prepared with whole milk	1 c	265	81	233	10	27	<1	10	5.4	2.4	0.7
	Milk Shakes:											
2020	Chocolate	1 c	166	72	211	6	34	3	6	3.8	1.8	0.2
2024	Vanilla	1 c	166	75	184	6	30	<1	5	3.1	1.4	0.2
	Milk Desserts:											
2062	Low-fat frozen dessert bars	1 ea	81	72	88	2	19	0	1	0.2	0.1	0.4
	Ice cream, vanilla (about 10% fat):											
2004	Hardened	1 c	132	61	265	5	31	1	15	9	3.9	0.6
2008	Soft serve	1 c	172	60	382	7	38	1	22	12.9	6	0.8
	Ice cream, rich vanilla (16% fat):											
2006	Hardened	1 c	148	57	369	5	33	0	24	15.3	6.6	1
2220	Ben & Jerry's	½ c	108		250	4	22	0	16	11		
	Ice milk, vanilla (about 4% fat):											
2009	Hardened	1 c	132	63	218	7	35	<1	5	3.4	1.1	0.2
2010	Soft serve (about 3.3% fat)	1 c	176	70	222	9	38	0	5	2.9	1.3	0.2
	Pudding, canned (5 oz can = .55 cup):											
2610	Chocolate	1 ea	142	69	197	4	33	1	6	1	2.4	2
2611	Tapioca	1 ea	142	74	169	3	28	<1	5	0.9	2.2	1.9
2612	Vanilla	1 ea	142	71	183	3	31	0	5	0.8	2.2	1.9
	Puddings, dry mix with whole milk:											
2604	Chocolate, regular, cooked	1 c	284	73	338	9	55	2	9	5.1	2.3	0.5
2606	Rice, cooked	1 c	288	72	348	9	60	<1	8	4.8	2.2	0.3
2607	Tapioca, cooked	1 c	282	74	324	8	55	0	8	4.9	2.2	0.3
2609	Vanilla, regular, cooked	1 c	280	75	314	8	52	<1	8	4.6	2	0.6
2011	Sherbet (2% fat)	1 c	198	66	285	2	60	7	4	2.3	1	0.2
	Frozen Yogurt, Low-Fat:											
2064	Cup	1 c	144	65	235	6	35	0	8	4.9	2.3	0.3
2082	Scoop	1 ea	79	74	78	4	16	0	<1	t	t	t
	Milk Substitutes:											
20440	Rice Milk	1 c	245	89	120	<1	25	0	2	0.2	1.3	0.3
20590	Rice/Soy Milk, blend	1 c	241	88	120	7	18	0	3	0.5		
20033	Soy Milk	1 c	245	89	120	9	11	3	5	0.5	0.8	2
7785	Edensoy	1 c	244	89	130	10	13	0	4	0.5	0.8	2.7
7804	Vanilla	1 c	244	87	150	6	23	0	3	0		
20693	Soy Dream	1 c	244		128	7	17	0	4	0.5		
483	Enriched	1 c	244		128	7	17	0	4	0.5		
20404	Veggie original	1 c	227		110	9	13		3	0		
20405	Chocolate	1 c	227		150	9	26		2	0		
	Yogurt:											
	Fat Free:											
2852	Strawberry, container	1 ea	227	86	120	8	22	0	0	0	0	0
2574	Vanilla	1 c	245	76	223	13	43	0	<1	0.3	0.1	t
2012	Plain	1 c	245	85	137	14	19	0	<1	0.3	0.1	t
	Low-Fat:											
2034	Fruit added with low-calorie sweetener	1 c	241	86	122	11	19	1	<1	0.2	t	t
2001	Fruit added	1 c	245	74	250	11	47	0	3	1.7	0.7	t
2000	Plain	1 c	245	85	154	13	17	0	4	2.4	1	0.1
2015	Vanilla or coffee flavor	1 c	245	79	208	12	34	0	3	2	0.8	t
2013	Whole Milk:	1 c	245	88	149	9	11	0	8	5.1	2.2	0.2
	EGGS											
	Raw, Large:											
19501	Whole, without shell	1 ea	50	76	74	6	<1	0	5	1.5	1.9	0.7
19506	White	1 ea	33	88	17	4	<1	0	<1	0	0	0

PAGE KEY: H–2 = Beverages H–4 = Dairy H–8 = Eggs H–10 = Fat/Oil H–12 = Fruit H–18 = Bakery
H–24 = Grain H–30 = Fish H–30 = Meats H–34 = Poultry H–36 = Sausage H–38 = Mixed/Fast H–42 = Nuts/Seeds H–44 = Sweets
H–46 = Vegetables/Legumes H–56 = Vegetarian H–60 = Misc H–60 = Soups/Sauces H–62 = Fast H–76 = Convenience H–78 = Baby foods

Chol (mg)	Calc (mg)	Iron (mg)	Magn (mg)	Pota (mg)	Sodi (mg)	Zinc (mg)	VT-A (RAE)	Thia (mg)	VT-E (mg α)	Ribo (mg)	Niac (mg)	V-B$_6$ (mg)	Fola (µg)	VT-C (mg)
14	406	4.87	119	731	267	4.12		0.41	5.4	0.48	5.48	0.53	118	31
9	407	4.83	112	755	268	4.14		0.4	5.3	0.42	5.47	0.52	118	31
0	13	0.48	15	130	53	0.17	1	0.04	0.02	0.04	0.42	0.03	11	0
26	260	0.56	40	456	159	1.09	69	0.14	0.16	0.49	0.69	0.12	24	0
26	339	3.76	45	578	231	1.17	904	0.76	0.16	1.32	11.08	1.01	19	32
7	63	0.15	20	159	104	0.21	17	0.11	0.07	0.19	1.1	0.09	10	1
32	310	0.24	45	485	209	1.14	87	0.21	0.32	0.64	1.38	0.17	21	1
22	188	0.51	28	332	161	0.68	43	0.1	0.18	0.41	0.27	0.08	8	1
18	203	0.15	20	289	136	0.6	61	0.07	0.1	0.3	0.31	0.09	8	1
1	81	0.04	9	107	44	0.26	38	0.03	0.07	0.11	0.06	0.03	3	1
58	169	0.12	18	263	106	0.91	156	0.05	0.4	0.32	0.15	0.06	7	1
157	225	0.36	21	304	105	0.89	279	0.08	1.05	0.31	0.16	0.08	15	1
136	173	0.5	16	232	90	0.7	269	0.06	0.75	0.25	0.12	0.07	12	0
75	100	0.36			60									0
33	153	0.11	18	275	98	0.96	182	0.05	0.12	0.23	0.13	0.04	7	0
21	276	0.11	25	389	123	0.93	51	0.09	0.11	0.35	0.21	0.08	11	2
4	128	0.72	30	256	183	0.6	14	0.04	0.41	0.22	0.49	0.04	4	3
1	119	0.33	11	136	226	0.38	<1	0.03	0.43	0.14	0.44	0.03	4	1
10	125	0.18	11	160	192	0.36	9	0.03	0	0.2	0.36	0.02	0	0
26	273	0.99	57	426	278	1.36	68	0.1	0.14	0.55	0.42	0.09	11	0
32	297	1.07	37	369	311	1.09	84	0.21	0.24	0.4	1.27	0.1	12	2
34	290	0.17	34	369	338	0.96	70	0.08	0.23	0.39	0.21	0.09	11	2
25	249	0.11	25	333	437	0.95	70	0.1	0.14	0.45	0.26	0.08	11	0
0	107	0.28	16	190	91	0.95	20	0.06	0.06	0.18	0.15	0.05	14	11
3	206	0.43	20	304	125	0.6	85	0.05	0.16	0.32	0.41	0.12	9	1
1	137	0.07	13	175	53	0.67	1	0.03	<.01	0.16	0.09	0.04	8	1
0	20	0.2	10	69	86	0.24	<1	0.08	1.76	0.01	1.91	0.04	91	1
0	13	1.08	40	270	85	0.9	0	0.09		0.1	0.4	0.08	26	
0	10	1.42	47	345	29	0.56	5	0.39	0.02	0.17	0.36	0.1	5	0
0	80	1.44	60	440	105	0.9	0	0.15		0.07	0.8	0.16	40	0
0	60	0.72	40	290	90	0.6	0	0.12		0.07	1.2	0.12	40	0
0	40	1.78	40	237	138	0.59	0	0.15		0.07	0.79	0.12	59	0
0	295	1.77	39	236	138	0.59		0.15	4.95	0.07	0.79	0.12	59	0
0	400			350	90									
0	401			351	130									
5	350	0		380	160		0							5
4	436	0.21	42	559	168	2.13	4	0.11	<.01	0.52	0.27	0.12	27	2
5	488	0.22	47	625	189	2.38	5	0.12	0	0.57	0.3	0.13	29	2
3	370	0.61	41	550	139	1.83		0.1	0.17	0.45	0.5	0.11	32	26
10	372	0.17	37	478	142	1.81	24	0.09	0.05	0.44	0.23	0.1	22	2
15	448	0.2	42	573	172	2.18	34	0.11	0.07	0.52	0.28	0.12	27	2
12	419	0.17	39	537	162	2.03	29	0.1	0.05	0.49	0.26	0.11	27	2
32	296	0.12	29	380	113	1.45	66	0.07	0.15	0.35	0.18	0.08	17	1
212	26	0.92	6	67	70	0.56	70	0.03	0.48	0.24	0.04	0.07	24	0
0	2	0.03	4	54	55	<.01	0	<.01	0	0.14	0.03	<.01	1	0

Table H–1

Food Composition

(Computer code number is for West Diet Analysis program)   (For purposes of calculations, use "0" for t, <1, <.1, <.01, etc.)

Computer Code Number	Food Description	Measure	Wt (g)	H₂O (%)	Ener (kcal)	Prot (g)	Carb (g)	Dietary Fiber (g)	Fat (g)	Fat Breakdown (g)		
										Sat	Mono	Poly
	EGGS—Continued											
19508	Yolk	1 ea	17	52	55	3	1	0	5	1.6	2	0.7
	Cooked:											
19509	Fried in margarine	1 ea	46	69	92	6	<1	0	7	2	2.9	1.2
19510	Hard-cooked, shell removed	1 ea	50	75	78	6	1	0	5	1.6	2	0.7
19511	Hard-cooked, chopped	1 c	136	75	211	17	2	0	14	4.4	5.5	1.9
19517	Poached, no added salt	1 ea	50	76	74	6	<1	0	5	1.5	1.9	0.7
19516	Scrambled with milk & margarine	1 ea	61	73	101	7	1	0	7	2.2	2.9	1.3
	Substitute, liquid:											
19525	Egg substitute, liquid:	½ c	126	83	106	15	1	0	4	0.8	1.1	2
19581	Egg Beaters, Fleischmann's	½ c	122		60	12	2	0	0	0	0	0
19552	Egg substitute, liquid, prepared	½ c	105	80	107	12	2	0	6	1.1	2.1	2.1
	FATS AND OILS											
	Butter:											
8000	Tablespoon:	1 tbs	14	16	100	<1	<1	0	11	5.7	4.7	0.4
8025	Unsalted	1 tbs	14	18	100	<1	<1	0	11	7.2	2.9	0.4
8001	Pat (about 1 tsp)	1 ea	5	16	36	<1	<1	0	4	2	1.7	0.1
8142	Whipped	1 tsp	3	16	22	<1	<1	0	2	1.5	0.7	t
	Fats, cooking:											
8004	Beef fat/tallow	1 c	205	0	1849	0	0	0	205	102.1	85.7	8.2
8005	Chicken fat	1 c	205	0	1845	0	0	0	205	61.1	91.6	42.8
8007	Vegetable shortening	1 tbs	13	0	115	0	0	0	13	3.2	5.8	3.4
	Margarine:											
8041	Imitation (about 40% fat), soft	1 tbs	14	58	48	<1	<1	0	5	1.1	2.2	1.9
90234	Regular, hard (about 80% fat)	1 tbs	14	16	101	<1	<1	0	11	2.2	5	3.6
8043	Regular, soft (about 80% fat)	1 tbs	14	16	100	<1	<1	0	11	1.9	4	4.8
8485	Saffola, unsalted	1 tbs	14	20	100	0	0	0	11	2	3	4.5
8486	Saffola, reduced fat	1 tbs	14	37	60	0	0	0	8	1.3	2.7	4.4
	Spread:											
8044	Hard (about 60% fat)	1 tbs	14	37	76	<1	0	0	9	2	3.6	2.5
8045	Soft (about 60% fat)	1 tbs	14	37	76	<1	0	0	9	1.8	4.4	1.9
8602	Touch of Butter (47% fat)	1 tbs	14		60	0	0	0	7	1.5		
	Oils:											
8084	Canola	1 tbs	14	0	124	0	0	0	14	1	8.2	4.1
8009	Corn	1 tbs	14	0	124	0	0	0	14	1.8	3.4	8.2
8008	Olive	1 tbs	14	0	124	0	0	0	14	1.9	10.3	1.4
8361	Olive, extra virgin	1 tbs	14	0	126	0	0	0	14	2	10.8	1.3
8026	Peanut	1 tbs	14	0	124	0	0	0	14	2.4	6.5	4.5
8010	Safflower	1 tbs	14	0	124	0	0	0	14	0.9	2	10.4
8012	Soybean	1 tbs	14	0	124	0	0	0	14	2	3.3	8.1
8028	Soybean/cottonseed	1 tbs	14	0	124	0	0	0	14	2.5	4.1	6.7
8011	Sunflower	1 tbs	14	0	124	0	0	0	14	1.4	2.7	9.2
	Salad Dressings/Sandwich Spreads:											
	Store Brand:											
8013	Blue Cheese	1 tbs	15	32	76	1	1	0	8	1.5	1.8	4.2
	French:											
8015	Regular	1 tbs	16	37	73	<1	2	0	7	0.9	1.3	3.4
8014	Low calorie	1 tbs	16	54	37	<1	5	<1	2	0.2	0.9	0.8
	Italian:											
8020	Regular	1 tbs	15	56	44	<1	2	0	4	0.7	0.9	1.9
8016	Low calorie	1 tbs	15	85	11	<1	1	0	1	t	0.3	0.3
	Ranch:											
8428	Regular	1 tbs	15		80	0	<1	0	8	1.2		
8465	Low calorie	1 tbs	14		22	<1	1	<1	2	0.3		
8022	Russian	1 tbs	15	34	74	<1	2	0	8	1.1	1.8	4.4
8021	Mayo Type	1 tbs	15	40	58	<1	4	0	5	0.7	1.4	2.7
	Mayonnaise:											
8032	Imitation, low calorie	1 tbs	15	63	35	<1	2	0	3	0.5	0.7	1.6
8046	Regular (soybean)	1 tbs	14	15	100	<1	1	0	11	1.7	2.7	6
8148	Regular, low calorie, low sodium	1 tbs	14	63	32	<1	2	0	3	0.5	0.6	1.5
8141	Salad dressing, low calorie, oil free	1 tbs	15	88	4	<1	1	<1	<1	t	0	t

PAGE KEY: H–2 = Beverages   H–4 = Dairy   H–8 = Eggs   H–10 = Fat/Oil   H–12 = Fruit   H–18 = Bakery
H–24 = Grain   H–30 = Fish   H–30 = Meats   H–34 = Poultry   H–36 = Sausage   H–38 = Mixed/Fast   H–42 = Nuts/Seeds   H–44 = Sweets
H–46 = Vegetables/Legumes   H–56 = Vegetarian   H–60 = Misc   H–60 = Soups/Sauces   H–62 = Fast   H–76 = Convenience   H–78 = Baby foods

Chol (mg)	Calc (mg)	Iron (mg)	Magn (mg)	Pota (mg)	Sodi (mg)	Zinc (mg)	VT-A (RAE)	Thia (mg)	VT-E (mg α)	Ribo (mg)	Niac (mg)	V-B₆ (mg)	Fola (µg)	VT-C (mg)
210	22	0.46	1	19	8	0.39	65	0.03	0.44	0.09	<.01	0.06	25	0
210	27	0.91	6	68	94	0.55	91	0.03	0.56	0.24	0.04	0.07	23	0
212	25	0.6	5	63	62	0.52	84	0.03	0.52	0.26	0.03	0.06	22	0
577	68	1.62	14	171	169	1.43	230	0.09	1.4	0.7	0.09	0.16	60	0
211	26	0.92	6	66	147	0.55	70	0.03	0.48	0.24	0.04	0.07	24	0
215	43	0.73	7	84	171	0.61	87	0.03	0.52	0.27	0.05	0.07	18	0
1	67	2.65	11	416	223	1.64	23	0.14	0.34	0.38	0.14	<.01	19	0
0	40	2.16		170	250	1.2				1.61	1.7		0.16	64	0
1	83	1.85	11	337	201	1.25			0.09	0.83	0.29	0.12	0.01	11	0
30	3	<.01	0	3	81	0.01	96	<.01	0.32	<.01	<.01	<.01	<1	0
30	3	<.01	0	3	2	0.01	96	<.01	0.32	<.01	<.01	<.01	<1	0
11	1	<.01	0	1	29	<.01	34	<.01	0.12	<.01	<.01	<.01	<1	0
7	1	<.01	0	1	25	<.01	21	<.01	0.07	<.01	<.01	<.01	<1	0
223	0	0	0	0	0	0	0	0	5.54	0	0	0	0	0
174	0	0	0	0	0	0	0	0	5.54	0	0	0	0	0
0	0	0	0	0	0	0	0	0	0.1	0	0	0	0	0
0	3	0	0	4	134	0	115	<.01	0.33	<.01	<.01	<.01	<1	0
0	4	<.01	0	6	132	0	115	<.01	1.26	<.01	<.01	<.01	<1	0
0	4	0	0	5	151	0	102	<.01	0.98	<.01	<.01	<.01	<1	0
0	0			0									0	
0	0			115									0	
0	3	0	0	4	139	0	102	<.01	0.7	<.01	<.01	<.01	<1	0
0	3	0	0	4	139	0	102	<.01	0.7	<.01	<.01	<.01	<1	0
0	0	0		0	110				1.27					0
0	0	0	0	0	0	0	0	0	2.39	0	0	0	0	0
0	0	0	0	0	0	0	0	0	2	0	0	0	0	0
0	0	0.09	0	0	<1	0	0	0	2.01	0	0	0	0	0
									1.74					
0	0	<.01	0	0	0	<.01	0	0	2.2	0	0	0	0	0
0	0	0	0	0	0	0	0	0	4.77	0	0	0	0	0
0	0	<.01	0	0	0	0	0	0	1.29	0	0	0	0	0
0	0	0	0	0	0	0	0	0	1.69	0	0	0	0	0
0	0	0	0	0	0	0	0	0	5.75	0	0	0	0	0
3	12	0.03	0	6	164	0.04	10	<.01	0.9	0.02	0.02	<.01	4	0
0	4	0.13	1	11	134	0.05	4	<.01	0.8	<.01	0.03	0	0	0
0	2	0.14	1	17	129	0.03	4	<.01	0.05	<.01	0.07	<.01	<1	0
0	1	0.09	0	7	248	0.02	<1	<.01	0.75	<.01	0	<.01	0	0
1	1	0.1	1	13	205	0.03	<1	0	0.03	<.01	0	0.01	0	0
5	0	0			105		0							0
3	12	0		16	120		0		0.7					0
3	3	0.09	0	24	130	0.06	2	<.01	0.6	<.01	0.09	<.01	2	1
4	2	0.03	0	1	107	0.03	10	<.01	0.31	<.01	<.01	<.01	1	0
4	0	0	0	2	75	0.02	0	0	0.3	0	0	0	0	0
5	3	0.07	0	5	80	0.02	12	0	0.73	0	<.01	0.08	1	0
3	0	0	0	1	15	0.02	1	0	0.53	<.01	0	<.01	<1	0
0	1	0.05	2	7	256	<.01	<1	<.01	<.01	<.01	<.01	<.01	<1	0

Table H–1

Food Composition

(Computer code number is for West Diet Analysis program)   (For purposes of calculations, use "0" for t, <1, <.1, <.01, etc.)

Computer Code Number	Food Description	Measure	Wt (g)	H₂O (%)	Ener (kcal)	Prot (g)	Carb (g)	Dietary Fiber (g)	Fat (g)	Fat Breakdown (g) Sat	Mono	Poly
	FATS AND OILS—Continued											
8034	Salad dressing, from recipe, cooked	1 tbs	16	69	25	1	2	0	2	0.5	0.6	0.3
53122	Tartar Sauce, low calorie	1 tbs	14	63	31	<1	2	<1	3	0.4	0.6	1.4
	Thousand Island	1 tbs	16									
8024	Regular	1 tbs	16	47	59	<1	2		6	0.8	1.3	2.9
8023	Low Calorie	1 tbs	15	61	31	<1	3	<1	2	0.1	1	0.4
8035	Vinegar and oil	1 tbs	16	47	72	0	<1	0	8	1.5	2.4	3.9
	Kraft, Deliciously Right:											
8563	1000 Island	1 tbs	16		34	0	3	0	2	0.2		
8574	Cucumber ranch	1 tbs	16		31	0	1	0	3	0.5		
	Wishbone:											
8442	Creamy Italian, lite	1 tbs	15	72	26	<1	2		2	0.4	0.9	0.7
8413	Italian, lite	1 tbs	16		6	0	1		<1	0	0.2	0.1
8427	Ranch, lite	1 tbs	15		50	0	2	0	4	0.8		
	FRUITS and FRUIT JUICES											
	Apples:											
	Fresh, w/peel:											
3000	2¾" diam (about 3/lb w/cores)	1 ea	138	86	72	<1	19	3	<1	t	t	t
3001	3¼" diam (about 2/lb w/cores)	1 ea	212	86	110	1	29	5	<1	t	t	0.1
3004	Slices	1 c	110	87	53	<1	14	1	<1	t	t	t
3005	Dried, sulfured	10 ea	64	32	156	1	42	6	<1	t	t	t
3576	Cup	¼ c	40	30	107	<1	26	3	<1	0		
3008	Juice, bottled or canned	1 c	248	88	117	<1	29	<1	<1	t	t	t
3147	Applesauce, sweetened	1 c	255	80	194	<1	51	3	<1	t	t	0.1
3006	Applesauce, unsweetened	1 c	244	88	105	<1	28	3	<1	t	t	t
	Apricots:											
3157	Fresh, w/o pits (about 12 per lb w/pits)	3 ea	105	86	50	1	12		<1	t	0.2	t
	Canned (fruit and liquid):											
3011	Heavy syrup	1 c	240	78	199	1	52	4	<1	t	t	t
3633	Halves	3 ea	120	78	100	1	26	2	<1	t	t	t
3151	Juice pack	1 c	244	87	117	2	30	4	<1	t	t	t
3152	Halves	3 ea	108	87	52	1	13	2	<1	t	t	t
3013	Dried, halves	10 ea	35	31	84	1	22	3	<1	t	t	t
3580	Cup	¼ c	40	30	107	1	25	2	<1	0		
3217	Dried, cooked, unsweetened, w/liquid	1 c	250	76	212	3	55		<1	t	t	t
3015	Nectar, canned	1 c	251	85	141	1	36	2	<1	t	t	t
	Avocados, raw, edible part only:											
3210	California	1 ea	173	72	289	3	15	12	27	3.7	17	3.5
3212	Florida	1 ea	304	79	365	7	24		31	6	16.8	5.1
3017	Mashed, fresh, average	1 c	230	73	368	5	20	15	34	4.9	22.5	4.2
	Bananas:											
	Fresh, w/o peel:											
3020	Whole, 8¾ long (175g w/peel)	1 ea	118	75	105	1	27	3	<1	0.1	t	t
3021	Slices	1 c	150	75	134	2	34	4	<1	0.2	t	0.1
3023	Dehydrated slices	½ c	50	3	173	2	44	5	1	0.3	t	0.2
	Berries:											
3024	Blackberries, raw	1 c	144	88	62	2	14		1	t	t	0.4
	Blueberries:											
3029	Fresh	1 c	145	84	83	1	21	3	<1	t	t	0.2
3232	Frozen, sweetened	10 oz	284	77	230	1	62	6	<1	t	t	0.2
3231	Frozen, thawed	1 c	230	77	186	1	50	5	<1	t	t	0.1
	Cranberries:											
3042	Juice cocktail, vitamin C added	1 c	253	86	144	0	36	<1	<1	t	t	0.1
3276	Juice, low calorie	1 c	237	95	45	<1	11	0	<1	0	0	0
3223	Cranberry-apple juice, vitamin C added	1 c	245	82	174	<1	44	<1	<1	0	0	0
3040	Sauce, canned, strained	1 c	277	61	418	1	108	3	<1	t	t	0.2
3865	Dried	⅓ c	40		120	0	33	2	0	0	0	0
	Raspberries:											
3131	Fresh	1 c	123	86	64	1	15	8	1	t	t	0.5

PAGE KEY: H–2 = Beverages   H–4 = Dairy   H–8 = Eggs   H–10 = Fat/Oil   H–12 = Fruit   H–18 = Bakery
H–24 = Grain   H–30 = Fish   H–30 = Meats   H–34 = Poultry   H–36 = Sausage   H–38 = Mixed/Fast   H–42 = Nuts/Seeds   H–44 = Sweets
H–46 = Vegetables/Legumes   H–56 = Vegetarian   H–60 = Misc   H–60 = Soups/Sauces   H–62 = Fast   H–76 = Convenience   H–78 = Baby foods

Chol (mg)	Calc (mg)	Iron (mg)	Magn (mg)	Pota (mg)	Sodi (mg)	Zinc (mg)	VT-A (RAE)	Thia (mg)	VT-E (mg α)	Ribo (mg)	Niac (mg)	V-B₆ (mg)	Fola (µg)	VT-C (mg)
9	13	0.08	1	19	117	0.06	8	<.01	0.13	0.02	0.04	<.01	3	0
3	1	0.06	0	4	82	0.02	1	<.01	0.84	<.01	<.01	<.01	<1	0
4	3	0.19	1	17	138	0.04	2	0.23	0.18	<.01	0.07	0	0	0
0	2	0.14	1	30	125	0.03	2	<.01	0.15	<.01	0.07	0	0	0
0	0	0	0	1	<1	0	0	0	0.74	0	0	0	0	0
5	0	0		29	165		0							0
0	0	0		10	248		0							0
0	0	0			148		0	0	0.56	0	0			0
0	1	0			255			0	0.24	0	0			0
2	0	0			120		0							0
0	8	0.17	7	148	1	0.06	4	0.02	0.25	0.04	0.13	0.06	4	6
0	13	0.25	11	227	2	0.08	6	0.04	0.38	0.06	0.19	0.09	6	10
0	6	0.08	4	99	0	0.06	2	0.02	0.06	0.03	0.1	0.04	0	4
0	9	0.9	10	288	56	0.13	0	0	0.34	0.1	0.59	0.08	0	2
0	9	0.67		178	251		<1							1
0	17	0.92	7	295	7	0.07	<1	0.05	0.02	0.04	0.25	0.07	0	2
0	10	0.89	8	156	8	0.1	3	0.03	0.54	0.07	0.48	0.07	3	4
0	7	0.29	7	183	5	0.07	2	0.03	0.51	0.06	0.46	0.06	2	3
0	14	0.41	10	272	1	0.21	101	0.03	0.93	0.04	0.63	0.06	9	10
0	22	0.72	17	336	10	0.26	149	0.05	1.44	0.05	0.9	0.13	5	7
0	11	0.36	8	168	5	0.13	74	0.02	0.72	0.03	0.45	0.06	2	4
0	29	0.73	24	403	10	0.27	207	0.04	1.46	0.05	0.84	0.13	5	12
0	13	0.32	11	178	4	0.12	92	0.02	0.65	0.02	0.37	0.06	2	5
0	19	0.93	11	407	4	0.14	63	<.01	1.52	0.03	0.91	0.05	4	0
0	20	1.56		520	1		16							6
0	48	2.35	28	1028	10	0.35	160	0.01	3.82	0.06	2.29	0.13	8	1
0	18	0.95	13	286	8	0.23	166	0.02	0.78	0.04	0.65	0.06	3	2
0	22	1.06	50	877	14	1.18	12	0.13	3.41	0.25	3.31	0.5	107	15
0	30	0.52	73	1067	6	1.22	21	0.06	8.09	0.16	2.04	0.24	106	53
0	28	1.26	67	1116	16	1.47	16	0.15	4.76	0.3	4	0.59	133	23
0	6	0.31	32	422	1	0.18	4	0.04	0.12	0.09	0.78	0.43	24	10
0	8	0.39	40	537	2	0.22	4	0.05	0.15	0.11	1	0.55	30	13
0	11	0.57	54	746	2	0.3	6	0.09	0.2	0.12	1.4	0.22	7	4
0	42	0.89	29	233	1	0.76	16	0.03	1.68	0.04	0.93	0.04	36	30
0	9	0.41	9	112	1	0.23	4	0.05	0.83	0.06	0.61	0.08	9	14
0	17	1.11	6	170	3	0.17	6	0.06	1.48	0.15	0.72	0.17	20	3
0	14	0.9	5	138	2	0.14	5	0.05	1.2	0.12	0.58	0.14	16	2
0	8	0.38	5	46	5	0.18	1	0.02	0	0.02	0.09	0.05	0	90
0	21	0.09	5	59	7	0.05	<1	0	0.12	<.01	<.01	<.01	0	76
0	12	0.29	5	69	17	0.44	<1	0.01	0	0.05	0.15	0.05	0	78
0	11	0.61	8	72	80	0.14	6	0.04	2.3	0.06	0.28	0.04	3	6
0	0	0			0		0							0
0	31	0.85	27	186	1	0.52	2	0.04	1.07	0.05	0.74	0.07	26	32

Table H–1

Food Composition

(Computer code number is for West Diet Analysis program)   (For purposes of calculations, use "0" for t, <1, <.1, <.01, etc.)

Computer Code Number	Food Description	Measure	Wt (g)	H₂O (%)	Ener (kcal)	Prot (g)	Carb (g)	Dietary Fiber (g)	Fat (g)	Fat Breakdown (g)		
										Sat	Mono	Poly
	FRUITS and FRUIT JUICES—Continued											
71120	Frozen, sweetened	10 oz	284	73	293	2	74	12	<1	t	t	0.3
3235	Cup, thawed measure	1 c	250	73	258	2	65	11	<1	t	t	0.2
	Strawberries:											
3134	Fresh, whole, capped	1 c	144	91	46	1	11	3	<1	t	t	0.2
3236	Frozen, sliced, sweetened	1 c	255	73	245	1	66	5	<1	t	t	0.2
3663	Breadfruit	1 c	220	71	227	2	60	11	1	0.1	t	0.1
	Cherries:											
3035	Sour, red pitted, canned water pack	1 c	244	90	88	2	22	3	<1	t	t	t
3036	Sweet, red pitted, raw	10 ea	68	82	43	1	11	1	<1	t	t	t
3862	Dried	¼ c	40		120	2	26	3	0	0	0	0
	Dates:											
3044	Whole, without pits	10 ea	83	21	234	2	62	7	<1	t	t	t
3043	Chopped	1 c	178	21	502	4	134	14	1	t	t	t
3162	Figs, dried	10 ea	190	30	473	6	121	19	2	0.3	0.3	0.7
	Fruit cocktail, canned, not drained:											
3045	Heavy syrup pack	1 c	248	80	181	1	47	2	<1	t	t	t
3164	Juice pack	1 c	237	87	109	1	28	2	<1	t	t	t
	Grapefruit:											
	Raw 3¾" diam (half w/rind = 241g)											
3818	Pink/red, half fruit, edible part	1 ea	123	88	52	1	13	2	<1	t	t	t
3047	White, half fruit, edible part	1 ea	118	90	39	1	10	1	<1	t	t	t
3050	Canned sections with light syrup	1 c	254	84	152	1	39	1	<1	t	t	t
	Juice:											
3051	Fresh, white, raw	1 c	247	90	96	1	23	<1	<1	t	t	t
3052	Canned, unsweetened	1 c	247	90	94	1	22	<1	<1	t	t	t
3165	Sweetened	1 c	250	87	115	1	28	<1	<1	t	t	t
3053	Prepared from concentrate	1 c	247	89	101	1	24	<1	<1	t	t	t
	Grapes, European (adherent skin):											
	Fresh:											
3055	Thompson seedless	10 ea	50	81	34	<1	9	<1	<1	t	t	t
3056	Tokay/Emperor, seeded types	10 ea	50	81	34	<1	9	<1	<1	t	t	t
3064	Juice, prepared from frozen, vit C added	1 c	250	87	128	<1	32	<1	<1	t	t	t
3062	Low calorie	1 c	253	84	154	1	38	<1	<1	t	t	t
3636	Jackfruit, fresh, sliced	1 c	165	73	155	2	40	3	<1	0.1	t	0.1
3065	Kiwi fruit, raw, peeled (88g w/peel)	1 ea	76	83	46	1	11	2	<1	t	t	0.2
	Lemons:											
	Fresh:											
3066	Without peel and seeds (about 4/lb)	1 ea	58	89	17	1	5	2	<1	t	t	t
	Juice:											
254	Fresh:	1 c	244	91	61	1	21	1	0	0	0	0
3068	Tablespoon	1 tbs	15	91	4	<1	1	<1	0	0	0	0
3069	Canned or bottled	1 tbs	15	92	3	<1	1	<1	<1	t	t	t
258	Frozen, single strength, Unsweetened:	1 c	244	92	54	1	16	1	1	0.1	t	0.2
3070	Tablespoon	1 tbs	15	92	3	<1	1	<1	<1	t	t	t
	Lime juice:											
3072	Fresh	1 tbs	15	90	4	<1	1	<1	<1	t	t	t
3073	Canned or bottled, unsweetened	1 c	246	93	52	1	16	1	1	t	t	0.2
3758	Pomelos, raw	1 ea	609	89	231	5	59	6	<1			
3221	Mangos, raw, edible part (300g w/skin & seeds)	1 ea	207	82	135	1	35	4	1	0.1	0.2	0.1
	Melons:											
	Raw, without rind and contents:											
3076	Cantaloupe, 5" diam (2⅓ lb whole w/refuse), orange flesh	½ ea	276	90	94	2	23		1	0.1	t	0.2
3081	Honeydew, 6" diam (5⅓ lb whole w/refuse), slice = ⅒ melon	1 pce	160	90	58	1	15	1	<1	t	t	t
3215	Nectarines, raw, w/o pits, 2" diam	1 ea	136	88	60	1	14	2	<1	t	0.1	0.2
	Oranges:											
	Fresh:											

PAGE KEY: H–2 = Beverages H–4 = Dairy H–8 = Eggs H–10 = Fat/Oil H–12 = Fruit H–18 = Bakery
H–24 = Grain H–30 = Fish H–30 = Meats H–34 = Poultry H–36 = Sausage H–38 = Mixed/Fast H–42 = Nuts/Seeds H–44 = Sweets
H–46 = Vegetables/Legumes H–56 = Vegetarian H–60 = Misc H–60 = Soups/Sauces H–62 = Fast H–76 = Convenience H–78 = Baby foods

Chol (mg)	Calc (mg)	Iron (mg)	Magn (mg)	Pota (mg)	Sodi (mg)	Zinc (mg)	VT-A (RAE)	Thia (mg)	VT-E (mg α)	Ribo (mg)	Niac (mg)	V-B$_6$ (mg)	Fola (µg)	VT-C (mg)
0	43	1.85	37	324	3	0.51	9	0.05	2.04	0.13	0.65	0.1	74	47
0	38	1.62	32	285	2	0.45	8	0.05	1.8	0.11	0.57	0.08	65	41
0	23	0.6	19	220	1	0.2	1	0.03	0.42	0.03	0.56	0.07	35	85
0	28	1.5	18	250	8	0.15	3	0.04	0.59	0.13	1.02	0.08	38	106
0	37	1.19	55	1078	4	0.26	0	0.24	0.22	0.07	1.98	0.22	31	64
0	27	3.34	15	239	17	0.17	93	0.04	0.56	0.1	0.43	0.11	20	5
0	9	0.24	7	151	0	0.05	2	0.02	0.05	0.02	0.1	0.03	3	5
0	20	0.36			5		5							0
0	32	0.85	36	544	2	0.24	<1	0.04	0.04	0.05	1.06	0.14	16	0
0	69	1.82	77	1168	4	0.52	1	0.09	0.09	0.12	2.27	0.29	34	1
0	308	3.86	129	1292	19	1.04	1	0.16	0.66	0.16	1.18	0.2	17	2
0	15	0.72	12	218	15	0.2	25	0.04	0.99	0.05	0.93	0.12	7	5
0	19	0.5	17	225	9	0.21	36	0.03	0.95	0.04	0.96	0.12	7	6
0	27	0.1	11	166	0	0.09	71	0.05	0.16	0.04	0.25	0.07	16	38
0	14	0.07	11	175	0	0.08	2	0.04	0.15	0.02	0.32	0.05	12	39
0	36	1.02	25	328	5	0.2	0	0.1	0.23	0.05	0.62	0.05	23	54
0	22	0.49	30	400	2	0.12	5	0.1	0.54	0.05	0.49	0.11	25	94
0	17	0.49	25	378	2	0.22	1	0.1	0.1	0.05	0.57	0.05	25	72
0	20	0.9	25	405	5	0.15	1	0.1	0.1	0.06	0.8	0.05	25	67
0	20	0.35	27	336	2	0.12	1	0.1	0.1	0.05	0.54	0.11	10	83
0	5	0.18	4	96	1	0.04	2	0.03	0.1	0.04	0.09	0.04	1	5
0	5	0.18	4	96	1	0.04	2	0.03	0.1	0.04	0.09	0.04	1	5
0	10	0.25	10	52	5	0.1	1	0.04	0	0.06	0.31	0.1	2	60
0	23	0.61	25	334	8	0.13	1	0.07	0	0.09	0.66	0.16	8	0
0	56	0.99	61	500	5	0.69	25	0.05	0.25	0.18	0.66	0.18	23	11
0	26	0.24	13	237	2	0.11	3	0.02	1.11	0.02	0.26	0.05	19	70
0	15	0.35	5	80	1	0.03	1	0.02	0.09	0.01	0.06	0.05	6	31
0	17	0.07	15	303	2	0.12	2	0.07	0.37	0.02	0.24	0.12	32	112
0	1	<.01	1	19	<1	<.01	<1	<.01	0.02	<.01	0.02	<.01	2	7
0	2	0.02	1	15	3	<.01	<1	<.01	0.02	<.01	0.03	<.01	2	4
0	20	0.29	20	217	2	0.12	2	0.14	0.2	0.03	0.33	0.15	24	77
0	1	0.02	1	13	<1	<.01	<1	<.01	0.01	<.01	0.02	<.01	2	5
0	1	<.01	1	16	<1	<.01	<1	<.01	0.02	<.01	0.02	<.01	1	4
0	30	0.57	17	184	39	0.15	2	0.08	0.3	<.01	0.4	0.07	20	16
0	24	0.67	37	1315	6	0.49	2	0.21	0.55	0.16	1.34	0.22	158	371
0	21	0.27	19	323	4	0.08	79	0.12	2.32	0.12	1.21	0.28	29	57
0	25	0.58	33	737	44	0.5	466	0.11	0.14	0.05	2.03	0.2	58	101
10	10	0.27	16	365	29	0.14	5	0.06	0.03	0.02	0.67	0.14	30	29
0	8	0.38	12	273	0	0.23	23	0.05	1.05	0.04	1.53	0.03	7	7

Table H–1

Food Composition

(Computer code number is for West Diet Analysis program)   (For purposes of calculations, use "0" for t, <1, <.1, <.01, etc.)

Computer Code Number	Food Description	Measure	Wt (g)	H₂O (%)	Ener (kcal)	Prot (g)	Carb (g)	Dietary Fiber (g)	Fat (g)	Fat Breakdown (g) Sat	Mono	Poly
	FRUITS and FRUIT JUICES—Continued											
3082	Whole w/o peel & seeds, 2⅜" diam (180g w/peel & seeds)	1 ea	131	87	62	1	15	3	<1	t	t	t
3083	Sections, without membranes	1 c	180	87	85	2	21	4	<1	t	t	t
	Juice:											
3090	Fresh, all varieties	1 c	248	88	112	2	26	<1	<1	t	t	t
3093	Canned, unsweetened	1 c	249	89	105	1	25	<1	<1	t	t	t
3480	Calcium fortified	1 c	247		110	1	27	0	0	0	0	0
3092	Chilled	1 c	249	88	110	2	25	<1	1	t	0.1	0.2
3091	Prepared from concentrate	1 c	249	88	112	2	27	<1	<1	t	t	t
20004	Prepared from dry crystals	1 c	248	87	122	0	31	<1	0	0	0	0
3170	Orange and grapefruit juice, canned	1 c	247	89	106	1	25	<1	<1	t	t	t
	Papayas:											
	Fresh:											
3172	½" slices	1 c	140	89	55	1	14	3	<1	t	t	t
3171	Whole, 3" diam by 5⅛"	1 ea	304	89	119	2	30	5	<1	0.1	0.1	t
3095	Nectar, canned	1 c	250	85	142	<1	36	2	<1	0.1	0.1	t
	Peaches:											
	Fresh:											
3096	Whole, 2" diam	1 ea	98	89	38	1	9		<1	t	t	t
3097	Sliced	1 c	170	89	66	2	16		<1	t	0.1	0.1
	Canned, not drained:											
	Heavy syrup pack:											
3098	Cup	1 c	262	79	194	1	52	3	<1	t	t	0.1
3099	Half	1 ea	98	79	73	<1	20	1	<1	t	t	t
	Juice pack:											
3175	Cup	1 c	248	87	109	2	29	3	<1	t	t	t
3176	Half	1 ea	98	87	43	1	11	1	<1	t	t	t
	Dried:											
3100	Uncooked	10 ea	130	32	311	5	80	11	1	0.1	0.4	0.5
3214	Cooked, w/fruit & liquid	1 c	258	78	199	3	51	7	1	t	0.2	0.3
	Frozen, sweetened:											
3234	10-oz package,vitamin C added	1 ea	284	75	267	2	68	5	<1	t	0.1	0.2
57481	Cup, vitamin C added	1 c	250	75	235	2	60	4	<1	t	0.1	0.2
3101	Nectar, canned	1 c	249	86	134	1	35	1	<1	t	t	t
	Pears:											
	Fresh, with skin, cored:											
3103	Bartlett, 2½" diam (about 2½/lb)	1 ea	166	84	96	1	26	5	<1	t	t	t
3105	Bosc, 2⅛" diam (about 3/lb)	1 ea	139	84	81	1	21	4	<1	t	t	t
3106	D'Anjou, 3" diam (about 2/lb)	1 ea	209	84	121	1	32	6	<1	t	t	t
	Canned, fruit and liquid:											
	Heavy syrup pack:											
3107	Cup	1 c	266	80	197	1	51	4	<1	t	t	t
3108	Half	1 ea	76	80	56	<1	15	1	<1	t	t	t
	Juice pack:											
3179	Cup	1 c	248	86	124	1	32	4	<1	t	t	t
3180	Half	1 ea	76	86	38	<1	10	1	<1	t	t	t
3109	Dried halves	10 ea	175	27	458	3	122	13	1	t	0.2	0.3
3110	Nectar, canned	1 c	250	84	150	<1	39	2	<1	t	t	t
	Pineapple:											
3111	Fresh chunks, diced	1 c	155	86	74	1	20	2	<1	t	t	t
	Canned, not drained:											
	Heavy syrup pack:											
3114	Crushed, chunks, tidbits	½ c	127	79	99	<1	26	1	<1	t	t	t
3115	Slices	1 ea	49	79	38	<1	10	<1	<1	t	t	t
	Juice pack:											
3183	Crushed, chunks, tidbits	1 c	250	84	150	1	39	2	<1	t	t	t
3184	Slices	1 ea	47	84	28	<1	7	<1	<1	t	t	t
3120	Juice, canned, unsweetened:	1 c	250	86	140	1	34	<1	<1	t	t	t
	Plantains, yellow fleshed, without peel:											
3195	Raw slices (whole=179g w/o peel)	1 c	148	65	181	2	47	3	1	0.2	t	0.1
3196	Cooked, boiled, sliced	1 c	154	67	179	1	48	4	<1	0.1	t	t

PAGE KEY: H–2 = Beverages   H–4 = Dairy   H–8 = Eggs   H–10 = Fat/Oil   H–12 = Fruit   H–18 = Bakery
H–24 = Grain   H–30 = Fish   H–30 = Meats   H–34 = Poultry   H–36 = Sausage   H–38 = Mixed/Fast   H–42 = Nuts/Seeds   H–44 = Sweets
H–46 = Vegetables/Legumes   H–56 = Vegetarian   H–60 = Misc   H–60 = Soups/Sauces   H–62 = Fast   H–76 = Convenience   H–78 = Baby foods

Chol (mg)	Calc (mg)	Iron (mg)	Magn (mg)	Pota (mg)	Sodi (mg)	Zinc (mg)	VT-A (RAE)	Thia (mg)	VT-E (mg α)	Ribo (mg)	Niac (mg)	V-B$_6$ (mg)	Fola (µg)	VT-C (mg)
0	52	0.13	13	237	0	0.09	14	0.11	0.24	0.05	0.37	0.08	39	70
0	72	0.18	18	326	0	0.13	20	0.16	0.32	0.07	0.51	0.11	54	96
0	27	0.5	27	496	2	0.12	25	0.22	0.1	0.07	0.99	0.1	74	124
0	20	1.1	27	436	5	0.17	22	0.15	0.5	0.07	0.78	0.22	45	86
0	300	0		430	15		0	0	0		0	0	40	78
0	25	0.42	27	473	2	0.1	10	0.28	0.47	0.05	0.7	0.13	45	82
0	22	0.25	25	473	2	0.12	12	0.2	0.5	0.04	0.5	0.11	110	97
0	126	0.02	2	60	10	0.02	191	0	0	0.22	2.54	0.25	0	73
0	20	1.14	25	390	7	0.17	15	0.14	0.35	0.07	0.83	0.06	35	72
0	34	0.14	14	360	4	0.1	77	0.04	1.02	0.04	0.47	0.03	53	87
0	73	0.3	30	781	9	0.21	167	0.08	2.22	0.1	1.03	0.06	116	188
0	25	0.85	8	78	12	0.38	45	0.02	0.6	<.01	0.38	0.02	5	8
0	6	0.24	9	186	0	0.17	16	0.02	0.72	0.03	0.79	0.02	4	6
0	10	0.42	15	323	0	0.29	27	0.04	1.24	0.05	1.37	0.04	7	11
0	8	0.71	13	241	16	0.24	45	0.03	1.28	0.06	1.61	0.05	8	7
0	3	0.26	5	90	6	0.09	17	0.01	0.48	0.02	0.6	0.02	3	3
0	15	0.67	17	317	10	0.27	47	0.02	1.22	0.04	1.44	0.05	7	9
0	6	0.26	7	125	4	0.11	19	<.01	0.48	0.02	0.57	0.02	3	4
0	36	5.28	55	1295	9	0.74	140	<.01	0.25	0.28	5.69	0.09	0	6
0	23	3.38	34	826	5	0.46	26	0.01	0.15	0.05	3.92	0.1	0	10
0	9	1.05	14	369	17	0.14	40	0.04	1.76	0.1	1.85	0.05	9	268
0	8	0.92	12	325	15	0.12	35	0.03	1.55	0.09	1.63	0.04	8	235
0	12	0.47	10	100	17	0.2	32	<.01	0.72	0.03	0.72	0.02	2	13
0	15	0.28	12	198	2	0.17	2	0.02	0.2	0.04	0.26	0.05	12	7
0	13	0.24	10	165	1	0.14	1	0.02	0.17	0.03	0.22	0.04	10	6
0	19	0.36	15	249	2	0.21	2	0.03	0.25	0.05	0.33	0.06	15	9
0	13	0.59	11	173	13	0.21	0	0.03	0.21	0.06	0.64	0.04	3	3
0	4	0.17	3	49	4	0.06	0	<.01	0.06	0.02	0.18	0.01	1	1
0	22	0.72	17	238	10	0.22	1	0.03	0.2	0.03	0.5	0.03	2	4
0	7	0.22	5	73	3	0.07	<1	<.01	0.06	<.01	0.15	0.01	1	1
0	60	3.68	58	933	10	0.68	<1	0.01	0.1	0.25	2.4	0.13	0	12
0	12	0.65	8	32	10	0.18	<1	<.01	0.12	0.03	0.32	0.04	2	3
0	20	0.43	19	178	2	0.16	5	0.12	0.03	0.05	0.76	0.17	23	56
0	18	0.48	20	132	1	0.15	1	0.11	0.01	0.03	0.36	0.09	6	9
0	7	0.19	8	51	<1	0.06	<1	0.04	<.01	0.01	0.14	0.04	2	4
0	35	0.7	35	305	2	0.25	5	0.24	0.02	0.05	0.71	0.18	12	24
0	7	0.13	7	57	<1	0.05	1	0.04	<.01	<.01	0.13	0.03	2	4
0	42	0.65	32	335	2	0.28	1	0.14	0.05	0.06	0.64	0.24	58	27
0	4	0.89	55	739	6	0.21	83	0.08	0.21	0.08	1.02	0.44	33	27
0	3	0.89	49	716	8	0.2	69	0.07	0.2	0.08	1.16	0.37	40	17

Table H–1

Food Composition

(Computer code number is for West Diet Analysis program) (For purposes of calculations, use "0" for t, <1, <.1, <.01, etc.)

Computer Code Number	Food Description	Measure	Wt (g)	H₂O (%)	Ener (kcal)	Prot (g)	Carb (g)	Dietary Fiber (g)	Fat (g)	Fat Breakdown (g) Sat	Mono	Poly
	FRUITS and FRUIT JUICES—Continued											
	Plums:											
3121	Fresh, medium, 2⅛" diam	1 ea	66	87	30	<1	8	1	<1	t	t	t
	Canned, purple, not drained:											
	Heavy syrup pack:											
3124	Cup	1 c	258	76	230	1	60	2	<1	t	0.2	t
3125	Each	3 ea	138	76	123	<1	32	1	<1	t	t	t
	Juice Pack:											
3185	Cup	1 c	252	84	146	1	38	2	<1	t	t	t
3186	Each	3 ea	138	84	80	1	21	1	<1	t	t	t
3197	Pomegranate, fresh	1 ea	154	81	105	1	26	1	<1	t	t	t
	Prunes:											
	Dried, pitted:											
3126	Uncooked (10 = 97g w/pits, 84g w/o pits)	10 ea	84	31	202	2	54	6	<1	t	t	t
3127	Cooked, unsweetened, fruit & liq, (250 g w/pits)	1 c	248	70	265	2	70	8	<1	t	0.3	t
3128	Juice, bottled or canned	1 c	256	81	182	2	45	3	<1	t	t	t
	Raisins, seedless:											
3130	Cup, not pressed down	1 c	145	15	434	4	115	5	1	t	t	t
3764	One packet, ½ oz	½ oz	14	15	42	<1	11	1	<1	t	t	t
3133	Rhubarb, cooked, added sugar	1 c	240	68	278	1	75	5	<1	t	t	t
	Tangerines, without peel and seeds:											
3138	Fresh (2⅜" whole) 116g w/refuse	1 ea	84	88	37	1	9	2	<1	t	t	t
3237	Canned, light syrup, fruit and liquid	1 c	252	83	154	1	41	2	<1	t	t	t
3140	Juice, canned, sweetened	1 c	249	87	124	1	30	<1	<1	t	t	t
	Watermelon, raw, without rind and seeds:											
3143	Piece, 1/16th wedge	1 pce	286	91	86	2	22	1	<1	t	0.1	0.1
3142	Diced	1 c	152	91	46	1	11	1	<1	t	t	t
	BAKED GOODS: BREADS, CAKES, COOKIES, CRACKERS, PIES											
	Bagels:											
42100	Cinnamon raisin, 3½" diam.	1 ea	71	32	195	7	39	2	1	0.2	0.1	0.5
42620	Dunkin Donuts	1 ea	125		340	11	72	4	1	0		
42000	Plain, enriched, 3½" diam.	1 ea	71	33	195	7	38	2	1	0.2	t	0.5
42619	Dunkin Donuts	1 ea	125		330	12	68	3	1	0		
42596	Oat Bran	1 ea	110	33	280	12	59	4	1	0.2	0.3	0.5
42617	Whole Wheat	1 ea	110	28	291	12	62	10	2	0.3	0.2	0.6
	Biscuits:											
42001	From home recipe	1 ea	60	29	212	4	27	1	10	2.6	4.2	2.5
42002	From mix	1 ea	57	29	191	4	28	1	7	1.6	2.4	2.5
	Bread:											
42672	Cornbread, 2.5 x 2.5 x 1.5" piece	1 pce	65	49	152	4	23	2	5	1.6	2.4	0.5
42015	Croissants, 4½ x 4 x 1¾"	1 ea	57	23	231	5	26	1	12	6.6	3.1	0.6
42148	Croutons, seasoned	½ c	20	4	93	2	13	1	4	1	1.9	0.5
42004	Crumbs, dry, grated (see 364, 365 for soft crumbs)	1 c	108	7	427	14	78	5	6	1.3	1.1	2.2
42052	Boston brown, canned, 3¼" slice	1 pce	45	47	88	2	19	2	1	0.1	t	0.3
	Cracked wheat (¼ cracked-wheat & ¾ enriched wheat flour):											
42042	Slice (18 per loaf)	1 pce	25	36	65	2	12	1	1	0.2	0.5	0.2
	French/Vienna, enriched:											
42044	Slice, 4¾ x 4½"	1 pce	25	34	68	2	13	1	1	0.2	0.3	0.2
42043	French, slice, 5 x 2"	1 pce	25	34	68	2	13	1	1	0.2	0.3	0.2
	French toast: see Mixed Dishes, and Fast Foods, #691											
42476	Honey wheatberry	1 pce	38		100	3	18	2	2	0	0.5	1
	Italian, enriched:											
42046	Slice, 4½ x 3¼ x ¾"	1 pce	30	36	81	3	15	1	1	0.3	0.2	0.4
	Mixed grain, enriched:											
42047	Slice (18 per loaf)	1 pce	26	38	65	3	12	2	1	0.2	0.4	0.2
42048	Slice, toasted	1 pce	24	32	65	3	12	2	1	0.2	0.4	0.2

PAGE KEY: H–2 = Beverages   H–4 = Dairy   H–8 = Eggs   H–10 = Fat/Oil   H–12 = Fruit   H–18 = Bakery
H–24 = Grain   H–30 = Fish   H–30 = Meats   H–34 = Poultry   H–36 = Sausage   H–38 = Mixed/Fast   H–42 = Nuts/Seeds   H–44 = Sweets
H–46 = Vegetables/Legumes   H–56 = Vegetarian   H–60 = Misc   H–60 = Soups/Sauces   H–62 = Fast   H–76 = Convenience   H–78 = Baby foods

Chol (mg)	Calc (mg)	Iron (mg)	Magn (mg)	Pota (mg)	Sodi (mg)	Zinc (mg)	VT-A (RAE)	Thia (mg)	VT-E (mg α)	Ribo (mg)	Niac (mg)	V-B$_6$ (mg)	Fola (µg)	VT-C (mg)
0	4	0.11	5	104	0	0.07	11	0.02	0.17	0.02	0.28	0.02	3	6
0	23	2.17	13	235	49	0.18	34	0.04	0.46	0.1	0.75	0.07	8	1
0	12	1.16	7	126	26	0.1	18	0.02	0.25	0.05	0.4	0.04	4	1
0	25	0.86	20	388	3	0.28	126	0.06	0.45	0.15	1.19	0.07	8	7
0	14	0.47	11	213	1	0.15	69	0.03	0.25	0.08	0.65	0.04	4	4
0	5	0.46	5	399	5	0.18	8	0.05	0.92	0.05	0.46	0.16	9	9
0	36	0.78	34	615	2	0.37	33	0.04	0.36	0.16	1.58	0.17	3	1
0	47	1.02	45	796	2	0.47	42	0.06	0.47	0.25	1.79	0.54	0	7
0	31	3.02	36	707	10	0.54	<1	0.04	0.31	0.18	2.01	0.56	0	10
0	72	2.73	46	1086	16	0.32	0	0.15	0.17	0.18	1.11	0.25	7	3
0	7	0.26	4	105	2	0.03	0	0.01	0.02	0.02	0.11	0.02	1	0
0	348	0.5	29	230	2	0.19	10	0.04	0.65	0.06	0.48	0.05	12	8
0	12	0.08	10	132	1	0.2	29	0.09	0.13	0.02	0.13	0.06	17	26
0	18	0.93	20	197	15	0.6	106	0.13	0.25	0.11	1.12	0.11	13	50
0	45	0.5	20	443	2	0.07	32	0.15	0.37	0.05	0.25	0.08	12	55
0	20	0.69	29	320	3	0.29	80	0.09	0.14	0.06	0.51	0.13	9	23
0	11	0.36	15	170	2	0.15	43	0.05	0.08	0.03	0.27	0.07	5	12
0	13	2.7	20	105	229	0.8	15	0.27	0.22	0.2	2.19	0.04	79	0
0	40	3.6			470		0							4
0	53	2.53	21	72	379	0.62	0	0.38	0.21	0.22	3.24	0.04	75	0
0	20	4.5			690		0							4
0	13	3.39	34	126	558	0.99	1	0.36	0.36	0.37	3.26	0.05	108	0
0	32	3.52	116	379	592	2.52		0.34	0.99	0.28	5.75	0.3	66	0
2	141	1.74	11	73	348	0.32	14	0.21	0.78	0.19	1.77	0.02	37	0
2	105	1.17	14	107	544	0.35	15	0.2	0.23	0.2	1.72	0.04	30	0
22	71	0.82	13	105	356	0.38		0.12	0.64	0.16	0.93	0.05	6	0
38	21	1.16	9	67	424	0.43	117	0.22	0.48	0.14	1.25	0.03	50	0
1	19	0.56	8	36	248	0.19	1	0.1	0.08	0.08	0.93	0.02	21	0
0	198	5.22	46	212	791	1.57	0	1.04	0.09	0.44	7.16	0.13	116	0
0	32	0.94	28	143	284	0.22	11	<.01	0.14	0.05	0.5	0.04	5	0
0	11	0.7	13	44	134	0.31	0	0.09	0.15	0.06	0.92	0.08	15	0
0	19	0.63	7	28	152	0.22	0	0.13	0.08	0.08	1.19	0.01	37	0
0	19	0.63	7	28	152	0.22	0	0.13	0.08	0.08	1.19	0.01	37	0
0	20	0.72			200		0	0.12	0.24	0.07	0.8			0
0	23	0.88	8	33	175	0.26	0	0.14	0.09	0.09	1.31	0.01	57	0
0	24	0.9	14	53	127	0.33	0	0.11	0.09	0.09	1.13	0.09	31	0
0	24	0.9	14	53	127	0.33	0	0.08	0.08	0.08	1.02	0.08	28	0

Table H–1

Food Composition (Computer code number is for West Diet Analysis program) (For purposes of calculations, use "0" for t, <1, <.1, <.01, etc.)

Computer Code Number	Food Description	Measure	Wt (g)	H_2O (%)	Ener (kcal)	Prot (g)	Carb (g)	Dietary Fiber (g)	Fat (g)	Fat Breakdown (g) Sat	Mono	Poly
	BAKED GOODS: BREADS, CAKES, COOKIES, CRACKERS, PIES—Continued											
	Oatmeal, enriched:											
42049	Slice (18 per loaf)	1 pce	27	37	73	2	13	1	1	0.2	0.4	0.5
42050	Slice, toasted	1 pce	25	31	73	2	13	1	1	0.2	0.4	0.5
42007	Pita pocket bread, enr, 6″ round	1 ea	60	32	165	5	33	1	1	t	t	0.3
	Pumpernickel (⅔ rye & ⅓ enr wheat flr):											
42006	Slice, 5 x 4 x ⅜″	1 pce	26	38	65	2	12	2	1	0.1	0.2	0.3
42054	Slice, toasted	1 pce	29	32	80	3	15	2	1	0.1	0.3	0.4
	Raisin, enriched:											
42051	Slice (18 per loaf)	1 pce	26	34	71	2	14	1	1	0.3	0.6	0.2
42055	Slice, toasted	1 pce	24	28	71	2	14	1	1	0.3	0.6	0.2
	Rye, light (⅓ rye & ⅔ enr wheat flr):											
42005	1-lb loaf	1 ea	454	37	1176	39	219	26	15	2.8	6	3.6
42005	Slice, 4¾ x 3¾ x 7/16″	1 pce	32	37	83	3	15	2	1	0.2	0.4	0.3
42056	Slice, toasted	1 pce	24	31	68	2	13	2	1	0.2	0.3	0.2
	Wheat (enr wheat & whole-wheat flour):											
42012	Slice (18 per loaf)	1 pce	25	37	65	2	12	1	1	0.2	0.4	0.2
42031	Slice, toasted	1 pce	23	32	65	2	12	1	1	0.2	0.4	0.2
	White, enriched:											
42138	Slice	1 pce	42	35	120	3	21	1	2	0.5	0.5	1.2
	Whole Wheat:											
42014	Slice (16 per loaf)	1 pce	28	38	69	3	13	2	1	0.3	0.5	0.3
42029	Slice, toasted	1 pce	25	30	69	3	13	2	1	0.3	0.5	0.3
	Bread stuffing, prepared from mix:											
42037	Dry type	1 c	200	65	356	6	43	6	17	3.5	7.6	5.2
	Cakes:											
	Prepared from mixes using enrich flour and veg shortening, w/frostings made from margarine:											
46004	Angel Food, 1/12 of cake	1 pce	28	33	72	2	16	<1	<1	t	t	0.1
46002	Boston cream pie, ⅛ of cake	1 pce	92	45	232	2	39	1	8	2.2	4.2	0.9
46005	Coffee Cake, ⅙ of cake	1 pce	56	30	178	3	30	1	5	1	2.2	1.8
46013	Devil's food, chocolate frosting, 1/16 of cake	1 pce	64	23	235	3	35	2	10	3.1	5.6	1.2
	Home recipes w/enrich flour:											
	Fruitcake, dark:											
46205	Piece, 1/32 of cake, ⅔″ arc	1 pce	43	25	139	1	26	2	4	0.5	1.8	1.4
46015	Sheet, plain, made w/margarine, uncooked white frosting, ⅑ of cake	1 pce	64	22	239	2	38	<1	9	1.5	3.9	3.3
	Commercial:											
49004	Cheesecake, 1/12 of cake	1 pce	80	46	257	4	20	<1	18	7.9	6.9	1.3
46016	Pound cake, 1/17 of loaf, 2″ slice	1 pce	28	25	109	2	14	<1	6	3.2	1.7	0.3
	Yellow, chocolate frosting, 2 layer:											
46012	Slice, 1/16 of cake	1 pce	64	22	243	2	35	1	11	3	6.1	1.4
	Snack:											
46011	Chocolate w/creme filling, Ding Dong	1 ea	50	20	188	2	30	<1	7	1.4	2.8	2.6
46008	Sponge cake w/creme filling,Twinkie	1 ea	43	20	157	1	27	<1	5	1.1	1.8	1.4
46001	Sponge cake, 1/12 of 12″ cake	1 pce	38	30	110	2	23	<1	1	0.3	0.4	0.2
44032	Chex party mix	1 c	43	4	183	5	28	2	7	2.4	3.9	1.1
	Chips:											
44061	Bagel	5 pce	70	3	298	6	52	4	7	1.3	2.1	3.4
42396	Bagel, onion garlic, toasted	1 oz	28		181	5	30	3	7	1.6	4.9	0
43703	Potato chips	20 pce	28		148	2	15	1	10	3		
4022	Baked	11 pce	28		109	2	23	2	1	0		
	Cookies made with enriched flour:											
	Brownies with nuts:											
47000	Commercial w/frosting, 1½ x 1¾ x ⅞″	1 ea	61	14	247	3	39	1	10	2.6	5.5	1.4
47214	Fat free fudge, Entenmann's	1 pce	40		110	2	27	1	0	0	0	0
	Chocolate chip cookies:											
47001	Commercial, 2¼″diam	4 ea	60	12	275	2	35	2	15	4.4	7.8	2.1
47002	Home recipe, 2¼″ diam	4 ea	64	6	312	4	37	2	18	5.2	6.6	5.4

PAGE KEY: H–2 = Beverages H–4 = Dairy H–8 = Eggs H–10 = Fat/Oil H–12 = Fruit H–18 = Bakery
H–24 = Grain H–30 = Fish H–30 = Meats H–34 = Poultry H–36 = Sausage H–38 = Mixed/Fast H–42 = Nuts/Seeds H–44 = Sweets
H–46 = Vegetables/Legumes H–56 = Vegetarian H–60 = Misc H–60 = Soups/Sauces H–62 = Fast H–76 = Convenience H–78 = Baby foods

Chol (mg)	Calc (mg)	Iron (mg)	Magn (mg)	Pota (mg)	Sodi (mg)	Zinc (mg)	VT-A (RAE)	Thia (mg)	VT-E (mg α)	Ribo (mg)	Niac (mg)	V-B₆ (mg)	Fola (µg)	VT-C (mg)
0	18	0.73	10	38	162	0.28	1	0.11	0.13	0.06	0.85	0.02	17	0
0	18	0.74	10	38	163	0.28	1	0.09	0.13	0.06	0.77	0.02	13	0
0	52	1.57	16	72	322	0.5	0	0.36	0.18	0.2	2.78	0.02	64	0
0	18	0.75	14	54	174	0.38	0	0.09	0.11	0.08	0.8	0.03	24	0
0	21	0.91	17	66	214	0.47	0	0.08	0.13	0.09	0.89	0.04	25	0
0	17	0.75	7	59	101	0.19	0	0.09	0.07	0.1	0.9	0.02	28	0
0	17	0.76	7	59	102	0.19	0	0.07	0.07	0.09	0.81	0.02	24	0
0	331	12.85	182	754	2996	5.18	2	1.97	1.5	1.52	17.27	0.34	499	2
0	23	0.91	13	53	211	0.36	<1	0.14	0.11	0.11	1.22	0.02	35	0
0	19	0.74	10	44	174	0.3	<1	0.09	0.09	0.08	0.9	0.02	25	0
0	26	0.83	12	50	132	0.26	0	0.1	0.07	0.07	1.03	0.02	23	0
0	26	0.83	12	50	132	0.26	0	0.08	0.07	0.06	0.93	0.02	19	0
1	24	1.25	8	61	151	0.27	9	0.17	0.36	0.16	1.51	0.02	38	0
0	20	0.92	24	71	148	0.54	<1	0.1	0.09	0.06	1.07	0.05	14	0
0	20	0.93	24	71	148	0.55	<1	0.08	0.08	0.05	0.97	0.05	10	0
0	64	2.18	24	148	1086	0.56	236	0.27	2.8	0.21	2.95	0.08	78	0
0	39	0.15	3	26	210	0.02	0	0.03	0.03	0.14	0.25	<.01	10	0
34	21	0.35	6	36	132	0.15	22	0.38	0.14	0.25	0.18	0.02	13	0
27	76	0.8	10	63	236	0.25	20	0.09	0.11	0.1	0.85	0.03	27	0
27	28	1.41	22	128	214	0.44	17	0.02	<.01	0.09	0.37	0.03	11	0
2	14	0.89	7	66	116	0.12	3	0.02	0.39	0.04	0.34	0.02	9	0
35	40	0.68	4	34	220	0.16	12	0.06	1.22	0.04	0.32	0.02	17	0
44	41	0.5	9	72	166	0.41	114	0.02	1.26	0.15	0.16	0.04	14	0
62	10	0.39	3	33	111	0.13	42	0.04	0.18	0.06	0.37	0.01	11	0
35	24	1.33	19	114	216	0.4	21	0.08	1.45	0.1	0.8	0.02	14	0
8	36	1.68	20	61	212	0.26	2	0.11	1.09	0.15	1.21	0.01	20	0
7	19	0.55	3	37	157	0.12	2	0.07	0.51	0.06	0.53	0.01	17	0
39	27	1.03	4	38	93	0.19	17	0.09	0.09	0.1	0.73	0.02	18	0
0	15	10.62	27	116	437	0.9	3	0.67	0.11	0.21	7.24	0.67	22	20
0	9	1.39	39	167	419	0.88		0.13	1.71	0.12	1.62	0.15	46	0
0	0	2.37			461		0	0.37	<.01	0.22	3.29			0
0	0	0			178		0							6
0	40	0.36			148		0							1
10	18	1.37	19	91	190	0.44	12	0.16	0.09	0.13	1.05	0.02	29	0
0	0	1.08		90	140		0		<.01					0
0	9	1.45	21	56	196	0.28	<1	0.07	1.74	0.12	0.97	0.1	23	0
20	25	1.57	35	143	231	0.6	92	0.12	1.84	0.11	0.87	0.05	21	0

Table H–1

Food Composition   (Computer code number is for West Diet Analysis program)   (For purposes of calculations, use "0" for t, <1, <.1, <.01, etc.)

Computer Code Number	Food Description	Measure	Wt (g)	H₂O (%)	Ener (kcal)	Prot (g)	Carb (g)	Dietary Fiber (g)	Fat (g)	Fat Breakdown (g) Sat	Mono	Poly
	BAKED GOODS: BREADS, CAKES, COOKIES, CRACKERS, PIES—Continued											
47013	From refrigerated dough, 2¼" diam	4 ea	64	13	284	3	39	1	13	4.3	6.7	1.4
47012	Fig bars	4 ea	64	16	223	2	45	3	5	0.7	1.9	1.8
47376	Fruit bar, no fat	1 ea	28		90	2	21	0	0	0	0	0
47324	Fudge, fat free, Snackwell	1 ea	16	14	53	1	12	<1	<1	t	t	t
47154	Nabisco Newtons, fat free, all flavors	1 ea	23		69	1	16		0	0	0	0
47003	Oatmeal raisin, 2⅝" diam	4 ea	60	6	261	4	41	2	10	1.9	4.1	3
47010	Peanut butter, home recipe, 2⅝" diam	4 ea	80	6	380	7	47	2	19	3.6	8.7	5.8
47006	Sandwich-type, all	4 ea	40	2	189	2	28	1	8	1.5	3.4	2.9
47007	Shortbread, commercial, small	4 ea	32	4	161	2	21	1	8	2	4.3	1
47004	Sugar from refrigerated dough,2" diam	4 ea	48	5	232	2	31	<1	11	2.8	6.2	1.4
47160	Vanilla sandwich, Snackwell's	2 ea	26	4	109	1	21	1	2	0.5	0.8	0.2
47008	Vanilla wafers	10 ea	40	5	176	2	29	1	6	1.5	2.6	1.6
	Crackers:											
43500	Cheese-enriched	10 ea	10	3	50	1	6	<1	3	0.9	1.2	0.2
43501	Cheese with peanut butter-enriched	4 ea	28	3	139	3	16	1	7	1.2	3.6	1.4
	Fat free-enriched:											
43596	Cracked pepper, Snackwell	1 ea	14	2	61	1	10	<1	2	0.3	0.6	0.2
43593	Wheat, Snackwell	7 ea	15	1	60	2	12	1	<1	0.1	0.1	0.1
43590	Whole wheat, herb seasoned	5 ea	14		50	2	11	2	0	0	0	0
43591	Whole wheat, onion	5 ea	14		50	2	11	2	0	0	0	0
43502	Graham-enriched	2 ea	14	4	59	1	11	<1	1	0.2	0.6	0.5
43509	Melba toast, plain-enriched	1 pce	5	5	20	1	4	<1	<1	t	t	t
44064	Rice cakes, unsalted-enriched	2 ea	18	6	70	1	15	1	1	0.1	0.2	0.2
43504	Rye wafer, whole grain	2 ea	22	5	73	2	18	5	<1	t	t	t
43506	Saltine-enriched	4 ea	12	4	52	1	9	<1	1	0.4	0.8	0.2
43586	Saltine, unsalted tops-enriched	2 ea	6		25	1	4	0	<1			
43543	Snack-type, round like Ritz-enriched	3 ea	9	4	45	1	5	<1	2	0.3	1	0.9
43584	Triscuits	7 ea	31	4	150	3	21		6	1	2	0.5
43558	Wasa extra crsip crispbread	1 pce	6	6	24	1	4	<1	<1	t	0.2	t
43508	Whole-wheat wafers	2 ea	8	3	35	1	5	1	1	0.3	0.5	0.5
	Pastry:											
	Danish:											
428	Round piece, plain, 4¼" diam, 1" high	1 ea	88	21	349	5	47	<1	17	3.5	10.6	1.6
45512	Ounce, plain	1 oz	28	21	111	2	15	<1	5	1.1	3.4	0.5
45513	Round piece with fruit	1 ea	94	29	335	5	45		16	3.3	10.1	1.6
42094	Pan dulce, sweet roll w/topping	1 ea	79	21	291	5	48	1	9	1.8	4	2.5
49009	Peach crisp, 3 x 3 piece	1 pce	139	73	165	1	30	2	5	0.8		
	Toaster:											
45504	Fortified (PopTarts)	1 ea	52	12	204	2	37	1	5	0.8	2.2	2
	Strudel:											
45647	Cream Cheese	1 ea	54		200	3	24	<1	10	3		
45648	French Toast	1 ea	54		200	3	24	<1	10	3		
	Doughnuts:											
45505	Cake type, plain, 3¼" diam	1 ea	47	21	198	2	23	1	11	1.7	4.4	3.7
45698	Sugared, Dunkin Donuts	1 ea	67		310	4	28	1	20	4		
45506	Yeast-leavened, glazed, 3 ¾"diam	1 ea	60	25	242	4	27	1	14	3.5	7.7	1.7
45708	Dunkin Donuts	1 ea	46		160	3	23	1	7	2		
	Muffins:											
	English:											
42059	Plain, enriched	1 ea	57	42	134	4	26	2	1	0.1	0.2	0.5
42061	Toasted	1 ea	52	37	133	4	26	2	1	0.1	0.2	0.5
42082	Whole wheat	1 ea	66	46	134	6	27	4	1	0.2	0.3	0.6
44504	Cornmeal	1 ea	50	30	160	4	25	1	5	1.4	2.6	0.6
44614	Banana nut	1 ea	95		340	6	53	2	12	3		
44622	Blueberry	1 ea	95		310	5	51	2	10	2.5		
44616	Chocolate chip	1 ea	95		400	5	63	2	16	6		
	Granola Bars:											
23104	Soft	1 ea	28	6	124	2	19	1	5	2	1.1	1.5
23059	Hard	1 ea	25	4	118	3	16	1	5	0.6	1.1	3
47294	Fat free, all flavors	1 ea	42		140	2	35	3	0	0	0	0

PAGE KEY: H–2 = Beverages H–4 = Dairy H–8 = Eggs H–10 = Fat/Oil H–12 = Fruit H–18 = Bakery
H–24 = Grain H–30 = Fish H–30 = Meats H–34 = Poultry H–36 = Sausage H–38 = Mixed/Fast H–42 = Nuts/Seeds H–44 = Sweets
H–46 = Vegetables/Legumes H–56 = Vegetarian H–60 = Misc H–60 = Soups/Sauces H–62 = Fast H–76 = Convenience H–78 = Baby foods

Chol (mg)	Calc (mg)	Iron (mg)	Magn (mg)	Pota (mg)	Sodi (mg)	Zinc (mg)	VT-A (RAE)	Thia (mg)	VT-E (mg α)	Ribo (mg)	Niac (mg)	V-B$_6$ (mg)	Fola (µg)	VT-C (mg)
15	16	1.44	15	115	134	0.32	12	0.12	1.48	0.12	1.26	0.02	36	0
0	41	1.86	17	132	224	0.25	6	0.1	0.42	0.14	1.2	0.05	22	0
0	0	0.36			95		0		<.01					0
0	3	0.29	5	26	71	0.08		0.02	<.01	0.02	0.26	<.01		0
				77										
20	60	1.59	25	143	323	0.52	86	0.15	1.5	0.1	0.76	0.04	18	0
25	31	1.78	31	185	414	0.66	110	0.18	3.04	0.17	2.81	0.07	44	0
0	10	1.55	18	70	242	0.32	<1	0.03	0.63	0.07	0.83	<.01	20	0
6	11	0.88	5	32	146	0.17	6	0.11	0.11	0.11	1.07	0.03	22	0
15	43	0.88	4	78	225	0.13	6	0.09	0.1	0.06	1.16	0.01	34	0
0	17	0.61	5	28	95	0.16		0.05		0.07	0.69	0.01		0
20	19	0.95	6	39	125	0.14	3	0.11	0.09	0.13	1.24	0.03	24	0
1	15	0.48	4	14	100	0.11	3	0.06	<.01	0.04	0.47	0.06	15	0
0	14	0.76	16	61	199	0.29		0.15	0.66	0.08	1.63	0.04	26	0
0	24	0.51	4	16	117	0.11		0.04	0	0.05	0.75	<.01	11	0
0	28	0.58	7	43	169	0.21		0.04		0.07	0.73	0.02		0
0	0				80									2
0	0	0			80									2
0	3	0.52	4	19	85	0.11	<1	0.03	0.05	0.04	0.58	<.01	6	0
0	5	0.18	3	10	41	0.1	0	0.02	0.02	0.01	0.21	<.01	6	0
0	2	0.27	24	52	5	0.54	<1	0.01	0.02	0.03	1.41	0.03	4	0
0	9	1.31	27	109	175	0.62	<1	0.09	0.18	0.06	0.35	0.06	10	0
0	14	0.65	3	15	156	0.09	0	0.07	0.01	0.06	0.63	<.01	15	0
0		0.36		5	50				0.1					
0	11	0.32	2	12	76	0.06	0	0.04	0.18	0.03	0.36	<.01	8	0
0		1.44		95	160									
0	5	0.19	3	17	38	0.08		0.03	0.08	0.02	0.2	<.01	1	0
0	4	0.25	8	24	53	0.17	0	0.02	0.07	<.01	0.36	0.01	2	0
27	37	1.8	14	96	326	0.48	5	0.26	0.79	0.19	2.2	0.05	55	3
9	12	0.57	4	31	104	0.15	2	0.08	0.25	0.06	0.7	0.02	17	1
19	22	1.4	14	110	333	0.48	25	0.29	0.85	0.21	1.8	0.06	31	2
26	13	1.84	9	57	75	0.35		0.23	1.22	0.21	2.02	0.03	19	0
0	23	0.95	13	198	69	0.2		0.05	2.46	0.05	1.05	0.03	9	5
0	14	1.81	9	58	218	0.34	150	0.15	0.46	0.19	2.05	0.2	40	0
10	0	1.08			220	0								0
10	0	1.08			220	0								0
17	21	0.92	9	60	257	0.26	18	0.1	0.91	0.11	0.87	0.03	24	0
0	0	1.08			380	0								2
4	26	1.22	13	65	205	0.46	2	0.22	0.21	0.13	1.71	0.03	29	0
0	0	0.36			200	0								1
0	99	1.42	12	75	264	0.4	0	0.25	0.18	0.16	2.21	0.02	54	0
0	98	1.41	11	74	262	0.4	0	0.2	0.17	0.14	1.98	0.02	45	0
0	175	1.62	47	139	420	1.06	<1	0.2	0.27	0.09	2.25	0.11	32	0
31	38	0.97	10	66	398	0.32	20	0.12	0.75	0.14	1.05	0.05	28	0
35	40	1.8			210	0								1
35	40	1.44			190	0								0
35	40	1.8			190	0								0
0	29	0.72	21	91	78	0.42	0	0.08	0.34	0.05	0.14	0.03	7	0
0	15	0.74	24	84	74	0.51	2	0.07	0.33	0.03	0.4	0.02	6	0
0	0	3.6			5									0

Table H–1

Food Composition

(Computer code number is for West Diet Analysis program)   (For purposes of calculations, use "0" for t, <1, <.1, <.01, etc.)

Computer Code Number	Food Description	Measure	Wt (g)	H₂O (%)	Ener (kcal)	Prot (g)	Carb (g)	Dietary Fiber (g)	Fat (g)	Fat Breakdown (g)		
										Sat	Mono	Poly
	BAKED GOODS: BREADS, CAKES, COOKIES, CRACKERS, PIES—Continued											
	Pancakes, 4" diam:											
45001	Plain, from home recipe	1 ea	38	53	86	2	11	1	4	0.8	0.9	1.7
45002	Plain, from mix; egg, milk, oil added	1 ea	38	53	74	2	14	<1	1	0.2	0.3	0.3
	Piecrust, enriched flour, vegetable shortening, baked:											
45501	Home recipe, 9" shell	1 ea	180	10	949	12	86	3	62	15.5	27.3	16.4
45503	From mix, 1 pie shell	1 ea	160	11	802	11	81	3	49	12.3	27.7	6.2
48014	Pie, cherry, commercial fried	1 ea	128	38	404	4	55	3	21	3.1	9.5	6.9
44015	Pretzels, thin twists, 3¼x 2¼ x ¼"	10 pce	60	3	229	5	48	2	2	0.4	0.8	0.7
	Rolls & buns, enriched, commercial:											
42157	Cloverleaf rolls, 2½" diam, 2"high	1 ea	28	32	84	2	14	1	2	0.5	1	0.3
42021	Hot dog buns	1 ea	40	35	112	4	20	1	2	0.4	0.4	0.8
42020	Hamburger buns	1 ea	43	35	120	4	21	1	2	0.5	0.5	0.8
42022	Hard roll, white, 3¾" diam, 2"high	1 ea	57	31	167	6	30	1	2	0.3	0.6	1
42034	Submarine rolls/hoagies, 11¼ x 3 x 2½"	1 ea	135	34	386	11	68	4	7	1.6	3.4	1.2
	Sports/fitness bar:											
	Clif Bars:											
4042	Chocolate brownie	1 ea	68	17	236	10	41	6	4	1		
62709	Chocolate chip	1 ea	68	15	238	10	42	5	4	1		
62711	Chocolate chip peanut	1 ea	68	15	241	12	39	5	5	1.1		
62716	Cookies and cream	1 ea	68	20	225	10	39	5	4	1.4		
62207	Forza energy bar	1 ea	70		231	10	45	4	1			
62205	Tiger sports bar	1 ea	65		260	11	33	2	9	1.9		
	Power bars:											
62275	Apple cinnamon	1 ea	65		230	10	45	3	2	0.5	1.5	0.5
62276	Banana	1 ea	65		230	9	45	3	2	0.5	1	0.5
62823	Chocolate harvest	1 ea	65		240	7	45	4	4	1		
62279	Malt nut	1 ea	65		230	10	45	3	2	0.5	1	1
62280	Mocha	1 ea	65		230	10	45	3	2	1	1	0.5
	Tortillas:											
42023	Corn, enriched, 6" diam	1 ea	26	44	58	1	12	1	1	t	0.2	0.3
42025	Flour, 10" diam	1 ea	72	27	234	6	40	2	5	1.3	2.7	0.8
42027	Taco shells	1 ea	14	6	66	1	9	1	3	0.5	1.3	1.2
	Waffles, 7" diam:											
45003	From home recipe	1 ea	75	42	218	6	25	1	11	2.1	2.6	5.1
45017	Whole grain, prepared from frozen	1 ea	39	43	105	4	13	1	4	1.2	1.8	1.1
	GRAIN PRODUCTS: CEREAL, FLOUR, GRAIN, PASTA and NOODLES, POPCORN											
	Grain:											
38070	Amaranth	1 c	195	10	729	28	129	30	13	3.2	2.8	5.6
	Barley:											
38002	Pearled, dry, uncooked	1 c	200	10	704	20	155	31	2	0.5	0.3	1.1
38003	Pearled, cooked	1 c	157	69	193	4	44	6	1	0.1	t	0.3
38072	Buckwheat, whole grain, dry	1 c	170	10	583	23	122	17	6	1.3	1.8	1.8
38027	Bulgar, dry, uncooked	1 c	140	9	479	17	106	26	2	0.3	0.2	0.8
38028	Bulgar, cooked	1 c	182	78	151	6	34	8	<1	t	t	0.2
38076	Couscous, cooked	1 c	157	73	176	6	36	2	<1	t	t	0.1
38329	Cracked wheat	1 c	120	10	407	16	87	14	2	0.4	0.3	0.9
38052	Millet, cooked	1 c	240	71	286	8	57	3	2	0.4	0.4	1.2
38064	Oat bran, dry	1/4 c	24	7	59	4	16	4	2	0.3	0.6	0.7
38079	Quinoa, dry	1 c	170	9	636	22	117	10	10	1	2.6	4
	Rice:											
38010	Brown, cooked	1 c	195	73	216	5	45	4	2	0.4	0.6	0.6
56998	Spanish, cooked	1 c	246		130	3	28	2	1	0		
	White, enriched, all types:											
38012	Regular/long grain, dry	1 c	185	12	675	13	148	2	1	0.3	0.4	0.3
38013	Regular/long grain, cooked	1 c	158	68	205	4	45	1	<1	0.1	0.1	0.1
38019	Instant, prepared without salt	1 c	165	76	162	3	35	1	<1	t	t	t
	Parboiled/converted:											
38015	Raw, dry	1 c	185	10	686	13	151	3	1	0.3	0.3	0.3

PAGE KEY: H–2 = Beverages H–4 = Dairy H–8 = Eggs H–10 = Fat/Oil H–12 = Fruit H–18 = Bakery
H–24 = Grain H–30 = Fish H–30 = Meats H–34 = Poultry H–36 = Sausage H–38 = Mixed/Fast H–42 = Nuts/Seeds H–44 = Sweets
H–46 = Vegetables/Legumes H–56 = Vegetarian H–60 = Misc H–60 = Soups/Sauces H–62 = Fast H–76 = Convenience H–78 = Baby foods

Chol (mg)	Calc (mg)	Iron (mg)	Magn (mg)	Pota (mg)	Sodi (mg)	Zinc (mg)	VT-A (RAE)	Thia (mg)	VT-E (mg α)	Ribo (mg)	Niac (mg)	V-B6 (mg)	Fola (µg)	VT-C (mg)
22	83	0.68	6	50	167	0.21	21	0.08	0.36	0.11	0.6	0.02	14	0
5	48	0.59	8	66	239	0.15	4	0.08	0.32	0.08	0.65	0.03	14	0
0	18	5.2	25	121	976	0.79	0	0.7	0.56	0.5	5.95	0.04	121	0
0	96	3.44	24	99	1166	0.62	0	0.48	8.83	0.3	3.8	0.09	112	0
0	28	1.56	13	83	479	0.29	12	0.18	0.55	0.14	1.82	0.04	23	2
0	22	2.59	21	88	1029	0.51	0	0.28	0.21	0.37	3.15	0.07	103	0
0	33	0.88	6	37	146	0.22	<1	0.14	0.09	0.09	1.13	0.02	27	0
0	55	1.33	8	38	192	0.26	0	0.16	0.03	0.13	1.66	0.03	44	0
0	59	1.43	9	40	206	0.28	0	0.17	0.03	0.14	1.79	0.03	48	0
0	54	1.87	15	62	310	0.54	0	0.27	0.24	0.19	2.42	0.02	54	0
0	188	4.28	27	190	756	0.84	0	0.65	0.62	0.42	5.31	0.06	36	0
0	271	5.75	122	257	149	3.65		0.39	20.18	0.29	3.54	0.42	84	66
0	265	5.22	96	206	76	3.45		0.35	20.26	0.28	3.49	0.39	86	66
0	265	5.37	111	305	274	3.51		0.4	20.34	0.3	6.09	0.41	95	66
0	279	5.23	103	212	179	3.21		0.35	20.26	0.27	3.45	0.43	85	66
0	300	6.3	160	220	65	5.25		1.5	18.35	1.7	20	2	400	60
0	557	5.01	186		139			2.37		1.1	5.57	1.11		11
0	300	6.3	140	110	90	5.25	0	1.5	18.35	1.7	20	2	400	60
0	300	6.3	140	200	90	5.25	0	1.5	18.35	1.7	20	2	400	60
0	150	2.7	60		80	2.25	0	0.75	18.34	0.85	10	1	200	60
0	300	6.3	140	110	90	5.25	0	1.5	18.35	1.7	20	2	400	60
0	300	6.3	140	145	90	5.25	0	1.5	18.35	1.7	20	2	400	60
046		0.36	17	40	42	0.24	<1	0.03	0.02	0.02	0.39	0.06	26	0
090		2.38	19	94	344	0.51	0	0.38	0.4	0.21	2.57	0.04	75	0
022		0.35	15	25	51	0.2	5	0.03	0.42	<.01	0.19	0.05	1	0
52	191	1.73	14	119	383	0.51	49	0.2	1.72	0.26	1.55	0.04	34	0
37	102	0.81	16	90	132	0.44		0.08	0.55	0.13	0.77	0.04	7	0
0	298	14.8	519	714	41	6.2	0	0.16	2.01	0.41	2.51	0.43	96	8
0	58	5	158	560	18	4.26	2	0.38	0.04	0.23	9.21	0.52	46	0
0	17	2.09	35	146	5	1.29	1	0.13	0.02	0.1	3.24	0.18	25	0
0	31	3.74	393	782	2	4.08	0	0.17	1.75	0.72	11.93	0.36	51	0
0	49	3.44	230	574	24	2.7	1	0.32	0.08	0.16	7.16	0.48	38	0
0	18	1.75	58	124	9	1.04	<1	0.1	0.02	0.05	1.82	0.15	33	0
0	13	0.6	13	91	8	0.41	0	0.1	0.2	0.04	1.54	0.08	24	0
0	41	4.67	166	486	6	3.53	0	0.54	0.3	0.26	7.64	0.41	53	0
0	7	1.51	106	149	5	2.18	<1	0.25	0.05	0.2	3.19	0.26	46	0
0	14	1.3	56	136	1	0.75	0	0.28	0.24	0.05	0.22	0.04	12	0
0	102	15.72	357	1258	36	5.61	0	0.34	8.28	0.67	4.98	0.38	83	0
0	20	0.82	84	84	10	1.23	0	0.19	0.06	0.05	2.98	0.28	8	0
0	0	0			1340		0							0
0	52	7.97	46	213	9	2.02	0	1.07	0.2	0.09	7.76	0.3	427	0
0	16	1.9	19	55	2	0.77	0	0.26	0.06	0.02	2.33	0.15	92	0
0	13	1.04	8	7	5	0.4	0	0.12	0.02	0.08	1.45	0.02	116	0
0	111	6.59	57	222	9	1.78	0	1.1	0.06	0.13	6.72	0.65	475	0

Table H–1

Food Composition   (Computer code number is for West Diet Analysis program)   (For purposes of calculations, use "0" for t, <1, <.1, <.01, etc.)

Computer Code Number	Food Description	Measure	Wt (g)	H₂O (%)	Ener (kcal)	Prot (g)	Carb (g)	Dietary Fiber (g)	Fat (g)	Fat Breakdown (g)		
										Sat	Mono	Poly
	GRAIN PRODUCTS: CEREAL, FLOUR, GRAIN, PASTA and NOODLES, POPCORN—Continued											
38016	Cooked	1 c	175	72	200	4	43	1	<1	0.1	0.1	0.1
38083	Sticky (Glutinous), cooked	1 c	174	77	169	4	37	2	<1	t	0.1	0.1
38021	Wild, cooked	1 c	164	74	166	7	35	3	1	t	t	0.3
38163	Rice and pasta (Rice-a-Roni), cooked	1 c	202	72	246	5	43	5	6	1.1	2.3	1.9
38034	Tapioca-pearl, dry	1 c	152	11	544	<1	135	1	<1	t	t	t
40002	Wheat, rolled, cooked	1 c	240	84	158	5	33	4	1	0.1	0.1	0.5
	Flour & Grain Fractions:											
38053	Buckwheat flour, dark	1 c	120	11	402	15	85	12	4	0.8	1.1	1.1
	Cornmeal:											
38059	Whole-ground, unbolted, dry	1 c	122	10	442	10	94	9	4	0.6	1.2	2
38004	Degermed, enriched, dry	1 c	138	12	505	12	107	10	2	0.3	0.6	1
38041	Degermed, enriched, baked	1 c	138	12	505	12	107	10	2	0.3	0.6	1
38056	Rye flour, medium	1 c	102	10	361	10	79	15	2	0.2	0.2	0.8
7502	Soy flour, low-fat	1 c	88	3	325	45	30	9	6	0.9	1.3	3.3
38024	Wheat bran, crude	1 c	58	10	125	9	37	25	2	0.4	0.4	1.3
38025	Wheat germ, raw	1 c	115	11	414	27	60	15	11	1.9	1.6	6.9
38026	Wheat germ, toasted	1 c	113	6	432	33	56	17	12	2.1	1.7	7.5
38055	Wheat germ, with brown sugar & honey	1 c	113	3	420	30	66	12	9	1.5	1.2	5.5
	Wheat flour:											
38030	All-purpose white flour, enriched	1 c	125	12	455	13	95	3	1	0.2	0.1	0.5
38033	Self-rising, enriched, unsifted	1 c	125	11	442	12	93	3	1	0.2	0.1	0.5
38032	Whole wheat, from hard wheats	1 c	120	10	407	16	87	15	2	0.4	0.3	0.9
	Breakfast Bars:											
47279	Store brand, fat free, all flavors	1 ea	38		110	2	26	3	0	0	0	0
	Snackwell:											
40255	Apple-cinnamon	1 ea	37	16	119	1	29	1	<1	t	t	0.1
40254	Blueberry	1 ea	37	16	121	1	29	1	<1	t	t	0.1
40253	Strawberry	1 ea	37	16	120	1	29	1	<1	t	t	0.1
	Breakfast cereals, hot, cooked: w/o salt added											
	Corn grits (hominy) enriched:											
38007	Regular/quick prep w/o salt, yellow:	1 c	242	85	143	3	31	1	<1	t	0.1	0.2
40089	Instant, prepared from packet, white	1 ea	137	82	93	2	21	1	<1	t	t	t
	Cream of wheat:											
40079	Regular, quick, instant	1 c	239	87	129	4	27	1	<1	t	t	0.3
40087	Mix and eat, plain, packet	1 ea	142	82	102	3	21	<1	<1	t	t	0.2
40006	Farina cereal, cooked w/o salt	1 c	233	88	112	3	24	1	<1	t	t	t
40014	Malt-O-Meal, cooked w/o salt	1 c	240	88	122	4	26	1	<1	0.1	t	t
40239	Maypo	1 c	216	83	153	5	29	4	2	0.4	0.6	0.5
	Oatmeal or rolled oats:											
40000	Cooked w/o salt, nonfortified	1 c	234	85	147	6	25	4	2	0.4	0.7	0.9
492	Plain, from packet, fortified	½ c	118	86	65	3	11	2	1	0.2	0.3	0.4
	Breakfast cereals, ready to eat:											
40095	All-Bran	1 c	62	3	161	8	46	20	2	0.3	0.5	1.3
40258	Alpha Bits	1 c	28	1	110	2	24	1	1	0.1	0.2	0.2
40098	Apple Jacks	1 c	33	3	129	1	30	1	1	0.1	0.2	0.3
40029	Bran Buds	1 c	90	3	225	6	72	39	2	0.4	0.4	1.1
40323	Bran Chex	1 c	49	2	156	5	39	8	1	0.2	0.3	0.7
40084	Honey BucWheat Crisp	1 c	38	5	147	4	31	3	1	0.2	0.2	0.5
40031	C.W. Post, with raisins	1 c	103	4	446	9	74	14	15	11	1.7	1.4
40032	Cap'n Crunch	1 c	37	2	148	2	31	1	2	0.6	0.4	0.3
40033	Cap'n Crunchberries	1 c	35	2	140	2	30	1	2	0.5	0.4	0.3
40034	Cap'n Crunch, peanut butter	1 c	35	2	146	2	28	1	3	0.7	1.4	0.8
40297	Cheerios	1 c	23	3	85	3	17	2	1	0.3	0.5	0.2
40102	Cocoa Krispies	1 c	41	3	156	1	36	1	1	0.8	0.2	t
40257	Cocoa Pebbles	1 c	32	3	127	1	28	1	1	1.2	0.1	t
40036	Corn Bran	1 c	36	2	121	2	31	6	1	0.3	0.3	0.4
40325	Corn Chex	1 c	28	2	104	2	24	1	<1	t	t	t
40195	Corn Flakes, Kellogg's	1 c	28	3	101	2	24	1	<1	t	t	t

PAGE KEY: H–2 = Beverages H–4 = Dairy H–8 = Eggs H–10 = Fat/Oil H–12 = Fruit H–18 = Bakery
H–24 = Grain H–30 = Fish H–30 = Meats H–34 = Poultry H–36 = Sausage H–38 = Mixed/Fast H–42 = Nuts/Seeds H–44 = Sweets
H–46 = Vegetables/Legumes H–56 = Vegetarian H–60 = Misc H–60 = Soups/Sauces H–62 = Fast H–76 = Convenience H–78 = Baby foods

Chol (mg)	Calc (mg)	Iron (mg)	Magn (mg)	Pota (mg)	Sodi (mg)	Zinc (mg)	VT-A (RAE)	Thia (mg)	VT-E (mg α)	Ribo (mg)	Niac (mg)	V-B₆ (mg)	Fola (µg)	VT-C (mg)
0	33	1.98	21	65	5	0.54	0	0.44	0.02	0.03	2.45	0.03	133	0
0	3	0.24	9	17	9	0.71	0	0.03	0.07	0.02	0.5	0.05	2	0
0	5	0.98	52	166	5	2.2	<1	0.09	0.39	0.14	2.11	0.22	43	0
2	16	1.9	24	85	1147	0.57	0	0.25	0.27	0.16	3.6	0.2	89	0
0	30	2.4	2	17	2	0.18	0	<.01	0	0	0	0.01	6	0
0	22	1.49	55	170	0	1.18	<1	0.17	0.58	0.13	2.13	0.17	34	0
0	49	4.87	301	692	13	3.74	0	0.5	0.38	0.23	7.38	0.7	65	0
0	7	4.21	155	350	43	2.22	13	0.47	0.51	0.25	4.43	0.37	30	0
0	7	5.7	55	224	4	0.99	15	0.99	0.21	0.56	6.95	0.35	322	0
0	7	5.7	55	224	4	0.99	28	0.74	0.23	0.48	6.6	0.27	52	0
0	24	2.16	76	347	3	2.03	0	0.29	0.81	0.12	1.76	0.27	19	0
0	165	5.27	202	2262	16	1.04	2	0.33	0.17	0.25	1.9	0.46	361	0
0	42	6.13	354	686	1	4.22	<1	0.3	0.86	0.33	7.88	0.76	46	0
0	45	7.2	275	1026	14	14.13	0	2.16	16.1	0.57	7.83	1.5	323	0
0	51	10.27	362	1070	5	18.84	6	1.89	18.07	0.93	6.32	1.11	398	7
0	56	9.1	307	1089	12	15.68	0	1.51	34.07	0.78	5.34	0.56	685	0
0	19	5.8	28	134	2	0.88	0	0.98	0.08	0.62	7.38	0.06	229	0
0	422	5.84	24	155	1588	0.78	0	0.84	0.06	0.52	7.29	0.06	245	0
0	41	4.66	166	486	6	3.52	1	0.54	0.98	0.26	7.64	0.41	53	0
0	20	0.72			25									1
0	17	5	6	68	103	3.88		0.39		0.44	5.2	0.52		0
0	14	4.83	5	44	107	3.85		0.39		0.44	5.2	0.52		0
0	14	4.82	6	47	102	3.83		0.39		0.44	5.2	0.52		2
0	7	1.45	12	51	5	0.17	5	0.2	0.05	0.13	1.75	0.05	80	0
0	8	7.96	10	38	288	0.18	<1	0.16	0.03	0.19	2.21	0.05	47	0
0	50	10.28	12	45	139	0.33	0	0.24	0.02	0	1.43	0.03	108	0
0	20	8.09	7	38	241	0.24	376	0.43	0.01	0.28	4.97	0.57	101	0
0	9	1.16	5	30	5	0.19	0	0.14	0.02	0.1	1.14	0.02	79	0
0	5	9.6	5	31	2	0.17	0	0.48	0.02	0.24	5.76	0.02	5	0
0	117	7.54	48	190	233	1.34	631	0.64	0.15	0.71	8.42	0.83	11	25
0	19	1.59	56	131	2	1.15	0	0.26	0.23	0.05	0.3	0.05	9	0
0	66	5.12	27	63	53	0.54	190	0.17	0.12	0.2	2.4	0.25	51	0
0	205	9.92	236	701	160	3.72	326	0.74	0.76	0.87	9.92	3.72	812	12
0	8	2.66	17	54	178	1.48		0.36	0.02	0.42	4.93	0.5	99	0
0	8	4.59	18	40	157	1.65	51	0.56	0.05	0.43	5.08	0.5	102	15
0	57	13.5	184	900	608	4.5	460	1.08	1.42	1.26	15.3	6.03	1210	18
0	29	13.99	69	216	345	6.48	5	0.64	0.56	0.26	8.62	0.88	173	26
0	54	10.86	43	142	361	0.68	914	0.9	8.99	1.03	12.06	1.88	11	36
0	50	16.38	74	261	161	1.64		1.34	0.72	1.54	18.13	1.85	364	0
0	6	7.07	21	74	277	5.87	3	0.58	0.34	0.66	7.83	0.78	576	0
0	7	6.65	20	72	245	5.54	2	0.55	0.26	0.63	7.38	0.74	545	0
0	4	6.42	24	83	260	5.35	3	0.53	0.42	0.61	7.27	0.71	545	0
0	77	6.21	31	74	209	2.88	115	0.29	0.08	0.33	3.84	0.38	153	5
0	53	6.15	16	66	251	1.97	202	0.49	0.25	0.57	6.56	0.66	135	20
0	4	1.99	12	47	173	1.65		0.41		0.47	5.52	0.55	110	0
0	26	11.1	19	75	309	5.5	3	0.18	0.24	0.62	7.34	0.73	533	0
0	93	8.4	8	23	269	3.5	141	0.35	0.05	0.4	4.68	0.47	187	6
0	2	8.4	3	25	203	0.08	150	0.36	0.04	0.43	5.01	0.5	102	6

Table H–1

Food Composition   (Computer code number is for West Diet Analysis program)   (For purposes of calculations, use "0" for t, <1, <.1, <.01, etc.)

Computer Code Number	Food Description	Measure	Wt (g)	H₂O (%)	Ener (kcal)	Prot (g)	Carb (g)	Dietary Fiber (g)	Fat (g)	Fat Breakdown (g) Sat	Mono	Poly
	GRAIN PRODUCTS: CEREAL, FLOUR, GRAIN, PASTA and NOODLES, POPCORN—Continued											
40263	Corn Flakes, Post Toasties	1 c	28	4	101	2	24	1	<1	0	t	t
40206	Corn Pops	1 c	31	3	118	1	28	<1	<1	t	t	t
40205	Cracklin' Oat Bran	1 c	65	3	266	5	46	8	9	2.7	5.4	1.4
40040	Crispy Wheat `N Raisins	1 c	43	7	143	3	35	4	1	0.2	t	0.3
40041	Fortified Oat Flakes	1 c	48	3	180	8	36	1	1	0.2	0.3	0.4
40202	40% Bran Flakes, Kellogg's	1 c	39	3	124	4	31	7	1	0.2	0.2	0.4
40259	40% Bran Flakes, Post	1 c	47	4	150	4	38	8	1	0.2	0.2	0.7
40218	Froot Loops	1 c	32	3	126	1	28	1	1	0.5	0.1	0.2
40217	Frosted Flakes	1 c	41	3	150	1	37	1	<1	t	t	0.1
40043	Frosted Mini-Wheats	1 c	55	6	187	5	44	6	1	0.2	0.1	0.6
40105	Frosted Rice Krispies	1 c	35	3	133	1	31	<1	<1	t	0.1	0.1
40266	Fruity Pebbles	1 c	32	3	128	1	28	<1	1	0.3	0.6	0.4
40299	Golden Grahams	1 c	39	3	145	2	32	1	1	0.2	0.5	0.5
40048	Granola, homemade	½ c	61	5	299	9	32	5	15	2.8	4.7	6.5
40197	Granola, low fat, commercial	½ c	45	4	165	4	36	2	2	0.7	1.1	0.4
40277	Grape Nuts	½ c	55	4	197	6	45	5	1	0.2	0.2	0.6
40265	Grape Nuts Flakes	1 c	39	3	142	4	32	3	1	0.2	0.3	0.6
40238	Heartland Natural with raisins	1 c	110	5	468	11	76	6	16	4	4.2	6.2
40112	Honey & Nut Corn Flakes	1 c	37	2	150	3	31	1	2	0.3	0.8	0.6
40051	Honey Nut Cheerios	1 c	33	2	123	3	26	2	1	0.3	0.4	0.5
40052	HoneyBran	1 c	35	2	119	3	29	4	1	0.3	t	0.3
40264	HoneyComb	1 c	22	2	87	1	20	1	<1	0.1	0.1	0.2
40054	King Vitaman	1 c	21	2	81	1	18	1	1	0.2	0.2	0.2
40010	Kix	1 c	19	2	72	1	16	1	<1	t	0.1	0.1
40011	Life	1 c	44	4	165	4	34	3	2	0.4	0.7	0.6
40300	Lucky Charms	1 c	32	2	122	2	27	2	1	0.3	0.3	0.3
40124	Mueslix Five Grain	1 c	82	8	289	6	63	6	5	0.7	2	1.8
40008	Nature Valley Granola	1 c	113	4	510	12	74	7	20	2.6	13.3	3.8
40115	Nutri Grain Almond Raisin	1 c	40	6	147	3	31	3	2	t	1	1.2
40281	100% Bran	1 c	66	3	178	8	48	20	3	0.6	0.6	1.9
40063	100% Natural cereal, plain	1 c	104	2	473	11	69	8	20	8.6	4.6	2
40064	100% Natural with apples & cinnamon	1 c	104	2	477	11	70	7	20	15.5	1.8	1.3
40065	100% Natural with raisins & dates	1 c	110	4	496	12	72	7	20	13.6	3.7	1.7
40216	Product 19	1 c	30	3	100	2	25	1	<1	t	0.1	0.2
40066	Quisp	1 c	30	2	122	1	26	1	2	0.5	0.3	0.2
40209	Raisin Bran, Kellogg's	1 c	61	8	195	5	47	7	2	0.3	0.3	0.9
512	Raisin Bran, Post	1 c	59	9	187	5	46	8	1	0.2	0.2	0.7
40117	Raisin Squares	1 c	71	10	239	7	56	7	1	0.2	0.2	0.6
40333	Rice Chex	1 c	33	3	124	2	28	<1	<1	0.1	t	t
40017	Rice Krispies, Kellogg's	1 c	28	2	111	2	25	<1	<1	t	t	t
40018	Rice, puffed	1 c	14	4	54	1	12	<1	<1	t	t	t
40022	Shredded Wheat	1 c	43	5	154	5	35	4	1	0.1	0.1	0.4
40211	Special K	1 c	31	3	117	7	22	1	<1	0.1	0.1	0.2
40261	Super Golden Crisp	1 c	33	2	123	2	30	<1	<1	t	t	0.1
40068	Honey Smacks	1 c	36	2	139	2	32	1	1	0.1	0.2	0.3
40070	Tasteeos	1 c	24	2	94	3	19	3	1	0.2	0.2	0.2
40021	Total, wheat, with added calcium	1 c	40	3	130	3	30	3	1	0.2	0.2	0.4
40306	Trix	1 c	28	2	109	1	25	1	1	0.2	0.6	0.3
40335	Wheat Chex	1 c	46	2	159	5	37	5	1	0.2	0.1	0.4
40023	Wheat cereal, puffed, fortified	1 c	12	4	44	2	9	1	<1	t	t	0.1
40307	Wheaties	1 c	29	3	103	3	23	3	1	0.2	0.3	0.3
7508	Natto	1 c	175	55	371	31	25	9	19	2.8	4.3	10.9
	Pasta:											
	Cellophane Noodles:											
38146	Cooked	1 c	190	79	160	<1	39	<1	<1	t	t	t
7196	Dry	1 c	140	13	491	<1	121	1	<1	t	t	t
38048	Chow Mein, dry	1 c	45	1	237	4	26	2	14	2	3.5	7.8
	Cooked:											
38092	Fresh	2 oz	57	69	75	3	14	1	1	t	t	0.2
38118	Linguini/Rotini	1 c	140	66	197	7	40	2	1	0.1	0.1	0.4

PAGE KEY: H–2 = Beverages   H–4 = Dairy   H–8 = Eggs   H–10 = Fat/Oil   H–12 = Fruit   H–18 = Bakery
H–24 = Grain   H–30 = Fish   H–30 = Meats   H–34 = Poultry   H–36 = Sausage   H–38 = Mixed/Fast   H–42 = Nuts/Seeds   H–44 = Sweets
H–46 = Vegetables/Legumes   H–56 = Vegetarian   H–60 = Misc   H–60 = Soups/Sauces   H–62 = Fast   H–76 = Convenience   H–78 = Baby foods

Chol (mg)	Calc (mg)	Iron (mg)	Magn (mg)	Pota (mg)	Sodi (mg)	Zinc (mg)	VT-A (RAE)	Thia (mg)	VT-E (mg α)	Ribo (mg)	Niac (mg)	V-B$_6$ (mg)	Fola (µg)	VT-C (mg)
0	1	5.4	4	33	266	0.13		0.38	0.67	0.43	5	0.5	100	0
0	5	1.92	2	26	120	1.52	151	0.37	0.03	0.43	4.99	0.5	102	6
0	27	2.4	80	292	186	2.02	298	0.5	0.91	0.57	6.7	0.66	133	21
0	0	5.85	33	178	197	5.85	117	0.58	0.27	0.67	7.83	0.78	157	0
0	68	13.73	58	228	220	2.54		0.62	0.34	0.72	8.45	0.86	169	0
0	21	24.18	55	230	279	20.48	307	2.11	36.14	2.3	26.91	2.73	542	81
0	26	12.69	101	290	344	2.35		0.59		0.67	7.83	0.78	157	0
0	25	4.51	9	35	151	1.5	155	0.38	0.15	0.42	4.99	0.51	100	15
0	2	5.94	3	30	196	0.07	212	0.49	0.02	0.62	6.64	0.66	134	8
0	18	15.95	65	187	6	1.76		0.41	0.3	0.46	5.39	0.54	108	0
0	4	2.1	9	33	254	0.32	176	0.46	0.02	0.49	5.95	0.6	234	7
0	2	2.13	6	35	187	1.78		0.44		0.5	5.93	0.59	118	0
0	455	5.85	11	64	349	4.88	195	0.49	0.14	0.55	6.51	0.65	130	8
0	48	2.59	107	328	13	2.51	1	0.45	3.59	0.18	1.29	0.19	51	1
0	19	1.35	34	135	111	2.84	169	0.28	3.78	0.32	3.74	1.48	302	3
0	19	15.36	55	169	336	1.14		0.36		0.4	4.74	0.47	95	0
0	15	10.89	40	133	188	1.61		0.5	0.1	0.57	6.72	0.67	135	0
0	66	4.01	141	415	226	2.83	3	0.32	0.77	0.14	1.54	0.2	44	1
0	4	3.03	3	40	249	0.26	152	0.26	0.09	0.3	3.37	0.33	74	10
0	110	4.95	35	101	296	4.12	165	0.41	0.59	0.47	5.51	0.55	220	7
0	16	5.56	46	150	202	0.9	463	0.46	0.81	0.52	6.16	0.63	23	19
0	4	2.05	8	26	163	1.14		0.28		0.32	3.79	0.38	76	0
0	3	6.09	18	58	176	2.62	210	0.26	1.41	0.3	3.5	0.35	280	8
0	95	5.13	5	22	169	2.38	101	0.24	0.04	0.27	3.17	0.32	127	4
0	154	12.31	42	125	226	5.68	1	0.55	0.24	0.64	7.57	0.76	572	0
0	107	4.8	17	61	217	4	160	0.4	0.1	0.45	5.34	0.53	213	6
0	67	8.94	82	369	107	7.46	747	0.75	8.94	0.84	9.84	0.99	197	1
0	85	3.53	107	375	183	2.27	0	0.35	7.97	0.12	1.25	0.16	17	0
0	122	1	9	143	142	2.72	0	0.28	4	0.32	3.64	0.36	80	0
0	46	8.12	312	652	457	5.74	0	1.58	1.53	1.78	20.92	2.11	47	63
2	124	2.68	115	515	50	2.45	1	0.33	3.34	0.28	2.09	0.19	37	1
0	157	2.89	72	514	52	2		0.33	0.73	0.57	1.87	0.11	17	1
0	160	3.12	124	538	47	2.11		0.31	0.77	0.65	2.09	0.16	45	0
0	5	18.09	16	50	207	15.3	225	1.5	20.13	1.71	20.01	2.07	400	61
0	3	5.51	16	57	222	4.59	12	0.46	0.2	0.52	6.12	0.61	467	3
0	29	4.64	83	372	362	1.55	155	0.39	0.48	0.44	5.18	0.52	104	0
0	27	10.8	88	357	360	2.25		0.38		0.42	5	0.5	100	0
0	28	19.88	56	342	4	1.99	0	0.5	0.37	0.57	6.67	0.67	134	0
0	110	9.9	10	32	311	4.12	165	0.41	0.02	0.47	5.51	0.55	220	7
0	5	0.7	12	27	206	0.46	371	0.52	0	0.59	6.92	0.69	88	15
0	1	0.4	4	16	1	0.15	0	0.06	0.02	0.04	0.49	0	22	0
0	16	1.81	57	155	4	1.42	0	0.11	0.23	0.12	2.26	0.11	22	0
0	9	8.37	19	61	224	0.9	230	0.53	7.07	0.59	7.13	1.98	400	21
0	7	2.08	20	48	51	1.75		0.43	0.12	0.5	5.81	0.59	116	0
0	9	0.47	21	54	67	0.47	204	0.5	0.18	0.58	6.66	0.68	135	8
0	11	6.86	26	71	183	0.69	318	0.31	0.08	0.36	4.22	0.43	85	13
0	1333	24	32	119	256	20	200	2	26.84	2.27	26.68	2.67	533	80
0	93	4.2	3	16	181	3.5	140	0.35	0.56	0.4	4.68	0.47	93	6
0	92	13.34	37	172	410	3.68	138	0.34	0.33	0.39	4.6	0.46	368	6
0	3	0.53	16	44	1	0.37	0	0.08	0	0.05	0.63	0.02	18	0
0	0	7.83	31	107	210	7.25	145	0.72	0.18	0.82	9.66	0.97	193	6
0	380	15.05	201	1276	12	5.3	0	0.28	0.02	0.33	0	0.23	14	23
0	13	0.86	3	3	8	0.2	0	0.04	0.06	0	0.06	0.02	1	0
0	35	3.04	4	14	14	0.57	0	0.21	0.18	0	0.28	0.07	3	0
0	9	2.13	23	54	198	0.63	<1	0.26	1.57	0.19	2.68	0.05	40	0
19	3	0.65	10	14	3	0.32	3	0.12	0.09	0.09	0.57	0.02	36	0
0	10	1.96	25	43	1	0.74	0	0.29	0.08	0.14	2.34	0.05	108	0

Table H–1

Food Composition

(Computer code number is for West Diet Analysis program)   (For purposes of calculations, use "0" for t, <1, <.1, <.01, etc.)

Computer Code Number	Food Description	Measure	Wt (g)	H₂O (%)	Ener (kcal)	Prot (g)	Carb (g)	Dietary Fiber (g)	Fat (g)	Fat Breakdown (g) Sat	Mono	Poly
	GRAIN PRODUCTS: CEREAL, FLOUR, GRAIN, PASTA and NOODLES, POPCORN—Continued											
38047	Egg noodles, cooked, enriched	1 c	160	69	213	8	40	2	2	0.5	0.7	0.7
	Macaroni, cooked:											
38102	Enriched	1 c	140	66	197	7	40	2	1	0.1	0.1	0.4
38110	Whole wheat	1 c	140	67	174	7	37	4	1	0.1	0.1	0.3
38117	Vegetable, enriched	1 c	134	68	172	6	36	6	<1	t	t	t
	Spaghetti:											
38121	With salt, enriched	1 c	140	66	197	7	40	2	1	0.1	0.1	0.4
38060	Whole-wheat, cooked	1 c	140	67	174	7	37	6	1	0.1	0.1	0.3
38062	Spinach noodles, dry	3½ oz	100	8	372	13	75	11	2	0.2	0.2	0.6
	Popcorn:											
44012	Air popped, plain	1 c	8	4	31	1	6	1	<1	t	t	0.2
44013	Popped in vegetable oil/salted	1 c	11	3	55	1	6	1	3	0.5	0.9	1.5
44014	Sugar-syrup coated	1 c	35	3	151	1	28	2	4	1.3	1	1.6
	MEATS: FISH AND SHELLFISH											
	Fish:											
17029	Bass, baked or broiled	4 oz	113	69	165	27	0	0	5	1.1	2.1	1.5
17031	Bluefish, baked or broiled	4 oz	113	63	180	29	0	0	6	1.3	2.6	1.5
17126	Catfish, breaded/flour fried	4 oz	113	49	329	21	14	<1	20	4.5	9.2	5.2
	Cod:											
17037	Baked	4 oz	113	76	119	26	0	0	1	0.2	0.1	0.3
17000	Batter fried	4 oz	113	67	197	20	8	<1	9	1.8	3.6	3
17001	Poached, no added fat	4 oz	113	77	116	25	0	0	1	0.1	0.1	0.3
17002	Fish sticks, breaded pollock	2 ea	56	46	152	9	13	1	7	1.8	2.8	1.8
17068	Flounder/sole, baked	4 oz	113	73	132	27	0	0	2	0.4	0.3	0.7
17071	Grouper, baked or broiled	4 oz	113	73	133	28	0	0	1	0.3	0.3	0.5
17007	Haddock, breaded, fried	4 oz	113	60	247	23	10	<1	12	2.6	5.3	3.7
	Halibut:											
17291	Baked	4 oz	113	72	158	30	0	0	3	0.5	1.1	1.1
17259	Smoked	4 oz	113		203	34	0	0	4	0.6	1.2	1.5
17044	Raw	4 oz	113	78	124	24	0	0	3	0.4	0.8	0.8
17012	Herring, pickled	4 oz	113	55	296	16	11	0	20	2.7	13.5	1.9
	Ocean Perch:											
17093	Baked/Broiled	4 oz	113	73	137	27	0	0	2	0.4	0.9	0.6
17015	Breaded/Fried	4 oz	113	58	255	23	10	<1	13	2.7	5.8	3.9
19025	Octopus, raw	4 oz	113	80	93	17	2	0	1	0.3	0.2	0.3
17096	Pollock, baked, broiled, or poached	4 oz	113	74	128	27	0	0	1	0.3	0.2	0.6
17099	Salmon, broiled or baked	4 oz	113	62	244	31	0	0	12	2.2	6	2.7
17060	Sardines, Atlantic, canned, drained, 2=24g	4 oz	113	60	235	28	0	0	13	1.7	4.4	5.8
17022	Snapper, baked or broiled	4 oz	113	70	145	30	0	0	2	0.4	0.4	0.7
19047	Squid, fried in flour	4 oz	113	65	198	20	9	0	8	2.1	3.1	2.4
17080	Surimi	4 oz	113	76	112	17	8	0	1	0.2	0.2	0.5
17065	Swordfish, raw	4 oz	113	76	137	22	0	0	5	1.2	1.7	1
17066	Swordfish, baked or broiled	4 oz	113	69	175	29	0	0	6	1.6	2.2	1.3
17082	Trout, baked or broiled	4 oz	113	70	170	26	0	0	7	1.8	2	2.1
	Tuna:											
	Light, canned, drained solids:											
17025	Oil pack	1 c	145	60	287	42	0	0	12	2.2	4.3	4.2
17027	Water pack	1 c	154	75	179	39	0	0	1	0.4	0.2	0.5
17085	Bluefin, fresh	4 oz	113	68	163	26	0	0	6	1.4	1.8	1.6
	Shellfish:											
	Clams:											
19128	Raw meat only	1 ea	145	82	107	19	4	0	1	0.1	0.1	0.4
19002	Canned, drained	1 c	160	64	237	41	8	0	3	0.3	0.3	0.9
19000	Steamed, meat only	10 ea	95	64	141	24	5	0	2	0.2	0.2	0.5
	Crab, meat only:											
19033	Blue crab, cooked	1 c	118	77	120	24	0	0	2	0.3	0.3	0.8
19037	Imitation, from surimi	4 oz	113	74	115	14	12	0	1	0.3	0.2	0.8
19006	Lobster meat, cooked w/moist heat	1 c	145	76	142	30	2	0	1	0.2	0.2	0.1

PAGE KEY: H–2 = Beverages H–4 = Dairy H–8 = Eggs H–10 = Fat/Oil H–12 = Fruit H–18 = Bakery
H–24 = Grain H–30 = Fish H–30 = Meats H–34 = Poultry H–36 = Sausage H–38 = Mixed/Fast H–42 = Nuts/Seeds H–44 = Sweets
H–46 = Vegetables/Legumes H–56 = Vegetarian H–60 = Misc H–60 = Soups/Sauces H–62 = Fast H–76 = Convenience H–78 = Baby foods

Chol (mg)	Calc (mg)	Iron (mg)	Magn (mg)	Pota (mg)	Sodi (mg)	Zinc (mg)	VT-A (RAE)	Thia (mg)	VT-E (mg α)	Ribo (mg)	Niac (mg)	V-B$_6$ (mg)	Fola (μg)	VT-C (mg)
53	19	2.54	30	45	11	0.99	10	0.3	0.26	0.13	2.38	0.06	102	0
0	10	1.96	25	43	1	0.74	0	0.29	0.08	0.14	2.34	0.05	108	0
0	21	1.48	42	62	4	1.13	<1	0.15	0.42	0.06	0.99	0.11	7	0
0	15	0.66	25	42	8	0.59	7	0.15	0.12	0.08	1.44	0.03	87	0
0	10	1.96	25	43	140	0.74	0	0.29	0.08	0.14	2.34	0.05	108	0
0	21	1.48	42	62	4	1.13	<1	0.15	0.42	0.06	0.99	0.11	7	0
0	58	2.13	174	376	36	2.76	23	0.37	0.64	0.2	4.55	0.32	48	0
0	1	0.21	10	24	<1	0.28	1	0.02	0.02	0.02	0.16	0.02	2	0
0	1	0.31	12	25	97	0.29	1	0.01	0.55	0.01	0.17	0.02	2	0
2	15	0.61	12	38	72	0.2	1	0.02	0.42	0.02	0.77	<.01	2	0
98	116	2.16	43	515	102	0.94	40	0.1	0.84	0.1	1.72	0.16	19	2
86	10	0.7	47	539	87	1.18	156	0.08	0.71	0.11	8.19	0.52	2	0
91	62	1.88	36	391	240	1.17		0.46	2.87	0.21	3.78	0.22	17	1
62	16	0.55	47	276	88	0.66	16	0.1	0.92	0.09	2.84	0.32	9	1
56	33	0.81	28	437	104	0.57		0.08	1.49	0.11	2.58	0.37	10	2
52	10	0.37	31	484	90	0.56	9	0.02	0.33	0.05	2.45	0.45	7	3
63	11	0.41	14	146	326	0.37	17	0.07	0.29	0.1	1.19	0.03	24	0
77	20	0.38	66	389	119	0.71	15	0.09	0.75	0.13	2.46	0.27	10	0
53	24	1.29	42	537	60	0.58	56	0.09	0.71	<.01	0.43	0.4	11	0
88	70	2.02	49	377	194	0.63		0.11	1.93	0.12	4.94	0.31	16	0
46	68	1.21	121	651	78	0.6	61	0.08	1.23	0.1	8.05	0.45	16	0
59	87	1.56	154	833		0.78	86	0.11	1.11	0.14	10.83	0.64	22	0
36	53	0.95	94	508	61	0.47	53	0.07	0.96	0.08	6.61	0.39	14	0
15	87	1.38	9	78	983	0.6	292	0.04	1.93	0.16	3.73	0.19	2	0
61	155	1.33	44	396	108	0.69	16	0.15	1.84	0.15	2.75	0.31	11	1
71	151	1.88	39	335	201	0.75		0.18	2.87	0.2	2.98	0.24	13	1
54	60	5.99	34	396	260	1.9	51	0.03	1.36	0.05	2.37	0.41	18	6
108	7	0.32	82	437	131	0.68	28	0.08	0.89	0.09	1.86	0.08	5	0
98	8	0.62	35	424	75	0.58	71	0.24	1.42	0.19	7.54	0.25	6	0
160	432	3.3	44	449	571	1.48	36	0.09	2.31	0.26	5.93	0.19	14	0
53	45	0.27	42	590	64	0.5	40	0.06	0.71	<.01	0.39	0.52	7	2
294	44	1.14	43	315	346	1.97	12	0.06	2.09	0.52	2.94	0.07	16	5
34	10	0.29	49	127	162	0.37	23	0.02	0.71	0.02	0.25	0.03	2	0
44	5	0.92	31	325	102	1.3	41	0.04	0.56	0.11	10.94	0.37	2	1
56	7	1.18	38	417	130	1.66	46	0.05	0.71	0.13	13.32	0.43	2	1
78	97	0.43	35	506	63	0.58	17	0.17	0.57	0.11	6.52	0.39	21	2
26	19	2.02	45	300	513	1.3	33	0.06	1.26	0.17	17.98	0.16	7	0
46	17	2.36	42	365	521	1.19	26	0.05	0.51	0.11	20.45	0.54	6	0
43	9	1.15	56	285	44	0.68	740	0.27	1.13	0.28	9.78	0.51	2	0
49	67	20.27	13	455	81	1.99	130	0.12	0.45	0.31	2.56	0.09	23	19
107	147	44.74	29	1005	179	4.37	290	0.24	0.99	0.68	5.37	0.18	46	35
64	87	26.56	17	597	106	2.59	162	0.14	1.86	0.4	3.19	0.1	28	21
118	123	1.07	39	382	329	4.98	2	0.12	2.17	0.06	3.89	0.21	60	4
23	15	0.44	49	102	950	0.37	23	0.04	0.11	0.03	0.2	0.03	2	0
104	88	0.57	51	510	551	4.23	38	0.01	1.45	0.1	1.55	0.11	16	0

Table H–1

Food Composition   (Computer code number is for West Diet Analysis program)   (For purposes of calculations, use "0" for t, <1, <.1, <.01, etc.)

Computer Code Number	Food Description	Measure	Wt (g)	H₂O (%)	Ener (kcal)	Prot (g)	Carb (g)	Dietary Fiber (g)	Fat (g)	Fat Breakdown (g) Sat	Mono	Poly
	MEATS: FISH AND SHELLFISH—Continued											
	Oysters:											
	Raw:											
19026	Eastern	1 c	248	85	169	17	10	0	6	1.9	0.8	2.4
578	Pacific	1 c	248	82	201	23	12	0	6	1.3	0.9	2.2
	Cooked:											
19009	Eastern, breaded, fried, medium	5 ea	73	65	144	6	8	<1	9	2.3	3.4	2.4
19008	Western, simmered	5 ea	125	64	204	24	12	0	6	1.3	1	2.2
	Scallops:											
19030	Breaded, cooked from frozen	6 ea	93	58	200	17	9	<1	10	2.5	4.2	2.7
19046	Imitation, from surimi	4 oz	113	74	112	14	12	0	<1	t	t	0.2
	Shrimp:											
19012	Cooked, boiled, 2 large=11g	16 ea	88	77	87	18	0	0	1	0.3	0.2	0.4
19016	Canned, drained	1/2 c	64	73	77	15	1	0	1	0.2	0.2	0.5
19014	Fried, 2 large=15g,breaded	12 ea	90	53	218	19	10	<1	11	1.9	3.4	4.6
19032	Raw, large, about 7g each	14 ea	98	76	104	20	1	0	2	0.3	0.2	0.7
19039	Imitation, from surimi	4 oz	113	75	114	14	10	0	2	0.3	0.2	0.8
	MEATS: BEEF, LAMB, AND PORK											
	Beef:											
10008	Corned, canned	4 oz	113	58	282	31	0	0	17	7	6.7	0.7
10009	Dried, cured	1 oz	28	54	43	9	1	0	1	0.3	0.2	t
	Ground beef, broiled, patty:											
10030	Extra lean, about 16% fat	4 oz	113	54	299	32	0	0	18	7	7.8	0.7
10032	Lean, 21% fat	4 oz	113	53	316	32	0	0	20	7.8	8.7	0.7
10002	Rib, whole, rstd, choice, ¼" trim	4 oz	113	46	425	25	0	0	35	14.2	15.2	1.3
	Roast:											
10001	Blade, chuck, lean, brsd, choice, ¼" trim	4 oz	113	55	297	35	0	0	16	6.3	7	0.5
10016	Bottom round, brsd, ¼" trim	4 oz	113	52	311	32	0	0	19	7.2	8.3	0.7
10014	Bottom round, lean, brsd, choice, ¼" trim	4 oz	113	57	249	36	0	0	11	3.6	4.7	0.4
10000	Pot, chuck arm, brsd, choice, ¼" trim	4 oz	113	47	393	30	0	0	29	11.5	12.5	1.1
10017	Round eye, rstd, choice, ¼" trim	4 oz	113	59	272	30	0	0	16	6.2	6.8	0.6
10013	Round eye, lean, rstd, choice, ¼" trim	4 oz	113	65	198	33	0	0	6	2.3	2.7	0.2
	Steak:											
10056	Rib, lean, brld, ¼" trim	4 oz	113	58	250	32	0	0	13	5.1	5.3	0.4
10005	Top sirloin, lean, brld, choice, ¼" trim	4 oz	113	62	228	34	0	0	9	3.5	3.9	0.4
10006	T-bone, brld, choice, ¼" trim	4 oz	113	51	364	26	0	0	28	11	12.7	1
10007	T-bone, lean, brld, choice, ¼" trim	4 oz	113	61	232	30	0	0	11	4.1	5.6	0.3
	Variety meats:											
10010	Liver, fried	4 oz	113	62	198	30	6	0	5	1.7	0.7	0.7
10011	Tongue, cooked	4 oz	113	58	314	22	0	0	25	9.2	11.4	0.7
	Lamb:											
	Chop:											
13508	Shoulder arm, brsd, choice, ¼" trim	1 ea	70	44	242	21	0	0	17	6.9	7.1	1.2
13509	Shoulder arm, lean, brsd, ¼" trim	1 ea	55	49	153	20	0	0	8	2.8	3.4	0.5
13512	Loin, brld, choice, ¼" trim	1 ea	64	52	202	16	0	0	15	6.3	6.2	1.1
13513	Loin chop, lean, brld, choice, ¼" trim	1 ea	46	61	99	14	0	0	4	1.6	2	0.3
13517	Cutlet, avg of lean cuts, cooked	4 oz	113	54	330	28	0	0	23	9.9	9.8	1.7
	Leg:											
13500	Whole, rstd, choice,¼" trim	4 oz	113	57	292	29	0	0	19	7.8	7.9	1.3
13501	Whole, lean, rstd, choice, ¼" trim	4 oz	113	64	216	32	0	0	9	3.1	3.8	0.6
	Rib:											
13618	Rstd, choice, ¼" trim	4 oz	113	48	406	24	0	0	34	14.4	14.1	2.5
13511	Lean, rstd, choice, ¼" trim	4 oz	113	60	262	30	0	0	15	5.4	6.6	1
	Shoulder:											
13502	Whole, rstd, choice, ¼" trim	4 oz	113	56	312	25	0	0	23	9.5	9.2	1.8
13503	Whole, lean, rstd, choice, ¼" trim	4 oz	113	63	231	28	0	0	12	4.6	4.9	1.1

PAGE KEY: H–2 = Beverages H–4 = Dairy H–8 = Eggs H–10 = Fat/Oil H–12 = Fruit H–18 = Bakery
H–24 = Grain H–30 = Fish H–30 = Meats H–34 = Poultry H–36 = Sausage H–38 = Mixed/Fast H–42 = Nuts/Seeds H–44 = Sweets
H–46 = Vegetables/Legumes H–56 = Vegetarian H–60 = Misc H–60 = Soups/Sauces H–62 = Fast H–76 = Convenience H–78 = Baby foods

Chol (mg)	Calc (mg)	Iron (mg)	Magn (mg)	Pota (mg)	Sodi (mg)	Zinc (mg)	VT-A (RAE)	Thia (mg)	VT-E (mg α)	Ribo (mg)	Niac (mg)	V-B$_6$ (mg)	Fola (µg)	VT-C (mg)
131	112	16.52	117	387	523	225.21	74	0.25	2.11	0.24	3.42	0.15	25	9
124	20	12.67	55	417	263	41.22	201	0.17	2.11	0.58	4.98	0.12	25	20
59	45	5.07	42	178	304	63.6	66	0.11	1.66	0.15	1.2	0.05	23	3
125	20	11.5	55	378	265	41.55	182	0.16	1.06	0.55	4.52	0.11	19	16
57	39	0.76	55	310	432	0.99	21	0.04	1.77	0.1	1.4	0.13	34	2
25	9	0.35	49	116	898	0.37	23	0.01	0.12	0.02	0.35	0.03	2	0
172	34	2.72	30	160	197	1.37	60	0.03	1.21	0.03	2.28	0.11	4	2
111	38	1.75	26	134	108	0.81	12	0.02	0.6	0.02	1.76	0.07	1	1
159	60	1.13	36	202	310	1.24	51	0.12	1.35	0.12	2.76	0.09	16	1
149	51	2.36	36	181	145	1.09	53	0.03	1.08	0.03	2.5	0.1	3	2
41	21	0.68	49	101	797	0.37	23	0.03	0.12	0.04	0.19	0.03	2	0
97	14	2.35	16	154	1137	4.03	0	0.02	0.17	0.17	2.75	0.15	10	0
22	1	0.81	6	81	781	1.11	0	0.02	0	0.06	0.93	0.07	2	0
112	10	3.13	28	417	93	7.27	0	0.08	0.2	0.36	6.61	0.36	12	0
114	14	2.77	27	394	101	7.01	0	0.07	0.23	0.27	6.75	0.34	12	0
96	12	2.61	21	334	71	5.92	0	0.08	0.27	0.19	3.8	0.26	8	0
120	15	4.16	26	297	80	11.61	0	0.09	0.16	0.32	3.02	0.33	7	0
108	7	3.53	25	319	56	5.55	0	0.08	0.21	0.27	4.21	0.37	11	0
108	6	3.91	28	348	58	6.19	0	0.08	0.2	0.29	4.61	0.41	12	0
112	11	3.45	21	275	67	7.57	0	0.08	0.26	0.27	3.54	0.32	10	0
81	7	2.07	27	406	67	4.87	0	0.09	0.23	0.18	3.92	0.4	7	0
78	6	2.2	31	446	70	5.36	0	0.1	0.12	0.19	4.24	0.43	8	0
90	15	2.9	31	445	78	7.9	0	0.11	0.16	0.25	5.42	0.45	9	0
101	12	3.8	36	455	75	7.37	0	0.15	0.16	0.33	4.84	0.51	11	0
77	9	3.4	24	311	77	4.73	0	0.1	0.25	0.24	4.37	0.37	8	0
67	7	4.14	29	370	87	5.77	0	0.12	0.16	0.28	5.23	0.44	9	0
431	7	6.97	25	397	87	5.91	8751	0.2	0.52	3.87	19.75	1.16	294	1
149	6	2.95	17	208	73	46.22	0	0.02	0.34	0.33	3.94	0.18	8	1
84	18	1.67	18	214	50	4.26	0	0.05	0.1	0.18	4.66	0.08	13	0
67	14	1.48	16	186	42	4.01	0	0.04	0.1	0.15	3.48	0.07	12	0
64	13	1.16	15	209	49	2.23	0	0.06	0.08	0.16	4.54	0.08	12	0
44	9	0.92	13	173	39	1.9	0	0.05	0.07	0.13	3.15	0.07	11	0
110	12	2.26	25	340	77	4.67	0	0.12	0.15	0.32	7.48	0.16	19	0
105	12	2.24	27	354	75	4.97	0	0.11	0.17	0.31	7.45	0.17	23	0
101	9	2.4	29	382	77	5.58	0	0.12	0.2	0.33	7.16	0.19	26	0
110	25	1.81	23	306	82	3.94	0	0.1	0.11	0.24	7.63	0.12	17	0
99	24	2	26	356	92	5.05	0	0.1	0.17	0.26	6.96	0.17	25	0
104	23	2.23	26	284	75	5.91	0	0.1	0.16	0.27	6.95	0.15	24	0
98	21	2.41	28	299	77	6.83	0	0.1	0.2	0.29	6.51	0.17	28	0

H Appendix

Table H–1

Food Composition

(Computer code number is for West Diet Analysis program)   (For purposes of calculations, use "0" for t, <1, <.1, <.01, etc.)

Computer Code Number	Food Description	Measure	Wt (g)	H₂O (%)	Ener (kcal)	Prot (g)	Carb (g)	Dietary Fiber (g)	Fat (g)	Sat	Mono	Poly
	MEATS—Continued											
	Variety meats:											
13624	Brains, pan-fried	4 oz	113	76	164	14	0	0	11	2.9	2.1	1.2
13507	Sweetbreads, cooked	4 oz	113	60	264	26	0	0	17	7.7	6.2	0.8
13527	Tongue, cooked	4 oz	113	58	311	24	0	0	23	8.8	11.3	1.4
	Pork:											
	Cured:											
12000	Bacon, medium slices	3 pce	19	12	103	7	<1	0	8	2.6	3.5	0.9
12009	Breakfast strips, cooked	2 pce	23	27	106	7	<1	0	8	2.9	3.8	1.3
12002	Canadian-style bacon	2 pce	47	62	87	11	1	0	4	1.3	1.9	0.4
	Ham, roasted:											
12211	Reg, 11% fat	4 oz	113	65	201	26	0	0	10	3.5	5	1.6
12212	Extra lean, 5% fat	4 oz	113	68	164	24	2	0	6	2	3	0.6
12209	Extra lean, 4% fat, cnd	4 oz	113	69	154	24	1	0	6	1.8	2.8	0.5
	Chop:											
12029	Whole loin, brsd	1 ea	89	58	213	24	0	0	12	4.5	5.4	1
12033	Whole loin, lean, brsd	1 ea	80	61	163	23	0	0	7	2.7	3.3	0.6
12192	Center loin, w/bone, brld	1 ea	82	58	197	24	0	0	11	3.9	4.8	0.8
12025	Center loin, lean, brld	1 ea	74	61	149	22	0	0	6	2.2	2.7	0.4
12044	Center loin, pan fried	1 ea	78	53	216	23	0	0	13	4.7	5.5	1.5
12040	Blade loin, lean, pan fried	1 ea	63	59	152	16	0	0	10	3.3	3.9	1.2
	Leg:											
12016	Ham, whole, rstd	4 oz	113	55	308	30	0	0	20	7.3	8.9	1.9
12017	Ham, rump, lean, rstd	4 oz	113	61	233	35	0	0	9	3.2	4.3	0.9
	Rib:											
12050	Center loin, w/bone, rstd	4 oz	113	56	288	31	0	0	17	6.7	7.9	1.4
12055	Center loin, lean, w/bone, rstd	4 oz	113	59	252	32	0	0	13	4.9	5.9	1
	Shoulder:											
12003	Picnic, brsd	4 oz	113	48	372	32	0	0	26	9.6	11.7	2.6
12004	Picnic, lean, brsd	4 oz	113	54	280	36	0	0	14	4.7	6.5	1.3
12010	Spareribs, brsd	4 oz	113	40	449	33	0	0	34	12.6	15.2	3.1
14004	Rabbit, roasted (1 cup meat=140g)	4 oz	113	61	223	33	0	0	9	2.7	2.5	1.8
	Veal:											
11519	Short ribs, rstd	4 oz	113	60	258	27	0	0	16	6.1	6.1	1.1
11500	Liver, brsd	4 oz	113	60	217	32	4	0	7	2.2	1.3	1.2
14013	Venison (deer meat), roasted	4 oz	113	65	179	34	0	0	4	1.4	1	0.7
	Chicken:											
15016	Canned, boneless chicken	4 oz	113	69	186	25	0	0	9	2.5	3.6	2
	Fried, batter dipped:											
15013	Breast	1 ea	280	52	728	70	25	1	37	9.9	15.3	8.6
15030	Drumstick	1 ea	72	53	193	16	6	<1	11	3	4.6	2.7
15036	Thigh	1 ea	86	52	238	19	8	<1	14	3.8	5.8	3.4
15034	Wing	1 ea	49	46	159	10	5	<1	11	2.9	4.4	2.5
	Fried, flour coated:											
15003	Breast	1 ea	196	57	435	62	3	<1	17	4.8	6.9	3.8
15057	Breast, without skin	1 ea	172	60	322	58	1	0	8	2.2	3	1.8
15007	Drumstick	1 ea	49	57	120	13	1	<1	7	1.8	2.7	1.6
15009	Thigh	1 ea	62	54	162	17	2	<1	9	2.5	3.6	2.1
15011	Thigh, without skin	1 ea	52	59	113	15	1	0	5	1.4	2	1.3
15029	Wing	1 ea	32	49	103	8	1	<1	7	1.9	2.8	1.6
15902	Patty, breaded, cooked	1 ea	75	49	213	12	11	<1	13	4.1	6.4	1.7
	Roasted:											
15000	All types of meat	1 c	140	64	266	41	0	0	10	2.9	3.7	2.4
15027	Dark meat	1 c	140	63	287	38	0	0	14	3.7	5	3.2
15032	Light meat	1 c	140	65	242	43	0	0	6	1.8	2.2	1.4
15004	Breast, without skin	1 ea	172	65	284	53	0	0	6	1.7	2.1	1.3
15035	Drumstick, without skin	1 ea	44	67	76	12	0	0	2	0.7	0.8	0.6
15156	Leg, without skin	1 ea	95	65	181	26	0	0	8	2.2	2.9	1.9
15010	Thigh	1 ea	62	59	153	16	0	0	10	2.7	3.8	2.1
15012	Thigh, without skin	1 ea	52	63	109	13	0	0	6	1.6	2.2	1.3
15006	Stewed, all types	1 c	140	67	248	38	0	0	9	2.6	3.3	2.2

Appendix **H**

PAGE KEY: H–2 = Beverages H–4 = Dairy H–8 = Eggs H–10 = Fat/Oil H–12 = Fruit H–18 = Bakery
H–24 = Grain H–30 = Fish H–30 = Meats H–34 = Poultry H–36 = Sausage H–38 = Mixed/Fast H–42 = Nuts/Seeds H–44 = Sweets
H–46 = Vegetables/Legumes H–56 = Vegetarian H–60 = Misc H–60 = Soups/Sauces H–62 = Fast H–76 = Convenience H–78 = Baby foods

Chol (mg)	Calc (mg)	Iron (mg)	Magn (mg)	Pota (mg)	Sodi (mg)	Zinc (mg)	VT-A (RAE)	Thia (mg)	VT-E (mg α)	Ribo (mg)	Niac (mg)	V-B₆ (mg)	Fola (µg)	VT-C (mg)
2309	14	1.9	16	232	151	1.54	0	0.12	1.73	0.27	2.79	0.12	6	14
452	14	2.4	21	329	59	3.03	0	0.02	0.78	0.24	2.89	0.06	15	23
214	11	2.97	18	179	76	3.38	0	0.09	0.36	0.47	4.17	0.19	3	8
21	2	0.27	6	107	439	0.66	2	0.08	0.06	0.05	2.11	0.07	<1	0
24	3	0.45	6	107	483	0.85	0	0.17	0.06	0.08	1.75	0.08	1	0
27	5	0.39	10	183	727	0.8	0	0.39	0.16	0.09	3.25	0.21	2	0
67	9	1.51	25	462	1695	2.79	0	0.82	0.35	0.37	6.95	0.35	3	0
60	9	1.67	16	324	1359	3.25	0	0.85	0.28	0.23	4.55	0.45	3	0
34	7	1.04	24	393	1283	2.52	0	1.17	0.29	0.28	5.53	0.51	6	0
71	19	0.95	17	333	43	2.12	2	0.56	0.21	0.23	3.93	0.33	3	1
63	14	0.9	16	310	40	1.98	2	0.53	0.17	0.21	3.67	0.31	3	0
67	27	0.66	20	294	48	1.85	2	0.87	0.27	0.24	4.3	0.35	5	0
61	23	0.63	20	278	44	1.76	1	0.85	0.31	0.23	4.1	0.35	4	0
72	21	0.71	23	332	62	1.8	2	0.89	0.2	0.24	4.37	0.37	5	1
52	14	0.67	16	230	49	2.44	1	0.46	0.14	0.23	2.8	0.26	3	1
106	16	1.14	25	398	68	3.34	3	0.72	0.25	0.35	5.17	0.45	11	0
108	8	1.29	33	442	73	3.4	3	0.91	0.46	0.4	5.56	0.38	3	0
82	32	1.06	24	476	52	2.33	2	0.82	0.41	0.34	6.91	0.37	3	0
80	29	1.11	25	494	53	2.41	2	0.86	0.55	0.36	7.25	0.39	3	0
123	20	1.82	21	417	99	4.72	3	0.61	0.34	0.35	5.89	0.4	5	0
129	9	2.2	25	458	115	5.62	2	0.68	0.33	0.41	6.71	0.46	6	0
137	53	2.09	27	362	105	5.2	3	0.46	0.38	0.43	6.19	0.4	5	0
93	21	2.57	24	433	53	2.57	0	0.1	0.96	0.24	9.53	0.53	12	0
124	12	1.1	25	333	104	4.62	0	0.06	0.4	0.31	7.89	0.28	15	0
577	7	5.77	23	372	88	12.69	23894	0.21	0.77	3.23	14.86	1.04	374	1
127	8	5.05	27	379	61	3.11	0	0.2	0.28	0.68	7.58	0.42	5	0
70	16	1.79	14	156	568	1.59	38	0.02	0.29	0.15	7.15	0.4	5	2
238	56	3.5	67	563	770	2.66	56	0.32	2.97	0.41	29.46	1.2	42	0
62	12	0.97	14	134	194	1.68	19	0.08	0.88	0.15	3.67	0.19	13	0
80	15	1.25	18	165	248	1.75	25	0.1	1.05	0.2	4.91	0.22	16	0
39	10	0.63	8	68	157	0.68	17	0.05	0.52	0.07	2.58	0.15	9	0
174	31	2.33	59	508	149	2.16	29	0.16	1.12	0.26	26.93	1.14	12	0
157	28	1.96	53	475	136	1.86	12	0.14	0.72	0.22	25.43	1.1	7	0
44	6	0.66	11	112	44	1.42	12	0.04	0.41	0.11	2.96	0.17	5	0
60	9	0.92	16	147	55	1.56	18	0.06	0.52	0.15	4.31	0.2	7	0
53	7	0.76	14	135	49	1.45	11	0.05	0.3	0.13	3.7	0.2	5	0
26	5	0.4	6	57	25	0.56	12	0.02	0.18	0.04	2.14	0.13	2	0
45	12	0.94	15	184	399	0.78	11	0.07	1.46	0.1	5.04	0.23	8	0
125	21	1.69	35	340	120	2.94	22	0.1	0.38	0.25	12.84	0.66	8	0
130	21	1.86	32	336	130	3.92	31	0.1	0.38	0.32	9.17	0.5	11	0
119	21	1.48	38	346	108	1.72	13	0.09	0.38	0.16	17.39	0.84	6	0
146	26	1.79	50	440	127	1.72	10	0.12	0.46	0.2	23.58	1.03	7	0
41	5	0.57	11	108	42	1.4	8	0.03	0.12	0.1	2.67	0.17	4	0
89	11	1.24	23	230	86	2.72	18	0.07	0.26	0.22	6	0.35	8	0
58	7	0.83	14	138	52	1.46	30	0.04	0.17	0.13	3.95	0.19	4	0
49	6	0.68	12	124	46	1.34	10	0.04	0.14	0.12	3.39	0.18	4	0
116	20	1.64	29	252	98	2.79	21	0.07	0.38	0.23	8.56	0.36	8	0

Table H–1

Food Composition   (Computer code number is for West Diet Analysis program)   (For purposes of calculations, use "0" for t, <1, <.1, <.01, etc.)

Computer Code Number	Food Description	Measure	Wt (g)	H₂O (%)	Ener (kcal)	Prot (g)	Carb (g)	Dietary Fiber (g)	Fat (g)	Fat Breakdown (g) Sat	Mono	Poly
	MEATS—Continued											
	Variety meats:											
15025	Gizzards, simmered	1 c	145	68	212	44	0	0	4	1	0.8	0.5
15024	Hearts, simmered	1 c	145	65	268	38	<1	0	11	3.3	2.9	3.3
15215	Liver, simmered: Ounce	3 oz	85	67	142	21	1	0	6	1.8	1.2	1.1
	Duck:											
16295	Whole, w/skin, rstd, about 2.7 cups	½ ea	382	52	1287	73	0	0	108	36.9	49.3	13.9
14000	Whole, w/o skin, rstd, about 1.5 cups	½ ea	221	64	444	52	0	0	25	9.2	8.2	3.2
14003	Goose, whole, w/skin, rstd, about 5.5 cups	½ ea	774	52	2361	195	0	0	170	53.2	79.3	19.5
	Turkey:											
	Roasted, meat only:											
16002	Dark meat	4 oz	113	63	211	32	0	0	8	2.7	1.9	2.4
16158	Light meat	4 oz	113	66	177	34	0	0	4	1.2	0.6	1
16000	All types, chopped or diced	1 c	140	65	238	41	0	0	7	2.3	1.4	2
16003	Ground, cooked	4 oz	113	59	266	31	0	0	15	3.8	5.5	3.6
16010	With gravy, frozen package	3 oz	85	85	57	5	4	0	2	0.7	0.8	0.4
16307	Patty, breaded, fried	2 oz	57	50	161	8	9	<1	10	2.7	4.3	2.7
16308	Roasted, from frozen, seasoned	4 oz	113	68	175	24	3	0	7	2.1	1.4	1.9
16008	Roll, light meat	1 pce	28	72	41	5	<1	0	2	0.6	0.7	0.5
	Lunchmeat:											
	Turkey:											
	Breast:											
13259	Barbecued, Louis Rich	2 oz	56	72	57	11	2		<1	0.2	0.2	0.1
13108	Hickory smoked, Louis Rich	1 pce	80	73	80	16	2	0	1	0		
13110	Honey roasted, Louis Rich	1 pce	80	73	80	16	3	0	1	0.5		
13109	Oven roasted, Louis Rich	1 pce	80		70	16	0	0	1	0		
13114	Fat Free	1 pce	28	76	24	4	1	0	<1	t	t	t
13020	Pastrami	2 pce	57	71	80	10	1	0	4	1	1.2	0.9
13144	Salami	1 pce	28	72	41	4	<1	0	3	0.8	0.9	0.7
13014	Ham	2 pce	57	72	72	10	1	<1	3	0.9	1.1	0.8
	Bologna:											
13002	Beef	1 pce	23	54	72	2	1	0	6	2.6	2.8	0.2
13176	Beef, light, Oscar Mayer	1 pce	28	65	56	3	2	0	4	1.6	2	0.1
13006	Beef & pork	1 pce	28	52	85	4	2	0	7	2.7	2.9	0.3
13218	Healthy Favorites	1 pce	23		22	4	1	0	<1	0		
13032	Pork	1 pce	23	61	57	4	<1	0	5	1.6	2.2	0.5
13174	Regular, light, Oscar Mayer	1 pce	28	65	57	3	2	0	4	1.6	2	0.4
13007	Turkey	1 pce	28	65	59	3	1	<1	4	1.2	1.9	1.1
13149	Turkey, Louis Rich	1 pce	56	67	115	6	1	0	10	2.9	3.6	2.6
	Chicken:											
13222	Chicken breast, Healthy Favorites	4 pce	52		40	9	1	0	0	0	0	0
	Beef:											
13039	Corned beef loaf, jellied	1 pce	28	69	43	6	0	0	2	0.7	0.8	t
	Ham:											
13057	Chopped ham, packaged	2 c	42	64	96	7	0	0	7	2.4	3.4	0.9
13220	Honey ham, Healthy Favorites	4 pce	52	73	55	9	2	0	1	0.4	0.6	0.1
13168	Oscar Mayer lower sodium ham	1 pce	21	73	23	3	1	0	1	0.3	0.4	t
13048	Mortadella lunchmeat	2 pce	30	52	93	5	1	0	8	2.9	3.4	0.9
13049	Olive loaf lunchmeat	2 pce	57	58	134	7	5	0	9	3.3	4.5	1.1
13051	Pickle & pimento loaf	2 pce	57	57	149	7	3	0	12	4.5	5.5	1.5
	Sausages:											
13035	Beerwurst/beer salami, beef	1 oz	28	57	77	4	1	<1	6	2.4	2.8	0.6
13031	Beerwurst/beer salami, pork	1 oz	28	61	67	4	1	0	5	1.8	2.5	0.7
13001	Berliner sausage	1 oz	28	61	64	4	1	0	5	1.7	2.2	0.4
13066	Braunschweiger sausage	2 pce	57	51	186	8	2	0	16	5.2	7.5	1.6
13036	Bratwurst-link	1 ea	70	51	226	10	2		19	7	9.3	2
13037	Cheesefurter/cheese smokie	2 ea	86	52	281	12	1	0	25	9	11.8	2.6
13070	Chorizo, pork & beef	1 ea	60	32	273	14	1	0	23	8.6	11	2.1
	Frankfurters:											
13008	Beef, large link, 8/package	1 ea	57	52	188	6	2	0	17	6.7	8.2	0.7
13010	Beef and pork, large link, 8/package	1 ea	45	56	137	5	1		12	4.8	6.2	1.2

Appendix **H**

PAGE KEY: H–2 = Beverages   H–4 = Dairy   H–8 = Eggs   H–10 = Fat/Oil   H–12 = Fruit   H–18 = Bakery
H–24 = Grain   H–30 = Fish   H–30 = Meats   H–34 = Poultry   H–36 = Sausage   H–38 = Mixed/Fast   H–42 = Nuts/Seeds   H–44 = Sweets
H–46 = Vegetables/Legumes   H–56 = Vegetarian   H–60 = Misc   H–60 = Soups/Sauces   H–62 = Fast   H–76 = Convenience   H–78 = Baby foods

Chol (mg)	Calc (mg)	Iron (mg)	Magn (mg)	Pota (mg)	Sodi (mg)	Zinc (mg)	VT-A (RAE)	Thia (mg)	VT-E (mg α)	Ribo (mg)	Niac (mg)	V-B₆ (mg)	Fola (µg)	VT-C (mg)
536	25	4.63	4	260	81	6.41	0	0.04	0.29	0.3	4.52	0.1	7	0
351	28	13.09	29	191	70	10.58	12	0.1	2.32	1.07	4.06	0.46	116	3
479	9	9.89	21	224	65	3.38	3384	0.25	0.7	1.69	9.39	0.64	491	24
321	42	10.31	61	779	225	7.11	241	0.66	2.67	1.03	18.43	0.69	23	0
197	27	5.97	44	557	144	5.75	51	0.57	1.55	1.04	11.27	0.55	22	0
704	101	21.9	170	2546	542	20.28	163	0.6	13.47	2.5	32.26	2.86	15	0
96	36	2.63	27	328	89	5.04	0	0.07	0.72	0.28	4.12	0.41	10	0
78	21	1.53	32	345	72	2.31	0	0.07	0.1	0.15	7.73	0.61	7	0
106	35	2.49	36	417	98	4.34	0	0.09	0.46	0.25	7.62	0.64	10	0
115	28	2.18	27	305	121	3.23	0	0.06	0.38	0.19	5.45	0.44	8	0
15	12	0.79	7	52	471	0.6	11	0.02	0.3	0.11	1.53	0.08	3	0
35	8	1.25	9	157	456	0.82	6	0.06	0.72	0.11	1.31	0.11	16	0
60	6	1.84	25	337	768	2.87	0	0.05	0.43	0.18	7.09	0.31	6	0
12	11	0.36	4	70	137	0.44	0	0.02	0.04	0.06	1.96	0.09	1	0
25	14	0.61	16		592	0.58	0							0
35	0	0.72			1060		0							0
35	0	0.72			940		0							0
35	0				910		0							0
9	3	0.31	8	57	334	0.24								0
31	5	0.95	8	148	596	1.23	0	0.03	0.13	0.14	2.01	0.15	3	0
21	11	0.35	6	60	281	0.65								0
41	5	1.33	13	164	635	1.48	4	0.02	0.36	0.08	1.21	0.12	4	0
13	7	0.25	3	40	248	2.09	3	<.01	0.08	0.02	0.58	0.04	2	3
12	4	0.34	4	44	322	0.53	0						4	0
17	24	0.34	5	88	206	0.64	7	0.06	0.1	0.05	0.71	0.08	2	0
8		0.18			255									
14	3	0.18	3	65	272	0.47	0	0.12	0.06	0.04	0.9	0.06	1	0
16	14	0.39	6	46	313	0.45	0	0.04		0.03	0.86	0.05	5	0
21	34	0.84	4	38	351	0.36	3	0.01	0.13	0.03	0.73	0.07	3	4
44	68	0.9	10	103	484	1.14	0	0.03		0.1	2.15	0.1		0
25		0.72			620									
13	3	0.57	3	28	267	1.15	0	0	0.05	0.03	0.49	0.03	2	0
21	3	0.35	7	134	576	0.81	0	0.27	0.1	0.09	1.63	0.15	<1	0
24	6	0.7	18	144	635	1.02	0							0
9	1	0.3	5	197	174	0.42	0							0
17	5	0.42	3	49	374	0.63	0	0.04	0.07	0.05	0.8	0.04	1	0
22	62	0.31	11	169	846	0.79	34	0.17	0.14	0.15	1.05	0.13	1	0
21	54	0.58	10	194	792	0.8	13	0.17	0.14	0.14	1.17	0.11	3	0
17	8	0.48	5	68	205	0.62	<1	0.07	0.05	0.05	0.83	0.06	1	0
17	2	0.21	4	71	347	0.48	0	0.16	0.06	0.05	0.91	0.1	1	0
13	3	0.32	4	79	363	0.69	0	0.11	0.06	0.06	0.87	0.06	1	0
103	5	6.38	6	113	661	1.6	2405	0.14	0.2	0.87	4.77	0.19	25	0
44	34	0.72	11	197	778	1.47	0	0.18	0.19	0.16	2.31	0.09	4	0
58	50	0.93	11	177	931	1.94	40	0.22	0	0.14	2.49	0.11	3	0
53	5	0.95	11	239	741	2.05	0	0.38	0.13	0.18	3.08	0.32	1	0
30	8	0.86	8	89	650	1.4	0	0.02	0.11	0.08	1.35	0.05	3	0
22	5	0.52	4	75	504	0.83	8	0.09	0.11	0.05	1.19	0.06	2	0

Table H–1

Food Composition

(Computer code number is for West Diet Analysis program)   (For purposes of calculations, use "0" for t, <1, <.1, <.01, etc.)

Computer Code Number	Food Description	Measure	Wt (g)	H₂O (%)	Ener (kcal)	Prot (g)	Carb (g)	Dietary Fiber (g)	Fat (g)	Fat Breakdown (g) Sat	Mono	Poly
	MEATS—Continued											
13012	Turkey frankfurter, 10/package	1 ea	45	63	102	6	1	0	8	2.7	2.5	2.2
13129	Turkey/chicken frank 8/pkg	1 ea	43	67	81	5	2	0	6	1.7	2.4	1.4
13043	Kielbasa sausage	1 pce	26	54	81	3	1	0	7	2.6	3.4	0.8
13044	Knockwurst sausage, link	1 ea	68	55	209	8	2	0	19	6.9	8.7	2
13021	Pepperoni	2 pce	11	31	51	2	<1	<1	4	1.8	2.1	0.3
13022	Polish	1 oz	28	53	91	4	<1	0	8	2.9	3.8	0.9
13024	Salami, pork and beef	2 pce	57	60	142	8	1	0	11	4.6	5.2	1.2
13026	Salami, pork and beef, dry	3 pce	30	35	125	7	1	0	10	3.7	5.1	1
13025	Salami, turkey	2 pce	57	54	126	8	11	<1	5	2	1.8	1.4
13029	Smoked link sausage, beef and pork	1 ea	68	54	218	8	2	0	20	6.6	8.3	2.7
13027	Smoked link sausage, pork	1 ea	68	39	265	15	1	0	22	7.7	10	2.6
13030	Summer sausage	2 pce	46	51	154	7	<1	0	14	5.5	6	0.6
13052	Turkey breakfast sausage	1 pce	28	60	64	6	0		5	1.6	1.8	1.2
13054	Vienna sausage, canned	2 ea	32	60	89	3	1	0	8	3	4	0.5
	Sandwich spreads:											
13034	Ham salad spread	2 tbsp	30	63	65	3	3	0	5	1.5	2.2	0.8
13056	Pork and beef	2 tbsp	30	60	70	2	4	<1	5	1.8	2.3	0.8
	MIXED DISHES											
15907	Almond Chicken	1 c	242	77	280	22	16	3	15	1.9	6.1	5.6
7084	Bean cake	1 ea	32	23	130	2	16	1	7	1	2.9	2.6
56124	Beef fajita	1 ea	223	65	399	23	36	3	18	5.5	7.6	3.5
56119	Beef flauta	1 ea	113	51	354	14	13	2	28	4.8	11.8	9.4
5513	Broccoli, batter fried	1 c	85	74	122	3	9	2	9	1.3	2.1	4.9
15903	Buffalo wings/spicy chicken wings	2 pce	32	53	98	8	<1	<1	7	1.8	2.7	1.8
56649	Cheeseburger deluxe	1 ea	219	52	563	28	38		33	15	12.6	2
56123	Chicken fajita	1 ea	223	65	363	20	44	5	12	2.3	5.5	3.1
56120	Chicken flauta	1 ea	113	55	330	13	12	2	26	4.2	10.7	9.3
50312	Chili con carne	½ c	127	77	128	12	11		4	1.7	1.7	0.3
15915	Chicken teriyaki, breast	1 ea	128	67	178	27	7	<1	4	0.9	1.1	0.9
45557	Chinese Pastry	1 oz	28	46	67	1	13	<1	2	0.2	0.5	0.8
56094	Chop suey with beef & pork	1 c	220	63	421	22	31		24	4.7	8.3	9.2
5461	Coleslaw	1 c	132	74	195	2	17		15	2.1	3.2	8.5
5366	Corn pudding	1 c	250	76	272	11	32	4	13	6.3	4.3	1.7
19539	Deviled egg (½ egg + filling)	1 ea	31	70	63	4	<1	0	5	1.2	1.7	1.5
	Egg Foo Yung Patty:											
56132	Meatless	1 ea	86	77	113	6	3	1	8	2	3.4	2.1
56290	With beef	1 ea	86	76	119	8	3	<1	8	2	2.9	2.2
56287	With chicken	1 ea	86	76	121	8	4	1	8	1.9	2.8	2.3
	Egg Roll:											
56110	Meatless	1 ea	64	70	101	3	10	1	6	1.2	2.9	1.3
57523	With Meat	1 ea	64	66	113	5	9	1	6	1.4	3	1.3
56003	Egg salad	1 c	183	57	584	17	3	0	56	10.5	17.4	23.9
56102	Falafel	1 ea	17	35	57	2	5		3	0.4	1.7	0.7
50182	Hot & Sour Soup (Chinese)	1 c	244	87	162	15	5	1	8	2.7	3.4	1.2
56659	Hamburger deluxe	1 ea	110	49	279	13	27		13	4.1	5.3	2.6
7081	Hummous/hummus	¼ c	62	65	110	3	12	2	5	0.7	3	1.3
16335	Kung Pao Chicken	1 c	162	54	431	29	11	2	31	5.2	13.9	9.7
	Lasagna:											
56108	With meat, homemade	1 pce	245	67	392	23	40		16	8	5.2	0.8
56071	Without meat, homemade	1 pce	218	69	306	16	40		10	5.6	2.5	0.6
56073	Chicken alfredo, w/broccoli	1 ea	340	75	389	24	41		14	6.7	5.5	0.8
57521	Lo mein, meatless	1 c	200	82	135	6	27	4	1	0.1	t	0.3
57522	Lo mein, with meat	1 c	200	72	283	20	21	3	14	2.6	4	6
56080	Moussaka (lamb & eggplant)	1 c	250	82	238	17	13	4	13	4.6		
5514	Mushrooms, batter fried	5 ea	70	63	156	2	11	1	12	1.5	3.6	6
56005	Potato salad with mayonnaise & eggs	½ c	125	76	179	3	14	2	10	1.8	3.1	4.7
56618	Pizza, combination, ½ of 12" round	1 pce	79	48	184	13	21		5	1.5	2.5	0.9
56619	Pizza, pepperoni, ½ of 12" round	1 pce	71	47	181	10	20		7	2.2	3.1	1.2
56098	Quiche Lorraine ⅛ of 8" quiche	1 pce	176	53	526	15	25	1	41	18.9	14.3	5.2
38067	Ramen noodles-cooked	1 c	227	86	154	3	20	1	7	1.7	1.2	3.3

PAGE KEY: H–2 = Beverages   H–4 = Dairy   H–8 = Eggs   H–10 = Fat/Oil   H–12 = Fruit   H–18 = Bakery
H–24 = Fish   H–30 = Meats   H–30 = Poultry   H–36 = Sausage   H–38 = Mixed/Fast   H–42 = Nuts/Seeds   H–44 = Sweets
H–46 = Vegetables/Legumes   H–56 = Vegetarian   H–60 = Misc   H–60 = Soups/Sauces   H–62 = Fast   H–76 = Convenience   H–78 = Baby foods

Chol (mg)	Calc (mg)	Iron (mg)	Magn (mg)	Pota (mg)	Sodi (mg)	Zinc (mg)	VT-A (RAE)	Thia (mg)	VT-E (mg α)	Ribo (mg)	Niac (mg)	V-B$_6$ (mg)	Fola (µg)	VT-C (mg)
48	48	0.83	6	81	642	1.4	0	0.02	0.28	0.08	1.86	0.1	4	0
40	56	0.94	10	69	488	0.8								0
17	11	0.38	4	70	280	0.53	0	0.06	0.06	0.06	0.75	0.05	1	0
41	7	0.45	7	135	632	1.13	0	0.23	0.39	0.1	1.86	0.12	1	0
13	2	0.16	2	35	197	0.3	0	0.06	0.03	0.03	0.6	0.04	1	0
20	3	0.4	4	66	245	0.54	0	0.14	0.06	0.04	0.96	0.05	1	0
37	7	1.52	9	113	607	1.22	0	0.14	0.13	0.21	2.03	0.12	1	0
24	2	0.45	5	113	558	0.97	0	0.18	0.08	0.09	1.46	0.15	1	0
45	42	0.88	15	226	619	1.77	1	0.24	0.14	0.17	2.27	0.24	6	12
39	8	0.51	9	122	619	0.86	9	0.13	0.09	0.07	2	0.11	1	0
46	20	0.79	13	228	1020	1.92	0	0.48	0.17	0.17	3.08	0.24	3	1
34	6	1.17	6	125	571	1.18	0	0.07	0.1	0.15	1.98	0.12	1	0
23	5	0.51	6	75	188	0.96	0	0.03	0.14	0.08	1.4	0.08	1	0
17	3	0.28	2	32	305	0.51	0	0.03	0.07	0.03	0.52	0.04	1	0
11	2	0.18	3	45	274	0.33	0	0.13	0.52	0.04	0.63	0.04	<1	0
11	4	0.24	2	33	304	0.31	8	0.05	0.52	0.04	0.52	0.04	1	0
40	69	1.97	60	549	526	1.62		0.08	3.8	0.2	9.48	0.44	26	7
0	3	0.67	6	58	1	0.16	0	0.07	1.24	0.05	0.55	0.02	9	0
45	84	3.76	38	479	316	3.52	22	0.39	1.74	0.3	5.4	0.38	23	27
37	51	1.87	28	313	68	3.45		0.06	4.65	0.13	1.88	0.24	10	19
15	66	0.98	20	242	64	0.38		0.08	2.85	0.13	0.73	0.11	36	53
26	5	0.4	6	59	25	0.57		0.01	0.27	0.04	2.06	0.13	1	0
88	206	4.66	44	445	1108	4.6	140	0.39	1.18	0.46	7.38	0.28	81	8
39	101	3.32	48	534	343	1.65		0.43	1.71	0.33	6.12	0.38	42	37
35	50	0.95	27	269	71	1.13		0.05	4.36	0.09	3.1	0.22	8	18
67	34	2.6	23	347	505	1.79	42	0.06	0.81	0.57	1.24	0.17	23	1
82	27	1.71	35	309	1683	1.96		0.08	0.35	0.19	8.75	0.47	12	3
0	6	0.18	7	25	3	0.16	<1	0.02	0.26	<.01	0.27	0.04	1	0
43	39	4.19	54	519	950	3.48		0.36	1.8	0.37	5.73	0.39	44	20
7	45	0.96	12	236	356	0.26	48	0.05	5.28	0.04	0.11	0.15	51	11
250	100	1.4	38	402	138	1.25	135	1.03	0.52	0.32	2.47	0.3	62	7
122	15	0.35	3	37	50	0.3	50	0.02	0.61	0.15	0.02	0.05	13	0
185	31	1.04	12	117	317	0.7		0.04	1.22	0.26	0.43	0.09	30	5
166	25	1.01	11	139	131	1.01		0.05	1.06	0.23	0.65	0.13	22	3
167	27	0.82	11	136	132	0.76		0.05	1.1	0.23	0.89	0.12	22	3
30	14	0.81	9	97	274	0.25		0.08	0.85	0.11	0.8	0.05	13	3
37	15	0.83	10	124	274	0.46		0.16	0.8	0.13	1.28	0.09	10	2
581	74	1.81	13	181	464	1.45	263	0.09	7.66	0.66	0.09	0.46	61	0
0	9	0.58	14	99	50	0.26	<1	0.02	0.19	0.03	0.18	0.02	16	0
34	29	1.9	29	384	1011	1.51		0.27	0.15	0.25	5	0.2	13	1
26	63	2.63	22	227	504	2.06	4	0.23	0.82	0.2	3.68	0.12	52	2
0	30	0.97	18	107	150	0.68	<1	0.06	0.46	0.03	0.25	0.25	37	5
64	49	1.96	63	428	907	1.5		0.15	3.9	0.15	13.23	0.59	43	8
58	270	3.07	50	460	391	3.33		0.24	1.16	0.34	4.2	0.25	20	14
32	265	2.35	44	373	365	1.82		0.23	1.09	0.28	2.51	0.17	17	14
55	265	3.82	64	759	840	3.7		0.29	3.45	0.39	5.07	0.32	28	13
0	46	2.03	33	386	564	0.92	65	0.23	0.35	0.24	2.82	0.19	48	12
42	29	2.07	42	332	142	1.83		0.41	2.06	0.28	5.02	0.36	53	11
97	75	1.74	40	565	460	2.57		0.16	0.98	0.31	4.14	0.23	46	6
2	15	1.22	7	154	112	0.42		0.11	2.34	0.26	2.25	0.04	8	1
85	24	0.81	19	318	661	0.39	40	0.1	2.33	0.08	1.11	0.18	9	12
21	101	1.53	18	179	382	1.11	58	0.21		0.17	1.96	0.09	32	2
14	65	0.94	9	153	267	0.52	53	0.13		0.23	3.05	0.06	37	2
221	231	1.88	24	239	221	1.5		0.26	2.02	0.49	2.01	0.1	19	1
0	13	0.39	10	49	802	0.18	2	0.02	2.34	0.01	0.25	0.01	3	0

Table H–1

Food Composition

(Computer code number is for West Diet Analysis program)   (For purposes of calculations, use "0" for t, <1, <.1, <.01, etc.)

Computer Code Number	Food Description	Measure	Wt (g)	H₂O (%)	Ener (kcal)	Prot (g)	Carb (g)	Dietary Fiber (g)	Fat (g)	Fat Breakdown (g)		
										Sat	Mono	Poly
	MIXED DISHES—Continued											
56302	Ravioli, meat	½ c	125	69	197	11	18	1	9	3	3.7	1
38145	Fried rice (meatless)	1 c	166	68	271	5	34	1	12	1.8	3.2	6.7
	Spaghetti (enriched) in tomato sauce:											
56097	With cheese, home recipe	1 c	250	77	260	9	37		9	2		
56100	With meatballs, home recipe	1 c	248	71	362	18	28	3	18	4.8		
56076	Spinach souffle	1 c	136	74	219	11	3		18	7.1	6.8	3.1
2995	Spring roll, vegetable	1 ea	63	49	158	4	20	1	7	0.9		
	Sushi:											
56313	Fish and vegetable	1 c	166	65	232	9	47	2	1	0.2	0.2	0.2
56314	Vegetable seaweed	1 c	166	71	194	4	43	1	<1	0.1	0.1	0.1
12900	Sweet & sour pork	1 c	226	77	231	15	25	2	8	2.1	3.2	2.3
15921	Sweet & sour chicken breast	1 ea	131	79	118	8	15	1	3	0.5	0.9	1.4
56916	Tabouli	1 c	160	77	199	3	16	4	15	2	10.8	1.4
	Thai dishes											
1984	Beef peanut satay, svg	1 ea	129	60	286	26	5	1	18	5.9		
1988	Drunkard noodles, svg	1 ea	366	80	344	5	51	4	14	1.8		
1994	Lemongrass vegetables, svg	1 ea	187	75	238	8	19	4	16	2.6		
1998	Peanut chicken, svg	1 ea	309	81	272	19	26	3	11	2.3		
2994	Spicy noodles, svg	1 ea	310	58	626	34	65	3	26	4.4		
2994	Sweet noodles, svg	1 ea	211	58	426	23	44	2	18	3		
56118	Three bean salad	1 c	150	81	140	4	15	5	8	1.1	1.7	4.4
56006	Waldorf salad	1 c	137	58	411	4	12	3	41	4.3		
56111	Wonton, meat filled	1 ea	19	45	55	3	5	<1	3	0.8	1.2	0.3
	FAST FOODS and SANDWICHES											
	(see end of this appendix for additional Fast Foods)											
	Burritos:											
66026	Beef & bean	1 ea	116	52	255	11	33	3	9	4.2	3.5	0.6
66025	Bean	1 ea	109	53	225	7	36	4	7	3.5	2.4	0.6
16255	Chicken con queso	1 ea	299		350	14	60	6	6	2.5		
	Cheeseburgers:											
66013	With bun, 4-oz patty	1 ea	166	51	417	21	35		21	8.7	7.8	2.7
66015	With bun, regular	1 ea	154	55	359	18	28		20	9.2	7.2	1.5
56668	Corndog	1 ea	175	47	460	17	56		19	5.2	9.1	3.5
15181	Chicken	1 ea	113		271	11	23	2	15			
66021	Enchilada	1 ea	163	63	319	10	29		19	10.6	6.3	0.8
66032	English muffin with egg, cheese, bacon	1 ea	146	57	308	18	28	2	13	5	5	1.7
	Fish sandwich:											
66010	Large, no cheese	1 ea	158	47	431	17	41	<1	23	5.2	7.7	8.2
66011	Regular, with cheese	1 ea	183	45	523	21	48	<1	29	8.1	8.9	9.4
	Hamburgers:											
56658	With bun, regular	1 ea	107	45	275	12	35	2	10	3.6	3.4	1
66006	With bun, 4-oz patty	1 ea	215	51	576	32	39		32	12	14.1	2.8
66004	Hotdog/frankfurter with bun	1 ea	98	54	242	10	18		15	5.1	6.9	1.7
	Lunchables:											
56938	Bologna & American cheese	1 ea	128		450	18	19	0	34	15		
56939	Ham & cheese	1 ea	128		320	22	19	0	17	8		
56930	Honey ham & Amer. w/choc pudding	1 ea	176		390	18	34	<1	20	9		
56931	Honey turkey & cheddar w/Jello	1 ea	163		320	17	27	<1	16	9		
56940	Pepperoni & American cheese	1 ea	128		480	20	19	0	36	17		
56936	Salami & American cheese	1 ea	128		430	18	18	0	32	15		
56937	Turkey & cheddar cheese	1 ea	128		360	20	20	1	22	11		
	SANDWICHES:											
	Avocado, chesse, tomato & lettuce:											
56021	On white bread, firm	1 ea	210	62	429	14	35	5	27	7.7		
56022	On part whole wheat	1 ea	201	63	402	14	30	6	27	7.8		
56023	On whole wheat	1 ea	214	63	424	15	33	8	28	8.2		
	Bacon, lettuce & tomato sandwich:											
56009	On white bread, soft	1 ea	124	52	318	10	29	2	18	4.1		
56011	On part whole wheat	1 ea	124	53	314	11	26	3	19	4.6		

PAGE KEY: H–2 = Beverages   H–4 = Dairy   H–8 = Eggs   H–10 = Fat/Oil   H–12 = Fruit   H–18 = Bakery
H–24 = Grain   H–30 = Fish   H–30 = Meats   H–34 = Poultry   H–36 = Sausage   H–38 = Mixed/Fast   H–42 = Nuts/Seeds   H–44 = Sweets
H–46 = Vegetables/Legumes   H–56 = Vegetarian   H–60 = Misc   H–60 = Soups/Sauces   H–62 = Fast   H–76 = Convenience   H–78 = Baby foods

Chol (mg)	Calc (mg)	Iron (mg)	Magn (mg)	Pota (mg)	Sodi (mg)	Zinc (mg)	VT-A (RAE)	Thia (mg)	VT-E (mg α)	Ribo (mg)	Niac (mg)	V-B$_6$ (mg)	Fola (µg)	VT-C (mg)
85	35	2.15	20	264	90	1.7		0.16	1.32	0.22	2.99	0.14	14	4
43	28	1.94	23	128	261	0.92		0.21	2.51	0.11	2.24	0.15	22	4
8	80	2.25	26	408		1.3		0.25	2.75	0.18	2.25	0.2	8	12
65	92	3.33	44	479	1133	3.4		0.25	2.46	0.3	4.38	0.29	68	16
184	230	1.35	38	201	763	1.29	267	0.09	1.22	0.3	0.48	0.12	80	
3	58	1.78	12	74	262	0.38		0.18	1.29	0.14	1.88	0.03	32	1
11	25	2.33	27	218	93	0.84		0.28	0.62	0.07	2.96	0.16	15	4
0	22	1.65	21	106	5	0.75	33	0.21	0.13	0.04	1.99	0.15	11	3
39	28	1.44	34	386	839	1.47		0.55	1.09	0.21	3.63	0.41	10	20
23	15	0.84	21	185	506	0.67		0.06	0.67	0.08	3.09	0.18	6	12
0	29	1.25	36	246	799	0.48	34	0.08	2.16	0.05	1.14	0.11	31	29
71	15	2.77	39	408	1173	5.81		0.1	1.23	0.24	4.75	0.41	16	3
0	37	1.77	37	524	1227	0.77		0.16	3.83	0.13	1.89	0.39	42	159
0	146	2.88	49	453	741	1.09		0.18	2.41	0.12	1.45	0.29	48	92
36	42	2.65	48	344	900	1.21		0.22	1.73	0.12	8.24	0.48	66	67
212	73	4.77	95	623	1660	2.31		0.26	1.97	0.32	9.47	0.55	90	20
144	49	3.25	65	424	1130	1.57		0.17	1.34	0.22	6.45	0.38	61	14
0	35	1.48	27	247	520	0.58		0.08	1.74	0.1	0.44	0.04	56	4
21	44	0.97	37	258	235	0.7		0.09	8.62	0.05	0.54	0.36	34	5
20	4	0.4	4	51	10	0.32		0.09	0.18	0.06	0.63	0.04	3	0
24	53	2.46	42	329	670	1.93	16	0.27	0.7	0.42	2.71	0.19	58	1
2	57	2.27	44	328	495	0.76	9	0.32	0.87	0.31	2.04	0.15	44	1
35	40	1.8			590									6
60	171	3.42	30	335	1051	3.49	71	0.35		0.28	8.05	0.18	61	2
52	182	2.65	26	229	976	2.62	82	0.32	1.34	0.23	6.38	0.15	65	2
79	102	6.18	18	262	973	1.31	60	0.28	0.7	0.7	4.16	0.09	103	0
53	60	2.71			738		0							0
44	324	1.32	51	240	784	2.51	99	0.08	1.47	0.42	1.91	0.39	65	1
250	161	2.6	25	212	777	1.66	188	0.53	0.6	0.48	3.55	0.16	73	2
55	84	2.61	33	340	615	1	33	0.33	0.87	0.22	3.4	0.11	85	3
68	185	3.5	37	353	939	1.17	130	0.46	1.83	0.42	4.23	0.11	92	3
30	127	2.74	24	254	539	2.27	4	0.29	0.43	0.24	3.95	0.12	52	2
103	92	5.55	45	527	742	5.8	2	0.34	1.61	0.41	6.73	0.37	84	1
44	24	2.31	13	143	670	1.98	0	0.24	0.27	0.27	3.65	0.05	48	0
85	300	2.7			1620									0
60	300	1.8			1770									
55	250	2.7			1540									
50	20	6			1360									
95	250	2.7			1840									
80	250	2.7			1740									
70	300	1.8			1650									
30	283	6.1	48	548	507	1.45		0.37	3.15	0.43	3.76	0.3	102	12
29	272	5.96	56	576	454	1.59		0.3	3.2	0.37	3.47	0.31	85	12
30	270	6.36	83	636	499	2.21		0.3	3.51	0.36	3.67	0.38	77	13
21	68	2.21	22	240	632	0.98		0.41	2.33	0.26	3.7	0.16	66	6
22	61	2.22	32	288	625	1.21		0.37	2.56	0.21	3.68	0.18	53	6

Table H–1

Food Composition

(Computer code number is for West Diet Analysis program)   (For purposes of calculations, use "0" for t, <1, <.1, <.01, etc.)

Computer Code Number	Food Description	Measure	Wt (g)	H₂O (%)	Ener (kcal)	Prot (g)	Carb (g)	Dietary Fiber (g)	Fat (g)	Fat Breakdown (g) Sat	Mono	Poly
	MIXED DISHES—Continued											
56010	On whole wheat	1 ea	137	52	339	12	29	5	20	4.9		
	Cheese, grilled:											
56013	On white bread, soft	1 ea	119	37	399	17	30	1	23	11.9		
56015	On part whole wheat	1 ea	119	37	402	18	26	2	25	13.1		
56014	On whole wheat	1 ea	132	38	432	20	30	4	27	13.8		
56000	Chicken fillet	1 ea	182	47	515	24	39		29	8.5	10.4	8.4
	Chicken salad:											
56017	On white bread, soft	1 ea	110	41	366	10	31	2	22	2.7		
56019	On part whole wheat	1 ea	110	41	369	11	27	3	24	3.1		
56018	On whole wheat	1 ea	123	41	399	13	32	5	26	3.4		
56020	Corned beef & swiss on rye	1 ea	156	47	427	28	22	6	26	9.5		
	Egg salad:											
56025	On white bread, soft	1 ea	117	43	379	9	31	1	24	3.8		
56027	On part whole wheat	1 ea	116	44	378	10	27	2	26	4.3		
56026	On whole wheat	1 ea	130	44	410	11	31	5	28	4.6		
	Ham:											
56032	On rye bread	1 ea	150	52	345	22	29	4	15	3.2		
56029	On white bread, soft	1 ea	157	52	365	24	30	2	16	3.3		
56031	On part whole wheat	1 ea	156	54	355	25	26	2	17	3.6		
56030	On whole wheat	1 ea	169	53	378	27	29	4	18	3.9		
	Ham & cheese:											
56035	On white bread, soft	1 ea	157	48	423	24	31	2	23	8.1		
56037	On part whole wheat	1 ea	156	49	417	25	26	2	24	8.7		
56036	On whole wheat	1 ea	170	48	446	27	30	4	25	9.2		
56033	Ham & swiss on rye	1 ea	150	48	386	22	30	4	19	6.5		
	Ham salad:											
56064	On white bread, soft	1 ea	131	47	361	10	37	1	19	4.2		
56066	On part whole wheat	1 ea	131	48	358	11	33	2	21	4.7		
56065	On whole wheat	1 ea	144	48	383	12	37	4	22	5		
56038	Patty melt: Ground beef & cheese on rye	1 ea	182	46	561	37	22	6	37	12.7		
	Peanut butter & jelly:											
56040	On white bread, soft	1 ea	101	27	348	11	47	3	14	2.7		
56042	On part whole wheat	1 ea	101	27	339	12	45	6	15	3		
56041	On whole wheat	1 ea	114	27	398	13	51	5	17	3.6		
	Roast beef:											
66003	On a bun	1 ea	139	49	346	22	33		14	3.6	6.8	1.7
56044	On white bread, soft	1 ea	157	46	405	29	35	1	16	2.9		
56046	On part whole wheat	1 ea	156	47	398	30	30	2	17	3.2		
56045	On whole wheat	1 ea	169	47	423	32	34	4	18	3.4		
	Tuna salad:											
56048	On white bread, soft	1 ea	122	46	326	13	35	1	14	1.9		
56050	On part whole wheat	1 ea	122	47	322	14	32	2	16	2.2		
56049	On whole wheat	1 ea	135	47	347	16	36	4	17	2.4		
	Turkey:											
56052	On white bread, soft	1 ea	156	54	346	24	30	1	14	1.9		
56054	On part whole wheat	1 ea	155	55	336	25	25	2	15	2.1		
56053	On whole wheat	1 ea	169	54	360	27	29	4	16	2.3		
	Turkey ham:											
56103	On rye bread	1 ea	150	60	280	21	20	6	14	2.5		
56104	On white bread, soft	1 ea	156	55	331	21	30	2	14	2.5		
56106	On part whole wheat	1 ea	156	56	346	21	25	2	18	3.6		
56105	On whole wheat	1 ea	169	56	344	24	29	4	15	3		
57531	Taco	1 ea	171	58	369	21	27		21	11.4	6.6	1
	Tostada:											
66017	With refried beans	1 ea	144	66	223	10	27		10	5.4	3.1	0.7
66018	With beans & beef	1 ea	225	70	333	16	30		17	11.5	3.5	0.6
56062	With beans & chicken	1 ea	156	70	242	19	16	3	11	4.5		
	NUTS, SEEDS and PRODUCTS											
	Almonds:											
4571	Dry roasted, salted	1 c	138	3	824	30	27	16	73	5.6	46.4	17.5
4503	Slivered, packed, unsalted	1 c	108	5	624	23	21	13	55	4.2	34.7	13.2

PAGE KEY: H–2 = Beverages H–4 = Dairy H–8 = Eggs H–10 = Fat/Oil H–12 = Fruit H–18 = Bakery
H–24 = Grain H–30 = Fish H–30 = Meats H–34 = Poultry H–36 = Sausage H–38 = Mixed/Fast H–42 = Nuts/Seeds H–44 = Sweets
H–46 = Vegetables/Legumes H–56 = Vegetarian H–60 = Misc H–60 = Soups/Sauces H–62 = Fast H–76 = Convenience H–78 = Baby foods

Chol (mg)	Calc (mg)	Iron (mg)	Magn (mg)	Pota (mg)	Sodi (mg)	Zinc (mg)	VT-A (RAE)	Thia (mg)	VT-E (mg α)	Ribo (mg)	Niac (mg)	V-B$_6$ (mg)	Fola (µg)	VT-C (mg)
23	51	2.54	61	346	690	1.87		0.37	2.91	0.2	3.97	0.25	45	7
53	407	2	26	162	1155	2.03		0.29	1.01	0.4	2.37	0.08	60	0
57	431	2	38	208	1197	2.37		0.24	1.15	0.37	2.25	0.1	46	0
60	440	2.33	68	264	1293	3.13		0.24	1.44	0.36	2.46	0.16	37	0
60	60	4.68	35	353	957	1.87	31	0.33		0.24	6.81	0.2	100	9
30	76	2.21	20	146	483	0.79		0.3	5.48	0.24	4.09	0.27	63	1
33	70	2.25	32	194	468	1.04		0.25	6.08	0.2	4.15	0.31	49	1
35	59	2.6	63	252	528	1.76		0.25	6.65	0.18	4.47	0.38	39	1
83	267	3.03	28	232	1470	3.59		0.2	2.59	0.33	2.76	0.18	32	0
154	87	2.36	18	122	500	0.76		0.31	4.24	0.38	2.45	0.21	74	0
166	80	2.38	29	165	479	0.99		0.25	4.65	0.34	2.3	0.24	60	0
177	71	2.75	60	222	543	1.71		0.26	5.17	0.33	2.54	0.32	52	0
50	54	2.89	40	390	1245	2.76		0.84	2.36	0.42	6.42	0.37	54	0
54	76	3.1	32	384	1237	2.61		0.9	2.52	0.44	6.83	0.39	60	0
56	68	3.12	43	438	1252	2.92		0.88	2.72	0.4	6.91	0.42	44	0
59	57	3.46	72	500	1336	3.66		0.9	3.06	0.39	7.29	0.49	34	0
64	246	2.82	33	329	1366	2.71		0.7	2.58	0.46	5.36	0.31	61	0
67	248	2.82	44	379	1388	3.03		0.67	2.78	0.42	5.35	0.34	46	0
71	247	3.17	74	441	1487	3.8		0.69	3.14	0.41	5.7	0.41	36	0
56	241	2.68	43	370	1397	3		0.65	2.47	0.45	5.06	0.29	55	0
29	72	2.25	21	167	935	1.06		0.55	3.34	0.28	3.69	0.18	59	0
30	65	2.26	32	213	950	1.31		0.52	3.67	0.23	3.65	0.21	44	0
32	54	2.59	62	269	1029	2.02		0.53	4.08	0.21	3.92	0.28	34	0
113	222	4.17	39	391	714	7.04		0.25	3.52	0.46	6.14	0.37	37	0
1	76	2.29	52	240	429	1.05		0.29	2.55	0.23	5.44	0.15	79	2
0	54	2.41	88	320	418	1.82		0.22	2.9	0.15	5.43	0.22	53	2
0	80	2.66	75	335	465	1.5		0.28	3.24	0.21	6.41	0.2	76	2
51	54	4.23	31	316	792	3.39	11	0.38	0.19	0.31	5.87	0.26	57	2
43	76	4.13	30	436	1606	3.73		0.35	3.39	0.36	6.78	0.4	67	0
45	67	4.21	41	493	1643	4.11		0.29	3.63	0.32	6.86	0.44	51	0
47	57	4.6	70	557	1743	4.9		0.29	4.01	0.3	7.24	0.51	42	0
13	76	2.41	25	168	589	0.68		0.3	2.77	0.24	5.89	0.13	63	1
13	69	2.44	36	215	578	0.9		0.25	3.07	0.19	6.07	0.16	48	1
14	59	2.79	67	273	642	1.6		0.25	3.46	0.17	6.47	0.23	38	1
43	72	2.19	31	307	1589	1.33		0.31	3.28	0.29	9.29	0.42	60	0
45	63	2.15	42	356	1625	1.56		0.25	3.52	0.24	9.52	0.45	45	0
47	53	2.46	71	417	1736	2.25		0.25	3.92	0.22	10.07	0.53	35	0
55	51	4.04	27	343	1179	2.94		0.22	2.86	0.33	4.32	0.31	29	0
53	77	4.22	29	351	1252	2.85		0.32	2.83	0.41	5.29	0.29	62	0
57	68	4.22	39	395	1264	3.12		0.26	1.23	0.36	5.16	0.34	46	0
58	59	4.72	69	466	1361	3.95		0.26	3.41	0.36	5.63	0.39	37	0
56	221	2.41	70	474	802	3.93	108	0.15	1.88	0.44	3.21	0.24	68	2
30	210	1.89	59	403	543	1.9	45	0.1	1.15	0.33	1.32	0.16	43	1
74	189	2.45	68	490	871	3.17	101	0.09	1.8	0.5	2.86	0.25	86	4
55	146	1.57	41	263	386	1.93		0.09	0.66	0.15	4.26	0.31	25	5
0	367	6.22	395	1029	468	4.89	<1	0.1	35.88	1.19	5.31	0.17	46	0
0	268	4.64	297	786	1	3.63	<1	0.26	27.94	0.88	4.24	0.14	31	0

Table H–1

Food Composition   (Computer code number is for West Diet Analysis program)   (For purposes of calculations, use "0" for t, <1, <.1, <.01, etc.)

Computer Code Number	Food Description	Measure	Wt (g)	H₂O (%)	Ener (kcal)	Prot (g)	Carb (g)	Dietary Fiber (g)	Fat (g)	Fat Breakdown (g) Sat	Mono	Poly
	NUTS, SEEDS and PRODUCTS—Continued											
4502	Whole, dried, unsalted	1 oz	28	5	162	6	6	3	14	1.1	9	3.4
4534	Almond butter:	1 tbs	16	1	101	2	3	1	9	0.9	6.1	2
4572	Salted	1 tbs	16	1	101	2	3	1	9	0.9	6.1	2
4750	Brazil nuts, dry (about 7)	1 c	140	3	918	20	17	10	93	21.2	34.4	28.8
	Cashew nuts:											
4519	Dry roasted, salted	1 oz	28	2	161	4	9	1	13	2.6	7.6	2.2
4621	Dry roasted, unsalted	1 oz	28	2	161	4	9	1	13	2.6	7.6	2.2
4596	Oil roasted, salted	1 oz	28	2	163	5	8	1	13	2.4	7.3	2.4
4622	Oil roasted, unsalted:	1 c	130	3	754	22	39	4	62	11	33.7	11.1
4622	Ounce	1 oz	28	3	162	5	8	1	13	2.4	7.3	2.4
4537	Cashew butter, unsalted	1 tbs	16	3	94	3	4	<1	8	1.6	4.7	1.3
4662	Salted	1 tbs	16	3	94	3	4	<1	8	1.6	4.7	1.3
4538	Chestnuts, European, roasted, (1 cup = approx 17 kernels)	1 c	143	40	350	5	76	7	3	0.6	1.1	1.2
	Coconut, raw:											
4508	Piece 2 x 2 x ½"	1 pce	45	47	159	1	7	4	15	13.4	0.6	0.2
4507	Shredded/grated, unpacked	½ c	40	47	142	1	6	4	13	11.9	0.6	0.1
	Coconut, dried, shredded/grated:											
4510	Unsweetened	1 c	78	3	515	5	18	13	50	44.6	2.1	0.6
4511	Sweetened	1 c	93	13	466	3	44	4	33	29.3	1.4	0.4
4559	Coconut milk, canned	1 c	226	73	445	5	6	3	48	42.7	2	0.5
4514	Filberts/hazelnuts, chopped	1 oz	28	5	176	4	5	3	17	1.2	12.8	2.2
	Mixed Nuts:											
4592	Dry roasted, salted	1 c	137	2	814	24	35	12	70	9.5	43	14.8
4593	Oil roasted, salted	1 c	142	2	876	24	30	13	80	12.4	45	18.9
4533	Oil roasted, unsalted	1 c	142	2	876	24	30	14	80	12.4	45	18.9
4762	Peanuts, oil roasted, salted	1 oz	28	1	168	8	4	3	15	2.4	7.3	4.3
4578	Pecan halves, dried, unsalted	1 oz	28	4	193	3	4	3	20	1.7	11.4	6.1
4583	Dry roasted, salted	¼ c	28	4	199	3	4	3	21	1.8	12.3	5.8
4554	Pine nuts/pinons, dried	1 oz	28	6	176	3	5	3	17	2.6	6.4	7.2
4520	Pistachios, dried, shelled	1 oz	28	4	156	6	8	3	12	1.5	6.5	3.8
4540	Dry roasted,salted,shelled	1 c	128	2	727	27	34	13	59	7.1	31	17.8
4522	Pumpkin kernels, dried, unsalted	1 oz	28	7	151	7	5	1	13	2.4	4	5.9
4625	Roasted, salted	1 c	227	7	1185	75	30	9	96	18.1	29.7	43.6
4524	Sesame seeds, hulled, dried	¼ c	38	5	225	8	6	5	21	2.9	7.9	9.1
8878	Soy nuts, BBQ	5 pce	28		119	12	9	4	4	1		
8877	Salted	5 pce	28		119	12	9	5	4	1		
	Sunflower seed kernels:											
4545	Dry	¼ c	36	5	205	8	7	4	18	1.9	3.4	11.8
4546	Oil roasted	¼ c	34	3	209	7	5	2	20	2	3.7	12.9
4532	Tahini (sesame butter)	1 tbs	15	3	91	3	3	1	8	1.2	3.2	3.7
44059	Trail Mix w/chocolate chips	1 c	146	7	707	21	66	8	47	8.9	19.8	16.5
4525	Black walnuts, chopped	1 oz	28	5	173	7	3	2	17	0.9	4.2	9.8
4556	English walnuts, chopped	1 oz	28	4	183	4	4	2	18	1.7	2.5	13.2
	SWEETENERS and SWEETS (see also Dairy (milk desserts) and Baked Goods)											
23000	Apple butter	2 tbs	36	56	62	<1	15	1	0	0	0	0
1124	Butterscotch topping	2 tbs	41	32	103	1	27	<1	<1	t	t	0
23069	Caramel topping	2 tbs	41	32	103	1	27	<1	<1	t	t	0
	Cake frosting, creamy vanilla:											
46009	Canned	2 tbs	39	15	164	0	26	<1	6	1.2	1.9	3.1
46018	From mix	2 tbs	39	12	161	<1	29	<1	5	0.7	1.5	1.1
	Candy:											
23405	Almond Joy candy bar	1 oz	28	8	134	1	17	1	8	4.9	1.5	0.3
23184	Butterscotch morsels	¼ c	43	8	243	0	27	0	12	10.6		
23015	Caramel, plain or chocolate	1 pce	10	8	38	<1	8	<1	1	0.7	t	t
90712	Chewing gum	1 pce	3	3	7	0	2		<1	t	t	t
25125	Sugarless	1 pce	3		6	0	2		0	0	0	0
	Chocolate:											
23016	Milk chocolate	1 oz	28	2	150	2	17	1	8	4	3.7	0.2

PAGE KEY: H–2 = Beverages H–4 = Dairy H–8 = Eggs H–10 = Fat/Oil H–12 = Fruit H–18 = Bakery
H–24 = Grain H–30 = Fish H–30 = Meats H–34 = Poultry H–36 = Sausage H–38 = Mixed/Fast H–42 = Nuts/Seeds H–44 = Sweets
H–46 = Vegetables/Legumes H–56 = Vegetarian H–60 = Misc H–60 = Soups/Sauces H–62 = Fast H–76 = Convenience H–78 = Baby foods

Chol (mg)	Calc (mg)	Iron (mg)	Magn (mg)	Pota (mg)	Sodi (mg)	Zinc (mg)	VT-A (RAE)	Thia (mg)	VT-E (mg α)	Ribo (mg)	Niac (mg)	V-B$_6$ (mg)	Fola (µg)	VT-C (mg)
0	69	1.2	77	204	<1	0.94	<1	0.07	7.24	0.23	1.1	0.04	8	0
0	43	0.59	48	121	2	0.49	0	0.02	3.25	0.1	0.46	0.01	10	0
0	43	0.59	48	121	72	0.49	<1	0.02	4.16	0.1	0.46	0.01	10	0
0	224	3.4	526	923	4	5.68	0	0.86	8.02	0.05	0.41	0.14	31	1
0	13	1.68	73	158	179	1.57	0	0.06	0.26	0.06	0.39	0.07	19	0
0	13	1.68	73	158	4	1.57	0	0.06	0.26	0.06	0.39	0.07	19	0
0	12	1.69	76	177	86	1.5	0	0.1	0.26	0.06	0.49	0.09	7	0
0	56	7.86	355	822	17	6.96	0	0.47	1.2	0.28	2.26	0.42	32	0
0	12	1.69	76	177	4	1.5	0	0.1	0.26	0.06	0.49	0.09	7	0
0	7	0.8	41	87	2	0.83	0	0.05	0.25	0.03	0.26	0.04	11	0
0	7	0.8	41	87	98	0.83	0	0.05	0.15	0.03	0.26	0.04	11	0
0	41	1.3	47	847	3	0.82	1	0.35	0.72	0.25	1.92	0.71	100	37
0	6	1.09	14	160	9	0.5	0	0.03	0.11	<.01	0.24	0.02	12	1
0	6	0.97	13	142	8	0.44	0	0.03	0.1	<.01	0.22	0.02	10	1
0	20	2.59	70	424	29	1.57	0	0.05	0.34	0.08	0.47	0.23	7	1
0	14	1.79	46	313	244	1.69	0	0.03	0.36	0.02	0.44	0.25	7	1
0	41	7.46	104	497	29	1.27	0	0.05	1.47	0	1.44	0.06	32	2
0	32	1.32	46	190	0	0.69	<1	0.18	4.21	0.03	0.5	0.16	32	2
0	96	5.07	308	818	917	5.21	<1	0.27	14.99	0.27	6.44	0.41	68	1
0	153	4.56	334	825	926	7.21	<1	0.71	10.22	0.32	7.19	0.34	118	1
0	153	4.56	334	825	16	7.21	1	0.71	8.52	0.32	7.19	0.34	118	1
0	17	0.43	49	203	90	0.92	0	0.02	1.94	0.02	3.87	0.13	34	0
0	20	0.71	34	115	0	1.27	1	0.18	0.39	0.04	0.33	0.06	6	0
0	20	0.78	37	119	107	1.42	2	0.13	0.36	0.03	0.33	0.05	4	1
0	2	0.86	66	176	20	1.2	<1	0.35	0.98	0.06	1.22	0.03	16	1
0	30	1.16	34	287	<1	0.62	8	0.24	0.64	0.04	0.36	0.48	14	1
0	141	5.38	154	1334	518	2.94	17	1.08	2.47	0.2	1.82	1.63	64	3
0	12	4.19	150	226	5	2.09	5	0.06	0	0.09	0.49	0.06	16	1
0	98	33.91	1212	1830	1305	16.89	43	0.48	0	0.72	3.95	0.2	129	4
0	50	2.96	132	155	15	3.9	1	0.27	0.1	0.03	1.78	0.06	36	0
0	59	1.07			415		0							0
0	59	1.07			148		0							0
0	42	2.44	127	248	1	1.82	1	0.82	12.42	0.09	1.62	0.28	82	1
0	19	2.28	43	164	1	1.77	<1	0.11	13.25	0.1	1.4	0.27	80	0
0	21	0.95	53	69	<1	1.57	<1	0.24	0.34	0.02	0.85	0.02	15	0
6	159	4.95	235	946	177	4.58	3	0.6	15.62	0.33	6.43	0.38	95	2
0	17	0.87	56	146	1	0.94	1	0.02	0.5	0.04	0.13	0.16	9	0
0	27	0.81	44	123	1	0.87	<1	0.1	0.2	0.04	0.32	0.15	27	0
0	5	0.11	2	33	5	0.02	<1	<.01	<.01	<.01	0.02	0.01	<1	0
0	22	0.08	3	34	143	0.08	11	<.01	0	0.04	0.02	<.01	1	0
0	22	0.08	3	34	143	0.08	11	<.01	0	0.04	0.02	<.01	1	0
0	1	0.06	0	13	72	0.03	0	<.01	0.82	0.12	0.09	0	3	0
0	2	<.01	1	4	44	<.01	31	<.01	0.76	<.01	<.01	<.01	0	0
1	18	0.36	18	71	40	0.22		<.01	<.01	0.04	0.13	0.02		0
0	0	0		80	45		0	0.03		0.03	0.03			0
1	14	0.01	2	21	24	0.04	<1	<.01	0.28	0.03	0.02	<.01	<1	0
0	0	0	0	0	<1	0	0	0	0	0	0	0	0	0
				0	0									
6	53	0.66	18	104	22	0.56	14	0.03	0.57	0.08	0.11	0.01	3	0

Table H–1

Food Composition

(Computer code number is for West Diet Analysis program)   (For purposes of calculations, use "0" for t, <1, <.1, <.01, etc.)

Computer Code Number	Food Description	Measure	Wt (g)	H₂O (%)	Ener (kcal)	Prot (g)	Carb (g)	Dietary Fiber (g)	Fat (g)	Fat Breakdown (g) Sat	Mono	Poly
	SWEETENERS and SWEETS (see also Dairy (milk desserts) and Baked Goods)—Continued											
23018	Milk chocolate with almonds	1 oz	28	2	147	3	15	2	10	4.8	3.8	0.6
23058	Milk chocolate with rice cereal	1 oz	28	2	139	2	18	1	7	4.4	2.4	0.2
23012	Semisweet chocolate chips	1 c	168	1	805	7	106	10	50	29.8	16.7	1.6
23057	Sweet Dark chocolate (candy bar)	1 ea	41	1	218	2	24	3	13	7.9	2.1	0.2
23024	Fondant candy, uncoated (mints, candy corn, other)	1 pce	16	7	60	0	15	0	<1	0	0	0
23409	Gumdrops	1 c	182	1	721	0	180	<1	0	0	0	0
23031	Hard candy-all flavors	1 pce	6	1	24	0	6	0	<1	0	0	0
23033	Jellybeans	10 pce	11	6	41	0	10	<1	<1	0	0	0
23025	Fudge, chocolate	1 pce	17	10	70	<1	13	<1	2	1	0.5	t
23046	M&M's plain chocolate candy	10 pce	7	2	34	<1	5	<1	1	0.9	0.2	t
23048	M&M's peanut chocolate candy	10 pce	20	2	103	2	12	1	5	2.1	2.2	0.8
23037	MARS almond bar	1 ea	50	4	234	4	31	1	12	3.6	5.3	2
23038	MILKY WAY candy bar	1 ea	60	6	254	3	43	1	10	4.7	3.6	0.4
23021	Milk chocolate-coated peanuts	1 c	149	2	773	20	74	7	50	21.8	19.3	6.5
23081	Peanut brittle, recipe	1 c	147	1	711	11	104	4	28	6.1	11.9	6.7
23036	Skor English toffee candy bar	1 ea	39	2	209	1	24	1	13	7.3	3.6	0.5
1131	Snickers candy bar (2.2oz)	1 ea	62	5	297	5	37	2	15	5.6	6.5	3
23086	Fruit Roll-Up (small)	1 ea	14	10	52	<1	12	<1	<1	t	0.2	t
23174	Fruit juice bar (2.5 fl oz)	1 ea	77	78	63	1	16	1	<1	t	0	t
23052	Gelatin dessert/Jello, prepared	½ c	135	84	84	2	19	0	0	0	0	0
23093	SugarFree	½ c	117	95	23	1	5	0	0	0	0	0
25001	Honey	1 tbs	21	17	64	<1	17	<1	0	0	0	0
23003	Jellies:	1 tbs	19	30	51	<1	13	<1	<1	t	t	t
23004	Packet	1 ea	14	30	37	<1	10	<1	<1	t	t	t
23005	Marmalade	1 tbs	20	33	49	<1	13	<1	0	0	0	0
23007	Marshmallows	1 ea	7	16	22	<1	6	<1	<1	t	t	t
23071	Marshmallow creme topping	2 tbs	38	20	122	<1	30	<1	<1	t	t	t
25003	Molasses	2 tbs	41	22	119	0	31	0	<1	t	t	t
23050	Popsicle/ice pops	1 ea	128	80	92	0	24	0	0	0	0	0
23171	Rice crispie bar	1 ea	28	13	107	1	20	<1	3	0.6	1.3	0.8
	Sugars:											
25005	Brown sugar	1 c	220	2	829	0	214	0	0	0	0	0
25006	White sugar, granulated	1 tbs	12	0	46	0	12	0	0	0	0	0
25007	Packet	1 ea	6	0	23	0	6	0	0	0	0	0
	Sweeteners:											
25038	Equal, packet	1 ea	1	12	4	<1	1	0	<1	t	t	t
25208	Sweet 'N Low, packet	1 ea	1		4	0	1	0	0	0	0	0
	Syrups											
23014	Chocolate, hot fudge type	2 tbs	43	22	150	2	27	1	4	1.7	1.7	0.1
23056	Thin type	2 tbs	38	29	93	1	25	1	<1	0.3	0.2	t
23042	Pancake table syrup (corn and maple)	2 tbs	40	38	94	0	25	<1	0	0	0	0
	VEGETABLES AND LEGUMES											
	Amaranth leaves:											
5375	Raw, chopped	1 c	28	92	6	1	1	<1	<1	t	t	t
5376	Raw, each	1 ea	14	92	3	<1	1	<1	<1	t	t	t
5377	Cooked	1 c	132	91	28	3	5	2	<1	t	t	0.1
6033	Arugula, raw, chopped	½ c	10	92	2	<1	<1	<1	<1	t	t	t
5000	Artichokes, cooked globe (300 g with refuse)	1 ea	120	84	60	4	13	6	<1	t	t	t
	Artichoke hearts:											
5192	Cooked from frozen	1 c	168	86	76	5	15	8	1	0.2	t	0.4
5191	Marinated	1 c	130		116	5	14	5	7	0		
7962	In water	½ c	100		38	2	6	0	0	0	0	0
	Asparagus, green:											
5003	Cooked from fresh, cuts and tips	½ c	90	93	20	2	4	2	<1	t	t	0.1
5004	Spears, ½" diam at base	4 ea	60	93	13	1	2	1	<1	t	t	t
5005	Cooked from frozen, cuts and tips	½ c	90	94	16	3	2	1	<1	t	t	0.2
5006	Spears, ½" diam at base	4 ea	60	94	11	2	1	1	<1	t	t	0.1
5007	Canned, spears, ½" diam at base	4 pce	72	94	14	2	2		<1	0.1	t	0.2

PAGE KEY: H–2 = Beverages H–4 = Dairy H–8 = Eggs H–10 = Fat/Oil H–12 = Fruit H–18 = Bakery
H–24 = Grain H–30 = Fish H–30 = Meats H–34 = Poultry H–36 = Sausage H–38 = Mixed/Fast H–42 = Nuts/Seeds H–44 = Sweets
H–46 = Vegetables/Legumes H–56 = Vegetarian H–60 = Misc H–60 = Soups/Sauces H–62 = Fast H–76 = Convenience H–78 = Baby foods

Chol (mg)	Calc (mg)	Iron (mg)	Magn (mg)	Pota (mg)	Sodi (mg)	Zinc (mg)	VT-A (RAE)	Thia (mg)	VT-E (mg α)	Ribo (mg)	Niac (mg)	V-B$_6$ (mg)	Fola (µg)	VT-C (mg)
5	63	0.46	25	124	21	0.38	12	0.02	1.26	0.12	0.21	0.01	4	0
5	48	0.21	14	96	41	0.31	17	0.02	0.56	0.08	0.13	0.02	4	0
0	54	5.26	193	613	18	2.72	0	0.09	0.39	0.15	0.72	0.06	5	0
2	12	0.87	13	206	2	<.01		0	0.08	<.01	0	0	0	0
0	0	<.01	0	1	3	0	0	<.01	0	<.01	<.01	<.01	0	0
0	5	0.73	2	9	80	0	0	0.01	0	0.02	0.02	<.01	0	0
0	0	0.02	0	0	2	<.01	0	<.01	0	<.01	<.01	<.01	0	0
0	0	0.01	0	4	6	<.01	0	<.01	0	<.01	<.01	<.01	0	0
2	8	0.3	6	22	8	0.19	7	<.01	0.03	0.01	0.03	<.01	1	0
1	7	0.08	2	14	4	0.08	2	<.01	0.08	0.01	0.01	<.01	<1	0
2	20	0.23	15	69	10	0.48	5	0.02	0.51	0.03	0.82	0.02	8	0
8	84	0.55	36	162	85	0.56	8	0.02	3.88	0.16	0.47	0.03	4	0
8	78	0.46	20	145	144	0.43	11	0.02	0.75	0.13	0.21	0.03	4	1
13	155	1.95	143	748	61	3.56	51	0.17	5.16	0.26	6.33	0.31	12	0
18	40	1.79	62	247	654	1.28	57	0.2	3.76	0.06	3.89	0.12	68	0
21	51	0.22	4	60	124	0.07		<.01	0.02	0.04	0.05	0.01	1	0
8	58	0.47	35	185	165	0.97	26	0.08	0.68	0.12	0.98	0.04	19	0
0	4	0.14	3	41	44	0.03	1	0.01	0.08	<.01	0.01	0.04	1	17
0	4	0.15	3	41	3	0.04	1	<.01	0.1	0.01	0.12	0.02	5	7
0	4	0.03	1	1	101	0.01	0	0	0	<.01	<.01	0	1	0
0	4	0.01	1	1	56	0	0	0	0	0	0	0	0	0
0	1	0.09	0	11	1	0.05	0	0	0	<.01	0.03	<.01	<1	0
0	1	0.04	1	10	6	<.01	<1	<.01	0	<.01	<.01	<.01	<1	0
0	1	0.03	1	8	4	<.01	<1	0	0	<.01	<.01	<.01	<1	0
0	8	0.03	0	7	11	<.01	1	<.01	0.01	<.01	0.01	<.01	2	1
0	0	0.02	0	0	6	<.01	0	<.01	0	<.01	<.01	<.01	<1	0
0	1	0.08	1	2	30	0.02	<1	<.01	0	<.01	0.03	<.01	<1	0
0	84	1.94	99	600	15	0.12	0	0.02	0	<.01	0.38	0.27	0	0
0	0	0	1	5	15	0.03	0	0	0	0	0	0	0	0
0	2	0.51	4	12	123	0.15		0.1	0.41	0.11	1.31	0.13	28	4
0	187	4.2	64	761	86	0.4	0	0.02	0	0.02	0.18	0.06	2	0
0	0	<.01	0	0	0	0	0	0	0	<.01	0	0	0	0
0	0	<.01	0	0	0	0	0	0	0	<.01	0	0	0	0
0	0	<.01	0	0	<1	0	0	0	0	0	0	0	0	0
0	0	0			0		0							
1	43	0.68	28	194	149	0.36	2	0.03	1.07	0.12	0.16	0.03	2	0
0	5	5.15	25	183	58	0.28		<.01	0.01	0.31	12.76	<.01	2	0
0	1	0.01	1	6	33	0.03	0	<.01	0	<.01	<.01	<.01	0	0
0	60	0.65	15	171	6	0.25	0	<.01	0.22	0.04	0.18	0.05	24	12
0	30	0.32	8	86	3	0.13	0	<.01	0.11	0.02	0.09	0.03	12	6
0	276	2.98	73	846	28	1.16	183	0.03	0.66	0.18	0.74	0.23	75	54
0	16	0.15	5	37	3	0.05	12	<.01	0.04	<.01	0.03	<.01	10	2
0	54	1.55	72	425	114	0.59	11	0.08	0.23	0.08	1.2	0.13	61	12
0	35	0.94	52	444	89	0.6	13	0.1	0.27	0.27	1.54	0.15	200	8
0	0	0			488		0							46
0	0	1.35			250		6							4
0	21	0.82	13	202	13	0.54	45	0.15	0.34	0.13	0.98	0.07	134	7
0	14	0.55	8	134	8	0.36	30	0.1	0.23	0.08	0.65	0.05	89	5
0	16	0.5	9	155	3	0.37	36	0.06	1.08	0.09	0.93	0.02	122	22
0	11	0.34	6	103	2	0.25	24	0.04	0.72	0.06	0.62	0.01	81	15
0	12	1.32	7	124	207	0.29	30	0.04	0.22	0.07	0.69	0.08	69	13

Table H–1

Food Composition

(Computer code number is for West Diet Analysis program)   (For purposes of calculations, use "0" for t, <1, <.1, <.01, etc.)

Computer Code Number	Food Description	Measure	Wt (g)	H₂O (%)	Ener (kcal)	Prot (g)	Carb (g)	Dietary Fiber (g)	Fat (g)	Fat Breakdown (g)		
										Sat	Mono	Poly
	VEGETABLES AND LEGUMES—Continued											
	Bamboo shoots:											
5401	Canned, drained slices	1 c	131	94	25	2	4		1	0.1	t	0.2
6736	Raw slices	1 c	151	91	41	4	8	3	<1	0.1	t	0.2
5249	Cooked slices	1 c	120	96	14	2	2	1	<1	t	t	0.1
	Beans (see also alphabetical listing this section):											
7034	Adzuki beans, cooked	½ c	115	66	147	9	28	8	<1	t		
7012	Black beans, cooked	½ c	86	66	114	8	20	7	<1	0.1	t	0.2
	Canned beans (white/navy):											
7004	With pork and tomato sauce	½ c	127	73	124	7	25	6	1	0.5	0.6	0.2
7023	With sweet sauce	½ c	130	71	144	7	27	7	2	0.7	0.8	0.2
56101	With frankfurters	½ c	130	69	185	9	20	9	9	3.1	3.7	1.1
	Lima beans:											
5247	Fordhooks, cooked from frozen	½ c	85	73	88	5	16	5	<1	t	t	0.1
5019	Baby, cooked from frozen	½ c	90	72	94	6	18	5	<1	t	t	0.1
7010	Cooked from dry, drained	½ c	94	70	108	7	20	7	<1	t	t	0.2
7352	Red Mexican, cooked f/dry	½ c	112	70	127	8	24	9	<1	t	t	0.2
	Snap bean/green string beans:											
5011	Cooked from fresh	½ c	63	89	22	1	5	2	<1	t	t	t
5013	Cooked from frozen	½ c	68	91	19	1	4	2	<1	t	t	t
5015	Canned, drained	½ c	68	93	14	1	3	1	<1	t	t	t
5194	Snap bean, yellow, cooked f/fresh	½ c	63	89	22	1	5	2	<1	t	t	t
	Bean sprouts (mung):											
5020	Raw	½ c	52	90	16	2	3	1	<1	t	t	t
5246	Cooked, stir fried	½ c	62	84	31	3	7	1	<1	t	t	t
5021	Cooked, boiled, drained	½ c	62	93	13	1	3	<1	<1	t	t	t
5197	Canned, drained	½ c	63	96	8	1	1	1	<1	t	t	t
	Beets:											
5022	Cooked from fresh, sliced or diced	½ c	85	87	37	1	8	2	<1	t	t	t
5023	Whole beets, 2" diam	2 ea	100	87	44	2	10	2	<1	t	t	t
5024	Canned, sliced or diced	½ c	79	91	24	1	6	1	<1	t	t	t
5310	Pickled slices	½ c	114	82	74	1	19	3	<1	t	t	t
5025	Beet greens, cooked, drained	½ c	72	89	19	2	4	2	<1	t	t	t
	Broccoli:											
5027	Broccoli, raw, spear	1 ea	31	89	11	1	2	1	<1	t	t	t
5029	Cooked from fresh, spears	1 ea	180	89	63	4	13	6	1	0.1	t	0.3
5028	Chopped	½ c	78	89	27	2	6	3	<1	t	t	0.1
5234	Cooked from frozen, spear, small piece	½ c	92	91	26	3	5	3	<1	t	t	t
5030	Chopped	½ c	92	91	26	3	5	3	<1	t	t	t
5679	Broccoflower-steamed	½ c	78	90	25	2	5	2	<1	t	t	t
5033	Brussels sprouts, cooked from fresh	½ c	78	89	28	2	6	2	<1	t	t	0.2
5035	Cooked from frozen	½ c	78	87	33	3	6	3	<1	t	t	0.2
5036	Cabbage, common, raw, chopped	1 c	70	92	17	1	4		<1	t	t	t
5038	Cooked, drained	1 c	150	94	33	2	7		1	t	t	0.3
	Cabbage, Chinese:											
5041	Bok Choy, raw, shredded	1 c	70	95	9	1	2	1	<1	t	t	t
5237	Cooked, drained	1 c	170	96	20	3	3		<1	t	t	0.1
5535	Kim chee style	1 c	150	92	31	2	6	2	<1	t	t	0.1
5040	Pe Tsai, raw, chopped	1 c	76	94	12	1	2		1	<1	t	t
5235	Cooked	1 c	119	95	17	2	3	2	<1	t	t	t
6766	Cabbage, red, raw, chopped	1 c	89	90	28	1	7	2	<1	t	t	t
5238	Cooked, drained	1 c	150	91	44	2	10	4	<1	t	t	t
5043	Cabbage, savoy, raw, chopped	1 c	70	91	19	1	4	2	<1	t	t	t
5044	Cooked	1 c	145	92	35	3	8	4	<1	t	t	t
5511	Capers	1 ea	5		0	<1	<1		<1			
	Carrots:											
5045	Raw, whole, 7½ x 1⅛"	1 ea	72	88	30	1	7	2	<1	t	t	t
5046	Grated	½ c	55	88	23	1	5	2	<1	t	t	t
5047	Cooked from raw	½ c	78	90	27	1	6	2	<1	t	t	t
5358	From frozen	½ c	73	90	27	<1	6	2	<1	t	t	0.2
5199	Canned	½ c	73	93	18	<1	4	1	<1	t	t	t

PAGE KEY: H–2 = Beverages   H–4 = Dairy   H–8 = Eggs   H–10 = Fat/Oil   H–12 = Fruit   H–18 = Bakery
H–24 = Grain   H–30 = Fish   H–30 = Meats   H–34 = Poultry   H–36 = Sausage   H–38 = Mixed/Fast   H–42 = Nuts/Seeds   H–44 = Sweets
H–46 = Vegetables/Legumes   H–56 = Vegetarian   H–60 = Misc   H–60 = Soups/Sauces   H–62 = Fast   H–76 = Convenience   H–78 = Baby foods

Chol (mg)	Calc (mg)	Iron (mg)	Magn (mg)	Pota (mg)	Sodi (mg)	Zinc (mg)	VT-A (RAE)	Thia (mg)	VT-E (mg α)	Ribo (mg)	Niac (mg)	V-B$_6$ (mg)	Fola (µg)	VT-C (mg)
0	10	0.42	5	105	9	0.85	1	0.03	0.83	0.03	0.18	0.18	4	1
0	20	0.76	5	805	6	1.66	2	0.23	1.51	0.11	0.91	0.36	11	6
0	14	0.29	4	640	5	0.56	0	0.02	0.8	0.06	0.36	0.12	2	0
0	32	2.3	60	612	9	2.04	<1	0.13	0.12	0.07	0.82	0.11	139	0
0	23	1.81	60	305	1	0.96	<1	0.21	0.07	0.05	0.43	0.06	128	0
9	71	4.17	44	381	559	7.44	5	0.07	0.69	0.06	0.63	0.09	29	4
9	79	2.16	44	346	437	1.95	1	0.06	0.7	0.08	0.46	0.11	48	4
8	62	2.25	36	306	559	2.43	5	0.08	0.6	0.07	1.17	0.06	39	3
0	26	1.55	36	258	59	0.63	8	0.06	0.25	0.05	0.91	0.1	18	11
0	25	1.76	50	370	26	0.5	7	0.06	0.58	0.05	0.69	0.1	14	5
0	16	2.25	40	478	2	0.89	0	0.15	0.17	0.05	0.4	0.15	78	0
0	42	1.87	48	371	6	0.87	<1	0.13	0.08	0.07	0.38	0.12	94	2
0	28	0.41	11	92	1	0.16	22	0.05	0.28	0.06	0.39	0.04	21	6
0	33	0.6	16	86	6	0.33	19	0.02	0.24	0.06	0.26	0.04	16	3
0	18	0.61	9	74	178	0.2	15	0.01	0.19	0.04	0.14	0.03	22	3
0	29	0.81	16	188	2	0.23	3	0.05	0.28	0.06	0.39	0.04	21	6
0	7	0.47	11	77	3	0.21	1	0.04	0.05	0.06	0.39	0.05	32	7
0	8	1.18	20	136	6	0.56	1	0.09	<.01	0.11	0.74	0.08	43	10
0	7	0.4	9	63	6	0.29	1	0.03	0.04	0.06	0.51	0.03	18	7
0	9	0.27	6	17	88	0.18	<1	0.02	0.03	0.04	0.14	0.02	6	0
0	14	0.67	20	259	65	0.3	2	0.02	0.03	0.03	0.28	0.06	68	3
0	16	0.79	23	305	77	0.35	2	0.03	0.04	0.04	0.33	0.07	80	4
0	12	1.44	13	117	153	0.17	1	<.01	0.02	0.03	0.12	0.05	24	3
0	13	0.47	17	169	301	0.3	1	0.01	0.15	0.05	0.29	0.06	31	3
0	82	1.37	49	654	174	0.36	276	0.08	1.3	0.21	0.36	0.1	10	18
0	15	0.23	7	98	10	0.13	10	0.02	0.24	0.04	0.2	0.05	20	28
0	72	1.21	38	527	74	0.81	176	0.11	2.61	0.22	1	0.36	194	117
0	31	0.52	16	229	32	0.35	76	0.05	1.13	0.1	0.43	0.16	84	51
0	47	0.56	18	166	22	0.28	52	0.05	1.21	0.07	0.42	0.12	28	37
0	30	0.56	12	131	10	0.26	52	0.05	1.21	0.07	0.42	0.12	52	37
0	25	0.55	16	251	18	0.39	3	0.06	0.23	0.07	0.59	0.14	38	49
0	28	0.94	16	247	16	0.26	30	0.08	0.34	0.06	0.47	0.14	47	48
0	20	0.37	14	226	12	0.19	36	0.08	0.4	0.09	0.42	0.23	79	36
0	33	0.41	10	172	13	0.13	6	0.04	0.1	0.03	0.21	0.07	30	23
0	46	0.26	12	146	12	0.14	10	0.09	0.18	0.08	0.42	0.17	30	30
0	74	0.56	13	176	46	0.13	156	0.03	0.06	0.05	0.35	0.14	46	32
0	158	1.77	19	631	58	0.29	360	0.05	0.15	0.11	0.73	0.28	70	44
0	145	1.28	27	375	995	0.36	213	0.07	0.24	0.1	0.75	0.34	88	80
0	59	0.24	10	181	7	0.17	12	0.03	0.09	0.04	0.3	0.18	60	21
0	38	0.36	12	268	11	0.21	57	0.05	0.14	0.05	0.6	0.21	63	19
0	40	0.71	14	216	24	0.2	50	0.06	0.1	0.06	0.37	0.19	16	51
0	63	0.99	26	393	12	0.38	3	0.11	0.18	0.09	0.57	0.34	36	16
0	24	0.28	20	161	20	0.19	35	0.05	0.12	0.02	0.21	0.13	56	22
0	44	0.55	35	267	35	0.33	64	0.07	0.15	0.03	0.03	0.22	67	25
0	2	0.05			105		1							0
0	24	0.22	9	230	50	0.17	433	0.05	0.48	0.04	0.71	0.1	14	4
0	18	0.16	7	176	38	0.13	331	0.04	0.36	0.03	0.54	0.08	10	3
0	23	0.27	8	183	45	0.16	671	0.05	0.8	0.03	0.5	0.12	11	3
0	26	0.39	8	140	43	0.26	607	0.02	0.74	0.03	0.3	0.06	8	2
0	18	0.47	6	131	177	0.19	407	0.01	0.54	0.02	0.4	0.08	7	2

Table H–1

Food Composition
<small>(Computer code number is for West Diet Analysis program)</small>   <small>(For purposes of calculations, use "0" for t, <1, <.1, <.01, etc.)</small>

Computer Code Number	Food Description	Measure	Wt (g)	H₂O (%)	Ener (kcal)	Prot (g)	Carb (g)	Dietary Fiber (g)	Fat (g)	Fat Breakdown (g) Sat	Mono	Poly
	VEGETABLES AND LEGUMES—Continued											
5226	Juice, canned	1 c	236	89	94	2	22	2	<1	t	t	0.2
5625	Cassava, cooked	1 c	137	59	221	2	53	2	<1	0.1	0.1	t
5049	Cauliflower, flowerets, raw:	½ c	50	92	12	1	3	1	<1	t	t	t
5051	Cooked from fresh, drained	½ c	62	93	14	1	3	2	<1	t	t	0.1
5053	From frozen, drained	½ c	90	94	17	1	3	2	<1	t	t	t
5055	Celery, raw, large outer stalk, 8 x 1½" (root end)	1 ea	40	95	6	<1	1		<1	t	t	t
5054	Diced	1 c	120	95	17	1	4		<1	t	t	t
5200	Celeriac/celery root, cooked	1 c	155	92	42	1	9	2	<1	t	t	0.2
5057	Chard, swiss, raw, chopped	1 c	36	93	7	1	1	1	<1	t	t	t
5059	Cooked	1 c	175	93	35	3	7	4	<1	t	t	t
5413	Chayote fruit, raw	1 ea	203	95	35	2	8	3	<1	t	t	0.1
5414	Cooked	1 c	160	93	38	1	8	4	1	0.1	t	0.3
	Chickpeas (see Garbanzo Beans #854)											
5061	Collards, cooked from raw	½ c	95	92	25	2	5	3	<1	t	t	0.2
5062	From frozen	½ c	85	88	31	3	6	2	<1	t	t	0.2
	Corn, yellow:											
5364	Cooked from frozen, on cob, 3½" long	1 ea	63	73	59	2	14	2	<1	t	0.1	0.2
5065	Kernels, cooked from frozen	½ c	82	77	66	2	16	2	1	t	0.2	0.3
5068	Canned, cream style	½ c	128	79	92	2	23	2	1	t	0.2	0.3
5067	Whole kernel, vacuum pack	½ c	105	77	83	3	20	2	1	t	0.2	0.2
	Cowpeas (see Black-eyed peas #814-816)											
5610	Cucumber, kim chee style	1 c	150	91	32	2	7	2	<1	t	t	t
5241	Dandelion greens, raw	1 c	55	86	25	1	5	2	<1	t	t	0.2
5242	Chopped, cooked, drained	1 c	105	90	35	2	7	3	1	0.2	t	0.3
5072	Eggplant, cooked	1 c	99	90	35	1	9	2	<1	t	t	t
5202	Endive, fresh, chopped	1 c	50	94	8	1	2		2	<1	t	t
856	Escarole/curly endive-chopped	1 c	50	94	8	1	2	2	<1	t	t	t
7001	Garbanzo beans (Chickpeas), cooked	1 c	164	60	269	15	45	12	4	0.4	1	1.9
7961	Grape leaf, raw:	1 ea	3	73	3	<1	1	<1	<1	t	t	t
7914	Cup	1 c	14	73	13	1	2	2	<1	t	t	0.1
7021	Great northern beans, cooked	1 c	177	69	209	15	37	12	1	0.2	t	0.3
5077	Jerusalem artichoke, raw slices	1 c	150	78	114	3	26	2	<1	0	t	t
5224	Jicama	1 c	120	90	46	1	11		<1	t	t	t
5075	Kale, cooked from raw	1 c	130	91	36	2	7	3	1	t	t	0.3
5076	From frozen	1 c	130	90	39	4	7	3	1	t	t	0.3
7292	Kidney beans, canned	1 c	256	77	218	13	40	16	1	0.1	t	0.5
5078	Kohlrabi, raw slices	1 c	135	91	36	2	8	5	<1	t	t	t
5079	Cooked	1 c	165	90	48	3	11	2	<1	t	t	t
5205	Leeks, raw, chopped	1 c	89	83	54	1	13	2	<1	t	t	0.1
5203	Cooked, chopped	1 c	104	91	32	1	8	1	<1	t	t	0.1
7006	Lentils, cooked from dry	1 c	198	70	230	18	40	16	1	0.1	0.1	0.3
5390	Lentils, sprouted, stir fried	1 c	124	69	125	11	26	5	1	t	0.1	0.2
5389	Raw	1 c	77	67	82	7	17	3	<1	t	t	0.2
	Lettuce:											
5082	Butterhead/Boston, head, 5" diameter	¼ ea	41	96	5	1	1	<1	<1	t	t	t
5081	Leaves, inner or outer	4 ea	30	96	4	<1	1	<1	<1	t	t	t
5083	Iceberg/crisphead, chopped	1 c	55	96	6	<1	1		<1	t	t	t
5085	Head, 6" diameter	1 ea	539	96	54	4	11		1	t	t	0.3
866	Wedge, ¼ head	1 ea	135	96	14	1	3		<1	t	t	t
5086	Looseleaf, chopped	½ c	28	95	4	<1	1	<1	<1	t	t	t
5088	Romaine, chopped	½ c	28	95	5	<1	1	1	<1	t	t	t
5089	Romaine, inner leaf	3 pce	30	95	5	<1	1	1	<1	t	t	t
5528	Luffa, cooked (Chinese okra)	1 c	178	90	57	3	13	4	<1	t	t	t
6778	Manioc, raw	1 c	206	60	330	3	78	4	1	0.2	0.2	t
	Mushrooms:											
5090	Raw, sliced	½ c	35	92	8	1	1	<1	<1	t	t	t
5092	Cooked from fresh, pieces	½ c	78	91	22	2	4	2	<1	t	t	0.1
1962	Stir fried, shitake slices	½ c	73	83	40	1	10	2	<1	t	t	t
5094	Canned, drained	½ c	78	91	20	1	4	2	<1	t	t	t
5612	Mushroom caps, pickled	8 ea	47	92	11	1	2	<1	<1	t	t	t

Chol (mg)	Calc (mg)	Iron (mg)	Magn (mg)	Pota (mg)	Sodi (mg)	Zinc (mg)	VT-A (RAE)	Thia (mg)	VT-E (mg α)	Ribo (mg)	Niac (mg)	V-B$_6$ (mg)	Fola (µg)	VT-C (mg)
0	57	1.09	33	689	68	0.42	2256	0.22	2.74	0.13	0.91	0.51	9	20
0	21	0.35	28	338	18	0.45	1	0.1	0.26	0.06	1.06	0.11	24	19
0	11	0.22	8	152	15	0.14	<1	0.03	0.04	0.03	0.26	0.11	28	23
0	10	0.2	6	88	9	0.11	1	0.03	0.04	0.03	0.25	0.11	27	27
0	15	0.37	8	125	16	0.12	<1	0.03	0.05	0.05	0.28	0.08	37	28
0	16	0.08	4	104	32	0.05	9	<.01	0.11	0.02	0.13	0.03	14	1
0	48	0.24	13	312	96	0.16	26	0.03	0.32	0.07	0.38	0.09	43	4
0	40	0.67	19	268	95	0.31	0	0.04	0.31	0.06	0.66	0.16	5	6
0	18	0.65	29	136	77	0.13	110	0.01	0.68	0.03	0.14	0.04	5	11
0	102	3.96	150	961	313	0.58	536	0.06	3.31	0.15	0.63	0.15	16	32
0	35	0.69	24	254	4	1.5	0	0.05	0.24	0.06	0.95	0.15	189	16
0	21	0.35	19	277	2	0.5	3	0.04	0.19	0.06	0.67	0.19	29	13
0	133	1.1	19	110	15	0.22	386	0.04	0.84	0.1	0.55	0.12	88	17
0	178	0.95	26	213	42	0.23	489	0.04	1.06	0.1	0.54	0.1	65	22
0	2	0.38	18	158	3	0.4	8	0.11	0.05	0.04	0.96	0.14	20	3
0	2	0.39	23	191	1	0.52	8	0.02	0.06	0.05	1.08	0.08	29	3
0	4	0.49	22	172	365	0.68	5	0.03	0.09	0.07	1.23	0.08	58	6
0	5	0.44	24	195	286	0.48	4	0.04	0.04	0.08	1.23	0.06	51	9
0	14	7.23	12	176	1532	0.76	25	0.04	0.24	0.04	0.69	0.16	34	5
0	103	1.7	20	218	42	0.23	136	0.1	2.63	0.14	0.44	0.14	15	19
0	147	1.89	25	244	46	0.29	521	0.14	3.57	0.18	0.54	0.17	14	19
0	6	0.25	11	122	1	0.12	2	0.08	0.41	0.02	0.59	0.09	14	1
0	26	0.42	8	157	11	0.4	54	0.04	0.22	0.04	0.2	<.01	71	3
0	26	0.42	8	157	11	0.4	54	0.04	0.22	0.04	0.2	<.01	71	3
0	80	4.74	79	477	11	2.51	2	0.19	0.57	0.1	0.86	0.23	282	2
0	11	0.08	3	8	<1	0.02	41	<.01	0.06	0.01	0.07	0.01	2	0
0	51	0.37	13	38	1	0.09	193	<.01	0.28	0.05	0.33	0.06	12	2
0	120	3.77	88	692	4	1.56	<1	0.28	0.53	0.1	1.21	0.21	181	2
0	21	5.1	26	644	6	0.18	2	0.3	0.28	0.09	1.95	0.12	20	6
0	14	0.72	14	180	5	0.19	1	0.02	0.55	0.03	0.24	0.05	14	24
0	94	1.17	23	296	30	0.31	885	0.07	1.1	0.09	0.65	0.18	17	53
0	179	1.22	23	417	20	0.23	956	0.06	1.2	0.15	0.87	0.11	18	33
0	61	3.23	72	658	873	1.41	0	0.27	1.54	0.23	1.17	0.06	131	3
0	32	0.54	26	472	27	0.04	3	0.07	0.65	0.03	0.54	0.2	22	84
0	41	0.66	31	561	35	0.51	3	0.07	0.86	0.03	0.64	0.25	20	89
0	53	1.87	25	160	18	0.11	74	0.05	0.82	0.03	0.36	0.21	57	11
0	31	1.14	15	90	10	0.06	2	0.03	0.63	0.02	0.21	0.12	25	4
0	38	6.59	71	731	4	2.51	1	0.33	0.22	0.14	2.1	0.35	358	3
0	17	3.84	43	352	12	1.98	2	0.27	0.11	0.11	1.49	0.2	83	16
0	19	2.47	28	248	8	1.16	2	0.18	0.07	0.1	0.87	0.15	77	13
0	14	0.51	5	98	2	0.08	68	0.02	0.07	0.03	0.15	0.03	30	2
0	10	0.37	4	71	2	0.06	50	0.02	0.05	0.02	0.11	0.02	22	1
0	11	0.19	4	84	5	0.09	9	0.02	0.02	0.01	0.07	0.03	31	2
0	108	1.89	43	819	49	0.86	86	0.2	0.16	0.11	0.67	0.25	302	21
0	27	0.47	11	205	12	0.22	22	0.05	0.04	0.03	0.17	0.06	76	5
0	10	0.24	4	54	8	0.05	104	0.02	0.08	0.02	0.1	0.03	11	5
0	9	0.27	4	69	2	0.06	81	0.02	0.04	0.02	0.09	0.02	38	7
0	10	0.29	4	74	2	0.07	87	0.02	0.04	0.02	0.09	0.02	41	7
0	112	0.8	101	573	9	0.98	52	0.23	1.23	0.1	1.55	0.33	81	29
0	33	0.56	43	558	29	0.7	2	0.18	0.39	0.1	1.76	0.18	56	42
0	1	0.18	3	110	1	0.18	0	0.03	<.01	0.15	1.35	0.04	6	1
0	5	1.36	9	278	2	0.68	0	0.06	<.01	0.23	3.48	0.07	14	3
0	2	0.32	10	85	3	0.97	0	0.03	0.01	0.12	1.1	0.12	15	0
0	9	0.62	12	101	332	0.56	0	0.07	<.01	0.02	1.24	0.05	9	0
0	2	0.51	5	140	2	0.28	0	0.03	0.05	0.16	1.42	0.03	6	1

Table H–1

Food Composition (Computer code number is for West Diet Analysis program) (For purposes of calculations, use "0" for t, <1, <.1, <.01, etc.)

Computer Code Number	Food Description	Measure	Wt (g)	H₂O (%)	Ener (kcal)	Prot (g)	Carb (g)	Dietary Fiber (g)	Fat (g)	Fat Breakdown (g) Sat	Mono	Poly
	VEGETABLES AND LEGUMES—Continued											
5096	Mustard greens, cooked from raw	½ c	70	94	10	2	1	1	<1	t	t	t
5097	From frozen	½ c	75	94	14	2	2		<1	t	t	t
7022	Navy beans, cooked from dry	1 c	182	63	258	16	48	12	1	0.3	t	0.4
	Okra, cooked:											
5098	From fresh pods	8 ea	85	93	19	2	4		<1	t	t	t
5100	From frozen slices	1 c	184	91	52	4	11	5	1	0.1	t	0.1
5644	Batter fried from fresh	1 c	92	67	175	2	14	2	13	1.7	3.1	7.1
	Onions, yellow/white:											
7499	Raw, chopped	½ c	80	89	34	1	8	1	<1	t	t	t
7808	Raw, sliced	½ c	58	89	24	1	6	1	<1	t	t	t
7812	Cooked, drained, chopped	½ c	105	88	46	1	11	1	<1	t	t	t
5113	Dehydrated flakes	¼ c	14	4	49	1	12	1	<1	t	t	t
5530	Onions, pearl, cooked	½ c	93	88	41	1	9	1	<1	t	t	t
5114	Spring/green onions, bulb and top, chopped	½ c	50	90	16	1	4	1	<1	t	t	t
5190	Onion rings, breaded, heated from frozen	2 ea	20	28	81	1	8	<1	5	1.7	2.2	1
5522	Palm hearts, cooked slices	1 c	146	70	150	4	39	2	<1	t	t	0.1
26012	Parsley, raw, chopped	½ c	30	88	11	1	2	1	<1	t	t	t
5212	Parsnips, sliced, cooked	½ c	78	80	55	1	13	3	<1	t	t	t
	Peas:											
7018	Black-eyed, cooked from dry, drained	½ c	86	70	100	7	18	6	<1	0.1	t	0.2
5213	From fresh, drained	½ c	82	75	80	3	17	4	<1	t	t	0.1
5115	From frozen, drained	½ c	85	66	112	7	20	5	1	0.1	t	0.2
5122	Edible pod peas, cooked	½ c	80	89	34	3	6	2	<1	t	t	t
5119	Green, canned, drained:	½ c	85	82	59	4	11	3	<1	t	t	0.1
5267	Unsalted	½ c	124	86	66	4	12	4	<1	t	t	0.2
5118	Green, cooked from frozen	½ c	80	80	62	4	11	4	<1	t	t	0.1
6836	Snow peas, raw	½ c	49	89	21	1	4	1	<1	t	t	t
5121	each	10 ea	34	89	14	1	3	1	<1	t	t	t
7020	Split, green, cooked from dry	½ c	98	69	116	8	21	8	<1	t	t	0.2
5123	Peas & carrots, cooked from frozen	½ c	80	86	38	2	8	2	<1	t	t	0.2
5281	Canned w/liquid	½ c	128	88	49	3	11	3	<1	t	t	0.2
	Peppers, hot:											
5063	Hot green chili, canned	½ c	68	92	14	1	3	1	<1	t	t	t
5400	Raw	1 ea	45	88	18	1	4	1	<1	t	t	t
5288	Hot red chili, raw, diced	1 tbs	9	88	4	<1	1	<1	<1	t	t	t
5293	Jalapeno, chopped, canned	½ c	68	89	18	1	3	2	1	t	t	0.3
	Peppers, sweet:											
6846	Green, whole, raw	1 ea	119	94	24	1	6	2	<1	t	t	t
5126	Cooked, chopped	½ c	68	92	19	1	5	1	<1	t	t	t
5128	Red, raw, chopped	½ c	75	92	20	1	5		<1	t	t	0.1
5294	Raw, each	1 ea	74	92	19	1	4		<1	t	t	0.1
5278	Cooked, chopped	½ c	68	92	19	1	5	1	<1	t	t	t
5441	Yellow, raw, whole	1 ea	186	92	50	2	12	2	<1	t	t	0.2
5442	Strips	10 pce	52	92	14	1	3	<1	<1	t	t	t
7013	Pinto beans, cooked from dry	½ c	85	64	119	8	21	7	1	t	t	0.2
5229	Poi - two finger	½ c	120	72	134	<1	33	<1	<1	t	t	t
	Potatoes:											
	Baked in oven, 4¾" x 2⅓" diam											
5129	Flesh only	1 ea	156	75	145	3	34	2	<1	t	t	t
5339	Skin only	1 ea	58	47	115	2	27	5	<1	t	t	t
	Baked in microwave, 4¾" x 2⅓" dm:											
5340	With skin	1 ea	202	72	212	5	49	5	<1	t	t	t
5345	Flesh only	1 ea	156	74	156	3	36	2	<1	t	t	t
5350	Skin only	1 ea	58	64	77	3	17	3	<1	t	t	t
	Boiled, about 2½" diam:											
5133	Peeled after boiling	1 ea	136	77	118	3	27	2	<1	t	t	t
5135	Peeled before boiling	1 ea	135	77	116	2	27	2	<1	t	t	t
5139	French fried, strips, oven heated	10 ea	50	35	166	2	20	2	9	3	5.7	0.7
5140	Hashed browns from frozen	1 c	156	56	340	5	44	3	18	7	8	2.1

PAGE KEY: H–2 = Beverages H–4 = Dairy H–8 = Eggs H–10 = Fat/Oil H–12 = Fruit H–18 = Bakery
H–24 = Grain H–30 = Fish H–30 = Meats H–34 = Poultry H–36 = Sausage H–38 = Mixed/Fast H–42 = Nuts/Seeds H–44 = Sweets
H–46 = Vegetables/Legumes H–56 = Vegetarian H–60 = Misc H–60 = Soups/Sauces H–62 = Fast H–76 = Convenience H–78 = Baby foods

Chol (mg)	Calc (mg)	Iron (mg)	Magn (mg)	Pota (mg)	Sodi (mg)	Zinc (mg)	VT-A (RAE)	Thia (mg)	VT-E (mg α)	Ribo (mg)	Niac (mg)	V-B$_6$ (mg)	Fola (µg)	VT-C (mg)
0	52	0.49	10	141	11	0.08	221	0.03	0.85	0.04	0.3	0.07	51	18
0	76	0.84	10	104	19	0.15	266	0.03	1.01	0.04	0.19	0.08	52	10
0	127	4.51	107	670	2	1.93	<1	0.37	0.73	0.11	0.97	0.3	255	2
0	65	0.24	31	115	5	0.37	12	0.11	0.23	0.05	0.74	0.16	39	14
0	177	1.23	94	431	6	1.14	31	0.18	0.59	0.23	1.44	0.09	269	22
2	61	1.26	36	190	122	0.5		0.18	3.04	0.14	1.44	0.12	38	10
0	18	0.15	8	115	2	0.13	<1	0.04	0.02	0.02	0.07	0.12	15	5
0	13	0.11	6	84	2	0.09	<1	0.03	0.01	0.01	0.05	0.09	11	4
0	23	0.25	12	174	3	0.22	<1	0.04	0.02	0.02	0.17	0.14	16	5
0	36	0.22	13	227	3	0.26	<1	0.07	0.03	0.01	0.14	0.22	23	10
0	20	0.22	10	154	3	0.2	0	0.04	0.12	0.02	0.15	0.12	14	5
0	36	0.74	10	138	8	0.2	25	0.03	0.28	0.04	0.26	0.03	32	9
0	6	0.34	4	26	75	0.08	2	0.06	0.14	0.03	0.72	0.02	13	0
0	26	2.47	15	2637	20	5.45	5	0.07	0.73	0.25	1.25	1.06	30	10
0	41	1.86	15	166	17	0.32	126	0.03	0.22	0.03	0.39	0.03	46	40
0	29	0.45	23	286	8	0.2	0	0.06	0.78	0.04	0.56	0.07	45	10
0	21	2.16	46	239	3	1.11	1	0.17	0.24	0.05	0.43	0.09	179	0
0	105	0.92	43	343	3	0.84	33	0.08	0.18	0.12	1.15	0.05	104	2
0	20	1.8	42	319	4	1.21	3	0.22	0.26	0.05	0.62	0.08	120	2
0	34	1.58	21	192	3	0.3	43	0.1	0.31	0.06	0.43	0.12	23	38
0	17	0.81	14	147	214	0.6	23	0.1	0.03	0.07	0.62	0.05	37	8
0	22	1.26	21	124	11	0.87	89	0.14	0.02	0.09	1.04	0.08	36	12
0	19	1.22	18	88	58	0.54	84	0.23	0.02	0.08	1.18	0.09	47	8
0	21	1.02	12	98	2	0.13	26	0.07	0.19	0.04	0.29	0.08	21	29
0	15	0.71	8	68	1	0.09	18	0.05	0.13	0.03	0.2	0.05	14	20
0	14	1.26	35	355	2	0.98	<1	0.19	0.03	0.05	0.87	0.05	64	0
0	18	0.75	13	126	54	0.36	374	0.18	0.42	0.05	0.92	0.07	21	6
0	29	0.96	18	128	333	0.74	370	0.09	0.24	0.07	0.74	0.11	23	8
0	5	0.34	10	127	798	0.12	24	0.01	0.47	0.03	0.54	0.1	7	46
0	8	0.54	11	153	3	0.14	27	0.04	0.31	0.04	0.43	0.13	10	109
0	1	0.09	2	29	1	0.02	4	<.01	0.06	<.01	0.11	0.05	2	13
0	16	1.28	10	131	1136	0.23	58	0.03	0.47	0.03	0.27	0.13	10	7
0	12	0.4	12	208	4	0.15	21	0.07	0.44	0.03	0.57	0.27	13	96
0	6	0.31	7	113	1	0.08	10	0.04	0.36	0.02	0.32	0.16	11	51
0	5	0.32	9	158	2	0.19	118	0.04	1.18	0.06	0.73	0.22	14	142
0	5	0.32	9	156	1	0.18	116	0.04	1.17	0.06	0.72	0.22	13	141
0	6	0.31	7	113	1	0.08	187	0.04	1.12	0.02	0.32	0.16	11	116
0	20	0.86	22	394	4	0.32	19	0.05	1.28	0.05	1.66	0.31	48	341
0	6	0.24	6	110	1	0.09	5	0.01	0.36	0.01	0.46	0.09	14	95
0	36	1.77	35	246	9	0.86	0	0.08	0.8	0.06	0.17	0.08	146	1
0	19	1.06	29	220	14	0.26	4	0.16	2.76	0.05	1.32	0.33	25	5
0	8	0.55	39	610	8	0.45	0	0.16	0.06	0.03	2.18	0.47	14	20
0	20	4.08	25	332	12	0.28	1	0.07	0.02	0.06	1.78	0.36	13	8
0	22	2.5	55	903	16	0.73	0	0.24	0.1	0.06	3.46	0.69	24	31
0	8	0.64	39	641	11	0.51	0	0.2	0.06	0.04	2.54	0.5	19	24
0	27	3.45	21	377	9	0.3	0	0.04	0.02	0.04	1.29	0.29	10	9
0	7	0.42	30	515	5	0.41	<1	0.14	0.01	0.03	1.96	0.41	14	18
0	11	0.42	27	443	7	0.36	<1	0.13	0.01	0.03	1.77	0.36	12	10
0	6	0.83	12	270	306	0.2	0	0.04	0.25	0.02	1.33	0.11	11	3
0	23	2.36	27	680	53	0.5	0	0.17	0.3	0.03	3.78	0.2	11	10

Table H–1

Food Composition   (Computer code number is for West Diet Analysis program)   (For purposes of calculations, use "0" for t, <1, <.1, <.01, etc.)

Computer Code Number	Food Description	Measure	Wt (g)	H₂O (%)	Ener (kcal)	Prot (g)	Carb (g)	Dietary Fiber (g)	Fat (g)	Fat Breakdown (g)		
										Sat	Mono	Poly
	VEGETABLES AND LEGUMES—Continued											
	Mashed:											
5137	Home recipe with whole milk	½ c	105	79	87	2	18	2	1	0.3	0.1	t
5272	Home recipe with milk and marg	½ c	105	75	119	2	18	2	4	1	1.8	1.3
5138	Prepared from flakes with milk and marg	½ c	110	76	124	2	17	3	6	1.6	2.5	1.7
	Potato products, prepared:											
5276	Au gratin from dry mix	½ c	123	79	114	3	16	1	5	3.2	1.4	0.2
5275	From home recipe, using butter	½ c	122	74	161	6	14	2	9	4.3	3.2	1.3
5271	Scalloped from dry mix	½ c	122	79	113	3	16	1	5	3.2	1.5	0.2
5270	From home recipe, using butter	½ c	123	81	106	4	13	2	5	1.7	1.7	0.9
	Potato Salad (see Mixed Dishes #715)											
5265	Potato Puffs, cooked from frozen	½ c	64	53	142	2	20	2	7	3.3	2.8	0.5
5396	Pumpkin, cooked from fresh, mashed	½ c	123	94	25	1	6	1	<1	t	t	t
5142	Canned	½ c	123	90	42	1	10	4	<1	0.2	t	t
5451	Radicchio, raw, shredded	½ c	20	93	5	<1	1	<1	<1	t	t	t
5452	Leaf	10 ea	80	93	18	1	4	1	<1	t	t	t
5143	Red radishes	10 ea	45	95	7	<1	2		<1	t	t	t
1793	Daikon radishes (Chinese) raw	½ c	44	95	8	<1	2	1	<1	t	t	t
7024	Refried beans, canned	½ c	126	76	118	7	20	7	2	0.6	0.7	0.2
7225	Rutabaga, cooked cubes	½ c	85	89	33	1	7	2	<1	t	t	t
5145	Sauerkraut, canned with liquid	½ c	118	93	22	1	5	3	<1	t	t	t
6857	Seaweed, kelp, raw	½ c	40	82	17	1	4	1	<1	t	t	t
5260	Seaweed, spirulina, dried	½ c	8	5	23	5	2	<1	1	0.2	t	0.2
5427	Shallots, raw, chopped	1 tbs	10	80	7	<1	2	<1	<1	t	t	t
5666	Snow Peas, stir fried	½ c	83	89	35	2	6	2	<1	t	t	t
7015	Soybeans, cooked from dry	½ c	86	63	149	14	9	5	8	1.1	1.7	4.4
7063	dry roasted	½ c	86	1	388	34	28	7	19	2.7	4.1	10.5
	Soybean products:											
7503	Miso	½ c	138	41	284	16	39	7	8	1.2	1.9	4.7
	Soy milk (see Dairy)											
	Tofu (soybean curd):											
	Silken:											
7540	Extra firm	½ c	126	88	69	9	3	<1	2	0.4	0.4	1.3
7542	Firm	½ c	126	87	78	9	3	<1	3	0.5	0.7	1.9
7500	Regular	½ c	124	87	76	8	2	<1	5	0.7	1	2.6
7721	Block	1 pce	116	85	88	9	2	<1	6	0.8	1.2	3.1
7720	Firm	1 pce	81	70	117	13	3	2	7	1	1.6	4
7519	Dried, frozen	1 pce	17	6	82	8	2	1	5	0.7	1.1	2.9
90039	Fried	1 pce	13	51	35	2	1	1	3	0.4	0.6	1.5
7541	Soft	½ c	124	89	68	6	4	<1	3	0.4	0.6	1.9
7518	Prepared with nigari	½ c	126	84	97	10	4	1	6	0.8	1.2	3.2
	Spinach:											
5146	Raw, chopped	1/2 c	15	91	3	<1	1	<1	<1	t	t	t
5147	Cooked, from fresh, drained	½ c	90	91	21	3	3	2	<1	t	t	t
5148	From frozen (leaf)	½ c	95	89	30	4	5	4	<1	t	0	0.2
5149	Canned, unsalted	½ c	107	92	25	3	4	3	1	t	t	0.2
	Spinach souffle (see Mixed Dishes)											
	Squash, summer varieties, cooked w/skin:											
5152	Varieties averaged	½ c	90	94	18	1	4	1	<1	t	t	0.1
5322	Crookneck	½ c	90	94	18	1	4	1	<1	t	t	0.1
5327	Zucchini	½ c	90	95	14	1	4	1	<1	t	t	t
	Squash, winter varieties, cooked:											
5303	Average of all varieties, baked, cubes	1 c	205	89	76	2	18	6	1	0.2	t	0.4
5314	Acorn, baked, mashed	½ c	123	83	69	1	18	5	<1	t	t	t
5316	Boiled, mashed	½ c	122	90	41	1	11	3	<1	t	t	t
	Butternut squash:											
5317	Baked, mashed	½ c	103	88	41	1	11	3	<1	t	t	t
5274	Cooked from frozen	½ c	120	88	47	1	12	3	<1	t	t	t
5453	Hubbard, baked, mashed	½ c	120	85	60	3	13	3	1	0.2	t	0.3
5454	Boiled, mashed	½ c	118	91	35	2	8	3	<1	t	t	0.2
5455	Spaghetti, baked or boiled	½ c	77	92	21	1	5	1	<1	t	t	t

PAGE KEY: H–2 = Beverages H–4 = Dairy H–8 = Eggs H–10 = Fat/Oil H–12 = Fruit H–18 = Bakery
H–24 = Grain H–30 = Fish H–30 = Meats H–34 = Poultry H–36 = Sausage H–38 = Mixed/Fast H–42 = Nuts/Seeds H–44 = Sweets
H–46 = Vegetables/Legumes H–56 = Vegetarian H–60 = Misc H–60 = Soups/Sauces H–62 = Fast H–76 = Convenience H–78 = Baby foods

Chol (mg)	Calc (mg)	Iron (mg)	Magn (mg)	Pota (mg)	Sodi (mg)	Zinc (mg)	VT-A (RAE)	Thia (mg)	VT-E (mg α)	Ribo (mg)	Niac (mg)	V-B$_6$ (mg)	Fola (µg)	VT-C (mg)
2	23	0.28	19	311	317	0.3	4	0.09	0.02	0.05	1.18	0.24	8	7
1	21	0.27	20	342	350	0.32	43	0.1	0.44	0.05	1.23	0.26	9	11
4	54	0.24	20	256	365	0.2	51	0.12	0.77	0.06	0.74	<.01	8	11
18	102	0.39	18	269	540	0.3	64	0.02	1.48	0.1	1.15	0.05	9	4
18	145	0.78	24	483	528	0.84	78	0.08	0.64	0.14	1.21	0.21	13	12
13	44	0.46	17	248	416	0.3	43	0.02	0.18	0.07	1.26	0.05	12	4
7	70	0.7	23	465	412	0.49	41	0.08	0.4	0.11	1.3	0.22	14	13
0	19	1	12	243	477	0.19	<1	0.13	0.03	0.05	1.38	0.15	11	4
0	18	0.7	11	283	1	0.28	308	0.04	0.98	0.1	0.51	0.05	11	6
0	32	1.71	28	253	6	0.21	957	0.03	1.3	0.07	0.45	0.07	15	5
0	4	0.11	3	60	4	0.12	<1	<.01	0.45	<.01	0.05	0.01	12	2
0	15	0.46	10	242	18	0.5	1	0.01	1.81	0.02	0.2	0.05	48	6
0	11	0.15	4	105	18	0.13	<1	<.01	0	0.02	0.11	0.03	11	7
0	12	0.18	7	100	9	0.07	0	<.01	0	<.01	0.09	0.02	12	10
10	44	2.09	42	336	377	1.47	0	0.03	0	0.02	0.4	0.18	14	8
0	41	0.45	20	277	17	0.3	<1	0.07	0.27	0.03	0.61	0.09	13	16
0	35	1.73	15	201	780	0.22	1	0.02	0.12	0.03	0.17	0.15	28	17
0	67	1.14	48	36	93	0.49	2	0.02	0.35	0.06	0.19	<.01	72	1
0	10	2.28	16	109	84	0.16	2	0.19	0.4	0.29	1.03	0.03	8	1
0	4	0.12	2	33	1	0.04	6	<.01	<.01	<.01	0.02	0.03	3	1
0	36	1.73	20	166	3	0.22	5	0.11	0.32	0.06	0.47	0.13	28	42
0	88	4.42	74	443	1	0.99	<1	0.13	0.3	0.25	0.34	0.2	46	1
0	120	3.4	196	1173	2	4.1	0	0.37	3.96	0.65	0.91	0.19	176	4
0	91	3.78	58	226	5033	4.58	6	0.13	0.01	0.34	1.19	0.3	46	0
0	39	1.5	34	194	79	0.76	0	0.1	0.18	0.04	0.3	0.01		0
0	40	1.3	34	244	45	0.77	0	0.13	0.24	0.05	0.31	0.01		0
0	138	1.38	33	149	10	0.79	<1	0.06	0.01	0.05	0.66	0.06	55	0
0	406	6.22	35	140	8	0.93	5	0.09	0.01	0.06	0.23	0.05	17	0
0	553	2.15	47	192	11	1.27	6	0.13	0.02	0.08	0.31	0.07	23	0
0	62	1.65	10	3	1	0.83	4	0.08	0.01	0.05	0.2	0.05	16	0
0	125	0.63	12	19	2	0.26	0	0.02	<.01	<.01	0.01	0.01	4	0
0	38	1.02	36	223	6	0.64	0	0.12	0.25	0.05	0.37	0.01		0
0	204	1.83	58	222	10	1.27	0	0.12	0.03	0.13	0.01	0.08	42	0
0	15	0.41	12	84	12	0.08	70	0.01	0.3	0.03	0.11	0.03	29	4
0	122	3.21	78	419	63	0.68	472	0.09	1.87	0.21	0.44	0.22	131	9
0	145	1.86	78	287	92	0.47	573	0.07	3.36	0.17	0.42	0.13	115	2
0	136	2.46	81	370	29	0.49	524	0.02	2.08	0.15	0.42	0.11	105	15
0	24	0.32	22	173	1	0.35	10	0.04	0.13	0.04	0.46	0.06	18	5
0	20	0.33	14	153	0	0.2	7	0.04	0.11	0.04	0.46	0.08	18	5
0	12	0.32	20	228	3	0.16	50	0.04	0.11	0.04	0.39	0.07	15	4
0	45	0.9	27	896	2	0.45	535	0.03	0.25	0.14	1.01	0.33	41	20
0	54	1.14	53	538	5	0.21	26	0.21	0.15	0.02	1.08	0.24	23	13
0	32	0.68	32	321	4	0.13	50	0.12	0.15	<.01	0.65	0.14	13	8
0	42	0.62	30	293	4	0.13	575	0.07	1.33	0.02	1	0.13	20	16
0	23	0.7	11	160	2	0.14	200	0.06	0.16	0.05	0.56	0.08	19	4
0	20	0.56	26	430	10	0.18	362	0.09	0.14	0.06	0.67	0.21	19	11
0	12	0.33	15	253	6	0.12	236	0.05	0.14	0.03	0.39	0.12	12	8
0	16	0.26	8	90	14	0.15	5	0.03	0.09	0.02	0.62	0.08	6	3

Table H–1

Food Composition

(Computer code number is for West Diet Analysis program)   (For purposes of calculations, use "0" for t, <1, <.1, <.01, etc.)

Computer Code Number	Food Description	Measure	Wt (g)	H₂O (%)	Ener (kcal)	Prot (g)	Carb (g)	Dietary Fiber (g)	Fat (g)	Fat Breakdown (g) Sat	Mono	Poly
	VEGETABLES AND LEGUMES—Continued											
5154	Succotash, cooked from frozen	½ c	85	74	79	4	17	3	1	0.1	0.1	0.4
	Sweet potatoes:											
5155	Baked in skin, peeled, 5 x 2" diam	1 ea	114	76	103	2	24	4	<1	t	t	t
5159	Boiled without skin, 5 x 2" diam	1 ea	151	80	115	2	27	4	<1	t	0	t
5166	Candied pieces, 2½ x 2"	1 pce	105	67	144	1	29	3	3	1.4	0.7	0.2
5162	Canned, solid pack	½ c	128	74	129	3	30	2	<1	t	t	0.1
5164	Vacuum pack, mashed	½ c	127	76	116	2	27	2	<1	t	t	0.1
5543	Taro shoots, cooked slices	1 c	140	95	20	1	4	1	<1	t	t	t
5544	Taro, tahitian, cooked slices	1 c	137	86	60	6	9	1	1	0.2	t	0.4
5445	Tomatillos, raw, each	1 ea	34	92	11	<1	2	1	<1	t	t	0.1
5444	Chopped	1 c	132	92	42	1	8	3	1	0.2	0.2	0.6
	Tomatoes:											
5169	Raw, whole, 2⅗" diam	1 ea	123	94	22	1	5	1	<1	t	t	0.1
6492	Whole, small	1 ea	62	94	11	1	2	1	<1	t	t	t
5170	Chopped	1 c	180	94	32	2	7	2	<1	t	t	0.2
5178	Cooked from raw	1 c	240	94	43	2	10	2	<1	t	t	0.1
5179	Canned, solids and liquid:	1 c	240	94	41	2	9	2	<1	t	t	0.1
6992	Unsalted	1 c	240	94	46	2	10	2	<1	t	t	0.1
5446	Sundried:	1 c	54	15	139	8	30	7	2	0.2	0.3	0.6
5447	Pieces	10 pce	20	15	52	3	11	2	1	t	t	0.2
5448	Oil pack, drained	10 pce	30	54	64	2	7	2	4	0.6	2.6	0.6
5188	Tomato juice, canned:	1 c	243	94	41	2	10	1	<1	t	t	t
5397	Unsalted	1 c	243	94	41	2	10	1	<1	t	t	t
	Tomato products, canned:											
5181	Paste-no added salt	1 c	262	74	215	11	50	12	1	0.3	0.2	0.4
5225	Puree-no added salt	1 c	250	88	95	4	22	5	1	t	t	0.2
5180	Sauce-with salt	1 c	245	89	78	3	18	4	1	t	t	0.2
5183	Turnips, cubes, cooked from fresh	1 c	156	94	34	1	8	3	<1	t	t	t
5185	Turnip greens, cooked from fresh, leaves and stems	1 c	144	93	29	2	6	5	<1	t	t	0.1
5186	From frozen, chopped	1 c	164	90	48	5	8	6	1	0.2	t	0.3
20080	Vegetable juice cocktail, canned	1 c	242	94	46	2	11	2	<1	t	t	t
5305	Vegetables, mixed, canned, drained	½ c	81	87	40	2	8	2	<1	t	t	t
5187	Frozen, cooked, drained	½ c	91	83	59	3	12	4	<1	t	t	t
5386	Water chestnuts, Chinese, raw	½ c	62	73	60	1	15	2	<1	t	t	t
5387	Canned, slices	½ c	70	86	35	1	9	2	<1	t	t	t
5388	Whole	4 ea	28	86	14	<1	3	1	<1	t	t	t
5222	Watercress, fresh, chopped	½ c	17	95	2	<1	<1	<1	<1	t	t	t
	VEGETARIAN FOODS:											
7509	Bacon strips, meatless	3 ea	15	49	46	2	1	<1	4	0.7	1.1	2.3
7038	Baked beans, canned	1/2 c	127	73	118	6	26	6	1	0.1	t	0.2
7526	Bakon Crumbles	1/4 c	7	8	31	2	2	1	2	0		
7723	Chicken, fillet	3 oz	85		90	15	8	4	2	0		
7557	Chili w/meat substitute	½ c	107	65	141	19	15	5	2	0.3	0.6	0.9
	Frankfurters:											
7550	Meatless	1 ea	51	58	102	10	4	2	5	0.8	1.2	2.6
8127	Deli	1 ea	76	72	80	16	4	1	0	0	0	0
8835	Tofu wiener	1 ea	38		45	9	2	0	<1	0		
8839	Jumbo	1 ea	76		100	16	7	2	2	0		
8840	Hot and spicy	1 ea	52		70	13	3	2	1	0		
	Garden Burger patties:											
7504	Regular	1 ea	71		110	6	16	3	3	1.5		
8811	Mushroom	1 ea	71		120	6	18	4	2	1		
8813	Roasted vegetable	1 ea	71		120	6	18	4	2	1		
7652	Veggie medley	1 ea	71		90	5	18	3	0	0	0	0
7505	Garden Sausage, patty	1 ea	71		83	8	3	3	6	0		
57433	Nuteena	1 ea	55	58	162	6	6	2	13	5.2	5.8	1.7
7556	Pot pie, meatless	1 ea	227	60	510	14	41	5	32	8.6	12.4	9.6
8831	Soyburger mix	⅓ c	55	73	60	10	4	3	0	0	0	0

PAGE KEY: H–2 = Beverages   H–4 = Dairy   H–8 = Eggs   H–10 = Fat/Oil   H–12 = Fruit   H–18 = Bakery
H–24 = Grain   H–30 = Fish   H–30 = Meats   H–34 = Poultry   H–36 = Sausage   H–38 = Mixed/Fast   H–42 = Nuts/Seeds   H–44 = Sweets
H–46 = Vegetables/Legumes   H–56 = Vegetarian   H–60 = Misc   H–60 = Soups/Sauces   H–62 = Fast   H–76 = Convenience   H–78 = Baby foods

Chol (mg)	Calc (mg)	Iron (mg)	Magn (mg)	Pota (mg)	Sodi (mg)	Zinc (mg)	VT-A (RAE)	Thia (mg)	VT-E (mg α)	Ribo (mg)	Niac (mg)	V-B6 (mg)	Fola (µg)	VT-C (mg)
0	13	0.76	20	225	38	0.38	8	0.06	0.15	0.06	1.11	0.08	28	5
0	43	0.79	31	542	41	0.36	1096	1.65	0.81	0.12	1.7	0.33	7	22
0	41	1.09	27	347	41	0.3	1190	0.08	1.42	0.07	0.81	0.25	9	19
8	27	1.19	12	198	74	0.16	219	0.02	3.99	0.04	0.41	0.04	12	7
0	38	1.7	31	269	96	0.27	968	0.03	0.35	0.12	1.22	0.3	14	7
0	28	1.13	28	396	67	0.23	507	0.05	1.27	0.07	0.94	0.24	22	34
0	20	0.57	11	482	3	0.76	4	0.05	1.4	0.07	1.13	0.16	4	26
0	204	2.14	70	854	74	0.14	121	0.06	3.7	0.27	0.66	0.16	10	52
0	2	0.21	7	91	<1	0.07	2	0.01	0.13	0.01	0.63	0.02	2	4
0	9	0.82	26	354	1	0.29	8	0.06	0.5	0.05	2.44	0.07	9	15
0	12	0.33	14	292	6	0.21	52	0.05	0.66	0.02	0.73	0.1	18	16
0	6	0.17	7	147	3	0.11	26	0.02	0.33	0.01	0.37	0.05	9	8
0	18	0.49	20	427	9	0.31	76	0.07	0.97	0.03	1.07	0.14	27	23
0	26	1.63	22	523	26	0.34	58	0.09	1.34	0.05	1.28	0.19	31	55
0	74	2.33	26	451	307	0.34	14	0.11	1.7	0.11	1.76	0.22	19	22
0	72	1.32	29	545	24	0.38	17	0.11	1.92	0.07	1.76	0.22	19	34
0	59	4.91	105	1851	1131	1.07	24	0.29	<.01	0.26	4.89	0.18	37	21
0	22	1.82	39	685	419	0.4	9	0.11	<.01	0.1	1.81	0.07	14	8
0	14	0.8	24	470	80	0.23	19	0.06	0.16	0.11	1.09	0.1	7	31
0	24	1.04	27	556	654	0.36	56	0.11	0.78	0.08	1.64	0.27	49	44
0	24	1.04	27	556	24	0.36	56	0.11	0.78	0.08	1.64	0.27	49	44
5	94	7.81	110	2657	257	1.65	199	0.16	11.27	0.4	8.06	0.57	31	57
5	45	4.45	58	1098	70	0.9	65	0.06	4.92	0.2	3.66	0.32	28	26
0	32	2.5	39	811	1284	0.49	42	0.06	5.1	0.16	2.39	0.24	22	17
0	51	0.28	14	276	25	0.19	0	0.04	0.03	0.04	0.47	0.1	14	18
0	197	1.15	32	292	42	0.2	549	0.06	2.71	0.1	0.59	0.26	170	39
0	249	3.18	43	367	25	0.67	882	0.09	4.36	0.12	0.77	0.11	64	36
0	27	1.02	27	467	653	0.48	189	0.1	12.1	0.07	1.76	0.34	51	67
0	22	0.85	13	236	121	0.33	471	0.04	0.28	0.04	0.47	0.06	19	4
0	23	0.75	20	154	32	0.45	195	0.06	0.4	0.11	0.77	0.07	17	3
0	7	0.04	14	362	9	0.31	0	0.09	0.74	0.12	0.62	0.2	10	2
0	3	0.61	4	83	6	0.27	0	<.01	0.35	0.02	0.25	0.11	4	1
0	1	0.24	1	33	2	0.11	0	<.01	0.14	<.01	0.1	0.04	2	0
0	20	0.03	4	56	7	0.02	40	0.02	0.17	0.02	0.03	0.02	2	7
0	3	0.36	3	26	220	0.06	1	0.66	1.03	0.07	1.13	0.07	6	0
0	64	0.37	41	376	504	1.78	6	0.19	0.67	0.08	0.54	0.17	30	4
0	16	0.28			163		0							0
0	80	1.8			170		0							1
0	54	4.38	36	366	355	1.27	37	0.12	1.28	0.07	1.22	0.15	82	6
0	17	0.92	9	76	219	0.61	0	0.56	0.98	0.61	8.16	0.5	40	0
0	40	0.72			590		0							1
0	20	2.16	8	90	240	0.6	0	0.15						0
0	20	4.5	16	170	480	3.75	0	0.38						0
0	20	3.6	16	135	400	3	38	0.3						0
20	60	0			560		0							0
20	100	0.36			370									0
15	100	0.36			330		0							0
0	40	0			280									0
0	33	1.19			198		0							0
0	9	0.27		166	119	0.46	0	0.1		0.35	1.04	0.45		0
20	68	2.96	32	378	486	1.08		0.82	4.23	0.44	5.17	0.41	58	10
0	40	2.7		240	270	3	0	0.22		0.14	3	0.2		0

Table H–1

Food Composition

(Computer code number is for West Diet Analysis program)   (For purposes of calculations, use "0" for t, <1, <.1, <.01, etc.)

Computer Code Number	Food Description	Measure	Wt (g)	H₂O (%)	Ener (kcal)	Prot (g)	Carb (g)	Dietary Fiber (g)	Fat (g)	Fat Breakdown (g)		
										Sat	Mono	Poly
	VEGETARIAN FOODS:—Continued											
7751	Griller patty	1 ea	113		150	21	11	6	4	0		
7562	Patty, with cheese	1 ea	135	51	308	20	30	4	12	3.6	3.6	3.6
7517	Soy protein isolate	1 oz	28	5	95	23	2	2	1	0.1	0.2	0.5
7564	Tempeh	1 c	166	60	320	31	16		18	3.7	5	6.4
7670	Vegan burger, patty	1 ea	78	71	83	13	7	4	<1	t	0.3	0.1
8842	Veggie slices, soy	1 pce	15		19	3	1	<1	<1	0		
8830	Veggie ground soy	⅓ c	55	73	60	10	4	3	0	0	0	0
7902	Vegetarian burger mix	½ c	47		170	8	30	5	3	0		
7511	Vegetarian sausage link	1 ea	25	50	64	5	2	1	5	0.7	1.1	2.3
7512	Patty	1 ea	38	50	98	7	4	1	7	1.1	1.7	3.5
	Vegetarian foods, Green Giant harvest burger											
7673	Italian, patty	1 ea	90		140	17	8	5	4	1.5	0.5	0.5
7674	Original, patty	1 ea	90	65	138	18	7	6	4	1	2.1	0.3
6658	Mix, frozen	⅔ c	54		90	15	7	3	0	0	0	0
7675	Southwestern, patty	1 ea	90		140	16	9	5	4	1.5	0	0.5
	Vegetarian Foods, Loma Linda											
7727	Chik nuggets, frozen	5 pce	85	47	245	12	13	5	16	2.5	4	8.8
7767	Fakin Bakin, bacon bits	1 tsp	3		12	1	1	0	1	0		
7860	Smokey tempeh strip	1 pce	19		27	3	2	<1	1	0.2		
7744	Franks, big, canned	1 ea	51	58	118	12	2		7	0.8	1.5	3.7
7997	Gimme Lean! Meatless ground beef	2 oz	57		70	9	8	1	0	0	0	0
7998	Meatless sausage	2 oz	57		70	9	8	1	0	0	0	0
8159	Italian sausage, link	1 ea	40	68	60	5	5	0	2	1		
7747	Linketts, canned	1 ea	35	60	72	7	3	2	4	0.7	1.2	2.5
57434	Redi-burger, patty	1 ea	85	59	172	16	7	5	10	1.5	2.4	5.8
7755	Swiss stake w/gravy, canned	1 pce	92	71	120	9	8	4	6	0.8	1.5	3.3
57435	Vege-Burger, patty	1 ea	55	71	66	11	3	2	2	0.4	0.6	0.5
7999	Light	3 oz	85	64	130	16	12	2	1	0		
	Vegetarian foods, Morningstar Farms:											
7766	Better-n-eggs	¼ c	57	88	26	5	<1	0	<1	t	0.1	0.1
57436	Breakfast links	2 pce	45	68	64	9	2	1	2	0.4	0.5	1.1
7752	Breakfast strips	2 pce	16	43	56	2	2	1	4	0.7	0.9	2.6
477	Buffalo wings	5 ea	85	50	204	13	18	3	9	1.2	2.2	4.8
7725	Burger crumbles, svg	1 ea	55	60	116	11	3	3	6	1.6	2.3	2.5
7726	Burger, spicy black bean	1 ea	78	60	115	12	15	5	1	0.2	0.2	0.4
62541	Chik nuggets	4 pce	86	53	183	14	17	4	7	0.6	1.1	2.7
7665	Chik pattie	1 ea	71	56	148	10	14	3	6	0.9	1.6	3.6
7724	Frank, deli	1 ea	57	52	141	13	5	3	8	1.1	2.5	4.2
7722	Garden vege pattie	1 ea	67	60	119	11	10	4	4	0.5	1.1	2.2
7746	Grillers	1 ea	64	56	139	15	5	2	6	1.1	1.5	3.1
62545	Griller Prime	1 ea	71	56	160	17	4	2	8	1	3.6	3.8
7789	Ground burger	½ c	55	71	62	10	4	1	<1	t	t	0.3
62359	Sausage breakfast patty	1 ea	38	54	79	10	4	2	3	0.5	0.7	1.3
	Vegetarian foods, Worthington:											
7634	Beef style, meatless, frzn	3 pce	55	58	113	9	4	3	7	1.2	2.7	2.6
7732	Burger, meatless, patty	¼ c	55	71	60	9	2	1	2	0.3	0.5	1.1
7610	Choplets, slices, canned	2 pce	92	72	93	17	3	2	2	0.9	0.3	0.3
7608	Corned beef style, meatless, frzn	4 pce	57	55	138	10	5	2	9	1.9	4.1	3.1
7607	Country stew, canned	1 c	240	81	208	13	20	5	9	1.6	2.3	4.8
7612	Numete, slices, canned	1 pce	55	58	132	6	5	3	10	2.4	4.4	2.7
7613	Prime stakes, slices, canned	1 pce	92	71	136	9	4	4	9	1.4	2.9	4.9
7617	Protose, slices, canned	1 pce	55	53	131	13	5	3	7	1	3	2.4
7606	Roast, dinner, meatless, frzn	1 ea	85	63	180	12	5	3	13	2.2	5	5.2
7619	Saucette links, canned	1 pce	38	62	86	6	2	1	6	1.1	1.6	3.8
7620	Savory slices, canned	1 pce	28	66	48	3	2	1	3	1.2	1.3	0.6
7735	Stakelets, frzn	1 pce	71	58	145	12	6	2	8	1.4	2.7	3.9
7625	Turkee slices, canned	1 pce	33	64	68	5	1	1	5	0.8	1.9	2.1
7782	Vegan meatless patty	1 ea	71		90	10	12	2	0	0	0	0
7765	Vita Burger chunks	¼ c	21	6	75	10	6	4	1	0.3	0.2	0.6

PAGE KEY: H–2 = Beverages   H–4 = Dairy   H–8 = Eggs   H–10 = Fat/Oil   H–12 = Fruit   H–18 = Bakery
H–24 = Grain   H–30 = Fish   H–30 = Meats   H–34 = Poultry   H–36 = Sausage   H–38 = Mixed/Fast   H–42 = Nuts/Seeds   H–44 = Sweets
H–46 = Vegetables/Legumes   H–56 = Vegetarian   H–60 = Misc   H–60 = Soups/Sauces   H–62 = Fast   H–76 = Convenience   H–78 = Baby foods

Chol (mg)	Calc (mg)	Iron (mg)	Magn (mg)	Pota (mg)	Sodi (mg)	Zinc (mg)	VT-A (RAE)	Thia (mg)	VT-E (mg α)	Ribo (mg)	Niac (mg)	V-B$_6$ (mg)	Fola (μg)	VT-C (mg)
0					475									
9	158	2.94	27	242	922	1.91		0.8	1.58	0.59	8.32	0.84	65	1
0	50	4.06	11	23	281	1.13	0	0.05	0	0.03	0.4	0.03	49	0
0	184	4.48	134	684	15	1.89	0	0.13	0.03	0.59	4.38	0.36	40	0
0	80	2.66	15	398	351	0.69		0.23	<.01	0.51	3.77	0.18	225	0
0	10	0.87	4	31	104	0.73	2	0.07						0
0	40	2.7		240	270	3	0	0.22		0.14	3	0.2		0
0	60	1.8			320									2
0	16	0.93	9	58	222	0.36	0	0.59	0.52	0.1	2.8	0.21	6	0
0	24	1.41	14	88	337	0.55	0	0.89	0.8	0.15	4.25	0.31	10	0
0	80	2.7			370	6.75	0	0.3		0.14	4	0.3		0
0	102	3.85	70	432	411	8.07		0.32	1.56	0.2	6.3	0.39	22	0
0	100	1.8			370	4.5	0	0.22		0.1	3	0.16		0
0	80	2.7			370	6.75	0	0.3		0.14	4	0.3		0
2	40	1.4		153	709	0.43	0	0.67		0.3	2.89	0.45		0
0	0	0			30		0							0
0	30	0.3			77									0
0	10	0.99		61	224	1.2		0.28		0.68	5.78	0.67		
0	40	1.81			241		0							2
0	40	1.81			292		0							2
0	20	1.08			160		0							2
1	4	0.39		29	160	0.46	0	0.13		0.22	0.64	0.29		0
1	12	1.06		121	455	1.1	0	0.14		0.3	1.9	0.51		0
2	24	0.31		225	433	0.41	0	1.25		0.65	5.41	1		0
0	8	0.5		30	114	0.58	0	0.2		0.25	0.78	0.31		0
0	110	3.96			410		0							0
1	24	0.83		60	98	0.6		0.05		0.36	0	0.11		0
1	9	1.77		46	355	0.35	0	5.43		0.15	2.5	0.38		0
0	3	0.33		16	228	0.06	0	0.67		0.05	0.75	0.08		0
2	39	2.75		299	628	0.65		0.31		0.26	2.82	0.37		0
0	40	3.2	1	89	238	0.82		4.96	0.35	0.18	1.49	0.27		0
1	56	1.84	44	269	499	0.93		8.06	0.36	0.14	0	0.21		0
1	37	1.95		297	632	0.83	0	1.28		0.29	2.64	0.24		0
1	24	2.31		174	514	0.58	0	1.39		0.23	2.26	0.18		0
1	22	0.77	5	63	545	0.48		0.18	1.59	0.03	0	0.01		0
1	48	1.21	29	180	382	0.58	134	6.47	0.55	0.1	0	0	59	0
2	22	2.5		122	269	0.67	0	11.81		0.2	4.07	0.48		0
1	25	1.43		159	365	0.87	0	0.23		0.15				0
0	21	0.62		135	262	0.72	0	0.35		0.15	6.39	0.36		0
1	18	1.92	1	102	259	0.37		5.38	0.3	0.13	1.84	0.19		0
0	4	2.63		44	624	0.22	0	0.89		0.34	6.46	0.56		0
0	4	1.73		25	269	0.38	0	0.13		0.1	1.96	0.24		0
0	6	0.37		40	500	0.65	0	0.05		0.06	0	0.06		0
1	6	1.17		58	524	0.26	0	10.61		0.07	1.36	0.3		0
2	51	5.09		270	826	1.03	108	1.85		0.29	4.22	0.86		0
0	10	1.12		155	272	0.56	0	0.08		0.06	0.54	0.2		0
2	12	0.38		82	445	0.38	0	0.12		0.13	1.98	0.38		0
0	1	1.84		50	283	0.7	0	0.18		0.13	1.34	0.24		0
2	36	2.87		38	566	0.64	0	2.13		0.26	6.02	0.6		0
1	9	1.15		25	205	0.26	0	0.59		0.08	0.1	0.13		0
0	0	0.47		14	179	0.08	0	0.08		0.06	0.48	0.1		0
2	49	0.99		95	484	0.5	0	1.51		0.12	3.1	0.26		0
1	3	0.48		16	203	0.11	0	1.13		0.05	0.39	0.09		0
0	40	4.5			230		0							0
0	27	1.83		499	353	4.79	0	1.23		0.17	3.06	0.12		0

Table H–1

Food Composition   (Computer code number is for West Diet Analysis program)   (For purposes of calculations, use "0" for t, <1, <.1, <.01, etc.)

Computer Code Number	Food Description	Measure	Wt (g)	H₂O (%)	Ener (kcal)	Prot (g)	Carb (g)	Dietary Fiber (g)	Fat (g)	Fat Breakdown (g) Sat	Mono	Poly
	MISCELLANEOUS											
28006	Baking powder, low sodium	1 tsp	5	6	5	<1	2	<1	<1	t	t	t
28003	Baking soda	1 tsp	5	0	0	0	0	0	0	0	0	0
26001	Basil, dried	1 tbsp	5	6	13	1	3	2	<1	t	t	0.1
38063	Carob flour	1 c	103	4	229	5	92	41	1	t	0.2	0.2
27000	Catsup	1 tbsp	15	70	14	<1	4	<1	<1	t	t	t
26027	Cayenne/red pepper	1 tbsp	5	8	16	1	3	1	1	0.2	0.1	0.4
26040	Celery seed	1 tsp	2	6	8	<1	1	<1	1	t	0.3	t
26002	Chili powder	1 tbsp	8	8	25	1	4	3	1	0.2	0.3	0.6
23010	Chocolate, baking, unsweetened	1 oz	28	1	140	4	8	5	15	9.1	4.5	0.4
	(For other chocolate items, see Sweeteners & Sweets)											
26038	Cilantro/Coriander, fresh	1 tbsp	1	92	0	<1	<1	<1	<1	t	t	t
26003	Cinnamon	1 tsp	2	10	5	<1	2	1	<1	t	t	t
30197	Cornstarch	1 tbsp	8	8	30	<1	7	<1	<1	t	t	t
26004	Curry powder	1 tsp	2	10	6	<1	1	1	<1	t	0.1	t
26021	Dill weed, dried	1 tbsp	3	7	8	1	2	<1	<1	t	t	t
26005	Garlic cloves	1 ea	3	59	4	<1	1	<1	<1	t	t	t
26007	Garlic powder	1 tsp	3	6	10	1	2	<1	<1	t	t	t
23009	Gelatin, dry, unsweetened: Envelope	1 ea	7	13	23	6	0	0	<1	t	t	t
26043	Ginger root, slices, raw	2 pce	5	79	4	<1	1	<1	<1	t	t	t
27004	Horseradish, prepared	1 tbsp	15	85	7	<1	2	<1	<1	t	t	t
7081	Hummous/hummus	1 c	246	65	435	12	49	10	21	2.8	12.1	5.1
27072	Mustard, country dijon	1 tsp	5		5	1	1	0	0	0	0	0
6313	Mustard, gai choy chinese	1 tbsp	16	97	1	<1	<1	<1	0	0	0	0
	Miso (see Vegetables and Legumes, Soybean products)											
27009	Olives, ripe, pitted	5 ea	22	80	25	<1	1	1	2	0.3	1.7	0.2
26008	Onion powder	1 tsp	2	5	7	<1	2	<1	<1	t	t	t
26009	Oregano, ground	1 tsp	2	7	6	<1	1	1	<1	t	t	0.1
26010	Paprika	1 tsp	2	10	6	<1	1	1	<1	t	t	0.2
26011	Parsley, freeze dried	¼ c	1	2	3	<1	<1	<1	<1	t	t	t
26016	Pepper, black	1 tsp	2	11	5	<1	1	1	<1	t	t	t
27012	Pickles, dill, medium, 3¾ x 1¼" diam	1 ea	65	92	12	<1	3		<1	t	t	t
27016	Sweet, medium	1 ea	35	65	41	<1	11	<1	<1	t	t	t
	Popcorn (see Grain Products)											
44006	Potato chips:	10 pce	20	2	107	1	11	1	7	2.2	2	2.4
44076	Unsalted	1 oz	28	2	150	2	15	1	10	3.1	2.8	3.4
26031	Sage, ground	1 tsp	1	8	3	<1	1	<1	<1	t	t	t
27020	Salsa, from recipe	1 tbsp	15	95	3	<1	1	<1	<1	t	t	t
26014	Salt	1 tsp	6	0	0	0	0	0	0	0	0	0
	Salt Substitutes:											
26090	Morton, salt substitute	1 tsp	6	0	0	0	<1		0	0	0	0
26048	Morton, light salt	1 tsp	6	0	0	0	<1		<1			
26098	Seasoned salt, no MSG	1 tsp	5	5	0	0	0	0	0	0	0	0
27007	Vinegar, cider	½ c	120	94	17	0	7	0	0	0	0	0
27148	Balsamic	1 tbsp	15	64	21	0	5	0	0	0	0	0
27154	Malt	1 tbsp	15	90	5	0	1	0	0	0	0	0
27158	Tarragon	1 tbsp	15	95	3	0	<1	0	0	0	0	0
27156	White wine	1 tbsp	15	89	5	0	2	0	0	0	0	0
28001	Yeast, baker's, dry, active, package	1 ea	7	8	21	3	3	1	<1	t	0.2	t
	SOUPS, SAUCES, AND GRAVIES											
	SOUPS, canned:											
	Unprepared, condensed:											
50650	Cream of celery	1 c	251	85	181	3	18	2	11	2.8	2.6	5
50654	Cream of chicken	1 c	251	82	233	7	19	1	15	4.2	6.6	3
50666	Cream of mushroom	1 c	251	81	259	4	19	1	19	5.1	3.6	8.9
50668	Onion	1 c	246	86	113	8	16	2	3	0.5	1.5	1.3
	Prepared w/equal volume of whole milk:											
50008	Clam chowder, New England	1 c	248	85	164	9	17	1	7	3	2.3	1.1
50015	Cream of celery	1 c	248	86	164	6	15	1	10	3.9	2.5	2.7

Appendix **H**

PAGE KEY: H–2 = Beverages   H–4 = Dairy   H–8 = Eggs   H–10 = Fat/Oil   H–12 = Fruit   H–18 = Bakery
H–24 = Grain   H–30 = Fish   H–30 = Meats   H–34 = Poultry   H–36 = Sausage   H–38 = Mixed/Fast   H–42 = Nuts/Seeds   H–44 = Sweets
H–46 = Vegetables/Legumes   H–56 = Vegetarian   H–60 = Misc   H–60 = Soups/Sauces   H–62 = Fast   H–76 = Convenience   H–78 = Baby foods

Chol (mg)	Calc (mg)	Iron (mg)	Magn (mg)	Pota (mg)	Sodi (mg)	Zinc (mg)	VT-A (RAE)	Thia (mg)	VT-E (mg α)	Ribo (mg)	Niac (mg)	V-B6 (mg)	Fola (µg)	VT-C (mg)
0	217	0.41	1	505	4	0.04	0	0	0	0	0	0	0	0
0	0	0	0	0	1368	0	0	0	0	0	0	0	0	0
0	106	2.1	21	172	2	0.29	23	<.01	0.37	0.02	0.35	0.12	14	3
0	358	3.03	56	852	36	0.95	1	0.05	0.65	0.47	1.95	0.38	30	0
0	3	0.08	3	57	167	0.04	7	<.01	0.22	0.07	0.23	0.02	2	2
0	7	0.39	8	101	2	0.12	104	0.02	1.49	0.05	0.44	0.12	5	4
0	35	0.9	9	28	3	0.14	<1	<.01	0.02	<.01	0.06	0.02	<1	0
0	22	1.14	14	153	81	0.22	119	0.03	2.32	0.06	0.63	0.29	8	5
0	28	4.87	92	232	7	2.7	0	0.04	0.11	0.03	0.38	<.01	8	0
0	1	0.02	0	5	<1	<.01	3	<.01	0.02	<.01	0.01	<.01	1	0
0	25	0.76	1	10	1	0.04	<1	<.01	0.02	<.01	0.03	<.01	1	1
0	0	0.04	0	0	1	<.01	0	0	0	0	0	0	0	0
0	10	0.59	5	31	1	0.08	1	<.01	0.44	<.01	0.07	0.02	3	0
0	54	1.46	14	99	6	0.1	9	0.01		<.01	0.08	0.05		2
0	5	0.05	1	12	1	0.03	0	<.01	<.01	<.01	0.02	0.04	<1	1
0	2	0.08	2	33	1	0.08	0	0.01	0.02	<.01	0.02	0.09	<1	1
0	4	0.08	2	1	14	<.01	0	<.01	0	0.02	<.01	<.01	2	0
0	1	0.03	2	21	1	0.02	0	<.01	0.01	<.01	0.04	<.01	1	0
0	8	0.06	4	37	47	0.12	<1	<.01	<.01	<.01	0.06	0.01	9	4
0	121	3.86	71	426	595	2.68	1	0.22	1.84	0.13	0.98	0.98	145	19
0				10	120									
5		0.07	1		14								4	
0	19	0.73	1	2	192	0.05	4	<.01	0.36	0	<.01	<.01	0	0
0	7	0.05	2	19	1	0.05	0	<.01	<.01	<.01	0.01	0.02	3	0
0	32	0.88	5	33	<1	0.09	7	<.01	0.38	<.01	0.12	0.02	5	1
0	4	0.47	4	47	1	0.08	53	0.01	0.6	0.03	0.31	0.08	2	1
0	2	0.54	4	63	4	0.06	32	0.01	0.06	0.02	0.1	0.01	2	1
0	9	0.58	4	25	1	0.03	<1	<.01	0.01	<.01	0.02	<.01	<1	0
0	6	0.34	7	75	833	0.09	6	<.01	0.06	0.02	0.04	<.01	1	1
0	1	0.21	1	11	329	0.03	3	<.01	0.03	0.01	0.06	<.01	<1	0
0	5	0.33	13	255	119	0.22	0	0.03	1.82	0.04	0.77	0.13	9	6
0	7	0.46	19	357	2	0.31	0	0.05	2.55	0.06	1.07	0.18	13	9
0	17	0.28	4	11	<1	0.05	3	<.01	0.07	<.01	0.06	0.03	3	0
0	1	0.05	1	23	1	0.02	3	<.01	0.04	<.01	0.06	0.01	2	2
0	1	0.02	0	0	2325	<.01	0							
33			0	3018	1									
2			4	1518	1182									
0					1583									
0	7	0.72	26	120	1	0	0	0	0	0	0	0	0	0
2		0.15	10	3		0.15			0.15	0.15			0	
2		0.15	14	4		0.15			0.15	0.15			2	
0		0.15	2	1		0.15			0.15	0.15			0	
1		0.15	12	1		0.15			0.15	0.15			0	
0	4	1.16	7	140	4	0.45	0	0.17	0	0.38	2.78	0.11	164	0
28	80	1.25	13	246	1900	0.3	55	0.06	3.49	0.1	0.67	0.03	5	1
20	68	1.2	5	176	1973	1.25	108	0.06	1.36	0.12	1.64	0.03	3	0
3	65	1.05	10	168	1737	1.18	20	0.06	2.03	0.17	1.62	0.03	8	2
0	54	1.35	5	138	2116	1.23	7	0.07	0.54	0.05	1.21	0.1	30	2
22	186	1.49	22	300	992	0.79	57	0.07	0.45	0.24	1.03	0.13	10	3
32	186	0.69	22	310	1009	0.2	114	0.07	0.97	0.25	0.44	0.06	7	1

Table H–1

Food Composition

(Computer code number is for West Diet Analysis program)   (For purposes of calculations, use "0" for t, <1, <.1, <.01, etc.)

Computer Code Number	Food Description	Measure	Wt (g)	H₂O (%)	Ener (kcal)	Prot (g)	Carb (g)	Dietary Fiber (g)	Fat (g)	Fat Breakdown (g) Sat	Mono	Poly
	SOUPS, SAUCES, AND GRAVIES—Continued											
50006	Cream of chicken	1 c	248	85	191	7	15	<1	11	4.6	4.5	1.6
50011	Cream of mushroom	1 c	248	85	203	6	15	<1	14	5.1	3	4.6
50026	Cream of potato	1 c	248	87	149	6	17	<1	6	3.8	1.7	0.6
50024	Oyster stew	1 c	245	89	135	6	10	0	8	5	2.1	0.3
50012	Tomato	1 c	248	85	161	6	22	3	6	2.9	1.6	1.1
	Prepared with equal volume of water:											
50000	Bean with bacon	1 c	253	84	172	8	23	9	6	1.5	2.2	1.8
50003	Beef noodle	1 c	244	92	83	5	9	1	3	1.1	1.2	0.5
50005	Chicken noodle	1 c	241	92	75	4	9	1	2	0.7	1.1	0.6
50020	Chicken rice	1 c	241	94	60	4	7	1	2	0.5	0.9	0.4
50007	Chili beef	1 c	250	85	170	7	21	10	7	3.4	2.8	0.3
50021	Clam chowder, Manhattan	1 c	244	92	78	2	12	1	2	0.4	0.4	1.3
50018	Cream of chicken	1 c	244	91	117	3	9	<1	7	2.1	3.3	1.5
50049	Cream of mushroom	1 c	244	90	129	2	9	<1	9	2.4	1.7	4.2
50009	Minestrone	1 c	241	91	82	4	11	1	3	0.6	0.7	1.1
50022	Onion	1 c	241	93	58	4	8	1	2	0.3	0.7	0.7
50025	Split pea & ham	1 c	253	82	190	10	28	2	4	1.8	1.8	0.6
50028	Tomato	1 c	244	90	85	2	17	<1	2	0.4	0.4	1
50014	Vegetable beef	1 c	244	92	78	6	10	<1	2	0.9	0.8	0.1
50013	Vegetarian vegetable	1 c	241	92	72	2	12	<1	2	0.3	0.8	0.7
50052	Ready to serve, Chunky chicken soup	1 c	251	84	178	13	17	2	7	2	3	1.4
	SOUPS, dehydrated:											
	Prepared with water:											
50032	Beef broth/bouillon	1 c	244	97	20	1	2	0	1	0.3	0.3	t
50034	Chicken broth	1 c	244	97	22	1	1	0	1	0.3	0.4	0.4
50037	Chicken noodle	1 c	252	94	58	2	9	<1	1	0.3	0.5	0.4
50036	Cream of chicken	1 c	261	91	107	2	13	<1	5	3.4	1.2	0.4
50040	Onion	1 c	246	96	27	1	5	1	1	0.1	0.3	t
50041	Split pea	1 c	255	87	125	7	21	3	1	0.4	0.7	0.3
50043	Tomato vegetable	1 c	253	94	56	2	10	1	1	0.4	0.3	t
	Unprepared, dry products:											
50051	Beef bouillon, packet	1 ea	6	3	14	1	1	0	1	0.3	0.2	t
50054	Onion soup, packet	1 ea	39	4	115	5	21	4	2	0.5	1.4	0.3
	SAUCES											
53097	From home recipe, lowfat cheese sauce	¼ c	61	74	81	6	4	<1	5	1.8	1.8	0.8
	Ready to serve:											
53396	Alfredo sauce, reduced fat	¼ c	69	74	144	5	9	0	9	6.2		
53000	Barbeque sauce	1 tbsp	16	81	12	<1	2	<1	<1	t	0.1	0.1
53355	Creole sauce	¼ c	62	89	25	1	4		1	t	0.2	0.3
53392	Pesto sauce	2 tbsp	16		83	2	1	<1	8	1.5		
53002	Soy sauce	1 tbsp	16	71	8	1	1	<1	<1	t	t	t
53349	Szechuan sauce	1 tbsp	16	71	21	<1	3	<1	1	0.1	0.3	0.4
53004	Teriyaki sauce	1 tbsp	18	68	15	1	3	<1	0	0	0	0
53011	Spaghetti with meat sauce	1 c	250	85	178	7	19	4	8	1.8	3.3	1.8
	GRAVIES											
53023	Beef	1 c	233	87	123	9	11	1	5	2.7	2.2	0.2
53022	Chicken	1 c	238	85	188	5	13	1	14	3.4	6.1	3.6
53026	Mushroom	1 c	238	89	119	3	13	1	6	1	2.8	2.4
	FAST FOOD RESTAURANTS											
	ARBY'S											
	Roast beef sandwiches:											
56336	Regular	1 ea	155		326	21	35	2	14	6.9		
56337	Junior	1 ea	89		200	11	23	1	8	3.4		
56338	Super	1 ea	254		467	23	50	3	22	8.3		
69056	Beef 'n cheddar	1 ea	194		451	23	42	2	23	8.8		
56341	Chicken breast sandwich	1 ea	204		539	24	46	2	29	4.9		
56342	Ham'n cheese sandwich	1 ea	169		338	23	35	1	13	4.5		
69048	Italian sub sandwich	1 ea	297		742	28	47	3	50	14.3		
56343	Turkey sandwich, deluxe	1 ea	218		292	26	37	3	6	0.6	2.5	2.6
69044	Turkey sub sandwich	1 ea	277		570	24	46	2	33	8.1		

Appendix **H**

PAGE KEY: H–2 = Beverages   H–4 = Dairy   H–8 = Eggs   H–10 = Fat/Oil   H–12 = Fruit   H–18 = Bakery
H–24 = Grain   H–30 = Fish   H–30 = Meats   H–34 = Poultry   H–36 = Sausage   H–38 = Mixed/Fast   H–42 = Nuts/Seeds   H–44 = Sweets
H–46 = Vegetables/Legumes   H–56 = Vegetarian   H–60 = Misc   H–60 = Soups/Sauces   H–62 = Fast   H–76 = Convenience   H–78 = Baby foods

Chol (mg)	Calc (mg)	Iron (mg)	Magn (mg)	Pota (mg)	Sodi (mg)	Zinc (mg)	VT-A (RAE)	Thia (mg)	VT-E (mg α)	Ribo (mg)	Niac (mg)	V-B₆ (mg)	Fola (µg)	VT-C (mg)
27	181	0.67	17	273	1047	0.67	179	0.07	0.25	0.26	0.92	0.07	7	1
20	179	0.6	20	270	918	0.64	35	0.08	1.24	0.28	0.91	0.06	10	2
22	166	0.55	17	322	1061	0.67	52	0.08	0.1	0.24	0.64	0.09	10	1
32	167	1.05	20	235	1041	10.34	56	0.07	0.49	0.23	0.34	0.06	10	4
17	159	1.81	22	449	744	0.3	64	0.13	1.24	0.25	1.52	0.16	17	68
3	81	2.05	46	402	951	1.04	46	0.09	0.76	0.03	0.57	0.04	33	2
5	15	1.1	5	100	952	1.54	7	0.07	0.68	0.06	1.07	0.04	20	0
7	17	0.77	5	55	1106	0.39	36	0.05	0.1	0.06	1.39	0.03	22	0
7	17	0.75	0	101	815	0.27	22	0.02	0.1	0.02	1.13	0.02	0	0
12	42	2.12	30	525	1035	1.4	75	0.06	1.52	0.08	1.07	0.16	18	4
2	27	1.63	12	188	578	0.98	56	0.03	0.34	0.04	0.82	0.1	10	4
10	34	0.61	2	88	986	0.63	163	0.03	0.2	0.06	0.82	0.02	2	0
2	46	0.51	5	100	881	0.59	15	0.05	0.95	0.09	0.72	0.01	5	1
2	34	0.92	7	313	911	0.75	118	0.05	0.07	0.04	0.94	0.1	36	1
0	27	0.67	2	67	1053	0.6	<1	0.03	0.19	0.02	0.6	0.05	14	1
8	23	2.28	48	400	1007	1.32	23	0.15	0.15	0.08	1.47	0.07	3	2
0	12	1.76	7	264	695	0.24	29	0.09	2.32	0.05	1.42	0.11	15	66
5	17	1.12	5	173	791	1.54	95	0.04	0.37	0.05	1.03	0.08	10	2
0	22	1.08	7	210	822	0.46	116	0.05	0.41	0.05	0.92	0.06	10	1
30	25	1.73	8	176	889	1	68	0.09	0.33	0.17	4.42	0.05	5	1
0	10	0.02	7	37	1362	0.07	0	<.01	0.1	0.02	0.36	0	0	0
0	15	0.07	5	24	1484	0	2	<.01	0.07	0.03	0.2	0	2	0
10	5	0.5	8	33	577	0.2	3	0.2	0.13	0.08	1.09	0.03	18	0
3	76	0.26	5	214	1185	1.57	21	0.1	0.6	0.2	2.61	0.05	5	1
0	12	0.15	5	64	849	0.05	0	0.03	0	0.06	0.48	0	2	0
3	20	0.94	43	224	1148	0.56	3	0.21	0.03	0.14	1.26	0.05	41	0
0	8	0.63	20	104	1146	0.18	10	0.06	0.35	0.05	0.79	0.05	10	6
1	4	0.06	3	27	1019	0	0	<.01	0.13	0.01	0.27	0.01	2	0
2	55	0.58	25	260	3493	0.23	<1	0.11	0.03	0.24	1.99	0.04	6	1
9	167	0.24	10	101	307	0.74		0.03	0.51	0.14	0.16	0.03	4	0
31	103	0	8	93	618		0			0.11	0		0	0
0	3	0.14	3	28	130	0.03	<1	<.01	<.01	<.01	0.14	0.01	1	1
0	35	0.31	9	187	339	0.1		0.03	0.61	0.02	0.53	0.07	9	0
4	65	0.09		26	137			0		0.02	0		2	0
0	3	0.32	5	29	914	0.06	0	<.01	0	0.02	0.54	0.03	3	0
0	2	0.12	2	13	218	0.02		<.01	0.07	<.01	0.1	<.01	1	0
0	4	0.31	11	40	690	0.02	0	<.01	0	0.01	0.23	0.02	4	0
15	54	2.1	43	742	982	1.27		0.13	2.98	0.12	3.46	0.31	25	19
7	14	1.63	5	189	1305	2.33	2	0.07	0.05	0.08	1.54	0.02	5	0
5	48	1.12	5	259	1373	1.9	2	0.04	0.31	0.1	1.05	0.02	5	0
0	17	1.57	5	252	1357	1.67	0	0.08	0.19	0.15	1.6	0.05	29	0
44	59	3.55	16	422	879	3.75	0							0
28	41	1.86	8	201	483	1.5		0.18		0.26	6.6	0.1	7	
47	62	3.73	25	533	1099	3.73		0.39		0.58	12.4	0.3	21	1
49	98	3.53		321	1146	2.94		0.42		0.63	9.8			1
88	78	1.77	30	330	1138	0.15		0.22		0.54	8.99	0.38	18	4
89	149	2.68	31	380	1441	0.89		0.82		0.37	7.75	0.31	26	1
114	238	2.57		565	2323			0.91		0.49	8.19			2
45	90	2.02	33	394	1157	1.69		0.09		0.46	17.22	0.58	22	1
91	181	0.33		500	1964			13.2		0.54	18.8			2

Table H–1

Food Composition   (Computer code number is for West Diet Analysis program)   (For purposes of calculations, use "0" for t, <1, <.1, <.01, etc.)

Computer Code Number	Food Description	Measure	Wt (g)	H₂O (%)	Ener (kcal)	Prot (g)	Carb (g)	Dietary Fiber (g)	Fat (g)	Fat Breakdown (g) Sat	Mono	Poly
	FAST FOOD RESTAURANTS—Continued											
	Milkshakes:											
2124	Chocolate	1 ea	340	71	411	9	72	0	14	6.9		
2125	Jamocha	1 ea	326	72	386	8	67	0	12	5.7		
2126	Vanilla	1 ea	312	72	369	8	65	0	12	5.5		
6431	Salad, roast chicken	1 ea	400		152	19	14	6	2	0		
	Source: Arby's											
	BOSTON MARKET											
	Chicken											
15247	Half chicken with skin, svg	1 ea	277		590	70	4	0	33	10		
15246	Dark meat, svg	1 ea	95		190	22	1	0	10	3		
15249	Dark meat with skin, svg	1 ea	125		320	30	2	0	21	6		
15244	White meat, svg	1 ea	140		170	33	2	0	4	1		
15245	White meat, with skin, svg	1 ea	152	64	280	40	2	0	12	3.5		
57530	Chicken pot pie	1 ea	425		780	32	61	4	46	13		
15248	Turkey breast, svg	1 ea	142		170	36	1	0	1	0.5		
	Sandwiches											
69084	Ham and turkey club	1 ea	266		430	29	64	4	6	2		
69081	Ham	1 ea	266		440	25	66	4	8	2.5		
69076	Chicken	1 ea	281	64	430	34	62	4	4	1		
69075	Chicken with cheese and sauce	1 ea	352		750	41	72	5	33	12		
69079	Turkey	1 ea	266		400	45	61	4	4	1		
69078	Turkey with cheese and sauce	1 ea	337		710	45	68	4	28	10		
69083	Turkey and ham with cheese and sauce	1 ea	379		890	47	79	4	43	20		
69074	Meatloaf	1 ea	351		690	40	86	6	21	7		
69082	Meatloaf with cheese	1 ea	383		860	46	95	6	33	16		
	Side dishes											
7393	BBQ baked beans	¾ c	201		270	8	48	12	5	2		
7387	New Potatoes	¾ c	131		130	3	25	2	2	0		
7390	Mashed potatoes	⅔ c	161		190	3	24	1	9	6		
28257	Red beans and rice	1 c	227		260	8	45	4	5	0		
7388	Whole kernel corn	¾ c	146		180	5	30	2	4	0.5		
7386	Steamed vegetables	⅔ c	105		35	2	7	3	<1	0		
52111	Caesar side salad	4 oz	113		199	7	7	1	17	4.5		
52103	Chunky chicken salad	¾ c	158		370	28	3	1	27	4.5		
52108	Mediterranean pasta salad	¾ c	129		170	4	16	2	10	2.5		
	Source: Boston Chicken, Inc.											
	BURGER KING											
56346	Croissant sandwich, egg, sausage & cheese	1 ea	176	45	575	22	30	1	41	15		
	Sandwiches:											
56354	Whopper	1 ea	270		660	28	51	4	38	11.7		
56355	Whopper with cheese	1 ea	294		757	33	53	4	46	16.5		
57002	BK broiler chicken sandwich	1 ea	248		529	29	50	3	24	4.8		
56352	Cheeseburger	1 ea	138		375	20	31	2	18	9.1		
56360	Chicken sandwich	1 ea	229		675	26	54	3	40	8.2		
56356	Double beef	1 ea	351		915	48	53	4	57	19.9		
56357	Double beef & cheese	1 ea	375	52	1012	53	55	4	64	24.8		
56353	Double cheeseburger with bacon	1 ea	218		649	40	34	2	39	19.1		
56351	Hamburger	1 ea	126		328	18	31	2	14	6.1		
56362	Ocean catch fish fillet	1 ea	255		688	23	65	4	37	13.6		
15158	Chicken tenders	1 ea	88		251	16	15	1	14	3.4		
6141	French fries (salted), svg	1 ea	116		357	4	46	4	18	5		
42429	French toast sticks, svg	1 ea	141		491	8	58	3	25	5.7		

Appendix **H**

PAGE KEY: H–2 = Beverages   H–4 = Dairy   H–8 = Eggs   H–10 = Fat/Oil   H–12 = Fruit   H–18 = Bakery
H–24 = Grain   H–30 = Fish   H–30 = Meats   H–34 = Poultry   H–36 = Sausage   H–38 = Mixed/Fast   H–42 = Nuts/Seeds   H–44 = Sweets
H–46 = Vegetables/Legumes   H–56 = Vegetarian   H–60 = Misc   H–60 = Soups/Sauces   H–62 = Fast   H–76 = Convenience   H–78 = Baby foods

Chol (mg)	Calc (mg)	Iron (mg)	Magn (mg)	Pota (mg)	Sodi (mg)	Zinc (mg)	VT-A (RAE)	Thia (mg)	VT-E (mg α)	Ribo (mg)	Niac (mg)	V-B_6 (mg)	Fola (µg)	VT-C (mg)
39	428	0.62	48	410	317	1.5		0.12		0.68	0.8	0.14	14	2
37	411	0.59	36	525	320	1.48		0.12		0.68	0.8	0.14	14	2
35	393	0.85	36	686	283	1.49		0.12		0.68	4	0.14	37	2
38	76	1.71		877	667			0.31		0.54	5.6			0
290	0	2.7			1010	0								0
115	0	1.08			440	0								0
155	0	1.8			500	0								0
85	0	0.72			480	0								0
175	0	1.08			510	0								0
135	40	3.6			1480									4
100	20	1.8			850	0								0
55	80	2.7			1330									9
45	80	1.8			1450									9
65	100	1.8			910									9
135	400	2.7			1860									9
60	100	2.7			1070									9
110	500	2.7			1390									9
150	800	4.5			2310									9
120	100	4.5			1610									15
165	400	4.5			2270									15
0	100	3.6			540									6
0	0	0.72			150	0								12
25	60	0.36			570									6
5	60	2.7			1050									12
0	0	0.36			170	10								5
0	20	0.36			35	200								21
15	199	1.08			448									12
120	20	0.72			800									4
10	60	1.08			490									9
219	173	3.11			1173									0
78	97	5.24			913									9
102	243	5.24			1349									9
101	58	3.46			1067									6
56	152	2.74			761									0
72	82	2.76			1360									0
149	149	7.16			1014									9
169	298	7.14			1448									9
128	266	4.79			1244									0
46	82	2.77			543									0
48	78	3.49			1163									0
34	0	0.41			606	0								0
0	20	0.71			684	0								15
0	76	2.27			554	0								0

Table H–1

Food Composition

(Computer code number is for West Diet Analysis program)   (For purposes of calculations, use "0" for t, <1, <.1, <.01, etc.)

Computer Code Number	Food Description	Measure	Wt (g)	H₂O (%)	Ener (kcal)	Prot (g)	Carb (g)	Dietary Fiber (g)	Fat (g)	Fat Breakdown (g)		
										Sat	Mono	Poly
	FAST FOOD RESTAURANTS—Continued											
56363	Onion rings, svg	1 ea	124		436	5	55	4	22	5.5		
2127	Milk shakes, chocolate	1 ea	284		315	9	57	3	6	3.6		
2129	Vanilla	1 ea	284		315	9	57	1	6	3.6		
48134	Fried apple pie	1 ea	113		340	2	52	1	14	3		
	Source: Burger King Corporation											
	CARL'S JR											
	Sandwiches:											
91413	Carl's catch	1 ea	201	51	510	18	50	1	27	7		
91407	Charbroiled BBQ Chicken	1 ea	199		280	25	37	2	3	1		
91411	Bacon swiss chicken	1 ea	291		720	32	66	3	36	10		
91410	Ranch crispy chicken	1 ea	266		730	29	77	4	34	7.1		
91406	Hamburger	1 ea	134		330	18	34	1	13	5		
91402	Famous star hamburger	1 ea	254		580	25	49	2	32	9		
91403	Super star hamburger	1 ea	345		790	42	49	2	46	14		
91404	Western bacon cheeseburger	1 ea	225		650	32	63	2	30	12		
91405	Western bacon double cheeseburger	1 ea	308		900	51	64	2	49	21		
91424	Charbroiled chicken salad to go	1 ea	350		200	25	12	3	7	3.5		
	Side dishes:											
91437	Bacon, 2 slices, svg	1 ea	9	15	50	3	0	0	4	1.5		
91418	Criss-cut french fries, svg	1 ea	139		410	5	43	4	24	5		
91414	French fries, svg	1 ea	92	36	290	5	37	3	14	3		
91415	Onion rings, svg	1 ea	127		430	7	53	3	21	5		
	Source: Carl Karcher Enterprises, Inc.											
	CHICK-FIL-A											
	Sandwiches:											
69153	Chargrilled chicken	1 ea	150		290	26	30	1	7	2.1		
69154	Chicken salad club	1 ea	232		413	34	36	2	15	5.7		
	Salads:											
52139	Carrot and raisin	1 ea	76		109	1	18	2	4	0.8		
69181	Chargrilled chicken caesar	1 ea	241		240	31	6	2	10	6		
52134	Chicken garden, charbroiled	1 ea	397	86	257	33	11	4	9	4.3		
52135	Chick-n-strips	1 ea	451		487	43	27	4	23	7.2		
52138	Cole slaw	1 ea	79		158	1	11	2	13	1.9		
69184	Chargrilled chicken cool wrap	1 ea	240		390	31	53	3	7	3		
69183	Chicken caesar cool wrap	1 ea	227		460	38	51	3	11	6		
69188	Chicken breast fillet, breaded	1 ea	105		230	23	10	0	11	2.5		
15263	Chicken nuggets, svg	1 ea	110	54	253	25	12	1	12	2.4		
15262	Chicken-n- strips, svg	1 ea	119	54	275	28	13	0	12	2.8		
50885	Hearty breast of chicken soup, svg	1 ea	215	88	100	9	13	1	2	0		
7973	Waffle potato fries, svg	1 ea	85	34	280	3	37	5	14	5		
46489	Cheesecake, svg	1 ea	88		322	6	28	2	20	11.4		
49134	Fudge nut brownie, svg	1 ea	74	13	330	4	45	2	15	3.5		
20601	Icedream, svg	1 ea	127		151	4	26	0	4	1.9		
48214	Lemon pie, svg	1 ea	99	39	280	6	45	3	9	3.1		
20602	Lemonade, svg	1 ea	255		170	0	41	0	<1	0		
	Source: Chick-Fil-A											
	DAIRY QUEEN											
	Ice cream cones:											
2144	Small vanilla	1 ea	142		230	6	38	0	7	4.5		
2143	Regular vanilla	1 ea	213		355	9	57	0	10	6.5		
2142	Large vanilla	1 ea	253		410	10	65	0	12	8		
2136	Chocolate dipped	1 ea	220		490	8	59	1	24	13		

PAGE KEY: H–2 = Beverages   H–4 = Dairy   H–8 = Eggs   H–10 = Fat/Oil   H–12 = Fruit   H–18 = Bakery
H–24 = Grain   H–30 = Fish   H–30 = Meats   H–34 = Poultry   H–36 = Sausage   H–38 = Mixed/Fast   H–42 = Nuts/Seeds   H–44 = Sweets
H–46 = Vegetables/Legumes   H–56 = Vegetarian   H–60 = Misc   H–60 = Soups/Sauces   H–62 = Fast   H–76 = Convenience   H–78 = Baby foods

Chol (mg)	Calc (mg)	Iron (mg)	Magn (mg)	Pota (mg)	Sodi (mg)	Zinc (mg)	VT-A (RAE)	Thia (mg)	VT-E (mg α)	Ribo (mg)	Niac (mg)	V-B₆ (mg)	Fola (µg)	VT-C (mg)
0	136	0			627	0								0
25	250	1.29			193									0
18	286	0			243									4
0	0	1.44			470									0
80	150	1.8			1030									2
60	80	2.7			830									5
75	250	3.6			1610									6
59	177	4.24			1436									6
45	60	3.6			480	0								2
70	100	4.5			910									6
130	100	5.4			910									9
80	200	4.5			1430									1
155	300	6.3			1770									1
75	200	1.8			440									27
10	0	0			140	0								0
0	20	1.8			950	0								12
0	0	1.08			170	0								21
0	20	0.36			700	0								0
62	83	1.86			1034	0								0
92	172	3.1			1573									4
0	17	0.3			75									3
85	350	0.36			1170									0
100	214	0.51			1042									9
122	215	1.55			974									9
15	30	0.27			135									20
70	200	3.6			1120									5
85	400	3.6			1540									0
60	40	1.08			990									0
68	39	1.05			1061	0								0
77	44	1.19			628	0								0
20	40	0			940	0								21
15	20	0			105	0								0
85	57	0			255									0
20	20	1.8			210									0
14	94	0.34			75									4
96	131	0			193									15
0	0	0.36			10	0								
20	200	1.08			115									1
32	269	1.94			172									3
40	350	1.8			200									2
30	250	1.8			190									2

Table H–1

Food Composition

(Computer code number is for West Diet Analysis program)   (For purposes of calculations, use "0" for t, <1, <.1, <.01, etc.)

Computer Code Number	Food Description	Measure	Wt (g)	H₂O (%)	Ener (kcal)	Prot (g)	Carb (g)	Dietary Fiber (g)	Fat (g)	Fat Breakdown (g) Sat	Mono	Poly
	FAST FOOD RESTAURANTS—Continued											
2154	Chocolate sundae	1 ea	234		400	8	71	0	10	6		
2131	Banana split	1 ea	369		510	8	96	3	12	8		
2151	Peanut buster parfait	1 ea	305		730	16	99	2	31	17		
2133	Buster bar	1 ea	149		450	10	41	2	28	12		
2135	Dilly bar	1 ea	85	55	210	3	21	0	13	7		
	Milkshakes:											
2152	Large	1 ea	461	72	600	13	101	<1	16	10	2	2
2145	Malted	1 ea	418	68	610	13	106	<1				
2351	Misty slush, small	1 ea	454		220	0	56	0	0	0	0	0
2359	Starkiss	1 ea	85		80	0	21	0	0	0	0	0
	Sandwiches:											
56372	Cheeseburger, double	1 ea	219		540	35	30	2	31	16		
56371	Single	1 ea	152	55	340	20	29	2	17	8		
56379	Chicken	1 ea	191		466	18	45	4	24	4.2		
69029	Chicken fillet, grilled	1 ea	184		310	24	30	3	10	2.5		
56381	Fish fillet sandwich	1 ea	170	58	370	16	39	2	16	3.5		
56382	With cheese	1 ea	184	56	420	19	40	2	21	6		
56368	Hamburger, single	1 ea	138	56	290	17	29	2	12	5		
56369	Hamburger, double	1 ea	212		440	30	29	2	22	10		
	Hotdogs:											
56374	Regular	1 ea	99	55	240	9	19	1	14	5		
56375	With cheese	1 ea	113		290	12	20	1	18	8	8	2
56376	With chili	1 ea	128		297	13	20	2	19	8.1		
6143	French fries, small	1 ea	112	41	347	4	42	3	18	3.5		
56383	Onion rings	1 ea	113		320	5	39	3	16	4		

Source: International Dairy Queen

	HARDEES'S											
	Sandwiches:											
69061	Frisco burger hamburger	1 ea	242		793	36	41	2	54	15.3		
56422	Chicken sandwich	1 ea	187		444	23	39	2	17	4.6		
6146	French fries, svg	1 ea	96		289	3	38	0	14	1.7		

Source: Hardees Food Systems

	IN-N-OUT BURGERS											
81110	Hamburger with spread	1 ea	243		390	16	39	3	19	5		
81111	Hamburger w/mustard and ketchup	1 ea	243		310	16	41	3	10	4		
81113	Cheeseburger with spread	1 ea	268		480	22	39	3	27	10		
81114	Cheeseburger w/mustard and ketchup	1 ea	268		400	22	41	3	18	9		
81115	Cheeseburger, protein style, without bun	1 ea	300		330	18	11	2	25	9		
81116	Double cheeseburger with spread	1 ea	328		670	37	40	3	41	18		
81117	Double cheeseburger with mustard and ketchup	1 ea	328		590	37	42	3	32	17		
81118	Double cheeseburger, protein style, without bun	1 ea	361		520	33	11	2	39	17		

Source: In-N-Out Burgers

	JACK IN THE BOX											
	Breakfast items:											
56430	Breakfast jack sandwich	1 ea	126		280	17	28	1	12	5		
56431	Sausage crescent	1 ea	181		660	20	37	0	48	15		
56432	Supreme crescent	1 ea	164		530	21	37	0	34	10		
	Sandwiches:											
69032	Bacon cheeseburger	1 ea	274		760	39	39	2	50	17		

PAGE KEY: H–2 = Beverages   H–4 = Dairy   H–8 = Eggs   H–10 = Fat/Oil   H–12 = Fruit   H–18 = Bakery
H–24 = Grain   H–30 = Fish   H–30 = Meats   H–34 = Poultry   H–36 = Sausage   H–38 = Mixed/Fast   H–42 = Nuts/Seeds   H–44 = Sweets
H–46 = Vegetables/Legumes   H–56 = Vegetarian   H–60 = Misc   H–60 = Soups/Sauces   H–62 = Fast   H–76 = Convenience   H–78 = Baby foods

Chol (mg)	Calc (mg)	Iron (mg)	Magn (mg)	Pota (mg)	Sodi (mg)	Zinc (mg)	VT-A (RAE)	Thia (mg)	VT-E (mg α)	Ribo (mg)	Niac (mg)	V-B$_6$ (mg)	Fola (µg)	VT-C (mg)
30	250	1.44			210									0
30	250	1.8			180									15
35	300	1.8			400									1
15	150	1.08			280									0
10	100	0.36			75									0
50	450	1.44		660	260			0.15		0.68	0.8			0
45				570										
0	0	0			20	0								0
0	0	0			10	0								0
115	250	4.5			1130									4
55	150	3.6			850									4
28	75	4.19			1016									2
50	200	2.7			1040	0								0
45	40	1.8		280	630	0		0.3		0.22	3			0
60	100	1.8		290	850			0.3		0.26	5			0
45	60	2.7			630									4
90	60	4.5			680									6
25	60	1.8			730									4
40	150	1.8		180	950			0.22		0.17	2			4
41	135	1.62			983									3
0	20	1.07			872	0								4
0	20	1.44			180	0								0
111					1192									
61					1081									
0					331									
40	40	3.6			640									15
35	40	3.6			720									15
60	200	3.6			1000									15
55	200	3.6			1080									15
60	200	1.08			720									18
120	350	5.4			1430									15
115	350	5.4			1510									15
120	350	1.08			1160									18
190	150	3.6		120	750									10
240	100	1.8		160	860									0
225	100	1.8		165	1060									4
135	250	4.5		530	1570			0.27		0.54	9.96	0.44		9

Table H–1

Food Composition   (Computer code number is for West Diet Analysis program)   (For purposes of calculations, use "0" for t, <1, <.1, <.01, etc.)

Computer Code Number	Food Description	Measure	Wt (g)	H₂O (%)	Ener (kcal)	Prot (g)	Carb (g)	Dietary Fiber (g)	Fat (g)	Sat	Mono	Poly
	FAST FOOD RESTAURANTS—Continued											
1655	Chicken sandwich	1 ea	164		400	15	38	3	21	3		
56443	Chicken supreme	1 ea	305		830	33	66	3	49	7		
56366	Double cheeseburger	1 ea	158		440	24	31	2	24	11		
69033	Grilled sourdough burger	1 ea	233		690	34	37	2	45	15		
56436	Jumbo jack burger	1 ea	271		550	27	43	2	30	10		
56437	Jumbo jack burger with cheese	1 ea	296		640	31	44	2	38	15		
56377	Tacos, regular	1 ea	90	66	170	7	12	2	10	3.5		
56378	Super	1 ea	138	63	270	12	19	4	17	6		
57014	Chicken teriyaki bowl	1 ea	502		670	26	128	3	4	1		
6150	French fries	1 ea	113	40	350	4	46	3	16	4		
6149	Hash browns	1 ea	57		170	1	14	1	12	2		
2164	Milkshake, strawberry	1 ea	382		640	10	85	0	28	15		
48135	Apple turnover	1 ea	107	40	340	4	41	2	18	4		

Source: Jack in the Box Restaurant, Inc

	KENTUCKY FRIED CHICKEN											
	Original Recipe chicken:											
15163	Side breast	1 ea	153		400	29	16	1	24	6	14.4	3.6
15165	Drumstick	1 ea	61	56	140	13	4	0	9	2	5.3	1.7
15166	Thigh	1 ea	91	54	250	16	6	1	18	4.5	10.2	3.4
15167	Wing	1 ea	47	48	140	9	5	0	10	2.5	5.8	1.7
	Hot & spicy chicken:											
15185	Center breast	1 ea	180	48	505	38	23	1	29	8		
15183	Thigh	1 ea	107		355	19	13	1	26	7		
15187	Wing	1 ea	55	37	210	10	9	1	15	4		
	Extra Crispy Recipe chicken:											
15169	Center breast	1 ea	168	48	470	39	17	1	28	8	16.7	3.3
15170	Drumstick	1 ea	67	48	195	15	7	1	12	3	7.4	1.6
15171	Thigh	1 ea	118		380	21	14	1	27	7	15.8	4.2
15172	Wing	1 ea	55	35	220	10	10	1	15	4	8.9	2.1
7139	Baked beans	½ c	167		203	6	35	6	3	1.1	1.4	0.7
42331	Buttermilk biscuit	1 ea	56		180	4	20	1	10	2.5	5.4	2.1
56451	Coleslaw, svg	1 ea	142		232	2	26	3	14	2	3.5	7
6152	Corn-on-the-cob	1 ea	162		150	5	35	2	2	0	0.6	0.9
15177	Chicken, hot wings, svg	1 ea	135		471	27	18	2	33	8		
56681	Macaroni & cheese, svg	1 ea	153		180	7	21	2	8	3		
56453	Mashed potatoes & gravy, svg	1 ea	136	82	120	1	17	2	6	1	3.6	1.4
56454	Potato salad	½ c	160		230	4	23	3	14	1.8	4.5	7.8
6188	Potato wedges, svg	1 ea	135		278	5	28	5	13	4		

Source: Kentucky Fried Chicken Corp

	McDONALD'S											
	Sandwiches:											
69010	Big Mac	1 ea	216		590	24	47	3	34	11		
69013	Filet-o-fish	1 ea	156	43	470	15	45	1	26	5		
69011	Quarter-pounder	1 ea	171		428	23	37	2	21	8		
69012	Quarter-pounder with cheese	1 ea	199		527	28	38	2	30	12.9		
15174	Chicken McNuggets	4 pce	71		187	10	13	1	11	2.5		
	Sauces (packet):											
27070	Hot mustard	1 ea	28		60	1	7	1	4	0		
53176	Barbecue	1 ea	28		45	0	10	0	0	0	0	0
53177	Sweet & sour	1 ea	28		50	0	11	0	0			
	Low-fat (frozen yogurt) milk shakes:											
2167	Chocolate	1 ea	295		360	11	60	1	9	6		
2168	Strawberry	1 ea	294	72	360	11	60	0	9	6		
2169	Vanilla	1 ea	293		360	11	59	0	9	6		
	Low-fat (frozen yogurt) sundaes:											

PAGE KEY: H–2 = Beverages H–4 = Dairy H–8 = Eggs H–10 = Fat/Oil H–12 = Fruit H–18 = Bakery
H–24 = Grain H–30 = Fish H–30 = Meats H–34 = Poultry H–36 = Sausage H–38 = Mixed/Fast H–42 = Nuts/Seeds H–44 = Sweets
H–46 = Vegetables/Legumes H–56 = Vegetarian H–60 = Misc H–60 = Soups/Sauces H–62 = Fast H–76 = Convenience H–78 = Baby foods

Chol (mg)	Calc (mg)	Iron (mg)	Magn (mg)	Pota (mg)	Sodi (mg)	Zinc (mg)	VT-A (RAE)	Thia (mg)	VT-E (mg α)	Ribo (mg)	Niac (mg)	V-B$_6$ (mg)	Fola (µg)	VT-C (mg)
40	100	2.7		200	770									5
65	200	3.6		250	2140									9
80	250	4.5		290	1100									1
105	200	4.5		480	1180									9
75	150	4.5		490	880									9
105	250	4.5		530	1340									9
15	100	1.08	40	235	390	1.38								0
30	200	1.44	49	365	630	1.8								2
15	100	4.5		620	1730									24
0	10	0.72		590	710		0							6
0	10	0.18		100	250		0							0
85	350	0		620	300									0
0	10	1.8		85	510									10
135	40	1.08			1116									1
75	20	0.72			422									1
95	20	0.72			747									1
55	20	0.36			414									1
162	60	1.08			1170									1
126	20	0.72			630									1
55	20	0.72			350									1
160	20	1.08			874									1
77	20	0.72			375									1
118	20	1.08			625									1
55	20	0.36			415									1
5	86	1.93			814									1
0	20	1.08			560									1
8	30	0.36			285									34
0	20	0.36			20	5								4
150	40	1.44			1230									1
10	150	0.36			860									1
1	20	0.36			440									1
15	20	2.7			540									1
5	20	1.79			744	5								1
85	300	4.5	46	455	1090	4.8		0.49	1.01	0.44	6.07	0.25	49	4
50	200	1.8	34	286	890	0.76			1.64					0
70	199	4.47	34	405	835	4.66		0.39	0.36	0.32	6.78	0.24	27	2
95	348	4.48			1303				0.81		6.78		33	2
35	9	0.7	16	199	355	0.65	0		0.93		4.87			0
5	7	0.72		27	240			<.01		<.01	0.14			0
0	3	0		45	250		0	<.01		<.01	0.15			4
0	2	0.14		7	140			0		<.01	0.07			0
40	350	0.72		543	250									1
40	350	0.72		542	180									6
40	350	0.36		533	250									1

Table H–1

Food Composition (Computer code number is for West Diet Analysis program) (For purposes of calculations, use "0" for t, <1, <.1, <.01, etc.)

Computer Code Number	Food Description	Measure	Wt (g)	H₂O (%)	Ener (kcal)	Prot (g)	Carb (g)	Dietary Fiber (g)	Fat (g)	Fat Breakdown (g) Sat	Mono	Poly
	FAST FOOD RESTAURANTS—Continued											
2170	Hot caramel	1 ea	182		360	7	61	0	10	6		
2171	Hot fudge	1 ea	179	59	340	8	52	1	12	9		
2172	Strawberry	1 ea	178		290	7	50		7	5		
2166	Vanilla	1 ea	90		150	4	23	0	4	3		
47147	Cookies, McDonaldland	1 ea	42	13	169	2	28	1	6	1.5		
47146	Chocolaty chip	1 ea	35		175	2	23	1	9	5		
42333	Muffin, apple bran, fat-free	1 ea	114		300	6	61	3	3	0.5		
48136	Pie, apple	1 ea	77		260	3	34	1	13	3.5		
	Breakfast items:											
42064	English muffin with spread	1 ea	63	33	189	5	30	2	6	2.4	1.5	1.3
69005	Egg McMuffin	1 ea	137		292	17	27	1	12	4.5		
45069	Hotcakes with marg & syrup	1 ea	228		600	9	104	0	17	3		
19579	Scrambled eggs	1 ea	102		160	13	1	0	11	3.5		
12230	Pork sausage	1 ea	43		170	6	0	0	16	5		
69006	Sausage McMuffin	1 ea	112		360	13	26	1	23	8		
69007	Sausage McMuffin with egg	1 ea	163	52	443	19	27	1	28	10.1		
42332	Biscuit with biscuit spread	1 ea	84	33	292	5	37	1	13	3		
69003	Biscuit with sausage	1 ea	127	36	465	11	34	1	32	9.1		
69004	Biscuit with sausage & egg	1 ea	178	49	538	18	34	1	36	11		
69002	Biscuit with bacon, egg, cheese	1 ea	157	44	496	21	32	1	32	10.3		
56479	Garden salad	1 ea	149		100	7	4	2	6	3		
	Source: McDonald's Corporation											
	PIZZA HUT											
	Pan pizza:											
56481	Cheese	2 pce	216		569	24	55	4	27	11.8		
56482	Pepperoni	2 pce	208		549	22	55	4	27	9.8		
56483	Supreme	2 pce	273		657	27	60	6	35	12.3		
56484	Super supreme	2 pce	286		680	28	60	6	36	12		
	Thin 'n crispy pizza:											
56485	Cheese	2 pce	174		409	20	45	4	18	10.2		
56486	Pepperoni	2 pce	168		394	19	44	4	19	8.3		
56487	Supreme	2 pce	232		496	24	46	4	26	11.9		
56488	Super supreme	2 pce	247		532	25	44	4	28	11.4		
	Hand tossed pizza:											
56489	Cheese	2 pce	216		489	24	57	4	20	10.2		
56490	Pepperoni	2 pce	208		502	23	50	4	23	10.8		
56492	Super supreme	2 pce	286		599	29	64	7	27	11.1		
56493	Pepperoni personal pan pizza	1 ea	255		615	26	69	5	28	10.9		
	Source: Pizza Hut.											
	SUBWAY											
	Deli style sandwich:											
69102	Ham	1 ea	171		253	13	42	4	5	1.8		
69103	Roast beef	1 ea	180		262	15	42	4	5	2.4		
69107	Tuna with light mayo	1 ea	178		350	14	38	3	17	4.8		
69101	Turkey	1 ea	180		262	15	43	4	4	1.8		
	Sandwiches, 6 inch:											
69117	Club on white bread	1 ea	246		309	23	44	4	6	1.9		
69113	Cold cut trio on white bread	1 ea	246		421	20	45	4	20	6.7		
69115	Ham on white bread	1 ea	232		290	18	46	4	5	1.5		
69139	Italian B.M.T. on white bread	1 ea	246		476	23	46	4	24	8.9		
69129	Meatball on white bread	1 ea	260		480	22	48	5	24	9.1		
69127	Melt with turkey, ham, bacon, cheese, on white bread	1 ea	251		397	23	46	4	16	5.8		
69121	Roast Beef on white bread	1 ea	232		303	20	47	4	5	2.1		

Appendix **H**

PAGE KEY: H–2 = Beverages   H–4 = Dairy   H–8 = Eggs   H–10 = Fat/Oil   H–12 = Fruit   H–18 = Bakery
H–24 = Grain   H–30 = Fish   H–30 = Meats   H–34 = Poultry   H–36 = Sausage   H–38 = Mixed/Fast   H–42 = Nuts/Seeds   H–44 = Sweets
H–46 = Vegetables/Legumes   H–56 = Vegetarian   H–60 = Misc   H–60 = Soups/Sauces   H–62 = Fast   H–76 = Convenience   H–78 = Baby foods

Chol (mg)	Calc (mg)	Iron (mg)	Magn (mg)	Pota (mg)	Sodi (mg)	Zinc (mg)	VT-A (RAE)	Thia (mg)	VT-E (mg α)	Ribo (mg)	Niac (mg)	V-B$_6$ (mg)	Fola (µg)	VT-C (mg)
35	250				180									1
30	250	0.72			170									1
30	200	0.36			95									1
20	100	0.36			75									1
0	15	1.33	8	46	184	0.29	0		0.74		1.5			0
25	12	0.9	15	89	106	0.25		0.09	0.58	0.1	0.92			0
0	100	1.44	20	117	380	0.5	0	0.22	0	0.22	2.01	0.04	8	1
0	20	1.08	7	63	200	0.21		0.18	1.38	0.11	1.42	0.03	8	24
13	103	1.59	13	69	386	0.42	32	0.25	0.13	0.32	2.61	0.04	57	1
237	201	2.72	24	199	796	1.56		0.49	0.85	0.44	3.32	0.15	33	1
20	100	4.5	28	292	770	0.55			1.23				<1	0
425	40	1.08	10	126	170	1.06	150	0.07	0.92	0.51	0.06	0.12	44	0
35	7	0.36	7	102	290	0.78	0	0.18	0.26	0.06	1.7	0.09		0
45	200	1.8	22	191	740	1.51			0.66				16	0
257	252	2.72	26	251	895	2.06			1.11				30	0
0	49	2.19	10	116	779	0.33			0.89		2.46		5	0
40	45	2.7	16	221	1055	1.08		0.51	1.13	0.31	4.19	0.13	5	0
269	88	2.97	21	283	1110	1.68			1.59		4.14		28	0
258	155	2.79	21	253	1456	1.69			1.54		3.43		31	0
75	150	1.08			120		75							15
20	393	3.53			1159									5
29	196	3.53			1197									5
41	308	3.69			1375									12
50	300	3.6			1560									12
20	409	2.95			1208									5
31	207	2.99			1265									5
40	297	3.57			1408									18
48	285	3.42			1596									17
20	408	2.93			1325									5
36	359	3.23			1417									43
44	333	3.99			1618									13
30	298	4.46			1419									6
12	72	4.34			927									14
18	72	6.44			787									14
26	159	3.81			879									13
18	72	4.29			870									14
34	58	5.21			1254									20
53	144	5.17			1608									23
25	60	3.6			1270									21
55	149	3.57			1885									24
50	136	4.89			1232									24
44	145	3.49			1677									23
21	63	6.58			951									22

H Appendix

Table H–1

Food Composition

(Computer code number is for West Diet Analysis program)   (For purposes of calculations, use "0" for t, <1, <.1, <.01, etc.)

Computer Code Number	Food Description	Measure	Wt (g)	H₂O (%)	Ener (kcal)	Prot (g)	Carb (g)	Dietary Fiber (g)	Fat (g)	Sat	Mono	Poly
	FAST FOOD RESTAURANTS—Continued											
69125	Roasted chicken breast on white bread	1 ea	246		334	24	49	5	5	2.1		
69147	Seafood and crab with light mayo on white bread	1 ea	246		396	15	50	5	15	4.3		
69119	Steak and cheese on white bread	1 ea	257		392	24	48	5	14	5		
69143	Tuna with light mayo on white bread	1 ea	246		434	19	44	4	21	5.8		
69111	Turkey on white bread	1 ea	232		293	19	48	4	5	1.6		
69137	Turkey and ham on white bread	1 ea	232		290	20	46	4	5	1.5		
69109	Veggie delite on white bread	1 ea	175		242	9	46	4	3	1.1		
	Salads:											
52124	B.M.T., classic italian	1 ea	331		302	17	12	3	20	8.6		
52115	Club	1 ea	331		154	17	12	3	4	1.5		
52120	Cold cut trio	1 ea	330		240	15	11	3	16	6.3		
52123	Ham	1 ea	316		120	12	12	3	3	1.1		
52129	Meatball	1 ea	345		319	17	17	4	20	9		
52131	Melt	1 ea	336		211	18	12	3	11	4.8		
52126	Roast beef	1 ea	316		131	13	11	3	3	1.6		
52119	Roasted chicken breast	1 ea	331		153	17	13	3	3	1.1		
52116	Seafood and crab with light mayo	1 ea	331		211	9	18	4	12	3.7		
52130	Steak and cheese	1 ea	342		195	18	13	4	9	3.8		
52118	Tuna with light mayo	1 ea	331		253	14	11	3	17	4.2		
52113	Veggie delite	1 ea	260		56	2	10	3	1	0		
	Cookies:											
47658	With M&M's	1 ea	48	5	235	2	32	1	11	4.3		
47659	Chocolate chunk	1 ea	48		235	2	32	1	11	4.3		
47656	Oatmeal raisin	1 ea	48		213	3	32	2	9	2.7		
47657	Peanut butter	1 ea	48	5	235	4	28	1	13	4.3		
47660	Sugar	1 ea	48		245	2	30	0	13	4.3		
47661	White chip macademia	1 ea	48		235	2	30	1	12	4.3		
	Source: Subway International											
	TACO BELL											
	Burritos:											
56519	Bean with red sauce	1 ea	198		370	14	55	8	10	3.5		
56688	Chicken burrito	1 ea	171		283	14	34	3	10	4.1		
56522	Supreme with red sauce	1 ea	255		452	19	52	7	19	8.2		
56691	7 layer burrito	1 ea	283		530	18	67	10	22	8		
	Tacos:											
56525	Soft taco	1 ea	90		191	9	19	2	9	4.1		
56526	Soft taco supreme	1 ea	142		276	12	23	3	15	7.4		
56689	Soft taco, chicken	1 ea	121		232	17	23	1	7	3.1		
56693	Soft taco, steak	1 ea	128		282	12	21	1	17	4.5		
56528	Tostada with red sauce	1 ea	177		260	11	30	7	10	4.2		
56531	Mexican pizza	1 ea	220		560	21	47	7	32	11.2		
56537	Taco salad with salsa	1 ea	539		799	31	74	13	42	15.2		
56533	Nachos, regular	1 ea	99	40	320	5	33	2	19	4.5		
56534	Nachos, bellgrande	1 ea	312		790	20	81	12	44	13.2		
56536	Pintos & cheese with red sauce	1 ea	120	69	169	9	19	6	7	3.3		
53186	Taco sauce, packet	1 ea	11		4	0	<1	0	0	0	0	0
45585	Cinnamon twists	1 ea	28		128	1	22	0	4	0.8		
	Source: Taco Bell Corporation											
	WENDY'S											
56571	Cheeseburger, bacon	1 ea	166		382	20	34	2	19	7		
69059	Chicken sandwich, grilled	1 ea	189		302	24	36	2	7	1.5		
	Baked potatoes:											
6167	Plain	1 ea	284		310	7	72	7	0	0	0	0

PAGE KEY: H–2 = Beverages H–4 = Dairy H–8 = Eggs H–10 = Fat/Oil H–12 = Fruit H–18 = Bakery
H–24 = Grain H–30 = Fish H–30 = Meats H–34 = Poultry H–36 = Sausage H–38 = Mixed/Fast H–42 = Nuts/Seeds H–44 = Sweets
H–46 = Vegetables/Legumes H–56 = Vegetarian H–60 = Misc H–60 = Soups/Sauces H–62 = Fast H–76 = Convenience H–78 = Baby foods

Chol (mg)	Calc (mg)	Iron (mg)	Magn (mg)	Pota (mg)	Sodi (mg)	Zinc (mg)	VT-A (RAE)	Thia (mg)	VT-E (mg α)	Ribo (mg)	Niac (mg)	V-B$_6$ (mg)	Fola (µg)	VT-C (mg)
47	63	5.63			1042									22
24	145	3.47			1235									23
35	151	8.13			1215									24
39	14	3.47			1148									23
21	63	3.76			1055									22
25	60	3.6			1220									21
0	63	3.8			538									22
59	108	1.16			1714									32
36	41	18.5			1141									31
57	157	1.88			1431									31
27	44	1.18			1170									33
55	100	1.79			1047									36
48	106	1.14			1490									32
22	44	1.97			787									33
49	44	1.18			874									33
26	105	1.14			1023									32
38	109	3.91			966									33
42	105	1.14			928									32
0	45	1.21			346									33
16	0	1.15			112		0							0
11	0	1.15			112		0							0
16	0	1.15			192		0							0
11	0	1.15			213									0
16	0	1.15			144		0							0
16	0	1.15			171									0
10	200	2.7			1200									5
31	138	1.86			876									6
41	206	2.78			1368									9
25	300	3.6			1360									5
23	91	1.64			564									2
42	159	1.91			668									5
37	122	1.32			672									1
30	101	1.45			655									4
16	156	1.5			739									5
46	356	3.67			1049									6
66	405	6.37			1689									21
5	80	0.72			530		0							0
35	203	2.74			1317									6
14	141	1.01			656									3
0	0	0			102									0
0	0	0.29			120		0							0
55	151	3.62		322	895									9
55	80	2.71		432	744									9
0	20	3.6		1190	25		0							36

Table H–1

Food Composition (Computer code number is for West Diet Analysis program) (For purposes of calculations, use "0" for t, <1, <.1, <.01, etc.)

Computer Code Number	Food Description	Measure	Wt (g)	H₂O (%)	Ener (kcal)	Prot (g)	Carb (g)	Dietary Fiber (g)	Fat (g)	Fat Breakdown (g)		
										Sat	Mono	Poly
	FAST FOOD RESTAURANTS—Continued											
56579	With bacon & cheese	1 ea	380		580	18	79	7	22	6		
56580	With broccoli & cheese	1 ea	411		480	9	81	9	14	3		
56582	With sour cream & chives	1 ea	439		521	10	103	10	8	5.6		
50311	Chili	1 ea	227		200	17	21	5	6	2.5		
2176	Frosty dairy dessert	1 ea	227		330	8	56	0	8	5		
	Source: Wendy's International											
	CONVENIENCE FOODS & MEALS											
	HAAGEN DAZS											
70642	Ice cream bar, vanilla almond	1 ea	87		304	5	21	1	22	11.5		
70645	Lemon sorbet	½ c	113		120	0	31	<1	0	0	0	0
70646	Raspberry	½ c	105		120	0	30	2	0	0	0	0
	Source: Pillsbury											
	HEALTHY CHOICE											
	Entrees:											
18825	Fish, lemon pepper	1 ea	303		320	14	50	5	7	2		
81039	Lasagna	1 ea	383		420	26	59	6	9	3		
11119	Meatloaf, traditional	1 ea	340	78	316	15	52	6	5	2.5	1.9	0.6
	Low-Fat ice cream:											
2184	Brownie	½ c	71		120	3	22	1	2	1	0.3	0.7
2185	Chocolate chip	½ c	71		120	3	21	1	2	1	1	0
2105	Cookie & cream	½ c	71		120	3	21	1	2	1	1	0
2123	Rocky road	½ c	71		140	3	28	1	2	1	1	0
	Source: ConAgra Frozen Foods, Omaha, NE											
	HEALTH VALLEY											
	Soups, fat-free:											
50355	Beef broth, no salt added	1 c	240		18	5	0	0	0	0	0	0
50366	Beef broth, w/salt	1 c	240		30	5	2	0	0	0	0	0
50363	Black bean & vegetable	1 c	240		110	11	24	12	0	0	0	0
50364	Chicken broth	1 c	240		30	6	0	0	0	0	0	0
50365	14 garden vegetable	1 c	240		80	6	17	4	0	0	0	0
50362	Lentil & carrot	1 c	240		90	10	25	14	0	0	0	0
50361	Split pea & carrot	1 c	240		110	8	17	4	0	0	0	0
50360	Tomato vegetable	1 c	240		80	6	17	5	0	0	0	0
	Source: Health Valley											
	LA CHOY											
83016	Egg rolls, mini, chicken, svg	1 ea	106		108	3	13	1	5	1.3		
83013	Egg rolls, mini, shrimp, svg	1 ea	106		98	3	14	1	3	0.8		
	Source: Beatrice/Hunt-Wesson											
	LEAN CUISINE											
	Dinners:											
56901	Baked cheese ravioli	1 ea	241		260	12	38	4	7	3.5	1.5	0.5
15964	Chicken chow mein	1 ea	255		240	14	37	3	3	1	1.5	0.5
56740	Lasagna	1 ea	291		293	19	37	4	8	4.4	2	1
56702	Macaroni & cheese	1 ea	255		261	13	38	2	6	3.6	1.3	0.4
56732	Spaghetti w/meatballs	1 ea	269	75	299	18	40	5	8	2.1	2.7	1.3
56734	French bread sausage pizza	1 ea	170		210	8	24	1	9	3.5		
	Source: Stouffer's Foods Corp, Solon OH											

PAGE KEY: H–2 = Beverages H–4 = Dairy H–8 = Eggs H–10 = Fat/Oil H–12 = Fruit H–18 = Bakery
H–24 = Grain H–30 = Fish H–30 = Meats H–34 = Poultry H–36 = Sausage H–38 = Mixed/Fast H–42 = Nuts/Seeds H–44 = Sweets
H–46 = Vegetables/Legumes H–56 = Vegetarian H–60 = Misc H–60 = Soups/Sauces H–62 = Fast H–76 = Convenience H–78 = Baby foods

Chol (mg)	Calc (mg)	Iron (mg)	Magn (mg)	Pota (mg)	Sodi (mg)	Zinc (mg)	VT-A (RAE)	Thia (mg)	VT-E (mg α)	Ribo (mg)	Niac (mg)	V-B₆ (mg)	Fola (µg)	VT-C (mg)
40	200	3.6		1410	950									42
5	200	4.5		1400	510									72
21	84	5.07		1731	56									51
35	80	1.8		470	870									2
35	300	1.08		590	200									0
74	123	0.59		180	66					0.15				0
0	0	0		30	5	0								4
0	0	0		56	0	0								2
30	20	1.08			480									30
35	150	3.6		500	580			0.3		0.26	2			6
37	48	2.24			459									55
5	100	0		268	55									0
5	100	0		240	50									0
5	100			254	90			0.03		0.15				0
5	100	0		168	60			0.03		0.15				0
0				196	74					0.98				
0	0	0			160	0								5
0	40	3.6			280									9
0	20	1.8			170	0								1
0	40	1.8			250									15
0	60	5.4			220									2
0	40	5.4			230									9
0	40	5.4			240									9
8	10	0.56			335									0
5	10	0.56			377									0
35	150	1.44	42	450	590	1.5		0.06		0.26	1.2	0.2	48	5
35	40	0.72	30	300	590	1.1		0.15		0.17	5			0
29	244	1.41	44	596	576	2.9		0.15		0.25	3	0.32		6
18	180	0.65		423	567	0		0.12		0.25	1.2			0
5	94	2.37		539	465									6
10	100	2.7	39	165	630	2.2		0.45		0.51	0.05	0.07		1

Table H–1

Food Composition   (Computer code number is for West Diet Analysis program)   (For purposes of calculations, use "0" for t, <1, <.1, <.01, etc.)

Computer Code Number	Food Description	Measure	Wt (g)	H₂O (%)	Ener (kcal)	Prot (g)	Carb (g)	Dietary Fiber (g)	Fat (g)	Fat Breakdown (g)		
										Sat	Mono	Poly
	CONVENIENCE FOODS & MEALS—Continued											
	TASTE ADVENTURE SOUPS											
50325	Black bean	1 c	242		807	51	148	36	4	0.9	0.3	1.5
50324	Curry lentil	1 c	241		795	66	135	71	3	0.4	0.5	1.2
50326	Lentil chili	1 c	242		411	24	75	15	2			
50323	Split pea	1 c	244		807	58	143	60	3	0.4	0.6	1.2
	Source: Taste Adventure Soups											
	BABY FOODS											
60465	Apple juice	½ c	125	88	59	0	15	<1	<1	t	t	t
60475	Applesauce, strained	1 tbsp	16	87	8	<1	2	<1	<1	t	t	t
60502	Carrots, strained	1 tbsp	14	92	4	<1	1	<1	<1	t	t	t
60563	Cereal, mixed, milk added	1 tbsp	15	75	17	1	2	<1	1	0.3		
60622	Cereal, rice, milk added	1 tbsp	15	75	17	1	3	<1	1	0.3		
60515	Chicken and noodles, strained	1 tbsp	16	86	11	<1	1	<1	<1	t	0.1	t
60603	Peas, strained	1 tbsp	15	88	6	1	1	<1	<1	t	t	t
60634	Teething biscuits	1 ea	11	6	43	1	8	<1	<1	0.2	0.2	t

Chol (mg)	Calc (mg)	Iron (mg)	Magn (mg)	Pota (mg)	Sodi (mg)	Zinc (mg)	VT-A (RAE)	Thia (mg)	VT-E (mg α)	Ribo (mg)	Niac (mg)	V-B$_6$ (mg)	Fola (µg)	VT-C (mg)
0	296	12.18	405	3521	1978	8.67	39	2.12	0.12	0.48	4.8	0.84	1043	1
0	140	22.04	256	2170	2182	8.51	48	1.12	0.72	0.6	6.33	1.25	1005	16
			1476	1016										
0	140	10.65	272	2324	1729	7.14		1.71	2.75	0.51	6.84	0.42	646	5

Chol (mg)	Calc (mg)	Iron (mg)	Magn (mg)	Pota (mg)	Sodi (mg)	Zinc (mg)	VT-A (RAE)	Thia (mg)	VT-E (mg α)	Ribo (mg)	Niac (mg)	V-B$_6$ (mg)	Fola (µg)	VT-C (mg)
0	5	0.71	4	114	4	0.04	1	<.01	0.75	0.02	0.1	0.04	0	72
0	1	0.04	0	11	<1	<.01	<1	<.01	0.03	<.01	<.01	<.01	<1	6
0	3	0.05	1	27	5	0.02	80	<.01	0.15	<.01	0.06	0.01	2	1
2	33	1.56	4	30	7	0.11	4	0.06		0.09	0.87	<.01	2	0
2	36	1.83	7	28	7	0.1	3	0.07		0.08	0.78	0.02	1	0
3	4	0.1	2	22	4	0.09	18	<.01	0.03	<.01	0.12	0.01	2	0
0	3	0.14	2	17	1	0.05	3	0.01	0.01	<.01	0.15	0.01	4	1
0	29	0.39	4	36	40	0.1	3	0.03	0.03	0.06	0.48	0.01	5	1

WHO: Nutrition Recommendations Canada: Choice System and Guidelines

This appendix first presents nutrition recommendations from the World Health Organization (WHO) and then provides details for Canadians on Canada's *Food Guide to Healthy Eating* and on the exchange system (called the choice system).

Nutrition Recommendations from WHO

The World Health Organization (WHO) has assessed the relationships between diet and the development of chronic diseases. Its recommendations include:

- Total energy: sufficient to support normal growth, physical activity, and healthy body weight (body mass index = 20 to 22).
- Total fat: 15 to 30 percent of total energy.
- Saturated fat: less than 10 percent of total energy.
- Total carbohydrate: 55 to 75 percent of total energy.
- Added sugars: less than 10 percent of total energy.
- Protein: 10 to 15 percent of total energy.
- Salt: less than 5 grams/day, preferably iodized.
- Fruit and vegetables: at least 400 grams (almost 1 pound) daily.
- Physical activity: one hour per day of moderate intensity on most days of the week.

Canada's *Food Guide to Healthy Eating*

Figure I-1 presents the 1992 Canada's *Food Guide to Healthy Eating*, which interprets Canada's *Guidelines for Healthy Eating* (see Table 2-1 on p. 42) for consumers and recommends a range of servings to consume daily from each of the four food groups. The following publications, which are available from Health Canada, through its website, explain how to use the *Guide: Using the Food Guide; Food Guide Facts: Background for Educators and Communicators; Canada's Food Guide to Healthy Eating—Focus on Preschoolers: Background for Educators and Communicators;* and *Canada's Food Guide to Healthy Eating—Focus on Children Six to Twelve Years: Background for Educators and Communicators.* Figure I-2 presents Canada's Physical Activity Guide.

Canada's *Guidelines for Healthy Eating* and Canada's *Food Guide to Healthy Eating* are being reviewed for consistency with the new Dietary Reference Intakes. Check the website for the Health Canada Office of Nutrition Policy and Promotion, **www.hc-sc.gc.ca/hpfb-dgpsa/onpp-bppn/**, for the status of the review.

I Appendix

FIGURE I-1 Canada's *Food Guide to Healthy Eating*

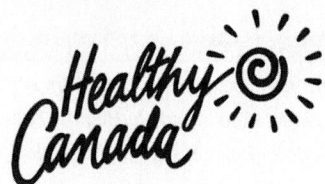

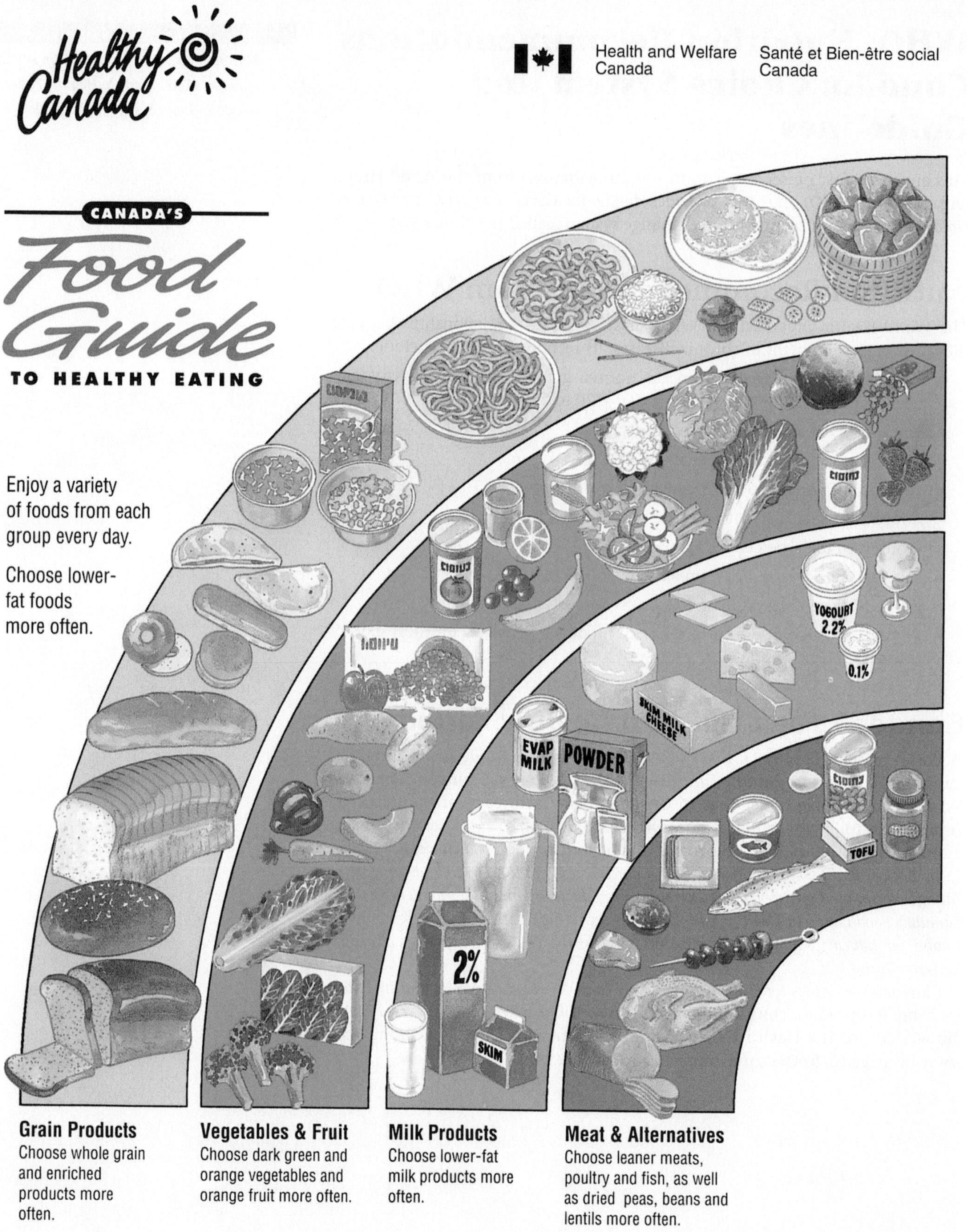

Healthy Canada
Health and Welfare Canada
Santé et Bien-être social Canada
CANADA'S
Food Guide
TO HEALTHY EATING
Enjoy a variety of foods from each group every day.
Choose lower-fat foods more often.
YOGOURT 2.2%
0.1%
SKIM MILK CHEESE
EVAP MILK
POWDER
2%
SKIM
TOFU
Grain Products
Choose whole grain and enriched products more often.
Vegetables & Fruit
Choose dark green and orange vegetables and orange fruit more often.
Milk Products
Choose lower-fat milk products more often.
Meat & Alternatives
Choose leaner meats, poultry and fish, as well as dried peas, beans and lentils more often.

FIGURE I-1 Canada's *Food Guide to Healthy Eating*—continued

CANADA'S

Food Guide

TO HEALTHY EATING

FOR PEOPLE FOUR YEARS AND OVER

Different People Need Different Amounts of Food

The amount of food you need every day from the 4 food groups and other foods depends on your age, body size, activity level, whether you are male or female and if you are pregnant or breast-feeding. That's why the Food Guide gives a lower and higher number of servings for each food group. For example, young children can choose the lower number of servings, while male teenagers can go to the higher number. Most other people can choose servings somewhere in between.

Grain Products
5-12 SERVINGS PER DAY

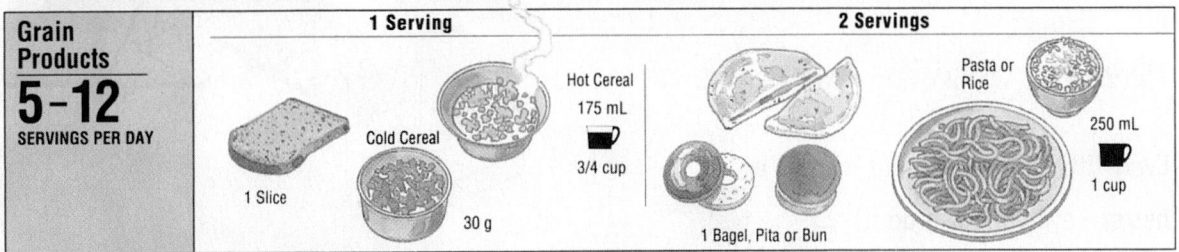

Vegetables & Fruit
5-10 SERVINGS PER DAY

Milk Products
SERVINGS PER DAY
Children 4–9 years: 2–3
Youth 10–16 years: 3–4
Adults: 2–4
Pregnant & Breast-feeding Women: 3–4

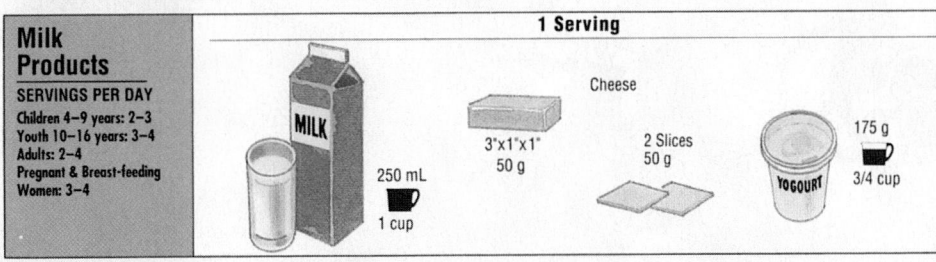

Other Foods

Taste and enjoyment can also come from other foods and beverages that are not part of the 4 food groups. Some of these foods are higher in fat or Calories, so use these foods in moderation.

Meat & Alternatives
2-3 SERVINGS PER DAY

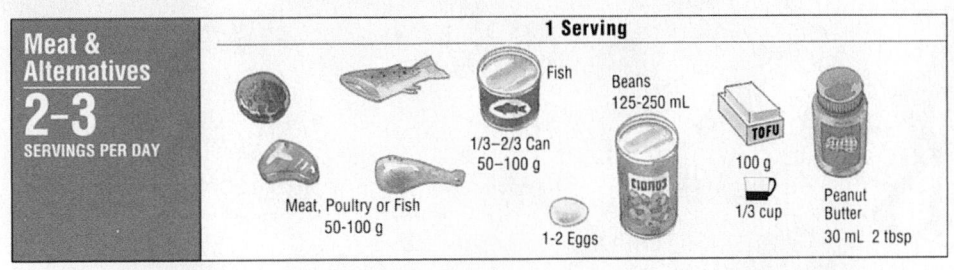

Enjoy eating well, being active and feeling good about yourself. That's VITALITÉ

© Minister of Supply and Services Canada 1992   Cat. No. H39-252/1992E   No changes permitted. Reprint permission not required.
ISBN 0-662-19648-1

FIGURE I-2 | Canada's Physical Activity Guide

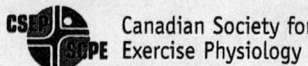

CANADA'S
Physical Activity Guide
to Healthy Active Living

Physical activity improves health.

Every little bit counts, but more is even better – everyone can do it!

Get active your way – build physical activity into your daily life...
- at home
- at school
- at work
- at play
- on the way
...that's active living!

TAKE THE STAIRS

Increase
Endurance
Activities

Increase
Flexibility
Activities

Increase
Strength
Activities

Reduce
Sitting for
long periods

Health Canada   Santé Canada

CSEP SCPE   Canadian Society for Exercise Physiology

FIGURE I-2   Canada's Physical Activity Guide—continued

Choose a variety of activities from these three groups:

Endurance

4-7 days a week
Continuous activities for your heart, lungs and circulatory system.

Flexibility

4-7 days a week
Gentle reaching, bending and stretching activities to keep your muscles relaxed and joints mobile.

Strength

2-4 days a week
Activities against resistance to strengthen muscles and bones and improve posture.

Starting slowly is very safe for most people. Not sure? Consult your health professional.

For a copy of the *Guide Handbook* and more information:
1-888-334-9769, or
www.paguide.com

Eating well is also important. Follow *Canada's Food Guide to Healthy Eating* to make wise food choices.

Get Active Your Way, Every Day–For Life!

Scientists say accumulate 60 minutes of physical activity every day to stay healthy or improve your health. As you progress to moderate activities you can cut down to 30 minutes, 4 days a week. Add-up your activities in periods of at least 10 minutes each. Start slowly... and build up.

Time needed depends on effort

Very Light Effort	Light Effort *60 minutes*	Moderate Effort *30-60 minutes*	Vigorous Effort *20-30 minutes*	Maximum Effort
• Strolling • Dusting	• Light walking • Volleyball • Easy gardening • Stretching	• Brisk walking • Biking • Raking leaves • Swimming • Dancing • Water aerobics	• Aerobics • Jogging • Hockey • Basketball • Fast swimming • Fast dancing	• Sprinting • Racing

Range needed to stay healthy

You Can Do It – Getting started is easier than you think

Physical activity doesn't have to be very hard. Build physical activities into your daily routine.

- Walk whenever you can – get off the bus early, use the stairs instead of the elevator.
- Reduce inactivity for long periods, like watching TV.
- Get up from the couch and stretch and bend for a few minutes every hour.
- Play actively with your kids.
- Choose to walk, wheel or cycle for short trips.

- Start with a 10 minute walk – gradually increase the time.
- Find out about walking and cycling paths nearby and use them.
- Observe a physical activity class to see if you want to try it.
- Try one class to start, you don't have to make a long-term commitment.
- Do the activities you are doing now, more often.

Benefits of regular activity:	Health risks of inactivity:
• better health • improved fitness • better posture and balance • better self-esteem • weight control • stronger muscles and bones • feeling more energetic • relaxation and reduced stress • continued independent living in later life	• premature death • heart disease • obesity • high blood pressure • adult-onset diabetes • osteoporosis • stroke • depression • colon cancer

 ACTIVE LIVING

No changes permitted. Permission to photocopy this document in its entirety not required.
Cat. No. H39-429/1998-1E ISBN 0-662-86627-7

 CANADA'S *Physical Activity Guide* to Healthy Active Living

Canada's Choice System for Meal Planning

The *Good Health Eating Guide* is the Canadian choice system of meal planning.[1] It contains several features similar to those of the U.S. exchange system including the following:

- Foods are divided into lists according to carbohydrate, protein, and fat content.
- Foods are interchangeable within a group.
- Most foods are eaten in measured amounts.
- An energy value is given for each food group.

Tables I-1 through I-8 present the Canadian choice system.

_____

[1]The tables for the Canadian choice system are adapted from the *Good Health Eating Guide Resource*, copyright 1994, with permission of the Canadian Diabetes Association.

TABLE I-1 Canadian Choice System: Starch Foods

1 starch choice = 15 g carbohydrate (starch), 2 g protein, 290 kJ (68 kcal)

Food	Measure	Mass (Weight)
Breads		
Bagels	½	30 g
Bread crumbs	50 mL (¼ c)	30 g
Bread cubes	250 mL (1 c)	30 g
Bread sticks	2	20 g
Brewis, cooked	50 mL (¼ c)	45 g
Chapati	1	20 g
Cookies, plain	2	20 g
English muffins, crumpets	½	30 g
Flour	40 mL (2½ tbs)	20 g
Hamburger buns	½	30 g
Hot dog buns	½	30 g
Kaiser rolls	½	30 g
Matzo, 15 cm	1	20 g
Melba toast, rectangular	4	15 g
Melba toast, rounds	7	15 g
Pita, 20 cm (8″) diameter	¼	30 g
Pita, 15 cm (6″) diameter	½	30 g
Plain rolls	1 small	30 g
Pretzels	7	20 g
Raisin bread	1 slice	30 g
Rice cakes	2	30 g
Roti	1	20 g
Rusks	2	20 g
Rye, coarse or pumpernickel	½ slice	30 g
Soda crackers	6	20 g
Tortillas, corn (taco shell)	1	30 g
Tortilla, flour	1	30 g
White (French and Italian)	1 slice	25 g
Whole-wheat, cracked-wheat, rye, white enriched	1 slice	30 g
Cereals		
Bran flakes, 100% bran	125 mL (½ c)	30 g
Cooked cereals, cooked	125 mL (½ c)	125 g
Dry	30 mL (2 tbs)	20 g

(continued on the next page)

TABLE I-1 — Canadian Choice System: Starch Foods—continued

1 starch choice = 15 g carbohydrate (starch), 2 g protein, 290 kJ (68 kcal)

Food	Measure	Mass (Weight)
Cornmeal, cooked	125 mL (½ c)	125 g
Dry	30 mL (2 tbs)	20 g
Ready-to-eat unsweetened cereals	125 mL (½ c)	20 g
Shredded wheat biscuits, rectangular or round	1	20 g
Shredded wheat, bite size	125 mL (½ c)	20 g
Wheat germ	75 mL (⅓ c)	30 g
Cornflakes	175 mL (⅔ c)	20 g
Rice Krispies	175 mL (⅔ c)	20 g
Cheerios	200 mL (¾ c)	20 g
Muffets	1	20 g
Puffed rice	300 mL (1¼ c)	15 g
Puffed wheat	425 mL (1⅔ c)	20 g
Grains		
Barley, cooked	125 mL (½ c)	120 g
Dry	30 mL (2 tbs)	20 g
Bulgur, kasha, cooked, moist	125 mL (½ c)	70 g
Cooked, crumbly	75 mL (⅓ c)	40 g
Dry	30 mL (2 tbs)	20 g
Rice, cooked, brown & white (short & long grain)	125 mL (½ c)	70 g
Rice, cooked, wild	75 mL (⅓ c)	70 g
Tapioca, pearl and granulated, quick cooking, dry	30 mL (2 tbs)	15 g
Couscous, cooked moist	125 mL (½ c)	70 g
Dry	30 mL (tbs)	20 g
Quinoa, cooked moist	125 mL (½ c)	70 g
Dry	30 mL (2 tbs)	20 g
Pastas		
Macaroni, cooked	125 mL (½ c)	70 g
Noodles, cooked	125 mL (½ c)	80 g
Spaghetti, cooked	125 mL (½ c)	70 g
Starchy Vegetables		
Beans and peas, dried, cooked	125 mL (½ c)	80 g
Breadfruit	1 slice	75 g
Corn, canned, whole kernel	125 mL (½ c)	85 g
Corn on the cob	½ medium cob	140 g
Cornstarch	30 mL (2 tbs)	15 g
Plantains	⅓ small	50 g
Popcorn, air-popped, unbuttered	750 mL (3 c)	20 g
Potatoes, whole (with or without skin)	½ medium	95 g
Yams, sweet potatoes (with or without skin)	½	75 g

Food	Choices per Serving	Measure	Mass (Weight)
Note: Food items found in this category provide more than 1 starch choice:			
Bran flakes	1 starch + ½ sugar	150 mL (⅔ c)	24 g
Croissant, small	1 starch + 1½ fats	1 small	35 g
Large	1 starch + 1½ fats	½ large	30 g
Corn, canned creamed	1 starch + ½ fruits and vegetables	12 mL (½ c)	113 g
Potato chips	1 starch + 2 fats	15 chips	30 g
Tortilla chips (nachos)	1 starch + 1½ fats	13 chips	20 g
Corn chips	1 starch + 2 fats	30 chips	30 g

(continued on the next page)

TABLE I-1   Canadian Choice System: Starch Foods—continued

1 starch choice = 15 g carbohydrate (starch), 2 g protein, 290 kJ (68 kcal)

Food	Choices per Serving	Measure	Mass (Weight)
Cheese twists	1 starch + 1½ fats	30 chips	30 g
Cheese puffs	1 starch + 2 fats	27 chips	30 g
Tea biscuit	1 starch + 2 fats	1	30 g
Pancakes, homemade using 50 mL (¼ c) batter (6″ diameter)	1½ starches + 1 fat	1 medium	50 g
Potatoes, french fried (homemade or frozen)	1 starch + 1 fat	10 regular size	35 g
Soup, canned* (prepared with equal volume of water)	1 starch	250 mL (1 c)	260 g
Waffles, packaged	1 starch + 1 fat	1	35 g

*Soup can vary according to brand and type. Check the label for Food Choice Values and Symbols or the core nutrient listing.

TABLE I-2   Canadian Choice System: Fruits and Vegetables

1 fruits and vegetables choice = 10 g carbohydrate, 1 g protein, 190 kJ (44 kcal)

Food	Measure	Mass (Weight)
Fruits (fresh, frozen, without sugar, canned in water)		
Apples, raw (with or without skin)	½ medium	75 g
Sauce unsweetened	125 mL (½ c)	120 g
Sweetened	see *Combined Food Choices (Table I-8)*	
Apple butter	20 mL (4 tsp)	20 g
Apricots, raw	2 medium	115 g
Canned, in water	4 halves, plus 30 mL (2 tbs) liquid	110 g
Bake-apples (cloudberries), raw	125 mL (½ c)	120 g
Bananas, with peel	½ small	75 g
Peeled	½ small	50 g
Berries (blackberries, blueberries, boysenberries, huckleberries, loganberries, raspberries)		
Raw	125 mL (½ c)	70 g
Canned, in water	125 mL (½ c), plus 30 mL (2 tbs) liquid	100 g
Cantaloupe, raw, with rind	¼ wedge	240 g
Cubed or diced	250 mL (1 c)	160 g
Cherries, raw, with pits	10	75 g
Raw, without pits	10	70 g
Canned, in water, with pits	75 mL (⅓ c), plus 30 mL (2 tbs) liquid	90 g
Canned, in water, without pits	75 mL (⅓ c), plus 30 mL (2 tbs) liquid	85 g
Crabapples, raw	1 small	55 g
Cranberries, raw	250 mL (1 c)	100 g
Figs, raw	1 medium	50 g
Canned, in water	3 medium, plus 30 mL (2 tbs) liquid	100 g
Foxberries, raw	250 mL (1 c)	100 g
Fruit cocktail, canned, in water	125 mL (½ c), plus 30 mL (2 tbs) liquid	120 g
Fruit, mixed, cut-up	125 mL (½ c)	120 g
Gooseberries, raw	250 mL (1 c)	150 g
Canned, in water	250 mL (1 c), plus 30 mL (2 tbs) liquid	230 g
Grapefruit, raw, with rind	½ small	185 g
Raw, sectioned	125 mL (½ c)	100 g
Canned, in water	125 mL (½ c), plus 30 mL (2 tbs) liquid	120 g

(continued on the next page)

TABLE I-2   Canadian Choice System: Fruits and Vegetables—continued

1 fruits and vegetables choice = 10 g carbohydrate, 1 g protein, 190 kJ (44 kcal)

Food	Measure	Mass (Weight)
Grapes, raw, slip skin	125 mL (½ c)	75 g
Raw, seedless	125 mL (½ c)	75 g
Canned, in water	75 mL (⅓ c), plus 30 mL (2 tbs) liquid	115 g
Guavas, raw	½	50 g
Honeydew melon, raw, with rind	½	225 g
Cubed or diced	250 mL (1 c)	170 g
Kiwis, raw, with skin	2	155 g
Kumquats, raw	3	60 g
Loquats, raw	8	130 g
Lychee fruit, raw	8	120 g
Mandarin oranges, raw, with rind	1	135 g
Raw, sectioned	125 mL (½ c)	100 g
Canned, in water	125 mL (½ c), plus 30 mL (2 tbs) liquid	100 g
Mangoes, raw, without skin and seed	⅓	65 g
Diced	75 mL (⅓ c)	65 g
Nectarines	½ medium	75 g
Oranges, raw, with rind	1 small	130 g
Raw, sectioned	125 mL (½ c)	95 g
Papayas, raw, with skin and seeds	¼ medium	150 g
Raw, without skin and seeds	¼ medium	100 g
Cubed or diced	125 mL (½ c)	100 g
Peaches, raw, with seed and skin	1 large	100 g
Raw, sliced or diced	125 mL (½ c)	100 g
Canned in water, halves or slices	125 mL (½ c), plus 30 mL (2 tbs) liquid	120 g
Pears, raw, with skin and core	½	90 g
Raw, without skin and core	½	85 g
Canned, in water, halves	1 half plus 30 mL (2 tbs) liquid	90 g
Persimmons, raw, native	1	30 g
Raw, Japanese	¼	50 g
Pineapple, raw	1 slice	75 g
Raw, diced	125 mL (½ c)	75 g
Canned, in juice, diced	75 mL (⅓ c), plus 15 mL (1 tbs) liquid	55 g
Canned, in juice, sliced	1 slice, plus 15 mL (1 tbs) liquid	55 g
Canned, in water, diced	125 mL (½ c), plus 30 mL (2 tbs) liquid	100 g
Canned, in water, sliced	2 slices, plus 15 mL (1 tbs) liquid	100 g
Plums, raw	2 small	60 g
Damson	6	65 g
Japanese	1	70 g
Canned, in apple juice	2, plus 30 mL (2 tbs) liquid	70 g
Canned, in water	3, plus 30 mL (2 tbs) liquid	100 g
Pomegranates, raw	½	140 g
Strawberries, raw	250 mL (1 c)	150 g
Frozen/canned, in water	250 mL (1 c), plus 30 mL (2 tbs) liquid	240 g
Rhubarb	250 mL (1 c)	150 g
Tangelos, raw	1	205 g
Tangerines, raw	1 medium	115 g
Raw, sectioned	125 mL (½ c)	100 g

(continued on the next page)

Appendix I

TABLE I-2 Canadian Choice System: Fruits and Vegetables—continued

1 fruits and vegetables choice = 10 g carbohydrate, 1 g protein, 190 kJ (44 kcal)

Food	Measure	Mass (Weight)
Watermelon, raw, with rind	1 wedge	310 g
Cubed or diced	250 mL (1 c)	160 g
Dried Fruit		
Apples	5 pieces	15 g
Apricots	4 halves	15 g
Banana flakes	30 mL (2 tbs)	15 g
Currants	30 mL (2 tbs)	15 g
Dates, without pits	2	15 g
Peaches	½	15 g
Pears	½	15 g
Prunes, raw, with pits	2	15 g
Raw, without pits	2	10 g
Stewed, no liquid	2	20 g
Stewed, with liquid	2, plus 15 mL (1 tbs) liquid	35 g
Raisins	30 mL (2 tbs)	15 g
Juices (no sugar added or unsweetened)		
Apricot, grape, guava, mango, prune	50 mL (¼ c)	55 g
Apple, carrot, papaya, pear, pineapple, pomegranate	75 mL (⅓ c)	80 g
Cranberry	see *Sugars (Table I-4)*	
Clamato	see *Sugars (Table I-4)*	
Grapefruit, loganberry, orange, raspberry, tangelo, tangerine	125 mL (½ c)	130 g
Tomato, tomato-based mixed vegetables	250 mL (1 c)	255 g
Vegetables (fresh, frozen, or canned)		
Artichokes, French, globe	2 small	50 g
Beets, diced or sliced	125 mL (½ c)	85 g
Carrots, diced, cooked or uncooked	125 mL (½ c)	75 g
Chestnuts, fresh	5	20 g
Parsnips, mashed	125 mL (½ c)	80 g
Peas, fresh or frozen	125 mL (½ c)	80 g
Canned	75 mL (⅓ c)	55 g
Pumpkin, mashed	125 mL (½ c)	45 g
Rutabagas, mashed	125 mL (½ c)	85 g
Sauerkraut	250 mL (1 c)	235 g
Snow peas	250 mL (1 c)	135 g
Squash, yellow or winter, mashed	125 mL (½ c)	115 g
Succotash	75 mL (⅓ c)	55 g
Tomatoes, canned	250 mL (1 c)	240 g
Tomato paste	50 mL (¼ c)	55 g
Tomato sauce*	75 mL (⅓ c)	100 g
Turnips, mashed	125 mL (½ c)	115 g
Vegetables, mixed	125 mL (½ c)	90 g
Water chestnuts	8 medium	50 g

*Tomato sauce varies according to brand name. Check the label or discuss with your dietitian.

TABLE I-3 — Canadian Choice System: Milk

Type of Milk	Carbohydrate (g)	Protein (g)	Fat (g)	Energy
Nonfat (0%)	6	4	0	170 kJ (40 kcal)
1%	6	4	1	206 kJ (49 kcal)
2%	6	4	2	244 kJ (58 kcal)
Whole (4%)	6	4	4	319 kJ (76 kcal)

Food	Measure	Mass (Weight)
Buttermilk (higher in salt)	125 mL (½ c)	125 g
Evaporated milk	50 mL (¼ c)	50 g
Milk	125 mL (½ c)	125 g
Powdered milk, regular	30 mL (2 tbs)	15 g
Instant	50 mL (¼ c)	15 g
Plain yogurt	125 mL (½ c)	125 g

Food	Choices per Serving	Measure	Mass (Weight)
Note: Food items found in this category provide more than 1 milk choice:			
Milk shake	1 milk + 3 sugars + ½ protein	250 mL (1 c)	300 g
Chocolate milk, 2%	2 milks 2% + 1 sugar	250 mL (1 c)	300 g
Frozen yogurt	1 milk + 1 sugar	125 mL (½ c)	125 g

TABLE I-4 — Canadian Choice System: Sugars

1 sugar choice = 10 g carbohydrate (sugar), 167 kJ (40 kcal)

Food	Measure	Mass (Weight)
Beverages		
Condensed milk	15 mL (1 tbs)	
Flavoured fruit crystals*	75 mL (⅓ c)	
Iced tea mixes*	75 mL (⅓ c)	
Regular soft drinks	125 mL (½ c)	
Sweet drink mixes*	75 mL (⅓ c)	
Tonic water	125 mL (½ c)	
*These beverages have been made with water.		
Miscellaneous		
Bubble gum (large square)	1 piece	5 g
Cranberry cocktail	75 mL (⅓ c)	80 g
Cranberry cocktail, light	350 mL (1⅓ c)	260 g
Cranberry sauce	30 mL (2 tbs)	
Hard candy mints	2	5 g
Honey, molasses, corn & cane syrup	10 mL (2 tsp)	15 g
Jelly bean	4	10 g
Licorice	1 short stick	10 g
Marshmallows	2 large	15 g
Popsicle	1 stick (½ popsicle)	
Powdered gelatin mix		
(Jello®) (reconstituted)	50 mL (¼ c)	
Regular jam, jelly, marmalade	15 mL (1 tbs)	
Sugar, white, brown, icing, maple	10 mL (2 tsp)	10 g
Sweet pickles	2 small	100 g
Sweet relish	30 mL (2 tbs)	

Food	Choices per Serving	Measures	Mass (Weight)
The following food items provide more than 1 sugar choice:			
Brownie	1 sugar + 1 fat	1	20 g
Clamato juice	1½ sugars	175 mL (⅔ c)	
Fruit salad, light syrup	1 sugar + 1 fruits & vegetables	125 mL (½ c)	130 g
Aero® bar	2½ sugars + 2½ fats	1 bar	43 g
Smarties®	4½ sugars + 2 fats	1 box	60 g
Sherbet	3 sugars + ½ fat	125 mL (½ c)	95 g

TABLE I-5   Canadian Choice System: Protein Foods

1 protein choice = 7 g protein, 3 g fat, 230 kJ (55 kcal)

Food	Measure	Mass (Weight)
Cheese		
Low-fat cheese, about 7% milk fat	1 slice	30 g
Cottage cheese, 2% milkfat or less	50 mL (¼ c)	55 g
Ricotta, about 7% milkfat	50 mL (¼ c)	60 g
Fish		
Anchovies	see *Extras* (Table I-7)	
Canned, drained (e.g., mackerel, salmon, tuna packed in water)	(⅓ of 6.5 oz can)	30 g
Cod tongues, cheeks	75 mL (⅓ c)	50 g
Fillet or steak (e.g., Boston blue, cod, flounder, haddock, halibut, mackerel, orange roughy, perch, pickerel, pike, salmon, shad, snapper, sole, swordfish, trout, tuna, whitefish)	1 piece	30 g
Herring	⅓ fish	30 g
Sardines, smelts	2 medium or 3 small	30 g
Squid, octopus	50 mL (¼ c)	40 g
Shellfish		
Clams, mussels, oysters, scallops, snails	3 medium	30 g
Crab, lobster, flaked	50 mL (¼ c)	30 g
Shrimp, fresh	5 large	30 g
Frozen	10 medium	30 g
Canned	18 small	30 g
Dry pack	50 mL (¼ c)	30 g
Meat and Poultry (e.g., beef, chicken, goat, ham, lamb, pork, turkey, veal, wild game)		
Back, peameal bacon	3 slices, thin	30 g
Chop	½ chop, with bone	40 g
Minced or ground, lean or extra-lean	30 mL (2 tbs)	30 g
Sliced, lean	1 slice	30 g
Steak, lean	1 piece	30 g
Organ Meats		
Hearts, liver	1 slice	30 g
Kidneys, sweetbreads, chopped	50 mL (¼ c)	30 g
Tongue	1 slice	30 g
Tripe	5 pieces	60 g
Soyabean		
Bean curd or tofu	½ block	70 g
Eggs		
In shell, raw or cooked	1 medium	50 g
Without shell, cooked or poached in water	1 medium	45 g
Scrambled	50 mL (¼ c)	55 g

Note: The following choices provide more than 1 protein exchange:

Food	Choices per Serving	Measures	Mass (Weight)
Cheese			
Cheeses	1 protein + 1 fat	1 piece	25 g
Cheese, coarsely grated (e.g., cheddar)	1 protein + 1 fat	50 mL (¼ c)	25 g

(continued on the next page)

TABLE I-5   Canadian Choice System: Protein Foods—continued

1 protein choice = 7 g protein, 3 g fat, 230 kJ (55 kcal)
Note: The following choices provide more than 1 protein exchange:

Food	Choices per Serving	Measures	Mass (Weight)
Cheese, dry, finely grated (e.g., parmesan)	1 protein + 1 fat	45 mL	15 g
Cheese, ricotta, high fat	1 protein + 1 fat	50 mL (¼ c)	55 g
Fish			
Eel	1 protein + 1 fat	1 slice	50 g
Meat			
Bologna	1 protein + 1 fat	1 slice	20 g
Canned lunch meats	1 protein + 1 fat	1 slice	20 g
Corned beef, canned	1 protein + 1 fat	1 slice	25 g
Corned beef, fresh	1 protein + 1 fat	1 slice	25 g
Ground beef, medium-fat	1 protein + 1 fat	30 mL (2 tbs)	25 g
Meat spreads, canned	1 protein + 1 fat	45 mL	35 g
Mutton chop	1 protein + 1 fat	½ chop, with bone	35 g
Paté	see *Fats and Oils (Table I-6)*		
Sausages, garlic, Polish or knockwurst	1 protein + 1 fat	1 slice	50 g
Sausages, pork, links	1 protein + 1 fat	1 link	25 g
Spareribs or shortribs, with bone	1 protein + 1 fat	1 large	65 g
Stewing beef	1 protein + 1 fat	1 cube	25 g
Summer sausage or salami	1 protein + 1 fat	1 slice	40 g
Wieners, hot dog	1 protein + 1 fat	½ medium	25 g
Miscellaneous			
Blood pudding	1 protein + 1 fat	1 slice	25 g
Peanut butter	1 protein + 1 fat	15 mL (1 tbs)	15 g

TABLE I-6 Canadian Choice System: Fats and Oils

1 fat choice = 5 g fat, 190 kJ (45 kcal)

Food	Measure	Mass (Weight)	Food	Measure	Mass (Weight)
Avocado*	⅛	30 g	Nuts (continued):		
Bacon, side, crisp*	1 slice	5 g	Sesame seeds	15 mL (1 tbs)	10 g
Butter*	5 mL (1 tsp)	5 g	Sunflower seeds		
Cheese spread	15 mL (1 tbs)	15 g	Shelled	15 mL (1 tbs)	10 g
Coconut, fresh*	45 mL (3 tbs)	15 g	In shell	45 mL (3 tbs)	15 g
Coconut, dried*	15 mL (1 tbs)	10 g	Walnuts	4 halves	10 g
Cream, Half and half			Oil, cooking and salad	5 mL (1 tsp)	5 g
(cereal), 10%*	30 mL (2 tbs)	30 g	Olives, green	10	45 g
Light (coffee), 20%*	15 mL (1 tbs)	15 g	Ripe black	7	57 g
Whipping, 32 to 37%*	15 mL (1 tbs)	15 g	Pâté, liverwurst,	15 mL (1 tbs)	15 g
Cream cheese*	15 mL (1 tbs)	15 g	meat spreads		
Gravy*	30 mL (2 tbs)	30 g	Salad dressing: blue,	10 mL (2 tsp)	10 g
Lard*	5 mL (1 tsp)	5 g	French, Italian,		
Margarine	5 mL (1 tsp)	5 g	mayonnaise,		
Nuts, shelled:			Thousand Island	5 mL (1 tsp)	5 g
Almonds	8	5 g	Salad dressing,	30 mL (2 tbs)	30 g
Brazil nuts	2	10 g	low-kcalorie		
Cashews	5	10 g	Salt pork, raw	5 mL (1 tsp)	5 g
Filberts, hazelnuts	5	10 g	or cooked*		
Macadamia	3	5 g	Sesame oil	5 mL (1 tsp)	5 g
Peanuts	10	10g	Sour cream		
Pecans	5 halves	5 g	12% milkfat	30 mL (2 tbs)	30 g
Pignolias, pine nuts	25 mL (5 tsp)	10 g	7% milkfat	60 mL (4 tbs)	60 g
Pistachios, shelled	20	10 g	Shortening*	5 mL (1 tsp)	
Pistachios, in shell	20	20 g			
Pumpkin and squash seeds	20 mL (4 tsp)	10 g			

*These items contain higher amounts of saturated fat.

TABLE I-7 Canadian Choice System: Extras

Extras have no more than 2.5 g carbohydrate, 60 kJ (14 kcal).

Vegetables 125 mL (½ c)

Artichokes

Asparagus

Bamboo shoots

Bean sprouts, mung or soya

Beans, string, green, or yellow

Bitter melon (balsam pear)

Bok choy

Broccoli

Brussels sprouts

Cabbage

Cauliflower

Celery

Chard

Cucumbers

Eggplant

Endive

Fiddleheads

Greens: beet, collard, dandelion, mustard, turnip, etc.

Kale

Kohlrabi

Leeks

Lettuce

Mushrooms

Okra

Onions, green or mature

Parsley

Peppers, green, yellow or red

Radishes

Rapini

Rhubarb

Sauerkraut

Shallots

Spinach

Sprouts: alfalfa, radish, etc.

Tomato wedges

Watercress

Zucchini

Free Foods (may be used without measuring)

Artificial sweetener, such as cyclamate or aspartame	Lime juice or lime wedges
	Marjoram, cinnamon, etc.
Baking powder, baking soda	Mineral water
Bouillon from cube, powder, or liquid	Mustard
	Parsley
Bouillon or clear broth	Pimentos
Chowchow, unsweetened	Salt, pepper, thyme
Coffee, clear	Soda water, club soda
Consommé	Soya sauce
Dulse	Sugar-free Crystal Drink
Flavorings and extracts	Sugar-free Jelly Powder
Garlic	Sugar-free soft drinks
Gelatin, unsweetened	Tea, clear
Ginger root	Vinegar
Herbal teas, unsweetened	Water
Horseradish, uncreamed	Worcestershire sauce
Lemon juice or lemon wedges	

Condiments

Food	Measure
Anchovies	2 fillets
Barbecue sauce	15 mL (1 tbs)
Bran, natural	30 mL (2 tbs)
Brewer's yeast	5 mL (1 tsp)
Carob powder	5 mL (1 tsp)
Catsup	5 mL (1 tsp)
Chili sauce	5 mL (1 tsp)
Cocoa powder	5 mL (1 tsp)
Cranberry sauce, unsweetened	15 mL (1 tbs)
Dietetic fruit spreads	5 mL (1 tsp)
Maraschino cherries	1
Nondairy coffee whitener	5 mL (1 tsp)
Nuts, chopped pieces	5 mL (1 tsp)
Pickles	
unsweetened dill	2
sour mixed	11
Sugar substitutes, granular	5 mL (1 tsp)
Whipped toppings	15 mL (1 tbs)

TABLE I-8 Canadian Choice System: Combined Food Choices

Food	Choices per Serving	Measure	Mass (Weight)
Angel food cake	½ starch + 2½ sugars	1/12 cake	50 g
Apple crisp	½ starch + 1½ fruits & vegetables + 1 sugar + 1–2 fats	125 mL (½ c)	
Applesauce, sweetened	1 fruits & vegetables + 1 sugar	125 mL (½ c)	
Beans and pork in tomato sauce	1 starch + ½ fruits & vegetables + ½ sugar + 1 protein	125 mL (½ c)	135 g
Beef burrito	2 starches + 3 proteins + 3 fats		110 g
Brownie	1 sugar + 1 fat	1	20 g
Cabbage rolls*	1 starch + 2 proteins	3	310 g
Caesar salad	2–4 fats	20 mL dressing (4 tsp)	
Cheesecake	½ starch + 2 sugars + ½ protein + 5 fats	1 piece	80 g
Chicken fingers	1 starch + 2 proteins + 2 fats	6 small	100 g
Chicken and snow pea Oriental	2 starches + ½ fruits & vegetables + 3 proteins + 1 fat	500 mL (2 c)	
Chili	1½ starches + ½ fruits & vegetables + 3½ protein	300 mL (1¼ c)	325 g
Chips			
Potato chips	1 starch + 2 fats	15 chips	30 g
Corn chips	1 starch + 2 fats	30 chips	30 g
Tortilla chips	1 starch + 1½ fats	13 chips	
Cheese twist	1 starch + 1½ fats	30 chips	30 g
Chocolate bar			
Aero®	2½ sugars + 2½ fats	bar	43 g
Smarties®	4½ sugars + 2 fats	package	60 g
Chocolate cake (without icing)	1 starch + 2 sugars + 3 fats	1/10 of a 8″ pan	
Chocolate devil's food cake (without icing)	2 starches + 2 sugars + 3 fats	1/12 of a 9″ pan	
Chocolate milk	2 milks 2% + 1 sugar	250 mL (1 c)	300 g
Clubhouse (triple-decker) sandwich	3 starches + 3 proteins + 4 fats		
Cookies			
Chocolate chip	½ starch + ½ sugar + 1½ fats	2	22 g
Oatmeal	1 starch + 1 sugar + 1 fat	2	40 g
Donut (chocolate glazed)	1 starch + 1½ sugars + 2 fats	1	65 g
Egg roll	1 starch + ½ protein + 1 fat		75 g
Four bean salad	1 starch + ½ protein + 1 fat	125 mL (½ c)	
French toast	1 starch + ½ protein + 2 fats	1 slice	65 g
Fruit in heavy syrup	1 fruits & vegetables + 1½ sugars	125 mL (½ c)	
Granola bar	½ starch + 1 sugar + 1–2 fats		30 g
Granola cereal	1 starch + 1 sugar + 2 fats	125 mL (½ c)	45 g
Hamburger	2 starches + 3 proteins + 2 fats	junior size	

*If eaten with sauce, add ½ fruits & vegetables exchange.

(continued on the next page)

TABLE I-8 Canadian Choice System: Combined Food Choices—continued

Food	Choices per Serving	Measure	Mass (Weight)
Ice cream and cone, plain flavour			
Ice cream	½ milk + 2–3 sugars + 1–2 fats		100 g
Cone	½ sugar		4 g
Lasagna			
Regular cheese	1 starch + 1 fruits & vegetables + 3 proteins + 2 fats	3″ × 4″ piece	
Low-fat cheese	1 starch + 1 fruits & vegetables + 3 proteins	3″ × 4″ piece	
Legumes			
Dried beans (kidney, navy, pinto, fava, chick peas)	2 starches + 1 protein	250 mL (1 c)	180 g
Dried peas	2 starches + 1 protein	250 mL (1 c)	210 g
Lentils	2 starches + 1 protein	250 mL (1 c)	210 g
Macaroni and cheese	2 starches + 2 proteins + 2 fats	250 mL (1 c)	210 g
Minestrone soup	1½ starches + ½ fruits & vegetables + ½ fat	250 mL (1 c)	
Muffin	1 starch + ½ sugar + 1 fat	1 small	45 g
Nuts (dry or roasted without any oil added):			
Almonds, dried sliced	½ protein + 2 fats	50 mL (¼ c)	22 g
Brazil nuts, dried unblanched	½ protein + 2½ fats	5 large	23 g
Cashew nuts, dry roasted	½ starch + ½ protein + 2 fats	50 mL (¼ c)	28 g
Filbert hazelnut, dry	½ protein + 3½ fats	50 mL (¼ c)	30 g
Macadamia nuts, dried	½ protein + 4 fats	50 mL (¼ c)	28 g
Peanuts, raw	1 protein + 2 fats	50 mL (¼ c)	30 g
Pecans, dry roasted	½ fruits & vegetables + 3 fats	50 mL (¼ c)	22 g
Pine nuts, pignolia dried	1 protein + 3 fats	50 mL (¼ c)	34 g
Pistachio nuts, dried	½ fruits & vegetables + ½ protein + 2½ fats	50 mL (¼ c)	27 g
Pumpkin seeds, roasted	2 proteins + 2½ fats	50 mL (¼ c)	47 g
Sesame seeds, whole dried	½ fruits & vegetables + ½ protein + 2½ fats	50 mL (¼ c)	30 g
Sunflower kernel, dried	½ protein + 1½ fats	50 mL (¼ c)	17 g
Walnuts, dried chopped	½ protein + 3 fats	50 mL (¼ c)	26 g
Perogies	2 starches + 1 protein + 1 fat	3	
Pie, fruit	1 starch + 1 fruits & vegetables + 2 sugars + 3 fats	1 piece	120 g
Pizza, cheese	1 starch + 1 protein + 1 fat	1 slice (⅛ of a 12″)	50 g
Pork stir fry	½ to 1 fruits & vegetables + 3 proteins	200 mL (¾ c)	
Potato salad	1 starch + 1 fat	125 mL (½ c)	130 g
Potatoes, scalloped	2 starches + 1 milk + 1–2 fats	200 mL (¾ c)	210 g
Pudding, bread or rice	1 starch + 1 sugar + 1 fat	125 mL (½ c)	
Pudding, vanilla	1 milk + 2 sugars	125 mL (½ c)	
Raisin bran cereal	1 starch + ½ fruits & vegetables + ½ sugar	175 mL (⅔ c)	40 g
Rice krispie squares	½ starch + 1½ sugars + ½ fat	1 square	30 g
Shepherd's pie	2 starches + 1 fruits & vegetables + 3 proteins	325 mL (1⅓ c)	
Sherbet, orange	3 sugars + ½ fat	125 mL (½ c)	
Spaghetti and meat sauce	2 starches + 1 fruits & vegetables + 2 proteins + 3 fats	250 mL (1 c)	
Stew	2 starches + 2 fruits & vegetables + 3 proteins + ½ fat	200 mL (¾ c)	
Sundae	4 sugars + 3 fats	125 mL (½ c)	
Tuna casserole	1 starch + 2 proteins + ½ fat	125 mL (½ c)	
Yogurt, fruit bottom	1 fruits & vegetables + 1 milk + 1 sugar	125 mL (½ c)	125 g
Yogurt, frozen	1 milk + 1 sugar	125 mL (½ c)	125 g

Glossary

Many medical terms have their origins in Latin or Greek. By learning a few common derivations, you can glean the meaning of words you have never heard of before. For example, once you know that "hyper" means above normal, "glyc" means glucose, and "emia" means blood, you can easily determine that "hyperglycemia" means high blood glucose. The derivations at left will help you to learn many terms presented in this glossary.

GENERAL

a- or *an-* = not or without
ana- = up
ant- or *anti-* = against
ante- or *pre-* or *pro-* = before
cata- = down
co- = with or together
bi- or *di-* = two, twice
dys- or *mal-* = bad, difficult, painful
endo- = inner or within
epi- = upon
extra- = outside of, beyond, or in addition
exo- = outside of or without
gen- or *-gen* = gives rise to, producing
homeo- = like, similar, constant unchanging state
hyper- = over, above, excessive
hypo- = below, under, beneath
in- = not
inter- = between, in the midst
intra- = within
-itis = infection or inflammation
-lysis = break
macro- = large or long
micro- = small
mono- = one, single
neo- = new, recent
oligo- = few or small
-osis or *-asis* = condition
para- = near
peri- = around, about
poly- = many or much
semi- = half
-stat or *-stasis* = stationary
tri- = three

BODY

angi- or *vaso-* = vessel
arterio- = artery
cardiac or *cardio-* = heart
-cyte = cell
enteron = intestine
gastro- = stomach
hema- or *-emia* = blood
hepatic = liver
myo- or *sarco-* = muscle
nephr- or *renal* = kidney
neuro- = nerve
osteo- = bone
pulmo- = lung
ure- or *-uria* = urine
vena = vein

CHEMISTRY

-al = aldehyde
-ase = enzyme
-ate = salt
glyc- or *gluc-* = sweet (glucose)
hydro- or *hydrate* = water
lipo- = lipid
-ol = alcohol
-ose = carbohydrate
saccha- = sugar

A

absorption: the uptake of nutrients by the cells of the small intestine for transport into either the blood or the lymph.

Acceptable Daily Intake (ADI): the estimated amount of a sweetener that individuals can safely consume each day over the course of a lifetime without adverse effect.

Acceptable Macronutrient Distribution Ranges (AMDR): ranges of intakes for the energy nutrients that provide adequate energy and nutrients and reduce the risk of chronic diseases.

acesulfame (AY-sul-fame) **potassium:** an artificial sweetener composed of an organic salt that has been approved for use in both the United States and Canada; also known as **acesulfame-K** because K is the chemical symbol for potassium.

acetaldehyde (ass-et-AL-duh-hide): an intermediate in alcohol metabolism.

acetyl CoA (ASS-eh-teel, or ah-SEET-il, coh-AY): a 2-carbon compound (**acetate,** or **acetic acid,** shown in Figure 5-1 on p. 142) to which a molecule of CoA is attached.

acid controllers: medications used to prevent or relieve indigestion by suppressing production of acid in the stomach; also called **H2 blockers.** Common brands include Pepcid AC, Tagamet HB, Zantac 75, and Axid AR.

acid-base balance: the equilibrium in the body between acid and base concentrations (see Chapter 12).

acidophilus (ASS-ih-DOF-ih-lus) **milk:** a cultured milk created by adding *Lactobacillus acidophilus,* a bacterium that breaks down lactose to glucose and galactose, producing a sweet, lactose-free product.

acidosis (assi-DOE-sis): above-normal acidity in the blood and body fluids.

acids: compounds that release hydrogen ions in a solution.

acne: a chronic inflammation of the skin's follicles and oil-producing glands, which leads to an accumulation of oils inside the ducts that surround hairs; usually associated with the maturation of young adults.

acupuncture (AK-you-PUNK-cher): a technique that involves piercing the skin with long thin needles at specific anatomical points to relieve pain or illness. Acupuncture sometimes uses heat, pressure, friction, suction, or electromagnetic energy to stimulate the points.

acute PEM: protein-energy malnutrition caused by recent severe food restriction; characterized in children by thinness for height (wasting).

adaptive thermogenesis: adjustments in energy expenditure related to changes in environment such as extreme cold and to physiological events such as overfeeding, trauma, and changes in hormone status.

added sugars: sugars and syrups used as an ingredient in the processing and preparation of foods such as breads, cakes, beverages, jellies, and ice cream as well as sugars eaten separately or added to foods at the table.

additives: substances not normally consumed as foods but added to food either intentionally or by accident.

adenomas (ADD-eh-NOH-mahz): cancers that arise from glandular tissues.

adequacy (dietary): providing all the essential nutrients, fiber, and energy in amounts sufficient to maintain health.

Adequate Intake (AI): the average daily amount of a nutrient that appears sufficient to maintain a specified criterion; a value used as a guide for nutrient intake when an RDA cannot be determined.

adipose (ADD-ih-poce) **tissue:** the body's fat tissue; consists of masses of triglyceride-storing cells.

adolescence: the period from the beginning of puberty until maturity.

adrenal glands: glands adjacent to, and just above, each kidney.

adrenocorticotropin (ad-REE-noh-KORE-tee-koh-TROP-in) or **ACTH:** a hormone, so named because it stimulates *(trope)* the adrenal cortex. The adrenal gland, like the pituitary, has two parts, in this case an outer portion *(cortex)* and an inner core *(medulla).* The realease of ACTH is mediated by **corticotropin-releasing hormone (CRH).**

adverse reactions: unusual responses to food (including intolerances and allergies).

aerobic (air-ROE-bic): requiring oxygen.

agribusiness: agriculture practiced on a massive scale by large corporations owning vast acreages and employing intensive technological, fuel, and chemical inputs.

AIDS (acquired immune deficiency syndrome): the end stage of HIV infection, in which severe complications are manifested. The cluster of mild symptoms that sometimes occurs early in the course of AIDS is called **AIDS-related complex (ARC).**

alcohol: a class of organic compounds containing hydroxyl (OH) groups.

alcohol abuse: a pattern of drinking that includes failure to fulfill work, school, or home responsibilities; drinking in situations that are physically dangerous (as in driving while intoxicated); recurring alcohol-related legal problems (as in aggravated assault charges); or continued drinking despite ongoing social problems that are caused by or worsened by alcohol.

alcohol dehydrogenase (dee-high-DROJ-eh-nayz): an enzyme active in the stomach and the liver that converts ethanol to acetaldehyde.

alcoholism: a pattern of drinking that includes a strong craving for alcohol, a loss of control and an inability to stop drinking once begun, withdrawal symptoms (nausea, sweating, shakiness, and anxiety) after heavy drinking, and the need for increasing amounts of alcohol in order to feel "high."

alcohol-related birth defects (ARBD): malformations in the skeletal and organ systems (heart, kidneys, eyes, ears) associated with prenatal alcohol exposure.

alcohol-related neurodevelopmental disorder (ARND): abnormalities in the central nervous system and cognitive development associated with prenatal alcohol exposure.

aldosterone (al-DOS-ter-own): a hormone secreted by the adrenal glands that regulates blood pressure by increasing the reabsorption of sodium by the kidneys. Aldosterone also regulates chloride and potassium concentrations.

alitame (AL-ih-tame): an artificial sweetener composed of two amino acids (alanine and aspartic acid); FDA approval pending.

alkalosis (alka-LOE-sis): above-normal alkalinity (base) in the blood and body fluids.

alpha-lactalbumin (lact-AL-byoo-min): a major protein in human breast milk, as opposed to **casein** (CAY-seen), a major protein in cow's milk.

alpha-tocopherol: the active vitamin E compound.

Alzheimer's disease: a degenerative disease of the brain involving memory loss and major structural changes in neuron networks; also known as *senile dementia of the Alzheimer's type (SDAT), primary degenerative dementia of senile onset,* or *chronic brain syndrome.*

amenorrhea (ay-MEN-oh-REE-ah): the absence of or cessation of menstruation. **Primary amenorrhea** is menarche delayed beyond 16 years of age. **Secondary amenorrhea** is the absence of three to six consecutive menstrual cycles.

amino acid pool: the supply of amino acids derived from either food proteins or body proteins that collect in the cells and circulating blood and stand ready to be incorporated in proteins and other compounds or used for energy.

amino acid scoring: a measure of protein quality assessed by comparing a protein's amino acid pattern with that of a reference protein; sometimes called **chemical scoring.**

amino (a-MEEN-oh) **acids:** building blocks of proteins. Each contains an amino group, an acid group, a hydrogen atom, and a distinctive side group, all attached to a central carbon atom.

ammonia: a compound with the chemical formula NH_3; produced during the deamination of amino acids.

amniotic (am-nee-OTT-ic) **sac:** the "bag of waters" in the uterus, in which the fetus floats.

amylase (AM-ih-lace): an enzyme that hydrolyzes amylose (a form of starch). Amylase is a *carbohydrase,* an enzyme that breaks down carbohydrates.

anabolic steroids: drugs related to the male sex hormone, testosterone, that stimulate the development of lean body mass.

anabolism (an-ABB-o-lism): reactions in which small molecules are put together to build larger ones. Anabolic reactions require energy.

anaerobic (AN-air-ROE-bic): not requiring oxygen.

anaphylactic (an-AFF-ill-LAC-tic) **shock:** a life-threatening whole-body allergic reaction.

anemia (ah-NEE-me-ah): literally, "too little blood." Anemia is any condition in which too few red blood cells are present, or the red blood cells are immature (and therefore large) or too small or contain too little hemoglobin to carry the normal amount of oxygen to the tissues. It is not a disease itself but can be a symptom of many different disease conditions, including any nutrient deficiencies, bleeding, excessive red blood cell destruction, and defective red blood cell formation.

anencephaly (AN-en-SEF-a-lee): an uncommon and always fatal type of neural tube defect; characterized by the absence of a brain.

angina (an-JYE-nah or AN-ji-nah): a painful feeling of tightness or pressure in and around the heart, often radiating to the back, neck, and arms; caused by a lack of oxygen to an area of heart muscle.

angiotensin (AN-gee-oh-TEN-sin): a hormone involved in blood pressure regulation. Its precursor protein is called *angiotensinogen;* is activated by **renin** (REN-in), an enzyme from the kidneys.

anions (AN-eye-uns): negatively charged ions.

anorexia (an-oh-RECK-see-ah) **nervosa:** an eating disorder characterized by a refusal to maintain a minimally normal body weight and a distortion in perception of body shape and weight.

antacids: medications used to relieve indigestion by neutralizing acid in the stomach.

antagonist: a competing factor that counteracts the action of another factor. When a drug displaces a vitamin from its site of action, the drug renders the vitamin ineffective and thus acts as a vitamin antagonist.

anthropometric (AN-throw-poe-MET-rick): relating to measurement of the physical characteristics of the body, such as height and weight.

antibodies: large proteins of the blood and body fluids, produced by the immune system in response to the invasion of the body by foreign molecules (usually proteins called *antigens*). Antibodies combine with and inactivate the foreign invaders, thus protecting the body.

antidiuretic hormone (ADH): a hormone produced by the pituitary gland in response to dehydration (or a high sodium concentration in the blood). It stimulates the kidneys to reabsorb more water and therefore prevents water loss in urine (also called **vasopressin**). (This ADH should not be confused with the enzyme alcohol dehydrogenase, which is also sometimes abbreviated ADH.)

antigens: substances that elicit the formation of antibodies or an inflammation reaction from the immune system. A bacterium, a virus, a toxin, and a protein in food that causes allergy are all examples of antigens.

antimicrobial agents: preservatives that prevent microorganisms from growing.

antioxidants: in the body, compounds that protect others from oxidation by being oxidized themselves, thereby decreasing the adverse effects of free radicals on normal physiological functions.

antioxidants: as an additive, preservatives that delay or prevent rancidity of fats in foods and other damage to food caused by oxygen.

antipromoters: factors that oppose the development of cancer.

antiscorbutic (AN-tee-skor-BUE-tik) **factor:** the original name for vitamin C.

anus (AY-nus): the terminal outlet of the GI tract.

appendix: a narrow blind sac extending from the beginning of the colon that stores lymph cells.

appetite: the integrated response to the sight, smell, thought, or taste of food that initiates or delays eating.

arachidonic (a-RACK-ih-DON-ic) **acid:** an omega-6 polyunsaturated fatty acid with 20 carbons and four double bonds; present in small amounts in meat and other animal products and synthesized in the body from linoleic acid.

arginine: a nonessential amino acid falsely promoted as enhancing the secretion of human growth hormone, the breakdown of fat, and the development of muscle.

aroma therapy: a technique that uses oil extracts from plants and flowers (usually applied by massage or baths) to enhance physical, psychological, and spiritual health.

arteries: vessels that carry blood from the heart to the tissues.

artesian water: water drawn from a well that taps a confined aquifer in which the water is under pressure.

arthritis: inflammation of a joint, usually accompanied by pain, swelling, and structural changes.

artificial colors: certified food colors added to enhance appearance. (*Certified* means approved by the FDA.) **Bleaching agents** may be used to whiten foods such as flour and cheese.

artificial fats: zero-energy fat replacers that are chemically synthesized to mimic the sensory and cooking qualities of naturally occurring fats, but are totally or partially resistant to digestion.

artificial flavors, flavor enhancers: chemicals that mimic natural flavors and those that enhance flavor.

artificial sweeteners: sugar substitutes that provide negligible, if any, energy; sometimes called **nonnutritive sweeteners.**

ascorbic acid: one of the two active forms of vitamin C (see Figure 10-15). Many people refer to vitamin C by this name.

–ase (ACE): a word ending denoting an enzyme. The word beginning often identifies the compounds the enzyme works on.

aspartame (ah-SPAR-tame or ASS-par-tame): an artificial sweetener composed of two amino acids (phenylalanine and aspartic acid); approved for use in both the United States and Canada.

atherosclerosis (ATH-er-oh-scler-OH-sis): a type of artery disease characterized by placques (accumulations of lipid-containing material) on the inner walls of the arteries (see Chapter 18).

atoms: the smallest components of an element that have all of the properties of the element.

ATP or **adenosine** (ah-DEN-oh-seen) **triphosphate** (try-FOS-fate): a common high-energy compound composed of a purine (adenine), a sugar (ribose), and three phosphate groups.

atrophic (a-TRO-fik) **gastritis** (gas-TRY-tis): chronic inflammation of the stomach accompanied by a diminished size and functioning of the mucous membrane and glands.

atrophy (AT-ro-fee): becoming smaller; with regard to muscles, a decrease in size (and strength) because of disuse, undernutrition, or wasting diseases.

autoimmune disorder: a condition in which the body develops antibodies to its own proteins and then proceeds to destroy cells containing these proteins. In type 1 diabetes, the body develops antibodies to its insulin and destroys the pancreatic cells that produce the insulin, creating an insulin deficiency.

autonomic nervous system: the division of the nervous system that controls the body's automatic responses. Its two branches are the **sympathetic** branch, which helps the body respond to stressors from the outside environment, and the **parasympathetic** branch, which regulates normal body activities between stressful times.

ayurveda (AH-your-VAY-dah): a traditional Hindu system of improving health by using herbs, diet, meditation, massage, and yoga to stimulate the body, mind, and spirit to prevent and treat disease.

B

balance (dietary): providing foods in proportion to each other and in proportion to the body's needs.

basal metabolic rate (BMR): the rate of energy use for metabolism under specified conditions— after a 12-hour fast and restful sleep, without any physical activity or emotional excitement, and in a comfortable setting. It is usually expressed as kcalories per kilogram body weight per hour.

basal metabolism: the energy needed to maintain life when a body is at complete digestive, physical, and emotional rest.

bases: compounds that accept hydrogen ions in a solution.

B-cells: lymphocytes that produce antibodies. B stands for *bursa*, an organ in the chicken associated with the first identification of the B-cells.

bee pollen: a product consisting of bee saliva, plant nectar, and pollen that supposedly aids in weight loss and boosts athletic performance. It does neither and may cause an allergic reaction in individuals sensitive to it.

beer: an alcoholic beverage brewed by fermenting malt and hops.

behavior modification: the changing of behavior by the manipulation of antecedents (cues or environmental factors that trigger behavior), the behavior itself, and consequences (the penalties or rewards attached to behavior).

belching: the expulsion of gas from the stomach through the mouth.

beriberi: the thiamin-deficiency disease.

beta-carotene (BAY-tah KARE-oh-teen): one of the carotenoids; an orange pigment and vitamin A precursor found in plants.

BHA and **BHT:** preservatives commonly used to slow the development of off-flavors, odors, and color changes caused by oxidation.

bicarbonate: a compound with the formula HCO_3 that results from the dissociation of carbonic acid; of particular importance in maintaining the body's acid-base balance. (Bicarbonate is also an alkaline secretion of the pancreas, part of the pancreatic juice.)

bifidus (BIFF-id-us, by-FEED-us) **factors:** factors in colostrum and breast milk that favor the growth of the "friendly" bacterium *Lactobacillus* (lack-toh-ba-SILL-us) *bifidus* in the infant's intestinal tract, so that other, less desirable intestinal inhabitants will not flourish.

bile: an emulsifier that prepares fats and oils for digestion; an exocrine secretion made by the liver, stored in the gallbladder, and released into the small intestine when needed.

binders: chemical compounds in foods that combine with nutrients (especially minerals) to form complexes the body cannot absorb. Examples include **phytates** (FYE-tates) and **oxalates** (OCK-sa-lates).

binge-eating disorder: an eating disorder whose criteria are similar to those of bulimia nervosa, excluding purging or other compensatory behaviors.

bioaccumulation: the accumulation of contaminants in the flesh of animals high on the food chain.

bioavailability: the rate at and the extent to which a nutrient is absorbed and used.

bioelectromagnetic medical applications: the use of electrical energy, magnetic energy, or both to stimulate bone repair, wound healing, and tissue regeneration.

biofeedback: the use of special devices to convey information about heart rate, blood pressure, skin temperature, muscle relaxation, and the like to enable a person to learn how to consciously control these medically important functions.

biofield therapeutics: a manual healing method that directs a healing force from an outside source (commonly God or another supernatural being) through the practitioner and into the client's body; commonly known as "laying on of hands."

biological value (BV): a measure of protein quality assessed by measuring the amount of protein nitrogen that is retained from a given amount of protein nitrogen absorbed.

biotechnology: the use of biological systems or organisms to create or modify products. Examples include the use of bacteria to make yogurt, yeast to make beer, and cross-breeding to enhance crop production.

bioterrorism: the intentional spreading of disease-causing microorganisms or toxins.

biotin (BY-oh-tin): a B vitamin that functions as a coenzyme in metabolism.

blind experiment: an experiment in which the subjects do not know whether they are members of the experimental group or the control group.

blood lipid profile: results of blood tests that reveal a person's total cholesterol, triglycerides, and various lipoproteins.

body composition: the proportions of muscle, bone, fat, and other tissue that make up a person's total body weight.

body mass index (BMI): an index of a person's weight in relation to height; determined by dividing the weight (in kilograms) by the square of the height (in meters).

bolus (BOH-lus): a portion; with respect to food, the amount swallowed at one time.

bomb calorimeter (KAL-oh-RIM-eh-ter): an instrument that measures the heat energy released when foods are burned, thus providing an estimate of the potential energy of the foods.

bone density: a measure of bone strength. When minerals fill the bone matrix (making it dense), they give it strength.

bone meal or **powdered bone:** crushed or ground bone preparations intended to supply calcium to the diet. Calcium from bone is not well absorbed and is often contaminated with toxic

materials such as arsenic, mercury, lead, and cadmium.

boron: a nonessential mineral that is promoted as a "natural" steroid replacement.

bottled water: drinking water sold in bottles.

botulism (BOT-chew-lism): an often fatal food-borne illness caused by the ingestion of foods containing a toxin produced by bacteria that grow without oxygen (see Chapter 19 for details).

bovine growth hormone (BGH): a hormone produced naturally in the pituitary gland of a cow that promotes growth and milk production; now produced for agricultural use by bacteria.

branched-chain amino acids: the essential amino acids leucine, isoleucine, and valine, which are present in large amounts in skeletal muscle tissue; falsely promoted as fuel for exercising muscles.

brewer's yeast: a preparation of yeast cells, containing a concentrated amount of B vitamins and some minerals; falsely promoted as an energy booster.

brown adipose tissue: masses of specialized fat cells packed with pigmented mitochondria that produce heat instead of ATP.

brown sugar: refined white sugar crystals to which manufacturers have added molasses syrup with natural flavor and color; 91 to 96% pure sucrose.

bulimia (byoo-LEEM-ee-ah) **nervosa:** an eating disorder characterized by repeated episodes of binge eating usually followed by self-induced vomiting, misuse of laxatives or diuretics, fasting, or excessive exercise.

C

caffeine: a natural stimulant found in many common foods and beverages, including coffee, tea, and chocolate; may enhance endurance by stimulating fatty acid release. High doses cause headaches, trembling, rapid heart rate, and other undesirable side effects.

calcitonin (KAL-see-TOE-nin): a hormone secreted by the thyroid gland that regulates blood calcium by lowering it when levels rise too high.

calcium: the most abundant mineral in the body; found primarily in the body's bones and teeth.

calcium rigor: hardness or stiffness of the muscles caused by high blood calcium concentrations.

calcium tetany (TET-ah-nee): intermittent spasm of the extremities due to nervous and muscular excitability caused by low blood calcium concentrations.

calcium-binding protein: a protein in the intestinal cells, made with the help of vitamin D, that facilitates calcium absorption.

calmodulin (cal-MOD-you-lin): an inactive protein that becomes active when bound to calcium. Once activated, it becomes a messenger that tells other proteins what to do. The system serves as an interpreter for hormone- and nerve-mediated messages arriving at cells.

calories: units by which energy is measured. Food energy is measured in **kilocalories** (1000 calories equal 1 kilocalorie), abbreviated **kcalories** or **kcal.** One kcalorie is the amount of heat necessary to raise the temperature of 1 kilogram (kg) of water 1°C. The scientific use of the term *kcalorie* is the same as the popular use of the term *calorie.*

cancers: diseases that result from the unchecked growth of malignant tumors.

capillaries (CAP-ill-aries): small vessels that branch from an artery. Capillaries connect arteries to veins. Exchange of oxygen, nutrients, and waste materials takes place across capillary walls.

carbohydrase (KAR-boe-HIGH-drase): an enzyme that hydrolyzes carbohydrates.

carbohydrate loading: a regimen of moderate exercise followed by the consumption of a high-carbohydrate diet that enables muscles to store glycogen beyond their normal capacities; also called **glycogen loading** or **glycogen super compensation.**

carbohydrates: compounds composed of carbon, oxygen, and hydrogen arranged as monosaccharides or multiples of monosaccharides. Most, but not all, carbohydrates have a ratio of one carbon molecule to one water molecule: $(CH_2O)_n$.

carbonated water: water that contains carbon dioxide gas, either naturally occurring or added, that causes bubbles to form in it; also called *bubbling* or *sparkling water*. Seltzer, soda, and tonic waters are legally soft drinks and are not regulated as water.

carbonic acid: a compound with the formula H_2CO_3 that results from the combination of carbon dioxide (CO_2) and water (H_2O); of particular importance in maintaining the body's acid-base balance.

carcinogens (car-SIN-oh-jenz): substances or agents that are capable of causing cancer.

carcinomas (KAR-see-NOH-mahz): cancers that arise from epithelial tissues.

cardiac output: the volume of blood discharged by the heart each minute; determined by multiplying the stroke volume by the heart rate. The stroke volume is the amount of oxygenated blood the heart ejects toward the tissues at each beat. Cardiac output (volume/minute) × stroke volume (volume/beat) = heart rate (beats/minute).

cardiorespiratory conditioning: improvements in heart and lung function and increased blood volume, brought about by aerobic training.

cardiorespiratory endurance: the ability to perform large-muscle, dynamic exercise of moderate-to-high intensity for prolonged periods.

cardiovascular disease (CVD): a general term for all diseases of the heart and blood vessels. Atherosclerosis is the main cause of CVD. When the arteries that carry blood to the heart muscle become blocked, the heart suffers damage known as **coronary heart disease (CHD).**

carnitine (CAR-neh-teen): a nonessential, nonprotein amino acid made in the body from lysine that helps transport fatty acids across the mitochondrial membrane. Carnitine supposedly "burns" fat and spares glycogen during endurance events, but in reality it does neither.

carotenoids (kah-ROT-eh-noyds): pigments commonly found in plants and animals, some of which have vitamin A activity. The carotenoid with the greatest vitamin A activity is beta-carotene.

carpal tunnel syndrome: a pinched nerve at the wrist, causing pain or numbness in the hand. It is often caused by repetitive motion of the wrist.

cartilage therapy: the use of cleaned and powdered connective tissue, such as collagen, to improve health.

catabolism (ca-TAB-o-lism): reactions in which large molecules are broken down to smaller ones. Catabolic reactions release energy.

catalyst (CAT-uh-list): a compound that facilitates chemical reactions without itself being changed in the process.

cataracts (KAT-ah-rakts): thickenings of the eye lenses that impair vision and can lead to blindness.

cathartic (ka-THAR-tik): a strong laxative.

cations (CAT-eye-uns): positively charged ions.

CD4+ T-lymphocytes: circulating white blood cells that contain the CD4+ protein on their surfaces and are a necessary component of the immune system.

CDC (Centers for Disease Control): a branch of the Department of Health and Human Services that is responsible for, among other things, monitoring foodborne diseases.

cell: the basic structural unit of all living things.

cell differentiation (DIF-er-EN-she-AY-shun): the process by which immature cells develop specific functions different from those of the original that are characteristic of their mature cell type.

cell membrane: the thin layer of tissue that surrounds the cell and encloses its contents; made primarily of lipid and protein.

cell salts: a preparation of minerals supposedly harvested from living cells, sold as a health-promoting supplement.

cellulite (SELL-you-light or SELL-you-leet): supposedly, a lumpy form of fat; actually, a fraud. Fatty areas of the body may appear lumpy when the strands of connective tissue that attach the skin to underlying muscles pull tight where the fat is thick. The fat itself is the same as fat anywhere else in the body. If the fat in these areas is lost, the lumpy appearance disappears.

central nervous system: the brain and spinal cord.

central obesity: excess fat around the trunk of the body; also called **abdominal fat** or **upper-body fat.**

certification: the process in which a private laboratory inspects shipments of a product for selected chemicals and then, if the product is free of violative levels of those chemicals, issues a guarantee to that effect.

cesarean section: a surgically assisted birth involving removal of the fetus by an incision into the uterus, usually by way of the abdominal wall.

CHD risk equivalents: disorders that raise the risk of heart attacks, strokes, and other complications associated with cardiovascular disease to the same degree as existing CHD. These disorders include symptomatic carotid artery disease, peripheral arterial disease, abdominal aortic aneurysm, and diabetes mellitus.

chelate (KEY-late): a substance that can grasp the positive ions of a metal.

chelation (kee-LAY-shun) **therapy:** the use of ethylene diamine tetraacetic acid (EDTA) to bind with metallic ions, thus healing the body by removing toxic metals.

chiropractic (KYE-roh-PRAK-tik): a manual healing method of manipulating the spine to restore health.

chloride (KLO-ride): the major anion in the extracellular fluids of the body. Chloride is the ionic form of chlorine, Cl_2; see Appendix B for a description of the chlorine-to-chloride conversion.

chlorophyll (KLO-row-fil): the green pigment of plants, which absorbs light and transfers the energy to other molecules, thereby initiating photosynthesis.

cholecystokinin (COAL-ee-SIS-toe-KINE-in), or **CCK:** a hormone produced by cells of the intestinal wall. Target organ: the gallbladder. Response: release of bile and slowing of GI motility.

cholesterol (koh-LESS-ter-ol): one of the sterols containing a four-carbon ring structure with a carbon side chain.

cholesterol-free: less than 2 mg cholesterol per serving and 2 g or less saturated fat and *trans* fat combined per serving.

choline (KOH-leen): a nitrogen-containing compound found in foods and made in the body from the amino acid methionine. Choline is part of the phospholipid lecithin and the neurotransmitter acetylcholine.

chromium picolinate (CROW-mee-um pick-oh-LYN-ate): a trace mineral supplement; falsely promoted as building muscle, enhancing energy, and burning fat. **Picolinate** is a derivative of the amino acid tryptophan that seems to enhance chromium absorption.

chromosomes: a set of structures within the nucleus of every cell that contains the cell's genetic material, DNA, associated with other materials (primarily proteins).

chronic diseases: diseases characterized by a slow progression and long duration. Examples include heart disease, cancer, and diabetes.

chronic PEM: protein-energy malnutrition caused by long-term food deprivation; characterized in children by short height for age (stunting).

chronological age: a person's age in years from his or her date of birth.

chylomicrons (kye-lo-MY-cronz): the class of lipoproteins that transport lipids from the intestinal cells to the rest of the body.

chyme (KIME): the semiliquid mass of partly digested food expelled by the stomach into the duodenum.

cirrhosis (seer-OH-sis): advanced liver disease in which liver cells turn orange, die, and harden, permanently losing their function; often associated with alcoholism.

clinically severe obesity: a BMI of 40 or greater or a BMI of 35 or greater with additional risk factors. A less preferred term used to describe the same condition is *morbid obesity*.

CoA (coh-AY): coenzyme A; the coenzyme derived from the B vitamin pantothenic acid and central to energy metabolism.

coenzyme Q10: a lipid found in cells (mitochondria) shown to improve exercise performance in heart disease patients, but not effective in improving the performance of healthy athletes.

coenzymes: complex organic molecules that work with enzymes to facilitate the enzymes' activity. Many coenzymes have B vitamins as part of their structures (Figure 10-1 on p. 325 in Chapter 10 illustrates coenzyme action).

colitis (ko-LYE-tis): inflammation of the colon.

collagen (KOL-ah-jen): the protein from which connective tissues such as scars, tendons, ligaments, and the foundations of bones and teeth are made.

colonic irrigation: the popular, but potentially harmful practice of "washing" the large intestine with a powerful enema machine.

colostrum (ko-LAHS-trum): a milklike secretion from the breast, present during the first day or so after delivery before milk appears; rich in protective factors.

complementary and alternative medicine: diverse medical and health care systems, practices, and products that are not currently considered part of conventional medicine; also called *adjunctive, unconventional,* or *unorthodox therapies.*

complementary medicine: an approach that uses alternative therapies as an adjunct to, and not simply a replacement for, conventional medicine.

complementary proteins: two or more dietary proteins whose amino acid assortments complement each other in such a way that the essential amino acids missing from one are supplied by the other.

complex carbohydrates (starches and fibers): polysaccharides composed of straight or branched chains of monosaccharides.

compound: a substance composed of two or more different atoms—for example, water (H_2O).

conception: the union of the male sperm and the female ovum; fertilization.

condensation: a chemical reaction in which two reactants combine to yield a larger product.

conditionally essential amino acid: an amino acid that is normally nonessential, but must be supplied by the diet in special circumstances when the need for it exceeds the body's ability to produce it.

conditioning: the physical effect of training; improved flexibility, strength, and endurance.

confectioners' sugar: finely powdered sucrose, 99.9% pure.

congregate meals: nutrition programs that provide food for the elderly in conveniently located settings such as community centers.

constipation: the condition of having infrequent or difficult bowel movements.

contaminants: substances that make a food impure and unsuitable for ingestion.

contamination iron: iron found in foods as the result of contamination by inorganic iron salts from iron cookware, iron-containing soils, and the like.

control group: a group of individuals similar in all possible respects to the experimental group except for the treatment. Ideally, the control group receives a placebo while the experimental group receives a real treatment.

conventional medicine: diagnosis and treatment of diseases as practiced by medical doctors (M.D.) and doctors of osteopathy (D.O.) and allied health professionals such as physical therapists and registered nurses; also called *allopathy; Western, mainstream, orthodox,* or *regular medicine;* and *biomedicine.*

cool-down: 5 to 10 minutes of light activity, such as walking or stretching, following a vigorous workout to return the body's core gradually to near-normal temperature.

Cori cycle: the path from muscle lactic acid (which travels to the liver) to glucose (which can travel back to the muscle); named after the scientist who elucidated this pathway.

corn sweeteners: corn syrup and sugars derived from corn.

corn syrup: a syrup made from cornstarch that has been treated with acid, high temperatures, and enzymes that produce glucose, maltose, and dextrins. See also *high-fructose corn syrup (HFCS).*

cornea (KOR-nee-uh): the transparent membrane covering the outside of the eye.

coronary heart disease (CHD): the damage that occurs when the blood vessels carrying blood to the heart (the **coronary arteries**) become narrow and occluded.

correlation (CORE-ee-LAY-shun): the simultaneous increase, decrease, or change in two variables. If A increases as B increases, or if A decreases as B decreases, the correlation is **positive.** (This does not mean that A causes B or vice versa.) If A increases as B decreases, or if A decreases as B increases, the correlation is **negative.** (This does not mean that A prevents B or vice versa.) Some third factor may account for both A and B.

cortical bone: the very dense bone tissue that forms the outer shell surrounding trabecular bone and comprises the shaft of a long bone.

coupled reactions: pairs of chemical reactions in which some of the energy released from the breakdown of one compound is used to create a bond in the formation of another compound.

covert (KOH-vert): hidden, as if under covers.

creatine (KREE-ah-tin): a nitrogen-containing compound that combines with phosphate to form the high-energy compound creatine phosphate (or phosphocreatine) in muscles. Claims that creatine enhances energy use and muscle strength need further confirmation.

creatine phosphate (also called phosphocreatine): a high-energy compound in muscle cells that acts as a reservoir of energy that can maintain a steady supply of ATP. CP provides the energy for short bursts of activity.

C-reactive protein (CRP): a protein produced during the acute phase of infection or inflammation that enhances immunity by promoting phagocytosis and activating platelets. Its presence may be used to assess a person's risk of an impending heart attack or stroke.

cretinism (CREE-tin-ism): a congenital disease characterized by mental and physical retardation and commonly caused by maternal iodine deficiency during pregnancy.

critical periods: finite periods during development in which certain events occur that will have irreversible effects on later developmental stages; usually a period of rapid cell division.

cross-contamination: the contamination of food by bacteria that occurs when the food comes into contact with surfaces previously touched by raw meat, poultry, or seafood.

crypts (KRIPTS): tubular glands that lie between the intestinal villi and secrete intestinal juices into the small intestine.

cyclamate (SIGH-kla-mate): an artificial sweetener that is being considered for approval in the United States and is available in Canada as a tabletop sweetener, but not as an additive.

cytokines (SIGH-toe-kines): special proteins that direct immune and inflammatory responses.

cytoplasm (SIGH-toh-plazm): the cell contents, except for the nucleus.

cytosol: the fluid of cytoplasm; contains water, ions, nutrients, and enzymes.

D

Daily Values (DV): reference values developed by the FDA specifically for use on food labels.

deamination (dee-AM-eh-NAY-shun): removal of the amino (NH_2) group from a compound such as an amino acid.

defecate (DEF-uh-cate): to move the bowels and eliminate waste.

deficient: the amount of a nutrient below which almost all healthy people can be expected, over time, to experience deficiency symptoms.

dehydration: the condition in which body water output exceeds water input. Symptoms include thirst, dry skin and mucous membranes, rapid heartbeat, low blood pressure, and weakness.

Delaney Clause: a clause in the Food Additive Amendment to the Food, Drug, and Cosmetic Act that states that no substance that is known to cause cancer in animals or human beings at any dose level shall be added to foods.

denaturation (dee-NAY-chur-AY-shun): the change in a protein's shape and consequent loss of its function brought about by heat, agitation, acid, base, alcohol, heavy metals, or other agents.

dental caries: decay of teeth.

dental plaque: a gummy mass of bacteria that grows on teeth and can lead to dental caries and gum disease.

desiccated liver: dehydrated liver powder that supposedly contains all the nutrients found in liver in concentrated form; possibly not dangerous, but has no particular nutritional merit and is considerably more expensive than fresh liver.

dextrose: an older name for glucose.

DHA, or docosahexaenoic (DOE-cossa-HEXA-ee-NO-ick) **acid:** an omega-3 polyunsaturated fatty acid with 22 carbons and six double bonds; present in fish and synthesized in limited amounts in the body from linolenic acid.

DHEA (dehydroepiandrosterone) and **androstenedione:** hormones made in the adrenal glands that serve as precursors to the male hormone testosterone; falsely promoted as burning fat, building muscle, and slowing aging. Side effects include acne, aggressiveness, and liver enlargement.

diabetes (DYE-uh-BEET-eez) **mellitus** (MELL-ih-tus or mell-EYE-tus): a disorder of carbohydrate metabolism characterized by altered glucose regulation and utilization, usually resulting from insufficient or ineffective insulin.

diarrhea: the frequent passage of watery bowel movements.

diet: the foods and beverages a person eats and drinks.

diet history: a record of eating behaviors and the foods a person eats.

dietary folate equivalents (DFE): the amount of folate available to the body from naturally occurring sources, fortified foods, and supplements, accounting for differences in the bioavailability from each source.

Dietary Reference Intakes (DRI): a set of nutrient intake values for healthy people in the United States and Canada. These values are used for planning and assessing diets and include:

• Estimated Average Requirements (EAR).

• Recommended Dietary Allowances (RDA).

• Adequate Intakes (AI).

• Tolerable Upper Intake Levels (UL).

digestion: the process by which food is broken down into absorbable units.

digestive enzymes: proteins found in digestive juices that act on food substances, causing them to break down into simpler compounds.

digestive system: all the organs and glands associated with the ingestion and digestion of food.

dioxins (dye-OCK-sins): a class of chemical pollutants created as by-products of chemical manufacturing, incineration, chlorine bleaching of paper pulp, and other industrial processes. Dioxins persist in the environment and accumulate in the food chain.

dipeptide (dye-PEP-tide): two amino acids bonded together.

disaccharides (dye-SACK-uh-rides): pairs of monosaccharides linked together. See Appendix C for the chemical structures of the disaccharides.

disordered eating: eating behaviors that are neither normal nor healthy, including restrained eating, fasting, binge eating, and purging.

dissociates (dis-SO-see-ates): physically separates.

distilled liquor or **hard liquor:** an alcoholic beverage made by fermenting and distilling grains; sometimes called *distilled spirits*.

distilled water: water that has been vaporized and recondensed, leaving it free of dissolved minerals.

diverticula (dye-ver-TIC-you-la): sacs or pouches that develop in the weakened areas of the intestinal wall (like bulges in an inner tube where the tire wall is weak).

diverticulitis (DYE-ver-tic-you-LYE-tis): infected or inflamed diverticula.

diverticulosis (DYE-ver-tic-you-LOH-sis): the condition of having diverticula. About one in every six people in Western countries develops diverticulosis in middle or later life.

DNA (deoxyribonucleic acid): the genetic material of cells necessary in protein synthesis; falsely promoted as an energy booster.

dolomite: a compound of minerals (calcium magnesium carbonate) found in limestone and marble. Dolomite is powdered and is sold as a calcium-magnesium supplement, but may be contaminated with toxic minerals, is not well absorbed, and interacts adversely with absorption of other esssential minerals.

double-blind experiment: an experiment in which neither the subjects nor the researchers know which subjects are members of the experimental group and which are serving as control subjects, until after the experiment is over.

Down syndrome: a genetic abnormality that causes mental retardation, short stature, and flattened facial features.

drink: a dose of any alcoholic beverage that delivers ½ oz of pure ethanol.

drug: a substance that can modify one or more of the body's functions.

drug history: a record of all the drugs, over-the-counter and prescribed, that a person takes routinely.

duodenum (doo-oh-DEEN-um, doo-ODD-num): the top portion of the small intestine (about "12 fingers' breadth" long in ancient terminology).

duration: length of time (for example, the time spent in each activity session).

dysentery (DISS-en-terry): an infection of the digestive tract that causes diarrhea.

dysphagia (dis-FAY-gee-ah): difficulty in swallowing.

E

eating disorders: disturbances in eating behavior that jeopardize a person's physical or psychological health.

eclampsia (eh-KLAMP-see-ah): a severe stage of preeclampsia characterized by convulsions.

edema (eh-DEEM-uh): the swelling of body tissue caused by excessive amounts of fluid in the interstitial spaces; seen in protein deficiency (among other conditions).

eicosanoids (eye-COSS-uh-noyds): derivatives of 20-carbon fatty acids; biologically active compounds that help to regulate blood pressure, blood clotting, and other body functions. They include *prostaglandins* (PROS-tah-GLAND-ins), *thromboxanes* (throm-BOX-ains), and *leukotrienes* (LOO-ko-TRY-eens).

electrolyte solutions: solutions that can conduct electricity.

electrolytes: salts that dissolve in water and dissociate into charged particles called ions.

electron transport chain: the final pathway in energy metabolism that transports electrons from hydrogen to oxygen and captures the energy released in the bonds of ATP.

element: a substance composed of atoms that are alike—for example, iron (Fe).

embolism (EM-boh-lizm): the obstruction of a blood vessel by an **embolus** (EM-boh-luss), or traveling clot, causing sudden tissue death.

embryo (EM-bree-oh): the developing infant from two to eight weeks after conception.

emergency kitchens: programs that provide prepared meals to be eaten on site; often called *soup kitchens*.

emerging risk factors: recently identified factors that enhance the ability to predict disease risk in an individual.

emetic (em-ETT-ic): an agent that causes vomiting.

empty-kcalorie foods: a popular term used to denote foods that contribute energy but lack protein, vitamins, and minerals.

emulsifier (ee-MUL-sih-fire): a substance with both water-soluble and fat-soluble portions that promotes the mixing of oils and fats in a watery solution.

emulsifiers and **gums:** as additives, thickening and stabilizing agents that maintain emulsions, foams, or suspensions or lend a desirable thick consistency to foods.

endoplasmic reticulum (en-doh-PLAZ-mic reh-TIC-you-lum): a complex network of intracellular membranes. The **rough endoplasmic reticulum** is dotted with ribosomes, where protein synthesis takes place. The **smooth endoplasmic reticulum** bears no ribosomes.

enemas: solutions inserted into the rectum and colon to stimulate a bowel movement and empty the lower large intestine.

energy: the capacity to do work. The energy in food is chemical energy. The body can convert this chemical energy to mechanical, electrical, or heat energy.

energy density: a measure of the energy a food provides relative to the amount of food (kcalories per gram).

energy-yielding nutrients: the nutrients that break down to yield energy the body can use.

enriched: the addition to a food of nutrients that were lost during processing so that the food will meet a specified standard.

enteropancreatic (EN-ter-oh-PAN-kree-AT-ik) **circulation:** the circulatory route from the pancreas to the intestine and back to the pancreas.

enzymes: proteins that facilitate chemical reactions without being changed in the process; protein catalysts.

EPA (Environmental Protection Agency): a federal agency that is responsible for, among other things, regulating pesticides and establishing water quality standards.

EPA, or **eicosapentaenoic** (EYE-cossa-PENTA-ee-NO-ick) **acid:** an omega-3 polyunsaturated fatty acid with 20 carbons and five double bonds; present in fish and synthesized in limited amounts in the body from linolenic acid.

epidemic (EP-ee-DEM-ick): the appearance of a disease (usually infectious) or condition that attacks many people at the same time in the same region.

epiglottis (epp-ee-GLOTT-iss): cartilage in the throat that guards the entrance to the trachea and prevents fluid or food from entering it when a person swallows.

epinephrine (EP-ih-NEFF-rin): a hormone of the adrenal gland that modulates the stress response; formerly called **adrenaline**. When administered by injection, epinephrine counteracts anaphylactic shock by opening the airways and maintaining heartbeat and blood pressure.

epithelial (ep-i-THEE-lee-ul) **cells:** cells on the surface of the skin and mucous membranes.

epithelial tissue: the layer of the body that serves as a selective barrier between the body's interior and the environment (examples are the cornea of the eyes, the skin, the respiratory lining of the lungs, and the lining of the digestive tract).

epoetin (eh-poy-EE-tin): a drug derived from the human hormone erythropoietin and marketed under the trade name Epogen; illegally used to increase oxygen capacity.

ergogenic (ER-go-JEN-ick) **aids:** substances or techniques used in an attempt to enhance physical performance.

erythrocyte (eh-RITH-ro-cite) **hemolysis** (he-MOLL-uh-sis): the breaking open of red blood cells (erythrocytes); a symptom of vitamin E–deficiency disease in human beings.

erythrocyte protoporphyrin (PRO-toe-PORE-ferin): a precursor to hemoglobin.

erythropoietin (eh-REE-throh-POY-eh-tin): a hormone that stimulates red blood cell production.

esophageal (ee-SOF-ah-GEE-al) **sphincter:** a sphincter muscle at the upper or lower end of the esophagus. The *lower esophageal sphincter* is also called the *cardiac sphincter.*

esophagus (ee-SOFF-ah-gus): the food pipe; the conduit from the mouth to the stomach.

essential amino acids: amino acids that the body cannot synthesize in amounts sufficient to meet physiological needs (see Table 6-1 on p. 182).

essential fatty acids: fatty acids needed by the body, but not made by it in amounts sufficient to meet physiological needs.

essential nutrients: nutrients a person must obtain from food because the body cannot make them for itself in sufficient quantity to meet physiological needs; also called **indispensable nutrients.** About 40 nutrients are currently known to be essential for human beings.

Estimated Average Requirement (EAR): the average daily amount of a nutrient that will maintain a specific biochemical or physiological function in half the healthy people of a given age and gender group.

Estimated Energy Requirement (EER): the average dietary energy intake that maintains energy balance and good health in a person of a given age, gender, weight, height, and level of physical activity.

estrogens: hormones responsible for the menstrual cycle and other female characteristics.

ethanol: a particular type of alcohol found in beer, wine, and distilled spirits; also called *ethyl alcohol* (see Figure H7-1 on p. 240). Ethanol is the most widely used—and abused—drug in our society. It is also the only legal, nonprescription drug that produces euphoria.

exchange lists: diet-planning tools that organize foods by their proportions of carbohydrate, fat, and protein. Foods on any single list can be used interchangeably.

exercise: planned, structured, and repetitive bodily movement that promotes or maintains physical fitness.

experimental group: a group of individuals similar in all possible respects to the control group except for the treatment. The experimental group receives the real treatment.

extra lean: less than 5 g of fat, 2 g of saturated fat and *trans* fat combined, and 95 mg of cholesterol per serving and per 100 g of meat, poultry, and seafood.

extracellular fluid: fluid outside the cells. Extracellular fluid includes two main components—the interstitial fluid and plasma. Extracellular fluid accounts for approximately one-third of the body's water.

F

fad diets: popular eating plans that promise quick weight loss. Most fad diets severely limit certain foods or overemphasize others (for example, never eat potatoes or pasta or eat cabbage soup daily).

faith healing: healing by invoking divine intervention without the use of medical, surgical, or other traditional therapy.

false negative: a test result indicating that a condition is not present (negative) when in fact it is present (therefore false).

false positive: a test result indicating that a condition is present (positive) when in fact it is not (therefore false).

famine: extreme scarcity of food in an area that causes starvation and death in a large portion of the population.

FAO (Food and Agriculture Organization): an international agency (part of the United Nations) that has adopted standards to regulate pesticide use among other responsibilities.

fat replacers: ingredients that replace some or all of the functions of fat and may or may not provide energy.

fat-free: less than 0.5 g of fat per serving (and no added fat or oil); synonyms include "zero-fat," "no-fat," and "nonfat."

fats: lipids that are solid at room temperature (70°F or 25°C).

fatty acid: an organic compound composed of a carbon chain with hydrogens attached and an acid group (COOH) at one end and a methyl group (CH$_3$) at the other end.

fatty acid oxidation: the metabolic breakdown of fatty acids to acetyl CoA; also called **beta oxidation.**

fatty liver: an early stage of liver deterioration seen in several diseases, including kwashiorkor and alcoholic liver disease. Fatty liver is characterized by an accumulation of fat in the liver cells.

fatty streaks: accumulations of cholesterol and other lipids along the walls of the arteries.

FDA (Food and Drug Administration): a part of the Department of Health and Human Services' Public Health Service that is responsible for ensuring the safety and wholesomeness of all dietary supplements and foods processed and sold in interstate commerce except meat, poultry, and eggs (which are under the jurisdiction of the USDA); inspecting food plants and imported foods; and setting standards for food composition and product labeling.

female athlete triad: a potentially fatal combination of three medical problems—disordered eating, amenorrhea, and osteoporosis.

fermentable: the extent to which bacteria in the GI tract can break down fibers to fragments that the body can use. Dietary fibers are fermented by bacteria in the colon to short-chain fatty acids, which are absorbed and metabolized by cells in the GI tract and liver (Chapter 5 describes fatty acids).

ferritin (FAIR-ih-tin): the iron storage protein.

fertility: the capacity of a woman to produce a normal ovum periodically and of a man to produce normal sperm; the ability to reproduce.

fetal alcohol effects (FAE): an older, less preferred, term used to describe both alcohol-related birth defects and neurodevelopmental disorders.

fetal alcohol syndrome (FAS): a cluster of physical, behavioral, and cognitive abnormalities associated with prenatal alcohol exposure, including facial malformations, growth retardation, and central nervous disorders.

fetal programming: the influence of substances during fetal growth on the development of diseases in later life.

fetus (FEET-us): the developing infant from eight weeks after conception until term.

fibers: in plant foods, the *nonstarch polysaccharides* that are not digested by human digestive enzymes, although some are digested by GI tract bacteria. Fibers include cellulose, hemicelluloses, pectins, gums, and mucilages and the nonpolysaccharides lignins, cutins, and tannins.

fibrocystic (FYE-bro-SIS-tik) **breast disease:** a harmless condition in which the breasts develop lumps, sometimes associated with caffeine consumption. In some, it responds to abstinence from caffeine; in others, it can be treated with vitamin E.

fibrosis (fye-BROH-sis): an intermediate stage of liver deterioration seen in several diseases, including viral hepatitis and alcoholic liver disease. In fibrosis, the liver cells lose their function and assume the characteristics of connective tissue cells (fibers).

filtered water: water treated by filtration, usually through *activated carbon filters* that reduce the lead in tap water, or by *reverse osmosis* units that force pressurized water across a membrane removing lead, arsenic, and some microorganisms from tap water.

fitness: the characteristics that enable the body to perform physical activity; more broadly, the ability to meet routine physical demands with enough reserve energy to rise to a physical challenge; or the body's ability to withstand stress of all kinds.

flavonoids (FLAY-von-oyds): yellow pigments in foods; phytochemicals that may exert physiological effects on the body.

flaxseed: the small brown seed of the flax plant; valued as a source of linseed oil, fiber, and omega-3 fatty acids.

flexibility: the capacity of the joints to move through a full range of motion; the ability to bend and recover without injury.

fluid balance: maintenance of the proper types and amounts of fluid in each compartment of the body fluids (see also Chapter 12).

fluorapatite (floor-APP-uh-tite): the stabilized form of bone and tooth crystal, in which fluoride has replaced the hydroxyl groups of hydroxyapatite.

fluorosis (floor-OH-sis): discoloration and pitting of tooth enamel caused by excess fluoride during tooth development.

folate (FOLE-ate): a B vitamin; also known as folic acid, folacin, or pteroylglutamic (tare-o-EEL-glue-TAM-ick) acid (PGA). The coenzyme forms are **DHF (dihydrofolate)** and **THF (tetrahydrofolate).**

follicle-stimulating hormone (FSH): a hormone that stimulates maturation of the ovarian follicles in females and the production of sperm in males. (The ovarian follicles are part of the female reproductive system where the eggs are produced.) The release of FSH is mediated by **follicle-stimulating hormone releasing hormone (FSH–RH).**

food allergy: an adverse reaction to food that involves an immune response; also called **food-hypersensitivity reaction.**

food aversions: strong desires to avoid particular foods.

food bank: a facility that collects and distributes food donations to authorized organizations feeding the hungry.

food chain: the sequence in which living things depend on other living things for food.

food cravings: strong desires to eat particular foods.

food frequency questionnaire: a checklist of foods on which a person can record the frequency with which he or she eats each food.

food group plans: diet-planning tools that sort foods of similar origin and nutrient content into groups and then specify that people should eat certain numbers of servings from each group.

food insecurity: limited or uncertain access to foods of sufficient quality or quantity to sustain a healthy and active life.

food insufficiency: an inadequate amount of food due to a lack of resources.

food intolerances: adverse reactions to foods that do not involve the immune system.

food pantries: programs that provide groceries to be prepared and eaten at home.

food poverty: hunger resulting from inadequate access to available food for various reasons, including inadequate resources, political obstacles, social disruptions, poor weather conditions, and lack of transportation.

food record: an extensive, accurate log of all foods eaten over a period of several days or weeks. A food record that includes associated information such as when, where, and with whom each food is eaten is sometimes called a **food diary.**

food recovery: collecting wholesome food for distribution to low-income people who are hungry.

food security: certain access to enough food for all people at all times to sustain a healthy and active life.

food substitutes: foods that are designed to replace other foods.

foodborne illness: illness transmitted to human beings through food and water, caused by either an infectious agent (foodborne infection) or a poisonous substance (food intoxication); commonly known as **food poisoning.**

foods: products derived from plants or animals that can be taken into the body to yield energy and nutrients for the maintenance of life and the growth and repair of tissues.

fortified: the addition to a food of nutrients that were either not originally present or present in insignificant amounts. Fortification can be used to correct or prevent a widespread nutrient deficiency or to balance the total nutrient profile of a food.

fossil fuels: coal, oil, and natural gas.

free: "nutritionally trivial" and unlikely to have a physiological consequence; synonyms include "without," "no," and "zero." A food that does not contain a nutrient naturally may make such a claim, but only as it applies to all similar foods (for example, "applesauce, a fat-free food").

free radicals: unstable and highly reactive atoms or molecules that have one or more unpaired electrons in the outer orbital (see Appendix B for a review of basic chemistry concepts).

frequency: the number of occurrences per unit of time (for example, the number of activity sessions per week).

fructose (FRUK-tose or FROOK-tose): a monosaccharide. Sometimes known as fruit sugar or **levulose,** fructose is found abundantly in fruits, honey, and saps.

fuel: compounds that cells can use for energy. The major fuels include glucose, fatty acids, and amino acids; other fuels include ketone bodies, lactic acid, glycerol, and alcohol.

functional foods: foods that contain physiologically active compounds that provide health benefits beyond their nutrient contributions; sometimes called *designer foods* or *nutraceuticals.* Functional foods may include whole foods, fortified foods, and modified foods.

G

galactose (ga-LAK-tose): a monosaccharide; part of the disaccharide lactose.

gallbladder: the organ that stores and concentrates bile. When it receives the signal that fat is present in the duodenum, the gallbladder contracts and squirts bile through the bile duct into the duodenum.

gastric glands: exocrine glands in the stomach wall that secrete gastric juice into the stomach.

gastric juice: the digestive secretion of the gastric glands of the stomach.

gastric-inhibitory peptide: a hormone produced by the intestine. Target organ: the stomach. Response: slowing of the secretion of gastric juices and of GI motility.

gastrin: a hormone secreted by cells in the stomach wall. Target organ: the glands of the stomach. Response: secretion of gastric acid.

gastroesophageal reflux: the backflow of stomach acid into the esophagus, causing damage to the cells of the esophagus and the sensation of heartburn. **Gastroesophageal reflux disease (GERD)** is characterized by symptoms of reflux occurring two or more times a week.

gastrointestinal (GI) tract: the digestive tract. The principal organs are the stomach and intestines.

gatekeepers: with respect to nutrition, key people who control other people's access to foods and thereby exert profound impacts on their nutrition. Examples are the spouse who buys and cooks the food, the parent who feeds the children, and the caregiver in a day-care center.

gelatin: a soluble form of the protein collagen, used to thicken foods; sometimes falsely promoted as a strength enhancer.

gene pool: all the genetic information of a population at a given time.

generally recognized as safe (GRAS): food additives that have long been in use and are believed safe. First established by the FDA in 1958, the GRAS list is subject to revision as new facts become known.

genetic engineering: the use of biotechnology to modify the genetic material of living cells so that they will produce new substances or perform new functions. Foods produced via this technology are called **genetically modified (GM)** or **genetically engineered (GE) foods.**

genome (GEE-nome): the full complement of genetic material (DNA) in the chromosomes of a cell. In human beings, the genome consists of 46 chromosomes. The study of genomes is called **genomics.**

gestation (jes-TAY-shun): the period from conception to birth. For human beings, the average length of a healthy gestation is 40 weeks. Pregnancy is often divided into thirds, called **trimesters.**

gestational diabetes: abnormal glucose tolerance during pregnancy.

ghrelin (GRELL-in): a protein produced by the stomach cells that enhances appetite and decreases energy expenditure.

ginseng: a plant whose extract supposedly boosts energy. Side effects of chronic use include nervousness, confusion, and depression.

gland: a cell or group of cells that secretes materials for special uses in the body. Glands may be **exocrine** (EKS-oh-crin) **glands,** secreting their materials "out" (into the digestive tract or onto the surface of the skin), or **endocrine** (EN-doe-crin) **glands,** secreting their materials "in" (into the blood).

gliomas (gly-OH-mahz): cancers that arise from glial cells of the central nervous system.

glucagon (GLOO-ka-gon): a hormone that is secreted by special cells in the pancreas in response to low blood glucose concentration and elicits release of glucose from liver glycogen stores.

glucocorticoids: hormones from the adrenal cortex that affect the body's management of glucose.

gluconeogenesis (gloo-co-nee-oh-GEN-ih-sis): the making of glucose from a noncarbohydrate source (described in more detail in Chapter 7).

glucose (GLOO-kose): a monosaccharide; sometimes known as blood sugar or **dextrose.**

glucose polymers: compounds that supply glucose, not as single molecules, but linked in chains somewhat like starch. The objective is to attract less water from the body into the digestive tract (osmotic attraction depends on the number, not the size, of particles).

glycemic index: a method of classifying foods according to their potential for raising blood glucose.

glycemic (gligh-SEEM-ic) **response:** the extent to which a food raises the blood glucose concentration and elicits an insulin response.

glycerol (GLISS-er-ol): an alcohol composed of a three-carbon chain, which can serve as the backbone for a triglyceride.

glycine: a nonessential amino acid, promoted as an ergogenic aid because it is a precursor of the high-energy compound creatine phosphate. Other amino acids commonly packaged for athletes that are equally useless include tryptophan, ornithine, arginine, lysine, and the branched-chain amino acids.

glycogen (GLY-co-gen): an animal polysaccharide composed of glucose; manufactured and stored in the liver and muscles as a storage form of glucose. Glycogen is not a significant food source of carbohydrate and is not counted as one of the complex carbohydrates in foods.

glycolysis (gligh-COLL-ih-sis): the metabolic breakdown of glucose to pyruvate. Glycolysis does not require oxygen (anaerobic).

goblet cells: cells of the GI tract (and lungs) that secrete mucus.

goiter (GOY-ter): an enlargement of the thyroid gland due to an iodine deficiency, malfunction of the gland, or overconsumption of a goitrogen. Goiter caused by iodine deficiency is **simple goiter.**

goitrogen (GOY-troh-jen): a substance that enlarges the thyroid gland and causes **toxic goiter.** Goitrogens occur naturally in such foods as cabbage, kale, brussels sprouts, cauliflower, broccoli, and kohlrabi.

Golgi (GOAL-gee) **apparatus:** a set of membranes within the cell where secretory materials are packaged for export.

good source of: the product provides between 10 and 19% of the Daily Value for a given nutrient per serving.

granulated sugar: crystalline sucrose; 99.9% pure.

growth hormone (GH): a hormone secreted by the pituitary that regulates the cell division and protein synthesis needed for normal growth. The release of GH is mediated by **GH-releasing hormone (GHRH).**

growth hormone releasers: herbs or pills that supposedly regulate hormones; falsely promoted as enhancing athletic performance.

guarana: a reddish berry found in Brazil's Amazon valley that is used as an ingredient in carbonated sodas and taken in powder or tablet form. Guarana is marketed as an ergogenic aid to enhance speed and endurance, an aphrodisiac, a "cardiac tonic," an "intestinal disinfectant," and a smart drug that supposedly improves memory and concentration and wards off senility. Because guarana contains seven times as much caffeine as its relative the coffee bean, there are concerns that high doses can stress the heart and cause panic attacks.

H

hard water: water with a high calcium and magnesium content.

hazard: a source of danger; used to refer to circumstances in which harm is possible under normal conditions of use.

Hazard Analysis Critical Control Points (HACCP): a systematic plan to identify and correct potential microbial hazards in the manufacturing, distribution, and commercial use of food products; commonly referred to as "HASS-ip."

HDL (high-density lipoprotein): the type of lipoprotein that transports cholesterol back to the liver from the cells; composed primarily of protein.

health claims: statements that characterize the relationship between a nutrient or other substance in a food and a disease or health-related condition.

health history: an account of a client's current and past health status and disease risks.

healthy: a food that is low in fat, saturated fat, cholesterol, and sodium and that contains at least 10% of the Daily Values for vitamin A, vitamin C, iron, calcium, protein, or fiber.

Healthy Eating Index: a measure developed by the USDA for assessing how well a diet conforms to the recommendations of the Food Guide Pyramid and the *Dietary Guidelines for Americans.*

Healthy People: a national public health initiative under the jurisdiction of the U.S. Department of Health and Human Services (DHHS) that identifies the most significant preventable threats to health and focuses efforts toward eliminating them.

heart attack: sudden tissue death caused by blockages of vessels that feed the heart muscle; also called **myocardial** (my-oh-KAR-dee-al) **infarction** (in-FARK-shun) or **cardiac arrest.**

heartburn: a burning sensation in the chest area caused by backflow of stomach acid into the esophagus.

heat stroke: a dangerous accumulation of body heat with accompanying loss of body fluid.

heavy metal: any of a number of mineral ions such as mercury and lead, so called because they are of relatively high atomic weight. Many heavy metals are poisonous.

Heimlich (HIME-lick) **maneuver (abdominal thrust maneuver):** a technique for dislodging an object from the trachea of a choking person (see Figure H3-2 on p. 95); named for the physician who developed it.

hematocrit (hee-MAT-oh-krit): measurement of the volume of the red blood cells packed by centrifuge in a given volume of blood.

heme (HEEM): the iron-holding part of the hemoglobin and myoglobin proteins. About 40% of the iron in meat, fish, and poultry is bound into heme; the other 60% is **nonheme** iron.

hemochromatosis (HE-moh-KRO-ma-toe-sis): a hereditary defect in iron absorption characterized by deposits of iron-containing pigment in many tissues, with tissue damage.

hemoglobin (HE-moh-GLOW-bin): the globular protein of the red blood cells that carries oxygen from the lungs to the cells throughout the body.

hemolytic (HE-moh-LIT-ick) **anemia:** the condition of having too few red blood cells as a result of erythrocyte hemolysis.

hemophilia (HE-moh-FEEL-ee-ah): a hereditary disease in which the blood is unable to clot because it lacks the ability to synthesize certain clotting factors. Hemophilia is caused by a genetic defect and has no relation to vitamin K.

hemorrhagic (hem-oh-RAJ-ik) **disease:** a disease characterized by excessive bleeding.

hemorrhoids (HEM-oh-royds): painful swelling of the veins surrounding the rectum.

hemosiderin (heem-oh-SID-er-in): an iron storage protein primarily made in times of iron overload.

hemosiderosis (HE-moh-sid-er-OH-sis): a condition characterized by the deposition of hemosiderin in the liver and other tissues.

hepatic vein: the vein that collects blood from the liver capillaries and returns it to the heart.

herbal (ERB-al) **medicine:** the use of plants to treat disease or improve health; also known as *botanical medicine* or *phytotherapy.*

herbal steroids or **plant sterols:** curious mixtures of herbs, "adaptogens," and "aphrodisiacs" that supposedly enhance hormone activity. Products marketed as herbal steroids include astragalus, damiana, dong quai, fo ti teng, ginseng root, licorice root, palmetto berries, sarsaparilla, schizardra, unicorn root, yohimbe bark, and yucca.

hGH (human growth hormone): a hormone produced by the brain's pituitary gland that regulates normal growth and development; also called *somatotropin.* Some athletes misuse this hormone to increase their height and strength.

hiccups (HICK-ups): repeated cough-like sounds and jerks that are produced when an involuntary spasm of the diaphragm muscle sucks air down the windpipe; also spelled *hiccoughs.*

high: 20% or more of the Daily Value for a given nutrient per serving; synonyms include "rich in" or "excellent source."

high fiber: 5 g or more fiber per serving. A high-fiber claim made on a food that contains more than 3 g fat per serving and per 100 g of food must also declare total fat.

high potency: 100% or more of the Daily Value for the nutrient in a single supplement and for at least two-thirds of the nutrients in a multinutrient supplement.

high-fructose corn syrup (HFCS): a syrup made from cornstarch that has been treated with an enzyme that converts some of the glucose to the sweeter fructose; made especially for use in processed foods and beverages, where it is the predominant sweetener. With a chemical structure similar to sucrose, HFCS has a fructose content of 42, 55, or 90%, with glucose making up the remainder.

high-quality proteins: dietary proteins containing all the essential amino acids in relatively the same amounts that human beings require. They may also contain nonessential amino acids.

high-risk pregnancy: a pregnancy characterized by indicators that make it likely the birth will be surrounded by problems such as premature delivery, difficult birth, retarded growth, birth defects, and early infant death.

histamine (HISS-tah-mean or HISS-tah-men): a substance produced by cells of the immune system as part of a local immune reaction to an antigen; participates in causing inflammation.

HIV (human immunodeficiency virus): the virus that causes AIDS. The infection progresses to become an immune system disorder that leaves its victims defenseless against numerous infections.

HMB (beta-hydroxy-beta-methylbutyrate): a metabolite of the branched-chain amino acid leucine. Claims that HMB increases muscle mass and strength are based on the results of two studies from the lab that developed HMB as a supplement.

homeopathy (hoh-me-OP-ah-thee): a practice based on the theory that "like cures like," that is, that substances that cause symptoms in healthy people can cure those symptoms when given in very dilute amounts.

homeostasis (HOME-ee-oh-STAY-sis): the maintenance of constant internal conditions (such as blood chemistry, temperature, and blood pressure) by the body's control systems. A homeostatic system is constantly reacting to external forces so as to maintain limits set by the body's needs.

honey: sugar (mostly sucrose) formed from nectar gathered by bees. An enzyme splits the sucrose into glucose and fructose. Composition and flavor vary, but honey always contains a mixture of sucrose, fructose, and glucose.

hormones: chemical messengers. Hormones are secreted by a variety of glands in response to altered conditions in the body. Each hormone travels to one or more specific target tissues or organs, where it elicits a specific response to maintain homeostasis.

hormone-sensitive lipase: an enzyme inside adipose cells that responds to the body's need for fuel by hydrolyzing triglycerides so that their parts (glycerol and fatty acids) escape into the general circulation and thus become available to other cells for fuel. The signals to which this enzyme responds include epinephrine and glucagon, which oppose insulin (see Chapter 4).

hunger: the painful sensation caused by a lack of food that initiates food-seeking behavior.

hydrochloric acid: an acid composed of hydrogen and chloride atoms (HCl), normally produced by the gastric glands.

hydrogenation (HIGH-dro-gen-AY-shun or high-DROJ-eh-NAY-shun): a chemical process by which hydrogens are added to monounsaturated or polyunsaturated fatty acids to reduce the number of double bonds, making the fats more saturated (solid) and more resistant to oxidation (protecting against rancidity). Hydrogenation produces *trans*-fatty acids.

hydrolysis (high-DROL-ih-sis): a chemical reaction in which a major reactant is split into two products, with the addition of a hydrogen atom (H) to one and a hydroxyl group (OH) to the other (from water, H_2O). (The noun is **hydrolysis;** the verb is **hydrolyze.**)

hydrophilic (high-dro-FIL-ick): a term referring to water-loving, or water-soluble, substances.

hydrophobic (high-dro-FOE-bick): a term referring to water-fearing, or non-water-soluble, substances; also known as **lipophilic** (fat loving).

hydrotherapy: the use of water (in whirlpools, as douches, or packed as ice, for example) to promote relaxation and healing.

hydroxyapatite (high-drox-ee-APP-ah-tite): crystals made of calcium and phosphorus.

hyperactivity: inattentive and impulsive behavior that is more frequent and severe than is typical of others a similar age; professionally called **attention-deficit/hyperactivity disorder (ADHD).**

hypertension: higher-than-normal blood pressure. Hypertension that develops without an identifiable cause is known as **essential** or **primary hypertension;** hypertension that is caused by a specific disorder such as kidney disease is known as **secondary hypertension.**

hyperthermia: an above-normal body temperature.

hypertrophy (high-PER-tro-fee): growing larger; with regard to muscles, an increase in size (and strength) in response to use.

hypnotherapy: a technique that uses hypnosis and the power of suggestion to improve health behaviors, relieve pain, and heal.

hypoallergenic formulas: clinically tested infant formulas that do not provoke reactions in 90% of infants or children with confirmed cow's milk allergy. Like all infant formulas, hypoallergenic formulas must demonstrate nutritional suitability to support infant growth and development. Extensively hydrolyzed and free amino acid–based formulas are examples.

hypoglycemia (HIGH-po-gligh-SEE-me-ah): an abnormally low blood glucose concentration.

hyponatremia (HIGH-poe-na-TREE-mee-ah): a decreased concentration of sodium in the blood.

hypothalamus (high-po-THAL-ah-mus): a brain center that controls activities such as maintenance of water balance, regulation of body temperature, and control of appetite.

hypothermia: a below-normal body temperature.

hypothesis (hi-POTH-eh-sis): an unproven statement that tentatively explains the relationships between two or more variables.

I

ileocecal (ill-ee-oh-SEEK-ul) **valve:** the sphincter separating the small and large intestines.

ileum (ILL-ee-um): the last segment of the small intestine.

imagery: a technique that guides clients to achieve a desired physical, emotional, or spiritual state by visualizing themselves in that state.

imitation foods: foods that substitute for and resemble another food, but are nutritionally inferior to it with respect to vitamin, mineral, or protein content. If the substitute is not inferior to the food it resembles and if its name provides an accurate description of the product, it need not be labeled "imitation."

immune system: the body's natural defense against foreign materials that have penetrated the skin or mucous membranes.

immunity: the body's ability to defend itself against diseases; see Chapter 18.

immunoglobulins (IM-you-noh-GLOB-you-linz): proteins capable of acting as antibodies.

impaired glucose tolerance: blood glucose levels higher than normal but not high enough to be diagnosed as diabetes; sometimes called **prediabetes.**

implantation: the stage of development in which the zygote embeds itself in the wall of the uterus and begins to develop; occurs during the first two weeks after conception.

indigestion: incomplete or uncomfortable digestion, usually accompanied by pain, nausea, vomiting, heartburn, intestinal gas, or belching.

indirect or **incidental additives:** substances that can get into food as a result of contact with foods during growing, processing, packaging, storing, cooking, or some other stage before the foods are consumed; sometimes called **accidental additives.**

initiators: factors that cause mutations that give rise to cancer.

inorganic: not containing carbon or pertaining to living things.

inosine: an organic chemical that is falsely said to "activate cells, produce energy, and facilitate exercise," but actually has been shown to reduce the endurance of runners.

inositol (in-OSS-ih-tall): a nonessential nutrient that can be made in the body from glucose. Inositol is a part of cell membrane structures.

insoluble fibers: indigestible food components that do not dissolve in water. Examples include the tough, fibrous structures found in the strings of celery and the skins of corn kernels.

insulin (IN-suh-lin): a hormone secreted by special cells in the pancreas in response to (among other things) increased blood glucose concentration. The primary role of insulin is to control the transport of glucose from the bloodstream into the muscle and fat cells.

insulin resistance: the condition in which a normal amount of insulin produces a subnormal effect, resulting in an elevated fasting glucose; a metabolic consequence of obesity that precedes type 2 diabetes.

integrated pest management (IPM): management of pests using a combination of natural and biological controls and few or no pesticides.

integrative medicine: an approach that incorporates alternative therapies into the practice of conventional medicine (similar to complementary medicine, but a closer relationship is implied).

intensity: the degree of exertion while exercising (for example, the amount of weight lifted or the speed of running).

intentional food additives: additives intentionally added to foods, such as nutrients, colors, and preservatives.

intermittent claudication (klaw-dih-KAY-shun): severe calf pain caused by inadequate blood supply. It occurs when walking and subsides during rest.

interstitial (IN-ter-STISH-al) **fluid:** fluid between the cells (intercellular), usually high in sodium and chloride. Interstitial fluid is a large component of extracellular fluid.

intra-abdominal fat: fat stored within the abdominal cavity in association with the internal abdominal organs, as opposed to the fat stored directly under the skin (subcutaneous fat).

intracellular fluid: fluid within the cells, usually high in potassium and phosphate. Intracellular fluid accounts for approximately two-thirds of the body's water.

intrinsic factor: a glycoprotein (a protein with short polysaccharide chains attached) manufactured in the stomach that aids in the absorption of vitamin B_{12}.

invert sugar: a mixture of glucose and fructose formed by the hydrolysis of sucrose in a chemical process; sold only in liquid form and sweeter than sucrose. Invert sugar is used as a food additive to help preserve freshness and prevent shrinkage.

ions (EYE-uns): atoms or molecules that have gained or lost electrons and therefore have electrical charges. Examples include the positively charged sodium ion (Na^+) and the negatively charged chloride ion (Cl^-). For a closer look at ions, see Appendix B.

iridology: the study of changes in the iris of the eye and their relationships to disease.

iron deficiency: the state of having depleted iron stores.

iron-deficiency anemia: severe depletion of iron stores that results in low hemoglobin and small, pale red blood cells. Anemias that impair hemoglobin synthesis are **microcytic** (small cell).

iron overload: toxicity from excess iron.

irradiation: sterilizing a food by exposure to energy waves, similar to ultraviolet light and microwaves.

irritable bowel syndrome: an intestinal disorder of unknown cause. Symptoms include abdominal discomfort and cramping, diarrhea, constipation, or alternating diarrhea and constipation.

J

jejunum (je-JOON-um): the first two-fifths of the small intestine beyond the duodenum.

K

kcalorie (energy) control: management of food energy intake.

kcalorie-free: fewer than 5 kcal per serving.

keratin (KARE-uh-tin): a water-insoluble protein; the normal protein of hair and nails. Keratin-producing cells may replace mucus-producing cells in vitamin A deficiency.

keratinization: accumulation of keratin in a tissue; a sign of vitamin A deficiency.

keratomalacia (KARE-ah-toe-ma-LAY-shuh): softening of the cornea that leads to irreversible blindness; seen in severe vitamin A deficiency.

keto (KEY-toe) **acid:** an organic acid that contains a carbonyl group (C=O).

ketone (KEE-tone) **bodies:** the product of the incomplete breakdown of fat when glucose is not available in the cells.

ketosis (kee-TOE-sis): an undesirably high concentration of ketone bodies in the blood and urine.

kwashiorkor (kwash-ee-OR-core, kwash-ee-or-CORE): a form of PEM that results either from inadequate protein intake or, more commonly, from infections.

L

lactadherin (lack-tad-HAIR-in): a protein in breast milk that attacks diarrhea-causing viruses.

lactase: an enzyme that hydrolyzes lactose.

lactase deficiency: a lack of the enzyme required to digest the disaccharide lactose into its component monosaccharides (glucose and galactose).

lactation: production and secretion of breast milk for the purpose of nourishing an infant.

lactic acid: a 3-carbon compound produced from pyruvate during anaerobic metabolism.

lactoferrin (lack-toh-FERR-in): a protein in breast milk that binds iron and keeps it from supporting the growth of the infant's intestinal bacteria.

lacto-ovo-vegetarians: people who include milk, milk products, and eggs, but exclude meat, poultry, fish, and seafood from their diets.

lactose (LAK-tose): a disaccharide composed of glucose and galactose; commonly known as milk sugar.

lactose intolerance: a condition that results from inability to digest the milk sugar lactose; characterized by bloating, gas, abdominal discomfort, and diarrhea. Lactose intolerance differs from milk allergy, which is caused by an immune reaction to the protein in milk.

lactovegetarians: people who include milk and milk products, but exclude meat, poultry, fish, seafood, and eggs from their diets.

large intestine or **colon** (COAL-un): the lower portion of intestine that completes the digestive process. Its segments are the ascending colon, the transverse colon, the descending colon, and the sigmoid colon.

larynx: the voice box (see Figure H3-1 on p. 94).

laxatives: substances that loosen the bowels and thereby prevent or treat constipation.

LDL (low-density lipoprotein): the type of lipoprotein derived from very-low-density lipoproteins (VLDL) as VLDL triglycerides are removed and broken down; composed primarily of cholesterol.

lean: less than 10 g of fat, 4.5 g of saturated fat and *trans* fat combined, and 95 mg of cholesterol per serving and per 100 g of meat, poultry, and seafood.

lean body mass: the weight of the body minus the fat content.

lecithin (LESS-uh-thin): one of the phospholipids. Both nature and the food industry use lecithin as an emulsifier to combine water-soluble and fat-soluble ingredients that do not ordinarily mix, such as water and oil.

legumes (lay-GYOOMS, LEG-yooms): plants of the bean and pea family, rich in protein compared with other plant-derived foods.

leptin: a protein produced by fat cells under direction of the *ob* gene that decreases appetite and increases energy expenditure; sometimes called the **ob protein.**

less: at least 25% less of a given nutrient or kcalories than the comparison food (see individual nutrients); synonyms include "fewer" and "reduced."

less cholesterol: 25% or less cholesterol than the comparison food (reflecting a reduction of at least 20 mg per serving), and 2 g or less saturated fat and *trans* fat combined per serving.

less fat: 25% or less fat than the comparison food.

less saturated fat: 25% or less saturated fat and *trans* fat combined than the comparison food.

let-down reflex: the reflex that forces milk to the front of the breast when the infant begins to nurse.

leukemias (loo-KEE-mee-ahz): cancers that arise from the white blood cells.

levulose: an older name for fructose.

life expectancy: the average number of years lived by people in a given society.

life span: the maximum number of years of life attainable by a member of a species.

light or **lite:** one-third fewer kcalories than the comparison food; 50% or less of the fat or sodium than in the comparison food; any use of the term other than as defined must specify what it is referring to (for example, "light in color" or "light in texture").

lignans: phytochemicals present in flaxseed, but not in flax oil, that are converted to phytosterols by intestinal bacteria and are under study as possible anticancer agents.

limiting amino acid: the essential amino acid found in the shortest supply relative to the amounts needed for protein synthesis in the body. Four amino acids are most likely to be limiting:
• Lysine.
• Methionine.
• Threonine.
• Tryptophan.

linoleic (lin-oh-LAY-ick) **acid:** an essential fatty acid with 18 carbons and two double bonds.

linolenic (lin-oh-LEN-ick) **acid:** an essential fatty acid with 18 carbons and three double bonds.

lipase (LYE-pase): an enzyme that hydrolyzes lipids (fats).

lipids: a family of compounds that includes triglycerides, phospholipids, and sterols. Lipids are characterized by their insolubility in water. (Lipids also include the fat-soluble vitamins, described in Chapter 11.)

lipoprotein lipase (LPL): an enzyme that hydrolyzes triglycerides passing by in the bloodstream and directs their parts into the cells, where they can be metabolized for energy or reassembled for storage.

lipoproteins (LIP-oh-PRO-teenz): clusters of lipids associated with proteins that serve as transport vehicles for lipids in the lymph and blood.

liver: the organ that manufactures bile. (The liver's many other functions are described in Chapter 7.)

longevity: long duration of life.

low: an amount that would allow frequent consumption of a food without exceeding the Daily Value for the nutrient. A food that is naturally low in a nutrient may make such a claim, but only as it applies to all similar foods (for example, "fresh cauliflower, a low-sodium food"); synonyms include "little," "few," and "low source of."

low birthweight (LBW): a birthweight of 5½ lb (2500 g) or less; indicates probable poor health in the newborn and poor nutrition status in the mother during pregnancy, before pregnancy, or both. Normal birthweight for a full-term baby is 6½ to 8¾ lb (about 3000 to 4000 g).

low cholesterol: 20 mg or less cholesterol per serving and 2 g or less saturated fat and *trans* fat combined per serving.

low fat: 3 g or less fat per serving.

low kcalorie: 40 kcal or less per serving.

low saturated fat: 1 g or less saturated fat and less than 0.5 g of *trans* fat per serving.

low sodium: 140 mg or less per serving.

low-risk pregnancy: a pregnancy characterized by indicators that make a normal outcome likely.

lumen (LOO-men): the space within a vessel, such as the intestine.

lutein (LOO-teen): a plant pigment of yellow hue; a phytochemical believed to play roles in eye functioning and health.

luteinizing (LOO-tee-in-EYE-zing) **hormone (LH):** a hormone that stimulates ovulation and the development of the corpus luteum (the small tissue that develops from a ruptured ovarian follicle and secretes hormones); so called because the follicle turns yellow as it matures. In men, LH stimulates testosterone secretion. The release of LH is mediated by **luteinizing hormone–releasing hormone (LH–RH).**

lycopene (LYE-koh-peen): a pigment responsible for the red color of tomatoes and other red-hued vegetables; a phytochemical that may act as an antioxidant in the body.

lymph (limf): a clear yellowish fluid that is almost identical to blood except that it contains no red blood cells or platelets. Lymph from the GI tract transports fat and fat-soluble vitamins to the bloodstream via lymphatic vessels.

lymphatic (lim-FAT-ic) **system:** a loosely organized system of vessels and ducts that convey fluids toward the heart. The GI part of the lymphatic system carries the products of fat digestion into the bloodstream.

lymphocytes (LIM-foh-sites): white blood cells that participate in acquired immunity; B-cells and T-cells.

lymphomas (lim-FOH-mahz): cancers that arise from lymph tissue.

lysosomes (LYE-so-zomes): cellular organelles; membrane-enclosed sacs of degradative enzymes.

M

ma huang: an evergreen plant derivative that supposedly boosts energy and helps with weight control. Ma huang contains ephedrine, a cardiac stimulant, and has been associated with high blood pressure, rapid heart rate, nerve damage, muscle injury, psychosis, stroke, and memory loss.

macrobiotic diets: extremely restrictive diets limited to a few grains and vegetables; based on metaphysical beliefs and not nutrition. A macrobiotic diet might consist of brown rice, miso soup, and sea vegetables, for example.

macular (MACK-you-lar) **degeneration:** deterioration of the macular area of the eye that can lead to loss of central vision and eventual blindness. The **macula** is a small, oval, yellowish region in the center of the retina that provides the sharp, straight-ahead vision so critical to reading and driving.

magnesium: a cation within the body's cells, active in many enzyme systems.

major minerals: essential mineral nutrients found in the human body in amounts larger than 5 g; sometimes called **macrominerals.**

malnutrition: any condition caused by excess or deficient food energy or nutrient intake or by an imbalance of nutrients.

maltase: an enzyme that hydrolyzes maltose.

maltose (MAWL-tose): a disaccharide composed of two glucose units; sometimes known as malt sugar.

mammary glands: glands of the female breast that secrete milk.

maple sugar: a sugar (mostly sucrose) purified from the concentrated sap of the sugar maple tree.

marasmus (ma-RAZ-mus): a form of PEM that results from a severe deprivation, or impaired absorption, of energy, protein, vitamins, and minerals.

margin of safety: when speaking of food additives, a zone between the concentration normally used and that at which a hazard exists. For common table salt, for example, the margin of safety is ⅕ (five times the amount normally used would be hazardous).

massage therapy: a healing method in which the therapist manually kneads muscles to reduce tension, increase blood circulation, improve joint mobility, and promote healing of injuries.

matrix (MAY-tricks): the basic substance that gives form to a developing structure; in the body, the formative cells from which teeth and bones grow.

matter: anything that takes up space and has mass.

Meals on Wheels: a nutrition program that delivers food for the elderly to their homes.

meat replacements: products formulated to look and taste like meat, fish, or poultry; usually made of textured vegetable protein.

meditation: a self-directed technique of relaxing the body and calming the mind.

melanomas (MEL-ah-NOH-mahz): cancers that arise from pigmented skin cells.

MEOS or **microsomal** (my-krow-SO-mal) **ethanol-oxidizing system:** a system of enzymes in the liver that oxidize not only alcohol, but also several classes of drugs.

metabolic syndrome: a combination of risk factors—insulin resistance, hypertension, abnormal blood lipids, and abdominal obesity—that greatly increase a person's risk of developing coronary heart disease; also called **Syndrome X, insulin resistance syndrome,** or **dysmetabolic syndrome.**

metabolism: the sum total of all the chemical reactions that go on in living cells. Energy metabolism includes all the reactions by which the body obtains and expends the energy from food.

metalloenzymes (meh-tal-oh-EN-zimes): enzymes that contain one or more minerals as part of their structures.

metallothionein (meh-TAL-oh-THIGH-oh-neen): a sulfur-rich protein that avidly binds with and transports metals such as zinc.

metastasize (me-TAS-tah-size): to spread from one part of the body to another.

MFP factor: a factor associated with the digestion of meat, fish, and poultry that enhances nonheme iron absorption.

micelles (MY-cells): tiny spherical complexes of emulsified fat that arise during digestion; most contain bile salts and the products of lipid digestion, including fatty acids, monoglycerides, and cholesterol.

microvilli (MY-cro-VILL-ee, MY-cro-VILL-eye): tiny, hairlike projections on each cell of every villus that can trap nutrient particles and transport them into the cells; singular **microvillus.**

milk anemia: iron-deficiency anemia that develops when an excessive milk intake displaces iron-rich foods from the diet.

milliequivalents (mEq): the concentration of electrolytes in a volume of solution. Milliequivalents are a useful measure when considering ions because the number of charges reveals characteristics about the solution that are not evident when the concentration is expressed in terms of weight.

mineral oil: a purified liquid derived from petroleum and used to treat constipation.

mineral water: water from a spring or well that typically contains 250 to 500 parts per million (ppm) of minerals. Minerals give water a distinctive flavor. Many mineral waters are high in sodium.

mineralization: the process in which calcium, phosphorus, and other minerals crystallize on the collagen matrix of a growing bone, hardening the bone.

minerals: inorganic elements. Some minerals are essential nutrients required in small amounts by the body for health.

mitochondria (my-toh-KON-dree-uh); singular **mitochondrion:** the cellular organelles responsible for producing ATP aerobically; made of membranes (lipid and protein) with enzymes mounted on them.

moderate exercise: activity equivalent to the rate of exertion reached when walking at a speed of 4 miles per hour (15 minutes to walk one mile).

moderation (dietary): providing enough but not too much of a substance.

moderation: in relation to alcohol consumption, not more than two drinks a day for the average-sized man and not more than one drink a day for the average-sized woman.

molasses: the thick brown syrup produced during sugar refining. Molasses retains residual sugar and other by-products and a few minerals; blackstrap molasses contains significant amounts of calcium and iron.

molecule: two or more atoms of the same or different elements joined by chemical bonds. Examples are molecules of the element oxygen, composed of two oxygen atoms (O_2), and molecules of the compound water, composed of two hydrogen atoms and one oxygen atom (H_2O).

molybdenum (mo-LIB-duh-num): a trace element.

monoglycerides: molecules of glycerol with one fatty acid attached. A molecule of glycerol with two fatty acids attached is a **diglyceride.**

monosaccharides (mon-oh-SACK-uh-rides): carbohydrates of the general formula $C_nH_{2n}O_n$ that consist of a single ring. See Appendix C for the chemical structures of the monosaccharides.

monosodium glutamate (MSG): a sodium salt of the amino acid glutamic acid commonly used as a flavor enhancer. The FDA classifies MSG as a "generally recognized as safe" ingredient.

monounsaturated fatty acid: a fatty acid that lacks two hydrogen atoms and has one double bond between carbons—for example, oleic acid. A **monounsaturated fat** is composed of triglycerides in which most of the fatty acids are monounsaturated.

more: at least 10% more of the Daily Value for a given nutrient than the comparison food; synonyms include "added" and "extra."

mouth: the oral cavity containing the tongue and teeth.

MSG symptom complex: an acute, temporary intolerance reaction that may occur after the ingestion of the additive MSG (monosodium glutamate). Symptoms include burning sensations, chest and facial flushing and pain, and throbbing headaches.

mucous (MYOO-kus) **membranes:** the membranes, composed of mucus-secreting cells, that line the surfaces of body tissues.

mucus (MYOO-kus): a slippery substance secreted by cells of the GI lining (and other body linings) that protects the cells from exposure to digestive juices (and other destructive agents). The lining of the GI tract with its coat of mucus is a **mucous membrane.** (The noun is **mucus;** the adjective is **mucous.**)

muscle dysmorphia (dis-MORE-fee-ah): a psychiatric disorder characterized by a preoccupation with building body mass.

muscle endurance: the ability of a muscle to contract repeatedly without becoming exhausted.

muscle strength: the ability of muscles to work against resistance.

muscular dystrophy (DIS-tro-fee): a hereditary disease in which the muscles gradually weaken. Its most debilitating effects arise in the lungs.

myoglobin: the oxygen-holding protein of the muscle cells.

N

NAD (nicotinamide adenine dinucleotide): the main coenzyme form of the vitamin niacin. Its reduced form is NADH.

narcotic (nar-KOT-ic): a drug that dulls the senses, induces sleep, and becomes addictive with prolonged use.

natural water: water obtained from a spring or well that is certified to be safe and sanitary. The mineral content may not be changed, but the water may be treated in other ways such as with ozone or by filtration.

naturopathic (NAY-chur-oh-PATH-ick) **medicine:** a system that taps the natural healing forces within the body by integrating several practices, including traditional medicine, herbal medicine, clinical nutrition, homeopathy, acupuncture, East Asian medicine, hydrotherapy, and manipulative therapy.

neotame (NEE-oh-tame): an artificial sweetener composed of two amino acids (phenylalanine and aspartic acid); approved for use in the United States.

net protein utilization (NPU): a measure of protein quality assessed by measuring the amount of protein nitrogen that is retained from a given amount of protein nitrogen eaten.

neural tube defects: malformations of the brain, spinal cord, or both during embryonic development that often result in lifelong disability or death. The two main types of neural tube defects are **spina bifida** (literally, "split spine") and **anencephaly** ("no brain").

neurons: nerve cells; the structural and functional units of the nervous system. Neurons initiate and conduct nerve impulse transmissions.

neuropeptide Y: a chemical produced in the brain that stimulates appetite, diminishes energy expenditure, and increases fat storage.

neurotransmitters: chemicals that are released at the end of a nerve cell when a nerve impulse arrives there. They diffuse across the gap to the next cell and alter the membrane of that second cell to either inhibit or excite it.

niacin (NIGH-a-sin): a B vitamin. The coenzyme forms are **NAD (nicotinamide adenine dinucleotide)** and **NADP (the phosphate form of NAD).** Niacin can be eaten preformed or made in the body from its precursor, tryptophan, one of the amino acids.

niacin equivalents (NE): the amount of niacin present in food, including the niacin that can theoretically be made from its precursor, tryptophan, present in the food.

niacin flush: a temporary burning, tingling, and itching sensation that occurs when a person takes a large dose of nicotinic acid; often accompanied by a headache and reddened face, arms, and chest.

night blindness: slow recovery of vision after flashes of bright light at night or an inability to see in dim light; an early symptom of vitamin A deficiency.

nitrites (NYE-trites): salts added to food to prevent botulism. One example is sodium nitrite, which is used to preserve meats.

nitrogen balance: the amount of nitrogen consumed (N in) as compared with the amount of nitrogen excreted (N out) in a given period of time.

nitrosamines (nye-TROHS-uh-means): derivatives of nitrites that may be formed in the stomach when nitrites combine with amines. Nitrosamines are carcinogenic in animals.

nonessential amino acids: amino acids that the body can synthesize (see Table 6-1 on p. 182).

nonnutrients: compounds in foods that do not fit within the six classes of nutrients.

nonpoint water pollution: water pollution caused by runoff from all over an area rather than from discrete "point" sources. An example is the pollution caused by runoff from agricultural fields.

nucleus: a major membrane-enclosed body within every cell, which contains the cell's genetic material, DNA, embedded in chromosomes.

nursing bottle tooth decay: extensive tooth decay due to prolonged tooth contact with formula, milk, fruit juice, or other carbohydrate-rich liquid offered to an infant in a bottle.

nutrient additives: vitamins and minerals added to improve the nutritive value of foods.

nutrient density: a measure of the nutrients a food provides relative to the energy it provides. The more nutrients and the fewer kcalories, the higher the nutrient density.

nutrients: chemical substances obtained from food and used in the body to provide energy, structural materials, and regulating agents to support growth, maintenance, and repair of the body's tissues. Nutrients may also reduce the risks of some diseases.

nutrition: the science of foods and the nutrients and other substances they contain, and of their actions within the body (including ingestion, digestion, absorption, transport, metabolism, and excretion). A broader definition includes the social, economic, cultural, and psychological implications of food and eating.

nutrition assessment: a comprehensive analysis of a person's nutrition status that uses health, socioeconomic, drug, and diet histories; anthropometric measurements; physical examinations; and laboratory tests.

nutrition screening: the use of preliminary nutrition assessment techniques to identify people who are malnourished or are at risk for malnutrition.

nutritional genomics: the science of how nutrients affect the activities of genes and how genes affect the activities of nutrients.

nutritive sweeteners: sweeteners that yield energy, including both sugars and sugar replacers.

O

octacosanol: an alcohol isolated from wheat germ; often falsely promoted as enhancing athletic performance.

oils: lipids that are liquid at room temperature (70°F or 25°C).

olestra: a synthetic fat made from sucrose and fatty acids that provides 0 kcalories per gram; also known as **sucrose polyester.**

omega: the last letter of the Greek alphabet (ω), used by chemists to refer to the position of the first double bond from the methyl (CH_3) end of a fatty acid.

omega-3 fatty acid: a polyunsaturated fatty acid in which the first double bond is three carbons away from the methyl (CH_3) end of the carbon chain.

omega-6 fatty acid: a polyunsaturated fatty acid in which the first double bond is six carbons from the methyl (CH_3) end of the carbon chain.

omnivores: people who have no formal restriction on the eating of any foods.

opportunistic infections: infections from microorganisms that normally do not cause disease in the general population but can cause great harm in people once their immune systems are compromised (as in HIV infection).

opsin (OP-sin): the protein portion of the visual pigment molecule.

oral rehydration therapy (ORT): the administration of a simple solution of sugar, salt, and water, taken by mouth, to treat dehydration caused by diarrhea.

organelles: subcellular structures such as ribosomes, mitochondria, and lysosomes.

organic: in agriculture, crops grown and processed according to USDA regulations defining the use of fertilizers, herbicides, insecticides, fungicides, preservatives, and other chemical ingredients.

organic: in chemistry, a substance or molecule containing carbon-carbon bonds or carbonhydrogen bonds. This definition excludes coal, diamonds, and a few carbon-containing compounds that contain only a single carbon and no hydrogen, such as carbon dioxide (CO_2), calcium carbonate ($CaCO_3$), magnesium carbonate ($MgCO_3$), and sodium cyanide ($NaCN$).

organic: on food labels, that at least 95% of the product's ingredients have been grown and processsed according to USDA regulations defining the use of fertilizers, herbicides, insecticides, fungicides, preservatives, and other chemical ingredients (see Chapter 19).

organic halogen: organic compounds containing one or more atoms of a halogen—fluorine, chlorine, iodine, or bromine.

orlistat (OR-leh-stat): a drug used in the treatment of obesity that inhibits the absorption of fat in the GI tract, thus limiting kcaloric intake.

ornithine: a nonessential amino acid falsely promoted as enhancing the secretion of human growth hormone, the breakdown of fat, and the development of muscle.

orthomolecular medicine: the use of large doses of vitamins to treat chronic disease.

oryzanol: a plant sterol that supposedly provides the same physical responses as anabolic steroids without the adverse side effects; also known as *ferulic acid, ferulate,* or *FRAC.*

osmosis: the movement of water across a membrane *toward* the side where the solutes are more concentrated.

osmotic pressure: the amount of pressure needed to prevent the movement of water across a membrane.

osteoarthritis: a painful, degenerative disease of the joints that occurs when the cushioning cartilage in a joint deteriorates; joint structure is damaged, with loss of function; also called **degenerative arthritis.**

osteomalacia (OS-tee-oh-ma-LAY-shuh): a bone disease characterized by softening of the bones. Symptoms include bending of the spine and bowing of the legs. The disease occurs most often in adult women.

osteoporosis (OS-tee-oh-pore-OH-sis): a disease in which the bones become porous and fragile due to a loss of minerals; also called **adult bone loss.**

overnutrition: excess energy or nutrients.

overt (oh-VERT): out in the open and easy to observe.

overweight: body weight above some standard of acceptable weight that is usually defined in relation to height (such as BMI).

ovum (OH-vum): the female reproductive cell, capable of developing into a new organism upon fertilization; commonly referred to as an egg.

oxaloacetate (OKS-ah-low-AS-eh-tate): a carbohydrate intermediate of the TCA cycle.

oxidants (OK-see-dants): compounds (such as oxygen itself) that oxidize other compounds. Compounds that prevent oxidation are called *antioxidants,* whereas those that promote it are called *prooxidants.*

oxidation (OKS-ee-day-shun): the process of a substance combining with oxygen; oxidation reactions involve the loss of electrons.

oxidative stress: a condition in which the production of oxidants and free radicals exceeds the body's ability to handle them and prevent damage.

oxytocin (OCK-see-TOH-sin): a hormone that stimulates the mammary glands to eject milk during lactation and the uterus to contract during childbirth.

oyster shell: a product made from the powdered shells of oysters that is sold as a calcium supplement, but is not well absorbed by the digestive system.

ozone therapy: the use of ozone gas to enhance the body's immune system.

P

pancreas: a gland that secretes digestive enzymes and juices into the duodenum.

pancreatic (pank-ree-AT-ic) **juice:** the exocrine secretion of the pancreas, containing enzymes for the digestion of carbohydrate, fat, and protein as well as bicarbonate, a neutralizing agent. The juice flows from the pancreas into the small intestine through the pancreatic duct. (The pancreas also has an endocrine function, the secretion of insulin and other hormones.)

pangamic acid: also called vitamin B_{15} (but not a vitamin, nor even a specific compound—it can be anything with that label); falsely claimed to speed oxygen delivery.

pantothenic (PAN-toe-THEN-ick) **acid:** a B vitamin. The principal active form is part of coenzyme A, called "CoA" throughout Chapter 7.

parathormone (PAIR-ah-THOR-moan): a hormone from the parathyroid glands that regulates blood calcium by raising it when levels fall too low; also known as **parathyroid hormone.**

pasteurization: heat processing of food that inactivates some, but not all, microorganisms in the food; not a sterilization process. Bacteria that cause spoilage are still present.

pathogens (PATH-oh-jens): microorganisms capable of producing disease.

PBB (polybrominated biphenyl) and **PCB (polychlorinated biphenyl):** toxic organic compounds used in pesticides, paints, and flame retardants.

peak bone mass: the highest attainable bone size and density for an individual, developed during the first three decades of life.

peer review: a process in which a panel of scientists rigorously evaluates a research study to assure that the scientific method was followed.

pellagra (pell-AY-gra): the niacin-deficiency disease.

pepsin: a gastric enzyme that hydrolyzes protein. Pepsin is secreted in an inactive form, pepsinogen, which is activated by hydrochloric acid in the stomach.

peptic ulcer: an erosion in the mucous membrane of either the stomach (a gastric ulcer) or the duodenum (a duodenal ulcer).

peptidase: a digestive enzyme that hydrolyzes peptide bonds. *Tripeptidases* cleave tripeptides; *dipeptidases* cleave dipeptides. *Endopeptidases* cleave peptide bonds within the chain to create smaller fragments, whereas *exopeptidases* cleave bonds at the ends to release free amino acids.

peptide bond: a bond that connects the acid end of one amino acid with the amino end of another, forming a link in a protein chain.

percent fat-free: may be used only if the product meets the definition of *low fat* or *fat-free* and must reflect the amount of fat in 100 g (for example, a food that contains 2.5 g of fat per 50 g can claim to be "95 percent fat free").

peripheral (puh-RIFF-er-ul) **nervous system:** the peripheral (outermost) part of the nervous system; the vast complex of wiring that extends from the central nervous system to the body's outermost areas. It contains both somatic and autonomic components.

peristalsis (per-ih-STALL-sis): wavelike muscular contractions of the GI tract that push its contents along.

pernicious (per-NISH-us) **anemia:** a blood disorder that reflects a vitamin B_{12} deficiency caused by lack of intrinsic factor and characterized by abnormally large and immature red blood cells. Other symptoms include muscle weakness and irreversible neurological damage.

persistence: stubborn or enduring continuance; with respect to food contaminants, the quality of persisting, rather than breaking down, in the bodies of animals and human beings.

pesticides: chemicals used to control insects, weeds, fungi, and other pests on plants, vegetables, fruits, and animals. Used broadly, the term includes herbicides (to kill weeds), insecticides (to kill insects), and fungicides (to kill fungi).

pH: the unit of measure expressing a substance's acidity or alkalinity. The lower the pH, the higher the H^+ ion concentration and the stronger the acid. A pH above 7 is alkaline, or base (a solution in which OH^- ions predominate).

phagocytes (FAG-oh-sites): white blood cells (neutrophils and macrophages) that have the ability to ingest and destroy foreign substances.

phagocytosis (FAG-oh-sigh-TOH-sis): the process by which phagocytes engulf and destroy foreign materials.

pharynx (FAIR-inks): the passageway leading from the nose and mouth to the larynx and esophagus, respectively.

phosphate pills: a product demonstrated to increase the levels of a metabolically important phosphate compound (diphosphoglycerate) in red blood cells and the potential of the cells to deliver oxygen to the body's muscle cells. However, it does not extend endurance or increase efficiency of aerobic metabolism and may cause calcium losses from the bones if taken in excess.

phospholipid (FOS-foe-LIP-id): a compound similar to a triglyceride but having a phosphate group (a phosphorus-containing salt) and choline (or another nitrogen-containing compound) in place of one of the fatty acids.

phosphorus: a major mineral found mostly in the body's bones and teeth.

photosynthesis: the process by which green plants use the sun's energy to make carbohydrates from carbon dioxide and water.

physical activity: bodily movement produced by muscle contractions that substantially increase energy expenditure.

physiological age: a person's age as estimated from her or his body's health and probable life expectancy.

phytic (FYE-tick) **acid:** a nonnutrient component of plant seeds; also called **phytate** (FYE-tate). Phytic acid occurs in the husks of grains, legumes, and seeds and is capable of binding minerals such as zinc, iron, calcium, magnesium, and copper in insoluble complexes in the intestine, which the body excretes unused.

phytochemicals (FIE-toe-KEM-ih-cals): nonnutrient compounds found in plant-derived foods that have biological activity in the body.

phytoestrogens: plant-derived compounds that have structural and functional similarities to human estrogen. Phytoestrogens include genistein, daidzein, and glycitein.

phytosterols: plant-derived compounds that have structural similarities to cholesterol and lower blood cholesterol by competing with cholesterol for absorption. Phytosterols include sterol esters and stanol esters.

pica (PIE-ka): a craving for nonfood substances. Also known as **geophagia** (gee-oh-FAY-gee-uh) when referring to clay eating and **pagophagia** (pag-oh-FAY-gee-uh) when referring to ice craving.

pigment: a molecule capable of absorbing certain wavelengths of light so that it reflects only those that we perceive as a certain color.

placebo (pla-see-bo): an inert, harmless medication given to provide comfort and hope; a sham treatment used in controlled research studies.

placebo effect: the result of expectations in the effectiveness of a medicine, even medicine without pharmaceutical effects.

placenta (plah-SEN-tuh): the organ that develops inside the uterus early in pregnancy, through which the fetus receives nutrients and oxygen and returns carbon dioxide and other waste products to be excreted.

plant-pesticides: pesticides made by the plants themselves.

plaque (PLACK): an accumulation of fatty deposits, smooth muscle cells, and fibrous connective tissue that develops in the artery walls in atherosclerosis. Plaque associated with atherosclerosis is known as **atheromatous** (ATH-er-OH-ma-tus) **plaque.**

platelets: tiny, disc-shaped bodies in the blood, important in blood clot formation.

point of unsaturation: the double bond of a fatty acid, where hydrogen atoms can easily be added to the structure.

polypeptide: many (ten or more) amino acids bonded together.

polysaccharides: compounds composed of many monosaccharides linked together. An intermediate string of three to ten monosaccharides is an **oligosaccharide.**

polyunsaturated fatty acid (PUFA): a fatty acid that lacks four or more hydrogen atoms and has two or more double bonds between carbons—for example, linoleic acid (two double bonds) and linolenic acid (three double bonds). A **polyunsaturated fat** is composed of triglycerides in which most of the fatty acids are polyunsaturated.

portal vein: the vein that collects blood from the GI tract and conducts it to capillaries in the liver.

post term (infant): an infant born after the 42nd week of pregnancy.

postpartum amenorrhea: the normal temporary absence of menstrual periods immediately following childbirth.

potassium: the principal cation within the body's cells; critical to the maintenance of fluid balance, nerve impulse transmissions, and muscle contractions.

precursors: substances that precede others; with regard to vitamins, compounds that can be converted into active vitamins; also known as **provitamins.**

preeclampsia (PRE-ee-KLAMP-see-ah): a condition characterized by hypertension, fluid retention, and protein in the urine; formerly known as *pregnancy-induced hypertension.*

preformed vitamin A: dietary vitamin A in its active form.

prehypertension: slightly higher-than-normal blood pressure, but not as high as hypertension.

prenatal alcohol exposure: subjecting a fetus to a pattern of excessive alcohol intake characterized by substantial regular use or heavy episodic drinking.

preservatives: antimicrobial agents, antioxidants, and other additives that retard spoilage or maintain desired qualities, such as softness in baked goods.

pressure ulcers: damage to the skin and underlying tissues as a result of compression and poor circulation; commonly seen in people who are bedridden or chairbound.

preterm (infant): an infant born prior to the 38th week of pregnancy; also called a **premature infant.** A **term** infant is born between the 38th and 42nd week of pregnancy.

primary deficiency: a nutrient deficiency caused by inadequate dietary intake of a nutrient.

probiotics: microbial food ingredients that are beneficial to health. Nondigestible food ingredients that encourage the growth of favorable bacteria are called **prebiotics.**

processed foods: foods that have been treated to change their physical, chemical, microbiological, or sensory properties.

progesterone: the hormone of gestation (pregnancy).

progressive overload principle: the training principle that a body system, in order to improve, must be worked at frequencies, durations, or intensities that gradually increase physical demands.

prolactin (pro-LAK-tin): a hormone secreted from the anterior pituitary gland that acts on the mammary glands to promote (*pro*) the production of milk (*lacto*). The release of prolactin is mediated by **prolactin-inhibiting hormone (PIH).**

promoters: factors that favor the development of cancer once it has started.

proof: a way of stating the percentage of alcohol in distilled liquor. Liquor that is 100 proof is 50% alcohol; 90 proof is 45%, and so forth.

prooxidants: substances that significantly induce oxidative stress.

protease (PRO-tee-ace): an enzyme that hydrolyzes protein.

protein digestibility: a measure of the amount of amino acids absorbed from a given protein intake.

protein digestibility–corrected amino acid score (PDCAAS): a measure of protein quality assessed by comparing the amino acid score of a food protein with the amino acid requirements of preschool-age children and then correcting for the true digestibility of the protein.

protein turnover: the degradation and synthesis of protein.

protein efficiency ratio (PER): a measure of protein quality assessed by determining how well a given protein supports weight gain in growing rats; used to establish the protein quality for infant formulas and baby foods.

protein-energy malnutrition (PEM), also called **protein-kcalorie malnutrition (PCM):** a deficiency of protein, energy, or both, including kwashiorkor, marasmus, and instances in which they overlap.

proteins: compounds composed of carbon, hydrogen, oxygen, and nitrogen atoms, arranged into amino acids linked in a chain. Some amino acids also contain sulfur atoms.

protein-sparing action: the action of carbohydrate (and fat) in providing energy that allows protein to be used for other purposes.

puberty: the period in life in which a person becomes physically capable of reproduction.

public water: water from a municipal or county water system that has been treated and disinfected.

purified water: water that has been treated by distillation or other physical or chemical processes that remove dissolved solids. Because purified water contains no minerals or contaminants, it is useful for medical and research purposes.

pyloric (pie-LORE-ic) **sphincter:** the circular muscle that separates the stomach from the small intestine and regulates the flow of partially digested food into the small intestine; also called *pylorus* or *pyloric valve.*

pyruvate (PIE-roo-vate): a 3-carbon compound that plays a key role in energy metabolism.

Q

qi gong (chée GUNG): a Chinese system that combines movement, meditation, and breathing techniques to enhance the flow of qi (vital energy) in the body.

quality of life: a person's perceived physical and mental well-being.

R

randomization (ran-dom-ih-zay-shun): a process of choosing the members of the experimental and control groups without bias.

raw sugar: the first crop of crystals harvested during sugar processing. Raw sugar cannot be sold in the United States because it contains too much filth (dirt, insect fragments, and the like). Sugar sold as "raw sugar" domestically has actually gone through over half of the refining steps.

Recommended Dietary Allowance (RDA): the average daily amount of a nutrient considered adequate to meet the known nutrient needs of practically all healthy people; a goal for dietary intake by individuals.

rectum: the muscular terminal part of the intestine, extending from the sigmoid colon to the anus.

reduced kcalorie: at least 25% fewer kcalories per serving than the comparison food.

reference protein: a standard against which to measure the quality of other proteins.

refined: the process by which the coarse parts of a food are removed. When wheat is refined into flour, the bran, germ, and husk are removed, leaving only the endosperm.

reflux: a backward flow.

relaxin: the hormone of late pregnancy.

remodeling: the dismantling and re-formation of a structure, as in the case of bone.

renin (REN-in): an enzyme from the kidneys that activates angiotensin.

rennin: an enzyme that coagulates milk; found in the gastric juice of cows, but not human beings.

replication (REP-lee-KAY-shun): repeating an experiment and getting the same results. The skeptical scientist, on hearing of a new, exciting finding, will ask, "Has it been replicated yet?" If it hasn't, the scientist will withhold judgment regarding the finding's validity.

requirement: the lowest continuing intake of a nutrient that will maintain a specified criterion of adequacy.

residues: whatever remains. In the case of pesticides, those amounts that remain on or in foods when people buy and use them.

resistant starches: starches that escape digestion and absorption in the small intestine of healthy people.

resting metabolic rate (RMR): similar to the basal metabolic rate (BMR), a measure of the energy use of a person at rest in a comfortable setting, but with less stringent criteria for recent food intake and physical activity. Consequently, the RMR is slightly higher than the BMR.

retina (RET-in-uh): the layer of light-sensitive nerve cells lining the back of the inside of the eye; consists of rods and cones.

retinoids (RET-ih-noyds): chemically related compounds with biological activity similar to that of retinol; metabolites of retinol.

retinol activity equivalents (RAE): a measure of vitamin A activity; the amount of retinol that the body will derive from a food containing preformed retinol or its precursor beta-carotene.

retinol-binding protein (RBP): the specific protein responsible for transporting retinol.

rheumatoid (ROO-ma-toyd) **arthritis:** a disease of the immune system involving painful inflammation of the joints and related structures.

rhodopsin (ro-DOP-sin): a light-sensitive pigment of the retina; contains the retinal form of vitamin A and the protein opsin.

riboflavin (RYE-boh-flay-vin): a B vitamin. The coenzyme forms are **FMN (flavin mononucleotide)** and **FAD (flavin adenine dinucleotide).**

ribose: a 5-carbon sugar falsely promoted as improving the regeneration of ATP and thereby the speed of recovery after high-power exercise.

ribosomes (RYE-boh-zomes): protein-making organelles in cells; composed of RNA and protein.

rickets: the vitamin D–deficiency disease in children characterized by inadequate mineralization of bone (manifested in bowed legs or knock-knees,

outward-bowed chest, and knobs on ribs). A rare type of rickets, not caused by vitamin D deficiency, is known as *vitamin D–refractory rickets.*

risk: a measure of the probability and severity of harm.

risk factor: a condition or behavior associated with an elevated frequency of a disease but not proved to be causal. Leading risk factors for chronic diseases include obesity, cigarette smoking, high blood pressure, high blood cholesterol, physical inactivity, and a diet high in saturated fats and low in vegetables, fruits, and whole grains.

RNA (ribonucleic acid): the genetic material of cells necessary for protein synthesis; falsely promoted as enhancing athletic performance.

royal jelly: the substance produced by worker bees and fed to the queen bee; falsely promoted as increasing strength and enhancing performance.

S

saccharin (SAK-ah-ren): an artificial sweetener that has been approved for use in the United States. In Canada, approval for use in foods and beverages is pending; currently available only in pharmacies and only as a tabletop sweetener, not as an additive.

safety: the condition of being free from harm or danger.

saliva: the secretion of the salivary glands. Its principal enzyme begins carbohydrate digestion.

salivary glands: exocrine glands that secrete saliva into the mouth.

salt: a compound composed of a positive ion other than H^+ and a negative ion other than OH^-. An example is sodium chloride ($Na^+ Cl^-$).

salt sensitivity: a characteristic of individuals who respond to a high salt intake with an increase in blood pressure.

sarcomas (sar-KOH-mahz): cancers that arise from muscle, bone, or connective tissues.

sarcopenia (SAR-koh-PEE-nee-ah): loss of skeletal muscle mass, strength, and quality.

satiating: having the power to suppress hunger and inhibit eating.

satiation (say-she-AY-shun): the feeling of satisfaction and fullness that occurs during a meal and halts eating. Satiation determines how much food is consumed during a meal.

satiety (sah-TIE-eh-tee): the feeling of fullness and satisfaction that occurs after a meal and inhibits eating until the next meal. Satiety determines how much time passes between meals.

saturated fat-free: less than 0.5 g of saturated fat and 0.5 g of *trans* fat per serving.

saturated fatty acid: a fatty acid carrying the maximum possible number of hydrogen atoms— for example, stearic acid. A **saturated fat** is composed of triglycerides in which most of the fatty acids are saturated.

scurvy: the vitamin C–deficiency disease.

secondary deficiency: a nutrient deficiency caused by something other than an inadequate intake such as a disease condition or drug interaction that reduces absorption, accelerates use, hastens excretion, or destroys the nutrient.

secretin (see-CREET-in): a hormone produced by cells in the duodenum wall. Target organ: the pancreas. Response: secretion of bicarbonate-rich pancreatic juice.

sedentary: physically inactive (literally, "sitting down a lot").

segmentation (SEG-men-TAY-shun): a periodic squeezing or partitioning of the intestine at intervals along its length by its circular muscles.

selenium (se-LEEN-ee-um): a trace element.

senile dementia: the loss of brain function beyond the normal loss of physical adeptness and memory that occurs with aging.

serotonin (SER-oh-TONE-in): a neurotransmitter important in sleep regulation, appetite control, and sensory perception among other roles. Serotonin is synthesized in the body from the amino acid tryptophan with the help of vitamin B_6.

set point: the point at which controls are set (for example, on a thermostat). The set-point theory that relates to body weight proposes that the body tends to maintain a certain weight by means of its own internal controls.

sibutramine (sigh-BYOO-tra-mean): a drug used in the treatment of obesity that slows the reabsorption of serotonin in the brain, thus suppressing appetite and creating a feeling of fullness.

sickle-cell anemia: a hereditary form of anemia characterized by abnormal sickle- or crescent-shaped red blood cells. Sickled cells interfere with oxygen transport and blood flow. Symptoms are precipitated by dehydration and insufficient oxygen (as may occur at high altitudes) and include hemolytic anemia (red blood cells burst), fever, and severe pain in the joints and abdomen.

simple carbohydrates (sugars): monosaccharides and disaccharides.

small intestine: a 10-foot length of small-diameter intestine that is the major site of digestion of food and absorption of nutrients. Its segments are the duodenum, jejunum, and ileum.

socioeconomic history: a record of a person's social and economic background, including such factors as education, income, and ethnic identity.

sodium: the principal cation in the extracellular fluids of the body; critical to the maintenance of fluid balance, nerve impulse transmissions, and muscle contractions.

sodium bicarbonate: baking soda; an alkaline salt believed to neutralize blood lactic acid and thereby to reduce pain and enhance possible workload. "Soda loading" may cause intestinal bloating and diarrhea.

sodium-free and **salt-free:** less than 5 mg of sodium per serving.

soft water: water with a high sodium or potassium content.

solanine (SOH-lah-neen): a poisonous narcotic-like substance present in potato peels and sprouts.

soluble fibers: indigestible food components that dissolve in water to form a gel. An example is pectin from fruit, which is used to thicken jellies.

solutes (SOLL-yutes): the substances that are dissolved in a solution. The number of molecules in a given volume of fluid is the **solute concentration.**

somatic (so-MAT-ick) **nervous system:** the division of the nervous system that controls the voluntary muscles, as distinguished from the autonomic nervous system, which controls involuntary functions.

somatostatin (GHIH): a hormone that inhibits the release of growth hormone; the opposite of **somatotropin (GH).**

sperm: the male reproductive cell, capable of fertilizing an ovum.

sphincter (SFINK-ter): a circular muscle surrounding, and able to close, a body opening. Sphincters are found at specific points along the GI tract and regulate the flow of food particles.

spina (SPY-nah) **bifida** (BIFF-ih-dah): one of the most common types of neural tube defects; characterized by the incomplete closure of the spinal cord and its bony encasement.

spirulina: a kind of alga ("blue-green manna") that supposedly contains large amounts of protein and vitamin B_{12}, suppresses appetite, and improves athletic performance. It does none of these things and is potentially toxic.

sports anemia: a transient condition of low hemoglobin in the blood, associated with the early stages of sports training or other strenuous activity.

spring water: water originating from an underground spring or well. It may be bubbly (carbonated), or "flat" or "still," meaning not carbonated. Brand names such as "Spring Pure" do not necessarily mean that the water comes from a spring.

starches: plant polysaccharides composed of glucose.

sterile: free of microorganisms, such as bacteria.

sterols (STARE-ols or STEER-ols): compounds containing a four-carbon ring structure with any of a variety of side chains attached.

stevia (STEE-vee-ah): a South American shrub whose leaves are used as a sweetener; sold in the United States as a dietary supplement that provides sweetness without kcalories.

stomach: a muscular, elastic, saclike portion of the digestive tract that grinds and churns swallowed food, mixing it with acid and enzymes to form chyme.

stools: waste matter discharged from the colon; also called **feces** (FEE-seez).

stress: any threat to a person's well-being; a demand placed on the body to adapt.

stress fractures: bone damage or breaks caused by stress on bone surfaces during exercise.

stress response: the body's response to stress, mediated by both nerves and hormones.

stressors: environmental elements, physical or psychological, that cause stress.

stroke: an event in which the blood flow to a part of the brain is cut off; also called **cerebrovascular accident (CVA).**

structure-function claims: statements that characterize the relationship between a nutrient or other substance in a food and its role in the body.

subclinical deficiency: a deficiency in the early stages, before the outward signs have appeared.

subjects: the people or animals participating in a research project.

successful weight-loss maintenance: achieving a weight loss of at least 10 percent of initial body weight and maintaining the loss for at least one year.

succinate: a compound synthesized in the body and involved in the TCA cycle; falsely promoted as a metabolic enhancer.

sucralose (SUE-kra-lose): an artificial sweetener approved for use in the United States and Canada.

sucrase: an enzyme that hydrolyzes sucrose.

sucrose (SUE-krose): a disaccharide composed of glucose and fructose; commonly known as table sugar, beet sugar, or cane sugar. Sucrose also occurs in many fruits and some vegetables and grains.

sudden infant death syndrome (SIDS): the unexpected and unexplained death of an apparently well infant; the most common cause of death of infants between the second week and the end of the first year of life; also called *crib death.*

sugar replacers: sugarlike compounds that can be derived from fruits or commercially produced from dextrose; also called **sugar alcohols** or **polyols.** Sugar alcohols are absorbed more slowly than other sugars and metabolized differently in the human body; they are not readily utilized by ordinary mouth bacteria. Examples are **maltitol, mannitol, sorbitol, xylitol, isomalt,** and **lactitol.**

sugar-free: less than 0.5 g of sugar per serving.

sulfites: salts containing sulfur that are added to foods to prevent spoilage.

sulfur: a mineral present in the body as part of some proteins.

superoxide dismutase (SOD): an enzyme that protects cells from oxidation. When it is taken orally, the body digests and inactivates this protein; it is useless to athletes.

supplements: pills, capsules, tablets, liquids, or powders that contain vitamins, minerals, herbs, or amino acids; intended to increase dietary intake of these substances.

sushi: vinegar-flavored rice and seafood, typically wrapped in seaweed and stuffed with colorful vegetables. Some sushi is stuffed with raw fish; other varieties contain cooked seafood.

sustainable: able to continue indefinitely. In Chapter 20, the term means using resources at such a rate that the earth can keep on replacing them and producing pollutants at a rate with which the environment and human cleanup efforts can keep pace, so that no net accumulation of pollution occurs.

sustainable agriculture: agricultural practices that use individualized approaches appropriate to local conditions so as to minimize technological, fuel, and chemical inputs.

synergistic (SIN-er-JIST-ick): multiple factors operating together in such a way that their combined effects are greater than the sum of their individual effects.

T

tagatose (TAG-ah-tose): a monosaccharide structurally similar to fructose that is incompletely absorbed and thus provides only 1.5 kcalories per gram; approved for use as a "generally recognized as safe" ingredient.

TCA cycle or **tricarboxylic** (try-car-box-ILL-ick) **acid cycle:** a series of metabolic reactions that break down molecules of acetyl CoA to carbon dioxide and hydrogen atoms; also called the **Kreb's cycle** after the biochemist who elucidated its reactions.

T-cells: lymphocytes that attack antigens. T stands for the thymus gland, where the T-cells are stored for a while.

tempeh (TEM-pay): a fermented soybean food, rich in protein and fiber.

teratogenic (ter-AT-oh-jen-ik): causing abnormal fetal development and birth defects.

testosterone: a steroid hormone from the testicles, or testes. The steroids, as explained in Chapter

5, are chemically related to, and some are derived from, the lipid cholesterol.

textured vegetable protein: processed soybean protein used in vegetarian products such as soy burgers; see also *meat replacements.*

theory: a tentative explanation that integrates many and diverse findings to further the understanding of a defined topic.

thermic effect of food (TEF): an estimation of the energy required to process food (digest, absorb, transport, metabolize, and store ingested nutrients); also called the **specific dynamic effect (SDE)** of food or the **specific dynamic activity (SDA)** of food. The sum of the TEF and any increase in the metabolic rate due to overeating is known as **diet-induced thermogenesis (DIT).**

thermogenesis: the generation of heat; used in physiology and nutrition studies as an index of how much energy the body is expending.

thiamin (THIGH-ah-min): a B vitamin. The coenzyme form is **TPP (thiamin pyrophosphate).**

thirst: a conscious desire to drink.

thrombosis (throm-BOH-sis): the formation of a **thrombus** (THROM-bus), or a blood clot, that may obstruct a blood vessel, causing gradual tissue death.

thyroid-stimulating hormone (TSH): a hormone secreted by the pituitary that stimulates the thyroid gland to secrete its hormones—thyroxine and triiodothyronine. The release of TSH is mediated by **TSH-releasing hormone (TRH).**

tocopherol (tuh-KOFF-er-ol): a general term for several chemically related compounds, one of which has vitamin E activity (see Appendix C for chemical structures).

tofu (TOE-foo): a curd made from soybeans, rich in protein and often fortified with calcium; used in many Asian and vegetarian dishes in place of meat.

Tolerable Upper Intake Level (UL): the maximum daily amount of a nutrient that appears safe for most healthy people and beyond which there is an increased risk of adverse health effects.

tolerance level: the maximum amount of a residue permitted in a food when a pesticide is used according to label directions.

toxicity: the ability of a substance to harm living organisms. All substances are toxic if high enough concentrations are used.

trabecular (tra-BECK-you-lar) **bone:** the lacy inner structure of calcium crystals that supports the bone's structure and provides a calcium storage bank.

trace minerals: essential mineral nutrients found in the human body in amounts smaller than 5 g; sometimes called **microminerals.**

trachea (TRAKE-ee-uh): the windpipe; the passageway from the mouth and nose to the lungs.

training: practicing an activity regularly, which leads to conditioning. (Training is what you do; conditioning is what you get.)

trans **fat-free:** less than 0.5 g of *trans* fat and less than 0.5 g of saturated fat per serving.

*trans***-fatty acids:** fatty acids with hydrogens on opposite sides of the double bond.

transamination (TRANS-am-ih-NAY-shun): the transfer of an amino group from one amino acid to a keto acid, producing a new nonessential amino acid and a new keto acid.

transferrin (trans-FAIR-in): the iron transport protein.

transient hypertension of pregnancy: high blood pressure that develops in the second half of

pregnancy and resolves after childbirth, usually without affecting the outcome of the pregnancy.

transient ischemic (is-KEY-mik) **attack (TIA):** a temporary reduction in blood flow to the brain, which causes temporary symptoms that vary depending on the part of the brain affected. Common symptoms include light-headedness, visual disturbances, paralysis, staggering, numbness, and inability to swallow.

travelers' diarrhea: nausea, vomiting, and diarrhea caused by consuming food or water contaminated by any of several organisms, most commonly, *E. coli, Shigella, Campylobacter jejuni,* and *Salmonella.*

triglycerides (try-GLISS-er-rides): the chief form of fat in the diet and the major storage form of fat in the body; composed of a molecule of glycerol with three fatty acids attached; also called **triacylglycerols** (try-ay-seel-GLISS-er-ols).

tripeptide: three amino acids bonded together.

tumor: a new growth of tissue forming an abnormal mass with no function; also called a **neoplasm** (NEE-oh-plazm). Tumors that multiply out of control, threaten health, and require treatment are **malignant** (ma-LIG-nant). Tumors that stop growing without intervention or can be removed surgically and pose no threat to health are **benign** (bee-NINE).

turbinado (ter-bih-NOD-oh) **sugar:** sugar produced using the same refining process as white sugar, but without the bleaching and anti-caking treatment. Traces of molasses give turbinado its sandy color.

24-hour recall: a record of foods eaten by a person for one 24-hour period.

type 1 diabetes: the less common type of diabetes in which the person produces no insulin.

type 2 diabetes: the more common type of diabetes in which the fat cells resist insulin.

type I osteoporosis: osteoporosis characterized by rapid bone losses, primarily of trabecular bone.

type II osteoporosis: osteoporosis characterized by gradual losses of both trabecular and cortical bone.

U

ulcer: an erosion in the topmost, and sometimes underlying, layers of cells in an area. See also *peptic ulcer.*

ultrahigh temperature (UHT) treatment: sterilizing a food by brief exposure to temperatures above those normally used.

umbilical (um-BILL-ih-cul) **cord:** the ropelike structure through which the fetus's veins and arteries reach the placenta; the route of nourishment and oxygen to the fetus and the route of waste disposal from the fetus. The scar in the middle of the abdomen that marks the former attachment of the umbilical cord is the **umbilicus** (um-BILL-ih-cus), commonly known as the "belly button."

undernutrition: deficient energy or nutrients.

underweight: body weight below some standard of acceptable weight that is usually defined in relation to height (such as BMI).

unsaturated fatty acid: a fatty acid that lacks hydrogen atoms and has at least one double bond between carbons (includes monounsaturated and polyunsaturated fatty acids). An **unsaturated fat**

is composed of triglycerides in which most of the fatty acids are unsaturated.

unspecified eating disorders: eating disorders that do not meet the defined criteria for specific eating disorders.

urea (you-REE-uh): the principal nitrogen-excretion product of protein metabolism. Two ammonia fragments are combined with carbon dioxide to form urea.

USDA (U.S. Department of Agriculture): the federal agency responsible for enforcing standards for the wholesomeness and quality of meat, poultry, and eggs produced in the United States; conducting nutrition research; and educating the public about nutrition.

uterus (YOU-ter-us): the muscular organ within which the infant develops before birth.

V

validity (va-LID-ih-tee): having the quality of being founded on fact or evidence.

variables: factors that change. A variable may depend on another variable (for example, a child's height depends on his age), or it may be independent (for example, a child's height does not depend on the color of her eyes). Sometimes both variables correlate with a third variable (a child's height and eye color both depend on genetics).

variety (dietary): eating a wide selection of foods within and among the major food groups.

vasoconstrictor (VAS-oh-kon-STRIK-tor): a substance that constricts or narrows the blood vessels.

vegans (VEE-guns, VAY-guns, or VEJ-ans): people who exclude all animal-derived foods (including meat, poultry, fish, eggs, and dairy products) from their diets; also called **pure vegetarians, strict vegetarians,** or **total vegetarians.**

vegetarians: a general term used to describe people who exclude meat, poultry, fish, or other animal-derived foods from their diets.

veins (VANES): vessels that carry blood to the heart.

very low sodium: 35 mg or less per serving.

villi (VILL-ee, VILL-eye): fingerlike projections from the folds of the small intestine; singular **villus.**

viscous: a gel-like consistency.

vitamin A: all naturally occurring compounds with the biological activity of retinol (RET-ih-nol), the alcohol form of vitamin A.

vitamin A activity: a term referring to both the active forms of vitamin A and the precursor forms in foods without distinguishing between them.

vitamin B$_{12}$: a B vitamin characterized by the presence of cobalt (see Figure 13-12 on p. 458). The active forms of coenzyme B$_{12}$ are **methylcobalamin** and **deoxyadenosylcobalamin.**

vitamin B$_6$: a family of compounds—pyridoxal, pyridoxine, and pyridoxamine. The primary active coenzyme form is **PLP (pyridoxal phosphate).**

vitamins: organic, essential nutrients required in small amounts by the body for health.

VLDL (very-low-density lipoprotein): the type of lipoprotein made primarily by liver cells to transport lipids to various tissues in the body; composed primarily of triglycerides.

VO$_2$max: the maximum rate of oxygen consumption by an individual at sea level.

vomiting: expulsion of the contents of the stomach up through the esophagus to the mouth.

W

waist circumference: an anthropometric measurement used to assess a person's abdominal fat.

warm-up: 5 to 10 minutes of light activity, such as easy jogging or cycling, prior to a workout to prepare the body for more vigorous activity.

wasting syndrome: an involuntary loss of more than 10% of body weight, common in AIDS and cancer.

water balance: the balance between water intake and output (losses).

water intoxication: the rare condition in which body water contents are too high in all body fluid compartments.

wean: gradually replacing breast milk with infant formula or other foods appropriate to an infant's diet.

weight training (also called **resistance training**): the use of free weights or weight machines to provide resistance for developing muscle strength and endurance. A person's own body weight may also be used to provide resistance as when a person does push-ups, pull-ups, or abdominal crunches.

well water: water drawn from ground water by tapping into an aquifer.

Wernicke-Korsacoff (VER-nee-key KORE-sah-kof) **syndrome:** a neurological disorder typically associated with chronic alcoholism and caused by a deficiency of the B vitamin thiamin; also called *alcohol-related dementia.*

wheat germ oil: the oil from the wheat kernel; often falsely promoted as an energy aid.

whey protein: a by-product of cheese production; falsely promoted as increasing muscle mass. Whey is the watery part of milk that separates from the curds.

white sugar: pure sucrose or "table sugar," produced by dissolving, concentrating, and recrystallizing raw sugar.

WHO (World Health Organization): an international agency concerned with promoting health and eradicating disease.

whole grain: a grain milled in its entirety (all but the husk), not refined.

wine: an alcoholic beverage made by fermenting grape juice.

X

xanthophylls (ZAN-tho-fills): pigments found in plants; responsible for the color changes seen in autumn leaves.

xerophthalmia (zer-off-THAL-mee-uh): progressive blindness caused by severe vitamin A deficiency.

xerosis (zee-ROW-sis): abnormal drying of the skin and mucous membranes; a sign of vitamin A deficiency.

Y

yogurt: milk fermented by specific bacterial cultures.

Z

zygote (ZY-goat): the product of the union of ovum and sperm; so-called for the first two weeks after fertilization.

Credits

4 © Michael Newman/PhotoEdit; 5 © Ariel Skelley/CORBIS; 6 © SW Productions/Index Stock Imagery; 7 © Photodisc Collection/Getty Images; 8 © Felicia Martinez/Photo Edit; 8 © Felicia Martinez /PhotoEdit; 8 © Thomas Harm, Tom Peterson/Quest Photographic Inc.; 8 © Tony Freeman/Photo Edit; 10 © Matthew Farruggio (both); 11 © Ron Chapple/Taxi/Getty Images; 13 © L. V. Bergman and Associates, Inc.; 13 © R. Benali/Gamma; 13 USDA Agricultural Research Service; 13 © 2001 PhotoDisc, Inc.; 15 © Craig M. Moore; 16 © 1999 PhotoDisc, Inc.; 20 © Tom & Dee Ann McCarthy/CORBIS; 23 © Burke-Triolo Productions/FoodPix/Getty Images; 25 © Roy Morsch/CORBIS; 30 © USDA, Agricultural Research Service; 33 © Marilyn Herbert Photography; 40 © Polara Studios Inc.; 41 © Polara Studios Inc.; 43 © Matthew Farruggio; 44–45 © Polara Studios Inc. (all); 47 © Becky Luigart-Stayner/Corbis; 47 © PhotoDisc Inc.; 47 © PhotoDisc Inc.; 50 Matthew Farruggio (all); 51 Photo reprinted with permission of the Pillsbury Company, 2001; 52 © Thomas Harm/Tom Peterson/Quest Photographic Inc.; 53 © Geri Engberg Photography; 54 © 1998 Photo Disc Inc.; 54 © 1998 Photo Disc Inc.; 54 © Michael Newman/PhotoEdit; 54 © Felicia Martinez/PhotoEdit; 57 © Bob Daemmrich Photography; 66 © Michael S. Yamashita/CORBIS; 68–69 Reprinted from *The Journal of the American Dietetic Association,* 102(4): 483–489, © 2002, with permission from The American Dietetic Association; 74 © Joe Pellegrini/FoodPix/Getty Images; 84 © Michael S. Yamashita/Corbis; 85 © Bill Crew/Super Stock; 85 © Don W. Fawcett; 94 © Ronnie Kaufman/CORBIS; 98 © Polara Studios Inc.; 107 © Wartenberg/Picture Press/Corbis; 109 © Polara Studios, Inc.; 110 © Banana Stock/SuperStock; 115 © Brian Leatart/FoodPix/Getty Images; 120 © Polara Studies, Inc.; 121 Matthew Farruggio; 124 © Rita Maas/The Image Bank/Getty Images; 133 Funnette Division, Hoechst Celanese Corp.; 138 © Craig Moore; 147 © Polara Studios Inc.; 149 Matthew Farruggio; 157 © Jim Cummins/Taxi/Getty Images; 158 © Bob Thomas/Stone/Getty Images; 164 © Polara Studios, Inc. (all); 166 © Polara Studios Inc.; 168 © 1998 PhotoDisc, Inc.; 173 © Philip Salaverry/FoodPix/Getty Images; 174 Matthew Farruggio; 175 Matthew Farruggio; 175 © www.comstock.com; 189 © Dr. Stanley Flegler/Visuals Unlimited; 192 © IT International eStock Photography/PhotoQuest; 194 © Polara Studios Inc.; 196 © Polara Studios Inc.; 197 AP/Wide World Photos; 198 AP/Wide World Photos; 199 © Paul A. Souders/Corbis; 200 © World Health Organization, Geneva; 202 © Courtesy National Cattlemen's Beef Association; 208 © Polara Studios, Inc.; 220 © Chris Cole/The Image Bank/ Getty Images; 223 © Jim Cummins/Taxi/Getty Images; 234 © George Shelley/Masterfile; 240 © 1998 PhotoDisc, Inc.; 240 © Polara Studios, Inc.; 241 Matthew Farruggio; 254 © Banana Stock, Ltd/Picture Quest; 254 © Creatas/PictureQuest; 254 © Benelux Press/Index Stock Imagery; 254 © Creatas/PictureQuest; 255 © Owen Franken/CORBIS; 255 © Polara Studios Inc. (both); 257 © Bob Torrez/Stone/Getty Images; 261 © Rick Schaff; 261 © Underwood & Underwood/Corbis; 261 AP/Wide World Photos; 262 © Lori Adamski Peek/Stone/Getty Images; 266 © Fitness & Wellness, Boise, Idaho; 266 © David Young-Wolff/PhotoEdit; 266 © Geri Engberg; 266 Photo Courtesy of Life Measurement, Inc.; 266 © Photo Courtesy of Hologic, Inc.; 268 © Photo by Joe Sampson, Courtesy of Jennifer Portnick; 272 © Geri Engberg; 273 Matthew Farruggio; 275 1998 Photo Disc Inc.; 283 © Courtesy Amgen, Inc.; 288 © Anne Dowie; 294 Matthew Farruggio (all); 296 © Mike Chew/CORBIS; 310 © Steve Niedorf Photography/The Image Bank/Getty Images; 311 © Reuters NewMedia Inc./CORBIS; 315 © Michael Newman/PhotoEdit; 322 Polara Studios, Inc.; 326 © NMSB/Custom Medical Stock Photo; 326 © Polara Studios Inc.; 329 © Polara Studios Inc.; 330 © Dr. M. A. Ansary/Photo Researchers, Inc.; 331 © Polara Studios Inc.; 335 © Polara Studios Inc.; 340 © Polara Studios Inc.; 342 © Carolina Biological/Visuals Unlimited (both); 347 © Custom Medical Stock Photo; 347 © Science Photo Library/Photo Researchers, Inc.; 351 © L. V. Bergman & Associates Inc.; 351 © Dr. P. Marazzi/Photo Researchers, Inc.; 352 © PhotoDisc, Inc.; 352 © Polara Studios, Inc.; 359 © J. Share/Stone/Getty Images; 364 © Anne Dowie; 371 © David Farr/Image Smythe (all); 372 © Ken Greer/Visuals Unlimited; 373 © 2002 Massachusetts Medical Society; 375 © Polara Studios Inc.; 377 Biophoto Associates/Photo Researchers Inc.; 377 © Photo Courtesy of Dr. Normal Carvalho at Children's Healthcare of Atlanta; 377 © Francisco Cruz/Super Stock; 378 © Fotographia/Corbis; 381 © Craig M. Moore; 383 © Simon Fraser/Photo Researchers, Inc.; 384 © Polara Studios Inc.; 389 © Nick Clements/Taxi/Getty Images; 392 © Jeffry Myers/Stock, Boston/PictureQuest; 395 © Michael Pole/CORBIS ; 403 © Craig M. Moore (both); 404 © Norbert Schaefer/CORBIS; 408 © BSIP Agency/Index Stock Imagery; 409 Matthew Farruggio (all); 411 © Polara Studios Inc.; 415 © David Dempster from J Bone Miner Res,1986 (both); 417 © Thomas Harm, Tom Peterson/Quest Photographic Inc.; 428 © Photo Disc Inc.; 429 Courtesy of Gjon Mili; 429 © David Dempster from J Bone Miner Res, 1986 (both); 432 © Rob Lewine/ CORBIS; 441 © Benjamin F. Fink Jr./Brand X Pictures/Getty Images; 443 © Dr. Gladden Willis/Visuals Unlimited (both); 445 © Craig M. Moore; 447 © Polara Studios Inc.; 449 © H. Sanstead, University of Texas at Galveston; 449 © Polara Studios Inc.; 452 Bob Daemmrich/The Image Works; 452 © Craig M. Moore; 456 © Dr. P. Marrazi/Science Photo Library/Photo Researchers Inc.; 465 © John E. Kelly/FoodPix/Getty Images; 467

© Craig M. Moore; 468 (broccoli sprouts) © Courtesy of Brassica Protection Products; (apples, soy, wine/grapes, strawberries, tea, tomatoes, blueberries) © 2001 PhotoDisc; (garlic, oranges) © EyeWire, Inc.; (flaxseed) Courtesy of Flax Council of Canada; (spinach) Photodisc/Getty Images, (chocolate) Matthew Farruggio; 469 © Craig M. Moore; 474 © Zoran Milich/Masterfile; 475 © Lori Adamski Peek/Stone/Getty Images; 476 © Photo Disc Inc.; 476 © David Hanover Photography; 476 © David Hanover Photography; 477 Ryan McVay/PhotoDisc/Getty Images; 477 © David Madison/Stone/Getty Images; 480 © Bob Winsett/CORBIS; 481 © Simon Watson/The Image Bank/Getty Images; 482 © David Young-Wolff/PhotoEdit; 483 © Bonnie Kamin/PhotoEdit, Inc.; 483 David Leahy/Taxi/Getty Images; 485 © 2001 PhotoDisc, Inc.; 486 © Jim Cummins/CORBIS; 489 © Bob Thoma/Stone/Getty Images; 490 © Kevin R. Morris/CORBIS; 491 Rick Mariani Photography/StockFood; 494 © Polara Studios Inc. (all); 495 Matthew Farruggio (all); 499 © Ellen Stagg/Stone/Getty Images; 509 (photos 1, 2, 3) © Petit Format/Nestle/Photo Researchers, Inc.; 509 (photo 4) © Anthony M. Vanelli; 511 © Lennart Nilsson/Albert Bonniers Förlag AB, from *A Child is Born,* Dell Publishing Co. (both); 513 © Jirtle and Waterland; 514 © Mug Shots/CORBIS; 517 © 2000 Tracy Fraukel/Image Bank/Getty Images; 517 © Rick Gomez/CORBIS; 524 © Nik Kleinberg/Stock, Boston; 529 © Jose I. Pelaez, Inc./CORBIS; 531 © Ariel Skelley/CORBIS; 532 © Ariel Skelley/CORBIS; 533 © Photo Disc Inc.; 539 © Streissguth, A. P./Landesman-Dwyer, S., Martin, J. C., & Smith, D. W.); 540 © 1995 George Steinmetz; 541 © 1995 George Steinmetz; 541 Matthew Farruggio; 544 © Mark Scott/Taxi/Getty Images; 547 © 2001 PhotoDisc, Inc.; 549 © RubberBall Productions/Getty Images; 550 © E. H. Gill/Custom Medical Stock Photo; 552 © Polara Studios Inc.; 553 © Donna Day/Stone/Getty Images; 554 © SW Productions/Index Stock Imagery; 554 Anthony M. Vanelli (both); 557 © 2001 PhotoDisc, Inc.; 558 © Tony Freeman/PhotoEdit; 561 Polara Studios, Inc.; 564 © Pat Doyle/CORBIS; 566 Mary Kate Denny/PhotoEdit; 567 © PhotoEdit; 570 © Bob Daemmrich/Stock Boston/Picture Quest; 571 © Steve Dunwell/Index Stock Imagery; 572 Ariel Skelley/CORBIS; 581 © Ross Whitaker/The Image Bank/Getty Images; 582 © Courtesy of Zeneca Pharmaceutical Division, Cheshire, England (both); 584 © Tom & Dee Ann McCarthy/CORBIS; 590 © R. W. Jones/Corbis; 590 © Geoff Manasse/Photodisc/PictureQuest; 590 © IT Stock Free/PictureQuest; 590 © IT Stock Free/PictureQuest; 590 © Ron Chapple/Thinkstock/PictureQuest; 593 Courtesy of Dr. William Evans; 594 © Bob Thomas/Stone/Getty Images; 595 © Kwame Zikomo/SuperStock; 601 © Deborah Davis/PhotoEdit; 602 © Richard Pasley/Stock Boston/Picture Quest; 603 Myrleen Ferguson Cate/PhotoEdit; 603 © Lonnie Duka/Index Stock Imagery; 604 © Elena Rooraid/PhotoEdit; 608 © David Woods/CORBIS; 617 © Syracuse Newspapers/David Lassman/The ImageWorks; 619 ©1998 PhotoDisc Inc.; 629 © Jim Cummins/Taxi/Getty Images; 630 © Michael Keller/Index Stock Imagery; 631 © Stefan Hallberg/Index Stock Imagery/PictureQuest; 635 © Ann Dowie; 638 © Kent Meireis/The Image Works; 640 © Polara Studios Inc.; 642 © Bruce Ayres/Stone/Getty Images; 647 © Art Montes De Oca/Taxi/Getty Images; 649 © Sheila Terry/Photo Researchers, Inc.; 649 © James Worrell/Time/Getty Images (others); 650 © Darrell Gulin/Stone/Getty Images; 657 Digital Imagery © 2001 PhotoDisc Inc.; 659 © Polara Studios Inc.; 661 © 2002 PhotoDisc/Getty Images; 662 © EyeWire Inc.; 663 David Chasey/Photodisc Red/Getty Images; 672 © Angelo Cavalli/The Image Bank/Getty Images; 673 Photodisc Blue/Getty Images; 673 Ariel Skelley/CORBIS; 674 United States Department of Agriculture; 675 Polara Studios Inc.; 676 © Polara Studios Inc.; 677 © Polara Studios Inc.; 679 © 1998 PhotoDisc, Inc.; 681 © 1998 PhotoDisc, Inc.; 686 Tony Freeman/PhotoEdit; 686 Smithsonian photo by Antonio Mortaner; 688 Photo courtesy of David Garvin, USDA-ARS; 689 Korengold/ZUMA Press. © 2003 by Glen Korengold; 695 Joseph Sohm; ChromoSohm Inc./CORBIS; 696 AP/Wide World Photos; 698 AP/Wide World Photos; 699 Photo by Chris Hondros/Getty Images; 700 © Alexandra Avakian/Contact Press/PictureQuest; 702 © 1998 PhotoDisc, Inc.; 704 © Elsa Peterson/Stock, Boston; 705 NASA; 708 © Jim Foster/CORBIS; 713 © Paul Barton/CORBIS; A-1 © Jay Ahrend/FoodPix/Getty Images

Chapter Opening Photos:

2 © Lew Robertson/FoodPix/Getty Images; 38 © Brian Hagiwara/FoodPix/Getty Images; 72 © Burke/Triolo Productions/FoodPix/Getty Images; 102 © Mary Jan Cardenas/The Image Bank/Getty Images; 140 © Luc Hautecoeur/Stone/Getty Images; 180 © James Jackson/Stone/Getty Images; 214 © Mark Thomas/FoodPix/Getty Images; 250 © Charles Thatcher/Stone/Getty Images; 278 © Ken Scott/Stone/Getty Images; 320 © Stephen Wilkes/The Image Bank/Getty Images; 366 © Paul Poplis/FoodPix/Getty Images; 394 © Jack Andersen/FoodPix/Getty Images ; 436 © Brian Hagiwara/FoodPix/Getty Images; 472 © Michael Pohuski/FoodPix/Getty Images; 506 © Wally Eberhart/Botanica/Getty Images; 542 © Simon Watson/FoodPix/Getty Images; 586 © Judd Pilossof/FoodPix/Getty Images; 612 © Tif Hunter/Stone/Getty Images; 656 © Leigh Beisch/FoodPix/Getty Images; 692 © Nino Mascardi/The Image Bank/Getty Images

Index

vitamin A and, 370
vitamin D and, 375–376, 376, 377f
zinc deficiency, 449
See also Fetal development
Growth charts, 545f, E-8f to E-13f
Growth hormone (GH), 190t, **A-4,** A-6
Growth hormone releasers, **500**
GTF (glucose tolerance factor), **456**
Guarana, **500**
Guidelines for Healthy Eating (Canada), 42t
Gums (fiber), 109, **677,** 678

H

Habit and food choice, 4, 298–299
See also Lifestyle choices
Hair, 422, E-17t
Hard liquor, **241,** 246t
Hard water, 241, **398**
Hazard, **657**
See also Safety
Hazard Analysis Critical Control Points (HACCP), **661**
HCl. *See* Hydrochloric acid
HDL (high-density lipoprotein), **154**
blood levels, 159, 622t
composition of, 155f
function of, 154
heart disease risk, 154, 622t, 623, 625
LDL ratio to, 154, 155, 623
levels, improving, 626t, 627, 628, 629
in obese children, 582
size and density of, 155f
Head circumference, E-7
Health
overview, 24–25
body composition and, 260–268, 261f, 263f, 265f, 267f
food choices for, 5
"It's All About You" program, 61
risks. *See* Disease risk and prevention
status. *See* Nutrition assessment; Nutrition status
strategies for older adults, 601t
websites, 643–644
Health Canada website, 34
Health care professionals
in alternative health, 654
certified lactation consultants, **531**
dietetic technician, **31,** 33
dietitian career descriptions, 32–33, 33t
guidelines for medications, 611
nutritionists, **31,** 32
physicians, 32
registered dietitian, **19,** 31
websites on, 34t, 37
Health claims, **60**
food choices and, 5
on food labels, 60, 60t
phytochemicals, 467
on supplement labels, 363–364
See also Information on nutrition, validity of
Health history, 20–21, 22f, E-1t, **E-2,** E-2t
Healthy Eating Index, **48,** 48f, 49
Healthy People 2010, **23**
overview of goals, 23, 23t
adolescent pregnancy, 527
AIDS/HIV infection, 616
alcohol, 241, 247, 249, 528
allergies, 561
antibiotic resistant *Salmonella,* 680
birthweight, 524
breastfeeding, 530
calcium, 415, 418
cancer, 636
cholesterol (blood), 623
Daily Food Guide, 47
diabetes, 632

eating disorders, 316
exercise, 476, 477, 478, 479, 565
fat intake, 163
fiber, 126
fish contaminants, 670
fluoridated water, 456
folate and neural tube defects, 512
food safety, 659, 662
growth retardation, 556
health behavior counseling, 25
heart disease, 620, 623
hunger, 694
hypertension, 629
iron deficiency, 442, 555
iron deficiency in pregnancy, 520
lead levels, 559
nutrition, 22, 23, 23t
nutrition counseling and services, 32, 299
obesity, 280, 562
older adults' diets, 602
osteoporosis, 418
overweight, 22t
pesticide exposure, 672
physician dietary counseling, 642
pregnancy, 523, 527, 528
school health education, 570
school nutrition, 569
smoking/tobacco use, 528, 574
sodium intake, 407
stroke, 620
toxic chemicals, 670
validity of Internet information, 30
water, safety of, 682
water-borne diseases, 681
website for, 26
website quality, 30
weight gain during pregnancy, 515
weight management, 299
Heart attacks, **620**
alcohol and, 246
blood cholesterol and, 13f
death rate, 621
mechanism of, 621
obesity and, 267–268
See also Cardiovascular disease (CVD)
Heartburn, **96**
cause of, 80, 98–99
during pregnancy, 522, 522t
preventing, 100t
Heart disease. *See* Cardiovascular disease
Heart rate, 478, 479
Heat energy
ATP and, 218
generation of, 256
kcalorie as unit of, **8–9,** 252
white *vs.* brown adipose tissue, 284
Heat stroke, **490**
Heavy metals, 350, 458–**459,** 668
See also specific metals
Height measurement, E-6, E-6f
Heimlich maneuver, 94, 95f, **96**
Helicobacter pylori, 99, 100, 341
Hematocrit, **442,** E-20, E-20t, E-21t
Heme iron, **440,** 440f
Hemicelluloses, 109, 109n, C-2f to C-3f
Hemochromatosis, **444**
Hemoglobin, **184**
function of, 184, 439
in iron deficiency, 442, E-20, E-20t, E-21t
iron in, 184f, 439
precursor, 442
in sickle-cell anemia, 187, 189, 189f
structure of, 184f
zinc and, 454

Hemolytic anemia, **381**
Hemophilia, **382**
Hemorrhagic disease, **382**
Hemorrhoids, **96,** 97, 125t, 522
Hemosiderin, **441**
Hemosiderosis, **444**
Heparin, 352n
Hepatic vein, **88,** 88f
Hepatitis, 651, 660t
Hepatitis A virus, 664
Herbal medicines/supplements, 650–653
for athletes, 503
dosage, 652–653
herbal medicines, **648,** 650–653, 652t, 653t, 659f
herb-drug interactions, 653t
lactation and, 534
"natural" *vs.* safe, 36f, 651–653
obesity treatment, 288, 289t
during pregnancy, 528
websites on, 654
Herbal sterols, **500**
Herbert, Victor, 33
Heredity. *See* Genetics/genes
Herpes, 203
Hesperidin, 344
Hexoses, **104**
hGH (human growth hormone), **499,** 503
Hiccups, **96,** 98
High blood pressure. *See* Hypertension
High-carbohydrate diets. *See* Diets, high-carbohydrate
High-density lipoprotein. *See* HDL
High potency, **359,** 363
High-quality protein, **194,** 195
High-risk pregnancy, **523–530,** 523t
Histamine, **350**
Historical data (in nutrition assessment), 20–21, 22f, 274–275, E-1t, E-1 to E-5, E-2t
HIV (human immunodeficiency virus), 534, 615–618, **616,** 616t
HMB (beta-hydroxy-beta-methylbutyrate), **500**
Homeopathy, **648**
Homeostasis, **89**
blood glucose, 116, 117f
calcium, 413
nervous system and, A-7 to A-8
See also Acid-base balance; Electrolytes
Homocysteine
alcohol abuse and, 245
Alzheimer's disease and, 600
B vitamins and, 628t
folate and, 338–339
heart disease and, 200, 338–339
Honey, **120**
botulism risk, 121n, 553, 665
laxative effect, 97
nutrients in, 121, 121t
Hormonal athletic supplements, 503, 503t
Hormones, **90,** A-3 to A-7, **A-4,** A-4f
blood glucose levels and, 116–117, 117f
blood pressure and, 399, A-6
bone remodeling, 413–414, 414f, 431, 431n
calcium balance and, 413, 414f, A-6
changes with age, 592, 592n
cholesterol in synthesis of, 149
enzymes *vs.,* 79
fat storage and, 413
functions of, 190, 190n, 190t
gastrointestinal, 89–90, 91, 150, A-6 to A-7
ghrelin as, **283–284**
as incidental food additives, 680
lactation and, 531, A-5
nervous system and, 89–90
osteoporosis and, 431–432, 431n
pregnancy, A-6

Aids to Calculation

Many mathematical problems have been worked out in the "How to" and "Nutrition Calculations" sections of the text. These pages provide additional help and examples.

Conversion Factors

A conversion factor is a fraction in which the numerator (top) and the denominator (bottom) express the same quantity in different units. For example, 2.2 pounds (lb) and 1 kilogram (kg) are equivalent; they express the same weight. The conversion factors used to change pounds to kilograms and vice versa are:

$$\frac{1 \text{ kg}}{2.2 \text{ lb}} \text{ and } \frac{2.2 \text{ lb}}{1 \text{ kg}}.$$

Because a conversion factor equals 1, measurements can be multiplied by the factor to change the *unit* of measure without changing the *value* of the measurement. To change one unit of measurement to another, use the factor with the unit you are seeking in the numerator (top) of the fraction.

Example 1 Convert the weight of 130 pounds to kilograms.

- Choose the conversion factor in which the kilograms are on top and multiply by 130 pounds:

$$\frac{1 \text{ kg}}{2.2 \text{ lb}} \times 130 \text{ lb} = \frac{130 \text{ kg}}{2.2} = 59 \text{ kg}.$$

Example 2 Consider a 4-ounce (oz) hamburger that contains 7 grams (g) of saturated fat. How many grams of saturated fat are contained in a 3-ounce hamburger?

- Because you are seeking grams of saturated fat, the conversion factor is:

$$\frac{7 \text{ g saturated fat}}{4 \text{ oz hamburger}}.$$

- Multiply 3 ounces of hamburger by the conversion factor:

$$3 \text{ oz hamburger} \times \frac{7 \text{ g saturated fat}}{4 \text{ oz hamburger}} = \frac{3 \times 7}{4} = \frac{21}{4} =$$

5 g saturated fat (rounded off).

Percentages

A percentage is a comparison between a number of items (perhaps the number of kcalories in your daily energy intake) and a standard number (perhaps the number of kcalories used for Daily Values on food labels). To find a percentage, first divide by the standard number and then multiply by 100 to state the answer as a percentage (*percent* means "per 100").

Example 3 Suppose your energy intake for the day is 1500 kcalories (kcal): What percentage of the Daily Value (DV) for energy does your intake represent? (Use the Daily Value of 2000 kcalories as the standard.)

- Divide your kcalorie intake by the Daily Value:

  1500 kcal (your intake) ÷ 2000 kcal (DV) = 0.75.

- Multiply your answer by 100 to state it as a percentage:

  0.75 × 100 = 75% of the Daily Value.

Example 4 Sometimes the percentage is more than 100. Suppose your daily intake of vitamin C is 120 milligrams (mg) and your RDA (male) is 90 milligrams. What percentage of the RDA for vitamin C is your intake?

120 mg (your intake) ÷ 90 mg (RDA) = 1.33.

1.33 × 100 = 133% of the RDA.

Example 5 Sometimes the comparison is between a part of a whole (for example, your kcalories from protein) and the total amount (your total kcalories). In this case, the total is the number you divide by. If you consume 60 grams (g) protein, 80 grams fat, and 310 grams carbohydrate, what percentages of your total kcalories for the day come from protein, fat, and carbohydrate?

- Multiply the number of grams by the number of kcalories from 1 gram of each energy nutrient (conversion factors):

$$60 \text{ g protein} \times \frac{4 \text{ kcal}}{1 \text{ g protein}} = 240 \text{ kcal}.$$

$$80 \text{ g fat} \times \frac{9 \text{ kcal}}{1 \text{ g fat}} = 720 \text{ kcal}.$$

$$310 \text{ g carbohydrate} \times \frac{4 \text{ kcal}}{1 \text{ g carbohydrate}} = 1240 \text{ kcal.}$$

- Find the total kcalories:

  $$240 + 720 + 1240 = 2200 \text{ kcal.}$$

- Find the percentage of total kcalories from each energy nutrient (see Example 3):

  Protein: $240 \div 2200 = 0.109 \times 100 = 10.9 =$ 11% of kcal.

  Fat: $720 \div 2200 = 0.327 \times 100 = 32.7 =$ 33% of kcal.

  Carbohydrate: $1240 \div 2200 = 0.563 \times 100 = 56.3 =$ 56% of kcal.

  Total: $11\% + 33\% + 56\% = 100\%$ of kcal.

In this case, the percentages total 100 percent, but sometimes they total 99 or 101 because of rounding—a reasonable estimate.

Ratios

A ratio is a comparison of two (or three) values in which one of the values is reduced to 1. A ratio compares identical units and so is expressed without units.

Example 6 Suppose your daily intakes of potassium and sodium are 3000 milligrams (mg) and 2500 milligrams, respectively. What is the potassium-to-sodium ratio?

- Divide the potassium milligrams by the sodium milligrams:

  $$3000 \text{ mg potassium} \div 2500 \text{ mg sodium} = 1.2.$$

The potassium-to-sodium ratio is 1.2:1 (read as "one point two to one" or simply "one point two"), which means there are 1.2 milligrams of potassium for every 1 milligram of sodium. A ratio greater than 1 means that the first value (in this case, potassium) is greater than the second (sodium). When the ratio is less than 1, the second value is larger.

Weights and Measures

LENGTH
1 meter (m) = 39 in.
1 centimeter (cm) = 0.4 in.
1 inch (in) = 2.5 cm.
1 foot (ft) = 30 cm.

TEMPERATURE

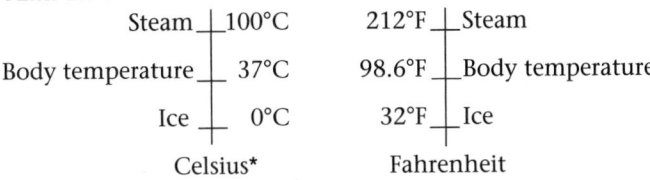

- To find degrees Fahrenheit (°F) when you know degrees Celsius (°C), multiply by 9/5 and then add 32.
- To find degrees Celsius (°C) when you know degrees Fahrenheit (°F), subtract 32 and then multiply by 5/9.

VOLUME
1 liter (L) = 1000 mL, 0.26 gal, 1.06 qt, or 2.1 pt.
1 milliliter (mL) = 1/1000 L or 0.03 fluid oz.
1 gallon (gal) = 128 oz, 8 c, or 3.8 L.
1 quart (qt) = 32 oz, 4 c, or 0.95 L.
1 pint (pt) = 16 oz, 2 c, or 0.47 L.
1 cup (c) = 8 oz, 16 tbs, about 250 mL, or 0.25 L.
1 ounce (oz) = 30 mL.
1 tablespoon (tbs) = 3 tsp or 15 mL.
1 teaspoon (tsp) = 5 mL.

WEIGHT
1 kilogram (kg) = 1000 g or 2.2 lb.
1 gram (g) = 1/1000 kg, 1000 mg, or 0.035 oz.
1 milligram (mg) = 1/1000 g or 1000 µg.
1 microgram (µg) = 1/1000 mg.
1 pound (lb) = 16 oz, 454 g, or 0.45 kg.
1 ounce (oz) = about 28 g.

ENERGY
1 kilojoule (kJ) = 0.24 kcal.
1 millijoule (mJ) = 240 kcal.
1 kcalorie (kcal) = 4.2 kJ.
1 g carbohydrate = 4 kcal = 17 kJ.
1 g fat = 9 kcal = 37 kJ.
1 g protein = 4 kcal = 17 kJ.
1 g alcohol = 7 kcal = 29 kJ.

*Also known as *centigrade*.

Daily Values for Food Labels

The Daily Values are standard values developed by the Food and Drug Administration (FDA) for use on food labels. The values are based on 2000 kcalories a day for adults and children over 4 years old. Chapter 2 provides more details.

Nutrient	Amount
Protein[a]	50 g
Thiamin	1.5 mg
Riboflavin	1.7 mg
Niacin	20 mg NE
Biotin	300 µg
Pantothenic acid	10 mg
Vitamin B_6	2 mg
Folate	400 µg
Vitamin B_{12}	6 µg
Vitamin C	60 mg
Vitamin A	5000 IU[b]
Vitamin D	400 IU[b]
Vitamin E	30 IU[b]
Vitamin K	80 µg
Calcium	1000 mg
Iron	18 mg
Zinc	15 mg
Iodine	150 µg
Copper	2 mg
Chromium	120 µg
Selenium	70 µg
Molybdenum	75µg
Manganese	2 mg
Chloride	3400 mg
Magnesium	400 mg
Phosphorus	1000 mg

[a]The Daily Values for protein vary for different groups of people: pregnant women, 60 g; nursing mothers, 65 g; infants under 1 year, 14 g; children 1 to 4 years, 16 g.
[b]Equivalent values for nutrients expressed as IU are: vitamin A, 1500 RAE (assumes a mixture of 40% retinol and 60% beta-carotene); vitamin D, 10 µg; vitamin E, 20 mg.

Food Component	Amount	Calculation Factors
Fat	65 g	30% of kcalories
Saturated fat	20 g	10% of kcalories
Cholesterol	300 mg	Same regardless of kcalories
Carbohydrate (total)	300 g	60% of kcalories
Fiber	25 g	11.5 g per 1000 kcalories
Protein	50 g	10% of kcalories
Sodium	2400 mg	Same regardless of kcalories
Potassium	3500 mg	Same regardless of kcalories

GLOSSARY OF NUTRIENT MEASURES

kcal: kcalories; a unit by which energy is measured (Chapter 1 provides more details).

g: grams; a unit of weight equivalent to about 0.03 ounces.

mg: milligrams; one-thousandth of a gram.

µg: micrograms; one-millionth of a gram.

IU: international units; an old measure of vitamin activity determined by biological methods (as opposed to new measures that are determined by direct chemical analyses). Many fortified foods and supplements use IU on their labels.
- For vitamin A, 1 IU = 0.3 µg retinol, 3.6 µg β-carotene, or 7.2 µg other vitamin A carotenoids.
- For vitamin D, 1 IU = 0.025 µg cholecalciferol.
- For vitamin E, 1 IU = 0.67 natural α-tocopherol (other conversion factors are used for different forms of vitamin E).

mg NE: milligrams niacin equivalents; a measure of niacin activity (Chapter 10 provides more details).
- 1 NE = 1 mg niacin.
    = 60 mg tryptophan (an amino acid).

µg DFE: micrograms dietary folate equivalents; a measure of folate activity (Chapter 10 provides more details).
- 1 µg DFE = 1 µg food folate.
    = 0.6 µg fortified food or supplement folate.
    = 0.5 µg supplement folate taken on an empty stomach.

µg RAE: micrograms retinol activity equivalents; a measure of vitamin A activity (Chapter 11 provides more details).
- 1 µg RAE = 1 µg retinol.
    = 12 µg β-carotene.
    = 24 µg other vitamin A carotenoids.

mmol: millimoles; one-thousanth of a mole, the molecular weight of a substance. To convert mmol to mg, multiply by the atomic weight of the substance.
- For sodium, mmol × 23 = mg Na.
- For chloride, mmol × 35.5 = mg Cl.
- For sodium chloride, mmol × 58.5 = mg NaCl.

Recommended Dietary Allowances (RDA) and Adequate Intakes (AI) for Vitamins

Age (yr)	Thiamin RDA (mg/day)	Riboflavin RDA (mg/day)	Niacin RDA (mg/day)[a]	Biotin AI (µg/day)	Pantothenic acid AI (mg/day)	Vitamin B_6 RDA (mg/day)	Folate RDA (µg/day)[b]	Vitamin B_{12} RDA (µg/day)	Choline AI (mg/day)	Vitamin C RDA (mg/day)	Vitamin A RDA (µg/day)[c]	Vitamin D AI (µg/day)[d]	Vitamin E RDA (mg/day)[e]	Vitamin K AI (µg/day)
Infants														
0–0.5	0.2	0.3	2	5	1.7	0.1	65	0.4	125	40	400	5	4	2.0
0.5–1	0.3	0.4	4	6	1.8	0.3	80	0.5	150	50	500	5	5	2.5
Children														
1–3	0.5	0.5	6	8	2	0.5	150	0.9	200	15	300	5	6	30
4–8	0.6	0.6	8	12	3	0.6	200	1.2	250	25	400	5	7	55
Males														
9–13	0.9	0.9	12	20	4	1.0	300	1.8	375	45	600	5	11	60
14–18	1.2	1.3	16	25	5	1.3	400	2.4	550	75	900	5	15	75
19–30	1.2	1.3	16	30	5	1.3	400	2.4	550	90	900	5	15	120
31–50	1.2	1.3	16	30	5	1.3	400	2.4	550	90	900	5	15	120
51–70	1.2	1.3	16	30	5	1.7	400	2.4	550	90	900	10	15	120
>70	1.2	1.3	16	30	5	1.7	400	2.4	550	90	900	15	15	120
Females														
9–13	0.9	0.9	12	20	4	1.0	300	1.8	375	45	600	5	11	60
14–18	1.0	1.0	14	25	5	1.2	400	2.4	400	65	700	5	15	75
19–30	1.1	1.1	14	30	5	1.3	400	2.4	425	75	700	5	15	90
31–50	1.1	1.1	14	30	5	1.3	400	2.4	425	75	700	5	15	90
51–70	1.1	1.1	14	30	5	1.5	400	2.4	425	75	700	10	15	90
>70	1.1	1.1	14	30	5	1.5	400	2.4	425	75	700	15	15	90
Pregnancy														
≤18	1.4	1.4	18	30	6	1.9	600	2.6	450	80	750	5	15	75
19–30	1.4	1.4	18	30	6	1.9	600	2.6	450	85	770	5	15	90
31–50	1.4	1.4	18	30	6	1.9	600	2.6	450	85	770	5	15	90
Lactation														
≤18	1.4	1.6	17	35	7	2.0	500	2.8	550	115	1200	5	19	75
19–30	1.4	1.6	17	35	7	2.0	500	2.8	550	120	1300	5	19	90
31–50	1.4	1.6	17	35	7	2.0	500	2.8	550	120	1300	5	19	90

NOTE: For all nutrients, values for infants are AI. The glossary on the inside back cover defines units of nutrient measure.

[a] Niacin recommendations are expressed as niacin equivalents (NE), except for recommendations for infants younger than 6 months, which are expressed as preformed niacin.

[b] Folate recommendations are expressed as dietary folate equivalents (DFE).

[c] Vitamin A recommendations are expressed as retinol activity equivalents (RAE).

[d] Vitamin D recommendations are expressed as cholecalciferol and assume an absence of adequate exposure to sunlight.

[e] Vitamin E recommendations are expressed as α-tocopherol.

Recommended Dietary Allowances (RDA) and Adequate Intakes (AI) for Minerals

Age (yr)	Sodium AI (mg/day)	Chloride AI (mg/day)	Potassium AI (mg/day)	Calcium AI (mg/day)	Phosphorus RDA (mg/day)	Magnesium RDA (mg/day)	Iron RDA (mg/day)	Zinc RDA (mg/day)	Iodine RDA (µg/day)	Selenium RDA (µg/day)	Copper RDA (µg/day)	Manganese AI (mg/day)	Fluoride AI (mg/day)	Chromium AI (µg/day)	Molybdenum RDA (µg/day)
Infants															
0–0.5	120	180	400	210	100	30	0.27	2	110	15	200	0.003	0.01	0.2	2
0.5–1	370	570	700	270	275	75	11	3	130	20	220	0.6	0.5	5.5	3
Children															
1–3	1000	1500	3000	500	460	80	7	3	90	20	340	1.2	0.7	11	17
4–8	1200	1900	3800	800	500	130	10	5	90	30	440	1.5	1.0	15	22
Males															
9–13	1500	2300	4500	1300	1250	240	8	8	120	40	700	1.9	2	25	34
14–18	1500	2300	4700	1300	1250	410	11	11	150	55	890	2.2	3	35	43
19–30	1500	2300	4700	1000	700	400	8	11	150	55	900	2.3	4	35	45
31–50	1500	2300	4700	1000	700	420	8	11	150	55	900	2.3	4	35 .	45
51–70	1300	2000	4700	1200	700	420	8	11	150	55	900	2.3	4	30	45
>70	1200	1800	4700	1200	700	420	8	11	150	55	900	2.3	4	30	45
Females															
9–13	1500	2300	4500	1300	1250	240	8	8	120	40	700	1.6	2	21	34
14–18	1500	2300	4700	1300	1250	360	15	9	150	55	890	1.6	3	24	43
19–30	1500	2300	4700	1000	700	310	18	8	150	55	900	1.8	3	25	45
31–50	1500	2300	4700	1000	700	320	18	8	150	55	900	1.8	3	25	45
51–70	1300	2000	4700	1200	700	320	8	8	150	55	900	1.8	3	20	45
>70	1200	1800	4700	1200	700	320	8	8	150	55	900	1.8	3	20	45
Pregnancy															
≤18	1500	2300	4700	1300	1250	400	27	12	220	60	1000	2.0	3	29	50
19–30	1500	2300	4700	1000	700	350	27	11	220	60	1000	2.0	3	30	50
31–50	1500	2300	4700	1000	700	360	27	11	220	60	1000	2.0	3	30	50
Lactation															
≤18	1500	2300	5100	1300	1250	360	10	14	290	70	1300	2.6	3	44	50
19–30	1500	2300	5100	1000	700	310	9	12	290	70	1300	2.6	3	45	50
31–50	1500	2300	5100	1000	700	320	9	12	290	70	1300	2.6	3	45	50

Tolerable Upper Intake Levels (UL) for Vitamins

Age (yr)	Niacin (mg/day)[a]	Vitamin B6 (mg/day)	Folate (μg/day)[a]	Choline (mg/day)	Vitamin C (mg/day)	Vitamin A (μg/day)[b]	Vitamin D (μg/day)	Vitamin E (mg/day)[c]
Infants								
0–0.5	—	—	—	—	—	600	25	—
0.5–1	—	—	—	—	—	600	25	—
Children								
1–3	10	30	300	1000	400	600	50	200
4–8	15	40	400	1000	650	900	50	300
9–13	20	60	600	2000	1200	1700	50	600
Adolescents								
14–18	30	80	800	3000	1800	2800	50	800
Adults								
19–70	35	100	1000	3500	2000	3000	50	1000
>70	35	100	1000	3500	2000	3000	50	1000
Pregnancy								
≤18	30	80	800	3000	1800	2800	50	800
19–50	35	100	1000	3500	2000	3000	50	1000
Lactation								
≤18	30	80	800	3000	1800	2800	50	800
19–50	35	100	1000	3500	2000	3000	50	1000

[a] The UL for niacin and folate apply to synthetic forms obtained from supplements, fortified foods, or a combination of the two.

[b] The UL for vitamin A applies to the preformed vitamin only.
[c] The UL for vitamin E applies to any form of supplemental α-tocopherol, fortified foods, or a combination of the two.

Tolerable Upper Intake Levels (UL) for Minerals

Age (yr)	Sodium (mg/day)	Chloride (mg/day)	Calcium (mg/day)	Phosphorus (mg/day)	Magnesium (mg/day)[d]	Iron (mg/day)[b]	Zinc (mg/day)	Iodine (μg/day)	Selenium (μg/day)	Copper (μg/day)	Manganese (mg/day)	Fluoride (mg/day)	Molybdenum (μg/day)	Boron (mg/day)	Nickel (mg/day)	Vanadium (mg/day)
Infants																
0–0.5	—[e]	—[e]	—	—	—	40	4	—	45	—	—	0.7	—	—	—	—
0.5–1	—[e]	—[e]	—	—	—	40	5	—	60	—	—	0.9	—	—	—	—
Children																
1–3	1500	2300	2500	3000	65	40	7	200	90	1000	2	1.3	300	3	0.2	—
4–8	1900	2900	2500	3000	110	40	12	300	150	3000	3	2.2	600	6	0.3	—
9–13	2200	3400	2500	4000	350	40	23	600	280	5000	6	10	1100	11	0.6	—
Adolescents																
14–18	2300	3600	2500	4000	350	45	34	900	400	8000	9	10	1700	17	1.0	—
Adults																
19–70	2300	3600	2500	4000	350	45	40	1100	400	10,000	11	10	2000	20	1.0	1.8
>70	2300	3600	2500	3000	350	45	40	1100	400	10,000	11	10	2000	20	1.0	1.8
Pregnancy																
≤18	2300	3600	2500	3500	350	45	34	900	400	8000	9	10	1700	17	1.0	—
19–50	2300	3600	2500	3500	350	45	40	1100	400	10,000	11	10	2000	20	1.0	—
Lactation																
≤18	2300	3600	2500	4000	350	45	34	900	400	8000	9	10	1700	17	1.0	—
19–50	2300	3600	2500	4000	350	45	40	1100	400	10,000	11	10	2000	20	1.0	—

[d] The UL for magnesium applies to synthetic forms obtained from supplements or drugs only.
[e] Source of intake should be from human milk (or formula) and food only.

NOTE: An Upper Limit was not established for vitamins and minerals not listed and for those age groups listed with a dash (—) because of a lack of data, not because these nutrients are safe to consume at any level of intake. All nutrients can have adverse effects when intakes are excessive.

SOURCE: Adapted with permission from the *Dietary Reference Intakes* series, National Academy Press. Copyright 1997, 1998, 2000, 2001, by the National Academy of Sciences. Courtesy of the National Academy Press, Washington, D.C.